Carpentry

This text is dedicated
to the memory of Gaspar J. Lewis.

Carpentry

THIRD EDITION

Gaspar Lewis
Floyd Vogt

Doug Holman, Contributor

DELMAR
THOMSON LEARNING™

Australia Canada Mexico Singapore Spain United Kingdom United States

NOTICE TO THE READER

Business Unit Director:
Alar Elken

Channel Manager:
Mona Caron

Executive Editor:
Sandy Clark

Marketing Coordinator:
Kasey Young

Acquisitions Editor:
Mark Huth

Executive Production Manager:
Mary Ellen Black

Developmental Editor:
Jeanne Mesick

Production Coordinator:
Toni Hansen

Editorial Assistant:
Dawn Daugherty

Project Editor:
Barbara L. Diaz

Executive Marketing Manager:
Maura Theriault

Art/Design Coordinator:
Rachel Baker

Printed in the United States of America
3 4 5 6 7 03 02 01

For more information, contact Delmar at 3 Columbia Circle, PO Box 15015, Albany, New York 12212-5015; or find us on the World Wide Web at: http://www.delmar.com.

Library of Congress Cataloging-in-Publication Data
Lewis, Gaspar J.
Carpentry / Gaspar Lewis, Floyd Vogt.
p. cm
Includes index.
ISBN 0-7668-1081-X (alk. paper)
1. Carpentry. I. Vogt, Floyd. II. Title.
TH5604.L44 2000
694—dc21 99-057913

Printed on acid-free paper.

CONTENTS

Preface

In 1995, only months after the second edition of *Carpentry* was released, Gaspar Lewis died. Mr. Lewis was an experienced carpenter and devoted author who was dedicated to assisting carpentry students in becoming competent carpenters with skills to grow into leadership positions. His efforts brought together a blend of old and new construction concepts, producing a textbook well suited for use by carpentry students in secondary and post-secondary vocational programs and on-the-job carpentry apprentices. The third edition continues this legacy while incorporating the changes that time brings to the construction industry.

As the succeeding author to Mr. Lewis, I am sincerely interested in maintaining his approach. My personal experience is that of growing up in a family-owned and -operated residential construction business and teaching carpentry at a two-year technical college in upstate New York. Preserving the heritage of pride in our trade and clarifying the learning process were two of my goals in revising this book for the third edition.

The basic approach of this book is to build a house. Each topic is presented in a fashion that follows the construction process. For example, when the roof is framed and the building enclosed, a quick set of stairs is needed. But later, after much of the finished material has been installed, a finished set of stairs is needed. The material in this book is organized to reflect this order of construction.

This book incorporates different methods of teaching to accomodate variations in the ways that different people learn. Each unit begins with a list of objectives that explains exactly what the student is expected to accomplish. The material is presented through the use of text information, graphic illustrations, and step-by-step procedures. This three-fold approach gives the student many opportunities to grasp the material. At the end of each unit is a "Building for Success" section designed to stimulate thought-provoking discussions and present issues that face today's carpenter. The student activity feature gives students the opportunity to apply their knowledge by conducting research, presenting new information, and solving problems. These activities may be done individually or within a group setting. Throughout the book are illustrated "Helpful Hints," which are techniques and tips to help make the job at hand easier, faster, and more accurate.

Changes and modifications from the previous edition were made to address changes in the methods and materials that have impacted the construction industry. Changes include an increased reliance on the calculator to simplify mathematical calculations, the addition of the speed square to the rafter chapters, and the use of the Pythagorean Theorem to facilitate laying out the location of a building from a plot plan and calculating the rafter lengths of a true gambrel roof. Some diagrams and illustrations were refined to improve the clarity of the presented material.

Modifications include adding and deleting information in the scaffolding and cabinet construction areas. The scaffolding section includes more information and assembly steps for steel-pipe scaffolding. The wood scaffolding section was reduced by excluding the step-by-step procedures for wood scaffold erection. These adjustments were made to reflect the changes in the safety requirements from organizations such as OSHA. The cabinet construction units were condensed to reflect the fact that the construction industry is relying more on outside woodworking shops to make custom cabinets rather than having the cabinets custom-built on site. Because many carpenters are involved with installing cabinets, the cabinet chapter in the third edition focuses more on cabinet and countertop installation.

Additional study material is provided for the student in the Workbook to supplement the text material. The Workbook correlates directly with the text, with exercises and suggested activities focusing on critical thinking, planning, preparation, safety, cutting, assembly, tool use, plan reading, communication, calculations, and site cleanup. An Instructor's Guide is available to assist the instructor with additional resource materials and ideas, answers to questions from the text and workbook, and transparency masters taken from the artwork of the text.

This book is dedicated to the memory of Gaspar Lewis. It is also dedicated to those who seek a life in carpentry and are willing to carry on the long legacy of pride in our profession while pursuing excellence.

Floyd H. Vogt
SUNY Delhi
Delhi, NY 13753
vogtfh@delhi.edu

Acknowledgments

This revision could not have been completed without the assistance of the many people who worked on this edition. Special thanks to Richard Harrington, SUNY Delhi, Delhi, New York; Steve McKeegan, P.C., SUNY Delhi, Delhi, New York; and Steve Munson, Munson's Building Supplies, Oneonta, New York, for their valuable input and contributions.

A special note of thanks is owed to Douglas A. Holman, Apprenticeship and Training Coordinator, East Tennessee Carpenters Joint Apprenticeship and Training Program, Knoxville, Tennessee, for his rewriting of the scaffolding section. Doug's contribution has surely made a difference in the text. We would also like to acknowledge the contrubutions of Larry Kness, Southeast Community College, Milford, Nebraska, who wrote the "Building for Success" features.

As always, our gratitude is extended to those who reviewed the revision and past edition for their contributions and suggestions on making a better text. Their contributions were invaluable. Special thanks to:

John E. Mackay
Training Coordinator
New Jersey Carpenters Technical
Training Centers
Kenilworth, New Jersey

Dave Rainforth
Southeast Community College
Milford, Nebraska

Kathy Swan
Carpenters Training Trust of
Western Washington
Renton, Washington

Rick Glanville
Camosun College
Victoria, British Columbia, Canada

Jim Loosle
Kearns High School
Taylorsville, Utah

Douglas A. Holman
East Tennessee Carpenters Joint
Apprenticeship and Training Program
Knoxville, Tennessee

Carl Gamarino
Industrial Technical Education Center
Brockport, Pennsylvania

Richard Cappelmann
Technical College of the Low Country
Beaufort, South Carolina

Terry Schaefer
Western Wisconsin Technical College
LaCrosse, Wisconsin

Larry Kness
Southeast Community College
Milford, Nebraska

Introduction

The history of carpentry goes back to 8000 B.C., when primitive people used stone axes to shape wood to build shelters. Stone Age Europeans built rectangular timber houses more than 100 feet long—proving the existence of carpentry even at this early date. The Egyptians used copper woodworking tools as early as 4000 B.C. By 2000 B.C. they had developed bronze tools and were proficient in the drilling, dovetailing, mitering, and mortising of wood.

In the Roman Empire, two-wheeled chariots, called *carpentum* in Latin, were made of wood. A person who built such chariots was called a *carpentarius,* from which the English word *carpenter* is derived. Roman carpenters handled iron adzes, saws, rasps, awls, gouges, and planes.

During the Middle Ages, most carpenters were found in larger towns where work was plentiful. They would also travel with their tools to outlying villages or wherever there was a major construction project in progress. By this time, they had many efficient, steel-edged hand tools. During this period skillful carpentry was required for the building of timber churches and castles.

In the twelfth century, carpenters banded together to form *guilds.* The members of the guild were divided into *masters,* **journeymen,** and **apprentices**. The master was a carpenter with much experience and knowledge who trained apprentices. The apprentice lived with the master and was given food, clothing, and shelter and worked without pay. After a period of five to nine years, the apprentice became a journeyman who could work for wages. Eventually, a journeyman could become a master. Guilds were the forerunners of the modern labor unions and associations.

Starting in the fifteenth century, carpenters used great skill in constructing the splendid buildings of the Renaissance period and afterward. With the introduction of the balloon frame in the early nineteenth century, more modern construction began to replace the slower mortise-and-tenon frame. In 1873, electric power was used for the first time to drive machine tools. The first electric hand drill was developed in 1917, and in 1925, electric portable saws were being used.

At present, many power tools are available to the carpenter to speed up the work. Although the volume of the carpenter's work has been reduced by the use of manufactured parts, some of the same skills carpenters used in years past are still needed for the intricate interior finish work in buildings.

Carpenters construct and repair structures and their parts using wood, plywood, and other building materials. They lay out, cut, fit, and fasten the materials to erect the framework and apply the finish. They build houses, factories, banks, schools, hospitals, churches, bridges, dams, and other structures. In addition to new construction, a large part of the industry is engaged in remodeling and repair of existing buildings.

The majority of workers in the construction industry are carpenters (Fig. I-1). They are the first trade workers on the job, laying out excavation and building lines. They take part in every phase of the construction, working below the ground, at ground level, or at great heights. They are the last to leave the job when they put the key in the lock.

Fig. I-1 Carpenters make up the majority of workers in the construction industry.

Specialization

In large cities, where there is a great volume of construction, carpenters tend to specialize in one area of the trade. They may be specialists in rough carpentry, which are called rough carpenters or framers (Fig. I-2). Rough carpentry does not mean that the workmanship is crude. Just as much care is taken in the rough work as in any other work. Rough carpentry will be covered eventually by the finish work or dismantled, as in the case of concrete form construction.

Finish carpenters specialize in applying exterior and interior finish, sometimes called trim (Fig. I-3). Other specialties are constructing concrete forms (Fig. I-4), laying finish flooring, building stairs, applying gypsum board (Fig. I-5), roofing, insulating, and installing suspended ceilings.

In smaller communities, where the volume of construction is lighter, carpenters perform tasks in all areas of the trade from the rough to the finish. The general carpenter needs a more complete knowledge of the trade than the specialist does.

Fig. I-2 Some carpenters specialize in framing.

Fig. I-3 Carpenters may be specialists in applying interior trim. (*Courtesy of Senco Products, Inc.*)

Fig. I-4 Erecting concrete formwork may be a specialty.

Fig. I-5 Some carpenters choose to do only drywall application.

Requirements

Carpenters need to know how to use and maintain hand and power tools. They need to know the kinds, grades, and characteristics of the materials with which they work—how each can be cut, shaped, and most satisfactorily joined. Carpenters must be familiar with the many different fasteners available and choose the most appropriate for each task.

Carpenters must know how to lay out and frame floors, walls, stairs, and roofs. They must know how to install windows and doors, and how to apply numerous kinds of exterior and interior finish. They must use good judgment to decide on proper procedures to do the job at hand in the most efficient and safest manner.

Carpenters must be in good physical condition because much of the work is done by hand and sometimes requires great exertion. They must lift large sheets of plywood, heavy wood timbers, and bundles of roof shingles; they also have to climb ladders and scaffolds.

An attitude of care and concern for the job and for other workers is vital. The carpenter must feel that nothing less than a first-class job is acceptable.

Training

Vocational training in carpentry is offered in many high schools for those who become seriously interested at an early age. For those who have completed high school, carpentry training programs are offered at many post-secondary vocational schools and community colleges. Most of these programs train students for duties up to and including that of foreman. Because there is so much more to learn, vocational school training should then be followed by an apprenticeship.

Some schools participate in programs in which students go to school part-time and work as on-the-job apprentices part-time. Upon graduation, they usually continue with the same employer as full-time apprentices.

Apprenticeship training programs (usually four years in length) are offered by the United Brotherhood of Joiners and Carpenters of America, a carpenters' union, and by contractors' associations, such as the National Association of Home Builders of the United States (NAHB) and the Associated General Contractors of America (AGC) in cooperation with the Bureau of Apprenticeship and Training of the U. S. Department of Labor (Fig. I-6). In Canada, apprenticeships are administered by the provincial governments with a final exam called the interprovincial examination written for national accreditation. Usually these organizations give apprenticeship credit for completion of previous vocational school training in carpentry, resulting in a shortened apprenticeship and a higher starting wage.

The apprentice (Fig. I-7), must be at least seventeen years of age, in some areas eighteen, and must learn the trade while working on the job with experienced journeymen. Under the apprenticeship agreement, basic standards provide, among other things, that there can be only a certain number of apprentices hired by a particular contractor in relation to the number of journeymen. This is to ensure that apprentices receive proper supervision and training on the job.

Basic standards also provide that there is a progressively increasing schedule of wages. Starting pay for the apprentice carpenter is usually about 50 percent of the journeyman carpenter's rate. Due to periodic increases about every six months, the apprentice should receive 95 percent of the journeyman's wage during the last six months of the apprenticeship.

Under the terms of agreement, the apprentice is required to attend classes for a certain number of hours each year. A minimum of 144 hours per year is normally considered necessary. These classes are usually held at local schools, twice a week for about thirty-six weeks during the school year, for the length of the apprenticeship. The ap-

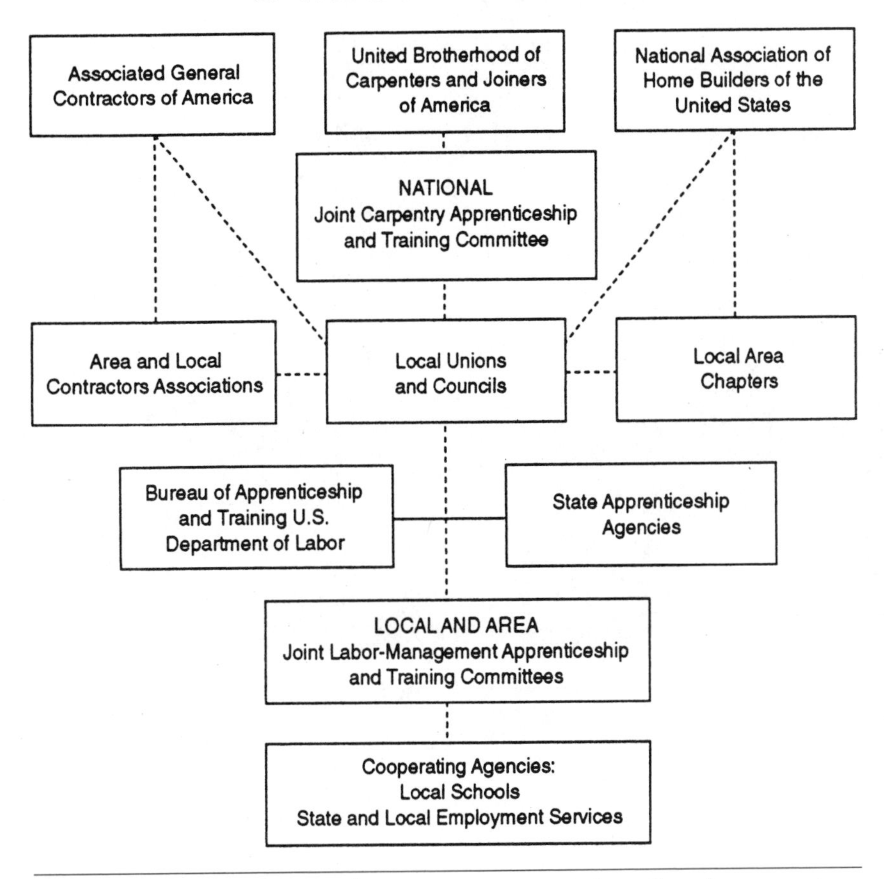

Fig. I-6 Industry, labor, and government work together in carpentry apprenticeship programs.

prentice becomes accepted as a journeyman carpenter when the training is completed and is awarded a Certificate of Completion of Apprenticeship. The newly graduated apprentice is now expected, within reason, to do jobs required of the journeyman.

Although it is in the best interests of the apprentice to be indentured, that is, to have a written contract with an organization, with conditions of the apprenticeship agreed upon, it is possible to learn the trade as a helper until enough skills have been acquired to demand the recognition and rewards of a journeyman carpenter. However, self-discipline is required to gain knowledge of both the practical and theoretical aspects of the trade. There may be time on the job to explain to a helper or apprentice how to do a certain task, but there usually is not time to explain concepts.

Many opportunities exist for the journeyman carpenter. Advancement depends on dependability, skill, productivity, and ingenuity, among other characteristics. Carpentry foremen, construction superintendents, and general contractors usually rise from

Fig. I-7 Many opportunities lie ahead for the apprentice carpenter.

Fig. I-8 Construction superintendents usually rise from the ranks of the carpenters. (*Courtesy of Brotherhood of Carpenters and Joiners of America*)

the ranks of the journeyman carpenters (Fig. I-8). Many who start as apprentice carpenters eventually operate their own construction firms. A survey revealed that 90 percent of the top officials (presidents, vice presidents, owners, and partners) of construction companies who replied began their careers as apprentices. Many of the project managers, superintendents, and craft supervisors employed by these companies also began as apprentices.

Summary

Carpentry is a trade in which there is a great deal of self-satisfaction, pride, and dignity associated with the work. It is an ancient trade and the largest of all trades in the building industry.

Skilled carpenters who have labored to the best of their ability can take pride in their workmanship whether the job was a rough concrete form or the finest finish in an elaborate staircase (Fig. I-9). At the end of each working day, carpenters can stand back and actually see the results of their labor. As the years roll by, the buildings that carpenters' hands had a part in creating still can be viewed with pride in the community.

Fig. I-9 After completing a complicated piece of work, such as this intricate staircase, carpenters can take pride in and view their accomplishment. (*Courtesy of L. J. Smith, Inc.*)

SECTION

Tools and Materials

UNIT

1

Wood and Lumber

An understanding of the nature of wood and the characteristics of lumber is necessary for all those who work with it. With this knowledge, the carpenter is able to protect lumber from decay, select it for appropriate use, work it with proper tools, and join and fasten it to the best advantage.

OBJECTIVES

After completing this unit, the student should be able to:

- name the parts of a tree trunk and state their function.
- describe methods of cutting the log into lumber.
- define hardwood and softwood, give examples of some common kinds, and list their characteristics.
- explain moisture content at various stages of seasoning, tell how wood shrinks, and describe some common lumber defects.
- state the grades and sizes of lumber and compute board measure.

UNIT CONTENTS

CHAPTER 1 WOOD

The carpenter works with wood more than any other material and must understand its characteristics in order to use it intelligently. Wood is a remarkable substance. It can be cut, shaped, or bent into just about any form. It is an efficient insulating material. It takes almost 6 inches of brick, 14 inches of concrete, or over 1,700 inches of aluminum to equal the insulating value of only 1-inch of wood.

There are many kinds of wood that vary in strength, workability, elasticity, color, grain, and texture. It is important to keep these qualities in mind when selecting wood. For instance, baseball bats, diving boards, and tool handles are made from hickory and ash because of their greater ability to bend without breaking (elasticity). Oak and maple are used for floors because of their hardness and durability. Redwood, cedar, cypress, and teak are used in exterior situations because of their resistance to decay. Cherry, mahogany, and walnut are typically chosen for their beauty.

With proper care, wood will last indefinitely. It is a material with beauty and warmth that has thousands of uses. Wood is one of our greatest natural resources. With wise conservation practices, wood will always be in abundant supply. It is fortunate that we have perpetually producing forests that supply this major building material to construct homes and other structures that last for hundreds of years. When those structures have served their purpose and are torn down, the wood used in their construction can be salvaged and used again (recycled) in new building, remodeling, or repair. Wood is biodegradable and when it is considered not feasible for reuse, it is readily absorbed back into the earth with no environmental harm.

Fig. 1-1 Redwood is often used for exterior trim and siding. (*Courtesy of California Redwood Association*)

Structure and Growth

Wood is made up of many hollow cells held together by a natural substance called *lignin.* The size, shape, and arrangement of these cells determine the strength, weight, and other properties of wood. Tree growth takes place in the **cambium layer,** which is just inside the protective shield of the tree called the *bark.* The tree's roots absorb water that passes upward through the **sapwood** to the leaves, where it is combined with carbon dioxide from the air. Sunlight causes these materials to change into food, which is then carried down and distributed toward the center of the trunk through the **medullary rays.**

As the tree grows outward from the **pith** (center), the inner cells become inactive and turn into **heartwood.** Heartwood is the central part of the tree and usually is darker in color and more durable than sapwood. The heartwood of cedar, cypress, and redwood, for instance, is extremely resistant to decay and is used extensively for outdoor furniture, patios, and exterior siding (Fig. 1-1). Used for the same purposes, sapwood decays more quickly.

Each spring and summer, a tree adds new layers to its trunk. Wood grows rapidly in the spring; it is rather porous and light in color. In summer, tree growth is slower; the wood is denser and darker, forming distinct rings. Because these rings are formed each year, they are called **annular rings** (Fig. 1-2). By counting the dark rings, the age of a tree can be determined. By studying the width of the rings, periods of abundant rainfall and sunshine or periods of slow growth can be discerned. Some trees, like the Douglas fir, grow rapidly to great heights and have very wide and pronounced annular rings. Mahogany, which grows in a tropical climate where the weather is more constant, has annular rings that are not so contrasting and sometimes are hardly visible.

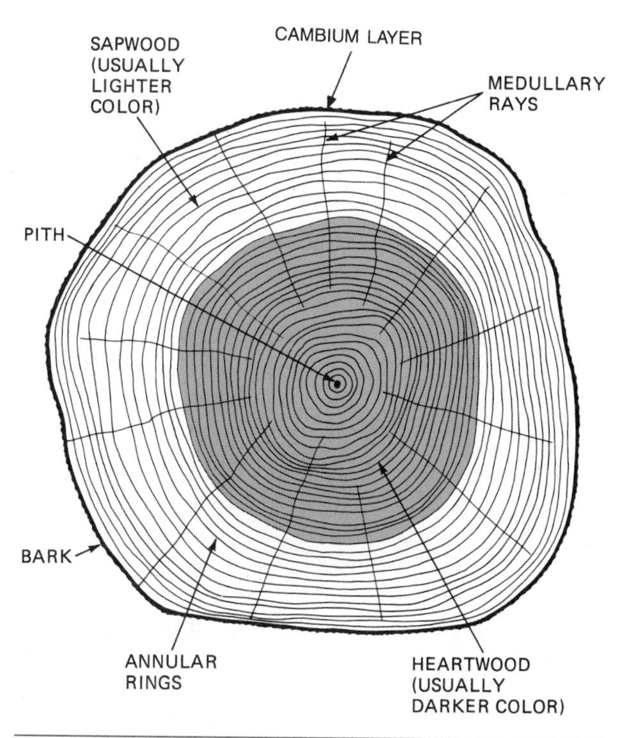

Fig. 1-2 A cross-section of a tree showing its structure. (*Courtesy of Western Wood Products Association*)

Hardwoods and Softwoods

Woods are classified as either hardwood or softwood. There are different methods of classifying these woods. The most common method of classifying wood is by its source. Hardwood comes from **deciduous** trees that shed their leaves each year. Softwood is cut from **coniferous,** or cone-bearing, trees, commonly known as evergreens (Fig. 1-3). In this method of classifying wood, some of the softwoods may actually be harder than the hardwoods. For instance, fir, a softwood, is harder and stronger than basswood, a hardwood. There are other methods of classifying hardwoods and softwoods, but this method is the one most widely used.

Some common hardwoods are ash, birch, cherry, hickory, maple, mahogany, oak, and walnut. Some common softwoods are pine, fir, hemlock, spruce, cedar, cypress, and redwood.

Wood may also be divided into two groups according to cell structure. *Open-grained* wood has large cells that show tiny openings or pores in the surface. In order to obtain a smooth finish, these pores must be filled with a specially prepared paste wood filler. Examples of open-grained wood are oak, mahogany, and walnut. All softwoods are *close-grained.* Some close-grained hardwoods are birch, cherry, maple, and poplar. (See Fig. 1-4 for common kinds of softwoods and their characteristics and Fig. 1-5 for hardwoods.)

Identification of Wood

Identifying different kinds of wood can be very difficult because some closely resemble each other. For instance, ash and white oak are hard to distinguish from each other, as are some pine, hemlock, and

Fig. 1-3 Hardwood is from broad-leaf trees, softwood from cone-bearing trees.

SOFTWOODS

KIND	COLOR	GRAIN	HARDNESS	STRENGTH	WORK-ABILITY	ELASTICITY	DECAY RESISTANCE	USES	OTHER
Red Cedar	Dark Reddish Brown	Close Medium	Soft	Low	Easy	Poor	Very High	Exterior	Cedar Odor
Cypress	Orange Tan	Close Medium	Soft to Medium	Medium	Medium	Medium	Very High	Exterior	
Fir	Yellow to Orange Brown	Close Coarse	Medium to Hard	High	Hard	Medium	Medium	Framing Millwork Plywood	
Ponderosa Pine	White with Brown Grain	Close Coarse	Medium	Medium	Medium	Poor	Low	Millwork Trim	Pine Odor
Sugar Pine	Creamy White	Close Fine	Soft	Low	Easy	Poor	Low	Patternmaking Millwork	Large Clear Pieces
Western White Pine	Brownish White	Close Medium	Soft to Medium	Low	Medium	Poor	Low	Millwork Trim	
Southern Yellow Pine	Yellow Brown	Close Coarse	Soft to Hard	High	Hard	Medium	Medium	Framing Plywood	Much Pitch
Redwood	Reddish Brown	Close Medium	Soft	Low	Easy	Poor	Very High	Exterior	Light Sapwood
Spruce	Cream to Tan	Close Medium	Medium	Medium	Medium	Poor	Low	Siding Subflooring	Spruce Odor

Fig. 1-4 Common kinds of softwood and their characteristics.

spruce. Not only are they the same color, but the grain pattern and weight are about the same. Only the most experienced workers are able to tell the difference.

It is possible to get some clues to identifying wood by studying literature, but the best way to learn the different kinds of wood is by working with them. Each time you handle a piece of wood, examine it. Look at the color and the grain; feel if it is heavy or light, if it is soft or hard; and smell it for a characteristic odor. Aromatic cedar, for instance, can always be identified by its pleasing moth-repelling odor, if for no other reason. After studying the characteristics of the wood, ask or otherwise find out the kind of wood you are holding, and remember it. In this manner, after a period of time, those kinds of wood that are used regularly on the job can be identified easily. Identification of kinds of wood that are seldom worked with can be accomplished in the same manner, but, of course, the process will take a little longer.

HARDWOODS

KIND	COLOR	GRAIN	HARDNESS	STRENGTH	WORK-ABILITY	ELASTICITY	DECAY RESISTANCE	USES	OTHER
Ash	Light Tan	Open Coarse	Hard	High	Hard	Very High	Low	Tool Handles Oars Baseball Bats	
Basswood	Creamy White	Close Fine	Soft	Low	Easy	Low	Low	Drawing Bds Veneer Core	Imparts No Taste Or Odor
Beech	Light Brown	Close Medium	Hard	High	Medium	Medium	Low	Food Containers Furniture	
Birch	Light Brown	Close Fine	Hard	High	Medium	Medium	Low	Furniture Veneers	
Cherry	Lt. Reddish Brown	Close Fine	Medium	High	Medium	High	Medium	Furniture	
Hickory	Light Tan	Open Medium	Hard	High	Hard	Very High	Low	Tool Handles	
Lauan	Lt. Reddish Brown	Open Medium	Soft	Low	Easy	Low	Low	Veneers Paneling	
Mahogany	Russet Brown	Open Fine	Medium	Medium	Excellent	Medium	High	Quality Furniture	
Maple	Light Tan	Close Medium	Hard	High	Hard	Medium	Low	Furniture Flooring	
Oak	Light Brown	Open Coarse	Hard	High	Hard	Very High	Medium	Flooring Boats	
Poplar	Greenish Yellow	Close Fine	Medium Soft	Medium Low	Easy	Low	Low	Furniture Veneer Core	
Teak	Honey	Open Medium	Medium	High	Excellent	High	Very High	Furniture Boat Trim	Heavy Oily
Walnut	Dark Brown	Open Fine	Medium	High	Excellent	High	High	High Quality Furniture	

Fig. 1-5 Common kinds of hardwood and their characteristics.

CHAPTER 2 LUMBER

Manufacture of Lumber

When logs arrive at the sawmill, the bark is removed first. Then a huge bandsaw slices the log into large planks, which are passed through a series of saws. The saws slice, edge, and trim them into various dimensions, and the pieces become **lumber.**

Once trimmed of all uneven edges, the lumber is stacked according to size and grade and taken outdoors where *sticking* takes place. Sticking is the process of restacking the lumber on small cross-sticks that allow air to circulate between the pieces. This air-seasoning process may take six months to two years due to the large amount of water found in lumber.

Following air-drying, the lumber is then dried in huge ovens. Once dry, the rough lumber is surfaced to standard sizes and shipped (Fig. 2-1).

The long, narrow surface of a piece of lumber is called its edge, the long, wide surface is termed its side, and its extremities are called ends. The distance across the edge is called its thickness, across its side is called its width, and from end to end is called its length (Fig. 2-2). The best-appearing side is called the face side, and the best-appearing edge is called the face edge.

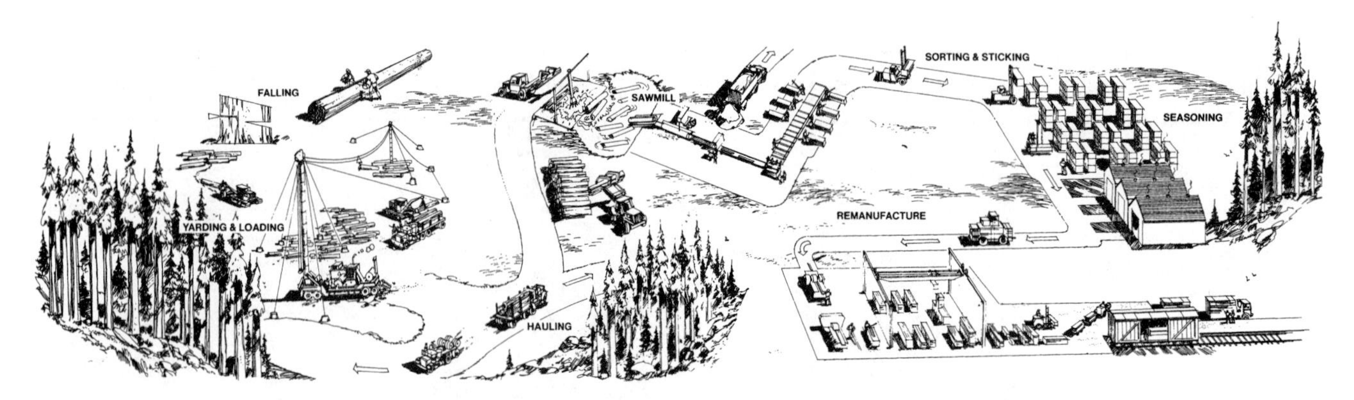

Fig. 2-1 Manufacture of lumber from forest to finished product. (*Courtesy of California Redwood Association*)

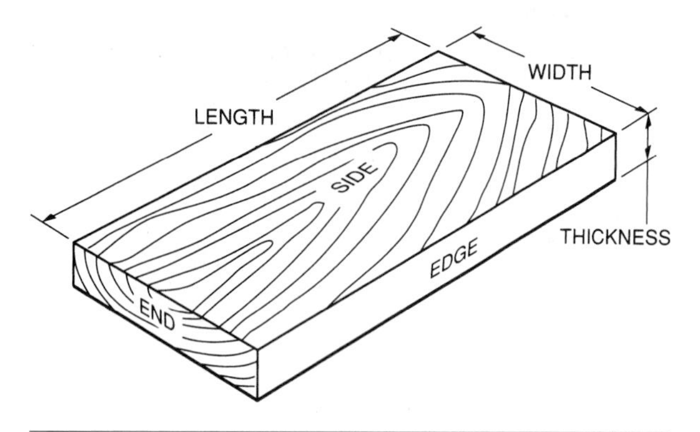

Fig. 2-2 Lumbar surfaces are distinguished by specific names.

In certain cases, the surfaces of lumber may acquire different names. For instance, the distance from top to bottom of a beam is called its depth, and the distance across its top or bottom may be called its width or its thickness. The length of posts or columns, when installed, may be called their height.

Plain-Sawed Lumber

A common way of cutting lumber is called the **plain-sawed** method, in which the log is cut tangent to the annular rings. This method produces a distinctive grain pattern on the wide surface (Fig. 2-3). This method of sawing is the least expensive and produces greater widths. However, plain-sawed lumber shrinks more during drying and warps easily. Plain-sawed lumber is sometimes called slash-sawed lumber.

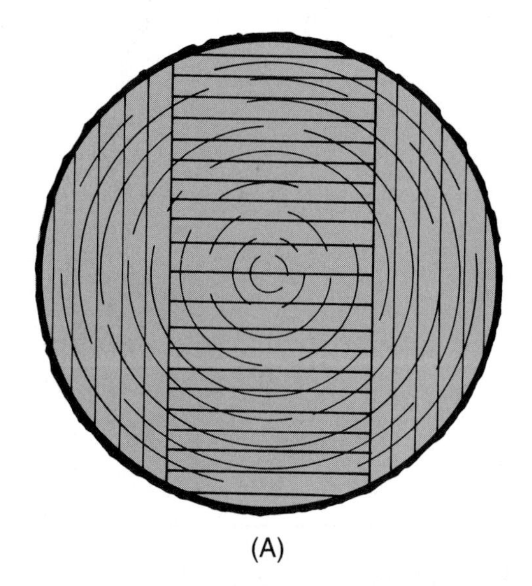

(A)

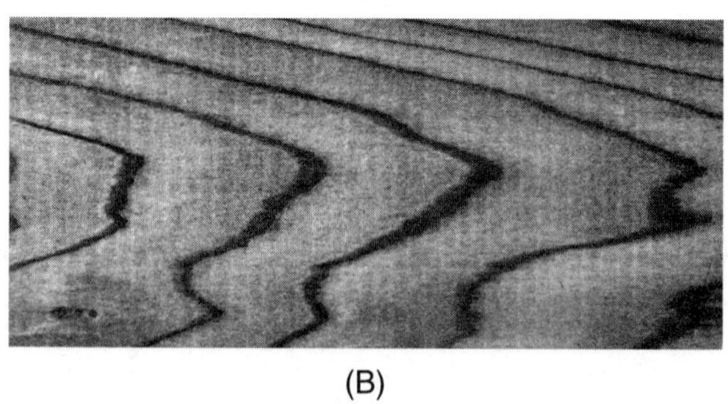

(B)

Fig. 2-3 A. Plain-sawed lumber; B. surface of plain-sawed lumber. (*Courtesy of California Redwood Association*)

Quarter-Sawed Lumber

Another method of cutting the log, called **quarter-sawing**, produces pieces in which the annular rings are at or almost at right angles to the wide surface. **Quarter-sawed** lumber has less tendency to warp and shrinks less and more evenly when dried. This type of lumber is durable because the wear is on the edge of the annular rings. Quarter-sawed lumber is frequently used for flooring.

A distinctive and desirable grain pattern is produced in some wood, such as oak, because the lumber is sawed along the length of the medullary rays. Quarter-sawed lumber is sometimes called vertical-grain or edge-grain (Fig. 2-4).

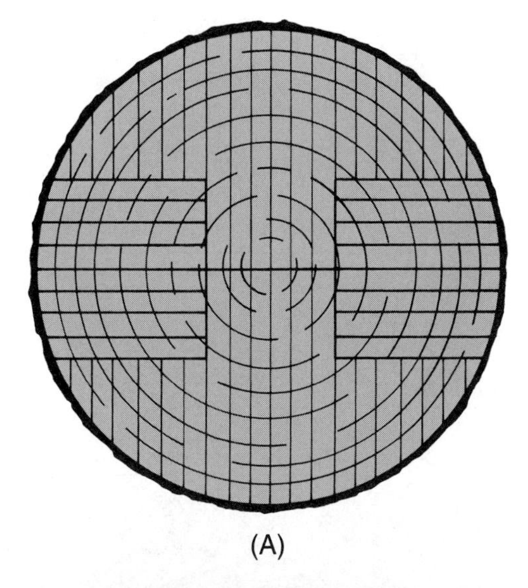

(A)

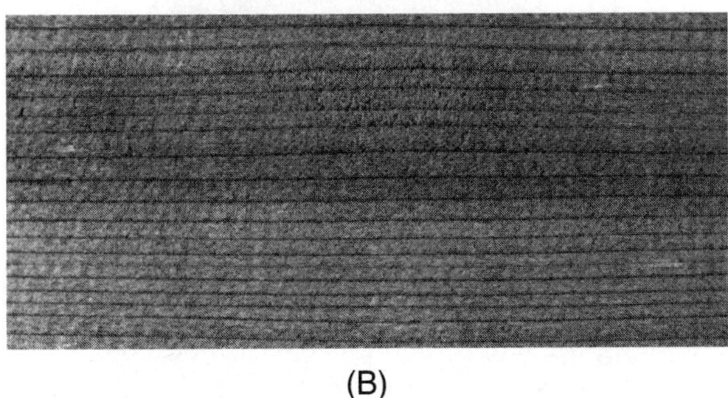

(B)

Fig. 2-4 A. Quarter-sawed lumber; B. Surface of quarter-sawed lumber. (*Courtesy of California Redwood Association*)

Combination Sawing

Most logs are cut into a combination of plain-sawed and quarter-sawed lumber. With computers and laser-guided equipment, the **sawyer** determines how to cut the log with as little waste as possible in the shortest amount of time to get the desired amount and kinds of lumber (Fig. 2-5).

Fig. 2-5 Combination-sawed lumber. (*Courtesy of Western Wood Products Association*)

Moisture Content and Shrinkage

When a tree is first cut down, it contains a great amount of water. Lumber, when first cut from the log, is called *green lumber* and is very heavy because most of its weight is water. A piece 2 inches thick, 6 inches wide, and 10 feet long may contain as much as 4 1/4 gallons of water weighing about 35 pounds (Fig. 2-6).

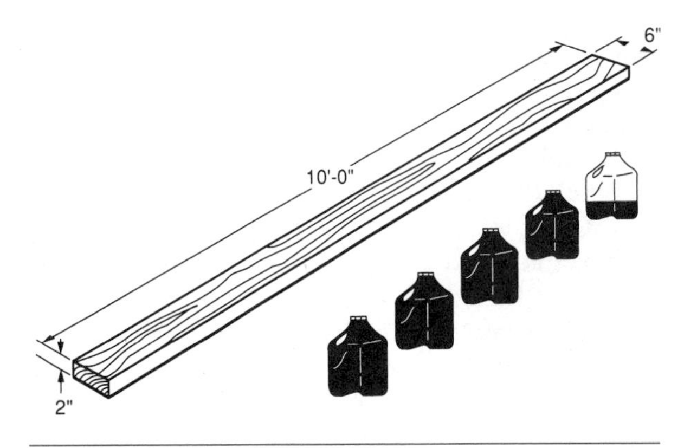

Fig. 2-6 Green lumber contains a large amount of water.

Green lumber should not be used in construction because it eventually will dry to the same moisture content as the surrounding air. As green lumber dries, it shrinks considerably and unequally because of the large amount of water that leaves it. When it shrinks, it usually warps, depending on the way it was cut from the log. The use of green lumber in construction results in cracked ceilings and walls, squeaking floors, sticking doors, and many other problems caused by shrinking and warping of the lumber as it dries. Therefore, lumber must be dried to a suitable degree before it can be surfaced and used.

Green lumber is also subject to decay. Decay is caused by *fungi*, low forms of plant life that feed on wood. This decay is commonly known as dry rot because it usually is not discovered until the lumber has dried. Decay will not occur unless wood moisture content is in excess of 19 percent. Wood construction maintained at moisture content of less than 20 percent will not decay. It is important that lumber with an excess amount of moisture be exposed to an environment that will allow the moisture to evaporate. Seasoned lumber must be protected to prevent the entrance of moisture that allows the growth of fungi and decay of the wood. (The subject of fungi and wood decay is discussed in detail in Chapter 33.)

Moisture Content

The **moisture content** (M.C.) of lumber is expressed as a percentage, and indicates how much of the weight of a wood sample is actually water. It is derived by determining the difference in the weight of the sample before and after it has been oven-dried and dividing that number by the dry weight. For example, if a wood sample weighs 16 ounces prior to baking and 13 ounces after baking, we assume that there were (16–13) 3 ounces of water in the wood. To determine the moisture content of the sample before baking, we divide the evaporated water weight by the dry weight and multiply by a hundred: $(3 \div 13) \times 100 = 23.0769$ percent. Thus, the moisture content of the sample before drying was roughly 23 percent. Lumber used for framing and exterior finish should have an M.C. that does not exceed 19 percent, preferably 15 percent. For interior finish, an M.C. of 10 to 12 percent is recommended.

Green lumber may have water in the hollow part of the wood cells as well as in the cell walls. When wood starts to dry, the water in the cell cavities, called *free water,* is first removed. When all of the free water is gone, the wood has reached the **fiber-saturation point;** about a third of the remaining weight is water. No noticeable shrinkage of wood takes place up to this point.

As wood continues to dry, the water in the walls of the cells is removed and the wood starts to shrink. It shrinks considerably from the size at its fiber-saturation point of approximately 30 percent M.C. to the desired percent of M.C. suitable for construction. Lumber at this stage is called *dry* or **seasoned** and now must be prevented from getting wet.

It is important to understand not only that wood shrinks as it dries, but also how it shrinks. So little shrinkage occurs along the length of lumber that it is not considered. Most shrinkage takes place in the direction of the annular rings, with more shrinkage taking place on the longer rings. When viewing plain-sawed lumber in cross-section (looking at the end), it can be seen that the piece warps as it shrinks because of the unequal length of the annular rings. A cross-section of quarter-sawed lumber shows annular rings of equal length. Therefore, although the piece shrinks somewhat, it shrinks evenly with no warp (Fig. 2-7). Wood warps as it dries according to the way it was cut from the tree. Cross-section views of the annular rings are different along the length of a piece of lumber; therefore, various kinds of warp result when lumber dries and shrinks.

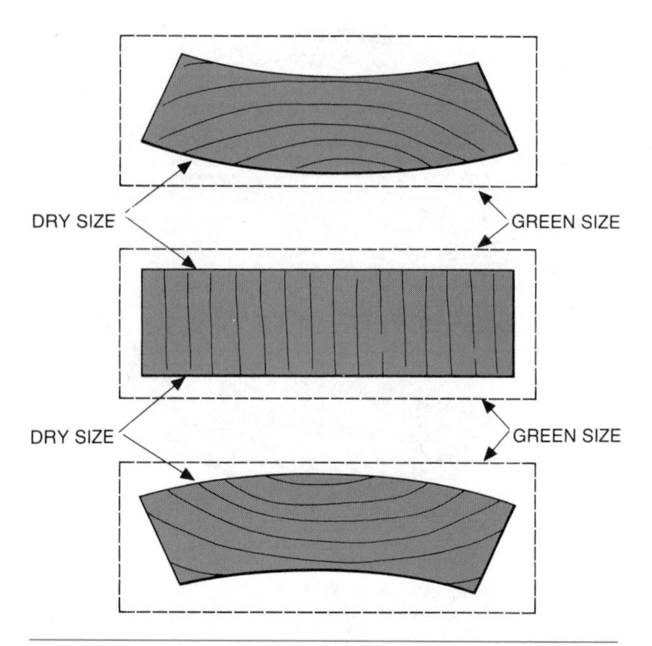

Fig. 2-7 Lumber shrinks in the direction of annular rings.

When the moisture content of lumber reaches that of the surrounding air (about 10 to 12 percent M.C.), it is at **equilibrium moisture content.** At this point, lumber shrinks or swells only slightly with changes in the moisture content of the air. Realizing that lumber undergoes certain changes when moisture is absorbed or lost, the experienced carpenter uses techniques to deal with this characteristic of wood (Fig. 2-8).

Drying Lumber

Lumber is either **air-dried, kiln-dried,** or a combination of both. In air-drying the lumber is stacked in piles with spacers, which are called stickers, placed between each layer to permit air to circulate through the pile (Fig. 2-9). Kiln-dried lumber is stacked in the same manner but is dried in buildings, called kilns, which are like huge ovens (Fig. 2-10).

Under carefully controlled temperatures, humidity, and air circulation, the removal of moisture takes place. First, the humidity level is raised and the temperature is increased; the humidity is then gradually decreased. Kiln-drying has the advantage of drying lumber in a shorter period of time, but is more expensive than air-drying.

The recommended moisture content for lumber to be used for exterior finish at the time of installation is 12 percent, except in very dry climates,

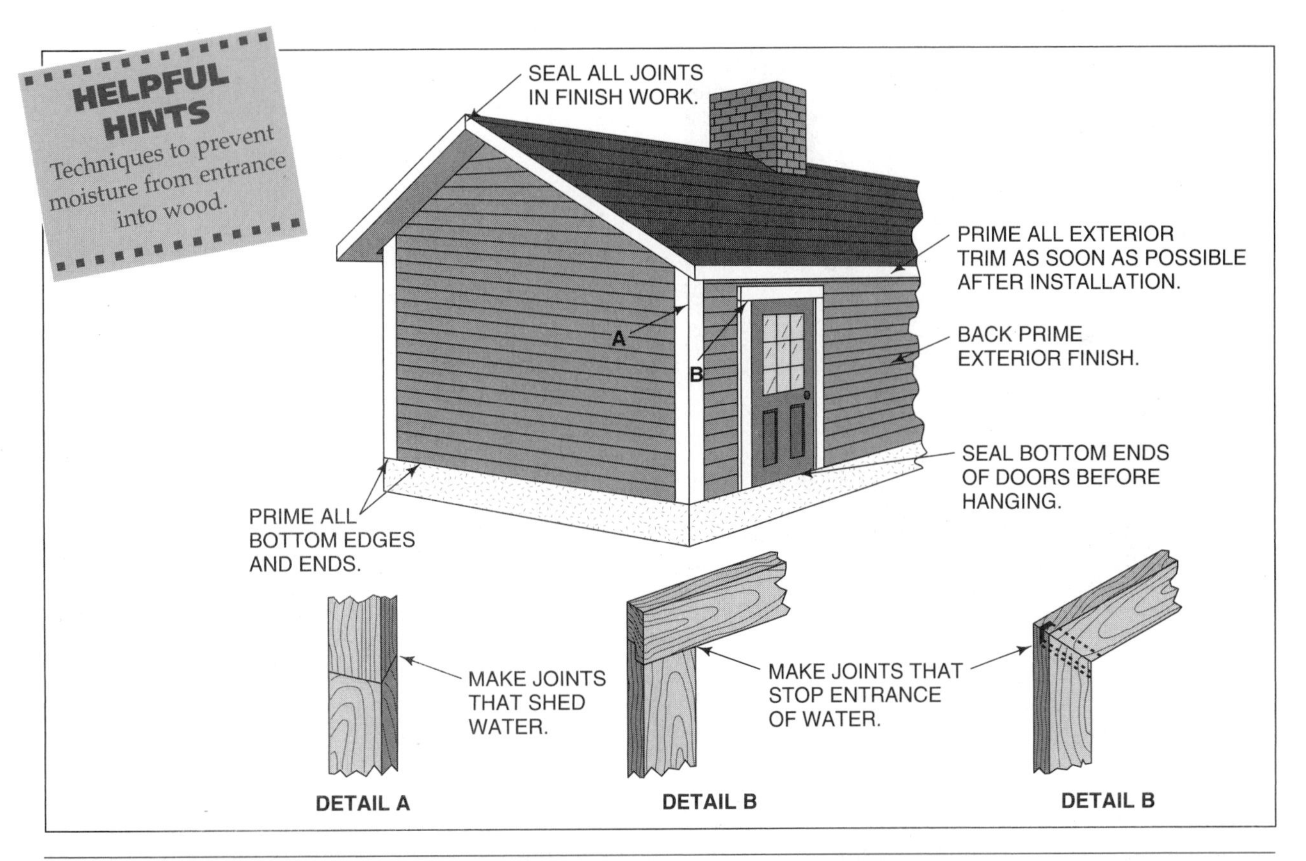

Fig. 2-8 Techniques to prevent water from getting in behind the wood surface.

Fig. 2-9 Drying lumber in the air. (*Courtesy of Western Wood Products Association*)

Fig. 2-10 Drying lumber in a kiln. (*Courtesy of Western Wood Products Association*)

where 9 percent is recommended. Lumber with low moisture content (8 to 10 percent) is necessary for interior trim and cabinet work.

The moisture content of lumber is determined by the use of a **moisture meter** (Fig. 2-11). Points on the ends of the wires of the meter are driven into the wood, and the moisture content is read off the meter.

Experienced workers know when lumber is green (because it is much heavier than dry lumber), and they can estimate fairly accurately the moisture content of lumber simply by lifting it.

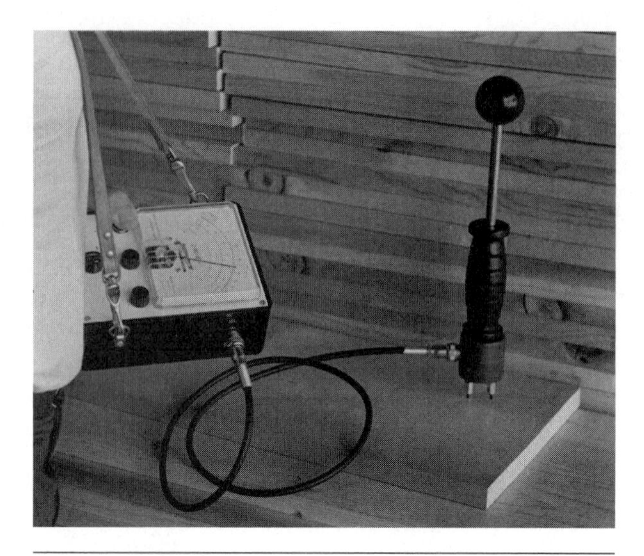

Fig. 2-11 Moisture meter. (*Courtesy of Moisture Register Company*)

Lumber is brought to the planer mill, where it is straightened, smoothed, and uniformly sized. This process may be done when the lumber is dry or green. Most construction lumber is surfaced on four sides (S4S) to standard thicknesses and widths. Some may be surfaced on only two sides (S2S) to required thicknesses.

Lumber Storage

Lumber should be delivered to the job site so materials are accessible in the proper sequence; that is, those that are to be used first are on the top and those to be used last are on the bottom.

Lumber stored at the job site should be adequately protected from moisture and other hazards. A common practice that must be avoided is placing unprotected lumber directly on the ground. Use short lengths of lumber running at right angles to the length of the pile and spaced close enough to keep the pile from sagging and coming into contact with the ground. The base on which the lumber is to be placed should be fairly level to keep the pile from falling over.

Protect the lumber with a tarp or other type of cover. Leave enough room at the bottom and top of the pile for circulation of air. A cover that reaches to the ground will act like a greenhouse, trapping ground moisture within the stack.

Keep the piles in good order. Lumber spread out in a disorderly fashion can cause accidents as well as subject the lumber to stresses that may cause warping.

Lumber Defects

A defect in lumber is any fault that detracts from its appearance, function, or strength. One type of defect is called a **warp.** Warps are caused by, among other things, drying lumber too fast, careless handling and storage, or surfacing the lumber before it is thoroughly dry. Warps are classified as **crooks, bows, cups,** and **twists** (Fig. 2-12).

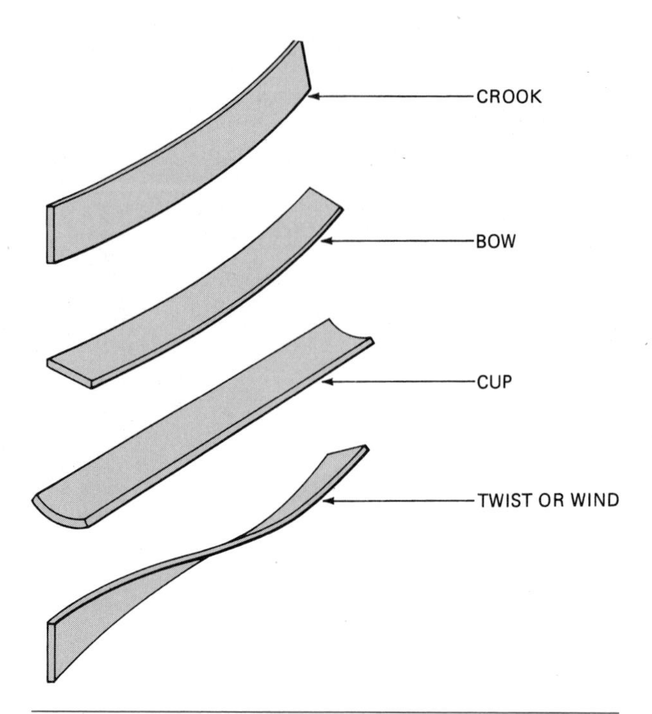

Fig. 2-12 Kinds of warp.

Splits in the end of lumber running lengthwise and across the annular rings are called *checks* (Fig. 2-13). Checks are caused by faster drying of the end than of the rest of the stock. Checks can be prevented to a degree by sealing the ends of lumber with paint, wax, or other material during the drying period. Cracks that run parallel to and between the annular rings are called **shakes** and may be caused by weather or other damage to the tree.

The *pith* is the spongy center of the tree. It contains the youngest portion of the lumber, called **juvenile wood.** Juvenile wood is the portion of wood that contains the first seven to fifteen growth rings. The wood cells in this region are not well aligned and are therefore unstable when they dry. They shrink in different directions, causing internal stresses. If a board has a high percentage of juvenile wood, it will warp and twist in remarkable ways. **Knots** are cross-sections of branches in the trunk of the tree. Knots are not necessarily defects unless they are loose or weaken the piece. **Pitch pockets** are small cavities that hold pitch, which sometimes oozes out. A **wane** is bark on the edge of lumber or the surface from which the bark has fallen. *Pecky* wood has small grooves or channels running with the grain. This is common in cypress. Pecky cypress is often used as an interior wall paneling when that effect is desired. Some other defects are *stains, decay,* and *wormholes.*

Fig. 2-13 A severe check in the end of a piece of lumber.

Lumber Grades and Sizes

Lumber grades and sizes are established by wood products associations of which many wood mills are members. Member wood mills are closely supervised by the associations to assure that standards are maintained. The grade stamp of the association is assurance that lumber grade standards have been met.

Member mills use the association grade stamp to indicate strict quality control. A typical grade stamp is shown in Figure 2-14 and shows the association trademark, the mill number, the lumber grade, the species of wood, and whether the wood was green or dry when it was planed.

Softwood Grades

The largest softwood association of lumber manufacturers is the Western Wood Products Association (WWPA), which grades lumber in three categories: **boards** (under 2 inches thick), *dimension* (2 to 4 inches thick), and **timbers** (5 inches and thicker). The board group is divided into boards, sheathing, and form lumber. The dimension group is divided into light framing, studs, structural light framing, and structural joists and planks. Timbers are divided into beams and stringers. The three main categories are further classified according to strength and appearance as shown in Figure 2-15.

Hardwood Grades

Hardwood grades are established by the National Hardwood Lumber Association. **Firsts and seconds** *(FAS)* is the best grade of hardwood and must yield about 85 percent clear cutting. Each piece must be at least 6 inches wide and 8 feet long. The next best grade is called *select.* For this, the minimum width is 4 inches and the minimum length is 6 feet. **No. 1 common** allows even narrower widths and shorter lengths, with about 65 percent clear cutting.

Lumber Sizes

Rough lumber that comes directly from the sawmill is close in size to what it is called, nominal size. There are slight variations to nominal size because of the heavy machinery used to cut the log into lumber. When rough lumber is planed, it is reduced in thickness and width to standard and uniform sizes. Its nominal size does not change even though the actual size does. Therefore, when *dressed* (surfaced), although a piece may be called a 2 × 4, its actual size is 1 1/2 inches (38 mm) by 3 1/2 inches (89 mm). The same applies to all surfaced lumber; the nominal size (what it is called) and the actual size are not the same. Figure 2-16 shows the standard nominal and dressed sizes of softwood lumber,

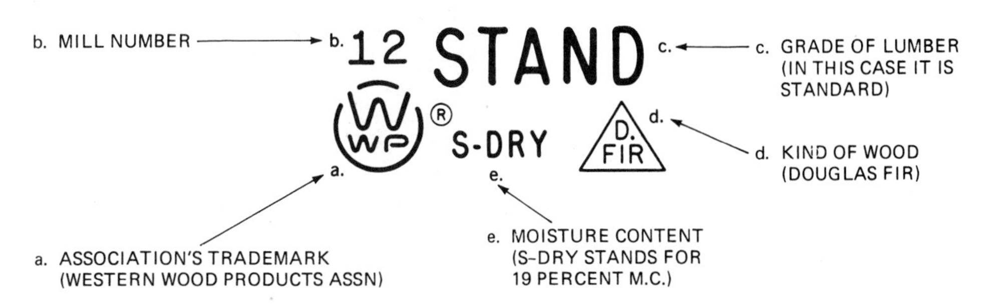

Fig. 2-14 Typical softwood lumber grade stamp. (*Courtesy of Western Wood Products Association*)

Grade Selector Charts

Boards

APPEARANCE GRADES	**SELECTS**	B & BETTER (IWP—SUPREME)* C SELECT (IWP—CHOICE) D SELECT (IWP—QUALITY)
	FINISH	SUPERIOR PRIME E
	PANELING	CLEAR (ANY SELECT OR FINISH GRADE) NO. 2 COMMON SELECTED FOR KNOTTY PANELING NO. 3 COMMON SELECTED FOR KNOTTY PANELING
	SIDING (BEVEL, BUNGALOW)	SUPERIOR PRIME
BOARDS SHEATHING & FORM LUMBER		NO. 1 COMMON (IWP—COLONIAL) NO. 2 COMMON (IWP—STERLING) NO. 3 COMMON (IWP—STANDARD) NO. 4 COMMON (IWP—UTILITY) NO. 5 COMMON (IWP—INDUSTRIAL)
		ALTERNATE BOARD GRADES SELECT MERCHANTABLE CONSTRUCTION STANDARD UTILITY ECONOMY

SPECIFICATION CHECK LIST

- ☐ Grades listed in order of quality.
- ☐ Include all species suited to project.
- ☐ Specify lowest grade that will satisfy job requirement.
- ☐ Specify surface texture desired.
- ☐ Specify moisture content suited to project.
- ☐ Specify (WWP) grade stamp. For finish and exposed pieces, specify stamp on back or ends.

Western Red Cedar

FINISH PANELING AND CEILING	CLEAR HEART A B
BEVEL SIDING	CLEAR — V.G. HEART A — BEVEL SIDING B — BEVEL SIDING C — BEVEL SIDING

*Idaho White Pine carries its own comparable grade designations.

Dimension/All Species 2″ to 4″ thick (also applies to finger-jointed stock)

LIGHT FRAMING 2″ to 4″ Thick 2″ to 4″ Wide	CONSTRUCTION STANDARD UTILITY	This category for use where high strength values are **NOT** required; such as studs, plates, sills, cripples, blocking, etc.
STUDS 2″ to 4″ Thick 2″ and Wider	STUD	An optional all-purpose grade. Characteristics affecting strength and stiffness values are limited so that the "Stud" grade is suitable for vertical framing members, including load bearing walls.
STRUCTURAL LIGHT FRAMING 2″ to 4″ Thick 2″ to 4″ Wide	SELECT STRUCTURAL #1&Btr.* NO. 1 NO. 2 NO. 3	These grades are designed to fit those engineering applications where higher bending strength ratios are needed in light framing sizes. Typical uses would be for trusses, concrete pier wall forms, etc.
STRUCTURAL JOISTS & PLANKS 2″ to 4″ Thick 5″ and Wider	SELECT STRUCTURAL #1&Btr.* NO. 1 NO. 2 NO. 3	These grades are designed especially to fit in engineering applications for lumber five inches and wider, such as joists, rafters and general framing uses.

*Douglas fir/Larch, Hem-Fir only (DF–L)

Timbers 5″ and thicker

BEAMS & STRINGERS 5″ and thicker Width more than 2″ greater than thickness	SELECT STRUCTURAL NO 1 NO. 2** NO. 3**
POSTS & TIMBERS 5″ x 5″ and larger Width not more than 2″ greater than thickness	SELECT STRUCTURAL NO. 1 NO. 2** NO. 3**

**Design values are not assigned.

Fig. 2-15 Softwood lumber grades. (*Courtesy of Western Wood Products Association*)

Standard Lumber Sizes / Nominal, Dressed, Based on WWPA Rules

Product	Description	Nominal Size: Thickness In.	Nominal Size: Width In.	Dressed Dimensions: Thicknesses and Widths In. Surfaced Dry	Dressed Dimensions: Thicknesses and Widths In. Surfaced Unseasoned	Lengths Ft.
DIMENSION	S4S ...... Other surface combinations are available. See "Abbreviations" below.	2 3 4	2 3 4 5 6 8 10 12 Over 12	1-1/2 2-1/2 3-1/2 4-1/2 5-1/2 7-1/2 9-1/2 11-1/2 3/4 off nominal	1-9/16 2-9/16 3-9/16 4-5/8 5-5/8 7-1/2 9-1/2 11-1/2 Off 1/2	6′ and longer, generally shipped in multiples of 2′
SCAFFOLD PLANK	Rough Full Sawn or S4S ...... (Usually shipped unseasoned)	1¼ & Thicker	8 and Wider	If Dressed refer to "DIMENSION" sizes		6′ and longer, generally shipped in multiples of 2′
TIMBERS	Rough or S4S ...... (Shipped unseasoned)	5 and Larger		1/2″ off nominal (S4S) See 3.20 of WWPA Grading Rules for Rough		6′ and longer, generally shipped in multiples of 2′
		Nominal Size		**Dressed Dimensions**		
		Thickness In.	**Width In.**	**Thickness In.**	**Width In.**	**Lengths Ft.**
DECKING	2″ Single T&G	2	5 6 8 10 12	1½	4 5 6¾ 8¾ 10¾	6′ and longer, generally shipped in multiples of 2′
	3″ and 4″ Double T&G	3 4	6	2½ 3½	5¼	
FLOORING	(D & M), (S2S & CM)	3/8 1/2 5/8 1 1¼ 1½	2 3 4 5 6	5/16 7/16 9/16 3/4 1 1¼	1⅛ 2⅛ 3⅛ 4⅛ 5⅛	4′ and longer, generally shipped in multiples of 2′
CEILING AND PARTITION	(S2S & CM)	3/8 1/2 5/8 3/4	3 4 5 6	5/16 7/16 9/16 11/16	2⅛ 3⅛ 4⅛ 5⅛	4′ and longer, generally shipped in multiples of 2′
FACTORY AND SHOP LUMBER	S2S	1 (4/4) 1¼ (5/4) 1½ (6/4) 1¾ (7/4) 2 (8/4) 2½ (10/4) 3 (12/4) 4 (16/4)	5″ and wider (except 4″ and wider in 4/4 No. 1 Shop and 4/4 No. 2 Shop, and 2″ and wider in 5/4 & Thicker No. 3 Shop)	3/4 (4/4) 1 5/32 (5/4) 1 13/32 (6/4) 1 19/32 (7/4) 1 13/16 (8/4) 2⅜ (10/4) 2¾ (12/4) 3¾ (16/4)	Usually sold random width	6′ and longer, generally shipped in multiples of 2′

ABBREVIATIONS

Abbreviated descriptions appearing in the size table are explained below.

S1S — Surfaced one side.
S2S — Surfaced two sides.
S4S — Surfaced four sides.
S1S1E — Surfaced one side, one edge.
S1S2E — Surfaced one side, two edges.
CM — Center matched.
D & M — Dressed and matched.
T & G — Tongue and grooved.
Rough Full Sawn — Unsurfaced green lumber cut to full specified size.

Product Classification

	thickness in.	width in.		thickness in.	width in.
board lumber	1″	2″ or more	**beams & stringers**	5″ and thicker	more than 2″ greater than thickness
light framing	2″ to 4″	2″ to 4″	**posts & timbers**	5″ x 5″ and larger	not more than 2″ greater than thickness
studs	2″ to 4″	2″ to 6″ 10′ and shorter	**decking**	2″ to 4″	4″ to 12″ wide
structural light framing	2″ to 4″	2″ to 4″	**siding**	thickness expressed by dimension of butt edge	
structural joists & planks	2″ to 4″	5″ and wider	**mouldings**	size at thickest and widest points	

Lengths of lumber generally are 6 feet and longer in multiples of 2′

Nailing Diagram

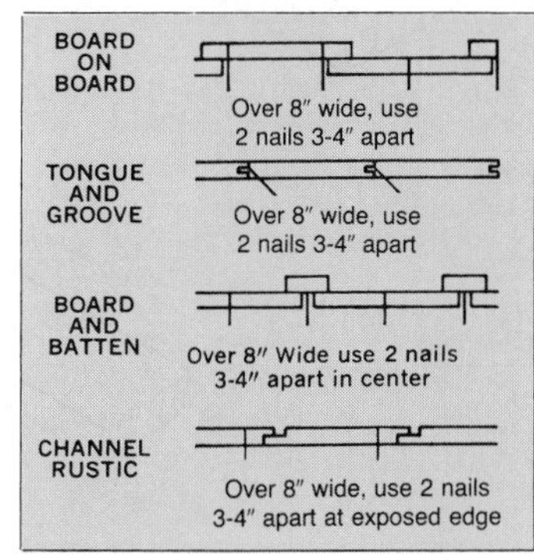

Fig. 2-16 Softwood lumber sizes. (*Courtesy of Western Wood Products Association*)

based on WWPA rules. Hardwood lumber is usually purchased in the rough and straightened, smoothed, and sized as needed.

Board Measure

Softwood lumber is usually purchased by specifying the number of pieces—thickness(") × width(") × length(') (i.e., 35–2" × 6" × 16')—in addition to the grade. Often, when no particular lengths are required, the thickness, width, and total number of linear feet (length in feet) are ordered. The length of the pieces then may vary and are called *random* lengths. Another method of purchasing softwood lumber is by specifying the thickness, width, and total number of **board feet.** Lumber purchased in this manner may also contain random lengths.

Hardwood lumber is purchased by specifying the grade, thickness, and total number of board feet. Large quantities of both softwood and hardwood lumber are priced and sold by the board foot.

A board foot is a measure of lumber. It is equivalent to a piece 1 inch thick, 12 inches wide, and 1 foot long. A piece of lumber 1 inch thick and 6 inches wide must be 2 feet long to equal 1 board foot. A piece 2 inches thick has twice as many board feet as a piece 1 inch thick of the same width and length (Fig. 2-17).

To calculate the number of board feet, use the formula: number of pieces × thickness in inches × width in inches × length in feet ÷ 12 = number of board feet. For example: 16 pieces × 2 inches × 4 inches × 8 feet ÷ 12 = 85 1/3 board feet.

$$\frac{\overset{4}{\cancel{16}} \times 2 \times 4 \times 8}{\underset{3}{\cancel{12}}} = \frac{256}{3} = 85\ 1/3 \text{ bd ft}$$

When dealing with dressed lumber, use the nominal dimensions in the calculations, not the actual dimensions.

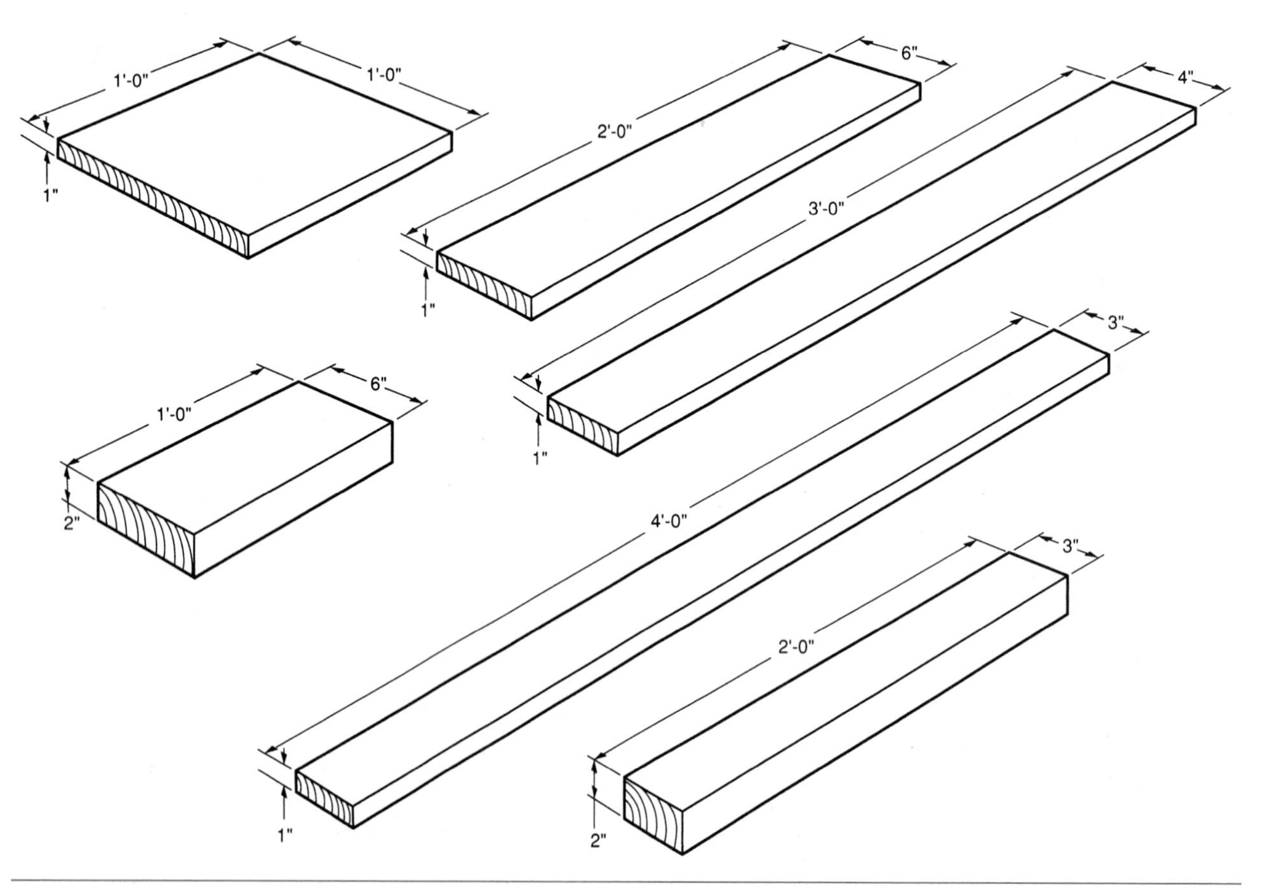

Fig. 2-17 Each piece contains one board foot.

Review Questions

Select the most appropriate answer.

1. The center of a tree is called the
a. heartwood. c. pith.
b. lignin. d. sapwood.

2. New wood cells of a tree are formed in the
a. heartwood. c. medullary rays.
b. bark. d. cambium layer.

3. Tree growth is faster in the
a. spring. c. fall.
b. summer. d. winter.

4. Quarter-sawed lumber is sometimes called
a. tangent-grained. c. vertical-grained.
b. slash-sawed. d. plain-sawed.

5. Lumber is called "green" when
a. it is stained by fungi.
b. the tree is still standing.
c. it is first cut from the log.
d. it has reached equilibrium moisture content.

6. Wood will not decay unless its moisture content is in excess of
a. 15 percent. c. 25 percent
b. 19 percent. d. 30 percent.

7. When all of the free water in the cell cavities of wood is removed and before water is removed from cell walls, lumber is at what is called
a. fiber-saturation point.
b. 30 percent moisture content.
c. equilibrium moisture content.
d. shrinkage commencement.

8. Air-dried lumber cannot be dried less than
a. 8 to 10 percent moisture content.
b. equilibrium moisture content.
c. 25 to 30 percent moisture content.
d. its fiber saturation point

9. A commonly used and abundant softwood is
a. ash. c. basswood.
b. fir. d birch.

10. One of the woods extremely resistant to decay is
a. pine. c. cypress.
b. spruce. d. hemlock.

BUILDING FOR SUCCESS

Why Construction Education?

You have probably made a decision to pursue a career in the carpentry field or at least have sensed the need to become better informed about a particular carpentry process. Regardless of the reason, you have taken it upon yourself to become better educated. This is a process that will never stop during your many years in the construction industry. As you set goals for your career, the process of obtaining those goals will undoubtedly involve some form of education.

Assuming that your interests have led you in the direction of a carpentry career, how have you decided to pursue that career? Planning for your career can begin at any time during your school years. Today many people change careers at some point in their lives. Some job changes may be made due to economic influences, new interests, or other opportunities where job prospects appear to be very good. Many times job changes are intentionally planned in order to gain the variety of work experiences necessary to reach an ultimate career goal.

There are a few very good publications available to help you undertake an intensive study of the specific jobs available in the construction arena. A full array of job descriptions can be found in resources such as the occupational outlooks and guidebooks at your school or local library. These guidebooks are quite complete and will provide you with the initial information to begin a more detailed research of any occupation. Your instructor or librarian can assist you in locating these guidebooks. The federal and state governments publish literature that addresses the many state or local job opportunities for each area of the country. Some of these informative publications are available free of charge; you may have to pay a small fee for others.

The planning process must begin with your sincere desire to investigate in depth the general career area. You must then begin to eliminate jobs that do not seem appealing. Once you have decided on one or two careers in which you have the most interest, you may intensify and focus your research. By looking at the exact job tasks that are conducted regularly in that line of work, you will be able to make the best decisions for your education.

Today's construction careers require more education and training than in past years. The competition for quality construction jobs is increasing rapidly. The economy, both nationwide and global, will have a tremendous impact on the progress and growth of every industry. Since the construction industry is the pacesetter for much of the nation during both prosperous economic times and recession, we can use it as one of the gauges for the country's economic wellness.

There are various ways to gain the education necessary to enter the career of your choice. Perhaps by using this text, you have entered one of those avenues. The specific job you have selected will reveal the type and amount of informal and formal education required for an entry-level job. If you are pursuing a professional career such as architect, engineer, or construction manager, a four-year degree may be the best avenue. If your concentration is more on the technical side, several options for entry-level jobs exist. One popular way to gain the technical skills required of carpenters, for example, is by way of the local carpenters' union or a professional construction organization. The Associated General Contractors, Associated Builders and Contractors, and the National Association of Home Builders are examples of associations that have excellent construction education opportunities. All of these professional associations have state or local chapters listed in most metropolitan area phone directories.

Other educational programs are offered through various post-secondary institutions such as trade schools, junior colleges, and community colleges. In each state and province there are institutions that offer a vast array of programs that prepare technicians for construction careers. In most cases these schools prepare their graduates for an entry-level job in two years or less. This option allows the graduates a very quick return on their education investment. Another advantage of institutional construction training programs across the country is that many offer some form of continuing education. There will always be a need to continue gaining new skills and knowledge for almost every trade or occupation where high technology has an impact.

Quality construction education programs must be interwoven with the basic academic curriculum. There can be no division between academics and the vocational/technical realm. The tie has become so necessary that it cannot be neglected in a quality construction curriculum. The basic skills that business and industry require are dependent on the proper blend of academic and specific technical courses. From now on carpenters and other skilled workers will be expected to perform not only technical tasks on the job, but leadership, business, and interpersonal communication tasks as well.

With the fast pace of computer technology, and the introduction of new products and assembly systems, the need for updating will continue. The time is past when only basic technical training and preparation for work will be adequate for a lifetime. Our global society has dictated that we should be fully committed to ongoing education and training programs not only to advance in our employment, but simply to maintain our jobs.

The need for continued education may be a good topic for discussion during the interview process. If the employer sees that continuing education is important for quality growth, he will be more willing to assist people in finding the time and financial resources to pursue additional training once they are employed.

As you select a career area and begin the necessary preparation for becoming gainfully employed, keep in mind the fact that the education process has only started and that it will continue at least until retirement. Now is the best time to enter into or continue your education for your desired career. Very seldom do the opportunities become more appropriate, convenient, or economical than at the present. Once you establish a family and make financial commitments, you will need to become the best trained or prepared worker possible.

The competition will continue to increase for better-paying jobs. There will be no reduction of necessary requirements such as technical skills, special abilities, knowledge, or other unique qualifications for the jobs of the twenty-first century. "This is the first day of the rest of your life." What decisions do you have to make to properly pursue a technical career area? Have you approached knowledgeable people for answers to your questions? Have you arrived at the best educational avenue for gaining the necessary entry-level job skills? By answering these questions, you will be well on your way to a more productive and meaningful career.

FOCUS QUESTIONS: For individual or group discussion

1. Do you know what your career goals are? Can you identify how you are going to reach those goals?
2. What might be some of the considerations that must be part of making a career decision?
3. How does education play a part in helping you attain your career goals?
4. Who do you think could be potential resources to work with as you plan your career? Are you acquainted with knowledgeable people that represent your area of interest?

UNIT

2

Engineered Panels

Growing concern over efficient use of forest resources led to the development of reconstituted wood in the form of large sheets, commonly called panels or boards. Since the advent of plywood, the first of these products, many more are now available for use in construction. The carpenter must know the different kinds, sizes, and recommended applications to use them correctly and to the best advantage.

OBJECTIVES

After completing this unit, the student should be able to describe the composition, kinds, sizes, grades, and several uses of:

- plywood
- oriented strand board
- composite panels
- particleboard
- hardboard
- medium-density fiberboard
- softboard

UNIT CONTENTS

CHAPTER 3 RATED PLYWOOD AND PANELS

The term *engineered panels* refers to man-made products in the form of large reconstituted wood sheets, sometimes called **panels** or **boards.** In some cases, the tree has been taken apart and its contents have been redistributed into sheet or panel form. The panels are widely used in the construction industry. They are also used in the aircraft, automobile, and boat-building industries, as well as in the making of road signs, furniture, and cabinets.

With the use of engineered panels, construction progresses at a faster rate because a greater area is covered in a shorter period of time (Fig. 3-1). These panels, in certain cases, present a more attractive appearance and give more protection to a surface than does solid lumber. It is important to know the kinds and uses of various engineered panels in order to use them to the best advantage.

APA-Rated Panels

Many mills belong to associations that inspect, test, and allow mills to stamp the product to certify that it conforms to government and industrial standards. The grade stamp assures the consumer that the product has met the rigid quality and performance requirements of the association.

The trademarks of the largest association of this type, the American Plywood Association (APA), appear only on products manufactured by APA member mills (Fig. 3-2). This association is concerned not only with quality supervision and testing of **plywood** (cross-laminated wood veneer) but also of *composites* (veneer faces bonded to reconstituted wood cores) and nonveneered panels, commonly called *oriented strand board* (Fig. 3-3).

Plywood

One of the most extensively used engineered panels is plywood. Plywood is a sandwich of wood. Most plywood panels are made up of sheets of veneer (thin pieces) called *plies.* These plies, arranged in layers, are bonded under pressure with glue to form

Fig. 3-1 Sheet material covers a greater area in a shorter period of time. (*Courtesy of American Plywood Association*)

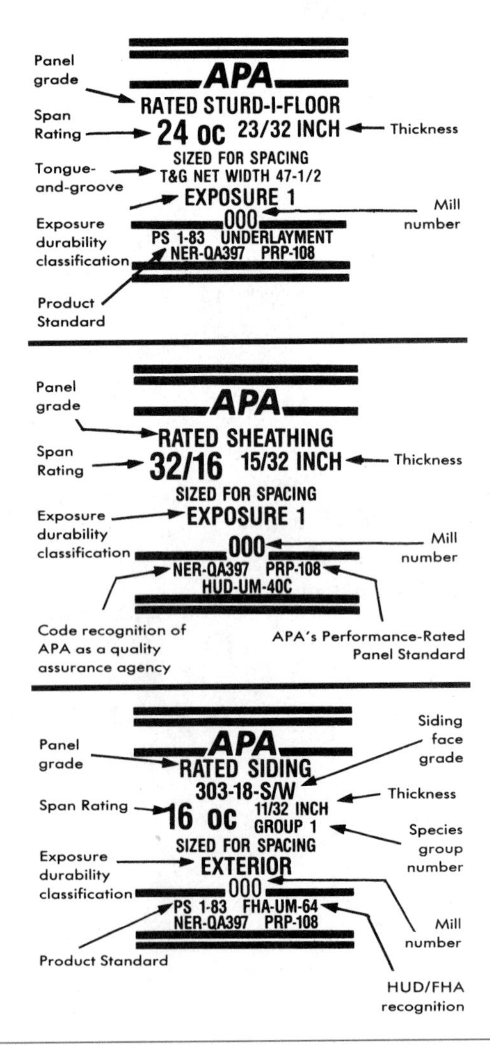

Fig. 3-2 The grade stamp is assurance of a high-quality, performance-rated panel. (*Courtesy of American Plywood Association*)

Fig. 3-3 APA performance-rated panels. (*Courtesy of American Plywood Association*)

a very strong panel. The plies are glued together so that the grain of each layer is at right angles to the next one. This cross-graining results in a sheet that is as strong as or stronger than the wood it is made from. Plywood usually contains an odd number of layers so that the face grain on both sides of the sheet runs in the direction of the long dimension of the panel (Fig. 3-4). Softwood plywood is commonly made with three, five, or seven layers. Because of its construction, plywood is more stable with changes of humidity and is resistant to shrinking and swelling.

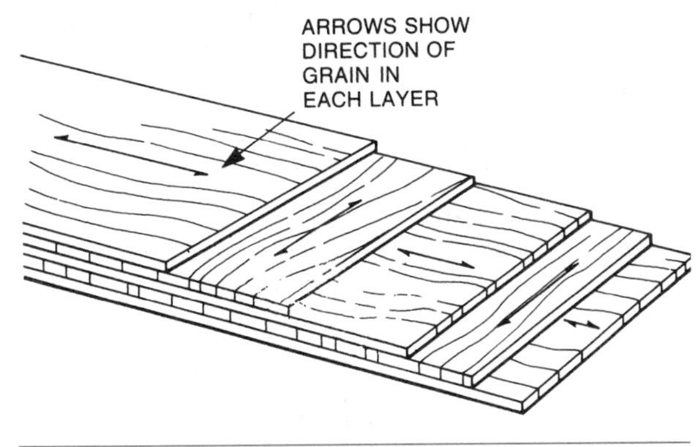

Fig. 3-4 In plywood construction, the grains of veneer plies are at right angles to each other.

Manufacture of Veneer Core Plywood. Specially selected "peeler logs" are mounted on a huge lathe in which the log is rotated against a sharp knife. As the log turns, a thin layer is peeled off like paper unwinding from a roll (Fig. 3-5). The entire log is used. The small remaining spindles are utilized for making other wood products.

The long ribbon of veneer is then cut into desired widths, sorted, and dried to a moisture content of 5 percent. After drying, the veneers are fed through glue spreaders that coat them with a uniform thickness. The veneers are then assembled to make panels (Fig. 3-6). Large presses bond the as-

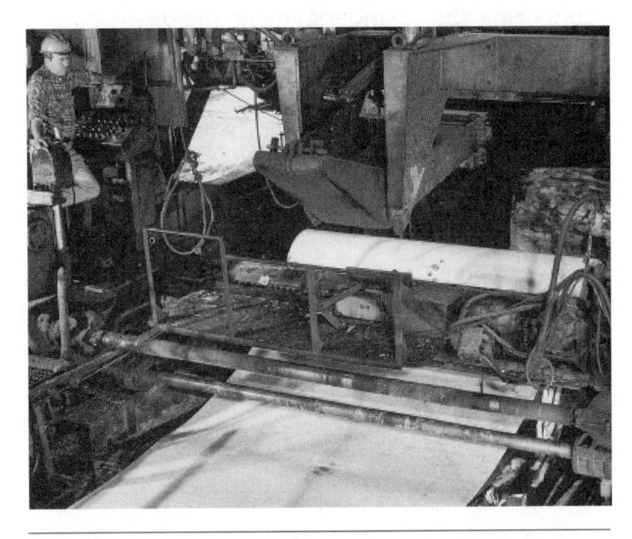

Fig. 3-5 The veneer is peeled from the log like paper unwinding from a roll. (*Courtesy of American Plywood Association*)

Fig. 3-6 Gluing and assembling plywood veneers into panels. (*Courtesy of American Plywood Association*)

sembly under controlled heat and pressure. From the presses, the panels are either left unsanded, touch-sanded or smooth-sanded, cut to size, inspected, and stamped.

Veneer Grades. In declining order, the letters *A, B, C plugged, C,* and *D* are used to indicate the appearance quality of panel veneers. Two letters are found in the grade stamp of veneered panels. One letter indicates the quality of one face, while the other letter indicates the quality of the opposite face. The exact description of these letter grades is shown in Figure 3-7. Panels with B-grade or better veneer faces are always sanded smooth. Some panels, such as APA-Rated Sheathing, are unsanded because their intended use does not require sanding. Other panels used for such purposes as subflooring and underlayment require only a touch-sanding to make the panel thickness more uniform.

Strength Grades. Softwood veneers are made of many different kinds of wood. These woods are classified in groups according to their strength (Fig. 3-8). Group 1 is the strongest. Douglas fir and southern pine are in Group 1 and are used to make most of the softwood plywood. The group number is also shown in the grade stamp.

Veneer Grades

A	Smooth, paintable. Not more than 18 neatly made repairs, boat, sled, or router type, and parallel to grain, permitted. Wood or synthetic repairs permitted. May be used for natural finish in less demanding applications.
B	Solid surface. Shims, sled or router repairs, and tight knots to 1 inch across grain permitted. Wood or synthetic repairs permitted. Some minor splits permitted.
C Plugged	Improved C veneer with splits limited to 1/8-inch width and knotholes or other open defects limited to 1/4 x 1/2 inch. Wood or synthetic repairs permitted. Admits some broken grain.
C	Tight knots to 1 - 1/2 inch. Knotholes to 1 inch across grain and some to 1 - 1/2 inch if total width of knots and knotholes is within specified limits. Synthetic or wood repairs. discoloration and sanding defects that do not impair strength permitted. Limited splits allowed. stiching permitted.
D	Knots and knotholes to 2 - 1/2 inch width across grain and 1/2 inch larger within specified limits. Limited splits are permitted. Stiching permitted. Limited to Exposure 1 or interior panels.

Fig. 3-7 Veneer letter grades define veneer appearance. (*Courtesy of American Plywood Association*)

Oriented Strand Board

Oriented strand board (OSB) is a nonveneered performance-rated structural panel composed of small oriented (lined up) strand-like wood pieces arranged in three to five layers with each layer at right angles to the other (Fig. 3-9). The cross-lamination of the layers achieves the same advantages of strength and stability as in plywood.

Manufacture of OSB. Logs from specially selected species are debarked and sliced into strands that are between 25/1000 and 30/1000 of an inch thick, 3/4 and 1 inch wide, and between 2 1/2 and 4 1/2 inches long. The strands are dried, loaded into a blender, coated with liquid resins, formed into a mat consisting of three or more layers of systematically oriented wood fibers, and fed into a press where, under high temperatures and pressure, they form a dense panel. As a safety measure, the panels have one side textured to help prevent slippage.

Nonveneered panels have previously been sold with such names as waferboard, structural particleboard, and others. At present almost all panels manufactured with oriented strands or wafers are called oriented strand board. Various manufacturers have their own particular brand names for OSB, such as Oxboard, Aspenite, and many others.

Composite Panels

Composite panels are manufactured by bonding veneers of wood to both sides of reconstituted wood panels. More efficient use of wood is thus allowed with this product while retaining the wood grain appearance on both faces of the panel.

Composite panels rated by the American Plywood Association are called *COM-PLY* and are manufactured in three or five layers. A three-layer panel has a reconstituted wood core with wood veneers on both sides. A five-layer panel has a wood veneer in the center as well as on both sides.

Performance Ratings

A Performance-Rated Panel meets the requirements of the panel's end use. The three end uses for which panels are rated are single-layer flooring, exterior siding, and sheathing for roofs, floors, and walls. Names given to designate end uses are *APA-Rated Sheathing, Structural I, APA-Rated Sturdi-I-Floor,* and *APA-Rated Siding* (Fig. 3-10). Panels are tested to meet standards in areas of resistance to moisture, strength, and stability.

CLASSIFICATION OF SPECIES				
Group 1	**Group 2**	**Group 3**	**Group 4**	**Group 5**
Apitong Beech American Birch Sweet Yellow Douglas Fir 1[(a)] Kapur Keruing Larch, Western Maple, Sugar Pine Caribbean Ocote Pine South Loblolly Longleaf Shortleaf Slash Tanoak	Cedar Port Orford Cypress Douglas Fir 2[(a)] Fir Balsam California Red Grand Noble Pacific Silver White Hemlock, Western Lauan Almon Bagtikan Mayapis Red Tangile White Maple, Black Mengkulang Meranti, Red[(b)] Mersawa Pine Pond Red Virginia Western White Spruce Black Red Sitka Sweetgum Tamarack Yellow- Poplar	Alder Red Birch Paper Cedar Alaska Fir Subalpine Hemlock, Eastern Maple Bigleaf Pine Jack Lodgepole Ponderosa Spruce Redwood Spruce Engelmann White	Aspen Bigtooth Quaking Cativo Cedar Incense Western Red Cottonwood Eastern Black (Western Poplar) Pine Eastern White Sugar	Basswood Poplar, Balsam

(a) Douglas Fir from trees grown in the states of Washington, Oregon, California, Idaho, Montana, Wyoming, and the Canadian Provinces of Alberta and British Columbia shall be classed as Douglas Fir No. 1. Douglas Fir from trees grown in the states of Nevada, Utah, Colorado, Arizona and New Mexico shall be classed as Douglas Fir No. 2.

(b) Red Meranti shall be limited to species having a specific gravity of 0.41 or more based on green volume and oven dry weight.

Fig. 3-8 Plywood is classified into groups according to strength and stiffness. *(Courtesy of American Plywood Association)*

Fig. 3-9 Oriented strand board being used for wall sheathing.

Exposure Durability

APA performance-rated panels are manufactured in three exposure durability classifications: *Exterior, Exposure 1,* and *Exposure 2.* Panels marked Exterior are designed for permanent exposure to the weather or moisture. Exposure 1 panels are intended for use where long delays in construction may cause the panels to be exposed to the weather before being protected. Panels marked Exposure 2 are designed for use when only moderate delays in providing protection from the weather are expected. The exposure durability of a panel may be found in the grade stamp.

Span Ratings

The span rating in the grade stamp on APA-Rated Sheathing appears as two numbers separated by a

Grade	Typical Trademark	Description
APA Rated Sheathing Typical Trademark	APA RATED SHEATHING 24/16 7/16 INCH SIZED FOR SPACING EXPOSURE 1 000 NER-QA397 PRP-108 HUD-UM-40C	Specially designed for subflooring and wall and roof sheathing. Also good for a broad range of other construction and industrial applications. Can be manufactured as plywood, as a composite, or as OSB. EXPOSURE DURABILITY CLASSIFICATIONS: Exterior, Exposure 1, Exposure 2. COMMON THICKNESSES: 5/16, 3/8, 7/16, 15/32, 1/2, 19/32, 5/8, 23/32, 3/4.
APA Structural I Rated Sheathing (c) Typical Trademark	APA RATED SHEATHING STRUCTURAL I 32/16 15/32 INCH SIZED FOR SPACING EXPOSURE 1 000 PS 1-83 C-D NER-QA397 PRP-108 APA RATED SHEATHING 32/16 15/32 INCH SIZED FOR SPACING EXPOSURE 1 000 STRUCTURAL I RATED DIAPHRAGMS-SHEAR WALLS PANELIZED ROOFS NER-QA397 PRP-108	Unsanded grade for use where shear and cross-panel strength properties are of maximum importance, such as panelixed roofs and diaphragms. Can be manufactured as plywood, as a composite, or as OSB. EXPOSURE DURABILITY CLASSIFICATIONS: Exterior, Exposure 1. COMMON THICKNESSES: 5/16, 3/8, 7/16, 15/32, 1/2, 19/32, 5/8, 23/32, 3/4.
APA Rated Sturd-I-Floor Typical Trademark	APA RATED STURD-I-FLOOR 20 oc 19/32 INCH SIZED FOR SPACING [illegible] EXPOSURE 1 000 NER-QA397 PRP-108 HUD-UM-40C	Specially designed as combination subfloor-underlayment. Provides smooth surface for application of carpet and pad and possesses high concentrated and impact load resistance. Can be manufactured as plywood, as composite, or as OSB. Available square edge or tonge-and-groove. EXPOSURE DURABILITY CLASSIFICATIONS: Exterior, Exposure 1, Exposure 2. COMMON THICKNESSES: 19/32, 5/8, 23/32, 3/4, 1, 1-1/8.
APA Rated Siding TYPICAL TRADEMARK	APA RATED SIDING 24 oc 19/32 INCH SIZED FOR SPACING EXTERIOR 000 NER-QA397 PRP-108 HUD-UM-40C APA RATED SIDING 303-18-S/W 16 oc 11/32 INCH GROUP 1 SIZED FOR SPACING EXTERIOR 000 PS 1-83 FHA-UM-64 NER-QA397 PRP-108	For exterior siding, fencing, etc. Can be manufactured as plywood, as a composite or as an overlaid OSB. Both panel and lap siding available. Special surface treatment such as V-groove, channel groove, deep groove (such as APA Texture 1-11), brushed, rough sawn, and overlaid (MDO) with smooth- or texture-embossed face. Span Rating (stud spacing for siding qualified for APA Sturd-I-Wall applications) and face grade classification (for veneer-faced siding) indicated in trademark. EXPOSURE DURABILITY CLASSIFICATION: Exterior. COMMON THICKNESSES: 11/32, 3/8, 7/16, 15/32, 1/2, 19/32, 5/8.

(a) Specific grades, thicknesses and exposure durability classifications may be in limited supply in some areas. Check with your supplier before specifying.

(b) Specify Performance Rated Panels by thickness and Span Rating. Span Ratings are based on panel strength and stiffness. Because these properties are a function of panel composition and configuration, as well as thickness, the same Span Rating may appear on panels of different thicknesses. Conversely, panels of the same thickness may be marked with different Span Ratings.

(c) All piles in Structural I plywood panels are special improved grades and panels marked PS 1 and limited to Group 1 species. Other panels marked Structural I rated qualify through special performance testing. Structural II plywood panels are also provided for, but rarely manufactured. Application recommendations for Structural II plywood are identical to those for APA-RATED SHEATHING plywood.

Fig. 3-10 Guide to APA Performance-rated panels (*Courtesy of American Plywood Association*)

slash, such as 32/16 or 48/24. The left number denotes the maximum recommended spacing of supports when the panel is used for roof or wall sheathing. The right number indicates the maximum recommended spacing of supports when the panel is used for subflooring. In both cases, the long dimension of the panel must be placed across three or more supports. A panel marked 32/16, for example, may be used for roof sheathing over rafters not more than 32 inches on center, or for subflooring over joists not more than 16 inches on center.

The span ratings on APA-Rated Sturd-I-Floor and APA-Rated Siding appear as a single number. APA-Rated Sturd-I-Floor panels are designed specifically for combined subflooring-underlayment applications and are manufactured with span ratings of 16, 20, 24, and 48 inches.

APA-Rated Siding is produced with span ratings of 16 and 24 inches. The rating applies to vertical installation of the panel. All siding panels may be applied horizontally direct to studs 16 to 24 inches on center provided horizontal joints are blocked.

CHAPTER 4 NONSTRUCTURAL PANELS

Plywood

All the rated products discussed in the previous chapter may be used for nonstructural applications. In addition, other plywood products, grade-stamped by the American Plywood Association, are available for nonstructural use. They include sanded and touch-sanded plywood panels (Fig. 4-1) and specialty plywood panels (Fig. 4-2).

Grade / Typical Trademark	Description
APA A-A Typical Trademark A-A•G-1•EXPOSURE 1-APA•000•PS 1-83	Use where appearance of both sides is important for interior applications such as built-ins, cabinets, furniture, partitions; and for exterior applications such as fences, signs, boats, shipping containers, tanks, ducts, etc. Smooth surfaces suitable for painting. EXPOSURE DURABILITY CLASSIFICATIONS: Interior, Exposure 1, Exterior. COMMON THICKNESSES: 1/4, 11/32, 3/8, 15/32, 1/2, 19/32, 5/8. 23/32, 3/4.
APA A-B Typical Trademark A-B•G-1•EXPOSURE 1-APA•000•PS 1-83	For use where appearance of one side is less important but where two solid srurfaces are necessary. EXPOSURE DURABILITY CLASSIFICATIONS: Interior, Exposure 1, Exterior. COMMON THICKNESSES: 1/4, 11/32, 3/8, 15/32, 1/2, 19/32, 5/8, 23/32, 3/4.
APA A-C Typical Trademark APA A-C GROUP 1 EXTERIOR 000 PS 1-83	For use where appearance of only one side is important in exterior applications, such as soffits, fences, farm buildings, etc.(f) EXPOSURE DURABILITY CLASSIFICATION: Exterior. COMMON THICKNESSES: 1/4, 11/32, 3/8, 15/32, 1/2, 19/32, 5/8, 23/32, 3/4.
APA A-D Typical Trademark APA A-D GROUP 1 EXPOSURE 1 000 PS 1-83	For use where appearance of only one side is important in interior applications, such as paneling, built-ins, shelving, partitions, flow racks, etc.(f) EXPOSURE DURABILITY CLASSIFICATIONS: Interior, Exposure 1. COMMON THICKNESSES: 1/4, 11/32, 3/8, 15/32, 5/8, 23/32, 3/4.
APA B-B Typical Trademark B-B•G-2•EXPOSURE 1-APA•000•PS 1-83	Utility panels with two solid sides. EXPOSURE DURABILITY CLASSIFICATIONS: Interior, Exposure 1, Exterior. COMMON THICKNESSES: 1/4, 11/32, 3/8, 15/32, 1/2, 19/32, 5/8, 23/32, 3/4.
APA B-C Typical Trademark APA B-C GROUP 1 EXTERIOR 000 PS 1-83	Utility panel for farm service and work buildings, boxcar and truck linings, containers, tanks, agricultural equipment, as a base for exterior coatings and other exterior uses or applications subject to high or continuous moisture.(f) EXPOSURE DURABILITY CLASSIFICATION: Exterior. COMMON THICKNESSES: 1/4, 11/32, 3/8, 15/32, 1/2, 19/32, 5/8, 23/32, 3/4.

Fig. 4-1 Guide to APA sanded and touched-sanded plywood panels. (*Courtesy of American Plywood Association*)

Grade	Trademark	Description
APA B-D Typical Trademark	APA B-D GROUP 2 EXPOSURE 1 000 PS 1-83	Utility panel for backing, sides of built-ins, industry shelving, slip sheets, separator boards, bins and other interior or protected applications.(f) EXPOSURE DURABILITY CLASSIFICATIONS: Interior, Exposure 1. COMMON THICKNESSES: 1/4, 11/32, 3/8, 15/32, 1/2, 19/32, 5/8, 23/32, 3/4.
APA Underlayment Typical Trademark	APA UNDERLAYMENT GROUP I EXPOSURE 1 000 PS 1-83	For application over structural subfloor. Provides smooth surface for application of carpet and pad and possesses high concentrated and impact load resistance. For areas to be covered with resilient flooring, specify panels with "sanded face."(e) EXPOSURE DURABILITY CLASSIFICATIONS: Interior, Exposure 1. COMMON THICKNESSES (d): 1/4, 11/32, 3/8, 15/32, 1/2, 19/32, 5/8, 23/32, 3/4.
APA C-C Plugged (g) Typical Trademark	APA C-C PLUGGED GROUP 2 EXTERIOR 000 PS 1-83	For use as an underlayment over structural subfloor, refrigerated or controlled-atmosphere storage rooms, pallet fruit bins, tanks, boxcar and truck floors and linings, open soffits, and other similar applications where continuous or severe moisture may be present. Provides smooth surface for application of carpet and pad and possesses high concentrated and impact load resistance. For areas to be covered with resilient flooring, specify panels with "sanded face."(e) EXPOSURE DURABILITY CLASSIFICATION: Exterior. COMMON THICKNESSES:(d) 11/32, 3/8, 15/32, 1/2, 19/32, 5/8, 23/32, 3/4.
APA C-D Plugged Typical Trademark	APA C-D PLUGGED GROUP 2 EXPOSURE1 000 PS 1-83	For open soffits, built-ins, cable reels, separator boards, and other interior or protected applications. Not substitute for Underlayment or APA Rated Sturd-I- Floor as it lacks their puncture resistance. EXPOSURE DURABILITY CLASSIFICATIONS: Interior, Exposure 1. COMMON THICKNESSES: 3/8, 15/32, 1/2, 19/32, 5/8, 23/32, 3/4.

(a) Specific plywood grades, thicknesses, and exposure durability classifications may be in limited supply in some areas. Check with your supplier before specifying.

(b) Sanded Exterior plywood panels, C-C Plugged, C-D Plugged, and Underlayment grades can also be manufactured in Structural I (all plies limited to Group 1 species).

(c) Some manufacturers also produce plywood panels with premium N-grade veneer on one or both faces. Available only by special order. Check with the manufacturer.

(d) Panels 1/2 inch and thicker are Span Rated and do not contain species group number in trademark.

(e) Also available in Underlayment A-C or Underlayment B-C grades, marked either "touch sanded" or "sanded face."

(f) For nonstructural floor underlayment, or other applications requiring improved inner ply construction, specify panels marked either "plugged inner plies" (may also be designated plugged crossbands under face or plugged crossbands or core); or "meets underlayment requirements."

(g) Also may be designated APA Underlayment C-C Plugged.

Fig. 4-1 *(Continued)*

Hardwood Plywood

Plywood is available with hardwood face veneers, of which the most popular are birch, oak, and lauan. Beautifully grained hardwoods are sometimes matched in a number of ways to produce interesting face designs. Hardwood plywood is used in the interior of buildings for such things as wall paneling, built-in cabinets, and fixtures. Much of this kind of plywood is manufactured with a lumber core (Fig. 4-3), although some are manufactured with a reconstituted wood core.

Particleboard

Particleboard is a reconstituted wood panel made of wood flakes, chips, sawdust, and planer shavings (Fig. 4-4). These wood particles are mixed with an adhesive, formed into a mat, and pressed into sheet form. The kind, size, and arrangement of the wood particles determine the quality of the board.

The highest-quality particleboard is made of large wood flakes in the center. The flakes become gradually smaller toward the surfaces where finer particles are found. This type of construction results in an extremely hard board with a very smooth surface. Softer and lower-quality boards contain the same size particles throughout. These boards usually have a rougher surface texture. In addition to the size, kind, and arrangement of the particles, the quality of the board is determined by the method of manufacture.

	Typical Trademark	Description
APA DECORATIVE Typical Trademark	APA DECORATIVE GROUP 2 INTERIOR 000 PS 1-83	Rough-sawn, brushed, grooved, or striated faces. For paneling, interior accent walls, built-ins, counter facing, exhibit displays. Can also be made by some manufacturers in Exterior for exterior siding, gable ends, fences and other exterior applications. Use recommendations for Exterior panels vary with the particular product. Check with the manufacturer. EXPOSURE DURABILITY CLASSIFICATONS: Interior, Exposure 1, Exterior. COMMON THICKNESSES: 5/16, 3/8, 1/2, 5/8.
APA HIGH DENSITY OVERLAY (HDO)(b) Typical Trademark	HDO•A-A•G-1•EXT-APA•000•PS 1-83	Has a hard semi-opaque resin-fiber overlay on both faces. Abrasion resistant. For concrete forms, cabinets. countertops, signs, tanks. Also available with skid-resistant screen-grid surface. EXPOSURE DURABILITY CLASSIFICATION: Exterior. COMMON THICKNESSES: 3/8, 1/2, 5/8, 3/4.
APA MEDIUM DENSITY OVERLAY (MDO)(b) Typical Trademark	APA M. D. OVERLAY GROUP 1 EXTERIOR 000 PS 1-83	Smooth, opaque, resin-fiber overlay on one or both faces. Ideal base for paint, both indoors and outdoors. For exterior siding, paneling, shelving, exhibit displays, cabinets, signs. EXPOSURE DURABILITY CLASSIFICATION: Exterior. COMMON THICKNESSES: 11/32, 3/8, 15/32, 1/2, 19/32, 5/8, 23/32, 3/4.
APA MARINE TYPICAL TRADEMARK	MARINE•A-A•EXT-APA•PS 1-83	Ideal for boat hulls. Made only with Douglas fir or western larch. Subject to special limitations on core gaps and face repairs. Also available with HDO faces. EXPOSURE DURABILITY CLASSIFICATION: Exterior. COMMON THICKNESSES: 1/4, 3/8, 1/2, 5/8, 3/4.
APA B-B PLYFORM CLASS I(b) Typical Trademark	APA PLYFORM B - B CLASS1 EXTERIOR 000 PS 1-83	Concrete form grades with high reuse factor. Sanded both faces and mill-oiled unless otherwise specified. Special restrictions on species. Also available in HDO for very smooth concrete finish, and with special overlays. EXPOSURE DURABILITY CLASSIFICATION: Exterior. COMMON THICKNESSES: 19/32, 5/8, 23/32, 3/4.
APA PLYRON Typical Trademark	PLYRON•EXPOSURE 1-APA•000	Hardboard face on both sides. Faces tempered, untempered, smooth, or screened. For countertops, shelving, cabinet doors, flooring. EXPOSURE DURABILITY CLASSIFICATIONS: Interior, Exposure 1, Exterior. COMMON THICKNESSES: 1/2, 5/8, 3/4.

(a) Specific plywood grades, thicknesses, and exposure durability classifications may be in limited supply in some areas. Check with your supplier before specifying.

(b) Can also be manufactured in Structural 1 (all plies limited to Group 1 species).

Fig. 4-2 Guide to APA specialty plywood panels. (*Courtesy of American Plywood Association*)

Particleboard Grades

The quality of particleboard is indicated by its density (hardness), which ranges from 28 to 55 pounds per cubic foot. Nonstructural particleboard is used in the construction industry for the construction of kitchen cabinets and countertops, and for the core of veneer doors and similar panels. *Duraflake, Novoply,* and *Tuf-Flake* are some brand names of particleboard.

Fiberboards

Fiberboards are manufactured as *high-density, medium-density,* and *low-density* boards.

Hardboards

High-density fiberboards are called *hardboards* and are commonly known by the trademark *Masonite* regardless of the manufacturer. The hardboard industry makes almost complete use of the great natural resource of wood by utilizing the wood chips and board trimmings, which were once considered waste.

Wood chips are reduced to fibers and water is added to make a soupy pulp. The pulp flows onto a traveling mesh screen where water is drawn off to form a mat. The mat is then pressed under heat to weld the wood fibers back together by utilizing lignin, the natural adhesive in wood.

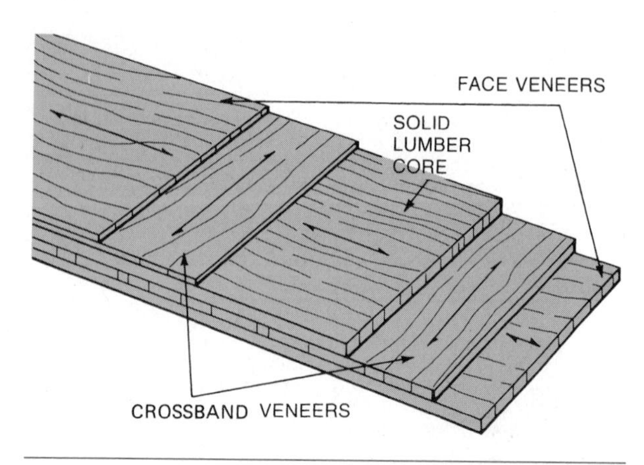

Fig. 4-3 Construction of lumber-core plywood.

Fig. 4-4 Particleboard is made from wood flakes, shavings, resins, and waxes. *(Courtesy of Duraflake Division, Williamette Industries, Inc.)*

Some panels are **tempered** (coated with oil and baked to increase hardness, strength, and water resistance). Carbide-tipped saws trim the panels to standard sizes.

Sizes of Hardboard. The most popular thicknesses of hardboard range from 1/8 to 3/8 inch. The most popular sheet size is 4 feet by 8 feet, although sheets may be ordered in practically any size.

Classes and Kinds of Hardboard. Hardboard is available in three different classes: tempered, standard, and service-tempered (Fig. 4-5). It may be obtained smooth-one-side (S1S) or smooth-two-sides (S2S). Hardboard is available in many forms, such as perforated, grooved, and striated.

Uses of Hardboard. Hardboard may be used inside or outside. It is widely used for exterior siding and interior wall paneling. It is also used extensively for cabinet backs and drawer bottoms. It can be used wherever a dense, hard panel is required (Fig. 4-6).

CLASS	SURFACE	NOMINAL THICKNESS
1 Tempered	S1S and S2S	1/8 1/4
2 Standard	S1S and S2S	1/8 1/4 3/8
3 Service-tempered	S1S and S2S	1/8 1/4 3/8

Fig. 4-5 Kinds and thicknesses of hardboard.

Because it is a wood-based product, hardboard can be sawed, routed, shaped, and drilled with standard woodworking tools. It can be securely fastened with glue, screws, staples, or nails.

Hardboard products are manufactured to comply with U.S. Department of Commerce Voluntary Product Standards. Most hardboard producers belong to the American Hardboard Association.

Medium-density Fiberboard

Commonly called *medium-density fiberboard (MDF)* is known by such names as *Medite, Baraboard,* and

Fig. 4-6 Hardboard is widely used for exterior siding and interior paneling.

Fibrepine, among others. It is manufactured in a manner similar to that used to make hardboard except that the fibers are not pressed as tightly together. The refined fiber produces a fine-textured, homogeneous board with an exceptionally smooth surface. Densities range from 28 to 65 pounds. It is available in thicknesses ranging from 3/16 to 1 1/2 inches and comes in widths of 4 feet and 5 feet. Lengths run from 6 to 18 feet.

MDF may be used for case goods, drawer parts, kitchen cabinets, cabinet doors, signs, and some interior wall finish.

Softboard

Low-density fiberboard is called *softboard.* Common brand names include *Temlok, Celotex,* and *Baracore.* Softboard is very light and contains many tiny air spaces because the particles are not compressed tightly.

The most common thicknesses range from 1/2 to 1 inch. The most common sheet size is 4 feet by 8 feet, although many sizes are available.

Uses of Softboard. Because of their lightness, softboard panels are used primarily for insulating or sound control purposes. They are used extensively as decorative panels in suspended ceilings and as ceiling tiles (Fig. 4-7). Much use is made of softboard for exterior wall sheathing. This type may be coated or impregnated with asphalt to protect it from moisture during construction.

Softboard panels can easily be cut with a knife, hand saw, or power saw. They cannot be hand-planed with any satisfactory results. Wide-headed nails, staples, or adhesive are used to fasten softboards in place, depending on the type and use.

Other

The preceding chapters have been limited to engineered wood panels and boards. There are many more products used in the construction industry besides those already mentioned. It is recommended that the student study *Sweet's Architectural File* to become better acquainted with the thousands of building material products on the market. This reference is well-known by architects, contractors and builders, and is revised and published annually. Sweet's may be found online at http://www.sweets.com. Two publications, *Products for General Building and Renovations* and *Products for Home building and Remodeling,* may be available at the school or city library.

Fig. 4-7 Softboards are used extensively for decorative ceiling panels. (*Courtesy of Armstrong World Industries, Inc.*)

Review Questions

Select the most appropriate answer.

1. Plywood usually contains
   a. three layers.
   b. four layers.
   c. an even number of layers.
   d. an odd number of layers.

2. Most softwood plywood is made of
   a. cedar.
   b. fir.
   c. western pine.
   d. spruce.

3. The best-appearing face veneer of a softwood plywood panel is indicated by the letter
   a. A.
   b. B.
   c. E.
   d. Z.

4. Which is a good selection of plywood for exterior wall sheathing?
   a. APA Structural Rated Sheathing, Exposure 1
   b. APA A-C, Exterior
   c. APA-Rated Sheathing, Exposure 2
   d. CD, Plugged, Exterior

5. A number "0" on the right of the slash in the Identification Index of Plywood means
   a. there are no knots in the inner plies.
   b. it should not be used for subflooring.
   c. it should not be used for roof sheathing.
   d. it should not be used for any structural purpose.

6. Particleboard not rated as structural may be used for
a. countertops.
b. subflooring.
c. wall sheathing.
d. roof sheathing.

7. Hardboard may be used in
a. interior applications only.
b. exterior and interior applications.
c. applications protected from moisture.
d. cabinet and furniture work only.

8. Much of the softboard is used in the construction industry for
a. underlayment for wall-to-wall rugs.
b. roof covering.
c. decorative ceiling panels.
d. interior wall finish.

9. For use as an underlayment over structural subflooring where continuous or severe moisture may be present, select
a. A-C, Group 1, Exterior plywood.
b. B-D, Group 2, Exposure 1 plywood.
c. Underlayment, Group 1, Exposure 1 plywood.
d. C-C Plugged, Group 2, Exterior plywood.

10. The recommended selection of plywood where the appearance of one side is important for interior applications such as built-ins and cabinets is
a. A-A, Group 1, Exposure 1 plywood.
b. M.D. Overlay, Group 1, Exterior plywood.
c. Decorative, Group 2, Interior plywood.
d. A-D, Group 2, Exposure 2 plywood.

BUILDING FOR SUCCESS

Listening Skills

In order to be effective, the carpenter has to be receptive to information, instruction, and details. These are criteria for quality work. Not only what we listen to, but *how* we listen forms the basis for our gaining knowledge and understanding.

Many of us have played the game in which a group of people sit in a circle and someone begins to relay a story or situation to the person at his or her side. As this message is being whispered into a person's ear, many times it becomes garbled or distorted as it is passed around the circle. Each person interprets what is heard and passes it on until the message comes back to the original communicator. The final report of what was received by the person who first stated the story usually has changed quite a bit. A variety of versions can be shared as to what people thought they heard. This exercise can help us understand the need for clear communication.

The world of construction is similar to other industries across the nation and calls for good listening skills. These skills must be utilized and nurtured daily in order for communication to be productive. Listening skills must be developed.

Just like the variety of readers that range from rapid scanners to detail seekers, people tend to listen in different ways. Usually the contractor has a daily orientation session with superintendents, supervisors, and foremen in order to provide clear understandings about the day's activities. This is where information is shared about the goals, work assignments, and schedules the company has planned. There may be discussion about the previous day's accomplishments and problems before the new day is brought into focus. Many times this information exchange occurs before the normal work day begins.

As these initial meetings end, the supervisors then relate the new information to each work crew. The leaders relate what they heard and brief the workers. It is imperative that the information does not change.

Listening at the construction site may be difficult due to noise created by construction equipment and assembly processes. Hearing protection may need to be worn in order to prevent hearing loss. Today much of the communication linking is conducted by mobile phones and radios in the field. These higher-technology forms of audio communication are almost a necessity and require constant access by each worker.

Assuming that the conditions are good for hearing, how does the carpenter develop good listening skills?

First, all workers must be open to listening. Being receptive to instruction and information is a necessity for learning to take place. A closed mind may be demonstrated when instruction is not being followed.

Second, we must not be selective listeners. Selective listening may influence whom we are willing to listen to and what portion of the total information we actually hear. Appropriate listening will include all information. As people study for construction work, they must develop the ability to listen to everything that is said.

Third, we must focus our attention on those who know the industry and serve as our advisors. Usually the builder or other leader will work directly with us to provide an immediate communication link. The best skills will be developed by listening to those with more experience. This textbook has been developed by an experienced professional who has a vast array of information that needs to be absorbed by each student. Some of the content of the text will be received through lectures or discussions led by the course instructor. Your full attention will be required in order for learning to take place.

The instruction being presented by each educational outreach using this text is also being reinforced with personal construction work experience. Most of the information in those work experiences will be communicated orally.

One example of good listening skills is demonstrated by a custom home builder who was "all ears" as a student. During the two years of preparation at Southeast Community College, Sam Manzitto, a custom home builder in Lincoln, Nebraska, was continually absorbing information that would be necessary in the home building industry. He had a desire to gain all of the technical data and information possible for the years ahead. He and his brother Mike now are recognized as excellent home builders in the Lincoln community. In 1995, Manzitto was recognized by the National Association of Home Builders as one of "America's Best Builders." His company was one of eleven to receive this prestigious award. If Manzitto's listening skills had been poor, he might not have even been in a position to be considered as a candidate for this honor.

How can we develop good listening skills or refine those we currently possess?

1. We should try to determine the speaker's motive, purpose, and perspective. Premature evaluation of what is being said could lead to the wrong assumption. This abruptness will be the breeding grounds for frustration, ill feelings, and errors.
2. Listen for what the speaker actually means. Just as we must sometimes read between the lines, we must try to concentrate on what is meant by the statements.
3. Listen for the speaker's outlook or attitude. Motives shape attitudes and attitudes are visible in behaviors. This may help us understand the speaker and see where the motivation is based.
4. Do not give negative feedback. Both verbal and nonverbal messages can be sent by the listener. This could have an effect on the speaker's delivery. For example, crossed arms and a stare at the ceiling will undoubtedly tell the speaker you are disagreeing even before hearing the entire message.
5. Record and use what has been said to help you make personal decisions. Most of what is said has value. The value may be new information, reinforcement, or verification for your decisions. Somehow, put it to use.

A key to good listening will continue to be how receptive the listener is to all information. A closed mind receives very little that is meaningful. You have the responsibility to sort through what is being said and apply the real substance to your life.

FOCUS QUESTIONS: For individual or group discussion

1. What might be some good reasons for developing good listening skills?
2. Do you personally feel you listen well? Why?
3. In the past, have you communicated with people who did not even hear you out? What were their reactions? What was yours?
4. How do you feel you can improve your own listening skills? Will those skills reflect your attitude toward the issue and person talking?

UNIT

3

Engineered Lumber Products

The modern lumber industry is faced with many problems. Ensuring the survival of endangered wildlife, making wise use of forest resources, and harvesting logs are some of the problems. Although the supply of *old-growth* trees available for harvesting is steadily diminishing, enough lumber must still be manufactured to meet increasing demands.

OBJECTIVES

After completing this unit, the student should be able to describe the manufacture, composition, uses, and sizes of:

- laminated veneer lumber
- parallel strand lumber and laminated strand lumber
- wood I-beams
- glue-laminated beams

UNIT CONTENTS

Old-growth trees are large, tall, tight-grain trees that take over two hundred years to mature. Most of the large dimension, structural, solid-sawn lumber comes from old-growth trees because not enough can be cut from smaller second- and third-growth trees. Yet it is not economical to plant slow-growing trees and wait for them to mature. Although the supply of old-growth trees is decreasing, the supply of smaller second- and third-growth trees is abundant and is actually increasing as the result of replanting and reforestation.

An inevitable result of the decreasing supply of large old-growth trees and the abundance of smaller trees is the development and use of **engineered lumber products.**

Engineered lumber products, commonly called ELPs, are reconstituted wood products and assemblies designed to replace traditional structural lumber. Engineered lumber products consume less wood and can be made from smaller trees. Traditional lumber processes typically convert 40 percent of a log to structural solid lumber. Engineered lumber processes convert up to 75 percent of a log into structural lumber. In addition, the manufacturing processes of engineered lumber consume far less energy than those of solid lumber. Also, some ELPs make use of abundant, fast-growing species not currently harvested for solid lumber.

The final engineered lumber product has greater strength and consequently can span greater distances. It is predicted that engineered lumber will be used more than solid lumber in the near future. It is important that present and future builders be thoroughly informed about engineered lumber products.

This unit describes how engineered lumber products are made, where they are used, and what sizes are available. Construction details for ELPs are shown in following units on floor, wall, and roof framing.

CHAPTER 5 LAMINATED VENEER LUMBER

Laminated Veneer Lumber

Laminated veneer lumber, commonly called LVL, is one of several types of engineered lumber products (Fig. 5-1). It was first used to make airplane propellers during World War II. The world's first commercially produced LVL for building construction was patented as MICRO-LAM laminated veneer lumber in 1970. Since then, because of increased demand for the product, several manufacturers are marketing it under similar names, such as GANG-LAM (Fig. 5-2), STRUCLAM, and VERSA-LAM. LVL is now widely used in wood frame construction.

Fig. 5-2 Laminated veneer lumber is marketed under several brand names. (*Courtesy of Louisiana-Pacific Corporation*)

Manufacture of LVL

Like plywood, LVL is a wood veneer product. The grain in each layer of veneer in LVL runs in the same direction, parallel to its length (Fig. 5-3). This

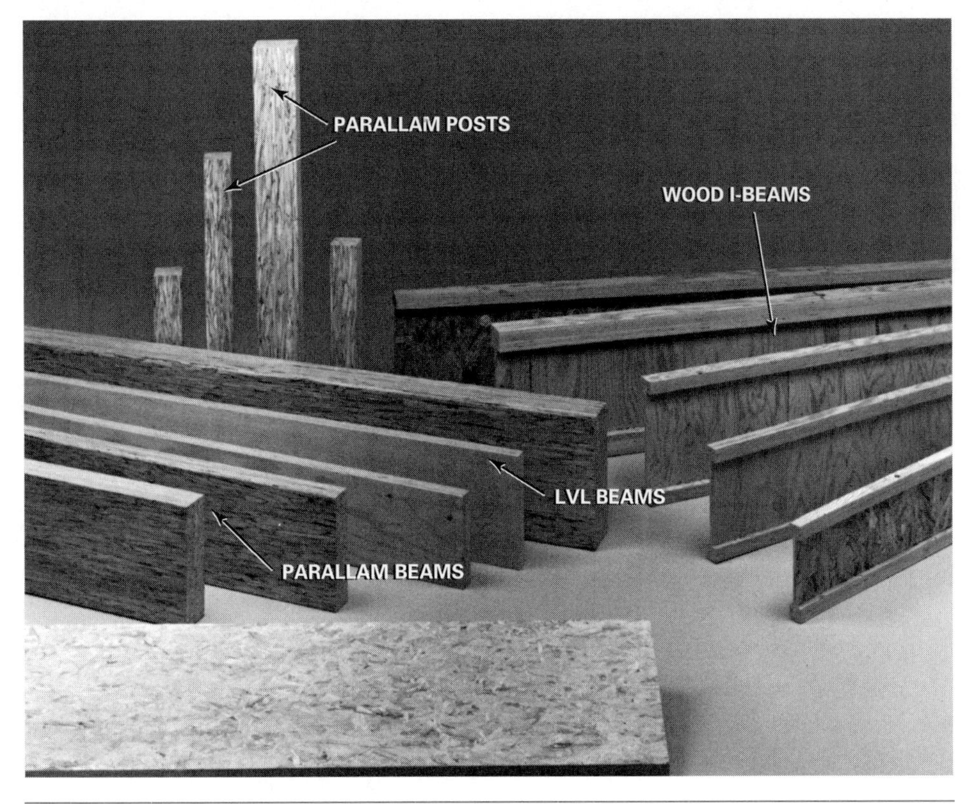

Fig. 5-1 There are several types of engineered lumber. (*Courtesy of Trus Joist MacMillan, Boise, Idaho*)

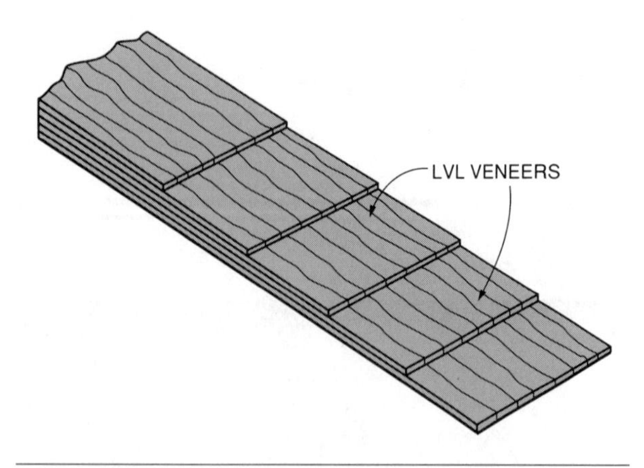

Fig. 5-3 In LVL, the grain in each layer of veneer runs in the same direction.

is unlike plywood, in which each layer of veneer is laid with the grain at right angles to each other.

Laminated veneer lumber is made from sheets of veneer peeled from logs, similar to the first step in the manufacture of plywood (see Fig. 3-5). Douglas fir or southern pine is used because of its strength. The veneer is peeled in widths of 27 inches or 54 inches and from 1/10 to 3/16 inch in thickness. It is then dried, cut into sheets, ultrasonically graded for strength, and sorted.

The veneer sheets are laid in a staggered pattern so that the ends overlap. They are then permanently bonded together with an exterior type adhesive in a continuous press under precisely controlled heat and pressure. Unlike plywood, the LVL veneers are *densified,* that is, the thickness is compressed and made more compact; fifteen to twenty layers of veneer make up a typical 1 3/4-inch thick beam. The edges of the bonded veneers are then edge trimmed to specified widths and end-cut to specified lengths (Fig. 5-4).

LVL Sizes

Laminated veneer lumber is manufactured up to 3 1/2 inches thick, 18 inches wide, and 80 feet long. The usual thickness are 1 1/2 inches and 1 3/4 inches. The 1 1/2 inch thickness is the same as nominal 2-inch-thick framing lumber, while doubling the 1 3/4-inch thickness equals the width of nominal 4-inch framing. LVL widths are usually 9 1/4, 11 1/4, 11 7/8, 14, 16, and 18 inches. LVL beams may be fastened together to make a thicker and stronger beam (Fig. 5-5).

Uses of LVL

Laminated veneer lumber is intended for use as high-strength, load-carrying beams to support the weight of construction over window and door openings, and in floor and roof systems of residential and light commercial wood frame construction (Fig. 5-6). Its relatively light weight makes it easy to handle and suitable for use in concrete forming in manufactured housing, and in many other special-

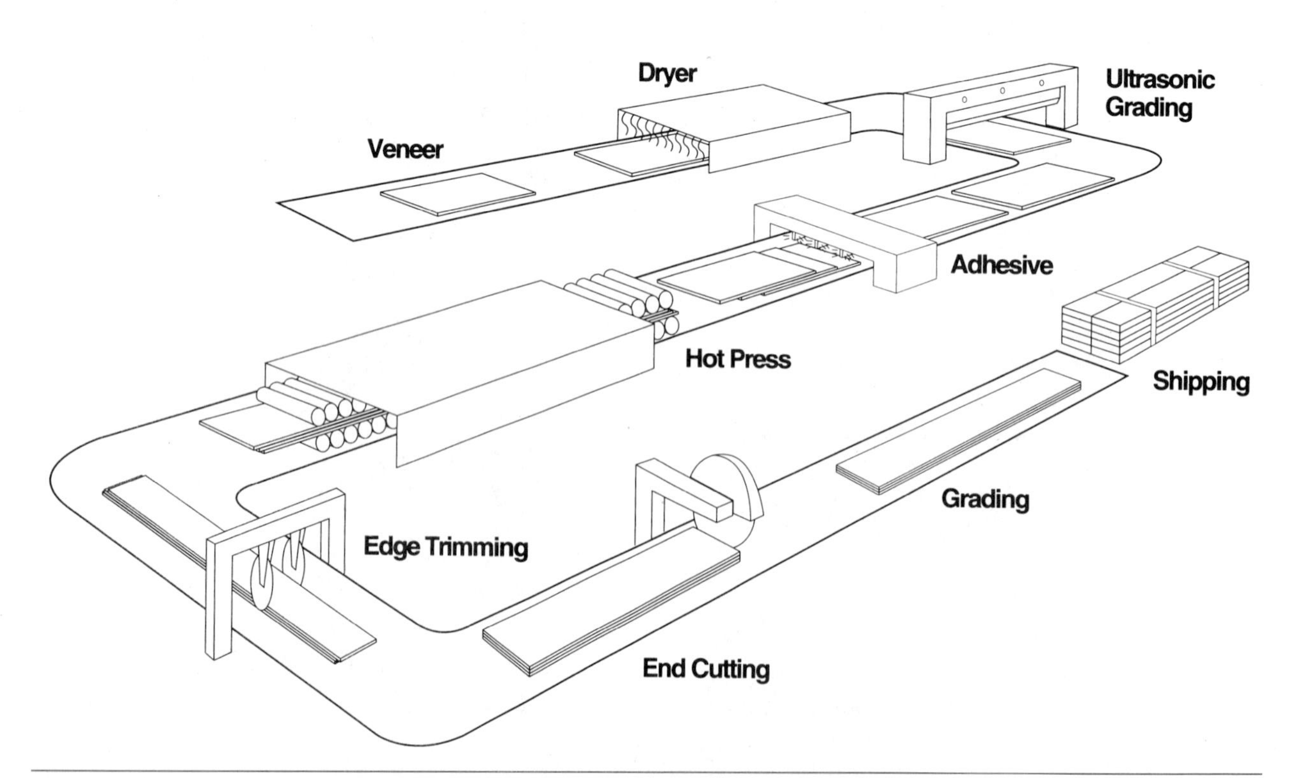

Fig. 5-4 LVL manufacturing process, simplified. (*Courtesy of Trus Joist MacMillan*)

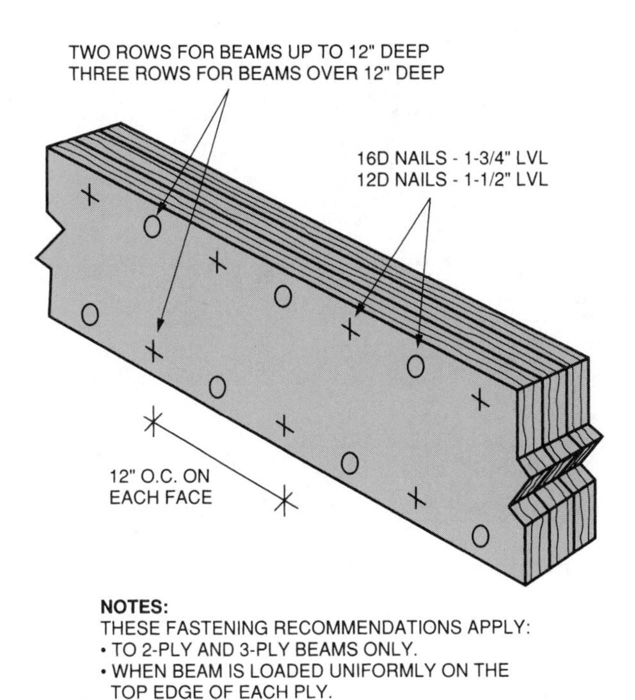

Fig. 5-5 Recommended nailing pattern for fastening LVL beams together. (*Courtesy of Louisiana-Pacific Corporation*)

ties where a light-weight, strong beam is required. It can be cut with regular tools and requires no special fasteners.

Fig. 5-6 LVL is designed to be used for load-carrying beams. (*Courtesy Trus Joist MacMillian*)

CHAPTER 6 PARALLEL STRAND LUMBER AND LAMINATED STRAND LUMBER

Parallel Strand Lumber

Parallel strand lumber (PSL) commonly known by its brand name, Parallam (Fig. 6-1), was developed, like all engineered lumber products, to meet the need of the building industry. PSL provides large dimension lumber (beams, planks, and posts). PSL also utilizes small-diameter, second-growth trees, thus protecting the diminishing supply of old-growth trees.

Manufacture of PSL

Parallam PSL is manufactured by peeling veneer of Douglas fir and southern pine from logs in much the same manner as for plywood and LVL. The veneer is then dried and clipped into *strands* (narrow strips) up to 8 feet in length and 1/8 or 1/10 inch in thickness. Small defects are removed and the strands are then coated with a waterproof adhesive. The oriented strands are fed into a rotary belt press and bonded using a patented microwave pressing process. The result is a continuous timber, up to 11 inches thick by 17 inches wide, which can then be factory-ripped into widths and thicknesses to fit builders' needs (Fig. 6-2). The four surfaces are sanded smooth before the product is shipped.

PSL Sizes

PSL comes in many thicknesses and widths and is manufactured up to 66 feet long. PSL is available in square and rectangular shapes for use as posts and beams. Beams are sold in convenient 1 3/4 inch thickness for installation of single and multiple laminations. Solid 3 1/2-inch thicknesses are compatible with 2 × 4 wall framing. A 2 11/16-inch thickness is also available. A list of beam and post sizes

Fig. 6-1 Parallel strand lumber is commonly called Parallam and is used as beams and posts to carry heavy loads. (*Courtesy of Trus Joist MacMillan*)

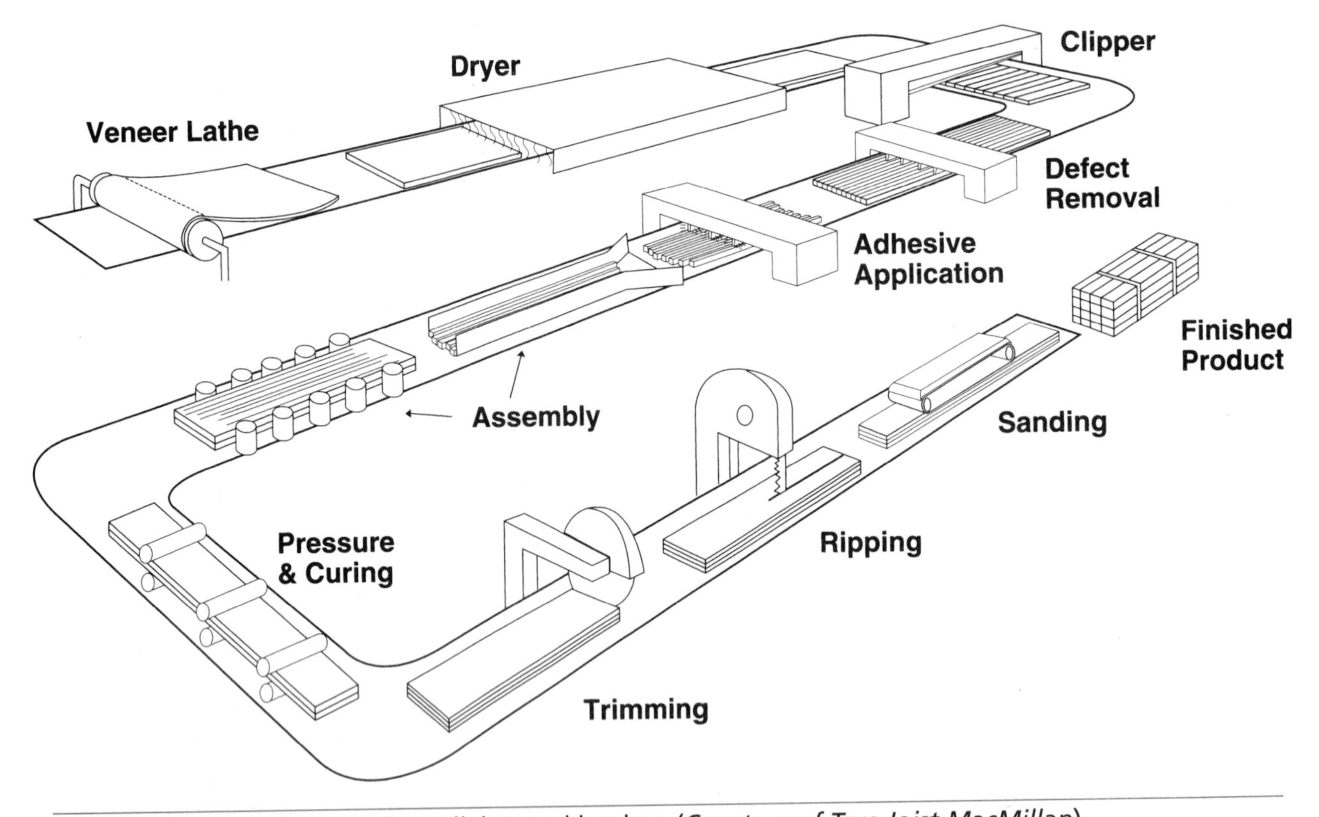

Fig. 6-2 The manufacture of parallel strand lumber. (*Courtesy of Trus Joist MacMillan*)

is shown in Figure 6-3. Also available is Parallam 269, which measures 2 11/16 inches thick.

Uses of PSL

Parallel strand lumber can be used wherever there is a need for a large beam or post. The differences between PSL and solid lumber are many. Solid lumber beams may have defects, like knots, checks, and shakes, which weaken them, while PSL is consistent in strength throughout its length. PSL is readily available in longer lengths and its surfaces are sanded smooth, eliminating the need to cover them by boxing the beams.

Laminated Strand Lumber

The registered brand name of **laminated strand lumber** (LSL) is *TimberStrand* (Fig. 6-4). While LVL and PSL are made from Douglas fir and southern

THICKNESS	WIDTH
3 1/2"	3 1/2"
3 1/2"	5 1/4"
3 1/2"	7"
5 1/4"	5 1/4"
5 1/4"	7"
7"	7"

PARALLAM COLUMN & POST SIZES

THICKNESS	DEPTHS
1 3/4"	7 1/4"
3 1/2"	9 1/4"
5 1/4"	11 1/4"
7"	11 1/2"
	11 7/8"
	12"
	12 1/2"
	14"
	16"
	18"

PARALLAM BEAM SIZES

Fig. 6-3 Post and beam sizes of parallel strand lumber.

Fig. 6-4 Laminated strand lumber is commonly called TimberStrand. (*Courtesy of Trus Joist MacMillan*)

pine, LSL can be made from very small logs of practically any species of wood; its strands are much shorter than those of parallel strand lumber. At present, LSL is being manufactured from surplus, overmature aspen trees that usually are not large, strong, or straight enough to produce ordinary wood products.

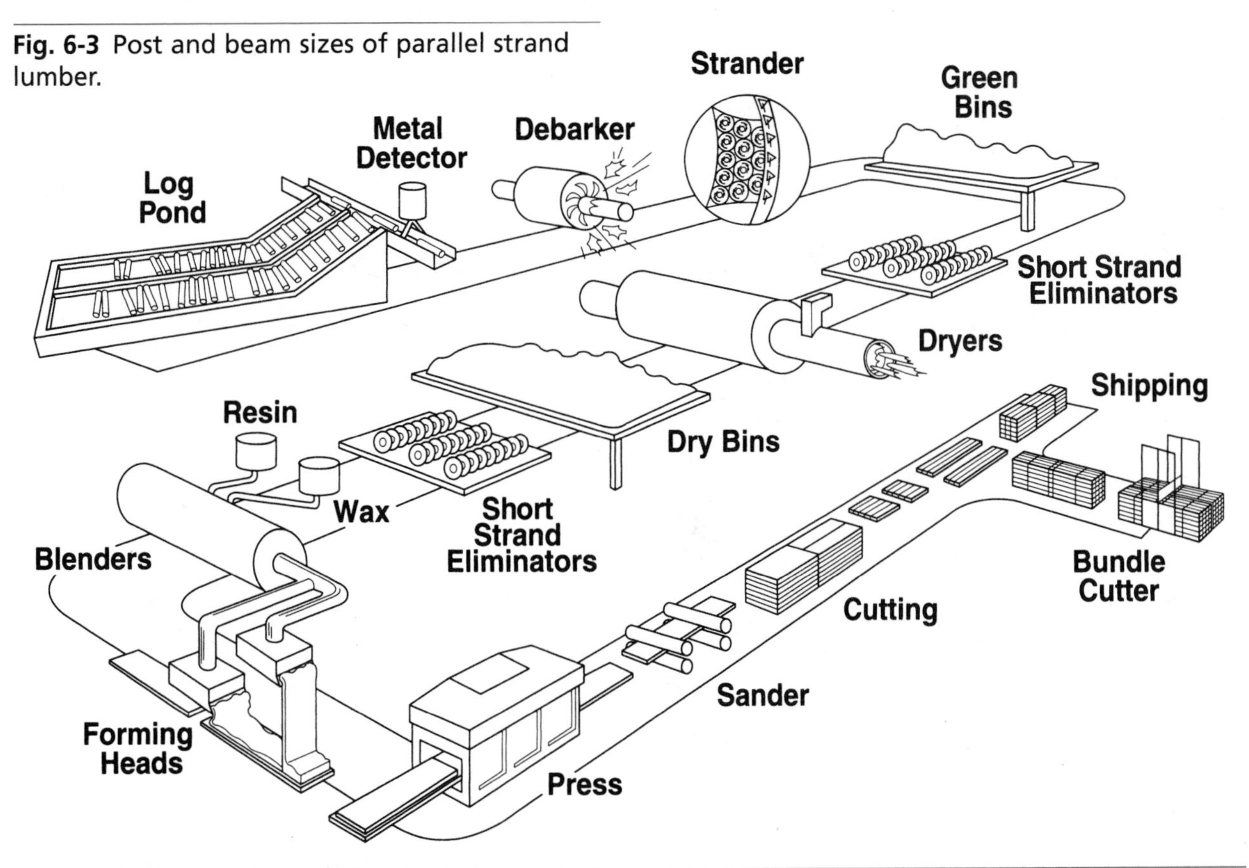

Fig. 6-5 Laminated strand lumber can be manufactured from practically any species of wood. (*Courtesy of Trus Joist MacMillan*)

Manufacture of LSL

The TimberStrand LSL manufacturing process begins by cleaning and debarking 8-foot aspen logs. The wood is then cut into strands up to 12 inches long, dried, and treated with a resin. The treated strands are aligned parallel to each other to take advantage of the wood's natural strength. The strands are pressed into solid *billets* (large blocks) up to 5 1/2 inches thick, 8 feet wide, and 35 feet long (Fig. 6-5). Scraps from the process fuel the furnace that provides heat and steam for the plant.

Sizes and Uses of LSL

The long billet is resawn and sanded to sizes as required by customers. It is used for a wide range of **millwork,** such as doors, windows, and virtually any product that requires high-grade lumber. It is also used for truck decks, manufactured housing, and some structural lumber, such as window and door headers.

One of the differences between PSL and LSL is the kind of wood from which they are manufactured. Because PSL is made from stronger wood and with longer strands, it is designed to carry heavy loads, while LSL is not.

CHAPTER 7 WOOD I-BEAMS

Wood I-Beams

There are several manufacturers of *wood I-beams,* and each has its own registered trademark or brand name, such as BCI joist, GNI joist, StrucJoist, TJI joist, and Wood I-Beam. The TJI joist was invented in 1969, as a substitute for solid lumber **joists** (structural members of a floor frame).

Fig. 7-1 Wood I-beams are available in many sizes. (*Courtesy of Louisiana-Pacific Corporation*)

Wood I-beams are engineered wood assemblies that utilize an efficient "I" shape, common in steel beams, which gives them tremendous strength in relation to their size and weight (Fig. 7-1). Consequently, they are able to carry heavy loads over long distances while using considerably less wood than solid lumber of a size necessary to carry the same load over the same span.

The flanges of the beam may be made of laminated veneer lumber or specially selected **finger-jointed** solid wood lumber (Fig. 7-2). The web of the beam may be made of plywood, laminated veneer lumber, or oriented strand board.

The Manufacture of Wood I-Beams

Regardless of who makes them, the manufacturing process of wood I-beams consists of gluing top and

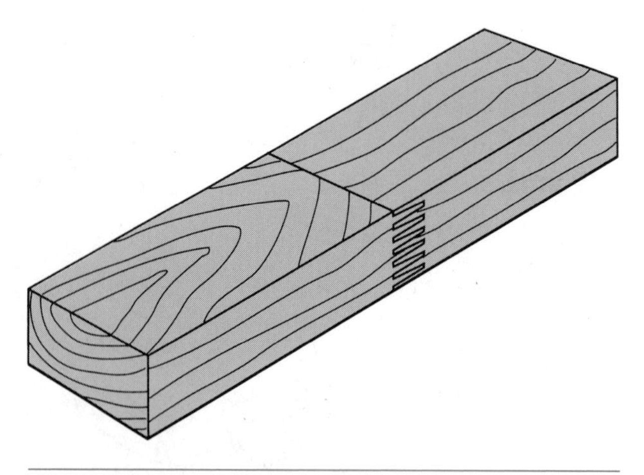

Fig. 7-2 Finger-joints are used to join the ends of short pieces of lumber to make a longer piece.

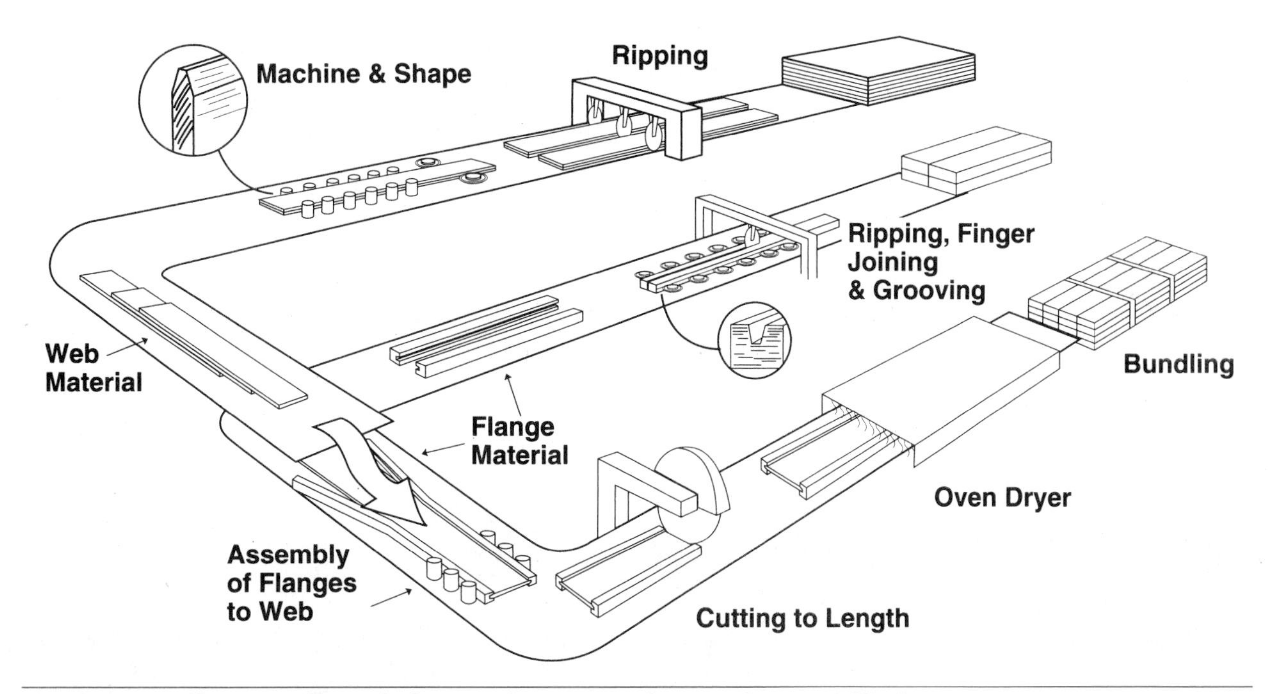

Fig. 7-3 Manufacture of wood I-beams. (*Courtesy of Trus Joist MacMillan*)

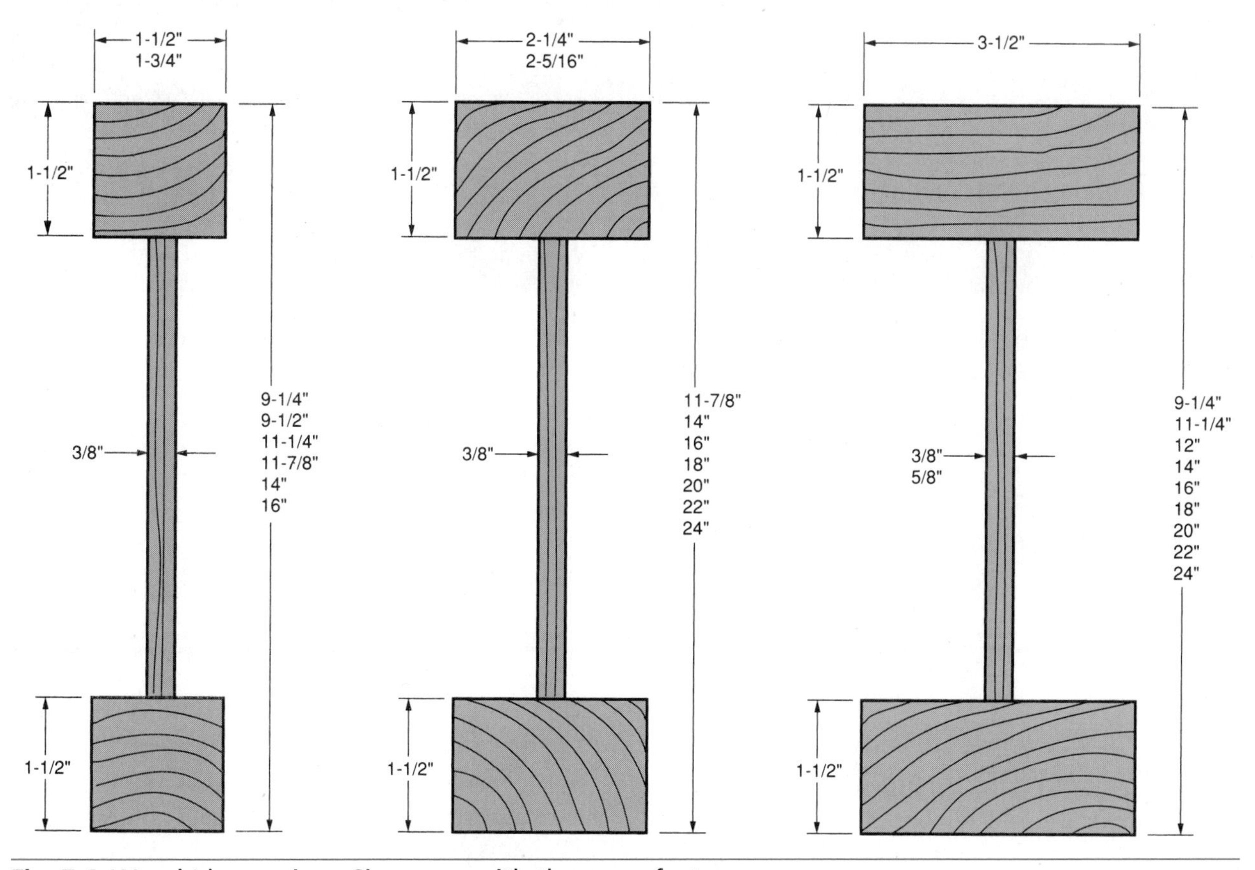

Fig. 7-4 Wood I-beam sizes. Sizes vary with the manufacturer.

bottom flanges to a connecting center web. First the web material is ripped to specified width and then the edges and ends are shaped for joining to flanges and adjacent web sections. The ends of the flange material are finger-jointed for gluing end to end. One side of the flange material is grooved to receive the beam's web. Flanges and webs are then assembled with waterproof glue by pressure-fitting the web into the flanges. The wood I-beam is end-trimmed and the adhesive cured in an oven or at room temperature (Fig. 7-3). As with most engineered wood products, wood I-beams are produced to approximate equilibrium moisture content.

Wood I-Beam Sizes

Wood I-beams may have webs of various thicknesses, and flanges may vary in thickness and width, depending on intended end use and the manufacturer. Beam depths are available from 9 1/4 to 30 inches (Fig. 7-4). Beams with larger webs and flanges are designed to carry heavier loads. Wood I-beams are available up to 80 feet long.

Uses of Wood I-Beams

Wood I-beams are intended for use in residential and commercial construction as floor joists, roof rafters, and **headers** for window, entrance door, and garage door openings (Fig. 7-5). Window and door headers are beams that support the load above wall openings.

Fig. 7-5 Wood I-beams are used as roof rafters as well as for floor joists and window and door headers. (*Courtesy of Louisiana-Pacific Corporation*)

CHAPTER 8 GLUE-LAMINATED LUMBER

Glue-Laminated Lumber

Glue-laminated lumber, commonly called *glulam*, is constructed of solid lumber glued together, side against side, to make beams and joists of large dimensions that are stronger than natural wood of the same size (Fig. 8-1). Even if it were possible, it would not be practical to make solid wood beams as large as most glulams. They are used for structural purposes, but are decorative as well and, in most cases, their surfaces are left exposed to show the natural wood grain.

Manufacture of Glulams

Glulam beams are made by gluing stacks of *lam* stock into large beams, headers, and columns (Fig. 8-2). Lams are the individual pieces in the glued-up stack. The lams, which are at approximate equilibrium moisture content, are glued together with exterior adhesives. Thus glulams are rated in terms of weather resistance.

Lam Layup. The lams are arranged in a certain way for maximum strength and minimum shrinkage.

Different grade lams are placed where they will do the most good under load conditions. High-grade *tension* lams are used in tension faces, and high-grade *compression* lams are used in compression faces.

Fig. 8-1 Glue-laminated lumber is commonly called *glulam.* (*Courtesy of American Plywood Association*)

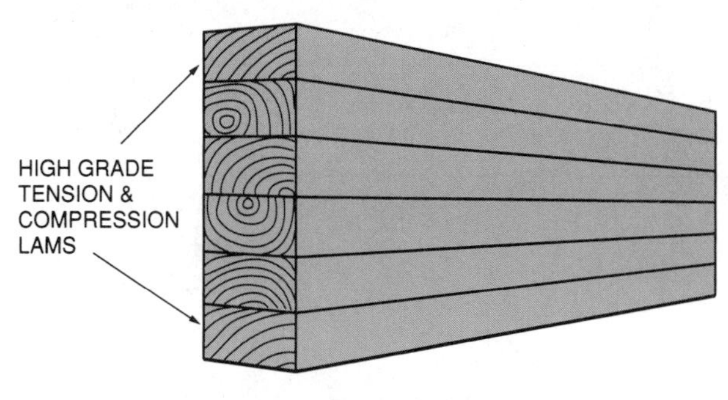

Fig. 8-2 Glulams are made by gluing stacks of solid lumber.

Tension is a force applied to a member that tends to increase along its length, while compression is a force tending to decrease along the length of the lam (Fig. 8-3). When a load is imposed on a glulam beam that is supported on both ends, the topmost lams are in compression while those at the bottom are in tension.

More economical lams are used in the lower-stressed middle sections of the glulam. The sequence of lam grades, from bottom to top of a glulam, is referred to as *lam layup,* and it is a vitally important factor in glulam performance. Because of this lam layup, glulam beams come with one edge stamped "TOP." *Always remember to install glulam beams with the "TOP" stamp facing toward the sky* (Fig. 8-4).

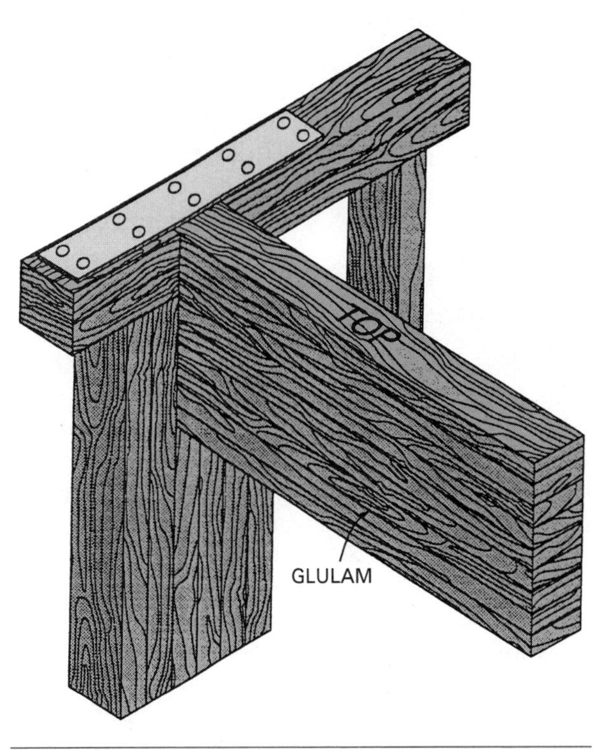

Fig. 8-4 Glulam beams must be installed with the edge stamped "TOP" pointed toward the sky.

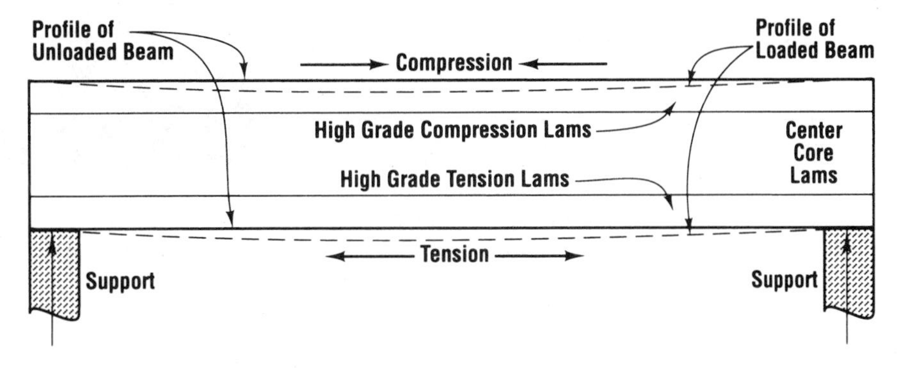

Fig. 8-3 The load on a beam places lams in tension and compression. (*Courtesy of Bohemia, Inc.*)

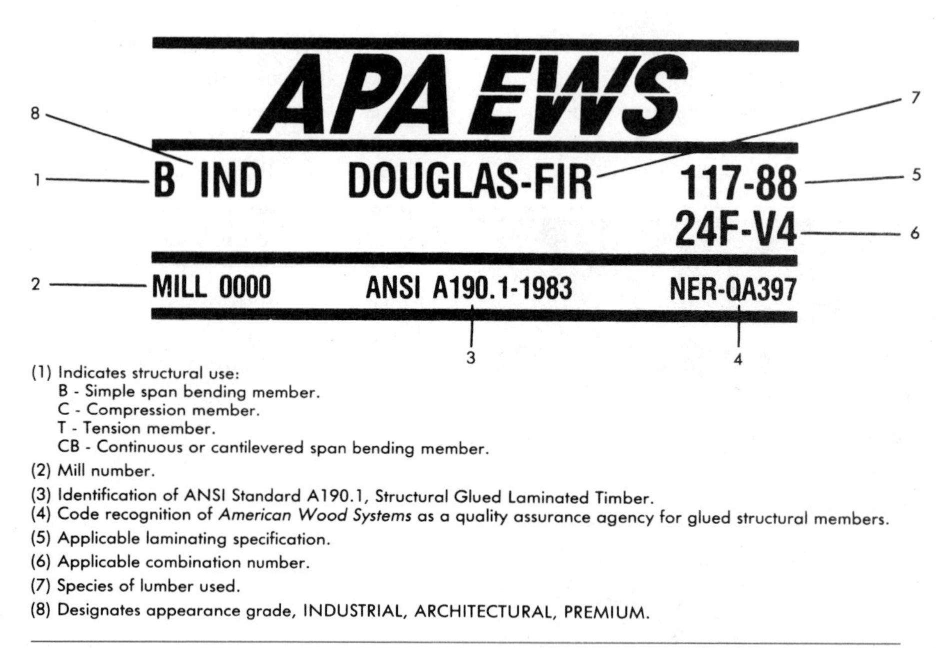

Fig. 8-5 The grade stamp assures that the beam has met all the necessary requirements. (*Courtesy of American Plywood Association*)

Glulam Grades, Sizes, and Uses

Grades. The American Plywood Association-Engineered Wood Systems trademark, APA-EWS (Fig. 8-5), appears on all beams manufactured by American Wood Systems member mills. The AWS is a related corporation of the APA. The trademark guarantees that the glulams meet all the requirements of the American National Standards Institute (ANSI).

Glulams are manufactured in three grades for appearance. The *industrial appearance grade* is used in warehouses, garages, and other structures in which appearance is not of primary importance or when the beams are not exposed. *Architectural appearance grade* is for projects where appearance is important, and the *premium appearance grade* is used when appearance is critical (Fig. 8-6). There is no difference in strength among the different appearance grades. All glulams with the same design values are rated for the same loadings, regardless of appearance grade.

Sizes. The dimensions of glulam beams are indicated by width, depth, and length (Fig. 8-7). Widths range from 2 1/2 to 8 3/4 inches, depths from 6 to 28 1/2 inches, and lengths are generally available from 10 to 40 feet in 2-foot increments (Fig. 8-8).

Uses. Various wood species are used to produce straight, curved, arched, and special shapes for all structures—from elegant homes and churches to large malls, warehouses, and civic centers. In all, the

Fig. 8-6 The appearance of an exposed beam is important. Some glulam beams are manufactured for appearance. (*Courtesy of Bohemia, Inc.*)

American Institute of Timber Construction (AITC) recognizes over one hundred glulam beam combinations, most of which are for specialized applica-

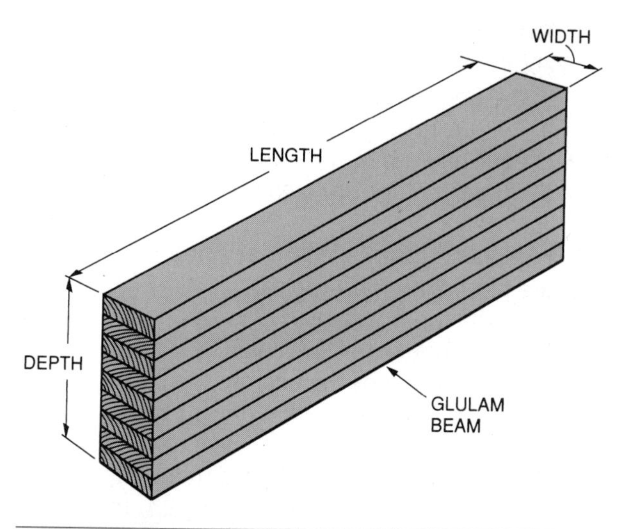

Fig. 8-7 The dimensions of glulam beam are width, depth, and length.

tions. However, five glulam beam combinations are the most commonly used and meet most consumer needs (Fig. 8-9).

Combination	Species	Use
24F-V4	Df/Df	Simple Span
24F-V8	Df/Df	Multi Span & Cantilever
24F-V5	Df/Hf	Simple Span
24F-V10	Df/Hf	Multi Span & Cantilever
Combination #1	Df/Df	Columns

Combination Number Table Notes

- The 24F means the beam's tension face is rated at 2,400 psi bending stress.
- V stands for visually graded laminations, and the 4, 5, 8, or 10 refer to the lamination lay up.
- The V4 and V8 combinations use only Douglas Fir. V5 and V10 combine Douglas Fir with Hemfir core lams.
- Combination #1 is for columns and uses only Douglas Fir L3 laminates.

Fig. 8-9 Only five glulam combinations, out of more than one hundred, are commonly used. (*Courtesy of Bohemia, Inc.*)

WIDTH	DEPTH IN INCHES															
	6	7 1/2	9	10 1/2	12	13 1/2	15	16 1/2	18	19 1/2	21	22 1/2	24	25 1/2	27	28 1/2
2 1/2"		*	*	*	*	*	*	*	*	*	*	*				
3 1/8"	*	*	*	*	*	*	*	*	*	*	*	*				
3 1/2"	*	*	*	*	*	*	*	*	*	*	*	*				
5 1/8"			*	*	*	*	*	*	*	*	*	*	*	*		
5 1/2"			*	*	*	*	*	*	*	*	*	*	*	*		
6 3/4"					*	*	*	*	*	*	*	*	*	*	*	*
8 3/4"					*	*	*	*	*	*	*	*	*	*	*	*
Sizes generally available from 10 to 40 feet long in 2'-0" increments																

Fig. 8-8 Glulam beam sizes.

Review Questions

Select the most appropriate answer.

1. Engineered lumber products are designed as replacements or substitutes for
a. sawn solid lumber.
b. second-growth lumber.
c. steel framing.
d. structural lumber.

2. Unlike plywood, the veneers of LVL are
a. deciduous.
b. densified.
c. diversified.
d. double-faced.

3. Laminated veneer lumber is manufactured in lengths up to
a. 40 feet. c. 80 feet.
b. 50 feet. d. 100 feet.

4. Parallel strand lumber is made from
a. Alaskan cedar and California redwood.
b. Douglas fir and southern pine.
c. Idaho pine and eastern hemlock.
d. Englemann spruce and western pine.

5. Parallel strand lumber is manufactured in lengths up to
a. 36 feet. c. 66 feet.
b. 50 feet. d. 80 feet.

6. The web of wood I-beams may be made of
a. hardboard. c. solid lumber.
b. particleboard. d. strand board.

7. The flanges of wood I-beams are generally made from
a. glue-laminated lumber.
b. laminated veneer lumber.
c. parallel strand lumber.
d. laminated strand lumber.

8. The lams used in glue-laminated lumber are at approximately
a. the fiber-saturation point.
b. 19 percent moisture content.
c. equilibrium moisture content.
d. 30 percent moisture content.

9. In glulam beams, specially selected tension lams are placed
a. at the top of the beam.
b. at the bottom of the beam.
c. in the center of the beam.
d. throughout the beam.

10. Although over one hundred different glulam combinations are recognized for use, the number of combinations commonly used are
a. five. c. twelve.
b. ten. d. eighteen.

BUILDING FOR SUCCESS

Characteristics of Construction Materials

Carpenters and other construction workers need to know many things about the materials they use. The most successful builders understand the properties, characteristics, uses, and limitations of the materials they install. For example, a basic requirement for selecting the correct species of wood for a framing job is to know how the wood will perform under loads and the span capability for that species.

Each construction job will vary and call for specific materials. It is true that the architect, engineer, and designer will possess a good understanding of material requirements for a job. The builder must also have some of the same information. The product manufacturer, lumber producer, and materials supplier will have access to that information, but it must be passed on to the builder. For the builder, this becomes a continual learning process and reinforces the need for maintaining a personal construction materials and products file.

High technology continues to create new products from existing or new materials at an accelerated pace. Economic and environmental factors have forced the manufacturing and fabricating industries to look at new ways of producing construction materials. This in turn obligates everyone to look at new or alternate methods of construction. The lumber industry has been drastically affected by environmental issues that show little or no chance for reversal. This turn of events has prompted new thinking and new products.

New or improved construction products such as structural and nonstructural wood panels, laminated veneer lumber, parallel and laminated strand lumber, wood I-beams, glue-laminated beams, and other preengineered products are now the norm in many areas of the country. These innovative products are allowing for the consumption of smaller trees with minimal waste. The use of traditional dimensional lumber will probably continue to decline in many areas. Environmental factors will continue to affect the products used in the conventional platform frame system. We must continually learn as much as possible about these new products as they become a part of our materials and products inventory. The carpenter of the twenty-first century will be expected to know more about construction materials and products in order to be effective.

With the information about construction materials presented in this text, manufacturers' data sheets, and other technical publications, construction students and workers will be able to assemble quality products. The selection and installation of construction products become less complex with a better understanding of each material's characteristics. For example, wood technology must be studied to gain a good working knowledge of defects and how those defects affect the appearance or performance of dimensional lumber and boards. Knowing how various species will perform and why will help us select which wood to use in every case. Softwoods and hardwoods possess many different characteristics that make one or the other more suitable for a certain job. Carpenters must know the machineability, strength, color, texture, decay resistance, and so forth of the woods they use.

A carpenter can read a grade stamp on lumber or sheet goods to determine if code requirements have been met. This in turn helps assure quality. Using inferior materials for a project will cause problems even before the project is completed.

More than one builder has had to correct errors after inappropriate lumber was installed in a frame system. In one incident, a carpenter installed floor joists with a number 3 grade when the specs called for a number 2 grade. The problem surfaced when a building inspector noticed the grade stamp, after the subflooring was in place and the walls had been erected. The most economical solution was to place an additional number 2 grade joist alongside the original, thus doubling the cost of the floor joists. By using lumber with the appropriate grade stamp, the code requirements were met. With a knowledgeable and alert carpenter that problem could have been avoided.

The utilization of building materials will be greatly enhanced by their proper placement and storage at the construction site. As the lumber, sheet goods, and preengineered products arrive at the building site, they must be properly stored until used. A proper base system such as pallets or support runners must be used to keep the materials off the ground and free of moisture. The location should be relatively flat and offer easy access for a forklift or loader. Additionally the material should be within easy reach of construction workers but not block vehicles from unloading other materials.

Polyethylene film or other impermeable barriers should be used to cover the materials and protect them from excess moisture. Air circulation is necessary to prevent condensation from forming on the underside of the barrier during long periods of storage.

Many construction sites need security to prevent pilferage and vandalism. Some estimates have placed theft and vandalism damage in the billions of dollars across the country. All of this adds to the cost of construction, which eventually is paid by the owners and their customers.

There will continue to be a need for carpenters to maintain a good working knowledge of the materials and products they install. By knowing the basic composition of a product, its structural rating, and its appropriate uses, the carpenter can help guarantee quality. Quality never seems to go out of style. People are continually looking for quality and expect it in the construction of their homes and places of business. A carpenter's problem-solving skills will be greatly enhanced by knowing as much as possible about building materials. The end result will normally reflect greater safety, quality construction, and profit for the builder.

FOCUS QUESTIONS: For individual or group discussion

1. What are some reasons for carpenters to understand the characteristics of the materials they install?
2. How can being able to fully read and interpret a lumber grade stamp help avoid potential problems while framing?
3. Assuming that someday you will be involved in the design and selection of construction materials, how do you think your decisions will be made? Will a good knowledge of material characteristics play a part? How?

UNIT

4

Fasteners

Many kinds of fasteners are used in the construction industry. It is important to know the best kind and size to use for each job.

OBJECTIVES

After completing this unit, the student should be able to name and describe the following commonly used fasteners and select them for appropriate use:

- nails
- screws and lag screws
- bolts
- solid and hollow wall anchors
- adhesives

UNIT CONTENTS

CHAPTER 9 NAILS, SCREWS, AND BOLTS

Nails, screws, and bolts are the most widely used of all fasteners. They come in many styles and sizes. The carpenter must know what is available and wisely select those most appropriate for fastening various materials under different conditions.

Nails

There are hundreds of kinds of nails manufactured for just about any kind of fastening job. They differ according to purpose, shape, material, coating, and in other ways. Nails are made of aluminum, brass, copper, steel, and other metals. Some nails are hardened so that they can be driven into masonry without bending. Only the most commonly used nails are described in this chapter (Fig. 9-1).

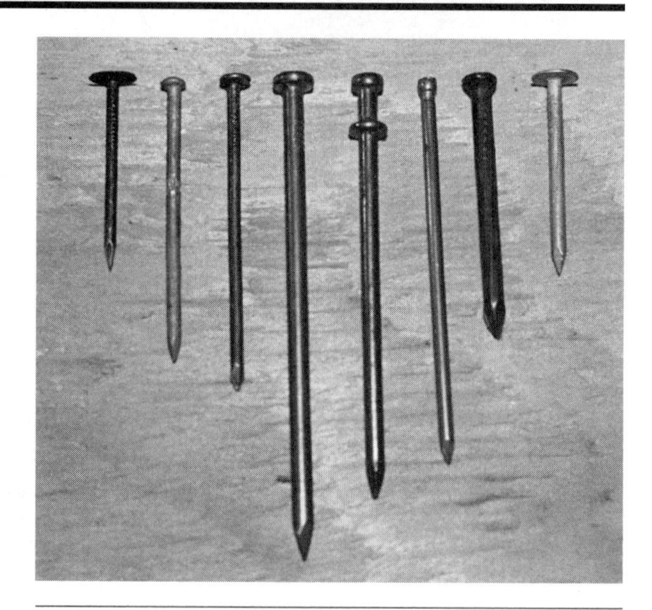

Fig. 9-1 Kinds of commonly used nails.

Uncoated steel nails are called *bright* nails. Various coatings may be applied to reduce corrosion, increase holding power, and enhance appearance. To prevent rusting, steel nails are coated with *zinc.* These nails are called **galvanized** nails. They may be coated by being dipped in molten zinc (*hot-dipped galvanized* nails), or they may be *electroplated* with corrosion-resistant metal (*plated* nails). Hot-dipped nails have a heavier coating than plated nails and are more resistant to rusting. Many manufacturers specify that their products be fastened with hot-dipped nails because of the heavier coating.

When fastening metal that is going to be exposed to the weather, use nails of the same material if possible. For example, when fastening aluminum, copper, or galvanized iron, use nails made of the same metal. Otherwise, a reaction with moisture and the two different metals, called **electrolysis,** will cause one of the metals to disintegrate over time.

When fastening some woods, such as cedar, redwood, and oak, exposed to the weather, use stainless steel nails. Otherwise, a reaction between the acid in the wood and bright nails causes dark, ugly stains to appear around the fasteners.

Nail Sizes

The sizes of some nails are designated by the **penny** system. The origin of this system of nail measurement is not clear. Although many people think it should be discarded, it is still used in the United States. Some believe it originated years ago when nails cost a certain number of pennies for a hundred of a specific length. Of course, the larger nails cost more per hundred than smaller ones, so nails that cost 8 pennies were larger than those that cost 4 pennies. The symbol for penny is *d;* perhaps it is the abbreviation for *denarius,* an ancient Roman coin.

In the penny system the shortest nail is 2d and is 1 inch long. The longest nail is 60d and is 6 inches long (Fig. 9-2). A sixpenny nail is written as 6d and is 2 inches long. Eventually, a carpenter can determine the penny size of nails just by looking at them.

The thickness or diameter of the nail is called its gauge. In the penny system, gauge depends on the kind and length of the nail. The gauge increases with the length of the nail. Long nails, 20d and over, are called **spikes.** The length of nails not included in the penny system is designated in inches and fractions of an inch, and the gauge may be specified by a number. As a general rule, select a nail that is three times longer than the material being fastened.

Kinds of Nails

Most nails, cut from long rolls of metal wire, are called *wire nails. Cut nails,* used only occasionally, are wedge-shaped pieces stamped from thin sheets of metal. The most widely used wire nails are the common, box, and finish nails (Fig. 9-3).

Common Nails. *Common nails* are made of wire, are of heavy gauge, and have a medium-sized head. They have a pointed end and a smooth shank. A barbed section just under the head increases the holding power of common nails.

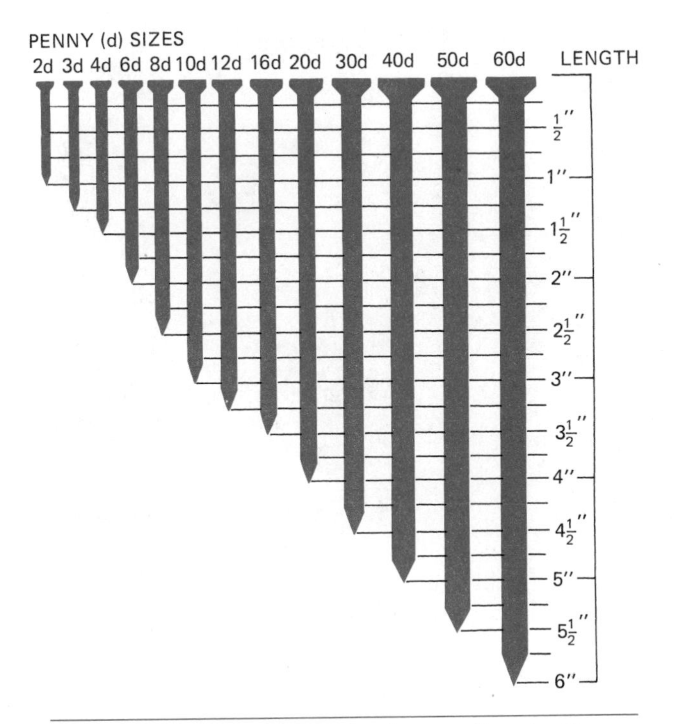

Fig. 9-2 Some nails are sized according to the penny system.

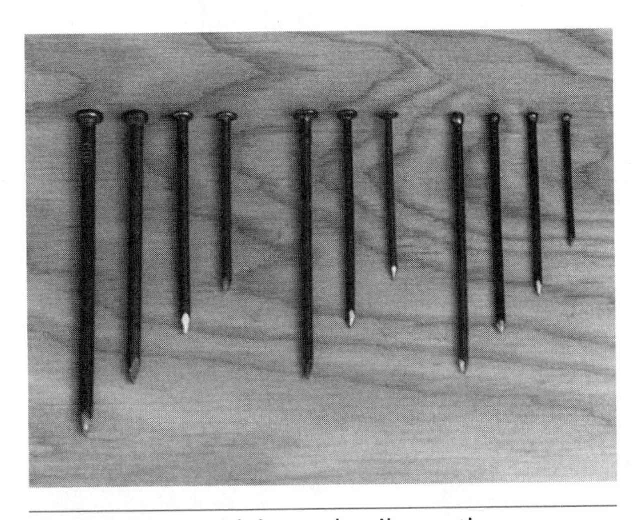

Fig. 9-3 Most widely used nails are the common, box, and finish.

Box Nails. Box nails are similar to common nails, except they are thinner. Because of their small gauge, they can be used close to edges and ends with less danger of splitting the wood. Many box nails are coated with resin cement to increase their holding power.

Finish Nails. Finish nails are of light gauge with a very small head. They are used mostly to fasten interior trim. The small head is sunk into the wood with a nail set and covered with a filler. The small head of the finish nail does not detract from the appearance of a job as much as would a nail with a larger head.

Casing Nails. *Casing nails* are similar to finish nails. Many carpenters prefer them to fasten exterior finish. The head is cone-shaped and slightly larger than that of the finish nail, but smaller than that of the common nail. The shank is the same gauge as that of the common nail.

Duplex Nails. On temporary structures, such as wood scaffolding and concrete forms, the **duplex nail** is often used. The lower head ensures that the piece is fastened tightly. The projecting upper head makes it easy to pry the nail out when the structure is dismantled (Fig. 9-4).

Brads. Brads are small finishing nails (Fig. 9-5). They are sized according to length in inches and

Fig. 9-4 Duplex nails are used on temporary structures.

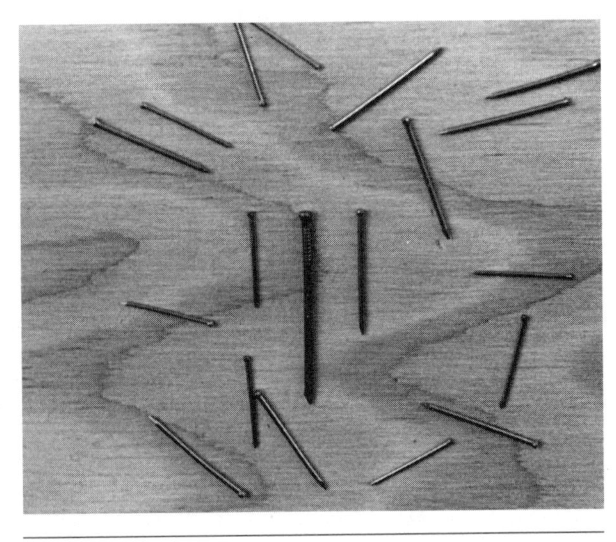

Fig. 9-5 Brads are small finishing nails.

gauge. Usual lengths are from 1/2 inch to 1 1/2 inches, and gauges run from #14 to #20. The higher the gauge number, the thinner the brad. Brads are used for fastening thin material, such as small molding.

Roofing Nails. *Roofing nails* are short nails of fairly heavy gauge with wide, round heads. They are used for such purposes as fastening roofing material and softboard wall sheathing. The large head holds thin or soft material more securely.

Some roofing nails are coated to prevent rusting. Others are made from noncorrosive metals such as aluminum and copper. The shank is usually barbed to increase holding power. Usual sizes run from 3/4 inch to 2 inches. The gauge is not specified when ordering.

Masonry Nails. *Masonry nails* may be cut nails or wire nails (Fig. 9-6). These nails are made from hardened steel to prevent them from bending when being driven into concrete or other masonry. The cut nail has a blunt point that tends to prevent splitting when it is driven into hardwood. Some masonry and flooring nails have round shanks of various designs for better holding power.

Miscellaneous Fasteners. Tacks are small pointed nails with a head (Fig. 9-7) used to fasten thin material such as wire-mesh screen. They are available in sizes 1 to 24, with the larger number designating the longer tack.

Staples are U-shaped fasteners (Fig. 9-8). They come in a number of sizes, which are designated by length. Two most popular sizes are 3/4 inch and 1 1/2 inches, with the gauge of the staple increasing with the length. They are usually galvanized and are frequently used to fasten wire fencing and heavy wire mesh. They can be driven individually with a hammer. Widely used in the industry are staples that come glued together in rows to be driven by hand (Fig. 9-9) or power staplers (Fig. 9-10).

Corrugated fasteners are thin, wrinkled metal pieces used to fasten joints (Fig. 9-11). They are used on light or rough work and on finish work where the fastener is not exposed. They are available in individual pieces to be driven with a hammer or in glued strips for use in power drivers. The usual widths are 1/2 inch and 5/8 inch.

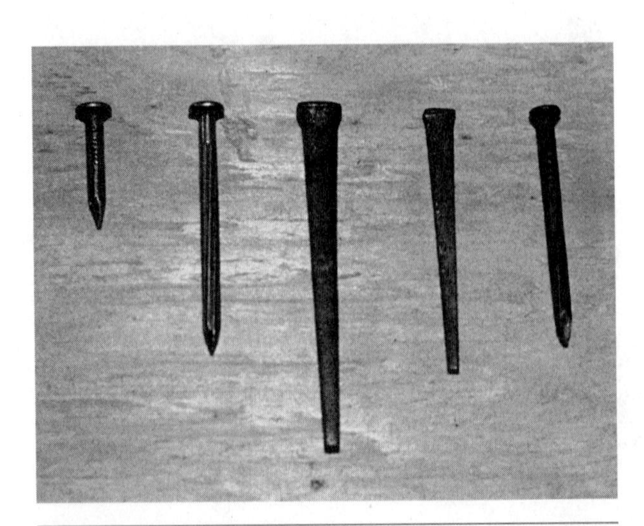

Fig. 9-6 Masonry nails are made of hardened steel.

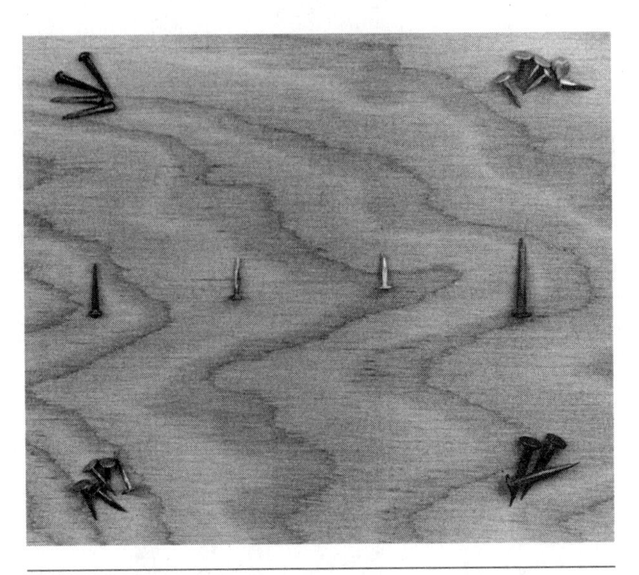

Fig. 9-7 Tacks are sometimes used to fasten screen mesh or other thin material.

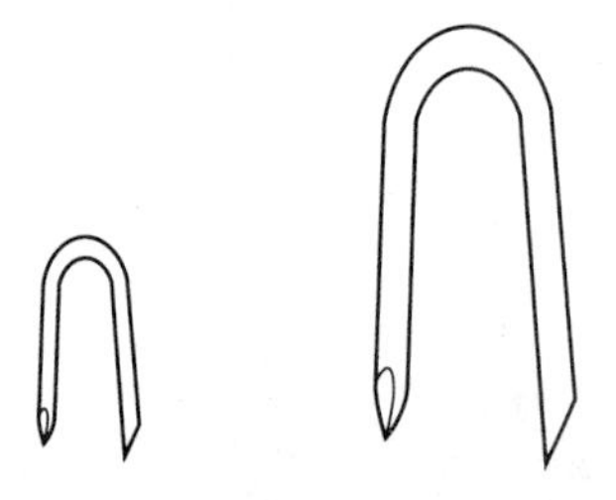

Fig. 9-8 Staples are U-shaped nails.

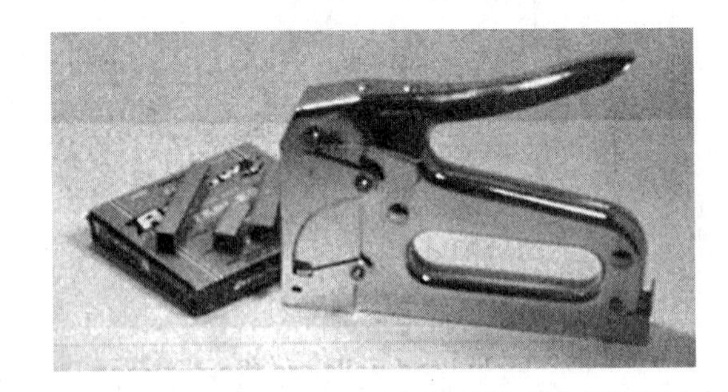

Fig. 9-9 Staples may come glued together in strips for use in power staplers.

Fig. 9-10 Using a power stapler to fasten plywood roof sheathing. (*Courtesy of American Plywood Association*)

Fig. 9-11 Corrugated fasteners are used to fasten joints. (*Courtesy of American Plywood Association*)

Screws

Wood screws are used when greater holding power is needed and when the work being fastened must at times be removed. For example, door hinges must be applied with screws because nails would pull loose after a while, and the hinges may, at times, need to be removed. Screws cost more than nails and require more time to drive. When ordering screws, specify the length, gauge, type of head, coating, kind of metal, and screwdriver slot.

Kinds of Screws

A wood screw is identified by the shape of the screwhead and screwdriver slot. For instance, a screw may be called a *flat head Phillips* or a *round head common* screw. Three of the most common shapes of screwheads are the *flat head, round head,* and *oval head.* Other shapes include the *truss head* and *fillister head.*

The pointed end of a screw is called the *gimlet point.* The threaded section is called the *thread.* The smooth section between the head and thread is called the *shank.* Screw lengths are measured from the point to that part of the head that sets flush with the wood when fastened (Fig. 9-12).

Screw Slots. A screwhead that is made with a straight, single slot is called a *common screw.* A **Phillips head** screw has a crossed slot. There are many other types of screwdriver slots, each with a different name (Fig. 9-13).

Sheet Metal Screws. Wood screws are threaded, in most cases, only partway to the head.

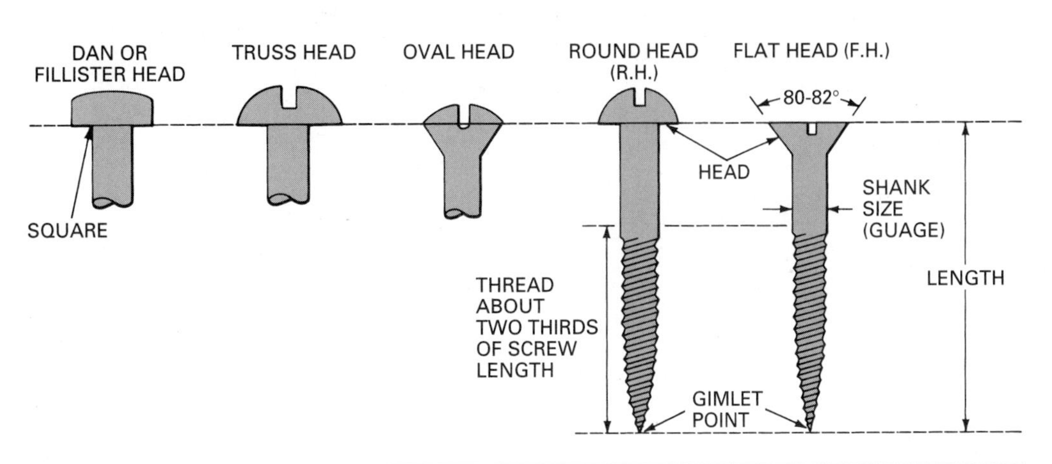

Fig. 9-12 Common kinds of screws and screw terms.

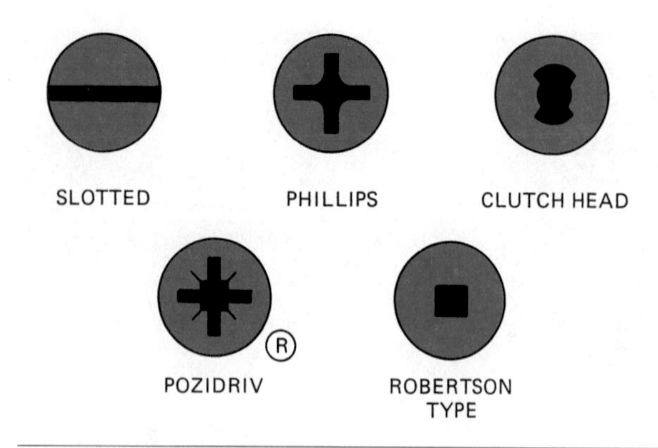

Fig. 9-13 Common kinds of screw slots.

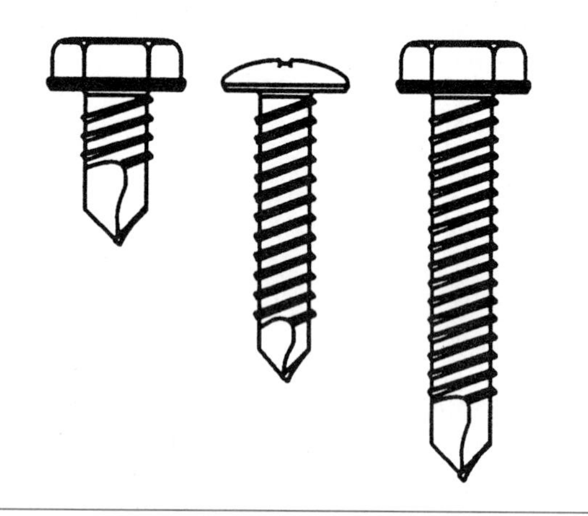

Fig. 9-14 The point of a self-drilling screw has cutting edges that drill a hole as the screw is driven.

The threads of sheet metal screws extend for the full length of the screw and are much deeper. Sheet metal screws are used for fastening thin metal. They are also recommended for fastening hardboard and particleboard because their deeper thread grabs better in softer and fiberless material.

Another type of screw, used with power screwdrivers, is the *self-drilling screw,* which is used extensively to fasten metal framing. This screw has a cutting edge on its point to eliminate predrilling a hole (Fig. 9-14). It is important that the drilling process be completed before the threading process begins. Drill points are available in various lengths and must be equal to the thickness of the metal being fastened.

Many other screws are available that are designed for special purposes. Like nails, screws come in a variety of metals and coatings. Steel screws with no coating are called *bright* screws.

Screw Sizes

Screws are made in many different sizes. Usual lengths range from 1/4 inch to 4 inches. Gauges run from 0 to 24 (Fig. 9-15). Unlike brads and some nails, the higher the gauge number, the greater the diameter of the screw. Screw lengths are not available in every gauge. The lower gauge numbers are for shorter, thinner screws. Higher gauge numbers are for longer screws.

Lag Screws

Lag screws (Fig. 9-16) are similar to wood screws except that they are larger and have a square or hex head designed to be turned with a wrench instead

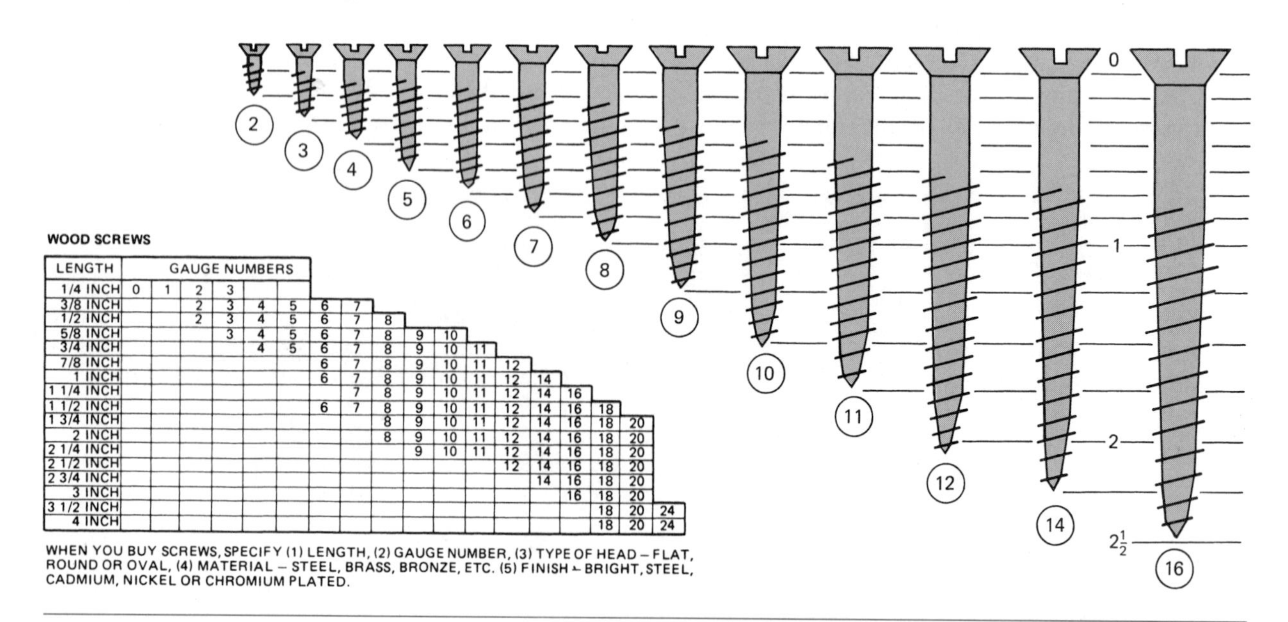

WOOD SCREWS

LENGTH	GAUGE NUMBERS																
1/4 INCH	0	1	2	3													
3/8 INCH			2	3	4	5	6	7									
1/2 INCH			2	3	4	5	6	7	8								
5/8 INCH				3	4	5	6	7	8	9	10						
3/4 INCH					4	5	6	7	8	9	10	11					
7/8 INCH							6	7	8	9	10	11	12				
1 INCH							6	7	8	9	10	11	12	14			
1 1/4 INCH								7	8	9	10	11	12	14	16		
1 1/2 INCH							6	7	8	9	10	11	12	14	16	18	
1 3/4 INCH									8	9	10	11	12	14	16	18	20
2 INCH									8	9	10	11	12	14	16	18	20
2 1/4 INCH										9	10	11	12	14	16	18	20
2 1/2 INCH													12	14	16	18	20
2 3/4 INCH														14	16	18	20
3 INCH															16	18	20
3 1/2 INCH																18	20	24
4 INCH																18	20	24

WHEN YOU BUY SCREWS, SPECIFY (1) LENGTH, (2) GAUGE NUMBER, (3) TYPE OF HEAD – FLAT, ROUND OR OVAL, (4) MATERIAL – STEEL, BRASS, BRONZE, ETC. (5) FINISH – BRIGHT, STEEL, CADMIUM, NICKEL OR CHROMIUM PLATED.

Fig. 9-15 Wood screw sizes.

Fig. 9-16 Lag screws are large screws with a square or hex head.

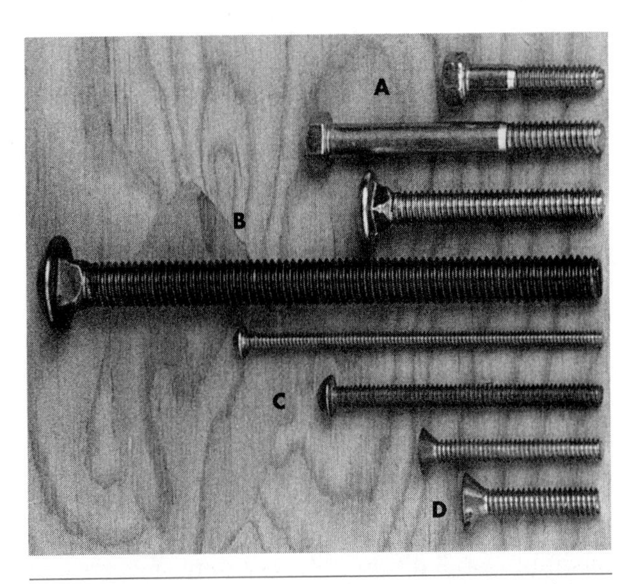

Fig. 9-17 Commonly used bolts include Machine (A), Carriage (B), Round-head stove (C), and Flat-head stove (D).

of a screwdriver. This fastener is used when great holding power is needed to join heavy parts and where a bolt cannot be used.

Lag screws are sized by diameter and length. Diameters range from 1/4 inch to 1 inch, with lengths from 1 inch to 12 inches and up. Shank and pilot holes to receive lag screws are drilled in the same manner as for wood screws. (See Chapter 13.) Place a flat washer under the head to prevent the head from digging into the wood as the lag screw is tightened down. Apply a little wax to the threads to allow the screw to turn more easily and to prevent the head from twisting off.

Bolts

Most bolts are made of steel. To retard rusting, galvanized bolts are used. As with nails and screws, they are available in different kinds of metal and coatings. Many kinds are used for special purposes, but only a few are generally used.

Kinds of Bolts

Commonly used bolts are the carriage, machine, and stove bolts (Fig. 9-17).

Carriage Bolts. The *carriage bolt* has a square section under its oval head. The square section is embedded in wood and prevents the bolt from turning as the nut is tightened.

Machine Bolts. The *machine bolt* has a square or hex head. This is held with a wrench to keep the bolt from turning as the nut is tightened.

Stove Bolts. *Stove bolts* have either round or flat heads with a screwdriver slot. They are usually threaded all the way up to the head. *Machine screws* are very similar to stove bolts.

Bolt Sizes

Bolt sizes are specified by diameter and length. Carriage and machine bolts range from 3/4 inch to 20 inches in length and from 3/16 to 3/4 inch in diameter. Stove bolts are small in comparison to other bolts. They commonly come in lengths from 3/8 inch to 6 inches and from 1/8 to 3/8 inch in diameter.

Drill holes for bolts the same diameter as the bolt. Use flat washers under the head (except for carriage bolts) and under the nut to prevent the nut from cutting into the wood and to distribute the pressure over a wider area. Apply a little light lubricating oil to the threads before the nut is turned. Use wrenches of the correct size to tighten the bolt. Be careful not to over-tighten carriage bolts. The head need only be drawn snug, not pulled below the surface.

CHAPTER 10 ANCHORS AND ADHESIVES

Anchors

Special kinds of fasteners used to attach parts to solid masonry and hollow walls and ceilings are called **anchors.** There are hundreds of types available. Those most commonly used are described in this chapter.

Solid Wall Anchors

Solid wall anchors may be classified as heavy-, medium-, or light-duty. Heavy-duty anchors are used to install such things as machinery, hand rails, dock bumpers, and storage racks. Medium-duty anchors may be used for hanging pipe and ductwork, securing window and door frames, and installing cabinets. Light-duty anchors are used for fastening such things as junction boxes, bathroom fixtures, closet organizers, small appliances, smoke detectors, and other lightweight objects.

Heavy-Duty Anchors. The *wedge anchor* (Fig. 10-1) is used when high resistance to pullout is required. The anchor and hole diameter are the same, simplifying installation. The hole depth is not critical as long as the minimum is drilled. Proper installation requires cleaning out the hole (Fig. 10-2).

The *self-drilling anchor* drills its own hole, usually with the aid of a rotary hammer drill. After using the anchor to drill the hole, remove the anchor. Clean both the hole and the anchor. Insert the external tapered plug into the toothed end of the anchor. Place the assembly, plug down, into the hole until it bottoms out. Drive the anchor over the plug and shear off the top.

The *stud anchor* (Fig. 10-3) and hole diameter are the same. This feature eliminates hole layout time because holes are drilled in masonry through the mounting holes in the fixture being installed. However, this anchor requires precise hole depth for proper expansion.

The *sleeve anchor* (Fig. 10-4) and its hole size are also the same, but the hole depth need not be exact. After inserting the anchor in the hole, it is expanded by tightening the nut.

The *drop-in anchor* (Fig. 10-5) consists of an expansion shield and a cone-shaped, internal expander plug. The hole must be drilled at least equal to the length of the anchor. A setting tool, supplied with the anchors, must be used to drive and expand the anchor. This anchor takes a machine screw or bolt.

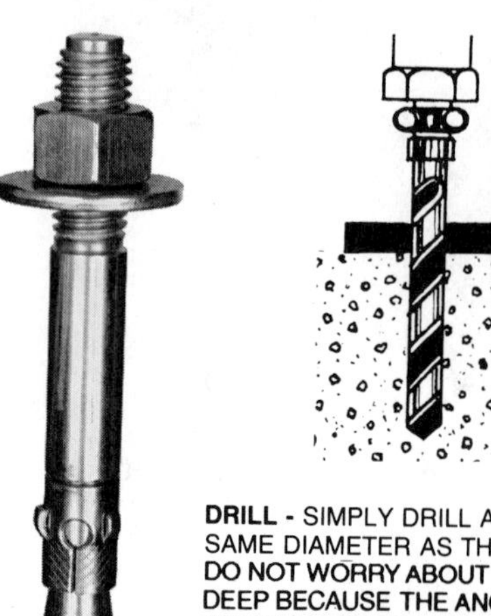

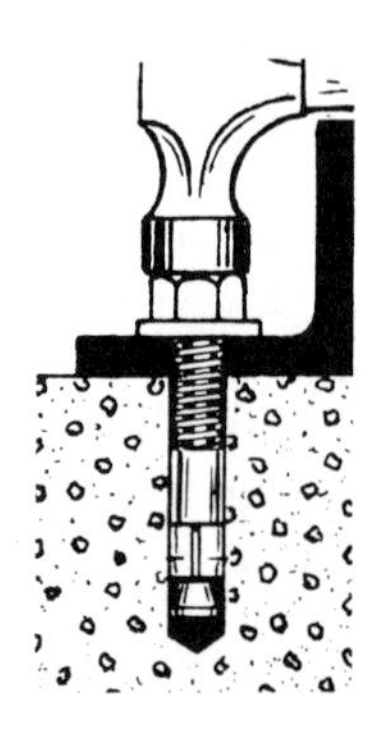

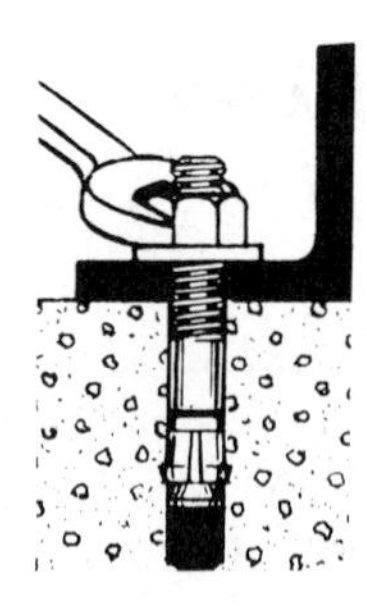

Fig. 10-1 The wedge anchor has high resistance to pullout. (*Courtesy of U.S. Anchor, Pompano Beach, Florida*)

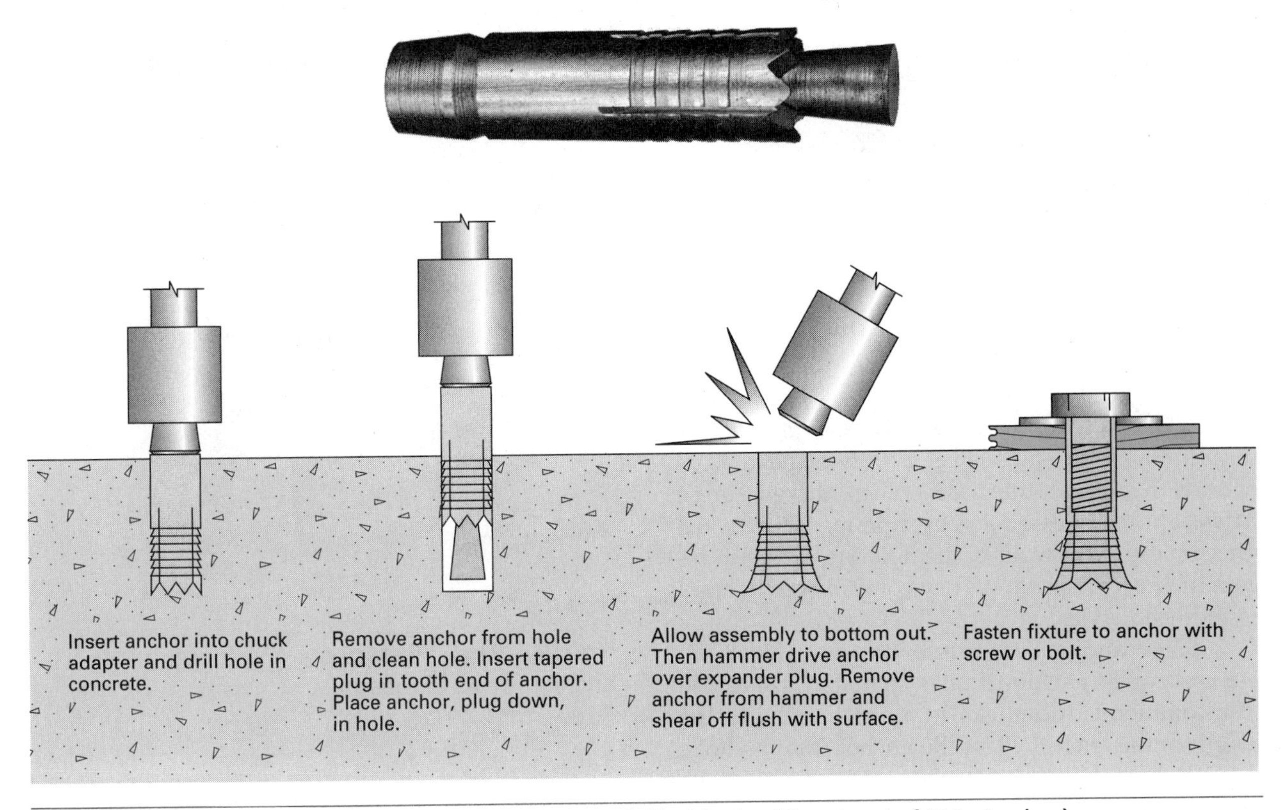

Fig. 10-2 The self-drilling anchor requires no predrilled hole. (*Courtesy of U.S. Anchor*)

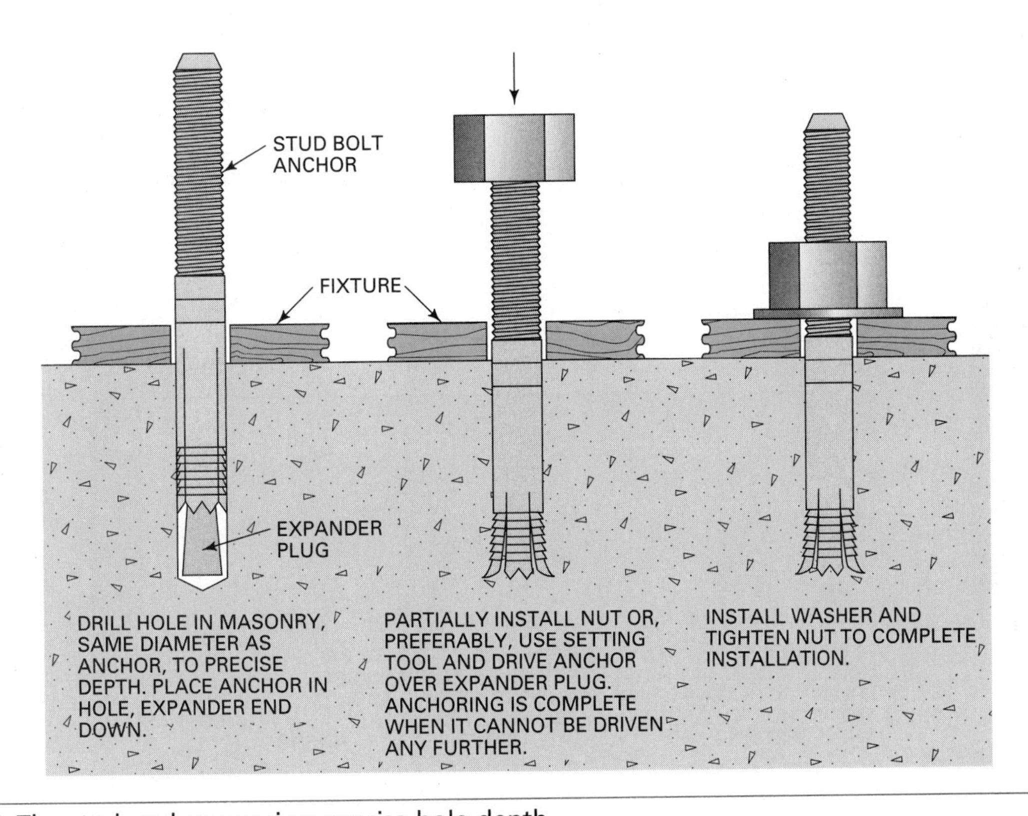

Fig. 10-3 The stud anchor requires precise hole depth.

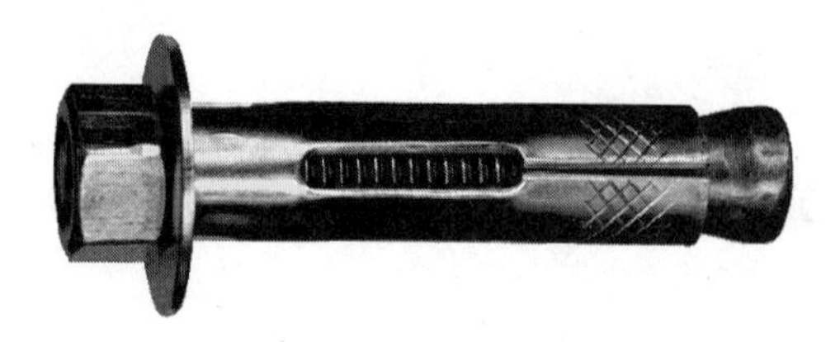

Fig. 10-4 Sleeve anchors eliminate the problem of exact hole depth requirements. (*Courtesy of U.S. Anchor*)

Medium-Duty Anchors. *Split fast anchors* (Fig. 10-6) are one-piece steel with two sheared expanded halves at the base. When driven, these halves are compressed and exert immense outward force on the inner walls of the hole as they try to regain their original shape. They come in both flat and round head styles.

Single and *double expansion anchors* (Fig. 10-7) are used with machine screws or bolts. Drill a hole of recommended diameter to a depth equal to the length of the anchor. Place the anchor into the hole, flush with or slightly below the surface. Position the object to be fastened and bolt into place. Once fastened, the object may be unbolted, removed, and refastened, if desired.

The *lag shield* (Fig. 10-8) is used with a lag screw. The shield is a split sleeve of soft metal, usually a zinc alloy. It is inserted into a hole of recommended diameter and a depth equal to the length of the shield plus 1/2 inch or more. The lag screw

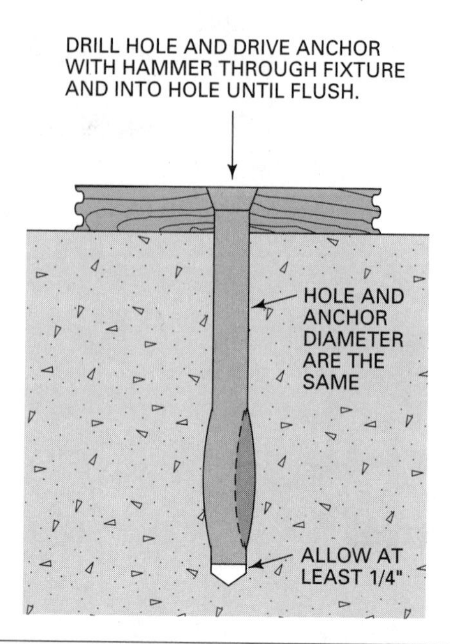

Fig. 10-6 The split fast is a one-piece, all-steel anchor for hard masonry.

length is determined by adding the length of the shield, the thickness of the material to be fastened, plus 1/4 inch. The tip of the lag screw must protrude from the bottom of the anchor to ensure proper expansion. As the fastener is threaded in, the shield expands tightly and securely in the drilled hole.

The *concrete screw* (Fig. 10-9) utilizes specially fashioned high and low threads that cut into a prop-

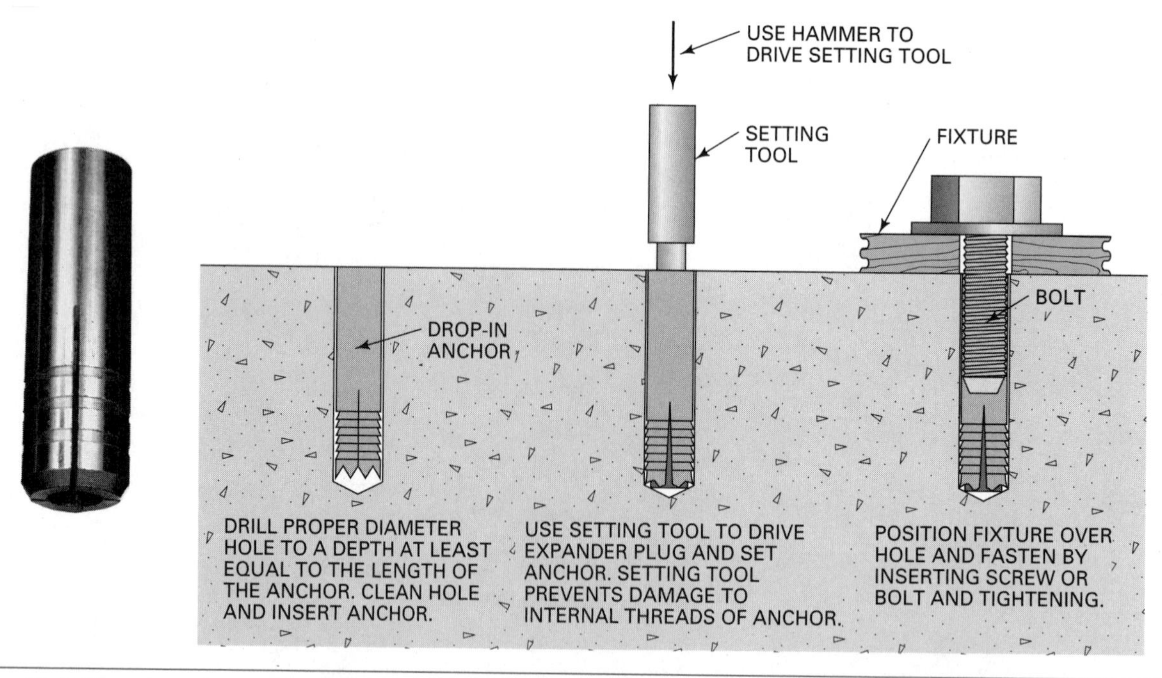

Fig. 10-5 The drop-in anchor is expanded with a setting tool. (*Courtesy of U.S. Anchor*)

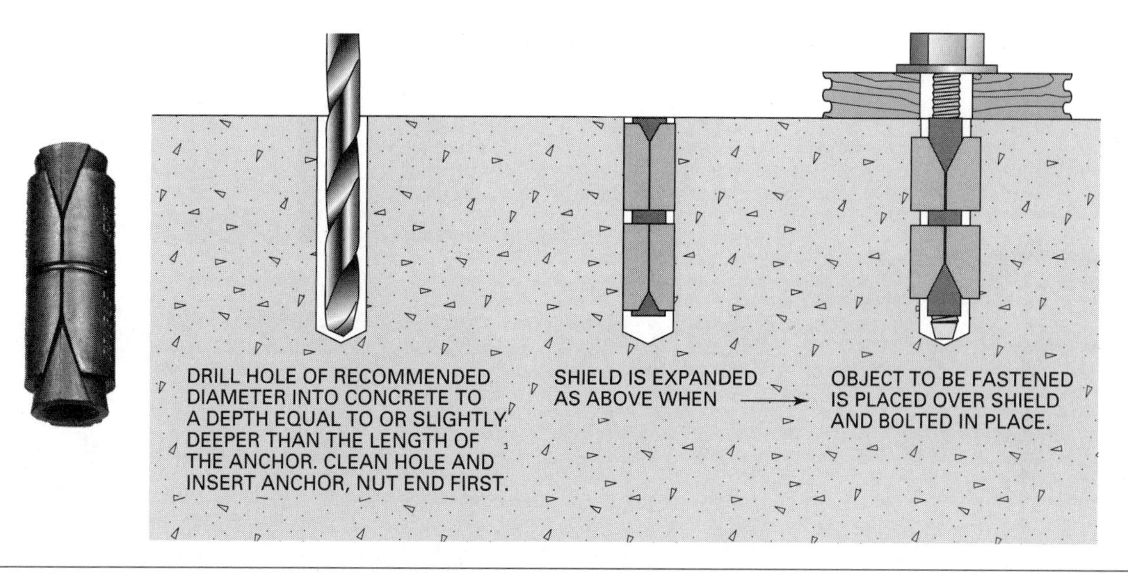

Fig. 10-7 Two opposing wedges of the double expansion anchor pull toward each other, expanding the full length of the anchor body. *(Courtesy of U.S. Anchor)*

Fig. 10-8 Lag shields are designed for light- to medium-duty fastening in masonry.

erly sized hole in concrete. Screws come in 3/16 and 1/4 inch diameters and up to 6 inches in length. The hole diameter is important to the performance of the screw. It is recommended that a minimum of 1 inch and a maximum of 1 3/4 inch embedment be used to determine the fastener length. The concrete screw system eliminates the need for plastic or lead anchors.

The *coil anchor* (Fig. 10-10) is a lag type screw that uses a wire coil around its threads. The hole is drilled the same diameter as the anchor. Anchor, washer, and coil are assembled, inserted, and tightened. The hole depth is determined by the anchor length minus the fixture thickness plus 1/4 inch minimum.

The *machine screw anchor* (Fig. 10-11) consists of two parts. A lead sleeve slides over a threaded, cone-shaped piece. Using a special setting punch that comes with the anchors, the lead sleeve is driven over the cone-shaped piece to expand the sleeve and hold it securely in the hole. A hole of recommended diameter is drilled to a depth equal to the length of the anchor.

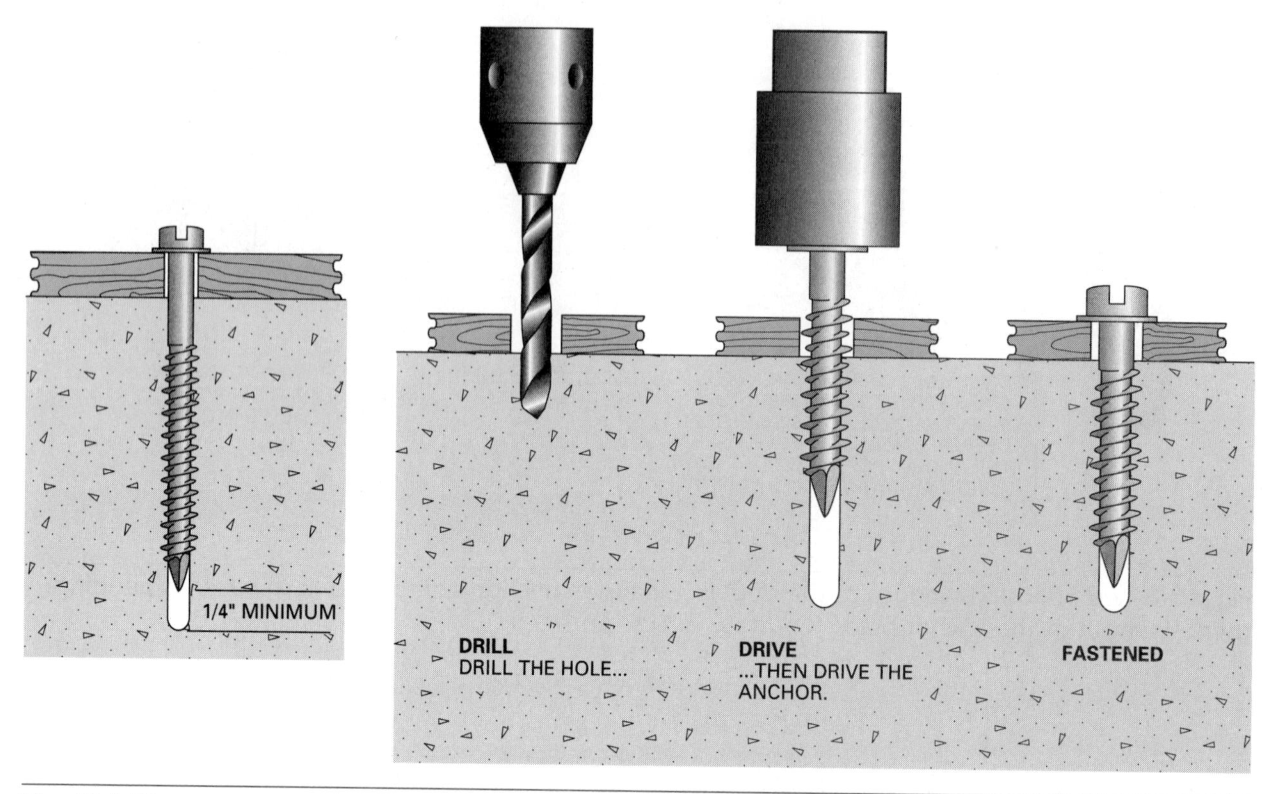

Fig. 10-9 The concrete screw system eliminates the need for an anchor when fastening into concrete.

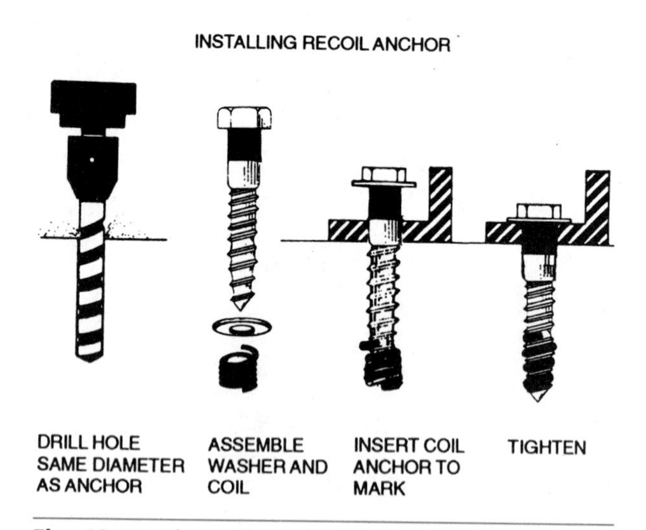

Fig. 10-10 The coil anchor utilizes a small metal coil around the threads of the screw. (*Courtesy of U.S. Anchor*)

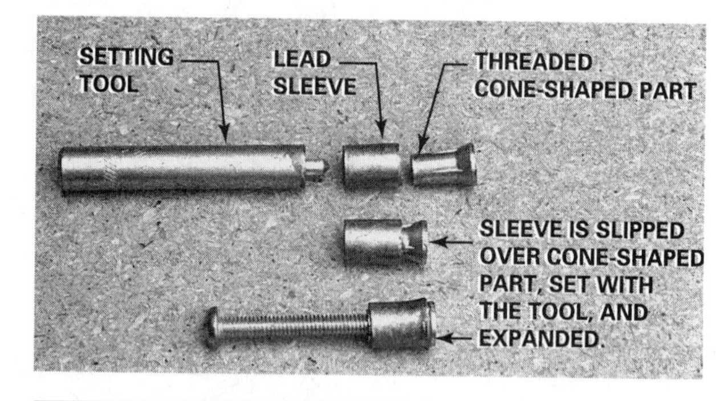

Fig. 10-11 Machine screw anchors are commonly called tamp-ins.

Light-Duty Anchors. Three kinds of *drive* anchors are commonly used for quick and easy fastening in solid masonry. The *hammer drive anchor* (Fig. 10-12) has a body of zinc alloy containing a steel expander pin. In the *aluminum drive anchor,* both the body and the pin are aluminum to avoid the corroding action of electrolysis. The *nylon nail anchor* utilizes a nylon body and a threaded steel expander pin. All are installed in a similar manner.

Lead and *plastic anchors,* also called *inserts* (Fig. 10-13), are commonly used for fastening lightweight fixtures to masonry walls. These anchors have an unthreaded hole into which a wood or sheet metal screw is driven. The anchor is placed into a hole of recommended diameter and 1/4 inch or more deeper than the length of the anchor. As the screw is turned, the threads of the screw cut into the soft material of the insert. This causes the insert to expand and tighten in the drilled hole. Ribs on the

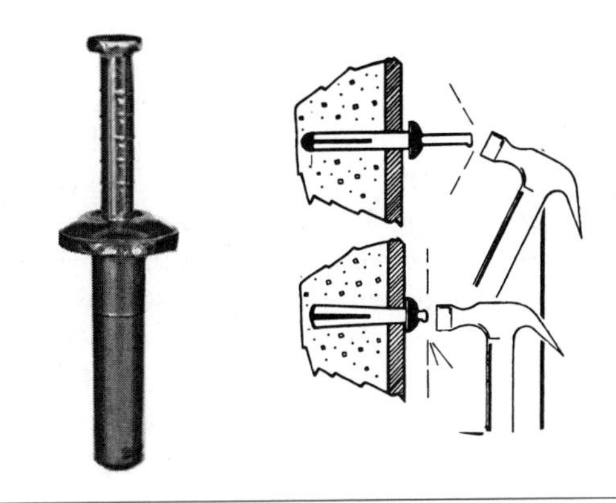

Fig. 10-12 Hammer drive anchors come assembled for quick and easy fastening. (*Courtesy of U.S. Anchor*)

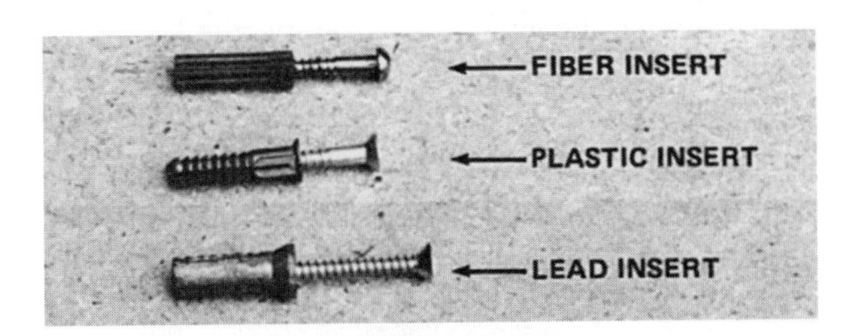

Fig. 10-13 Lead and plastic anchors or inserts are used for light-duty fastening.

sides of the anchors prevent them from turning as the screw is driven.

Chemical Anchoring Systems. Threaded studs, bolts, and re-bar (concrete reinforcing rod) may be anchored in solid masonry with a chemical bond using an *epoxy resin compound.* Two types of systems commonly used are the *epoxy injection* (Fig. 10-14) and *chemical capsule* (Fig. 10-15).

In the injection system, a dual cartridge is inserted into a tool similar to a caulking gun. The chemical is automatically mixed as it is dispensed. Small cartridges are available to accurately dispense epoxy from an ordinary caulking gun.

Chemical capsules contain the exact amount of all chemicals needed for one installation. Each capsule is marked with the appropriate hole size to be used. Drill holes according to diameter and depth indicated on each capsule. It is important to thoroughly clean and clear the hole of all concrete dust before inserting the capsule.

With all solid masonry anchors, follow the specifications in regard to hole diameter and depth, minimum embedment, maximum fixture thickness, and allowable load on anchor.

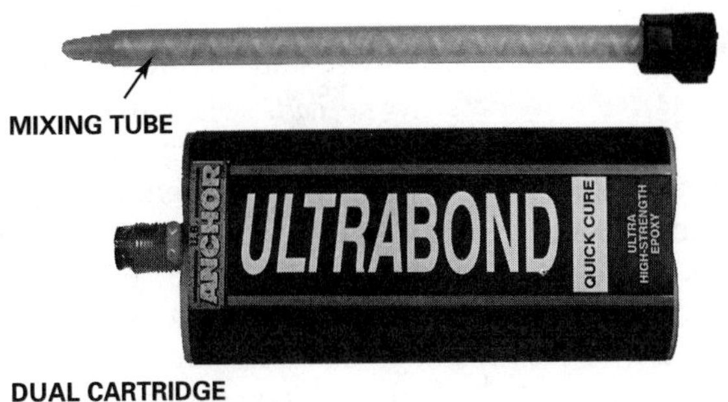

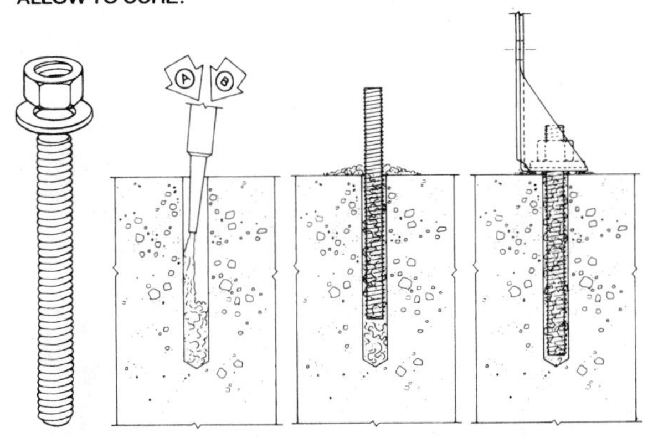

Fig. 10-14 The epoxy injection system is designed for high-strength anchoring. (*Courtesy of U.S. Anchor*)

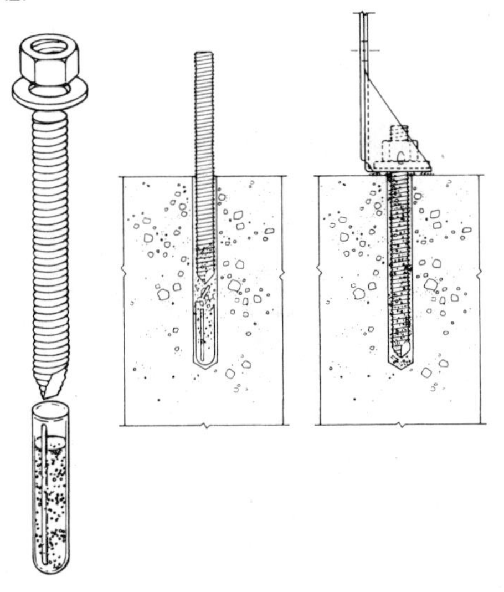

Fig. 10-15 The chemical capsule anchoring system provides for easy, premeasured application.

Hollow Wall Fasteners

Toggle Bolts. *Toggle bolts* (Fig. 10-16) may have a wing or a tumble toggle. The wing toggle is fitted with springs, which cause it to open. The tumble toggle falls into a vertical position when passed through a drilled hole in the wall. The hole must be drilled large enough for the toggle of the bolt to slip through. A disadvantage of using toggle bolts is that, if removed, the toggle falls off inside the wall.

Plastic Toggles. The *plastic toggle* (Fig. 10-17) consists of four legs attached to a body that has a hole through the center and fins on its side to prevent turning during installation. The legs collapse to allow insertion into the hole. As sheet metal screws are turned through the body, they draw in and expand the legs against the inner surface of the wall.

Expansion Anchors. Hollow wall *expansion anchors* are commonly called *molly screws* (Fig. 10-18). The anchor consists of an expandable sleeve, a machine screw, and a fiber washer. The collar on the outer end of the sleeve has two sharp prongs that grip into the surface of the wall material. This prevents the sleeve from turning when the screw is tightened to expand the anchor. After expanding the sleeve, the screw is removed, inserted through the part to be attached, and then screwed back into the anchor. Some types require that a hole be drilled, while other types have pointed ends that may be driven through the wall material.

Installed fixtures may be removed and refastened or replaced by removing the anchor screw without disturbing the anchor. Anchors are manufactured for various wall board thicknesses. Make sure to use the right size anchor for the wall thickness in which the anchor is being installed.

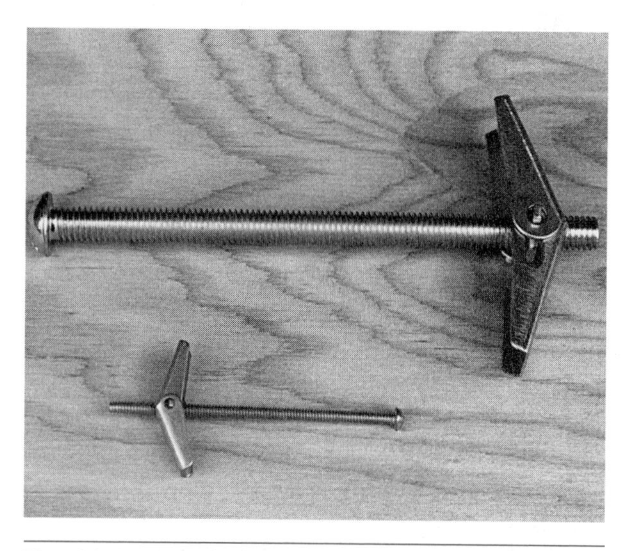

Fig. 10-16 Toggle bolts are used for fastening in hollow walls.

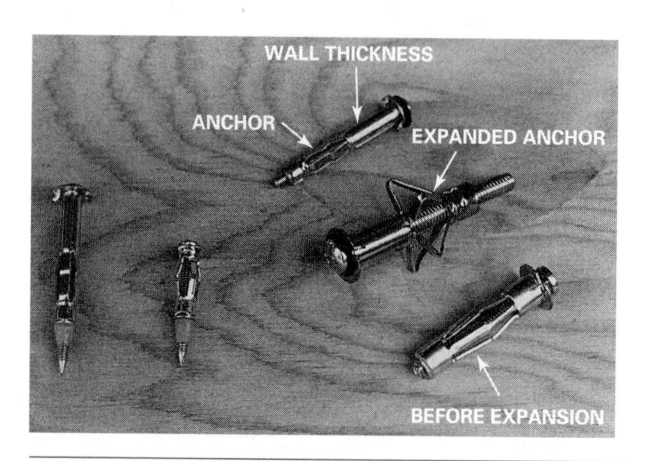

Fig. 10-18 Hollow wall expansion anchors are commonly called molly screws.

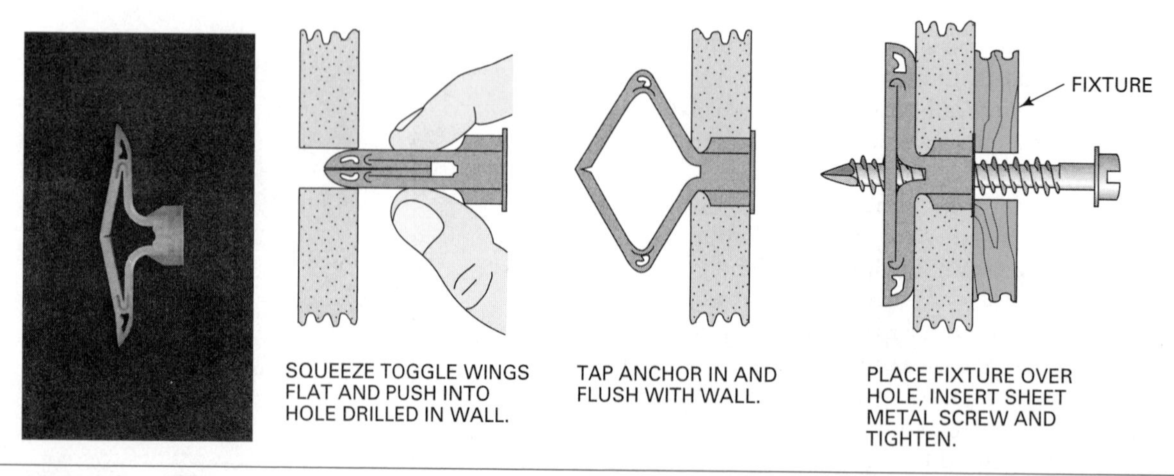

Fig. 10-17 The plastic toggle is a unique removable and reusable hollow wall anchor. (*Courtesy of U.S. Anchor*)

Conical Screws. The deep threads of the *conical screw* anchor (Fig. 10-19) resist stripping out when screwed into gypsum board, strand board, and similar material. After the plug is seated flush with the wall, the fixture is placed over the hole and fastened by driving a screw through the center of the plug.

Universal Plugs. The *universal plug* (Fig. 10-20) is made of nylon and is used for a number of hollow wall and some solid wall applications. A hole of proper diameter is drilled. The plug is inserted, and the screw is driven to draw or expand the plug.

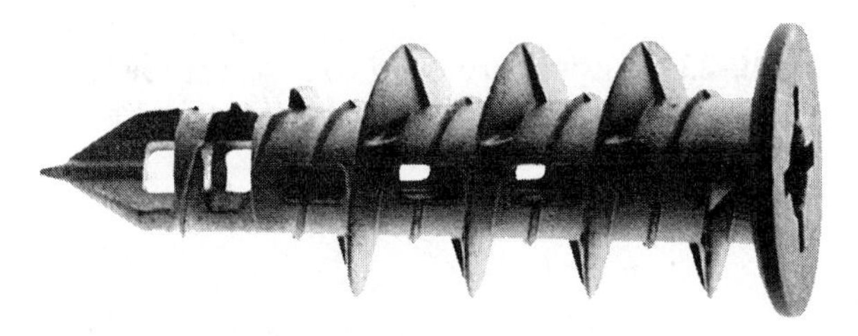

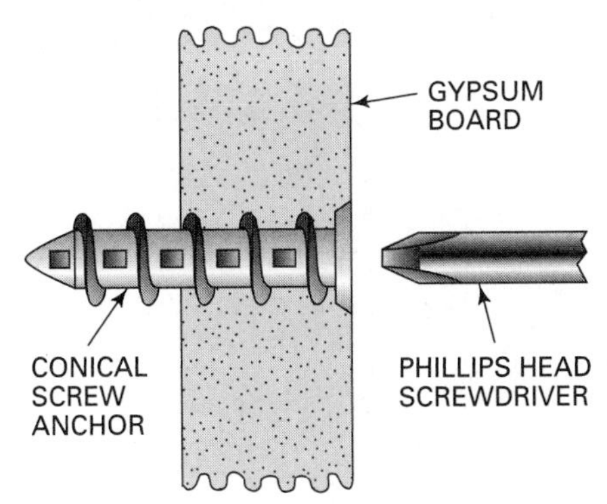

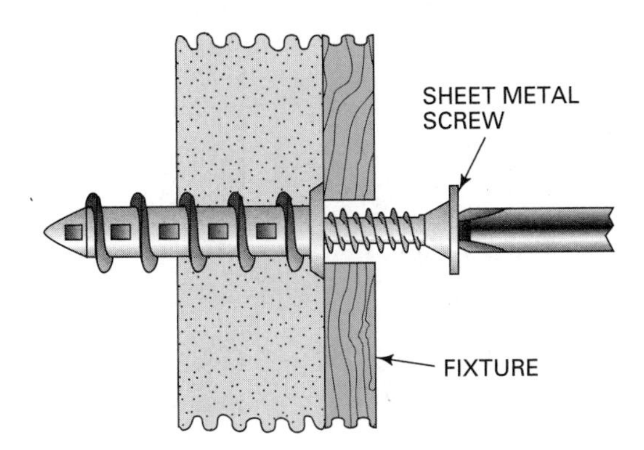

Fig. 10-19 The conical screw anchor is a self-drilling, hollow wall anchor for lightweight fastenings. (*Courtesy of U.S. Anchor*)

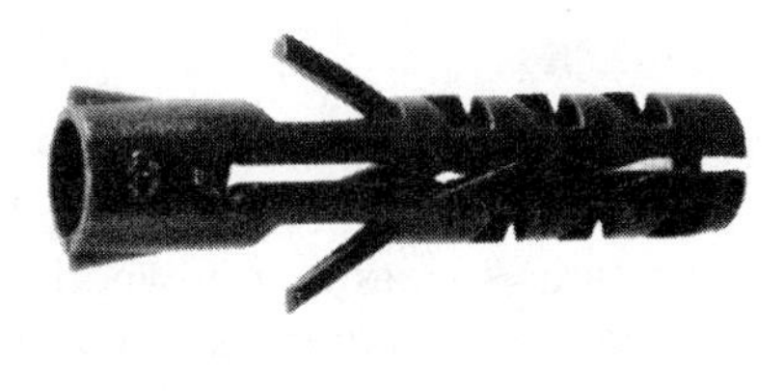

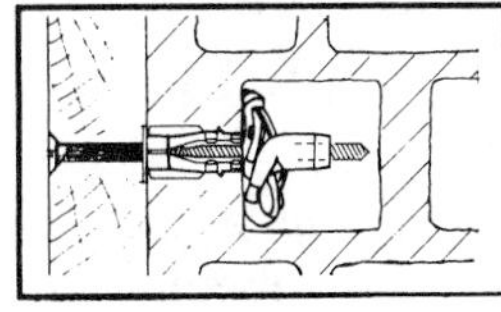

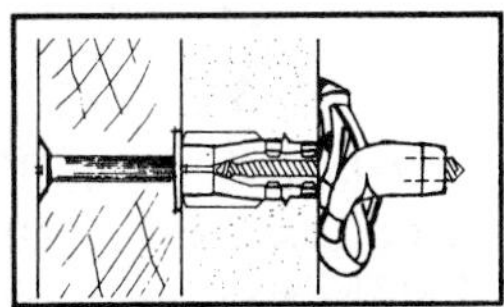

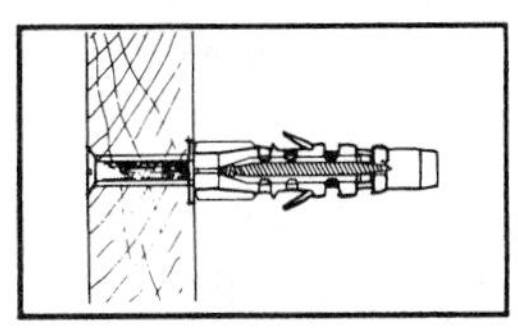

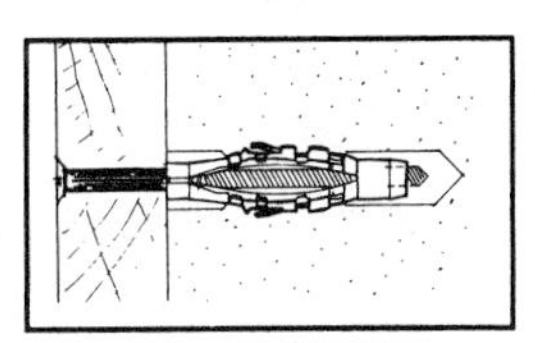

Fig. 10-20 The universal plug is used for many types of hollow wall fastening. (*Courtesy of U.S. Anchor*)

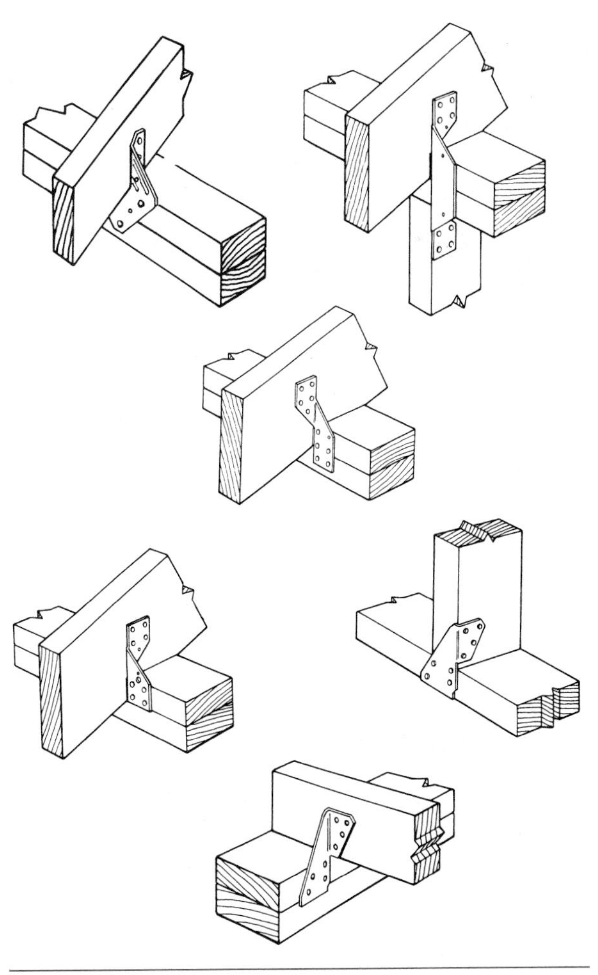

Fig. 10-21 Framing ties and anchors are manufactured in many unique shapes. (*Courtesy of Simpson Strong-Tie Co.*)

Connectors

Widely used in the construction industry are devices called *connectors.* They are metal pieces formed into various shapes to join wood to wood, or wood to concrete or other masonry. They are called specific names depending on their function.

Wood-to-Wood. *Framing anchors* and *seismic* and *hurricane ties* (Fig. 10-21) are used to join parts of a wood frame. *Post* and *column caps* and *bases* are used at the top and bottom of those members (Fig. 10-22). *Joist hangers* and *beam hangers* are available in many sizes and styles (Fig. 10-23). It is important to use the proper style, size, and quantity of nails in each hanger.

Wood-to-Concrete. Some wood-to-concrete connectors are *sill anchors, anchor bolts,* and *holdowns* (Fig. 10-24). A *girder hanger* and a *beam seat* (Fig. 10-25) make beam-to-foundation wall connections. *Post bases* come in various styles. They are used to anchor posts to concrete floors or footings.

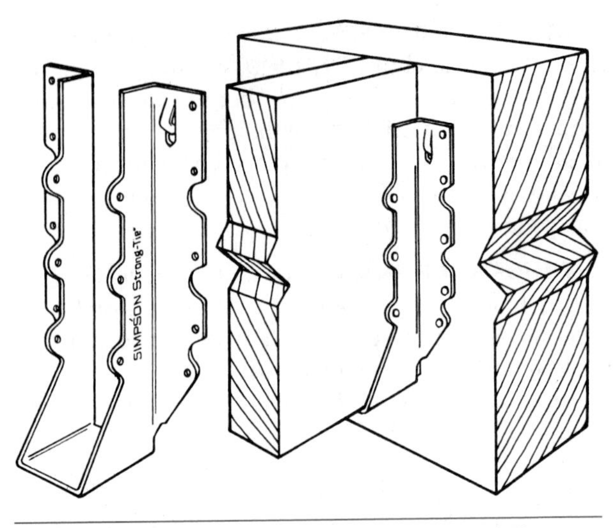

Fig. 10-23 Hangers are used to support joists and beams. (*Courtesy of Simpson Strong-Tie Co.*)

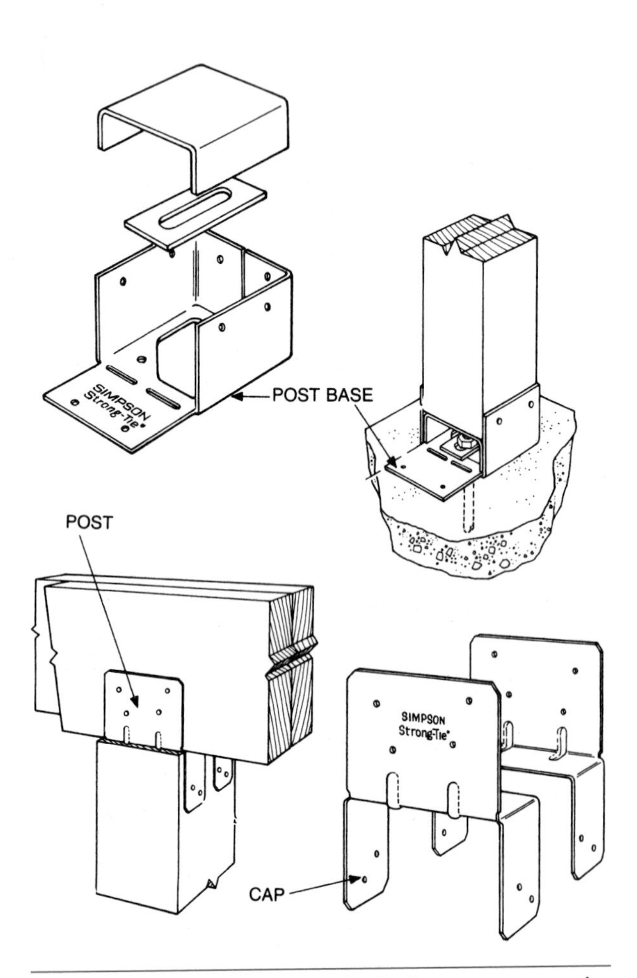

Fig. 10-22 Caps and bases help fasten tops and bottoms of posts and columns. (*Courtesy of Simpson Strong-Tie Co.*)

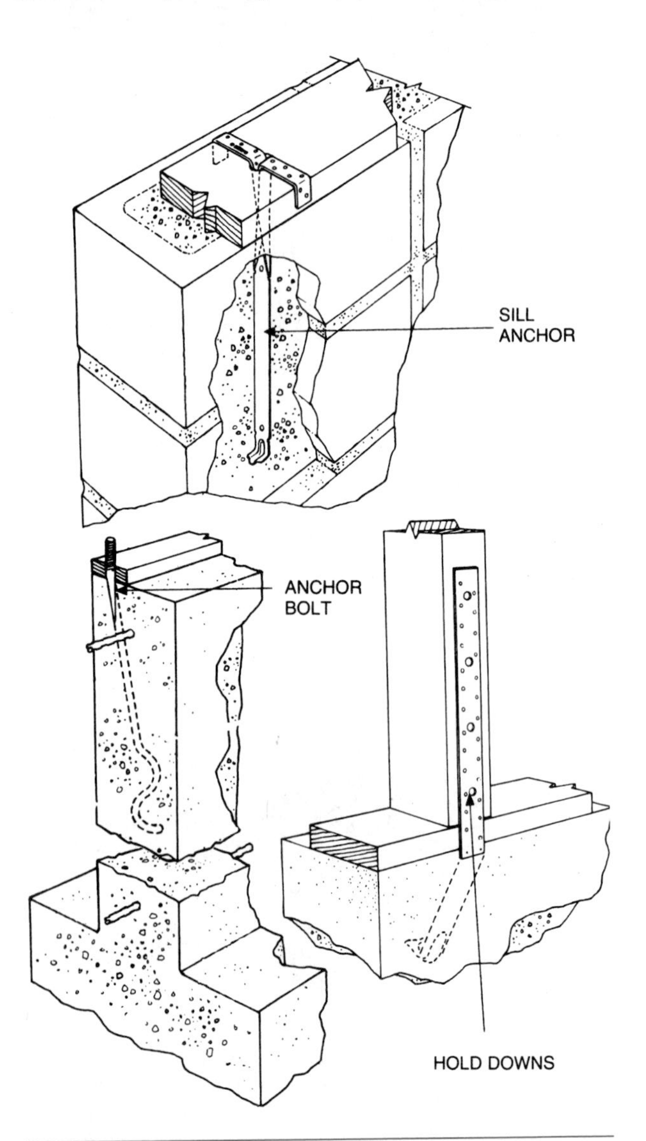

Fig. 10-24 Sill anchors, anchor bolts, and holdowns connect frame members to concrete. (*Courtesy of Simpson Strong-Tie Co.*)

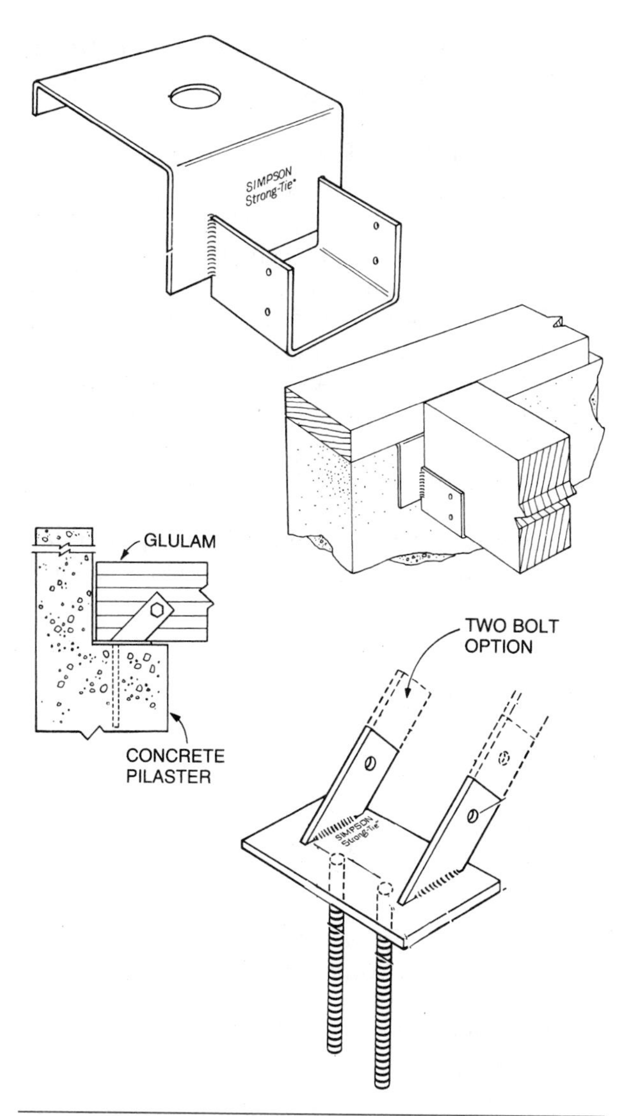

Fig. 10-25 Girder and beam seats provide support from concrete walls. (*Courtesy of Simpson Strong-Tie Co.*)

Many other specialized connectors are used in frame construction. Some are described in the framing sections of this book.

Adhesives

The carpenter seldom uses any glue in the frame or exterior finish. Glue is used on some joints and other parts of the interior finish work. A number of **mastics** (heavy, paste-like adhesives) are used throughout the construction process.

Glue

White and Yellow Glue. Most of the glue used by the carpenter is the so-called white glue or yellow glue. The white glue is *polyvinyl acetate;* the yellow glue is *aliphatic resin.* Neither type is resistant to moisture. Both are fast-setting, so joints should be made quickly after applying the glue. They are available under a number of trade names and are excellent for joining wood parts not subjected to moisture.

Urea Resin Glue. When a moisture-resistant glue is needed, *urea resin,* commonly called *plastic resin,* glue is used. This type comes in powder form and is mixed with water to the consistency of heavy cream. The glue sets slowly, and the joint must remain under pressure for eight hours or more, depending on the temperature.

Resorcinol Resin Glue. For a completely waterproof joint, *resorcinol resin* glue is used. A dry powder is mixed with a liquid part to make this glue. Pieces glued with this substance must be clamped under pressure for twelve to twenty-four hours.

Note: If the work being glued, no matter what type, is to receive a penetrating stain, it is extremely important that all excess glue be removed from the surface to be stained. A clean cloth, dampened with water, is recommended for this purpose. If excess glue is allowed to dry on the surface of the wood, it seals the surface. This prevents the stain from penetrating, and results in a blotchy stain job. Not so much care is needed to remove excess glue if the work is eventually to be painted or given a clear finish.

Contact Cement. Contact cement is so named because pieces coated with it bond on contact and need not be clamped under pressure. It is extremely important that pieces are positioned accurately before contact is made. Contact cement is widely used to apply plastic laminates for kitchen countertops. It is also used to bond other thin or flexible material that otherwise might require elaborate clamping devices.

Mastics

There are several types of mastics used throughout the construction trades. They come in cans or cartridges used in hand or air guns. With these adhesives, the bond is made stronger and fewer, if any, fasteners are needed.

Construction Adhesive. One type of mastic is called *construction adhesive.* It is used in a glued floor system, described in a following unit on floor framing. It can be used in cold weather, even on wet or frozen wood. It is also used on stairs to increase stiffness and eliminate squeaks.

Panel Adhesive. *Panel adhesive* (Fig. 10-26) is used to apply such things as wall paneling, foam

Fig. 10-26 Applying panel adhesive to stud with a caulking gun.

insulation, gypsum board, and hardboard to wood, metal, and masonry. It is usually dispensed with a caulking gun.

Troweled Mastics. Other types of mastics may be applied by hand for such purposes as installing vinyl base, vinyl floor tile, or ceramic wall tile. A notched trowel is usually used to spread the adhesive. The depth and spacing of the notches along the edges of the trowel determine the amount of adhesive left on the surface.

It is important to use a trowel with the correct notch depth and spacing. Failure to follow recommendations will result in serious consequences. Too much adhesive causes the excess to squeeze out onto the finished surface. This leaves no alternative but to remove the applied pieces, clean up, and start over. Too little adhesive may result in loose pieces.

Review Questions

Select the most appropriate answer.

1. The length of an eight penny nail is
a. 1 1/2 inches.
b. 2 inches.
c. 2 1/2 inches.
d. 3 inches.

2. Fasteners coated with zinc to retard rusting are called
a. coated.
b. dipped.
c. electroplated.
d. galvanized.

3. Brads are
a. types of screws.
b. small box nails.
c. small finishing nails.
d. kinds of stove bolts.

4. When a moisture-resistant glue is required use
a. aliphatic resin.
b. plastic resin.
c. polyvinyl acetate.
d. yellow glue.

5. Many carpenters prefer to use casing nails to fasten
a. interior finish.
b. exterior finish.
c. door casings.
d. roof shingles.

6. The blunt point on the end of a cut nail helps
a. drive the nail straight.
b. prevent splitting the wood.
c. hold the fastened material more securely.
d. start the nail in the material.

7. On temporary structures, such as wood scaffolding and concrete forms, use
a. common nails.
b. duplex nails.
c. galvanized nails.
d. spikes.

8. As a general rule, how should the length of a nail compare to the thickness of the material being fastened?
a. The same.
b. Twice as long.
c. 2 1/2 times as long.
d. 3 times as long.

9. The name of a one-piece, all steel, hammer-driven anchor is
a. wedge anchor.
b. drop-in anchor.
c. split fast anchor.
d. expansion anchor.

10. For a waterproof bond, use
a. aliphatic resin glue.
b. urea resin glue.
c. resorcinol resin glue.
d. polyvinyl acetate glue.

BUILDING FOR SUCCESS

Fasteners: So What's the Connection?

Along with each new engineered construction product, manufacturers have designed sophisticated fasteners. The traditional nail, screw, and bolt have all been improved and complemented in many ways. Holding power for fasteners and corrosion resistance have been increased. These three mainstays are still being used to securely attach components in the assembly process. Mechanical and automated devices such as electric and pneumatic systems have greatly increased the speed of construction during the past decade. A vast array of anchors and adhesives have made the assembling of materials and products easier.

Along with each new fastener has come a need to assure proper installation. Manufacturers' recommendations become the basis for numerous building code requirements. Testing for public safety by both manufacturers and independent testing agencies has helped ensure proper and safe installation procedures. With adherence to these recommended fastening procedures, the carpenter will assemble a safe and long-lasting product. In all too many cases, if proper fastening procedures are neglected, a tragedy can occur.

A number of years ago a carpenter was installing the track and other hardware for an overhead garage door. On this type of door, the track and mounting brackets for the door suspension system are lagged into the door trimmer and header. The pilot holes for the lag screws are to be drilled to the root diameter of the lag. In this case, the pilot holes greatly exceeded the root diameter. Some were driven in with a hammer, not torqued in with the required wrench. The lag screws holding the cable connecting S-hook and bracket pulled out under tension when the door was in the closed position. At the time, the carpenter had his arm over the track and received the full impact of the S-hook and mounting bracket as they pulled loose. His upper arm was severely damaged in the process.

A mistake of this magnitude should not have happened. It was costly both physically and financially for the carpenter. An informed and knowledgeable carpenter should always know the capabilities of fasteners and their proper installation requirements. It is the worker's responsibility to select the correct fastener for the job and to install it properly. Fasteners, whether anchors, bolts, nails, screws, mastics, or adhesives, are to be used with care. Directions are to be followed to avoid tragedies.

Students, apprentices, journeymen, and other skilled workers must "make the proper use connection." Safety must connect with common sense each time fasteners are used. Shortcuts may appear to save time, but usually affect quality and worker safety. There can be no substitute for safe and proper installation procedures in today's construction world.

FOCUS QUESTIONS: For individual or group discussion

1. What might be sound reasons for selecting and installing the proper fastener on the job?
2. What can be learned from the accident described in this study?
3. What are some suggestions for selecting and installing fasteners so they do not cause problems?
4. What are some good ideas to implement if you do not have the proper fastener for the job at hand?

UNIT

5

Hand Tools

Even though there has been a substantial increase in the use of power tools, there are still many occasions when hand tools must be used. There may be times when it is easier and faster to use hand tools than power tools.

Knowing how to choose the proper hand tool, how to use it skillfully, and how to keep it in good working condition are essential to the carpentry trade. A student should not underestimate the importance of hand tools and neglect their proper use and care. Carpenters are expected to have their own hand tools and keep them in working condition.

Buy tools of good quality. Inferior tools sometimes cannot be brought to a sharp edge and dull rapidly. Keep hand tools sharp and in good condition. If they get wet on the job, dry them as soon as possible, and coat them with a little light oil to prevent them from rusting.

OBJECTIVES

After completing this unit, the student should be able to:

- identify and describe the hand tools that are commonly used by the carpenter
- use each of the hand tools in a safe and appropriate manner
- sharpen and maintain hand tools in suitable working condition

UNIT CONTENTS

CHAPTER 11 LAYOUT TOOLS

Layout Tools

Many layout tools must be used by the carpenter. They are used, among other things, to measure distances, lay out lines and angles, test for depths of cuts, and set vertical and level pieces.

Measuring Tools

The ability to take measurements quickly and accurately must be mastered early in the carpenter's training. Practice reading the rule or tape to gain skill in fast and precise measuring.

Most industrialized countries use the metric system of measure. Linear metric measure centers on the meter, which is slightly larger than a yard. Smaller parts of a meter are denoted by the prefix *deci-* (1/10), which is used intsead of *feet. Centi-* (1/100) and *milli-* (1/1000) are used instead of inches and fractions. The prefix *kilo-* represents 1000 times larger and is used instead of *miles.* For example, in metric measure a 2 × 4 is 39 mm (millimeter) × 89 mm. The metric system is easier to use than the English system because all measurements are in decimal form and there are no fractions.

Rules and Tapes. *Rules* used in construction in the United States are divided into feet, inches, and usually 16ths of an inch. Most rules and tapes used by the carpenter have clearly marked increments of 16 inches to help in laying out spaced framing members.

Some carpenters prefer the *6-foot folding rule* (Fig. 11-1) while others use the *pocket tape* (Fig. 11-2). The folding rule sometimes has a metal extension on one end for taking inside measurements. Oil the joints occasionally to prevent breaking the rule when opening and closing it (Fig. 11-3). Pocket tapes are available in 6- to 30-foot lengths. A hook on the end slides back and forth slightly to compensate for the thickness of its metal when taking outside and inside measurements.

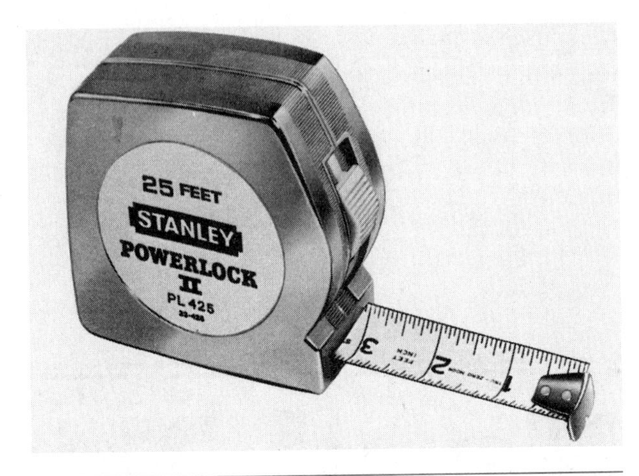

Fig. 11-2 Pocket tape. *(Courtesy of Stanley Tools)*

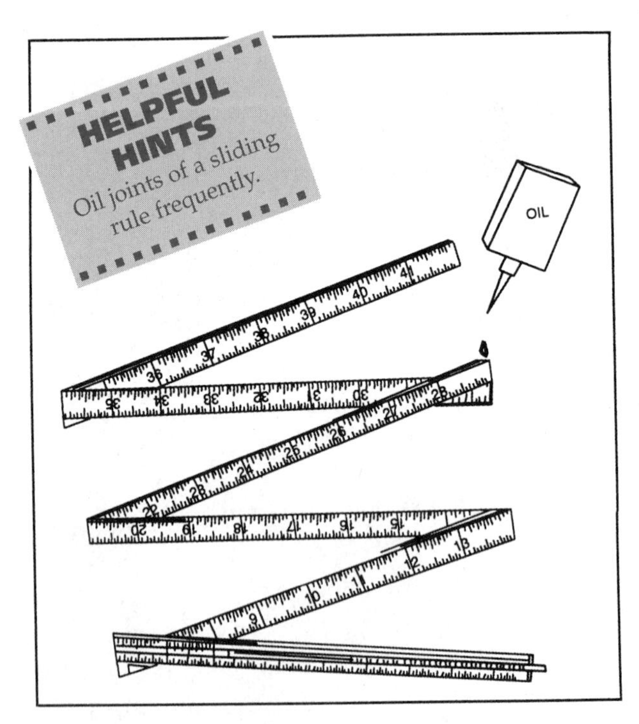

Fig. 11-3 All tools require some form of preventive maintenance.

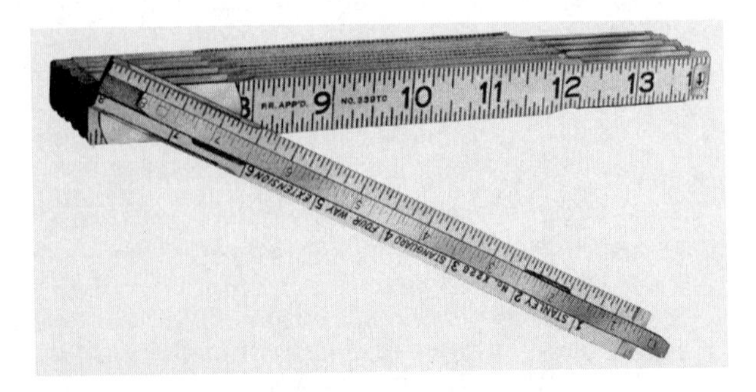

Fig. 11-1 Six-foot folding rule. *(Courtesy of Stanley Tools)*

Steel tapes of 50- and 100-foot lengths are commonly used to lay out longer measurements. The end of the tape has a steel ring with a folding hook attached. The hook may be unfolded to go over the edge of an object. It may also be left in the folded position and the ring placed over a nail when extending the tape. Remember to place the nail so that the *outside* of the ring, which is the actual end of the tape, is to the mark (Fig. 11-4). Rewind the tape when not using it. If the tape is stepped on when extended, it may snap or become kinked. Keep it out of water. If it gets wet, dry it while rewinding.

Slip-sticks. A pair of *slip-sticks* is a helpful and accurate device for laying out stock that has to be fitted between two surfaces. In some cases, it is better than a folding rule or tape. For instance, the distance from floor to ceiling is too great for a rule, and a tape is not rigid enough to stand straight for that height without support. Slip-sticks can be made from two narrow strips of a convenient length of scrap lumber, more than half the distance to be determined. Hold the slip-sticks together. Slide them until the end of one touches a surface while the end of the other touches the opposite surface (Fig. 11-5). Hold the pieces without moving them and lay them on the stock to be marked for length. This is an accurate and convenient method of laying out the length of wall studs, closet shelves, baseboard, and other trim.

Squares

The carpenter has the use of a number of different kinds of squares to lay out for square and other angle cuts.

Combination Squares. The *combination square* (Fig. 11-6) consists of a movable blade, 1 inch wide and 12 inches long, which slides along the body of the square. It is used to lay out or test 90- and 45-degree angles. Hold the body of the square against the edge of the stock and mark along the blade (Fig. 11-7). It can function as a depth gauge to lay out or test the depth of **rabbets, grooves,** and **dadoes.** It can also be used with a pencil as a marking gauge to draw lines parallel to the edge of a board. Drawing lines in this manner is called *gauging* lines. Lines may also be gauged by holding the pencil and riding the finger along the edge of the board. Finger-gauging takes practice, but once mastered saves a lot of time. Be sure to check the edge of the wood for slivers first.

Speed Squares. Some carpenters prefer to use a triangular-shaped square known by the brand name *Speed Square* (Fig. 11-8). Speed Squares are made of one-piece plastic and aluminum alloy and are available in two sizes. They can be used to lay out 90- and 45-degree angles and as guides for portable power saws. A degree scale allows angles to be laid out; other scales may be used to lay out rafters.

Framing Squares. The *framing square,* often called the *steel square* (Fig. 11-9), is an L-shaped tool made of thin steel or aluminum. The longer of the two legs is called the *blade* or *body* and is 2 inches (50 mm) wide and 24 inches (600 mm) long. The shorter leg is called the *tongue* and it is 1 1/2 inches (38 mm) wide and 16 inches (400 mm) long. The outside corner is called the *heel.*

The side that has the manufacturer's name stamped on it is called the *face* side. A number of different tables are stamped on both sides of the square. Most are rarely used except for the *rafter table,* found on the face side of the blade, which is used to find the length of several kinds of rafters.

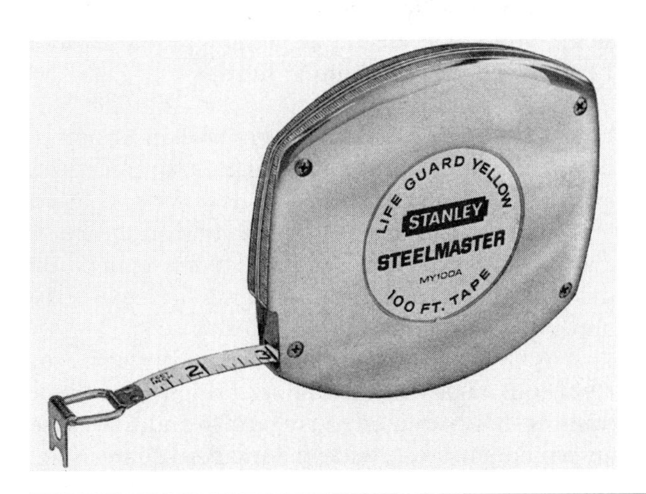

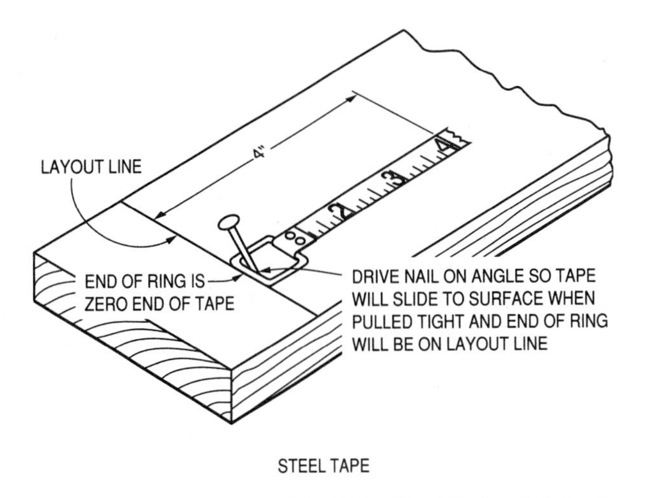

Fig. 11-4 Steel tape. (*Courtesy of Stanley Tools*)

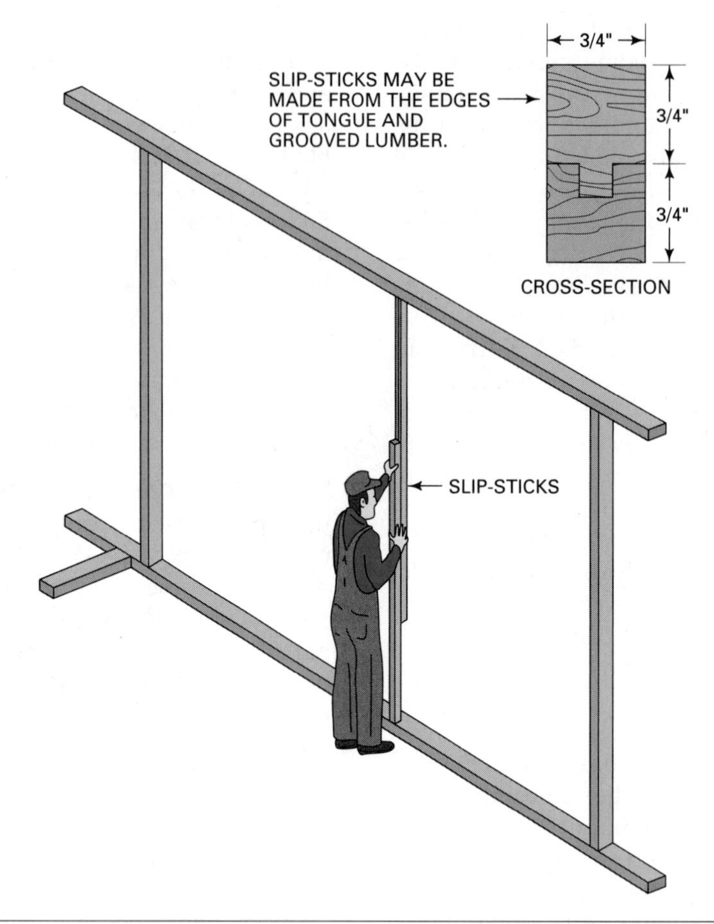

Fig. 11-5 Using slip-sticks is a practical and accurate way to lay out inside dimensions.

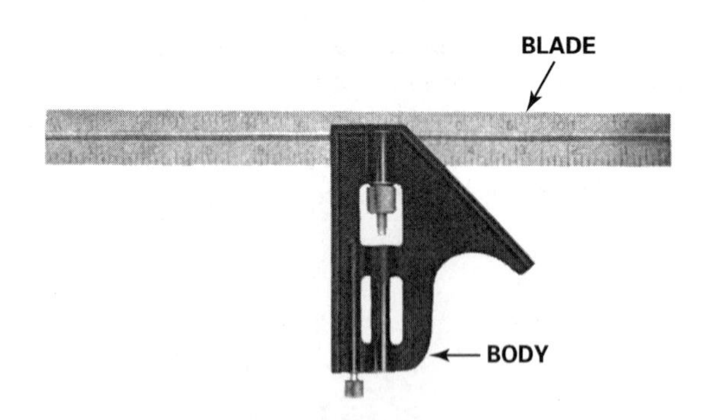

Fig. 11-6 The body and blade of a combination square are adjustable.

(How to use this table will be explained in the following units on roof framing.) On the same side of the square, on the tongue, can be found the **octagon scale**, which is used to lay out eight-sided timbers from square ones.

On the back side of the square, the *Essex board foot table* is used to calculate the number of board feet in lumber. The *brace table* is used to figure the length of diagonal braces. The *hundredths scale,* consisting of an inch divided into one hundred parts, is used to find 1/100ths of an inch. This scale may be used to convert fractions to decimals and vice versa.

On the face side, the edges are divided into inches which are graduated into 1/8ths on the inside and 1/16ths on the outside. The edges on the back are divided into inches and 1/12ths on the outside, while one inside edge is graduated into 1/16ths and the other into 1/10ths.

The framing square has been used for centuries. Entire books have been written about it. Based on the use of the right-angle triangle, many layout techniques have been devised and used throughout the years. Although those techniques, and the scales, tables, and graduations retained on the square, are now rarely used, the carpenter once depended on them.

At present, the framing square is useful for laying out angles for roof rafters, bridging, and stair framing. It is also used to lay out 90- and 45-degree angles on stock too wide for smaller squares (Fig. 11-10).

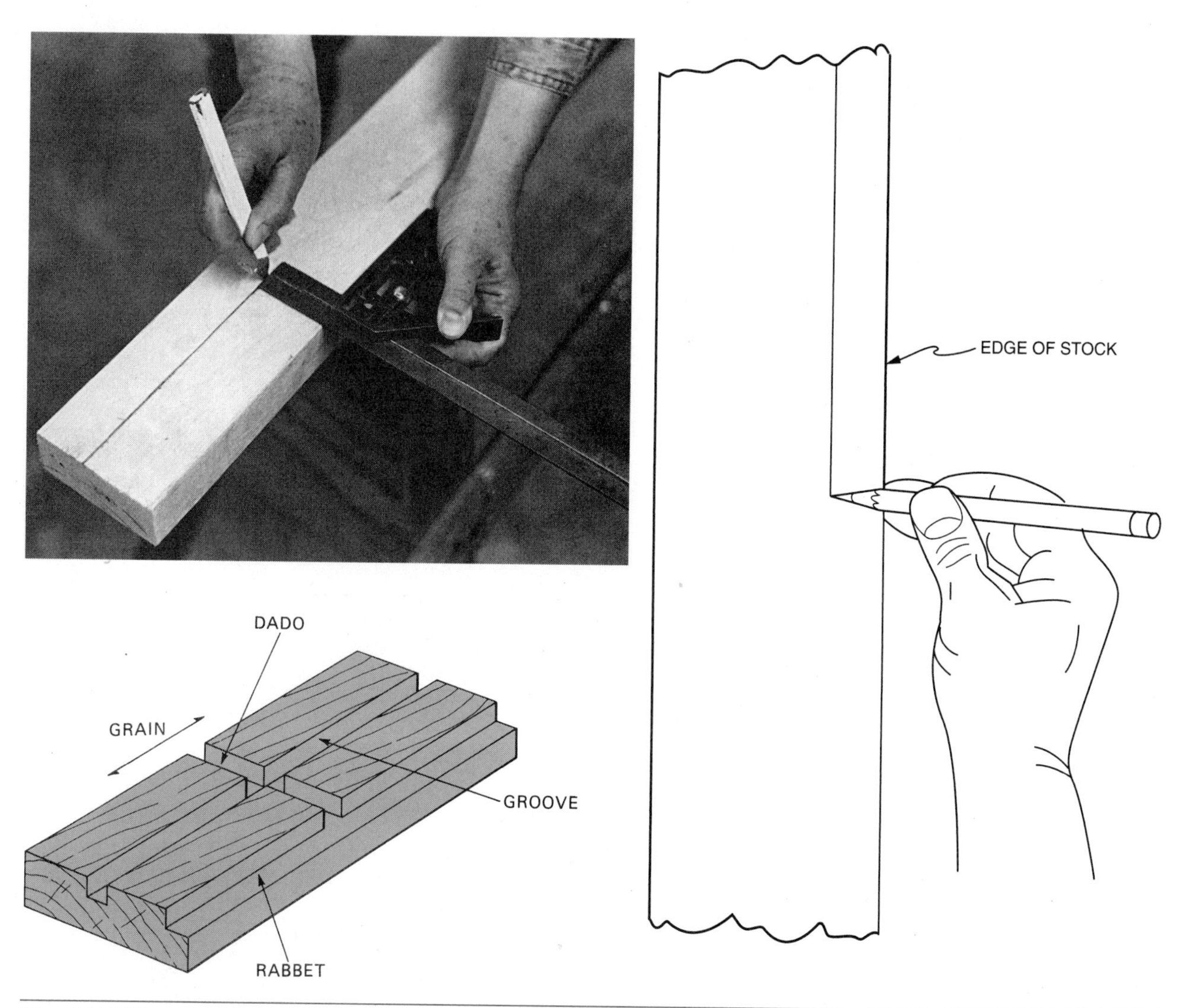

Fig. 11-7 The combination square is useful for squaring and as a marking gauge. A pencil held in one hand is a quick way to draw a parallel line. Check the wood first to reduce the potential for splinters.

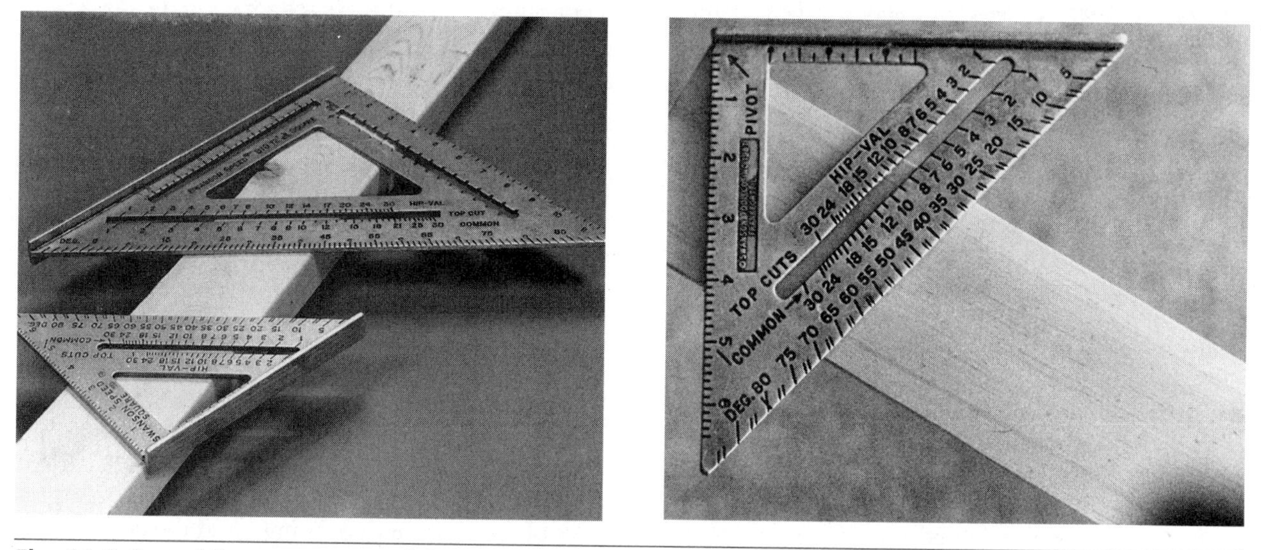

Fig. 11-8 Speed Squares are used for layout of rafters and other angles.

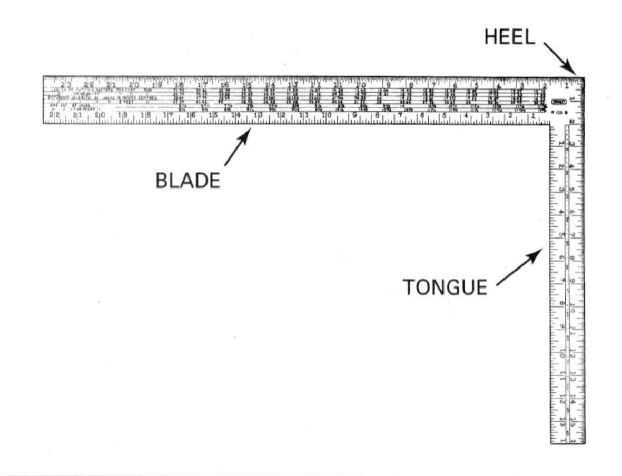

Fig. 11-9 Framing square. (*Courtesy of Stanley Tools*)

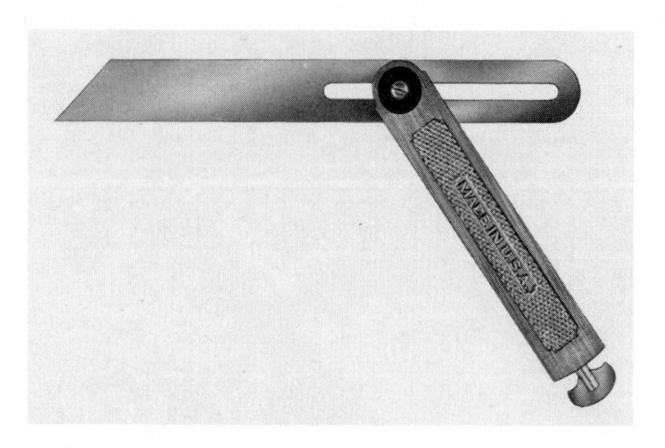

Fig. 11-11 Sliding T-bevel. (*Courtesy of Stanley Tools*)

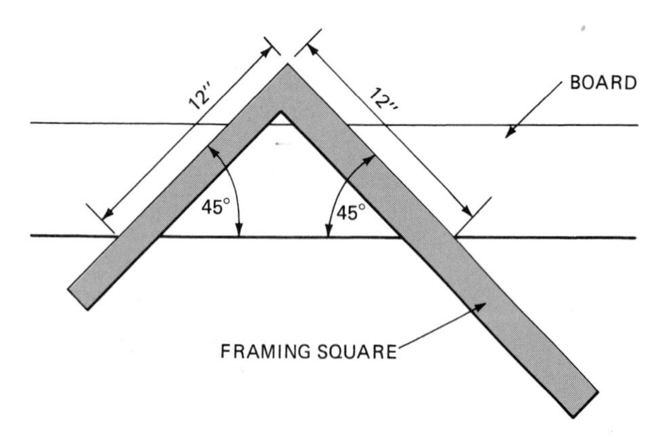

Fig. 11-10 Laying out 45-degree angles with a framing square.

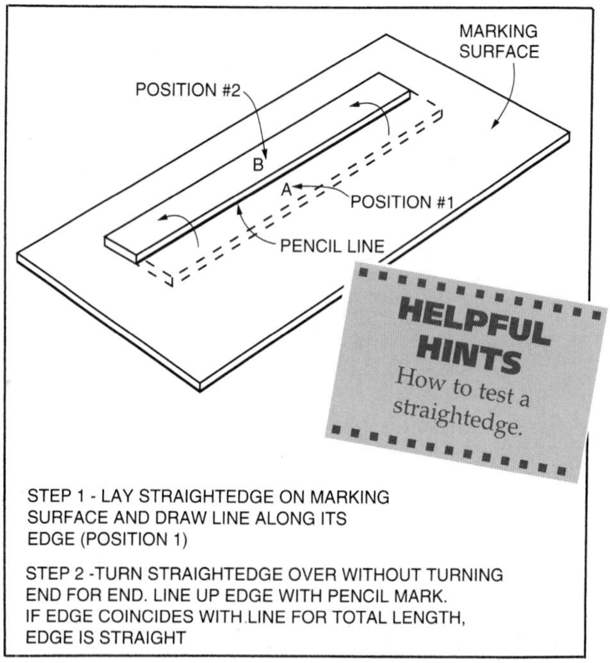

MARKING SURFACE
POSITION #2
B
A
POSITION #1
PENCIL LINE

HELPFUL HINTS
How to test a straightedge.

STEP 1 - LAY STRAIGHTEDGE ON MARKING SURFACE AND DRAW LINE ALONG ITS EDGE (POSITION 1)

STEP 2 -TURN STRAIGHTEDGE OVER WITHOUT TURNING END FOR END. LINE UP EDGE WITH PENCIL MARK. IF EDGE COINCIDES WITH LINE FOR TOTAL LENGTH, EDGE IS STRAIGHT

Fig. 11-12 Testing the straightness of a straightedge by flipping over on a line.

Sliding T-Bevels. The *sliding T-bevel*, sometimes called a *bevel square* or just a **bevel** (Fig. 11-11), consists of a body and a sliding blade that can be turned to any angle and locked in position. It is used to lay out or test angles other than those laid out with squares. The body of the tool is held against the edge of the stock, and the angle is laid out by marking along the blade.

Straightedges

A **straightedge** can be made of metal or wood. It can have any thickness, width, or length, as long as the size is convenient for its intended use and it has at least one edge that is absolutely straight from one end to the other. To determine if it is straight, sight along the edge. Another way is to lay the piece on its side and mark along its edge from one end to the other. Turn the piece over. Hold each end on the line just marked, and mark another line. If both lines coincide, the edge is straight (Fig. 11-12).

Straightedges are useful for many purposes. The framing square, the blade of the combination square, the back of a saw, or a yardstick could be used as a straightedge for drawing short, straight lines. Other larger straightedges may be used for leveling over long distances, or testing the straightness of the sides of door frames and the edges of intersecting wall studs.

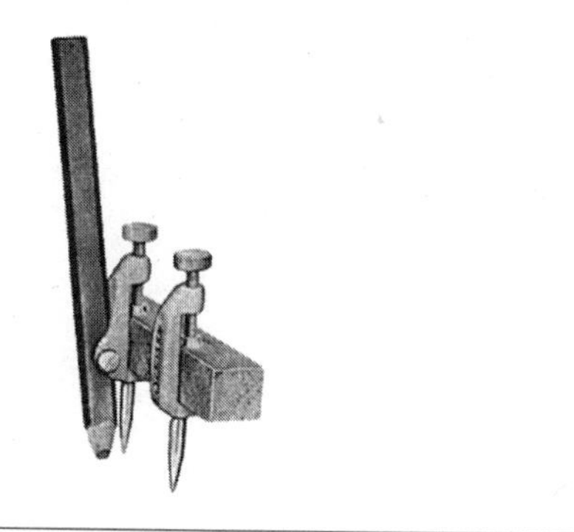

Fig. 11-13 Trammel points are used to lay out arcs of large diameter.

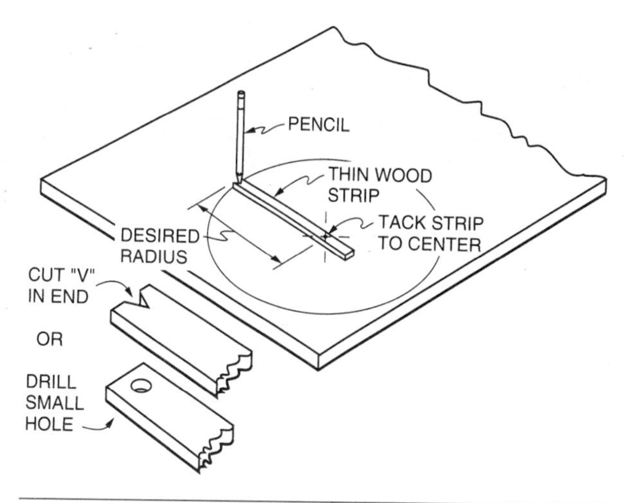

Fig. 11-14 A thin strip of wood can be used to lay out circles or arcs.

Trammel Points. A pair of tools called *trammel points* may be used to draw circles or parts of circles, called arcs (Fig. 11-13), which may be too large for a compass. They can be clamped to a strip of wood any distance apart according to the desired radius of the circle to be laid out. One trammel point can be set on the center while the other, which may have a pencil attached, is swung to lay out the circle or arc.

In place of trammel points, the same kinds of layouts can be made by using a thin strip of wood with a brad or small finish nail through it for a center point. Measure from the end of the strip a distance equal to the desired radius. Drive the brad through the strip until the point comes through. Set the point of the brad on the center, and hold a pencil against the end while swinging the strip to form the circle or arc (Fig. 11-14). To keep the pencil from slipping, a small V may be cut on the end of the strip or a hole may be drilled near the end to insert the pencil. Make sure measurements are taken from the bottom of the V or the center of the hole.

Levels

In construction, the term **level** is used to indicate that which is *horizontal,* and the term **plumb** is used to mean the same as *vertical.* An important point to remember is that level and plumb lines, or objects, must also be straight throughout their length or height. Parts of a structure may have their end points level or plumb with each other. If they are not straight in between, however, they are not level or plumb for their entire length (Fig. 11-15).

Carpenter's Levels. The *carpenter's level* (Fig. 11-16) is used to test both level and plumb sur-

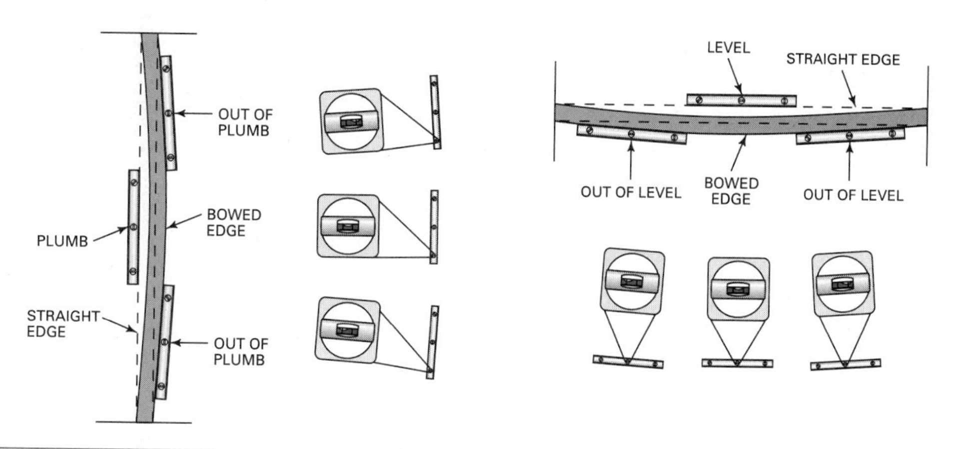

Fig. 11-15 To be level or plumb for their entire length, pieces must be straight from end to end.

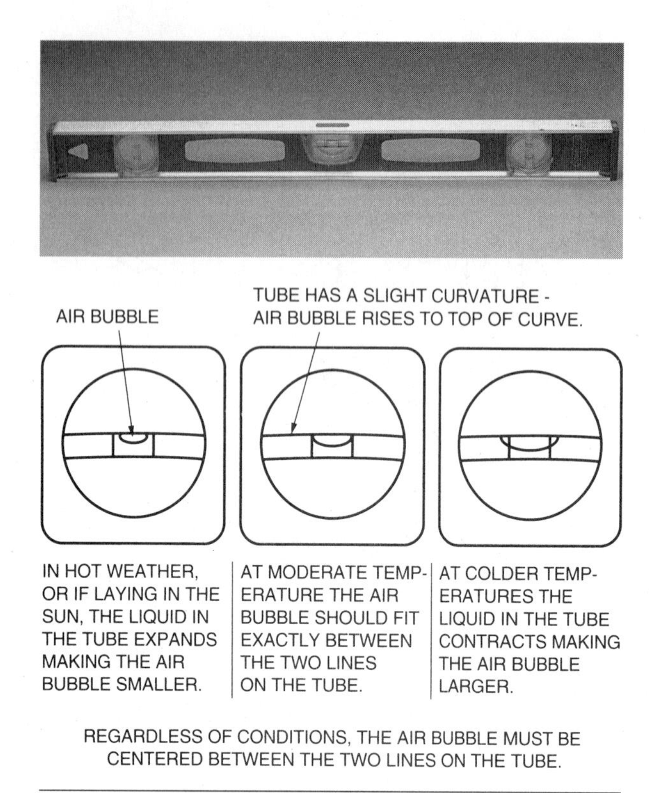

Fig. 11-16 The bubble size of a carpenters level can be affected by temperature.

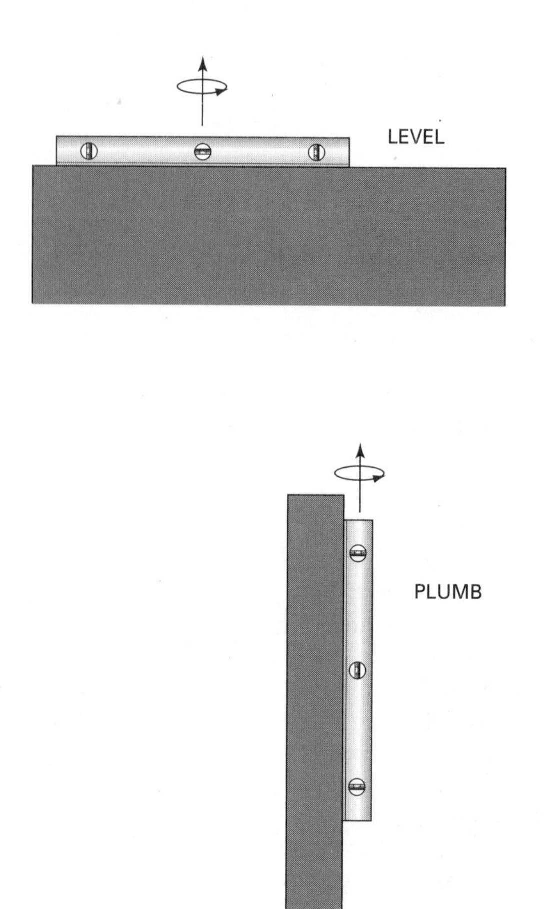

Fig. 11-17 Technique for checking to see if a level is accurate.

faces. Accurate use of the level depends on accurate reading. The air bubble in the slightly crowned glass tube of the level must be exactly centered between the lines marked on the tube. The tubes of a level are oriented in two directions for testing level and plumb. The number of tubes in a level depend on the level length and manufacturer.

Levels are made of wood or metal, usually aluminum. Most carpenters prefer the metal type in lengths of 24 to 28 inches. Lengths of 48, 72, and 78 inches are also available.

Note: Care must be taken not to drop the level because this could break the glass or disturb the accuracy of the level. To check a level for accuracy, place it on a nearly level or plumb object that is firm. Note the exact position of the level on the object. Read the level carefully and remember where the bubble is located within the lines on the bubble tube. Rotate the level along its vertical axis and reposition it in the same place on the object (Fig. 11-17). If the bubble reads the same as the previous measurement, then the level is accurate.

Line Levels. The *line level* (Fig. 11-18) consists of one glass tube encased in a metal sleeve with hooks on each end. The hooks are attached to a stretched line, which is then moved up or down until the bubble is centered. However, this is not an accurate method and gives only an approximate levelness. Care must be taken that the level be attached close to the center of the suspended line because the weight of the level causes the line to sag. If the line level is off center to any great degree, the results are very faulty.

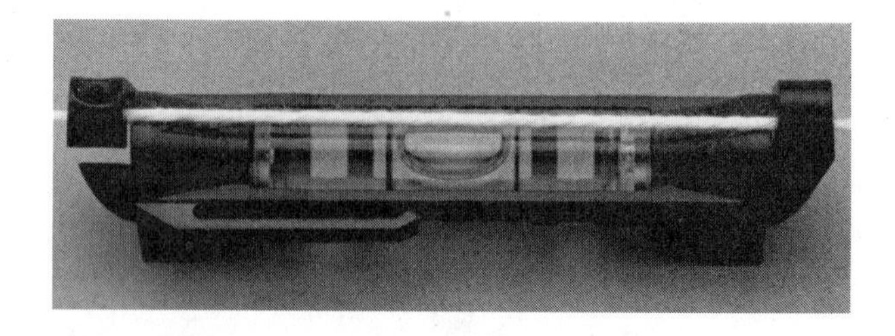

Fig. 11-18 Line level.

Plumb Bobs. The **plumb bob** (Fig. 11-19) is very accurate and is used frequently for testing and establishing plumb lines. Suspended from a line, the plumb bob hangs absolutely vertical when it stops swinging. However, it is difficult to use outside when the wind is blowing. Plumb bobs come in several different weights. Heavy plumb bobs stop swinging more quickly than lighter ones. Some have hollow centers that are filled with heavy metal to increase the weight without enlarging the size.

The plumb bob is useful for quick and accurate plumbing of posts, studs, door frames, and other vertical members of a structure (Fig. 11-20). It can be suspended from a great height to establish a point that is plumb with another. Its only limitation is the length of the line.

Chalk Lines

Long straight lines are laid out by using a *chalk line.* A line coated with chalk dust is stretched tightly between two points and snapped against the surface (Fig. 11-21). The chalk dust is dislodged from the line and remains on the surface.

Widely used at present is the *chalk line reel,* also called a *chalk box* (Fig. 11-22). The box is filled with chalk dust that comes in a number of colors. The most popular colors are blue, yellow, red, and white. The dust saturates the line, which is on a reel inside the box. The line is ready to be snapped when it is pulled out of the box. After several snaps, the line will need more chalk and must be reeled in to be recoated with chalk. Shaking the box helps recoat the line.

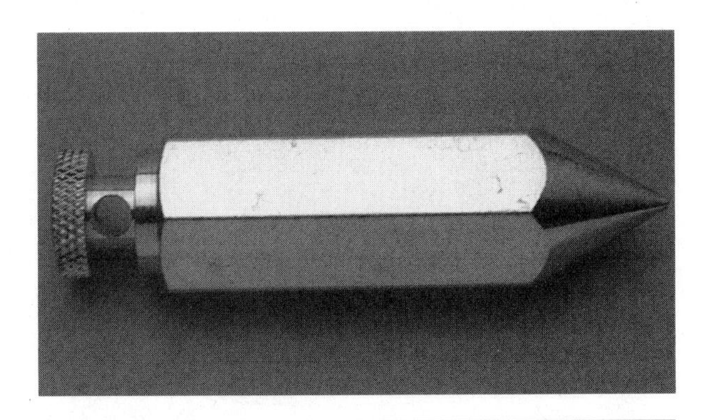

Fig. 11-19 Plumb bob.

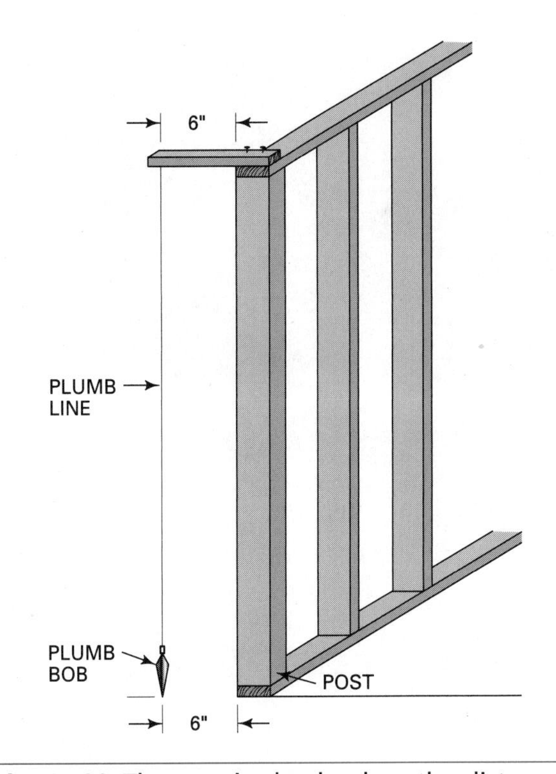

Fig. 11-20 The post is plumb when the distance between it and the plumb line is the same.

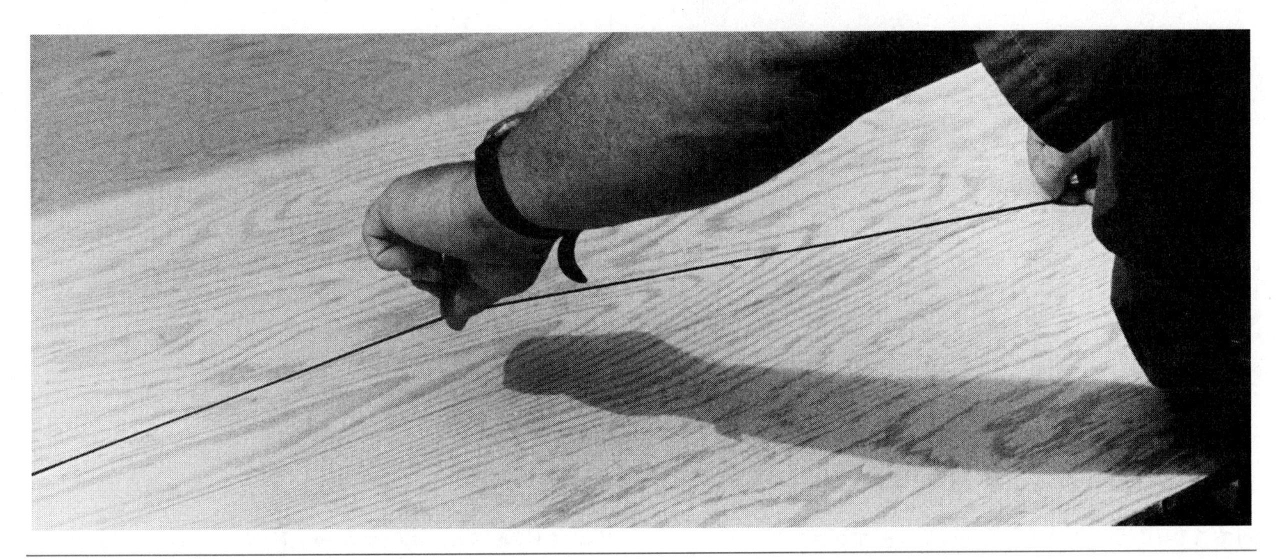

Fig. 11-21 Snapping a chalk line.

Chalk Line Techniques. When unwinding and chalking the line, keep it off the surface until snapped. Otherwise many lines will be made on the surface, and this could be confusing. Make sure lines are stretched tight in order to snap a straight and true line before snapping. Sight long lines by eye for straightness to make sure there is no sag in the line. If there is a sag, take it out by supporting the line near the center. Over long distances, stretch and fasten the line at both ends. Press the center of the line to the deck and snap the line on both sides of the center. Keep the line from getting wet. A wet line is practically useless.

Wing Dividers

Wing dividers can be used as a compass to lay out circles and arcs and as dividers to space off equal distances. However, this tool is used mainly for **scribing** and is often called a *scriber.* Scribing is the technique of laying out stock to fit against an irregular surface (Fig. 11-23). For easier and more accurate scribing, heat and bend the end of the solid metal leg outward (Fig. 11-24). Pencils are usually used in place of the interchangeable steel marking

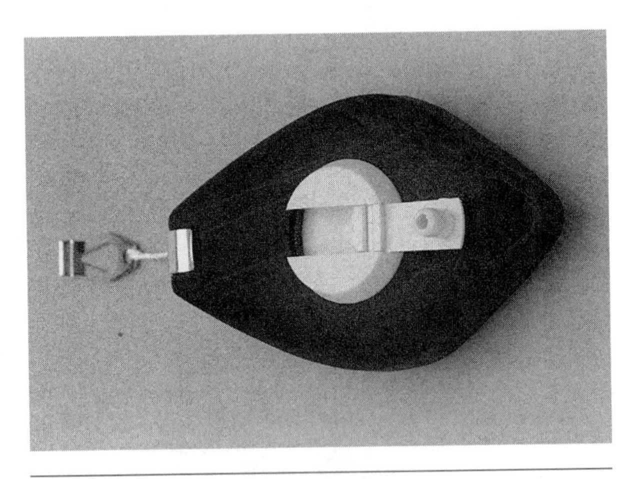

Fig. 11-22 Chalk line reel.

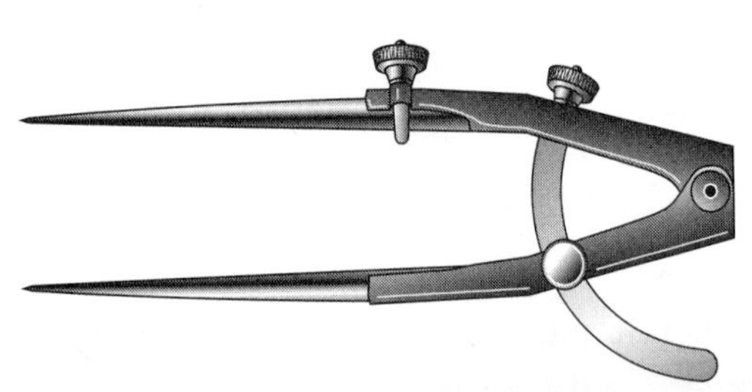

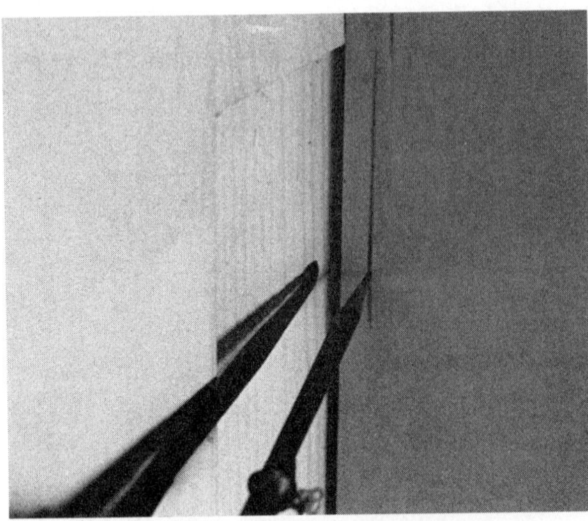

Fig. 11-23 Scribing is laying out a piece to fit against an irregular surface.

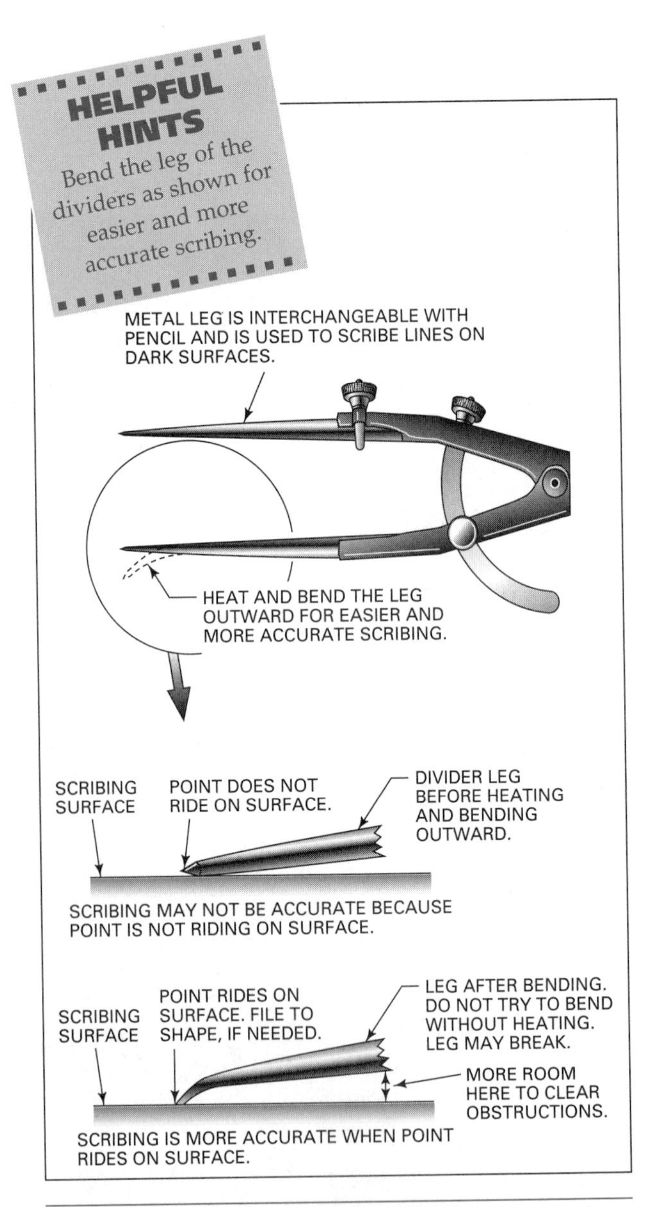

Fig. 11-24 Adjusting one of the metal legs of a scriber makes it a more accurate tool.

leg. Use pencils with hard lead that keep their points longer.

Butt Markers

Butt markers (Fig. 11-25) are available in three sizes. They are often used to mark hinge gains. The marker is laid on the door edge at the hinge location and tapped with a hammer to outline the cutout for the hinge.

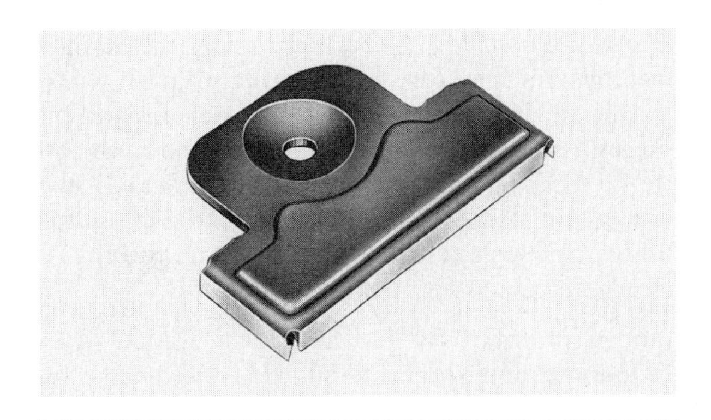

Fig. 11-25 Butt hinge markers come in three sizes.

CHAPTER 12 BORING AND CUTTING TOOLS

Boring Tools

The carpenter is often required to cut holes in wood and metal. Boring denotes cutting larger holes in wood. Drilling is often thought of as making holes in metal or smaller holes in wood. Boring tools include those that actually do the cutting and those used to turn the cutting tool. See also Saws, Drills, and Drivers, p. 112.

Bit Braces

The *bit brace* (Fig. 12-1) is used to hold and turn auger bits to bore holes in wood. It may also be used to hold screwdriver bits to drive large screws. Its size is determined by its *sweep* (the diameter of the circle made by its handle). Sizes range from 8 to 12 inches. Most bit braces come with a ratchet that can be used when there is not enough room to make a complete turn of the handle.

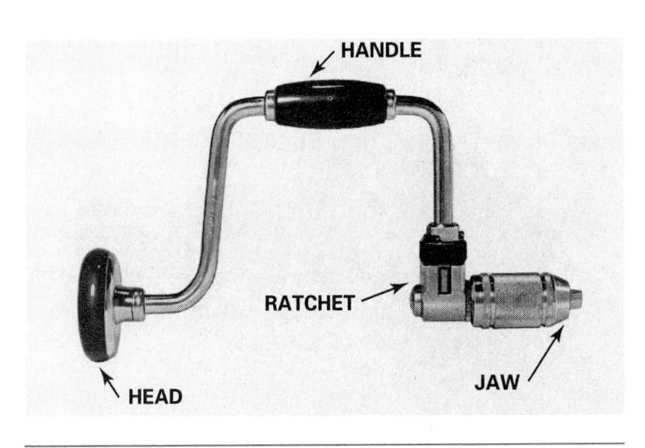

Fig. 12-1 Bit brace. (*Courtesy of Stanley Tools*)

Bits

Auger Bits. *Auger bits* (Fig. 12-2) are available with coarse or fine *feed screws.* Bits with coarse feed

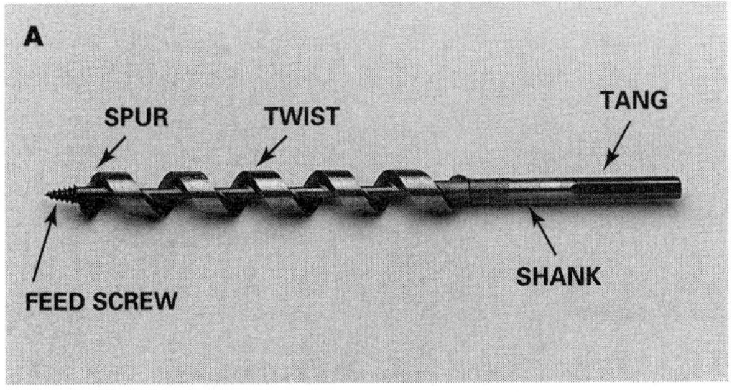

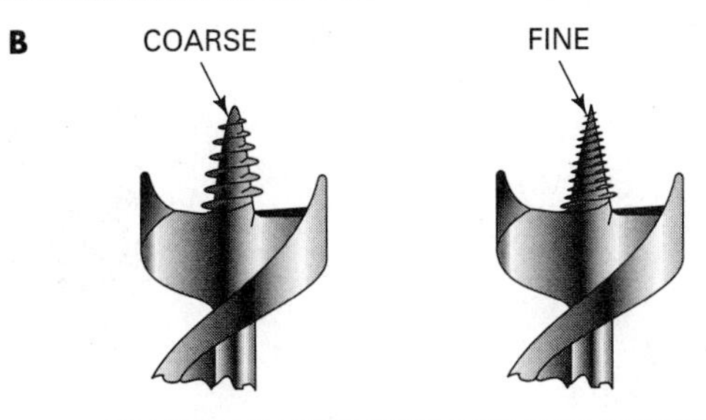

Fig. 12-2 A. Auger bit. (*Courtesy of Stanley Tools*) B. Coarse and fine feed screws.

screws are used for fast boring in rough work. Fine feed bits are used for slower boring in finish work. As the bit is turned, the feed screw pulls the bit through the wood so little or no pressure on the bit is necessary. The *spurs* score the outer circle of the hole in advance of the *cutting lips.* The cutting lips lift the chip up and through the twist of the bit.

Auger Bit Sizes. A full set of auger bits ranges in sizes from 1/4 to 1 inch, graduated in 1/16-inch increments. The bit size is designated by the number of 1/16-inch increments in its diameter. For instance, a #12 bit has 12 sixteenths. Therefore, it will bore a 3/4-inch diameter hole.

Expansive Bits. To bore holes over 1 inch in diameter, the carpenter usually uses an *expansive bit* (Fig. 12-3). With two interchangeable and adjustable cutters, holes up to 3 inches in diameter may be bored. This tool is handy for boring holes in doors for locksets when power tools are not available. Usually these locksets require a hole of 2 1/8 inches in diameter.

Boring Techniques. To avoid splintering the back side of a piece when boring all the way through, stop when the point comes through. Finish by boring from the back side. This is especially important when both sides are exposed to view as in the case of a door. Care must be taken not to strike any nails or other objects that might cause blunting and shortening of the spurs. If the spurs become too short, the auger bit is ruined.

Sharpening Bits. Auger bits are sharpened with an auger bit file. File across the inside of the spurs only enough to bring them to a sharp edge. Excessive filing of the spurs reduces their height, eventually making the bit useless. Do not file on the outside because this will reduce the diameter of the circle being scored by the spurs.

File the cutting lips on the top side and on the original bevel only until the edge is sharp (Fig. 12-4).

Drills

Twist Drills. *Twist drills* range in size from 1/16 to 1/4 inch in increments of 1/64 inch. These drills are particularly useful for drilling holes for screws and are also used in power drills (Fig. 12-5).

Countersinks. The countersink (Fig. 12-6) may be turned in a power drill. It forms a recess for

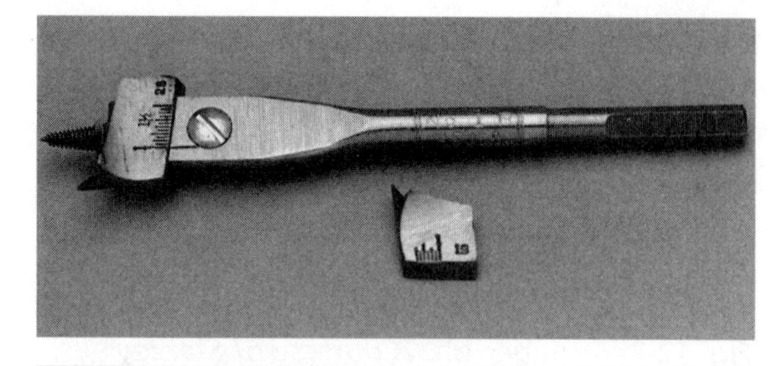

Fig. 12-3 Expansive bit.

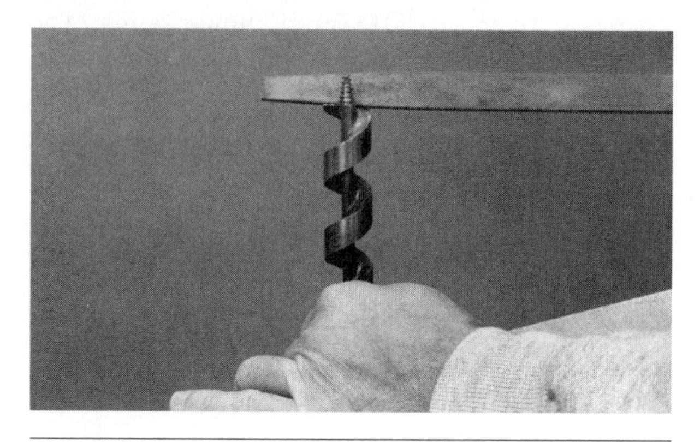

Fig. 12-4 Sharpening an auger bit.

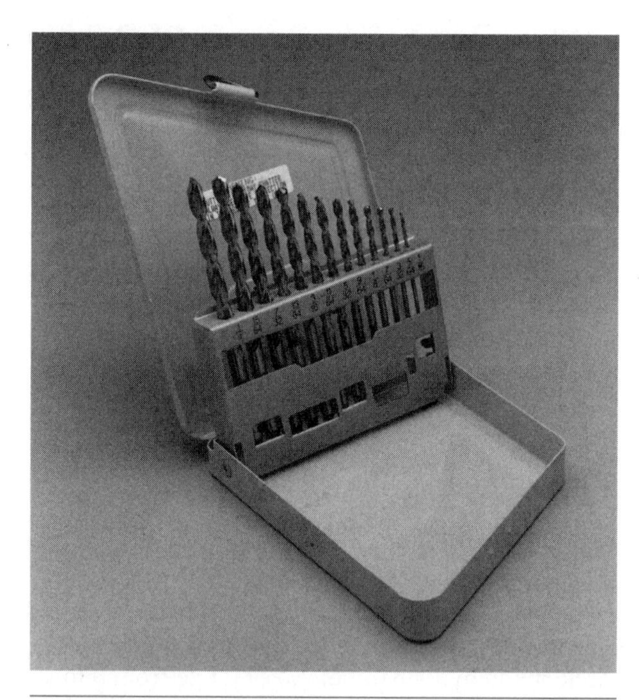

Fig. 12-5 A set of twist drills from 1/16 to 1/4 inch by in increments of sixty-fourths. *(Courtesy of Stanley Tools)*

Fig. 12-6 Countersink boring bit.

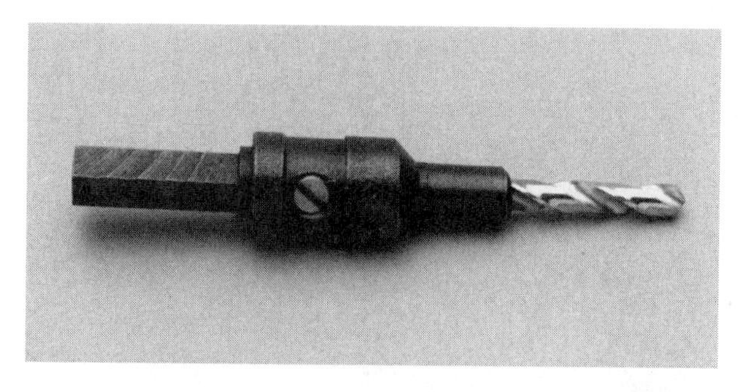

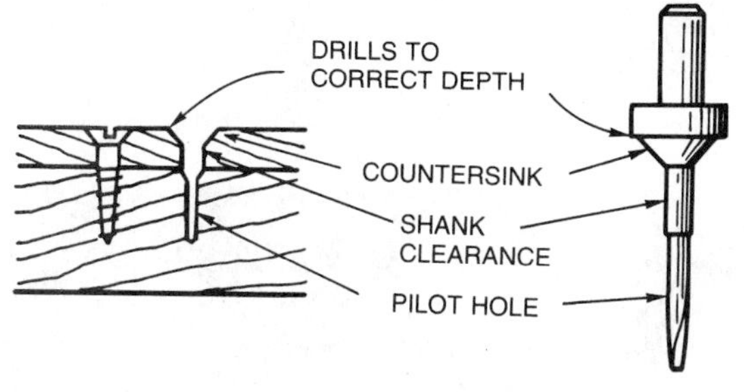

Fig. 12-7 Combination drill and countersink.

a flat head screw to set flush with the surface of the material in which it is driven.

Combination Drills. *Combination drills and countersinks* are used to drill shank and pilot holes for screws and countersink in one operation (Fig. 12-7).

Cutting Tools

The carpenter uses many kinds of cutting tools and must know which to select for each job, as well as how to use, sharpen, and maintain them.

Edge-cutting Tools

Wood Chisels. The *wood chisel* (Fig. 12-8), is used to cut recesses in wood for such things as door hinges and locksets and to make joints.

Chisels are sized according to the width of the blade and are available in widths of 1/8 inch to 2 inches. Most carpenters can do their work with a set consisting of chisels that are 1/4, 1/2, 3/4, 1, and 1 1/2 inches in size.

Firmer chisels have long, thick blades and are used on heavy framing. *Butt* chisels are short, with a thinner blade. They are preferred for finish work.

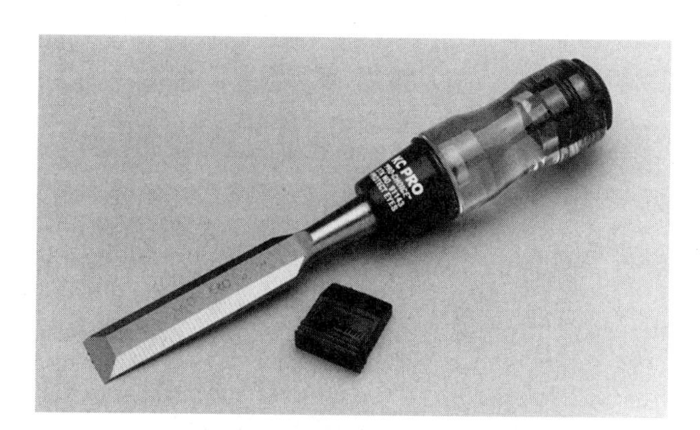

Fig. 12-8 Wood chisel.

CAUTION Improper use of chisels has caused many accidents. When using chisels, keep both hands behind the cutting edge at all times (Fig. 12-9). When not in use, the cutting edge should be shielded (Fig. 12-10). Never put or carry chisels or other sharp or pointed tools in pockets.

Bench Planes. *Bench planes* (Fig. 12-11) come in several sizes. They are used for smoothing rough surfaces and bringing work down to the desired size. Large planes are used on work like door edges to produce a straight surface over a long distance. Long planes will bridge hollows in a surface and cut high spots better than short plane (Fig. 12-12). Small planes are more easily used for shorter work.

Fig. 12-9 Keep both hands in back of the chisel's cutting edge.

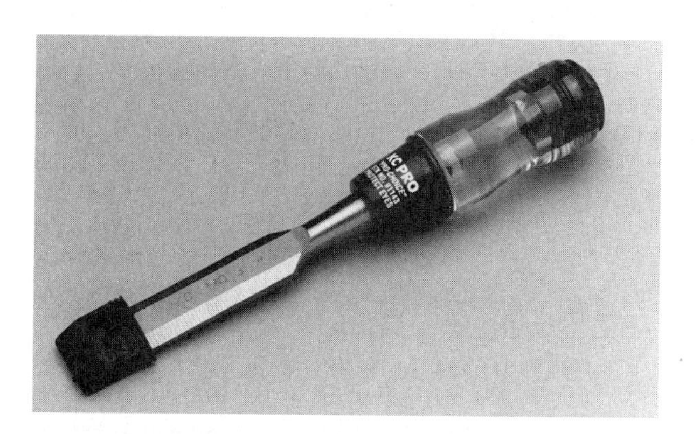

Fig. 12-10 Keep the chisel edge shielded when it is not being used.

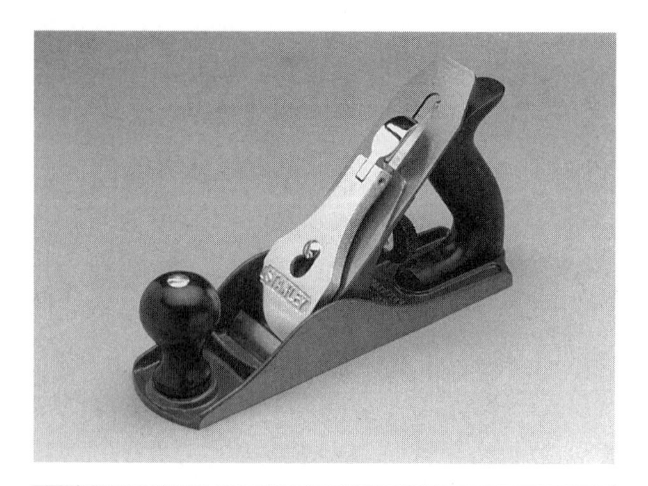

Fig. 12-11 The jack plane is a general purpose plane.

Fig. 12-12 Longer planes bridge hollows to plane long, straight edges.

Bench planes are given names according to their length. The longest is called the **jointer.** In declining order are the *fore, jack,* and *smooth* planes. It is not necessary to have all the planes. The jack plane is 14 inches long and of all the bench planes is considered the best for all-around work.

Block Planes. *Block planes* are small planes designed to be held in one hand. They are often used to smooth the edges of short pieces and for trimming end grain to make fine joints (Fig. 12-13).

Block planes are designed differently than bench planes. On bench planes, the cutting edge bevel is on the bottom side. On block planes, it is on the top. In addition, the bench plane iron has a plane iron cap attached to it, while the block plane iron has none (Fig. 12-14). Practice assembling and adjusting both types until proficient to produce a fine shaving.

Unlike bench planes, block planes are available with their blades set at a high angle or at a low angle. Most carpenters prefer the low-angle block plane because it seems to have a smoother cutting action and because it fits into the hand more comfortably.

Using Planes. When planing, have the stock securely held against a stop. Always plane with the grain. When starting, push forward while applying pressure downward on the **toe** (front). When the **heel** (back) clears the end, apply pressure downward on both ends while pushing forward. When the opposite end is approached, relax pressure on the toe and continue pressure on the heel until the

Fig. 12-13 Block plane is small and often has a low blade angle. (*Courtesy of Stanley Tools.*)

cut is complete (Fig. 12-15). This method prevents tilting the plane over the ends of the stock and helps ensure a straight, smooth edge.

Sharpening Chisels and Plane Irons. The bevel of the wood chisel or plane iron is shaped by a grinding wheel. The bevel should have a concave surface, which is called a *hollow grind.* In order to obtain a hollow grind, a grinding attachment may be used.

CAUTION Use safety goggles or otherwise protect your eyes when grinding.

If a grinding attachment is not available, hold the chisel or plane iron by hand on the tool rest at the proper angle. A general rule is that the width of the bevel is approximately twice the thickness of the blade (Fig. 12-16). Move the blade up on the grinding for a longer bevel and down for a shorter bevel.

Let the index finger of the hand holding the chisel ride against the outside edge of the tool rest

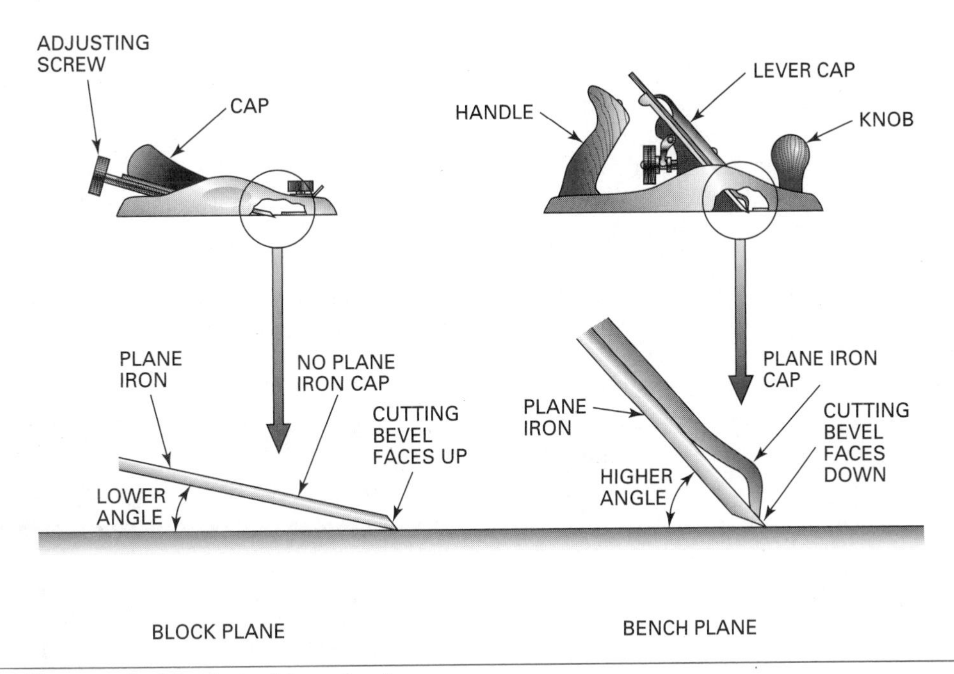

Fig. 12-14 Difference in block and bench planes.

Fig. 12-15 Correct method of planing edges.

as the chisel is moved back and forth across the revolving wheel. Dip the blade in water frequently to prevent overheating. Do not move the position of your index finger, making sure that the tool can be replaced on the wheel at exactly the same angle to obtain a smooth hollow to the bevel. Grind the chisel or plane iron until an edge is formed (Fig. 12-17). A burr or *wire-edge* will be formed on the edge on the flat side. This can be felt by lightly rubbing your thumb along the flat side toward the edge.

Grinding does not sharpen the tool; it only shapes it. To produce a keen edge, it must be **whetted** (sharpened) using an oilstone or waterstone. Hold the tool on a well-oiled stone so that the bevel rests flat on it. Move the tool back and forth across the stone for a few strokes. Then, make a few strokes with the flat side of the chisel or plane iron held absolutely flat on the stone. Continue whetting in this manner until as keen an edge as possible is obtained. To obtain a keener edge, repeat the procedure on a finer stone or on a piece of leather. The edge is sharp when, after having whetted the bevel and before turning it over, no wire edge can be felt on the flat side (Fig. 12-18).

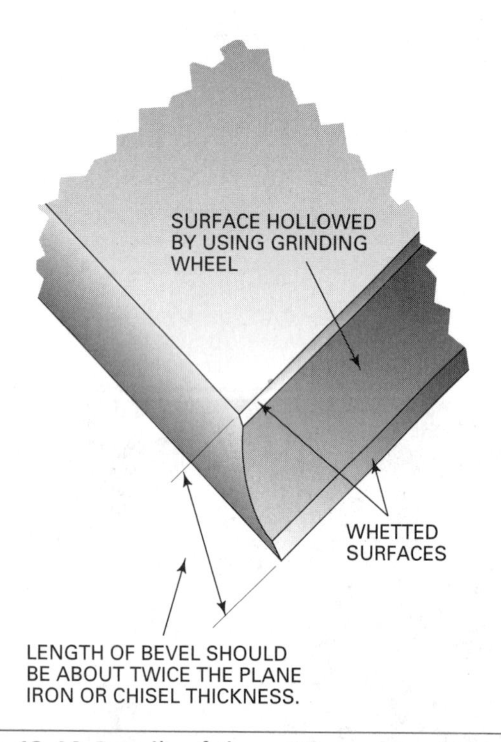

Fig. 12-16 Details of the cutting edge of chisels and planes.

Fig. 12-17 Grinding a cold chisel.

STEP BY STEP
PROCEDURES

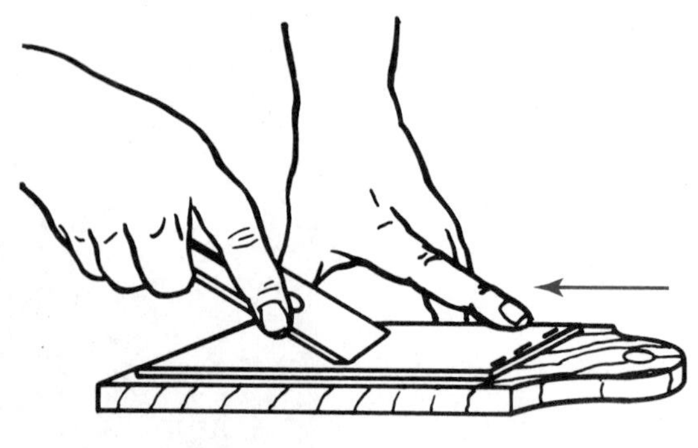

Fig. 12-18 Whetting a plane iron.

Chisels and plane irons do not have to be ground each time they need sharpening. Grinding is necessary only when the bevel has lost its concave shape by repeated whettings, the edge is badly nicked, or the bevel has become too short and blunt. The edge of a blade may be whetted many times before it needs grinding.

Snips. *Straight tin snips* (Fig. 12-19) are generally used to cut straight lines on thin metal, such as roof flashing and metal roof edging. Three styles of *aviation snips* (Fig. 12-20) are available for straight metal cutting and for left and right curved cuts. The color of the handles denotes the differences in the design of the snips. Yellow handles are for straight cuts, green are for cutting curves to the right, and red are for cutting to the left.

Fig. 12-19 Metal shears.

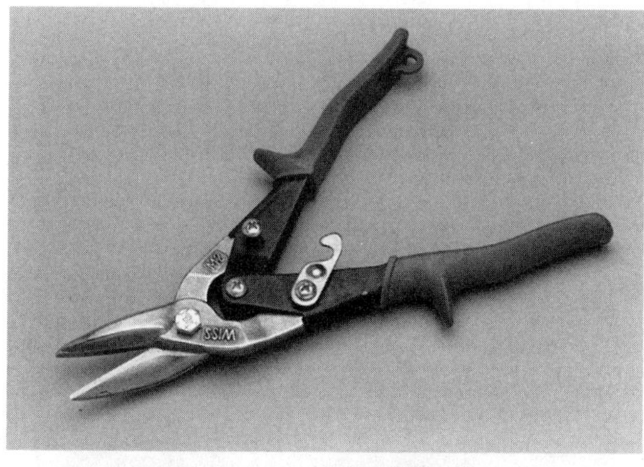

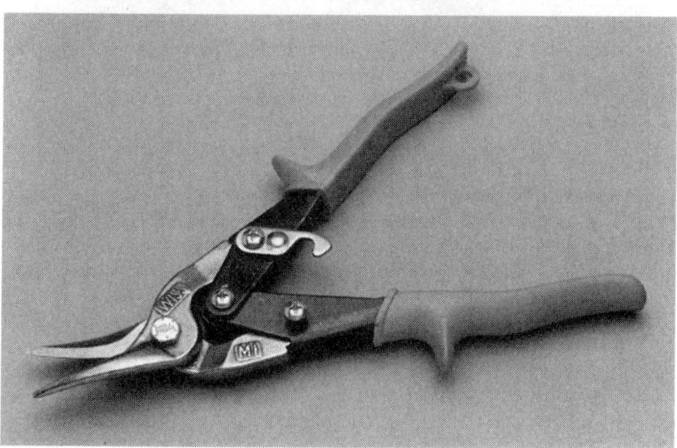

Fig. 12-20 Right- and left-cutting aviation snips.

Hatchets. For wood shingling of side walls and roofs, among other purposes, the *shingling hatchet* is used. In addition to the shingling hatchet, many carpenters carry a slightly heavier hatchet for such uses as pointing stakes or otherwise tapering rough stock. A special drywall hatchet is also used for the installation of gypsum board (Fig. 12-21).

CAUTION When using hatchets for driving fasteners, make sure there are no workers in the path of the backswing.

Knives. A carpenter usually has a jackknife of good quality. The jackknife is used mostly for sharpening pencils and for laying out recessed cuts for some types of finish hardware, such as door hinges. The jackknife is used for laying out this type of work because a finer line can be obtained with it than with a pencil. In addition to marking, it also scores the layout line that is helpful when chiseling the recess (Fig. 12-22).

The *utility knife* (Fig. 12-23) is frequently used for such things as cutting gypsum board and softboards. Replacement blades are carried inside the handle.

Scrapers. The *hand scraper* (Fig. 12-24) is very useful for removing old paint, dried glue, pencil, crayon, and other marks from wood surfaces. The scraper blades are reversible, removable, and replaceable. They dull quickly, but can be easily sharpened by filing on the bevel and against the cutting edge (Fig. 12-25).

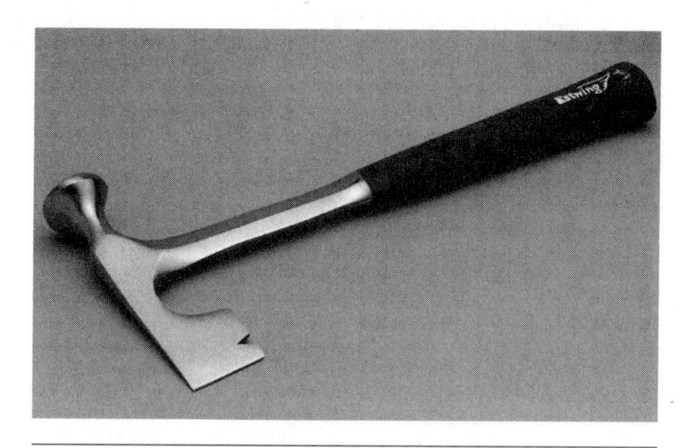

Fig. 12-21 Shingling hatchet.

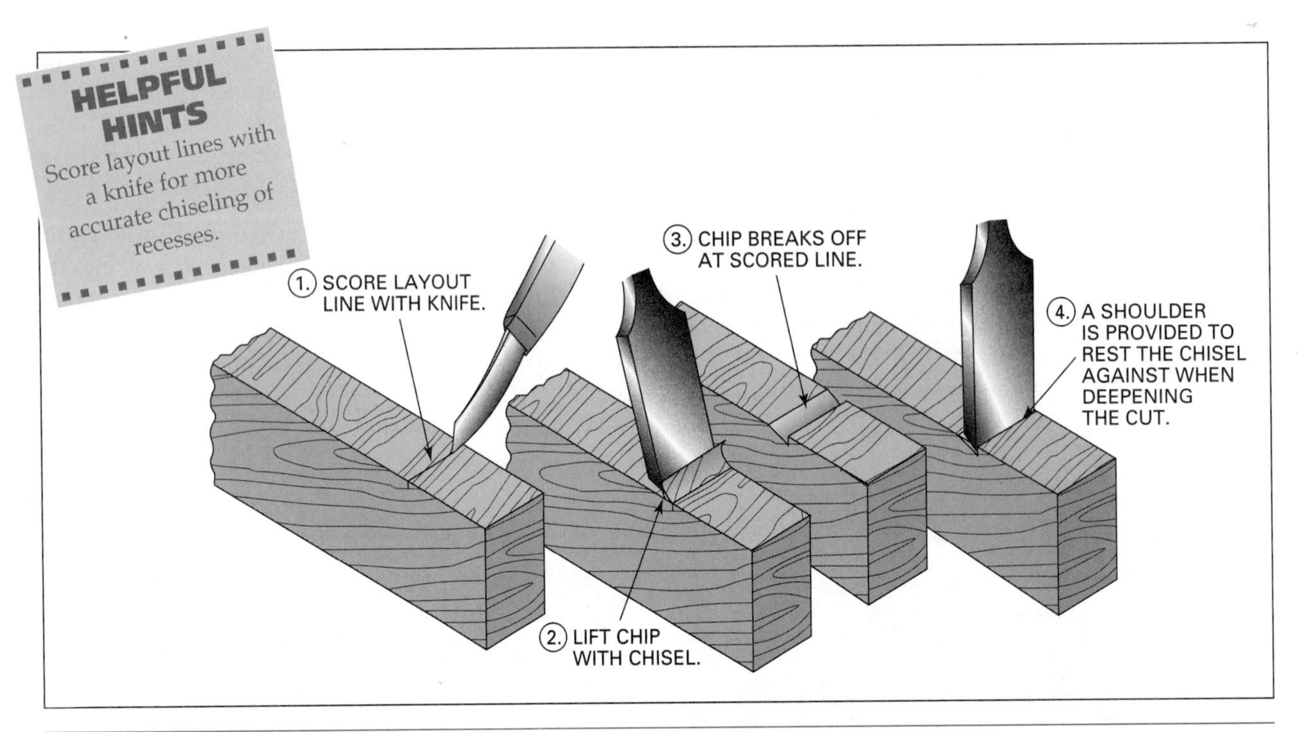

Fig. 12-22 Technique for making a square cut in wood using a wood chisel.

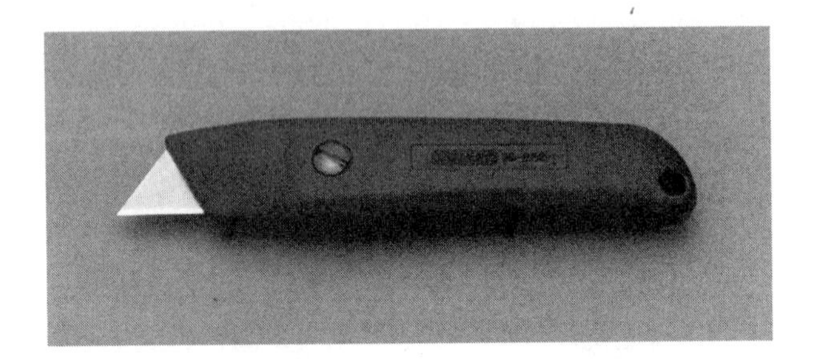

Fig. 12-23 Utility knife.

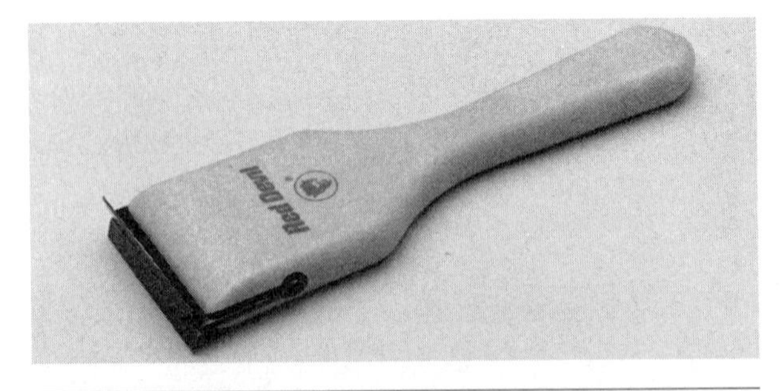

Fig. 12-24 Hand scraper. (*Courtesy of Stanley Tools*)

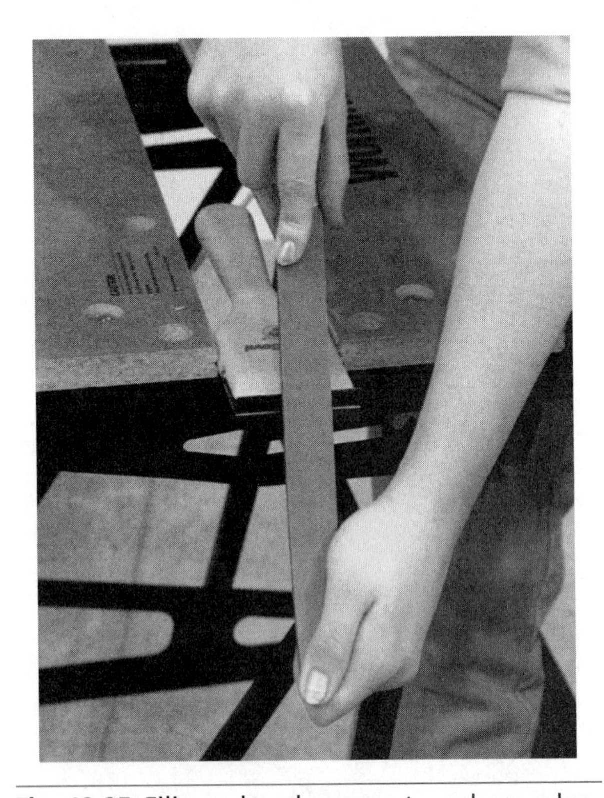

Fig. 12-25 Filing a hand scraper to a sharp edge.

CAUTION Be careful when filing so that the filing hand does not come in contact with the cutting edge. Also, care should be taken not to hollow out the center of the blade or the outside corners will dig into the work.

Tooth-cutting Tools

The carpenter uses several kinds of saws to cut wood, metal, and other material. Each one is designed for a particular purpose.

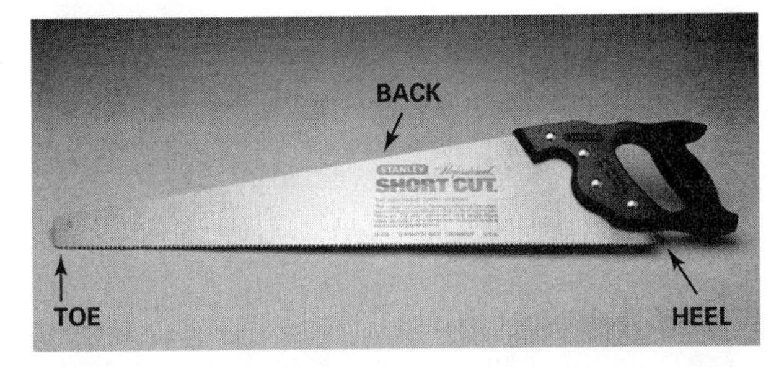

Fig. 12-26 Handsaws are still useful on the jobsite. Some handsaws are made with deeper teeth, which are designed to cut in both directions. *(Courtesy of Stanley Tools)*

Handsaws

Handsaws (Fig. 12-26) used to cut across the grain of lumber are called **crosscut** *saws.* To cut with the grain, *ripsaws* are used. The difference in the cutting action is in the shape of the teeth. The crosscut saw has teeth shaped like knives. These teeth cut through the wood fibers to give a smoother action and surface when cutting across the grain. The ripsaw has teeth shaped like rows of tiny chisels that cut the wood ahead of them. This results in a faster sawing action (Fig. 12-27). Another design to handsaw teeth, called a *shark tooth saw,* makes the teeth longer and able to cut in both directions of blade travel. To keep the saw from binding, the teeth are *set,* that is, alternately bent, to make the saw cut or *kerf* wide enough to give clearance for the blade.

The carpenter ordinarily uses two crosscut saws. A 7-point or 8-point (number of tooth points to the inch) saw is used to cut across the grain of framing and other rough lumber. A 10-point or 11-point saw is used for fine crosscuts on finish work. The number of points is usually stamped on the saw blade at the heel.

Using Handsaws. Stock is handsawed with the face side up because the back side is splintered, along the cut, by the action of the saw going through the stock. This is not important on rough

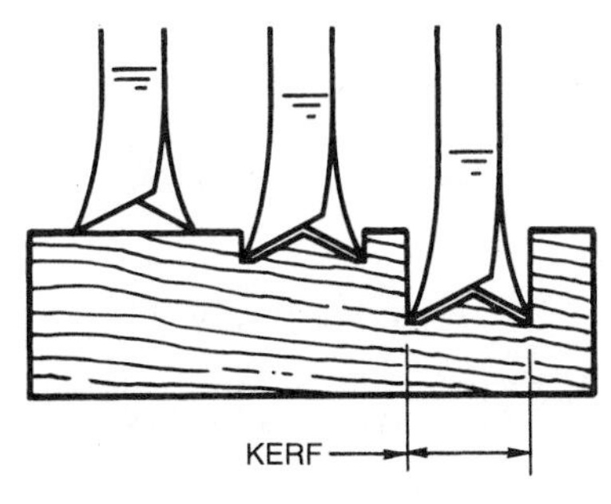

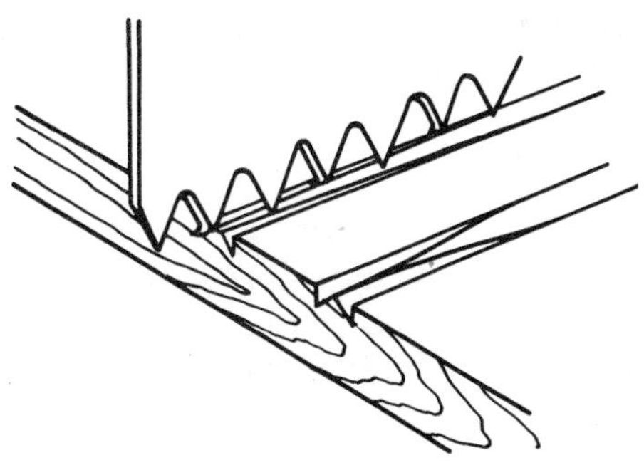

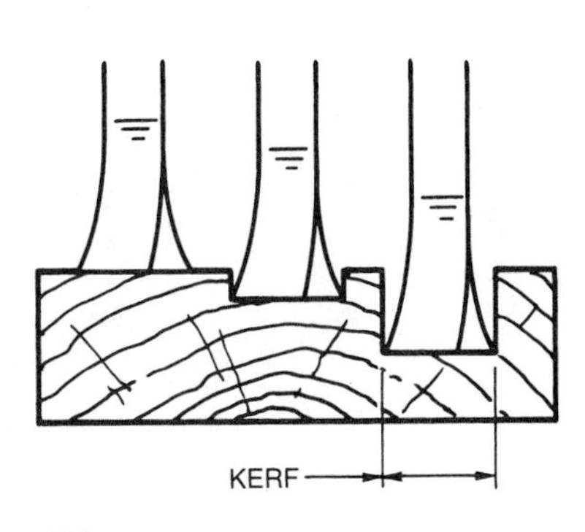

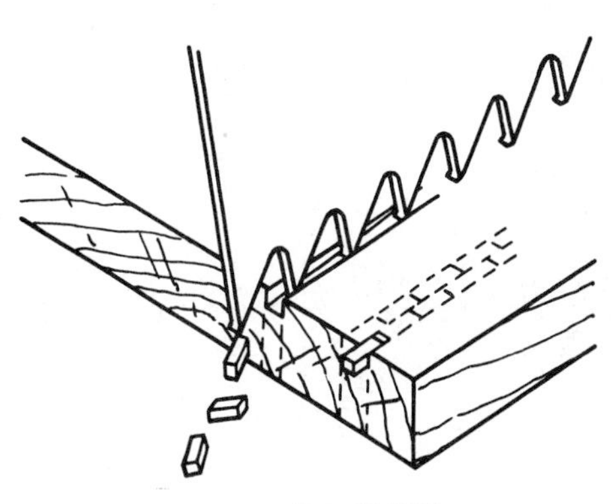

Fig. 12-27 Cutting action of rip- and crosscut saws. *(Courtesy of Disston)*

work. However, on finish work, it is essential to identify the face side of a piece and to make all layout lines and saw cuts with the face side up. It is possible, if no attention is given, to cut one end from the face side and the other end from the back side. Then, no matter which way the piece is installed, one splintered end will be exposed to view.

The saw cut is made on the waste side of the layout line by cutting away part of the line and leaving the rest. This becomes difficult and takes practice, especially when it is important to make thin layout lines rather than broad, heavy ones. Press the blade of the saw against the thumb when starting a cut. Make sure the thumb is above the teeth; steady it with the index finger, with the rest of the hand on the work. Move the thumb until the saw is where desired and start the cut on the downstroke (Fig. 12-28). Move the hand away when the cut is deep enough. When handsawing, hold crosscut saws at about a 45-degree angle. Hold ripsaws, with the handle higher, at about 60 degrees.

CAUTION Do not use a ripsaw for cutting across the grain. It can jump at the start of the cut, possibly causing injury.

Most carpenters prefer to have their saws set and filed by sharpening shops. Sharpening handsaws requires special tools, much skill, and experience to do a professional job.

Special Purpose Saws

Compass and Keyhole Saws. The *compass saw* is used to make circular cuts in wood. The *keyhole saw* is similar to the compass saw except its blade is narrower for making curved cuts of smaller diameter (Fig. 12-29). To start the saw cut, a hole needs to be bored (except in soft material, when the point of the saw blade can be pushed through).

Coping Saws. The *coping saw* (Fig. 12-30) is used primarily to cut molding to make coped joints. A *coped joint* is made by cutting and fitting the end of a molding against the face of a similar piece. (Coping is explained in detail in a following unit.) The coping saw is also used to make any small, irregular curved cuts in wood or other soft material.

Use coping saw blades with fine teeth, and install with the teeth pointing away from the handle. Cut on the downstroke with the handle above the work and the layout line.

Hacksaws. *Hacksaws* (Fig. 12-31) are used to saw metal. Hacksaw blades are available with 18, 24, and 32 points to the inch. Coarse-toothed blades are used for fast cutting in thick metal. Fine-toothed blades are used for smooth cutting of thin metal. At least three teeth of the blade should be in contact with thin metal or cutting will be difficult. Make sure that blades are installed with the teeth pointing away from the handle.

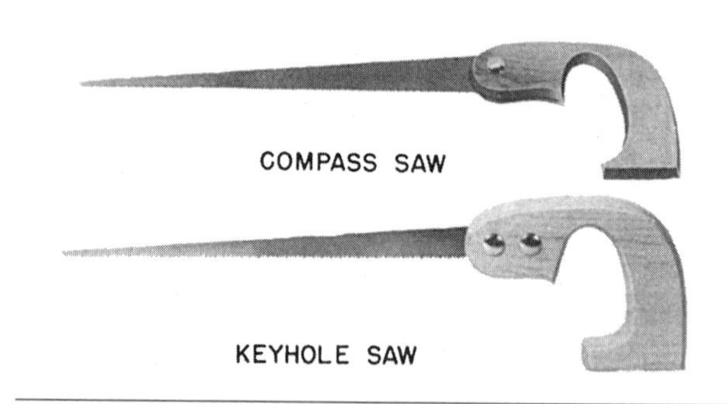

Fig. 12-29 Compass and keyhole saws.

Fig. 12-28 Starting a cut with a handsaw.

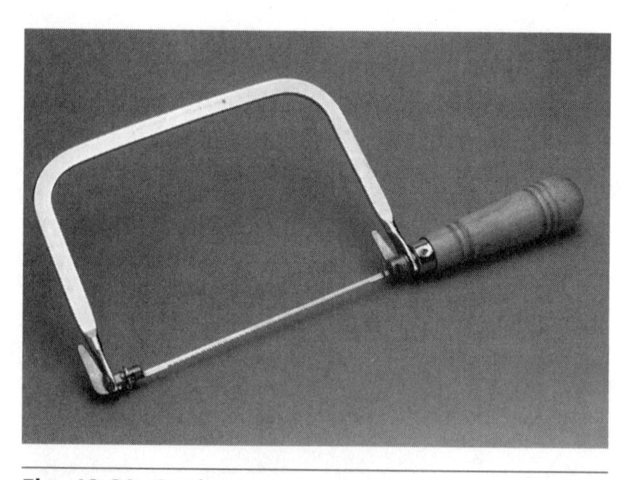

Fig. 12-30 Coping saw.

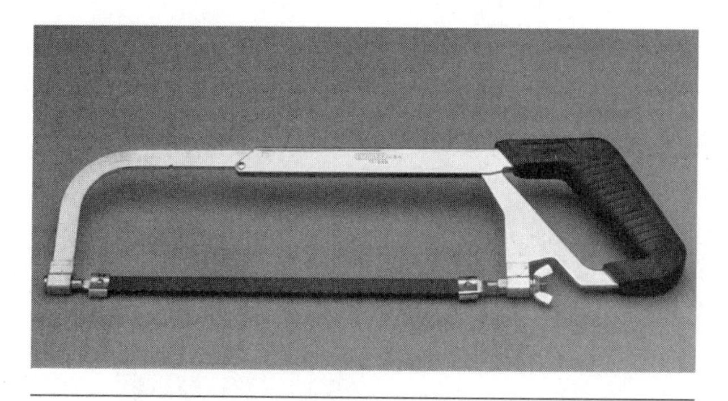

Fig. 12-31 Hacksaw.

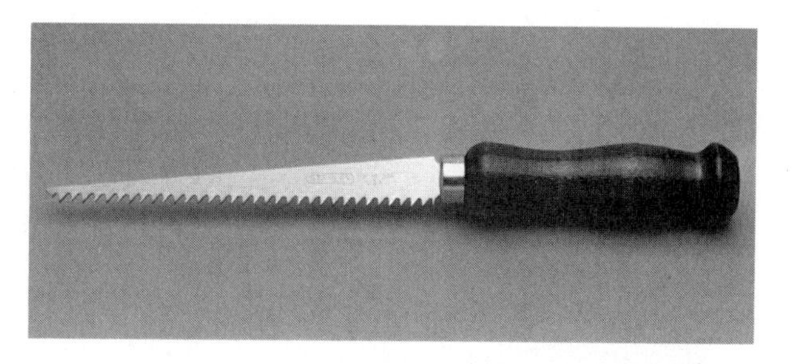

Fig. 12-32 Wallboard saws.

Wallboard Saws. The *wallboard saw* (Fig. 12-32) is similar to the compass saw but is designed especially for gypsum board. The point is sharpened to make self-starting cuts for electric outlets, pipes, and other projections. Another type with a handsaw handle is also used frequently. (Their use is discussed in detail in a following unit on drywall construction.)

Miter Boxes. The *miter box* (Fig. 12-33) is used to cut angles of various degrees on finish lumber by swinging the saw to the desired angle and locking it in place. These cuts are called **miters.** The joint between the pieces cut at these angles is called a **mitered joint.**

A mitered joint is made by cutting each piece at half the angle at which it is to be joined to another piece (Fig. 12-34). For instance, if two pieces are to be joined with a mitered joint at 90 degrees to each other, each piece is cut at a 45-degree angle.

The miter box has built-in stops to locate the saw to cut 90, 67 1/2, 60, and 45 degree angles, which are commonly used miter angles. The back saw in a miter box should only be used in the miter box and for no other purpose.

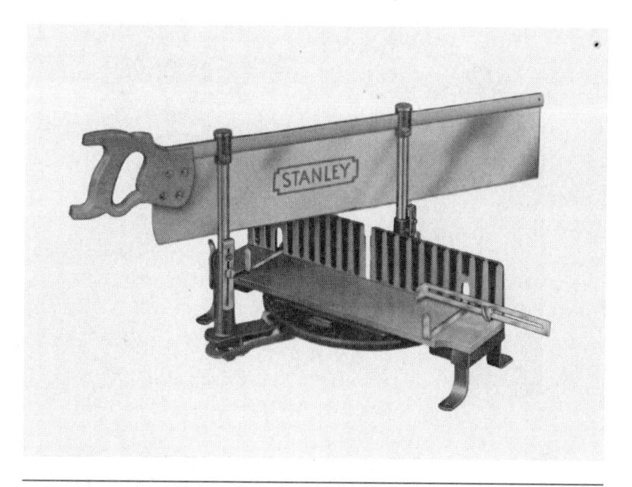

Fig. 12-33 A handsaw miter box. (*Courtesy of Stanley Tools*)

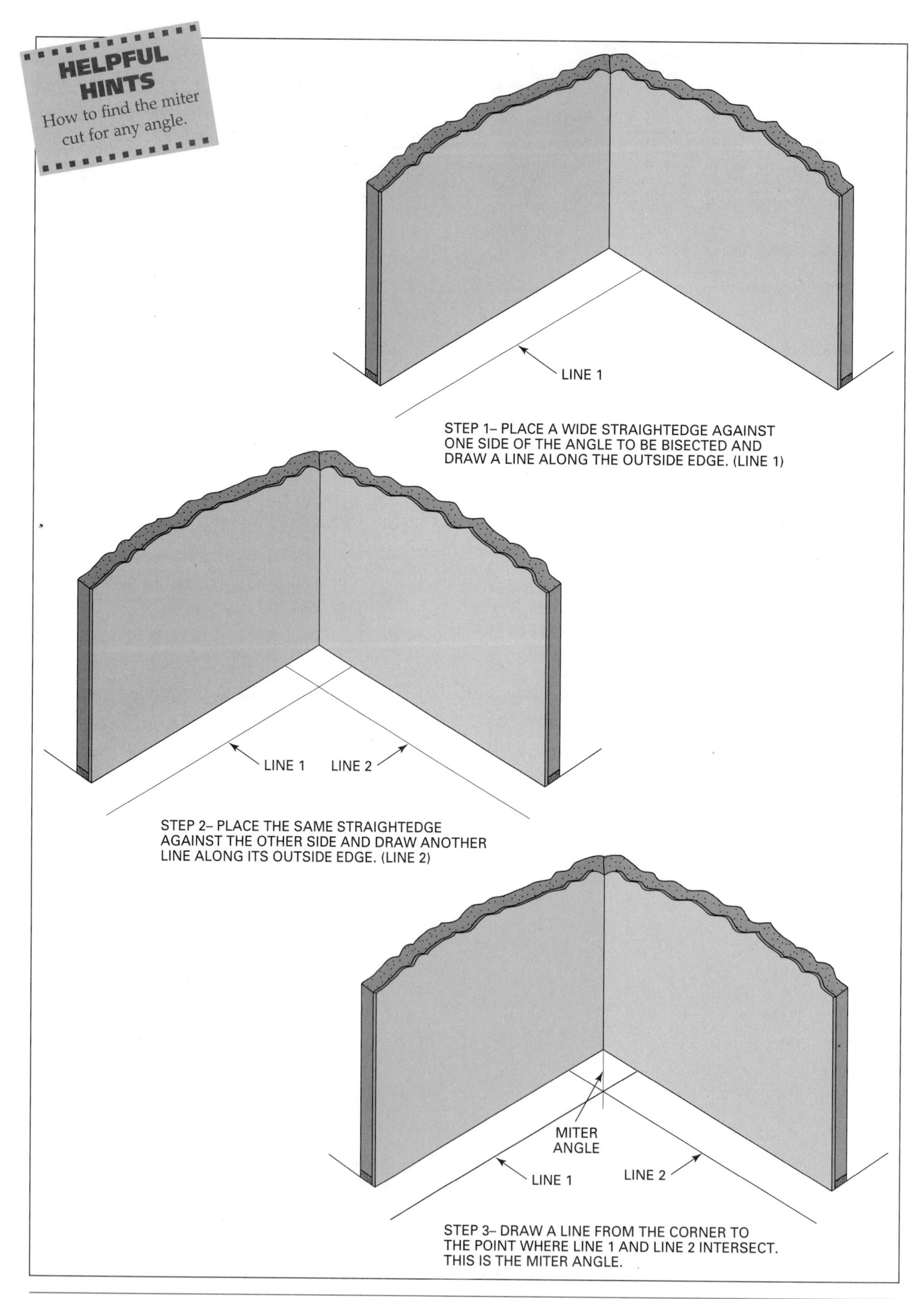

Fig. 12-34 Technique for finding the miter angles for wall intersections of any angle.

CHAPTER 13 FASTENING AND DISMANTLING TOOLS

Discussed in this chapter are those tools used to drive nails and turn screws and other fasteners. Tools used to clamp, hold, pry, and dismantle workpieces are also included.

Fastening Tools

The carpenter must decide which fastening tool to select and be able to use it competently and safely for the job at hand.

Hammers

The carpenter's *claw hammer* is available in a number of styles and weights. The claws may be straight or curved. Head weights range from 7 to 32 ounces. Most popular for general work is the 16-ounce, curved claw hammer (Fig. 13-1). For rough work, a 20- or 22-ounce *framing hammer* (Fig. 13-2), is often used. This has a longer handle and may have a straight or curved claw. In some areas, a 28- or 32-ounce framing hammer is preferred for extra driving power.

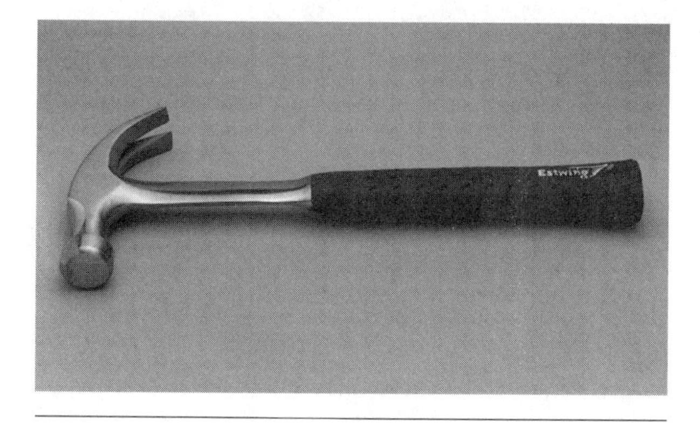

Fig. 13-1 Curved claw hammer.

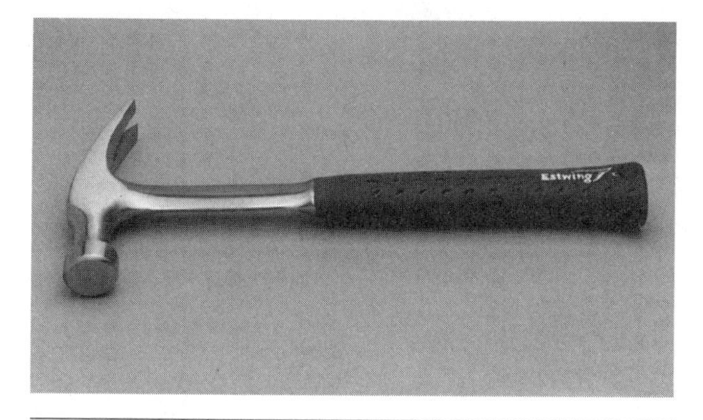

Fig. 13-2 Framing hammer.

Nail Sets. *Nail sets* (Fig. 13-3) are used to set nail heads below the surface. The most common sizes are 1/32, 2/32, and 3/32 inch. The size refers to the diameter of the tip. The surface of the tip is concave to prevent it from slipping off the nail head. If the tip becomes flattened, the nail set has lost its usefulness.

CAUTION Do not set hardened nails or hit the tip of the nail set with a hammer. This will cause the tip to flatten and may produce chips of flying metal. If it is necessary to drive hardened steel nails, wear eye protection (Fig. 13-4). Do not hit the side of the nail set for setting nails in tongue and groove flooring.

Nailing Techniques. Hold the hammer firmly, close to the end of the handle, and hit the nail squarely. If the hammer frequently glances off the nail head, try cleaning the hammer face (Fig. 13-5). As a general rule, use nails that are three times longer than the thickness of the material being fastened. To swing a hammer, the entire arm and shoulder are used. It is important to use the wrist. During the latter part of the swing, as the hammer nears the nail, the wrist is rotated quickly, giving more speed to the hammerhead. This increased speed generates more nail-driving force, all with less arm effort.

Toenailing is the technique of driving nails at an angle to fasten the end of one piece to another (Fig. 13-6). It is used when nails cannot be driven into the end, called *face nailing*. Toenailing generally uses smaller nails than face nailing and offers greater withdrawal resistance of the pieces joined.

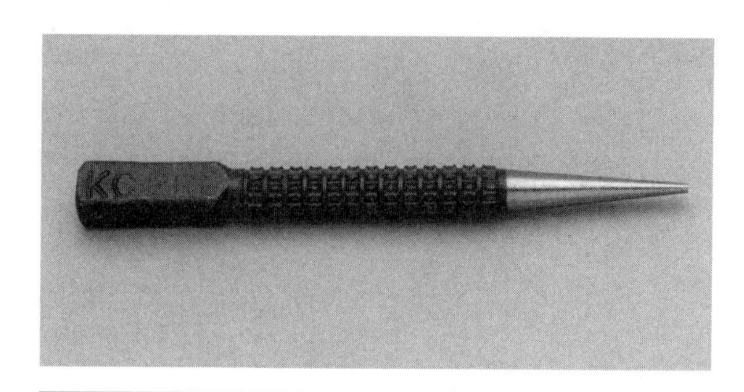

Fig. 13-3 Nail set.

Start the nail about 3/4 to 1 inch from the end and at an angle of about 30 degrees from the surface.

Drive finish nails almost home. Then set the nail below the surface with a nail set to avoid making hammer marks on the surface. Set finish nails at least 1/8 inch deep so the filler will not fall out.

In hardwood, or close to edges or ends, drill a hole slightly smaller than the nail shank to prevent the wood from splitting or the nail from bending. If a twist drill of the desired size is not available, cut the head off a finish nail of the same gauge and use it for making the hole.

Fig. 13-4 Wear eye protection when driving nails.

Fig. 13-6 Toenailing is the technique of driving nails at an angle.

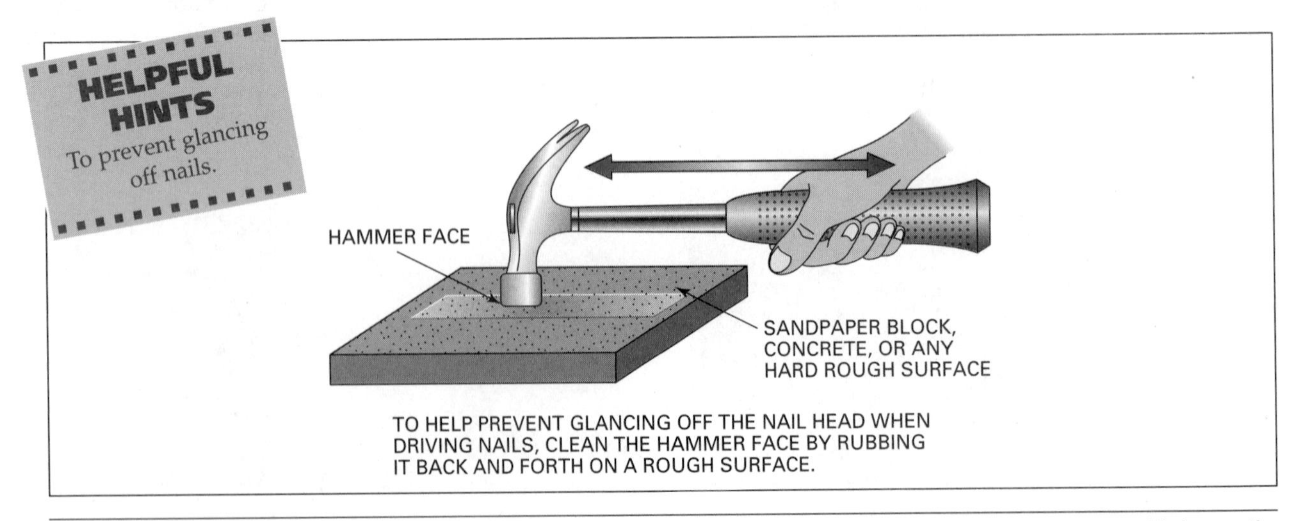

Fig. 13-5 Roughing up the hammerhead face helps keep the hammerhead from glancing off the nail.

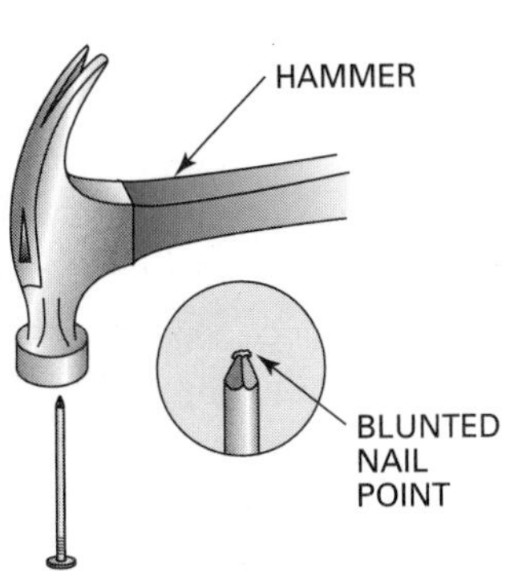

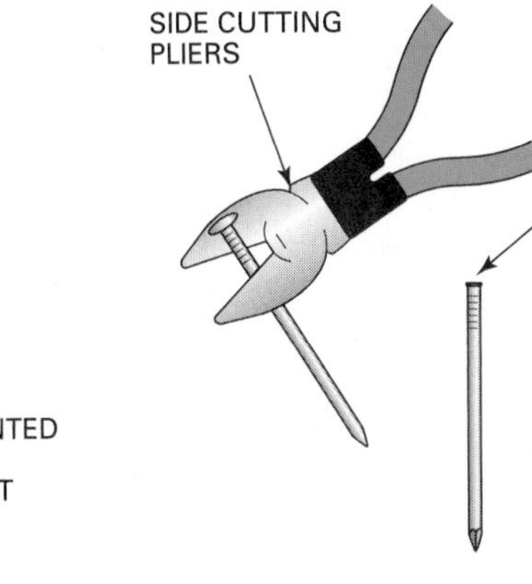

Fig. 13-7 Methods to avoid splitting wood.

Blunting or cutting off the point of the nail also helps prevent splitting the wood (Fig. 13-7). The point spreads the wood fibers as the nail is driven, while the blunt end pushes the fibers ahead of it and reduces the possibility of splitting. A little *paraffin* applied to the nail shank makes driving the nail easier. The end of a used candle is a handy thing to carry in a tool box for this and other lubricating purposes.

Holding the nail tightly with the thumb and as many fingers as possible while driving the nail in hardwood helps prevent bending the nail. Of course, hold the nail in this manner only as long as possible. Be careful not to glance the hammer off the nail and hit the fingers.

When nailing along the length of a piece, stagger the nails from edge to edge, rather than in a straight line. This avoids splitting and provides greater strength (Fig. 13-8). Drive nails at an angle into end grain for greater holding power. When fastening pieces side to side, nails are driven at an angle for greater strength. In addition, this may keep the nail points from protruding (Fig. 13-9) if using 10d or 3-inch nails to fasten 2-inch nominal stock together.

Fig. 13-8 Stagger nails for greater strength and to avoid splitting the stock.

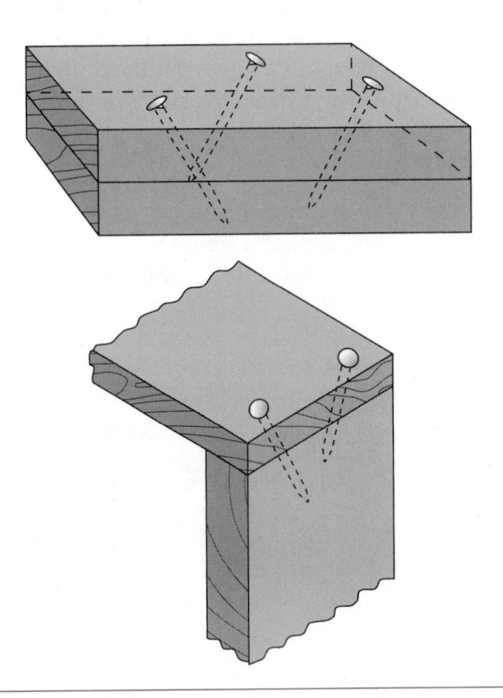

Fig. 13-9 Driving nails at an angle increases holding power.

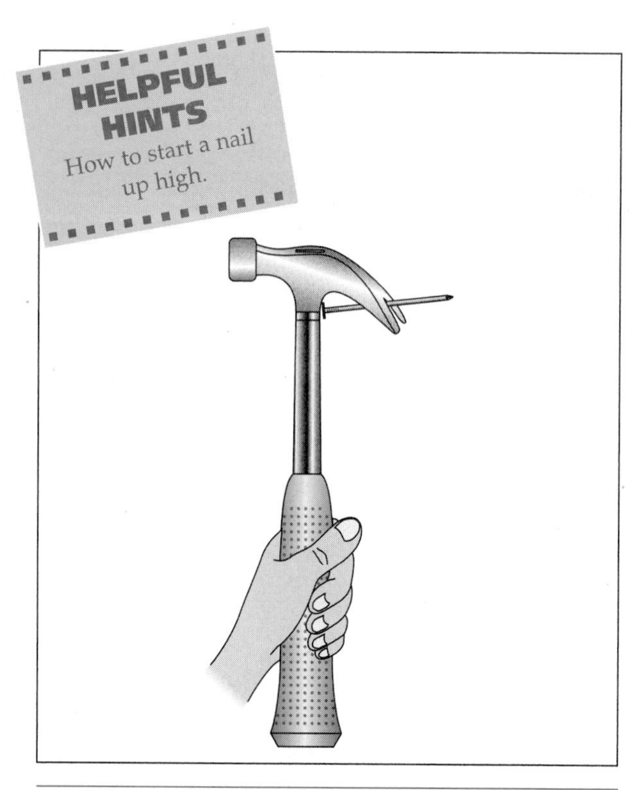

Fig. 13-10 Method of starting a nail with one hand.

When it is necessary to start a nail higher than you can hold it, use the nail starter located in the head of many framing hammers. If your hammer does not have one, press the nail tightly between the claws of the hammer, with the head of the nail against the handle (Fig. 13-10). Turn the claws of the hammer toward the surface. Reach up and swing the hammer to start and hold the nail in the stock. Pull the hammer claws away from the nail, turn the hammer around, and drive the nail home.

Screwdrivers

Screwdrivers are manufactured to fit all types of screw slots. The carpenter generally uses only the *slotted* screwdriver (Fig. 13-11), which has a straight tip to drive common screws, and the *Phillips* screwdriver, which has a cross-shaped tip (Fig. 13-12). Other screwdrivers include the Robertson screwdriver, which has a squared tip (see Fig. 9-13).

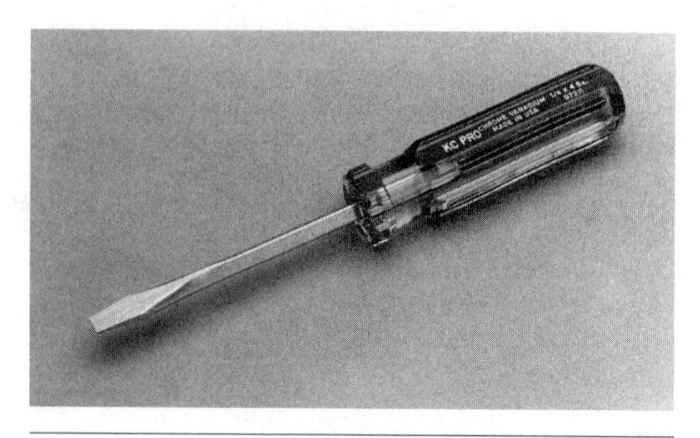

Fig. 13-11 Slotted screwdriver.

Screwdriver Sizes. Slotted screwdrivers are sized by the length of the blade and by the type. Lengths generally run from 3 to 12 inches. Phillips screwdrivers are sized by their length and point size. Commonly used sizes are lengths that run the same as common screwdriver and points that come in numbers 0, 1, 2, 3, and 4. The higher number indicates a point with a larger diameter.

Screwdrivers should fit snugly, without play, into the slot of the screw being driven. The screwdriver tip should not be wider than the screwhead, nor should it be too narrow (Fig. 13-13). The correct size screwdriver helps assure that the screw will be driven without slipping out of the slot. When seated, the screwdriver slot should look the same as before the screw was driven, with no burred edges.

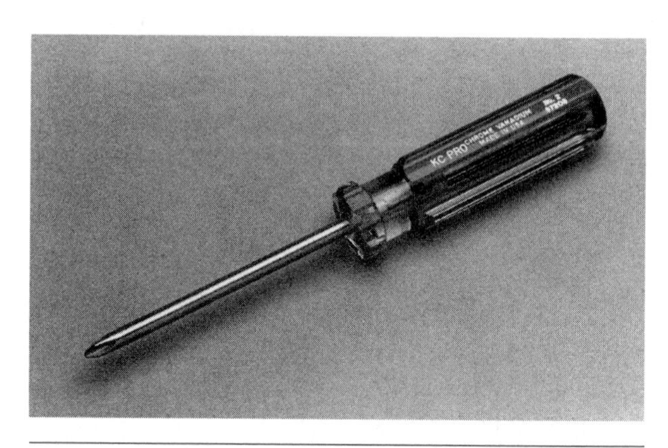

Fig. 13-12 Phillips screwdriver.

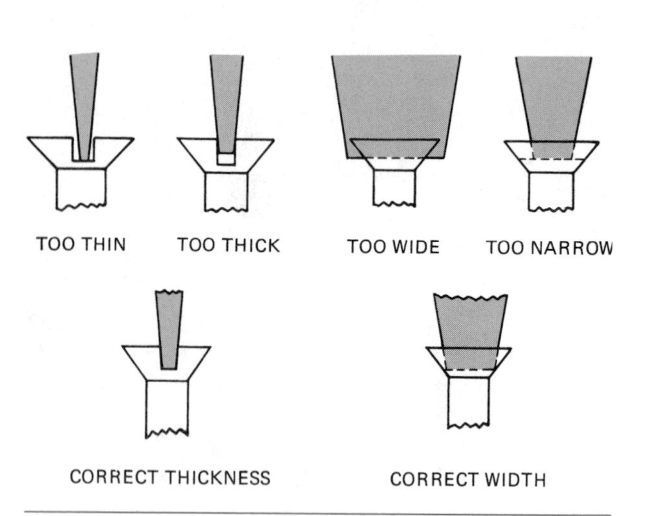

Fig. 13-13 Select the correct size screwdriver for the screw being driven.

Screwdriver Bits. *Screwdriver bits* (Fig. 13-14) are available in many shapes and sizes to accomodate a variety of screws. They are designed to drive a screw using a drill or screw gun.

Fig. 13-14 Screwgun drive bits for various screw head styles. (*Courtesy of Stanley Tools*)

Screwdriving Techniques. If possible, select screws so that two-thirds of their length penetrates the piece in which they are gripping. In preparation for driving a screw, a *shank* hole and a *pilot hole* must be drilled. In addition, if the screw has a flat head, the shank hole must be countersunk so that the screwhead will be flush with or slightly below the surface when driven (Fig. 13-15).

To select a drill for the shank hole, hold the drill bit against the shank of the screw and determine by eye if it is the same size. For the pilot hole, hold the drill bit against the threaded portion of the screw so that the drill covers the solid center section but leaves the threads visible.

Select drills with great care. Smaller drills may be used for a pilot hole in softwoods. Some may advocate that in softwood no pilot hole is necessary, but it is wise to drill them anyhow. It does not take that much more time, and the screw can be driven straight and more easily. Without a pilot hole, the screw may follow the grain and go in at an undesirable angle.

In hardwoods, if the pilot hole is too small or not deep enough, difficulty may be encountered in driving the screw. This causes slipping and damage

STEP BY STEP
PROCEDURES

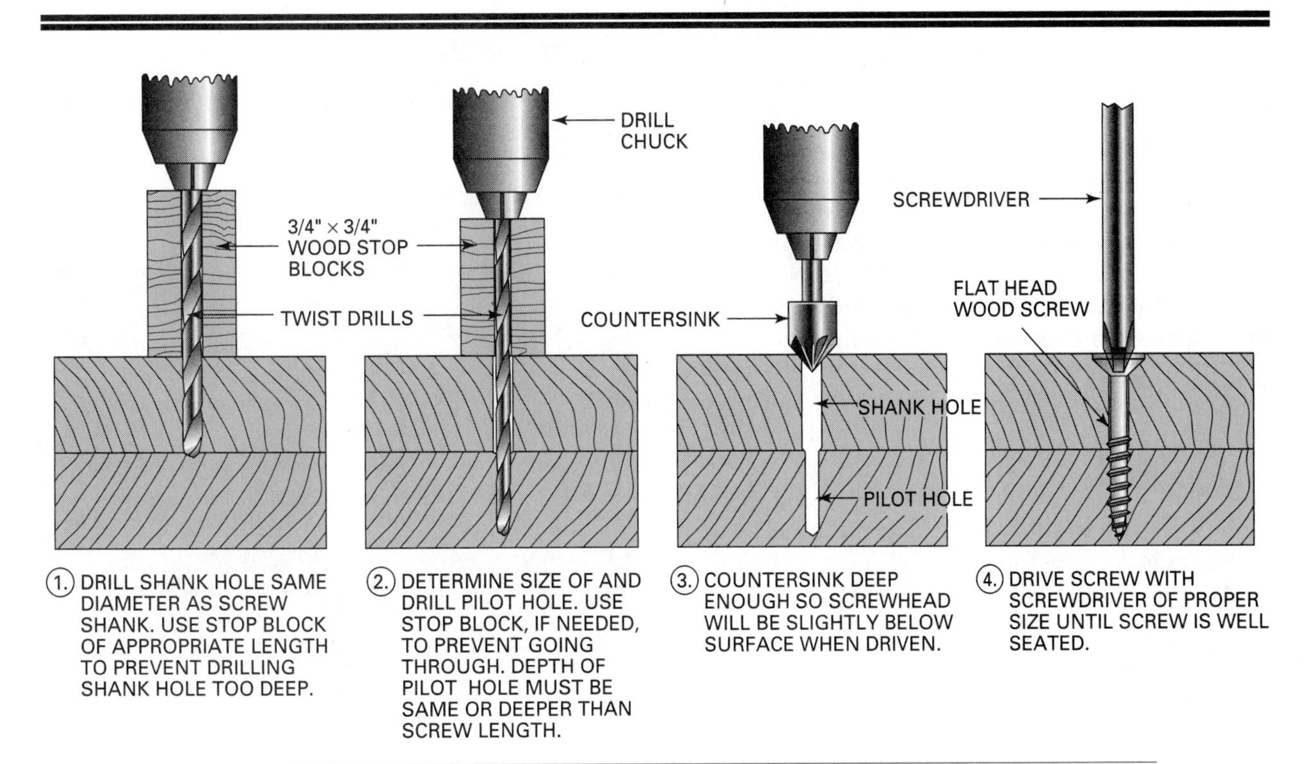

Fig. 13-15 Preparation needed to drive a screw.

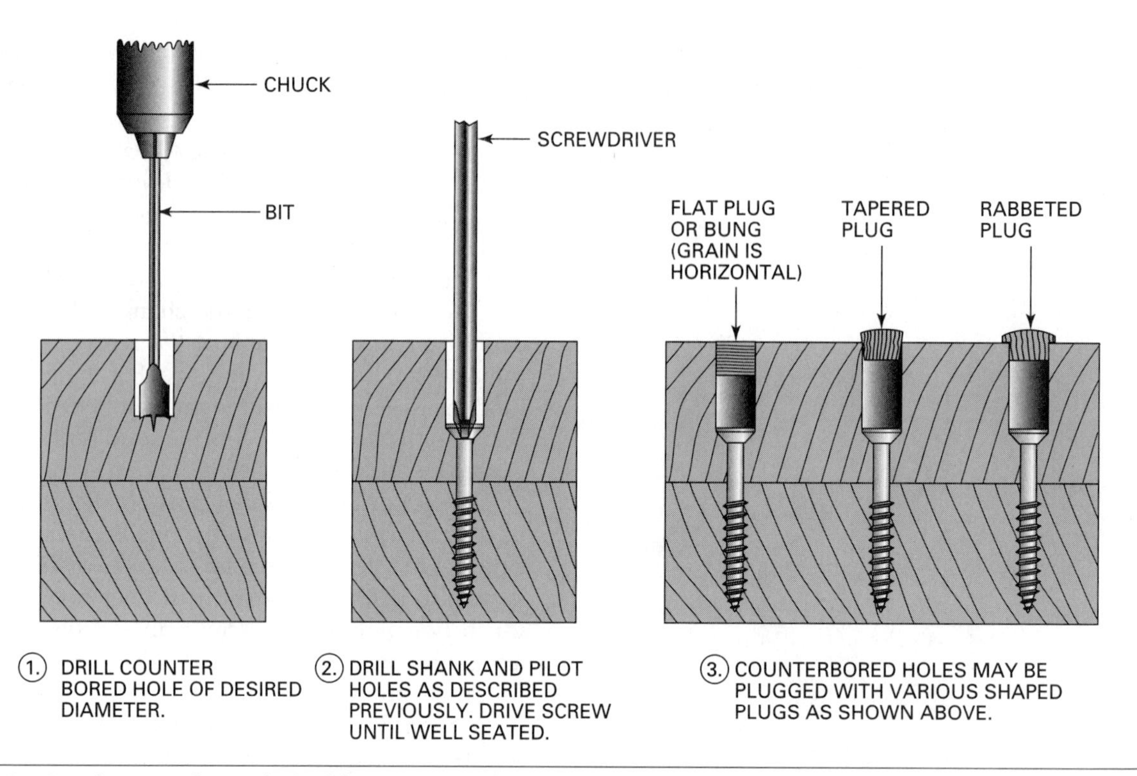

Fig. 13-16 It is sometimes desirable to counterbore when driving screws.

to the screw slot. Also, if too much pressure is applied when driving the screw, the head may be twisted off. This is particularly true when driving screws of soft metal, such as aluminum or brass. It might be wise to first drive a steel screw, remove it, and then drive the screw of softer metal. Rub some wax (paraffin) on the threads of the screw to make driving easier. Remember that if the pilot hole is too large, the screw will not grip.

Use a stop when drilling the shank hole to make sure it will not be drilled too deep. This will prevent it from going through the material when drilling the pilot hole. A simple stop can be made by drilling a hole lengthwise through a piece of nominal 1 × 1 stock, cutting it to the desired length, and inserting it on the twist drill against the chuck.

If the material to be fastened is thick, the screw may be set below the surface by **counterboring** to gain additional penetration without resorting to a longer screw. To set the screwhead below the surface, bore the counterbored hole first. Use a stop block to assure desired depth. The diameter of the hole should be equal to or slightly larger than the diameter of the screwhead. Next, drill the shank hole; then, the pilot hole (Fig. 13-16). Drilling shank and pilot holes first leaves no stock for the center point of the bit used to make the counterbored hole.

Dismantling Tools

Dismantling tools are used to take down staging and scaffolding, concrete forms, and other temporary structures. In addition, they are used for tearing out sections of a building when remodeling. Carpenters must be skilled in the use of the tools, and in the work, so that the dismantled members are not damaged any more than necessary.

Hammers

In addition to fastening, hammers are often used for pulling nails to dismantle parts. To increase leverage and make nail pulling easier, place a small block of wood under the hammer head (Fig. 13-17).

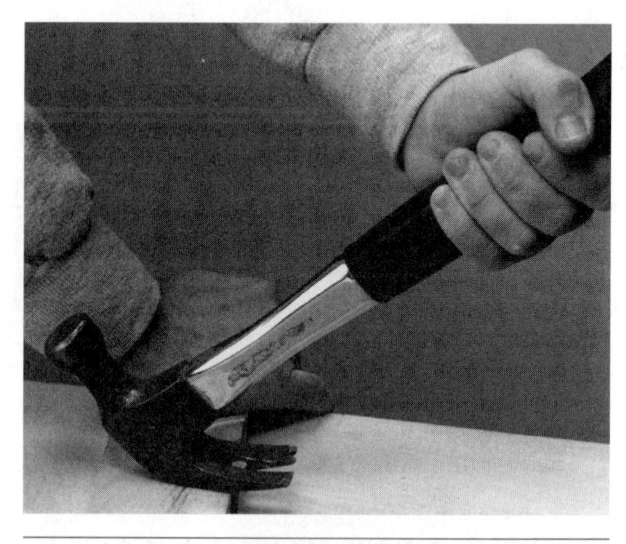

Fig. 13-17 Pull a nail more easily by placing a block of wood under the hammer.

Bars and Pullers

The *wrecking bar* (Fig. 13-18) is used to withdraw spikes and to pry when dismantling parts of a structure (Fig. 13-19). They are available in lengths from 12 to 36 inches, with the 30-inch size preferred for construction work.

Carpenters need a small *flat bar,* similar to that shown in Fig. 13-20, to pry small work and pull small nails. To extract nails that have been driven home (all the way in) a *nail claw,* commonly called a *cat's paw,* is used (Fig. 13-21).

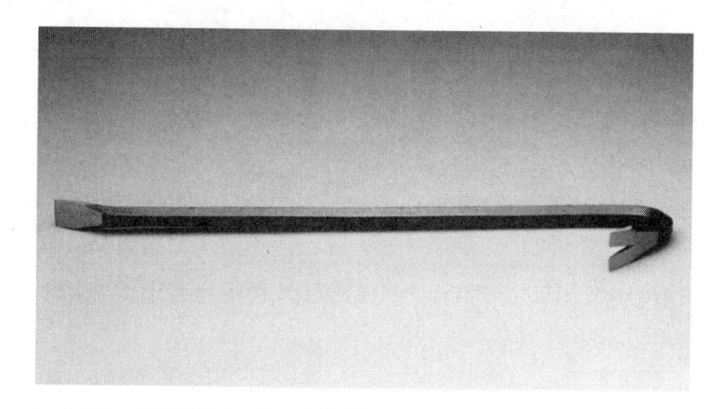

Fig. 13-18 Wrecking bar.

Fig. 13-19 Using a wrecking bar to pry stock loose.

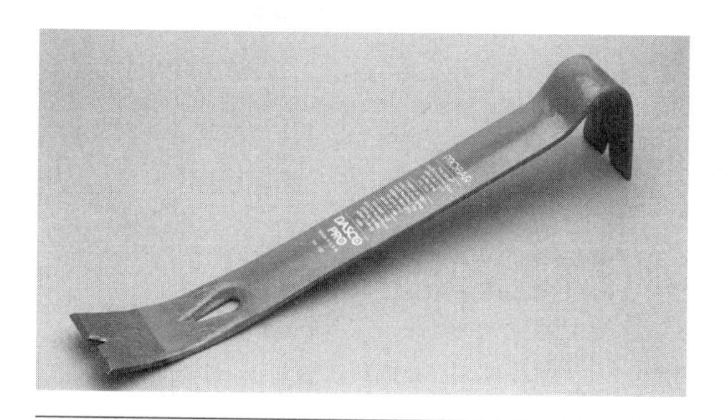

Fig. 13-20 Flat bar.

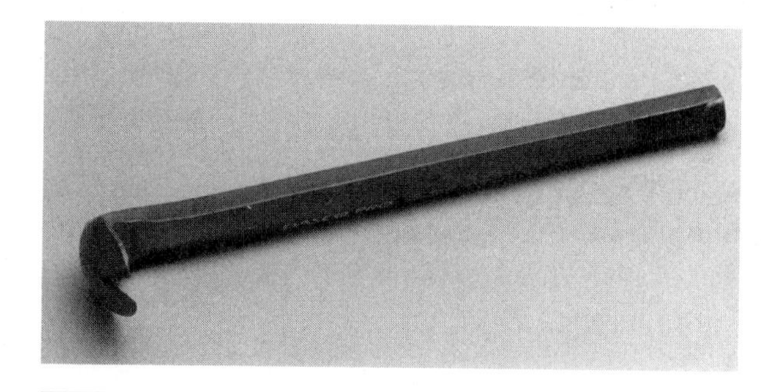

Fig. 13-21 Nail claw.

Holding Tools

To turn nuts, lag screws, bolts, and other objects, an *adjustable wrench* is often used (Fig. 13-22). The wrench is sized by its overall length. The 10-inch adjustable wrench is the one most widely used.

For extracting, turning, and holding objects, a pair of pliers is often used. Many kinds are manufactured, but the *combination pliers* (Fig. 13-23) is designed for general use.

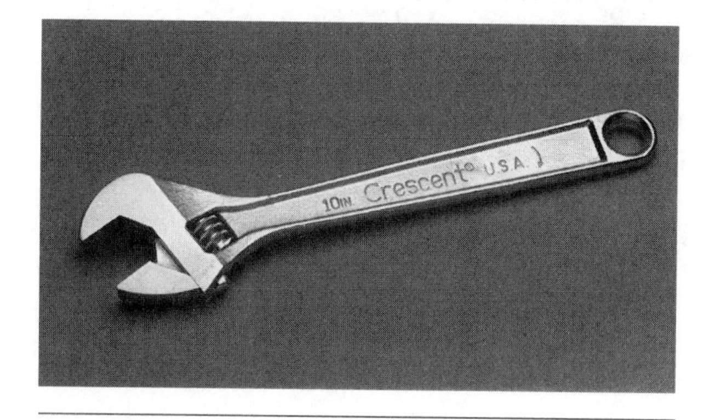

Fig. 13-22 Adjustable wrench.

Fig. 13-23 Combination pliers.

C clamps (Fig. 13-24) are useful for holding objects together while they are being fastened, holding temporary guides, applying pressure to glued joints, and many other purposes. The size is designated by the throat opening. C clamps are available in a number of styles. The carpenter will find a few 3- or 4-inch light C clamps to be very useful.

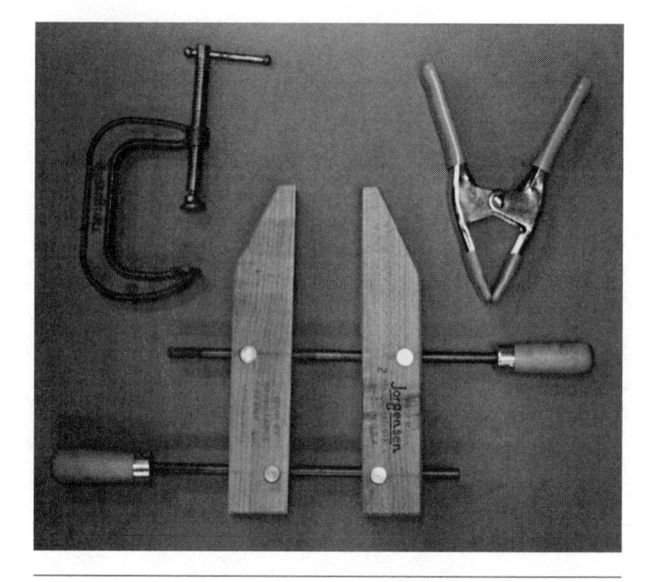

Fig. 13-24 C Clamp, wood screw, and spring clamp.

Review Questions

Select the most appropriate answer.

1. When stretching a steel tape to lay out a measurement, place the ring on a nail so the
a. one-inch mark is on the starting line.
b. end of the tape is on the starting line.
c. inside of the ring is on the starting line.
d. outside of the ring is on the starting line.

2. In construction, the term *plumb* means
a. horizontal.
b. level.
c. straight.
d. vertical.

3. In addition to laying out circles and equal spaces, the wing dividers may be used for
a. scaling.
b. scoring.
c. scribing.
d. squaring.

4. A #8 auger bit will bore a hole
a. 1/4 inch in diameter.
b. 1/2 inch in diameter.
c. 3/4 inch in diameter.
d. 1 inch in diameter.

5. Countersinks are tools used to
a. make sink cutouts in countertops.
b. drive large screws.
c. form recesses for flat head screws.
d. set nails below the surface.

6. The longest of the bench planes is called a
a. fore plane.
b. jack plane.
c. jointer plane.
d. smooth plane.

7. For finish work, the carpenter may use a crosscut saw with
a. 5 1/2 points.
b. 7 to 8 points.
c. 11 to 12 points.
d. 16 to 18 points.

8. The kind of joint made by cutting one piece of molding to fit against the irregular face of a similar piece is made by using a
a. chisel.
b. coping saw.
c. compass saw.
d. countersink.

9. To prevent a nail set from slipping off the nail head, its tip is
a. flat and smooth.
b. convex.
c. concave.
d. checkered.

10. Grind the bevel on a chisel so it
a. is hollow.
b. has a wire-edge.
c. is sharp.
d. is smooth.

BUILDING FOR SUCCESS

Tool Care and Maintenance

It is possible to make some assumptions about a craftsperson's quality of work by inspecting his or her tools and equipment. In order to do quality work, skilled workers must properly maintain their entire tool inventory. It is essential for a good carpenter, electrician, mason, or plumber to develop a regular tool maintenance program. Sharp tools are actually safer than dull ones because they require less effort to use.

All hand and power tools are to be properly used and cared for, regardless of the owner. The contractor's tools and equipment used by the worker on a daily basis must carry the same value as those of the worker. A carpenter must be ready to perform maintenance on the job if any equipment malfunctions or needs repair. Tasks such as honing a cutting edge, lubricating, adjusting, mending, and repairing can be completed successfully on the job. The majority of required maintenance will be conducted after work and on Saturdays.

An ongoing tool maintenance program may be a one-person operation or require the services of a professional repair person. Regardless of the situation, the scheduling of routine or special repair is required. An efficient tool maintenance program must begin with the proper storing and use of all hand and power tools. The tool inventory must be immediately accessible to be functional. A nail gun will not function properly if it is tossed into the back of a pickup.

An experienced craftsperson knows which tools call for daily, weekly, or monthly maintenance. The most frequently used tools, such as chisels, saws, planes, and router bits, may require daily inspection. The habit of storing tools in a protected manner such as tool boxes, bags, cases, or on shelves will prolong their service life.

The craftsperson who knows building materials, tools, and equipment will select the best tool for the job. When the hand or power tool is needed, it should be in proper working order. Potential accidents resulting from dull or malfunctioning equipment will be reduced. If a tool must be forced to perform, it has not been properly maintained.

When properly trained, a carpenter can sharpen a noncarbide saw blade at the job site in just a few minutes. A simple homemade sharpening jig, a sawhorse, and two C clamps will quickly bring a blade back into use. A plane iron can be honed to regain a sharp cutting edge in a very short time. At times, electric cords can be mended or replaced on the job. Blades and spare parts such as brushes can be carried in a tool box.

As repairs are made on tools and equipment, caution should be taken to assure the worker's safety. All power tools must be disconnected from their power source before any attempt is made to analyze the problem. When the problem appears to be beyond the repair capability of the worker, the equipment must be taken to an experienced tool repair facility.

One of the best assurances for having hand and power tools ready for use is to create and use an effective maintenance program. The following suggestions will help guarantee that hand and power tools are properly cared for.

1. Protect your tools and equipment during daily transport and use.
2. To prolong the life of your equipment, store it properly in tool boxes, carrying cases, racks, shelving, or special containers.
3. Carry an inventory of the most common replacement parts, such as cords, brushes, batteries, and gas cylinders (nail guns) for on-the-job repairs.
4. Avoid leaving tools and equipment in unprotected areas, on floors, on the ground (in water or mud), on ladders or scaffolding, on roofs, or near heavy traffic areas.
5. Sideline tools that have malfunctioned and schedule them for immediate repair.
6. Have a maintenance system in operation at the job site and schedule regular daily, weekly or monthly maintenance.
7. Establish and use as needed the services of a reputable maintenance service center.
8. Read and keep the service/use manual that accompanies each hand and power tool in your inventory. File each manual where it is easily accessible.
9. Clean, check for malfunctions, and properly store your equipment on a daily basis.

10. Schedule a brief period of time at the end of each week to perform maintenance in preparation for the next work week.

Remember that the tools and equipment in your inventory play a major role in generating your income. Do not assume that they will continue to operate without regular care and maintenance.

FOCUS QUESTIONS: For individual or group discussion

1. In what ways can a personal tool maintenance process benefit you as a carpenter or other skilled worker?
2. What are some of the outcomes of not having a tool or machine maintenance program in place?
3. Assuming that you are working out of a vehicle, how would you transport your tools and machines that are used practically every day?
4. Do you know where you can take tools and machines for repair within your community? How do you locate and make initial contact with a tool service center?

UNIT 6

Portable Power Tools

Portable power tools are widely used in the construction industry. They enable the carpenter to do more work in shorter time with less effort. However, they can cause serious injury if operated improperly. Most accidents are caused by a complete disregard for common sense safety precautions. Be aware of the dangers of operating portable power tools, and take steps to avoid these dangers and prevent accidents. Always wear eye protection when using power equipment.

OBJECTIVES

After completing this unit, the student should be able to:

- state general safety rules for operating portable power tools
- identify, describe, and safely use the following portable power tools: circular saws, saber saws, reciprocating saws, drills, hammer-drills, screwdrivers, planes, routers, sanders, staplers, nailers, and powder-actuated drivers

UNIT CONTENTS

GENERAL SAFETY RULES

CAUTION Have a complete understanding of a tool before attempting to operate it.

Electrical shock has caused many fatal accidents. Make sure the tool is properly grounded. A *ground fault circuit interrupter* (GFCI) must be in the circuit being used. Even though the circuit is grounded, it is still possible for an operator of a portable power tool to be electrocuted should a short circuit occur without a GFCI in the circuit. The GFCI will trip at about 5 milliamperes, turning off the power before any chance of electrical shock or electrocution occurs.

Do not allow your attention to be distracted when operating power tools.

Never use a power tool that has a worn or frayed cord.

Use the proper size extension cord. Do not use one any longer than necessary to prevent excessive voltage drop.

Be careful to place extension cords where they will not be damaged, cut, caught in the tool, or tripped on.

Unplug tools when making adjustments or changing cutters.

Do not wear loose clothing that might become caught in the tool.

Wear eye protection during all operations that cause particles to fly.

Make sure the material being worked on is securely held and supported. Use sharp tools and all safety guards supplied.

Stay alert and develop an attitude of care and concern for yourself and other workers.

CHAPTER 14 SAWS, DRILLS, AND DRIVERS

Saws

The carpenter uses several kinds of portable saws for crosscutting and ripping, for making circular cuts, and for cutting openings in floors, walls, and ceilings.

Electric Circular Saws

Commonly called the *skilsaw,* the portable electric circular saw (Fig. 14-1) is used by the carpenter more than any other portable power tool. The circular saw blade is driven by an electric motor. The saw has a base that rests on the work to be cut. A handle with a trigger switch is provided for the operator to control the tool. The saw is adjustable for depth of cut. A retractable safety guard is provided over the blade, extending under the base.

The base may be tilted for making *bevel* and *miter* cuts. Bevel cuts are those at an angle through the thickness and with the grain or along the length. Edge or end miters are angle cuts through the thickness across the grain. Flat miters are cuts at an angle across the width. Compound miters are angle cuts across the width and also through the thickness (Fig. 14-2).

Fig. 14-1 Using a portable electric circular saw to cut compound angles.

Saws are manufactured in many styles and sizes. They are available from very light- to heavy-duty models ranging from 1/6 hp to 1 1/2 hp. The

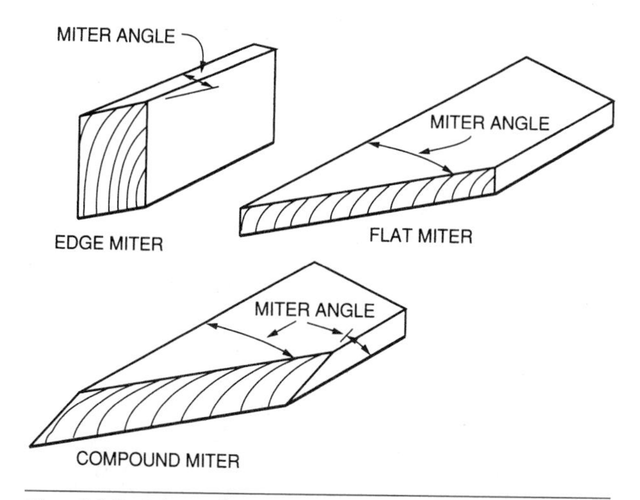

Fig. 14-2 Edge, flat, and compound miters.

size is determined by the diameter of the blade, which ranges from 4 1/2 to 16 inches. The handle and switch may be located on the top or in back. The blade may be driven directly by the motor or through a worm gear (Fig. 14-3).

The forward end of the base is notched in two places to serve as guides for following layout lines. One notch is used to follow layout lines when the base is tilted to 45 degrees and the other when the base is not tilted.

CAUTION Make sure the saw blade is installed with the teeth pointing in the correct direction. The teeth of the saw blade projecting below the base should be pointing away from the operator.

To loosen the bolt that holds the blade in place, first disconnect the saw by unplugging it, and then insert a nail under the base and through the hole provided in the blade. Turn the bolt in the same direction as the rotation of the blade. To tighten, turn the bolt in a direction opposite to the rotation of the blade. Most saws have an arrangement that locks the spindle while loosening or tightening the bolt.

On most models an adjustable attachment that fits into the base is used for ripping narrow pieces parallel with the edge. The saw may also be guided by tacking or clamping a straightedge to the material and running the edge of the saw base against it. Allowance must be made for the distance from the saw blade to the edge of the saw base when positioning the straightedge.

Circular Saw Blades. Circular saw blades are available in a number of styles. The shape and number of teeth per inch determine their cutting action. Carbide-tipped blades are used more than high-speed steel blades. They stay sharper longer when cutting material that dulls ordinary blades quickly. (More complete information on saw blades can be found in Unit 7.)

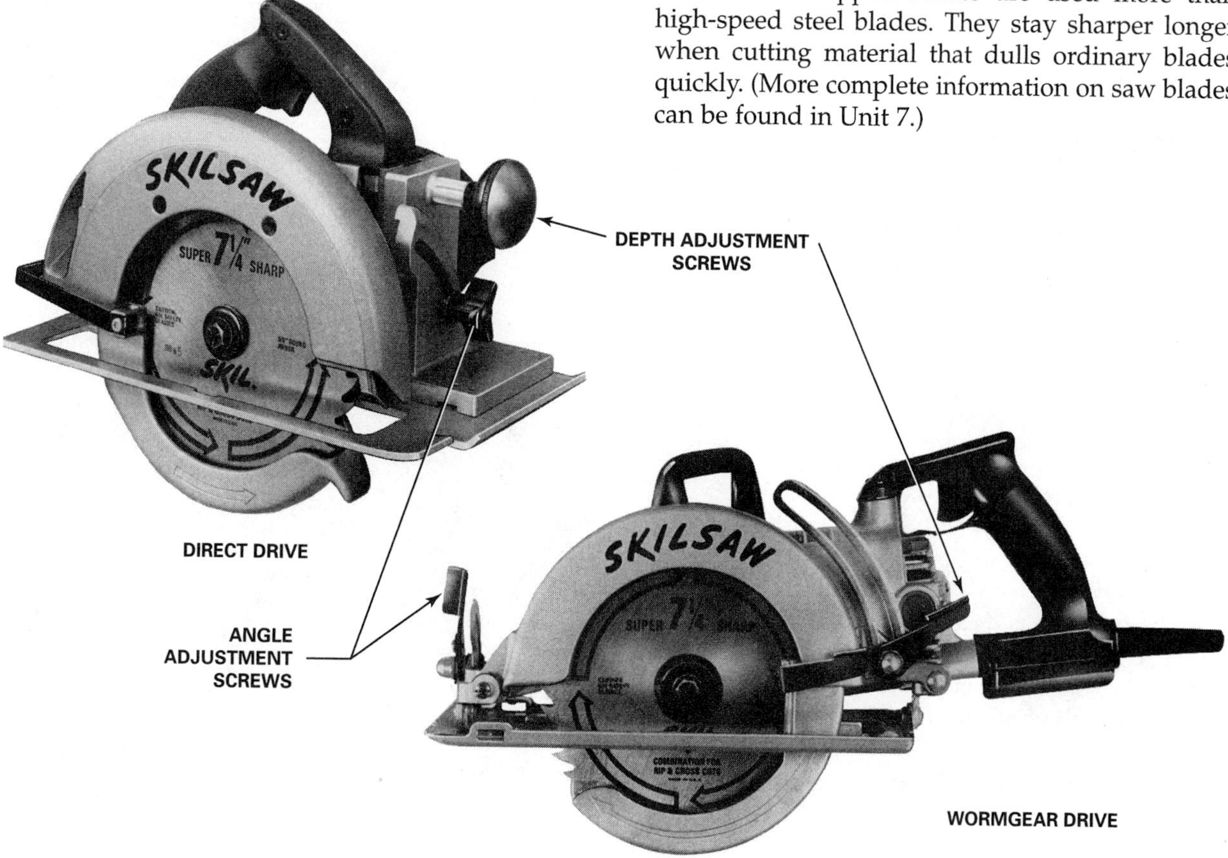

Fig. 14-3 Direct drive and worm gear drive portable electric circular saws. (*Courtesy of Skil Power Tools*)

Fig. 14-4 Saw cuts are made over the end of supports so the waste will fall clear and not bind the blade.

Using the Portable Circular Saw. Safe and efficient cutting follows an established method:

- Make sure the work is securely held and that the waste will fall clear and not bind the saw blade (Fig. 14-4).
- Adjust the depth of cut so that the blade just cuts through the work. Never expose the blade any more than is necessary (Fig. 14-5).

CAUTION Make sure the guard operates properly. Be aware that the guard may possibly stick in the open position. Never wedge the guard back in an open position.

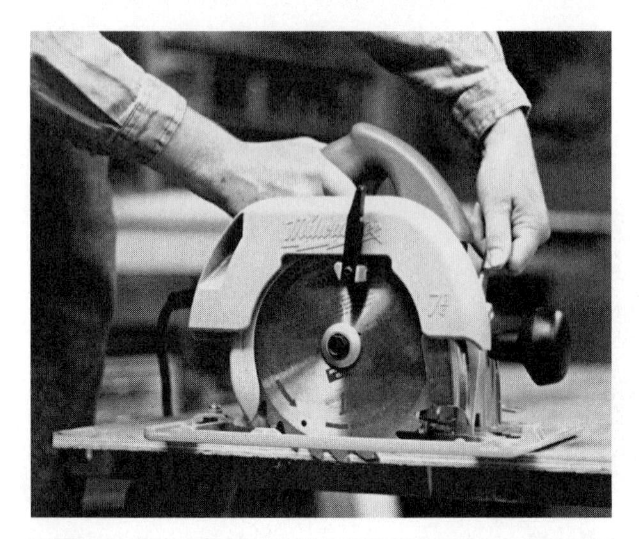

Fig. 14-5 The blade of the saw is adjusted for depth only enough to cut through the work.

- Mark the stock. Put on safety goggles. Rest the forward end of the base on the work. With the blade clear of the material, start the saw.
- Advance the saw into the work when it has reached full speed. Make sure to observe the line to be followed. With the saw cut in the waste, cut as close to the line as possible for a short distance. Then observe the notch in the forward end of the base. Follow the line by using the notch on the saw base.
- Follow the line closely.

CAUTION Any deviation from the line may cause the saw to bind and kick back. Do not force the saw forward. In case the saw does bind, stop the motor and bring the saw back to where it will run free. Continue following the line closely.

- Near the end of the cut, the forward end of the base will go off the work. Guide the saw by observing the line at the saw blade and finish the cut. The saw may also be guided by watching the layout line at the saw cut for the whole length. Let the waste drop clear and release the switch.

CAUTION Keep the saw clear of your body until the saw blade has completely stopped.

- When starting cuts across stock at an angle, it may be necessary to retract the guard by hand. A handle is provided for this purpose (Fig. 14-6). Release the handle after the cut has been started and continue as above.

Fig 14-6 Retracting the guard of the portable circular saw by hand.

- Compound miter cuts may be made by cutting across the stock at an angle with the base tilted.

Portable circular saws cut on the upstroke. The saw blade rotates upward through the material. As the teeth of the saw blade come through the top surface, splintering of the stock occurs at the layout line. The severity of the splintering depends on the kind of blade used, kind and thickness of the material being cut, and other factors. More splintering occurs when cutting across grain than with the grain.

On *finish work,* that is, work that will ultimately be exposed to view, any splintering along the cut is unacceptable. One way to prevent this is to mark layout lines on, and make cuts from, the back side. If it is not possible to cut from the back side, or if both sides may be exposed to view, mark the layout lines on the face side. Then score along the layout lines with a sharp knife. Make the cuts just outside the scored lines.

Making Plunge Cuts. Many times it is necessary to make internal cuts in the material such as for sinks in countertops or openings in floors and walls. To make these cuts with a portable electric circular saw, the saw must be plunged into the material:

- Accurately lay out the cut to be made. Adjust the saw for depth of cut.
- Wearing eye protection, hold the guard open and tilt the saw up with the front edge of the base resting on the work. Have the saw blade over, and in line with, the cut to be made (Fig. 14-7).
- Make sure the teeth of the blade are clear of the work and start the saw. Lower the blade slowly into the work, following the line carefully, until the entire base rests squarely on the material.
- Advance the saw into the corner. Release the switch and wait until the saw stops before removing it from the cut.
- Reverse the direction and cut into the corner. Again, wait until the saw stops before removing it from the cut.
- Proceed in like manner to cut the other sides. Finish the cut into the corners with a handsaw or saber saw.

Saber Saws

The *saber saw* (Fig. 14-8) is sometimes called a *jigsaw* or *bayonet saw.* It is widely used to make curved cuts. The teeth of the blade point upward when installed, so the saw cuts on the upstroke. To produce a splinter-free cut on the face side, it is best to cut on the side opposite the face of the work, if possible, or to score with a knife along the layout line.

There are many styles and varieties of saber saws. The length of the stroke along with the amperage of the motor determine its size and quality. Strokes range from 1/2 to 1 inch. The longest stroke is the best for faster and easier cutting. Some saws can be switched from straight up-and-down strokes to several orbital (circular) motions to provide the most effective cutting action for various materials (Fig. 14-9). The base of the saw may be tilted to make bevel cuts.

Fig. 14-7 Making a plunge cut with a portable circular saw. First retract the guard, place the front edge of the saw base on the material and then pivot, running the saw slowly into the material.

Fig. 14-8 The jigsaw may be used either in straight or orbital cutting actions.

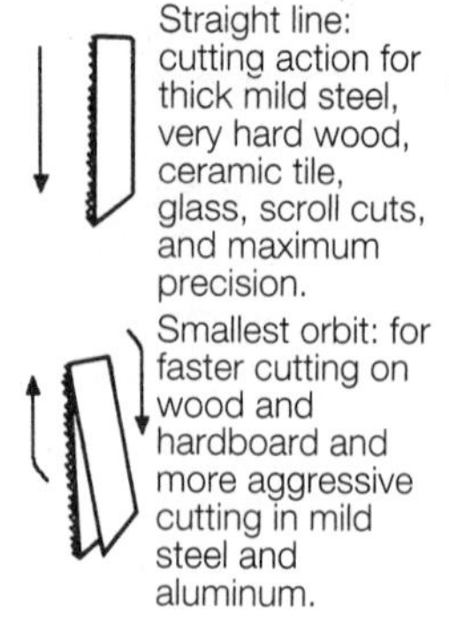

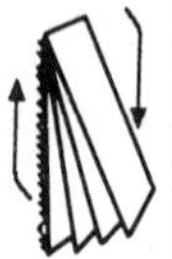

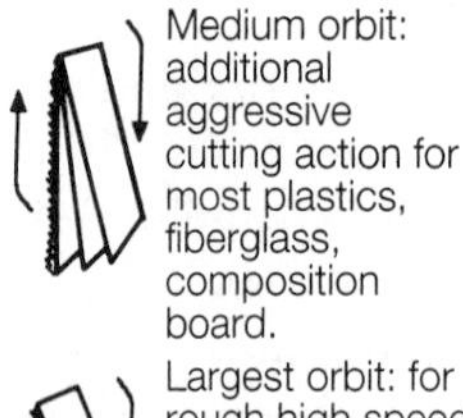

Fig. 14-9 Guide to the selection of saber saw blade action.

Blade Selection. Many blades are available for fine or coarse cutting in wood or metal (Fig. 14-10). Wood cutting blades have teeth that are from 6 to 12 points to the inch. Blades with coarse teeth (less points to the inch) cut faster, but rougher. Blades with more teeth to the inch may cut slower, but produce a smoother cut surface. They do not splinter the work as much. Proper blade selection and use will produce the best cuts. The blade will last longer and the saber saw will not be subjected to harmful abuse by applying excessive pressure when cutting.

Using the Saber Saw. Follow a safe and established procedure:

- Outline the cut to be made. Secure the work either by hand, tacking, clamping, or by some other method.
- Using eye protection, hold the base of the saw firmly on the work. With the blade clear, pull the trigger.

APPLICATION	PART NUMBER	TEETH PER INCH	OVERALL LENGTH (INCHES)	WIDTH (INCHES)
Metal cutting (High speed steel)				
Non-ferrous metal cutting ¼" to ¾" thick	**94612**	8	3⅝	5/16
Metal cutting over ⅛" thick	**94613**	12	2¾	5/16
Metal cutting up to ⅛" thick	**94614**	21	2¾	5/16
Metal cutting up to 1/16" thick	**94615**	36	2¾	5/16
Wood cutting (Taper ground, high speed steel)				
Fast, smooth curve cutting in wood up to 2" thick	**94618**	6	3⅝	¼
Fast, smooth cutting in wood up to 2" thick	**94619**	6	3⅝	5/16
Very smooth curve cutting in wood up to 1" thick	**94622**	10	3⅝	¼
Very smooth cutting in wood up to 1" thick	**94623**	10	3⅝	5/16
Wood cutting (Alternate set—ground, high carbon steel)				
Fast, medium curve cutting in wood up to 2" thick	**94627**	6	3⅝	¼
Fast, medium cutting in wood up to 2" thick	**94628**	6	3⅝	5/16
Wood cutting (Alternate set—high carbon steel)				
Fast, rough cutting in wood up to 2" thick	**94631**	6	3⅝	5/16
Medium cutting in wood up to 2" thick	**94634**	8	3⅝	5/16
General purpose (Alternate set—high speed steel)				
General purpose cutting in wood, plastic, metal, etc.	**94637**	12	3⅝	5/16
Wood scroll cutting (Taper ground, high carbon steel)				
Smooth, intricate scroll cuts in wood up to 1¾" thick	**94640**	20	2¾	3/16

Fig. 14-10 Saber saw blade selection guide.

- Push the saw into the work, following the line closely. Make the saw cut into the waste, and cut as close to the line as possible without completely removing it.
- Keep the saw moving forward, holding the base down firmly on the work. Turn the saw as necessary in order to follow the line to be cut. Feeding the saw into the work as fast as it will cut, but not forcing it, finish the cut. Keep the saw clear of your body until it has stopped.

Making Plunge Cuts. Plunge cuts may be made with the saber saw in a manner similar to that used with the circular saw:

- Tilt the saw up on the forward end of its base with the blade in line and clear of the work (Fig. 14-11).
- Start the motor, holding the base steady. Very gradually and slowly, lower the saw until the blade penetrates the work and the base rests firmly on it.

CAUTION Hold the saw firmly to prevent it from jumping when the blade makes contact with the material and to make a successful plunge cut.

- Cut along the line into the corner. Back up for about an inch, turn the corner, and cut along the other side and into the corner.
- Continue in this manner until all the sides of the opening are cut.

CAUTION Make sure the tool comes to a complete stop before withdrawing it from the material being cut.

- Turn the saw around and cut in the opposite direction to cut out the corners.

Reciprocating Saws

The **reciprocating saw** (Fig. 14-12), sometimes called a *sawzall,* is used primarily for *roughing in* work. This work consists of cutting holes and openings for such things as pipes, heating and cooling ducts, and roof vents. It can be likened to a powered compass saw.

Most models have a variable speed of from 0 to 2,400 strokes per minute. Like saber saws, some models may be switched to several *orbital* cutting strokes from a straight back and forth (reciprocal) cutting action. Ordinarily, the orbital cutting mode is used for fast cutting in wood. The reciprocating stroke is used for cutting metal.

Reciprocating Saw Blades. Common blade lengths run from 4 to 12 inches. They are available for cutting practically any type of material. Blades are available to cut wood, metal, plaster, fiberglass, ceramics, and other material.

Using the Reciprocating Saw. The reciprocating saw is used in a manner similar to the saber saw. The difference is that the reciprocating saw is heavier and more powerful. It can be used

Fig. 14-11 Making an internal cut by plunging the saber saw.

Fig. 14-12 Using the reciprocating saw to cut an opening in the subfloor.

more efficiently to cut through rough, thick material, such as walls when remodeling. With a long blade, it can be used to cut flush with a floor or along the side of a stud.

To use the saw, lines must be laid out and followed. The base or shoe of the saw is held firmly against the work whenever possible. To make cutouts, first drill a hole in the material. Then insert the blade, start the motor, and follow the layout lines. The blades can be reversed for cutting in confined areas.

Drills and Drivers

Portable power drills, manufactured in a great number of styles and sizes, are widely used to drill holes and drive fasteners in all kinds of construction materials. Battery-powered cordless models are used because of their convenience or when electrical power is not available.

Drills

The drills used in the construction industry are classified as light-duty or heavy-duty. Light-duty drills usually have a *pistol-grip* handle. Heavy-duty drills may have a *spade-shaped* or *D-shaped* handle (Fig. 14-13).

Drill Sizes. The size of a drill is determined by the capacity of the *chuck,* that is, its maximum opening. The chuck is that part of the drill that holds the cutting tool. The most popular sizes for light-duty models are 1/4 and 3/8 inch. Heavy-duty drills have a 1/2-inch chuck or larger.

Drill Speed and Rotation. Most drills have *variable speed* and *reversible* controls. Speed of rotation can be controlled from 0 to maximum RPM (revolutions per minute) by varying the pressure on the trigger switch. Slow speeds are desirable for drilling larger holes or for driving fasteners. Faster speeds are used for drilling smaller holes. A reversing switch changes direction of the rotation for removing screws or withdrawing bits and drills from holes.

Bits and Twist Drills. Twist drills are used in electric drills to make small holes in wood or metal (Fig. 14-14). For larger holes in wood, plastics, and composition materials, a variety of wood-cutting bits are used.

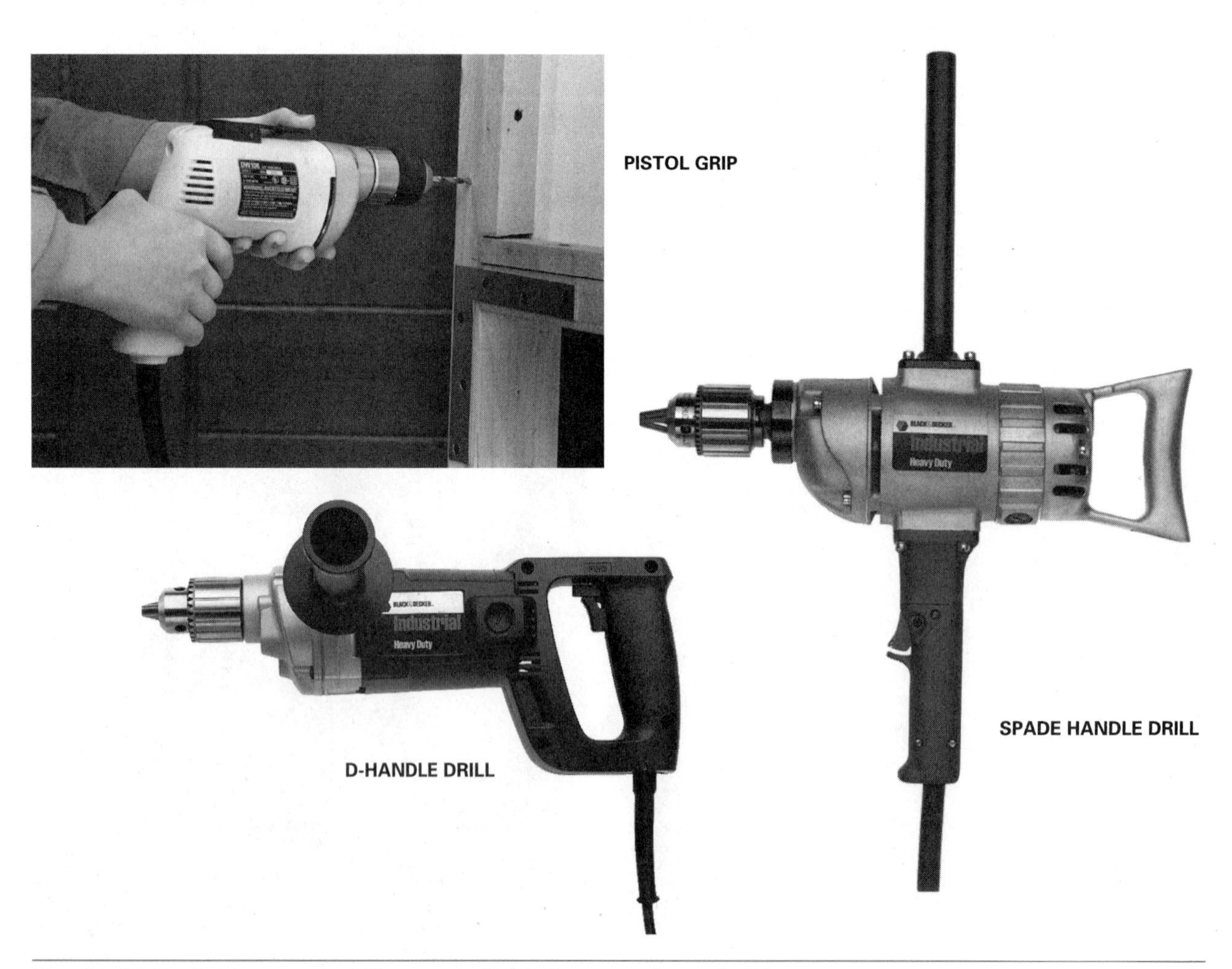

Fig. 14-13 Portable power drills are available in a number of styles.

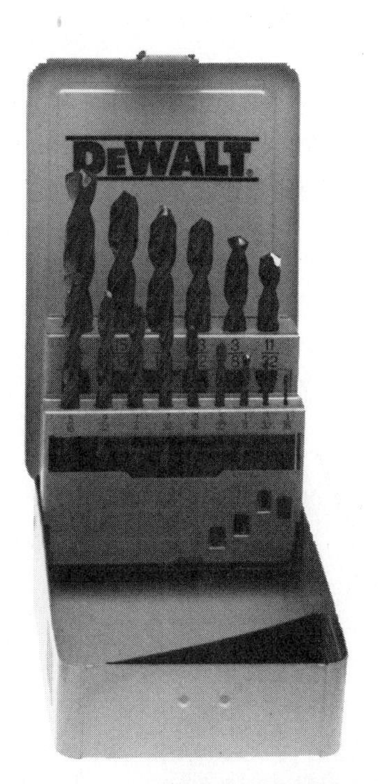

Fig. 14-14 Twist drills are used to drill small holes. (*Courtesy of DeWalt Industrial Tool Company*)

For boring holes in rough work, the *spade or paddle bit* is commonly used. For a hole with a cleaner edge in finish work, the *Power Bore* bit may be used (Fig. 14-15). Notice that none of these types has a center point that is threaded.

CAUTION Never use bits with threaded center points in electric drills when boring deep holes. A threaded center will draw the bit into the work. This makes it difficult to withdraw the bit and may cause the operator to lose control of the drill.

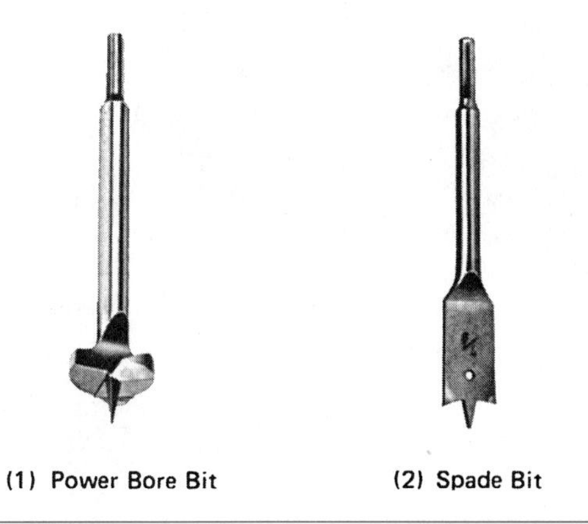

Fig. 14-15 Spade and Power Bore bits are used to drill larger holes in wood and similar material.

Other Drill Accessories. Occasionally, carpenters may use *hole saws.* These saws cut holes through thin material from 5/8 inch to 6 inches in diameter. One disadvantage of the hole saw is that a hole cannot be made partially through the material. Only the circumference is cut and the waste is not expelled.

The *combination drill and countersink,* discussed in Chapter 12, is frequently used with power drills for fast drilling, countersinking, and counterboring in one operation when needed.

Masonry drill bits have carbide tips for drilling holes in concrete, brick, tile, and other masonry. They are frequently used in portable power drills. They are more efficiently used in *hammer-drills.*

Using Portable Electric Drills. Select the proper size bit or twist drill and insert it. Tighten the chuck with the chuck key or by holding the chuck of a *keyless* chuck. For accuracy, holes in wood are center-punched. Holes in metal must be center-punched because the drill will wander off center.

CAUTION Hold small pieces securely by clamping or other means. When drilling through metal, especially, the drill has a tendency to hang up when it penetrates the underside. If the piece is not held securely, the hangup will cause the piece to rotate with the drill. It could then hit anything in its path and possibly cause serious injury to a person before power to the drill can be shut off.

Place the bit on the center of the hole to be drilled. Hold the drill at the desired angle (Fig. 14-16) and start the motor. Apply pressure as required, but do not force the bit. Drill into the stock, being careful not to wobble the drill. Failure to hold the drill steady may result in breakage of small twist drills.

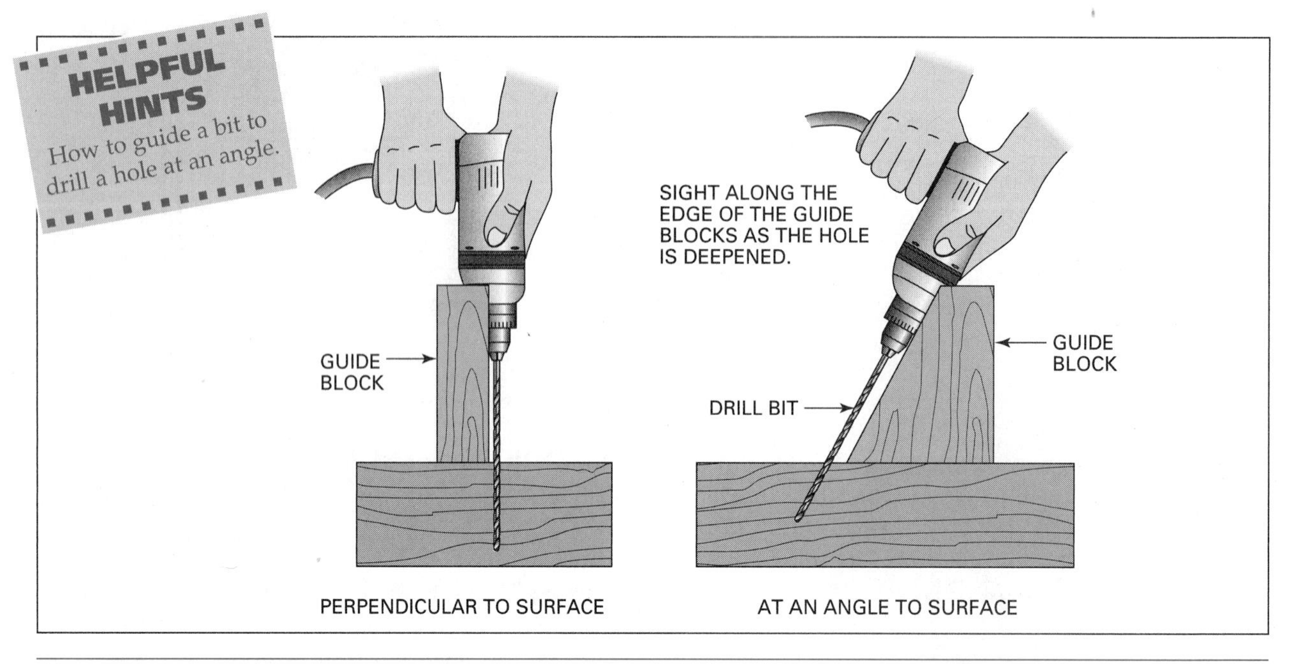

Fig. 14-16 Techniques for drilling a hole where the angle of the hole is accurately reproducible.

CAUTION Remove the bit from the hole frequently to clear the chips. Failure to do this may result in the drill binding and twisting from the operator's hands. Be ready to instantly release the trigger switch if the drill does bind.

Hammer-drills

Hammer-drills (Fig. 14-17) are similar to other drills. However, they can be changed to hammers as they drill, quickly making holes in concrete or other masonry. Some models deliver as much as 50,000 hammer blows per minute. Most popular are the 3/8- and 1/2-inch sizes.

A depth stop is usually attached to the side of the hammer-drill. It can be converted to a conventional drill by a quick-change mechanism. Most models have a variable speed of from 0 up to 2,600 rpm. The hammer-drill has the same type chuck and is used in the same manner as conventional portable power drills.

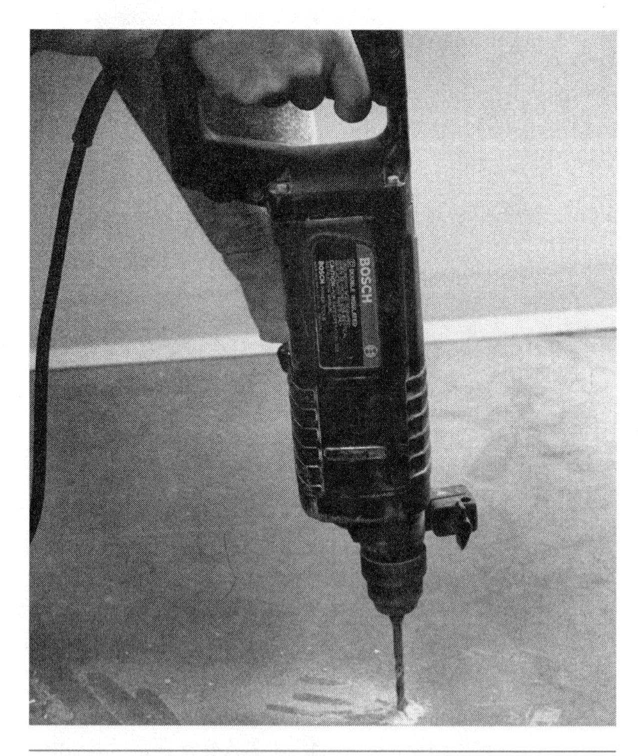

Fig 14-17 The hammer-drill is used to make holes in concrete.

Screwguns

Screwguns or *drywall drivers* (Fig. 14-18) are used extensively for fastening gypsum board to walls and ceilings with screws. They are similar in appearance to the light-duty drills, except for the chuck. They have a pistol-type grip for one-hand operation and controls for varying the speed and reversing the rotation.

The chuck is made to receive special screwdriver bits of various shapes and sizes. A screwgun has an adjustable nosepiece, which surrounds the bit. When the forward end of the nosepiece touches the surface, the clutch is separated and the bit stops turning. Adjusting the nosepiece makes variations in the screw depth.

Accessories are available for driving hex head as well as screws with slotted heads. Hex head nut-

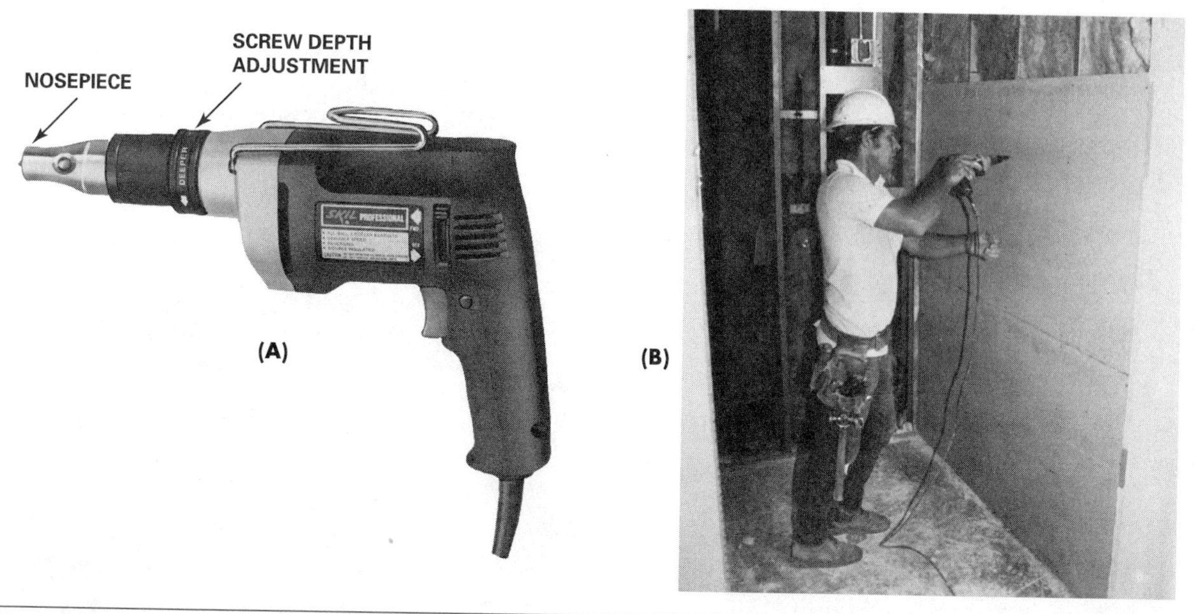

Fig. 14-18 The drywall driver is used to fasten wallboard with screws. (A) (*Courtesy of Skil Power Tools*), (B) (*Courtesy of U.S. Gypsum Corporation*)

setters are available in magnetic or nonmagnetic styles, in sizes ranging from 3/16 to 3/8 inch. Screwdriver bits include slotted, Phillips, Pozidrive, Torx, clutch head, and Robertson.

Cordless Tools

Cordless power tools are widely used due to their convenience, strength and durability (Fig. 14-19). The tools' power source is a removable battery usually attached to the handle of the tool. The batteries range in voltage from 4 to 24 volts and, in general, the higher the voltage, the stronger the tool.

Cordless drills come with variable-speed reversing motors and a positive clutch, which can be adjusted to drive screws to a desired torque and depth. Cordless circular saws can cut dimension lumber and are very handy as trim saws.

Downtime is practically eliminated with spare batteries and improvements in the chargers, which

Fig. 14-19 Cordless tools and battery charger. (*Courtesy of DeWalt Industrial Tool Company*)

Fig. 14-20 Cordless tool battery charger. (*Courtesy of DeWalt Industrial Tool Co.*)

can quick charge in as little as fifteen minutes (Fig. 14-20). Improved chargers can charge batteries of various voltages. They stop charging when the voltage level is reached, before the battery begins to heat. This prolongs the life of the battery (Fig. 14-21).

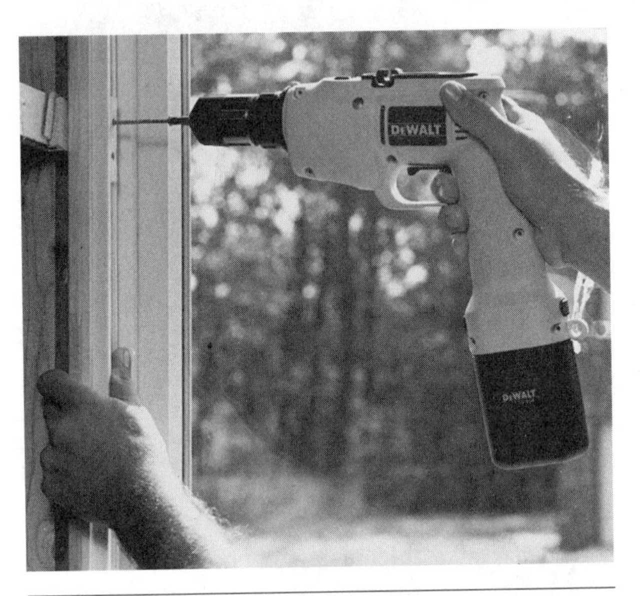

Fig. 14-21 Cordless screwdriver being used to install a window. (*Courtesy of DeWalt Industrial Tool Co.*)

CHAPTER 15 PLANES, ROUTERS, AND SANDERS

Portable Power Planes

Portable power planes make some planing jobs much easier for the carpenter. Planing the ends and edges of hardwood doors and stair treads, for example, takes considerable effort with hand planes, even with razor-sharp cutting edges.

Jointer Planes

The *jointer plane* is used primarily to smooth and straighten long edges, such as fitting doors in openings (Fig. 15-1). It is manufactured in lengths up to 18 inches. The electric motor powers a cutter head that may measure up to 3 3/4 inches wide. The planing depth, or the amount that can be taken off with one pass, can be set for 0 up to 1/8 inch.

An adjustable fence allows planing square, beveled edges to 45 degrees, or **chamfers** (Fig. 15-2). A rabbeting guide is used to cut rabbets up to 7/8 inch deep.

Block Planes

The *block plane* is about 7 inches long. It is operated with one hand. It is used to smooth the ends and edges of small pieces of stock (Fig. 15-3). Like the jointer plane, a motor powers the cutter head. The toe of the shoe can be adjusted for depth of cut. The fence may be used for guiding the block plane in the same manner as for jointer planes.

Abrasive Planes

At least one manufacturer has replaced the cutter head of the block plane with a heavy-duty sleeve. This tool is called an *abrasive plane*. Its maximum depth of cut is 1/64 inch. The plane leaves the sur-

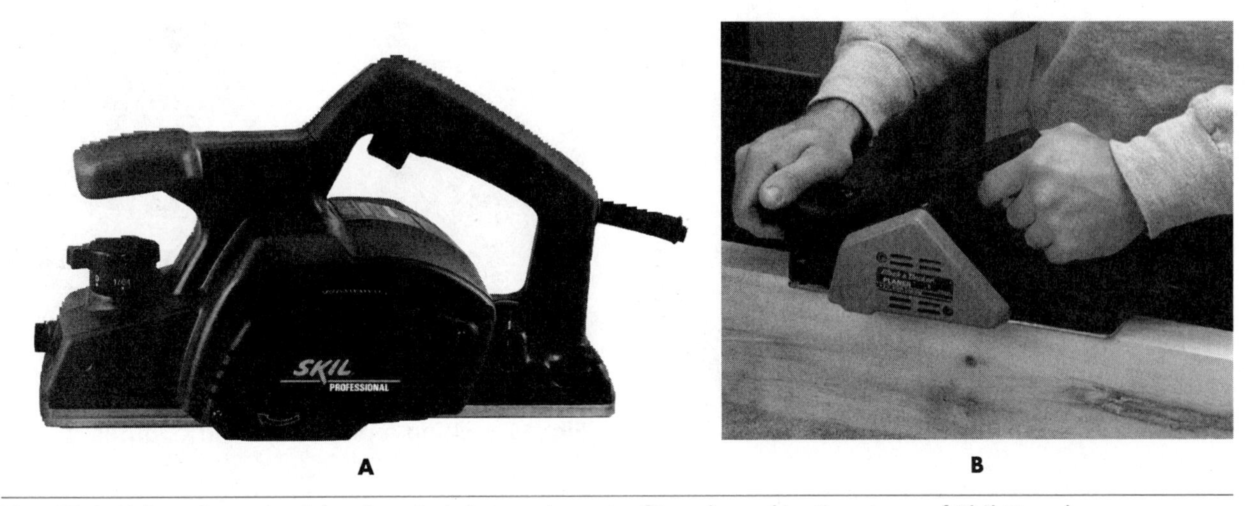

Fig. 15-1 Using the portable electric jointer plane to fit a door. (A. *Courtesy of Skil Corp.*)

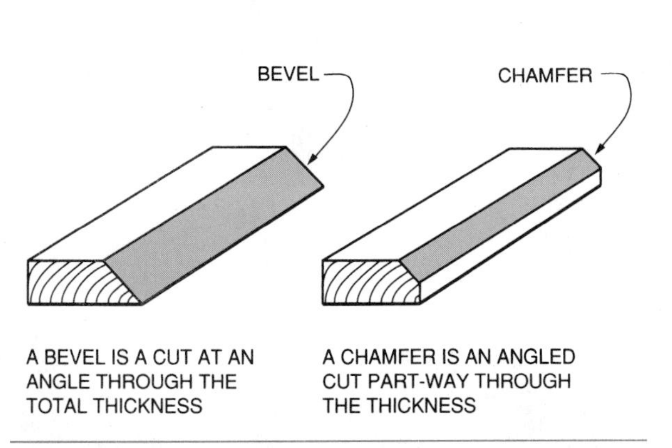

Fig. 15-2 A bevel and chamfer.

Fig. 15-3 Using a power block plane requires only one hand.

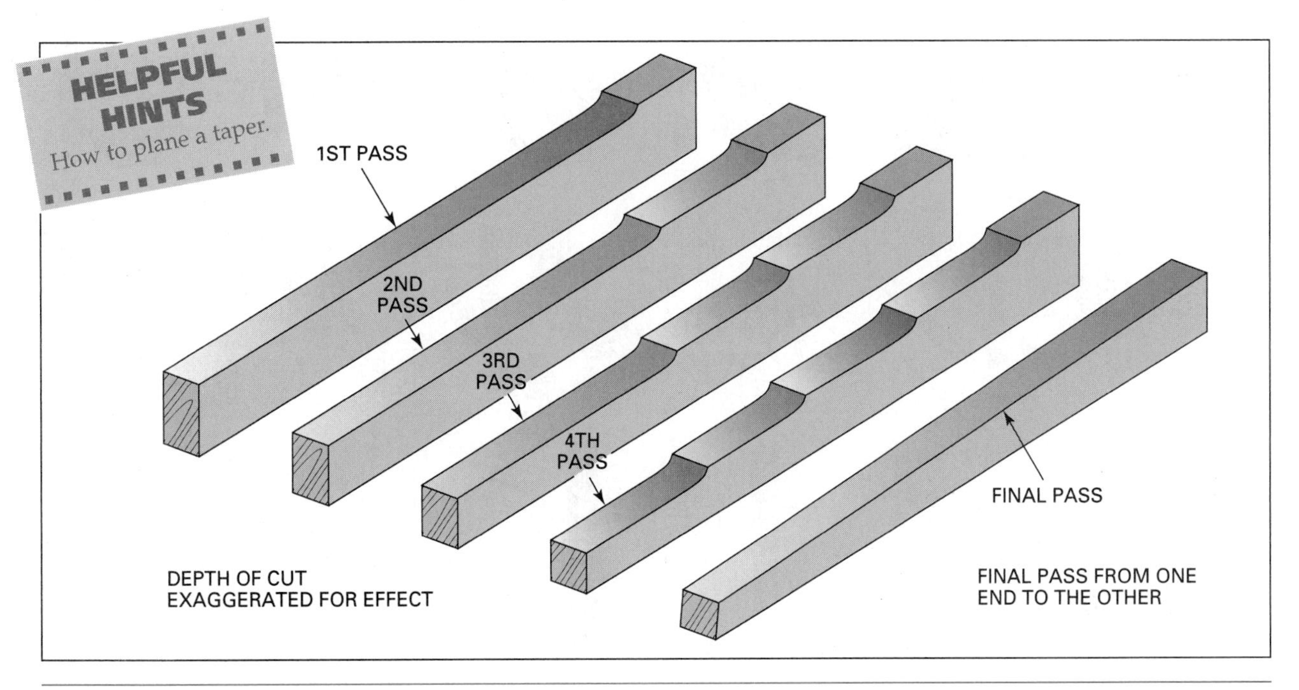

Fig. 15-4 Technique for planing a taper.

face sanded and prepared for finishing. With carbide abrasive sleeves, the tool can be used to sand masonry and metal, in addition to wood and plastics.

Operating Power Planes

CAUTION Extreme care must be taken when operating power planes. There is no retractable guard, and the high-speed cutterhead is exposed on the bottom of the plane. Keep the tool clear of your body until it has completely stopped. Keep extension cords clear of the tool.

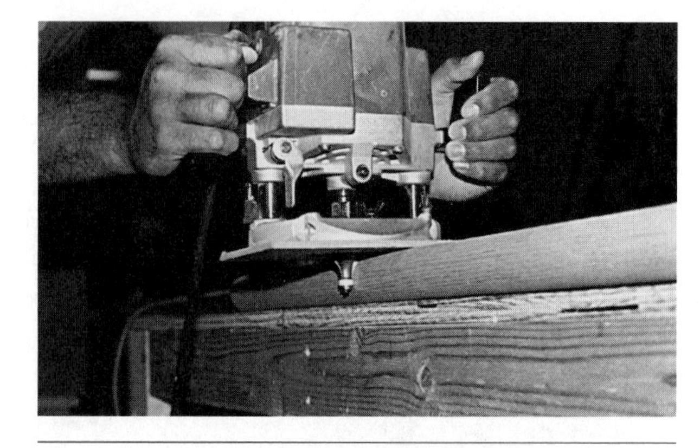

Fig. 15-5 Using a portable electric router.

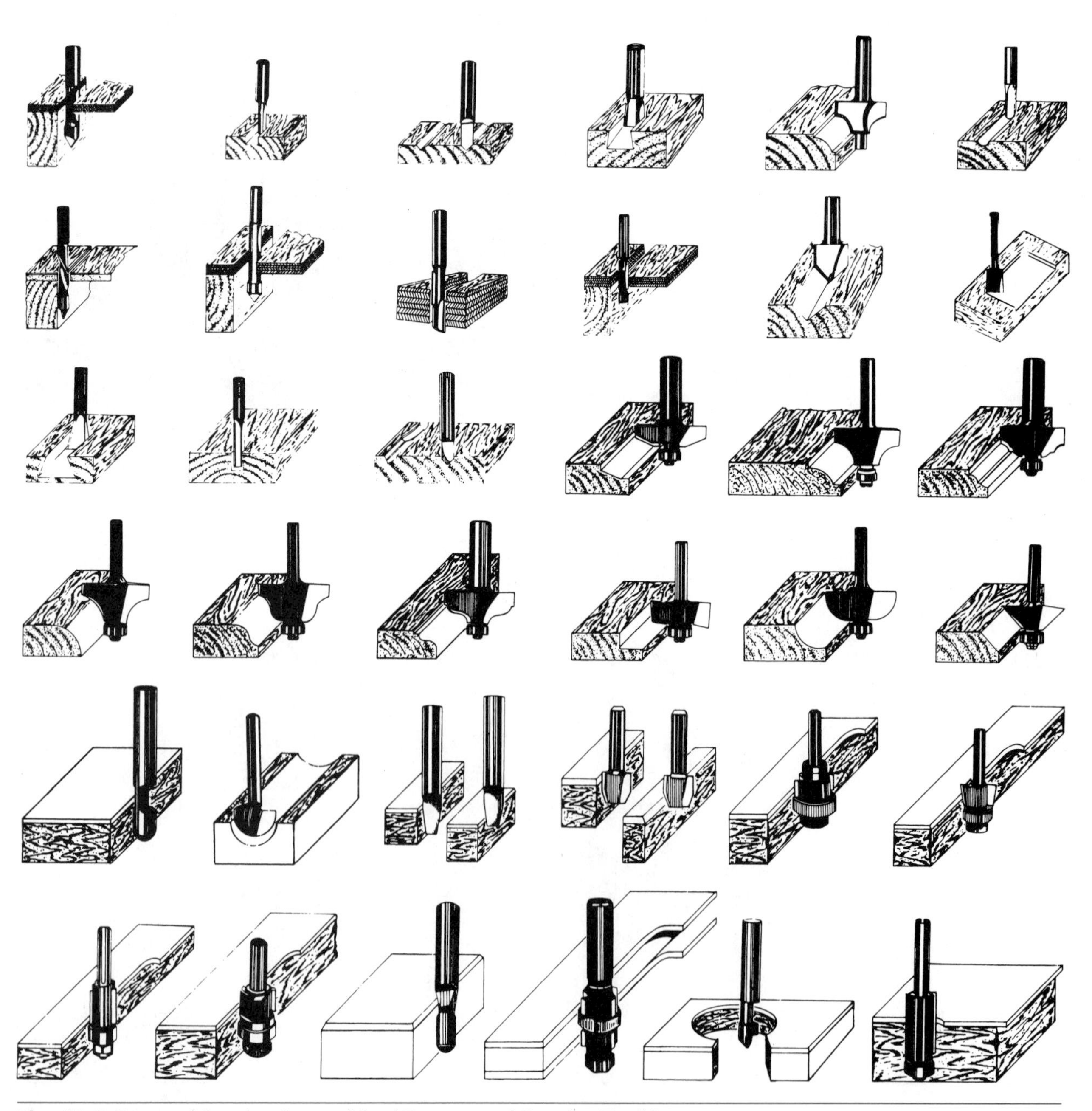

Fig. 15-6 Router bit selection guide. (*Courtesy of Stanley Tools*)

- Set the side guide to the desired angle, and adjust the depth of cut.
- Hold the toe (front) firmly on the work, with the plane cutterhead clear of the work.
- Start the motor. With steady, even pressure make the cut through the work for the entire length. Guide the angle of the cut by holding the guide against the side of the stock. Apply pressure to the toe of the plane at the beginning of the cut. Apply pressure to the heel (back) at the end of the cut to prevent tipping the plane over the ends of the work.
- To plane a **taper**, that is, to take more stock off one end than the other, make a number of passes. Each pass should be shorter than the preceding one. Lift the plane clear of the stock at the end of the pass. Make the last pass completely from one end to the other (Fig. 15-4 page 123).

Portable Electric Routers

One of the most versatile portable tools used in the construction industry is the *router* (Fig. 15-5). It is available in many models, ranging from 1/4 hp to over 3 hp with speeds of from 18,000 to 30,000 rpm. These tools have high-speed motors that enable the operator to make clean, smooth-cut edges.

The motor powers a chuck in which cutting bits of various sizes and shapes are held (Fig. 15-6). An adjustable base is provided to control the depth of cut. A trigger or toggle switch controls the motor.

The router is used to make many different cuts, including grooves, dadoes, rabbets, and **dovetails** (Fig. 15-7). It is also used to shape edges and make cutouts, such as for sinks in countertops. It is extensively used with accessories, called **templates**, to cut recesses for hinges in door edges and door frames. When operating the router, it is important to be mindful of the bit at all times. Watch what you are cutting and keep the router moving. Stalling the movement of the router will cause the bit to burn or melt the material.

Laminate Trimmers. A light-duty specialized type of router is called a *laminate trimmer.* It is used almost exclusively for trimming the edges of *plastic laminates* (Fig. 15-8). Plastic laminate is a thin, hard material used primarily as a decorative covering for kitchen and bathroom cabinets and countertops. (The installation of this material is described in a following chapter.)

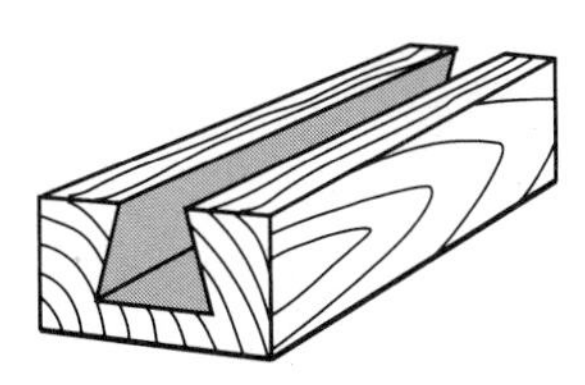

Fig. 15-7 A dovetail cut is easily made with a router.

Fig. 15-8 The laminate trimmer is used to trim the edges of plastic laminates. (*Courtesy of Rockwell, International*)

Guiding the Router

Controlling the sideways motion of the router is accomplished by the following methods:

- By using a router bit with a pilot (guide) (Fig. 15-9). The pilot may be solid and rotate with the bit or may have ball-bearing pilots. These guide the router along the uncut portion of the material being routed.

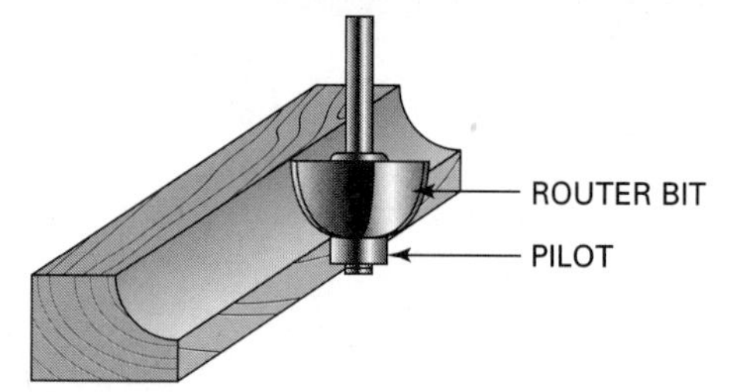

Fig. 15-9 Guiding the router with a pilot on the bit for a guide.

- By guiding the edge of the router base against a straightedge (Fig. 15-10). Be sure to keep the router tight to the straightedge, and do not rotate the router during the cut, as its base may not be centered.

Fig. 15-11 A guide attached to the base of the router rides along the edge of the stock and controls the sideways motion of the router. (*Courtesy of Black & Decker*)

- By using an adjustable guide attached to the base of the router (Fig. 15-11). The guide rides along the edge of the stock. Make sure the edge is in good condition.
- By using a template (pattern) with template guides attached to the base of the router (Fig. 15-12). This is the method widely used for cutting recesses for door hinges. (It is explained in detail in a following unit on installing doors.)

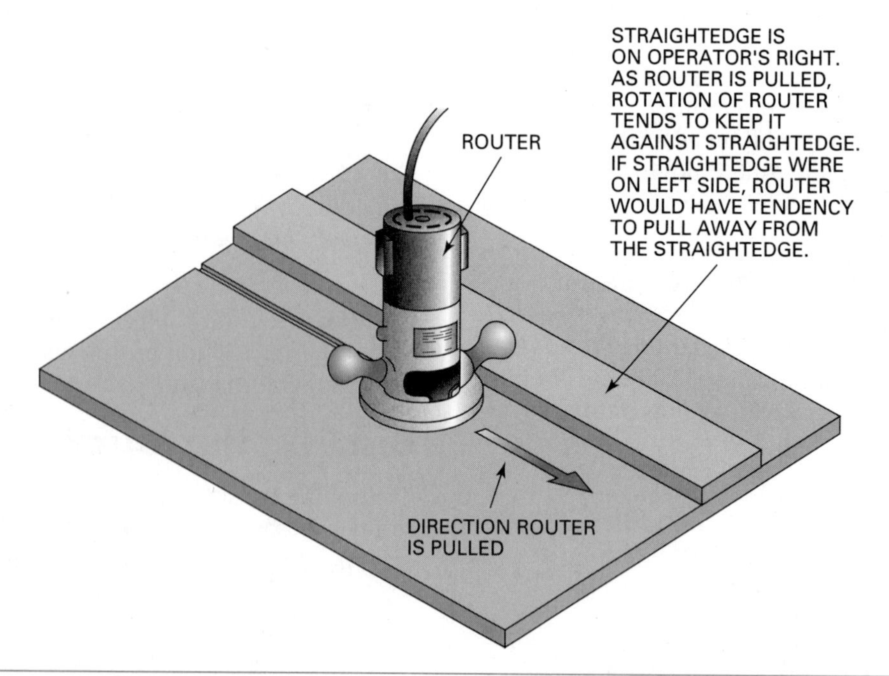

Fig. 15-10 Using a straightedge to guide the router.

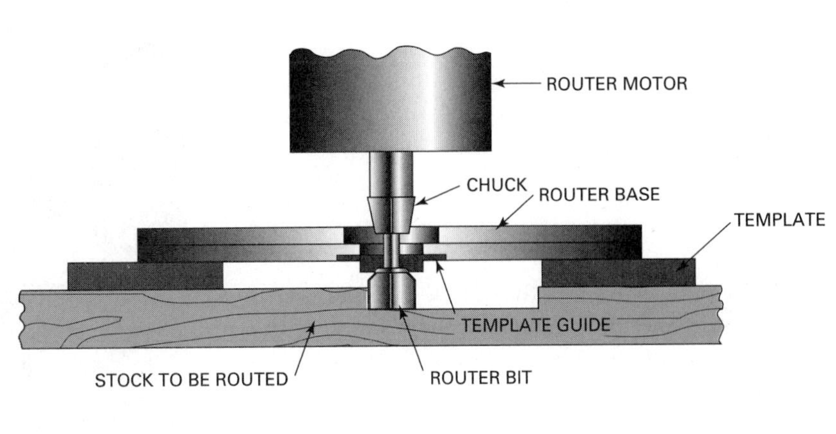

Fig 15-12 Guiding the router by means of a template and template guide.

Fig. 15-13 Freehand routing. (*Courtesy of Black & Decker*)

- By freehand routing, in which the sideways motion of the router is controlled by the operator only (Fig. 15-13).
- To make *circular* cuts, remove the subbase. Replace it, using the same screwholes, with a custom-made one in which one side extends to any desired length. Along a centerline make a series of holes to fasten the newly made subbase to the center of the desired arc (Fig. 15-14).

Using the Router

Before using the router, make sure power is disconnected. Follow the method outlined:

- Select the correct bit for the type of cut to be made.

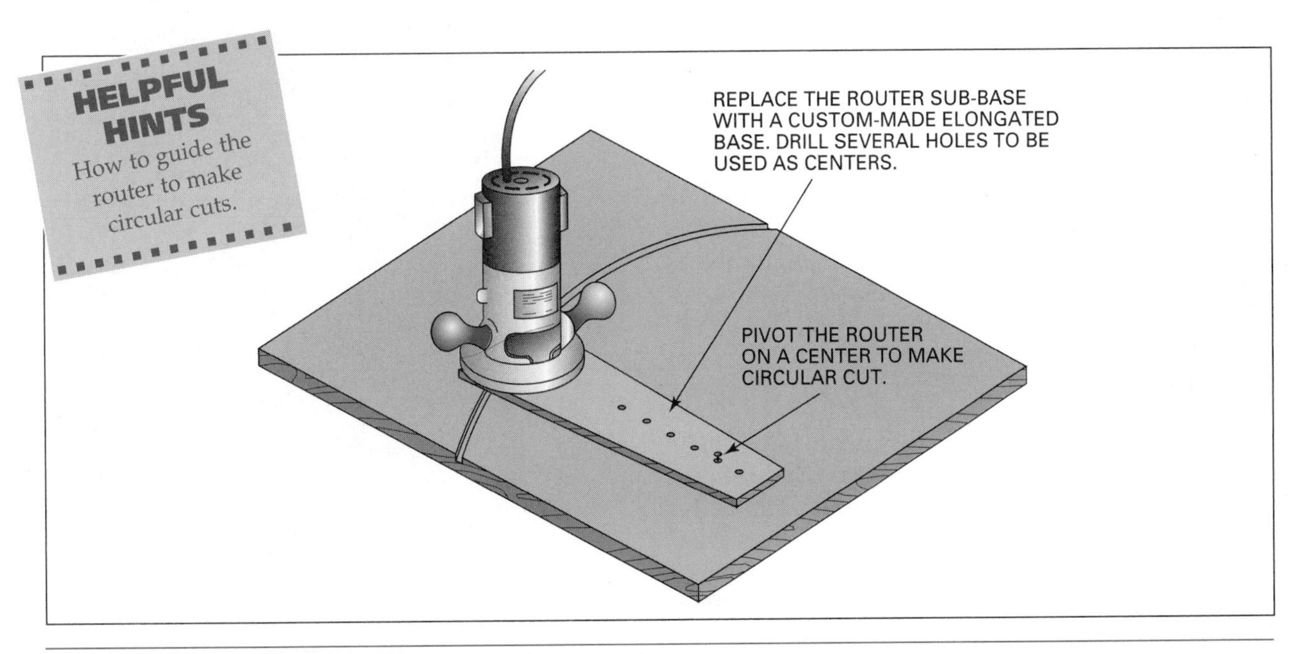

Fig. 15-14 Technique for making arcs using a router.

- Insert the bit into the chuck. Make sure the chuck grabs at least 1/2 inch of the bit. Adjust the depth of cut.
- Control the sideways motion of the router by one of the methods previously described.
- Clamp the work securely in position. Plug in the cord.
- Lay the base of the router on the work with the router bit clear of the work. Start the motor.
- Advance the bit into the cut, pulling the router in a direction that is against the rotation of the bit. On outside edges and ends, the router is moved counterclockwise around the piece. When making internal cuts, the router is moved in a clockwise direction.

CAUTION Finish the cut, keeping the router clear of your body until it has stopped. Be aware that the router bit is unguarded.

Portable Electric Sanders

Interior trim, cabinets, and other finish should be sanded before any finishing coats of paint, stain, polyurethane, or other material is applied. It is shoddy workmanship to coat finish work without sanding. In too many cases, expediency seems to take precedence over quality. Trim needs to be sanded because the grain probably has been *raised.* This happens because the stock has been exposed to moisture in the air between the time it was planed and the time of installation. Also, rotary planing of lumber leaves small ripples in the surface. Although hardly visible before a finish is applied, they become very noticeable later.

Some finishing coats require more sanding than others. If a penetrating stain is to be applied, extreme care must be taken to provide a surface that is evenly sanded with no cross-grained scratches. If paints or transparent coatings are to be applied, surfaces need not be sanded as thoroughly. Portable sanders make sanding jobs less tedious.

Portable Electric Belt Sanders

The *belt sander* is used frequently for sanding cabinetwork and interior finish (Fig. 15-15). The size of the belt determines the size of the sander. Belt widths range from 2 to 4 inches. Belt lengths vary from 21 to 24 inches. The 3 inch by 21 inch belt sander is a popular, lightweight model. Some sanders have a bag to collect the sanding dust. Remember to wear eye protection.

Fig. 15-15 Using a portable electric belt sander.

Installing Sanding Belts. Sanding belts are usually installed by retracting the forward roller of the belt sander. An arrow is stamped on the inside of some sanding belts to indicate the direction in which the belt should run. The sanding belt should run *with,* not against, the lap of the joint (Fig. 15-16). Sanding belts joined with butt joints may be installed in either direction. Install the belt over the rollers. Then release the forward roller to its operating position.

The forward roller can be tilted to keep the sanding belt centered as it is rotating. Stand the sander on its back end. Hold it securely and start it. Turn the adjusting screw one way or the other to track the belt and center it on the roller (Fig. 15-17).

Fig. 15-16 Sanding belts should be installed in the proper direction. Use the arrow on the back of the belt as a guide.

Fig. 15-17 The belt should be centered on its rollers by using the tracking screw.

Fig. 15-18 Hold the belt sander with both hands so the edge of the sanding belt can be seen clearly.

Using the Belt Sander. There has been probably more work ruined by improper use of the portable belt sander than any other tool. It is wise to practice on scrap stock until enough experience in its use will assure an acceptable sanded surface. Care must be taken to sand squarely on the sander's pad. Allowing the sander to tilt sideways or to ride on either roller results in a gouged surface.

CAUTION Make sure the switch of the belt sander is off before plugging the cord into a power outlet. Some trigger switches can be locked in the "ON" position. If the tool is plugged in when the switch is locked in this position, the sander will travel at high speed across the surface. This could cause damage to the work and/or injury to anyone in its path.

- Secure the work to be sanded. Make sure the belt is centered on the rollers and is tracking properly.
- Hold the tool with both hands so that the edge of the sanding belt can be clearly seen (Fig. 15-18).
- Start the machine. Place the pad of the sander flat on the work. Pull the sander back and lift it just clear of the work at the end of the stroke.

CAUTION Be careful to keep the electrical cord clear of the tool. Because of the constant movement of the sander, the cord may easily get tangled in the sander if the operator is not alert.

- Bring the sander forward. Continue sanding using a skimming motion that lifts the sander just clear of the work at the end of every stroke. Sanding in this manner prevents overheating the sander, the belt, and the material being sanded. It allows debris to be cleared from the work. The operator can also see what has been done.
- Do not sand in one spot too long. Be careful not to tilt the sander in any direction. Always sand with the pad flat on the work. Do not exert excessive pressure. The weight of the sander is enough. Always sand with the grain to produce a smooth finish.
- Make sure the sander has stopped before setting it down. It is a good idea to lay it on its side to prevent accidental traveling.

Finishing Sanders

The finishing sander or palm sander (Fig. 15-19) is used for the final sanding of interior work. These tools are manufactured in many styles and sizes. They are available in cordless models.

Finishing sanders either have an orbital motion, an oscillating (straight back and forth) motion, or a combination of motions controlled by a switch. The orbital motion has faster action, but leaves scratches across the grain. The *random orbital* sander reduces this problem with a design that randomly moves the center of the rotating paper at high speed. This allows the paper to sand in all directions at once. The straight line motion is slower, but produces no cross-grain scratches on the surface.

Most sanders take 1/4 or 1/2 sheet of sandpaper. It is usually attached to the pad by some type of friction or spring device. Some sanding sheets come precut with an adhesive backing for easy attachment to the sander pad.

Fig. 15-19 A portable electric finishing sander.

To use the finishing sander, proceed as follows:

- Select the desired grit sandpaper. Attach it to the pad, making sure it is tight. A loose sheet will tear easily.
- Start the motor and sand the surface evenly, *slowly* pushing and pulling the sander with the grain. Let the action of the sander do the work. Do not use excessive pressure as this may overload the machine and burn out the motor. Always hold the sander flat on its pad.

Coated Abrasives

Sandpaper is the common name for *coated abrasives.* Sandpaper does not use sand as an abrasive. Nor is its backing necessarily paper. It comes in the form of sheets, belts, discs, and sleeves, among others. The quality of the coated abrasive depends on the kind of abrasive, the backing on which the abrasive is coated, and the type of adhesive used to secure the abrasive.

Abrasives. The quality of the abrasive is determined by the length of time it is able to retain its sharp cutting edges. *Flint* and *garnet* are natural minerals used as abrasives. Although sandpaper made with flint is less expensive, it does not last as long as garnet. Synthetic (man-made) abrasives include *aluminum oxide* and *silicon carbide.* Sandpaper coated with aluminum oxide is probably the most widely used for wood.

Grits. Sandpaper *grit* refers to the size of the abrasive particles. Sandpaper with large abrasive

DESCRIPTION	GRIT NO.	0 SERIES
VERY FINE	400	10/0
	360	—
	320	9/0
	280	8/0
FINE	240	7/0
	220	6/0
	180	5/0
	150	4/0
MEDIUM	120	3/0
	100	2/0
COARSE	80	1/0
	60	1/2
	50	1
	40	1 -1/2
VERY COARSE	36	2
	30	2 - 1/2
	24	3
	20	3 - 1/2
	16	4
	12	4 - 1/2

Fig. 15-20 Grits of coated abrasives.

particles is considered coarse. Small abrasive particles are used to make fine sandpaper.

Sandpaper grits are designated either by a grit number or by an equivalent "0" series number. The grit numbers range from No. 12 (coarsest) to No. 600 (finest). The "0" series number 10/0 is equivalent to a 400 grit number. A 1/0 is the same as 80 grit (Fig. 15-20). Commonly used grits are 60 or 80 for rough sanding, and 100 or 120 for finish sanding.

Sand with a coarser grit until a surface is uniformly sanded. Do not switch to a finer grit too soon. Do not use worn or clogged abrasives. Their use causes the surface to become glazed or burned.

CHAPTER 16 FASTENING TOOLS

Portable fastening tools included in this chapter are those called **pneumatic** (powered by compressed air) and *powder-actuated* (drive fasteners with explosive powder cartridges). They are widely used throughout the construction industry for practically every fastening job from foundation to roof.

Pneumatic Staplers and Nailers

Pneumatic staplers and *nailers* are commonly called *guns* (Fig. 16-1). They are used widely for quick fastening of framing, subfloors, wall and roof sheathing, roof shingles, exterior finish, and interior trim. A number of manufacturers make a variety of models in several sizes for special fastening jobs. For instance, a *framing gun,* although used for many fastening jobs, is not used to apply interior trim.

Older models required frequent oiling, either by hand or by an oiler installed in the air line. With improvements in design, some newer models require no oiling at all.

Remember to wear eye protection. It is also recommended to wear ear protection. Prolonged exposure to loud noises will damage the ear.

Nailing Guns

The heavy-duty *framing gun* (Fig. 16-2) drives smooth-shank, headed nails up to 3 1/4 inches, ring-shank nails 2 3/8 inches and screw shank nails 3 inches to fasten all types of framing. A light-duty version (Fig. 16-3) drives smooth-shank nails up to 2 3/8 inches and ring-shank nails up to 1 3/4 inches to fasten light framing, subfloor, sheathing, and similar components of a building. Nails come glued

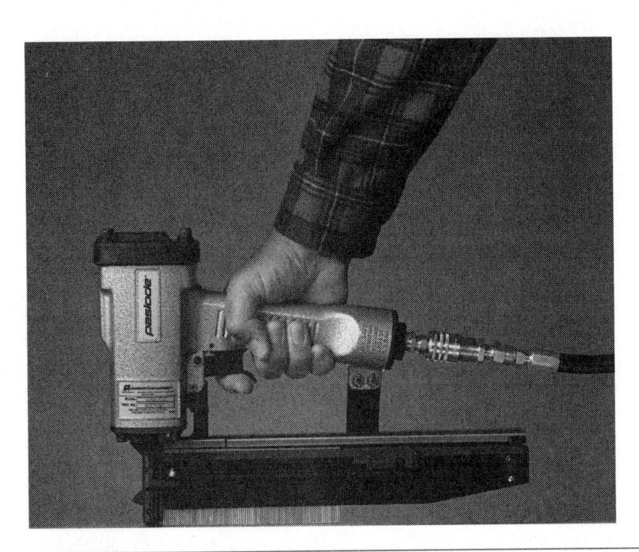

Fig. 16-1 Pneumatic nailers and staplers are widely used to fasten building parts. (*Courtesy of ITW Paslode*)

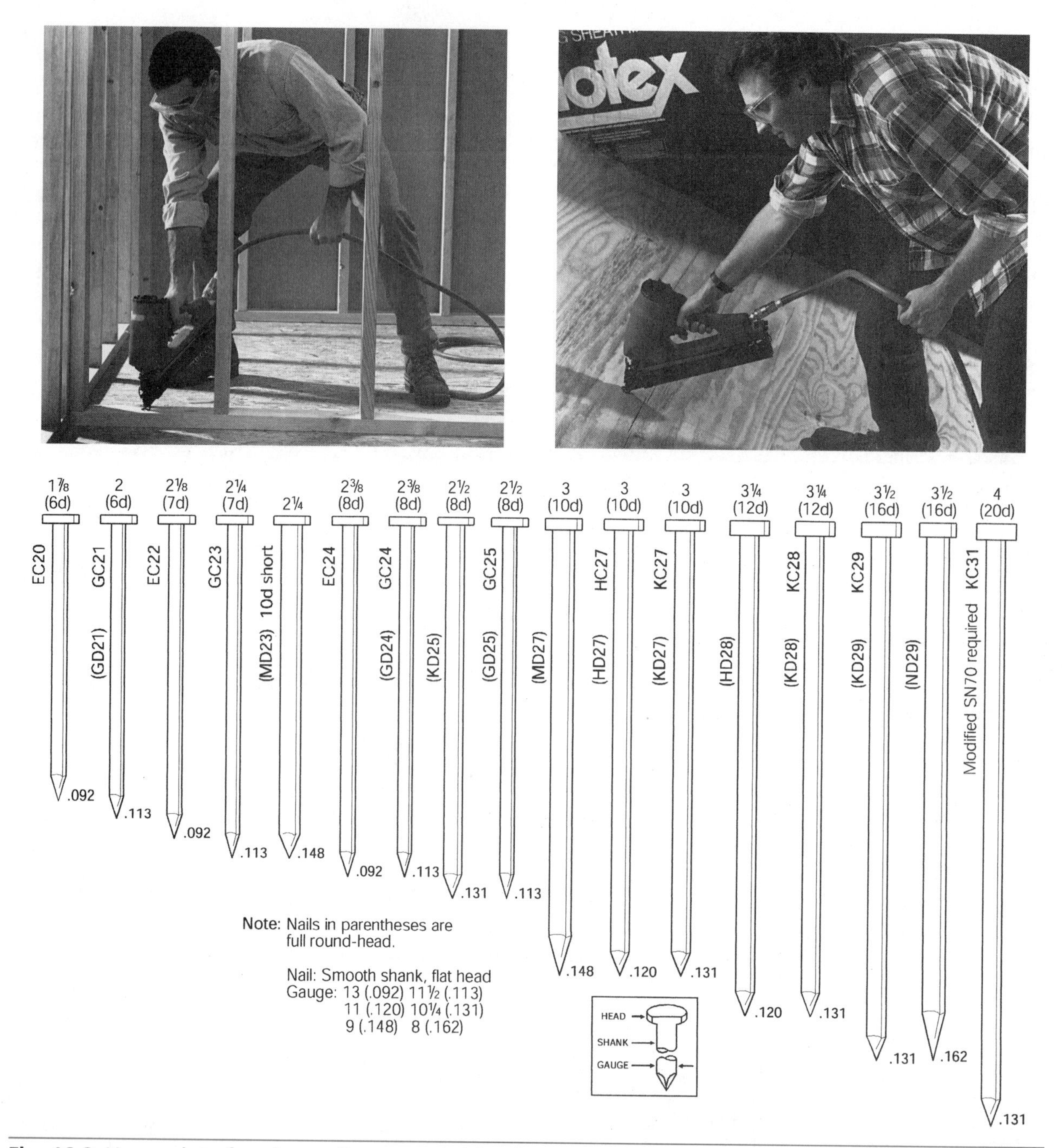

Fig. 16-2 Heavy-duty framing nailers are used for floor, wall, and roof framing. (*Courtesy of Senco Products, Inc.*)

in strips for easy insertion into the magazine of the gun (Fig. 16-4).

The *finish nailer* (Fig. 16-5) drives finish nails from 1 to 2 inches long. It can be used for the application of practically all kinds of exterior and interior finish work. It sets or flush drives nails as desired. A nail set is not required, and the possibility of marring the wood is avoided.

The *brad nailer* (Fig. 16-6) drives both slight-headed and medium-headed brads ranging in length from 1/2 inch to 1 5/8 inches. It is used to fasten small moldings and trim, cabinet door frames and panels, and other miscellaneous finish carpentry.

The *roofing nailer* (Fig. 16-7) is designed for fastening asphalt and fiberglass roof shingles. It drives five different sizes of wide, round-headed roofing nails from 7/8 to 1 3/4 inches. The nails come in coils of 120 (Fig. 16-8), which are easily loaded in a nail canister.

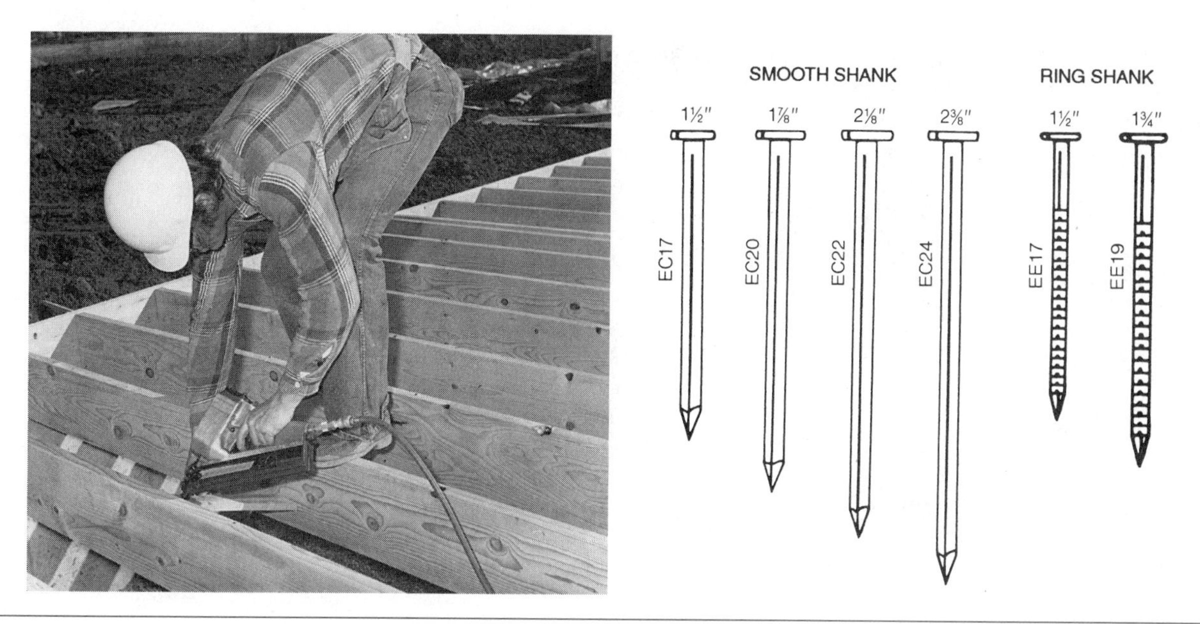

Fig. 16-3 A light-duty nailer is used to fasten light framing, subfloors, and sheathing. (*Courtesy of Senco Products, Inc.*)

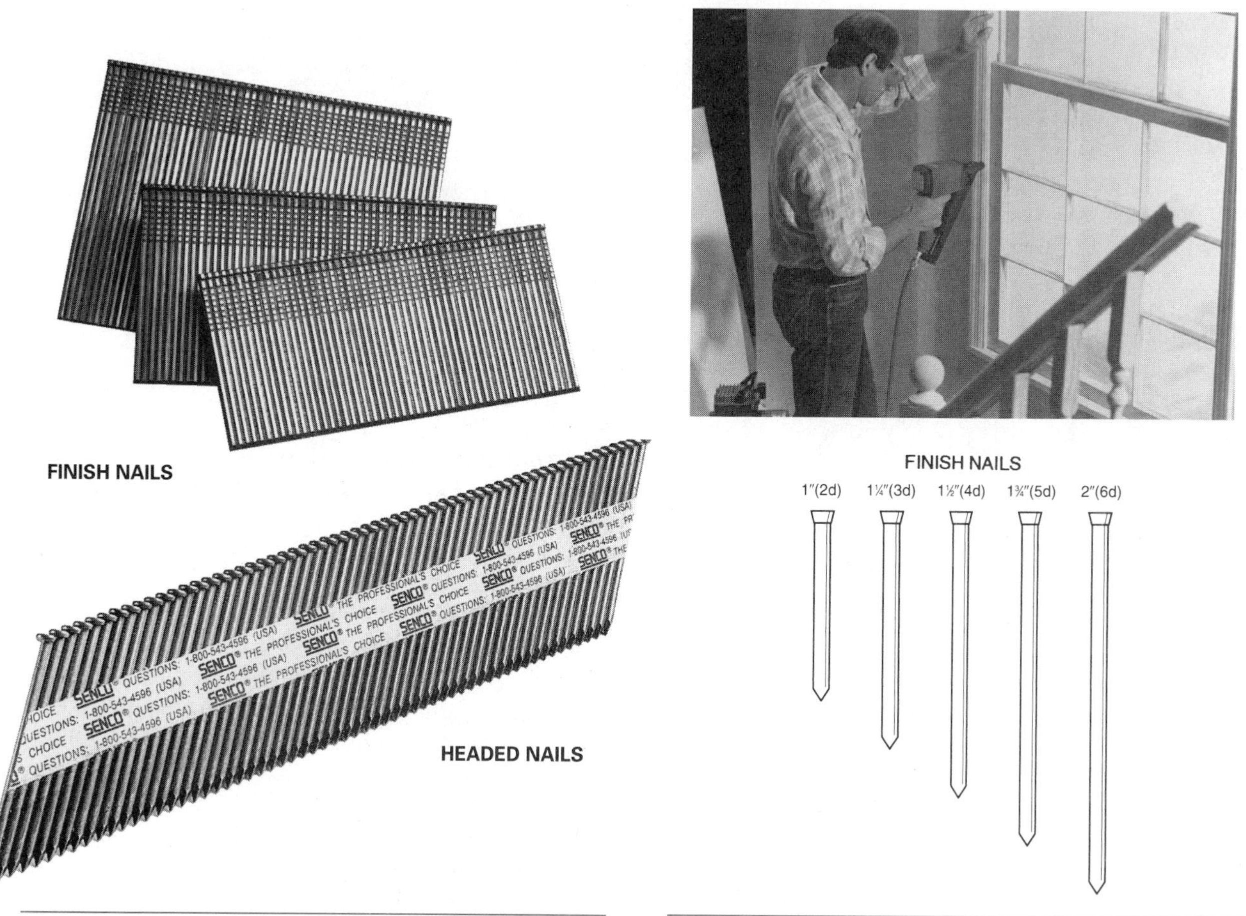

Fig. 16-4 Both headed and finish nails used in nailing guns come glued together in strips. (*Courtesy of Senco Products, Inc.*)

Fig. 16-5 The finish nailer is used to fasten all kinds of interior trim. (*Courtesy of Senco Products, Inc.*)

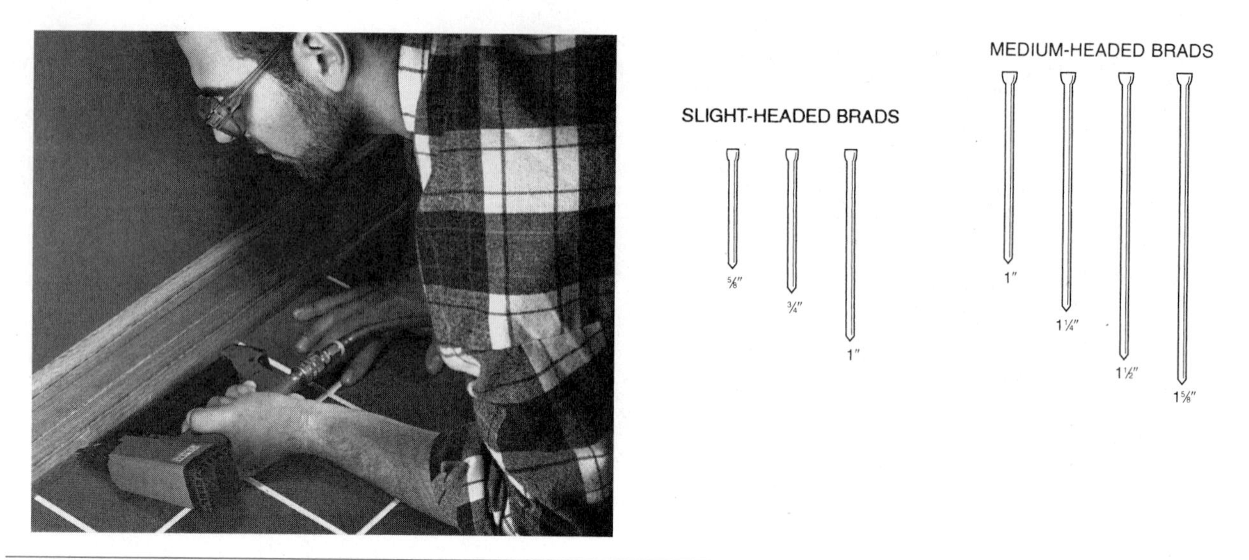

Fig. 16-6 A light-duty brad nailer is used to fasten thin molding and trim. (*Courtesy of Senco Products, Inc.*)

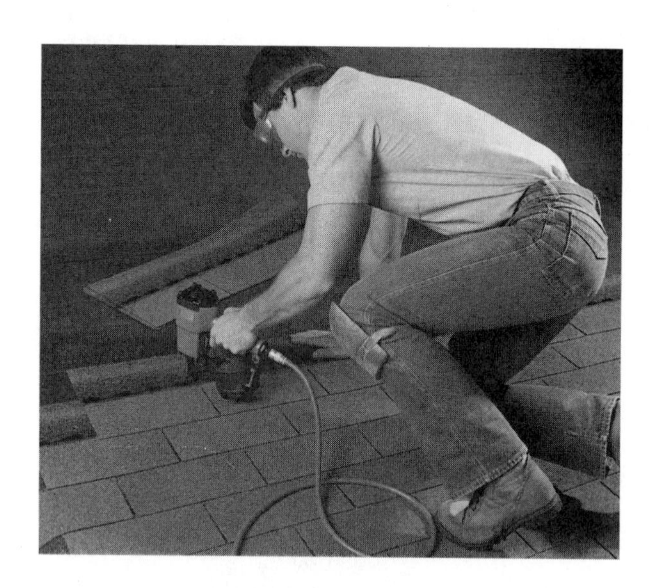

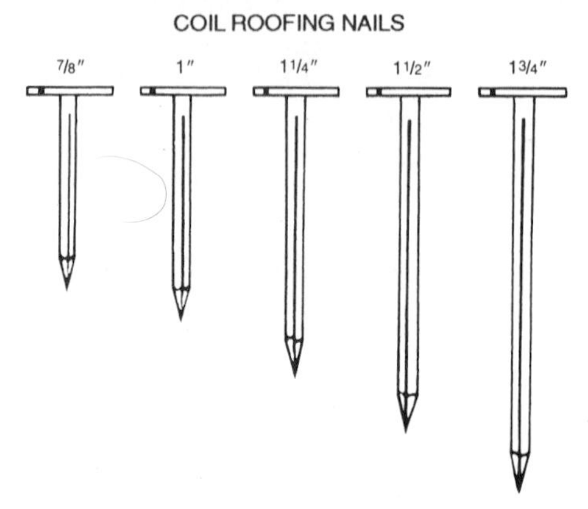

Fig. 16-7 A coil roofing nailer is used to fasten asphalt roof shingles. (*Courtesy of Senco Products, Inc.*)

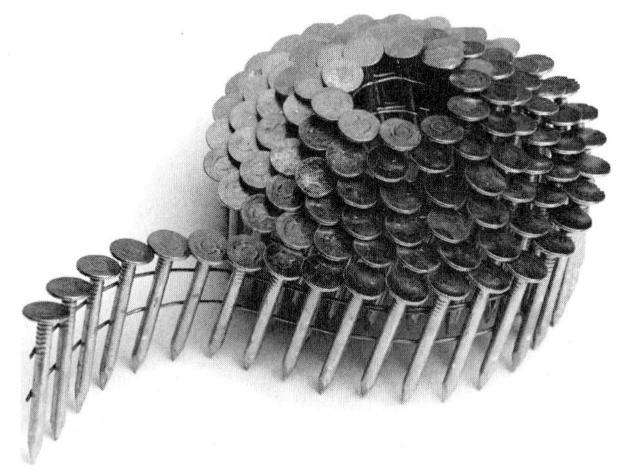

Fig. 16-8 Roofing nails come in coils for use in the roofing nailer. (*Courtesy of Senco Products, Inc.*)

Staplers

Like nailing guns, *staplers* are manufactured in a number of models and sizes. Although they differ somewhat, they are classified as light-, medium-, and heavy-duty.

A popular tool is the *roofing stapler* (Fig. 16-9), which is used to fasten roofing shingles. It comes in several models and drives 1-inch wide-crown staples in lengths from 3/4 inch to 1 1/2 inches. It can also be used for fastening other material, such as lath wire, insulation, furniture, and cabinets. The

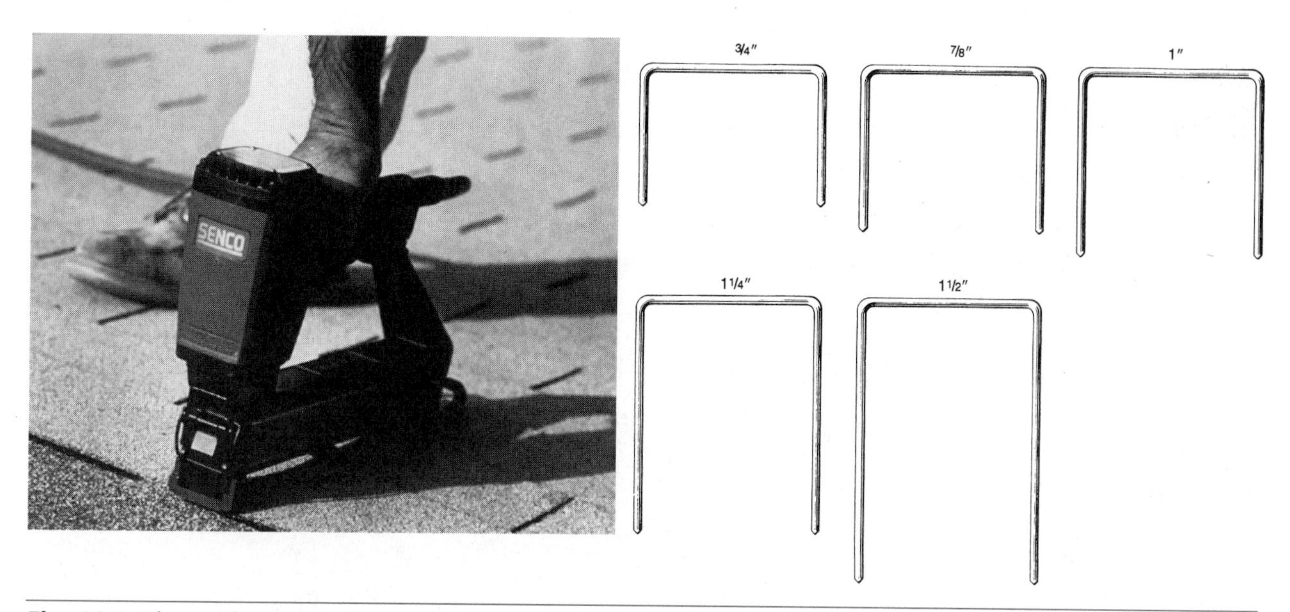

Fig. 16-9 The wide-crown stapler, being used to fasten roof shingles, may be used to fasten a variety of materials. (*Courtesy of Senco Products, Inc.*)

staples, like nails, come glued together in strips (Fig. 16-10) for quick and easy reloading. Most stapling guns can hold up to 150 staples.

No single model stapler drives all widths and lengths of staples. Models are made to drive narrow-crown and intermediate-crown, in addition to wide-crown staples. A popular model drives 3/8-inch intermediate-crown staples from 3/4 inch to 2 inches in length.

Cordless Guns

Conventional pneumatic staplers and nailers are powered by compressed air. The air is supplied by an air compressor (Fig. 16-11) through long lengths of air hose stretched over the construction site. The development of *cordless* nailing and stapling guns (Fig. 16-12) eliminates the need for air compressors and hoses. The cordless gun utilizes a disposable fuel cell. A battery and spark plug power an internal combustion engine that forces a piston down to drive the fastener. Another advantage is the time saved in setting up the compressor, draining it at the end of the day, coiling the hoses, and storing the equipment.

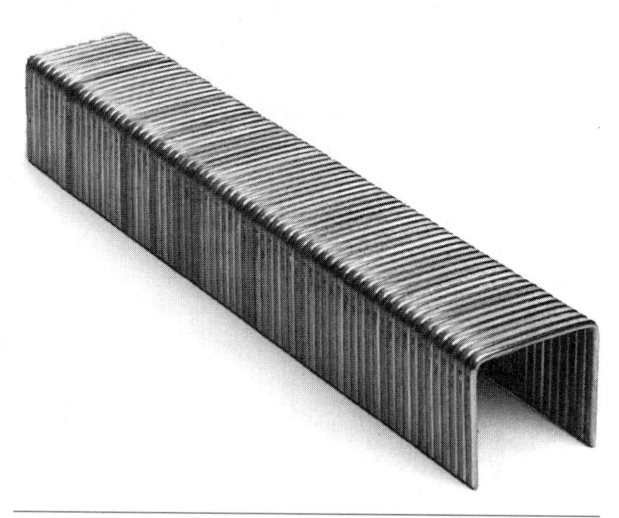

Fig. 16-10 Staples, like nails, come glued together in strips for use in stapling guns. (*Courtesy of Senco Products, Inc.*)

The *cordless framing nailer* (Fig. 16-13) drives nails from 2 to 3 1/4 inches in length. Each fuel cell will deliver energy to drive about 1,200 nails. The battery will last long enough to drive about 4,000 nails before recharging is required.

The *cordless finish nailer* (Fig. 16-14) drives finish nails from 3/4 to 2 1/2 inches in length. It will drive about 2,500 nails before a new fuel cell is needed and about 8,000 nails before the battery has to be recharged.

The *cordless stapler* (Fig. 16-15) drives intermediate-crown staples from 3/4 inch to 2 inches in length. It will drive about 2,500 staples with each fuel cell and about 8,000 staples with each charge of the battery.

Using Staplers and Nailers

Because of the many designs and sizes of staplers and nailers, you should study the manufacturer's directions and follow them carefully. Use the right nailer or stapler for the job at hand. Make sure all safety devices are working properly and always wear eye protection. A work contact element allows

Fig. 16-11 The development of cordless nailers and staplers eliminates the need for air compressors.

the tool to operate only when this device is firmly depressed on a work surface and the trigger is pulled, promoting safe tool operation.

- Load the magazine with the desired size staples or nails.

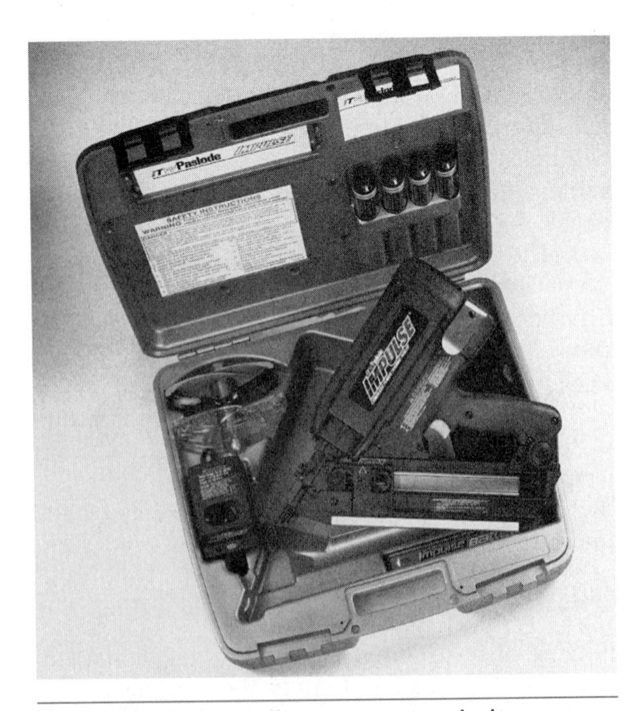

Fig 16-12 Each cordless gun comes in its own case with battery, battery charger, safety glasses, instructions, and storage for fuel cells. *(Courtesy of ITW Paslode)*

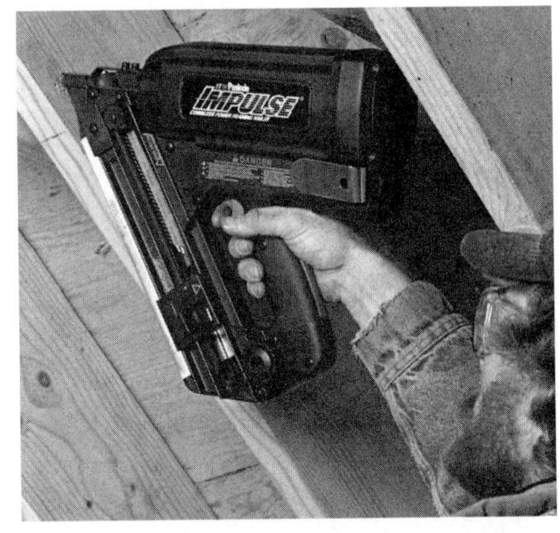

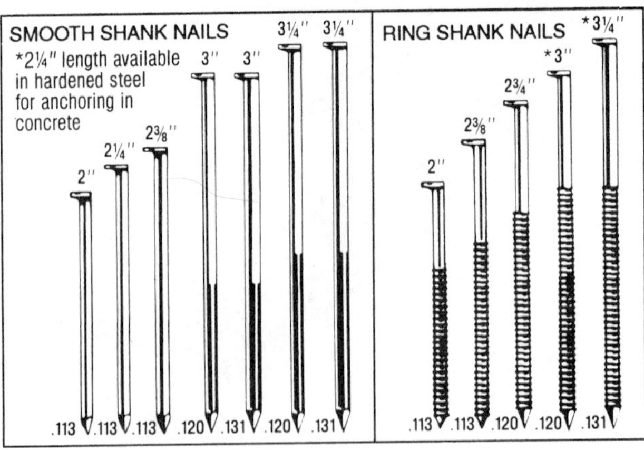

Fig. 16-13 The cordless framing nailer eliminates the nuisance of bothersome air hoses. *(Courtesy of ITW Paslode)*

APPLICATION	DIMENSION
PANELING, BED MOLDING	3/4" X 16 GA.
PANELING, CAP/SHOE MOLDING	1" X 16 GA.
CAP/SHOE MOLDING	1 1/4" X 16 GA.
BASEBOARD, CROWN, CHAIR	1 1/2" X 16 GA.
BASEBOARD, CROWN, CHAIR	1 3/4" X 16 GA.
BASEBOARD, CROWN, CASINGS	2" X 16 GA.
CASINGS	2 1/4" X 16 GA.
CASINGS	2 1/2" X 16 GA.

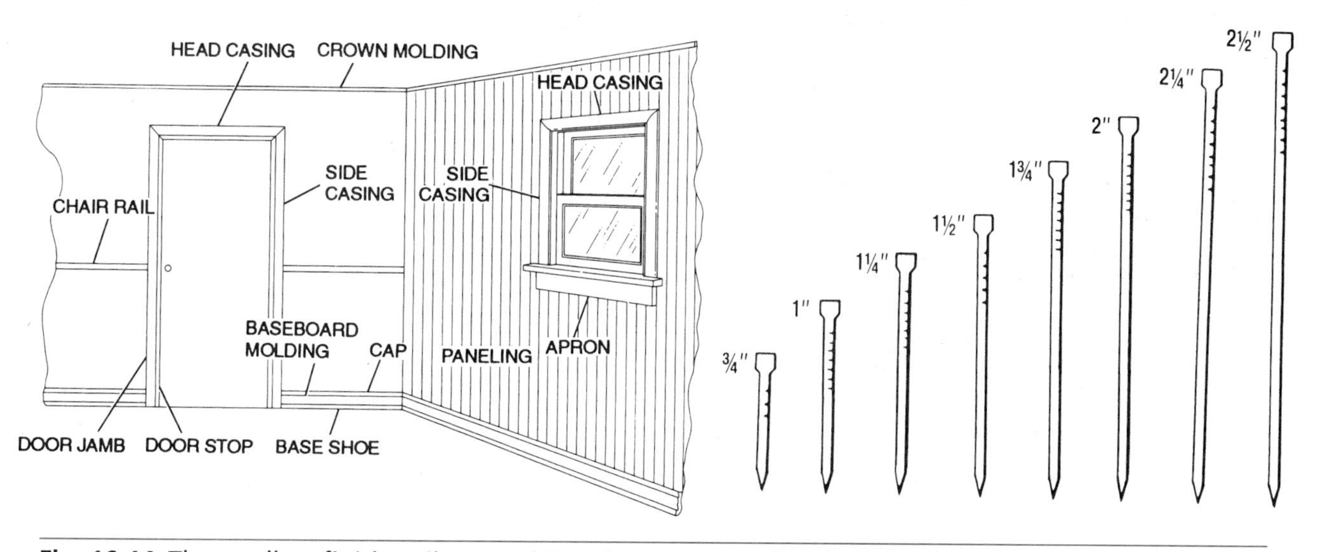

Fig. 16-14 The cordless finish nailer can drive about 2,500 nails with one fuel cell. (*Courtesy of ITW Paslode*)

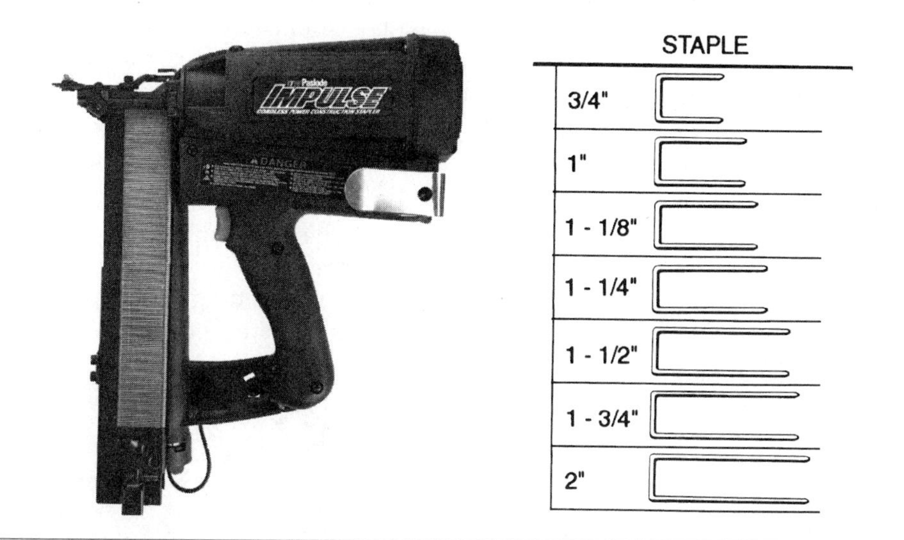

Fig 16-15 The cordless stapler can drive 2500 intermediate-crown staples from 3/4 inch up to 2 inches without changing fuel cells. (*Courtesy of ITW Paslode*)

- Connect the air supply to the tool. For those guns that require it, make sure there is an oiler in the air supply line, adequate oil to keep the gun lubricated during operation, and an air filter to keep moisture from damaging the gun. Use the recommended air pressure.

CAUTION Exceeding the recommended air pressure may cause damage to the gun or burst air hoses, possibly causing injury to workers.

- Press the trigger and tap the nose of the gun to the work. When the trigger is depressed, a fastener is driven each time the nose of the gun is tapped to the work. The fastener may also be safely driven by first pressing the nose of the gun to the surface and then pulling the trigger.
- Upon completion of fastening, disconnect the air supply.

CAUTION Never leave an unattended gun with the air supply connected. Always keep the gun pointed toward the work. Never point it at other workers or fire a staple except into the work. A serious injury can result from horseplay with the tool.

Powder-actuated Drivers

Powder-actuated drivers (Fig. 16-16) are used to drive specially designed pins into masonry or steel. They are used in a manner similar to firing a gun. Powder charges of various strengths drive the pin when detonated.

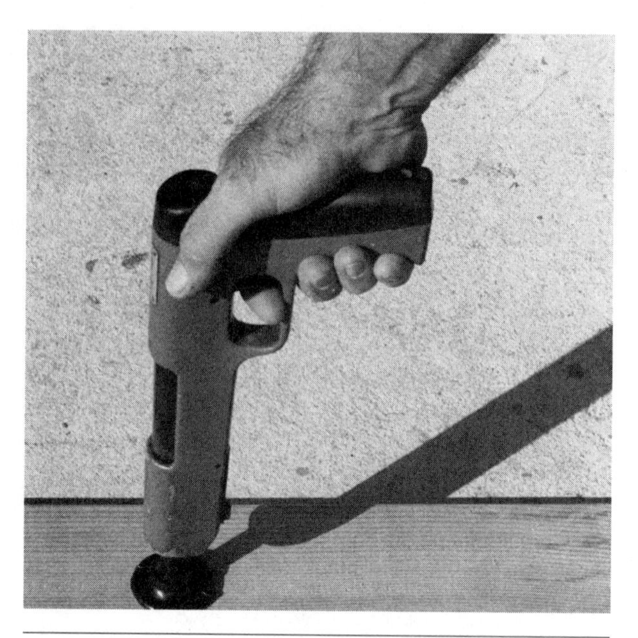

Fig. 16-16 Powder-actuated drivers are used for fastening into masonry or steel.

Drivepins

Drivepins are available in a variety of sizes. Three styles are commonly used. The *headed* type is used for fastening material. The *threaded* type is used to bolt an object after the pin is driven. The *eyelet* type is used when attachments are to be made with wire.

Powder Charges

Powder charges are color-coded according to strength. Learn the color codes for immediate recognition of the strength of the charge. A stronger charge is needed for deeper penetration or when driving into hard material. The strength of the charge must be selected with great care.

Because of the danger in operating these guns, many states require certification of the operator. Certificates may be obtained from the manufacturer's representative after a brief training course.

Using Powder-actuated Drivers

- Study the manufacturer's directions for safe and proper use of the gun. Use eye and ear protection.
- Make sure the drivepin will not penetrate completely through the material into which it is driven. This has been the cause of fatal accidents.
- To prevent ricochet hazard, make sure the recommended shield is in place on the nose of the gun. A number of different shields are available for special fastening jobs.
- Select the proper fastener for the job. Consult the manufacturer's drivepin selection chart to determine the correct fastener size and style.
- Select a powder charge of necessary strength. Always use the weakest charge that will do the job. Load the driver with the pin first and the cartridge second.
- Keep the tool pointed at the work. Wear safety goggles. Press hard against the work surface, and pull the trigger. The resulting explosion drives the pin. Eject the spent cartridge.

CAUTION If the gun does not fire, hold it against the work surface for at least 30 seconds. Then remove the cartridge according to the manufacturer's directions. Do not attempt to pry out the cartridge with a knife or screwdriver; most cartridges are rim-fired and could explode.

Review Questions

Select the most appropriate answer.

1. To prevent electrical shock, make sure the tool is properly
a. connected.
b. grounded.
c. insulated.
d. wired.

2. The guard of the portable electric saw should never be
a. lubricated.
b. adjusted.
c. retracted by hand.
d. wedged open.

3. The saber saw is used primarily for making
a. curved cuts.
b. holes.
c. internal cuts.
d. straight cuts.

4. The reciprocating saw is most similar to a
a. hammer-drill.
b. powered compass saw.
c. saber saw.
d. vibrating sander.

5. The size of an electric drill is determined by its
a. ampere rating.
b. horsepower.
c. weight.
d. chuck capacity.

6. In power drills, never use bits with
a. a double twist.
b. no center.
c. a single spur.
d. a threaded center.

7. To prevent a drill bit from binding in a hole
a. clear chips frequently.
b. drill at a slow speed.
c. drill at a high speed.
d. use bits with a twist.

8. The amount of screw tightening in the positive clutch model electric screwdriver is dependent on the
a. adjustment of clutch torque.
b. release of pressure.
c. size of the screw being driven.
d. ampere rating of the screwdriver.

9. The portable electric jointer plane is well suited to
a. nose the edge of closet shelves.
b. plane miter joints to fit.
c. trim end grain of stair treads.
d. plane doors to fit in openings.

10. When using the router to shape external edges and ends, the router is pulled in a
a. direction with the grain.
b. clockwise direction.
c. counterclockwise direction.
d. direction with the rotation of the bit.

BUILDING FOR SUCCESS

The Importance of Following Instructions

How many times have we all rushed through something only to find that the results did not meet our expectations? Perhaps the time required to read and follow instructions would have produced better results.

Our nation's construction workforce continues to be very active and geared to be time conscious. We move through a work day at rates that minimize time for planning, thinking, and organizing properly. Some workers believe that taking time to create a plan of action only slows them down. A short pause to assess the situation will usually help us obtain better results and meet our expected goals.

Almost everyone has purchased or received as a gift something that had instructions for assembling or processing. The instructions were supplied by the manufacturer to assist each person in utilizing the product. Some instructions are lengthy; others are brief but adequate to meet the objectives.

Instructions should focus on the goals and lead users through the assembly process successfully.

The key to following any instructions is to believe they are necessary for success. With that understanding, people will adequately reach their goals.

As a society, we tend to be more audio- or visual-oriented than reading-oriented. For the carpenter, reading skills are crucial to following blueprints, codes, and specifications. Each carpenter is reading, listening to, or observing instructions daily.

People must believe that they can benefit from following the instructions. As instructions are followed, a learning process takes place. By repeating the procedures at regular intervals, strict reliance on the written instructions is lessened. Once the procedures are memorized or become automatic, reliance for adherence is transferred more to memory.

How should we follow instructions?

1. Observe, listen to, and read instructions before attempting to follow through on the process. Work with an experienced person whenever possible.
2. Prepare the necessary tools and materials called for in instructions. This may call for preparation a few days in advance.
3. Begin with the proverbial "step one," then proceed sequentially through to the next step. Do not skip steps.
4. Ask questions along the way that will help complete a step or procedure.
5. Repeat any step that did not bring the expected results. The next step will be based on completion of the previous step.
6. As the project or process is completed, maintain the instructions for future reference. If the instructions include maintenance procedures, they must be forwarded to whomever has that responsibility.

A carpenter working with a quality product will want to assure the new homeowner that the installation instructions have been followed. Deviation from the manufacturer's instructions may very well affect the warranty of the product. For example, the application of "Colorlok" siding (Masonite Corporation) will call for joint moldings, 1/8-inch expansion gaps at the ends where siding pieces butt against trim and corners, and nailing to be on each stud 5/8 inch down from the top edge of each course. If the expansion gap is not allowed, buckling will occur. If joint moldings are not installed, weathering can take place at the ends of each exposed piece of siding. If the nails are exposed below the buttline of the succeeding course, undue weather of the painted surface will shorten the life of the product.

As the result of improper installation, the carpenter may be responsible for product failure, not the manufacturer. Any disregard for instructions could affect the carpenter's profit margin on a job.

In conclusion, a carpenter must be prepared to observe all instructions to avoid possible setbacks and financial losses. The attitude for success understands the need for following instructions.

FOCUS QUESTIONS: For individual or group discussion

1. Who will be affected by the decision to follow or disregard instructions? How can they be affected?
2. What might be some good reasons to develop the habit of following directions?
3. What might be the attitudes possessed by workers who disregard installation instructions?
4. In what way can a skilled worker such as a carpenter maintain a volume of the manufacturer's installation instructions?

UNIT

7

Stationary Power Tools

Many kinds of stationary power woodworking tools are used in wood mills and cabinet shops for specialized work. On the building site, usually only the radial arm, table, and miter saws are available for use by carpenters. These are not heavy-duty machines. They must be light enough to be transported from one job to another. These saws are ordinarily furnished by the contractor.

OBJECTIVES

After completing this unit, the student should be able to:

- describe different types of circular saw blades and select the proper blade for the job at hand
- describe, adjust, and operate the radial arm saw and table saw safely to crosscut lumber to length, rip to width, and make miters, compound miters, dadoes, grooves, and rabbets
- operate the table saw in a safe manner to taper rip and to make cove cuts
- describe, adjust, and operate the power miter saw to crosscut to length, making square and miter cuts safely and accurately

UNIT CONTENTS

GENERAL SAFETY RULES

- Be trained and competent in the use of stationary power tools before attempting to operate them without supervision.
- Make sure power is disconnected when making adjustments to machines.
- Make sure saw blades are sharp and suitable for the operation. Ensure that safety guards are in place and that all guides are in proper alignment and secured.
- Wear eye protection and appropriate, properly fitted clothing.
- Keep the work area clear of scrap that might present a tripping hazard.
- Keep stock clear of saw blades before starting a machine.
- Do not allow your attention to be distracted while operating power tools.
- Turn off the power and make sure a machine has stopped before leaving the area.

Safety precautions that apply to specific operations are given when those operations are described in this unit.

CHAPTER 17 CIRCULAR SAW BLADES

All the stationary power tools described in this unit use circular saw blades. To ensure safe and efficient saw operation, it is important to know which type to select for a particular purpose.

Circular Saw Blades

The more teeth a saw blade has, the smoother the cut. However, a fine-toothed blade does not cut as fast. This means that the stock must be fed more slowly. A coarse-toothed saw blade leaves a rough surface, but cuts more rapidly. Thus, the stock can be fed faster by the operator.

If the feed is too slow, the blade may overheat. This will cause it to lose its shape and wobble at high speed. This is a dangerous condition that must be avoided. An overheated saw will probably start to bind in the cut, possibly causing kickback and serious operator injury.

The same results occur when trying to cut material with a dull blade. Always use a sharp blade. Use fine-toothed blades for cutting thin, dry material, and coarse-toothed blades for cutting heavy, rough lumber.

Types of Circular Saw Blades

Saw blades are classified into two general groups. In one group, the saw blade is made with high-speed steel. In the other group, a steel blade has *tungsten carbide cutting tips.* Tungsten carbide is such a hard metal that it is sharpened with diamond-impregnated grinding wheels. The carbide-tipped blades stay sharp longer. They are used for cutting material that contains adhesives, or other foreign material, such as particleboard, plywood, hardboard, plastics, and engineered lumber that rapidly dull ordinary steel blades.

In both groups of saw blades, the number and shape of the teeth vary to give different cutting actions according to the kind, size, and condition of the material to be cut. Most of the saw blades being used today are carbide-tipped. However, they have not entirely replaced high-speed steel blades.

High-speed Circular Saw Blades

High-speed steel blades are classified as *rip, crosscut,* or *combination* blades (Fig. 17-1). They may be given other names, such as plywood, panel, or flooring blades, but they are still in one of the three classifications just named.

Ripsaws. The ripsaw blade (Fig. 17-2) usually has less teeth than crosscut or combination blades. As in the hand ripsaw, every tooth of the circular ripsaw blade is filed or ground at right angles to the face of the blade. This produces teeth with a cutting edge all the way across the tip of the tooth. These teeth act like a series of small chisels that cut and clear the stock ahead of them. Look for these cutting edges to identify a ripsaw.

Fig. 17-1 High-speed circular saw blades.

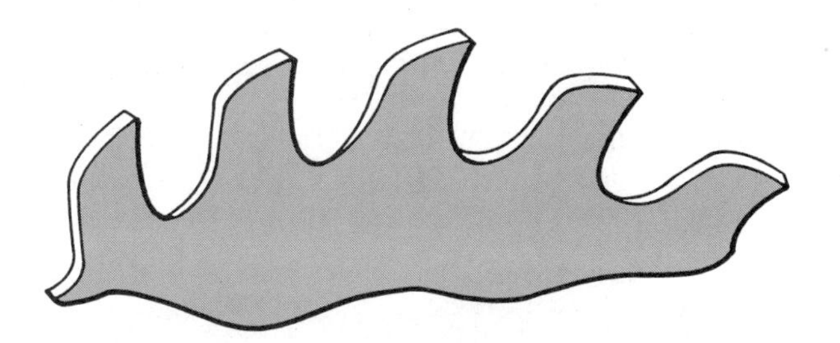

Fig. 17-2 The teeth of a ripsaw have square edge cutting tips.

Use a ripsaw for cutting solid lumber with the grain when a smooth edge is not necessary. Also, use a ripsaw when cutting unseasoned or green lumber and lumber of heavy dimension with the grain.

Crosscut Saws. The teeth of a crosscut circular saw blade are shaped like those of the crosscut handsaw (Fig. 17-3). The sides of the teeth are alternately filed or ground on a bevel. This produces teeth that come up to points instead of edges. Look for these beveled sides and points on the teeth to determine a crosscut blade.

Crosscut teeth slice through the wood fibers smoothly. A crosscut blade is an ideal blade for cutting across the grain of solid lumber. It also cuts plywood satisfactorily with little splintering of the cut edge.

Fig. 17-3 The teeth of a crosscut blade have beveled sides and pointed tips.

Fig. 17-4 The teeth of a combination blade.

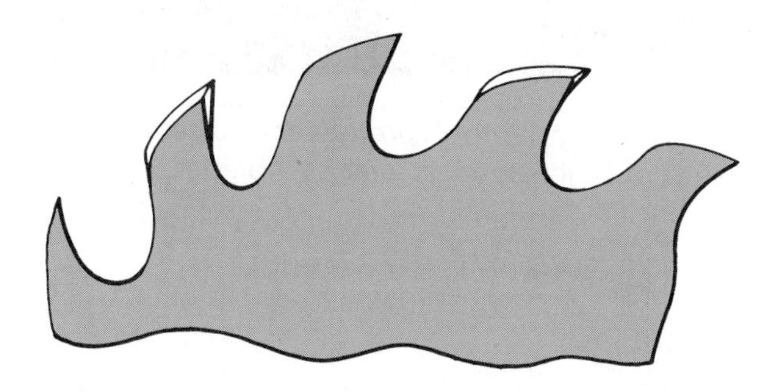

Fig. 17-5 The teeth of a chisel-point combination blade.

Combination Saws. The combination blade is used when a variety of ripping and crosscutting is to be done. It eliminates the need to change blades for different operations.

There are several types of combination blades. One type has groups of teeth around its circumference (Fig. 17-4). The leading tooth in each group is a rip tooth. The ones following are crosscut teeth.

The *chisel-point* combination blade (Fig. 17-5) is generally available with from twenty to forty teeth per blade. The number of teeth depends on the size of the blade. Each tooth is the same shape. Ripsaw teeth are given a slight alternate bevel to produce combination crosscut and rip cutting action.

Providing Clearance in the Saw Cut

All saw blades must have some provision for clearance in the saw cut. Without this clearance, the saw blade, no matter how sharp, will bind in the cut and overheat. The teeth of the blades just described are set in a manner similar to that used for handsaws to provide this clearance. Special circular saw blade setting tools are needed for this purpose (Fig. 17-6).

Another type of blade provides this clearance by being **taper ground.** A *taper ground blade* is thicker at the tips of the teeth and is thinner toward its center (Fig. 17-7). Its teeth have no set. The taper of the blade provides the clearance in the saw cut. A taper ground blade can be identified by the ridge of thicker metal around the arbor hole.

Fig. 17-6 Special tools are needed to set the teeth of high-speed circular saw blades.

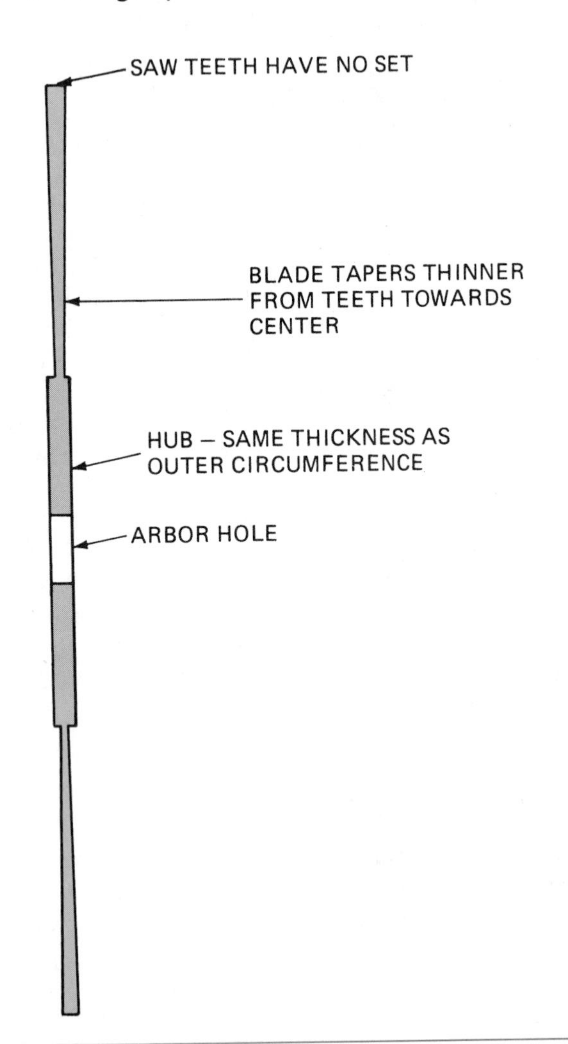

Fig. 17-7 Cross-section of a taper ground circular saw planer blade.

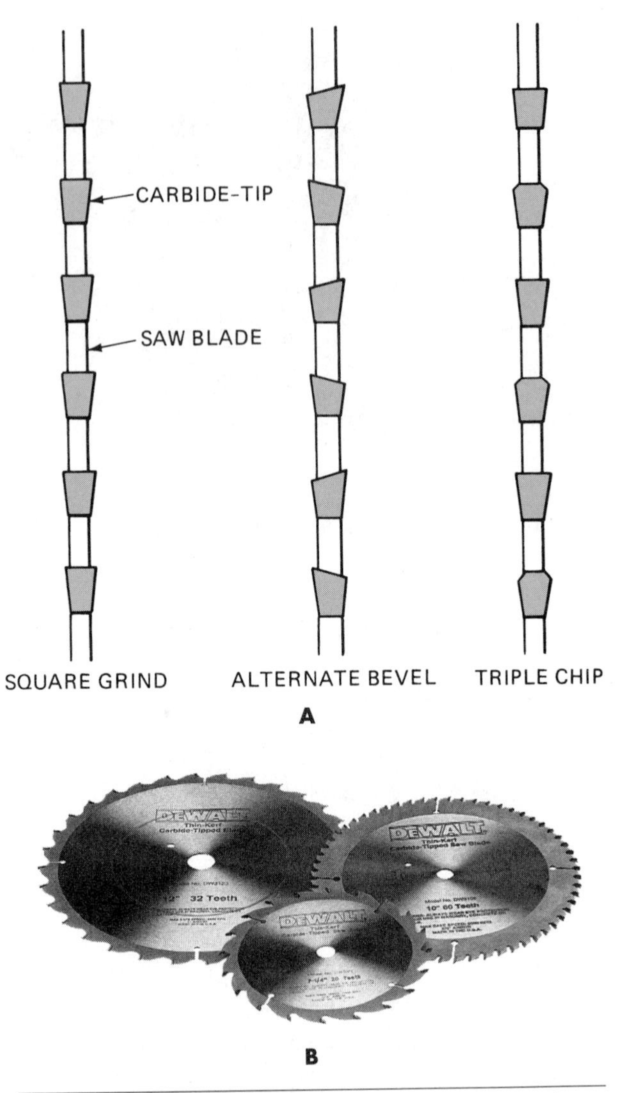

Fig. 17-8 Three main styles of carbide-tipped saw blades are commonly used. (*B. Courtesy of Dewalt*)

The reason for using a taper ground blade is to obtain an extremely smooth cut surface. Taper ground blades usually are a combination type. They are called *planer* blades. Use planer blades only on straight, dry lumber of relatively small dimension. Taper ground blades do not provide much clearance. The slightest twist in the stock will cause it to bind against the blade. Binding causes overheating and dangerous conditions. Binding also produces black, burned areas on the cut surface.

Carbide-tipped Blades

The carbide tips welded on a circular saw blade are very hard. They are also very *brittle*. For this reason, extreme care must be taken when handling carbide-tipped blades. Be careful not to drop the blades. Some of the tips could shatter like glass. Although the tips are shaped in many ways for various purposes, three main styles of teeth are commonly used (Fig. 17-8).

Square Grind. The *square grind* is similar to the rip teeth in a steel blade. It is used primarily to cut solid wood with the grain. It can also be used on composition boards when the quality of the cut surface is not important.

Alternate Top Bevel. The *alternate top bevel* grind is used with excellent results for crosscutting solid lumber, plywood, hardboard, particleboard, fiberboard, and other wood composite products. It can also be used for ripping operations. However, the feed is slower than square or combination grinds.

Triple Chip. The *triple chip grind* is designed for cutting brittle material without splintering or chipping the surface. It is particularly useful for cutting an extremely smooth edge on plastic laminated material, such as countertops. It can also be used like a planer blade to produce a smooth cut surface on straight, dry lumber of small dimension.

Combination. Carbide-tipped saws also come with a combination of teeth. The leading tooth in each set is square ground. The following teeth are ground at alternate bevels. These saws are ideally suited for ripping and crosscutting solid lumber and all kinds of engineered wood products. Because of their versatility, these saw blades are probably the most widely used by carpenters.

Carbide-tipped teeth are not set. The carbide tips are slightly thicker than the saw blade itself. Therefore, they provide clearance for the blade in the saw cut. In addition, the sides of the carbide tips are slightly beveled back to provide clearance for the tip.

Removing and Replacing Circular Saw Blades

Saw arbor shafts may have a right- or left-hand thread for the nut, depending on which side of the blade it is located. No matter what direction the arbor shaft is threaded, the arbor nut is loosened in the same direction in which the saw blade rotates (Fig. 17-9). The arbor nut is always tightened against the rotation of the saw blade. This design prevents the arbor nut from loosening during operation.

Fig. 17-9 The saw arbor nut is loosened in the same direction in which the blade rotates. Always unplug the saw before touching blade or cutters. (*Courtesy of DeWalt*)

CHAPTER 18 RADIAL ARM AND MITER SAWS

The radial arm saw and the miter saw are similar in purpose. While many other operations can be performed with the radial arm saw, it is used primarily for crosscutting framing members. The miter saw is specifically designed to crosscut interior and exterior trim.

Radial Arm Saws

The *radial arm saw* (Fig. 18-1) is ideally suited for crosscutting operations. The stock remains stationary while the saw moves across it. The operator does not have to push or pull the stock through the blade. The end of long lengths are easily cut squared or mitered. The cut is made above the work and all layout lines are clearly visible. The radial arm saw may also be used for ripping operations. In many cases, this saw is brought to the job site when construction begins.

The size of the radial arm saw is determined by the diameter of the largest blade that can be used. The arm of the saw moves horizontally in a complete circle (Fig. 18-2). The motor unit tilts to any desired angle and also rotates in a complete circle. The depth of cut is controlled by raising or lowering the arm. This flexibility allows practically any kind of cut to be made. Extreme care should be employed to follow all set-up procedures exactly. Remember to wear eye and ear protection.

Fig. 18-1 The radial arm saw is well suited for crosscutting operations. (*Courtesy of DeWalt*)

Fig. 18-2 The arm of the radial saw moves horizontally in a complete circle. (*Courtesy of Delta International*)

Crosscutting

For straight crosscutting, make sure the arm is at right angles to the fence. Adjust the depth of cut so that the teeth of the saw blade are about 1/16 inch below the surface of the table. With the saw all the way back and all guards in place, hold the stock against the fence. Make the cut by bringing the saw forward and cutting to the layout line (Fig. 18-3).

CAUTION When holding the stock on the left side of the blade, hold with the left hand and pull the saw with the right hand. When holding the stock on the right side of the blade, hold with the right hand and pull the saw with the left hand. Do not cross your hands over each other.

Fig 18-3 Straight crosscutting using the radial arm saw. (*Courtesy of Rockwell, International*)

When crosscutting stock thicker than the capacity of the saw, cut through half the thickness. Then turn the stock over and make another cut.

CAUTION Do not pull the saw quickly through the work. Forcing the saw may cause it to jam in the work or ride over it. This could cause injury to the operator. Pull the saw slowly through the work or even hold it back somewhat, especially when cutting thick material. If the stock is bowed, crosscut with the crown of the bow down on the table. If the cut is made with the crown up, the saw will probably bind as the stock flattens out on the table surface.

When the cut is complete, return the saw to the starting position, behind the fence. Turn off the power. As a safety precaution, all saws should automatically return to the retracted rest position if the operator should let go of the saw.

Cutting Identical Lengths

When many pieces of identical length need to be cut, clamp or otherwise fasten a *stop block* to the table in the desired location. Cut one end of the stock. Slide the cut end against the stop block. Then cut the other end (Fig. 18-4). Continue in this manner until the desired number of pieces is cut.

CAUTION Be sure to keep your hand out of the line of travel of the saw in case the saw should jump into the stock toward the operator. Duller blades tend to jump forward into the cut more than sharp ones.

Fig. 18-4 Cutting identical lengths.

Be careful not to bring the stock against the stop block with too much force. Doing so will cause the stop block to move. Any movement results in pieces of unequal length. It is helpful, when cutting many identical lengths, to mark the location of the stop block. Occasionally, look to see if it has moved.

Be careful that no sawdust or wood chips are trapped between the stock and the stop block. This results in pieces shorter than desired. A stop block with a rabbeted end helps prevent sawdust or chips from being trapped between the stop block and the piece being cut (Fig. 18-5).

CAUTION Replace the fence of the radial arm saw when saw cuts in it become too wide or too close together. Wide spaces in the fence allow small cut ends to shoot out toward the rear. The pieces can bounce back and cause injury to the operator.

Mitering

Flat miters. To cut a *flat miter*, the arm of the saw is rotated to the right or left to the desired angle. Usually the arm is swung to the right for more advantageous miter cutting, where the motor does not obscure the view or get in the way. The *miter latch* locates the arm in the 90- or 45- degree angle position.

For any other angle, the arm may be located by viewing the degree scale. This method may not be accurate, however. The pointer of the scale may have been bent or moved. In addition, the gradations on the scale are so fine that it is difficult to rely on the accuracy of the setting. Test pieces of known accuracy should be used to adjust the saw to cut miters other than 45 degrees.

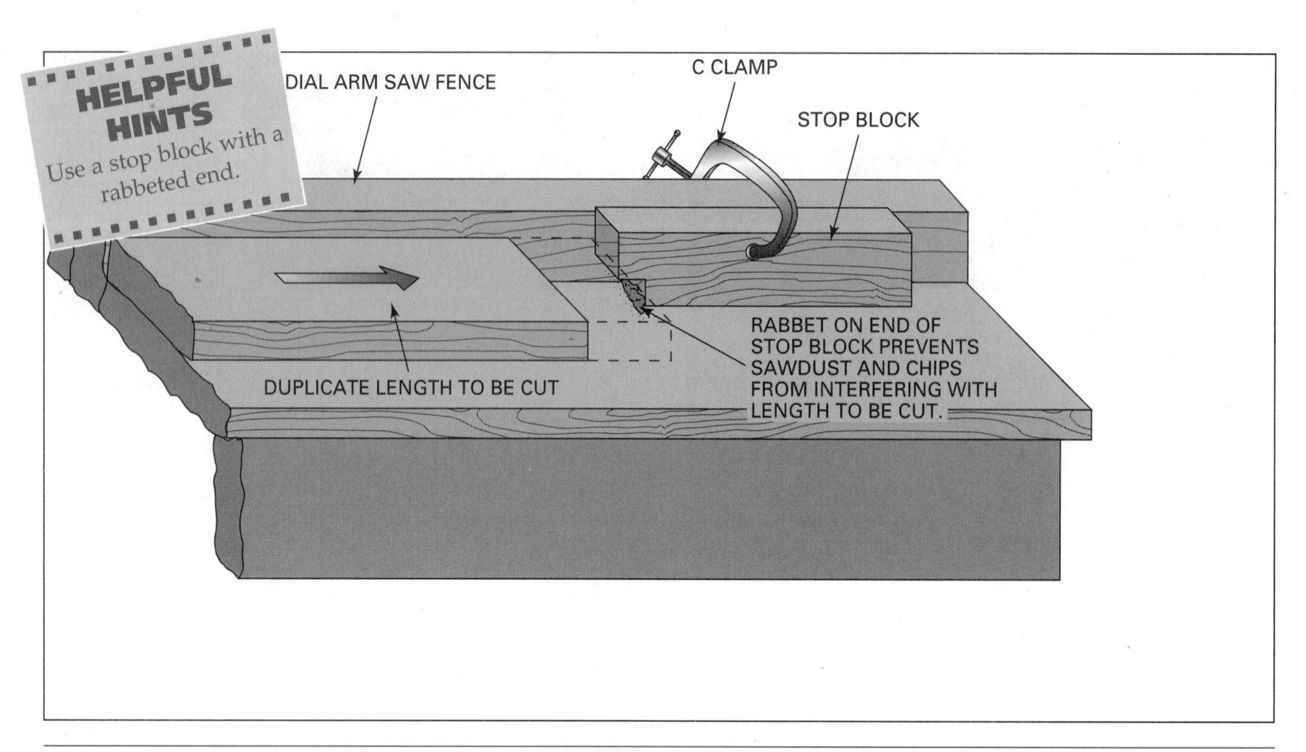

Fig. 18-5 Technique for making a fence stop where sawdust will not interfere with the operation.

Fig. 18-6 Making miter cuts.

In both cases, the arm is locked in position by the *miter clamp* after being adjusted. Miter cuts are made in the same manner as straight crosscutting (Fig. 18-6).

Mitering Jig. Flat miters may also be cut with the use of a *mitering jig.* One type of mitering jig consists of a piece of plywood, or similar material, to which strips are fastened at 45-degree angles to its edge. The jig is clamped to the radial arm table surface.

The pieces are cut by holding them at a 45-degree angle against the strips with the radial arm saw in its straight crosscutting position (Fig. 18-7). Using this jig allows both right and left miters, such as for door and window casings, to be cut without swinging the arm of the saw. Mitering jigs may also be constructed to cut miters other than 45 degrees.

End Miters. An *end miter* is cut in the same manner, except that the saw blade is tilted to the desired angle (Fig. 18-8). Stops locate the blade angle at 90, 45, and 0 degrees. Any other angle is located by viewing the angle degree scale or by using test blocks. When the blade is tilted to the desired angle, it is locked in position with the clamp.

Making Compound Miters. *Compound miters* are frequently made on some types of roof rafters and moldings. The compound miter is made by swinging the arm and also tilting the motor to the desired angles (Fig. 18-9).

Ripping

Although it is possible to rip lumber using the radial arm saw, it should be avoided if a table saw can be used. The setup for a radial arm saw to rip must be followed very carefully to protect the operator and the material being cut.

For ripping operations, the arm is locked in the straight crosscutting position at right angles to the fence. The motor unit is rotated horizontally to either the in-rip or the out-rip position (Fig. 18-10). The out-rip position is used to rip wider material than the in-rip position. The ripping width may be measured with a rule or tape from the fence to the

STEP BY STEP

PROCEDURES

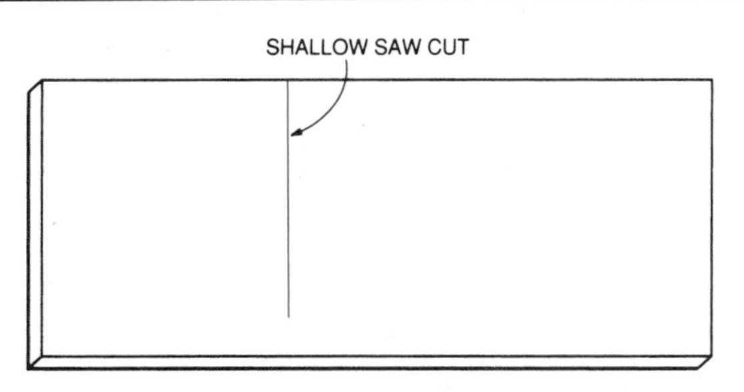

1. USE 1/2" PLYWOOD AS WIDE & LONG AS THE RADIAL ARM SAW TABLE. PLACE ONE EDGE AGAINST THE FENCE & MAKE A SHALLOW CUT ABOUT 1/16" DEEP PART-WAY ACROSS THE PLYWOOD. REMOVE IT FROM THE SAW.

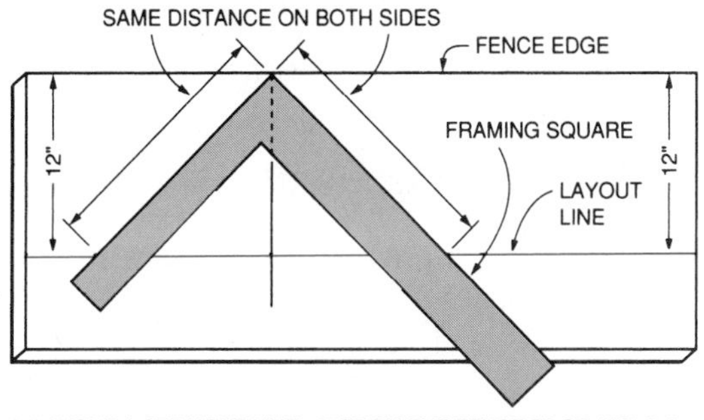

2. LAYOUT A STRAIGHT LINE 12" FROM THE FENCE EDGE. HOLD A FRAMING SQUARE SO ITS HEEL IS CENTERED ON THE SAW CUT, FLUSH WITH THE FENCE EDGE, & WITH EQUAL DISTANCES TO THE LAYOUT LINE ALONG BOTH TONGUE & BLADE. MARK LAYOUT LINES ALONG THE OUTSIDE EDGES OF THE SQUARE.

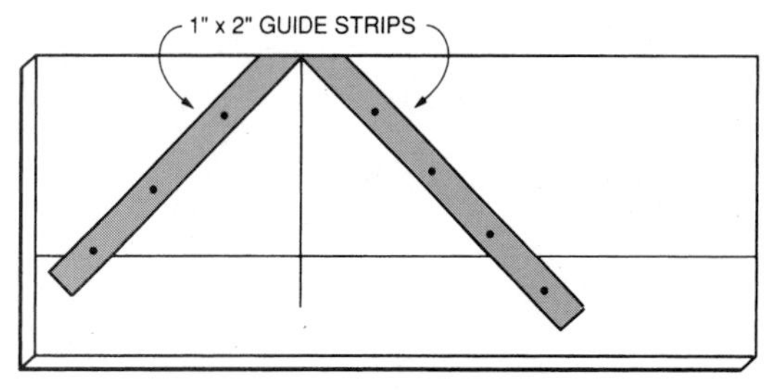

3. ATTACH 1" x 2" GUIDE STRIPS SO THEIR INSIDE EDGES ARE TO THE LAYOUT LINES.

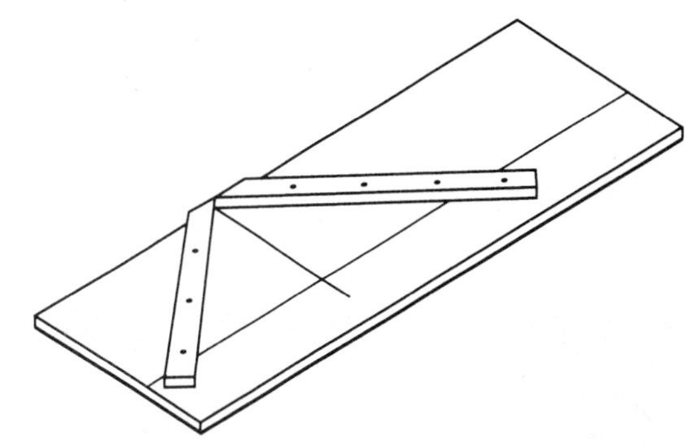

4. THE COMPLETED MITER JIG. TO USE, PLACE ON TABLE AGAINST FENCE SO SAW CUT LINES UP WITH SAW BLADE. CLAMP OR, OTHERWISE FASTEN, TO TABLE SURFACE. PIECES TO BE MITERED ARE HELD AGAINST INSIDE EDGES OF GUIDE STRIPS.

Fig. 18-7 Technique for making and using miter jig for a radial arm saw.

Fig. 18-8 Cutting an end miter.

Fig. 18-9 Making a compound miter cut.

blade. It may also be set by reading the rip scale built in on the arm of the saw (Fig. 18-11).

Lock the saw carriage in the desired position by tightening the *rip lock* against the side of the arm. Lower the motor until the saw teeth are just below the table surface.

Adjust the safety guard so that the in-feed end almost touches the material to be cut. This is important because the guard holds the stock to the table and prevents the saw from picking up the stock during the rip and binding the motor. It also keeps most of the sawdust from flying at the operator.

Lower the kickback assembly so that the kickback fingers are about 1/8 inch lower than the material to be ripped. Using the fence as a guide, feed the stock evenly into the saw blade (Fig. 18-12).

Fig. 18-10 In-rip and out-rip positions of the radial arm saw.

Fig. 18-11 The rip scale indicates the width of cut for both the in-rip and out-rip positions.

CAUTION Do not feed stock from the kickback end of the guard. ***Feed the stock against the rotation of the blade.*** Feeding stock in the wrong direction may cause it to be pulled from the operator's hands and through the saw with great force. This could cause serious or fatal injury to anyone in its path.

Fig. 18-12 Using the radial arm saw to rip lumber (note direction of feed in relation to blade rotation).

Fig. 18-13 Ripping stock at a bevel.

Do not force the stock into the saw. Feed it with a continuous motion. When ripping narrow stock, use a *push stick* on the end of the stock to push it between the saw blade and the fence.

Bevel Ripping. *Bevel ripping* is done in the same manner as straight ripping, except that the blade is tilted to the desired angle (Fig. 18-13). A V-groove may be made in this position by cutting partway through the stock. Then turn it around, end for end, and run it through again. The depth, width, and angle of the groove may be varied with different adjustments.

Dadoing, Grooving, and Rabbeting

Dadoes, grooves, and *rabbets* can be cut using a single saw blade. A *dado set* is commonly used to make them faster, with one pass of the stock through the cutting tool. One type of dado head consists of two outside circular saw blades with several *chippers* of different thicknesses placed in between (Fig. 18-14). Most dado sets make cuts from 1/4 to 13/16 inch wide. Wider cuts are obtained by making two or more passes through the saw.

When installing this type of dado set, make sure the tips of the chippers are opposite the gullets and not against the side of the blade. Chipper tips are *swaged* (made wider than the body of the chipper) to assure a clean cut across the width of the dado or groove.

Another type is the one-unit *adjustable dado head* (Fig. 18-15), commonly called a *wobbler head.* This type can be adjusted, for width of dado, without removing it from the saw arbor. Both types are available in high-speed steel or carbide-tipped styles.

Make dadoes and end rabbets in the same manner as for crosscutting. Edge rabbets and grooves are made with the saw in the ripping position. V-grooves can be made by tilting the saw blade (Fig. 18-16).

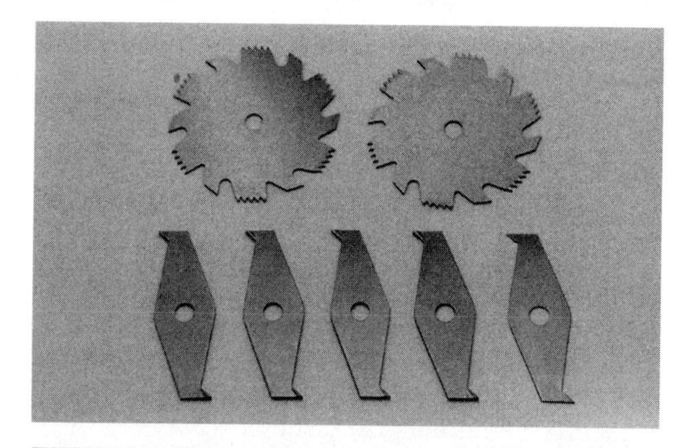

Fig. 18-14 The dado set consists of two blades and several chippers of different thicknesses from 1/16 to 1/4 inch.

Fig. 18-15 The one-unit adjustable dado head.

Fig. 18-16 Cutting dadoes and grooves.

Fig. 18-17 The power miter saw is widely used to cut miters on interior trim. (*Courtesy of Skil Power Tools*)

Fig. 18-18 The blade of the compound miter saw may be tilted. (*Courtesy of DeWalt Industrial Company*)

Miter Saw

The *power miter saw* is also called a *power miter box*. Usual sizes, designated by the blade diameter, are 10 and 12 inches (Fig. 18-17). It is not considered a portable power tool. However, it is light enough to be moved around wherever needed on the job.

A circular saw blade, mounted above the base, is pushed down using a chopping action to make the cut. The saw blade may be turned 45 degrees to the right or left. Positive stops are located at 90, 67 1/2, 60, and 45 degrees. Other angles are cut by reading the degree scale and locking in the desired position.

In some models, the blade can be tilted to make compound miter cuts. The 12-inch *compound miter saw* can make compound miters in crown molding up to 6 5/8 inches wide and other compound cuts up to 45 degrees by 45 degrees (Fig. 18-18). This saw, although used mostly for finish work, can also easily cut a 2 × 8 at 90 degrees and a 2 × 6 or 4 × 4 at 45 degrees.

The compound miter cuts on crown molding may be made with the stock flat on the saw table when the saw is adjusted to the compound angles using the preset stops on the saw. Check the manufacturer's instructions for the saw to determine crown molding cutting procedures.

To make cuts, the stock is held firmly on the table and against the fence. The saw is adjusted to the desired angle and brought down through the stock. A retractable blade guard provides operator protection and should be in place when the saw is being used (Fig. 18-19).

Fig. 18-19 Using the compound miter saw to miter a wide molding.

CHAPTER 19 TABLE SAWS

The table saw is one of the most frequently used woodworking power tools. In many cases, it is brought to the job site when the interior finish work begins. It is a useful tool because so many kinds of work can be performed with it. Common table saw operations, with different jigs to aid the process, are discussed in this chapter.

Table Saws

The size of the *table saw* (Fig. 19-1) is determined by the diameter of the saw blade. It may measure up to 16 inches or more. A commonly used table saw on the construction site is the 10-inch model.

The blade is adjusted for depth of cut and tilted up to 45 degrees by means of handwheels. A *rip fence* guides the work during ripping operations. A **miter gauge** is used to guide the work when cutting square ends and miters. The miter gauge slides in grooves in the table surface. It may be turned and locked in any position up to 45 degrees.

A guard should always be placed on the blade to protect the operator. Exceptions to this include some table saw operations like dados, rabbets, and cuts where the blade does not penetrate the entire stock thickness. A general rule is if the guard can be used for any operation on the table saw, it should be used.

Fig. 19-1 The 10-inch contractor's saw is frequently used on the job site.

Ripping Operations

For *ripping* operations, the table saw is easier to use and safer than the radial arm saw. To rip stock to width, measure from the *rip fence* to the point of a saw tooth set closest to the fence. Lock the fence in place. Adjust the height of the blade to about 1/4 inch above the stock to be cut. With the stock clear of the blade, turn on the power.

CAUTION Stand to the left of the blade. Never stand directly in back of the saw blade. Make sure no one else is in line with the saw blade in case of kickback. Be prepared to turn the saw off at any moment should the blade bind.

Hold the stock against the fence with the left hand. Push the stock forward with the right hand, holding the end of the stock (Fig. 19-2). As the end approaches the saw blade, let it slip through the left hand. Remove the left hand from the work. Push the end all the way through the saw blade with the right hand, if the stock is of sufficient width (at least 5 inches wide). If the stock is not wide enough, use a *push stick* (Fig. 19-3).

CAUTION Make sure the stock is pushed all the way through the saw blade. Leaving the cut stock between the fence and a running saw blade may cause a kickback, injuring anyone in its path. Use a push stick when ripping narrow pieces. Do not pick small waste pieces or strips from the saw table when the saw is running. Remove them with a stick or wait until the saw has stopped. Always use the rip fence for ripping operations. Never make freehand cuts. Never reach over a running saw blade.

A

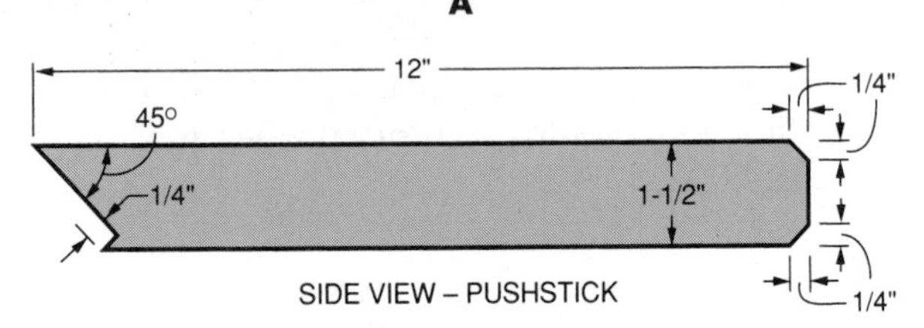

NOTE: PUSH STICK MUST BE THIN ENOUGH TO PASS BETWEEN THE RIP FENCE AND SAW BLADE WITH CLEARANCE.

B

Fig. 19-3 A. Use a push stick to rip narrow pieces (guard has been removed for clarity). B. Push stick design.

Bevel Ripping

Ripping on a *bevel* is done in the same manner as straight ripping except that the blade is tilted. For safer and more efficient bevel ripping, tilt the blade away from the fence and not toward it (Fig. 19-4).

Fig. 19-2 Using the table saw to rip lumber (guard has been removed for clarity).

Fig. 19-4 When bevel ripping, the blade is tilted (guard has been removed for clarity).

Taper Ripping

Tapered pieces (one end narrower than the other) can be made with the table saw by using a *taper ripping jig*. The jig consists of a wide board with the length and amount of taper cut out of one edge. The other edge is held against the rip fence. The stock to be tapered is held in the cutout of the jig. The taper is cut by holding the stock in the jig as both are passed through the blade (Fig. 19-5).

Fig. 19-5 Using a taper ripping jig to cut identical wedges (guard has been removed for clarity).

By using taper ripping jigs, small tapered pieces, such as wedges for concrete forms, or large ones, such as rafter tails, may be cut according to the design of the jig. A handle on the jig makes it safer to use. Also, if the jig is the same thickness as the stock to be cut, then the cutout section of the jig can be covered with a thin strip of wood to prevent the stock from flying out of the jig and back toward the operator.

Fig. 19-7 Cutting a groove with a dado head installed on the table saw.

Rabbeting and Grooving

Making *rabbets* and narrow *grooves* may be done using a single saw blade in a ripping fashion. To make a rabbet, usually two settings of the rip fence and two passes through the saw blade are required (Fig. 19-6). For narrow grooves, one or more passes are required. Move the rip fence slightly with each pass until the desired width of groove is obtained.

To make these cuts with fewer passes, a dado head is usually used (Fig. 19-7).

Crosscutting Operations

For most crosscutting operations, the *miter gauge* is used. To cut stock to length with both ends squared, first check the miter gauge for accuracy. Hold a

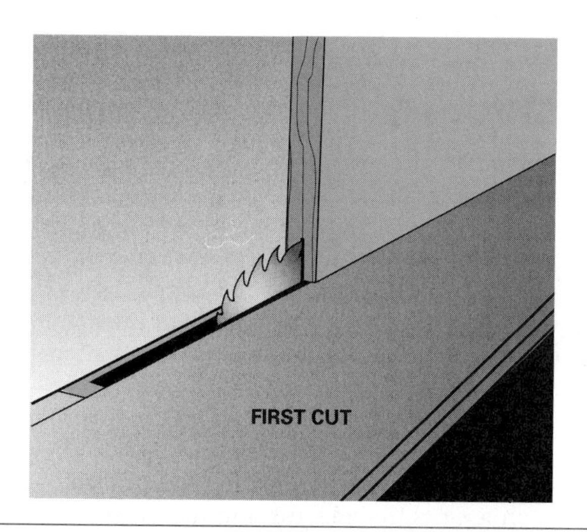

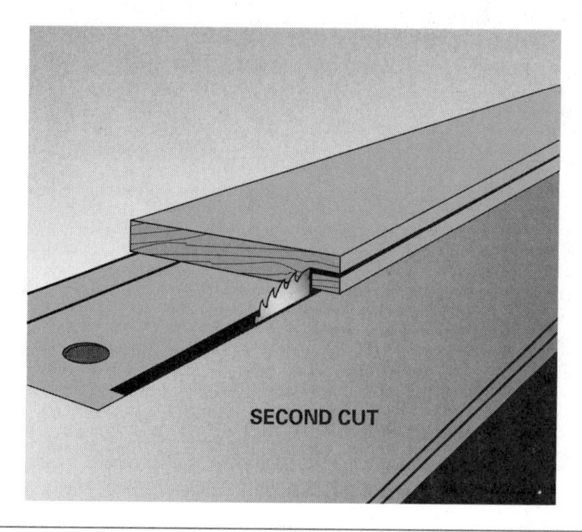

Fig. 19-6 Rabbeting an edge using a single saw blade by making two passes.

framing square against it and the side of the saw blade. Usually the miter gauge is operated in the left-hand groove. The right-hand groove is used only when it is more convenient.

Square one end of the stock by holding the work firmly against the miter gauge with one hand while pushing the miter gauge forward with the other hand. Measure the desired distance from the squared end. Mark on the front edge of the stock. Repeat the procedure, cutting to the layout line (Fig. 19-8).

Cutting Identical Lengths

When a number of identical lengths need to be cut, first square one end of each piece. Clamp a stop to an *auxiliary wood fence* installed on the miter gauge. Place the square end of the stock against the stop block. Then make the cut. Slide the remaining stock across the table until its end comes in contact with the stop block. Then make another cut. Continue in this manner until the desired number of pieces is cut (Fig. 19-9).

When short pieces of identical length are to be cut, a block is clamped to the rip fence in a location so that once the cut is made, there is clearance between the cut piece and the rip fence. The fence is adjusted so that the desired distance is between the face of the stop block and saw blade.

Square one end of the stock. Slide the squared end against the stop block. Then make a cut. Continue making cuts in this manner until the desired number of pieces is obtained (Fig. 19-10).

Do not use the miter gauge and the rip fence together on opposite sides of the blade unless a stop block is used. Pieces cut off between the blade and the fence can easily bind and be hurled across the room or at the operator.

Fig. 19-8 Using the miter gauge as a guide to crosscut (guard has been removed for clarity).

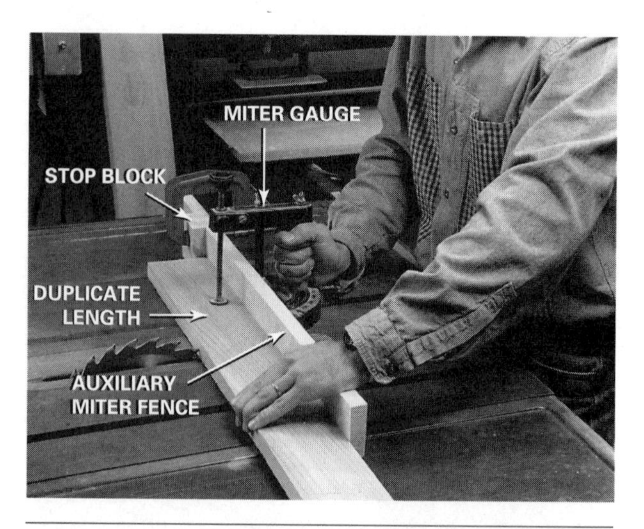

Fig. 19-9 Cutting identical lengths using a stop block on an auxiliary fence of the miter gauge (guard has been removed for clarity).

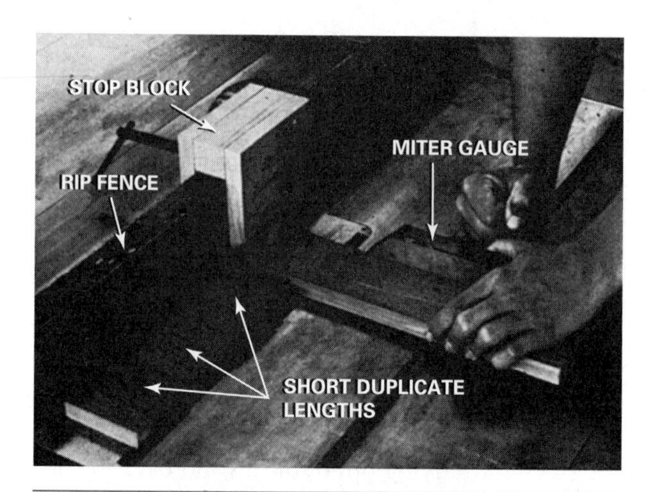

Fig. 19-10 Using the rip fence as a stop to cut short, identical lengths (guard has been removed for clarity).

Mitering

Flat miters are cut in the same manner as square ends, except the miter gauge is turned to the desired angle. *End miters* are made by adjusting the miter gauge to a square position and making the cut with the blade tilted to the desired angle.

Compound miters are cut with the miter gauge turned and the blade tilted to the desired angles (Fig. 19-11).

A *mitering jig* similar to that used on the radial arm saw can be used with the table saw. Figure 19-12 shows the construction and use of the jig. Use of such a jig eliminates turning the miter gauge each time for left- and right-hand miters.

Fig. 19-11 Making a compound miter cut (guard has been removed for clarity).

Dadoing

Dadoing is done in a similar manner as crosscutting except, with the use of a dado set (Fig. 19-13). The dado set is only used to cut *partway* through the stock thickness.

Table Saw Aids

A very useful aid is an *auxiliary fence.* A straight piece of 3/4-inch plywood about 12 inches wide and as long as the ripping fence is screwed or bolted to the metal fence. When cuts must be made close to the fence, the use of an auxiliary wood fence

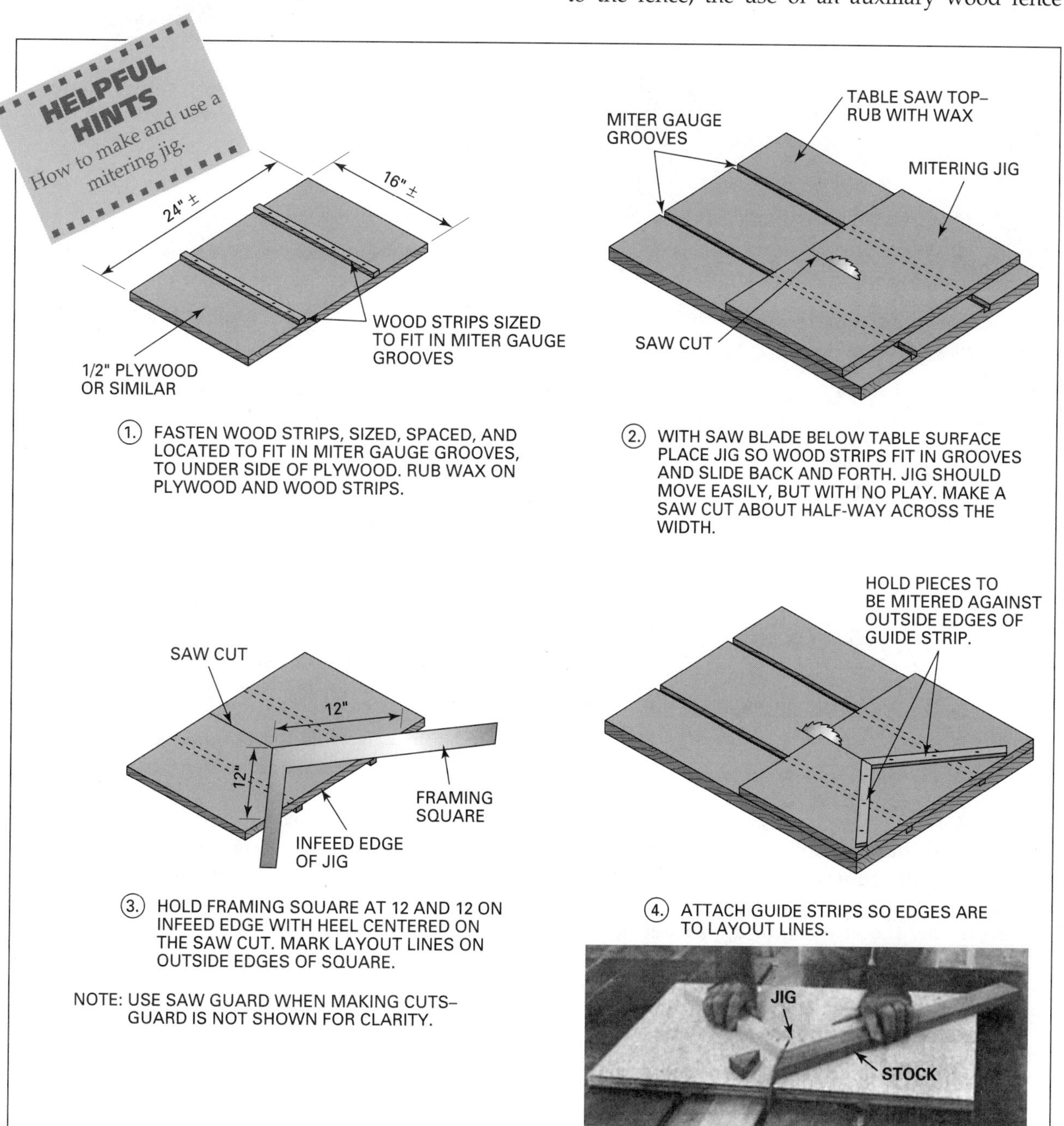

Fig. 19-12 Making a table saw jig to cut 45-degree miters.

Fig. 19-13 Cutting dadoes using a dado head.

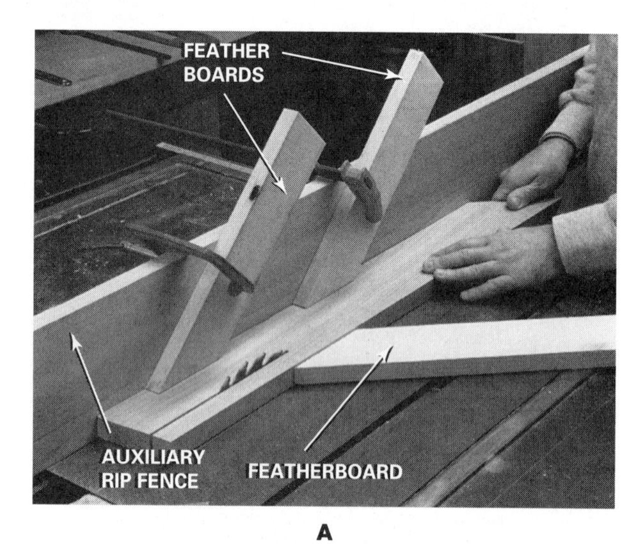

A

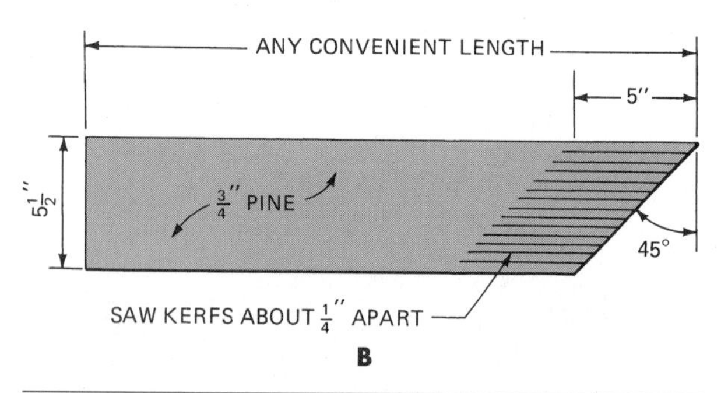

B

Fig. 19-14 A. Feather boards are useful aids to hold work during table saw operations (guard has been removed for clarity). Note that the locations of the feather boards do not cause the saw blade to bind. B. Feather board design.

Fig. 19-15 An auxiliary table top serves many useful purposes (guard has been removed for clarity).

prevents the saw blade from cutting into the metal fence. Also, the additional height provided by the fence gives a broader surface to steady wide work when its edge is being cut. The auxiliary wood fence also provides a surface on which to clamp feather boards. An auxiliary fence is also useful when attached to the miter gauge.

Feather boards are useful aids to hold work against the fence as well as down on the table surface during ripping operations (Fig. 19-14). Feather boards may be made of 1 × 6 nominal lumber, with one end cut at a 45-degree angle. Saw cuts are made in this end about 1/4 inch apart. This gives the end some spring and allows pressure to be applied to the piece being ripped.

An *auxiliary table top* serves many useful purposes. With the saw blade below the table surface, a wide piece of thin plywood, or similar material, as long as the saw table top is clamped in place after the rip fence is adjusted (Fig. 19-15). The saw motor is started. The blade is slowly raised to the desired height through the auxiliary top. The top prevents narrow rippings from slipping through the space between the saw blade and the table insert. It also prevents thin material from slipping under the rip fence. Use of the auxiliary table top can prevent accidents and serious injury.

Review Questions

Select the most appropriate answer.

1. The most frequently used blade in general carpentry is the
a. combination planer blade.
b. combination carbide-tipped blade.
c. square-grind carbide-tipped blade.
d. ripsaw blade.

2. A high-speed steel circular saw blade that has no set is the
a. combination blade.
b. crosscut blade.
c. ripsaw blade.
d. taper ground blade.

3. The alternate top bevel grind, carbide-tipped circular saw blade is designed to
a. rip solid lumber.
b. crosscut solid lumber.
c. cut in either direction of the grain.
d. cut green lumber of heavy dimension.

4. Saw arbor nuts are loosened
a. clockwise.
b. with the rotation of the blade.
c. counterclockwise.
d. against the rotation of the blade.

5. The saw designed primarily for the cutting of interior trim is a
a. power miter box.
b. radial arm saw.
c. table saw.
d. all of the above.

6. When using the radial arm saw for ripping operations, the in-feed end of the blade guard is lowered close to the work to
a. protect the operator from the saw blade.
b. prevent sawdust from flying out.
c. hold the work down on the table surface.
d. all of the above.

7. The table saw guide used for cutting with the grain is called
a. rip fence.
b. miter gauge.
c. tilting arbor.
d. ripping jig.

8. The tool designed to hold a piece of stock safely while being cut with a table saw is a
a. miter gauge.
b. push stick.
c. feather board.
d. all of the above.

9. A dado is a wide cut partway through the thickness of the material and
a. across the grain.
b. with the grain.
c. in either direction of the grain.
d. close to the edge.

10. When using the table saw for ripping operations, use a push stick if the ripped width is narrower than
a. 3 inches.
b. 4 inches.
c. 5 inches.
d. 6 inches.

BUILDING FOR SUCCESS

Implementing a Safety Program

At the job site or in the school construction lab, no objective should neglect to take into consideration the safety and welfare of workers and students. Completion dates, schedules, quality, and profits are important goals in the construction industry. However, none should be attained at the expense of people's lives. The highest priority on the job has to be an effective plan to ensure worker safety. The safety program has to be ongoing from day one to be effective. The responsibility of this program is everyone's. Both skilled and unskilled men and women who subject themselves to numerous hazards and potential accidents while working deserve the best protection available.

In most cases the contractor has informed his or her employees of appropriate job safety. Acceptable safety practices are reinforced by insurance companies and local, state, and national agencies such as the Occupational Safety and Health Administration (OSHA).

A company safety program must be well researched, prepared, and presented to the employees so they see it is for their benefit. Implementation must become part of everyone's daily tasks. The basic premise should be to greatly reduce the probability of construction-related accidents. Safety research should be conducted using all available resources, including statistics depicting the types and causes of construction accidents. The collected information should then be used to structure and implement the safety program.

Much research information may be obtained from insurance companies, labor statistics, OSHA, and the National Safety Council. A desired result of the research is to prepare a safety program that the employees will accept. The focus will center on the workers' welfare. Realistically this concentration should be a high priority with all workers.

Employees have to be involved from the beginning. The success of the program depends on how much the workers themselves believe in staying healthy. It has to be a program that automatically kicks in on a daily basis. A safe working environment has to be important to every worker or student on the job. This calls for influence on the part of instructors or contractors. They will have to "sell" the safety plan to employees. Almost everyone has listened to radio station "WIIFM" (What's in it for me?). That will be the key for acceptance. This connection has to be made if the workers are to accept the plan as a part of their job and security.

The annual number of injuries and fatalities in the construction industry share the lead with those in mining and farming. The very nature of these professions exposes people to a high risk for accidents. Each trade represented in construction has specific high-risk tasks that endanger workers. The overall safety program should become an element in all construction jobs.

Discussions, videos, speakers, demonstrations, and literature can be a vital part of an effective safety program. Each student or worker can be given a part in the presentation, application, or feedback report of a company safety program. As potential accidents are discovered and studied by workers, attention can be drawn to accident prevention.

Hazards and dangerous working conditions can be eliminated from the work site through employee involvement. Potential subjects for good safety programs are numerous due to the multitude of construction related jobs. Concentrations can be on excavating, shoring, hauling, lifting, cutting, climbing, driving, assembling, electrical grounding, or myriad other activities. Ideally the studies should center on tasks that are carried out by the people on a daily basis, then go on to other more infrequent tasks.

OSHA guidelines are usually used to create a safety program that informs workers and keeps a safety program ongoing. The program should have the safety of the workers in mind. Many construction companies take pride in the length of time between accidents on the job. An incentive program helps everyone become strongly committed to worker safety.

A good safety program will answer the important questions of Who, Why, What, When, Where, and How (in no particular order). To answer these questions: (1) identify the need and target population (Who); (2) justify the safety program with reason (Why); (3) relate the safety program concept (What); (4) establish a time frame for the program (When); (5) list the various work sites for implementation (Where); and (6) clarify the implementation process for everyone (How). Then be prepared to document, report, and reinforce the use of the safety program. A successful safety program will always involve the target population in the planning process. Ownership at this point is extremely important if goals are to be reached. The business owners and/or managers may see how a good safety record can ensure staying on schedule or assure profits for the company, but the employees may view it differently. The workers will be interested in assisting the company with their goals of quality and profits. They will become more highly motivated, however, by a safety program that centers on their health and welfare. That's when the buy-in takes place.

As employers reflect on the time and cost of an effective safety program, they will see and feel the benefits in satisfied workers, lower insurance rates, fewer accidents, less downtime, and healthier work climates. The employees in turn will feel they are a valued part of "their" company.

FOCUS QUESTIONS: For individual or group discussion

1. What might be the reasons for implementing a safety program at school or on the job?
2. In order for a safety program to be effective, people must see how it can benefit them. How could you as a contractor persuade them to buy in to this type of program?
3. As you introduce a safety program, how might you answer the questions Who, Why, What, When, Where, and How?

UNIT

8

Blueprints and Building Codes

The ability to interpret blueprints and to understand building codes, zoning ordinances, building permits, and inspection procedures is necessary for promotion on the job. With this ability, the carpenter has the competence needed to handle jobs of a supervisory nature, including those of foreman, superintendent, and contractor.

OBJECTIVES

After completing this unit, the student should be able to:

- describe and explain the function of the various kinds of drawings contained in a set of blueprints
- demonstrate how specifications are used
- read and use an architect's scale
- identify various types of lines and read dimensions
- identify and explain the meaning of symbols and abbreviations used on blueprints
- read and interpret plot, foundation, floor, and framing plans
- locate and explain information found in exterior and interior elevations
- identify and utilize information from sections and details
- use schedules to identify and determine location of windows, doors, and interior finish
- define and explain the purpose of building codes and zoning laws
- explain the requirements for obtaining a building permit and the duties of a building inspector

UNIT CONTENTS

CHAPTER 20 UNDERSTANDING BLUEPRINTS

Blueprints

An *architect* designs buildings, supervises construction, and manages projects. The *drafter* makes working drawings from the architect's designs. These drawings use various kinds of lines, dimensions, notes, symbols, and abbreviations to describe a structure to be built.

Standards for architectural drawings are more flexible than, for instance, machine drafting. Therefore, the method of drawing, lettering, and dimensioning may vary slightly with the drafter.

Many copies of the drawings must be made for use by clients, contractors, and subcontractors, for bidding purposes, for obtaining building permits, and for use on the job.

Making Blueprints

Original black line drawings are made with graphite or ink on translucent paper, cloth, or polyester film. Copies are usually made with a *diazo* printer. They are called diazo or *ozalid* prints. The copying process utilizes an ultraviolet light that passes through the paper, but not the lines. The light exposes a chemical on the copy paper, except where the lines are drawn. The copy paper is then exposed to ammonia vapor. This vapor develops the remaining chemical into blue, black, or brown lines against a white background. The diazo printing process is faster and less expensive than older methods.

Blueprinting is an older method of making copies. It uses a water wash to produce prints with white lines against a dark blue background. Although these are true blueprints, the process is seldom used today. The word *blueprint* remains in use, however, to mean any copy of the original drawing.

Kinds of Drawings

Different kinds of drawings are required for the construction of a building. These are called **plans, elevations, sections,** and *details.* When put together, they constitute a set of blueprints.

Types of Views

It may be necessary to use several views to describe an object with clarity. *Pictorial drawings* are usually three-dimensional *perspective* or **isometric** views. Pictorials do not give much information for building.

Multiview, also called *orthographic,* drawings are two-dimensional. They are used to convey most of the information needed for construction. They constitute the bulk of a set of blueprints (Fig. 20-1).

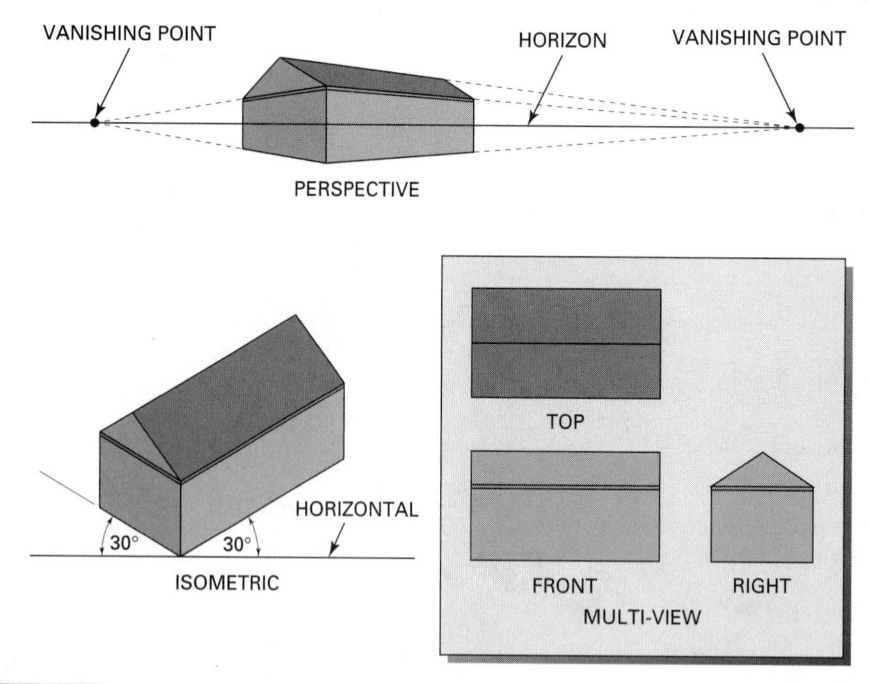

Fig. 20-1 Different views are used in architectural drawings.

Fig. 20-2 A perspective view is used for a presentation drawing.

The lines in a perspective diminish in size as they converge toward vanishing points on a line called a *horizon.* In an isometric drawing, the horizontal lines are drawn at 30 degree angles. All lines are drawn to actual scale. They do not diminish or converge as in perspective drawings.

In multiview projections, views of the front, right side, left side, and rear are called *elevations.* Views from the top are called *plans,* of which there are several kinds. The bottom view is never used in architectural drawings.

Presentation Drawings

A *presentation drawing* is usually a perspective type (Fig. 20-2). It shows the building from a desirable vantage point to display its most interesting features. Walks, streets, shrubs, trees, vehicles, and even people may be drawn. Many times, the drawing is colored for greater eye appeal. Presentation drawings provide no information for construction purposes. They are usually shown to clients to let them see how the completed building will look.

Plans

- The *floor plan* is a drawing of the structure viewed from above. Imagine a horizontal cut made about 4 to 5 feet above the floor. After the material above the cut is removed, the floor plan remains (Fig. 20-3).

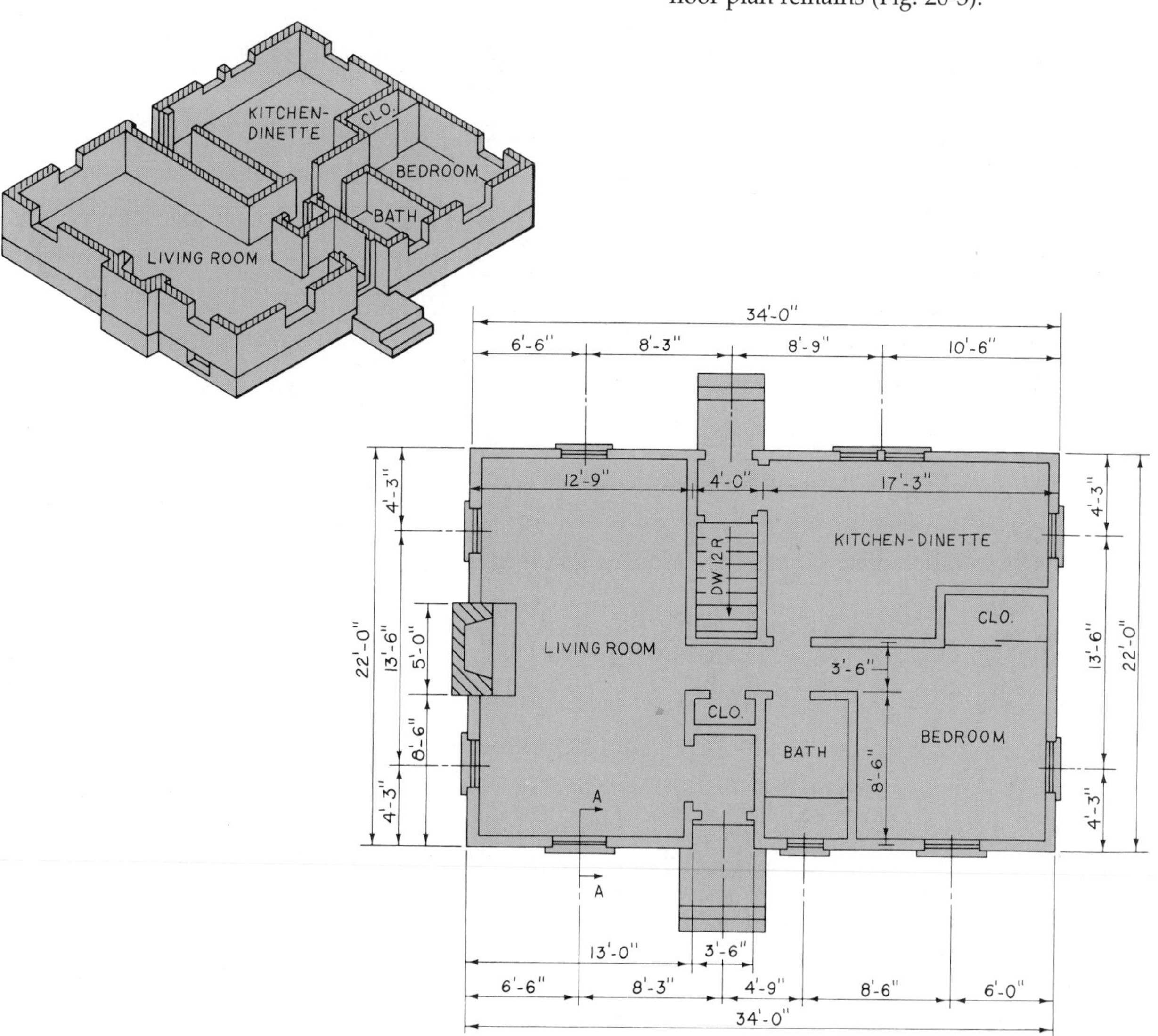

Fig. 20-3 A floor plan is a horizontal cut through the building.

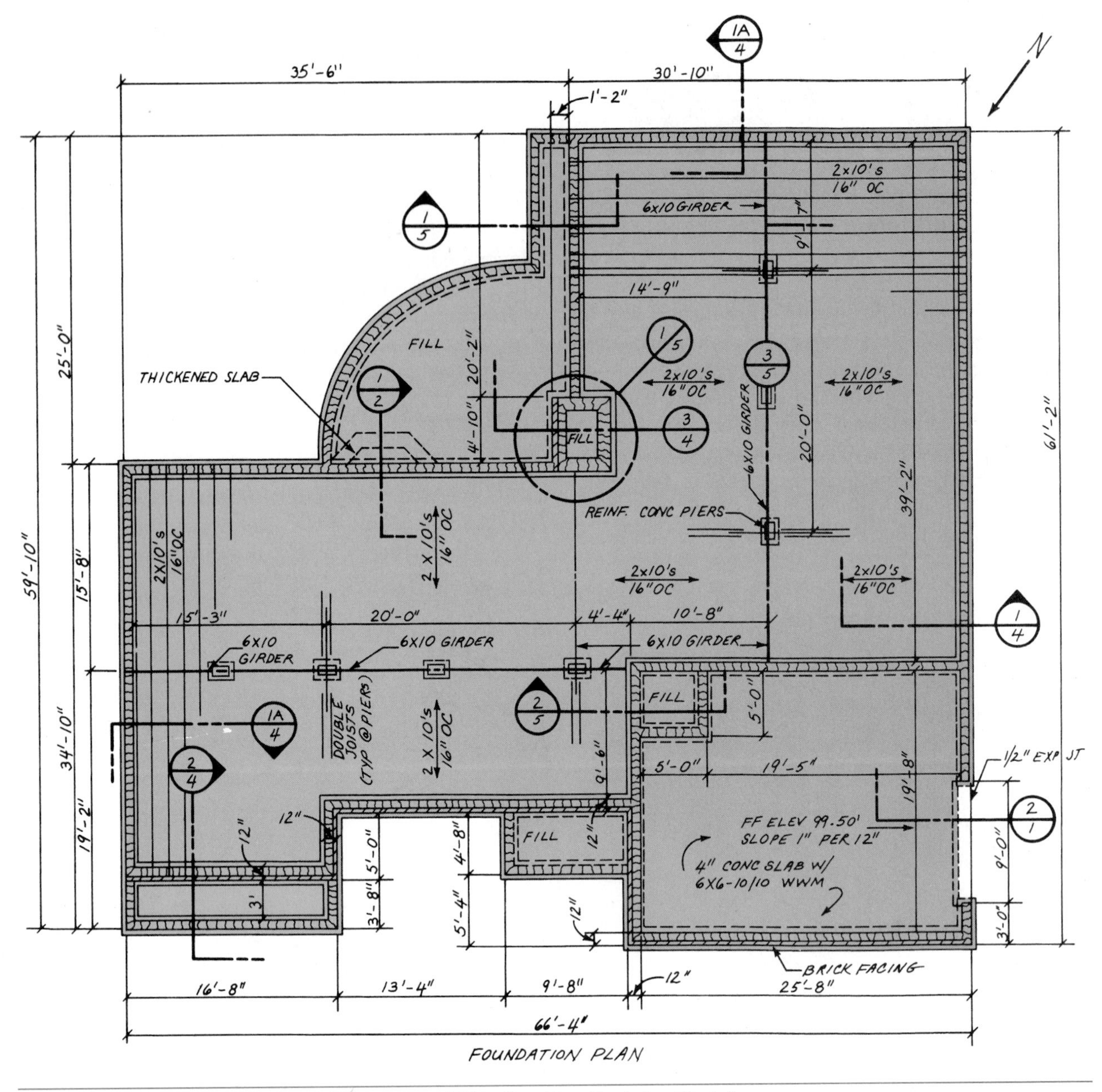

Fig. 20-4 A foundation plan. (*Courtesy of Pinellas Voc-Tech Institute*)

- The *foundation plan* (Fig. 20-4) shows a horizontal cut through the walls. It is similar to a floor plan. It shows the shape and dimensions of the foundation, among other things.
- The *plot plan* (Fig. 20-5) shows information about the lot, such as the location of the building, walks, and driveways. The drawing simulates a view looking down from a considerable height. It is made at a small scale because of the relatively large area it represents.
- *Framing plans* are not always found in a set of prints. They include floor framing and roof framing. They show the direction and spacing of the framing members (Fig. 20-6.)

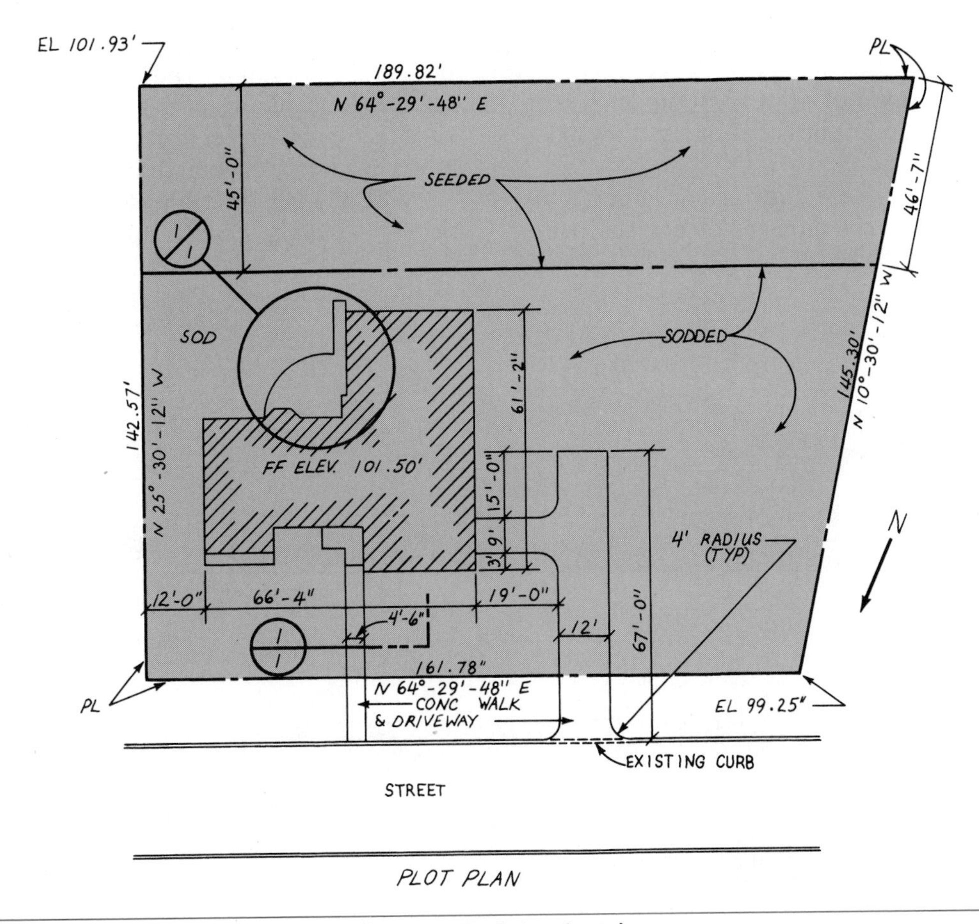

Fig. 20-5 A plot plan. (*Courtesy of Pinellas Voc-Tech Institute*)

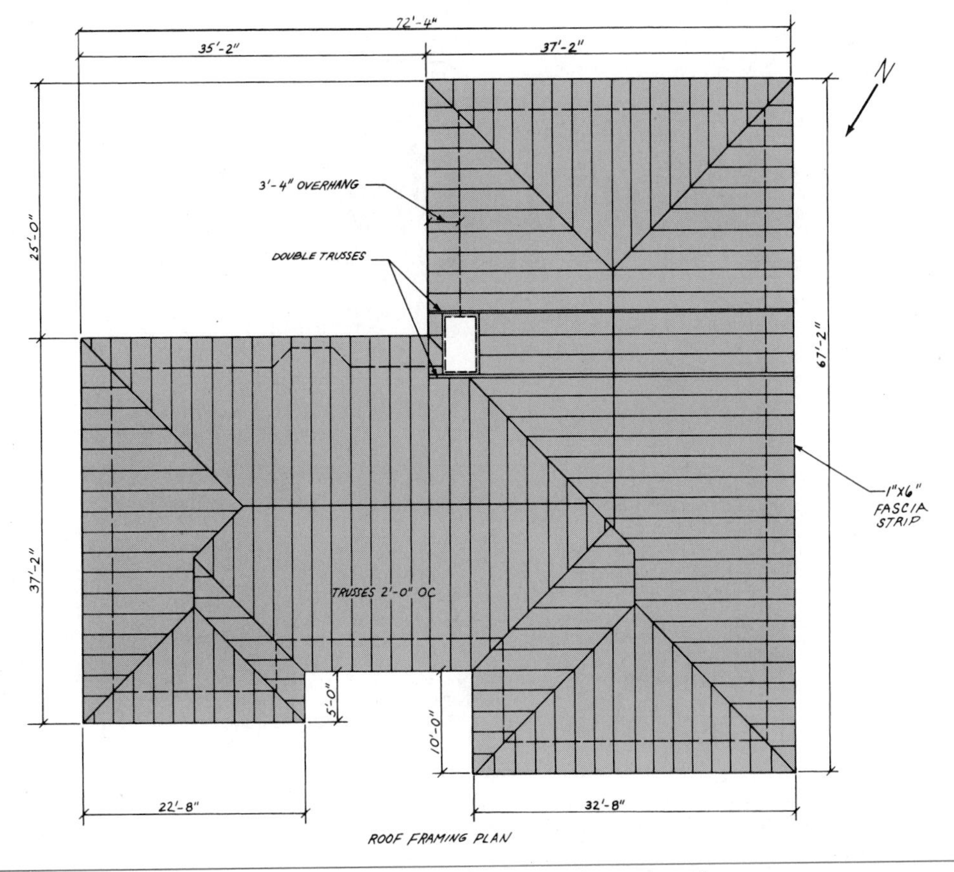

Fig. 20-6 A roof framing plan. (*Courtesy of Pinellas Voc-Tech Institute*)

Elevations

Elevations are a group of drawings (Fig. 20-7) that show the shape and finishes of all sides of the exterior of a building.

Interior elevations are drawings of certain interior walls. The most common are kitchen and bathroom wall elevations. They show the design and size of cabinets built on the wall (Fig. 20-8). Other walls that have special features, such as a fireplace, may require an elevation drawing.

Occasionally found in some sets of prints are *framing elevations.* Similar to framing plans, they

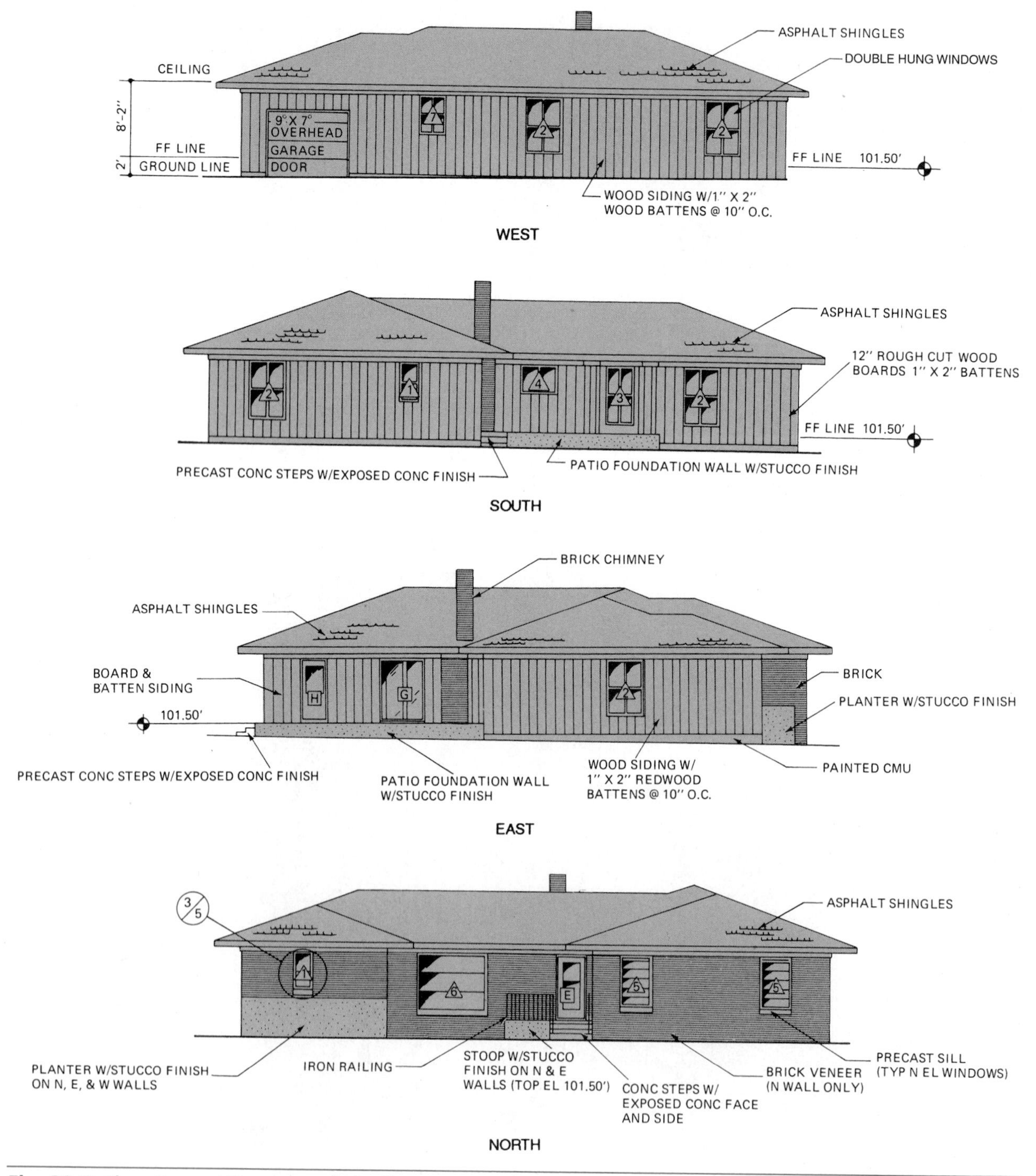

Fig. 20-7 Elevations. (*Courtesy of Pinellas Voc-Tech Institute*)

show the spacing, location, and sizes of wall framing members. No further description of framing drawings is required to be able to interpret them.

Sections

A *section* is a drawing that shows a vertical cut through all or part of a construction (Fig. 20-9). A section reference line is found in the plans or elevations to identify the section being viewed.

Details

To make parts of the construction more clear, it is usually necessary to draw *details.* Details are small parts drawn at a very large scale, even full-size. The detail shown in Figure 20-10 shows a portion of the construction more clearly.

Other Drawings

Drawings relating to electrical work, plumbing, heating, and ventilating may be on separate sheets in a set of prints. For smaller projects, separate plans are not always needed. All necessary information can be usually found on the floor plan.

The carpenter is responsible for building to accommodate wiring, pipes, and ducts. He or she should be able to read these plans with some degree of proficiency to understand the work involved.

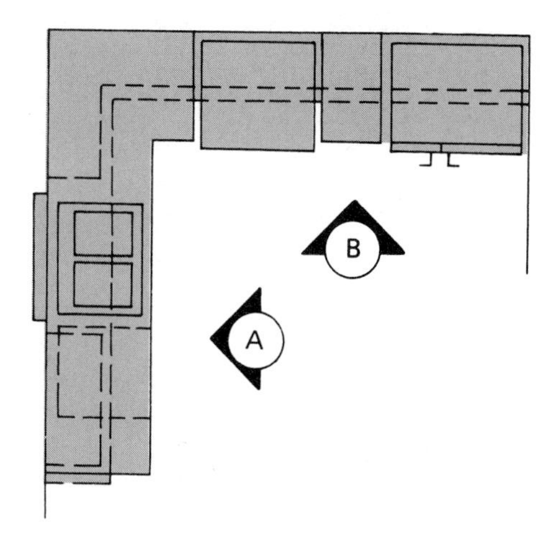

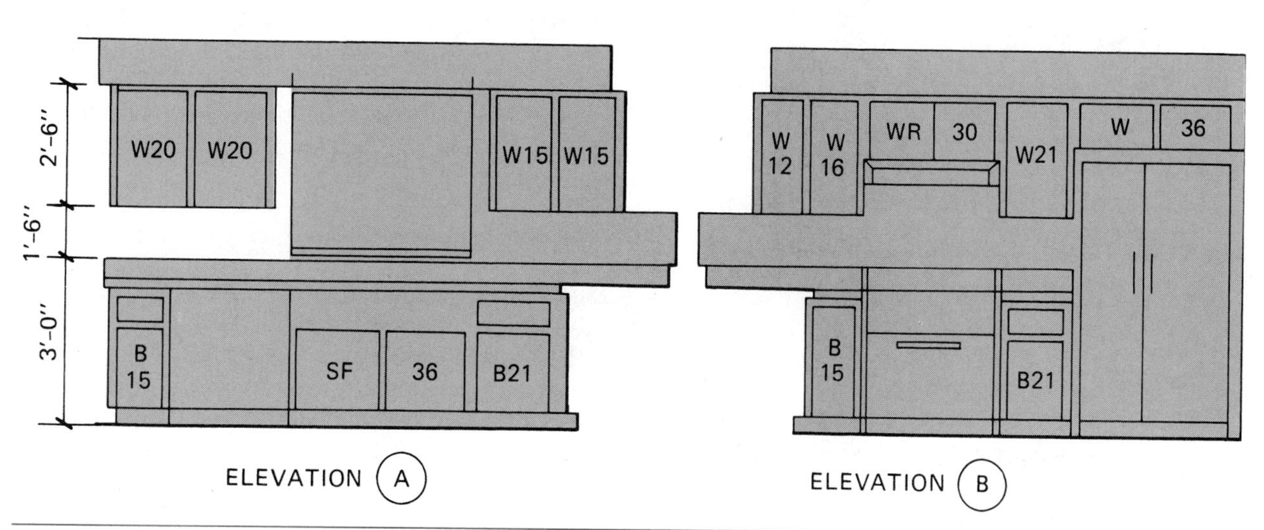

Fig. 20-8 Interior wall elevations. (*Courtesy of Pinellas Voc-Tech Institute*)

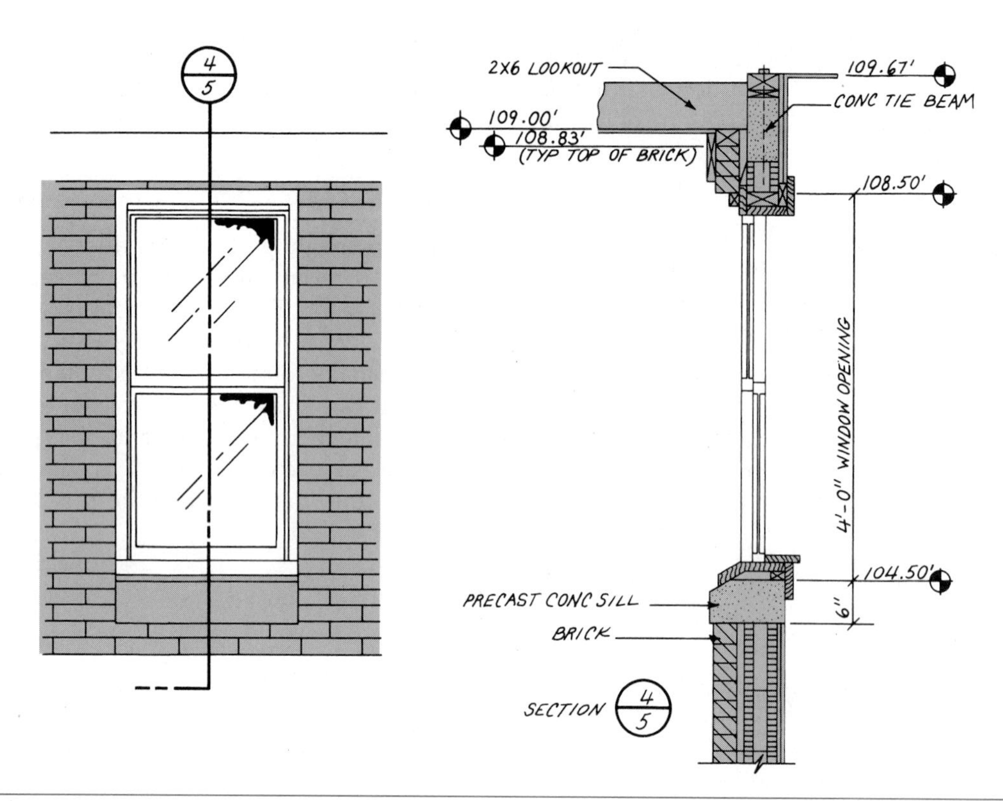

Fig. 20-9 A section is a view of a vertical cut through part of the construction. (*Courtesy of Pinellas Voc-Tech Institute*)

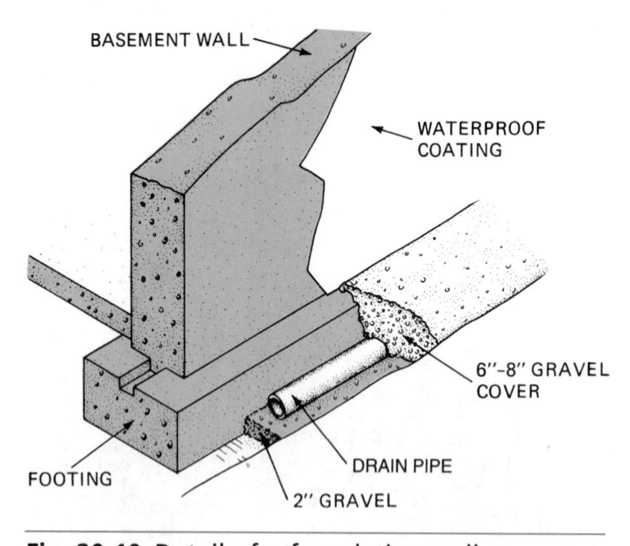

Fig. 20-10 Detail of a foundation wall.

Schedules

Besides drawings, printed instructions are included in a set of drawings. *Window schedules* and *door schedules* (Fig. 20-11) give information about the location, size, and kind of windows and doors to be installed in the building. Each of the different units is given a number or letter. A corresponding number or letter is found on the floor plan to show the location of the unit. Windows may be identified by letters and doors by numbers. The letters and numbers may be framed with various geometric figures, such as circles and triangles.

A *finish schedule* (Fig. 20-12) may also be included in a set of prints. This schedule gives information on the kind of finish material to be used on the floors, walls, and ceilings of the individual rooms.

Window, door, and finish schedules are used for easy understanding and to conserve space on floor plans.

Specifications

Specifications, commonly called *specs,* are written to give information that cannot be completely provided in the drawings or schedules. They supplement the working drawings with more complete descriptions of the methods, materials, and quality of construction. If there is a conflict, the specifications take precedence over the drawings. Any conflict should be pointed out to the architect so corrections can be made.

WINDOW SCHEDULE (SIZE OF OPENING FOR FRAME)				
TYPE	HEIGHT	WIDTH	STYLE	MATERIAL
1	4°	2°	DOUBLE HUNG	VINYL
2	6°	4°	DOUBLE HUNG	VINYL
3	6°	3°	DOUBLE HUNG	VINYL
4	3°	4°	CASEMENT	VINYL
5	5°	3°	AWNING	VINYL
6	6°	8°	AWNING	VINYL
7	4°	3°	DOUBLE HUNG	VINYL

DOOR SCHEDULE			
TYPE	HEIGHT	STYLE	MATERIAL
A	$2^8 \times 6^8$	H.C.	WOOD
B	(2)$2^\circ \times 6^8$	H.C. DOUBLE (LOUVER)	WOOD
C	(2)$2^\circ \times 6^8$	S.C. DOUBLE (SWINGING)	WOOD
D	(2)$2^\circ \times 6^8$	S.C. FRENCH DOORS	WOOD & GLASS
E	$3^\circ \times 6^8$	S.C.	VINYL
F	$3^\circ \times 6^8$	H.C.	VINYL
G	$5^\circ \times 6^8$	DOUBLE SLIDING GLASS	GLASS & ALUM.
H	$3^\circ \times 6^8$	S.C. (ONE - LIGHT)	VINYL & GLASS

Fig. 20-11 A typical window and door schedule. (*Courtesy of Pinellas Voc-Tech Institute*)

FINISH SCHEDULE							
ROOM	WALLS	PAINT COLORS	BASE	FLOOR	CEILING	CORNICE	REMARKS
LIV. RM.	DRY WALL	BONE	WOOD	OAK	PLASTER	WOOD	BOOKCASE
DIN. RM.	"	"	"	"	"	PICT. MLDG	CLIPBD.
KITCHEN	"	EGG SHELL	TILE	VINYL	"	——	——
HALL	"	"	WOOD	OAK	"	WOOD	SEE DTL.
ENTRY	"	"	"	"	"	——	——

Fig. 20-12 A typical finish schedule.

The amount of detail contained in the specs will vary, depending on the size of the project. On small jobs, they may be written by the architect. On larger jobs, a *specifications writer,* trained in the construction process, may be required. For complex commercial projects, a specifications guide, used by spec writers, has been developed by the Construction Specifications Institute (CSI). The guide has sixteen major divisions, each containing a number of subdivisions (Fig. 20-13.)

Using the specification guide, under Division 6—WOOD AND PLASTICS, Section 06200, an example of the content and the manner in which specifications are written is shown in Figure 20-14.

For some light commercial and residential construction, many sections of the spec guide would not apply. A shortened version is then used, eliminating divisions 12,000, 13,000, and 14,000. On simpler plans, notations made on the same sheets as

DIVISION 1 – GENERAL REQUIREMENTS
01010 SUMMARY OF WORK
01100 ALTERNATIVES
01150 MEASUREMENT & PAYMENT
01200 PROJECT MEETINGS
01300 SUBMITTALS
01400 QUALITY CONTROL
01500 TEMPORARY FACILITIES & CONTROLS
01600 MATERIAL & EQUIPMENT
01700 PROJECT CLOSEOUT

DIVISION 2 – SITE WORK
02010 SUBSURFACE EXPLORATION
02100 CLEARING
02110 DEMOLITION
02200 EARTHWORK
02250 SOIL TREATMENT
02300 PILE FOUNDATIONS
02350 CAISSONS
02400 SHORING
02500 SITE DRAINAGE
02550 SITE UTILITIES
02600 PAVING & SURFACING
02700 SITE IMPROVEMENTS
02800 LANDSCAPING
02850 RAILROAD WORK
02900 MARINE WORK
02950 TUNNELING

DIVISION 3 – CONCRETE
03100 CONCRETE FORMWORK
03150 FORMS
03200 CONCRETE REINFORCEMENT
03250 CONCRETE ACCESSORIES
03300 CAST-IN-PLACE CONCRETE
03350 SPECIALLY FINISHED (ARCHITECTURAL) CONCRETE
03360 SPECIALLY PLACED CONCRETE
03400 PRECAST CONCRETE
03500 CEMENTITIOUS DECKS
03600 GROUT

DIVISION 4 – MASONRY
04100 MORTAR
04150 MASONRY ACCESSORIES
04200 UNIT MASONRY
04400 STONE
04500 MASONRY RESTORATION & CLEANING
04550 REFRACTORIES

DIVISION 5 – METALS
05100 STRUCTURAL METAL FRAMING
05200 METAL JOISTS
05300 METAL DECKING
05400 LIGHTGAGE METAL FRAMING
05500 METAL FABRICATIONS
05700 ORNAMENTAL METAL
05800 EXPANSION CONTROL

DIVISION 6 – WOOD & PLASTICS
06100 ROUGH CARPENTRY
06130 HEAVY TIMBER CONSTRUCTION
06150 TRESTLES
06170 PREFABRICATED STRUCTURAL WOOD
06200 FINISH CARPENTRY
06300 WOOD TREATMENT
06400 ARCHITECTURAL WOODWORK
06500 PREFABRICATED STRUCTURAL PLASTICS
06600 PLASTIC FABRICATIONS

DIVISION 7 – THERMAL & MOISTURE PROTECTION
07100 WATERPROOFING
07150 DAMPPROOFING
07200 INSULATION
07300 SHINGLES & ROOFING TILES
07400 PREFORMED ROOFING & SIDING
07500 MEMBRANE ROOFING
07570 TRAFFIC TOPPING
07600 FLASHING & SHEET METAL
07800 ROOF ACCESSORIES
07900 SEALANTS

DIVISION 8 – DOOR & WINDOWS
08100 METAL DOORS & FRAMES
08200 WOOD & PLASTIC DOORS
08300 SPECIAL DOORS
08400 ENTRANCES & STOREFRONTS
08500 METAL WINDOWS
08600 WOOD & PLASTIC WINDOWS
08650 SPECIAL WINDOWS
08700 HARDWARE & SPECIALTIES
08800 GLAZING
08900 WINDOW WALLS/CURTAIN WALLS

DIVISION 9 – FINISHES
09100 LATH & PLASTER
09250 GYPSUM WALLBOARD
09300 TILE
09400 TERRAZZO
09500 ACOUSTICAL TREATMENT
09540 CEILING SUSPENSION SYSTEMS
09550 WOOD FLOORING
09650 RESILIENT FLOORING
09680 CARPETING
09700 SPECIAL FLOORING
09760 FLOOR TREATMENT
09800 SPECIAL COATINGS
09900 PAINTING
09950 WALL COVERING

DIVISION 10 – SPECIALTIES
10100 CHALKBOARDS & TACKBOARDS
10150 COMPARTMENTS & CUBICLES
10200 LOUVERS & VENTS
10240 GRILLES & SCREENS
10260 WALL & CORNER GUARDS
10270 ACCESS FLOORING
10280 SPECIALTY MODULES
10290 PEST CONTROL
10300 FIREPLACES
10350 FLAGPOLES
10400 IDENTIFYING DEVICES
10450 PEDESTRIAN CONTROL DEVICES
10500 LOCKERS
10530 PROTECTIVE COVERS
10550 POSTAL SPECIALTIES
10600 PARTITIONS
10650 SCALES
10670 STORAGE SHELVING
10700 SUNCONTROLDEVICES(EXTERIOR)
10750 TELEPHONE ENCLOSURES
10800 TOILET & BATH ACCESSORIES
10900 WARDROBE SPECIALTIES

DIVISION 11 – EQUIPMENT
11050 BUILT-IN MAINTENANCE EQUIPMENT
11100 BANK & VAULT EQUIPMENT
11150 COMMERCIAL EQUIPMENT
11170 CHECKROOM EQUIPMENT
11180 DARKROOM EQUIPMENT
11200 ECCLESIASTICAL EQUIPMENT
11300 EDUCATIONAL EQUIPMENT
11400 FOOD SERVICE EQUIPMENT
11480 VENDING EQUIPMENT
11500 ATHLETIC EQUIPMENT
11550 INDUSTRIAL EQUIPMENT
11600 LABORATORY EQUIPMENT
11630 LAUNDRY EQUIPMENT
11650 LIBRARY EQUIPMENT
11700 MEDICAL EQUIPMENT
11800 MORTUARY EQUIPMENT
11830 MUSICAL EQUIPMENT
11850 PARKING EQUIPMENT
11860 WASTE HANDLING EQUIPMENT
11870 LOADING DOCK EQUIPMENT
11880 DETENTION EQUIPMENT
11900 RESIDENTIAL EQUIPMENT
11970 THEATER & STAGE EQUIPMENT
11990 REGISTRATION EQUIPMENT

DIVISION 12 – FURNISHINGS
12100 ARTWORK
12300 CABINETS & STORAGE
12500 WINDOW TREATMENT
12550 FABRICS
12600 FURNITURE
12670 RUGS & MATS
12700 SEATING
12800 FURNISHING ACCESSORIES

DIVISION 13 – SPECIAL CONSTRUCTION
13010 AIR SUPPORTED STRUCTURES
13050 INTEGRATED ASSEMBLIES
13100 AUDIOMETRIC ROOM
13250 CLEAN ROOM
13350 HYPERBARIC ROOM
13400 INCINERATORS
13440 INSTRUMENTATION
13450 INSULATED ROOM
13500 INTEGRATED CEILING
13540 NUCLEAR REACTORS
13550 OBSERVATORY
13600 PREFABRICATED STRUCTURES
13700 SPECIAL PURPOSE ROOMS & BUILDINGS
13750 RADIATION PROTECTION
13770 SOUND & VIBRATION CONTROL
13800 VAULTS
13850 SWIMMING POOLS

DIVISION 14 – CONVEYING SYSTEMS
14100 DUMBWAITERS
14200 ELEVATORS
14300 HOISTS & CRANES
14400 LIFTS
14500 MATERIAL HANDLING SYSTEMS
14570 TURNTABLES
14600 MOVING STAIRS & WALKS
14700 TUBE SYSTEMS
14800 POWERED SCAFFOLDING

DIVISION 15 - MECHANICAL
15010 GENERAL PROVISIONS
15050 BASIC MATERIALS & METHODS
15180 INSULATION
15200 WATER SUPPLY & TREATMENT
15300 WASTE WATER DISPOSAL & TREATMENT
15400 PLUMBING
15500 FIRE PROTECTION
15600 POWER OR HEAT GENERATION
15650 REFRIGERATION
15700 LIQUID HEAT TRANSFER
15800 AIR DISTRIBUTION
15900 CONTROLS & INSTRUMENTATION

DIVISION 16 - ELECTRICAL
16010 GENERAL PROVISIONS
16100 BASIC MATERIALS & METHODS
16200 POWER GENERATION
16300 POWER TRANSMISSION
16400 SERVICE & DISTRIBUTION
16500 LIGHTING
16600 SPECIAL SYSTEMS
16700 COMMUNICATIONS
16850 HEATING & COOLING
16900 CONTROLS & INSTRUMENTATION

Fig. 20-13 CSI format for specifications.

the drawings may take the place of specifications. The ability to read notations and specifications accurately is essential in order to conform to the architect's design.

Blueprint Language

Carpenters must be able to read and understand the combination of lines, dimensions, symbols, and no-

DIVISION 6—WOOD AND PLASTICS

Section 06200—FINISH CARPENTRY

General: This section covers all finish woodwork and related items not covered elsewhere in these specifications. The contractor shall furnish all materials, labor, and equipment necessary to complete the work, including rough hardware, finish hardware, and specialty items.

Protection of Materials: All millwork (finish woodwork*) and trim is to be delivered in a clean and dry condition and shall be stored to insure proper ventilation and protection from dampness. Do not install finish woodwork until concrete, masonry, plaster, and related work is dry.

Materials: All materials are to be the best of their respective kind. Lumber shall bear the mark and grade of the association under whose rules it is produced. All millwork shall be kiln dried to a maximum moisture content of 12%.

1. Exterior trim shall be select grade white pine, S4S.
2. Interior trim and millwork shall be select grade white pine, thoroughly sanded at the time of installation.

Installation: All millwork and trim shall be installed with tight fitting joints and formed to conceal future shrinkage due to drying. Interior woodwork shall be mitered or coped at corners (cut in a special way to form neat joints*). All nails are to be set below the surface of the wood and concealed with an approved putty or filler.

*(explanations in parentheses have been added to aid the student.)

Fig. 20-14 Sample specifications following the CSI format. (*From Huth,* Construction Technology, *1989, by Delmar*)

tations on the drawings. Only then can they build exactly as the architect has designed the construction (Fig. 20-15). No deviation from the blueprints may be made without the approval of the architect.

Scales

It would be inconvenient and impractical to make full-sized drawings of a building. Therefore, they are made *to scale.* This means that each line in the drawing is reduced proportionally to a size that clearly shows the information and can be handled conveniently. Not all drawings in a set of prints, or even on the same page, are drawn at the same scale. The scale of a drawing is stated in the title block of the page. It can also be directly below a drawing.

Fig. 20-15 The carpenter must be able to read blueprints.

Architect's Scale. The triangular *architect's scale* is commonly used to scale lines when making drawings (Fig. 20-16). Both sides of the three edges are divided into different size gradations. The side

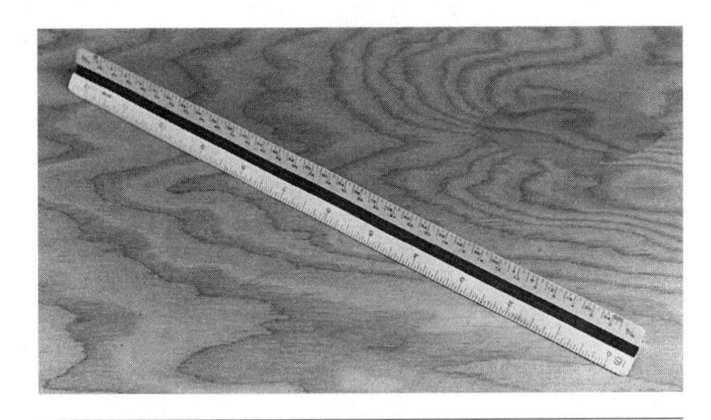

Fig. 20-16 The architect's scale.

of one edge is divided into 1-inch and 1/16-inch increments. Along each edge of the remaining five sides are two scales. One scale is twice as large as the other. One is read from left to right. The other scale is read from right to left.

The scales are paired as follows:

3″ = 1′-0″ and 1 1/2″ = 1′-0″
1″ = 1′-0″ and 1/2″ = 1′-0″
3/4″ = 1′-0″ and 3/8″ = 1′-0″
1/4″ = 1′-0″ and 1/8″ = 1′-0″
3/16″ = 1′-0″ and 3/32″ = 1′-0″

When reading a scale in one direction, be careful not to confuse it with the one running in the opposite direction. On the end of each scale, a space representing one foot is divided into inches. In some cases, it is divided into fractions of an inch. This space is used when scaling off fractions of a foot (Fig. 20-17).

Commonly Used Scales. Probably the most commonly used scale found on blueprints is 1/4 inch equals 1 foot. This is indicated as 1/4″ = 1′-0″. It is often referred to as a "quarter-inch scale." This

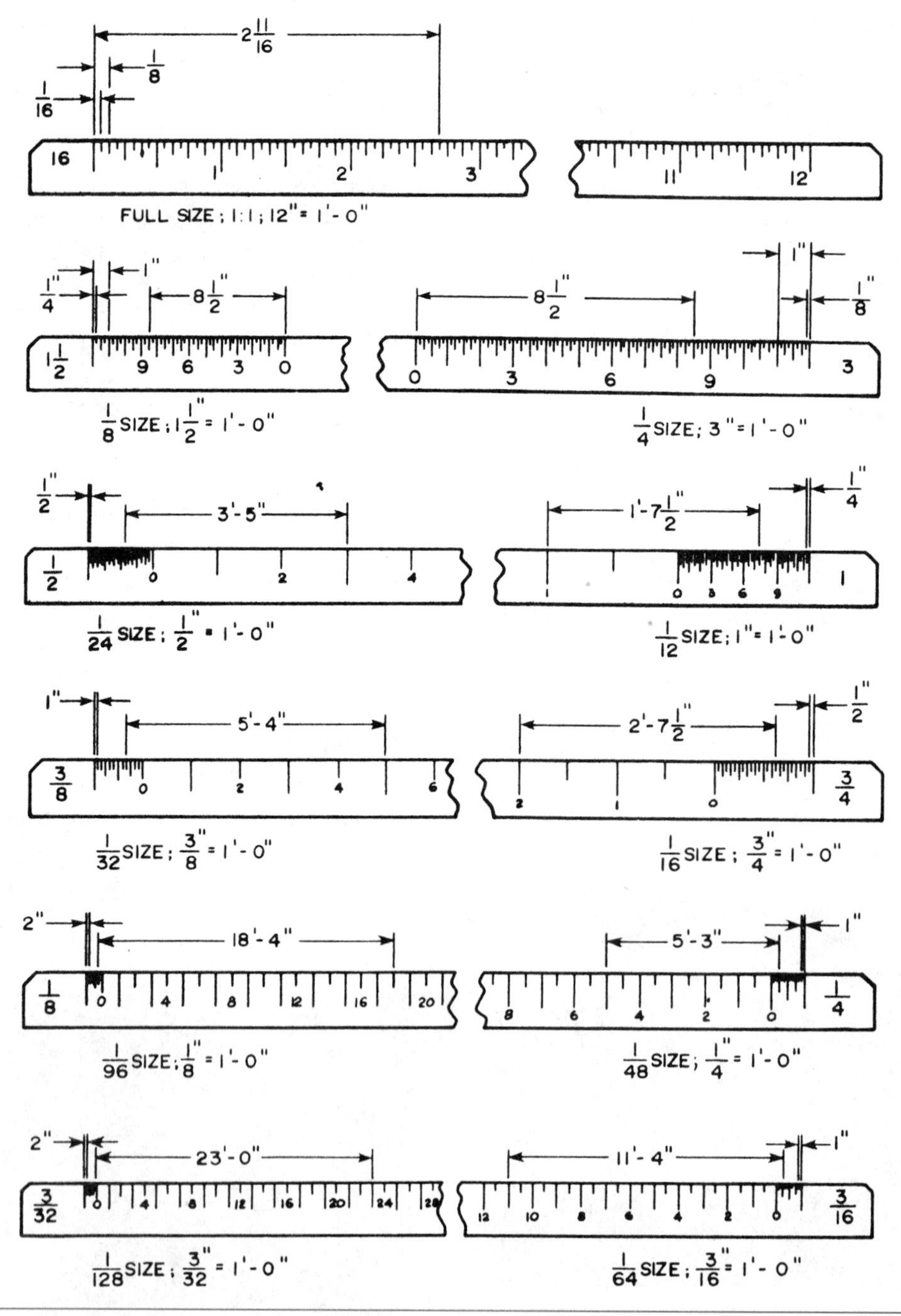

Fig. 20-17 Various architect's scales. (*From Jefferis and Madsen,* Architectural Drafting and Design,*1991, by Delmar*)

means that every 1/4 inch on the drawing will equal one foot in the building. Floor plans and exterior elevations for most residential buildings are drawn at this scale.

To show the location of a building on a lot and other details of the site, the architect may use a scale of 1/16″ = 1′-0″. This reduces the size of the drawing to fit it on the paper. To show certain details more clearly, larger scales of 1 1/2″ = 1′-0″ or 3″ = 1′-0″ are used. Complicated details may be drawn full size or half size. Other scales are used when appropriate. Views showing the elevation of interior walls are often drawn at 1/2″ = 1′-0″ or 3/4″ = 1′-0″.

Drawing blueprints to scale is important. The building and its parts are shown in true proportion, making it easier for the builder to visualize the construction. However, *the use of a scale rule to determine a dimension should be a last resort.* Dimensions on blueprints should be determined either by reading the dimension or adding and subtracting other dimensions to determine it. The use of a scale rule to determine a dimension results in inaccuracies.

Types of Lines

Some lines in an architectural drawing look darker than others. They are broader so they stand out clearly from other lines. This variation in width is called *line contrast.* This technique, like all architectural drafting standards, is used to make the drawing easier to read and understand (Fig. 20-18).

- Lines that outline the object being viewed are broad, solid lines called *object lines.*
- To indicate an object not visible in the view, a *hidden line* consisting of short, fine, uniform dashes is used. Hidden lines are used only when necessary. Otherwise the drawing becomes confusing to read.

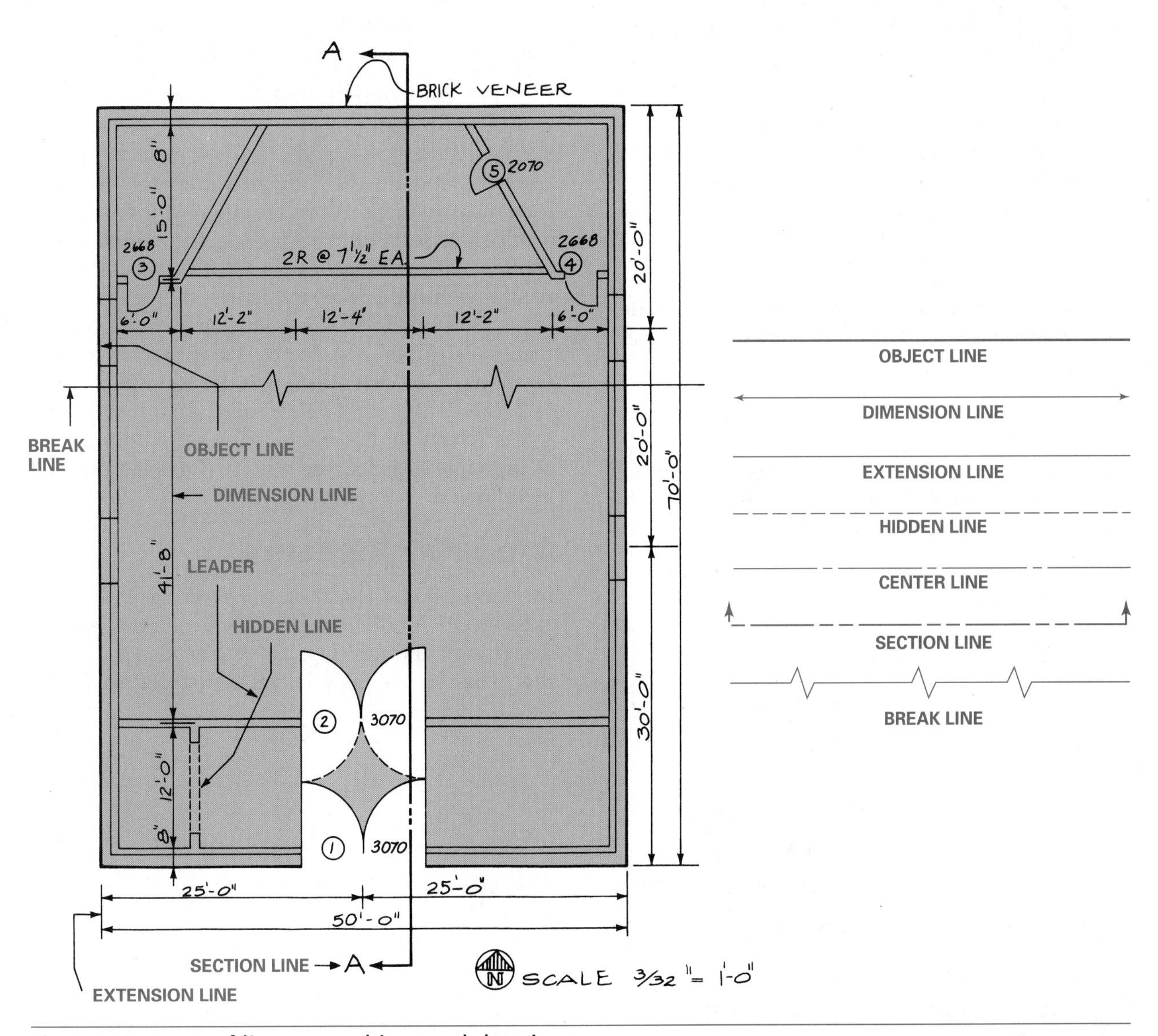

Fig. 20-18 Types of lines on architectural drawings.

- *Centerlines* are indicated by a fine, long dash, then a short dash, then a long dash, and so on. They show the centers of doors, windows, partitions, and similar parts of the construction.
- A *section reference* or *cutting-plane line* is, sometimes, a broad line consisting of a long dash followed by two short dashes. At its ends are arrows. The arrows show the direction in which the cross-section is viewed. Letters identify the cross-sectional view of that specific part of the building. More elaborate methods of labeling section reference lines are used in larger, more complicated sets of plans (Fig. 20-19). The sectional drawings may be on the same page as the reference line or on other pages.
- A *break line* is used in a drawing to terminate part of an object that, in actuality, continues. It can only be used when there is no change in the drawing at the break. Its purpose is to shorten the drawing to utilize space.
- A *dimension line* is a fine, solid line used to indicate the location, length, width, or thickness of an object. It is terminated with arrowheads, dots, or slashes (Fig. 20-20.)
- *Extension lines* are fine, solid lines projecting from an object to show the extent of a dimension.
- A *leader line* is a fine solid line. It terminates with an arrowhead, and points to an object from a notation.

Dimensions

Dimension lines on a blueprint are generally drawn as continuous lines. The dimension appears above and near the center of the line. All dimensions on vertical lines should appear above the line when the print is rotated 1/4 turn clockwise. Extension lines are drawn from the object so that the end point of the dimension is clearly defined. When the space is too small to permit dimensions to be shown clearly, they may be drawn as shown in Figure 20-21.

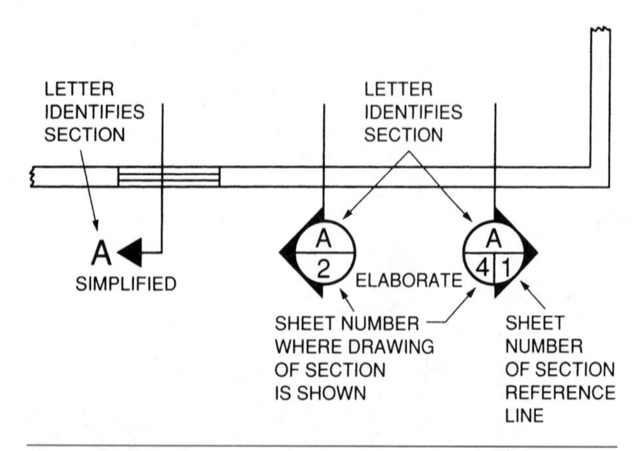

Fig. 20-19 Several ways of labeling section reference lines.

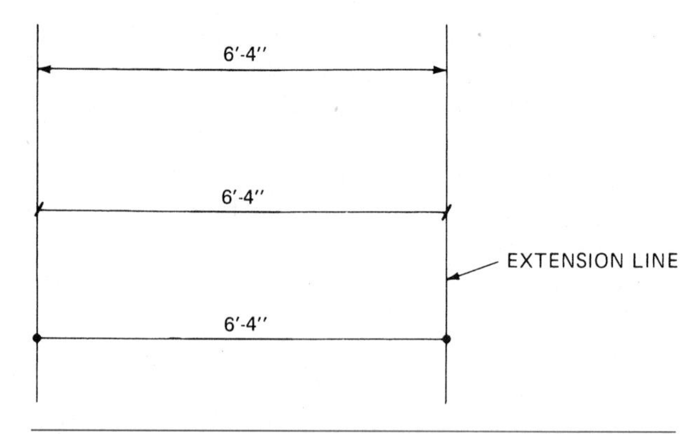

Fig. 20-20 Several methods of terminating dimension lines.

Kinds of Dimensions. Dimensions on architectural blueprints are given in feet and inches, such as 3'-6", 4'-8", and 13'-7". A dash is always used to separate the foot measurement from the inch measurement. When the dimension is a whole number of feet with no inches, the dimension is written with zero inches, as 14'-0". The use of the dash prevents mistakes in reading dimensions.

Dimensions of 1 foot and under are given in inches, as 10", 8", and so on. Dimensions involving fractions of an inch are shown, for example, as 1'-0 1/2", 2'-3 3/4", or 6 1/2". If there is a difference between a written dimension and a scaled dimension of the same distance, the written dimension should be followed.

Modular Measure

In recent years, *modular measurement* has been used extensively. A grid with a unit of 4 inches is used in designing buildings (Fig. 20-22). The idea is to draw the plans to use material manufactured to fit the

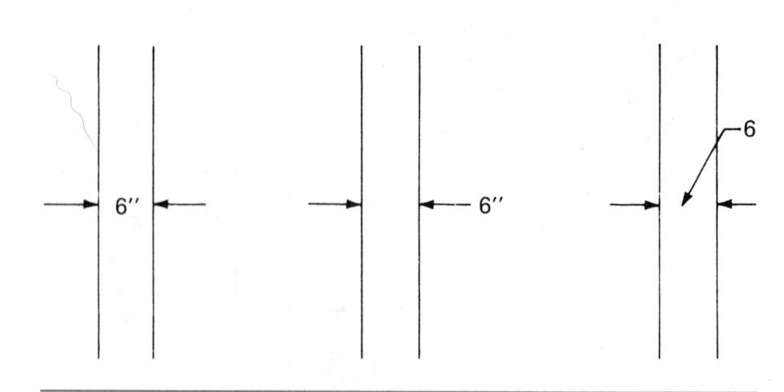

Fig. 20-21 Dimensioning small spaces.

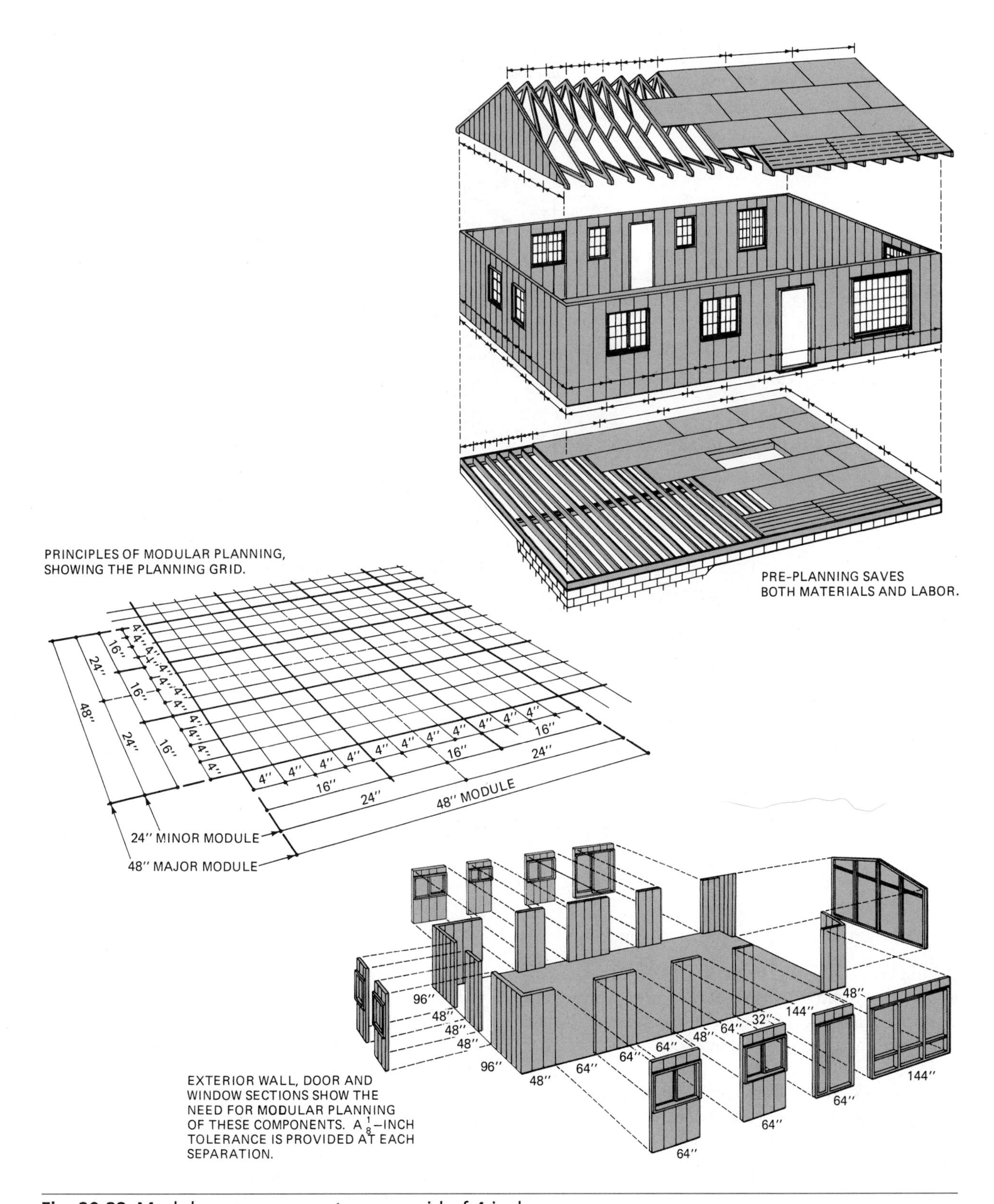

Fig. 20-22 Modular measurement uses a grid of 4 inches.

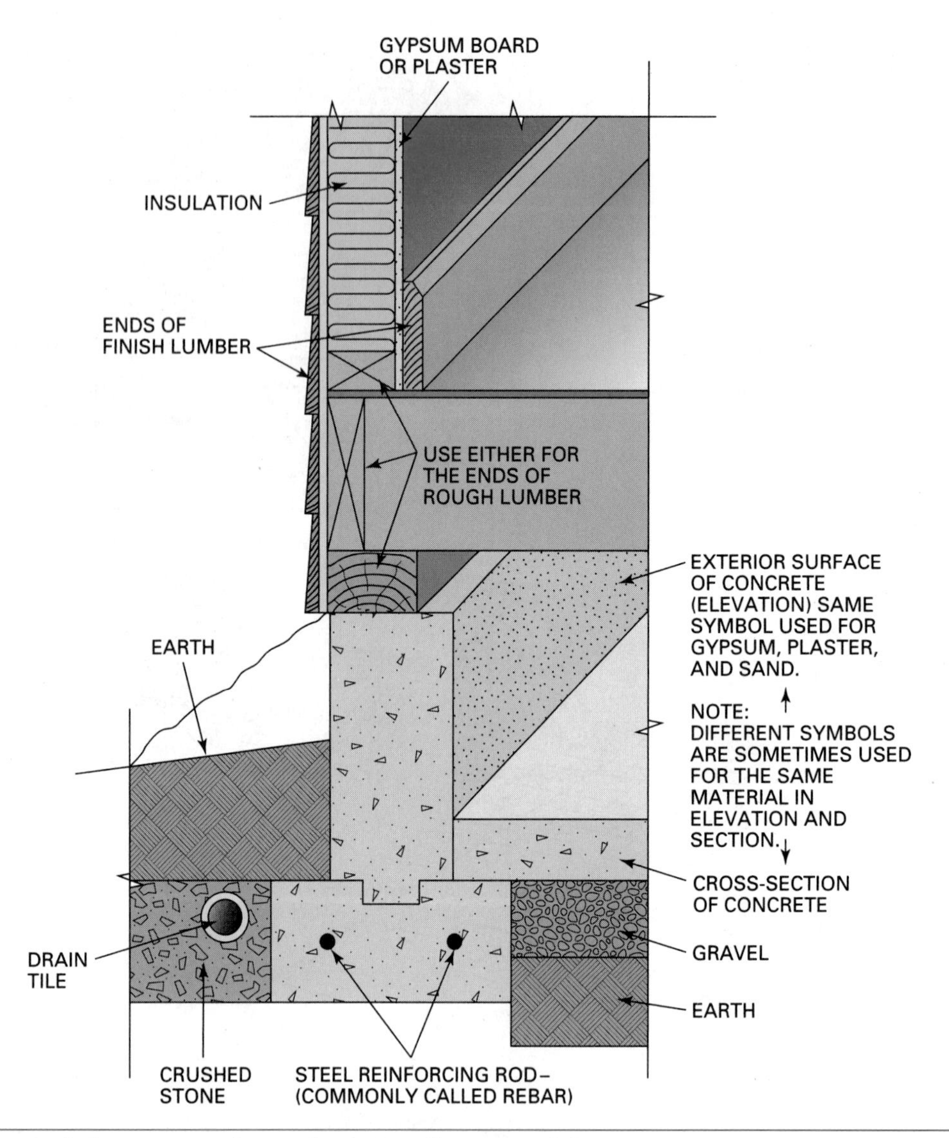

Fig. 20-23 Symbols for commonly used construction materials.

grid spaces. Drawing plans to a modular measure enables the builder to use manufactured component parts with less waste, such as 4 x 8 sheet materials and manufactured wall, floor, and roof sections that fit together with greater precision.

The spacing of framing members and the location and size of windows and doors adhering to the concept of modular measurement cut down cost and conserve materials.

Symbols

Symbols are used on drawings to represent objects in the building, such as doors, windows, cabinets, plumbing, and electrical fixtures. Others are used in regard to the construction, such as for walls, stairs, fireplaces, and electrical circuits. They may be used for identification purposes, such as those used for section reference lines. The symbols for various construction materials, such as lumber, concrete, sand, and earth (Fig. 20-23), are used when they make the drawing easier to read. (More detailed illustration, description, and use of architectural symbols are presented in following units where appropriate.)

Abbreviations

Architects find it necessary to use abbreviations on drawings to conserve space. Only capital letters, such as DR for door, are used. Abbreviations that make an actual word, such as FIN. for finish, are followed by a period. Several words may use the same abbreviation, such as W for west, width, or with. The *location* of these abbreviations is the key to their meaning. A list of commonly used abbreviations is shown in Figure 20-24.

Access Panel	AP
Acoustic	ACST
Acoustical Tile	AT
Aggregate	AGGR
Air Conditioning	AIR COND
Aluminum	AL
Anchor Bolt	AB
Angle	
Apartment	APT
Approximate	APPROX
Architectural	ARCH
Area	A
Area Drain	AD
Asbestos	ASB
Asbestos Board	AB
Asphalt	ASPH
Asphalt Tile	AT
Basement	BSMT
Bathroom	B
Bathtub	BT
Beam	BM
Bearing Plate	BRG PL
Bedroom	BR
Blocking	BLKG
Blueprint	BP
Boiler	BLR
Book Shelves	BK SH
Brass	BRS
Brick	BRK
Bronze	BRZ
Broom Closet	BC
Building	BLDG
Building Line	BL
Cabinet	CAB
Calking	CLKG
Casing	CSG
Cast Iron	CI
Cast Stone	CS
Catch Basin	CB
Cellar	CEL
Cement	CEM
Cement Asbestos Board	CEM AB
Cement Floor	CEM FL
Cement Mortar	CEM MORT
Center	CTR
Center to Center	C to C
Center Line	or CL
Center Matched	CM
Ceramic	CER
Channel	CHAN
Cinder Block	CIN BL
Circuit Breaker	CIR BKR
Cleanout	CO
Cleanout Door	COD
Clear Glass	CL GL
Closet	C, CL, or CLO
Cold Air	CA
Cold Water	CW
Collar Beam	COL B
Concrete	CONC
Concrete Block	CONC B
Concrete Floor	CONC FL
Conduit	CND
Construction	CONST
Contract	CONT
Copper	COP
Counter	CTR
Cubic Feet	CU FT
Cut Out	CO
Detail	DET
Diagram	DIAG
Dimension	DIM
Dining Room	DR
Dishwasher	DW
Ditto	DO
Double-Acting	DA
Double Strength Glass	DSG
Down	DN
Downspout	DS
Drain	D or DR
Drawing	DWG
Dressed and Matched	D & M
Dryer	D
Electric Panel	EP
End to End	E to E
Excavate	EXC
Expansion Joint	EXP JT
Exterior	EXT
Finish	FIN
Finished Floor	FIN FL
Firebrick	FBRK
Fireplace	FP
Fireproof	FPRF
Fixture	FIX
Flashing	FL
Floor	FL
Floor Drain	FD
Flooring	FLG
Fluorescent	FLUOR
Flush	FL
Footing	FTG
Foundation	FND
Frame	FR
Full Size	FS
Furring	FUR
Galvanized Iron	GI
Garage	GAR
Gas	G
Glass	GL
Glass Block	GL BL
Grille	G
Gypsum	GYP
Hardware	HDW
Hollow Metal Door	HMD
Hose Bib	HB
Hot Air	HA
Hot Water	HW
Hot Water Heater	HWH
I Beam	I
Inside Diameter	ID
Insulation	INS
Interior	INT
Iron	I
Jamb	JB
Kitchen	K
Landing	LDG
Lath	LTH
Laundry	LAU
Laundry Tray	LT
Lavatory	LAV
Leader	L
Length	L, LG, or LNG
Library	LIB
Light	LT
Limestone	LS
Linen Closet	L CL
Lining	LN
Living Room	LR
Louver	LV
Main	MN
Marble	MR
Masonry Opening	MO
Material	MATL
Maximum	MAX
Medicine Cabinet	MC
Minimum	MIN
Miscellaneous	MISC
Mixture	MIX
Modular	MOD
Mortar	MOR
Moulding	MLDG
Nosing	NOS
Obscure Glass	OBSC sL
On Center	OC
Opening	OPNG
Outlet	OUT
Overall	OA
Overhead	OVHD
Pantry	PAN
Partition	PTN
Plaster	PL or PLAS
Plastered Opening	PO
Plate	PL
Plate Glass	PL GL
Platform	PLAT
Plumbing	PLBG
Plywood	PLY
Porch	P
Precast	PRCST
Prefabricated	PREFAB
Pull Switch	PS
Quarry Tile Floor	QTF
Radiator	RAD
Random	RDM
Range	R
Recessed	REC
Refrigerator	REF
Register	REG
Reinforce or Reinforcing	REINF
Revision	REV
Riser	R
Roof	RF
Roof Drain	RD
Room	RM or R
Rough	RGH
Rough Opening	RO
Rubber Tile	R TILE
Scale	SC
Schedule	SCH
Screen	SCR
Scuttle	S
Section	SECT
Select	SEL
Service	SERV
Sewer	SEW
Sheathing	SHTHG
Sheet	SH
Shelf and Rod	SH & RD
Shelving	SHELV
Shower	SH
Sill Cock	SC
Single Strength Glass	SSG
Sink	SK or S
Soil Pipe	SP
Specification	SPEC
Square Feet	SQ FT
Stained	STN
Stairs	ST
Stairway	STWY
Standard	STD
Steel	ST or STL
Steel Sash	SS
Storage	STG
Switch	SW or S
Telephone	TEL
Terra Cotta	TC
Terrazzo	TER
Thermostat	THERMO
Threshold	TH
Toilet	T
Tongue and Groove	T & G
Tread	TR or T
Typical	TYP
Unfinished	UNF
Unexcavated	UNEXC
Utility Room	URM
Vent	V
Vent Stack	VS
Vinyl Tile	V TILE
Warm Air	WA
Washing Machine	WM
Water	W
Water Closet	WC
Water Heater	WH
Waterproof	WP
Weather Stripping	WS
Weephole	WH
White Pine	WP
Wide Flange	WF
Wood	WD
Wood Frame	WF
Yellow Pine	YP

Fig. 20-24 Commonly used abbreviations.

CHAPTER 21 FLOOR PLANS

The chapters in this unit are based on a set of drawings for a two-story, three-bedroom residence (Fig. 21-1). Part of the basement is an activity room.

Floor Plans

Floor plans (Fig. 21-2) contain a substantial amount of information. They are used more than any other kind of drawing. After consideration of many factors that determine the size and shape of the building, floor plans are drawn first. Others, such as the foundation plan and elevations, are derived from it. They are generally drawn at a scale of 1/4″ = 1′-0″ or 1:50. A separate plan is made for each floor of buildings with more than one story.

Floor Plan Symbols

In order to make the plan as uncluttered as possible, numerous *symbols* are used. Recognition of commonly used symbols makes it easier to read the floor plan as well as other plans that use the same symbols. (Symbols used in elevation and section drawings are different from plan symbols. They are described in following chapters.)

Door Symbols. Symbols for exterior doors are drawn with a line representing the outside edge of the sill. Interior door symbols show no sill line. The symbols in Figure 21-3 identify the *swing* and show on which side of the opening to hang the door.

Similarly, exterior *sliding* door symbols show the sill line. The symbols for interior sliding doors, called *bypass* doors, show none. **Pocket doors** slide inside the wall (Fig. 21-4).

Bifold and *accordian* doors open to almost the full width of a closet opening. They are used when complete access to the closet is desired. The sections or panels of the doors are clearly seen in the symbols (Fig. 21-5).

Window Symbols. The inside and outside lines of window symbols represent the edges of the window sill. In between, other lines are drawn for the panes of glass. A window with a fixed, single

Fig. 21-1 The plans for this residence are used throughout this unit. (*From Jefferis and Madsen,* Architectural Drafting & Design, *1991, by Delmar.*)

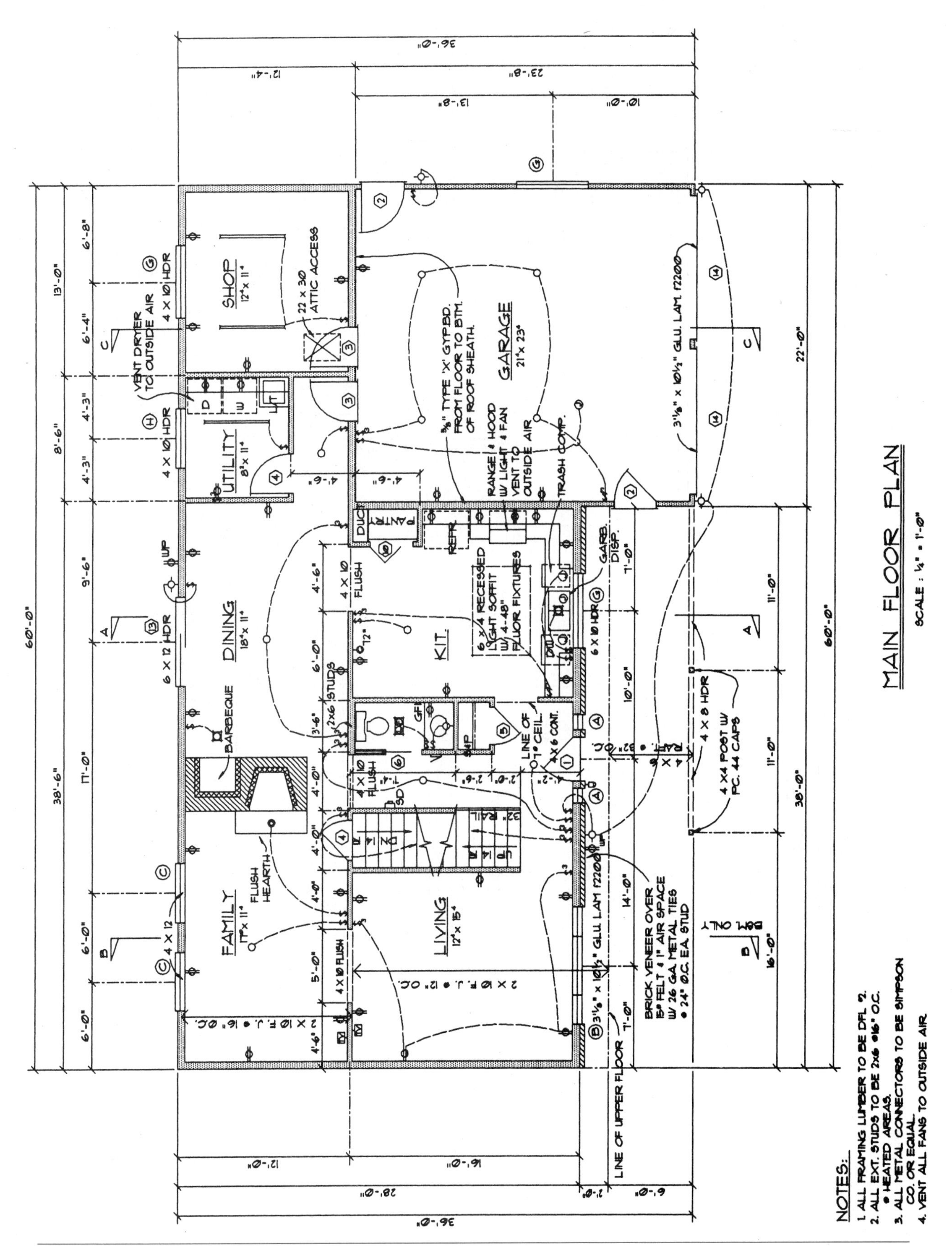

Fig. 21-2 First and second floor plans. (*From Jefferis and Madsen,* Architectural Drafting & Design, *1991, by Delmar.*)

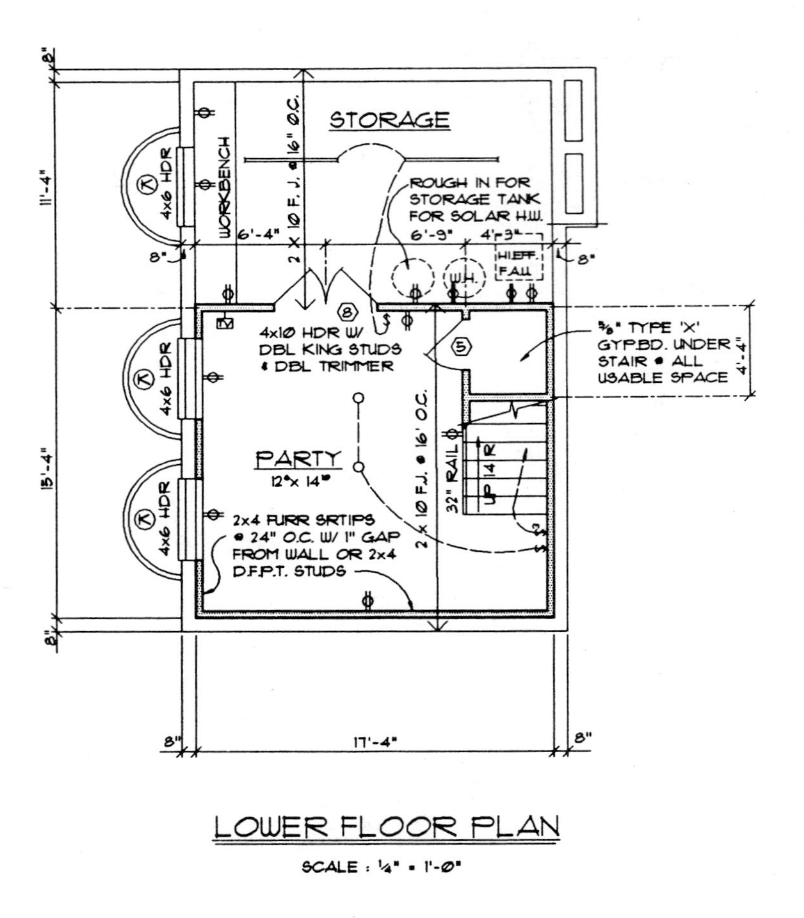

WINDOW SCHEDULE

SYM.	SIZE	MODEL	ROUGH OPEN	QUAN
A	$1^{0} \times 5^{0}$	JOB-BUILT		2
B	$8^{0} \times 5^{0}$	W4N5 CSM	8'-2¾"x5'-5⅞"	1
C	$4^{0} \times 5^{0}$	W2N5 CSM	4'-2¼"x5'-5⅞"	2
D	$4^{0} \times 3^{6}$	W2N3 CSM	4'-2¼"x3'-5½"	2
E	$3^{6} \times 3^{6}$	2N3 CSM	3'-5¼"x3'-5½"	2
F	$6^{0} \times 4^{0}$	G64 SLDG	6'-0½"x4'-0½"	1
G	$5^{0} \times 3^{6}$	G536 SLDG	5'-0½"x3'-6½"	4
H	$4^{0} \times 3^{6}$	G436 SLDG	4'-0½"x3'-6½"	1
J	$4^{0} \times 2^{0}$	A41 AWN	4-0½"x2'-0⅝"	3
K	$4^{0} \times 2^{0}$	G42 SLDG	4'-0½"x2'-0½"	3

DOOR SCHEDULE

SYM.	SIZE	TYPE	QUAN
1	$3^{0} \times 6^{8}$	S.C. R.P. METAL INSULATED	1
2	$3^{0} \times 6^{8}$	S.C.-FLUSH-METAL INSUL	2
3	$2^{8} \times 6^{8}$	S.C-SELF CLOSING	2
4	$2^{8} \times 6^{8}$	H.C.	5
5	$2^{6} \times 6^{8}$	H.C.	5
6	$2^{6} \times 6^{8}$	POCKET	2
7	$2^{4} \times 6^{8}$	POCKET	1
8	PR $2^{6} \times 6^{8}$	H.C.	1
9	$5^{0} \times 6^{8}$	BI-PASS	2
10	$3^{0} \times 6^{8}$	BI-FOLD	1
11	$4^{0} \times 6^{8}$	BI-FOLD	1
12	$2^{0} \times 6^{0}$	SHATTER PROOF	1
13	$6^{0} \times 6^{8}$	WOOD FRAME-TEMP. SLDG. GL.	1
14	$9^{0} \times 7^{0}$	OVERHEAD GARAGE	2

Fig. 21-2 *(continued)*

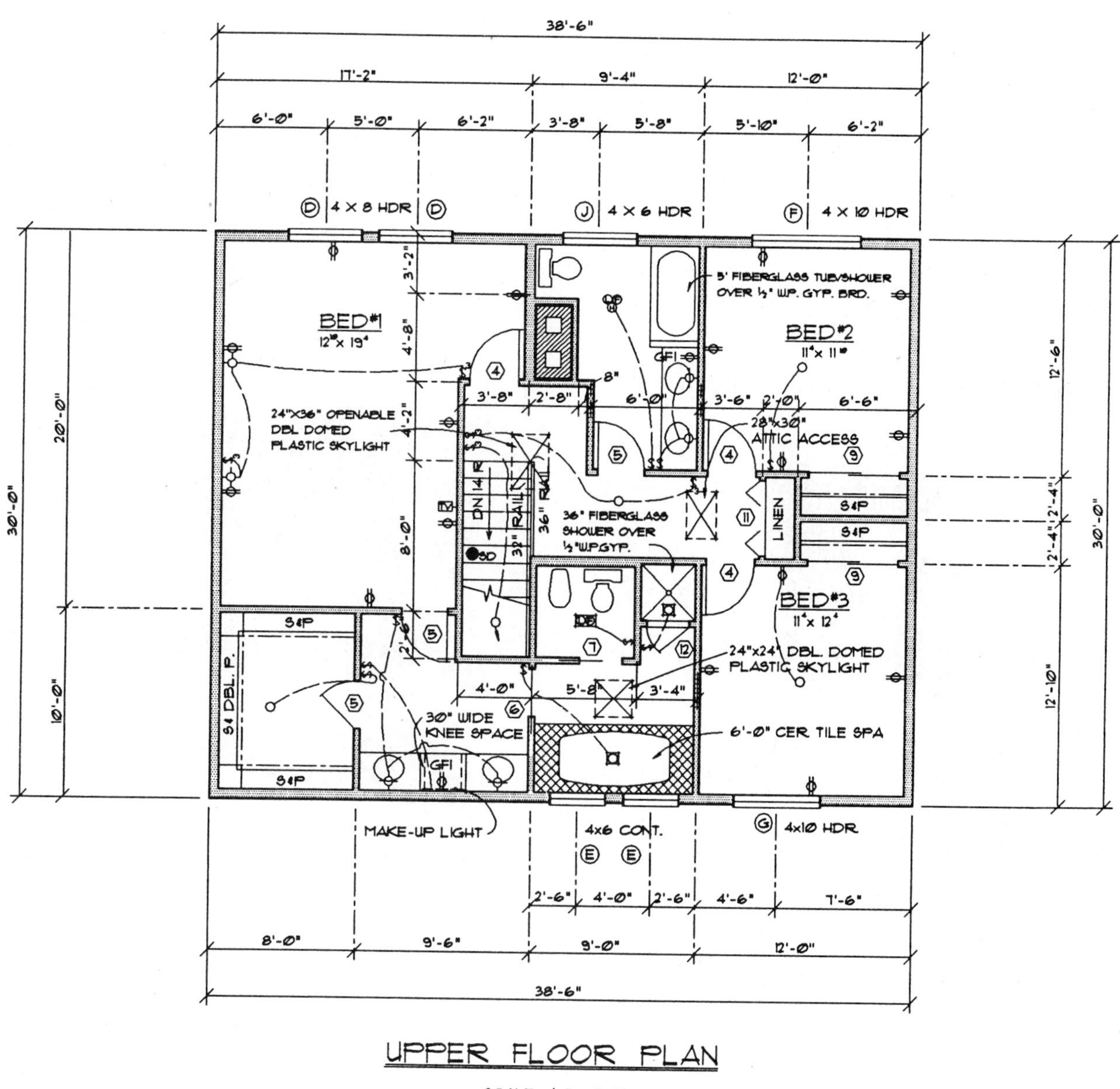

Fig. 21-2 (*continued*)

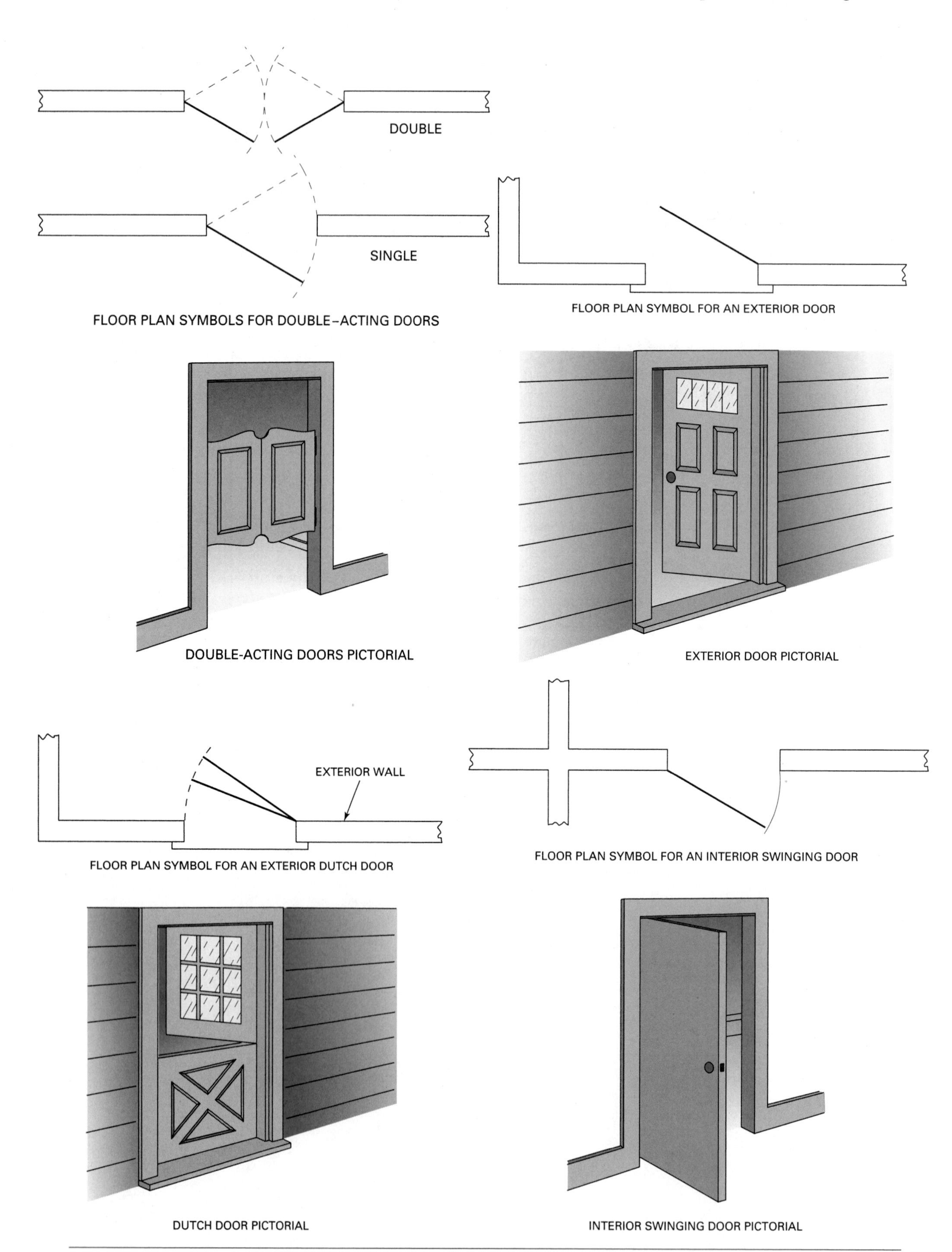

Fig. 21-3 Floor plan symbols for exterior and interior swinging doors.

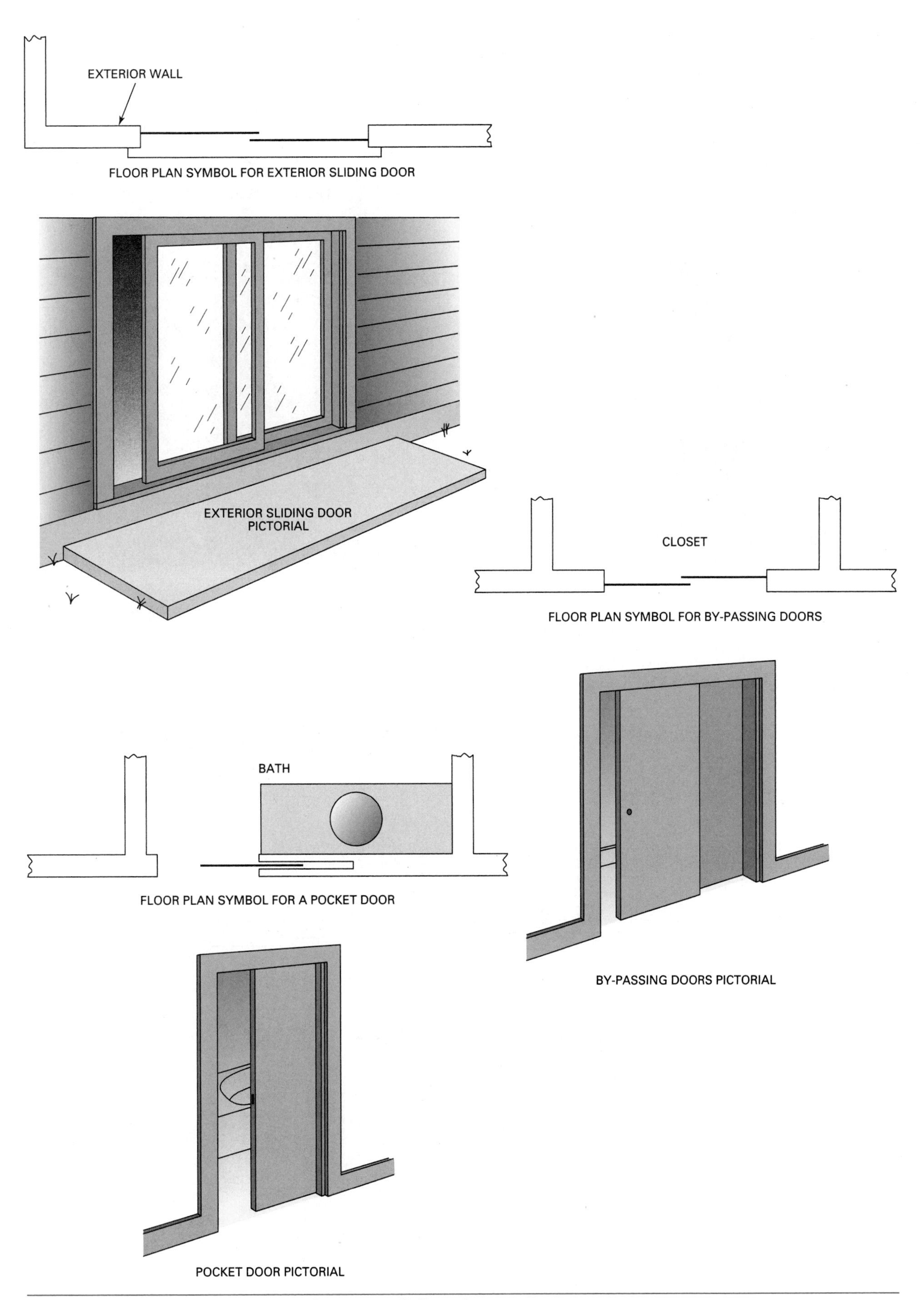

Fig. 21-4 Symbols for exterior and interior sliding doors.

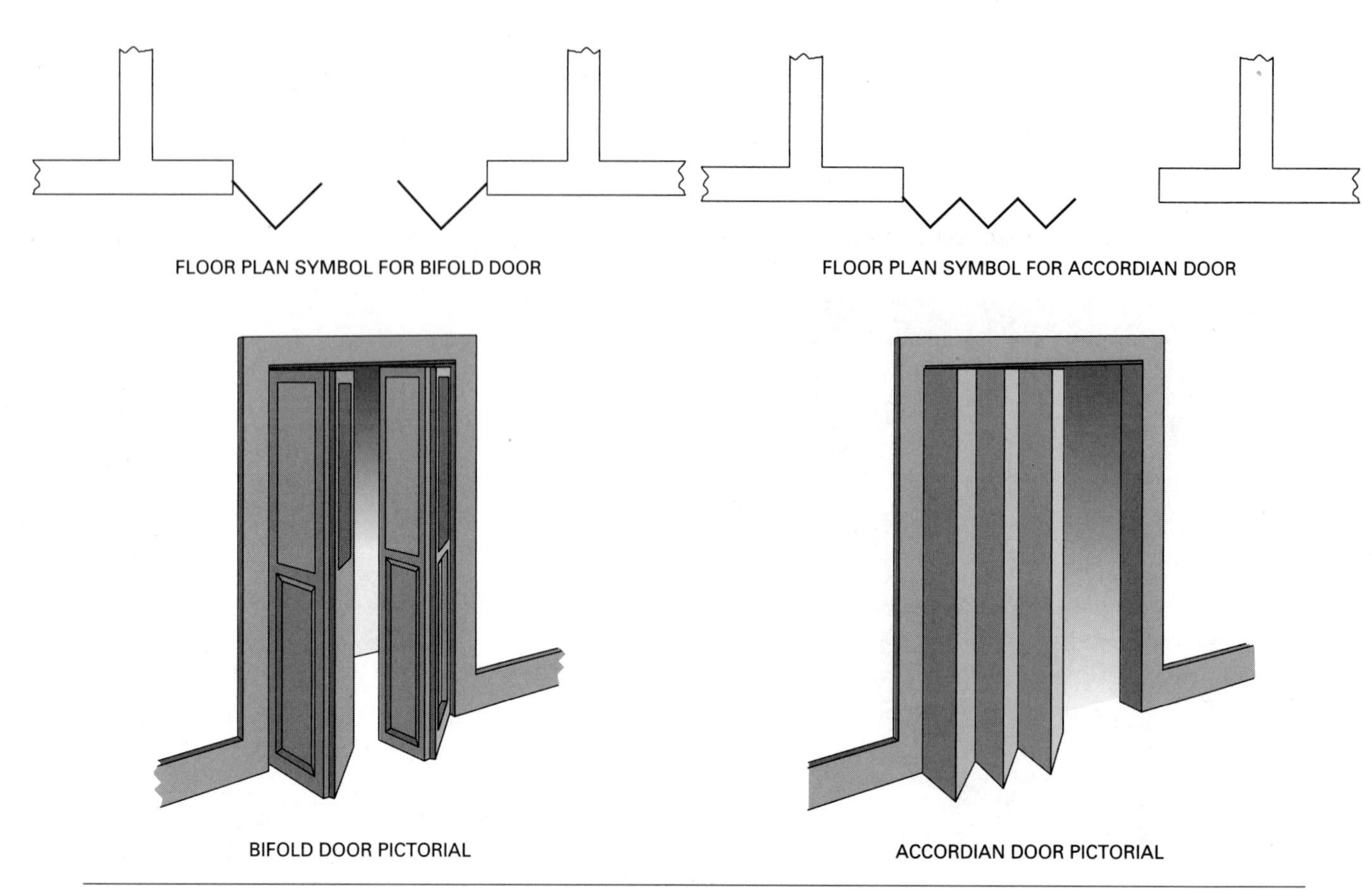

Fig. 21-5 Bifold and accordian doors are used on closets and wardrobes.

sash is indicated by one line. Because the **double-hung window** has two sash that slide vertically, its symbol shows two lines (Fig. 21-6).

The **casement window,** which swings outward, is depicted by a symbol similar to that for a swinging door. They may be shown having two or

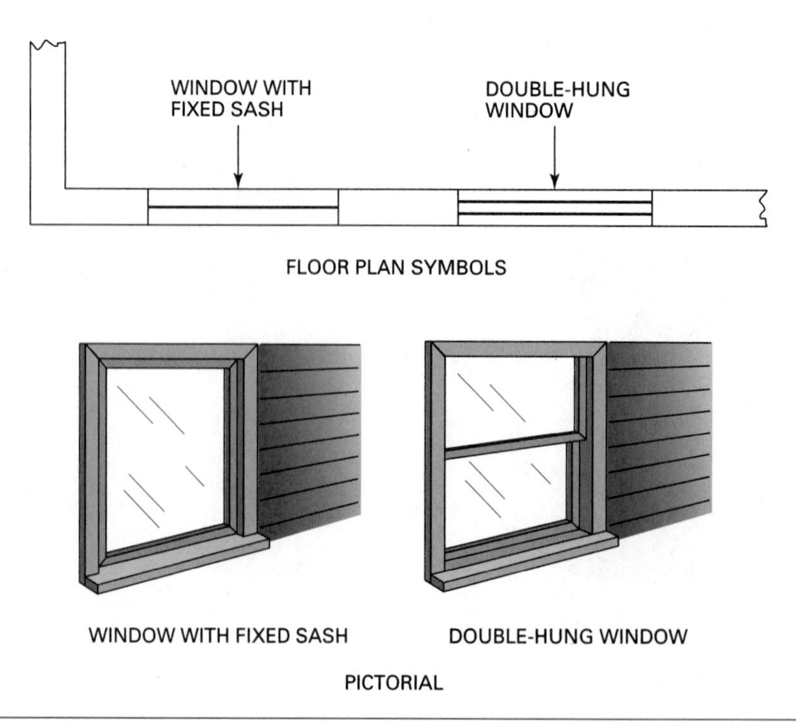

Fig. 21-6 Fixed sash and double-hung window symbols.

more units in each window (Fig. 21-7). An **awning window** is similar to a casement except it swings outward from the top. Its open position is indicated by dashed lines (Fig. 21-8).

The symbol for a *sliding window* is similar to that for a sliding door except for a line indicating the inside edge of the sill (Fig. 21-9).

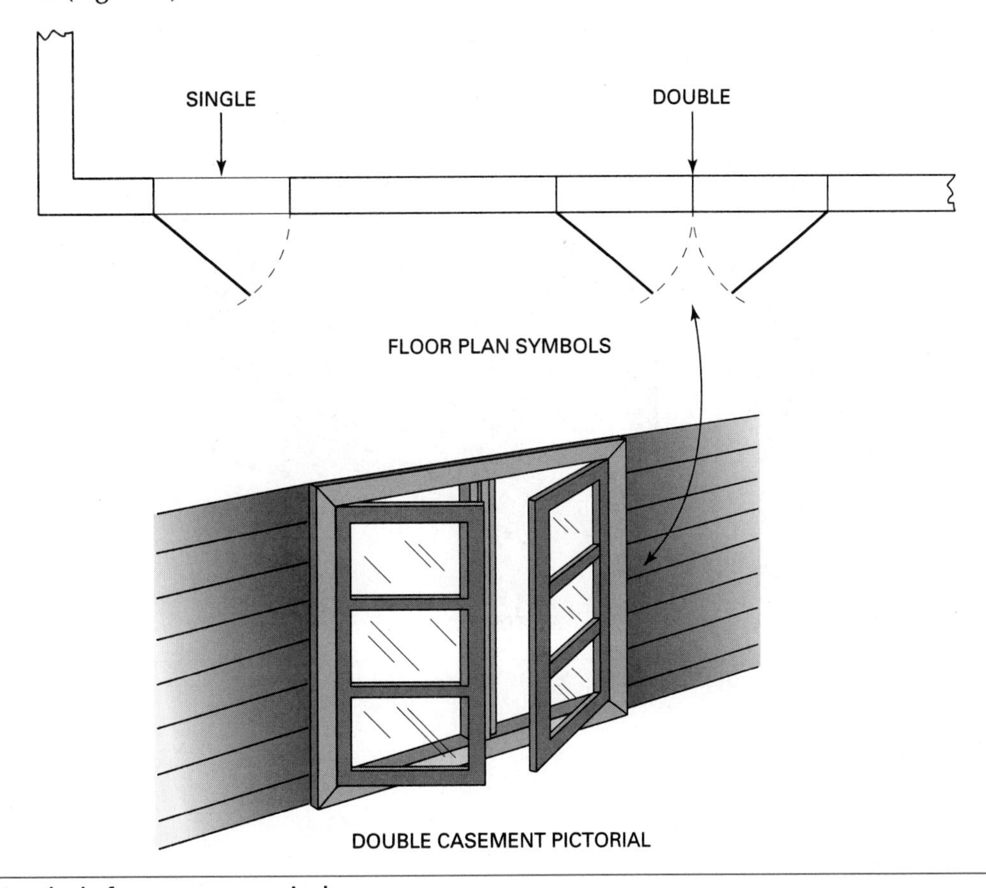

Fig. 21-7 Symbols for casement windows.

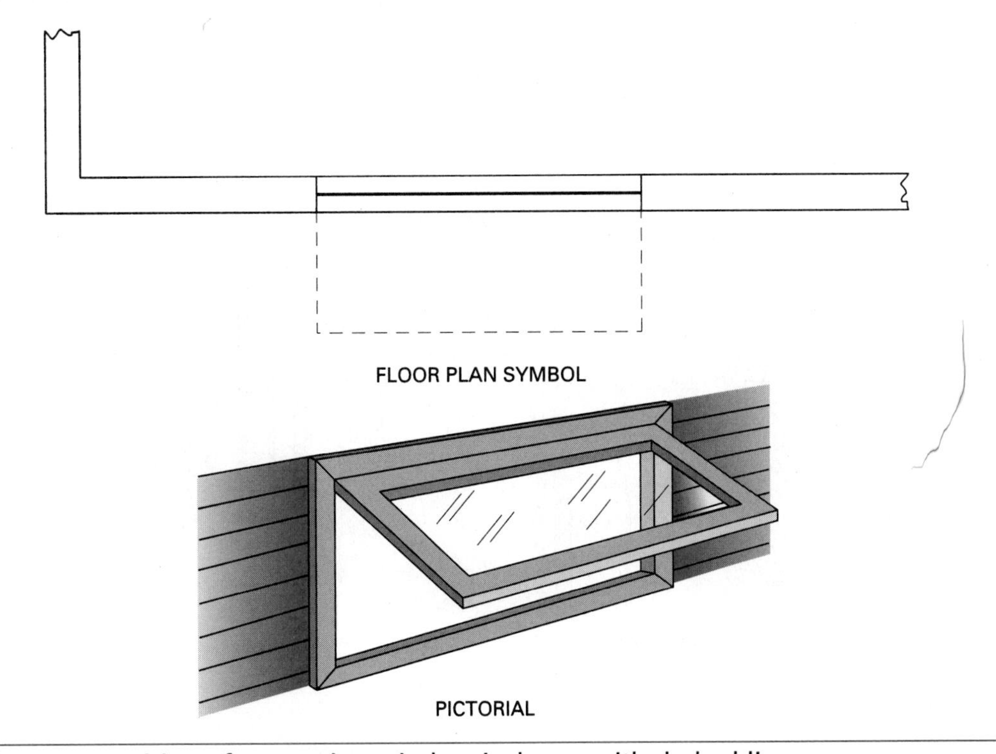

Fig. 21-8 The open position of an awning window is shown with dashed lines.

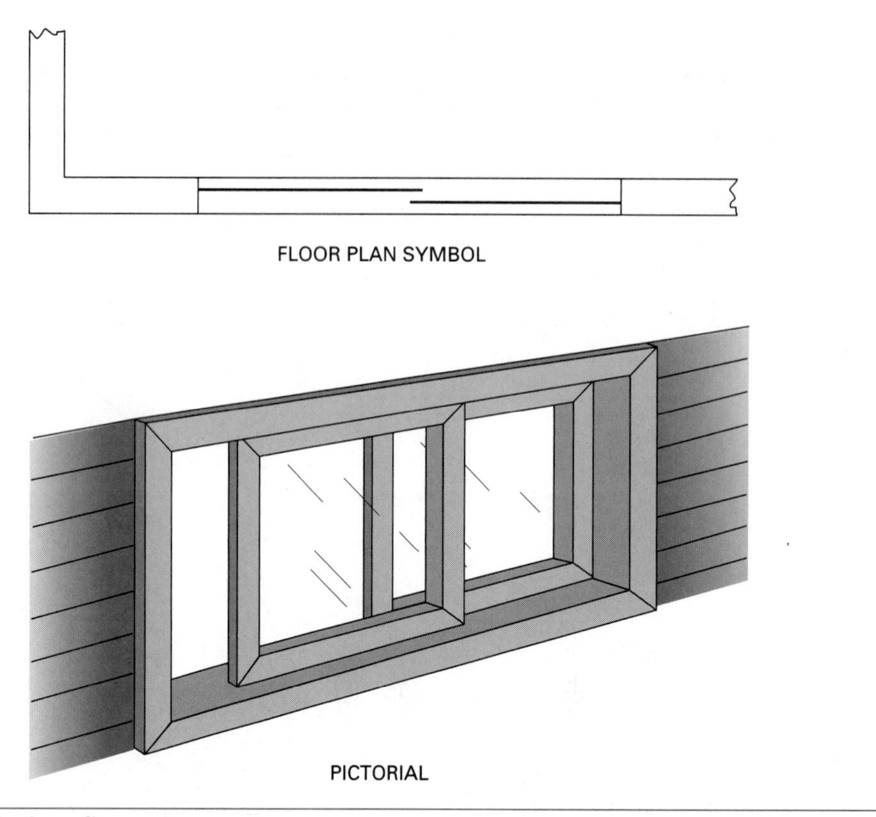

Fig. 21-9 Sliding window floor plan symbol.

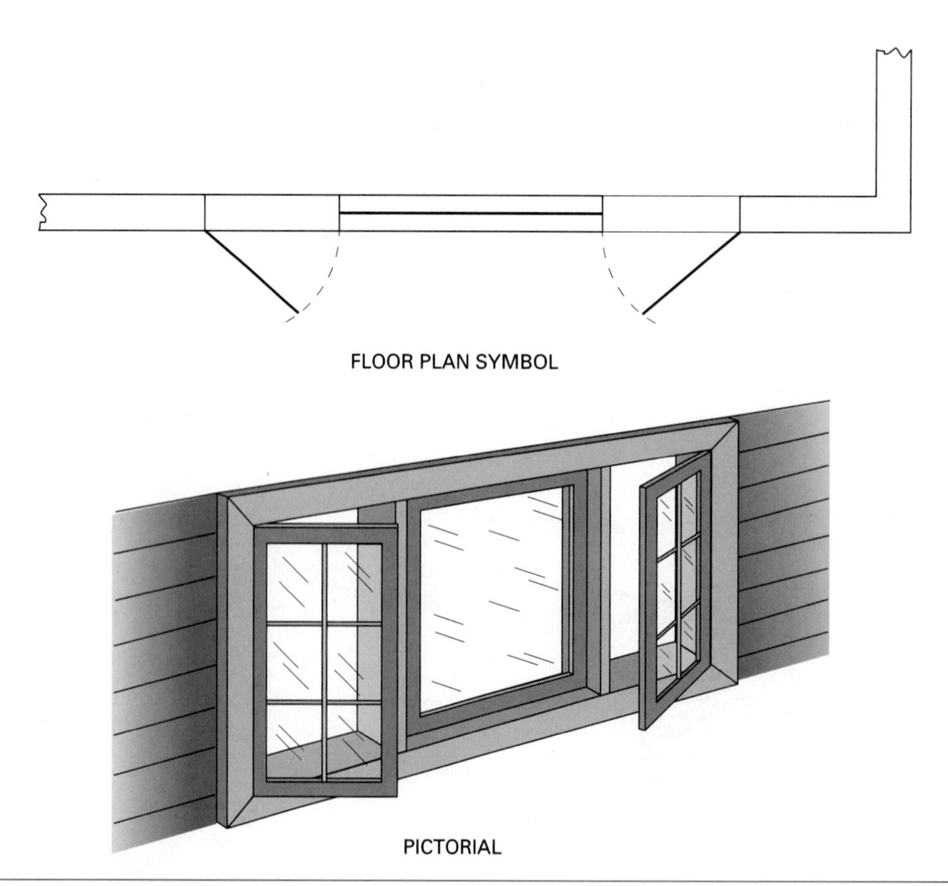

Fig. 21-10 A window consisting of a fixed sash and casement units.

Different kinds of windows may be used in combination. Figure 21-10 shows a window with a fixed sash, with casements on both sides. Main entrances may consist of a door with a **sidelight** on one or both sides (Fig. 21-11).

Close to the window and door symbols are letters and numbers that identify the units in the window and door schedules.

Structural Members. Openings without doors in interior walls for passage from one area to another are indicated by dashed lines. Notations are given if the opening is to be *cased* (trimmed with molding) and if the top is to be *arched* or treated in any other manner (Fig. 21-12).

The location of garage door *headers* may be shown by a series of dashes. Their size is usually indicated with a notation (Fig. 21-13). Window and other headers are shown in the same manner.

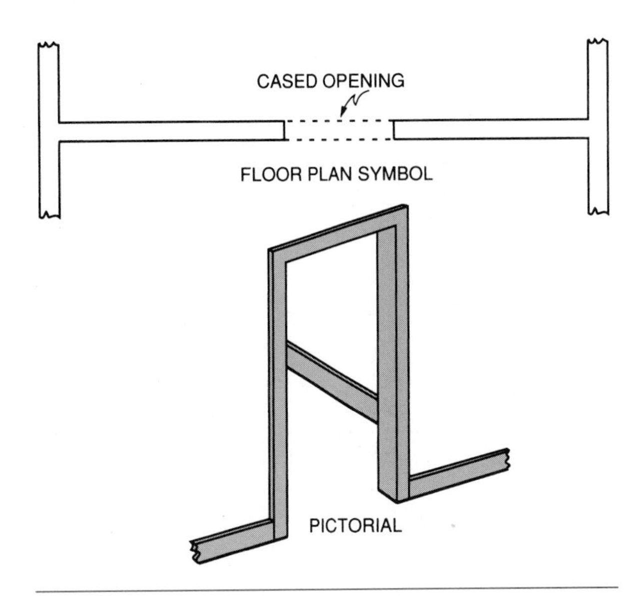

Fig. 21-12 Dashed lines indicate an interior wall opening without a door.

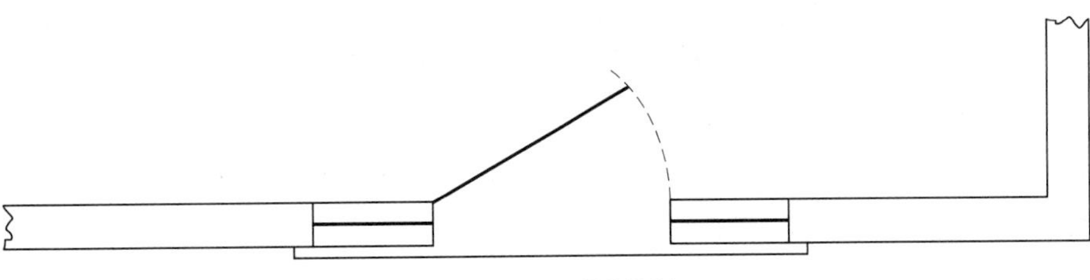

Fig. 21-11 An entrance door flanked by sidelights.

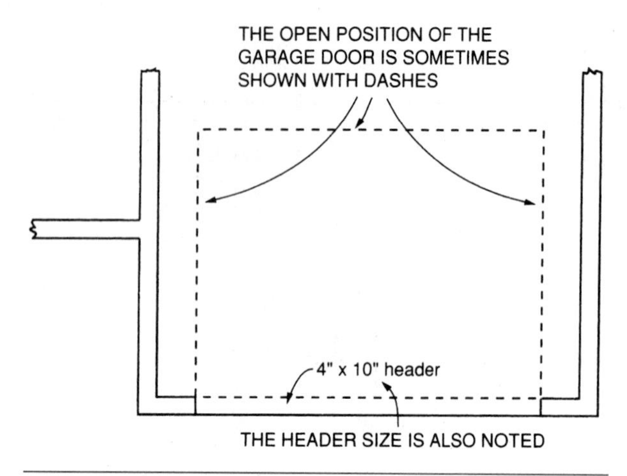

Fig. 21-13 The symbol for a garage door header is a dashed line.

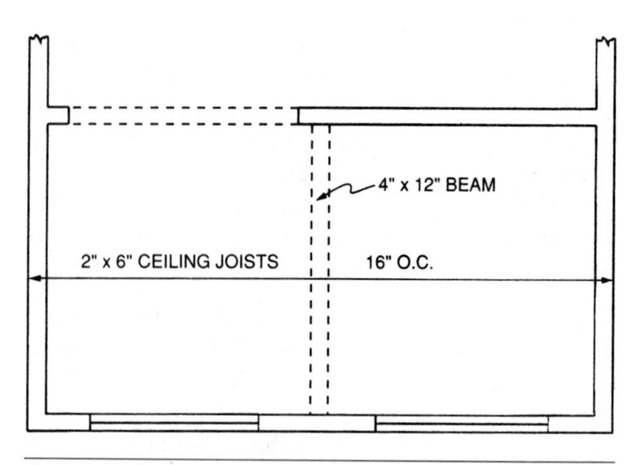

Fig. 21-14 Symbols for ceiling beams and ceiling joists.

Ceiling beams above the cutting plane of the floor, which support the ceiling joists, are also represented by a series of dashes. *Ceiling joists* or trusses are identified in the floor plan with a double-ended arrow showing the direction in which they run. Their size and spacing are noted alongside the arrow (Fig. 21-14).

Kitchen, Bath, and Utility Room.

The location of *bathroom* and *kitchen fixtures,* such as sinks, tubs, refrigerators, stoves, washers, and dryers, are shown by obvious symbols, abbreviations, and notations. The extent of the *base cabinets* is indicated by a line indicating the edge of the countertop. Objects such as dishwashers, trash compactors, and lazy susans are shown by a dashed line and notations or abbreviations. The *upper cabinets* are symbolized by dashed lines (Fig. 21-15).

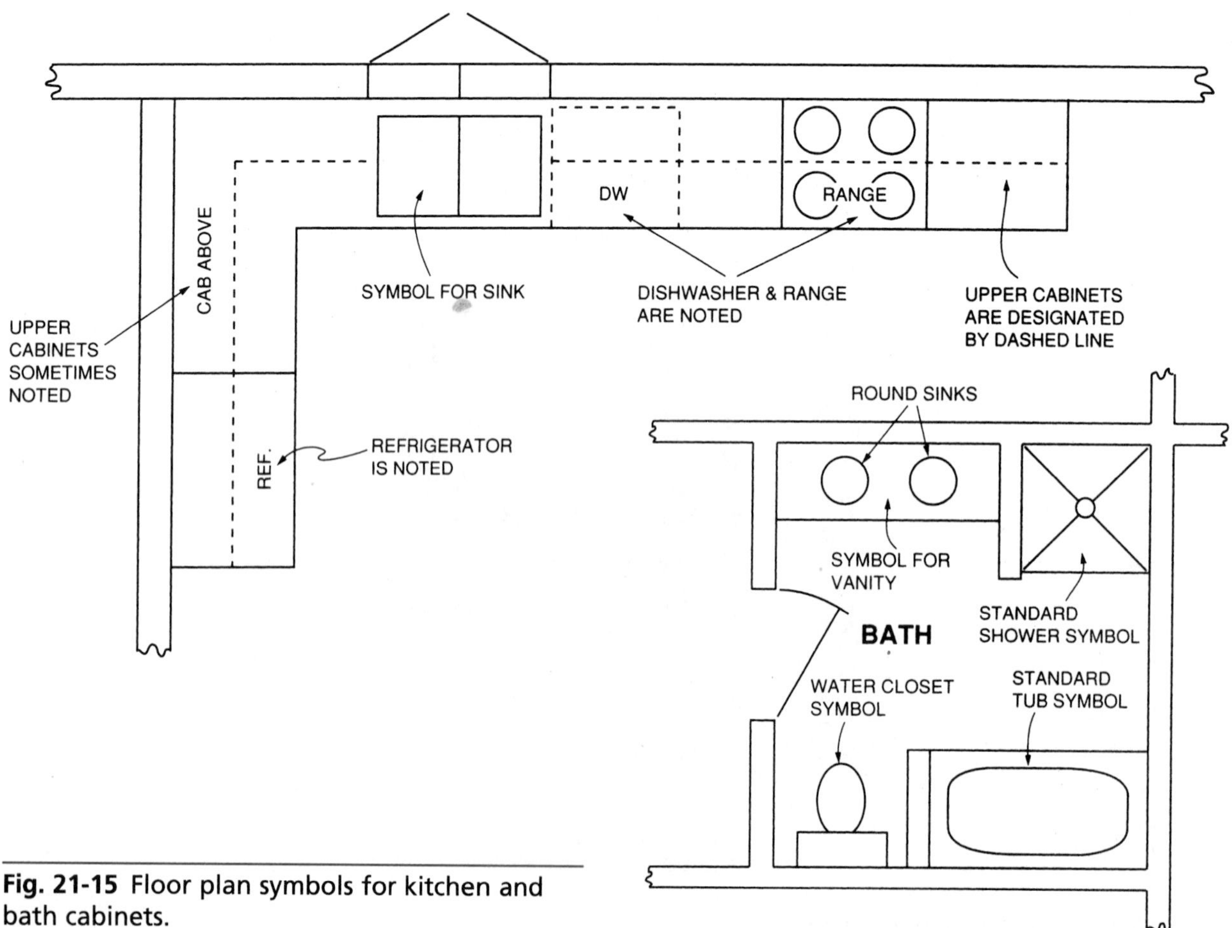

Fig. 21-15 Floor plan symbols for kitchen and bath cabinets.

Other Floor Plan Symbols. The floor plan also shows the location of *stairways.* Lines indicate the outside edges of the **treads.** Also shown is the direction of travel and the number of **risers** vertical distance from tread to tread) in the staircase (Fig. 21-16).

The location and style of *chimneys, fireplaces,* and **hearths** are shown by the use of appropriate symbols. Fireplace dimensions are generally not given. The size of the chimney flue and the hearth is usually indicated. The fireplace material may be shown by symbols according to the material specified (Fig. 21-17). The kind of material may also be identified with a notation.

An *attic access,* also called a **scuttle,** is usually located in a closet or hall. It is outlined with dashed

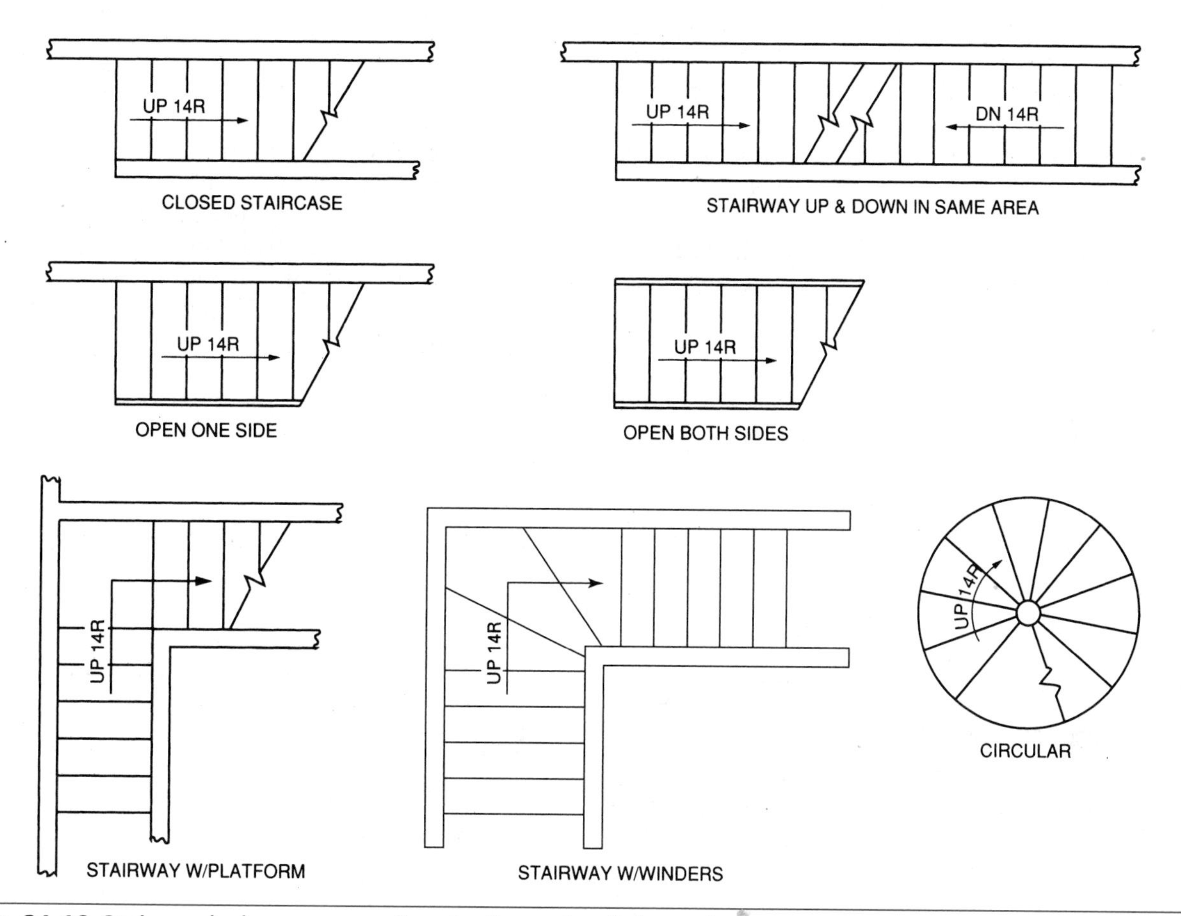

Fig. 21-16 Stair symbols vary according to the style of the staircase.

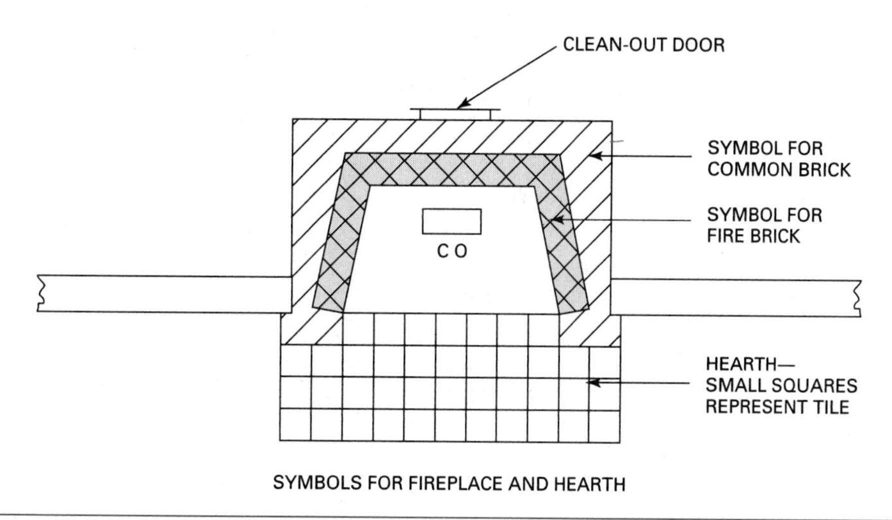

Fig. 21-17 Symbols for fireplace and hearth.

lines. It may also be identified by a notation (Fig. 21-18).

The floor plan symbol for a *floor drain* is a small circle, square, or circle within a square. The slope of the floor is shown by straight lines from the corners of the floor to the center of the drain. Floor drains are appropriately installed in utility rooms where washers, dryers, laundry tubs, and water heaters are located (Fig. 21-19).

The location of outdoor water faucets, called *hose bibbs,* is shown by a symbol (Fig. 21-20) projecting from exterior walls where desired. For clarity, the symbol is labeled.

If not distracting, *electrical wiring* may be shown on the floor plan of simpler structures by the use of curved, dashed lines running to *switches, outlets,* and *fixtures* (Fig. 21-21). The symbols for these electrical components are shown in Figure 21-22. Complex buildings require separate electrical plans, as well as prints for plumbing and for heating and ventilation.

Dimensions

Dimensions are placed and printed in a manner to make them as easy to read as possible. Read dimensions carefully. A mistake in reading a dimension early in the construction process could have serious consequences later.

Exterior Dimensions. On floor plans, the *overall* dimensions of the building are found on the outer dimension lines. In a *wood frame,* the dimensions are to the outside face of the frame. *Concrete block* walls are dimensioned to their outside face. *Brick veneer* walls are dimensioned to the outside face of the wood frame, with an added dimension and notation for the veneer (Fig. 21-23).

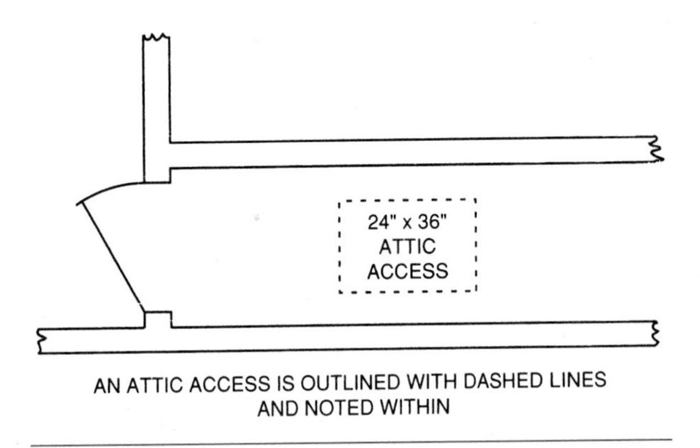

Fig. 21-18 Attic access is outlined with dashed lines.

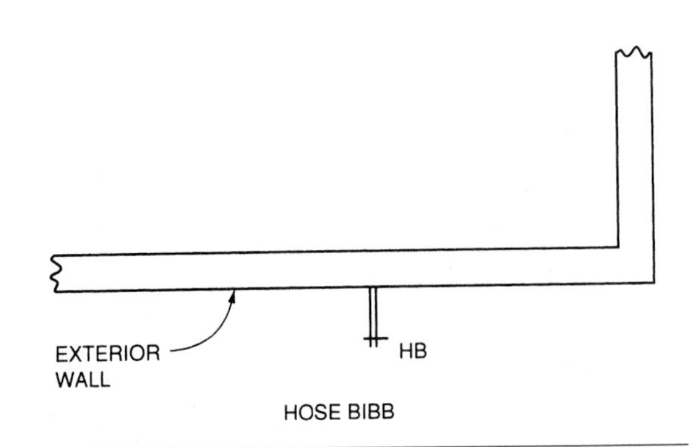

Fig. 21-20 Hose bibb.

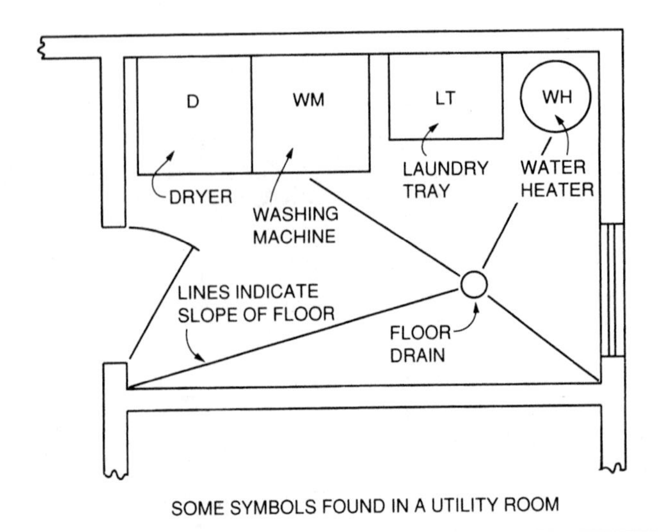

Fig. 21-19 Some symbols found in a utility room.

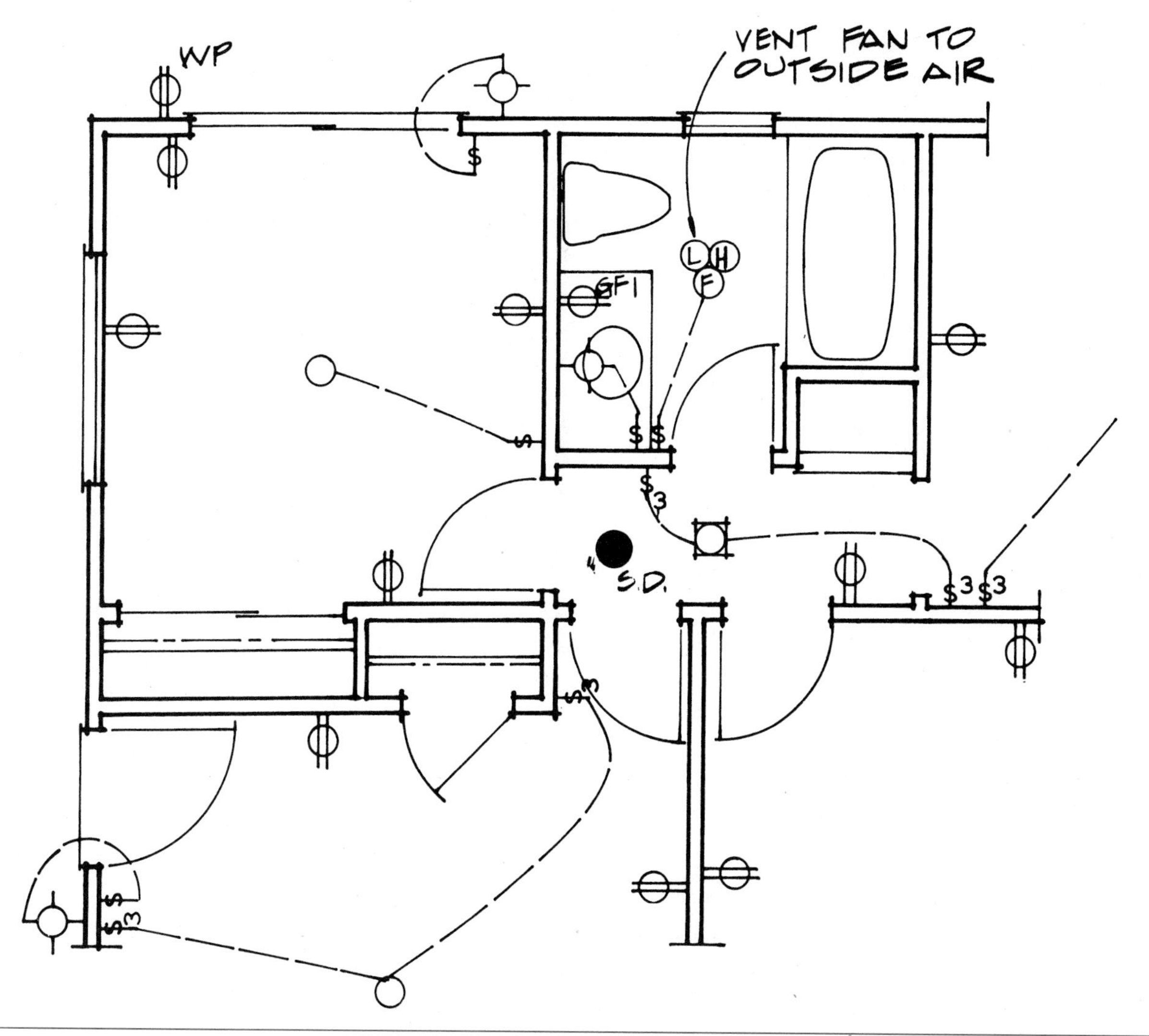

Fig. 21-21 Part of a typical electrical floor plan. (*From Jefferis and Madsen,* Architectural Drafting & Design, *1991, by Delmar.*)

Dimension lines closest to the exterior walls are used to show the location of *windows* and *doors.* In a wood frame, they are dimensioned to their centerline. In concrete block walls, the dimensions are to the edges of the openings and also show the opening width.

The second dimension line from the exterior wall is used to locate the centerline of *interior partitions,* which intersect the exterior wall.

Interior Dimensions. Dimensions are given from the outside of a wood frame or from the inside of concrete block walls, to the centerlines of **partitions.** Interior *doors* and other openings are dimensioned to their centerline similar to exterior walls (Fig. 21-24).

Not all interior dimensions are required. Some may be assumed. For instance, it can be clearly seen if a door is centered between two walls of a closet or hallway (Fig. 21-25).

If a door is to be located a minimum distance from a wall, place the door opening near enough to the wall so that any door **casing** later applied can be scribed (fitted) to the wall. Or place the opening far enough away to allow convenient finishing (painting, wallpapering) between the casing and the wall (Fig. 21-26). A small space (1 inch or less) between the outside edge of side casing and a wall is very difficult to finish.

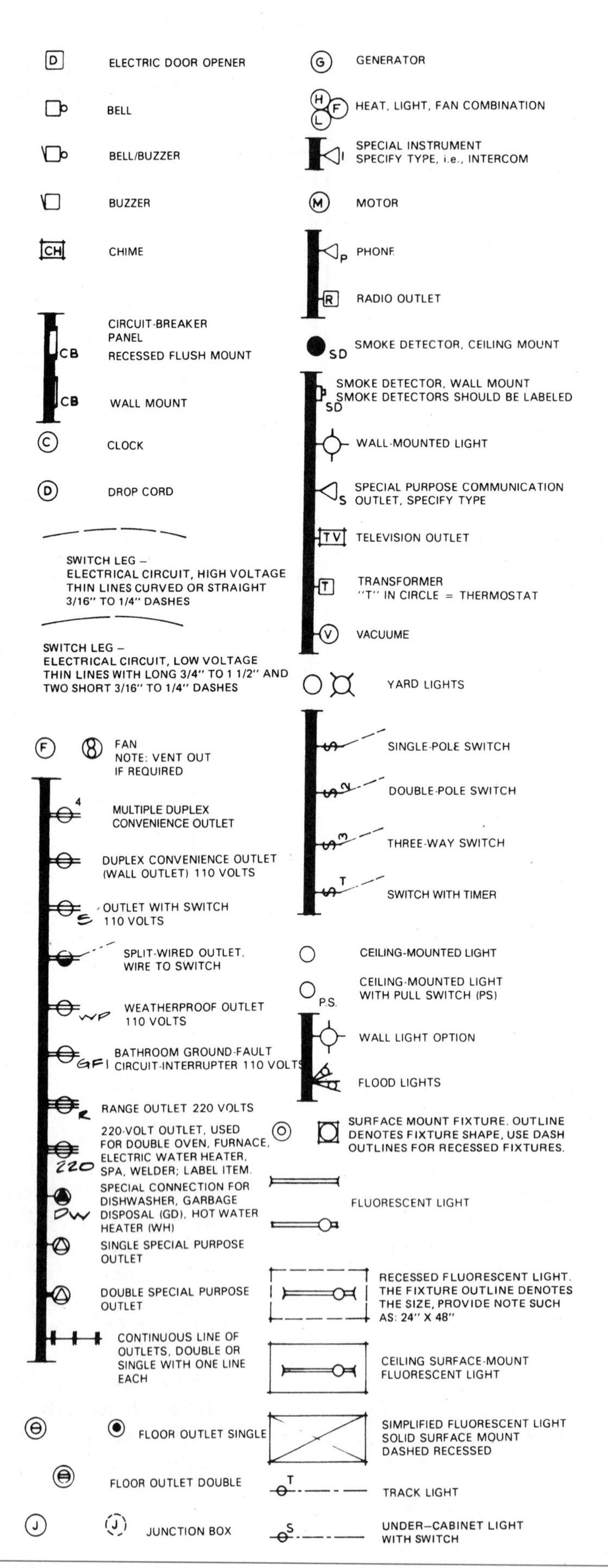

Fig. 21-22 Electrical symbols. (*From Jefferis and Madsen,* Architectural Drafting & Design, *1991, by Delmar.*)

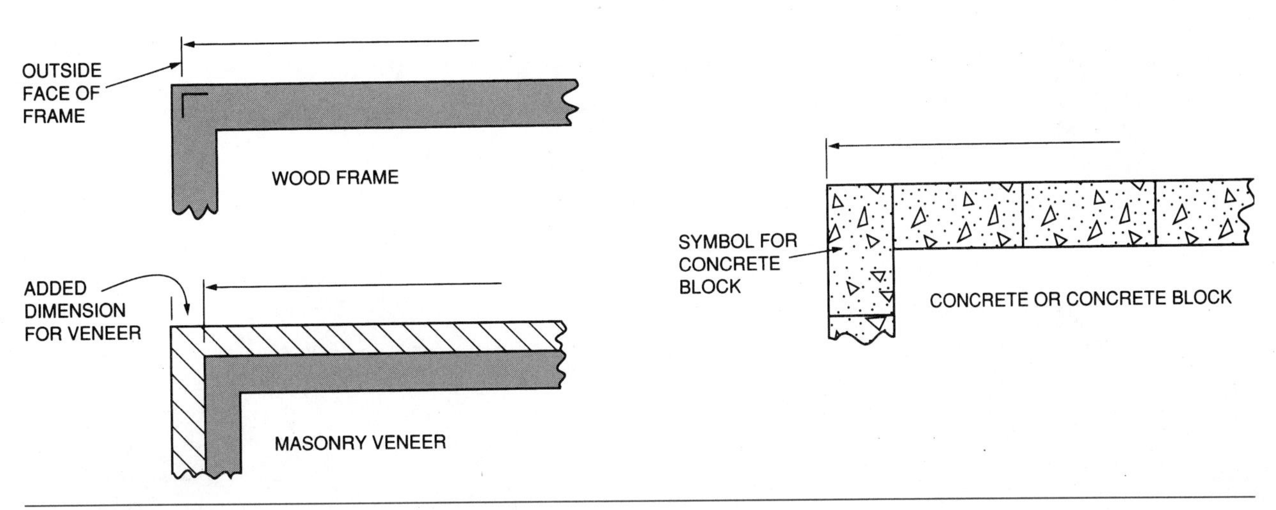

Fig. 21-23 Overall dimensions.

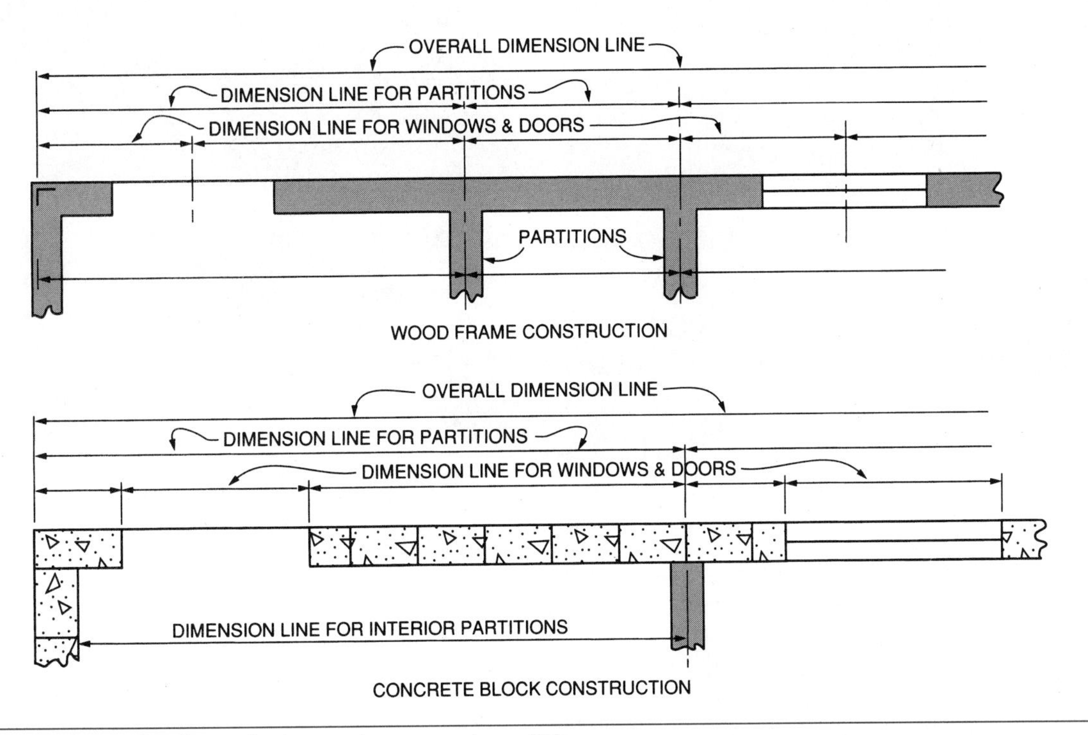

Fig. 21-24 Dimensioning windows, doors, and partitions.

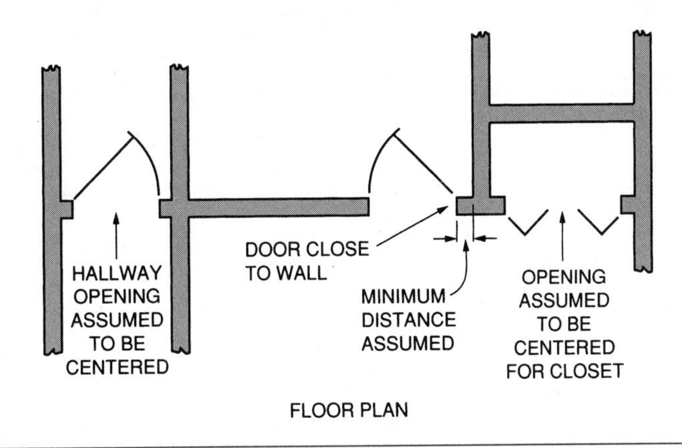

Fig. 21-25 Some dimensions are obvious and are not given.

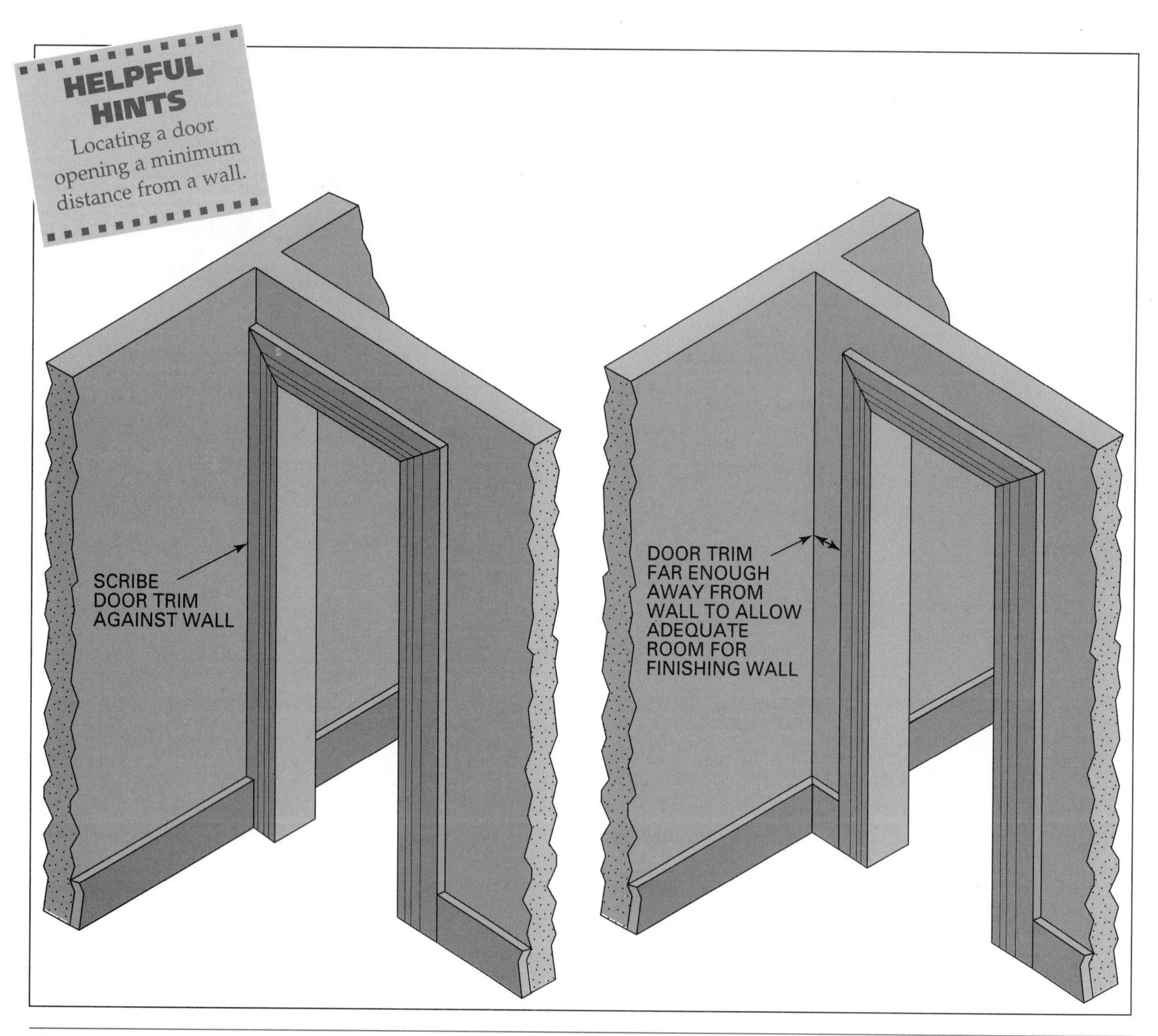

Fig. 21-26 Locating a door to allow room for finish.

CHAPTER 22 SECTIONS AND ELEVATIONS

Sections

Floor plans are views of a horizontal cut. *Sections* show *vertical* cuts called for on the floor, framing, and foundation plans (Fig. 22-1). Sections provide information not shown on other drawings. The number and type of section drawings in a set of prints depend on what is required for a complete understanding of the construction. They are usually drawn at a scale of 3/8″ = 1′-0″ or 1:25.

Kinds of Sections

Full sections cut across the width or through the length of the entire building (Fig. 22-2). For a small residence, only one full section may be required to fully understand the construction. Commercial structures may require several full and many partial sections for complete understanding.

A *partial section* shows the vertical relationship of the parts of a small portion of the building.

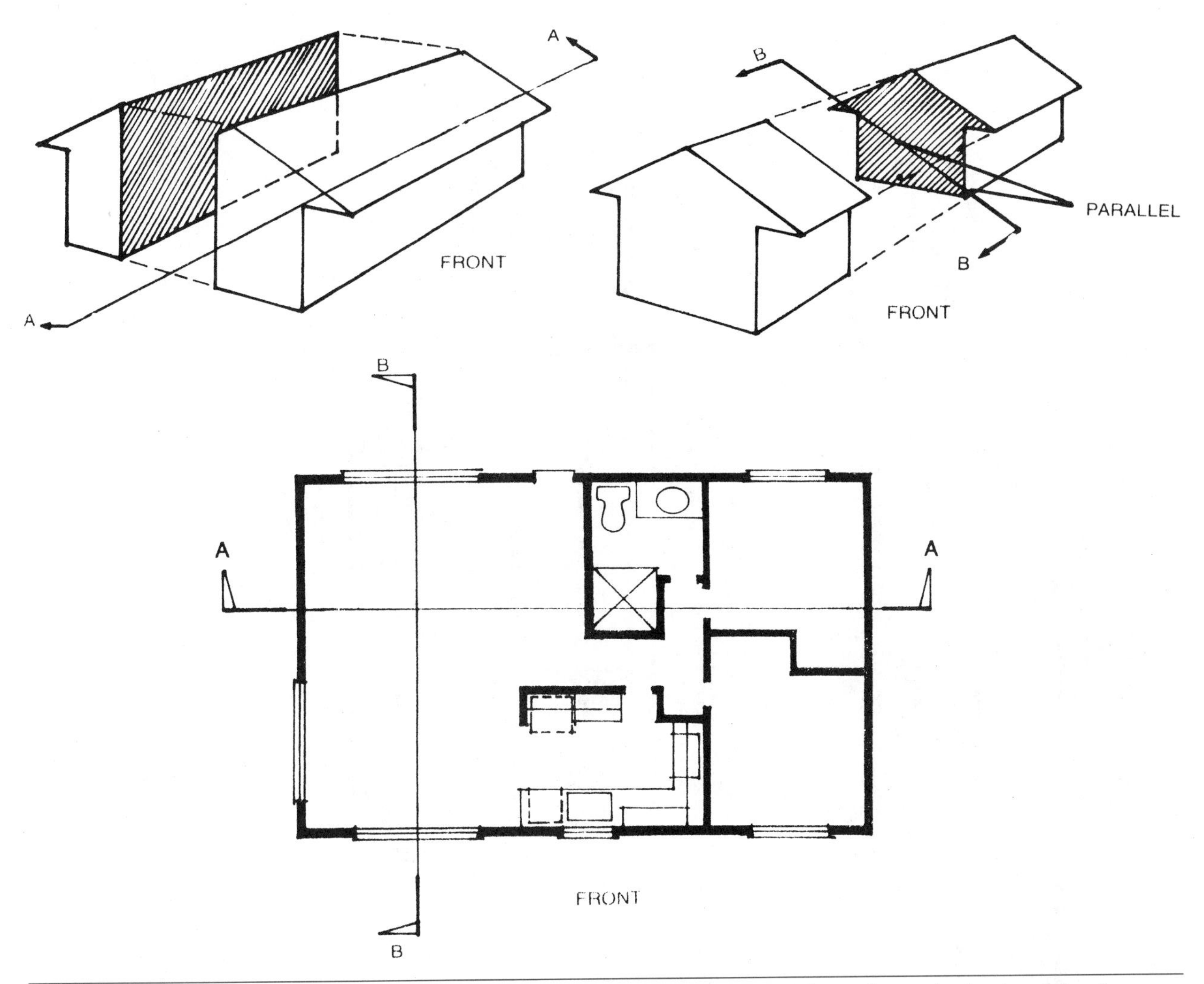

Fig. 22-1 Sections are views of vertical cutting planes across the width or through the length of a building. Section reference lines identify the location of the section and the direction from which it is being viewed. (*From Jefferis and Madsen,* Architectural Drafting & Design, *1991, by Delmar.*)

Partial sections through exterior walls are often used to give information about materials from foundation to roof (Fig. 22-3). They are drawn at a larger scale. The information given in one wall section does not necessarily apply for all walls. It may not apply, in fact, for all parts of the same wall. The wall section, or any other section being viewed, applies only to that part of the construction located by the section reference lines. Because the construction changes throughout the building, many section views are needed to provide clear and accurate information.

An enlargement of part of a section is required when enough information cannot be given in the space of a smaller scale drawing. These large-scale drawings are called *details.* An example is the foundation detail shown in Figure 22-4.

Reading Sections

From a typical exterior wall section of a residence, shown in Figure 22-3, much necessary information is provided. If a desired dimension is not given, enough others are provided to calculate it. Starting from the bottom and working upward to the roof, the following is an example of data that may be read from a full cross-section.

Foundation

- The height, width, and shape of the foundation **footings,** including the footing **keyway.** The material used for the footing, and the location and size of any steel reinforcement. The location of the finished grade and the depth of the footings below the grade line.

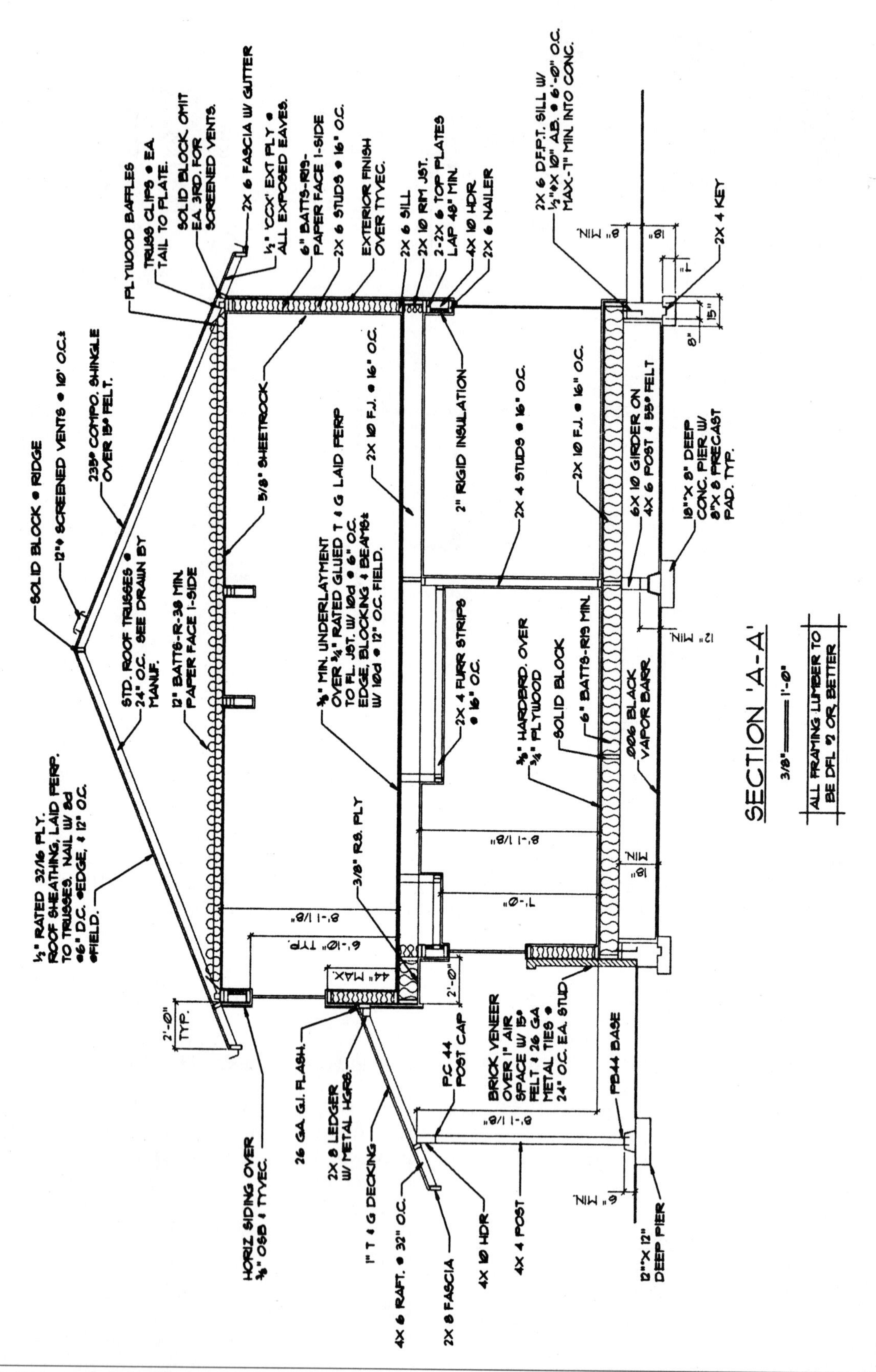

Fig. 22-2 A typical full cross-section of a residence. (*From Jefferis and Madsen,* Architectural Drafting & Design, *1991, by Delmar.*)

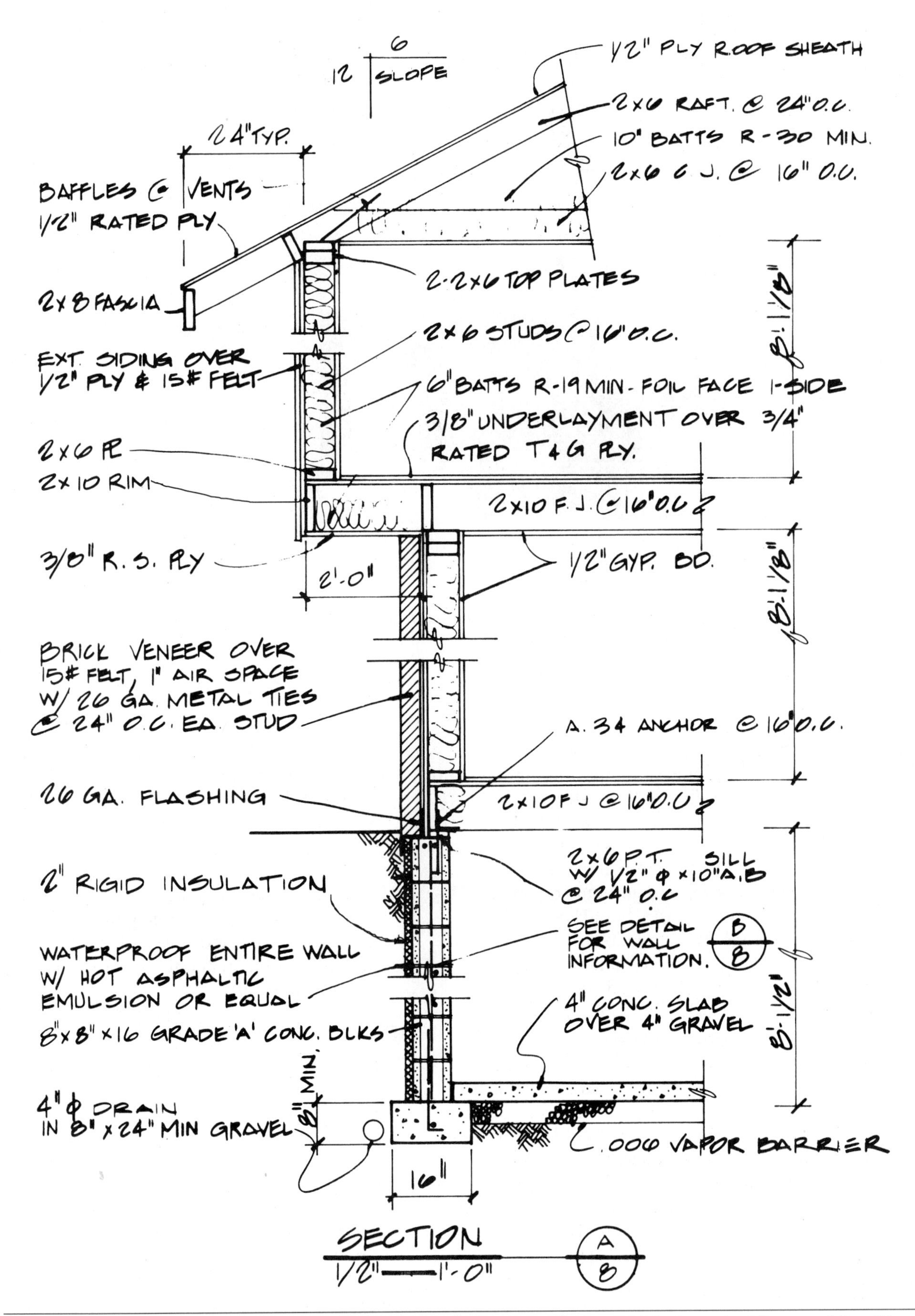

Fig. 22-3 A partial section through an exterior wall. (*From Jefferis and Madsen,* Architectural Drafting & Design, *1991, by Delmar.*)

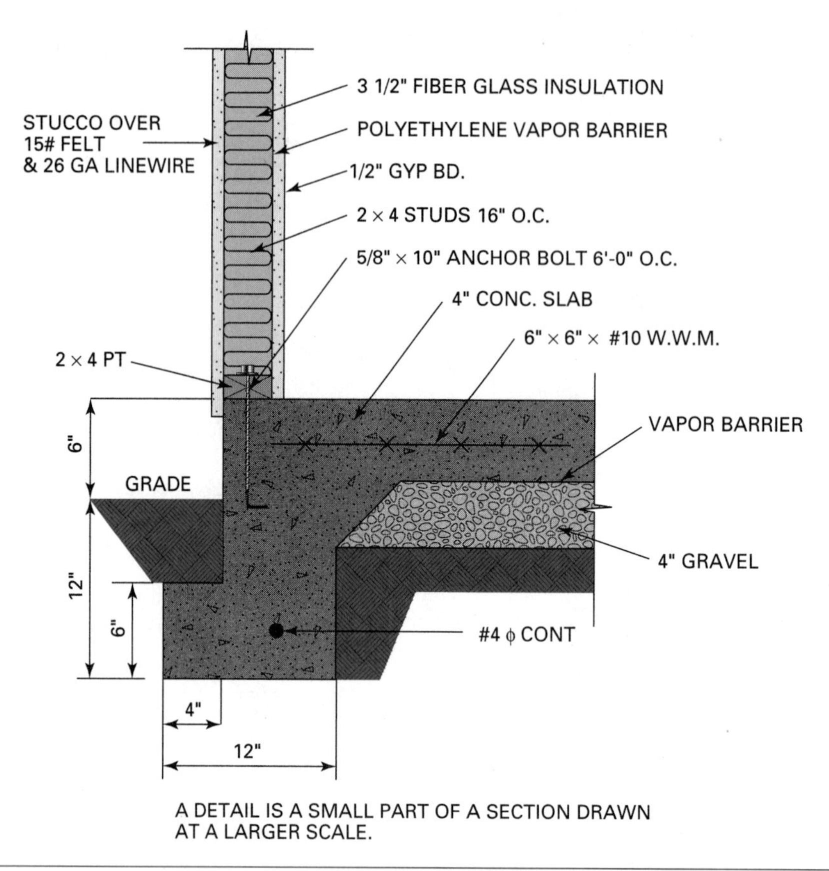

Fig. 22-4 A detail is a small part of a section drawn at a large scale.

- The height and thickness of the foundation wall and the material used to construct it. The shape, size, and location of a wall ledge for brick veneer. The location and size of reinforcing rods and anchor bolts in the foundation wall. The type, size, and location of below grade foundation wall insulation. The kind and surface coverage of foundation wall waterproofing. The location of foundation drain pipe and the kind and amount of drain bed material.
- The thickness of the basement floor slab. The floor slab material and the size, gauge, and kind of slab reinforcement. The thickness and kind of material directly under the slab. The kind, size, and location of a vapor barrier. Column footings material, size, and depth below grade. The location, size, and spacing of steel reinforcing rods in the column footings. The type and size of columns are given, but the number and spacing are found in a longitudinal section or a foundation plan.

Floor

- The thickness, width, location, and kind of sill plate. The width, depth, location, and kind of girder. The kind, thickness, width, length, and spacing of floor joists. The header joist or rim joist material, thickness, and width. The size, type, location, and kind of bridging. The kind and thickness of the subfloor and finish floor. The kind, size, and location of floor insulation. The type, location, and spacing of framing anchors, connectors, and hold-downs. The location, shape, and material used for flashings and termite shields.

Walls

- The kind, location, thickness, and width of sole and top plates. The kind, width, and spacing of the studs. The type and thickness of wall insulation. The thickness and width of certain window and door opening headers, and their height from the finish floor. The kind and thickness of exterior wall sheathing, and finish, including type of building paper or housewrap. The location of certain partitions. The kind and thickness of interior wall finish. The type, shape, and size of the base and base shoe. The finish floor to ceiling height.

Ceiling and Roof

- The kind, width, thickness, length, and spacing of ceiling joists. The kind, size, and spacing of ceiling **furring strips.** The type and thickness of ceiling insulation and ceiling finish. The kind size, spacing, and pitch of roof rafters or trusses. The thickness and width of the **ridgeboard.** The size and location of collar ties. The type and thickness of roof sheathing and the type of roofing felt and roof finish. The amount of roof overhang. The kind, thickness, and width of **cornice** finish; **soffit, fascia,** and **frieze.** The type and location of soffit and roof vents, including insulation baffles for air circulation. Kind and location of framing ties and connectors.

Elevations

Elevations are orthographic drawings. They are usually drawn at the same scale as the floor plan. They show each side of the building as viewed from the outside. Generally four elevations, one for each side, are included in a set of drawings. They are titled Front, Rear, Left Side, and Right Side. They may also be titled according to the compass direction that they face, for instance, North, South, East, and West. From the exterior elevations, the general shape and design of the building can be determined (Fig. 22-5).

Symbols. Elevation symbols are different than floor plan symbols for the same object. In elevation

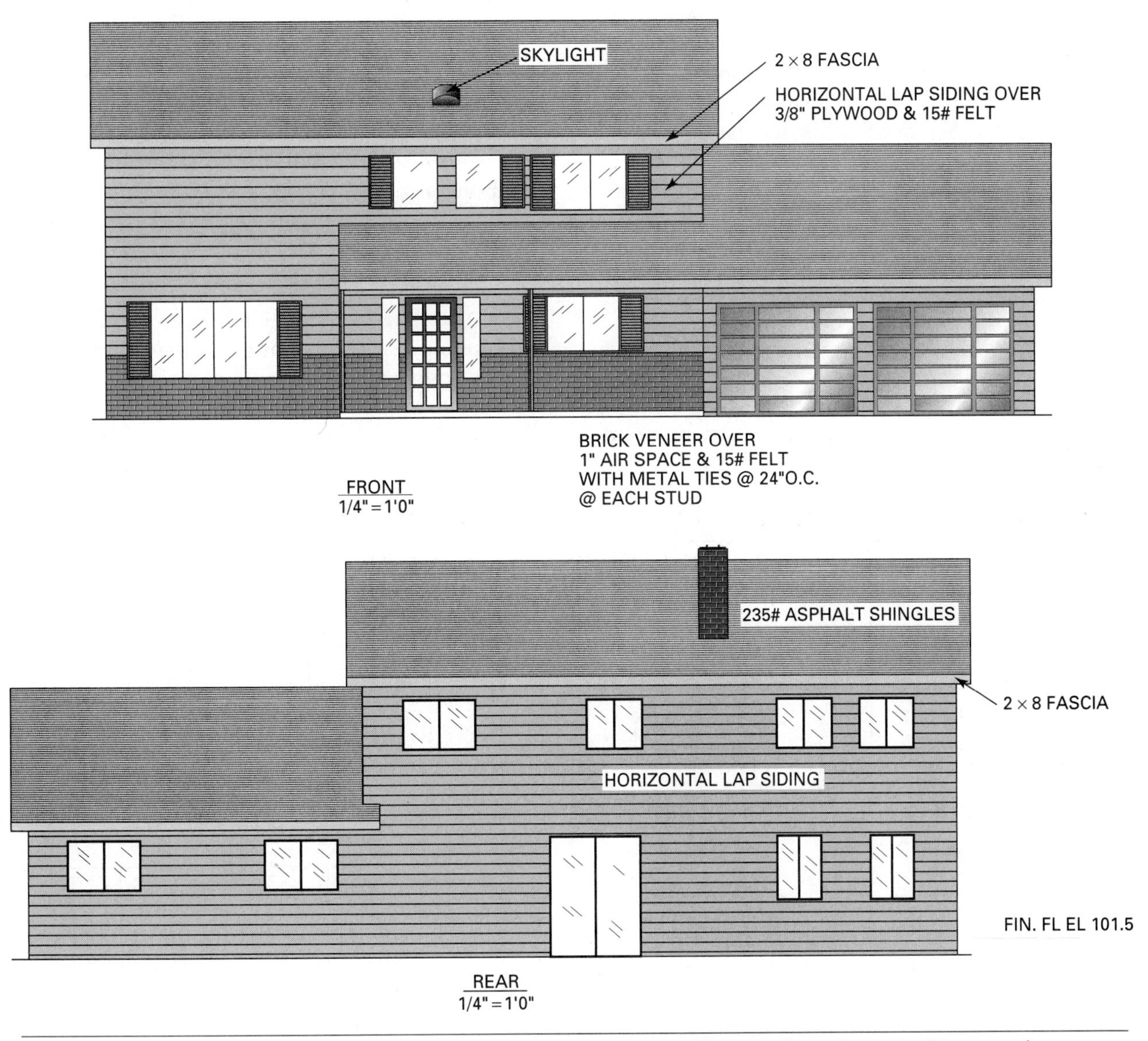

Fig. 22-5 Elevations show the exterior of a building. (*From Jefferis and Madsen,* Architectural Drafting & Design, *1991, by Delmar.*)

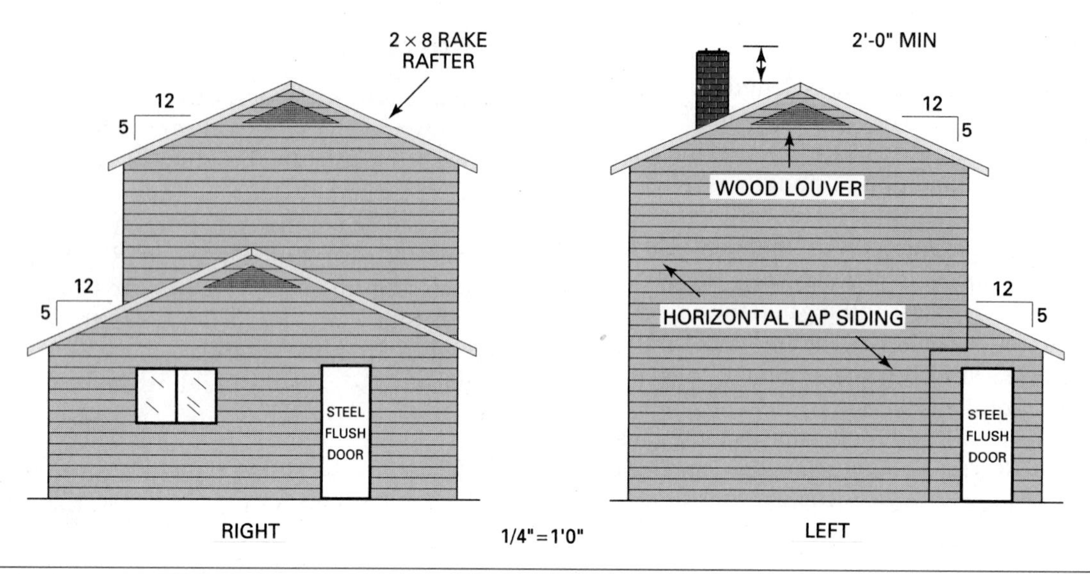

Fig. 22-5 *(continued)*

drawings, the symbols represent, as closely as possible, what they would appear to the eye. To make the drawing more clear, the symbols are usually identified with a notation.

Interpreting Elevations

- The location of any steps, porches, **dormers**, **skylights**, and chimneys, although not dimensioned, can be seen.
- Foundation footings and walls below the grade level may be shown with hidden lines.
- The kind and size of exterior siding, railings, entrances, and special treatment around doors and windows are shown (Fig. 22-6).

The elevations show the windows and doors in their exact location. Other openings, such as vents and **louvers**, are shown in place. Their style and size are identified by appropriate symbols and notations (Fig. 22-7).

The type of roofing material, the roof pitch, and the cornice style may also be determined from the exterior elevations (Fig. 22-8).

Dimensions. In relation to other drawings, elevations have few dimensions. Some dimensions usually given are floor to floor heights, distance from grade level to finished floor, height of window openings from the finished floor, and distance from the ridge to the top of the chimney.

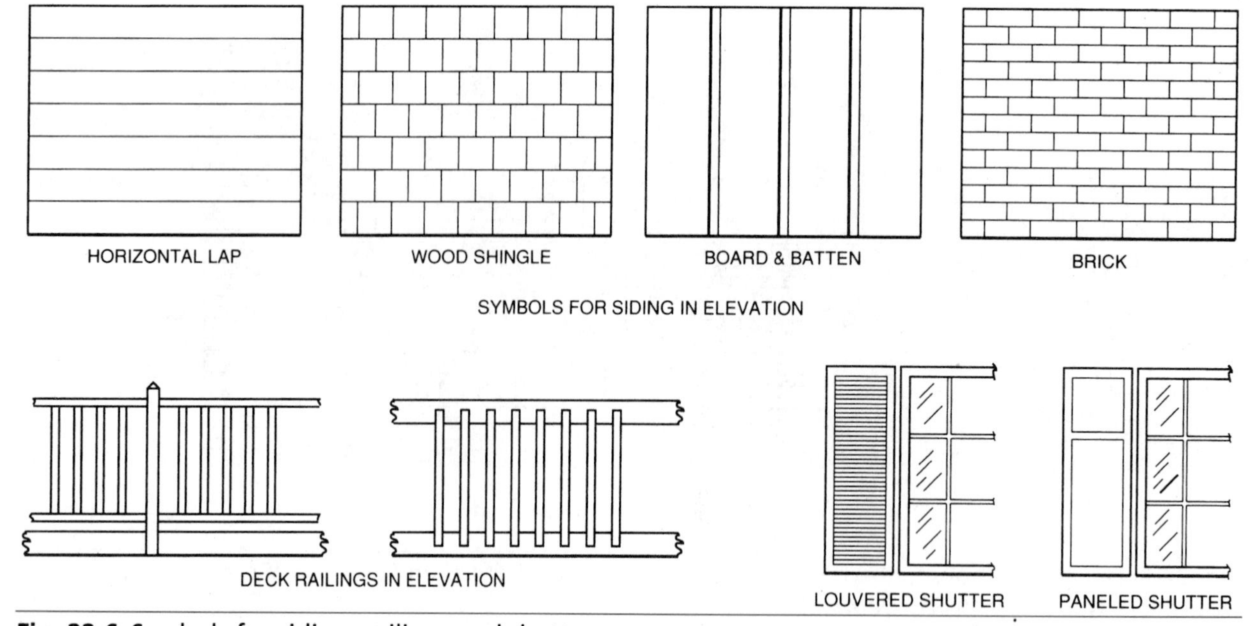

Fig. 22-6 Symbols for siding, railings and shutters.

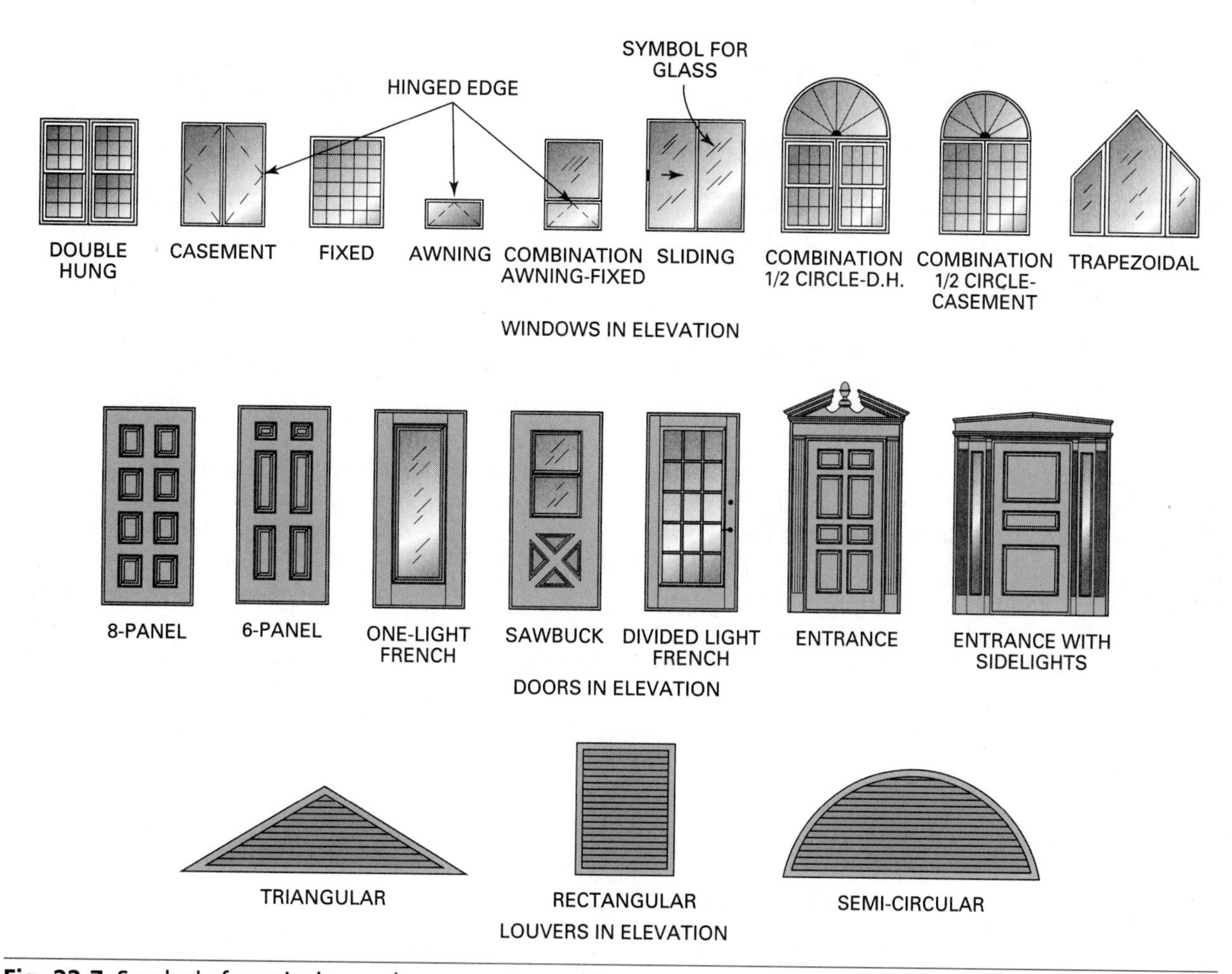

Fig. 22-7 Symbols for windows, doors, vents, and louvers.

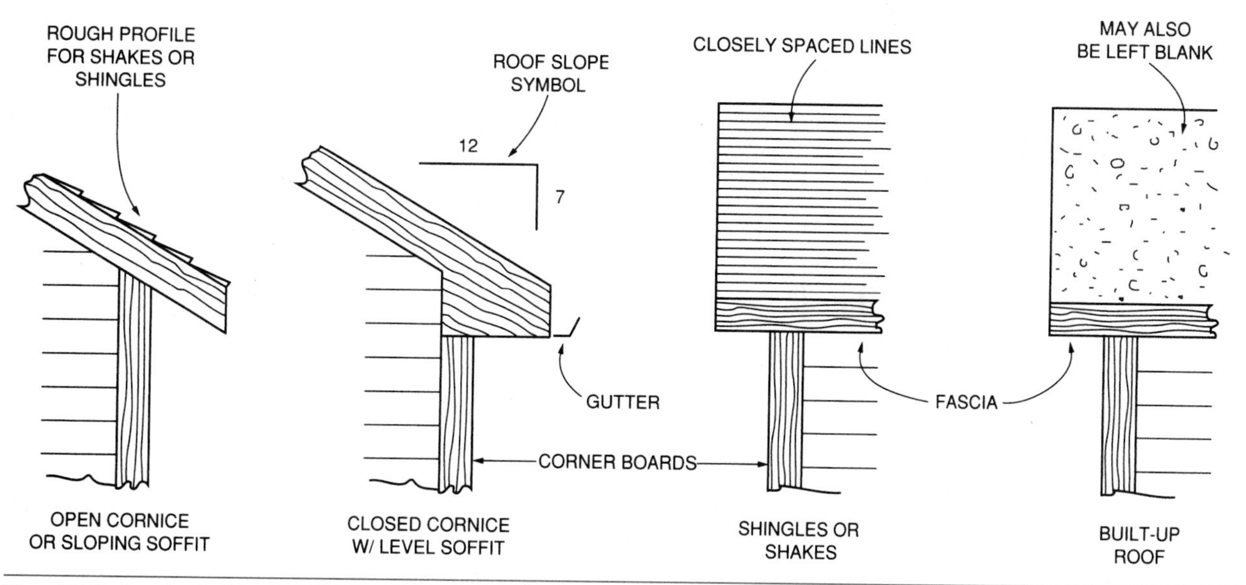

Fig. 22-8 Symbols for roofing, roof slope, and cornice style.

A number of other things may be shown on exterior elevations, depending on the complexity of the structure. Little information is given in elevations that cannot be seen in more detail in plans and sections. However, elevations serve an important purpose in making the total construction easier to visualize.

CHAPTER 23 PLOT AND FOUNDATION PLANS

Plot Plans

A **plot plan** is a map of a section of land used to show the proposed construction (Fig. 23-1). Depending on the size, the scale of plot plans may vary from 1″ = 10′ to 1″ = 200′ or 1:100 to 1:250. It is a required drawing when applying for a permit to build in practically every community. It is a necessary drawing to plan construction that may be affected by various features of the land. The plan must show compliance with zoning and health regulations. Although plot plan requirements may vary with localities, certain items in the plan are standard.

Property Lines

The property line *measurements* and *bearings*, known as **metes and bounds**, show the shape and size of the parcel. They are standard in every plot plan.

Measurements. The boundary lines are measured in *feet, yards, rods, chains,* or *meters.*

3 feet equals one yard.

16 1/2 feet or 5 1/2 yards equal one rod.

66 feet or 22 yards or 4 rods equal one chain.

Parts of measurement units are expressed as decimals. For instance, a boundary line dimension is expressed as 100.50 feet, not 100 feet, 6 inches.

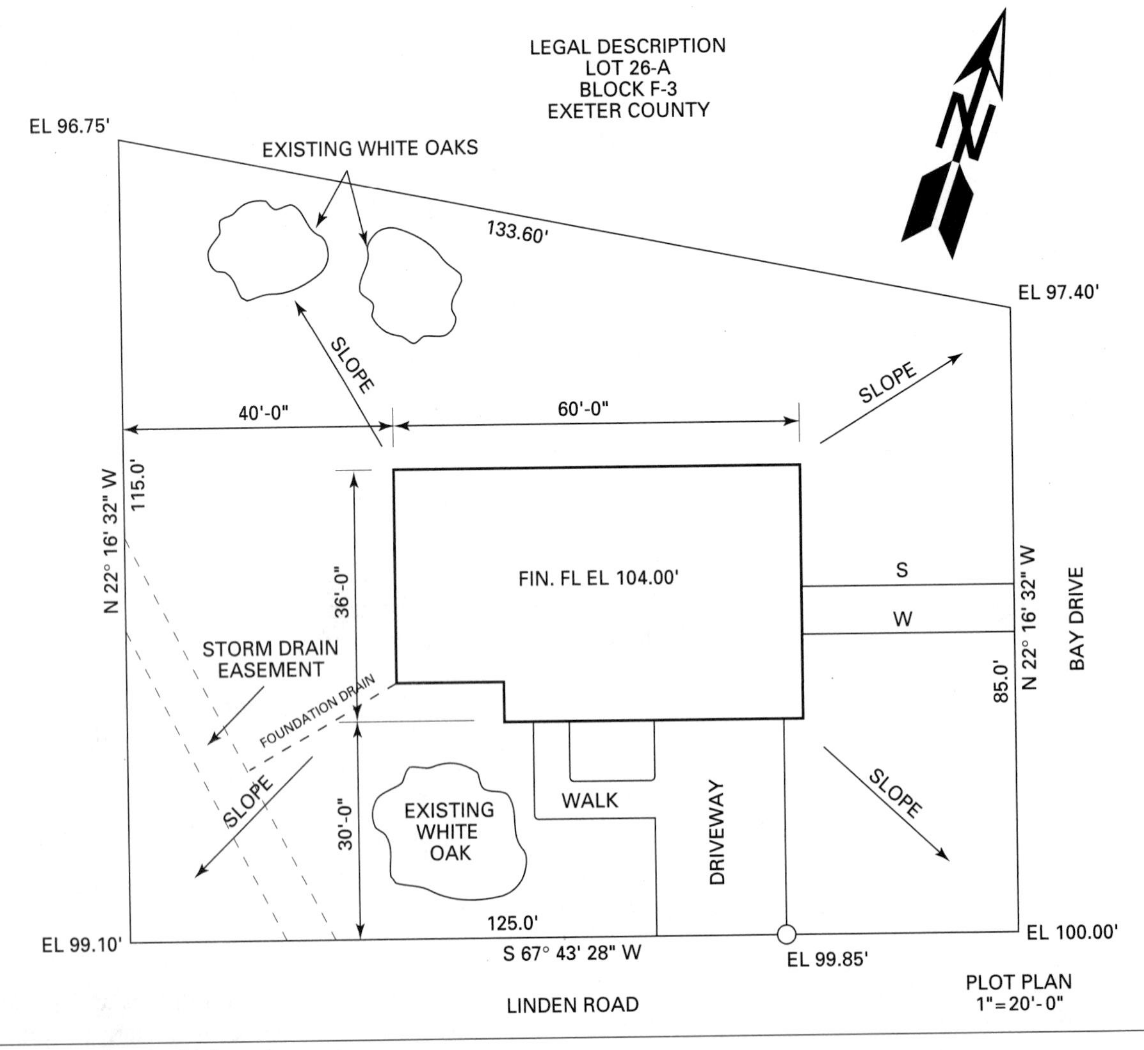

Fig. 23-1 A typical plot plan.

It is shown centered on, close to, and inside the line.

North. The North compass direction is clearly marked on every plot plan. In a clear space, an arrow of any style labeled with the letter "N" is pointed in the north direction (Fig. 23-2).

Bearings. In addition to the length of the boundary line, its *bearing* is shown. The bearing is a compass direction given in relation to a *quadrant* of a circle. There are 360 degrees in a circle and 90 degrees in each quadrant. Degrees are divided into *minutes* and *seconds*.

One degree equals 60 minutes (60′)

One minute equal 60 seconds (60″)

The boundary line bearing is expressed as a certain number of degrees clockwise or counterclockwise from either North or South. For instance, a bearing may be shown as N 30° E, N 60° W, S 45° E, or S 75° W (Fig. 23-3). No bearings begin with East or West as a direction. The bearing is shown centered close to and outside the boundary line opposite its length.

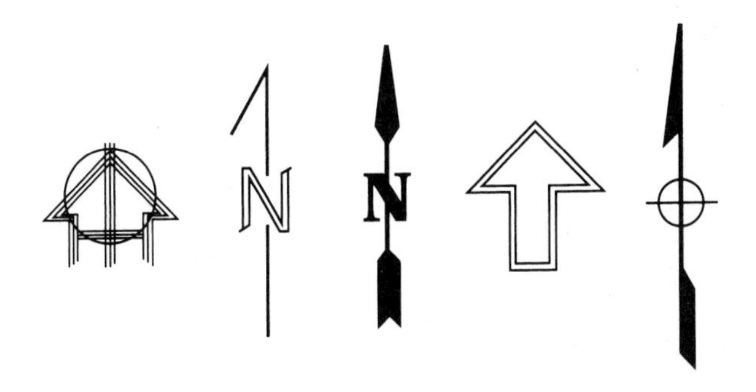

Fig. 23-2 Typical North direction symbols.

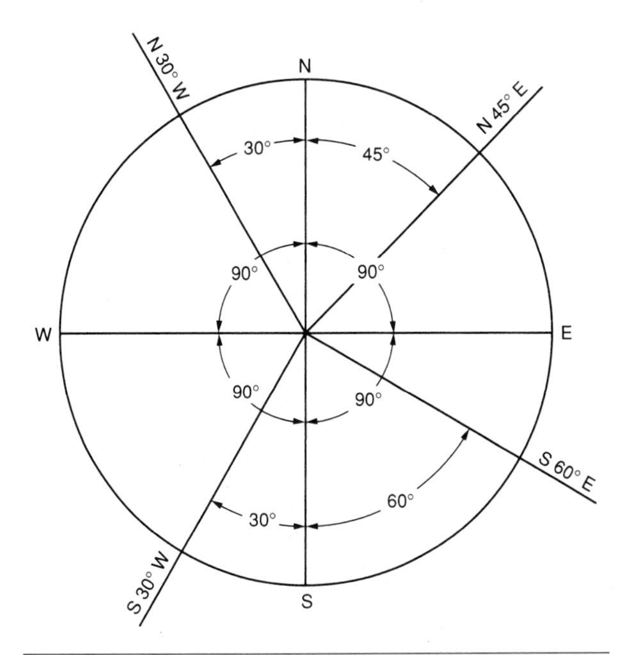

Fig. 23-3 Method of indicating property line bearings.

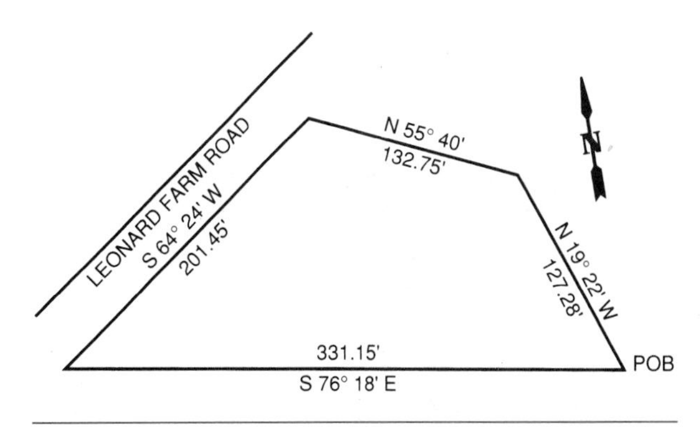

Fig. 23-4 Measurements, bearings, and legal description of a parcel of land.

Point of Beginning. An object that is unlikely to be moved easily, such as a large rock, tree, or iron rod driven into the ground, is used for a *point of beginning*. To denote this point on the plot plan, one corner of the lot may be marked with the abbreviation POB. It is from this point that the lot is laid out and drawn (Fig. 23-4).

Topography

Topography is the detailed description of the land surface. It includes any outstanding physical features and differences in *elevation* of the building site. Elevation is the height of a surface above sea level. It is expressed in feet and 1/100 of a foot or meters and decimals.

Contour Lines. Contour lines are irregular, curved lines connecting points of the same elevation of the land. The vertical distance between contour lines is called the *contour interval*. It may vary depending on how specifically the contour of the land needs to be shown on the plot plan.

When contour lines are close together, the slope is steep. Widely spaced contour lines indicate a gradual slope. At intervals, the contour lines are broken and the elevation of the line inserted in the space (Fig. 23-5). On some plans, dashed contour lines indicate the existing grade and solid lines depict the new grade. Topography is not always a requirement on plot plans. This is especially true for sites where there is little or no difference in elevation of the land surface. The slope of the finished grade may be shown by arrows instead of contour lines (Fig. 23-6).

Elevations

The height of several parts of the site and the construction are indicated on the plot plan. It is neces-

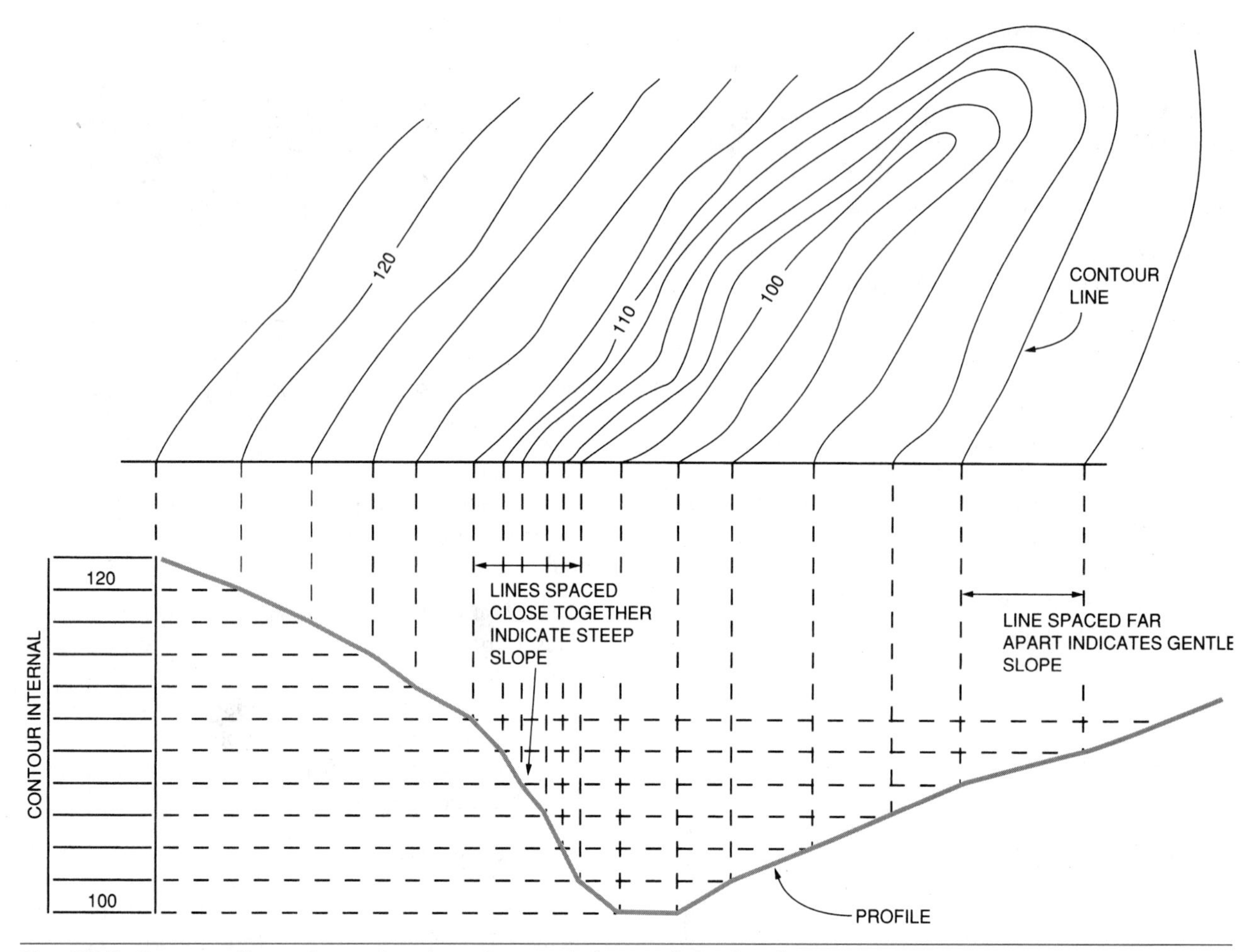

Fig. 23-5 Contour lines show the elevation and slope of the land.

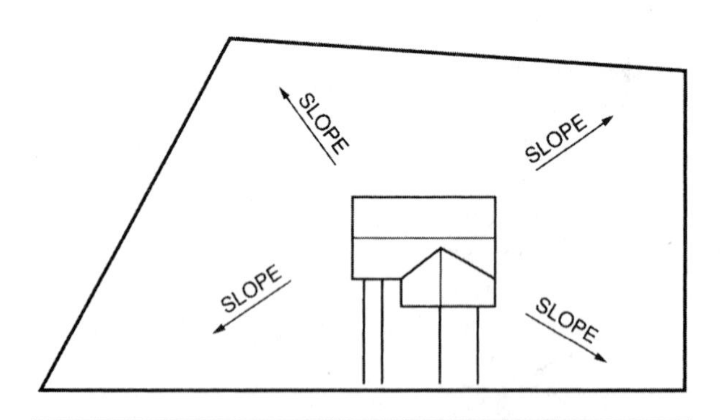

Fig. 23-6 Arrows are sometimes used in place of contour lines to show the slope of the land.

sary to know these elevations for grading the lot and construction of the building and accessories.

Benchmark. Before construction begins, a reference point, called a **benchmark,** is established on or close to the site. It is used for conveniently determining differences in elevation of various land and building surfaces.

The benchmark is established on some permanent object, which will not be moved or destroyed, at least until the construction is complete. It may be the actual elevation in relation to sea level. It may also be given an arbitrary elevation of 100.00 feet. All points on the lot, therefore, would either be above, level with, or below the benchmark (Fig. 23-7). The location of the benchmark is clearly shown on the plot plan with the abbreviation BM.

Finish Floor. The elevation of finished floor levels is clearly shown and noted on the plot plan. Building lines are laid out to the same height as the top of the foundation wall before any floors are built. Therefore, the distance from the finish floor to the top of the foundation wall must be found. It is then subtracted from the finish floor elevation to determine the elevation of the top of the foundation wall. This distance can be calculated with information from an exterior wall section (Fig. 23-8).

Converting Decimals to Fractions. Calculations, especially those dealing with eleva-

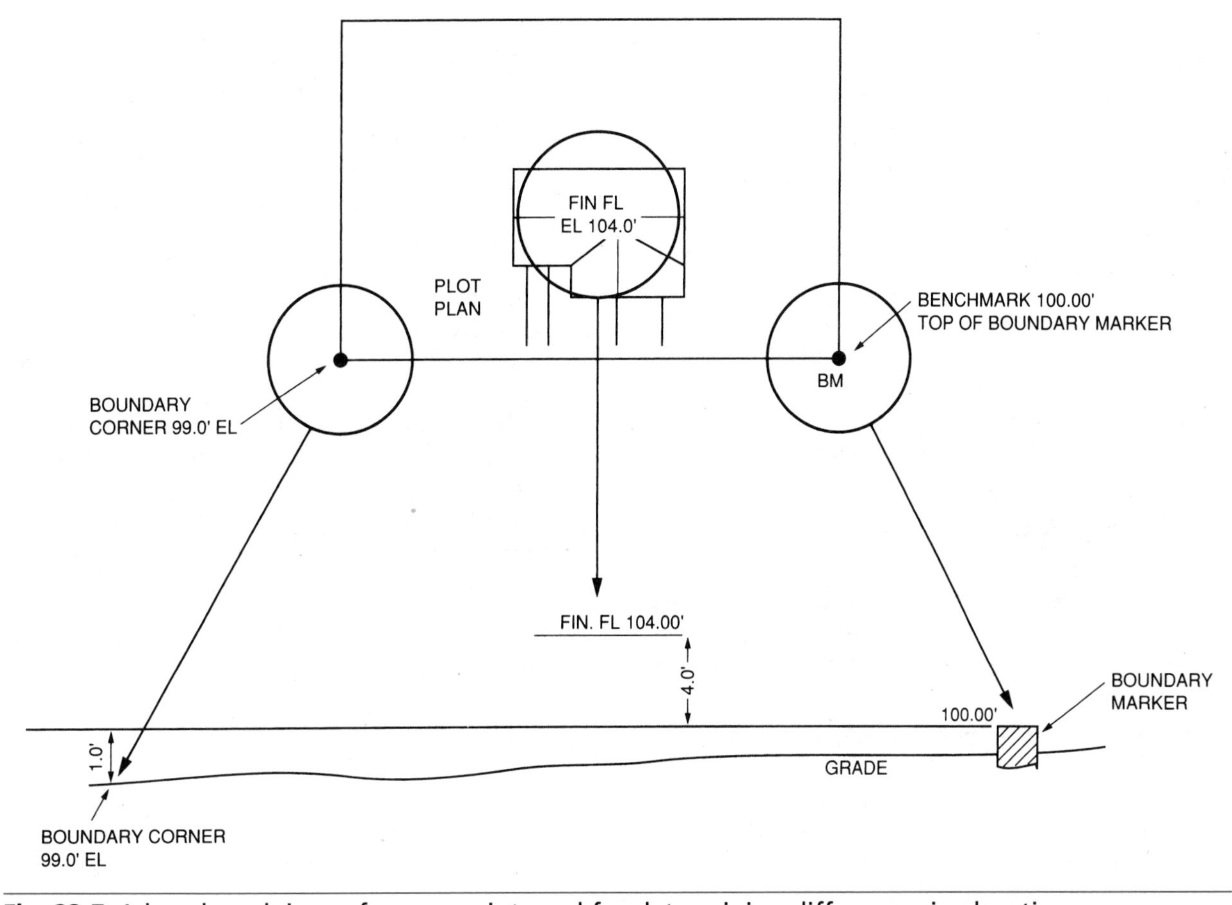

Fig. 23-7 A benchmark is a reference point used for determining differences in elevation.

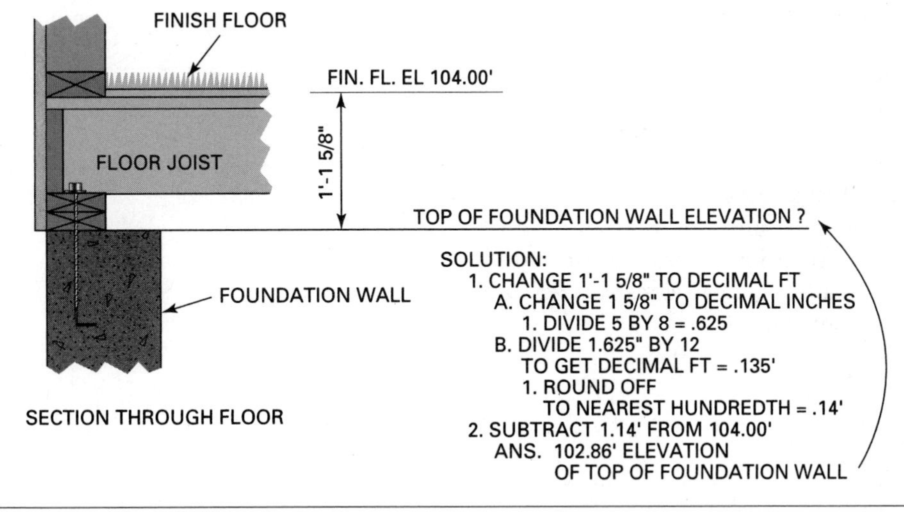

Fig. 23-8 The distance from the finish floor elevation to the top of the foundation wall must be calculated.

tions and roof framing, require the carpenter to convert decimals of a foot to feet, inches, and 16ths of an inch as found on the rule or tape. To convert, use the following method:

- Multiply a decimal of a foot by 12 (the number of inches in a foot) to get inches.
- Multiply any remainder decimal of an inch by 16 (the number of 16ths in an inch) to get 16ths of an inch. Round off any remainder to the nearest 16th of an inch.
- Combine whole feet, whole inches, and 16ths of an inch to make the conversion.

Example: Convert the finish floor elevation of 104.65 feet to feet, inches, and 16ths of an inch.

1. Multiply .65 ft. × 12 = 7.80 inches
2. Multiply .80 inches × 16 = 12.8 16ths of an inch.
3. Round off 12.8 16ths of an inch to 13 16ths of an inch.
4. Combine feet, inches, and sixteenths = 104′ – 7 13/16″.

It is more desirable to remember the method of conversion rather than use conversion tables. Reliance on conversion tables requires access to the tables and knowledge of their use, encourages dependence on them, and results in helplessness without them.

Other Elevations

- In addition to contour lines, the elevation of each corner of the property is noted on the plot plan (Fig. 23-9). The top of one of the boundary corner markers makes an excellent benchmark.
- Existing and proposed roads adjacent to the property are shown as well as any **easements.** Easements are right-of-way strips running through the property. They are granted for various purposes, such as access to other property, storm drains, or utilities. Elevations of a street at a driveway and at its centerline are usually required. One of these is sometimes used as the benchmark.

The Structure

The shape and location of the building are shown. Distances, called *setbacks,* are dimensioned from the boundary lines to the building.

The shape, width, and location of patios, walks, driveways, and parking areas are also found on plot plans. Details of their construction are found in another drawing.

A plot plan may also show any *retaining walls.* These walls are used to hold back earth to make more level surfaces instead of steep slopes.

Utilities. The water supply and public sewer connections are shown by noted lines from the structure to the appropriate boundary line. If a private sewer disposal system is planned, it is shown on the plot plan (Fig. 23-10). Although there are many kinds of sewer disposal systems, those most commonly used consist of a septic tank and leaching field. There are usually strict regulations in regard to location and construction.

The location of gas lines is shown, if applicable. Sometimes the location of the nearest utility pole is given. It is shown using a small solid circle as a symbol and is noted. Foundation drain lines leading to a storm drain, drywell, or other drainage may be shown and labeled.

Landscaping. The location and kind of existing and proposed trees are shown on the plot plan.

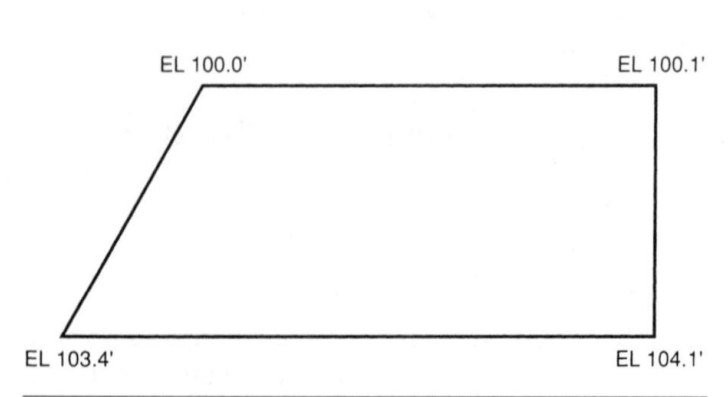

Fig. 23-9 The elevations of property line corners are indicated on the plot line.

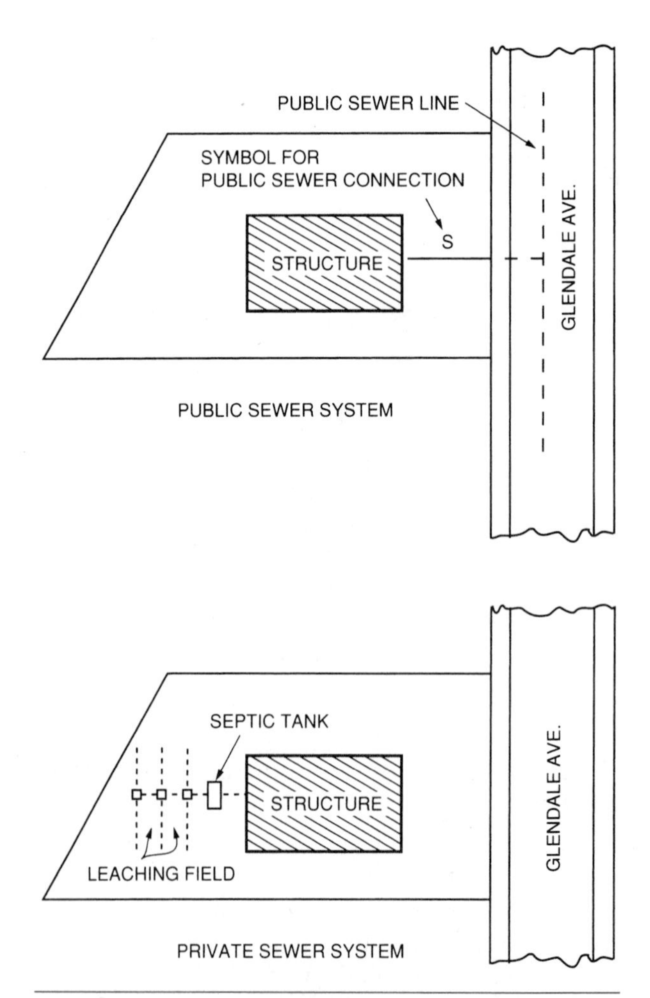

Fig. 23-10 Sewage disposal systems are shown on the plot plan.

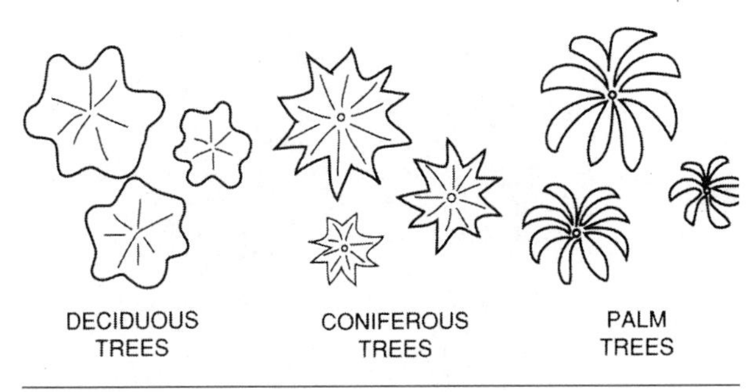

Fig. 23-11 Various symbols are used to indicate different kinds of trees.

Existing trees are noted, whether they are to be saved or removed. Those that are to be saved are protected by barriers during the construction process. Various symbols are used to show different kinds of trees (Fig. 23-11).

Identification. Included on the plot plan are the name and address of the property owner, the title and scale of the drawing, and a legal description of the property (Fig. 23-12).

Foundation Plans

The foundation plan is drawn at the same scale as the floor plan. It is a view from above of a horizontal cut through the foundation. Great care must be taken when reading the foundation plan so no mistakes are made. A mistake in the foundation affects the whole structure, and generally requires adjustments throughout the construction process.

Two commonly used types of foundations are those having a crawl space, or basement below grade, and those with a concrete slab floor at grade level (Fig. 23-13).

Crawl Space and Basement Foundations

The **crawl space** is the area enclosed by the foundation between the ground and the floor above. A minimum distance of 18 inches from the ground to the floor and 12 inches from the ground to the bottom of any beam is required. The ground is covered with a plastic sheet, called a **vapor barrier,** to prevent moisture rising from the ground from penetrating into the floor frame above.

A foundation enclosing a *basement* is similar to that of a crawl space except the walls are higher, windows and doors may be installed, and a concrete floor is provided below grade. The basement may be used for additional living area, garage, utility room, or workshop.

Reading Plans. Whether the foundation supports a floor using closely spaced floor joists or more widely spaced post-and-beam construction, the information given in the foundation plan is similar. A typical foundation plan is shown in Figure 23-14. Its usual components are as follows:

- The inside and outside of the foundation wall are clearly outlined. Dashed lines on both sides of the wall show the location of the foundation footing. The type, size, and spacing of anchor bolts are shown by a notation.

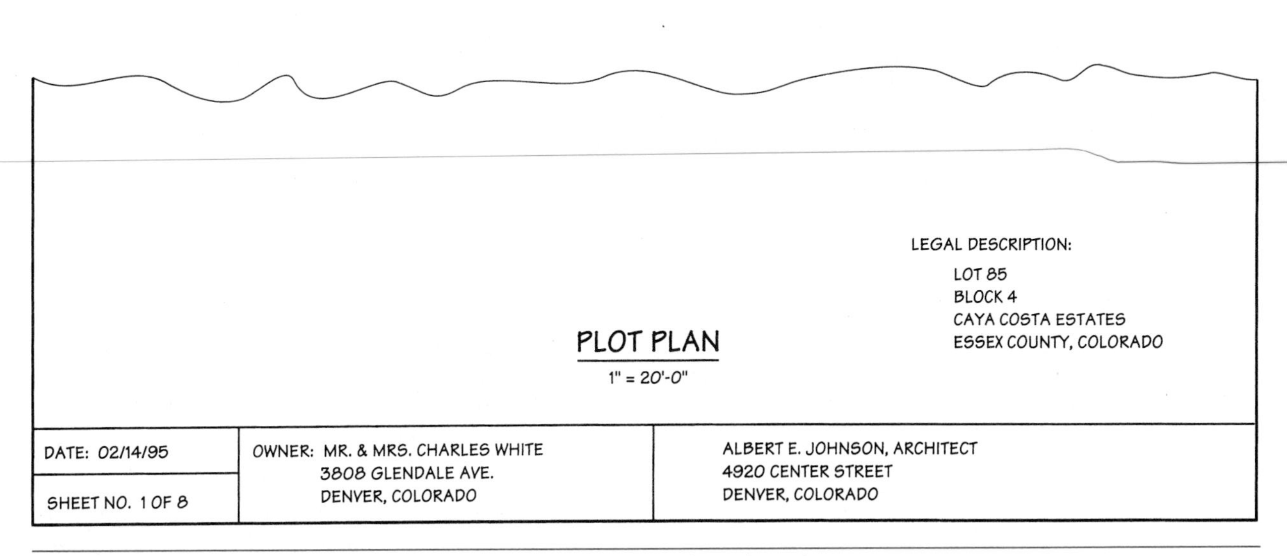

Fig. 23-12 Certain identification items are needed on the plot plan.

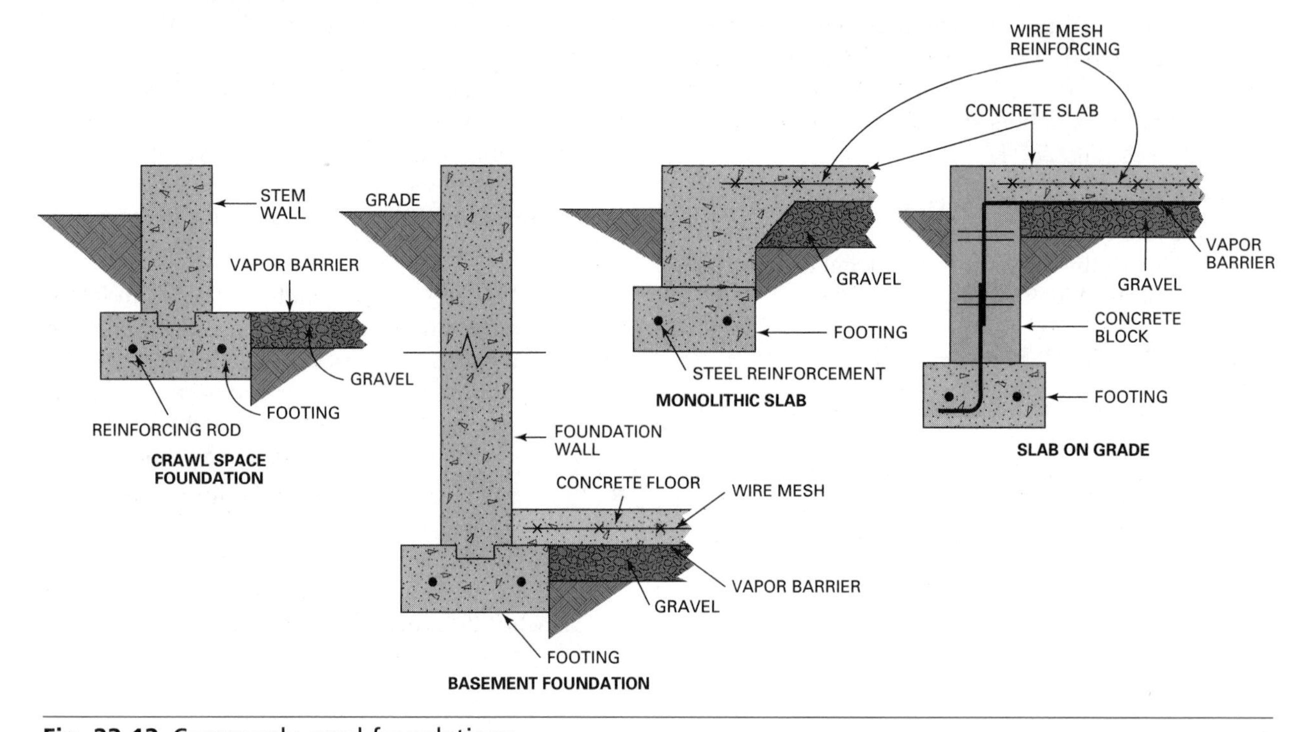

Fig. 23-13 Commonly used foundations.

- Wall openings for windows, doors, or crawl space access and vents are shown with appropriate symbols and noted. Small retaining walls of concrete or metal, called **areaways**, to hold earth away from windows that are below grade may be shown (Fig. 23-15).
- Walls for *stoops* (platforms for entrances) are shown. A notation is made in regard to the material with which to fill the enclosed area and cap the surface.
- Other footings shown by dashed lines include those for chimneys, fireplaces, and columns or posts. Columns or posts support girders (shown by a series of long and short dashes directly over the center of the columns). A recess in the foundation wall, called a **pocket**, to support the ends of the girder is shown. Notations are made to identify all of these items (Fig. 23-16).
- Floor joist direction is shown by a line with arrows on both ends similar to those shown in the floor plan for ceiling joists. A notation gives the size and spacing of the joists.
- The composition, thickness, and underlying material of the basement floor or crawl space surface are noted.

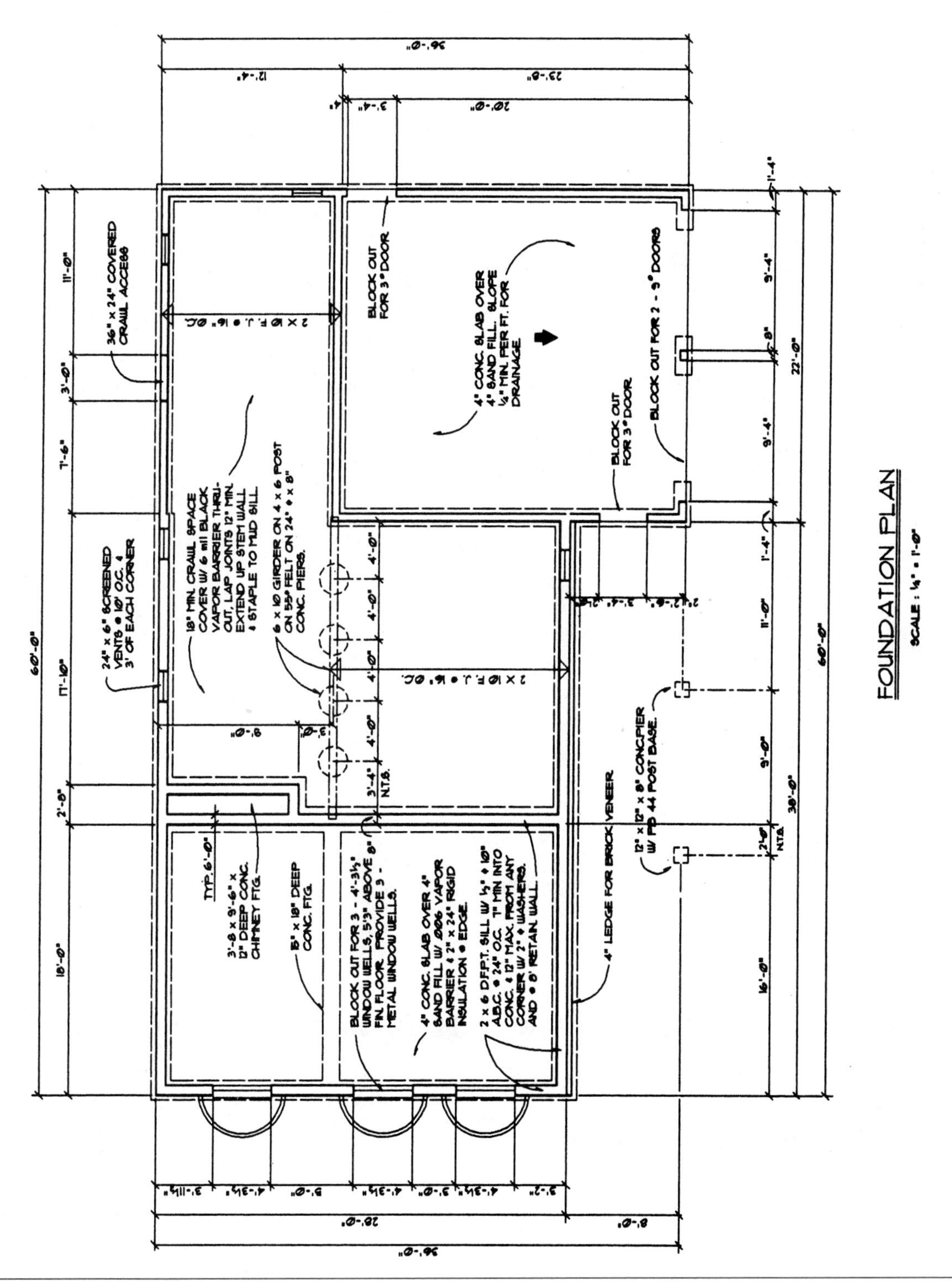

Fig. 23-14 A foundation plan for a partial basement. (*From Jefferis and Madsen,* Architectural Drafting & Design, *1991, by Delmar.*)

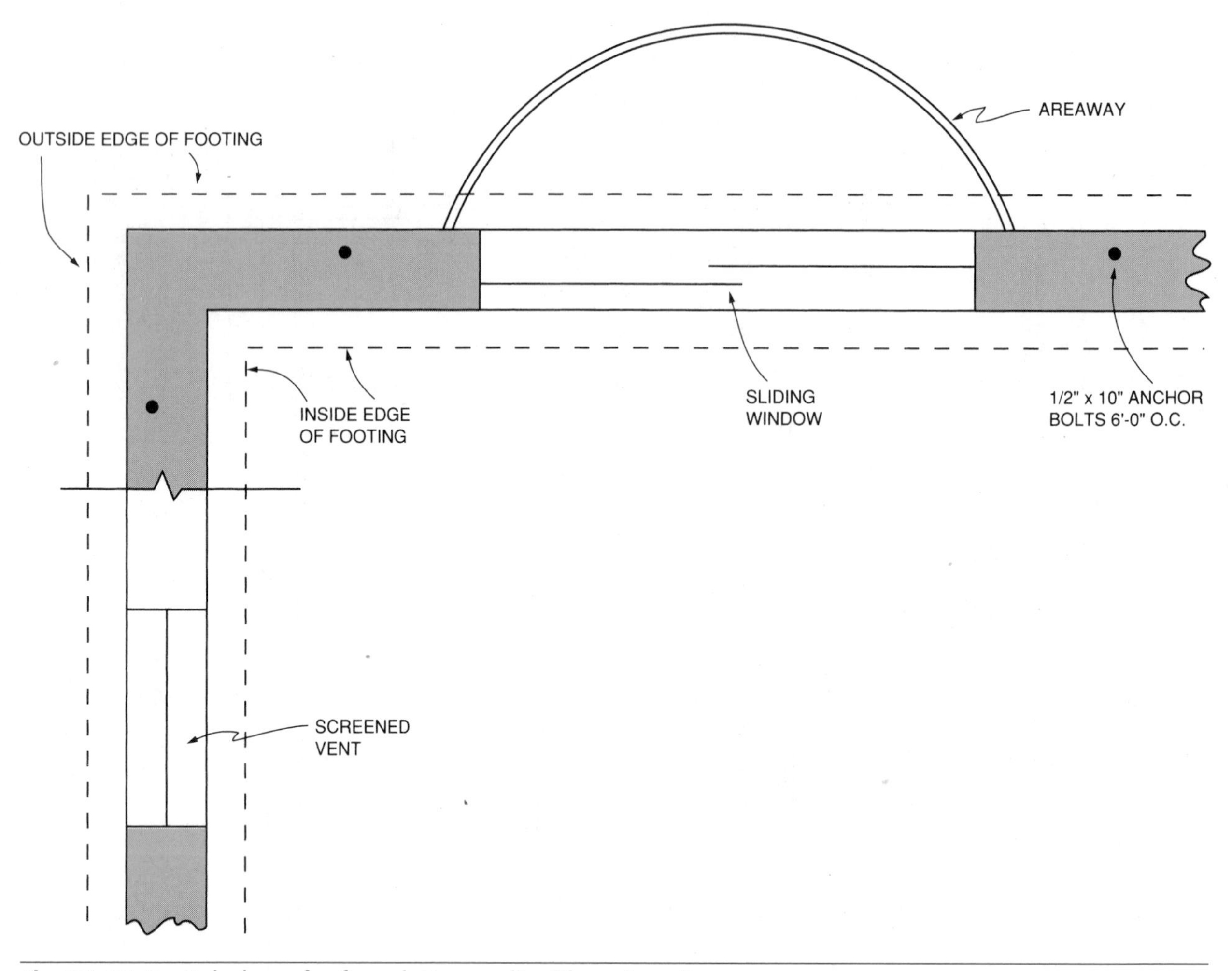

Fig. 23-15 Partial plan of a foundation wall with various items.

- The location of a stairway to the basement is shown with the same symbol as used in the floor plan.
- Although it may be stated in the specifications, the strength of concrete used for various parts of the foundation, and the wood type and grade, in addition to size, may be specified by a notation.
- Plans for foundations with basements may show the location, with appropriate notations, of furnaces and other items generally found in the floor plan if the basement is used as part of the living area.

Dimensions. It is important to understand how parts of the foundation are dimensioned. Foundation walls are dimensioned face to face. Interior footings, columns, posts, girders, and beams are dimensioned to their centerline (Fig. 23-17).

Slab-on-Grade Foundation

The *slab-on-grade foundation* is used in many residential and commercial buildings. It takes less labor and material than foundations to support beam and joist floor framing. Often the concrete for the footing, foundation, and slab can be placed at the same time. There are several kinds, but slab-on-grade foundation plans (Fig. 23-18) show common components:

- The shape and size of the slab are shown with solid lines, as are patios and similar areas. Changes in floor level, such as for a fireplace hearth or a sunken living area, are indicated by solid lines. A notation gives the depth of the recess.
- Exterior and interior footing locations ordinarily below grade are indicated with dashed lines. Footings outside the slab and ordinarily above grade are shown with solid lines.

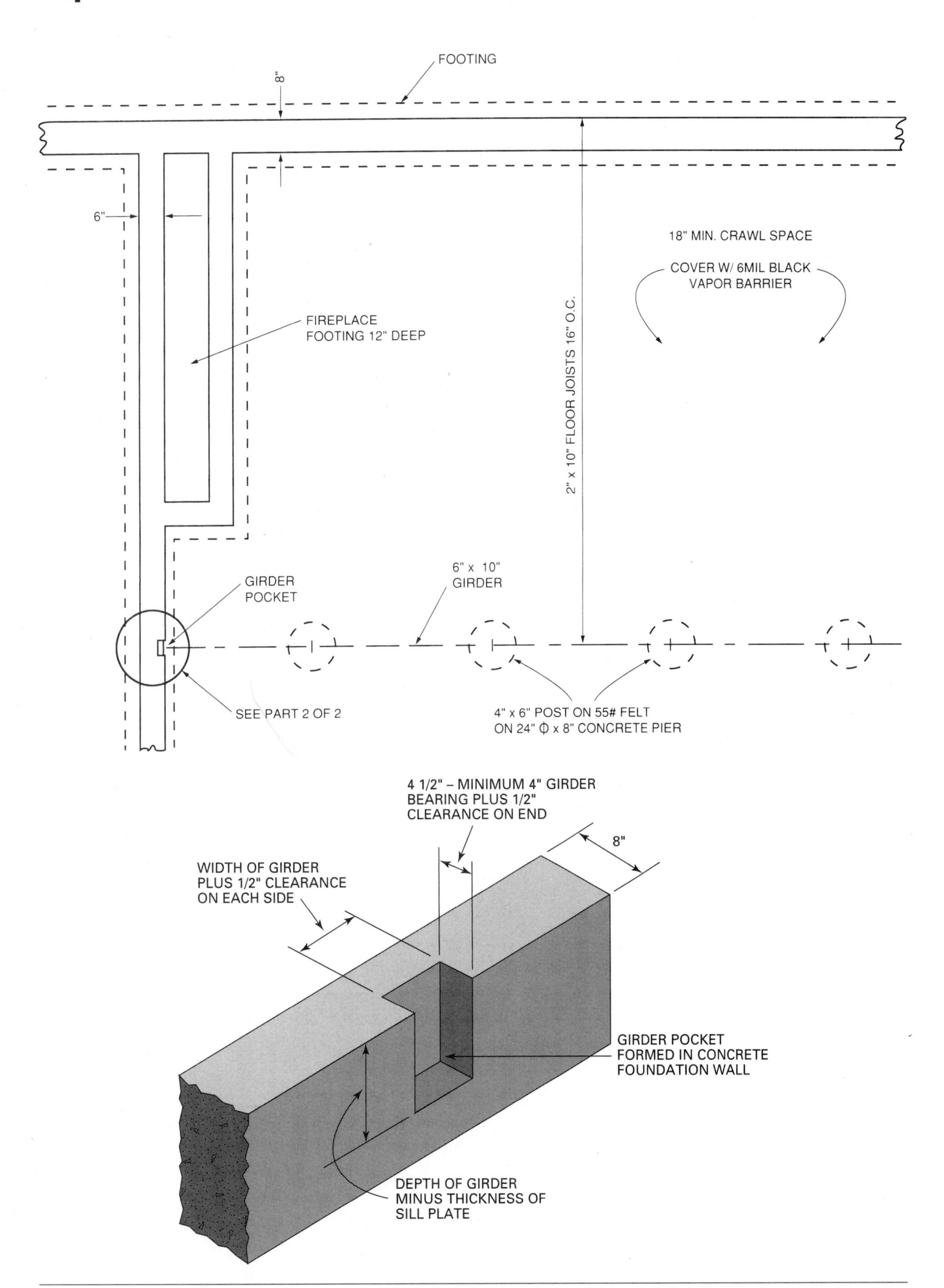

Fig. 23-16 Enlarged view of foundation plan showing girder and column details.

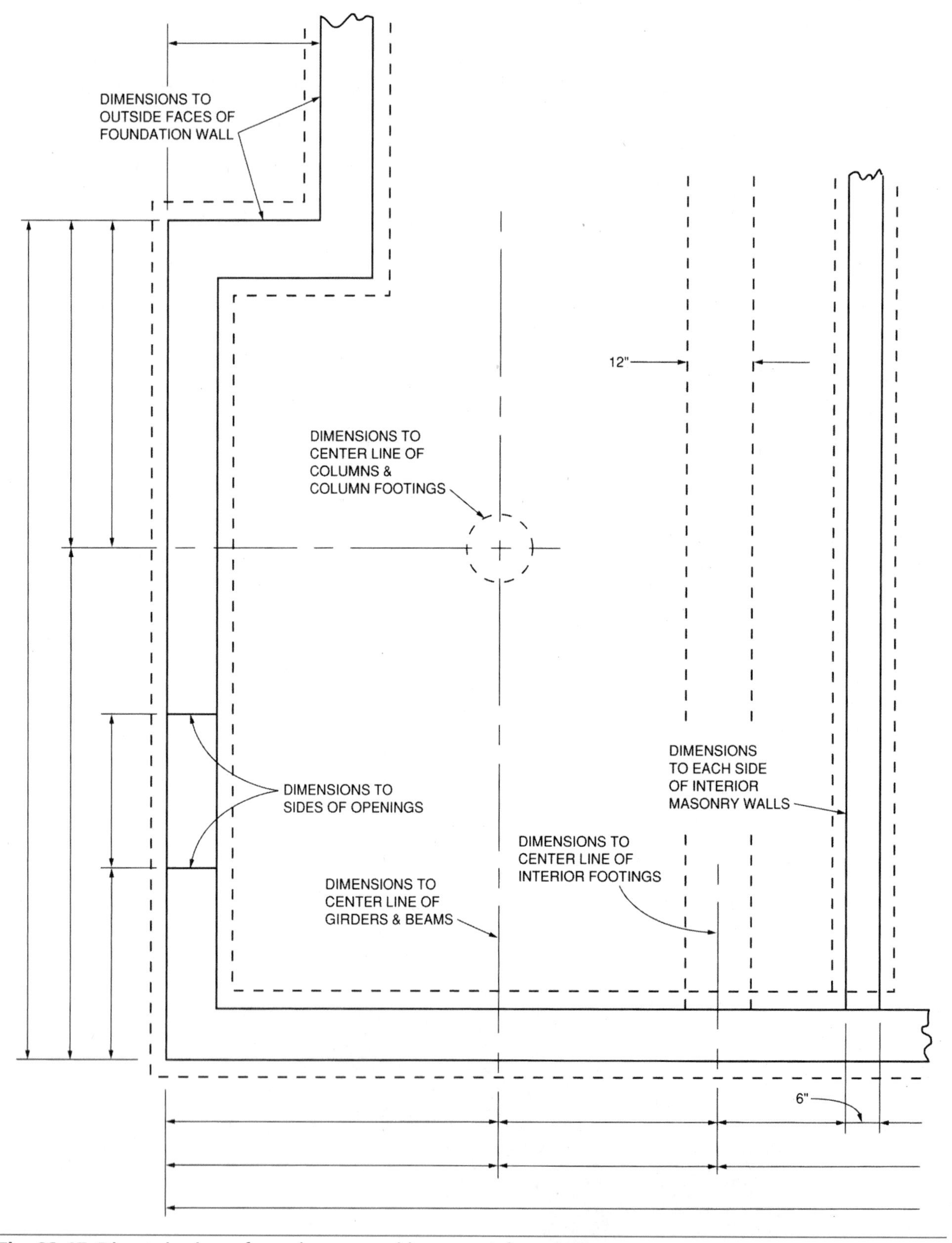

Fig. 23-17 Dimensioning of crawl space and basement foundations.

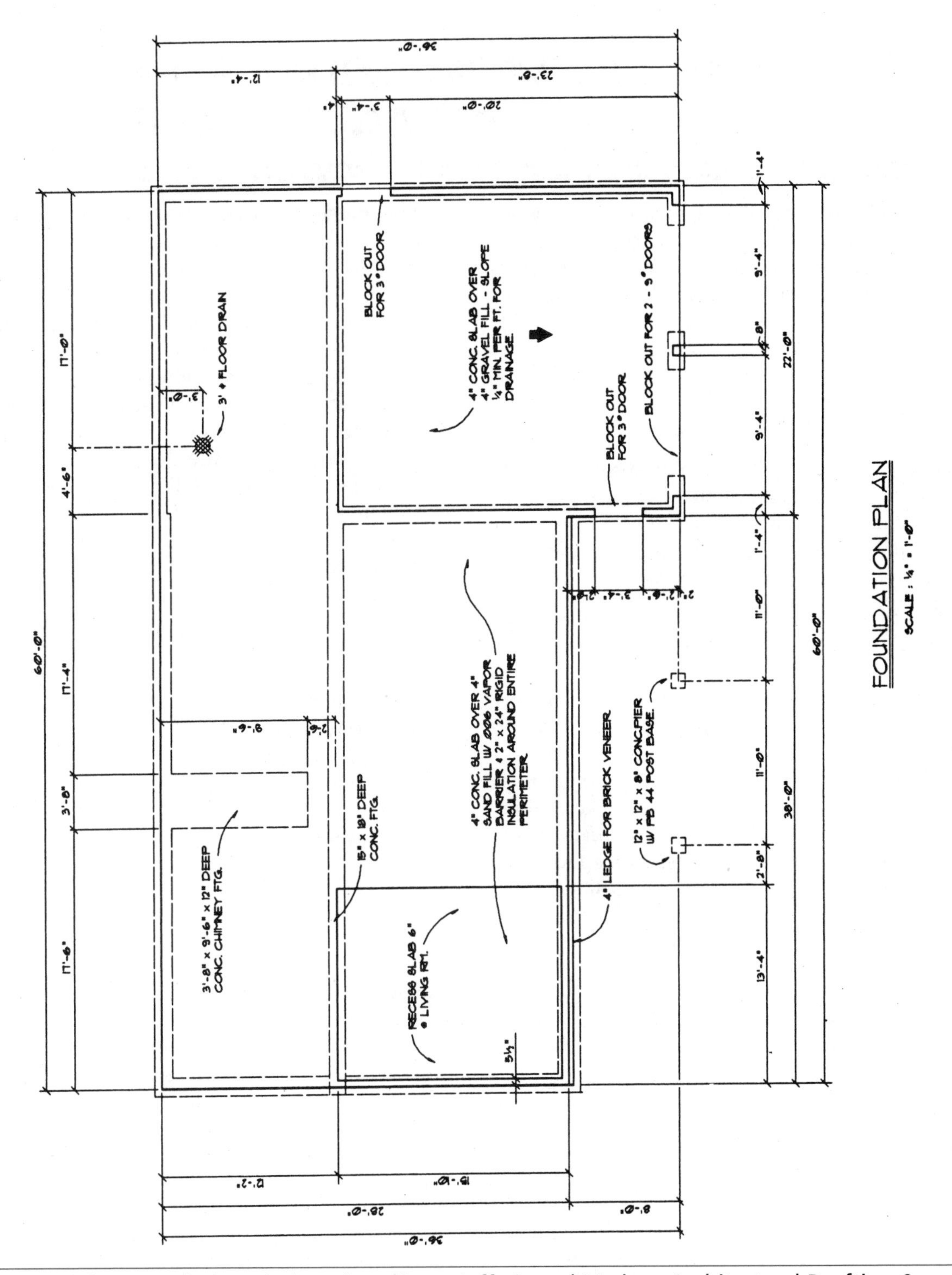

Fig. 23-18 Slab-on-grade foundation plan. (*From Jefferis and Madsen,* Architectural Drafting & Design, *1991, by Delmar.*)

- Appropriate symbols and notations are used for fireplaces, floor drains, and ductwork for heating and ventilation. Note: Blueprints for commercial work usually have separate drawings for electrical, plumbing, and mechanical work.
- Notations are made for the slab thickness, wire mesh reinforcing, fill material, and vapor barrier under the slab.
- Interior footing, mudsill, reinforcing steel, and anchor bolt size, location, and spacing are written.

Dimensions. Overall dimensions are to the outside of the slab. Interior piers are located to their centerline. Door openings are dimensioned to their sides.

CHAPTER 24 BUILDING CODES AND ZONING REGULATIONS

Cities and towns have laws governing many aspects of new construction and remodeling. These laws protect the consumer and the community. Codes and regulations provide for safe, properly designed buildings in a planned environment. Contractors and carpenters should have knowledge of zoning regulations and building codes.

Zoning Regulations

Zoning regulations deal, generally speaking, with keeping buildings of similar size and purpose in areas for which they have been planned. They also regulate the space in each of the areas. The community is divided into areas called *zones,* shown on *zoning maps.*

Zones. The names given to different zones vary from community to community. The zones are usually abbreviated with letters or a combination of letters and numbers. A large city may have thirty or more zoning districts.

There may be several *single family residential zones.* Some zones have less strict requirements than others. Other areas may be zoned as *multifamily residential.* They may be further subdivided into areas according to the number of apartments. Other residential zones may be set aside for *mobile home parks,* and those that allow a combination of *residences, retail stores, and offices.*

Other zones may be designated for the *central business district,* various kinds of *commercial districts* and different *industrial* zones.

Lots. Zoning laws regulate buildings and building sites. Most cities specify a *minimum lot size* for each zone and a *maximum ground coverage* by the structure. The *maximum height* of the building for each zoning district is stipulated. A *minimum lot width* is usually specified as well as *minimum yards.*

Minimum yard refers to the distance buildings must be kept from property lines. These distances are called *setbacks.* They are usually different for front, rear, and side.

Some communities require a certain amount of landscaped area, called *green space,* to enhance the site. In some residential zones, as much as half the lot must be reserved for green space. In a central business area, only 5 to 10 percent may be required.

In most zones, off-street parking is required. For instance, in single-family residential zones, room for two parking spaces on the lot is required.

Nonconforming Buildings. Because some cities were in existence before the advent of zoning laws, many buildings and businesses may not be in their proper zone. They are called *nonconforming.* It would be unfair to require that buildings be torn down, or to stop businesses, in order to meet the requirements of zoning regulations.

Nonconforming businesses or buildings are allowed to remain. However, restrictions are placed on rebuilding. If partially destroyed, they may be allowed to rebuild, depending on the amount of destruction. If 75 percent or more is destroyed, they are not usually allowed to rebuild in the same manner or for the same purpose in the same zone.

Any hardships imposed by zoning regulations may be relieved by a *variance.* Variances are granted by a Zoning Board of Appeals within each community. A public hearing is held after a certain period of time. The general public, and, in particular, those abutting the property are notified. The petitioner must prove certain types of hardship specified in the zoning laws before the zoning variance can be granted.

Building Codes

Building codes regulate the design and construction of buildings by establishing minimum safety standards. They prevent such things as roofs being ripped off by high winds, floors collapsing from inadequate support, buildings settling because of a poor foundation, and tragic deaths from fire due to lack of sufficient exits from buildings. In addition to building codes, other codes govern the mechanical, electrical, and plumbing trades.

Some communities have no building codes. Some write their own. Some have codes, but exempt residential construction. Some have adopted one of three national model building codes. Some use one of the national codes supplemented with their own.

It is important to have a general knowledge of the building code used by a particular community. Construction superintendents and contractors must have extensive knowledge of the codes.

National Building Codes. The *Basic National Building Code* (BNBC) is used primarily in the Northeast and Midwest. The *Uniform Building Code* (UBC) has been adopted in the West and Southwest. The *Standard Building Code* is used by states in the Southeast.

Many communities use the Council of American Building Officials (CABO) One- and Two-Family Dwelling Code for residential work. They then keep one of the national codes for commercial work. For residential work, other communities may use *Dwelling Construction under the Uniform Building Code.*

In Canada, the *National Building Code* sets the minimum standard. Some provinces augment this code with more stringent requirements and publish the combination as a *Provincial Building Code.* A few cities have charters, which allow them to publish their own building codes.

Use of Residential Codes. In addition to structural requirements, major areas of residential codes include:

- Exit facilities, such as doors, halls, stairs, and windows as emergency exits, and smoke detectors.
- Room dimensions, such as ceiling height and minimum area.
- Light, ventilation, and sanitation, such as window size and placement, maximum limits of glass area, fans vented to the outside, requirements for baths, kitchens, and hot and cold water.

Use of Commercial Codes. Codes for commercial work are much more complicated than those for residential work. The structure must first be defined for code purposes. In order to define the structure, six classifications must be used.

1. The *occupancy group* classifies the structure by how and whom it will be used. The classification is designated by a letter, such as R, which includes not only single-family homes but apartments and hotels.
2. The size and location of the building.
3. The type of construction. Five general types are given numbers 1 through 5. Types 1 and 2 require that all structural parts be noncombustible. Construction in types 3, 4, or 5 can be made of either masonry, steel, or wood.
4. The floor area of the building.
5. The height of the building. Zoning regulations may also affect the height.
6. The number of people who will use the building, called the *occupant load,* determines such things as the number and location of exits.

Once the structure is defined, the code requirements may be studied.

Building Permits

A *building permit* is needed before construction can begin. Application is made by the contractor to the office of the local building official. The building permit application form (Fig. 24-1) requires a general description of the construction, legal description and location of the property, estimated cost of construction, and information about the applicant.

Drawings of the proposed construction are submitted with the application. The type and kind of drawings required depend on the complexity of the building. For commercial work, usually five sets of plot plans and two sets of other drawings are required. The drawings are reviewed by the building inspection department. If all is in order, a permit (Fig. 24-2) is granted upon payment of a fee. The fee is usually based on the estimated cost of the construction. Electrical, mechanical, plumbing, water, and sewer permits are usually obtained by subcontractors. The permit card must be displayed on the site in a conspicuous place until the construction is completed.

Inspections

Building inspectors visit the job site to perform code inspections at various intervals. These inspections may include:

1. A *foundation inspection* takes place after the trenches have been excavated and forms erected and ready for the placement of concrete. No reinforcing steel or structural framework of any part of any building may be covered without an inspection and a release.
2. A *frame inspection* takes place after the roof, framing, fire blocking, and bracing are in place, and all concealed wiring, pipes, chimneys, ducts, and vents are complete.
3. The *final inspection* occurs when the building is finished. A Certificate of Occupancy or Completion is then granted.

It is the responsibility of the contractor to notify the building official when the construction is ready for a scheduled inspection. If all is in order, the inspector signs the permit card in the appropriate space and construction continues. If the inspector finds a code violation, it is brought to the attention of the contractor or architect for compliance.

These inspections ensure that construction is proceeding according to approved plans. They also make sure construction is meeting code requirements. This protects the future occupants of the building and the general public. In most cases, a good rapport exists between inspectors and builders, enabling construction to proceed smoothly and on schedule.

CITY OF ST. PETERSBURG
APPLICATION FOR BUILDING PERMIT

RADON GAS FEE ______

FOR OFFICE USE ONLY

Permit Type ______ Permit # ______

Permit Class of Work ______ Log # ______

Permit Use Code ______ Issue Date ______

Lot ______ Block ______ Sub ______ Permit Cost ______

Fire Zone: IN ______ OUT ______ Zone: ______ T.I.F. Due (Y/N or NA) ______

Utility Notification 1. FL Power ______ 2. Peoples Gas ______ 3. Water Dept. ______

B of A (Y/N) ______ Case No.
E.D.C. (Y/N) ______ Case No.
C.R.A. (Y/N) ______ Case No.
H.P.C. (Y/N) ______ Case No.

NOTE: Items with* must be entered in computer.

*Plat Page ______ *Sec ______ *Township ______ *Range ______ Zone ______

*Dept of Commerce Code ______ *Const. Type ______ Protected ______ Unprotected ______

*Additional Permits Required:
Building ______ Plumbing ______ No. of W.C. ______ No. of Meters ______
Electrical ______ Mechanical ______ Gas ______ Fire Sprk. ______ Landscape ______
Park/Paving ______ Total Spaces ______ Handicap ______

*Flood Zone ______ *Setbacks: Front ______ Left Side ______ Right Side ______
Rear ______ Other Requirements ______

Threshold Building YES ______ NO ______

Special Notes/Comments to Inspector: ______

APPLICANT PLEASE FILL OUT THIS SECTION

JOB ADDRESS ______ Suite or Apt. No. ______

CONTRACTOR ______ Cert./Reg. No. ______ Telephone ______

PROPERTY OWNER'S Name ______ Address ______

City ______ State ______ Zip ______ Telephone ______

Building Description: Total Sq. Ft. ______ Estimated Job Value ______
(LF-SF or Dimensions ______ Building Use ______
Valuation of Work ______ Former Use ______
No. of Units ______ No. of Suites ______ No. of Stories ______

Special Notes or Comments: ______

PHONE 893-7386 FOR ALL INSPECTIONS

HCS-12 Rev. 6-1-88

(OVER)

Fig. 24-1 A form used to apply for a building permit.

CITY OF ST. PETERSBURG
DEPARTMENT OF HOUSING & CONSTRUCTION SERVICES

BUILDING PERMIT

THIS PERMIT BECOMES INVALID IF NO INSPECTIONS HAVE BEEN MADE DURING ANY 3 MONTH PERIOD.

Flood Elevation - ________ Lowest Floor Minimum Required

- ☐ New Construction
- ☐ Grounds Improvements
- ☐ Utility Building
- ☐ Reroofing
- ☐ Moving
- ☐ Fences
- ☐ Pool
- ☐ Other ________________
- ☐ Siding
- ☐ Walls

Permit No. ________________ (ZONE) ________________

Job Address ________________________________

Lot ________ **Blk.** ________ **Sub.** ________________

Date ________________________________

This permit covers building construction only. Additional permits are required for electric, plumbing, gas and/or mechanical installations.

BUILDING			ELECTRICAL			PLUMBING			MECHANICAL-GAS		
Type of Inspection	Date	Inspector	Type of Inspection	Date	Inspector	Type of Inspection	Date	Inspector	Type of Inspection	Date	Inspector

NOTE: Building, Electrical, Plumbing and Mech/Gas Inspections shall be dated and initialed by inspectors before walls and ceilings are covered.

BUILDING OK TO COVER		ELECTRICAL OK TO COVER		PLUMBING OK TO COVER		MECH/GAS OK TO COVER	
Date	Inspector	Date	Inspector	Date	Inspector	Date	Inspector

NOTE: This card shall remain posted at the job site until all final inspections have been dated and initialed by inspectors.

BUILDING FINAL OK		ELECTRICAL FINAL OK		PLUMBING FINAL OK		MECH/GAS FINAL OK	
Date	Inspector	Date	Inspector	Date	Inspector	Date	Inspector

THIS CARD MUST BE POSTED IN AN EASILY SEEN LOCATION.

For INSPECTIONS, call 893-7386. ■ For other information, call 893-7388.

Fig. 24-2 A building permit. Communities use different kinds of forms.

Review Questions

Select the most appropriate answer.

1. A view looking from the top downward is called
   a. an elevation.
   b. a perspective.
   c. a plan.
   d. a section.

2. A view showing a vertical cut through the construction is called
   a. an elevation.
   b. a detail.
   c. a plan.
   d. a section.

3. The most commonly used scale for floor plans is
   a. 1/4" = 1'-0".
   b. 3/4" = 1'-0".
   c. 1 1/2" = 1'-0".
   d. 3" = 1'-0".

4. To determine a missing dimension,
   a. use the scale rule.
   b. calculate it.
   c. read the specifications.
   d. find a typical wall section.

5. Centerlines are indicated by a
   a. series of short, uniform dashes.
   b. a series of long and short dashes.
   c. a long dash followed by two short dashes.
   d. a solid, broad, dark line.

6. Which of the dimensions below would be found on a blueprint?
   a. 3'.
   b. 3 ft.
   c. 3'-0".
   d. 36".

7. On what drawing would the setback of the building from the property lines be found?
   a. Floor plan.
   b. Plot plan.
   c. Elevation.
   d. Foundation plan.

8. To find out which edge of doors are to be hinged, look at the
   a. elevations.
   b. framing plan.
   c. floor plan.
   d. wall section.

9. The elevation of the finished first floor is usually found on the
   a. plot plan.
   b. floor plan.
   c. foundation plan.
   d. framing plan.

10. An exterior wall stud height can best be determined from the
   a. floor plan.
   b. framing elevation.
   c. wall section.
   d. specifications.

BUILDING FOR SUCCESS

The Role of Construction Documents

The construction student who desires to excel as a carpenter and eventually as a builder will need to be proficient in interpreting plans and reading specifications. Units 8 and 9 have provided essential information about reading plans, codes, and specifications. With this information, you will want to combine your basic understanding of how all of this documentation fits together. All of the information contained in the construction documents will be implemented, at various intervals, either before, during, or after the construction process is completed.

Construction documents vary. However, they normally consist of at least the working drawings and specifications. Part of the package may also involve local codes, ordinances, bidding information and forms, conditions of the contract, and any other pertinent agreements. The complexity of the project will often dictate the exact makeup of the document package. Generally residential projects do not contain the extensive documentation that commercial construction projects do. The full purpose of this information is to assure that the construction process is completed as proposed. Additionally, having a structure that is safe for occupancy and completed on schedule are important goals included in the document group. By law the contracts become binding, and may require legal counsel to clarify discrepancies or disputes.

During the construction planning process, the builder will use the blueprints, specifications, codes, and local regulations to prepare an accurate cost estimate of the job and materials needed to complete the project, The cost estimate in turn will be submitted for consideration by the client. Until the contractor submits the bid, the opportunity exists for questions to be answered that will affect his or her estimated bid. These questions may center on a variety of things, such as materials, assemblies, time schedules, responsibilities of all parties, or other factors affecting the project.

Throughout the construction process, everyone involved will be required to adhere to the construction documents and all agreements. Once construction is underway, the client, contractor, subcontractors, inspectors, and architects will refer to the documents for accuracy and accountability. The outcomes must be as proposed in the beginning. Payment for services will be made by the owner only if these services comply with the stipulations contained in the construction documents.

The successful conclusion of the project will be based on how well the construction requirements were met. Without a clear understanding of the stipulations surrounding the construction project, many setbacks will occur. Any delays or unexpected deviations from the proposed plan will greatly affect the schedules and profit margin for the builder. Additionally, deviations not planned will affect the relationship of the builder, subcontractors, and client.

Construction documents serve as the focal point for the entire construction process. Therefore, they must be understood by everyone. In many cases involving disputes, the specifications and contract will have precedence over the blueprints. Historically, successful construction projects have been associated with clear understandings centered on the construction document—*before* the construction began.

FOCUS QUESTIONS: For individual or group discussion

1. What do you think are included in the term *construction documents?*
2. What bearing should these various written stipulations and drawings have on the outcome of a new custom-built home?
3. How can these documents play a role in the settling of disputes that may come up during or following the construction of a new commercial structure such as a hotel?
4. Is there a need to maintain the construction documents for future reference? Think of a case in which there was structural failure that affected people's lives.

STUDENT ACTIVITIES for Section 1: Tools and Materials

Unit 1 Wood and Lumber

1. Using a tree cross-section, identify the various parts. Explain the function of each in the growth process.
2. Identify and present eight wood defects. Explain how these defects affect the wood's use and grading.
3. Present and explain a lumber grade stamp in detail.

4. Identify and show the difference between softwood lumber that is plain-sawed and quarter-sawed.
5. Using boards, define and show the types of warp.
6. Calculate the quantity of board feet and cost of the following dimensional lumber:
   a. 14 – 2″ × 10″ × 16′ @ $525/MBF
   b. 84 – 1″ × 8″ × 14′ @ $754/MBF
   c. 450 – 2″ × 4″ × 8′ @ $485/MBF

Unit 2 Engineered Panels

1. Select a plywood or composite panel grade stamp. Interpret the stamp markings. Then state the possible uses for that panel in construction.
2. Select three nonstructural panels. State their manufacturing processes and uses in construction.
3. Using a residential blueprint, calculate the number of sheets of 4 × 8 floor, wall, and roof sheathing needed to cover the exterior of the house. (Make no deductions for openings.)

Unit 3 Engineered Lumber

1. Select a laminated veneer lumber (LVL) product. Explain the manufacturing process. Using an example from the manufacturer's product guide, explain how the sizing is determined for a floor beam product.
2. Using a manufacturer's guide and design manual (such as Boise Cascade, BCI Joist; Louisiana Pacific, Inner Seal I, Joists 23 series, or Truss Joist MacMillan), explain how an "I" joist system can be sized for a residential floor and roof. Explain how to use the span charts with an example house plan.
3. Using the same product guide and manual, explain the process of web hole location and sizing in the "I" joist products.
4. Explain in detail the fastening systems for "I" floor and roof assemblies by using framing connectors. Give examples of proper sizing, placing, and nailing with the connectors.

Unit 4 Fasteners

1. Prepare and present a display board showing a large variety of nails used in residential and light commercial construction. Explain the names, available sizes, various surface treatments, uses, approximate cost, and in what quantities they may be purchased.
2. Prepare and present a display board showing a large variety of screws, bolts, and other construction fasteners. Include the same information as in the nail display.
3. Create a display board assortment of the various construction adhesives. Exhibit products, empty containers, product labels, literature, or samples that illustrate a large array available for use. Classify the adhesives as mastics, glues, and so forth. State their safe uses.

Unit 5 Hand Tools

1. Orally present the fractions of an inch, by 16ths (begin with 1/16, 1/8, 3/16, etc.)
2. Using a grinder and a hone, demonstrate the proper procedure for sharpening a chisel and plane iron.
3. Using a block plane, a combination square, and a piece of softwood approximately 1″ × 4″ × 12″, with a 4- to 5-degree bevel on one edge, reestablish a square edge to replace the bevel.
4. Using a hammer, fasten together two pieces of 2″ × 4″ × 2′, with two 3 inch box nails. Set the nails flush with the surface. Do not leave any hammer marks. Then fasten a small 1-inch board (fascia) to the surface of the 2″ × 4″ assembly, using 2-1/2-inch box nails. Leave no hammer marks.
5. On a piece of 2″ × 4″ lay out a 90 degree angle across the surface and on the edge. Using an 8- to 10-pt. hand saw, make a perpendicular cut through the lumber. Repeat the procedure using a 30 degree angle across the surface of the lumber. Cut to the line.

Unit 6 Portable Power Tools

The following exercises should only be attempted after proper instruction has been given by the instructor or qualified tool representative. Use safety glasses or goggles at all times.

1. Using saw horses and a portable electric saw, safely cut the following on the marked lines.
   a. Using a 6-foot 2″ × 4″, cut off four or five one-inch marked increments at 90 degrees, freehand. Then cut four or five using a saw protractor. Repeat the process, making a 45 degree compound miter across the end.

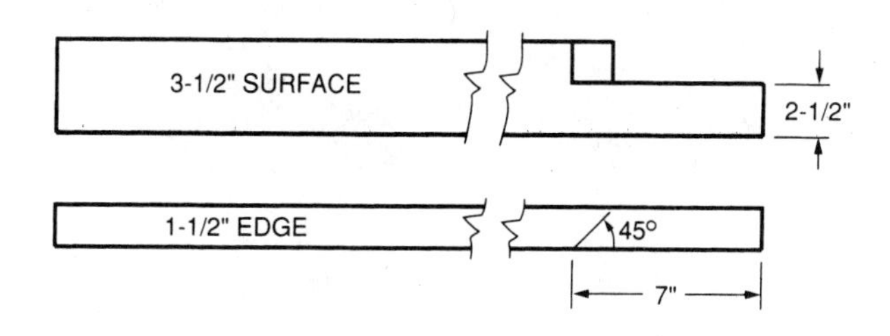

b. Using the rip guide and protractor, make a notch in the 2″ × 4″ as shown. Do not overcut.

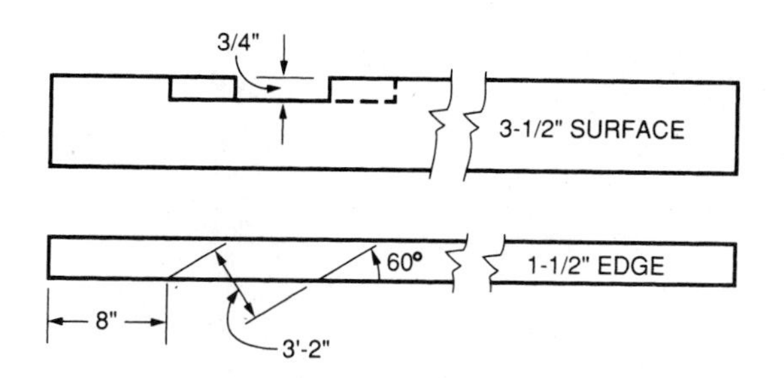

c. Cut a notch in the lumber as shown. A hammer and chisel may be used to complete the notch.

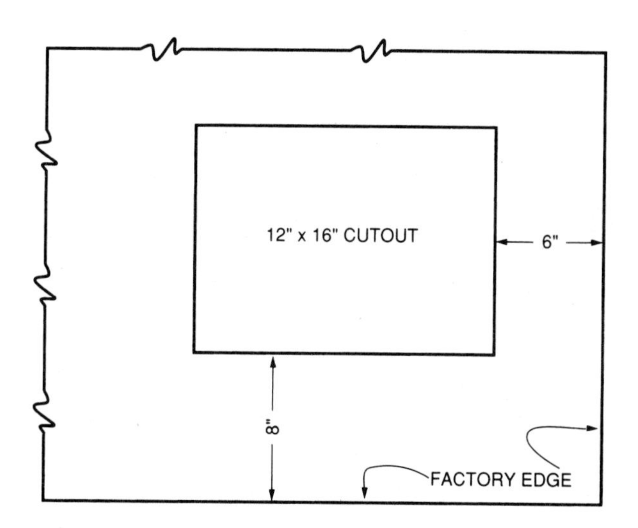

d. Using a piece of 3/4-inch plywood or waferboard, make a plunge cut as shown.

2. Using a portable electric jointer plane, and a 6-foot 2″ × 4″, plane a 3 degree bevel on the edge. The completed bevel should show no variations from a straight line or 3 degree angle. Check with a straightedge and a sliding T-bevel.

3. Select any three router bits. Safely demonstrate the setup and cutting process on a piece of plywood or lumber. Relate the names, type of cut made, and where they would be used in carpentry.

4. Demonstrate the safe procedure for setting up a portable compressor and pneumatic nailer for nailing dimensional lumber and sheet materials. Include: compressor/nailer preparation, airline connections, safe nailing procedures, disconnections, and proper storage.

5. Following a demonstration and training course by a certified field representative, and personal experience, demonstrate the safe and proper use of a powder-actuated driver fastening system. Explain and demonstrate how to safely fasten wood to concrete and wood to steel. Explain the powder charge color-coded system.

Unit 7 Stationary Power Tools

Upon receiving safe tool operation instruction by an instructor, complete the following.

1. Properly change the blade on a table saw and radial arm saw.

2. Using the appropriate saw (table saw, radial arm saw, or power miter box) make the following cuts: crosscut, rip, miter, compound miter, dado, plough, rabbet, bevel rip, taper rip, cove cut, duplicate length cuts.

3. Using the jointer, cut a 90-degree edge joint, end grain joint, taper cut, and 45-degree edge bevel joint.

Unit 8 Blueprints and Building Codes

Using a full set of residential blueprints and specifications (if included), a building code, architect's scale, and calculator, assemble a take-off list of construction materials for the concrete (foundation, floors, driveway, and walks), frame, exterior, and interior finish. The completed list should contain quantities and volumes of the construction materials. (This estimating exercise can be divided into various individual assignments.)

SECTION 2

Rough Carpentry

Building Layout

Before construction begins, lines must be laid out show the location and elevation of the building foundation. Accuracy in laying out these lines is essential in order to comply with local zoning ordinances. In addition, accurate layout lines provide for a foundation that is level and to specified dimensions. Accuracy in the beginning makes the work of the carpenter and other construction workers easier later.

OBJECTIVES

After completing this unit, the student should be able to:

- establish level points across a building area using a water level and by using a carpenter's hand spirit level in combination with a straightedge.
- accurately set up and use the builder's level, transit-level, and laser level for leveling, determining and establishing elevations, and laying out angles.
- lay out building lines by using the Pythagorean Theorem method for squaring corners and check the layout for accuracy.
- build batter boards and accurately establish layout lines for a building using building layout instruments,

UNIT CONTENTS

CHAPTER 25 LEVELING AND LAYOUT TOOLS

Building layout requires leveling lines as well as laying out various angles over the length and width of the structure. The carpenter must be able to set up, adjust, and use a variety of leveling and layout tools.

Leveling Tools

Several tools, ranging from simple to state-of-the-art, are used to level the layout. More sophisticated leveling and layout tools, although preferred, are not always available.

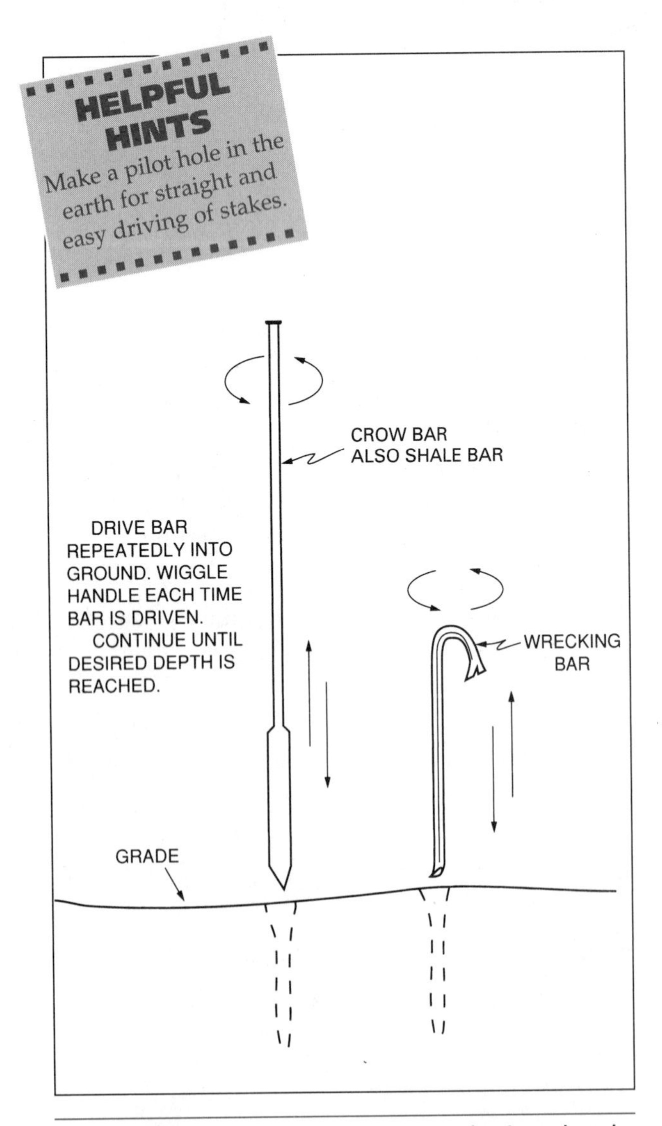

Fig. 25-1 Methods of starting a stake into hard ground.

Levels and Straightedges

If no other tools are available, a *carpenter's hand level* and a long *straightedge* may be used to level across the building area. Select an 8- to 10-foot length of lumber. Make sure it has a straight edge and is wide enough that it will not sag when placed on edge and supported only on its ends. (Use the method previously described in this book to determine if the edge is straight.)

Hold the straightedge on edge. Place one end on the surface to be leveled and the other end on the top of a stake driven in the ground so that the straightedge is level. Determine levelness by holding the spirit level against the bottom edge.

Move the straightedge and place the beginning end on the driven stake. Place the other end on another driven stake until the straightedge is again level. Stakes can be driven straight with relative ease if a pilot hole is first made in the earth (Fig. 25-1). Continue moving the straightedge from stake to stake until the desired distance is leveled (Fig. 25-2). This is an accurate, although time-consuming, method of leveling over a long distance. It can be done by one person.

If you want to level to the corners of building layouts, start by driving a stake near the center so its top is to the desired height. Level from the center stake to each corner in the manner described (Fig. 25-3).

Water Levels

A *water level* is a very accurate tool, dating back centuries. It is used for leveling from one point to another. Its accuracy, within a pencil point, is based on the principle that water seeks its own level (Fig. 25-4).

One commercial model consists of 50 feet of small diameter, clear vinyl tubing and a small tube storage container. A built-in reservoir holds the colored water that fills the tube. One end is held to the starting point. The other end is moved down until the water level is seen and marked on the surface to be leveled (Fig. 25-5).

Although highly accurate, the water level is somewhat limited by the length of the plastic tube. However, extension tubings are available. Also, though slightly inconvenient, the water level may be moved from point to point.

A water level is accurate only if both ends of the tube are open to the air and there are no air bubbles in the length of the tube. Because both ends

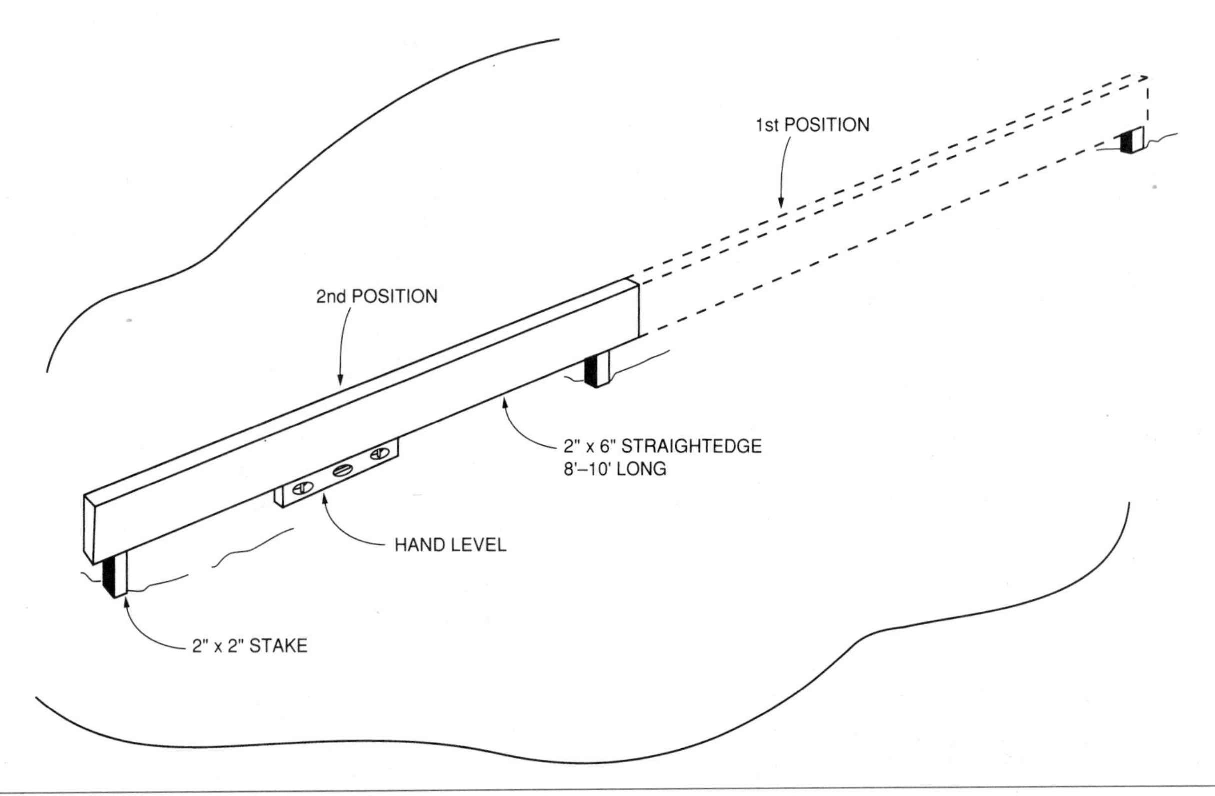

Fig. 25-2 Leveling with a straightedge from stake to stake.

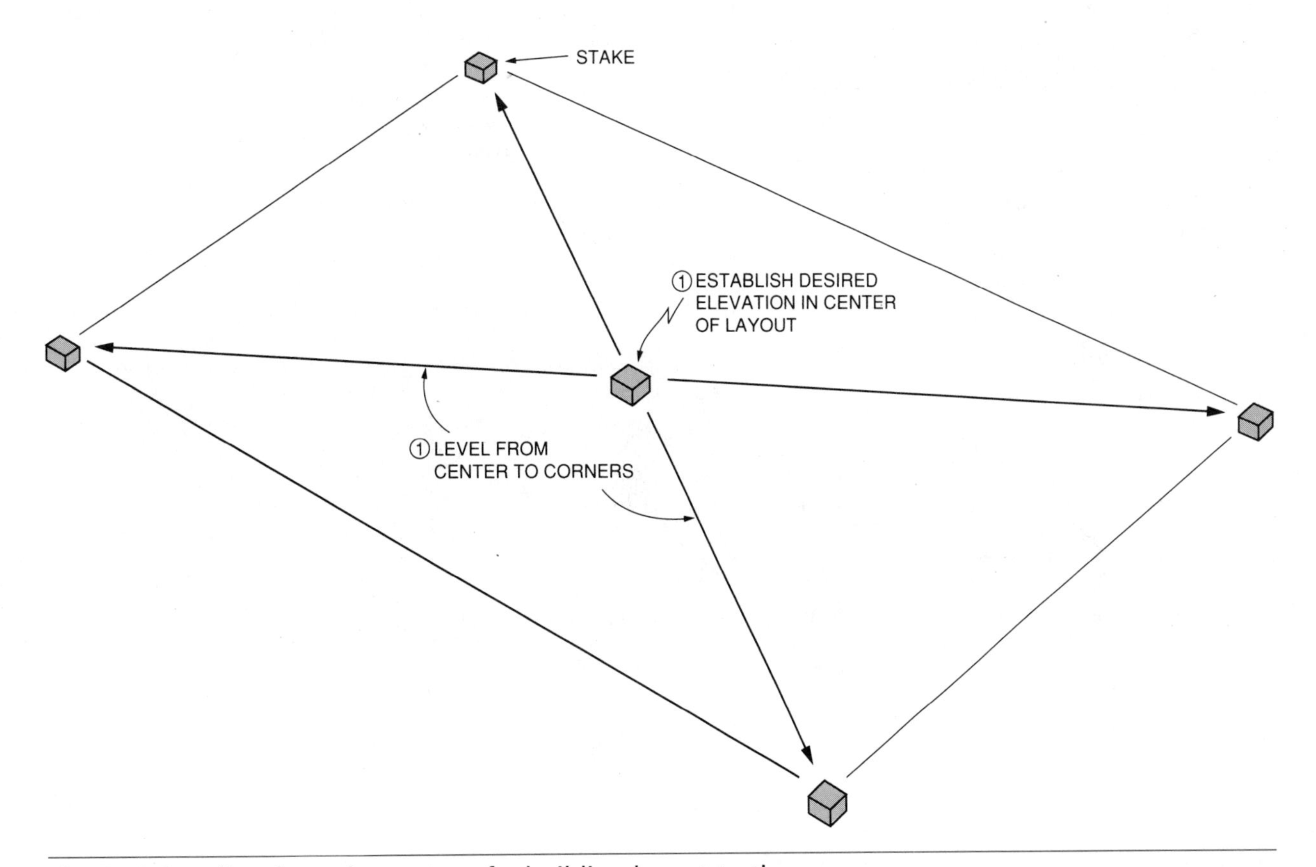

Fig. 25-3 Leveling from the center of a building layout to the corners.

must be open, there may occasionally be some loss of liquid. However, this is replenished by the reservoir. Any air bubbles can be easily seen with the use of colored water.

In spite of these drawbacks, the water level is an extremely useful, inexpensive, simple tool for leveling from room to room, where walls obstruct views, down in a hole, or around obstructions. Another advantage is that leveling can be done by one person.

Optical Levels

The most commonly used optical instruments for leveling, plumbing, and angle layout are the *builder's level* and the *transit-level.*

Builder's Levels

The *builder's level* (Fig. 25-6) consists of a *telescope* to which a *spirit level* is mounted. The telescope is fixed in a horizontal position. It can rotate 360 degrees for measuring horizontal angles but cannot be tilted up or down.

Transit-levels

The *transit-level* (Fig. 25-7) is similar to the builder's level. However, its telescope can be moved up and down 45 degrees in each direction. This feature enables it to be used more effectively than the builder's level.

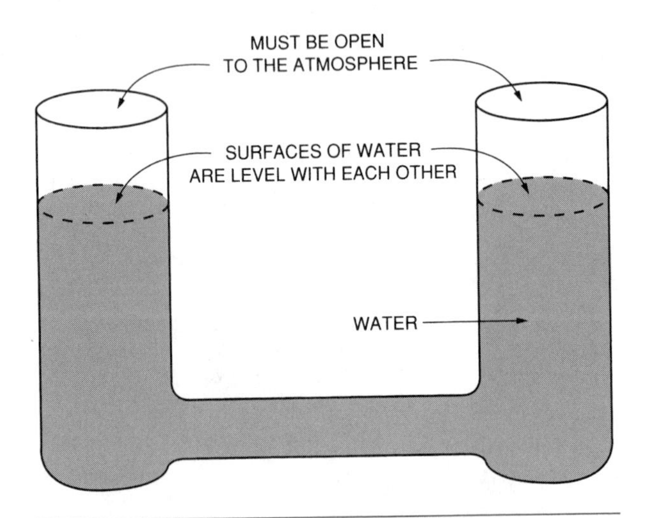

Fig. 25-4 Water seeks its own level. Both ends of the water level must be open to the atmosphere.

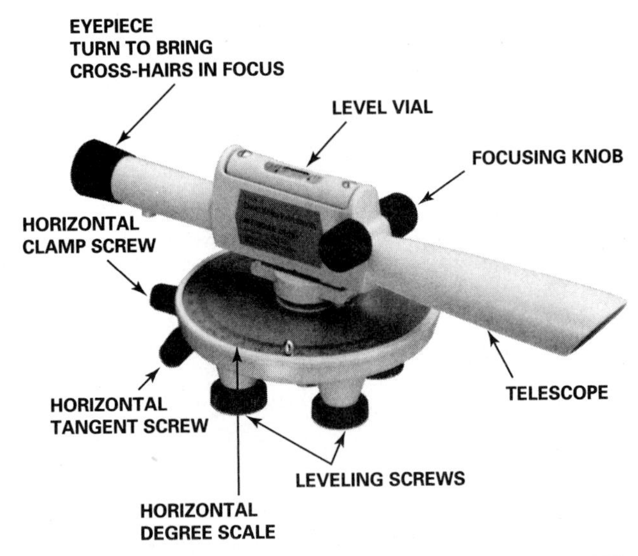

Fig. 25-6 The builder's level. (*Courtesy of David White*)

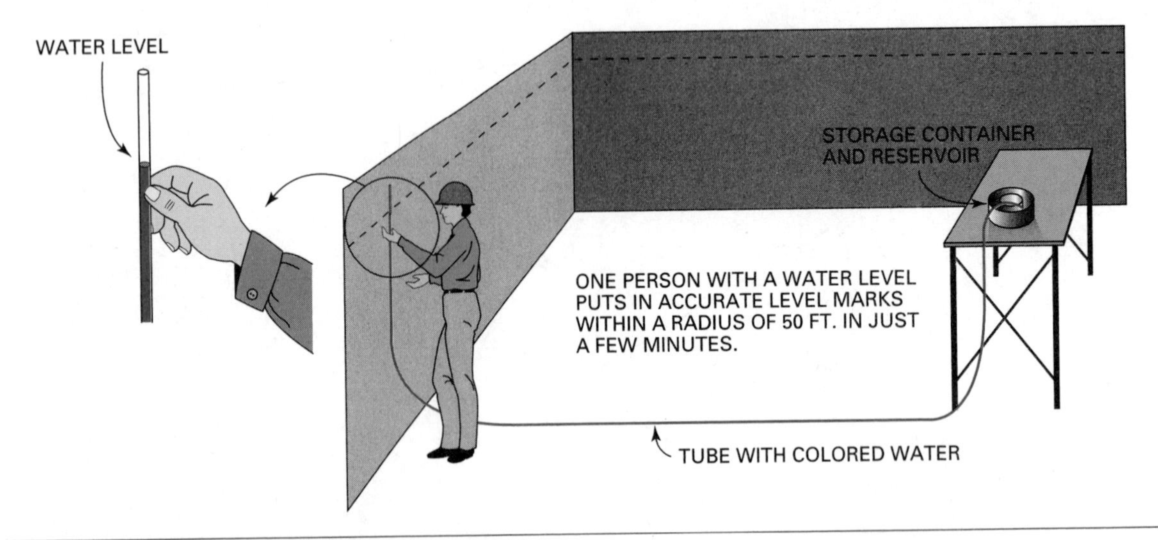

Fig. 25-5 The water level is a simple tool to use.

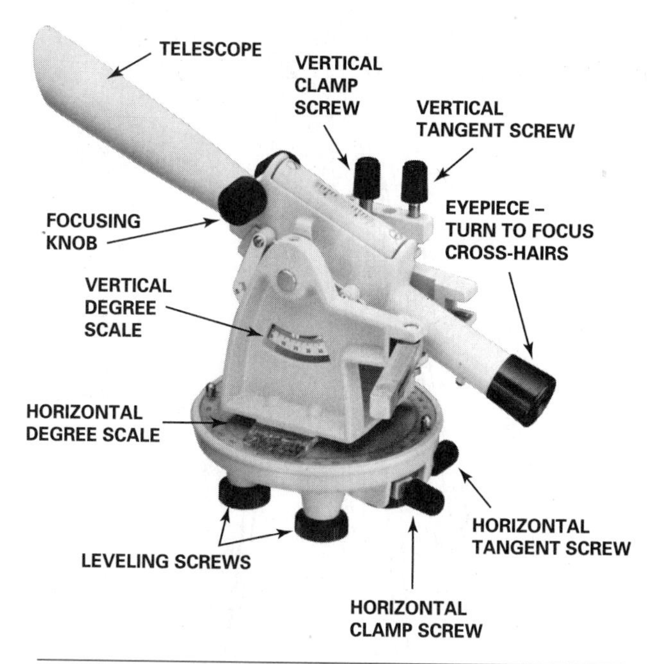

Fig. 25-7 The telescope of the transit-level may be moved up and down 45 degrees each way. (*Courtesy of David White*)

Automatic Levels

Automatic levels and *automatic transit-levels* (Fig. 25-8) are similar to those previously described except that they have an internal *compensator.* This compensator uses gravity to maintain a true level line of sight. Even if the instrument is jarred, the line of sight stays true because gravity does not change.

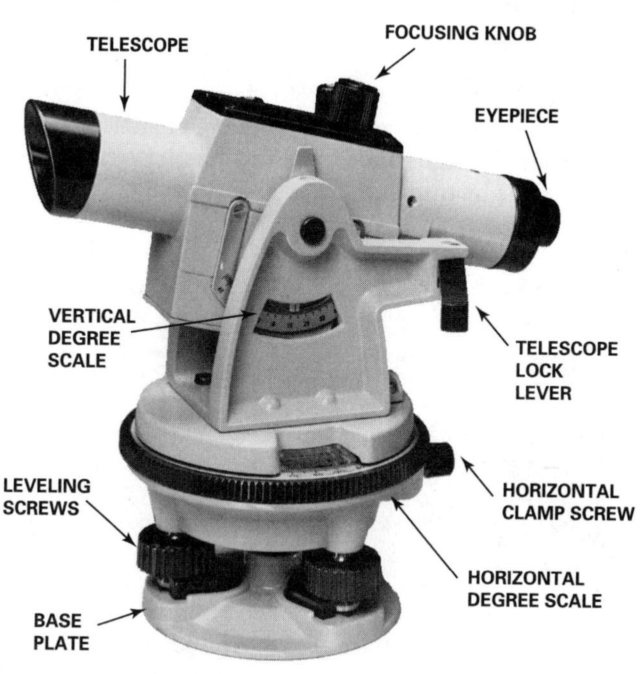

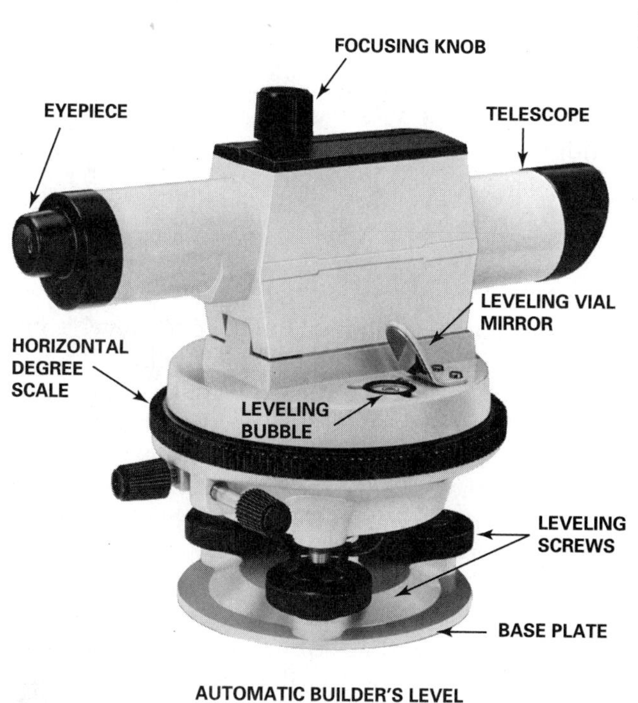

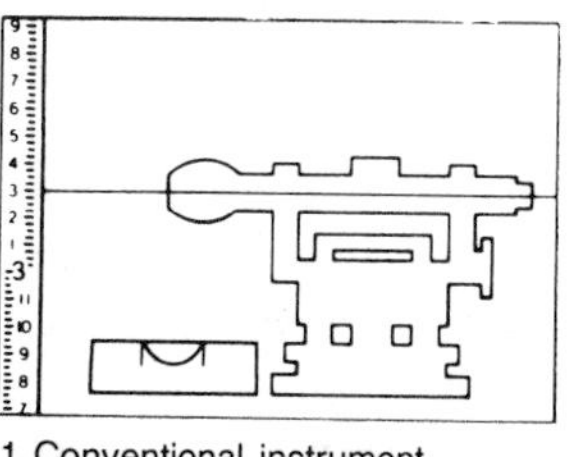

1 Conventional instrument correctly leveled. Rod reading is 3′-3″.

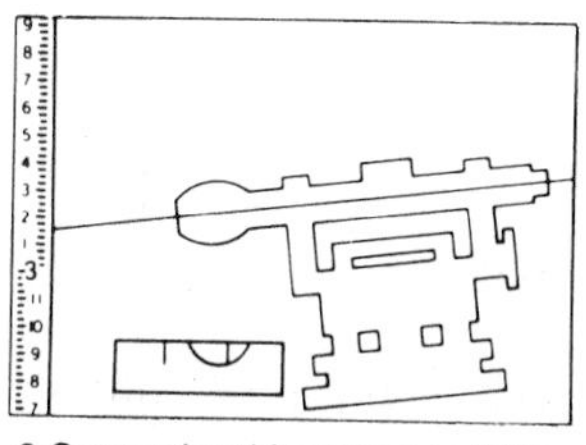

2 Conventional instrument slightly out of level. Vial bubble is off center and incorrect rod reading is 3′-1 1/2″.

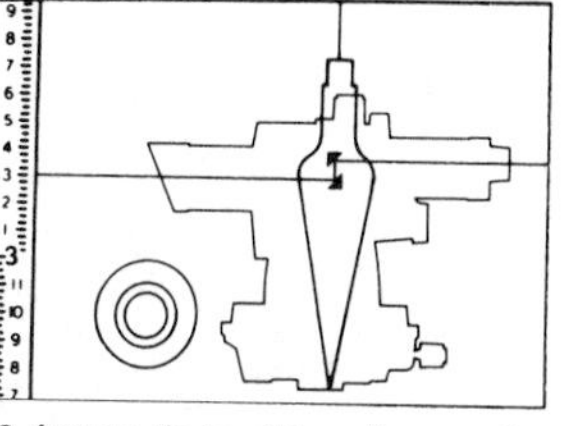

3 Automatic level-transit correctly leveled. Rod reading is 3′-3″.

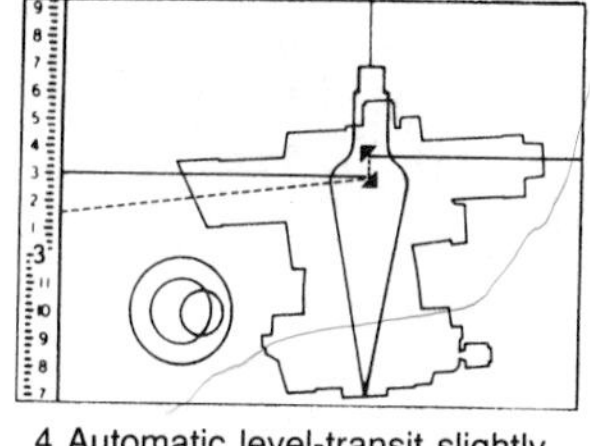

4 Automatic level-transit slightly out of level. Circular bubble is off center, but the compensator corrects for the variation from level and maintains a correct rod reading of 3′-3″.

THE CORRECTION RANGE ± 15 MINUTES OF ANGLE. THIS AMOUNTS TO AN ERROR CORRECTION CAPABILITY OF UP TO 5 INCHES ON A LEVELING ROD READING AT 100 FEET.

Fig. 25-8 Automatic levels and transit-levels. (*Courtesy of David White*)

There are many models of leveling instruments available. To become familiar with more sophisticated levels, study the manufacturers' literature. No matter what type of level is used, the basic procedures are the same.

Using Optical Levels

Before the level can be used, it must be placed on a *tripod* or some other solid support and leveled.

Setting Up and Adjusting the Level

The telescope is adjusted to a level position by means of four *leveling screws* that rest on a *base leveling plate.* In higher-quality levels, the base plate is part of the instrument. In less expensive models, the base plate is part of the tripod.

Open and adjust the legs of the tripod to a convenient height. Spread the legs of the tripod well apart, and firmly place its feet into the ground.

CAUTION On a smooth surface it is essential that the points on the feet hold without slipping. Make small holes or depressions for the tripod points to fit into. Or, insert screw eyes at the lower inside of the tripod legs and attach wire or light chain to the three screw eyes. (Fig. 25-9).

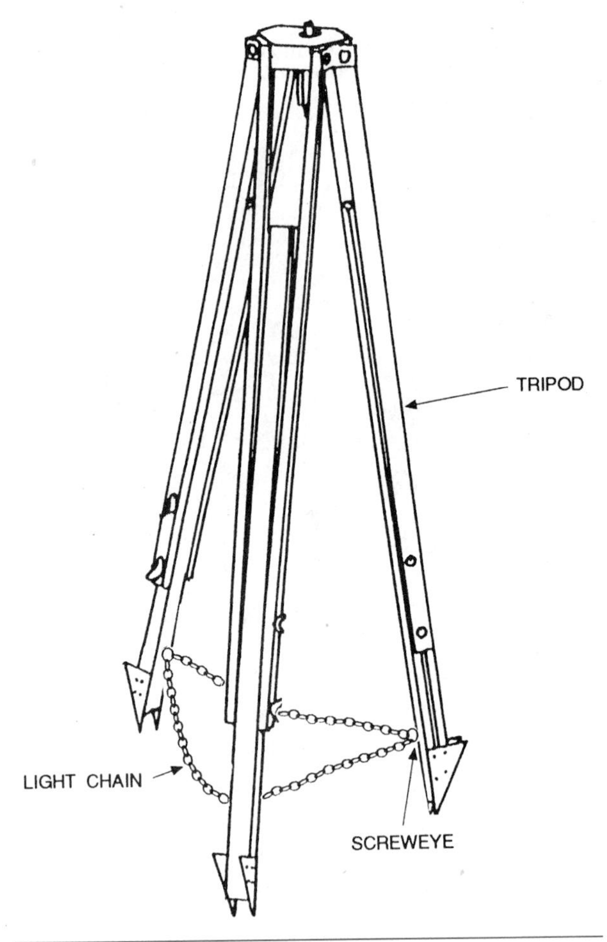

Fig. 25-9 Make sure the feet of the tripod do not slip on smooth or hard surfaces.

When set up, the top of the tripod should be as level as possible. Sight by eye and tighten the tripod wing nuts. With the top of the tripod close to level, adjustment of the instrument is made easier.

Lift the instrument from its case by the frame (do not hold on to the telescope to do so). Note how it is stored so it can be replaced in the case in the same position. Make sure the horizontal clamp screw is loose so the telescope revolves freely. While holding onto the frame, secure the instrument to the tripod. (The attaching mechanism varies according to the manufacturer and the quality of the instrument.)

CAUTION Care must be taken not to damage the instrument. Never use force on any parts of the instrument. All moving parts turn freely and easily by hand. Excessive pressure on the leveling screws may damage the threads of the base plate. Unequal tension on the screws will cause the instrument to wobble on the base plate resulting in leveling errors. Periodically use a toothbrush dipped in light instrument oil to clean and lubricate threads of the adjusting screws.

Accurate leveling of the builder's level is important. Line up the telescope directly over two opposite leveling screws. Turn the screws in opposite directions with forefingers and thumbs. Move the thumbs toward or away from each other, as the case may be, to center the bubble in the spirit level (Fig. 25-10). The bubble will always move in the same direction as your left thumb is moving.

Rotate the telescope 90 degrees over the other two opposite leveling screws and repeat the procedure. Make sure each of the screws has the same, but not too much, tension. Return to the original position, check, and make minor adjustments. Continue adjustments until the bubble remains exactly centered when the instrument is revolved in a complete circle.

CAUTION Do not leave a set-up instrument unattended near moving equipment.

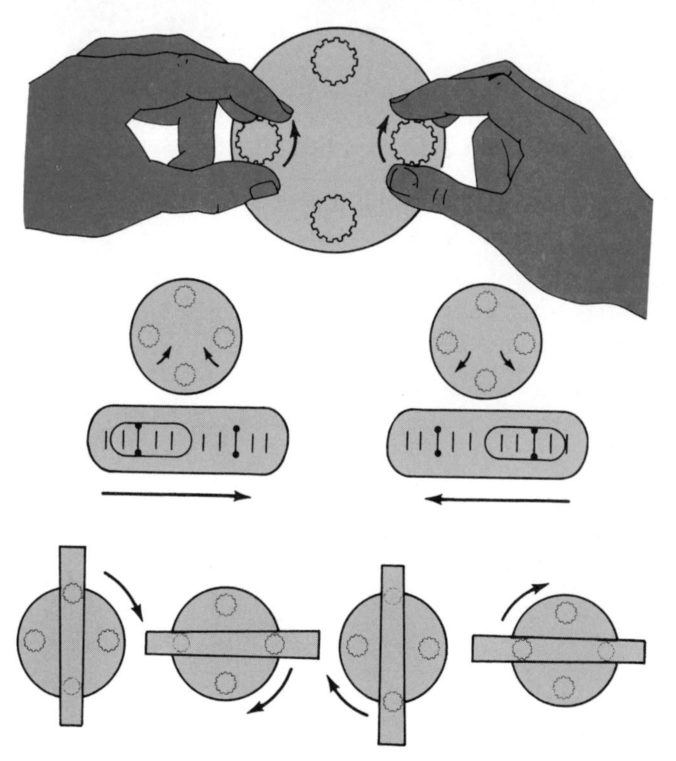

Fig. 25-10 A. Level the instrument by moving thumbs toward or away from each other. B. The instrument is level when the bubble remains centered as the telescope is revolved in a complete circle.

Sighting the Level

To sight an object, rotate the telescope and sight over its top, aiming it at the object. Look through the telescope. Focus it by turning the focusing knob one way or the other, until the object becomes clear. Keep both eyes open. This eliminates squinting, does not tire the eyes, and gives the best view through the telescope (Fig. 25-11).

CAUTION **If the lenses need cleaning, dust them with a soft brush or rag. Do *not* rub the dirt off. Rubbing may scratch the lens coating.**

When looking into the telescope, vertical and horizontal *cross-hairs* are seen. They enable the target to be centered properly (Fig. 25-12). The

Fig. 25-11 When sighting through the instrument, keep both eyes open. The worker is focusing the instrument.

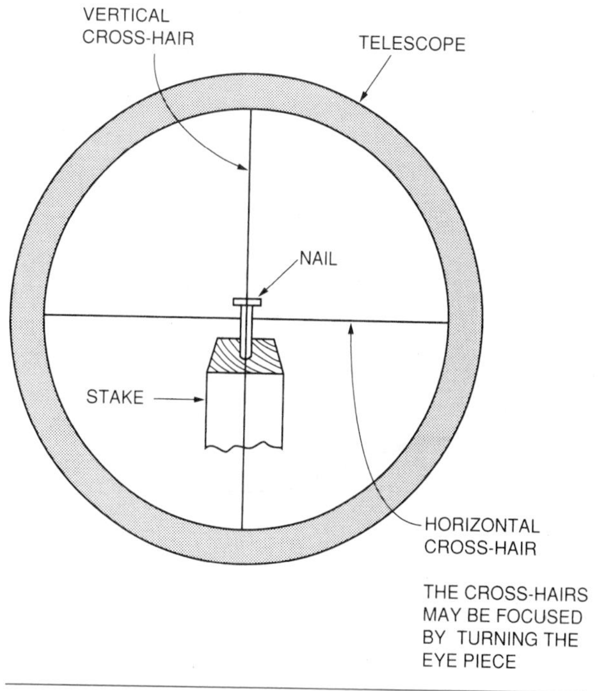

Fig. 25-12 When looking in the telescope, vertical and horizontal cross-hairs are seen.

cross-hairs can be brought into focus by turning the eyepiece one way or the other. Center the cross-hairs on the object by moving the telescope left or right. A fine adjustment can be made by tightening the horizontal clamp screw and turning the horizontal tangent screw one way or the other. The horizontal cross-hair is used for reading elevations. The vertical cross-hair is used when laying out angles and aligning vertical objects.

Leveling

When the instrument is leveled, a given point on the line of sight is exactly level with any other point Any line whose points are the same distance below or above the line of sight is also level (Fig. 25-13). To level one point with another, a helper must hold a *target* on the point to be leveled. A reading is taken. The target is then moved to selected points that are brought to the same elevation by moving those points up or down to get the same reading.

Targets. A *folding rule* is often used as a target (Fig. 25-14). The end of the rule is placed on the point to be leveled. The rule is then moved up or down until the same mark is read on the rule as was read at the starting point.

A tape may also be used as a target. Because of its flexibility, it may need to be backed up by a strip of wood to hold it rigid (Fig. 25-15). It is then used in the same manner as the folding rule.

The simplest target is a plain 1 × 2 strip of wood. The end of the stick is held on the starting point of desired elevation. The line of sight is marked on the stick. The end of the stick is then placed on top of various points. They are moved up or down to bring the mark to the same height as the line of sight (Fig. 25-16). A stick of practically any

Fig. 25-14 Leveling the top of a concrete form by sighting at a folding rule.

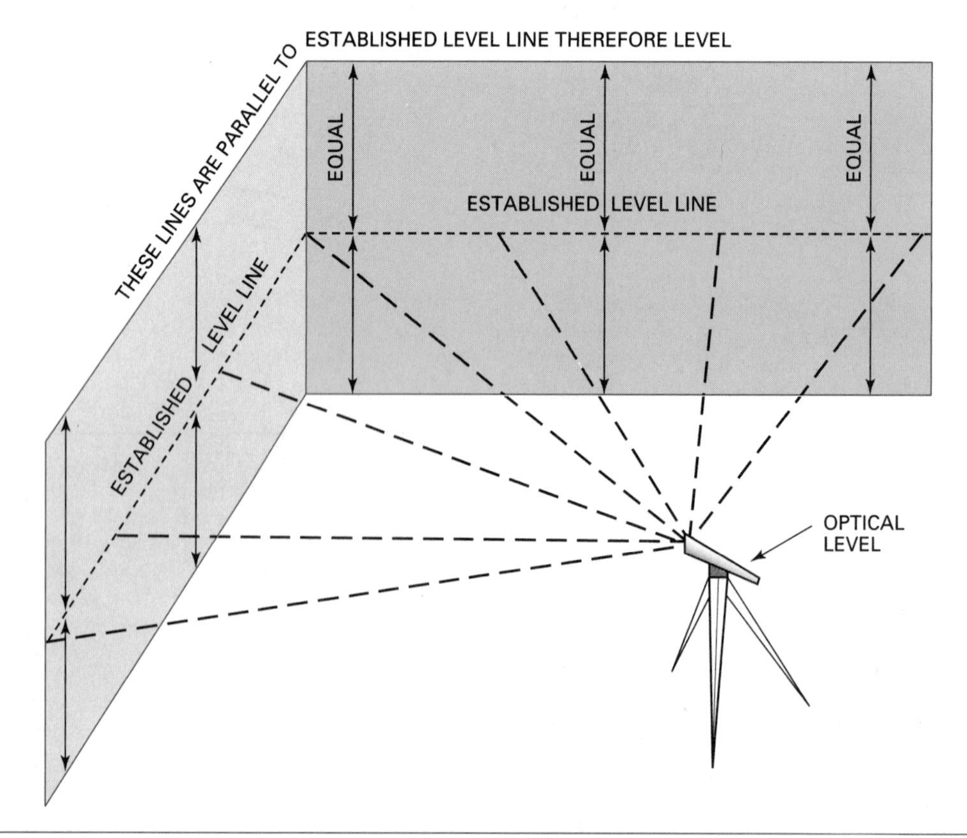

Fig. 25-13 Any line parallel to the established level line is also level.

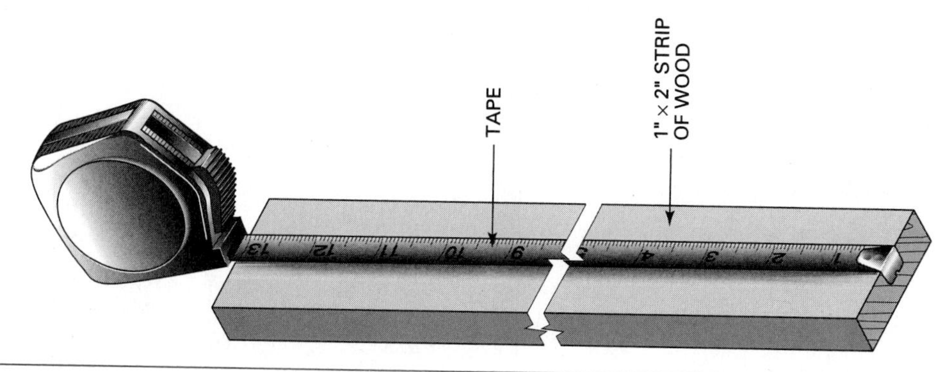

Fig. 25-15 A tape may need to be backed up by a strip of wood if used as a target.

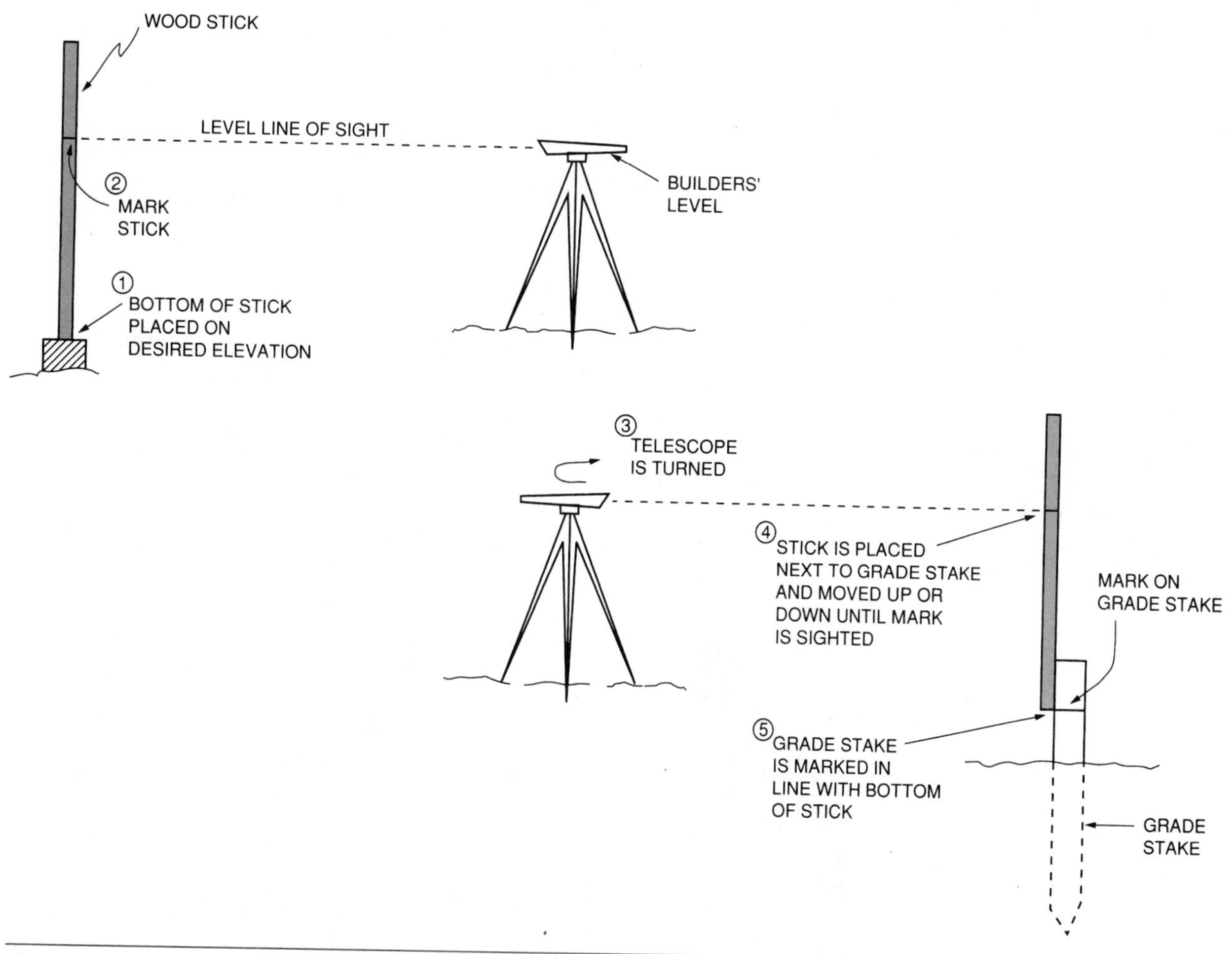

Fig. 25-16 Using a stick for a target.

length can be used. If a greater length is required, two or more strips may be fastened together.

Cut the stick to a length so that the mark to be sighted is a noticeable distance off center of its length. It is immediately noticeable if the stick is inadvertently turned upside down (Fig. 25-17). If, for some reason, it is not desirable to cut the stick, clearly mark the top and bottom ends.

Leveling Rods. For longer sightings, the *leveling* rod is used because of its clearer graduations. A variety of rods are manufactured of wood or fiberglass for several leveling purposes. They are made with two or more sections that extend easily and lock into place. Rods vary in length—from two-section rods extending 9′-0″ up to seven-section rods extending 25′-0″.

HELPFUL HINTS

Keep the stick an acceptable length so the mark is not close to being centered on its length.

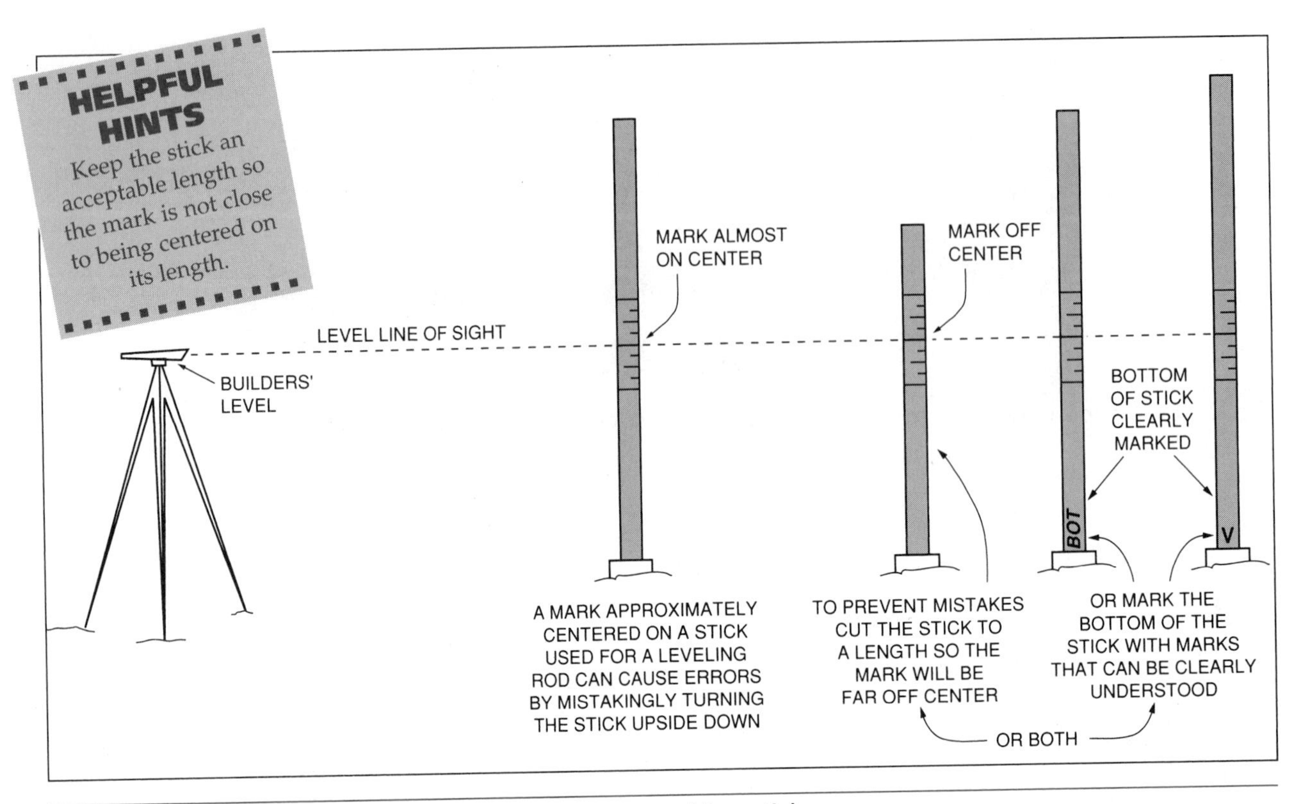

Fig. 25-17 Techniques for creating an easy-to-use marking stick.

The builder's rod has feet, inches, and 8ths of an inch. The graduations are 1/8 inch wide and 1/8 inch apart. The engineer's rod is very similar yet the scale is slightly different. It is in feet, tenths, and hundredths of a foot. Instead of inches, the number markings represent a tenth of a foot. The smaller graduations are 1/100 of a foot wide and 1/100 foot apart, which is slightly smaller than 1/8 inch. They are both designed for easy reading. An oval-shaped, red and white, movable target is available to fit on any rod for easy readings (Fig. 25-18). Other types of rods are graduated in feet, 10ths, and 100ths of a

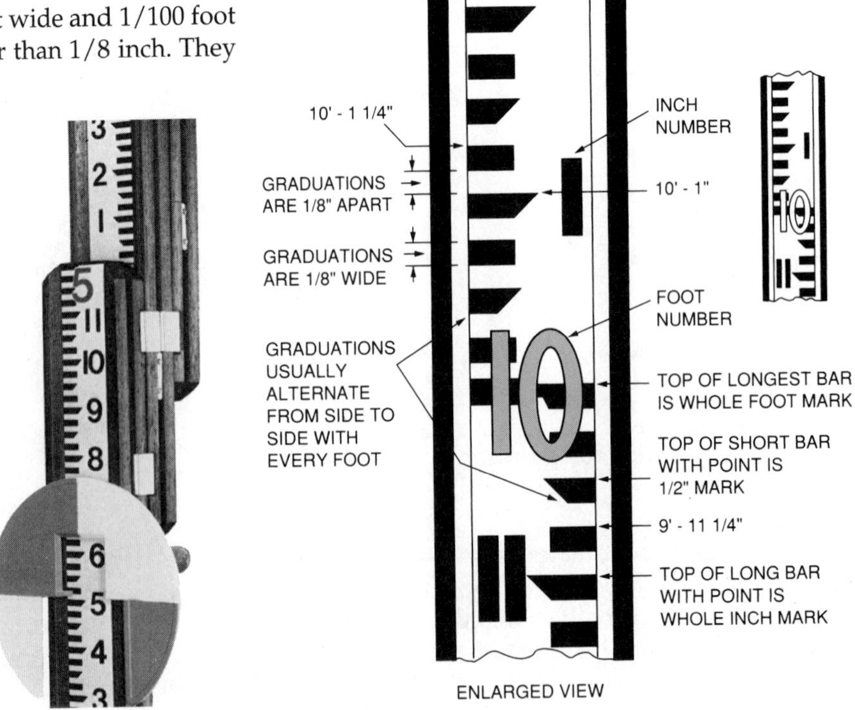

Fig. 25-18 The builder's leveling rod is marked in feet, inches, and eighths of an inch. (*Courtesy of David White*)

foot for surveying work. Some rods have metric gradations.

Communication. A responsible rod operator holds the rod vertical and faces the instrument so it can be read with ease and accuracy. Sighting distances are not usually over 100 to 150 feet. Sometimes voice commands cannot be used. Hand signals are then given to the rod operator to move the target as desired by the instrument operator. Usually appropriate hand signals are given even when distances are not great. Shouting on the job site is unnecessary, unprofessional, and creates more confusion than it eliminates.

Establishing Elevations

Many points on the job site, such as the depth of excavations, the height of foundation footing and walls, and the elevation of finish floors, are required to be set at specified elevations or grades. These elevations are established by starting from the *benchmark.* The benchmark is a point of designated elevation. The instrument operator records elevations and rod readings in a notebook to make calculations.

Height of the Instrument (HI). When it is necessary to set a point at some definite elevation, determine the *height of the instrument* (HI). To find HI, place the rod on the benchmark and add the reading to the elevation of the benchmark (Fig. 25-19). For instance, if the benchmark has an elevation of 100.00 feet and the rod reads 5′-8″, then the HI is 105′-8″.

Grade Rod. What must be read on the rod when its base is at the desired elevation is called the *grade rod.* This is found by subtracting the desired elevation from the height of the instrument (HI). For instance, if the elevation to be established is 102′-0″, subtract it from 105′-8″ (HI) to get 3′-8″ (the grade rod). The rod operator places the rod at the desired point. He or she then moves it up or down, at the direction of the instrument operator, until the

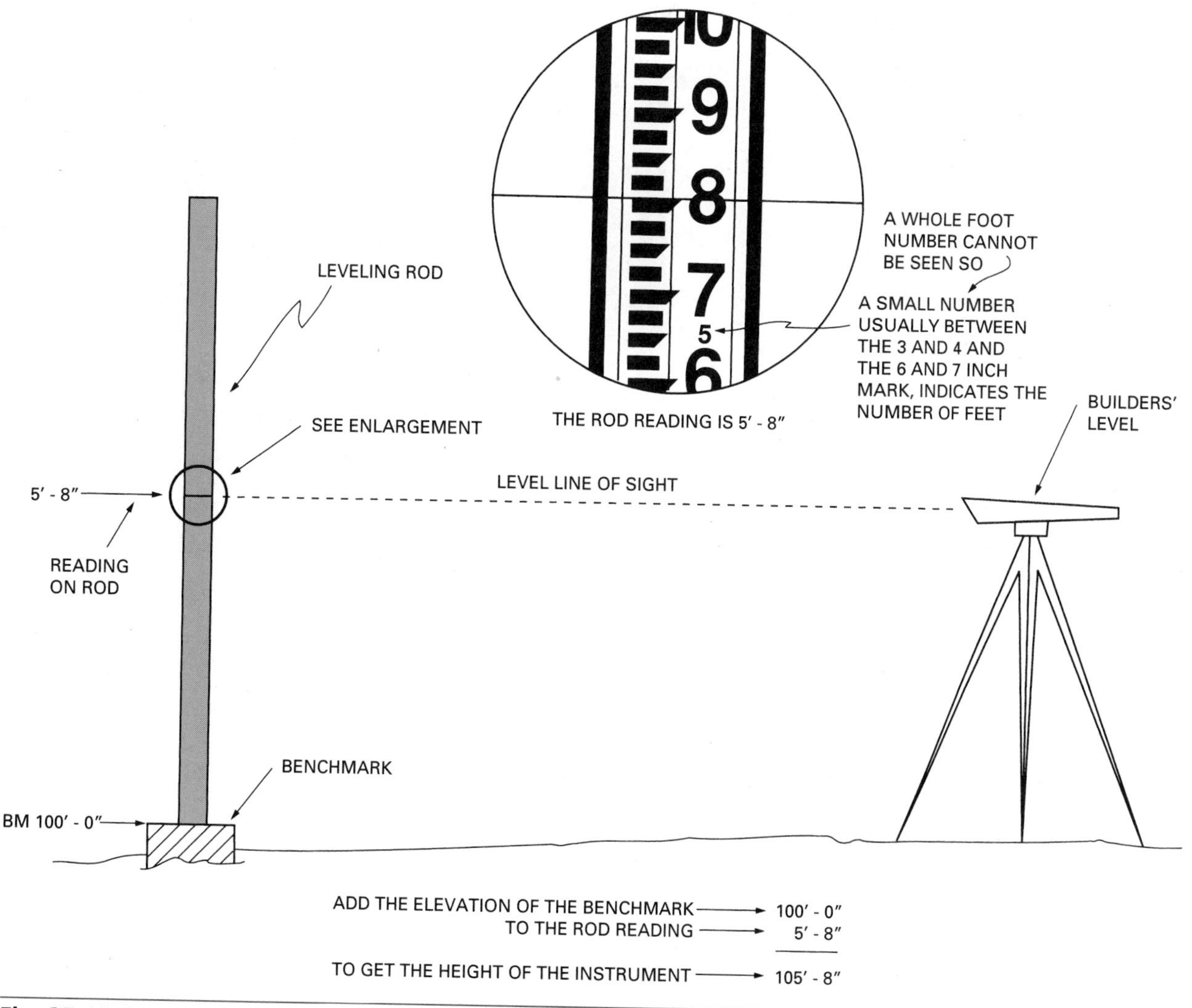

Fig. 25-19 Determining the height of the instrument (HI).

grade rod of 3′-8″ is read on the rod. The base of the builder's rod is then at the desired elevation (Fig. 25-20). A mark, drawn at the base of the rod on a stake or other object, establishes the elevation.

Base plates for bridges and large buildings must be accurately set to specified elevations. The concrete pier or foundation is formed slightly below grade. Four wedges, one on each corner, are placed upon the concrete foundation. The base plate is then lowered into place, resting on the wedges. The grade rod is calculated, and the rod is placed on one corner. The wedge at that corner is slowly driven until the correct reading and elevation are obtained. This procedure is repeated at the four corners and then rechecked. The space between the top of the foundation and the bottom of the base plate is then filled with concrete grout.

Determining Differences in Elevation

Differences in elevation need to be determined for such things as grading driveways, sidewalks, and parking areas, laying out drainage ditches, plotting contour lines, and estimating cut and fill requirements. The difference in elevation of two or more points is easily determined with the use of the builder's level or transit-level.

Determining Difference In Elevation Requiring Only One Setup. To find the difference in elevation between two points, set up the instrument about midway between them. Place the rod on the first point. Take a reading, and record it. This is called a *backsight.* All backsights are *plus (+) sights.*

Place the rod on the second point. Take a reading, and record it. This is called a *foresight.* All forward sights are *minus (−) sights.*

Add the plus and minus sights to get the difference in elevation (Fig. 25-21). For any sight, the plus or minus is reversed if readings are taken with the rod upside down (Fig. 25-22).

Determining Difference in Elevation Requiring More Than One Setup. Sometimes the difference in elevation of two points is too great or the distance is too far apart. Then it is necessary to make more than one instrument setup to determine the difference. The procedure is similar to a series of one setup operations until the final point is reached.

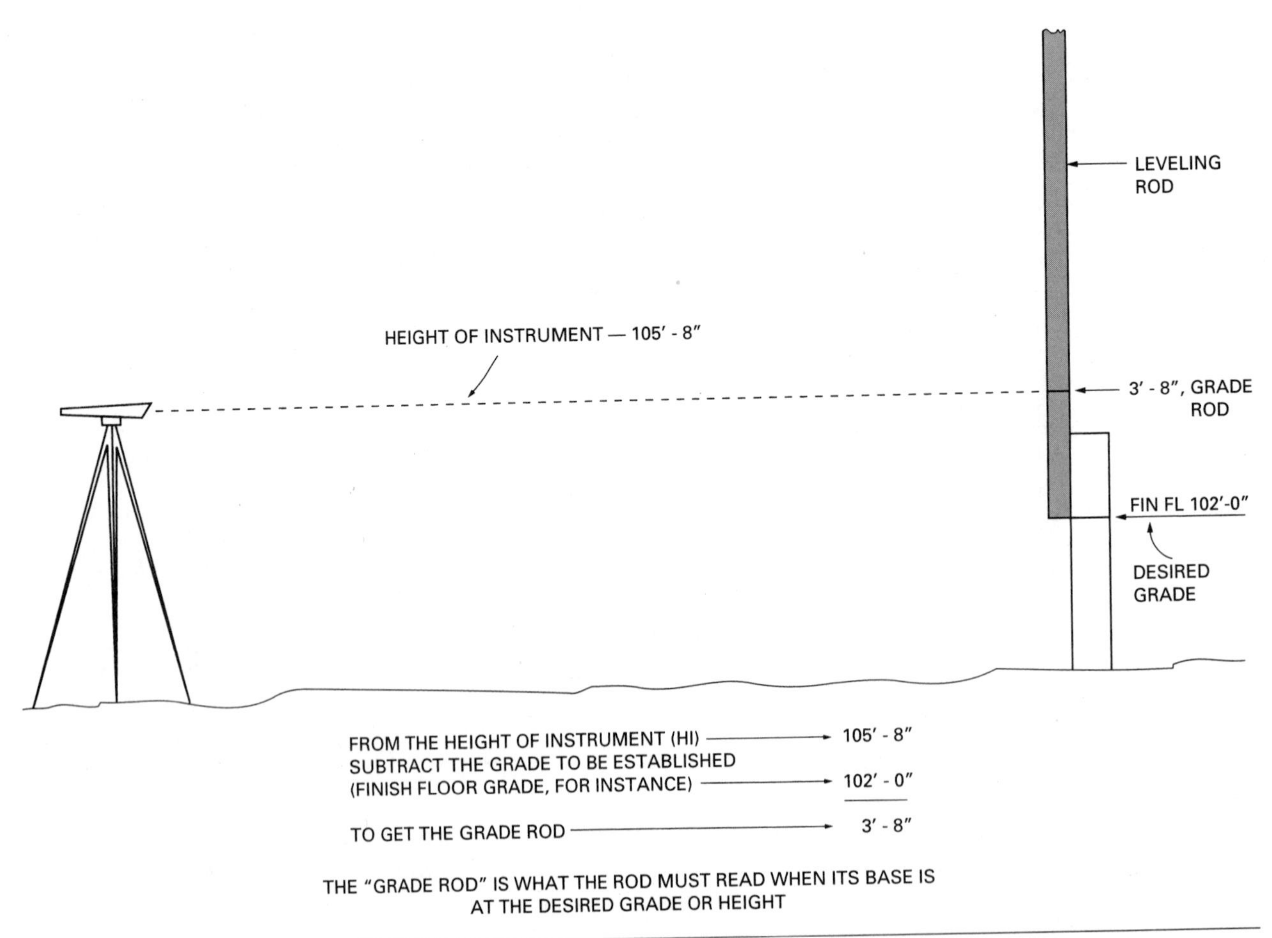

Fig. 25-20 Calculating the grade rod and establishing a desired elevation.

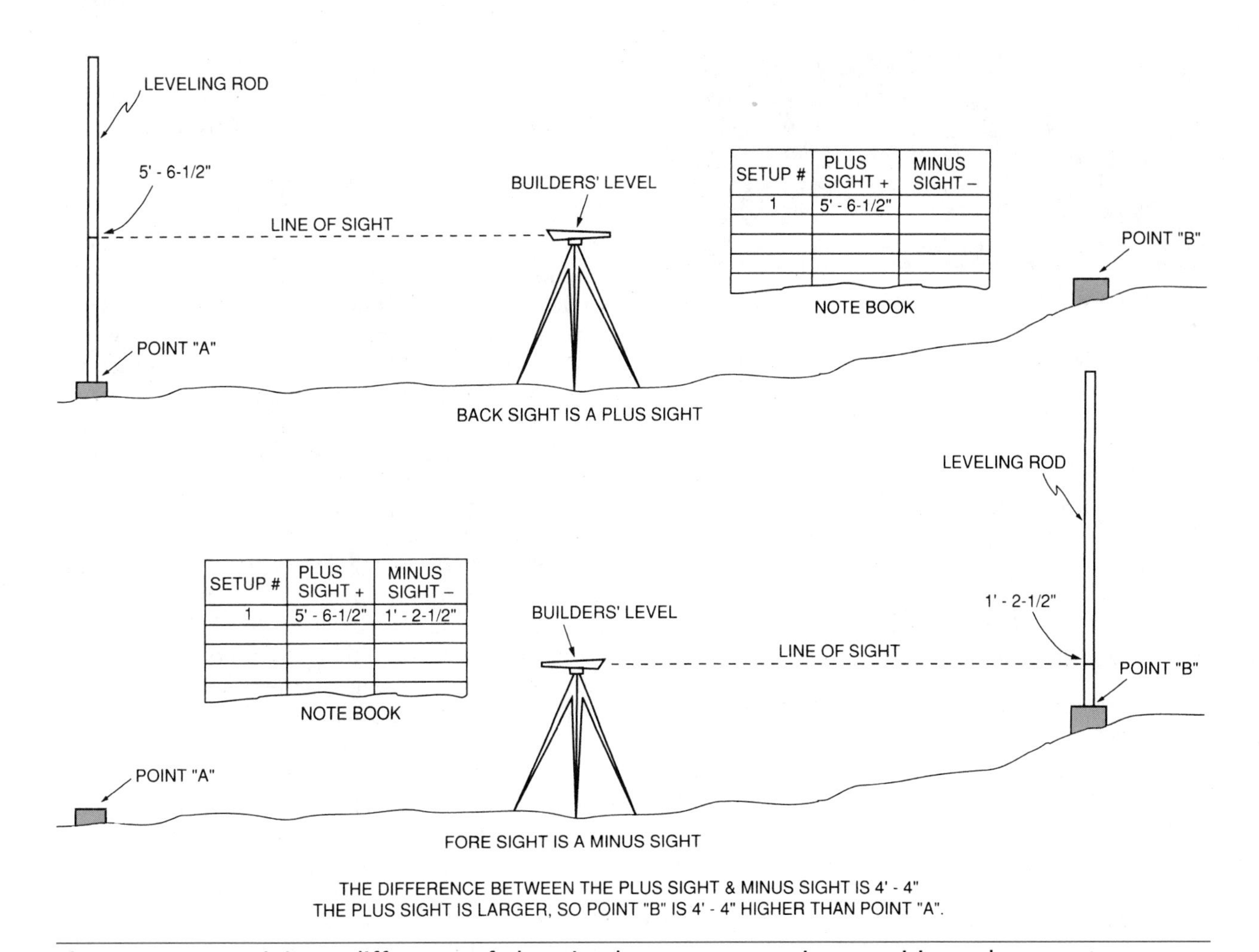

Fig. 25-21 Determining a difference of elevation between two points requiring only one setup.

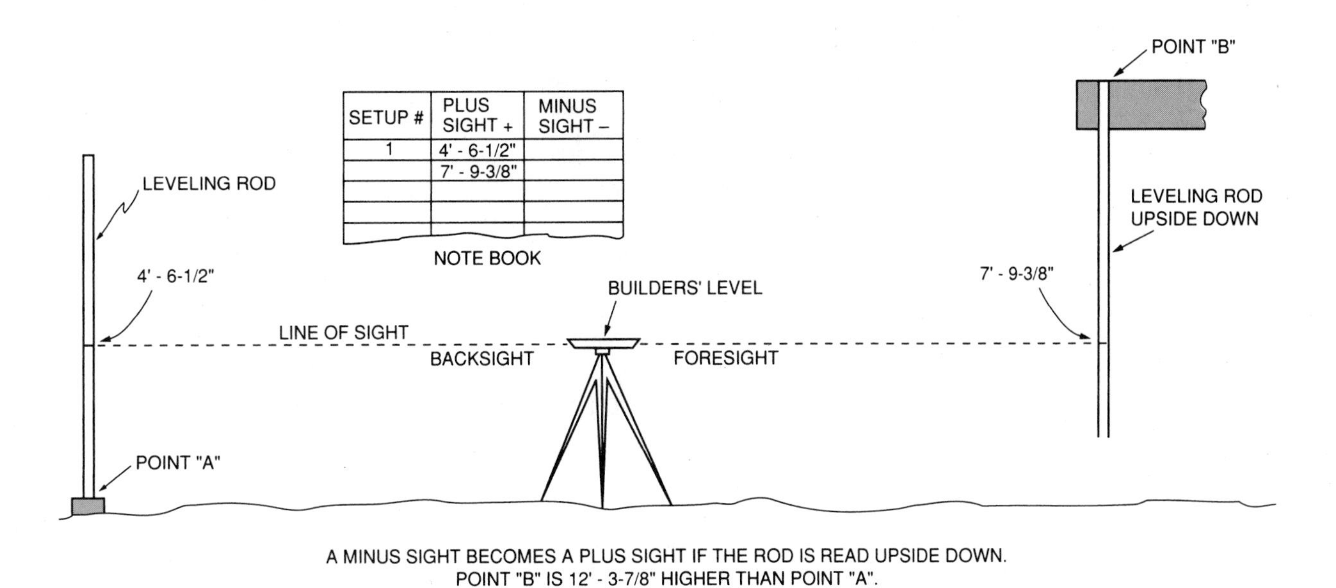

Fig. 25-22 Plus and minus sightings are reversed for any readings taken with the rod upside down.

CAUTION When carrying a tripod-mounted instrument, handle with care. Carry it in an upright position. Do not carry it over the shoulder or in a horizontal position. Be careful when going through buildings or close quarters not to bump the instrument.

Record all backsights as plus sights and all foresights as minus sights, unless the rod is upside down when read (Fig. 25-23). Find the sum of all minus sights and all plus sights. The difference between them is the difference in elevation of the beginning and ending points. If the sum of the plus sights is larger, then the end point is higher than the starting point. If the sum of the minus sights is larger, then the end point is lower than the starting point.

Measuring and Laying Out Angles

To measure or lay out angles, the instrument must be set over a point on the ground. A hook, centered below the instrument, is provided for suspending a plumb bob. The plumb bob is used to place the level directly over a particular point. In more sophisticated instruments, a built-in *optical plumb* allows the operator to sight to a point below, exactly plumb with the center of the instrument. This enables quick and accurate setups over a point (Fig. 25-24).

Setting Up over a Point. Suspend the plumb bob from the instrument. Secure it with a slip knot. Move the tripod and instrument so that the plumb bob appears to be over the point.

Press the legs of the tripod into the ground. Lower the plumb bob by moving the slip knot until it is about 1/4 inch above the point on the ground. The final centering of the instrument can be made by loosening any two adjacent leveling screws and slowly shifting the instrument until the plumb bob is directly over the point (Fig. 25-25). Retighten the same two leveling screws that were previously loosened, and level the instrument. Shift the instrument on the base plate until the plumb bob is directly over the point. Check the levelness of the instrument. Adjust, if necessary.

Circle Scale and Index. A *horizontal circle scale* (outside ring) is divided into 90-degree quadrants. A pointer, or *index,* turns with the telescope. The circle scale remains stationary and indicates the number of degrees the telescope is turned. When desired, the horizontal circle may be turned by hand for setting to zero degrees, no matter which way the telescope is pointing. Starting at zero and rotating the telescope on it, any horizontal angle can be easily measured (Fig. 25-26).

Reading the Horizontal Vernier. For more precise readings, the *horizontal vernier* is used to read minutes of a degree (Fig. 25-27). The

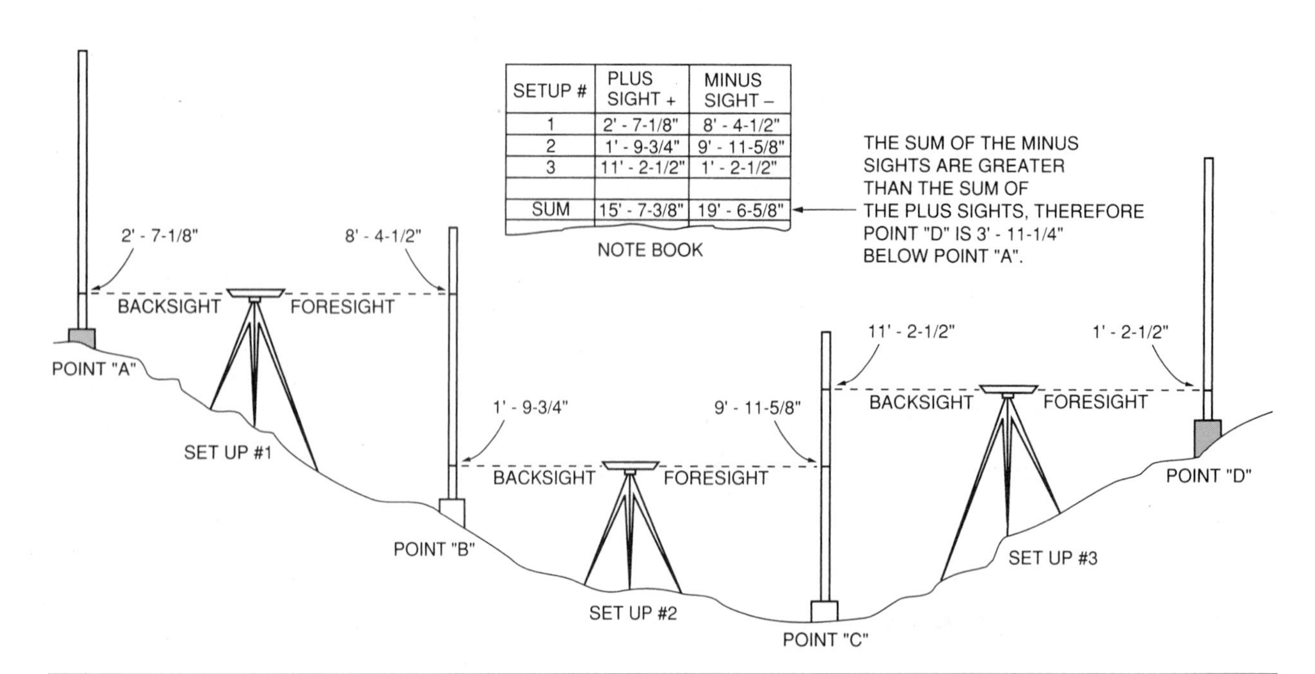

Fig. 25-23 Determining a difference of elevation between two points requiring more than one setup.

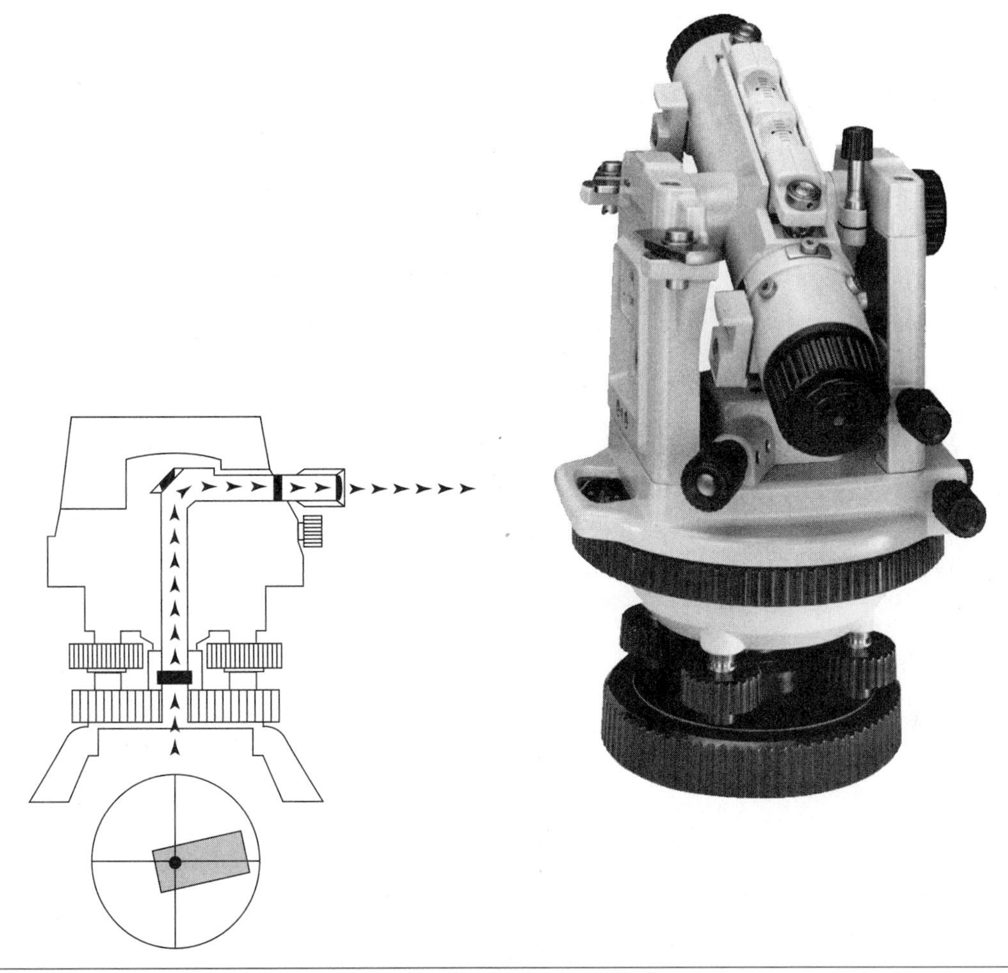

Fig. 25-24 Some instruments have a device called an optical plumb for setting the instrument directly over a point. (*Courtesy of David White*)

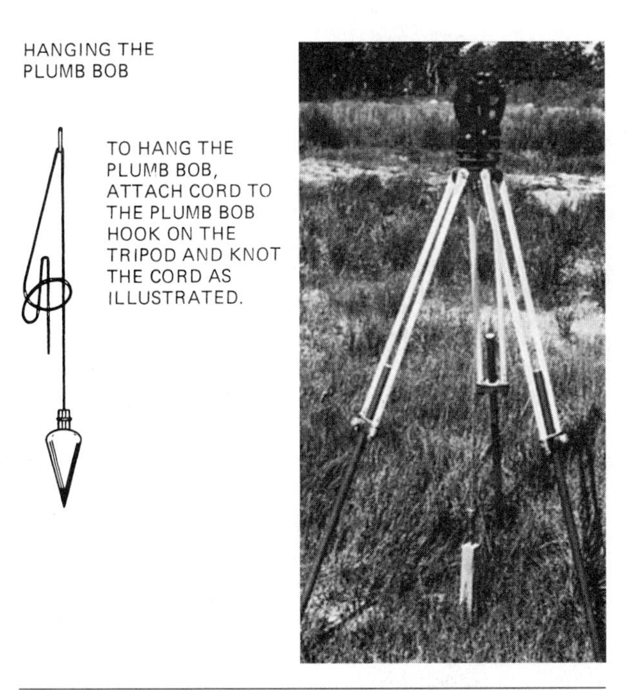

Fig. 25-25 To locate the instrument directly over a point, a plumb bob is suspended from the level.

vernier is actually two verniers, one on each side of the vernier zero index. This makes it possible to read any angle, whether turned to the right or to the left.

The vernier scale turns with the telescope. If the eye end of the telescope is turned to the left (clockwise), the vernier scale on the left side of the zero index is used. If the eye end of the telescope is turned to the right (counterclockwise), the vernier scale on the right side of the zero index is used. Use either the left or right vernier scale according to the direction in which the eye end of the telescope is turned when measuring or laying out angles (Fig. 25-28).

If the circle index zero lines up between degree marks on the circle scale, notice the last degree mark passed. On the vernier scale, locate a vernier mark that coincides with a circle mark. Add the reading in minutes to the degrees noted on the circle scale (Fig. 25-29). The carpenter is not usually concerned with any finer readings.

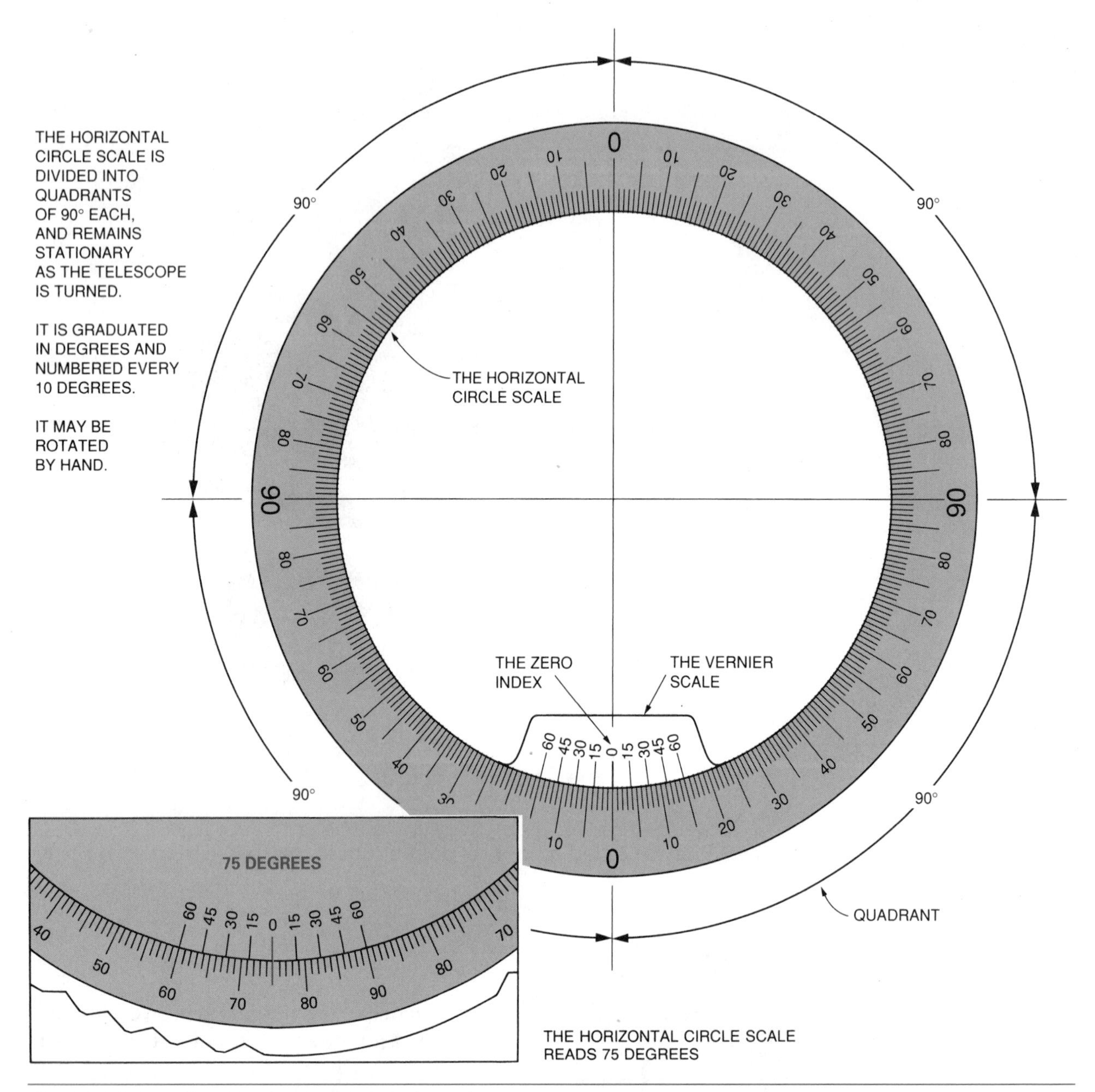

Fig. 25-26 Reading the horizontal circle scale.

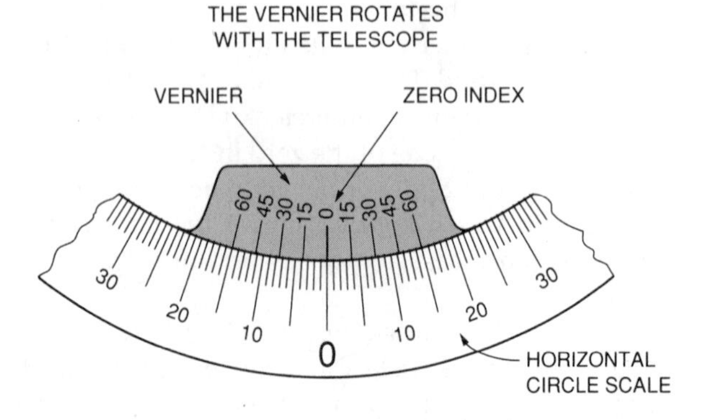

Fig. 25-27 The horizontal vernier scale is used for reading minutes of a degree.

Measuring a Horizontal Angle. After leveling the instrument over the point of an angle, called its *vertex,* loosen the horizontal clamp screw. Rotate the instrument until the vertical cross-hair is nearly in line with a distant point on one side of the angle. Tighten the clamp screw. Then turn the tangent screw to line up the vertical cross-hair exactly with the point.

If the point is above or below the line of sight, and a transit-level is not available, sight a straightedge held plumb from the point. Or, sight the line of a plumb bob line suspended over the point (Fig. 25-30). If using a transit-level, release the locking lever and tilt the telescope to sight the point.

By hand, turn the horizontal circle scale to zero. Loosen the clamp screw. Swing the telescope until the vertical cross-hair lines up with a point on the

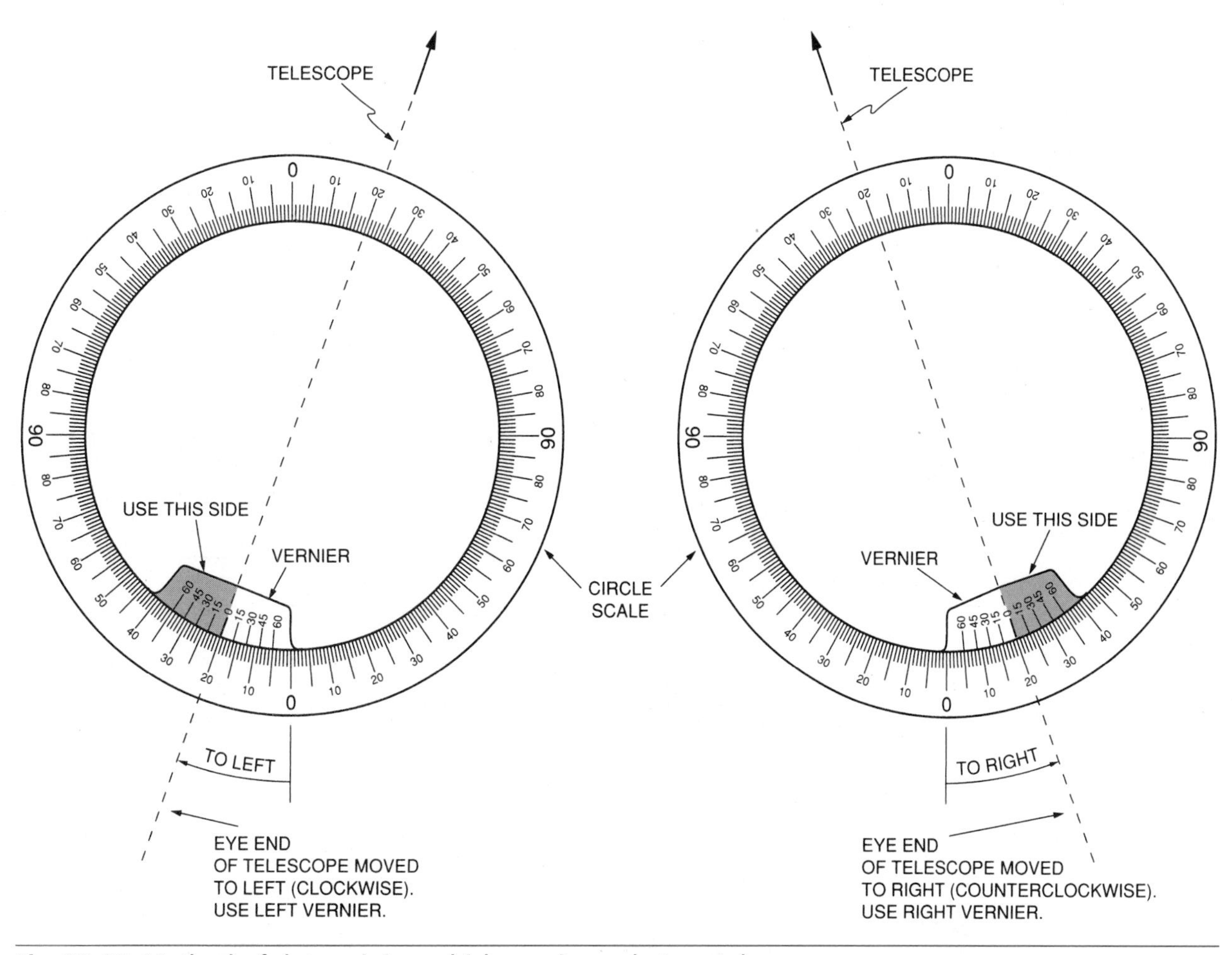

Fig. 25-28 Method of determining which vernier scale to read.

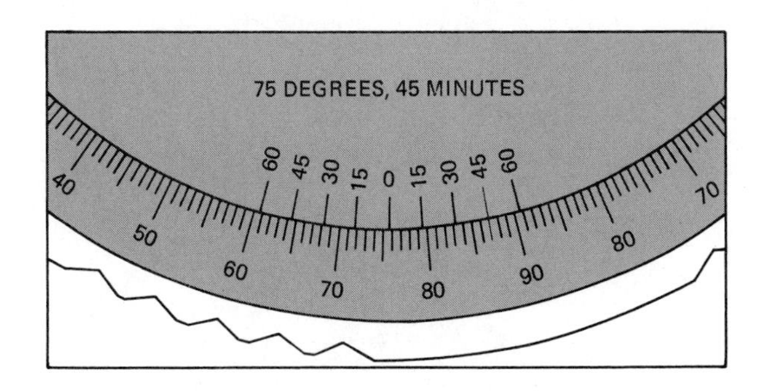

Fig. 25-29 Add the reading in minutes from the vernier scale to the reading in degrees from the circle scale.

other side of the angle. Tighten the horizontal clamp screw. Then turn the tangent screw for a fine adjustment, if necessary. Read the degrees on the circle scale and minutes on the vernier scale (Fig. 25-31).

Laying Out a Horizontal Angle. Center and level the instrument over the vertex of the angle to be laid out. Sight the telescope on a distant point on one side of the angle. Tighten the horizontal clamp, and set the circle scale to zero.

Fig. 25-30 Locating a point below the line of sight. A. Using a hand level and straightedge B. Using a plumb bob

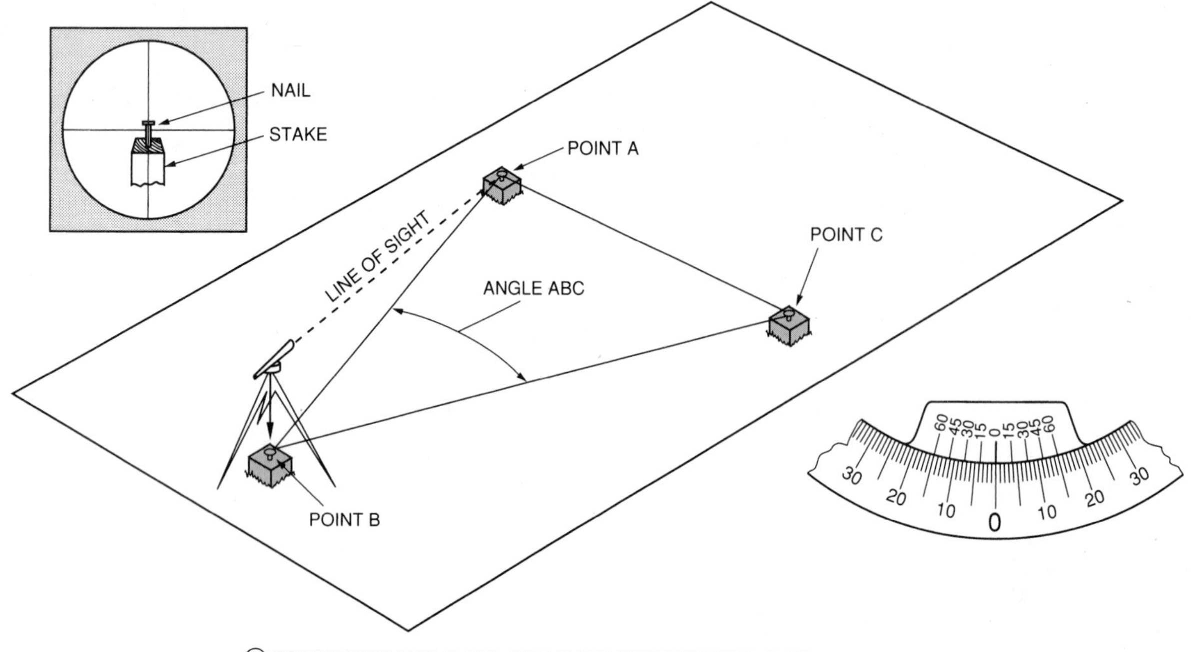

① TO MEASURE ANGLE ABC, SET UP THE TRANSIT-LEVEL OVER POINT B AND SIGHT TO POINT A. TIGHTEN HORIZONTAL CLAMP SCREW. BY HAND, SET THE HORIZONTAL CIRCLE SCALE TO ZERO.

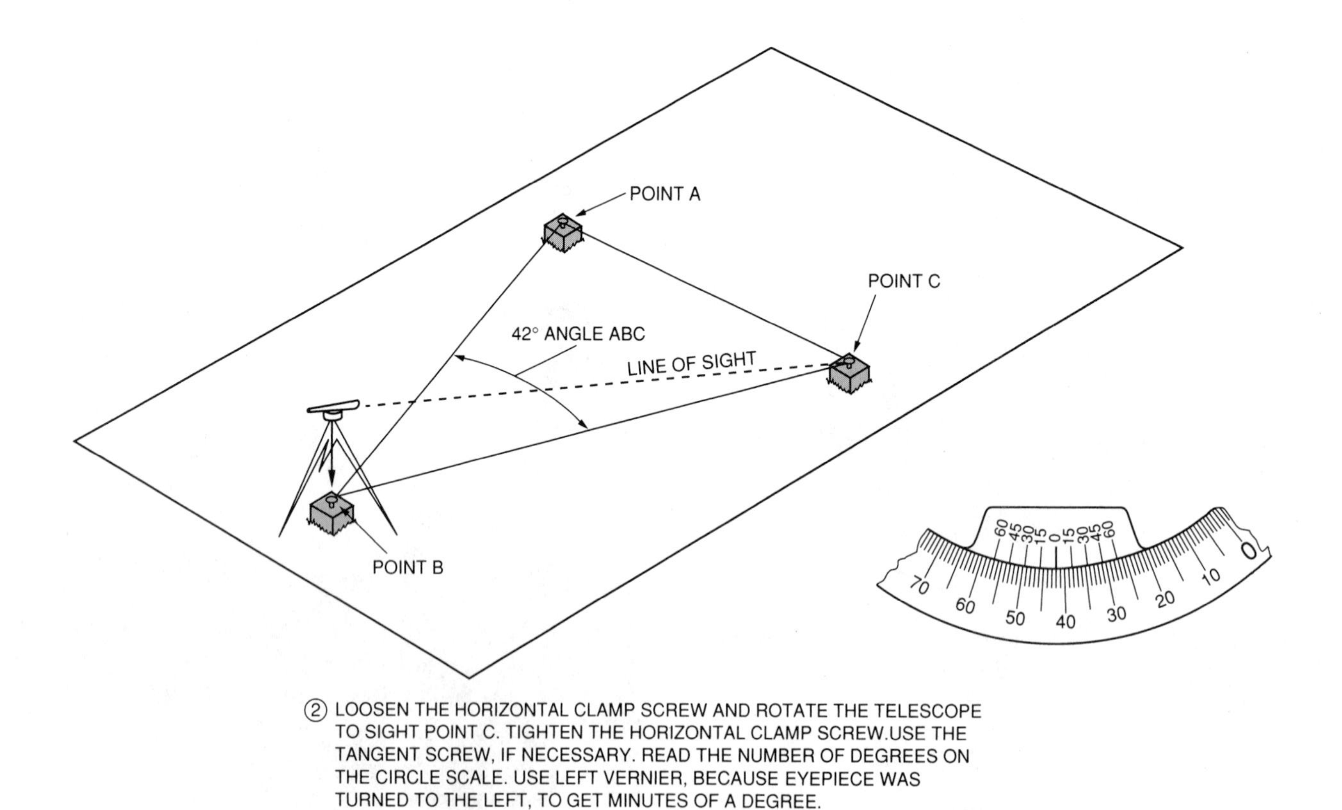

② LOOSEN THE HORIZONTAL CLAMP SCREW AND ROTATE THE TELESCOPE TO SIGHT POINT C. TIGHTEN THE HORIZONTAL CLAMP SCREW.USE THE TANGENT SCREW, IF NECESSARY. READ THE NUMBER OF DEGREES ON THE CIRCLE SCALE. USE LEFT VERNIER, BECAUSE EYEPIECE WAS TURNED TO THE LEFT, TO GET MINUTES OF A DEGREE.

Fig. 25-31 Measuring an angle.

Loosen the clamp. Then turn the telescope until close to the desired number of degrees. Tighten the clamp. Use the tangent screw and make a fine adjustment until the index reads exactly the desired number of degrees. Sight the vertical cross-hair to lay out the other side of the angle (Fig. 25-32).

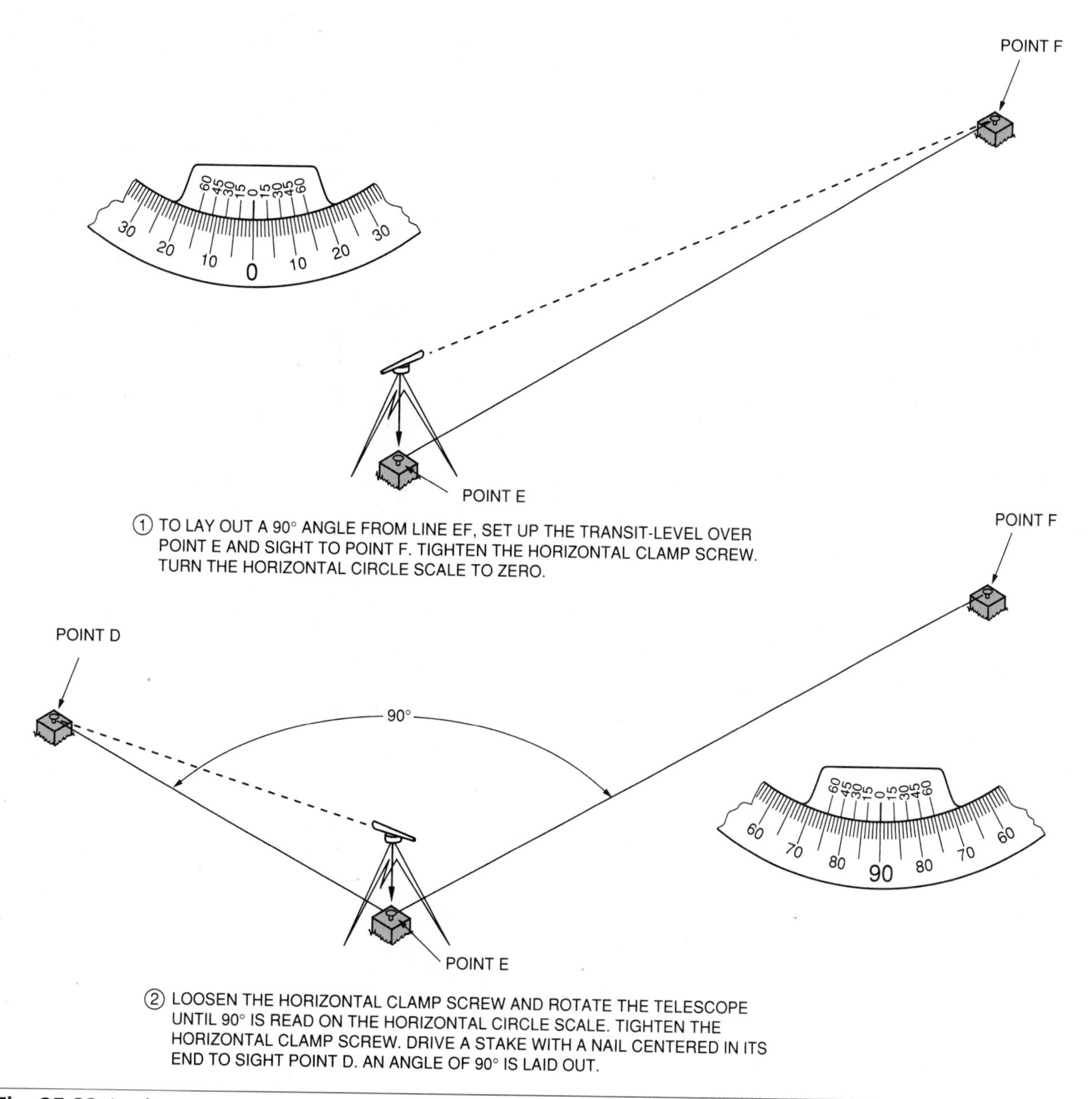

Fig. 25-32 Laying out an angle.

Special Uses of the Transit-level

The transit-level can be used efficiently for all the operations described previously in this chapter. In addition, it is a valuable tool for such things as measuring vertical angles, setting points in a line, and plumbing posts, columns, walls, and similar objects.

Measuring Vertical Angles. The *vertical arc* scale is attached to the telescope. It measures vertical angles to 45 degrees above and below the horizontal. By tilting and rotating the telescope, set the horizontal cross-hair on the points of the vertical angle being measured. Tighten the vertical clamp. Then turn the tangent screw for a fine adjustment to place the cross-hair exactly on the point. Vertical angles are read by means of the vertical arc scale and the obvious vernier similar to the reading of horizontal angles (Fig. 25-33).

Setting Points in a Straight Line. It may be necessary to set points in a straight line, such as on a property boundary line. Set up the instrument over the one point. Rotate and tilt the telescope to sight the other point fairly close. Tighten the horizontal clamp. Then turn the tangent screw until the vertical cross-hair is exactly on the far point. With the horizontal clamp tight, release the lock levers and depress the telescope, as required, to sight points between the corners and along the boundary line (Fig. 25-34).

If it is necessary to continue in a straight line beyond the far point, move and set up the instrument over the far point. Sight back to the first point. Tighten the horizontal clamp. Then turn the horizontal circle scale to zero. Loosen the horizontal clamp. Turn the telescope 180 degrees, and tighten the horizontal clamp again. Depress or raise the telescope as needed to sight additional points in a straight line (Fig. 25-35).

Plumbing Posts and Columns. To *plumb* posts, columns, walls, and similar objects, first set up the instrument away from the object at a distance equal to about the object's height. Sight so the vertical cross-hair lines up exactly with one edge of the object at its base. Tighten the horizontal

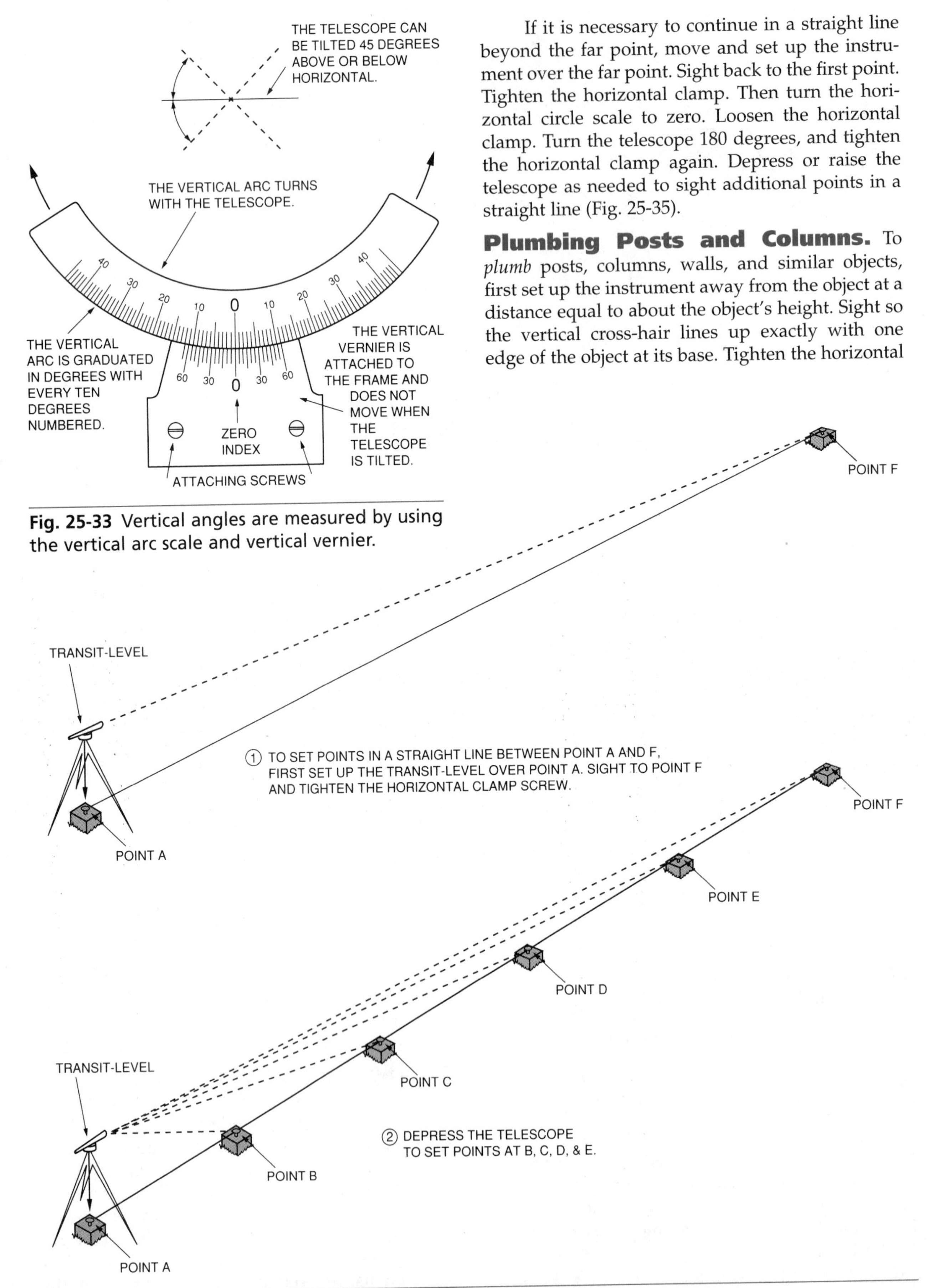

Fig. 25-33 Vertical angles are measured by using the vertical arc scale and vertical vernier.

Fig. 25-34 Setting points in a straight line.

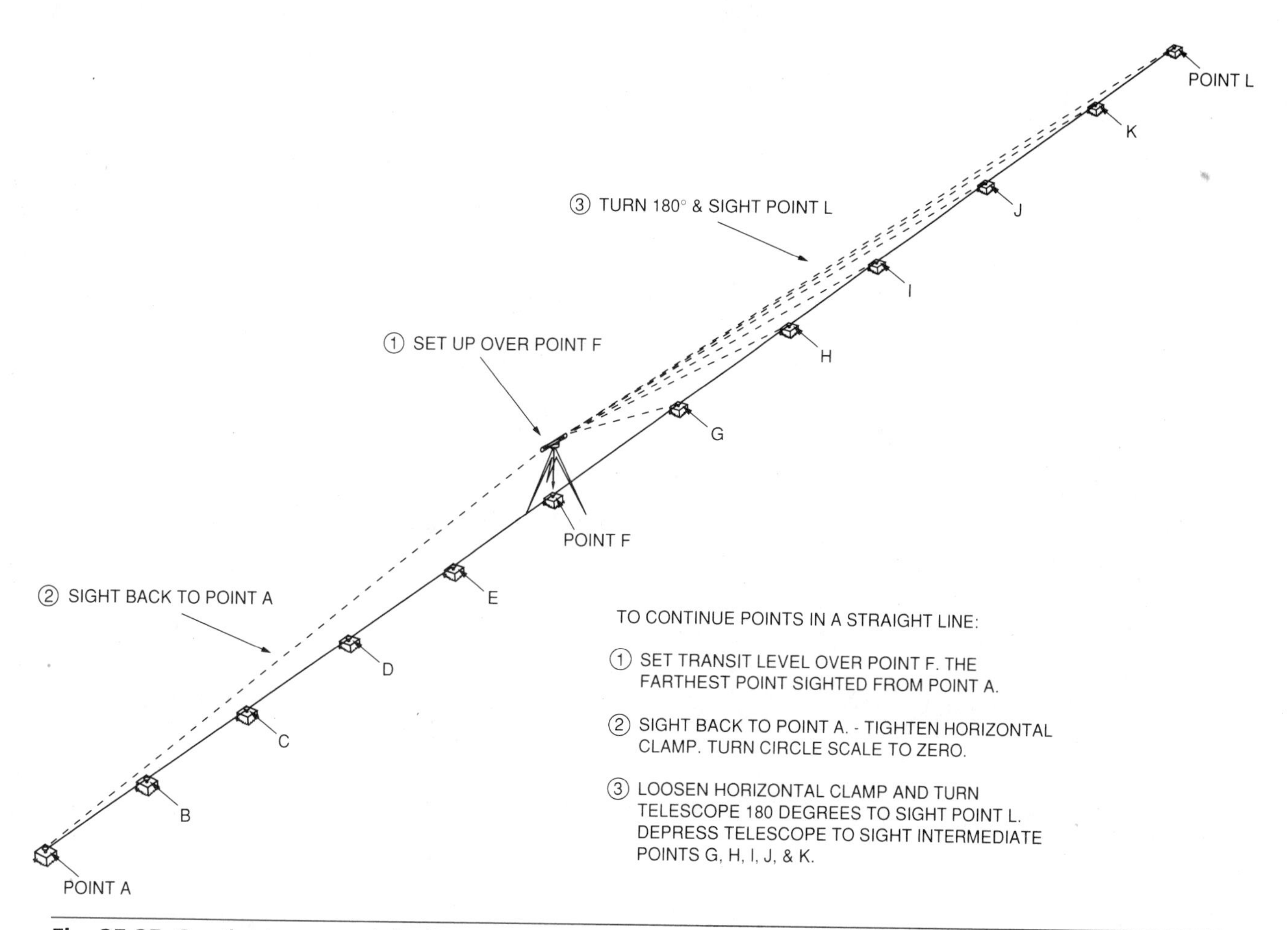

Fig. 25-35 Continuing a straight line.

clamp. Then raise the telescope to sight the same edge near the top of the object. If plumb, the same edge will line up exactly as it did at the base. If it does not line up, move the top of the object until it does and brace in position.

To complete the plumbing process, move and set the instrument up at a position about 90 degrees from the first position. Then repeat the procedure (Fig. 25-36).

Storing the Instrument. If the instrument gets wet, dry it before returning it to its case. Keep it in its carrying case when it is not being used or when being transported in a vehicle over long distances.

Laser Levels

A **laser** is a device in which energy is released in a narrow beam of light. The light beam is absolutely straight. Unless interrupted by an obstruction or otherwise disturbed, the light beam can be seen for a long distance.

Lasers have many applications in space, medicine, agriculture, and engineering. The *laser level* has been developed for the construction industry to provide more efficient layout work (Fig. 25-37).

Kinds and Uses of Laser Levels

Several manufacturers make laser levels in a number of different models. The least expensive models are the least sophisticated. A low-price unit is leveled and adjusted manually. More expensive ones are automatically adjusted to and maintained in level. Power sources include four C-cell batteries, a detachable, rechargeable battery pack, a 12-volt battery, or an AC/DC converter for 110 or 220 volts.

Establishing and Determining Elevations. A simple, easy-to-use model is mounted on a tripod or solid flat surface and leveled like manual or automatic optical instruments. The laser is turned on. It will emit a red beam, usually 3/8 inch in diameter. The beam rotates through a full 360 degrees, creating a level *plane* of light. As it rotates, it establishes equal points of elevation over the entire job site, similar to a line of sight being rotated by the telescope of an optical instrument (Fig. 25-38).

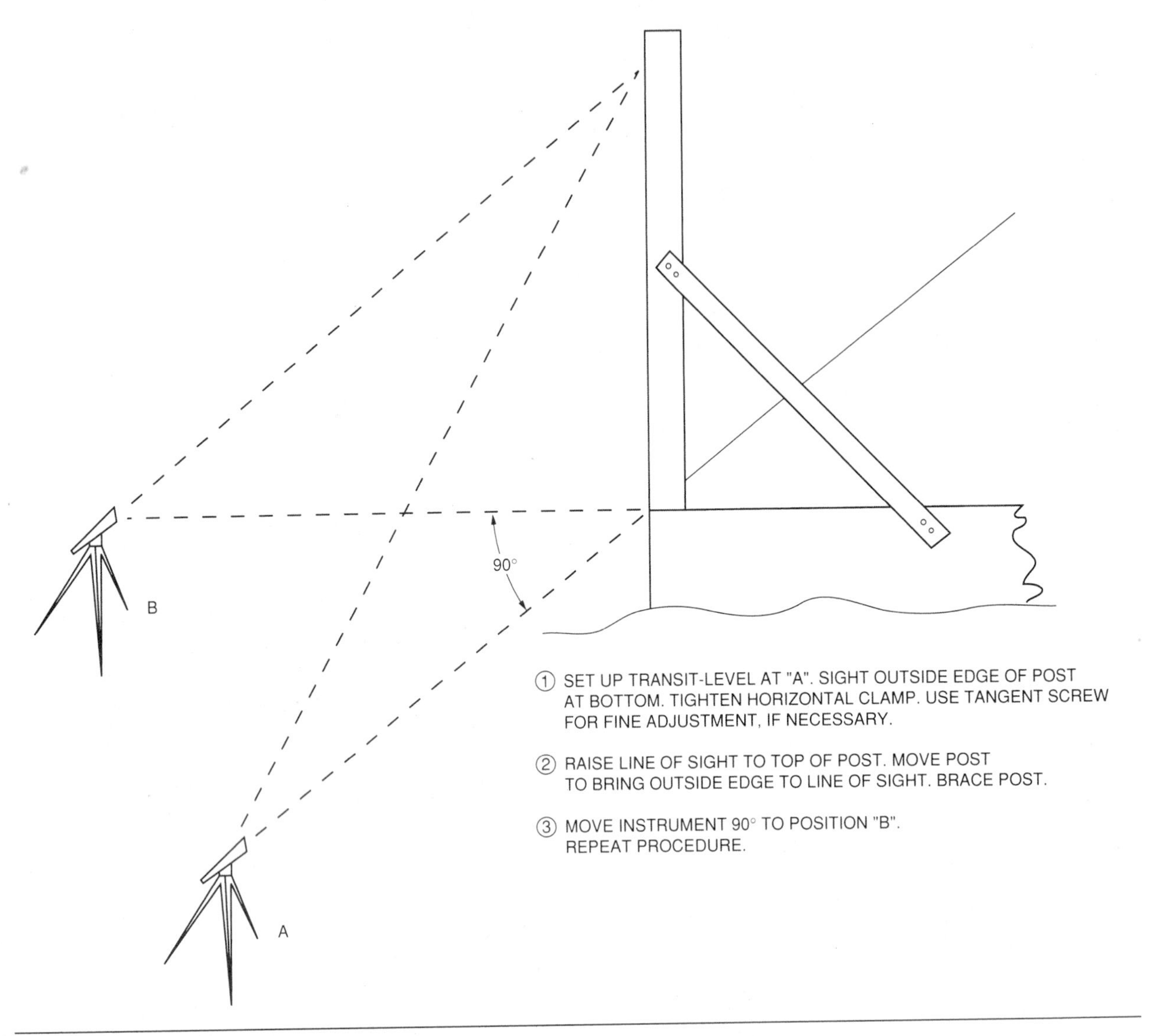

Fig. 25-36 Plumbing a post.

Fig. 25-37 Laser levels have been developed for use in the construction industry. (*Courtesy of Laser Alignment, Inc., and Spectra-Physics*)

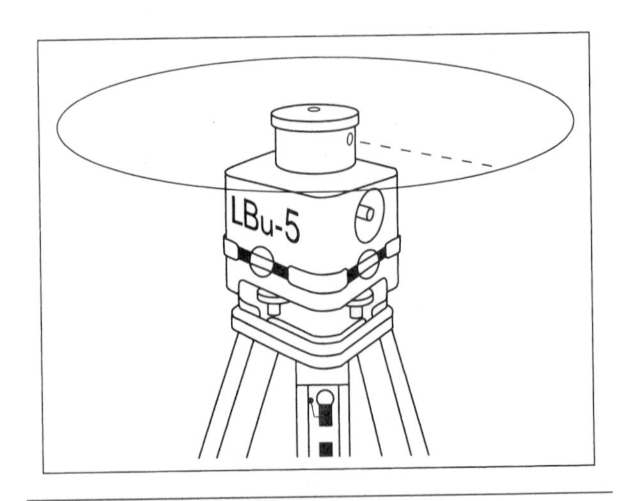

Fig. 25-38 The laser beam rotates 360 degrees, creating a level plane of light. (*Courtesy of Laser Alignment, Inc.*)

Depending on the quality of the instrument, the laser head may rotate at various revolutions per second (RPS), up to 40 RPS. Its quality also determines its working range. This may vary from a 75- to 1,000-foot radius.

Some laser beams cannot be seen by the human eye. Those that can are difficult to see outdoors in bright sunlight. To detect the beam, a battery powered electronic *sensor* target, also called a *receiver* or *detector,* is attached to the leveling rod or stick. Most sensors have a visual display with a selectable audio to indicate when it is close to or on the beam (Fig. 25-39). In addition to electronic sensor targets, specially designed targets are used for interior work, such as installing ceiling grids and leveling computer floors.

The procedures for establishing and determining elevations with laser levels are similar to those with optical instruments. To establish elevations, the sensor is attached to the grade rod. The grade rod is moved up or down until the beam indicates that the base of the rod is at grade.

To determine elevations, the base of the rod is held to the elevation to be determined. The sensor is moved up or down on the rod until the display indicates it is centered on the beam. The reading from the rod is then recorded.

HIGH .... FAST BEEPING
ON-GRADE ...... SOLID TONE
LOW .... SLOW BEEPING

Fig. 25-39 An electronic target senses the laser beam. An audio provides tones to match the visual display. *(Courtesy of Laser Alignment, Inc.)*

One of the great advantages of using laser levels is that, in most cases, only one person is needed to do the operations (Fig. 25-40). Another advantage is that certain operations are accomplished more easily in less time.

Special Horizontal Operations. For leveling *suspended ceiling grids,* the laser level is mounted to an adjustable grid mount bracket. Once the first strip of angle trim has been installed, the laser and bracket can be attached easily and clamped into place.

The unit is leveled. The height of the laser beam is then adjusted for use with a special type magnetic or clip-on target. Magnetic targets are used on steel grids. Clip-on targets are used on aluminum or steel. Once the laser is set up, the ceiling grid is quickly and easily leveled using the rotating beam as a reference and viewing the beam through the

Fig. 25-40 When using laser levels, only one person is required for leveling operations. *(Courtesy of Laser Alignment, Inc.)*

target (Fig. 25-41). Sprinkler heads, ceiling outlets, and similar objects can be set in a similar manner.

For leveling *raised access computer floors,* the laser unit is mounted on a *mini tripod* and leveled. Adjust the special computer floor target to the elevation of the beam by placing it on a floor pedestal set at the desired elevation. Move the target to each floor pedestal and set to the correct height indicated when the beam hits the target (Fig. 25-42). In addition to precise leveling jobs, like computer floors, European-style home and office cabinets and similar objects can also be leveled.

Special Vertical Operations. Most laser levels are designed to work laying on their side using special brackets or feet. The unit is placed on the floor and leveled. *To lay out a partition,* rotate the head of the laser downward. Position the laser beam over one end of the partition. Align the beam toward the far point by turning the head toward and adjusting to the second point. Recheck the alignment and turn on to rotate the beam. The beam will be displayed as continuous straight and plumb lines on the floor, ceiling, and walls to both align and plumb the partition at the same time (Fig. 25-43). Mounted on its side on a tripod, the laser can be used for aligning and plumbing tilt-ups, curtain walls, and a variety of similar work (Fig. 25-44).

Fig. 25-41 Leveling suspended ceiling grids. The beam is viewed through the target. (*Courtesy of Spectra-Physics*)

Fig. 25-42 Leveling pedestals for a computer floor. (*Courtesy of Spectra-Physics*)

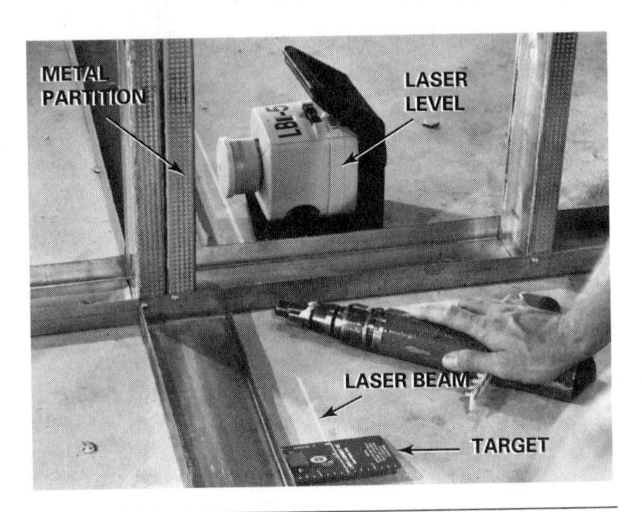

Fig. 25-43 Laid on its side, the laser level is used to lay out and align partitions and walls. (*Courtesy of Laser Alignment, Inc.*)

Fig. 25-44 On exterior applications, the laser unit is mounted on its side on a tripod to plumb and align formwork and similar objects. (*Courtesy of Laser Alignment, Inc.*)

Layout Operations. Some laser levels emit a plumb line of light projecting upward from the top at a right angle (90 degrees) to the plane of the rotating beam. The plumb reference beam allows one person to lay out 90-degree cross-walls or building lines. Set the unit on its side as for laying out a partition. The reference beam establishes a 90-degree corner with the rotating laser beam (Fig. 25-45).

Plumbing Operations. The vertical beam provides a ready reference for plumbing posts, columns, elevator shafts, slip forming, and wherever a plumb reference beam is required.

Mount the laser unit on a tripod over a point of known offset from the work to be plumbed. Suspend a plumb bob from the center of the tripod directly over the point. Apply power to the unit. The vertical beam that is projected is ready for use as a reference. Move the top of the object until it is offset from the beam the same distance as the bottom point.

The heads on some laser units can be tilted so the rotating beam produces a plane of light at an angle to the horizontal. Such units are used for laying out slopes. Other units are manufactured for special purposes such as pipelaying and tunnel guidance. Marine laser units, with ranges up to 10 miles, are used in port and pier construction and offshore work.

Laser Safety

With a little common sense, the laser can be used safely. All laser instruments are required to have warning labels attached (Fig. 25-46). The following are safety precautions for laser use:

- Only trained persons should set up and operate laser instruments.
- Never stare directly into the laser beam or view it with optical instruments.
- When possible, set the laser up so it is above or below eye level.
- Turn the laser off when not in use.
- Do not point the laser at others.

Fig. 25-45 A reference beam at right angles to the plane of the rotating beam is used for 90-degree layout work. (*Courtesy of Spectra-Physics*)

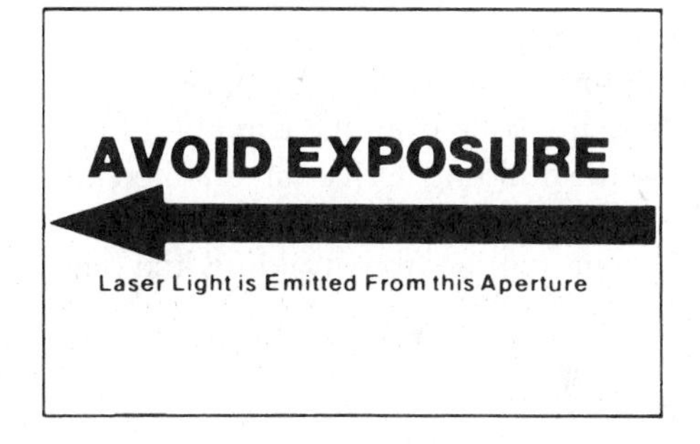

APERTURE LABEL

WARNING LABEL

Fig. 25-46 Warning labels must be attached to every laser instrument. (*Courtesy of Laser Alignment, Inc.*)

CHAPTER 26 LAYING OUT FOUNDATION LINES

Laying Out Foundation Lines

Before any layout can be made, the carpenter must determine the dimensions of the building and its location on the site from the plot plan. It is usually the carpenter's responsibility to lay out building lines.

Staking the Building

Find the survey rods that mark the corners of the property. Do not guess where the property lines are. Stretch and secure lines between each corner, laying out all the property boundary lines.

Measure in on each side from the front property line the specified front setback. Drive stakes on each end. Stretch a line between the stakes to mark the front line of the building (Fig. 26-1).

Measure back from the front building line the width of the building. Drive stakes. Then lay out the rear building line in a manner similar to that used when laying out the front building line (Fig. 26-2).

Along the front building line, measure in from the side property line the specified side setback of the building. Drive a stake. Drive a nail in the top of the stake to the exact side setback. From this nail, measure the dimension of the building along the front building line. Drive another stake. Drive a nail in the top of the stake marking the exact length of the building (Fig. 26-3).

Set up a level, for 90-degree layout, over the nail in one of the stakes marking the front corner of the building. A transit-level or a laser level with a reference beam is the preferred choice. Sight along the front property line to the nailhead in the stake of the opposite front corner.

Swing the telescope 90 degrees. Sight, locate, and drive a stake for the rear corner of the building. Drive a nail in the top of the stake marking the rear corner exactly (Fig. 26-4).

After squaring the corner, locate the stake for the other corner of the building by measuring along the established rear building line the same distance as in the front. Locate a stake at that point. Drive a

STEP BY STEP PROCEDURES

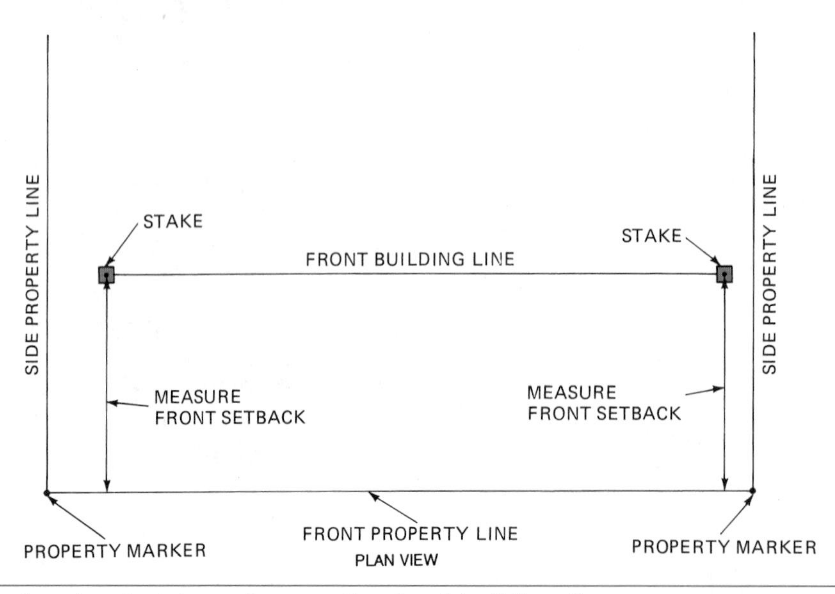

Fig. 26-1 Step 1—Locate, stretch, and secure the front building line.

STEP BY STEP

PROCEDURES

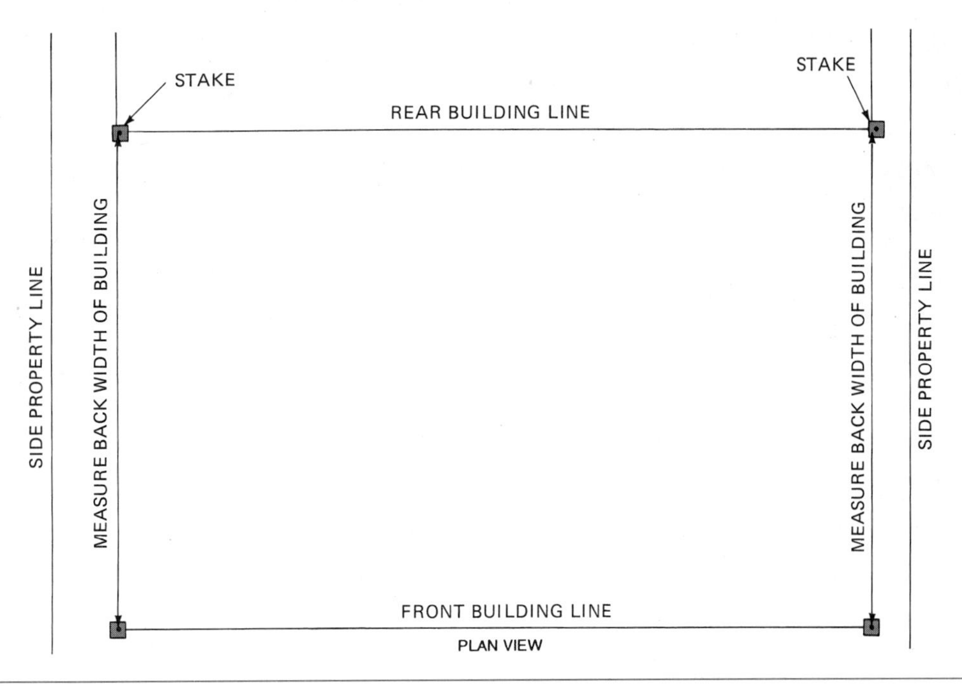

Fig. 26-2 Step 2—Locate, stretch, and secure the rear building line.

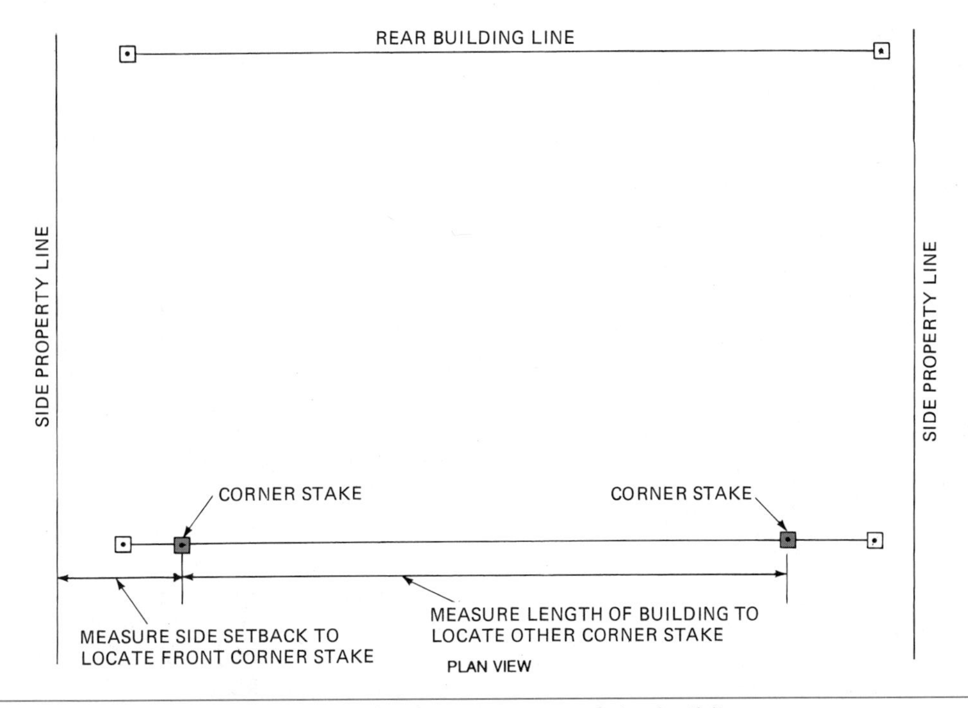

Fig. 26-3 Step 3—Drive stakes to lay out the front corners of the building.

STEP BY STEP
PROCEDURES

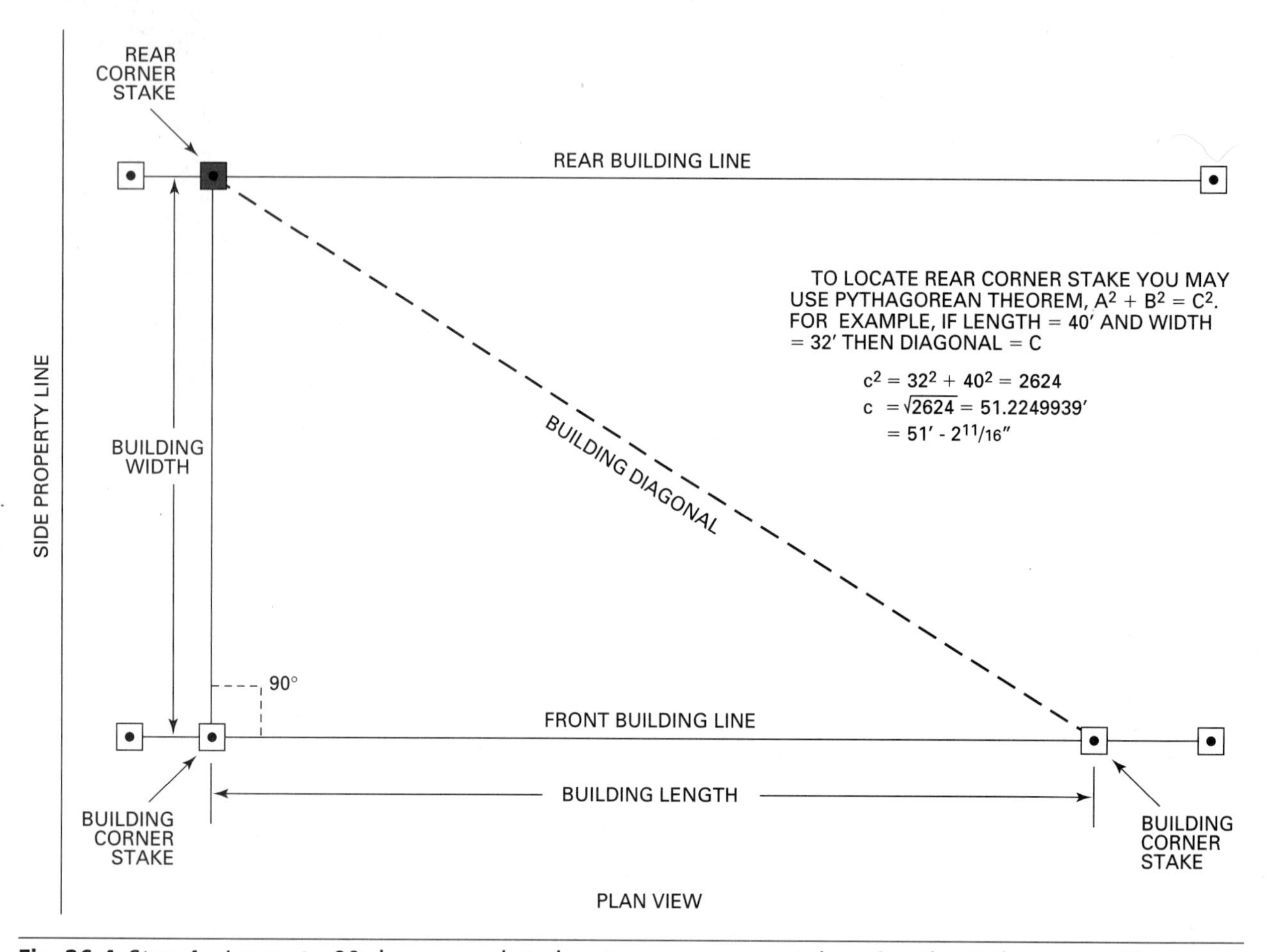

Fig. 26-4 Step 4—Lay out a 90-degree angle to locate a rear corner stake using the Pythagorean Theorem.

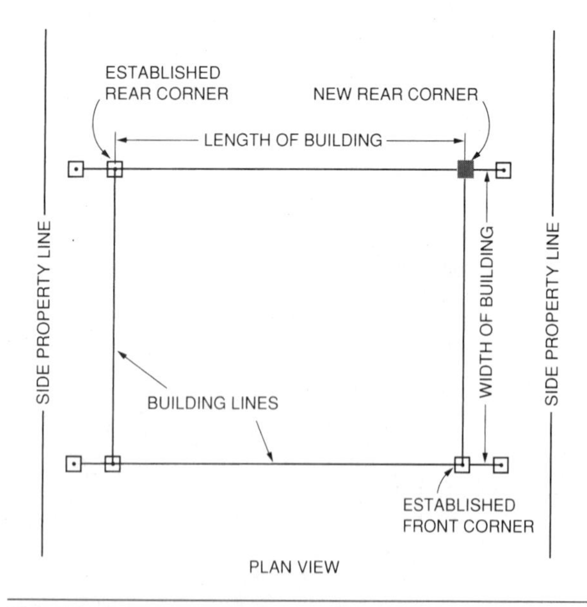

Fig. 26-5 Step 5—Measure from established front and rear corners to locate the other rear corner of the building.

nail in its top to mark exactly the other rear corner (Fig. 26-5).

Check the accuracy of the work by measuring diagonally from corner to corner. The diagonal measurements should be the same. Make adjustments if the measurements differ (Fig. 26-6).

All measurements must be made on the level. If the land slopes, the tape is held level with a plumb bob suspended from it (Fig. 26-7).

Irregular-shaped buildings are laid out using the fundamental principles outlined above. All corner stakes are located by measuring from the established front and side building lines (Fig. 26-8).

Stakes, other than those used for the front corners, may have to be moved slightly for accuracy. Also, staking the building is a temporary aid for erecting **batter boards.** Consequently, mark the corners by driving a long spike into the ground through a small square of thin cardboard (Fig. 26-9).

STEP BY STEP

PROCEDURES

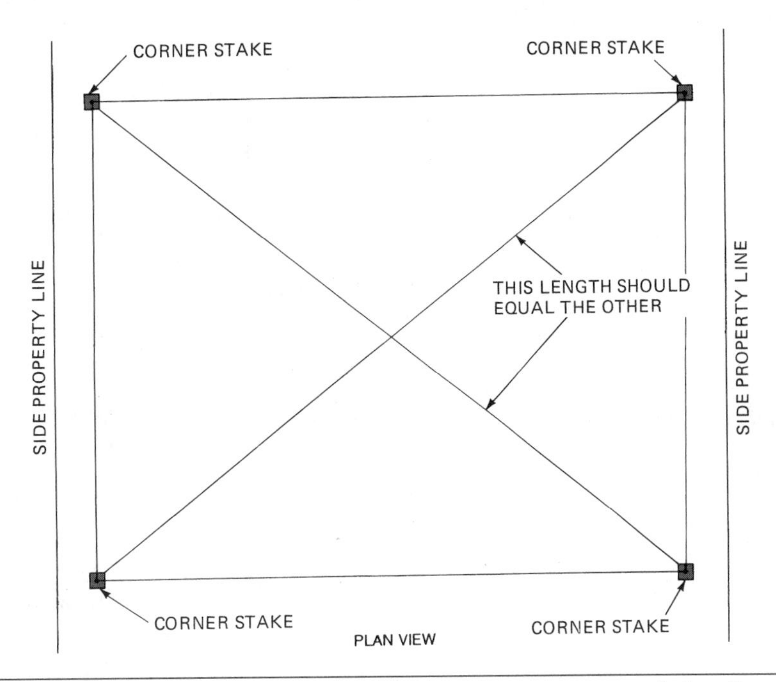

Fig. 26-6 Step 6—If the length and width measurements are accurate and the diagonal measurements are equal, then the corners are square.

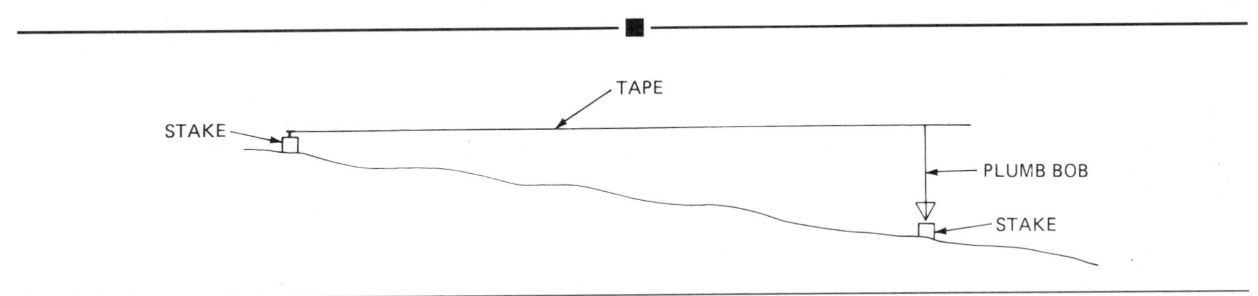

Fig. 26-7 For building layouts on sloping land, measurements must be taken on the level.

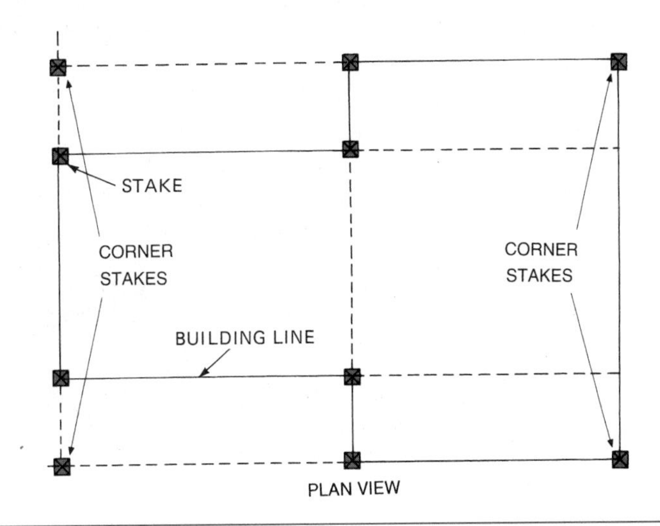

Fig. 26-8 Locate and drive corner stakes first, then intermediate ones, when laying out an irregular building.

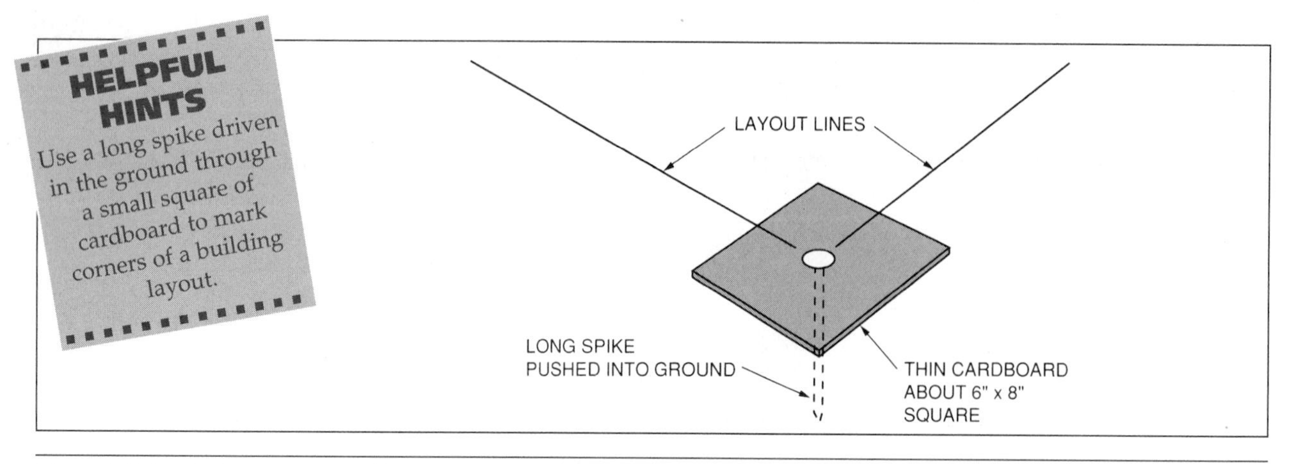

Fig. 26-9 Technique for marking the ground location of future stakes.

Erecting Batter Boards

Batter boards are wood frames to which building layout lines are secured. Batter boards consist of horizontal members, usually 1 × 6, called **ledgers.** These are attached to 2 × 4 stakes driven into the ground. The ledgers are fastened in a level position to the stakes, usually at the same height as the **foundation wall** (Fig. 26-10). Batter boards are built in the same way for both residential or commercial construction.

Batter boards are erected in such a manner that they will not be disturbed during excavation. Drive 2 × 4 stakes into the ground a minimum of 4 feet outside the building lines at each corner to construct right angle batter boards. When setting batter boards for large construction, make sure to leave room for heavy excavating equipment to operate without disturbing them. Straight batter boards are used for long setbacks. In loose soil or when stakes are higher than 3 feet, they must be braced (Fig. 26-11).

Set up the builder's level about center on the building location. Sight to the benchmark, and record the sighting. Determine the difference between the benchmark sighting and the height of the ledgers. Sight and mark each corner stake at the specified elevation. Attach ledgers to the stakes so that the top edge of each ledger is on the mark. Brace the batter boards for strength, if necessary.

Stretch lines between batter boards directly over the nailheads in the original corner stakes. Locate the position of the lines by suspending a plumb bob directly over the nailheads. When the lines are accurately located, make a saw cut on the

Fig. 26-10 Batter boards are erected for a large commercial building.

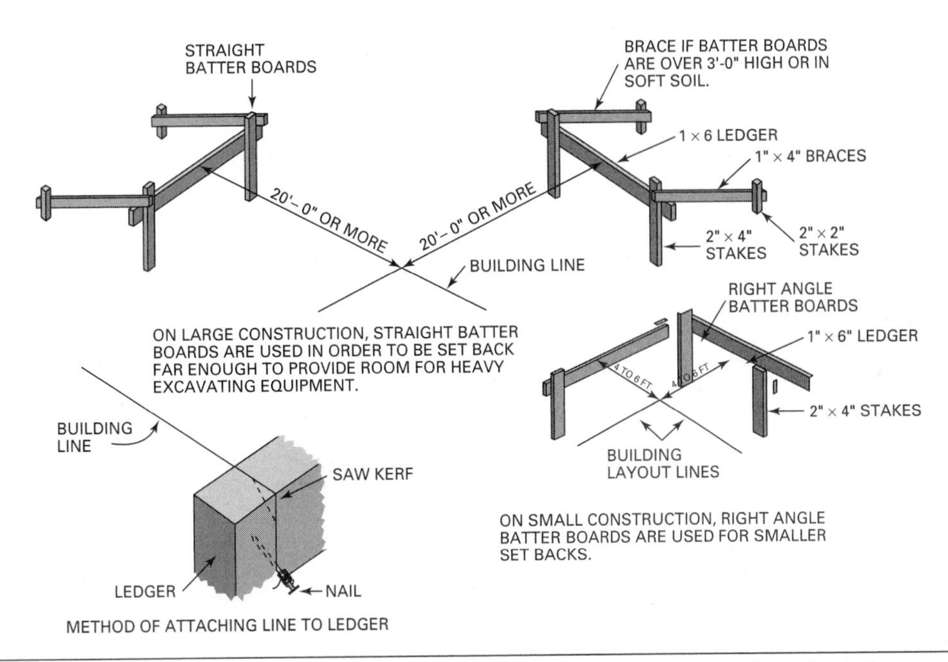

Fig. 26-11 Batter boards are placed back far enough so they will not be disturbed during excavation operations.

outside corner of the top edge of the ledger. This prevents the layout lines from moving when stretched and secured. Be careful not to make the saw cut below the top edge (Fig. 26-12).

Check the accuracy of the layout by again measuring the **diagonals** to see if they are equal. If not, make the necessary adjustment until they are equal (Fig. 26-13).

Saw cuts are also often made on batter boards to mark the location of the foundation footing. The footing width usually extends outside and inside of the foundation wall.

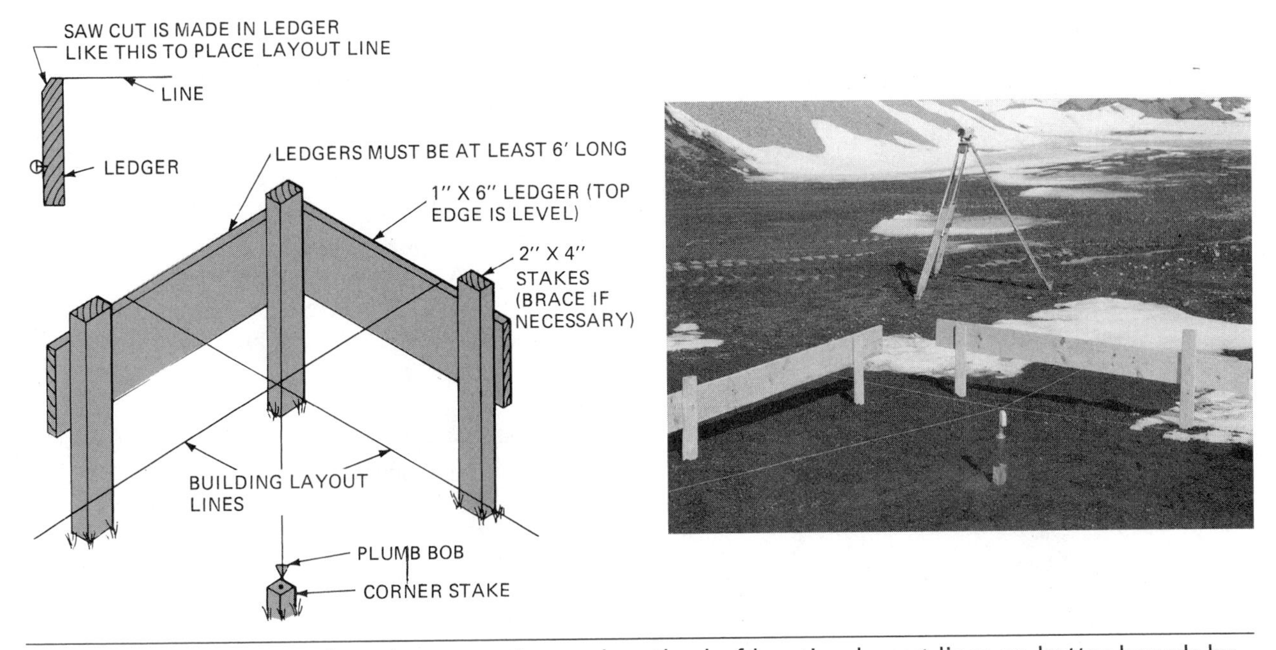

Fig. 26-12 Typical batter board construction and method of locating layout lines on batter boards by suspending a plumb bob directly over corner stakes.

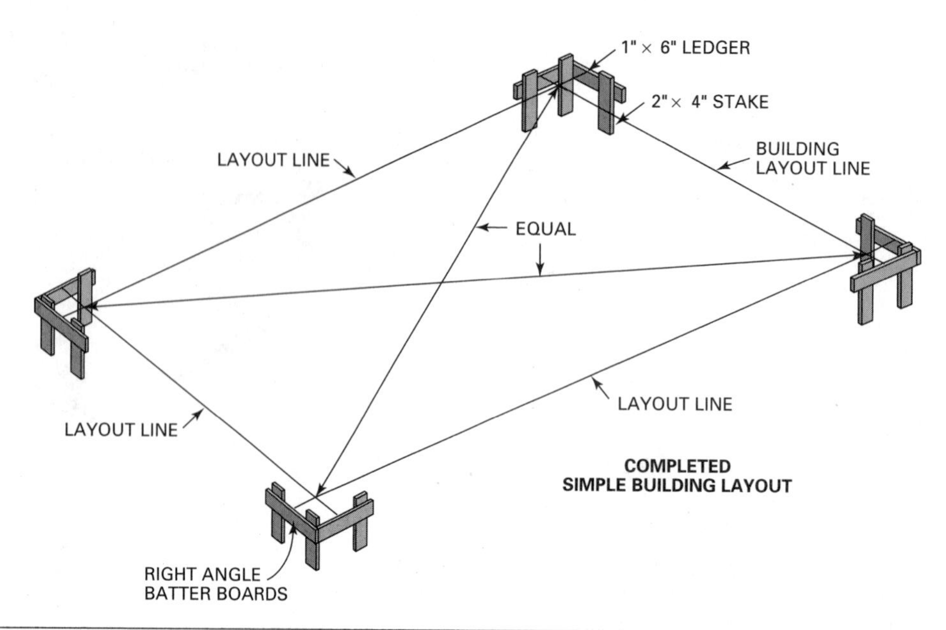

Fig. 26-13 Check the length and width and then the diagonals for accuracy of the building layout.

Review Questions

Select the most appropriate answer.

1. The location of the building on the lot is determined from the
a. foundation plan. c. architect.
b. floor plan. d. plot plan.

2. The builder's level is ordinarily used for
a. laying out straight lines.
b. reading elevations.
c. plumbing walls.
d. all of these.

3. When sighting objects through the telescope of the builder's level
a. use a sun shade.
b. keep both eyes open.
c. close one eye.
d. have the cross-hairs centered on the object.

4. The horizontal cross-hair is used for
a. laying out angles.
b. laying out straight lines.
c. plumbing walls and posts.
d. reading elevations.

5. The graduations of a leveling rod are
a. 1/8 inch wide and 1/8 inch apart.
b. 1/4 inch wide and 1/4 inch apart.
c. 1/16 inch wide and 1/16 inch apart.
d. 1/8 inch wide and 1/4 inch apart.

6. If the rod reading at point A is 6′-0″ and the reading at point B is 3′- 6″ then
a. the elevation at point B is 9′-6″.
b. point A is higher than point B.
c. point A has an elevation of 2′-4″.
d. point B is higher than point A.

7. The reference for establishing elevations on a construction site is called
a. starting point. c. benchmark.
b. reference point. d. sight mark.

8. Starting from point A, backsight rod readings are 36″, 48″, and 59″. Forward sight rod readings to point B are 28″, 42″, and 60″.
a. point A is 13″ higher than point B.
b. point A is 13″ lower than point B.
c. point A is 23″ higher than point B.
d. point A is 23″ lower than Point B.

9. To read minutes of a degree, when laying out horizontal angles with the builder's level, use the
a. base plate. c. circle scale.
b. index. d. vernier scale.

10. The diagonal of a rectangle whose dimensions are 30 feet × 40 feet is:
a. 45 feet. c. 60 feet.
b. 50 feet. d. 70 feet.

BUILDING FOR SUCCESS

Building Site Layout

Accurate building site layout requires placing the structure in the correct location and being in compliance with all local zoning regulations. The establishment of any required elevations will be determined in the layout process. An adequate working knowledge of site layout, instrument use, calculations, site plan reading, and local zoning ordinances should be possessed by the carpenter.

This phase of the construction process calls for workers to use their acquired skills in math, vernier and elevation reading, staking, and batter board construction. The layout must be squared. Then batter boards must be marked for building lines, footing lines, and excavation lines. If the process is followed as explained in Unit 9, the results will be accurate.

Site layout can pose many challenges for the carpenter. As site plans are read and interpreted, the layout person will be solving math problems in the process. Once the site layout is underway, it is wise to recheck the measurements. This will minimize errors. This is time to check for any easements that have been established by the local authorities. With each additional site layout, the need for extensive double checking will diminish.

Site layout is most convenient with two or three people. One person working alone must improvise throughout the process for accurate results. The site layout process takes on the characteristics of surveying and calls for a team approach to obtain the best results. Accuracy in the layout process also depends on how well the equipment is maintained and transported. Abused instruments will not provide accurate readings. Care should be taken to wipe or clean the instruments, leveling rods, tapes, tripods, and other tools after each layout is complete. The storing and transporting of the equipment should be done with care.

Weather variations will call for a variety of methods to be used during the site layout process. For example, steel stakes may be more appropriate as batter boards when the ground is frozen. Excessive winds may affect the string lines and plumb bobs. Differing temperatures and climatic conditions may affect the layout instruments. A warm builder's level or transit-level may have to adjust to the cold temperatures in the winter. Expansions and contractions in the metal housing can cause instruments to give incorrect readings. Rain, snow, and wind will interfere with the layout process. Rain may cause delays that last for days. Instruments should be covered if exposed to rain during use. Snow may have to be removed from the site in order to expose the soil. Wind may cause string lines to vibrate extensively.

Some of the more important aspects of accurate building site layout are:

1. Follow the site plan precisely. Confirm all measurements, setbacks, utility placements, and easements.
2. Properly interpret all local zoning ordinances. Ask local authorities for clarification.
3. Rely on the most recent survey to locate the property lines. If any property lines cannot be located, consult the surveyor.
4. Be sure the builder's level or transit-level is not damaged. This could give inaccurate readings.
5. Recheck measurements, elevation readings, and calculations as needed to confirm accuracy.
6. Construct durable batter boards and notch all necessary lines (excavation, footing, foundation), for future reference.

As experience is gained in site layout, confidence is increased. That confidence will in turn, allow each person to reach a high level of accuracy and efficiency.

FOCUS QUESTIONS: For individual or group discussion

1. What might be some reasons for accurate site layout?
2. Once the site layout process is understood and practiced, what would be a good procedure to follow?
3. What could be some of the results of an inaccurate building site layout? How could they affect the preestablished time schedule and goals of the project?

UNIT

10

Concrete Form Construction

The construction of concrete forms is the responsibility of the carpenter. The forms must meet specified dimensions and be strong enough to withstand tremendous pressure. If a form fails to hold, costly labor and materials are lost. Time must be spent in dismantling and cleanup. The formwork must be erected all over again. Cleanup is difficult because the concrete has probably set up by that time. If reinforcing rods are used in the concrete the cleanup and rebuilding are even more costly. The carpenter must know that the forms will hold before the concrete is placed.

OBJECTIVES

After completing this unit, the student should be able to:

- describe the composition of concrete and factors affecting its strength, durability, and workability.
- explain the reasons for reinforcing concrete and describe the materials used.
- job-mix a batch of concrete and explain the method of and reasons for making a slump test.
- explain techniques used for the proper placement and curing of concrete.
- construct forms for footings, slabs, walks, and driveways.
- construct concrete forms for foundation walls.
- construct concrete column forms.
- lay out and build concrete forms for stairs.
- estimate quantities of concrete.

UNIT CONTENTS

CHAPTER 27 CHARACTERISTICS OF CONCRETE

Understanding the characteristics of concrete is essential for the construction of reliable concrete forms, the correct handling of freshly mixed material, and the final quality of hardened concrete.

Concrete

Concrete is the most widely used of all building materials. It can be formed into practically any shape for construction of buildings, bridges, dams, and roads. Improvements over the years have created a product that is strong, durable, and versatile (Fig. 27-1).

Composition of Concrete

Concrete is a mixture of *portland cement,* fine and coarse *aggregate,* water, and various *admixtures.* Aggregates are fillers, usually sand, gravel, or stone. Admixtures are materials added to the mix to achieve certain desired qualities.

Portland Cement. In 1824, Joseph Aspdin, an Englishman, developed an improved type of cement. He named it *portland cement* because it produced a concrete that resembled stone found on the Isle of Portland, England. Portland cement is a fine gray powder. When mixed with water it forms a paste. A chemical reaction, called *hydration,* takes places that causes the cement to set. It then hardens to bind the aggregates together into a strong, durable, and watertight mass called concrete.

When moisture is present, hydration and, consequently, hardening of the concrete, continues for many years. Rapid evaporation or freezing of the water in concrete can cause the cement hydration process and the hardening to stop.

Portland cement is usually sold in paper bags that hold one cubic foot. Each bag weighs 94 pounds or 40 kilograms. The cement can still be used even after a long period of time as long as it remains dry. Eventually, however, it will absorb moisture from the air. It should not be used if it contains lumps that cannot be broken up easily. This condition is known as prehydration.

Types of Portland Cement. New developments have produced several types of portland cement. Type I is the familiar gray kind most generally purchased and used. Type I A is an *air-entraining* cement that is more resistant to freezing and thawing. Types II through V have special uses, such as for low or high heat generation, high early strength, and resistance to severe frost. White portland cement, used for decorative purposes, differs only in color from gray cement. There are many other types, each with its specific purpose—ranging from underwater use to sealing oil wells.

Water. Any water used for drinking is suitable for use to make concrete. Other water may be used but should be tested first to make sure it is acceptable.

The amount of water used to make concrete in relation to the amount of portland cement is called

Fig. 27-1 Concrete is the most widely used material in the construction industry.

the *water-cement ratio.* The amount of water used largely determines the quality of the concrete. A common practice of adding more water to a mix with the proper ratio to make it more fluid and to allow it to flow easier alters the water-cement ratio and weakens the concrete.

Aggregates. Aggregates have no cementing value of their own. They serve only as a filler but are important ingredients. They constitute from 60 to 80 percent of the concrete volume.

Fine aggregate consists of particles 1/4 inch or less in diameter. Sand is the most commonly used fine aggregate. *Coarse* aggregate, usually gravel or crushed stone, comes usually in sizes of 3/8, 1/2, 3/4, 1, 1 1/2, and 2 inches. When strength is the only consideration, 3/4 inch is the optimum size for most aggregates. Large-size aggregate uses less water and cement. Therefore, it is more economical. However, the maximum aggregate size must not exceed the following:

- 1/5 the smallest dimension of the unreinforced concrete.
- 3/4 the clear spacing between reinforcing bars or between the bars and the form.
- 1/3 the depth of unreinforced slabs on the ground (Fig. 27-2).

Fine and coarse particles must be in a proportion that allows the finer particles to fill the spaces between the larger particles. The aggregate should be clean and free of dust, loam, clay, or vegetable matter.

Air. An important advance was made with the development of *air-entrained* concrete. It is produced by using air-entraining portland cement or an **admixture.** The intentionally made air bubbles are very small. Billions of them are contained in a cubic yard of concrete.

The introduction of air into concrete was designed to increase its resistance to freezing and thawing. It also has other benefits: less water and sand are required, the workability is improved, separation of the water from the paste is reduced, and the concrete can be finished sooner and is more watertight than ordinary concrete. Air-entrained concrete is now recommended for almost all concrete projects.

Admixtures. Admixtures are available to quicken or retard setting time, develop early strength, inhibit corrosion, retard moisture, control bacteria and fungus, improve pumping, and color concrete, among many other purposes.

Mixing Concrete

Concrete mixtures are designed to achieve the desired qualities of strength, durability, and workability in the most economical manner. The mix will

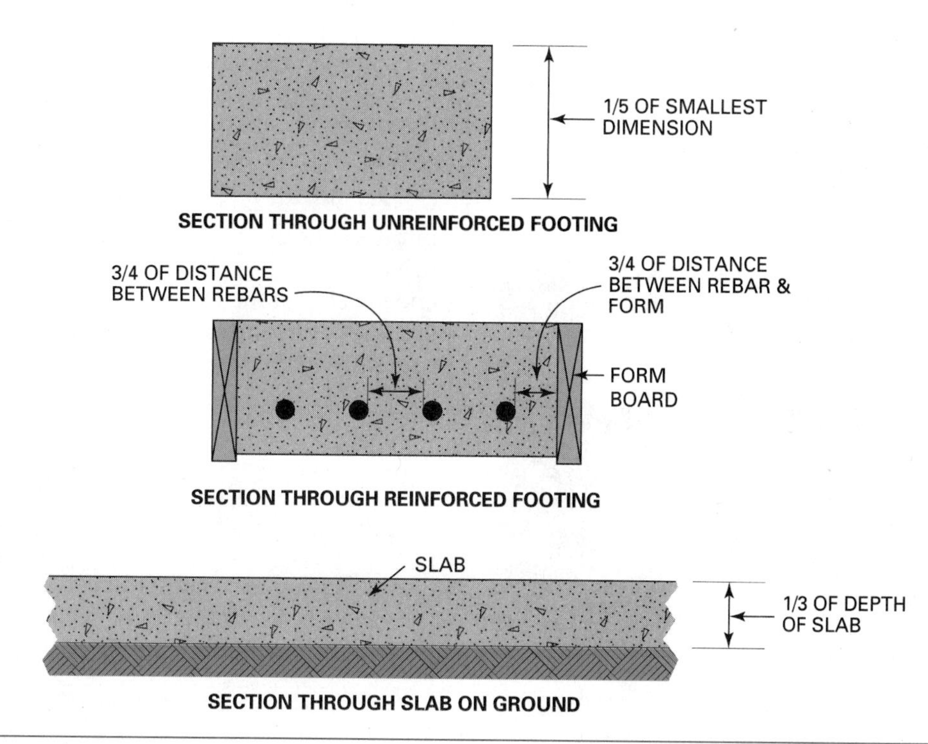

Fig. 27-2 The maximum aggregate size allowed in the concrete mixture.

vary according to the strength and other desired qualities, plus other factors such as the method and time of curing.

For very large jobs, such as bridges and dams, concrete mixing plants may be built on the job site. This is because freshly mixed concrete is needed on a round-the-clock basis. The construction engineer calculates the proportions of the mix and supervises tests of the concrete.

Ready-Mixed Concrete. Most concrete used in construction is *ready-mixed* concrete. It is sold by the cubic yard or cubic meter. There are 27 cubic feet to the cubic yard. The purchaser usually specifies the amount and the desired strength of the concrete. The concrete supplier is then responsible for mixing the ingredients in the correct proportion to yield the desired strength.

Sometimes the purchasers specify the proportion of the ingredients. They then assume responsibility for the design of the mixture.

The ingredients are accurately measured at the plant, often with computerized equipment. The mixture is delivered to the job site in *transit-mix trucks* (Fig. 27-3). The truck contains a large revolving drum, capable of holding from 1 to 10 cubic yards. There is a separate water tank with a water measuring device. The drum rotates to mix the concrete as the truck is driven to the construction site.

Specifications require that each batch of concrete be delivered within 1 1/2 hours after water has been added to the mix. If the job site is a short distance away, water is added to the cement and aggregates at the plant. If the job site is farther away, the proper amount of water is added to the dry mix from the water tank on the truck as it approaches or arrives at the job.

Job-Mixed Concrete. A small job may require that the concrete be mixed on the site either by hand or with a powered concrete mixer. All materials must be measured accurately. For measuring purposes, remember that a cubic foot of water weighs about 62 1/2 pounds and contains approximately 7 1/2 gallons.

All ingredients should be thoroughly mixed according to the proportions shown in Figure 27-4. A little water should be put in the mixer before the dry materials are added. The water is then added uniformly while the ingredients are mixed from one to three minutes.

Fig. 27-3 An over-the-cab, front-end discharge design of concrete trucks makes it easier to deliver ready-mix concrete.

Concrete Reinforcement

Concrete has high *compressive strength.* This means that it resists being crushed. It has, however, low *tensile* strength. It is not as resistant to bending or pulling apart. Steel bars, called **rebars,** are used in concrete to increase its tensile strength. Concrete is then called *reinforced concrete.*

Rebars. Rebars used in construction are usually *deformed.* Their surface has ridges that increase the bond between the concrete and the steel. They come in standard sizes, identified by numbers that indicate the diameter in eighths. For instance, a #6 rebar has a diameter of 6/8 or 3/4 inch (Fig. 27-5). Metric rebars are measured in millimeters.

The size, location, and spacing of rebars are determined by concrete engineers and shown on the plans. The iron worker places the rebars inside the form before the concrete is placed. Rebars in bridges and roads that experience a salty environment, such as sea water or salting during the winter months, have an epoxy coating. This protects the rebar from rusting, and thus helps prevent premature concrete failure. Cutting, bending, placing, and tying rebars require workers who have been trained in that trade.

Wire Mesh. Welded wire mesh is used to reinforce concrete floor slabs resting on the ground, driveways, and walks. It is identified by the gauge and spacing of the wire. Common gauges are #6, #8, and #10. The wire is usually spaced to make 6-inch squares (Fig. 27-6).

WATER GALS	PORTLAND CEMENT BAGS	AGGREGATE (FINE) CU. FT.	AGGREGATE (COARSE) CU. FT.
5 1/2	1	3	4
5	1	2 1/4	3
4	1	2	2 1/4

TRIAL BATCHES FOR SLUMP & WORKABILITY

Fig. 27-4 Formulas for several concrete mixtures.

BAR #	BAR Ø INCHES	METRIC SIZES (MM)	BAR WEIGHT LBS PER 100 LIN FT.
2	1/4	6	17
3	3/8	10	38
4	1/2	13	67
5	5/8	16	104
6	3/4	19	150
7	7/8	22	204
8	1	26	267

Fig. 27-5 Numbers and sizes of commonly used reinforcing steel bars.

MESH SIZE INCHES	MESH GAUGE	MESH WEIGHT LBS PER 100 SQ. FT.
6×6	#6	42
6×6	#8	30
6×6	#10	21

Fig. 27-6 Size, gauge, and weight of commonly used welded wire mesh.

Welded wire mesh is laid in the slab above the vapor barrier, if used, before the concrete is placed. It is spliced by lapping one full square plus 2 inches.

Placing Concrete

Sawdust, nails, and other debris should be removed from the forms. The inside surfaces are brushed or sprayed with oil to make form-removal easier. Also, before concrete is placed, the forms and subgrade are moistened with water. This is done to prevent rapid absorbing of water from the concrete.

Concrete is *placed,* not poured. Water should never be added so that concrete flows into forms without working it. Adding water alters the water-cement ratio on which the quality of the concrete depends.

Slump Test. Slump tests are made by supervisors on the job to determine the consistency of the concrete. The concrete sample for a test should be taken just before the concrete is placed. The *slump cone* is first dampened. It is placed on a flat surface and filled to about one-third of its capacity with the fresh concrete. The concrete is then *rodded* by moving a metal rod up and down twenty-five times over the entire surface. Two more approximately equal layers are added to the cone. Each layer is rodded in a similar manner with the rod penetrating the layer below. The excess is **screeded** from the top. The cone is removed by carefully lifting it vertically in 3 to 7 seconds. The cone is gently placed beside the concrete, and the amount of slump measured (Fig. 27-7). The test should be completed within 2 to 3 minutes.

Changes in slump should be corrected immediately. The table shown in Figure 27-8 shows recommended slumps for various types of construction. Concrete with a slump greater than 6 inches should not be used unless a slump-increasing admixture has been added.

Placing In Forms. The concrete truck should get as close as possible. Concrete is placed by chutes where needed. It should not be pushed or dragged any more than necessary. It should not be dropped more than 4 to 6 feet, and should be dropped vertically and not angled. Drop chutes should be used in high forms to prevent the buildup of dry concrete on the side of the form or reinforcing bars above the level of the placement. Drop chutes also prevent separation caused by concrete striking and bouncing off the side of the form.

Fig. 27-7 A slump test shows the wetness or dryness of a concrete mix.

TYPES OF CONSTRUCTION	SLUMP IN INCHES	
	MAXIMUM	MINIMUM
REINFORCED FOUNDATION WALLS & FOOTINGS	3	1
PLAIN FOOTINGS, CAISSONS & SUBSTRUCTURE WALLS	3	1
BEAMS & REINFORCED WALLS	4	1
BUILDING COLUMNS	4	1
PAVEMENTS & SLABS	3	1
HEAVY MASS CONCRETE	2	1

Fig. 27-8 Recommended slumps for different kinds of construction.

RATE (FT/HR)	(PSF)			
	50° F		70° F	
	COLUMNS	WALLS	COLUMNS	WALLS
1	330	330	280	280
2	510	510	410	410
3	690	690	540	540
4	870	870	660	660
5	1050	1050	790	790
6	1230	1230	920	920
7	1410	1410	1050	1050
8	1590	1470	1180	1090
9	1770	1520	1310	1130
10	1950	1580	1440	1170

PRESSURE INCREASES AS PLACEMENT RATE INCREASES.
PRESSURE INCREASES AT LOWER TEMPERATURES.

Pumps are used on large jobs that need concrete continuously over long distances or to heights up to 500 feet.

Rate of Placement. Concrete must not be placed at a rapid rate, especially in high forms. The amount of pressure at any point on the form is determined by the height and weight of the concrete above it. Pressure is not affected by the thickness of the wall (Fig. 27-9).

A slow rate of placement allows the concrete nearer the bottom to begin to harden. Once concrete hardens it cannot exert more pressure on the forms even though liquid concrete continues to be placed

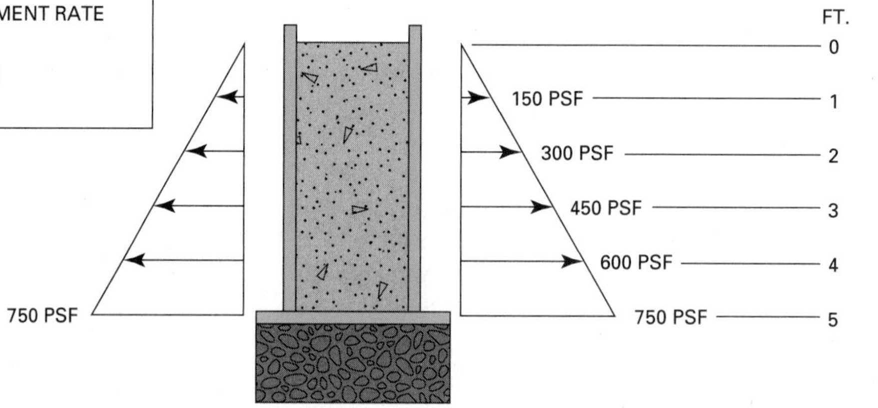

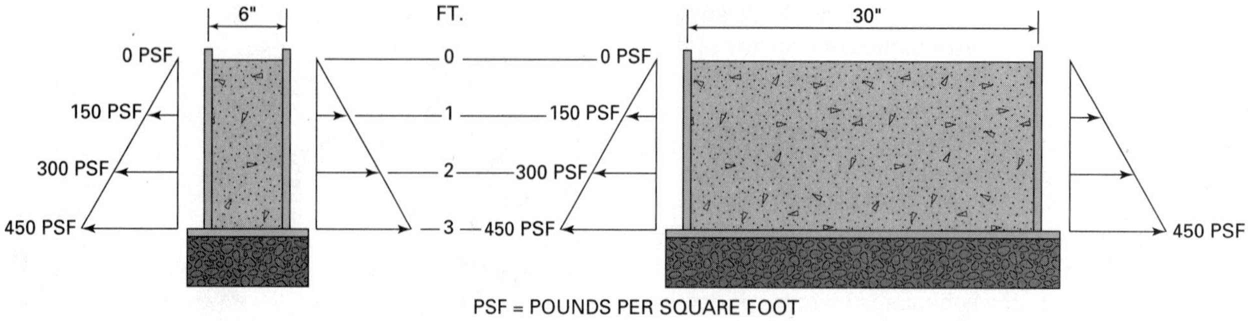

FT.
0
150 PSF 1
300 PSF 2
450 PSF 3
600 PSF 4
750 PSF 750 PSF 5

6"
FT.
0 PSF 0 0 PSF
150 PSF 1 150 PSF
300 PSF 2 300 PSF
450 PSF 3 450 PSF
30"
450 PSF
PSF = POUNDS PER SQUARE FOOT

Fig. 27-9 The height of concrete being poured affects the amount of pressure against the forms. Pressure is not affected by the thickness of the wall.

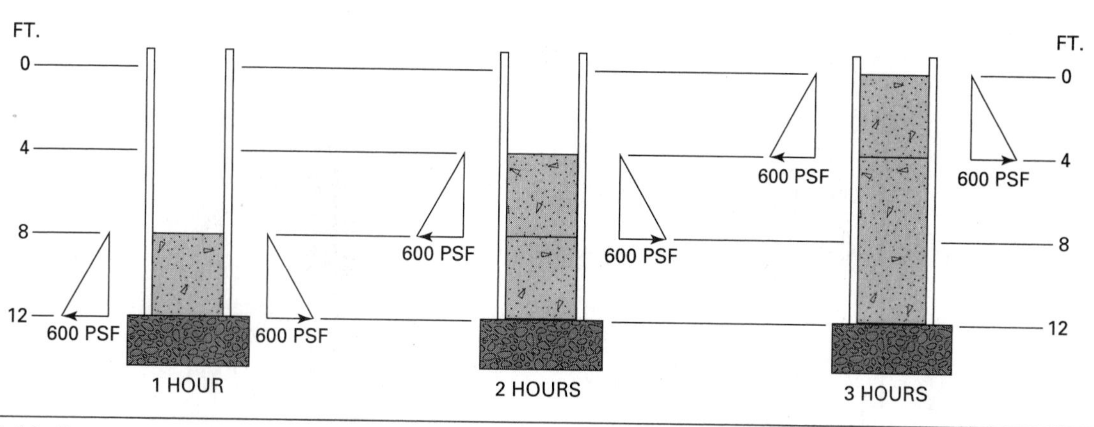

Fig. 27-10 Once concrete sets, it cannot exert more pressure on the form even though liquid concrete continues to be placed above it.

above it (Fig. 27-10). The use of stiff concrete with a low slump, which acts less like a liquid, will transmit less pressure. Rapid placing leaves the concrete in the bottom still in a fluid state. It will exert great lateral pressure on the forms at the bottom. This may cause the form ties to fail or the form to deflect excessively.

Concrete should be placed in forms in layers of not more than 12 to 18 inches thick. Each layer should be placed before setting occurs in the previous layer. The layers should be thoroughly consolidated (Fig. 27-11). The rate of concrete placement in high forms should be carefully controlled.

Consolidation of Concrete. To eliminate voids or honeycombs in the concrete, it should be thoroughly worked by hand spading or vibrated after it goes into the form. Vibrators make it possible to use a stiff mixture that would be difficult to consolidate by hand.

An immersion type vibrator, called a *spud vibrator,* has a metal tube on its end. This tube vibrates at a rapid rate. It is commonly used in construction to vibrate and consolidate concrete. Vibration makes the concrete more fluid and able to move, allowing trapped air to escape. This will prevent the formation of air pockets, honeycombs, and cold joints. The operator should be skilled in the use of the vibrator, keeping it moving up and down, uniformly vibrating the entire pour. Over-vibrating should not be done as vibrating increases the lateral pressure on the form (Fig. 27-12).

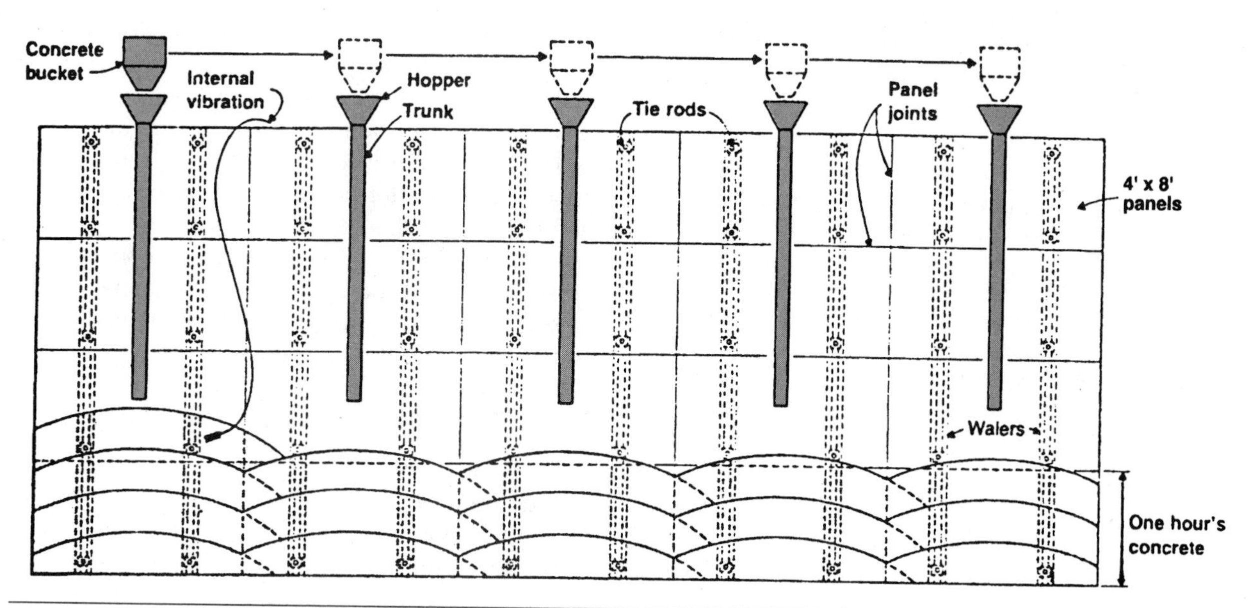

Fig. 27-11 A safe, consistent pour rate is accomplished using drop chutes to place concrete in approximately 12- to 18-inch internally vibrated layers.

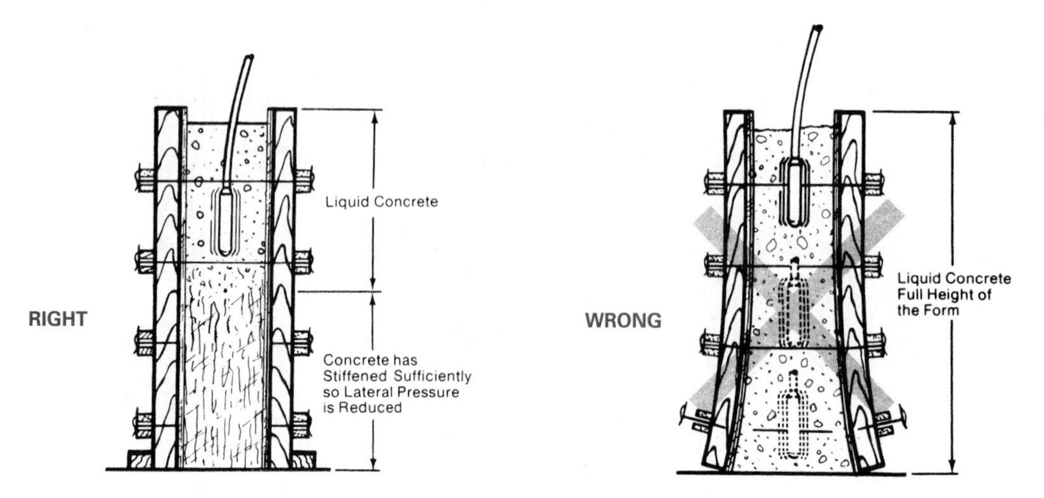

Fig. 27-12 Avoid excessive vibration of concrete. (*Courtesy of Dayton Superior*)

Curing Concrete

Concrete hardens and gains strength because of a chemical reaction called hydration. All the desirable properties of concrete are improved the longer this process takes place. A rapid loss of water from fresh concrete can stop the hydration process too soon, weakening the concrete. Curing prevents loss of moisture, allowing the process to continue so that the concrete can gain strength.

Concrete is *cured* either by keeping it moist or by preventing loss of its moisture for a period of time. For instance, if moist-cured for seven days, its strength is up to about 60% of full strength. A month later the strength is 95%, and up to full strength in about three months. Air-cure will reach only about 55% after three months and will never attain design strength. In addition, a rapid loss of moisture causes the concrete to shrink, resulting in cracks. Curing should be started as soon as the surface is hard enough to resist marring.

Methods of Curing. Flooding or constant sprinkling of the surface with water is the most effective method of curing concrete. Curing can also be accomplished by keeping the forms in place, covering the concrete with burlap, straw, sand, or other material that retains water, and wetting it continuously.

In hot weather, the main concern is to prevent rapid evaporation of moisture. Sunshades or windbreaks may need to be erected. The formwork may be allowed to stay in place or the concrete surface may be covered with plastic film or other waterproof sheets. The edges of the sheets are overlapped and sealed with tape or covered with planks. Liquid curing chemicals may be used to seal in moisture and prevent evaporation. However, manufacturers' directions for their use should be carefully followed.

Curing Time. Concrete should be cured for as long as practical. The curing time depends on the temperature. At or above 70 degrees, curing should take place at least three days. At or above 50 degrees, concrete is cured at least five days. Near freezing, there is practically no hydration and concrete takes considerably longer to gain strength. There is no strength gain while concrete is frozen. When thawed, hydration resumes with appropriate curing. If concrete is frozen within the first twenty-four hours after being placed, permanent damage to the concrete is almost certain. In cold weather, *accelerators* that shorten the setting time are sometimes used. Protect concrete from freezing for at least four days after being placed by providing insulation or artificial heat, if necessary.

Forms may be removed after the concrete has set and hardened enough to maintain its shape. This time will vary depending on the mix, temperature, humidity, and other factors.

Precautions Using Concrete

Avoid prolonged contact with fresh concrete or wet cement because of possible skin irritation. Wear protective clothing when working with newly mixed concrete. Wash skin areas that have been exposed to wet concrete as soon as possible. If any material containing cement gets into the eyes, flush immediately with water and get medical help.

CHAPTER 28 FORMS FOR FOOTINGS, SLABS, WALKS, AND DRIVEWAYS

Concrete for footings, slabs, walks, and driveways is placed directly on the soil. The supporting soil must be suitable for the type of concrete work placed upon it.

Footing Forms

The *footing* for a foundation provides a base on which to spread the load of a structure over a wider area of the soil. For foundation walls, the most typical type is a *continuous* or *spread* footing. To provide support for columns and posts, *pier* footings of square, rectangular, circular, or tapered shape are used. Sometimes it is not practical to excavate deep enough to reach loadbearing soil. Then **piles** are driven and capped with a *grade beam* (Fig. 28-1).

Continuous Wall Footings

In most cases, the footing is formed separately from the foundation wall. In residential construction, often the footing width is twice the wall thickness.

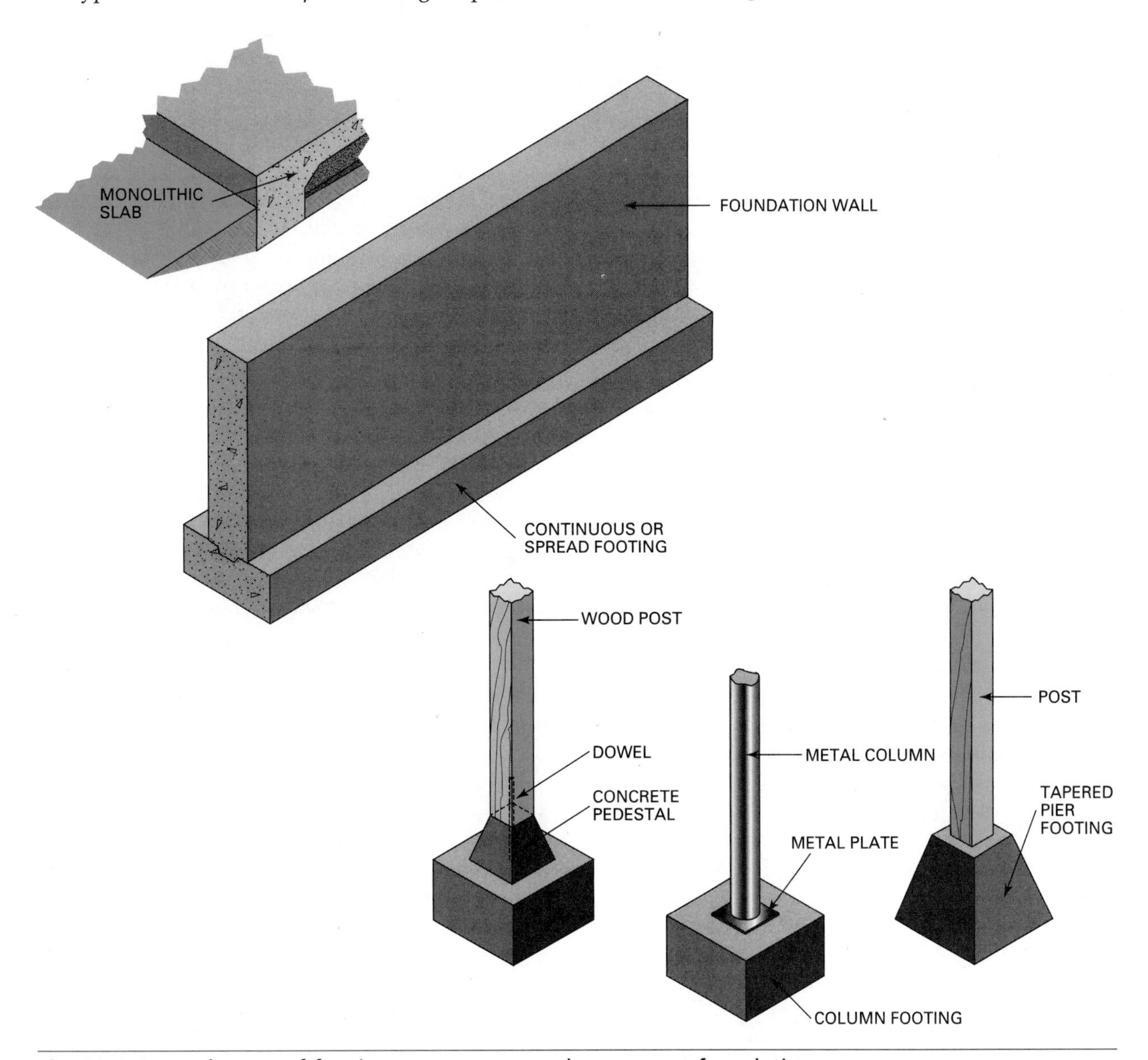

Fig. 28-1 Several types of footings are constructed to support foundations.

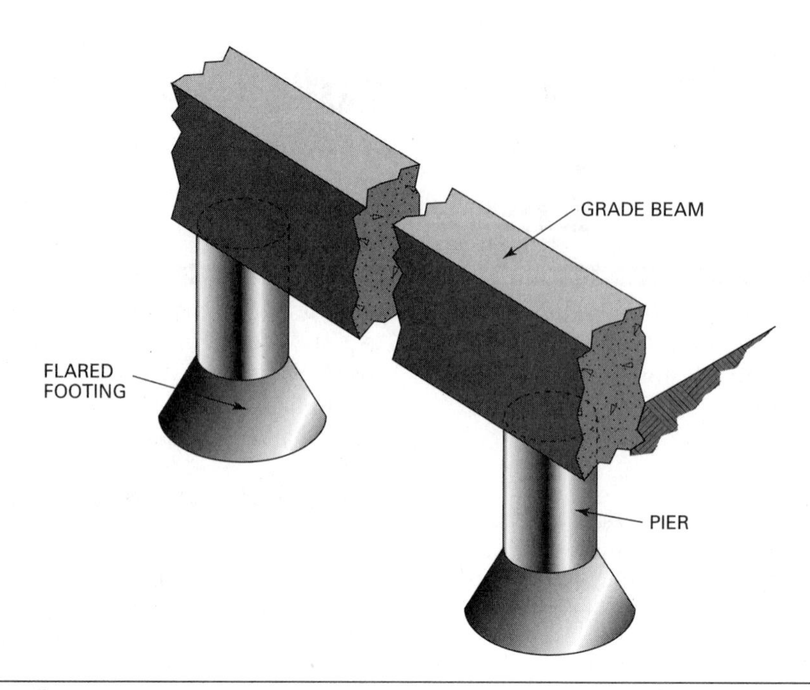

Fig. 28-1 (Continued)

The footing depth is often equal to the wall thickness (Fig. 28-2). However, to be certain, consult local building codes.

For larger buildings, architects or engineers design the footings to carry the load imposed upon them. Usually these footings are strengthened by reinforcing rods of specified size and spacing.

Frost Line. Footings must be located below the **frost line.** The frost line is the point below the surface to which the ground freezes in winter. Because water expands when frozen, foundations whose footings are above the frost line will heave and buckle when the ground freezes. In extreme northern climates, footings must be placed as much as 6 feet below the surface (Fig. 28-3). In tropical climates, footings only need to reach solid soil, with no consideration given to frost.

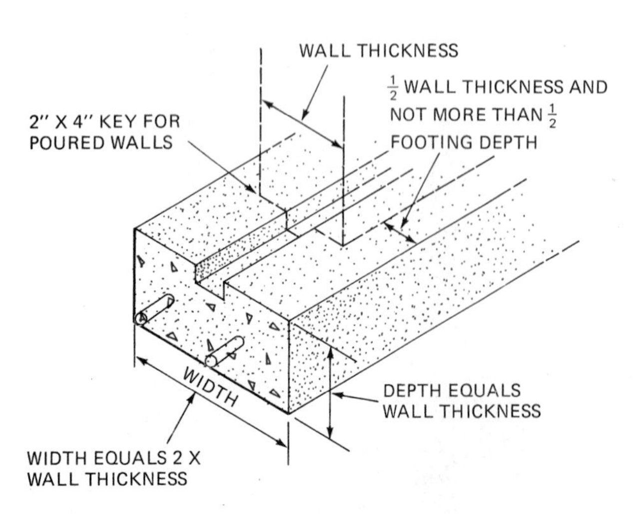

Fig. 28-2 Typical footing for residential construction.

In areas where the soil is stable, footings may be *trench poured* where no formwork is necessary. A trench is dug to the width and depth of the footing. The concrete is carefully placed in the trench (Fig. 28-4). In other cases, forms need to be built for the footing.

Locating Footings

To locate the footings, stretch lines on the batter boards in line with the outside of the footing. Suspend a plumb bob from the batter board lines at each corner. Drive stakes. Stretch lines between the stakes to the correct elevation of the top of the footing (Fig. 28-5).

Building Wall Footing Forms

When building wall footing forms, erect the outside form first. Usually 2×8 lumber is used for the side of the form, and stakes are used to hold the sides in position. Fasten the sides by driving nails through the stakes. Use duplex nails for easy removal. Steel stakes, spreaders, and braces are manufactured for use in building footing forms and other edge formwork. They come with nail holes for easy fastening (Fig. 28-6).

Keep the top inside corner of the form as close as possible to the line *without touching it*. Be sure the form does not touch the line. If the form touches the line at any point, the line is moved and is no longer

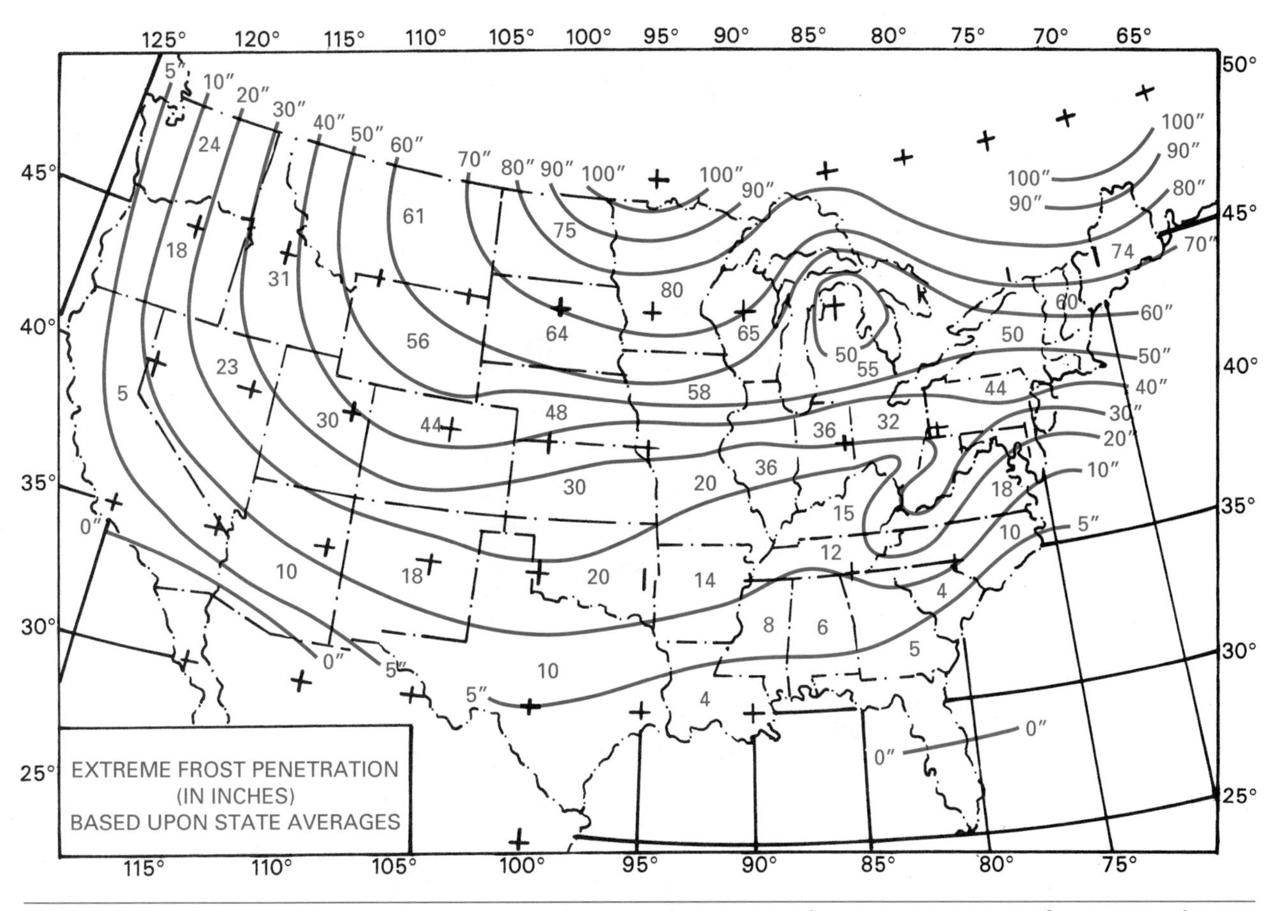

Fig. 28-3 Frost line penetration in the United States. (*Courtesy of U.S. Department of Commerce*)

Fig. 28-4 In stable soils, no footing formwork is necessary.

straight. Form the outside of the footing in this manner all around. Space stakes 4 to 6 feet apart or as necessary to hold the form straight.

Before erecting the inside forms, cut a number of **spreaders**. These are nailed to the top edges of the form. They tie the two sides together and keep them the correct distance apart. Erect the inside forms in a manner similar to that used in erecting the outside forms. Place stakes for the inside forms opposite those holding the outside form. Level across from the outside form to determine the height of the inside form. Fasten the spreaders across the form at intervals necessary to hold the form the correct distance apart.

Brace the stakes where necessary to hold the forms straight. In many cases, no bracing is necessary. Footing forms are sometimes braced by shoveling earth or placing large stones against the outside of the forms. Figure 28-7 shows the steps in the construction of a typical wall footing form.

Keyways. A keyway is formed in the footing by pressing 2 × 4 lumber into the fresh concrete (Fig. 28-8). The keyway form is beveled on both edges for easy removal after the concrete has set. The purpose of a keyway is to provide a lock between the footing and the foundation wall. This

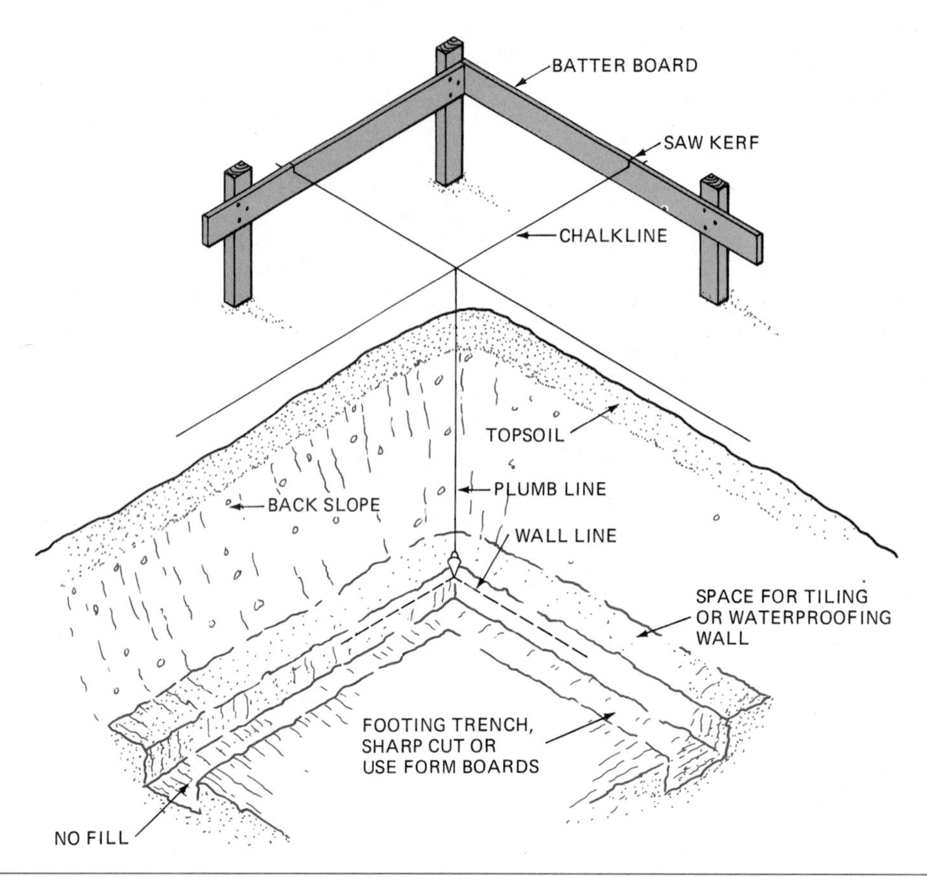

Fig. 28-5 The footing is located by suspending a plumb bob from the batter board lines.

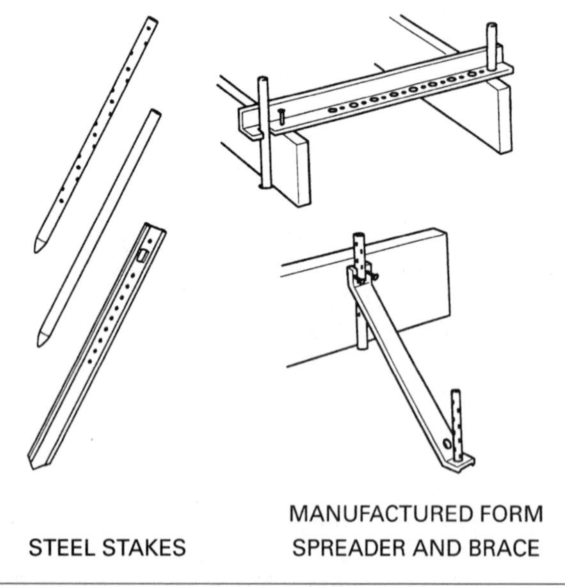

Fig. 28-6 Steel stakes, spreaders, and braces are manufactured for use in building footing forms and other edge formwork. (*Courtesy of Symons Corporation*)

joint helps the foundation wall resist the pressure of the back-filled earth against it. It also helps to prevent seepage of water into the basement. In some cases, where the design of the keyway is not so important, 2 × 4 pieces are not beveled on the edges, but are pressed into the fresh concrete at an angle.

Stepped Wall Footings

When the foundation is to be built on sloped land, it is sometimes necessary to *step* the footing. The footing is formed at different levels, to save material. In building stepped footing forms, the thickness of the footing must be maintained. The vertical and horizontal footing distances are adjusted so that a whole number of blocks or concrete forms can easily be placed into that section of the footing without cutting. The vertical part of each step should not exceed the footing thickness. The horizontal part of the step must be at least twice the vertical part (Fig. 28-9). Vertical boards are placed between the forms to retain the concrete at each step.

Column Footings

Concrete for footings, supporting columns, posts, fireplaces, chimneys, and similar objects is usually placed at the same time as the wall footings. The size and shape of the column footing vary according to what it has to support. The dimensions are determined from the foundation plan.

STEP BY STEP

PROCEDURES

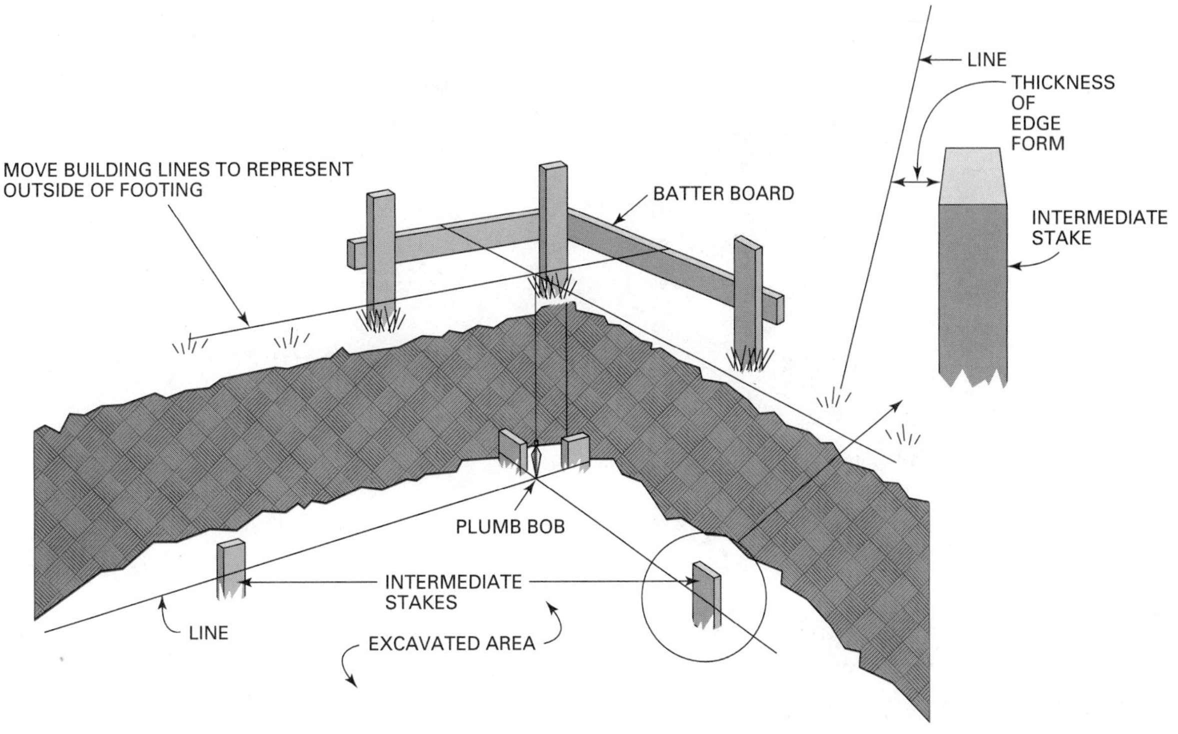

STEP 1 – SUSPEND PLUMB BOB TO LOCATE OUTSIDE OF FOOTING. DRIVE STAKES AT EACH CORNER & STRETCH LINES FROM CORNER TO CORNER AT ELEVATION OF TOP OF FOOTING.

STEP 2 – DRIVE INTERMEDIATE STAKES OUTSIDE OF STRETCHED LINE BY THE THICKNESS OF THE EDGE FORM.

STEP 3 – SNAP A CHALK LINE ON THE INSIDE FACE OF ALL STAKES. THIS LINE REPRESENTS THE TOP OF THE FORM.

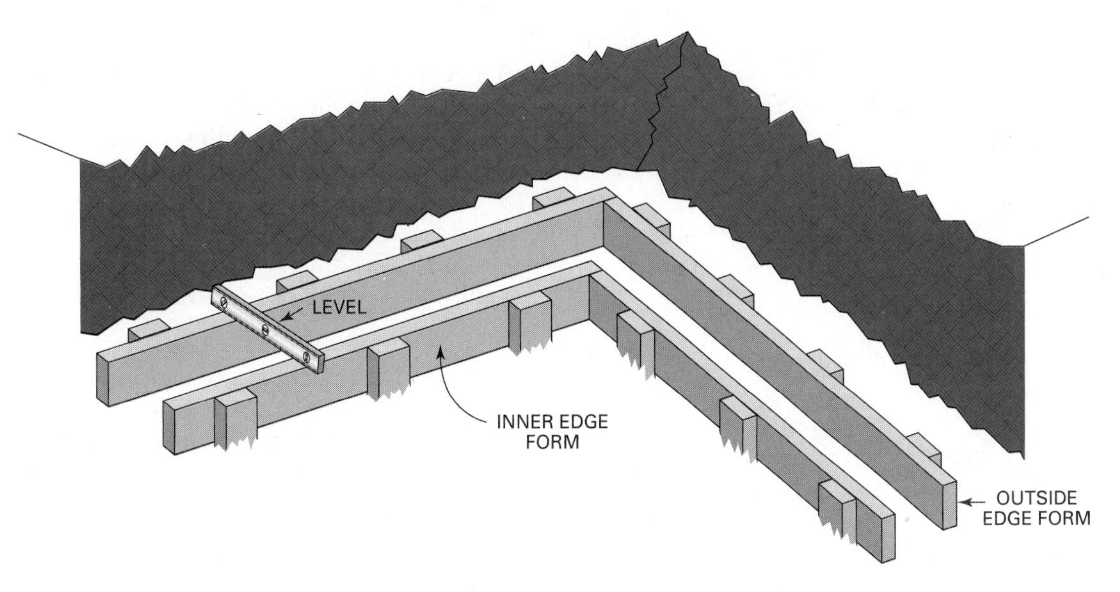

STEP 4 – FASTEN OUTSIDE EDGE FORM TO STAKES, KEEPING TOP EDGE TO CHALK LINE.

STEP 5 – LOCATE STAKES FOR INNER EDGE FORM. FASTEN INNER EDGE FORM TO STAKES. LEVEL ACROSS FROM OUTSIDE FORM FOR PROPER HEIGHT.

Fig. 28-7 Steps in the construction of a form for a typical continuous footing.

STEP BY STEP

P R O C E D U R E S

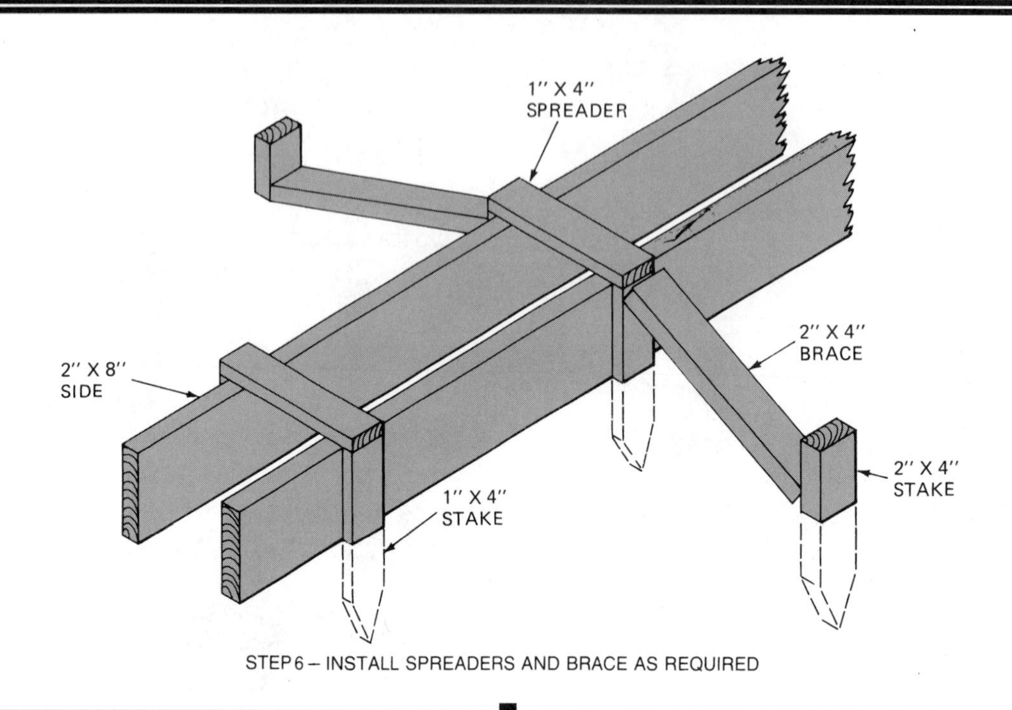

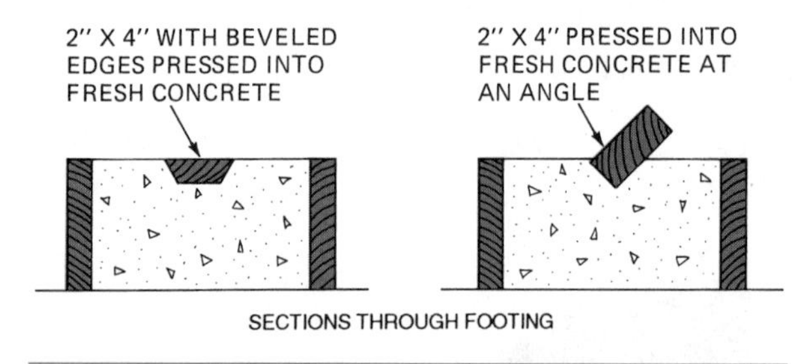

Fig. 28-8 Methods of forming keyways in the footing.

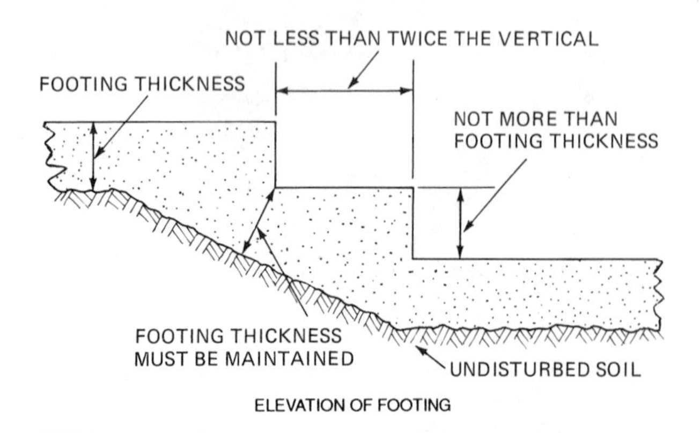

Fig. 28-9 Dimensions of a stepped footing.

In residential construction, the forms for this type of footing are usually built by spiking 2 × 8 pieces together in square, rectangular or tapered shapes to the specified size (Fig. 28-10).

Measurements are laid out on the wall footing forms to locate the column footings. Lines are stretched from opposite sides of the wall footing forms to locate the position of the forms. They are laid in position corresponding to the stretched lines (Fig. 28-11). Stakes are driven. Forms are usually fastened in a position so that the top edges are level with the wall footing forms.

Forms for Slabs

Building forms for slabs, walks, and driveways is similar to building continuous footing forms. The sides of the form are held in place by stakes driven into the ground. Forms for floor slabs are built level. Walks and driveways are formed to shed water. Usually 2 × 4 or 2 × 6 lumber is used for the sides of the form.

Slab-on-Grade

In warm climates, where frost penetration into the ground is not very deep, little excavation is necessary. The first floor may be a concrete slab placed

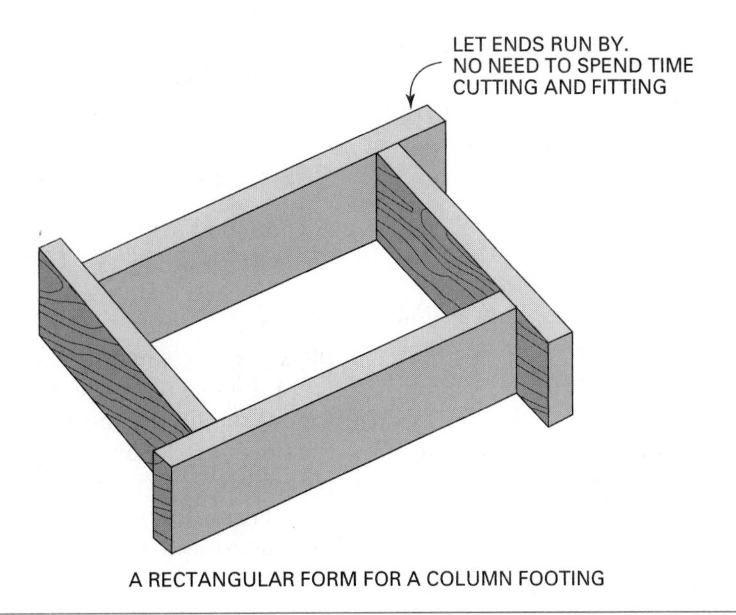

Fig. 28-10 Construction of pier forms.

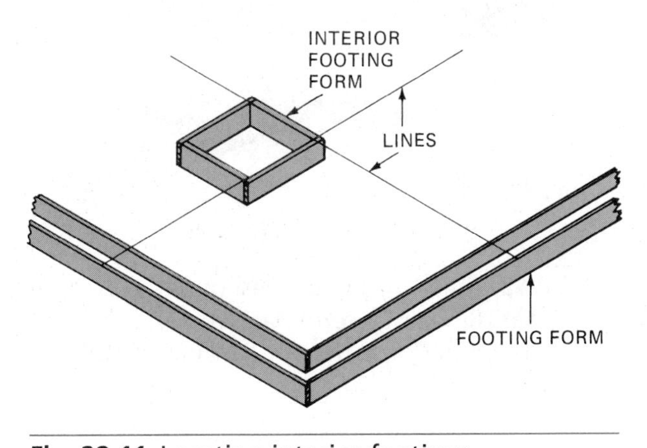

Fig. 28-11 Locating interior footings.

directly on the ground. This is commonly called *slab-on-grade* construction (Fig. 28-12). With improvements in the methods of construction, the need for lower construction costs, and the desire to give the structure a lower profile, slabs-on-grade are being used more often in all climates.

Basic Requirements

The construction of concrete floor slabs should meet certain basic requirements:

1. The finished floor level must be high enough so that the finish grade around the slab can be sloped away for good drainage. The top of the

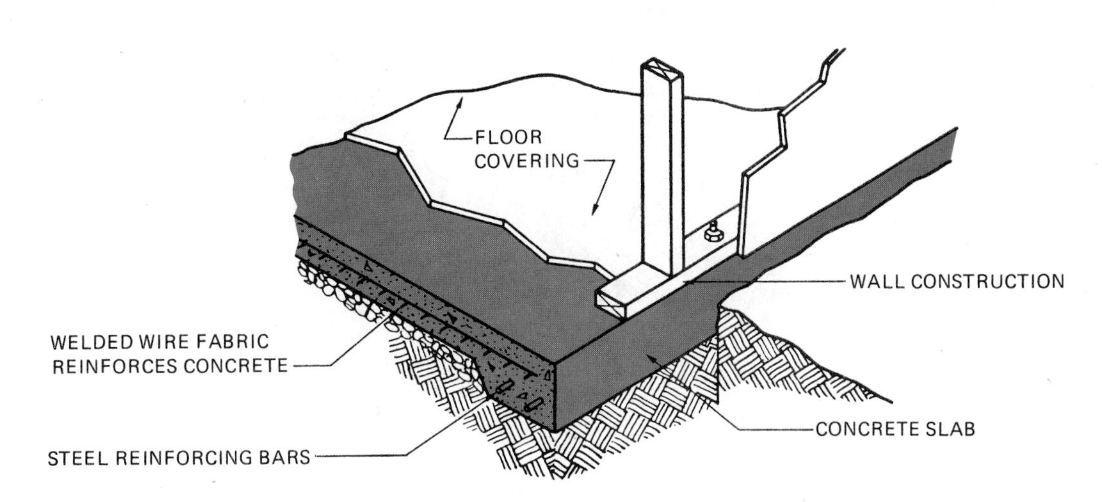

Fig. 28-12 Slab-on-grade foundation. (*From Mark W. Huth,* Understanding Construction Drawings, *©1983 by Delmar*)

slab should be no less than 8 inches above the finish grade.

2. All topsoil in the area in which the slab is to be placed must be removed. A base for the slab consisting of 4 to 6 inches of gravel, crushed stone, or other approved material must be well compacted in place.
3. The soil under the slab may be treated with chemicals for control of termites, but caution is advised. Such treatment should be done only by those thoroughly trained in the use of these chemicals.
4. All water and sewer lines, heating ducts, and other utilities that are to run under the slab must be installed.
5. A vapor barrier must be placed under the concrete slab to prevent soil moisture from rising through the slab. The vapor barrier should be a heavy plastic film, such as 6-mil polyethylene or other material having equal or superior resistance to the passage of vapor. It should be strong enough to resist puncturing during the placing of the concrete. Joints in the vapor barrier must be lapped at least 4 inches and sealed. A layer of sand may be applied to protect the membrane during concrete placement.
6. Where necessary, to prevent heat loss through the floor and foundation walls, a waterproof, rigid insulation is installed around the perimeter of the slab.
7. The slab should be reinforced with 6 × 6 inch, #10 welded wire mesh, or by other means to provide equal or superior reinforcing. The concrete slab must be at least 4 inches thick and *haunched* (made thicker) under loadbearing walls (Fig. 28-13).

Monolithic Slabs

A combined slab and foundation is called a *monolithic slab* (Fig. 28-14). This type of slab is also referred to as a *thickened edge slab.* It consists of a shallow footing around the perimeter. The perimeter is placed at the same time as the slab. The slab and footing make up a one-piece integral unit. The bottom of the footing must be at least one foot below the finish grade, unless local building codes dictate otherwise.

Forms for monolithic slabs are constructed using stakes and edge form boards, plank, or steel

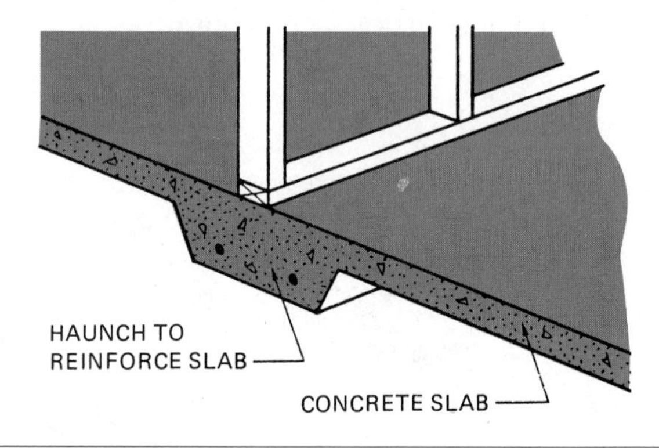

Fig. 28-13 The slab is haunched under load-bearing walls. (*From Mark W. Huth, Understanding Construction Drawings, ©1983 by Delmar*)

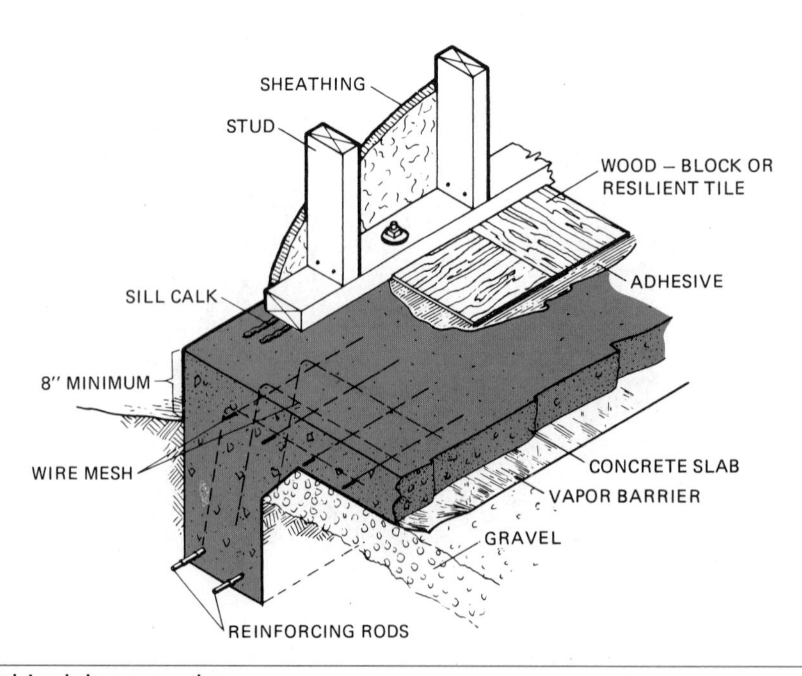

Fig. 28-14 Monolithic slab-on-grade.

manufactured especially for forming slabs and similar objects.

- Locate corner stakes by plumbing down from batter board lines. Drive the stakes outside of the building line by the thickness of the edge form material. Stakes should be long enough to be driven below the edge thickness of the slab.
- Stretch lines at desired elevation between the corner stakes. Drive intermediate stakes. Snap lines from corner stakes on to intermediate stakes. Fasten edge form boards to the tops and to the snapped lines.
- Tie all corners and joints together. Sight the forms for straightness. Brace by methods described earlier for continuous footing forms (Fig. 28-15).
- Excavate around the perimeter for the thickened edge and in the interior for any load bearing walls. Temporary screeds will need to be placed when the area is ready for concrete.

Independent Slabs

In areas where the ground freezes to any appreciable depth during winter, the footing for the walls of the structure must extend below the frost line. If slab-on-grade construction is desired in these areas, the concrete slab and foundation wall may be separate. This type of slab-on-grade is called an *independent slab.* It may be constructed in a number of ways according to conditions (Fig. 28-16).

When concrete blocks are used for the foundation wall, special blocks are used for the top course. They act as a form for the concrete slab. The top of

STEP BY STEP
PROCEDURES

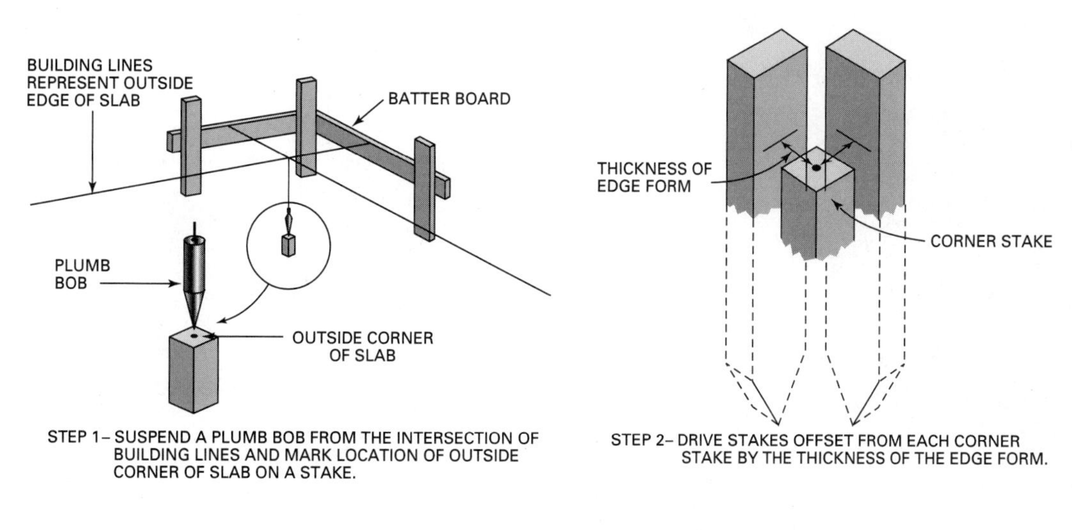

STEP 1 – SUSPEND A PLUMB BOB FROM THE INTERSECTION OF BUILDING LINES AND MARK LOCATION OF OUTSIDE CORNER OF SLAB ON A STAKE.

STEP 2 – DRIVE STAKES OFFSET FROM EACH CORNER STAKE BY THE THICKNESS OF THE EDGE FORM.

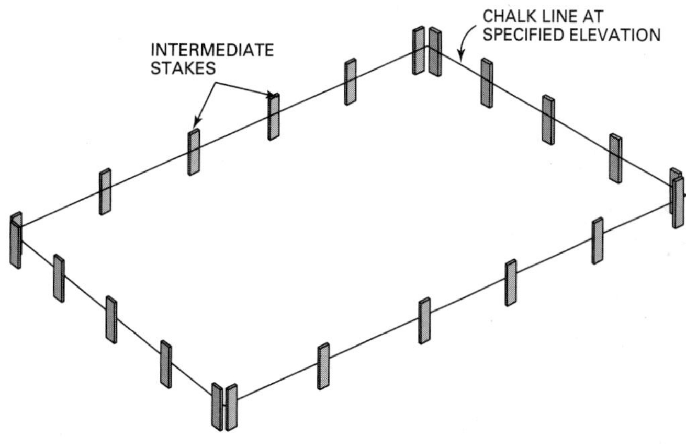

STEP 3 – STRETCH LINES FROM CORNERS AND DRIVE INTERMEDIATE STAKES. SNAP A LINE ON THE STAKES AT THE SPECIFIED ELEVATION.

Fig. 28-15 Steps in forming a monolithic slab.

STEP BY STEP
PROCEDURES

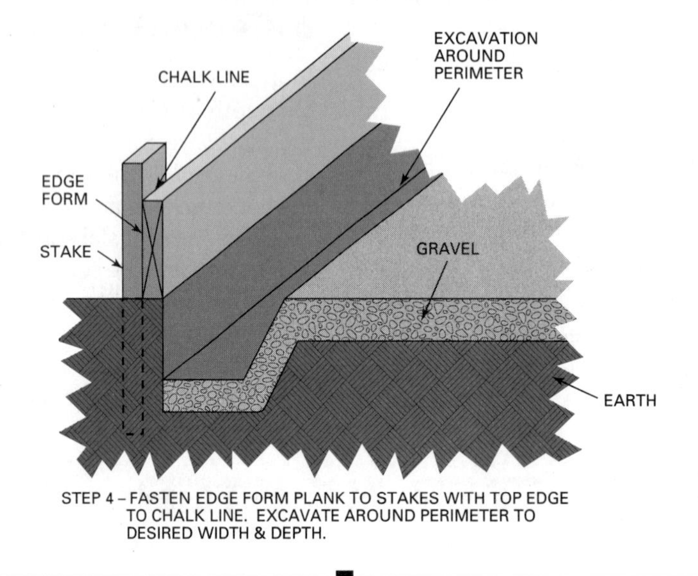

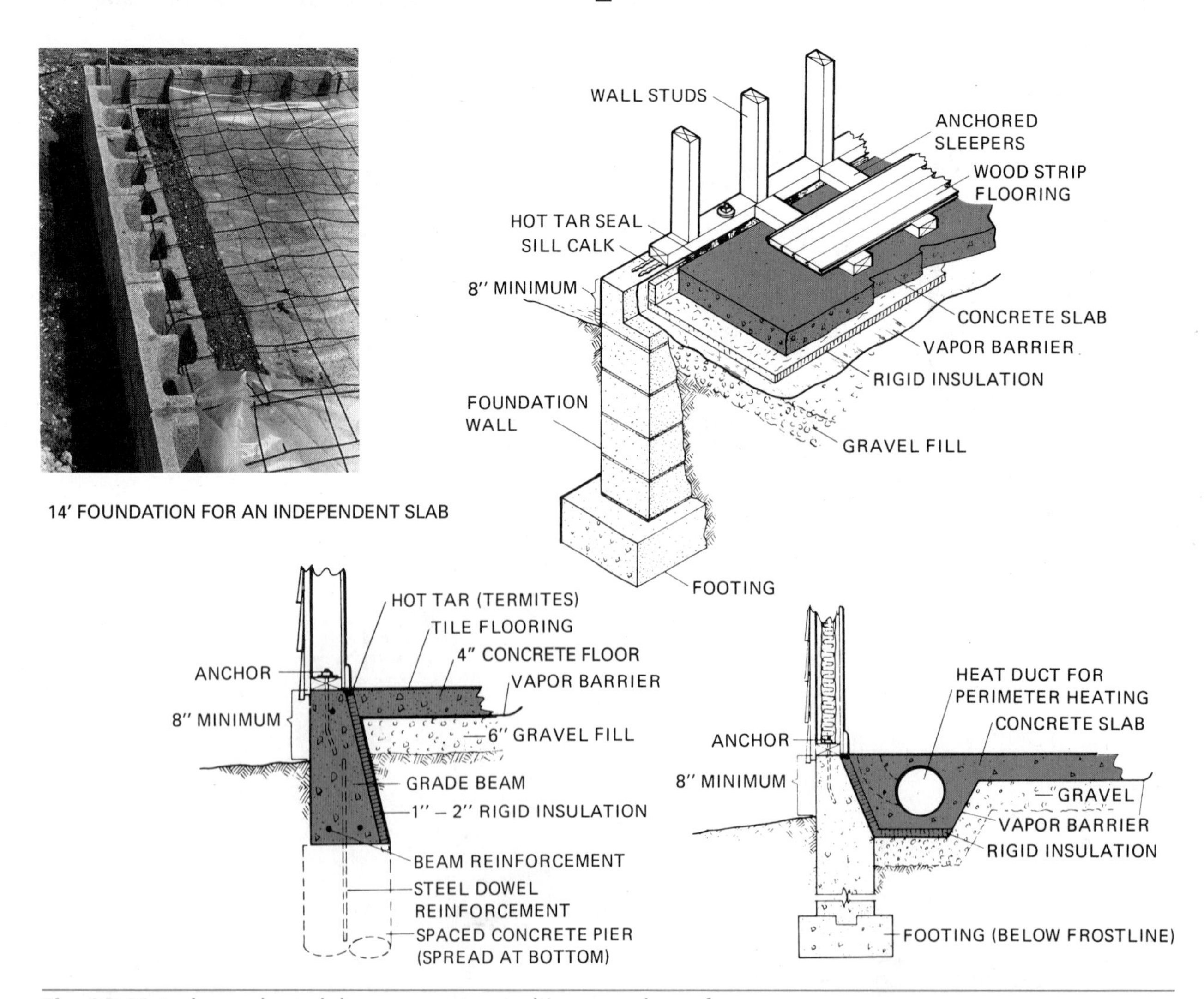

Fig. 28-16 Independent slabs are constructed in a number of ways.

a solid concrete wall may be formed with a shoulder for the same purpose.

If the frost line is not too deep, the footing and wall may be an integral unit and placed at the same time. In colder climates, the foundation wall and footing are formed and placed separately. The wall may be set on piles or on a continuous footing, the forming of which has been described previously.

Slab Insulation Requirements

Rigid insulation is placed between the foundation wall and the slab edge to provide an insulation joint. Insulation also prevents heat loss.

Perimeter insulation is required to prevent heat loss, except in warm climates. Two general rules to follow when determining insulation requirements are:

1. When average winter low temperatures are 0°F and higher, the **R factor** should be about 2.0. The depth of the insulation or the width under the slab should not be less than 1 foot.
2. When average winter temperatures are lower than 0°F, the R factor should be about 3.0 without floor heating. The depth or width of the insulation should be no less than 2 feet.

Forms for Walks and Driveways

Forms for walks and driveways are usually built so water will drain from the surface of the concrete. In these cases, grade stakes must be established and grade lines carefully followed (Fig. 28-17).

Usually a combination of 1 × 4 boards and 2 × 4 stakes are used or a combination of 2 × 4 edging and

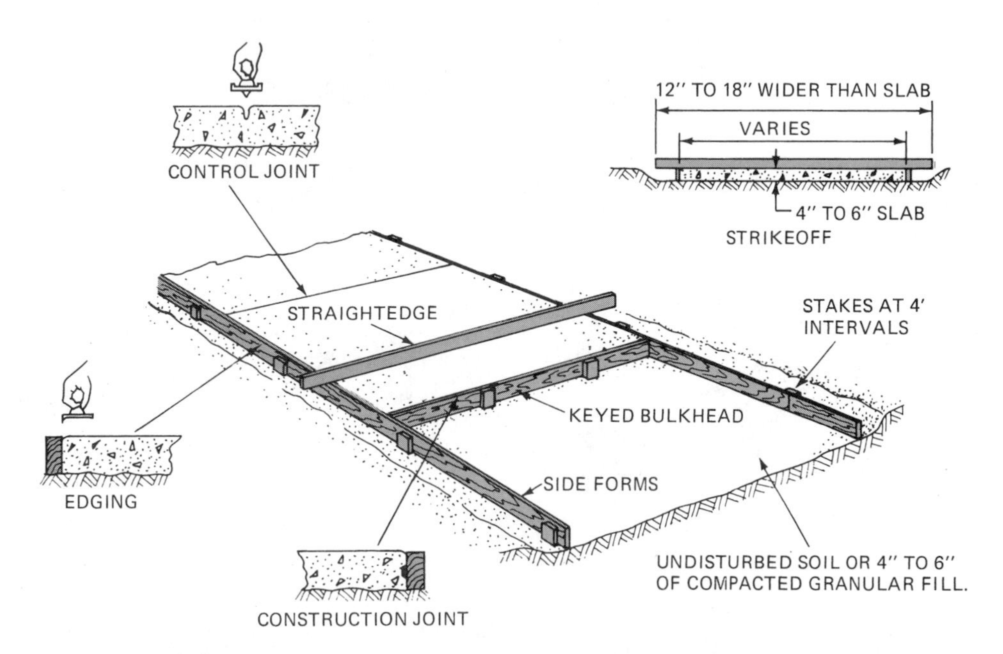

Fig. 28-17 Forms for walks are built so water drains from the surface.

1 × 4 stakes. Establish the grade of the walk or driveway on stakes at both ends. Stretch lines tightly between the end stakes. Drive intermediate stakes. Fasten the edge pieces to the stakes following the line in a manner similar to making continuous footing forms.

Forms for Curved Walks and Driveways

In many instances, walks and driveways are curved. To form the curve, 1/4-inch plywood or hardboard is used for small-radius curves. If using plywood, install it with the grain vertical for easier bending without breaking. Wetting the stock sometimes helps the bending process.

For curves of a long radius, 1 × 4 lumber may be used and satisfactorily bent if the curve is not too tight. Lumber may also be curved by making saw **kerfs** (cuts) spaced close together. The lumber is then bent until the saw kerfs close (Fig. 28-18). A method of determining the spacing of the saw kerfs to obtain a desired curve is shown in Figure 28-19.

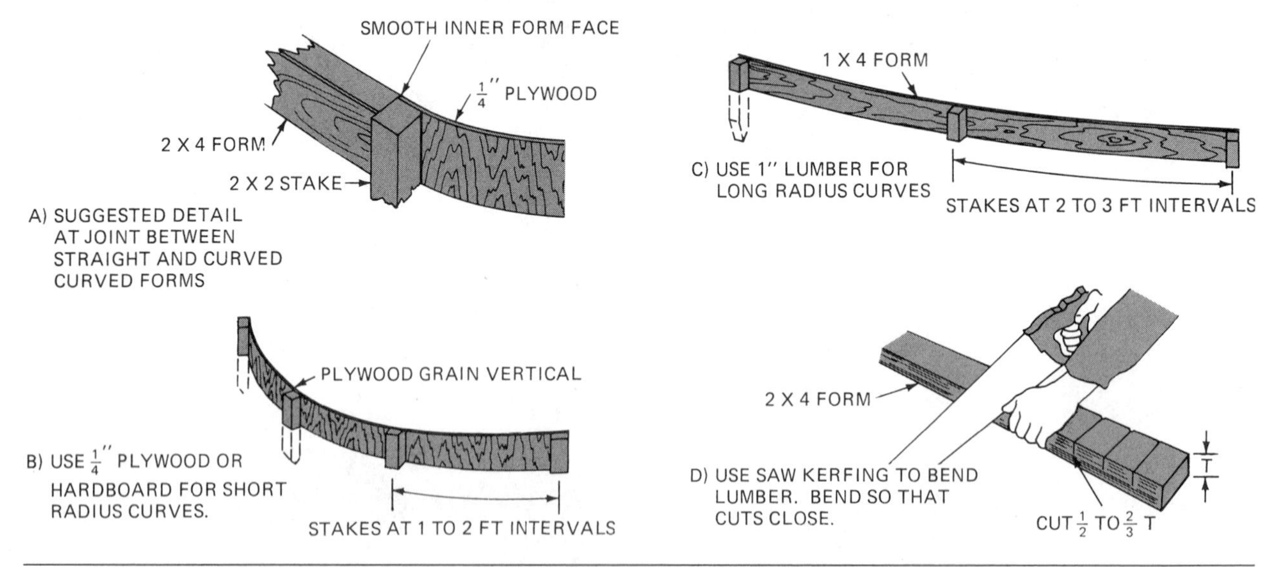

Fig. 28-18 Forming for curved walks and driveways.

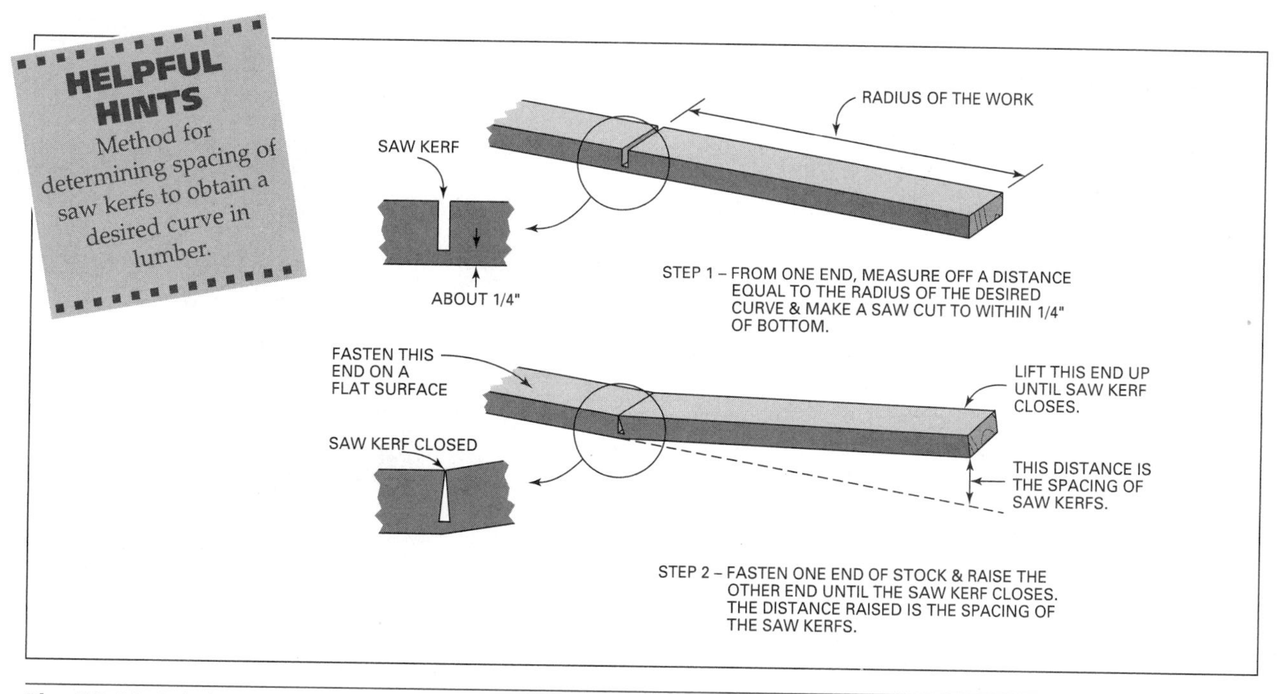

Fig. 28-19 Fabricating a curved form board using dimension lumber to fit a specific radius.

CHAPTER 29 WALL AND COLUMN FORMS

Foundation walls and columns are usually formed by using *panels* rather than building forms in place, piece by piece. Panel construction simplifies the erection and stripping of formwork.

Wall Forms

Wall Form Components

Various kinds of panels and panel systems are used. Some concrete panel systems are manufactured of wood, aluminum, or steel. Specially designed hardware is used for joining, spacing, aligning, and bracing the panels. Figure 29-1 shows the components of one system used for light formwork. (Other forming systems are described later in this chapter.)

Form Panels. Panels built of special form plywood backed by 2 × 4 **studs** (vertical framing members) are often used. They can be purchased or built on the job. They are placed side by side to form the inside and outside of the foundation walls. A usual panel size is 48 inches wide by 8 feet high. Narrower panels of several widths are used for fillers when the end space is too narrow for a standard size panel.

Snap Ties. Snap ties hold the wall forms together at the desired distance apart. They support both sides against the lateral pressure of the concrete (Fig. 29-2). These ties reduce the need for external bracing and greatly simplify the erection of wall forms. The design of a form for a particular job, including the spacing of the ties and studs, is decided by a construction engineer.

These ties are called snap ties because projecting ends are snapped off slightly inside the concrete after removal of the forms. A special snap tie wrench is used to break back the ties (Fig. 29-3). The small remaining holes are easily filled.

Because of the great variation in the size and shape of concrete forms, there are a large number of styles used. For instance, *flat ties* of various styles are used with some manufactured panels. For heavier formwork, *coil ties* and reuseable *coil bolts* are used (Fig. 29-4). For each kind of tie, there are also several sizes and styles. There are hundreds of kinds of form hardware. To become better acquainted with form hardware, study manufacturers' catalogs.

Walers. The snap ties run through and are wedged against form members called **walers.** Walers are doubled 2 × 4 pieces with spaces between them. They may be horizontal or vertical. Walers are spaced at right angles to the panel frame members. The number and spacing depend on the

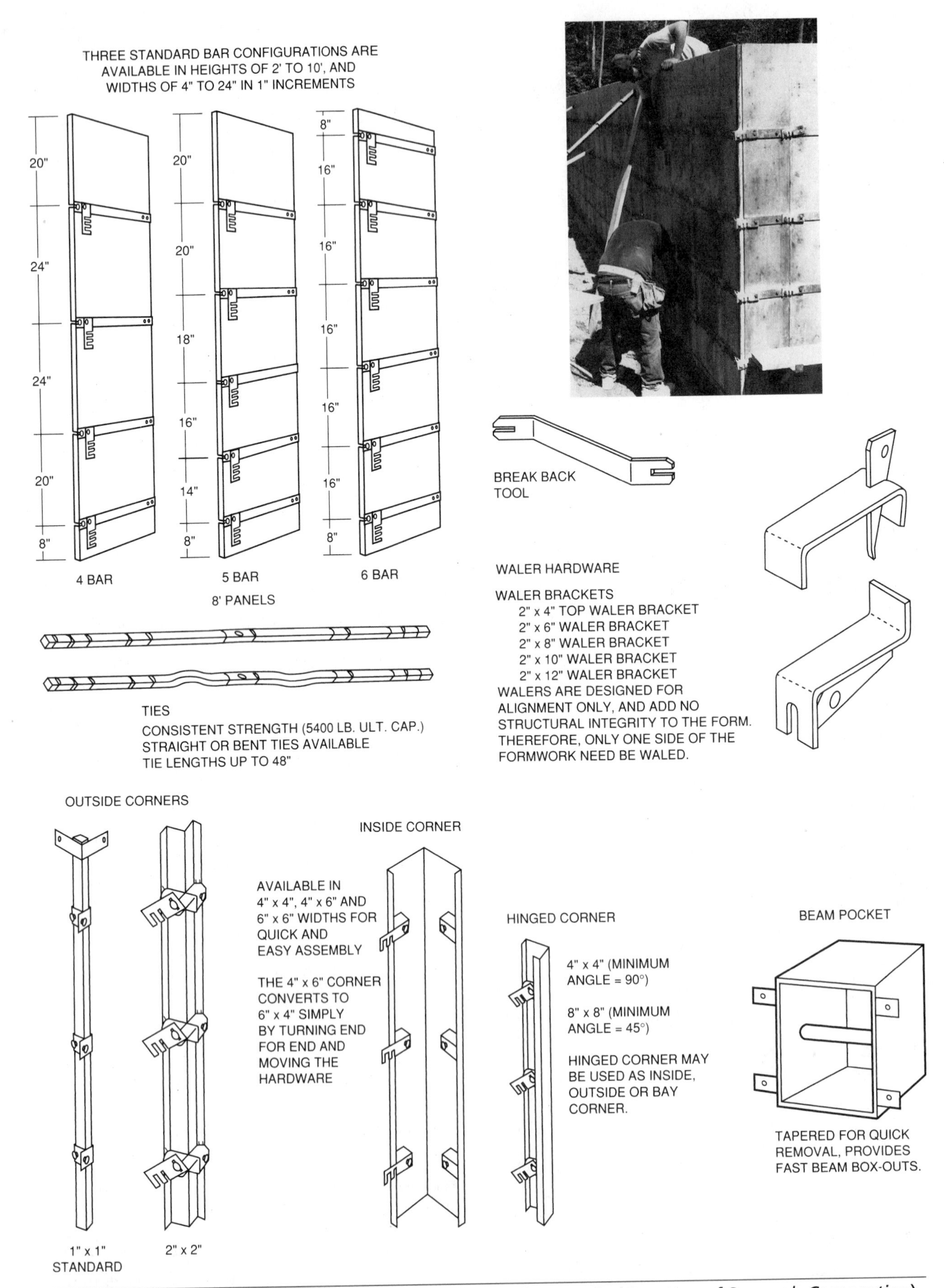

Fig. 29-1 Components of a panel system used for light formwork. (*Courtesy of Symonds Corporation*)

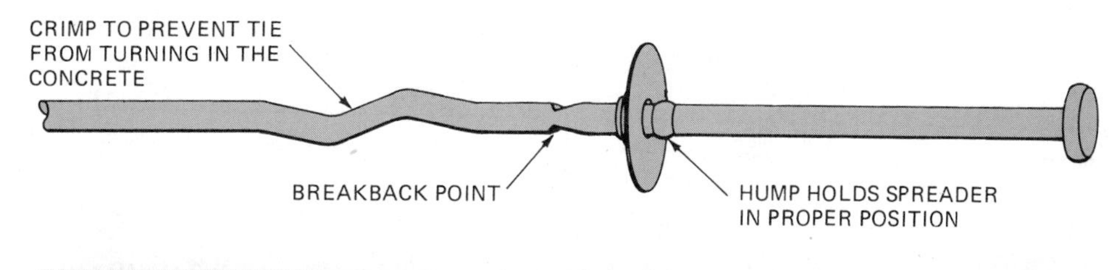

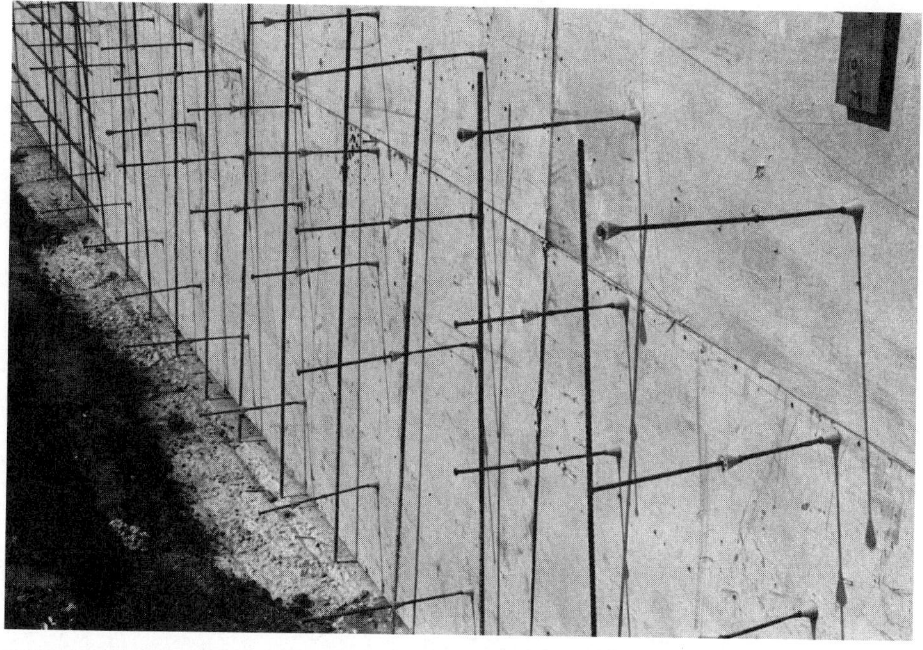

Fig. 29-2 A snap tie for light formwork. (*Courtesy of Dayton Superior*)

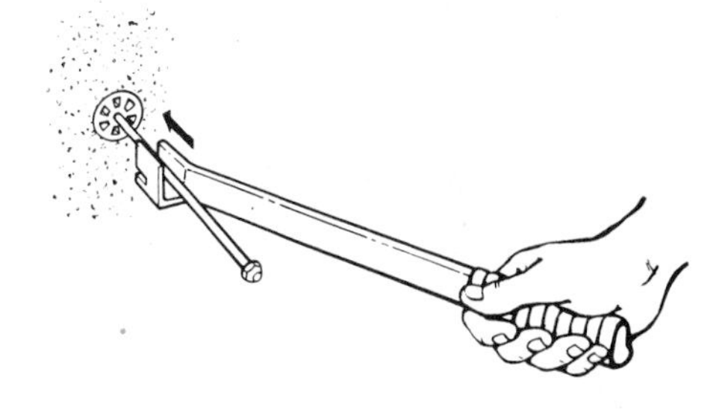

1. SLIDE THE SNAP TIE WRENCH UP AGAINST THE TIE SO THAT THE FRONT OF THE WRENCH IS TOUCHING THE CONCRETE.

2. KEEPING THE FRONT OF THE WRENCH TIGHT AGAINST THE CONCRETE, PUSH THE HANDLE END TOWARDS THE CONCRETE WALL SO THAT THE TIE IS BENT OVER AT APPROXIMATELY A 90° ANGLE.

3. ROTATE THE WRENCH AND TIE END 1/4 TO 1/2 TURN BREAKING OFF THE TIE END.

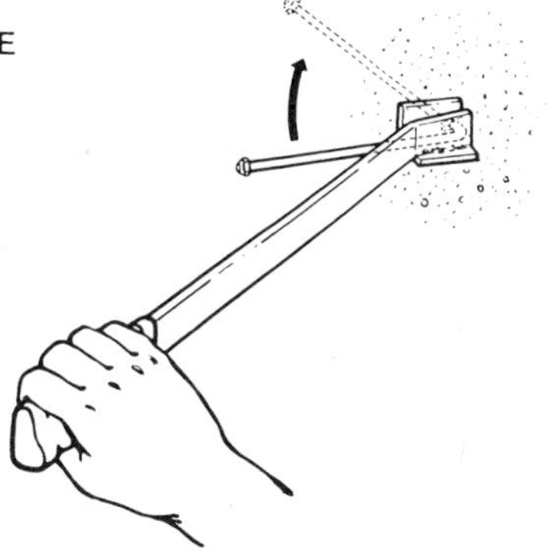

Fig. 29-3 Method of breaking back snap ties. (*Courtesy of Dayton Superior*)

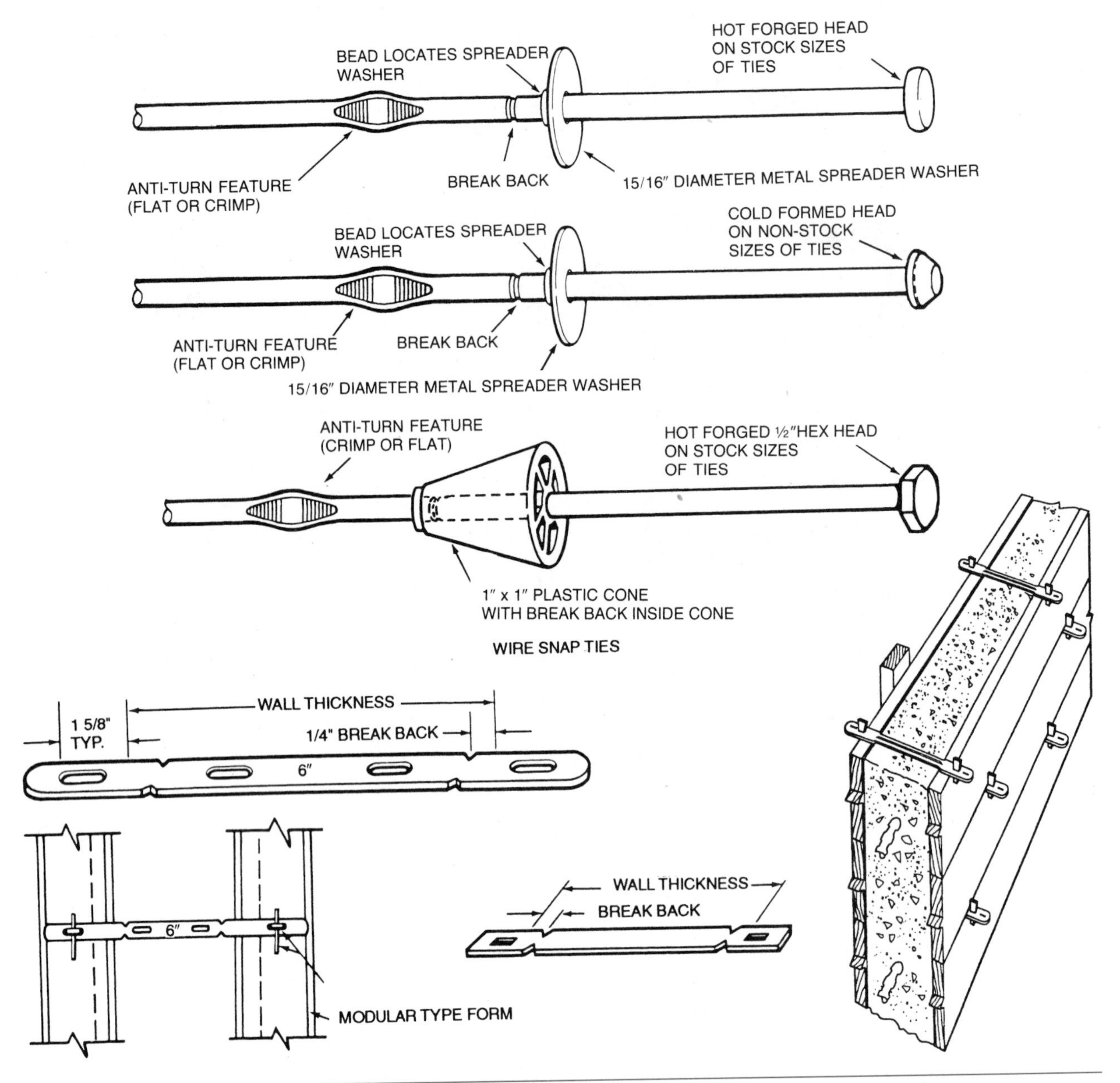

Fig. 29-4 A large variety of snap ties are manufactured. (*Courtesy of Dayton Superior*)

pressure exerted on the form. Figure 29-5 shows the construction of a typical wall form.

The vertical spacing of the **snap ties** and walers depends on the height of the concrete wall. The vertical spacing is closer together near the bottom. This is because there is more lateral pressure from the concrete there than at the top (Fig. 29-6).

For low wall forms less than 4 feet in height, the panel may be laid horizontal with vertical walers spaced as required (Fig. 29-7).

Erecting Wall Forms

After the concrete placed in the footing has hardened sufficiently (usually at least 3 days), the forms are removed and cleaned. The salvaged forms are reused.

Locating the Forms. Lines are stretched on the batter boards in line with the outside of the foundation wall. A plumb bob is suspended from the layout lines to the footing. Marks that are plumb with the layout lines are placed on the footing at each corner. A chalk line is snapped on the top of the footing between the corner marks outlining the outside of the foundation wall.

Installing Plates. Panels can be set directly on the concrete footing to the chalk line or on 2 × 4 or 2 × 6 lumber plates. Plates are

STEP BY STEP
PROCEDURES

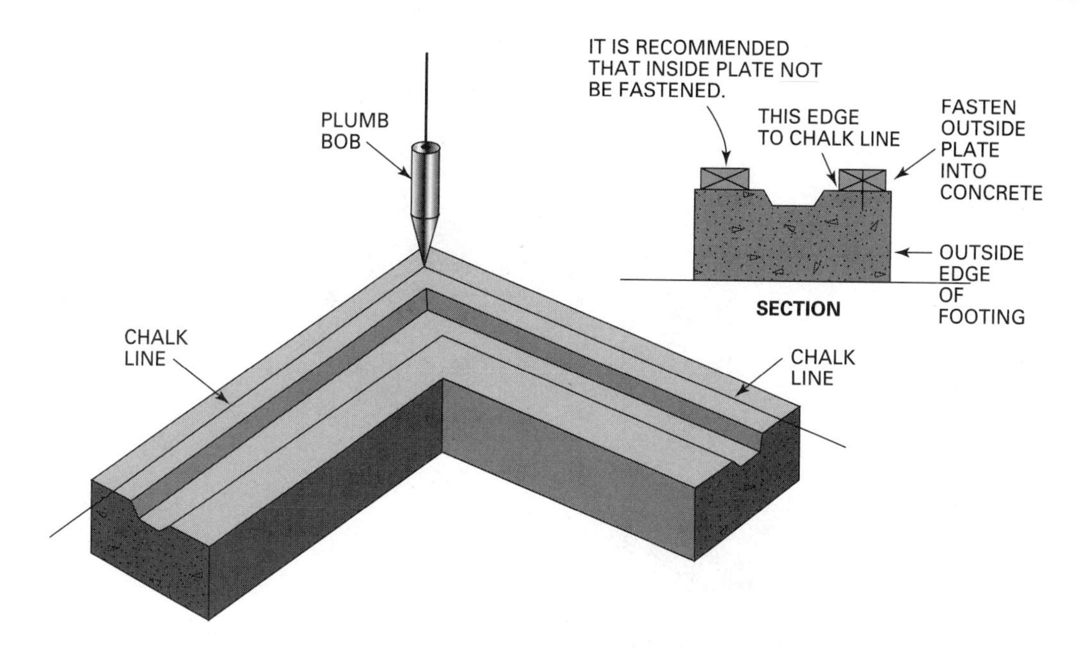

STEP 1 – SUSPEND PLUMB BOB FROM BUILDING LINES & MARK CORNERS OF WALL. SNAP CHALK LINE ON TOP OF FOOTING BETWEEN CORNERS.

STEP 2 – APPLY PLATES AROUND PERIMETER OF FOOTING.

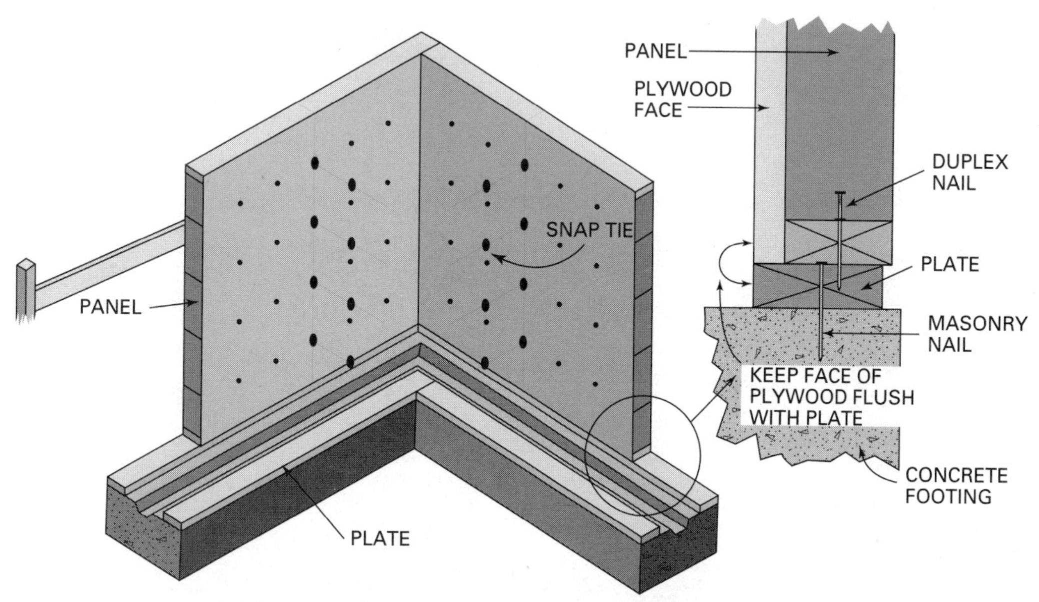

STEP 3 – START IN CORNER AND PLACE PANELS ON PLATE. FASTEN PANELS TO PLATE WITH DUPLEX NAILS. KEEP FACE OF PANELS FLUSH WITH INSIDE EDGE OF PLATE. INSERT SNAP TIES BETWEEN PANELS. BRACE PANELS WITH TEMPORARY BRACE.

Fig. 29-5 Construction of a typical wall form using plywood panels.

STEP BY STEP
PROCEDURES

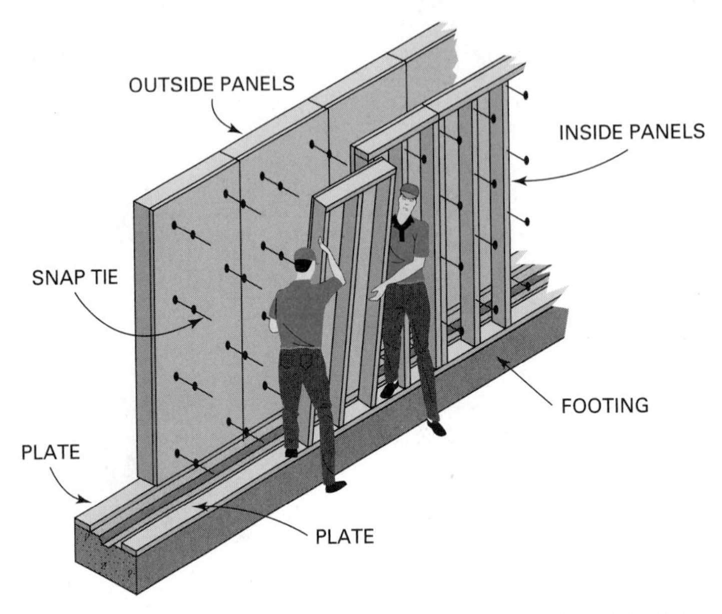

STEP 4 – INSERT REMAINDER OF TIES THROUGH OUTSIDE PANELS. INSTALL INSIDE PANELS OVER SNAP TIE ENDS. TILT TOP OF PANEL AND INSERT SNAP TIE ENDS STARTING FROM BOTTOM.

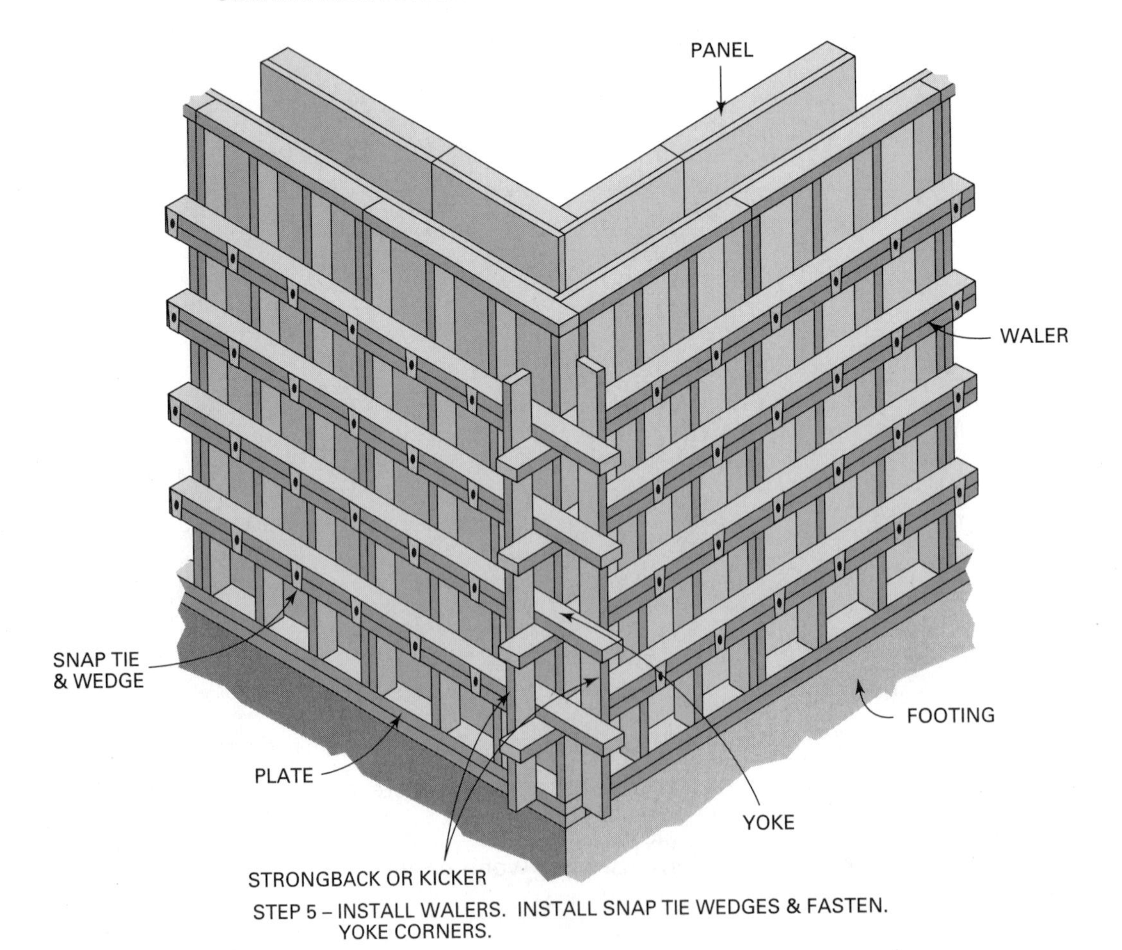

STEP 5 – INSTALL WALERS. INSTALL SNAP TIE WEDGES & FASTEN. YOKE CORNERS.

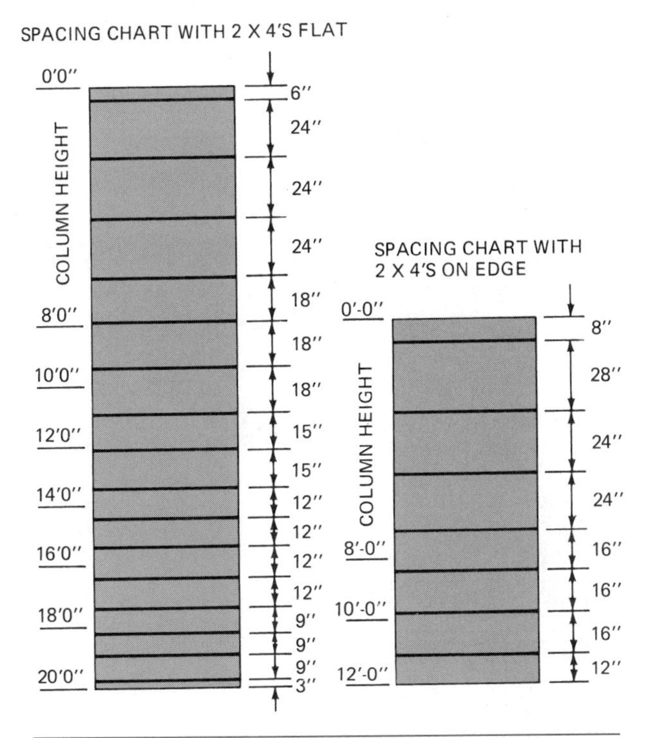

Fig. 29-6 Horizontal stiffeners are placed closer together near the bottom than they are at the top. Dimensions are for purposes of illustration only.

Fig. 29-7 When constructing low wall forms, the panels are laid horizontally.

recommended because they provide a positive on-line wall pattern. They also level out rough areas on the footing. Plates function to locate the position and size of pilasters, changes in wall thickness, and corners (Fig. 29-8). Secure plates to the footing using masonry nails or pins driven by a powder-actuated tool.

Erecting Panels. Stack the number of panels necessary to form the inside of the wall in the center of the excavation. Lay the panels needed for the outside of the wall around the walls of the excavation. The face of all panels should be oiled or treated with a chemical releasing agent. This provides a smooth face to the hardened concrete and makes stripping of the forms easy.

Erect the outside wall forms first. Set all corner panels in place by nailing into the plate with duplex nails. Make sure the corners are plumb by testing with a hand level.

Fill in between the corners with panels, keeping the same width panels opposite each other. Place snap ties in the dadoes between panels as work progresses. Tie panels together by driving U-shaped clamps over the edge 2×4s or by spiking them together with duplex nails. Use filler panels as necessary to complete each wall section. Brace the wall temporarily as needed.

Placing Snap Ties. After the panels for the outside of the wall have been erected, place snap ties in the intermediate holes. Be careful not to leave out any snap ties. Erect the panels for the inside of the wall. Keep joints between panels opposite to those for the outside of the wall. Insert the other end of the snap ties between panels and in intermediate holes as panels are erected (Fig. 29-9). If the concrete is to be reinforced, the rebars are tied in place before the inside panels are erected.

Installing Walers. When all panels are in place, install the walers. Let the snap ties come through them and wedge into place. Care must be taken when installing and driving snap tie wedges (Fig. 29-10). Let the ends of the walers extend by the corners of the formwork. Reinforce the corners with vertical 2×4s as shown in step 5 of Figure 29-5. This is called *yoking* the corners.

Forming Pilasters. The wall may be formed at intervals for the construction of **pilasters** (thickened portions of the wall). They may strengthen the wall or provide support for beams. They may be constructed on the inside or outside of the wall. They can be formed with job-built panels, making inside and outside corners in the usual manner. In the pilaster area longer snap ties are necessary (Fig. 29-11).

Straightening and Bracing. Brace the walls inside and outside as necessary to straighten them. Wall forms are easily straightened by sighting by eye along the top edge from corner to corner. Another method of straightening is by stretching a line from corner to corner at the top of the form over two blocks of the same thickness. Move the forms until a test block of equal thickness passes just under the line (Fig. 29-12).

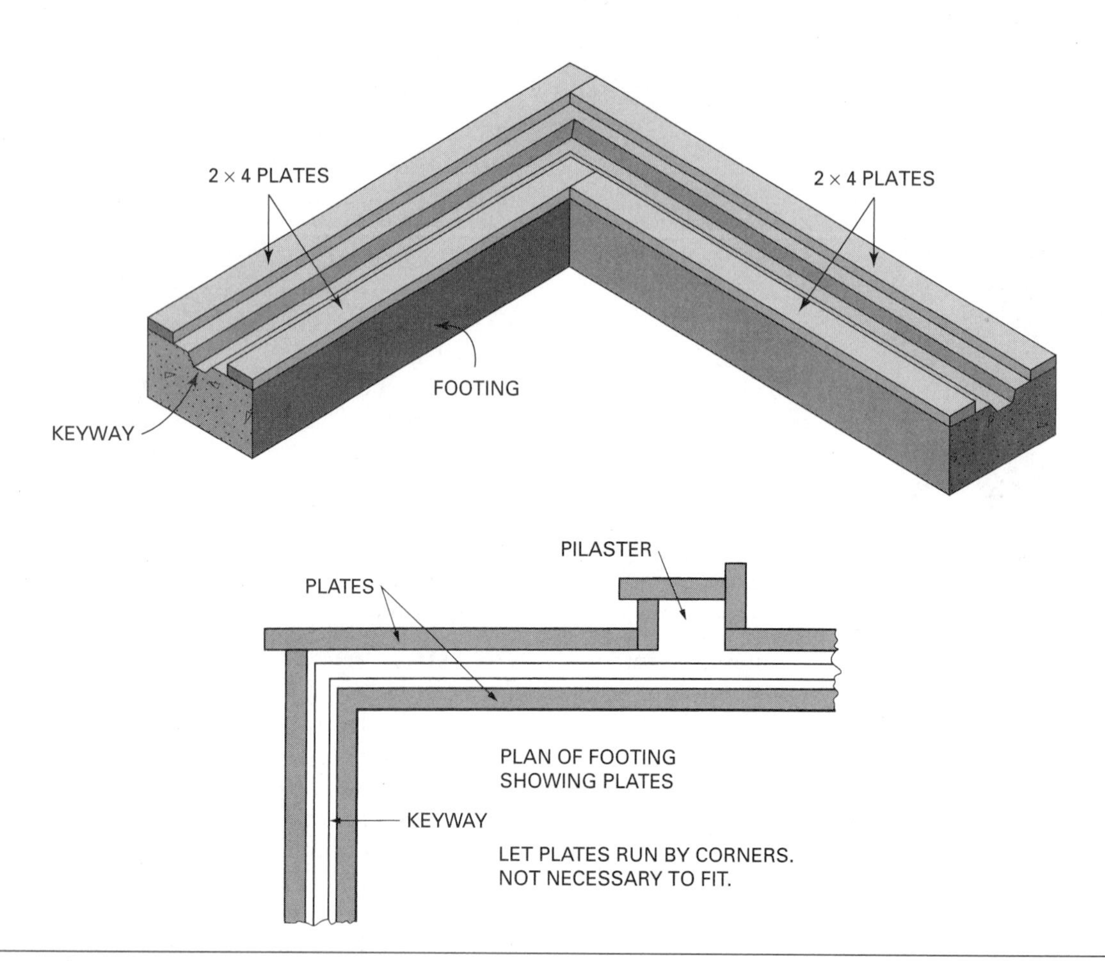

Fig. 29-8 Plates should be used when erecting wall forms.

Fig. 29-9 Erecting the walls for the inside of the form.

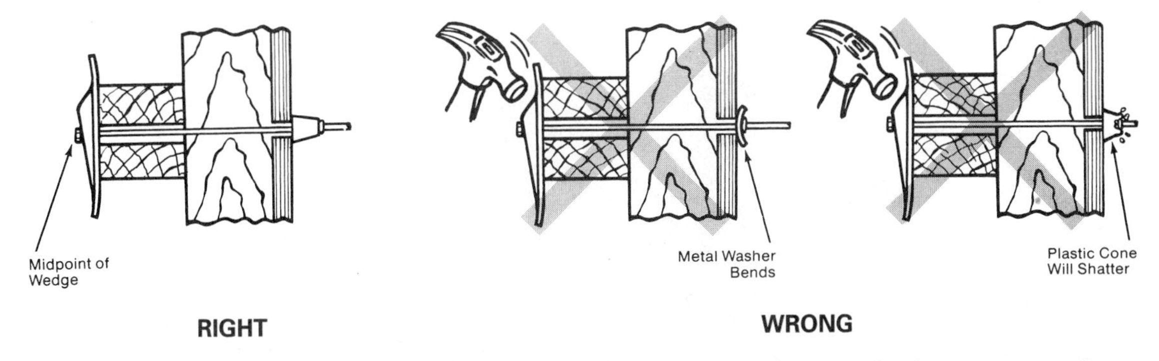

Do not attempt to draw up warped wales with a wedge or overtighten the wedge in any manner. Over-tightening will cause the metal spreader washers to bend out of shape or will shatter the plastic cones. resulting in a decreased and incorrect wall thickness.

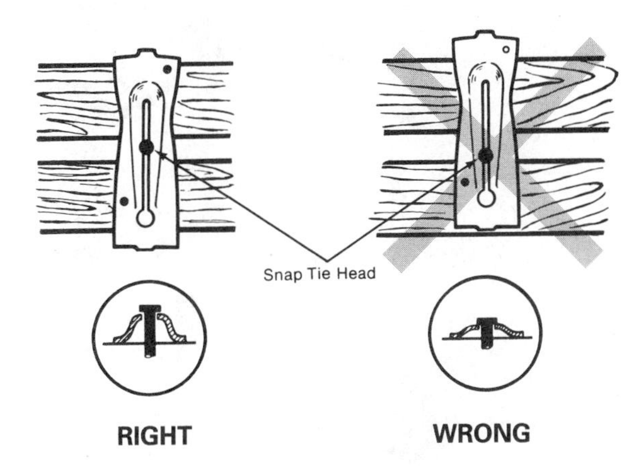

The optimimum wedge position is when the snap tie head is at the midpoint of the wedge. You may place the snap tie head higher on the wedge, as long as it is not overtightened. However, the snap tie head must not be positioned lower than the midpoint, as this will place it on a section of the wedge that has not been designed to carry the rated load.

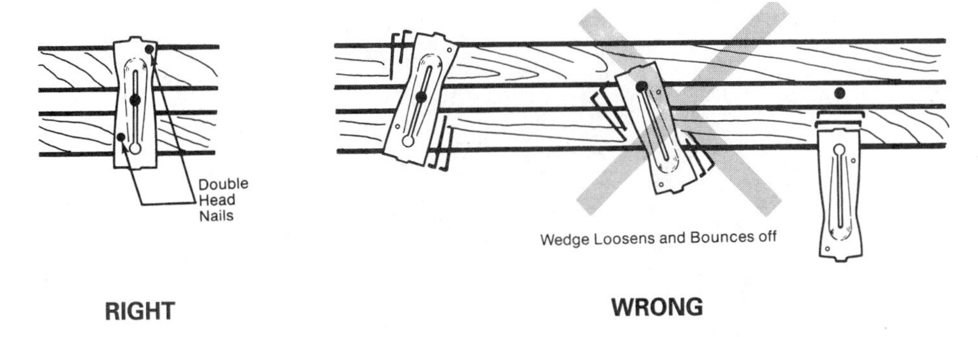

As the concrete is consolidated, internal vibrations may cause the steel wedges that have not been nailed into position to loosen, bounce around and eventually fall off resulting in premature form failure.

Fig. 29-10 Care must be taken when installing snap tie wedges.

Braces are cut with square ends and spiked into place against **strongbacks.** Strongbacks are placed across walers at right angles wherever braces are needed. The sharp corner of the square ends helps hold the braces in place. It also allows easy prying with a bar to tighten the braces and move the forms (Fig. 29-13). There is no need to make a bevel cut on the ends of braces. A special adjustable form brace and aligner are sometimes used for positioning and holding wall forms (Fig. 29-14).

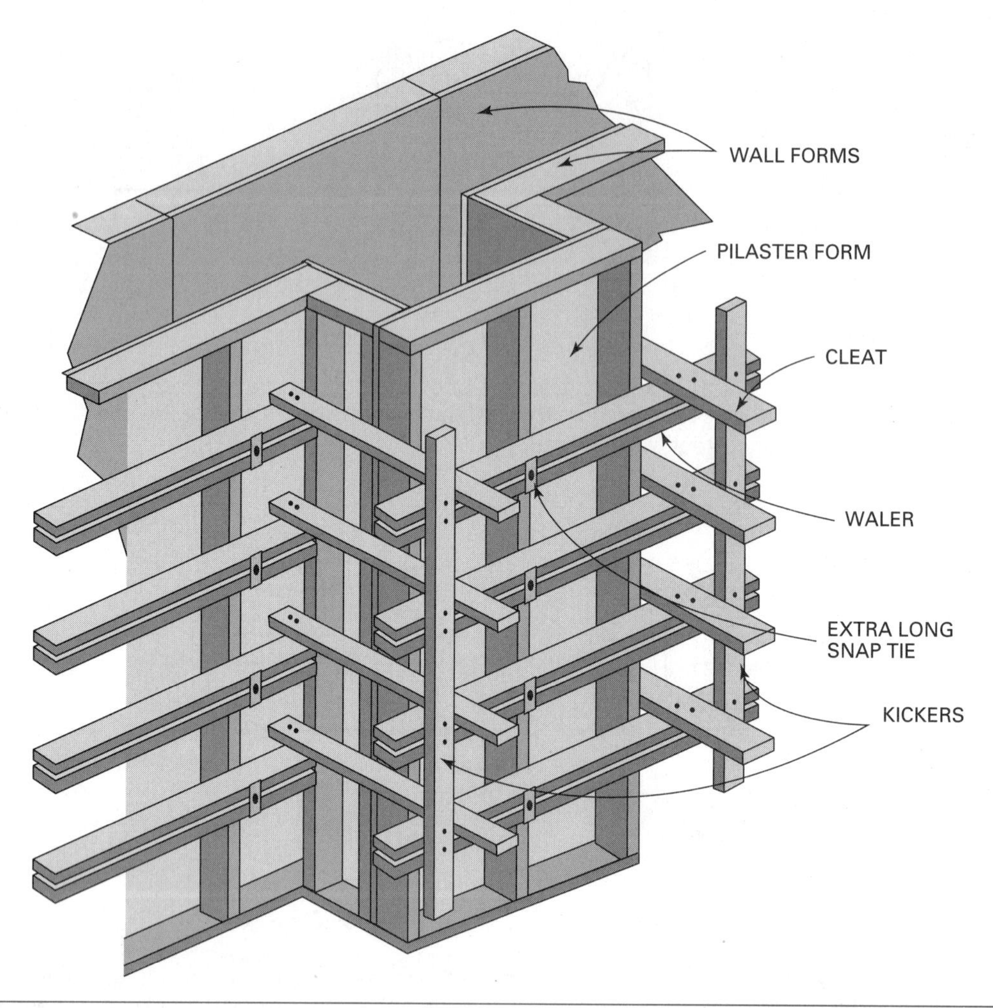

Fig. 29-11 Formwork for a pilaster.

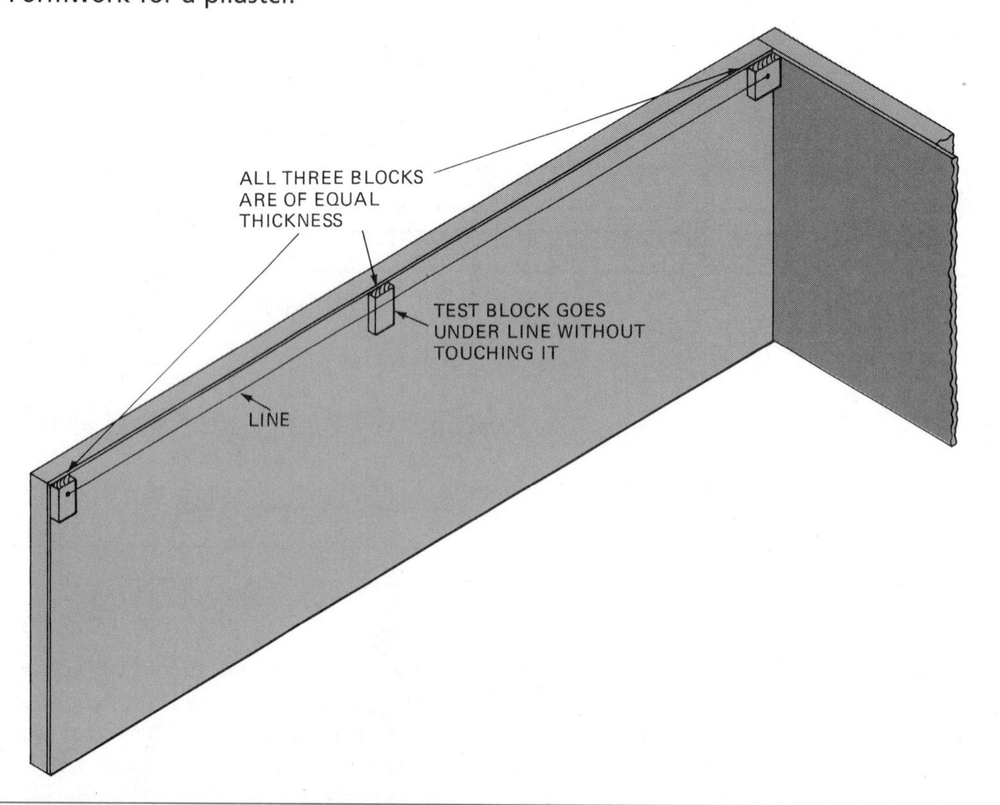

Fig. 29-12 Straightening the wall form with a line and test block.

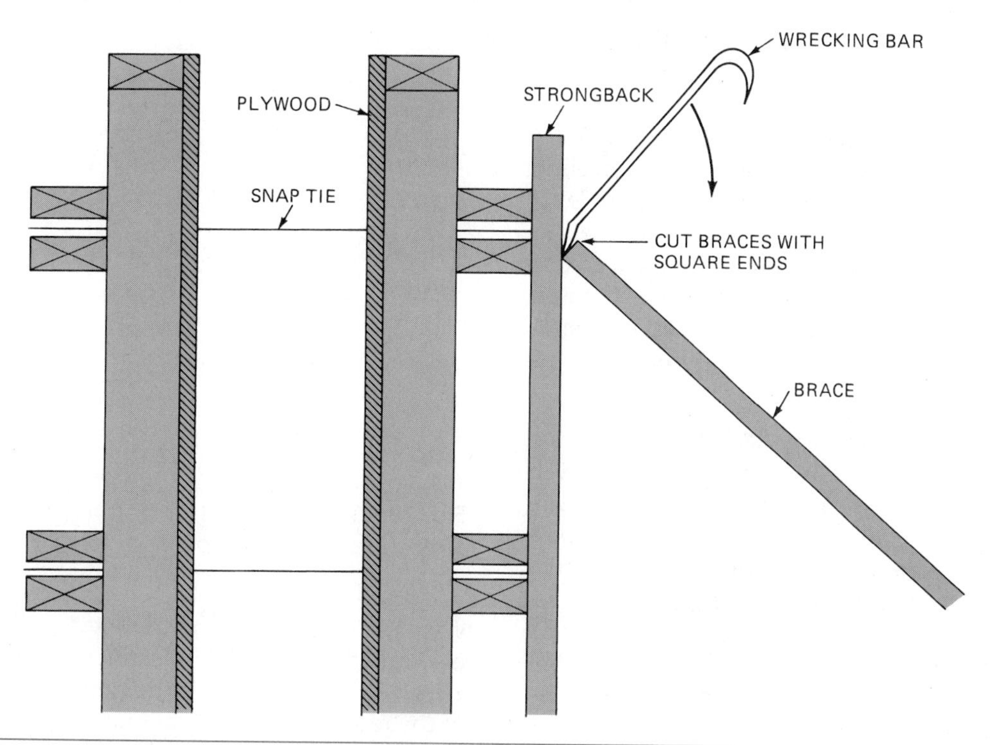

Fig. 29-13 Cut braces with square ends. This allows for easy prying when straightening the formwork.

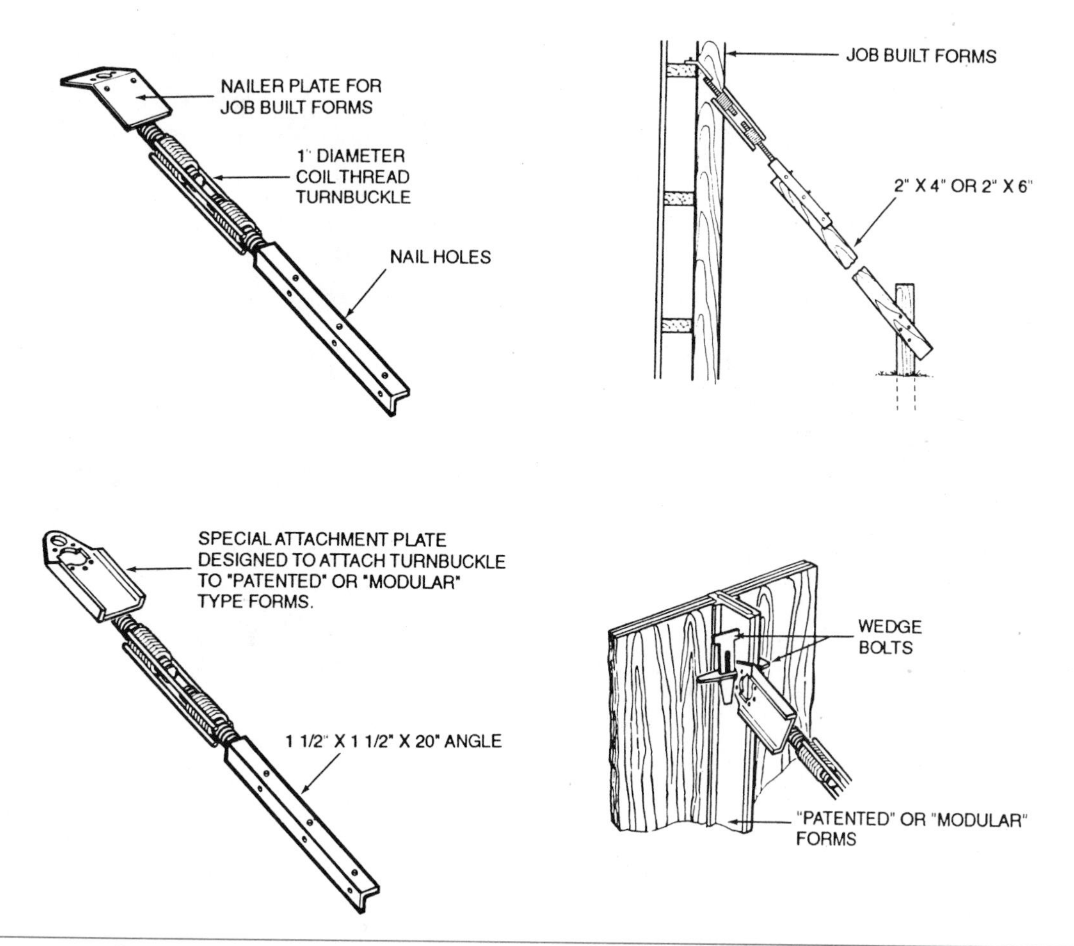

Fig. 29-14 Manufactured wall brace and aligner. (*Courtesy of Dayton Superior*)

Leveling. After the wall forms have been straightened and braced, chalk lines are snapped on the inside of the form for the height of the foundation wall. Grade nails may be driven partway in at intervals along the chalk line as a guide for leveling the top of the wall. If the tops of the panels are level with each other, a short piece of stock notched at both ends can be run along the panel tops to screed the concrete. Another method is to fasten strips on the inside walls along the chalk line and use a similar screeding board, notched to go over the strips (Fig. 29-15).

Setting Anchor Bolts. As soon as the wall is screeded, anchor bolts are set in the fresh concrete. A number of various styles and sizes are manufactured (Fig. 29-16). The type is usually specified on the foundation plan. Care must be taken to set the anchor bolts at the correct height and at specified locations. Bolts should be spaced 6 to 8 feet apart and between 6 and 12 inches from each end of the sill plates. An anchor bolt *template* is sometimes used to accurately place the bolts (Fig. 29-17).

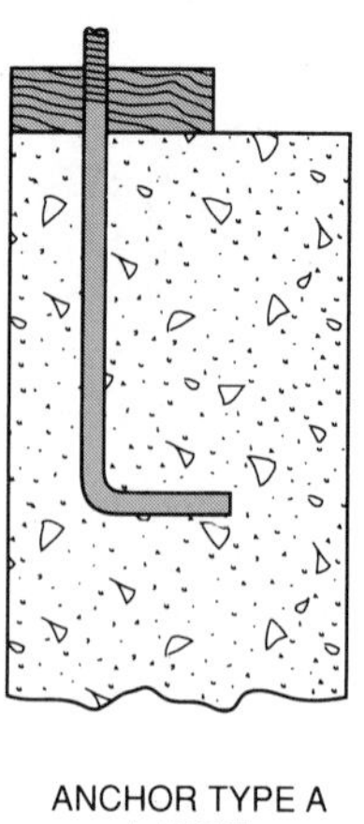

Fig. 29-16 Typical anchor bolt. (*Courtesy of Simpson Strong-Tie Co.*)

Openings in Concrete Walls

In many cases, openings must be formed in concrete foundation walls for such things as windows, doors, ducts, pipes, and beams. The forms used for providing the larger openings are called *blockouts.*

Constructing Blockouts. Blockouts are also called **bucks.** The blockout is usually made of 2-inch stock. Its width is the same as the thickness of the foundation wall. Nailing blocks or strips are often fastened to the outside of the bucks. These are beveled on both edges to lock them into the

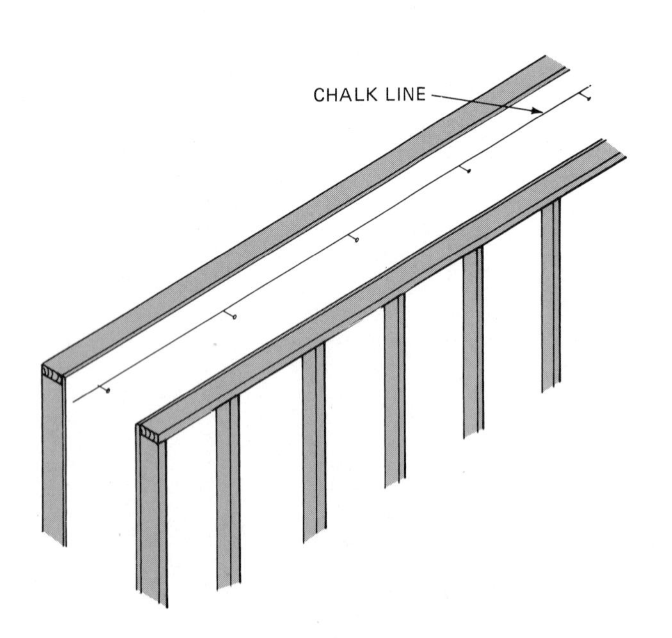

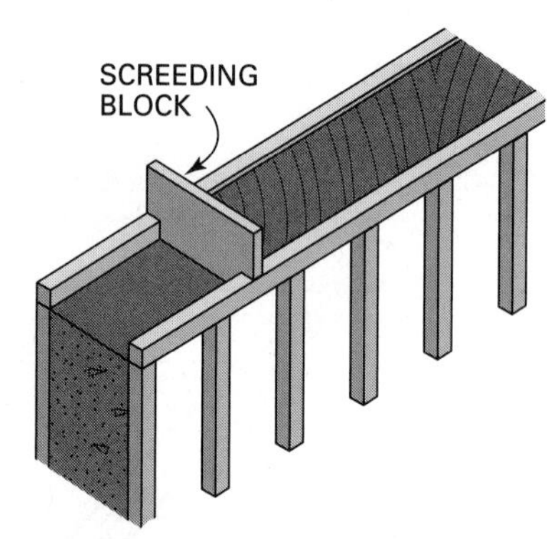

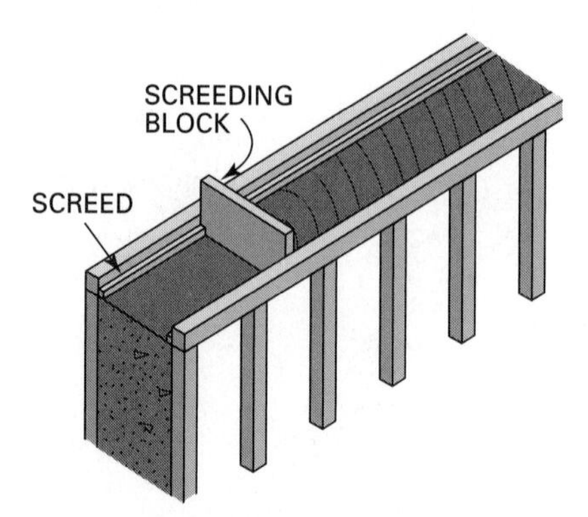

Fig. 29-15 Nails are driven partway in along a chalk line as a guide to screed the top of the wall. Walls are screeded with other methods.

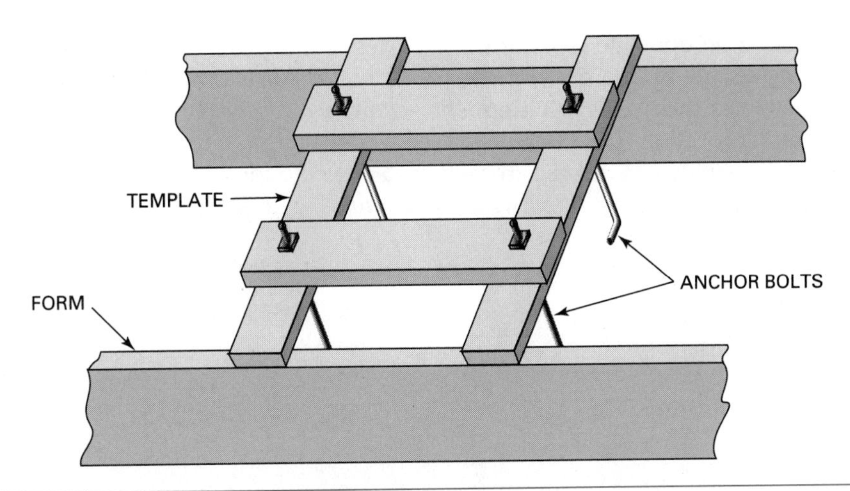

Fig. 29-17 Templates are sometimes used to place anchor bolts in fresh concrete accurately.

concrete when the form is stripped. They provide for the fastening of window and door frames in the openings. Intermediate pieces may be necessary in bucks for large openings to withstand the pressure of the concrete against them (Fig. 29-18).

Blockouts are made to the specified dimension. They are installed against the inside face of the outside panels. Duplex nails through the outside panels hold the blockouts in place. The inside wall panels are then installed against the other side of the blockouts. Nails are driven through the inside wall panels to secure the bucks on the inside.

Girder Pockets. Recesses in the top of the foundation wall are called *girder pockets.* They are sometimes required to receive the ends of girders (beams). A box of the size needed is made and fastened to the inside at the specified location (Fig. 29-19). Because it is near the top of the form, not much pressure from the liquid concrete is exerted against it.

Column Forms

To form columns, like all other kinds of formwork, as much use as possible is made of panels, manufactured or job-built, to simplify erection and stripping. Columns may be formed in square, rectangular, circular, or a number of other shapes.

Erecting Column Forms

The following is a procedure for forming a square or rectangular column. Stretch lines and mark the location of the column on the footing. Fasten two

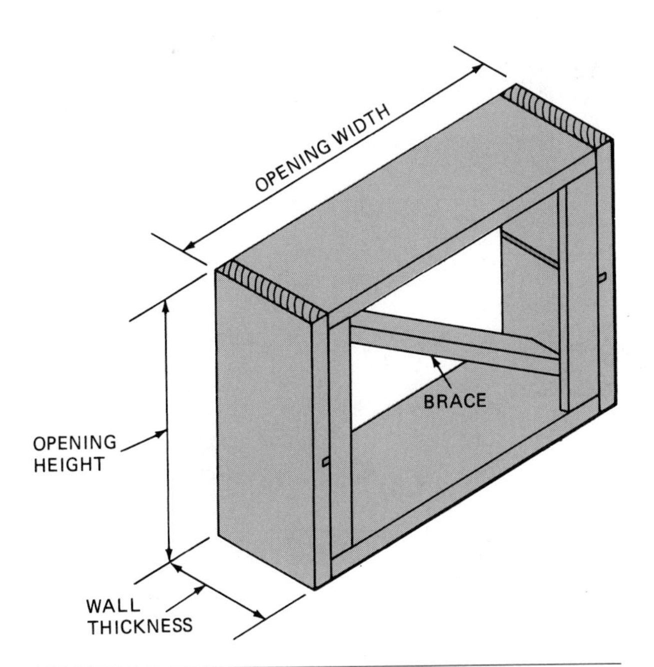

Fig. 29-18 Construction of a typical window blockout.

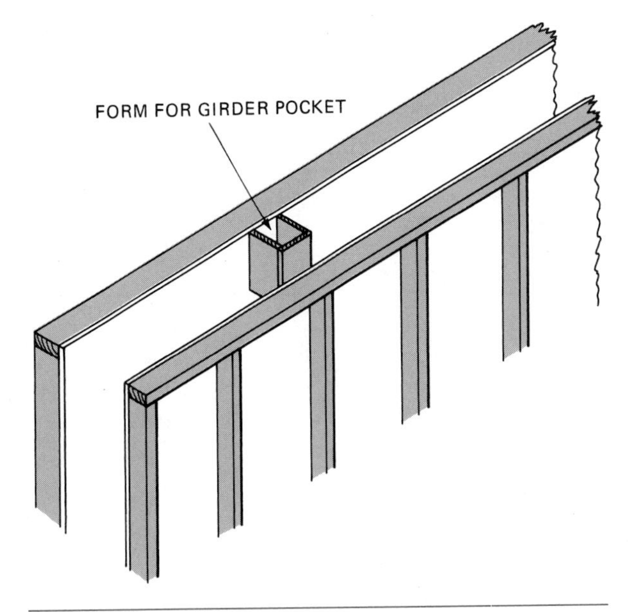

Fig. 29-19 A small box is used to form a girder pocket in a foundation wall.

2 × 4 pieces to the footing on opposite sides of the column and outside of the line by the panel thickness. Fasten the other two to the overlapping ends of the first pair in a similar location (Fig. 29-20).

Build and erect two panels to form the thickness and height of the column. A **cove** molding of desired size may be fastened on the edges of this panel to form a radius to the corners of the concrete column. A **quarter-round** molding may be used for a cove shape. Triangular-shaped strips of wood can be used to form a chamfer on the corners of the column (Fig. 29-21).

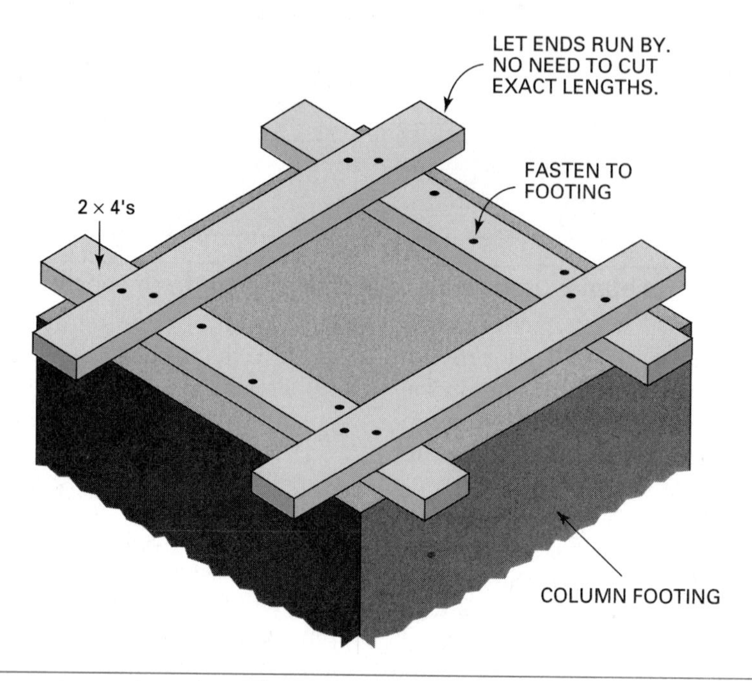

Fig. 29-20 A yoke is constructed on the column footing to start the forming of a column.

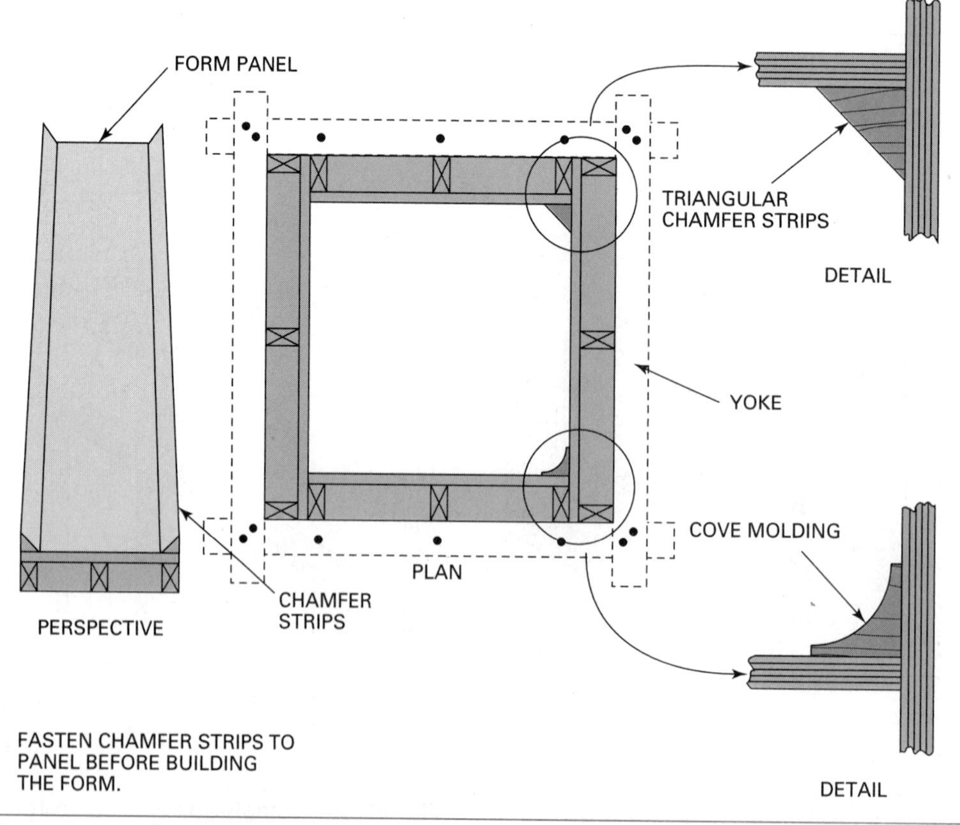

Fig. 29-21 The corners of a concrete column may be formed in several ways.

The face of the column may be decorated with **flutes** by fastening vertical strips of **half-round** molding spaced on the panel faces. In addition, *form liners* are often used. They provide various wood, brick, stone, and many other textures in the face of the concrete wall or column.

Build panels for the opposite sides of the column to overlap the previously built panels. Plumb and nail the corners together with duplex nails. Install 2 × 4 *yokes* around the column forms, letting their ends extend beyond the corners. Spike them together where they overlap. Yoke the column closer together at the bottom. The number and spacing of yokes depend on the height of the column. These details are specified by the form designer. Install vertical 2 × 4s between the overlapping ends of the yokes (Fig. 29-22). Brace the formwork securely to hold it plumb. Triangular strips placed in the corners of the form are sometimes used to bevel the column corners.

Concrete Forming Systems

A concrete forming system consists of manufactured items for concrete form construction. Among these are specially designed panels of steel or a combination of steel and wood. The panels may be tied together with metal wedges (Fig. 29-23). Much forming system hardware is specially designed for use with these systems.

Small column forms can be erected quickly with less material. The corners are wedged and act as yokes (Fig. 29-24). Large columns can be formed in much less time because less material is needed (Fig. 29-25). Circular columns and *column capitals* can be formed with reusable fiber glass and steel forms or with tubular fiber forms that are stripped by peeling them off (Fig. 29-26).

Walers are easily installed on wall forms when the forming system hardware is used (Fig. 29-27).

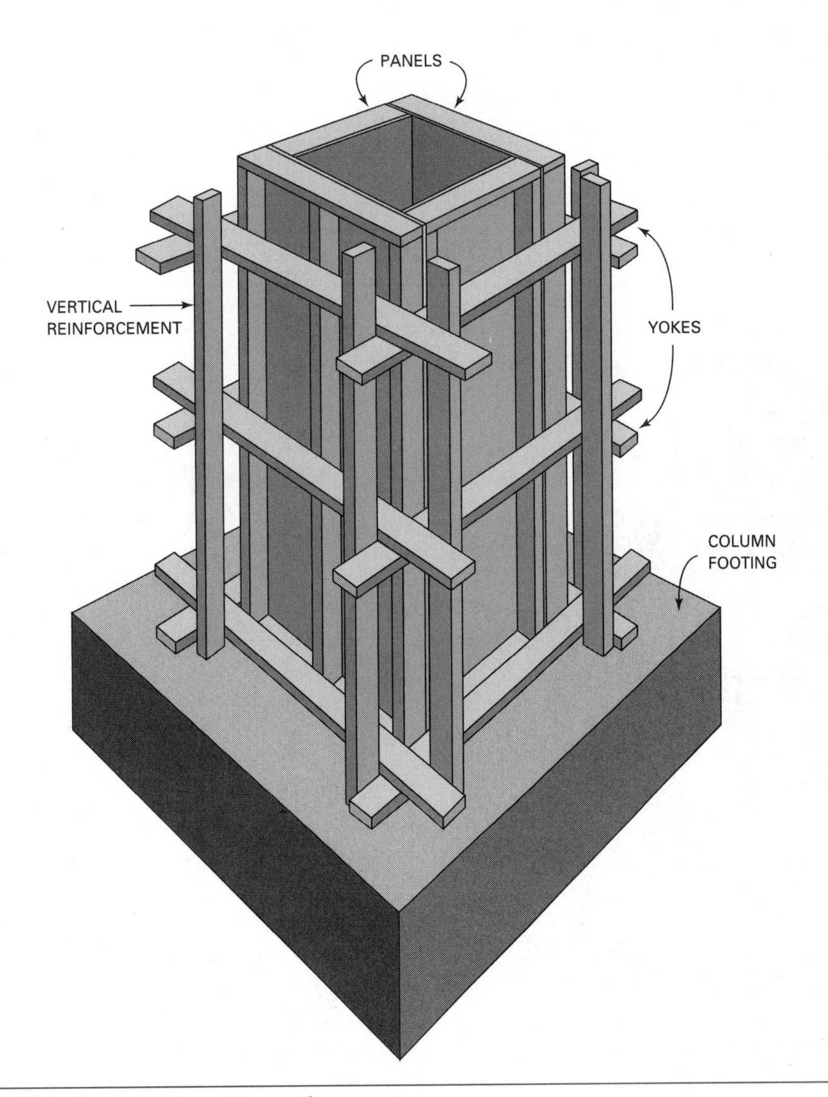

Fig. 29-22 A completed concrete column form.

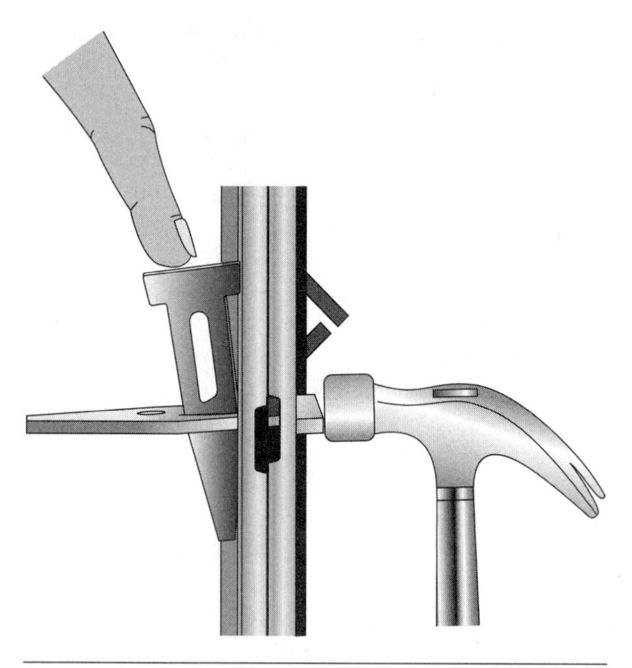

Fig. 29-23 Only a hammer is needed to tie panels or corners together when metal wedges are used. Tapping to the side as the wedge is pushed down adequately tightens the form, leaving it also loose enough to be easily removed later. (*Courtesy of Symons Corporation*)

Fig. 29-25 Large rectangular column form with walers and strongbacks attached. The forming system eliminates the need for yoking the corners. (*Courtesy of Symons Corporation*)

Fig. 29-24 Yokes are not necessary to form columns when forming systems are used. (*Courtesy of Symons Corporation*)

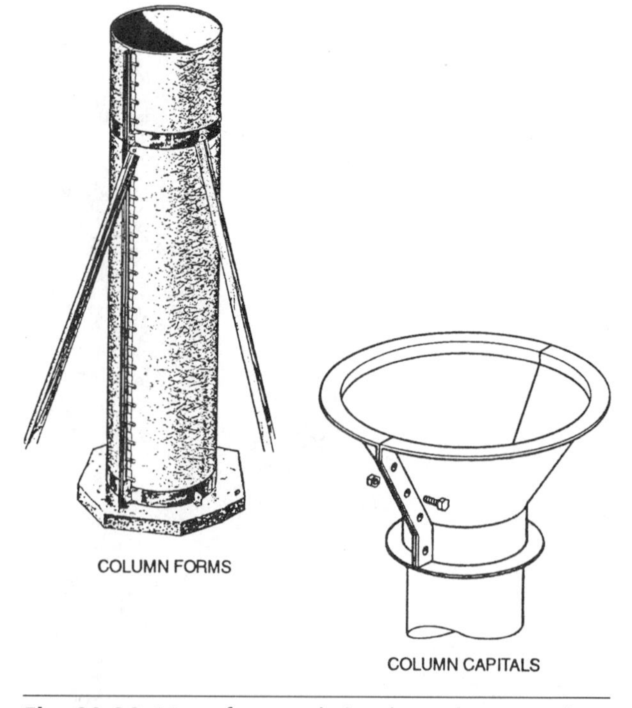

Fig. 29-26 Manufactured circular column and column capital forms need only to be set in place and braced. (*Courtesy of Symons Corporation*)

Adjustable braces are available. Few are required to hold the wall straight (Fig. 29-28). Steel inside corners are used with wall panels for fast and easier forming of the inside corner (Fig. 29-29).

Other accessories are available, such as molded beam forms of reinforced fiberglass (Fig. 29-30) and Domeforms® for forming beams and slabs (Fig. 29-31).

Forming systems require a considerable initial outlay of funds. However, the investment is saved many times over because of the reduced labor costs and reusability of the forming systems components. These systems are also available for rent. A number of forming systems are available for various types of work. Systems are made for light residential work up to heavy-duty concrete formwork. With forming systems, fewer ties and materials are used.

Fig. 29-27 Walers are easily installed on forming systems when special hardware is used. (*Courtesy of Symons Corporation*)

Fig. 29-28 Adjustable braces are available with forming systems. (*Courtesy of Symons Corporation*)

Fig. 29-29 Steel inside corners of a forming system make fast erection for formwork possible. (*Courtesy of Symons Corporation*)

Fig. 29-30 Slab and beam formed with a molded fiberglass beam form. (*Courtesy of Symons Corporation*)

Fig. 29-31 Domeforms® of fiberglass or steel form beams and floors easily. (*Courtesy of Symons Corporation*)

For special jobs, manufacturers also make custom-made formwork. The time saved by using reuseable custom-made formwork on a job is well worth the cost (Fig. 29-32).

Fig. 29-32 Custom-made framework was reused for all seventy floors of this high-rise building. (*Courtesy of Symons Corporation*)

CHAPTER 30 CONCRETE STAIR FORMS

Stair Forms

It may be necessary to refer to chapters 47 and 48 on stair framing for definition of stair terms, types of stairs, stair layout, and methods of stair construction.

Concrete stairs may be suspended or supported by earth (Fig. 30-1). Each type may be constructed between walls or have open ends.

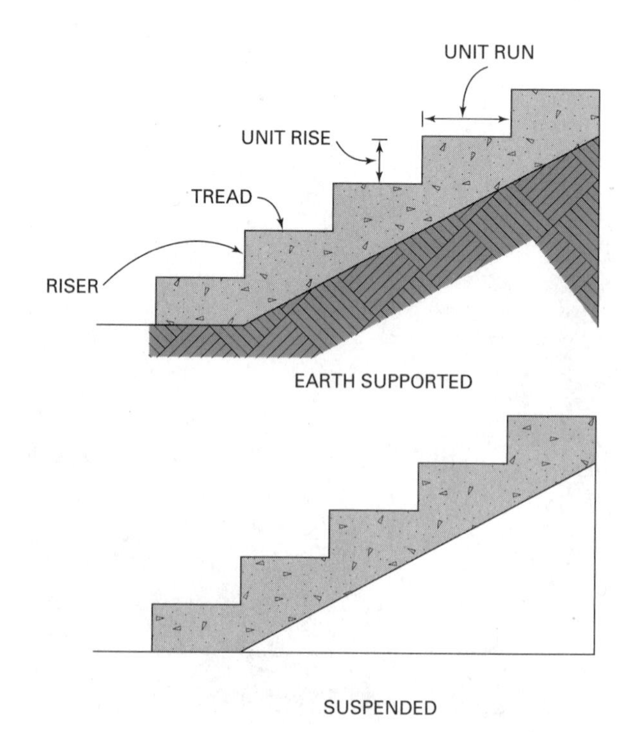

Fig. 30-1 Types of concrete stairs. Stairs may be formed with both ends closed or open, or with one end closed and the other end open.

Forms for Earth Supported Stairs

Before placing concrete, stone, gravel, or other suitable fill is graded to provide proper thickness to the stairs. It should not be overly thick, or concrete will be wasted. It may be necessary to lay out the stairs before the supporting material is placed.

Forming between Existing Walls. When earth-supported stairs are formed between two existing walls, the **rise** and **run** of each step are laid out on the inside of the existing walls. *Rise* is the vertical distance that a step will rise. It is the height of each step, which is also called the *riser. Run* is the horizontal distance of a step. It is roughly the width of each step, which is called the tread.

Planks are ripped to width to correspond to the height of each riser. The planks are then beveled on their bottom edge and wedged securely in position to the layout lines. The plank is wedged and secured in place with its inside face to the riser layout line. The top and bottom edges are to tread layout lines.

Beveling the bottom edge of the plank permits the mason to trowel the entire surface of the tread.

Fig. 30-2 Stairs formed between two existing walls. *(Courtesy of Portland Cement Association)*

Otherwise, the bottom edge of the riser form will leave its impression in the concrete tread. After the riser planks are secured in position, they are braced from top to bottom between the ends. This keeps them from bowing outward due to the pressure of the concrete (Fig. 30-2).

Forming Stairs with Open Ends. In cases where the ends of the stairs are to be open, panels are erected on each end. It does no harm if the panels are larger than needed. The panels must be plumb. The distance between them must be the desired width of the stairs. They are then braced firmly in position.

The risers and treads are laid out on the inside surfaces of the panels. **Cleats** (short strips of wood) are fastened at each riser location. Allowances must be made for the thickness of the riser form board. Screws or duplex nails should be used to make stripping the form easier. The space should then be filled with the supporting material to the proper level. Any necessary reinforcing should be installed (Fig. 30-3).

The boards or planks to form the risers are ripped to width and cut to length. The bottom edge, opposite the face side, is beveled to permit finishing of the total tread width. They are then fastened with duplex nails through the side panels in a position

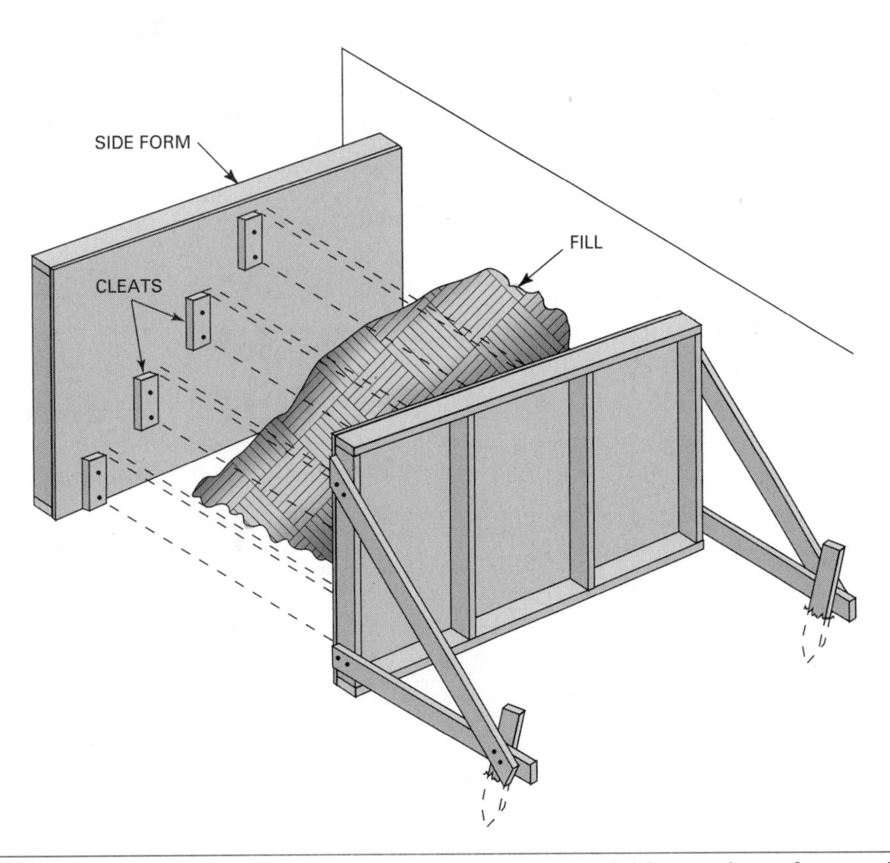

Fig. 30-3 End panels are braced firmly in position. The stairs are laid out, cleats fastened, and the space filled with supporting material.

against the cleats, with the top and bottom edges to the tread layout lines. The riser forms are then braced from top to bottom at intervals between the two ends (Fig. 30-4).

Forms for Suspended Stairs

Forms for suspended stairs are more difficult to build. Instead of earth support, a form needs to be built on the bottom to support the stair slab. With proper design and reinforcement, the stairs are strong enough to support themselves in addition to the weight of the traffic.

Forming Suspended Stairs between Existing Concrete Walls. A cross-section of the formwork for suspended stairs between existing walls, with the form members identified, is shown in Figure 30-5. First, lay out the treads, risers, and stair slab bottom on both of the walls between which the stairs run. This can be done by using a hand level, rule, and chalk line (Fig. 30-6).

The form for the bottom of the stair slab is then laid out. Allow for the thickness of the plywood deck, the width of the supporting joists, and

Fig. 30-4 Typical form construction for concrete stairs having open ends. (*Courtesy of Portland Cement Association*)

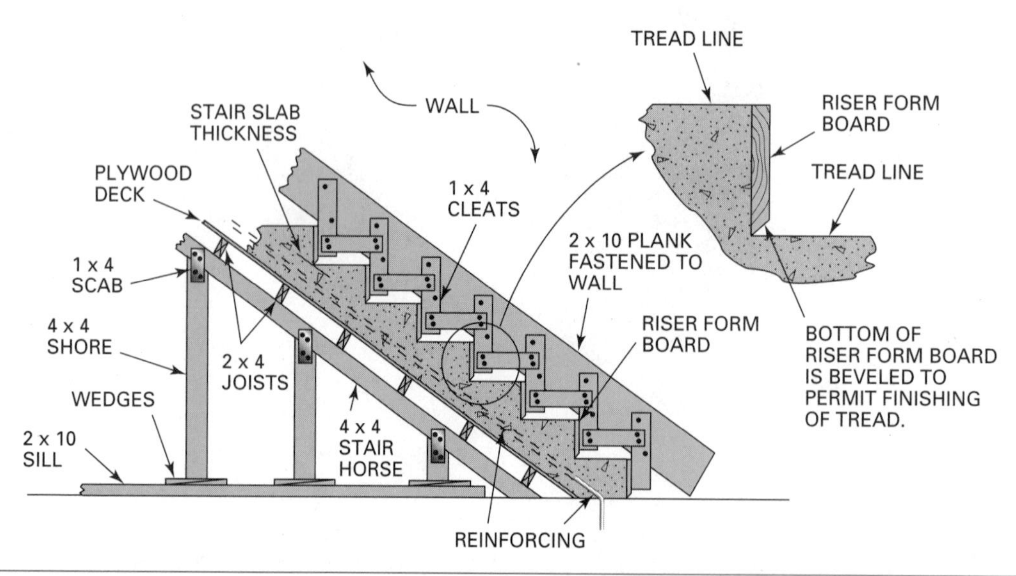

Fig. 30-5 Cross-section of formwork for suspended stairs between walls.

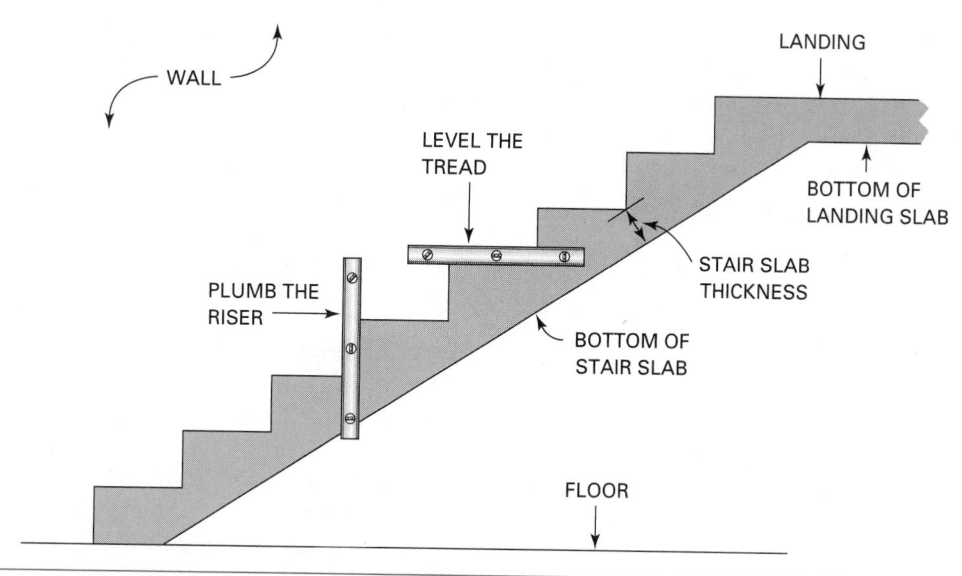

Fig. 30-6 Make the layout for suspended stairs on the inside of the walls.

the depth of the *horses*. Snap a line on both walls to indicate their bottom and also the top of the supporting *shores*.

Allowing for a plank sill and wedges at the bottom end of the shores, cut them to length with the correct angle at the top. They should be cut a little short to allow for wedging because of the unevenness of the concrete floor. Install them at specified intervals on top of the plank sill. Brace in position, and place and fasten the horses on top with **scabs** or **gusset** plates. Scabs are short lengths of narrow boards fastened across a joint to strengthen it. Install joists at right angles across the horses at the specified spacing (Fig. 30-7). Fasten the form plywood in position on top of the joists. Use only as many fasteners as needed for easier stripping later. Wedge the shoring as necessary to bring the surface of the plywood to the layout line. Fasten the wedges in place so they will not move. The plywood should now be oiled to facilitate stripping. Install the reinforcing.

Rip the riser planks to width. Cut to length, allowing for wedges. Bevel the outside bottom edge of the treads. This allows for easy leveling of the concrete at each tread. Install the riser form planks to the layout lines. Wedge in position.

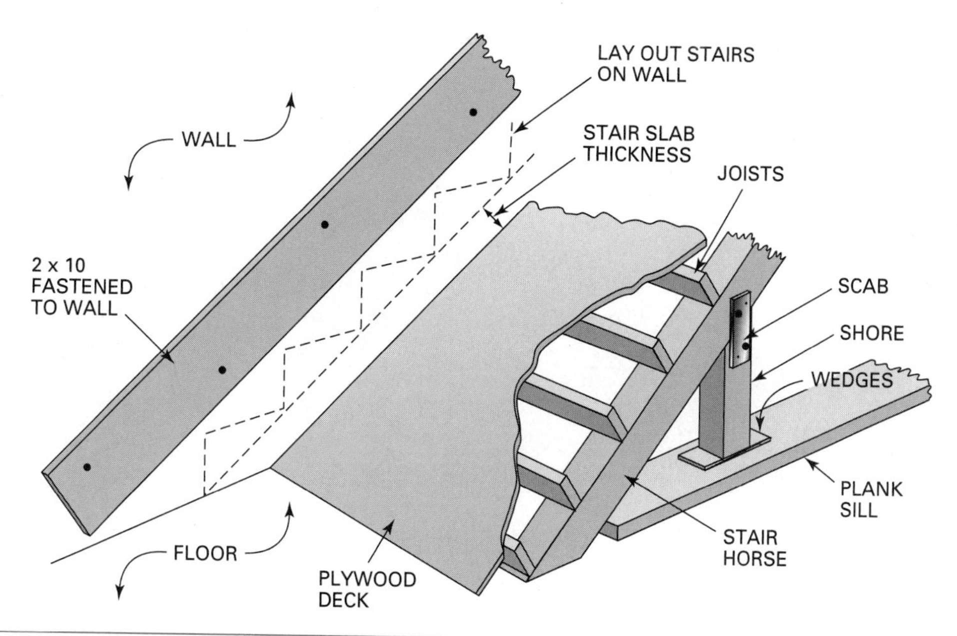

Fig. 30-7 The framework for the bottom of the suspended stairs.

Another method of securing riser boards is by fastening planks on the walls about an inch above the stair layout. Project the riser layout lines upward on the planks. Fasten cleats to the plank. Then brace the cleats and fasten the riser board to them.

Forming Suspended Stairs with Open Ends. A completed form for suspended stairs with open ends is shown in Figure 30-8. Because there are no walls upon which to lay out the stairs, the layout must be made on the side panels used for forming the ends of the stairs.

Laying Out the Side Forms. Lay out the risers, treads, and stair slab bottom on the left and right side form panels. First, measure and snap a line up from the bottom a distance equal to the thickness of the slab. The intersection of the treads and risers lies at intervals along this line.

Make a *pitch board* by laying out the width of the tread and the height of the riser on a thin piece of plywood. Mark tread and riser locations by using the pitch board held to the slab line just snapped (Fig. 30-9).

Laying Out the Stair Horses. Measure down from the bottom of the side to form a distance equal to the combined joist and plywood deck thickness. Snap a line parallel to the bottom of the side form to locate the top of the stair horse. Extend the bottom floor line and the top plumb line on the side forms to the snapped line. Mark the length and cuts of the stair horse (Fig. 30-10). Cut one stair horse for each side to length at the angles indicated.

Installing Horses and Joists. Temporarily support and brace the horses in position. Measure, cut, and install shores in a manner similar to that of closed stairs. Brace shores and horses firmly in position. Remove temporary supports and braces. Fasten joists at designated intervals to the horses, leaving an adequate and uniform overhang on both sides. Fasten decking to joists using a minimum number of nails.

Installing Side and Riser Forms. Snap lines on the deck for both sides of the stairs. Stand the side forms up with their inside face to the chalk line. Fasten them through the deck into the joists with duplex nails. Brace the side forms at intervals so they are plumb for their entire length (Fig. 30-11). Apply form oil to the deck before reinforcing bars are installed.

Rip the riser boards to width and crosscut them to length. Install them on the riser layout line. Fasten them with duplex nails through the side forms into their ends. Install cleats against the riser boards on both ends for additional support. Install intermediate braces to the riser form boards from top to bottom, if needed.

Economy and Conservation in Form Building

Economical concrete construction depends on the reuse of forms. Forms should be designed and built

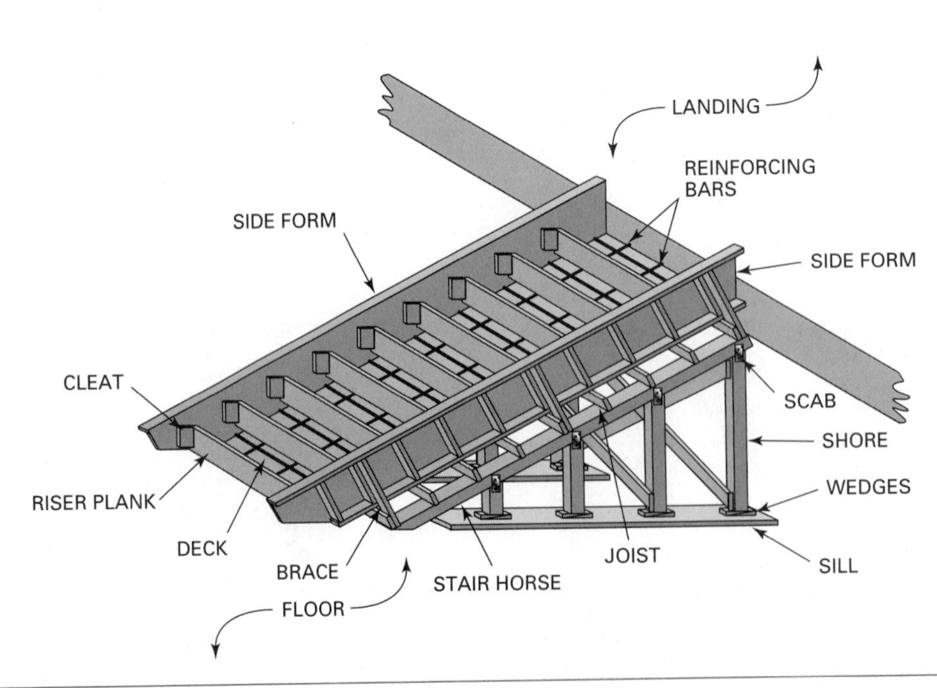

Fig. 30-8 A completed form for suspended stairs with open ends.

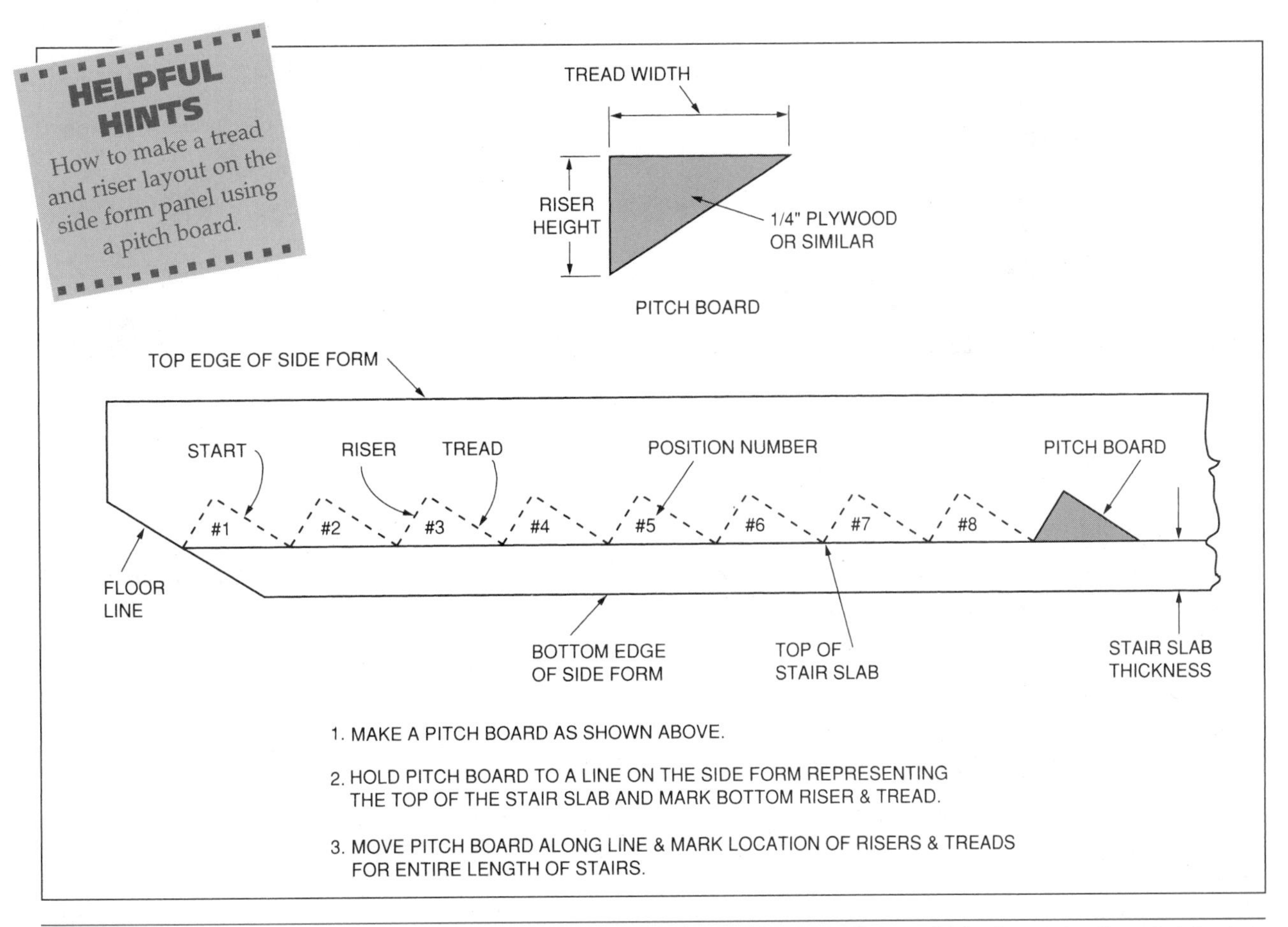

Fig. 30-9 Technique to lay out a set of stairs. A *pitch board* is a small wood block cut to the stair's rise and run.

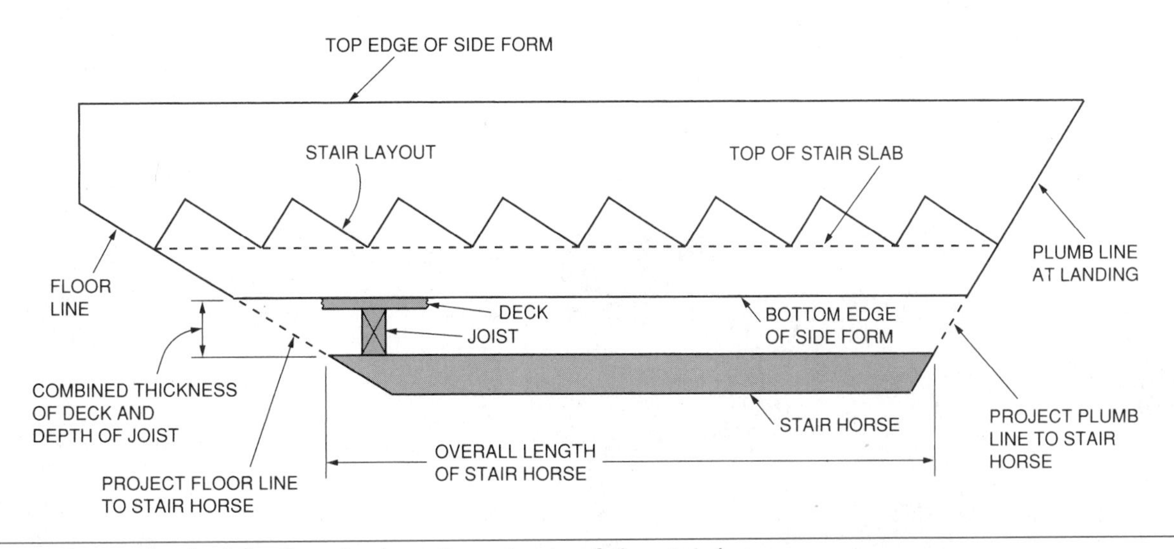

Fig. 30-10 Method of finding the length and cuts of the stair horses.

to facilitate stripping and reuse. Use panels to build forms whenever possible. Use only as many nails as necessary to make stripping forms easier.

Care must be taken when stripping forms to prevent damage to the panels so they can be reused. Stripped forms should be cleaned of all adhering concrete and stacked neatly.

CAUTION **Remove all protruding nails to eliminate the danger of stepping on or brushing against them.**

Long lengths of lumber can often be used without trimming. Random length walers can extend

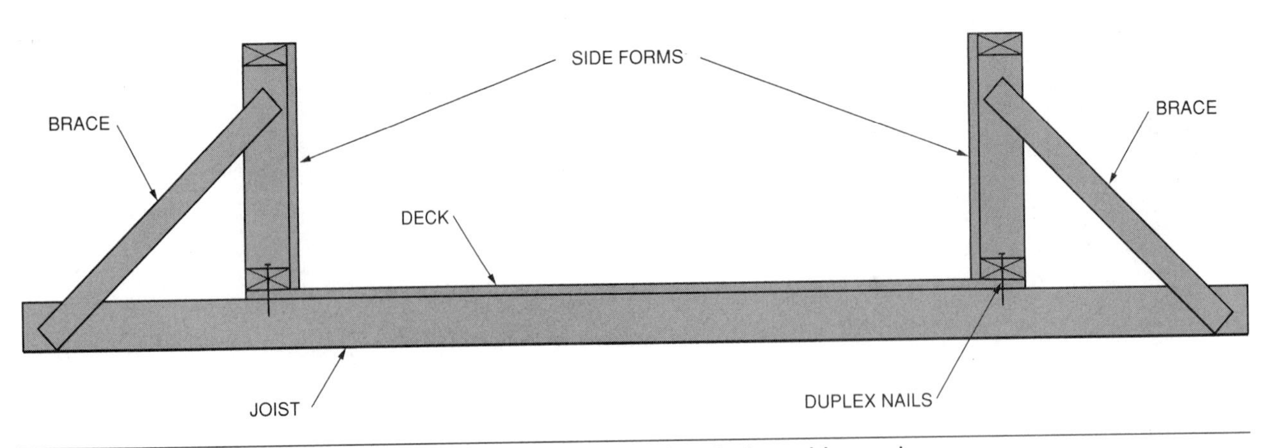

Fig. 30-11 Cross-section of the side form installed on the deck and braced.

beyond the forms. There is no need to spend a lot of time cutting lumber to exact length. The important thing is to form the concrete to specified dimensions without spending too much time in unnecessary fitting.

Estimating Concrete

Sometimes it is the carpenter's responsibility to order concrete for the job. Ready-mix concrete is sold by the cubic yard or cubic meter. To determine the number of cubic yards of concrete needed for a job, find the number of cubic feet and divide by 27. For example, a wall 8 inches thick, 8 feet high, and 36 feet long requires 7.1 cubic yards of concrete:

$$2/3 \times 8 \times 36 \div 27 = 7.1$$

In this formula, always change all dimensions to feet or fractions of a foot. In the above example, the 8-inch thickness is changed to 2/3 of a foot.

Unit Summary

Concrete can be formed in practically as many shapes as the human mind can conceive (Fig. 30-12). Specialized formwork such as for lift slabs, and tiltup construction, which were illustrated in Figure 27-1, is required for more efficient concrete construction. In addition, large crane-handled *gang, flying, slip, room tunnel, bridge,* and other special forms allow fast setting and quick stripping (Fig. 30-13).

Information in this unit is provided to enable the student to build several basic types of concrete forms. The knowledge of concrete form construction acquired from this unit can be used as a base for building more complicated concrete formwork.

Fig. 30-12 Concrete can be formed in many complex and pleasing shapes. (*Courtesy of American Plywood Association*)

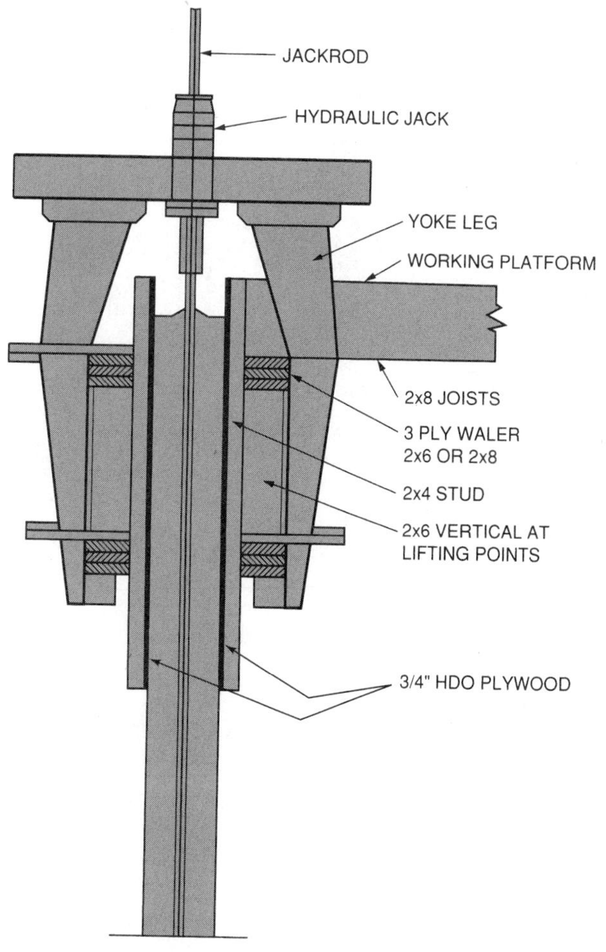

TWO FORMS, EACH CAPABLE OF FORMING A FOLDED PLATE 150 × 30 FEET, DEVELOPED THE ROOF SYSTEM FOR THIS 300 × 480 AIR TERMINAL.

THE WORK SCHEDULE REQUIRED 17 REUSES OF THE SAME FORM IN A WELL-PLANNED SERIES OF POURS. THE SAME 3/4-INCH PLYFORM SERVED THROUGH THE SERIES OF POURS, THEN WAS SALVAGED FOR USE IN OTHER PHASES OF THE JOB.

Fig. 30-13 Many methods are used to place and shape concrete. (*Courtesy of American Plywood Association*)

Review Questions

Select the most appropriate answer.

1. The inside surfaces of forms are oiled to
a. protect the forms from moisture.
b. prevent the loss of moisture from concrete.
c. strip the forms more easily.
d. prevent honeycombs in the concrete.

2. Rapid placing of concrete
a. prevents adequate vibrating.
b. may burst the forms.
c. separates the aggregate.
d. causes voids and honeycombs.

3. Unless footings are placed below the frost line,
a. the foundation will settle.
b. the foundation will heave and crack.
c. excavation is difficult in winter.
d. problems with water and moisture will result.

4. Spreaders for footing forms are used
a. to allow easy placement of the concrete.
b. to prevent forcing the form out of alignment.
c. because they are easier to fasten.
d. because they maintain the footing width.

5. The sides of residential slab forms are usually made with:
a. 2 × 4s and 2 × 6s.
b. 2 × 4s and 2 × 8s.
c. 2 × 6s and 2 × 8s.
d. 2 × 8s and 2 × 10s.

6. The horizontal surface of a stepped footing must be at least
a. 4 feet.
b. twice the vertical distance.
c. the vertical distance.
d. the thickness of the footing.

7. The usual size of a standard wall form panel is
a. 4 feet by 8 feet.
b. 3 feet by 8 feet.
c. 32 inches by 8 feet.
d. 2 feet by 8 feet.

8. Reinforcing the walers at the corners with vertical 2 × 4s is called
a. double locking the corners.
b. interlocking the corners.
c. yoking the corners.
d. tying the corners.

9. To screed the top of a foundation wall
a. space nails along a chalk line.
b. vibrate the top surface so it will flow level.
c. use a chalk line only.
d. add enough water to the concrete so it will flow level.

10. When erecting concrete formwork
a. drive extra nails for added strength.
b. drive all nails home.
c. use as few nails as possible.
d. use duplex nails throughout.

BUILDING FOR SUCCESS 

Quality Concrete Forms

Undoubtedly, the carpenter will continue to play a critical role in the forming and placing of concrete. Carpenters may be assisted by concrete workers, concrete finishers, or masons. However, the carpenter usually initiates, constructs, and disassembles concrete forms. Many small residential builders, remodelers, and some light commercial contractors do most of their own concrete work. Their lead carpenter generally will have extensive experience in building concrete forms of various types. Unit 10 has provided many of the important aspects and form building methods for footings, walls, columns, and floor systems. These methods and construction systems continue to be the most preferred in the industry.

"Hands-on" exercises help construction students develop the most meaningful skills. By digging a short footing by hand, the skill or "art" of using a shovel will be learned. Small form construction produces the skills and knowledge needed to build large ones. Understanding the basics of form building is a prerequisite for quality form construction. When the forms are plumb, level, square, inclined, or contoured properly, the correct size as called for on the prints, and in the right location, the criteria for quality forms have been met.

The site plan, footing and foundation plans, and perhaps an elevation plan will provide the necessary measurements and information for the forms. This is where excellent blueprint reading and measuring skills are a must. Accuracy will be the final goal in form construction. Much of the flatwork, such as sidewalks, driveways, and patios, will require certain inclines for proper drainage. This in turn calls for varying elevations to be maintained. Predetermined drops per lineal foot are required to divert water to designated areas. This will call for the carpenter to accurately read a builder's level, spirit level, layout lines, and grade stakes.

At times there is a tendency for the form carpenter to disregard accuracy and detail. This has proven to be a disaster for many concrete jobs. If water does not drain away from the structure, a multitude of complications surface soon after. The key to quality is attention to detail, code requirements, specifications, and plans. Shortcuts or unapproved alterations affect the appearance and safety of a job. For example, if the amount of shoring, bracing, and reinforcing of forms is minimized, there is no guarantee the forms will properly contain the concrete placed in them. The exterior stair form that does not allow water to drain from the treads will allow for ice to form and become potentially dangerous.

Today's regulations, uniform codes, ADA requirements, and other special considerations require a carpenter to be knowledgeable about all applicable requirements during the planning and construction process. Among other things, an acceptable concrete job must be well planned, meet its intended use, and be safe for people to use.

The versatility in skills required of a carpenter will continue indefinitely. Basic carpentry skills must be combined with problem solving abilities and communication skills. Good form carpenters must place as much emphasis on the quality of their work as trim carpenters. There can be little allowance for error or disregard for exactness anywhere in quality construction. After the project has been completed, people may not even notice how much care, talent, and skill went into forming the walls, columns, stairs, or floors. But when everyone can ascend or descend stairs with no difficulty, water drains away properly, and there are no slips or falls due to poor construction, then people notice and appreciate quality.

Quality work may go unnoticed, but a product lacking quality will become the center of attention. If form carpenters set quality work as a goal, they will not need to look for employment. Quality work is always recognized by builders throughout the country, and they are willing to pay for it. Probably the most important aspect of quality work is the pride one takes in a job well done.

FOCUS QUESTIONS: For individual or group discussion

1. How can you identify a properly formed concrete job?
2. What are some problems that develop when water does not flow in the designated direction on concrete flatwork?
3. How can you draw a correlation between concrete forming and trim carpentry?
4. If you were supervising the forming of a concrete patio slab that is attached to a house, what are some of the forming considerations that assure accuracy?

From rough work to a *smooth finish.*

The following pages outline the many different stages for residential construction. Each photo represents an important process for building a house. Photos have been cross-referenced to units throughout.

The prospective buyers choose the site with the assistance of a general contractor, real estate agent, architect, or engineer.

The architectural team draws up plans to meet state and local regulations.

The architect reviews plans and schedules with the clients.

The architect applies for the necessary permits.

The excavator uses heavy equipment to clear the land.

refer to
UNIT
9

A general contractor, or engineer, lays out the project.

Installation of septic, well, and other utilities will begin.

Heavy equipment begins the excavation process.

The general contractor, or mason contractor, pours the footings.

These photos show a block foundation and a poured concrete foundation.

The excavator backfills after the foundation is poured.

Carpenters, or framers, apply decking material to floor joists.

This photo shows the carpenter, or framer, raising an exterior wall.

This photo shows interior partitions, headers, rafters, staircase, etc.

At this stage, the carpenter applied the sheathing to exterior walls and roof, and completed the rakes and fascias.

The building inspector inspects all phases of construction.

When the building is ready for roofing, the general contractor, or roofing contractor, applies the shingles (cedar, slate, asphalt, etc.).

The builder applied exterior (rigid) insulation to increase the R-value, and to allow for different building applications.

Carpenters install double hung windows.

The builder applies vinyl, aluminum, or cedar siding.

Plumbers, or HVAC contractors, install HVAC (Heating, Ventilation and Air Conditioning). In this picture, the plumber is installing heating and cooling ducts.

At this point in the construction process, the plumber installs rough plumbing.

The electrician installs the rough electrical work.

The insulation contractor, or general contractor, insulates the ceiling.

A drywall contractor, or general contractor, installs drywall, which is commonly referred to as sheetrock.

The OSHA inspector examines the building site to ensure the safety of the work force.

The worker tapes joints after drywall is installed.

The finish carpenter installs the cabinets.

The hardwood flooring installer sands the flooring to prepare the wood for finishing.

This photo shows a variety of finished floors.

refer to UNIT 28

This photo shows a plumber setting a sink, which is part of the finish plumbing process.

Bath fixtures, showers, tubs, and toilets are all part of finish plumbing.

The electrician completes the finish electrical work at this time.

The trimmer installs the baseboard.

The land-scapers plant shrubs and trees to add a finishing touch.

CARPET
RAILING
CAP
DW
FOYER
QUARRY TILE
CARPET
DINING

UNIT

11

Floor Framing

Wood frame construction is used for residential and light commercial construction for important reasons of economy, durability, and variety. The cost for wood frame construction is generally less than for other types of construction. Fuel and air-conditioning expenses are reduced because wood frame construction provides better insulation.

Wood frame homes are very durable. If properly maintained, a wood frame building will last indefinitely. Many existing wood frame structures are hundreds of years old.

Because of the ease with which wood can be cut, shaped, fitted, and fastened, many different architectural styles are possible. In addition to single-family homes, wood frame construction is used for all kinds of lowrise buildings, such as apartments, condominiums, offices, motels, warehouses, and manufacturing plants.

OBJECTIVES

After completing this unit, the student should be able to:

- describe platform, balloon, and post-and-beam framing, and identify framing members of each.
- describe several energy and material conservation framing methods.
- build and install girders, erect columns, and lay sills.
- lay out and install floor joists.
- frame openings in floors.
- lay out, cut, and install bridging.
- apply subflooring.
- describe how termites and fungi destroy wood and state some construction techniques used to prevent destruction by wood pests.

UNIT CONTENTS

CHAPTER 31 TYPES OF FRAME CONSTRUCTION

There are several methods of framing a building. Some types are rarely used today but exist so knowledge of them is necessary when remodeling. Other types are relatively new and knowledge is not so widespread. Some wood frames are built using a combination of types. New designs utilizing engineered lumber are increasing the height and width to which wood frame structures can be built.

Platform Frame Construction

The **platform frame,** sometimes called the *western* frame, is most commonly used in residential construction (Fig. 31-1). In this type of construction, the floor is built and the walls are erected on top of it. When more than one story is built, the second-floor platform is erected on top of the walls of the first story.

A platform frame is easier to erect. At each floor level a flat surface is provided on which to work. A common practice is to assemble wall framing units on the floor and then tilt the units up in place.

Effects of Shrinkage

Lumber shrinks mostly across width and thickness. A disadvantage of the platform frame is the relatively large amount of settling caused by the shrinkage of the large number of horizontal load-bearing frame members. However, because of the equal amount of horizontal lumber, the shrinkage is more or less equal throughout the building. To reduce shrinkage, only framing lumber with the proper moisture content should be used.

Balloon Frame Construction

In the **balloon frame,** the wall *studs* and first-floor *joists* rest on the **sill.** The second-floor joists rest on a 1 × 4 **ribbon** that is cut in flush with the inside edges of the studs (Fig. 31-2). This type of construction is now rarely used, but a substantial number of structures built with this type of frame are still in use.

Effects of Shrinkage

There is so little shrinkage in lumber from one end to the other that it is not taken into consideration. In the balloon frame, settling caused by shrinkage of lumber is held to a minimum in the exterior walls. This is because the studs are continuous from sill to top **plate.** To prevent unequal settling of the frame due to shrinkage, the studs of **bearing partitions** rest directly on the *girder.*

This is a preferred frame when the exterior wall finish is to be brick **veneer** or stucco. There is little movement between the wood frame and the masonry veneer. Consequently, there are less cracks in the wall finish (Fig. 31-3).

In any type of frame, the closer the moisture content of lumber is to the *equilibrium moisture content,* the less effect shrinkage will have on the structure. The use of engineered lumber, manufactured with low moisture content, reduces shrinkage in the frame. Solid sawn lumber used for framing members should be, at least, S-Dry or, preferably at equilibrium moisture content of 8 to 12 percent, depending on the area.

Firestops

In a balloon frame, the studs run from sill to plate. **Firestops** must be installed in the walls in several locations. A firestop is an approved material used in the space between frame members to prevent the spread of fire for a certain period of time. In a wood frame, a firestop in a wall might consist of dimension lumber blocking between studs. In the platform frame, the wall plates act as firestops.

Firestops must be installed in the following locations:

- In all stud walls, partitions, and furred spaces at ceiling and floor levels.
- Between stair *stringers* at the top and bottom. (Stringers are stair framing members. They are sometimes called stair *horses.*)
- Around chimneys, fireplaces, vents, pipes, at ceiling and floor levels with noncombustible material.
- The space between floor joists at the sill and girder.
- All other locations as required by building codes (Fig. 31-4).

Post-and-Beam Frame Construction

The *post-and-beam* frame uses fewer but larger pieces. Large timbers, widely spaced, are used for joists, posts, and **rafters. Matched boards** (tongue

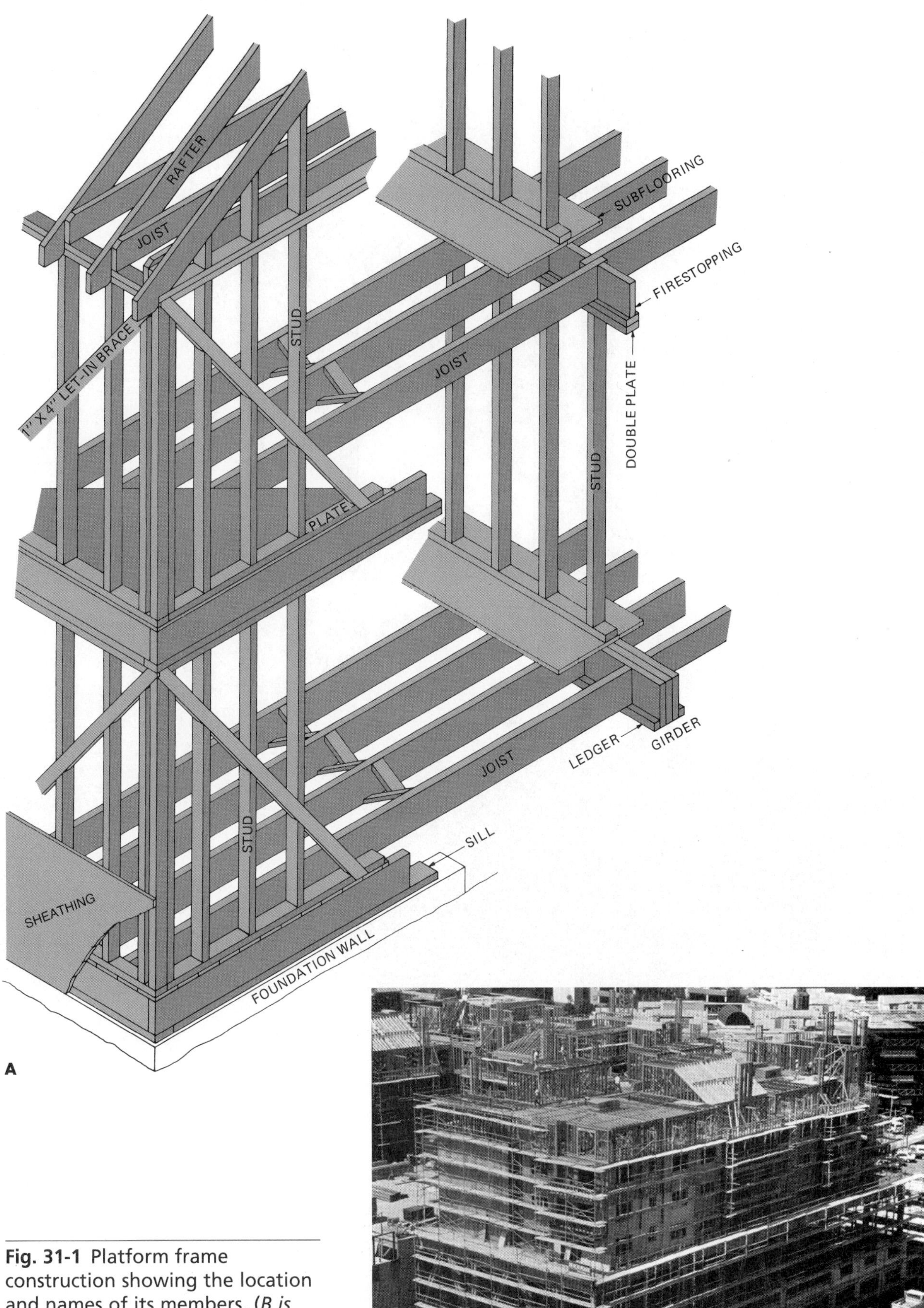

Fig. 31-1 Platform frame construction showing the location and names of its members. (*B is courtesy of Western Wood Products Association*)

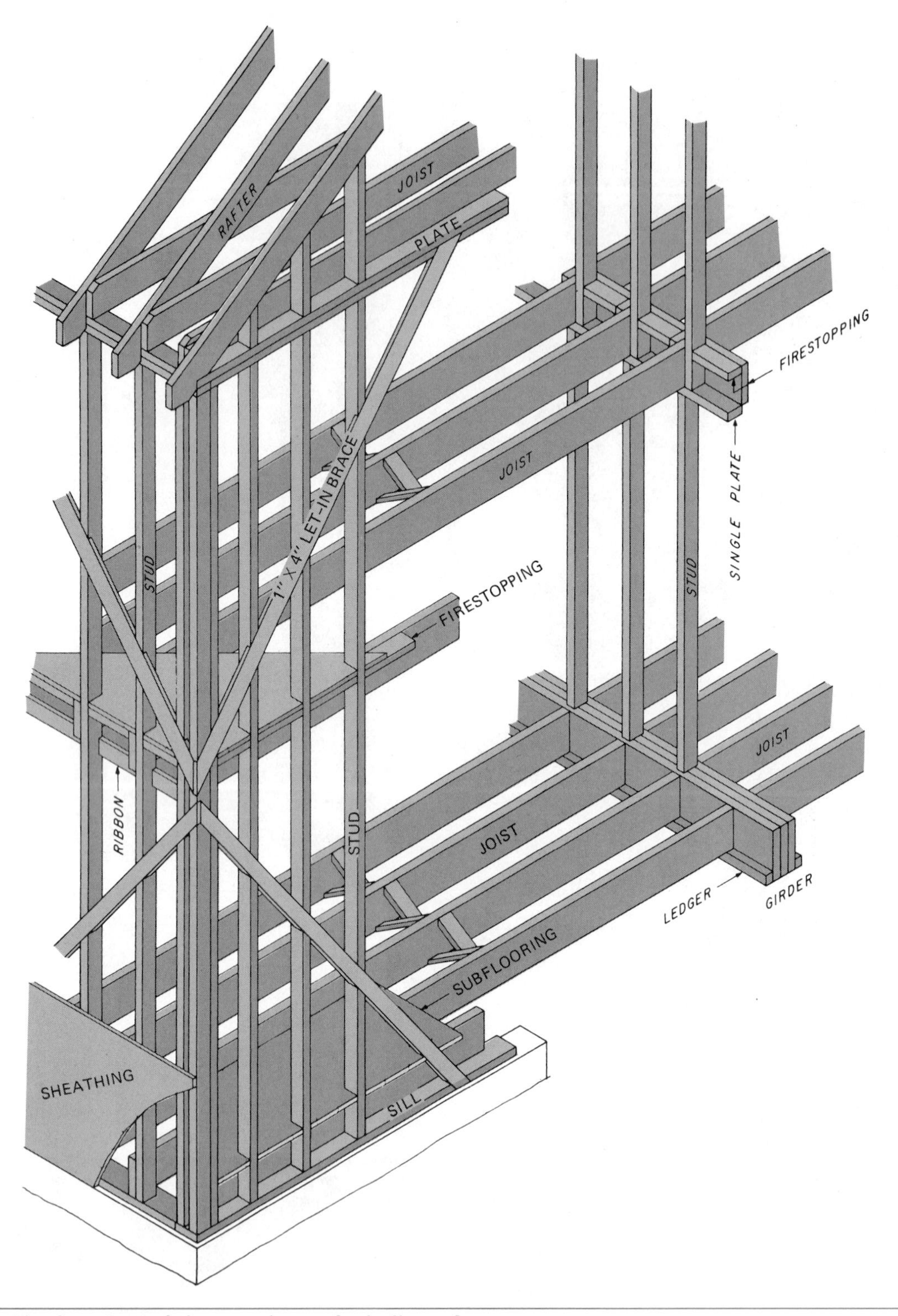

Fig. 31-2 The location of the members of a balloon frame.

and grooved) are often used for floors and roof sheathing (Fig. 31-5).

Floors

APA Rated Sturd-I-Floor 48 on center (OC), which is 1 3/32 inches thick, may be used on floor joists that are spaced 4 feet on center instead of matched boards (Fig. 31-6, p. 316). In addition to being nailed, the plywood panels are glued to the floor beams with construction adhesive applied with caulking guns. The use of matched planks allows the floor beams to be more widely spaced.

Walls

Exterior walls of a post-and-beam frame may be constructed with widely spaced posts. This allows

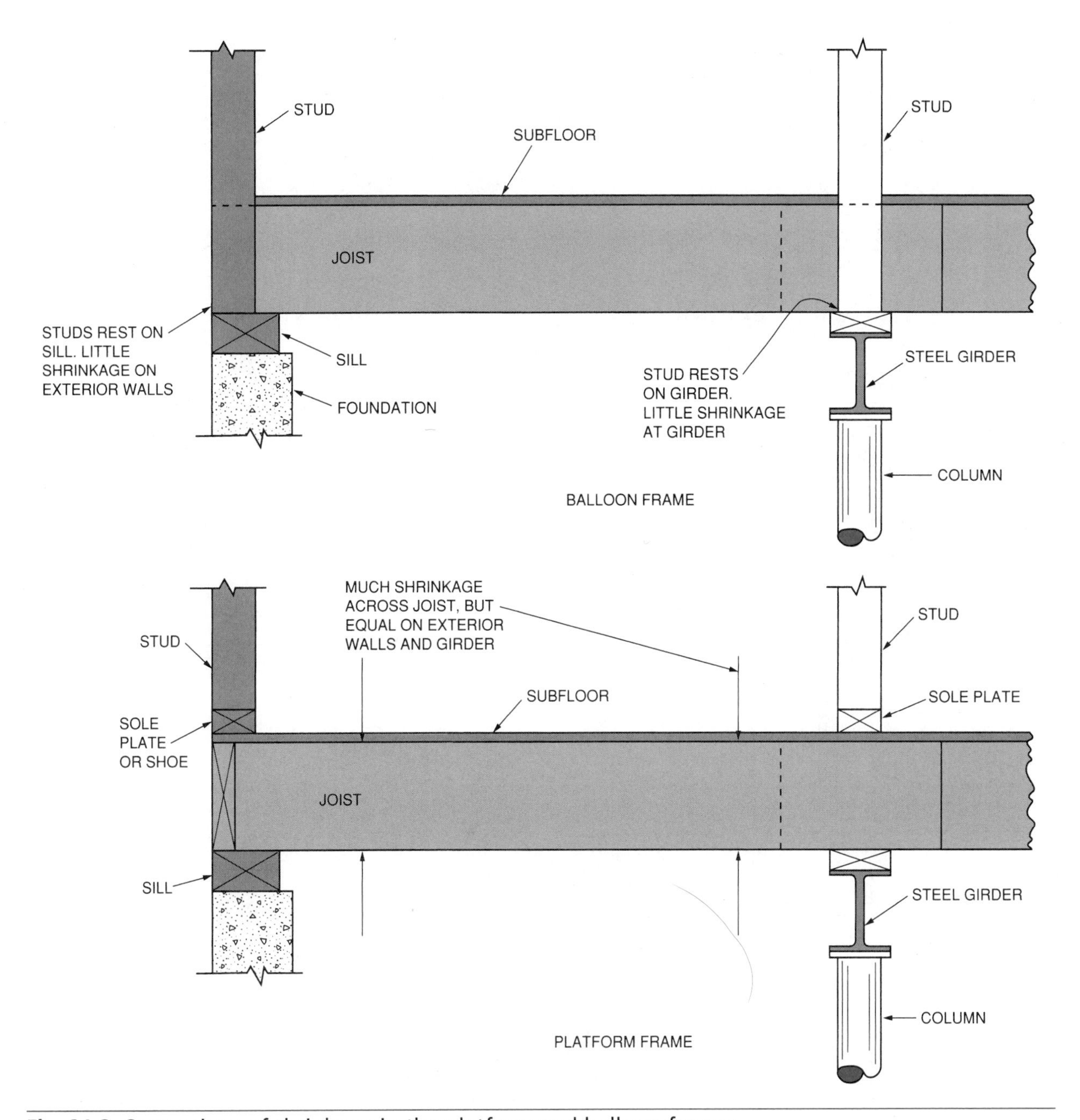

Fig. 31-3 Comparison of shrinkage in the platform and balloon frames.

wide expanses of glass to be used from floor to ceiling. Usually some sections between posts in the wall are studded at close intervals, as in platform framing. This provides for door openings, fastening for finish, and wall **sheathing.** In addition, close spacing of the studs permits the wall to be adequately braced (Fig. 31-7).

Roofs

The post-and-beam frame roof is widely used. The exposed roof beams and sheathing on the underside are attractive. Usually the bottom surface of the roof planks is left exposed to serve as the finished ceiling. Roof planks come in 2-, 3-, and 4-inch nominal thicknesses. Some are **end** matched as well as **edge** matched. Some buildings may have a post-and-beam roof, while the walls and floors may be conventionally framed.

The post-and-beam roof may be constructed with a *longitudinal* frame. The beams run parallel to the ridge beam. Or they may have a *transverse* frame. The beams run at right angles to the **ridge** beam similar to roof rafters.

The ridge beam and longitudinal beams, if used, are supported at each end by posts in the end walls. They must also be supported at intervals along their length. This prevents the side walls from spreading and the roof from sagging (Fig. 31-8). One of the disadvantages of a post-and-beam roof is

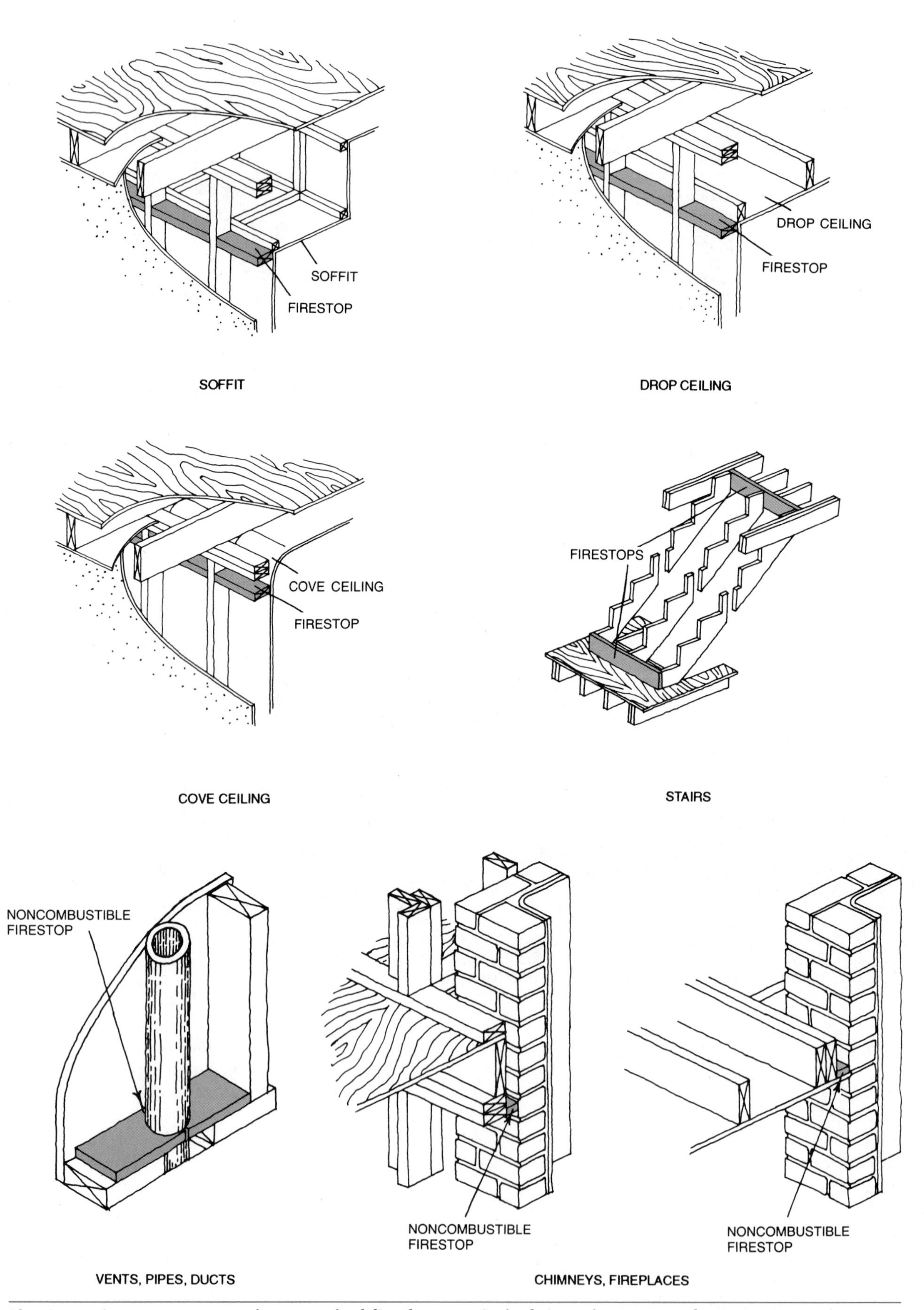

Fig. 31-4 Firestops prevent the spread of fire for a period of time. (*Courtesy of Western Wood Products Association*)

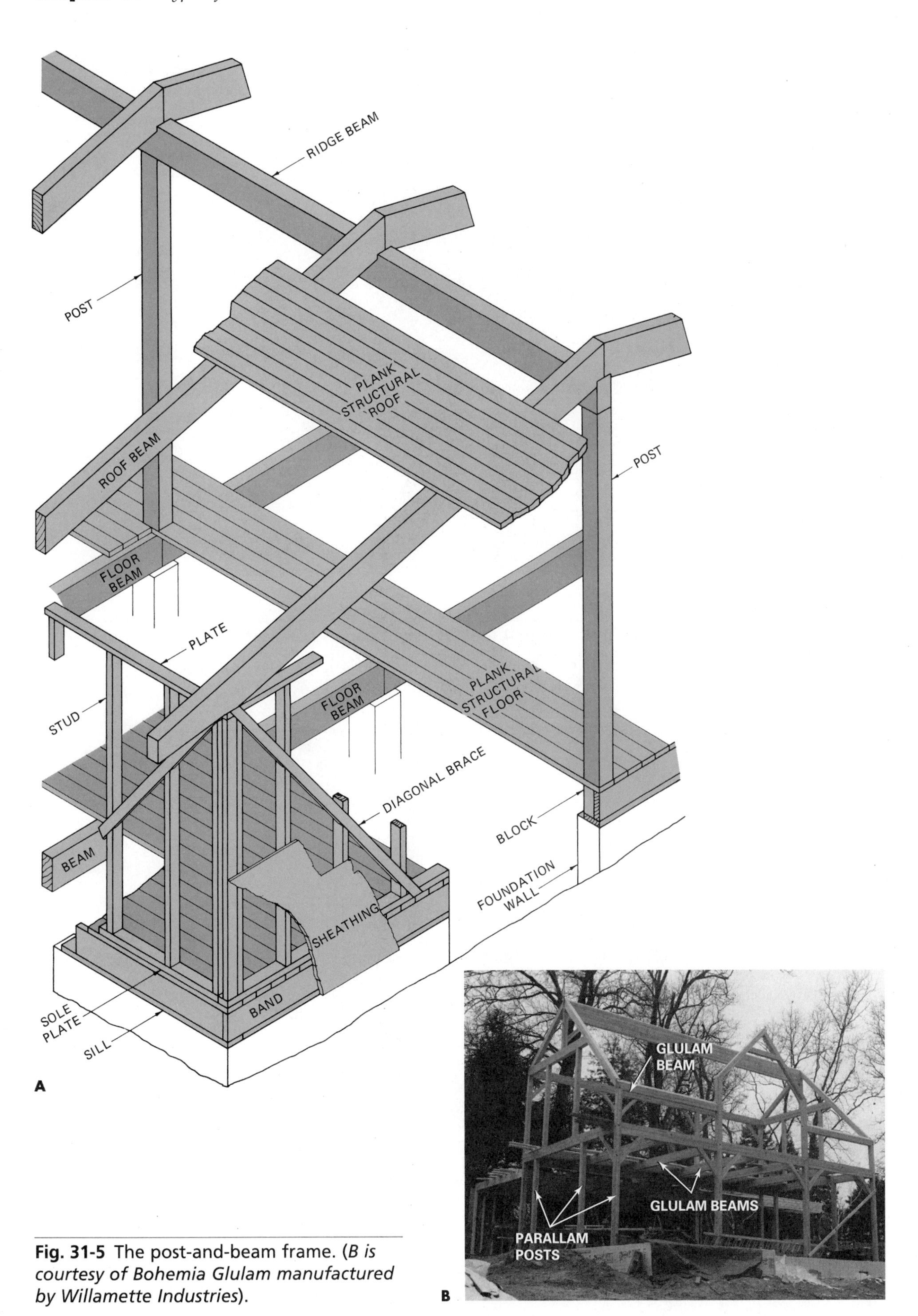

Fig. 31-5 The post-and-beam frame. (*B is courtesy of Bohemia Glulam manufactured by Willamette Industries*).

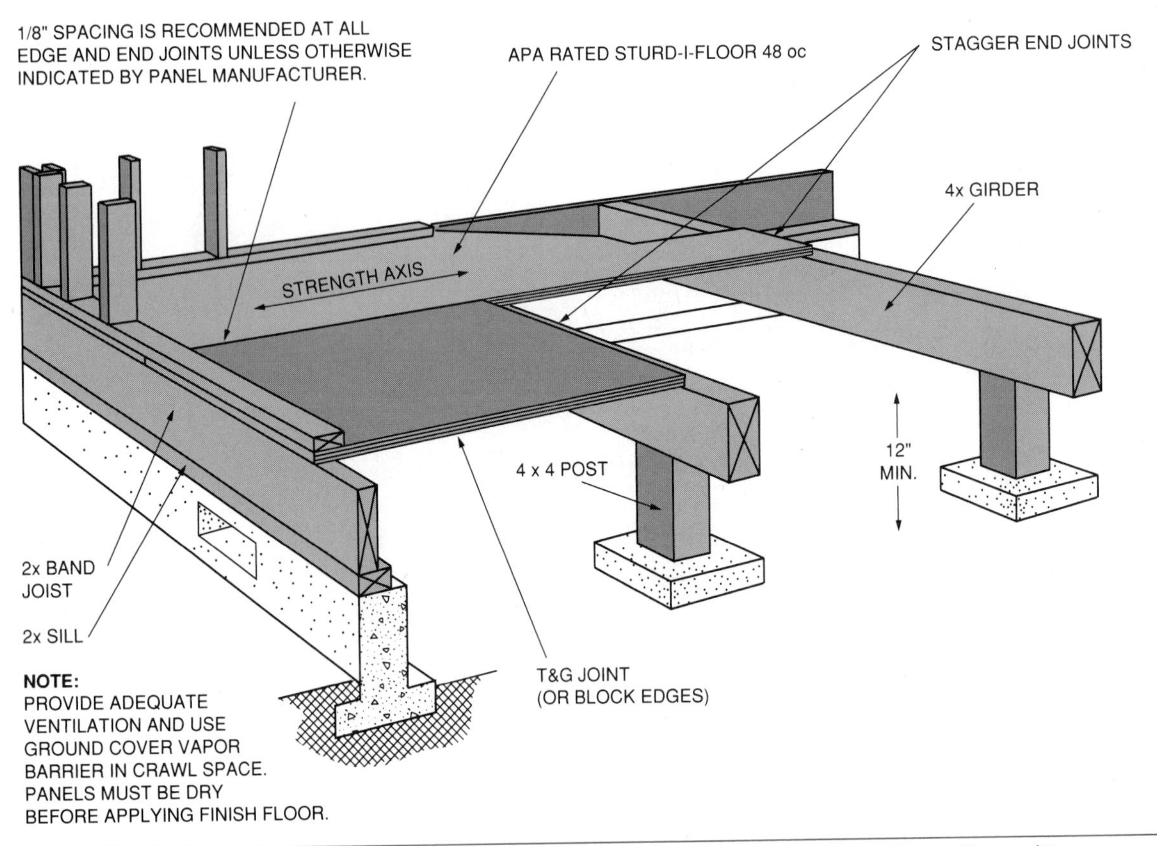

Fig. 31-6 Floor beams are sometimes spaced 4 feet OC when panels are used for a floor. (*Courtesy of American Plywood Association*)

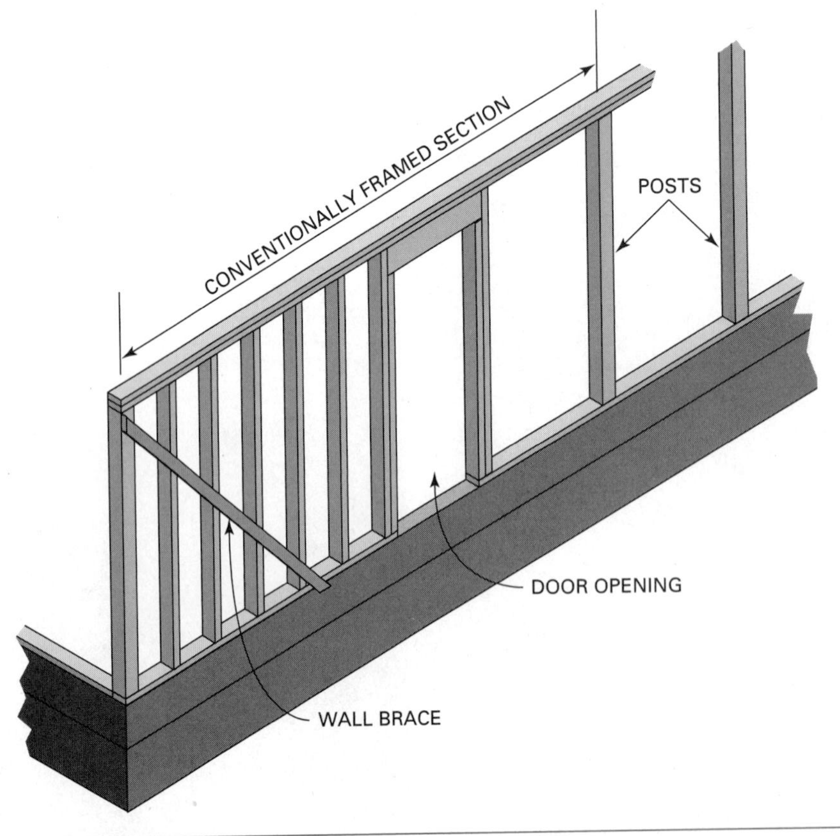

Fig. 31-7 Sections of the exterior walls of a post-and-beam wall may need to be conventionally framed.

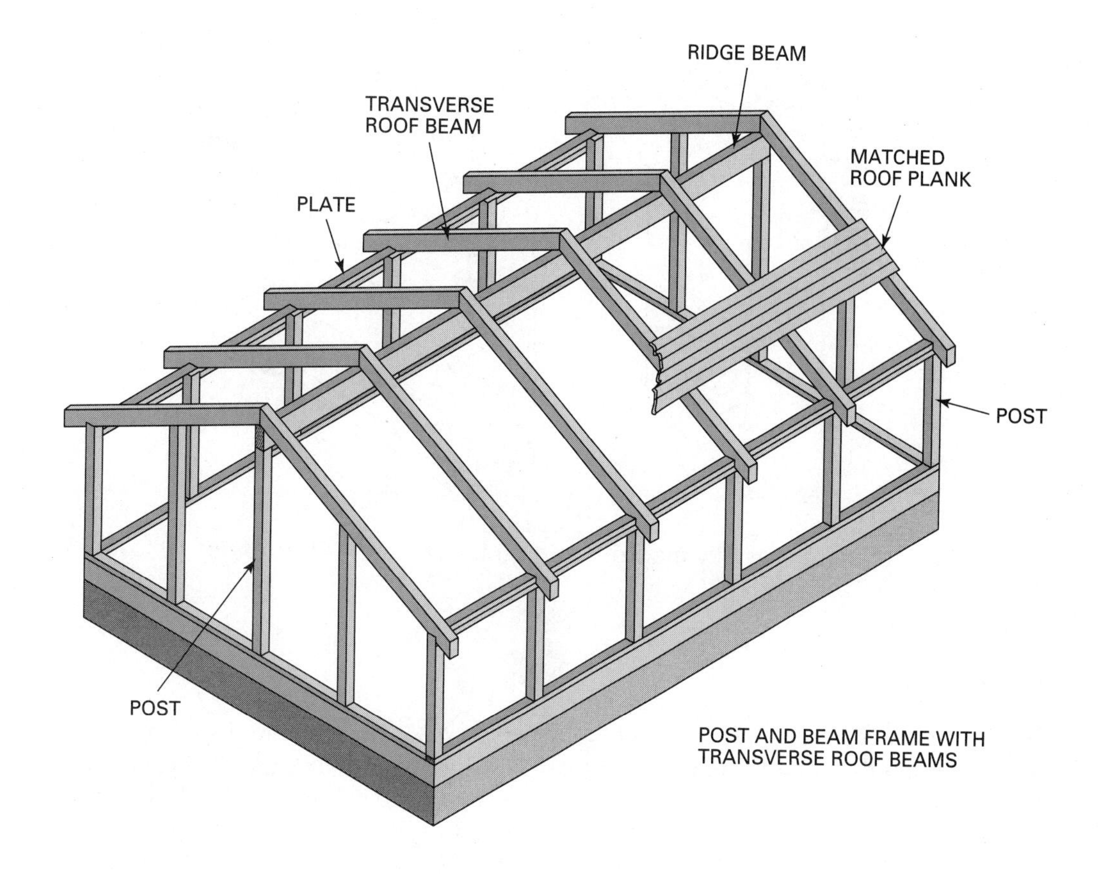

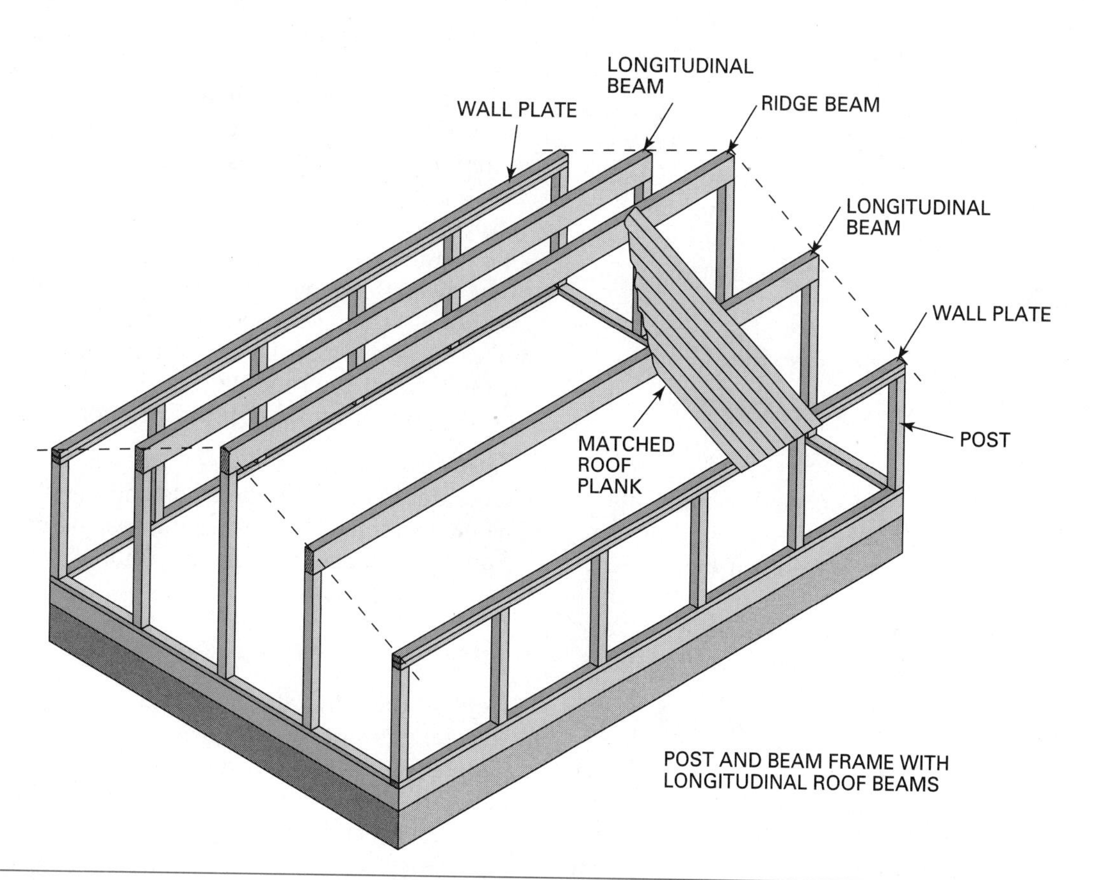

Fig. 31-8 Longitudinal and transverse post-and-beam roofs.

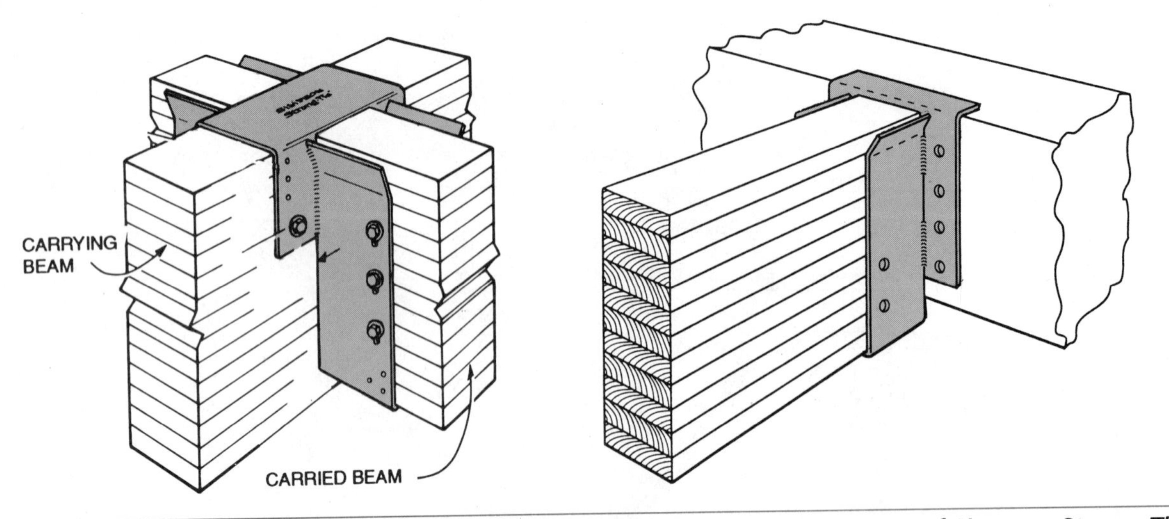

Fig. 31-9 Metal connectors are specially made to join glulam beams. (*Courtesy of Simpson Strong-Tie Company*)

that interior partitions and other interior features must be planned around the supporting roof beam posts.

The roof is insulated on top of the deck in order not to spoil the appearance of the exposed beams and deck on the underside. A vapor barrier is first applied. The insulation is installed before the roof covering is put on (See Fig. 49-19).

Because of the fewer number of pieces used, a well-planned post-and-beam frame saves material and labor costs. Care must be taken when erecting the frame to protect the surfaces and make well-fitting joints on exposed posts and beams. Glulam beams are well suited for and frequently used in post-and-beam construction. A number of metal connectors are used to join members of the frame (Fig. 31-9).

Energy and Material Conservation Framing Methods

There has been much concern and thought about conserving energy and materials in building construction. Several systems have been devised that differ from conventional framing methods. They conserve energy and use less material and labor. Check state and local building codes for limitations.

The 24-Inch Module Method

One of the new methods uses plywood over lumber framing spaced on a 24-inch module. All framing, floors, walls, and roofs are spaced 24 inches **on center (OC)**. Joists, studs, and rafters line up with each other (Fig. 31-10).

Floors. For maximum savings, a single layer of 3/4-inch tongue-and-grooved plywood is used over joists spaced 24 inches OC. In-line floor joists are used to make installation of the plywood floor easier (Fig. 31-11). The use of adhesive when fastening the plywood floor is recommended. Gluing increases stiffness and prevents squeaky floors (Fig. 31-12).

Walls. Studs up to 10 feet long with 24-inch spacing can be used in single-story buildings. Stud height for two-story buildings should be limited to 8 feet. A single layer of plywood acts as both sheathing and exterior siding. In this case, the plywood must be at least 1/2-inch thick (Fig. 31-13). If two-layer construction is used, 3/8-inch plywood is acceptable.

Wall openings are planned to be located so that at least one side of the opening falls on the module. Whenever possible, window and door sizes are selected so that the rough opening width is a multiple of the module (Fig. 31-14). Also, locate partitions at normal wall stud positions if possible.

Roofs. Rafters or trusses are spaced 24 inches on center in line with the studs under 3/8-inch plywood sheathing. Increasing the spacing of the roof framing members and using a thinner roof sheathing conserves material.

House Depths

House depths that are not evenly divisible by four waste floor framing and sheathing. Lumber for floor joists is produced in increments of 2 feet. A 25-foot house depth uses the same total linear footage of standard floor joist lengths as a house depth of 28 feet. Full 48-inch-wide plywood panels can be used without ripping for 24-, 28-, and 32-foot house depths.

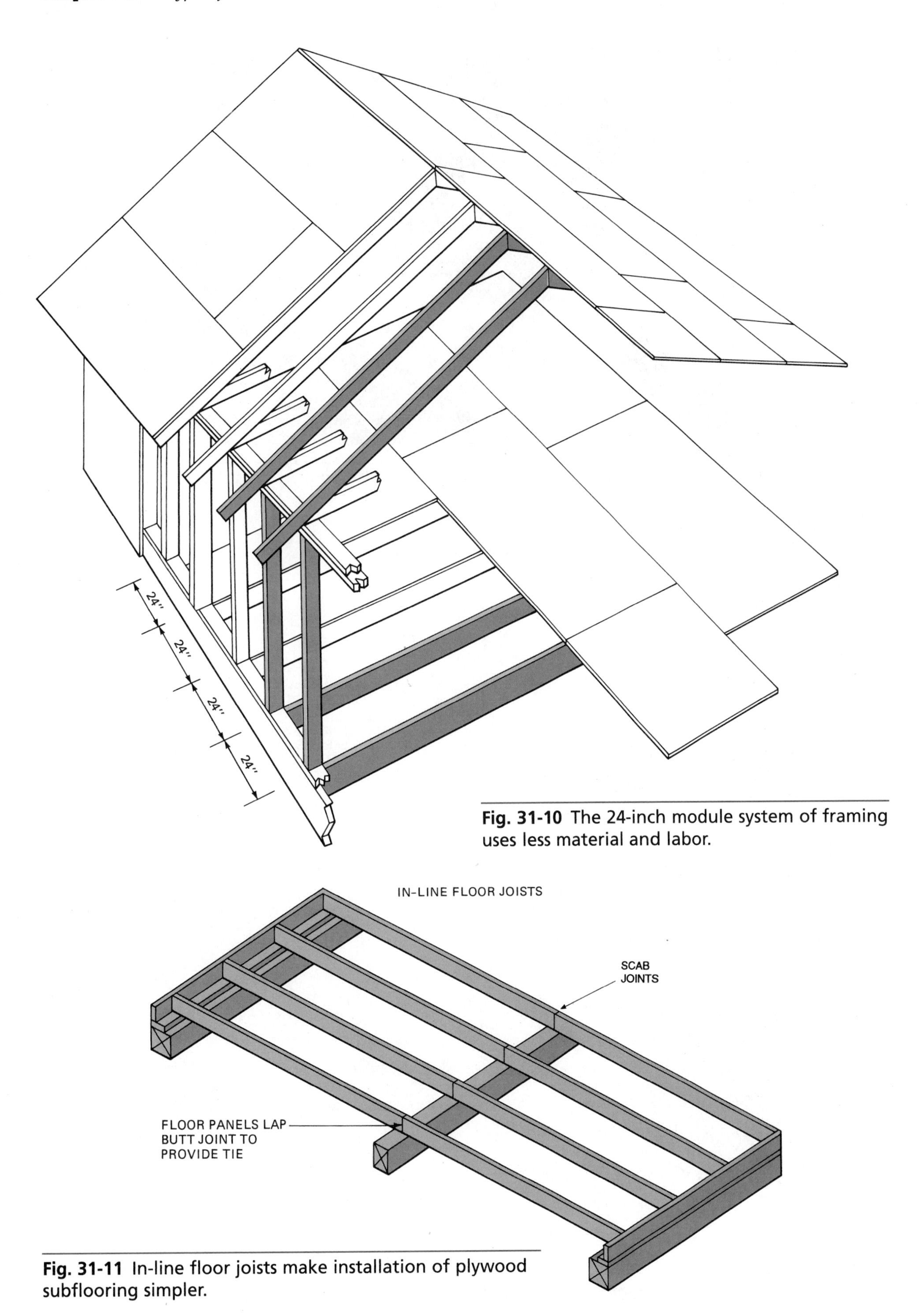

Fig. 31-10 The 24-inch module system of framing uses less material and labor.

Fig. 31-11 In-line floor joists make installation of plywood subflooring simpler.

Fig. 31-12 Using adhesive when fastening subflooring makes the floor frame stiffer and stronger. (*Courtesy of American Plywood Association*)

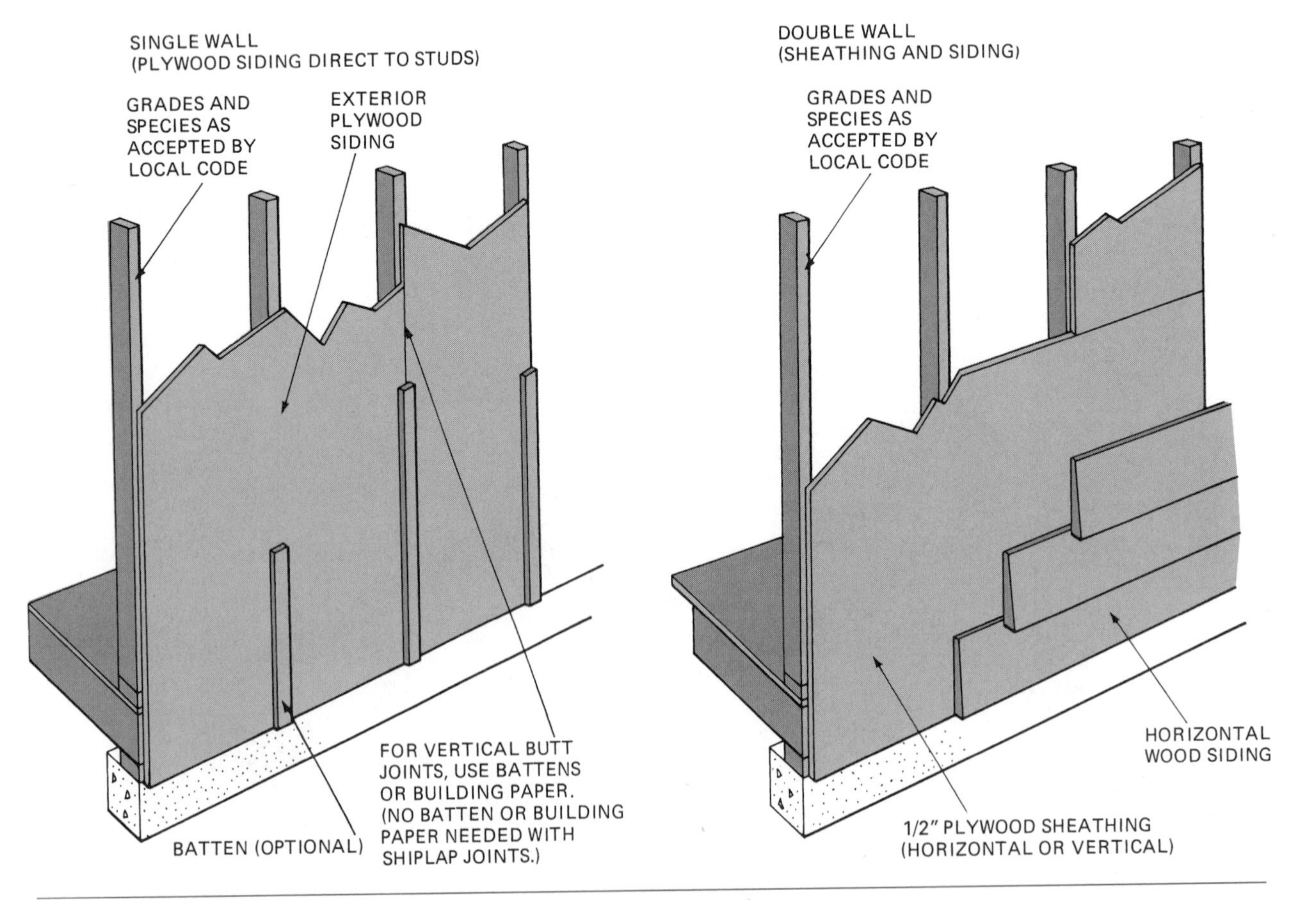

Fig. 31-13 Single-layer and double-layer exterior wall covering.

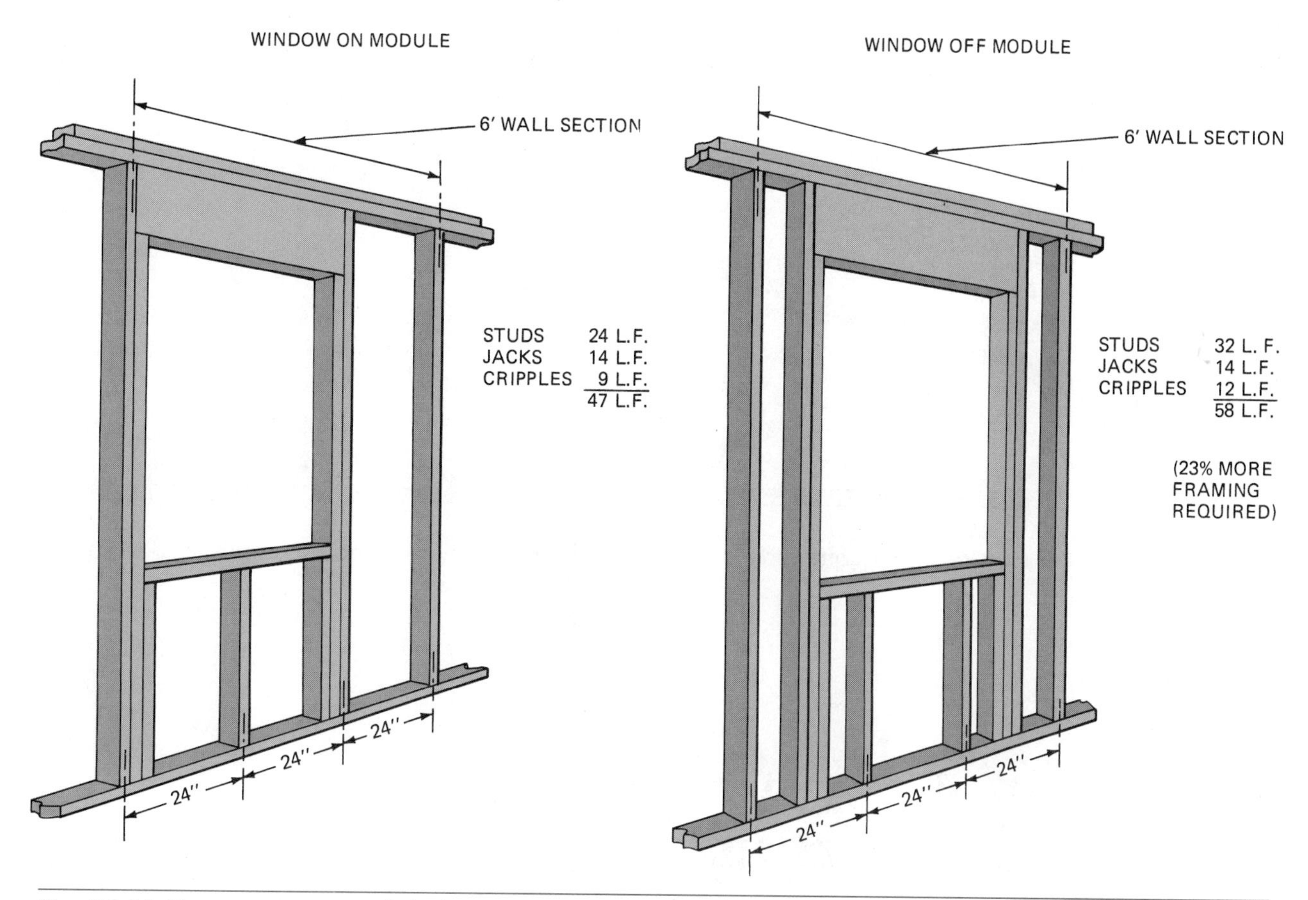

Fig. 31-14 To conserve materials, locate wall openings so they fall on the spacing module.

Reducing the Clear Span of Floor Joists

The need to select large-size floor joists can be avoided by reducing the clear span. This can be done by using wider sill plates and center-bearing plates (Fig. 31-15). When the use of wider plates makes it possible to use smaller joists, the added cost of the plates is little when compared to the savings achieved.

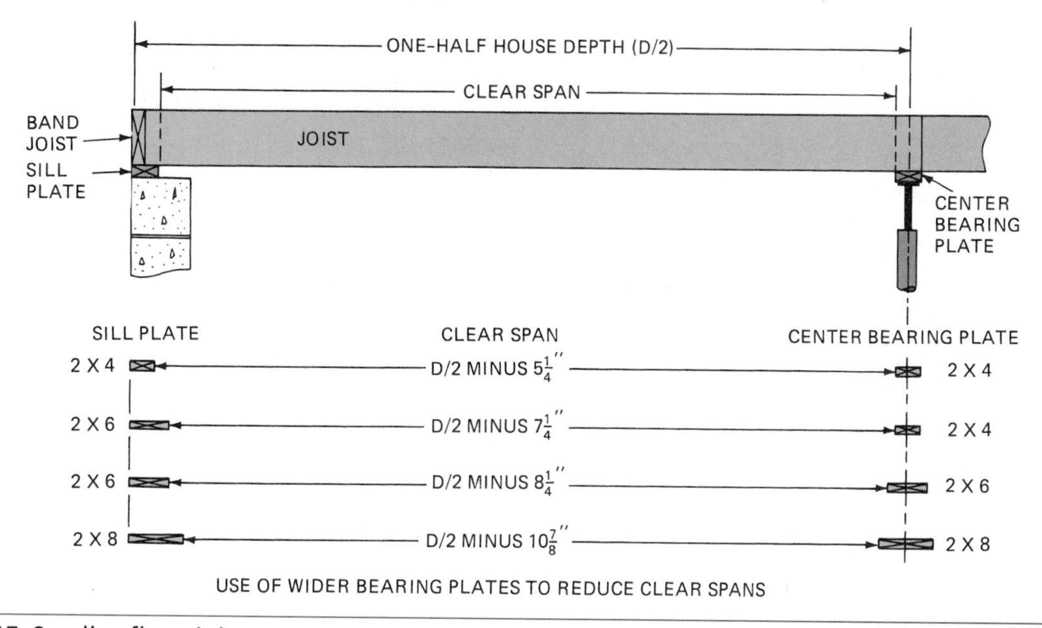

Fig. 31-15 Smaller floor joists can sometimes be used if the clear span is reduced. (*Courtesy of Southern Pine Marketing Council*)

The Arkansas System

An energy-saving construction system developed by the Arkansas Power and Light Company uses 2 × 6 wall studs spaced 24 inches CIC. This permits using full 6-inch insulation in the exterior walls. A modified truss accommodates 12 inches of insulation in the ceiling without compressing it at the eaves (Fig. 31-16).

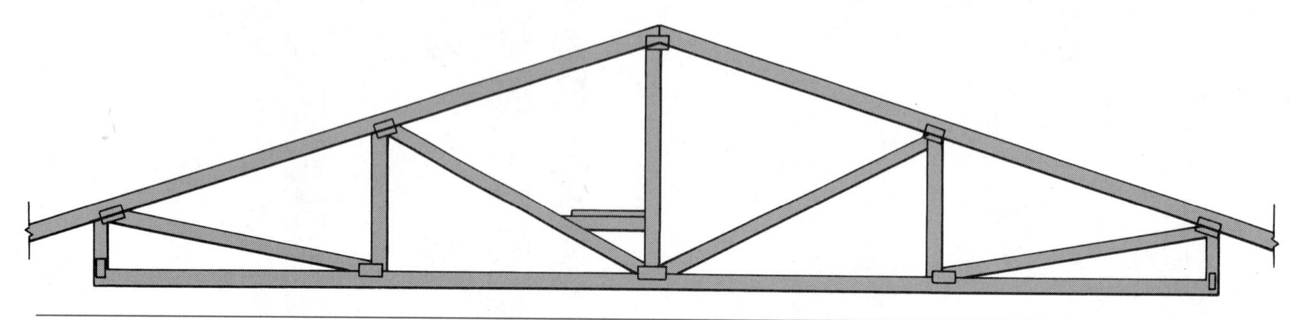

Fig. 31-16 Modified truss design accommodates 12-inch ceiling insulation without compressing at eaves.

CHAPTER 32 LAYOUT AND CONSTRUCTION OF THE FLOOR FRAME

A floor frame consists of members fastened together to support the loads a floor is expected to bear. The floor frame is started after the foundation has been placed and has hardened. A straight and level floor frame makes it easier to frame and finish the rest of the building.

Because platform framing is used more than any other type, this chapter describes how to lay out and construct its floor frame. The knowledge gained in this chapter can be used to lay out and construct any type of floor frame.

Description and Installation of Floor Frame Members

In the usual order of installation, the floor frame consists of *girders, posts* or *columns, sill plates, joists,* **bridging,** and *subflooring* (Fig. 32-1).

Description of Girders

Girders are heavy beams that support the inner ends of the floor joists. Several types are commonly used.

Kinds of Girders. Girders may be made of solid wood or built up of three or more 2-inch planks. Laminated veneer lumber or glulam beams may also be used as girders. Sometimes, wide flange, I-shaped steel beams are used (Fig. 32-2).

Determining the Size of Girders and Other Structural Members. The size of solid, built-up, engineered lumber, or steel girders, or any structural component, is best determined by professional architects or engineers. There are many factors to consider when determining structural lumber or steel sizes. Only people specifically trained should select the size of a structural member. Estimating or guessing the size can have serious consequences. If undersize, the member could sag. Or, worse, it could fail and possibly cause the section of building it supports to collapse. If oversize, it would not be economical and might take up more room than necessary. If the blueprints do not specify the kind or size, have the design checked by a professional engineer.

Built-Up Girders. If built-up girders of solid lumber are used, a minimum of three members are fastened together with three 3 1/2 inch or 16d nails at each end. The other nails are staggered not farther than 32 inches apart from end to end. Sometimes 1/2-inch bolts are required. Applying

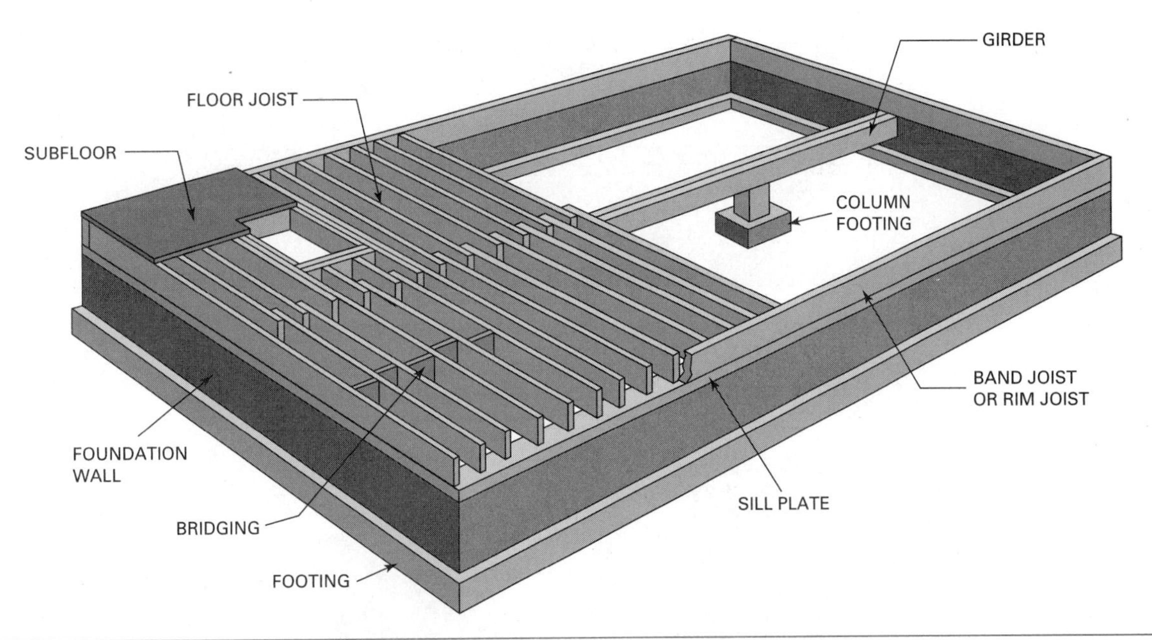

Fig. 32-1 A floor frame of platform construction. (*From* Architectural Drafting & Design, *Delmar*)

Fig. 32-2 A large glulam beam is used for a girder. (*Courtesy of Trus Joist MacMillan*)

glue between the pieces makes the bond stronger. Laminated veneer lumber may also be built up for use as girders (Fig. 32-3). The end joints of both types are placed directly over supports.

Girder Location. The ends of the girder are usually supported by a *pocket* formed in the foundation wall. The pocket should provide at least a 4-inch bearing for the girder. It should be wide enough to provide 1/2-inch clearance on both sides and the end. This allows any moisture to be evaporated by circulation of air. Thus no moisture will get into the girder, which would cause decay of the timber.

The pocket is formed deep enough to provide for shimming the girder to its designated height. An iron bearing plate is *grouted* in under the girder while it is supported temporarily (Fig. 32-4). **Grouting** is the process of filling in the small space between the iron plate and the bottom of the pocket with a thick paste of portland cement. Wood **shims** are not suitable for use under girders. The weight imposed on them compresses the wood, causing the girder to sink below its designated level.

Installing Girders

Steel girders usually come in one piece and are set in place with a crane. Wood girders are usually built-up and erected in sections. Start by building one section. Set one end in the pocket in the foundation wall. Place and fasten the other end on a braced temporary support. Continue building and erecting sections until the girder is completed to the opposite pocket. A solid wood girder is installed in the same manner as a built-up girder. Half lap joints are made directly over posts or columns.

Sight the girder by eye from one end to the other. Place wedges under the temporary supports to straighten the girder. Permanent posts or columns are usually installed after the girder has some weight imposed on it by the *floor joists.* Temporary posts should be strong enough to support the weight imposed on them until permanent ones are installed.

Sills

Sills , also called *mudsills* or *sill plates,* are horizontal members of a floor frame. They lie directly on the foundation wall and provide a bearing for *floor joists.* It is sometimes required that the sill be made

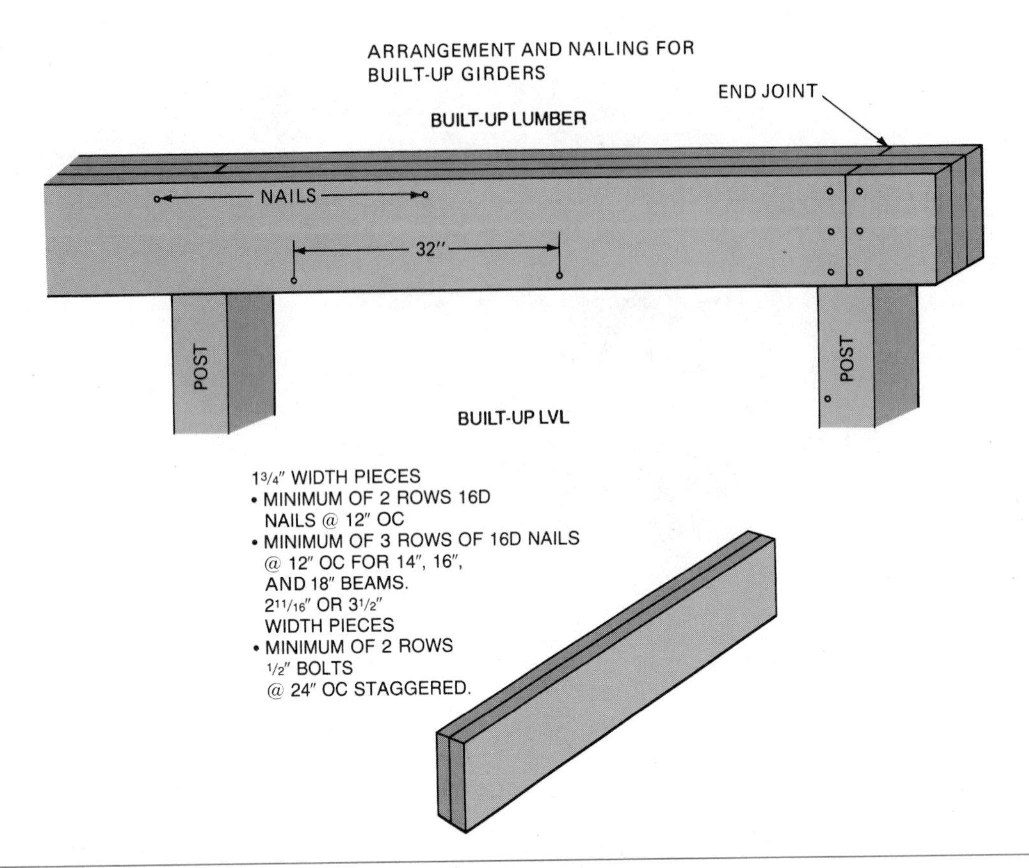

Fig. 32-3 Spacing of fasteners and joints of built-up girders.

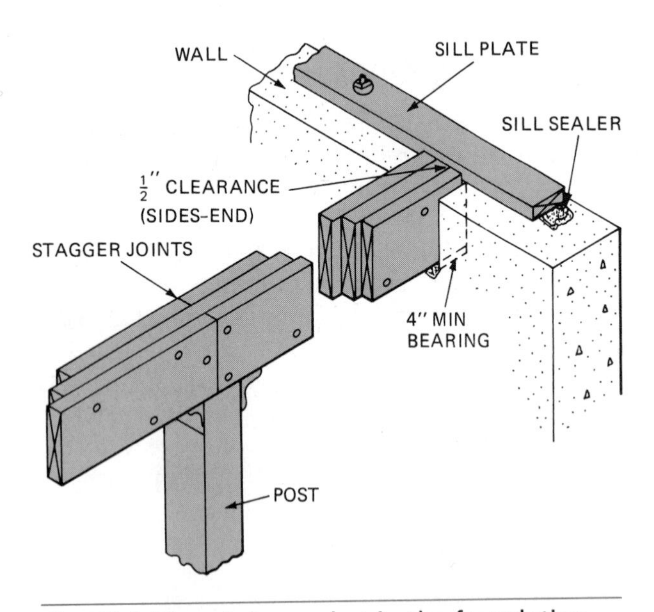

Fig. 32-4 The girder pocket in the foundation wall should be large enough to provide air space around the end of the girder.

with a decay-resistant material such as redwood, black locust, cedar, or pressure-treated lumber. Sills may consist of single 2 × 6, 4 × 6, or, in many cases, a doubled 2 × 6 solid lumber (Fig. 32-5).

The sill is attached to the foundation wall with anchor bolts. Their size, type, and spacing are specified on the blueprints. To take up irregularities between the foundation wall and the sill, a *sill sealer* is used.

The sill sealer should be an insulating material used to seal against drafts, dirt, and insects. It comes 6 inches wide and in rolls of 50 feet. It compresses when the weight of the structure is upon it.

Installing Sills

Sills must be installed so they are straight, level, and to the specified dimension of the building. The level of all other framing members depends on the care taken with the installation of the sill.

Sometimes the outside edge of the sill is flush with the outside of the foundation wall. Sometimes it is set in the thickness of the wall sheathing, depending on custom or design. In the case of brick-veneered exterior walls, the sill plate may be set back even farther (Fig. 32-6). Remove washers and nuts from the anchor bolts. Snap a chalk line on the top of the foundation wall in line with the inside edge of the sill.

Locating Anchor Bolts. Cut the sill sections to length. Hold the sill in place against the anchor bolts. Square lines across the sill on each

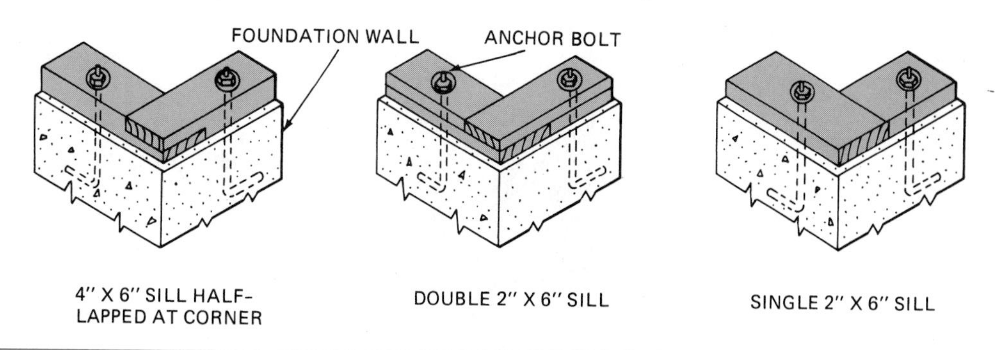

Fig. 32-5 Sill details at the corner. A bolt should be located 6 to 12 inches from the ends of each sill.

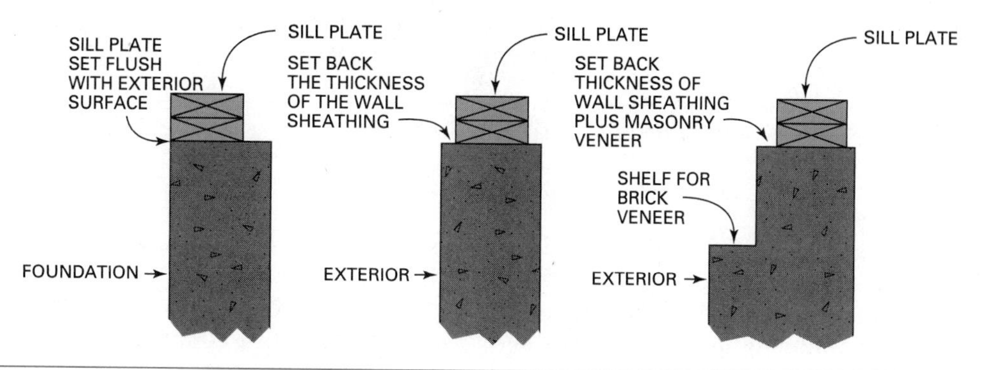

Fig. 32-6 The setback of the sill plate on the foundation wall may vary.

side of the bolts. Measure the distance from the center of each bolt to the chalk line. Transfer this distance at each bolt location to the sill by measuring from the inside edge (Fig. 32-7).

Bore holes in the sill for each anchor bolt. Bore the holes at least 1/8 -inch oversize to allow for adjustments. Place the sill sections in position over the anchor bolts after installing the sill sealer. The

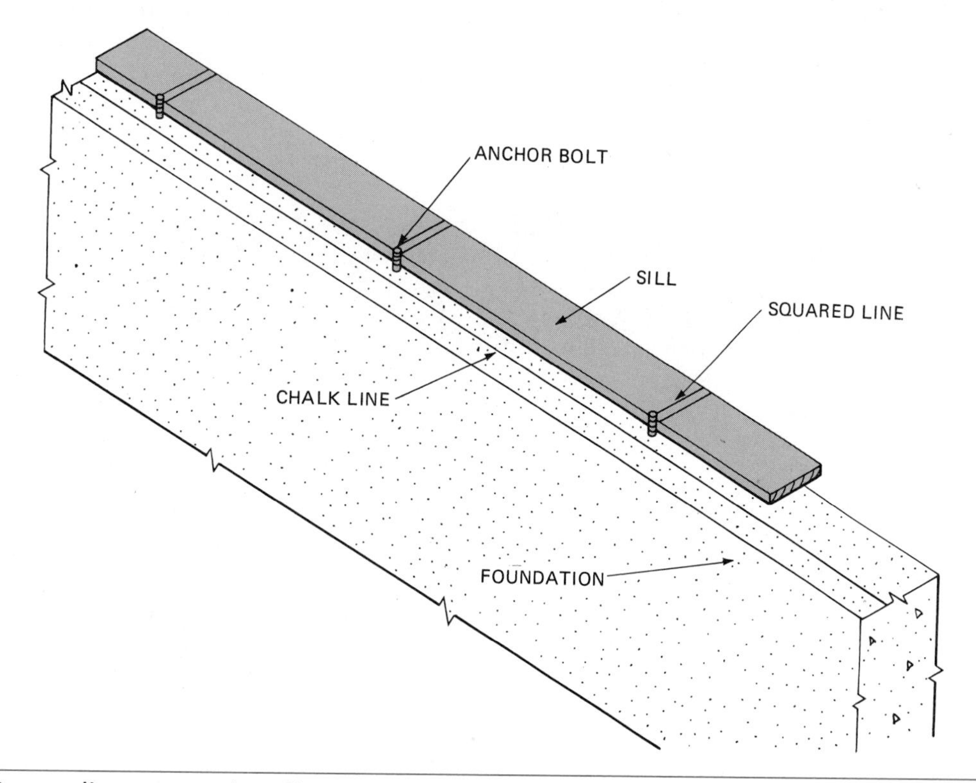

Fig. 32-7 Square lines across the sill to locate the center of anchor bolts.

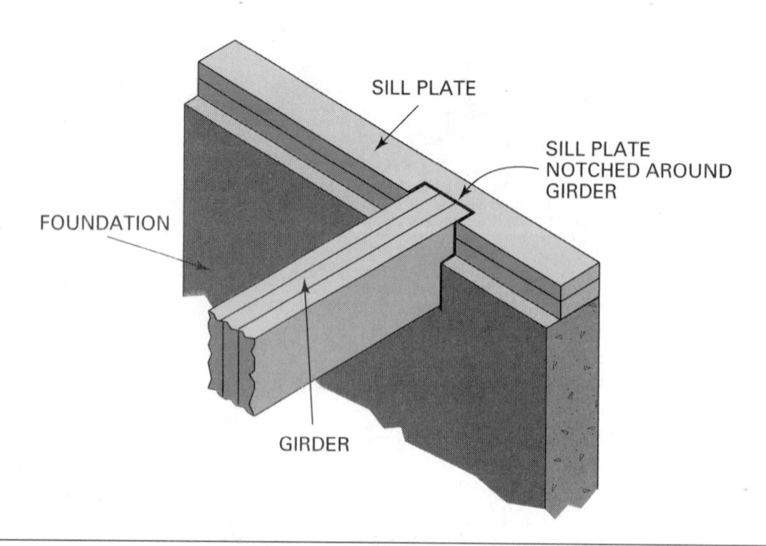

Fig. 32-8 Notch the sill around the end of the girder, if necessary.

inside edges of the sill sections should be on the chalk line. Replace the nuts and washers. Be careful not to overtighten the nuts, especially if the concrete wall is still **green** (not thoroughly dry and hard). This may crack the wall.

If the inside edge of the sill plate comes inside the girder pocket, notch the sill plate around the end of the girder (Fig. 32-8). Raise the ends of the girder so it is flush with the top of the sill plate.

Floor Joists

Floor joists are horizontal members of a frame. They rest on and transfer the load to sills and girders. In residential construction, nominal 2-inch thick lumber placed on edge has been traditionally used. Wood I-beams, with lengths up to 80 feet, are being specified more often today (Fig. 32-9). In commercial work, steel or a combination of steel and wood trusses is frequently used.

Joists are generally spaced 16 inches OC in conventional framing. They may be spaced 12 or 24 inches OC, depending on the type of construction and the load. The size of floor joists should be determined from the construction drawings.

Joist Framing at the Sill. Joists should rest on at least 1 1/2 inches of bearing on wood and 3 inches on **masonry.** In platform construction, the ends of floor joists are capped with a *band joist,* also

Fig. 32-9 In addition to solid lumber, wood I-beams are used for floor joists. (*Courtesy of Trus Joist MacMillan*)

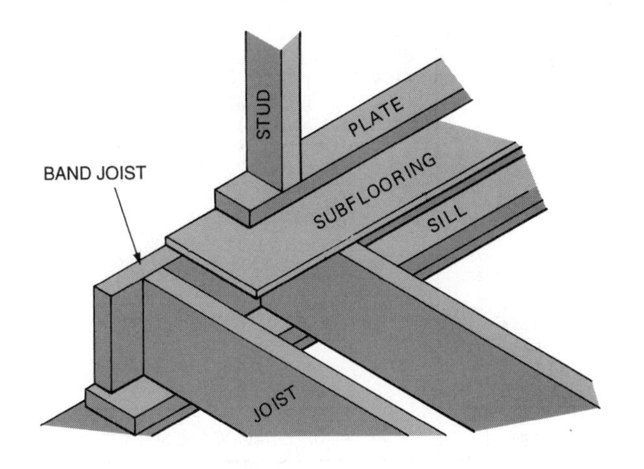

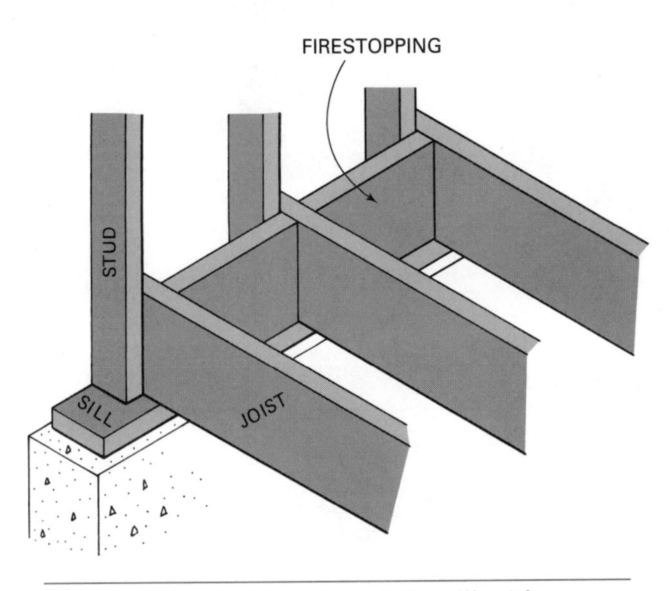

Fig. 32-10 Typical framing at the sill with lumber joists.

called a *rim joist,* **box header,** or **joist header.** In a balloon frame, joists are cut flush with the outside edge of the sill (Fig. 32-10). The use of wood I-beams requires sill construction as recommended by the manufacturer for satisfactory performance of the frame (Fig. 32-11).

Joist Framing at the Girder. If joists are lapped over the girder, the minimum amount of lap is 4 inches and the maximum overhang is 12 inches. There is no need to lap wood I-beams. They come in lengths long enough to span the building. However, they may need to be supported by girders depending on the span and size of the I-beam.

Sometimes, to gain more headroom, joists may be framed into the side of the girder. There are a number of ways to do this (Fig. 32-12). Joist hangers must be used to support wood I-beams. **Web stiffeners** should be applied to the beam ends if the hanger does not reach the top flange of the beam.

Notching and Boring of Joists

Notches in the bottom or top of sawn lumber floor joists should not exceed one-sixth of the joist depth. Notches should not be located in the middle one-third of the joist span. Notches on the ends should not exceed one-fourth of the joist depth.

Holes bored in joists for piping or wiring should not be larger than one-third of the joist depth. They should not be closer than 2 inches to the top or bottom of the joist (Fig. 32-13).

Some wood I-beams are manufactured with 1 1/2-inch perforated knockouts in the web at approximately 12 inches on center along its length. This allows easy installation of wiring and pipes. To cut other size holes in the web, consult the manufacturer's specifications guide. Do not cut or notch the flanges of wood I-beams.

Laying Out Floor Joists

The locations of floor joists are marked on the sill plate. A squared line marks the side of the joist. An X to one side of the line indicates on which side of the line the joist is to be placed (Fig. 32-14).

Floor joists must be laid out so that the ends of *plywood subfloor* sheets fall directly on the center of floor joists. Start the joist layout by measuring the joist spacing from the end of the sill. Measure back one-half the thickness of the joist. Square a line across the sill. This line indicates the side of the joist closest to the corner. Place an X on the side of the line on which the joist is to be placed (Fig. 32-15).

From the squared line, measure and mark the spacing of the joists along the length of the building. Place an X on the same side of each line as for the first joist location.

When measuring for the spacing of the joists, use a tape stretched along the length of the building. Most tapes have prominent markings for 16-inch spacing. Using a tape in this manner is more accurate. Measuring and marking each space individually with a rule or pocket tape generally causes a gain in the spacing. If the spacing is not laid out accurately, the plywood subfloor may not fall in the center of some floor joists. Time will then be lost either cutting the plywood back or adding strips of lumber to the floor joists (Fig. 32-16).

Laying Out Floor Openings. After marking floor joists for the whole length of the building, study the plans for the location of floor openings. Mark the sill plate, where joists are to be doubled, on each side of large floor openings.

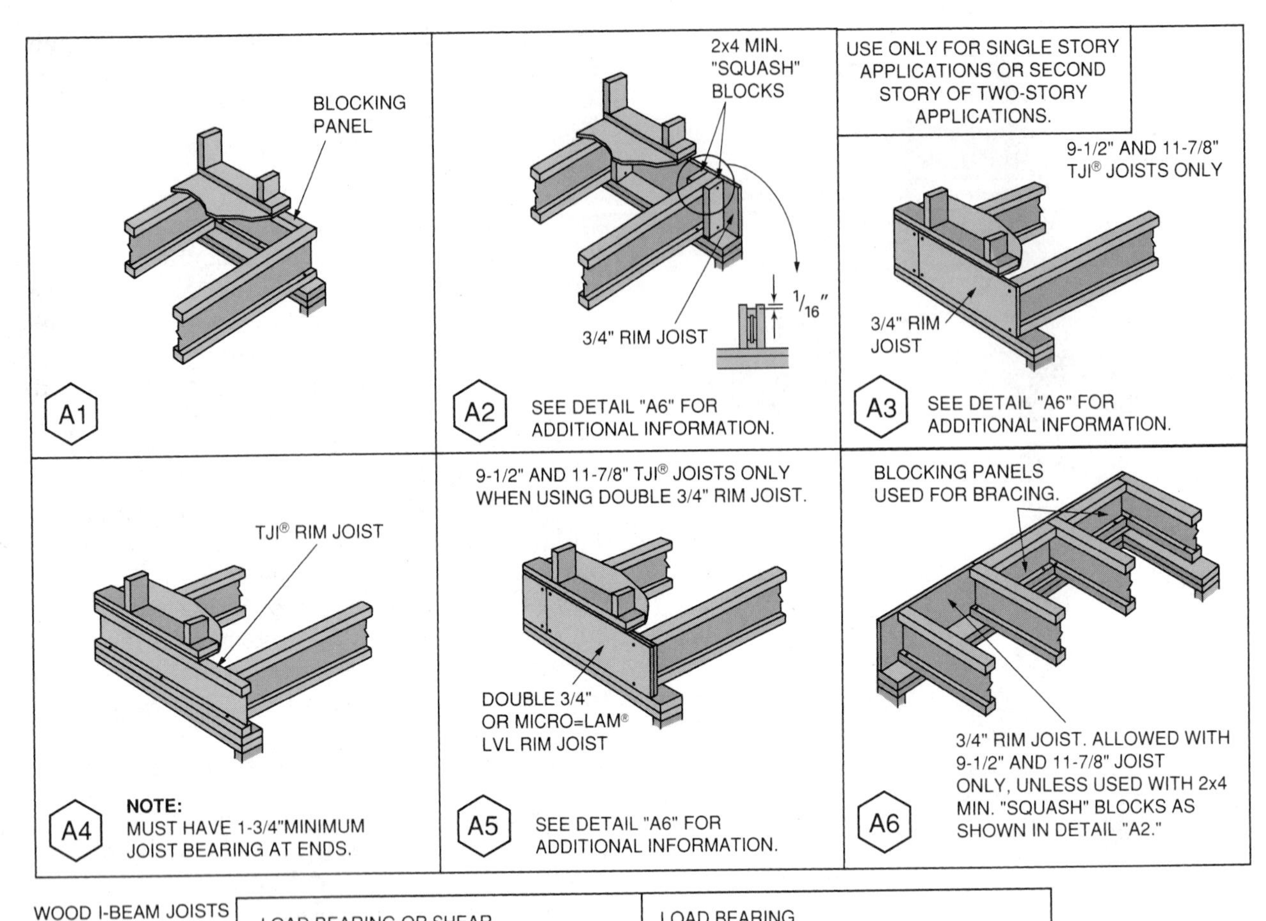

LOAD BEARING OR SHEAR WALL ABOVE (MUST STACK OVER WALL BELOW).
BLOCKING PANEL
GIRDER
NOTE:
WEB STIFFENERS MAY BE REQUIRED.

LOAD BEARING WALL ABOVE (MUST STACK OVER WALL BELOW).
2x4 MIN. "SQUASH" BLOCKS
GIRDER
NOTE:
WEB STIFFENERS MAY BE REQUIRED.
1/16"
SQUASH BLOCK DETAIL

NOTE:
WEB STIFFENERS MAY BE REQUIRED.

GIRDER
TOP MOUNT HANGER
FACE MOUNT HANGER
NOTE:
WEB STIFFENERS ARE REQUIRED IF THE SIDES OF THE HANGER DO NOT LATERALLY SUPPORT THE JOIST TOP FLANGE.

CHECK MANUFACTURERS' APPLICATION RECOMMENDATIONS

Fig. 32-11 Wood I-beam framing details at the sill.

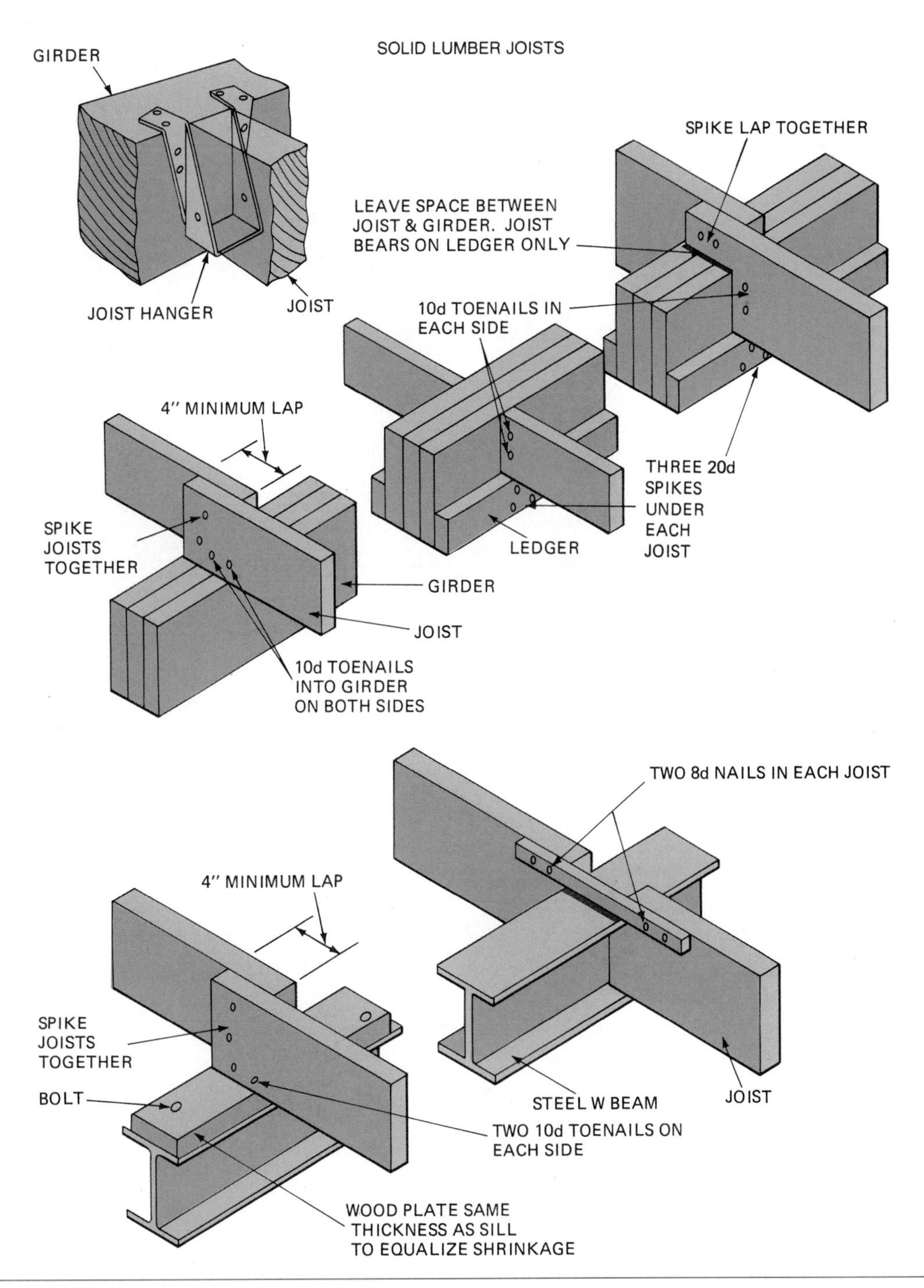

Fig. 32-12 Joist framing details at the girder.

Identify the layout marks that are not for full-length floor joists. Shortened floor joists at the ends of floor openings are called **tail joists.** They are usually identified by changing the *X* to a *T*. Lay out for partition supports, or wherever doubled floor joists are required (Fig. 32-17). Check the mechanical drawings to make adjustments in the framing to allow for the installation of mechanical equipment.

Lay out the floor joists on the girder and on the opposite wall. If the joists are in-line, *X*s are made on the same side of the mark on both the girder and the sill plate on the opposite wall. If the joists are lapped, an *X* is placed on both sides of the mark at the girder and on the opposite side of the mark on the other wall (Fig. 32-18).

Installing Floor Joists

Stack the necessary number of full-length floor joists at intervals along both walls. Each joist is

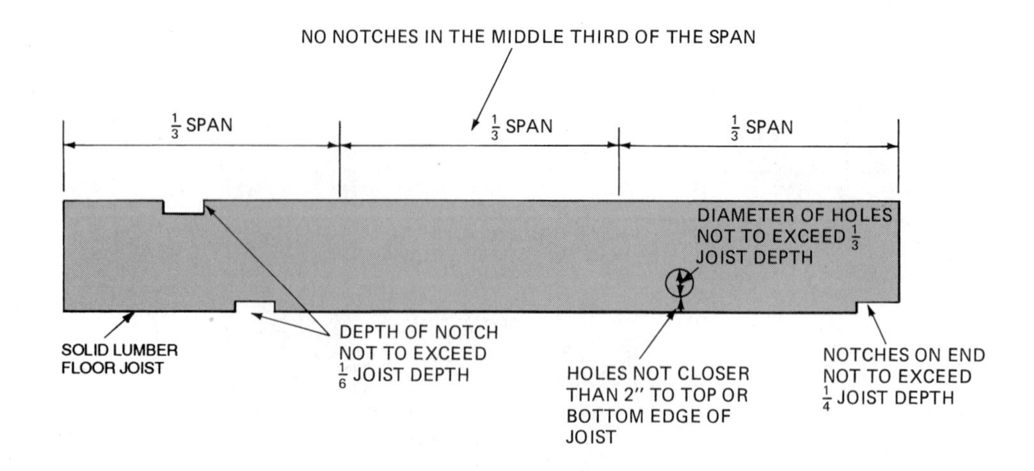

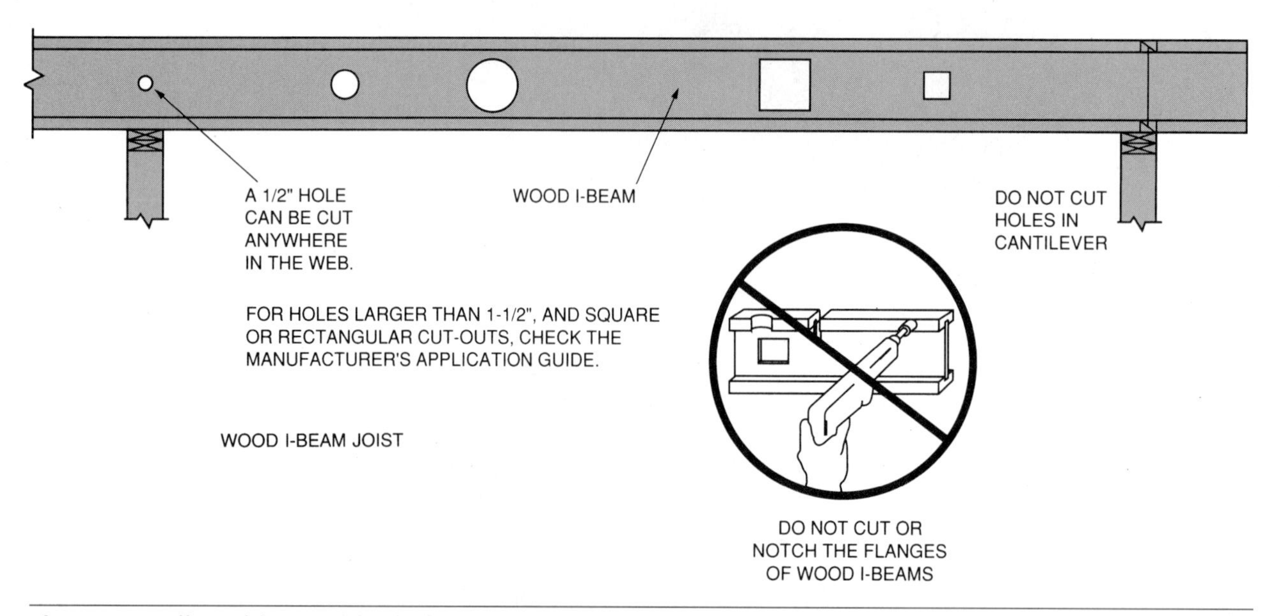

Fig. 32-13 Allowable notches, holes, and cutouts in floor joists.

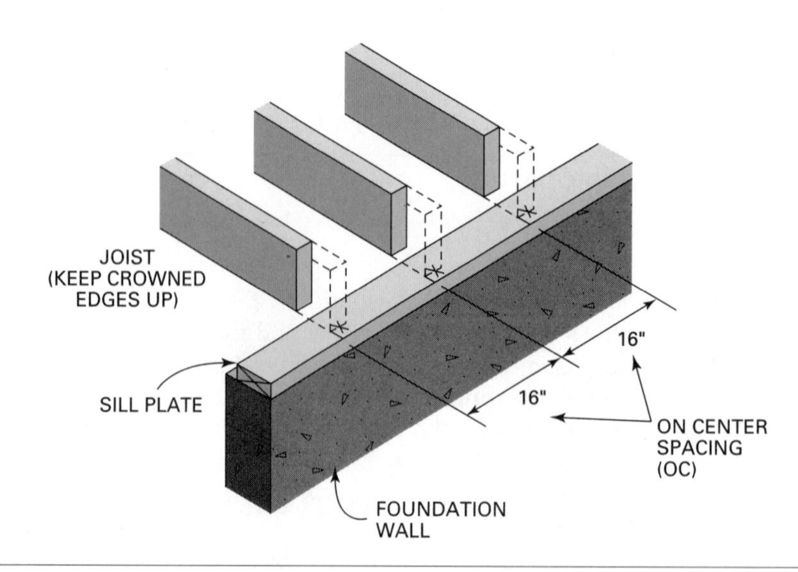

Fig. 32-14 A line marks this side of a joist. An *X* tells on which side of the line it is placed.

STEP BY STEP
PROCEDURES

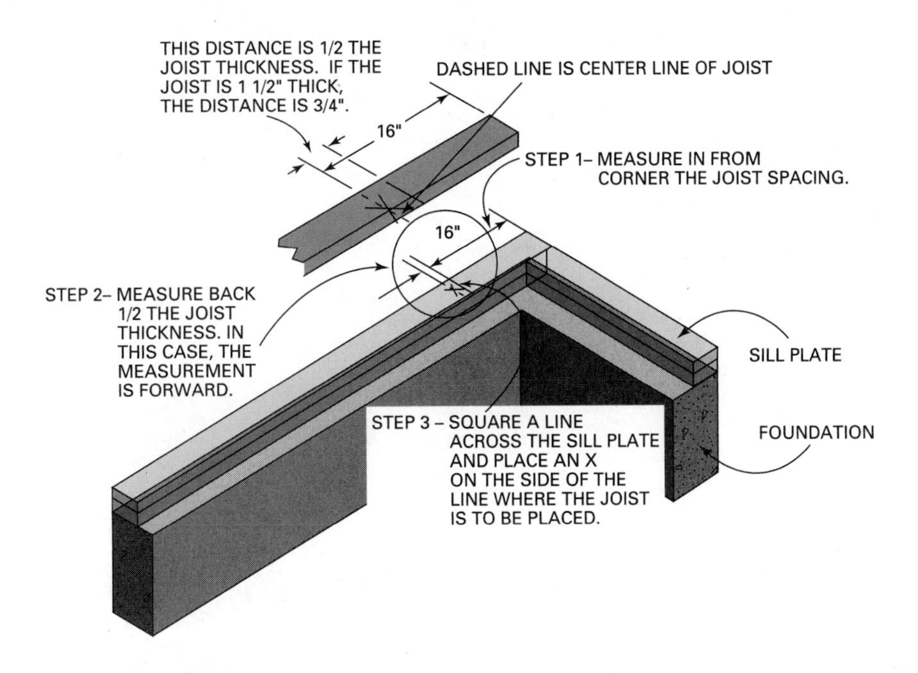

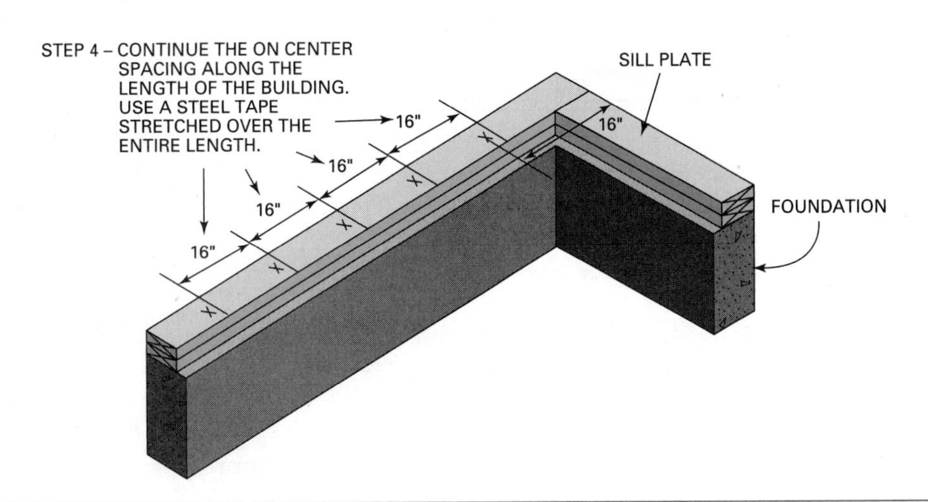

Fig. 32-15 Laying out the location of floor joists on the sill plate.

carefully sighted along its length by eye. Any joist with a severe crook or other warp should not be used (Fig. 32-19). Joists are installed with the crowned edge up.

Keep the end of the floor joist in from the outside edge of the sill plate by the thickness of the band joist. Toenail the joists to the sill and girder with 10d or 3-inch common nails. Spike the joists together if they lap at the girder (Fig. 32-20). When all floor joists are in position, they are sighted by eye from end to end and straightened. They may be held straight by strips of 1 × 3s tacked to the top of the joists about in the middle of the joist span.

Wood I-beams are installed using standard tools. They can be easily cut to any required length at the job site. A minimum bearing of 1 3/4 inches is

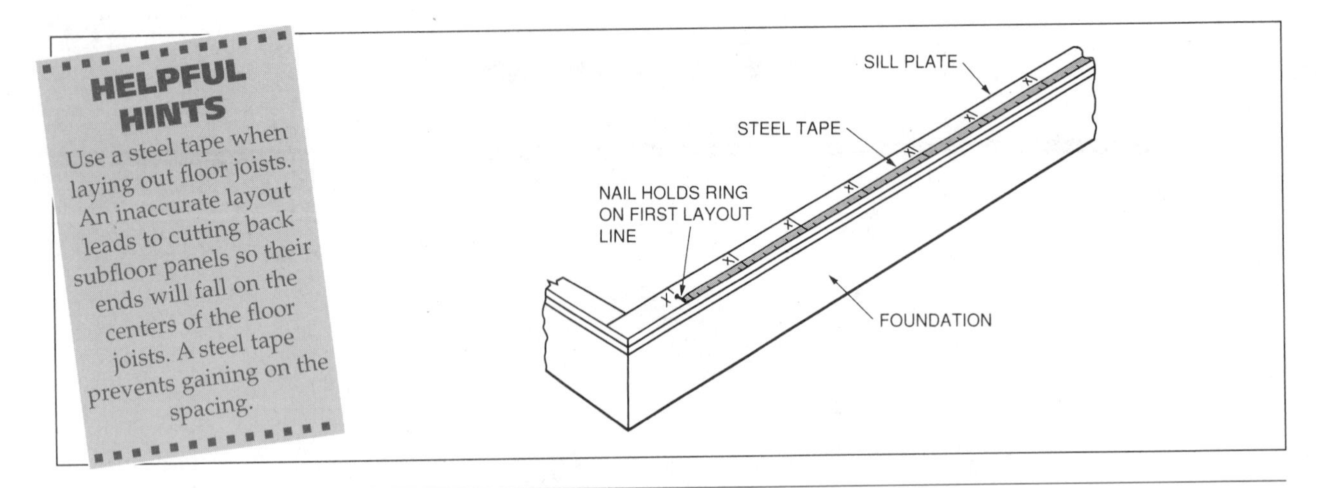

Fig. 32-16 It is important to use a steel tape for layout.

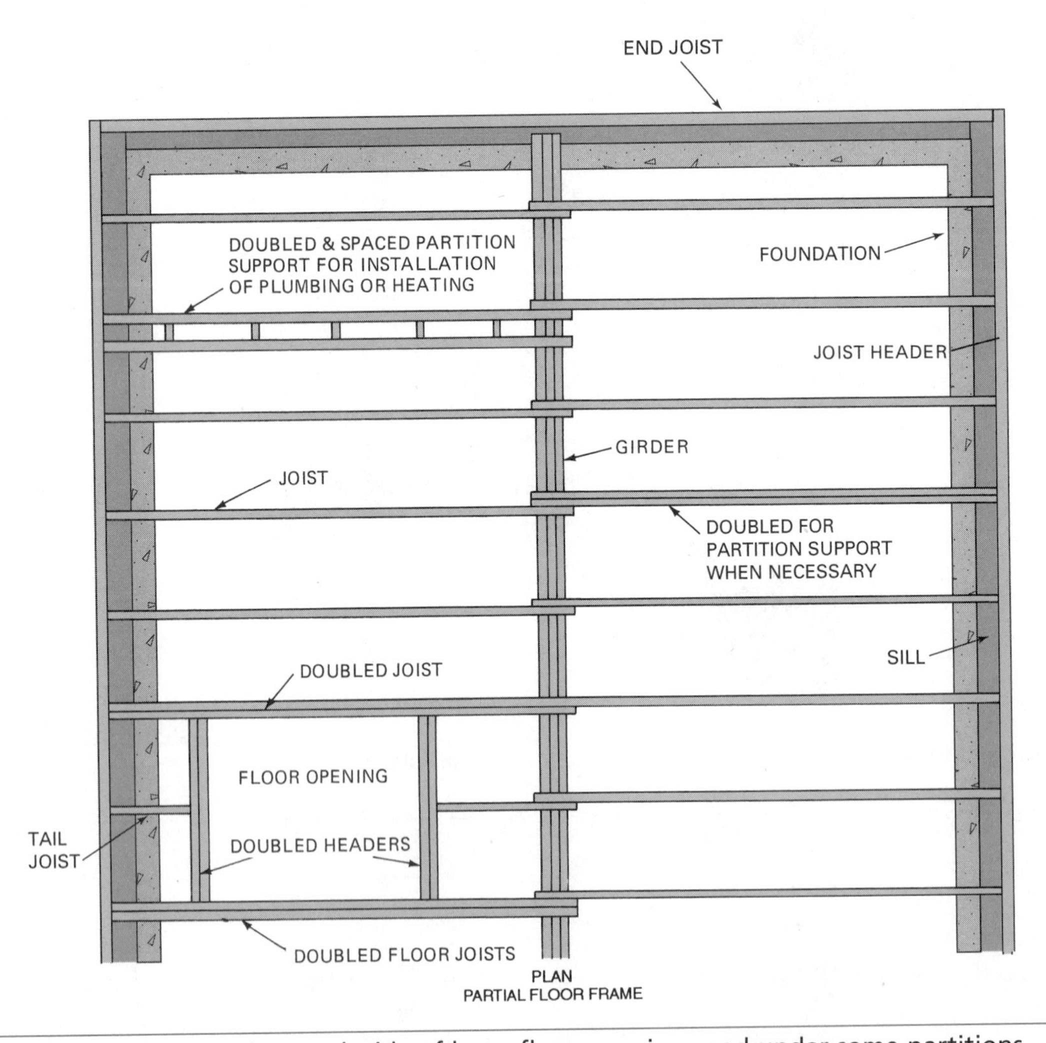

Fig. 32-17 Joists are doubled on each side of large floor openings and under some partitions.

required at joist ends and 3 1/2 inches over the girder. The wide, straight wood flanges on the joist make nailing easier, especially with pneumatic framing nailers (Fig. 32-21). Nail joists at each bearing with one 8d or 10d nail on each side. Keep nails at least 1 1/2 inches from the ends to avoid splitting.

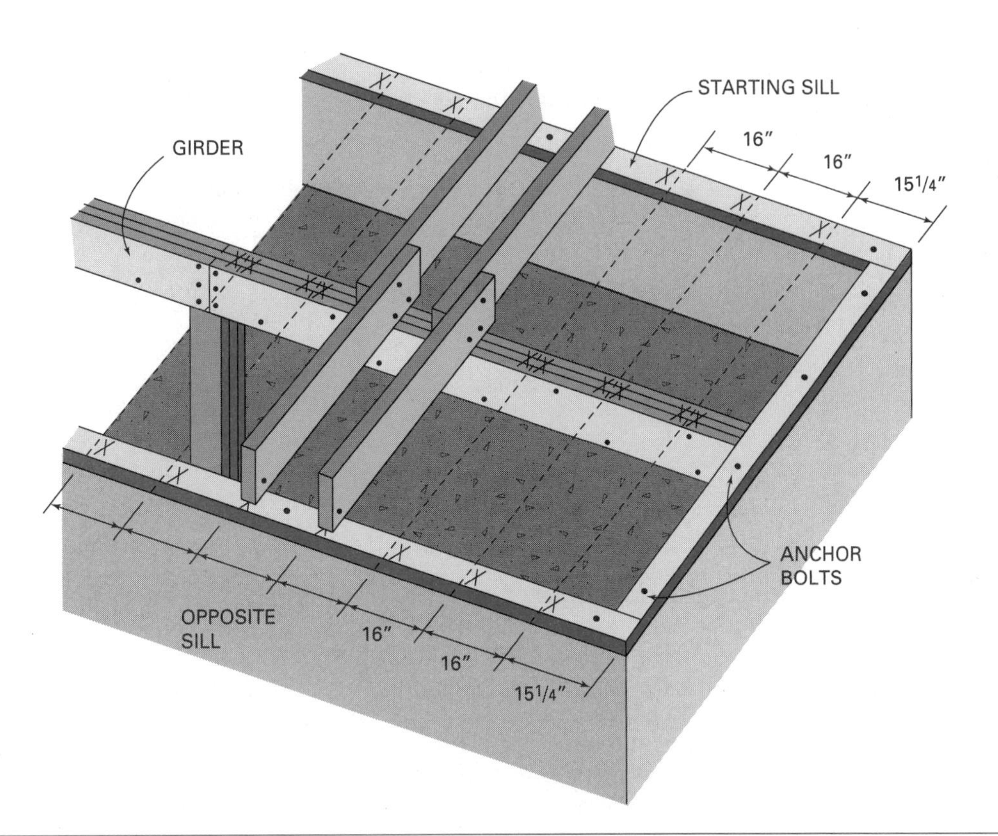

Fig. 32-18 Lay out lines for floor joists are the same whether marked on the girder or sills. Only the *X*s are different, being placed on different sides of the line depending on where the joist is to be located.

Fig. 32-19 Stack floor joists against both walls. Sight each joist. Install joists with the crowned edge up.

Fig. 32-20 A floor frame with joists lapped over the girder.

Fig. 32-21 Installing wood I-beam floor joists. Notice how extremely long lengths are easily handled. (*Courtesy of Trus Joist MacMillan*)

Doubling Floor Joists

For added strength, doubled floor joists must be securely fastened together. Their top edges must be even. In most cases, the top edges do not lie flush with each other. They must be brought even before they can be spiked together.

To bring them flush, toenail down through the top edge of the higher one, at about the center of their length. At the same time squeeze both together

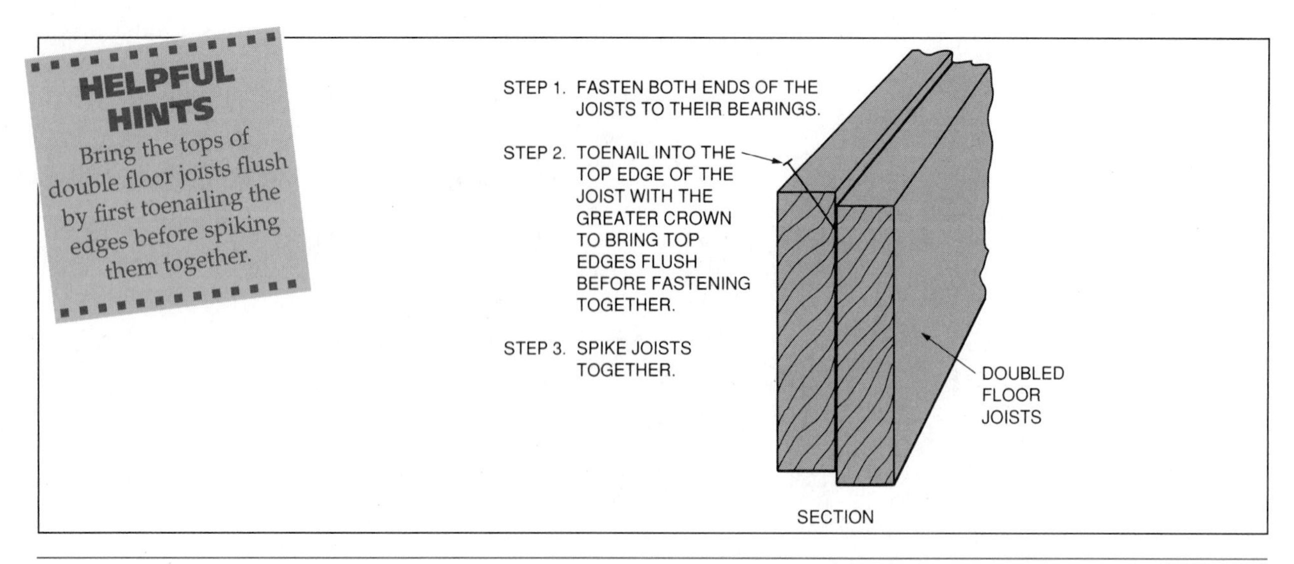

Fig. 32-22 Technique for aligning the top edges of dimension lumber.

tightly by hand. Use as many toenails as necessary, spaced where needed, to bring the top edges flush (Fig. 32-22). Usually no more than two or three nails are needed. Then, fasten the two pieces securely together. Drive nails from both sides, staggered from top to bottom, about 2 feet apart. Angle nails slightly so they do not protrude.

Framing Floor Openings

Large openings in floors should be framed before floor joists are installed. This is because room is needed for end nailing. To frame an opening in a floor, first fasten the **trimmer joists** in place. Trimmer joists are full-length joists that run along the inside of the opening. Mark the location of the *header joists* on the trimmers. Header joists are members of the opening that run at right angles to the floor joists.

Cut four *header joists* to length by taking the measurement at the sill between the trimmers. Taking the measurement at the sill where the trimmers are fastened, rather than at the opening, is standard practice. A measurement between trimmers taken at the opening may not be accurate. There may be bow in the trimmer joists (Fig. 32-23).

Place two headers, one for each end of the opening, on the sill between the trimmers. Transfer the layout of the tail joists on sill to the headers. Fasten the first header on each end of the opening in

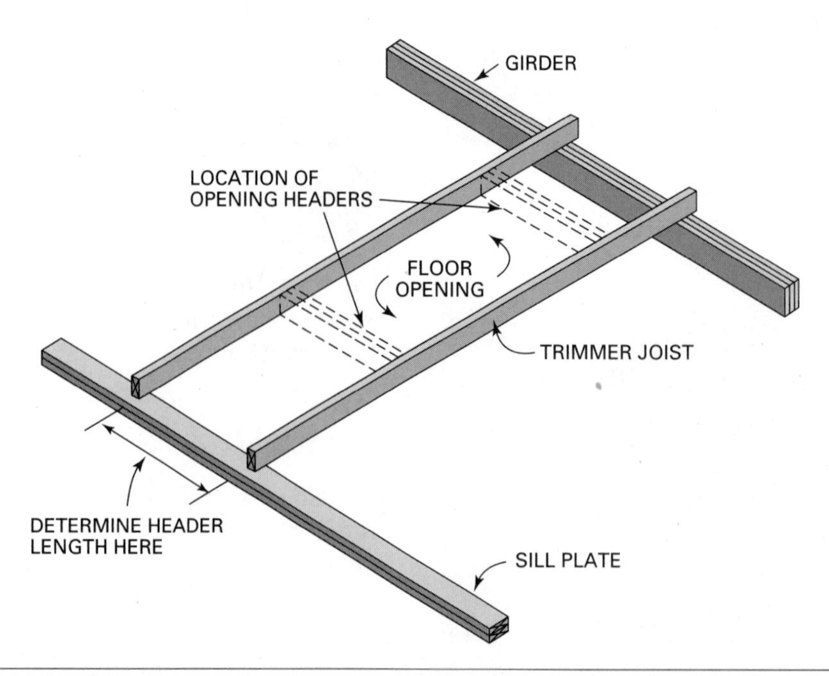

Fig. 32-23 The length of the header for the opening should be determined at the sill plate.

position by driving nails through the side of the trimmer into the ends of the headers. Be sure the first header is the header that is farthest from the floor opening. Fasten the tail joists in position. Double up the headers. Finally, double up the trimmer joists. Figure 32-24 shows the sequence of operations used to frame a floor opening. This particular sequence allows you to end nail the members rather than toenailing them. End nails are sufficient for headers up to 6 feet. Use joist hangers on headers

STEP BY STEP

PROCEDURES

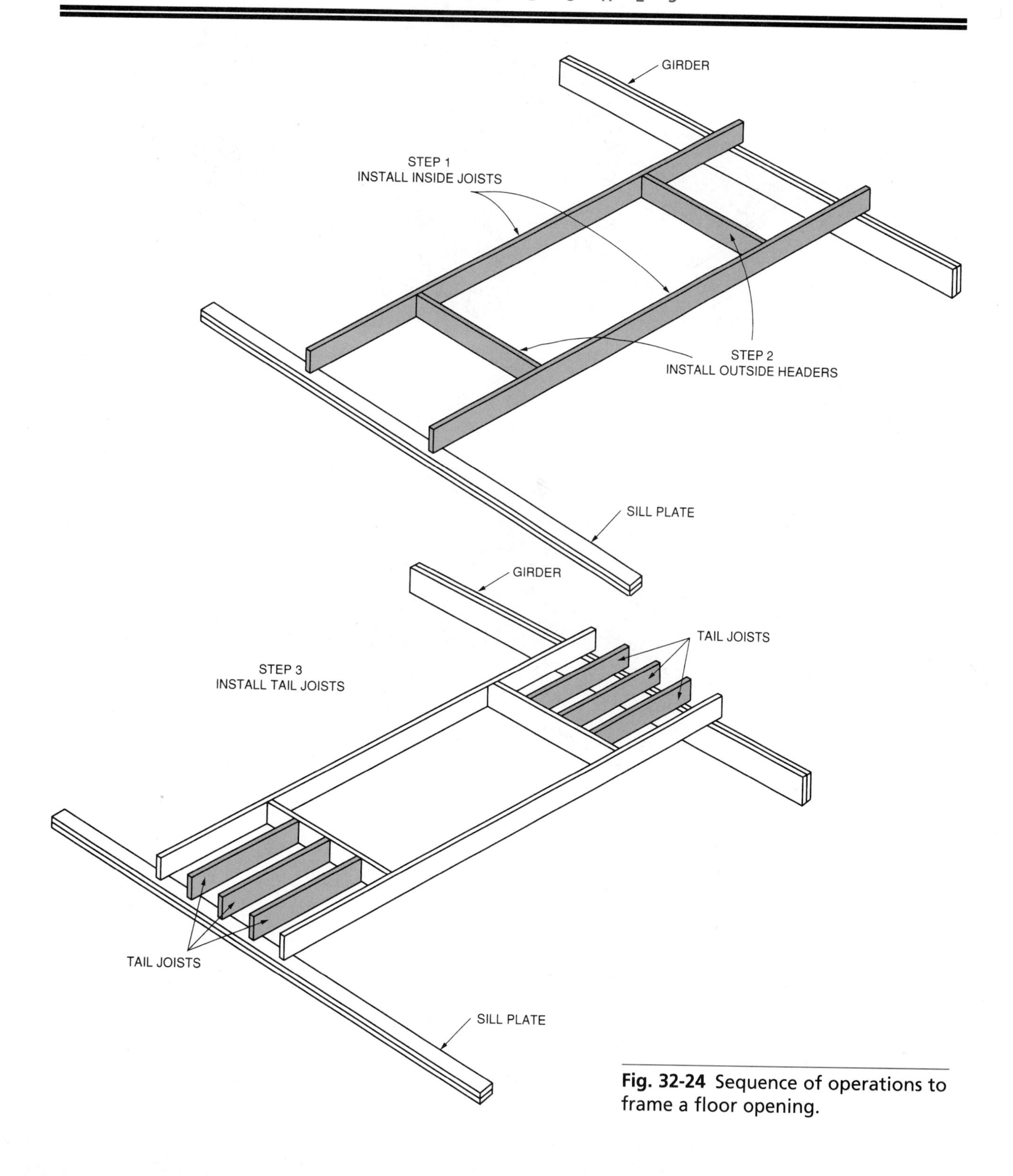

Fig. 32-24 Sequence of operations to frame a floor opening.

STEP BY STEP
PROCEDURES

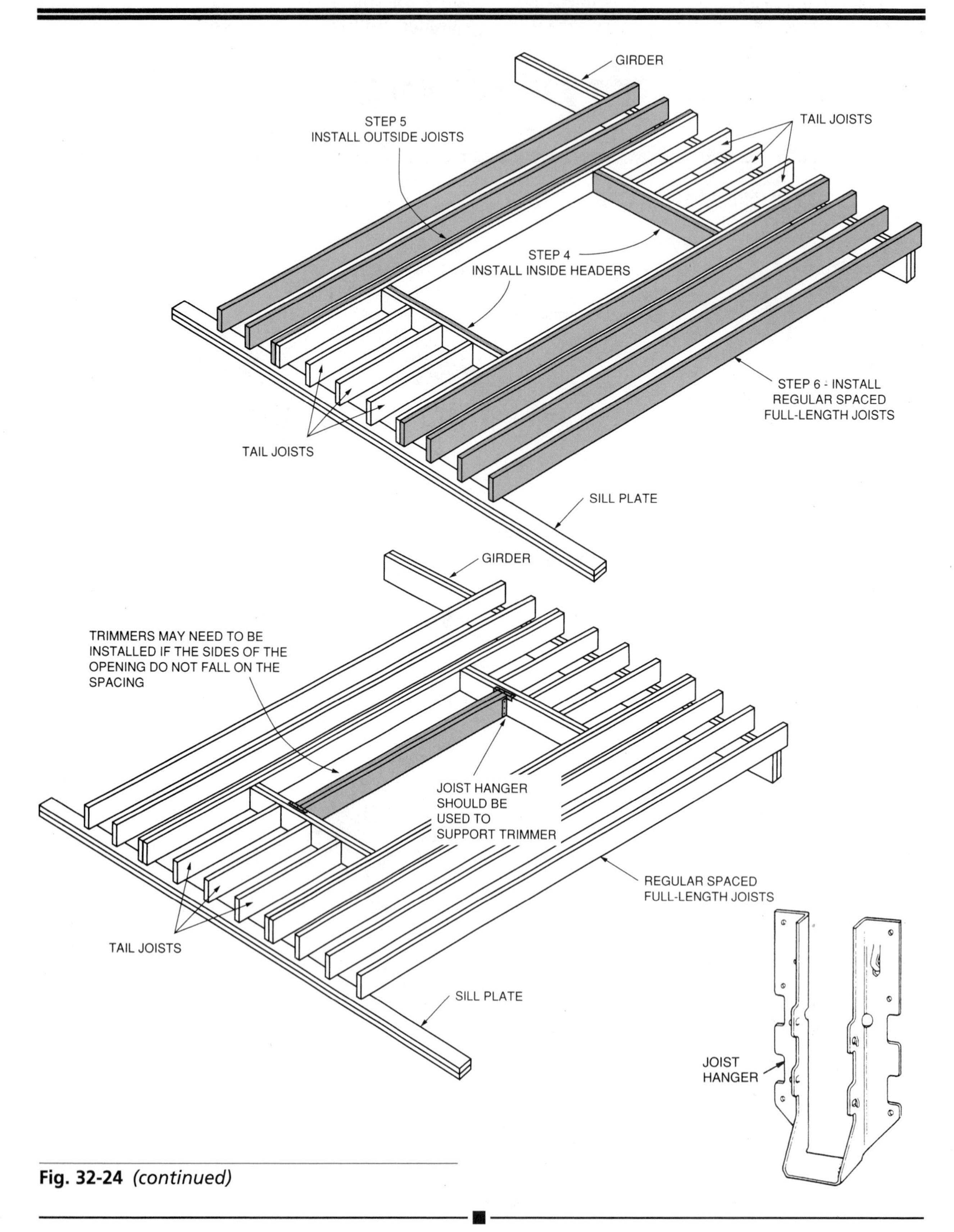

Fig. 32-24 *(continued)*

over 6 feet long and on tail joists over 12 feet long. Joist hangers must be installed with the proper amount, size, and type of nails. Roofing nails are not acceptable for joist hangers.

Installing the Band Joist. After all the openings have been framed and all floor joists are fastened, install the band joist. This closes in the ends of the floor joists. Band joists may be lumber of the same size as the floor joists. They also may be a single or double layer of 3/4-inch laminated veneer lumber.

Fasten the band joist into the end of each floor joist. If wood I-beams are used as floor joists, drive one nail into the top and bottom flange. The band joist is also toenailed to the sill plate at about 16-inch intervals.

Bridging

Bridging is installed in rows between floor joists at intervals not exceeding 8 feet. For instance, floor joists with spans 8 to 16 feet need one row of bridging near the center of the span. Its purpose is to distribute a concentrated load on the floor over a wider area. Although codes may not require bridging, many builders install it because of customary practice.

Bridging may be solid wood, wood cross-bridging, or metal cross-bridging (Fig. 32-25). Usually solid wood bridging is the same size as the floor joists. It is installed in an offset fashion to permit end nailing.

Wood cross-bridging should be at least nominal 1 × 3 lumber with two 6d nails at each end. It is placed in double rows that cross each other in the joist space.

Metal cross-bridging is available in different lengths for particular joist size and spacing. It is usually made of 18-gauge steel, and is 3/4 inch wide. It comes in a variety of styles. It is applied in a way similar to that used for wood cross-bridging.

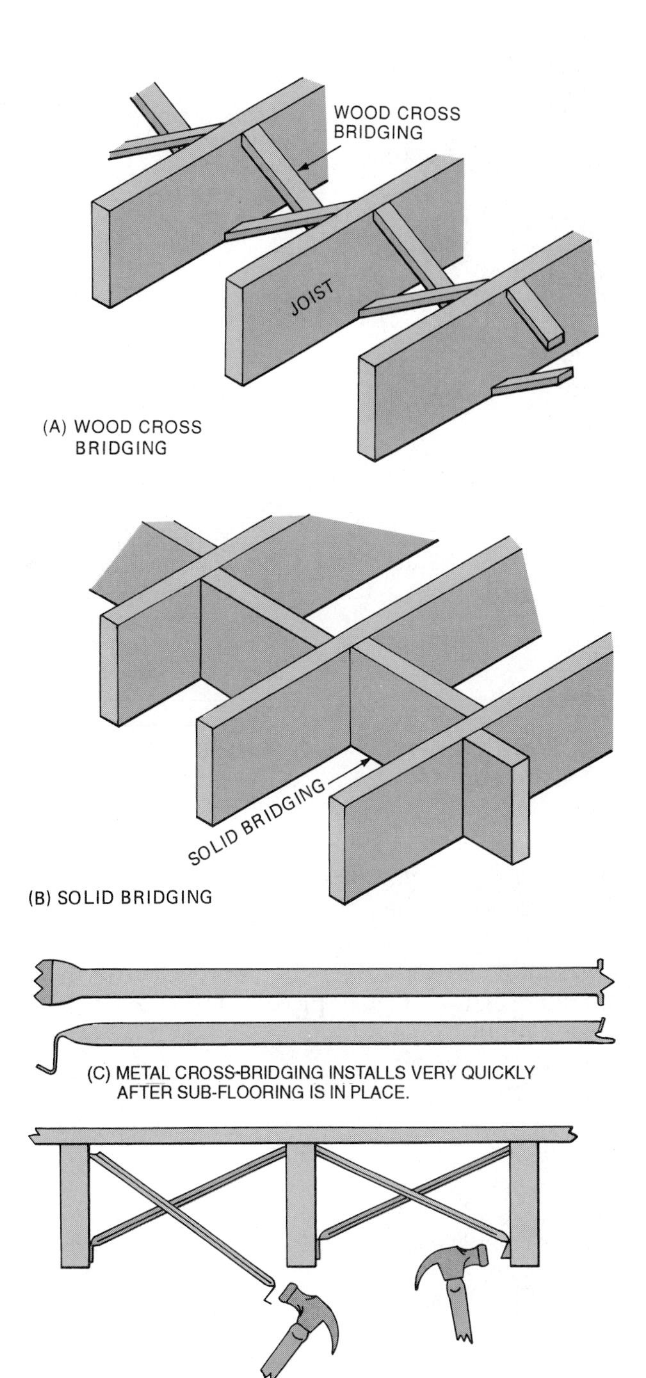

Fig. 32-25 Types of bridging.

Laying Out and Cutting Wood Cross-Bridging

Wood cross-bridging may be laid out using a framing square. Determine the actual distance between floor joists and the actual depth of the joist. For example, 2 × 10 floor joists 24 inches OC measure 22 1/2 inches between them. The actual depth of the joist is 9 1/4 inches.

Hold the framing square on the edge of a piece of bridging stock. Make sure the 9 1/4-inch mark of the tongue lines up with the upper edge of the stock. Also make sure the 22 1/2-inch mark of the blade lines up with the lower edge of the stock. Mark lines along the tongue and blade across the stock.

Rotate the square, keeping the same face up. Mark along the tongue (Fig. 32-26). The bridging may then be cut using a power miter box or radial arm saw. Tilt the blade and use a stop set to cut duplicate lengths.

Installing Bridging

Determine the centerline of the bridging. Snap a chalk line across the tops of the floor joist from one end to the other. Square down from the chalk line to the bottom edge of the floor joists on both sides.

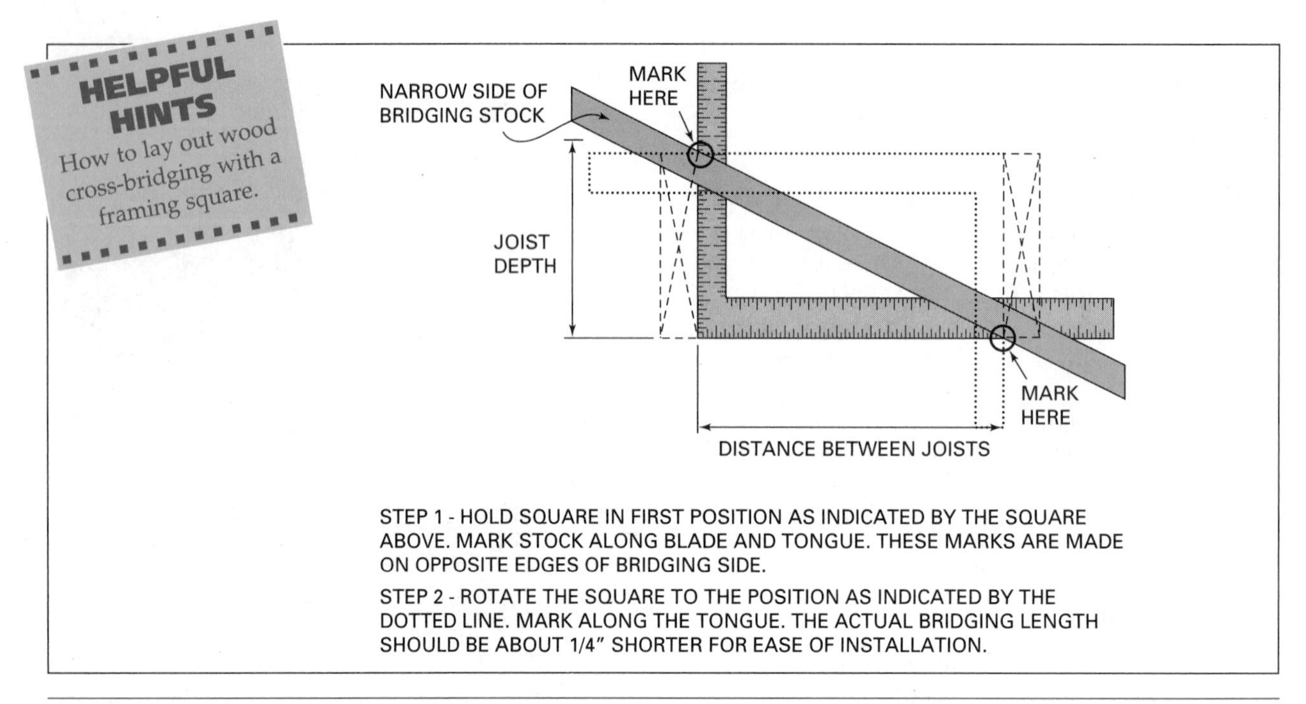

Fig. 32-26 Proper layout of cross-bridging using a framing square to mark the bridging on its opposite edges.

Solid Wood Bridging. To install solid wood bridging, cut the pieces to length. Install pieces in every other joist space on one side of the chalk line. Fasten the pieces by spiking through the joists into their ends. Keep the top edges flush with the floor joists. Install pieces in the remaining spaces on the opposite side of the line in a similar manner (Fig. 32-27).

Wood Cross-Bridging. To install wood cross-bridging, start two 6d nails in one end of the bridging. Fasten it flush with the top of the joist on one side of the line. Nail only the top end. The bottom ends are not nailed until the subfloor is fastened down.

Within the same joist cavity or space, fasten another piece of bridging to the other joist. Make sure

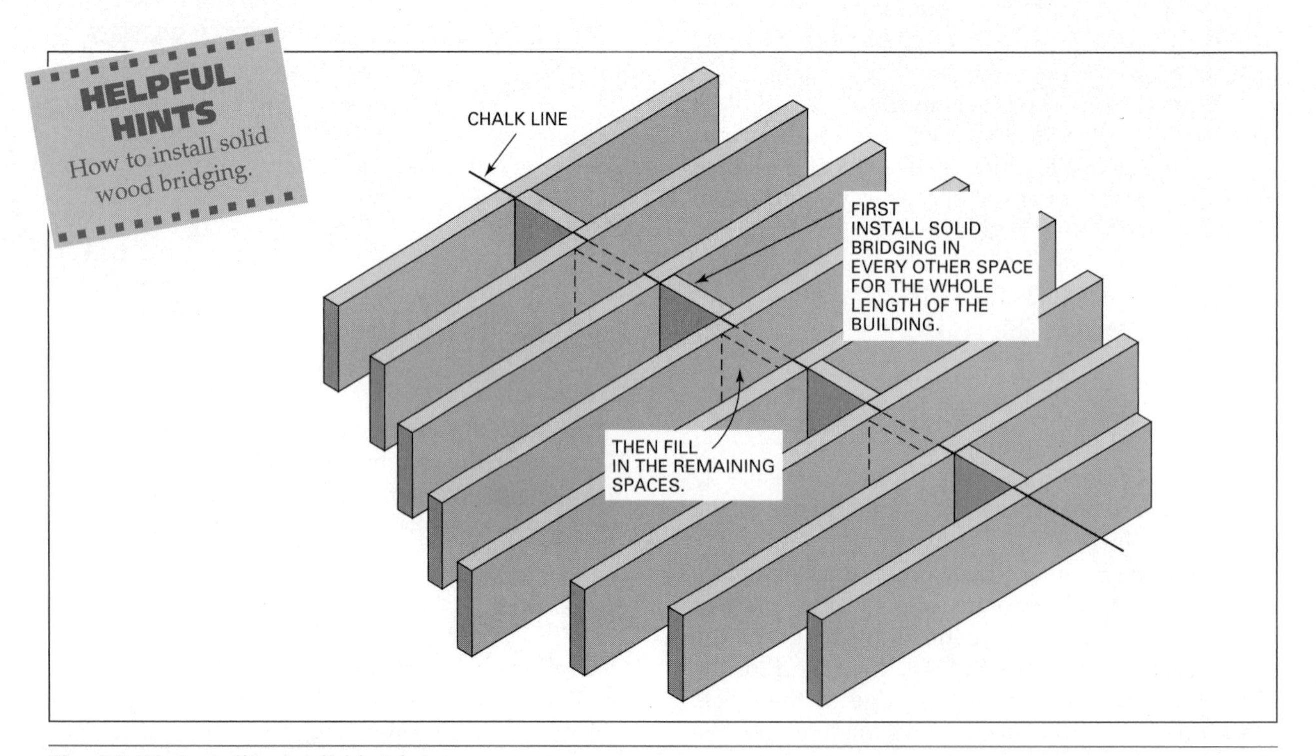

Fig. 32-27 Installing solid bridging.

STEP BY STEP

PROCEDURES

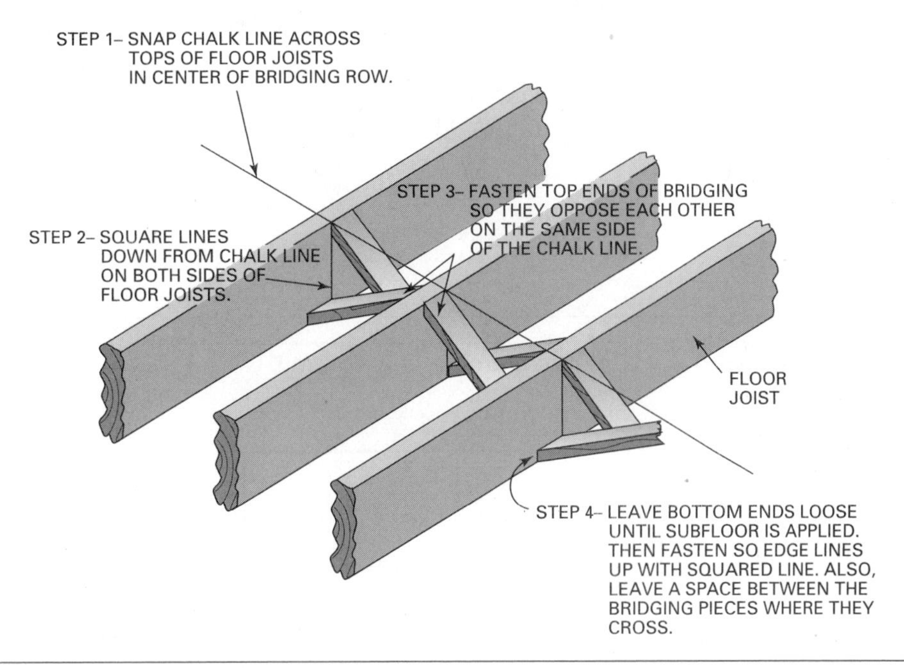

Fig. 32-28 Installing cross-bridging.

it is flush with top of the joist and positioned on the other side of the chalk line. Also leave a space between the bridging pieces where they form the X to minimize floor squeaks. Continue installing bridging in the other spaces, but alternate so that the top ends of the bridging pieces are opposite each other where they are fastened to same joist (Fig. 32-28).

Metal Cross-Bridging. Metal cross-bridging is fastened in a manner similar to that used for wood cross-bridging. The method of fastening may differ according to the style of the bridging. Usually the bridging is fastened to the top of the joists through predrilled holes in the bridging. Because the metal is thin, nailing to the top of the joists does not interfere with the subfloor.

Some types of metal cross-bridging have steel prongs that are driven into the side of the floor joists. This bridging can be installed from below to layout lines made previous to the installation of the subfloor.

Columns

Girders may be supported by framed walls, wood posts, or steel columns. Usually, columns of steel pipe filled and reinforced with concrete are used. These are commonly called *lally columns* (Fig. 32-29). Metal plates are used at the top and bottom of

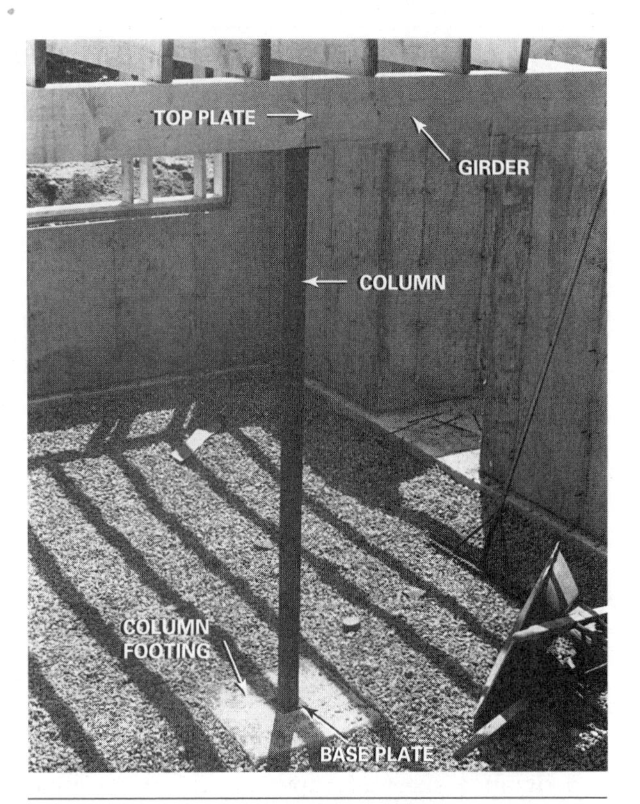

Fig. 32-29 Columns of steel pipe filled and reinforced with concrete are commonly called lally columns.

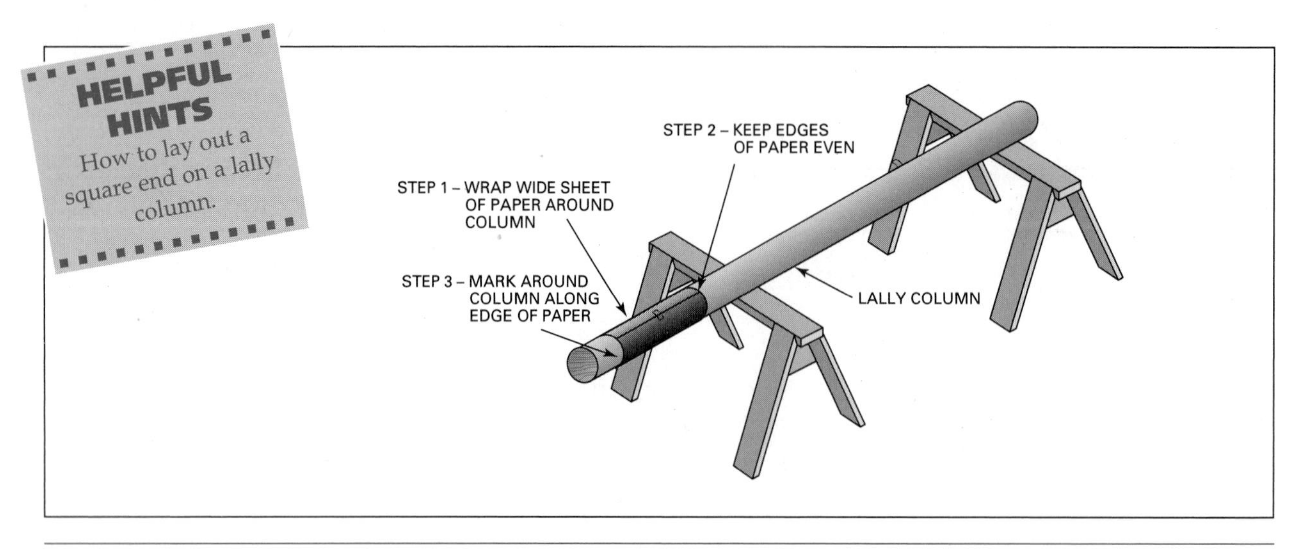

Fig. 32-30 Technique for laying out a square cut on a lally column.

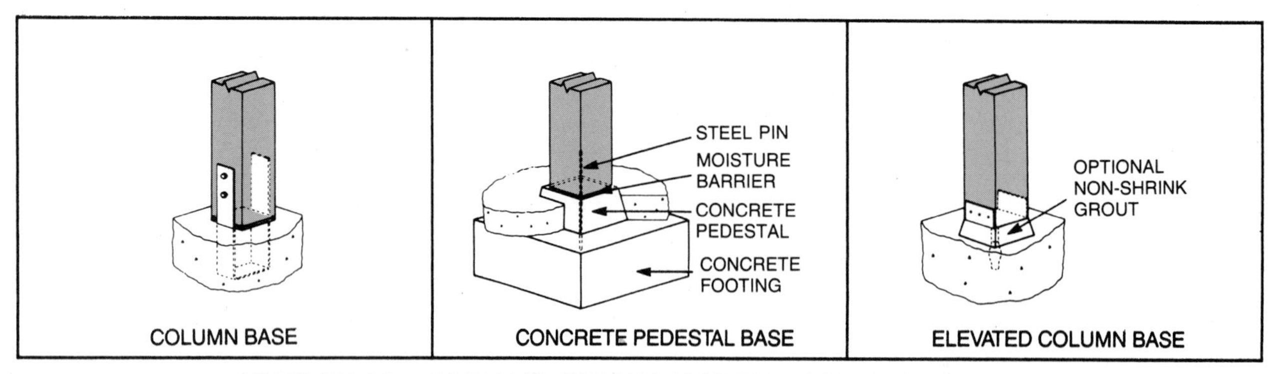

Fig. 32-31 The bottom of wood posts usually rests on a pedestal-type footing to keep them above the grade.

the columns to distribute the load over a wider area. The plates have predrilled holes so that they may be fastened to the girder. Notched sections prevent the columns from slipping off the plates. Column size should be determined from the blueprints.

Installing Columns

After the floor joists are installed and before any more weight is placed on the floor, the temporary posts supporting the girder are replaced with permanent posts or columns. Straighten the girder by stretching a line from end to end. Measure accurately from the column footing to the bottom of the girder. Hold a strip of lumber on the column footing. Mark it at the bottom of the girder. Transfer this mark to the column. Deduct the thickness of the top and bottom plate.

To mark around the column so it has a square end, wrap a sheet of paper around it. Keeping the edges even, mark along the edge of the paper (Fig. 32-30).

Cut through the metal along the line using a hacksaw. Snap the column waste off by hitting the end with a hammer. Trim off any protruding concrete. Install the columns in a plumb position under the girder and centered on the footing. Fasten the top plates to the girder with lag screws. If the girder is steel, then holes must be drilled. The plates are then bolted to the girder, or they may be welded to the girder. The bottoms of the columns are held in place when the finish concrete basement floor is placed around them. If there is to be no floor, then the bottom plate must be anchored to the footing.

Wood posts are installed in a similar manner, except their bottoms are placed on a pedestal footing (Fig. 32-31).

Subflooring

Subflooring is used over joists to form a working platform. This is also a base for finish flooring, such as hardwood flooring, or underlayment for carpet or resilient tiles. APA-Rated Sheathing Exposure 1 is generally used for subflooring in a two-layer floor system. Specifications for selecting and fastening panels is shown in Figure 32-32. APA-Rated

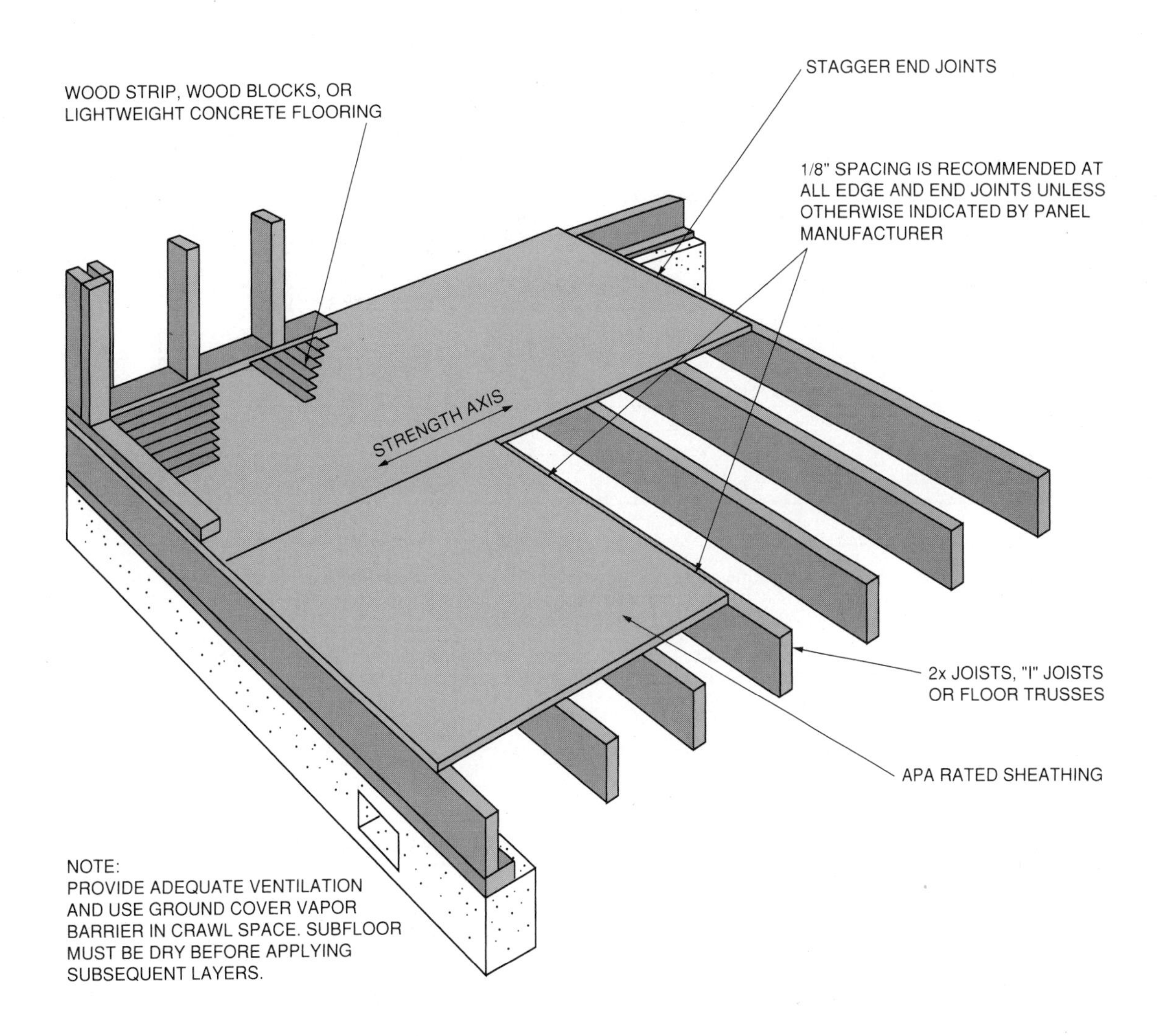

APA Panel Subflooring (APA RATED SHEATHING)[(a)]

Panel Span Rating	Minimum Panel Thickness (in.)	Maximum Span (in.)	Nail Size & Type[(e)]	Maximum Nail Spacing (in.)	
				Supported Panel Edges	Intermediate Supports
24/16	7/16	16	6d common	6	12
32/16	15/32	16[(b)]	8d common[(c)]	6	12
40/20	19/32	20[(b)(d)]	8d common	6	12
48/24	23/32	24	8d common	6	12
60/32	7/8	32	8d common	6	12

(a) For subfloor recommendations under ceramic tile and under gypsum concrete, contact manufacturerer of floor topping.

(b) Span may be 24 inches if 3/4-inch wood strip flooring is installed at right angles to joists.

(c) 6d common nail permitted if panel is 1/2 inch or thinner.

(d) Span may be 24 inches if a minimum 1-1/2 inches of lightweight concrete is applied over panels.

(e) Other code-approved fasteners may be used.

Fig. 32-32 APA Panel Subflooring selection and fastening guide. (*Courtesy of American Plywood Association*)

APA Rated Sturd-I-Floor (a)

Span Rating (Maximum Joist Spacing) (in.)	Minimum Panel Thickness(b) (in.)	Fastening: Glue-Nailed(c)			Fastening: Nailed-Only		
		Nail Size and Type	Maximum Spacing (in.)		Nail Size and Type	Maximum Spacing (in.)	
			Supported Panel Edges	Intermediate Supports		Supported Panel Edges	Intermediate Supports
16	19/32	6d ring- or screw-shank (d)	12	12	6d ring- or screw-shank	6	12
20	19/32	6d ring- or screw-shank (d)	12	12	6d ring- or screw-shank	6	12
24	23/32	6d ring- or screw-shank (d)	12	12	6d ring- or screw-shank	6	12
	7/8	8d ring- or screw-shank (d)	6	12	8d ring- or screw-shank	6	12
32	7/8	8d ring- or screw-shank (d)	6	12	8d ring- or screw-shank	6	12
48	1-3/32	8d ring- or screw-shank (e)	6	(f)	8d ring- or screw-shank (e)	6	(f)

(a) Special conditions may impose heavy traffic and concentrated loads that require construction in excess of the minimums shown.

(b) Panels in a given thickness may be manufactured in more than one Span Rating. Panels with a Span Rating greater than the actual joist spacing may be substituted for panels of the same thickness with a Span Rating matching the actual joist spacing. For example, 19/32-inch-thick Sturd-I-Floor 20 oc may be substituted for 19/32-inch-thick Sturd-I-Floor 16 oc over joists 16 inches on center.

(c) Use only adhesives conforming to APA Specification AFG-01, applied in accordance with the manufacturer's recommendations. If OSB panels with sealed surfaces and edges are to be used, use only solvent-based glues; check with panel manufacturer.

(d) 8d common nails may be substituted if ring- or screw-shank nails are not available.

(e) 10d common nails may be substituted with 1-1/8-inch panels if supports are well seasoned.

(f) Space nails maximum 6 inches for 48-inch spans and 12 inches for 32-inch spans.

Fig. 32-33 APA-Rated Sturdi-I-Floor selection and fastening guide. (*Courtesy of American Plywood Association*)

Sturd-I-Floor panels are used when a single-layer subfloor and under-layment system is desired. Blocking is required under the joints of these panels unless tongue-and-groove edges are used. Figure 32-33 is a selection and fastening guide for APA Sturd-I-Floor panels.

Applying Plywood Subflooring

Starting at the corner from which the floor joists were laid out, measure in 4 feet. It should be noted here that tongue-and-groove plywood subfloor is only 47 1/2" wide. Snap a line across the tops of the floor joists from one end to the other. Start with a full panel. Fasten the first row to the chalk line (Fig. 32-34) and align the joists to the correct spacing before nailing the panel. Leave a 1/16-inch space at all panel end joints to allow for expansion.

Start the second row with a half-sheet to stagger the end joints. Continue with full panels to finish the row. Leave a 1/8-inch space between panel edges. All end joints are made over joists.

Continue laying and fastening plywood sheets in this manner until the entire floor is covered (Fig. 32-35). Leave out sheets where there are to be openings in the floor. Snap chalk lines across the edges and ends of the building. Trim overhanging plywood with a portable electric circular saw.

Estimating Materials

Floor Joists. To determine the number of floor joists needed in a floor frame, divide the length of the building by the spacing and add one. Multiply by the number of rows of floor joists. Add the number needed for doubling and for band joists.

Bridging. To determine the total **linear feet** of wood cross-bridging needed, multiply the length of the building by 3 for each row of bridging. Linear feet is a measurement of length.

Fig. 32-34 Applying a rated panel subfloor.

STEP BY STEP PROCEDURES

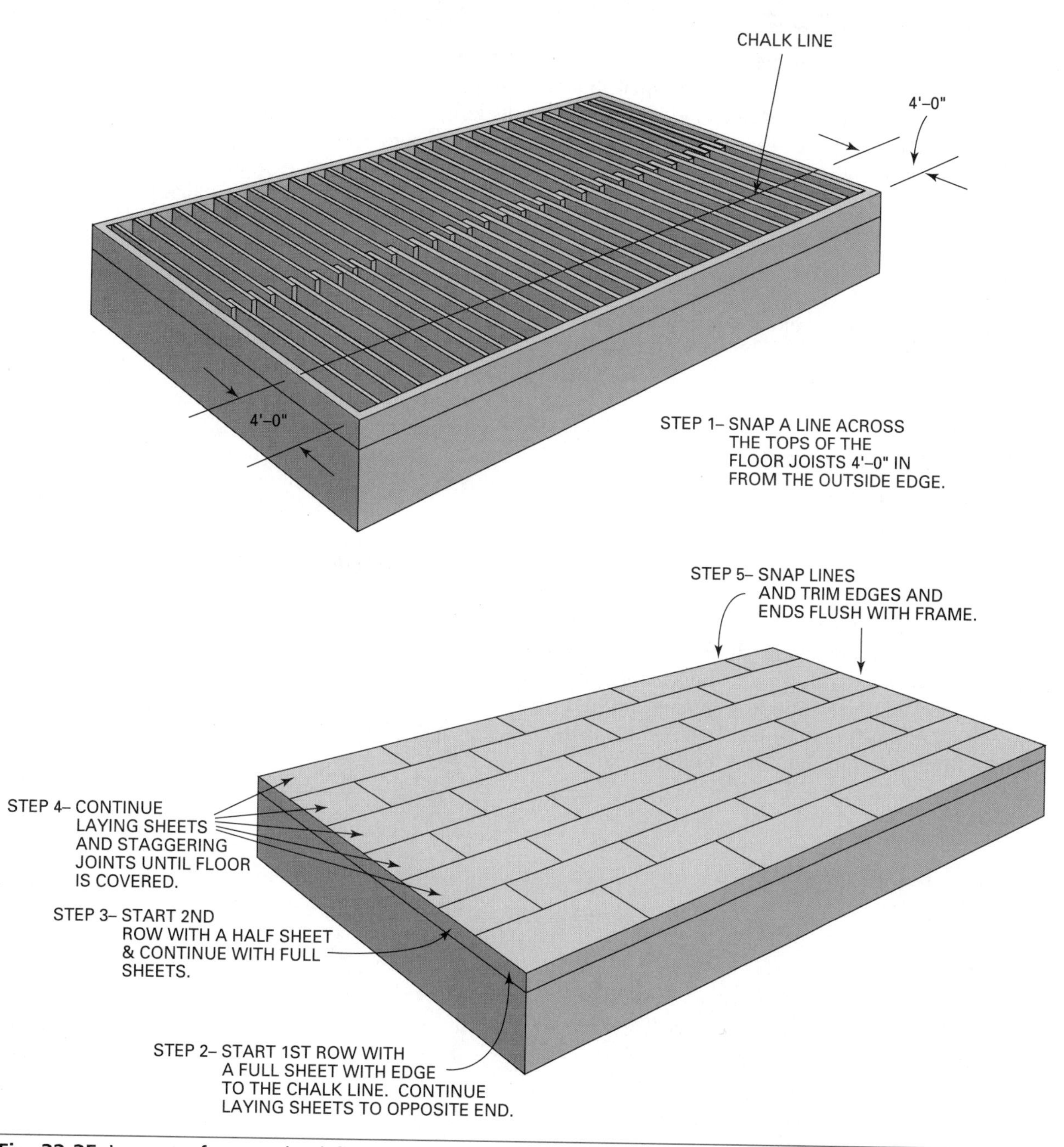

Fig. 32-35 Layout of a panel subfloor.

Panel Subfloor. To determine the number of rated panels of subflooring required, divide the floor area by 32. A rectangular floor area is found by multiplying its width by its length. The area of other shapes can be found by using appropriate formulas.

CHAPTER 33 CONSTRUCTION TO PREVENT TERMITE ATTACK

Of all the destructive wood pests, *termites* are the most common. They cause tremendous economic loss annually. They attack wood throughout most of the country, but they are more prevalent in the warmer sections (Fig. 33-1). Buildings should be designed and constructed to minimize termite attack.

Termites

Termites play a beneficial role in their natural habitat. They break down dead or dying plant material to enrich the soil. However, when termites feed on wood structures, they become pests.

Kinds of Termites

There are three kinds of termites: *drywood, dampwood,* and *subterranean.*

Drywood Termites. Drywood termites enter a building usually through attic or foundation vents. They attack sheathing and structural members. They also infest wood door and window frames and furniture. However, the colonies are small. Unchecked, they cause less damage to buildings than other kinds of termites. Treatment is usually tent fumigation of the entire building with a toxic gas.

Region I Very Heavy
Region II Moderate to Heavy
Region III Slight to Moderate
Region IV None to Slight

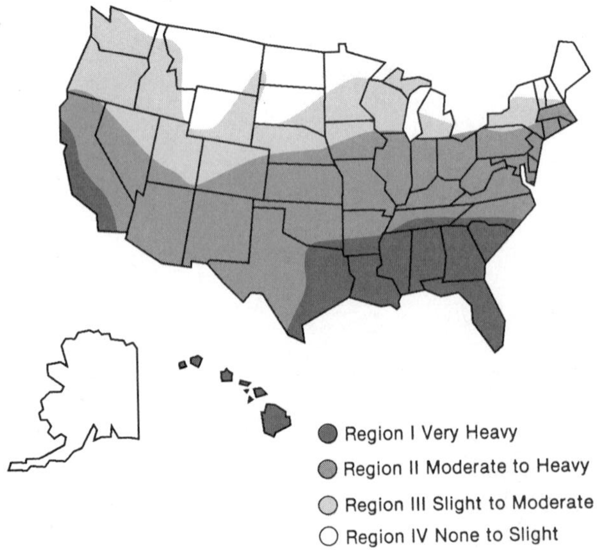

Fig. 33-1 Degree of subterranean termite hazard in the United States.

Dampwood Termites. Dampwood termites infest wood kept wet by lawn sprinklers, leaking toilets, showers, pipes, roofs, and other places where wood is damp. They are capable of doing great damage to a structure if undetected. Entrance into the building is usually where a continual source of moisture keeps wood wet. Discovery of a leak sometimes reveals a dampwood termite infestation. Once the wood is dry, dampwood termites will leave.

Subterranean Termites. Subterranean termites live in the ground. They are the most destructive species because they have such large underground colonies (Fig. 33-2). For protection against drying out, they must stay in close contact with the soil and its moisture. Above ground, they build earthen *shelter tubes* to protect themselves from the drying effects of the air. These tubes are usually built over a surface for support. Termites

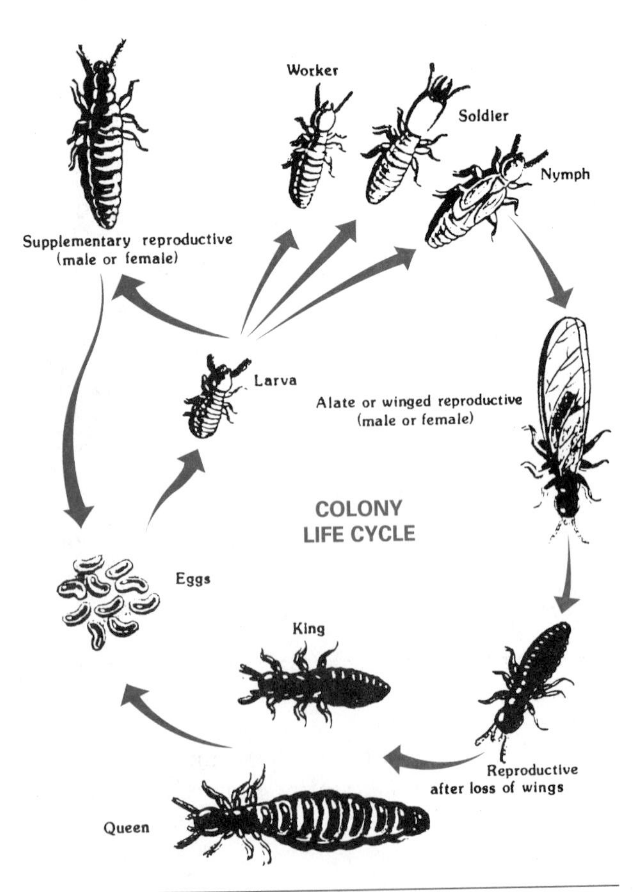

Fig. 33-2 Typical subterranean termite life cycle.

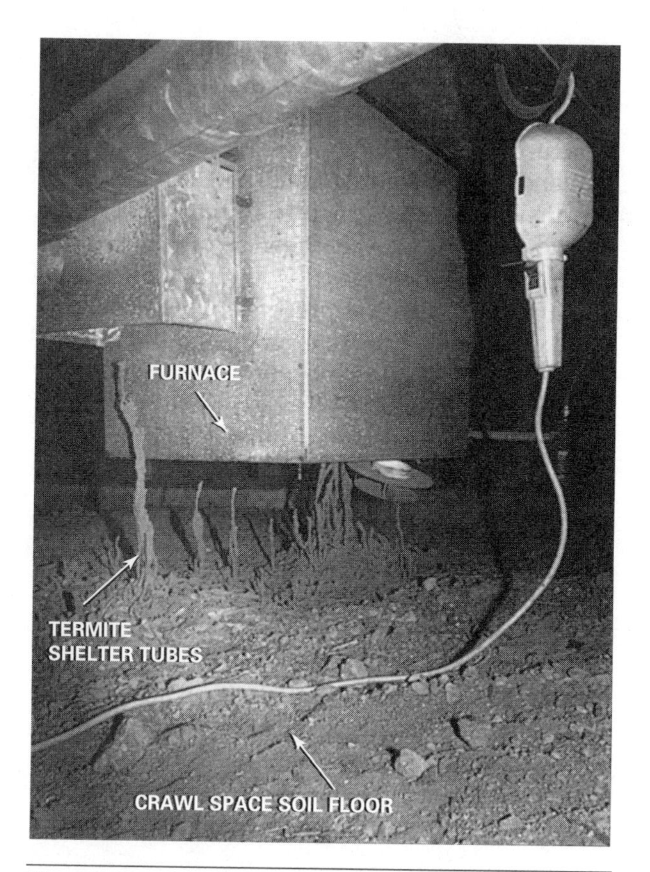

Fig. 33-3 Subterranean termites can build unsupported shelter tubes as high as 12 inches. Heat attracts them in their attempt to reach wood. (*Courtesy of* The Termite Report, *Pear Publishing; Don Pearman, photographer*)

can build unsupported tubes as high as 12 inches in their effort to reach wood (Fig. 33-3).

Treatment for subterranean termites generally consists of correcting conditions favorable to infestation, installation of ground-to-wood termite barriers, and chemical treatment of the foundation and soil. Chemicals should only be applied by trained technicians.

Techniques to Prevent Termites

Protection against subterranean termites should be considered during planning and construction of a building. Improper design and poor construction practices could lead to termite infestation after completion of the building. Preventative efforts in the planning stage and during construction may save the future owner much anxiety and expense. All the techniques used for the prevention of termite attack are based on keeping the wood in the structure dry (equilibrium moisture content) and making it as difficult as possible for termites to get to the wood. In lumber, a moisture content below 20 percent also prevents the growth of fungi, which cause wood to rot.

The Site

All tree stumps, roots, branches, and other wood debris should be removed from the building site. Do not bury it on the site. Footing and wall form plank, boards, stakes, spreaders, and scraps of lumber should be removed from the area before backfilling around the foundation. Lumber scraps should not be buried anywhere on the building site. None should be left on the ground beneath or around the building after construction is completed.

The site should be graded to slope away from the building on all sides. The outside finished grade should always be equal to or below the level of the soil in crawl spaces. This ensures that water is not trapped underneath the building (Fig. 33-4).

Chemical treatment of the soil before construction is one of the best methods of preventing termite attack. This should be a supplement to, and not a substitute for proper building practices.

Perforated drain pipe should be placed around the foundation, alongside the footing. This will drain water away from the foundation (Fig. 33-5). The foundation drain pipe should be sloped so water can drain to a lower elevation or a **dry well** some distance from the building. A dry well is a pit in the ground filled with stone to absorb the water from the drain pipe. Gutters and downspouts should be

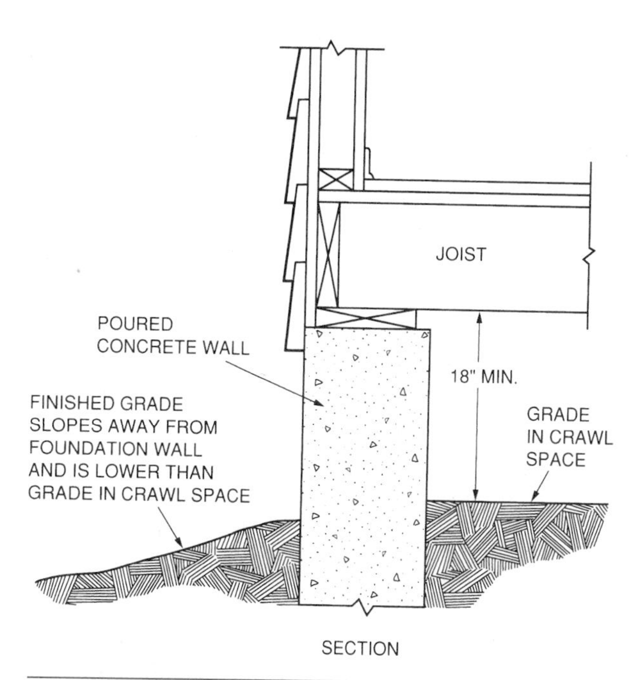

Fig. 33-4 The finished grade in the crawl space and around the outside of the foundation.

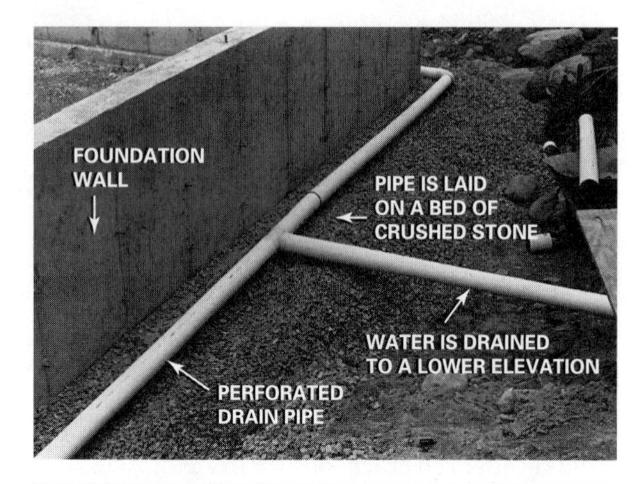

Fig. 33-5 Perforated drain pipe is placed alongside the foundation footing to drain water away from the building.

installed to lead roof water away from the foundation. Downspouts should be connected to a separate drain pipe to facilitate moving the water quickly.

Crawl Spaces

Solid concrete foundation walls should be properly reinforced and cured to prevent the formation of cracks. Cracks as little as 1/32-inch wide permit the passage of termites.

Concrete block walls should be either capped with a minimum of 4 inches of reinforced concrete or the top **course** filled completely with concrete (Fig. 33-6).

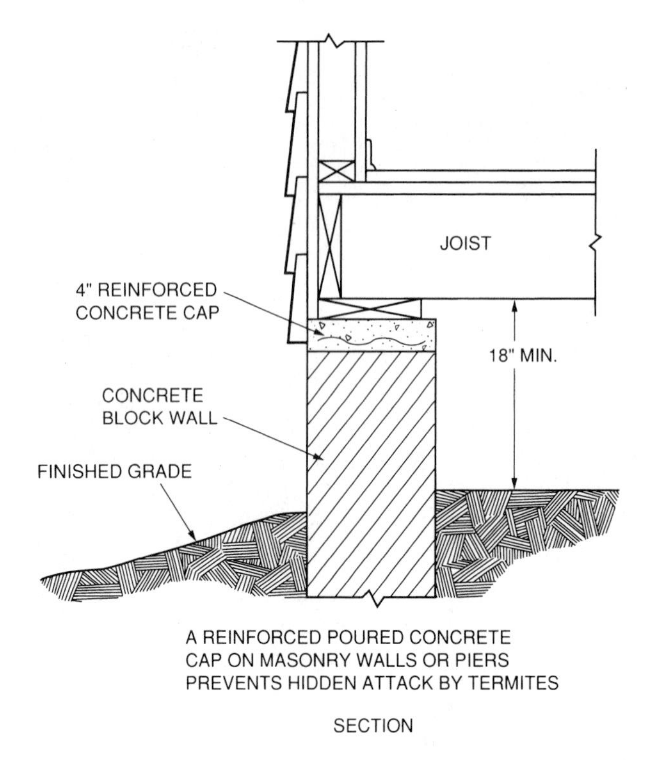

Fig. 33-6 The top course of concrete block walls should be capped or filled.

Air should be circulated in crawl spaces by means of ventilators placed to leave no pockets of stagnant air. In general, the total area of ventilation openings should be equal to 1/150th of the ground area of the crawl space. Shrubbery should be kept away from openings to permit free circulation of air. There should be access to the crawl space for inspection of inner wall surfaces for termite tubes.

In crawl spaces and other concealed areas, clearance between the bottom of floor joists and the ground should be at least 18 inches and at least 12 inches for beams and girders (Fig. 33-7).

Keep all plumbing and electrical conduits clear of the ground in crawl spaces. Suspend them from girders or joists. Do not support them with wood blocks or stakes in the ground. The soil around pipes extending from the ground to the wood above should be treated with chemicals.

Slab-on-Grade

Slab-on-grade is one of the most susceptible types of construction to termite attack. Termites gain access to the building over the edge of the slab, or through isolation joints, openings around plumbing, and cracks in the slab. Termite infestations in this type of construction are difficult to detect and control. For slab-on-grade construction, it is important to have the soil treated with chemicals before placing the concrete slab.

The monolithic slab provides the best protection against termites. The floor and footing are placed in one continuous operation, eliminating joints that permit hidden termite entry (Fig. 33-8). Proper curing of the slab helps eliminate the development of cracks through which termites can gain access to the wood above.

One type of independent slab extends completely across the top of the foundation. This prevents hidden termite entry. The lower edge of the slab should be open to view from the outside.

The top of the slab should be at least 8 inches above the grade (Fig. 33-9).

Independent slabs that rest either part-way on or against the side of the foundation wall are the least reliable. Termites may gain hidden access to the wood through isolation joints (Fig. 33-10). Fill the spaces around isolation joints, pipes, conduit, ducts, or steel columns with hot roofing-grade coaltar pitch.

Exterior Slabs. Spaces beneath concrete slabs for porch floors, entrance platforms, and similar units against the foundation should not be filled. Leave them open with access doors for inspection. If this cannot be done and spaces must be filled, have the soil treated for termites by a professional.

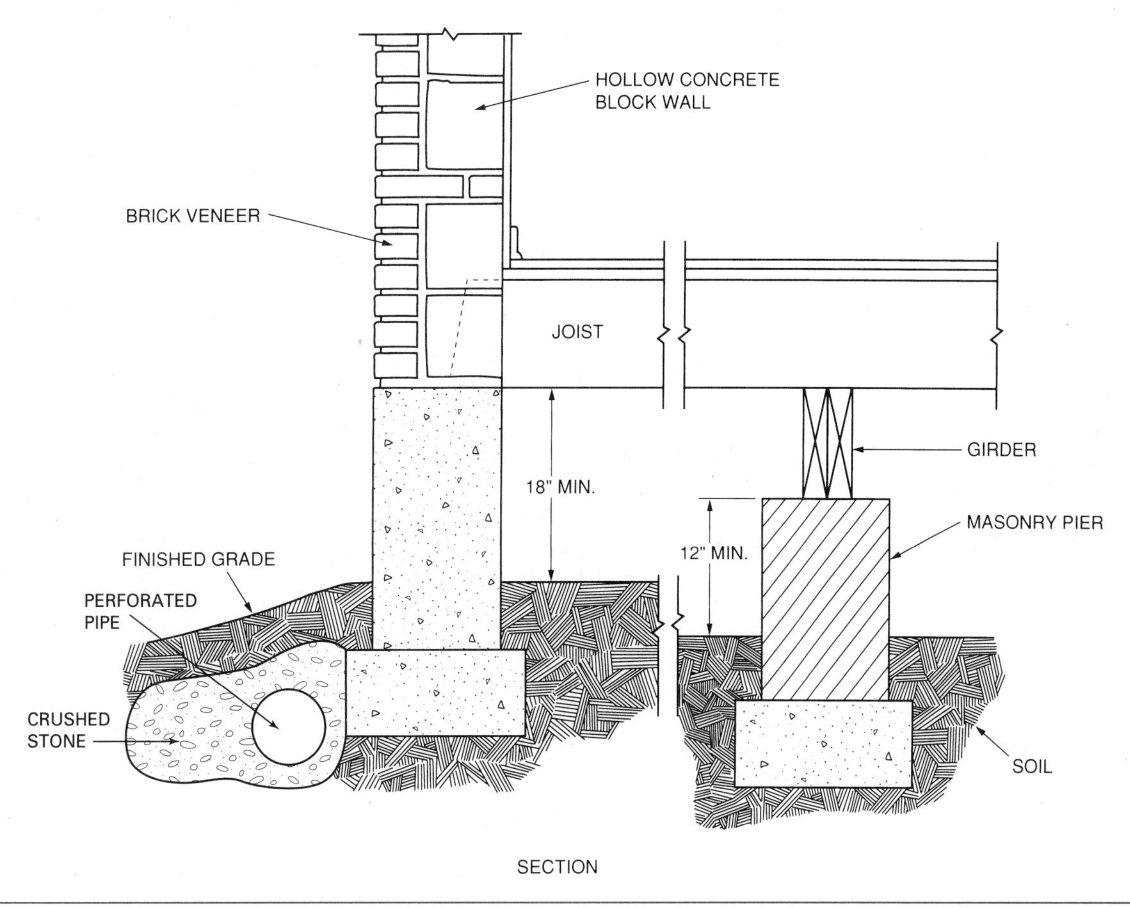

Fig. 33-7 Provide adequate clearance between wood and soil in crawl spaces.

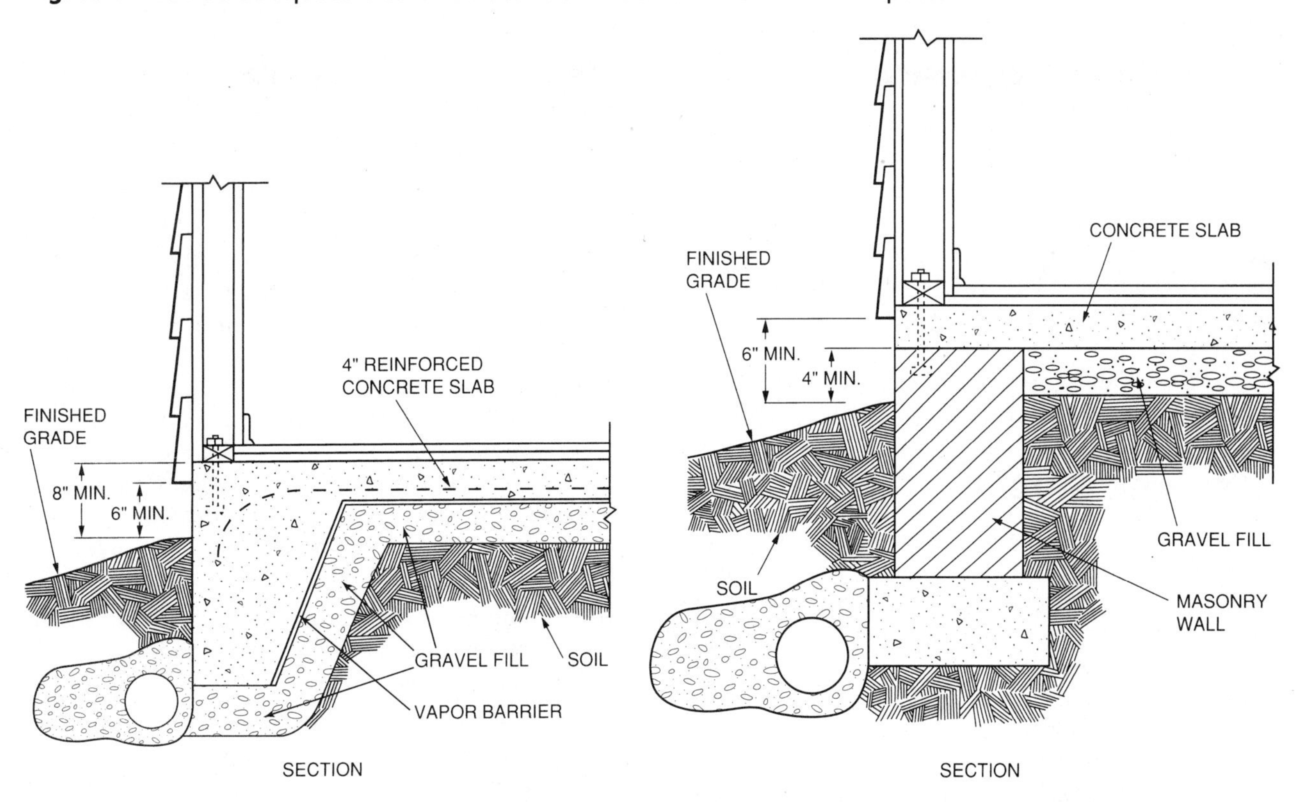

Fig. 33-8 In slab-on-grade construction, the monolithic slab provides the best protection against termites.

Fig. 33-9 An independent slab that extends across the top of the foundation wall prevents hidden termite attack.

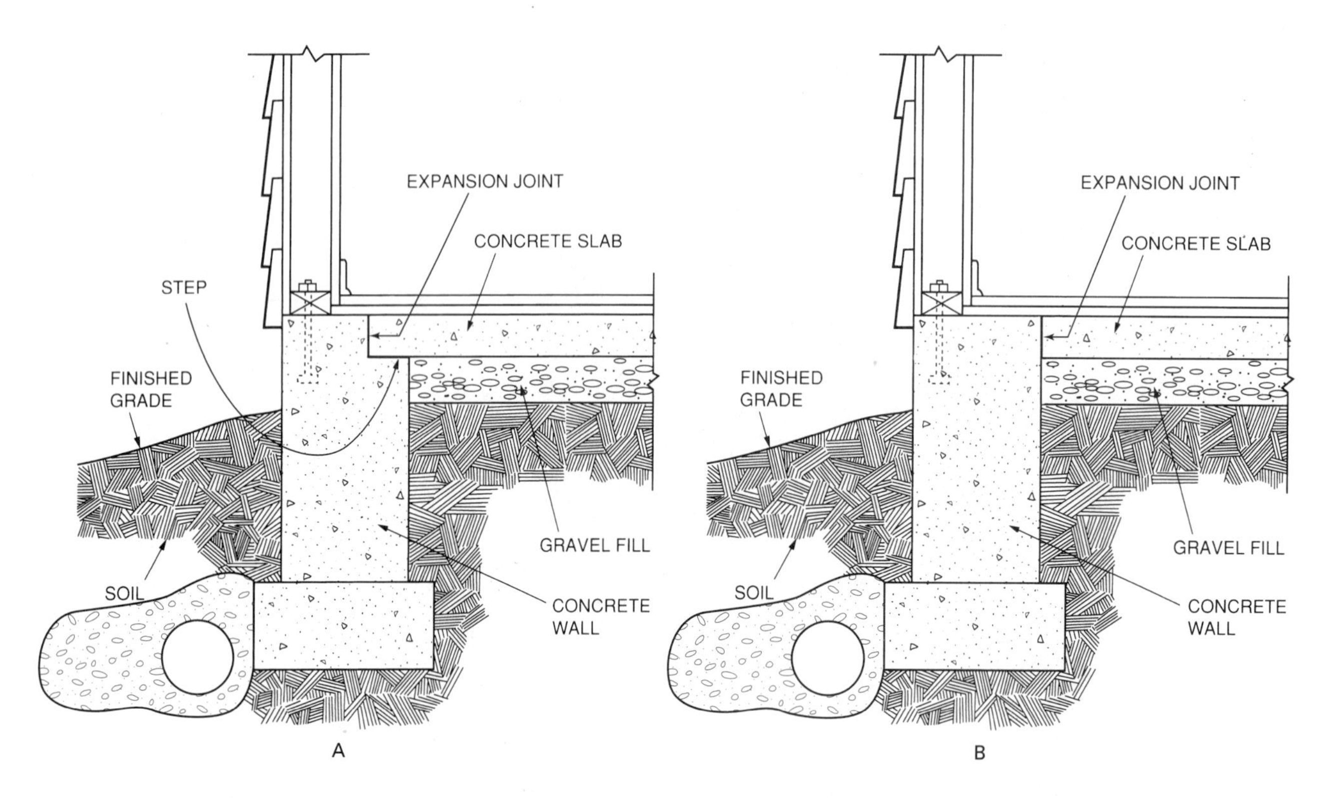

Fig. 33-10 Slab construction where termites may obtain access to wood through isolation joints.

Exterior Woodwork

Wall siding usually extends no more than 2 inches below the top of foundation walls. It should be at least 6 inches above the finished grade.

Porch supports should be placed not closer than 2 inches from the building to prevent hidden access by termites. Wood steps should rest on a concrete base that extends at least 6 inches above the ground.

Door jambs, posts, and similar wood parts should never extend into or through concrete floors.

Termite Shields

If **termite shields** are properly designed, constructed, installed, and maintained, they will force termites into the open. This will reveal any tubes constructed around the edge and over the upper surface of the shield (Fig. 33-11). However, research has shown that termite shields have not been effective in preventing termite infestations. Because of improper installation and infrequent inspection, they are not presently recommended by government agencies for detection and prevention of termite attack. However, check local building codes that may mandate their use.

Use of Pressure-Treated Lumber

Lumber in which preservatives are forced into the wood's cells under pressure is commonly called **pressure-treated** *lumber* (Fig. 33-12). Although other grades are available, two are generally used. *Above ground* is used for sill plates, joists, girders, decks, and similar members. *Ground contact* is suitable for contact with soil or fresh water. Typical grade stamps are shown in Figure 33-13. Special grades are manufactured for saltwater immersion and wood foundations.

Check building codes for requirements concerning the use of pressure-treated lumber. Generally, building codes require the use of pressure-treated lumber for the following structural members:

- Wood joists or the bottom of structural floors without joists that are located closer than 18 inches to exposed soil.
- Wood girders that are closer than 12 inches to exposed soil in crawl spaces or unexcavated areas.
- Sleeper, sills, and foundation plates on a concrete or masonry slab that is in direct contact with the soil.

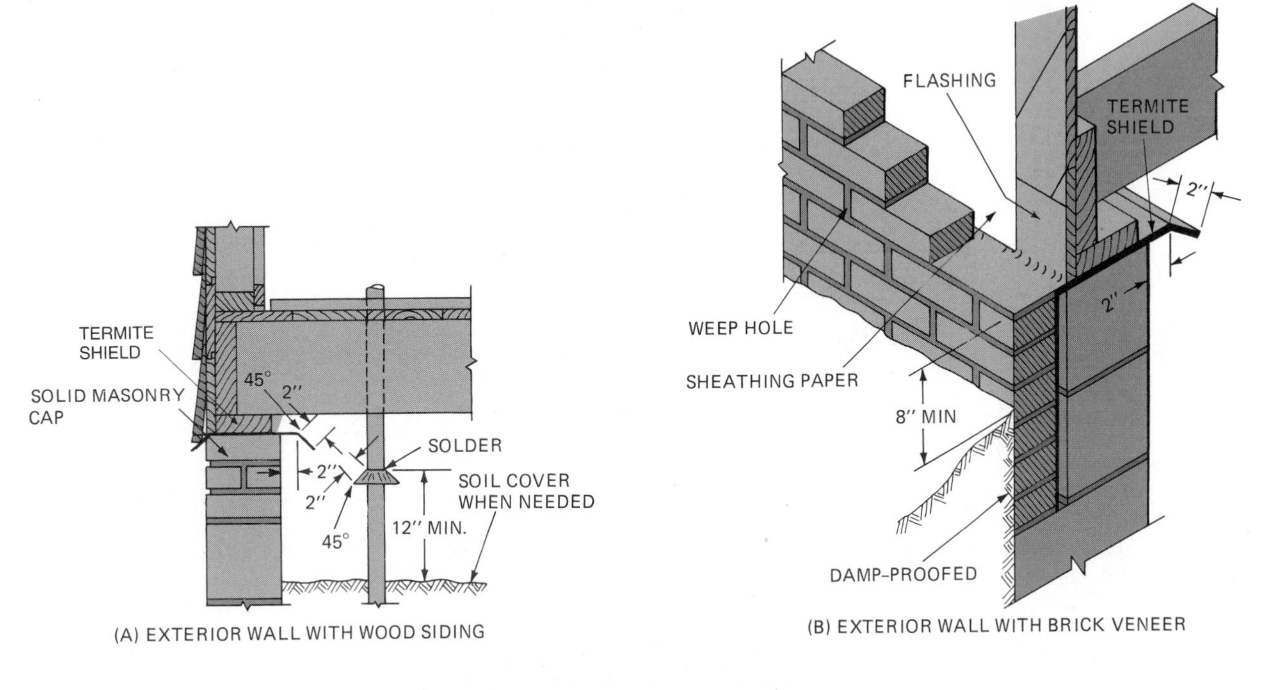

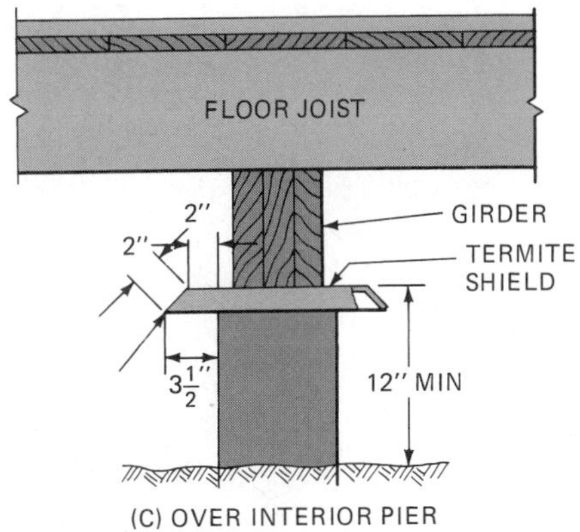

Fig. 33-11 Installation of termite shields.

Fig. 33-12 Preservatives are forced into lumber under pressure in large cylindrical tanks. (*Courtesy of Southern Pine Marketing Council*)

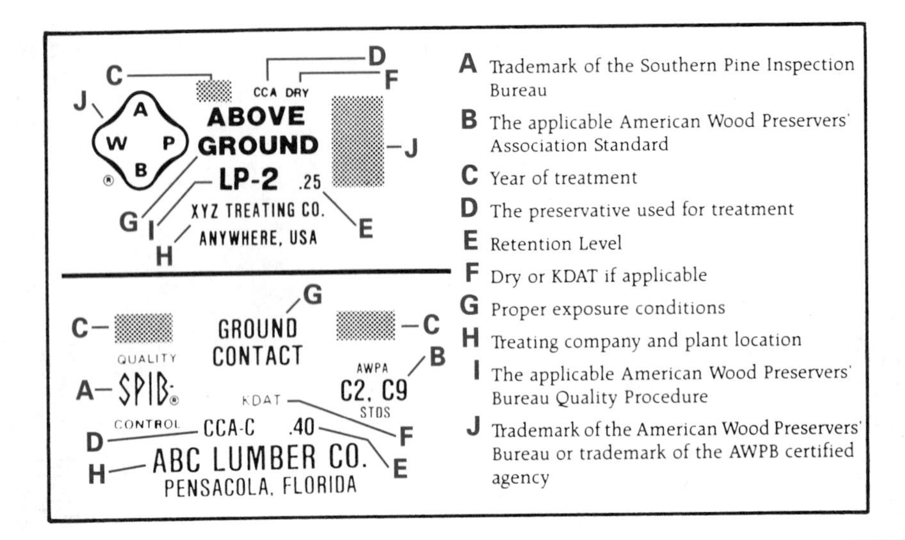

Fig. 33-13 Typical grade stamps for pressure-treated lumber. (*Courtesy of Southern Pine Marketing Council*)

Termites generally will not eat treated lumber. They will tunnel over it to reach untreated wood. Their shelter tubes then may be exposed to view and their presence easily detected upon inspection.

Follow these safety rules when handling pressure-treated lumber.

- Wear eye protection and a dust mask when sawing or machining treated wood (Fig. 33-14).
- When the work is completed, wash areas of skin contact thoroughly before eating or drinking.
- Clothing that accumulates sawdust should be laundered separately from other clothing and before reuse.
- Dispose of treated wood by ordinary trash collection or burial. Do not burn treated wood. The chemical retained in the ash could pose a health hazard.

Fig. 33-14 Know the safety precautions when handling pressure-treated lumber. (*Courtesy of Southern Pine Marketing Council*)

Review Questions

Select the most appropriate answer.

1. A platform frame is easy to erect because
a. only one-story buildings are constructed with this type of frame.
b. each platform may be constructed on the ground.
c. at each level a flat surface is provided on which to work.
d. less framing members are required.

2. One of the advantages of the balloon frame is that
a. the bottom plates act as firestops.
b. there is little vertical shrinkage in the frame.
c. the second-floor joists rest on a ribbon instead of a plate.
d. it is stronger, stiffer, and more resistant to lateral pressures.

3. A heavy beam that supports the inner ends of floor joists is called a
a. pier.
b. girder.
c. stud.
d. sill.

4. That member of a floor frame that rests directly on the foundation wall is called a
a. pier.
b. girder.
c. stud.
d. sill.

5. To mark a square end on a round column
a. use a square.
b. measure down from the existing end.
c. use a pair of dividers.
d. wrap a piece of paper around it.

6. Anchor bolts are spaced a maximum of
a. 6 feet OC.
b. 8 feet OC.
c. 10 feet OC.
d. 12 feet OC.

7. To protect against termites, keep wood in crawl spaces and other concealed areas at least
a. 8 inches above the ground.
b. 12 inches above the ground.
c. 18 inches above the ground.
d. 24 inches above the ground.

8. When floor joists rest on wood they should have at least
a. 4 inches of bearing.
b. 3 1/2 inches of bearing.
c. 2 1/2 inches of bearing.
d. 1 1/2 inches of bearing.

9. Floor joists should have a minimum lap of
a. 2 inches.
b. 4 inches.
c. 6 inches.
d. 8 inches.

10. It is extremely important when installing floor joists to
a. fasten them securely with enough nails.
b. fasten the girder end first.
c. have the crowned edges up.
d. notch and size the ends for any differences in width.

■ BUILDING FOR SUCCESS 

Sound Work Ethics: Obsolete or Essential?

Builders have a tremendous responsibility to maintain integrity in their businesses. Contractors and construction workers are highly visible in the public's eye. Their work rarely goes unnoticed. Few "professional" critics will be capable of doing better work than the worker. But the worker is the one in the spotlight and under inspection.

A builder's clientele today exhibits the same characteristics as those in the past. They desire quality work at reasonable prices, utilizing quality products. People desiring a new or remodeled home today typically do more homework when selecting a contractor. They are concerned about environmental issues that affect housing, financing, costs, and legalities. But as all of these issues become settled, today's clients still desire to know if they can trust the builder to give them the quality for which they are paying.

How does a builder or construction worker send the message of quality and trust? For the answer, we must return to the basics of long-term success.

Everyone desires to be treated fairly in their transactions with other people. They need to treat others as they themselves wish to be treated. This simple formula has traditionally set in motion the values and ethics necessary to demonstrate honesty and good will.

There are basic rules of conduct that are recognized by every class, group, and culture. These rules or principles are the guidelines for meaningful accomplishments, cooperation, satisfaction, and success. Any deviation from the accepted ethics could very well bring disaster.

Rules of conduct demonstrate the integrity of each person who adopts them. Integrity is defined in part, as "having moral principles, honesty, wholeness, or the quality of being undivided and incorruptible." A person's performance at work and at home should be transparent and require no clarification to show integrity. Integrity requires convictions and standards of ethical performance. In construction, when the specifications for the project are ignored, there can be no assurance of quality. Ignoring the requirements for quality negates integrity. It takes courage to adhere to specs, codes, and sound construction principles. That courage is the mainstay for exhibiting personal integrity.

The construction industry has building codes, state statutes, and local regulations that must be learned and adhered to daily. Every national, state, and local building organization implements a code of ethics for its membership. Individuals who desire to belong to and remain in good standing with an organization must be knowledgeable of the code and willing to apply those ethics on every job. This willingness becomes the basis for voluntary compliance.

Why adhere to a code of ethics? Because integrity must be a visual part of quality workmanship. Without sound work ethics, little credibility can be associated with a person, company, or enterprise. People that do not trust a builder or craftsman will not initiate any transactions with them.

One of the most noticeable characteristics of someone that exhibits sound work ethics is hard, dedicated work. Technology has alleviated many of the extremely difficult, laborious, and dangerous jobs that were once commonplace. Many physically demanding tasks have been minimized by new procedures that allow people to work smarter and more efficiently. Construction workers can learn to properly blend new technology, improved application methods, and honest work to accomplish a task. Although it may not be appealing to many coming into the workforce, meaningful labor and hard work remain a key ingredient to success. We should strive to work smarter, more diligently, and with purpose. Hard work always involves learning along the way. This learning translates into experience and wisdom. In the end, the skilled worker will generate more satisfaction and perhaps more profit. Good work habits are as hard to break as bad ones.

Remember, "the greatest ability may be dependability" to a builder. A successful construction business is based on people who work well on a regular basis. Punctuality on a regular basis is interpreted as proof that the employee values the working relationship with the employer. An employer strongly desires dependable employees.

As the costs of quality construction continue to rise at an accelerated rate, people become even more cautious about spending their savings or making long-term financial commitments. The competition for builders and skilled workers will also continue to increase. In turn, this challenges everyone to demonstrate credibility. The person who desires a career in a construction-related business today and well into the twenty-first century will have to compete for it. A sense of credibility will continue to be a survival technique. By believing in a sound code of ethics, and building a business on fairness and good moral principles, the chances of success are greatly increased.

Sound work ethics will never become obsolete because they are the basis for success. It has been said that "what goes around, comes around." Eventually we will be treated as we treat others. Builders or skilled workers will never have to be ashamed of their work if they perceive their jobs as serving others. Developing and maintaining sound work ethics in the beginning brings about the characteristics of integrity for a lifetime.

The following examples of sound work ethics and conduct reflect the characteristics needed by every worker who desires credibility.

- A day's work for a day's pay.
- Avoid shortcuts that save time but reduce quality.
- Give the customer the quality you yourself desire.
- Clearly state in writing the work to be done by all parties.
- Do not deceive or misrepresent anyone.
- Answer any questions honestly.
- Be straightforward in responding to the concerns of clients.
- Conduct your business in a professional manner.
- Charge your clients fairly.
- Assume your responsibility to your customers, community, and country.
- Pledge your support to the overall success of the project.

With the implementation of sound work ethics and principles, a construction worker will automatically send a message of integrity to the business world. This in turn will bring in a clientele that is comfortable and trusting.

The business world today does not appear to attach less value to sound ethics and principles than in the past. Therefore, each person entering the construction industry should be encouraged to learn and practice good work ethics. This will assure a successful and enjoyable career.

FOCUS QUESTIONS: For individual or group discussion

1. Describe in your own words what you feel are work ethics.
2. From your personal experiences, have any of these characteristics been observable in skilled workers or builders you have known?
3. What would be some outcomes of not expecting a skilled worker to have sound work ethics?
4. How might you personally go about acquiring or building on your present sound work ethics?

UNIT 12

Exterior Wall Framing

Exterior walls must be constructed to the correct height, corners braced plumb, walls straightened from corner to corner, and window and door openings framed to specified size. The techniques described in this unit will enable the apprentice carpenter to frame exterior walls with competence.

OBJECTIVES

After completing this unit, the student should be able to:

- identify and describe the function of each part of the wall frame.
- determine the length of exterior wall studs and the size of rough openings, and lay out a story pole.
- build corner posts and partition intersections, and describe several methods of forming them.
- lay out the wall plates.
- construct and erect wall sections to form the exterior wall frame.
- plumb, brace, and straighten the exterior wall frame.
- apply wall sheathing.
- describe the construction of wood foundation walls.

UNIT CONTENTS

CHAPTER 34 EXTERIOR WALL FRAME PARTS

The wall frame consists of a number of different parts. The student should know the name, function, location, and usual size of each member. Sometimes the names given to certain parts of a structure may differ according to the geographical area. For that reason, some members may be identified with more than one term.

Parts of an Exterior Wall Frame

An exterior wall frame consists of *plates, studs, headers, trimmers, corner posts, partition intersections, ribbons,* and *braces* (Fig. 34-1).

Plates

The top and bottom horizontal members of a wall frame are called *plates.* The bottom member is called a **soleplate.** It is also referred to as the *bottom plate* or *shoe.* The top members are called *top plates.* They usually consist of doubled 2-inch stock. In a balloon frame, the soleplate is not used. Instead, the studs rest directly on the sill plate.

Studs

Studs are vertical members of the wall frame. They run full-length between plates. *Jack studs* or *trimmers* are shortened studs that line the sides of an opening. They extend from the bottom plate up to the top of the opening. *Cripple* studs are shorter members above and below an opening, which extend from the top or the bottom plates to the opening.

Studs are usually 2 × 4s, but 2 × 6s are used when 6-inch insulation is desired in exterior walls. Studs are usually spaced 16 or 24 inches OC.

Headers

Headers or **lintels** run at right angles to studs. They form the top of window, door, and other wall openings, such as fireplaces. Headers must be strong enough to support the load above the opening. The depth of the header depends on the width of the opening. As the width of the opening increases, so must the strength of the header. Check drawings, specifications, codes or manufacturers' literature for header size.

Kinds of Headers. Solid or built-up lumber may be used for headers. For 4-inch walls, two pieces of 2-inch lumber with 1/2-inch plywood or strand board sandwiched in between them gives the header the full 3 1/2-inch thickness of the wall. In 6-inch walls, three pieces of 2-inch lumber with two pieces of 1/2-inch plywood or strand board in between makes up the 5 1/2-inch wall thickness (Fig. 34-2).

Much engineered lumber is now being used for window and door opening headers. Figures 5-6,

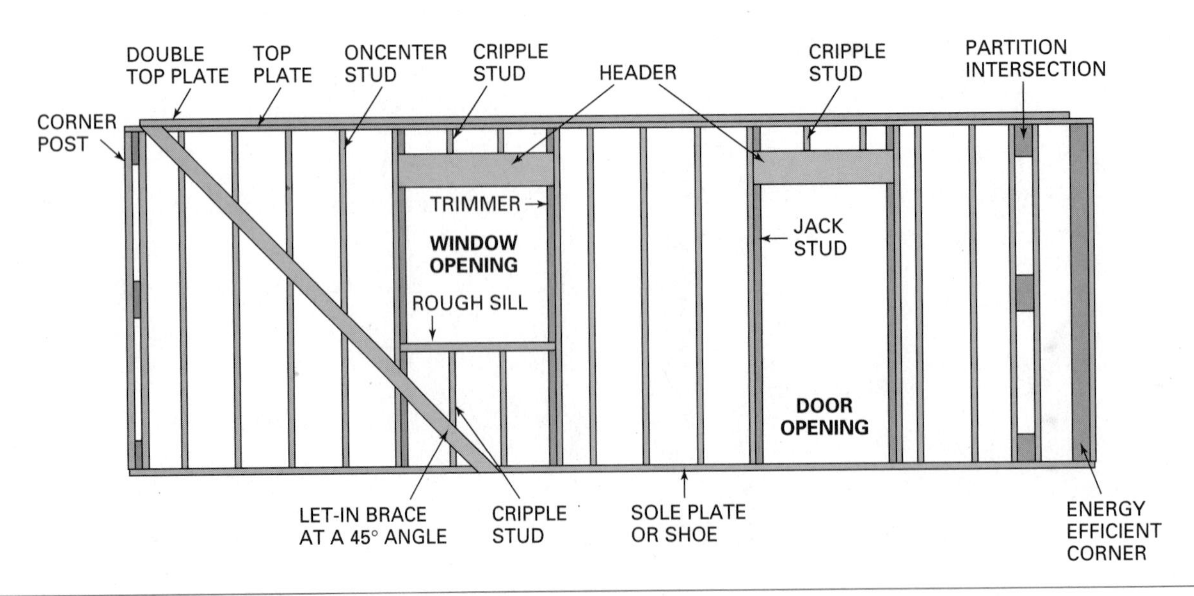

Fig. 34-1 Parts of an exterior wall frame.

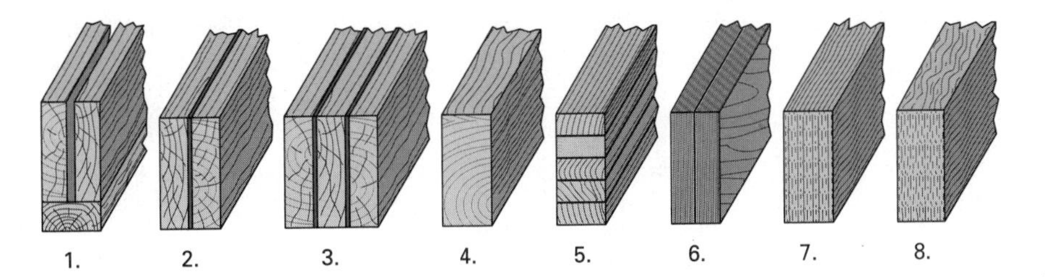

1. A BUILT-UP HEADER WITH A 2 × 4 OR 2 × 6 LAID FLAT ON THE BOTTOM.
2. A BUILT-UP HEADER WITH A 1/2" SPACER SANDWICHED IN BETWEEN.
3. A BUILT-UP HEADER FOR A 6" WALL.
4. A HEADER OF SOLID SAWN LUMBER.
5. GLULAM BEAMS ARE OFTEN USED FOR HEADERS.
6. A BUILT-UP HEADER OF LAMINATED VENEER LUMBER.
7. PARALLEL STRAND LUMBER MAKES EXCELLENT HEADERS.
8. LAMINATED STRAND LUMBER IS USED FOR LIGHT DUTY HEADERS.

Fig. 34-2 Several types of solid and built-up headers are used.

6-1, and 8-1 show the use of laminated veneer lumber, parallel strand lumber, and glulam beams as opening headers. In addition, a header called a *1.2E Light Duty Header* manufactured of laminated strand lumber is now available (Fig. 34-3). The use of engineered lumber permits the spanning of wide openings, such as double garage door openings. Such openings otherwise might need intermediate supports (Fig. 34-4).

In many buildings, when the opening must be supported without increasing the header size, the top of a wall opening may be trussed to provide support (Fig. 34-5). However, when the opening is fairly close to the top plate, the depth of the header is increased. This completely fills the space between the plate and the top of the opening. In this case, the same size header is usually used for all wall openings, regardless of the width of the opening (Fig. 34-6). This eliminates the need to install short cripple studs above the header.

Rough Sills

Forming the bottom of a window opening, at right angles to the studs are members called *rough sills.* They usually consist of a single 2-inch thickness. They carry little load. However, many carpenters prefer to use a double 2-inch thickness rough sill. This provides more surface on which to fasten window trim in the later stages of construction.

Trimmers

Trimmers are shortened studs that support the headers. They are fastened to the studs on each side of

Fig. 34-3 A 1.2E Light Duty Header of laminated strand lumber is used in a window opening. (*Courtesy of Trus Joist MacMillan*)

Fig. 34-4 Laminated veneer lumber is used for a garage door opening header. (*Courtesy of Trus Joist MacMillan*)

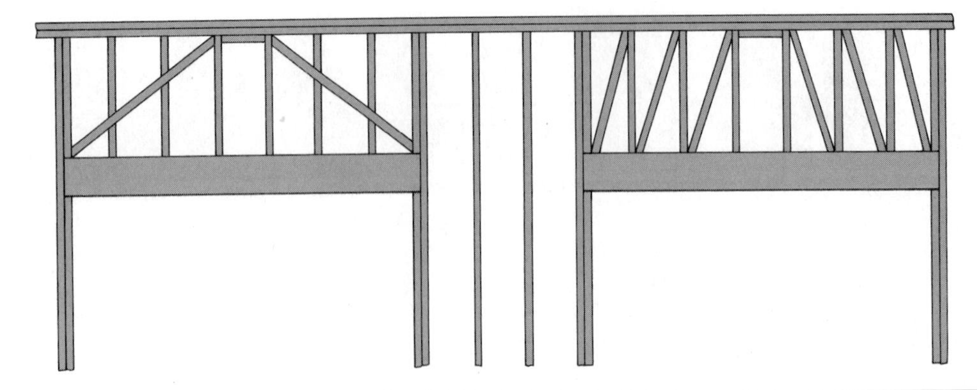

Fig. 34-5 Sometimes the top of a wall opening is trussed instead of increasing the header depth. Two methods of trussing an opening are shown.

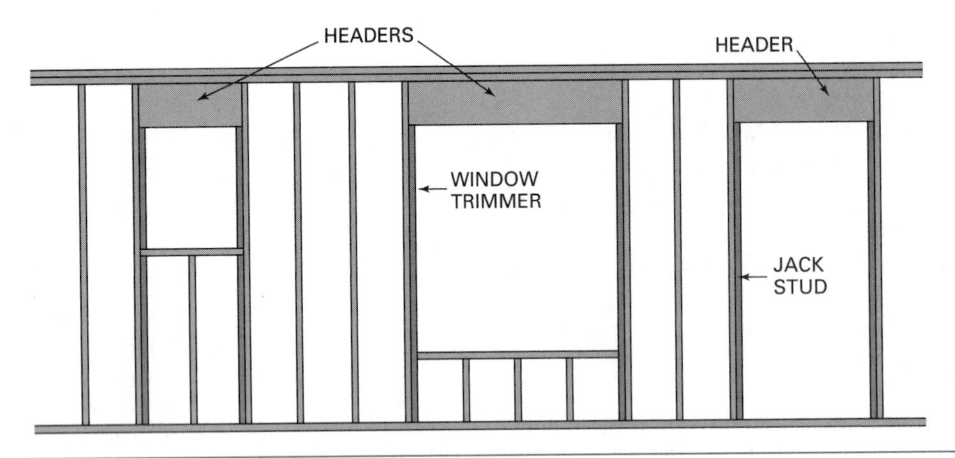

Fig. 34-6 It is common practice to use the same header depth. Completely fill the space between the opening and the top plate for all openings regardless of the opening width.

the opening. In window openings, they may fit snugly between the header and rough sill. Some codes, however, require that the trimmers run full length (Fig. 34-7). In door openings, the trimmers (also called *jack studs* and *liners*) fit between the header and the sole plate.

Some carpenters prefer to cut out the soleplate between the studs of a door opening. This lets the bottom end of the trimmers rest on the subfloor (Fig. 34-8). However, cutting out the soleplate may make it difficult to build wall sections on the subfloor and erect them into place. In some cases, door trimmers are left out until the wall section is erected. The soleplate is then cut out and trimmers installed.

Corner Posts

Corner posts are the same length as studs. They are constructed in a manner that provides an outside and an inside corner on which to fasten the exterior and interior wall coverings. Corner posts are built in a number of ways. Most of the time, three studs are used. Sometimes a solid 4 × 6 is used along with a single 2 × 4 (Fig. 34-9). Parallel strand lumber is used frequently for corner posts.

Partition Intersections

Wherever interior partitions meet an exterior wall, extra studs need to be put in the exterior wall. This provides wood for fastening the interior wall covering in the corner. In most cases, the *partition intersection* is made of two studs nailed to the edge of 2 × 4 blocks about a foot long. One block is placed at the bottom, one at the top, and one about center on the studs.

Another method is to maintain the regular spacing of the studs. Blocking is installed between them wherever partitions occur. The block is set back from the inside edge of the stud the thickness of a board. A 1 × 6 board is then fastened vertically on the inside of the wall so that it is centered on the partition.

Another method is to spike a continuous 2 × 8 backer to a full length stud. The edges of the backer

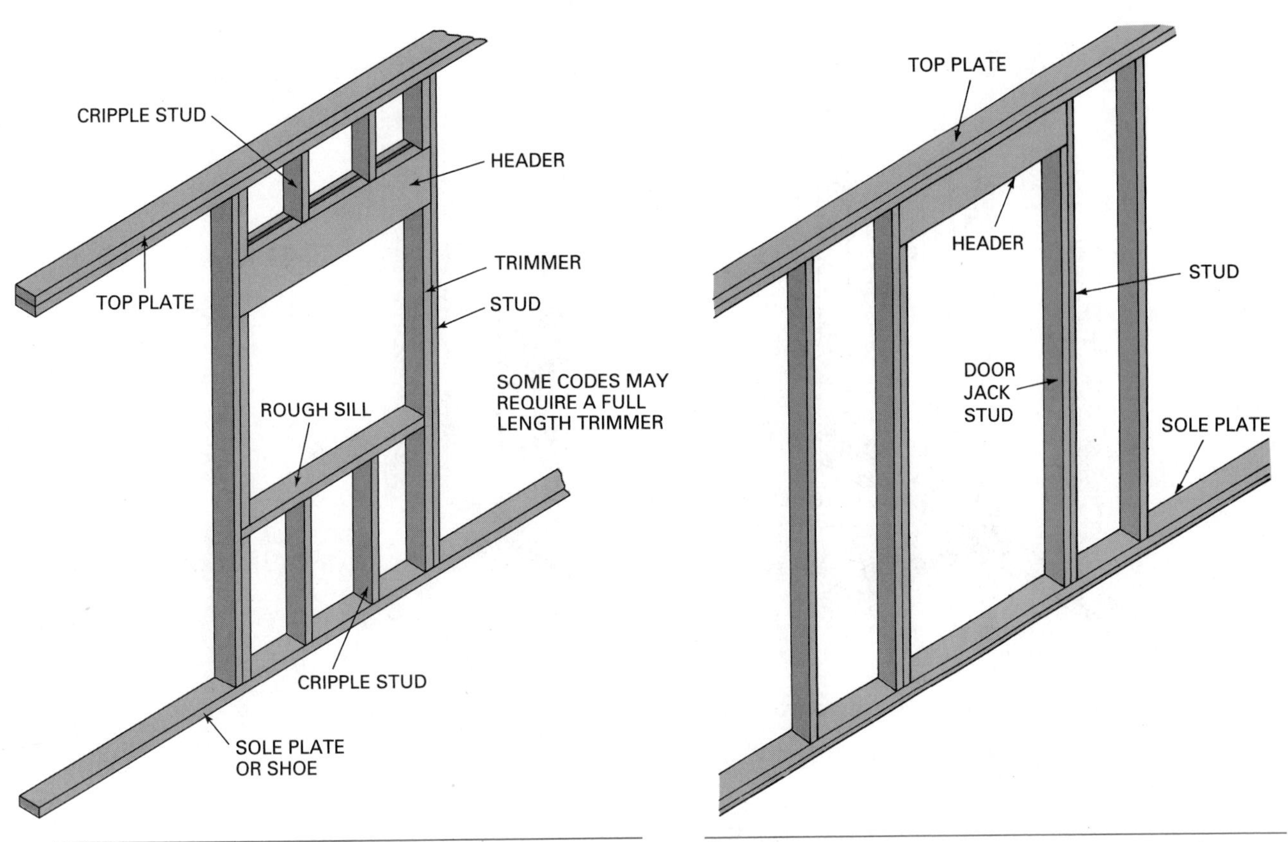

Fig. 34-7 Typical framing for a window opening.

Fig. 34-8 Typical framing for a door opening.

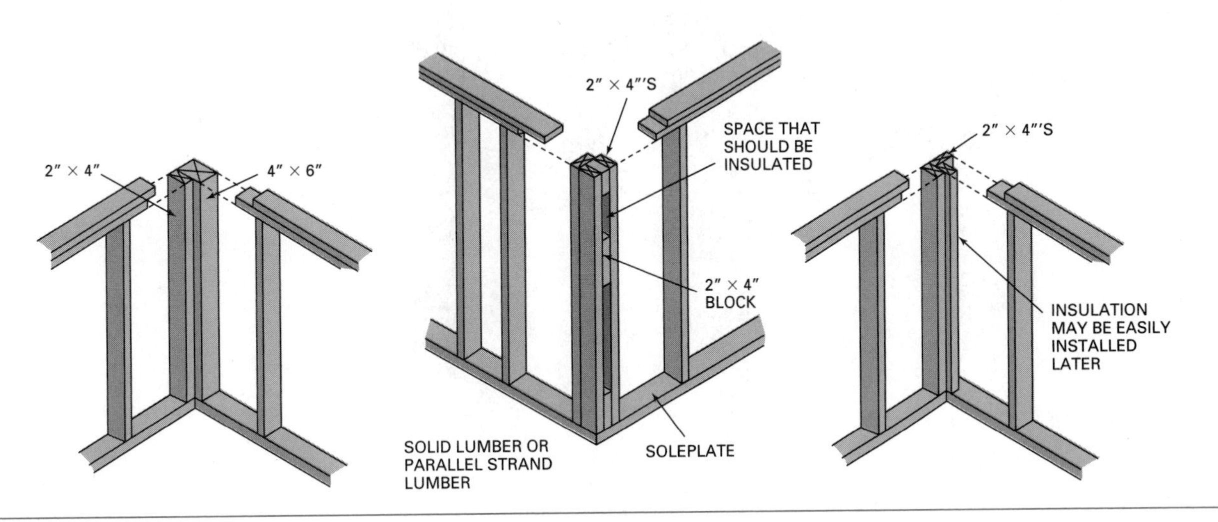

Fig. 34-9 Several methods of making corner posts are used.

project an equal distance beyond the edges of the stud (Fig. 34-10).

Ribbons

Ribbons are horizontal members of the exterior wall frame in balloon construction. They are used to support the second-floor joists. The inside edge of the wall studs is notched so that the ribbon lays flush with the edge (Fig. 34-11). Ribbons are usually made of 1 × 4 stock. Notches in the stud should be made carefully so the ribbon fits snugly in the notch. This prevents the floor joists from settling. If the notch is cut too deep the stud will be unnecessarily weakened.

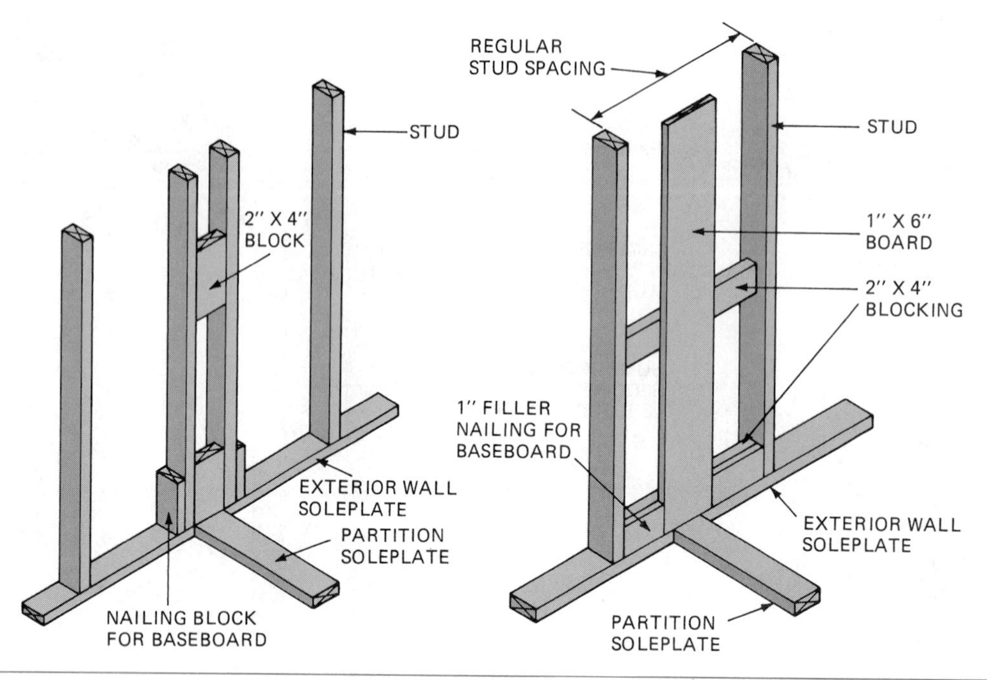

Fig. 34-10 Partition intersections are constructed in several ways.

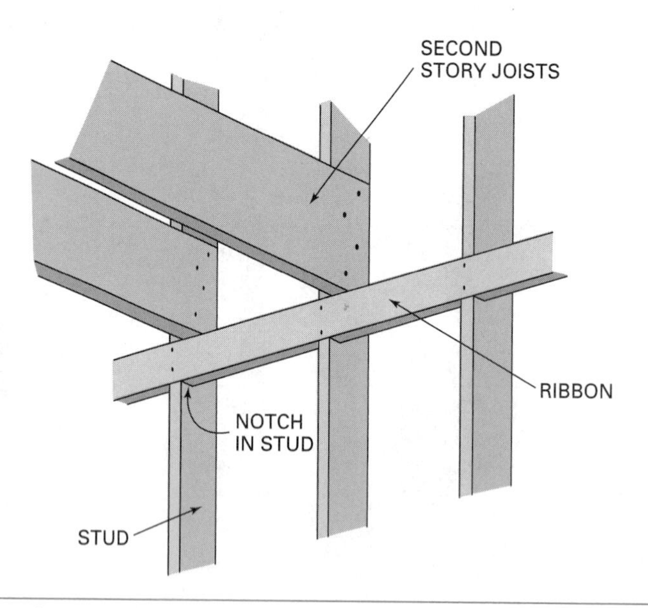

Fig. 34-11 Ribbons are used to support floor joists in a balloon frame.

Corner Braces

Generally, no wall bracing is required if rated panel wall sheathing is used. In other cases, such as when insulating board sheathing is used, walls are braced with metal wall bracing. They come in gauges of 22 to 16 in flat, T- or L-shapes. They are about 1 1/2" wide and run diagonally from the top to the bottom plates. They are nailed to the stud edges before the sheathing is applied. The T- and L-shapes require a saw kerf in the stud to allow them to lay flat when installed. Nominal 1 × 4 continuous diagonal strips of lumber may also be used. These are set into the face of the studs, top plate, and soleplate at each corner of the building. Or, 2 × 4s are cut in between the studs (Fig. 34-12).

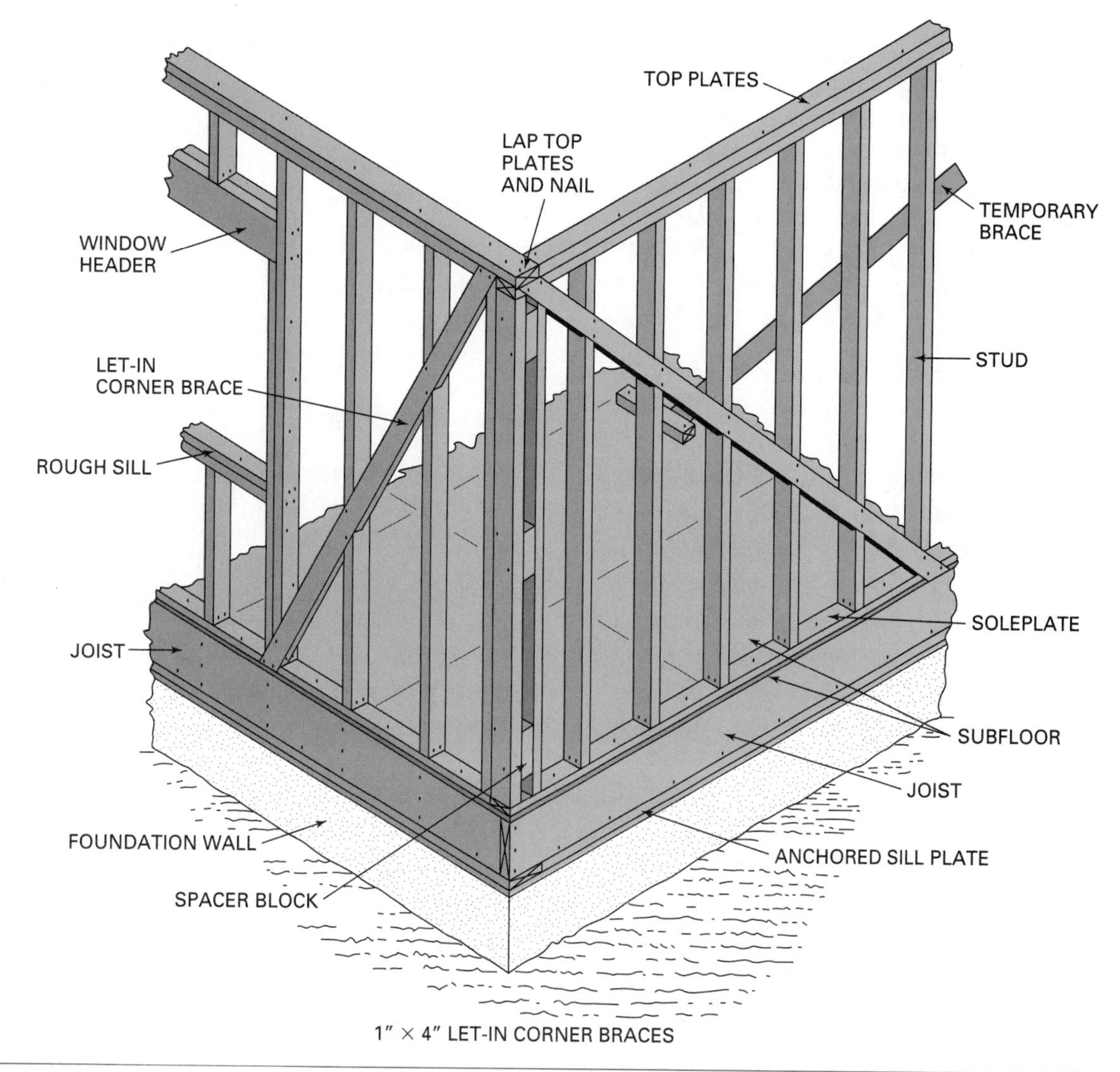

Fig. 34-12 Studs and plates may be notched to receive the let-in corner braces.

CHAPTER 35 FRAMING THE EXTERIOR WALL

Careful construction of the wall frame makes application of the exterior and interior finish easier. It also reduces problems for those who apply it later. This chapter shows how to frame and erect exterior walls in a proficient manner.

Exterior Wall Framing

Plans or blueprints usually indicate the height from finish floor to ceiling. This dimension is found in a wall section. It is needed to determine the length of wall studs.

Determining the Length of Studs

The stud length must be calculated so that, after the wall is framed, the distance from finish floor to ceiling will be as specified in the drawings. To determine the stud length, the thickness of the finish floor and the ceiling thickness below the joist must be known.

To the specified floor-to-ceiling height, add the thickness of the finish floor and the combined thickness of the ceiling below the ceiling joist. Include the thickness of *furring strips* in the ceiling, if used. Deduct the total thickness of the top plates and the soleplates to find the length of the stud (Fig. 35-1). (This applies only to platform frame construction.)

For example, in the plans the finish floor-to-ceiling height is found to be 7′-9″. Add 3/4 inch for the finish floor and 1/2 inch for the ceiling thickness below the joist. The total is 7′-10 1/4″. Deduct the combined thickness of the top plates and soleplates (4 1/2 inches). This leaves a stud length of 7′-5 3/4″.

For balloon frame construction, find the height from the sill to the top of the top plates by adding:

- finish floor to ceiling heights of all the stories,
- the thickness of both ceilings from the ceiling finish to the joists,
- the thickness of both finish floors and subfloors,
- and the combined width of all floor joists.

Then deduct the total thickness of the top plates (Fig. 35-2).

The location of the ribbon that supports the second-floor joists in a balloon frame is found by adding:

- the specified first floor to ceiling height,
- the thickness of the first-floor ceiling,
- finish floor of the first floor,
- subfloor of the first floor,
- and the width of the first-floor joists.

Measure this distance from the bottom end of the stud to the top edge of the ribbon.

Story Pole

The **story pole** is usually a straight strip of 1 × 2 or 1 × 3 lumber. Its length is the distance from the subfloor to the bottom of the joist above. Lines are squared across its width, indicating the height of the soleplates and top plates, headers, and rough sills for all openings.

The story pole is used to determine the length of studs, jack studs, trimmers, and cripple studs even

SECOND FLOOR JOIST
THICKNESS OF TOP PLATE
THICKNESS OF CEILING FINISH BELOW JOIST
STUD LENGTH
STUD
FINISH FLOOR TO CEILING HEIGHT
THICKNESS OF SOLEPLATE
THICKNESS OF FINISH FLOOR
SUBFLOOR
FIRST FLOOR JOIST
JOIST HEADER
SILL
FOUNDATION
(1) ADD THICKNESS OF CEILING FINISH AND FINISH FLOOR TO FINISH FLOOR TO CEILING HEIGHT
(2) SUBTRACT THE COMBINED THICKNESSES OF THE PLATES TO FIND THE STUD LENGTH

Fig. 35-1 Finding the length of a stud in an exterior wall frame of platform construction.

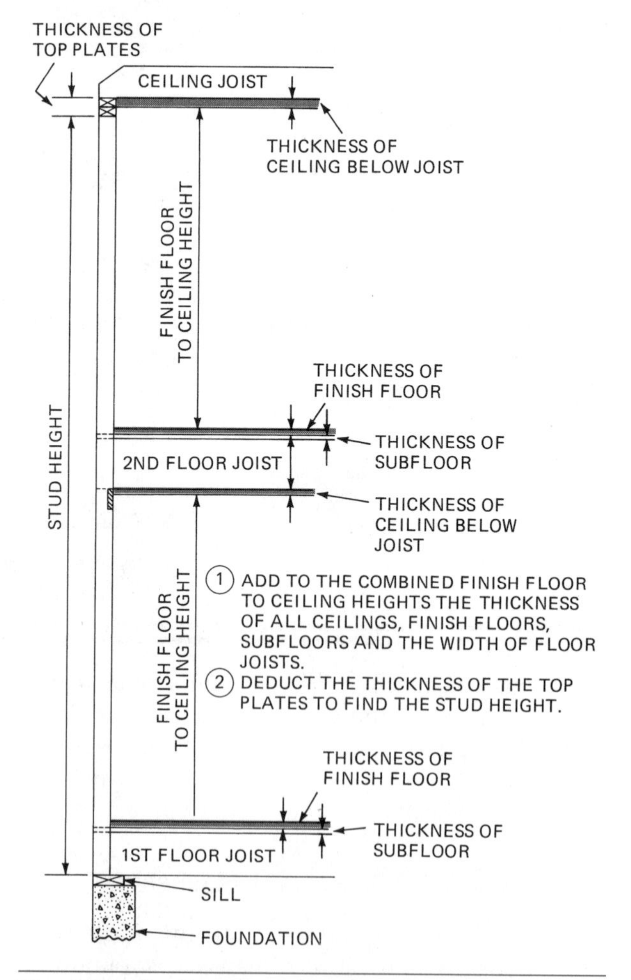

Fig. 35-2 Finding the length of a stud in a balloon frame.

if opening heights differ. The thickness and height of rough sills and the location of headers are also indicated and identified on the story pole (Fig. 35-3).

A different story pole is made for each floor of the building. As framing progresses, the story pole is used to test the accurate location of framing members in the wall.

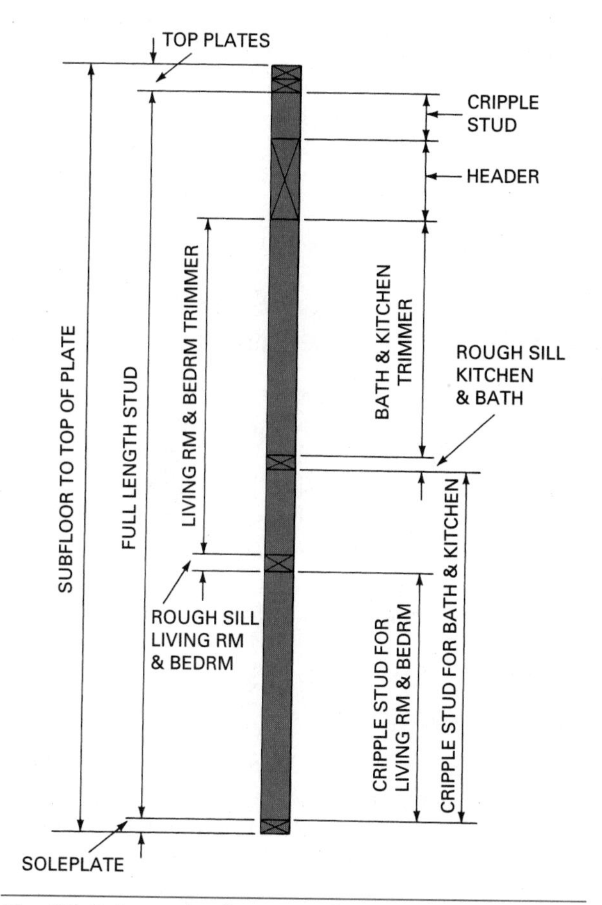

Fig. 35-3 Layout of a typical story pole.

Determining the Size of Rough Openings

A **rough opening** is an opening framed in the wall in which to install doors and windows. The width and height of rough openings are not usually indicated in the plans. It is the carpenter's responsibility to determine the rough opening size for the particular unit from the information given in the door and window schedule. The door and window schedule contains the kind, style, manufacturer's model number, size of each unit, and rough opening dimensions.

Rough Opening Sizes for Exterior Doors

The rough opening for an exterior door must be large enough to accommodate the door, door frame, and space for shimming the frame to a level and plumb position. Usually 1/2 inch is allowed for shimming, at the top and both sides, between the door frame and the rough opening. The amount allowed for the door frame itself depends on the thickness of the door frame beyond the door.

Care must be taken not to make the rough opening oversize. If the opening is made too large, the window or door finish may not cover it.

The sides and top of a door frame are called **jambs.** Jambs may vary in thickness. Sometimes *rabbeted* wood jambs are used. The rabbet is that part of the jamb that the door stops up against. At other times nominal 1- or 2-inch lumber is used for

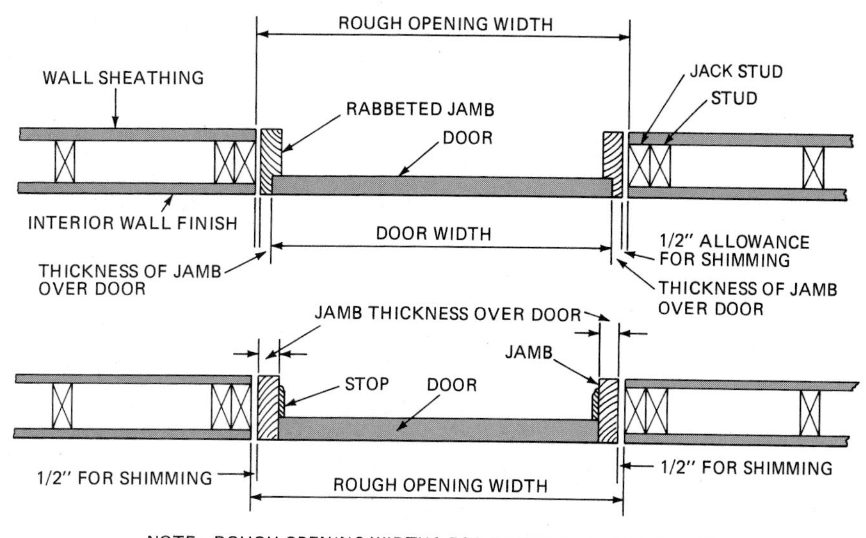

Fig. 35-4 Determinining the rough opening width of a door opening.

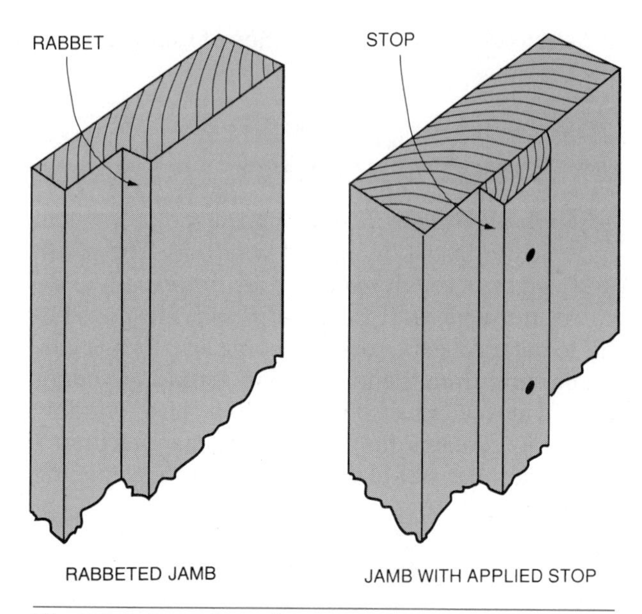

Fig. 35-4 (*continued*)

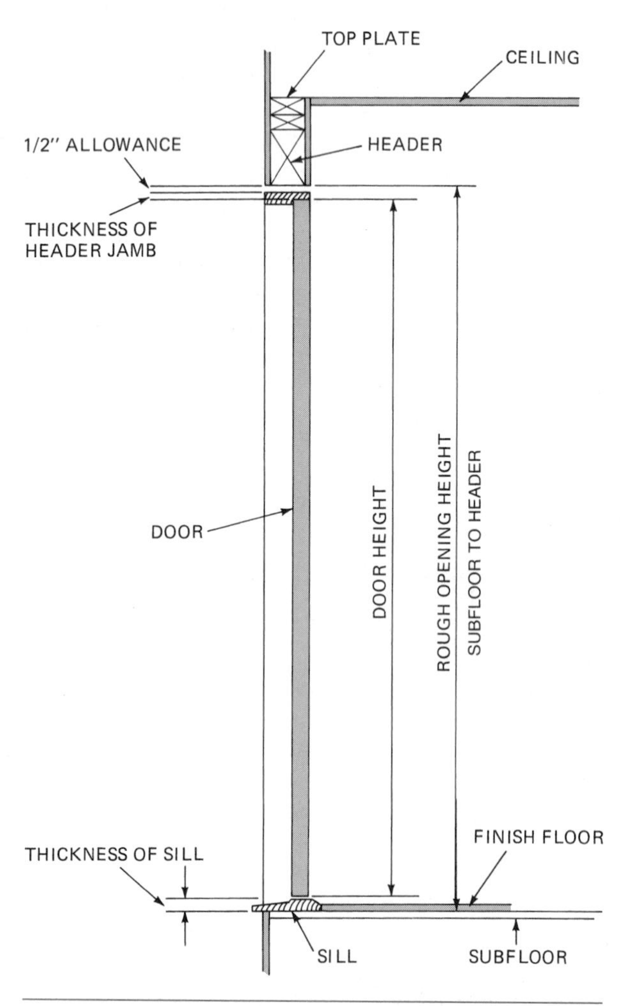

Fig. 35-5 Determining the rough opening height of an exterior door opening.

the jamb. A separate **stop** is applied (Fig. 35-4). Steel jambs have the door stop built-in similar to a rabbeted wood jamb.

The bottom member of the door frame is called a *sill*. Sills may be hardwood, metal, or a combination of wood and metal. The type of sill and its thickness must be known in order to figure the rough opening height (Fig. 35-5).

An exterior door may come *prehung* in its frame and the entire unit set in the opening. At other times, only the frame is set in the opening. The door is then hung in the frame at a later stage in the construction. In either case, the rough opening must be calculated in the same manner.

Rough Opening Sizes for Interior Doors

Rough openings for interior doors are found in a manner similar to that used with the exterior doors. Interior jambs are typically 3/4 inch in thickness with 1/2 inch on either side. These dimensions add up to 2 1/2 inches. The rough opening width is found by adding 2 1/2 inches to the width of the door. Rough opening height is found in the same manner as with an exterior door.

Rough Opening Sizes for Windows

Many kinds of windows are manufactured by a number of firms. Because of the number of styles, sizes, and variety of construction methods, it is best to consult the manufacturer's catalog to obtain the rough opening sizes. These catalogs show the style and size of the window unit. They also give the rough opening (RO) for each unit (Fig. 35-6). Catalogs are available from the lumber company that sells the windows.

Determining Wall Type

Before layout can begin, the carpenter must first determine if the wall to be laid out is load-bearing or non-load-bearing. It should be noted here that exterior walls are referred to as *walls* and interior walls are referred to as *partitions*. The load-bearing walls (LBW) usually are built first. They support the ceiling joist and rafters and typically run the length of the building. Non-load-bearing walls (NLBW) are end walls and run parallel with the joists. Interior partitions are also load and non-load-bearing. They are load-bearing partitions (LBP) if they run perpendicular to the joists, and are non-load-bearing partitions (NLBP) if they run parallel with the joists (Fig. 35-7).

Each type of wall has a slightly different layout characteristic. It is important to remember that

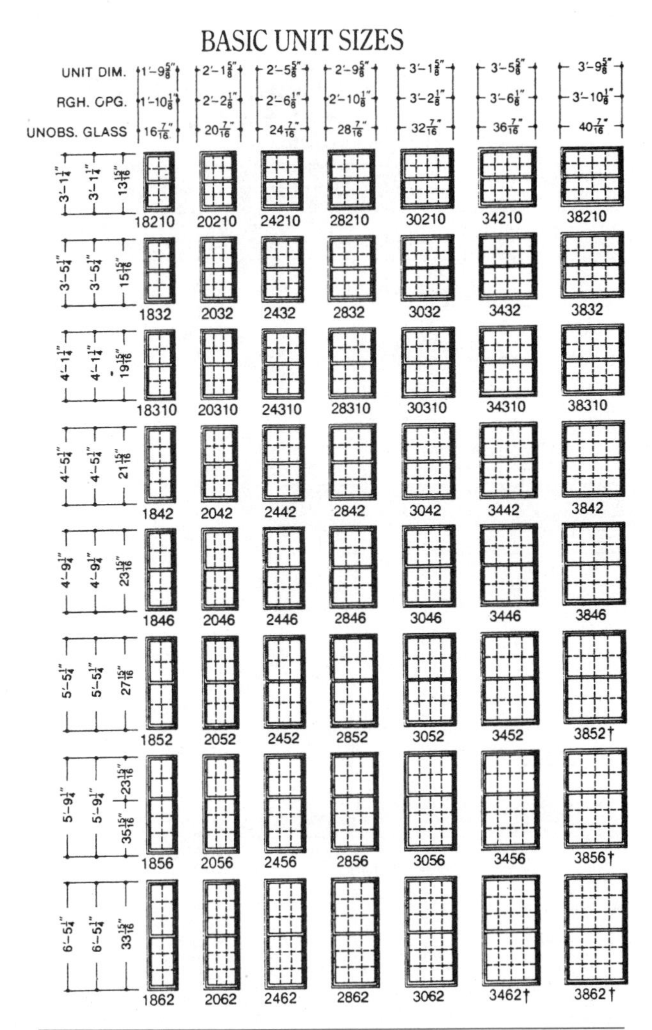

Fig. 35-6 Sample of a manufacturer's catalog showing rough opening sizes for window units. (*Courtesy of Andersen Corporation*)

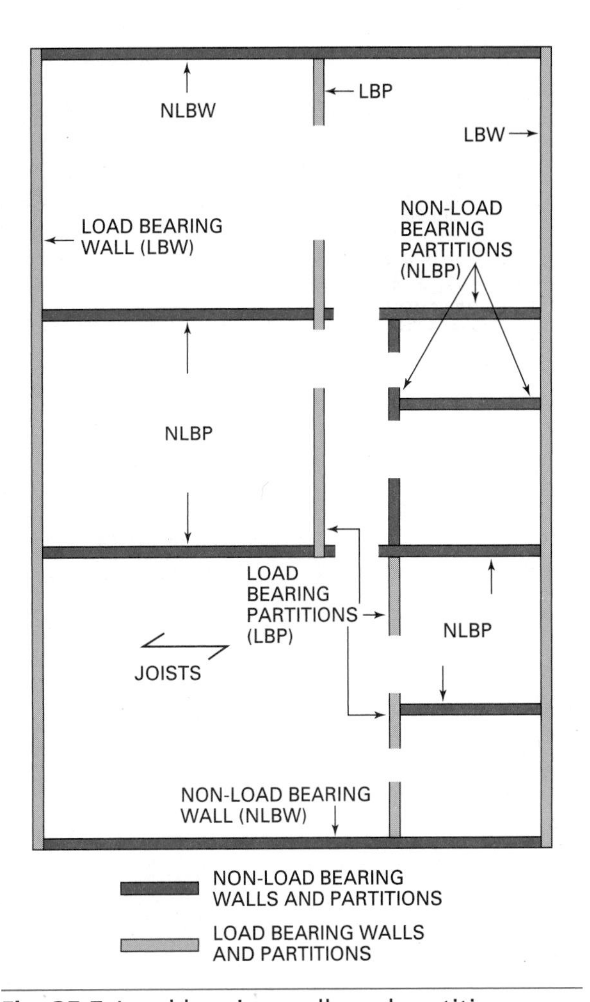

Fig. 35-7 Load-bearing walls and partitions run perpendicular to the joists. Non-load-bearing walls and partitions run parallel with the joists. Some minor non-load-bearing partitions may also run perpendicular.

all centerline dimensions for openings are measured from the building line, which is on the outside edge of the exterior framing. Layout must take this fact into account (Fig. 35-8). Figure 35-9 notes the similarities and differences of laying out walls and partitions.

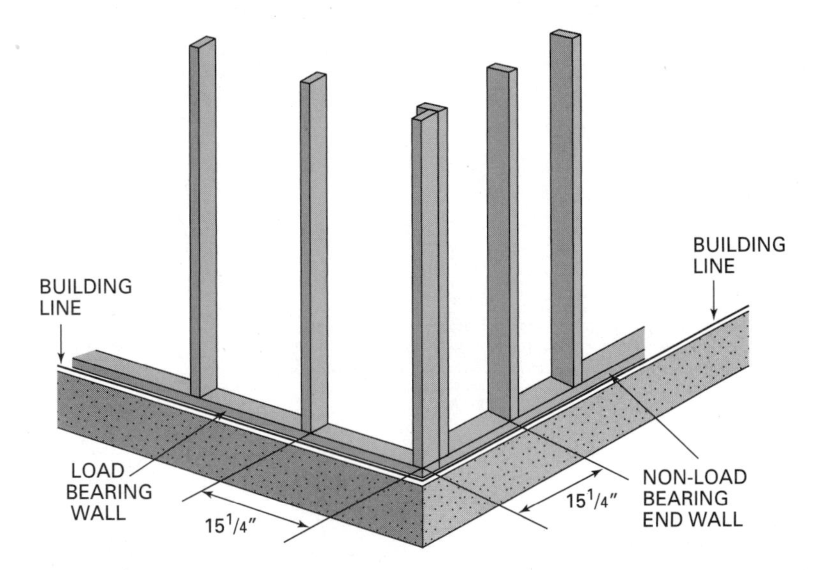

Fig. 35-8 Layout for on-center studs and centerlines for openings are measured from the building line.

Layout Variations for Walls and Partitions

	Measure to O.C. Studs	Measure to Centerlines of Openings
Load bearing wall (LBW)	from end of plate	from end of plate
Non-load bearing wall (NLBW)	include width of abutting wall and sheathing thickness	include width of abutting wall
Load bearing partition (LBP)	include width of abutting wall	include width of abutting wall
Non-load bearing partition (NLBP)	from end of plate	include width of abutting wall

1. LOAD BEARING WALL

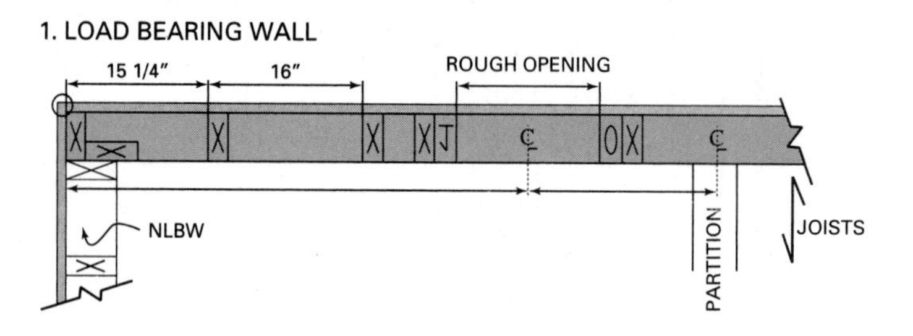

WALLS

2. NON-LOAD BEARING WALL

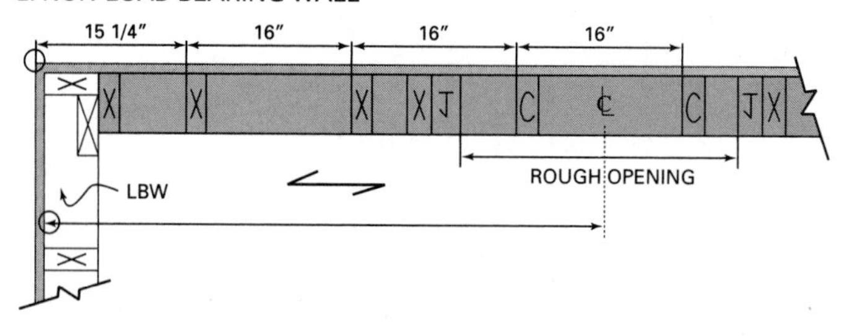

3. LOAD BEARING PARTITION

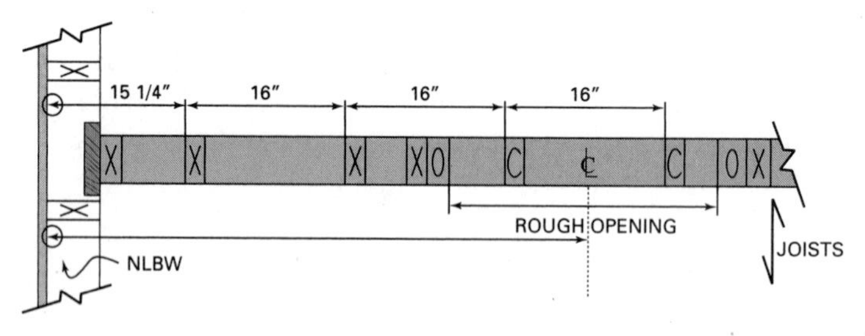

PARTITIONS

4. NON-LOAD BEARING PARTITION

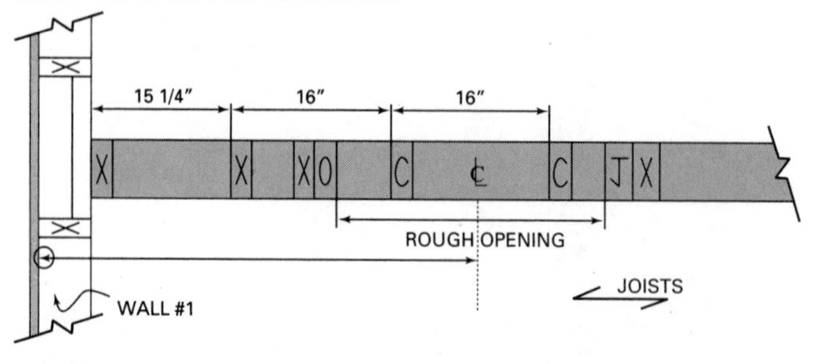

Fig. 35-9 Similarities and differences in wall layout.

Start the wall plate layout from the same corner that was used to start the layout for the floor frame.

Laying Out the Plates

To lay out the plates, measure in on the subfloor, at the corners, the thickness of the wall. Snap lines on the subfloor between the marks. *Tack* the soleplate in position so its inside edge is to the chalk line. Do not drive nails home (they have to be pulled later). Use only as many as are needed to hold the piece in place. Plan soleplate lengths so that joints between them fall in the center of a full-length stud for the convenient erection of wall sections later.

Wall Openings. From the blueprints, determine the centerline dimension of all the openings in the wall. Lay these out on the soleplate. Measure in each direction from the centerline one-half the width of the rough opening. Recheck the rough opening measurement to be sure it is correct. Square lines at these points across the soleplate. Mark a *T* for trimmer, or *O* for other, on the side of the line away from the center. To distinguish openings, a *T* may be used for window trimmers, a *J* for door jacks, and a *C* for cripple studs. It makes little difference what marks are used as long as the builder understands what they mean.

From the squared lines, measure back the stud thickness. Square lines across. Mark *X*s on the side of the line away from center for the full-length studs on each side of the openings (Fig. 35-10).

Partition Intersections

On blueprints, interior partitions usually are dimensioned to their centerline. Mark on the soleplate the centerline of all partitions intersecting the wall. From the centerlines, measure in each direction one-half the partition stud thickness. Square lines across the soleplate. Mark *X*s on the side of the lines away from center for the location of partition intersection studs (Fig. 35-11).

Studs and Cripple Studs

After all openings and partitions have been laid out, start laying out all full-length studs and cripple studs. Proceed in the same manner and from the same end as laying out floor joists. This keeps studs directly in line with the joists below.

Measure in from the outside corner the regular stud spacing. From this mark, measure in one direction or the other, one-half the stud thickness. Square a line across the soleplate. Place an *X* on the side of the line where the stud will be located.

Stretch a tape along the length of the soleplate from this first stud location. Square lines across the soleplate at each specified stud spacing. Place *X*s on the same side of the line as the first line.

Where openings occur, mark a *C* on the side of the line to indicate the location of cripple studs (Fig. 35-12). All regular and cripple studs should line up with the floor joists below. When laying out the opposite wall, start from the same end as the first wall to keep all framing lined up.

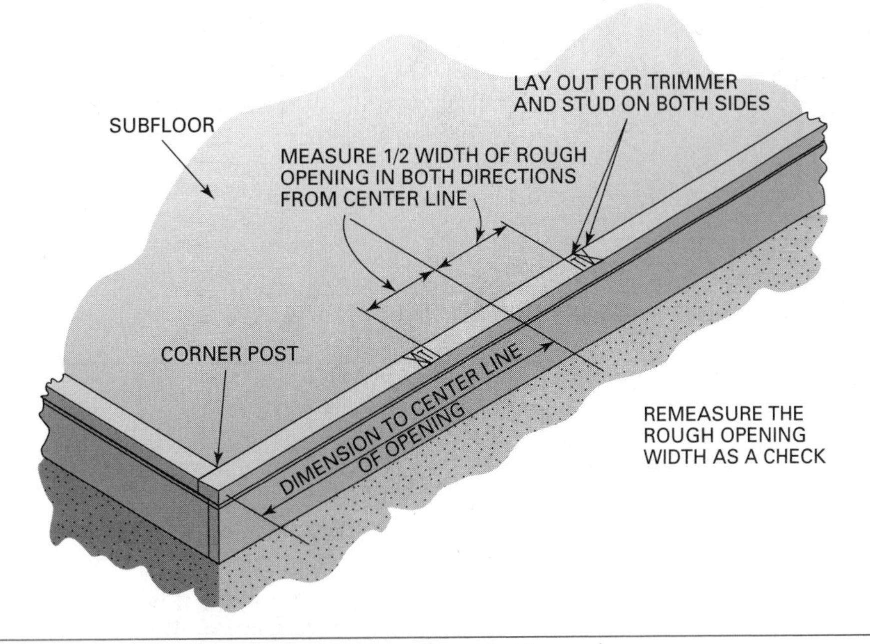

Fig. 35-10 Laying out a rough opening width on the soleplate.

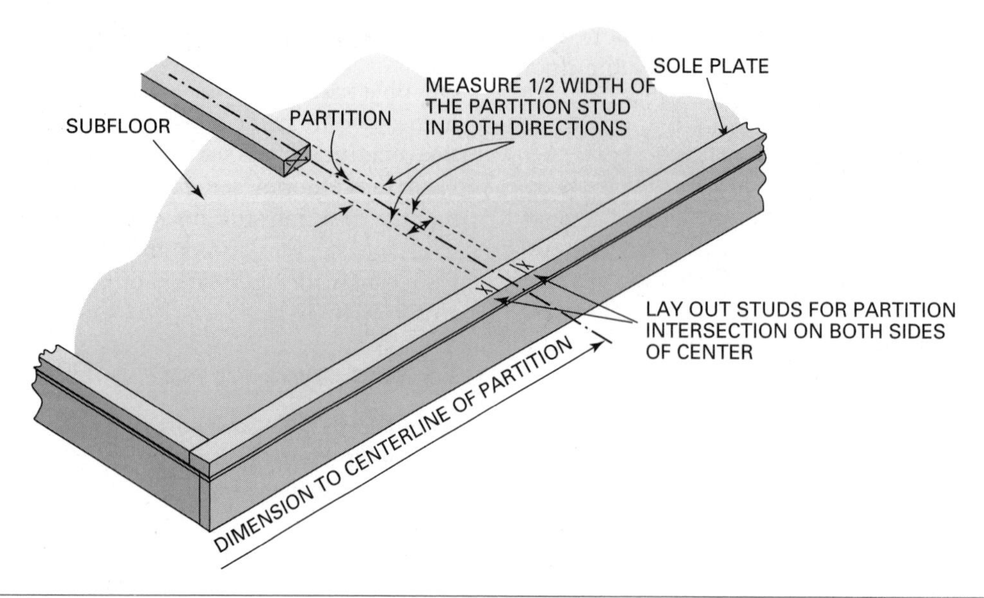

Fig. 35-11 Laying out a partition intersection on the soleplate.

STEP BY STEP
PROCEDURES

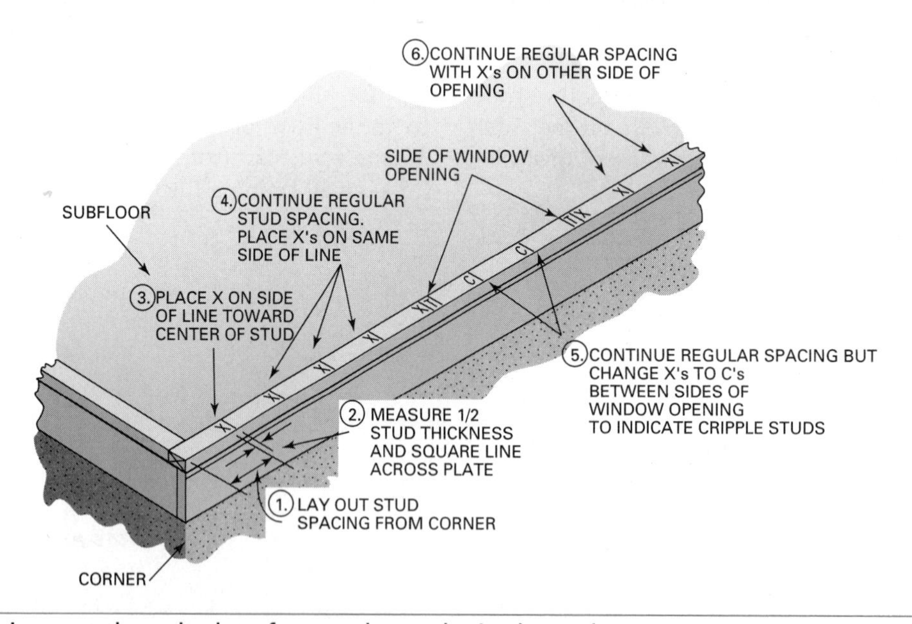

Fig. 35-12 Laying out the soleplate for regular and cripple studs.

Laying Out the Top Plate

Layouts for the floor joists, exterior walls, ceiling joists, and rafters all start at the same end of the building. For example always work from the north end toward the south end. To layout walls, use straight lengths of lumber for the top plate. Place a length alongside the soleplate. With one end flush with the end of the soleplate, mark and cut the other end of the top plate. Make sure it is centered on a stud and lines up with a joint in the soleplate (Fig. 35-13). Cut enough top plates for the length of the wall. Tack the top plate in position alongside the soleplate. Transfer the marks on the soleplate to the top plate (Fig. 35-14).

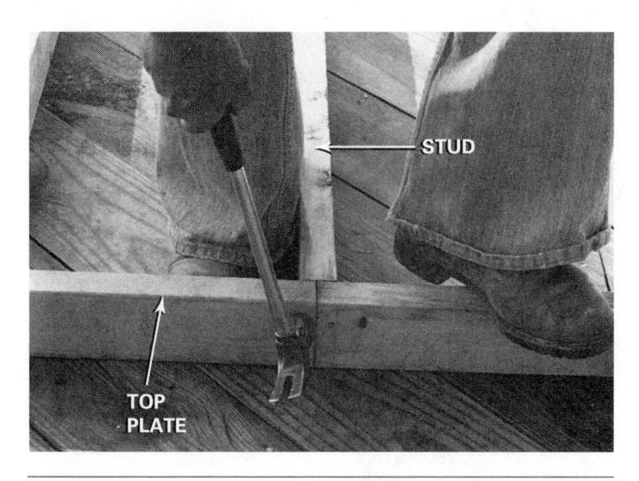

Fig. 35-13 Joints in the top plate should fall on the center of a stud.

Assembling and Erecting Wall Sections

The usual method of framing the exterior wall is to precut the wall frame members, assemble the wall frame on the subfloor, and erect the frame. With a small crew and without special equipment, the walls are raised section by section. When the frame is erected, the corners are plumbed and braced. Then the walls are straightened between corners. They are also braced securely in position. To prevent problems with the installation of the finish work later, it is important to keep the edges of the frame members flush wherever they join each other.

Precutting Wall Frame Members

Full-length Studs. To assemble a wall section, first cut the needed number of full-length studs to length. A radial arm or power miter saw is an effective tool for cutting studs and other framing to length. Set a stop the desired distance from the saw blade to cut duplicate lengths. If these saws are not available, a jig can be made for a portable electric circular saw, commonly known as a *skilsaw,* to cut duplicate lengths of framing (Fig. 35-15). Reject any studs that are severely warped. They may be cut into shorter lengths for blocking when making corner posts and partition intersections or between studs. It should be noted that precut studs that are already cut for length are available from most suppliers.

Corner Posts and Partition Intersections. Make up the necessary number of corner posts and partition intersections. Corners may be made by nailing two full-length studs together where one stud is rotated at a right angle. This detail allows for more insulation in the corner, thereby making the building more energy-efficient. Corners may also be made by nailing together two full-length studs with short blocks of the same material laid flat between them (Fig. 35-16).

Partition intersections may be made using ladder blocking, which again allows for more insulation in the wall. In this case no extra framing layout is needed because the ladder blocking is installed between on-center studs. Another method also

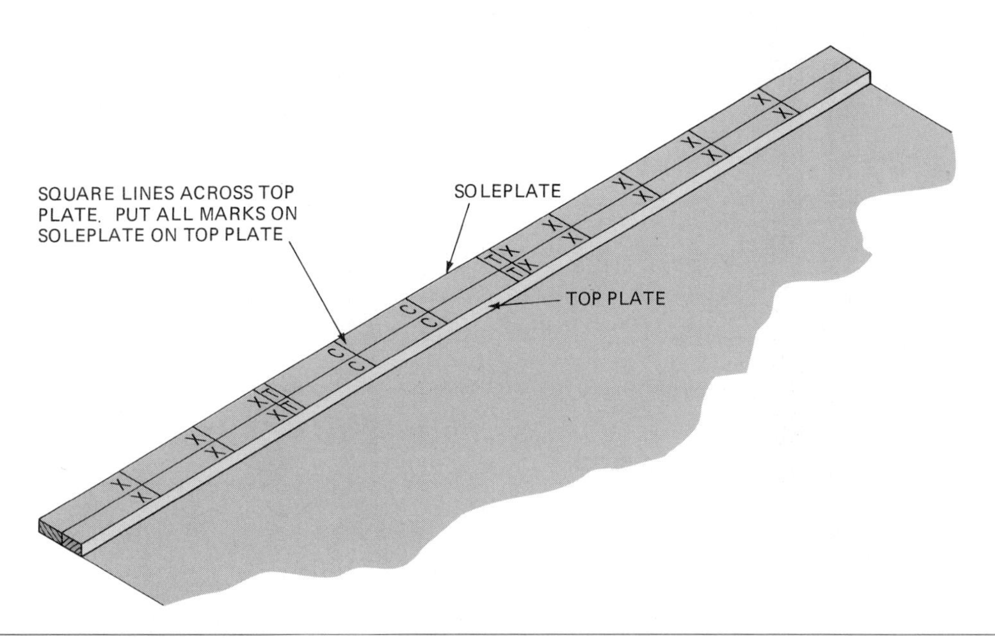

Fig. 35-14 Transferring the layout of a soleplate to the top plate.

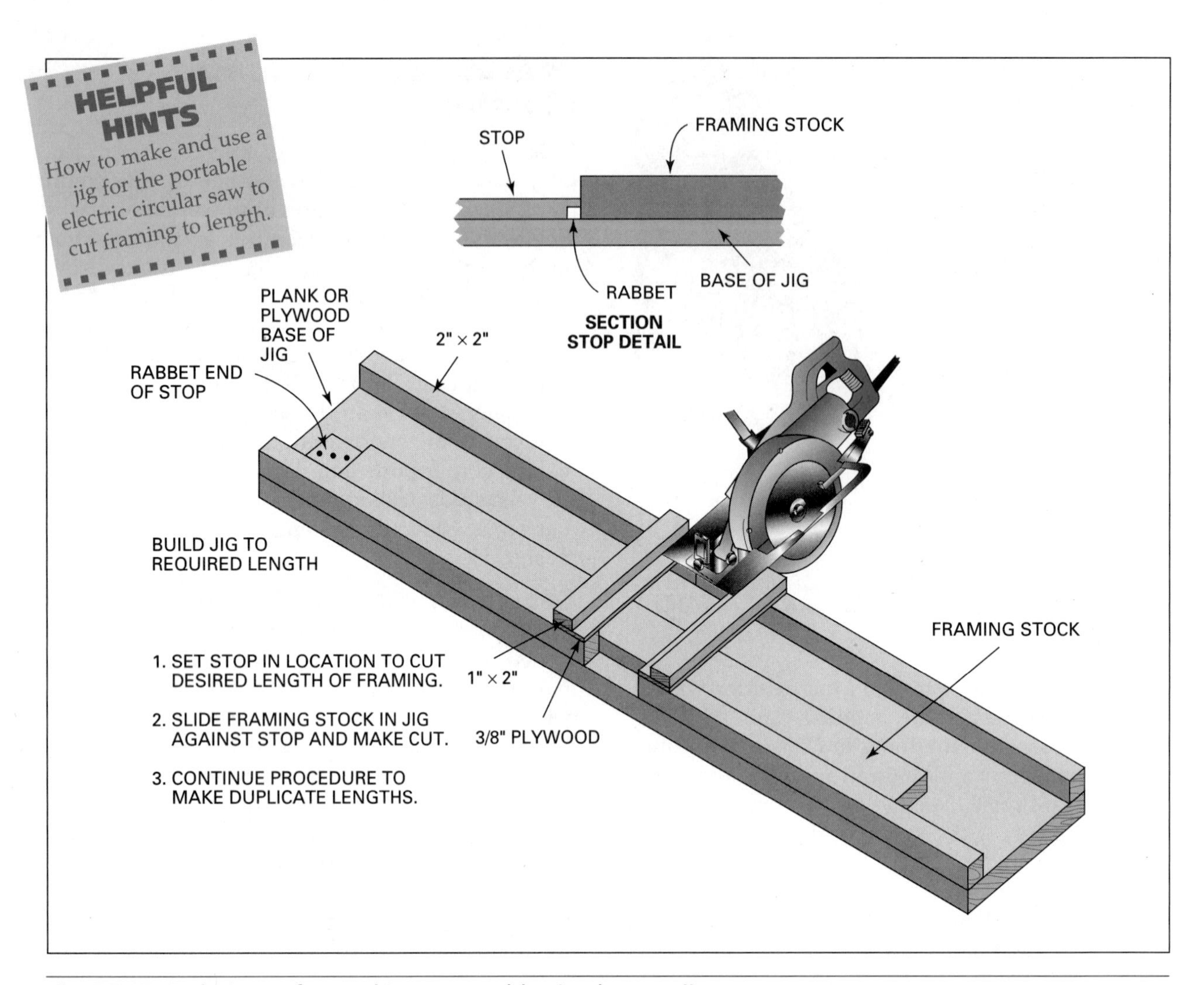

Fig. 35-15 Techniques for making a portable circular saw jig.

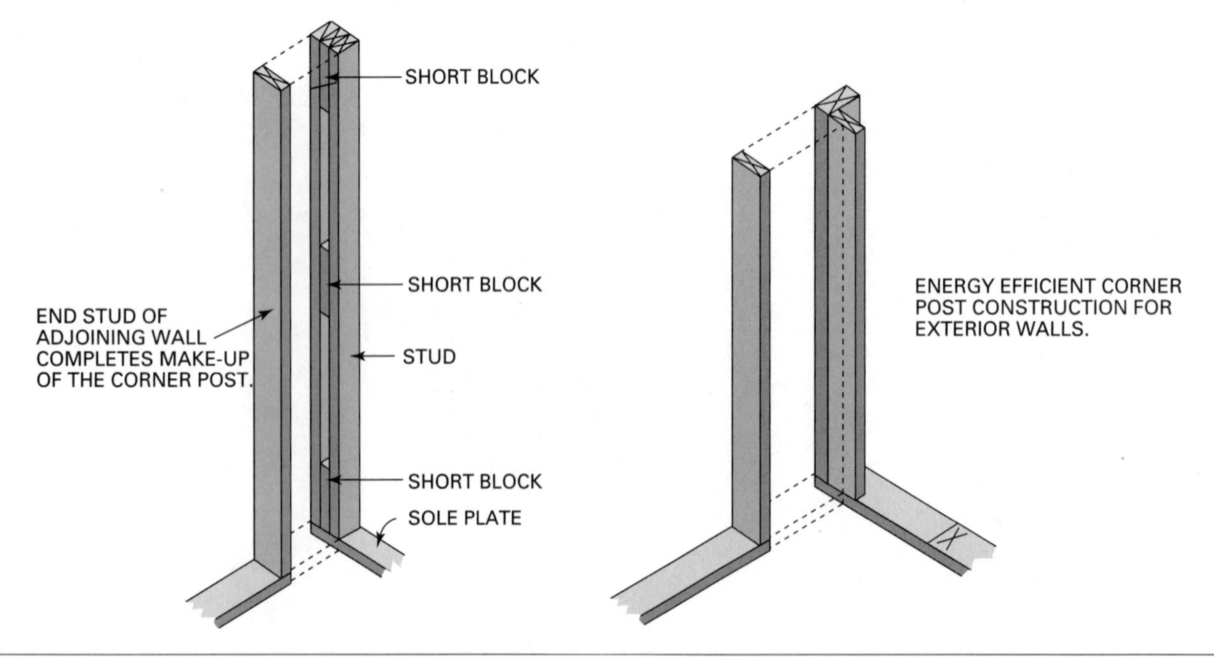

Fig. 35-16 Construction of a corner post.

allows for insulation easily to be installed later. A 2 × 6 is nailed at right angles to a stud. A third method is similar to the corner post except for the way the blocks are placed. They are placed in a similar location between two full-length studs, yet with blocks on their edge (Fig. 35-17).

Headers, Rough Sills, Jack Studs, and Trimmers. Cut all headers and rough sills. Their length can be determined from the layout on the plates.

Make a story pole. From it determine the length of all trimmers, jack studs, and cripple studs.

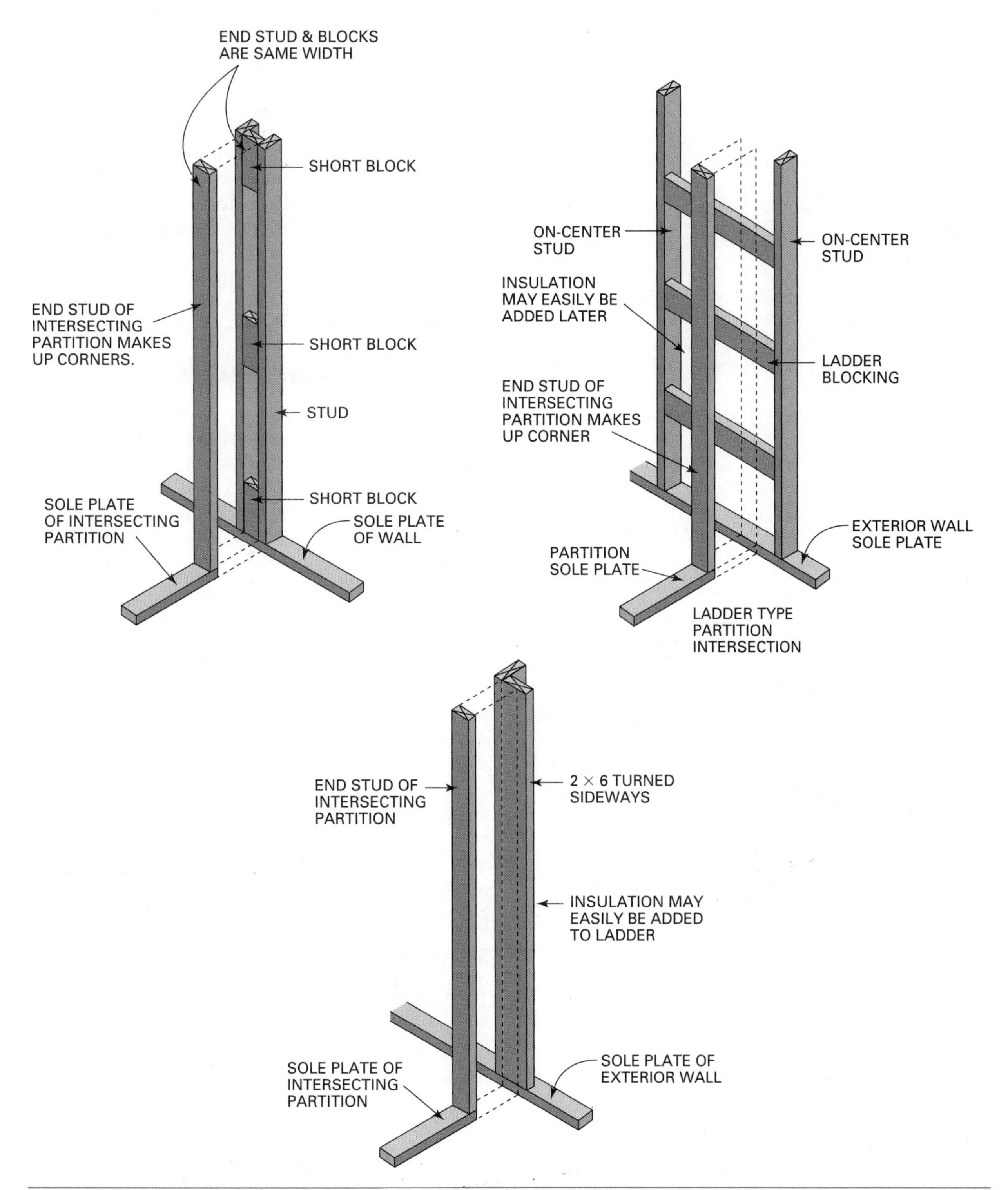

Fig. 35-17 Construction of a partition intersection.

Cut them accordingly. It may be necessary to place identifying marks on headers, rough sills, jacks, and trimmers if rough openings are different sizes. This will assist in locating the window or door unit to be placed in each rough opening.

Assembling Wall Sections

Separate the top plate from the sole plate. Stand them on edge. Place all full-length studs, corner posts, and partition intersections in between them. To avoid a mistake, be careful not to turn one of the plates around. Be certain that the layout lines on top and bottom plates line up.

Very few studs are absolutely straight from end to end. Carefully sight each full-length stud. It will be difficult for those who apply the interior finish if no attention is paid to the manner in which studs are installed in the wall. A stud that is installed with its crowned edge out next to one with its crowned edge in will certainly present problems later.

Fasten each stud, corner post, and partition intersection in the proper position by driving two 16d nails through the plates into the ends of the members. A pneumatic framing nailer makes the work easier.

Using the story pole, mark the position of all headers and rough sills on the full-length studs on each side of each opening. Fasten the headers and rough sills in position by nailing through the studs into their ends. If headers butt the top plate, nail through the plate into the header.

Fasten the rough sills in position to the trimmers and then fasten this assembly to the sole plate and header. Also fasten the trimmers to the full-length studs by driving the nails at a slight angle. Next, fasten the cripple studs as needed between the plates and sills or headers.

It may be necessary to toenail the ends of some members with 2 1/2-inch nails if nails cannot be driven in their ends. When doubling up studs, toenail their edges flush before nailing them together.

Bracing End Sections

If plywood sheathing is to be used, then no extra bracing is needed, because the plywood provides ample rigidity to the wall frame. If let-in braces are to be used, it is much easier to notch out the studs and plates and install the braces while the wall section is lying on the subfloor. Some builders nail the wall sheathing to the wall before erecting the wall. Before installing wall sheathing or laying out for the brace notches, the wall section should be square. To do this, snap a chalk line on the subfloor where the inside edge of the wall plate will rest. Align the bottom plate to the line and adjust the ends of the plate into their proper position (Fig. 35-18). Toenail the sole plate to the subfloor with 16d nails spaced about every 6 to 8 feet along what will be the top side of the sole plate when the wall is in its final position. Measure the diagonals from corner to corner both ways. If they are equal, the section is square. Toenail the top plate to the subfloor using one or two nails. The section is now ready for sheathing or bracing.

If let-in bracing is used, tack the brace in position against the studs and plates. If the brace is to be at a 45 degree angle, measure in from the corner the same distance as the height of the wall.

Mark the studs and plates along each edge of the brace. Remove the brace. Mark the depth of cut on both sides of each member. Use a portable electric circular saw with the blade set for the depth of the notch. Make multiple saw cuts between the layout lines. Use a wood chisel or the claw of a straight claw and trim the remaining waste to form the notch. Fasten the brace in the notches using two 8d common nails in each framing member (Fig. 35-19).

Erecting Wall Sections

To erect the wall, remove the toenails from the top plate while leaving the toenails in the bottom plate. The bottom toenails will remain until after the section is erected; they will help keep the frame in position while the frame is raised. Lift the wall section into place, plumb, and temporarily brace. After checking to be sure that the sole plate is on the chalk line, nail the sole plate to the band or floor joists below every 16 inches along the length. Sections are fastened together by nailing the top plates to the projecting end of the studs of the adjoining wall sections. In corners, fasten end studs together to complete the construction of the corner post. A completed corner post provides surfaces to fasten both exterior and interior wall finish.

Brace each section temporarily as erected. Fasten one end of a board to a 2 × 4 block that has been nailed to the subfloor and the other end to the side of a stud (Fig. 35-20). Either do not drive nails all the way home or use duplex head nails in temporary braces so they can be removed later.

Plumbing and Bracing the Corner

If the walls have not been previously braced while being framed on the subfloor, or if no permanent bracing is required, all corners must be plumbed and temporarily braced. Install braces for both sides of each corner. Fasten the top end of the brace to the top plate near the corner post on the inside of the wall.

Temporary braces are fastened to the inside of the wall. They can remain in position until

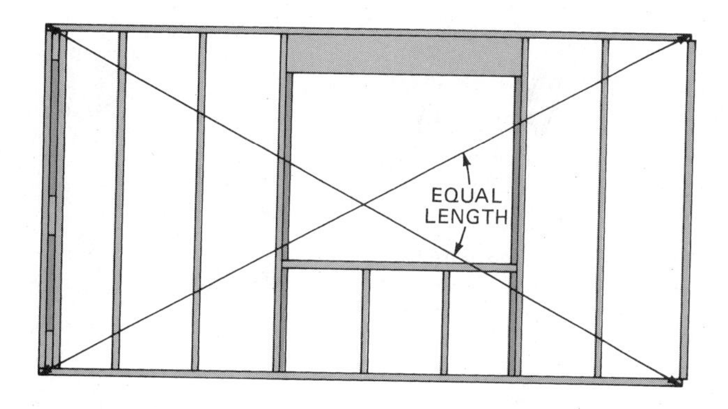

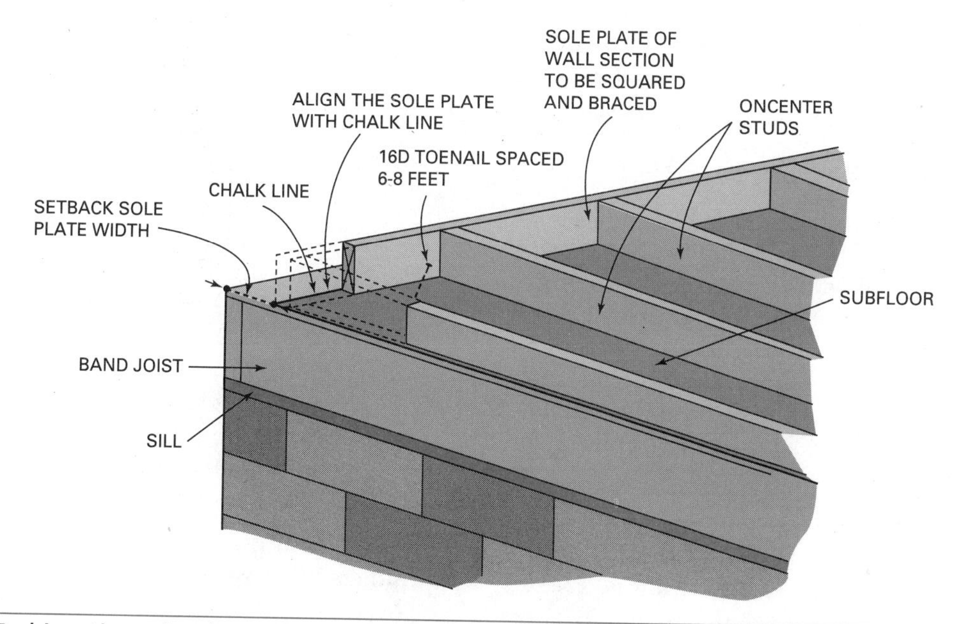

Fig. 35-18 Tacking the sole plate of a wall section to the deck helps in squaring the wall. When studs are equal, the plates are equal, and the diagonal measurements of a wall section are equal, the section is square.

Fig. 35-19 A let-in corner brace.

35-20 Raising an exterior wall section. The sheathing has already been applied.

the exterior wall sheathing is applied. Sometimes they remain until it is absolutely necessary to remove them for the application of the interior wall finish. Care must be taken not to let the ends of the braces extend beyond the corner post or top plate. This might interfere with the application of wall sheathing or subsequent ceiling or roof framing.

Method of Plumbing Corner Posts. Plumb the corner post and fasten the bottom end of the brace. Use duplex head nails or drive nails only partway in for easy removal. For accurate plumbing of the corner posts, use a 6-foot level with accessory aluminum blocks attached to each end. The blocks keep the level from resting against the entire surface of the corner post. This prevents any bow or irregularity in the surface from affecting the accurate reading of the level.

An alternate method is to use the carpenter's hand level in combination with a long straightedge. On the ends of the straightedge, two blocks of equal thickness are fastened to keep the edge of the straightedge away from the surface of the corner post (Fig. 35-21). A plumb bob, transit-level, or laser level may also be used to plumb corner posts.

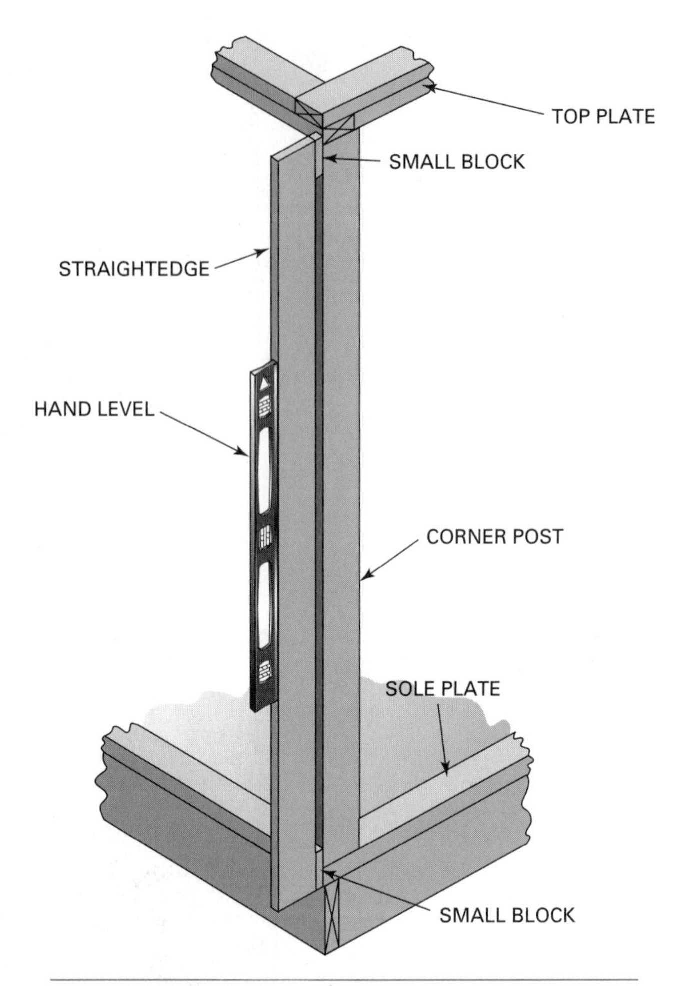

Fig. 35-21 Using a straightedge and level to plumb corner posts.

Doubling the Top Plate

If the complete wall is to be raised in one piece instead of in sections, the top plate can be doubled while the wall is still on the subfloor. Recess or extend the doubled top plate to make a lap joint at the corners and intersections. (Fig. 35-22). Be sure to nail the doubled top plate into the top plate such that nails are located above the studs. This will ensure that any holes drilled for wiring or plumbing into the top plates will not hit a nail.

Using Wall Jacks. Usually, if the full length of the wall is framed in one piece, the wall sheathing is also applied before the wall is raised. Sometimes even the windows, doors, and exterior wall finish are applied. When this is done, the wall is very heavy. Wall jacks have to be used to raise the wall into position (Fig. 35-23).

If the wall is erected in sections, after all the sections are raised, the top plate is doubled. First, toenail into the edges to bring them flush. Then spike them together with 16d nails staggered from edge to edge about every 16 inches. Stagger the joints in the top member as far as possible away from those in the bottom member. Nail only over the studs.

Partition Intersections. Where partitions intersect the exterior wall, a space is left for the top plate of the partition to lap the plate of the exterior wall (Fig. 35-24). Lapping the plates of the interior partitions with those of the exterior walls ties them together. This makes a more rigid structure.

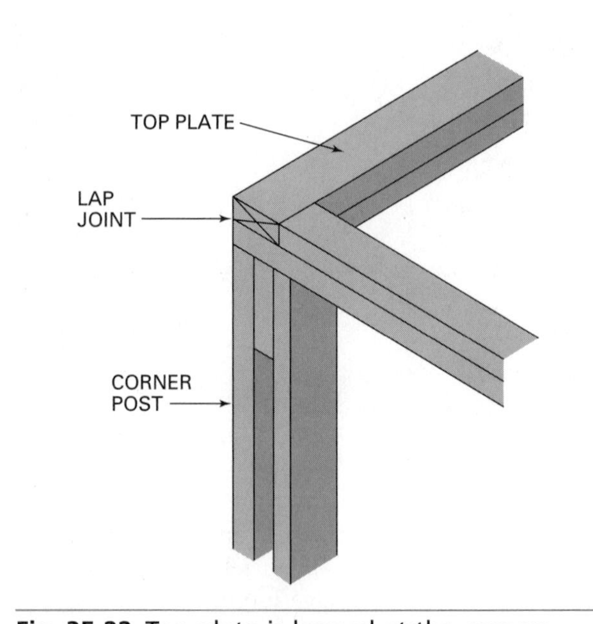

Fig. 35-22 Top plate is lapped at the corners.

Straightening the Walls

After the corner posts have been plumbed and the top plates doubled, the tops of the walls must be straightened and braced. To straighten each wall, use three blocks of 1-inch lumber of exactly the

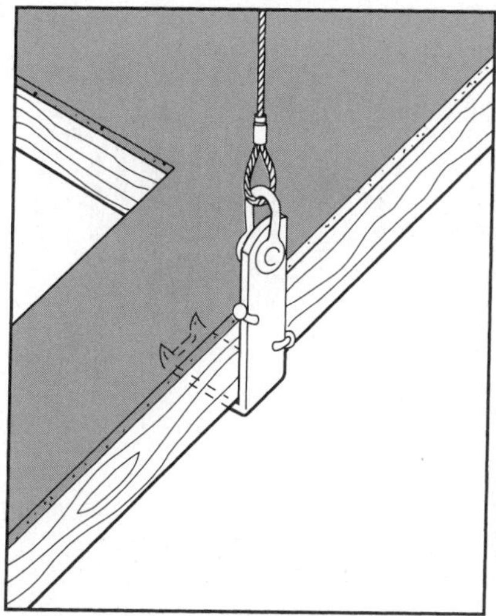

1. THE LIFTING BRACKETS ARE PLACED UNDER THE TOP PLATE OR HEADER AND SECURELY NAILED. THE BOTTOM OF THE WALL IS TOE-NAILED TO THE SUBFLOOR, OR OTHERWISE SECURED TO PREVENT IT FROM SLIPPING DURING THE LIFT.

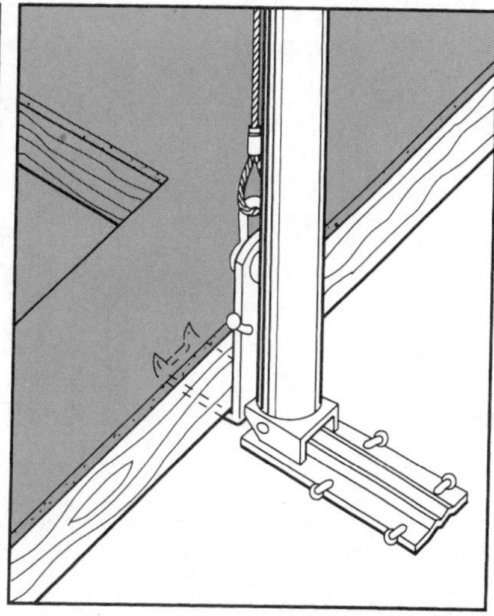

2. THE JACKS ARE PLACED UPRIGHT, FLUSH AGAINST THE LIFTING BRACKET. THE HINGED FOOT IS SECURELY NAILED TO THE SUB-FLOOR NEAR A FLOOR JOIST.

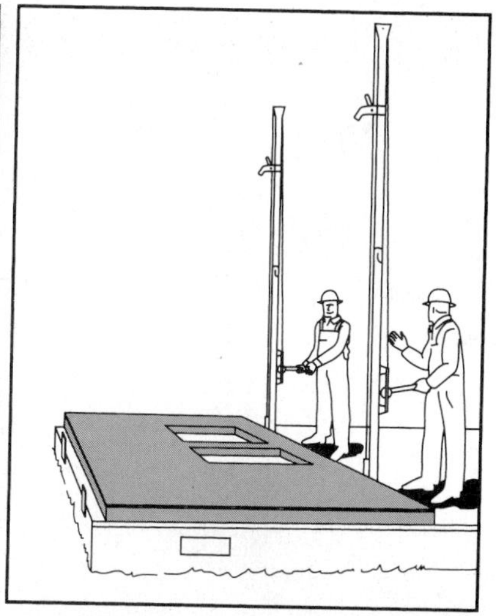

3. THE LIFT BEGINS, BOTH JACKS BEING OPERATED IN UNISON.
• IF ONE PERSON IS RAISING THE WALL, HE SIMPLY MOVES FROM ONE JACK TO THE OTHER AVOIDING EXCESSIVE TWISTING OF THE WALL.

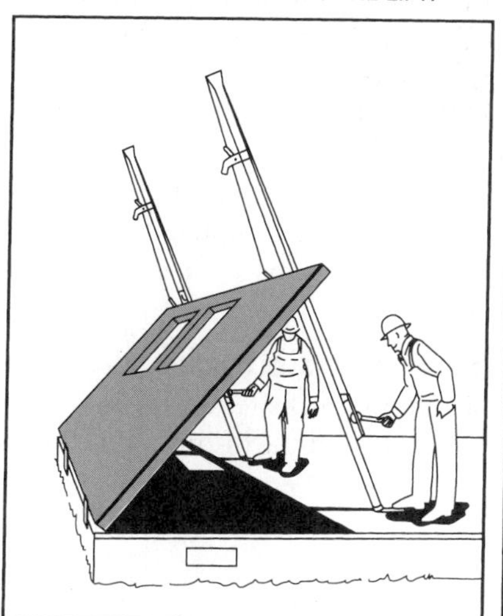

4. THE JACKS REMAIN IN CONTACT WITH THE TOP EDGE OF THE WALL DURING THE ENTIRE LIFT.

5. WHEN THE WALL REACHES THE UPRIGHT POSITION, IT IS STOPPED AND HELD FIRMLY BY THE PRESET WALL STOPS, WHICH ARE ADJUSTABLE.

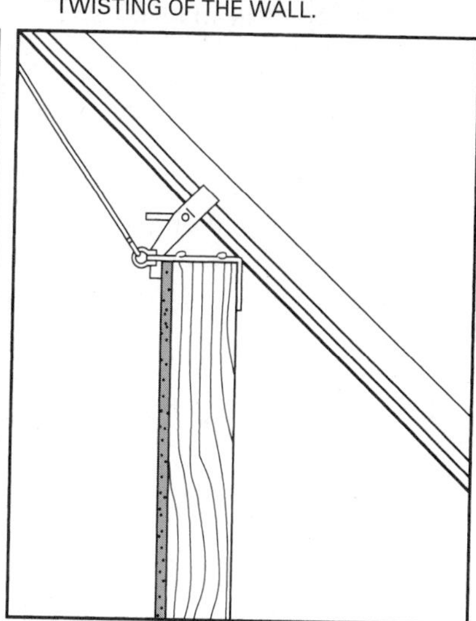

6. THE WALL JACKS PROVIDE A SUBSTANTIAL BRACE UNTIL THE WALL CAN BE BRACED IN A CONVENTIONAL MANNER.

Fig. 35-23 A small crew can easily erect a heavy wall using wall jacks. (*Courtesy of Proctor Products Company*)

same thickness. To assure equal thickness of the blocks, cut them from the same piece. Fasten two blocks to the inside of the top plate at each end.

Stretch a line tightly over and between the two corner blocks. Use the third block as a test block. Move the wall in or out at about 8- or 10-foot intervals until the test block just clears the line when held against the top plate (Fig. 35-25). It is important that the block does not move the line. It must come as close as possible *without touching*. If allowed to touch, the line is no longer straight.

Combination Wall Aligner and Brace. It may be difficult to hold the wall in alignment while bracing it. Either a manufactured, combination wall aligner and brace or a job-built type is then used. A manufactured type is similar to that used on concrete forms. A few turns on the

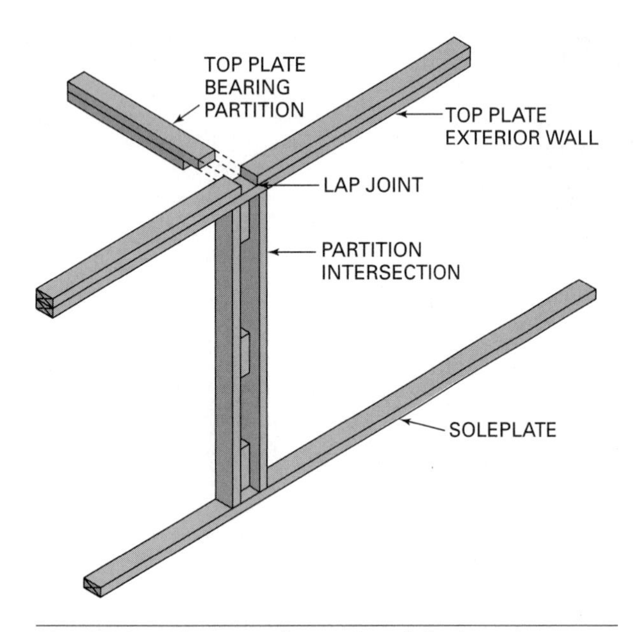

Fig. 35-24 The top plate of weight-bearing partitions laps the exterior wall.

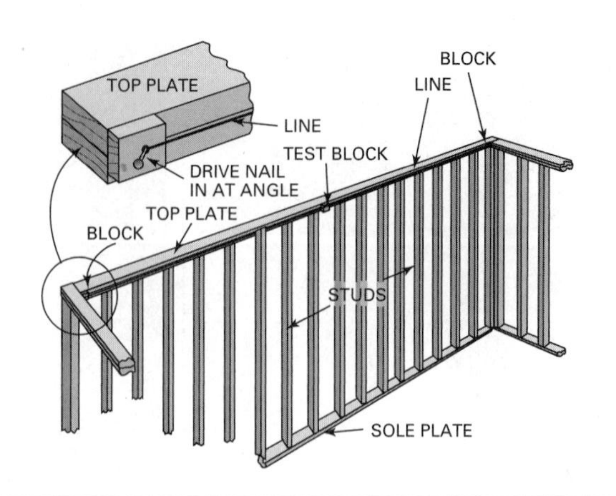

Fig. 35-25 Straightening the wall with a stretched line and test block.

adjusting screw moves the wall in or out as desired (Fig. 35-26). Use as many braces as necessary to straighten the wall along its entire length.

A job-built combination wall aligner and brace is called a *spring brace.* A spring brace consists of a 1 × 8 board about 12 or 14 feet long. The top end is securely fastened, with four or five 8d common nails driven home, to the underside of the top plate. The bottom end is fastened to the subfloor in the same manner so the top of the wall has to be brought in about an inch or less.

Spring the board upward by installing a piece of the same material in a vertical position between it and the subfloor about midway on the brace (Fig. 35-27). Within limits, the more the brace is sprung upward, the more the wall is brought inward. Bring the wall inward so the test block just fits between the stretched line and the top plate without touching the line. When the wall is straight, fasten the uprights by nailing through the brace and toenailing to the subfloor.

Installing Baseboard Blocking. Sometimes short 2 × 4 blocks are fastened against both sides of the corner posts and partition intersections

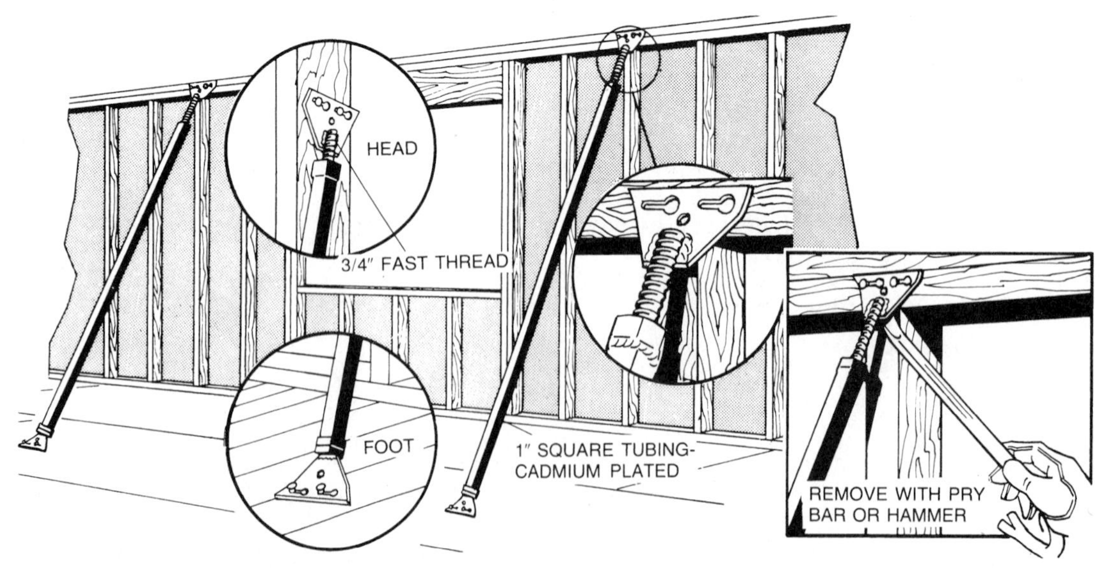

Fig. 35-26 A manufactured adjustable wall brace aligns and braces the wall. (*Courtesy of Proctor Products Company*)

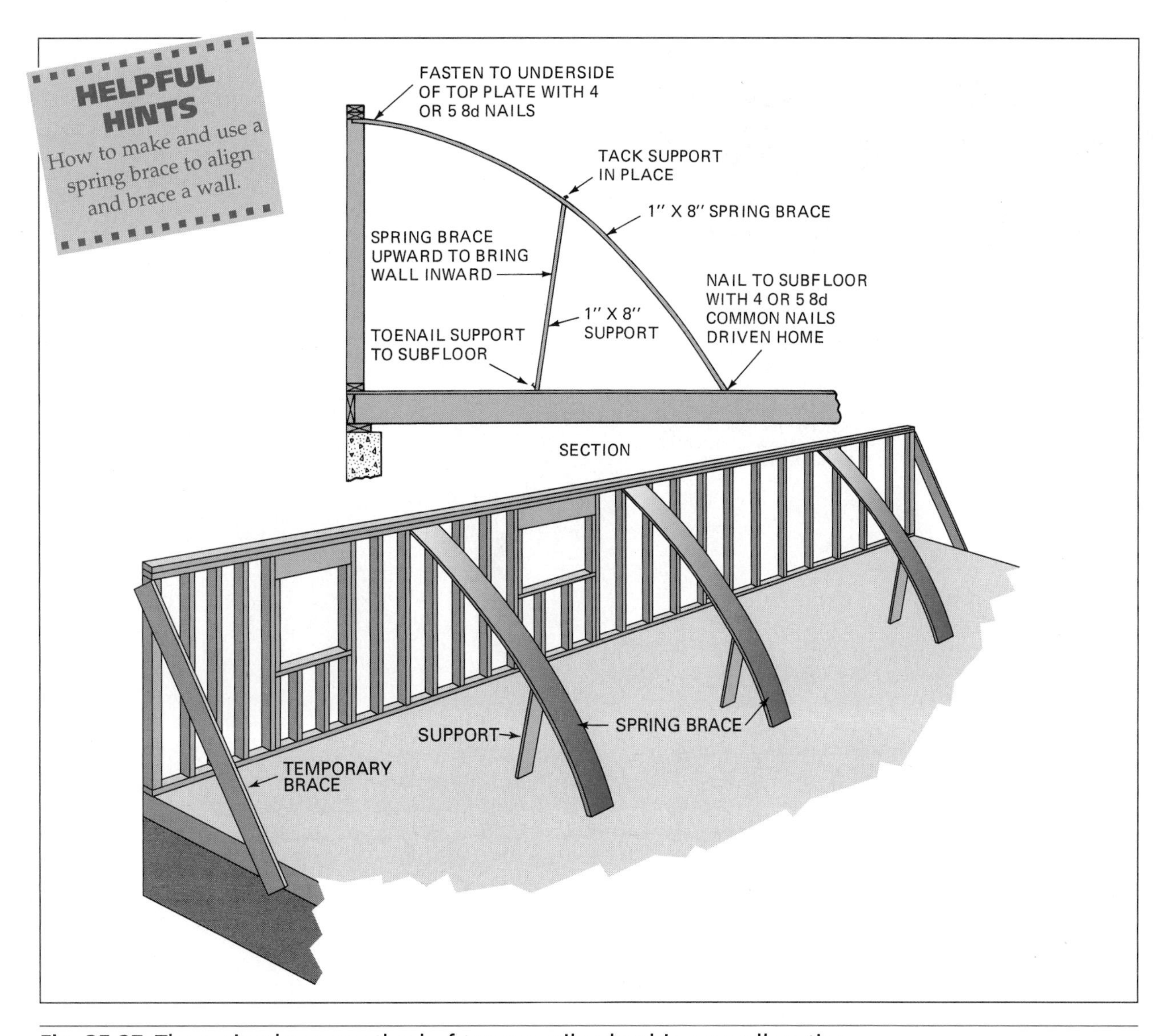

Fig. 35-27 The spring brace method of temporarily plumbing a wall section.

and down on the sole plate to provide fastening for the ends of the baseboard (Fig. 35-28).

Wall Sheathing

Wall sheathing covers the exterior walls. It may consist of boards, rated panels, fiberboard, **gypsum board,** or rigid foam board.

Boards. Boards are seldom used because of the high cost of material and labor in comparison to other kinds of wall sheathing. If used, they may be applied diagonally or horizontally. Diagonal sheathed walls require no other bracing. They make the frame stiffer and stronger than boards applied horizontally. Use two 2-1/2 inch or 8d common nails at each stud for 6- and 8-inch boards. Use three nails at each stud for 10- and 12-inch boards. End joints must fall over the center of studs. They must be staggered so no two successive end joints fall on the same stud.

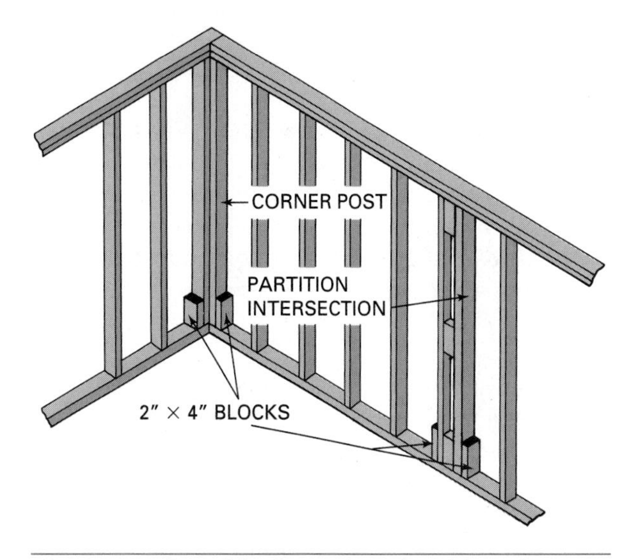

Fig. 35-28 Short blocks against the sides of corner posts and partition intersections provides fastening support for the ends of the baseboard.

APA Panel Wall Sheathing (a)
(APA RATED SHEATHING panels continuous over two or more spans.)

Panel Span Rating	Maximum Stud Spacing (in.)	Nail size (b) (c)	Maximum Nail spacing (in.)	
			Supported Panel Edges	Intermediate Supports
12/0, 16/0, 20/0 or Wall- 16 oc	16	6d for panels 1/2" thick or less; 8d for thicker panels	6	12
24/0, 24/16, 32/16 or Wall- 24 oc	24	6d for panels 1/2" thick or less; 8d for thicker panels	6	12

(a) See requirements for nailable panel sheathing when exterior covering is to be nailed to sheathing.

(b) Common, smooth, annular, spiral-thread, or galvanized box.

(c) Other code-approved fasteners may be used.

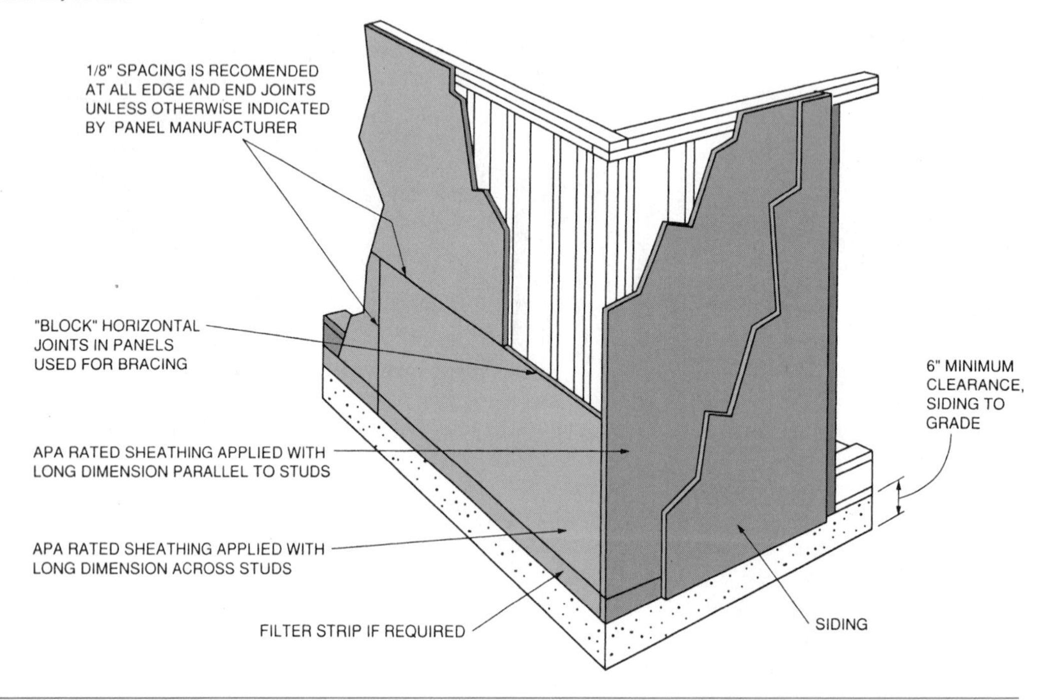

Fig. 35-29 Selection and fastening guide for APA-rated panel wall sheathing. (*Courtesy of American Plywood Association*)

Rated Panels. American Plywood Association-rated wall sheathing panels may be applied horizontally or vertically. There is no need for corner braces. A minimum 3/8-inch thickness is recommended when the sheathing is to be covered by exterior siding. Greater thicknesses are recommended when the sheathing also acts as the exterior finish siding.

Use 2-inch or 6d nails spaced 6 inches apart on the edges and 12 inches apart on intermediate studs for panels 1/2 inch thick or less. Use 2-1/2-inch or 8d nails for thicker sheathing panels (Fig. 35-29).

Rated sheathing panels are often used in combination with fiberboard, rigid foam, or gypsum board. When panels are applied vertically on both sides of the corner, no other corner bracing is necessary (Fig. 35-30).

Fig. 35-30 When plywood is used at the corners in combination with fiberboard sheathing, no additional corner bracing is required.

Other Sheathing Panels. *Fiberboard sheathing,* commonly called insulation board, is a soft, synthetic sheet material. Common sheet sizes are 2 × 8 and 4 × 8 and 4 × 9. The edges and ends may be **matched** (tongue-and-groove). Standard thicknesses are 1/2 inch and 25/32 inch. Most sheets are coated with asphalt for water resistance.

Fiberboard sheathing is fastened with roofing nails spaced 3 inches apart around the edges and 6 inches apart in the center. Fasteners must be kept a minimum of 3/8 inch from the edges. It may be applied vertically or horizontally.

Gypsum sheathing consists of a treated gypsum filler between sheets of water-resistant paper. Usually 1/2 inch thick, 2 feet wide, and 8 feet long, the sheets have matched edges to provide a tighter wall. Because of the soft material, galvanized wall board nails must be used to fasten gypsum sheathing.

Space the nails about 4 inches around the edges and 8 inches in the center. Gypsum board sheathing is used when a more fire-resistant sheathing is required.

Rigid foam sheathing is used when greater insulation is required (Fig. 35-31). A disadvantage of using fiberboard, gypsum board, or rigid foam sheathing is that the exterior wall siding that covers it cannot be nailed to it. Either the siding must be fastened to the studs or special fasteners must be used.

Application of Sheathing Panels. Sheathing panels are applied in a manner similar to applying subfloor. Snap a line 4 feet up across the wall. Fasten the first row of horizontally applied panels so their top edges are to the line. Start the next row with a half-panel to stagger the joints. It may be helpful to drive either some wood shingle tips or nails at intervals between the foundation and the sill plate. Let them project somewhat so the first row of panels can rest on them (Fig. 35-32). The same technique may be used to support the bottom ends of vertically applied panels. Panels should be installed with tight-fitting seams and as few seams as possible. This will make the house more airtight.

Estimating Materials for Exterior Walls

To estimate the amount of material needed for exterior walls, first determine the total linear feet of exterior wall. Then, figure one stud for every linear foot of wall, if spaced 16 inches on center. This allows for the extras needed for corner posts, partition intersections, trimmers, door jacks, and blocking openings. For 24-inch spacing, divide the total linear feet of wall by two. Allow an extra number for openings, corners, and partition intersections.

For plates, multiply the total linear feet of wall by three (one sole plate and two top plates). Add 5 percent for waste in cutting.

For headers and rough sills, calculate the size for each opening. Add together the material needed for different sizes.

For wall sheathing, first find the total area to be covered. Multiply the total linear feet of wall by the wall height to find the total number of square feet of wall area. Deduct the area of any large openings. Disregard small openings.

For board sheathing, add a percentage for waste to the area to be covered to find the total number of board feet. Add 20 percent for 6-inch boards, 15 percent for 8-inch boards, and 10 percent for 10-inch boards. These large percentages are not actually waste in cutting, but are mostly the difference between the nominal and actual sizes of the boards.

To find the number of sheathing panels, divide the total wall area to be covered by the number of square feet in each sheet. For instance, if a panel

Fig. 35-31 Rigid foam insulation is used for exterior wall sheathing. Insulation board sheathing is applied before the walls are erected. *(Courtesy of Dow Chemical Corporation)*

HELPFUL HINTS

Drive wood shingle tips or spikes between the foundation and the sill plate. Let them project so the first row of sheathing panels can rest on them. Remove when the panel is fastened.

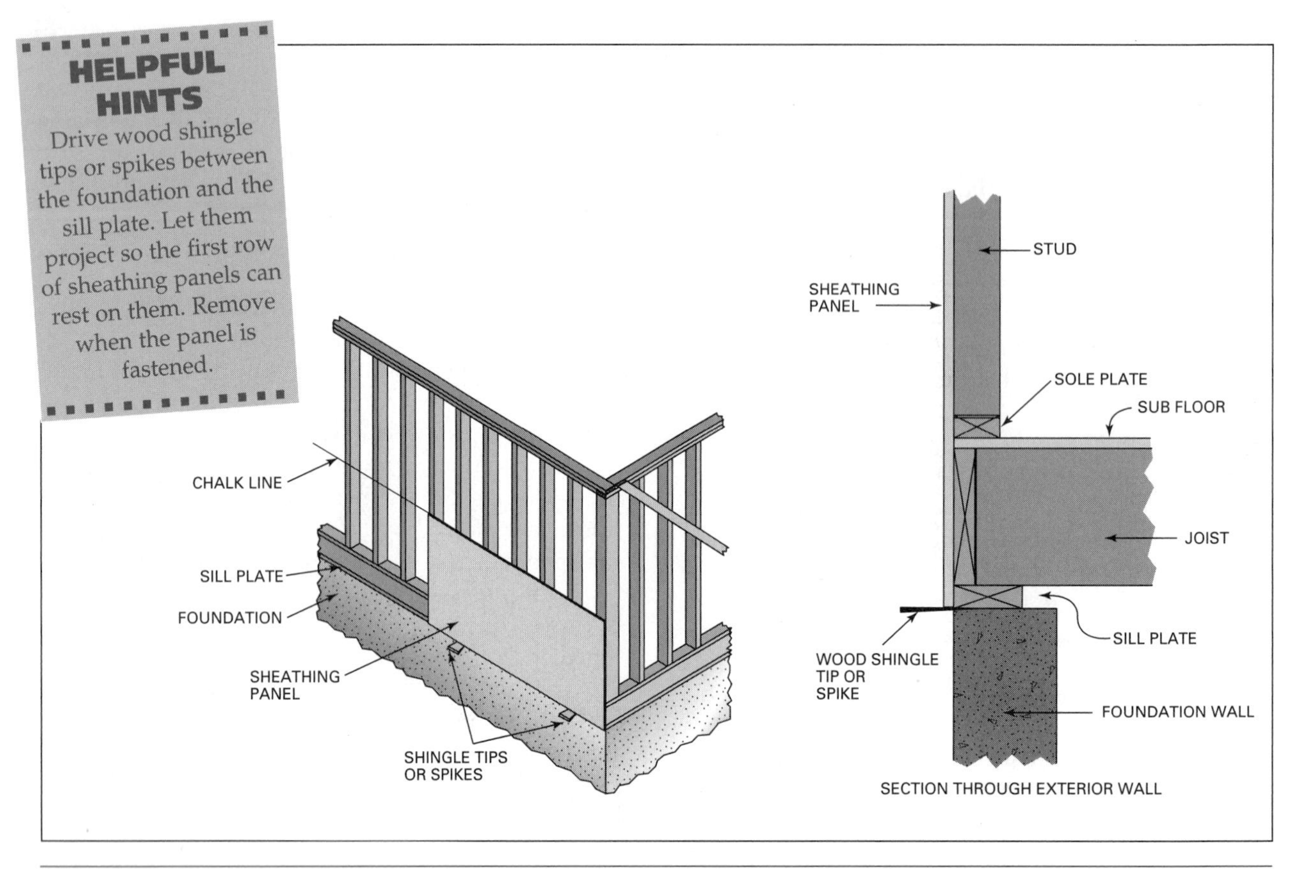

Fig. 35-32 Technique for supporting a piece of wall sheathing before nailing. Care should be taken not to injure the sill sealer.

measures 4 feet by 8 feet, it contains 32 square feet. Divide the wall area to be covered by 32 to find the number of panels required. Add about 5 percent for waste. If the answer does not come out even, round it up to the next whole panel.

Summary of Ways to Assemble and Erect Exterior Walls

There are a number of ways to assemble and erect exterior walls. The method described in this unit is probably the most common.

Another method is to fasten the sole plate securely. Nail all full-length studs to the top plate. Then, raise the sections. Toenail the studs to the sole plate. In this method, the rough openings are framed after the wall is erected. Then the sheathing is applied. This method is used when the framing crew is small.

Still another method consists of framing and sheathing the entire wall on the subfloor. Sometimes the windows and doors are installed and even the wall siding applied. A disadvantage to this method is that the wall is very heavy to raise into position and may require special raising equipment. Sometimes complete wall sections with doors, windows, and siding installed are prefabricated in the shop, transported to the job site, and erected in position.

All-Weather Wood Foundation

A wood foundation built of pressure-preservative-treated lumber and plywood is called the *All Weather Wood Foundation (AWWF)* (Fig. 35-33). It can be used to support light-frame buildings, such as houses, apartments, schools, and office buildings. It is accepted by major model building codes and by federal agencies such as the VA and FHA.

AWWF is mentioned in this chapter because much of the system is composed of wood exterior walls. The same techniques are used in their assembly and erection as described in this chapter.

For complete design and construction recommendations on the All-Weather Wood Foundation system, contact the American Forest & Paper Association and the Southern Forest Products Association.

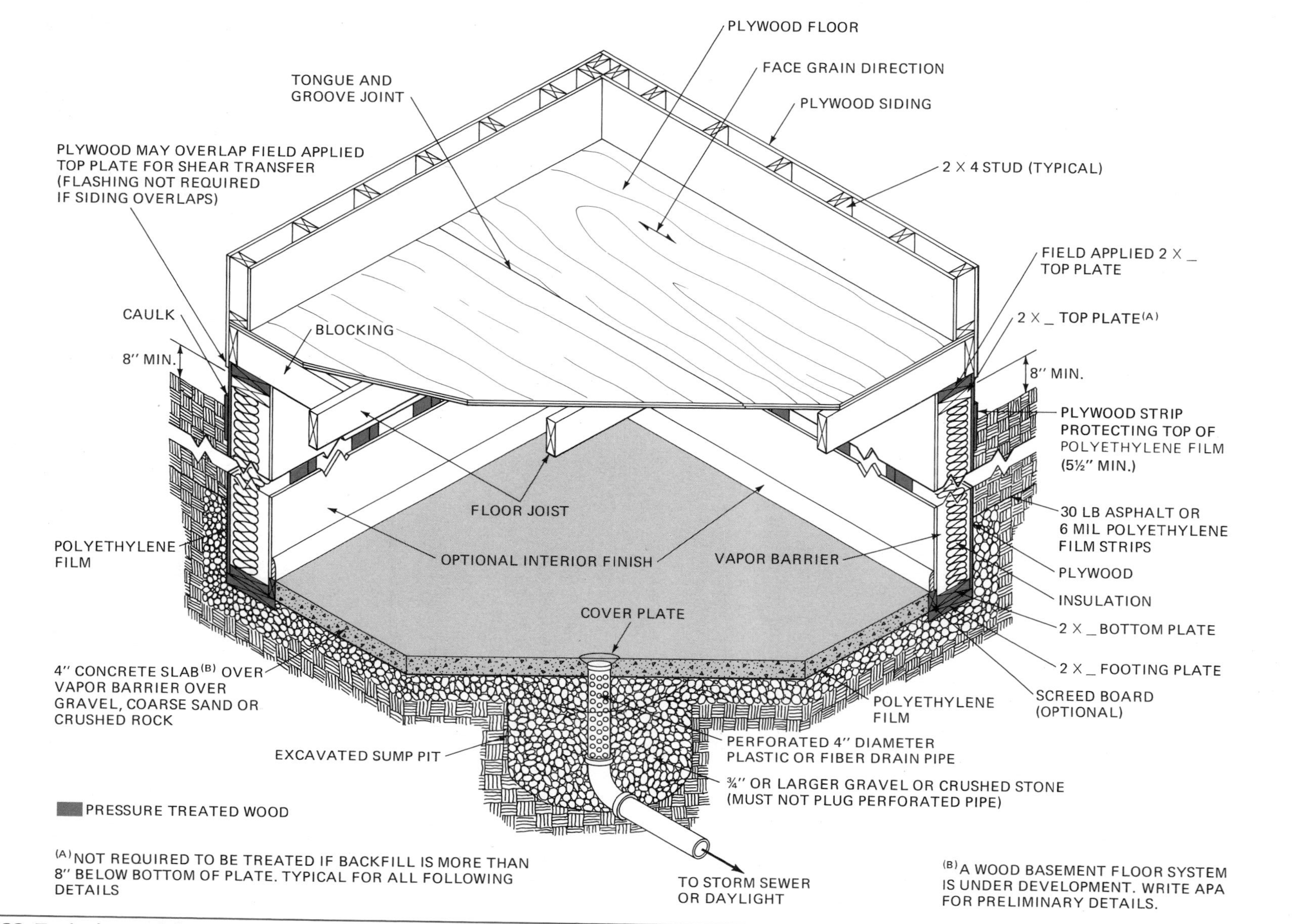

Fig. 35-33 Typical wood foundation (*Courtesy of American Plywood Association*)

Review Questions

Select the most appropriate answer.

1. The top and bottom horizontal members of a wall frame are called
a. headers. c. trimmers.
b. plates. d. sills.

2. The horizontal wall member supporting the load over an opening is called a
a. header. c. plate.
b. rough sill. d. truss.

3. Shortened studs above and below openings are called
a. shorts. c. cripples.
b. lame. d. stubs.

4. The finish floor to ceiling height in a platform frame is specified to be 7'-10". The finish floor is 3/4 inch thick and the ceiling material below the joist is 1 1/2 inches thick. A single sole plate and a double top plate are used, each of which has an actual thickness of 1 1/2 inches. What is the stud length?
a. 7'-6 3/4" c. 7'-9 1/4"
b. 7'-7 3/4" d. 7'-11 1/2"

5. A rabbeted jamb is 3/4 inch thick. Allowing 1/2 inch on each side for shimming the frame, what is the rough opening width for a door that is 2'-8" wide?
a. 2'-9 1/2" c. 2'-11 1/2"
b. 2'-10 1/2" d. 3'-0 1/2"

6. A story pole shows
a. the length of headers.
b. the length of rough sills.
c. the length of trimmers.
d. the width of the rough opening.

7. When laying out plates for walls and partitions, measurements for centerlines of openings start from the
a. end of the plate.
b. outside edge of the abutting wall.
c. building line.
d. nearest intersecting wall.

8. When assembling wall sections, studs are laid on the floor with the crowned edge
a. up. c. in either direction.
b. down. d. alternated.

9. Exterior walls are straightened by
a. sighting by eye.
b. using a line stretched between two blocks and testing with another block.
c. using a plumb bob dropped to the soleplate at intervals along the wall.
d. by sighting along the length of the wall using a builder's level.

BUILDING FOR SUCCESS

Job Site Housekeeping

Traditionally, one of the most overlooked components of a successful contracting company is maintaining a clean, functional, and safe work environment. The amount of productive work accomplished at a job site is directly related to the structure and operation of the work activities. Job site housekeeping plays an important part in this structure.

One of the common goals shared by all builders is a safe working environment. This environment is conducive to high production, efficiency, and generally more profit. With more federal, state, and local regulations for construction sites, builders cannot afford to be ignorant of or neglect those regulations. Insurance companies will continue to have major input into the safety affairs of all businesses. Job site housekeeping will continue to be a focus of attention for insurance companies, fire marshals, and many governmental agencies that look after the welfare of workers.

Traditionally, skilled workers have shared a sense of concern for the welfare of each person on the crew. They understand the need to maintain a clean and functional work site. Usually each person looks out for any job site hazards that could lead to accidents. Corrections are made once the hazard has been identified. With experience, the skill of spotting potential accidents becomes automatic.

Construction work has not changed drastically over the years. Production creates waste and by-products that must be disposed of regularly. In a very short time work locations can become cluttered and in a state of disarray. Waste and debris, in turn, become the basis for unsafe working conditions. Also, debris that is carried off the job site by wind will cause adverse public relations with neighbors. A plan to properly store, remove, or in some manner discard the waste must be established.

The ability to maintain a clean and safe working area has to begin with an understanding of the problem and how it is created. This is the appropriate time to analyze how much of the problem is simply neglect of good housekeeping.

There are many things to consider in the establishment of a safe work environment. One of the first considerations is helping the entire workforce understand what a safe environment is and how it relates to them personally. They must buy into the program or it is doomed to fail. Primarily, the longevity in their job will be affected by job site working conditions. They must see that fewer accidents occur at a clean and well-organized job site. As they ask the question, "What's in it for me?" evidence must be presented to demonstrate how they will be in a "win-win" situation.

Second, each worker must be willing to enforce sound housekeeping principles daily. Building materials must be properly stored so they do not interfere with the construction activities. Appropriate storage locations must be chosen that are out of the way but yet convenient.

Third, time has to be set aside each day for site cleanup. This activity has to be included in the bid or estimate for the construction project. If done on a daily basis, each person can carry out a housekeeping assignment in minimal time.

Fourth, housekeeping efforts must be visible and show the workers proof that the cleanup tasks have been beneficial and perhaps even profitable. An incentive program may help employees, subcontractors, and others see and feel the benefits. One proven benefit has been recognition by the company of individuals who have no or minimal accidents during a certain time period. Another is company recognition for traditionally clean and safe working environments by the community, professional construction organizations, or insurance companies. This, in turn, may have an impact on hiring quality personnel for a particular company. Many potential employees are concerned about the safety record or safe working conditions promoted by a company.

A possible employment situation involving job site housekeeping can be presented by the following example. In a typical custom cabinet shop each worker produces an abundance of wood waste daily. Debris left on the floor can cause walking hazards that turn into accidents. A buildup of wood by-products such as sawdust, wood chips, and shavings can become a fire hazard. If the fire marshal or insurance agent stops by unexpectedly, violations may cause production to halt until they are corrected. The cluttered or hazardous work site may be the grounds for insurance premium increases.

Trips, falls, and other accidents can be minimized or eliminated with good housekeeping habits. The time needed to daily clean up the work site will be small, if an organized plan is implemented. The regular elimination of waste can be completed in 10 minutes or less when specific cleanup jobs are scheduled.

Historically, workers are more productive and have a better sense of belonging to a company that has their welfare at heart. People tend to take pride in the fact that they can work for hundreds or thousands of hours without an accident. The vast majority of construction workers who are not involved in accidents in the workplace are working in a situation where good job site housekeeping is practiced. This practice is an integral part of the company safety program for employees. The end results of good housekeeping means many things, but characteristically they will mean a longer and safer construction career.

FOCUS QUESTIONS: For individual or group discussion

1. How would you explain to your employees the need for and benefits of a good housekeeping program in your place of business?
2. What do you feel might be the biggest obstacles to implementing an efficient, ongoing housekeeping program at a construction site?
3. How would you structure a housekeeping program for a construction company, custom cabinet shop, or factory-built housing organization?
4. What representatives from business and industry might be resources for ideas and support in establishing a good housekeeping structure within your company's safety program?

UNIT

13

Interior Rough Work

Interior rough work is constructed in the inside of a structure and later covered by some type of finish work. The interior rough work described in this unit includes the installation of *partitions, ceiling joists, furring strips, backing and blocking, plaster bases,* and *plaster grounds*. The term *rough work* does not imply that the work is crude. It is a kind of work that will eventually be covered by other material. Careful construction of the rough frame makes application of the finish work less complicated.

OBJECTIVES

After completing this unit, the student should be able to:

- assemble, erect, brace, and straighten bearing partitions.
- determine and make rough openings for interior doors.
- lay out, cut, and install ceiling joists.
- lay out and erect nonbearing partitions and install backing in walls for fixtures.
- apply plaster grounds and plaster bases.
- describe various components of light-gauge steel framing.
- lay out and frame light-gauge steel interior partitions.

UNIT CONTENTS

CHAPTER 36 INTERIOR PARTITIONS AND CEILING JOISTS

Partitions and *ceiling joists* constitute some of the interior framing. Ceiling joists tie the exterior side walls together and support the ceiling finish.

Interior partitions supporting a load are called *bearing partitions.* Partitions that merely divide the area into rooms are called *nonbearing partitions* (Fig. 36-1).

Bearing Partitions

Bearing partitions support the inner ends of ceiling joists or, in multi-story wood frame structures, the inner ends of joists are used to frame the floor above. They are placed directly over the girder or the bearing partition in the lower level. If several bearing partitions are used on the same floor, supported girders or walls are placed directly under each.

Bearing partitions are erected in a manner similar to exterior walls. A doubled top plate is usually required because of the weight on the wall. Headers over openings must be strong enough to support the load imposed on them. Common practice is to erect only bearing partitions at this stage. The framing of nonbearing partitions may be left until later so the roof can be made tight as soon as possible. Otherwise, non-load-bearing partitions may be framed the same as bearing walls and partitions.

If *roof trusses* are used, bearing partitions are not needed on single-story buildings or on the top floor of multi-story buildings. The load is carried by the trusses and transmitted to the exterior walls. (Details on roof trusses are given in a later unit.)

Layout and Framing of Bearing Partitions

From the floor plan, determine the location of the bearing partitions. Dimensions usually are given to the centerline. Measure half of the thickness of the partition from the centerline. Snap a chalk line on the subfloor.

Lay the edge of the soleplate to the chalk line, tacking it in position. Lay out the openings, partition intersections, and all on-center and cripple studs. On-center studs should line up with floor and ceiling joists. Lay the top plate next to the soleplate. Transfer the layout of the soleplate to the top plate. All joints in the plates should come in the center of a stud.

Frame, erect, straighten, and temporarily brace the bearing partition in the same manner as for exterior walls. Double the top plate, letting it lap the plate of the exterior wall (Fig. 36-2).

Rough Opening Sizes for Interior Doors

The rough opening width for an interior door opening is found by adding to the door width twice the

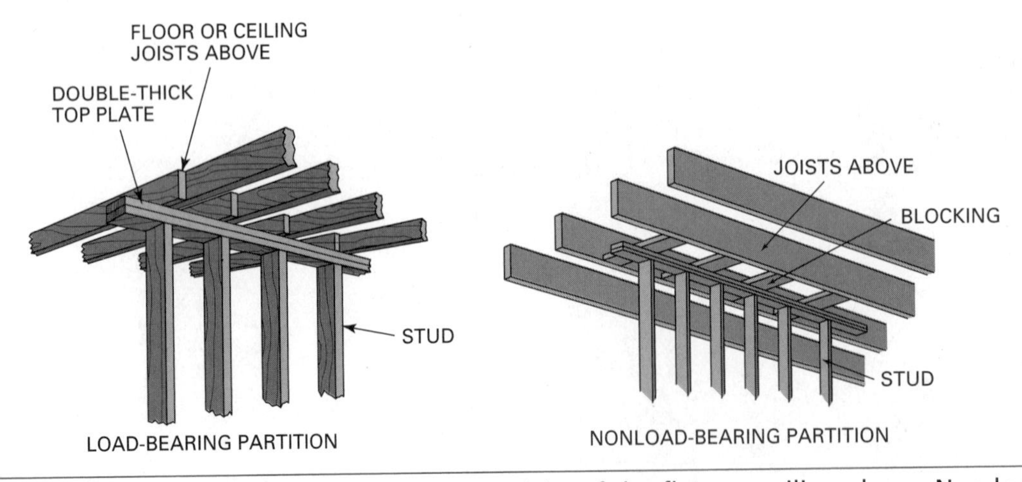

Figure 36-1 Load-bearing partitions support the weight of the floor or ceiling above. Non-load-bearing partitions merely divide an area into rooms.

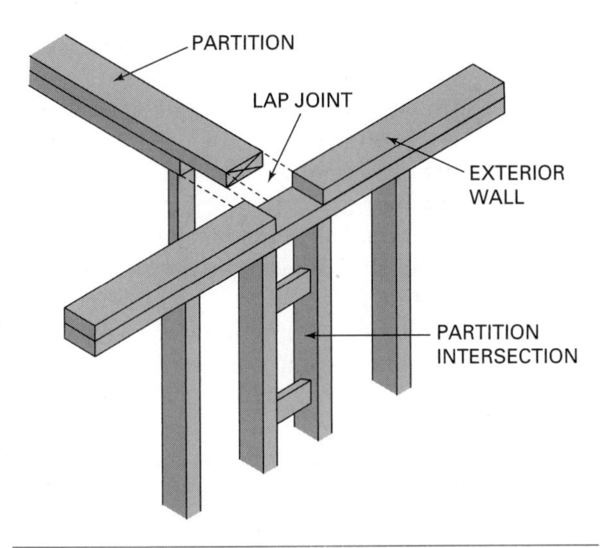

Fig. 36-2 The double top plates of walls and partitions overlap.

thickness of the jamb stock and one inch for shimming (1/2 inch for each side). The rough opening height is found by adding the following:

- the thickness of the finish floor
- 1/2 inch for clearance between the finish floor and the bottom of the door
- the height of the door, which is usually 6'-8" tall
- the thickness of the head jamb
- 1/2-inch clearance between the head jamb and the rough header.

Usually no threshold is used under interior doors. However, in case a threshold is used, substitute its thickness for the 1/2-inch clearance allowed under the door (Fig. 36-3).

Ceiling Joists

Ceiling joists generally run from the exterior walls to the bearing partition across the width of the building. Construction design varies according to geographic location, traditional practices, and the size and style of the building. The size of ceiling joists is based on the span, spacing, load, and the kind and grade of lumber used. Determine the size and spacing from the plans or from local building codes.

Methods of Installing Joists

In a conventionally framed roof, the *rafters* and the ceiling joists form a triangle. Framing a triangle is a common method of creating a strong and rigid building. The weight of the roof and weather are transferred from the roof to the exterior walls (Fig. 36-4). The rafters are located over the studs and the ceiling joists are fastened to the side of the rafters (Fig. 36-5).

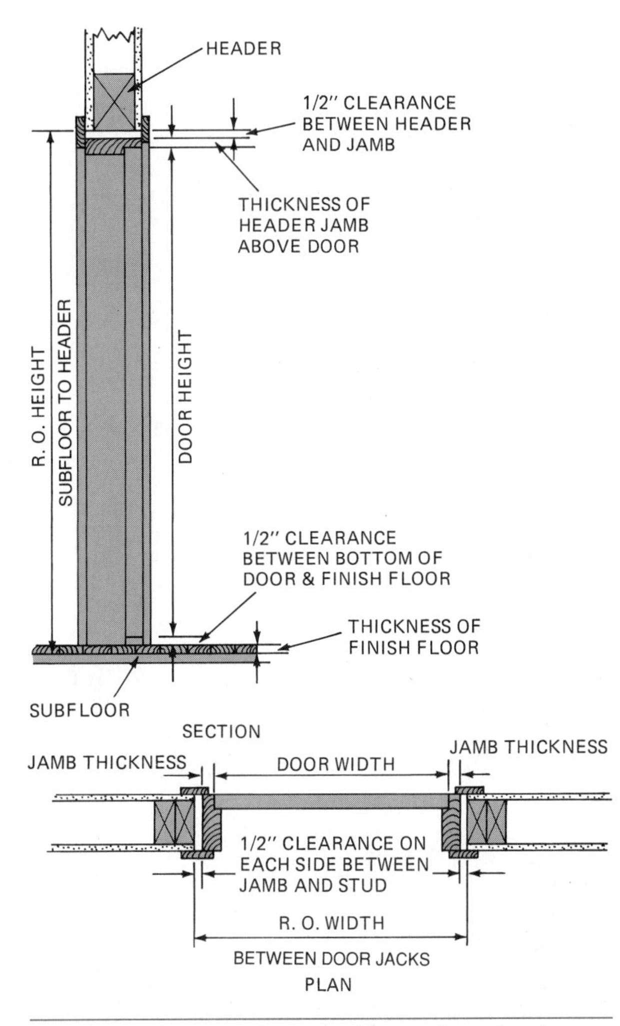

Fig. 36-3 Figuring the rough opening size for an interior door.

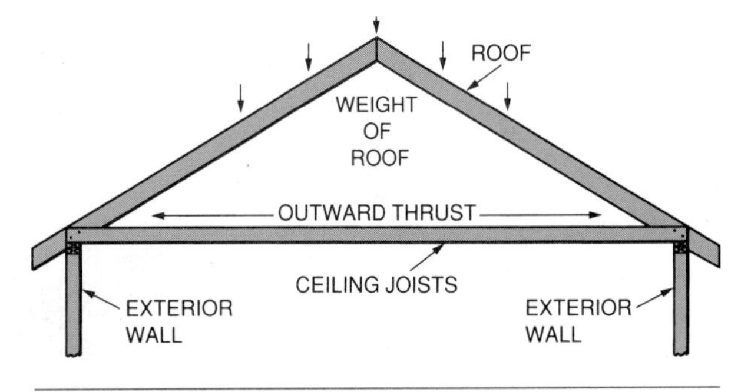

Fig. 36-4 The weight of the roof exerts pressures that tend to thrust the walls outward. Ceiling joists tie the frame together into a triangle, which resists the outward thrust.

This binds the rafters and ceiling joists together into a rigid triangle and keeps the walls from spreading outward due to the weight of the roof.

Ceiling joists may be made from engineered lumber and purchased in long lengths so that the

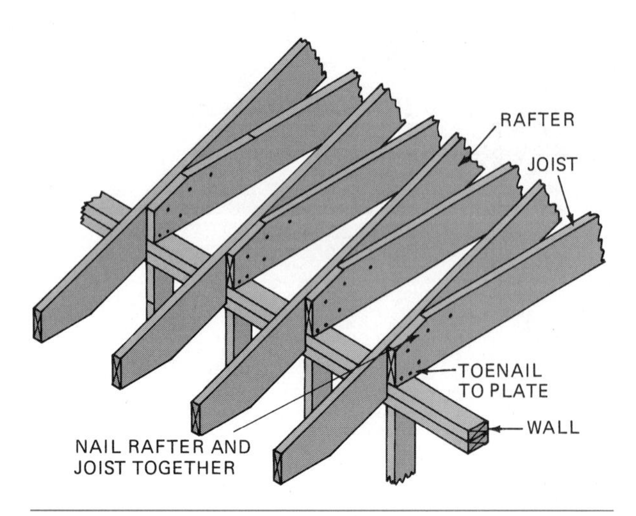

Fig. 36-5 Ceiling joists are located so they can be fastened to the side of rafters.

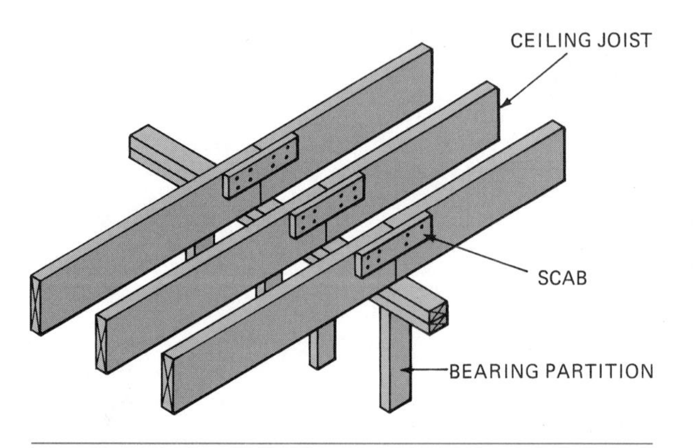

Fig. 36-6 The joint of in-line ceiling joists must be scabbed at the bearing partition.

rafter-ceiling joist triangle is easily formed. Typically, though, the ceiling joist lengths are half of the building width and therefore must be joined over a beam or bearing partition.

Sometimes the ceiling joists are installed in-line. Their ends butt each other at the centerline of the bearing partition. The joint must be *scabbed* to tie the joints together (Fig. 36-6). *Scabs* are short boards fastened to the side of the joist and centered on the joint. They should be a minimum of 24 inches long. The location for the ceiling joists on each exterior wall is on the same side of the rafter location when joists are in-line.

Another method of joining ceiling joists is to lap them over a bearing partition in the same manner as for floor joists (see Fig. 32-18). This puts a stagger in the line of the ceiling joist and consequently in the rafters as well (Fig. 36-7). This stagger is most visible at the ridgeboard. The layout lines for rafters and ceiling joists are measured from the outside end wall onto the top plate (Fig. 36-8). This measurement is exactly the same, with the only difference being the side of the line on which the ceiling joists and rafters are placed.

Cutting the Ends of Ceiling Joists

The ends of ceiling joists on the exterior walls usually project above the top of the rafter. This is especially true when the roof has a low slope. These ends must be cut to the slope of the roof, flush with or slightly below the top edge of the rafter.

Lay out the cut, using a framing square. Cut one joist for a pattern. Use the pattern to mark the

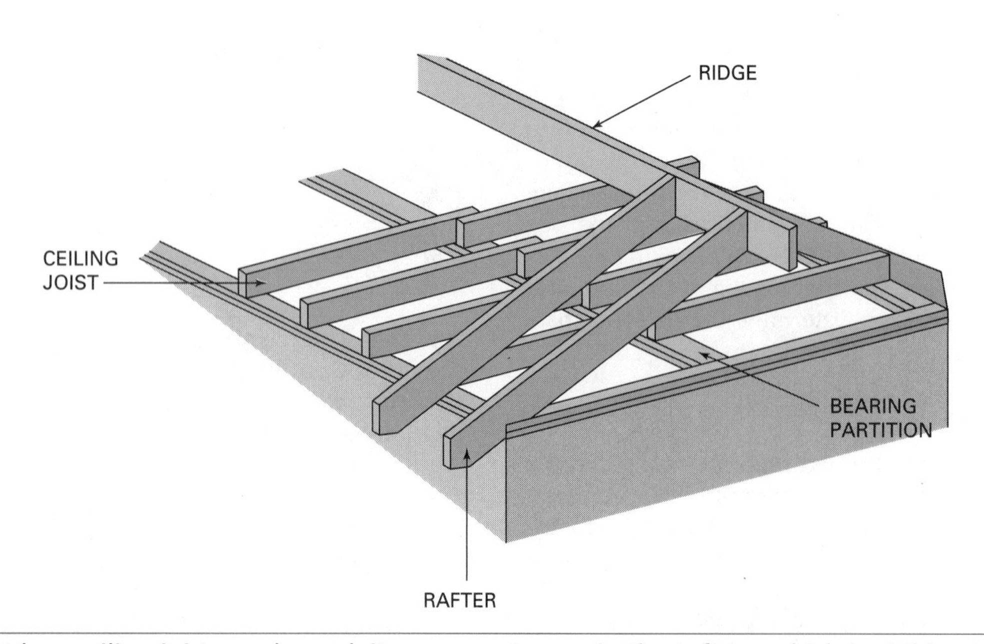

Fig. 36-7 When ceiling joists are lapped, it causes a stagger in the rafters, which is visible at the ridge.

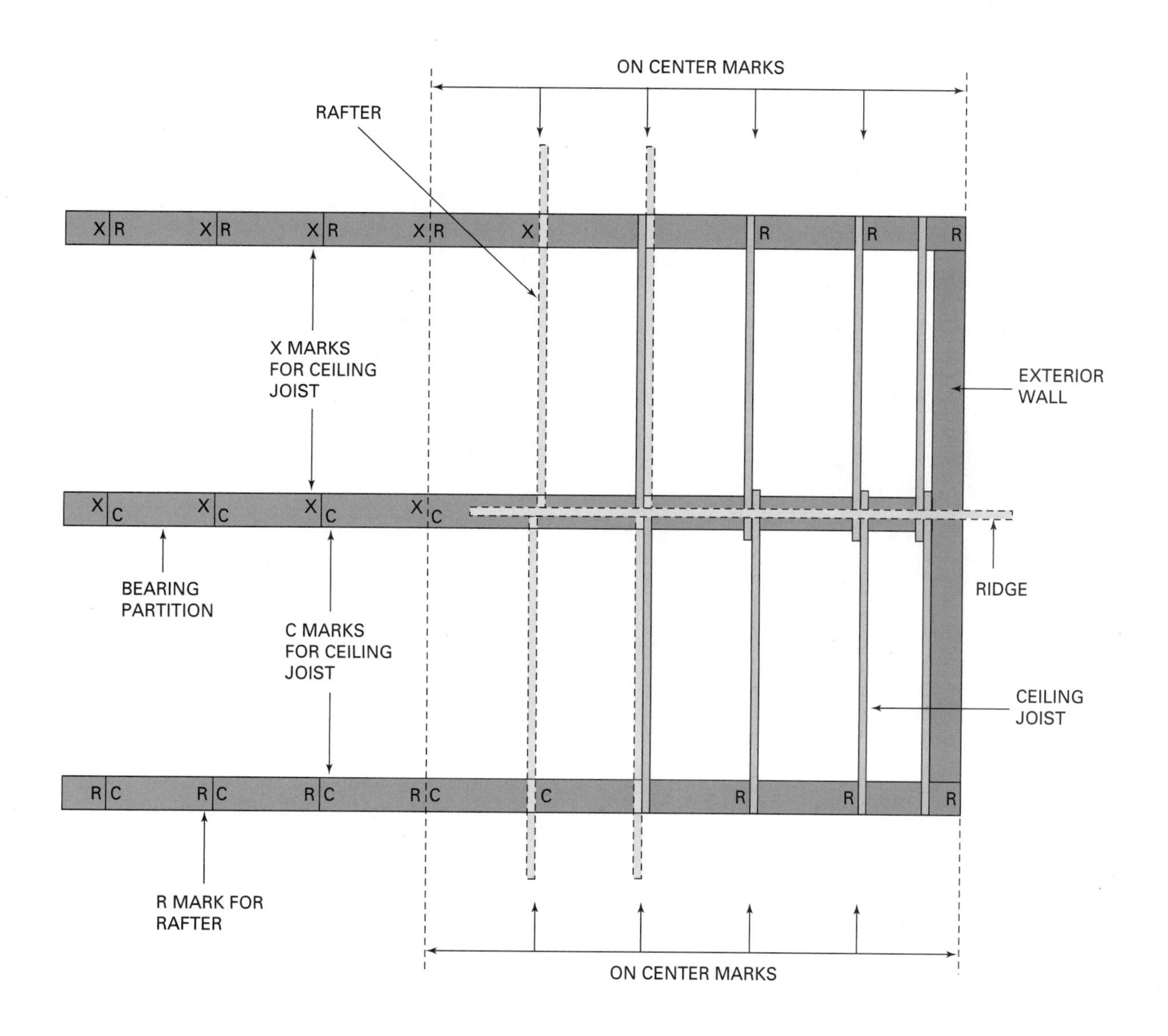

Fig. 36-8 Layout lines on all plates are the same measurements; only the position of the rafters and ceiling joists vary.

rest. Make sure when laying out the joists that you sight each for a crown. Make the cut on the crowned edge so that edge is up when the joists are installed. Cut the taper on the ends of all ceiling joists before installation. Make sure the length of the taper cut does not exceed three times the depth of the member. Also make sure that the end of the joist remaining after cutting is at least half the member's width (Fig. 36-9).

Stub Joists

Usually, ceiling joists run parallel to the end walls and in the same direction as the roof rafters. When low-pitched hip roofs are used, *stub joists,* which run at right angles to the end wall, must be installed. The use of stub joists allows clearance for the bottom edge of the rafters that run at right angles to the end wall (Fig. 36-10). The stub joists should be laid out so their centerlines coincide with those of any ceiling *furring strips,* if used. This provides a backing and a nailing surface for the ends of the furring strips. (Description of and application instructions for furring strips are given later in this chapter.)

Framing Ceiling Joists to a Beam

In many cases, the bearing partition does not run the length of the building because of large room areas. Some type of beam is then needed to support

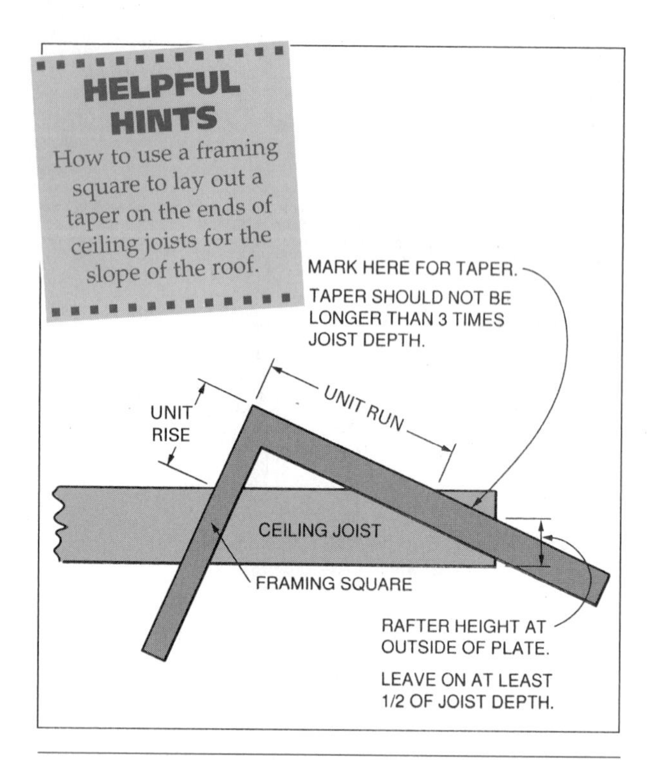

Fig. 36-9 Using a framing square to mark the slope of a tapered cut on a ceiling joist.

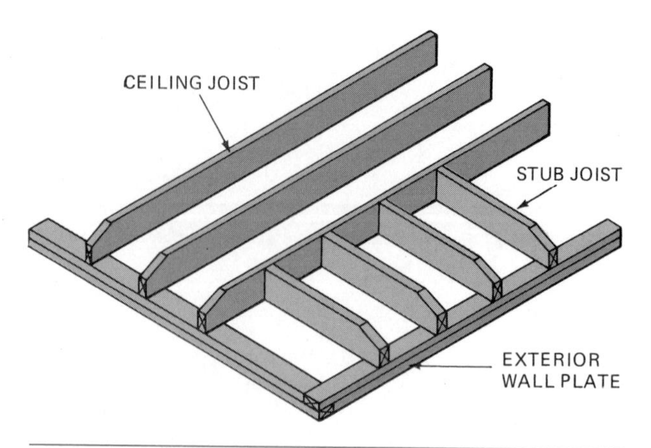

Fig. 36-10 Stub joists are used for low-pitched hip roofs.

the inner ends of the ceiling joists in place of the supporting wall. Similar in purpose and design to a girder, the beam may be of built-up, solid lumber or engineered lumber.

If the beam is to project below the ceiling, it is installed in the same manner as a header for an opening. The joists are then installed over the beam in the same manner as over the bearing partition (Fig. 36-11).

If the beam is to be raised in order to make a flush and continuous ceiling below, then the ends of the beam are set on the top of the bearing partition and the exterior end wall. The joists are butted to the beam. They may be supported by a ledger strip or by metal **joist hangers** (Fig. 36-12).

Openings

Openings in ceiling joists may need to be made for such things as chimneys, attic access (scuttle), or disappearing stairs. Large openings are framed in the same manner as for floor joists. For small openings, there is no need to double the joists or headers (Fig. 36-13).

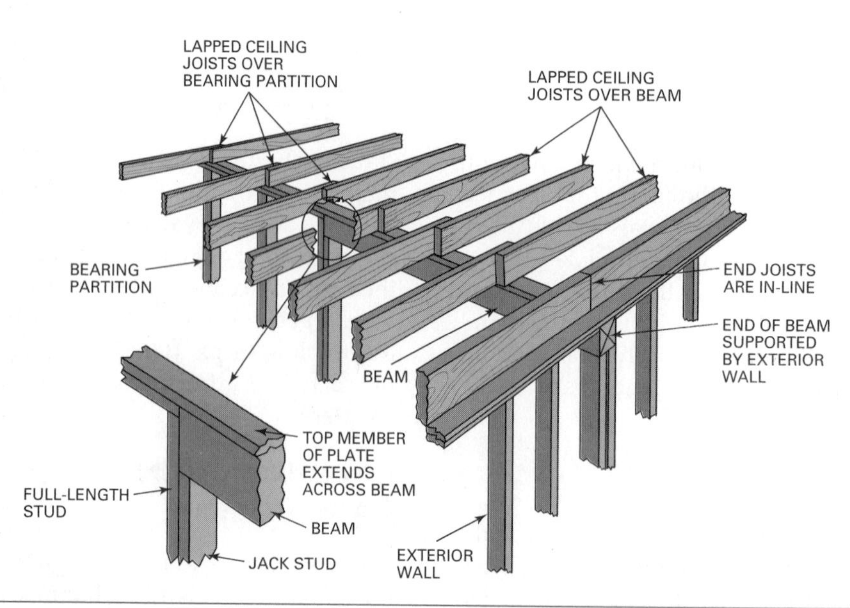

Fig. 36-11 Framing ceiling joists over a beam.

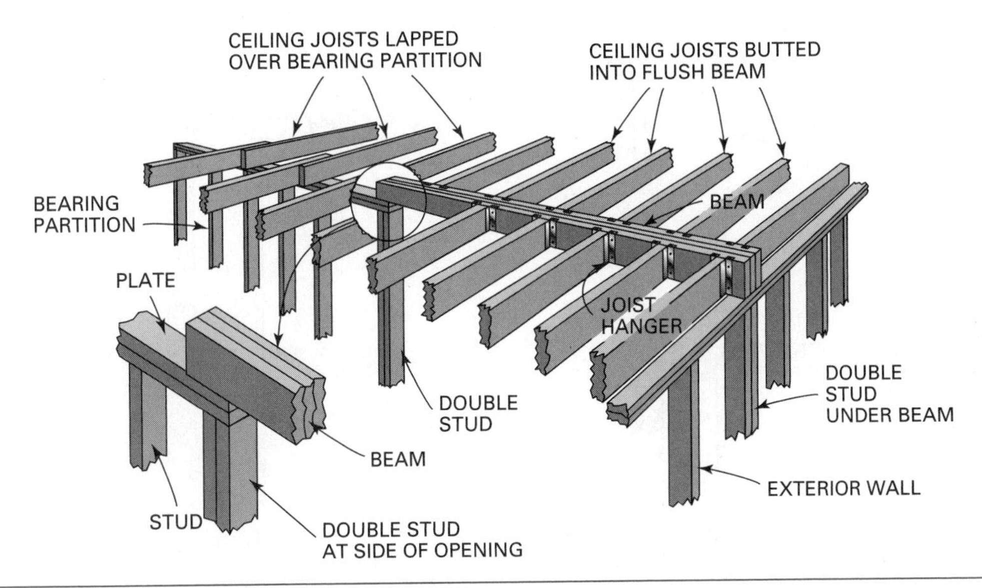

Fig. 36-12 Framing ceiling joists into a beam for a flush ceiling.

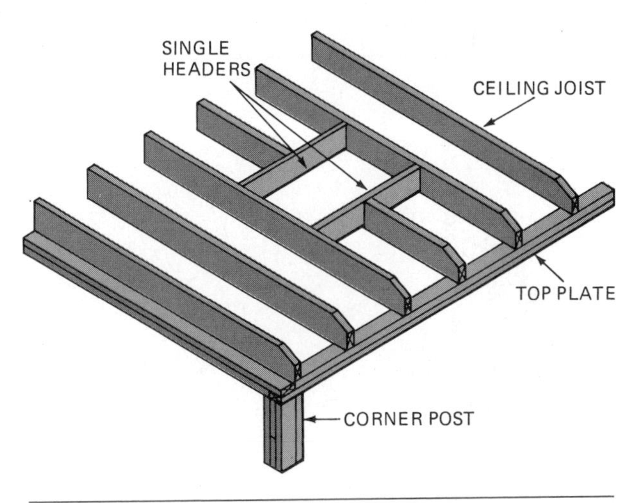

Fig. 36-13 Joists and headers need not be doubled for small ceiling openings.

Layout and Spacing of Ceiling Joists

Roof rafters rest on the plate directly over the regularly spaced studs in the exterior wall. Ceiling joists are installed against the side of the rafters and fastened to them. Spacing of the ceiling joists and rafters should be the same so they can be tied together at the plate line.

- Start the ceiling joist layout from the same corner of the building where as the floor joists and wall stud were laid out. Square up from the same side of each regularly spaced stud or cripple stud in the exterior wall and across the top of the plate. Mark an *R* on the side of the line over top of the stud for the rafter and an *X* or a *C* on the other side of the line for the location of the ceiling joists.
- Layout lines on the bearing partition and on the opposite exterior wall are the same layout as on the first wall. These layout lines should all be the same distance from the end wall. The only difference is the location of the marks for rafters and ceiling joists. They vary according to the construction method used.
- On the opposite exterior wall, place an *R* and a *C* or an *X* on either side of the line but opposite from the first exterior wall. This will allow for the stagger and lap of rafters and ceiling joist shown in Fig. 36-8.
- When ceiling joists are lapped, place the *C* and *X* on either side of the line and in line with its counterpart on the exterior wall (Fig. 36-14).
- When the ceiling joists are continuous or butted together, the marks are located on the same side of the line as the ceiling joist marks on the exterior walls, and the rafters are also on the same side of the line (Fig. 36-14).

The layouts for various methods of installing ceiling joists are shown in Fig. 36-14.

Installing Ceiling Joists

The ceiling joists on each end of the building are placed so the outside face is flush with the inside of the wall. They are installed in-line with their inner ends butting each other regardless of how the other joists are laid out. This will allow for easy installation of the **gable end** studs (Fig. 36-15).

In addition to other functions, these end joists provide fastening for the ends of furring strips or

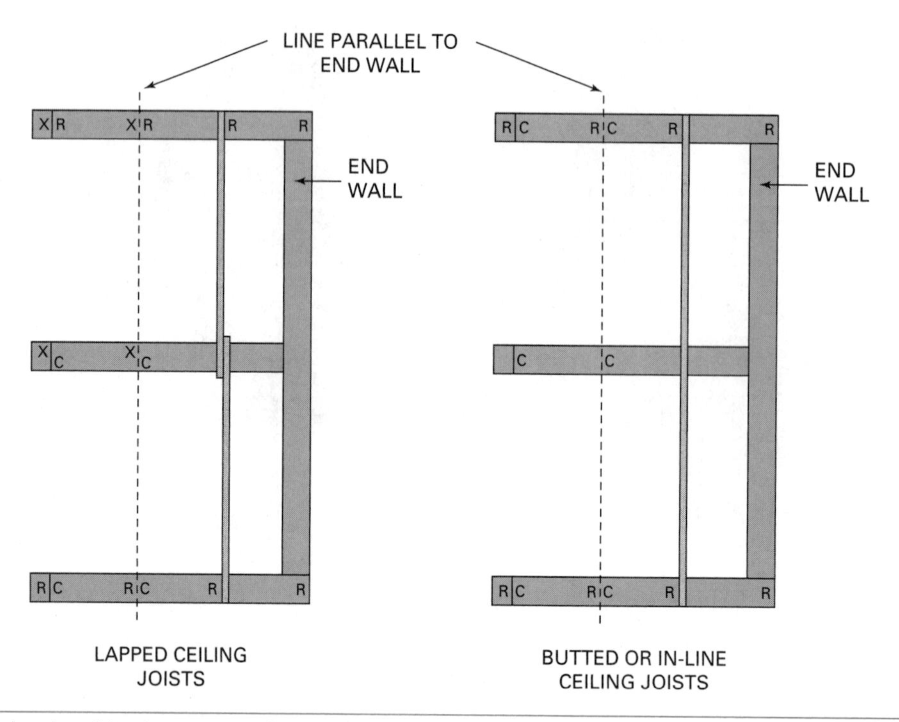

Fig. 36-14 Methods of laying out ceiling joists.

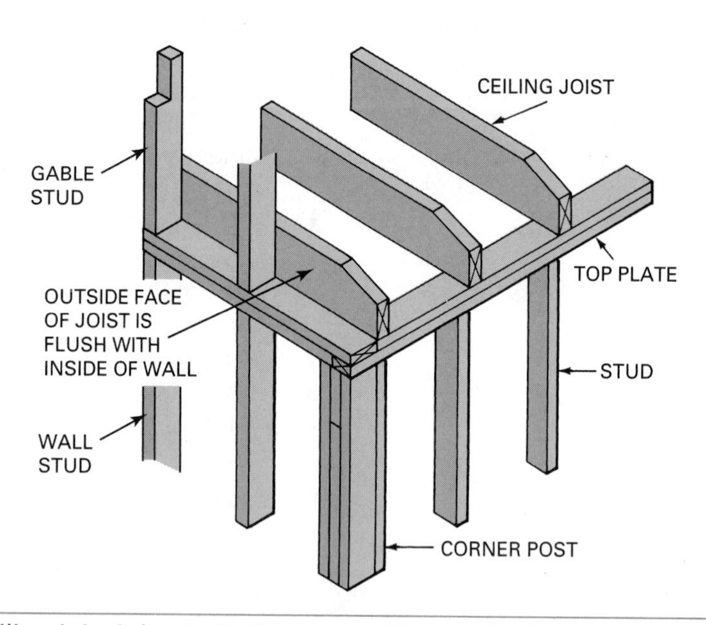

Fig. 36-15 The end ceiling joist is located with its outside face flush with the inside wall.

for the ceiling finish, if furring strips are not used. All other joists are fastened in position with their sides to the layout lines. If the outside ends of the joists have not been tapered, sight each joist for a crown. Install each one with the crowned edge up. If the outside ends have been tapered, install the ceiling joists with the cut edge up. Reject any badly warped joists.

Toenail joists with at least two 10d nails into the plates to the layout lines. Remember to keep joists on the correct side of the lines. Nail lapped joists together with at least four 10d nails, two from each side.

Nonbearing Partitions

Nonbearing partitions carry no load. They divide the floor area into rooms. Openings may be framed with single studs and headers (Fig. 36-16). Because the wall carries no load, headers are usually 2 × 4s.

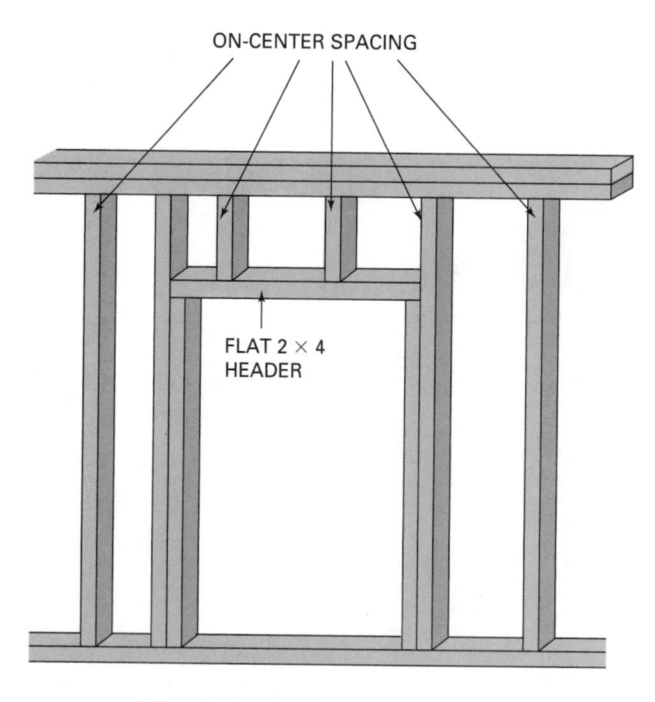

Fig. 36-16 A method for framing a header in a nonload bearing wall or partition.

Nonbearing partitions may be built after the ceiling joists are installed.

Installing Soleplates

The location of the partitions is determined from the floor plans. Dimensions are usually given to their centerline. Locate each partition on the subfloor. Snap chalk lines to indicate each side of the partition. Mark the end of each partition and all openings.

Fasten the soleplates for all partitions to the subfloor. Straighten the soleplates as necessary by toenailing 8d common nails into their edges. When possible, nail into the floor joists below, with one nail into each floor joist staggered along the length of the soleplate. In case partitions run between and parallel to the joists, fasten the soleplates by toenailing through both edges into the subfloor about 16 inches apart. Cut all soleplates to length. Leave spaces for door openings.

Locating Top Plates

The position of the top plate must be marked on the blocking or furring strips above. Hang a plumb bob from the furring strip so it lines up with the edge of the soleplate at one end of the partition. Mark the ceiling at this point (Fig. 36-17). Do the same at the other end of the partition. Snap a chalk line between

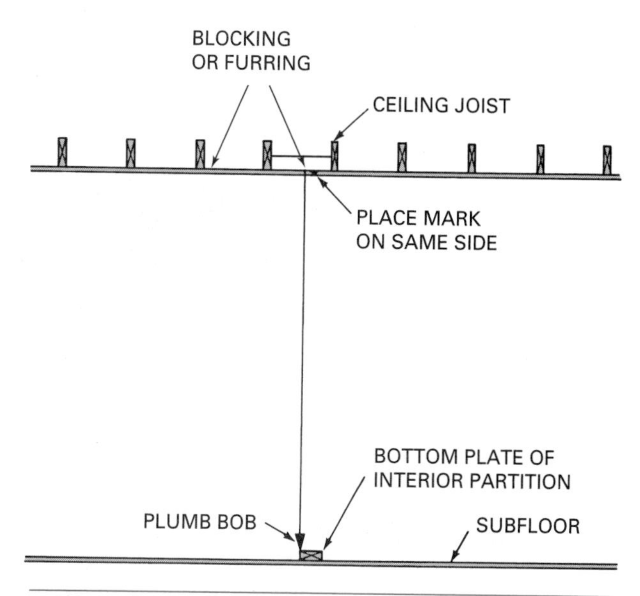

Fig. 36-17 Hang a plumb bob from the ceiling to locate the top plate of interior partitions.

the marks. Locate the top plates for all other partitions in the same manner.

Laying Out the Plates

Cut the top plate to run from one end of the partition to the other. Although the soleplate has been cut out for openings, the top plate runs uncut for the total length of the partition, except if the opening goes all the way to the ceiling.

Tack the top plate on top of the soleplate so that it can be removed later. Lay out the plates in a manner similar to exterior walls. Use the specified stud spacing (Fig. 36-18). Build corner posts and partition intersections as necessary. Find the stud length. Cut the needed number of full length studs. The stud length for nonbearing partitions is different than in bearing partitions or exterior walls.

To find the stud length for the nonbearing partition where furring strips are applied to the ceiling

Fig. 36-18 Laying out the plates for interior partitions.

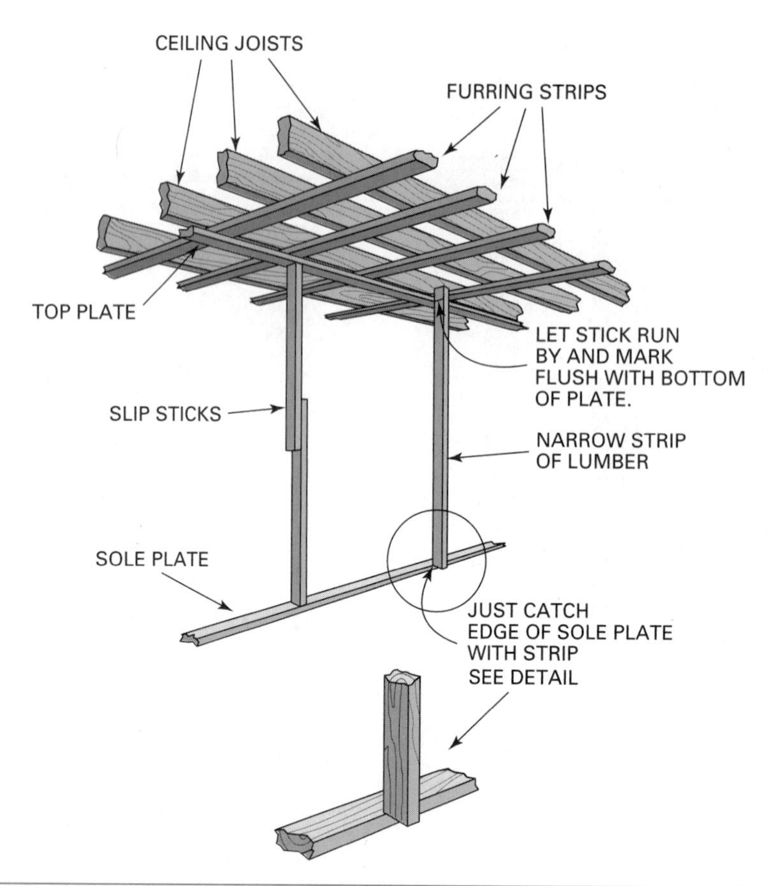

Fig. 36-19 Finding the length of a stud in a nonload-bearing partition when framing to a furred ceiling.

first, lay two thicknesses of plate material on the subfloor. Measure up to the furring strip (Fig. 36-19). Instead of actually measuring, a pair of slip sticks is useful for determining the length. Or, hold a strip of lumber on top of the bottom plate material. Let the strip run by the top plate. Mark the strip in line with the bottom surface of the top plate.

Constructing Nonbearing Partitions

Separate the top plate from the soleplate. Place all full-length studs, corner posts, and partition intersections on the floor. Make sure their crowned edges run in the same direction, either all up or all down, and their ends are against the fastened soleplate. Spike the top plate to these members to the layout lines.

Raise the section into position. Toenail the studs to the soleplate. Use four 8d, common nails in each stud, two on each side. Fasten the top plate into position to the chalk line by toenailing 8d, common nails through the edges of the plate into the furring strips or spike to the blocking..

Straighten the end stud that butts against another wall by toenailing through its edge into the center block of the partition intersection. Test for straightness using a long straightedge against its edge (Fig. 36-20).

Construct all nonbearing partitions in a similar manner. For ease of erecting, construct the longer partitions first. Then construct shorter cross-partitions, such as for closets. There is no hard and fast rule for constructing partitions. Some carpenters prefer to erect them in sections by nailing both top and bottom plates to the studs and raising them in position. Others prefer to first fasten both plates in place and then install studs in between. It matters little which method is used as long as the partitions are carefully laid out and constructed.

Bathroom and kitchen walls sometimes must be made thicker to accommodate plumbing. Sometimes 2×6 plates and studs are used or a double 2×4 partition is erected. Still another method is to use 2×6 plates with 2×4 studs, where studs are set to alternate sides. This allows fiberglass insulation to be woven between the studs for increased soundproofing. If the wall thickness needs to be increased only slightly, furring strips may be added to the edges of the studs and plates.

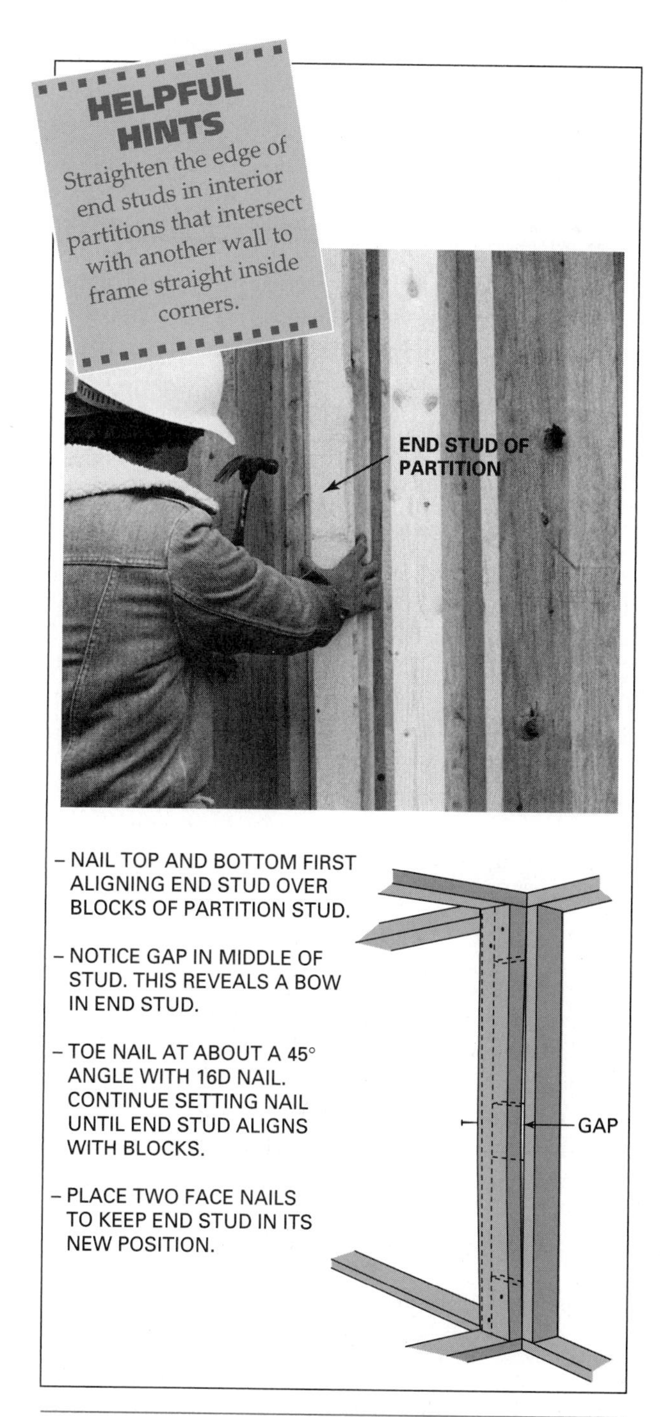

Fig. 36-20 Technique for straightening a crown stud.

Openings

Besides door openings, the carpenter must frame openings for heating and air-conditioning ducts, medicine cabinets, electrical panels, and similar items. If the items do not come in a stud space, the stud must be cut and a header installed. When ducts run through the floor, the soleplate and subfloor must be cut out (Fig. 36-21). The **reciprocating saw** is a useful tool for making these cuts.

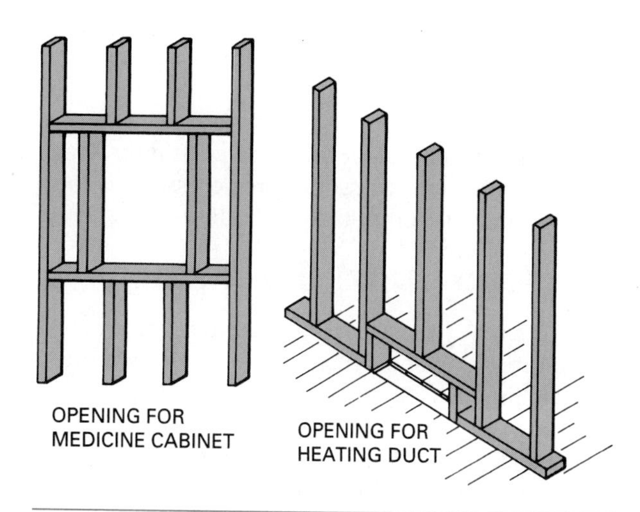

Fig. 36-21 Miscellaneous openings in interior partitions.

CHAPTER 37 BACKING, BLOCKING, AND BASES

This chapter deals with **backing, blocking,** and **plaster** bases. A *backing* is, ordinarily, a short block of lumber placed in floor, wall, and ceiling cavities to provide fastening for various parts and fixtures.

Blocking is installed for different purposes, such as providing support for parts of the structure, weathertightness, and firestopping. Sometimes blocking serves as backing.

Several types of *plaster bases* are used to provide a surface for plaster to be spread and on which to adhere. Various accessories are necessary to complete the installation of the plaster base.

Backing and Blocking

There are many places in the structure where blocking and backing should be placed. It would be unusual to find directions for the placement of backing and blocking in a set of plans. It is the wise builder who installs them, much to the delight of those who must install fixtures and finish of all types in later stages of construction.

Backing

Unit 12 recommended installing short blocks with their ends against the sole plate and their sides against the studs on inside corners and on both sides of door openings. This provides more fastening surface for the ends of **baseboard** (Fig. 37-1).

Much backing is needed in bathrooms. Plumbing rough-in work varies with the make and style of plumbing fixtures. The experienced carpenter will obtain the rough-in schedule from the plumber. He or she then installs backing in the proper location for such things as bathtub faucets, showerheads, lavatories, and water closets (Fig. 37-2). Backing should also be installed around the top of the bathtub.

In the kitchen, backing should be provided for the tops and bottoms of wall cabinets and for the tops of base cabinets. If the ceiling is to be built down to form a *soffit* at the tops of wall cabinets, backing should be installed to provide fastening for the soffit (Fig. 37-3).

A homeowner will appreciate the thoughtfulness of the builder who provides backing in appropriate locations in all rooms for the fastening of curtain and drapery hardware (Fig. 37-4).

Blocking

Some types of blocking have already been described in earlier units. See Chapter 32 for the installation of solid lumber blocking used for bridging.

When floor panels are used as a combination subfloor and **underlayment** (under carpet and pad), the panel edges must be tongue-and-groove or

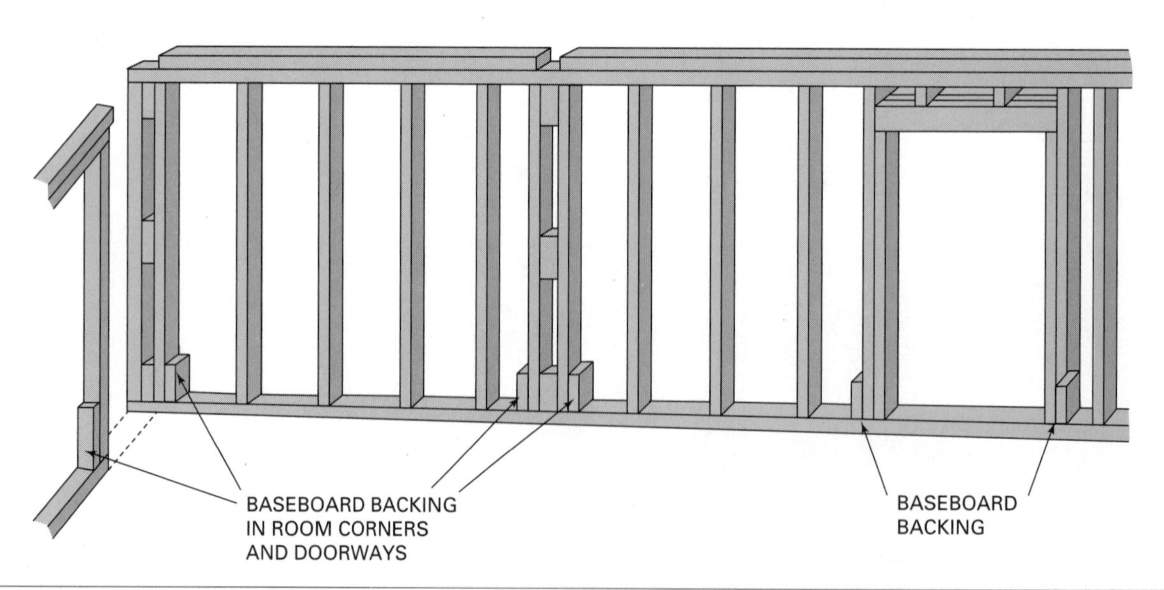

Fig. 37-1 Backing is installed at corners and door openings for baseboard.

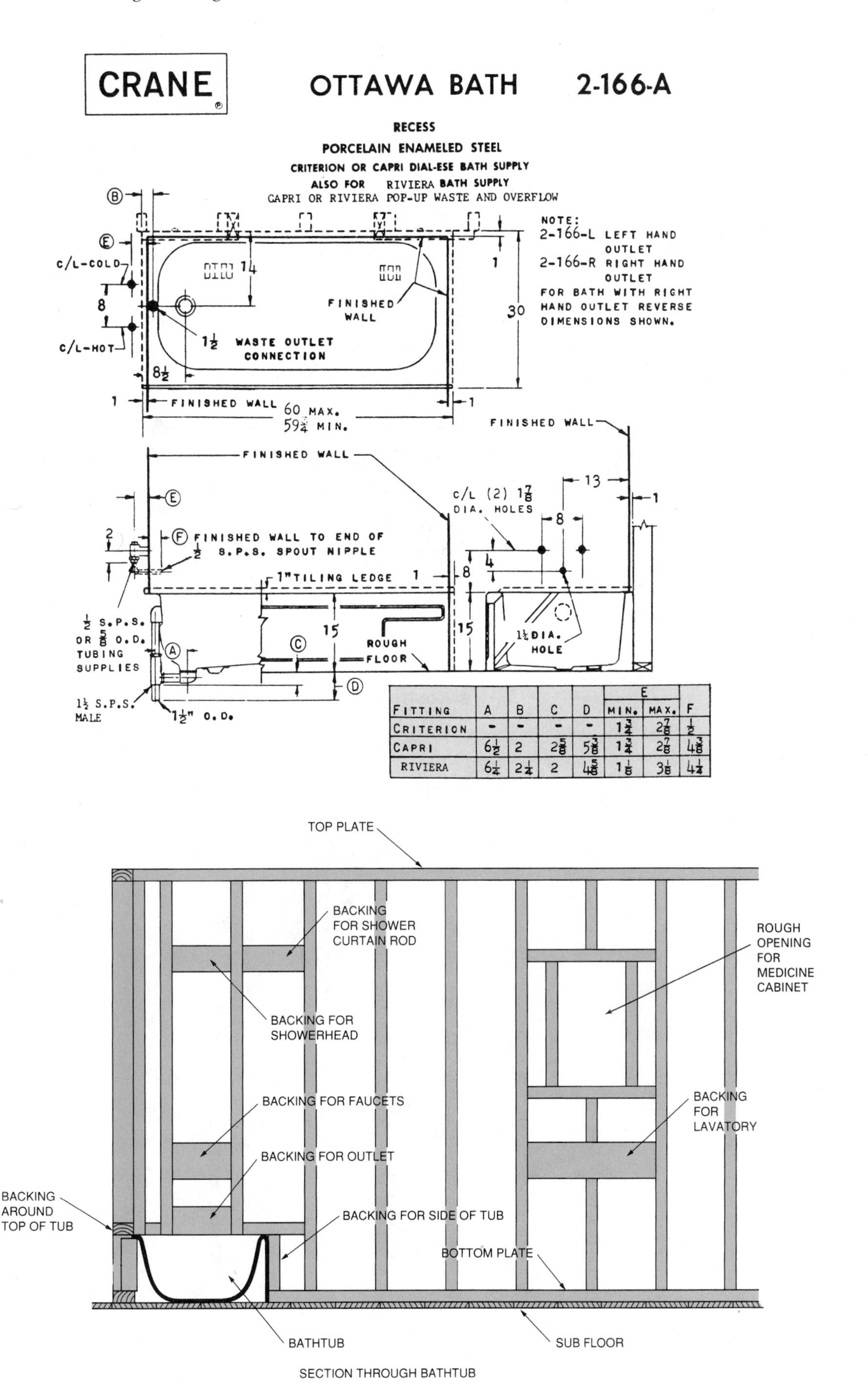

FITTING	A	B	C	D	E MIN.	E MAX.	F
CRITERION	-	-	-	-	1¾	2⅞	½
CAPRI	6½	2	2⅝	5⅜	1¾	2⅞	4⅜
RIVIERA	6¼	2¼	2	4⅝	1⅛	3⅛	4¼

Fig. 37-2 Considerable backing is needed in bathrooms.

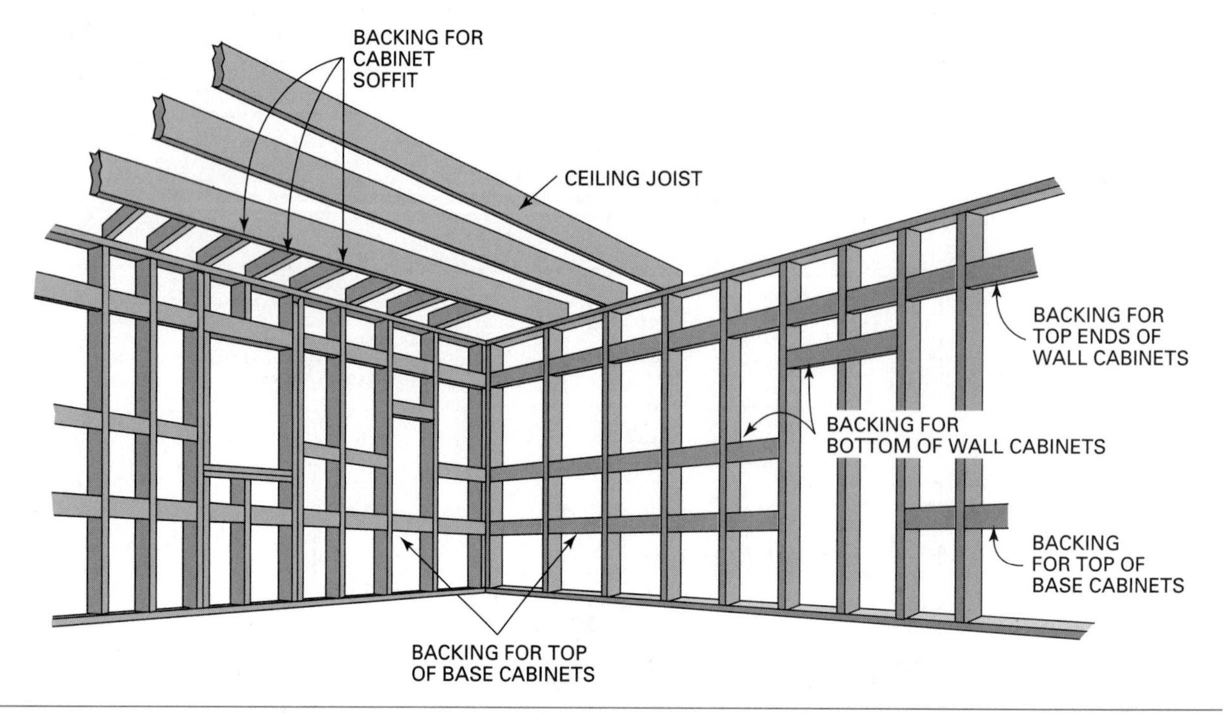

Fig. 37-3 Location and purpose of backing in kitchens.

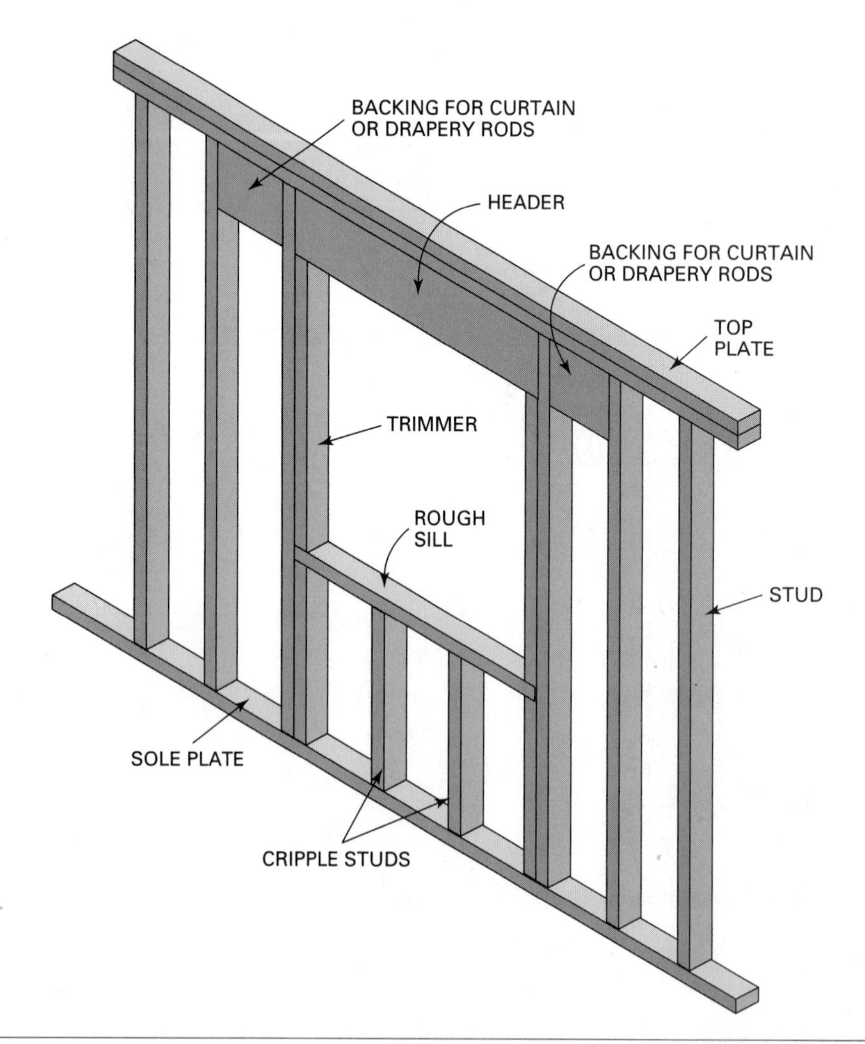

Fig. 37-4 An experienced builder will install backing in all rooms for curtain and drapery hardware.

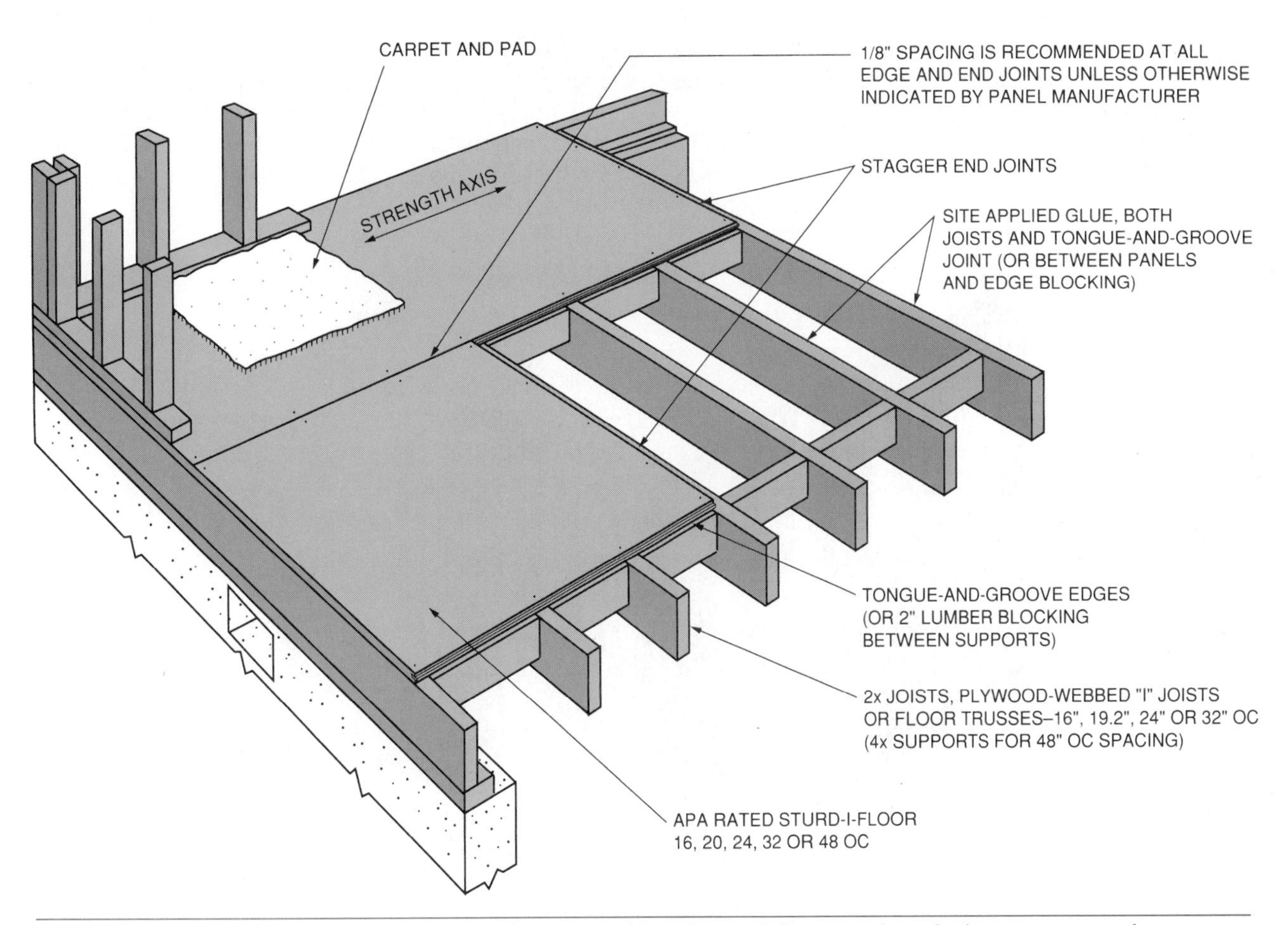

Fig. 37-5 The edges of floor panels used as a combination subfloor and underlayment must be tongue-and-groove or supported by blocking.

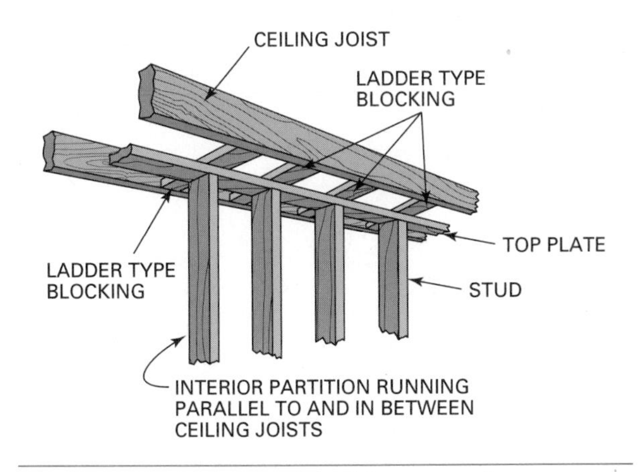

Fig. 37-6 Ladder-type blocking must be provided for some interior partitions not supported by furring strips.

supported on 2-inch lumber blocking installed between joists (Fig. 37-5).

When ceiling furring strips are not used, ladder-type blocking is needed between ceiling joists to support the top ends of partitions that run parallel to and between joists (Fig. 37-6).

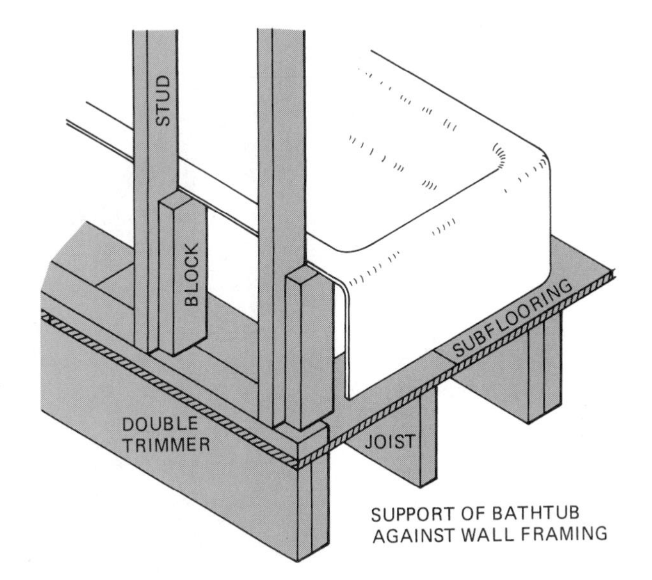

Fig. 37-7 Supporting the inner edge of a bathtub with blocking.

Blocks are sometimes installed to support bathtubs (Fig. 37-7). There is a negligible amount of shrinkage when the blocks are placed on end.

Wall blocking is required for weathertightness and support at the horizontal edges of wall sheathing panels permanently exposed to the weather. In addition, blocking is required behind all plywood seams in a wood foundation. When used for these purposes, they must be installed in a straight line (Fig. 37-8).

Blocking between studs is required in walls over 8′-1″ high. The purpose is to stiffen the studs and strengthen the structure. The required blocking also functions as a firestop in stud spaces. Blocking for these purposes may be installed in staggered fashion (Fig. 37-9).

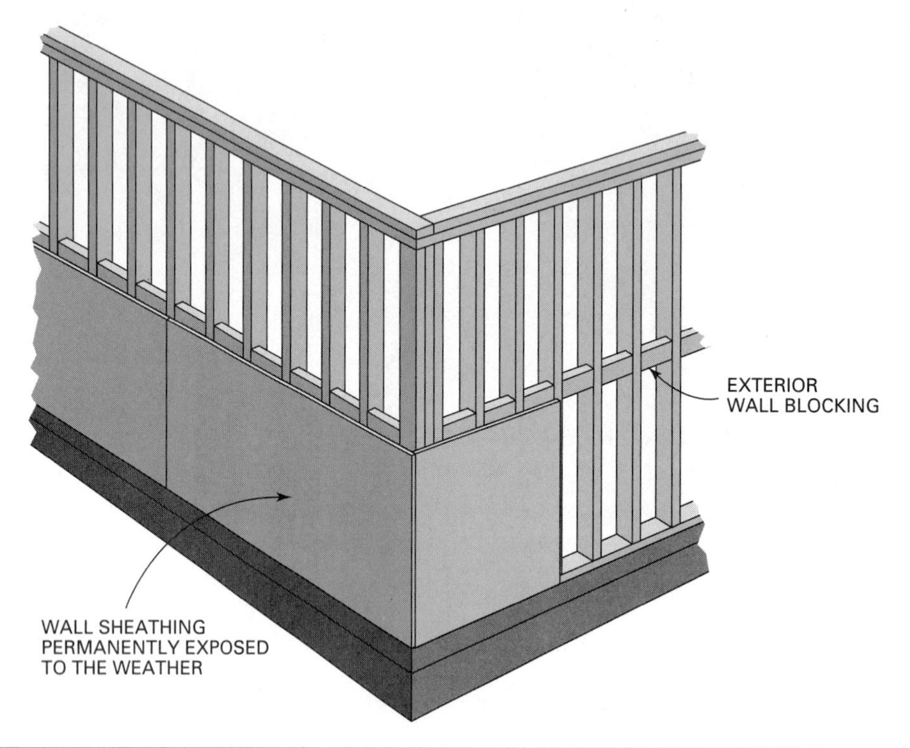

Fig. 37-8 Blocking for wall sheathing must be installed in a straight line.

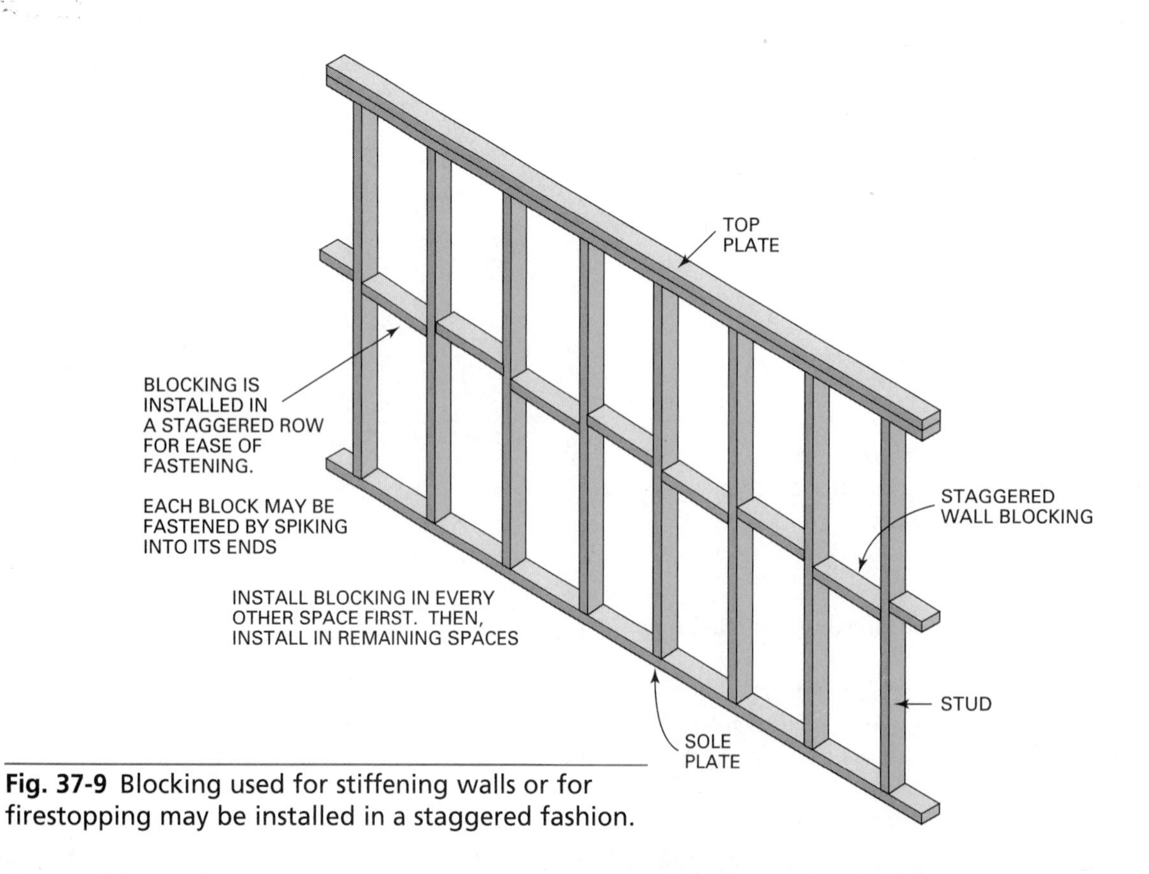

Fig. 37-9 Blocking used for stiffening walls or for firestopping may be installed in a staggered fashion.

Installing Backing and Blocking

Install blocking in a staggered row. Fasten by nailing through the studs into each end of each block in the same manner as staggered solid wood bridging described previously. Installing blocking in a straight line is more difficult. The ends of some pieces must be toenailed (Fig. 37-10). Snap a line across the framing. Square lines in from the chalk line on the sides of the studs.

It may be helpful to fasten short cleats on one side of each stud to support the blocking while the end is being toenailed. If cleats cannot be used or are not desired, start the toenails in the end of the block before positioning it (Fig. 37-11). Backing may

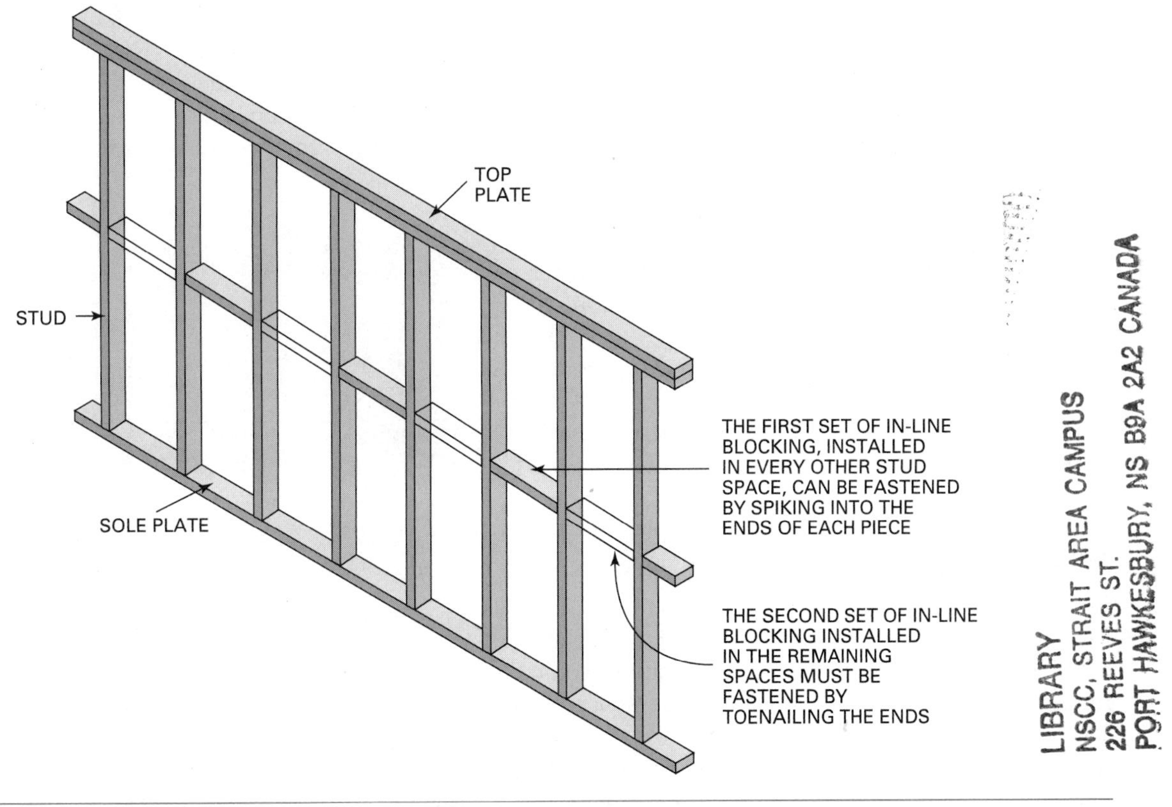

Fig. 37-10 Straight in-line blocking requires that the ends of some pieces be fastened by toenailing.

HELPFUL HINTS

Oil joints of a sliding rule frequently.

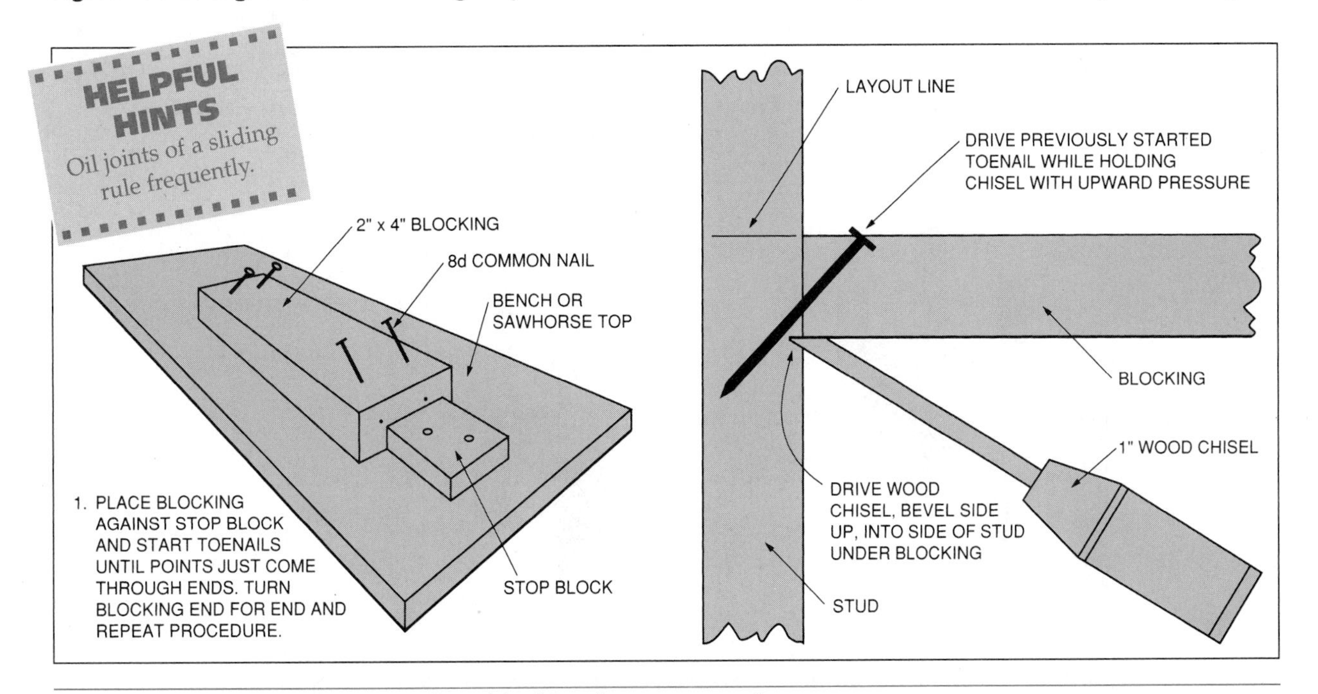

Fig. 37-11 Technique for toenailing blocking between studs.

also be installed in a continuous length by notching the studs and fastening into its edges (Fig. 37-12). If installing blocking between studs, fasten pieces in every other space first. Then go back and fill in. This prevents gaining on the stud space and bowing the studs, especially if the blocking is precut.

Plaster Bases

Plaster bases of some type must be fastened to the wall studs and ceiling joists or furring trips for the application of *plaster.*

Plaster

Plaster has been used for many years for wall and ceiling finish. It is made from a white mineral called **gypsum.** Gypsum is mined from the earth in rock form. The rock is crushed and ground to a powder form. It is then mixed with sand and a binder to make plaster.

The plaster powder is mixed with water to form a paste. When spread thinly on a flat surface, it dries to form a hard and smooth surface. The surface on which the plaster is spread is called a *plaster base* (Fig. 37-13).

Lath

In the past, thin wood strips called wood **lath** were used as a plaster base. They are still found in many older buildings and can still be purchased in some lumber yards. They were applied with a space between each strip that served to form a *key* or a lock for the plaster. Wood lath is no longer used, except perhaps in remodeling to match the existing construction. At the present time, *gypsum* lath, commonly called *rock* lath, and *metal* lath are used.

Gypsum Lath. Gypsum lath is a thin board consisting of a gypsum core sandwiched between a special paper covering. On the face side, the outer layers of a multi-layer paper absorb moisture from the plaster quickly to form a tight bond. The inner layers are treated to resist moisture and prevent softening of the gypsum core.

For most jobs, regular rock lath is used. When additional fire resistance is required, firecode rock lath, whose core contains special mineral materials, may be used.

Gypsum lath is made in several sizes. For framing spaced 16 inches on center, a 3/8-inch thickness may be used. A 1/2-inch thickness is also available. The maximum frame spacing is dependent on the thickness and type of lath used. The

Fig. 37-13 Applying plaster to a gypsum plaster base. *(Courtesy of U.S. Gypsum Company)*

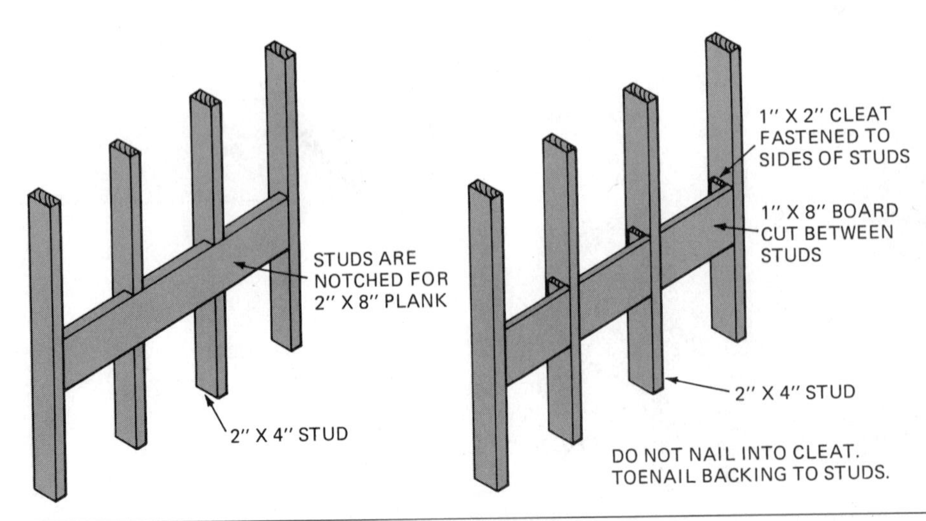

Fig. 37-12 Method of installing continuous backing for plumbing and other type fixtures.

standard width is 16 inches. The standard length is 48 inches. A 24 × 96 gypsum lath is available in 1/2-inch thickness only.

Metal Lath. Metal lath is formed from sheet steel that has been slit and expanded to form numerous small openings to key the plaster. The metal is expanded in several forms, such as diamond mesh, ribbed, or flat ribbed for various purposes (Fig. 37-14). The diamond mesh design is a general all-purpose type of metal lath. It is available in a *self-furring* type. It has dimple indentations spaced closely each way. It is used as an exterior stucco base, for column fireproofing, and for replastering over old surfaces. Metal lath is available in widths of 27 inches and in lengths of 96 inches. It is coated with black asphaltum paint or is galvanized.

Metal lath, commonly called *wire lath,* is used extensively in commercial construction. On large jobs it is installed by specialists called *lathers.* On smaller jobs, the carpenter may be required to apply either gypsum or metal lath. In residential construction, metal lath is recommended for use on ceilings that have a living area above. In shower and bathtub areas, where excessive moisture is present, only galvanized metal lath is recommended for a plaster base.

Metal lath requires three coats of plaster: a **scratch coat,** a brown coat, and a finish coat. Gypsum lath requires only two coats: the brown coat and the finish coat. Each coat must be partially dry before the next coat is applied. The combined thickness of the scratch and brown coat is 11/16 inch. The thickness of the finish coat is 1/16 inch (Fig. 37-15). Total thickness of the lath and plaster combined is 3/4 inch.

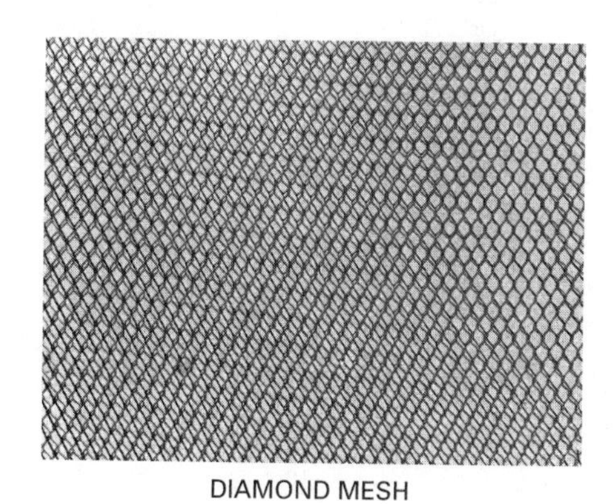

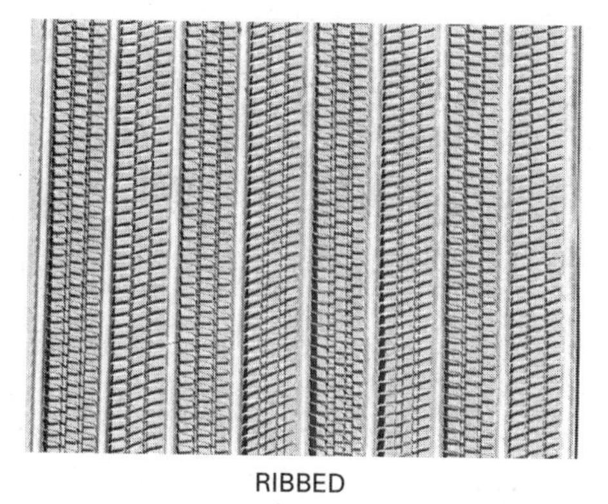

Fig. 37-14 Types of metal lath.

Applying Gypsum Lath

Gypsum lath is applied with its face out and long dimension across framing members. End joints should fall between framing members. They should be staggered in successive *courses* (rows) with none occurring in line with the sides of openings. Where end joints occur between framing, secure with clips.

Attach gypsum lath to framing with nails, screws, or staples (Fig. 37-16). Drive five fasteners into each framing member across the width of the

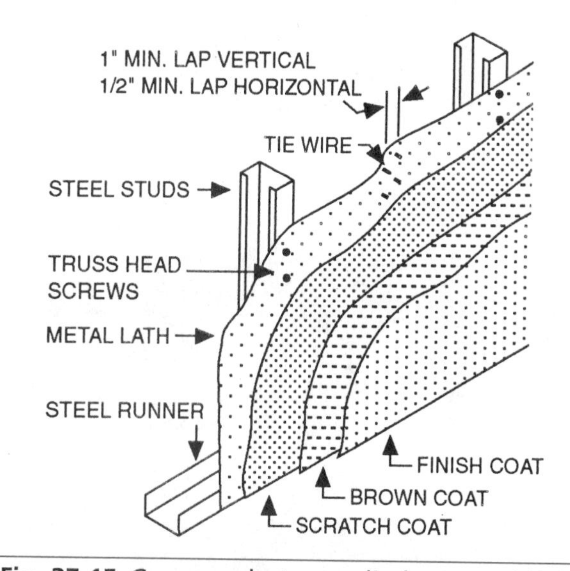

Fig. 37-15 Gypsum plaster applied to metal lath.

Fig. 37-16 Gypsum lath is applied with its end joints staggered. (*Courtesy of U.S. Gypsum Corporation*)

sheet, not less than 3/8 inch from ends and edges. For 24-inch wide, use six fasteners in each framing member. Be careful not to let the fastener heads cut into the face paper.

Gypsum lath may be cut to length by scoring one side with a utility knife. Bend and break the sheet along the scored line. Then score the back side on the bend. Snap off the waste (Fig. 37-17). A compass saw or wallboard saw is used to make cutouts for pipes or electrical outlets.

Reinforcing Plaster Bases

To prevent plaster cracks, certain sections are reinforced with expanded metal. Fasten small strips diagonally across the upper corners of all window and door openings (Fig. 37-18). Inside corners at the intersection of walls and ceilings are reinforced with metal lath called **cornerite.** These are strips bent at a 90-degree angle to fit in the corners. They project about 2 inches on each side (Fig. 37-19).

Corner beads of expanded metal are applied on all exterior corners. Use a straightedge. Install them level and plumb. Plumb corner beads from the plaster **grounds** (strips of wood used as guides to control thickness of plaster) installed at the base. Where the corner beads meet at the corner of an opening, make sure the bead of each is exactly flush (Fig. 37-20).

Applying Metal Lath

Metal lath is applied so that its edges and ends lap a minimum of 1 inch. No extra reinforcement is necessary except using corner beads at the exterior corners. Fasten the expanded metal to metal framing with tie wire and to wood framing with fasteners that catch two strands of the lath and spaced about 6 inches on center into each framing member. Cut expanded metal where necessary using tin snips.

CAUTION Handle the cut sheets carefully. The sharp ends of the metal can cut deeply.

Plaster Grounds

Plaster grounds are strips of wood or metal used as guides by the plasterer to control the thickness of

Fig. 37-17 Cutting gypsum board lath.

37-19 Interior corners are reinforced with wire lath commonly called cornerite.

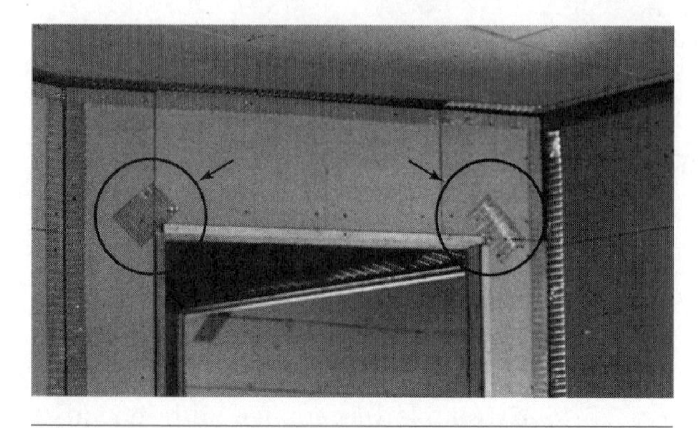

Fig. 37-18 Reinforce the tops of windows and door openings with small strips of metal lath.

Fig. 37-20 Exterior corners are reinforced with corner bead.

the plaster. The thickness of the ground is the total thickness of the lath and plaster combined. Recommended thickness is 3/4 inch.

Grounds are located at the base of walls and around door and window openings (Fig. 37-21). They are placed in a position so the ground will later be covered by the interior trim. They are also installed wherever it is imperative that a straight and even plaster line be provided for high-quality application of interior trim. Therefore, plaster grounds must be installed in a straight and true line.

Usually 3/4 × 2 1/2 -inch strips are used at the base. Wider strips may be used if the baseboard will cover them. Two narrow strips may be used, one at the floor line and one just below the top of the baseboard. Around door and window openings and at other locations, 3/4 × 1-inch strips usually are used.

In some cases, the inside edges of the window frame act as grounds when the jamb is the proper width. In most cases, the grounds are fastened permanently in position. They are left in place after the plaster is applied. However, they may be installed temporarily, especially around door openings. They are removed when the plastering is completed (Fig. 37-22). The carpenter is responsible for the installation of plaster grounds unless a specialist installs them.

Fig. 37-21 Plaster grounds serve as a guide to control the thickness of plaster.

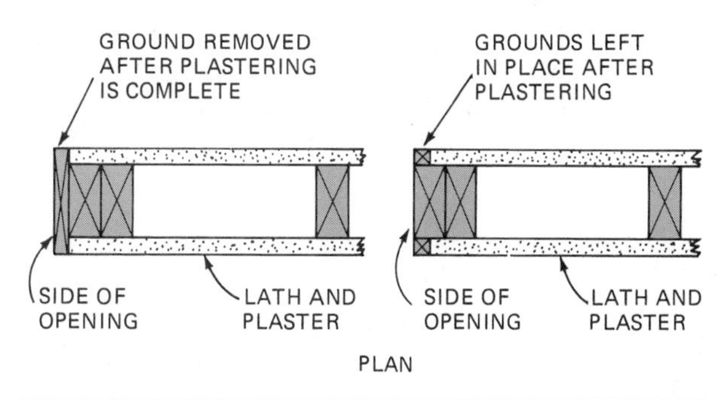

Fig. 37-22 Grounds may be installed temporarily or permanently around openings.

CHAPTER 38 STEEL FRAMING

Light steel framing is used for structural framing and interior non-load-bearing partitions. Specialists in light steel are required when framing is extensive. However, carpenters often frame interior partitions and apply *furring channels* of steel. This chapter, therefore, is limited to their installation.

The strength of steel framing members of the same design and size may vary with the manufacturer. The size and spacing of steel framing members should be determined from the drawings or by those qualified to do so.

Interior Steel Framing

The framing of steel interior partitions is quite similar to the framing of wood partitions. Different kinds of fasteners are used. Some special tools may be helpful.

Steel Framing Components

All steel framing components are coated with material to resist corrosion. The main parts of an interior steel framing system are *studs, track, channels,* and various *accessories.*

Studs. For interior non-load-bearing applications, *studs* are manufactured from 25-, 22-, and 20-gauge steel. The stud *web* has punchouts at intervals through which to run pipes and conduit. Studs come in widths of 1 5/8, 2 1/2, 3 5/8, 4, and 6 inches, with 1 and 1 1/4 leg thickness. Studs are available in stock lengths of 8, 9, 10, 12, and 16 feet

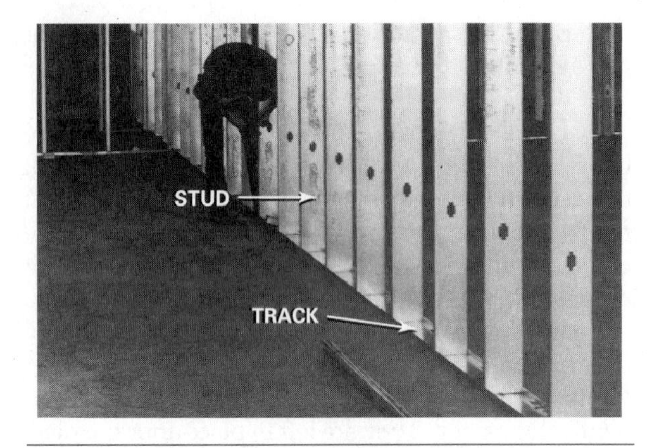

Fig. 38-1 Steel studs for light framing come in several widths, lengths, and gauges.

(Fig. 38-1). Custom lengths up to 28 feet are also available.

Track. The top and bottom horizontal members of a steel-framed wall are called *runners* or *track*. They are installed on floors and ceilings to receive the studs. They are manufactured in gauges, widths, and leg thicknesses to match studs (Fig. 38-2). Track is available in standard lengths of 10 feet.

Channels. Steel *cold-rolled channels* (CRC) are formed from 16-gauge steel. They are available in several widths. They come in lengths of 10, 16, and 20 feet. Channels are used in suspended ceilings and walls. When used for lateral bracing of walls, the channel is inserted through the stud punchouts. It is fastened with welds or clip angles to the studs (Fig. 38-3).

Furring Channels. *Furring channels* or hat track are hat-shaped pieces made of 20- and 25-gauge steel. Their overall cross-section size is 7/8 inch by 2 9/16 inches. They are available in lengths of 12 feet (Fig. 38-4). Furring channels are applied to

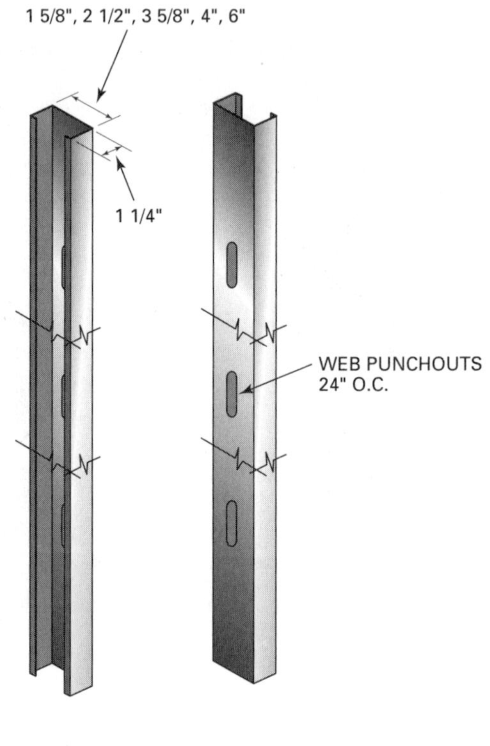

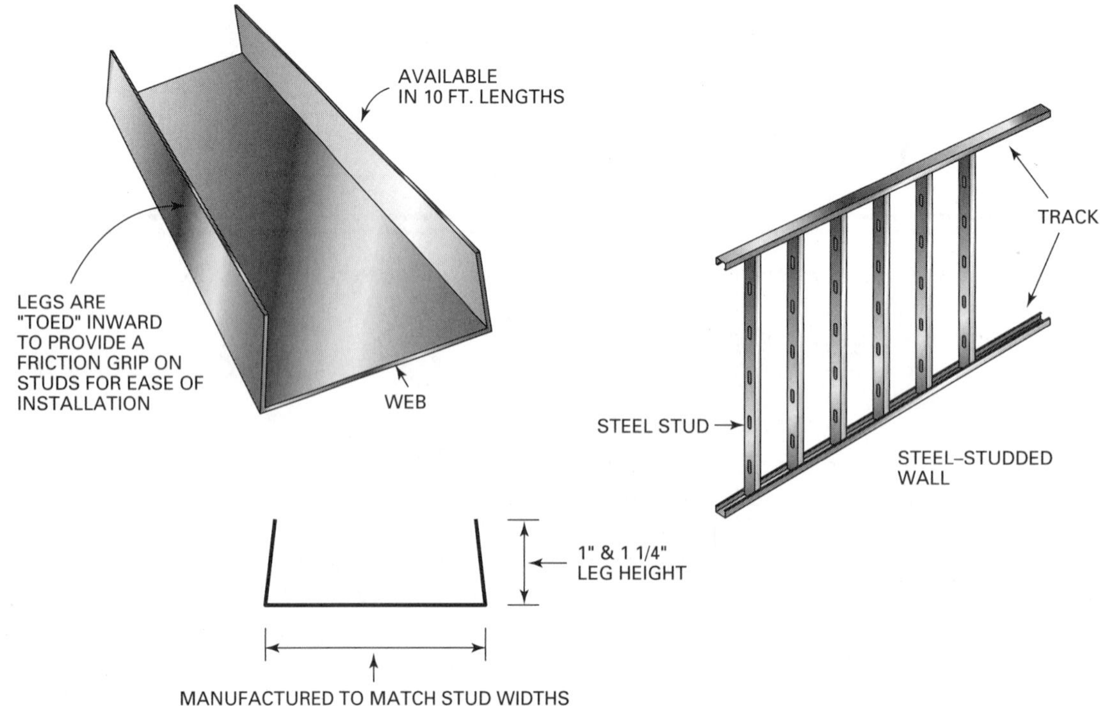

Fig. 38-2 Cold-rolled channels called *track* are the top and bottom plates of steel-stud walls.

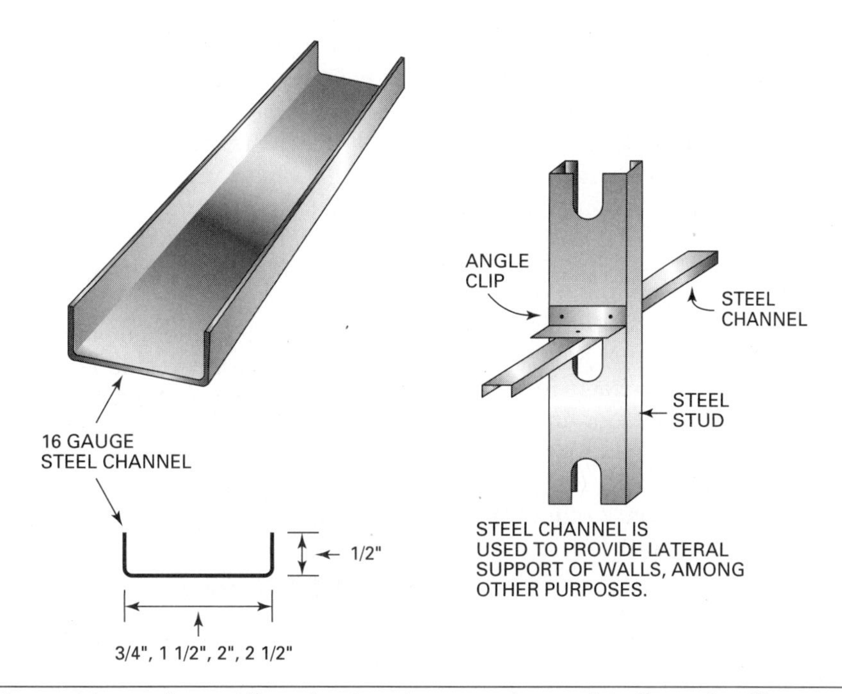

Fig. 38-3 Steel channel is used to stiffen the framing members of walls and ceilings.

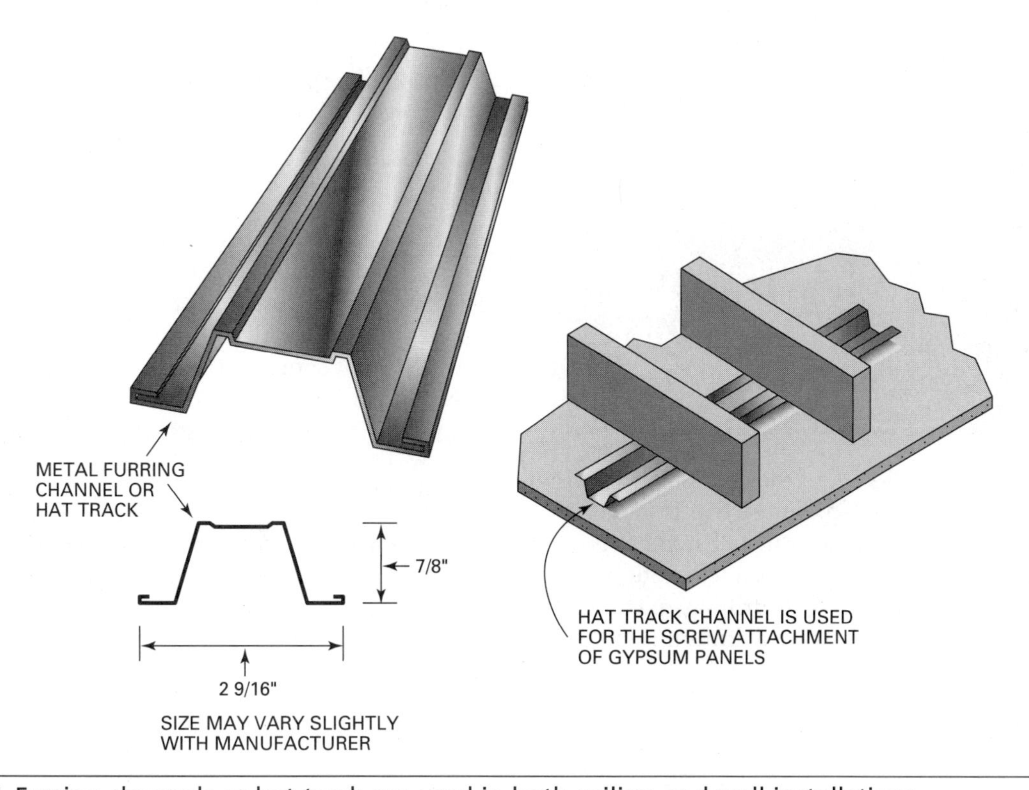

Fig. 38-4 Furring channels or hat track are used in both ceiling and wall installations.

walls and ceilings for the screw attachment of gypsum panels and plaster bases. Framing members may exceed spacing limits for various coverings. Furring can then be installed to meet spacing requirements and provide necessary support for the surfacing material.

Layout and Framing of Steel Partitions

Lay out steel-framed partitions as you would wood-framed partitions. Snap chalk lines on the floor. Plumb up from partition ends. Snap lines on the ceiling. Make sure that partitions will be plumb. Using a laser level is an efficient way to lay out floor and ceiling lines for partitions. See Figure 25-43.

Lay out the stud spacing and the wall opening on the bottom track. The top track is laid out after the first stud away from the wall is plumbed and fastened.

Installing Track. Fasten track to floor and ceiling so one edge is to the chalk line. Make sure both floor and ceiling track are on the same side of the line. Leave openings in floor track for door frames. Allow for the width of the door and thickness of the door frame. Track are usually fastened into concrete with powder-driven fasteners (see Fig. 16-16). Stub concrete nails or masonry screws may also be used. Fasten into wood with 1 1/4-inch oval head screws.

Attach the track with two fasteners about 2 inches from each end and a maximum of 24 inches on center in between. At corners, extend one track to the end. Then butt or overlap the other track (Fig. 38-5). It is not desirable or necessary to make mitered joints.

To cut metal framing to length, tin snips may be used on 25-gauge steel. Using tin snips becomes difficult on thicker metal. A power miter box, commonly called a *chop saw,* with a metal-cutting saw blade is the preferred tool.

CAUTION The sharp ends of cut metal can cause a serious injury. A pointed end presents an even greater danger. Avoid miter cuts on thin metal. Do not leave short ends of cut metal scattered around the job site. Dispose of them in a container as you cut them.

Installing Studs. Cut the necessary number of full-length studs needed. For ease of installation, cut them about 1/4 inch short. Install studs at partition intersections and corners. Fasten to bottom and top track. Use 3/8-inch self-drilling pan head screws or crimp together using a crimping tool designed for steel studs. If moisture may be present where a stud butts an exterior wall, place a strip of asphalt felt between the stud and the wall.

Place the first stud in from the corner between track. Fasten the bottom in position at the layout line. Plumb the stud. Using a magnetic level can be very helpful. Clamp the top end when plumb, and fasten to the top track. Lay out the stud spacing on the top track from this stud (Fig. 38-6).

Place all full-length studs in position between track with the open side facing in the same direction. The web punchouts should be aligned vertically. This provides for lateral bracing of the wall and the running of plumbing and wiring (Fig. 38-7). Fasten all studs except those on each side of door openings securely to top and bottom track.

Wall Openings

The method of framing door openings depends on the type of door frame used. A one-piece metal door frame must be installed before the gypsum board is applied. A three-piece, knocked-down frame is set in place after the wall covering is applied (Fig. 38-8).

Framing a Door Opening for a Three-Piece Frame. First, place full-length studs on each side of the opening in a plumb position. Fasten securely to the bottom and top plates. Cut a length of track for use as a header. Cut it 6 inches longer than the width of the opening. Cut the flanges in appropriate places. Bend the web to fit over the studs on each side of the opening. Fasten the fabricated header to the studs at the proper height. Install jack studs over the opening in positions that continue the regular stud spacing (Fig. 38-9). Window openings are framed in the same manner. However, a rough sill is installed at the bottom of the opening. Cripples are placed both above and below.

Framing a Door Opening for a One-Piece Frame. Place the studs on each side of the opening. Do not fasten to the track. Set the one-piece door frame in place. Level the door

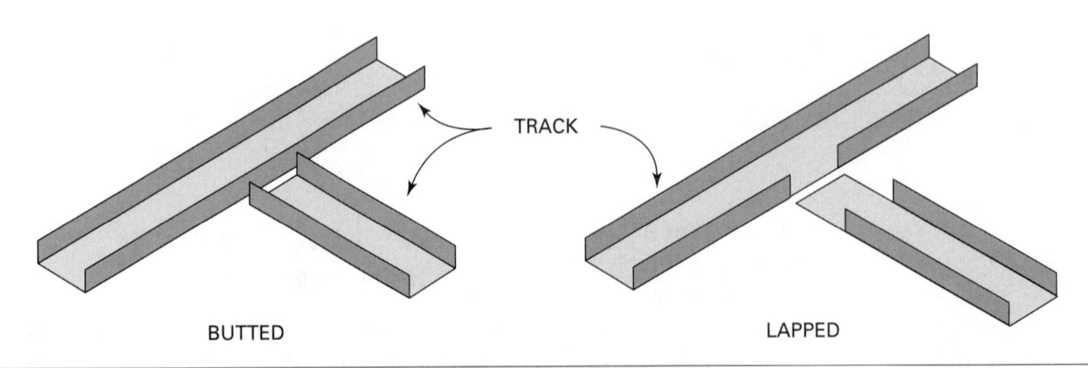

Fig. 38-5 Track may be butted or overlapped at intersections.

STEP BY STEP

PROCEDURES

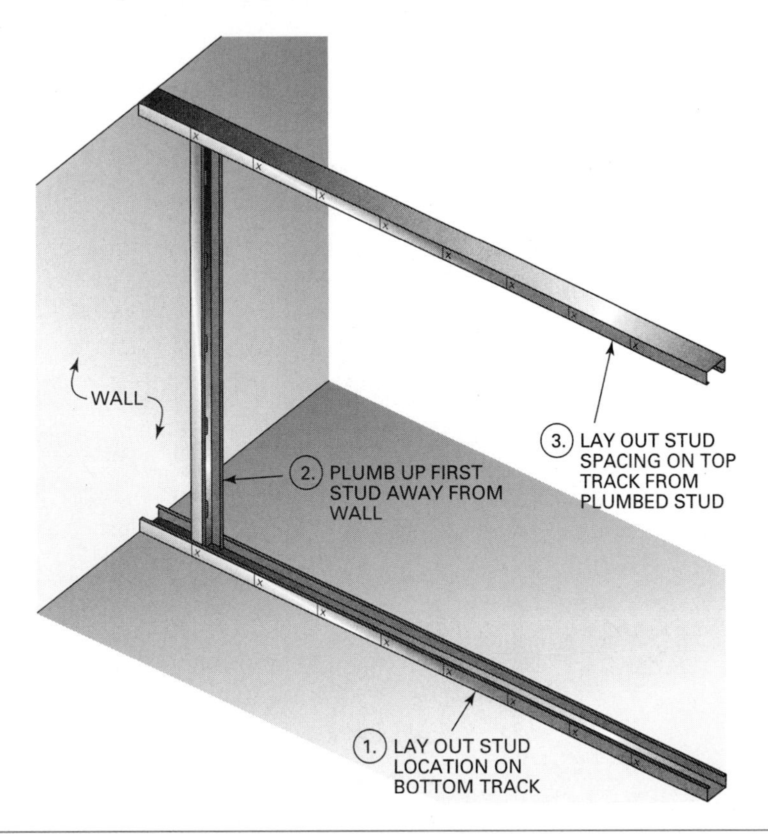

Fig. 38-6 Lay out the top track from the plumbed end stud.

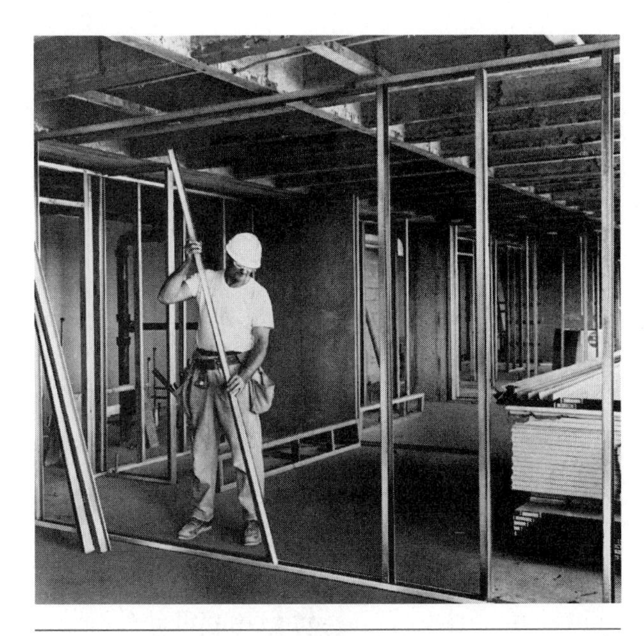

Fig. 38-7 Installing steel studs between top and bottom track. (*Courtesy of U.S. Gypsum Corporation*)

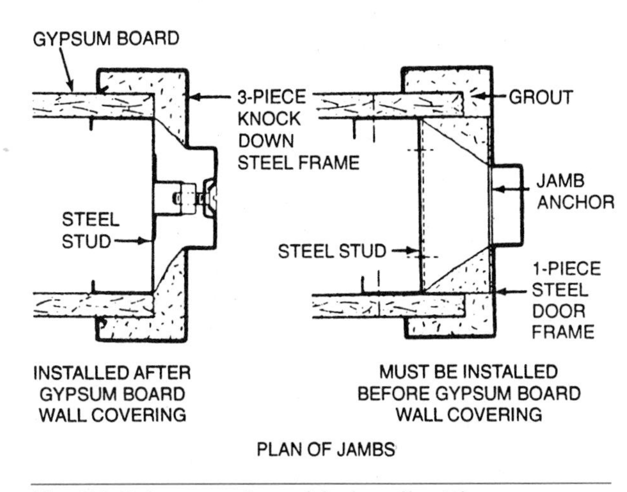

Fig. 38-8 In a steel-studded wall, either one-piece or three-piece, knocked-down metal door frames are used. (*Courtesy of U.S. Gypsum Corporation*)

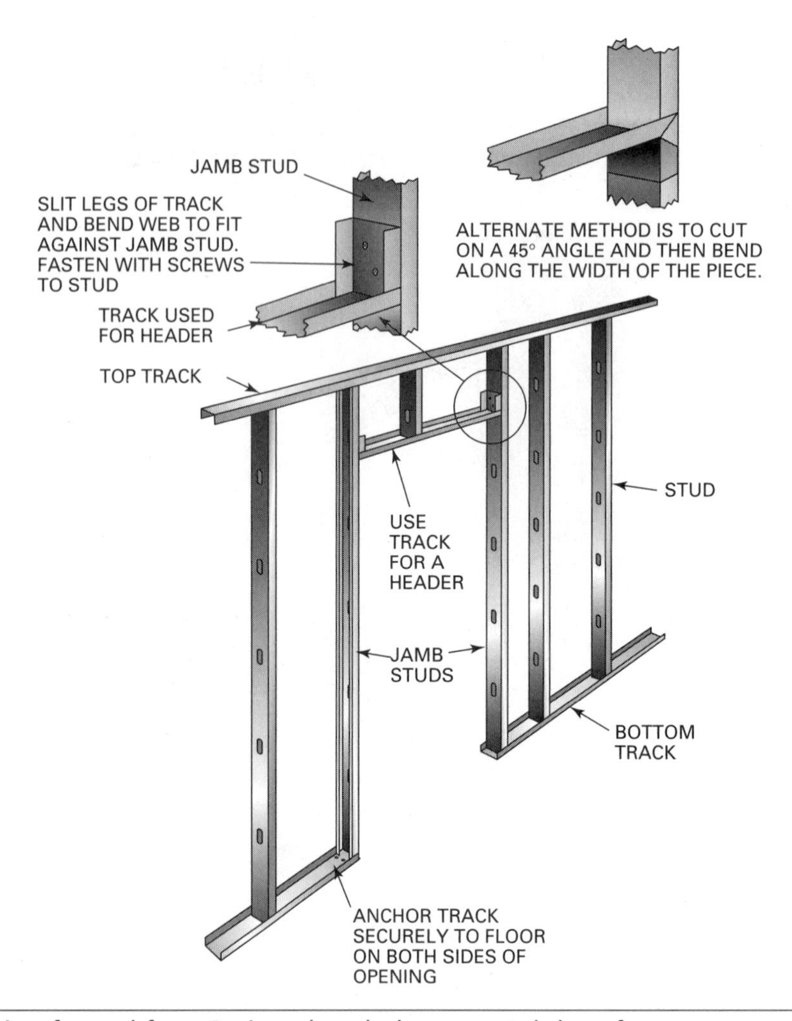

Fig. 38-9 An opening framed for a 3-piece, knock-down, metal door frame.

frame header by shimming under a jamb, if necessary. Fasten the bottom ends of the door jambs to the floor in the proper location. Fasten the studs to the door jambs. Then, fasten the studs to the bottom track. Plumb the door frame. Clamp the stud to the top track, and fasten with screws. Install header and jack studs in the same manner as described previously (Fig. 38-10).

The steel framing described above is suitable for average weight doors up to 2′-8″ wide. For wider and heavier doors, the framing should be strengthened by using 20-gauge steel framing. Also, double the studs on each side of the door opening (Fig. 38-11.)

Chase Walls

A **chase wall** is made by constructing two closely spaced, parallel walls for the running of plumbing, heating and cooling ducts, and similar items. They are constructed in the same manner as described previously. However, the spacing between the outside edges of the wall frames must not exceed 24 inches.

The studs in each wall should be installed with the flanges running in the same direction. They should be directly across from each other. The walls should be tied to each other either with pieces of 12-inch-wide gypsum board or short lengths of steel stud. If the wall studs are not opposite each other, install lengths of steel stud horizontally inside both walls. Tie together with shorter lengths of stud material spaced 24 inches on center (Fig. 38-12). Wall ties should be spaced 48 inches on center vertically.

Installing Metal Furring

Metal furring may be used on ceilings applied at right angles to joists. They may be applied vertically or horizontally to masonry walls. Space metal furring channels a maximum of 24 inches on center.

STEP BY STEP

PROCEDURES

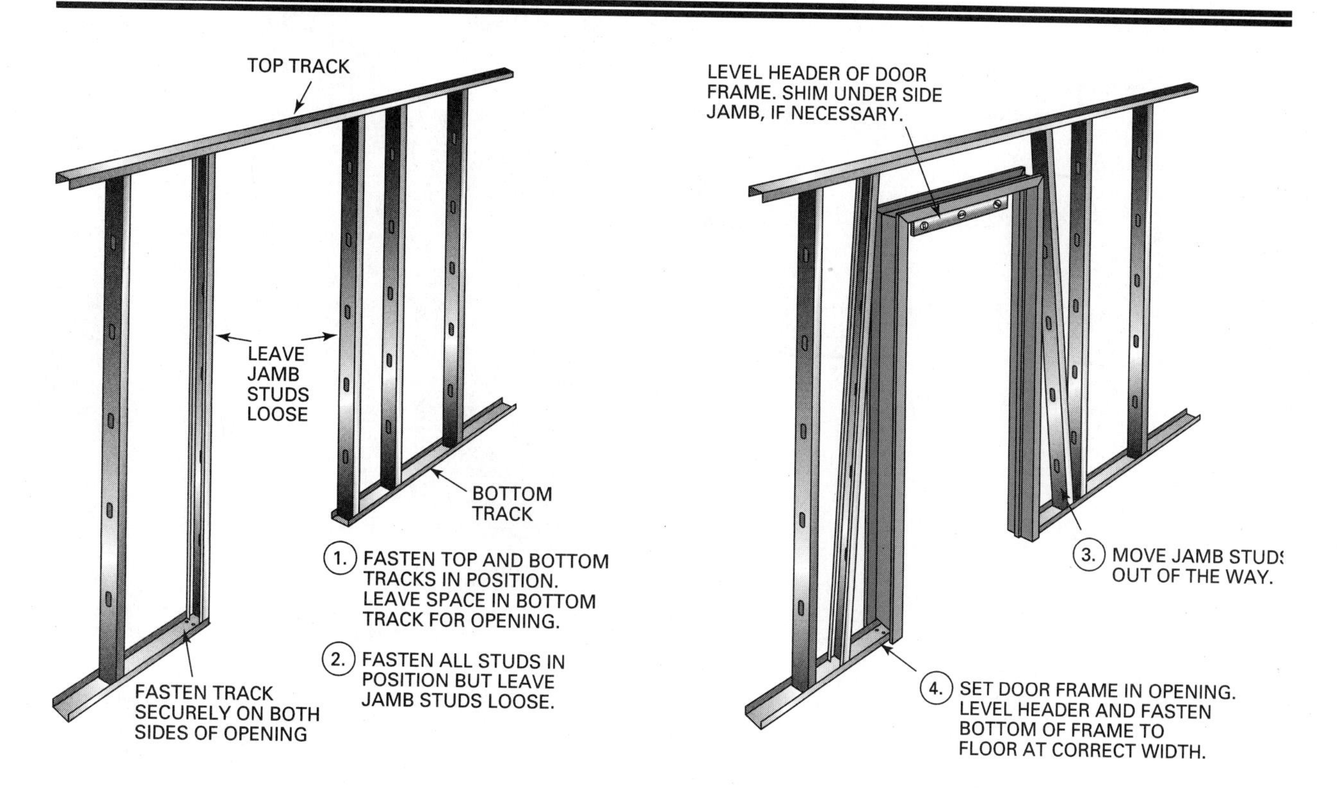

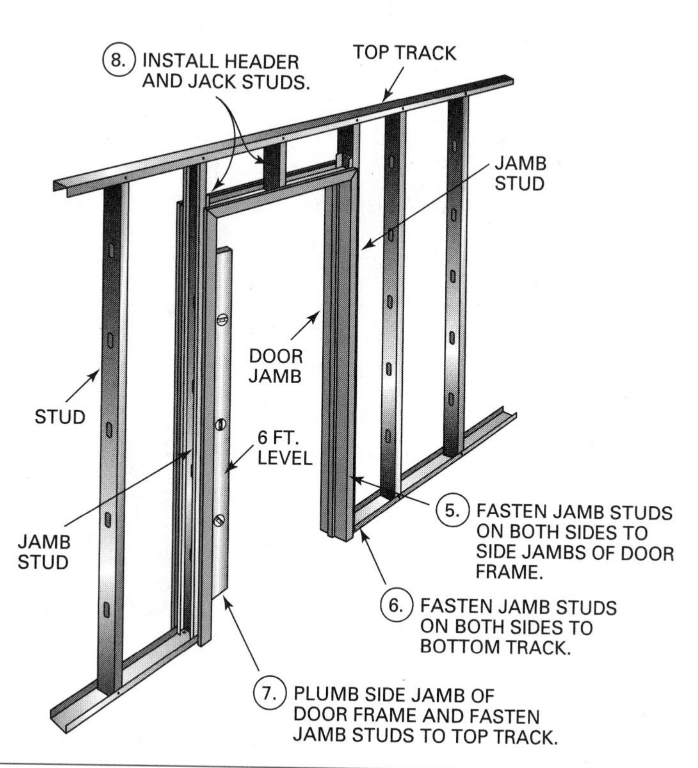

Fig. 38-10 Method of framing the opening for and setting a one-piece door frame.

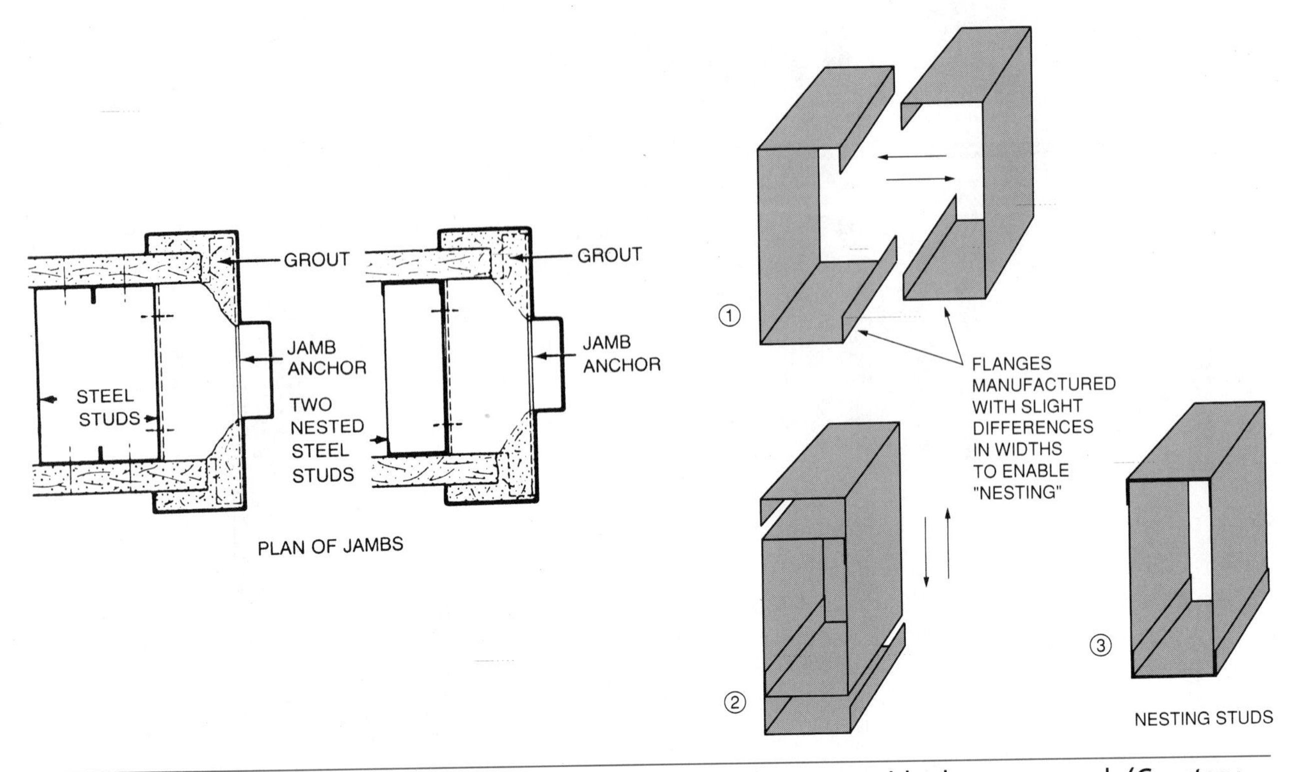

Fig. 38-11 The rough opening must be strengthened when heavy or wide doors are used. (*Courtesy of U.S. Gypsum Corporation*)

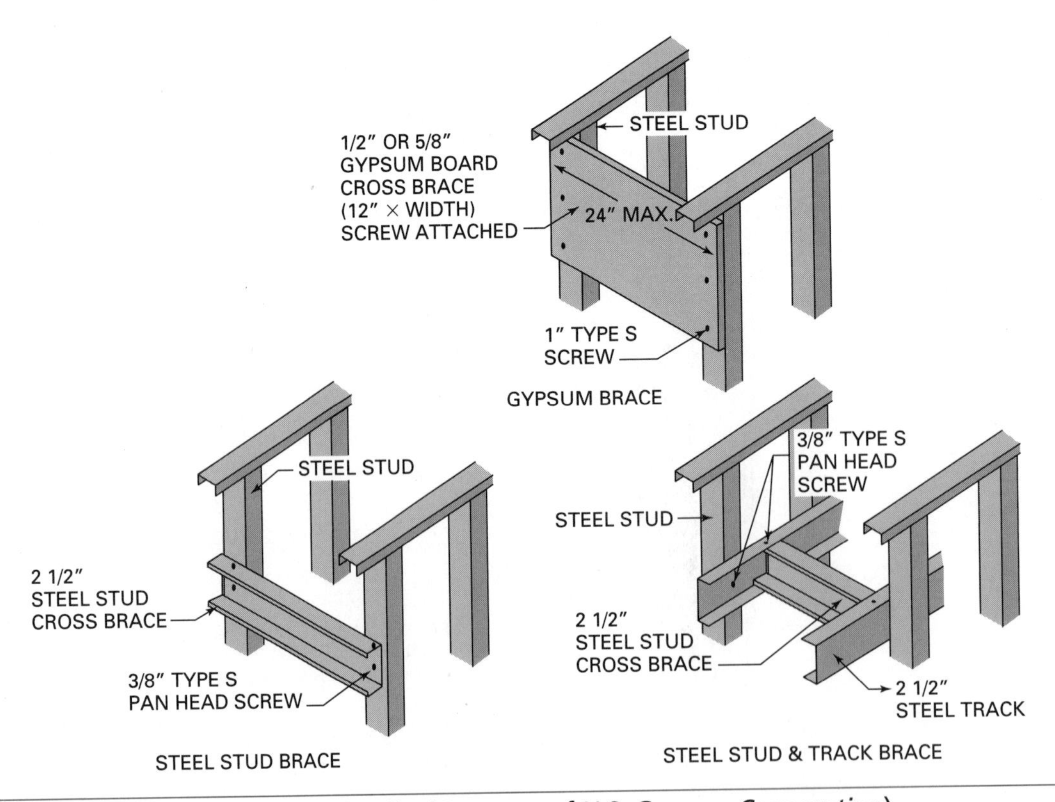

Fig. 38-12 Chase wall construction details. (*Courtesy of U.S. Gypsum Corporation*)

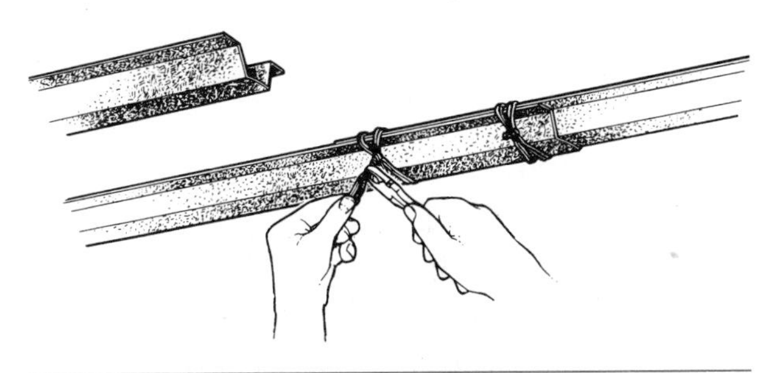

Fig. 38-13 Method of splicing furring channels. (*Courtesy of U.S. Gypsum Corporation*)

Ceiling Furring. Metal furring channels may be attached directly to structural ceiling members or suspended from them. For direct attachment, saddle tie with double-strand 18-gauge wire to each member. Leave a 1-inch clearance between ends of furring and walls. Metal furring channels may be spliced. Overlap the ends by at least 8 inches. Tie each end with wire (Fig. 38-13). Steel studs may be used with their open side up for furring when supporting framing is widely spaced. Several methods of utilizing metal furring channels or steel studs in suspended ceiling applications are shown in Figure 38-14.

Wall Furring. Vertical application of steel furring channels is preferred. Secure the channels by staggering the fasteners from one side to the other not more than 24 inches on center (Fig. 38-15). For horizontal application on walls, attach furring channels not more than 4 inches from the floor and ceiling. Fasten in the same manner as vertical furring.

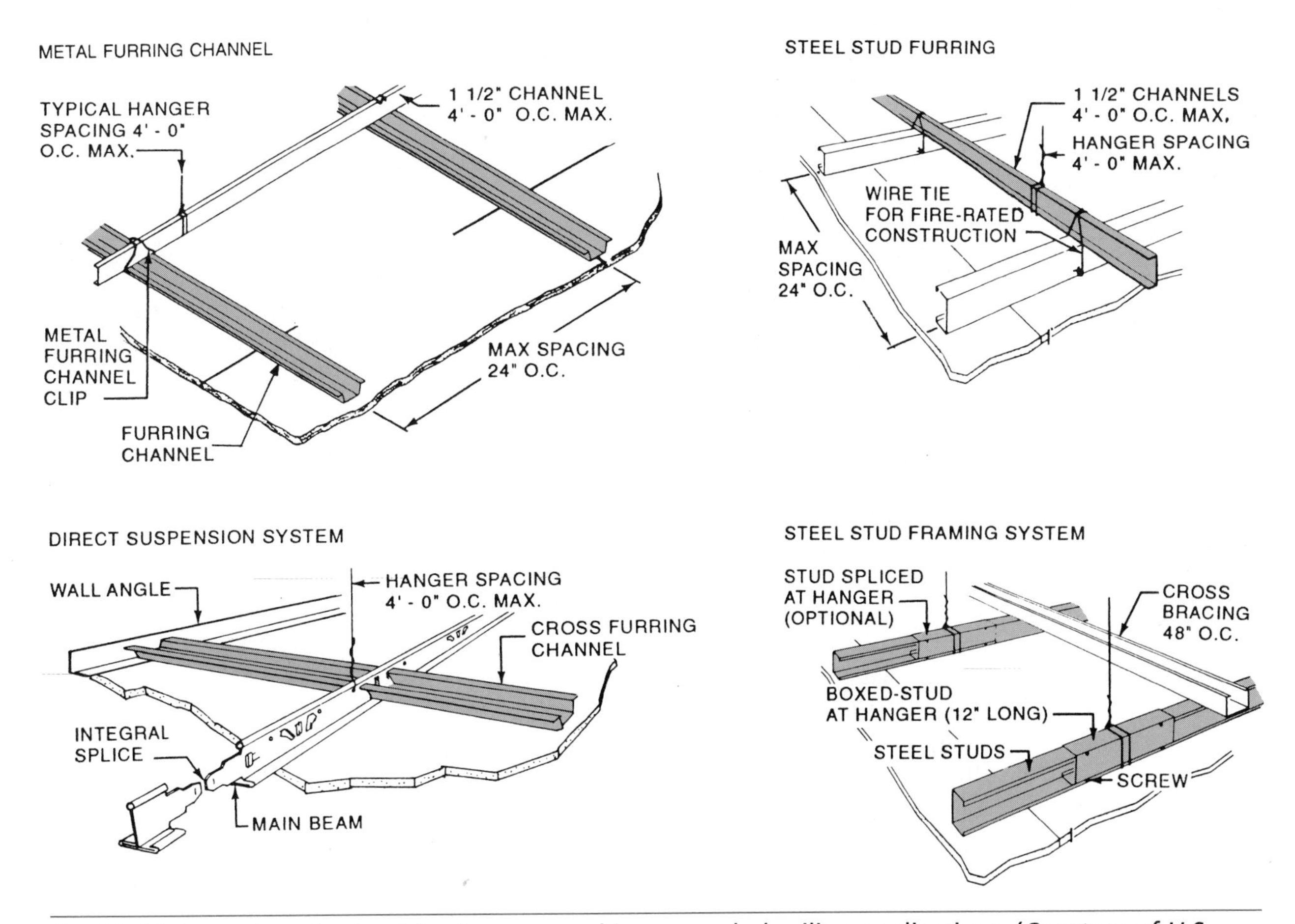

Fig. 38-14 Several methods of furring are used in suspended ceiling applications. (*Courtesy of U.S. Gypsum Corporation*)

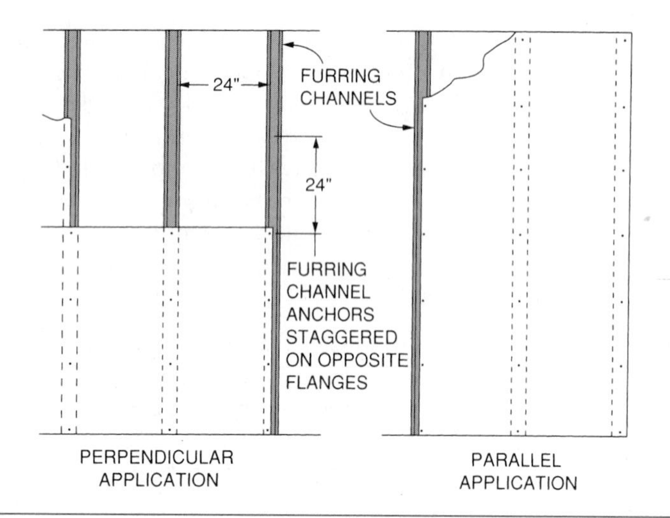

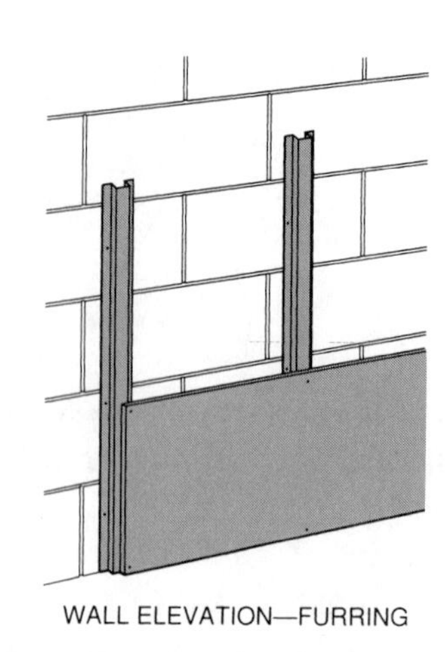

Fig. 38-15 Furring channels are attached directly to masonry walls. (*Courtesy of U.S. Gypsum Corporation*)

Review Questions

Select the most appropriate answer.

1. Bearing partitions
   a. have a single top plate.
   b. carry no load.
   c. are constructed like exterior walls.
   d. are erected after the roof is tight.

2. The top plate of the bearing partition
   a. laps the plate of the exterior wall.
   b. is a single member.
   c. butts the top plate of the exterior wall.
   d. is applied after the ceiling joists are installed.

3. What is the rough opening height of a door opening for a 6′-8″ door if the finish floor is 3/4-inch thick, 1/2-inch clearance is allowed between the door and the finish floor, and the jamb thickness is 3/4 inch?
   a. 6′-9″.
   b. 6′-9 1/2″.
   c. 6′-10″.
   d. 6′-10 1/2″.

4. Ceiling joists usually are installed
   a. with their end joints lapped at the bearing partition.
   b. with the joint over the bearing partition scabbed.
   c. by being fastened to the same side of the rafter.
   d. with blocks placed between them at the bearing partition.

5. The ends of ceiling joists are cut to the pitch of the roof
   a. for easy application of the roof sheathing.
   b. so they will not project above the rafters.
   c. so their crowned edges will be up.
   d. after they are fastened in position.

6. Stub joists
   a. run at right angles to regular ceiling joists.
   b. are used on low-pitched common roofs.
   c. are short pieces cut between regular joists.
   d. span the bearing partition and exterior wall.

7. When working with steel framing it can be noted that
   a. special fasteners are needed.
   b. top plates are usually doubled.
   c. studs are also called track.
   d. stud location is not that important.

8. The top plate of non-bearing partitions is located by
   a. measuring from the exterior walls the same distance as the soleplate.
   b. by plumbing with a transit from the soleplate.
   c. by hanging a plumb bob from the ceiling to the soleplate.
   d. any of the above methods.

9. The end stud of partitions that butt against another wall
   a. must be straightened.
   b. must be fastened securely to the wall.
   c. is left out until the wall is erected.
   d. must not be fastened in its center.

10. Gypsum board lath, commonly called rock lath, must be fastened to each framing member with at least
   a. three nails.
   b. four nails.
   c. five nails.
   d. six nails.

BUILDING FOR SUCCESS 

Job Planning to Avoid Stress

Someday you will have the opportunity to structure projects on the job. Many construction jobs involve tasks beyond the capabilities of one person. It is irrelevant whether the tasks are of a technical or business nature. The planning process is critical in order to establish some control over the outcome of the project. Undoubtedly there will be unexpected interruptions, stress, and people problems to overcome along the way. It is important that we know what we can control or change, and what we cannot.

Influences on job planning in construction are many, but include weather, customers, employees, financial lenders, schedules, subcontractors, material suppliers, and families. Identifying which elements can be controlled or at least predicted will help in the planning process.

Job planning involves developing a realistic time frame and sequence of tasks that culminate in the anticipated outcome. Most of the factors listed previously will affect job planning. Some will be more controllable and predictable. Others will remain relatively unpredictable. Weather, for example, can be forecast well at times. With enough notice, schedules can be made to coincide with favorable weather conditions. At other times inclement weather can cause undue hardships during the construction process. The best advice is to pay close attention to the forecasts, and plan the work accordingly.

Customers are sometimes unpredictable. Good communication skills will play an important role in working well with customers. Understand what they want and expect. In return, be clear about what you can and cannot do in relation to their desires. It is essential not to misunderstand or to give false impressions about what is to be done.

Employees will be a key in job planning. Make sure the technical and business skills of employees can deliver the expected results. Subcontractors may be required if the builder's own workforce is small or lacks skills. Subcontractors can be quite varied in their interpretations of what is expected on a particular job. They will often be involved in a variety of projects at the same time. They must clearly understand the schedule for the entire project. They must also understand what is to be done by what dates.

Local lending agencies are closely tied with their own lenders and national influences such as the prime interest rate and the local economy. At times, it may be difficult for you or your customers to borrow the necessary money to finance the construction project.

Material suppliers may cause delays. Usually control or scheduling here is not a major long-term problem. At times backorders cause delays in construction. This may require that a subcontractor return at an inconvenient time days or weeks down the road.

Personal family situations will often affect the job planning process. Time has to be taken for builders and employees to fulfill family obligations. Sometimes family members assist with the business.

Stress in the job planning process can be decreased by understanding what can be controlled or influenced and what cannot. Avoiding job-related stress can be accomplished by not worrying about uncontrollable items. Be as flexible as possible. Do not try to control every aspect of a construction project. Breakdowns and delays will happen. People sometimes will not show up for work or will fail to give notice of an absence.

Some good ways to reduce the level of stress include the following:

1. Learn when to temporarily walk away from a stressful situation. This gives you time to regenerate and address it later.
2. Seek others you trust to help solve the problem.
3. Understand what you can realistically accomplish in a day's work. Know your limits.
4. Plan either short or extended vacations. Allow time to concentrate on things you enjoy, especially your family.
5. Reduce large problems into manageable steps that move toward solutions. Complete the job one step at a time.
6. Understand and plan for the things you can personally do, then involve others with the balance of the job tasks.
7. Schedule time for a hobby.

Each person who has worked in construction could add many meaningful suggestions to this list. The important thing to remember is knowing your capabilities, what the job calls for, and how to be flexible in solving problems. Plan some form of task analysis to help reach the expected outcomes.

FOCUS QUESTIONS: For individual or group discussion

1. What might be some good reasons to plan for a particular job in the construction process?
2. What job-related tasks can lead to stress?
3. How can a worker deal with stress on the job?
4. Why is it important to know what we can control or influence as we make job-related decisions?
5. How would you go about planning a job where people, tasks, schedules, and the weather play major roles? How would a person's flexibility play a part in the planning process?

UNIT 14

Scaffolds, Ladders, and Horses

Scaffolding or *staging* are temporary working platforms. They are constructed at convenient heights above the floor or ground. They help carpenters perform their jobs quickly and safely. This unit includes information on the safe erection of various kinds of working platforms and the building of several construction aids.

OBJECTIVES

After completing this unit, the student should be able to:

- name the parts of wood single-pole and double-pole scaffolding.
- erect and dismantle metal scaffolding in accordance with recommended, safe procedures.
- build safe staging using roof brackets.
- safely set up, use, and dismantle pump jack scaffolding.
- describe the safe use of ladders, ladder jacks, trestles, and horses.
- build a ladder and sawhorse.

UNIT CONTENTS

CHAPTER 39 WOOD, METAL, AND PUMP JACK SCAFFOLDS

Scaffolds are an essential component of construction, as they allow work to be performed at various elevations. However, they also can create one of the most dangerous working environments. The United States Occupational Safety and Health Administration (OSHA) reports that in construction, falls are the number one killer, and 40 percent of those injured in falls had been on the job less than one year. A recent survey of scaffold accidents summarizes the problem (Fig. 39-1). A scaffold fatality and catastrophe investigation conducted by the OSHA revealed that the largest percentage, 47 percent, was due to equipment failure. In most instances, OSHA found the equipment did not just break; it was broken due to improper use and erection. Failures at the anchor points, allowing either the scaffold parts or its anchor points to break away, were often involved in these types of accidents. Other factors were improper, inadequate, and improvised construction, and inadequate fall protection. The point of this investigation is that accidents do not just happen; they are caused.

OSHA regulations on the fabrication of frame scaffolds are found in the Code of Federal Regulations 1926.450, 451, 452. These regulations should be thoroughly understood before any scaffold is erected and used. Furthermore, safety codes that are more restrictive than OSHA, such as those in Canada, California, Michigan, and Washington, should be consulted.

Scaffolds must be strong enough to support workers, tools, and materials. They must also provide an extra safety margin. The standard safety margin requirement is that all scaffolds must be capable of supporting at least four times the maximum intended load.

Those who erect scaffolding must be familiar with the different types and construction methods of scaffolding to provide a safe working platform for all workers. The type of scaffolding depends on its location, the kind of work being performed, the distance above the ground, and the load it is required to support. No job is so important as to justify risking one's safety and life. All workers deserve to be able to return to their families without injury.

The regulations on scaffolding enforced by OSHA make it clear that before erecting or using a scaffold, the worker must be trained about the hazards surrounding the use of such equipment. OSHA has not determined the length of training that should be required. Certainly that would depend on the expertise of the student in training.

Employers are responsible for ensuring that workers are trained to erect and use scaffolding. One level of training is required for workers, such as painters, to work from the scaffold. A higher level of training is required for workers involved in erecting, disassembling, moving, operating, repairing, maintaining, or inspecting scaffolds.

The employer is required to have a *competent person* to supervise and direct the scaffold erection.

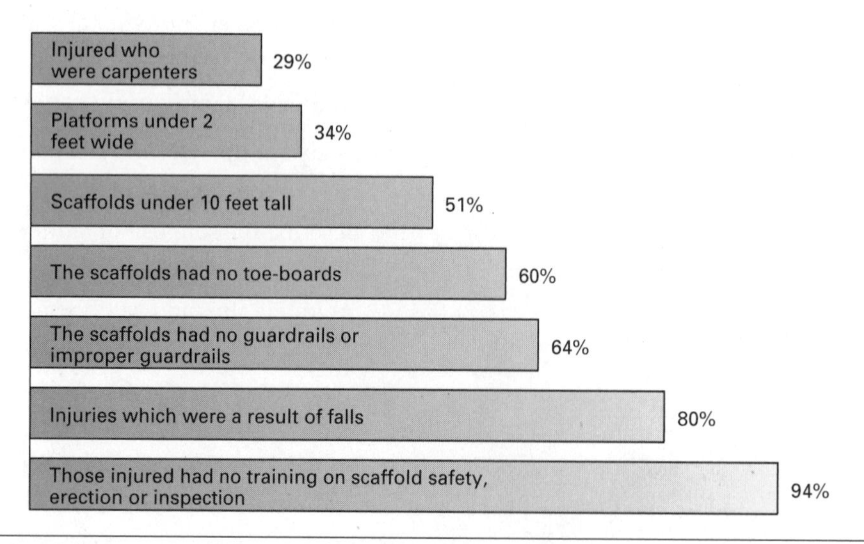

Fig. 39-1 Recent accident statistics involving scaffolding.

This individual must be able to identify existing and predictable hazards in the surroundings or working conditions that are unsanitary, hazardous, or dangerous to employees. This person also has authorization to take prompt corrective measures or eliminate such hazards. A competent person has the authority to take corrective measures and stop work if need be to ensure that scaffolding is safe to use.

Wood Scaffolds

Wood scaffolds are *single-pole* or *double-pole.* They are used when working on walls. The single-pole is used when it can be attached to the wall and does not interfere with the work (Fig. 39-2). The double-pole is used when the scaffolding must be kept clear of the wall for the application of materials or for other reasons (Fig. 39-3). Wood scaffolds are designated as light-, medium-, or heavy-duty, according to the loads they are required to support.

Fig. 39-2 A light-duty single-pole scaffold. Guardrails are required when the scaffolding is over 10 feet in height.

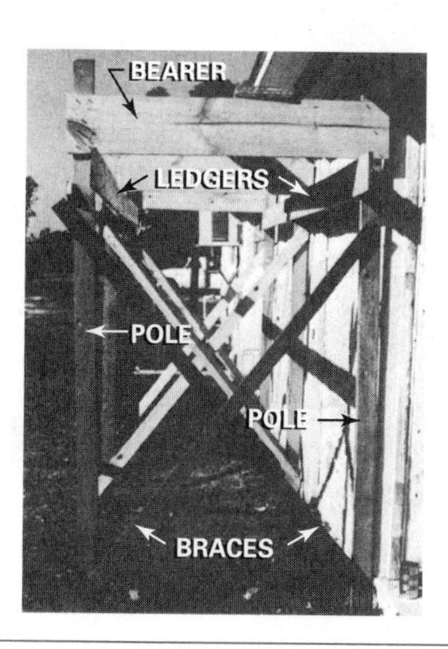

Fig. 39-3 View from the end of a double-pole wood scaffold showing cross braces from inner to outer poles.

Scaffolding Terms

Poles. The vertical members of a scaffold are called *poles.* All poles should be set plumb. They should bear on a footing of sufficient size and strength to spread the load. This prevents the poles from settling (Fig. 39-4). If wood poles need to be spliced for additional height, the ends are squared so the upper pole rests squarely on the lower pole. The joint is scabbed on at least two adjacent sides. Scabs should be at least 4 feet long. The scabs are fastened to the poles so they overlap the butted ends equally (Fig. 39-5).

Bearers. Bearers or *putlogs* are horizontal load-carrying members. They run from building to pole in a single-pole staging. Bearers run from pole to pole at right angles to the wall of the building in a double-pole staging. They are set with their greatest dimension vertical. They must be long enough to project a few inches outside the staging pole.

When placed against the side of a building, bearers must be fastened to a notched *wall ledger.* At each end of the wall, bearers are fastened to the corners of the building (Fig. 39-6).

Ledgers. *Ledgers* run horizontally from pole to pole. They are parallel with the building and support the bearers. Ledgers must be long enough to

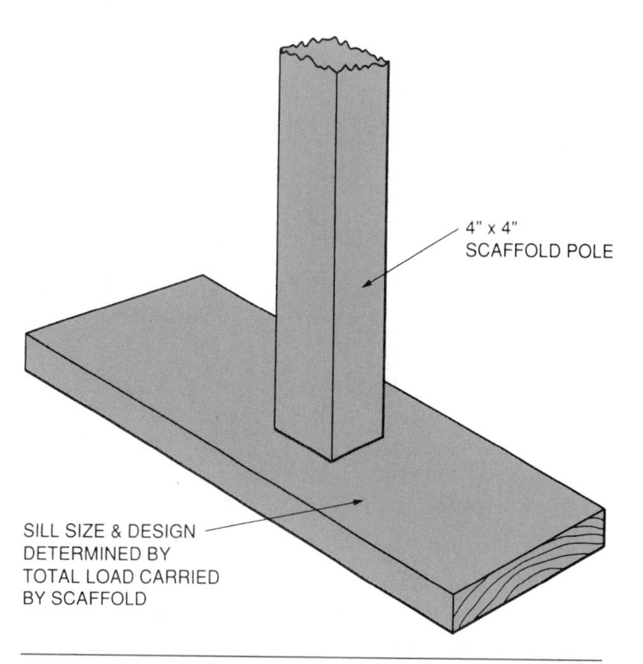

Fig. 39-4 The bottom end of scaffold poles are set on footings or pads to prevent them from sinking into the ground.

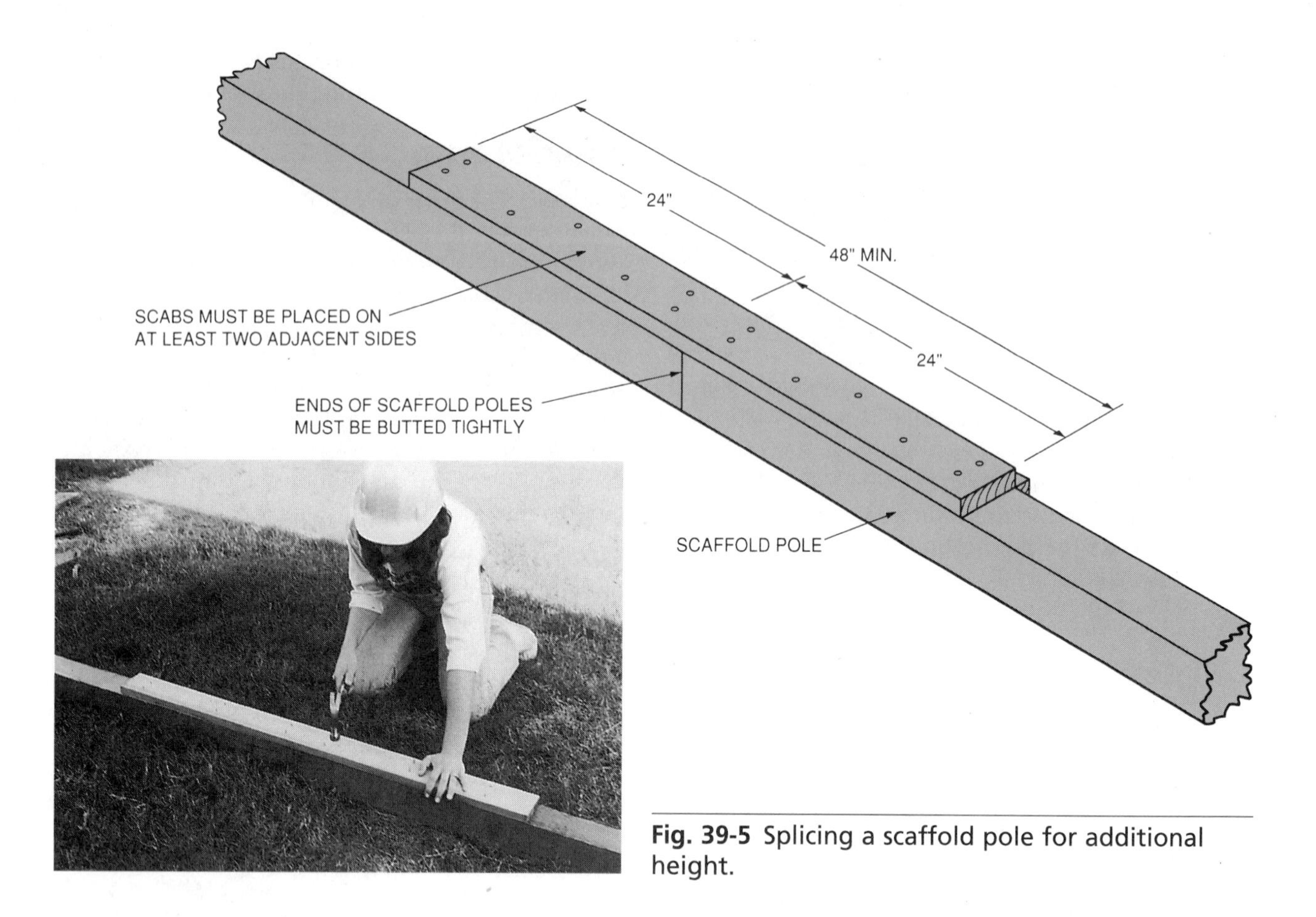

Fig. 39-5 Splicing a scaffold pole for additional height.

STEP BY STEP

PROCEDURES

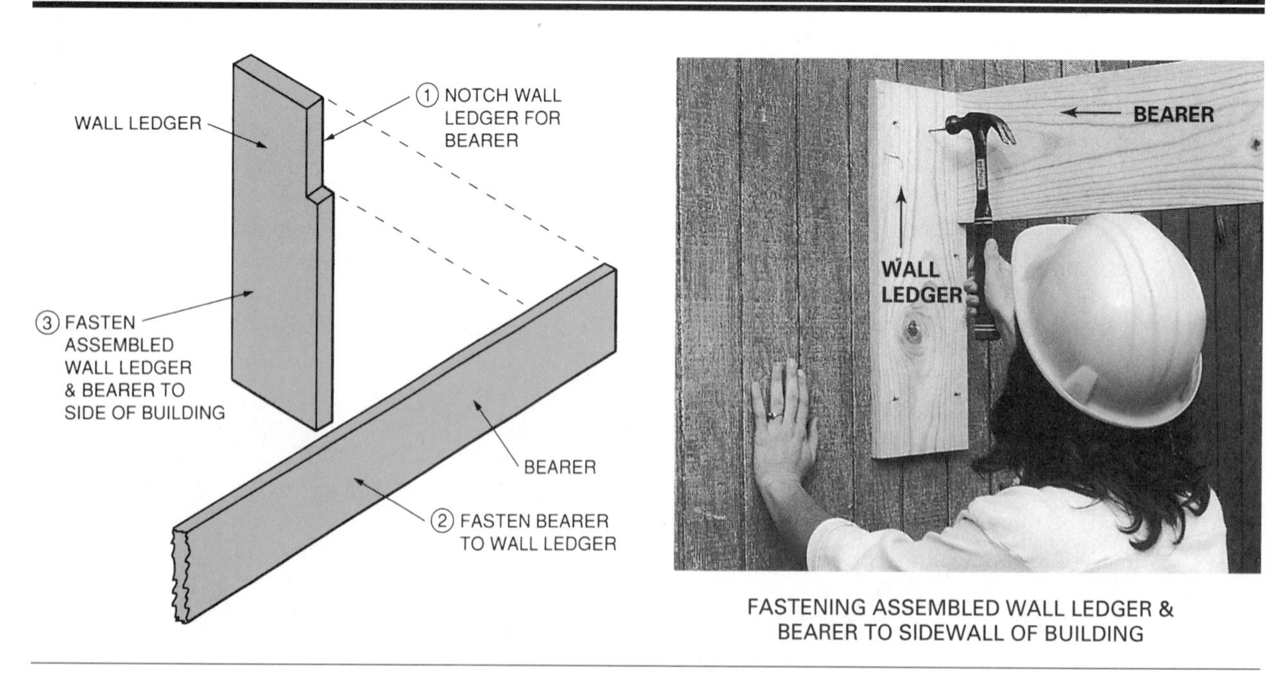

Fig. 39-6 Bearers must be fastened in a notch of a wall ledger for placement against the side of a building.

extend over two pole spaces. They must be overlapped at the pole and not spliced between them (Fig. 39-7).

Braces. *Braces* are diagonal members. They stiffen the scaffolding and prevent the poles from moving or buckling. Full diagonal face bracing is applied across the entire face of the scaffold in both directions. On medium- and heavy-duty double-pole scaffolds, the inner row of poles is braced in the same manner. Cross-bracing is also provided between the inner and outer sets of poles on all double-pole scaffolds. All braces are spliced on the poles (Fig. 39-8).

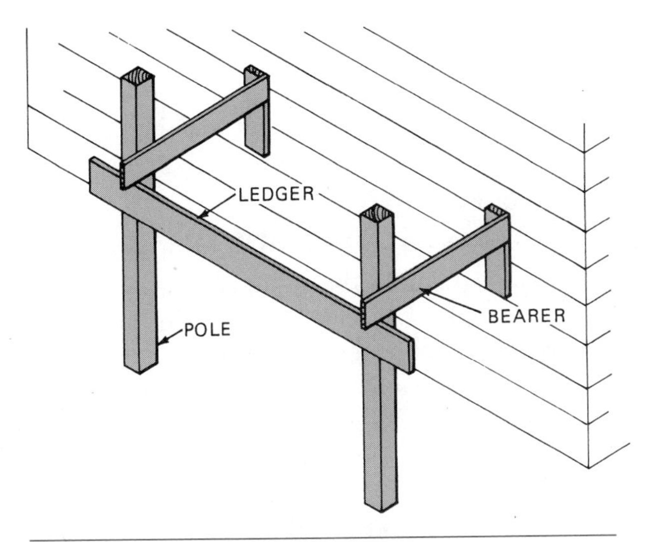

Fig. 39-7 Ledgers run horizontally from pole to pole and support the bearers.

Plank. *Staging planks* rest on the bearers. They are laid with the edges close together so the platform is tight. There should be no spaces through which tools or materials can fall. All planking should be scaffold grade or its equivalent. Planking should have the ends banded with steel to prevent excessive checking.

Overlapped planks should extend at least 6 inches beyond the bearer. Where the end of planks butt each other to form a flush floor, the butt joint is placed at the centerline of the pole. Each end rests on separate bearers. The planks are secured to prevent movement. End planks should not overhang the bearer by more than 12 inches (Fig. 39-9).

Guardrails. *Guardrails* are installed on all open sides and ends of scaffolds that are more than 10 feet in height. The top rail is usually of 2 × 4 lumber. It is fastened to the poles 38 to 45 inches (0.97–1.1 m) above the working platform. A middle rail of 1 × 6 lumber and a toeboard with a minimum height of 4 inches are also installed (Fig. 39-10).

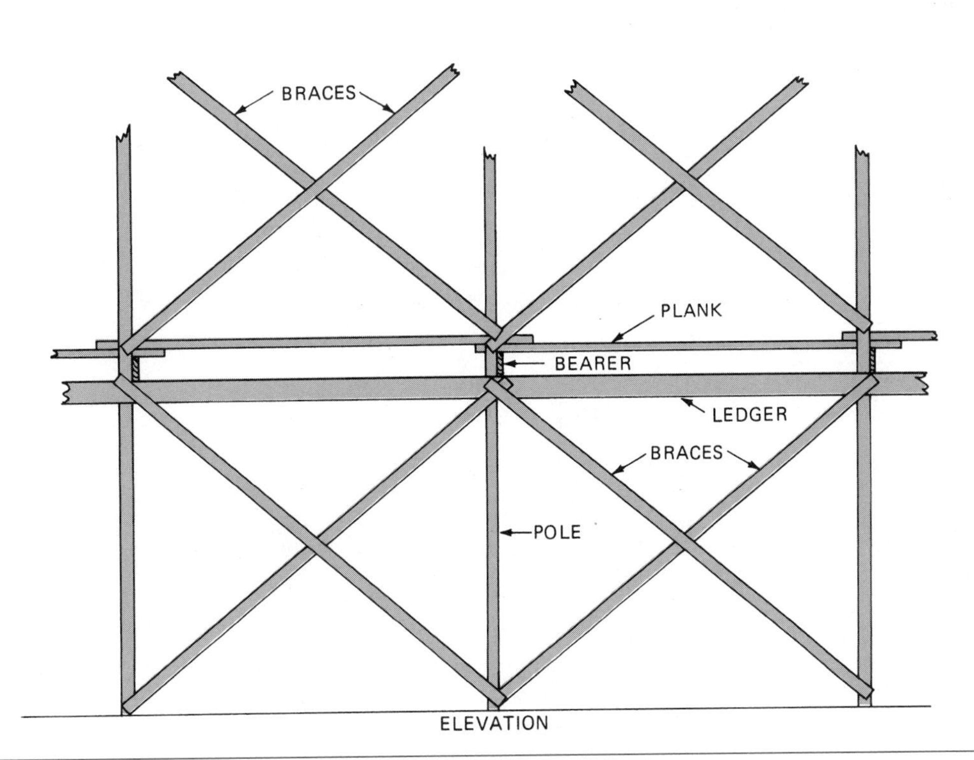

Fig. 39-8 Diagonal bracing is applied across the entire face of the scaffold.

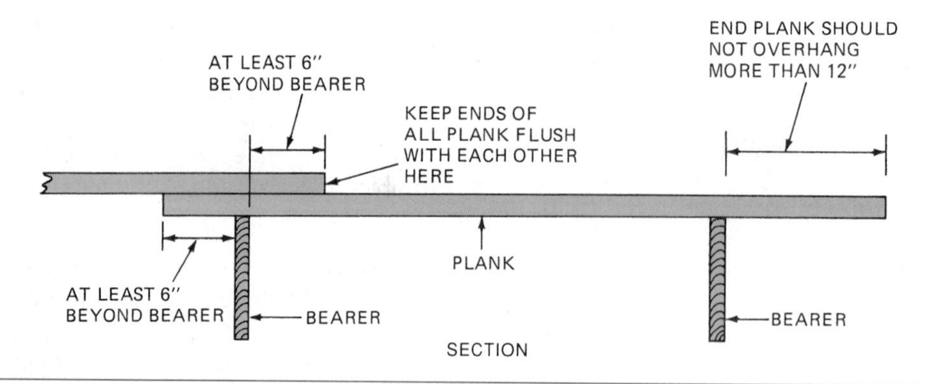

Fig. 39-9 Recommended placement of scaffold plank.

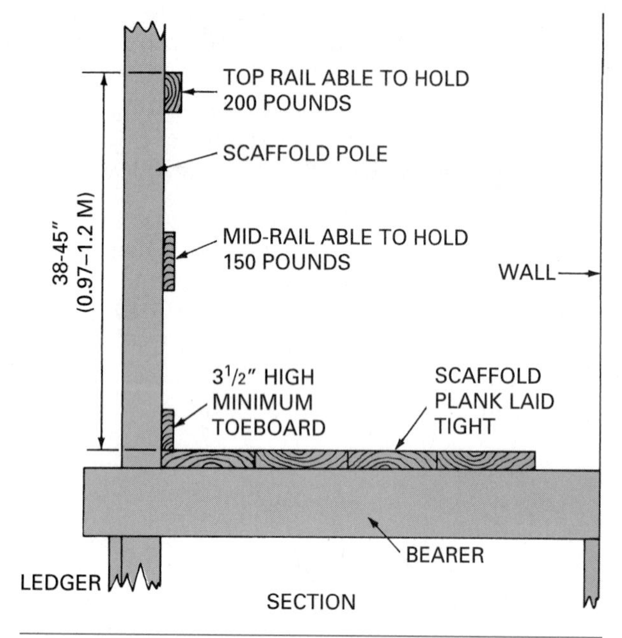

Fig. 39-10 Typical wood guardrail specifications.

Size and Spacing of Scaffold Members

Carpenters must know the minimum sizes of lumber to use and the maximum spacing of the members in order to build a safe scaffold. See Figure 39-11 for size and spacing recommendations of scaffold members.

Metal Scaffolding

Metal Tubular Frame Scaffold

Metal tubular frame scaffolding consists of manufactured *end frames* with folding *cross braces, adjustable screw legs, baseplates, platforms* and *guardrail* hardware (Fig. 39-12). They are erected in sections that consist of two end frames and two cross braces and typically come in 5 × 7 modules. Frame scaffolds are easy to assemble, which can lead to carelessness. Because untrained erectors may think scaf-

WOOD SCAFFOLDING				
MINIMUM SIZES AND MAXIMUM SPACING OF MEMBERS				
	LIGHT DUTY		MEDIUM DUTY	HEAVY DUTY
SCAFFOLD HEIGHT	20 FT	60 FT	60 FT	60 FT
Poles	2"×4"	4"×4"	4"×4"	4"×6"
Pole spacing	6'-0"	10'-0"	8'-0"	6'-0"
Scaffold width	5'-0"	5'-0"	5'-0"	5'-0"
Bearers to 3'-0"	2"×4"	2"×4"	2"×10" or 3"×4"	2"×10" or 3"×5"
Bearers to 5'-0"	2"×6" or 3"×4"	2"×6" or 3"×4"	2"×10" or 3"×4"	2"×10" or 3"×5"
Ledgers	1"×4"	1 1/4"×9"	2"×10"	2"×10"
Plank	1 1/4"×9"	2"×10"	2"×10"	2"×10"
Braces	1"×4"	1"×4"	1"×4"	2"×4"
Vertical distance between stages	7'-0"	9'-0"	7'-0"	6'-6"

Fig. 39-11 Minimum sizes and maximum spacing of wood scaffold members.

Fig. 39-12 A typical metal tubular frame scaffold.

folds are just stacked up, serious injury and death can result from a lack of training.

End frames consist of posts, horizontal bearers, and intermediate members. End frames come in a number of styles depending on the manufacturer. Frames can be wide or narrow, and some are designed for rolling tower scaffolds, while other frames have an access ladder built into the end frame (Fig. 39-13).

Cross braces rigidly connect one scaffold member to another member. Cross braces connect the bottoms and tops of frames. This diagonal bracing keeps the end frames plumb and provides the rigidity that allows them to attain their designed strength. The braces are connected to the end frames using a variety of locking devices (Fig. 39-14).

OSHA regulations require the use of baseplates on all supported or ground-based scaffolds (Fig. 39-15) in order to transfer the load of scaffolding, material, and workers to the supporting surface. It is extremely important to distribute this load over an area large enough to reduce the pounds per square foot load on the ground. If the scaffold sinks

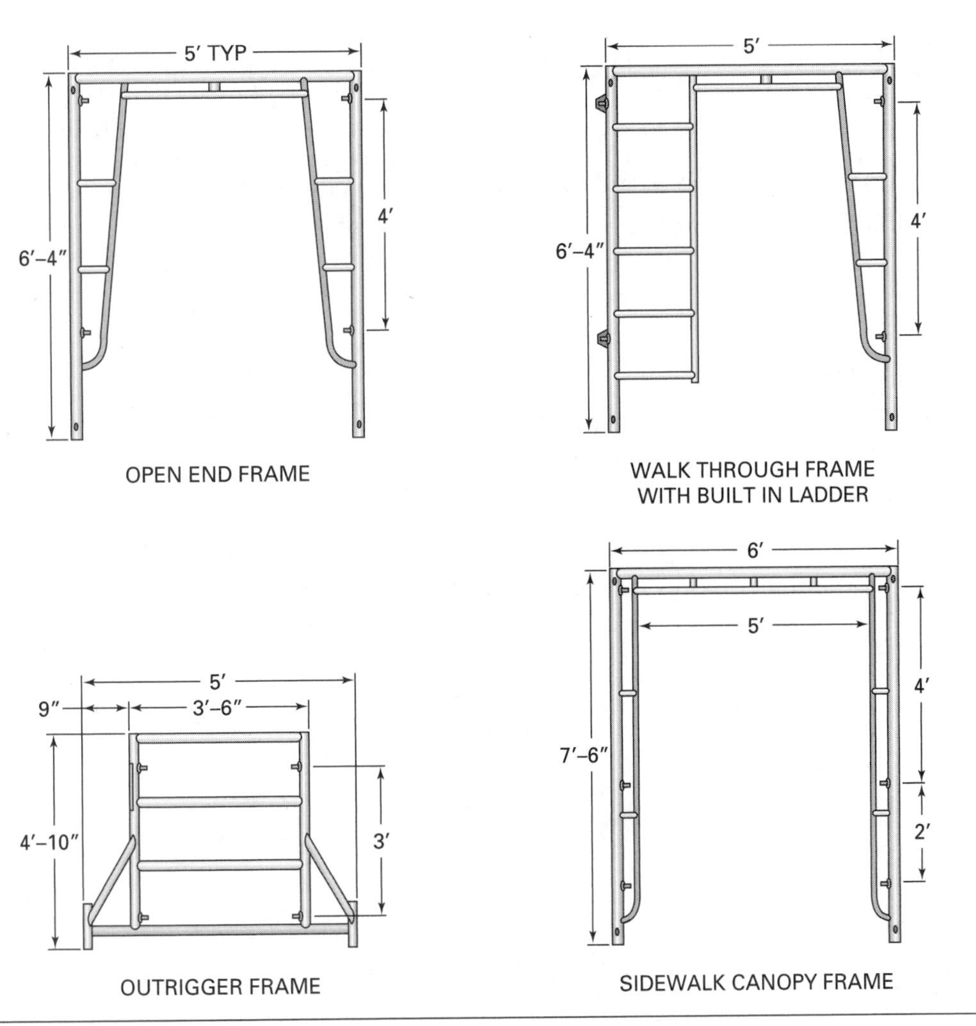

Fig. 39-13 Four examples of typical metal tubular end frames.

into the ground when it is being used, accidents could occur. Therefore, baseplates should sit on and be nailed to a *mud sill* (Fig. 39-16). A mud still is typically a 2 × 10 (5 × 25 cm) board approximately 18 to 24 inches (45–60 cm) long. On soft soil it may need to be longer and/or thicker.

To level an end frame while erecting a frame scaffold, screwjacks may be used. At least one-third of the screwjack must be inserted in the scaffold leg. Lumber may be used to crib up the legs of the scaffold (Fig. 39-17). Cribbing height is restricted to equal the length of the mud sill. Therefore, using a 19-inch (48 cm) long, 2 × 10 (5 × 25 cm) mud sill, the crib height is limited to 19 inches (48 cm). OSHA also prohibits the use of concrete blocks to level scaffolding.

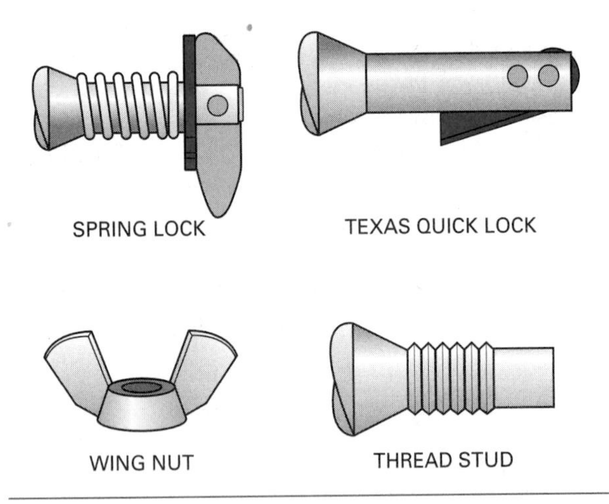

Fig. 39-14 Locking devices used to connect cross braces to end frames.

A guardrail system is a vertical fall-protection barrier consisting of, but not limited to, toprails, midrails, toeboards, and posts (Fig. 39-18). It prevents employees from falling off a scaffold platform or walkway. A guardrail system is required when the working height is 10 feet or more. Guardrail systems must have a top rail capacity of 200 pounds applied downward or horizontally. The top rail must be between 38 and 45 inches (0.97–1.2 m) above the work deck, with the midrail installed midway between the upper guardrail and the platform surface. The midrail must have a capacity of 150 pounds applied downward or horizontally.

If workers are on different levels of the scaffold, toeboards must be installed as an overhead protection for lower-level workers. Toeboards are typically 1 × 4-inch boards installed under the midrail at the platform. If materials or tools are stacked up higher than the toeboards, screening must be installed. Moreover, all workers on the scaffold must wear hard hats.

Coupling pins are used to stack the end frames on top of each other (Fig. 39-19). They have holes in them that match the holes in the end-frame legs; these holes allow locking devices to be installed. Workers must ensure the coupling pins are designed for the scaffold frames in use.

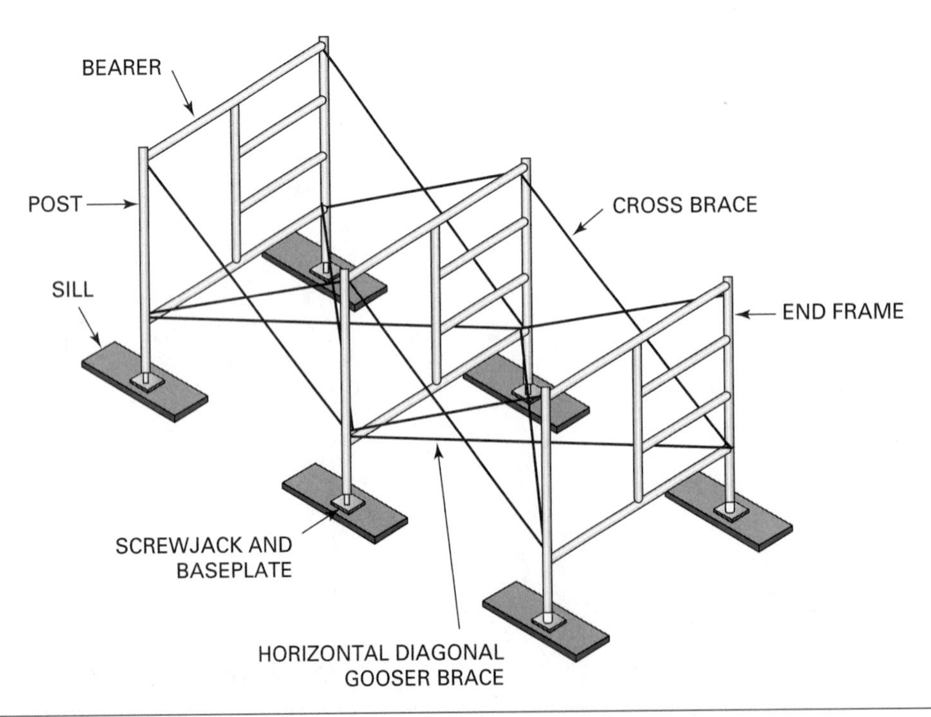

Fig. 39-15 Typical baseplate setup for a metal tubular frame scaffold.

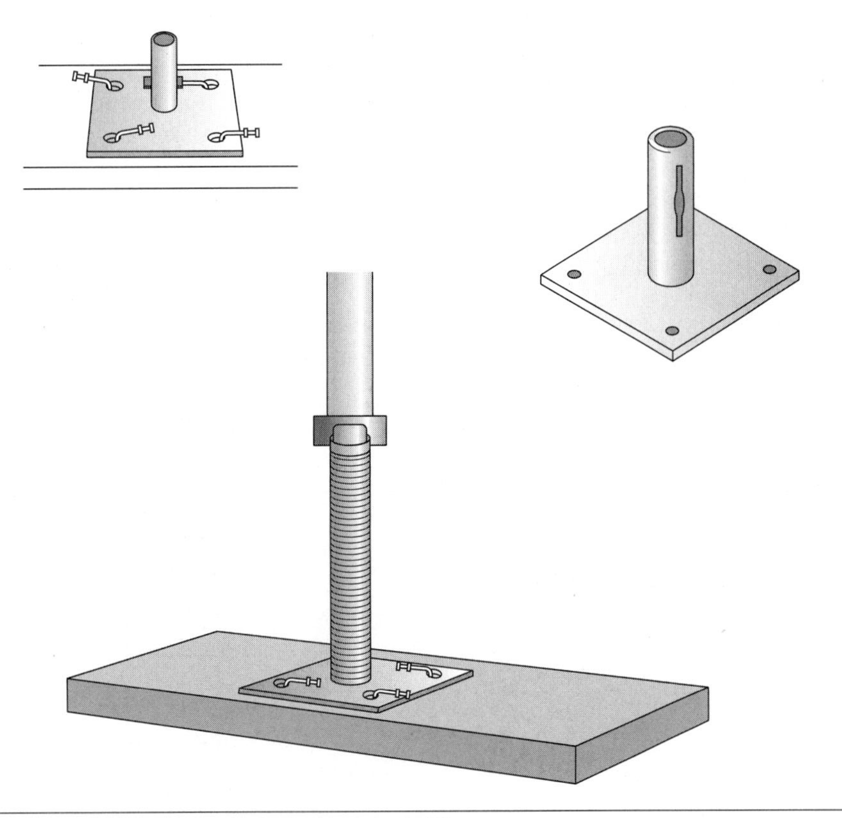

Fig. 39-16 Baseplates should be nailed to the mud sill.

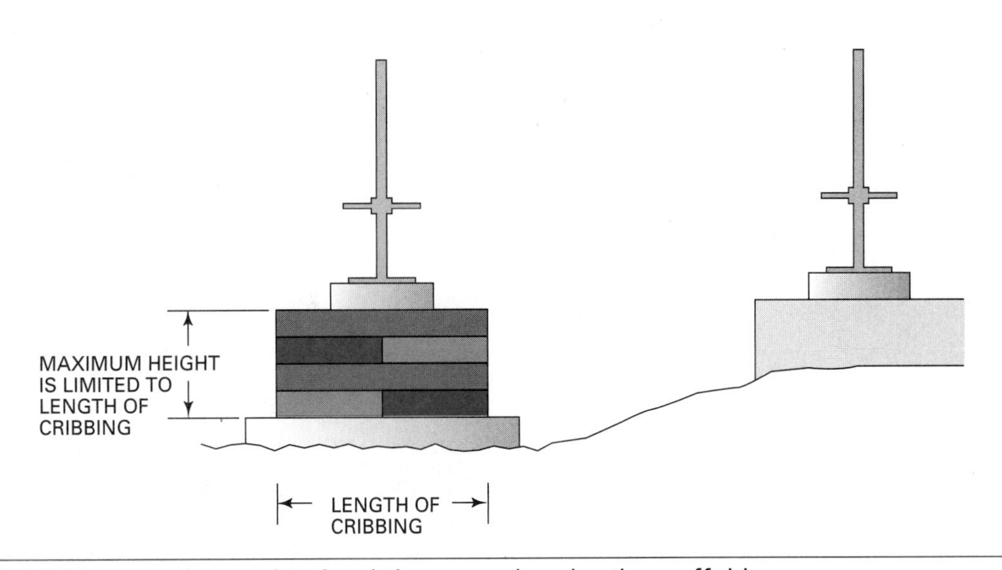

Fig. 39-17 Cribbing may be used to level the ground under the scaffold.

The scaffold end frames and platforms must have uplift protection installed when a potential for uplift exists. Installing locking devices through the legs of the scaffold and the coupling pins provides this protection (Fig. 39-20). If the platforms are not equipped with uplift protection devices, they can be tied down to the frames with number nine steel-tie wire.

OSHA requires safe access onto the scaffold for both erectors and users of the scaffolds. Workers can climb end frames only if they meet OSHA regulations. Frames may only be used as a ladder if they are designed as such. Frames meeting such design guidelines must have level horizontal members that are parallel and are not more than 16 3/4 inches (43 cm) apart vertically (Fig. 39-21). Scaffold erectors

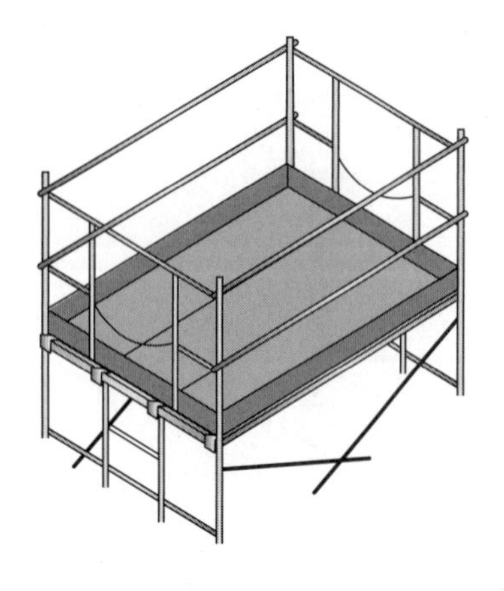

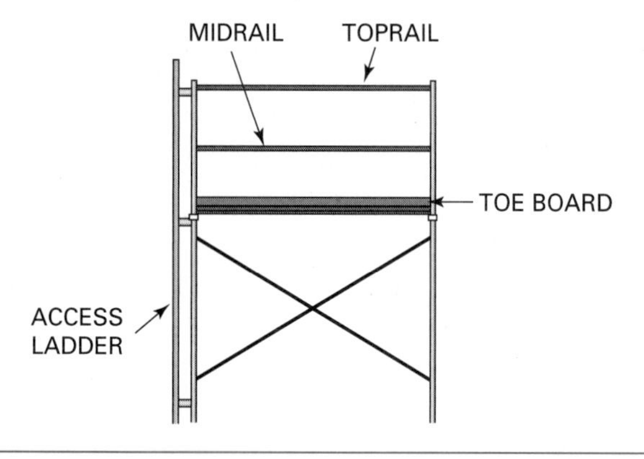

Fig. 39-18 Typical guardrail system for a metal tubular frame scaffold.

may climb end-frame rungs that are spaced up to 22 inches (46 cm). Platform planks should not extend over the end frames where end-frame rungs are used as a ladder access point. The cross braces should never be used as a means of access or egress. Attached ladders and stair units may be used (Fig. 39-22). A rest platform is required for every 35 feet (10.7 m) of ladder.

Manually propelled mobile scaffolds use wheels or casters in the place of baseplates (Fig. 39-23). Casters have a designed load capacity that should never be exceeded. Mobile scaffolds made from tubular end frames must use diagonal horizontal

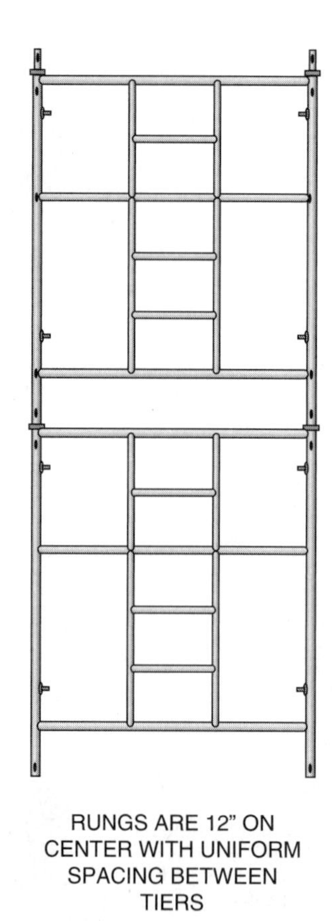

Fig. 39-21 The rings of an end frame designed for a scaffold user access ladder must be spaced no more than 16 3/4 inches (43 cm) apart.

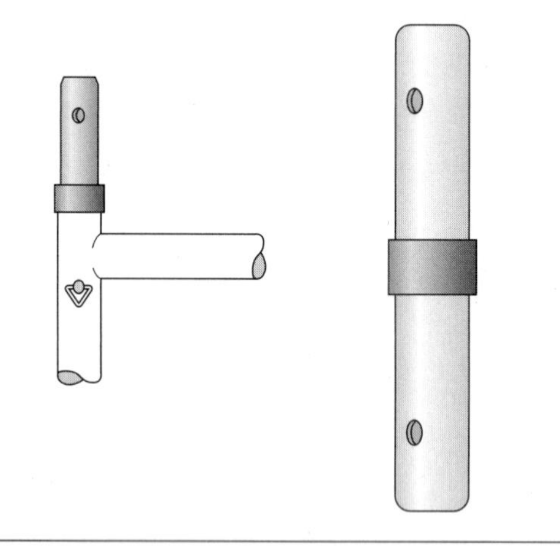

Fig. 39-19 Coupler pins to join end frames.

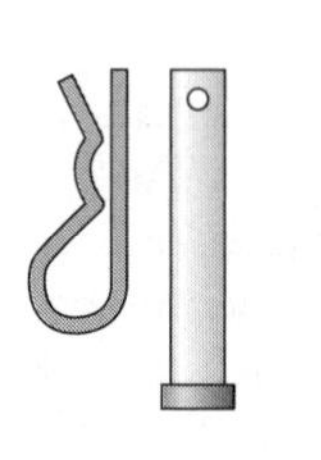

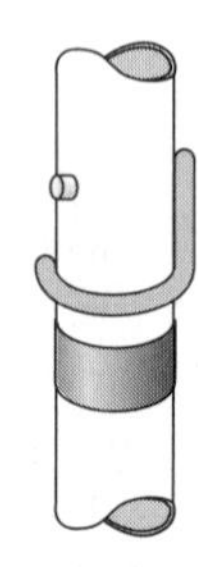

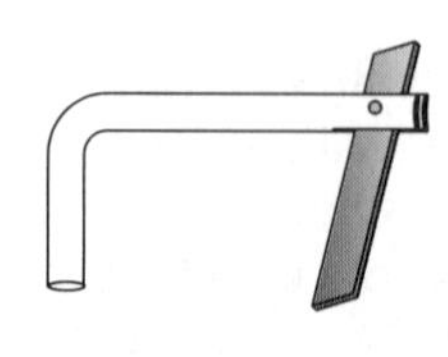

Fig. 39-20 Coupler locking devices to prevent scaffold uplift.

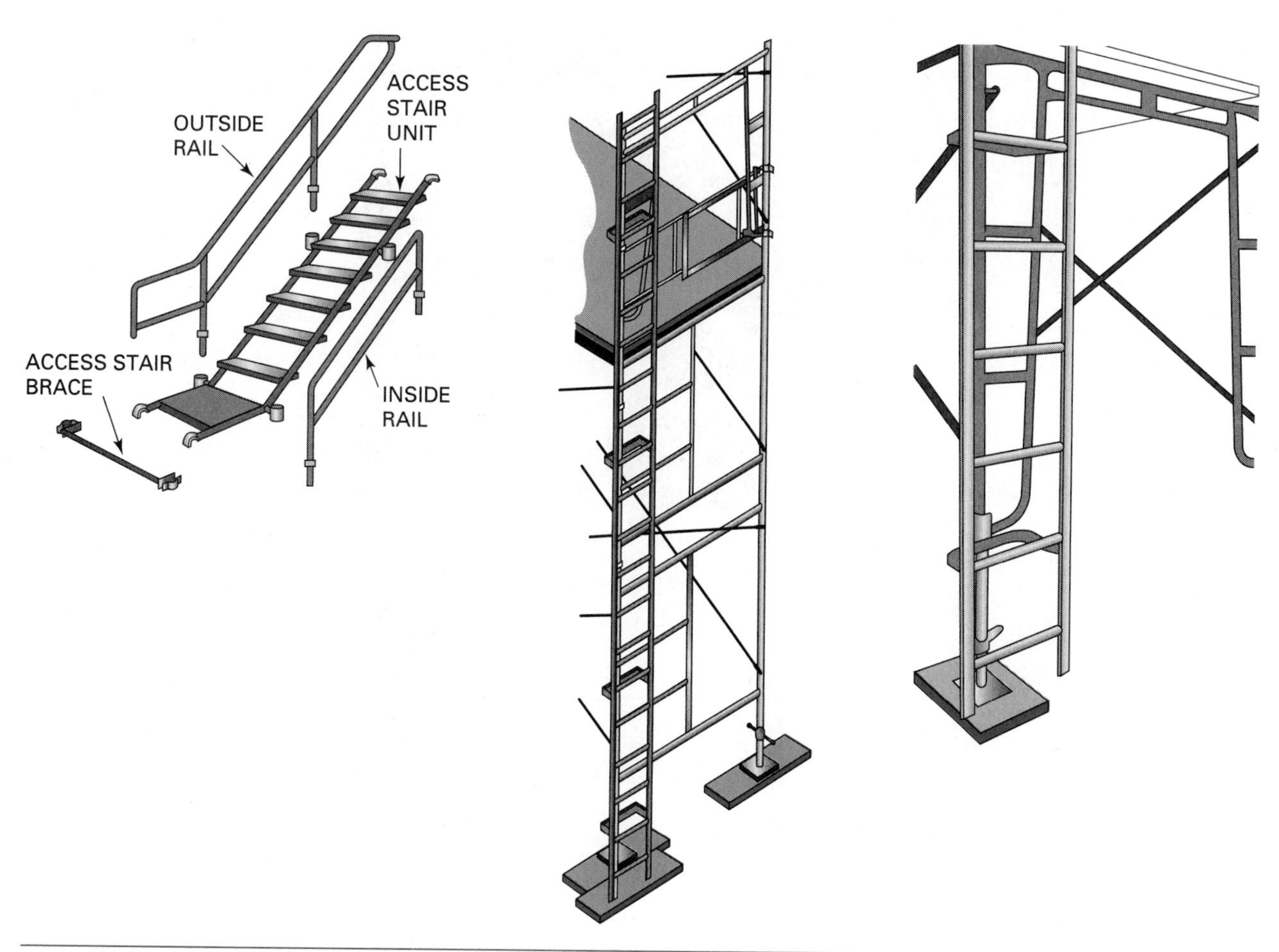

Fig. 39-22 Typical access ladder and stairway.

Fig. 39-23 Caster that replaces a baseplate to transform a metal tubular frame scaffold into a mobile scaffold.

braces, or gooser braces (see Fig. 39-15) to keep the mobile tower frame square.

Side brackets are light-duty (35 pounds per square foot maximum) extension pieces used to increase the working platform (Fig. 39-24). They are designed to hold personnel only and are not to be used for material storage. When side brackets are used, the scaffold must have tie-ins, braces, or outriggers to prevent the scaffold from tipping.

Hoist arms and wheel wells are sometimes attached to the top of the scaffold to hoist scaffold parts to the erector or material to the user of the scaffold (Fig. 39-25). The load rating of these hoist arms and wheel wells are typically no more than 100 pounds. The scaffold must be secured from overturning at the level of the hoist arm, and workers should never stand directly under the hoist arm when hoisting a load. They should stand a slight distance away, but not too far to the side, as this will increase the lateral or side loading force on the scaffold.

Scaffold Inspection

Almost half of all scaffold accidents, according to the U.S. Bureau of Labor Statistics, involve defective scaffolds or defective scaffold parts. This statistic means ongoing visual inspection of scaffold parts must play a major role in safe scaffold erection and use. OSHA requires that a competent person inspect all scaffolds at the beginning of every work shift.

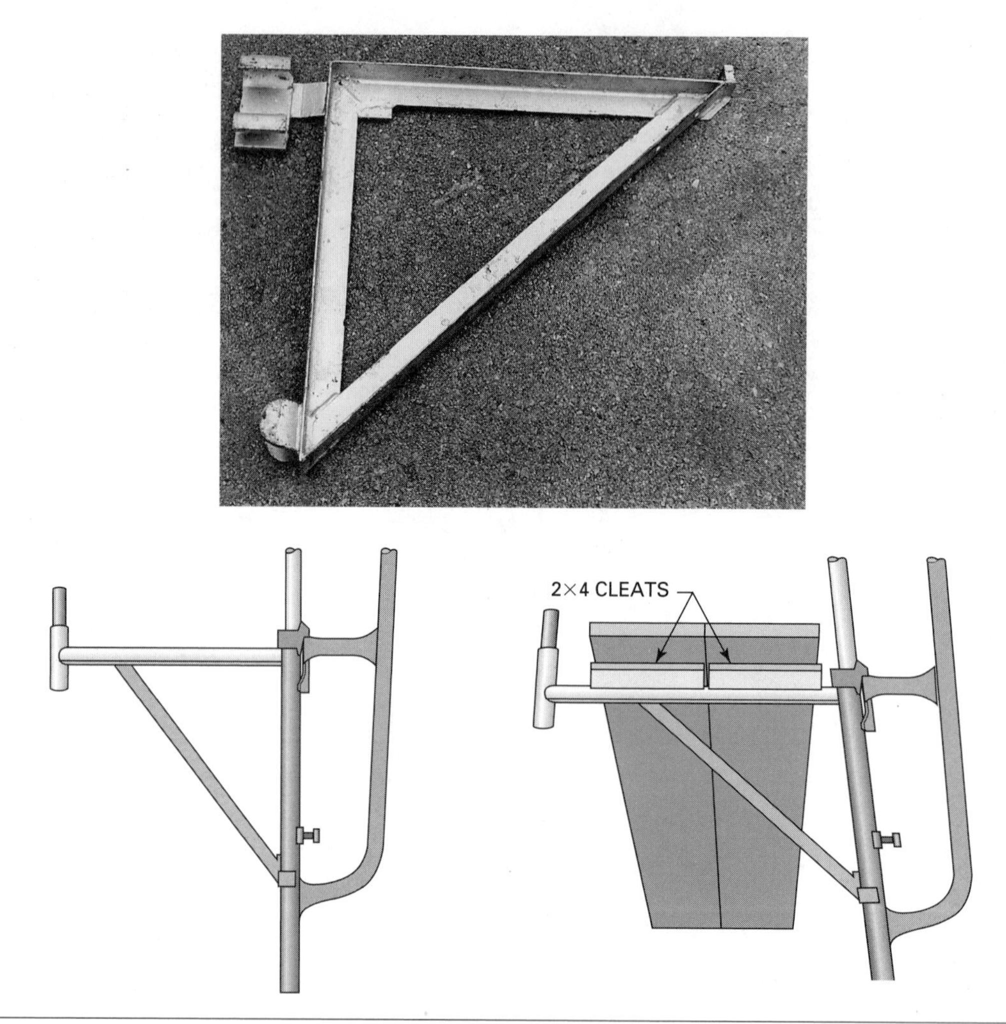

Fig. 39-24 Side brackets used to extend a scaffold work platform. These brackets should only be used for workers and never for material storage.

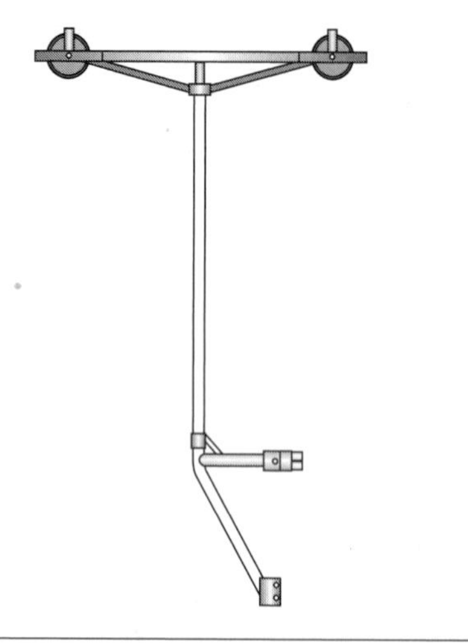

Fig. 39-25 Hoist that attaches to the top of a scaffold used to raise material and equipment.

Visual inspection of scaffold parts should take place at lease five times: before erection, during erection, during scaffold use, during dismantling, and before scaffold parts are put back in storage. All damaged parts should be red-tagged and removed from service and then repaired or destroyed as required. Things to look for during the inspection process include the following:

- Broken and excessively rusted welds
- Split, bent, or crushed tubes
- Cracks in the tube circumference
- Distorted members
- Excessive rust
- Damaged brace locks
- Lack of straightness
- Excessively worn rivets or bolts on braces
- Split ends on cross braces
- Bent or broken clamp parts
- Damaged threads on screwjacks
- Damaged caster brakes
- Damaged swivels on casters
- Corrosion of parts
- Metal fatigue caused by temperature extremes
- Leg ends filled with dirt or concrete

Scaffold Erection Procedure

The first thing to be done during the erection procedure is to inspect all scaffold components delivered to the job site. Defective parts must not be used.

The foundation of the scaffold must be stable and sound, able to support the scaffold and four times the maximum intended load without settling or displacement. Always start erecting the scaffold at the highest elevation, which will allow the scaffold to be leveled without any excavating by installing cribbing, screwjacks, or shorter frames under the regular frames. The scaffold must always be level and plumb. Lay out the baseplates and screwjacks on mud sills so the guardrails and end frames with cross braces may be properly installed (Fig. 39-26).

Stand one of the end frames up and attach the cross braces to each side, making sure the correct length cross braces have been selected for the job. Connect the other end of the braces to the second end frame. All scaffold legs must be braced to at least one other leg (Fig. 39-27). Make sure that all brace connections are secure. If any of these mechanisms are not in good working order, replace the frame with one that has properly functioning locks.

Use a level to plumb and level each frame (Fig. 39-28). Remember that OSHA requires that all tubular welded frame scaffolds be plumb and level. Adjust screwjacks or cribbing to level the scaffold. As each frame is added, keep the scaffold bays square with each other. Repeat this procedure until the scaffold run is erected. Remember, if the first level of scaffolding is plumb and level, the remaining levels will be more easily assembled.

The next step is to place the planks on top of the end frames. All planking must meet OSHA

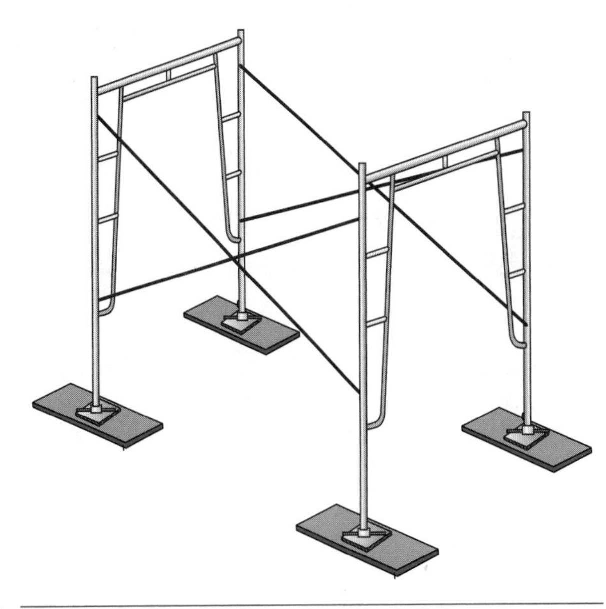

Fig. 39-27 Cross braces connect end frames, keeping them rigid and plumb.

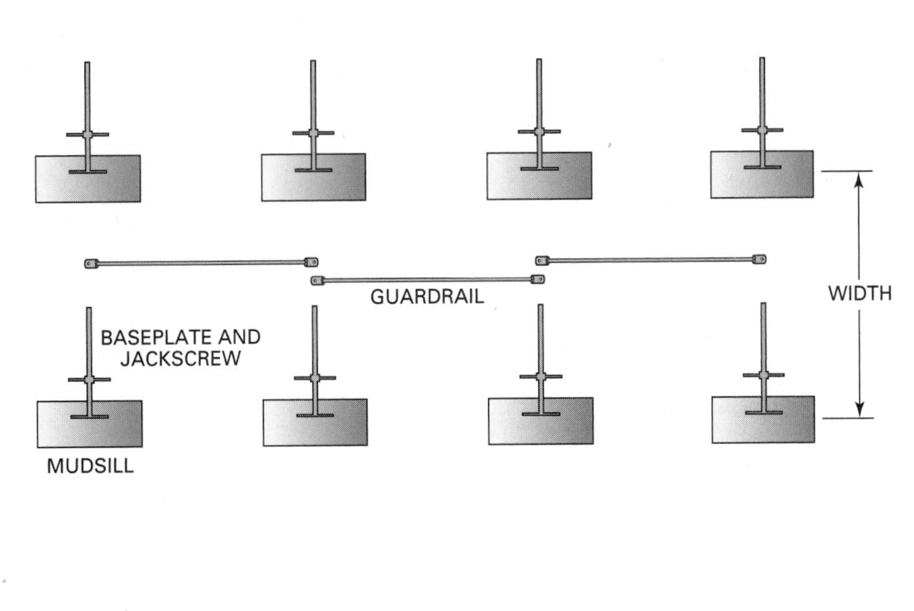

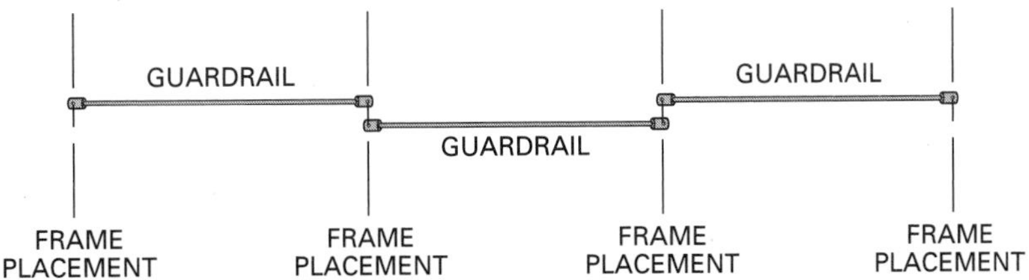

Fig. 39-26 During scaffold erection, the baseplates are spaced according to guardrail braces.

requirements and be in good condition. If planks that do not have hooks are used, they must extend over their end supports by at least 6 inches (15 cm) and not more than 12 inches (30 cm). A cleat should be nailed to both ends of wood planks to prevent plank movement (Fig. 39-29). Platform laps must be at least 12 inches, and all platforms must be secured from movement. Hooks on planks also have uplift protection installed on the ends.

It is a good practice to plank each layer fully as the scaffold is erected. If the deck is only to be used for erecting, then a minimum of two planks can be used. However, full decking is preferred, as it is a safer method for the erector.

Before the second lift is erected, the erector must provide an access ladder. Access may be on the end frame, if it is so designed, or an attached ladder. If the ladder is bolted to a horizontal member, the bolt must face downward (Fig. 39-30). Next,

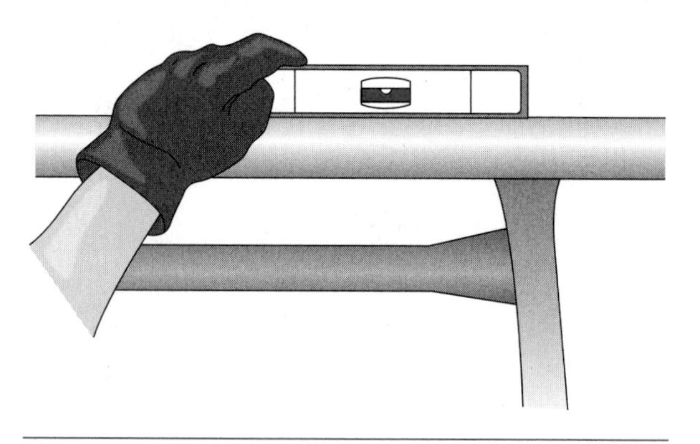

Fig. 39-28 End frames must be level.

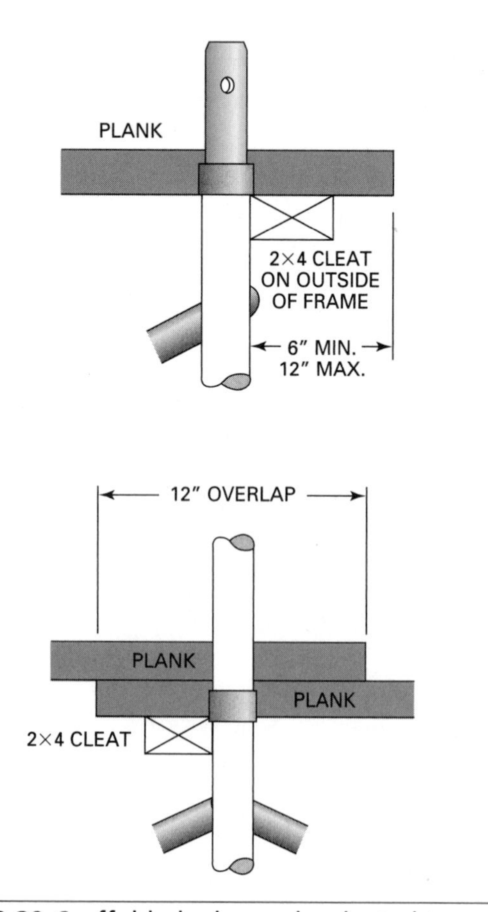

Fig. 39-29 Scaffold planks can be cleated to prevent them from moving.

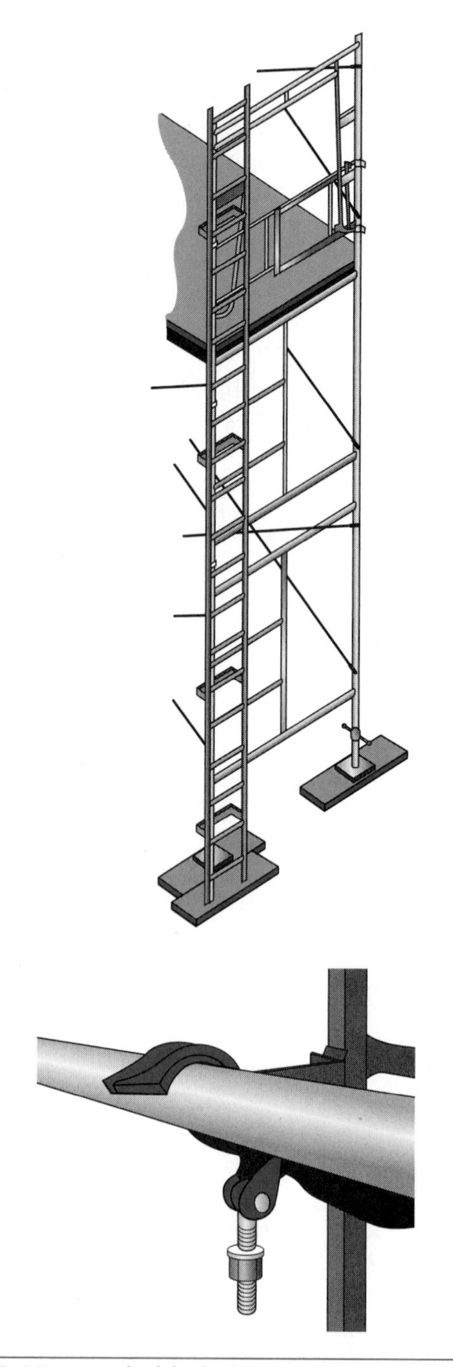

Fig. 39-30 Attachable ladders are connected so that the bolt is on the bottom.

the second level of frames may be hung temporarily over the ends of the first frames and then installed onto the coupling pins of the first-level frames (Fig. 39-31). Special care must be taken to ensure proper footing and balance when lifting and placing frames. OSHA requires erector fall protection—a full body harness attached to a proper anchor point on the structure—when it is feasible and not a greater hazard to do so.

After the end frames have been set in place and braced, they should have uplift protection pins installed through the legs and coupling pins. Wind, side brackets, and wheel wells can cause uplift so it is a good practice to pin all scaffold legs together.

The remaining scaffolding is erected in the same manner as the first. Remember, all work platforms must be fully decked and have a guardrail system or personal fall-arrest system installed before it can be turned over to the scaffold users. If the scaffold is higher than four times its minimum base dimension, it must be restrained from tipping by guying, tying, bracing, or equivalent means (Fig. 39-32). The scaffold is not allowed to tip into or away from the structure.

After the scaffold is complete, it is inspected again to make sure that it is plumb, level, and square before turning it over for workers to use. The inspection should also include checking that all legs are on base plates or screwjacks and mud sills (if required), ensuring the scaffolding is properly braced with all brace connections secured, and making sure all tie-ins are properly placed and secured,

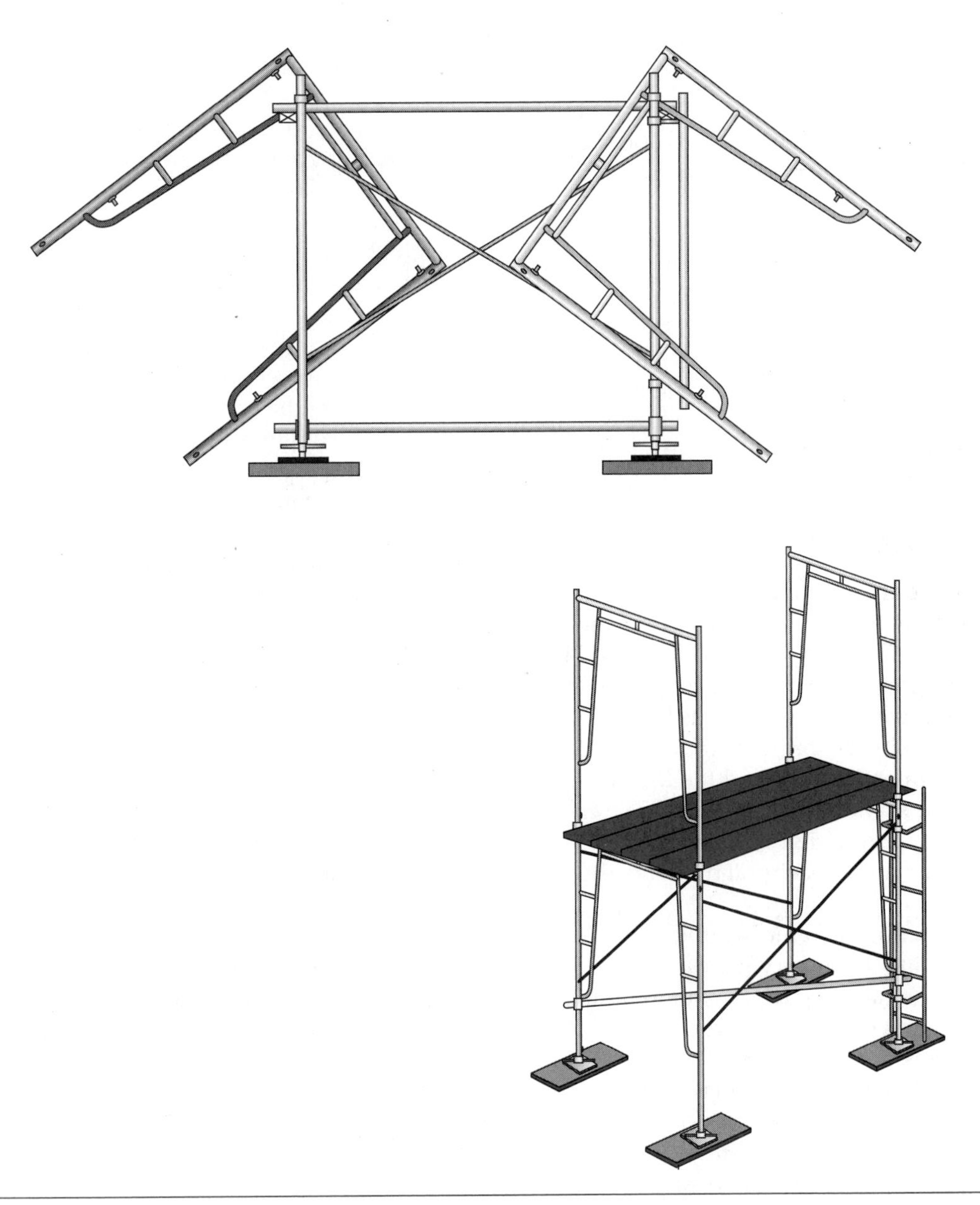

Fig. 39-31 Steps in setting the second lift of end frames.

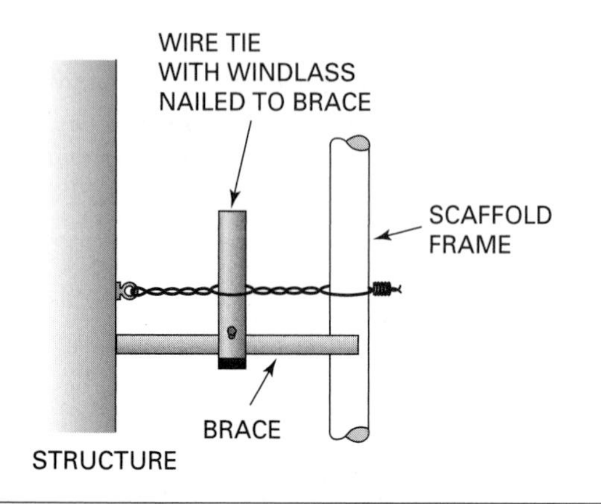

Fig. 39-32 A wire tie and a windlass are used to secure the scaffold tightly to the brace from the building.

both at the scaffold and at the structure. All platforms must be fully planked with proper decking and in good shape. Toeboards and/or screening should be installed as needed. Check that end and/or side brackets are fully secured and that any overturning forces are compensated for. All access units are inspected to ensure they are correctly installed and ladders and stairs are secured. Again, workers on the scaffold must wear hard hats.

After the scaffolding passes all inspections, it is ready to be turned over to the workers. Remember that this scaffolding must be inspected by a competent person at the beginning of each work shift and after any occurrence, such as a high wind or a rainstorm, which could affect its structural integrity.

Scaffold Capacity

All scaffolds and their components must be capable of supporting, without failure, their own weight and at least four times the maximum intended load applied or transmitted to them. Erectors and users of scaffolding must never exceed this safety factor.

Erectors and users of the scaffold must know the maximum intended load and the load-carrying capacities of the scaffold they are using. The erector must also know the design criteria, maximum intended load-carrying capacity, and intended use of the scaffold.

When erecting a frame scaffold, the erector should know the load-carrying capacities of its components. The rated leg capacity of a frame may never be exceeded on any leg of the scaffold. Also, the capacity of the top horizontal member of the end frame, called the bearer, may never be exceeded. Remember, it is possible to overload the bottom legs of the scaffold without overloading the bearer or top horizontal member of any frame. It is also possible to overload the bearer or top horizontal member of the frame scaffold and not overload the leg of that same scaffold. Erectors must pay careful attention to the load capacities of all scaffold components.

If the scaffold is covered with weatherproofing plastic or tarps, the lateral pressure applied to the scaffold will dramatically increase. Consequently, the number of tie-ins attached to prevent overturning must be increased. Additionally, any guy wires added for support will increase the downward pressure and weight of the scaffold.

OSHA regulations state that supported scaffolds with a ratio larger than four-to-one (4:1) of the height to narrow base width must be restrained from tipping by guying, tying, bracing, or equivalent means. Guys, ties, and braces must be installed at locations where horizontal members support both inner and outer legs. Guy, ties, and braces must be installed according to the scaffold manufacturer's recommendations or at the closest horizontal member to the 4:1 height. For scaffolds greater than 3-feet (0.91 m) wide, the vertical locations of horizontal members are repeated every 26 feet (7.9 m). The top guy, tie, or brace of completed scaffolds must be placed no further than the 4:1 height from the top. Such guys, ties, and braces must be installed at each end of the scaffold and at horizontal intervals not to exceed 30 feet (9.1 m). The tie or standoff should be able to take pushing and pulling forces so the scaffold does not fall into or away from the structure.

The supported scaffold poles, legs, post, frames, and uprights should bear on baseplates, mud sills, or other adequate, firm foundation. Because the mud sills have more surface area than baseplates, sills distribute loads over a larger area of the foundation. Sills are typically wood and come in many sizes. Erectors should choose a size according to the load and the foundation strength required. Mud sills made of 2 × 10-inch (5 × 25 cm) full thickness or nominal lumber should be 18 to 24 inches (46 to 60 cm) long and centered under each leg (Fig. 39-33).

The loads exerted onto the legs of a scaffold are not equal. Consider a scaffold with two loads on two adjacent platforms (Fig. 39-34). Half of load A is carried by end frame #1 and the other half is carried by #2. Half of load B is carried by end frames #2 and #3. End frame #2 carries two half loads or one full load, which is twice the load of end frames #1 and #3. At no time should the manufacturer's load rating for their scaffolding be exceeded.

Scaffold Platforms

The scaffolding's work area must be fully planked between the front uprights and the guardrail sup-

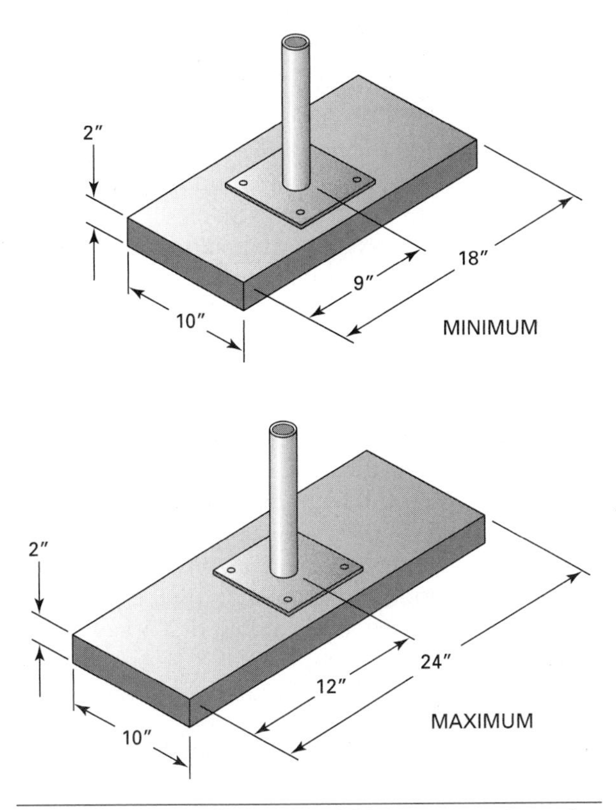

Fig. 39-33 Baseplates should be centered on the mud sills.

ports in order for the user to work from the scaffold. The planks should not have more than a 1-inch gap between them unless it is necessary to fit around uprights such as a scaffold leg. If the platform is planked as fully as possible, the remaining gap between the last plank and the uprights of the guardrail system must not exceed 9 1/2 inches.

Scaffold platforms must be at least 18 inches wide with a guardrail system in place. In areas where they cannot be 18 inches wide, they will be as wide as is feasible. The platform is allowed to be as much as 14 inches away from the face of the work.

Planking for the platforms, unless cleated or otherwise restrained by hooks or equivalent means, should extend over the centerline of their support at least 6 inches (15 cm) and no more than 12 inches (30 cm). If the platform is overlapped to create a long platform, the overlap shall occur only over supports and should not be less than 12 inches unless the platforms are nailed together or otherwise restrained to prevent movement.

When fully loaded with personnel, tools, and/or material, the wood plank used to make the platform must never deflect more than 1/60th of its span. In other words, a 2 × 10-inch (5 × 25 cm) plank that is 12 feet (3.7 m) long and is sitting on two end frames spaced 10 feet (3 m) apart should

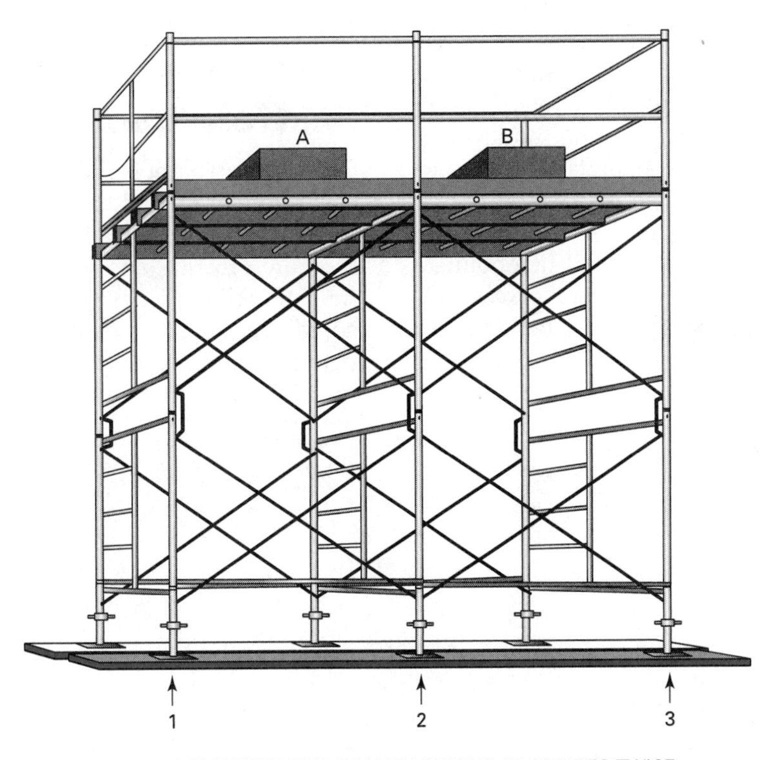

Fig. 39-34 The inner end frames, such as #2, often carry twice the load of the end frames located at the end of the scaffold.

not deflect more than 2 inches (5 cm) or 1/60th of the span.

Solid sawn wood planks should be scaffold-grade lumber as set out by the grading rules for the species of lumber being used. A recognized lumber-grading association, such as the Western Wood Products Association (WWPA) or the National Lumber Grades Authority (NLGA), establishes these grading rules. A grade should be stamped on the scaffold grade plank, indicating that it meets OSHA and industry requirements for scaffold planks. Two of the most common wood species used for scaffold planks are southern yellow pine and Douglas fir. OSHA does not require wood scaffold planks to bear grade stamps. The erector may use "equivalent" planks, which are determined equivalent by visually inspecting or test loading the wood plank in accordance with grading rules.

Scaffold platforms are usually rated for the intended load. Light-duty scaffolds are designed at 25 pounds per square foot, medium-duty scaffolds are rated at 50 pounds per square foot, and heavy duty at 75 pounds per square foot. The maximum span of a plank is tabulated in Figure 39-35. Using this chart, the maximum load that could be put on a nominal thickness plank (1 1/2 inch or 3.8 cm) with a span of 7 feet (2.1 m) is 25 pounds per square foot. Note that a load of 50 pounds per square foot would require a span of no more than 6 feet (1.8 m).

Fabricated planks and platforms are often used in lieu of solid sawn wood planks. These planks and platforms include fabricated wood planks that use a pin to secure the lumber sideways, oriented strand board planks, fiberglass composite planks, aluminum-wood decked planks, and high-strength galvanized steel planks. The loading of fabricated planks or platforms should be obtained from the manufacturer and never exceeded. Scaffold platforms must be inspected for damage before each use.

Scaffold Access

A means of access must be provided to any scaffold platform that is 2 feet above or below a point of access. Such means include a hook-on or attachable ladder, a ramp, or a stair tower and are determined by the competent person on the job.

If a ladder is used, it should extend 3 feet above the platform and be secured both at the top and bottom. Hook-on and attachable ladders should be specifically designed for use with the type of scaffold used, have a minimum rung length of 11 1/2 inches, and have uniformly spaced rungs with a maximum spacing between rung length of 16 3/4 inches. Sometimes a stair tower can be used for access to the work platform, usually on larger jobs (Fig. 39-36). A ramp can also be used as access to the scaffold or the work platform. When using a ramp, it is important to remember that a guardrail system or fall protection is required at 6 feet above a lower level.

The worker using the scaffold can sometimes access the work platform using the end frames of the scaffold itself. According to regulations, the end frame must be specifically designed and constructed for use as ladder rungs. The rungs can run up the center or to one side of the end frame; some have the rungs all the way across the end frame. Scaffold users should never climb any end frame unless the manufacturer of that frame designated it to be used for access.

Scaffold Use

Scaffolds must not be loaded in excess of their maximum intended load or rated capacities, whichever is less. Workers must know the capacity of scaffolds they are erecting and/or using. Before the beginning of each work shift, or after any occurrence that could affect a scaffold's structural integrity the competent person must inspect all scaffolds on the job.

	Maximum permissible plank span			
Maximum intended load	Full thickness, undressed lumber		Nominal thickness lumber	
Lbs/sq ft	Feet	Meters	Feet	Meters
25	10	3	8	2.4
50	8	2.4	6	1.8
75	6	1.8	-----	-----

Fig. 39-35 Maximum spacing of planks based on the load rating of the scaffold.

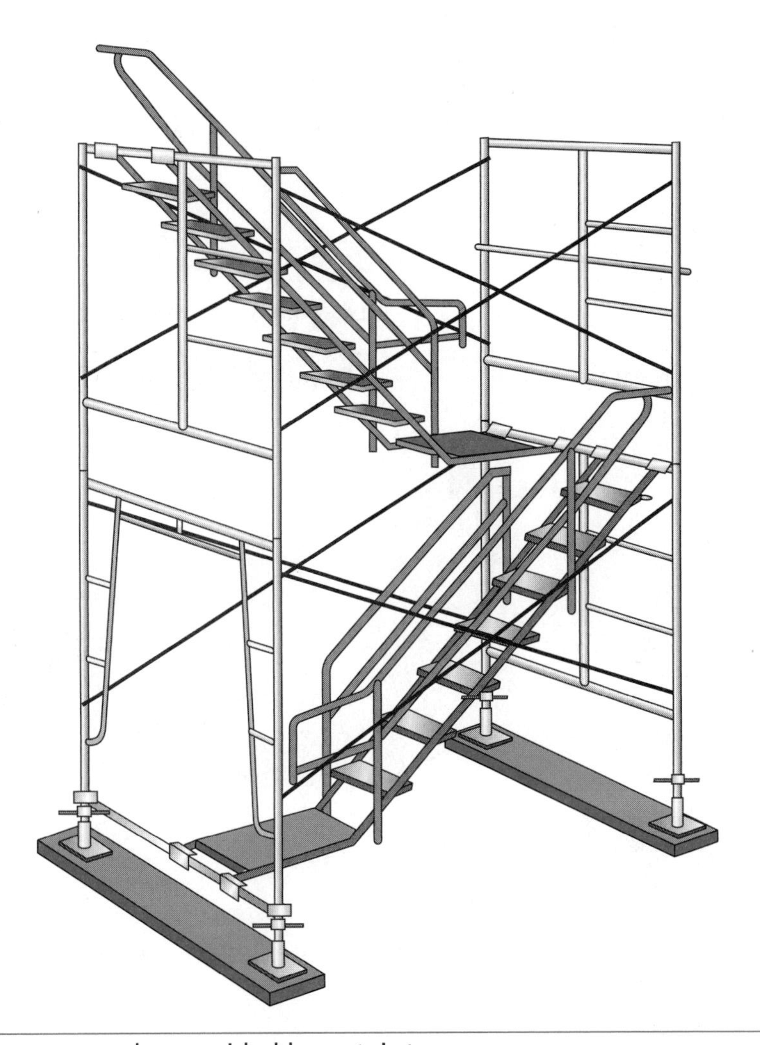

Fig. 39-36 Scaffold access may be provided by a stair tower.

Employees must not work on scaffolds covered with snow or ice except to remove the snow or ice. Generally, work on or from scaffolds is prohibited during storms or high winds. Debris must not be allowed to accumulate on the platforms. Makeshift scaffold devices, such as boxes or barrels, must not be used on the scaffold to increase workers' working height. Step ladders should not be used on the scaffold platform unless they are secured according to OSHA regulations.

Fall Protection

Current OSHA standards on scaffolding require fall protection when workers are working at heights above 10 feet. This regulation applies to both the user of the scaffold and the erector or dismantler of the scaffold. These regulations allow the employer the option of a guardrail system or a personal fall protection system. The fall protection system most often used is a complete guardrail system. A guardrail system has a toprail 38–45 inches (0.97 m–1.2 m) above the work deck, with a midrail installed midway between the toprail and the platform. The work deck should also be equipped with a toeboard. These requirements are for all open sides of the scaffold, except for those sides of the scaffold that are within 14 inches of the face of the building.

A typical personal fall protection system consists of five related parts: the harness, lanyard, lifeline, rope grab, and anchor (Fig. 39-37). The failure of any one part means failure of the system. Therefore, constant monitoring of a lifeline system is a critical responsibility. It is easy for a system to lose its integrity almost immediately, even on first use.

OSHA recognizes that sometimes fall protection may not be possible for erectors. As the scaffold increases in length, the personal fall-arrest system may not be feasible because of its fixed anchorage and the need for employees to traverse the entire length of the scaffold. Additionally, fall protection may not be feasible due to the potential for lifelines to become entangled or to create a tripping hazard

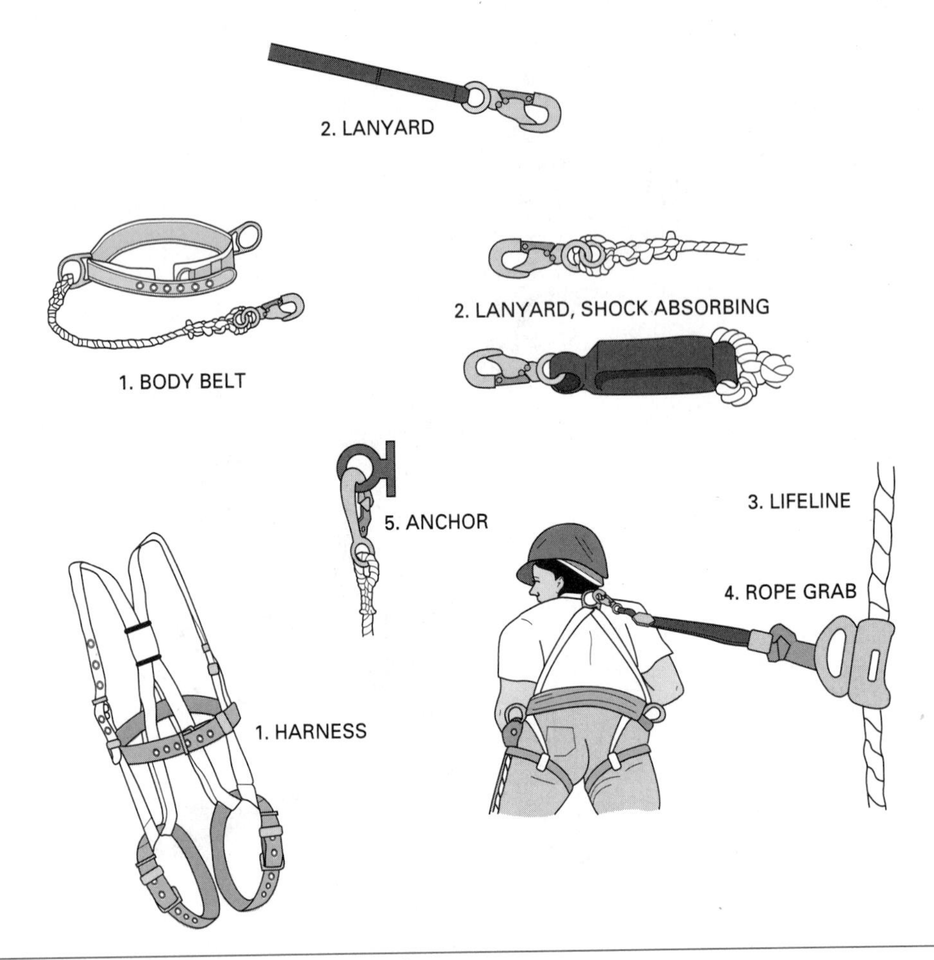

Fig. 39-37 Components of a personal protection system.

for erectors or dismantlers as they traverse the scaffold. Do not use the scaffold components as an anchor point of the fall-protection harness. OSHA puts the responsibility of when to use fall protection, both for the user of the scaffold and the erector, on the competent person.

Falling Object Protection

According to industry standards and OSHA requirements, workers must wear hard hats during the process of erecting a scaffold. In addition to hard hats, protection from potential falling objects may be required. When material on the scaffold could fall on workers below, some type of barricade must be installed to prevent that material from falling. OSHA lists toeboards as part of the falling object protection for the workers below the scaffold. The toeboard can serve two functions: it keeps material on the scaffold and keeps the workers on the scaffold platform if they happen to slip.

Dismantling Scaffolds

Many guidelines and rules for erection also apply to scaffold dismantling. However, dismantling requires additional precautions to ensure the scaffold will come down in a controlled, safe, and logical manner. Important factors to consider include the following:

1. Check every scaffold before dismantling. Any loose or missing ties or bracing must be corrected.
2. If a hoist is to be used to lower the material, the scaffold must be tied to the structure at the level of the hoist arm to dispel any overturning effect of the wheel and rope.
3. The erector should be tied off for fall protection, as required by the regulations, unless it is infeasible or a greater hazard to do so.
4. Start at the top and work in reverse order, following the step-by-step procedures for erection. Leave the work platforms in place as long as possible.
5. Do not throw planks or material from the scaffold. This practice will damage the material and presents overhead hazards for workers below.
6. Building tie-ins and bracing can only be removed when the dismantling process has

reached that level or location on the scaffold. An improperly removed tie can cause the entire scaffold to overturn.

7. Remove the ladders or the stairs only as the dismantling process reaches that level. Never climb or access the scaffold by using the cross braces.
8. As the scaffold parts come off the scaffold, they should be inspected for any wear or damage. If a defective part is found, it should be tagged for repair and not used again until inspected by the competent person.
9. Dismantled parts and materials should be organized, stacked, and placed in bins or racks out of the weather.
10. Secure the disassembled scaffold equipment to ensure that no unauthorized, untrained employees use it. All erectors must be trained, experienced, and under the supervision and direction of a competent person.
11. Always treat the scaffold components as if a life depended on them, as the next time the scaffold is erected, someone's life will be depending on it being sound.

Mobile Scaffolds

The rolling tower, or mobile scaffold, is widely used for small jobs, generally not more than 20 feet in height (Fig. 39-38). The components of the mobile scaffold are the same as those for the stationary frame scaffold, with the addition of casters and horizontal diagonal bracing. There are additional restrictions on rolling towers as well.

The height of a rolling tower must never exceed four times the minimum base dimension. For example, if the frame sections are 5 × 7, the rolling tower can only be 20 feet high. If the tower exceeds this height-to-base ratio, it must be secured to prevent overturning. When outriggers are used on a mobile tower, they must be used on both sides.

Casters on mobile towers must be locked with positive wheel swivel locks or the equivalent to prevent movement of the scaffold while it is stationary. Casters typically have a load capacity of 600 pounds each, and the legs of a frame scaffold can hold 2,000–3,000 pounds each. Care must be taken not to overload the casters.

Never put a cantilevered work platform, side bracket, or well wheels on the side or end of a mobile tower. Mobile towers can tip over if used incorrectly. Mobile towers must have horizontal, diagonal, or gooser braces at the base to prevent racking of the tower during movement (Fig. 39-39). Metal hook planks also help prevent racking if they are secured to the frames.

The force to move the scaffold should be applied as close to the base a practicable, but not more than 5 feet (1.5 m) above the supporting surface. The casters must be locked after each movement before beginning work again. Employees are not allowed to ride on rolling tower scaffolds during movement unless the height-to-base width ratio is two-to-one or less. Before the scaffold is moved, each employee on the scaffold must be made aware

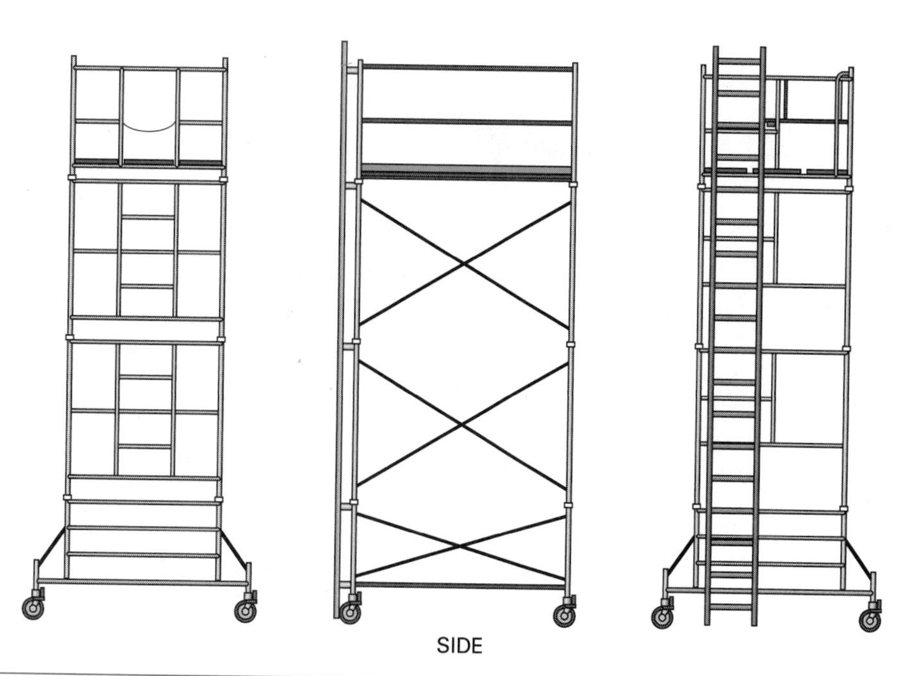

Fig. 39-38 Typical setup for a mobile scaffold.

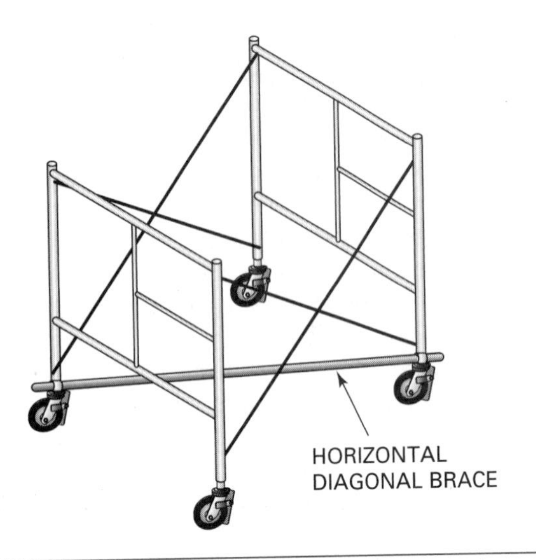

Fig. 39-39 The horizontal diagonal brace (or gooser) is used to keep the tower square when it is rolled.

of the move. Caster and wheel stems shall be pinned or otherwise secured in scaffold legs or adjustment screws. The surface that the mobile tower rolls on must be free of holes, pits, and obstructions and must be within 3 degrees of level. Only use a mobile scaffold on firm floors.

Pump Jack Scaffolds

Pump jack scaffolds consist of 4 × 4 poles, a pump jack mechanism, and metal braces for each pole (Fig. 39-40). The braces are attached to the pole at intervals and near the top. The arms of the bracket extend from both sides of the pole at 45-degree angles. The arms are attached to the sidewall or roof to hold the pole steady.

The scaffold is raised by pressing on the foot pedal of the pump jack (Fig. 39-41). The mechanism has brackets on which to place the scaffold plank. Other brackets hold a guardrail or platform. Reversing a lever allows the staging to be pumped downward.

Fig. 39-41 Pump jacks are raised by pressing the foot lever.

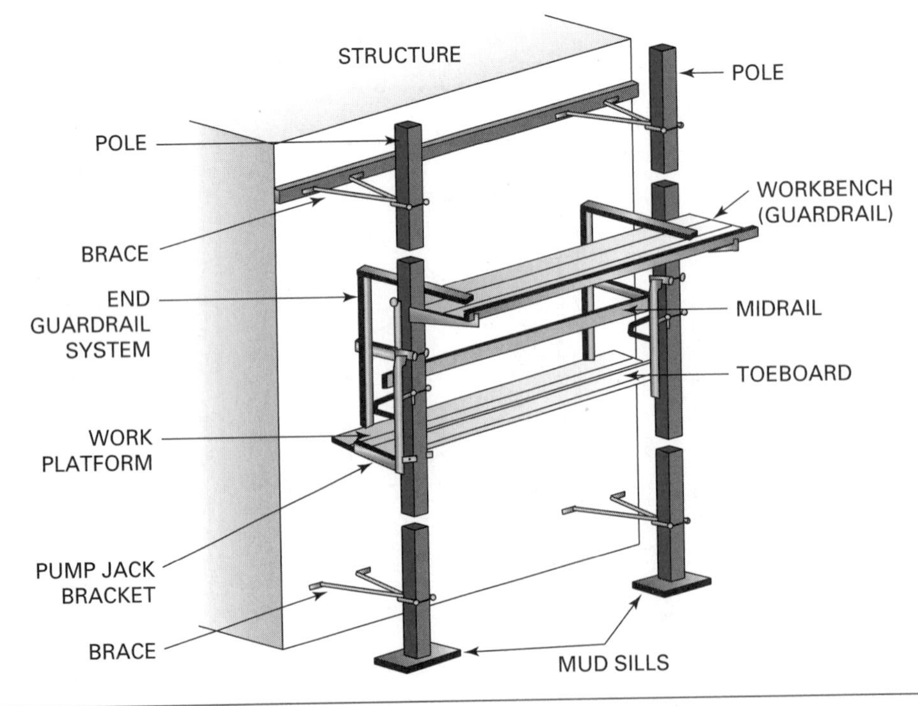

Fig. 39-40 Components of a pump jack system.

Pump jack scaffolds are used widely for siding, where staging must be kept away from the walls, and when a steady working height is desired. However, pump jack scaffolds have their limitations. They should not be used when the working load exceeds 500 pounds. No more than two persons are permitted at one time between any two supports. Wood poles must not exceed 30 feet in height. Braces must be installed at a maximum vertical spacing of not more than 10 feet.

In order to pump the scaffold past a brace location, temporary braces are used. The temporary bracing is installed about 4 feet above the original bracing. Once the scaffold is past the location of the original brace, it can be reinstalled. The temporary brace is then removed.

Wood pump jack poles are constructed of two 2 × 4s nailed together. The nails should be 3-inch or 10d, and no less than 12 inches apart, staggered uniformly from opposite outside edges.

CHAPTER 40 BRACKETS, HORSES, AND LADDERS

Roofing Brackets

Roofing brackets are used when the pitch of the roof is too steep for carpenters to work on without slipping (Fig. 40-1). Usually any roof with more than 4 on 12 slope requires roof brackets. Roofing brackets are made of wood or metal. Some are adjustable for roofs of different pitches.

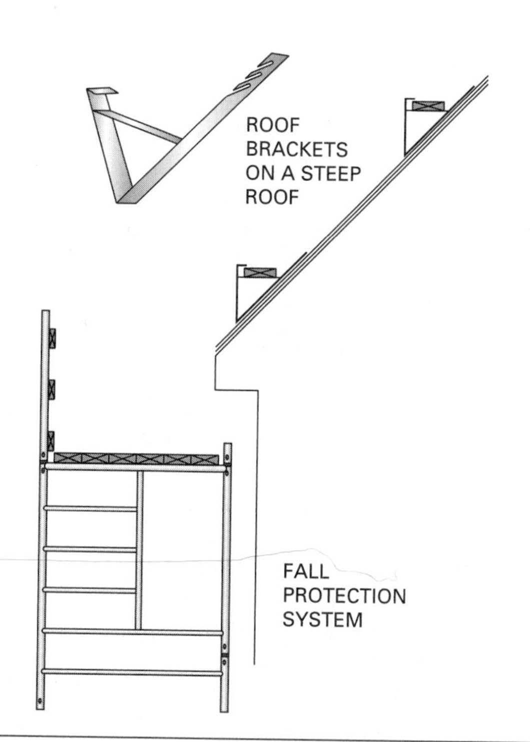

Fig. 40-1 Roof brackets are used when the roof pitch is too steep for carpenters to work on without slipping.

A metal plate at the top of the bracket has three slots in which to drive nails to fasten the bracket to the roof. The bottom of the slot is round and large enough to slip over the nailhead. This enables removal of the bracket from the fasteners without pulling the nails. The bracket is simply tapped upward from the bottom, and then lifted over the nailheads. The nails that remain are then driven home.

Applying Roof Brackets

Roof brackets are usually used when the roof is being shingled. On steep-pitched roofs, brackets are necessary when shingling. They keep the worker from slipping and also hold the roofing materials. Apply roof brackets in rows. Space them out so that they can be reached without climbing off the roof bracket staging below.

On asphalt-shingled roofs, place the brackets at about 6 to 8 foot (1.8–2.4 m) horizontal intervals. The top end of the bracket should be just below the next course of shingles. Nail the bracket over a joint or cutout in the tab of the shingle course below. No joint or cutout in the course above should fall in line with the nails holding the bracket. Otherwise, the roof will leak. Use three 3 1/4-inch or 12d common nails driven home. Try to get at least one nail into a rafter.

Open the brackets so the top member is approximately level or slightly leaning toward the roof. Place staging plank on the top of the brackets. Overlap them as in wall scaffolds. Keep the inner edges against the roof for greater support. A toeboard made of 1 × 6 or 1 × 8 lumber is usually placed flat on the roof with its bottom edge on top of the brackets. This protects the new roofing from

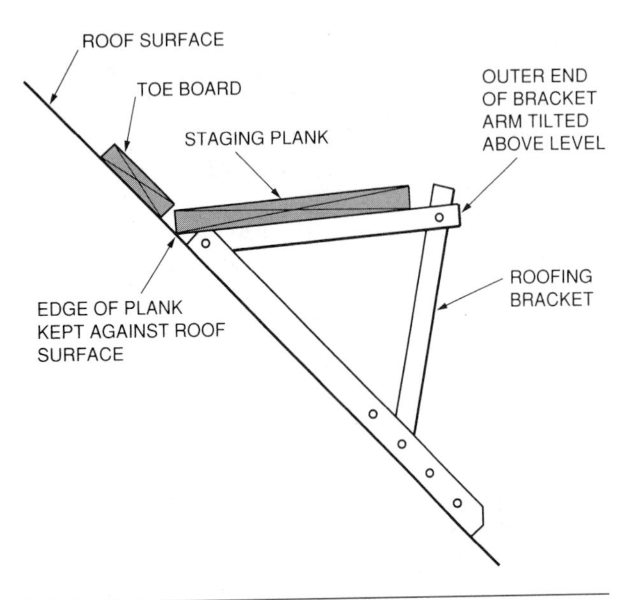

Fig. 40-2 The placement of a toeboard and plank used on roof brackets.

the workers' toes when the roofing has progressed that far (Fig. 40-2). After the shingles are applied, the bracket is tapped on the bottom upward along the slope of the roof to release it from the nails. Raise the shingle and drive the nails home so they do not stick up and damage the shingles.

Trestles

Trestles are used when a low scaffold on a level surface is desired. A trestle is a low working platform supported by a bearer with spreading legs at each end. This type of scaffold is used frequently in the building interior by lathers, plasterers, and drywall applicators working on ceilings.

Trestle Jacks

Manufactured metal trestle supports, called *trestle jacks*, are available for frequent and prolonged use. Trestle jacks are adjustable in height at about 3-inch intervals. They are clamped to a 2-inch wood support on which the scaffold planks are placed. The size of the support depends on how far apart the trestle jacks are placed. Metal braces hold the trestle rigid (Fig. 40-3).

Sawhorses are used for trestle supports when occasional use is required and their height meets the requirements of the job (Fig. 40-4). For light-duty work, horses and trestle jacks should not be spaced more than 8 feet apart. Do not use horses that have become weak or defective. The horizontal bearing member of horses used for trestle support should be at least a 3 × 4. The horse legs should be made of nominal 1 1/4 × 5 stock.

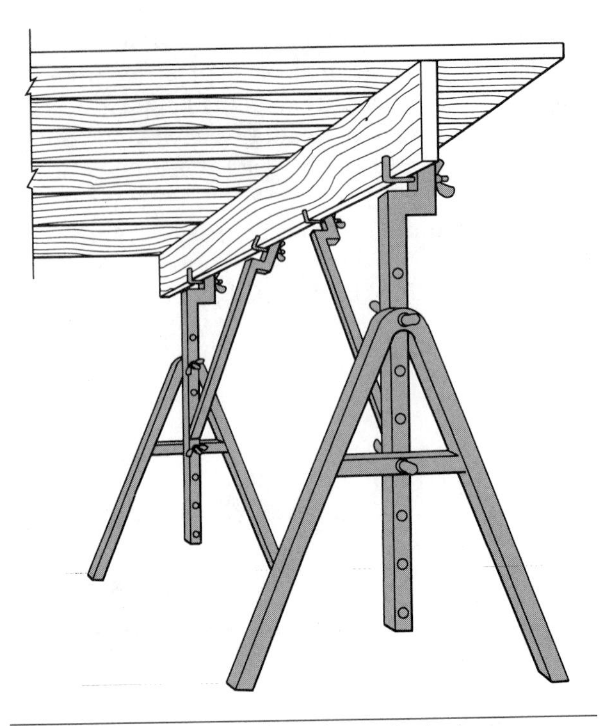

Fig. 40-3 Adjustable metal trestle jacks are manufactured for prolonged and repetitive use.

If a horse scaffold is arranged in tiers, no more than two tiers should be used. The legs of the horses in the upper tier should be nailed to the planks. Each tier should be cross-braced.

Ladders

Carpenters must often use *ladders* to work from or to reach working platforms above the ground. Most commonly used ladders are the *stepladder* and the *extension ladder.* They are usually made of wood, aluminum, or fiberglass. Make sure all ladders are in good condition before using them.

Extension Ladders

To raise an extension ladder, place its feet against a solid object. Pick up the other end. Walk forward under the ladder, pushing upward on each rung until the ladder is upright (Fig. 40-5). With the ladder vertical and close to a wall, extend the ladder by pulling on the rope with one hand while holding the ladder upright with the other. Raise the ladder to the desired height. Make sure the spring-loaded hooks are over the rungs on both sides.

Lean the top of the ladder against the wall. Move the base out until the distance from the wall is about 1/4 the vertical height. This will give the proper angle to the ladder. The proper angle for climbing the ladder can also be determined, as shown in Figure 40-6.

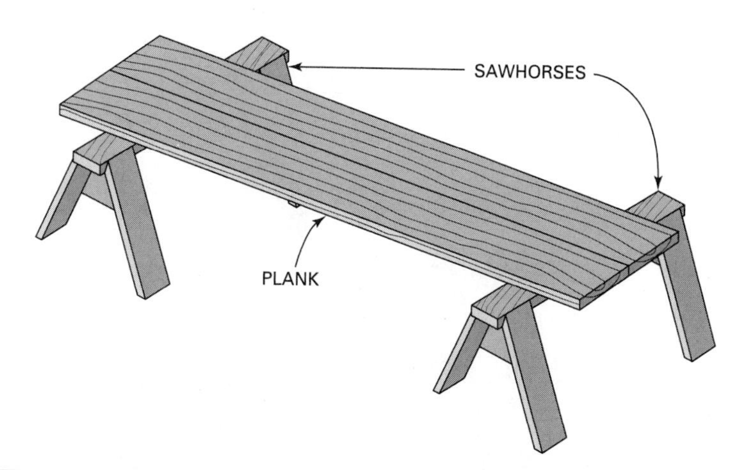

Fig. 40-4 Sawhorses are sometimes used as supports for a trestle scaffold.

Fig. 40-5 Raising an extension ladder.

If the ladder is used to reach a roof or working platform, it must extend above the top support by at least 3 feet.

> **CAUTION** Be careful of overhead power lines when using metal ladders. Metal ladders conduct electricity. Contact with power lines could result in electrocution.

When the ladder is in position, shim one leg, if necessary, to prevent wobbling, and secure the top of the ladder to the building. Face the ladder when climbing. Grasp the side rails or rungs with both hands (Fig. 40-7).

Stepladders

When using a stepladder, open the legs fully so the brackets are straight and locked. Make sure the ladder does not wobble. If necessary, place a shim under the leg to steady the ladder. Never work above the second step from the top. Do not use the ledge in back of the ladder as a step. The ledge is used to hold tools and materials only. Move the ladder as necessary to avoid overreaching. Make sure all materials and tools are removed from the ladder before moving it.

Ladder Jacks

Ladder jacks are metal brackets installed on ladders to hold scaffold plank. At least two ladders and two jacks are necessary for a section. Ladders should be heavy-duty, free from defects, and placed no more than 8 feet apart. They should have devices to keep them from slipping.

The ladder jack should bear on the side rails in addition to the ladder rungs. If bearing on the rungs only, the bearing area should be at least 10 inches on each rung. No more than two persons should occupy any 8 feet of ladder jack scaffold at any one time. The platform width must not be less than 18 inches. Planks must overlap the bearing surface by at least 10 inches (Fig. 40-8).

Construction Aids

Sawhorses, work stools, ladders, and other construction aids are sometimes custom-built by the carpenter on the job or in the shop.

Sawhorses

Sawhorses are used on practically every construction job. They support material that is being laid out or cut to size. Unless they are being used as supports for a trestle scaffold, sawhorses are usually made

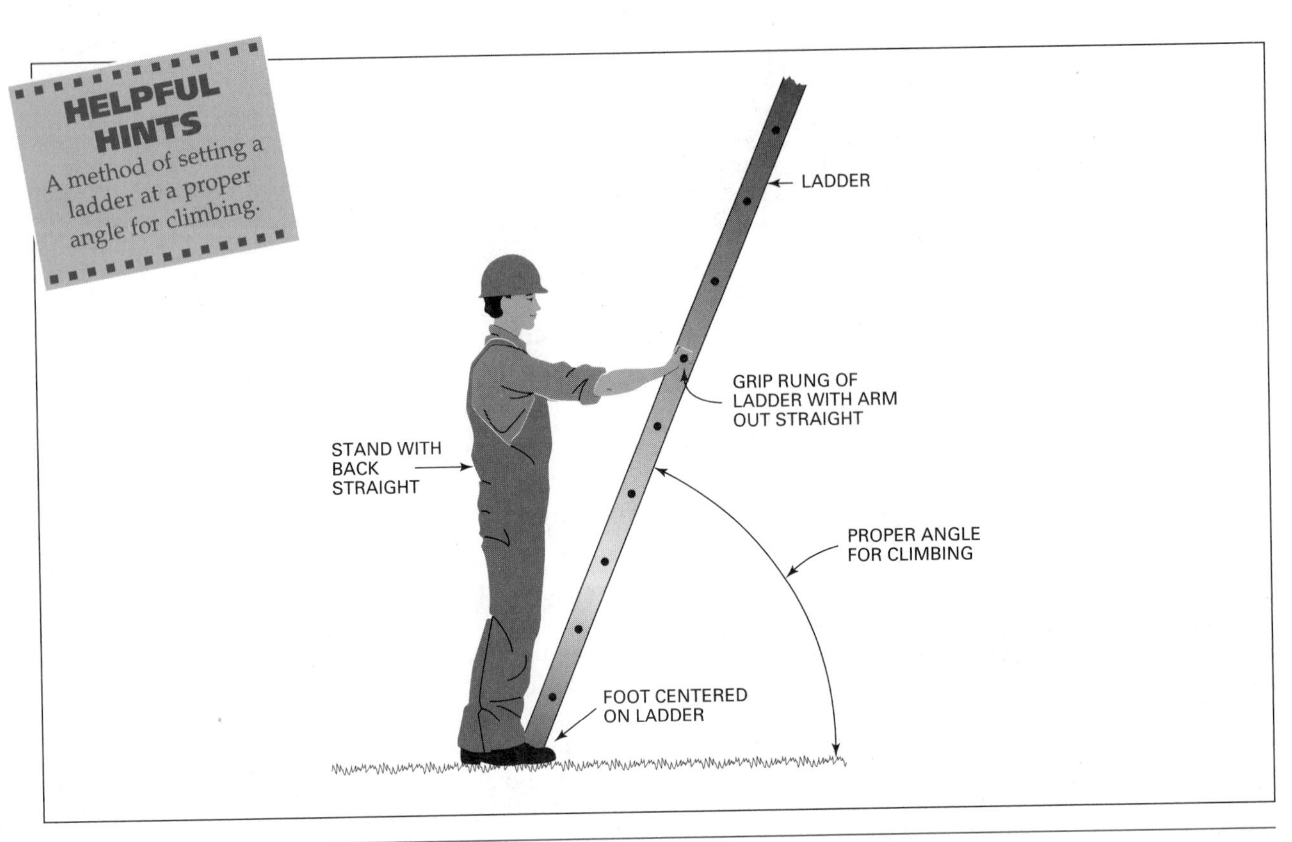

Fig. 40-6 Technique for finding the proper ladder angle before climbing.

Fig. 40-7 Face the ladder when climbing. Hold on with both hands.

Fig. 40-8 Ladder jacks are used to support scaffold plank for short-term, light repair work.

with a 2 × 4 or 2 × 6 top, 1 × 6 legs, and 3/8 or 1/2 inch plywood leg braces.

Sawhorses are constructed in a number of ways according to the preference of the individual. However, they should be of sufficient width, a comfortable working height, and light enough to be moved easily from one place to another. A typical sawhorse is 36 inches wide with 24-inch legs (Fig. 40-9). A tall person may wish to make the leg 26 inches long.

STEP BY STEP

PROCEDURES

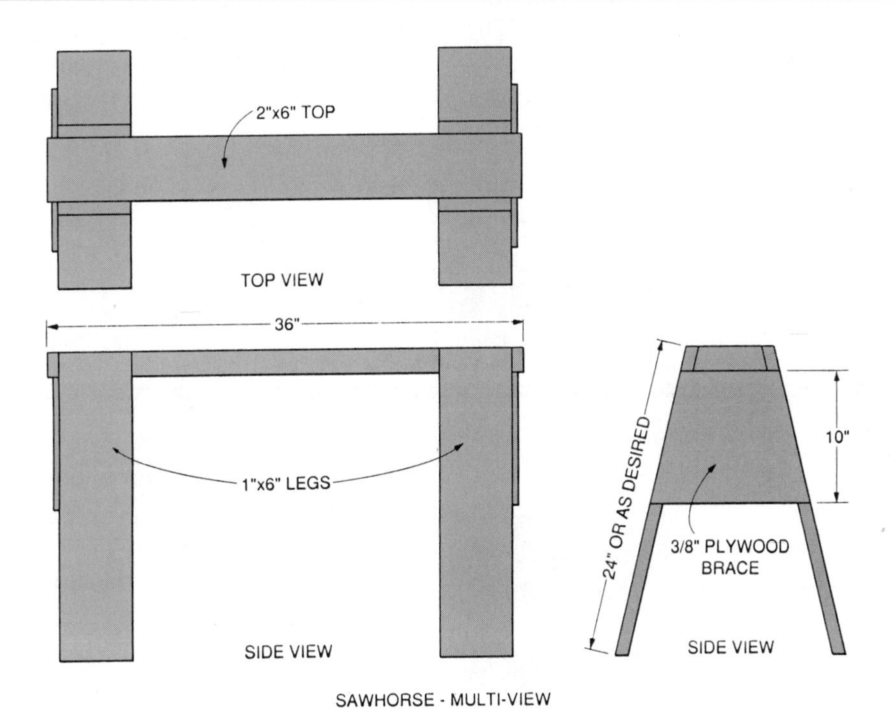

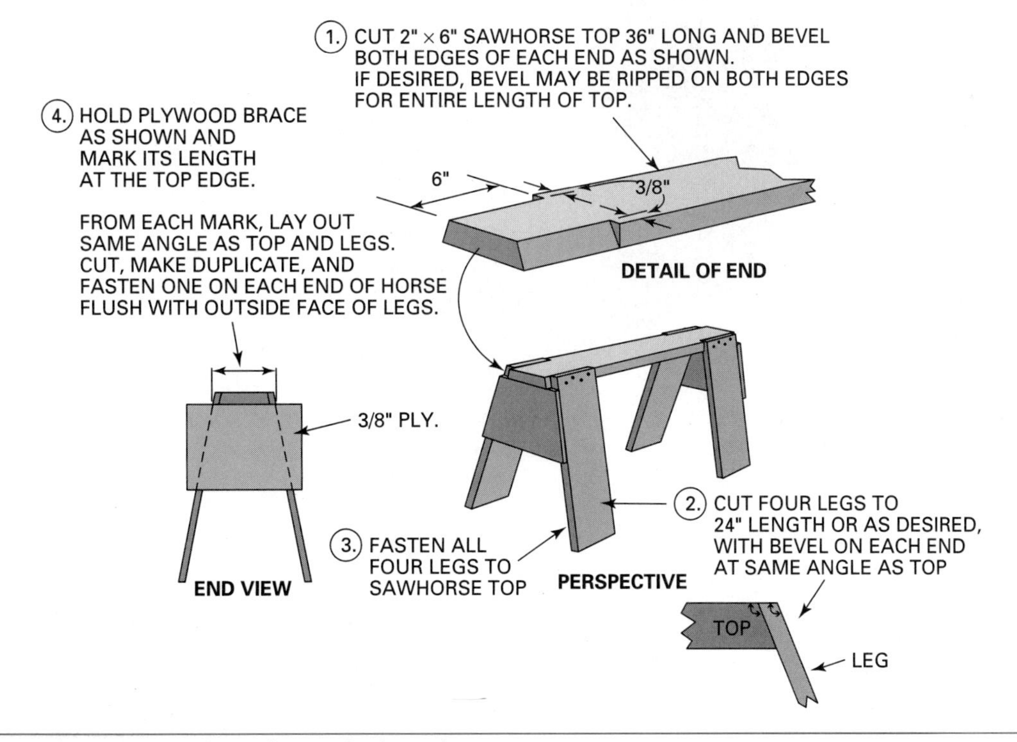

Fig. 40-9 Construction of a typical sawhorse.

Job-Made Ladders

At times it is necessary to build a ladder on the job. These are usually short, straight ladders no more than 24 feet in length. The side rails are made of clear, straight-grained 2 × 4 stock spaced 15 to 20 inches apart. *Cleats* or *rungs* are cut from 2 × 4 stock and inset into the edges of the side rails not more than 3/4 inch. Filler blocks are sometimes used on the rails between the cleats. Cleats are uniformly spaced at 12 inches top to top (Fig. 40-10).

Scaffold Safety

The safety of those working at a height depends partly on properly constructed scaffolds. Those who have the responsibility of constructing scaffolds must be thoroughly familiar with the sizes, spacing, and fastening of scaffold members and other scaffold construction techniques.

STEP BY STEP

PROCEDURES

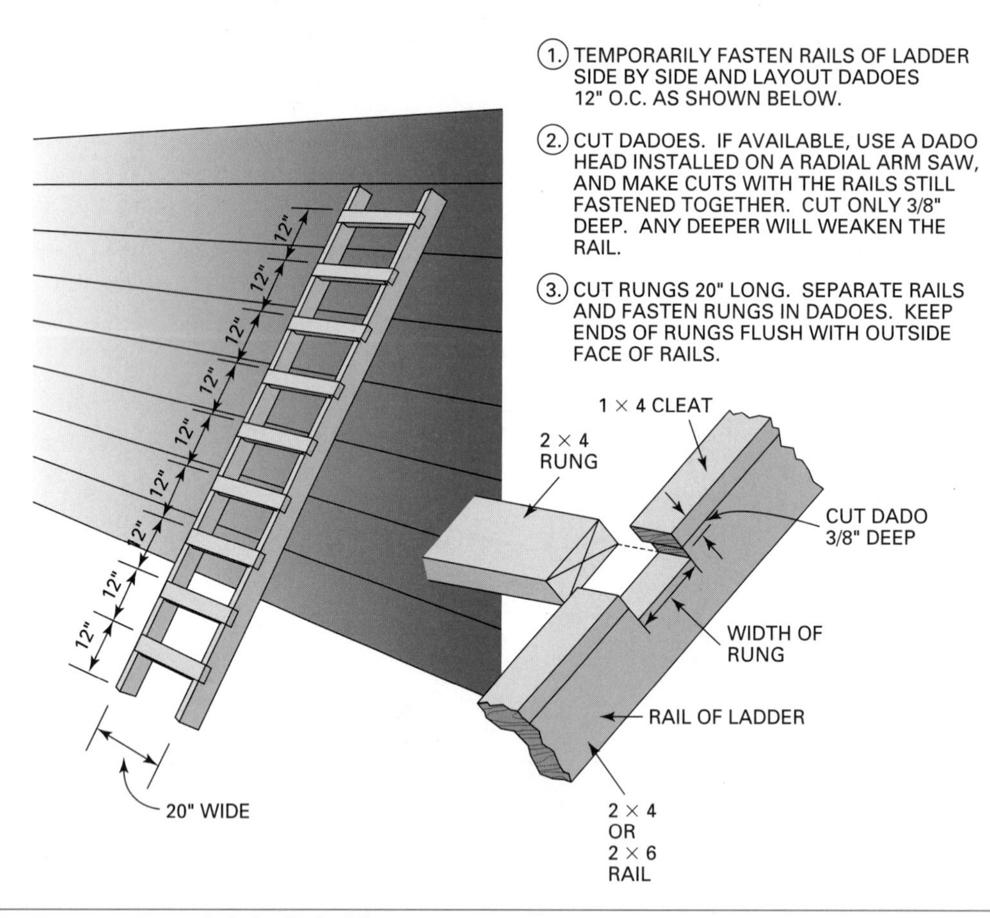

Fig. 40-10 Constructing a job-built ladder.

Review Questions

Select the most appropriate answer.

1. The vertical members of a scaffold are called
a. columns. c. poles.
b. piers. d. uprights.

2. Bearers support
a. ledgers. c. rails.
b. plank. d. braces.

3. Scaffold planks should be at least
a. 2 × 6. c. 2 × 10.
b. 2 × 8. d. 2 × 12.

4. Overlapped planks should extend beyond the bearer at least __ inches and no more than __ inches.
a. 3, 6. c. 6, 8.
b. 3, 8. d. 6, 12.

5. Metal tubular frame scaffolding is held rigidly plumb by
a. end frames. c. cross braces.
b. goosers. d. cribbing.

6. The part of a scaffold that protects workers below from falling objects is a
a. toe board. c. top rail.
b. midrail. d. posts.

7. The workers allowed to climb an access ladder for a metal tubular scaffold that has its rungs spaced 18 inches apart are the scaffold
a. users only.
b. erectors only.
c. erector and dismantlers only.
d. all of the above.

8. Wood scaffold planks, when loaded, should deflect no more than
a. 1/6 of the span.
b. 1/16 of the span.
c. 1/20 of the span.
d. 1/60 of the span.

9. The height of a mobile scaffold must not exceed the minimum base dimension by
a. three times. c. five times.
b. four times. d. six times.

10. Guardrails must be installed on all scaffolds more than
a. 10 feet in height.
b. 16 feet in height.
c. 20 feet in height.
d. 24 feet in height.

BUILDING FOR SUCCESS 

The Construction Team Networking Process

The concept of teamwork has always been at the heart of the construction process. A network implies the use of interconnected or interrelated groups to accomplish tasks. A team network will have common goals. These shared goals become the desired outcomes that formulate and direct the team's activities. Successful teamwork outcomes are reflected in a project that is completed on or ahead of schedule, within budget, and with no or minimal worker injuries. Satisfaction for the client and all interested parties will also be achieved with the proper use of a team-centered structure.

Effective teamwork has many characteristics, but some of the most important are coordination, cooperation, scheduling, problem solving, and quality assurance. If these are implemented successfully, the goals are most generally met.

Coordination is the process of harmonizing the various people and jobs on a construction project. This action is directed through a team facilitator who might be an architect, general contractor, or field superintendent. Good coordinators have a clear picture of the entire project. They must understand the skilled trades involved, the building materials/processes, and job scheduling.

Cooperation is necessary for a team to accomplish its goals. Good interaction and communication will ensure cooperation. People find it very difficult to cooperate without trust. Flexibility is a part of cooperation and must also be possessed by team members. This helps ensure that job tasks are completed, even by alternate routes.

Good scheduling is vital if deadlines are to be met. The project coordinator has the responsibility of implementing the best delivery times for materials and the optimal time to move people on and off the work site. Planning these sequential moves will demand experience and flexibility.

Problem solving becomes the responsibility of everyone involved in the construction process. Even the best planning by experienced people will be plagued at times by the unexpected. Typical examples of the unexpected are inclement weather, delivery delays, accidents, and disagreements. As problems arise, teamwork will be implemented to solve them. Each person is called upon to share expertise and ideas in finding solutions. A problem solving process should be agreed upon by all parties in advance. Usually the general contractor has implemented a procedure based on experience.

Quality assurance is a desired outcome with the construction team. Ongoing monitoring is required by everyone to maintain product quality. The client will become more aware of job and material quality as the construction progresses. At each level of construction, the skilled workers, technicians, and laborers must be committed to achieving the highest possible standards. The key to a successful team network concept is a well prepared plan. The need for such a plan must be integrated into everyone's thinking from the beginning. People who do not buy into the team network process will cause breakdowns. These will ultimately cost the construction process time and money. Losses of this type are first felt in the profit margin. The worst possible consequence could be the lack of coordination and scheduling that might cause personal injuries to workers.

In order to ensure the welfare of the construction workers and the general public, the team network has to be accepted by everyone at the job site. Efficiency will be an outcome when full acceptance of teamwork is gained. Negative factors such as hostilities, down time, and accidents will decrease when the team network concept is accepted and implemented by everyone involved in the project. Many times the common denominator is the profit margin. All builders and construction workers will understand that teamwork will assure greater chances for a profit to be maintained on each project.

FOCUS QUESTIONS: For individual or group discussion

1. The team network process in construction has its base established in working together to reach goals. What might be some good reasons to consider communicating or networking well with others in the construction industry?
2. What do you feel are necessary components in a network system approach to construction? What understandings have to be present for this system to be functional?
3. Can other components become part of a network system when considering all of the varied types of trades represented in the world of construction? What are they?
4. By concentrating on any one of the team network components, how can you defend its need?

UNIT

15

Roof Framing

The ability to lay out rafters and frame all types of roofs is an indication of an experienced carpenter. On most jobs, the boss lays out the different rafters, while workers make duplicates of them. Those persons aspiring to supervisory positions in the construction field must know how to frame various kinds of roofs.

OBJECTIVES

After completing this unit, the student should be able to:

- describe several roof types and define roof framing terms.
- describe the members of gable, gambrel, hip, intersecting, and shed roofs.
- lay out a common rafter and erect gable, gambrel, and shed roofs.
- lay out and install gable studs.
- lay out the members of and frame hip and equal-pitch intersecting roofs.
- erect a trussed roof.
- apply roof sheathing.

UNIT CONTENTS

CHAPTER 41 ROOF TYPES AND TERMS

Knowledge of diverse roof types and the terms used in their construction are essential to framing roofs.

Roof Types

Several roof styles are in common use. These roofs are described in the following material and are illustrated in Figure 41-1.

- The most common roof style is the **gable roof.** Two sloping roof surfaces meet at the top. They form triangular shapes at each end of the building called *gables.*
- The **shed roof** slopes in one direction. It is sometimes referred to as a *lean-to.* It is commonly used on additions to existing structures. It is also used extensively on contemporary homes.
- The **hip roof** slopes upward to the ridge from all walls of the building. This style is used when the same overhang is desired all around the building. The hip roof eliminates maintenance of gable ends.
- An **intersecting roof** is required on buildings that have wings. Where two roofs intersect, valleys are formed. This requires several types of rafters.
- The **gambrel roof** is a variation of the gable roof. It has two slopes on each side instead of one. The lower slope is much steeper than the upper slope. It is framed somewhat like two separate gable roofs.
- The **mansard roof** is a variation of the hip roof. It has two slopes on each of the four sides. It is framed somewhat like two separate hip roofs.
- The *butterfly roof* is an inverted gable roof. It is used on many modern homes. It resembles two shed roofs with their low ends placed against each other.
- Other roof styles are a combination of the styles just mentioned. The shape of the roof can be one of the most distinctive features of a building.

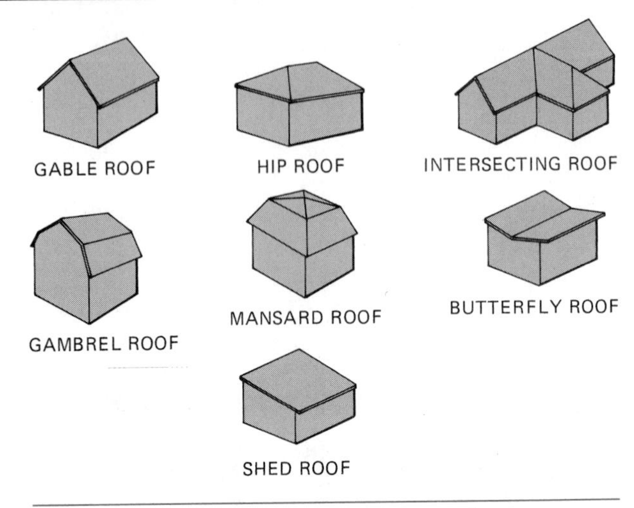

Fig. 41-1 Several roof styles are used for residential buildings.

Roof Framing Terms

It is important for the carpenter who wants to become proficient in roof framing to be familiar with roof framing terms. The terms are defined in the following material and illustrated in Figure 41-2.

- The *span* of a roof is the horizontal distance covered by the roof. This is the width of the building measured from the outer faces of the frame.
- A *rafter* is one of the sloping members of a roof frame. It supports the roof covering.
- The *total run* of a rafter is the horizontal distance over which the rafter rises. This is 1/2 the span, except in roofs of unequal pitch.
- The *unit run* is the same for all rafters that run at right angles to the plate. These rafters are called **common** *rafters.* The unit run of hip and valley rafters is longer because they usually project into the building at a 45-degree angle with the plates. This unit of run is the diagonal of a square whose sides are equal to the unit run of a common rafter. Some rafters project into the building at odd angles to the plates and thus have varying units of run.
- The *ridge* is the uppermost horizontal line of the roof.
- The **unit rise** is the number of inches the roof rises vertically for every unit run. In drawings, the slope or *cut* of a roof is indicated by a symbol. This symbol shows the unit rise per unit of run (4 and 12, 6 and 12, or 7 and 12, for example) as shown in Figure 22-8.
- The *total rise* is the total vertical distance that the roof rises. Total rise is found by multiplying the unit rise by the total run of the rafter.

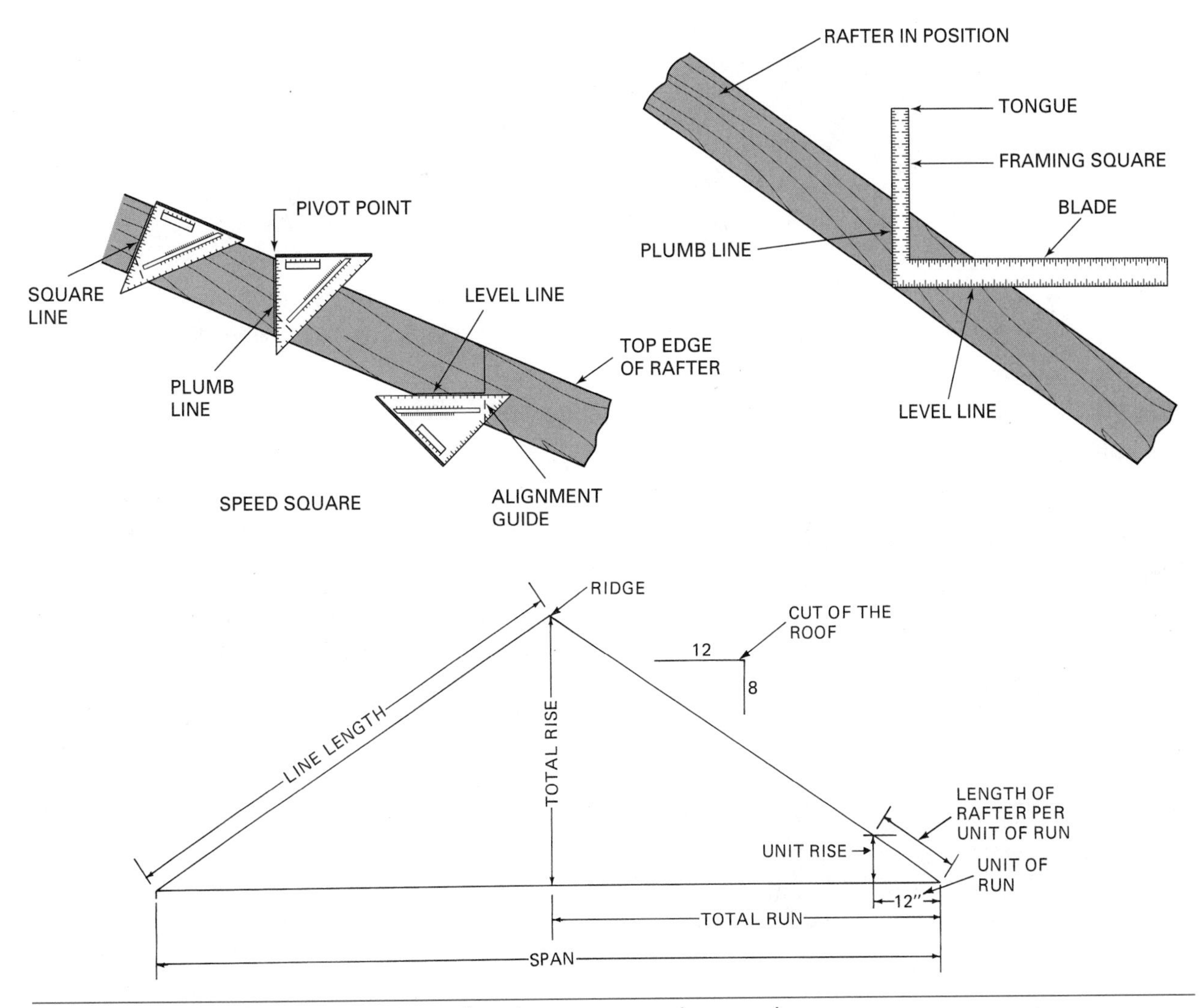

Fig. 41-2 Roof framing terms using a framing square and a speed square.

- The **line length** of a rafter is the *hypotenuse* (longest side) of a right triangle formed with its base as the total run and its other leg as the total rise. The line length gives no consideration to the thickness or width of the framing stock.
- The **pitch** is the amount of incline of a roof. This is usually expressed as a fraction. The pitch is found by dividing the total rise by the span. For example, if the span of the building is 32 feet and total rise is 8 feet, then 8 divided by 32 equals 0.25, which equals 1/4. The roof is said to be 1/4 pitch.
- A *plumb line* is any line on the rafter that is vertical when the rafter is in position. Always mark along the tongue of the framing square when laying out plumb cuts on rafters. With a speed square, always mark along the edge of the square where the inch ruler is located.
- A *level line* is any line on the rafter that is horizontal when the rafter is in position. Always mark along the blade of the framing square when laying out level cuts on rafters. With a speed square, always mark the long edge of the square where the degree scale is located after lining up the alignment guide with the plumb line (Fig. 41-3).

Calculating Rough Rafter Lengths

It is necessary to determine the length of lumber to order for use as rafters. The rough length of a rafter from plate to ridge can be calculated using the line length method. Let the blade of the square represent the total run. Let the tongue of the square represent the total rise. Use a scale of one inch to one foot. Measure the distance from blade to tongue from a

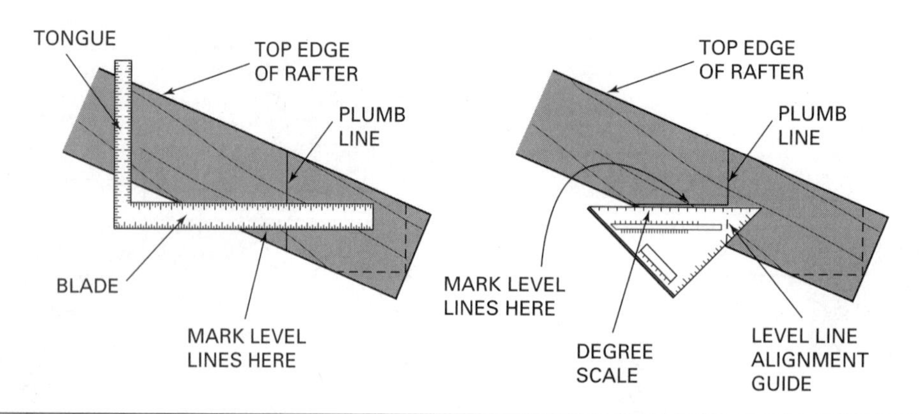

Fig. 41-3 Techniques for marking a level line on a rafter using a framing square and a speed square.

41-4 Scaling the rough length of a common rafter.

given total rise and total run to find the length of the rafter (Fig. 41-4). Add extra if the rafter overhangs the wall of the building.

EXAMPLE Assume the total rise of a rafter is 9 feet. Assume the total run is 12 feet. Locate 9 inches on the tongue of the square and 12 inches on the blade. Measure the distance between. This will be 15 inches. Using a scale of one inch equals one foot, the rough rafter length from plate to ridge is 15 feet. Standard stock length of 16 feet would have to be used for the rafter with no overhang.

CHAPTER 42 GABLE AND GAMBREL ROOFS

The *gable roof* is the style most commonly used. A *gambrel roof* is like a gable roof with two different slopes.

Framing the Gable Roof

An equal-pitched gable roof is a roof that has an equal slope on both sides intersecting the ridge in the center of the span (Fig. 42-1). This roof is the simplest to frame. Only one type of rafter, the *common rafter,* needs to be laid out.

Gable roofs may have different slopes on each side, such as the *saltbox* roof (Fig. 42-2). The principles for framing both equal or unequal-pitched gable roofs are the same.

Laying Out a Common Rafter

The *common rafter* extends at right angles from the plate upward to the ridge. It is called a common rafter because it is common to all types of roofs. It is used as a basis for laying out other kinds of rafters.

The *ridge,* although not absolutely necessary, simplifies the erection of the roof. It provides a means

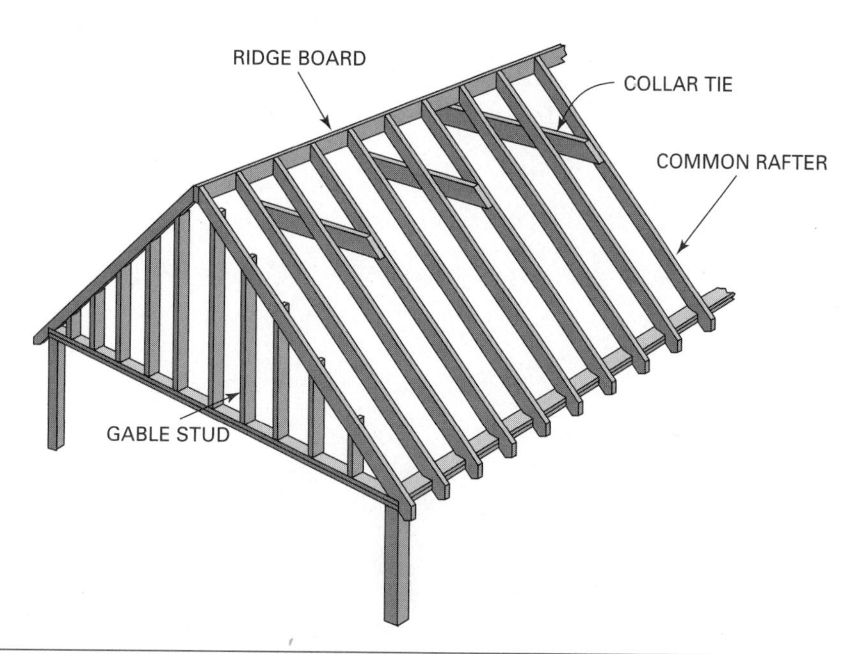

Fig. 42-1 The gable roof is framed with common rafters, ridge board, collar ties, and gable studs.

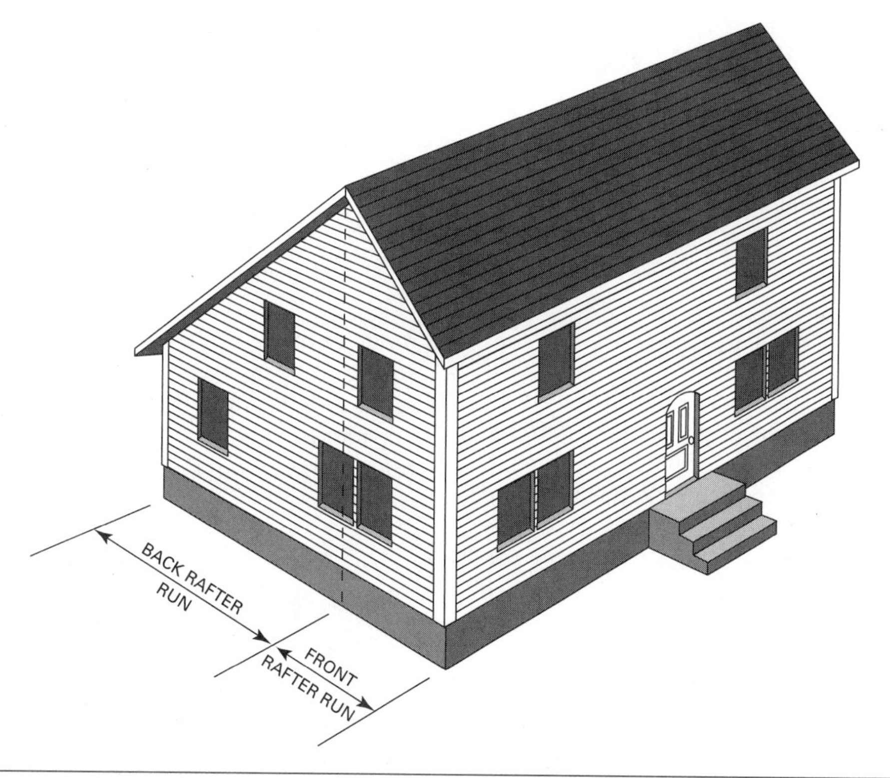

Fig. 42-2 The ridge of the saltbox roof is off-center.

of tying the rafters together before the roof is sheathed. Erecting a roof frame is called *raising* the roof.

Cuts on the Common Rafter

The common rafter requires layouts for making several cuts. The cut at the top is called the *plumb cut* or *ridge cut*. It fits against the ridge board. The cut at the plate, called the **bird's mouth** or *seat cut*, consists of a plumb line and a level line layout. It fits against the top and outside edge of the wall plate. The distance between the ridge cut and the seat cut, called the *line length* or *theoretical length*, must be determined. At the bottom end of the rafter, a **tail cut** is made on the rafter *tail* or *overhang* that extends beyond the building (Fig. 42-3).

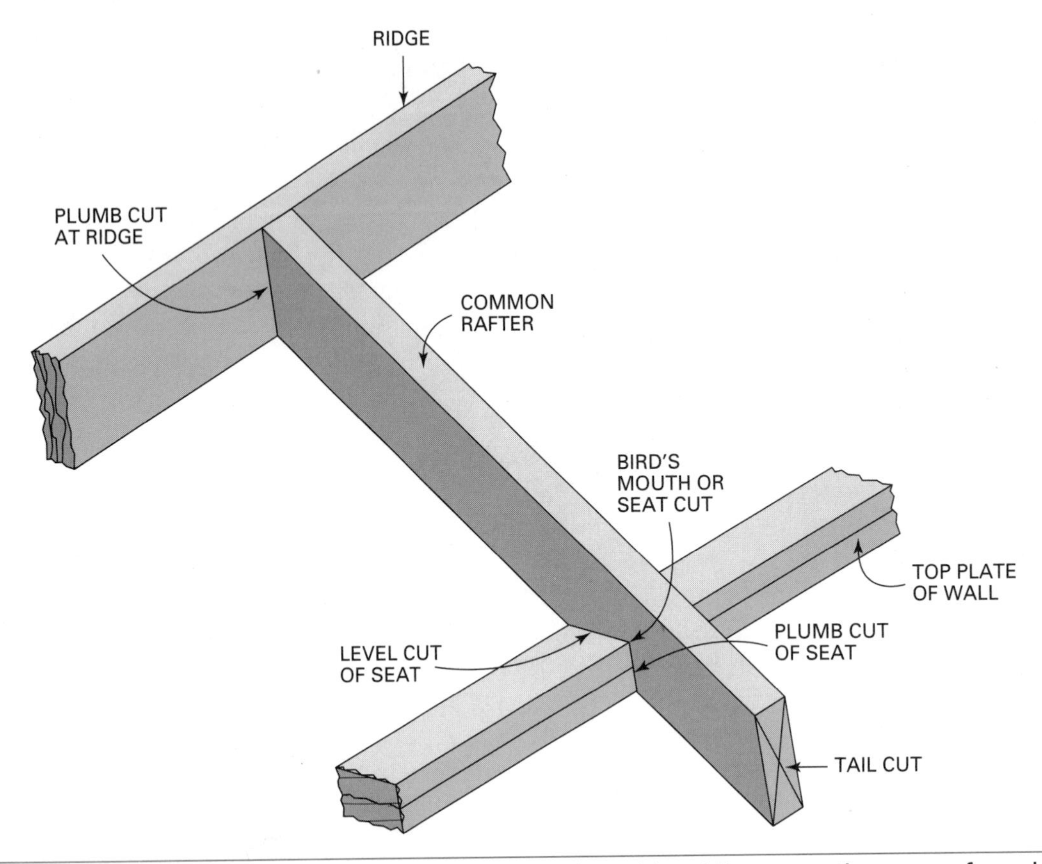

Fig. 42-3 The cuts of the common rafter. The seat plumb and level lines together are referred to as the bird's mouth or seat cut.

Common Rafter Pattern

Lay a piece of rafter stock across two sawhorses. Sight the stock along the edge for straightness. Select the straightest piece possible because it will be used as a pattern to mark the required number of rafters for the roof. If there is any slight crook in the edge of the pattern, the carpenter stands on the side that has the crowned edge. This will be the top edge of the rafter when it is installed.

Laying Out the Ridge Cut

Place the square down on the side of the stock at its left end. Hold the tongue of the square with the left hand and the blade with the right hand. Move the square until the outside edge of the tongue and the edge of the stock line up with the specified rise in inches. Make sure the blade of the square and the edge of the stock line up with 12 inches. Mark along the outside edge of the tongue for the plumb cut at the ridge.

When using a speed square, place the pivot point of the square on the top edge of the rafter. Rotate square with the pivot point touching the rafter. Looking in the rafter scale window, align the edge of the rafter with the number that corresponds to the rise per unit of run desired. Mark the plumb line along the edge of the square that has the inch ruler marked on it (Fig. 42-4).

Fig. 42-4 Laying out the plumb cut of the common rafter at the ridge.

EXAMPLE Assume the cut of the roof is 6 and 12. Hold the square so the 6-inch mark on the tongue and the 12-inch mark on the blade line up with the top edge of the rafter stock. Mark along the outside edge of the tongue of the square. Make a second plumb line for practice using a speed square. Place the speed square on the rafter stock and rotate the square, keeping the pivot point touching the edge of the rafter. Align the edge of the rafter with the number 6 on the common scale in the rafter scale window. Mark the plumb line along the edge of the square that has the ruler marked on it.

Finding the Length of a Common Rafter

The distance between the plumb cut at the ridge and the seat cut can be determined by the *step-off method* or by using a calculator and the *rafter tables.*

The Step-Off Method. The step-off method is based on the unit of run (12 inches for the common rafter). The rafter stock is stepped-off for each unit of run until the desired number of units or parts of units are stepped-off.

First, lay out the ridge cut. Mark where the blade intersects with the top edge of the rafter. Hold the square at the same angle. Shift it until the tongue lines up with this mark. Move the square in a similar manner until the total run of the rafter is laid out (Fig. 42-5). Mark a plumb line along the tongue of the square at the last step. This line will be parallel to the ridge cut.

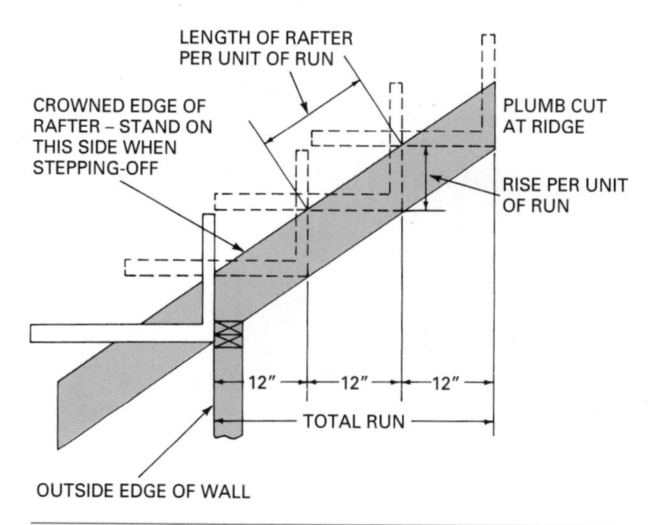

Fig. 42-5 Stepping-off the common rafter. This diagram shows three steps in a step-off.

EXAMPLE If the rafter has a total run of 16 feet, the square is moved 16 times.

Stepping-Off a Fractional Part of a Unit of Run. First, step-off for the total whole units of run. Move the square as if to step-off one more time. Instead of marking at the blade edge of the square, measure the fractional part of the unit of run along the blade. Mark it on the rafter. Holding the square in the same position, move it so the outside of the tongue lines up with the mark. Lay out a plumb line along the tongue of the square (Fig. 42-6).

EXAMPLE If the rafter has a total run of 16 feet, 8 inches, step off 16 times. Hold the square for the 17th step. Mark along the blade at the 8-inch mark. With the square held at the same angle, move it to the mark. Lay out the plumb cut of the bird's mouth by marking along the tongue.

Layout Aids for the Framing Square

Several aids may be used to keep the square at the same angle for each step during the step-off process.

Framing Square Gauges. *Framing square gauges* may be attached to the square. They are used as stops against the top edge of the rafter. These gauges are attached to the tongue of the square for the desired rise per foot of run and to the blade at the unit of run (Fig. 42-7).

Straightedge. A small straightedge clamped across the tongue and blade of the square at the desired location will ensure that the square is held at the same angle for every step. The square is moved by sliding the straightedge against the top edge of the rafter stock (Fig. 42-8).

Calculating the Common Rafter Length

Calculating a rafter using a calculator and the rafter tables can be faster and more accurate than using the step-off method. Rafter tables give information for finding the length and cuts on various kinds of rafters. They come in booklet form. They are also stamped on one side of most framing squares (Fig. 42-9). On the square, the inch marks indicate the rise of the rafter per unit of run. Directly below inch marks 2 through 18 are the length of several kinds of rafters in inches per unit of run. The common rafter length per unit of run can also be found using the Pythagorean Theorem, $A^2 + B^2 = C^2$ (see Fig. 26-4).

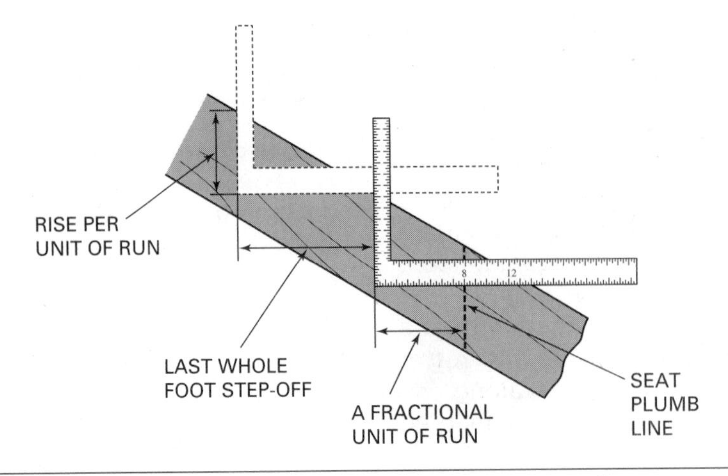

Fig. 42-6 Laying out a fractional part of the total run of a common rafter.

Fig. 42-7 Framing square gauges are attached to the square to hold it in the same position for every step-off.

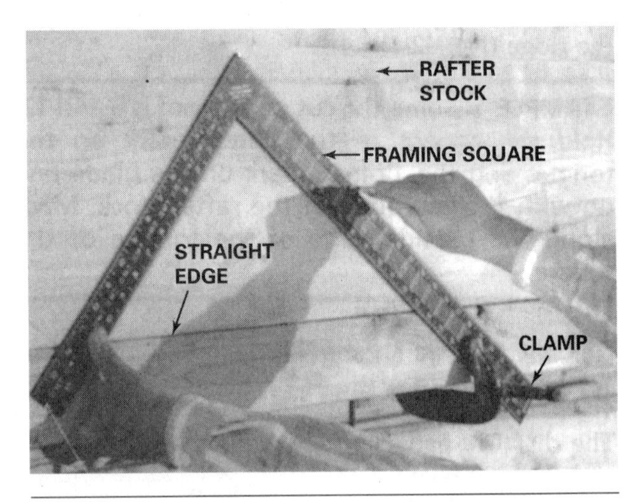

Fig. 42-8 A straightedge clamped to the square helps the step-off procedure.

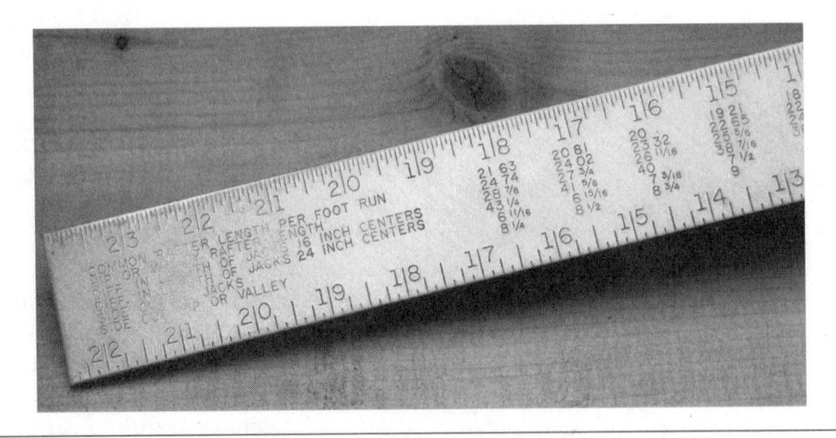

Fig. 42-9 Rafter tables are found on the framing square.

EXAMPLE Suppose the rise of the rafter per unit of run is 8 inches. Using the Pythagorean Theorem substitute 8 in for the A and 12 for the B. C is then the length per unit of run. To solve, square 8 giving 64 and square 12 giving 144. Adding these numbers gives 208. C is found by taking the square root of 208, which equals 14.422205. When rounded to the nearest hundredth, we have 14.42 inches.

Using Rafter Tables. To find the line length of a common rafter, first find the length per foot of run in the top line of the rafter table. This number will be under the inch mark that corresponds to pitch of the roof being framed. Multiply the length per foot of run found in the table by the number of units of run.

EXAMPLE Find the line length of a common rafter with a rise of 8 inches per foot of run for a building 28 feet wide. Reading below the 8-inch mark on the square, the length of the common rafter per foot of run is found to be 14.42 inches. Multiplying 14.42 inches by 14, which is the run of the rafter, gives 201.88 inches. Dividing by 12 to change inches to feet gives 16.823 feet. Changing 0.823 foot to inches by multiplying by 12 gives 9.876 inches. Changing 0.876 inch to 16ths of an inch by multiplying by 16, gives slightly over 14, making the total length of the rafter 16′-9 7/8″.

Laying Out the Common Rafter Length

Mark a plumb cut at the ridge. From the ridge cut, lay out, along the top edge of the rafter, the length of the rafter as determined by calculations. Mark the length and make a plumb line at the seat. At this point, the rafter length is theoretical because the rafter must be shortened due to the thickness of the ridge (Fig. 42-10).

Laying Out the Seat Cut of the Rafter

The *seat cut* or *bird's mouth* of the rafter is a combination of a level cut and a plumb cut. The level cut rests on top of the wall plate. The plumb cut fits snugly against the outside edge of the wall.

On the plumb line of the seat cut, which has already been laid out, measure down from the top edge of the rafter a distance that will leave sufficient stock to ensure enough strength to support any overhang. Mark on the plumb line. (This is usually a minimum of 2/3 the width of the stock.) Hold the square in the same position as for stepping-off. Slide it along the stock until the blade lines up with the mark. Mark along the blade for the level cut of the seat. With a speed square, hold the alignment guide of the square in line with the plumb line previously drawn. Mark along the long edge of the square to achieve level lines for seat cuts (Fig. 42-11). On roofs with moderate slopes, the length of the level cut of the seat is usually the width of the wall plate. For steep roofs, the level cut is shorter. Otherwise too much stock would be cut out of the rafter, weakening it.

Overhang of the Common Rafter

In most plans, the rafter projection is given in terms of a level measurement. To lay out the overhang,

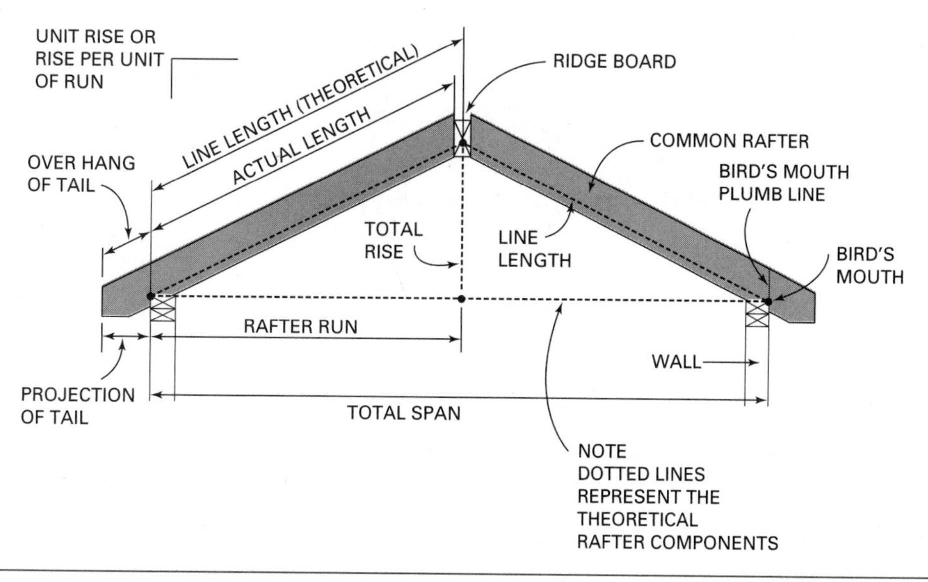

Fig. 42-10 Terms, components and concepts of rafter framing.

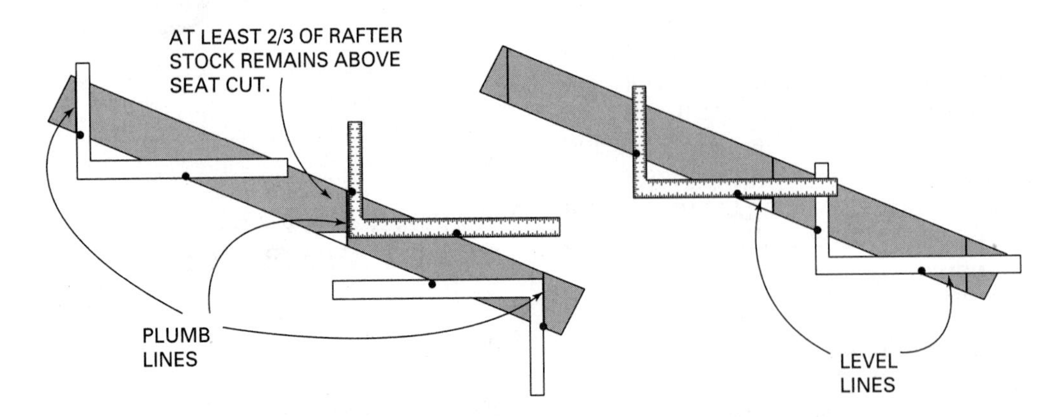

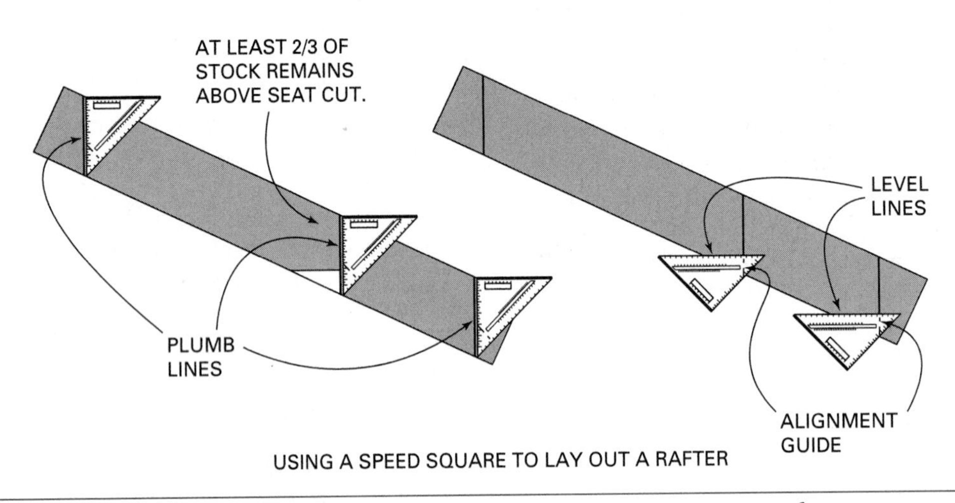

Fig. 42-11 Techniques for using a framing and speed square to lay out a rafter.

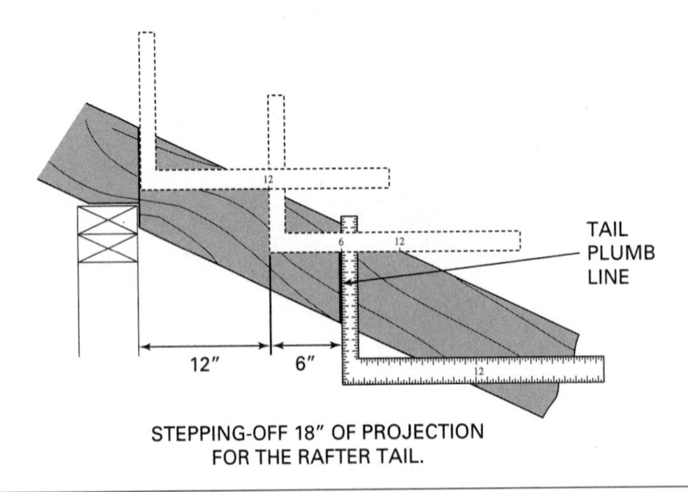

Fig. 42-12 Laying out the overhang of the common rafter with a projection of 18 inches.

simply step off the necessary amount of run from the plumb line of the seat cut (Fig. 42-12). The *tail cut* is the cut at the end of the rafter tail. It may be a plumb cut, level cut, a combination of cuts, or a square cut (Fig. 42-13). Sometimes the rafter tails are left to run *wild.* This means they are slightly longer than needed. They are cut off in a straight line after the roof frame is erected.

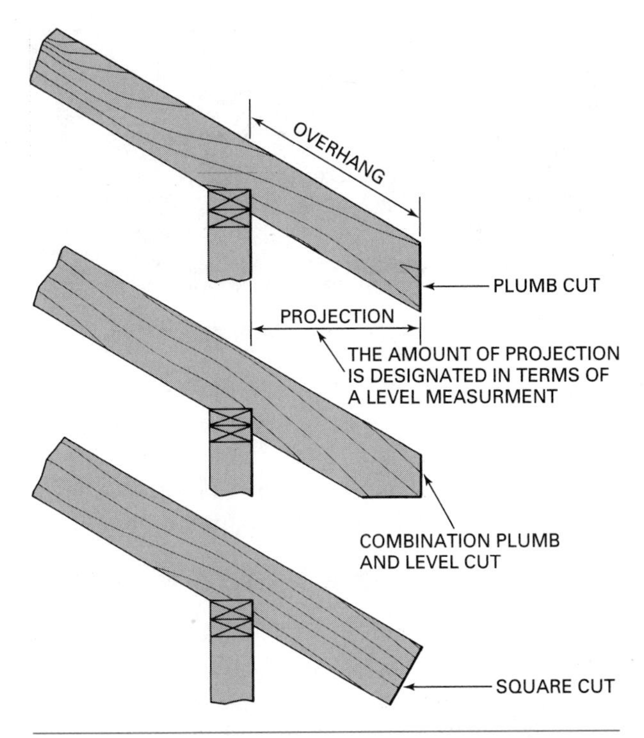

Fig. 42-13 Various tail cuts of the common rafter.

Shortening the Common Rafter

The length of the rafter at this point is from the outside of the wall to the centerline of the ridge (its line length). Therefore, the rafter must be shortened to allow for the thickness of a ridge board, if one is used (Fig. 42-14). If no ridge board is used, no shortening is necessary.

To shorten the rafter, measure back, at a right angle from the plumb line at the ridge, a distance equal to 1/2 the thickness of the ridge board. Lay out another plumb line at this point for the actual length of the rafter.

Note: Shortening is always measured at right angles to the ridge cut, regardless of the slope of the roof.

Cutting the Required Number of Rafters

To find the number of rafters required for a gable roof, divide the length of the building by the spacing

STEP BY STEP PROCEDURES

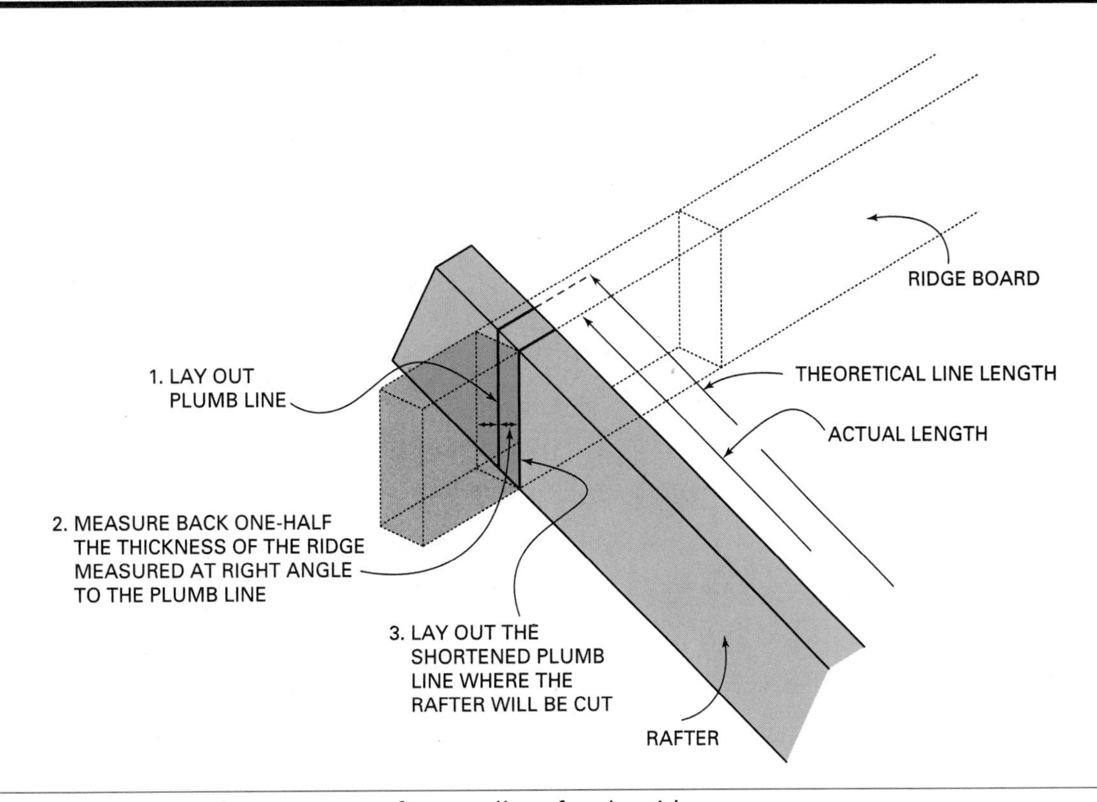

Fig. 42-14 Shortening the common rafter to allow for the ridge.

of the rafters. Add one, as a starter, and then multiply the total by two.

EXAMPLE A building is 42 feet long. The rafter spacing is 16 inches OC. Divide 42 by 1 1/3 (16 inches divided by 12 equals 1 1/3 feet) to get 31 1/2 spaces. Change 31 1/2 to the next whole number to get 32. Add 1 to make 33. Multiply by 2. This shows that sixty-six rafters are needed.

Pile the required number of rafter stock with the crowned edges all facing in the same direction. Lay the pattern on each piece so the crowned edge of the stock will be the top edge of the rafter. Mark and make the necessary cuts. Usually a skilsaw or radial arm saw is used to make these cuts.

Fig. 42-15 Wood I-beams frequently are used for roof rafters. (*Courtesy of Trus Joist MacMillan*)

CAUTION On some cuts that are at a sharp angle with the edge of the stock (i.e., the level cut of the seat), the guard of the skilsaw may not retract. In this case, retract the guard by hand until the cut is made a few inches into the stock. Never wedge the guard in an open position to overcome this difficulty.

Wood I-Beam Roof Details

In addition to solid lumber, wood I-beams may be used for rafters (Fig. 42-15). Some roof framing details for wood I-beams are shown in Figure 42-16. To determine sizes for various spans and more specific information, consult the manufacturer's literature.

Laying Out the Ridge Board

Lay a sufficient number of boards to be used for a ridge alongside the wall plate. Transfer the rafter layout from the plate to the ridge board. Joints in the ridge board are centered on a rafter. The total length of the ridge board should be the same as the length of the building, unless there is an overhang at the gable ends. If there is an overhang, add the necessary amount on both ends.

Erecting the Gable Roof Frame

Construct a scaffold, if necessary, in the center of the building at a convenient height from the ridge (usu-

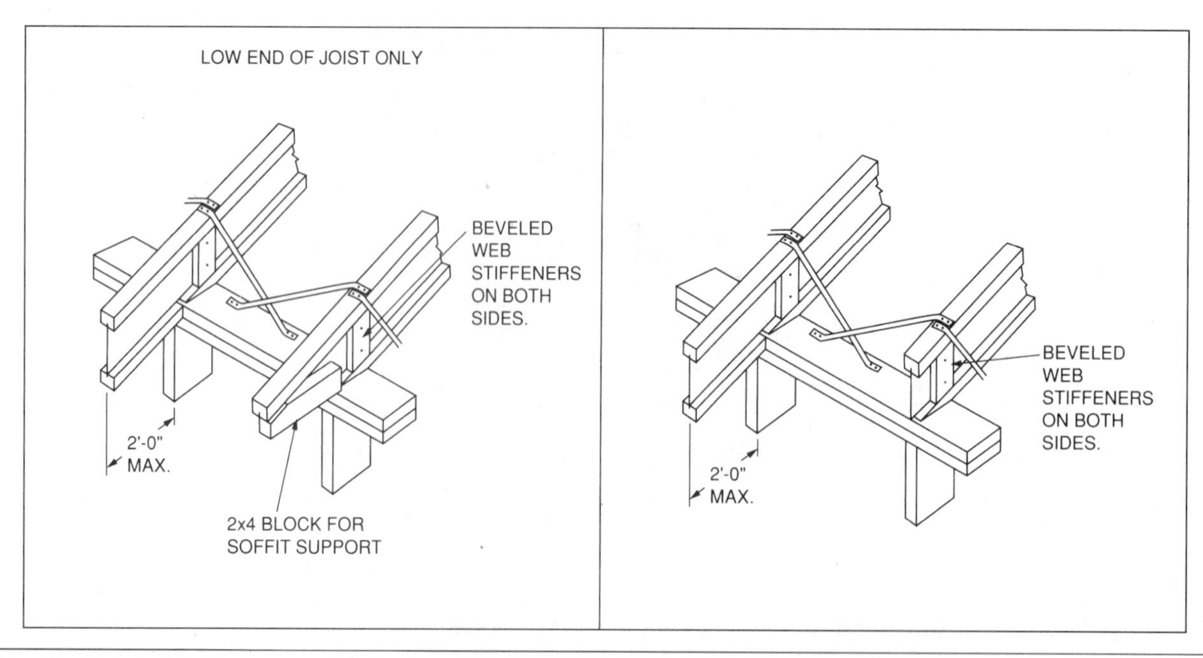

Fig. 42-16 Wood I-beam roof-framing details. (*Courtesy of Trus Joist MacMillan*)

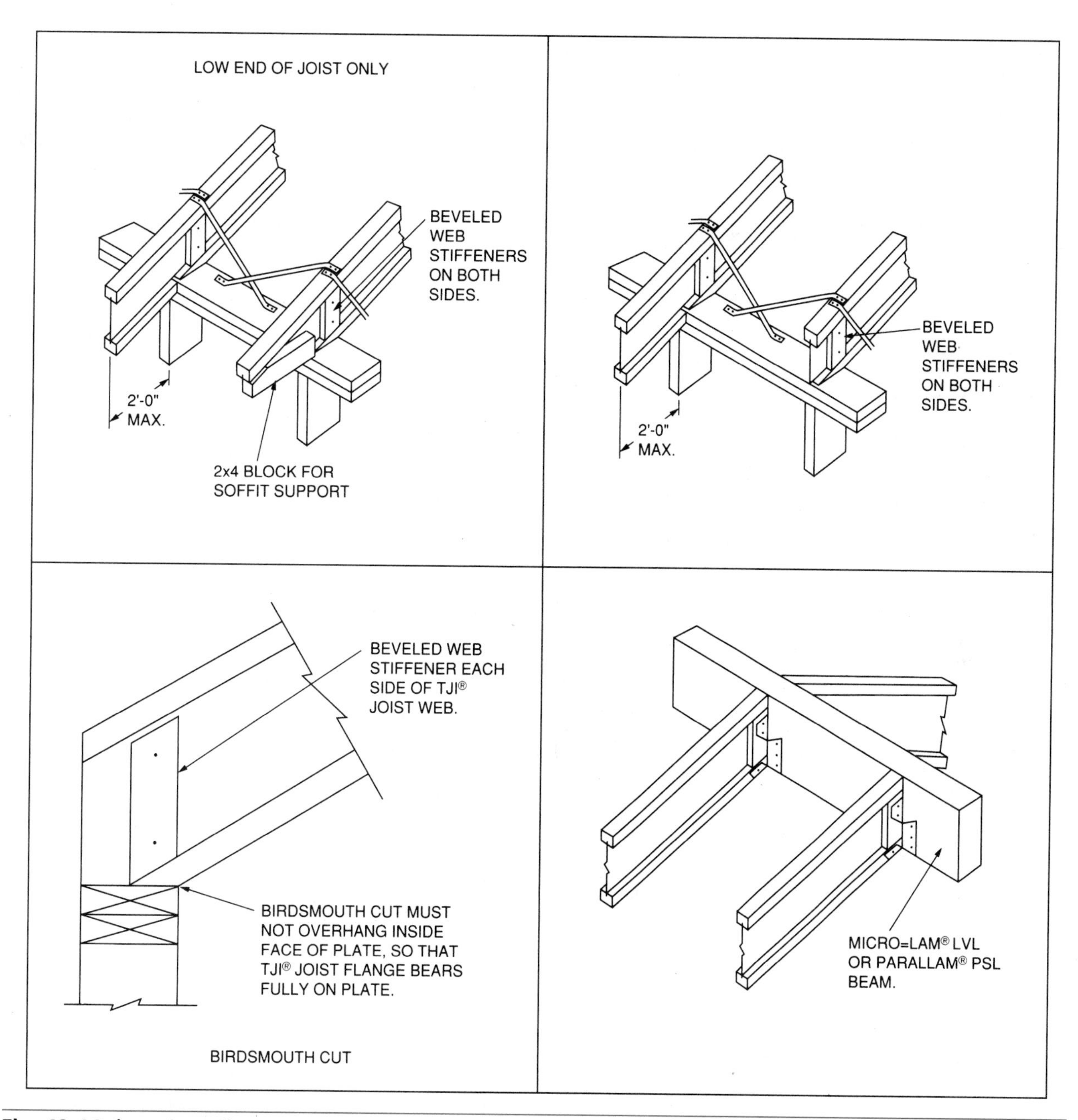

Fig. 42-16 (*continued*)

ally about 5 feet below the ridge). Lay the ridge board on top of the scaffold in the same direction it was laid out. Be careful not to turn the pieces around, end for end.

Lean the cut rafters against the building. Space them along each side, with the ridge cut up. Select four straight pieces for the end rafters, commonly called the **rake** rafters. One carpenter on each side of the building and one at the ridge are needed to raise the roof with efficiency.

Fasten a rafter to each end of the first section of ridge board. Raise the ridge board and two rafters into position. Fasten the rafters at the seat. Install two opposing rafters on the other side. Fasten the opposing rafters to the ridge board by nailing through the board at an angle into the cut end of the rafter. Drive the nail home with a nail set if necessary. Keep the bottom edge of the rafters and ridge board flush with each other. This will allow for greater airflow through the ridge vent and better support of the rafter.

Plumb and brace this section. Plumb the section from the end of the ridge board to the outside edge of the plate at the end of the building. Brace the section temporarily from attic floor to ridge. Raise all other sections in a similar manner.

Install the rest of the rafters. It is best to fasten one rafter and then the opposing one. Do not install

STEP BY STEP

PROCEDURES

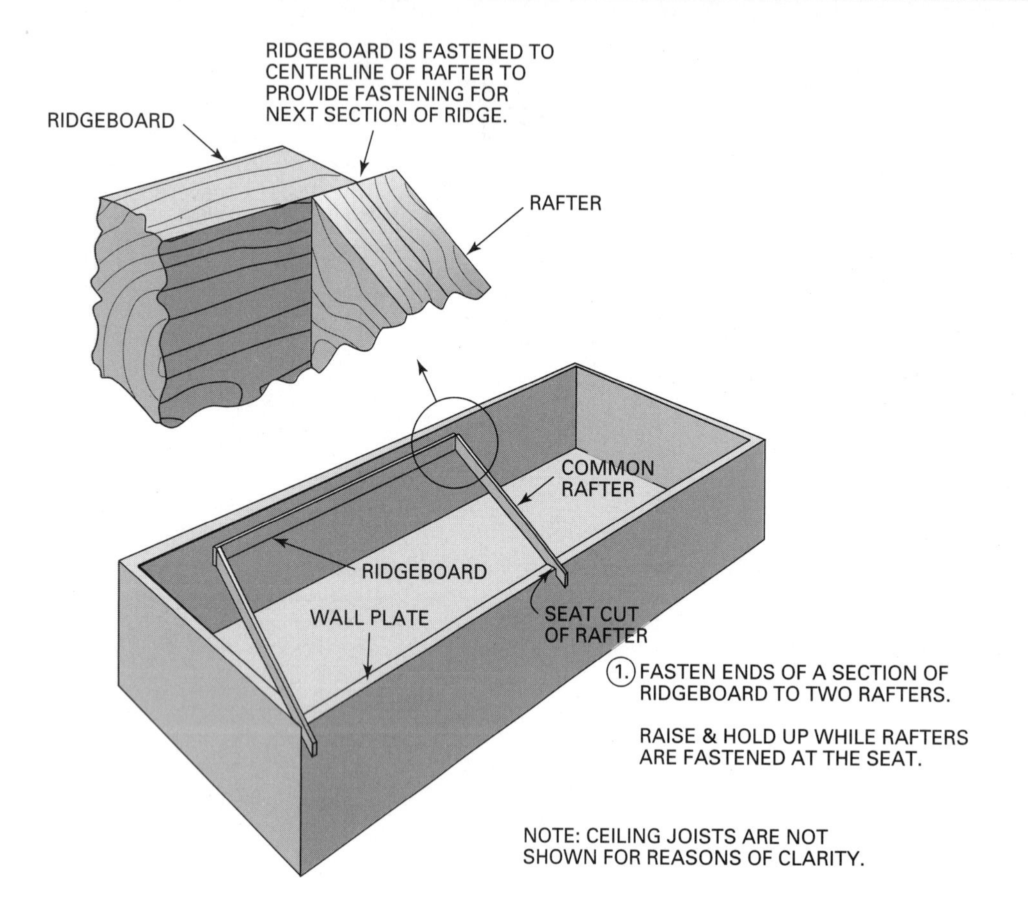

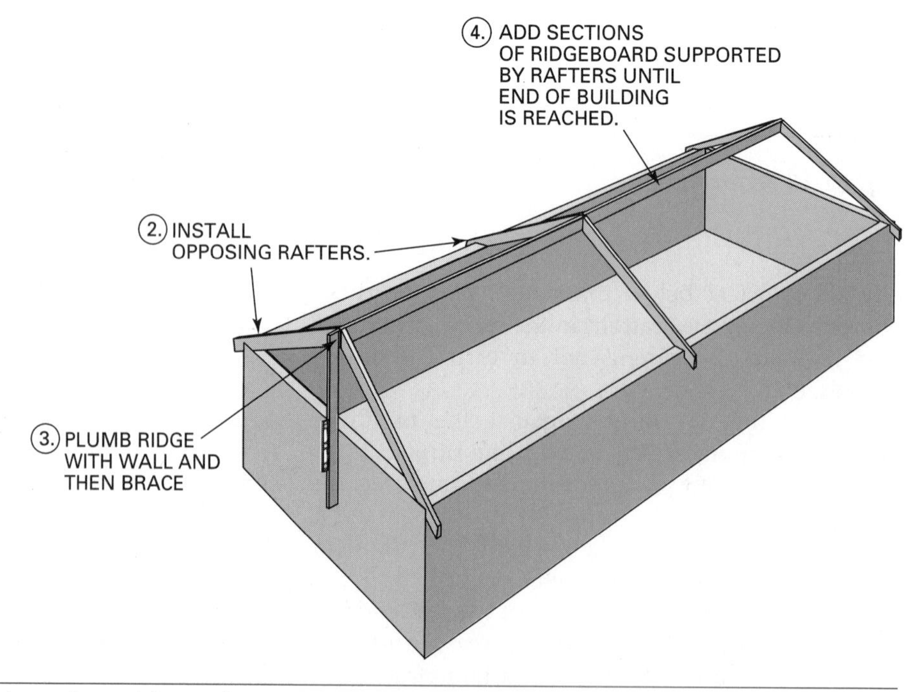

Fig. 42-17 Steps in the erection of a gable roof.

STEP BY STEP
PROCEDURES

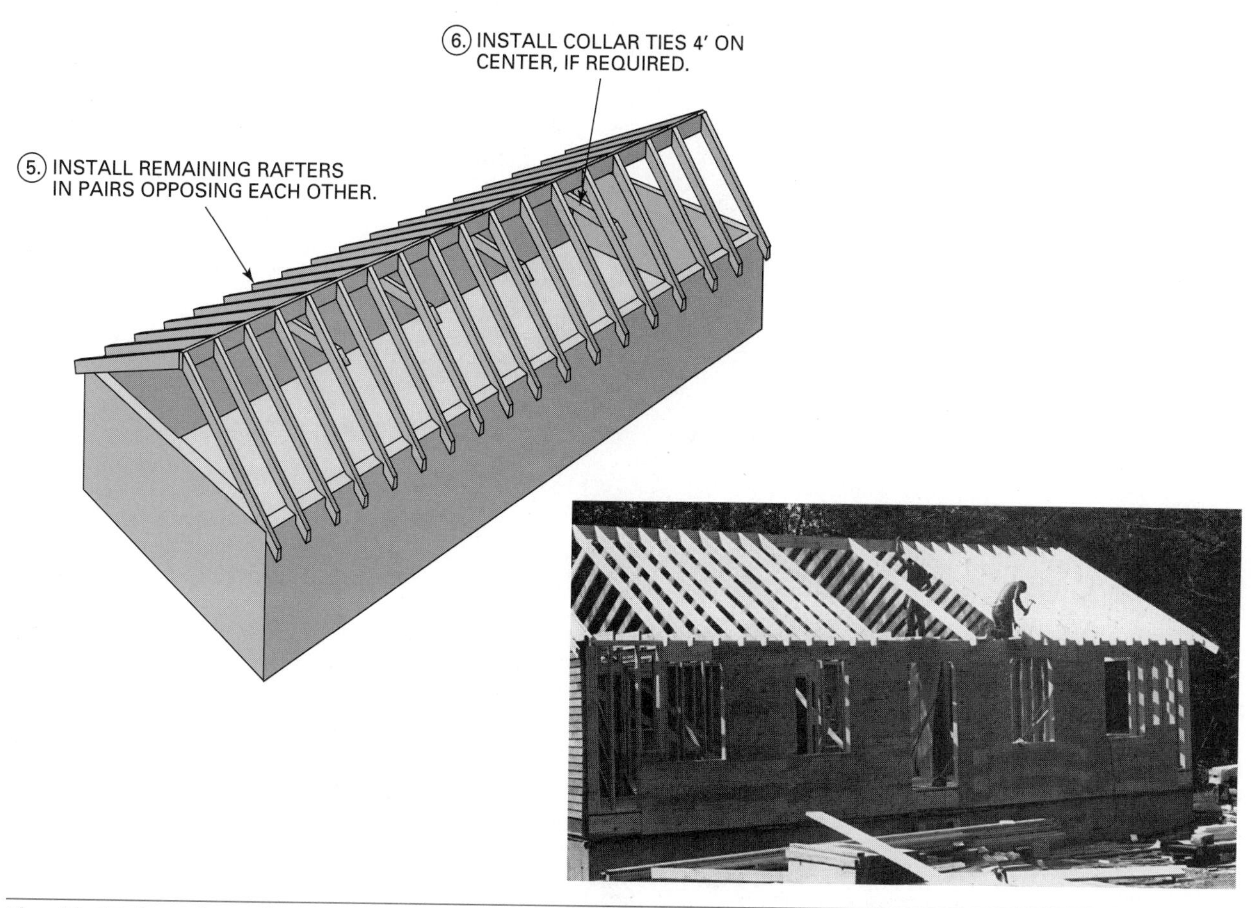

Fig. 42-17 *(continued)*

a number of rafters on one side only. Sight along the top edge of the ridge board for straightness as framing progresses. If not straight, make adjustments when installing the remaining rafters. In addition to fastening rafters to the wall plate, fasten them to the sides of ceiling joists.

Collar Ties. Collar ties are horizontal members fastened to opposing pairs of rafters, which effectively reduce the span of the rafter. Install collar ties to every third rafter pair or as required by drawings or codes. The length of a collar tie varies, but they are usually about 1/3 to 1/2 of the building span (Fig. 42-17).

Plumbing and Bracing. Check the roof for plumb again. Brace the roof frame securely with braces fastened to the under edges of the rafters running at a 45-degree angle from plate to ridge. Applying the brace to the underside of the roof frame allows the roof to be sheathed without disturbing the braces.

Rafter Tails. If the rafter tails have not previously been cut, measure and mark the end rafters for the amount of overhang. This is usually a level measurement from the outside of the wall studs to the tail plumb line (Fig. 42-13). Plumb the marks up to the top edge of the rafters. Stretch a line between these two marks across the top edges of all the rafters. Using a level, plumb down on the side of each rafter from the chalk line. Using a skilsaw, cut each rafter. Start the cut from the top edge and follow the line carefully.

If the tail cut is a combination of plumb and level cuts, make the plumb cuts first. Snap a line across the cut ends where specified. Level in from the chalk line. Make the level cuts with a skilsaw. Work from the outside toward the wall (Fig. 42-18). It should be noted here that it is usually easier to cut the rafter tails before the rafter is installed.

STEP BY STEP

PROCEDURES

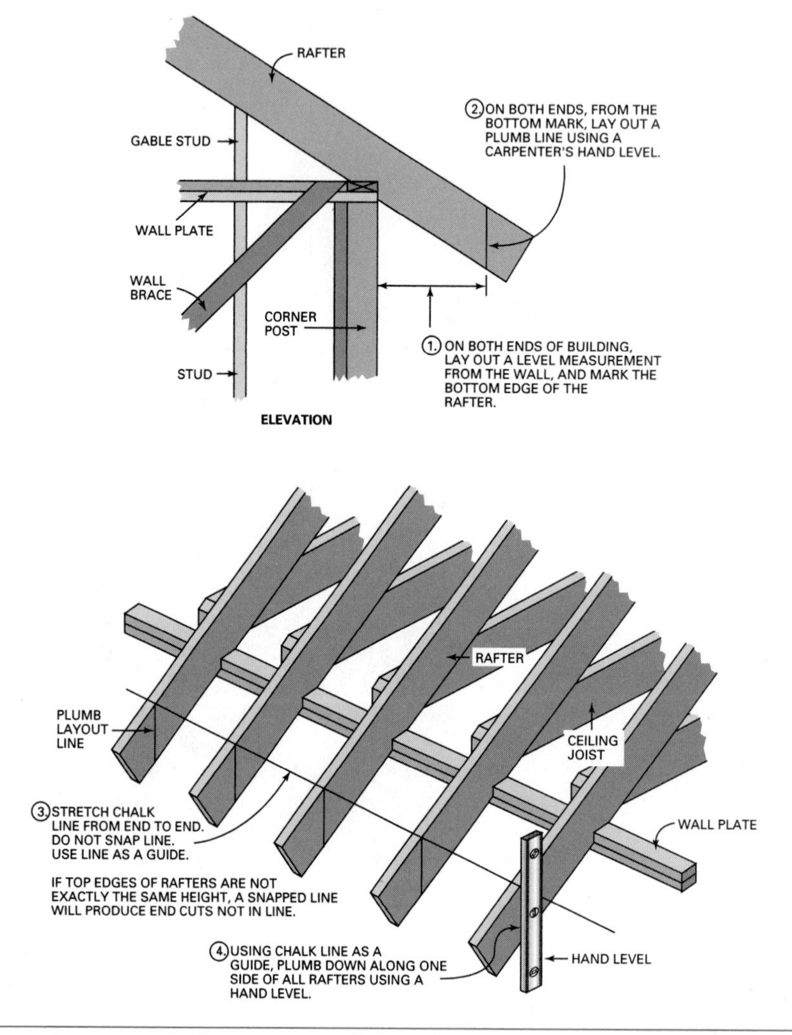

Fig. 42-18 Laying out the tail cut of common rafters after they are installed.

Constructing a Rake Overhang

If an overhang is specified at the rakes, horizontal structural members called lookouts must be installed. They support the rake rafter as shown in Figure 42-19.

Installing Gable Studs

The triangular areas formed by the rake rafters and the wall plate at the ends of the building are called *gables.* They must be framed with studs. These studs are called *gable studs.* The bottom ends are cut square. They fit against the top of the wall plate. The top ends fit snugly against the bottom edge and inside face of the end rafter. They are cut off flush with the top edge of the rafter (Fig. 42-20).

Laying Out Gable Studs

The end wall plate is laid out for the location of the gable studs. Studs should be positioned directly above the end wall studs. This allows for easier installation of the wall sheathing. Square a line up from the wall studs over to the top of the wall plate. With a straight edge and level, plumb these marks up to the end rafters for lay out of the top end of the gable end studs.

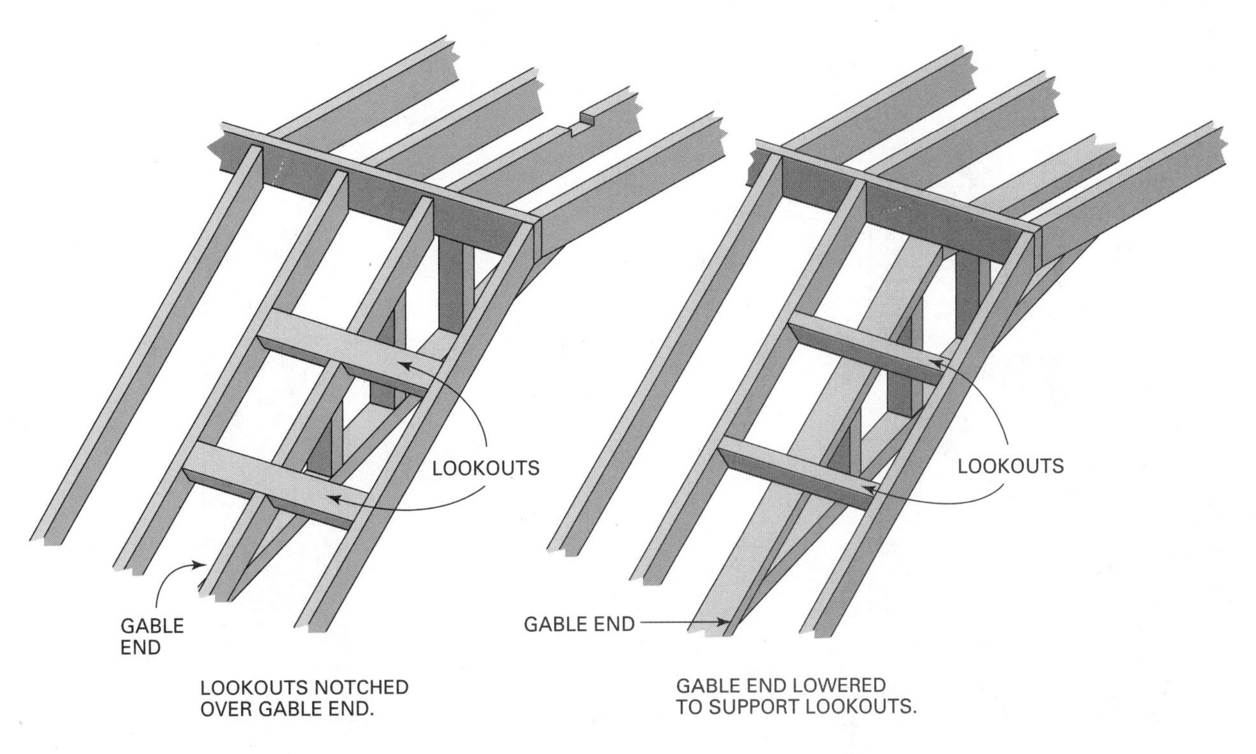

Fig. 42-19 Lookouts support the rake overhang.

Finding the Length of Gable Studs

One method of finding the length and cut of the stud is to stand it up plumb on the mark at the plate and on the inside face of the rafter. Mark it along the bottom and top edge of the rafter (Fig. 42-21). Remove the stud. Use a scrap piece of rafter stock to mark the depth of cut on the stud. Mark and cut all studs in a similar manner.

The studs are fastened by toenailing to the plate and by nailing through the rafter into the edge of the stud. Care must be taken when installing gable studs not to force the end rafters up. This creates a crown in them. Sight the top edge of the end rafters for straightness as gable studs are installed. After all gable studs are installed, the end ceiling joist is nailed to the inside edges of the studs.

Laying Out Gable Studs by a Common Difference

Gable studs that are spaced equally have a common difference in length. Each stud is shorter than the next one by the same amount (Fig. 42-22). Once the length of the first stud and the common difference are known, gable studs can be laid out easily and cut on the ground.

To find the common difference in the length of gable studs, multiply the spacing in units of run by the unit rise of the roof.

The quickest method of finding the length of the first stud is to place it in position. Mark it as described previously. Find the length of the longest gable stud. Mark the common difference successively along its length for a pattern to lay out all gable studs.

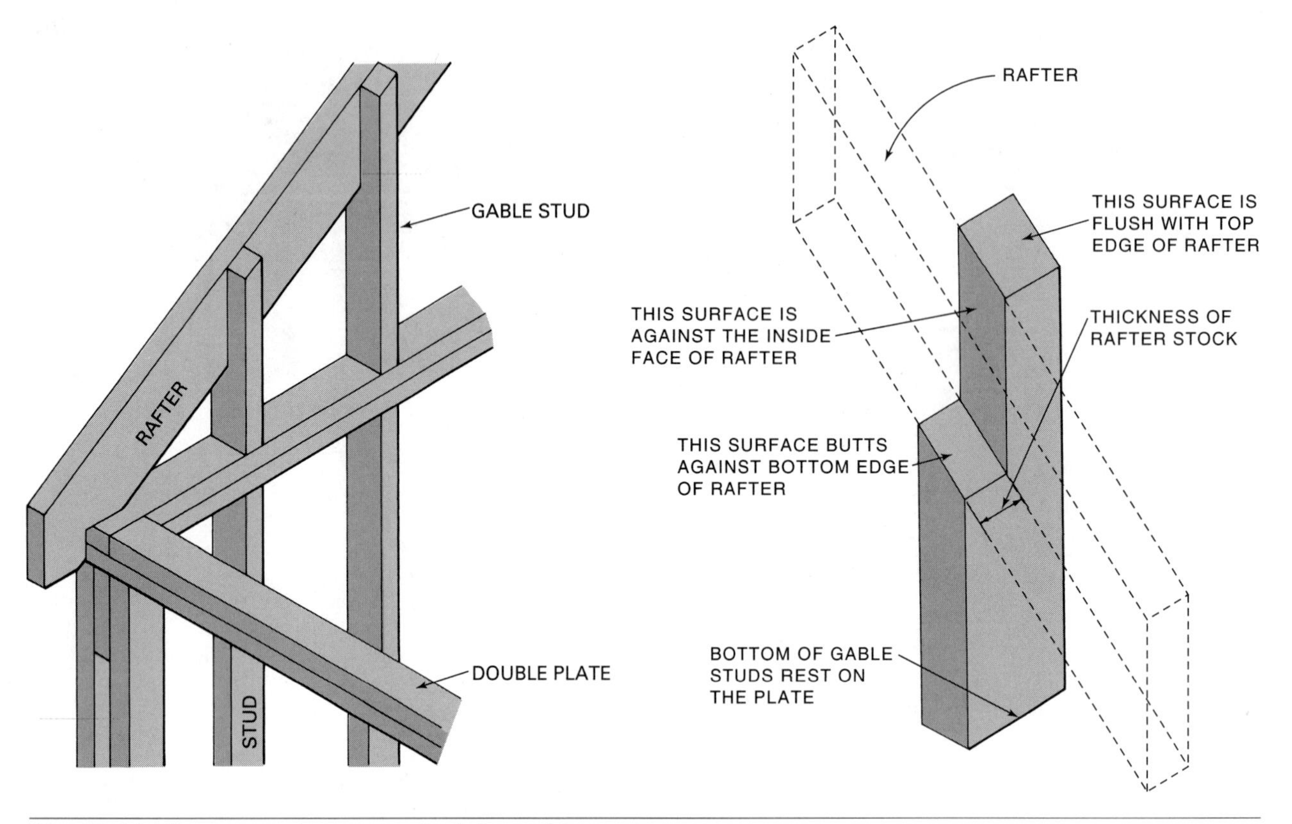

Fig. 42-20 The cut at the top end of the gable stud.

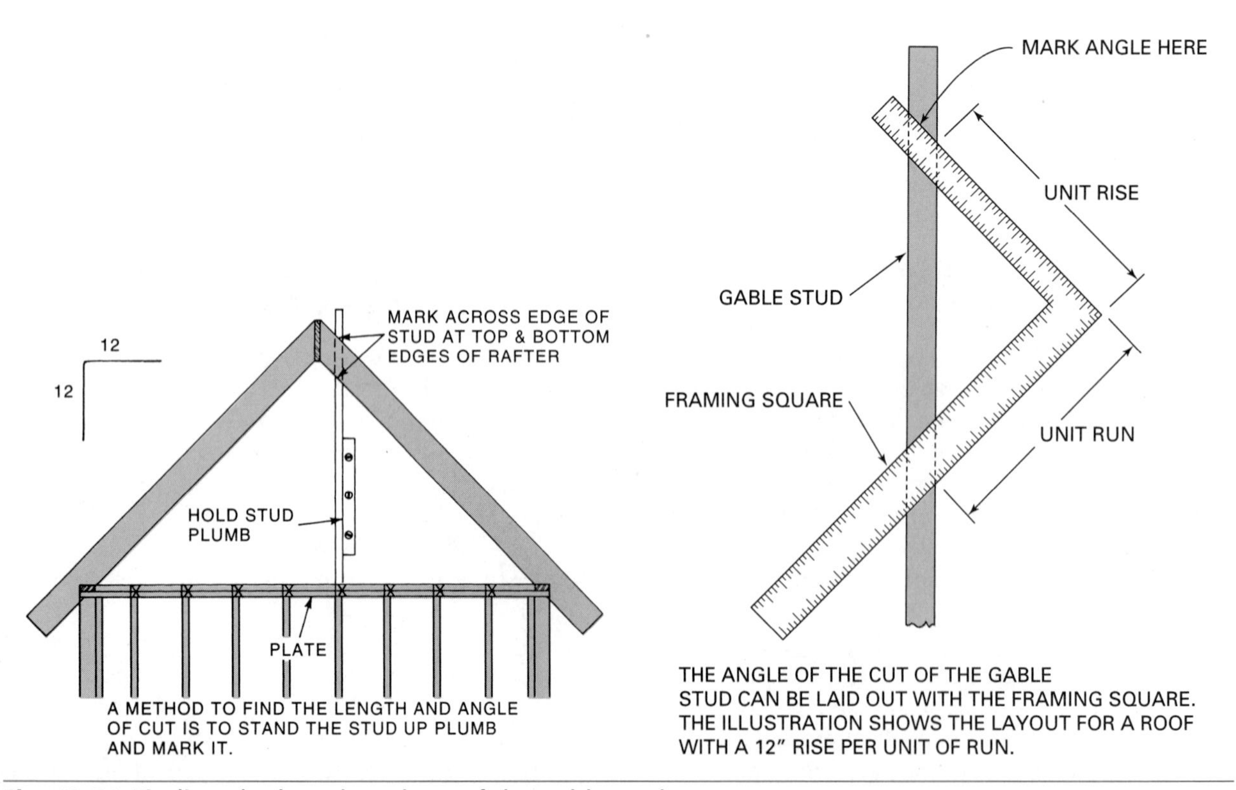

Fig. 42-21 Finding the length and cut of the gable stud.

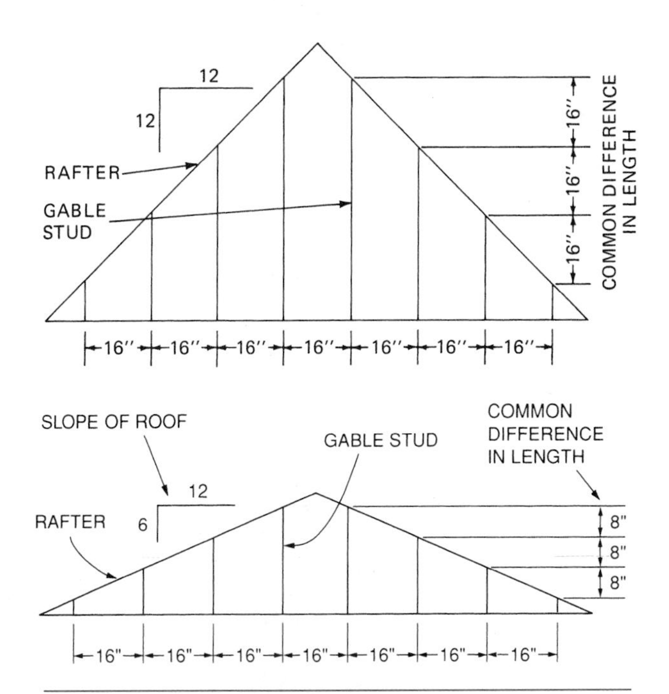

Fig. 42-22 Gable studs spaced equally have a common difference in length.

EXAMPLE Gable studs are spaced 16 inches OC. The roof rises 6 inches per foot of run. Change 16 inches to 1 1/3 feet. Multiply 1 1/3 by 6 to get 8. The common difference in length for gable end studs here is 8 inches.

Framing a Gambrel Roof

The two slopes of a *true* gambrel roof are chords of a semicircle whose diameter is the width of the building (Fig. 42-23). In order to frame a true gambrel roof, the rafter lengths may be calculated using the Pythagorean Theorem or determined from a full scale layout on a large, flat surface, such as the subfloor. From the layout, rafter lengths and angles can be determined.

The calculated method involving the Pythagorean Theorem may at first appear intimidating, but it really is only repetitive steps of the same process. Before calculations can begin, some information must be obtained from the plans or the architect. The measurements needed are the building span and the horizontal distances that the side rafter intersections are from the building's center and building line (Fig. 42-24). It is helpful to note that the total rise of the roof is equal to one-half the span. This happens because the height is also a radius of the semicircle. The distance from the side-rafter intersection to the center of the building is also a radius (Fig. 42-25). Note that there are only two different rafters because they come in pairs. This is because a gambrel roof is symmetrical. Bringing all of this information together shows that right triangles are

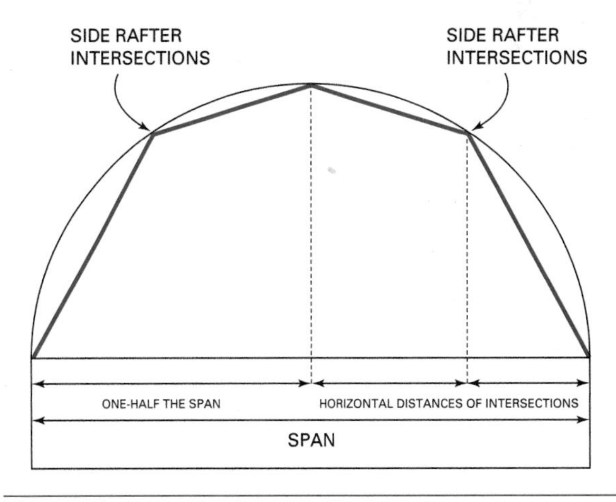

Fig. 42-24 Horizontal distances of intersection and span are needed before gambrel calculations may begin.

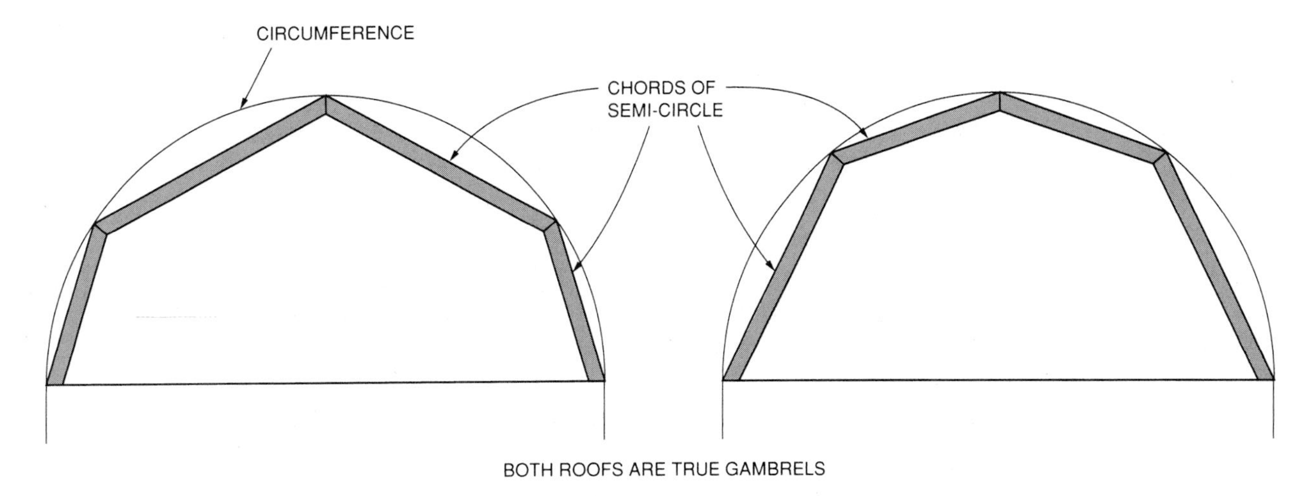

Fig. 42-23 The slopes of a true gambrel roof form the chords of a semicircle.

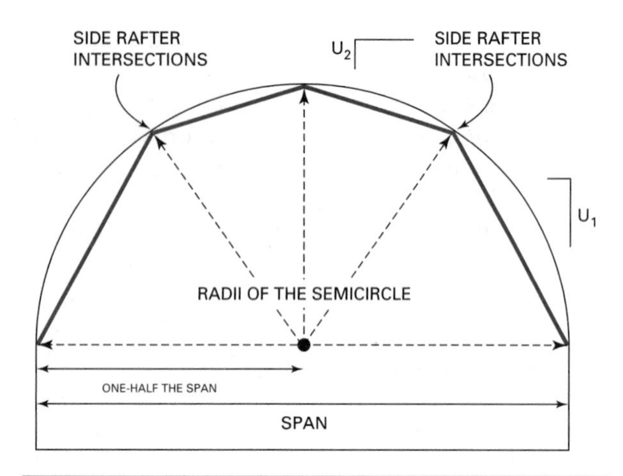

Fig. 42-25 Rafter intersections are located one radius from the center of the span.

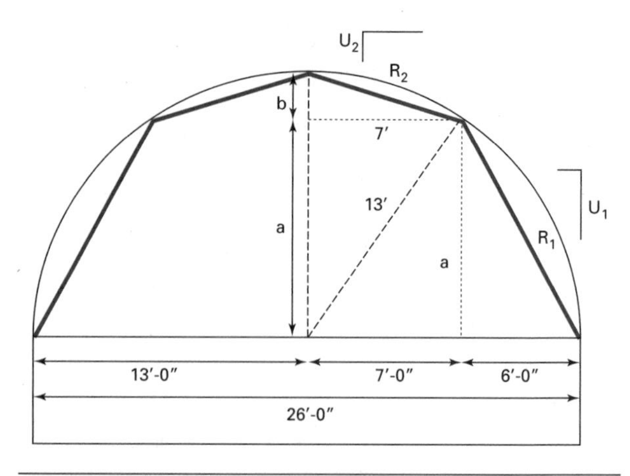

Fig. 42-27 Starting information to a gambrel calculation.

formed and the Pythagorean Theorem can be used to solve for rafter lengths (Fig. 42-26).

EXAMPLE: Find the lengths of the rafters labeled R_1 and R_2 for a gambrel roof system, given that the span is 26 feet and the horizontal distances of the side intersections are 6 and 7 feet (Fig. 42-27).

1. $a^2 + b^2 = c^2$ therefore $a^2 = c^2 - b^2$
2. $a^2 = 13^2 - 7^2 = 169 - 49 = 120$
   $a = \sqrt{120} = 10.954'$
3. $R_1 = \sqrt{a^2 + 6^2} = \sqrt{10.954^2 + 6^2} = \sqrt{120 + 36}$
   $R_1 = \sqrt{156} = 12.490' = 12'\text{-}5\ 7/8''$
4. $b = \text{radius} - a = 13' - 10.954' = 2.046'$
5. $R_2 = \sqrt{b^2 + 7^2} = \sqrt{2.046^2 + 7^2} = \sqrt{4.186 + 49}$
   $R_2 = \sqrt{53.186} = 7.293' = 7'\text{-}3\ 1/2''$

If the rafters are to be framed to a **purlin knee wall** as shown in Figure 42-28, the unit rise for each rafter may be found by dividing the total rise of the rafter, in inches, by its total run. For example in Figure 42-27, U_1 is found by first converting (a) to inches, 10.954 feet times 12 = 131.448 inches, then dividing it by 6 to get 21.908 or 21 15/16 inches unit rise. U_2 is found similarly, (b) divided by 7, 2.046 feet times 12 = 24.552 inches then divide by 7 to get 3.507 or 3 1/2 inches unit rise.

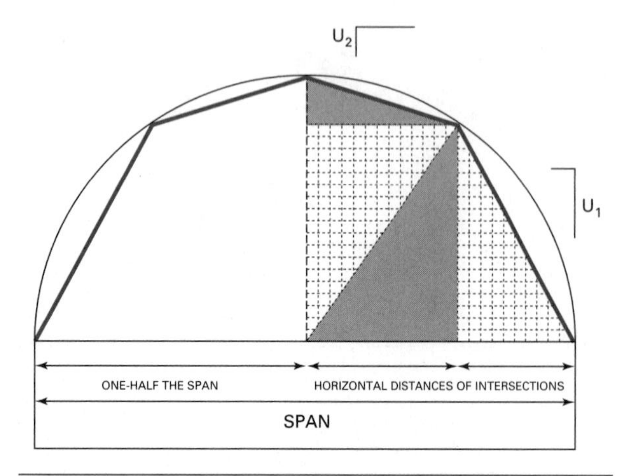

Fig. 42-26 Right triangles are formed by horizontal and vertical lines and radii of semicircle.

Usually the slope of gambrel rafters are given in the drawings. The slopes, then, may not actually be chords of a semicircle (Fig. 42-29). In this case, the rafter lengths and cuts can also be found with a full-size layout. They are usually laid out with a framing square, however, similar to the layout of gable roof rafters.

Laying Out Gambrel Roof Rafters

Determine the construction at the intersection of the two slopes. Usually gambrel roof rafters meet at a continuous member, similar to the ridge, called a *purlin.* Any structural member that runs at right angles to and supports rafters is called a purlin. Sometimes, the rafters of a gambrel roof meet at the top of a *knee wall.* In the lower and steeper slope, the rafters may be sized as wall members rather than roof members if the rafter is within 30 degrees of vertical (Fig. 42-28).

Determine the run and rise of the rafters for both slopes. Find their line length in the same way as for rafters in a gable roof. Lay out plumb lines at the top and bottom of rafters for each slope. Make the seat cut on the lower rafter. Notch all rafters where they intersect at the purlin (Fig. 42-30). Because of the steep slope of the lower roof, the level cut of the seat cannot be made the full width of the plate. At least 2/3 of the width of the rafter stock must remain.

Erecting the Gambrel Roof

If a purlin is used, fasten a rafter to each end of a section of purlin. Raise the assembly. Fasten the lower end of the rafters against ceiling joists and on the top plate of the wall. Brace the section under

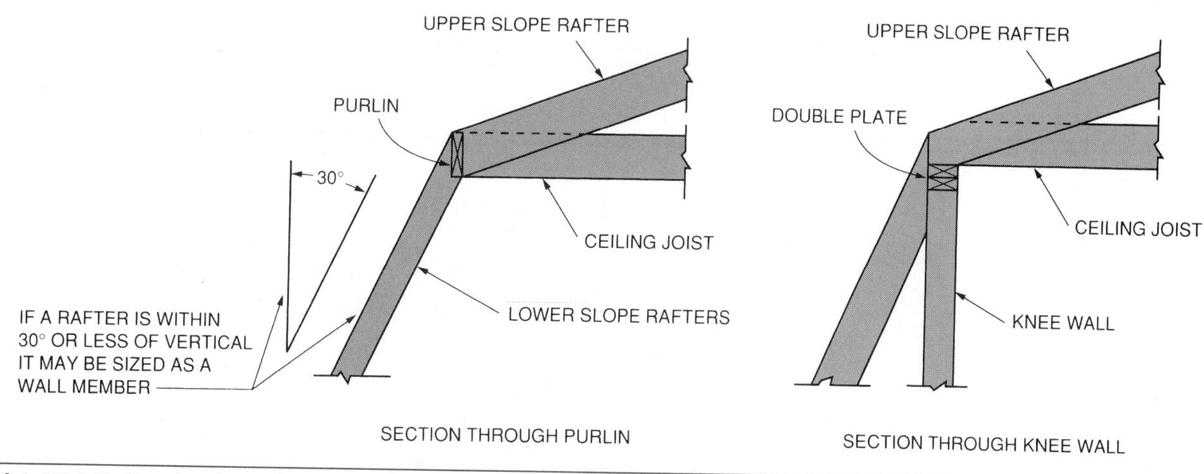

Fig. 42-28 Gambrel roof rafters intersect each other at a purlin or knee wall.

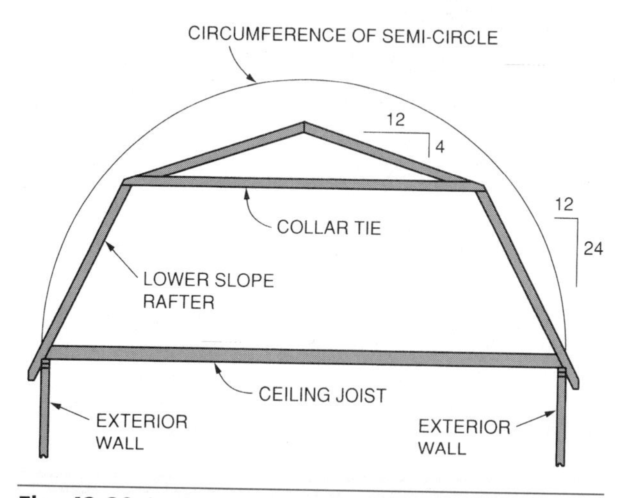

Fig. 42-29 In many cases, the slope of the gambrel rafters is given in the drawings.

each end of the purlin. Continue framing sections until the other end of the building is reached. Place temporary and adequate bracing under the purlin where needed. If a knee wall is used, build, straighten, and brace the wall.

Frame the roof by installing both lower and upper slope rafters opposite each other. This maintains equal stress on the frame. Plumb and brace the roof after a few rafters have been installed. Sight along the ridge and purlin or knee wall for straightness as framing progresses. After all rafters have been erected, install collar ties with their bottom edge resting on the purlin.

The open ends formed by the gambrel roof are framed with studs in the same manner as installing studs in gable ends.

STEP BY STEP

PROCEDURES

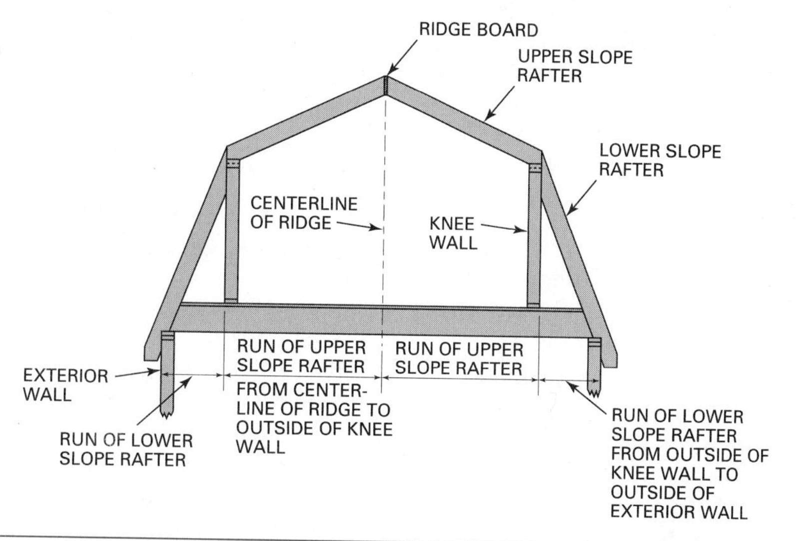

Fig. 42-30 Layout of the gambrel roof rafters for the lower and upper slopes.

STEP BY STEP
PROCEDURES

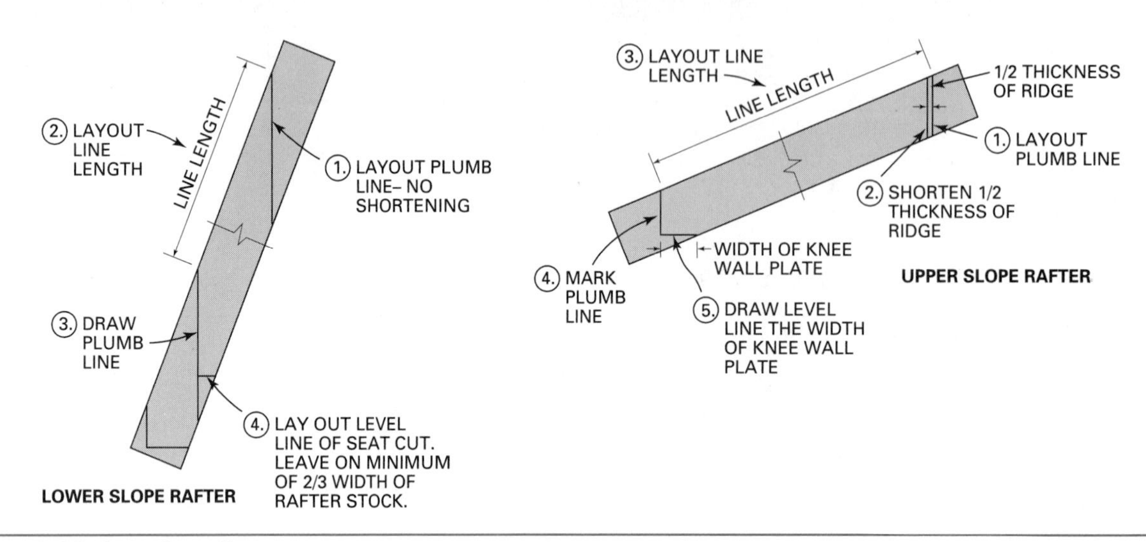

Fig. 42-30 *(continued)*

CHAPTER 43 HIP ROOFS

The hip roof is a little more complicated to frame than the gable roof. Two additional kinds of rafters need to be laid out.

Framing the Hip Roof

To frame the hip roof it is necessary to lay out, not only common rafters and a ridge, but also **hip rafters** and *hip jack* rafters (Fig. 43-1). Hip rafters are required where the slopes of the hip roof meet. The hip jack rafter runs at right angles from the plate to the hip rafter. Several lengths of hip jacks are required.

Run of the Hip Rafter

Because the hip rafter runs at a 45-degree angle from the plate, the amount of horizontal distance it covers (total run) is greater than that of the common rafter. The hip rafter must rise the same amount as the common rafter with the same number of steps (units of run) to meet at the ridge. Therefore, the unit of run for the hip rafter is increased.

The total run of the hip is the diagonal of a square formed by the total run of adjacent common rafters and the exterior walls. When the unit of run of the common rafter is 12 inches, the unit of run of the hip rafter becomes the diagonal of a 12 inch square, which is 16.97 inches or, 17 inches, for all practical purposes (Fig. 43-2).

EXAMPLE If the slope of a hip roof is an 8-inch rise per foot of run, the common rafter is stepped-off by holding the square at 8 and 12. The hip rafter would be laid out by holding the square at 8 and 17. The same number of steps are taken as for the common rafter.

Estimating the Rough Length of the Hip Rafter

The rough length of the hip rafter can be found in the manner previously described for the common rafter. Let the tongue of the square represent the length of common rafter. Let the blade of the square represent its total run. Measure across the square the distance between the two points. A scale of one inch equals one foot gives the length of the hip rafter from plate to ridge (Fig. 43-3).

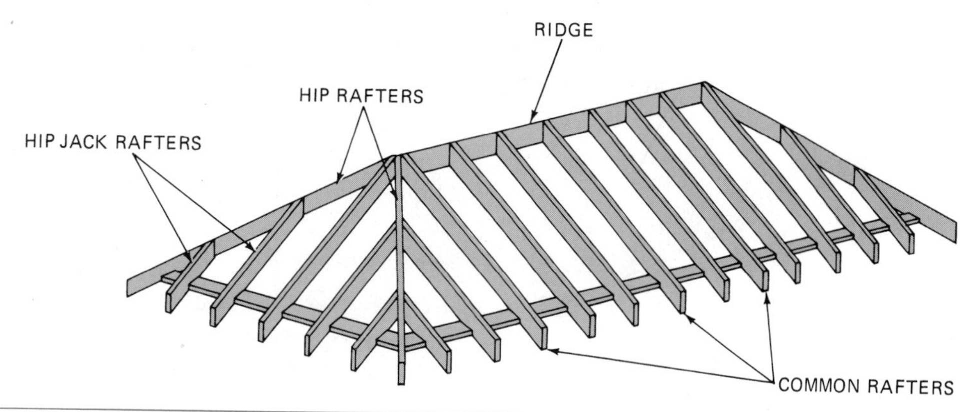

Fig. 43-1 Members of a hip roof frame.

STEP BY STEP
PROCEDURES

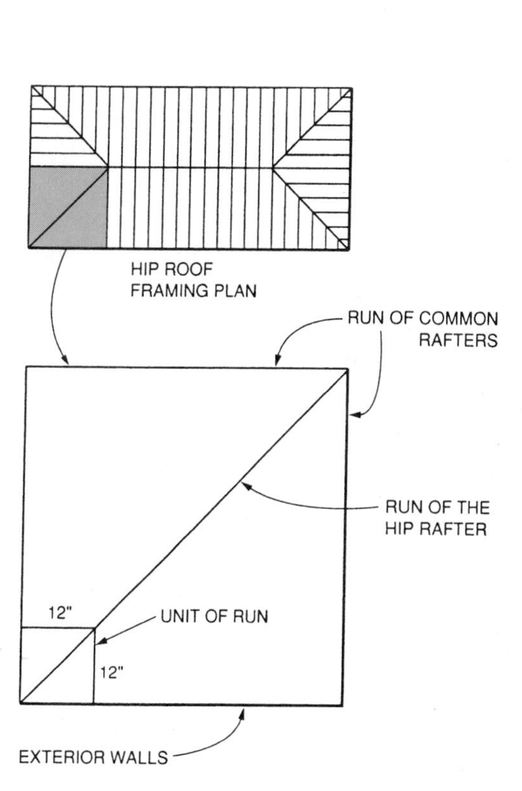

① THE EXTERIOR WALLS OF THE BUILDING AND THE RUN OF COMMON RAFTERS FORM A SQUARE. THE RUN OF THE HIP RAFTER IS THE DIAGONAL OF THE SQUARE.

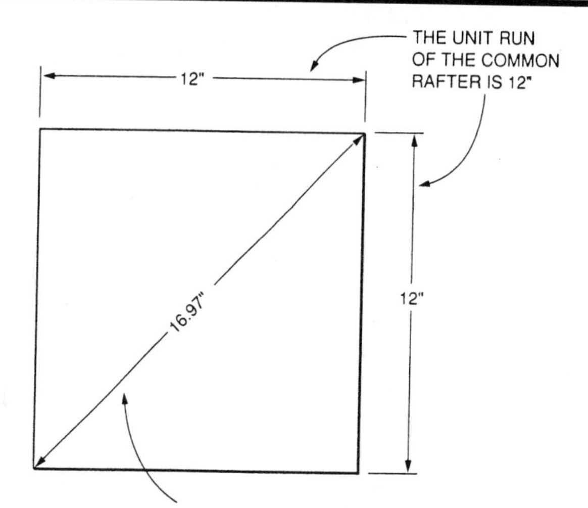

② THE UNIT OF RUN OF THE HIP RAFTER IS THE DIAGONAL OF A 12" SQUARE, WHICH IS 16.97" OR 17" FOR STEPPING OFF PURPOSES. THE DIAGONAL OF ANY SQUARE (RECTANGLE) CAN BE FOUND BY USING THE PYTHAGOREAN THEOREM.

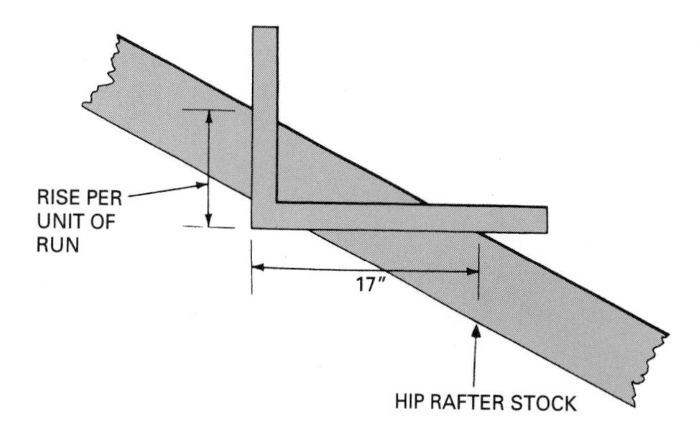

③ WHEN STEPPING OFF, HOLD THE FRAMING SQUARE AT 17" FOR THE RUN OF THE HIP RAFTER.

Fig. 43-2 The unit of run of the hip rafter is 16.97 inches, which is rounded to 17 inches for layout.

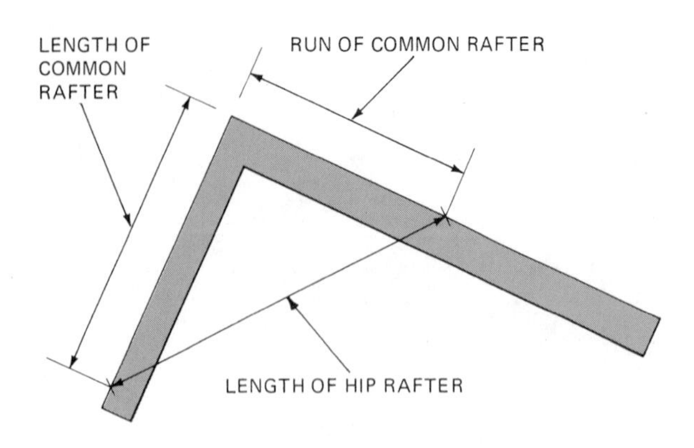

Fig. 43-3 Scaling across the framing square to estimate the rough length of the hip rafter.

EXAMPLE The common rafter length is 15 feet. Its total run is 12 feet. A measurement across these points scales off to 19'-2 1/2". It will be necessary to order 20-foot lengths for the hip rafters unless extra is needed for overhang.

Laying Out the Hip Rafter

The steps to lay out a hip rafter are similar to those for a common rafter. To lay out the hip rafter, make the cut at the ridge. Lay out its length. Make the seat cut. Lay out the length of the tail. Make the tail cut.

Ridge Cut of the Hip. The ridge cut of a hip rafter is a compound angle called a **cheek cut** or *side cut*. A *single cheek* cut or a *double cheek* cut may be made on the hip rafter according to the way it is framed at the ridge (Fig. 43-4).

Shortening the Hip. The rafter should be shortened before the cheek cuts are laid out. The amount of shortening depends on how it is framed at the ridge. When framed against the ridge, the hip rafter is shortened by half the 45-degree thickness of the ridge board. When framed against common rafters, the hip is shortened half the 45-degree thickness of the common rafter (Fig. 43-5).

To lay out the ridge cut, select a straight length of stock for a pattern. Lay it across two sawhorses. Mark a plumb line near the left end. Hold the tongue of the square at the rise and the blade of the square at 17 inches, the unit of run for the hip rafter. Shorten the rafter by measuring at right angles from the plumb line. Lay out another plumb line at that point (Fig. 43-6).

Laying Out Cheek Cuts. To complete the layout of the ridge cut, mark lines for a single or double cheek cut as required. The method of laying out cheek cuts shown in Figure 43-7 gives accurate results regardless of the pitch of the roof.

The two bottom lines of the rafter tables on the framing square can be also be used to find the angle of side cuts of hip and jack rafters. A measurement is given under the inch marks for various slopes. The square is held with the measurement at the tongue and 12 inches at the blade across the top edge of the rafter. The angle is laid out by marking along the blade (Fig. 43-8).

Laying Out the Hip-Rafter Length

The length of the hip rafter is laid out in a manner similar to that used for the common rafter. It may be found by the step-off method, rafter tables, or calculating. Remember always to start any layout for length from the original plumb line before any shortening.

Stepping-Off the Hip. In the step-off method, the number of steps for the hip rafter is the same as for the common rafter in the same roof. The same rise is used, but the unit of run for the hip is 17 not 12. For example, for a roof with a rise of 6 inches per foot of run, the square is held at 6 and 12 for the common rafter, and 6 and 17 for the hip rafter of the same roof (Fig. 43-9).

If the total run of the rafter contains a fractional part, the same fractional part of 17 must be used to determine the extra length of the hip rafter.

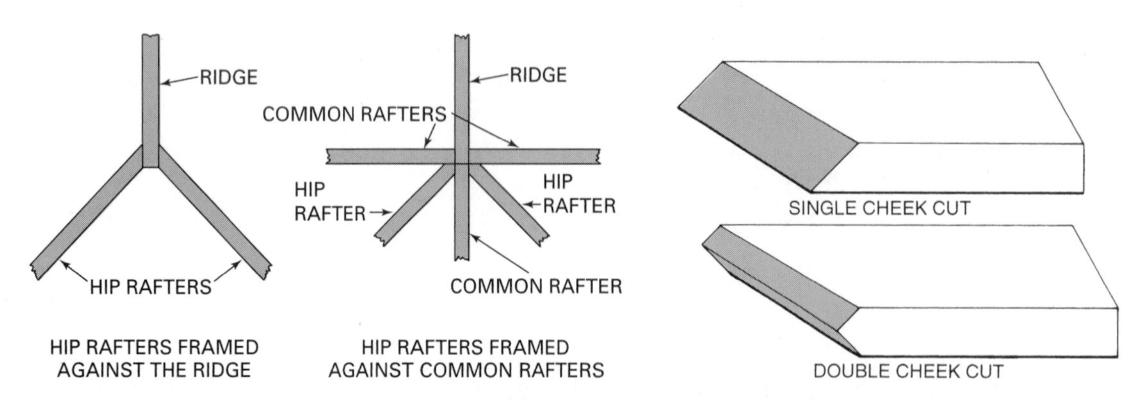

Fig. 43-4 Single or double cheek cuts are used depending on the method of framing the hip rafter at the ridge.

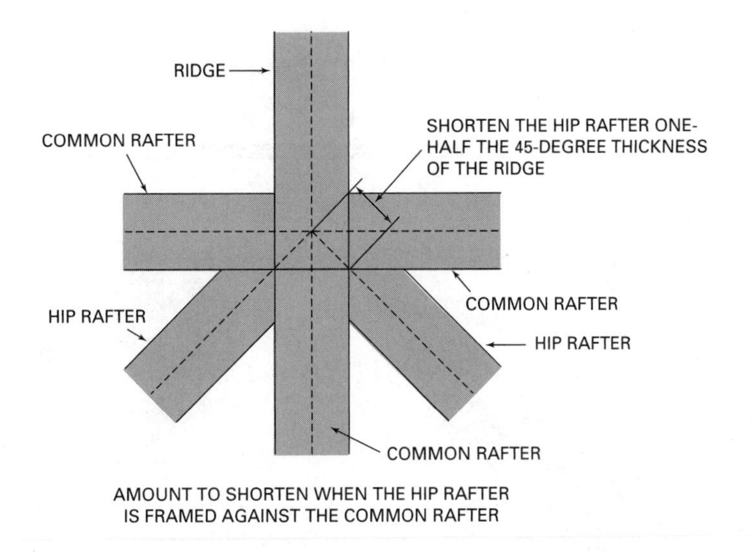

Fig. 43-5 Amount to shorten the hip rafter for both methods of framing.

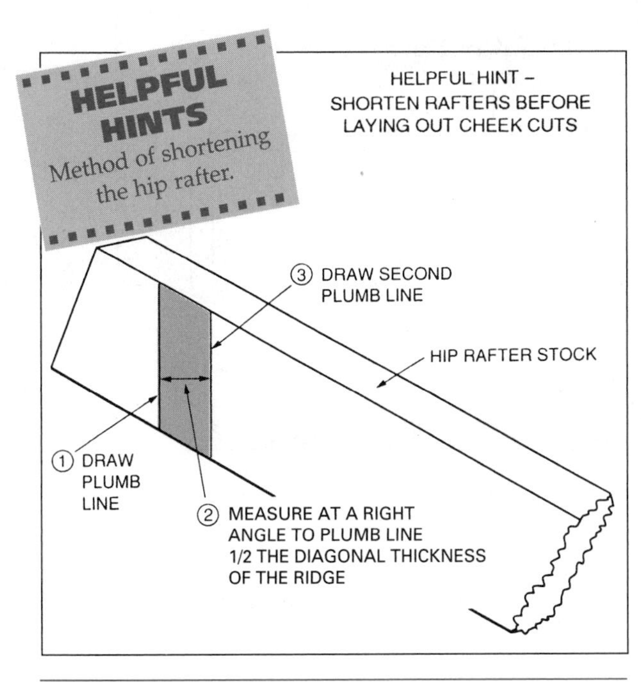

Fig. 43-6 Technique to shorten a hip rafter to allow for the ridge.

EXAMPLE If the total run of the rafter is 15′-6″, step-off 15 times. Use 17 as the unit of run. The 6 inches is 1/2 foot. One-half of 17 is 8 1/2 inches. Measure and mark this dimension along the blade of the square when held at the last step. Move the square until the tongue lines up with the mark. Mark along the tongue for the plumb line of the seat cut, which is the outside edge of the wall plate (Fig. 43-10).

Using the Rafter Tables. Finding the length of the hip using the tables found on the framing square is similar to finding the length of the common rafter. However, the figures from the second line are used. These figures give the length of the hip rafter in inches for every foot of run for the common rafter. It is only necessary to multiply this figure by the number of feet of run. The figure found in the tables may also be found by calculating as described previously for common rafters.

STEP BY STEP

PROCEDURES

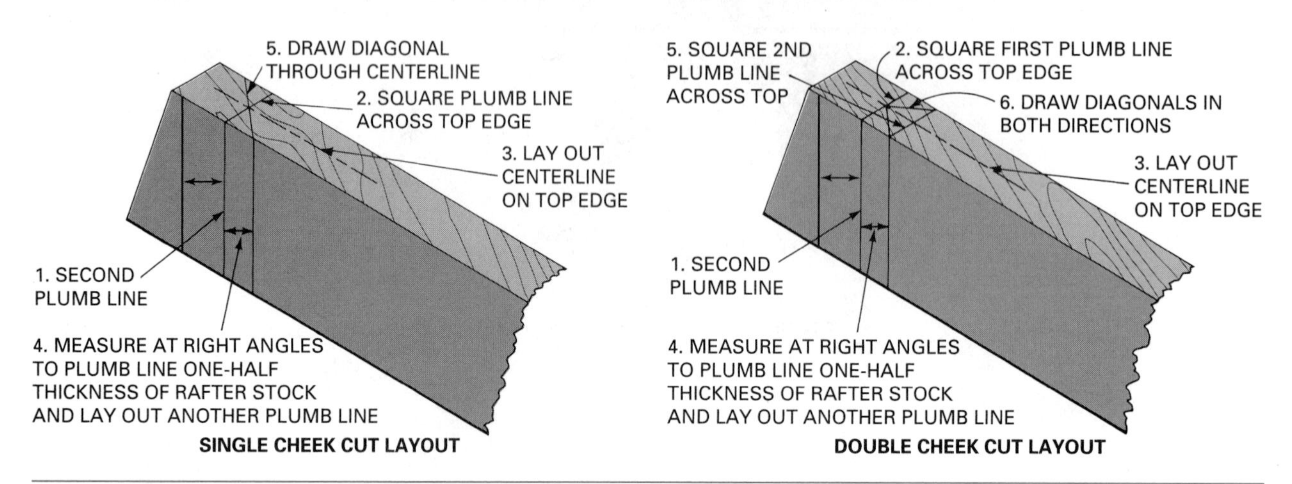

Fig. 43-7 Laying out single and double cheek cuts.

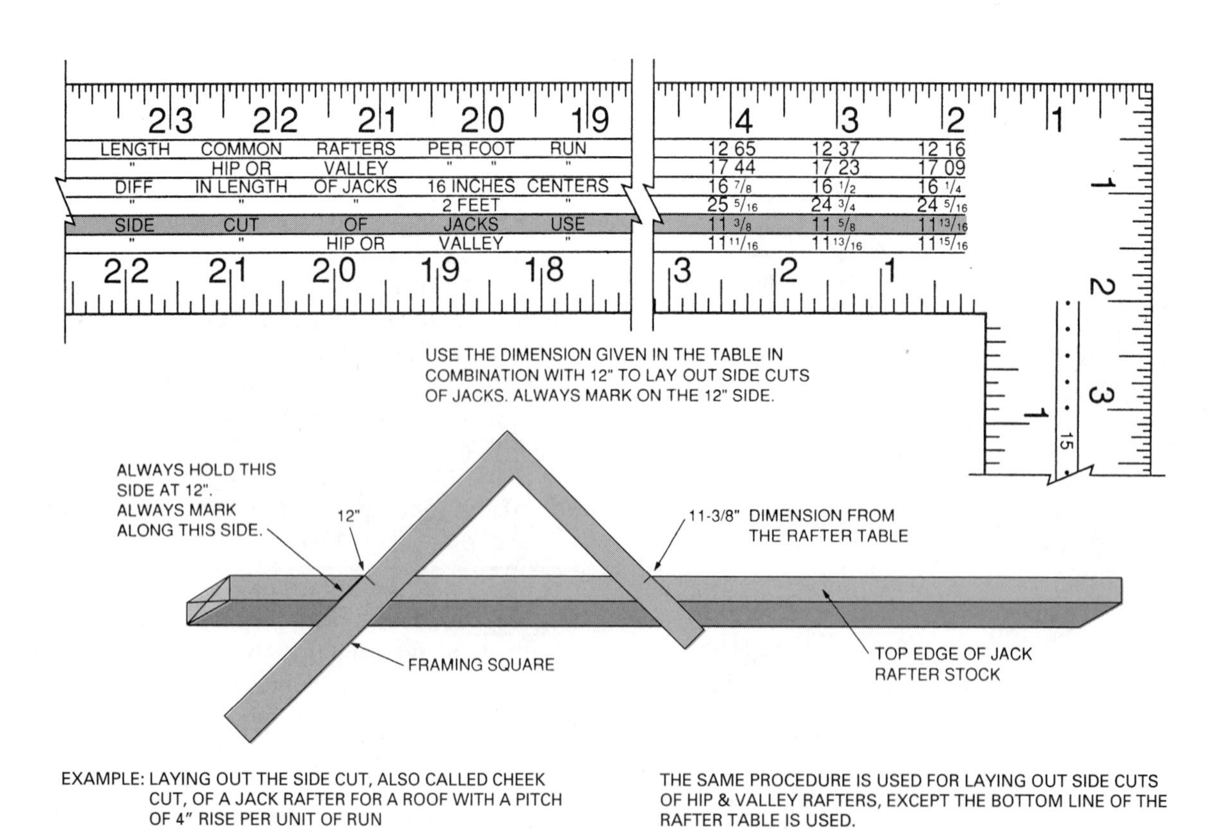

Fig. 43-8 The rafter tables can be used to lay out the side cut of hip and jack rafters.

STEP BY STEP
PROCEDURES

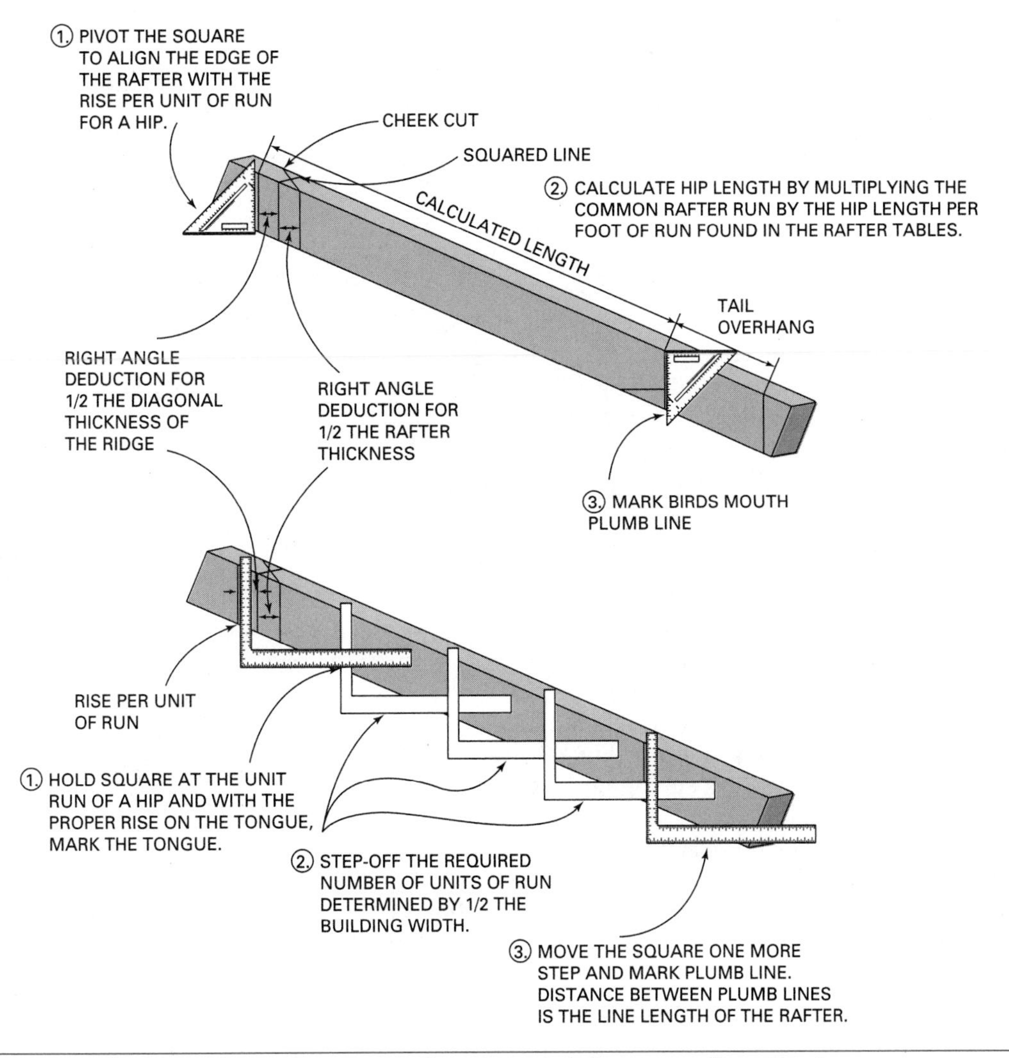

Fig. 43-9 Layout of a hip rafter using a speed square and the step-off method.

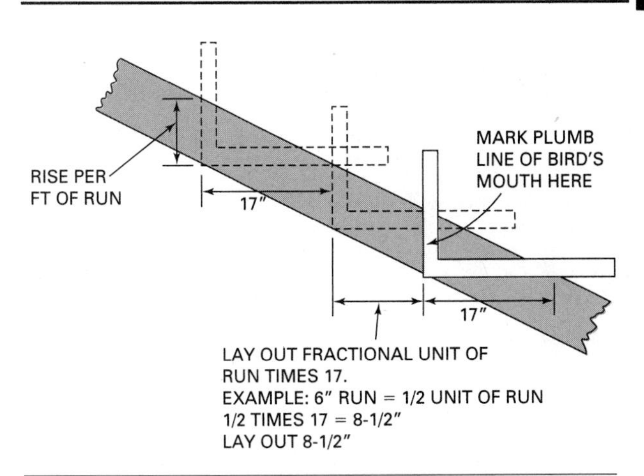

Fig. 43-10 Stepping-off a fractional part of a foot of run on the hip rafter.

EXAMPLE Find the length of a hip rafter for a roof with a rise of 8 inches per unit of run with a total number of units of 14. On the square below the 8-inch mark, it is found that the length of the hip rafter per unit of run is 18.76 inches. This multiplied by 14 equals 262.64 inches. Dividing by 12 gives 21.887 feet. Multiplying 0.887 foot by 12 gives 10.64 inches. Change 0.64 inch to 5/8 inch by multiplying by 16. The total length of the hip rafter is 21′-10 5/8″. It should be noted that using the metric system is much easier.

If the total run of the rafter contains a fractional part of a foot, multiply the figure in the tables by the whole number of feet plus the fractional part changed to a decimal. For instance, if the total run of the rafter is 15′-6″, multiply the figure in the tables by 15 1/2 changed to 15.5.

Lay out the length obtained by either method along the top edge of the hip rafter stock from the original plumb line at the ridge before any shortening.

Laying Out the Seat Cut of the Hip Rafter

Like the common rafter, the seat cut of the hip rafter is a combination of plumb and level cuts. The height above the bird's mouth on a hip rafter is the same as the common rafter. To locate where the seat level line should be, first measure down along the plumb line from the top of the common rafter and mark that same measurement on the hip. The plumb line is usually made square across the bottom edge of the hip rafter. No consideration needs to be given to fitting it around the corner of the wall. When making the level cut of the seat, consideration must be given to **dropping the hip** rafter.

Backing and Dropping the Hip. Because the hip rafter is at the intersection of the side and end slopes of the roof, the top outside corners of the rafter project above the plane of the roof. When the corners of the rafter are beveled flush with the roof, it is called **backing the hip** (Fig. 43-11). When the level line of the seat is cut higher, it is called *dropping* the hip (Fig. 43-12).

Dropping the hip is much easier and is more frequently used. The procedure for determining the amount of drop is shown in Figure 43-12. If you want to back the hip, make the seat cut of the hip

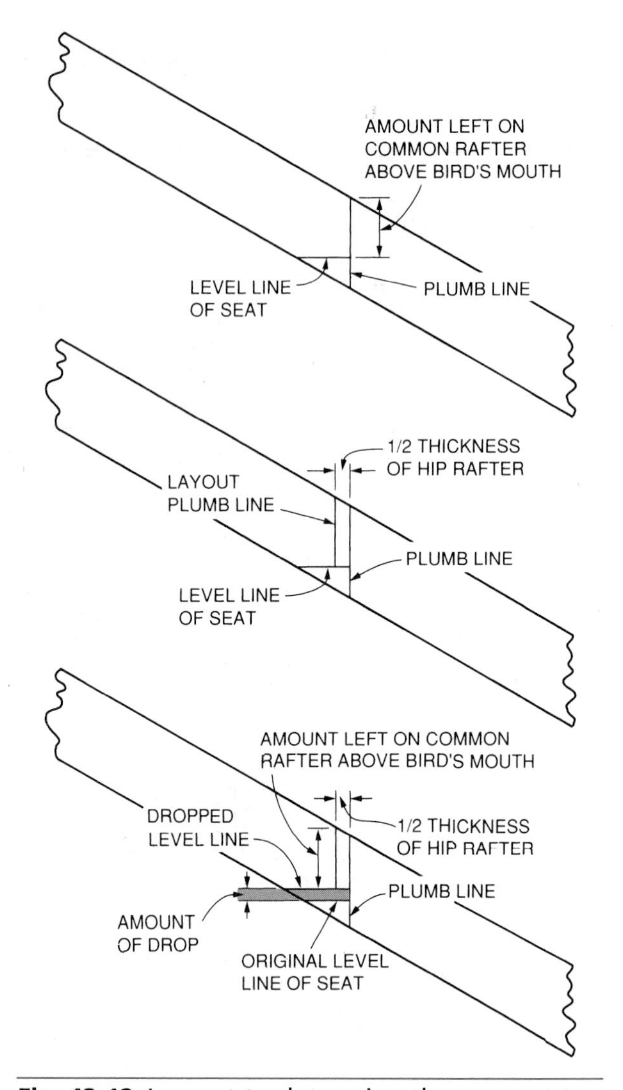

Fig. 43-12 Lay out to determine the amount to drop the hip rafter.

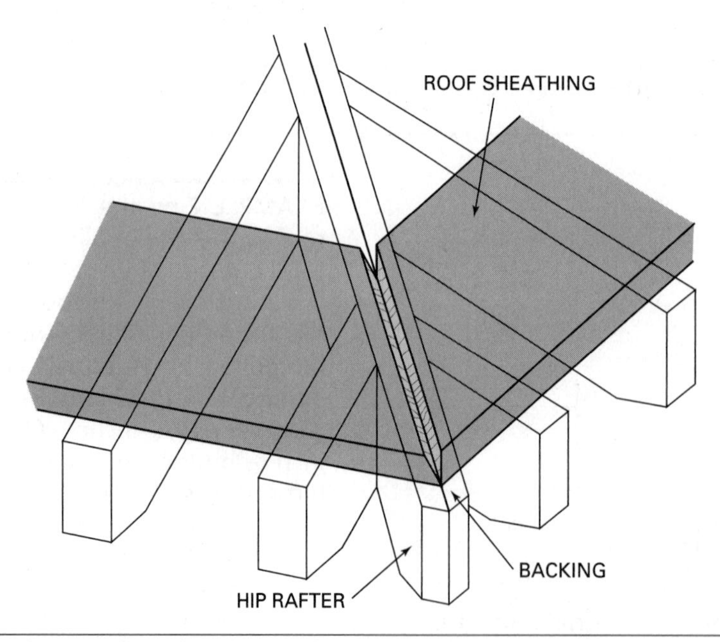

Fig. 43-11 Backing the hip to allow for the roof sheathing.

rafter so the same amount of stock is left on above it as the common rafter. The amount of backing, then, is the same as the amount of drop.

Laying Out the Tail of the Hip Rafter

If you wish to lay out the tail of the hip rafter, multiply its length per unit of run by the number of units of run of the common rafter projection to find its total length. Lay out the length along its top edge from the plumb line at the seat. Draw another plumb line. The tail cut of the hip rafter is usually a double cheek cut (Fig. 43-13).

Laying Out the Hip Jack Rafter

The *hip jack rafter* runs at right angles from the plate to the hip rafter. Its seat cut and overhang are the same as that of the common rafter. It is actually a shortened common rafter. It has the same unit of run, 12 inches (Fig. 43-14).

Finding the Length of Hip Jack Rafters

Hip jack rafters are framed in pairs against the side of the hip rafter. Each set is shorter than the next set by the same amount. This is called the *common difference* (Fig. 43-15). The common difference is found

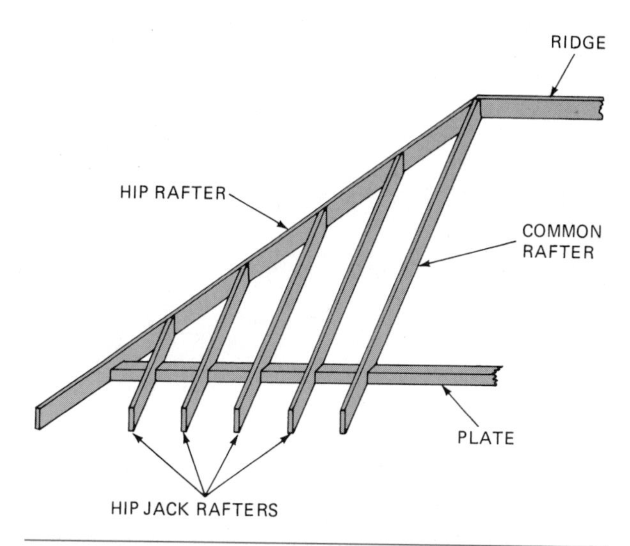

Fig. 43-14 Hip jack rafters are shortened common rafters.

in the rafter tables on the framing square for jacks 16 and 24 inches OC.

Once the length of the longest jack is determined, the length of all others can be found by making each set shorter by the common difference. To find the length of the longest jack, its total run must be known.

Finding the Total Run of the Longest Jack Rafter. The total run of any pair of jack rafters and the outside edge of the wall

STEP BY STEP

PROCEDURES

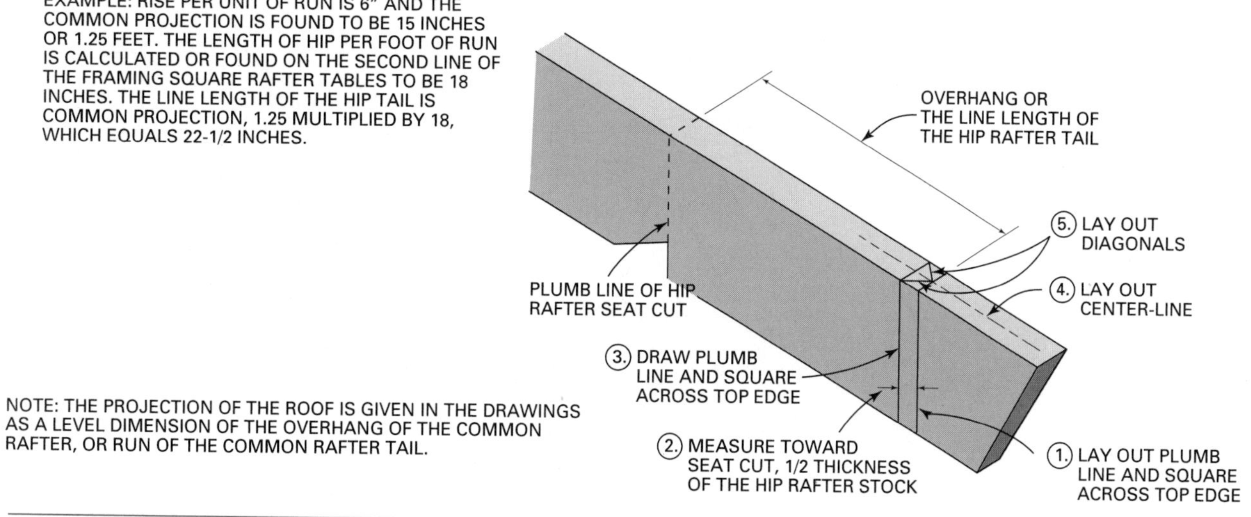

Fig. 43-13 Laying out the tail of the hip rafter.

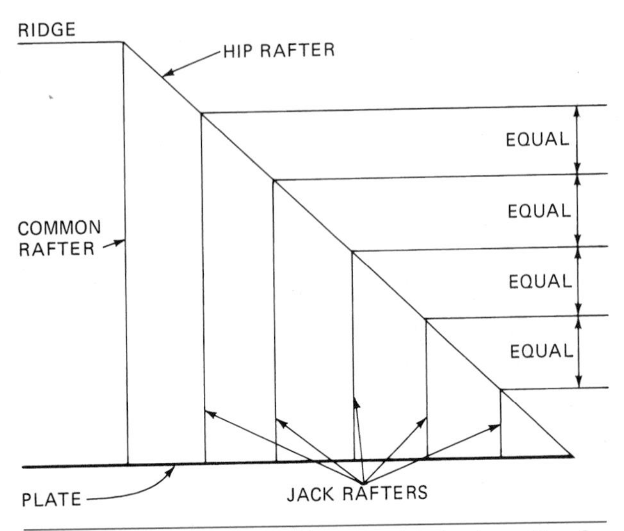

Fig. 43-15 Common difference in the length of hip jack rafters.

plate form a square. Because all sides of a square are equal, the total run of any hip jack rafter is equal to its distance from the corner of the building. At the plate, measure from the corner of the building to the centerline of the longest jack to find its total run (Fig. 43-16).

Find the length of the longest hip jack rafter by stepping-off. Or, use the rafter tables in the same way as for common rafters except that the total run of the jack as found above is used to determine the number of steps. Lay out the seat cut and tail exactly the same as for the common rafter.

Shortening the Hip Jack Rafter. Because the hip jack rafter meets the hip rafter at a 45-degree angle, it must be shortened half the 45-degree angle thickness of the hip rafter stock. Measure the distance at right angles to the original plumb line. Draw another plumb line.

Laying Out the Cheek Cut of the Hip Jack Rafter. The hip jack rafter has a single cheek cut. Square the shortened plumb line across the top of the rafter. Draw an intersecting centerline along the top edge. Measure back at right angles from the last plumb line a distance equal to half the jack rafter stock. Lay out another plumb line. On the top edge, draw a diagonal from the last plumb line through the intersection of the center and the second plumb line (Fig. 43-17). The direction of the diagonal depends on which side of the hip the jack rafter is framed.

Determine the common difference of hip jack rafters from the rafter tables under the inch mark that coincides with the slope of the roof being framed. The common difference may also be calculated. For rafters spaced 24 inches OC, the common difference is equal to the common rafter length for two units of run. For rafters spaced 16 inches OC, the common difference is equal to the common rafter length for 1 1/3 units of run (Fig. 43-18).

Once the common difference is determined, measure the distance successively along the top edge of the pattern for the longest jack rafter. This pattern is then used to cut all the jack rafters necessary to frame the roof. Cut *pairs* with cheek cuts going in opposite directions.

Finding the Length of the Hip Roof Ridge

The total run of the hip rafter is the diagonal of a square whose sides are equal to the run of the common rafter. Because the run of the common rafter is equal to half the width of the building, the *line length* of the ridge is found by subtracting the width of the building from the length of the building (Fig. 43-19).

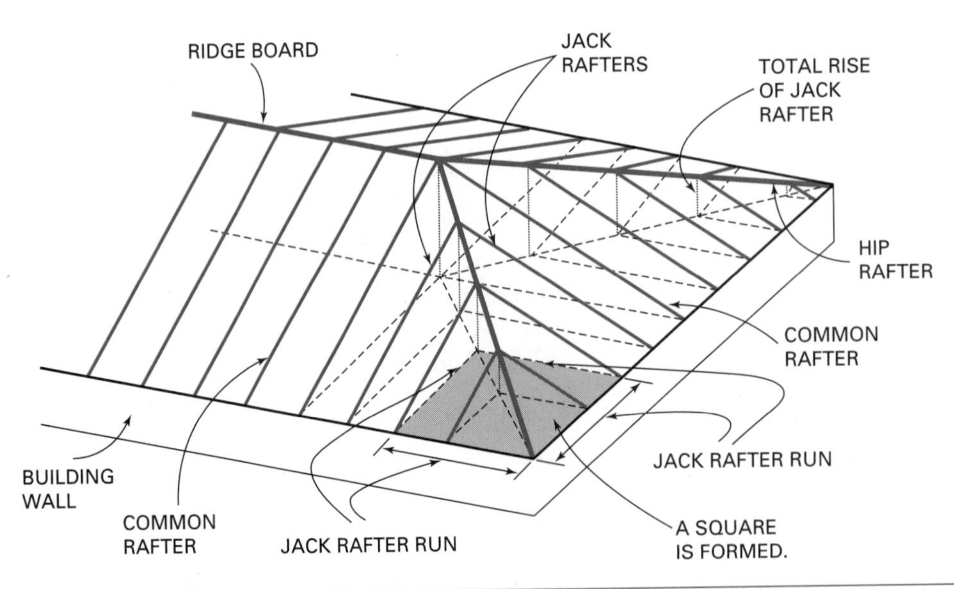

Fig. 43-16 The total run of any hip jack rafter is equal to its distance from the corner of the building.

STEP BY STEP
PROCEDURES

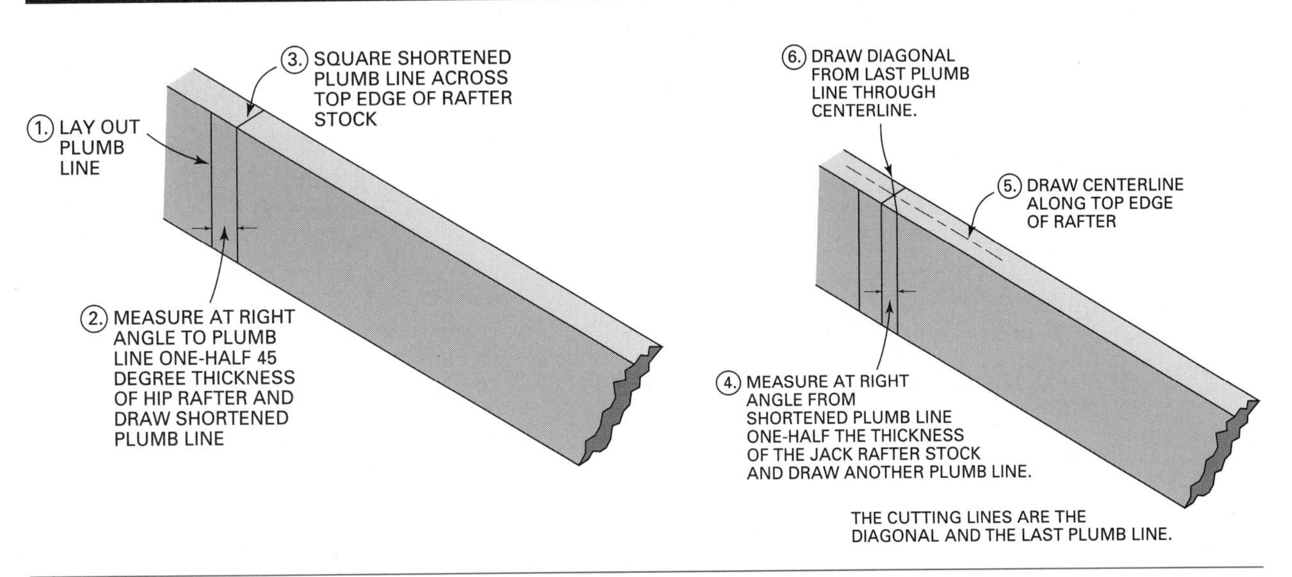

Fig. 43-17 Steps in the layout of the cheek cut of the hip jack rafter.

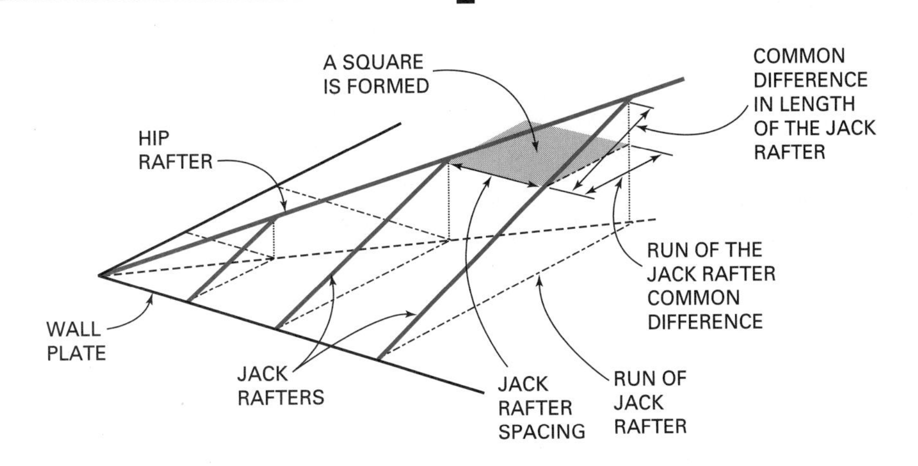

Fig. 43-18 Determining the common difference in hip jack rafters by calculating.

However, the *actual length* of the hip roof ridge must be cut longer than the line length. The amount of increase depends on the construction. If the hip rafters are framed against the ridge, increase the length at *each* end by half the 45-degree thickness of the hip rafter plus half the ridge board. If the hip rafters are framed against the common rafters, the line length of the ridge is increased at *each* end half the thickness of the common rafter stock (Fig. 43-20).

Raising the Hip Roof

Transfer the common rafter layout at the wall plate to the ridgeboard in a manner similar to the layout of the gable roof ridge. Erect the common

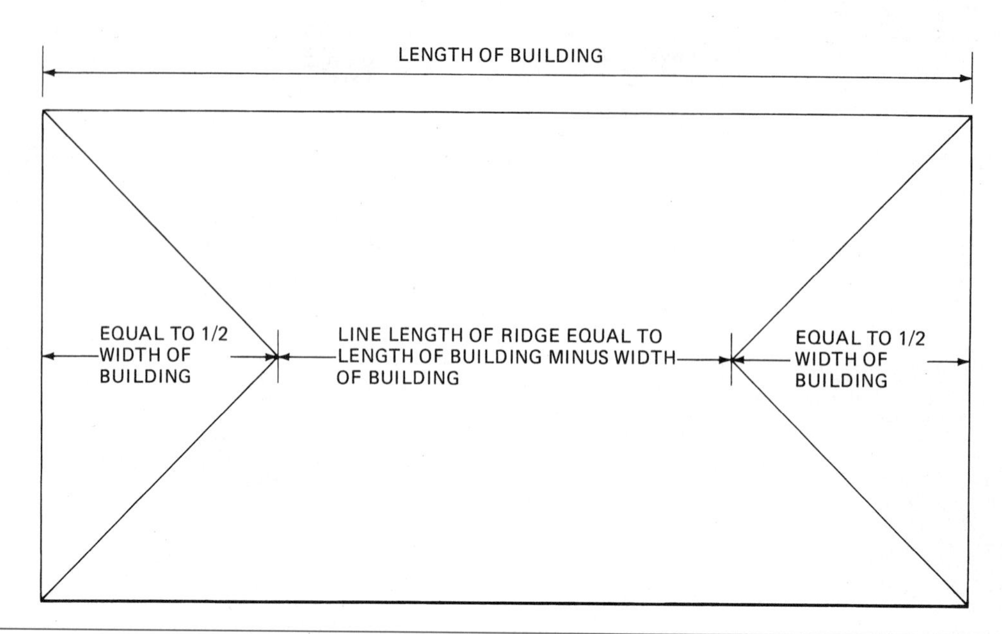

Fig. 43-19 Determining the line length of the hip roof ridge.

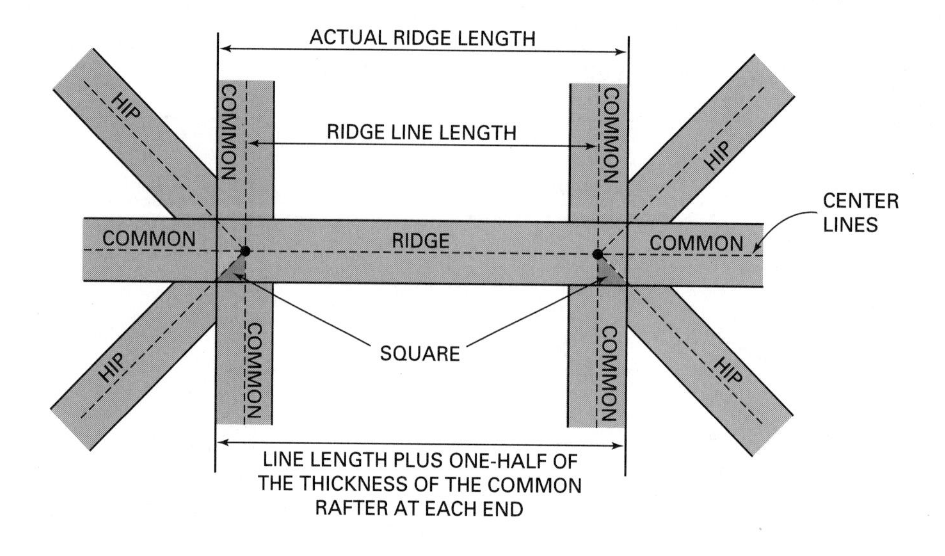

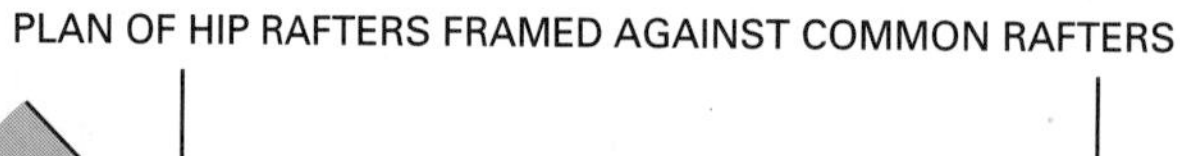

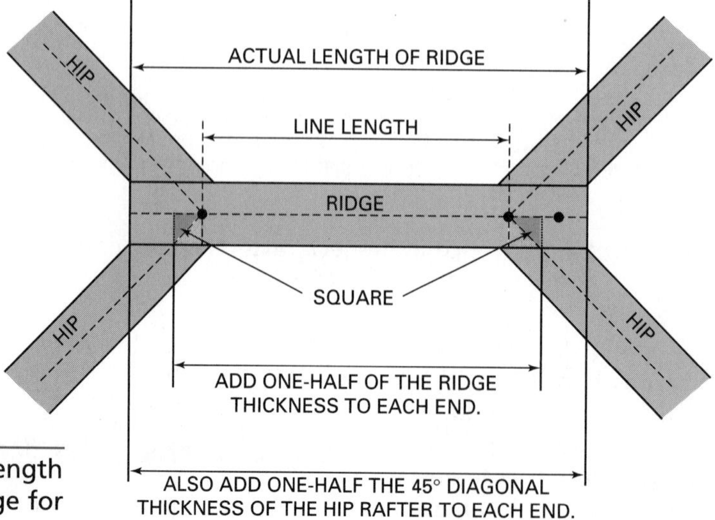

Fig. 43-20 Actual length of the hip roof ridge for both methods of framing the hip rafter.

roof section in the same manner as described previously. If the hip rafters are framed against the ridge, install them next. If they are framed against the common rafters, install the commons against the end of the ridge. Then install the hip rafters.

Fasten jack rafters to the plate and to hip rafters in pairs. As each pair of jacks is fastened, sight the hip rafter along its length. Keep it straight. Any *bow* in the hip is straightened by driving the jack a little tighter against the bowed-out side of the hip as the roof is framed (Fig. 43-21).

STEP BY STEP
PROCEDURES

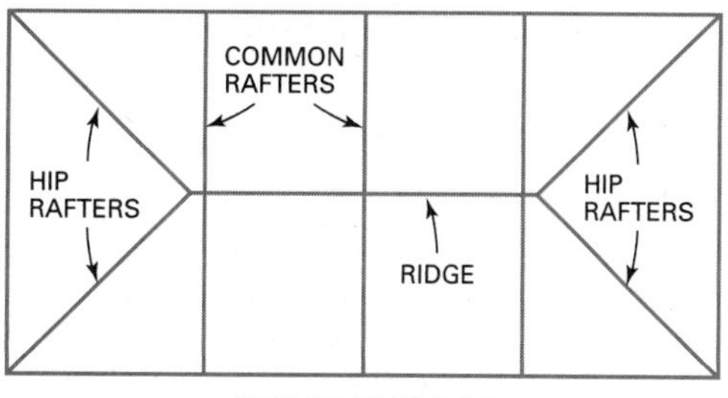

1. ERECT THE RIDGE WITH ONLY AS MANY COMMON RAFTERS AS NEEDED. THEN INSTALL THE HIP RAFTERS. THE HIP RAFTERS EFFECTIVELY BRACE THE ASSEMBLY.

NOTE: THIS PROCEDURE SHOWS THE HIPS FRAMED TO THE RIDGE. IF HIPS ARE FRAMED TO THE COMMON RAFTERS, INSTALL THE COMMONS TO THE END OF THE RIDGE BEFORE THE HIPS.

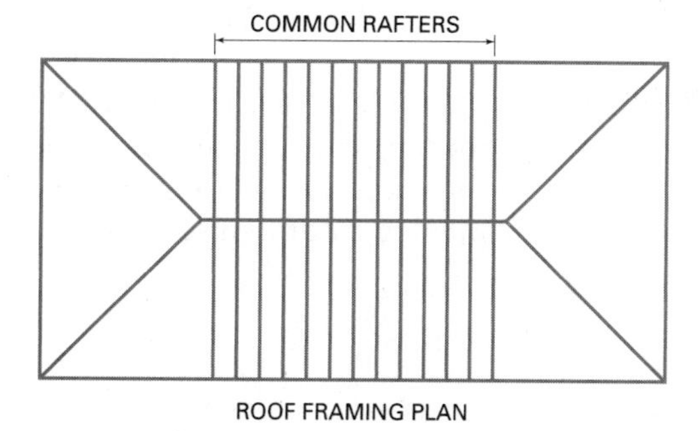

2. INSTALL THE REMAINING COMMON RAFTERS IN PAIRS OPPOSING EACH OTHER. SIGHT THE TOP EDGE OF THE RIDGE FOR STRAIGHTNESS AS FRAMING PROGRESSES.

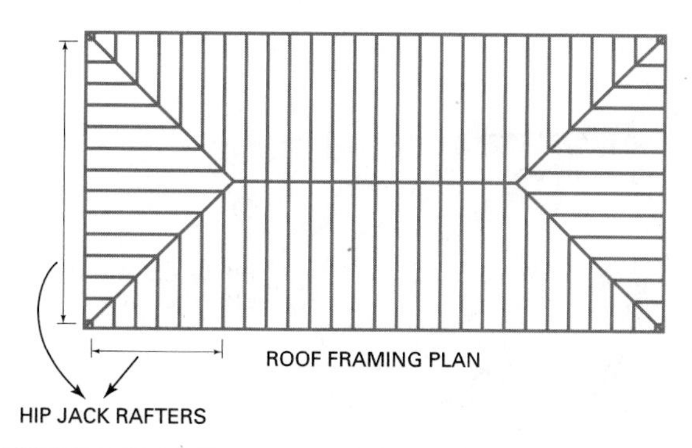

3. INSTALL THE HIP JACK RAFTERS IN PAIRS OPPOSING EACH OTHER. SIGHT THE HIP FOR STRAIGHTNESS AS JACKS ARE INSTALLED. IT IS BEST TO INSTALL A PAIR OF JACKS ABOUT HALF-WAY UP THE HIP TO STRAIGHTEN IT. THE HIP CAN BE STRAIGHTENED BY DRIVING THE JACK, ON THE CROWN SIDE OF BOW, DOWN TIGHTER.

Fig. 43-21 Steps in the erection of the hip roof frame.

CHAPTER 44 THE INTERSECTING ROOF

Buildings of irregular shape, such as L-, H-, or T-shaped, require a roof for each section. These roofs meet at an intersection, where **valleys** are formed. They are called *intersecting roofs*. The roof may be a gable, a hip, or a combination of types.

The Intersecting Roof

The intersecting roof requires the layout and installation of several kinds of rafters not previously described.

- **Valley rafters** form the intersection of the slopes of two roofs. If the heights of both roofs are different, two kinds of valley rafters are required.
- The *supporting valley rafter* runs from the plate to the ridge of the main roof.
- The *shortened valley rafter* runs from the plate to the supporting valley rafter.
- **Valley jack rafters** run from the ridge to the valley rafter.
- The **valley cripple jack rafter** runs between the supporting and shortened valley rafter.
- The *hip-valley cripple jack rafter* runs between a hip rafter and a valley rafter (Fig. 44-1).

Confusion concerning the layout of so many different kinds of rafters can be eliminated by remembering the following.

- The length of any kind of rafter can be found by its run.
- The amount of shortening is always measured at right angles to the plumb cut.
- Hip and valley rafters are similar. Common and cripple rafters are similar.
- The method previously described for laying out cheek cuts on rafters for any pitch.

Layout of Supporting Valley Rafters

The layout of valley rafters is similar to that of hip rafters. The unit of run for both is 17 inches. The total run of the supporting valley is the run of the common rafter of the main roof, called the *major span.* Its line length is found either by the step-off method or by the rafter tables in the same manner as for hip rafters. A single cheek cut is made at the ridge. The rafter is shortened by half the 45-degree thickness of the ridge board (Fig. 44-2).

Seat Cut of the Valley Rafter

The seat cut of the valley rafter must be extended to clear the inside corner of the wall plate. Measure down from the top edge of the rafter, along the plumb line at the plate, a distance equal to the amount left on the common rafter. Draw a level line at this point to make a bird's mouth. From the plumb line, measure back at a right angle towards the tail, a distance equal to one-half the valley rafter thickness. Draw a new plumb line and extend the previously drawn seat cut line (Fig. 44-3).

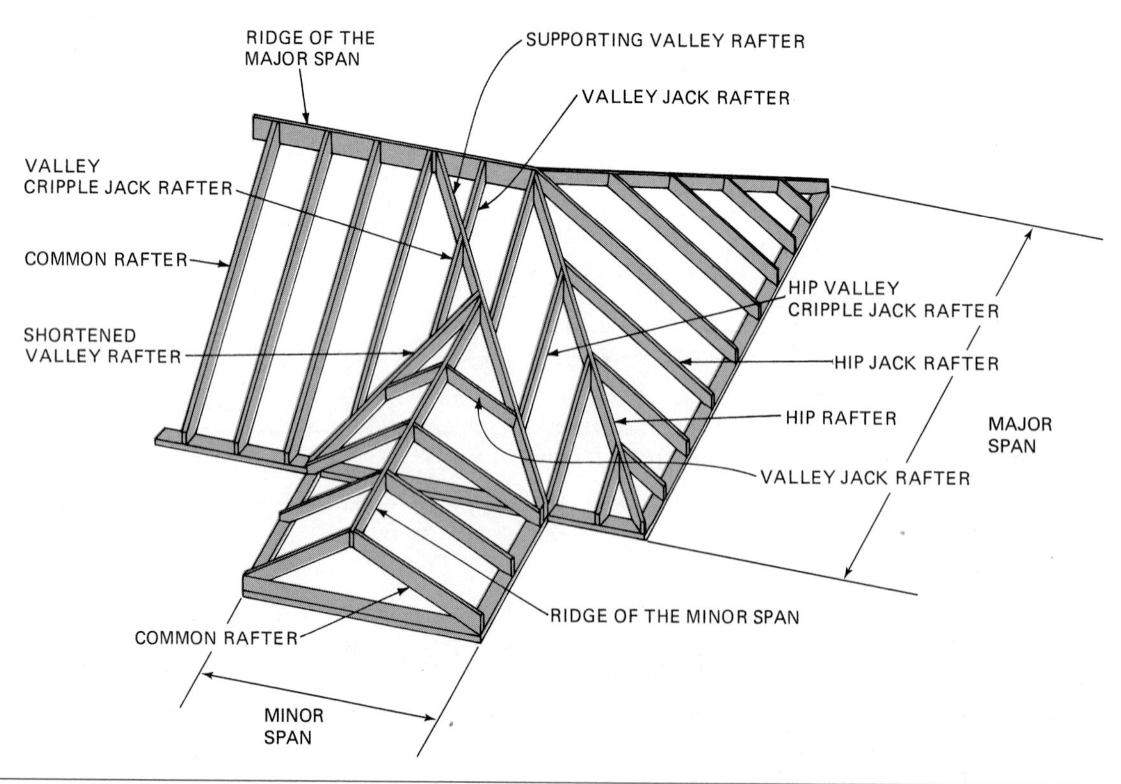

Fig. 44-1 Members of the intersecting roof frame.

STEP BY STEP

PROCEDURES

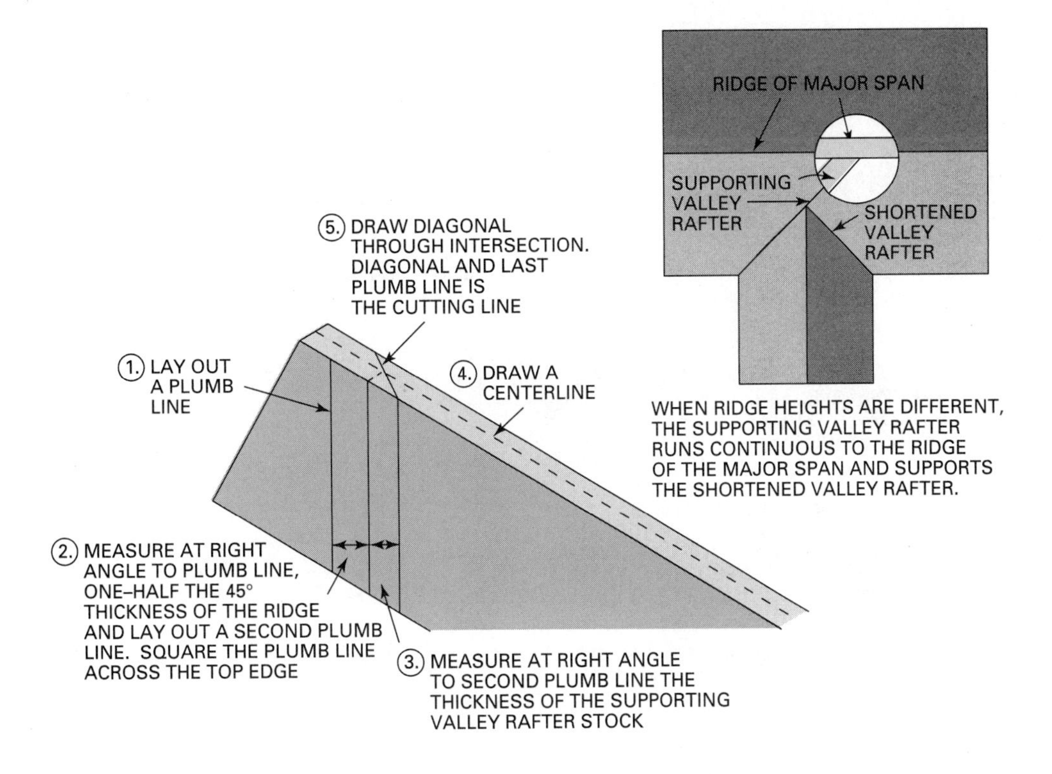

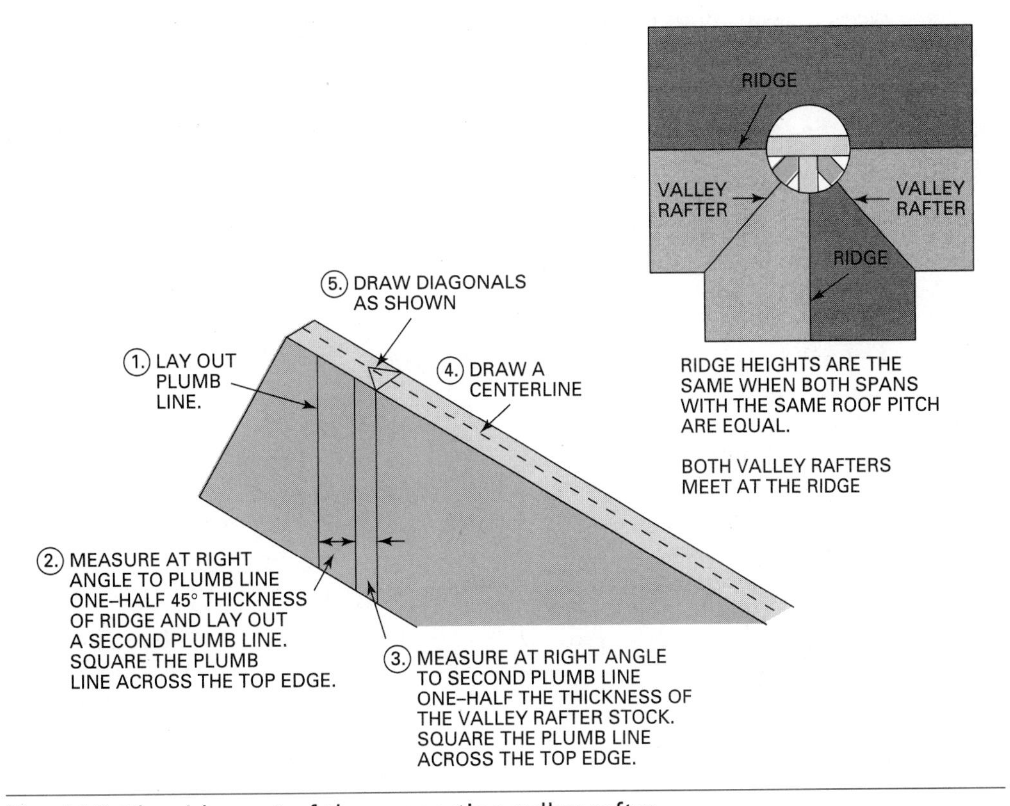

Fig. 44-2 The ridge cut of the supporting valley rafter.

STEP BY STEP
PROCEDURES

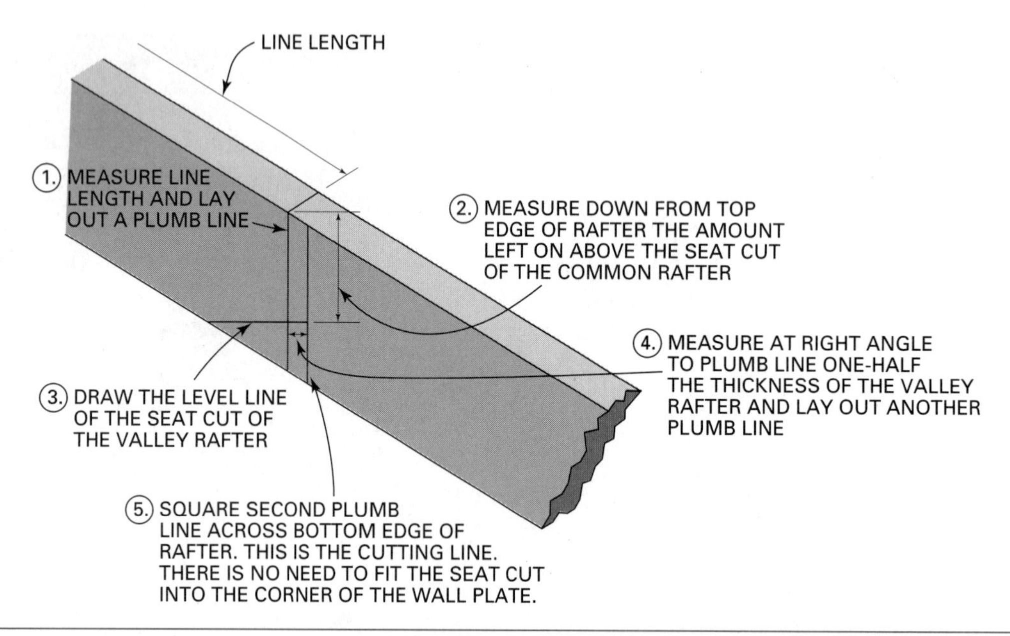

Fig. 44-3 The layout of the valley rafter seat cut.

The valley rafter is not dropped like the hip. Its top corners do not project above the slope of the roof. However, one edge of the supporting valley, from the shortened valley rafter to the ridge, needs backing.

Overhang and Tail Cut of the Valley Rafter

The length of the valley rafter overhang is found in the same manner as that used for the hip rafter. It may be stepped-off as shown in Fig. 43-10 or calculated as shown in Fig. 43-13. Step-off the same number of times, from the plumb line of the seat cut, as stepped off for the common rafter. However, use a unit run of 17 instead of 12.

The tail cut of the valley rafter is a double cheek cut. It is similar to that of the hip rafter, but angles inward instead of outward. From the tail plumb line, measure at right angles to the plumb line toward the tail, one-half the thickness of the rafter stock. Lay out another plumb line. Square both plumb lines across the top of the rafter. Draw diagonals from the center (Fig. 44-4).

Layout of the Shortened Valley

The length of the shortened valley is found by using the run of the common rafter of the smaller roof,

STEP BY STEP
PROCEDURES

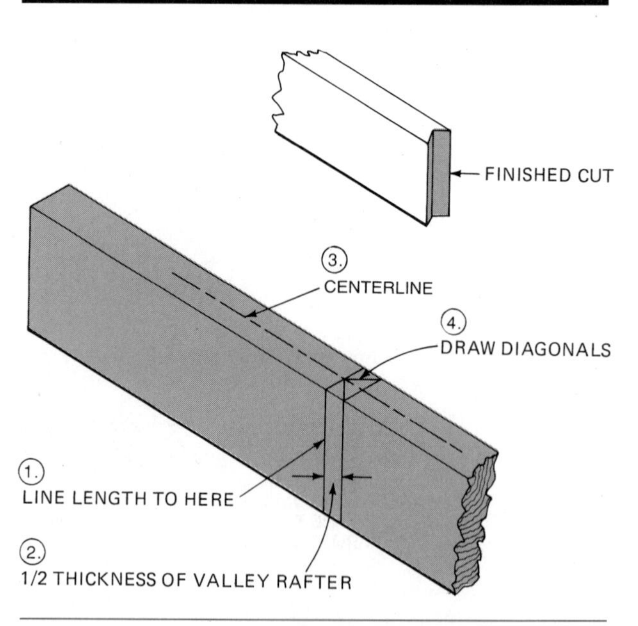

Fig. 44-4 The layout of the valley rafter tail cut.

called the *minor span*. Its seat and tail are laid out the same as for the supporting valley.

However, the plumb cut at the top end is different from that of the supporting valley. Because the two valley rafters meet at right angles, the cheek cut of the shortened valley is a square cut. To shorten this rafter, measure back, at right angles from the plumb line at the top end, a distance that is half the thickness of the supporting valley rafter stock. Lay out another plumb line (Fig. 44-5).

Layout of the Valley Jack

The length of the valley jack can be found if its total run is known. The total run of any valley jack rafter is equal to the run of the common rafter minus the horizontal distance that it is located from the corner of the building (Fig. 44-6). Remember that all jack rafters are shortened common rafters and that the unit of run is 12.

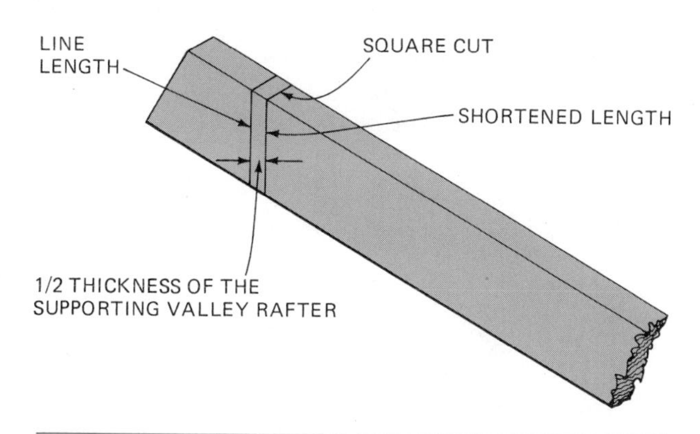

Fig. 44-5 Plumb cut layout on the top end of the shortened valley rafter.

The ridge cut of the valley jack is the same as a common rafter. It is shortened in the same way. The cheek cut against the valley rafter is a single cheek cut. The valley jack is shortened at this end by half the 45-degree angle thickness of the valley rafter stock. The layout of all other valley jack rafters is made by making each set shorter by the common difference found in the rafter tables.

Layout of the Valley Cripple Jack

As stated before, the length of any rafter can be found if its total run is known. The run of the valley cripple jack is always twice its horizontal distance from the intersection of the valley rafters (Fig. 44-7). Use the common rafter tables or step-off in a manner similar to that used for common rafters to find its length. Shorten each end by half the 45-degree thickness of the valley rafter stock (Fig. 44-8). Noting that they angle in opposite directions, make a single cheek cut.

Hip-Valley Cripple Jack Rafter Layout

Determine the length of the rafter by first finding its total run. The run of a hip-valley cripple jack rafter is equal to the plate line distance between the seat cuts of the hip and valley rafters (Fig. 44-9). Lay out the length using the cut of the common rafter. All hip-valley cripple jacks cut between the same hip and valley rafters are the same length. On each end, shorten by half the 45-degree thickness of the hip and valley rafters. Make single cheek cuts (Fig. 44-10).

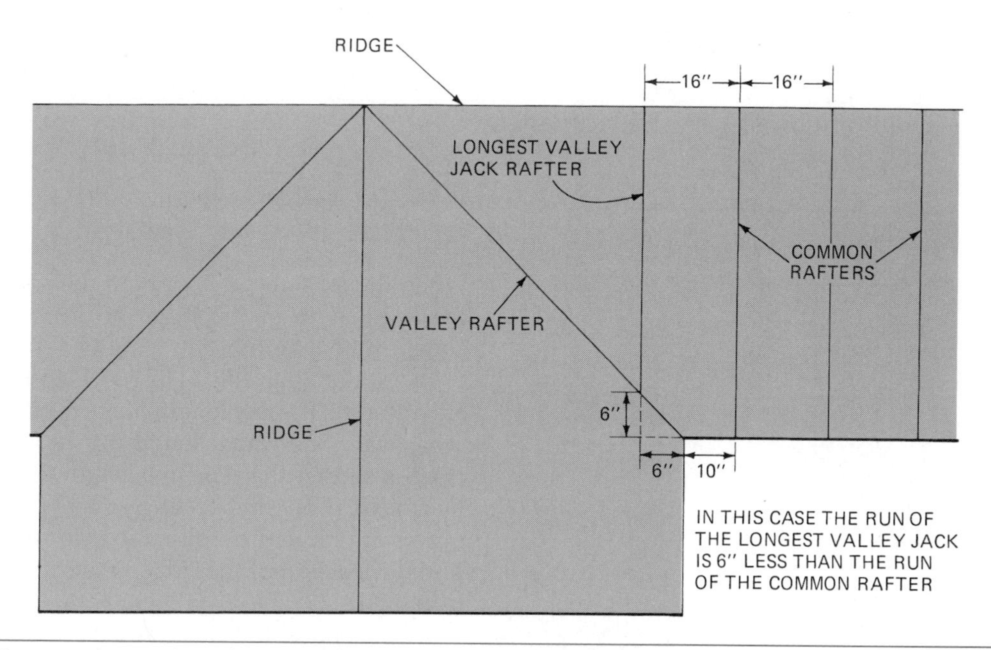

Fig. 44-6 How to determine the run of the longest valley jack rafter.

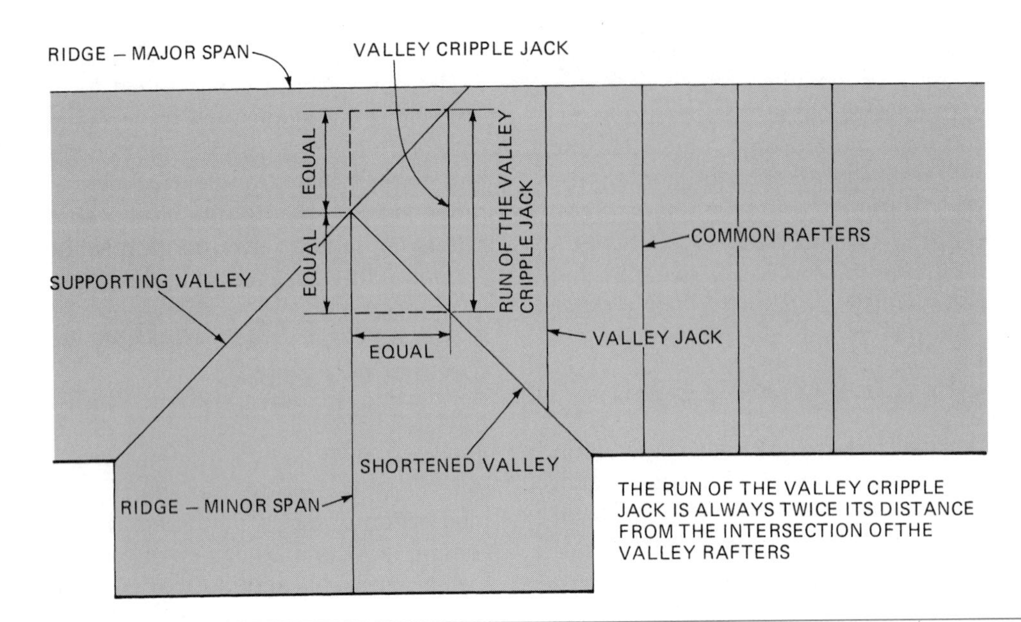

Fig. 44-7 Determining the run of the valley cripple jack rafter.

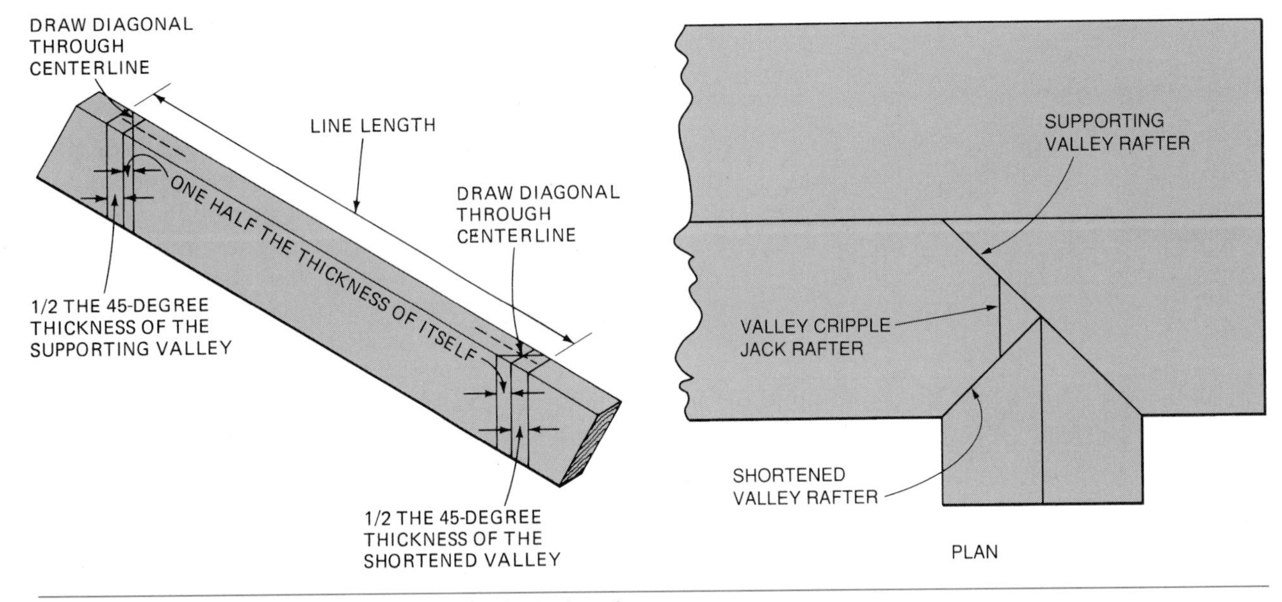

Fig. 44-8 The layout of the valley cripple jack rafter.

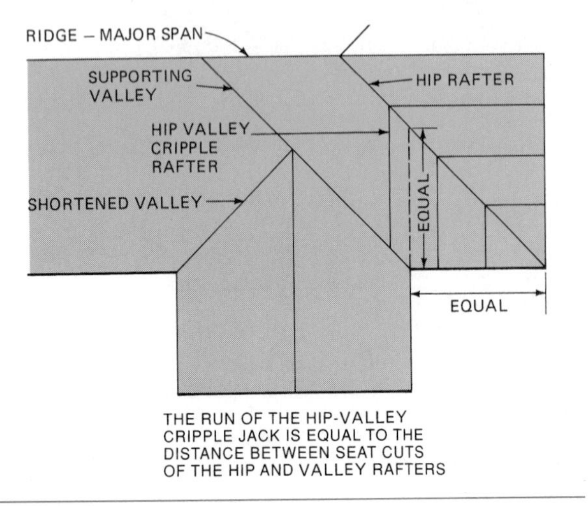

Fig. 44-9 Finding the total run of the hip-valley cripple jack rafter.

Ridge Length of the Minor Span

When the ridge heights of both roofs are the same, the line length of the ridge of the minor span is equal to the length of the wing plus half the width of the minor span (Fig. 44-11). When ridge heights are different, the line length of the ridge is found the same way. The actual length of the ridge in both cases is shorter than the line length to allow for the thicknesses of intersecting members. The end rafters are located by plumbing up from the plate. The surplus end of the ridge is cut off after the roof is framed.

The end of the ridge at the intersection of the valleys is simply a square cut. It is not necessary to make the ridge fit perfectly into the intersection.

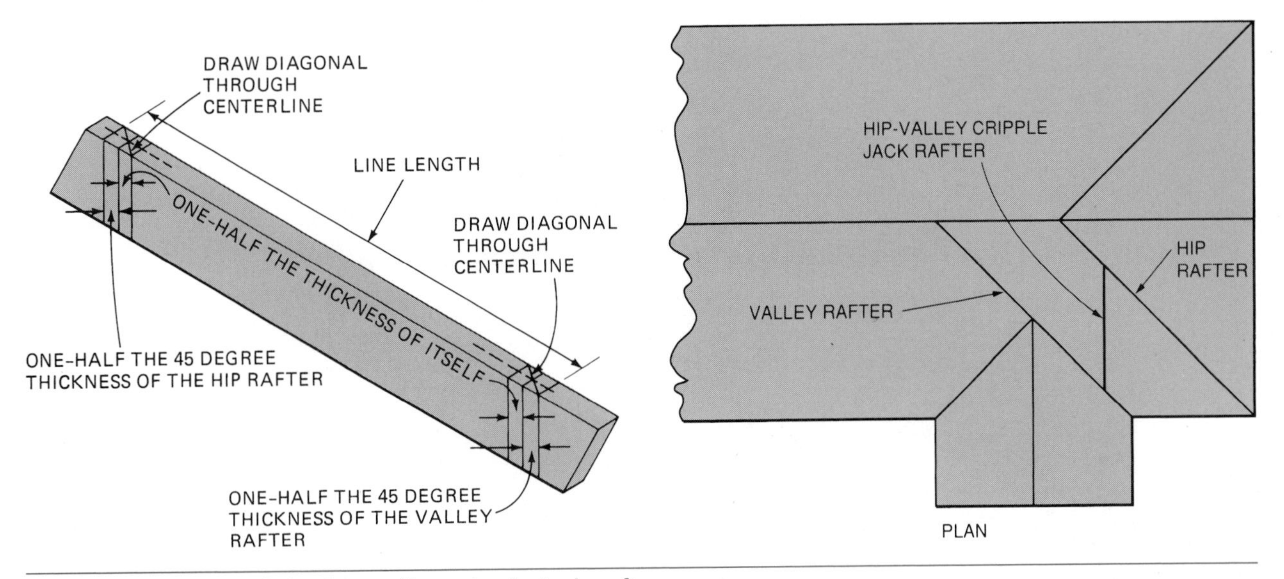

Fig. 44-10 Layout of the hip-valley cripple jack rafter.

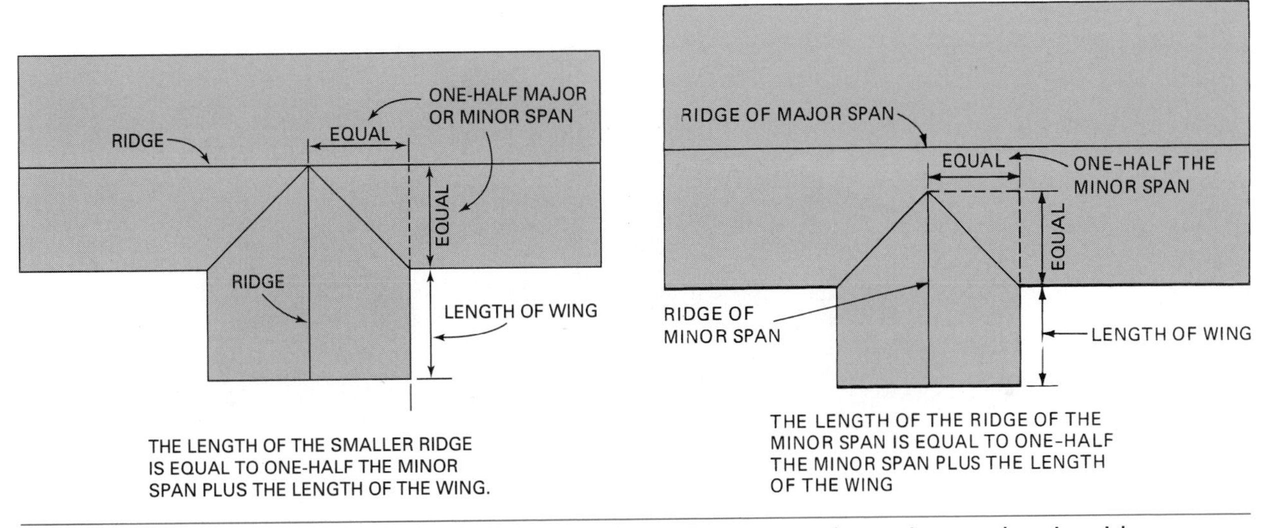

Fig. 44-11 Finding the line length of the ridge of the minor span when minor and major ridge heights are the same and when they are different.

Erecting the Intersecting Roof

The intersecting roof is raised by framing opposing members of the main span first. Then, the valley rafters are installed. To prevent bowing the ridge of the main span, install rafters to oppose the valley rafters. Install common and jack rafters in sets opposing each other. Sight members of the roof as framing progresses, keeping all members in a straight line.

CHAPTER 45 SHED ROOFS, DORMERS, AND OTHER ROOF FRAMING

The shed roof slopes in only one direction. It is relatively easy to frame. A shed roof may be freestanding or one edge may rest on an exterior wall while the other edge butts against an existing wall (Fig. 45-1). Shed and other type roofs are also used on *dormers*. A dormer is a framed projection above the plane of the roof. It contains one or more windows for the purpose of providing light and ventilation or for enhancement of the exterior design (Fig. 45-2).

Fig. 45-1 A building with shed roofs butting each other.

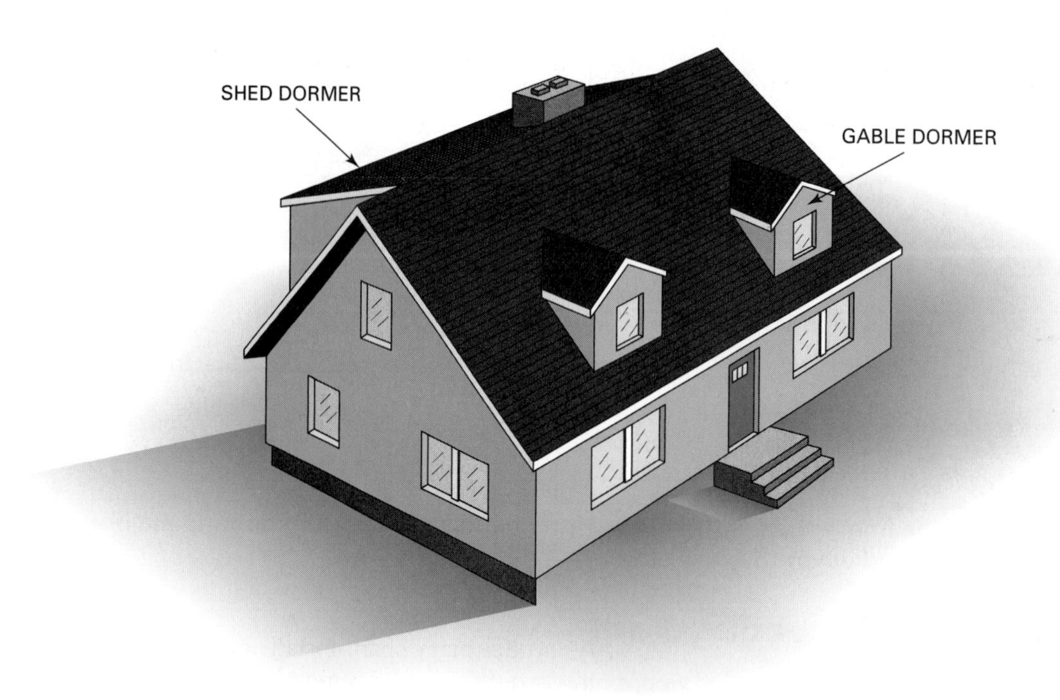

Fig. 45-2 A roof with gable and shed dormers.

Framing a Shed Roof

A *shed roof rafter* runs at a right angle to the plate line, similar to a common rafter. However, it may require two seat cuts instead of one. The unit of run is 12 inches, the same as the common rafter. The total run of the rafter, if both ends rest on exterior walls, is the width of the building minus the thickness of one of the walls (Fig. 45-3). For shed roofs that butt against an existing wall, the rafters are laid out the same as common rafters.

Layout of the Shed Roof Rafter

To lay out the shed roof rafter, mark a plumb line and a level line for a seat cut at one end. From the plumb line, step-off the rafter. Or, find its length us-

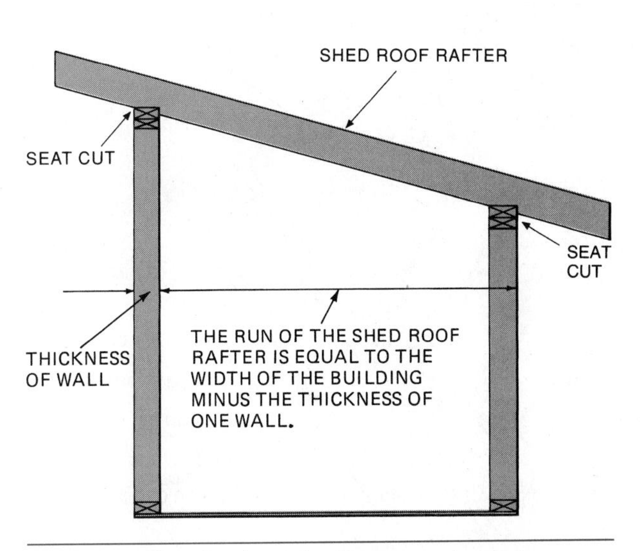

Fig. 45-3 The shed roof rafter may have two seat cuts.

ing the common rafter tables. Lay out another seat cut. Leave enough beyond the seat cut on both ends for overhangs at top and bottom. The overhang may be trimmed after the rafters have been installed, as described previously for common rafters. If preferred, the overhangs can be laid out and cut before installation (Fig. 45-4).

Shed roofs are framed by toenailing the rafters into the plate at the designated spacing. It is important that the plumb cut of the seat be kept snug up against the walls. In addition to nailing the rafters, metal *framing anchors* or *ties* may be required, not only for shed roofs, but for all types of framed roofs.

Dormers

Usually dormers have either gable or shed roofs. A gable dormer roof is framed similar to an intersecting gable roof (Fig. 45-5). In most cases, the shed dormer roof extends to the ridge of the main roof to gain enough incline (Fig. 45-6).

STEP BY STEP

PROCEDURES

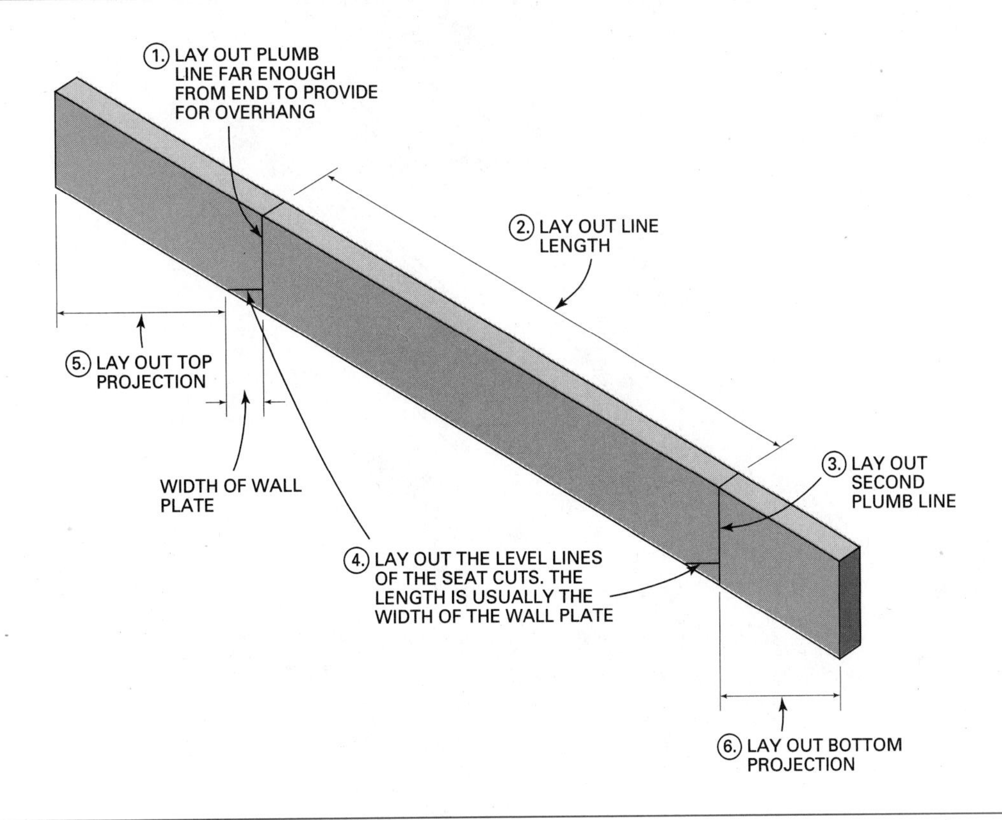

Fig. 45-4 Steps in the layout of a shed roof rafter.

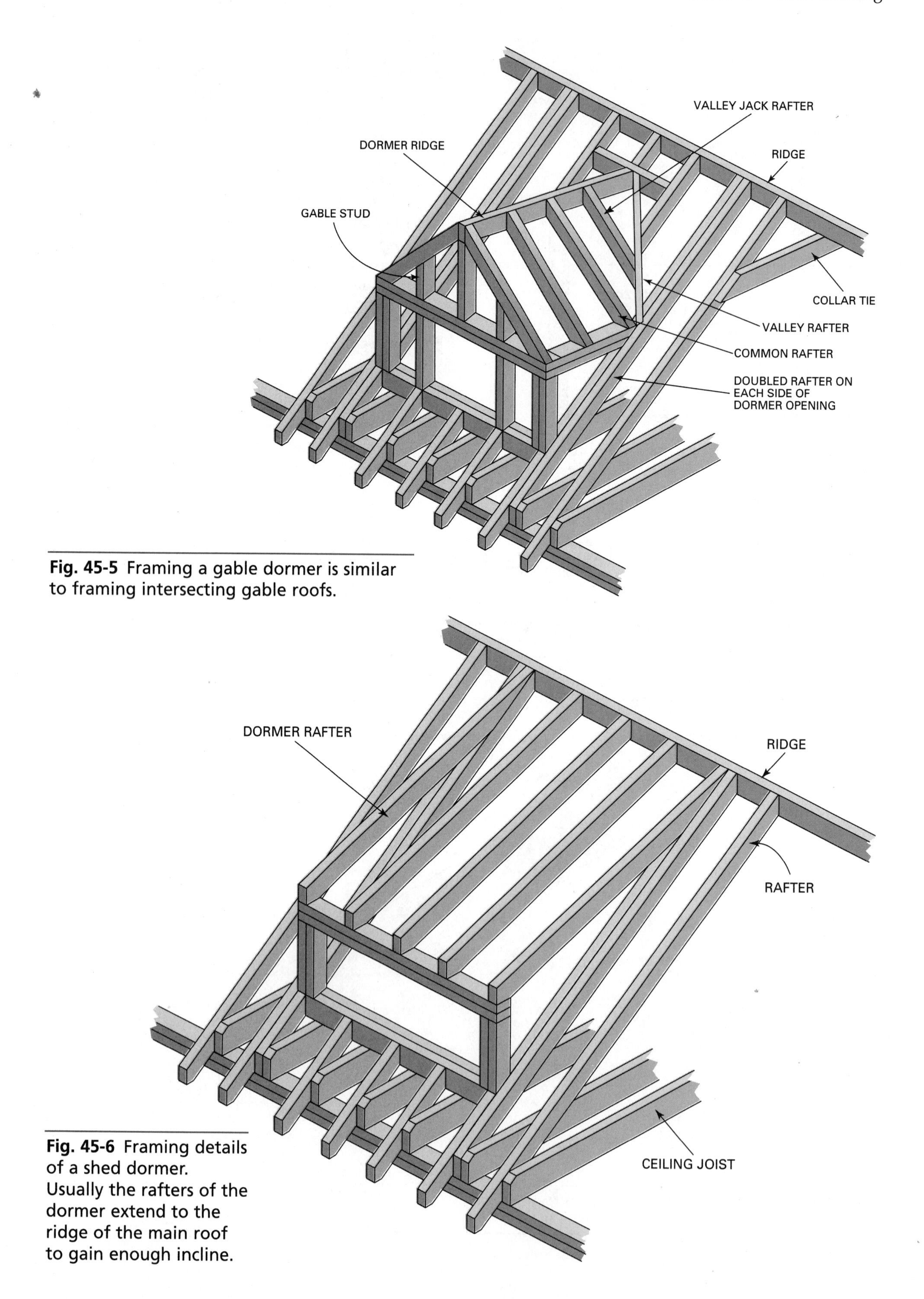

Fig. 45-5 Framing a gable dormer is similar to framing intersecting gable roofs.

Fig. 45-6 Framing details of a shed dormer. Usually the rafters of the dormer extend to the ridge of the main roof to gain enough incline.

When framing openings for dormers, the rafters on both sides of the opening are doubled. Some dormers have their front walls directly over the exterior wall below. When dormers are framed with their front wall partway up the main roof, top and bottom headers of sufficient strength must be installed.

Other Roof Framing

There are a number of other roof framing problems related in some way to the framing of roofs previously described. Solutions to some of the most commonly encountered problems are given.

Fitting a Valley Jack Rafter to Roof Sheathing

Some intersecting roofs are built after the main roof has been framed and sheathed. In this type of construction, the valley rafters are eliminated and common rafters are used full-length, making the main roof easier to frame. However, an extensive repair job will be required if the sheathing in the valley deteriorates because of a leak in the roof. Steps to lay out the seat cut of the valley jack rafter to fit against roofs of the same and different inclines are shown in Figure 45-7.

Layout of the Ridge Cut against Roof Sheathing

The layout of the ridge of the intersecting gable roof that fits against the roof sheathing is shown in Figure 45-8.

Cut of the Shed-Roof Rafter against a Roof

The top ends of shed-roof rafters are occasionally fitted against a roof of a different and steeper incline. The cuts can be laid out by using a framing square as outlined in Figure 45-9.

STEP BY STEP

PROCEDURES

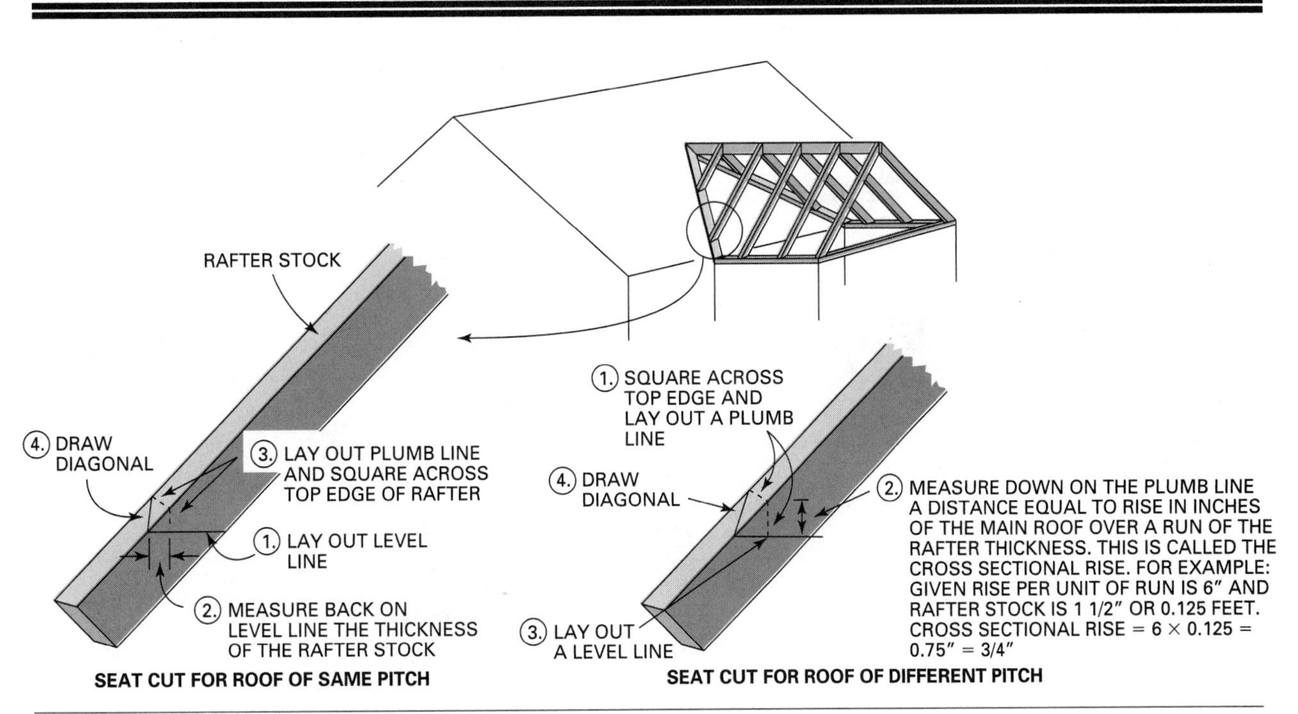

Fig. 45-7 Laying out the seats of valley jack rafters that fit against roofs of the same and different pitches.

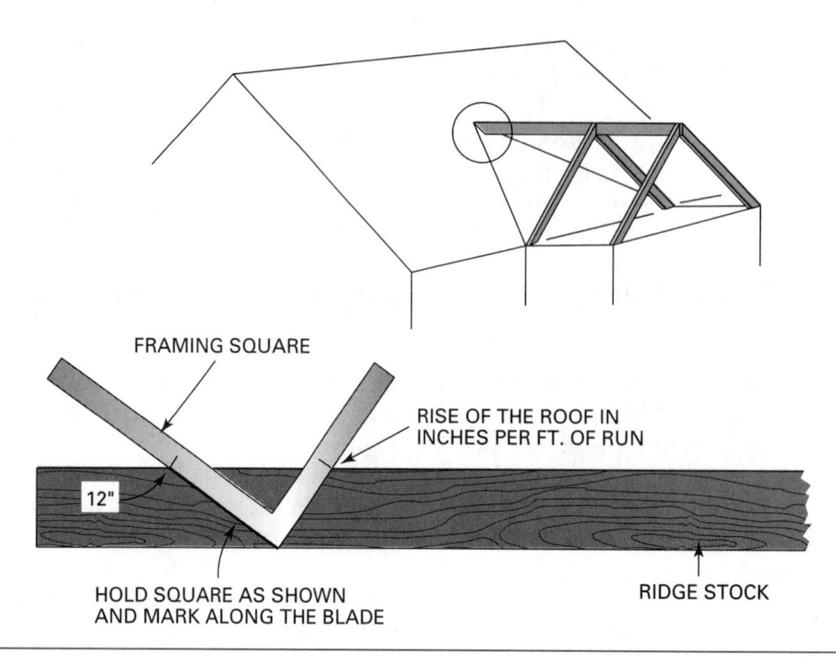

Fig. 45-8 Layout of a ridge that fits against roof sheathing.

STEP BY STEP PROCEDURES

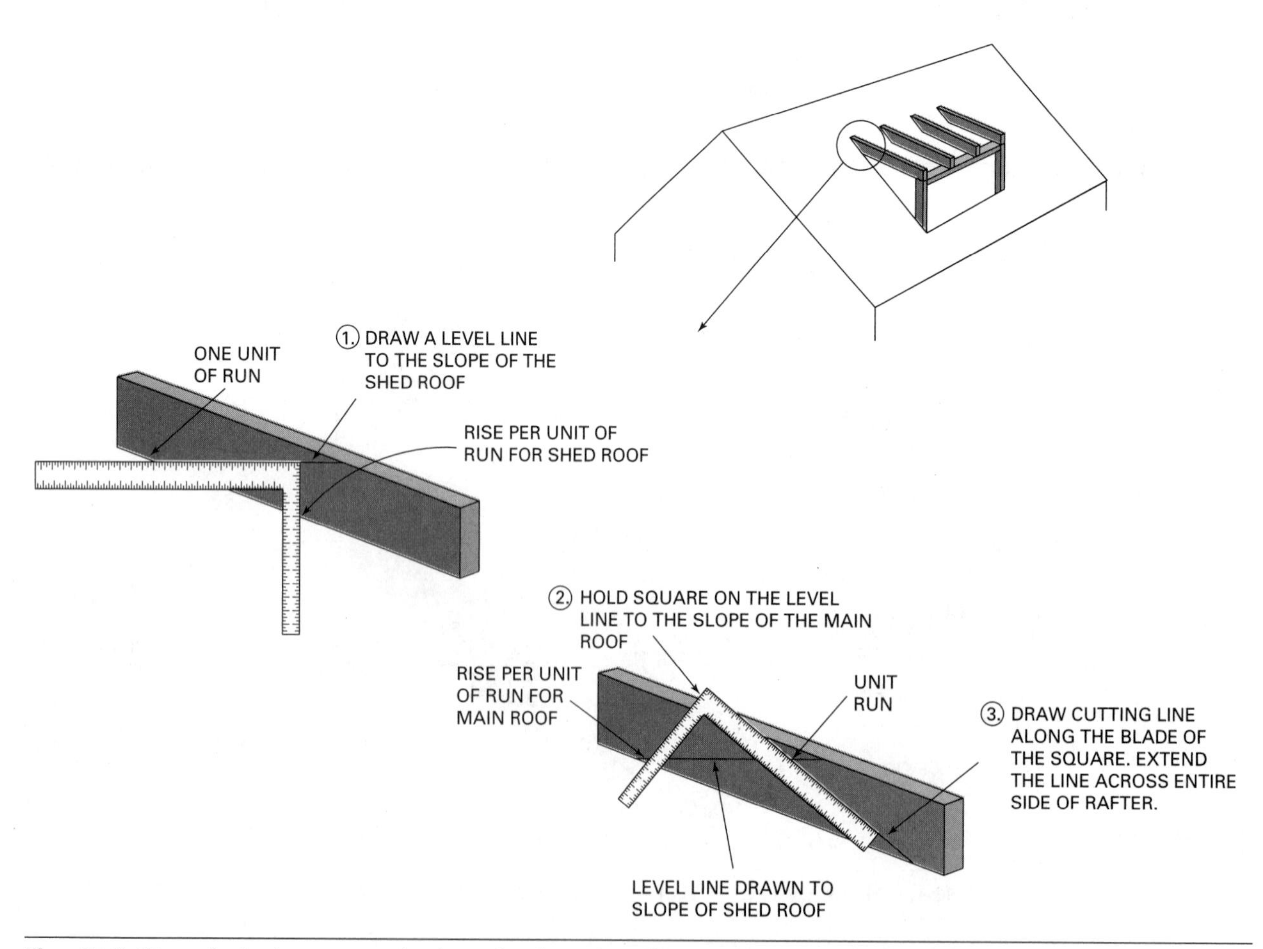

Fig. 45-9 Steps in laying out a shed roof rafter that fits against a roof of a different pitch.

CHAPTER 46 TRUSSED ROOFS

Roof Trusses

Roof trusses are extensively used in residential and commercial construction (Fig. 46-1). Because of their design, they can support the roof over a wide span. In residential construction, the use of roof trusses eliminates the need for load-bearing partitions below. The roof is also framed in much less time. However, because of their design, much usable attic space is lost.

Truss Construction

A roof truss consists of upper and lower *chords* and diagonals called *web members.* The upper chords act as rafters. The lower chords serve as ceiling joists. Joints are fastened securely with metal or wood gusset plates (Fig. 46-2).

Fig. 46-1 Trusses are used extensively for roof framing.

Fig. 46-2 The members of roof trusses are securely fastened with metal gussets.

Truss Design

Most trusses are made in fabricating plants. They are transported to the job site. Trusses are designed by engineers to support prescribed loads. Trusses may also be built on the job, but approved designs must be used. The carpenter should not attempt to design a truss. Approved designs and instructions for job-built trusses are available from the American Plywood Association and the Truss Plate Institute.

The most common truss design for residential construction is the *Fink* truss (Fig. 46-3). Other truss shapes are designed to meet special requirements (Fig. 46-4).

Erecting a Trussed Roof

Carpenters are more involved in the erection than the construction of trusses. The erection and bracing of a trussed roof is a critical stage of construction. Failure to observe recommendations for erection and bracing could cause a collapse of the structure. This could result in loss of life or serious injury, not to mention the loss of time and material.

The recommendations contained herein are technically sound. However, they are not the only method of bracing a roof system. They serve only as a guide. The builder must take necessary precautions during handling and erection to ensure that trusses are not damaged, which might reduce their strength.

Small trusses, which can be handled by hand, are placed upside down, hanging on the wall plates, toward one end of the building. Trusses for wide spans require the use of a crane to lift them into position. One at a time the truss is lifted, fastened, and braced in place. The end truss is installed first by swinging it up into place and bracing it securely in a plumb position.

Bracing the First Truss

All other trusses are tied to the first truss. Therefore, the bracing system depends, to a great extent, on how well the first truss is braced. One satisfactory method is for the first truss to be braced to stakes driven and securely anchored in the ground. The ground braces should be located directly in line with all rows of top chord continuous lateral bracing

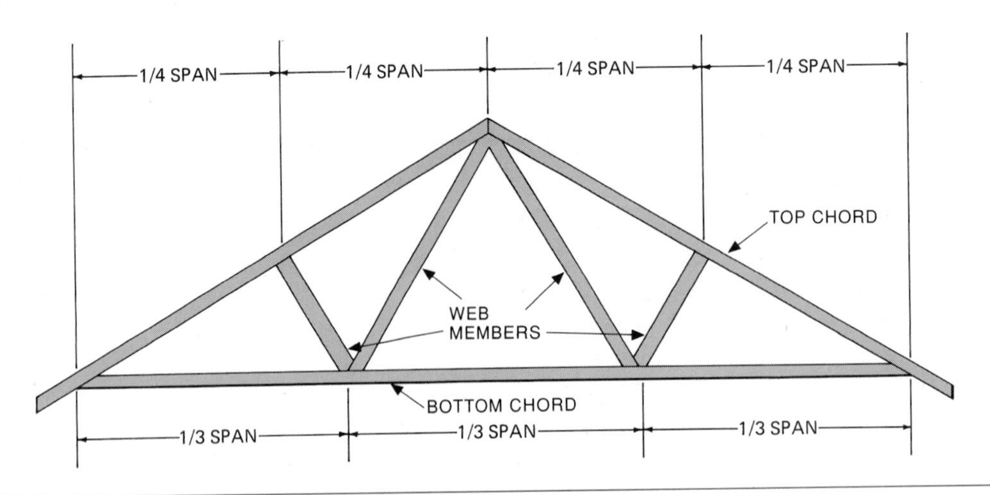

Fig. 46-3 The Fink truss is widely used in residential construction.

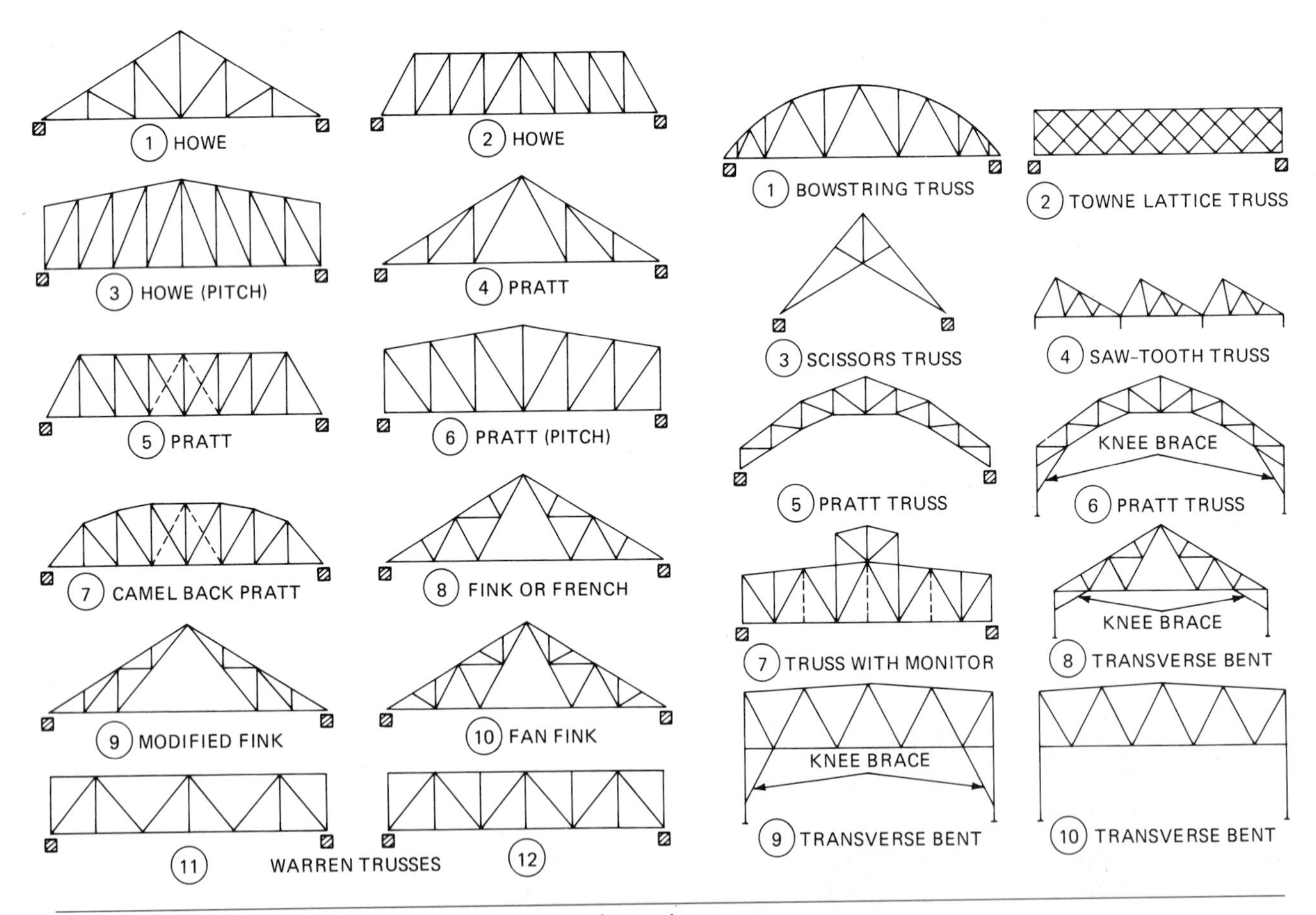

Fig. 46-4 Truss shapes are designed for special requirements.

(Fig. 46-5). The first truss is fastened to the plate and temporarily to the end bracing. Do not nail scabs to the end of the building to brace the first truss. These scabs can break off or pull out, causing a collapse of the roof.

Temporary Bracing of Trusses

As trusses are set in place, they are nailed to the plate. Metal framing ties are usually applied (Fig. 46-6).

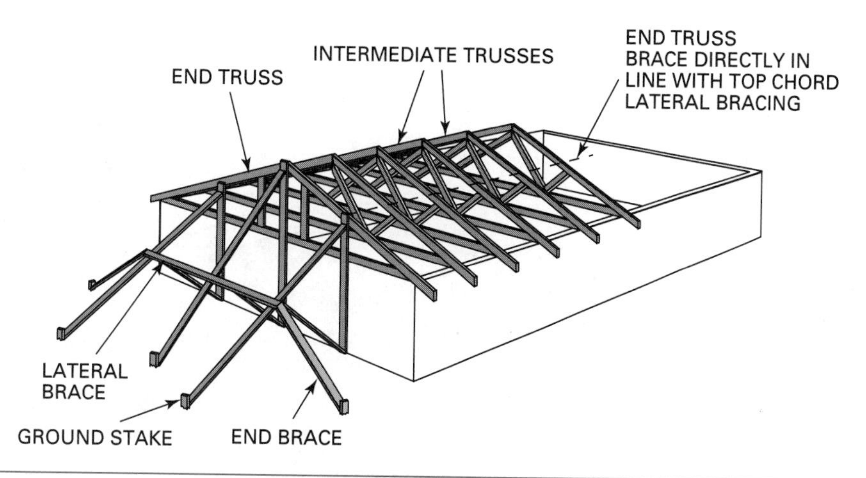

Fig. 46-5 The end truss must be well braced before the erection of other trusses.

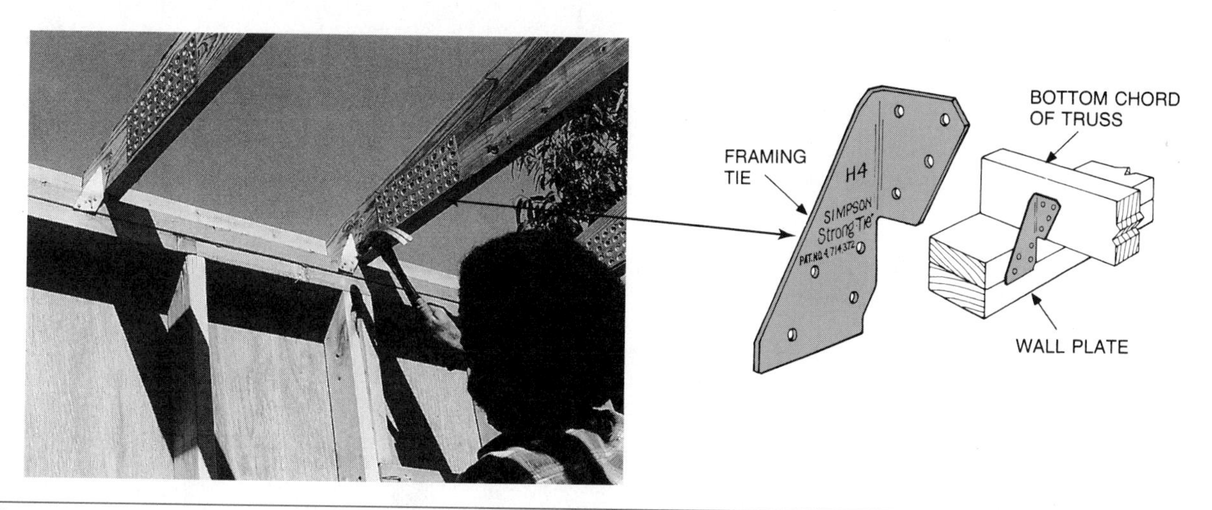

Fig. 46-6 Metal framing ties are used to fasten trusses to the wall plate.

Sufficient temporary bracing must be applied. Temporary bracing should be no less than 2 × 4 lumber as long as practical, with a minimum length of 8 feet. The 2 × 4s should be fastened with two 16d duplex nails at every intersection. Exact spacing should be maintained as bracing is installed. Adjusting the spacing later can cause trusses to collapse if a key brace is unfastened at the wrong time.

Temporary bracing must be applied to three planes of the truss assembly: the top chord or sheathing plane, the bottom chord or ceiling plane, and vertical web plane at right angles to the trusses.

Bracing the Plane of the Top Chord. Continuous lateral bracing should be installed within 6 inches of the ridge and at about 8- to 10-feet intervals between the ridge and wall plate. Diagonal bracing should be set at approximately 45-degree angles between the rows of lateral bracing. It forms triangles that provide stability to the plane of the top chord (Fig. 46-7). If possible, the bracing should be placed on the underside of the top chord. Then, it will not have to be removed as the roof sheathing is applied.

Bracing the Web Member Plane. Temporary bracing in the plane of the web members are diagonals placed at right angles to the trusses from top to bottom chords (Fig. 46-8). They usually become permanent braces of the web member plane.

Bracing the Plane of the Bottom Chord. In order to maintain the proper spacing on the bottom chord, continuous lateral bracing for the full length of the building must be applied. The bracing should be nailed to the top of the bottom chord at intervals no greater than 8 to 10 feet along the width of the building.

Diagonal bracing should be installed, at least, at each end of the building (Fig. 46-9). In most cases,

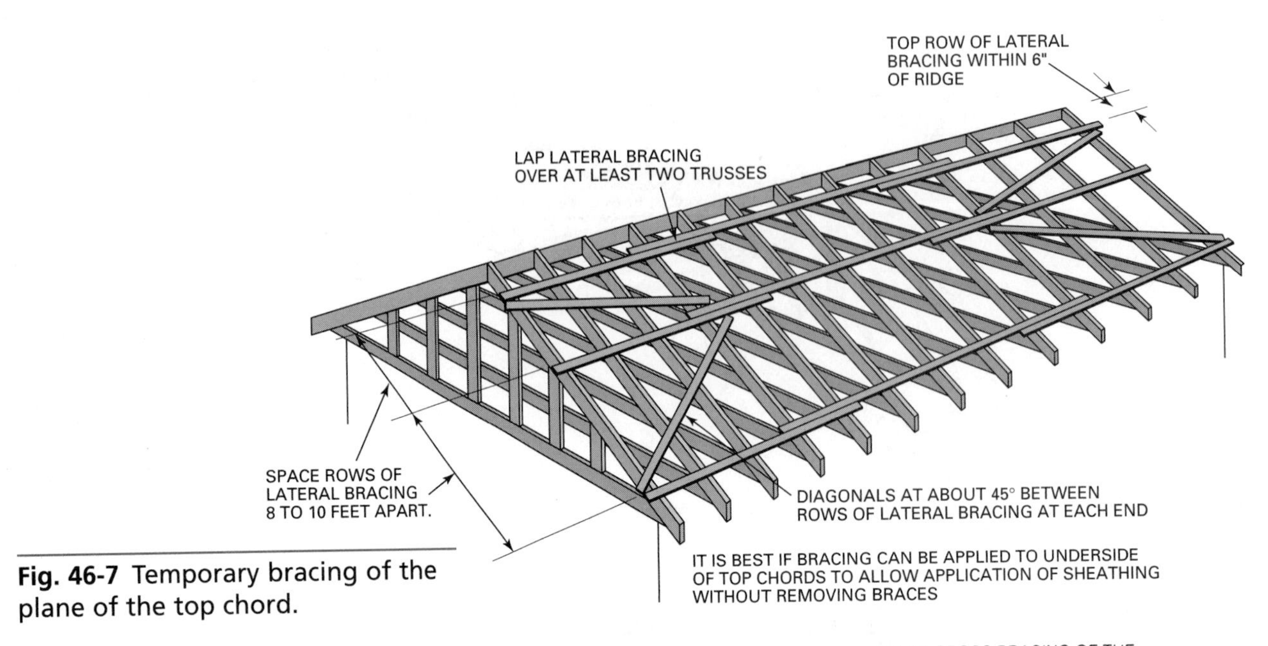

Fig. 46-7 Temporary bracing of the plane of the top chord.

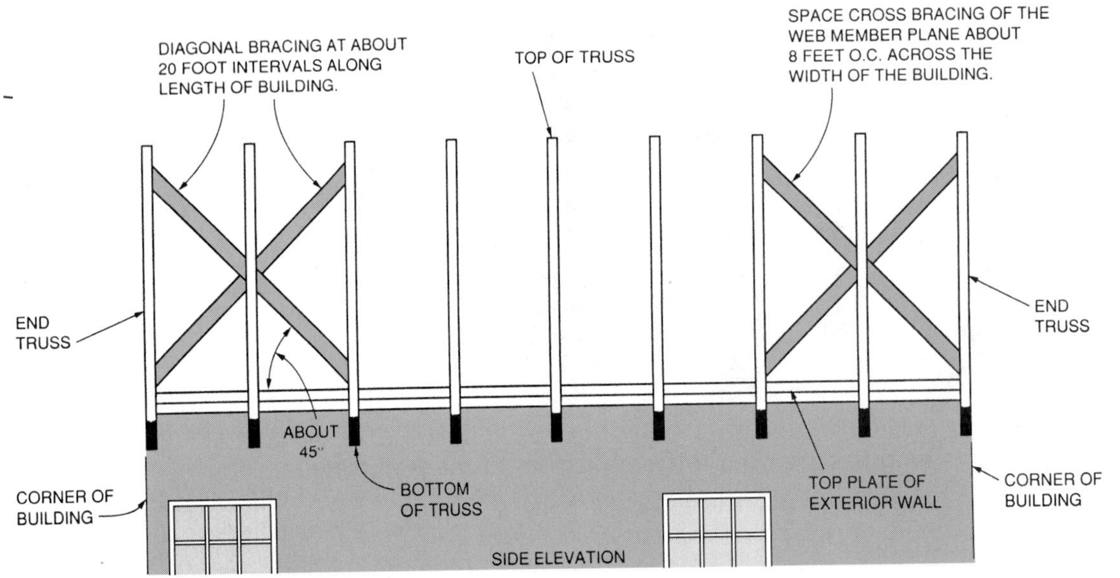

Fig. 46-8 Bracing of the web member plane prevents lateral movement of the trusses.

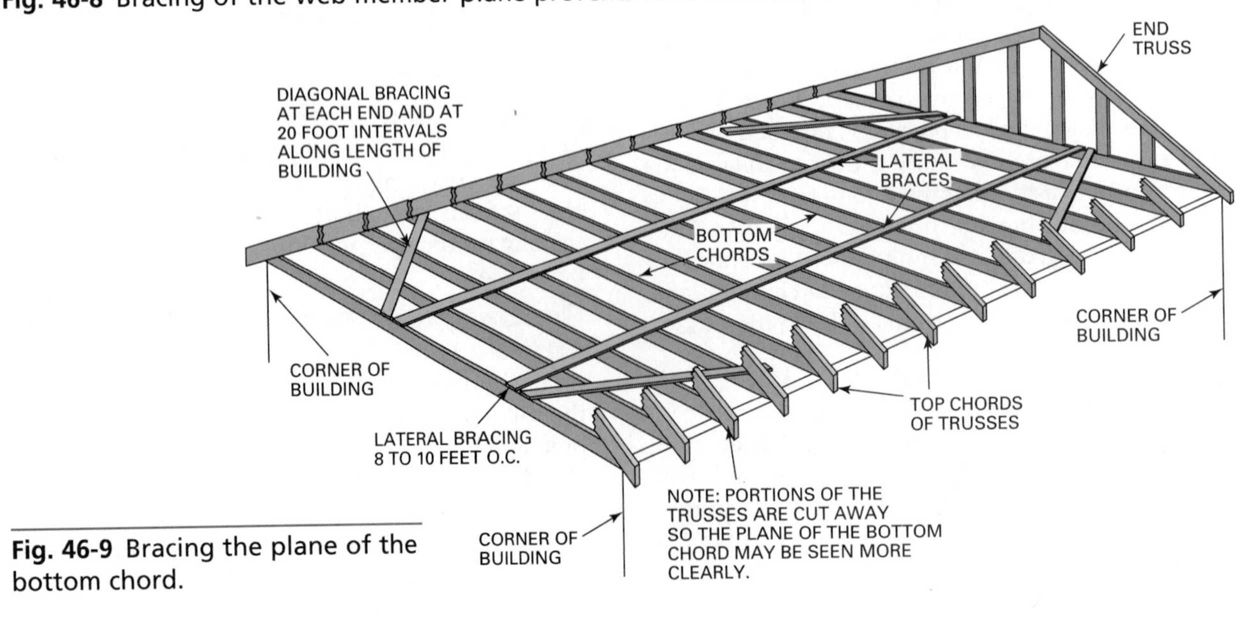

Fig. 46-9 Bracing the plane of the bottom chord.

temporary bracing of the plane of the bottom chord is left in place as permanent bracing.

Permanent Bracing

Permanent bracing is designed by the architect or engineer for the structural safety of a building. It is installed at the appropriate time as specified.

Roof Sheathing

Roof sheathing is applied after the roof frame is complete. Sheathing gives rigidity to the roof frame. It also provides a nailing base for the roof covering. Rated panels of plywood and strand board are commonly used to sheath roofs. Solid lumber sheathing boards are used only occasionally.

Lumber Sheathing

Lumber sheathing, if used, is usually square-edge or matched 1 × 8 or 1 × 10 boards. They are laid close together when a solid and continuous base is needed, such as for asphalt shingles. The boards may be spaced when the roof covering is wood shingles, shakes, or other stiff roofing. Spaced sheathing is usually 1 × 4 strips with the on center spacing the same as the shingle exposure. Joints between boards must be made on a rafter and staggered. Each board must bear on at least two rafters.

Plank decking is used in post and beam construction where the roof supports are spaced farther apart. Plank roof sheathing may be of 2-inch nominal thickness or greater, depending on the span. Usually both edges and ends are matched. The plank roof often serves as the finish ceiling for the rooms below (Fig. 46-10).

Panel Sheathing

Plywood and other rated panel roof sheathing is laid with the face grain running across the rafters for greater strength. End joints are made on the rafters and staggered. Nails are spaced 6 inches apart on the ends and 12 inches apart on intermediate supports (Fig. 46-11).

Adequate blocking, tongue and grooved edges, or other suitable edge support such as *panel clips* must be used when spans exceed the indicated value of the plywood roof sheathing. Panel clips are small metal pieces shaped like a capital H. They are used between adjoining edges of the plywood between rafters (Fig. 46-12). Two panel clips are used for 48-inch or greater spans. One clip is used for lesser spans.

The ends of sheathing are allowed to slightly and randomly overhang at the *rakes* until sheathing is completed. A chalk line is snapped between the ridge and plate at each end of the roof. The sheathing ends are trimmed to the line with a skilsaw.

Fig. 46-11 Sheathing a trussed roof with plywood.

Fig. 46-10 Matched plank being used to cover a flat roof of post-and-beam construction. (*Courtesy of Georgia Pacific*)

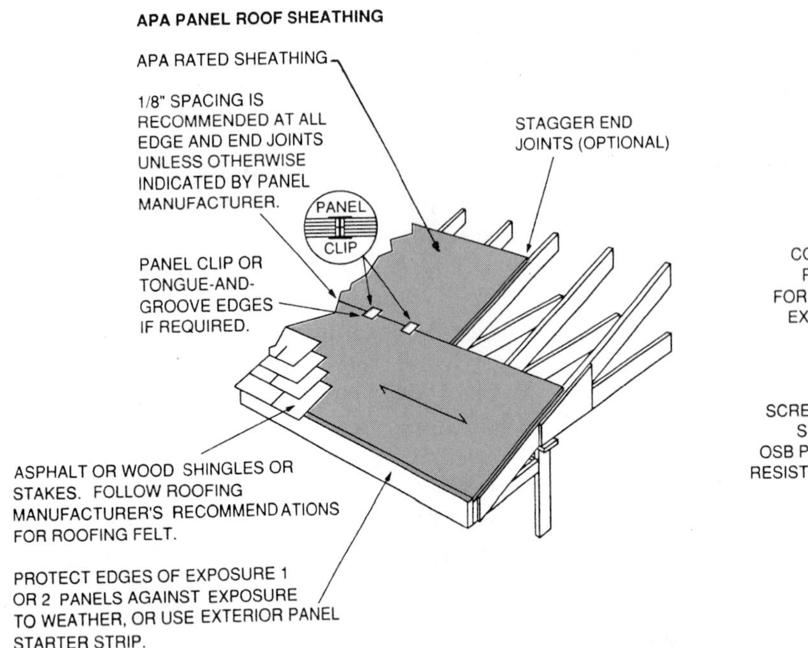

Recommended Minimum Fastening Schedule for APA Panel Roof Sheathing (Increased nail schedules may be required in high wind zones.)

Panel Thickness (b) (in.)	Nailing (c) (d)		
	Size	Maximum Spacing (in.) Panel Edges	Maximum Spacing (in.) Intermediate
5/16 - 1	8d	6	12 (a)
1 - 1/8	8d or 10d	6	12 (a)

(a) For spans 48 inches or greater, space nails 6 inches at all supports.
(b) For stapling asphalt shingles to 5/16-inch and thicker panels, use staples with a 15/16-inch minimum crown width and a 1-inch leg length. Space according to shingle manufacturer's recommendations.
(c) Use common smooth or deformed shank nails with panels to 1 inch thick. For 1-1/8-inch panels, use 8d ring- or screw-shank or 10d common smooth-shank nails.
(d) Other code-approved fasteners may be used.

Fig. 46-12 Recommendations for the application of APA panel roof sheathing. (*Courtesy of American Plywood Association*)

Review Questions

Select the most appropriate answer.

1. The minimum amount of stock left above the seat cut of the common rafter to ensure enough strength to support the overhang is usually
a. one-quarter of the rafter stock width.
b. one-half of the rafter stock width.
c. two-thirds of the rafter stock width.
d. three-quarters of the rafter stock width.

2. What is the line length of a common rafter from the centerline of the ridge to the plate with a rise of 5 inches per foot of run, if the building is 28′-0″ wide?
a. 14′-8 1/2″ c. 15′-6″
b. 15′-2″ d. 16′-2″

3. The common difference in the length of gable studs spaced 24 inches OC for a roof with a pitch of 8 inches rise per foot of run is
a. 8 inches. c. 16 inches.
b. 12 inches. d. 20 inches.

4. How much is laid out on the blade of the square for 6 inches of run when stepping-off a hip rafter?
a. 6 inches. c. 8 1/2 inches.
b. 6/17 foot. d. 17 inches minus 6 inches.

5. Using the rafter tables, the line length of a hip rafter with a rise of 6 inches per foot of run and a total run of 12 feet is
a. 16′-6″. c. 18′-0″.
b. 17′-4″. d. 18′-8″.

6. The unit of run of any jack rafter is
a. 12 inches. c. 17 inches.
b. 16 inches. d. the same as hip and valley rafters.

7. The length of any rafter in a roof of specified pitch can be found if
a. its total rise is known.
b. its total run is known.
c. the rise per foot of run is known.
d. its length per foot of run is known.

8. The total run of any hip jack rafter is equal to
a. its distance from the corner.
b. one-third of its total rise.
c. its length minus the total rise.
d. its common difference in length.

9. The total run of the shortened valley rafter is equal to
a. the total run of the minor span.
b. the total run of the major span.
c. the total run of the common rafter.
d. the total run of the hip rafter.

10. The total run of the hip-valley cripple jack rafter is equal to
a. one-half the run of the hip jack rafter.
b. one-half the run of the longest valley jack rafter.
c. the distance between bird's mouths of the hip and valley rafters.
d. the difference in run between the supporting and shortened valley rafters.

BUILDING FOR SUCCESS

Communication Skills in Construction

The world of construction has always relied on strong oral and written communication skills to relay information. Today we communicate electronically more than ever before. Yet the most effective exchange of information is still based on good speaking, writing, and listening skills.

Along with the personal computer, fax machines, fiber optics, and telecommunications, there is still a need to develop strong oral and written communication skills. These skills form the basis for using high-technology achievements successfully.

Oral communication has always been important in providing the desired "personal touch" for customers. The vast majority of construction projects require good oral and written interaction both in the office and at the job site. As we become a more service-oriented nation, we must continue to value the power of the spoken and written word in conducting business, To enhance the success of the project, the builder, sub-contractor, or technician on the job must maintain exactness and clarity when dealing with clients.

The diverse populations that make up the nation's workforce today bring new dimensions to effective communication. Construction personnel represent more cultures, education levels, and diversities than ever before. This alone calls for a greater need to promote clear and concise oral and written communication at the workplace. These new dimensions call for increased abilities in reading, writing, and listening.

Technicians today are more involved in the planning, coordinating, and marketing as well as the construction of a project. This will continue to require more verbal and writing skills. Legalities, warranties, quality assurances, and guarantees have taken on expanded roles. Interpreting codes, specifications, statutes, and other stipulations demands better reading and listening skills. Mistakes due to errors in reading or interpretation can be costly and even endanger lives. All workers must develop a proficiency in effective communication.

One of the biggest challenges to young people entering the industry today appears to be in the area of using appropriate language. The use of inappropriate or abusive language in the business world can seriously restrict the success of the project. Associated with this problem is the concern for forming appropriate business attitudes. To illustrate this concern, we need to look at the issue from the client's viewpoint.

Let's take, for example, the family that is having a new house built in a community. Excluding a business venture, this will be the largest investment they make in their lifetime. They are envisioning a new abode where they will be rearing their children, sharing family experiences, and possibly spending their retirement years. They take it seriously. Beginning with the planning phase, their expectation level is high.

Now enter the builders and workers representing the various trades. These representatives from the construction industry have the responsibility of being professional. They should have the trust of the customer as part of the construction package. An agreement has been reached. The builder and the team are obligated to provide the best quality for the price being paid. This must be communicated.

The builder must demonstrate professional attitudes and use professional language. If the builder or construction representative uses abusive or profane language on the job site, it sends a negative message to the client. The message will say, "This project is not important; it's just another routine job." That, in turn, will probably tell the client that the quality is less than what should be expected.

Workers with the proper attitude and command of the language will not spit chewing tobacco on the subfloor or present themselves vocally as crude or uncaring. We must remember that clients will advertise for the builder whether the builder likes it or not. They will share with others how they perceived the builder's professionalism and work ethics in addition to the quality of work. Not all communication is conducted orally or in writing. We also communicate who we are by our performance. People read each other continually.

We must ask ourselves, "How can I improve my communication skills?" All people should take personal inventories to see how well equipped they are. Some questions and suggestions follow that may help students determine their own strengths or weaknesses in personal communication skills.

- Can I list, in writing, my career goals?
- How much do these goals involve oral and written communications?
- What is my current skill level in communicating?
- Do I listen, write, and speak well? Is my vocabulary acceptable?
- What does my past communication performance in school indicate?
- Do I understand what personal communication skills will be demanded of me in a construction career?
- Do I understand how communication skills will describe my professional image, either positively or negatively?
- Do I know someone I can talk to that knows what communication skills will help me reach my goals?
- Do I know how these skills can be obtained?
- Once I understand what good communication skills will be required of me to be successful, will I desire to obtain those skills?

An important recommendation to any student pursuing a career in construction is to talk to knowledgeable people who can provide accurate information about the necessary communication requirements for the desired job. Trade representatives, counselors, parents, teachers, and the CEOs of construction businesses will be the people with that knowledge.

Someone once said, "It is not so much what we know as how well we use what we know." A personal vocabulary does not have to equal the volume of a dictionary. But words must be selected carefully. A good policy may be one that says, "If you would not write it and sign it, do not say it."

In conclusion, the element of effective communication must never be diminished or disregarded. It is a vital part of good business ethics. it has become even more important as we strive for success in the workplace.

FOCUS QUESTIONS: For individual or group discussion

1. Address each question presented in the preceding study.
2. What would be the consequences of not developing good communication skills as you pursue a career in construction?

UNIT

16

Stair Framing

A great degree of skill is necessary for the design and construction of staircases. The carpenter must know how to lay out and build well-designed staircases with first-class workmanship. Staircases must be comfortable to use and safe. This unit deals with stair framing. A later unit describes the application of stair finish.

OBJECTIVES

After completing this unit, the student should be able to:

- describe several stairway designs.
- define terms used in stair framing.
- determine the rise and tread run of a stairway.
- determine the length of and frame a stairwell.
- lay out a stair carriage and frame a straight stairway.
- lay out and frame a stairway with a landing.
- lay out and frame a stairway with winders.
- lay out and frame basement stairs.

UNIT CONTENTS

CHAPTER 47 STAIRWAYS AND STAIRWELLS

In addition to the staircase, consideration must be given to the layout and framing of the **stairwell.** The stairwell is the opening in the floor through which a person must pass when climbing and going down the stairs (Fig. 47-1). The stairwell is framed at the same time as the floor.

Stairway Design

Stairs, stairway, and *staircase* are terms used to designate one or more flights of steps leading from one level of a structure to another. Stairs are further defined as *finish or service* stairs. Finish stairs extend from one habitable level of a house to another. Service stairs extend from a habitable to a nonhabitable level. Stairways in residential construction should be at least 36 inches wide-preferably even wider. This allows the passage of two persons at a time and for the moving of furniture (Fig. 47-2). Many codes restrict the maximum height of a single flight of stairs to 12 feet.

Types of Stairways

A *straight* stairway is continuous from one floor to another. There are no turns or landings. *Platform* stairs have intermediate landings between floors. Platform-type stairs sometimes change direction at the landing. An L-type platform stairway changes direction 90 degrees. A U-type platform stairway changes direction 180 degrees.

Platform stairs are installed in buildings in which there is a higher floor-to-floor level. They also provide a temporary resting place. They are a safety feature in case of falls. The landing is usually constructed at about the middle of the staircase.

A *winding* staircase gradually changes direction as it ascends from one floor to another. In many cases, only a part of the staircase winds. Winding stairs may solve the problem of a shorter straight horizontal run. However, their use is not recommended. They are relatively difficult to construct. They pose a danger because of their tapered treads, which are very narrow on one end (Fig. 47-3).

Stairways constructed between walls are called *closed* stairways. Closed stairways are more economical to build. However, they add little charm or beauty to a building. Stairways that have one or both sides open to a room are called *open* stairways. One side of the staircase may be closed while the other side is open for all or part of the flight. Usually,

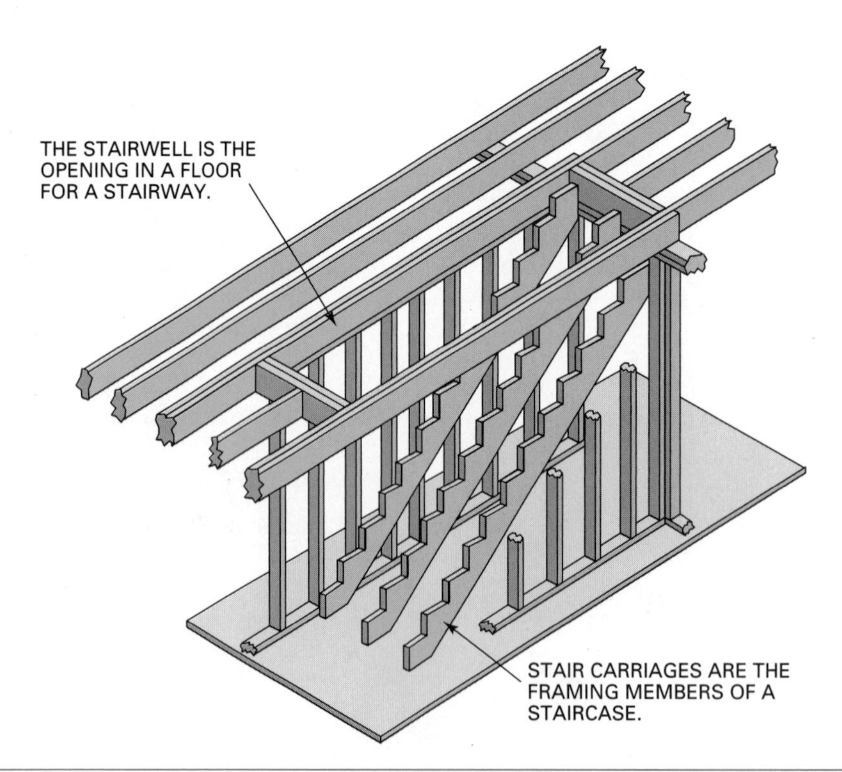

Fig. 47-1 Frame for stairs and stairwell.

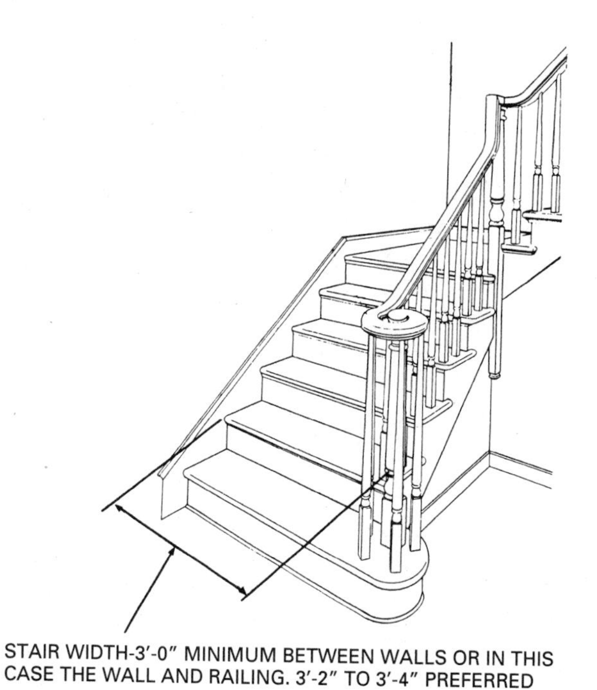

Fig. 47-2 Recommended stair widths as measured between the walls or railing.

a staircase is the outstanding feature of an entrance. It provides beauty and grace to a room. A staircase generally affects the character of the entire interior.

Stair Framing Terms

The terms used in stair framing are defined in the following material. Figure 47-4 illustrates the relationship of the various terms to each other and to the total staircase.

- The *total rise* of a stairway is the vertical distance between finish floors.
- The *total run* is the total horizontal distance that the stairway covers.
- A *rise* is the vertical distance from one step to another. A *riser* is the finish material that covers this distance.
- The *tread run* is the horizontal distance between the faces of the risers. A *tread* is the horizontal member on which the feet are placed when climbing or descending the stairs. The **nosing** is that part of the tread that extends beyond the face of the riser.

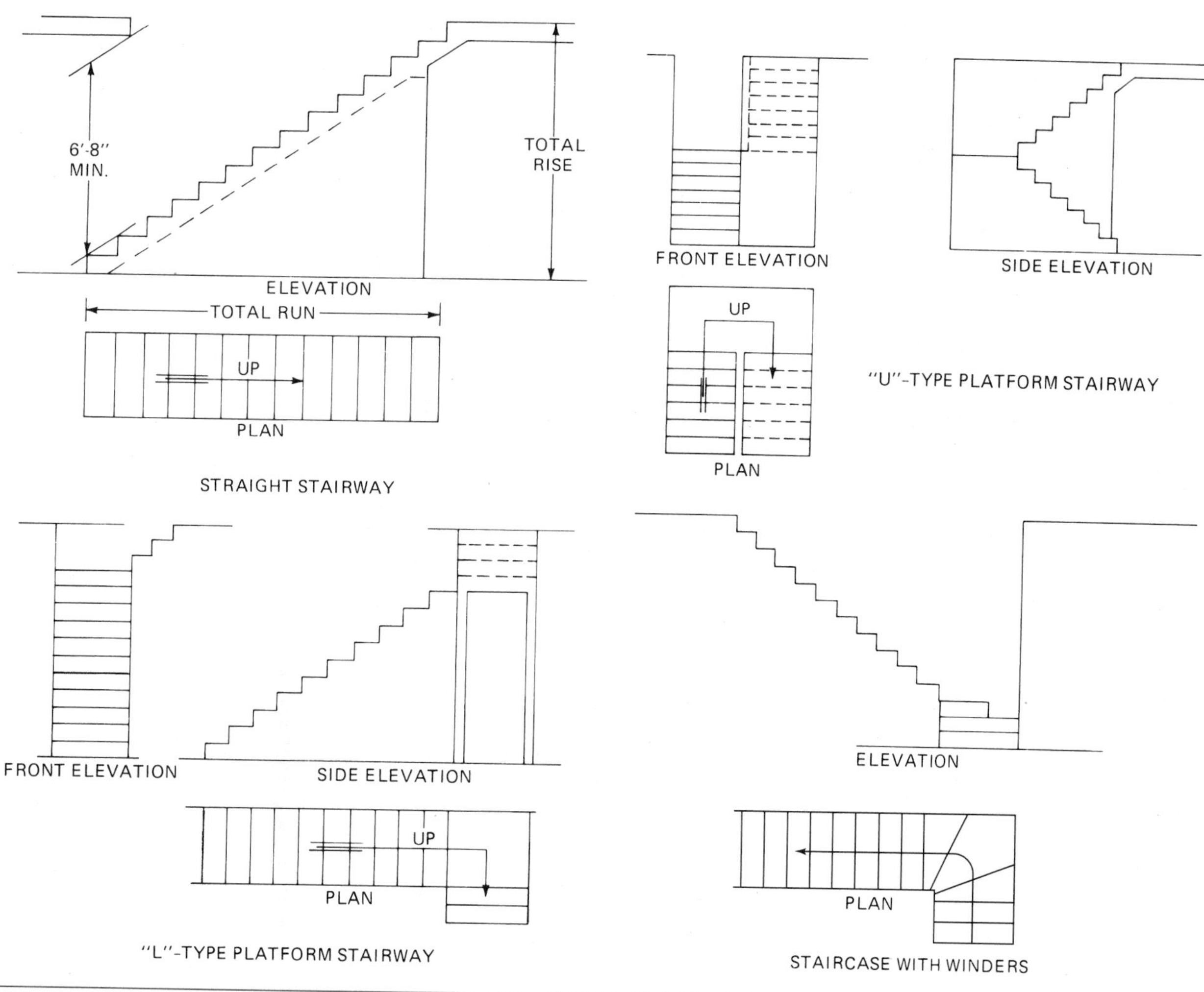

Fig. 47-3 Various types of stairways.

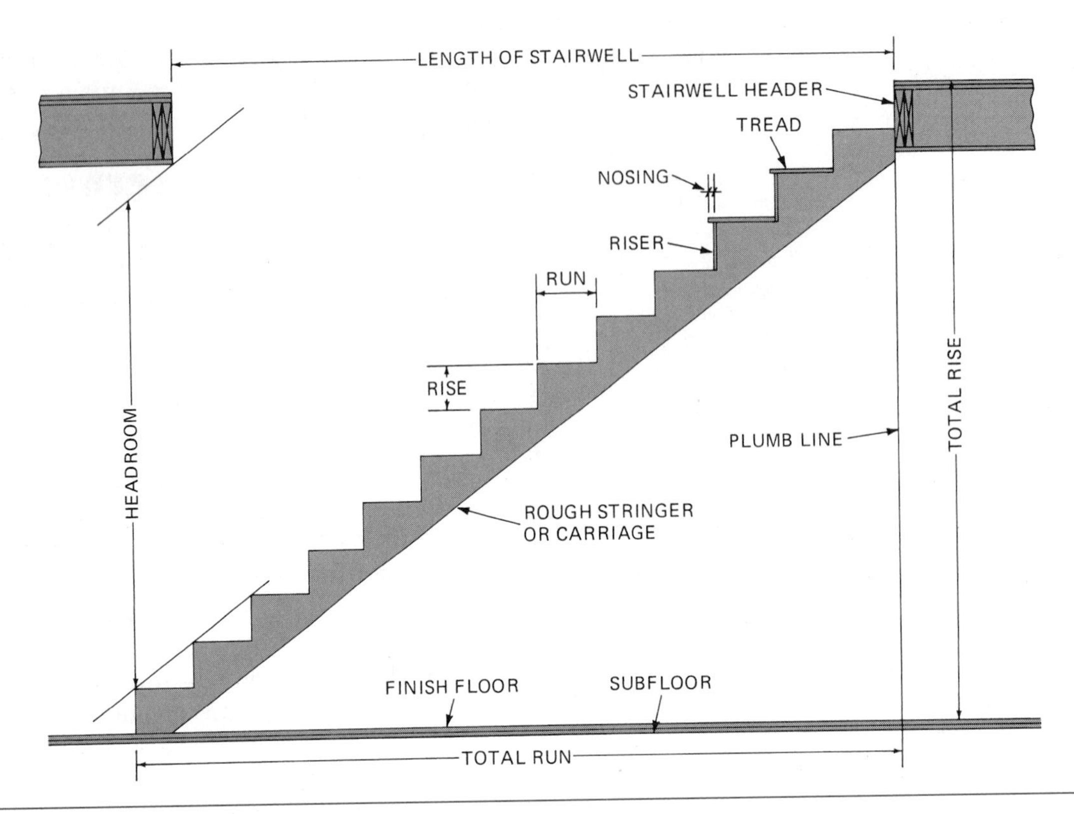

Fig. 47-4 Stair framing terms.

- A **stair carriage** is sometimes called a **stair horse** or **rough stringer.** It is usually a nominal 2 × 10 or 2 × 12 framing member that supports the treads and risers.
- A *stairwell* is an opening in the floor for the stairway to pass through. It provides adequate headroom for persons using the stairs.

Methods of Stair Construction

There are two principal methods of stair construction. The **housed finished stringer** is usually fabricated in a shop. Framing stairs on the job usually requires using *stair carriages.*

The Housed Stringer Staircase

For the housed stringer staircase, the framing crew frames only the stairwell. The staircase is installed when the house is ready for finishing. Dadoes are routed into the sides of the finish stringer. They *house* (make a place for) and support the risers and treads (Fig. 47-5).

Occasionally, the finish carpenter builds a housed stringer staircase on the job site. A router and jig are used to dado the stringers. Then, the treads, risers, and other stair parts are cut to size. The staircase is then assembled. Stair carriages are not required when the housed finish stringer method of construction is used. (Building housed finished stringer stairs is described in more detail in Chapter 81.)

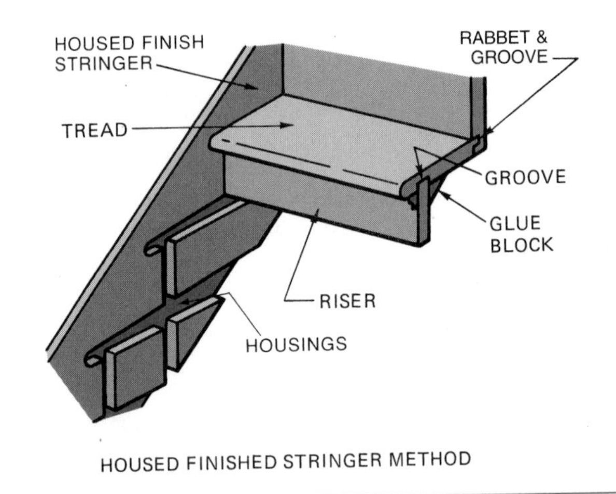

Fig. 47-5 A housed finish stringer method of stair construction.

The Job-Built Staircase

The *job-built* staircase uses stair carriages that are installed when the structure is being framed. The carriage is laid out for cutouts. Risers and treads are fastened to the cutouts (Fig. 47-6). Temporary rough treads are installed for easy access to upper levels

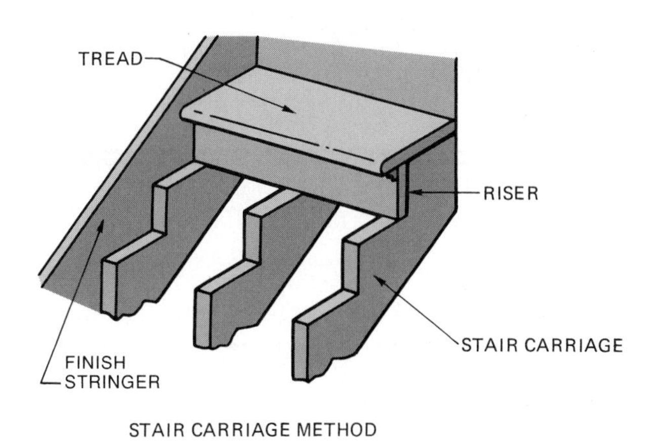

Fig. 47-6 Job-built stairs usually are framed using stair carriages.

until the stairs are ready for the finish treads and other stair trim. The layout for both housed stringers and stair carriages is made in a similar manner.

Determining the Rise and Tread Run

Staircases must be constructed at a proper angle for maximum ease in climbing and for safe descent. The relationship of the rise and run determines this angle (Fig. 47-7). The preferred angle is between 30 and 35 degrees. Stairs with a slope of less than 20 degrees waste a lot of valuable space. Stairs with a slope that is excessively steep (50 degrees or over) are difficult to climb and dangerous to descend.

Some building codes specify that the height of a riser shall not exceed 8 1/4 inches and that the width of a tread not be less than 9 inches. However, a rise of between 7 and 7 1/2 inches and a tread run of between 10 and 12 inches is recommended. The rise should be determined first.

Determining the Riser Height

To determine the individual rise, the total rise of the stairway must be known. Divide the total rise by 8 inches. (This measurement is more than is recommended for each rise.) The result is a usually a whole number and a fraction. The next largest whole number may be the total number of risers required.

For example, a total rise of 8'-10" (106 inches) is divided by 8 inches. The result is 13 1/4. The next largest whole number is 14. This is the number of risers that may be desired.

Dividing the total rise by the number of risers gives the height of each rise. For example, dividing the total rise (106 inches) by the number of risers (14) gives a unit rise of 7 9/16 inches.

To decrease the riser height, increase the number of risers by one. Divide the total rise by that number. For example, increasing 14 by 1 and dividing 106 inches by 15 equals slightly over 7 1/16 inches.

Another method of determining the riser height does not involve mathematics. Stand a *story rod* (usually a 1 × 2 strip of lumber) vertically with one end on the floor. Mark on it the finish floor to finish floor height (total rise of the stairs). Set the dividers at a comfortable rise. Divide the rod into equal spaces, adjusting the dividers slightly as necessary (Fig. 47-8).

Determining the Tread Run

The tread run (width) is measured from the face of one riser to the next. It does not include the nosing (Fig. 47-9).

To find the tread run, apply the following rule.

- The sum of one rise and one tread should equal between 17 and 18. For example, if the riser height is 7 3/8 inches, then the minimum

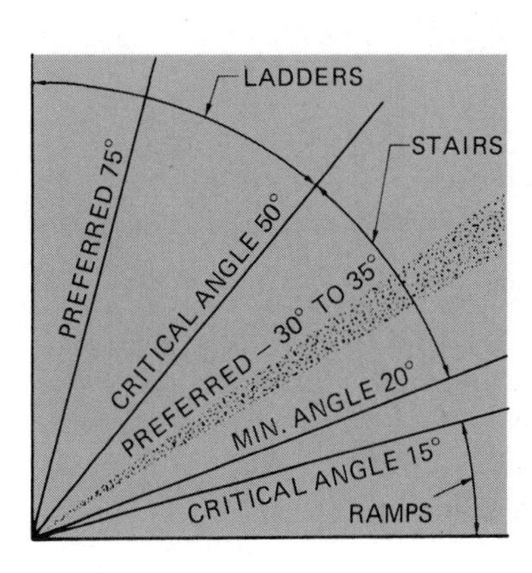

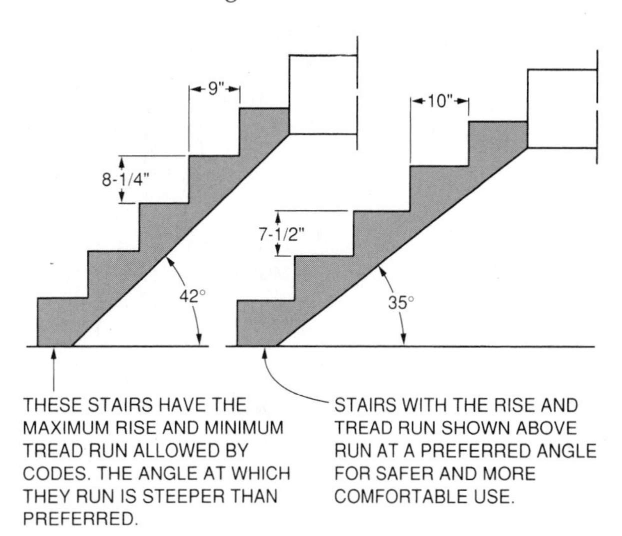

Fig. 47-7 Recommended angles for stairs.

STEP BY STEP
PROCEDURES

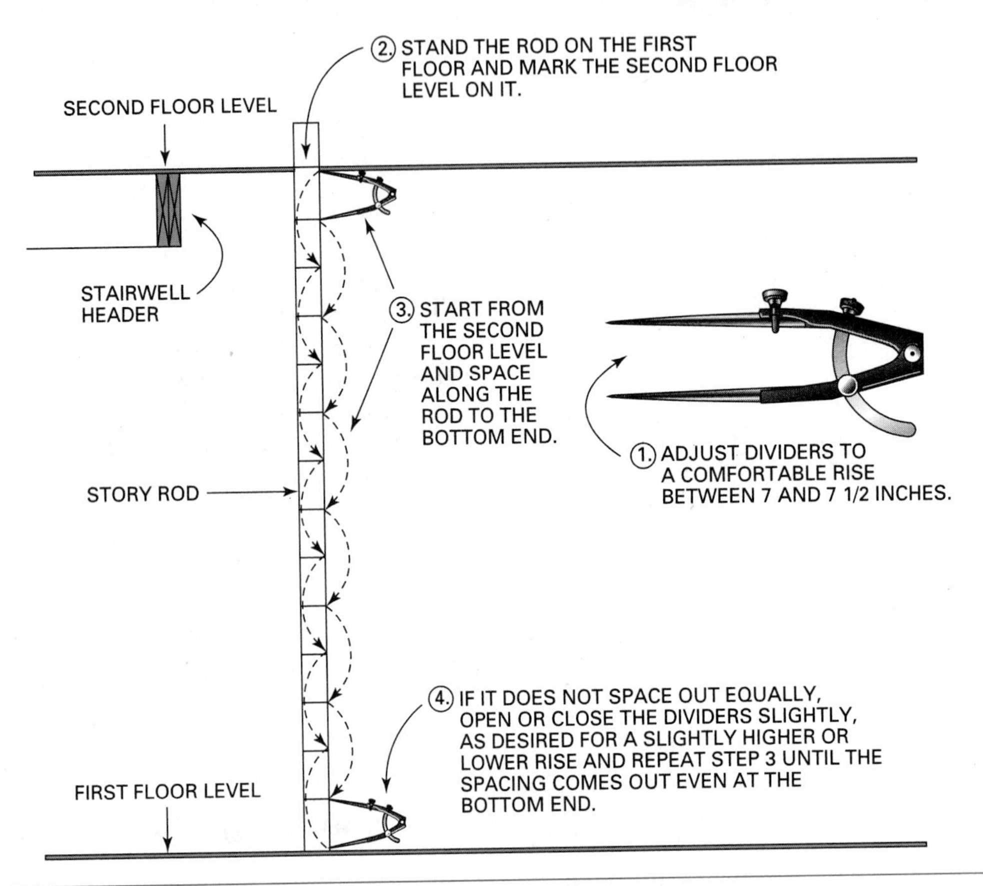

Fig. 47-8 The riser can be laid out on a story pole using dividers.

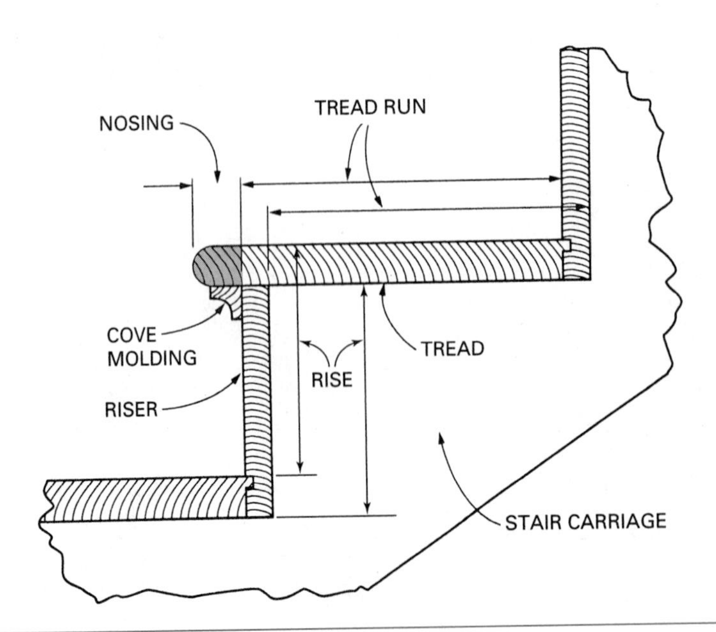

Fig. 47-9 The tread run does not include the nosing.

tread width may be 17 inches minus 7 3/8 inches. This equals 9 5/8 inches. The maximum tread width may be 18 inches minus 7 3/8 inches. This equals 10 5/8 inches.

Another formula to find the tread width is found in many building codes. It states that the sum of two risers and one tread shall not be less than 24 inches nor more than 25 inches (Fig. 47-10). With this formula, a rise of 7 3/8 inches calls for a minimum tread width of 9 1/4 inches and maximum of 10 1/4 inches.

Decreasing the rise increases the run of the stairs. This uses up more space. Increasing the riser height decreases the run. This makes the stairs steeper and more difficult to climb. The carpenter must use good judgment and adapt the rise and run dimensions to the space in which the stairway is to be constructed in conformance with the building code. A riser height of 7 1/2 inches and a tread run of 10 inches makes a safe, comfortable stairway.

Framing a Stairwell

A stairwell is framed in the same manner as any large floor opening. Unit 11, Floor Framing, describes methods of framing floor openings. Several methods of framing stairwells are illustrated in Figure 47-11.

Determining the Size of the Stairwell

Stairwell Width. The width of the stairwell depends on the width of the staircase. The drawings show the finish width of the staircase. However, the stairwell must be made wider than the staircase to allow for wall and stair finish (Fig. 47-12). Extra width will be required for a handrail and other finish parts of an open staircase that makes a U-turn on the landing above. The carpenter must be able to determine the width of the stairwell by studying the size, type, and placement of the stair finish before framing the stairs.

Length at the Stairwell. The length of the stairwell depends on the slope of the stairway. Stairs with a low angle require a longer stairwell to provide adequate *headroom.* Headroom is the vertical distance measured from the outside corner of a tread and riser to the ceiling above (Fig. 47-13). Most building codes require a minimum of 6′-8″ for headroom. However, 7′-0″ headroom or even more is preferred.

To find the minimum length of the stairwell, add the thickness of the floor assembly above (including subfloor, floor joists, and ceiling finish) to the desired headroom. Divide by the riser height.

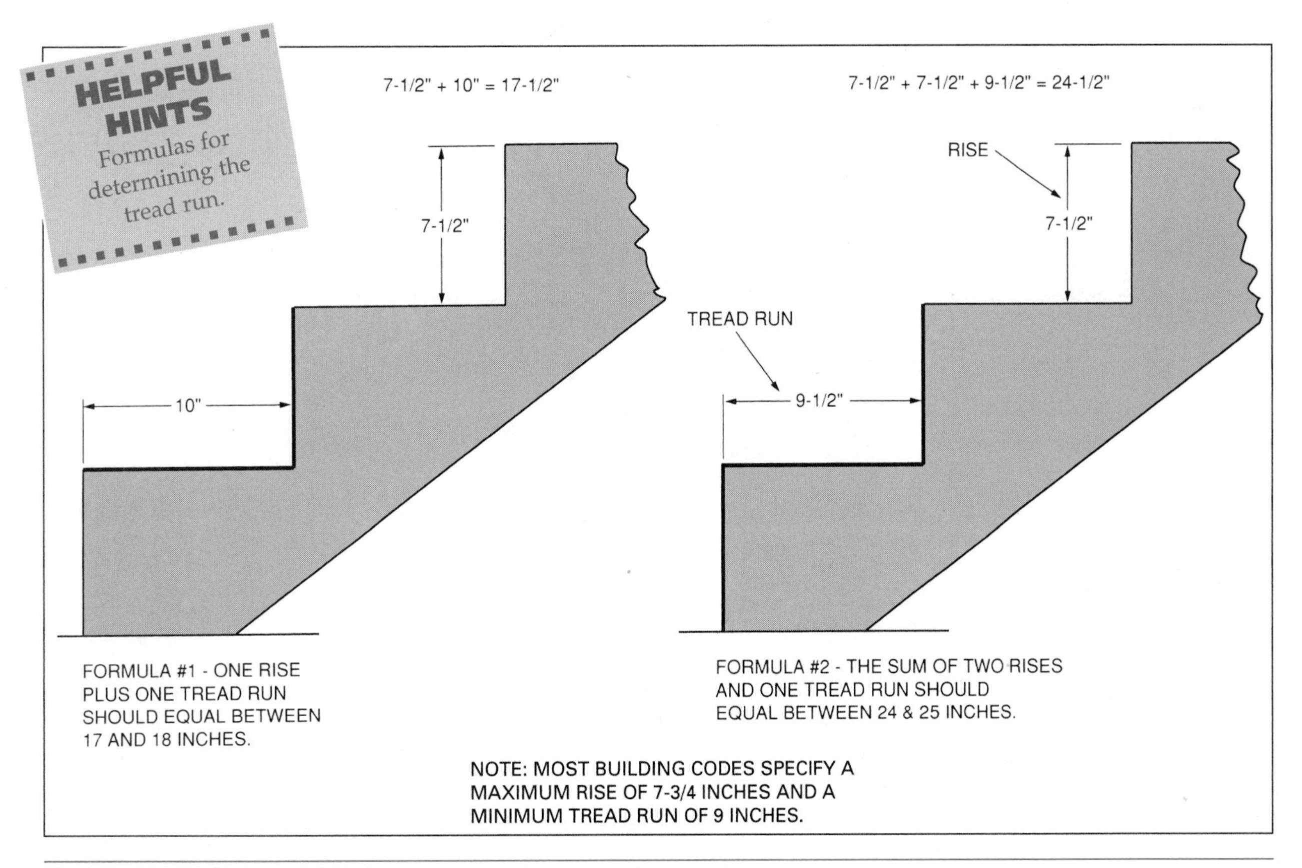

Fig. 47-10 Techniques for determining the run of a tread.

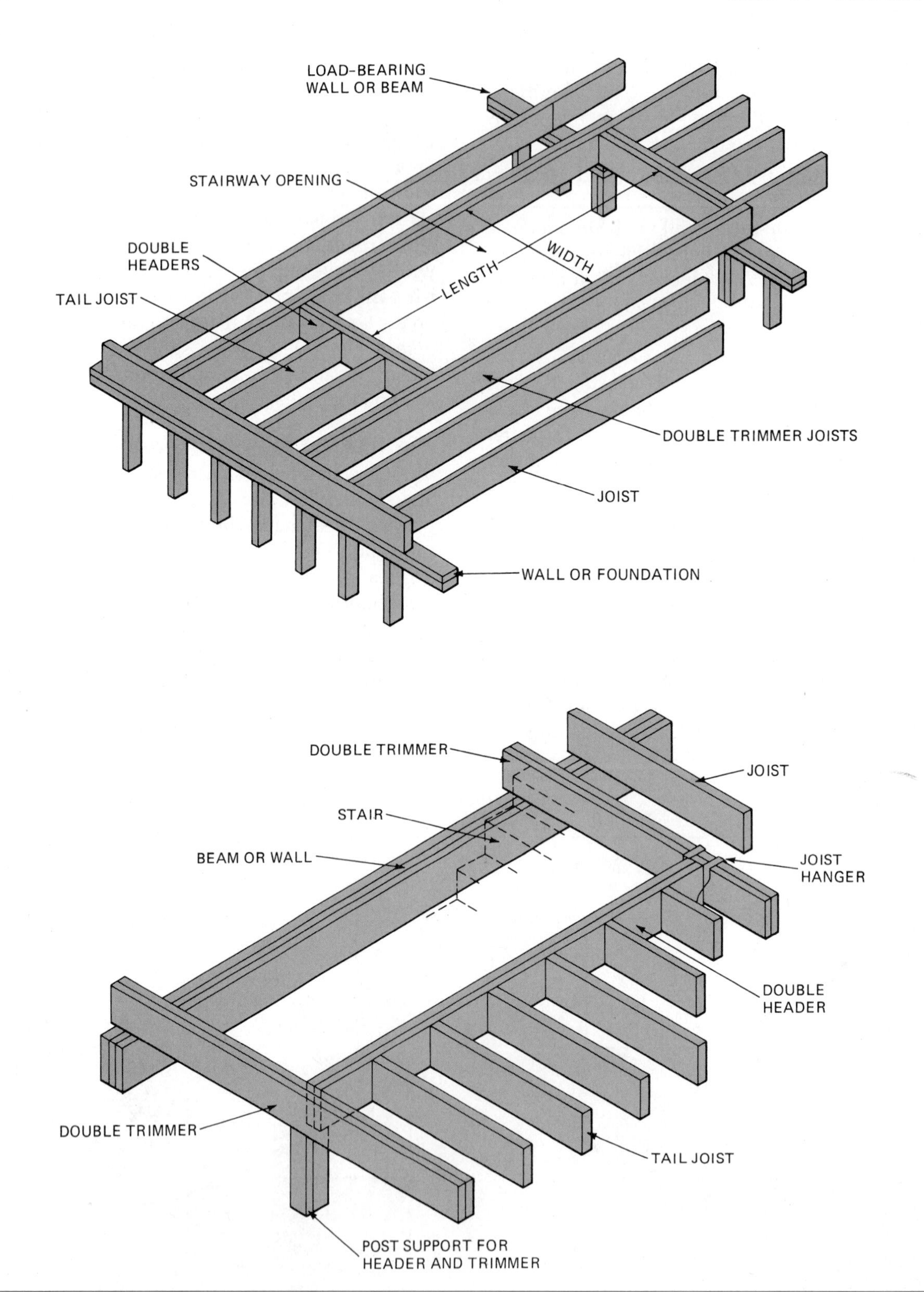

Fig. 47-11 Methods of framing stairwells.

The result will probably be a whole number and a fraction. Use the next largest whole number. Multiply by the tread width to find the length of the stairwell.

For example, a stairway has a riser height of 7 1/2 inches, a tread width of 10 inches, and a total thickness of the floor above of 11 3/4 inches. When 11 3/4 inches is added to the desired headroom of 84 inches, the total is 95 3/4 inches. Dividing this total by 7 1/2 equals 12.77. Changing 12.77 to 13 and multiplying by 10 inches equals 130 inches or 10′-10″, the minimum length of the stairwell (Fig. 47-14).

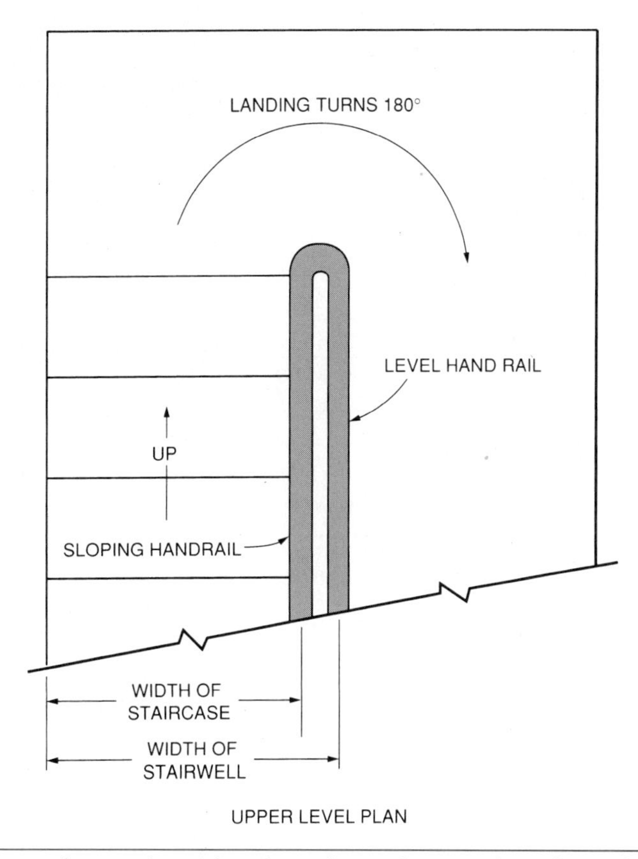

Fig. 47-12 The stairwell must be made wider than the staircase. This allows for wall and stair finish.

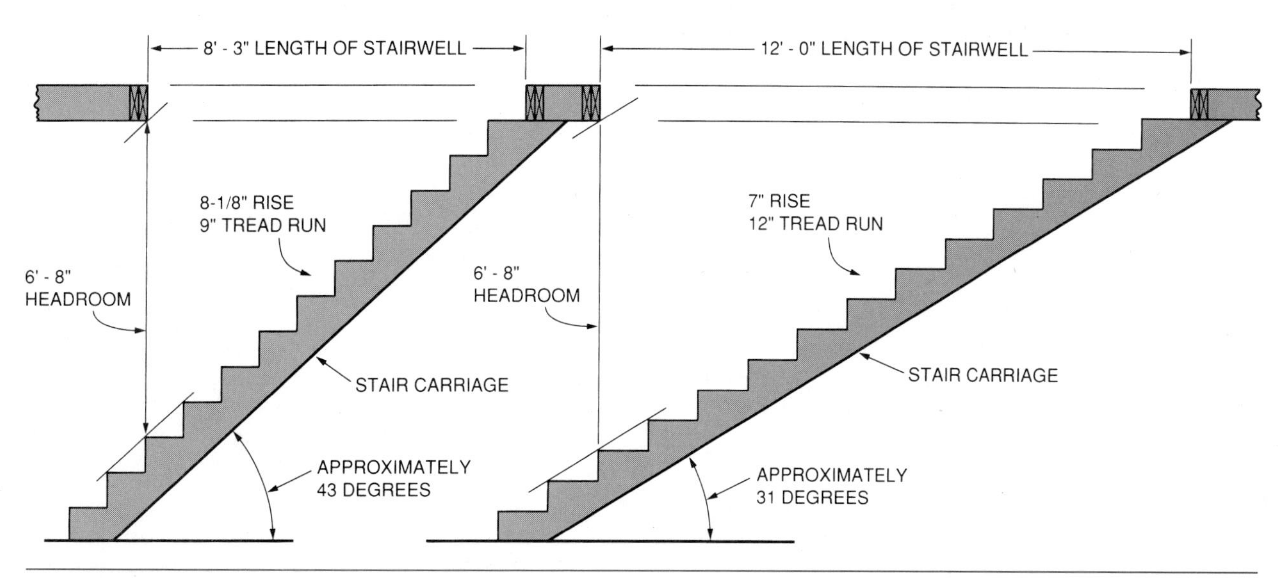

Fig. 47-13 Low-angle stairs require longer stairwells to provide adequate headroom.

This length is correct if the header of the stairwell acts as the top riser. If the carriage is framed flush with the top of the opening, add another tread width to the length of the stairwell. A longer stairwell will provide more headroom. Additional headroom can be obtained by framing the header above the low end of the staircase at the same angle as the stairs (Fig. 47-15).

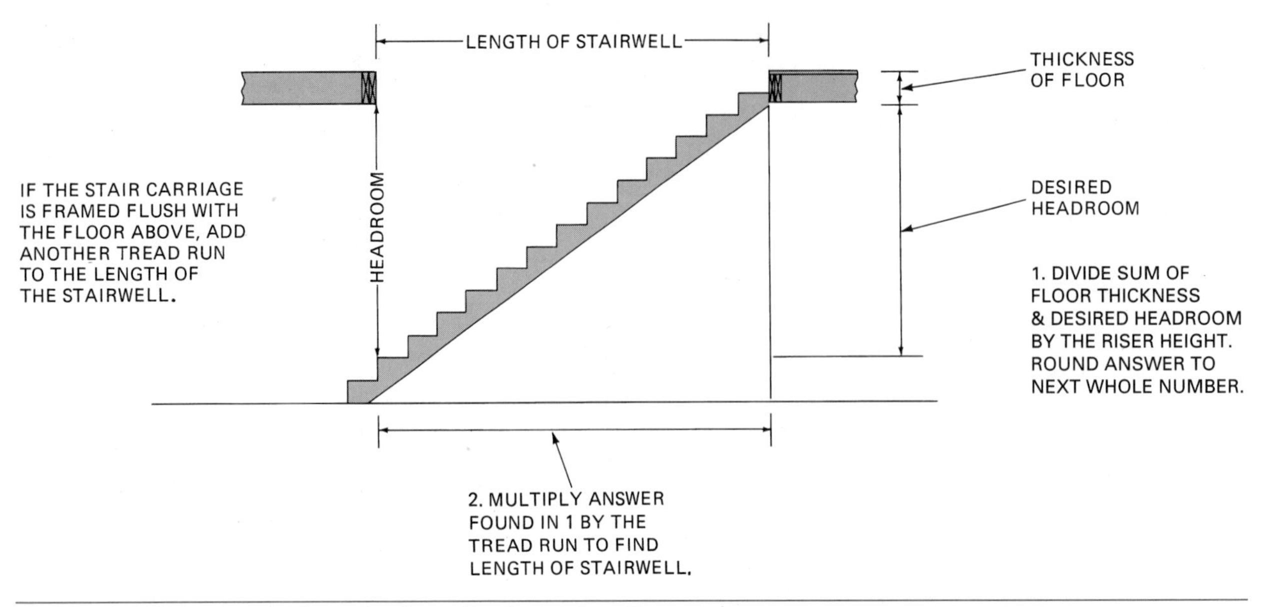

Fig. 47-14 How to calculate the length of a stairwell.

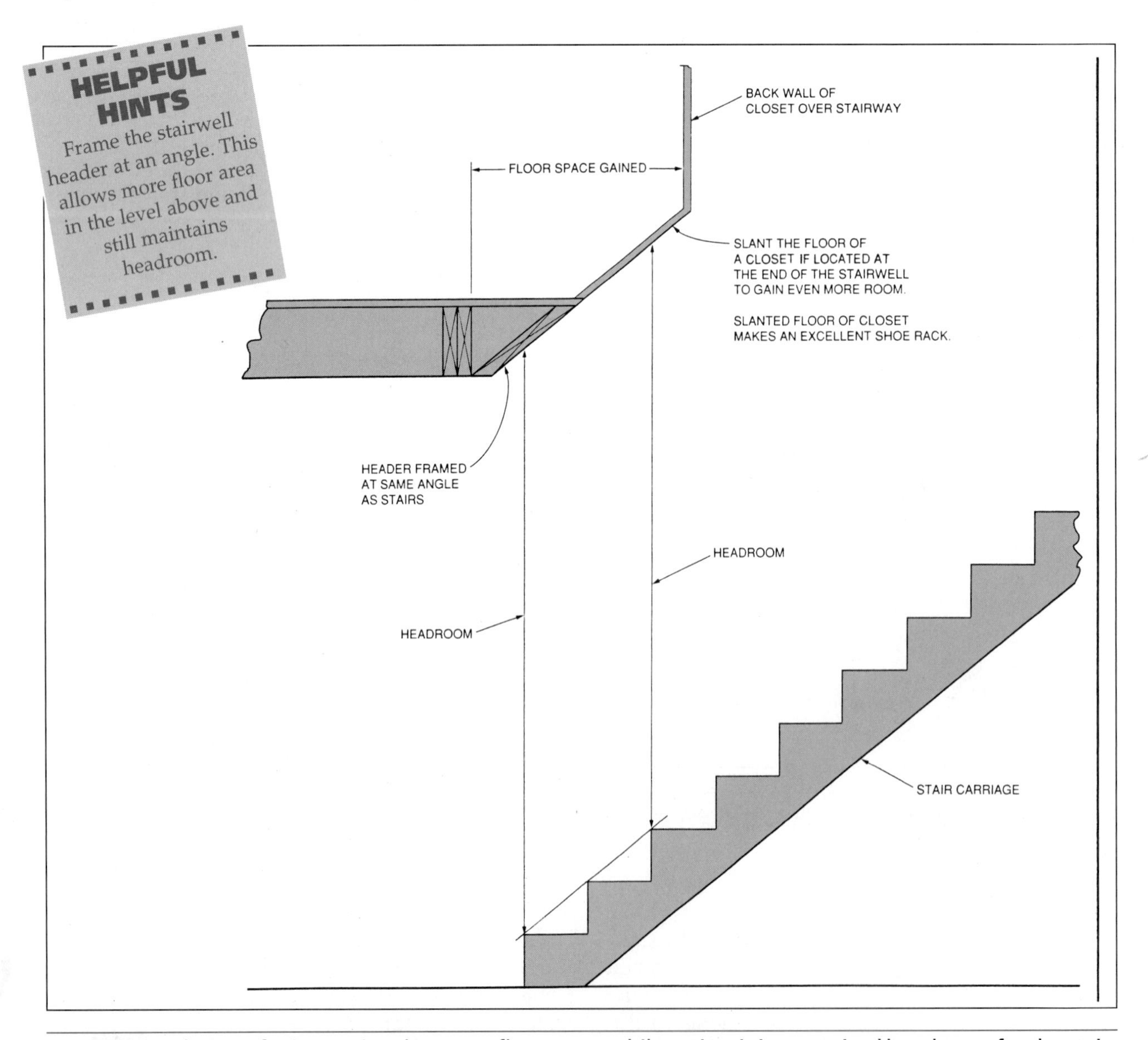

Fig. 47-15 Technique for increasing the upper floor space while maintaining required headroom for the stair.

CHAPTER 48 STAIR LAYOUT AND CONSTRUCTION

All stairs are laid out in approximately the same way. The use, location, and cost of stairs determine the way they are built. Regardless of the kind of stairs, where they are located, or how much they cost, care should be taken in their layout and construction.

Laying Out a Stair Carriage

When laying out stair carriages, make sure that every riser height will be the same. Also make sure all tread widths will be equal when the staircase is finished. It is very dangerous to descend a flight of stairs and unexpectedly step down a longer or shorter distance than the previous step. It is also dangerous for a person accustomed to stepping on treads of a certain width suddenly to step on a narrower or wider one when descending stairs. Stairs that are not laid out and constructed properly could cause a fatal accident.

Determining the Rough Length of a Stair Carriage

The length of lumber needed for the stair carriage is often determined using the Pythagorean Theorem. It also can be found by scaling across the framing square. Use the edge of the square that is graduated in 12ths of an inch. Mark the total rise on the tongue. Then mark the total run on the blade. Scale off in between the marks (Fig. 48-1).

For example, a stairway has a total rise of 8′-9″ and a total run of 12′-3″. Scaling across the square between 8–9/12ths and 12–3/12ths at a scale of 1″ = 1′-0″ reads a little over 15. A 16-foot length of lumber is needed for the stair carriage.

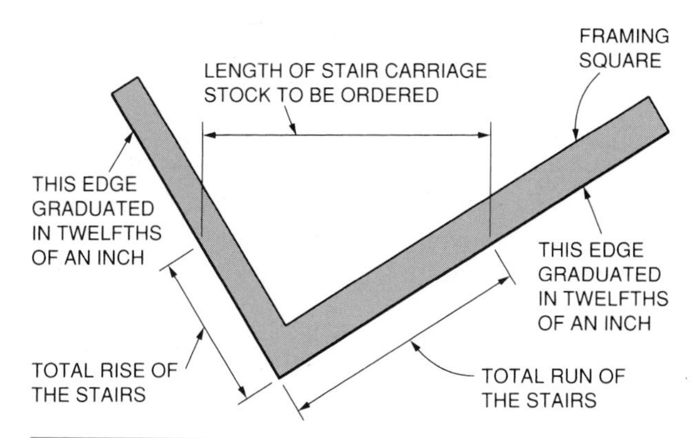

Fig. 48-1 Scale across the framing square to find the rough length of a stair carriage.

Stepping-Off the Stair Carriage

Place the stair carriage stock on a pair of sawhorses. Sight the stock for a crowned edge. Stand on the side of the crowned edge when laying out. This will be the top edge of the carriage. Set gauges on the framing square with the rise on the tongue and the tread run on the blade. Or, use a pitch block. Lines laid out along the tongue will be plumb lines. Those laid out along the blade will be level lines when the stair carriage is in its final position. Step-off the necessary number of times, marking along the outside of the tongue and blade. These lines are the back sides of the finish risers and undersides of the finish treads (Fig. 48-2). Lay out a level line at the top of the carriage for the tread. Lay out a level line on the bottom where the carriage sits on the floor.

Equalizing the Bottom Riser

A certain amount may have to be cut off the bottom end of the stair carriage. This will make the bottom riser equal in height to all the other risers when the staircase is finished. The thickness of the finish floor and the finish stair treads must be known.

- If the tread stock is thicker than the finish floor, and the carriage rests on the subfloor, then the first riser is made less by the difference.
- If the carriage rests on the finish floor, the first riser is made less by the thickness of the tread stock.
- If the bottom of the carriage rests on the subfloor, and the finish floor and tread stock are the same thickness, the height of the first rise is the same as all the rest. Nothing more is cut off the bottom end of the stair carriage.

Lay out a level line where the carriage will rest on the floor, according to the conditions. The first riser should be the same as all the rest when the staircase is complete. Cutting the bottom end off to equalize the bottom riser is called *dropping* the stair carriage (Fig. 48-3).

Equalizing the Top Riser

No amount is cut from the level line at the top of the carriage. The top riser is equalized by fastening the carriage a certain distance below the subfloor of

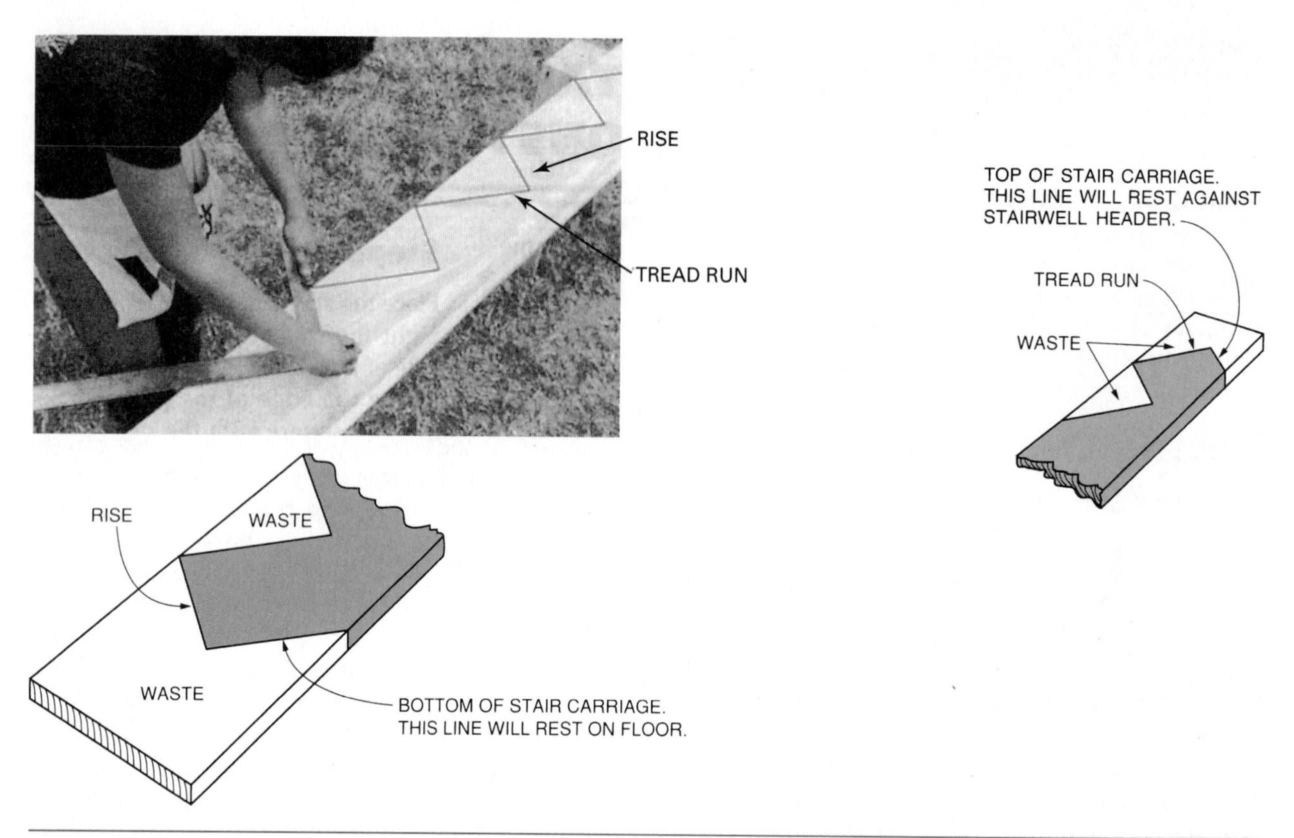

Fig. 48-2 Laying out a stair carriage by stepping-off with a framing square.

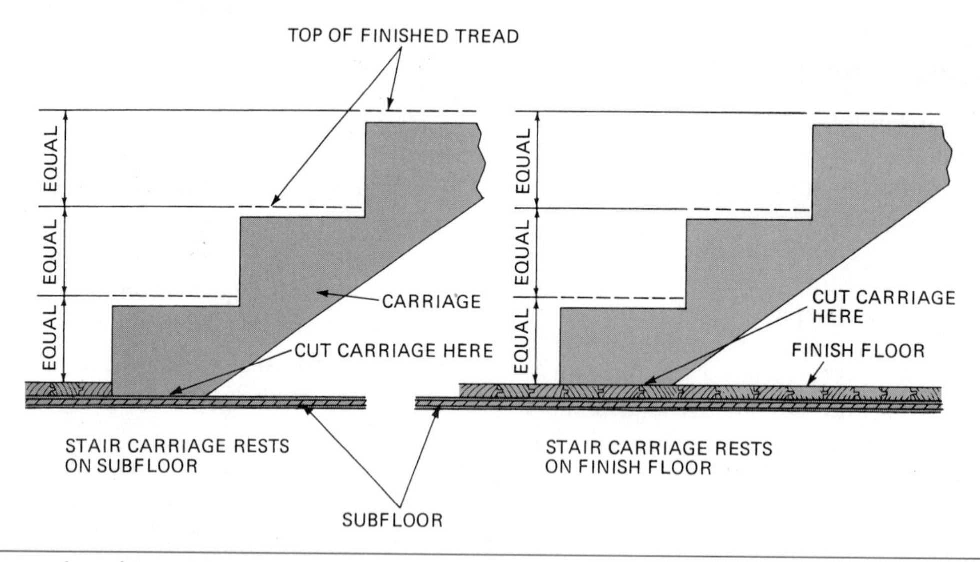

Fig. 48-3 Dropping the stair carriage to equalize the first riser.

the top landing. This distance depends on the thickness of the stair tread and the finish floor above.

- With the header of the stairwell acting as the top riser, if the stair tread thickness is the same as the finish floor, the carriage is fastened a riser height below the subfloor.
- If the stair tread is thicker than the finish floor, fasten the carriage a distance of a riser, plus the difference, below the subfloor of the top landing.

For example, if the tread stock is 1 1/16 inches and the finish floor above is 3/4 inch, the difference is 5/16 inch more than the riser height below the top of the subfloor above (Fig. 48-4). Whatever conditions may be encountered, take steps to equalize the top and bottom risers for the construction of a safe staircase.

Width of the Top Tread

The width of the top finished tread must be the same as all others in the staircase. If the carriage is

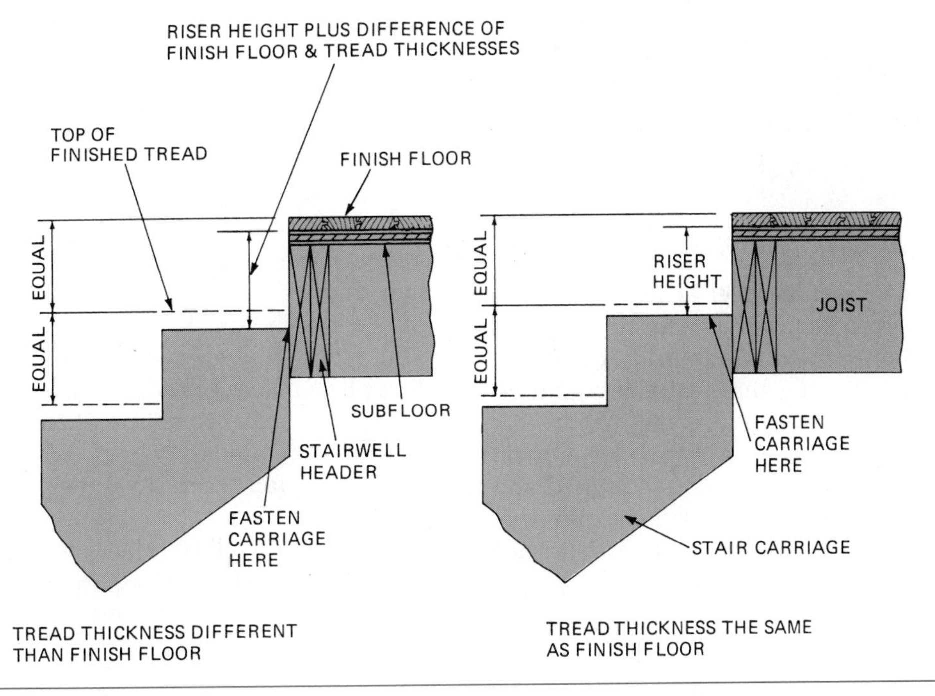

Fig. 48-4 Equalizing the top riser.

framed against the stairwell *header,* and the same thickness of finish is applied to the header as the thickness of the riser stock, then the tread cut at the top of the carriage is made the same as all others. Measure the tread run from the top riser. Lay out a plumb line.

If the carriage is framed to the stairwell in a different manner, allowances must be made so that the top finished tread width will be the same as all others in the staircase (Fig. 48-5).

Cutting the Stair Carriages

After the first carriage is laid out, cut it. Follow the layout lines carefully because this will be a pattern for others.

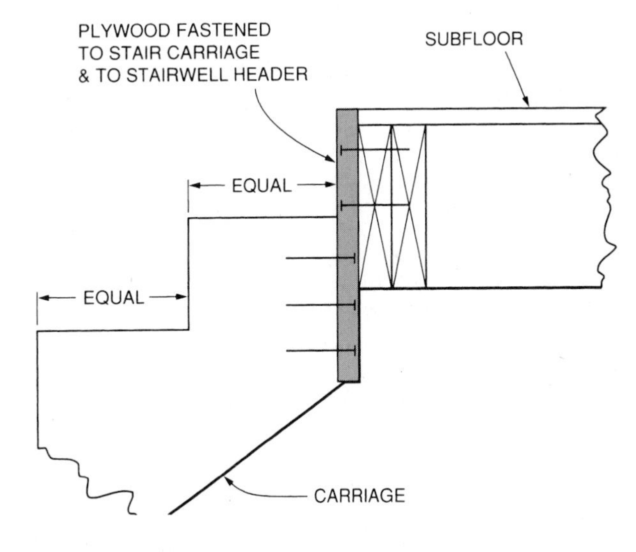

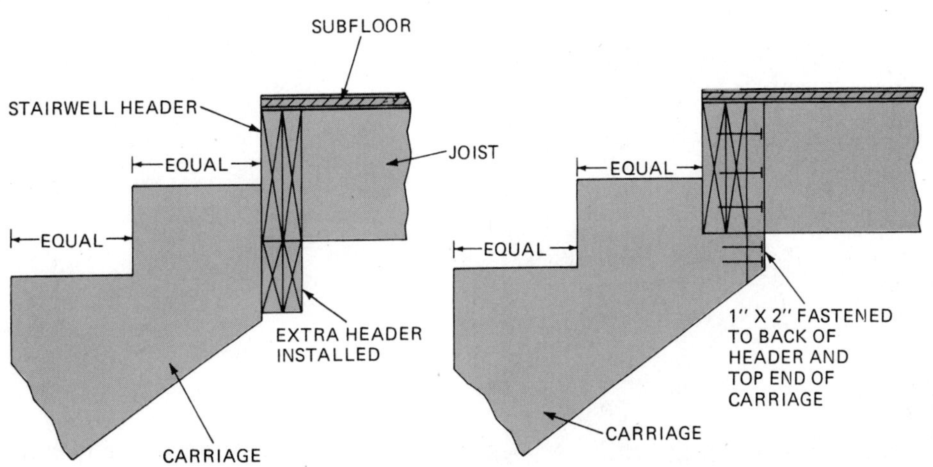

Fig. 48-5 Methods of framing the stair carriage to the stairwell header.

CAUTION When making a cut at a sharp angle to the edge, the guard of the saw may not retract. Retract the guard by hand until the cut is made a few inches into the stock. Then release the guard and continue the cut. Never wedge the guard in an open position.

Finish the cuts at the intersection of the riser and tread run with a handsaw. Use the first carriage as a pattern. Lay out and cut as many other carriages as needed. Usually, three carriages are used for residential staircases of average width. For wider stairs, the number of carriages depends on such factors as whether or not risers are used and the thickness of the tread stock. Check the drawings or building code for the spacing of carriages for wider staircases.

Built-Up Stair Carriage

In order to conserve materials, the triangular blocks cut out of the stair carriage may be fastened to the edge of a fairly clear, straight-grained 2 × 4 to form a *built-up stair carriage* (Fig. 48-6). The blocks are fastened by using the first stair carriage as a pattern. Built-up stair carriages are usually used as intermediate carriages in the stairway.

Framing a Straight Stairway

If the stairway is either completely closed or closed on one side, the walls must be prepared before the stair carriages are fastened in position.

Preparing the Walls of the Staircase

Gypsum board (drywall) is sometimes applied to walls before the stair carriages are installed against them. This procedure saves time. It eliminates the need to cut the drywall around the cutouts of the stair carriage. This method also requires no blocking between the studs to fasten the ends of the drywall panels. However, the time saved may be offset by some disadvantages.

Blocking between the studs in back of the stair carriage provides backing for fastening the stair trim. Lack of it may cause difficulty for those who apply the finish. Also, any repair or remodeling work will be more difficult if gypsum board is applied before the stair carriages. In addition, drywall is usually not applied until after the framing is complete. Workers are deprived of the stair frame for access to upper levels until that time.

If the drywall is to be applied after the stairs are framed, blocking is required between studs in back of the stair carriage to fasten the ends of the gypsum board. Snap a chalk line along the wall sloped at the same angle as the stairs. Be sure the top of the blocking is sufficiently above the stair carriage to be useful. Install 2 × 6 or 2 × 8 blocking, on edge, between and flush with the edges of the studs. Their top edge should be to the chalk line (Fig. 48-7).

Housed staircases may be installed after the walls are finished. Sometimes they are installed before the walls are finished. Then, they must be furred out away from the studs. This allows the wall finish to extend below the top edge of the finished stringer.

Fig. 48-6 Making a built-up stair carriage.

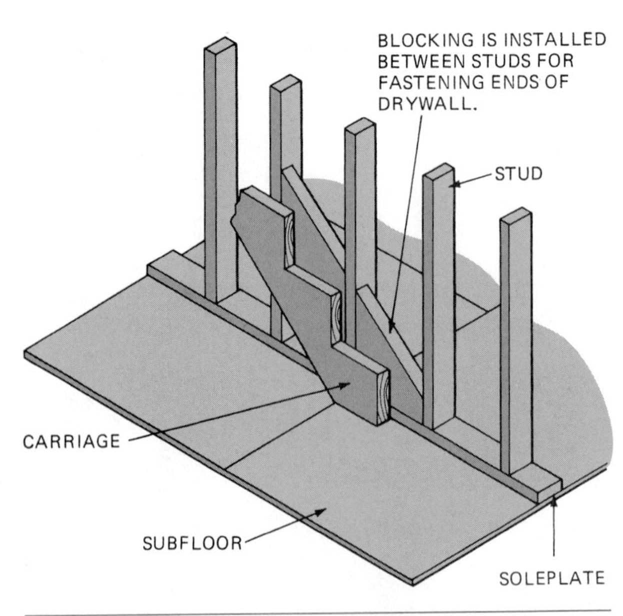

Fig. 48-7 Preparation of the wall for application of gypsum board before stair carriages are installed.

Installing the Stair Carriage

When installing the stair carriage, fasten the first carriage in position on one side of the stairway. Attach it at the top to the stairwell header. Make sure the distance from the subfloor above to the top tread run is correct.

Fasten the bottom end of the stair carriage to the soleplate of the wall and with intermediate fastenings into the studs. Drive spikes near the bottom edge of the carriage. This prevents splitting the triangular sections.

Fasten a second carriage on the other wall in the same manner as the first. If the stairway is to be open on one side, fasten the carriage at the top and bottom of the staircase. The location of the stair carriage on the open end of a stairway is in relation to the position of the **handrail**. First, determine the location of the centerline of the handrail. Then, position the stair carriage on the open side of a staircase. Make sure its outside face will be in a line plumb with the centerline of the handrail when it is installed (Fig. 48-8).

Fasten intermediate carriages at the top into the stairwell header and at the bottom into the subfloor. Test the tread run and riser cuts with a straightedge placed across the outside carriages (Fig. 48-9). About halfway up the flight, or where necessary, fasten a temporary riser board. This straightens and maintains the spacing of the carriages (Fig. 48-10).

If a wall is to be framed under the stair carriage at the open side, fasten a bottom plate to the subfloor plumb with the outside face of the carriage. Lay out the studs on the plate. Cut and install studs under the carriage in a manner similar to that used to install gable studs. Be careful to keep the carriage straight. Do not crown it up in the center (Fig. 48-11). Install rough lumber treads on the carriages until the stairway is ready for the finish treads.

Laying Out and Framing a Stairway with a Landing

A stair landing is an intermediate platform between two flights of stairs. A landing is designed for changing the direction of the stairs and as a resting

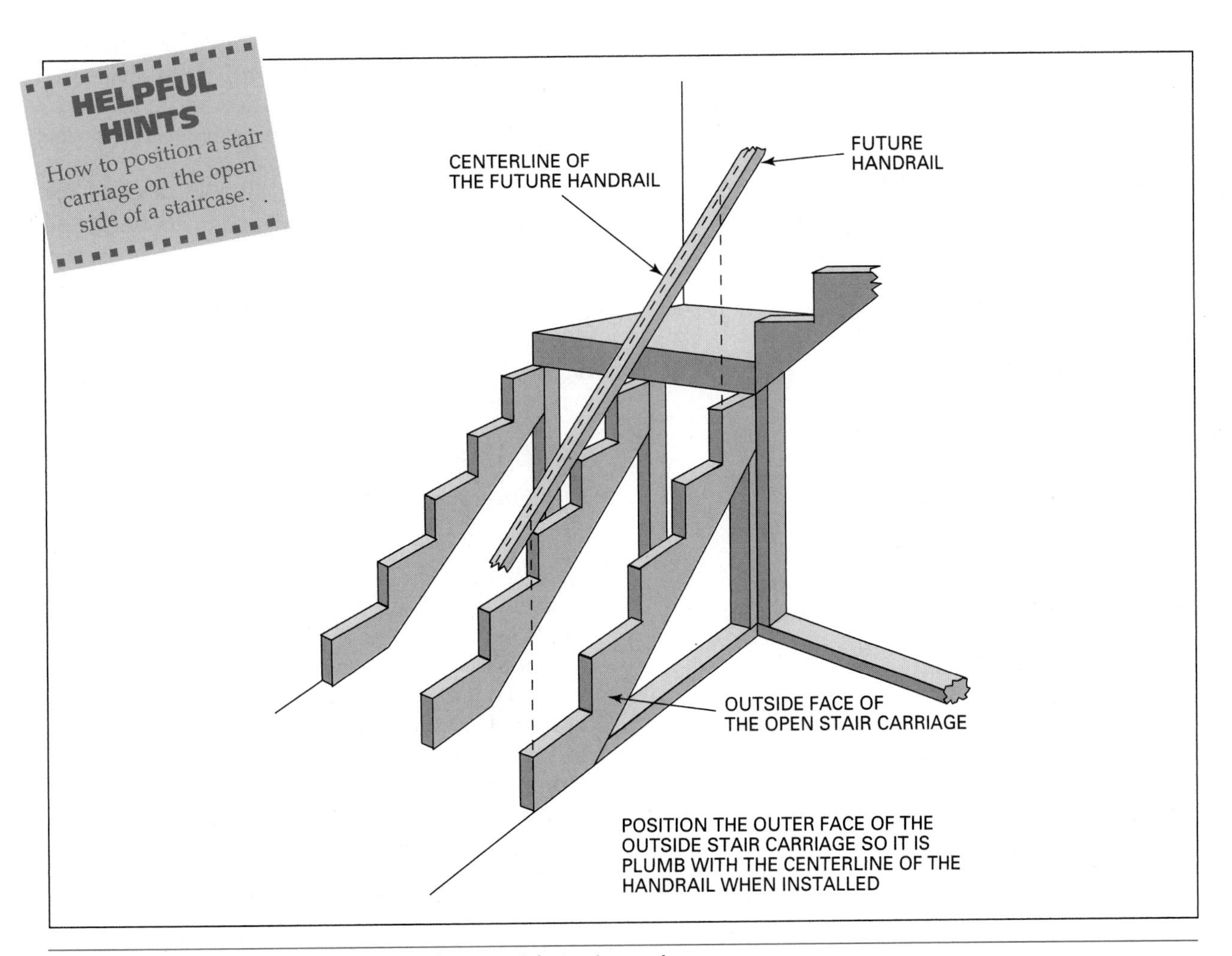

Fig. 48-8 Techniques for locating the outside stair carriage.

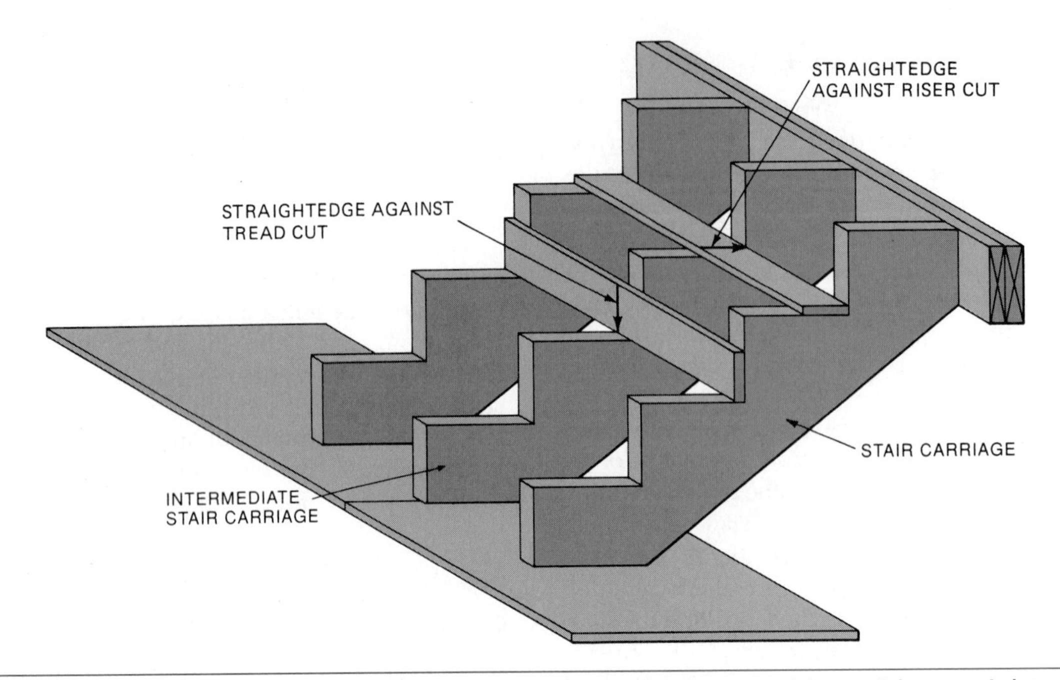

Fig. 48-9 Check the position of tread and riser cuts on intermediate carriages with a straightedge.

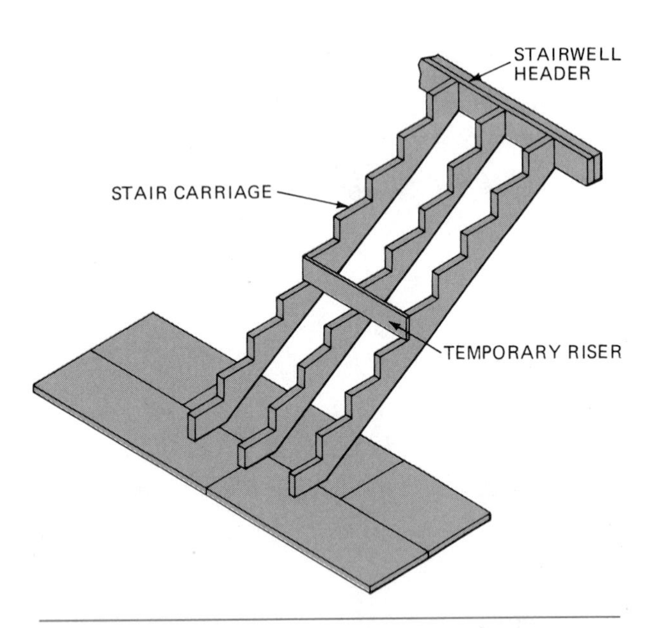

Fig. 48-10 Fasten a temporary riser about halfway up the flight. This straightens and maintains the carriage spacing.

place for long stair runs. The landing usually is floored with the same materials as the main floors of the structure. Some codes require that the minimum length of a landing be not less than 2′-6″. Other codes require the minimum dimension to be the width of the stairway. L-type stairs may have the landing near the bottom or the top. U-type stairs usually have the landing about midway on the flight. Many codes state that no flight of stairs shall have a vertical rise of more than 12 feet. Therefore, any staircase running between floors with a vertical distance of more than 12 feet must have at least one intermediate landing or platform.

Platform stairs are built by first erecting the platform. The finish floor of the platform should be the same height as if it were a finish tread in the staircase. This allows an equal riser height for both flights (Fig. 48-12). The stairs are then framed to the platform as two straight flights. Either the stair carriage or the housed stringer method of construction may be used (Fig. 48-13).

Laying Out Winding Stairs

Winding stairs change direction without a conventional landing. They often will allow for two extra risers in the space normally occupied by the landing. This is particularly useful for stairwells that are too small for normal stair construction. **Winders,** however, are not recommended for safety reasons. Some codes state that they may be used in individual dwelling units, if the required tread width is provided along an arc. This is called the *line of travel.* It is a certain distance from the side of the stairway where the treads are narrower. The same code also states that in no case shall any width of tread be less than 6 inches at any point (Fig. 48-14). Some other codes may permit a narrower tread. Check the building code for the area in which the work is being performed.

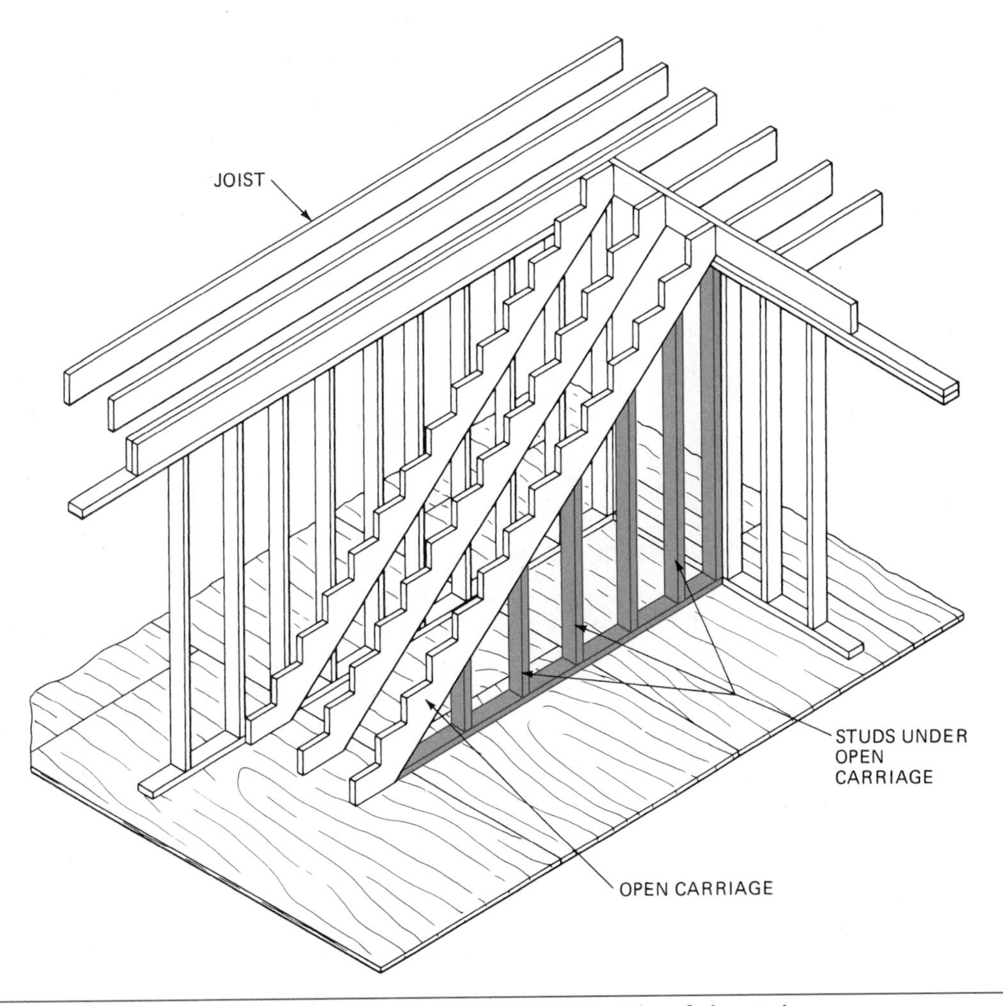

Fig. 48-11 Install studs under the stair carriage on the open side of the stairway.

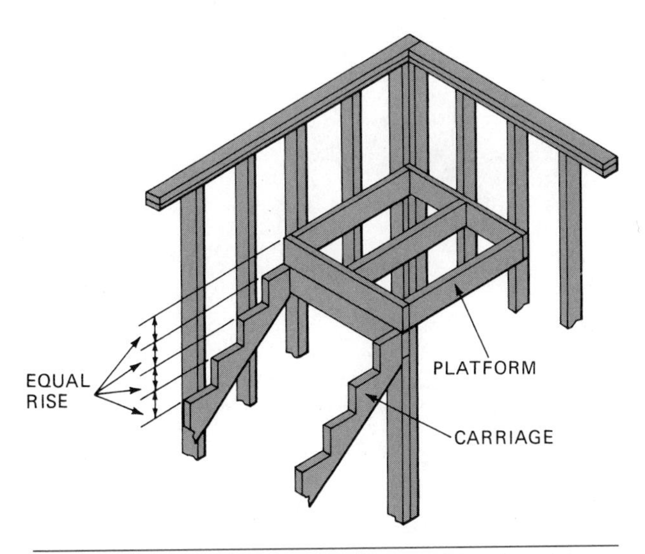

Fig. 48-12 The platform is located so that its top is in line with one of the tread runs.

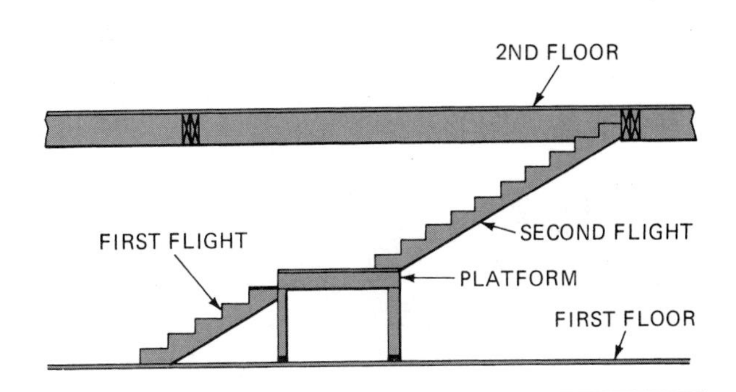

Fig. 48-13 Stair carriages are framed to the platform in a way similar to that used for two straight flights.

To lay out a winding turn of 90 degrees, draw a full-size winder layout on the floor directly below where the winders are to be installed. On a closed staircase, the walls at the floor line represent its sides. For stairs open on one side, lay out lines on the floor representing the outside of the staircase. The wall represents the inside. Swing an arc, showing the line of travel. Use the corner of the wall or

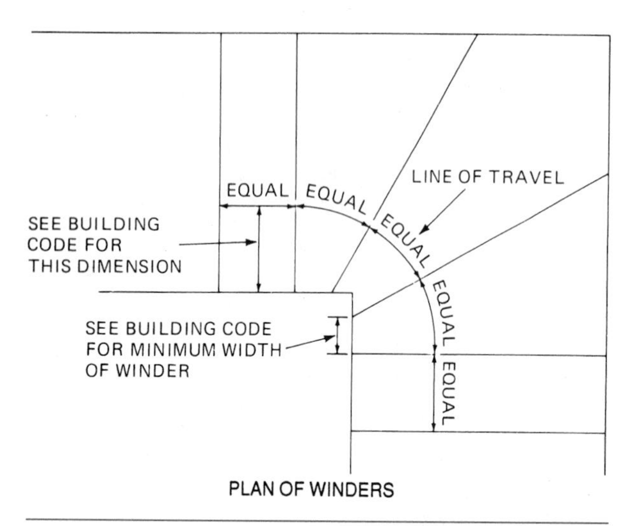

Fig. 48-14 Specifications for winding stairs.

layout lines as center. The radius of the arc may be 12 to 18 inches, as the codes permit. Fifteen inches is usually used.

From the same center, lay out the width of the narrow end of the treads in both directions. Square lines from the end points to the opposite side of the staircase. Divide the arc into equal parts. Project lines from the narrow end of the tread, through the intersections at the arc, to the wide end at the wall. These lines represent the faces of the risers. Draw lines parallel to these to indicate the riser thickness. Plumb these lines up the wall to intersect with the tread run for each winder (Fig. 48-15).

The cuts on the stair carriage for the winding steps are obtained from the full-size layout. Lay out and cut the carriage. Fasten it to the wall. Install rough treads until the stairs are ready for finishing.

If one side of the staircase is to be open, a **newel post** is installed. Then, the risers are mitered to or **mortised** into the post (Fig. 48-16). A **mortise** is a rectangular cavity in which the riser is inserted. Newel posts are part of the stair finish. They are described in more detail in Chapters 80 and 82.

STEP BY STEP

PROCEDURES

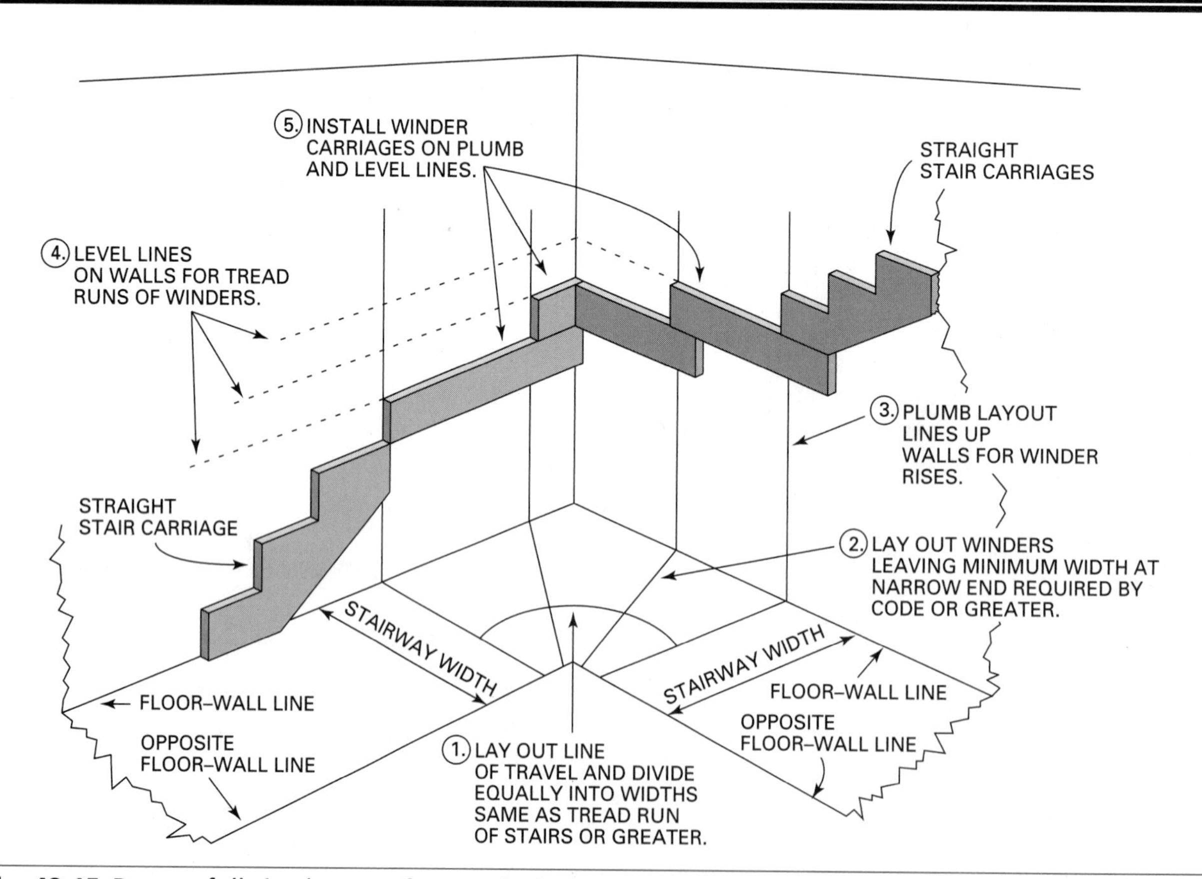

Fig. 48-15 Draw a full-size layout of a set of winders on the floor to find the rise and tread run of the stair carriage.

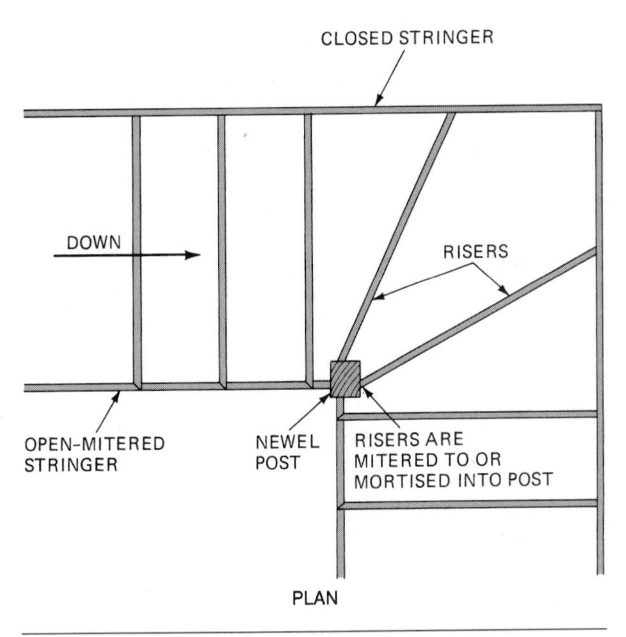

Fig. 48-16 On the open side, winder risers are mitered against or mortised into a newel post.

Framing Basement Stairs

Basement stairs are sometimes built without riser boards. Two carriages are used. Nominal 2 × 10 treads are cut in between them. The carriages are not always cut out, like those previously described. They may be dadoed to receive the treads. An alternative method is to fasten *cleats* to the carriages to support the treads (Fig. 48-17).

Lay out the carriages in the usual manner. Cut the bottoms on a level line to fit the floor. Cut the tops along a plumb line to fit against the header of the stairwell. "Drop" the carriages as necessary to provide a starting riser with a height equal to the rest of the risers.

Dadoed Carriages

If the treads of rough stairs are to be dadoed into the carriages, lay out the thickness of the tread on the stringer below the layout line. The top of the tread

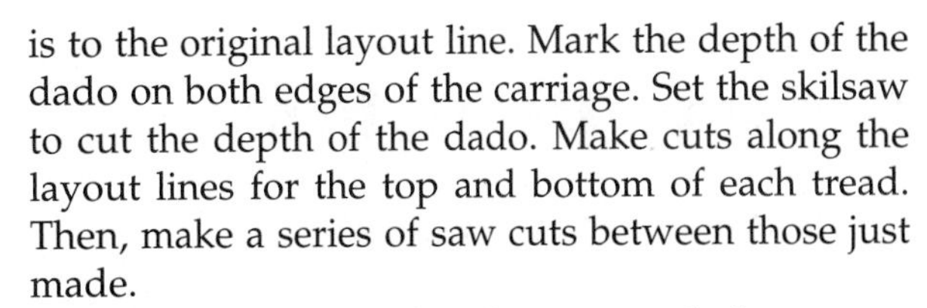
is to the original layout line. Mark the depth of the dado on both edges of the carriage. Set the skilsaw to cut the depth of the dado. Make cuts along the layout lines for the top and bottom of each tread. Then, make a series of saw cuts between those just made.

Chisel from both edges toward the center. Remove the excess to make the dado. Nail all treads into the dadoes. Assemble the staircase. Fasten the assembled staircase in position. Locate the top tread at a height to obtain an equal riser at the top. The lower end of a basement stairway is sometimes anchored by installing a **kicker** plate, which is fastened to the floor (Fig. 48-18).

Cleated Carriages

If the treads are to be supported by cleats, measure down from the top of each tread a distance equal to its thickness. Draw another level line. Fasten 1 × 3 cleats to the carriages. Make sure their top edges are to the bottom line. Fasten the carriages in position. Cut the treads to length. Install the treads between the carriages so the treads rest on the cleats. Fasten the treads by nailing the stair carriage into their ends.

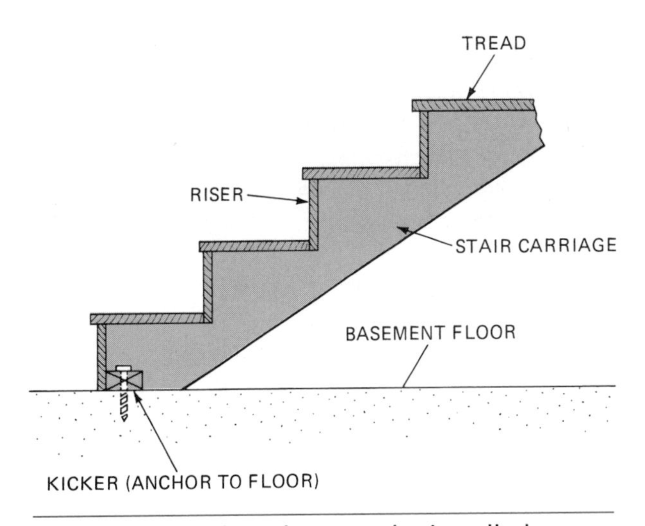

Fig. 48-18 A kicker plate may be installed to fasten the bottom end of basement stairs.

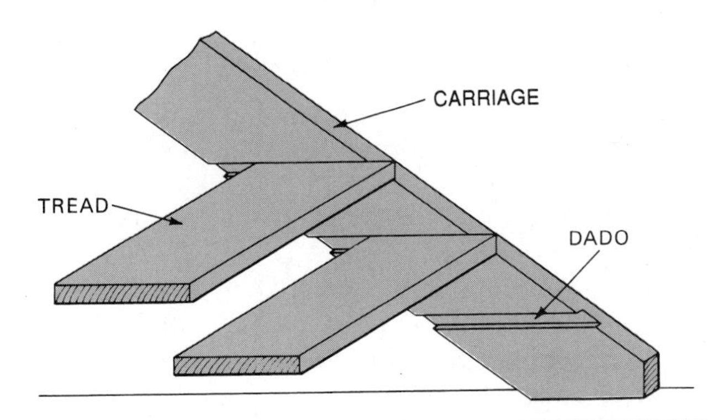

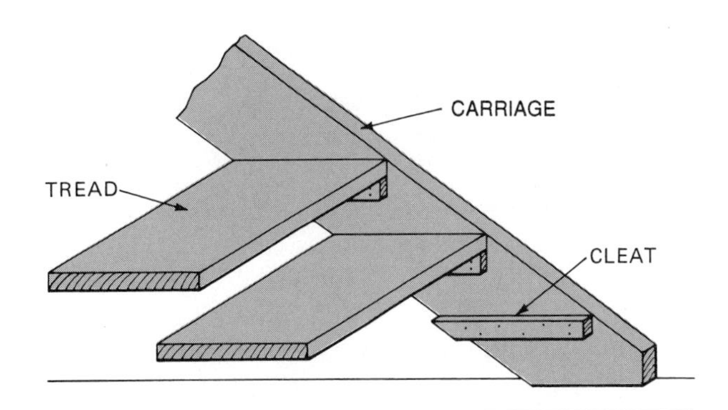

Fig. 48-17 Basement and service stair carriages may be dadoed or cleated to support the treads.

Review Questions

Select the most appropriate answer.

1. Stairways in residential construction should have a minimum width of
a. 30 inches. c. 36 inches.
b. 32 inches. d. 40 inches.

2. A nosing is
a. a horizontal member of a step.
b. a vertical member of a step.
c. the top molded edge of a finish stringer.
d. the overhang of the tread beyond the riser.

3. The maximum riser height stated in this unit is
a. 8 1/4 inches. c. 7 1/2 inches.
b. 7 3/4 inches. d. 7 1/4 inches.

4. The riser height for a stairway with a total vertical rise of 8′-9″ is
a. 8 inches. c. 7 5/8 inches.
b. 7 3/4 inches. d. 7 1/2 inches.

5. Applying the rules stated in this unit, what is the tread run if the riser height in a stairway is 7 1/4 inches?
a. 8 3/4 to 9 3/4 inches
b. 9 to 10 inches
c. 9 3/4 to 10 3/4 inches
d. 10 to 11 inches

6. Many building codes specify a minimum tread width of
a. 8 1/2 inches. c. 9 1/2 inches.
b. 9 inches. d. 10 inches.

7. A flight of stairs has a riser height of 7 1/2 inches and a tread run of 10 inches. The total thickness of the upper floor is 10 inches. What is the length of the stairwell if the stairwell header acts as the top riser and the desired minimum headroom is 6′-8″?
a. 9′-0″ c. 10′-0″
b. 9′-6″ d. 10′-2″

8. Most building codes specify a minimum head room clearance of
a. 6′-6″. c. 7′-0″.
b. 6′-8″. d. 7′-6″.

9. The stair carriage of a stairway with a riser height of 7 1/2 inches rests on the finish floor. What is the riser height of the first step if the tread thickness is 3/4 inch?
a. 6 3/4 inches c. 7 3/4 inches
b. 7 1/2 inches d. 8 1/4 inches

10. A stairway has a riser height of 7 1/2 inches. The tread stock thickness is 1 1/16 inches. The finish floor thickness of the upper floor is 3/4 inch. What distance down from the top of the upper subfloor must the rough carriage be fastened if the stairwell header acts as the top riser?
a. 7 3/16 inches c. 7 1/2 inches
b. 7 3/8 inches d. 7 13/16 inches

BUILDING FOR SUCCESS

Construction Associations and the College Student

How can a professional organization benefit college students who desire careers in construction?

This question is asked annually by many colleges and universities across the nation. The following true account is presented as a possible answer to this question. Although it is only one community college's response to involvement beyond the classroom, it may motivate other colleges as well.

In 1988 Larry Kness, the program chair for the building construction program at Southeast Community College (SCC), Milford, Nebraska, was looking for ways to provide more career information to students. Classrooms, textbooks, and labs were providing the essential opportunities and experiences for students in the construction program. The educational experiences were geared to prepare competent, skilled technicians to work in residential and light commercial construction. The program had proven to be exceptional in its capacity to send well-trained carpenters, masons, drafters, and estimators into the construction world. However, there was a need to make students more aware of the business aspect of construction.

General education courses were providing the basics in business, communications, accounting, finance, and math, but direct exposure to builders was lacking. The need for more student exposure to the local, state, and national construction world loomed as the missing link. After the need was identified, the program chair responded to an invitation from the Lincoln Home Builders Association (LHBA) to form a National Association of Home Builders (NAHB) student chapter on the Milford campus. LHBA offered to become the sponsor of the college student chapter. Through the combined efforts of builders, educators, and students, a new relationship was formed to promote quality construction education.

The idea was presented to the building construction students, who overwhelmingly accepted the proposal. Chuck Brazie, executive vice president of LHBA, initially proposed the formation of the chapter. Nadine Condello, the new CEO of LHBA, became the coordinator. She proved to be the catalyst who finalized the chapter formation at the national level. Condello demonstrated the necessary vision needed to make the student outreach a success.

Many local association members had been graduates of SCC's construction program, were employers of graduates, or were continuing supporters of the program. This membership and the board of directors were instrumental in the entire effort to create a local student chapter.

This new educational outreach opened the door to more student involvement in the construction industry. New opportunities to visit construction sites, listen to guest speakers, attend local NAHB meetings, and participate in student chapter competitions at the national level were introduced or expanded.

Students were provided an opportunity to look inside a national organization that has a goal of promoting quality construction businesses nationally. Avenues for meeting local, state, and national construction representatives became more readily available. The student chapter was introduced to new challenges and rewards through student involvement.

Participation in local activities with LHBA provided a means of generating finances to fund trips to the national LHBA convention. (This convention participation will be presented in the case study centering on the local LHBA activities.)

As we studied the opportunities available for chapter involvement at the national level, we saw dozens of other two-year and four-year colleges and universities actively participating with the NAHB. We saw what other schools were doing to supplement their construction programs. We were excited to belong to an organization that included many highly respected construction programs from colleges and universities such as Texas A&M, Purdue, Penn State, BYU, Snow College (Utah), Hinds Community College (Mississippi), and Ohio State. These schools represented a vast array of institutions that radiated quality. They became examples of excellence as we pursued more involvement.

Annually the NAHB student chapter community participates in national competitions such as construction marathons, essays on affordable housing, and "Outstanding Student Chapter." Through open competition, any chapter may compete with others within their classification (two-year or four-year schools) from across the country for national honors. The question we had to answer was, Can we realistically compete with the proven leaders across the nation?

After looking at the competition process, our student chapter and sponsoring association felt we should enter the "race" and learn how to compete at the national level. Our application for consideration as "Outstanding Student Chapter" was evaluated as a worthwhile competitor. We were notified that we had placed as one of the three finalists. We were invited to attend the national convention in Atlanta for the announcement of the winners. Our chapter set a goal to attend the national convention held in February 1991. At that convention we were personally awarded the honor of outstanding Student Chapter for 1990. Second place was awarded to Texas A&M. Third place went to Purdue University. This was the first time that a two-year construction program had won the top honor.

What began as an experiment to assist students in becoming better informed about their chosen industry has become an ongoing educational experience for our student chapter members. The potential financial awards, competitions, and attendance at national conventions are just a few of the benefits for students belonging to a college student chapter. The challenge for students to become more knowledgeable about the construction industry is an important part of belonging to a student chapter. This example of Southeast Community College and the NAHB is offered as proof to all that many opportunities exist for colleges to provide meaningful education beyond the classroom by way of a construction-related student chapter.

FOCUS QUESTIONS: For individual or group discussion

1. Considering your own educational institution, what might be some benefits of establishing a construction-related student chapter?
2. How could you initiate a charter chapter? Whom could you contact?
3. Do you feel that there would be institutional, faculty, industry, and student support to maintain an ongoing chapter?

UNIT 17

Insulation and Ventilation

Thermal insulation prevents the loss of heat in buildings during cold seasons. It also resists the passage of heat into air-conditioned buildings in hot seasons. *Acoustical* insulation reduces the passage of sound from one area to another. Adequate *ventilation* must be provided in a building. This evaporates moisture that may be formed in the space between thermal insulation and the cold surface.

OBJECTIVES

After completing this unit, the student should be able to:

- describe how insulation works and define insulating terms and requirements.
- describe the commonly used insulating materials and state where insulation is placed.
- properly install various kinds of insulation.
- state the purpose of and install vapor barriers.
- describe various methods of construction to reduce the transmission of sound.
- explain the need for ventilating a structure, describe types of ventilators, and state the minimum recommended sizes.

UNIT CONTENTS

CHAPTER 49 THERMAL AND ACOUSTICAL INSULATION

Most materials used in construction have some insulating value (Fig. 49-1). Even the dead air spaces between studs resist the passage of heat. After insulating material is installed, however, the stud space has many times the resistance to the passage of heat and sound than the air space alone had.

How Insulation Works

Air is an excellent insulator if confined to a small space and kept still. The **insulation** becomes more effective as the air spaces become smaller and greater in number. Millions of tiny air cells, trapped in its unique cellular structure, make wood a natural and effective insulating material. Most insulating materials are manufactured by trapping dead air in tiny spaces to provide resistance to the flow of heat and sound.

Thermal Insulation

Among the materials used for insulating are glass fibers, mineral fibers (rock), organic fibers (paper), and plastic foam. Aluminum foil is also used. It works by reflecting heat instead of blocking its passage.

Resistance Value of Insulation

The **R-value** of insulation is a number that indicates its measured resistance to the flow of heat. The higher the R-value, the more efficient the material is in retarding the passage of heat. Insulation is rated according to its efficiency. Its R-value is clearly printed on the wrapper (Fig. 49-2).

THERMAL PROPERTIES OF VARIOUS BUILDING MATERIALS, PER INCH OF THICKNESS	
Material	Thermal Resistance R
Wood	1.25
Air Space	0.97
Cinder Block	0.28
Common Brick	0.20
Face Brick	0.11
Concrete (Sand and Gravel)	0.08
Stone (Lime or Sand)	0.08
Steel	0.0032
Aluminum	0.00070

Fig. 49-1 Insulating values of various building materials per inch of thickness.

Fig. 49-2 Insulation is rated according to its thermal resistance. It is stamped with an R-value.

Insulation Requirements

The rising costs of energy coupled with the need to conserve it have resulted in higher R-value recommendations for new construction than in previous years. Average winter low-temperature zones of the United States are shown in Figure 49-3. This information is used to determine the R-value of insulation installed in walls, ceilings, and floors. Insulation requirements vary according to the average low temperature (Fig. 49-4).

In warmer climates, less insulation is needed to conserve energy and provide comfort in the cold season. However, air-conditioned homes should receive more insulation in walls, ceilings, and floors. This assures economy in the operation of air-conditioning equipment during the hot season.

Comfort and operating economy are dual benefits. Insulating for maximum comfort automatically provides maximum economy of heating and cooling operations. It also reduces the initial costs of heating and cooling equipment because smaller units may be adequate.

Where Heat is Lost

The amount of heat lost from the average house varies with the type of construction. The principal areas and approximate amount of heat loss for a

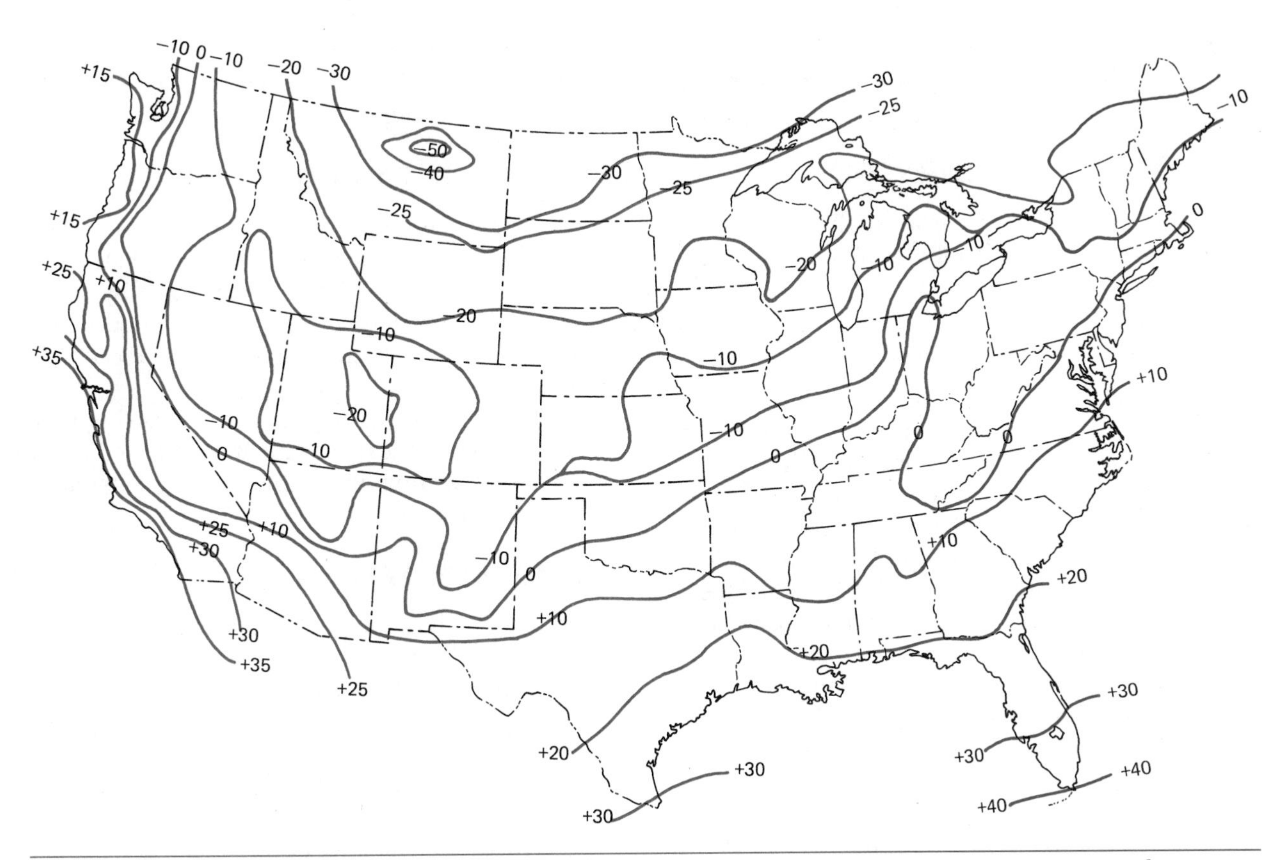

Fig. 49-3 Average low-temperature zones of the United States. (*Courtesy U.S. Department of Agriculture, Forest Service*)

RECOMMENDED R-VALUES OF INSULATION			
AVERAGE LOW TEMPERATURE Degrees Fahrenheit	CEILINGS	WALLS	FLOORS
0 & BELOW	38	17	19
0 TO +10	30	17	19
+10 TO +40	19	11	0

Fig. 49-4 Insulation requirements for various climates

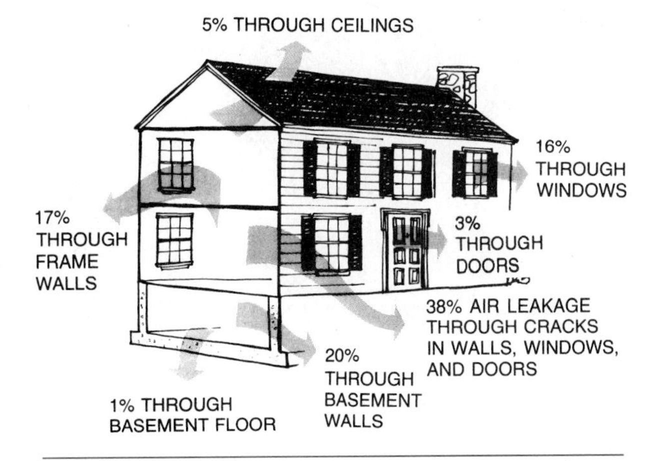

Fig. 49-5 Areas and amounts of heat loss for a typical house with moderate insulation. (*Courtesy of Dow Chemical Company*)

typical house with moderate insulation are shown in Figure 49-5.

Houses of different architectural styles vary in their heat loss characteristics. A single-story house, for example, contains less exterior wall area than a two-story house. It also has a proportionately greater ceiling area. Greater heat loss through floors is experienced in homes erected over concrete slabs or unheated crawl spaces unless these areas are well insulated.

The transfer of heat through uninsulated ceilings, walls, or floors can be reduced almost any desired amount by installing insulation. Maximum quantities in these areas can cut heat losses by up to 90 percent. The use of 2 × 6 studs in exterior walls permits installation of 6-inch-thick insulation. This achieves an R-19 value, or, with improved insulation, an R-21 value.

Windows and doors are generally sources of great heat loss. By reducing the heat transfer through other sections of the house, discomfort caused by these sources can be offset. **Weatherstripping** around windows and doors reduces heat loss. Heat loss through glass surfaces can be reduced

50 percent or more by installing double- or triple-glazed windows or by adding **storm sash** and storm doors.

Types of Insulation

Insulation is manufactured in a variety of forms and types. These are commonly grouped as *flexible, loose-fill, rigid, reflective,* and *miscellaneous.*

Flexible Insulation

Flexible insulation is manufactured in *blanket* and *batt* form. Blanket insulation (Fig. 49-6) comes in rolls. Widths are suited to 16- and 24-inch stud and joist spacing. The usual thickness is from 1 inch to 6 1/4 inches. The body of the blanket is made of fluffy rock or glass wool, wood fiber, or cotton. Organic insulation materials are treated to make them resistant to fire, decay, insects, and vermin. Most blanket insulation is faced with asphalt-laminated Kraft paper or aluminum foil with flanges on both edges for fastening to studs or joists. The facing of the blanket serves as a vapor barrier. It resists the movement of water vapor. It should always face the warm side of the wall.

Batt insulation (Fig. 49-7) is made of the same material as blanket insulation. It comes in thicknesses up to 12 inches. Widths are for standard stud and joist spacing. Lengths are either 48 or 93 inches. Batts come faced with Kraft paper or unfaced.

Fig. 49-6 Blanket insulation comes in rolls.

Fig. 49-7 Batt insulation is made up to 12 inches thick.

Loose-fill Insulation

Loose-fill insulation is usually composed of materials in bulk form. It is supplied in bags or bales. It is placed by pouring, blowing, or packing by hand. Materials include rock or glass wool, wood fibers, shredded redwood bark, cork, wood pulp products, and vermiculite.

Loose-fill insulation is suited for use between ceiling joists in unheated attics. It is also used in the sidewalls of existing houses that were not insulated during construction.

Rigid Insulation

Rigid insulation is usually a fiber or foamed plastic material in sheet or board forms (Fig. 49-8). The material is available in a wide variety of sizes, with widths up to 4 feet and lengths up to 12 feet. The most common types are made from polystyrene, polyurethane, glass fibers, processed wood, and sugar cane or other fibrous vegetable materials.

Structural insulating boards come in densities ranging from 15 to 31 pounds per cubic foot. They are used as building boards, roof decking, sheathing, and wallboard. Their primary purpose is structural. They also serve as insulation. In house construction, perhaps the most common forms are sheathing and decorative coverings in sheet or tile forms.

Reflective Insulation

Reflective insulation usually consists of outer layers of aluminum foil bonded to inner layers of various materials for added strength to resist heat flow.

Fig. 49-8 Rigid insulation being installed between floor joists over a crawl space.

Reflective insulation should be installed facing an air space with a depth of 3/4 inch or more.

Miscellaneous Insulation

Foamed-in-place insulation is sometimes used. A urethane foam is produced by mixing two chemicals together. It is injected into place. It expands on contact with the surface.

Sprayed insulation is usually inorganic fibrous material. It is blown against a surface that has been primed with an adhesive coating. It is often left exposed for acoustical as well as insulating properties.

Other types of insulating material, such as lightweight **vermiculite** and *perlite aggregates,* are sometimes used in plaster to reduce heat flow.

Where and How to Insulate

To reduce heat loss, all walls, ceilings, roofs, and floors that separate heated from unheated spaces should be insulated. Insulation should be placed in all outside walls and in the ceiling. In houses with unheated crawl spaces, it should be placed between the floor joists. Great care should be exercised when installing insulation. Voids in insulation of only 5 percent can create overall efficiency reduction of 25 percent. A good general rule is to be neat. If insulation looks neat, it will probably perform well. Pay attention to details around outlets, pipes, and any obstructions. Make the insulation conform to irregularities by cutting and piecing without bunching or squeezing. Keep the natural fluffiness of the insulation at all times.

A ground cover of roll roofing or plastic film such as polyethylene should be placed on the soil of crawl spaces. This decreases the moisture content of the space as well as of the wood members. Provision should also be made for ventilation of unheated areas. (The use of vapor retarders, the cause of moisture condensation, and the need for ventilation are discussed in greater detail later in this unit.)

In houses with flat or low-pitched roofs, insulation should be used in the ceiling area only if sufficient space is left above the insulation for ventilation. Insulation is used along the perimeter of houses built on slabs when required (Fig. 49-9).

Installing Flexible Insulation

Cut the batts or blankets with a knife. Make lengths slightly oversize so stapling flanges can be formed at the top and bottom. Measure out from one wall a distance equal to the desired lengths of insulation. Draw a line on the floor. Roll out the material from the wall. Compress the insulation with a straightedge on the line. Cut it with a sharp knife (Fig. 49-10). Cut the necessary number of lengths.

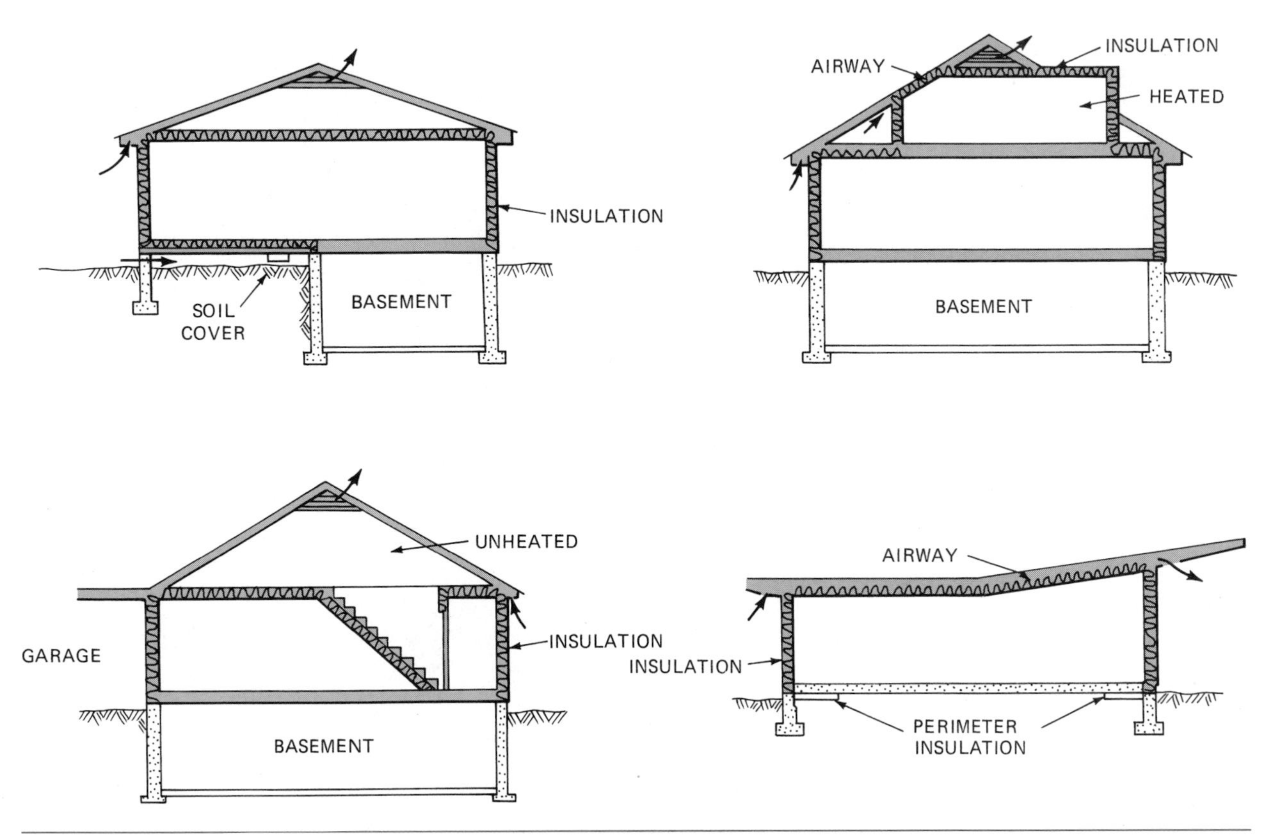

Fig. 49-9 Placement of insulation in various types of structures.

Fig. 49-10 Compress flexible insulation with a straightedge. Cut it with a sharp knife.

Fig. 49-11 Use a hand stapler to fasten wall insulation in place.

Place the batts or blankets between the studs. Staple the flanges of the vapor retarder either to the sides or to the inside edges of the studs as well as the top and bottom plates. Use a hand stapler to fasten the insulation in place (Fig. 49-11).

Fill any spaces around windows and doors with spray-can foam. Non-expanding foam will fill the voids with an airtight seal and protect the house from air leakage. After the foam cures, flexible insulation may be added to fill the remaining space. Do not pack the insulation tightly. Squeezing or compressing it reduces its effectiveness. Use narrow strips of vapor retarder material along the top, bottom, and sides of wall openings. This protects these areas (Fig. 49-12). Ordinarily these are not adequately covered by the retarder on the blanket or batt. If the insulation has no covering or stapling tabs, it is friction-fitted between the framing members.

Ceiling Insulation

Ceiling insulation is installed by stapling it to the ceiling joists or by friction-fitting it between them. If furring strips have been applied to the ceiling joists, the insulation is simply laid on top of the furring strips. In most cases, the use of unfaced insulation is recommended. This makes it easier to determine proper fit of the insulation as well as lowering the cost of materials.

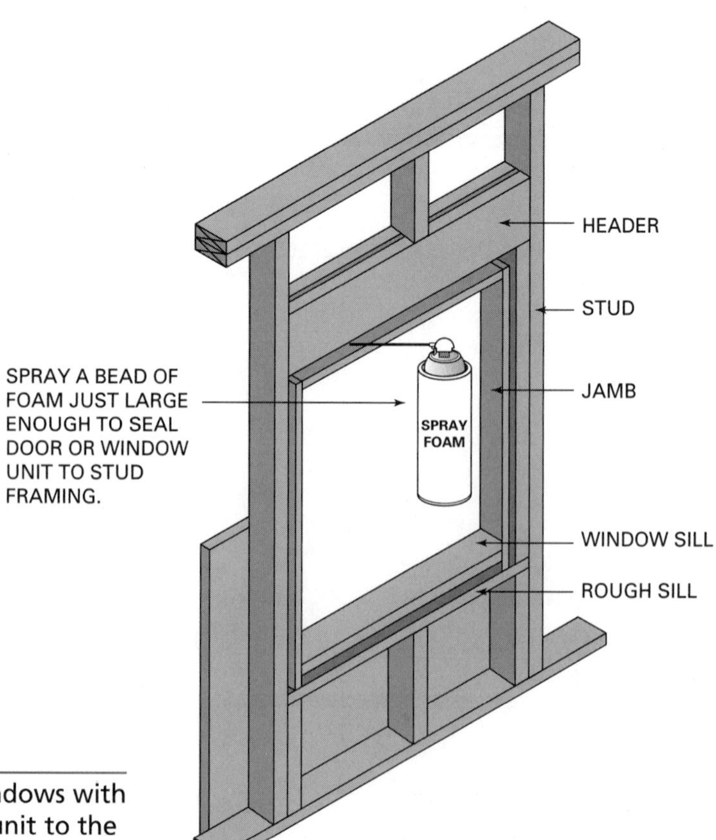

Fig. 49-12 Fill spaces around doors and windows with spray-foam insulation to seal the window unit to the rough opening.

Extend the insulation across the top plate. However, because the eaves are vented, care must be taken not to block the flow of air into the attic. It may be necessary to compress the insulation against the top of the wall plate to permit the free flow of air. An air-insulation dam should be installed to protect the insulation from air movement that reduces the insulation's R-value and performance (Fig. 49-13).

Insulating Floors over Crawl Spaces

Flexible insulation installed between floor joists over crawl spaces may be held in place by wire mesh or pieces of heavy-gauge wire wedged between the joists (Fig. 49-14).

Insulating Masonry Walls

Masonry walls require furring strips for one method of insulation. Fasten 1 × 2 furring strips 16 or 24 inches OC. Insulate as for wood frame walls (Fig. 49-15).

Installing Loose-fill Insulation

Loose-fill insulation is poured by hand or blown into place with special equipment. Level the surface of the material between ceiling joists to the desired depth (Fig. 49-16).

CAUTION Use a respirator to prevent inhaling fine dust particles when installing insulation.

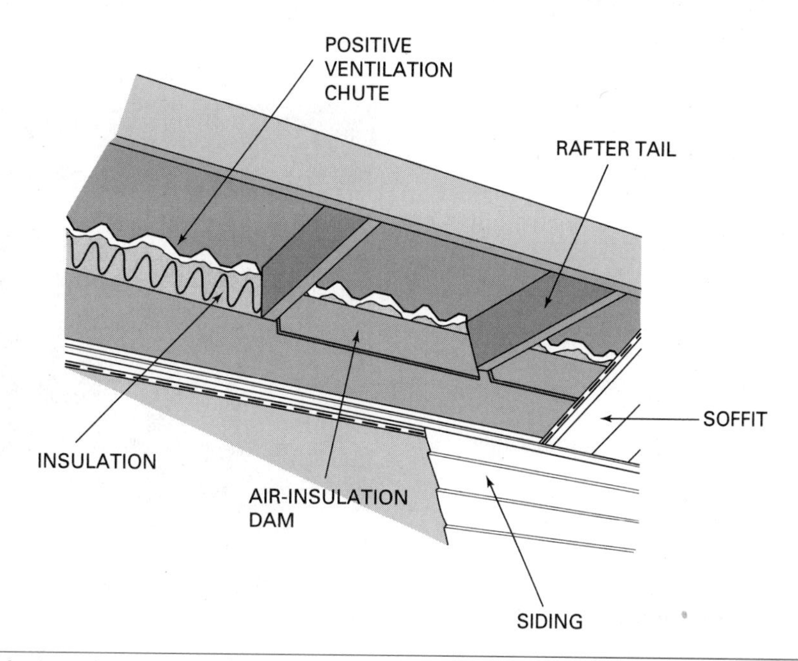

Fig. 49-13 Air-insulation dams protect the insulation from air infiltration.

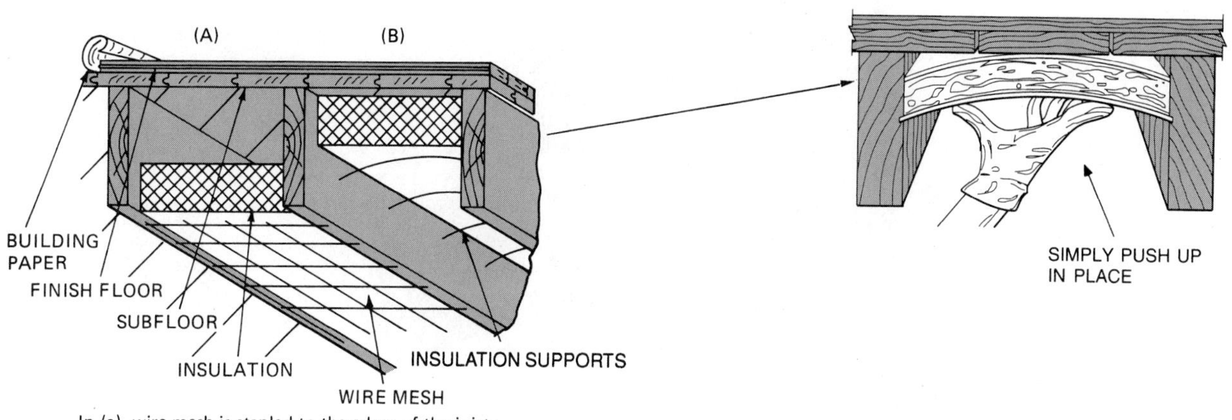

Fig. 49-14 Methods of installing insulation between floor joists in a crawl space.

Installing Rigid Insulation

Rigid insulation is easily cut with a knife or saw. It is usually applied by friction-fitting between the framing members. However, it may be applied to masonry walls with special adhesives (Fig. 49-17). Polystyrene foam plastic, commonly called Styrofoam, has excellent insulating qualities. It is also waterproof. Therefore, it can also act as a vapor retarder. It frequently is used to insulate existing structures before new siding is applied (Fig. 49-18). It is also used extensively on roofs either below or above the roof sheathing (Fig. 49-19).

Fig. 49-17 Extruded polystyrene rigid insulation may be applied to the exterior of basement walls before backfilling in regions where termites are not a problem.

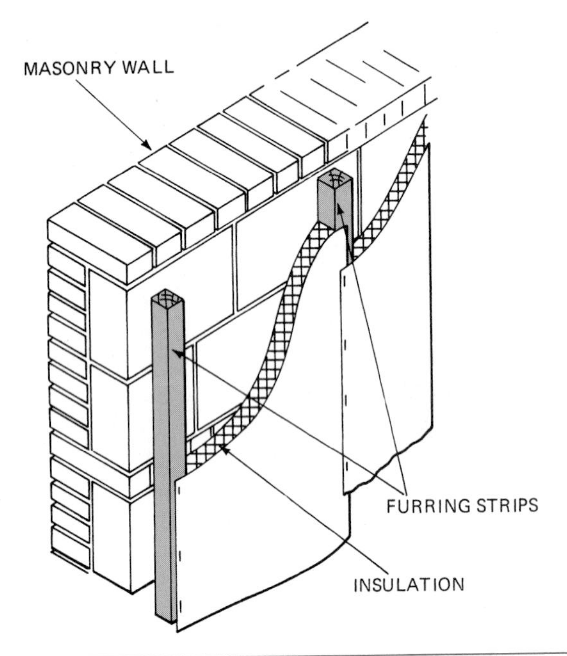

Fig. 49-15 Method of insulating masonry walls.

Fig. 49-18 Rigid insulation applied over old, existing siding.

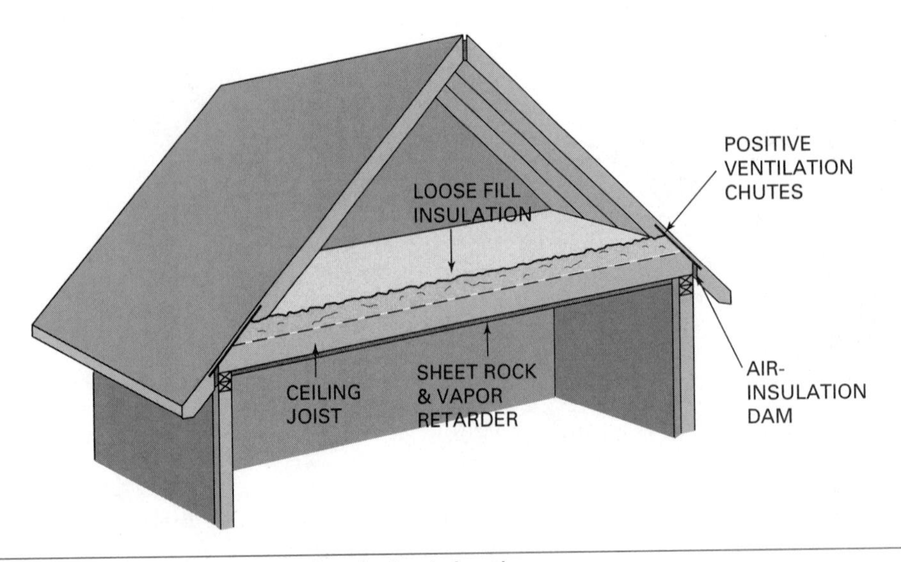

Fig. 49-16 Level loose-fill insulation to the desired depth.

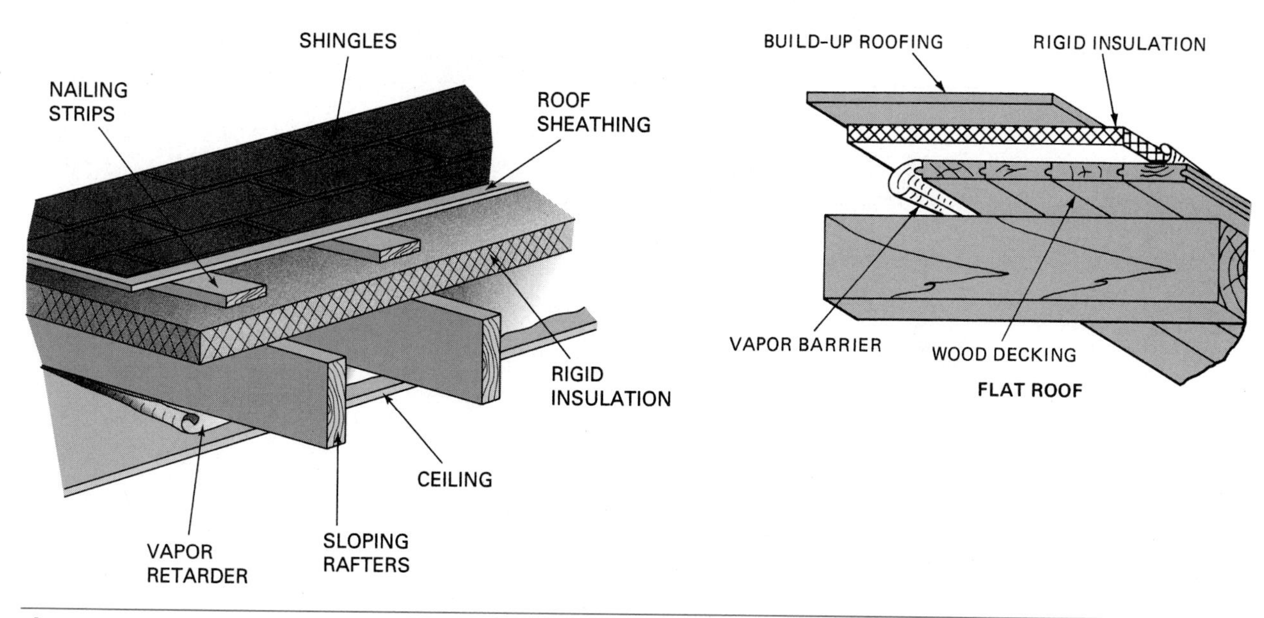

Fig. 49-19 Method of installing rigid insulation on roofs.

Installing Reflective Insulation

Reflective insulation usually is installed between framing members in the same manner as blanket insulation. It is attached to either the face or side of the studs. However, an air space of at least 3/4 inch must be maintained between its surface and the inside edge of the stud.

Properly Installed Insulation

Insulation that is not properly installed can render the material useless. The carpenter must be aware of the consequences of improper application and not take the installation of insulation lightly. The carpenter must know how to install insulation and vapor retarders in the recommended fashion and how to provide proper and adequate ventilation. Follow the manufacturer's directions carefully when installing any type of insulation.

Acoustical Insulation

Acoustical or *sound* insulation resists the passage of noise through a building section. In the past, the reduction of sound transfer between rooms was important in offices, apartments, and motels. Now, more attention is being paid to insulating for sound in homes. Excessive noise is not only annoying but harmful. It not only causes fatigue and irritability, it can damage the sensitive hearing nerves of the inner ear.

Sound insulation between active areas and quiet areas of the home is desirable. Sound insulation between the bedroom area and the living area is important. Bathrooms also should be insulated. Sound control is an important part of residential and commercial construction.

Sound Transmission

Noises such as a loud radio, barking dog, or shouting create sound waves. These waves radiate outward from the source until they strike a wall, floor, or ceiling. These surfaces then begin to vibrate by the pressure of the sound waves in the air. Because the wall vibrates, it conducts sound to the other side in varying degrees, depending on the wall construction.

Sound Transmission Class. The resistance of a building section, such as a wall, to the passage of sound is rated by its *Sound Transmission Class* (STC). The higher the number, the better the sound barrier. The approximate effectiveness of walls with varying STC numbers is shown in Figure 49-20.

Sound travels readily through the air and through some materials. When airborne sound strikes a wall, the studs act as conductors unless

25	Normal speech can be understood quite easily
30	Loud speech can be understood fairly well
35	Loud speech audible but not intelligible
42	Loud speech audible as a murmur
45	Must strain to hear loud speech
48	Some loud speech barely audible
50	Loud speech not audible

This chart from the Acoustical and Insulating Materials Association illustrates the degree of noise control achieved with barriers having different STC numbers.

Fig. 49-20 Approximate effectiveness of walls with varying STC ratings.

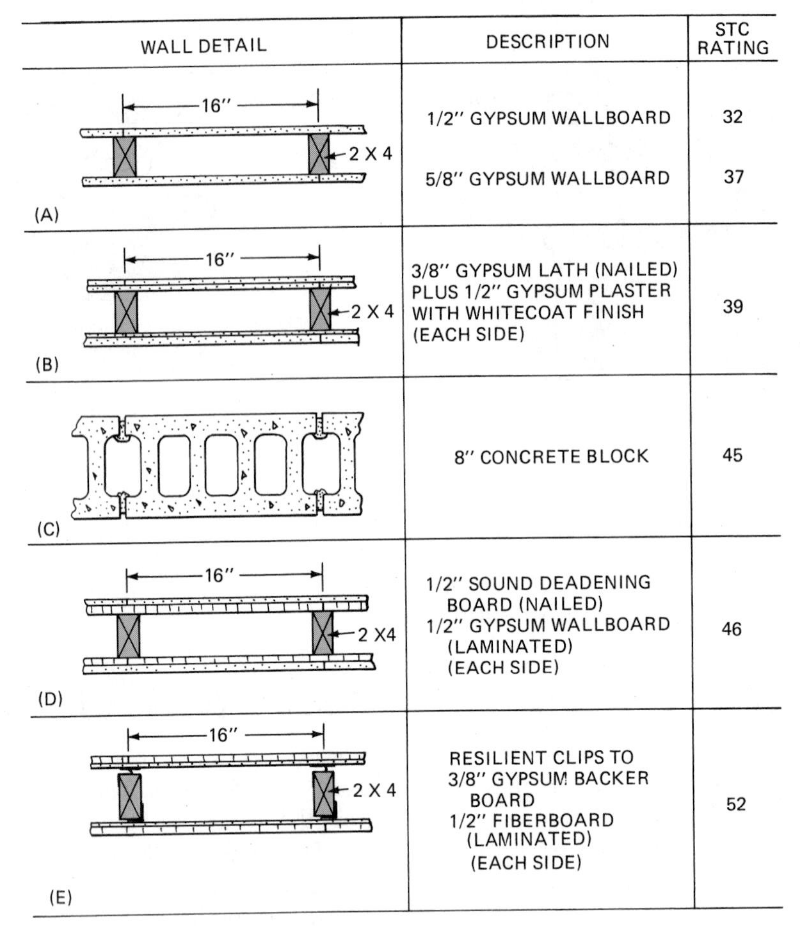

WALL DETAIL	DESCRIPTION	STC RATING
(A) 16" 2 X 4	1/2" GYPSUM WALLBOARD	32
	5/8" GYPSUM WALLBOARD	37
(B) 16" 2 X 4	3/8" GYPSUM LATH (NAILED) PLUS 1/2" GYPSUM PLASTER WITH WHITECOAT FINISH (EACH SIDE)	39
(C)	8" CONCRETE BLOCK	45
(D) 16" 2 X4	1/2" SOUND DEADENING BOARD (NAILED) 1/2" GYPSUM WALLBOARD (LAMINATED) (EACH SIDE)	46
(E) 16" 2 X 4	RESILIENT CLIPS TO 3/8" GYPSUM BACKER BOARD 1/2" FIBERBOARD (LAMINATED) (EACH SIDE)	52

SOUND INSULATION OF SINGLE WALLS

WALL DETAIL	DESCRIPTION	STC RATING
(A) 16" 2 X 4	1/2" GYPSUM WALLBOARD	
(B) 2 X 4	5/8" GYPSUM WALLBOARD (DOUBLE LAYER EACH SIDE)	45
(C) 2 X 4 BETWEEN OR "WOVEN"	1/2" GYPSUM WALLBOARD 1-1/2" FIBROUS INSULATION	49
(D) 2 X 4	1/2" SOUND DEADENING BOARD (NAILED) 1/2" GYPSUM WALLBOARD (LAMINATED)	50

SOUND INSULATION OF DOUBLE WALLS

Fig. 49-21 Sound transmission rating of various types of wall construction.

they are separated in some way from the covering material. Electrical outlet boxes placed back to back in a wall easily transmit sound. Faulty construction, such as poorly fitted doors, often allows sound to pass through. Therefore, good, airtight construction practices aid in controlling sound.

Wall Construction

A wall that provides sufficient resistance to airborne sound should have an STC rating of 45 or greater. At one time, the resistance usually was provided only by double walls, which resulted in increased costs. However, a system of using sound-deadening *insulating board* with a gypsum board outer covering has been developed. This system provides good sound resistance. Resilient steel channels placed at right angles to the studs isolate the gypsum board from the stud. Figure 49-21 shows various types of wall construction and their STC rating.

Floor and Ceiling Construction

Sound insulation between an upper floor and the ceiling below involves not only the resistance of airborne sounds, but also that of *impact noise.* Impact noise is caused by such things as dropped objects, footsteps, or moving furniture. The floor is vibrated by the impact. Sound is then radiated from both sides of the floor. Impact noise control must be considered as well as airborne sounds when constructing floor sections for sound insulation.

An *Impact Noise Rating* shows the resistance of various types of floor-ceiling construction to impact noises. The higher the positive value of the INR, the more resistant to impact noise transfer. Figure 49-22 shows various types of floor-ceiling construction with their STC and INR ratings.

Sound Absorption

The amount of noise in a room can be minimized by the use of *sound absorbing materials.* Perhaps the most commonly used material is **acoustical tile** made of **acoustical board.** These are most often used in the ceiling where they are not subjected to damage. The tiles are soft and are made of wood fiber or similar materials. The tile surface consists of small holes or fissures or a combination of both (Fig. 49-23). These holes or fissures act as sound traps. The sound waves enter, bounce back and forth, and finally die out. (A more complete description and method of installing ceiling tile can be found in a later unit.)

DETAIL	DESCRIPTION	ESTIMATED VALUE	
		STC RATING	APPROX. INR
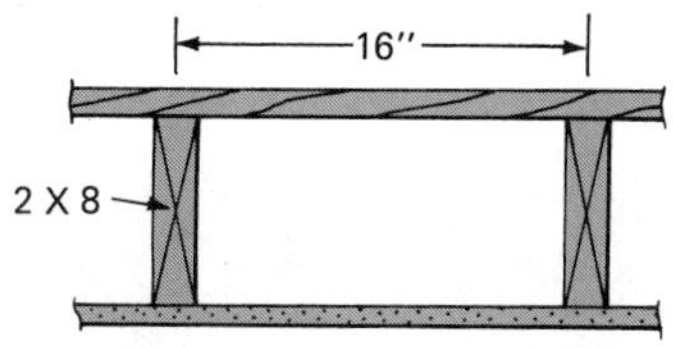 (A) 16" 2 X 8	FLOOR 7/8" T. & G. FLOORING CEILING 3/8" GYPSUM BOARD	30	−18
 (B) 2 X 8	FLOOR 3/4" SUBFLOOR 3/4" FINISH FLOOR CEILING 3/4" FIBERBOARD	42	−12
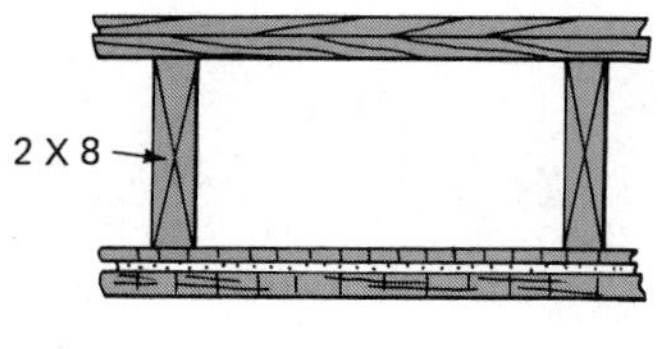 (C) 2 X 8	FLOOR 3/4" SUBFLOOR 3/4" FINISH FLOOR CEILING 1/2" FIBERBOARD LATH 1/2" GYPSUM PLASTER 3/4" FIBERBOARD	45	−4

RELATIVE IMPACT AND SOUND TRANSFER IN FLOOR-CEILING COMBINATIONS (2- BY 8-IN. JOISTS)

Fig. 49-22 Floor and ceiling construction rated for STC and INR.

		ESTIMATED VALUE	
DETAIL	DESCRIPTION	STC RATING	APPROX. INR
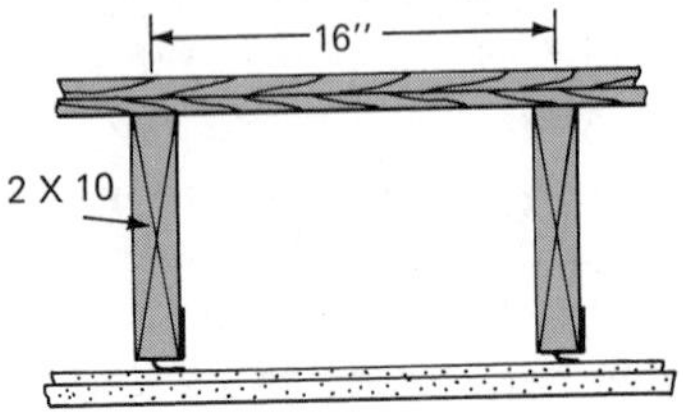 16" 2 X 10 (A)	FLOOR 3/4" SUBFLOOR (BUILDING PAPER) 3/4" FINISH FLOOR CEILING GPYSUM LATH AND SPRING CLIPS 1/2" GYPSUM PLASTER	52	−2
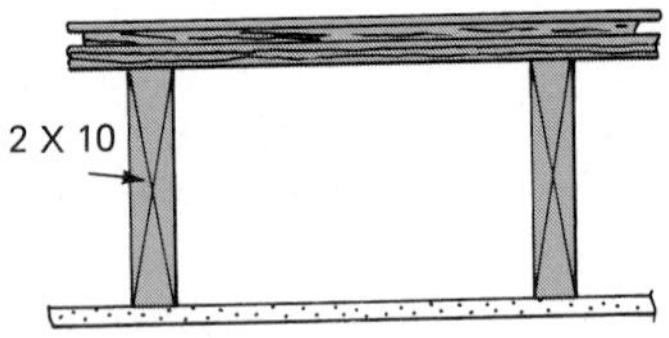 2 X 10 (B)	FLOOR 5/8" PLYWOOD SUBFLOOR 1/2" PLYWOOD UNDERLAYMENT 1/8" VINYL-ASBESTOS TILE CEILING 1/2" GYPSUM WALLBOARD	31	−17
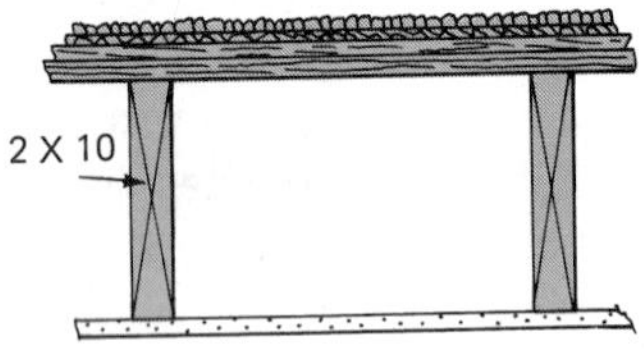 2 X 10 (C)	FLOOR 5/8" PLYWOOD SUBFLOOR 1/2" PLYWOOD UNDERLAYMENT FOAM RUBBER PAD 3/8" NYLON CARPET CEILING 1/2" GYPSUM WALLBOARD	45	+5

RELATIVE IMPACT AND SOUND TRANSFER IN FLOOR-CEILING COMBINATIONS (2- BY 10-IN. JOISTS)

Fig. 49-22 *(continued)*

Fig. 49-23 Sound-absorbing ceiling tile.

CHAPTER 50 CONDENSATION AND VENTILATION

Energy use and costs are reduced when insulation is installed in buildings. Savings in energy costs are realized when buildings are well insulated. The effectiveness of this insulation is reduced if moisture is trapped in the insulation. Problems of moisture migration into the insulation spaces are solved through the practice of airtight construction, the use of vapor retarders and adequate ventilation.

Condensation

Warm air holds more moisture than cold air. The warmer the air, the more moisture it can hold. As moisture-laden air is cooled, it reaches the **dew point.** At this point it is completely saturated with moisture. Fog is air that has reached its dew point. Any further lowering of the air temperature causes it to lose some of its moisture as **condensation** on the cold surface. Evidence of this can be seen as dew on grass in the morning when warm humid air has cooled off at night. The moisture it has lost is condensed on the ground.

Moisture in Buildings

Moisture in the form of water vapor comes from many activities. It is produced mainly by cooking, taking showers and baths, washing and drying clothes, and using a humidifier. The water contained in the warmer air tries to penetrate the insulation layer in its effort to equalize with the cooler, drier air.

In an insulated building, water vapor strives to move through the walls, ceilings, and floors. It may condense under certain conditions when it comes in contact with a cold surface (Fig. 50-1). In a wall, this contact is made on the inside surface of the exterior wall sheathing. In a crawl space, condensation can form on the floor frame and subfloor. In attics, the ceiling joists and roof frame can become saturated with moisture. Condensation of water vapor in walls, attics, roofs, and floors leads to serious problems.

Problems Caused by Condensation

- If insulation absorbs condensed water vapor, the dead air spaces in the insulation may become filled with water. This renders the insulation useless.
- Condensed moisture absorbed by the wood frame raises the moisture content of the wood. This improves the conditions for fungi to grow and decay the wood.
- If moisture is absorbed by the sheathing, it is eventually absorbed by the exterior trim and siding. This causes the exterior paint to blister and peel. Excessive condensation of water vapor may even damage the interior finish.

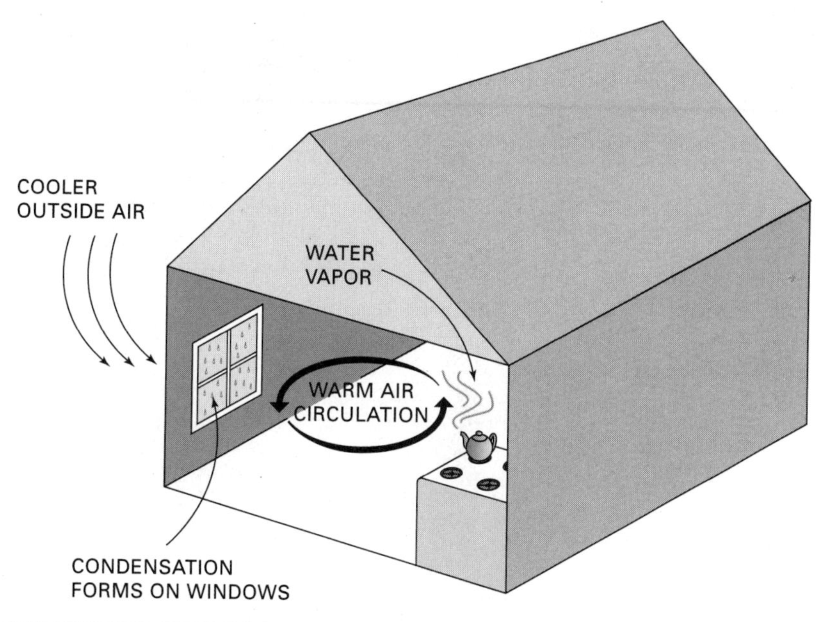

Fig. 50-1 Moisture in warm air condenses when it comes in contact with a cold surface.

- A warm attic that is inadequately ventilated may cause the formation of ice dams at the **eaves.** After a heavy snowfall, heat causes the snow next to the roof to melt. Water running down the roof freezes on the cold roof overhang, forming an ice dam. This causes the water to back up on the roof, under the shingles, and into the walls and ceiling (Fig. 50-2).

Prevention of Vapor Condensation

Logically, to prevent the condensation of moisture in a building, steps should be taken to reduce the effect of the moisture.

This can be accomplished in two ways. Either reduce the production of moisture within the house or reduce the moisture migration into the insulation layer. Sources of moisture within the house include cooking, bathing, washing, and damp basements. Check the exhaust piping of clothes dryers making sure it is not constricted and is working properly. Installing ventilation fans, and using them daily, can exhaust the moisture-laden air to the outside.

Airtight construction techniques are the best way to reduce moisture migration into the insulation layer. Virtually all the moisture migration occurs by air leakage into the insulation layer. Thus, if there is an airtight seal between the warmer and cooler air, moisture migration is stopped (Fig. 50-3).

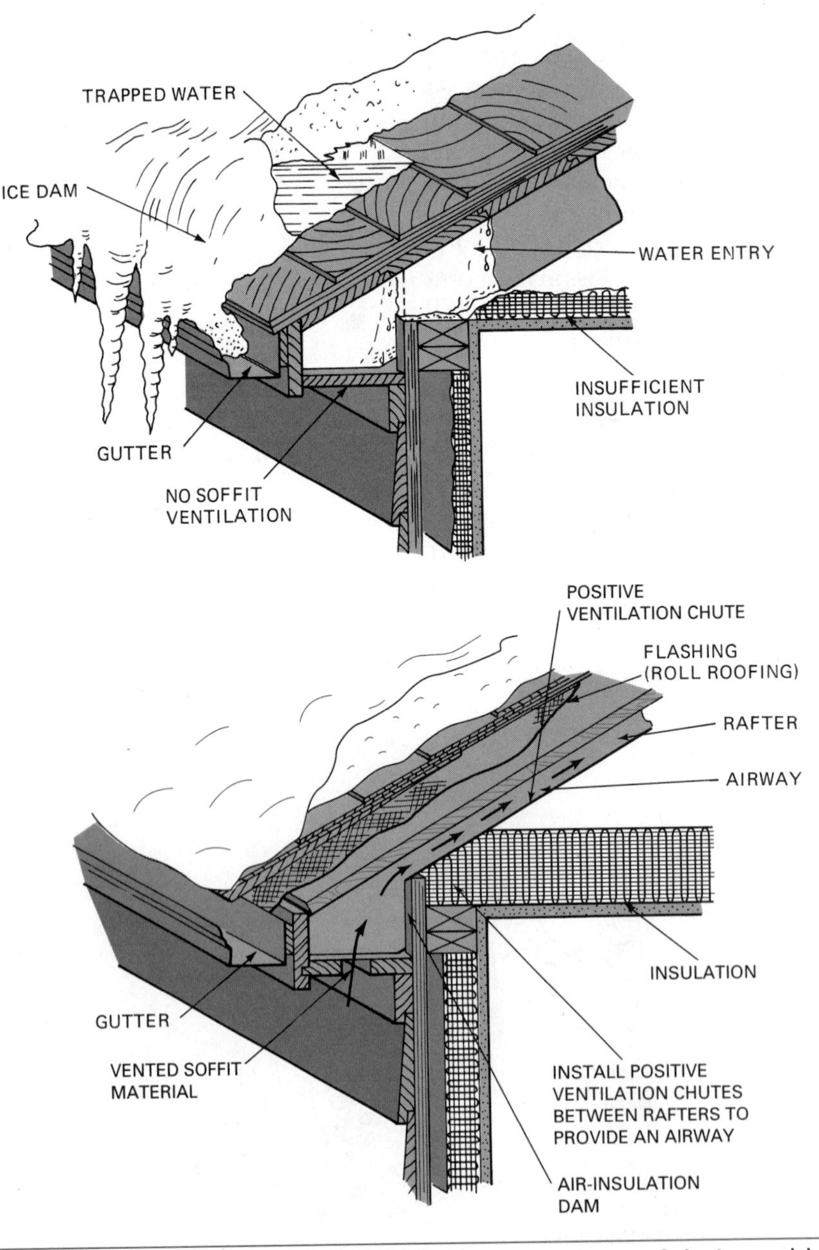

Fig. 50-2 A warm attic may cause the formation of ice dams on the roof during cold weather. A properly constructed and ventilated attic will keep the snow from melting and forming ice dams.

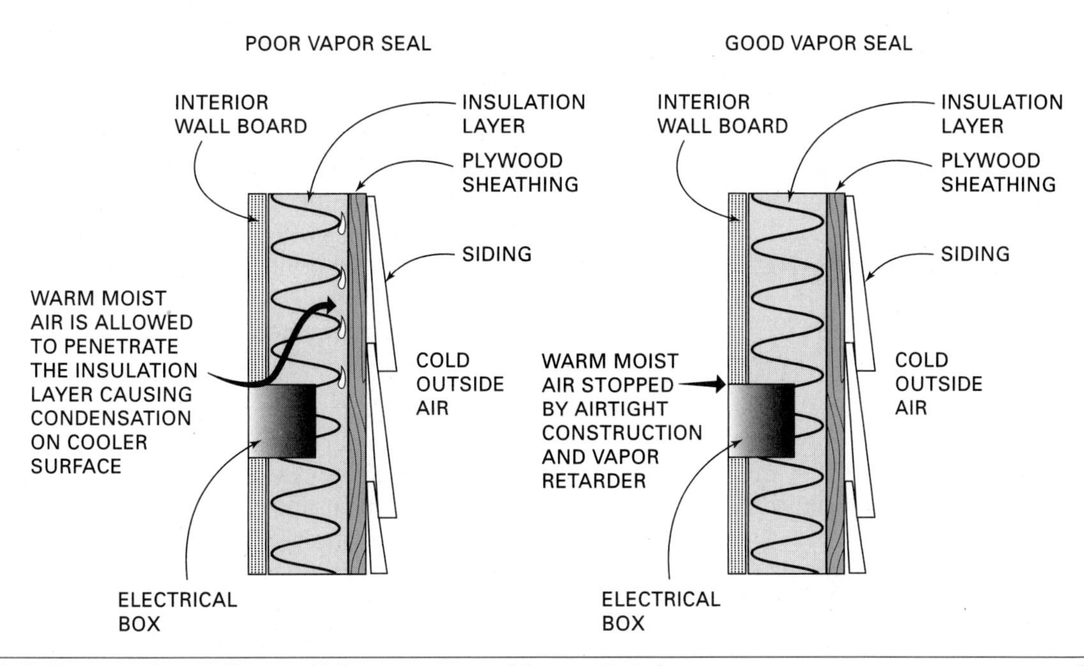

Fig. 50-3 An exterior wall should be constructed in an airtight manner to prevent water vapor from entering the insulation layer.

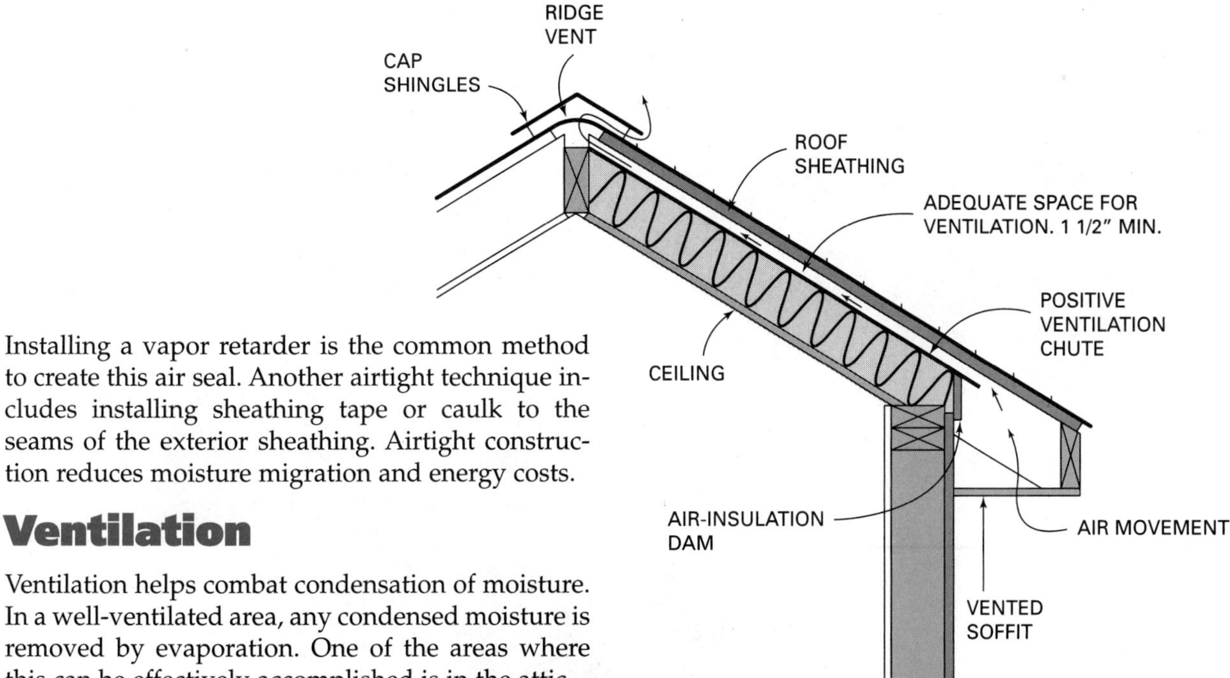

Fig. 50-4 Methods of providing adequate ventilation when the entire rafter cavity is insulated.

Installing a vapor retarder is the common method to create this air seal. Another airtight technique includes installing sheathing tape or caulk to the seams of the exterior sheathing. Airtight construction reduces moisture migration and energy costs.

Ventilation

Ventilation helps combat condensation of moisture. In a well-ventilated area, any condensed moisture is removed by evaporation. One of the areas where this can be effectively accomplished is in the attic.

With a well-insulated ceiling and adequate ventilation, attic temperatures are low. Melting of snow on the roof over the attic space can be eliminated. Also, roof shingles will stay cooler in the summer months and thus last longer.

In crawl spaces under floors, ventilation can easily be provided to evaporate any condensation of moisture. Here the vapor retarder should be installed on top of the ground to inhibit moisture from getting into the floor system above.

On roofs where the ceiling finish is attached to the roof rafters, insulation is usually installed between the rafters. An adequate air space of at least 1 1/2 inches must be maintained between the insulation and the roof sheathing. The air space must be amply vented with air inlets in the soffit and outlets at the ridge (Fig. 50-4). Failure to do so may result in reduced shingle life, formation of ice dams at the eaves, and possible decay of the roof frame. Application of a vapor retarder on the heated-in-winter side of the insulation helps prevent any condensation of moisture in the rafter space by blocking the passage of vapor.

Types of Ventilators

There are many types of ventilators. Their location and size are factors in providing adequate ventilation.

Ventilating Gable Roofs. Triangularly shaped louver vents are sometimes installed in both end walls of a gable roof. They come in various shapes and sizes and are installed as close to the roof peak as possible.

The best way to vent an attic is with continuous ridge and soffit vents (Fig. 50-5). Each rafter cavity is vented from soffit to ridge. The roof sheathing is cut back about 1 inch from the ridge on each side, and the vent material is nailed over this slot. Cap shingles then can be nailed directly to the vent. Perforated material or screen vents are installed in the soffits to provide the entry point for the ventilation. Positive-ventilation chutes should be installed to prevent any air obstructions by the ceiling insulation near the eaves (Fig. 50-6). This system can adequately vent the attic space of a house that is up to 50 feet wide.

Ventilating Hip Roofs. Hip roofs should have additional continuous venting along each hip rafter. This allows each hip-jack rafter cavity to be vented. When cutting a 2-1/2-inch wide slot for the vents, it is recommended to leave a 1-foot section of sheathing uncut between every 2 feet of slot section (Fig. 50-7). This allows for adequate ventilation

Fig. 50-5 Ridge and soffit vents work together to provide adequate attic ventilation.

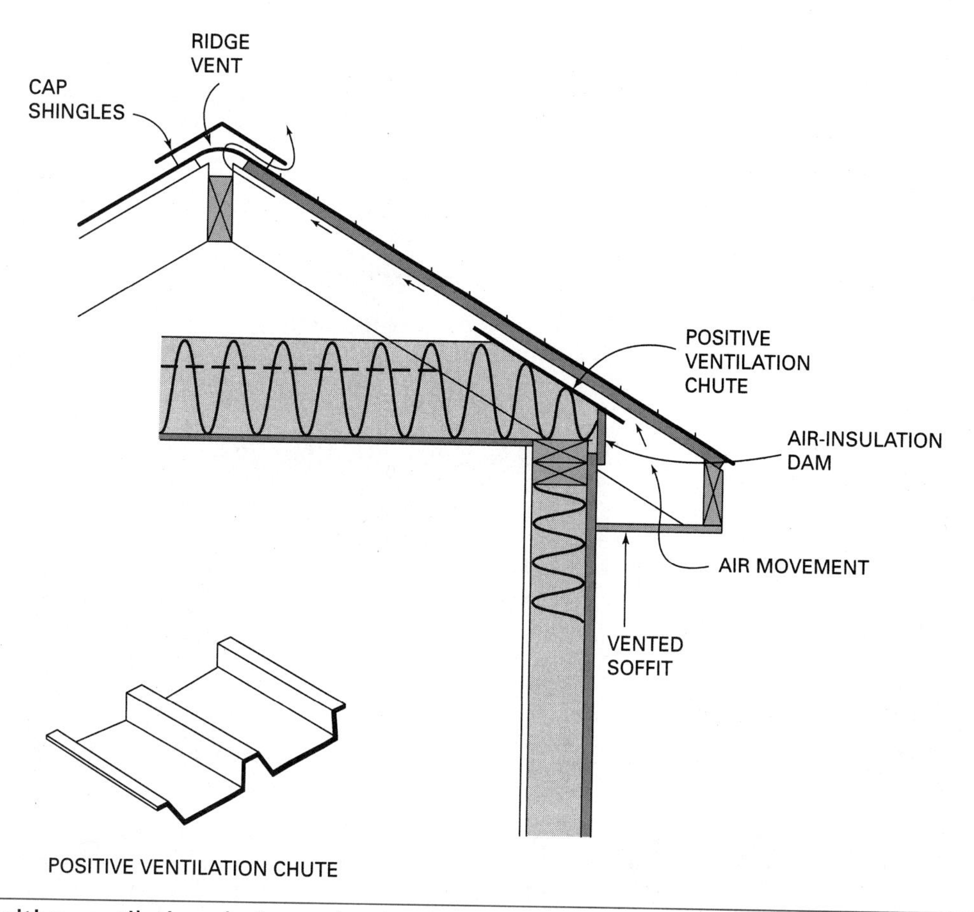

Fig. 50-6 Positive ventilation chutes maintain air space between insulation and rafters.

while maintaining the integrity of the hip roof structure.

Ventilating Crawl Spaces. For crawl spaces, usually rectangular metal or plastic louvered ventilators are used. Some are made to fit into a space made by leaving out a concrete block in the foundation. Vents should be placed near each corner of the crawl space foundation and in intermediate equally spaced locations (Fig. 50-8). The use of a ground cover moisture barrier is recommended under most conditions. It protects wood framing members from ground moisture. It also allows the use of a smaller number of ventilators.

Determining Ventilator Size

The location of ventilators is a factor in their efficiency. Ventilators must be of adequate size.

Size of Attic Ventilators. The minimum free-air area for attic ventilators is based on the ceiling area of the rooms below. The free-air area for the openings should be 1/150th of the ceiling area. For example, if the ceiling area is 1,200 square feet, the minimum total free-air area of the ventilators should be 8 square feet.

Crawl Space Ventilators. The total free-air area of crawl space ventilators should be 1/1600th of the ground area when a vapor retarder ground cover is used. For example, if a crawl space measures 50 feet by 100 feet, its ground area is 5,000 square feet. The free-air area of ventilators must be a little over 3 square feet.

The free-air area of a ventilator is the area through which air is able to pass. The actual area must be increased to allow for any restrictions such as louvers or wire screen mesh. Use as coarse a screen as conditions permit. Lint and dirt tend to clog fine-mesh screens. When painting, be careful not to contact the screen and clog the mesh with paint. Ventilators should be designed to keep the weather out. They should not be closed. In cold climates, it is especially important to leave them open in the winter as well as during the summer.

Description of Vapor Retarder

The most commonly used material for a vapor retarder is *polyethylene film.* It is a transparent plastic sheet. It comes in rolls of usually 100 feet in length and 10, 12, 14, 16, and 20 feet in width. The

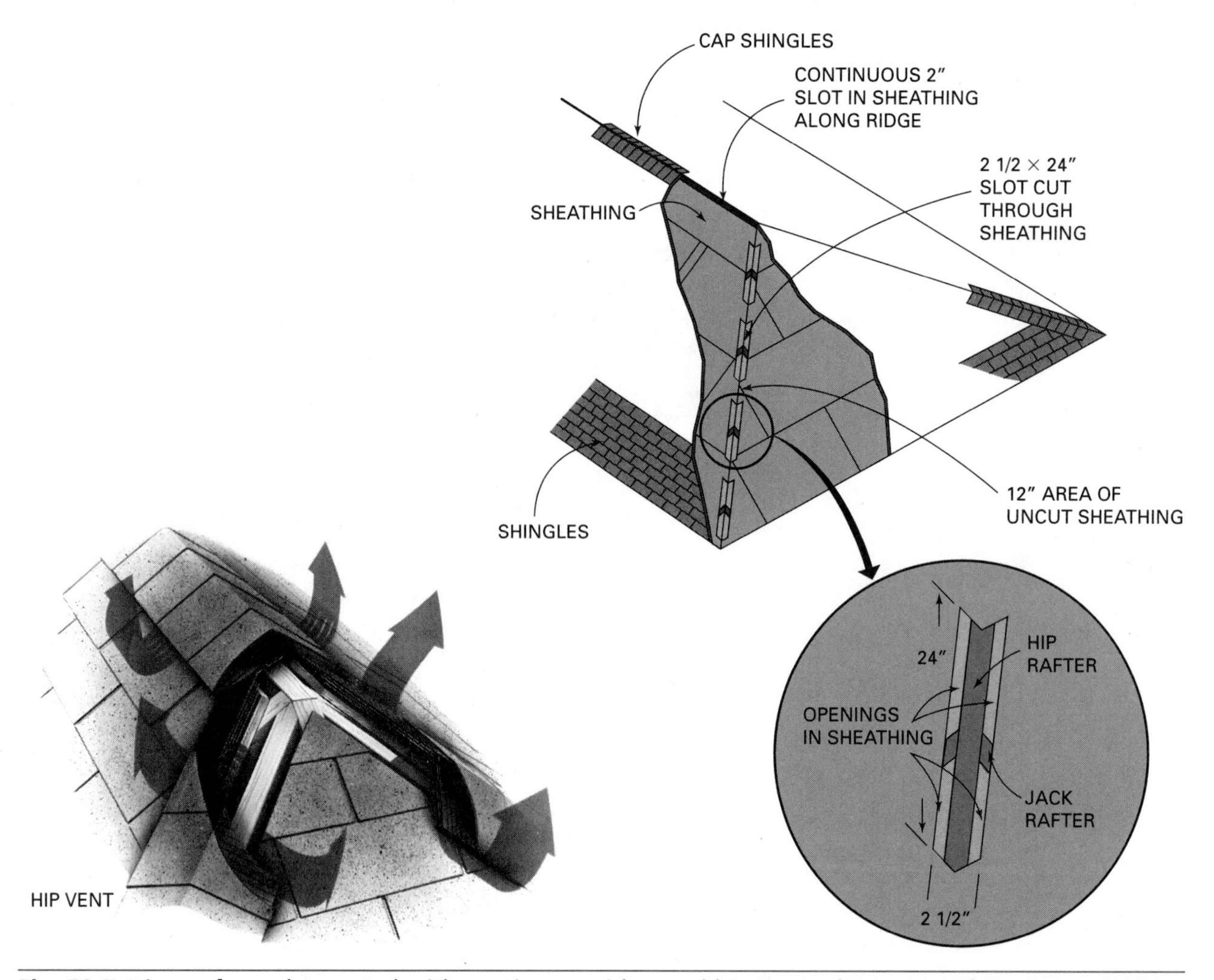

Fig. 50-7 Hip roofs can be vented with continuous ridge and hip vents. (*Courtesy of CorAvent, Inc.*)

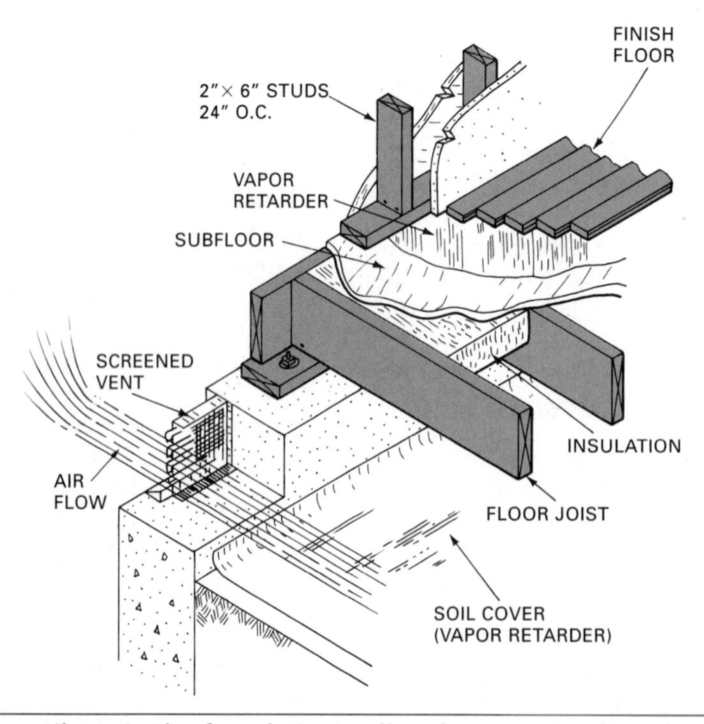

Fig. 50-8 Crawl space ventilator in the foundation wall and vapor retarders placed on the ground.

most commonly used thicknesses are 4 mil and 6 mil.

Other materials used as vapor retarders are *asphalt-laminated papers and aluminum foil.* Most blanket and batt insulations are manufactured with a vapor retarder on one side.

Installing a Polyethylene Vapor Retarder

Unroll a length long enough to cover the wall. Make sure to allow a few extra inches in length. Unfold the section. Staple it along the top plate, letting the rest drape down to the floor. Smooth out wrinkles. Staple about 6 to 12 inches apart to every stud (Fig. 50-9). Cut the material. Carefully fit it around all openings. Lap all joints. Repair any tears with tape. Cut off the excess at the floor line.

Some contractors apply the film over all openings. They then cut the film after the finish is applied. This assures a more positive seal. The retarder is fitted tightly around outlet boxes. A ribbon of sealing compound around outlets or switch boxes minimizes vapor penetration.

Fig. 50-9 Installing polyethylene film vapor retarder.

Polyethylene is applied over insulation having no vapor retarder facing. It also ensures excellent protection over any type of insulation.

Review Questions

Select the most appropriate answer.

1. Of these common building materials listed below, the most efficient insulator is
a. cinder block.
b. concrete.
c. stone.
d. wood.

2. The insulating term *R-value* is defined as the measure of
a. resistance of a material to the flow of heat.
b. the heat loss through a building section.
c. the conductivity of a material.
d. the total heat transfer through a building section.

3. To protect the insulation layer from air leakage and moisture migration a vapor retarder should be installed on the
a. warm side.
b. cold side.
c. inside.
d. outside.

4. Reflective insulation usually consists of
a. foamed plastic.
b. glass wool.
c. aluminum foil.
d. vermiculite.

5. Reflective insulation is effective only when
a. it is placed on the warm side.
b. it comes in contact with the wall covering.
c. it faces an air space of at least 3/4 inch deep.
d. it is used under the exterior finish.

6. Moisture migration into the insulation layer can be stopped by
a. installing a vapor retarder.
b. placing sheathing tape on the seams of exterior sheathing.
c. airtight construction techniques.
d. all of the above.

7. Squeezing or compressing flexible insulation tightly into spaces
a. reduces its effectiveness.
b. increases its efficiency because more insulation can be installed.
c. is necessary to hold it in place.
d. helps prevent air leakage by sealing cracks.

8. When insulation is placed between roof framing members, there should be an air space between the insulation and the roof sheathing of at least
a. 3/4 inch.
b. 1 inch.
c. 1 1/2 inches.
d. 2 inches.

9. Ventilators
a. are closed in winter to conserve heat.
b. are left open year round.
c. are closed in stormy weather.
d. are closed when the house is to be left vacant for long periods of time.

10. The best method of venting an attic space is with
a. gable vents.
b. hip vents.
c. roof windows.
d. continuous ridge and soffit vents.

BUILDING FOR SUCCESS

Nontraditional Workers in the Workplace

Across the country today, the traditional white male "turf" is being shared by more women and other minority groups. This demands fresh perspectives and increased flexibility. There will be a need for more understanding, demonstrations of respect, and cooperation by everyone.

All ethnic and diversity groups in the workplace must be accepted. Acceptance must be authentic and demonstrated if workers are to be successful. This acceptance is based on the understanding of a person's qualifications to do the job. These include, but are not limited to, technical skills, talents, abilities, knowledge, and experience.

Competition for jobs must be clearly centered on identifiable criteria that avoid discrimination and/or harassment. The criteria must focus on what is needed to do the job satisfactorily. Performance on the job should be evaluated by a person's ability to perform, and not whether the worker is male, female, white, of color, or disabled.

Nontraditional workers in the construction workforce today represent a variety of cultures and ethnic groups. We must all continue to learn more about each co-worker. Common goals must be shared by all workers regardless of their culture. There will always be a need for good company orientation programs that help introduce each worker to a compatible work environment. Supervisors who work with a specific ethnic group other than their own may need to learn another language for effective communication. There will continue to be a need to gain a better understanding of other customs and cultural traditions that can cause misunderstandings on the job. With the workforce gradually incorporating more females into its ranks, there may be a need to entertain short or extended maternity leaves. Flexibility has always played an important part in adjusting to new developments in the work world.

Increased ethnic and cultural diversity groups entering the workforce will influence the number of skilled and unskilled workers. The underlying criteria for determining who will obtain and maintain quality jobs in construction must center on the skills and abilities of the workers. Obtaining and retaining those jobs should be the responsibility and goal of every construction student, apprentice, and trainee nationwide. Good education and skill development will be the center of attention for those offering training programs. The person who is best prepared with the necessary technical skills, business skills, and abilities will generally gain an edge when seeking employment. Additionally, the person who has a good understanding of how to be productive as a member of a team that includes nontraditional workers will be more successful than others.

The avenue for obtaining the necessary skills and abilities may not be as important as learning the right technology and developing the required skills that accompany that technology. There are many educational opportunities available to people who desire construction careers. Universities, colleges, trade unions, professional construction organizations, and construction companies offer quality preparation and training. Competence in applying technical skills will be a must. So will competence in communicating, reading, teamwork, social skills, and self-management.

The best job offer may take people out of their familiar environments and place them in one that has a major population that represents another culture or diverse group. This will take a good understanding of nontraditional workers and flexibility. We must be prepared to work alongside those representing other ethnic groups. The nontraditional worker will continue to be an important part of the skilled workforce. All successful carpenters, plumbers, electricians, and masons that represent a vast array of cultures and ethnic groups will possess many similar technical skills. They will also need the capability to establish common goals and work alongside many nontraditional workers.

FOCUS QUESTIONS: For individual or group discussion

1. What is your current understanding of the nation's workforce in reference to nontraditional workers?
2. What impact is a diversified workforce having in your area of desired employment?
3. What elements of the population do you think will be represented in your technical specialty during the next decade? Will this population representation show a contrast from the present-day workforce?
4. How can you plan to be part of a workforce that promotes teamwork in problem solving? What would be some examples of new skills and understandings that would help bring success to your job?

STUDENT ACTIVITIES for Section 2: Rough Carpentry

Unit 9 Building Layout

1. By using the two site plans, lay out the buildings, erect batter boards, and identify and mark the building line, footing line, and digging line. Set up the builder's level in the site. Take elevation readings at each corner of the lot and building.

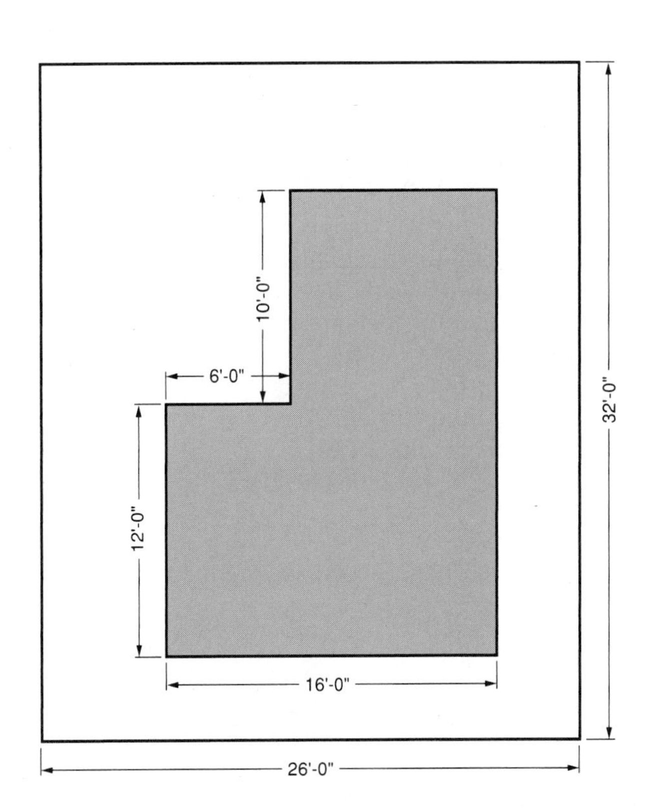

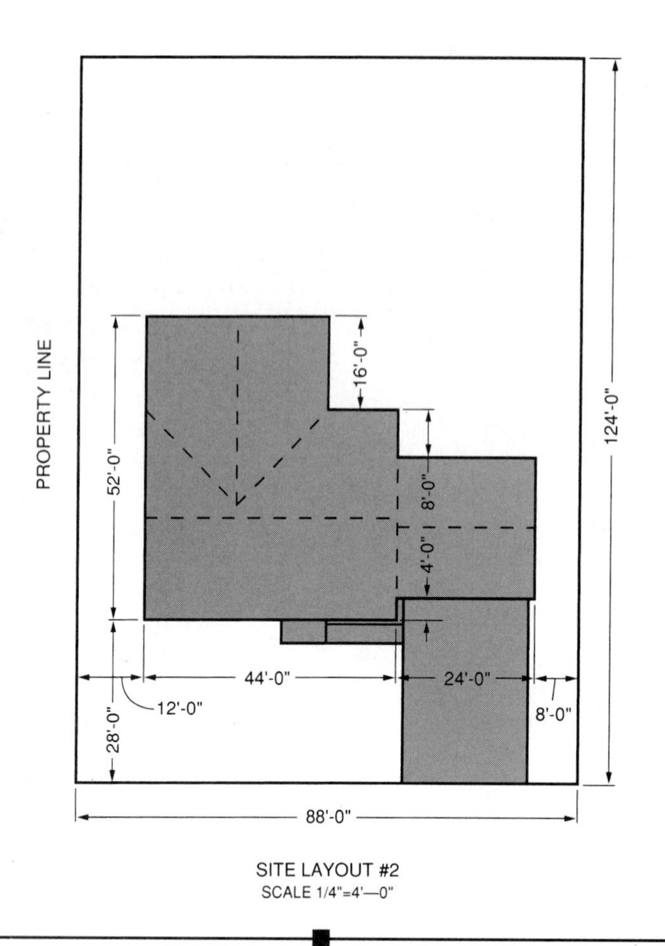

SITE LAYOUT #2
SCALE 1/4"=4'—0"

Unit 10 Concrete and Concrete Forms

1. Lay out and form a 4-inch patio slab, 12′ × 16′, with a slope of 1 inch across the 12-foot dimension. Use 1″ × 4″ form lumber. Check the slope with a straightedge and level or a builder's level.

Unit 11 Floor Framing

Using the drawings, complete the following framing activities by selecting all the needed tools and materials. Have the instructor check your work for accuracy, then make any necessary corrections.

1. Using the drawing of the box header, measure and lay out the header to receive floor joists. Materials required: 1 ea.
2″ × 4″ × 12′(box header).

Layout criteria:

1. Floor joists to be at 16 inches OC.
2. Mark end of the box header at 12′ × 0″ for cutting.
3. Make all changes in the joist positions to allow for stair trimmers and double joists under the walls.

4. Draw the accepted symbols or markings on the lumber.
a. on center
b. floor joist
c. trimmer
d. wall

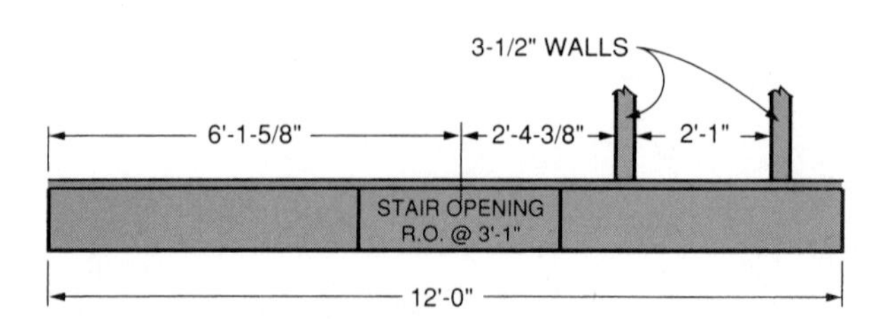

BOX HEADER

Unit 12 Wall Framing

Materials required: 1 ea. 2″ × 4″ × 12′, 1 ea. 2″ × 4″ × 8′.

Layout Criteria (platform frame):

1. Studs to be at 16 inches OC.
2. Window rough openings located as shown.
3. Single trimmers at opening sides.
4. Locate intersecting walls as shown.
5. Calculate header lengths.
6. Mark end of the wall plate at 12 feet for cutting.
7. Mark all changes in the stud positions to allow for openings, trimmers, and walls.

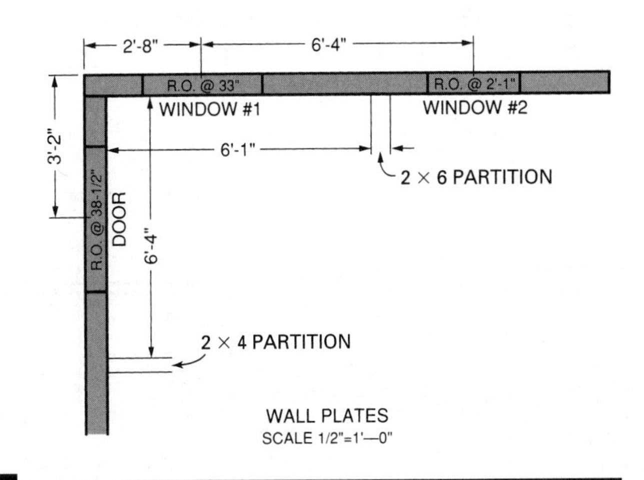

Unit 15 Roof Framing

Solve the following problems by finding the dimensions and calculations needed in the roof frame layout.

1. Given: 6 on 12 roof slope, 16 inches OC rafter spacing. Find the theoretical lengths of entire roof frame, with rafter lengths to plate and to the tail end.

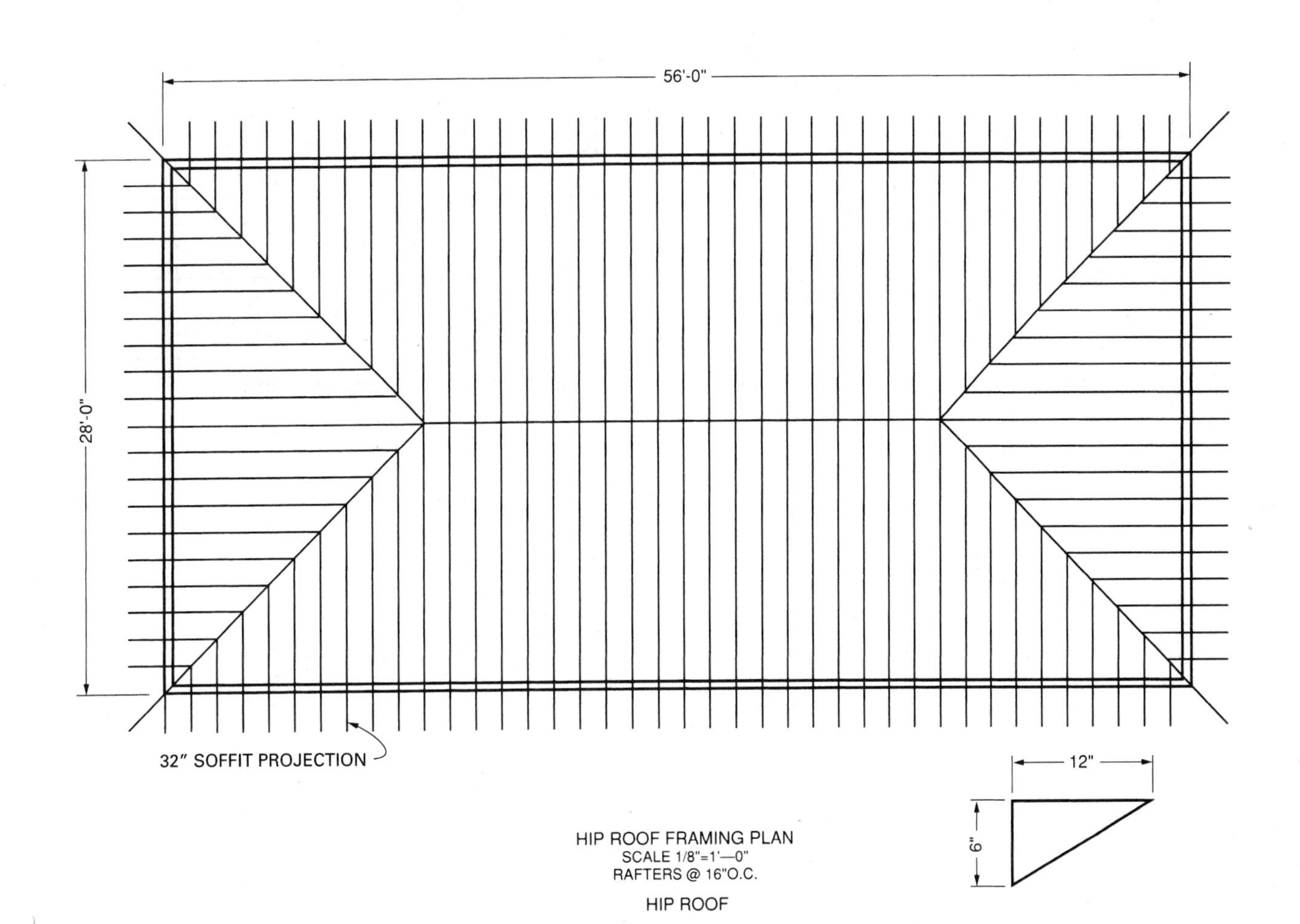

HIP ROOF FRAMING PLAN
SCALE 1/8"=1'—0"
RAFTERS @ 16"O.C.

HIP ROOF

2. Given: 5 on 12 roof slope, 16 inches OC rafter spacing. Find the theoretical lengths of entire roof frame of roofs A and B, to plate and tail end. Roof B may be shortened 4 feet to 6 feet to create cripple jacks

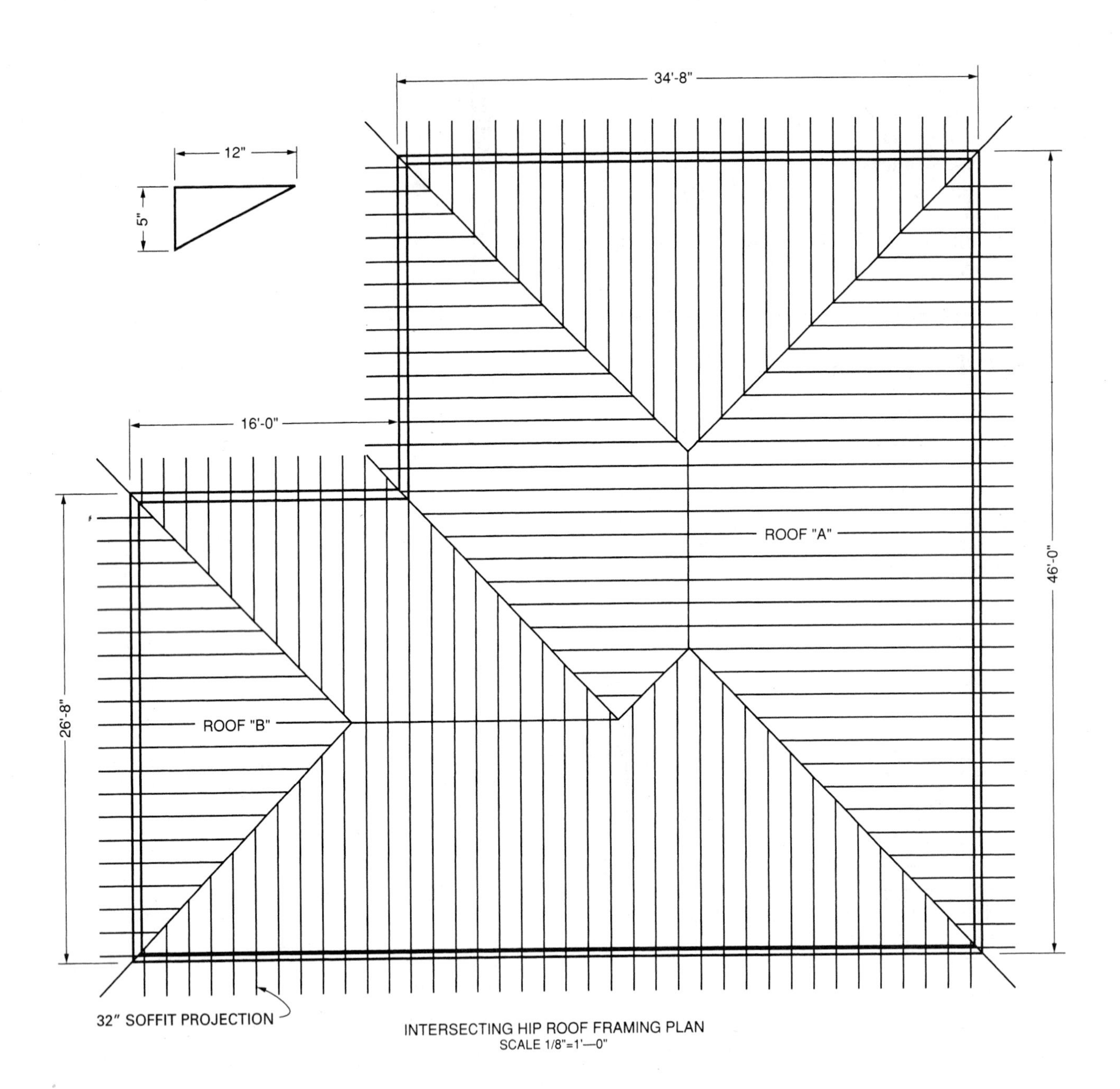

INTERSECTING HIP ROOF

3. Substitute new dimensions and slope in roof plan #2 for additional practice.

4. Lay out the end cut for a 2″ × 6″ ceiling joist having a 7 on 12 slope cut at 4 inches above the plate.

Unit 16 Stair Framing

1. Solve for the missing dimensions in the stair problem. Total rise, unit rise, total run, unit run, and headroom.

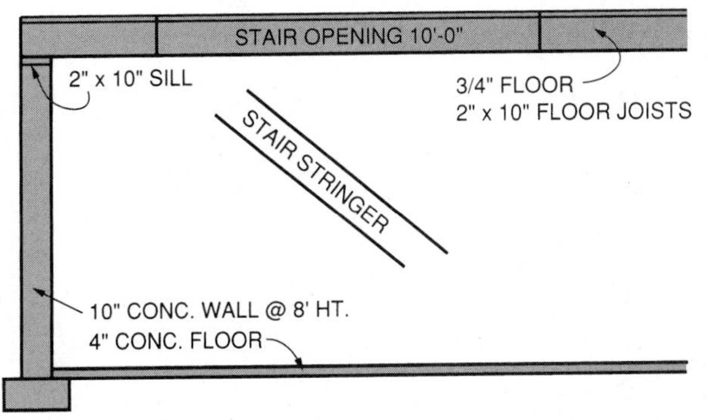

STAIR SECTION
SCALE 3/8"=1'—0"

2. Lay out and cut the stair stringer in the drawing.

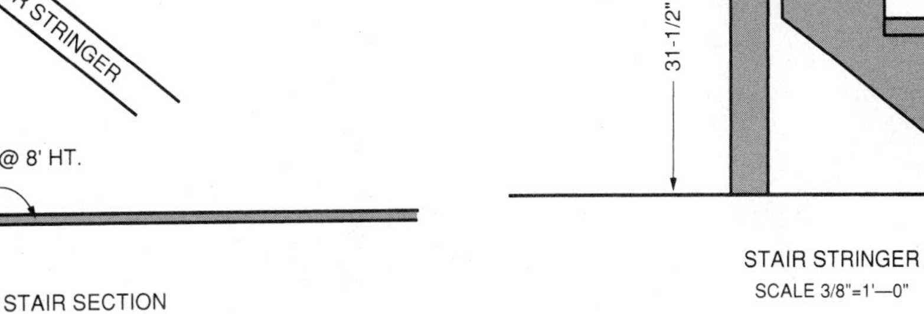

STAIR STRINGER
SCALE 3/8"=1'—0"

Unit 17 Insulation and Ventilation

1. Make a display board of the insulation used in residential and light commercial construction. List the manufacturer, product name, and costs.

2. Make a display of selected roof and soffit venting systems. List the manufacturer, product name, and costs.

SECTION

3

Exterior Finish

UNIT

18

Roofing

Materials used to cover a roof and make it tight are part of the exterior finish called *roofing*. Roofing adds beauty to the exterior and protects the interior. Before roofing is applied, the roof deck must be securely fastened. There must be no loose or protruding nails. All cornice trim should be in place and primed. Properly applied roofing gives years of dependable service.

OBJECTIVES

After completing this unit, the student should be able to:

- define roofing terms.
- describe and apply roofing felt, organic or fiber glass asphalt shingles, and roll roofing.
- describe various grades and sizes of wood shingles and shakes and apply them.
- flash valleys, sidewalls, chimneys, and other roof obstructions.
- estimate needed roofing materials.

UNIT CONTENTS

CHAPTER 51 ASPHALT SHINGLES

Asphalt shingles are the most commonly used roof covering for residential and light commercial construction. They are designed to provide protection from the weather for a period ranging from 20 to 30 years. They are available in many colors and styles.

Roofing Terms

An understanding of the terms most commonly used in connection with roofing is essential for efficient application of roofing material.

- A **square** is the amount of roofing required to cover 100 square feet of roof surface. There are usually three bundles of shingles per square or about 80 three-tab shingle strips (Fig. 51-1). One square of shingles can weigh between 235 and 325 pounds.
- *Deck* is the wood roof surface to which roofing materials are applied.
- *Coverage* is the number of overlapping layers of roofing and the degree of weather protection offered by roofing material. Roofing may be called single, double, or triple coverage, for example.
- A *shingle butt* is the bottom exposed edge of a shingle.
- *Courses* are horizontal rows of shingles or roofing.
- **Exposure** is the distance between courses of roofing measured from butt to butt. It is the amount of roofing in each course exposed to the weather (Fig. 51-2).

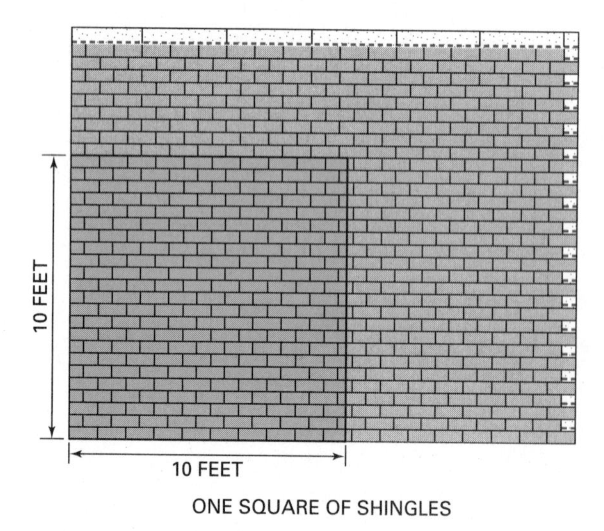

Fig. 51-1 One square of shingles will cover 100 square feet.

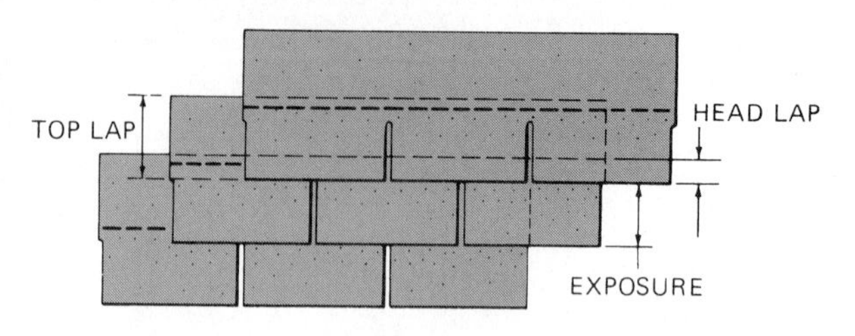

Fig. 51-2 Asphalt strip exposure and lap.

- The *top lap* is the height of the shingle or other roofing minus the exposure. In roll roofing this is also known as the **selvage.**
- The *head lap* is the distance from the butt of an overlapping shingle to the tip of a shingle two courses below measured up the slope.
- *End lap* is the horizontal distance that the ends of roofing in the same course overlap each other.
- **Flashing** are strips of thin sheet metal. They are usually made of lead, zinc, copper, or aluminum. They may also be strips of roofing material used to make watertight joints on a roof. Metal flashing comes in rolls of various widths. They are cut to the desired length.
- *Asphalt cements* and *coatings* are manufactured to various consistencies depending on the purpose for which they are to be used. *Cements* are classified as *plastic, lap,* and *quick-setting.* They will not flow at summer temperatures. They are used as adhesives to bond asphalt roofing products and flashings. They are usually troweled on the surface. *Coatings* are usually thin enough to be applied with a brush. They are used to resurface old roofing or metal that has become weathered.

Electrolysis is a reaction that occurs when unlike metals come in contact with water. This contact causes one of the metals to erode. A simple way to prevent the disintegration caused by electrolysis is to secure metal roofing material with fasteners of the same metal.

Preparing the Deck

A metal **drip edge** or a single course of wood shingles is installed along the roof edges. The metal drip edge is usually made of aluminum or galvanized iron. It is applied along the lower edge of the roof (Fig. 51-3). The drip edge or wood shingle course is

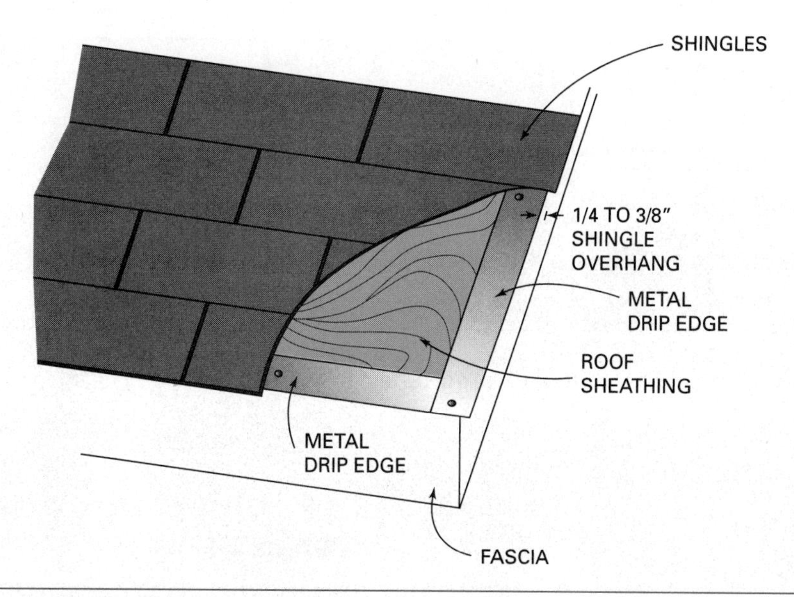

Fig. 51-3 Metal drip edge may be used to support the shingle edge overhang.

used to support the asphalt shingle overhang. Otherwise the shingles would droop from the heat of the sun.

Install the metal drip edge by using roofing nails of the same metal spaced 8 to 10 inches along its inner edge. Lap end joints by about 2 inches.

A wood shingle **starter course** serves the same purpose as the metal drip edge. Apply a shingle with the desired overhang at each end of the roof. Stretch a line between them. Install temporary shingles at intermediate points. Fasten the line to their **butts** to provide a straight line from one end of the building to the other (Fig. 51-4).

Apply wood shingles along the edge of the roof. Their butts should come as close to the line as possible without touching it. Sight each shingle for a bow. Fasten it with the crown of the bow up. Use two 3d galvanized nails halfway up the shingle about 1/2 inch in from each edge. Also fasten two nails through each shingle into the top edge of the fascia. Continue applying shingles until the course is completed.

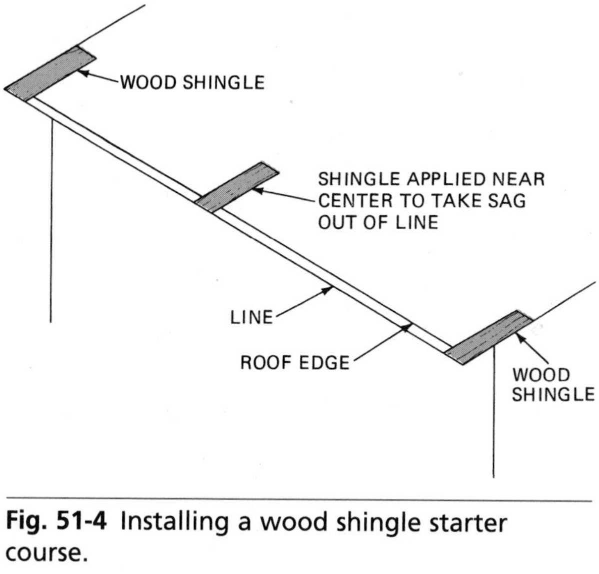

Fig. 51-4 Installing a wood shingle starter course.

Underlayment

The deck should next be covered with an asphalt shingle *underlayment.* The underlayment protects the sheathing from moisture until the roofing is applied. It also gives additional protection to the roof afterward. Use an underlayment that allows the passage of water vapor. This prevents moisture or frost from accumulating between the underlayment and the deck.

Asphalt Felts

Asphalt felts consist of heavy felt paper saturated with asphalt or coal tar. They are usually made in various weights of pounds per square (Fig. 51-5). Asphalt felt comes in 36-inch wide rolls. The rolls are 72 or 144 feet long covering 2 or 4 square. Usually the lightest-weight felt is used as an underlayment for asphalt shingles.

Apply a layer of asphalt felt underlayment over the deck. Lay each course of felt over the lower course at least 2 inches. Make any end laps at least 4 inches. Lap the felt 6 inches from both sides over all hips and ridges (Fig. 51-6).

Nail or staple through each lap and through the center of each layer about 16 inches apart. Roofing nails driven through the center of metal discs or specially designed, large head felt fasteners

SATURATED FELT	APPROX. WEIGHT PER ROLL	APPROX. WEIGHT PER SQUARE	SQUARES PER ROLL	ROLL LENGTH	ROLL WIDTH	SIDE OR END LAPS	TOP LAP	EXPOSURE
	60 #	15 #	4	144'	36"	4"	2"	34"
	60 #	30 #	2	72'	36"	TO		
	60 #	60 #	1	36'	36"	6"		

Fig. 51-5 Sizes and weights of asphalt-saturated felts.

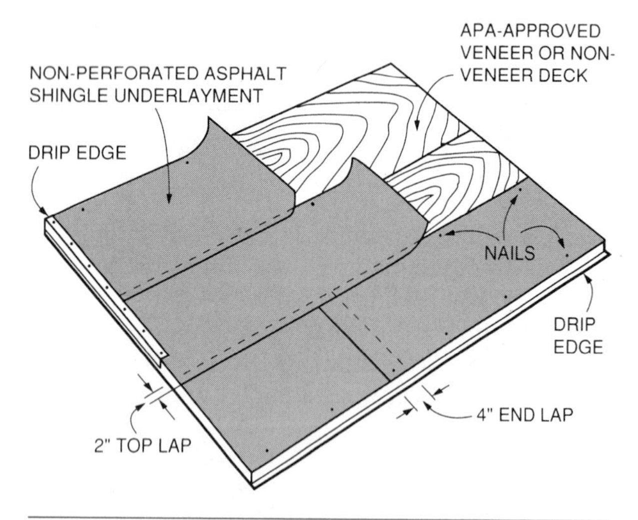

Fig. 51-6 Application of asphalt shingle underlayment felt. (*Courtesy of Asphalt Roofing Manufacturers' Association*)

hold the underlayment securely in strong winds until shingles are applied (Fig. 51-7). A metal drip edge should be installed along the rakes after the application of the underlayment.

Fig. 51-7 An asphalt felt underlayment is applied to the roof deck before shingling.

Low Slope Underlayment Application

If asphalt shingles are to be used on slopes less than 4 inches per foot, down to 2 inches per foot, special application of the underlayment is required. Lay a 19-inch course of felt along the eaves. This course should completely cover the drip edge or starter course. Then lay full-width sheets, each overlapping the preceding course by 19 inches and exposing 17 inches of the underlying sheet (Fig. 51-8). Use only enough fasteners to hold the felt securely to the deck until the shingles are applied.

The underlayment plies are then cemented together to form an *eaves flashing*. The eaves flashing should extend up the roof at least 24 inches beyond the inside wall line of the building. The layers are cemented together using a continuous layer of plastic asphalt cement. The cement should be applied uniformly with a notched trowel at a rate of two gallons per 100 square feet.

Kinds of Asphalt Shingles

Two types of *asphalt shingles* are manufactured. *Organic* shingles have a base made of heavy asphalt-saturated paper felt coated with additional asphalt. *Fiber glass* shingles have a base mat of glass fibers. The mat does not require the saturation process and only requires an asphalt coating. Both kinds of shingles are surfaced with selected mineral granules. The asphalt coating provides weatherproofing qualities. The granules protect the shingle from the sun, provide color, and protect against fire.

Fiber glass-based asphalt shingles have an Underwriters Laboratories Class A fire resistance rating. The Class A rating is the highest standard for resistance to fire. The Class C rating for organic shingles, while not as high, will meet most residential building codes.

Organic and fiber glass shingles come in a wide variety of colors, shapes, and weights (Fig. 51-9). They are applied in the same manner. Shingle qual-

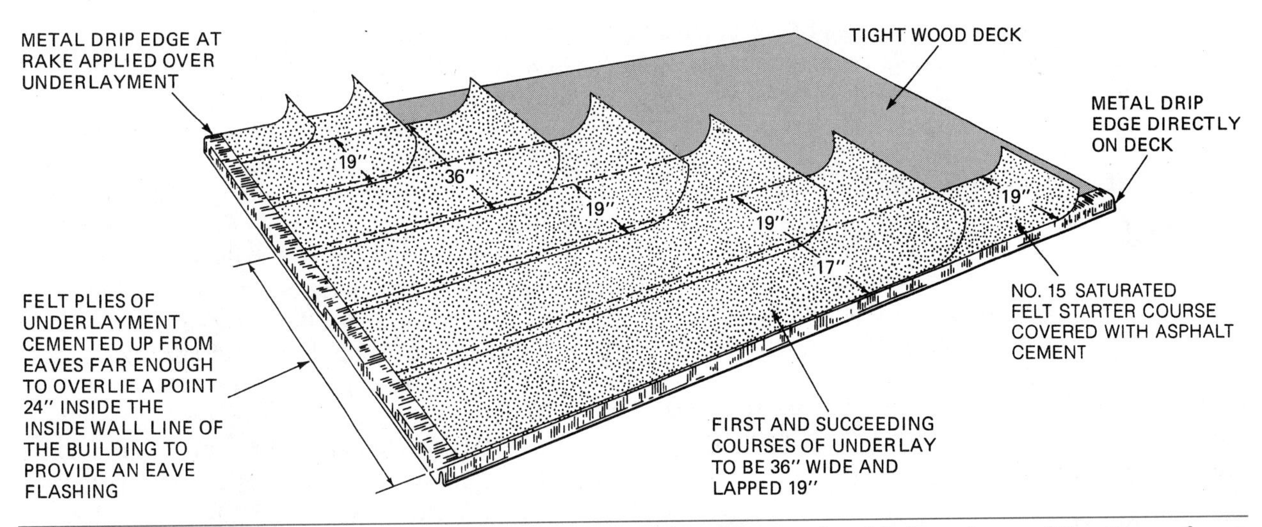

Fig. 51-8 Method of applying underlayment prior to shingling roofs with low slopes. (*Courtesy of Asphalt Roofing Manufacturers' Association*)

PRODUCT	Configuration	Per Square			Size		Exposure	Under-writers' Listing
		Approx. Shipping Weight	Shingles	Bundles	Width	Length		
Wood Appearance Strip Shingle More Than One Thickness Per Strip Laminated or Job Applied	Various Edge, Surface Texture & Application Treatments	285# to 390#	67 to 90	4 or 5	11-1/2" to 15"	36" or 40"	4" to 6"	A or C – Many Wind Resistant
Wood Appearance Strip Shingle Single Thickness Per Strip	Various Edge, Surface Texture & Application Treatments	Various 250# to 350#	78 to 90	3 or 4	12" or 12-1/4"	36" or 40"	4" to 5-1/8"	A or C – Many Wind Resistant
Self-Sealing Strip Shingle	Conventional 3 Tab	205#-240#	78 or 80	3	12" or 12-1/4"	36"	5" or 5-1/8"	A or C – All Wind Resistant
	2 or 4 Tab	Various 215# to 325#	78 or 80	3 or 4	12" or 12-1/4"	36"	5" or 5-1/8"	
Self-Sealing Strip Shingle No Cut Out	Various Edge and Texture Treatments	Various 215# to 290#	78 to 81	3 or 4	12" or 12-1/4"	36" or 36-1/4"	5"	A or C – All Wind Resistant
Individual Lock Down Basic Design	Several Design Variations	180# to 250#	72 to 120	3 or 4	18" to 22-1/4"	20" to 22-1/2"	–	C – Many Wind Resistant

Fig. 51-9 Asphalt shingles are available in a wide variety of sizes, shapes, and weights. (*Courtesy of Asphalt Roofing Manufacturers' Association*)

ity is generally determined by the weight per square. Most asphalt shingles are manufactured with factory-applied adhesive. This increases their resistance to the wind.

Applying Asphalt Shingles

Before applying strip shingles, make sure that the roof deck is properly prepared. The underlayment and drip edge should be applied. Asphalt roofing products become soft in hot weather. Be careful not to damage them by digging in with heavy shoes during application or by unnecessary walking on the surface after application.

Roof Slope

The slope of a roof should not be less than 4 inches per foot when conventional methods of asphalt shingle application are used. On lower roof slopes, down to 2 inches rise per foot of run, *self-sealing* asphalt shingles are used. If "free tab" strip shingles are used, cement each tab down with a spot of quick-setting asphalt cement about 1 1/2 inches in diameter (Fig. 51-10). A double coverage felt underlayment for low slopes must also be provided as previously described.

Asphalt Shingle Layout

On small roofs, strip shingles are applied by starting from either rake. On long buildings, a more accurate vertical alignment is ensured by starting at the center and working both ways (Fig. 51-11). Mark the center of the roof at the eaves and at the ridge. Snap a chalk line between the marks. Snap a series of chalk lines 6 inches apart on each side of the centerline if the shingle tab cutouts are to break on the halves. Snap lines 4 inches apart if the cutouts are to break on the thirds. When applying the shingles, start the course with the end of the shingle to the vertical chalk line. Start succeeding courses in the same manner. Break the joints as necessary. Pyramid the shingles up in the center. Work both ways toward the rakes.

If it is decided to start at the rakes and cutouts are to break on the halves, start the first course with a whole tab. The second course is started with a shingle from which 6 inches have been cut; the third course, with a strip from which the entire first tab is removed; the fourth, with one and one-half tabs removed, and so on (Fig. 51-12). These starting strips are precut for faster application. Waste from these strips is used on the opposite rake.

If the cutouts are to break on the thirds, cut the starting strip for the second course by removing 4 inches. Remove 8 inches from the strip for the third course, and so on (Fig. 51-13).

Cut the shingles by scoring them on the back side with a utility knife. Use a square as a guide for the knife. Bend the shingle. It will break on the scored line.

The layouts may have to be adjusted so that tabs on opposite rakes will be of approximately equal widths. No rake tab should be less than 3 inches in width.

Fig. 51-11 On long roofs, start shingling in the center. Work toward the rakes.

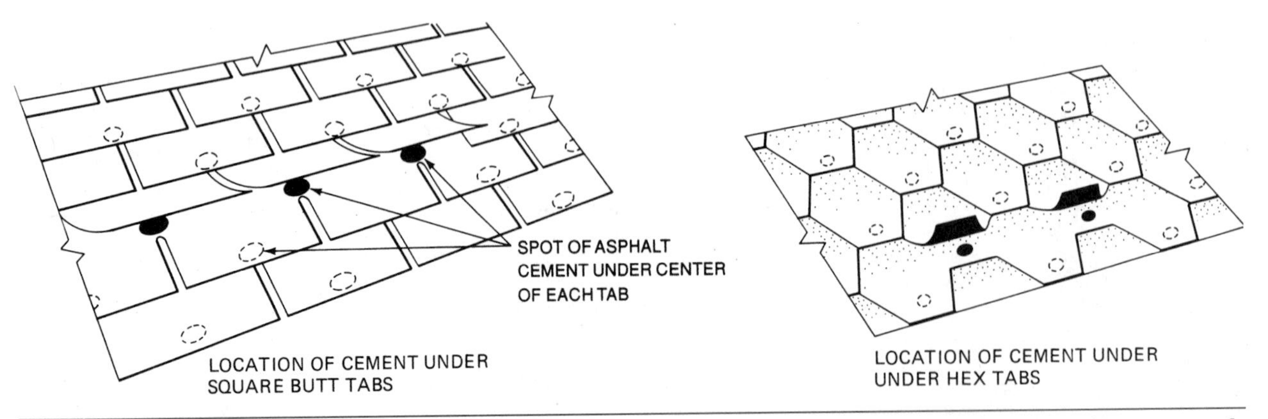

Fig. 51-10 On low slopes, use self-sealing shingles or spot free tabs with asphalt cement. (*Courtesy of Asphalt Roofing Manufacturers' Association*)

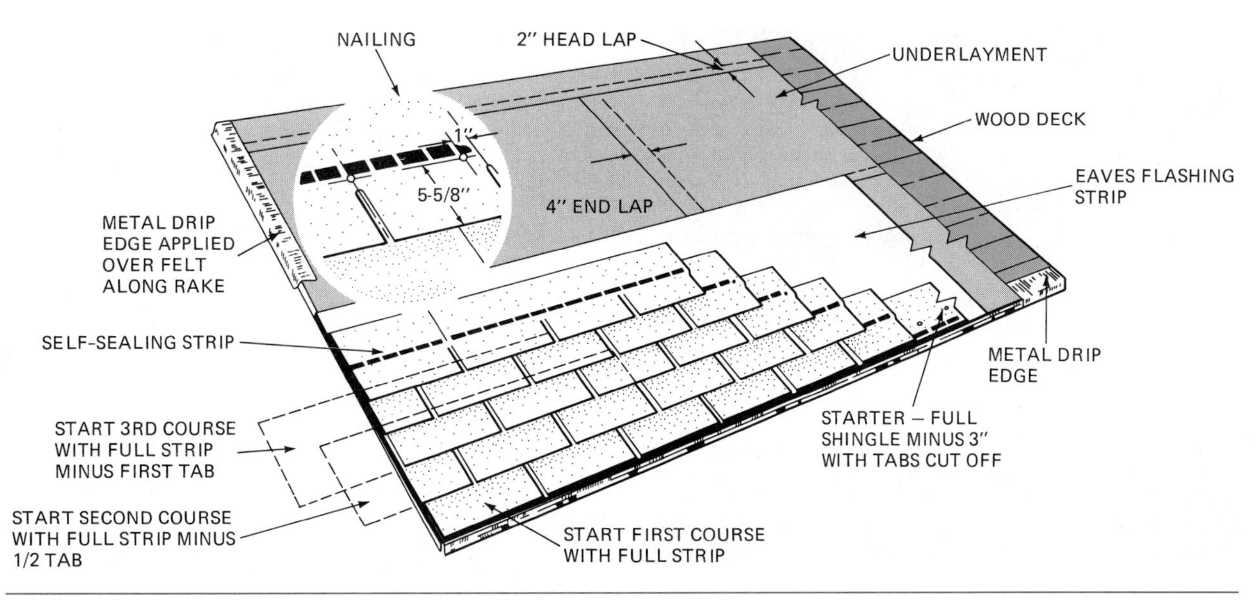

Fig. 51-12 Layout to stagger cutouts on the halves starting from the rake.

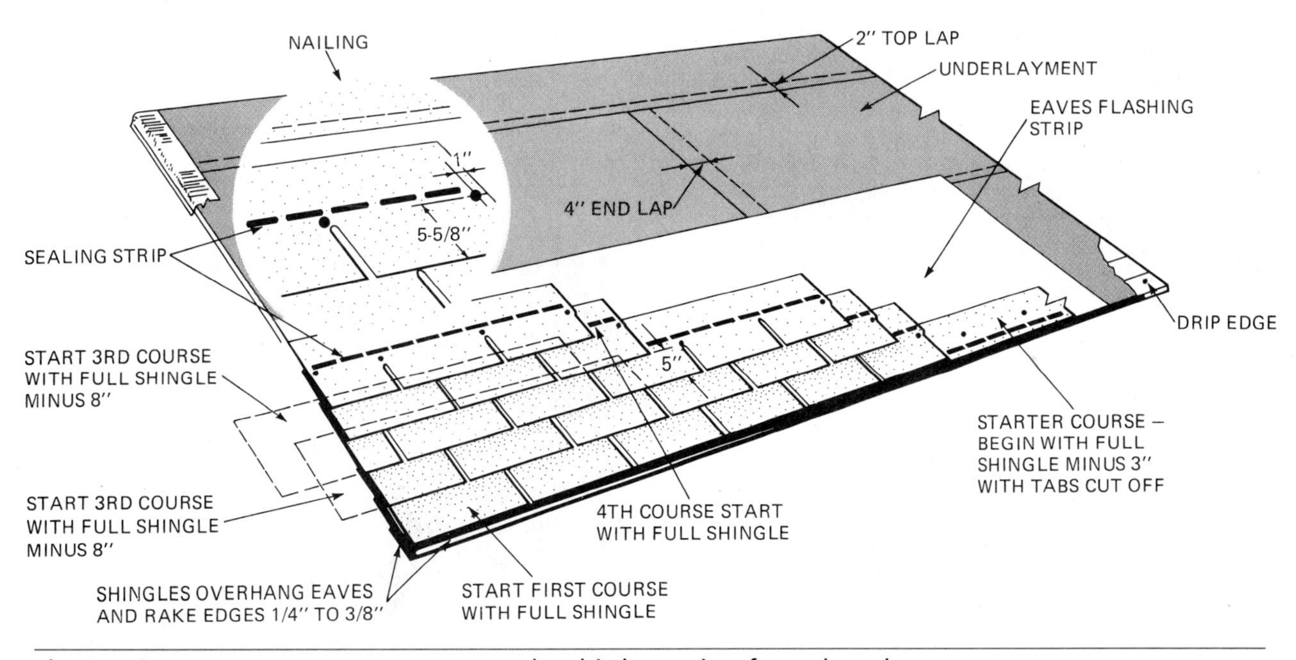

Fig. 51-13 Layout to stagger cutouts on the thirds starting from the rake.

Starter Course of Asphalt Shingles

The *starter course* backs up and fills in the spaces between tabs of the first regular course of shingles. Cut a minimum of 3 inches off the butt of the shingles needed for the starter course. Apply them with the factory-applied adhesive along the roof edge. Start the course so that no end joint will fall in line with an end joint or tab cutout of the regular first course of shingles. If the rake trim projects beyond the fascia, a shingle tab about 4 or 5 inches wide is installed to cover the top edge of the rake trim (Fig. 51-14). A wood shingle should first be placed under the tab to give it support.

Fastening Asphalt Shingles

Selecting suitable fasteners, using the recommended number, and putting them in the right places are important steps in the application of asphalt shingles. Lay the first regular course of shingles on top of the starter course. Keep their bottom edges flush with each other. Use a minimum of four fasteners in each strip shingle. Do not nail into or above the factory-applied adhesive (Fig. 51-15).

The fastener length should be sufficient to penetrate the sheathing at least 3/4 inch, or through approved panel sheathing. Roofing nails should be 11- or 12-gauge galvanized or aluminum with barbed shanks. They should have 3/8 to 7/16 inch

Fig. 51-14 A shingle tab is cut to cover the projecting edge of the rake trim.

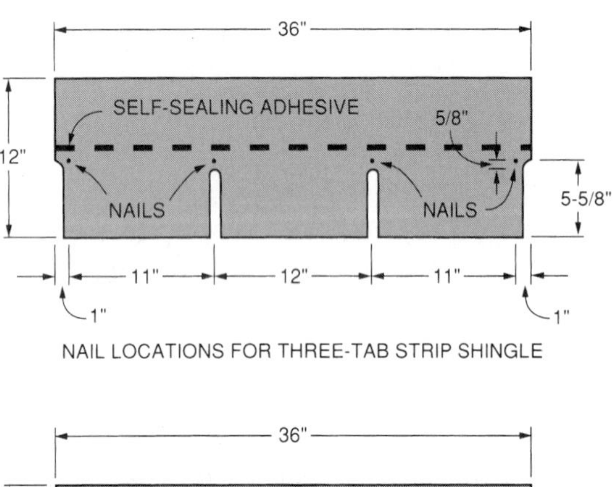

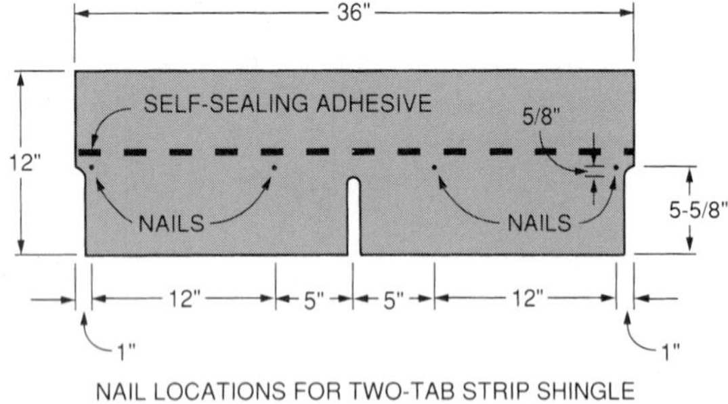

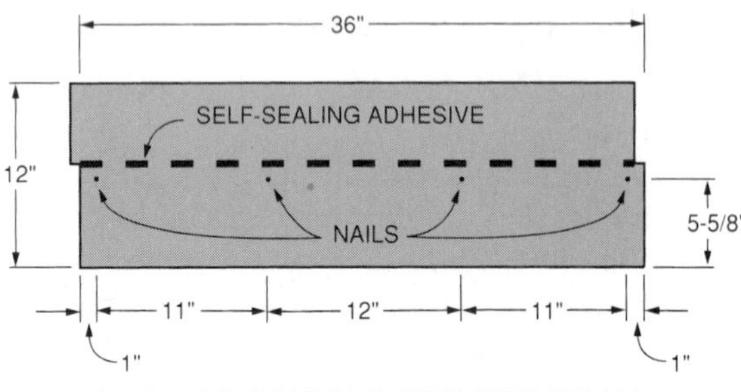

Fig. 51-15 Recommended fastener locations for asphalt strip shingles. (*Courtesy of Asphalt Roofing Manufacturers' Association*)

heads. Roofing nails may be driven by hand or with power nailers. Power-driven staples may be used in place of nails. However, their use is limited to shingles with factory-applied adhesive (Fig. 51-16). Staples must be at least 16-gauge. They should have a minimum crown width of 15/16 inch.

Align each shingle of the first course carefully. Fasten each shingle from the end nearest the shingle just laid. This prevents buckling. Drive fasteners straight so that the nail heads or staple crowns will not cut into the shingles. The entire crown of the staple or nail head should bear tightly against the shingle. It should not penetrate its surface (Fig. 51-17). Continue applying shingles until the first course is complete. Both ends of the course should be flush with the rake drip edge.

Shingle Exposure

The maximum *exposure* of asphalt shingles to the weather depends on the type of shingle. Recommended maximum exposures range from 4 to 6 inches. Most commonly used asphalt shingles have a maximum exposure of 5 inches. Less than the maximum recommended exposure may be used, if desired.

Fig. 51-16 Pneumatic staplers and nailers are often used to fasten asphalt shingles.

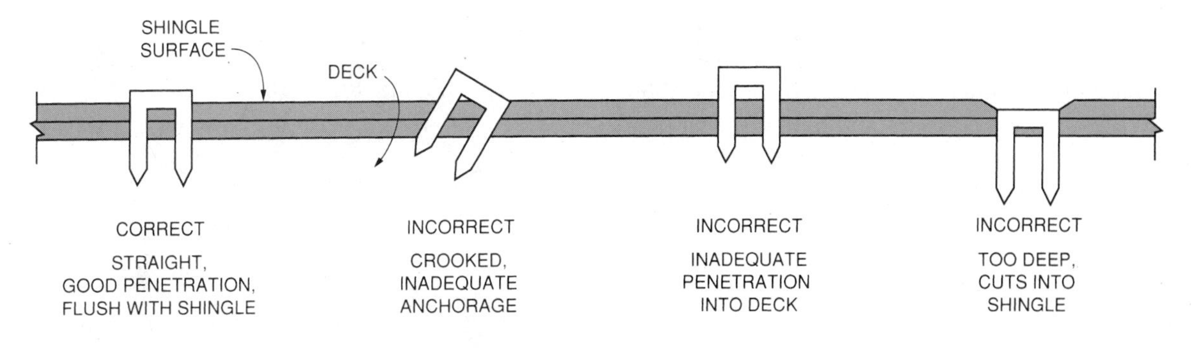

Fig. 51-17 It is important to fasten asphalt shingles correctly. (*Courtesy of Asphalt Roofing Manufacturers' Association*)

Laying Out Shingle Courses

When laying out shingle courses, space the desired exposure up each rake from the top edge of the first course of shingles. Snap lines across the roof for five or six courses. When snapping a long line, it may be necessary to thumb the line down against the roof at about center. Then snap the lines on both sides of the thumb.

Lay succeeding courses so their top edges are to the chalk line (Fig. 51-18). Start each course so the cutouts are staggered in the desired manner. Continue snapping lines and applying courses until a point 3 or 4 feet below the ridge is reached. Some carpenters snap a line to straighten out the course after using the top of the cutout for the shingle exposure for a number of courses.

Spacing Shingle Courses to the Ridge

The last course of shingles should be exposed below the *ridge cap* by about the same amount as all the other shingle courses. Cut a full tab from an asphalt strip. Center it on the ridge. Then bend it over the ridge. Do this at both ends of the building. Mark the bottom edges on the roof. It may be necessary, when reroofing older buildings, to mark and space up near the center also, if the ridge is sagged.

Measure up 2 inches. Snap a line between the marks. This line will be the top edge of the next to last course of shingles. It should be about 3 1/2 inches down from the ridge. Divide the distance between this line and the top of the last course of shingles applied into spaces as close as possible to the exposure of previous courses. Do not exceed the maximum exposure (Fig. 51-19). Snap lines across the roof and shingle up to the ridge.

The line for the last course of shingles is snapped on the face of the course below. Lay out the exposure from the bottom edge of the course on both ends of the roof. Snap a line across the roof. Fasten the last course of shingles by placing their butts to the line. Bend the top edges over the roof. Fasten their top edges to the opposite slope.

Applying the Ridge Cap

Cut hip and ridge shingles from shingle strips to make approximately 12 × 12-inch squares. Cut

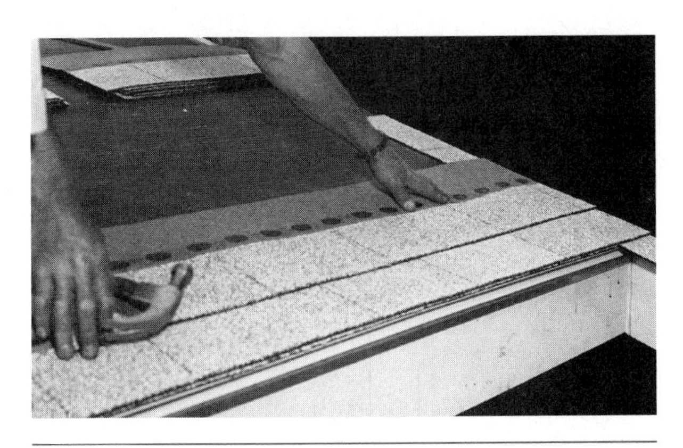

Fig. 51-18 Lay shingles so their top edge is to the chalk line.

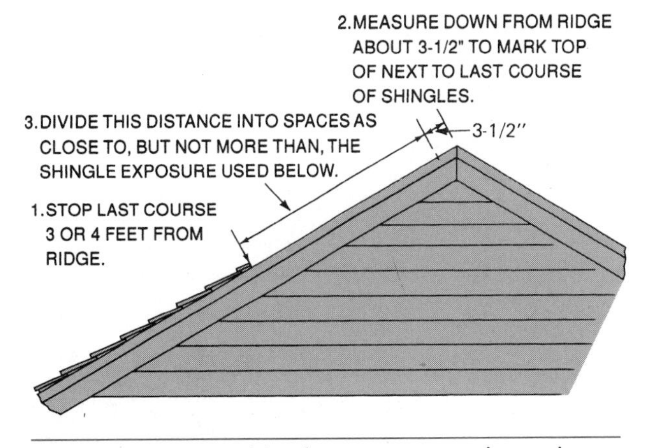

Fig. 51-19 Space shingle courses evenly to the ridge.

shingles from the top of the cutout to the top edge on a slight taper. The top edge should be narrower than the bottom edge (Fig. 51-20). Cutting the shingles in this manner keeps the top half of the shingle from protruding when it is bent over the ridge.

The ridge cap is applied after both sides of the roof have been shingled. At each end of the roof, center a shingle on the ridge. Bend it over the ridge. Mark its bottom edge on the front slope or the one most visible. Snap a line between the marks.

Beginning at the bottom of a hip or at one end of the ridge, apply the shingles over the hip or ridge. Expose each 5 inches. In cold weather, ridge cap shingles may have to be warmed in order to prevent cracking when bending them over the ridge. On the ridge, shingles are started from the end away from prevailing winds. The wind should blow over the shingle butts, not against them. Keep one edge, from the butt to the start of the tapered cut, to the chalk line. Secure each shingle with one fastener on each side, 5 1/2 inches from the butt and one inch up from each edge (Fig. 51-21). Apply the cap across the ridge until 3 or 4 feet from the end. Then space the cap to the end in the same manner as spacing the shingle course to the ridge. The last ridge shingle is cut to size. It is applied with one fastener on each side of the ridge. The two

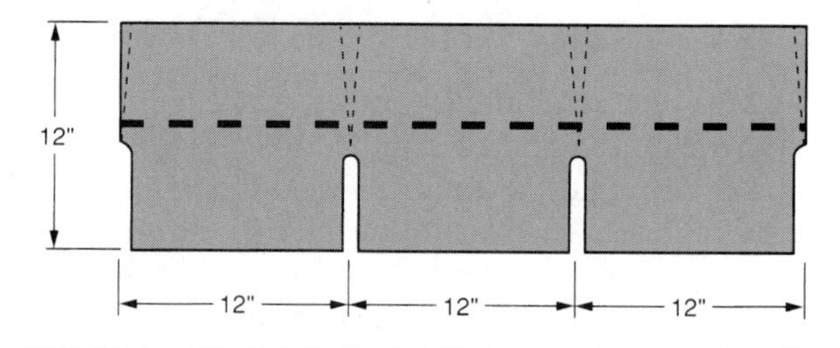

Fig. 51-20 Hip and ridge shingles are cut from strip shingles. (*Courtesy of Asphalt Roofing Manufacturers' Association*)

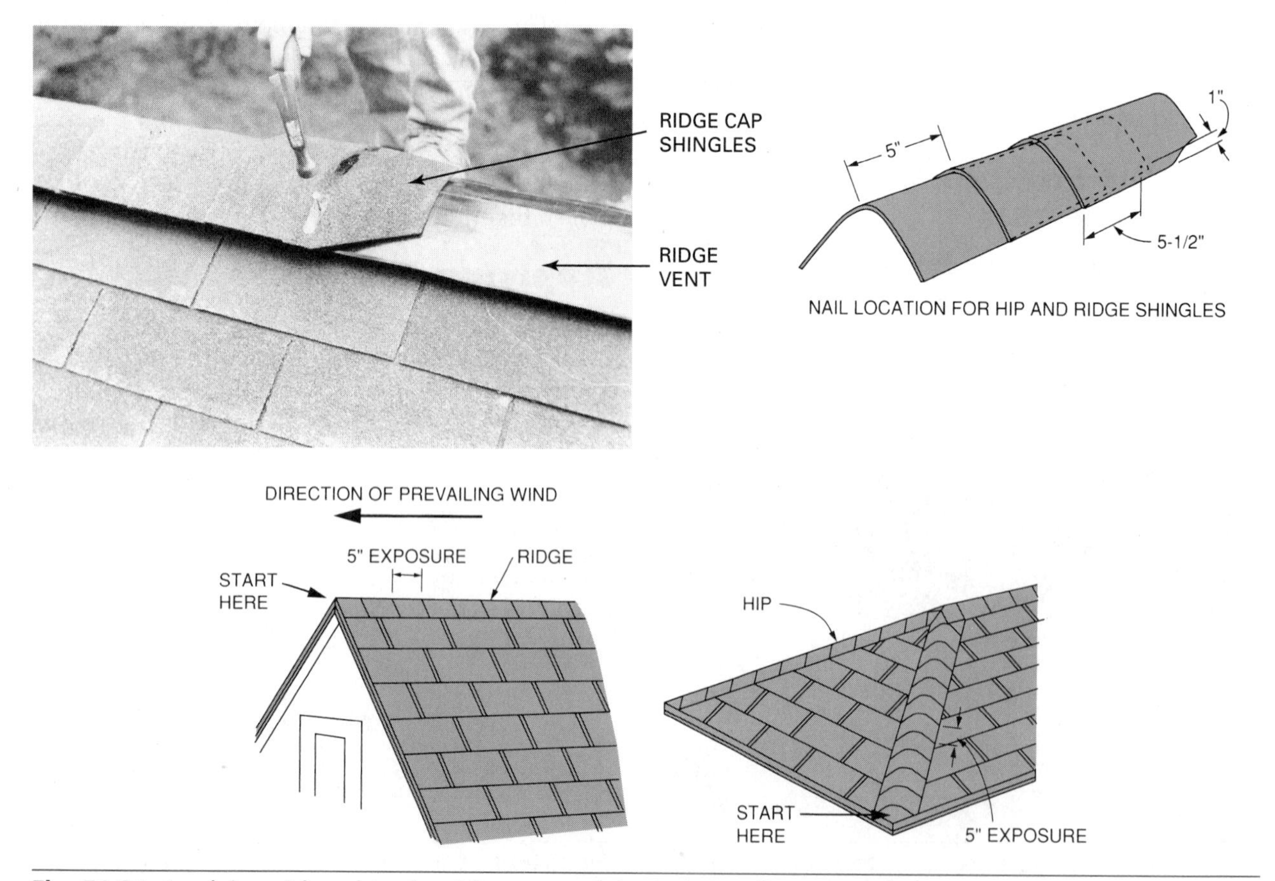

Fig. 51-21 Applying ridge shingles. (*Courtesy of Asphalt Roofing Manufacturers' Association*)

fasteners are covered with asphalt cement to prevent leakage.

Steep Slope Application

Steep roof slopes, like those on lower *mansard roof* slopes, reduce the effectiveness of factory-applied adhesive. This is especially true in cold or shaded areas. The maximum slope recommended for normal asphalt shingle application is 60 degrees or 21 inches rise per foot of run.

On steeper slopes, apply spots of quick-setting asphalt plastic cement about one inch in diameter under all shingle tabs immediately after installation. Apply one spot of cement under each tab of shingles with three or more tabs. Use two spots of cement under tabs of shingles with two tabs. For shingles with no cutouts, use three spots of cement.

The steep slope roof must be provided with through ventilation to evaporate moisture in the air behind the roof sheathing.

CHAPTER 52 ROLL ROOFING

Roll roofing can be used on roof slopes as little as one inch rise per foot of run. On steeper roofs, roll roofing is used when economy is the major factor and appearance is not so important.

Types of Roll Roofing

Roll roofing is made of the same materials as asphalt shingles. Various types are made with a base sheet of organic felt or glass fibers in a number of weights (Fig. 52-1). Some types are applied with exposed or concealed fasteners. They have a top lap of 2 to 4 inches. A concealed-nail type, called *double coverage* roll roofing, has a top lap of 19 inches. All kinds come in rolls that are 36 inches wide.

Roll roofing is recommended for use on roofs with slopes less than 4 inches rise per foot. However,

TYPICAL ASPHALT ROLLS

1	2		3	4		5		6	7
PRODUCT	Approximate Shipping Weight		Sqs. Per Package	Length	Width	Side or End Lap	Top Lap	Exposure	Underwriters' Listing
	Per Roll	Per Sq.							
 MINERAL SURFACE ROLL	75# to 90#	75# to 90#	One	36′ 38′	36″ 36″	6″	2″ 4″	34″ 32″	C
 MINERAL SURFACE ROLL DOUBLE COVERAGE	55# to 70#	55# to 70#	One Half	36′	36″	6″	19″	17″	C

Fig. 52-1 Types of roll roofing.

the exposed-nail type should not be used on pitches less than 2 inches rise per foot. Roll roofing applied with concealed nails and having a top lap of at least 3 inches may be used on pitches as low as 1 inch per foot. The exposed fastener type is rarely used. Only the concealed fastener type is recommended for use and described in this chapter. Use the same type and length of nails as for asphalt shingles.

General Application Methods

Apply all roll roofing when the temperature is above 45 degrees Fahrenheit. This prevents cracking the coating. Cut the roll into 12- to 18-foot lengths. Spread in a pile on a smooth surface to allow them to flatten out.

Use only the lap or quick-setting cement recommended by the manufacturer. Store cement in a warm place until ready for use. To warm it rapidly, place the unopened container in hot water.

CAUTION These materials are flammable. Never warm them over an open fire or place them in direct contact with a hot surface.

Apply roll roofing only on a solid, smooth, well-seasoned deck. Make sure the area below has sufficient ventilation to prevent the deck from absorbing condensation. This would cause the roofing to warp and buckle. A felt underlayment is not used with roll roofing.

Applying Roll Roofing with Concealed Fasteners

Apply 9-inch wide strips of the roofing along the eaves and rakes overhanging about 3/8 inch. Fasten the strips with two rows of nails one inch from each edge. Space them about 4 inches apart.

Apply the first course with its edges and ends flush with the strips. Secure the upper edge with nails staggered about 4 inches apart. Do not fasten within 18 inches of the rake edge. Apply cement only to that part of the edge strips covered by the first course. Press the lower edge and rake edges of the first course firmly in place over the edge strips. Finish nailing the upper edge out to the rakes. Apply succeeding courses in like manner. Make all end laps 6 inches wide. Apply cement the full width of the lap.

After all courses are in place, lift the lower edge of each course. Apply the cement in a continuous layer over the full width and length of the lap. Press the lower edges of the upper courses firmly into the cement. A small bead should appear along the entire edge of the sheet (Fig. 52-2). Care must be taken to apply the correct amount of cement.

Covering Hips and Ridges

Cut strips of 12 × 36 roofing. Bend the pieces lengthwise through their centers. Snap a chalk line on both sides down 5 1/2 inches from the hip or ridge. Apply cement from one line over the top to the other line.

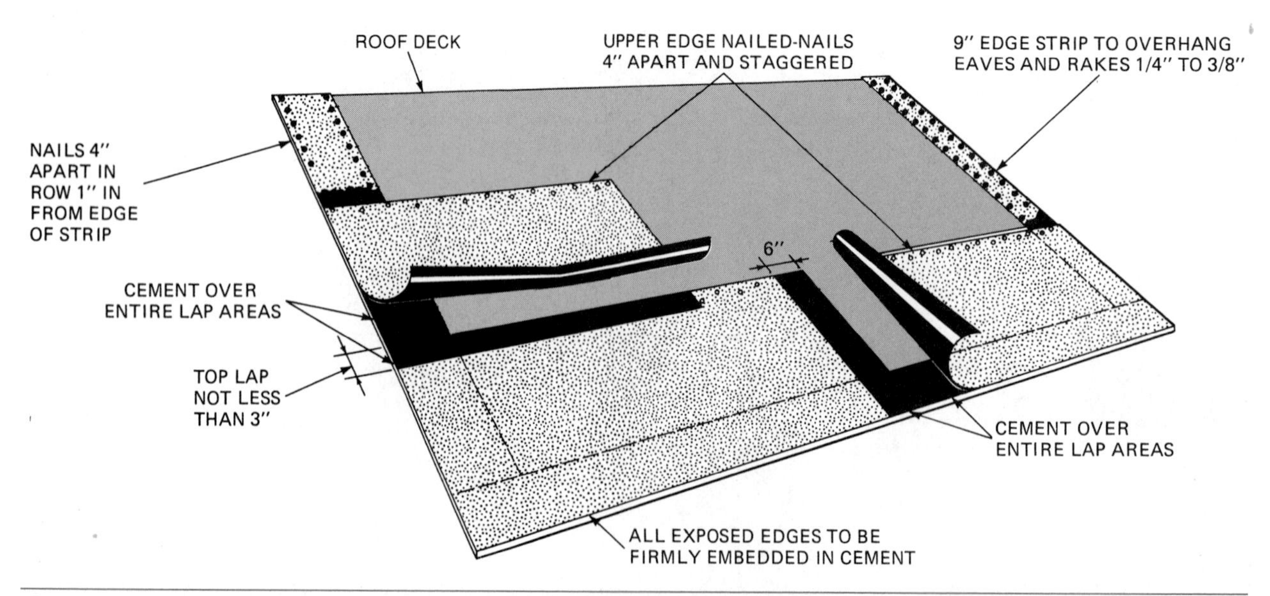

Fig. 52-2 Recommended procedure for applying concealed-nail type roll roofing.

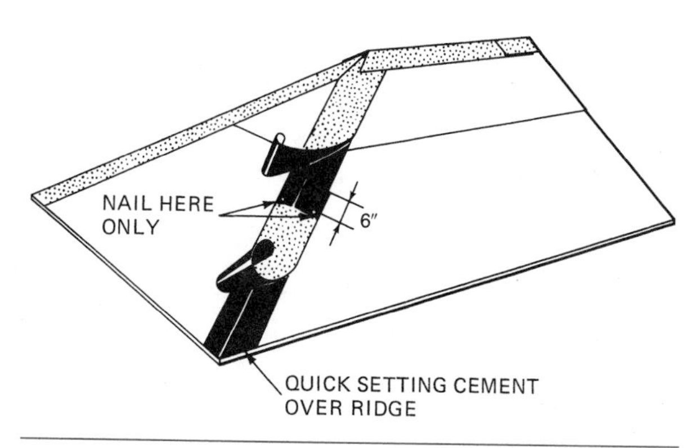

Fig. 52-3 Finishing hips and ridges using concealed-nail type roll roofing.

Fit the first strip over the hip or ridge. Press it firmly into place. Start at the lower end of a hip and at either end of a ridge. Lap each strip 6 inches over the preceding one. Nail each strip only on the end that is to be covered by the overlapping piece. Spread cement on the end of each strip that is lapped before the next one is applied. Continue in this manner until the end is reached (Fig. 52-3).

Applying Double Coverage Roll Roofing

Cut the 19-inch *selvage* portion from enough double coverage roll roofing to cover the length of the roof. Save the surfaced portion for the last course. Apply the selvage portion parallel to the eaves. It should overhang the drip edge by 3/8 inch. Secure it to the roof deck with three rows of nails. Place the top row 4 1/2 inches below the upper edge. Put the bottom row one inch above the bottom edge. Place the other row halfway between. Place the nails in the upper and middle rows slightly staggered about 12 inches apart. Place the nails in the lower row about 6 inches apart and slightly staggered. Nail along rakes in the same manner.

Apply the first course. Secure it with two rows of nails in the selvage portion. Place one row about 4 3/4 inches below the upper edge. Put the second row about 8 1/2 inches below the first. Space the nails about 12 inches apart in each row and stagger them (Fig. 52-4).

Apply succeeding courses in the same manner. Lap the full width of the 19-inch selvage each time. Make all end laps 6 inches wide. End laps are made in the manner shown in Figure 52-5. Stagger end laps in succeeding courses.

Lift and roll back the surface portion of each course. Starting at the bottom, apply cement to the entire selvage portion of each course. Apply it to within 1/4 inch of the surfaced portion. Press the overlying sheet firmly into the cement. Apply pressure over the entire area using a light roller to ensure adhesion between the sheets at all points.

Apply the remaining surfaced portion left from the first course as the last course. This type roofing may also be applied in like manner parallel to the rakes (Fig. 52-6). Hips and ridges are covered in the same manner shown in Figure 52-7.

It is important to follow specific application instructions because of differences in the manufacture of roll roofing. Some instructions call for hot asphalt. Others call for cold cement. Others give the option of either. Specific requirements for quantities and types of adhesive must be followed.

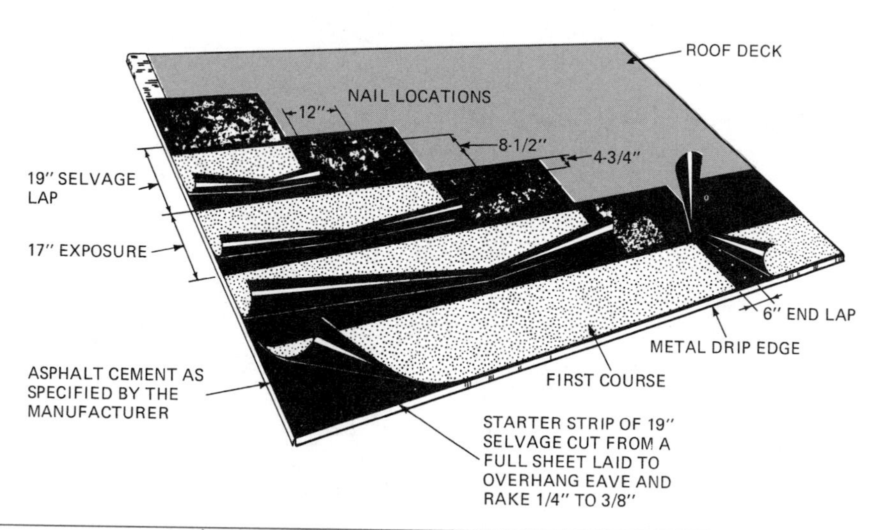

Fig. 52-4 Method of applying double coverage roll roofing.

STEP BY STEP

PROCEDURES

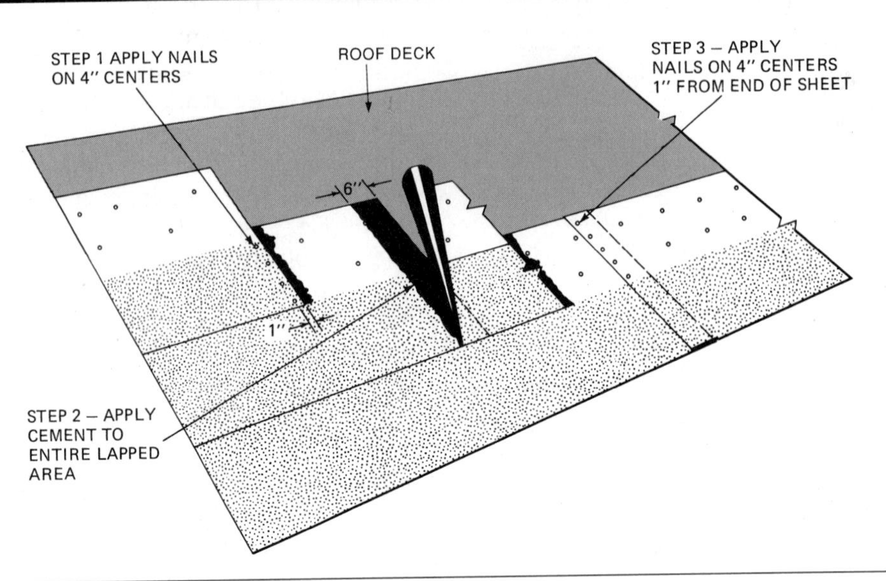

Fig. 52-5 Method of making end laps on double coverage roll roofing.

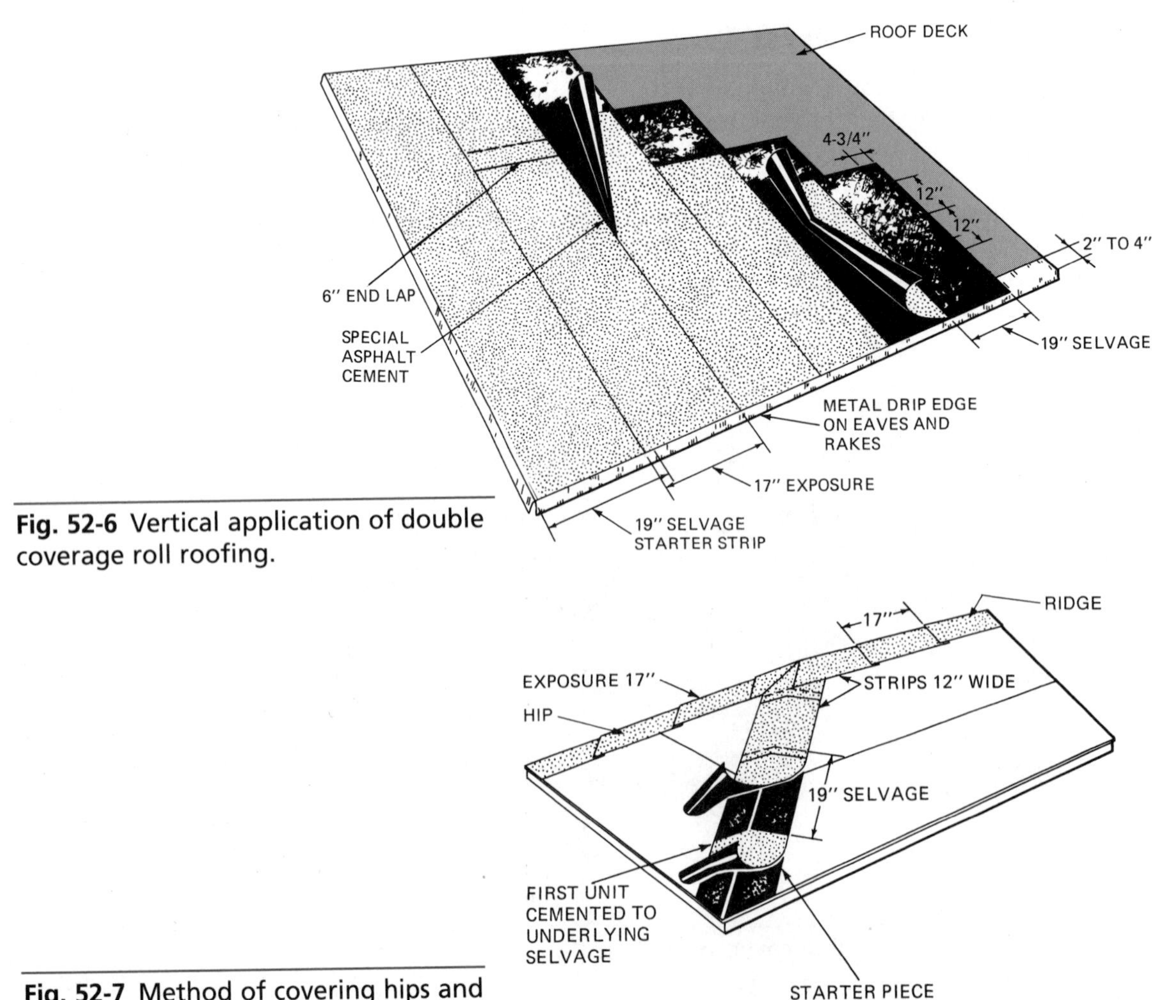

Fig. 52-6 Vertical application of double coverage roll roofing.

Fig. 52-7 Method of covering hips and ridges with double coverage roofing.

CHAPTER 53 WOOD SHINGLES AND SHAKES

Wood shingles and *shakes* are used extensively to cover roofs (Fig. 53-1). Most shingles and shakes are produced from western red cedar. Cedar logs are first cut into desired lengths. They are then split into sections from which shingles and shakes are sawn or split. All shingles are sawed. Most shakes are split. Shingles, therefore, have a relatively smooth surface. Most shakes have at least one highly textured, natural grain surface. Most shakes also have thicker butts than shingles (Fig. 53-2).

Fig. 53-1 Wood shingles and shakes are used extensively for roof covering.

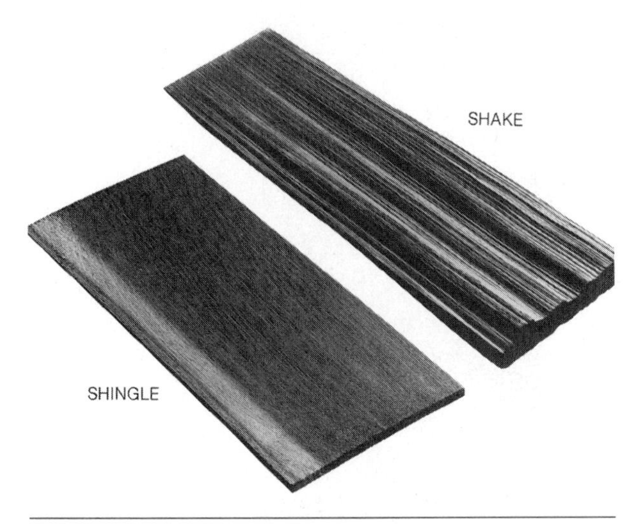

Fig. 53-2 All shingles are sawed from the log. Most shakes are split and have a rough surface. (*Courtesy of Cedar Shake and Shingle Bureau*)

Most wood shingles and shakes are produced by mills that are members of the Cedar Shake and Shingle Bureau. Their product label (Fig. 53-3) is assurance that the products meet quality standards.

Description of Wood Shingles and Shakes

Wood shingles are available for use on roofs in four standard grades. Shakes are manufactured by different methods to produce four types. Both shingles and shakes may be treated to resist fire or premature decay in areas of high humidity (Fig. 53-4).

Sizes and Coverage

Shingles come in lengths of 16, 18, and 24 inches. The butt thickness increases with the length. Shakes are available in lengths of 15, 18, and 24 inches. Their butt thicknesses are from 3/8 to 3/4 inch. The 15-inch length is used for starter and finish courses. Figure 53-5 shows the sizes and the amount of roof area that one square, laid at various exposures, will cover.

Maximum Exposures

The area covered by one square of shingles or shakes depends on the amount they are exposed to the weather. The maximum amount of shingle exposure depends upon the length and grade of the shingle or shake and the pitch of the roof. Shakes are not generally applied to roofs with slopes of less than 4 inches rise per foot. Shingles, with reduced exposures, may be used on slopes down to 3 inches rise per foot. Figure 53-6 shows the maximum recommended roof exposure for wood shingles and shakes.

Fig. 53-3 The product label assures that quality requirements are met. (*Courtesy of Cedar Shake and Shingle Bureau*)

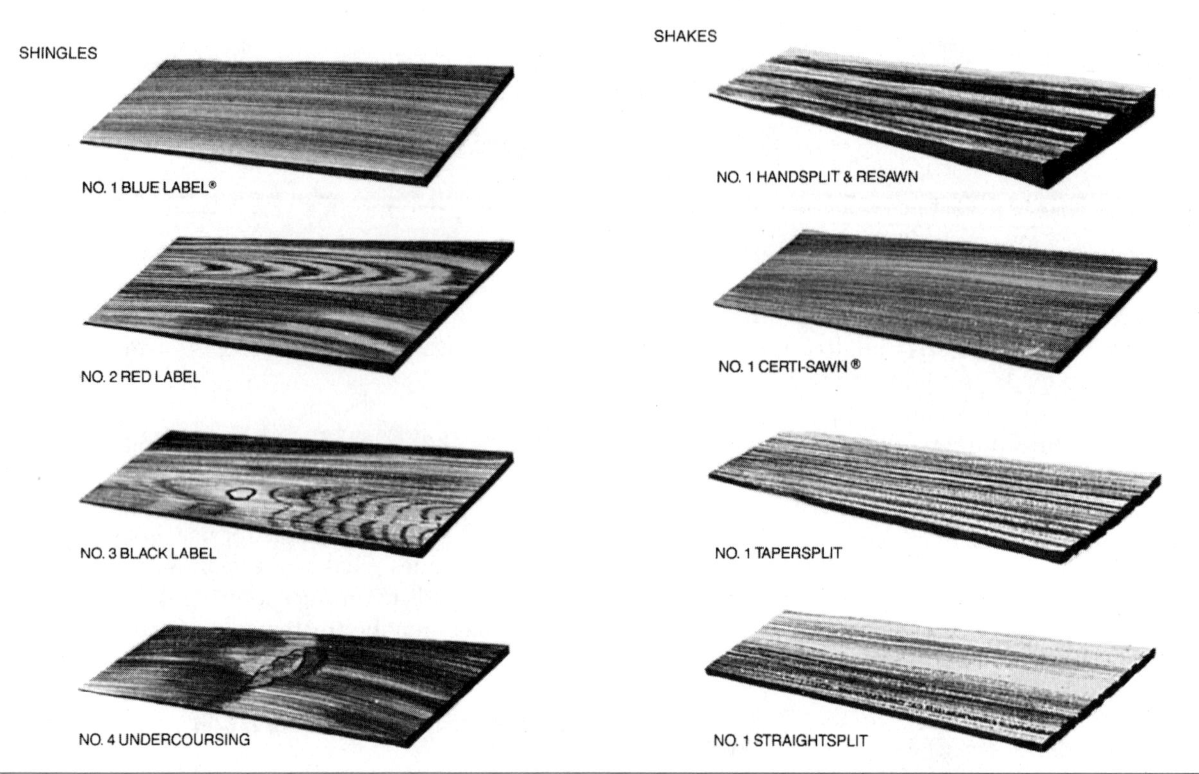

Fig. 53-4 Description of grades and kinds of wood shingles and shakes. (*Courtesy of Cedar Shake and Shingle Bureau*)

COVERAGE AND EXPOSURE TABLES

Shingle Coverage Table

LENGTH AND THICKNESS	Approximate coverage of one square (4 bundles) of shingles based on following weather exposures								
	3½"	4"	4½"	5"	5½"	6"	6½"	7"	7½"
16" × 5/2"	70	80	90	100*	—	—	—	—	—
18" × 5/2¼"	—	72½	81½	90½	100*	—	—	—	—
24" × 4/2"	—	—	—	—	73½	80	86½	93	100*
NOTE: *Maximum exposure recommended for roofs.									

Shake Coverage Table

SHAKE TYPE, LENGTH AND THICKNESS	Approximate coverage (in sq. ft.) of one square, when shakes are applied with an average ½" spacing, at following weather exposures, in inches (d):				
	5	5½	7½	8½	10
18" × ½" Handsplit-and-Resawn Mediums (a)	—	55(b)	75(c)	—	—
18" × ¾" Handsplit-and-Resawn Heavies (a)	—	55(b)	75(c)	—	—
18" × ⅝" Tapersawn	—	55(b)	75(c)	—	
24" × ⅜" Handsplit	50(e)	—	75(b)	—	—
24" × ½" Handsplit-and-Resawn Mediums	—	—	75(b)	85	100(c)
24" × ¾" Handsplit-and-Resawn Heavies	—	—	75(b)	85	100(c)
24" × ⅝" Tapersawn	—	—	75(b)	85	100(c)
24" × ½" Tapersplit	—	—	75(b)	85	100(c)
18" × ⅜" Straight-Split	—	65(b)	90(c)	—	—
24" × ⅜" Straight-Split	—	—	75(b)	85	100(c)
15" Starter-Finish course	Use supplementary with shakes applied not over 10" weather exposure.				

(a) 5 bundles will cover 100 sq. ft. roof area when used as starter-finish course at 10" weather exposure; 7 bundles will cover 100 sq. ft. roof area at 7½" weather exposure; see footnote (d).
(b) Maximum recommended weather exposure for 3-ply roof construction.
(c) Maximum recommended weather exposure for 2-ply roof construction.
(d) All coverage based on an average ½" spacing between shakes.
(e) Maximum recommended weather exposure.

Fig. 53-5 Tables show the sizes and coverage of wood shingles and shakes. (*Courtesy of Cedar Shake and Shingle Bureau*)

Shingle Exposure Table

PITCH	Maximum exposure recommended for roofs								
	Length								
	No. 1 Blue Label			No. 2 Red Label			No. 3 Black Label		
	16"	18"	24"	16"	18"	24"	16"	18"	24"
3/12 to 4/12	3 3/4"	4 1/4"	5 3/4"	3 1/2"	4"	5 1/2"	3"	3 1/2"	5"
4/12 and steeper	5"	5 1/2"	7 1/2"	4"	4 1/2"	6 1/2"	3 1/2"	4"	5 1/2"

Shake Exposure Table

PITCH	Maximum exposure recommended for roofs	
	Length	
	18"	24"
4/12 and steeper	7 1/2"	10"(a)
(a) 24" x 3/8" handsplit shakes limited to 5" maximum weather exposure, per UBC.		

Formula for calculating material at reduced exposures:

square footage ÷ reduced coverage = total material required

e.g. You are estimating a roof that measures 3200 square feet (32 squares). You have decided to put 16" shingles at 4" exposure. A 4-bundle square at 4" exposure covers 80 square feet.

3200 ÷ 80 = 40 squares of material

Fig. 53-6 Maximum exposures of wood shingles and shakes for various roof pitches. (*Courtesy of Cedar Shake and Shingle Bureau*)

Sheathing and Underlayment

Shingles and shakes may be applied over spaced or solid roof sheathing. *Spaced sheathing* is usually 1 × 4 or 1 × 6 boards. *Solid sheathing* is usually APA-rated panels of plywood or strand board. It may be required in regions subject to frequent earthquakes or under treated shingles and shakes. It is also recommended for use with shakes in areas where wind-driven snow is common.

Spaced Sheathing

Solid wood sheathing is applied from the eaves up to a point that is plumb with a line 12 to 24 inches inside the wall line. An eaves flashing is installed, if required. Spaced sheathing may then be used above the solid sheathing to the ridge.

For shingles, either 1 × 4 or 1 × 6 spaced sheathing may be used. Space 4-inch boards the same amount as the shingles are exposed to the weather. In this method of application, each course of shingles is nailed to the center of the board. If 6-inch boards are used, they are spaced two exposures. Two courses of shingles are nailed to the same board when courses are exposed up to, but not exceeding, 5 1/2 inches. For shingles with greater exposures, the sheathing is spaced a distance of one exposure (Fig. 53-7).

In shake application, spaced sheathing is usually 1 × 6 boards spaced the same distance, on

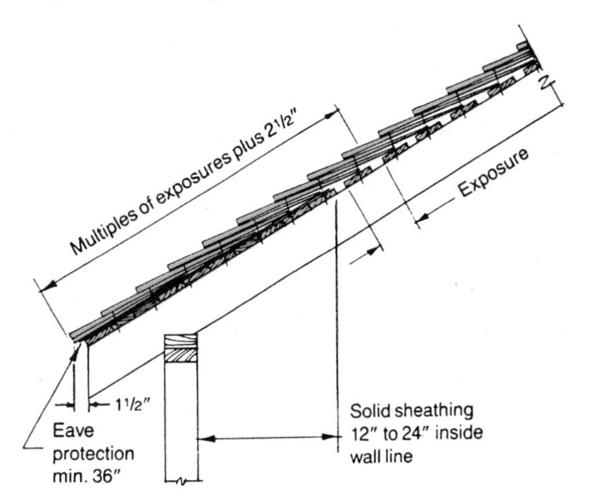

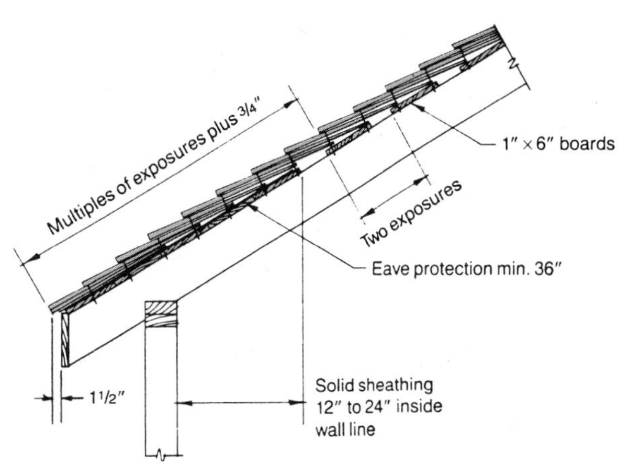

Fig. 53-7 Application of wood shingles on spaced sheathing. (*Courtesy of Cedar Shake and Shingle Bureau*)

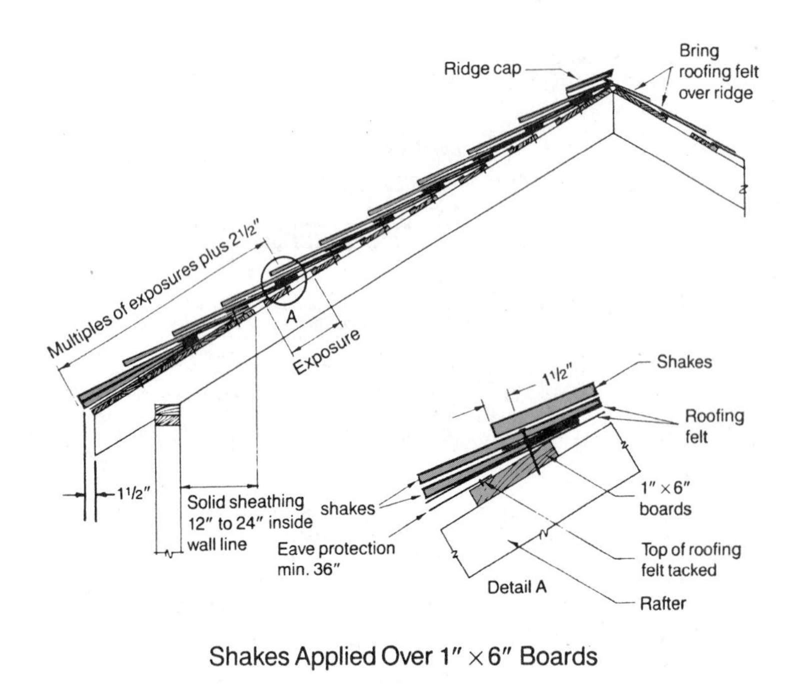

Fig. 53-8 Method of applying shakes on spaced sheathing. (*Courtesy of Cedar Shake and Shingle Bureau*)

center, as the shake exposure (Fig. 53-8). The spacing should never be more than 7 1/2 inches for 18-inch shakes and 10 inches for 24-inch shakes installed on roofs.

Underlayment

No underlayment is required under wood shingles. A breather-type roofing felt may be used over solid or spaced sheathing.

Application Tools and Fasteners

Shingles and shakes are usually applied with a shingling hatchet. Recommendations for the type and size of fasteners should be closely followed.

Shingling Hatchet

A *shingling hatchet* (Fig. 53-9), should be lightweight. It should have both a sharp *blade* and a *heel.* A *sliding gauge* is sometimes used for fast and accurate checking of shingle exposure. The gauge permits laying several shingle courses at a time without snapping a chalk line. A power nailer may be used. However, shingles and shakes often need to be trimmed or split with the hatchet. More time may be lost than gained by using a power nailer.

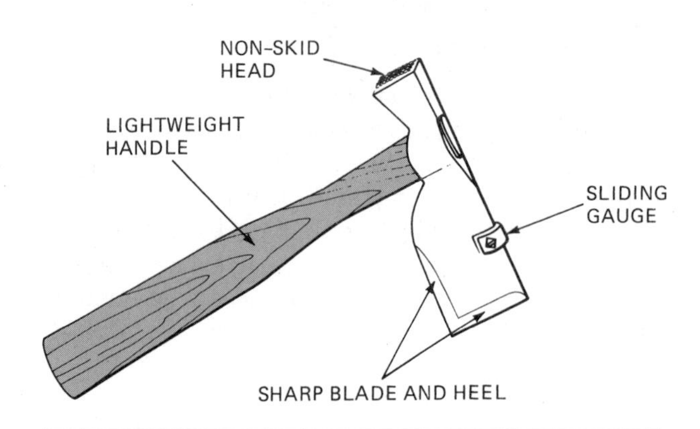

Fig. 53-9 A shingling hatchet commonly is used to apply wood shingles and shakes.

Fasteners

Apply each shingle with only two corrosion-resistant nails, such as stainless steel, hot-dipped galvanized, or aluminum. Box nails usually are used because their smaller gauge minimizes splitting. Minimum nail lengths for shingles and shakes are shown in Figure 53-10. Staples should be 16-gauge aluminum or stainless steel with a minimum crown of 7/16 inch. The staple should be driven with its crown across the grain of the shingle or shake. Staples should be long enough to penetrate the sheathing at least 1/2 inch.

Applying Wood Shingles

Install the first layer of a starter course of wood shingles as described in a previous chapter. If a gutter is used, overhang the shingles plumb with the center of the gutter. If no gutter is installed, over-

Type of Shingle or Shake	Nail Type and Minimum Length	
Shingles—New Roof	**Type**	**(in)**
16" and 18" Shingles	3d Box	1¼
24" Shingles	4d Box	1½
Shakes—New Roof	**Type**	**(in)**
18" Straight-Split	5d Box	1¾
18" and 24" Handsplit--and-Resawn	6d Box	2
24" Tapersplit	5d Box	1¾
18" and 24" Taper-sawn	6d Box	2

Fig. 53-10 Recommended nail types and sizes for wood shingles and shakes. (*Courtesy of Cedar Shake and Shingle Bureau*)

hang the starter course 1 1/2 inches beyond the fascia. Space adjacent shingles between 1/4 and 3/8 inch apart. Place each fastener about 3/4 inch in from the edge of the shingle and not more than one inch above the exposure line. Drive fasteners flush with the surface. Make sure that the head does not crush the surface.

Apply another layer of shingles on top of the first layer of the starter course. The starter course may be tripled, if desired, for appearance. This procedure is recommended in regions of heavy snowfall.

Lay succeeding courses across the roof. Apply several courses at a time. Stagger the joints in adjacent courses at least 1 1/2 inches. There should be no joint in any three adjacent courses in alignment. Joints should not line up with the centerline of the heart of wood or any knots or defects. **Flat grain** or *slash grain* shingles wider than 8 inches should be split in two before fastening. Trim shingle edges with the hatchet to keep their butts in line (Fig. 53-11).

After laying several courses, snap a chalk line to straighten the next course. Proceed shingling up the roof. When 3 or 4 feet from the ridge, check the distance on both ends of the roof. Divide the distance as close as possible to the exposure used. A full course should show below the ridge cap. Shingle tips are cut flush with the ridge. Tips can be easily cut across the grain with a hatchet if cut on an angle to the side of the shingle.

On intersecting roofs, stop shingling the roof a few feet away from the valley. Select and cut a shingle at a proper taper. Apply it to the valley. Do not break joints in the valley. Do not lay shingles with their grain parallel to the valley centerline. Work back out. Fit a shingle to complete the course (Fig. 53-12).

Hips and Ridges

After the roof is shingled, 4- to 5-inch wide hip and ridge caps are applied. Measure down on both sides of the hip or ridge at each end for a distance equal to the exposure. Snap a line between the marks.

Lay a shingle so its bottom edge is to the line. Fasten with two nails. Use longer nails so that they penetrate at least 1/2 inch into or through the sheathing. With the hatchet, trim the top edge flush with the opposite slope.

Lay another shingle on the other side with the butt even and its upper edge overlapping the first shingle laid. Trim its top edge at a bevel and flush with the side of the first shingle. Double this first set of shingles, alternating the joint. Apply succeeding layers of cap, with the same exposure as used on the roof. Alternate the overlap of each layer (Fig. 53-13). Space the exposure when nearing the end so all caps are about equally exposed.

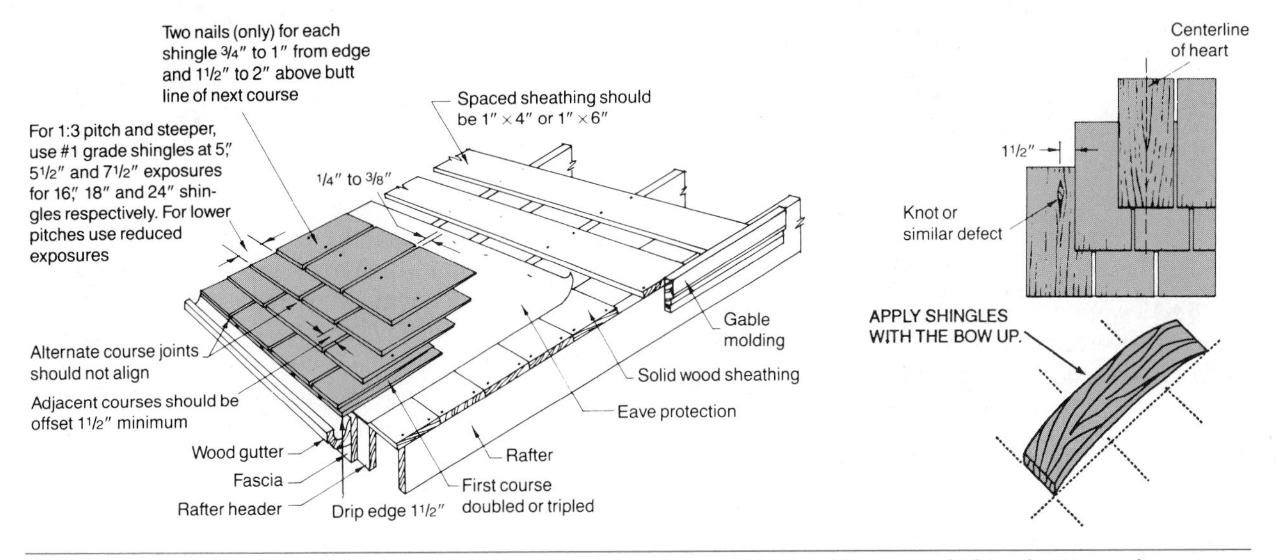

Fig. 53-11 Details of wood shingle application. (*Courtesy of Cedar Shake and Shingle Bureau*)

STEP BY STEP

PROCEDURES

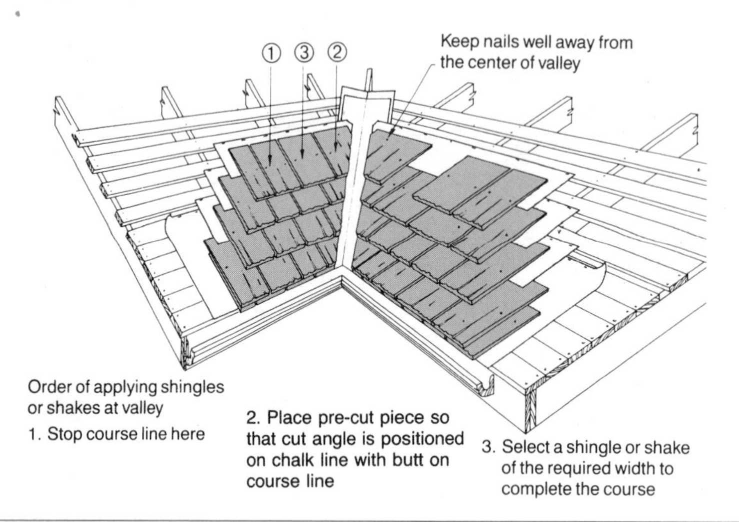

Fig. 53-12 Method of applying shingles or shakes along an open valley. (*Courtesy of Cedar Shake and Shingle Bureau*)

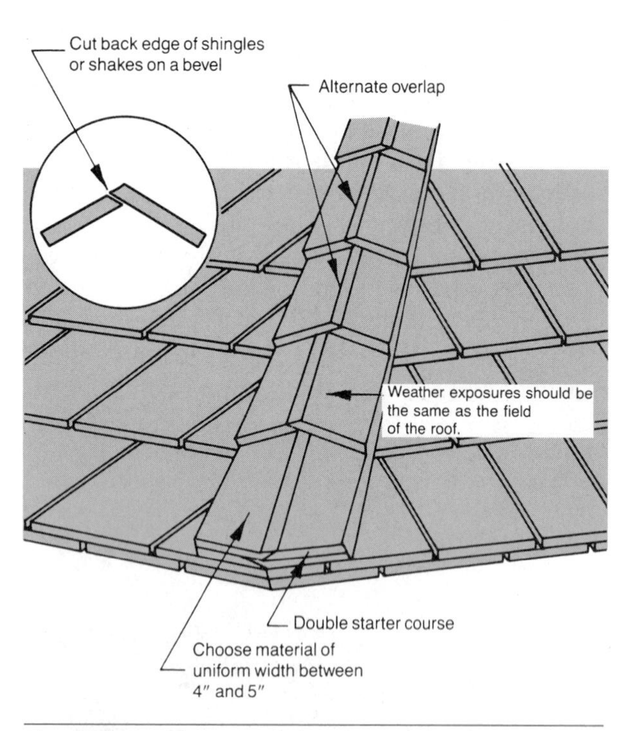

Fig. 53-13 When applying hip and ridge shingles, alternate the overlap. (*Courtesy of Cedar Shake and Shingle Bureau*)

Fig. 53-14 Three or four shake courses at a time are carried across the roof. (*Courtesy of Cedar Shake and Shingle Bureau*)

Applying Wood Shakes

Shakes are applied in much the same manner as shingles (Fig. 53-14). Mark the handle of the shingling hatchet at 7 1/2 and 10 inches from the top of the head. These are the exposures that are used most of the time when applying wood shakes.

An underlayment of felt is applied over a starter course of one or two layers of shingles or shakes that are overlaid with a course of shakes to be exposed. Butts of the starter course should project 1 1/2 inches beyond the fascia.

Interlayment

Next, lay an 18-inch wide strip of #30 roofing felt over the starter course. Its bottom edge should be positioned at a distance equal to twice the shake exposure above the butt line of the first course of shakes. For example, 24-inch shakes, laid 10 inches to the weather, would have felt applied 20 inches above the butts of the shakes. The felt will cover the top 4 inches of the shakes. It will extend up 14 inches on the sheathing. The top edge must rest on the spaced sheathing, if used.

Nail only the top edge of the felt. Fasten successive strips on their top edge only. Their bottom edges should be one shake exposure from the bottom of the previous strip. It is important to lay the felt straight. It serves as a guide for applying shakes. After the roof is felted, the tips of the shakes are tucked under the felt. The bottom should be exposed by the distance of twice the exposure (Fig. 53-15).

Apply the second and successive courses with joints staggered and fasteners placed the same as for shingles. The spacing between shakes should be at least 3/8 inch, but not more than 5/8 inch. Lay straight-split shakes with their smooth end toward the ridge (Fig. 53-16).

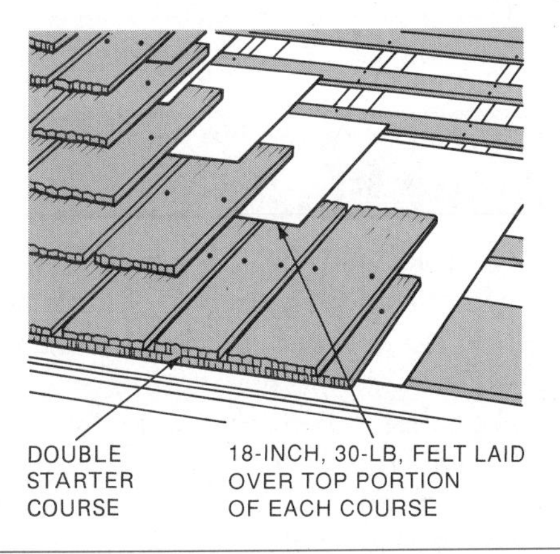

Fig. 53-15 An interlayment of felt is required when laying shakes.

Maintaining Shake Exposure

There is a tendency to angle toward the ground. Therefore, check the exposure regularly with the hatchet handle. An easy way to be sure of correct exposure is to look through the joint between the edges of the shakes in the course below the one being nailed. The bottom edge of the felt will be visible. The butt of the shake being nailed is positioned directly above it (Fig. 53-17).

Fig. 53-17 The tip of the shake is inserted between the layers of interlayment. Its butt is lined up with the bottom edge of another layer sighted between the edge joints of the course below. (*Courtesy of Cedar Shake and Shingle Bureau*)

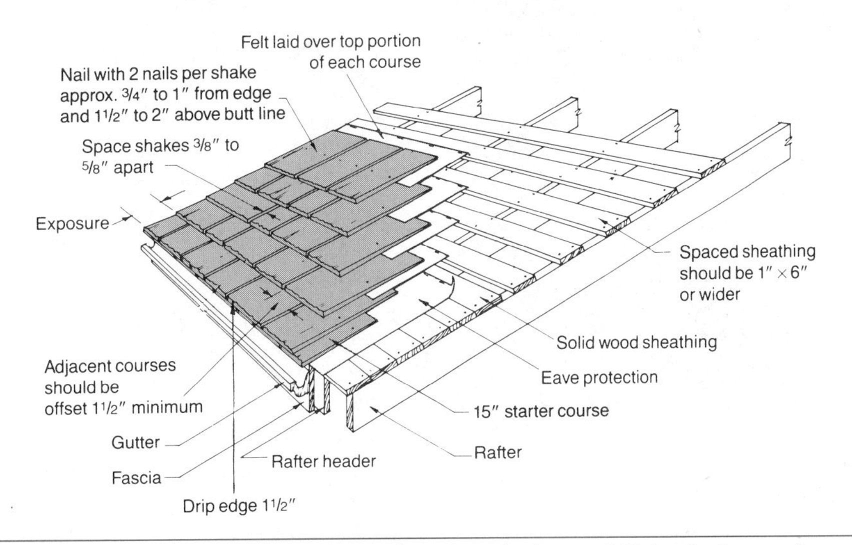

Fig. 53-16 Shake application details. (*Courtesy of Cedar Shake and Shingle Bureau*)

Ridges and Hips

Adjust the exposure so that tips of shakes in the next-to-last course just come to the ridge. Use economical 15-inch starter-finish shakes for the last course. They save time by eliminating the need to trim shake tips at the ridge. Cap ridges and hips with shakes in the same manner as that used with wood shingles.

Estimating Roofing Materials

Find the area of the roof in square feet. Divide the total by 100 to determine the number of squares needed. Add about 5 to 10 percent for waste. A simple roof with no dormers, valleys, or other obstructions requires less allowance for waste. A complicated roof requires more.

For wood shingles, add one square for every 240 linear feet of starter course. Add one square of shakes for 120 linear feet of starter course.

Add one extra square of shingles for every 100 linear feet of valleys and about two squares for shakes.

Add an extra bundle of shakes or shingles for every 16 feet of hip and ridge to be covered.

Figure 2 pounds of nails per square at standard exposure.

Remember that a square of roofing will cover 100 square feet of roof surface only when applied at standard exposures. Allow proportionally more material when these exposures are reduced.

CHAPTER 54 FLASHING

Flashing is a material used in various susceptible locations. It prevents water from entering a building (Fig. 54-1). The words *flash, flashed,* and *flashing* are also used as verbs to describe the installation of the material. Various kinds of flashing are applied at the eaves, valleys, chimneys, vents, and other roof projections. They prevent leakage at the intersections.

Fig. 54-1 Flashings are used to seal against leakage where the roof butts against adjoining walls and at other intersections. (*Courtesy of Cedar Shake and Shingle Bureau*)

Kinds of Flashing

Flashing material may be sheet copper, zinc, aluminum, galvanized iron, or mineral-surfaced asphalt roll roofing. Copper and zinc are high-quality flashing materials, but they are more expensive. Roll roofing is less expensive. Colors that match or contrast with the roof covering can be used. If properly applied, roll roofing used as a valley flashing will outlast the main roof covering. Sheet metal, especially copper, may last longer. However, it is good practice to replace all flashing when reroofing.

Eaves Flashing

Whenever there is a possibility of ice dams forming along the eaves and causing a backup of water, an *eaves flashing* is needed. For a slope of at least 4 inches rise per foot of run, install a course of 36-inch wide, smooth- or mineral-surface roll roofing, 50 pounds per square or heavier over the underlayment and drip edge. On lower slopes protection against water leakage from ice dams is gained by applying the underlayment as described previously and shown in Figure 51-8.

A course of a special 36-inch wide, self-adhering eaves flashing, called an *ice and water shield,* may also be used. Apply the flashing. Let it overhang the drip edge by 1/4 to 3/8 inch. The flashing should extend up the roof far enough to cover a point at least 12 inches inside the wall line of the building. If the overhang of the eaves requires that

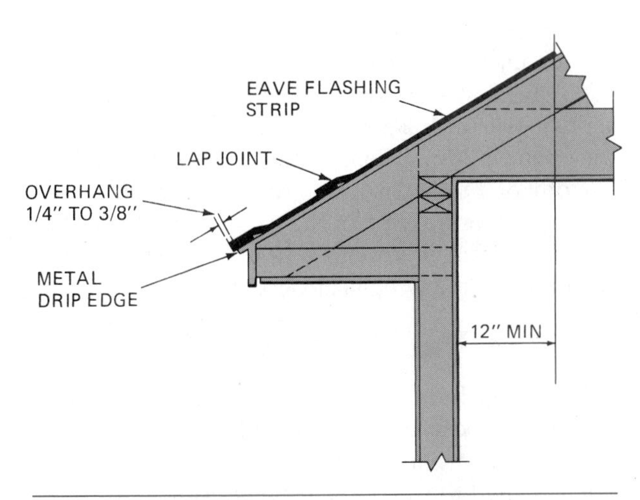

Fig. 54-2 An eaves flashing is installed if there is danger of ice dams forming along the eaves.

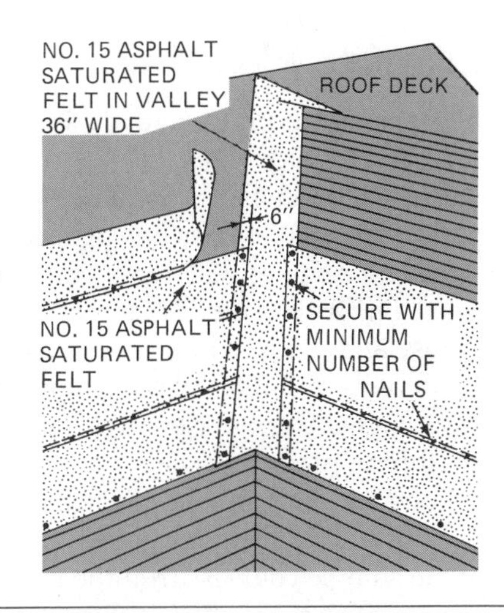

Fig. 54-3 A felt underlay is applied in the valley before flashing.

the flashing be wider than 36 inches, the necessary horizontal lap joint is cemented and located on the portion of the roof that extends outside the wall line (Fig. 54-2).

Flashing Valleys

Roof valleys are especially vulnerable to leakage. This is because of the great volume of water that flows down through them. Valleys must be carefully flashed according to recommended procedures. Valleys are flashed in two ways: *open* or *closed.*

Open Valley Flashing

Apply a 36-inch wide strip of asphalt felt centered in the **open valley.** Fasten it with only enough nails along its edges to hold it in place. Let the courses of felt underlayment applied to the roof overlap the valley underlayment by not less than 6 inches. Seat the felt well into the valley. Be careful not to cause any break in the felt. The eave flashing, if required, is then applied (Fig. 54-3).

Using Roll Roofing Flashing

Lay an 18-inch wide layer of mineral-surfaced roll roofing centered in the valley. Its mineral-surfaced side should be down. Use only enough nails spaced 1 inch in from each edge to hold the strip smoothly in place. Press the roofing firmly in the bottom of the valley when nailing the opposite edge. On top of the first strip, lay a 36-inch wide strip with its surfaced side up. Center it in the valley. Fasten it in the same manner as the first strip.

Snap a chalk line on each side of the valley. Use them as guides for trimming the ends of the shingle courses. These lines are spaced 6 inches apart at the ridge. They are spread 1/8 inch per foot as they approach the roof edge. Thus, a valley 16 feet long will be 8 inches wide at the eaves.

The upper corner of each end asphalt shingle is clipped. This helps keep water from entering between the courses. The roof shingles are cemented to the valley flashing with plastic asphalt cement (Fig. 54-4).

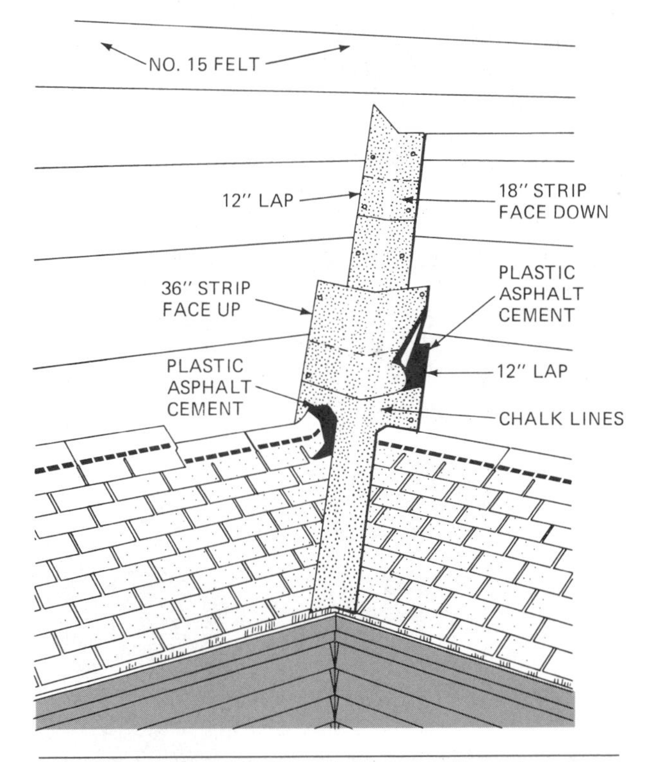

Fig. 54-4 Method of applying a roll roofing open valley flashing.

Using Metal Flashing

Prepare the valley with underlayment in the same manner as described previously. Lay a strip of sheet metal flashing centered in the valley. The metal should extend at least 10 inches on each side of the valley centerline for slopes with a 6-inch rise or less and 7 inches on each side for a steeper slope. Carefully press and form it into the valley. Fasten the metal with nails of similar material spaced close to its outside edges. Use only enough fasteners to hold it smoothly in place.

Snap lines on each side of the valley, as described previously. Use them as guides for cutting the ends of the shingle courses. Trim the last shingle of each course to fit on the chalk line. Clip 1 inch from its upper corner at a 45-degree angle. To form a tight seal, cement the shingle to the metal flashing with a 3-inch width of asphalt plastic cement (Fig. 54-5).

If a valley is formed by the intersection of a low-pitched roof and a much steeper one, a 1 inch high, crimped standing seam should be made in the center of the metal flashing of an open valley. The seam will keep heavy rain water flowing down the steeper roof from overrunning the valley and possibly being forced under the shingles of the lower slope.

Closed Valley Flashing

Closed valleys are those where the roof shingles meet in the center of the valley and completely cover the valley flashing. Using a closed valley protects the valley flashing. It thus adds to the weather resistance at very vulnerable points. Several methods are used to flash closed valleys.

The first step for any method is to apply the asphalt felt underlayment as previously described for open valleys. Then, center a 36-inch width of smooth or mineral surface roll roofing, 50-pound per square, or heavier, in the valley over the underlayment. Form it smoothly in the valley. Secure it with only as many nails as necessary.

Woven Valley Method

Valleys are commonly flashed by applying asphalt shingles on both sides of the valley and alternately weaving each course over and across the valley. This is called the *woven valley* (Fig. 54-6).

Fig. 54-6 Valleys are often flashed by weaving shingles together.

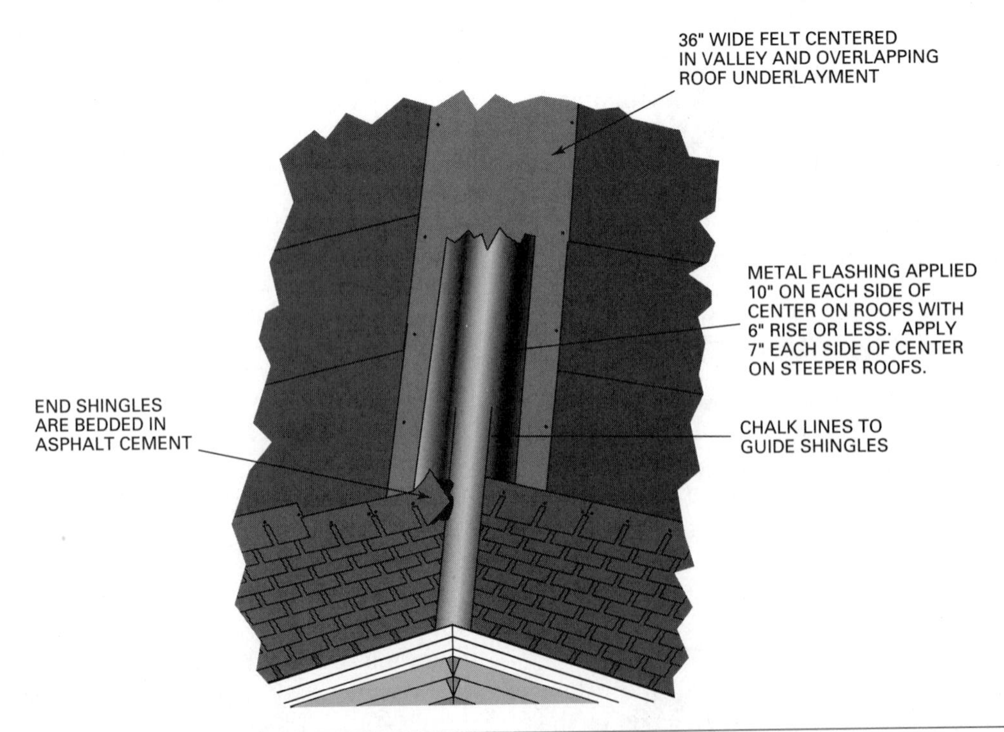

Fig. 54-5 Procedure for flashing an open valley with metal.

Lay the first course of shingles along the edge of one roof up to and over the valley for a distance of at least 12 inches. Lay the first course along the edge of the adjacent roof. Extend the shingles over the valley on top of the previously applied shingles.

Succeeding courses are then applied. Weave the valley shingles alternately, first on one roof and then on the other. When weaving the shingles, make sure they are pressed tightly into the valley. Also make sure no nail is closer than 6 inches to the valley centerline. Fasten the end of the woven shingle with two nails (Fig. 54-7). Most carpenters prefer to cover each roof area with shingles to a point approximately 3 feet from the valley. They weave the valley shingles in place later.

No end joints should occur in or near the center of the valley. Therefore, it may be necessary to occasionally cut a strip short that would otherwise end near the center. Continue from this cut end with a full-length strip over the valley.

Closed Cut Valley Method

Apply the shingles to one roof area. Let the end shingle of every course overlap the valley by at least 12 inches. Make sure no end joints occur at or near the center of the valley. Place fasteners no closer than 6 inches from the center of the valley. Form the end shingle of each course snugly in the valley. Secure its end with two fasteners.

Snap a chalk line along the center of the valley on top of the overlapping shingles. Apply shingles to the adjacent roof area. Cut the end shingle to the chalk line. Clip the upper corner of each shingle as described previously for open valleys. Bed the end that lies in the valley in about a 3-inch wide strip of asphalt cement. Make sure that no fasteners are located closer than 6 inches to the valley centerline (Fig. 54-8).

Step Flashing Method

Step flashings are individual metal pieces tucked between courses of shingles (see Fig. 54-1). When applying step flashings in valleys, first estimate the number of shingle courses required to reach the ridge. Using tin snips, cut a piece of metal flashing for each course of shingles. Each piece should be at least 18 inches wide for slopes with a 6-inch rise or greater and 24 inches wide for slopes with less pitch. The height of each piece should be at least 3 inches more than the shingle exposure.

Prepare the valley with underlayment as described previously. Snap a chalk line in the center of the valley. Apply the starter course on both roofs. Trim the end shingle of each course to the chalk line. Fit and form the first piece of flashing to the valley. Trim its bottom edge flush with the drip edge. Fasten it in the valley over the first layer of the starter course. Use fasteners of like material to prevent electrolysis. Fasten the upper corners of the flashing only.

Apply the first regular course of shingles to both roofs on each side of the valley. Trim the valley shingles so their ends lay on the chalk line. Bed them in plastic asphalt cement. Do not drive nails through the metal flashing.

Apply the next piece of flashing in the valley over the first course of shingles. Keep its bottom edge about 1/2 inch above the butts of the next

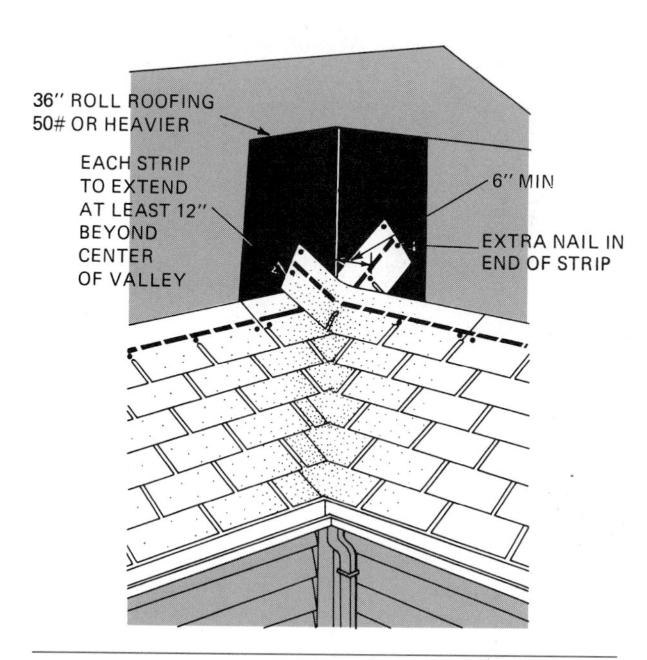

Fig. 54-7 Specifications for a woven valley flashing.

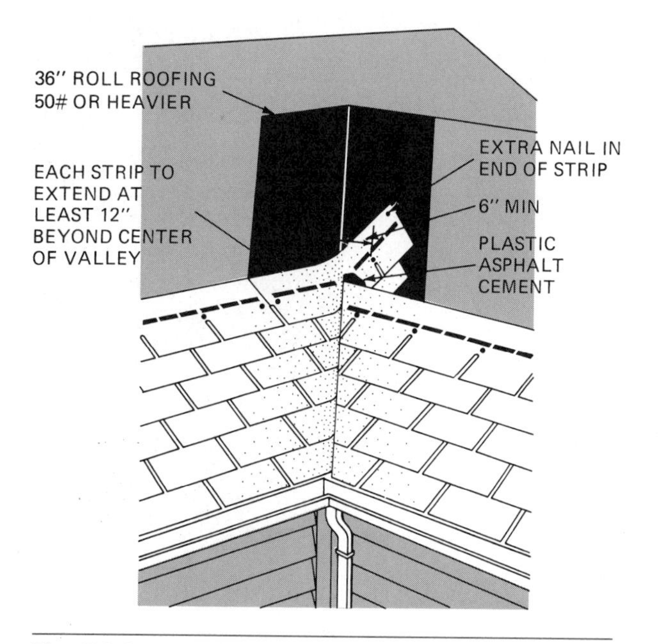

Fig. 54-8 Recommendations for applying a closed cut valley flashing.

course of shingles. Apply the second course of shingles in the same manner as the first. Secure a flashing over the second course. Apply succeeding courses and flashings in this manner (Fig. 54-9). Remember, a flashing is placed over each course of shingles. Do not leave any flashings out. When the valley is completely flashed, no metal flashing surface is exposed. If the valley does not extend all the way to the ridge of the main roof, a *lead saddle* is applied over the ridge of the minor roof (Fig. 54-10).

Flashing against a Wall

When roof shingles butt up against a vertical wall, the joint must be made watertight. The usual method of making the joint tight is with the use of metal step flashings.

The flashings are purchased or cut about 8 inches in width. They are bent at right angles in the center so they will lay about 4 inches on the roof and extend about 4 inches up the sidewall. The length of the flashings is about 3 inches more than the exposure of the shingles. When used with shingles exposed 5 inches to the weather, they are made 8 inches in length. Cut and bend the necessary number of metal flashings.

The roofing is applied and flashed before the siding is applied to the vertical wall. First, apply an underlayment of asphalt felt to the roof deck. Turn the ends up on the vertical wall by about 3 to 4 inches.

Apply the first layer of the starter course, working toward the vertical wall. Fasten a metal flashing, on top of the first layer of shingles. Its bottom edge should be flush with the drip edge. Use

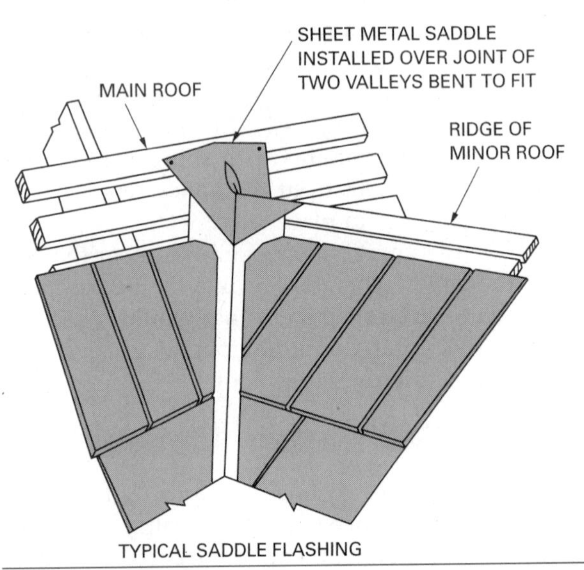

Fig. 54-10 A saddle is installed over the ridge of a minor roof where it intersects with the main roof.

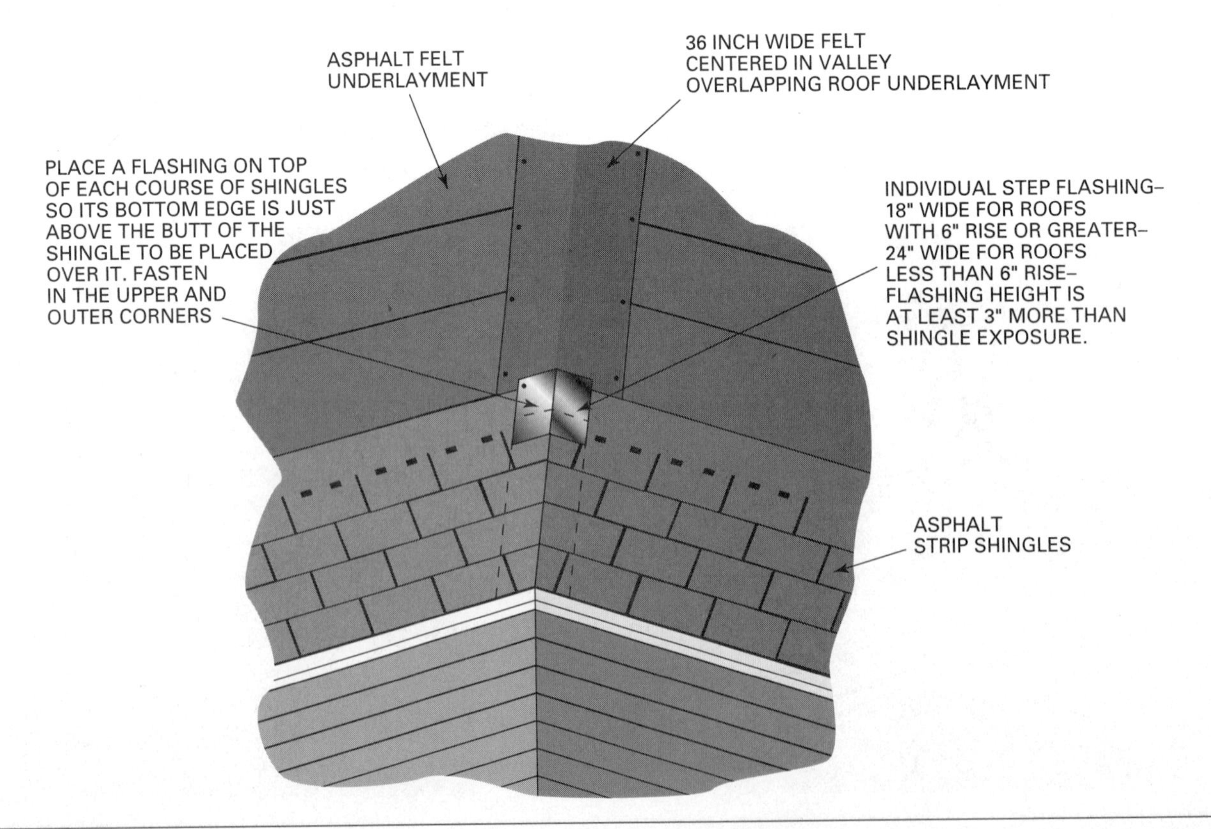

Fig. 54-9 Specifications for applying metal step flashings in a valley.

one fastener in each top corner. Lay the first regular course with its end shingle over the flashing and against the sheathing of the sidewall. Do not drive any fasteners through the flashings. It is usually not necessary to bed the shingles to the flashings with asphalt cement. The step flashing holds down the end of the shingle below it.

Apply a flashing over the first course and against the wall. Keep its bottom edge at a point that will be about 1/2 inch above the butt of the next course of shingles. Continue applying shingles and flashings in this manner until the ridge is reached (Fig. 54-11). Some carpenters prefer to cut the shingles back if a tab cutout occurs over a flashing. This prevents metal from being exposed to view.

Flashing a Chimney

In many cases, especially on steep pitch roofs, a **cricket** or **saddle** is built between the upper side of the chimney and roof deck. The cricket, although not a flashing in itself, prevents accumulation of water behind the chimney (Fig. 54-12).

Flashings are installed by *brick masons* who build the chimney. The upper ends of the flashing are bent around and **mortared** in between the courses of brick as the chimney is built. The flashings are long enough to be bent at and over the roof sheathing for tucking between shingles. These flashings are usually in place before the carpenter applies the roof covering.

The underlayment is applied and tucked under the existing flashings. The shingle courses are brought up to the chimney. They are applied under the flashing on the lower side of the chimney. This is called the *apron flashing.* Shingles are tucked under the **apron.** The top edge of the shingles is cut as necessary, until the shingle exposure shows below the apron. The apron is then pressed into place on top of the shingles in a bed of plastic cement. Its projecting ends are carefully and gently formed up around the sides of the chimney and under the lowest side flashings.

Along the sides of the chimney, the flashings are tucked in between the shingles in the same manner as in flashing against a wall. No nails are used in the flashings. The standing portions of the *side*

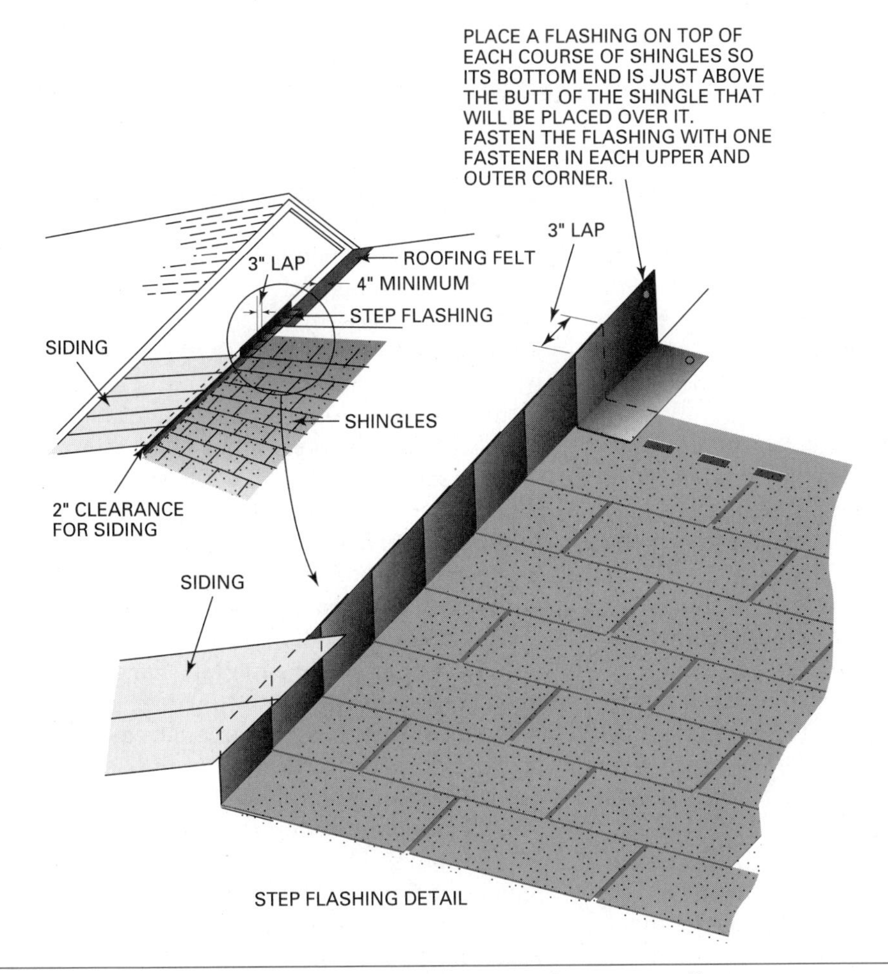

Fig. 54-11 Using metal step flashing where a roof joins a wall.

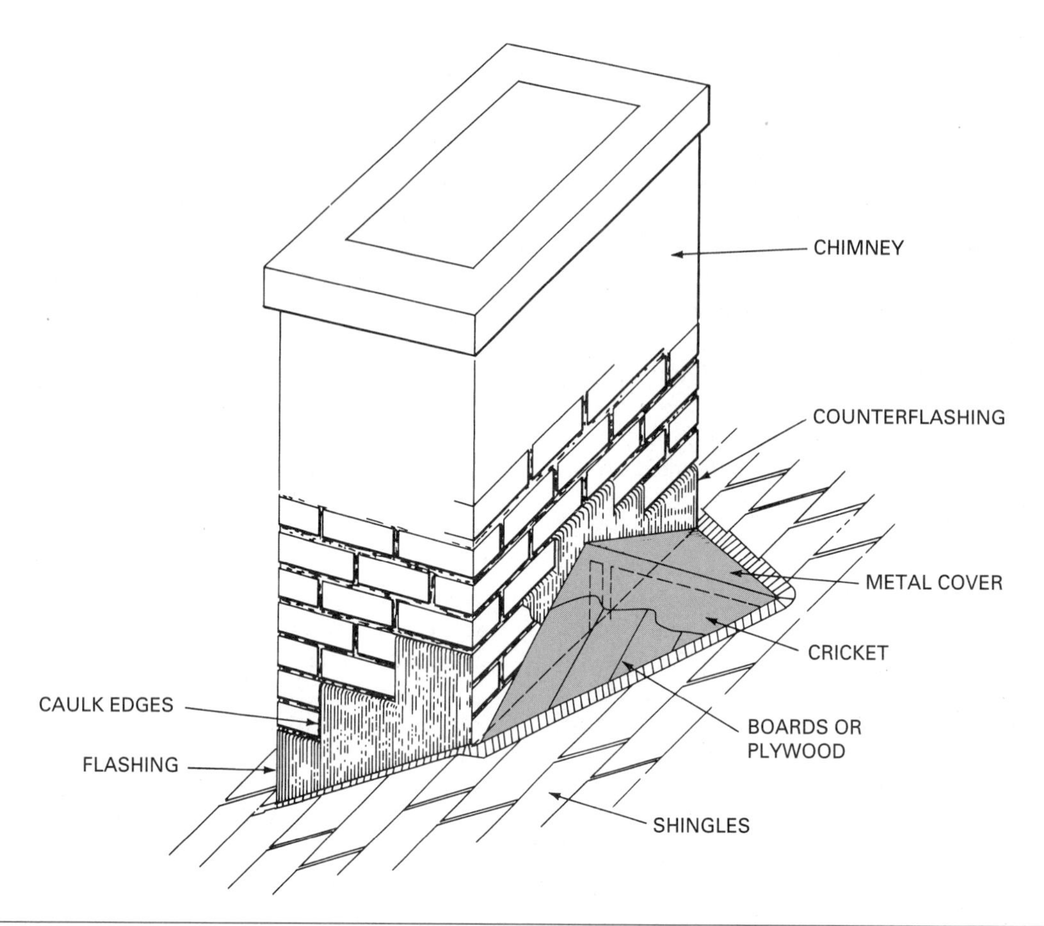

Fig. 54-12 A cricket is built to prevent the accumulation of water behind the chimney.

flashings are bedded to the chimney with asphalt cement. The roof portion is bedded to the shingle. The projecting edges of the lowest side flashings are carefully formed around the corner. They are folded against the low side on the chimney. The top edges of the highest side flashings are also folded around the corner and under the *head flashing* on the upper side of the chimney.

The head flashing is cemented to the roof. Shingles are applied over it. They are bedded to it with asphalt cement. Its projecting ends are also carefully formed around the corner on top of the side flashings.

The projecting ends of chimney flashings are carefully formed and folded around the corners of the chimney. Gently and carefully tap the metal with a hammer handle. Care must be taken not to break through the flashings. Chimneys may be flashed by other methods and materials, other than described above, depending on the custom of certain geographical areas (Fig. 54-13).

Other rectangular roof obstructions, such as skylights, are flashed in a similar manner. The carpenter usually applies the flashings to these obstructions.

Flashing Vents

Flashings for round pipes, such as *stack vents* for plumbing systems and *roof ventilators,* usually come as *flashing collars* made for various roof pitches. They fit around the stack. They have a wide flange on the bottom that rests on the roof deck. All joints in the flashing are soldered to make them watertight. The top end sometimes is bent and fitted inside the top end of the stack. Or, it may be sealed to the side with a lead ring and asphalt cement. The flashing is installed over the stack vent, with its flange on the roof sheathing. It is fastened in place with one fastener in each upper corner.

Shingle up to the lower end of the stack vent flashing. Lift the lower part of the flange. Apply shingle courses under it. Cut the top edge of the shingles, where necessary, until the shingle exposure, or less, shows below the lower edge of the flashing. Apply asphalt cement under the lower end of the flashing. Press it into place on top of the shingle courses.

Apply shingles around the stack and over the flange. Do not drive nails through the flashing. Bed shingles to the flashing with asphalt cement, where necessary (Fig. 54-14).

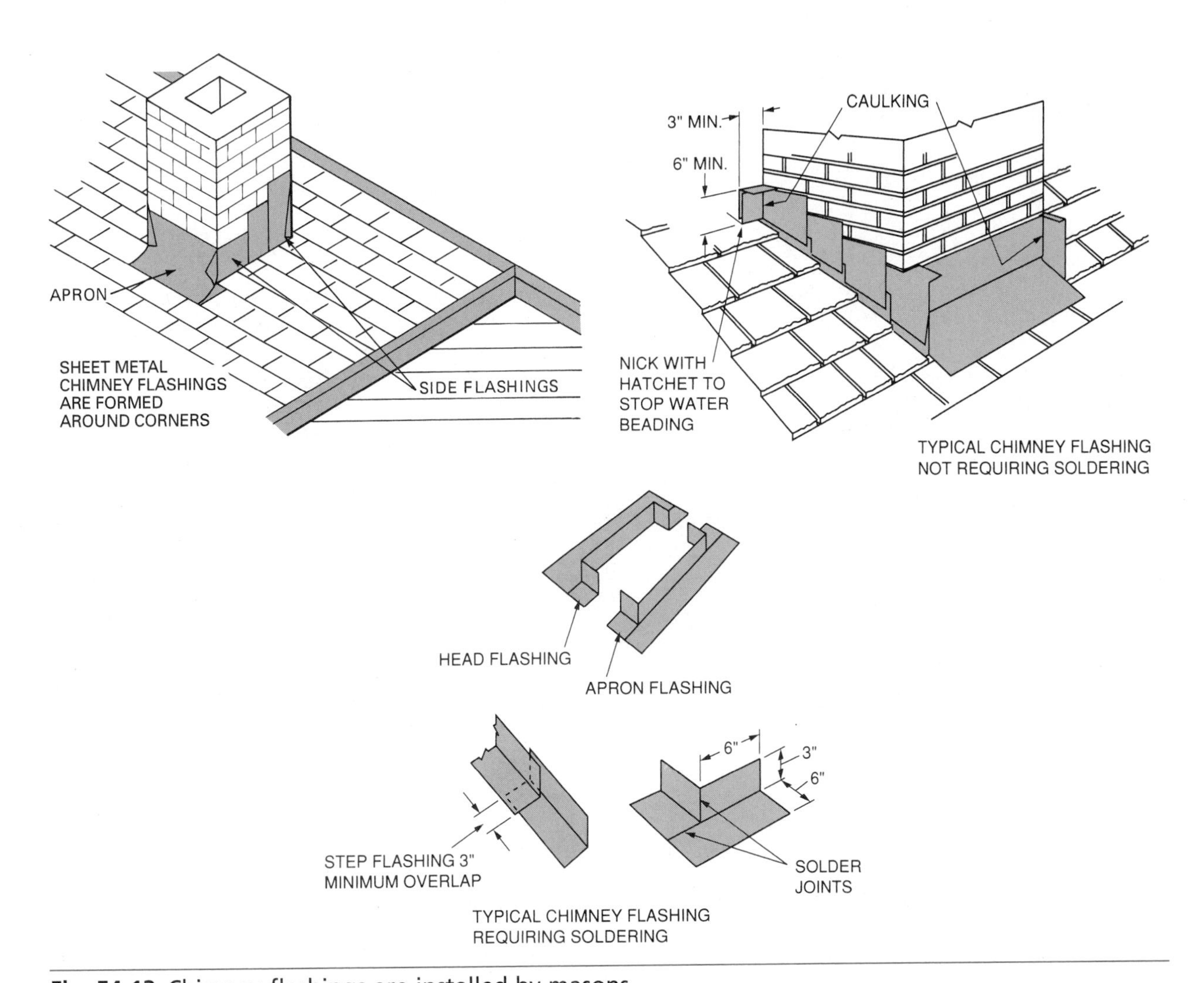

Fig. 54-13 Chimney flashings are installed by masons.

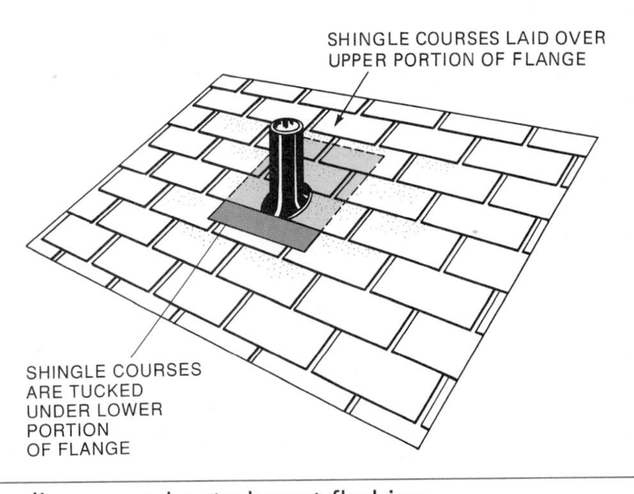

Fig. 54-14 Method of shingling around a stack vent flashing.

REVIEW QUESTIONS

Select the most appropriate answer.

1. A square is the amount of roofing required to cover
a. 50 square feet. c. 150 square feet.
b. 100 square feet. d. 200 square feet.

2. One roll of #15 asphalt felt will cover
a. 1 square. c. 3 square.
b. 2 square. d. 4 square.

3. When applying asphalt felt on a roof deck as underlayment, lap each course over the lower course by at least
a. 2 inches. c. 4 inches.
b. 3 inches. d. 6 inches.

4. Under no circumstances should asphalt strip shingles be used on slopes lower than
a. 1 inch rise per foot of run.
b. 2 inches rise per foot of run.
c. 3 inches rise per foot of run.
d. 4 inches rise per foot of run.

5. When applying asphalt shingles, it is recommended that no rake tab be less than
a. 2 inches in width. c. 4 inches in width.
b. 3 inches in width. d. 5 inches in width.

6. For slopes with a 6-inch rise per foot of run or less, metal valley flashings should extend on each side of the valley centerline by at least
a. 6 inches. c. 10 inches.
b. 8 inches. d. 12 inches.

7. When flashing a valley by weaving shingles, do not locate any nails closer to the valley centerline than
a. 6 inches. c. 10 inches.
b. 8 inches. d. 12 inches.

8. Flashings about 8 inches wide are used when flashing a roof that butts against a vertical wall. They are bent so that
a. 3 inches lays on the wall and 5 inches lays on the roof.
b. 4 inches lays on the wall and 4 inches lays on the roof.
c. 5 inches lays on the wall and 3 inches lays on the roof.
d. 2 inches lays on the wall and 6 inches lays on the roof.

9. A built-up section between the roof and the upper side of a chimney is called a
a. cricket. c. furring.
b. dutchman. d. counterflashing.

10. Concealed-nail roll roofing may be used on roofs with slopes as low as
a. 1 inch rise per foot of run.
b. 2 inches rise per foot of run.
c. 3 inches rise per foot of run.
d. 4 inches rise per foot of run.

11 Most wood shingles and shakes are made from
a. cypress. c. eastern white cedar.
b. redwood. d. western red cedar.

12. The longest available length of wood shingles and shakes is
a. 16 inches. c. 24 inches.
b. 18 inches. d. 28 inches.

13. The maximum exposure for No. 1, 16-inch wood shingles laid on roofs at least as steep as 4 in 12 is
a. 4 1/2 inches. c. 5 1/2 inches.
b. 5 inches. d. 6 inches.

14. Wood shingles normally overhang the fascia by
a. 3/8 inch. c. 1 1/2 inches.
b. 1 inch. d. 2 inches.

15. It is important to lay interlayment straight prior to applying shakes
a. so there are no wrinkles in the felt.
b. to obtain the proper lap.
c. for ease in nailing.
d. because the felt serves as a guide.

BUILDING FOR SUCCESS

What Employers are Looking For

It has been said, "There are three kinds of people: those who make things happen, those who watch things happen, and those who wonder what happened." More likely than not, we all have been in each of these categories at one time or another. In preparing for immediate and long-term employment, each person should desire to make positive things happen.

Study and careful preparations will make a difference in a person's employment opportunities. We have the responsibility to make positive things happen to ourselves and to people around us. Contractors and construction related companies are always looking for people who take the initiative to make a company or business progressive. That will not change as we continue into the twenty-first century.

Everyone entering a construction-related career must clearly understand the demands placed on technically skilled workers. As in the past, various construction trades call for workers to perform technical tasks on a daily basis. Additional business skills are increasingly important. Skilled workers, such as carpenters, will be expected to possess many social, personal management, and problem-solving skills. They have to be able to work with minimal supervision. More and more industries are calling for both new and current skilled workers to be more adaptable, autonomous, and willing to learn on the job.

Workplace basics are changing and affecting larger groups of our nation's population. Many employers desire each worker to respond well to new technology, assure quality control, be involved in participative management, and acquire new skills quickly. Due to intensive competition, the time to adjust and continue producing quality products has been shortened. Therefore, all industries must respond to new demands with a workforce that is flexible, well educated, and progressive in its thinking. The most successful companies understand the need to assist in the training and development of their employees. More contractors today look for both local and national training opportunities. They want their employees to obtain new skills that will support the company as they attain their goals. This is one of the many criteria the new construction graduates must look for in the job search process. A construction-related business that desires to be on the cutting edge in competition will have some element of employee development in place. These companies should be a high priority in the job selection process.

Learning on the job will continue to be a requirement for all skilled workers well into the next century. Employees at all levels must become proficient in observing, listening, and solving problems. Innovative ideas, reinforced with strong interpersonal communication skills, will play a larger role in the progress of a construction business. Teamwork will continue to increase as problems are addressed. Learning to learn begins in a person's early years and must continue throughout the employment years. These new skills will become the foundation on which more refined skills will be developed on the job.

Many construction leaders, managers, and supervisors have diversified educational backgrounds. Some arrived at their leadership positions through on-the-job training, apprenticeship programs, trade schools, community colleges, or four-year construction management programs. They possessed many of the basic reading, writing, and computation skills needed for the job. They also have accumulated an abundance of additional necessary technical and business skills while employed. Good communication skills, involving observing, listening, writing, and speaking, are essentials for all skilled workers who desire to excel. Students who do not place great importance on communication and social skills will not readily advance in their employment

Economics in the workplace will continue to call for creative thinking and problem-solving skills. Competition for a share of the market, quality products, and cost efficiency will be driving forces for each construction-related business. Both management and skilled workers will have excellent opportunities for input into a company's creative efforts. Through constant observations of trends and developments, employees must be ready to offer new ideas that will help make their employers pace-setters or strong competitors.

As students prepare for careers in construction, they must set goals that are challenging, realistic, and attainable. Most goals will involve education, training, and work experience. The selected program of study should contain a curriculum that will, as a minimum, prepare each person for the entry-level job. This becomes the first educational step that can be combined with other training opportunities and work experiences that lead to the next step. The entire sequence may take years to complete. All things of quality take investment of time, planning, and commitment to reach completion.

Each person has the opportunity to make things happen with each situation. No two people will have the same circumstances, backgrounds, or resources. What is done with abilities and desire is the key to success. Reasonable questions that must be asked are: Where do you want to go? How can you get there? Do you want to watch things happen around you? Do you want to forever wonder what happened? Or do you want to make positive things happen in your life? Ask those around you who have insight into the construction industry or career counseling to assist you in making these challenging decisions. That may possibly be the first step to a rewarding career.

FOCUS QUESTIONS: For individual or group discussion

1. What are some skills you feel an employer in construction will desire of you?
2. What ongoing changes in business and industry will affect your job? How well you can advance? How will new technology, trends, economic factors, and societal developments affect your opportunities for obtaining a job in construction?
3. How do you think being diversified in technical and business skills could help assure you of being considered for the job you desire?

UNIT

19

Windows

Windows are normally installed prior to the application of exterior *siding.* Care must be taken to provide easy-operating, weathertight, attractive units. Quality workmanship results in a more comfortable interior, saves energy by reducing fuel costs, minimizes maintenance, gives longer life to the units, and makes application of the exterior siding easier.

OBJECTIVES

After completing this unit, the student should be able to:

- describe the most popular styles of windows and name their parts.
- select and specify desired sizes and styles of windows from manufacturers' catalogs.
- install various types of windows in an approved manner.
- cut glass and glaze a sash.

UNIT CONTENTS

CHAPTER 55 WINDOW TERMS AND TYPES

Wood *windows* are one of many types of *millwork* (Fig. 55-1). Millwork is a term used to describe products, such as windows, doors, and cabinets, fabricated in woodworking plants that are used in the construction of a building. Windows are usually fully assembled and ready for installation when delivered to the construction site. Windows are also made of aluminum and steel. Windows made with exposed wood parts encased in vinyl are called *vinyl-clad* windows. The names given to various parts of a window are the same, in most cases, regardless of the window type.

Parts of a Window

When shipped from the factory, the window is complete except for the interior trim. It is important that the installer know the names, location, and functions of the parts of a window in order to understand, or to give, instructions concerning them.

The Sash

The **sash** is a frame in a window that holds the glass. The type of window is generally determined by the way the sash operates. The sash may be installed in a fixed position, move vertically or horizontally, or swing outward or inward.

Sash Parts. Vertical edge members of the sash are called **stiles.** Top and bottom horizontal members are called **rails.** The pieces of glass in a sash are called **lights.** There may be more than one light of glass in a sash. Small strips of wood that divide the glass into smaller lights are called **muntins.** Muntins may divide the glass into rectangular, diamond, or other shapes (Fig. 55-2).

Many windows come with false muntins called *grilles.* Grilles do not actually separate or support the glass. They are applied as an overlay to simulate small lights of glass. They are made of wood or plastic. They snap in and out of the sash for easy cleaning of the lights (Fig. 55-3). They may also be preinstalled between the layers of glass in double- or triple-glazed windows.

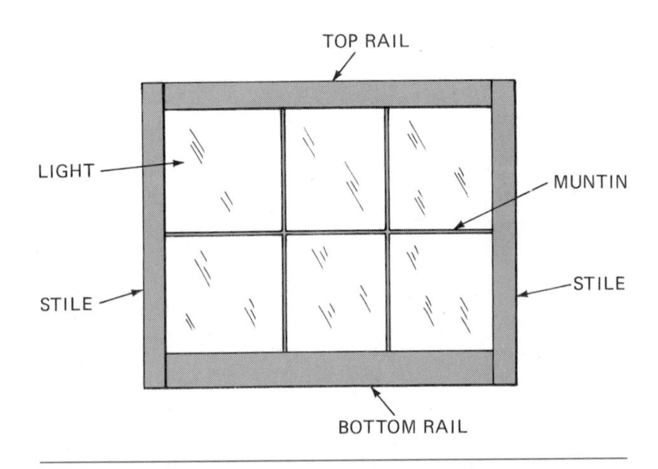

Fig. 55-2 A sash and its parts.

Fig. 55-1 Windows of many types and sizes are fully assembled in millwork plants and ready for installation. (*Courtesy of Andersen Windows, Inc.*)

Fig. 55-3 Removable grilles simulate true divided-light muntins. (*Courtesy of Andersen Windows, Inc.*)

Window Glass

Several qualities and thicknesses of sheet glass are manufactured for **glazing** and other purposes. The installation of glass in a window sash is called glazing. *Single strength* (SS) glass is about 3/32-inch thick. It is used for small lights of glass. For larger lights, *double strength* (DS) glass about 1/8-inch thick may be used. *Heavy sheet* glass about 3/16 inch thick is also manufactured (Fig. 55-4). Many other kinds of glass are made for use in construction.

Safety Glass

Most windows are not glazed with safety glass. If broken, they could fragment and cause injury. Care must be taken to handle windows in a manner to prevent breaking the glass. Some codes require a type of *safety glass* in windows with low sill heights or located near doors. Skylights and roof windows are generally required to be glazed with safety glass.

Safety glass is constructed, treated, or combined with other materials to minimize the possibility of injuries resulting from contact with it. Several types of safety glass are manufactured.

Laminated glass consists of two or more layers of glass with inner layers of transparent plastic bonded together. *Tempered glass* is treated with heat or chemicals. When broken at any point, the entire piece immediately disintegrates into a multitude of small granular pieces. *Wired glass* consists of a single sheet with wire mesh imbedded in the glass. *Transparent plastic* is also used for safety glazing material.

Insulated Glass

To help prevent heat loss, and to avoid condensation of moisture on glass surfaces, **insulated glass,** or *thermal pane windows,* are used frequently in place of single-thickness glass.

Insulated glass consists of two or three (generally two) layers of glass separated by a sealed air space 3/16 to 1 inch in thickness (Fig. 55-5). Moisture is removed from the air between the layers before the edges are sealed. To raise the R-value of insulated glass, the space between the layers is filled with *argon* gas. Argon conducts heat at a lower rate than air. Additional window insulation may be provided with the use of *removable glass panels* or *combination storm sash.*

Solar Control Glass. The R-value of windows may also be increased by using special *solar-control* insulated glass, called *high performance* or *LoE* glass. LoE is an abbreviation for **low emissivity.** It is

Fig. 55-5 Cutaway of insulated glass used to increase the R-value in a window.

Sheet		Range			Approx. Oz./Sq. ft.
Picture		.043-.053	AA, A, B	50[1]	9-11
		.054-.069 and	AA, A	70[1]	12-14 and
		.070-.080	B	80[1]	15-17
Window	SS	.085-.101	AA, A, B	120[1]	19.5
	DS	.115-.134		140[1]	26.0
Heavy Sheet	3/16	.182-.205	AA, A, B	84x120	40.0

[1] United inches (sum of width plus length.)

Fig. 55-4 Kinds and sizes of glass commonly used in residential construction.

used to designate a type of glazing that reflects heat back into the room in winter and blocks heat from entering in the summer (Fig. 55-6). An invisible, thin, metallic coating is bonded to the air space side of the inner glass. This lets light through, but reflects heat.

The Window Frame

The sash is hinged to, slides, or is fixed in a *window frame.* The frame usually comes with the exterior trim applied. It consists of several distinct parts (Fig. 55-7).

The Sill

The bottom horizontal member of the window frame is called a *sill.* It is set or shaped at an angle to shed water. Its bottom side usually is grooved so a weathertight joint can be made with the wall siding.

Jambs

The vertical sides of the window frame are called *side jambs.* The top horizontal member is called a *head jamb.*

Extension Jambs. The inside edge of the jamb should be flush with the finished interior wall surface when the window is installed. In some cases, windows can be ordered with jamb widths for standard wall thicknesses. In other cases, jambs are made narrow. **Extension jambs** are then provided with the window unit. The extensions are cut to width to accommodate various wall thicknesses.

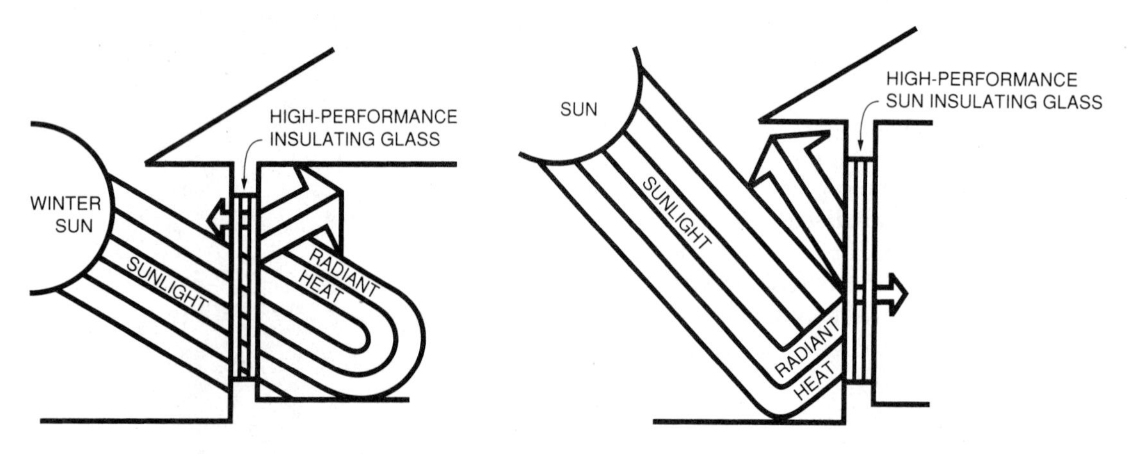

Fig. 55-6 Special solar-control glass is used in windows to help keep heat in during cold weather and out during hot weather. (*Courtesy of Andersen Windows, Inc.*)

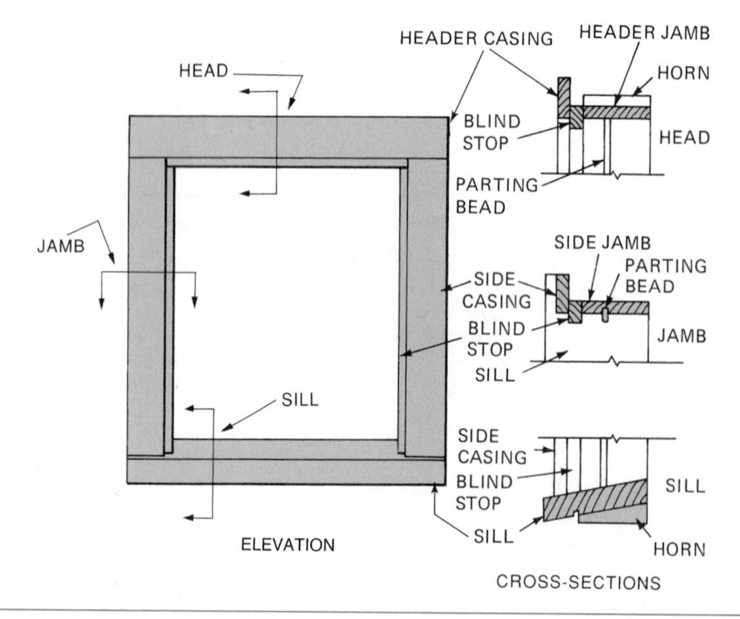

Fig. 55-7 A window frame consists of parts with specific terms.

They are applied to the inside edge of the jambs of the window frame (Fig. 55-8). The extension jambs are installed at a later stage of construction when the interior trim is applied. They should be stored for safekeeping until needed.

Blind Stops

Blind stops are sometimes applied to certain types of window frames. They are strips of wood attached to the outside edges of the jambs. Their inside edges

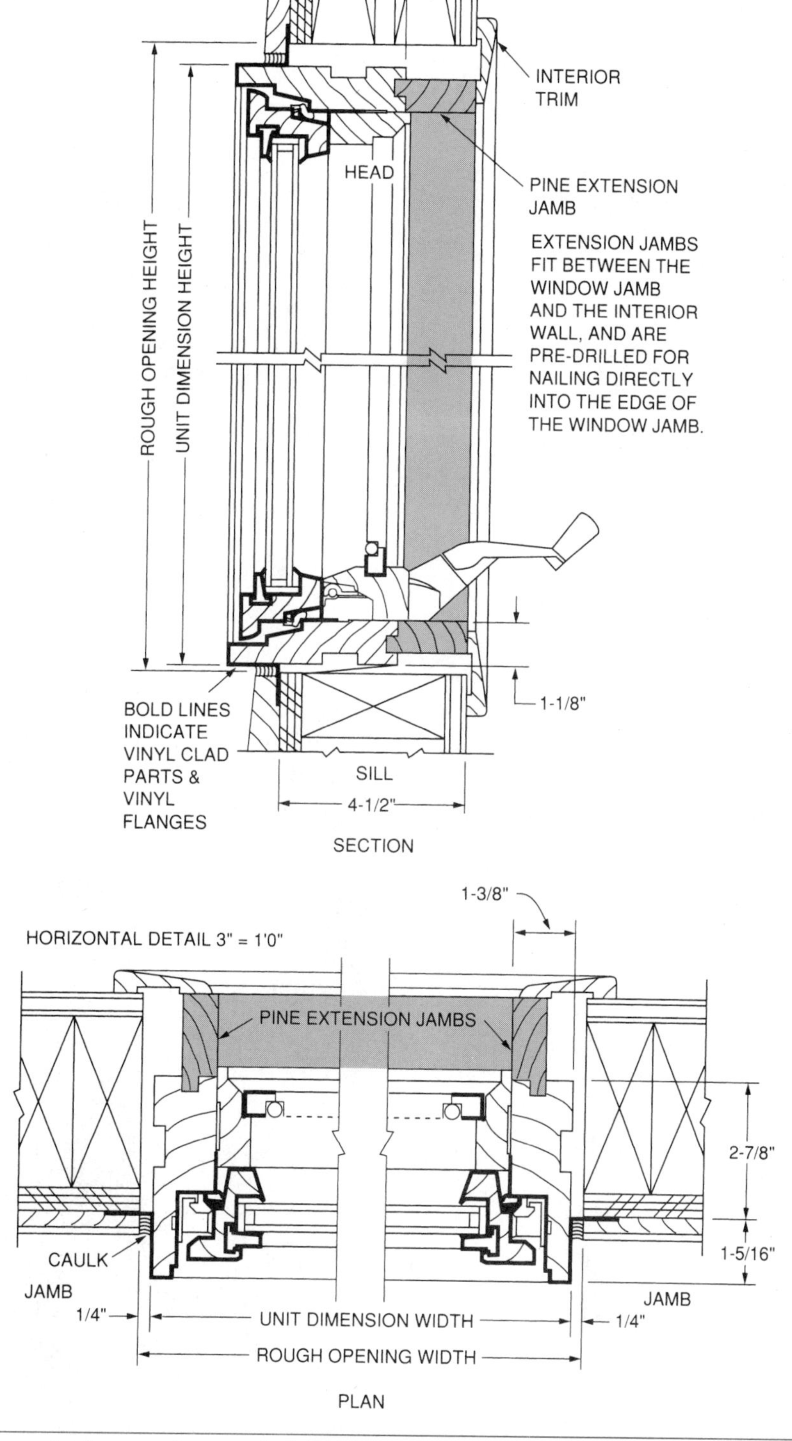

Fig. 55-8 To compensate for varying wall thicknesses, extension jambs are provided with some window units.

project about 1/2 inch inside the frame. They provide a weathertight joint between the outside casings and the frame. They also act as stops for screens and storm sash. They make the outer edge of the channel for top sash of *double-hung windows.*

Casings

Window units usually come with *exterior casings* applied. The side members are called *side casings.* In most windows, their lower ends are cut at a bevel and rest on the sill. The top member is called the *head casing.* On flat casings, a weathertight *rabbeted* or *tongue-and-grooved* joint is made between them. When molded casings are used, the mitered joints at the head are usually bedded in compound (Fig. 55-9).

When windows are installed or manufactured, side by side, in multiple units, a **mullion** is formed where the two side jambs are joined together. The casing covering the joint is called a *mullion casing.*

The Drip Cap

A **drip cap** comes with some windows. It is applied on the top edge of the head casing. It is shaped with a sloping top to carry rain water out over the window unit. Its inside edge is rabbeted. The wall siding is applied over it to make a weathertight joint.

Window Flashing

In some cases, a *window flashing* is also provided. This is a piece of metal as long as the head casing, which is also called a *drip cap.* It is bent to fit over the head casing and against the exterior wall (Fig. 55-10). The flashing prevents the entrance of water at this point. Flashings are usually made of aluminum or zinc. The vinyl flanges of vinyl-clad wood windows are usually formed as an integral part of the window. No additional head flashings are required.

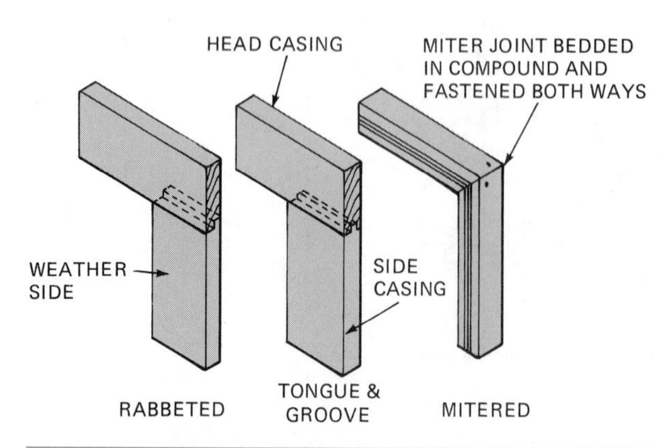

Fig. 55-9 A weathertight joint is made between side and header casings.

Making a Flashing on the Job. If window flashings are not provided with the unit, they can be made on the job site. Cut a length of sheet zinc or aluminum from a roll of the desired width. The metal should be wide enough to extend up the wall about 2 inches, over the top edge of the header casing, and about 1/4-inch down over its face side.

Tack the metal along its top edge to a plank about 2 inches in from the edge of the stock. Bend the metal by tapping the overhanging edge with a short 2 × 4 block (Fig. 55-11). The other small bend is made when the flashing is installed.

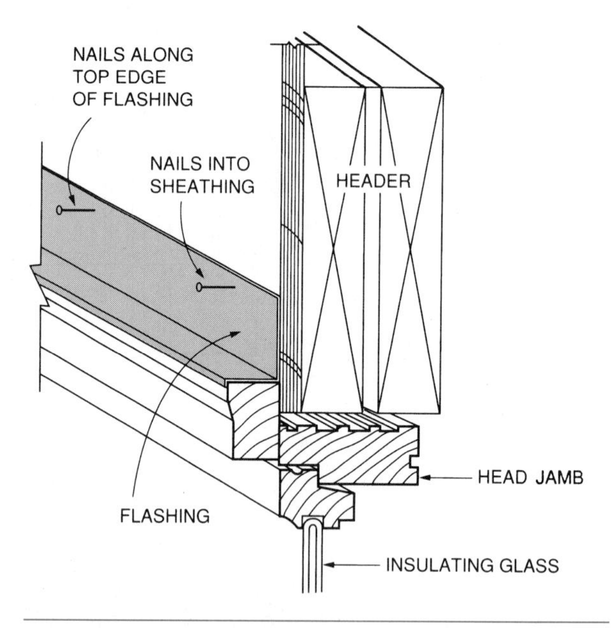

Fig. 55-10 A window flashing covers the top edge of the header casing. It extends up the sidewall above the window.

Fig. 55-11 Bending a window flashing on the job site.

Protective Coatings

Most window units are *primed with* a first coat of paint applied at the factory. Vinyl-clad wood windows are designed to eliminate painting. However, they are available in only a few colors. In case units need priming, this should be done before installation. Store the units under cover and protected from the weather until installed. Additional protective coats should be applied as soon as practical.

Screens

Most manufacturers provide screens as optional accessories for all kinds of windows. On outswinging and sliding windows, the screens are attached to the inside of the frame. On double-hung windows they are mounted on the outside of the frame.

The screen mesh is usually plastic or aluminum. Bronze is used in more expensive screens. Screens may be manufactured specifically for various types of windows. They are mounted in place with hardware already installed at the factory.

Wood screens are also available in various sizes. They may be installed full or half size. Half screens slide in tracks installed on the inside edges of the window's side casings. Figure 55-12 shows how a window screen with a wood frame is made.

STEP BY STEP
PROCEDURES

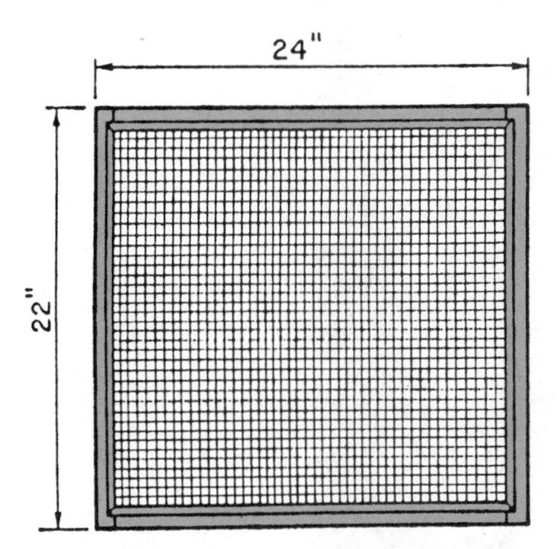

$\frac{1}{4}$" X $\frac{3}{4}$" SCREEN MOULDING

WIRE MESH

$\frac{3}{4}$"

$1\frac{1}{2}$"

SECTION OF SCREEN

MORTISE

TENON

RAIL

JOINT DETAIL

STILE

1. TO MAKE THE SCREEN SHOWN ABOVE
   a.) CUT 2 STILES 22" LONG
   b.) CUT RAILS 24" LONG
2. MAKE THROUGH MORTISE & TENON JOINTS BY:
   a.) MORTISE BOTH ENDS OF THE STILES
   b.) TENON BOTH ENDS OF THE RAILS
3. ASSEMBLE THE FRAME WITH GLUE & SCREWS IN EACH JOINT.
   a.) SQUARE THE FRAME BY MAKING DIAGONAL DISTANCES EQUAL

Fig. 55-12 Instructions for making a wood-framed window screen.

Types of Windows

Common types of windows are *fixed, single-* or *double-hung, casement, sliding, awning, hopper,* and *jalousie* windows.

Fixed Windows

Fixed windows consist of a frame in which a sash is fitted in a fixed position. They are manufactured in many shapes (Fig. 55-13).

Oval and *circular* windows are usually installed as individual units. *Elliptical, half rounds,* and *quarter rounds* are widely used in combination with other types (Fig. 55-14). In addition, fixed windows are manufactured in *geometric* shapes (*squares, rectangles, triangles, parallelograms, diamonds, trapezoids, pentagons, hexagons* and *octagons*). They may be assembled or combined with other types of windows in a great variety of shapes (Fig. 55-15). In addition to factory-assembled units, lengths of the frame stock may be purchased for cutting and assembling odd shape or size units on the job.

Arch windows have a curved top or head that make them well-suited to be joined in combination with a number of other types of windows or doors (Fig. 55-16). All of the windows mentioned come in a variety of sizes. With so many shapes and sizes, hundreds of interesting and pleasing combinations can be made.

Fig. 55-13 Fixed windows are often used in combination with other types of windows. Here, fixed, half-round windows are used above casement windows. (*Courtesy of Andersen Windows, Inc.*)

Fig. 55-14 Elliptical, half round, and quarter round windows. (*Courtesy of Andersen Windows, Inc.*)

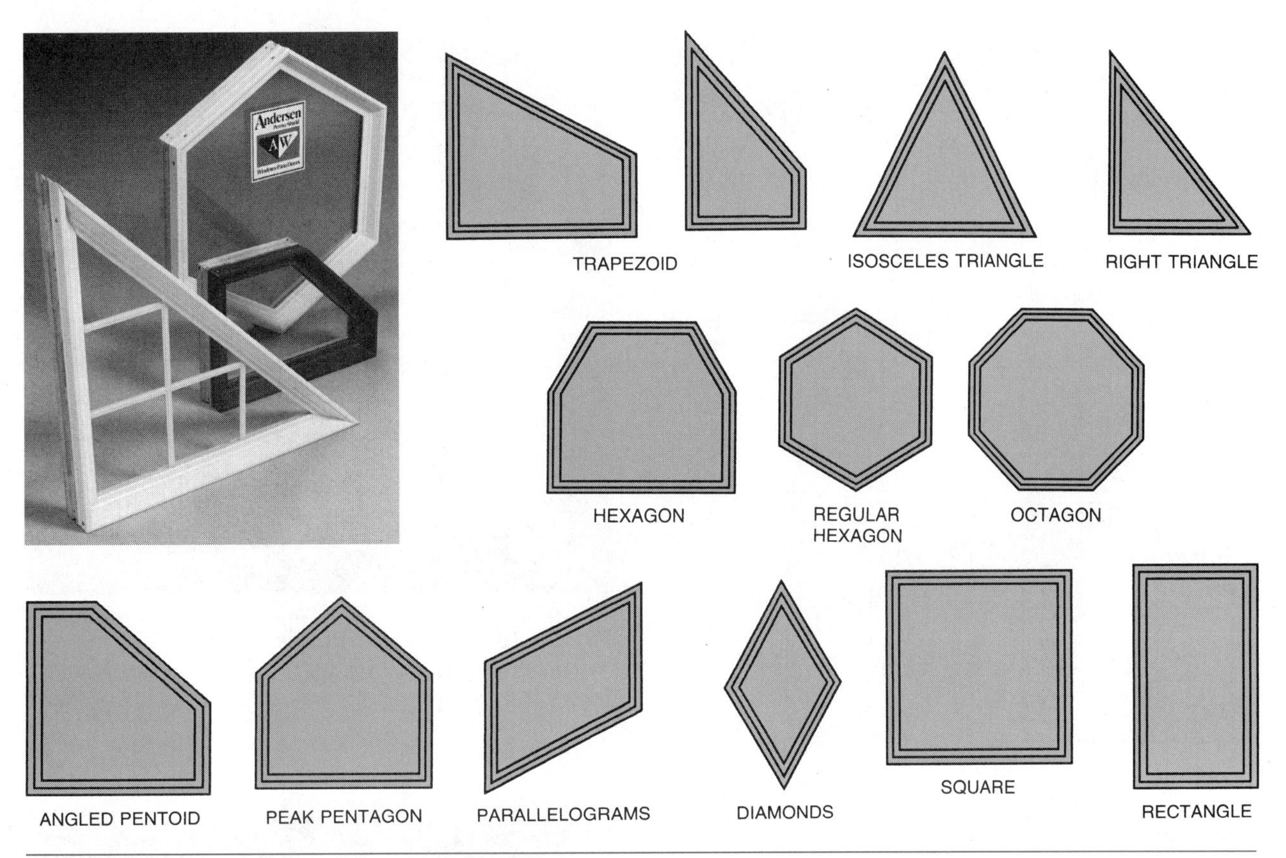

Fig. 55-15 Fixed windows are also made in many geometric forms.

Fig. 55-16 Arch windows have curved head jambs and casings. *(Courtesy of Andersen Windows, Inc.)*

Single- and Double-Hung Windows

The *double-hung window* consists of an upper and a lower sash that slide vertically by each other in separate channels of the side jambs (Fig. 55-17). The *single-hung* window is similar except the upper sash is fixed. A strip separating the sash is called a **parting strip.**

In most units, sash slide in metal channels that are installed in the frames. Each sash is provided

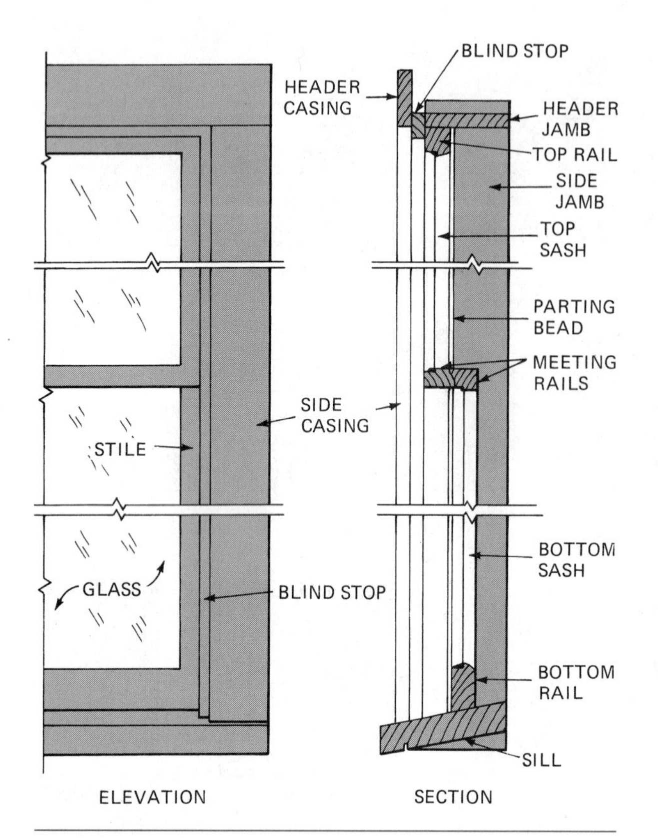

Fig. 55-17 The double-hung window and its parts.

Fig. 55-18 Double-hung windows are combined in this bay window unit. (*Courtesy of Andersen Windows, Inc.*)

Fig. 55-19 Casement windows are combined in this bow window unit. (*Courtesy of Andersen Windows, Inc.*)

with *springs,* **sash balances,** or *compression weatherstripping* to hold it in place in any position. Compression weatherstripping prevents air infiltration, provides tension, and acts as a counterbalance. Some types provide for easy removal of the sash for painting, repair, and cleaning.

When the sash are closed, specially shaped *meeting rails* come together to make a weathertight joint. *Sash* locks located at this point not only lock the window, but draw the rails tightly together. Other hardware consists of *sash* lifts that are fastened to the bottom rail of the bottom sash. They provide an uplifting force to make raising the sash easier and help keep the sash in the position it is placed.

Double-hung windows can be arranged in a number of ways. They can be installed side by side in multiple units or in combination with other types. Figure 55-18 shows them used in a **bay window** unit. Detachable vinyl grille simulates muntins.

Casement Windows

The *casement window* consists of sash hinged at the side. It swings outward by means of a crank or lever. Most casements swing outward. The inswinging type is very difficult to make weathertight. An advantage of the casement type is that the entire sash can be opened for maximum ventilation. Figure 55-19 shows the use of casement windows in a *bow window* unit.

Sliding Windows

Sliding windows have sash that slide horizontally in separate tracks located on the header jamb and sill (Fig. 55-20). When a window-wall effect is desired,

Fig. 55-20 The sash in sliding windows move horizontally by each other. (*Courtesy of Andersen Windows, Inc.*)

many units can be placed side by side. Most units come with all necessary hardware applied.

Awning and Hopper Windows

An *awning window* unit consists of a frame in which a sash hinged at the top swings outward by means of a crank or lever. A similar type, called the **hopper window**, is hinged at the bottom and swings inward.

Each sash is provided with an individual frame so that many combinations of width and height can be used. These windows are often used in combination with other types (Fig. 55-21).

Jalousie Windows

A **jalousie window** consists of a metal frame that holds a series of horizontal glass slats. The glass slats open and close together when operated by a single crank (Fig. 55-22). The use of jalousie windows is limited to porches and breezeways or to warm climates. They are not very weathertight.

Fig. 55-21 Awning windows are often used in stacks or in combination with other types of windows. (*Courtesy of Andersen Windows, Inc.*)

Fig. 55-22 A jalousie window has many horizontal glass slats that operate similar to a venetian blind.

Skylight and Roof Windows

Skylights provide light only. *Roof windows* contain operating sash to provide light and ventilation (Fig. 55-23). One type of roof window comes with a tilting sash that allows access to the outside surface for cleaning. Special flashings are used when multiple skylights or roof windows are ganged together.

Fig. 55-23 Skylights and roof windows are made in a number of styles and sizes. (*Courtesy of Andersen Windows, Inc.*)

CHAPTER 56 WINDOW INSTALLATION AND GLAZING

There are numerous window manufacturers that produce hundreds of kinds, shapes, and sizes of windows. Because of the tremendous variety and design differences, follow the manufacturer's instructions closely to ensure a correct installation. Directions in this unit are basic to most window installations. They are intended as a guide to be supplemented by procedures recommended by the manufacturer.

Selecting and Ordering Windows

The builder must study the plans to find the type and location of the windows to be installed. The floor plan shows the location of each unit. Each unit is usually identified by a number or a letter next to the window symbol.

Those responsible for designing and drawing plans for building or selecting windows must be aware of, and comply with, building codes that set certain standards in regard to windows. Most codes require minimum areas of natural light. Codes also require minimum ventilation by windows unless provided by other means. Some codes stipulate minimum window sizes in certain rooms for use as emergency egress.

Outswinging windows, such as awning and casement windows, should not swing out over decks, patios, and similar areas unless they are high enough to permit persons to travel under them. When lower, the projecting sash could cause serious injury.

Window Schedule

The numbers or letters found in the floor plan identify the window in more detail in the *window schedule*. This is, usually, part of a set of plans (Fig. 56-1). This schedule normally includes window style, size, manufacturer's name, and unit number. Rough opening sizes may or may not be shown.

Manufacturers' Catalogs

Sometimes a window schedule is not included. Units are identified only by the manufacturer's name and number on the floor plan. In order to gain more information, the builder must refer to the window *manufacturer's catalog*.

The catalog usually includes a complete description of the manufactured units and optional accessories, such as insect screens, glazing panels, and grilles. For a particular window style, the catalog typically shows overall unit dimensions, rough opening widths and heights, and glass sizes of man-

WINDOW SCHEDULE				
IDENT.	QUAN.	MANUFACTURER	SIZE	REMARKS
A	6	MORGAN	2'-8" X 3'-10"	D.H. SINGLE
B	3	MORGAN	2'-8" X 3'-10"	D.H. TRIPLE
C	2	MORGAN	3'-4" X 3'-10"	D.H. SINGLE
D	1	ANDERSEN	CW24	CASEMENT SINGLE
E	1	ANDERSEN	C34	CASEMENT TRIPLE
F	1	ANDERSEN	C23	CASEMENT DOUBLE

Fig. 56-1 Typical window schedule found in a set of plans.

ufactured units. Large-scale, cross-section details of the window unit also usually are included so the builder can more clearly understand its construction (Fig. 56-2).

Order window units giving the type and identification letters and/or numbers found in the window schedule or manufacturer's catalog. The size of all existing rough openings should be checked to make sure they correspond to the size given in the catalog before windows are ordered.

Installing Wood Windows

All rough window openings should be prepared to ensure weathertight window installations. Two methods of preparation are generally used.

Applying Felt Strips

Prepare the openings in wood frame walls by applying strips of #15 asphalt felt about 8 inches wide against the wall sheathing on each side of the window openings. Let each end of the strip project above and below the opening about 4 inches. Fasten the strips along each edge about 12 inches apart. Do not fasten the lower end below the window opening (Fig. 56-3).

Housewraps and Building Paper

Exterior walls are sometimes covered with a building paper prior to the application of **siding.** This prevents the infiltration of air into the structure. Yet, it also allows the passage of water vapor to the outside. In place of building paper, exterior walls may be covered with a type of air infiltration barrier commonly called **housewrap.**

Housewrap is a very thin, tough plastic material. It is used to cover the sheathing on exterior walls for the same purpose as building paper (Fig. 56-4). Housewraps are commonly known by the brand names of Typar and Tyvek. They are more resistant than building paper to air leakage and are virtually tearproof. Yet they are also breathable to allow water vapor to escape. Building paper only comes in 36-inch wide rolls. Housewrap rolls are 1.5, 3, 4.5, 5, 9, and 10 feet wide.

Building paper must be applied after windows and doors are installed. If installed before, there may be a prolonged period before it is covered by siding. During this time it could get wet and deteriorate or be torn by high winds. Therefore rough openings are prepared with strips of felt because a flashing must extend under the window casing.

Housewrap is designed to survive prolonged periods of exposure to the weather. It can and usually is applied immediately after framing is completed, but before doors and windows have been installed. The wrap, then, serves also as a flashing for the sides of windows and doors. It is not necessary to apply felt flashing strips to the sides of rough openings. Housewrap gets its name because it is completely wrapped around the building. It covers corners, window and door openings, plates, and sills.

Applying Housewrap. Begin at the corner holding the roll vertically on the wall. Unroll it a short distance. Make sure the roll is plumb. Secure the sheet to the corner, leaving about a foot extending beyond the corner to overlap later. Continue to unroll. Make sure the sheet is straight, with no buckles. Fasten every 12 to 18 inches (Fig. 56-5).

Unroll directly over window and door openings and around the entire perimeter of the building. Overlap all joints by at least 3 inches. Secure them with a special housewrap tape. On horizontal joints, the upper layer should overlap the lower layer.

Make cuts in the housewrap from corner to corner of rough openings. Fold the triangular flaps

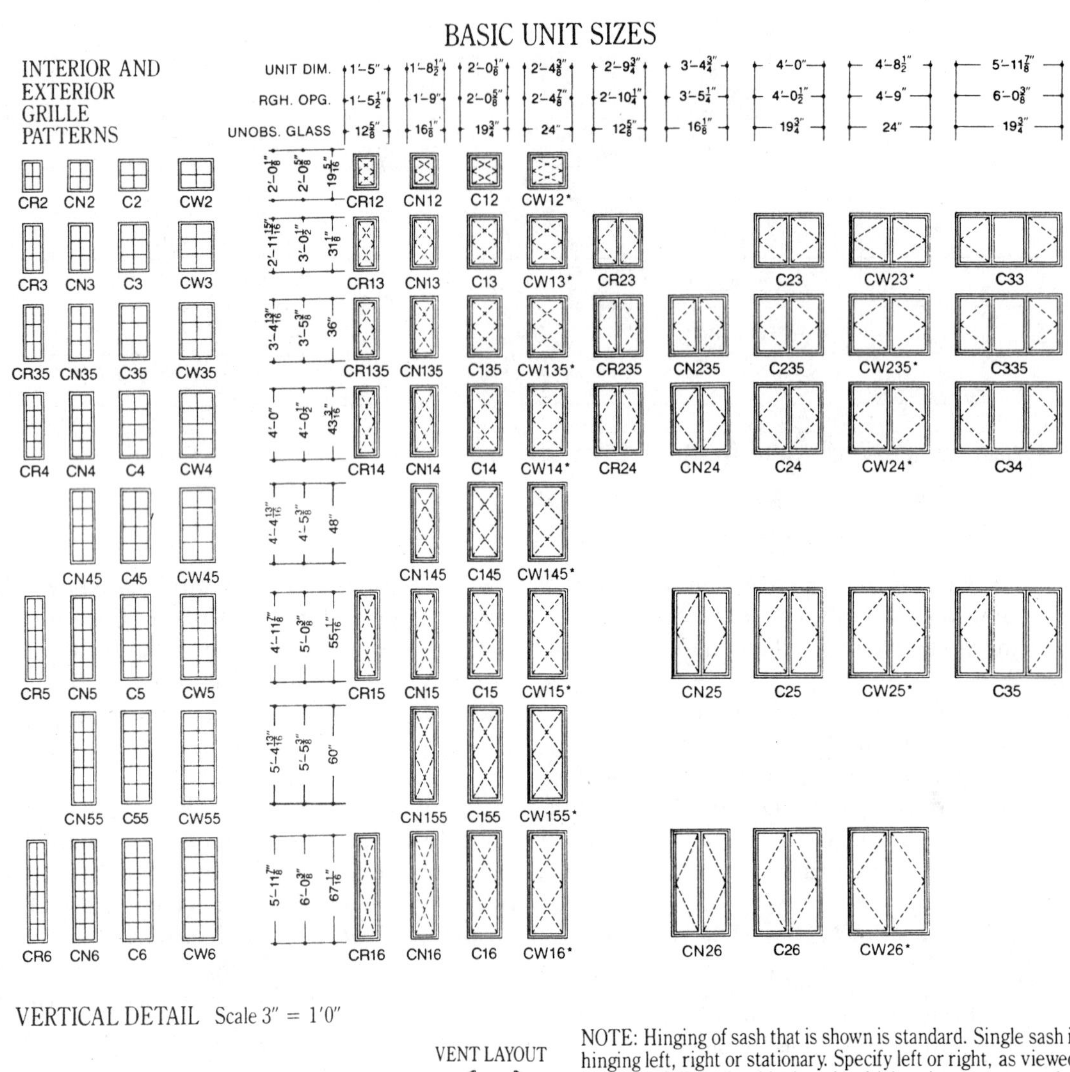

VERTICAL DETAIL Scale 3″ = 1′0″

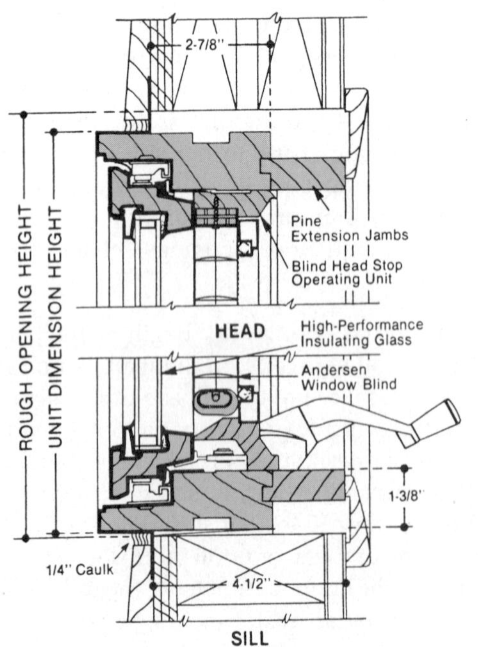

VENT LAYOUT

L.H. R.H.

NOTE: Hinging of sash that is shown is standard. Single sash indicates hinging left, right or stationary. Specify left or right, as viewed from the outside. For other hinging of multiple units contact your local Andersen supplier.

*CW series units (except CW2 and CW3 height) open to 20″ clear opening width using sill hinge control bracket. Bracket can be pivoted allowing for cleaning position.

CW series units are also available with a 22″ clear opening width. Please contact distributor for availability.

When ordering be sure to specify color desired.

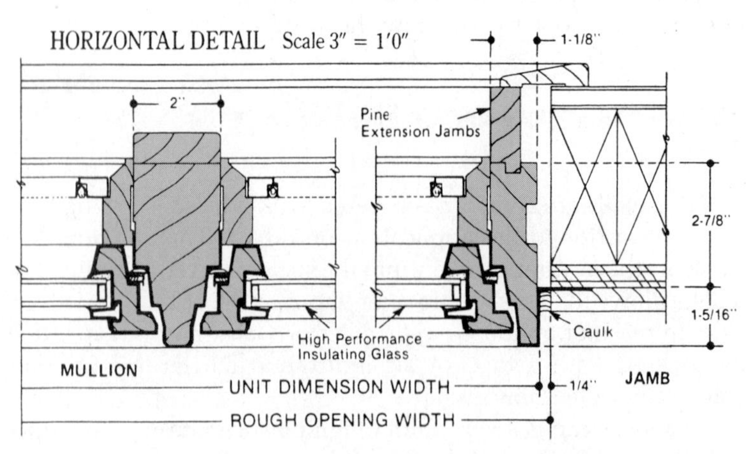

Fig. 56-2 Typical page from a window manufacturer's catalog. (*Courtesy of Andersen Windows, Inc.*)

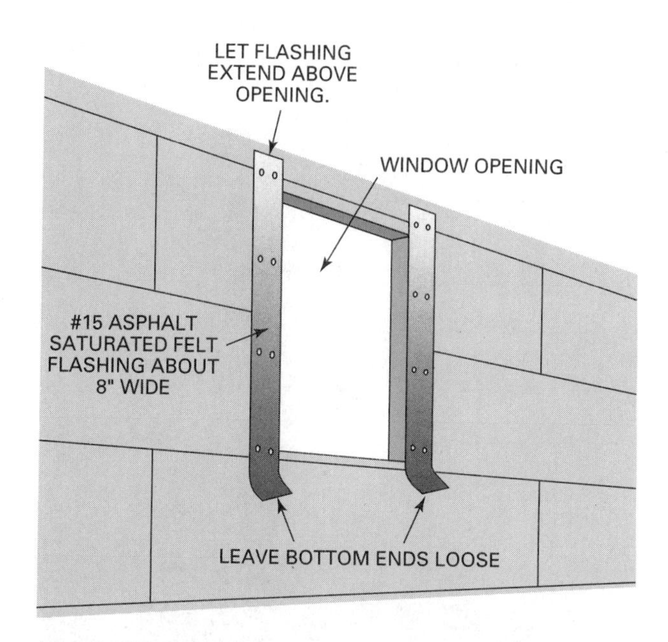

Fig. 56-3 Flash each side of the window opening with felt strips prior to installing the window.

Fig. 56-5 Housewrap being applied to a sidewall. (*Courtesy of Reemay, Inc.*)

Fig. 56-4 Housewrap is widely used as an air infiltration barrier on sidewalls. (*Courtesy of Reemay, Inc.*)

in and around the opening (Fig. 56-6). Secure the flaps on the inside with fasteners spaced about every 6 inches.

CAUTION Housewraps are slippery. They should not be used in any application where they will be walked on.

Establishing Window Height

Most of the windows in each story of a building are set so their head casings lie in a straight line at the same level, even if some of the windows are shorter than others. Snap a chalk line all around the building to the height of the head casing. Windows are then set by lining the top edges of the head casing to the chalk line.

Another way of establishing this height is to use a story pole. Lay out the height of the head jamb from the subfloor on the story pole. When setting windows, hold the story pole with one end against the subfloor. Line up the head jamb of the window with the layout line on the story pole (Fig. 56-7).

Installing Windows

Remove all protection blocks from the window unit. Cut off any **horns** left on the unit. Horns are extensions of the side jambs above the head jamb and be-

Fig. 56-6 Housewrap is cut and folded in wall openings. (*Courtesy of Reemay, Inc.*)

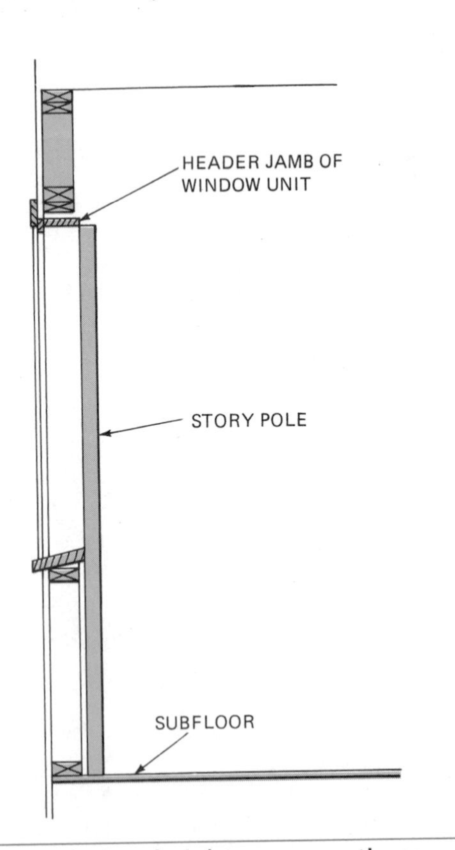

Fig. 56-7 Window heights are sometimes established using a story pole.

low the sill. The extensions or horns protect the units during storage, shipment to the dealer, and delivery to the construction site.

Do not remove any diagonal braces applied at the factory. Close and lock the sash. If windows are stored inside, they can easily be moved through the openings and set in place (Fig. 56-8). It is important to center the unit in the opening on the rough sill with the window casing overlapping the wall sheathing.

CAUTION Have sufficient help when setting large units. Handle them carefully to avoid damaging the unit or breaking the glass. Broken glass can cut through protective clothing and cause serious injury.

Place a level on the window sill. If not level, determine which side of the window is the highest.

Fig. 56-8 Windows can be easily installed, through the opening, from the inside.

Check to see if the high side of the unit is at the desired height. If not, bring it up by shimming with a wood **shingle tip** between the rough sill and the bottom end of the window's side jamb (Fig. 56-9). Remove the window unit from the opening and caulk the backside of the casing or nailing flange. This will seal the unit to the sheathing. Tack through the lower end of the casing into the sheathing on the high side. Shim the other side of the window in the same manner, so the window sill is level. Tack the lower end of the casing on that side.

On wide windows with long sills, shim at intermediate points so that the sill is perfectly straight and level with no sag. Use either a long level or a shorter level in combination with a straightedge. Also, sight the sill by eye from end to end to make sure it is straight.

Plumb the ends of the side jambs with a level. Tack the top ends of the side casings. Straighten the side jambs between sill and head jamb. Tack through the side casings at intermediate points. Straighten and tack the header casing.

Check the joint between sash and jamb. Make sure the sash operates properly. If not, make necessary adjustments. Then fasten the window permanently in place. Use galvanized casing or common nails spaced about 16 inches apart. Keep nails about 2 inches back from the ends of the casings to avoid splitting. Nail length depends on the thickness of the casing. Nails should be long enough to penetrate the sheathing and into the framing members. Set the nails so they can be puttied over later.

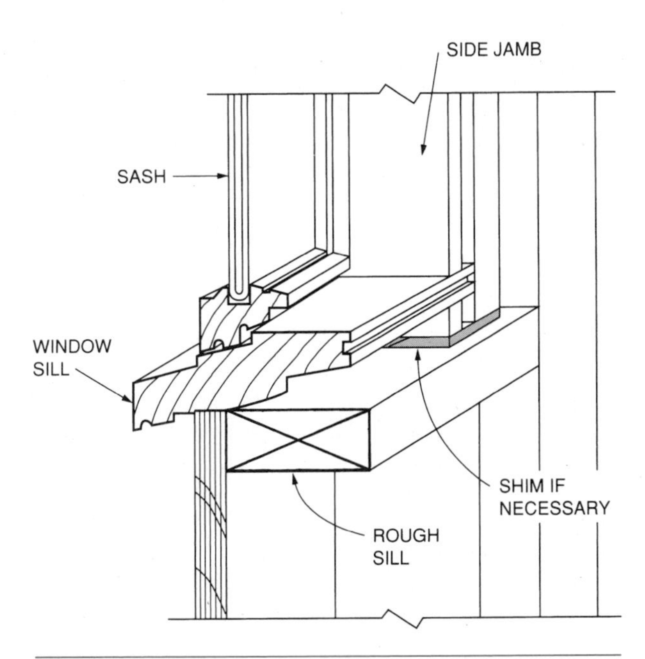

Fig. 56-9 Shim under the side jambs to level the window unit. (*Courtesy of Andersen Windows, Inc.*)

Vinyl-clad windows have a vinyl nailing flange. Large-head roofing nails are driven through the flange instead of through the casing (Fig. 56-10).

The installation of wood windows in masonry and brick veneer walls is similar to that of frame walls. Adequate clearance should be left for caulking around the entire perimeter between the window and masonry (Fig. 56-11).

Flashing the Head Casing

To flash the head casing, cut the flashing with tin snips. Its length should be equal to the overall width of the window. Do not let the ends project beyond the side casings. This will make the application of siding difficult. If the flashing must be applied in more than one piece, lap the joint about 3 inches. Place the flashing firmly on top of the head casing. Secure with fasteners along its top edge and into the wall sheathing (Fig. 56-12).

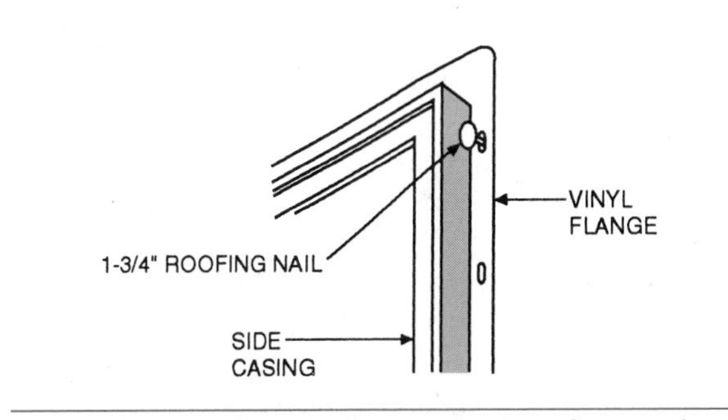

Fig. 56-10 Roofing nails are used to fasten the flanges of vinyl-clad windows. (*Courtesy of Andersen Windows, Inc.*)

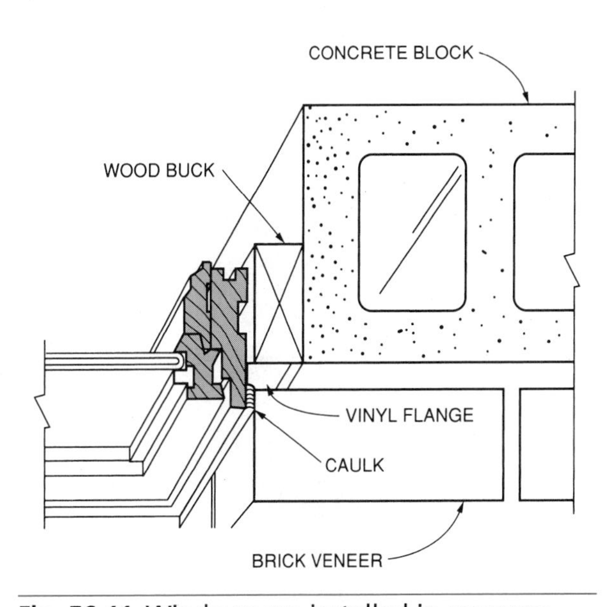

Fig. 56-11 Windows are installed in masonry openings against wood bucks. (*Courtesy of Andersen Windows, Inc.*)

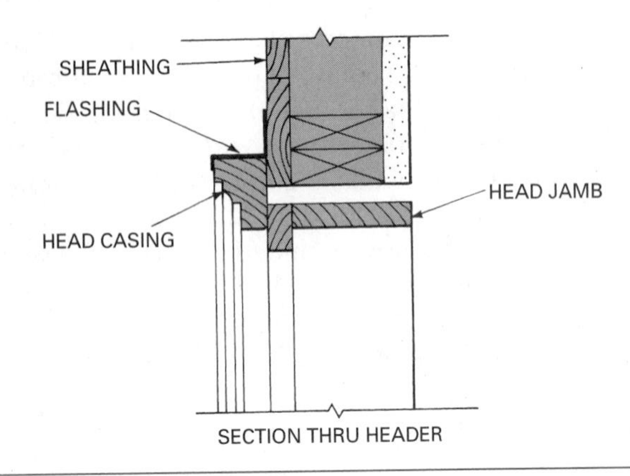

Fig. 56-12 A window flashing is used over the top of the head casing.

Flashing the Sill

If the sides of the window openings have been previously flashed with felt strips, cut a strip of felt about 8 inches wide. It should be long enough to extend to the outside edges of the side flashings. Tuck the edge of the strip under the loose ends of the felt flashings on each side and well into the groove in the bottom of the window sill. Fasten the strip. Make sure to leave a few inches of the bottom edge loose so that building paper can be applied under it later (Fig. 56-13).

If housewrap is used, apply a strip of it in the same manner and well up in the siding groove in the bottom of window sill. It must, however, be applied on top of housewrap already installed. Seal all overlapping edges and ends with housewrap tape.

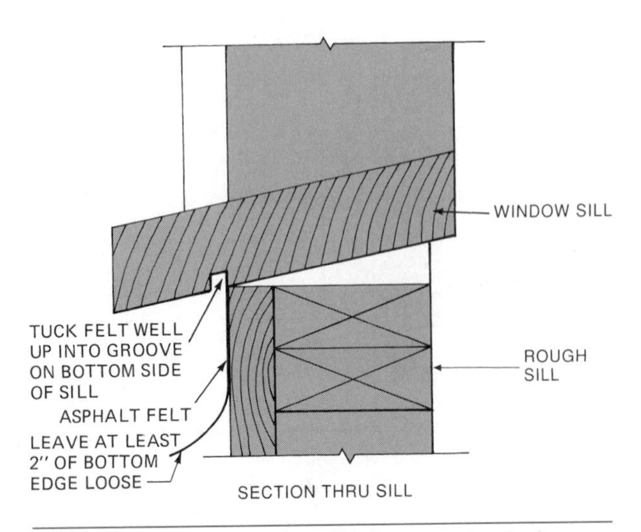

Fig. 56-13 Flashing the sill of a wood window with asphalt felt.

Metal Windows

Metal windows are available in the same styles as wood windows. The shape and sizes of the parts vary with the manufacturer and the intended use.

In frame construction, if metal windows are used, they may be set in a wood frame. The frame is then installed in the same manner as for wood windows. They may also be set in the opening with their flanges overlapping the siding or sheathing. **Caulking** is applied under the flanges. The unit is then screwed to the wall (Fig. 56-14).

In masonry construction, wood *bucks* are fastened to the sides of the opening. Metal windows are installed against them. The bottom usually rests against a rabbet of a preformed concrete sill. The flanges on the two sides and top are bedded in caulking. They are then screwed to the bucks. **Mortar** is applied to cover the flanges and beveled to the outside corner of the masonry opening (Fig. 56-15).

Carefully follow the installation directions provided with the units, whether wood or metal.

Glazing

The art of cutting and installing lights of glass in sash and doors is called *glazing.* Those who do such work are called **glaziers.** Sometimes the carpenter may have to replace a light of glass.

Fig. 56-14 Installing a metal awning window.

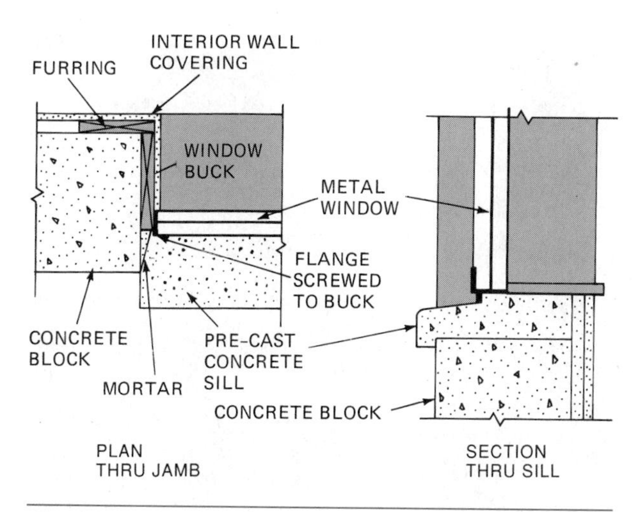

Fig. 56-15 Method of setting a metal window in a masonry opening.

Glazing Materials

Sash are made so that the glass is usually held in place with **glazing points** and **glazing compound** (Fig. 56-16). Glazing points are small triangular or diamond-shaped pieces of thin metal. They are driven into the sash parts to hold the glass in place. Glazing compound is commonly called *putty.* It is used to cover and seal around the edges of the light.

A light of glass is installed with its convex side, or crown of its bow, up in a thin bed of compound against the rabbet of the opening. Glass set in this manner is not as apt to break when installed.

Installing Glass

To replace or **glaze** a light of glass, first remove the broken glass.

CAUTION Use heavy gloves to handle the broken glass. Broken glass edges are sharp and can cut easily.

Clean all compound and glazier points from the rabbeted section of the sash. Apply a thin bed of glazing compound to the surface of the rabbet on which the glass will lay. Lay the glass in the sash with its crowned side up. Carefully seat the glass in the bed by moving it back and forth slightly.

Fasten the glass in place with glazier points. Slide the driver along the glass. Do not lift the driver off the glass. Special glazier point driving tools prevent glass breakage. If a driving tool is not available, drive the points with the side of a chisel or a putty knife.

Lay a bead of compound on the glass along the rabbet on top of the light of glass. Trim the compound at a bevel by drawing the putty knife along. One edge should be flush with the inside edge of the glass opening. The other edge of the compound should feather to the outside edge of the opening. Prime the compound as soon as possible after glazing. Lap the paint about 1/16th inch onto the glass to make a seal against the weather.

Cutting Glass

Sometimes it may be necessary to cut a light of glass to size so it will fit in the opening. Lay the glass on a clean, smooth surface. Brush some mineral spirits along the line of cut. Hold a straightedge on the line. Draw a glass cutter with firm pressure along the straightedge to make a clean, uniformly scored line (Fig. 56-17).

Do not go over the scored line. This will dull the glass cutter. The line must be scored along the whole length the first time with no skips. Otherwise, the glass may not break where desired. Practice on scrap pieces to become proficient in making clean breaks.

Move the glass so the scored line is even with the edge of the workbench. Apply downward pressure on the overhanging glass. If the glass is properly scored, it will break along the scored line (Fig. 56-18).

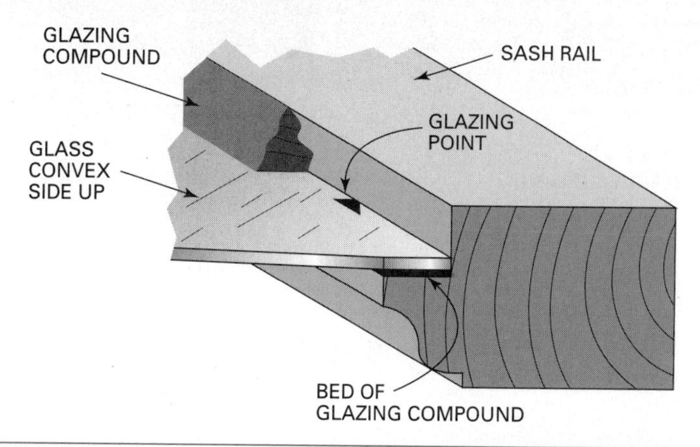

Fig. 56-16 Lights of glass are secured with glazing points and glazing compound.

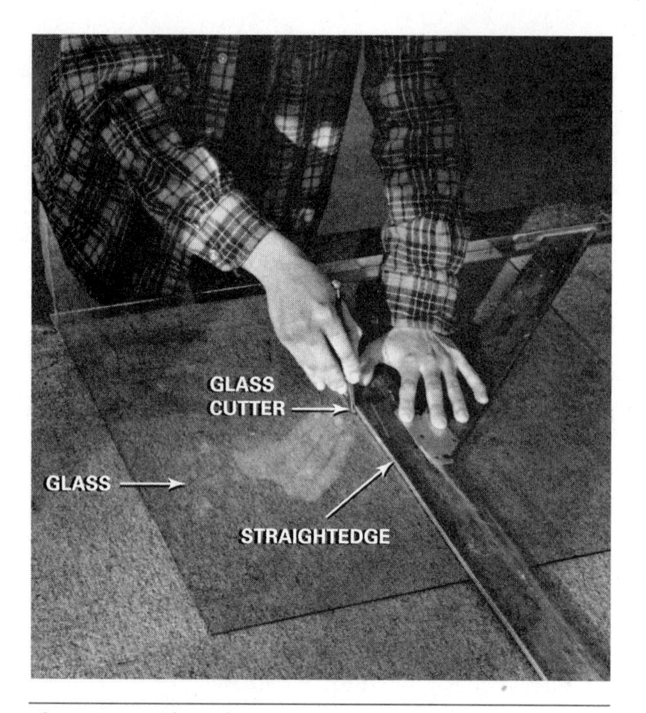

Fig. 56-17 The glass must be scored with a single stroke of the glass cutter.

Fig. 56-18 If the glass is scored correctly, it will break easily and cleanly along the scored line.

Review Questions

Select the most appropriate answer.

1. A frame in a window that holds the glass is called a
a. light. c. sash.
b. mullion. d. stile.

2. Small wood strips that divide the glass into smaller lights are called
a. mantels. c. mullions.
b. margins. d. muntins.

3. When windows are installed in multiple units, the joining of the side jambs forms a
a. mantel. c. mullion.
b. margin. d. muntin.

4. A window that consists of an upper and a lower sash that slide vertically is called a
a. casement window.
b. double-hung window.
c. hopper window.
d. sliding window.

5. A window that has a sash hinged on one side and swings outward is called
a. an awning window.
b. a casement window.
c. a double-hung window.
d. a hopper window.

6. The difference between a hopper and an awning window is that the hopper window
a. swings inward instead of outward.
b. swings outward instead of inward.
c. is hinged at the top rather than at the bottom.
d. is hinged on the side rather than on the bottom.

7. Before setting a window in an opening
a. flash all sides of the opening.
b. flash the bottom.
c. flash two sides.
d. flash the top.

8. Horns on windows are
a. placed on the header casing.
b. extensions of the side jambs.
c. moldings applied to a flat casing.
d. extensions of the header jamb.

9. The art of cutting and installing glass is called
a. gauging.
b. glazing.
c. gouging.
d. grouting.

10. Glass is installed
a. with its crowned side up.
b. with its concave side up.
c. with either side up.
d. on the inside face of the sash.

BUILDING FOR SUCCESS

The Building Inspector: Friend or Foe?

A career in construction will introduce you to a variety of skilled and professional trade representatives. Among these representatives will be building inspectors. In most cases, inspectors will be engaged by municipalities. They serve their communities as quality control agents who ensure public safety during the construction of buildings. A primary function of local building inspectors is to examine and evaluate construction. They determine if it is in compliance with the approved standards. Codes become the criteria by which the plans, specifications, and entire construction process is gauged.

Building inspectors have the responsibility of determining if a building or structure is safe for occupancy, meets minimum code requirements, and is safe from fire.

The working relationship among contractors, skilled workers, and building inspectors must be one of mutual respect, courtesy, and confidence. It is imperative that building contractors and subcontractors clearly understand the building inspector's role and how the inspector can assist in the building process. Everyone involved in construction should strive to establish good working relations with each other.

Trust can be built between inspectors and builders when both understand the inspector's function and purpose. This trust is based on how well inspectors know the adopted codes, evaluate construction projects, and relate their assessments to the builders. Equally important is the builder's history of performing quality work. Together these professionals can set the pace for meaningful and safe construction projects.

The local building standards should relate to the building's structural integrity, life safety, and environmental health capabilities. The pertinent community ordinances are reflected along with the local building codes. These codes will have many parts, possibly consisting of building, electrical, fire prevention, health, housing, life safety, mechanical, and plumbing requirements. Additionally, the local zoning ordinances will play a major role in the inspector's job. Workable codes should be performance codes rather than specification codes. As long as the materials and workmanship perform to code standards, many systems can be used. New technology, new products, climatic conditions, and environmental issues will continue to influence local codes and regulations.

Inspections will generally occur in a specific order. For example, site inspection may be followed by the inspections of footings and foundation forms, reinforcing steel, embedded electrical, mechanical, and plumbing, or steel reinforcing before the concrete is placed. It is the builder's responsibility to plan for and schedule the appropriate inspection throughout the construction process. Assisting the inspector as needed during the inspection process will establish the best working relationships. Partial inspections may be needed so the construction assembly process can remain on schedule.

Inspection records will be required at all construction sites to document properly each inspection until the project is completed. Occasionally, corrections will have to be made when violations or omissions have taken place. Adequate time will be given to make the necessary adjustments. Quick attention by the contractor generally will result in remaining on schedule and keep unforeseen expenses to a minimum.

This may be a time when all parties will need to implement their best communication and business skills to avoid frustrations and misunderstandings. The outcomes of these visits by the building inspector will form the basis for future cooperation. The inspector must remain impartial in interpreting the relevant codes and ordinances. The builder needs to recognize the infraction, make the necessary changes, and maintain the integrity of the relationship.

The building inspector has an extremely important role in the construction process. There will always be a need for someone to assure adequate and safe construction methods for the general populace. Occupants of buildings have great concern for a safe and well-built structure.

The building inspector is generally looked upon in a favorable manner by quality builders. They have learned to respect and trust the integrity of the position. Contractors can assure their customers a well-built home, office building, or place of business. The inspector's documentation verifies that the community's public safety requirements have been met throughout the construction process. In the vast majority of cases, the building inspector is a builder's friend and adviser in the construction process. The working relationships involving the inspector should remain favorable for a lifetime.

FOCUS QUESTIONS: For individual or group discussion

1. How would you define the role and purpose of the local building inspector?
2. How do you think a building inspector could help you establish and maintain a reputable business?
3. What might be some reasons to clearly understand the role and mission of the building inspector?
4. What could be some possible consequences of not cooperating with inspectors and the agency they represent?
5. What do you envision as the best working relationship between the contractor and the building inspection official?

UNIT

20

Exterior Doors

Exterior doors, like windows, are manufactured in millwork plants in a wide range of styles and sizes. Many entrance doors come pre-hung in frames, complete with exterior casings applied, and ready for installation. In other cases, the door frame must be constructed and set in the rough opening. The door is then fitted and hinged to it.

OBJECTIVES

After completing this unit, the student should be able to:

- name the parts of, build, and set an exterior door frame.
- describe the standard designs and sizes of exterior doors and name their parts.
- fit and hang an exterior door.
- install locksets in doors.

UNIT CONTENTS

CHAPTER 57 DOOR FRAME CONSTRUCTION AND INSTALLATION

Careful construction and installation of the door frame makes it easier to fit and hang the door afterward. In addition, it contributes to smooth operation of the door and protection against the weather for many years.

Parts of an Exterior Door Frame

Terms given to members of an exterior door frame are the same as similar members of a window frame. The bottom member is called a *sill* or *stool.* The vertical side members are called *side jambs.* The top horizontal part is a *head jamb.* The exterior door trim may consist of many parts to make an elaborate and eye-appealing entrance or a few parts for a more simple doorway. The *door casings,* if not too complex, are usually applied to the door frame before it is set (Fig. 57-1). When more intricate trim is specified, it is usually applied after the frame is set (Fig. 57-2).

Sills

In residential construction, door frames usually are designed and constructed for entrance doors that swing inward. Codes require that doors swing outward in buildings used by the general public. The shape of a wood door sill for an inswinging door is different from that for an outswinging door (Fig. 57-3).

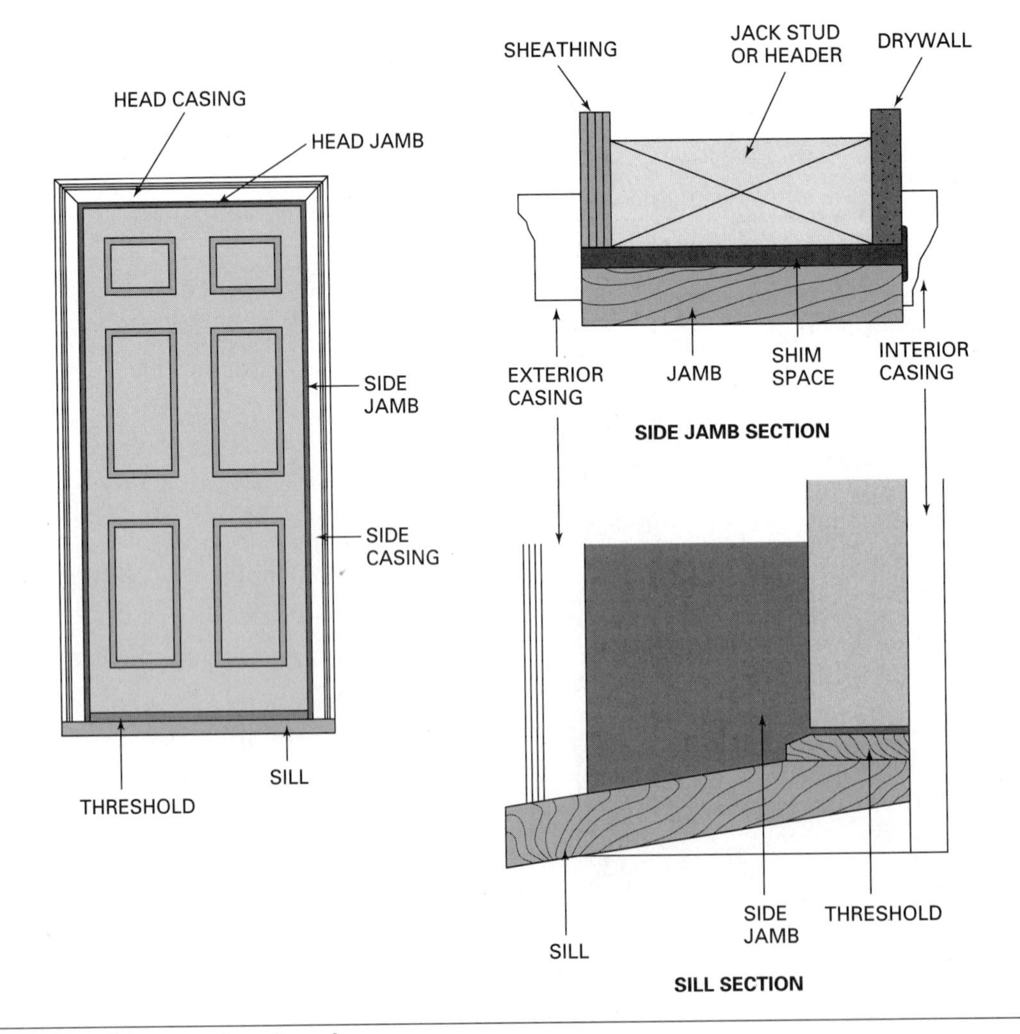

Fig. 57-1 Parts of an exterior door frame.

Fig. 57-2 Elaborate entrance trim is available in knocked-down form, ready for assembly. (*Courtesy of Morgan Manufacturing*)

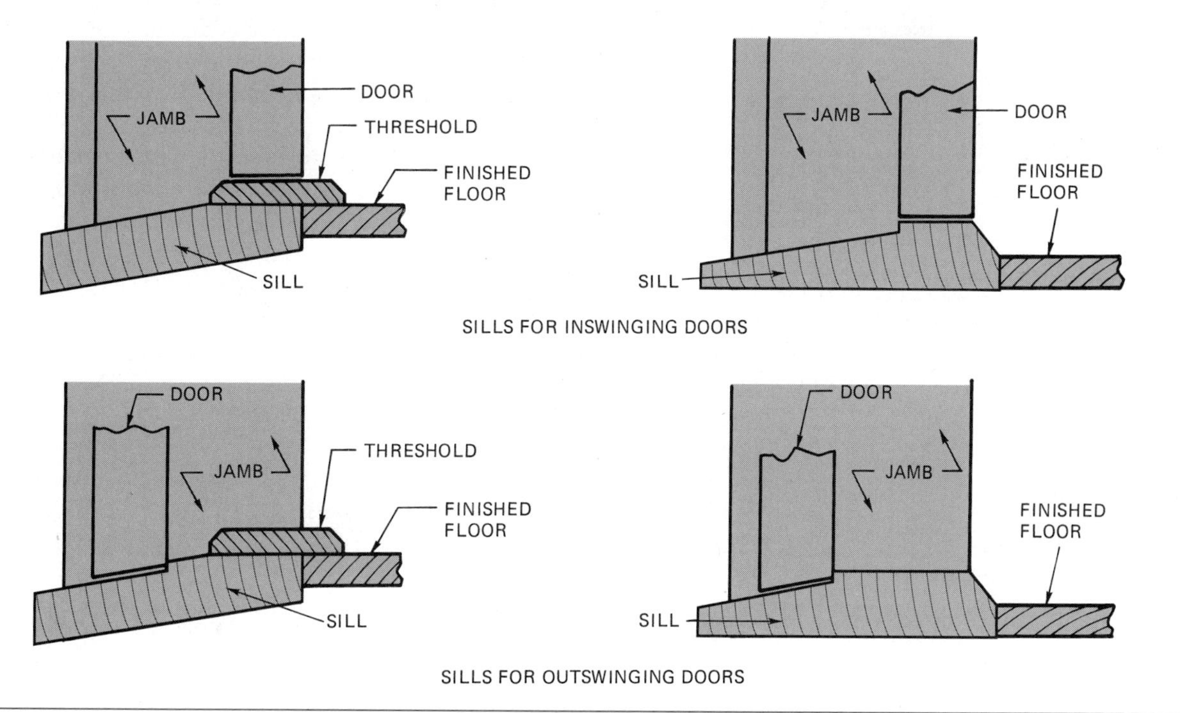

Fig. 57-3 Wood sill shapes vary according to the swing of the door.

In addition to wood, extruded aluminum sills of many styles are manufactured for both inswinging and outswinging doors. They usually come with vinyl inserts to weatherstrip the bottom of the door. Some are adjustable for exact fitting between the sill and door (Fig. 57-4).

Jambs

Side and header jambs are the same shape. Jambs may be square edge pieces of stock to which *door stops* are later applied. Or, they may be *rabbeted jambs,* with *single* or *double* rabbets. On double-rabbeted jambs, one rabbet is used as a stop for the main door. The other is used as a stop for storm and screen doors (Fig. 57-5). Several jamb widths are available for different wall thicknesses. For walls of odd thicknesses, jambs, except double-rabbeted ones, may be ripped to any desired width.

Exterior Casings

Exterior casings may be plain square-edge stock. Moldings are sometimes applied around the outside edges of the casings. This is done to improve the appearance of the entrance. Because the main entrance is such a distinctive feature of a building, the exterior casing may be enhanced on the job with *fluted,* or otherwise shaped, pieces and appropriate caps and moldings applied (Fig. 57-6). Flutes are narrow, closely spaced, concave grooves that run parallel to the edge of the trim. In addition, ornate main entrance trim may be purchased in knocked-down form. It is then assembled at the job site (Fig. 57-7).

Fig. 57-4 Some metal sills are adjustable for exact fitting at the bottom of the door.

Making a Door Frame

One of the most common types of door frames consists of single-rabbeted jambs with an oak sill for an inswinging door. (Instructions for building and setting this frame are given in following paragraphs. The specifics of the instructions may not apply to other types of door frames, but the basics are similar.) If the construction and setting of the door frame described herein are mastered, competence can be gained easily with other types of door frames.

A more accurately sized door frame can be constructed if the door itself is available for measurement. Many doors are manufactured slightly over the nominal size. Some are manufactured slightly undersize. Making the frame too small will result in excessive fitting of the door. Unacceptable wide joints between door and frame are the result of making the frame too big.

Determining Jamb Width

The first step in making any door frame is to determine the jamb width. The jamb width is equal to the overall wall thickness from the outside surface of the wall sheathing to the inside surface of interior wall covering. Because the interior wall covering probably has not been applied at this point, its planned thickness must be known.

Rip the jamb stock so its width corresponds to the wall thickness. Rip the edge opposite the rabbet. Smooth and slightly **back-bevel** the ripped

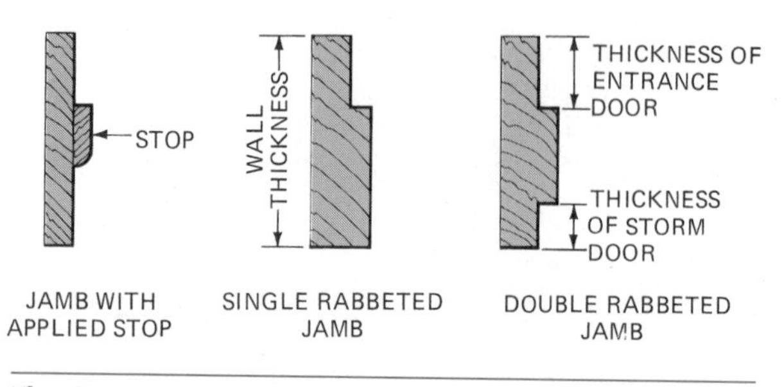

Fig. 57-5 Door jamb cross-sections may be square-edge, single-rabbeted, or double-rabbeted.

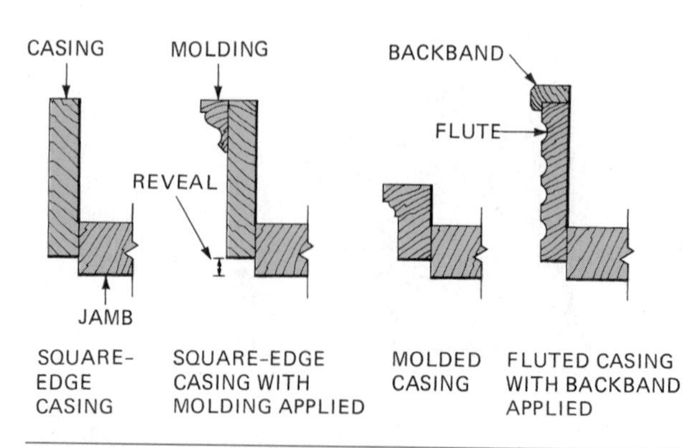

Fig. 57-6 Exterior door casings may be enhanced by applying moldings and by shaping.

Fig. 57-7 A few samples of manufactured entrance door trim. Many other styles are available. (*Courtesy of Morgan Manufacturing*)

edge with a jack plane (Fig. 57-8). Back-bevel the rabbeted edge, also, if this has not already done at the mill. The reason for back-beveling is to assure a tight fit between the edges of the jamb and door casings when they are applied. Slightly round over the sharp outside corners that will be left exposed.

Dadoing the Side Jambs

The side jambs are usually *dadoed* to receive the ends of the header jamb and the sill. Lay the side jambs on a pair of sawhorses, side by side. Make sure their outside edges are against each other and their ends flush.

Dadoing for the Header Jamb. Square a line across both pieces about 1 inch down from the top end. To lay out the width of the dado, hold a piece of jamb stock with its end on the squared line. Mark its overall thickness toward the bottom end of the side jambs (Fig. 57-9). Mark the depth of the dado on the edge of the jamb opposite the rabbet.

To make the dado, first cut along both layout lines with a fine-toothed hand crosscut saw to the depth of the dado. Be careful when making the bottom cut not to score the face of the rabbet. If scored, it will be exposed when the frame is assembled. Chisel in from both edges to make the dado. The bottom surface of the dado should lie in a straight line from edge to edge with no crown. A slight hollow in the surface is not objectionable (Fig. 57-10).

Dadoing for the Sill. Place the scrap piece of jamb stock, previously used in the dado, with its rabbeted side toward the bottom end of the side jamb. From the rabbet of the scrap piece, lay out a distance equal to the door height. Allow an additional 3/16 inch for top and bottom joints, along the side jamb. Square a line across the face of the rabbet.

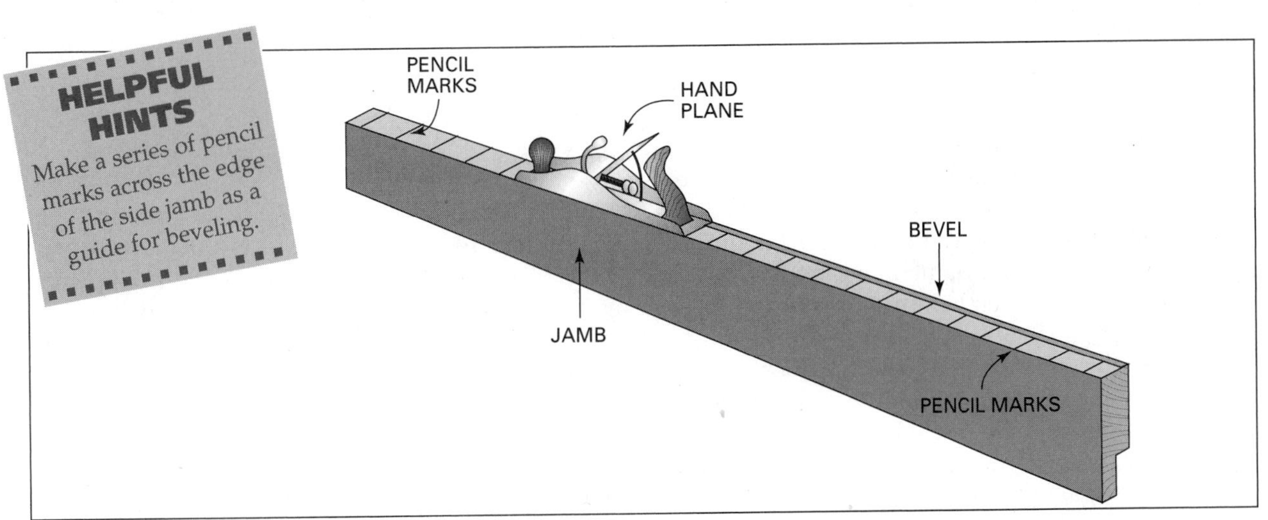

Fig. 57-8 Technique for determining the amount of material that is removed while planing.

Fig. 57-9 A scrap piece of jamb stock is used to lay out the width of the dado for the head jamb.

Fig. 57-11 Mark both sides of the scrap sill stock to lay out the dado for the sill.

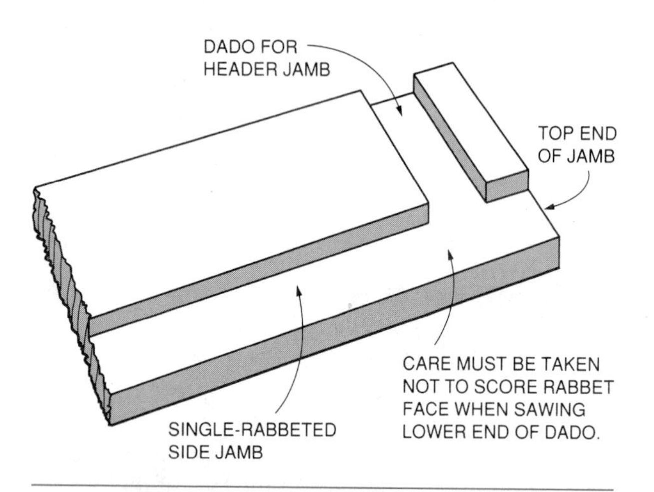

Fig. 57-10 A completed dado to receive the head jamb of the door frame.

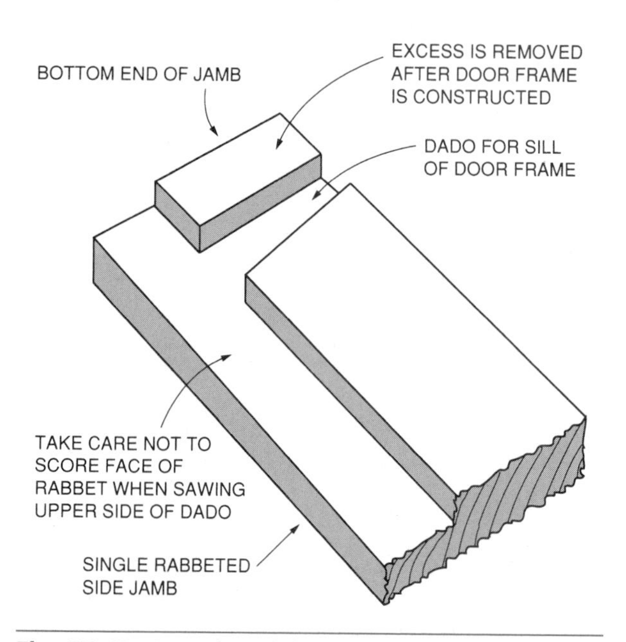

Fig. 57-12 Completed dado to receive the sill of the door frame.

Hold a piece of scrap sill stock so the short straight surface, on its face side, lines up with the mark. The top of the *chamfer* on the inside edge of the sill should line up with the inside edge of the side jamb. Mark the side jamb along both sides of the scrap sill piece when held in this position (Fig. 57-11).

Place the side jambs back together again. Transfer the layout for the sill dado from one side jamb to the other. Dado the side jambs for the sill in a manner similar to that for the header jamb (Fig. 57-12). Again, be careful not to score the face of the rabbet when making the upper saw cut.

Sill and Head Jamb

Cut the head jamb so it is square on both ends. Its length should be the width of the door plus 3/16 inch. The extra 3/16 inch allows for a 3/32-inch joint between door and frame on each side.

Lay out the same length on the sill stock length. Square lines across the face of the sill. From the squared lines, measure toward each end a distance of 5/16 inch more. Square across only on the chamfer. Notch both ends of the sill as shown in Figure 57-13. The projecting ends of the sill eventually butt against edges of the interior casing when applied.

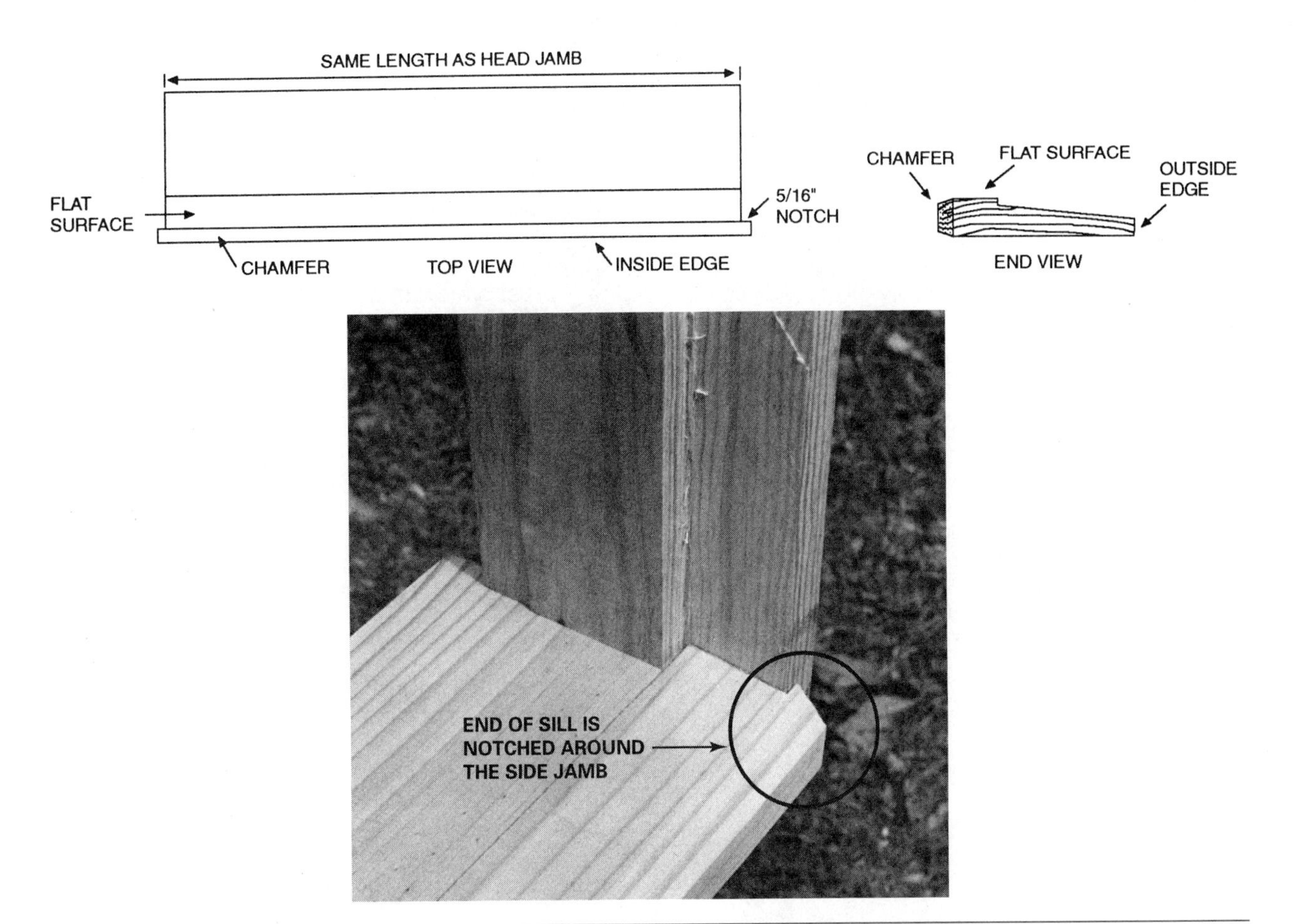

Fig. 57-13 The ends of the sill are notched so the inside edge of interior casings will later butt against them.

Assembling the Door Frame

Although some have different-shaped jambs and sills, all door frames are assembled in much the same manner. Sand any layout lines from the side jambs. Stand the header and sill on end. Place the side jamb on them with their ends inserted in the dadoes. Nail through the side jambs into the ends of the sill and header jamb. To assure a tight fit on the face side, drive a chisel between the back sides of both sill and header jamb and the shoulders of the dado before nailing. This will drive the sill and header jamb tight against the dado shoulders on the exposed side. Drive three 8d, common nails through the side jambs into the end of both pieces. Turn the assembly over. Install the other side jamb in the same manner. Keep the edges of the side and head jamb flush with each other. The notches in the sill should fit tightly against the inside edges of the side jambs (Fig. 57-14).

Applying the Exterior Casing

The exterior casing is applied so its inside edge is set back 1/2 inch from the inside face of the jambs. This setback is called a **reveal.** It is used as a stop for outswinging screen or storm doors (see Fig. 57-6).

Joints between side and head door casings are made similar to those on windows. This prevents wind and rain from entering the wall cavity. If square-edged casings are used, a *blind rabbet* makes a weathertight joint between them (Fig. 57-15).

Applying Square Edge Casings. Rabbet the top end of both side casings. Apply both side casings with noncorroding fasteners spaced about 12 inches apart into the edge of the side jamb. The casings should be positioned so the top ends and inside edges will allow a 1/2-inch reveal on the jambs. Let the bottom ends project below the sill to the bottom edge of the wall sheathing or any desired amount.

Cut the head casing to a length equal to the distance between the outside edges of the side casings. Make a blind rabbet on both ends to fit over the rabbeted ends of the side casings.

Applying Molded Casings. Molded casings are applied in the same manner as square edge casings. However, a *plain miter* or a *splined miter* joint is made at the head. Plain mitered joints are usually caulked with compound and fastened together. A splined miter joint is used when exterior door casings are subjected to severe weather. It is

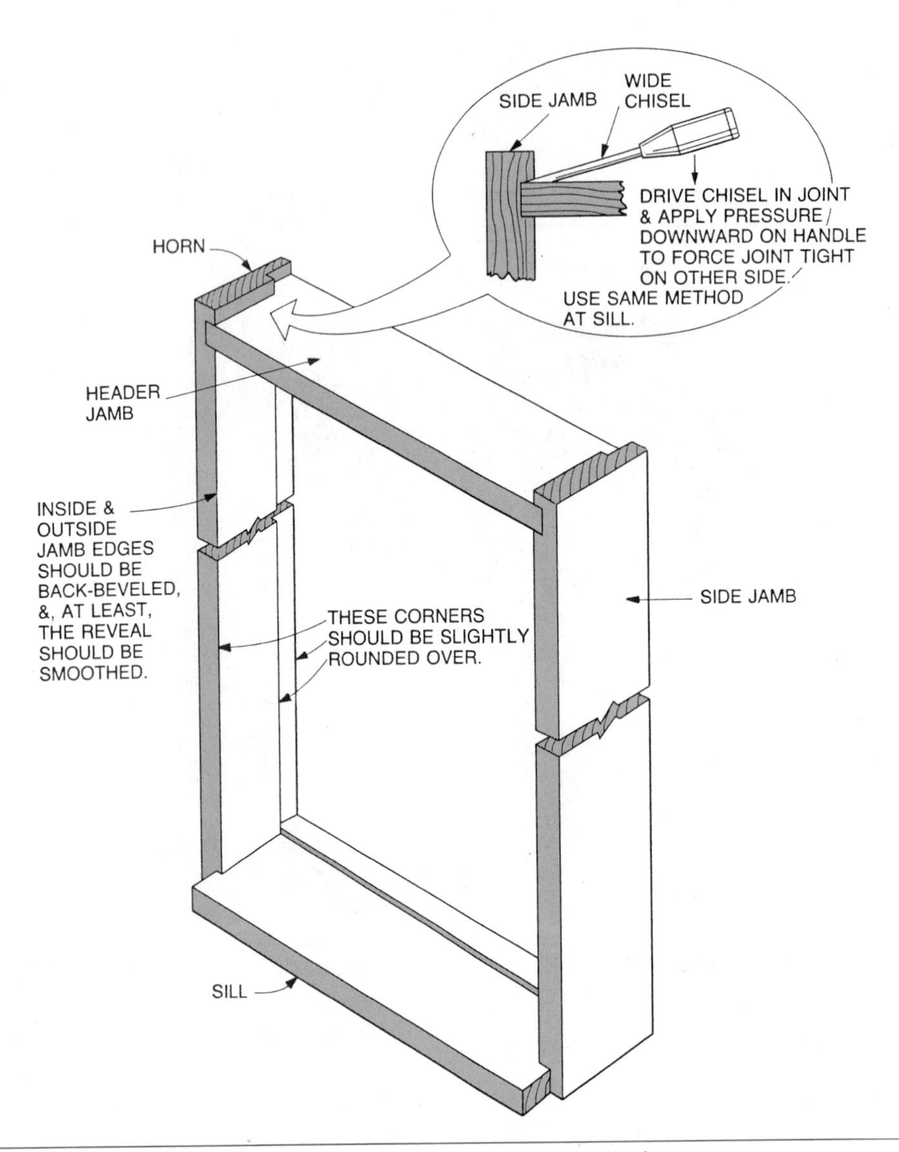

Fig. 57-14 The assembled frame before exterior casings are applied.

Fig. 57-15 A blind rabbet joint is used between exterior side and head door casings.

made by inserting a thin strip of wood or metal, known as a **spline,** across the joint into the ends of the miters (Fig. 57-16).

Setting the Door Frame

When setting a door frame, cut off the horns that project beyond the sill and header jamb. Flash both sides of the rough door opening with saturated felt in the same manner as for windows, except if housewrap has been used. Set the door frame in the opening with the sill on the subfloor (Fig. 57-17). Center the bottom of the frame in the opening. Level the sill by shimming under the side jambs, if necessary. Tack the lower end of both side casings to the wall.

The side jambs must be plumbed next. A 26- or 28-inch carpenter's level may not be accurate when plumbing the sides because of any bow that may be in the jambs. Use either a 6-foot level with blocks at-

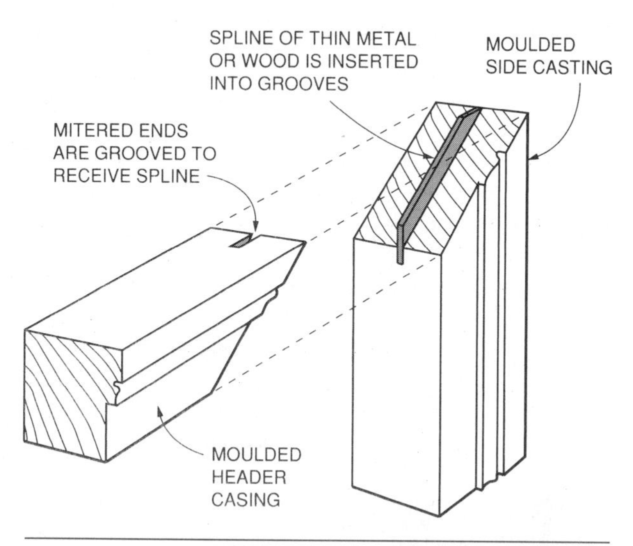

Fig. 57-16 A splined miter sometimes is used to join molded exterior door casings.

Fig. 57-17 Setting an exterior door frame.

tached to each end, a 6-foot straightedge in combination with a carpenter's level, or a plumb bob.

To use the plumb bob, measure from the side jamb a short distance on the header jamb. Hang the plumb bob from the mark to the sill. Move the top of the frame sideways, one way or the other, until the same distance is obtained at the sill (Fig. 57-18). When the side jambs are plumb, tack the top of the side casings to the wall.

Straighten the jambs between the top and bottom using a long straightedge placed against their sides. Tack the side casings to the wall at intermediate points when they are straight. Make sure the head jamb is straight. Tack the header casing to the wall.

Flash under the sill with felt before driving nails home. Felt flashing can then be installed under both the side casings and the side felt flashings. Check all parts again for level and plumb. Then drive and set all nails. Apply any decorative molding or other trim, if specified. Install the door flashing on the top edge of the head casing.

Setting Door Frames in Masonry Walls

In commercial construction, exterior wood or metal door frames are sometimes set in place before masonry walls are built. The frames must be set and firmly braced in a level and plumb position. The head jamb is checked for level. The bottom ends of

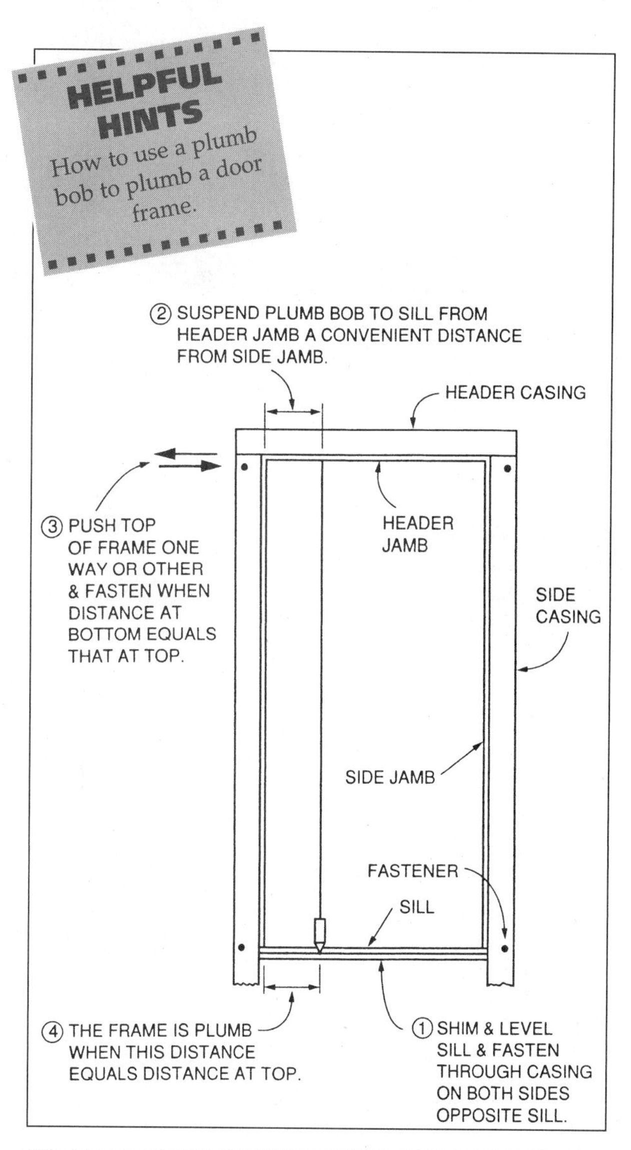

Fig. 57-18 Method of using a plumb bob.

the side jambs are secured in place. It may be necessary to shim one or the other side jamb in order to level the head jamb. The side jambs are then plumbed in a sideways direction. They are braced in position. Then, the frame is plumbed and braced at a right angle to the wall (Fig. 57-19).

Finally, the frame is checked to see if it has a **wind.** A wind is a twist in the door frame caused when the side jambs do not line up vertically with each other. No matter how carefully the side jambs of a door frame are plumbed, it is always best to check the frame to see if it has a wind.

One method of checking the door frame for a wind is to stand to one side. Sight through the door frame to see if the outer edge of one side jamb lines up with the inner edge of the other side jamb. If they do not line up, the frame is in wind. Make adjustments until they do. One way of making the adjustment is to plumb and brace one side at a right angle to the wall. Then sight, line up, and brace the other side jamb (Fig. 57-20).

Some workers check for a wind by stretching two lines diagonally from the comers of the frame. If both lines meet accurately at their intersections, the frame does not have a wind (Fig. 57-21).

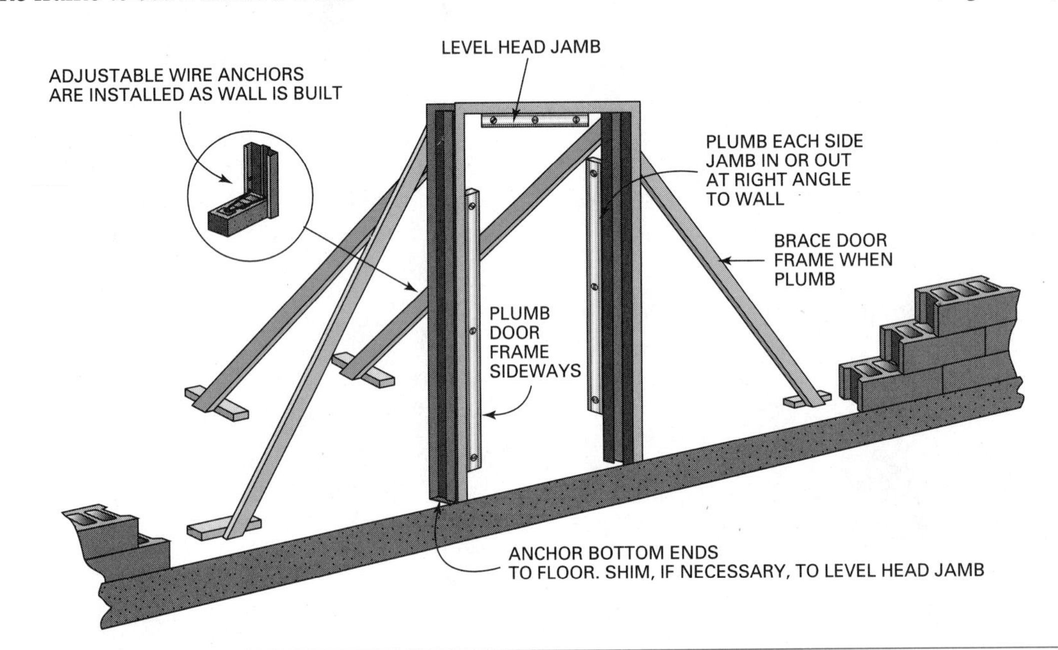

Fig. 57-19 Installation of an exterior door frame in a masonry wall.

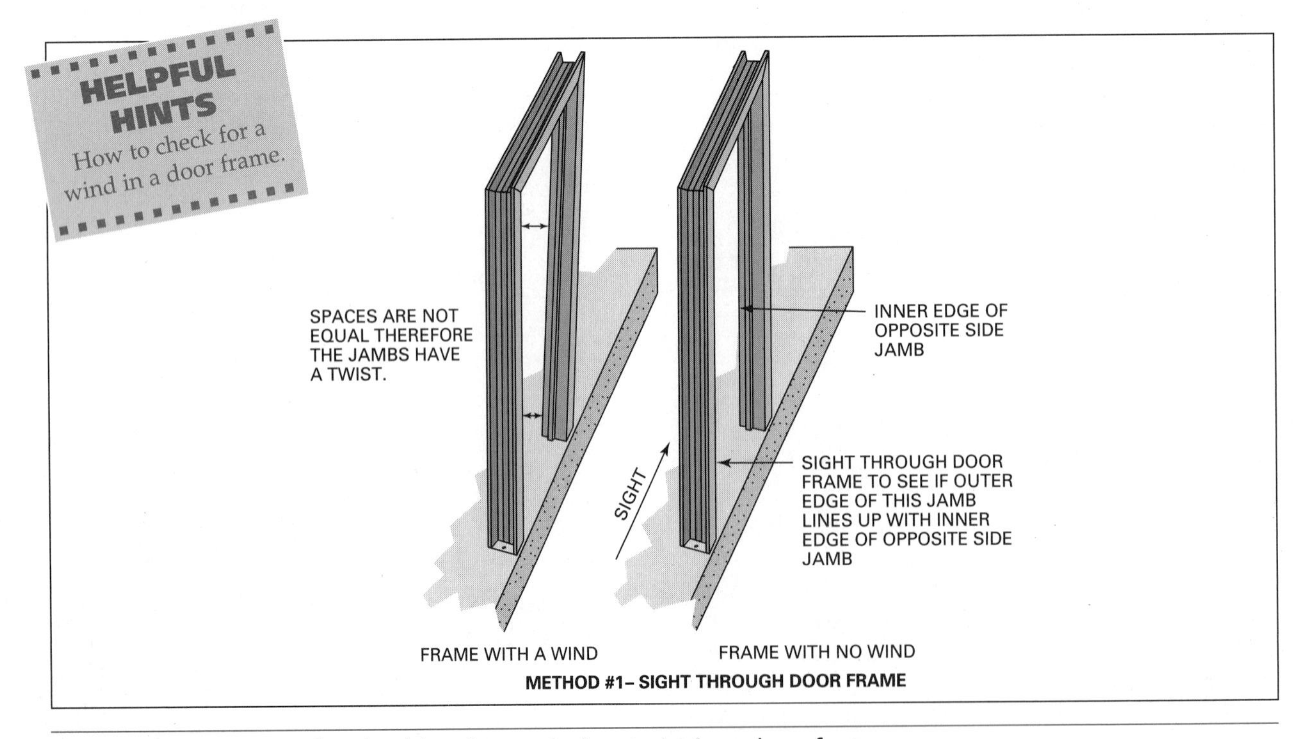

Fig. 57-20 Technique for checking for a wind or twist in a door frame.

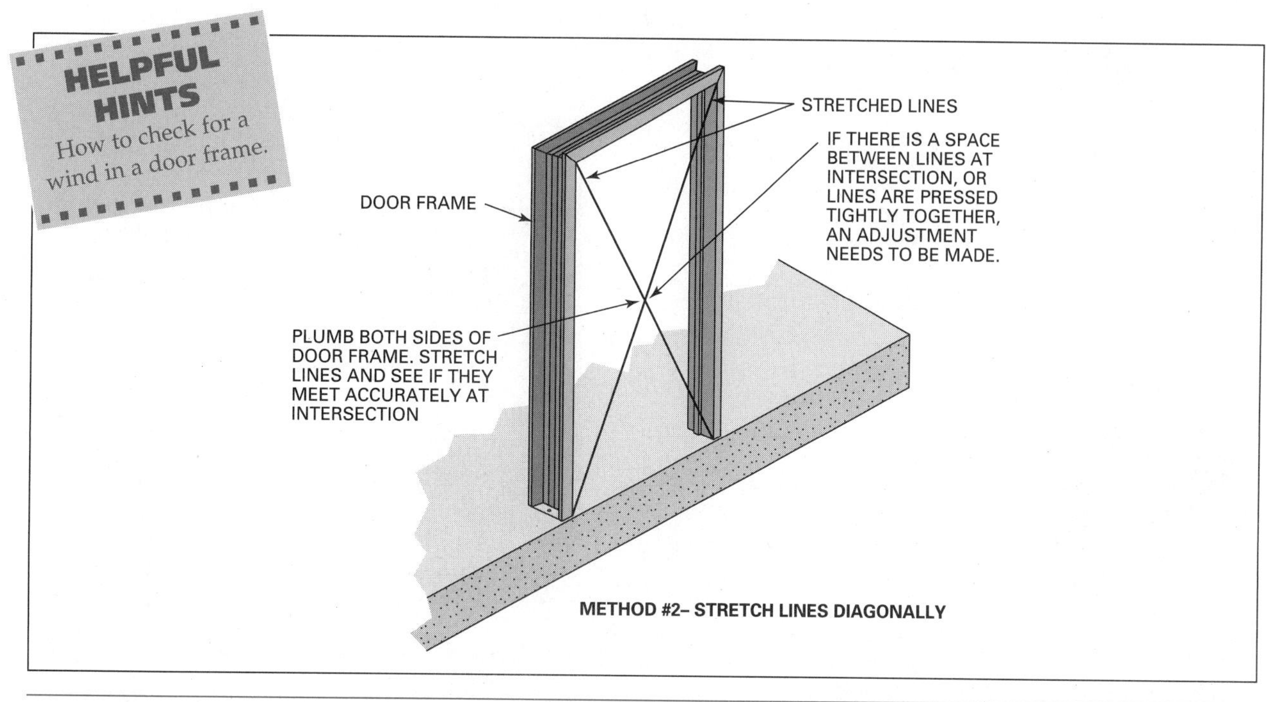

Fig. 57-21 Another technique for checking for a wind or twist in a door frame.

CHAPTER 58 DOOR FITTING AND HANGING

Fitting and hanging of wood doors is still an important part of the carpentry trade despite the increasing use of prehung door units. There are many situations in new construction and in remodeling work that require fitting and hanging doors on the job. It is helpful to be able to identify door styles, know their sizes, and understand their construction.

Door Styles and Sizes

Exterior *flush* and *panel* doors are available in many styles. Architects have many choices when designing entrances.

Flush Doors

An exterior flush door has a smooth, flat surface of wood veneer or metal. It has a framed, *solid core* of staggered wood blocks or composition board. Wood *core blocks* are inserted in appropriate locations in composition cores. They serve as *backing* for door locks (Fig. 58-1). Openings may be cut in flush doors either in the factory or on the job. Lights of various kinds and shapes are installed in them. Molding of various shapes may be applied in many designs to make the door more attractive.

Panel Doors

Panel doors are classified by one large manufacturer as *high-style, panel, sash, fire, insulated, French, Dutch,* and *ventilating* doors. *Sidelights,* although not actually doors, constitute part of some entrances. They are fixed in the door frame on one or both sides of the door (Fig. 58-2). *Transoms* are similar to sidelights. When used, they are installed above the door.

Panel Door Styles

High-style doors, as the name implies, are highly crafted designer doors. They may have a variety of cut-glass designs. **Panel doors** are made with raised panels of various shapes. Sash doors have panels of tempered or insulated glass that allow the passage of light. Fire doors are used where required by codes. These doors prevent the spread of fire for a certain period of time. Insulated doors have thicker panels with LoE or argon-filled insulated glass.

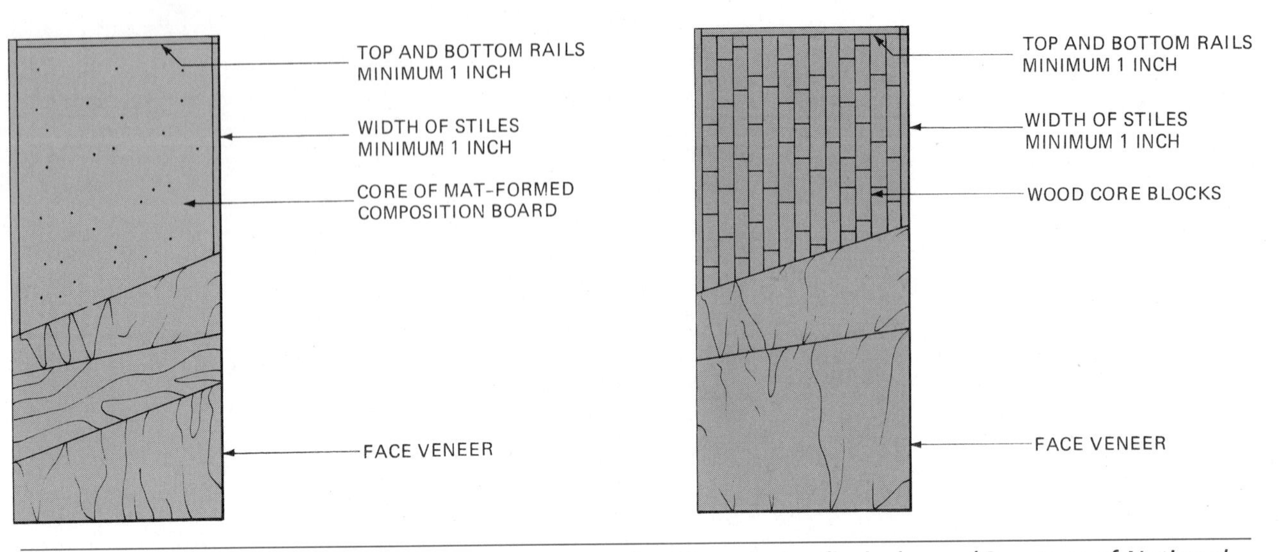

Fig. 58-1 Solid wood or composition cores are required in exterior flush doors. (*Courtesy of National Wood Window and Door Association*)

Fig. 58-2 Sidelights are installed on one or both sides of the main entrance door.

French doors may contain one, five, ten, twelve, fifteen, or eighteen lights of glass for the total width and length inside the frame. A Dutch door consists of top and bottom units, hinged independently of each other. Ventilating doors are made with operating sash and screen combinations (Fig. 58-3).

Parts of a Panel Door

A panel door consists of a frame that surrounds panels of solid wood and glass, or louvers. Some door parts are given the same terms as a window sash.

Fig. 58-3 Several kinds of exterior doors are made in many designs. (*Courtesy of Morgan Manufacturing*)

The outside vertical members are called *stiles.* Horizontal frame members are called *rails.* The *top rail* is generally the same width as the stiles. The *bottom rail* is the widest of all rails. A rail situated at lockset height, usually 38 inches from the finish floor to its center, is called the *lock rail.* Almost all other rails are called *intermediate rails. Mullions* are vertical members between rails dividing panels in a door.

The molded shape on the edges of stiles, rails, mullions, and bars, adjacent to panels, is called the **sticking**. The name is derived from the molding machine, commonly called a **sticker**, used to shape the parts. Several shapes are used to *stick* frame members.

Bars are narrow horizontal or vertical rabbeted members. They extend the total length or width of a

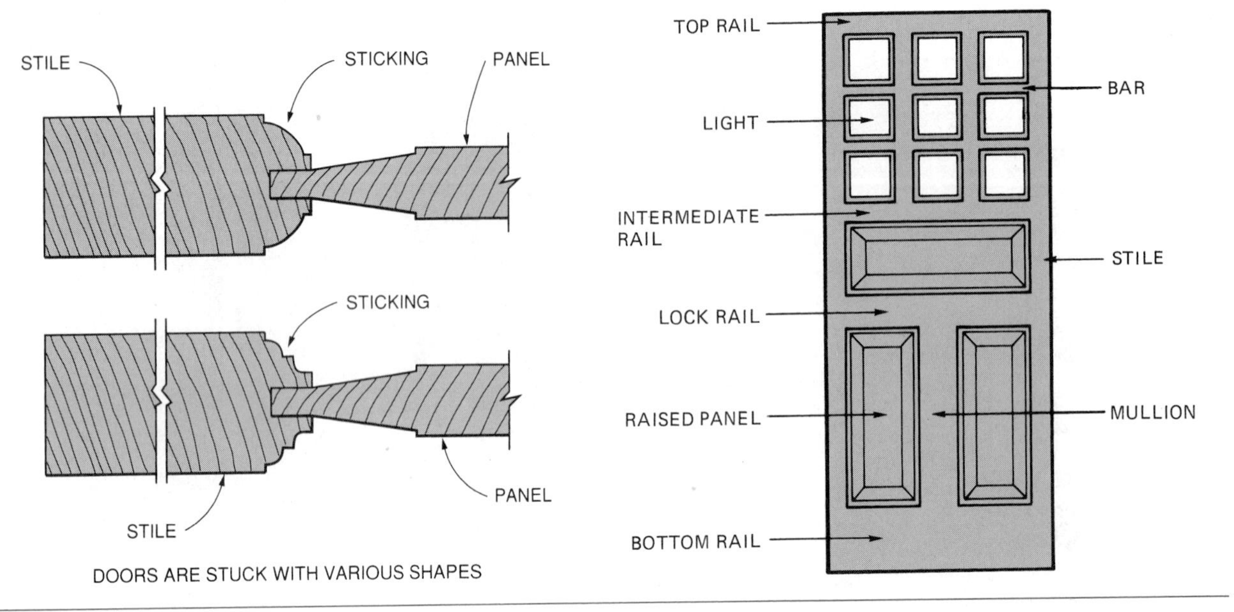

Fig. 58-4 The parts of an exterior paneled door.

glass opening from rail to rail or from stile to stile. Door *muntins* are short members, similar to and extending from bars to a stile, rail, or another bar. Bars and muntins divide the overall length and width of the glass area into smaller lights.

Panels fit between and are usually thinner than the stiles, rails, and mullions. They may be *raised* on one side or on both sides for improved appearance (Fig. 58-4).

Exterior Door Sizes

Practically all exterior entrance doors are manufactured in a thickness of 1 3/4 inches, in widths of 2′-8″ and 3′-0″, and in heights of 6′-8″ and 7′-0″. Doors that are less than 6′-8″ in height and less than 2′-8″ in width are not generally used in entrances. A few styles are produced in widths of 2′-6″ and 2′-10″ and in heights of 6′-6″, 6′-10″, and 8′-0″. Some styles, such as French, panel, and sash doors, are available in narrower widths, such as 1′-6″, 2′-0″, and 2′-4″, when double doors are used, for instance. Sidelights are made in both 1 3/8- and 1 3/4-inch thickness, and 1/2 inch greater in height than the entrance door. This allows for fitting the bottom against the sloping sill of the door frame. Auxiliary doors, such as combination storm and screen doors are made in 1 1/8-inch thickness.

There may be styles and sizes of stock doors other than those described above. Door manufacturers range from those that employ a few workers and serve only a local market to large manufacturers that ship their doors all over the world. It is not possible or desirable to describe all their products here. Study door manufacturers' literature to become better informed about door styles and sizes.

Check local building codes for entrance door requirements. Many codes specify minimum sizes, number, kind, and location of entrance doors.

Prehung Doors

Most doors come already fitted, and hinged in the door frame with the *lockset* and outside casings installed. A *prehung* door unit is set in the rough opening similar to setting door frames.

Installing a Prehung Door Unit

Working from the outside, center the unit in the opening with the casings removed. The door should be in a closed position with factory-installed spacers between door and frame still in place. Level the sill. Shim between the side jambs and the jack stud at the bottom. Tack through the side jambs and shims into the stud at the bottom using finishing nails. Plumb the side jambs. Shim at the top and tack. Shim at intermediate points along the side jamb. Fasten through the shims (Fig. 58-5).

Make sure the spacers between the door and the jamb are in place to maintain the proper joint between the door and jamb. Also make sure that the outside edge of the jamb is flush with the wall sheathing. Open the door to make sure it operates properly. Make any necessary adjustments. Drive additional nails as required. Set them all. Do not make any hammer marks on the finish. Drive the nail until it is almost flush. Then use a nail set to drive it the rest of the way (Fig. 58-6). Some installers prefer to leave the exterior casing on. They fasten the

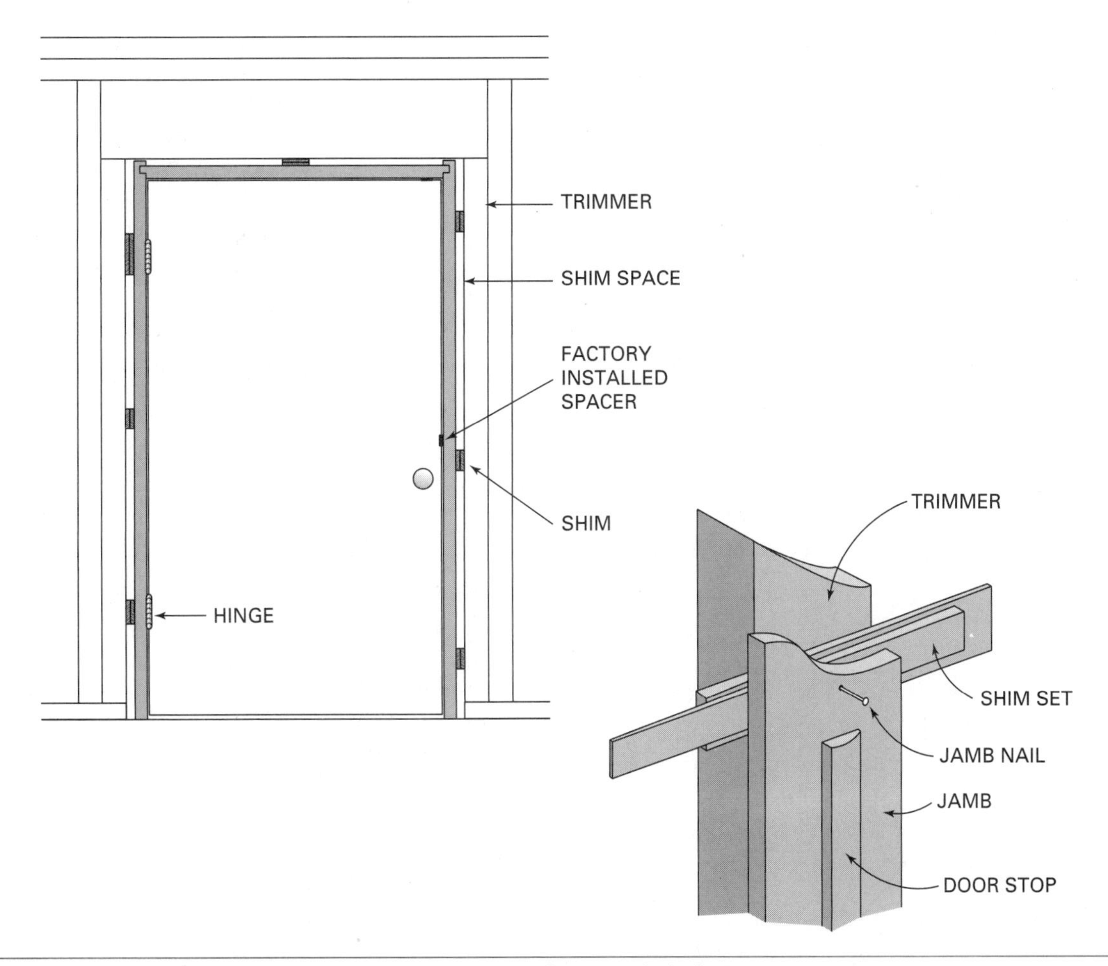

Fig. 58-5 The door remains in the prehung door frame while being set.

STEP BY STEP

PROCEDURES

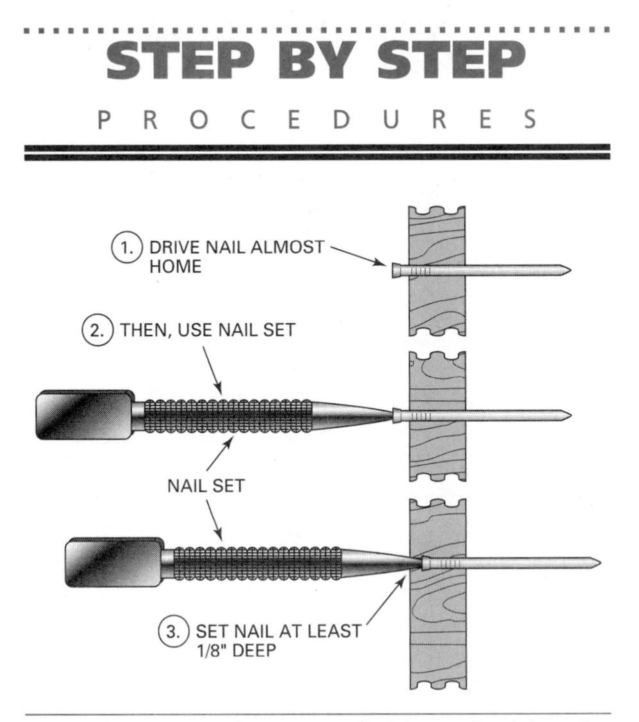

Fig. 58-6 To avoid marring finish work, drive the nail until almost home. Then use a nail set.

unit through the casing first. Then, they go inside to install the shims and additional fastening.

Fitting the Door

The first step in fitting a door is to determine the side that will close against the stops on the door frame. If the design of the door permits that either side may be used toward the stops, sight along the door stiles from top to bottom to see if the door is bowed. Hardly any doors are perfectly straight. Most are bowed to some extent.

Determining Stop Side of Door

The door should be fitted to the opening so that the hollow side of the bow will be against the door stops. Hanging a door in this manner allows the top and bottom of the closed door to come up tight against stops. The center comes up tight when the door is latched. Also, the door will not rattle.

If no attention is paid to which side the door stops against, then the reverse may happen. The door will come up against the stop at the center and away from the stop at the bottom and top (Fig. 58-7).

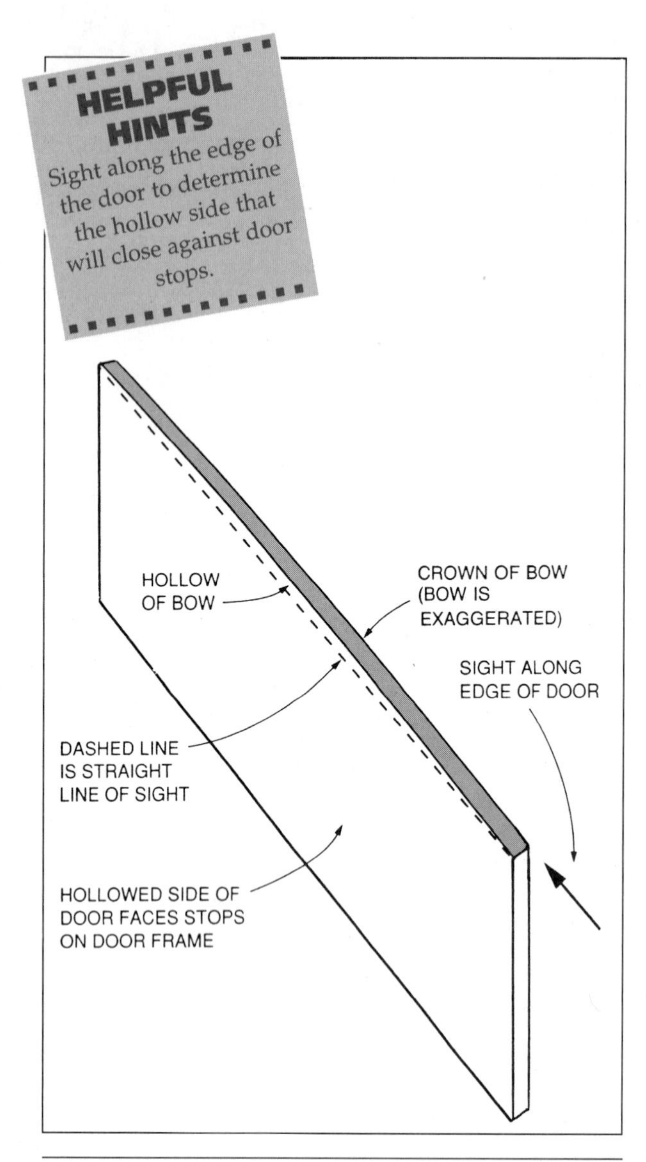

Fig. 58-7 Method of determining which side of the door will rest against the stop when the door is closed.

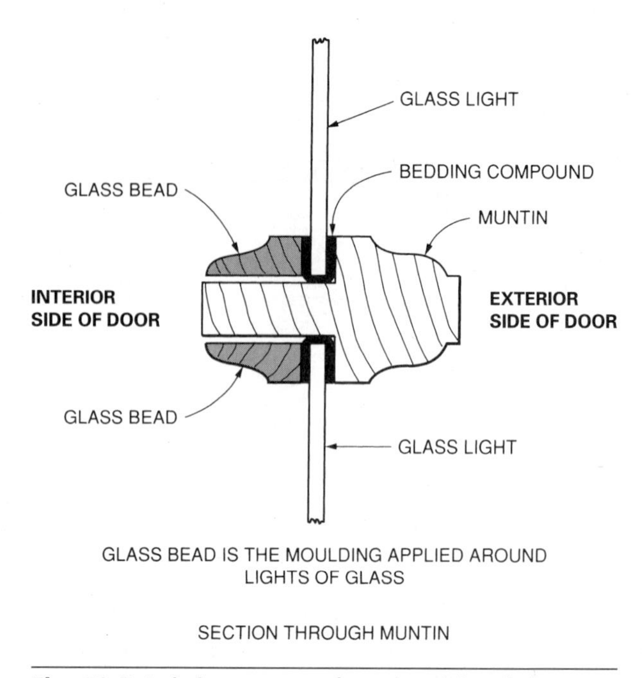

Fig. 58-8 It is important that the side of doors containing glass lights face in the direction recommended by the manufacturer.

Determining Exposed Side of Sash Doors

It is important to hang exterior doors containing lights of glass with the proper side exposed to the weather. This prevents wind-driven rainwater from seeping through joints. Manufacturers clearly indicate, with warning labels glued to the door, which side should face the exterior. Any door warranty is voided if the door is improperly hung. Do not hang exterior doors with the removable **glass bead** facing outward. Glass bead is small molding used to hold lights of glass in the opening. The bead can be identified by holes made when fasteners of the bead were set (Fig. 58-8).

Some doors are manufactured with *compression glazing*. This virtually eliminates the possibility of any water seeping through the joints. When LoE insulating glass is used in a door, it is especially important to have the door facing in the direction indicated by the manufacturer.

Determining the Swing of the Door

Determine the swing of the door from the plans. A door is designated as being *right-hand* or *left-hand,* depending on the direction it swings. The designation is determined by standing on the side of the door that swings away from the viewer. It is a right-hand door if the hinges are on the viewer's right. It is a left-hand door if the hinges are on the left (Fig. 58-9). Lightly mark the door to identify the hinge edge and the hinge pin side of the door.

Place the door on sawhorses, with its face side up. Cut any *horns* from the stiles. Horn refers to the extra stile material that extends above or below the rails. Measure carefully the width and height of the door frame. The frame should be level and plumb, but this may not be the case and should not be taken for granted.

The process of fitting a door into a frame is called *jointing. The door must be carefully jointed. An even joint* of approximately 3/32 inch must be made between the door and the frame on all sides. A wider joint of approximately 1/8 inch must be made to allow for swelling of the door and frame in extremely damp weather.

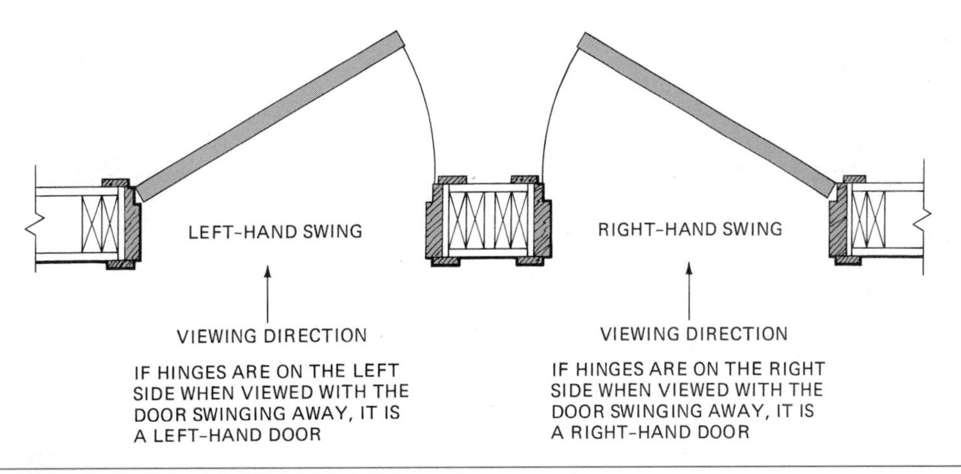

Fig. 58-9 Determining the swing of a door.

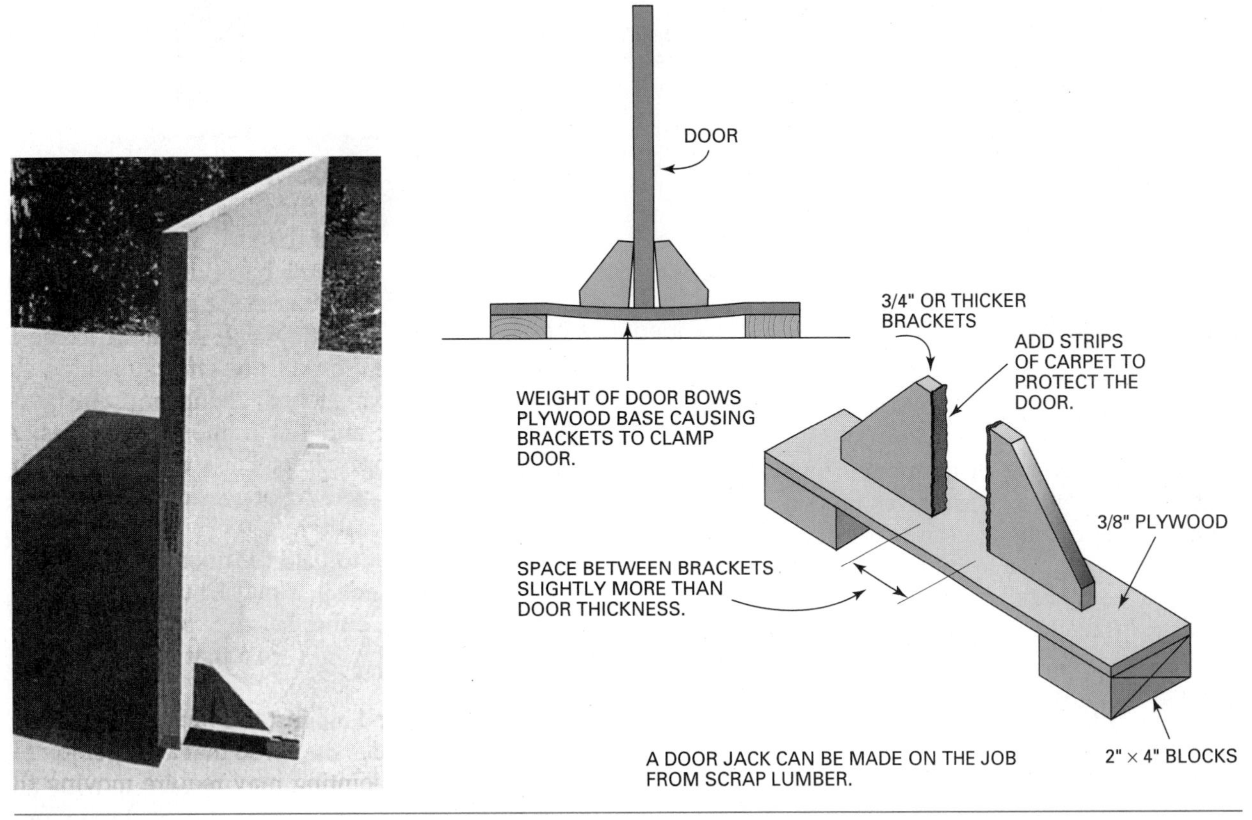

Fig. 58-10 When fitting a door, hold it in a door jack.

Use a *door jack* to hold the door steady. A manufactured or job-made jack may be used (Fig. 58-10). Use a *jointer plane,* either hand or power. Plane the edges and ends of the door so that it fits snugly in the door frame.

Fit the top end against the header jamb. Then fit the bottom against the sill so that a proper joint is obtained. Careful jointing may require moving the door in and out of the frame several times.

Next, joint the hinge edge against the side jamb. Finally, plane the lock edge so the desired joint is obtained on both sides. The lock edge must also be planed on a *bevel.* This is so the back corner does not strike the jamb when the door is opened. The bevel is determined by making the same joint between the back corner of the door and the side jamb when the door is slightly open, as when the door is closed (Fig. 58-11).

Extreme care must be taken when fitting doors not to get them undersize. The door can always be jointed a little more. However, it cannot be made wider or longer. Do not cut more than 1/2 inch total from the width of a door. Cut no more than 2 inches from its height. Cut equal amounts from

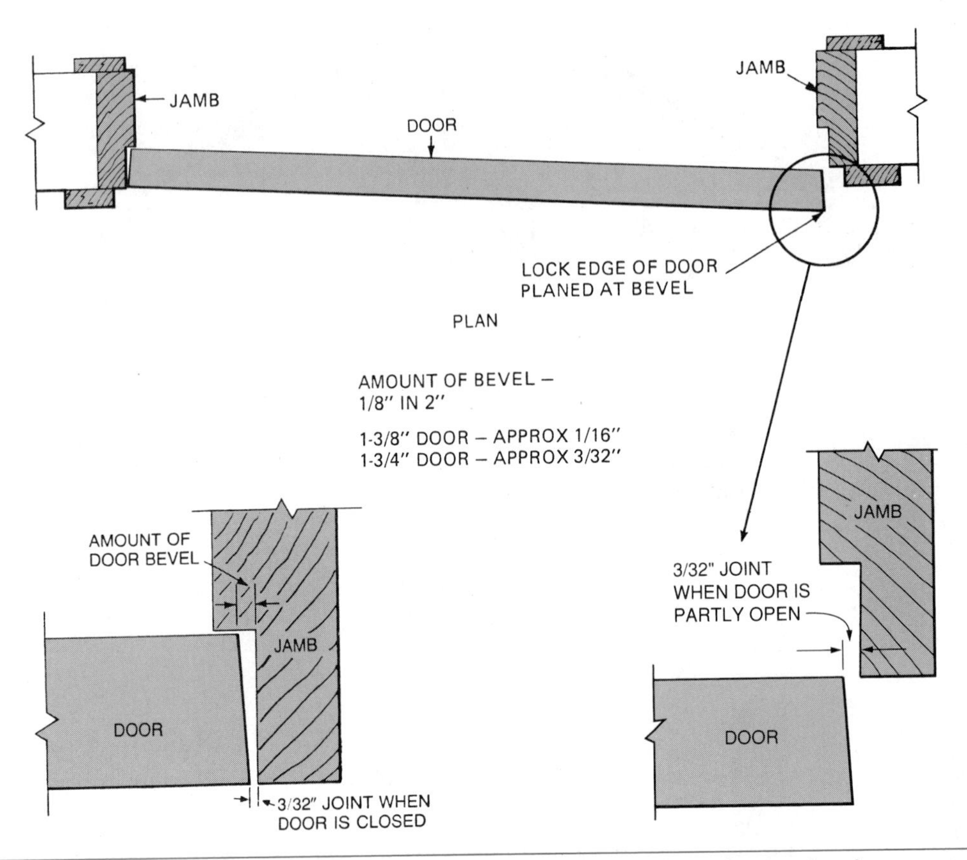

Fig. 58-11 The lock edge of a door must be planed at a bevel to clear the jamb when opened.

ends and edges when approaching maximum amounts. Check the fit frequently by placing the door in the opening, even if this takes a little extra time. Most entrance doors are very expensive. Care should be taken not to ruin one. Speed will come with practice. Handle the door carefully to avoid marring it or other finish. After the door is fitted, *ease* all sharp corners slightly with a block plane and sandpaper. To ease sharp corners means to round them over slightly.

Hanging the Door

On swinging doors, the *loose-pin* type *butt hinge* is ordinarily used. The pin is removed. Each *leaf* of the hinge is applied separately to the door and frame. The door is hung by placing it in the opening. The pins are inserted to rejoin the separated hinge leaves. Extreme care must be taken so that the hinge leaves line up exactly on the door and frame. Use three 4 × 4 hinges on 1 3/4-inch doors 7′-0″, or less in height. Use four hinges on doors over 7′-0″ in height.

The hinge leaves are recessed in flush with the door edge and only partway across. The recess for the hinge is called a *hinge gain* or *hinge mortise*. Hinge gains are only made partway across the edge of the door. This is so that the edge of the hinge is not exposed when the door is opened. The remaining distance, from the edge of the hinge to the side of the door, is called the **backset** of the hinge. Butt hinges must be wide enough so that the pin is located far enough beyond the door face to allow the door to clear the door trim when fully opened (Fig. 58-12).

Location and Size of Door Hinges

On paneled doors, the top hinge is usually placed with its upper end in line with the bottom edge of the top rail. The bottom hinge is placed with its lower end in line with the top edge of the bottom rail. The middle hinge is centered between them.

On flush doors, the usual placement of the hinge is approximately 9 inches down from the top and 13 inches up from the bottom, as measured to the center of the hinge. A middle hinge is centered between the two (Fig. 58-13).

Laying Out Hinge Locations

Place the door in the frame. Shim the top and bottom so the proper joint is obtained. Place shims between the lock edge of the door and side jamb of the frame. The hinge edge should be tightly against the door jamb.

Use a sharp knife. Mark across the door and jamb at the desired location for one end of each

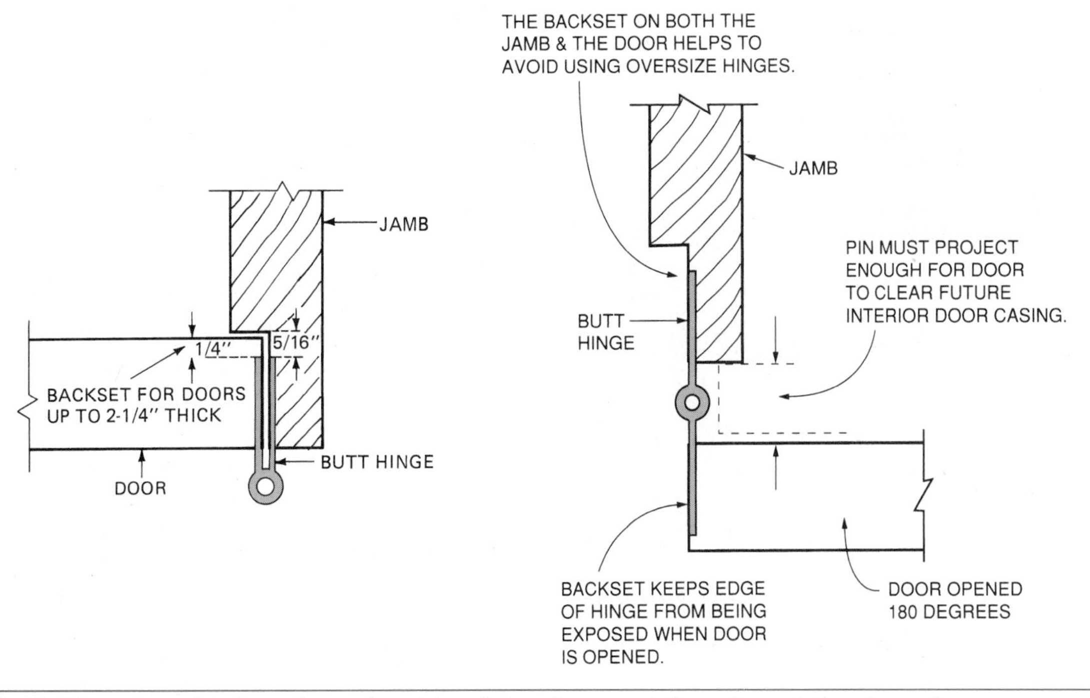

Fig. 58-12 Hinges are backset from the side of the door and the edge of the door stop.

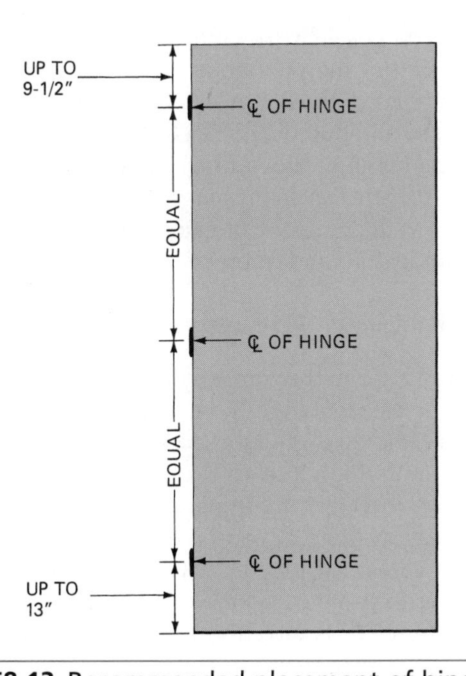

Fig. 58-13 Recommended placement of hinges on flush doors.

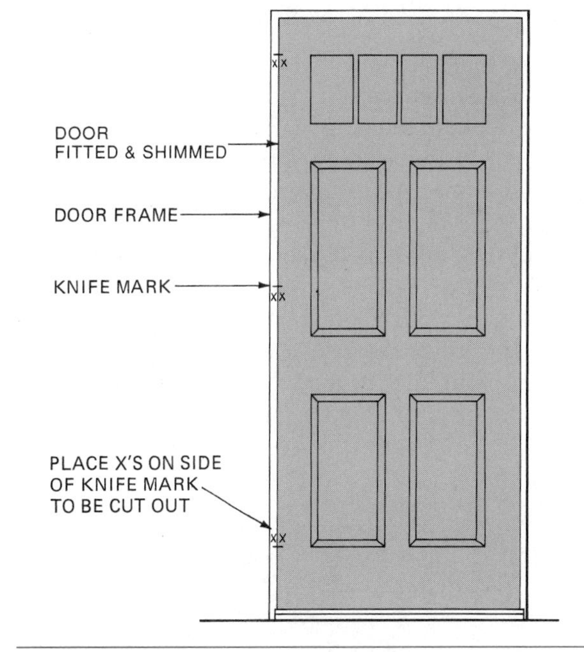

Fig. 58-14 Mark the location of hinges on the door and frame with a sharp knife.

hinge. A knife is used because it makes a finer line than a pencil. The marks on the door and jamb should not be any longer than the hinge thickness. Place a small X, with a pencil, on both the door and the jamb. This indicates on which side of the knife mark to cut the gain for the hinge (Fig. 58-14). Care must be taken to cut hinge gains on the same side of the layout line on both the door and the door frame.

Remove the door from the frame. Place it in the door jack with its hinge edge up in order to lay out and cut the hinge gains.

Laying Out the Hinge Gain

The first step in laying out a hinge gain is to mark its ends. Place a hinge leaf on the door edge with its end

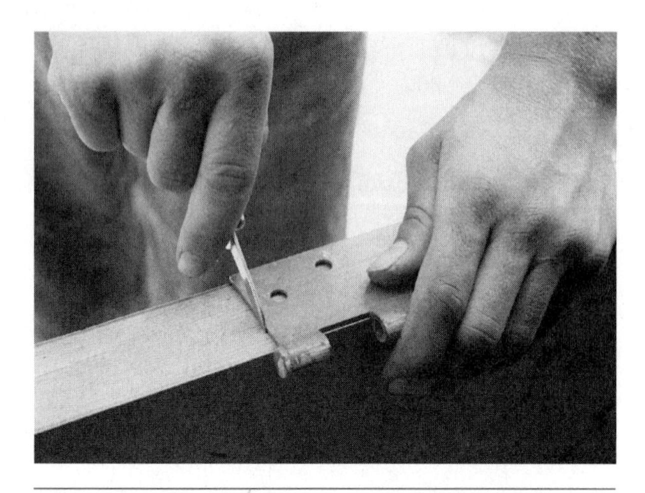

Fig. 58-15 The ends of hinge gains are laid out by scoring with a knife along a hinge leaf.

on the knife mark previously made. With the *barrel* of the leaf against the side of the door, hold the leaf firmly. Score a line along one end with a sharp knife. Then, tap the other end, until the leaf just covers the line. Score a line along the other end (Fig. 58-15). Score only partway across the door edge.

Using a Butt Gauge

The *butt gauge* contains two sliding rods that can be secured with set screws. One rod is adjusted to lay out the depth of the hinge gain. The other rod is used to lay out the backset of the hinge on both the door and door frame. Adjust this rod for the desired backset and the other rod for the depth of the hinge gain. Lay out the backset between the two end marks previously made (Fig. 58-16). Use the other rod and lay out the gain depth. Scoring lines between and not beyond the ends of the gain is a sign of a professional. Scored layout lines beyond the gain will remain exposed on the door after the gain is cut.

Fig. 58-16 Laying out the backset of the hinge gain using a butt gauge.

Cutting Hinge Gains

Use a sharp knife to deepen the scored backset line. It may be necessary to draw the knife along the line several times to score to the bottom of the gain. Do not use a chisel for scoring. Using a chisel will likely split the edge of the door (Fig. 58-17).

With the bevel of the chisel down, cut a small chip from each end of the gain. The chips will break off at the scored end marks (see Fig. 12-22). With the flat of the chisel against the shoulders of the gain, cut down to the bottom of the gain.

Make a series of small chisel cuts along the length of the gain. Brush off the chips. Then, with the flat of the chisel down, pare and smooth the excess down to the depth of the gain. Be careful not to slip and cut off the backset (Fig. 58-18).

After the gain is made, the hinge leaf should press-fit into it. It should be flush with the door edge. If the hinge leaf is above the surface, deepen the gain until the leaf lies flush. If the leaf is below the surface, it may be shimmed flush with thin pieces of cardboard from the hinge carton.

Applying Hinges

Press the hinge leaf in the gain. Drill pilot holes for the screws. Center the pilot holes carefully on the countersunk holes of the hinge leaf. A **centering punch** is often used for this purpose. Drilling off-center will cause the hinge to move from its position when the screw is driven. Fasten the hinge leaf with screws provided with the hinges. Cut gains and apply all hinge leafs on the door and frame in the same manner.

Shimming Hinges

Hang the door in the frame by inserting a pin in the barrels of the top hinge leaves first. Insert the other pins. Try the swing of the door. If the door binds against or is too close to the jamb on the hinged side, shim between the hinge leaves and the gain. Use a narrow strip of cardboard on the side of the screws nearest the pin. This will move the door toward the lock side of the door frame. If the door binds against or is too close to the jamb on the lock side, apply shims in the same manner. However, apply them on the opposite side of the hinge screws (Fig. 58-19). Check the bevel on the lock edge of the door. Plane to the proper bevel, if necessary. Ease all sharp, exposed corners.

Other Methods of Hanging Doors

Other methods are used to lay out and cut hinge gains. They may be faster and more efficient if the necessary tools and equipment are available.

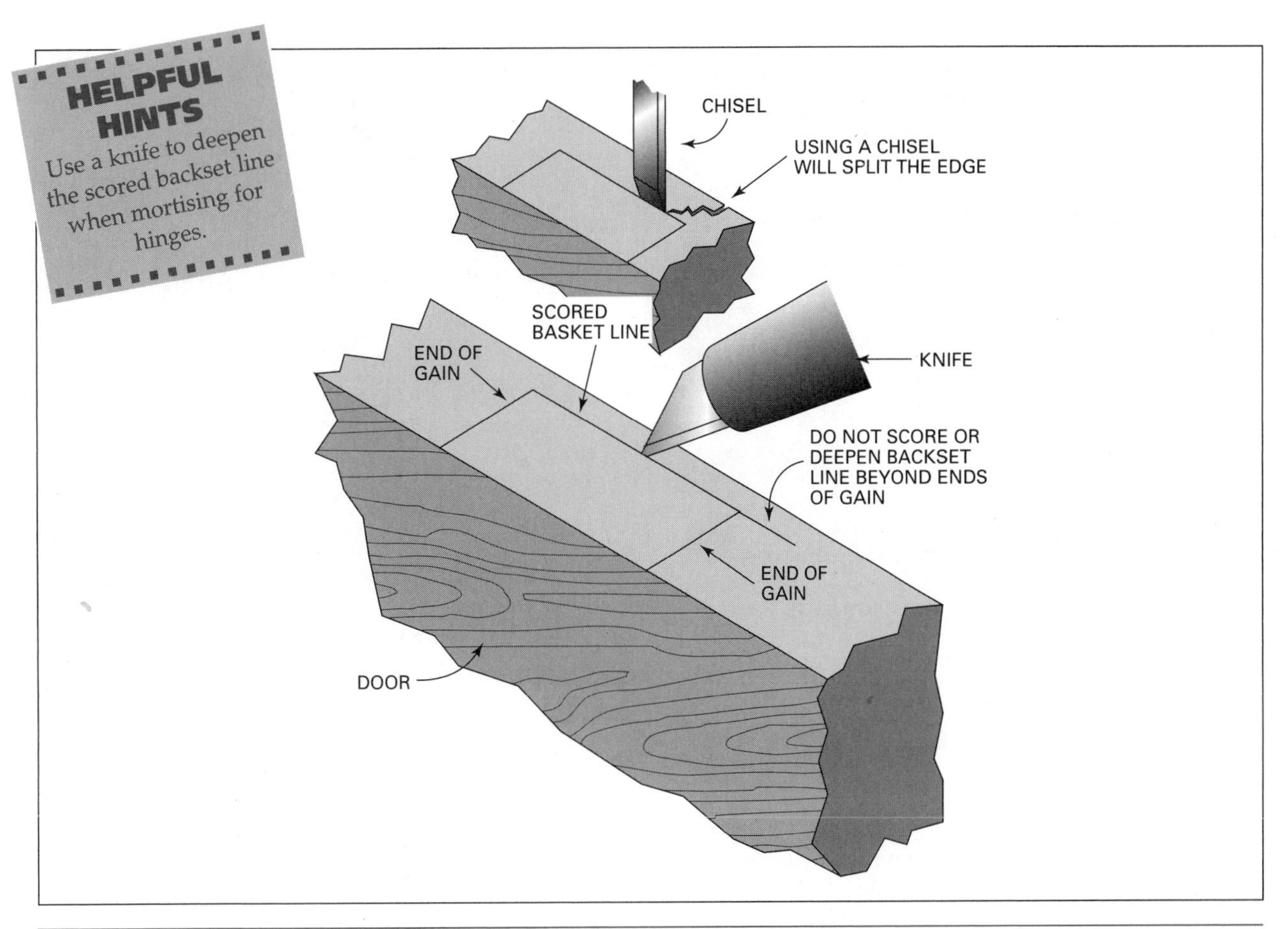

Fig. 58-17 A utility knife is useful in deepening the cut that is parallel to the grain.

Fig. 58-18 Chiseling out the hinge gain.

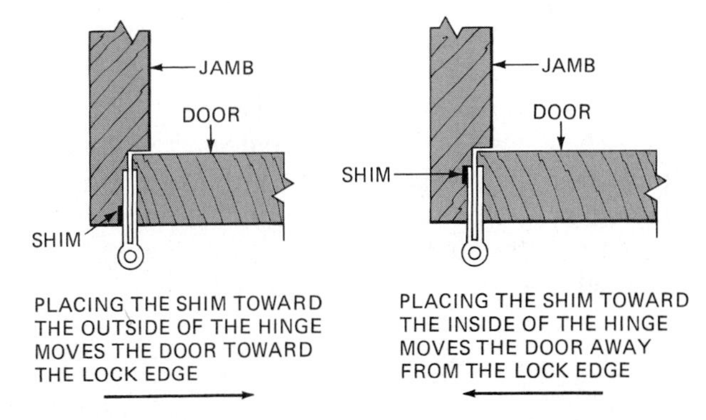

Fig. 58-19 Shimming the hinges will move the door toward or away from the lock edge.

Using Butt Hinge Markers

Instead of laying out hinge gains with a butt gauge and knife, *butt hinge markers* of several sizes are often used (see Fig. 11-26). With markers, location of the hinge leaves must still be marked on the door and jamb, as described. The width and length of the hinge gain are outlined by simply placing the marker in the proper location on the door edge. Tap it with a hammer (Fig. 58-20). However, the depth of the gain must be scored with a butt gauge or some other gauging method. The gain is then chiseled out in the manner previously described.

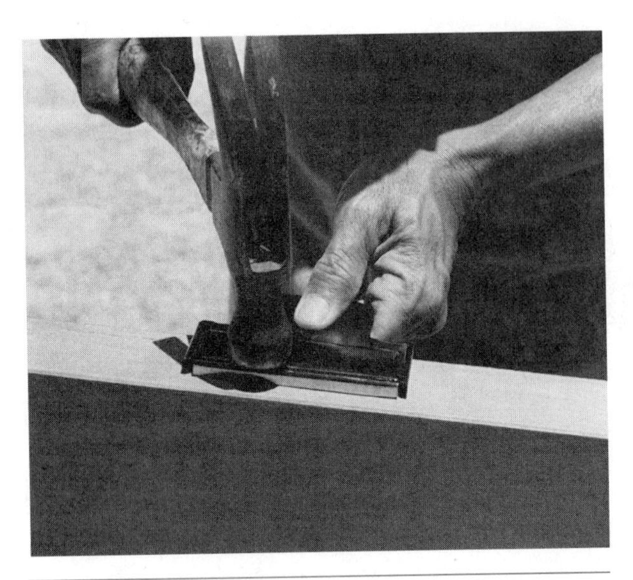

Fig. 58-20 Sometimes butt hinge markers are used to lay out hinge gains.

Fig. 58-21 A butt hinge template is used on both door and frame when routing hinge gains.

Using a Butt Hinge Template and Router

A *butt hinge template* fixture or jig and portable electric router are usually used when many doors need to be hung. Most hinge routing jigs contain three adjustable templates secured on a long rod. The templates are positioned and tacked in place with pins. The jig is used on both the door and frame as a guide to rout hinge gains. The templates are adjustable for different size hinges, hinge locations, and door thicknesses. An attachment positions the hinge template jig to provide the required joint between the top of the door and the frame.

The gains are cut using a router with a special hinge mortising bit. The bit is set to cut the gain to the required depth. A template guide is attached to the base of the router. It rides against the jig template when routing the hinge gain (Fig. 58-21).

CAUTION Care must be taken not to clip the template with a rotating router bit when lifting the router from the template. It is best to let the router come to a stop before removing it. Both the template and bit could be damaged beyond repair.

Butt hinges with rounded corners are used in routed gains. Hinges with square corners require that the rounded corner of a routed gain be chiseled to a square corner.

Other Exterior Doors

Other exterior doors, such as swinging *double doors*, are fitted and hung in a similar manner to single doors. Allowance must be made, when fitting swinging double doors, for an **astragal** between them for weathertightness (Fig. 58-22). An astragal is a **molding** that is rabbeted on both edges. It is de-

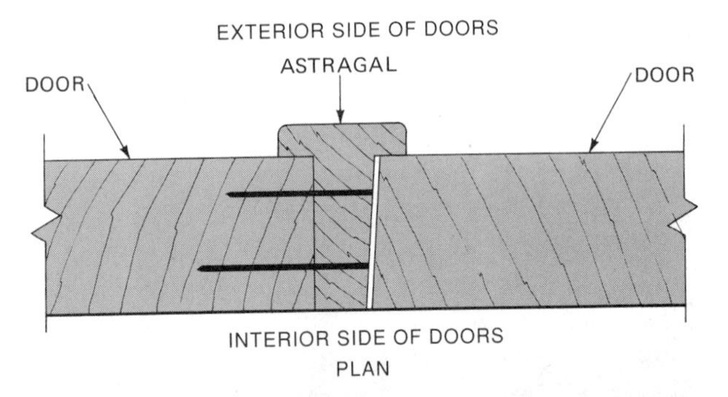

Fig. 58-22 An astragal is required between double doors for weathertightness. (*Courtesy of Andersen Corporation*)

signed to cover the joint between double doors. One edge has a square rabbet. The other has a beveled rabbet to allow for the swing of one of the doors.

Patio Doors

Patio door units normally consist of two or three sliding or swinging glass doors completely assembled in the frame (Fig. 58-23). Instructions for assembly are included with the unit. They should be followed carefully. Installation of patio door frames is similar to setting frames for swinging doors. After the frame is set, the doors are installed using special hardware supplied with the unit.

In sliding patio door units, one door is usually stationary. In two-door swinging units, either one or the other door is the swinging door. In three-door units, the center door is usually the swinging door.

Garage Doors

Overhead garage doors come in many styles, kinds, and sizes. Two popular kinds used in residential construction are the *one-piece* and the *sectional* door. The rigid one-piece unit swings out at the bottom. It then slides up and overhead. The sectional type has hinged sections. These sections move upward on rollers and turn to a horizontal position overhead. A *rolling steel door,* used mostly in commercial construction, consists of narrow metal slats that roll up on a drum installed above the opening.

Special hardware, required for all types, is supplied with the door. Equipment is available for power operation of garage doors, including remote control. Also supplied are the manufacturers' directions for installation. These should be followed carefully. There are differences in the door design and hardware of many manufacturers.

Fig. 58-23 Two or three doors usually are used in sliding- or swinging-type patio door units. (*Courtesy of Andersen Corporation*)

CHAPTER 59 INSTALLING EXTERIOR DOOR LOCKSETS

After the door has been fitted and hung in the frame, the *lockset* and other door hardware is installed. A large variety of locks are available from several large manufacturers in numerous styles and qualities, providing a wide range of choices.

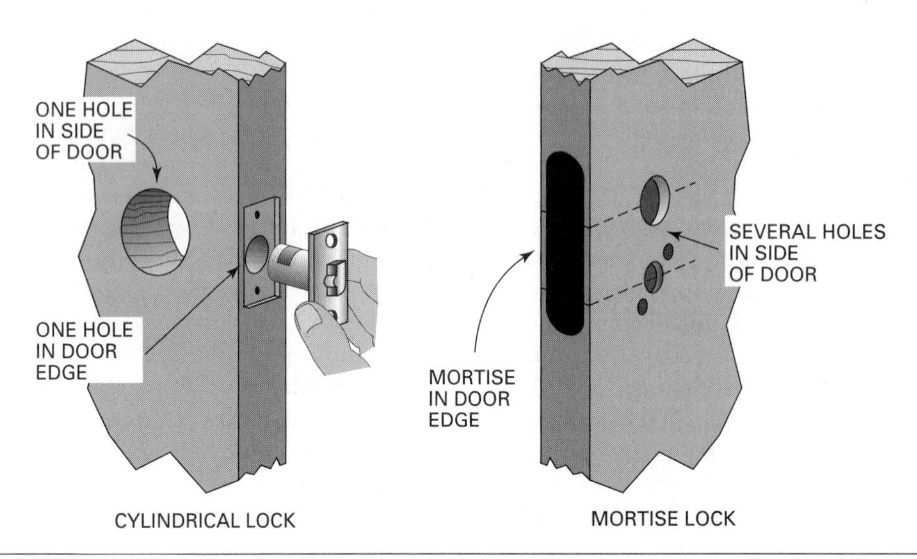

Fig. 59-1 The method of preparing the door for cylindrical locks differs from that for mortise locks.

Kinds of Locksets

Locksets can be grouped into two broad categories: *cylindrical* and *mortise* types. A distinguishing difference is the basic shape. Consequently, the preparation of the door to receive the lockset varies (Fig. 59-1).

Cylindrical Lockset

Cylindrical locksets are often called *key-in-the-knob locksets* (Fig. 59-2). They are the most commonly used type in both residential and commercial construction. This is primarily because of the ease and speed of installing them. They may be obtained in several groups, from light-duty residential to extra-heavy-duty commercial and industrial applications.

In place of knobs, *lever handles* are provided on locksets likely to be used by handicapped persons or in other situations where a lever is more suited than a knob (Fig. 59-3). **Deadbolt** *locks* are used for both primary and auxiliary locking of doors in residential and commercial buildings (Fig. 59-4). They provide additional security. They also make an attractive design in combination with grip *handle* locksets or latches (Fig. 59-5).

Fig. 59-2 Cylindrical locksets are the most commonly used type in both residential and commercial construction. (*Courtesy of Schlage Lock Co.*)

Fig. 59-3 Locks with lever handles are used when difficulty in turning knobs is expected. (*Courtesy of Schlage Lock Co.*)

Fig. 59-4 Deadbolt locks are used primarily as auxiliary locks for added security. (*Courtesy of Schlage Lock Co.*)

An *interconnected* lock combines deadbolt locking with a standard key-in-knob set (Fig. 59-6). By turning the knob on the inside both the locks are retracted for emergency exiting.

Mortise Locksets

The *mortise* lockset (Fig. 59-7) is more commonly used in commercial construction, such as hospitals, schools, office and similar buildings, where extra-heavy-duty locks of high quality are required. Mortise locks are generally not used in residential construction. This is because of the high cost of the lock and the time required to install it.

Lock Trim and Finish

The kind of trim and finish are factors, in addition to the kind and style, which determine the quality of a lockset. **Escutcheons** are decorative plates of various shapes that are installed between the door and the lock handle or knob. Locksets and escutcheons are available in various metals and finishes. More expensive locksets and trim are made of brass, bronze, or stainless steel. Less expensive ones may be of steel that is plated or coated with a finish. It is important that conditions and usage be taken into consideration when selecting locksets. This is especially the case in areas that have a humid climate or are near salt water.

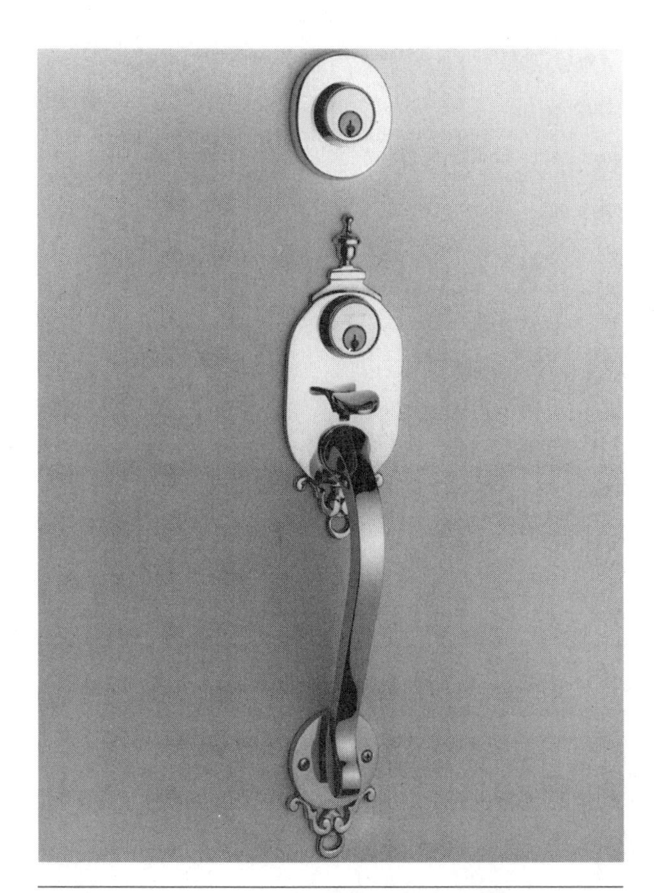

Fig. 59-5 A grip-handle lockset combines well with a deadbolt lock. (*Courtesy of Schlage Lock Co.*)

Fig. 59-6 The interconnected lock combines a deadbolt with a standard latch lockset. (*Courtesy of Schlage Lock Co.*)

Fig. 59-7 Mortise locks are generally used in commercial construction. (*Courtesy of Schlage Lock Co.*)

Installing Cylindrical Locksets

To install a cylindrical lockset, first check the contents and read the manufacturer's directions carefully (Fig. 59-8). There are so many kinds of locks

HOW TO INSTALL THIS A SERIES LOCK

MANUFACTURED BY SCHLAGE LOCK COMPANY

Part of worldwide Ingersoll-Rand

1. MARK DOOR

Mark height line (center line of latchbolt) on edge of door. Suggested height from floor 38″. Mark center point of door thickness. Position center line of template on height line. Hold in place and mark center point for 2⅛″ hole.

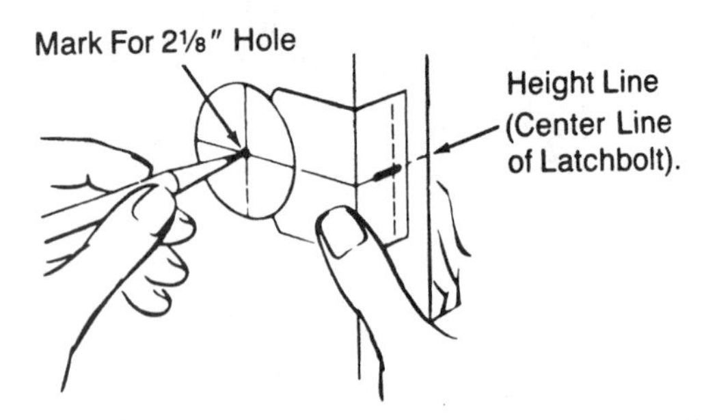

Note: "A" series locks are normally extended for doors as thick as 1⅞″. For doors over 1⅞″ thick, mark hole for latch ¾″ from outside face of door.

2. BORE TWO HOLES

Bore a 2⅛″ hole from both sides of door (to avoid splintering or otherwise damaging door).

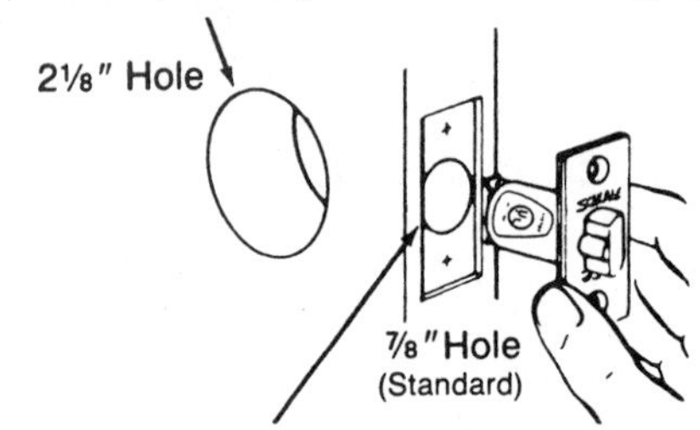

Bore a ⅞″ or 1″ hole, depending on latch housing diameter, straight into door edge to intersect with center of 2⅛″ hole.

Use latchfront faceplate to pattern for cutout. Front should fit flush with door surface. Install latch with screws provided.

2a. CIRCULAR LATCH INSTALLATION

For circular latch installation. Drill 1″ diameter latchbolt hole.

Insert latch partially into latchbolt hole. Line up beveled face of latchbolt with edge of jamb. Push latch into hole as far as it will go.

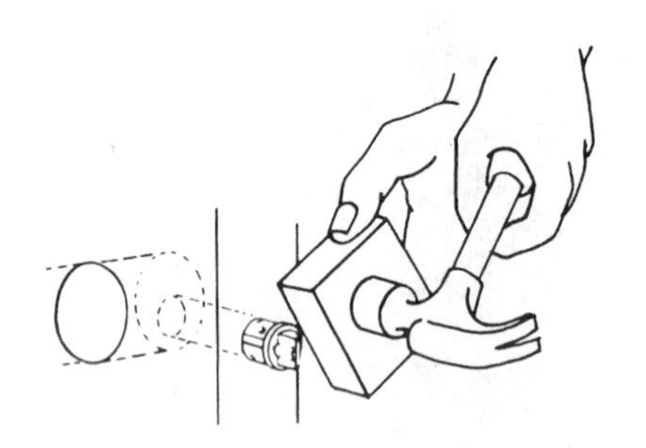

Place wooden block against bolt. Apply enough force to depress bolt. Tap block with hammer or mallet to drive latch into hole. Surface of latch faceplate should be flush with edge of door.

3. INSTALL STRIKE

Mark vertical line and height line on jamb exactly opposite center point of latch hole. Clean out hole and install strike.

Step 1. FOR T STRIKE bore two ⅞″ holes, ¹¹⁄₁₆″ deep in frame on vertical line ⅜″ above and below height line.

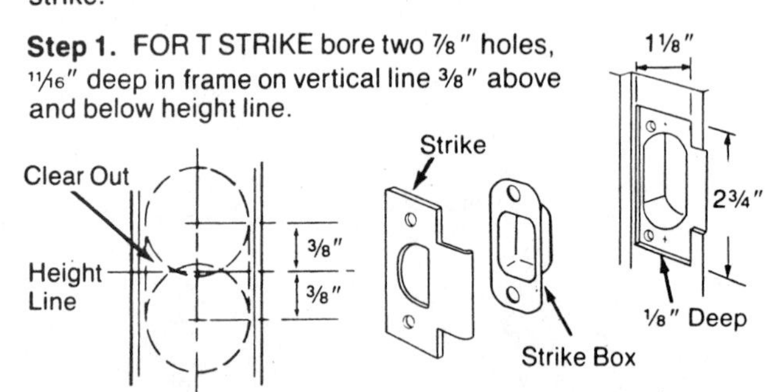

Step 2. FOR FULL LIP STRIKE mark screw holes for strike so that screws lie on same vertical center line as latch screws. Cut out frame providing for clearance of latch bolt and strike tongue and install strike.

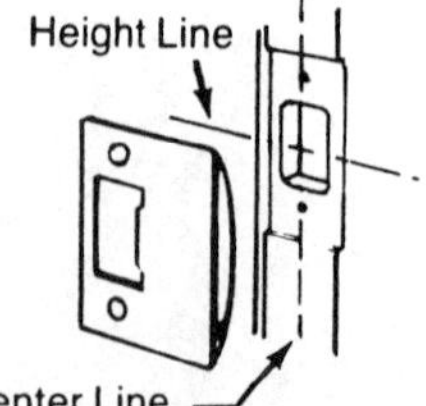

4. REMOVE INSIDE TRIM

Depress knob catch, slide knob off spindle and remove appropriate rose design as described.

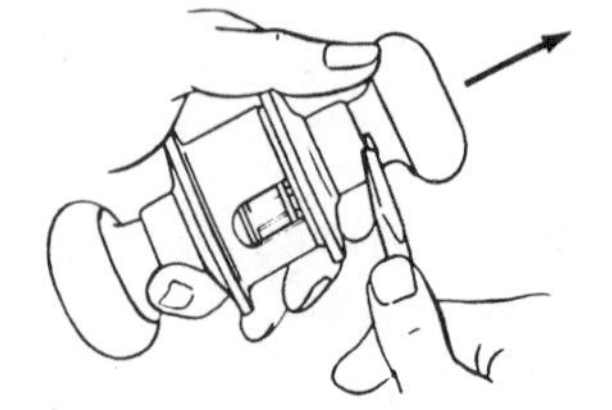

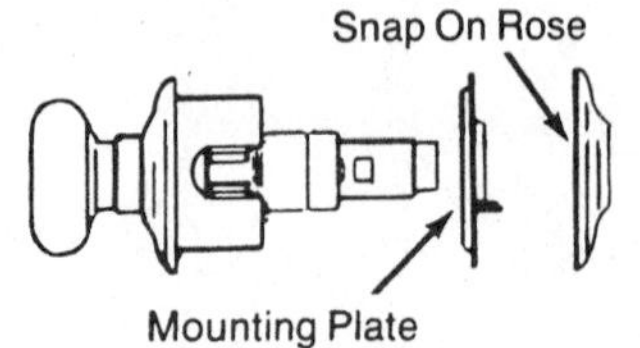

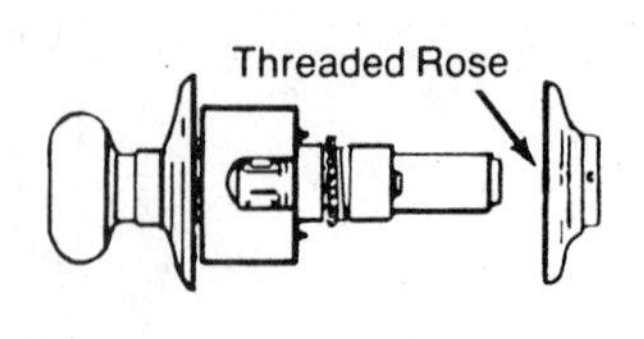

Fig. 59-8 Directions for installation are provided with the lockset. (*Courtesy of Schlage Lock Co.*)

5. ADJUST ROSE

Rotate Rose 1/16" Short of Housing for 1 3/8" Thick Door.

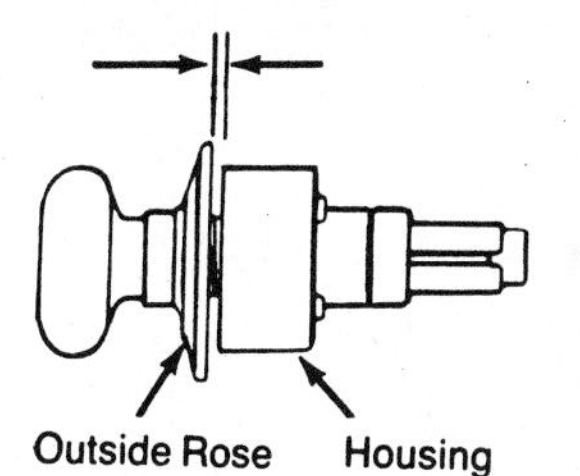

Rotate Out To 3/16" For 1 7/8" Door. This Is The Maximum Adjustment.

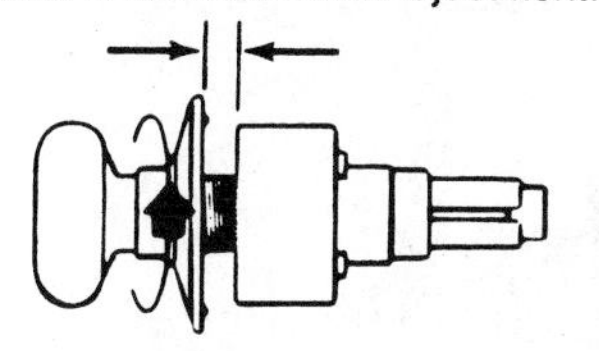

6. INTERLOCK UNITS

Latch unit must be in place before installing lock. Be sure lock housing engages with latch prongs and retractor interlocks with latch bar.

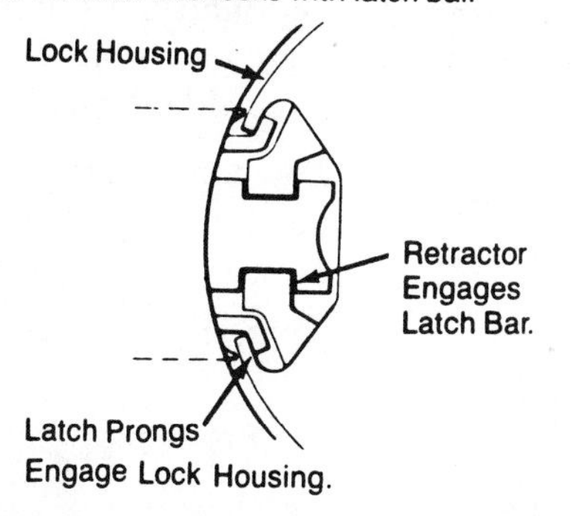

For proper installation, deadlocking plunger on latchbolt must stop against strike, preventing forcing when door is closed.

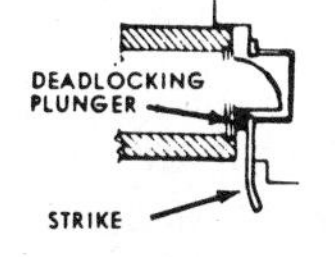

Caution—Do Not Attempt to Mount Lock Unit with Door Closed.

7. ATTACH TRIM & ROSE

Step 1: SNAP ON ROSE. Slip mounting plate over spindle and fasten securely with two machine screws. Snap rose over spring clip on mounting plate.

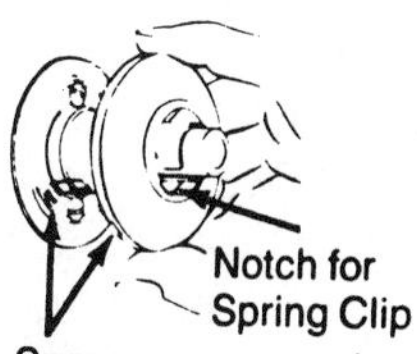

Snap rose over spring clip slot to remove rose

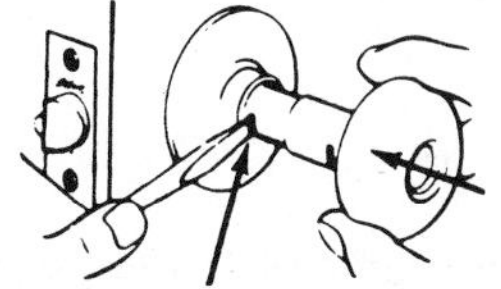

Depress Knob Catch Slide knob all the way onto spindle so catch engages into slot.

Step 2: THREADED ROSE
Slip rose over spindle and screw onto threaded hub. Turn clockwise and tighten with spanner wrench

8. TO CHANGE LOCK HAND

Pin tumbler cylinders are factory assembled in knobs for right or left hand doors as ordered. If necessary to change the hand of a lock so that cylinder is right side up, see following instructions.

8a. STANDARD CYLINDER

Insert pointed end of spanner wrench, or small nail, in the outside knob sleeve on side facing latchbolt. Exert pressure and turn key slowly until knob catch depresses; then pull off knob. Turn knob over. With key partially removed from cylinder, replace knob onto spindle. Slide knob up to knob catch. Insert key and turn key one-quarter turn to right, depress knob catch and push knob into position.

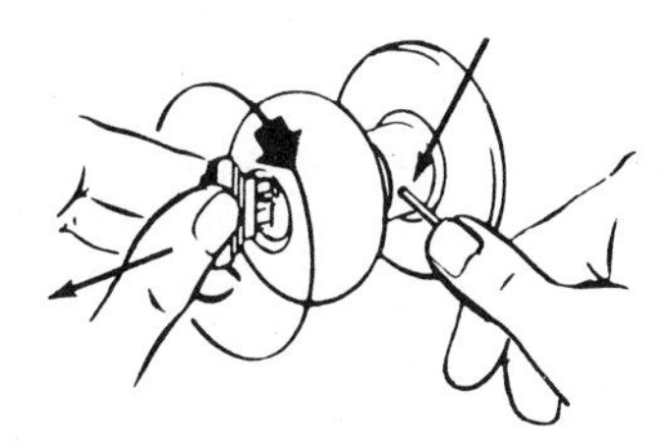

8b. INTERCHANGEABLE CORE CYLINDER

Follow steps in 8A for removing knob. CORE MUST NOW BE REMOVED FROM KNOB BEFORE REPLACING ON LOCK. Insert Control key and turn 15° to the right until action stops. Pulling on key will now extract core from knob.

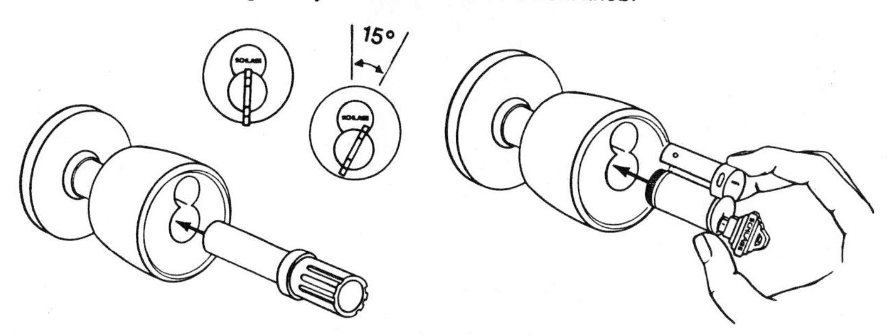

Turn knob over and replace knob on spindle. Slide knob up to knob catch. Insert long end of installation tool into knob until it engages driver. Turn tool one-quarter turn to right, depress knob catch and push knob into position. With control key fully inserted, push core into knob housing, rotate key 15° to the right and insert core completely. Rotate key back to upright and remove control key.

SCHLAGE
Part of worldwide Ingersoll-Rand

P509-403 **Rev. 11/90**

Fig. 59-8 (*continued*)

manufactured that the mechanisms vary greatly. The directions included with the lockset must be followed carefully.

However, there are certain basic procedures. Open the door to a convenient position. Wedge the bottom to hold it in place (Fig. 59-9). Measure up, from the floor, the recommended distance to the centerline of the lock. This is usually 36 to 40 inches. At this height, square a light line across the edge and stile of the door.

Marking and Boring Holes

Position the center of the paper template supplied with the lock on the squared lines. Lay out the centers of the holes that need to be bored (Fig. 59-10). It is important that the template be folded over the high corner of the beveled door edge. The distance from the door edge to the center of the hole through the side of the door is called the *backset* of the lock. Usual backsets are 2 3/8 inches for residential and 2 3/4 inches for commercial. Make sure the backset is marked correctly before boring the hole. One hole must be bored through the side and one into the edge of the door. The manufacturers' directions specify the hole sizes.

The hole through the side of the door should be bored first. Stock for the center of the boring bit is lost if the hole in the edge of the door is bored first. The hole is usually 2 1/8 inch in diameter. It can be bored with hand tools, using an expansion bit in a bit brace. However, it is a difficult job. If using hand tools, bore from one side until only the point of the bit comes through. Then bore from the other side to avoid splintering the door.

Using a Boring Jig. A *boring jig* is frequently used. It is clamped to the door to guide power-driven *multispur* bits. With a boring jig, holes can be bored completely through the door from one

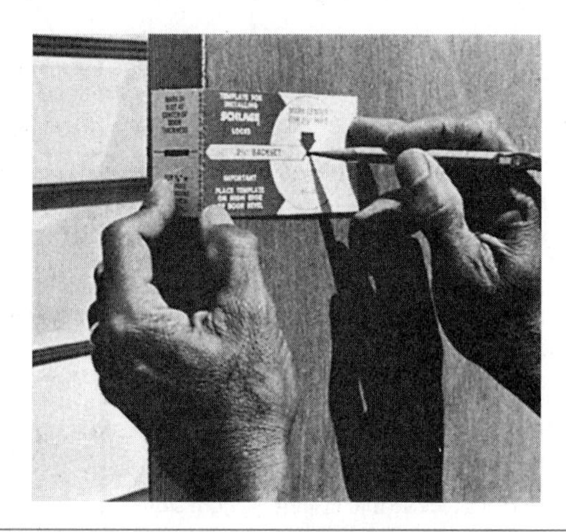

Fig. 59-10 Using a template to lay out the centers of the holes for a lockset.

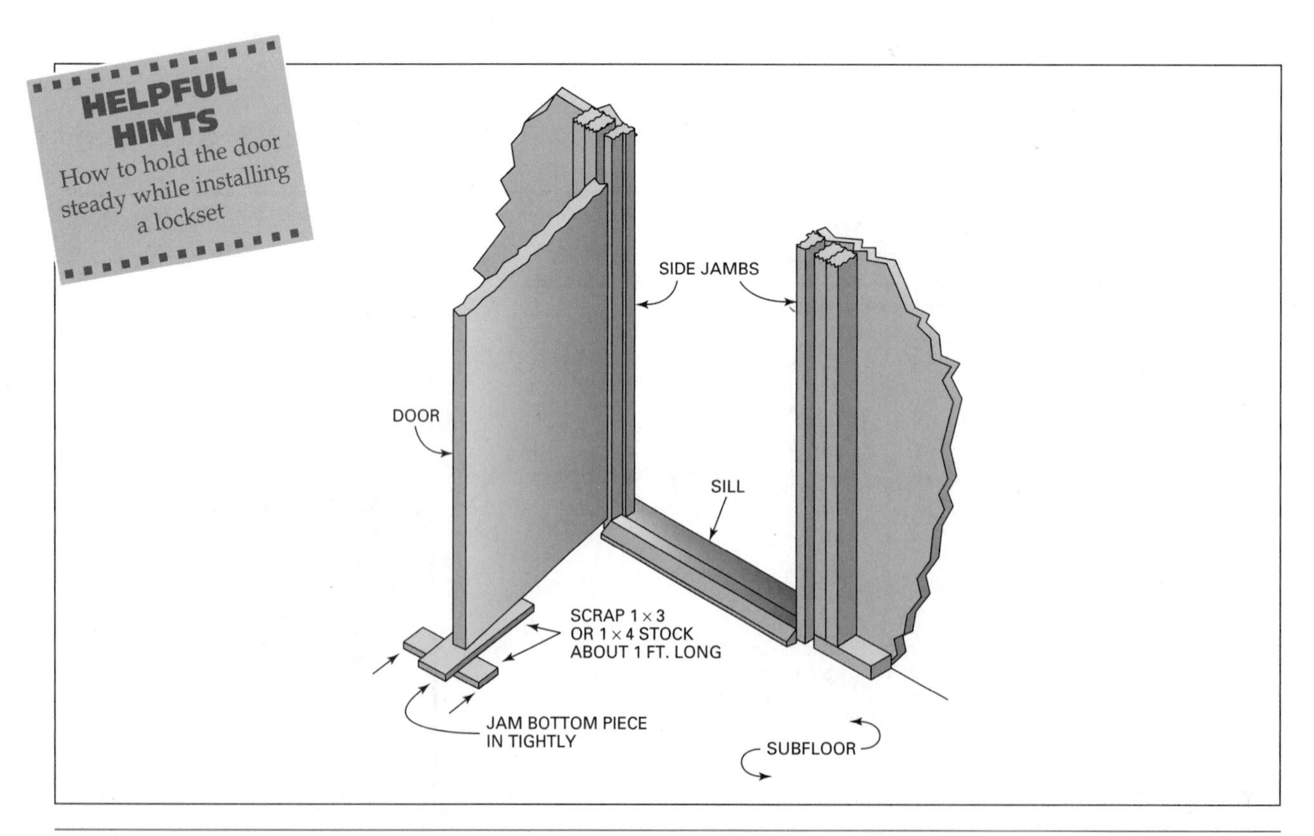

Fig. 59-9 A door may be shimmed from the floor to hold it plumb during installation.

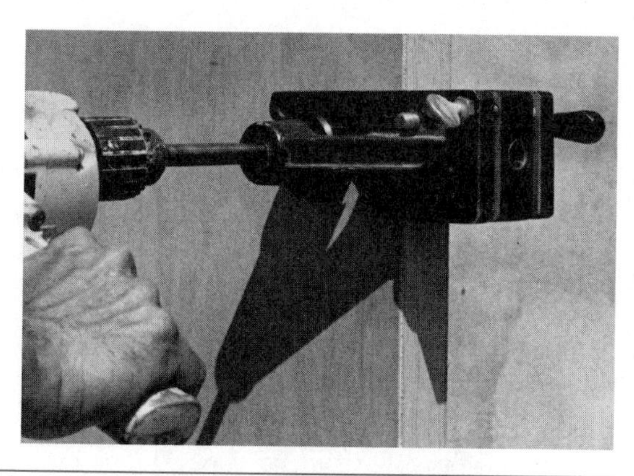

Fig. 59-11 Boring jigs are frequently used to guide bits when boring holes for locksets.

side. The clamping action of the jig prevents splintering (Fig. 59-11).

After the holes are bored, insert the latch in the hole bored in the door edge. Hold it firmly and score around its faceplate with a sharp knife. Remove the latch unit. Deepen the vertical lines with the knife in the same manner as with hinges. Do not use a chisel along these lines. This may split out the edge of the door. Then, chisel out the recess so that the faceplate of the latch lays flush with the door edge.

Faceplate markers, if available, may be used to lay out the mortise for the latch faceplate. A marker of the appropriate size is held in the bored latch hole and tapped with a hammer (Fig. 59-12). Complete the installation of the lockset by following specific manufacturers' directions.

Installing the Striker Plate

The *striker plate* is installed on the door jamb so when the door is closed it latches tightly with no play. If the plate is installed too far out, the door will not close tightly against the stop. It will then rattle. If the plate is installed too far in, the door will not latch.

To locate the striker plate in the correct position, place it over the latch in the door. Close the door snugly against the stops. Push the striker plate in against the latch. Draw a vertical line on the face of the plate flush with the outside face of the door (Fig. 59-13).

Open the door. Place the striker plate on the jamb. The vertical line, previously drawn on it, should be in line with the edge of the jamb. Center the plate on the latch. Hold it firmly while scoring a line around the plate with a sharp knife. Chisel out the mortise so the plate lies flush with the jamb. Screw the plate in place. Chisel out the center to receive the latch.

Fig. 59-12 Using a faceplate marker.

Mortise Lock

Mortise locks are so-called because the edge of the door must be mortised to the depth of the lock. The mortise is made by boring a series of holes along the centerline of the door edge. The mortise is then shaped with a chisel to receive the lock (Fig. 59-14). A *mortising jig* can be used to guide bits when boring holes for mortises (Fig. 59-15).

After the mortise is made for the body of the lock, a gain is chiseled in the edge of the door for the faceplate. When mortise locks are installed, some of the holes in the side of the door are not bored completely through. Care must be taken to read the manufacturers' directions carefully. A door installation would be unacceptable if a hole was bored through when it was supposed to be bored only partway through.

After the door is fitted, hung, and locked, remove all hardware. Prime the door and all exposed parts of the door frame. Replace the door and hardware after the prime coat is dry.

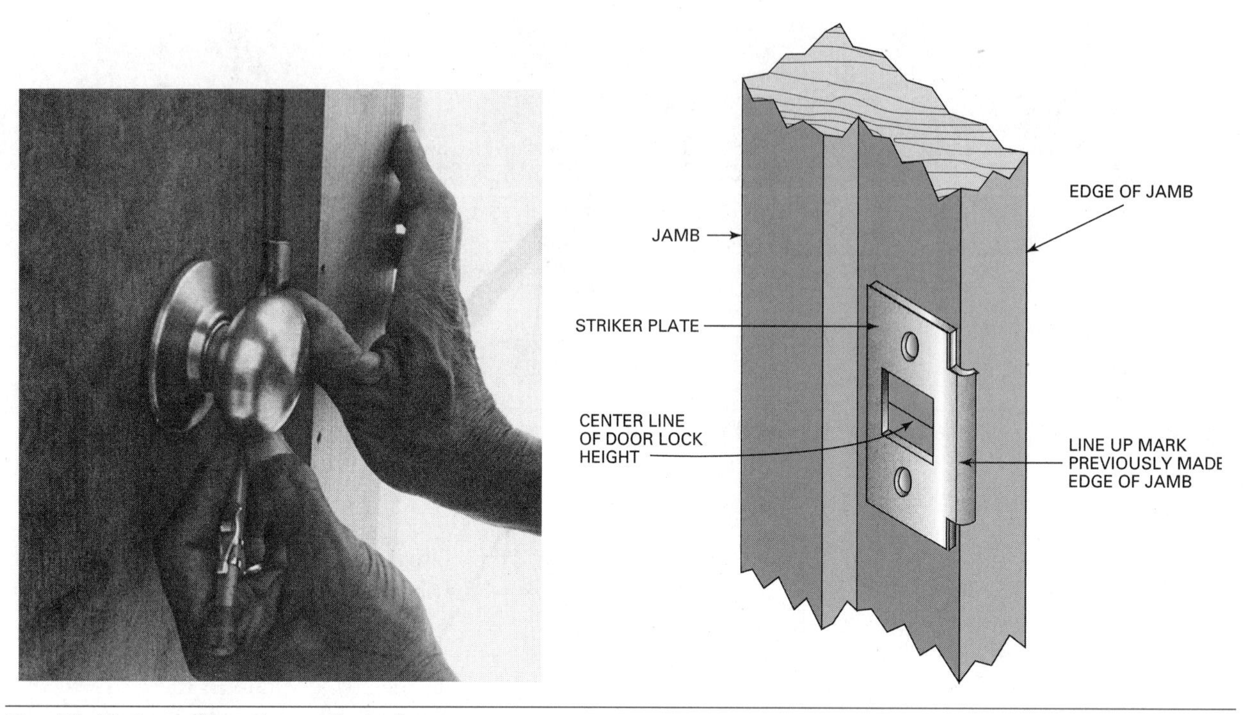

Fig. 59-13 Installing the striker plate.

Fig. 59-14 A series of holes is bored in the door edge to install a mortise lockset.

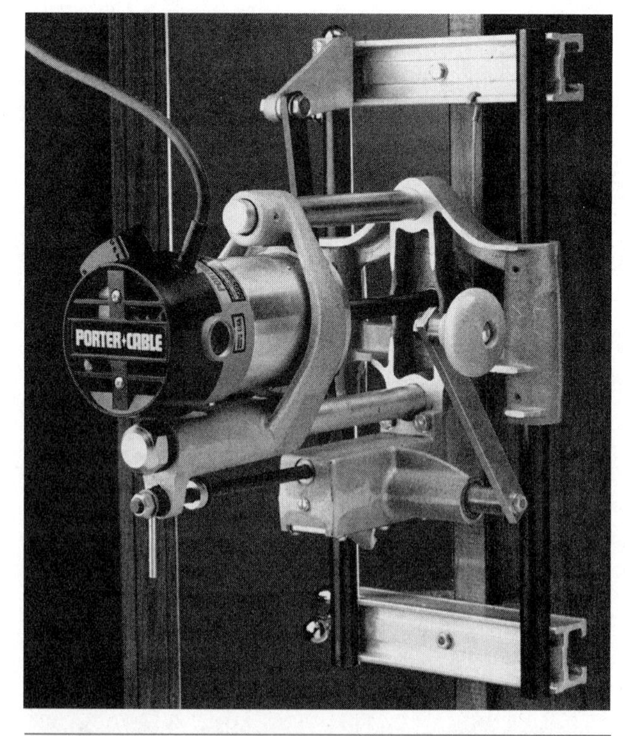

Fig. 59-15 A mortising jig can be used to bore holes when making mortises for locks. (*Courtesy of Porter Cable Co.*)

Review Questions

Select the most appropriate answer.

1. The standard thickness of exterior wood doors in residential construction is
a. 1 3/8 inches. c. 1 3/4 inches.
b. 1 1/2 inches. d. 2 1/4 inches.

2. The recommended minimum width of exterior entrance doors is
a. 3′-0″. c. 2′-6″.
b. 2′-8″. d. 2′-4″.

3. The height of exterior entrance doors in residential construction is generally not less than
a. 7′-0″. c. 6′-8″.
b. 6′-10″. d. 6′-6″.

4. The thickness of combination storm and screen auxillary doors is usually
a. 3/4 inch. c. 1 3/8 inches.
b. 1 1/8 inches. d. 1 1/2 inches.

5. A narrow member dividing the glass in a door into smaller lights is called a
a. bar. c. rail.
b. mullion. d. stile.

6. The setback of the door casing on the edge of the door jamb is called a
a. gain. c. rabbet.
b. backset. d. reveal.

7. Projections of the stiles of a door beyond the top and bottom rails are called
a. horns. c. casings.
b. ears. d. rabbets.

8. Before hanging a door, sight along its length for a bow. The hollow side of the bow should
a. face the outside.
b. face the door stops.
c. face the inside.
d. be straightened with a plane.

9. The joint between the door and door frame should be close to
a. 3/32 inch. c. 1/4 inch.
b. 3/64 inch. d. 3/16 inch.

10. On paneled doors, the top end of the top hinge is usually placed
a. in line with the bottom edge of the top rail of the door.
b. 7 inches down from the top end of the door.
c. not more than 10 3/4 inches down from the top end of the door.
d. in line with the top edge of an intermediate rail.

BUILDING FOR SUCCESS

Preparing for the Job Search

During the final months of your education, your attention will gradually be drawn toward finding gainful employment. Securing employment in an area related to your studies will most likely be where you concentrate your job-seeking efforts. But there may be a need to remain open to other job opportunities that might appear to be somewhat removed from your original employment plans.

For the apprentice carpenter, employment normally accompanies the education and training experience. Apprenticeship programs are one of the more secure avenues to gainful employment in the construction industry. Regardless of the avenue through which you obtained your training and education, you should feel competent and prepared for full-time construction work.

In addition to technical skills, what is needed to help students in their quests for specific jobs? Generally, most competitors will possess the necessary technical skills called for in the job. But what other skills, attitudes, and abilities will employers be looking for? How can students be sure they have the additional prerequisites an employer may be looking for?

Before the actual job search is under way, each student or graduate must do some self-assessment. Once people understand more about themselves and compare that data to the desired job, better job decisions may be made. If the position of self-employment in a construction-related business is the ultimate career goal for you, a quick review of your skills, abilities, financial status, and experience will indicate how soon that may become reality. At the very least, you will identify certain things that will have to be accomplished in a given time frame in order for you to become the owner of your own business. Personal work experience, must be obtained first, before self-employment can be considered. The amount of experience and the time it takes depend on the individual.

An assessment of each job with which you have an interest must be conducted. This will begin a process of elimination. The job's work conditions, specific technical tasks, climatic conditions, associated stresses, potential dangers, time demands, travel obligations, and potential wages will all help you target a desired job.

Included in the assessment of the job should be the satisfaction element. Usually jobs that bring minimal satisfaction, even if accompanied by larger salaries, will cause people to reconsider their decisions in time. It is commonplace today to see people change jobs for reasons such as advancement opportunities or more money. Other reasons center on increased stress and dissatisfaction.

In much of the business world today, there is an increased emphasis on developing a personal portfolio. Portfolios enable people to market their skills to a variety of businesses. This makes them more mobile in nature and able to work with more than one employer or client. This is more apt to be an avenue for the four-year college graduate, It may also be a consideration for the technician who has more management or business desires in mind. Many technically skilled workers are continuing to make themselves available to employers in the traditional way by contracting with only one employer.

As students prepare for graduation, they should further assess their personal skills and abilities. Important questions include the following.

1. Do I like to lead others or follow others' instructions? Why?
2. Do I like to make decisions regarding the planning of a project? Why?
3. Do I feel confident that I can perform the technical tasks that the job demands? Why?
4. Do I work well with others as a team player?
5. Do I feel more confident with maximum or minimal supervision and assistance?
6. Will I enjoy performing the same job tasks on a continual basis?
7. Do I understand the physical demands of the job?
8. Will I be satisfied with the starting wage, knowing my present financial obligations?
9. Can I demonstrate to my employer my capabilities and understand a good performance will be necessary for pay raise considerations?
10. Am I prepared to assume the job responsibilities?

The construction education program you are currently experiencing should use these and other questions as part of your job search orientation. Each job, once identified, can be looked at in greater depth with the assistance of the instructor or others with trade experience. The learning outcomes from personal and job assessment discussions will help point each person in the best direction for employment.

The following is an illustration of how an in-depth job assessment helped one student select a job that was related to construction but not involved in the actual building process. This young man had a desire to work with residential construction. He did not desire to be a frame or trim carpenter. He was completing a program of study in building construction at a community college. However, he was open to other ideas for employment. His interests were greater in associating with people, drafting, estimating, cabinet making, and sales. After looking closely at the retail building materials business, he decided to pursue that line of work. His construction background and education gave him the skills needed to work in the construction materials sales business. Prior to graduation, he interviewed with a major building materials company. He was offered a job in the location of his choice. In his later years as an educator in construction, he helped form a building materials program of study at a community college.

Preparing for the job search must precede the actual job search. It should be an ongoing process beginning at least, with the technical training and education that follow high school. An earlier preliminary evaluation of employment opportunities in high school would be even more beneficial for responsible decisions. Field trips throughout elementary and secondary school years have exposed many people to a variety of job opportunities. Individuals that can offer assistance in preparing for a job search are parents, teachers, trade representatives, counselors, clergy, and members of technical advisory teams. By seeking the assistance of people who understand various types of construction-related jobs, each student can better decide which job will be appealing. The avenue of education and training pursued by each person can be a result of spending time with one or all of those people.

FOCUS QUESTIONS: For individual or group discussion

1. Conduct a self-analysis by answering the questions stated in this study.
2. Record your responses to the self-analysis and the job analysis for guidance purposes.
3. Make a list of the potential job opportunities that appear to be worthy of your further research. Narrow this list down to a few that surface as the best for you.
4. With the assistance of your instructor, begin the job search.

UNIT

21

Siding and Cornice Construction

The exterior finish work is the major visible part of the architectural design of a building. Because the exterior is so prominent, it is important that all finish parts be installed straight and true with well-fitted joints.

The portion of the finish that covers the vertical area of a building is the siding. Siding does not include masonry covering, such as stucco or brick veneer. Siding is used extensively in both residential and commercial construction.

That area where the lower portion of the roof, or eaves, overhangs the walls is called the cornice. Variations in cornice design and detail can set the appearance of one building apart from another.

OBJECTIVES

After completing this unit, the student should be able to:

- describe the shapes, sizes, and grades of various siding products.
- install corner boards and prepare sidewalls for siding.
- apply horizontal and vertical lumber siding.
- apply plywood and hardboard panel and lap siding.
- apply wood shingles and shakes to sidewalls.
- apply aluminum and vinyl siding.
- estimate required amounts of siding.
- describe various types of cornices and name their parts.
- install gutters and downspouts.

UNIT CONTENTS

CHAPTER 60 WOOD SIDING TYPES AND SIZES

Siding is manufactured from *solid lumber, plywood, hardboard, aluminum, concrete,* and *vinyl.* It comes in many different patterns. Prefinished types eliminate the need to refinish for many years, if at all. Siding may be applied horizontally, vertically, or in other directions, to make many interesting designs (Fig. 60-1).

Wood Siding

The natural beauty and durability of solid wood have long made it an ideal material for siding. The *Western Wood Products Association (WWPA)* and the *California Redwood Association (CRA)* are two major organizations whose member mills manufacture siding and other wood products. They have to meet standards supervised by their associations. Grade stamps of the WWPA and CRA and other associations of lumber manufacturers are placed on siding produced by member mills. Grade stamps assure the consumer of a quality product that complies with established standards (Fig. 60-2). WWPA member mills produce wood siding from species such as *fir, larch, hemlock, pine, spruce,* and *cedar.* Most *redwood* siding is produced by mills that belong to the CRA.

Wood Siding Grades

Grain. Some siding is available in *vertical grain, flat grain,* or *mixed grain.* In vertical grain siding the annual growth rings, viewed in cross-section, must form an angle of 45 degrees or more with the surface of the piece. All other lumber is classified as flat grain (Fig. 60-3). Vertical grain siding is the highest

Fig. 60-2 Association trademarks assure that the product has met established standards of quality. (*Courtesy of Western Wood Products Association; Courtesy of California Redwood Association*)

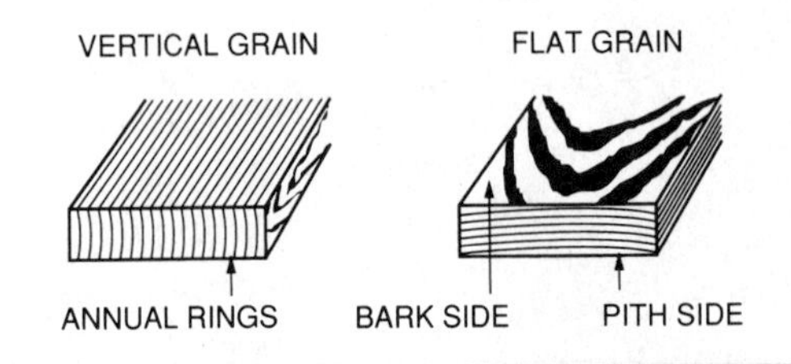

Fig. 60-3 In some species, siding is available in vertical, flat, or mixed grain. (*Courtesy of California Redwood Association*)

Fig. 60-1 Wood siding is used extensively in both residential and commercial construction. (*Courtesy of California Redwood Association*)

quality. It warps less, takes and holds finishes better, has less defects, and is easier to work.

Surface Texture. Sidings are manufactured with *smooth, rough,* or *saw-textured* surfaces. Saw-textured surface finishes are obtained by resawing in the mill. They generally hold finishes longer than smooth surfaces.

WWPA Grades. Grades published by the WWPA for siding products are shown in Figure 60-4. Siding graded as *premium* has fewer defects such as knots and pitch pockets. The highest premium grade is produced from clear, all-heart lumber. *Knotty* grade siding is divided into #1, #2, and #3 *common.* The grade depends on the type and number of knots and other defects.

CRA Grades. There are over thirty grades of redwood lumber. The best grades are grouped in a category called *architectural.* They are used for high-quality exterior and interior uses, including siding (Fig. 60-5).

Siding Patterns and Sizes

The names, descriptions, and sizes of siding patterns are shown in Figure 60-6. Some patterns can only be used for either horizontal or vertical applications. Others can be used for both. *Drop* and *tongue-and-grooved* sidings are manufactured in a variety of patterns other than shown. *Bevel* siding, more commonly known as *clapboards,* is a widely used kind (Fig. 60-7).

General Categories (Note that there are additional grades for bevel pattern)		GRADES		
		Western Species		Cedar
		Selects	Finish	Western & Canadian
ALL PATTERNS	PREMIUM GRADES	C Select D Select	Superior Prime	Clear Heart A Grade B Grade
Additional Grades for Bevel Patterns	Premium		Superior Bevel	Clear VG Heart A Bevel B Bevel Rustic C Bevel
	Knotty		Prime Bevel	Select Knotty Quality Knotty
ALL PATTERNS	KNOTTY GRADES	Commons	Alternate Boards	
		#2 Common #3 Common	Select Merch. Construction Standard	Select Knotty Quality Knotty

Fig. 60-4 Western Wood Products Association grade rules for siding products. (*Courtesy of Western Wood Products Association*)

Panel and Lap Siding

Most *panel* and *lap* siding is manufactured from plywood and hardboard. It comes in a variety of sizes, patterns, and surface textures. Some hardboard siding comes prefinished in a variety of colors. Or, it is primed for the later application of finish coats of any color.

Panel Siding

Plywood panel siding, manufactured by APA member mills, is known as *APA 303* siding. It is produced in a number of surface textures and patterns (Fig. 60-8). It comes in several thicknesses, 4-foot widths, and lengths of 8, 9, and 10 feet. It is usually applied vertically. It may also be applied horizontally.

Hardboard panel siding is available with similar surface treatments and sizes. In addition, some hardboard siding, for horizontal application only, is manufactured 12 inches wide and 16 feet long. It comes in a *drop lap* design of two 6-inch courses or three 4-inch courses (Fig. 60-9). Most panel siding is shaped with *shiplapped* edges for weathertight joints.

GRADE	DESCRIPTION
CLEAR ALL HEART	A SUPERIOR GRADE FOR FINE SIDINGS AND ARCHITECTURAL USES. IT IS ALL HEARTWOOD AND THE GRADED FACE OF EACH PIECE IS FREE OF KNOTS.
CLEAR	SIMILAR IN QUALITY TO CLEAR ALL HEART, EXCEPT THAT IT INCLUDES SAPWOOD IN VARYING AMOUNTS. SOME BOARDS MAY HAVE ONE OR TWO SMALL, TIGHT KNOTS ON THE GRADED FACE.
HEART B	AN ECONOMICAL ALL-HEART-WOOD GRADE CONTAINING A LIMITED NUMBER OF TIGHT KNOTS AND CHARACTERISTICS NOT PERMITTED IN CLEAR OR CLEAR ALL HEART. IT IS GRADED ON ONE FACE ON ONE EDGE.
B GRADE	AN ECONOMICAL GRADE CONTAINING A LIMITED NUMBER IF TIGHT KNOTS WITH SAPWOOD ACCENTING THE HEARTWOOD.

REDWOOD GRADES ARE ESTABLISHED BY THE REDWOOD INSPECTION SERVICE

Fig. 60-5 Redwood siding grades. (*Courtesy of California Redwood Association*)

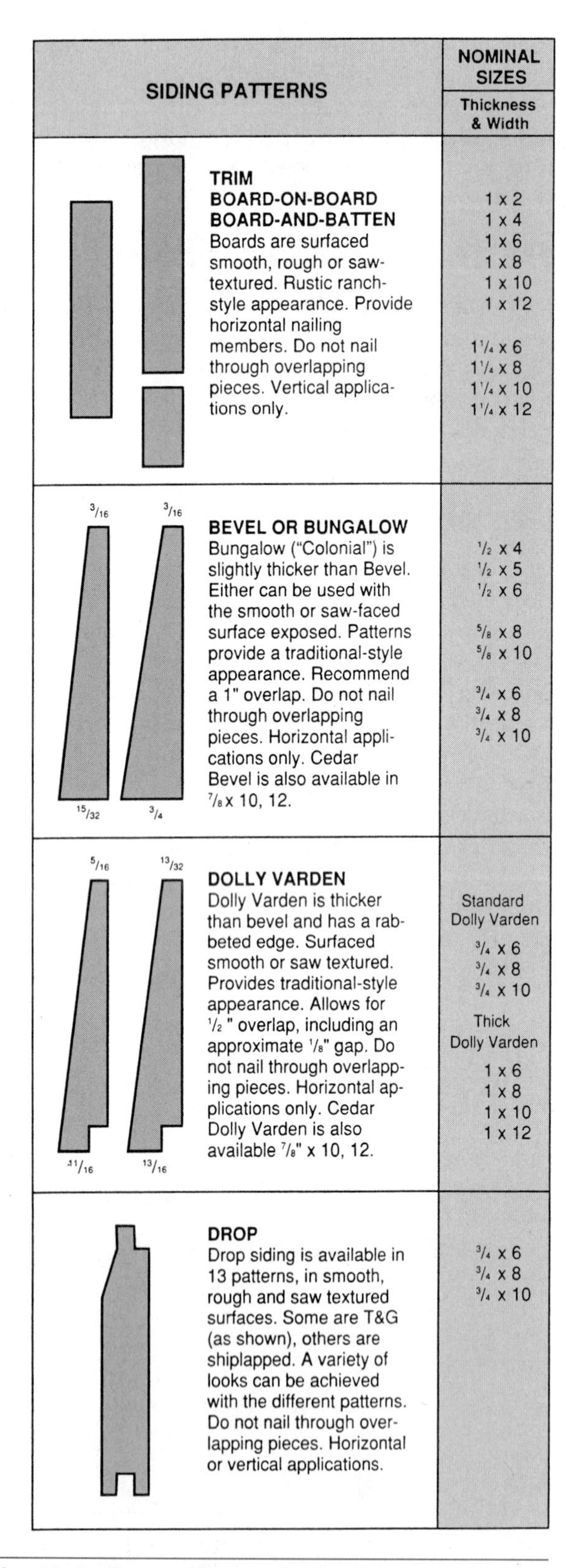

SIDING PATTERNS	NOMINAL SIZES Thickness & Width
TRIM BOARD-ON-BOARD BOARD-AND-BATTEN Boards are surfaced smooth, rough or saw-textured. Rustic ranch-style appearance. Provide horizontal nailing members. Do not nail through overlapping pieces. Vertical applications only.	1 x 2 1 x 4 1 x 6 1 x 8 1 x 10 1 x 12 $1\frac{1}{4}$ x 6 $1\frac{1}{4}$ x 8 $1\frac{1}{4}$ x 10 $1\frac{1}{4}$ x 12
BEVEL OR BUNGALOW Bungalow ("Colonial") is slightly thicker than Bevel. Either can be used with the smooth or saw-faced surface exposed. Patterns provide a traditional-style appearance. Recommend a 1" overlap. Do not nail through overlapping pieces. Horizontal applications only. Cedar Bevel is also available in $\frac{7}{8}$x 10, 12.	$\frac{1}{2}$ x 4 $\frac{1}{2}$ x 5 $\frac{1}{2}$ x 6 $\frac{5}{8}$ x 8 $\frac{5}{8}$ x 10 $\frac{3}{4}$ x 6 $\frac{3}{4}$ x 8 $\frac{3}{4}$ x 10
DOLLY VARDEN Dolly Varden is thicker than bevel and has a rabbeted edge. Surfaced smooth or saw textured. Provides traditional-style appearance. Allows for $\frac{1}{2}$" overlap, including an approximate $\frac{1}{8}$" gap. Do not nail through overlapping pieces. Horizontal applications only. Cedar Dolly Varden is also available $\frac{7}{8}$" x 10, 12.	Standard Dolly Varden $\frac{3}{4}$ x 6 $\frac{3}{4}$ x 8 $\frac{3}{4}$ x 10 Thick Dolly Varden 1 x 6 1 x 8 1 x 10 1 x 12
DROP Drop siding is available in 13 patterns, in smooth, rough and saw textured surfaces. Some are T&G (as shown), others are shiplapped. A variety of looks can be achieved with the different patterns. Do not nail through overlapping pieces. Horizontal or vertical applications.	$\frac{3}{4}$ x 6 $\frac{3}{4}$ x 8 $\frac{3}{4}$ x 10
TONGUE & GROOVE Tongue & groove siding is available in a variety of patterns. T&G lends itself to different effects aesthetically. Sizes given here are for Plain Tongue & Groove. Do not nail through overlapping pieces. Vertical or horizontal applications.	1 x 4 1 x 6 1 x 8 1 x 10 Note: T&G patterns may be ordered with $\frac{1}{4}$, $\frac{3}{8}$ or $\frac{7}{16}$" tongues. For wider widths, specify the longer tongue and pattern.
CHANNEL RUSTIC Channel Rustic has $\frac{1}{2}$" overlap (including an approximate $\frac{1}{8}$" gap) and a 1" to $1\frac{1}{4}$" channel when installed. The profile allows for maximum dimensional change without adversely affecting appearance in climates of highly variable moisture levels between seasons. Available smooth, rough or saw textured. Do not nail through overlapping pieces. Horizontal or vertical applications.	$\frac{3}{4}$ x 6 $\frac{3}{4}$ x 8 $\frac{3}{4}$ x 10
LOG CABIN Log Cabin siding is $1\frac{1}{2}$" thick at the thickest point. Ideally suited to informal buildings in rustic settings. The pattern may be milled from appearance grades (Commons) or dimension grades (2x material). Allows for $\frac{1}{2}$" overlap, including an approximately $\frac{1}{8}$" gap. Do not nail through overlapping pieces. Horizontal or vertical applications.	$1\frac{1}{2}$ x 6 $1\frac{1}{2}$ x 8 $1\frac{1}{2}$ x 10 $1\frac{1}{2}$ x 12

Fig. 60-6 Names, descriptions, and sizes of natural wood siding patterns. (*Courtesy of Western Wood Products Association*)

Fig. 60-7 Bevel siding is commonly known as clapboards. (*Courtesy of California Redwood Association*)

BRUSHED

Brushed or relief-grain surfaces accent the natural grain pattern to create striking textured surfaces. Generally available in 11/32", 3/8", 1/2", 19/32" and 5/8" thicknesses. Available in redwood, Douglas fir, cedar, and other species.

KERFED ROUGH-SAWN

Rough-sawn surface with narrow grooves providing a distinctive effect. Long edges shiplapped for continuous pattern. Grooves are typically 4" o.c. Also available with grooves in multiples of 2" o.c. Generally available in 11/32", 3/8", 1/2", 19/32" and 5/8" thicknesses. Depth of kerfgroove varies with panel thickness.

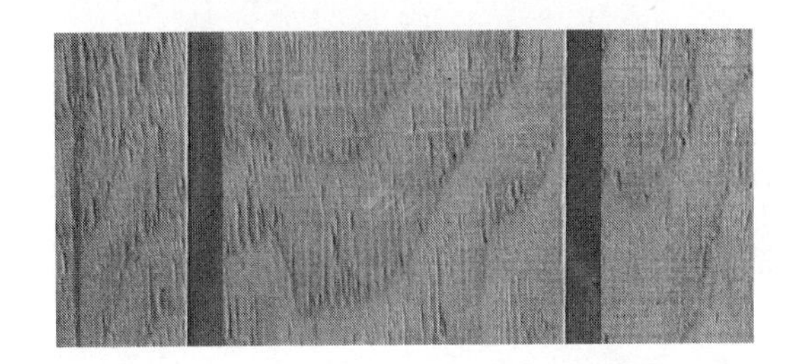

APA TEXTURE 1-11

Special 303 Siding panel with shiplapped edges and parallel grooves 1/4" deep, 3/8" wide; grooves 4" or 8" o.c. are standard. Other spacings sometimes available are 2", 6" and 12" o.c., check local availability. T 1-11 is generally available in 19/32" and 5/8" thicknesses. Also available with scratch-sanded, overlaid, rough-sawn, brushed and other surfaces. Available in Douglas fir, cedar, redwood, southern pine, other species.

ROUGH-SAWN

Manufactured with a slight, rough-sawn texture running across panel. Available without grooves, or with grooves of various styles; in lap sidings, as well as in panel form. Generally available in 11/32", 3/8", 1/2", 19/32" and 5/8" thicknesses. Rough-sawn also available in Texture 1-11, reverse board-and-batten (5/8" thick), channel groove (3/8" thick), and V-groove (1/2" or 5/8" thick). Available in Douglas fir, redwood, cedar, southern pine, other species.

CHANNEL GROOVE

Shallow grooves typically 1/16" deep, 3/8" wide, cut into faces of 3/8" thick panels, 4" or 8" o.c. Other groove spacings available. Shiplapped for continuous patterns. Generally available in surface patterns and textures similar to Texture 1-11 and in 11/32", 3/8" and 1/2" thicknesses. Available in redwood, Douglas fir, cedar, southern pine and other species.

REVERSE BOARD-AND-BATTEN

Deep, wide grooves cut into brushed, roughsawn, coarse sanded or other textured surfaces. Grooves about 1/4" deep, 1" to 1-1/2" wide, spaced 8", 12" or 16" o.c. with panel thickness of 19/32" and 5/8". Provides deep, sharp shadow lines. Long edges shiplapped for continuous pattern. Available in redwood, cedar, Douglas fir, southern pine and other species.

Fig. 60-8 APA 303 plywood panel siding is produced in a wide variety of sizes, surface textures, and patterns. (*Courtesy of American Plywood Association*)

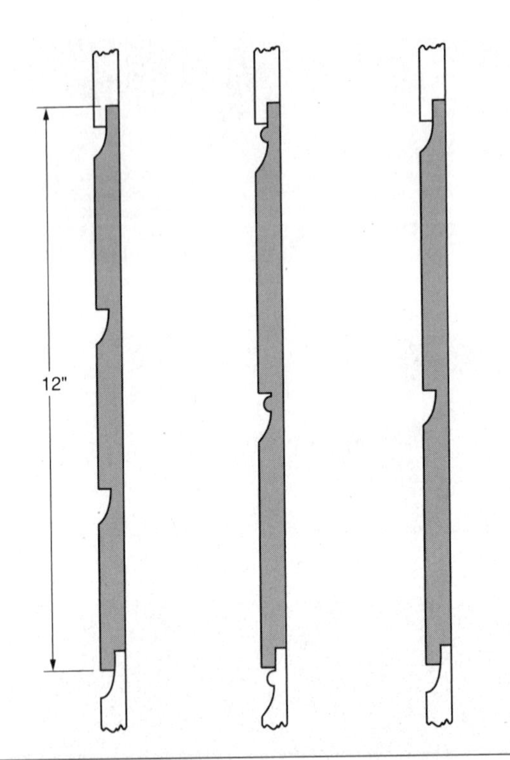

Fig. 60-9 Shapes of horizontally applied hardboard panel siding. Care should be taken not to over-drive the nails.

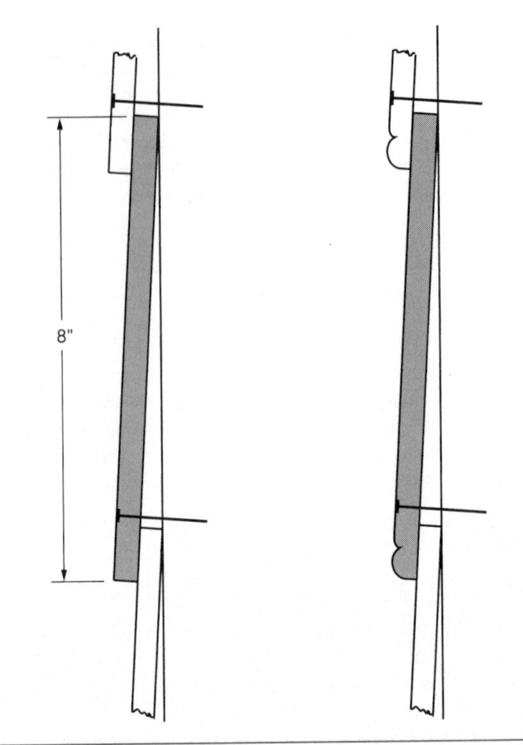

Fig. 60-10 Shapes and sizes of lapped hardboard siding. Care should be used not to over-drive the nails.

Lap Siding

Lap siding is manufactured with rough-sawn, weathered wood grain, or other embossed surface textures. Some surfaces are *grooved* or *beaded* with square or beveled edges. It comes in thicknesses from 7/16 to 19/32 inch, widths of 6, 8, and 12 inches, and lengths of 16 feet. Its long length makes installation go more quickly. One type of hardboard lap siding comes with a *self-aligning spline,* inserted in the back side, for faster installation (Fig. 60-10).

Estimating Siding

First, calculate the wall area by multiplying its length by its height. To calculate the area of a gable end, multiply the length by the height. Then divide the result by two. Add the square foot areas together with the areas of other parts that will be sided, such as dormers, bays, and porches.

Because windows and doors will not be covered, their total surface areas must be deducted. Multiply the width by the height (large openings only). Subtract the results from the total area to be covered by siding (Fig. 60-11).

The amount of siding to order depends on the kind of siding used. When calculating, use the *factor* for either linear feet or board feet, according to the way it is sold (Fig. 60-12). (Coverage factors for other kinds of siding, such as wood shingles, can be found in the appropriate chapters.)

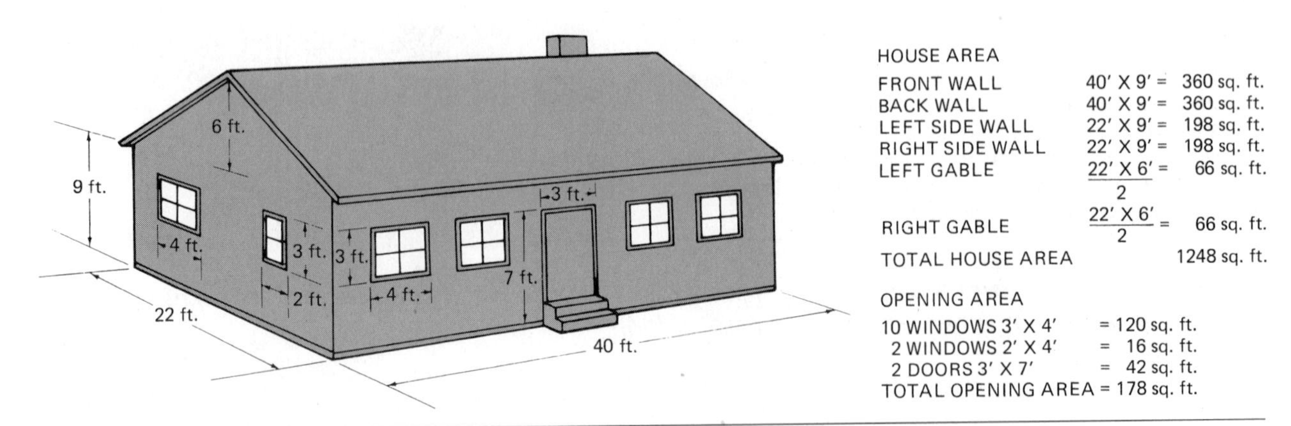

Fig. 60-11 Estimating the area to be covered by siding.

Pattern	Nominal Width	Width		Factor for Lineal Feet	Factor for Board Feet
		Dressed	Exposed Face		
Bevel & Bungalow	4	3½	2½	4.8	1.60
	6	5½	4½	2.67	1.33
	8	7¼	6¼	1.92	1.28
	10	9¼	8¼	1.45	1.21
Dolly Varden	4	3½	3	4.0	1.33
	6	5½	5	2.4	1.2
	8	7¼	6¾	1.78	1.19
	10	9¼	8¾	1.37	1.14
	12	11¼	10¾	1.12	1.12
Drop, T&G & Channel Rustic	4	3⅜	3⅛	3.84	1.28
	6	5⅜	5⅛	2.34	1.17
	8	7⅛	6⅞	1.75	1.16
	10	9⅛	8⅞	1.35	1.13
Log Cabin	6	5 7/16	4 15/16	2.43	2.43
	8	7⅛	6⅝	1.81	2.42
	10	9⅛	8⅝	1.39	2.32
Boards	2	1½	The exposed face width will vary depending on size selected and on how the boards-and-battens or boards-on-boards are applied.		
	4	3½			
	6	5½			
	8	7¼			
	10	9¼			

COVERAGE ESTIMATOR

Fig. 60-12 Multiply the area to be covered by either the linear feet or board feet factor to find the total footage required. (*Courtesy of Western Wood Products Association*)

CHAPTER 61 APPLYING VERTICAL AND HORIZONTAL WOOD SIDING

The method of siding application varies with the type. This chapter describes the application procedures for the most commonly used kinds of *solid* and *engineered wood* siding (Fig. 61-1).

Preparation for Siding Application

Before siding is applied, it must be determined how it will be *stopped* or treated at the foundation, eaves, and corners. The installation of various kinds of exterior wall trim may first be required.

Foundation Trim

In most cases, no additional trim is applied at the foundation. The siding is started so that it extends slightly below the top of the foundation. However, a **water table** may be installed for appearance. It sheds water a little farther away from the foundation. The water table usually consists of a board and a drip cap installed around the perimeter. Its bottom edge is slightly below the top of the foundation. The siding is started on top of the water table.

Eaves Treatment

At the eaves, the siding may end against the bottom edge of the frieze. The width of the frieze may vary. It is necessary to know its width when laying out horizontal siding courses. The siding may also terminate against the soffit, if no frieze is used. The joint between is then covered by a cornice molding. The size of the molding must be known to plan exposures on courses of horizontal siding.

Rake Trim

At the rakes, the siding may be applied under a furred-out rake fascia. When the rake overhangs the

Fig. 61-1 Applying solid redwood bevel siding. (*Couresy of California Redwood Association*)

sidewall, the siding may be fitted against the rake frieze. When fitted against the rake soffit the joint is covered with a molding (Fig. 61-2).

Gable End Treatment

Sometimes a different kind of siding is used on the gable ends than on the sidewalls below the plate. The joint between the two types must be weather-tight. One method of making the joint is to use a drip cap and flashing between the two types of material.

In another method, the plate and studs of the gable end are extended out from the wall a short distance. This allows the gable end siding to overlap the siding below (Fig. 61-3). Furring strips may also be used on the gable end framing in place of extending the gable plate and studs.

Treating Corners

One method of treating corners is with the use of **corner boards**. Horizontal siding may be mitered around exterior corners. Or, metal corners may be used on each course of siding. In interior corners, siding courses may butt against a square corner board or against each other (Fig. 61-4). The thickness of corner boards depends on the type of siding

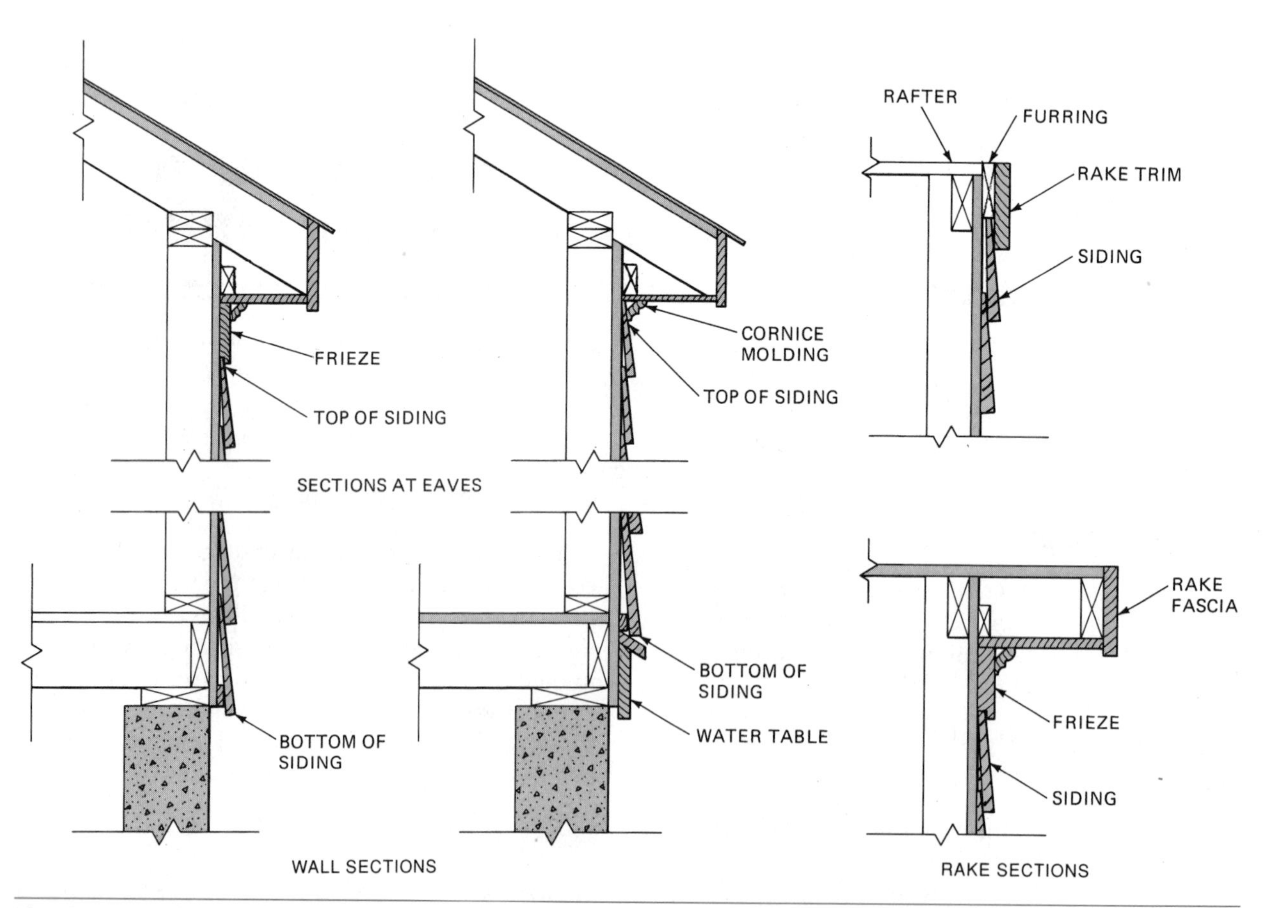

Fig. 61-2 Methods of stopping siding at foundation, eaves, and rake areas.

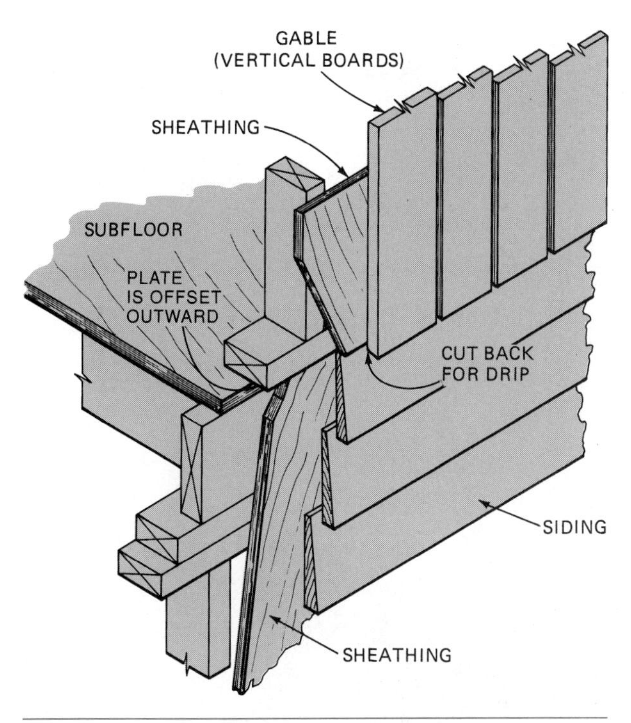

Fig. 61-3 The gable end may be constructed to extend over the siding on the wall below.

used. The corner boards should be thick enough so that the siding does not project beyond the face of the corner board.

The width of the corner boards depends on the effect desired. However, one of the two pieces, making up an outside corner, should be narrower than the other by the thickness of the stock. Then, after the wider piece is fastened to the narrower one, the same width is exposed on both sides of the corner. The joint between the two pieces should be on the side of the building that is least viewed.

Installing Corner Boards

Before installing corner boards, flash both sides of the corner. Install a strip of #15 felt vertically on each side. One edge should extend beyond the edge of the corner board at least 2 inches. Tuck the top ends under any previously applied felt.

With a sharp plane, slightly back-bevel one edge of the narrower of the two pieces that make up the outside corner board. This assures a tight fit between the two boards. Cut, fit, and fasten the narrow piece. Start at one end and work toward the bottom. Make sure to keep the beveled edge flush with the corner. Fasten with galvanized or other noncorroding nails spaced about 16 inches apart along its inside edge.

Cut, fit, and fasten the wider piece to the corner in a similar manner. Make sure its outside edge is flush with the surface of the narrower piece. The outside row of nails is driven into the edge of the narrower piece. Plane the outside edge of the wide piece wherever necessary to make it come flush. Slightly round over all sharp exposed corners by planing a small chamfer and sanding. Set all nails so they can be filled over later. Make sure a tight joint is obtained between the two pieces (Fig. 61-5).

Corner boards may also be applied by fastening the two pieces together first. Then install the assembly on the corner.

Applying Building Paper

A description of the kinds and purposes of building paper has been previously given in Chapter 56. A *breatheable* type building paper should be applied to sidewalls if housewrap has not already been applied. Building paper is not designed for prolonged exposure to the weather. It generally is not applied to the total wall area at one time, but only as the application of siding progresses.

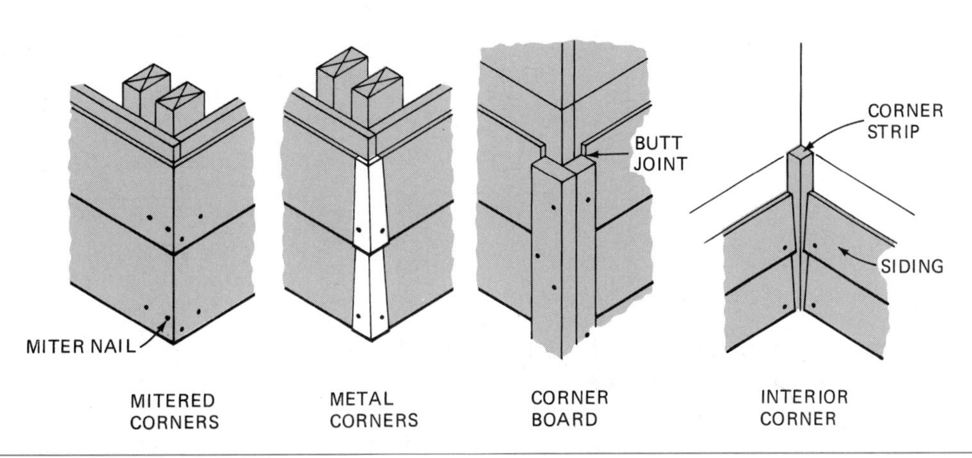

Fig. 61-4 Methods of returning and stopping courses of horizontal siding.

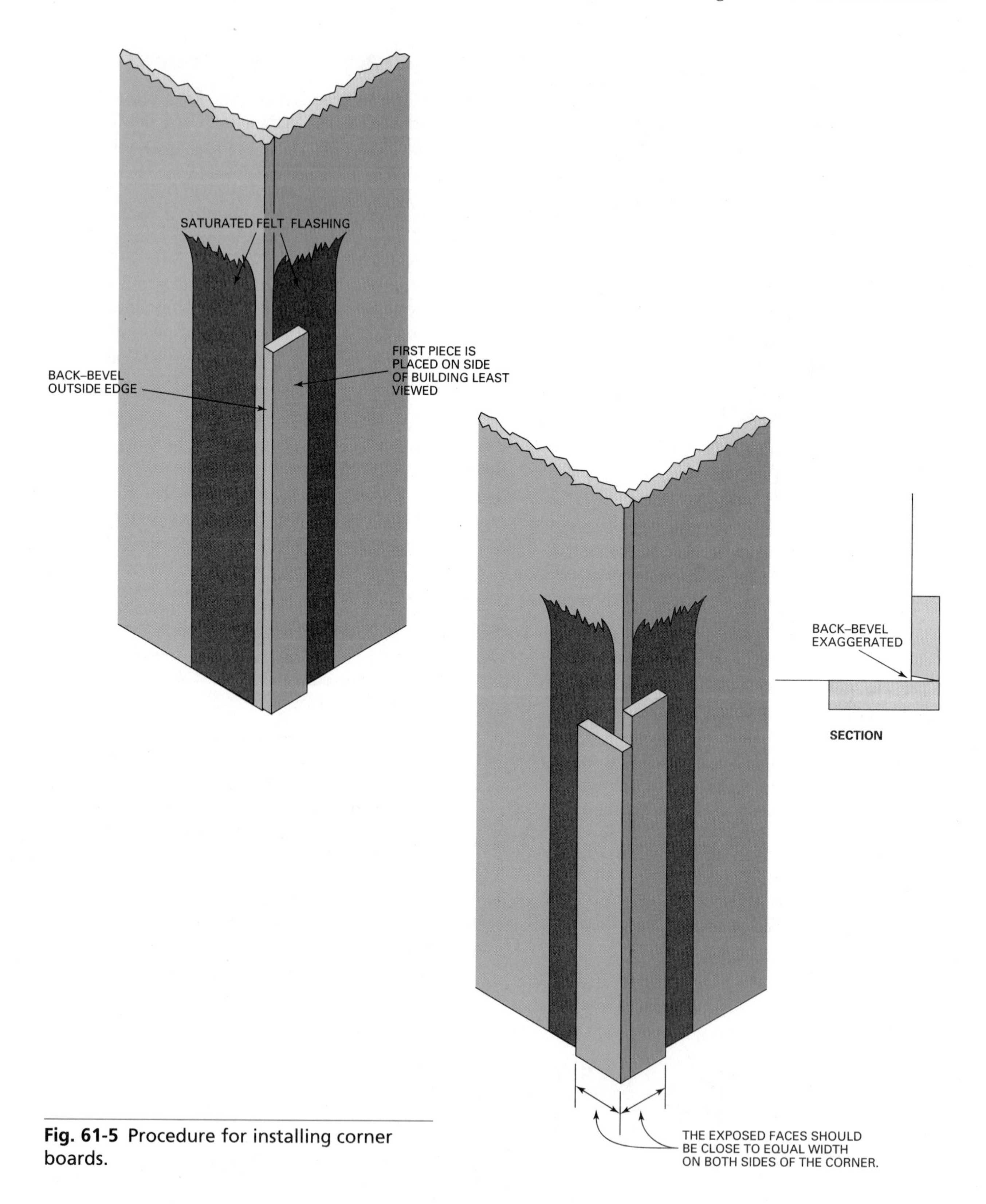

Fig. 61-5 Procedure for installing corner boards.

Apply the paper horizontally. Start at the bottom of the wall. Make sure the paper lies flat with no wrinkles. Fasten it in position. If nailing, use large-head roofing nails. Fasten in rows near the bottom, the center, and the top about 16 inches apart. Each succeeding layer should lap the lower layer by about 4 inches.

The sheathing paper should lap over any flashing applied at the sides and tops of windows and doors and at corner boards. It should be tucked under any flashings applied under the bottoms of windows or frieze boards. In any case, all laps should be over the paper below.

Installing Horizontal Wood Siding

One of the important differences between *bevel siding* and other types with *tongue-and-groove, shiplap,* or *rabbeted* edges is that exposure of courses of bevel siding can be varied somewhat. With other types, the amount exposed to the weather is constant with every course and cannot vary.

The ability to vary the exposure is a decided advantage. It is desirable from the standpoint of appearance, weathertightness, and ease of application to have a full course of horizontal siding exposed above and below windows and over the tops of doors (Fig. 61-6). The exposure of the siding may vary gradually up to a total of 1/2 inch over the entire wall, but the width of each exposure should not vary more than 1/8 inch from its neighbor.

Determining Siding Exposure

To determine the siding exposure so that it is about equal both above and below the window sill, divide the overall height of the window frame by the amount of exposure.

Dividing the distance by the recommended exposure results in the number of courses between the top and the bottom of the window. For example, if the overall height of a window is 64 inches and 8-inch siding is used, divide the height by 7, the maximum allowed exposure. You will need slightly more than nine courses. To obtain the exposure, divide 64 by 9. The result is 7 1/9 inches or a *slack* 7 1/8 inches.

The next step is to determine the exposure distance from the bottom of the window sill to about one inch below the bottom of the foundation wall. If this distance is 36 inches, then five courses are required at approximately 7 3/16 inches *strong*. The term *strong* is used to mean slightly more, about 1/32 inch, than the measurement indicated. The term *slack* means slightly less. Thus, the exposure above and below the window sill is almost the same. The difference is not noticeable to the eye. The same procedure is used to lay out courses above the window. The layout lines are then transferred to a story pole. The story pole is used to lay out the courses all around the building (Fig. 61-7). Because of varying window heights, it may not

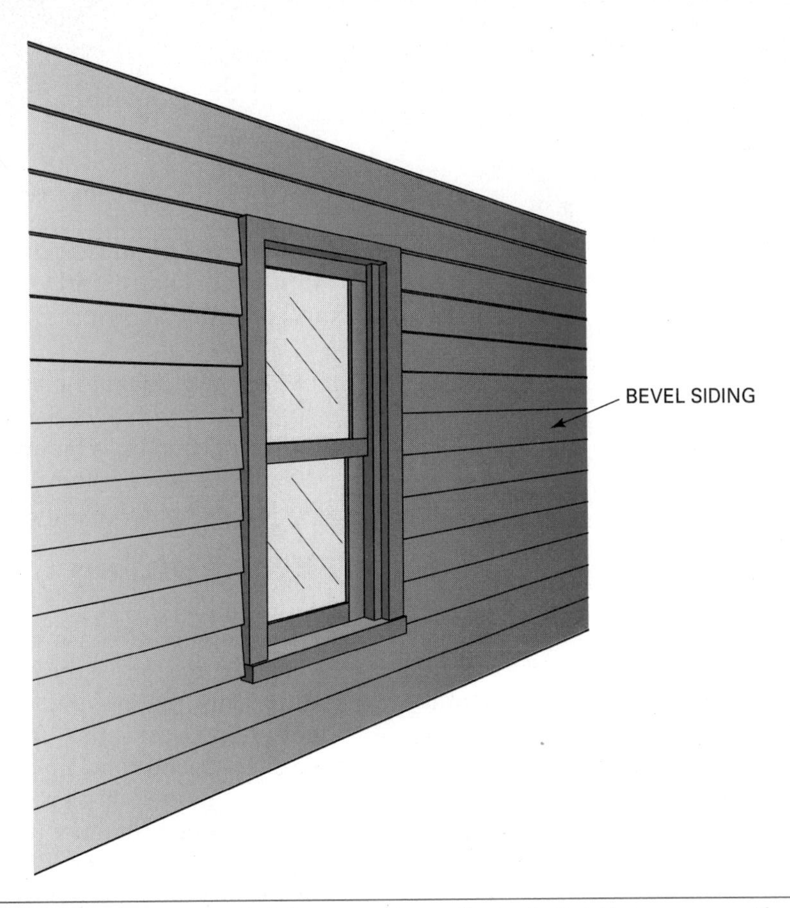

Fig. 61-6 The exposure of each course of siding may be adjusted slightly to ensure there is a full exposure of siding above and below windows and doors.

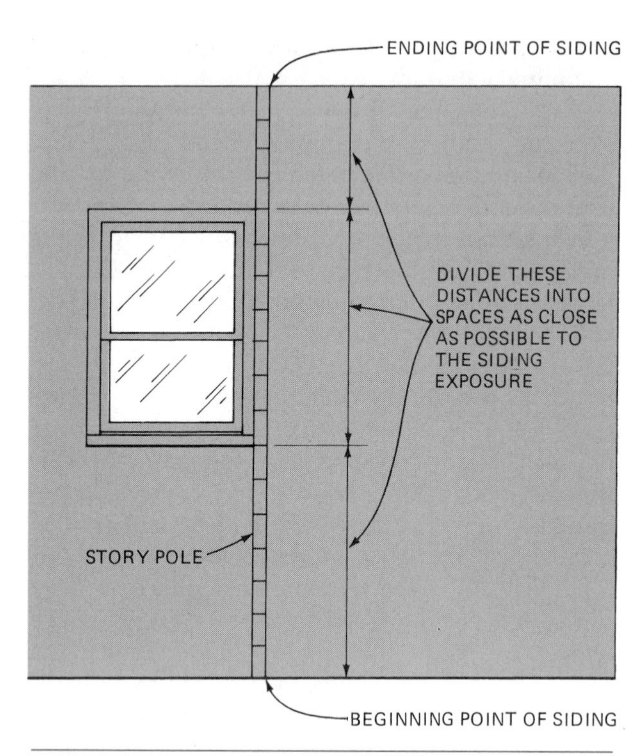

Fig. 61-7 Laying out a story pole for courses of horizontal siding.

always be possible to lay out full siding courses above and below every window or door.

Instead of calculating the number and height of siding courses mathematically, *dividers* may be used to space off the distances.

Starting Bevel Siding from the Top

Another advantage of using bevel siding is that application may be made starting at the top and working toward the bottom if more convenient. With this method, a number of chalk lines may be snapped without being covered by a previous course. This saves time. Any scaffolding already erected may be used and then dismantled as work progresses toward the bottom.

Starting Siding from the Bottom

Most horizontal siding, however, is usually started at the bottom. For bevel siding, a *furring strip* of the same thickness and width of the siding *headlap* is first fastened along the bottom edge of the sheathing (Fig. 61-8). For the first course, a line is snapped on the wall at a height that is in line with the *top edge* of the first course of siding.

For siding with constant exposure, the only other lines snapped are across the tops of wide entrances, windows, and similar objects to keep the courses in alignment. For lap siding, with exposures

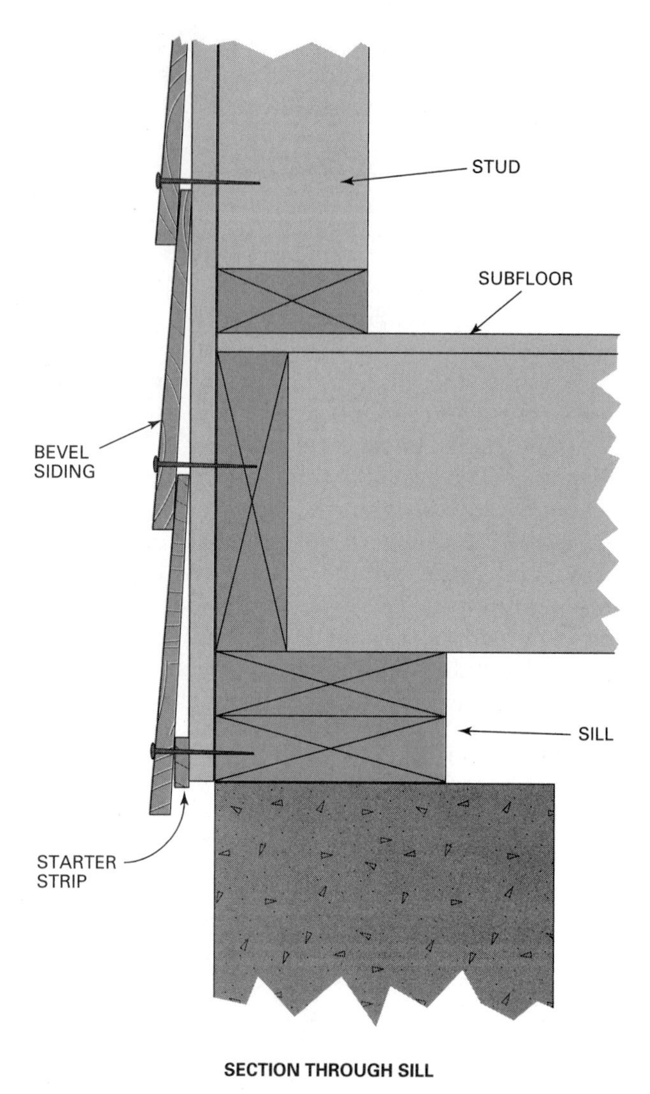

Fig. 61-8 For bevel siding, a strip of wood of the same thickness and width of the headlap is used as a starter strip.

that may vary, lines are snapped for each successive course in line with their bottom edge. Stagger joints in adjacent courses as far apart as possible. A small piece of felt paper often is used behind the butt seams to ensure the weathertightness of the siding.

Fitting Siding

When applying a course of siding, start from one end and work toward the other end. With this procedure, only the last piece will have to be fitted. Tight-fitting butt joints must be made between pieces. If an end joint must be fitted, use a block plane to trim the end as needed. When a piece has to be fitted between other pieces, measure carefully. Cut it just a trifle strong. Place one end in position. Bow the piece outward, position the other end, and snap into place (Fig. 61-9).

A **preacher** is often used for accurate layout of siding where it butts against corner boards, casings,

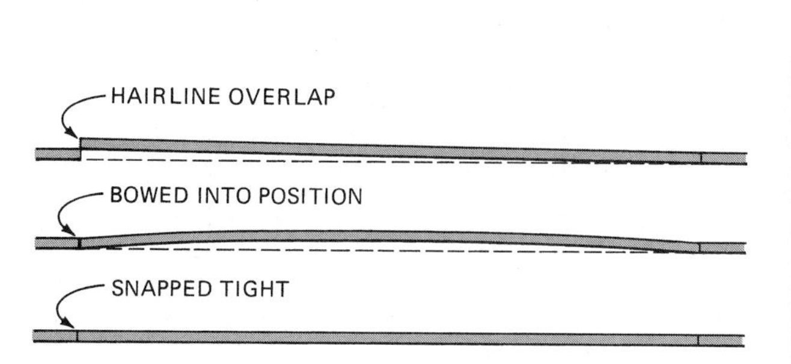

Fig. 61-9 Measure and cut individual pieces of bevel siding carefully for tight fits. (*Courtesy of California Redwood Association*)

HELPFUL HINTS

Use a preacher for accurate markings of bevel siding to fit against corner boards and casings.

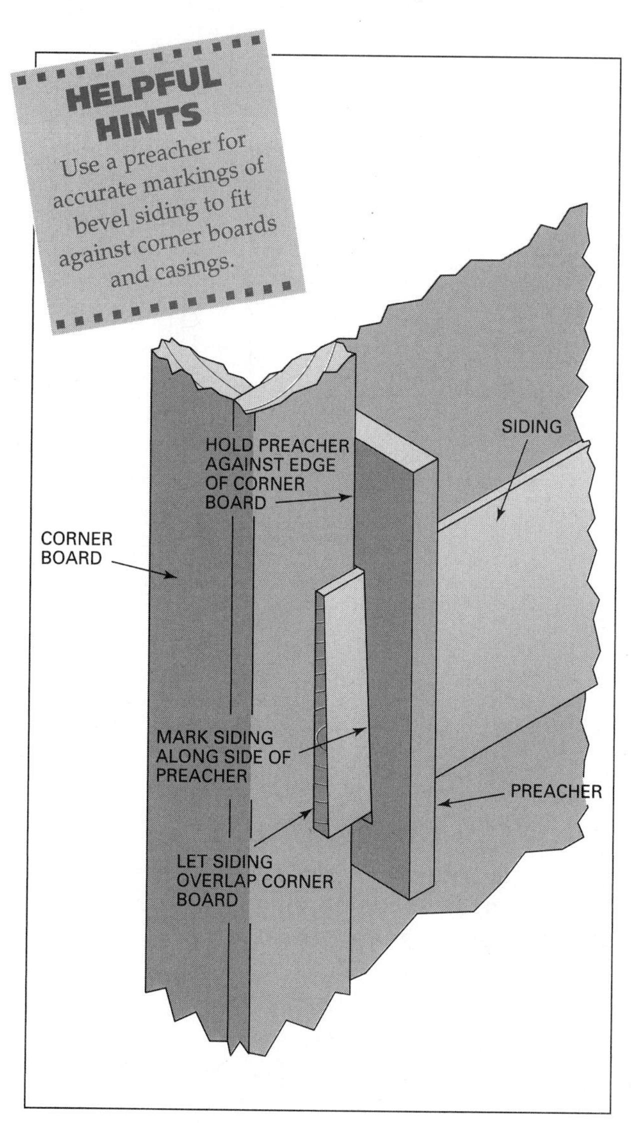

Fig. 61-10 A preacher may be used for accurately marking a siding piece for length.

and similar trim. The siding is allowed to overlap the trim. The preacher is held against the trim. A line is then marked on the siding along the face of the preacher (Fig. 61-10).

When fitting siding under windows, make sure the siding fits snugly in the groove on the underside of the window sill for weathertightness (Fig. 61-11).

Fastening Siding

Siding is fastened to each bearing stud or about every 16 inches. On bevel siding, fasten through the butt edge just above the top edge of the course

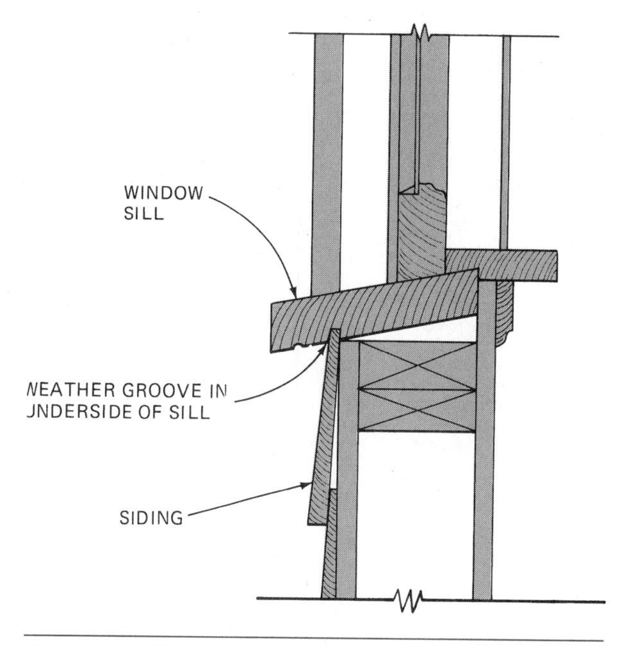

Fig. 61-11 Fit siding snugly in the weather groove on the underside of the window sill.

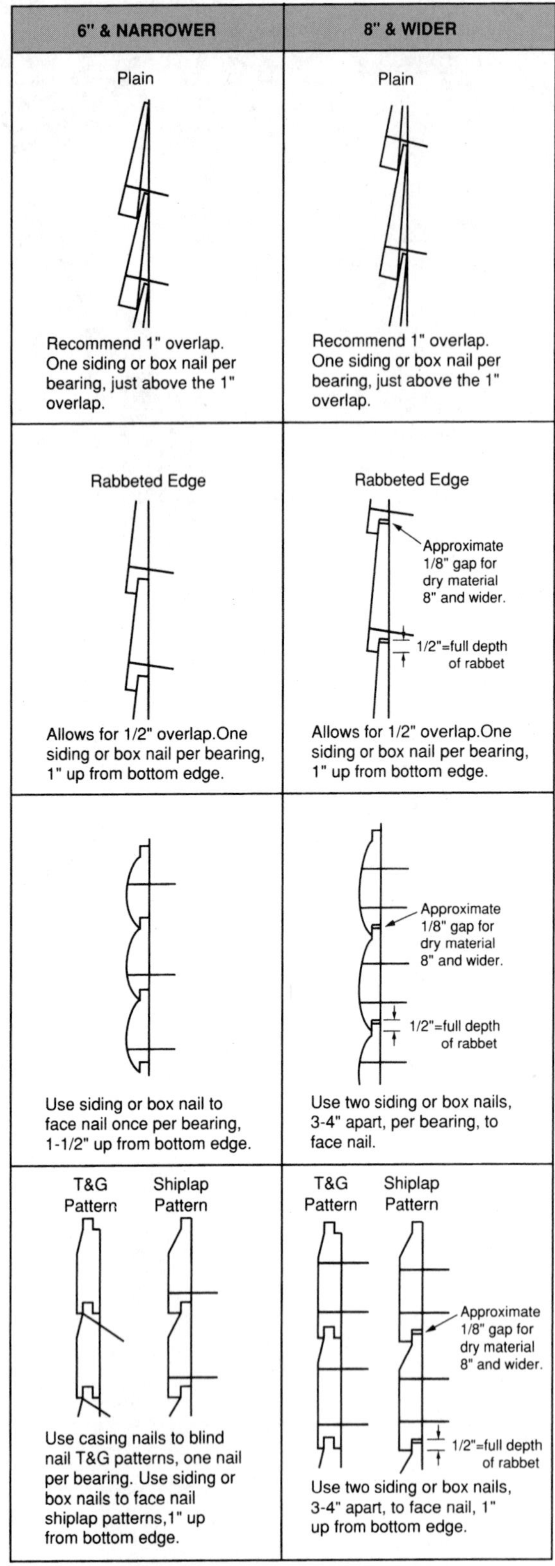

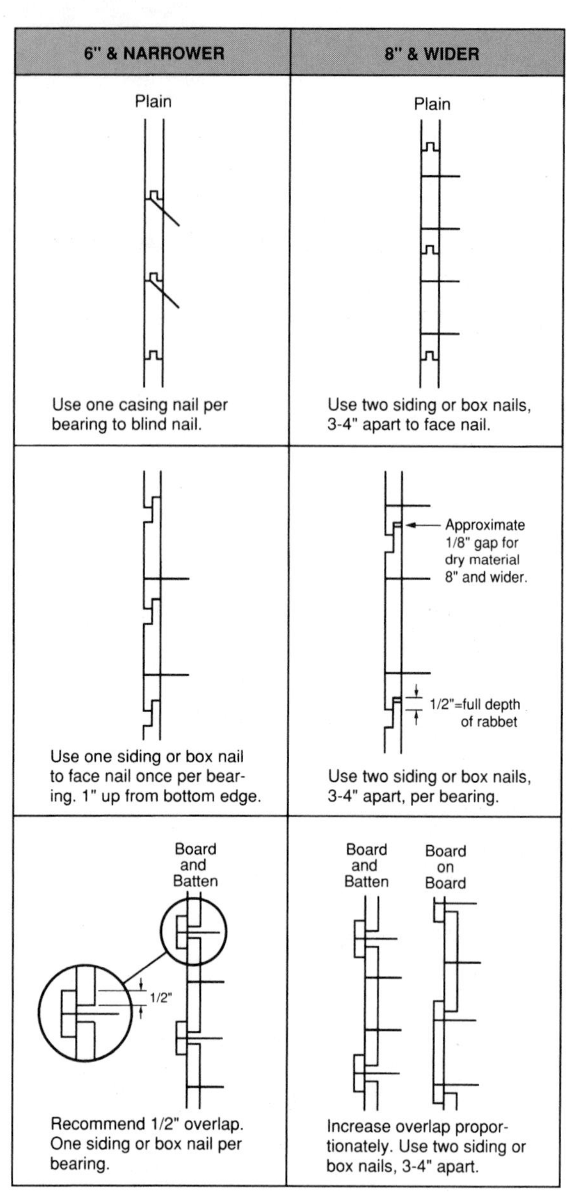

Fig. 61-12 Location and number of fasteners recommended for wood siding. (*Courtesy of Western Wood Products Association*)

below. Do not fasten through the lap. This prevents splitting of the siding that might be caused by slight swelling or shrinking due to moisture changes. Care must be taken to fasten as low as possible to avoid splitting the siding in the center. The location and number of fasteners recommended for siding are shown in Figure 61-12.

Vertical Application of Wood Siding

Bevel sidings are designed for horizontal applications only. *Board on board* and *board and batten* are applied only vertically. Almost all other patterns may be applied in either direction. The following paragraphs describe the vertical application of *tongue-and-groove* siding. Other edge-shaped siding is installed in a comparable way.

Installing Vertical Tongue-and-Groove Siding

Corner boards usually are not used when wood siding is applied vertically. The siding boards are fitted around the corner (Fig. 61-13). Rip the grooved edge from the starting piece. Slightly back-bevel the ripped edge. Place it vertically on the wall with the beveled edge flush with the corner similar to making a corner board.

The tongue edge should be plumb, the bottom end should be about 1 inch below the sheathing. The top end should butt or be tucked under any trim above. *Face nail* the edge nearest the corner. **Blind nail** into the tongue edge. Nails should be placed from 16 to 24 inches apart. *Blocking* must be provided between studs if siding is applied directly to the frame.

Fasten a temporary piece on the other end of the wall projecting below the sheathing by the same amount. Stretch a line to keep the bottom ends of other pieces in a straight line (Fig. 61-14). Apply succeeding pieces by toenailing into the tongue edge of each piece.

Fig. 61-13 Vertical tongue-and-groove siding needs little accessory trim, such as corner boards. (*Courtesy of California Redwood Association*)

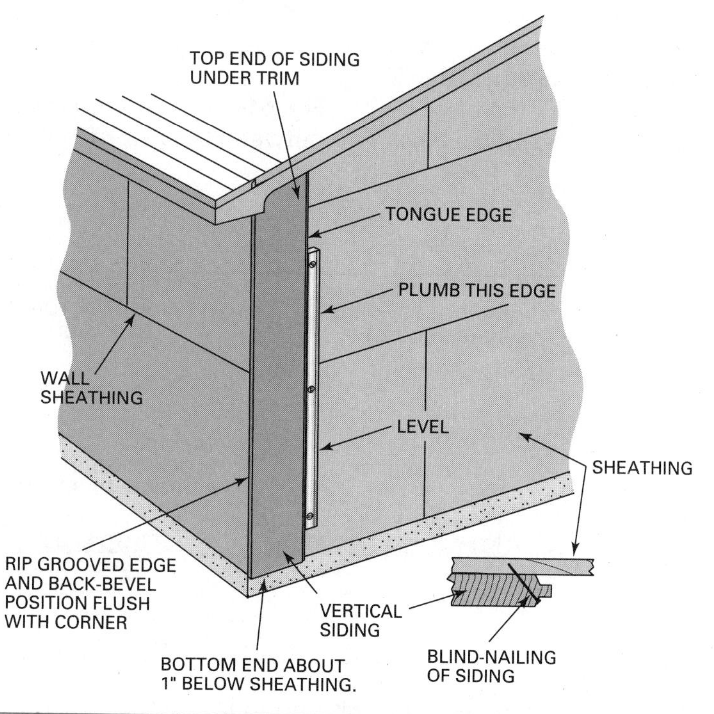

Fig. 61-14 Starting the application of vertical board siding.

Make sure the edges between boards come up tight. If they do not come up tight by nailing alone, drive a chisel, with its beveled edge side against the tongue, into the sheathing. Use it as a pry to force the board up tight. When the edge comes up tight, fasten it close to the chisel. If this method is not successful, toenail a short block of the siding with its grooved edge into the tongue of the board (Fig. 61-15). Drive the nail home until it forces the board up tight. Drive nails into the siding on both sides of the scrap block. Remove the scrap block.

Continue applying pieces in the same manner. Make sure to keep the bottom ends in a straight line. Avoid making horizontal joints between lengths. If joints are necessary, use a mitered or rabbeted joint for weathertightness (Fig. 61-16).

Fitting around Doors and Windows

Vertical siding is fitted tightly around window and door casings with different methods than those used for horizontal siding.

Approaching a Wall Opening. When approaching a door or window, cut and fit the piece just before the one to be fitted against the casing. Then remove it. Set it aside for the time being.

Cut, fit, and tack the piece to be fitted against the casing in place of the next to last piece. Level from the top of the window casing and the bottom of the sill. Mark the piece.

To lay out the piece so it will fit snugly against the side casing, first cut a scrap block of the siding material, about 6 inches long. Remove the tongue from one edge. Be careful to remove all of the tongue, but no more. Hold the block so its grooved edge is against the side casing and the other edge is on top of the siding to be fitted. Mark the piece by holding a pencil against the outer edge of the block while moving the block along the length of the side casing (Fig. 61-17).

Cut the piece, following the layout lines carefully. When laying out to fit against the bottom of the sill, make allowance to rabbet the siding to fit in the weathergroove on the bottom side of the window sill. Place and fasten the pieces in position.

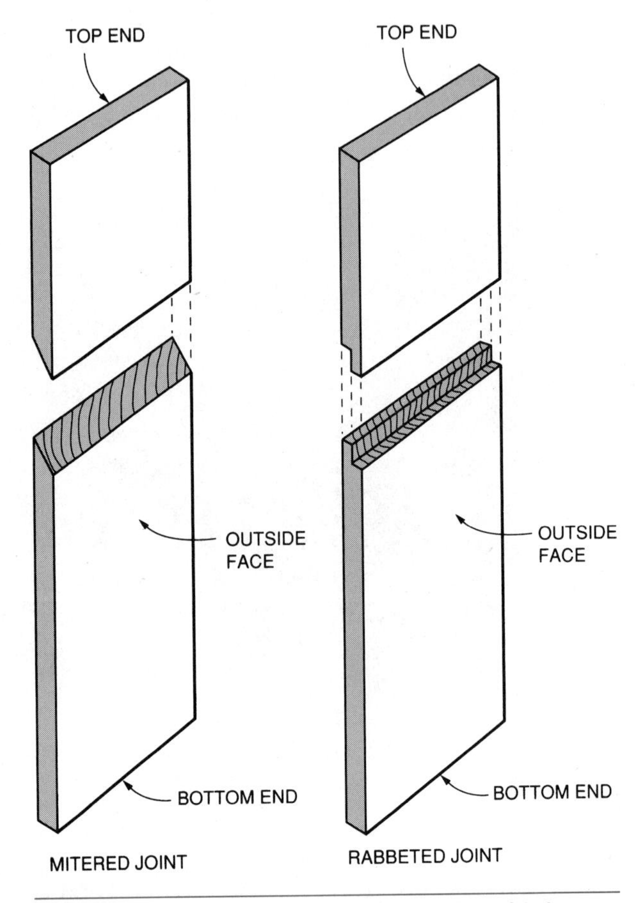

Fig. 61-16 Use mitered or rabbeted end joints between lengths of vertical board siding.

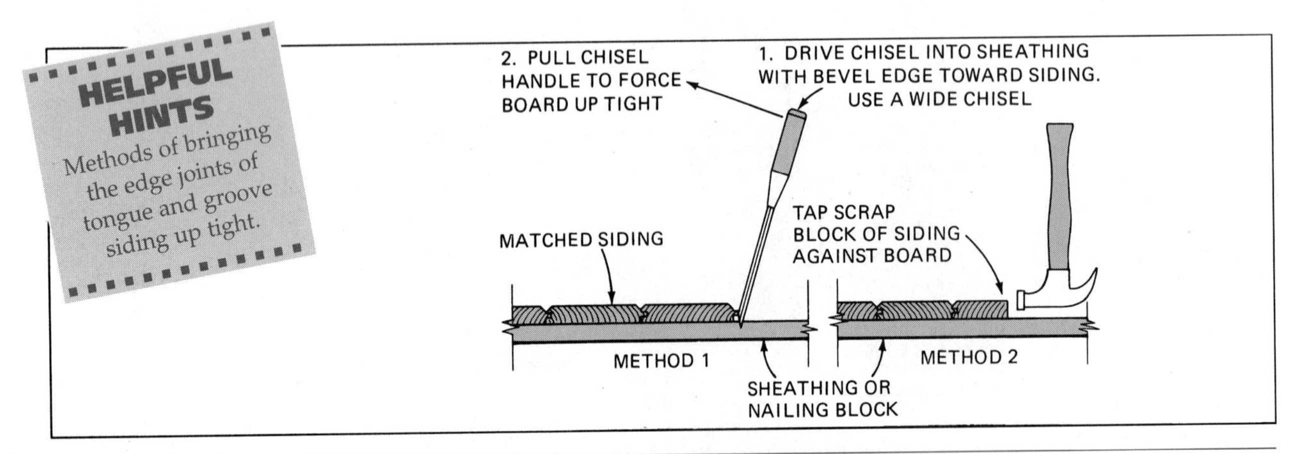

Fig. 61-15 Techniques to tighten the joints of boards during installation.

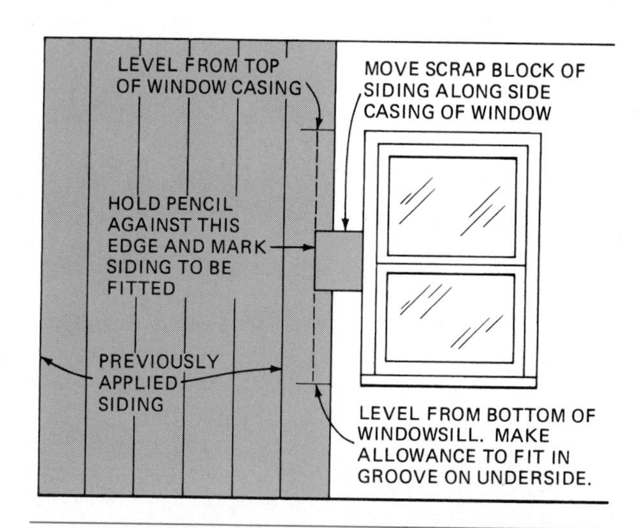

Fig. 61-17 Method of fitting vertical board siding when approaching a window casing.

Continue to apply the siding with short lengths across the top and bottom of the window. Each length under a window must be rabbeted at the top end to fit in the weather groove at the sill.

Leaving a Wall Opening. A full length must also be fitted to the casing on the other side of the window. To mark the piece, first tack a short length of scrap siding above and below the window and against the last pieces of siding installed. Tack the length of siding to be fitted against these blocks. Level from the top and bottom of the window. Mark the piece for the horizontal cuts.

To lay out the piece for the vertical cut that fits against the side casing, use the same block with the tongue removed, as used previously. Hold the grooved edge against the side casing. With a pencil against the other edge, ride the block along the side casing while marking the piece to be fitted (Fig. 61-18).

Remove the piece and the scrap blocks from the wall. Carefully cut the piece to the layout lines. Then fasten in position. Continue applying the rest of the siding until you are almost to the other end of the wall.

Method of Ending Vertical Siding

The last piece of vertical siding should be close to the same width of previously installed pieces. If siding is installed in *random* widths, plan the application. The width of the last piece should be equal, at least, to the width of the narrowest piece. It is not good practice to allow vertical siding to end with a narrow sliver.

Stop several feet from the end. Space off, and determine the width of the last piece. If it will not be a satisfactory width, install narrower or wider pieces for the remainder, as required. It may be necessary to rip available siding to narrower widths and reshape the grooves (Fig. 61-19).

When the corner is reached, the board is ripped to width along its tongue edge. It is slightly back-beveled for the first piece on the next wall to butt against. When the last corner is reached, the board is ripped in a similar manner. However, it is smoothed to a square edge to fit flush with the surface of the first piece installed. All exposed sharp corners should be eased.

Installing Panel Siding

Plywood, hardboard, and other panel siding is usually installed vertically. It can be installed horizontally, if desired. Lap siding panels are ordinarily applied horizontally.

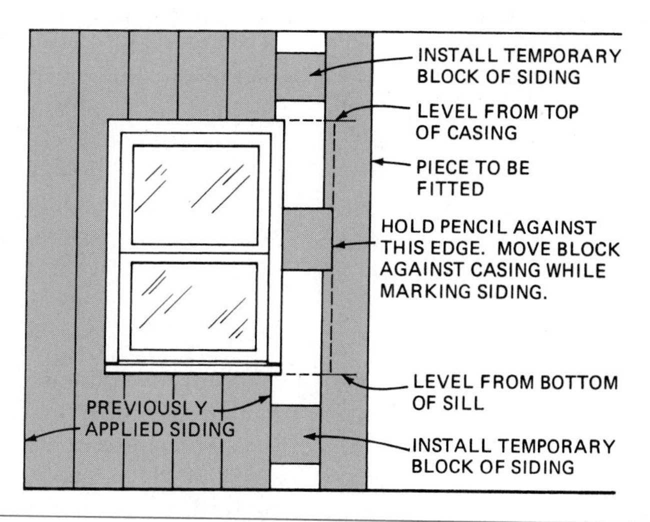

Fig. 61-18 Laying out vertical board siding to fit against a window casing, when leaving it.

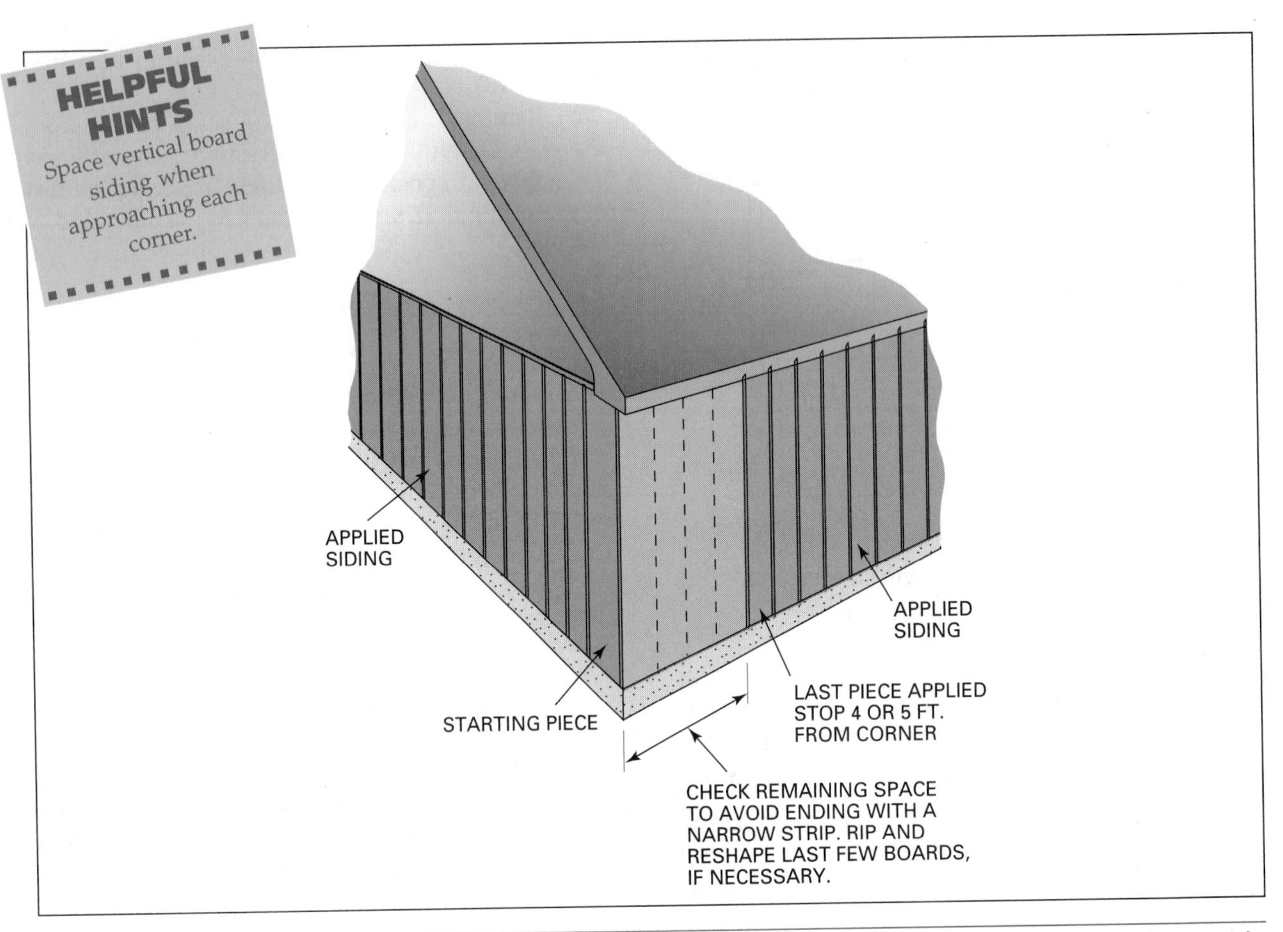

Fig. 61-19 Joints in vertical siding may be adjusted slightly to ensure the corner piece is nearly as wide as the others are.

Installing Vertical Panel Siding

Start a vertical panel so it is plumb, with one edge squared and flush with the starting corner. The inner edge should fall on the center of a stud. Fasten panels, of 1/2-inch thickness or less, with 6d *siding nails.* Use 8d siding nails for thicker panels. Fasten panel edges about every 6 inches and about every 12 inches along intermediate studs.

Apply successive sheets. Leave a 1/8-inch space between panels. Panels must be installed with their bottom ends in a straight line. There should be a minimum of 6 inches above the finished grade line.

Installing Horizontal Panel Siding

Mark the height of the first course of horizontal panel siding on both ends of the wall. Snap a chalk line between marks. Fasten a full-length panel with its top edge to the line, its inner end on the center of a stud, and its outer end flush with the corner. Fasten in the same way as for vertical panels.

Apply the remaining sheets in the first course in like manner. Trim the end of the last sheet flush with the corner. Start the next course so joints will line up with those in the course below.

Both vertical and horizontal panels may be applied to sheathing or directly to studs if backing is provided for all joints (Fig. 61-20).

Carefully fit and caulk around doors and windows. It is important that horizontal butt joints be either *offset and lapped, rabbeted,* or *flashed* (Fig. 61-21). Vertical joints are either shiplapped or covered with **battens.**

Applying Lap Siding Panels

Panels of lap siding are applied in much the same manner as wood bevel siding with some exceptions. First, install a strip, of the same thickness and width of the siding headlap, along the bottom of the wall. Determine the height of the top edge of

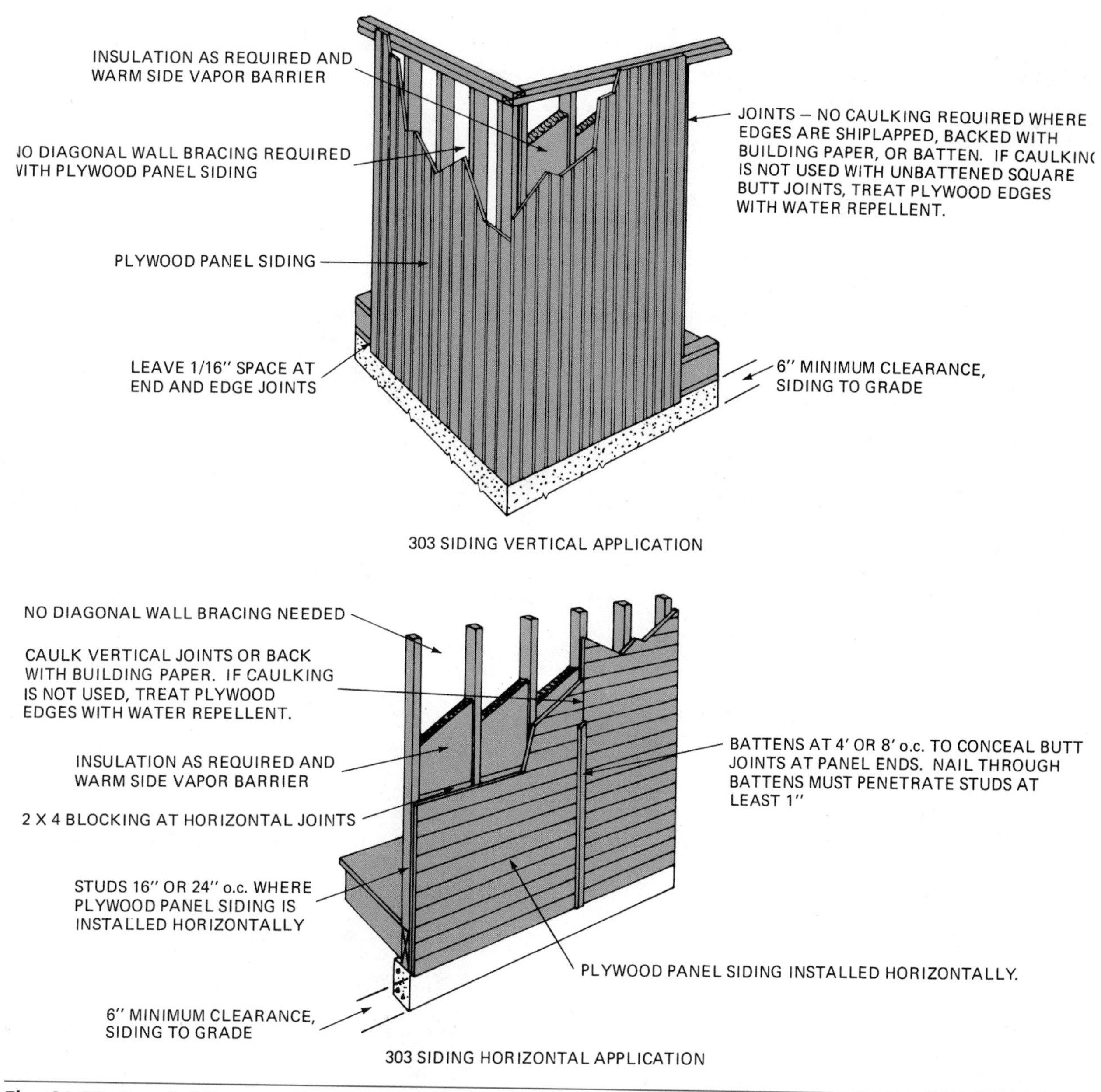

Fig. 61-20 Panel siding may be applied vertically or horizontally to sheathing or directly to studs. (*Courtesy of American Plywood Association*)

the first course. Snap a chalk line across the wall. Apply the first course with its top edge to the snapped line.

When applied over nailable sheathing, space nails 8 inches apart in a line about 3/4 inch above the bottom edge of the siding. When applied directly to framing, fasten at each stud location. A 1/8-inch, caulked joint between the ends of siding and trim is recommended. Joints between ends of siding may also be flashed with a narrow strip of #15 felt centered behind the joint and backed with a wood shingle wedge (Fig. 61-22). In some types of lap panel siding, a 1/8-inch gap is allowed between the ends. A special joint molding is inserted in the gap (Fig. 61-23).

Corners. Individual metal or plastic outside corners may be installed as work progresses. They can also be installed after the entire wall is sided. Applied corners are used because composition material does not miter well. Somewhat better success may be had when mitering plywood at outside corners. Slide the corners up into place with the top slipping under the course above. When the corner is in position, fasten it through the prepunched holes (Fig. 61-24). Lap siding may also be finished with corner boards.

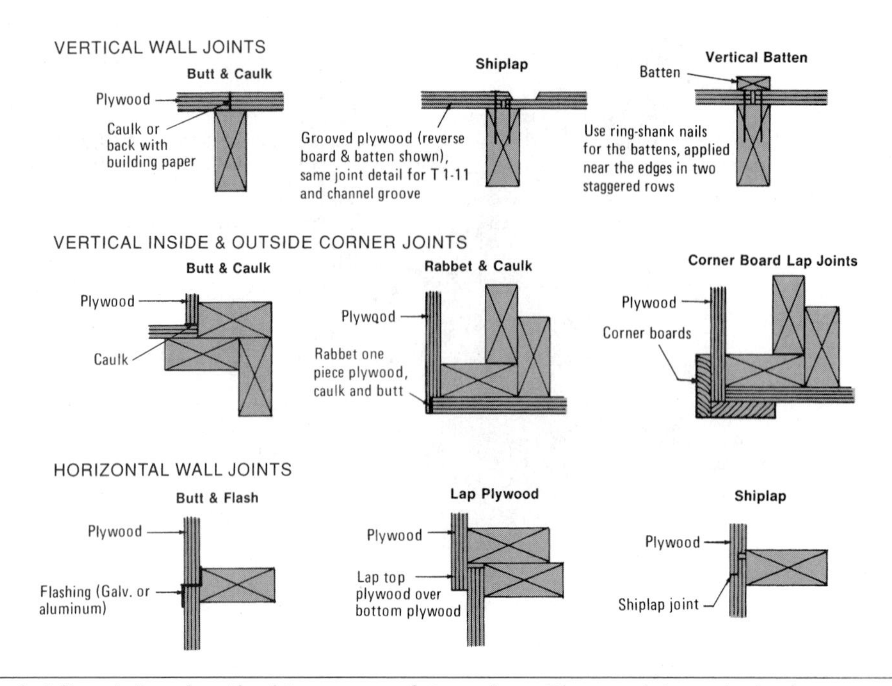

Fig. 61-21 Panel siding joint details. (*Courtesy of American Plywood Association*)

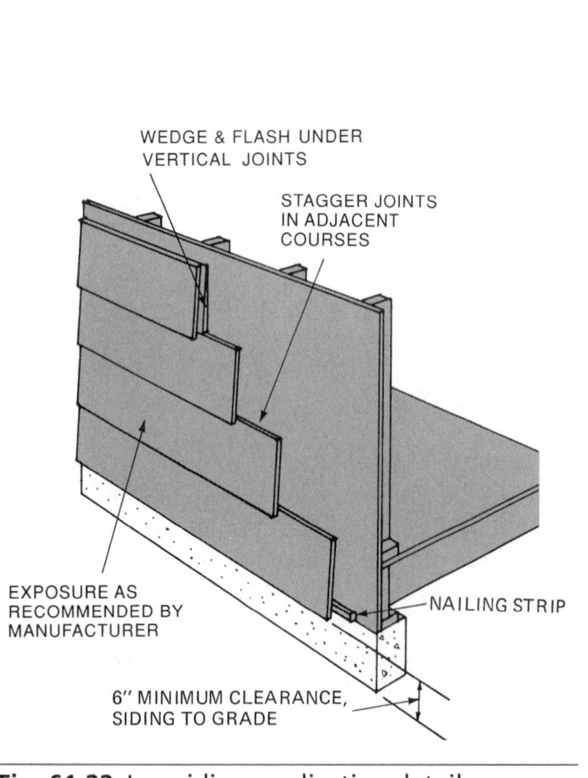

Fig. 61-22 Lap siding application details.

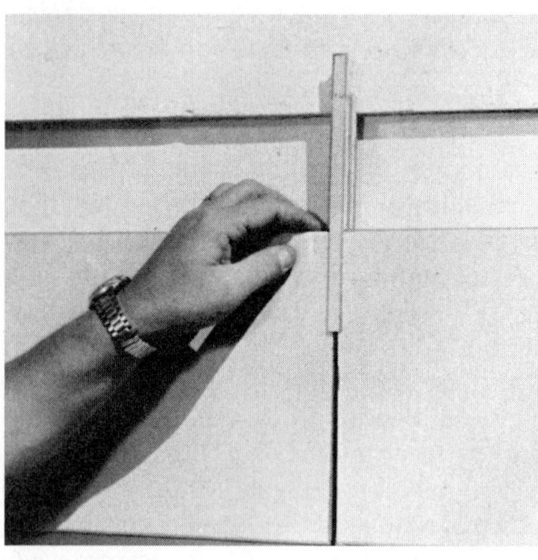

Fig. 61-23 A special molding is sometimes used on end joints of lap siding.

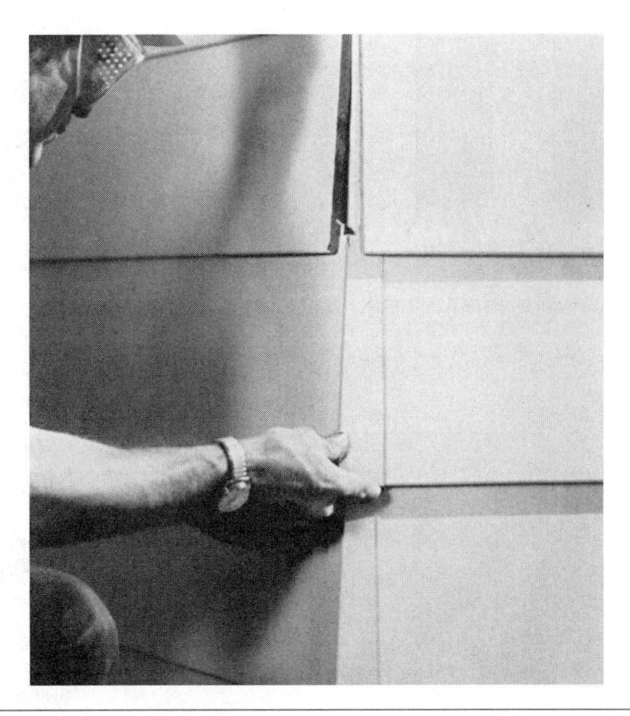

Fig. 61-24 Individual metal trim being applied to corners of lap siding.

CHAPTER 62 WOOD SHINGLE AND SHAKE SIDING

Wood shingles and *shakes* frequently are used for siding, as well as roofing (Fig. 62-1). Those previously described in Chapter 53 for roofing may also be applied to sidewalls.

Sidewall Shingles and Shakes

Some kinds of shingles and shakes are designed for sidewall use only (Fig. 62-2). *Rebutted* and *rejointed* ones are machine-trimmed with parallel edges and square butts for sidewall application. Rebutted and rejointed machine-grooved, sidewall shakes have **striated** faces.

Special *fancy butt* shingles are available in a variety of designs. They provide interesting patterns, in combination with square butts or other types of siding. Fancy butt shingles were used widely in the nineteenth century on Victorian style buildings. They are once again becoming popular (Fig. 62-3).

Fig. 62-1 Wood shingles and shakes frequently are used as siding. (*Courtesy of Western Wood Products Association*)

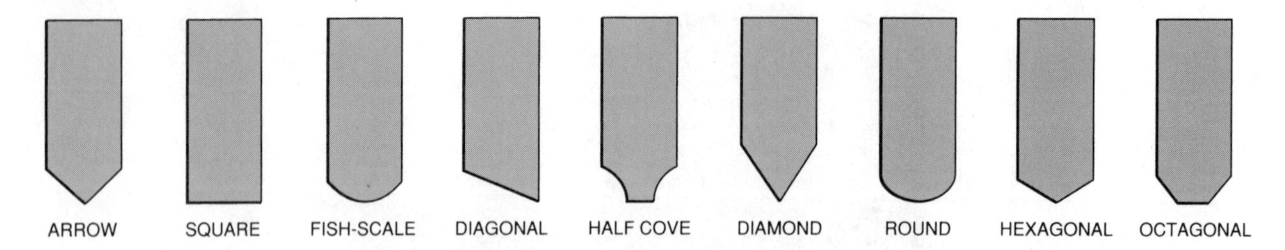

FANCY-BUTT RED CEDAR SHINGLES. NINE OF THE MOST POPULAR DESIGNS ARE SHOWN. FANCY-BUTT SHINGLES CAN BE CUSTOM PRODUCED TO INDIVIDUAL ORDERS.

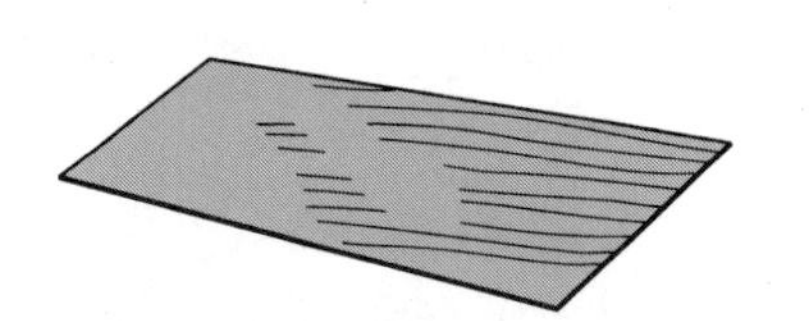

REBUTTED AND REJOINTED. MACHINE TRIMMED FOR PARALLEL EDGES WITH BUTTS SAWN AT RIGHT ANGLES. FOR SIDEWALL APPLICATION WHERE TIGHTLY FITTING JOINTS ARE DESIRED.

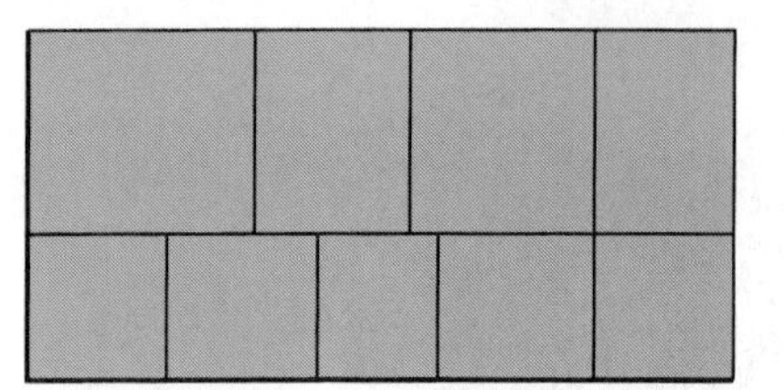

PANELS. WESTERN RED CEDAR SHINGLES ARE AVAILABLE IN 4- AND 8-FOOT PANELIZED FORM.

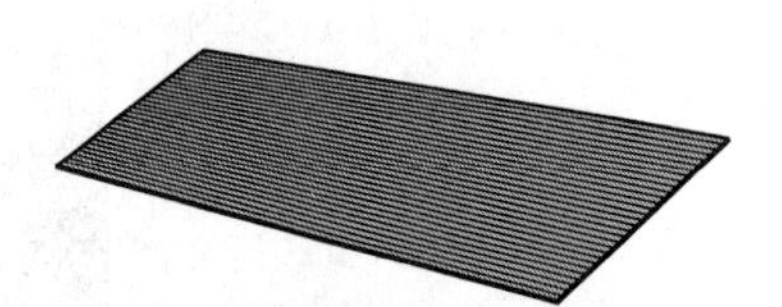

MACHINE GROOVED. MACHINE-GROOVED SHAKES ARE MANUFACTURED FROM SHINGLES AND HAVE STRIATED FACES AND PARALLEL EDGES. USED DOUBLE-COURSED ON EXTERIOR SIDEWALLS.

Fig. 62-2 Some wood shingles and shakes are made for sidewall application only. (*Courtesy of Cedar Shake and Shingle Bureau*)

Fig. 62-3 Fancy butt shingles are still used to accent sidewalls with distinctive designs. (Courtesy of Andersen Corporation)

Red cedar shingles, factory-applied on 4- and 8-foot panels, are available in several styles and exposures. Panels come with square butt shingles applied in straight or staggered lines and with a number of fancy butt designs.

Applying Wood Shingles and Shakes

Wood shingles and shakes may be applied to sidewalls in either single-layer or double-layer courses. In *single coursing,* shingles are applied to walls with a single layer in each course, in a way similar to roof application. However, greater exposures are allowed on sidewalls than on roofs (Fig. 62-4).

In *double coursing,* two layers are applied in one course. Consequently, even greater weather exposures are allowed. Double coursing is used when wide courses with deep, bold shadow lines are desired (Fig. 62-5).

Applying the Starter Course

The *starter course* of sidewall shingles and shakes is applied in much the same way as the starter course on roofs. A double layer is used for single-course

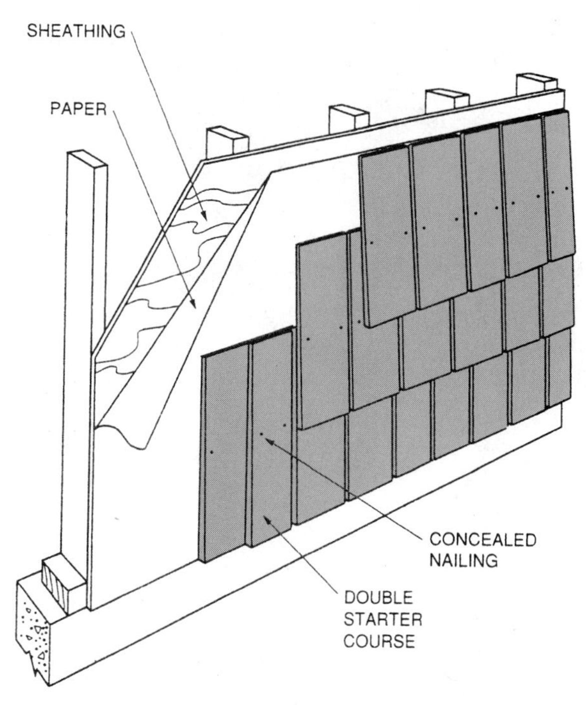

Fig. 62-4 Single-coursed shingle wall application is similar to roof application. However, greater weather exposures are allowed. (*Courtesy of Cedar Shake and Shingle Bureau*)

applications. A triple layer is used for triple coursing. Less expensive *undercourse* shingles are used for underlayers.

Fasten a shingle on both ends of the wall with its butt about 1 inch below the top of the founda-

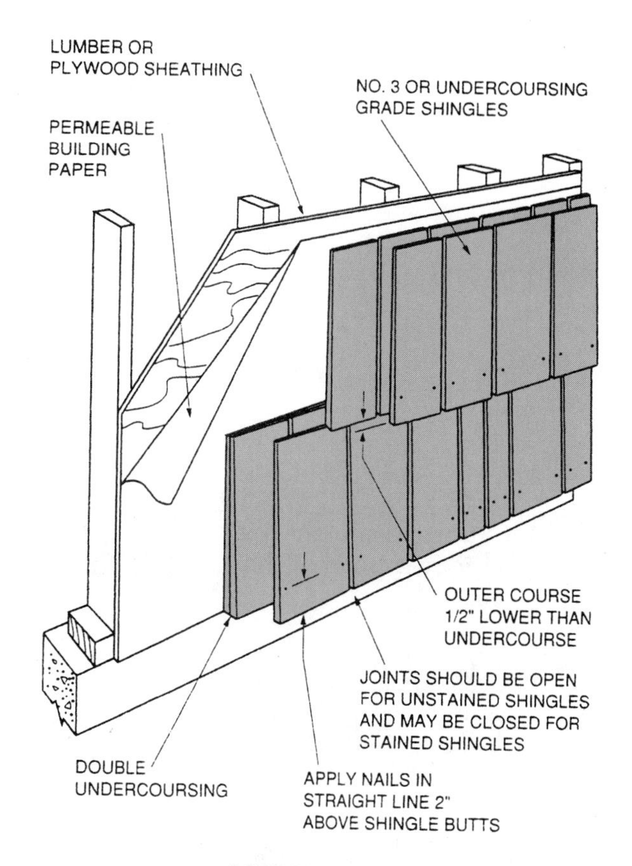

Shingle Length	Max. Weather Exposure Single Course	Double Course
16"	7 1/2"	12"
18"	8 1/2"	14"
24"	11 1/2"	16"

Shake Length & Type	Max. Weather Exposure Single course	Double course
16" Centigroove	7 1/2"	12"
18" Centigroove	8 1/2"	14"
18" resawn	8 1/2"	14"
24" resawn	11 1/2"	18"
18" taper-sawn	8 1/2"	14"
24" tapersplit	11 1/2"	18"
24" taper-sawn	11 1/2"	18"
18" straight-split	8 1/2"	16"

Fig. 62-5 Double-coursed shingles, with two layers in each row, permit even greater exposures when wide courses are desired. (*Courtesy of Cedar Shake and Shingle Bureau*)

tion. Stretch a line between them at the butts. Sight the line for straightness. Fasten additional shingles at necessary intervals. Attach the line to their butts to straighten it (Fig. 62-6). Even a tightly stretched line will sag in the center over a long distance.

Apply a single course of shingles so the butts are as close to the chalk line as possible without touching it. Remove the line. Apply another course on top of the first course. Offset the joints in the outer layer at least 1 1/2 inches from those in the

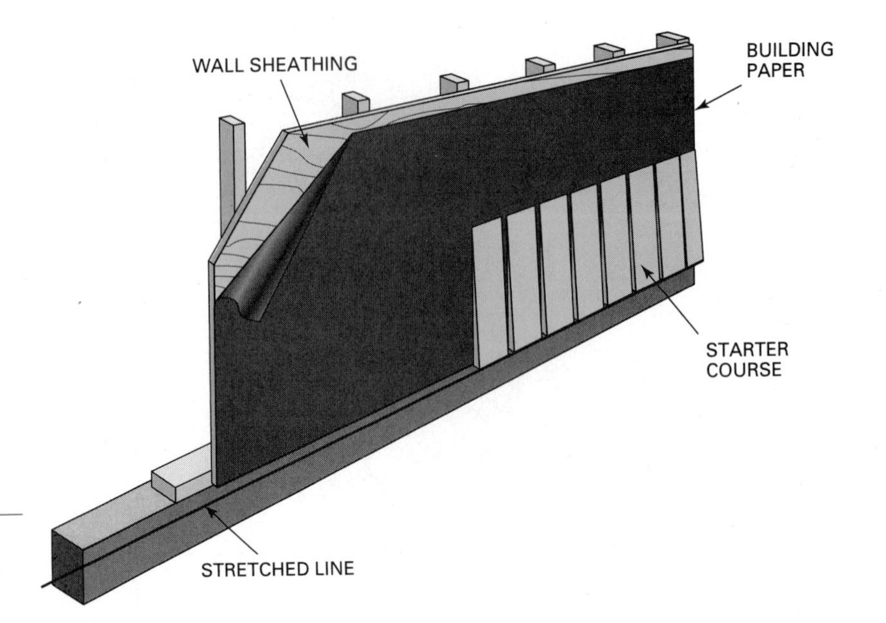

Fig. 62-6 Stretch a straight line as a guide for the butts of the starter course.

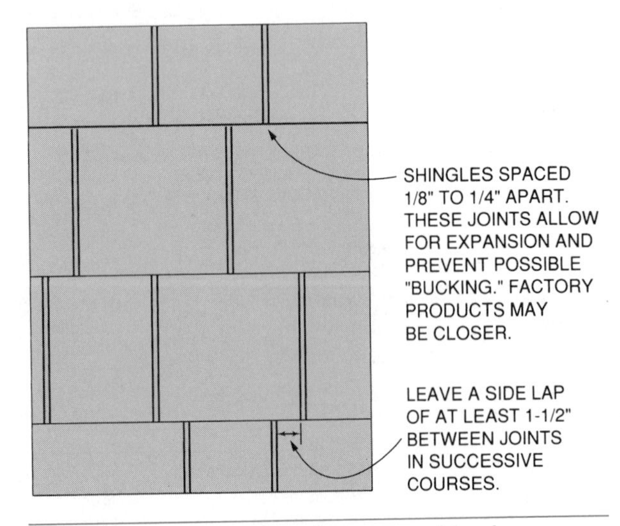

Fig. 62-7 Stagger joints between shingle courses. Space untreated shingles. (*Courtesy of Cedar Shake and Shingle Bureau*)

bottom layer. Untreated shingles should be spaced 1/8 to 1/4 inch apart to allow for swelling and to prevent buckling. Shingles can be applied close together if factory-primed or if treated soon after application (Fig. 62-7).

Single Coursing

A story pole may be used to lay out shingle courses in the same manner as with horizontal wood siding. Snap a chalk line across the wall, at the shingle butt line, to apply the first course. Using only as many finish nails as necessary, tack 1 × 3 straightedges to the wall with their top edges to the line. Lay individual shingles with their butts on the straightedge (Fig. 62-8). Use a shingling hatchet to trim and fit the edges, if necessary. Butt ends are not trimmed. If rebutted and rejointed shingles are used, no trimming should be necessary.

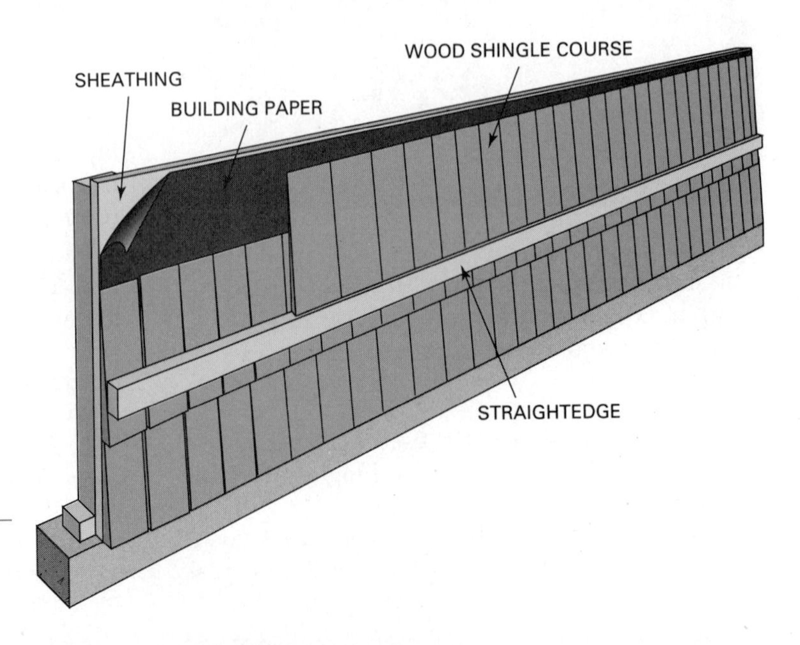

Fig. 62-8 Rest shingle butts on a straightedge when single-coursing on sidewalls.

At times it may be necessary to fit a shingle between others in the same course. Tack the next to last shingle in place with one nail. Slip the last shingle under it. Score along the overlapping edge with the hatchet. Cut along the scored line. Fasten both shingles in place (Fig. 62-9).

Fasteners. Each shingle, up to 8 inches wide, is fastened with two nails or staples about 3/4 inch in from each edge. On shingles wider than 8 inches, drive two additional nails about 1 inch apart near the center. Fasteners should be hot-dipped galvanized, stainless steel, or aluminum. They should be driven about 1 inch above the butt line of the next course. Fasteners must be long enough to penetrate the sheathing by at least 1/2 inch.

Staggered and Ribbon Coursing. An alternative to straight line courses are *staggered* and *ribbon coursing*. In staggered coursing, the butt lines of the shingles are alternately offset below, but not above, the horizontal line. Maximum offsets are one inch for 16- and 18-inch shingles and 1 1/2 inches for 24-inch shingles.

In ribbon coursing, both layers are applied in straight lines. The outer course is raised about 1 inch above the inner course (Fig. 62-10).

Shingle Panels. Shingles in *panelized* form are applied in the same manner as horizontal panel siding, described in the previous chapter (Fig. 62-11). The end shingles of each course in the panel are offset for staggered joints. They match up with adjacent panels.

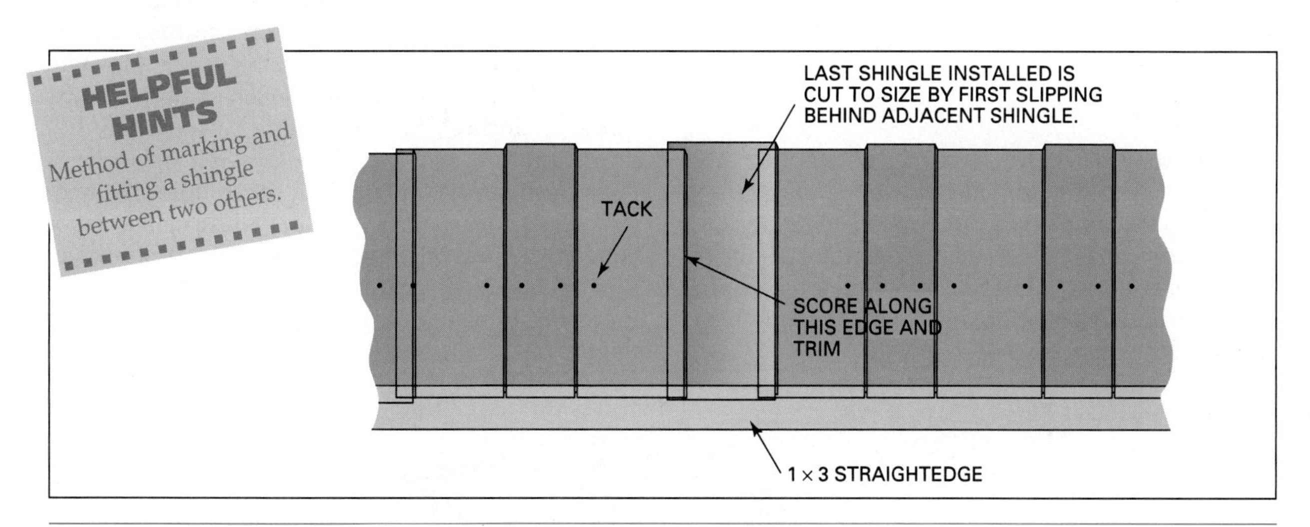

Fig. 62-9 Technique for cutting last shingle to the proper size.

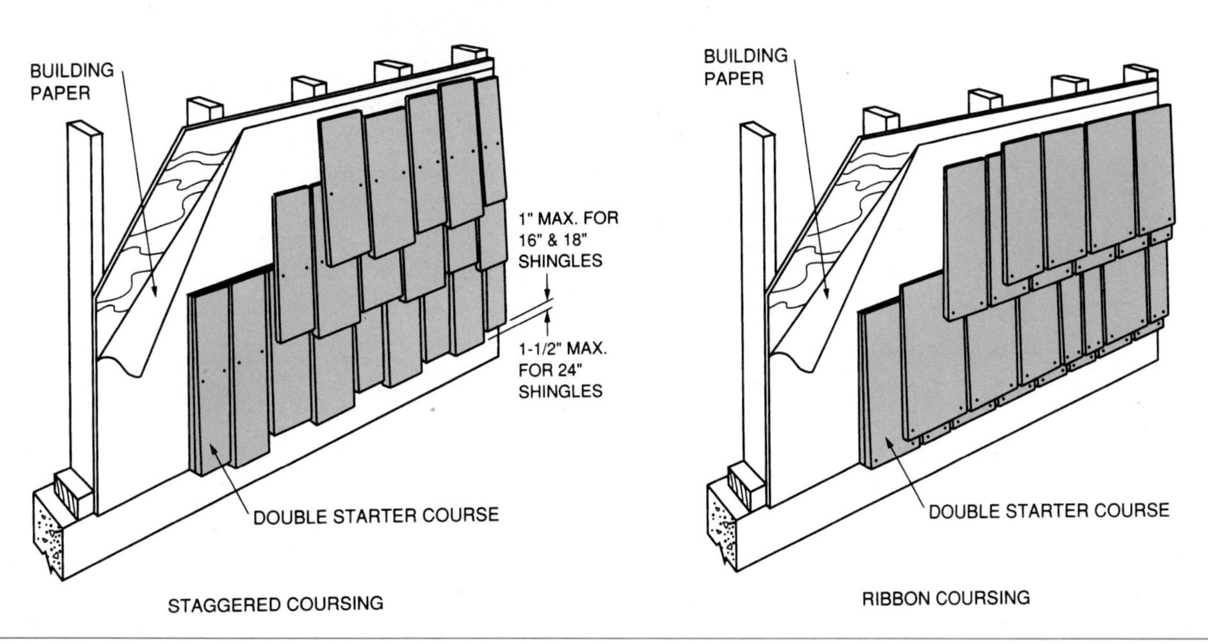

Fig. 62-10 Staggered and ribbon courses are alternatives to straight-line courses.

Fig. 62-11 Wood shingles are made in panel form for faster application. (*Courtesy of Cedar Shake and Shingle Bureau*)

Corners

Shingles may be butted to corner boards like any horizontal wood siding. On outside corners, they may also be applied by alternately overlapping each course in the same manner as in applying a wood shingle ridge. Inside corners may be woven by alternating the corner shingle first on one wall and then the other (Fig. 62-12).

Double Coursing

When double coursing, the starter course is tripled. The outer layer of the course is applied 1/2 inch lower than the inner layer. For ease in application, use a rabbeted straightedge or one composed of two pieces with offset edges (Fig. 62-13).

Fastening. Each inner layer shingle is applied with one fastener at the top center. Each outer course shingle is face-nailed with two 5d galvanized box or special 14-gauge shingle nails. The fasteners are driven about 2 inches above the butts, and about 3/4 inch in from each edge.

Estimating Shingle Siding

The number of squares of shingles needed to cover a certain area depends on how much they are exposed to the weather. One square of shingles will cover 100 square feet when 16-inch shingles are exposed 5 inches, 18-inch shingles exposed 5 1/2 inches, and 24-inch shingles exposed 7 1/2 inches.

A square of shingles will cover more area with greater exposures and less area with smaller exposures. After calculating the wall area as shown in Figure 60-11, determine the amount of shingles needed by using the coverage tables shown in Figure 62-14.

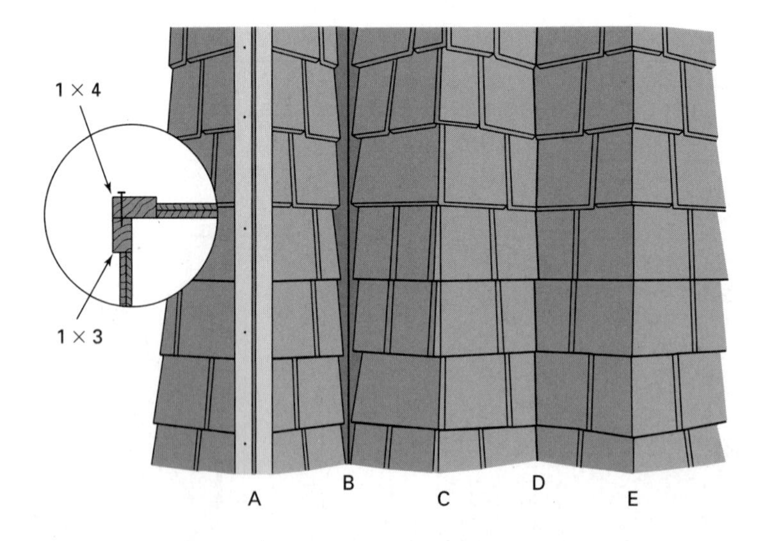

A) SHINGLES BUTTED AGAINST CORNER BOARDS
B) SHINGLES BUTTED AGAINST SQUARE WOOD STRIP ON INSIDE CORNER, FLASHING BEHIND
C) LACED OUTSIDE CORNER
D) LACED INSIDE CORNER WITH FLASHING BEHIND
E) MITERED CORNER

Fig. 62-12 Wood shingle corner details. (*Courtesy of Cedar Shake and Shingle Bureau*)

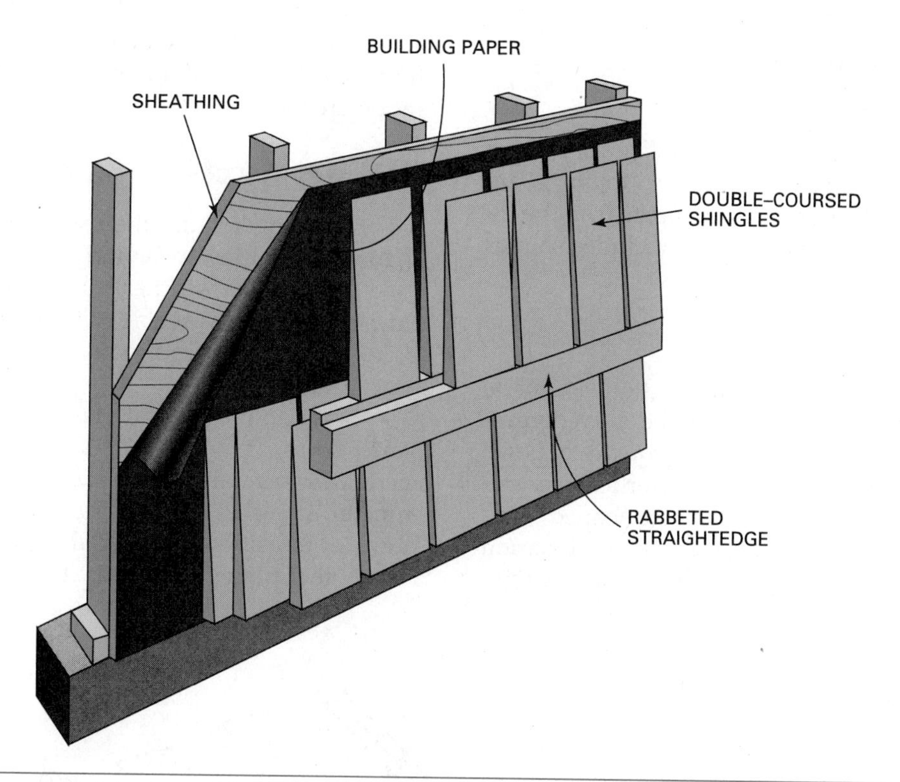

Fig. 62-13 Use a straightedge with rabbeted edges, or one made of two pieces with offset edges, for double-coursed application.

LENGTH AND THICKNESS	Approximate coverage of one square (4-bundle roof-pack) of shingles at indicated weather exposures:													
	3½″	4″	4½″	5″	5½″	6″	6½″	7″	7½″	8″	8½″	9″	9½″	10″
16″ × 5/2″	70	80	90	100	110	120	130	140	150	160	170	180	190	200
18″ × $5/2^{1/4}$″	—	72½	81½	90½	100	109	118	127	136	145½	154½	163½	172½	181½
24″ × 4/2″	—	—	—	—	73½	80	86½	93	100	106½	113	120	126½	133

LENGTH AND THICKNESS	10½″	11″	11½″	12″	12½″	13″	13½″	14″	14½″	15″	15½″	16″
16″ × 5/2″	210	220	230	240	—	—	—	—	—	—	—	—
18″ × $5/2^{1/4}$″	191	200	209	218	227	236	245½	254½	—	—	—	—
24″ × 4/2″	140	146½	153	160	166½	173	180	186½	193	200	206½	213

Fig. 62-14 Sidewall shingle coverage at various exposures. (*Courtesy of Cedar Shake and Shingle Bureau*)

CHAPTER 63 ALUMINUM AND VINYL SIDING

Except for the material, *aluminum* and *vinyl* siding systems are similar. Aluminum siding is finished with a baked-on enamel. In vinyl siding, the color is in the material itself. Both kinds are manufactured with inter-locking edges, for horizontal and vertical applications. Descriptions and instructions are given here for vinyl siding systems, much of which can be applied to aluminum systems.

Siding Panels and Accessories

Siding systems are composed of *siding panels* and several specially shaped *moldings.* Moldings are used on various parts of the building, to trim the installation. In addition, the system includes shapes for use on *soffits.*

Siding Panels

Siding panels, for horizontal application, are made in 8- and 12-inch widths. They come in configurations to simulate one, two, or three courses of bevel or drop siding. Panels designed for vertical application come in 12-inch widths. They are shaped to resemble boards. They can be used in combination with horizontal siding. Vertical siding panels with solid surfaces may also be used for soffits. For ventilation, *perforated* soffit panels of the same configuration are used (Fig. 63-1).

Siding System Accessories

Siding systems require the use of several specially shaped accessories. *Inside* and *outside corner posts* are used to provide a weather-resistant joint to corners. Corner posts are available with channels of appropriate widths to accommodate various configurations of siding.

Some other accessories include *J-channels, starter strips,* and *undersill/finish trim. J-channels* are made with several opening sizes. They are used in a number of places such as around doors and windows, at transition of materials, against soffits, and many other places (Fig. 63-2). The majority of vinyl

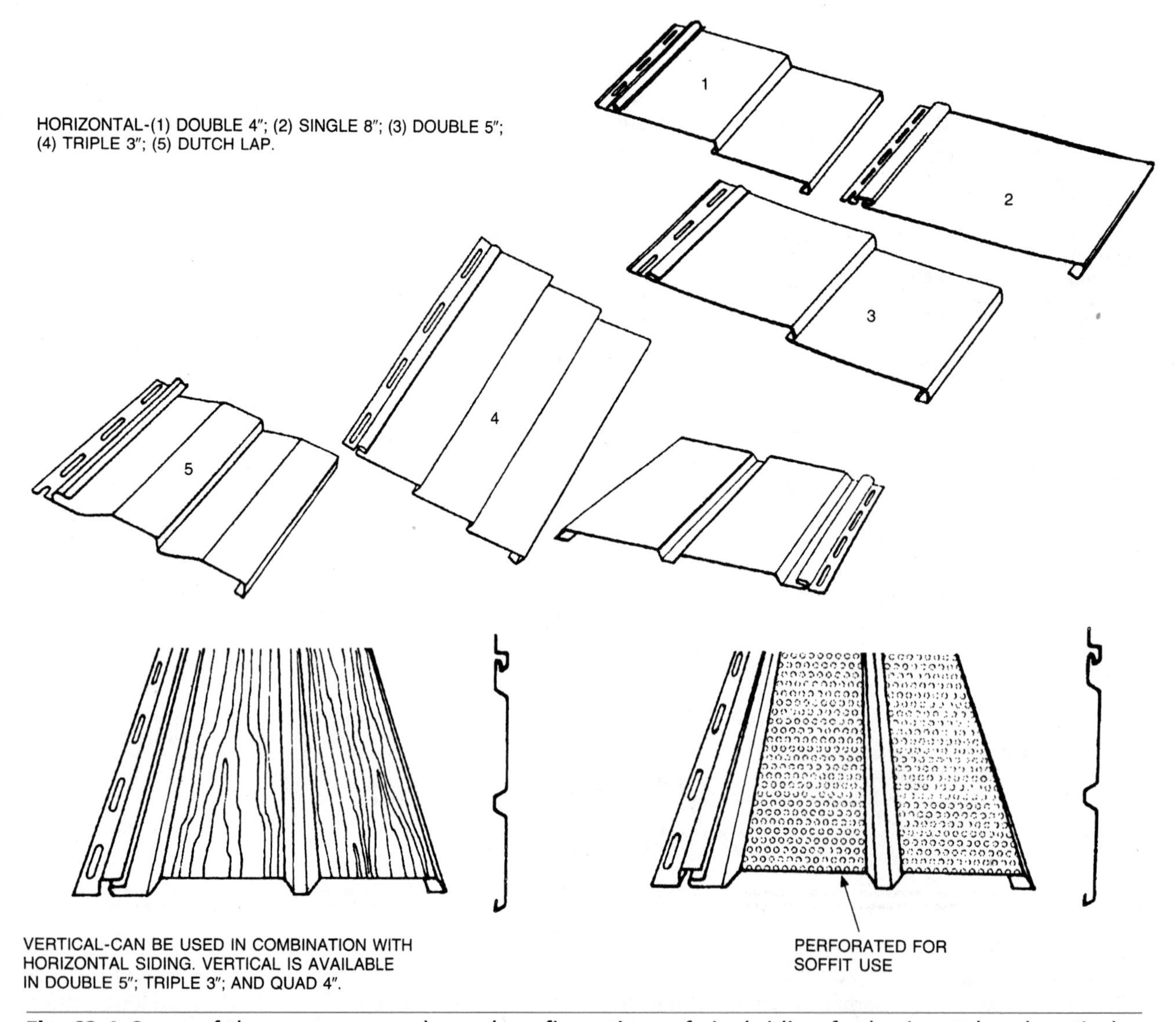

Fig. 63-1 Some of the most commonly used configurations of vinyl siding for horizontal and vertical application. (*Courtesy of Vinyl Siding Institute*)

siding panels and accessories are manufactured in 12′-6″ lengths.

Applying Horizontal Siding

The siding may expand and contract as much as 1/4 inch over a 12′-6″ length with changes in temperature. For this reason, it is important to center fasteners in the slots. Do not drive them too tightly. There should be about 1/32 inch between the head of the fastener, when driven, and the siding (Fig. 63-3). Space fasteners 16 inches apart for horizontal siding and 6 to 12 inches apart for accessories unless otherwise specified by the manufacturer.

Applying the Starter Strip

Snap a level line to the height of the starter strip all around the bottom of the building. Fasten the strips

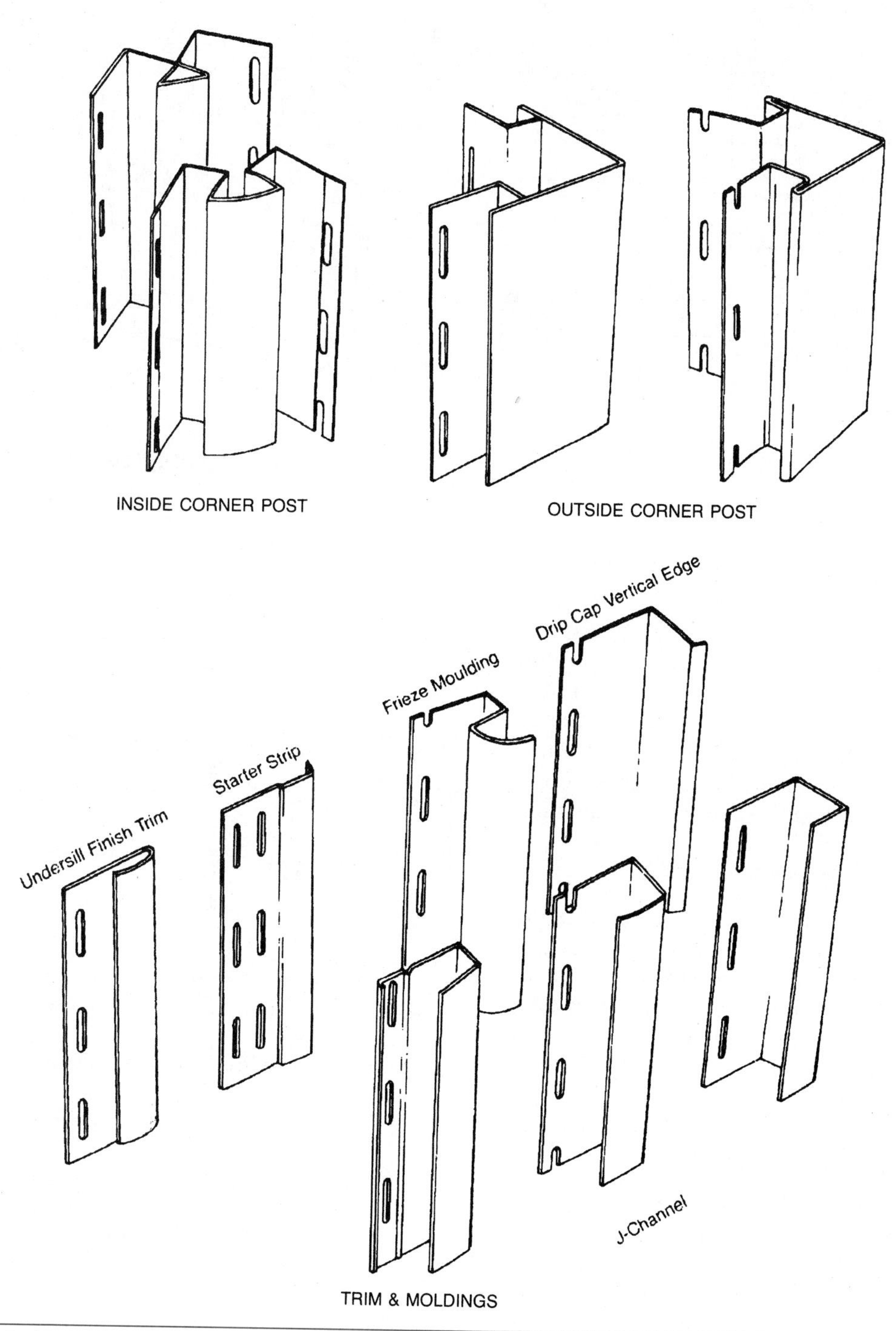

Fig. 63-2 Various accessories are used to trim a vinyl siding installation. (*Courtesy of Georgia-Pacific Corporation*)

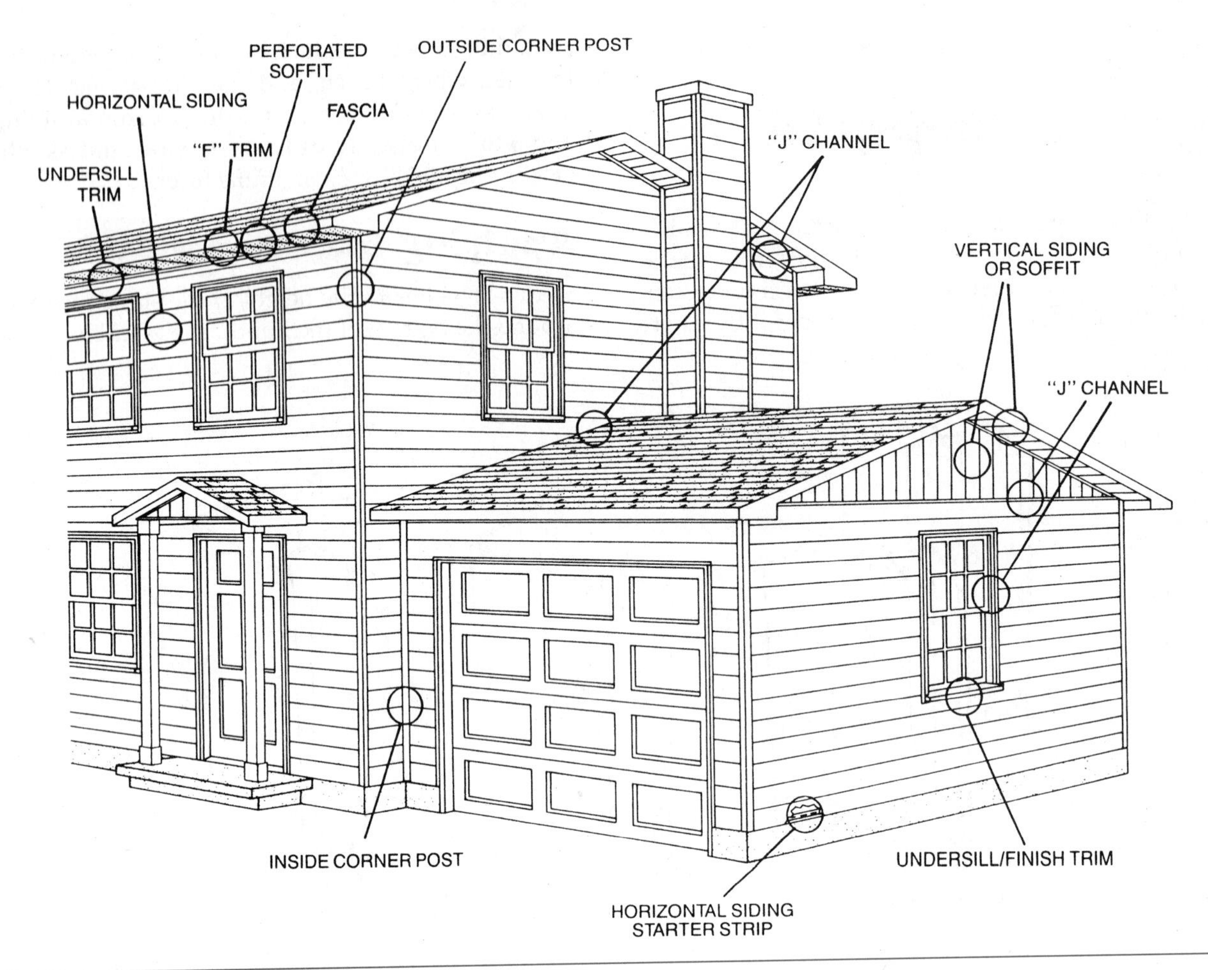

Fig. 63-2 *(continued)*

to the wall with their top edges to the chalk line. Leave a 1/4-inch space between them and other accessories to allow for expansion (Fig. 63-4). Make sure the starter strip is applied as straight as possible. It controls the straightness of entire installation.

Installing Corner Posts

Corner posts are installed in corners 1/4 inch below the starting strip and 1/4 inch from the top. Attach the posts by fastening in the top of the upper slot on each side. The posts will hang on these fasteners. The rest of the fasteners should be centered on the slots. Make sure the posts are straight and true from top to bottom (Fig. 63-5).

Installing J-Channel

Install J-channel across the top and along the sides of window and door casings. It may also be installed under windows or doors with the *undersill* nailed inside of the channel. To miter the corners, cut all pieces to extend, on both ends, beyond the casings and sills a distance equal to the width of the channel face. On both ends of the side J-

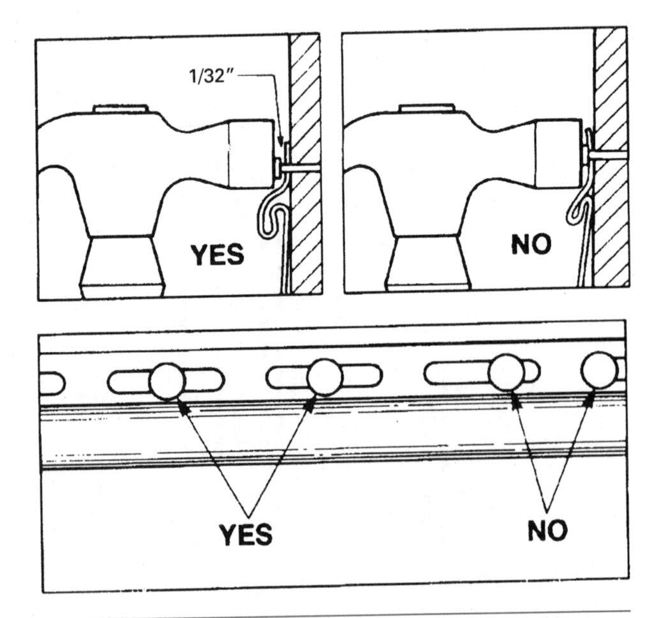

Fig. 63-3 Fasten siding to allow for expansion and contraction.

channels, cut a 3/4-inch notch out of the bottom. Fasten in place. On both ends of the top and bottom channels, make 3/4-inch cuts. Bend down the

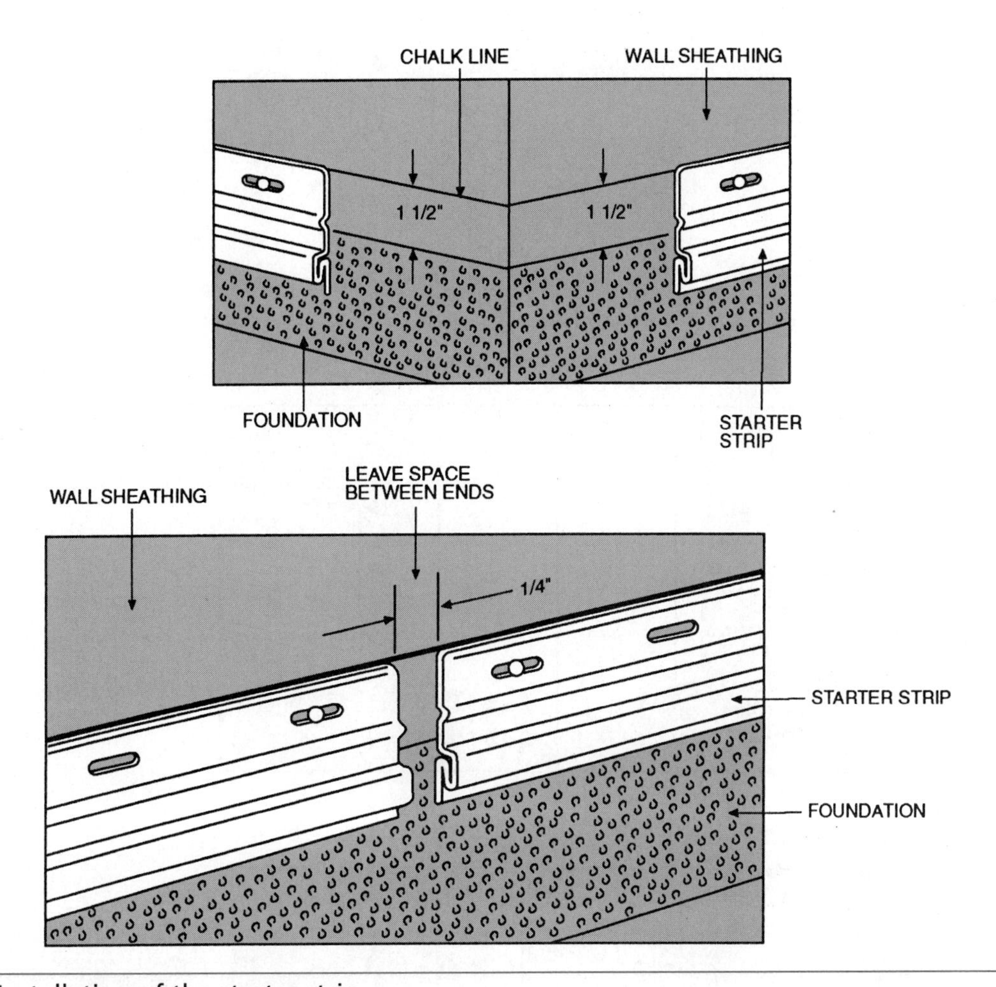

Fig. 63-4 Installation of the starter strip.

tabs and miter the faces. Install them so the mitered faces are in front of the faces of the side channels (Fig. 63-6).

Installing Siding Panels

Snap the bottom of the first panel into the starter strip. Fasten it to the wall. Start from a back corner, leaving a 1/4-inch space in the corner post channel. Work toward the front with other panels. Overlap each panel about 1 inch. The exposed ends should face the direction from which they are least viewed (Fig. 63-7).

Install successive courses by interlocking them with the course below and staggering the joints between courses. Use tin snips, hacksaw, utility knife, or power saw with an abrasive wheel or fine-tooth circular blade. Reverse the blade if a power saw is used, for smooth cutting through the vinyl.

Fitting Around Windows. Plan so there will be no joints under a window. Hold the siding panel under the sill. Mark the width of the cutout, allowing 1/4-inch clearance on each side. Mark the height of the cutout, allowing 1/4-inch clearance below the sill.

Make vertical cuts with tin snips. Score the horizontal layout lines with a utility knife or scoring tool. Bend the section to be removed back and forth until it separates. Using a special *snaplock punch,* punch the panel 1/4 inch below the cut edge at 6-inch intervals to produce *raised lugs* facing outward. Install the panel under the window and up in the undersill trim. The raised lugs cause the panel to fit snugly in the trim (Fig. 63-8).

Panels are cut and fit over windows in the same manner as under them. However, the lower portion is cut instead of the top. Install the panel by placing it into the J-channel that runs across the top of the window (Fig. 63-9).

Installing the Last Course under the Soffit

The last course of siding panels under the soffit is installed in a manner similar to fitting under a window.

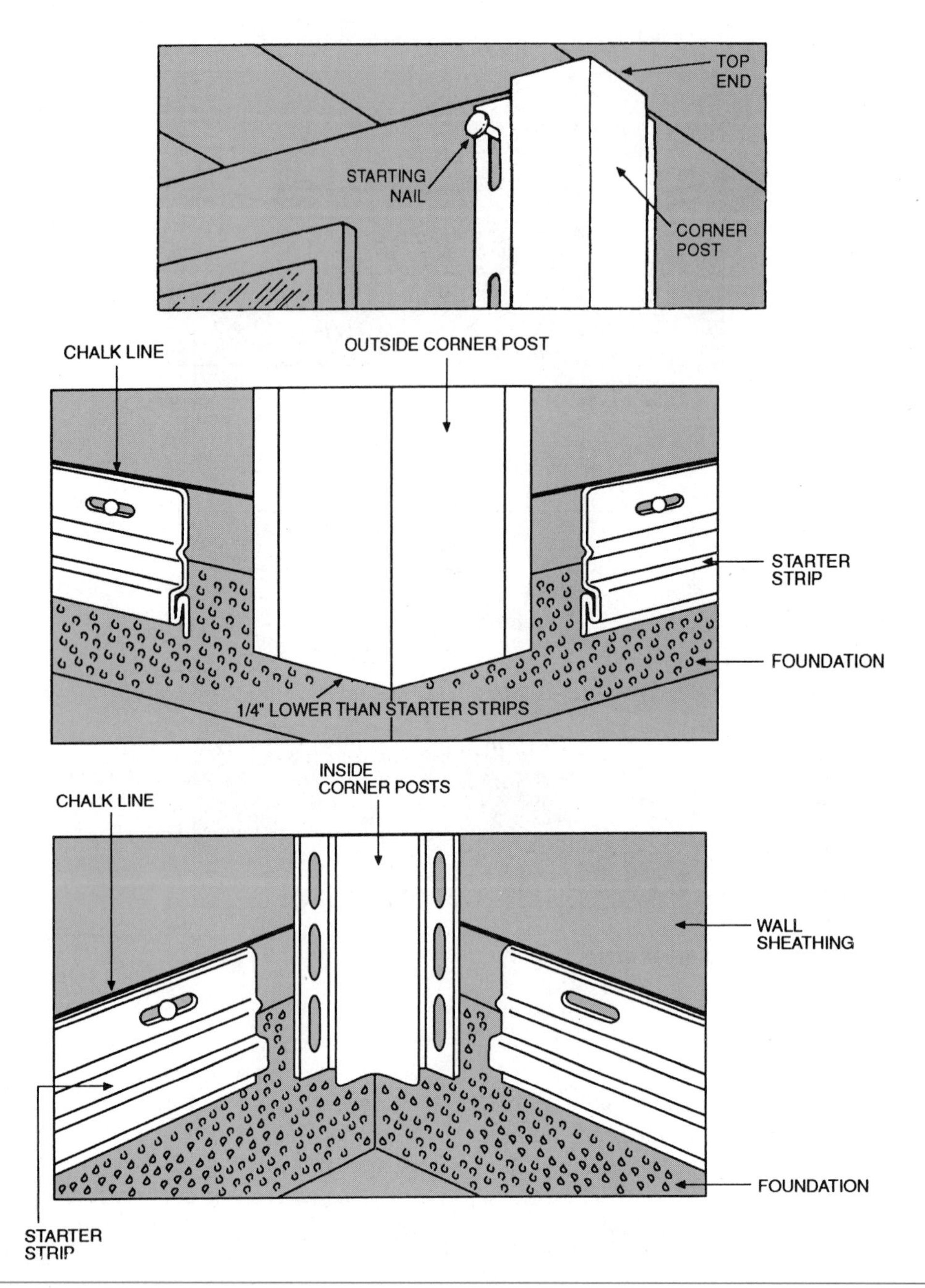

Fig. 63-5 Inside and outside corner posts are installed in a similar way.

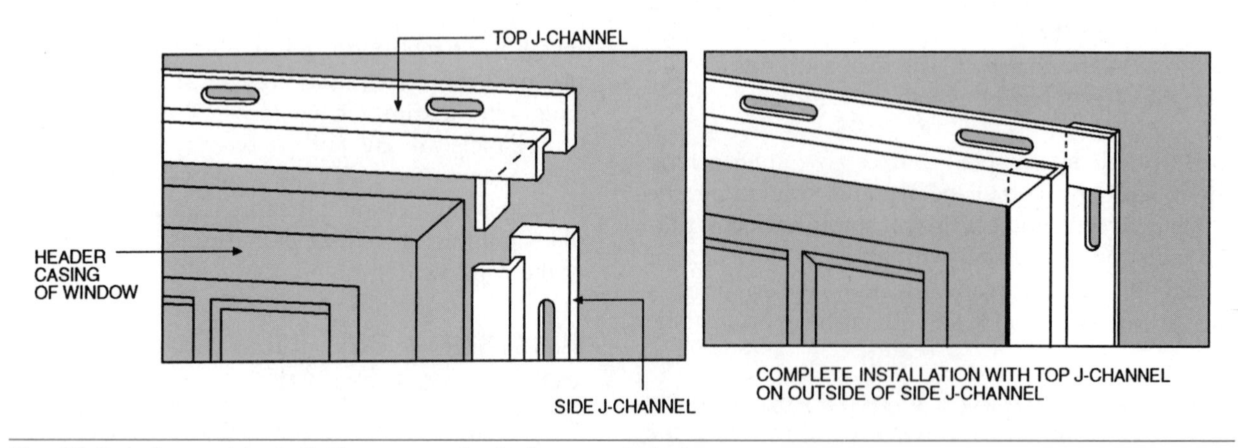

Fig. 63-6 J-channel installation around windows and doors.

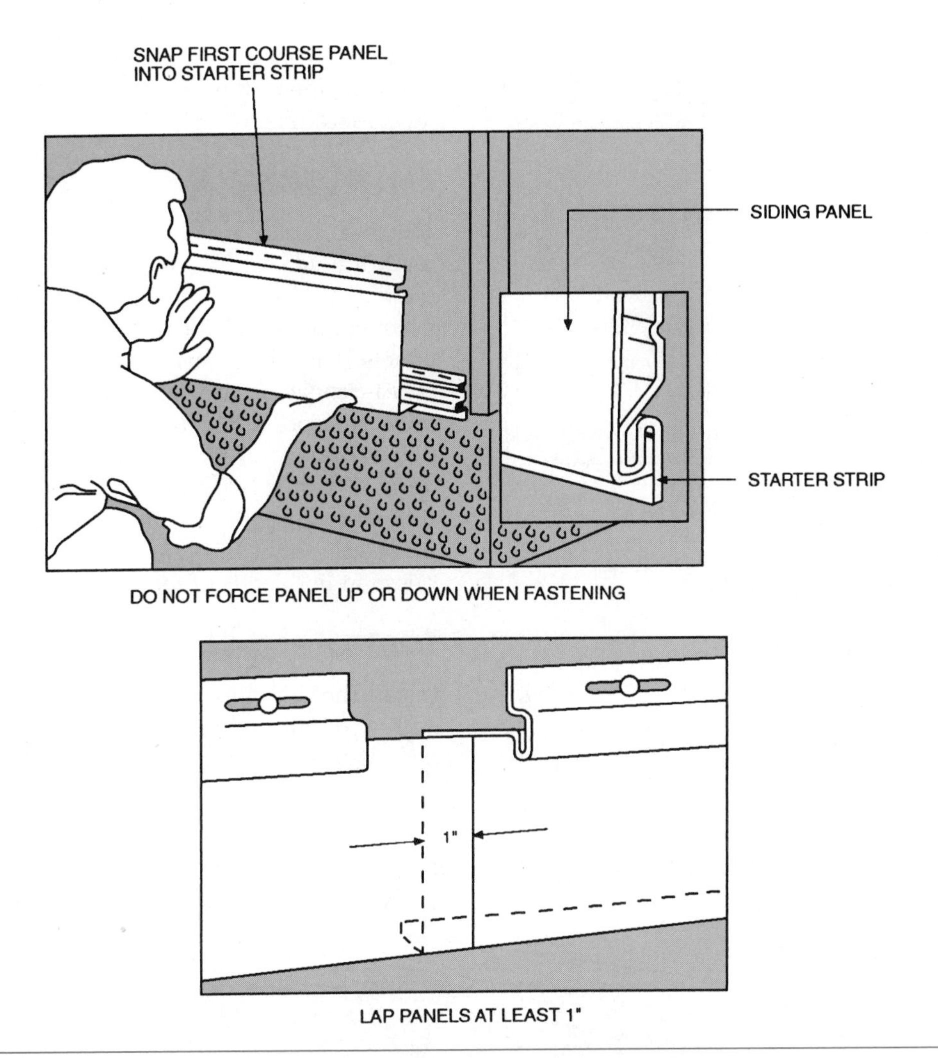

Fig. 63-7 Installing the first course of horizontal siding panels.

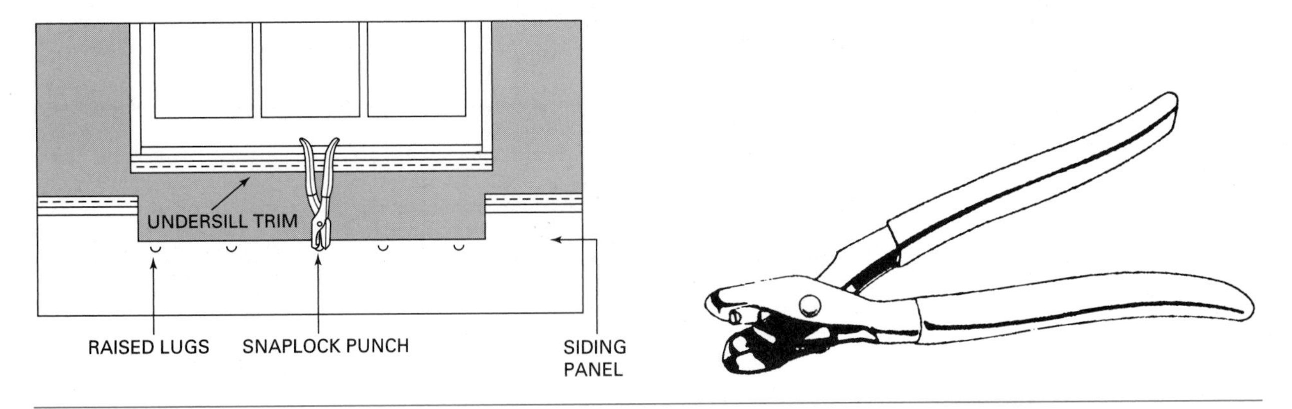

Fig. 63-8 Method of fitting a panel under a window.

An *undersill trim* is applied on the wall and up against the soffit. Panels in the last course are cut to width. Lugs are punched along the cut edges. The panels are then snapped firmly into place into the undersill trim (Fig. 63-10).

Gable End Installation

The rakes of a gable end are first trimmed with J-channels. The panel ends are inserted into the channel with 1/4-inch expansion gap. Make a pattern

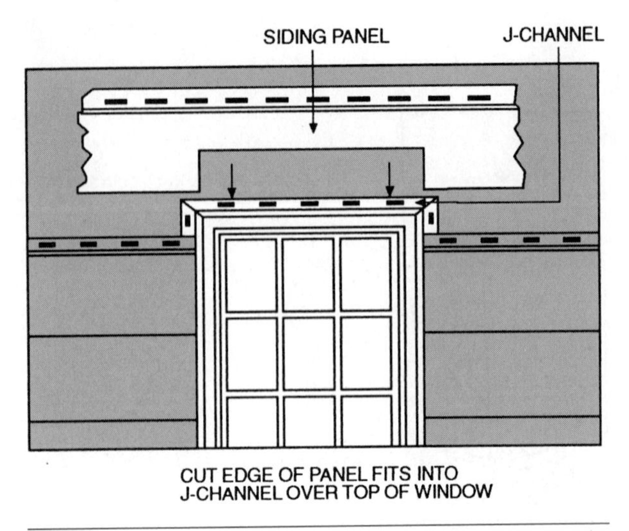

Fig. 63-9 Fitting a panel over a window.

for cutting gable end panels at an angle where they intersect with the rake. Use two scrap pieces of siding to make the pattern. Interlock one piece with an installed siding panel below. Hold the other piece on top of it and against the rake. Mark along the bottom edge of the slanted piece on the face of the level piece (Fig. 63-11).

Applying Vertical Siding

The installation of vertical siding is similar to that for horizontal siding with a few exceptions. The method of fastening is the same. However, space fasteners about 12 inches apart for vertical siding panels. The starter strip is different. It may be *1/2-inch J-channel* or *drip cap* flush with and fitted into the corner posts (Fig. 63-12). Around windows and doors, under soffits, against rakes, and other locations, 1/2-inch J-channel is used. One of the major differences is that a vertical layout should be planned so that the same and widest panel is exposed at both ends of the wall.

Installing the First Panel

Install the first vertical panel plumb on the starter strip with one edge into the corner post. Allow 1/4 inch at top and bottom. Place the first nails in the

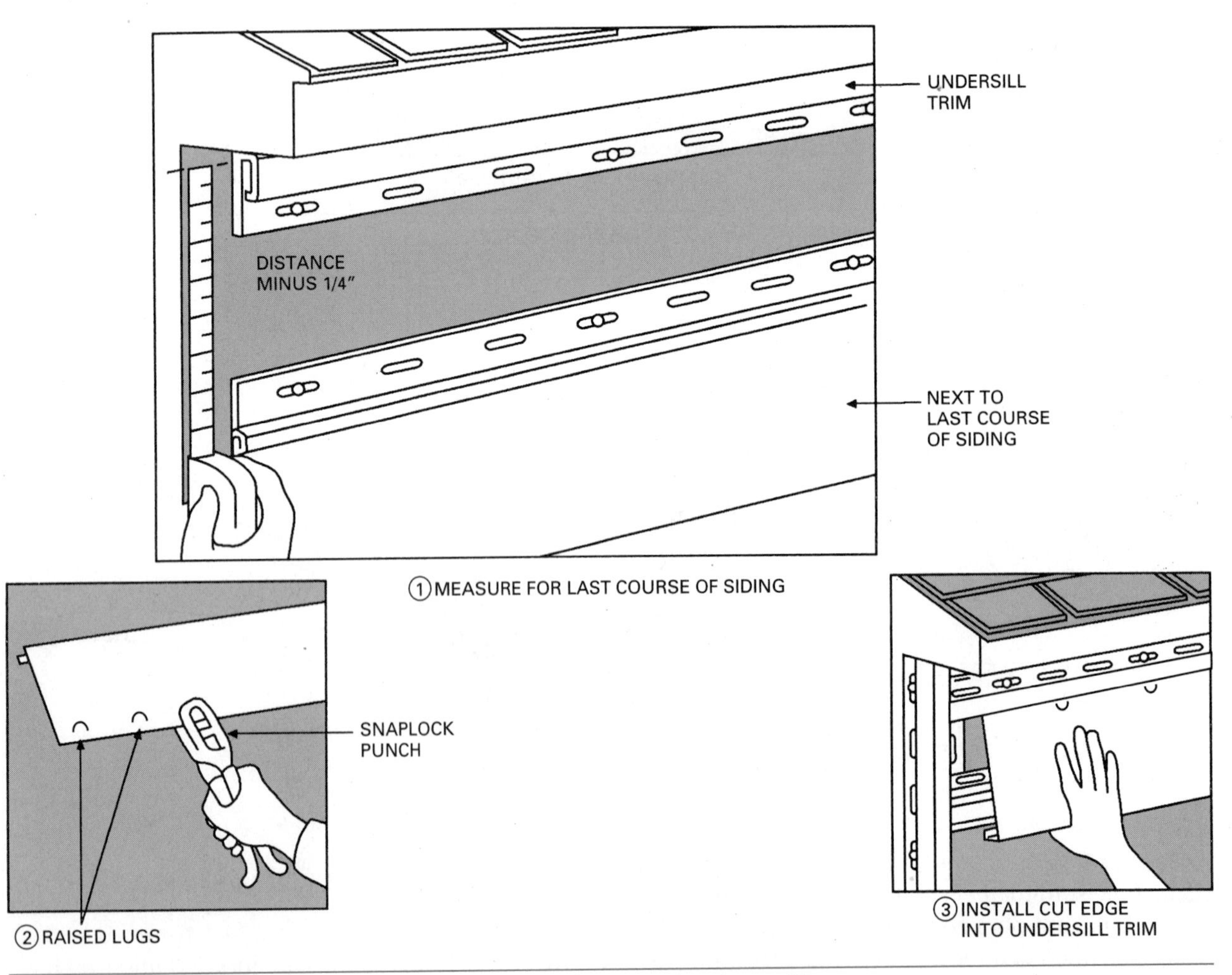

Fig. 63-10 Fitting the last course of horizontal siding under the soffit.

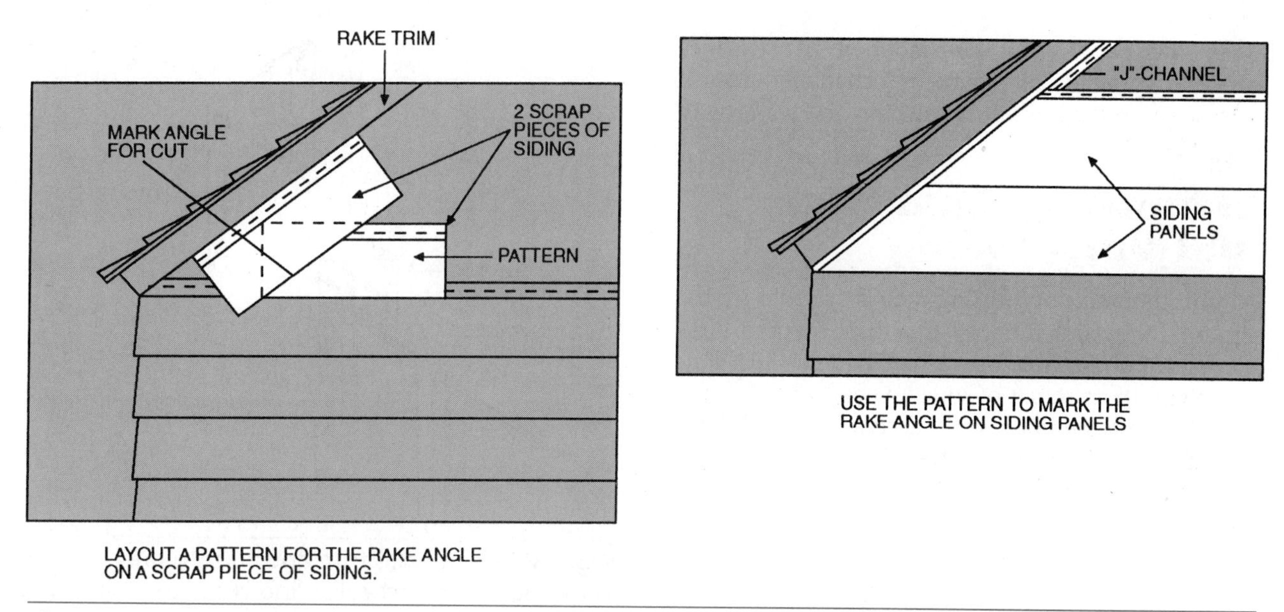

Fig. 63-11 Fitting horizontal siding panels to the rakes.

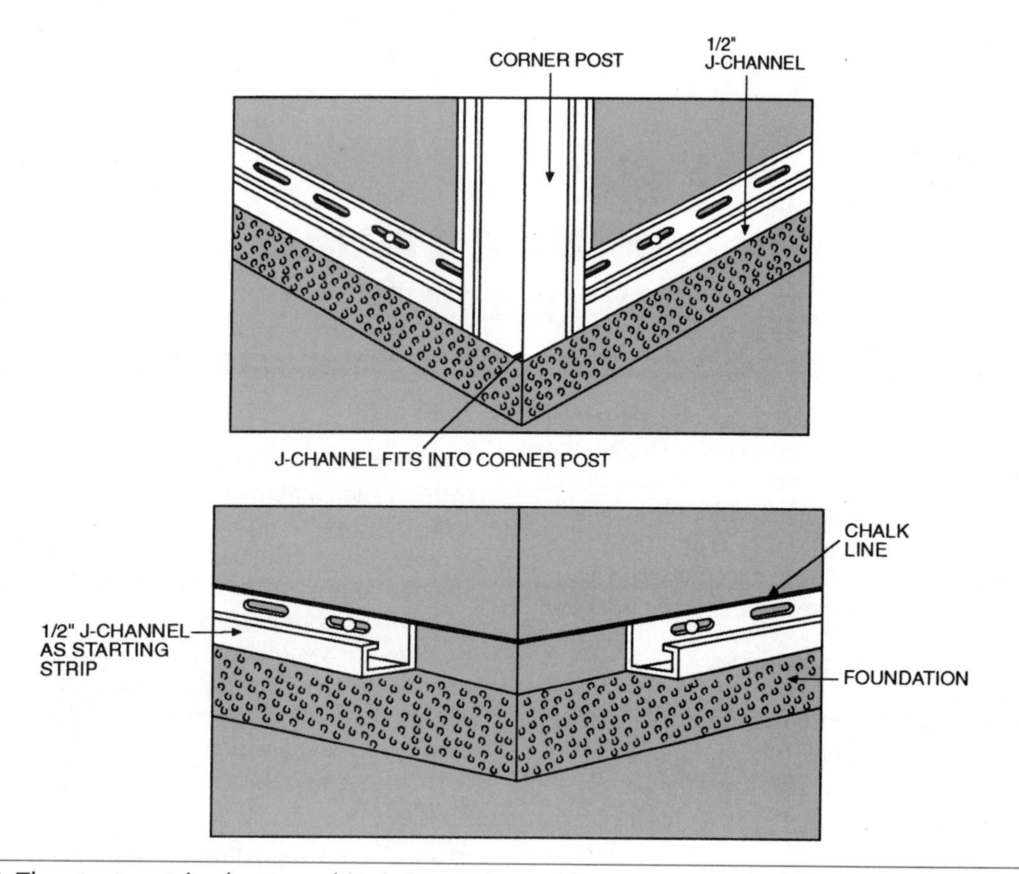

Fig. 63-12 The starter strip shape and its intersection with corner posts is different for vertical application of vinyl siding.

uppermost end of the top nail slots to hold it in position.

The edge of the panel may need to be cut in order for equal widths to be exposed on both ends of the wall. If the panel is cut on the flat surface, place a piece of undersill trim backed by furring into the channel of the corner post. Punch lugs along the cut edge of the panel at 6-inch intervals.

Snap the panel into the undersill trim. Edges of vertical panels cut to fit in J-channels around windows and doors are treated in the same way (Fig. 63-13).

Estimating Aluminum and Vinyl Siding

Aluminum and vinyl siding panels are sold by the square. Determine the wall area to be covered. Add 10 percent of the area for waste. Divide by 100. This gives you the number of squares needed.

Become familiar with accessories and how they are used. Measure the total linear feet required for each item.

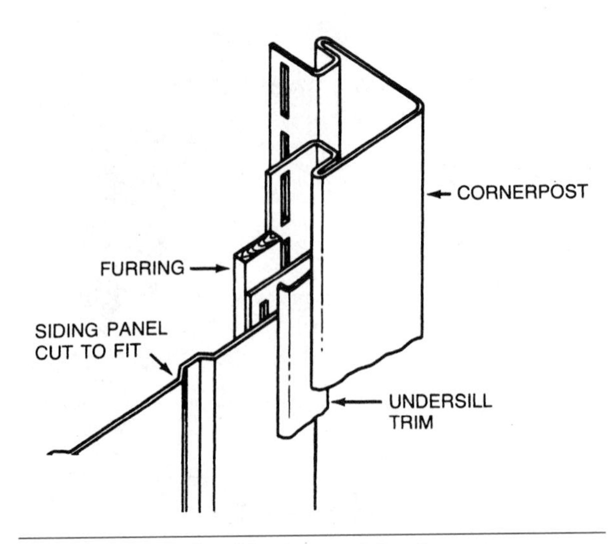

Fig. 63-13 Undersill trim and furring are required when vertical siding is cut to fit into corner posts and J-channel on the sides of windows and doors. (*Courtesy of Vinyl Siding Institute*)

CHAPTER 64 CORNICE TERMS AND DESIGN

Cornice terms from earlier times still remain in use. However, cornice design has changed considerably. In earlier times, cornices were very elaborate and required much time to build. Now, cornice design, in most cases, is much more simplified. Only occasionally is a building designed with an ornate cornice similar to those built in years gone by.

Parts of a Cornice

Several finish parts are used to build the cornice. In some cases, additional framing members are required.

Subfascia

The *subfascia* is sometimes called the *false fascia* or *rough fascia* (Fig. 64-1). It is a horizontal framing member fastened to the rafter tails. It provides an even, solid, and continuous surface for the attachment of other cornice members. When used, the subfascia is generally a nominal 1- or 2-inch-thick piece. Its width depends on the slope of the roof, the tail cut of the rafters, or the type of cornice construction.

Soffit

The finished member on the underside of the cornice is called a **plancier** and is often referred to as a *soffit* (Fig. 64-2). Soffit material may include solid lumber, plywood, strand board, fiberboard or corrugated aluminum and vinyl panels. Soffits should be perforated or constructed with screen openings to allow for ventilation of the rafter cavities. Soffits may be fastened to the bottom edge of the rafter tails to the slope of the roof. The soffit is an ideal location for the placement of attic ventilation.

Fascia

The *fascia* is fastened to the subfascia or to the ends of the rafter tails. It may be a piece of lumber grooved to receive the soffit. It also may be made from bent aluminum and vinyl material used to wrap the subfascia. Fascia provides a surface for the attachment of a **gutter**. The fascia may be built up from one or more members to enhance the beauty of the cornice.

The bottom edge of the fascia usually extends below the soffit by 1/4 to 3/8 inch. The portion of

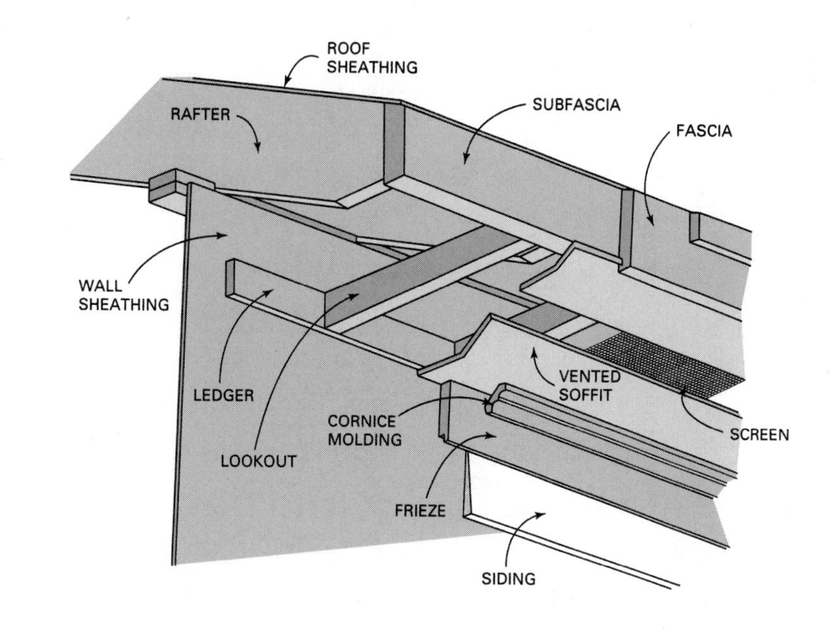

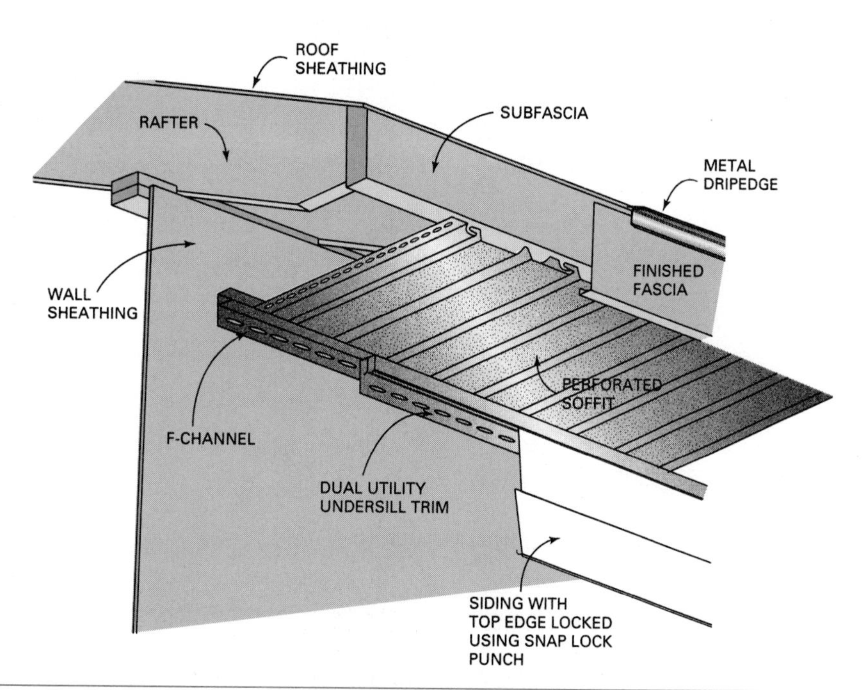

Fig. 64-1 The subfascia is a framing member that is fastened to the ends of the rafter tails.

the fascia that extends below the soffit is called the **drip**. The drip is necessary to prevent rainwater from being swept back against the walls of the building. In addition, a drip makes the cornice more attractive.

Frieze

The *frieze* is fastened to the sidewall with its top edge against the soffit. Its bottom edge is sometimes *rabbeted* to receive the sidewall finish. In other cases, the frieze may be furred away from the sidewall to allow the exterior wall *siding* to extend above and behind its bottom edge. However, the frieze is not always used. The sidewall finish may be allowed to come up to the soffit. The joint between the siding and the soffit is then covered by a molding.

Cornice Molding

The *cornice molding* is used to cover the joint between the frieze and the soffit. If the frieze is not used, the cornice molding covers the joint between the siding and the soffit.

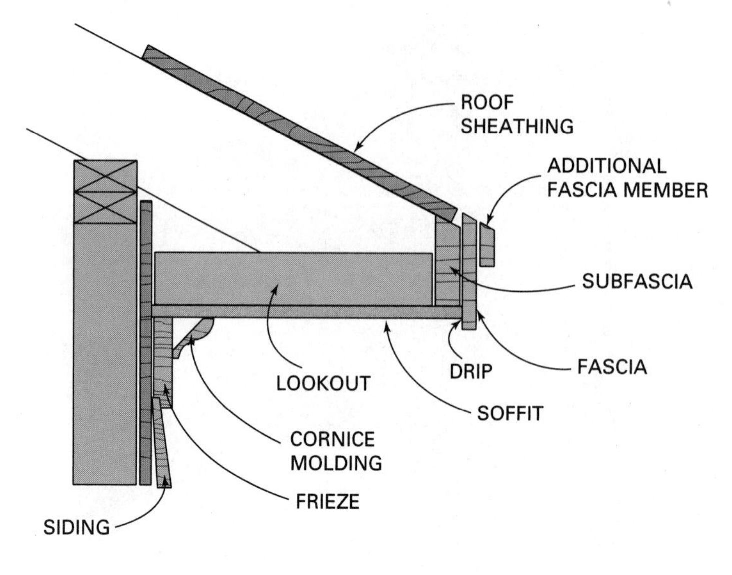

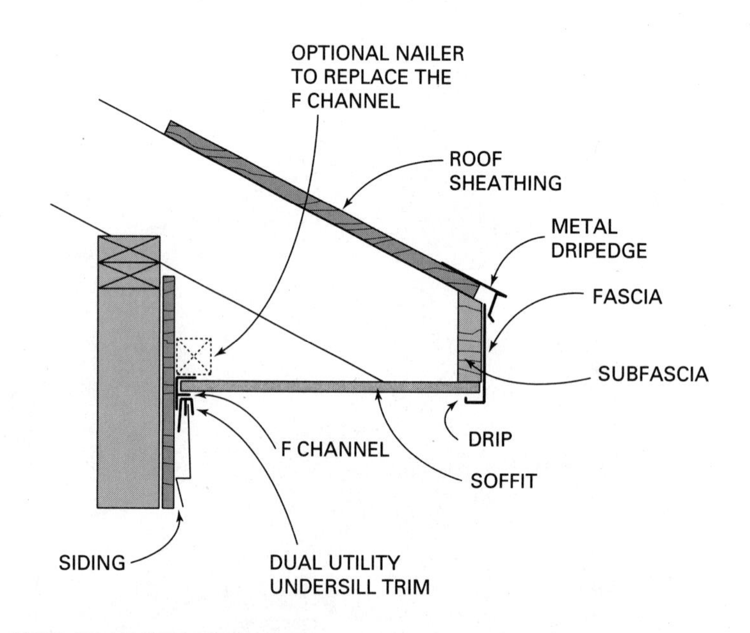

Fig. 64-2 The soffit is the bottom finish member of the cornice; through it, ventilation is provided to the attic.

Lookouts

Lookouts are framing members, usually 2 × 4 stock used to provide a fastening surface for the soffit. They run horizontally from the end of the rafter to the wall, adding extra strength to larger overhangs. Lookouts may be installed at every rafter or spaced 48 inches OC, depending on the material being used for the soffit (Fig. 64-2).

Cornice Design

Cornices are generally classified into three main types: *box, snub,* and *open* (Fig. 64-3).

The Box Cornice

The *box cornice* is probably most common. It gives a finished appearance to this section of the exterior. Because of its overhang, it helps protect the sidewalls from the weather. It also provides shade for windows.

Box cornices may be designated as narrow or wide. They may be constructed with level or sloping soffits. A *narrow box cornice* is one in which the cuts on the rafters serve as nailing surfaces for the cornice members. A *wide box cornice* may be constructed with a level or sloping soffit. A wide, level soffit requires the installation of lookouts.

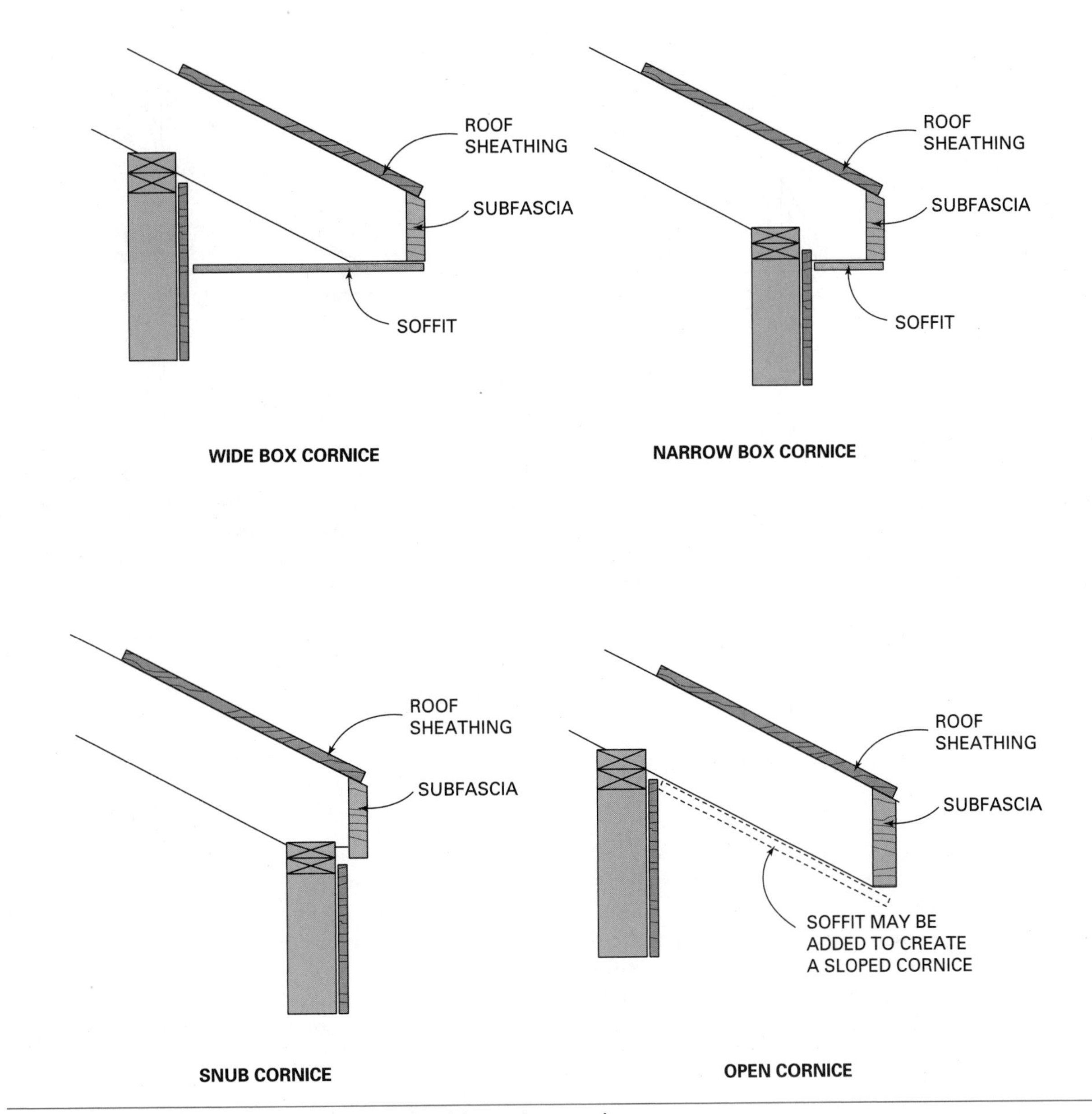

Fig. 64-3 The cornice may be constructed in various styles.

The Snub Cornice

The *snub cornice* does not provide as attractive an appearance. Nor does it give as much protection to the sidewalls of a building because of its small overhang. There is no rafter projection beyond the wall. The snub cornice is chosen primarily to cut down the cost of material and labor. A snub cornice is frequently used on the *rakes* of a gable end in combination with a boxed cornice on the sides of a building.

The Open Cornice

The *open cornice* has no soffit. It is used when it is desired to expose the rafter tails. It is often used when the rafters are large, laminated or solid beams with a wide overhang that exposes the roof decking on the bottom side. Open cornices give contemporary or rustic design to post-and-beam framing. They provide protection to sidewalls at low cost. This cornice might also be used for conventionally framed buildings for reasons of design and to reduce costs.

By adding a soffit, a *sloped cornice* is created. The soffit is installed directly to the underside of the rafter tails. This is sometimes done to simplify the cornice detail when there is also an overhang over the gable end of the building.

Rake Cornices

The *main cornice* is constructed on the rafter tails where they meet the walls of a building. On buildings with hip or mansard roofs, the main cornice, regardless of the type, extends around the entire building.

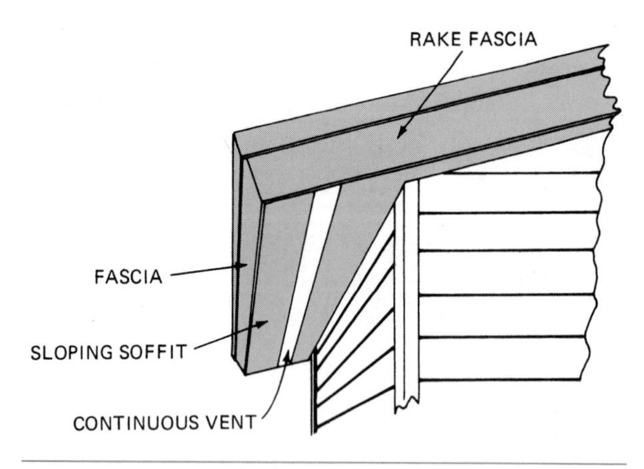

Fig. 64-4 A boxed cornice with a sloping soffit may be returned up the rakes of a gable roof.

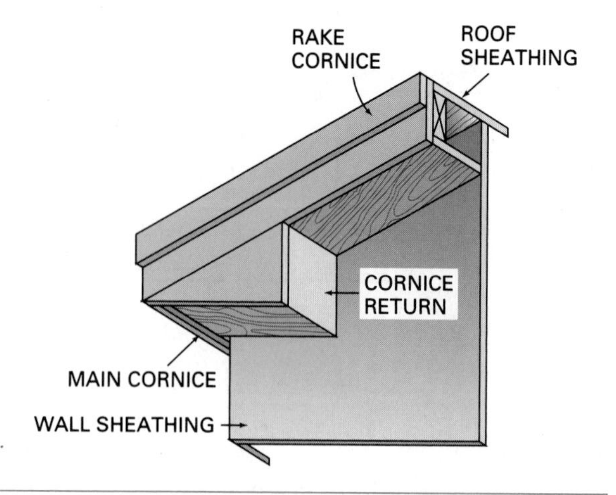

Fig. 64-6 A boxed main cornice with a level soffit may be changed to the angle of the roof with a simple cornice return.

On buildings with gable roofs, a boxed main cornice with a sloping soffit, attached to the bottom edge of the rafter tails, may be returned up the rakes to the ridge. The cornice that runs up the rakes is called a *rake cornice* (Fig. 64-4).

Cornice Returns. A main cornice with a horizontal soffit attached to level lookouts may, at times, be terminated at each end wall against a snub rake cornice (Fig. 64-5). At other times, a *cornice return* must be constructed to change the level box cornice to the angle of the roof.

A main cornice with a level soffit may also be *returned upon itself.* That is, the main cornice is mitered at each end. It is turned 90 degrees toward and beyond the corner as much as it overhangs on the side of the building. This short section on each end of a gable roof is called a *cornice return.* The cornice return provides a *stop* for the *rake cornice.* It adds to the design of the building at this point (Fig. 64-6). However, cornice returns of this type are rarely built today. A large amount of labor is involved in their construction. The main cornice is returned on the rakes of the gable end in a much more simple fashion as described.

Practical Tips for Installing Cornices

Install cornice members in a straight and true line with well fitting and tight joints. Do not dismiss the use of hand tools for cutting, fitting, and fastening exterior trim.

Many times a hand tool is a better choice than a power tool. For example, a sharp handsaw can do

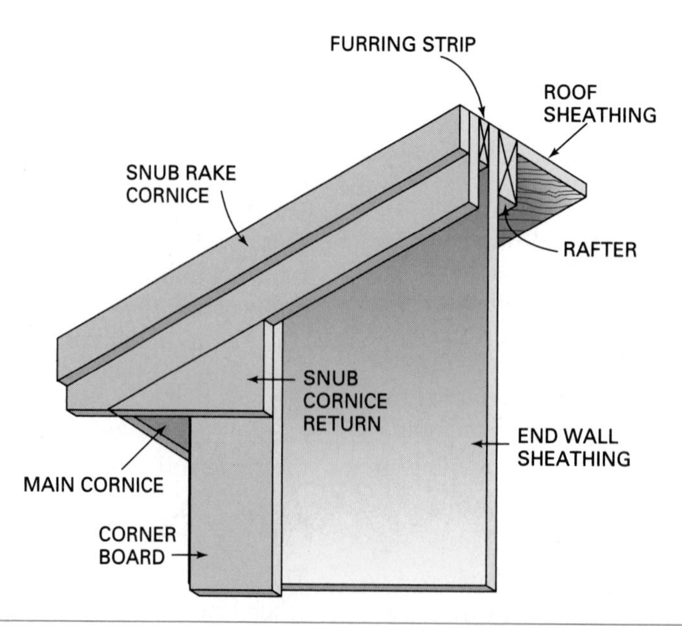

Fig. 64-5 A boxed main cornice with a level soffit may be terminated at the gable ends against the rake cornice.

some jobs faster and as neatly as a power saw, yet is much quieter. Joints may be fitted by hand planing or by running a handsaw cut through the assembled joint, when possible.

All fasteners should be noncorrosive. Stainless steel fasteners offer the best protection against corrosion. Hot-dipped galvanized fasteners provide better protection than plated ones. Fasteners used in wood should be well set. They should be puttied to conceal the fastener. This also prevents the heads from corroding. Setting and concealing fasteners in exterior trim is a mark of quality.

Paint or otherwise seal and protect all exterior trim as soon as possible after installation. Properly installed and protected wood exterior trim will last indefinitely.

CHAPTER 65 GUTTERS AND DOWNSPOUTS

A *gutter* is a shallow trough or conduit set below the edge of the roof along the fascia. It catches and carries off rainwater from the roof. A **downspout,** also called a **conductor,** is a rectangular or round pipe. It carries water from the gutter downward and away from the foundation (Fig. 65-1).

Gutters

Gutters or *eavestrough* may be made of wood, galvanized iron, aluminum, copper, or vinyl. Copper gutters require no finishing. Vinyl and aluminum gutters are prefinished and ready to install. Wood and galvanized metal gutters need several protective coats of finish after installation.

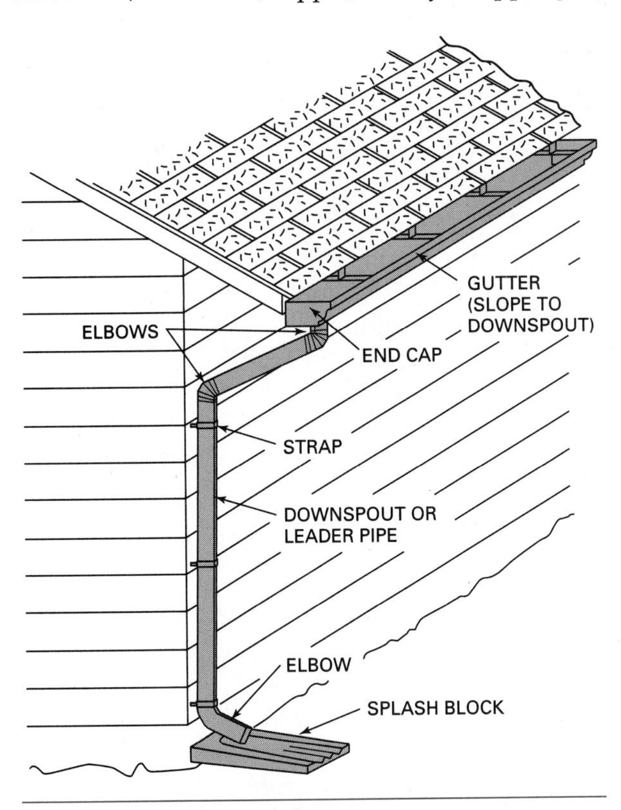

Fig. 65-1 Gutters and downspouts are an important system for conducting water away from the building. Otherwise, it might accumulate in crawl spaces or basements.

The size of a gutter is determined by area of the roof from which it handles the water runoff. Under ordinary conditions, 1 square inch of gutter cross-section is required for every 100 square feet of roof area. For instance, a 4 × 5 inch gutter has a cross-section area of 20 square inches. It is, therefore, capable of handling the runoff from 2,000 square feet of roof surface.

Wood Gutters

Wood gutters usually are made of Douglas fir, Western red cedar, or California redwood. They come in sizes of 3 × 4, 4 × 5, 4 × 6, and 5 × 7, and lengths of over 30 feet. Although wood gutters are not used as extensively as in the past, they enhance the cornice design. When properly installed and maintained, they will last as long as the building itself.

Metal Gutters

Metal gutters are made in rectangular, beveled, ogee, or semicircular shapes (Fig. 65-2). They come in a variety of sizes, from 2 1/2 inches to 6 inches in height and from 3 inches to 8 inches in width. Stock

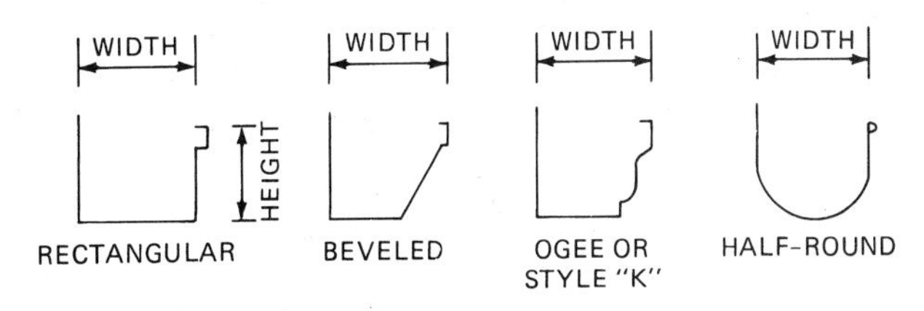

Fig. 65-2 Metal gutters are available in several shapes.

lengths run from 10 to 40 feet. *Forming machines* are often brought to the job site to form the gutters to practically any desired length.

Besides straight lengths, metal gutter systems have components comprised of *inside* and *outside corners, joint connectors, outlets, end caps,* and others. *Metal brackets* or *spikes and ferrules* are used to support the gutter sections on the fascia (Fig. 65-3).

Laying Out the Gutter Position

Gutters should be installed with a slight pitch to allow water to drain toward the downspout. A gutter of 20 feet or less in length may be installed to slope in one direction with a downspout on one end. On longer buildings, the gutter is usually crowned in the center. This allows water to drain to both ends.

On both ends of the fascia, mark the location of the bottom side of the gutter. The top outside edge of the gutter should be in relation to a straight line projected from the top surface of the roof. The height of the gutter depends on the pitch of the roof (Fig. 65-4).

Stretch a chalk line between the two marks. Move the center of the chalk line up enough to give the gutter the proper pitch. Thumb the line down. Snap it on both sides of the center. For a slope in one

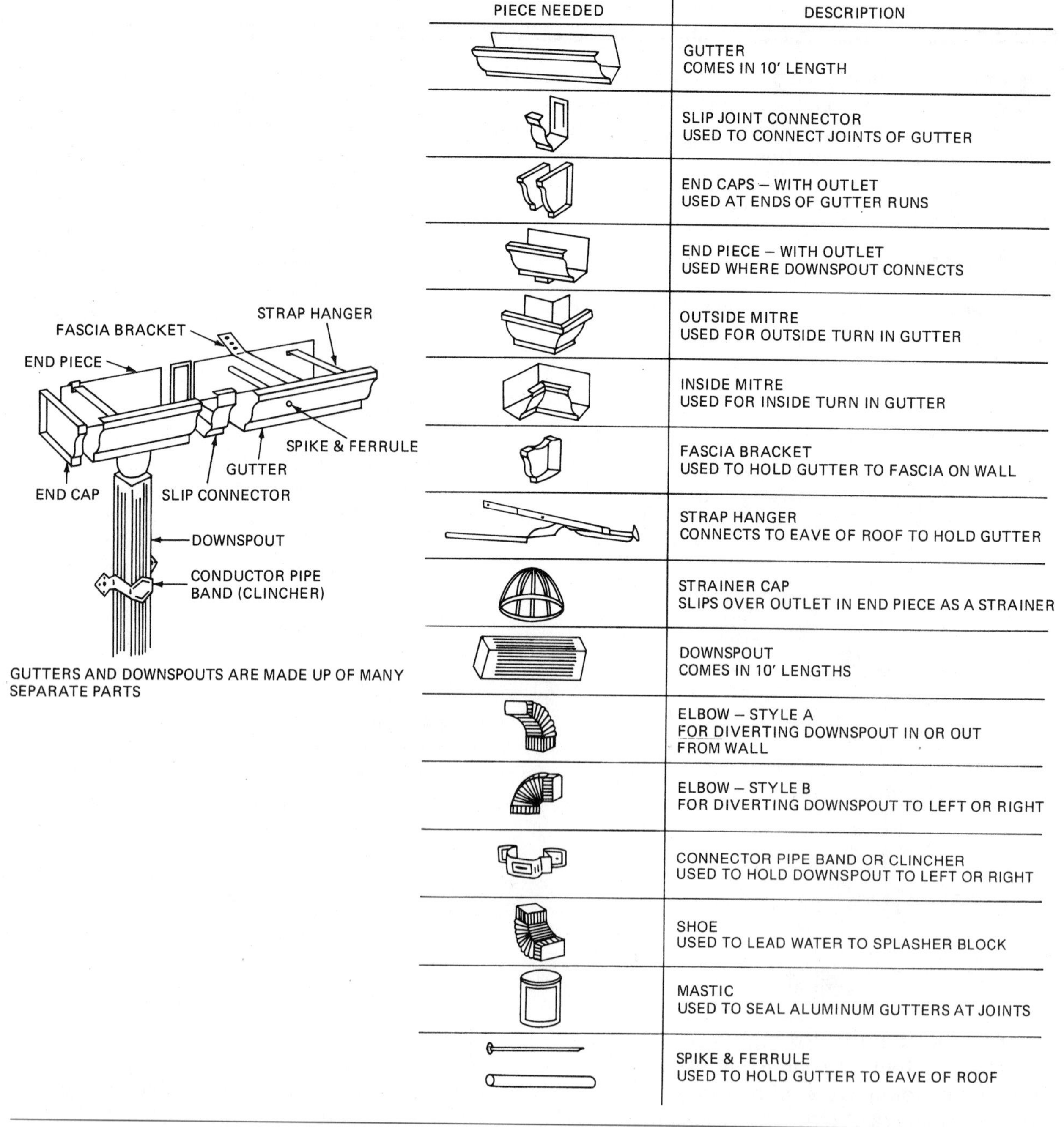

Fig. 65-3 Components of a metal gutter system.

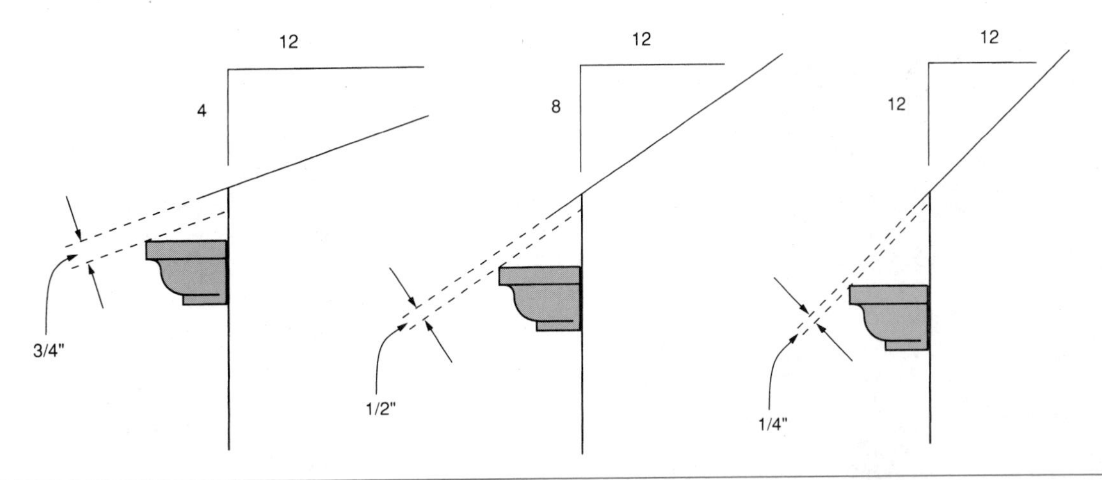

Fig. 65-4 The height of the gutter on the fascia is in relation to the pitch of the roof.

direction only, snap a straight line lower on one end than the other to obtain the proper pitch. It is important to install gutters on the chalk line. This avoids any dips that may prevent complete draining of the gutter.

Installing Metal and Vinyl Gutters

Fasten the gutter brackets to the chalk line on the fascia with screws. All screws should be made of stainless steel or other material that is corrosion-resistant. Aluminum brackets may be spaced up to 30 inches OC. Steel brackets may be spaced up to 48 inches OC. Install the gutter sections in the brackets. Use slip-joint connectors to join the sections. Apply the recommended gutter sealant to connectors before joining.

Locate the outlet tubes as required, keeping in mind that the downspout should be positioned plumb with the building corner and square with the building. Join with a connector. Add the end cap. Use either inside or outside corners where gutters make a turn.

Vinyl gutters and components are installed in a manner similar to metal ones.

Installing Downspouts

Metal or vinyl downspouts are fastened to the wall of the building in specified locations with aluminum straps. Downspouts should be fastened at the top and bottom and every 6 feet in between for long lengths.

The connection between the downspout and the gutter is made with 45-degree elbows and short straight lengths of downspout (Fig. 65-5). The connection will depend on the offset of the gutter from the downspout. Because water runs downhill, care should be taken when putting the downspout pieces together. The downspout components are assembled where the upper piece is inserted into the lower one (Fig. 65-6). This makes the joint lap such that the water cannot escape until it reaches the

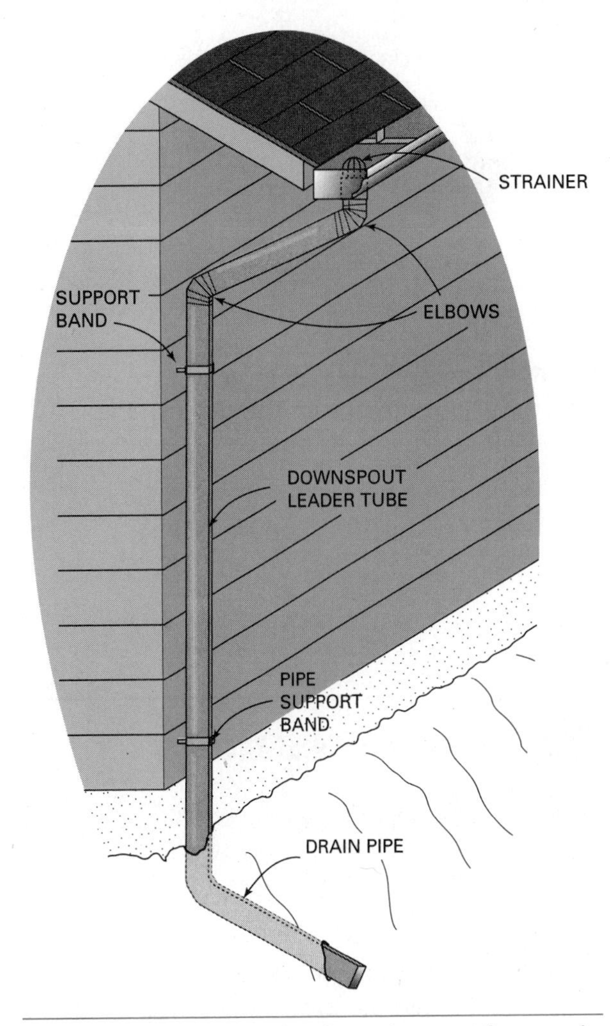

Fig. 65-5 Downspout leader tubes are fastened in place with pipe bands. Elbows are used to change the leader tube's direction. Water should be directed away from the building.

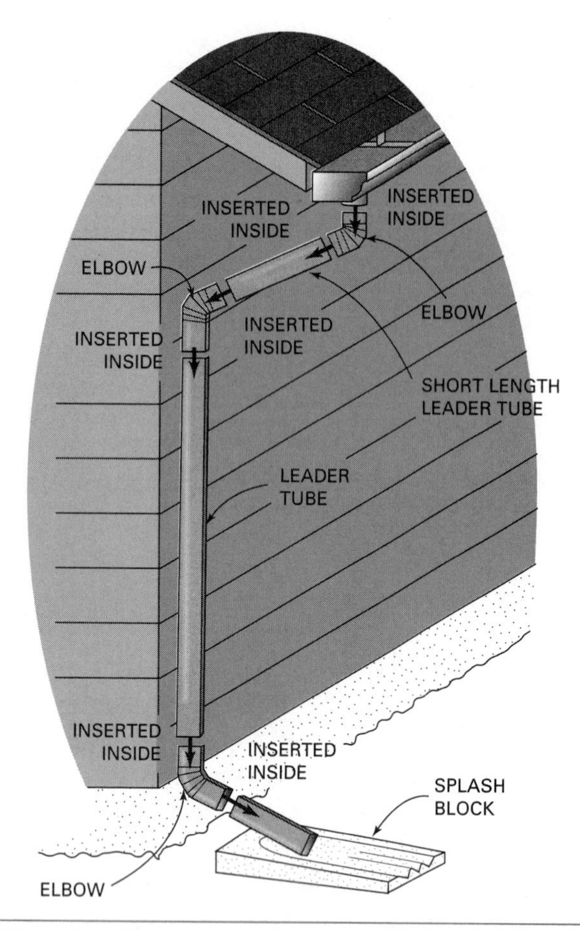

Fig. 65-6 Upper gutter components are inserted into lower ones to ensure the downspout doesn't leak.

bottom-most piece. An elbow, called a *shoe,* should be used with a splash block at the bottom of the downspout. This leads water away from the foundation. An alternate method is to connect the downspout with underground piping that carries the water away from the foundation. The piping can be connected to storm drains and drywells or piped to the surface elsewhere. Storm water as found in gutter downspouts should never be connected to footing or foundation drains. Strainer caps should be placed over gutter outlets if water is conducted by this alternate method. Leaves and other debris that fall into gutters flow into the drainage system and can cause clogging problems.

Review Questions

Select the most appropriate answer.

1. Bevel siding is applied
a. horizontally.
b. vertically.
c. horizontally or vertically.
d. horizontally, vertically, or diagonally.

2. A type of bevel siding, whose exposure cannot be changed, has
a. tongue-and-grooved edges.
b. rabbeted butt edges.
c. beveled back sides to lie flat against studs.
d. a convex shape.

3. A particular advantage of bevel siding over other types of horizontal siding is that
a. it comes in a variety of prefinished colors.
b. it has a constant weather exposure.
c. the weather exposure can be varied slightly.
d. application can be made in any direction.

4. When applying horizontal siding, it is desirable to
a. maintain exactly the same exposure with every course.
b. apply full courses above and below windows.
c. work from the top down.
d. use a water table.

5. In order to lay out a story pole for courses of horizontal siding, which of the following must be known?
a. the width of windows and doors
b. the kind and size of finish at the eaves and foundation
c. the location of windows, doors, and other openings
d. the length of the wall to which siding is to be applied
e. the size of corner boards

6. When installing aluminum or vinyl siding, drive nails
a. tightly against the flange.
b. loosely against the flange.
c. with large heads.
d. colored the same as the siding.

7. To allow for expansion when installing solid vinyl starter strips, leave a space between the ends of, at least,
a. 1/8 inch. c. 3/8 inch.
b. 1/4 inch. d. 1/2 inch.

8. The exterior trim that extends up the slope of the roof on a gable end is called the
a. box finish. c. return finish.
b. rake finish. d. snub finish.

9. A member of the cornice fastened in the vertical position to the rafter tails is called the
a. drip. c. soffit.
b. fascia. d. frieze.

10. Care should be taken when installing gutters so that
a. they are level with the fasica.
b. downspouts are in the center.
c. downspouts leader tubes are connected to the foundation drains.
d. parts are installed with the idea that water runs downhill.

BUILDING FOR SUCCESS

Employer/Employee Relationships

Each person pursuing employment in the construction industry should be aware of the need to establish good working relationships with employers. Goals will have to be set as each skilled worker becomes involved with an employer, customers, and peers. Careful thought and preparation will help make each worker more successful.

The reasons for forming good working relationships with others in the construction field need to be identified. Each day as tasks are carried out, meaningful relationships help the flow and progress of the work. The cooperation that is established in the work process will build friendships. In many cases, people work alongside their peers and employers for hours. What happens in these situations affects the amount of job stress a person has to deal with from day to day. This underscores the importance of team effort in solving problems on the job.

Other reasons for developing good working relationships on the job are centered on job longevity, advancement, customer service, production, and enjoyment. A career of framing wood or steel buildings on a crew will be more enjoyable when people like each other and are compatible. How and to what extent people advance depend on their working relationships. Customers see each worker as representing the business through personal contacts. They are either pleased or displeased by that one-on-one relationship. Production on the job is affected by relationships. Low production may mean that there are problems among workers on the job. Friction in personalities or work habits will create situations that distract from quality work. One of the primary reasons a person takes a job is for enjoyment. That criterion is one of the most important in selecting the job. Without job enjoyment, work is not rewarding and fulfilling. It will be only a matter of time before someone will leave a job if there is no work enjoyment.

How does one go about establishing good working relationships on the job? This is a good time to look at yourself and see how you work and what you expect on the job. Personalities will be one of the first things people notice when placed alongside someone in an employment situation. Compatibility will be established or rejected almost immediately. Usually people will expose their skills, talents, personality traits, and work habits in a very short time on the job. Many times the job itself calls for certain types of actions, skills, and abilities and sets the pace to which other people can react. How you present yourself and react to others will set the stage for working relationships.

Generally, people who are knowledgeable, friendly, productive, and use their skills well develop good working relationships with employers and peers. Being honest with people will be one good way of telling others what you are like. How can each person benefit from establishing good relationships with others? The dividends are there from the beginning in the form of a productive and safe working environment. Accidents on the job can often be traced to poor working relationships with others.

Some of the continuing benefits that come from having good working relationships are seen as characteristics of trustworthiness, dependability, concern for the progress of the company, and less stress in the workplace. These basic characteristics will be reflected in the attitudes and performance of each skilled worker on the job.

In looking at a way to hear why employers are concerned with good working relationships, try asking them. Contact one or more employers in construction-related businesses. Ask them to share their opinions with you. The information gained from this type of conversation will provide invaluable assistance in making a good job choice.

FOCUS QUESTIONS: For individual or group discussion

1. What might be some reasons for establishing good work relations with an employer?
2. How could good working relationships benefit you as a carpenter, drafter, estimator, or any other skilled worker?
3. What would you expect from a person who will not make an effort to work well with others?
4. Do you have a good idea of how to establish good working relationships with an employer and peers in a job situation? What would be your approach?
5. How would you help someone establish a good working relationship with you?

UNIT

22

Decks, Porches, and Fences

Among the final steps in finishing the exterior is the building of porches, decks, fences, and other accessory structures. Plans may not always show specific construction details. Therefore, it is important to know some of the techniques used to build these structures.

OBJECTIVES

After completing this unit, the student should be able to:

- describe the construction of and kinds of materials used in decks and porches.
- lay out and construct footings for decks and porches.
- install supporting posts, girders, and joists.
- apply decking in the recommended manner and install flashing, for an exposed deck, against a wall.
- construct deck stairs and railings.
- describe several basic fence styles.
- design and build a straight and sturdy fence.

CONTENTS

CHAPTER 66 DECK AND PORCH CONSTRUCTION

Wood porches and decks are built to provide outdoor living areas for various reasons, in both residential and commercial construction (Fig. 66-1). The construction of both is similar. However, a porch is covered by a roof. Its walls may be enclosed with wire mesh screens for protection against insects. With screen and storm window combinations, glass replaces screen to keep the porch comfortably warm in the cold months.

Deck Materials

Decking materials must be chosen for strength and durability, as well as appearance and resistance to decay. Redwood, cedar, and pressure-treated southern yellow pine are often used as decking boards. Other decking materials available include Timber Tech® and Trex®. These decking products are made from a mixture of plastic and sawdust. They are cut, fit, and fastened in the same manner as wood and have the added benefit of being made mostly from recycled material.

If not specified, the kind, grade, and sizes of material must be selected before building a deck. Also, the size and kind of fasteners, connectors, anchors, and other hardware must be determined.

Lumber

All lumber used to build decks should be *pressure-treated* with preservatives or be from a decay-resistant species, such as *redwood* or *cedar.* Remember, it is the *heartwood* of these species that is resistant to decay, not the *sapwood.* Either *all-heart* or *pressure-treated* lumber should be used wherever there is a potential for decay. This is essential for posts that are close to the ground and other parts subject to constant moisture. (A description of pressure-treated lumber and its uses can be found in Chapter 33.)

Lumber Grades. For pressure-treated southern pine and western cedar, #2 grade is structurally adequate for most applications. Appearance can be a deciding factor when choosing a grade. If a better appearance is desired, higher grades should be considered.

A grade called *Construction Heart* is the most suitable and most economical grade of California redwood for deck posts, beams, and joists. For decking and rails, a grade called *Construction Common* is recommended. Better appearing grades are available. However, they are more expensive.

Two grades of redwood, called *Redwood Deck Heart* and *Redwood Deck Common,* are manufactured especially for exterior walking surfaces. Two grades of decking, *Standard* and *Premium,* are also available in pressure-treated southern pine. Special decking grades of western cedar may also be obtained. The lumber grade or special purpose is shown in the grade stamp (Fig. 66-2).

Lumber Sizes. Specific sizes of supporting posts, girders, and joists depend on the spacing and

Fig. 66-1 Wood decks are built in many styles. This multilevel deck blends well with the landscape. (*Courtesy of California Redwood Association*)

Fig. 66-2 Grade stamps show the grade and special purpose of lumber used in deck construction. (*Courtesy of California Redwood Association; Courtesy of Southern Forest Products Association*)

height of supporting posts and the spacing of girders and joists. In addition, the sizes of structural members depend on the type of wood used and the weight imposed on the members. Too many factors prohibit generalization about sizes of structural members. Check with local building officials or with a professional to determine the sizes of structural members for specific deck construction. Determining sizes with incomplete information may result in failure of undersize members. Unnecessary expense is incurred with the use of oversize members.

Fasteners

All nails, fasteners, and hardware should be stainless steel, aluminum, or hot-dipped galvanized. Electroplated galvanizing is not acceptable because the coating is too thin. In addition to corroding and eventual failure, poor quality fasteners will react with substances in decay-resistant woods and cause unsightly stains.

Building a Deck

Most wood decks consist of *posts,* set on *footings,* supporting a platform of *girders* and *joists* covered with *deck boards. Posts, rails,* **balusters,** and other special parts make up the railing (Fig. 66-3). Other parts, such as shading devices, privacy screens, benches, and planters, lend finishing touches to the area.

Installing a Ledger

If the deck is to be built against a building, a *ledger,* usually the same size as the joists, is nailed or bolted to the wall for the entire length of the deck (Fig. 66-4). The ledger acts as a beam to support joists that run at

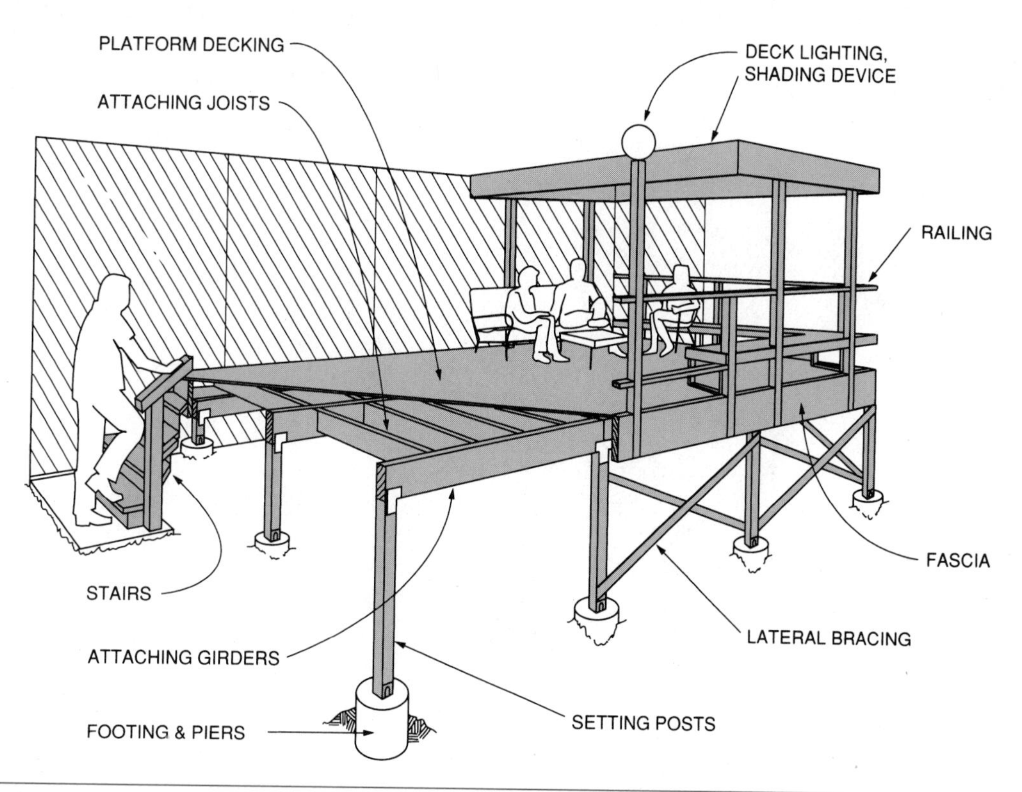

Fig. 66-3 The components of a deck.

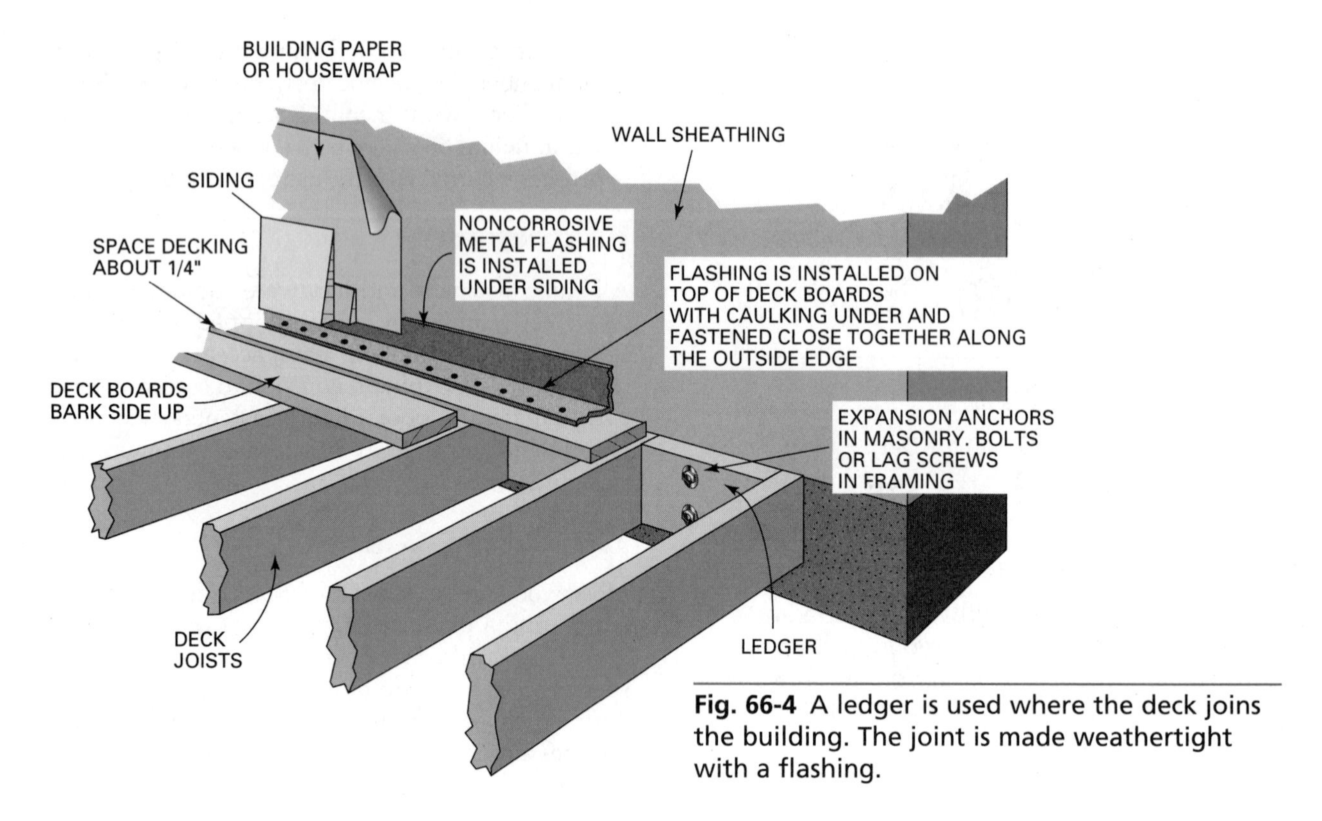

Fig. 66-4 A ledger is used where the deck joins the building. The joint is made weathertight with a flashing.

right angles to the wall. It is installed to a level line. Its top edge is located to provide a comfortable step down from the building after the decking is applied. The ledger height may be used as a benchmark for establishing the elevations of supporting posts and girders.

After the deck is applied, a flashing is installed under the siding and on top of the deck board. Caulking is applied between the deck and the flashing. The flashing is then fastened, close to and along its outside edge, with nails spaced closely together. The outside edge of the flashing should extend beyond the ledger.

Footing Layout and Construction

Footings for the supporting posts must be accurately located. To determine their location, erect batter boards and stretch lines in a manner previously described in Chapter 26. All footings require digging a hole and filling it with concrete (Fig. 66-5).

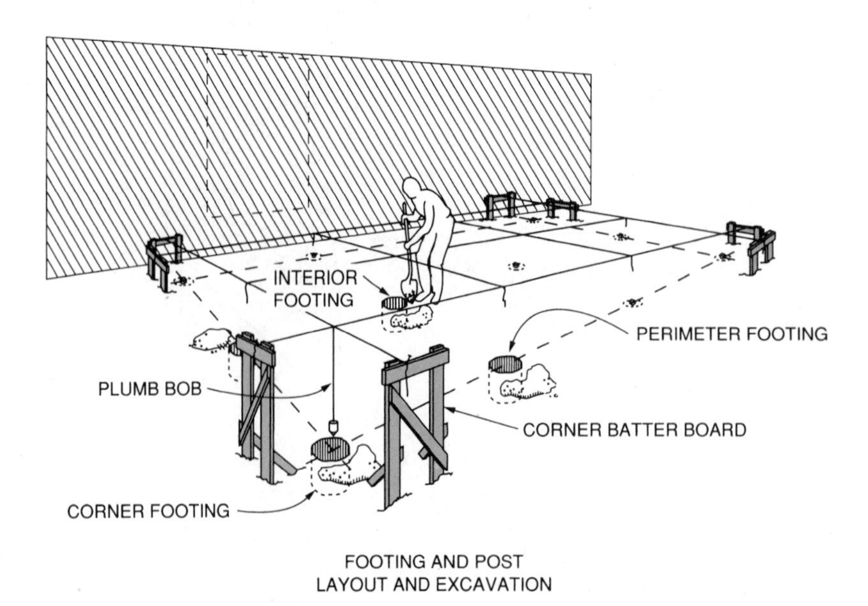

Fig. 66-5 Footing and post layout and excavation. (*Courtesy of Simpson Strong-Tie Company*)

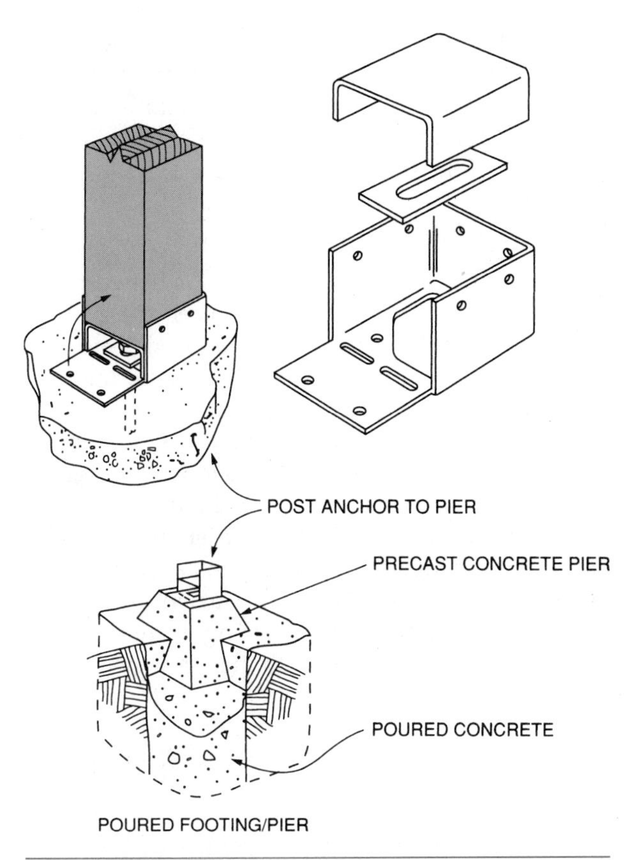

Fig. 66-6 Precast concrete pier embedded in the footing. (*Courtesy of Simpson Strong-Tie Company*)

Footing Size and Style

In stable soil and temperate climate, the footing width is usually made twice the width of the post it is supporting. The footing depth reaches undisturbed soil, at least 12 inches below grade. In cold climates the footing should extend below the frost line.

There are several footing styles that are commonly used. One method is to partially fill the footing hole with concrete to within a few inches from the top. Set a precast pier 2 inches into the wet concrete. After the concrete has set, attach a post anchor to the top of the precast pier (Fig. 66-6). Align piers and anchors with the layout lines.

The top of the footing may be brought above grade with the use of a wood box or fiber-tube form. Place concrete in the footing hole. Bring it to the top of the form. Set post anchors while the concrete is still wet (Fig. 66-7).

Erecting Supporting Posts

All supporting posts are set on footings. They are then plumbed and braced. Cut posts, for each footing, a few inches longer than their final length. Tack the bottom of each post to the anchor. Brace them in a plumb position in both directions (Fig. 66-8).

When all posts are plumbed and braced, the tops must be cut level to the proper height. From

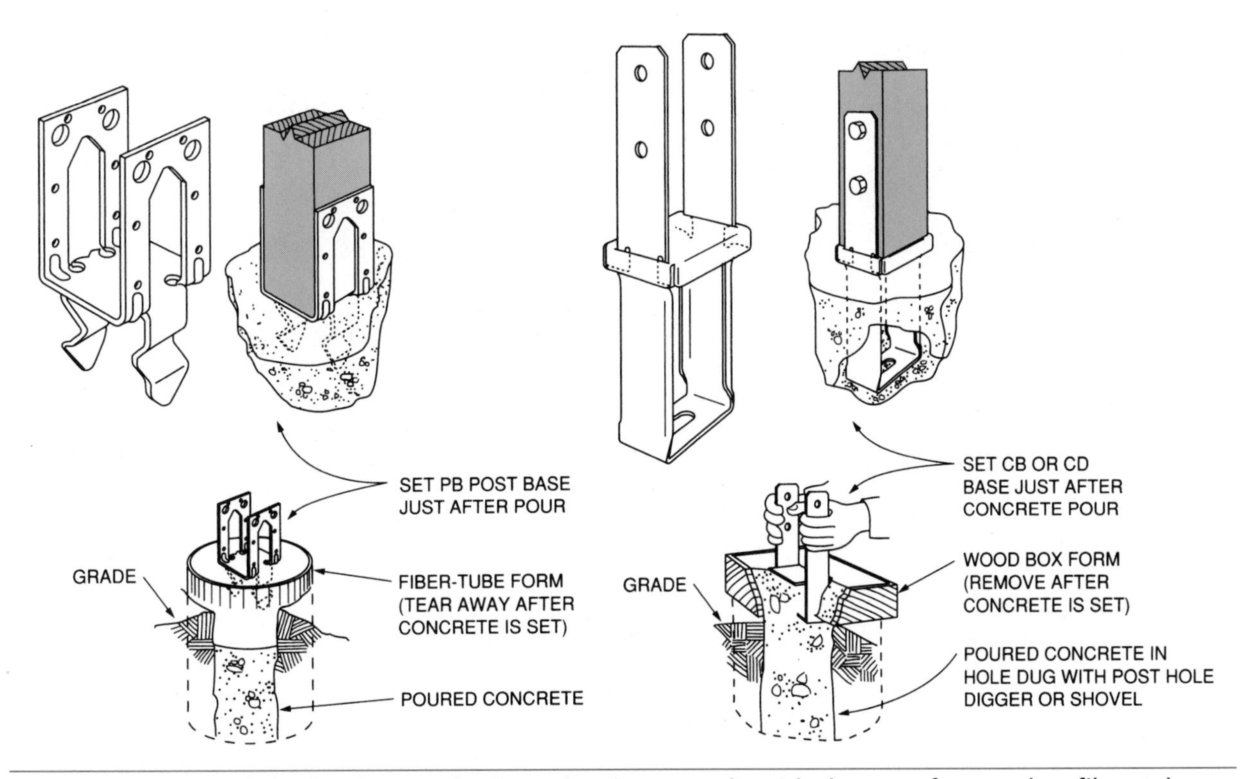

Fig. 66-7 The top of the footing can be brought above grade with the use of a wood or fiber-tube form. (*Courtesy of Simpson Strong-Tie Company*)

STEP BY STEP
PROCEDURES

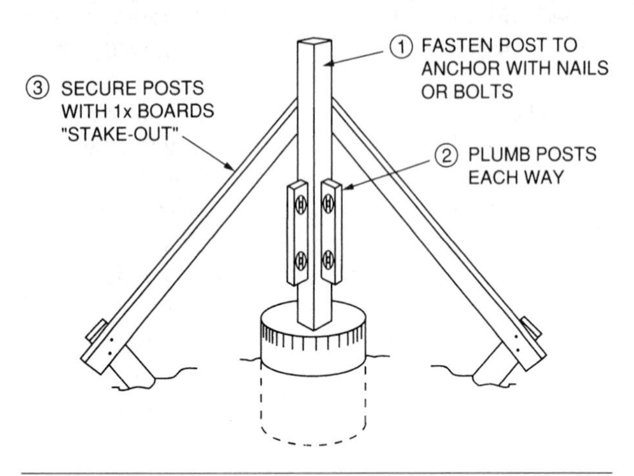

Fig. 66-8 Supporting posts are cut longer than required and braced plumb on footings. (*Courtesy of Simpson Strong-Tie Company*)

the height of the deck, deduct the deck thickness and the depth of the girder. Mark on a corner post. Mark the other posts by leveling from the first post marked. Mark each post completely around using a square. Cut the tops with a portable circular saw (Fig. 66-9).

Installing Girders

Install the girders on the posts using post and beam metal connectors. The deck should slope slightly, about 1/8 inch per foot, away from the building. The size of the connector will depend on the size of the posts and girders. Install girders with the crowned edge up. Any splice joints should fall over the center of the post (Fig. 66-10).

Installing Joists

Joists may be placed over the top or between the girders. When joists are hung between the girders, the overall depth of the deck is reduced. This provides more clearance between the frame and the ground. For decking run at right angles, joists may be spaced 24 inches on center. Joists should be spaced 16 inches on center for diagonal decking.

Lay out and install the joists in the same manner as previously described in earlier chapters. Use appropriate hangers if joists are installed between girders (Fig. 66-11). When joists are installed over girders, use recommended framing anchors. Make sure all joists are installed with their crowned edges facing upward.

Bracing Supporting Posts

If the deck is 4 feet or more above the ground, the supporting posts should be braced in a manner similar to that shown in Figure 66-12. Use minimum 1 × 6 braces for heights up to 8 feet. Use minimum 1 × 8 braces for higher decks applied continuously around the perimeter.

Applying Deck Boards

Specially shaped *radius edge decking* is available in both pressure-treated and natural decay-resistant lumber. It is usually used to provide the surface and

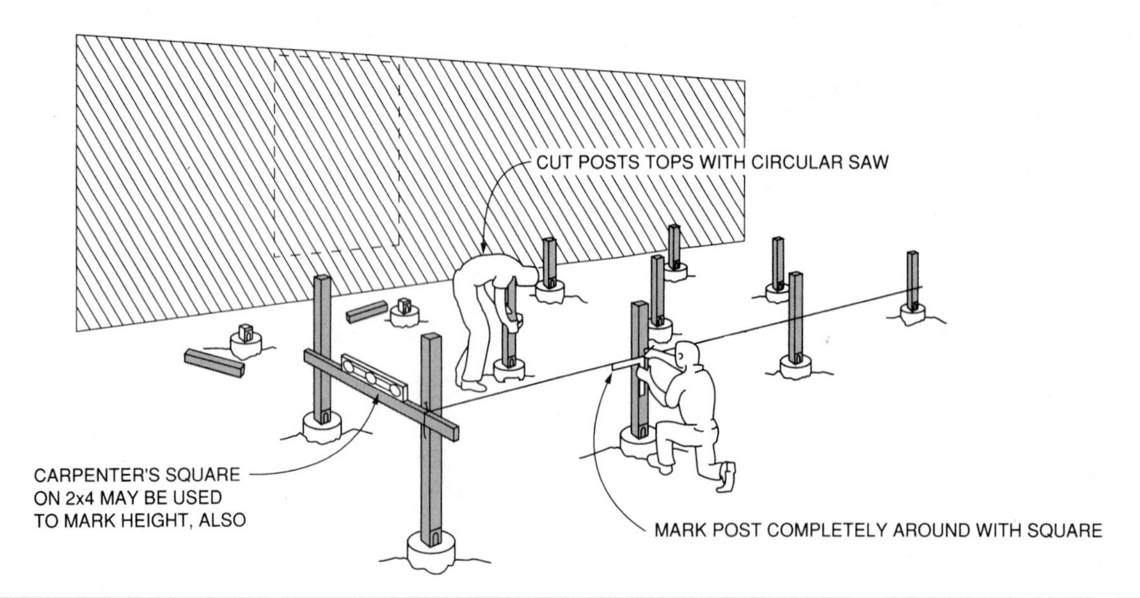

Fig. 66-9 The post height is determined and the tops cut with square ends level with each other. (*Courtesy of Simpson Strong-Tie Company*)

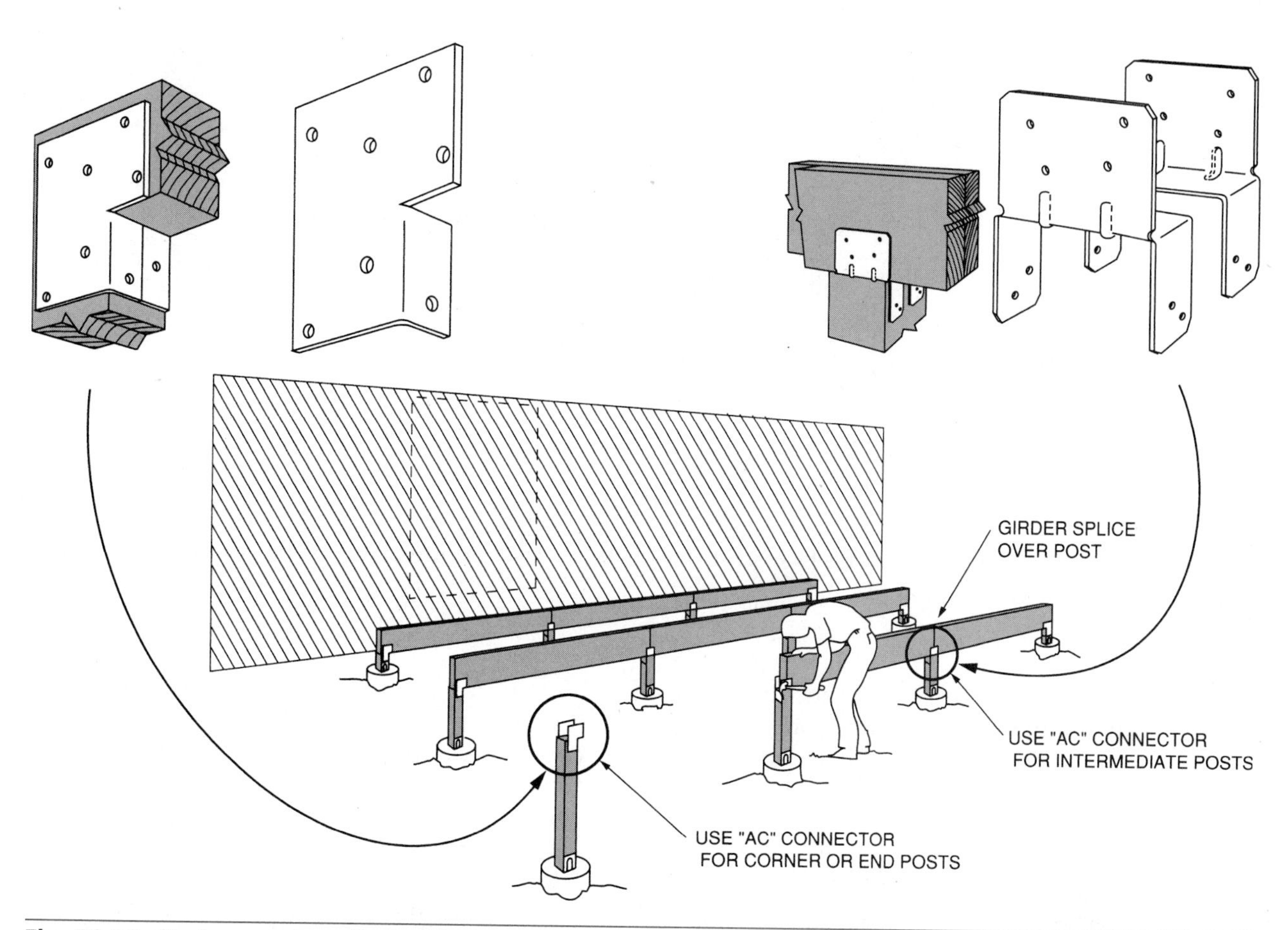

Fig. 66-10 Girders are installed with their crowned edges up on top of supporting posts. (*Courtesy of Simpson Strong-Tie Company*)

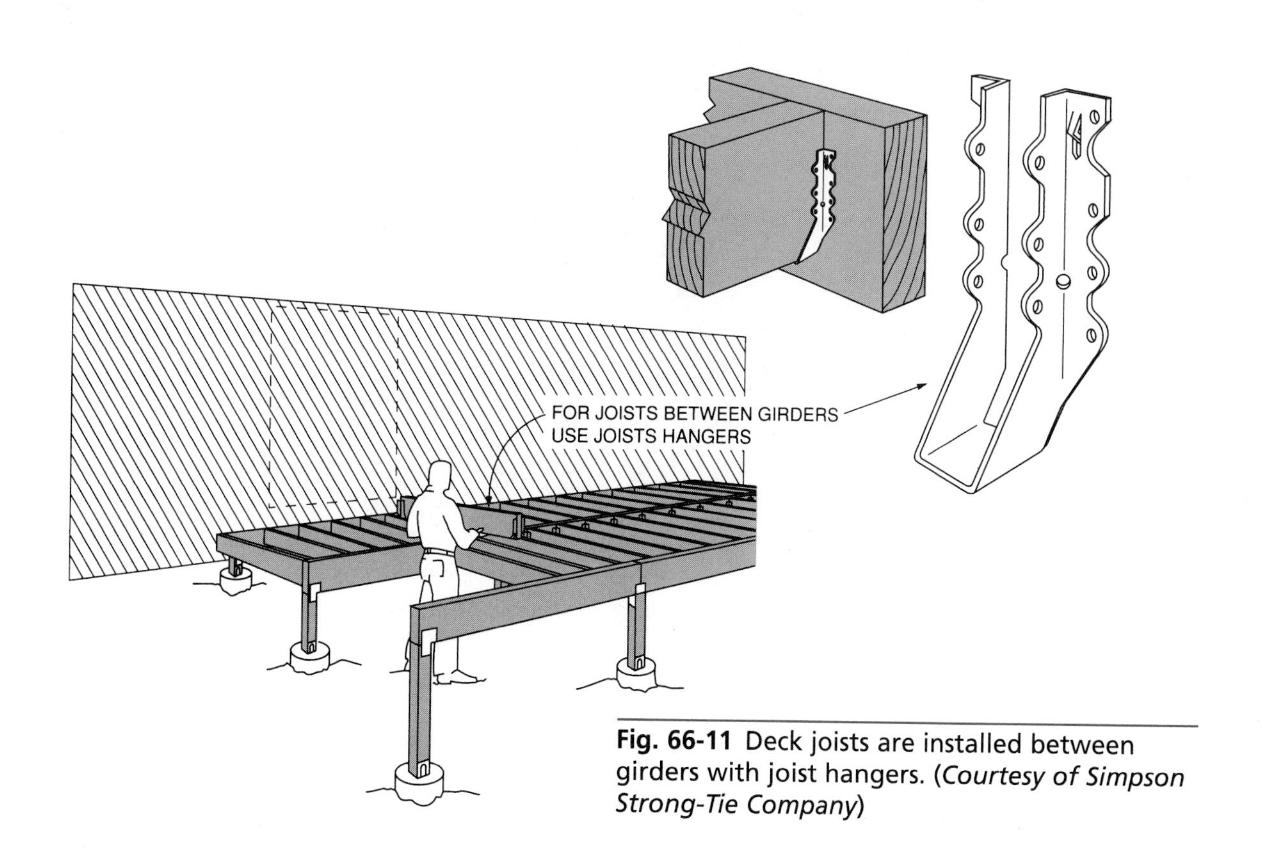

Fig. 66-11 Deck joists are installed between girders with joist hangers. (*Courtesy of Simpson Strong-Tie Company*)

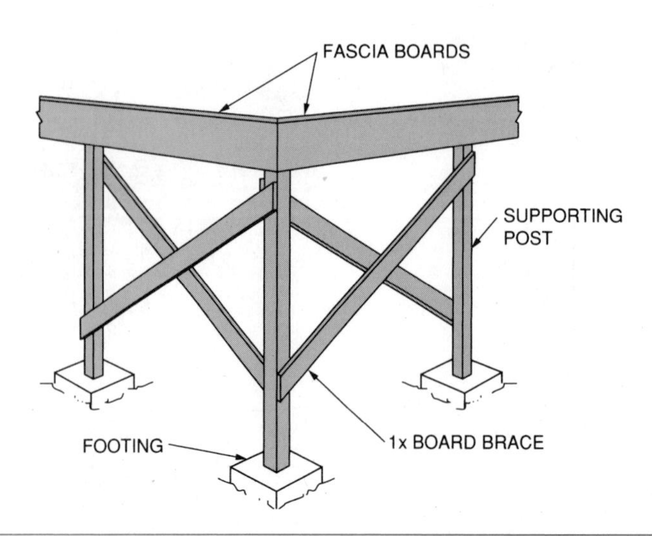

Fig. 66-12 Supporting posts must be braced if the deck is 4 feet or more above the ground. (*Courtesy of Simpson Strong-Tie Company*)

walking area of the deck. Dimension lumber of 2-inch thickness and widths of 4 and 6 inches is also widely used. Lay boards with the bark side up to minimize cupping (Fig. 66-13).

Boards are usually laid parallel with the long dimension of the deck. Deck boards usually do not come longer than 16 feet. It may be desirable to lay the boards parallel to the short dimension, if their length will span it, to eliminate end joints in the decking. Boards may also be laid in a variety of patterns including diagonal, herringbone, and parquet. Make sure the supporting structure has been designed and built to accommodate the design.

Much care should be taken with the application of deck boards. Snap a straight line as a guide to apply the starting row. Start at the outside edge if the deck is built against a building. A ripped and narrower ending row of decking is not so noticeable against the building as it is on the outside edge of the deck.

Straighten boards as they are installed. Maintain about a 1/4-inch space between them. If the decking boards are wet, as with most pressure-treated boards, they will shrink as they dry. Nailing them tight together is the preferred method as the 1/4" space will appear when the lumber reaches equilibrium moisture content. If the deck boards do

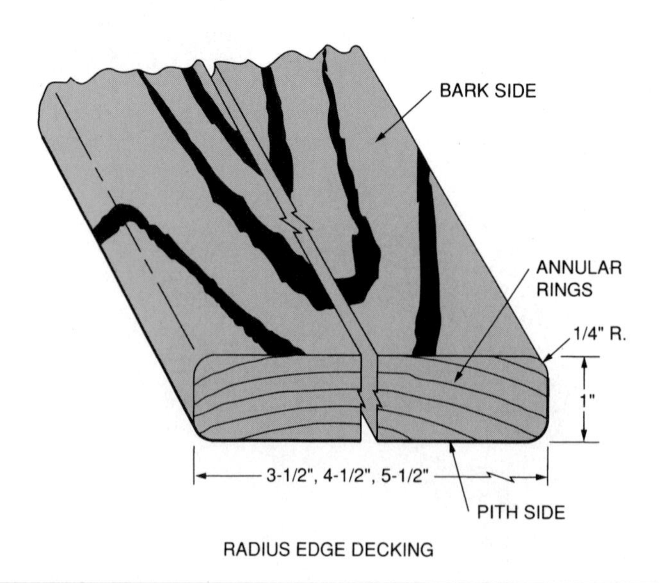

Fig. 66-13 Special radius edge decking is used and applied with the bark side up to resist cupping. (*Courtesy of Southern Pine Marketing Council*)

not span the entire width of the deck, cut the boards so their ends are centered over joists. Make tight-fitting end joints. Stagger them between adjacent rows. Predrill holes for fasteners to prevent splitting the ends. Let the end of each row overhang the deck. Use two screws or nails in each joist. Drive nails at an angle. Set the heads below the surface. This will keep the nails from working loose and the heads from staining the surface. When approaching the end, it is better to increase or decrease the spacing of the last six or seven rows of decking. This way you will end with a row that is nearly equal in width to all the rest, rather than with a narrow strip. When all the deck boards are laid, snap lines and cut the overhanging ends. Apply a preservative to them (Fig. 66-14).

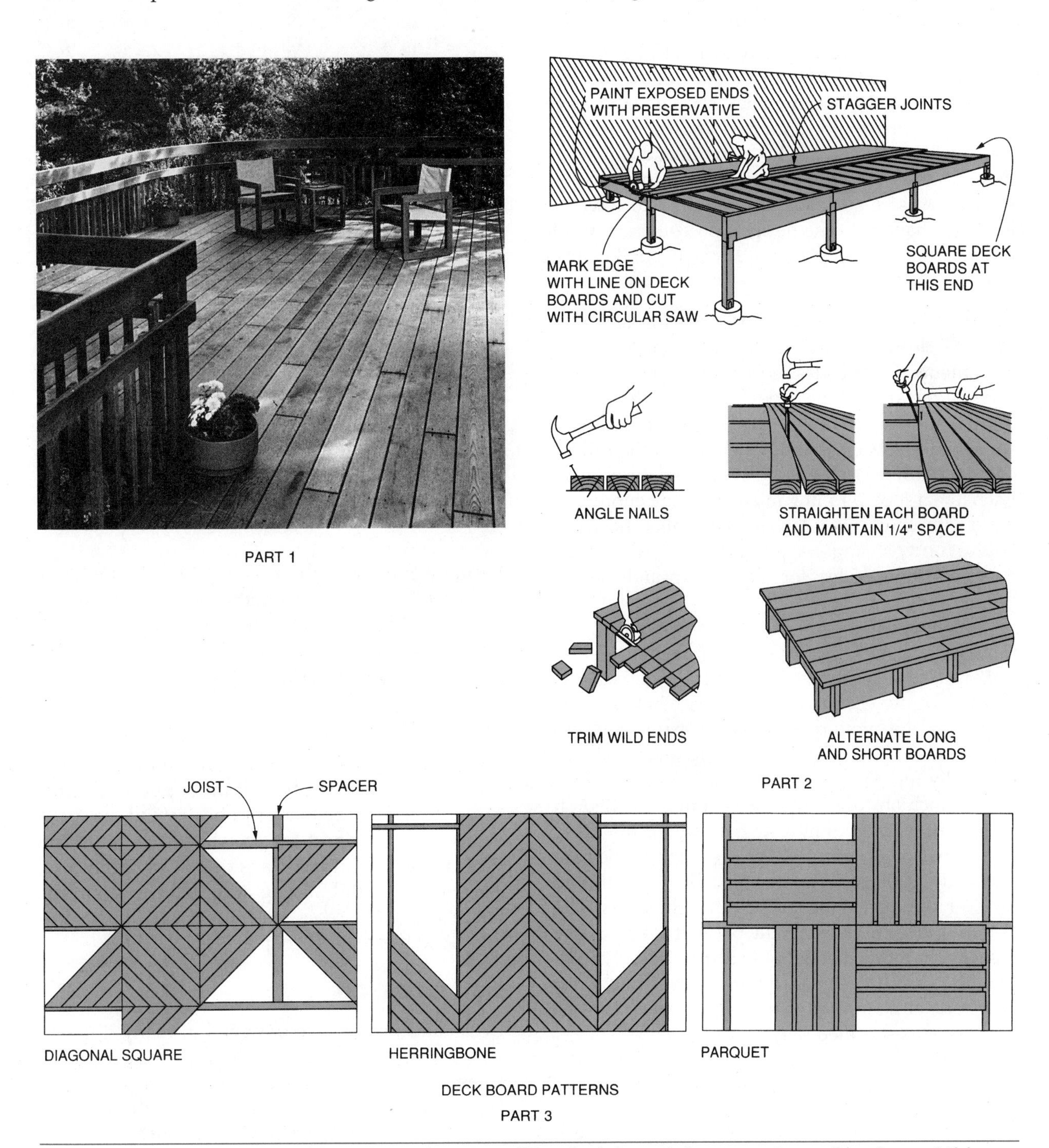

Fig. 66-14 Much care should be taken when installing deck boards because the surface is so visible. *(Part 1, Courtesy of California Redwood Association; Part 2, Courtesy of Simpson Strong-Tie Company; Part 3, Courtesy of Southern Pine Marketing Council)*

Applying Trim

A fascia board may be fastened around the perimeter of the deck. Its top edge should be flush with the deck surface. The fascia board conceals the cut ends of the decking and the supporting members below. The fascia board is optional.

Stairs and Railings

Stairs. Most decks require at least one or two steps leading to the ground. To protect the bottom ends of the stair carriage, they should be treated with preservative and supported by an above-grade concrete pad (Fig. 66-15). Stair layout and construction are described in Chapter 48. Stairs with more than two risers are generally required to have at least one handrail. The design and construction of the stair handrail should conform to that of the deck railing.

Rails. There are numerous designs for deck railings. All designs must conform to certain code requirements. Most codes require at least a 36-inch-high railing around the exposed sides, if the deck is more than 30 inches above the ground. In addition, some codes specify that no openings in the railing should allow a 4-inch sphere to pass through it. Each linear foot of railing must be strong enough to resist a pressure of 20 lbs./sq. ft. applied horizontally at a right angle against the top rail. Check local building codes for deck stair and railing requirements.

Railings may consist of *posts,* top, bottom, and intermediate *rails,* and *balusters.* Sometimes *lattice work* is used to fill in the rail spaces above the deck. It is frequently used to close the space between the deck and the ground. Posts, rails, balusters, and other deck parts are manufactured in several shapes especially for use on decks (Fig. 66-16).

Stanchions or posts are sometimes notched on their bottom ends to fit over the edge of the deck. They are usually spaced about 4 feet apart. They are fastened with lag screws or bolts. The top rail may go over the tops or be cut between the posts. The bottom rail is cut between the posts. It is kept a few inches (no more than 8) above the deck. The remaining space may be filled with intermediate rails, balusters, lattice work, or other parts in designs as desired or as specified (Fig. 66-17).

Deck Accessories

There are many details that can turn a plain deck into an attractive and more comfortable living area. *Shading structures* are built in many different designs. They may be completely closed in or spaced to provide filtered light and air circulation. *Benches* partially or entirely around the deck may double as a place to sit and act as a railing (Fig. 66-18). Bench seats should be 18 inches from the deck. Make allowance for cushion thickness, if used. The width of the seat should be from 18 to 24 inches.

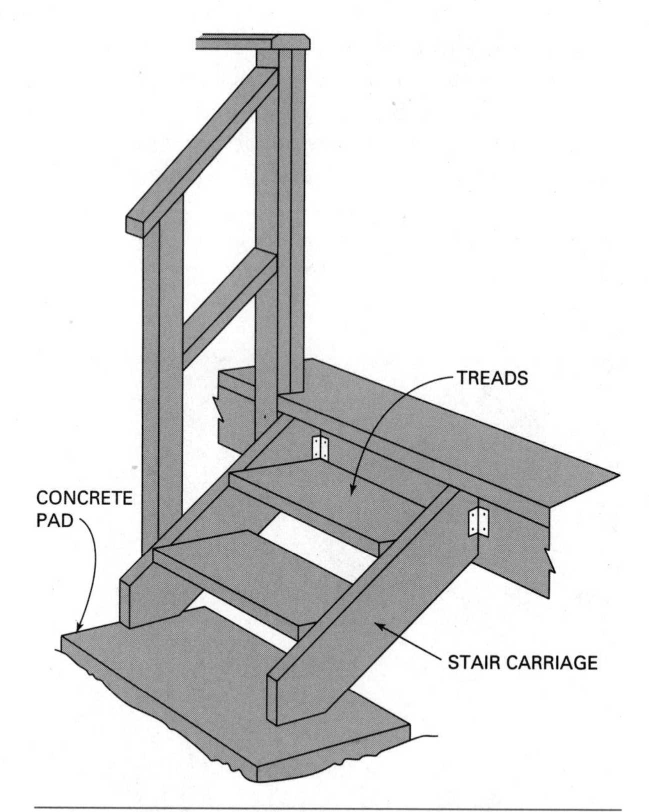

Fig. 66-15 Stairs for decks are usually constructed with a simple basement or utility design. *(Courtesy of Simpson Strong-Tie Company)*

Porches

The porch deck is constructed in a similar manner to an open deck. Members of the supporting structure may need to be increased in size and the spans decreased to support the weight of the walls and roof above. Work from plans drawn by professionals or check with building officials before starting. Porch walls and roofs are framed and finished as described in previous chapters (Fig. 66-19).

Summary

Decks and porches are designed in many different ways. There are other ways to construct them besides the procedures described in this chapter. However, making the layout, building the supporting structure, applying the deck, constructing the railing and stairs, and, in the case of a porch, building the walls and roof are basic steps that can be applied to the construction of practically any deck or porch.

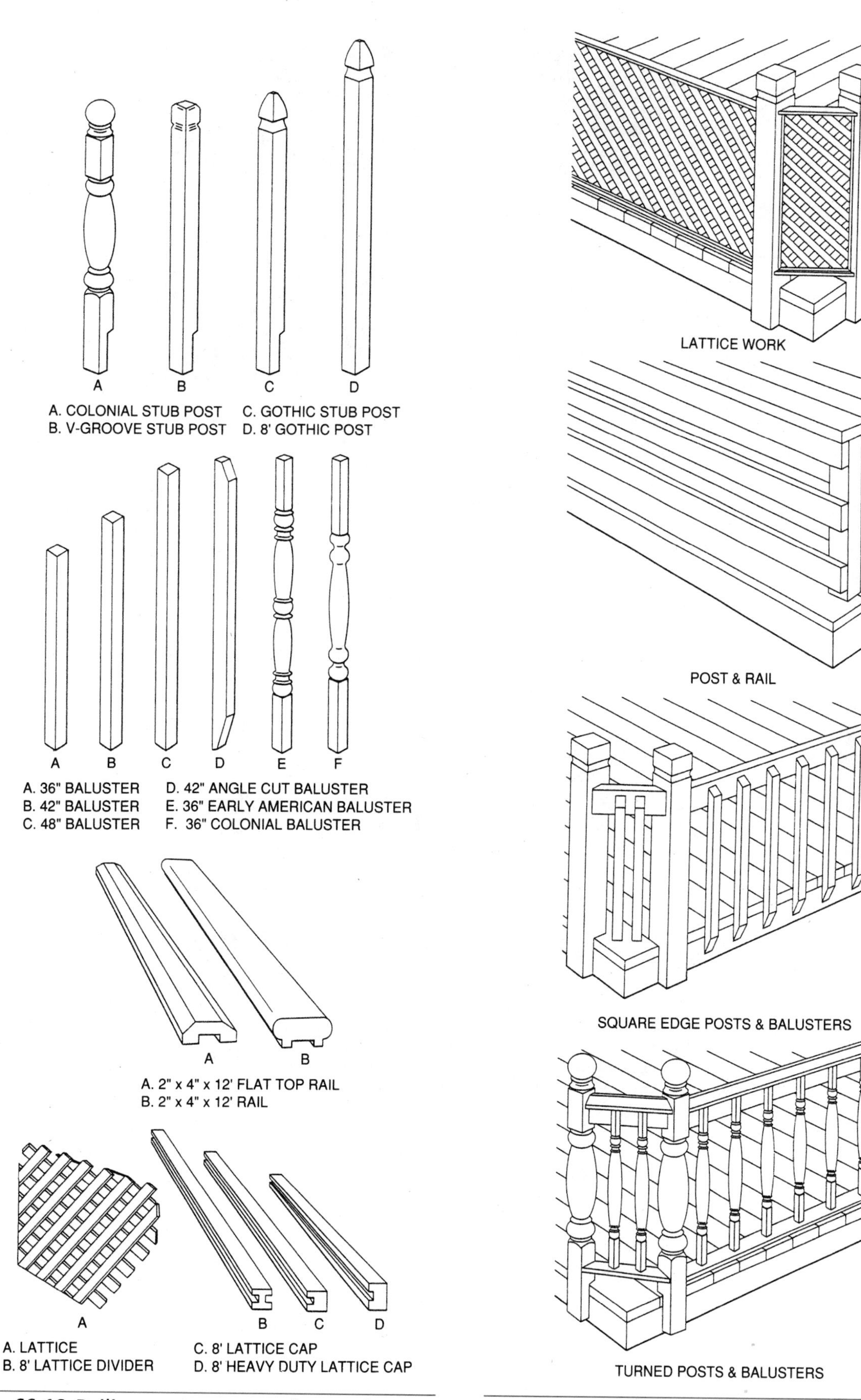

Fig. 66-16 Railing parts are manufactured in several shapes.

Fig. 66-17 The deck railing may be constructed with various kinds of parts in a number of designs.

Fig. 66-18 A plain deck can be made more attractive and comfortable with benches and shading devices. (*Courtesy of California Redwood Association*)

Fig. 66-19 A porch is composed of a deck enclosed by walls and a roof. (*Courtesy of Southern Pine Marketing Council*)

CHAPTER 67 FENCE DESIGN AND ERECTION

A superior fence combines utility and beauty (Fig. 67-1). It may define space, create privacy, provide shade or shelter, screen areas from view, or form required barriers around swimming pools and other areas for the protection of small children.

The design of a fence is often the responsibility of the builder. With creativity and imagination, he or she can construct an object of beauty and elegance that also fulfills its function. The objective of this chapter is to create an awareness of the importance of design by showing several styles of fences and the methods used to erect them.

Fence Material

Fences are not supporting structures. Lower, knotty grades of lumber may be used to build them. This is not the case if appearance is a factor and you want to show the natural grain of the wood. Many kinds of softwood may be used as long as they are shielded from the weather by protective coatings of paint, stains, or preservatives. However, for wood posts that are set into the ground and other parts that may be subjected to constant moisture, pressure-treated or all-heart, decay-resistant lumber must be used.

The same type of fasteners that are used on exposed decks should be used to build fences. Inferior hardware and fasteners will corrode and cause unsightly stains when in contact with moisture.

Fence Design

Fences consist of posts, rails, and fence boards. Fences may be constructed in almost limitless designs. Zoning regulations sometimes restrict their height or placement. Often the site will affect the design. For example, the fence may have to be stepped like a staircase on a steep slope (Fig. 67-2). Fences on property lines can be designed to look attractive from either side. Fence designs may block wind and sunlight. Fence boards may be spaced in many attractive patterns. Most fences are constructed in several basic styles, each of which can be designed in numerous ways. Provisions should be made in the fence design to drain water from any area where it may otherwise be trapped.

Picket Fence

The *picket* fence is commonly used on boundary lines or as barriers for pets and small children. Usually not more than 4 feet high, the pickets are spaced to provide plenty of air and also to conserve material. The tops of the pickets may be shaped in

Fig. 67-1 A fence can serve its intended purpose and also enhance the surroundings. (*Courtesy of California Redwood Association*)

Fig. 67-2 Sometimes the site may affect the design of a fence, such as on this steep slope. (*Courtesy of California Redwood Association*)

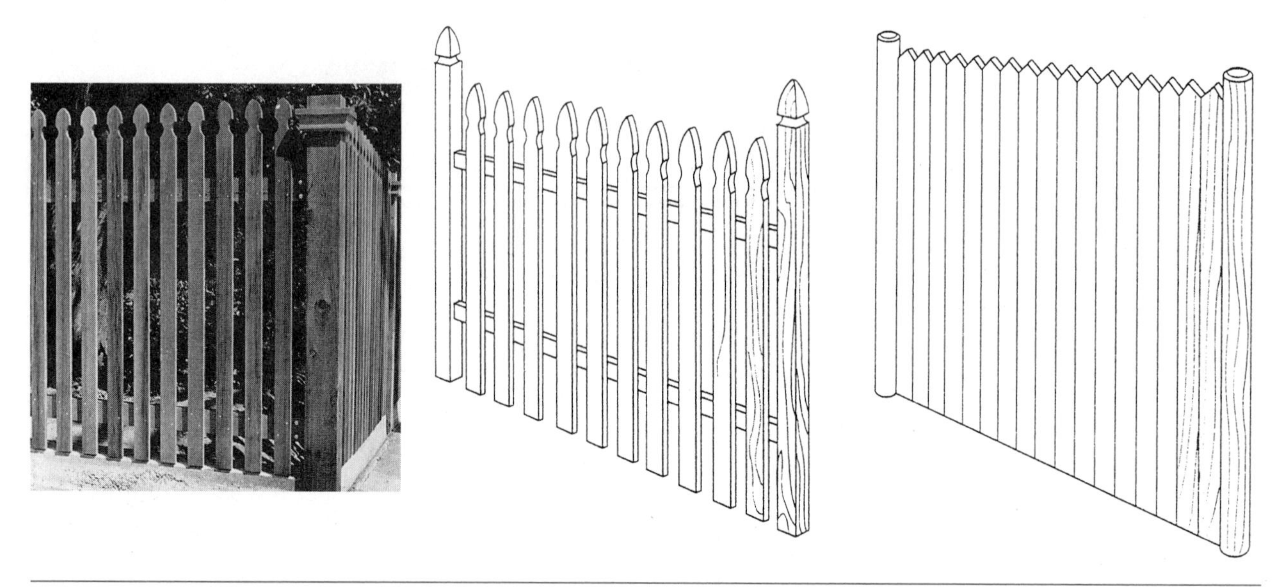

Fig. 67-3 The picket fence is constructed in many styles. (*Courtesy of California Redwood Association*)

various styles. Or, they may be cut square, with ends exposed or capped with a molding. The pickets may be applied with their tops in a straight line or in curves between posts. When pickets are applied with their edges tightly together, the assembly is called a *stockade* fence. Stockade fences are usually higher and are used when privacy is desired (Fig. 67-3).

Board on Board Fence

The *board on board* fence is similar to a picket fence. However, the boards are alternated from side to side so that the fence looks the same from both sides. The boards may vary in height and spacing according to the degree of privacy and protection from wind and sun desired. The tops or edges of the boards may be shaped in many different designs (Fig. 67-4).

Fig. 67-4 A board on board fence is similar to a picket fence. However, it looks the same on both sides.

Lattice Fence

This fence gets its name from narrow strips of wood called **lattice**. Strips are spaced by their own width. Two layers are applied at right angles to each other, either diagonally or in a horizontal and vertical fashion, to form a lattice work panel. Panels of various sizes can be prefabricated and installed between posts and rails, similar to a lattice work deck railing as shown in Figure 66-17.

Panel Fence

The *panel* fence creates a solid barrier with boards or panels fitted between top and bottom rails (Fig. 67-5). Fence boards may be installed diagonally or in other appealing designs. Alternating the panel design provides variety and adds to the visual appeal of the

Fig. 67-5 The panel fence restricts air. However, it provides shade and privacy. (*Courtesy of California Redwood Association*)

Fig. 67-6 Solid panels are combined with lattice work panels for an attractive design. (*Courtesy of California Redwood Association*)

Fig. 67-7 The louvered fence provides privacy and lets air through. (*Courtesy of California Redwood Association*)

fence. A small space should be left between panel boards to allow for swelling in periods of high humidity. In many cases, two or more basic styles may be combined to enhance the design. Figure 67-6 shows lattice panels combined with board panels.

Louvered Fence

The *louvered* fence is a panel fence with vertical boards set at an angle (Fig. 67-7). The fence permits the flow of air through it and yet provides privacy. This fence is usually used around patios and pools. It is not used as a barrier along property lines.

Post and Rail Fence

The *post and rail* fence is a basic and inexpensive style normally used for long boundaries (Fig. 67-8). Designs include two or more square edge or round rails, of various thicknesses, widths, and diameters, cut between the posts or fastened to their edges. Most post and rail fence designs have large openings. They are not intended as barriers.

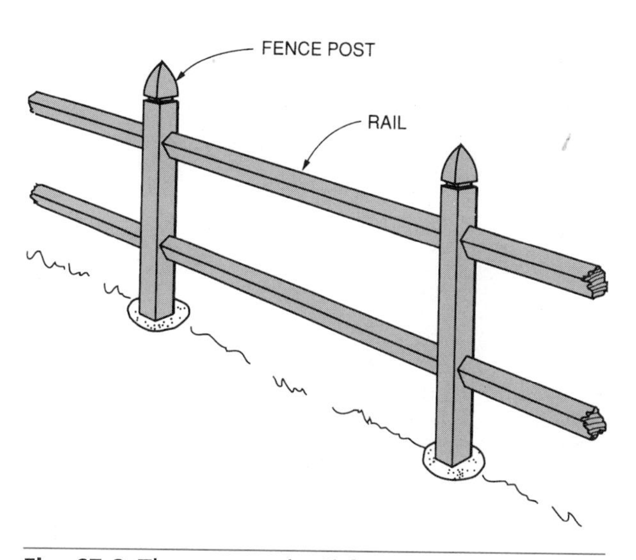

Fig. 67-8 The post and rail fence is an inexpensive design for long boundaries.

Fence Post Design

Fence posts are usually wood or iron. Wood posts are usually 4 × 4. Larger sizes may be used, depending on the design. The post tops may be shaped in various ways to enhance the design of the fence (Fig. 67-9). In order to conserve material and reduce expenses, a 4 × 4 post may be made to appear larger by applying furring and then boxing it in with 1-inch boards. The top may then be capped with shaped members and molding in various designs to make an attractive fence post (Fig. 67-10).

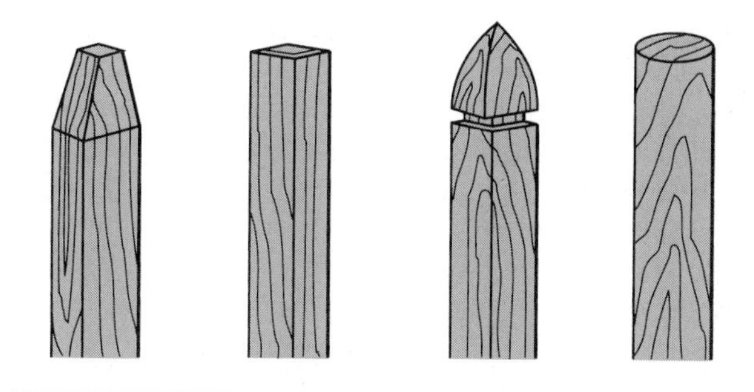

Fig. 67-9 Fence post tops are shaped in various ways to enhance the design of the fence.

Iron posts may be pipe or solid rod of from one inch to 1 1/4 inches in diameter for fences 3 to 4 feet tall. Larger diameters should be used for higher

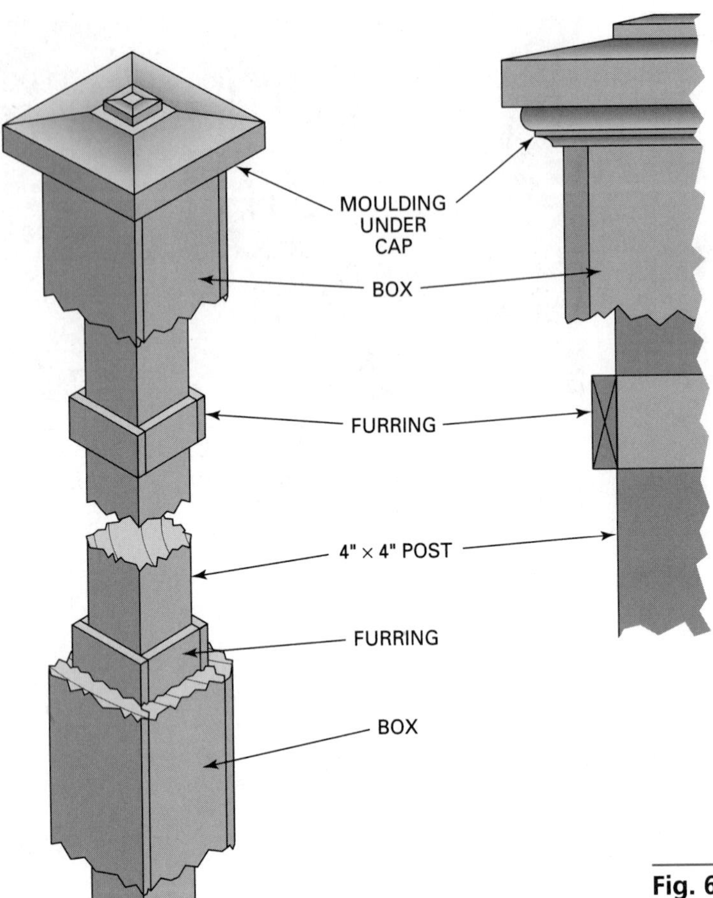

Fig. 67-10 Method of boxing and capping a wood post.

fences. Iron fence posts may be boxed in to simulate large wood posts. The tops are then usually capped with various shaped members similar to boxed wood posts (Fig. 67-11). Iron posts should be galvanized or otherwise coated to resist corrosion.

Building a Fence

The first step in building a fence is to set the fence posts. Locate and stretch a line between the end posts. If it is not possible to stretch a line because of steep sloping land, set up a transit-level to lay out a straight line. (See Chapter 26, Laying Out Foundation Lines.) If the fence is to be built on a property line, make sure that the exact locations of the boundary markers are known.

Setting Posts

Posts are generally placed about 8 feet apart, to their center lines. Mark the post locations with stakes along the fence line. Dig holes about 10 inches in diameter with a post hole digger. The depth of the hole depends on the height of the fence. Higher fences require that posts be set deeper (Fig. 67-12). The bottom of the hole should be filled with gravel or stone. This provides drainage and helps eliminate moisture to extend the life of the post.

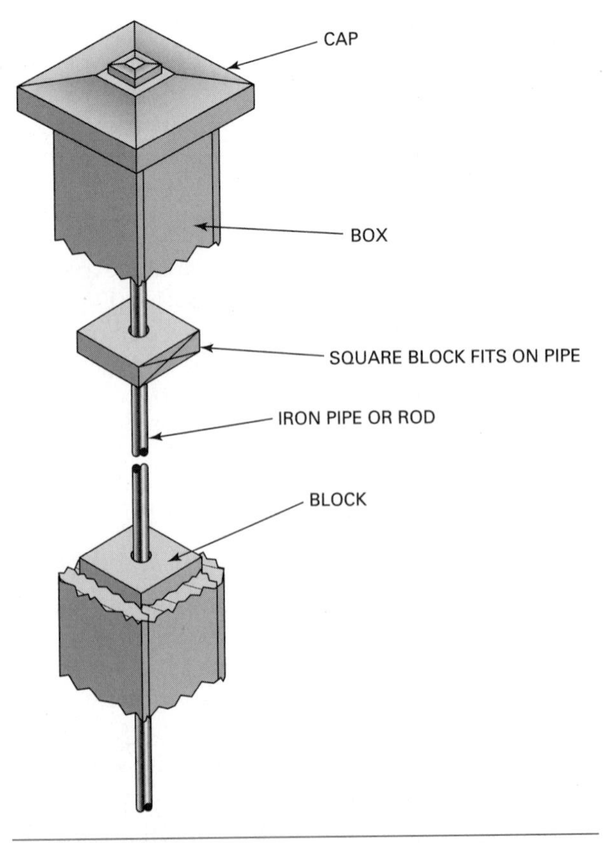

Fig. 67-11 Method of boxing and capping an iron fence post.

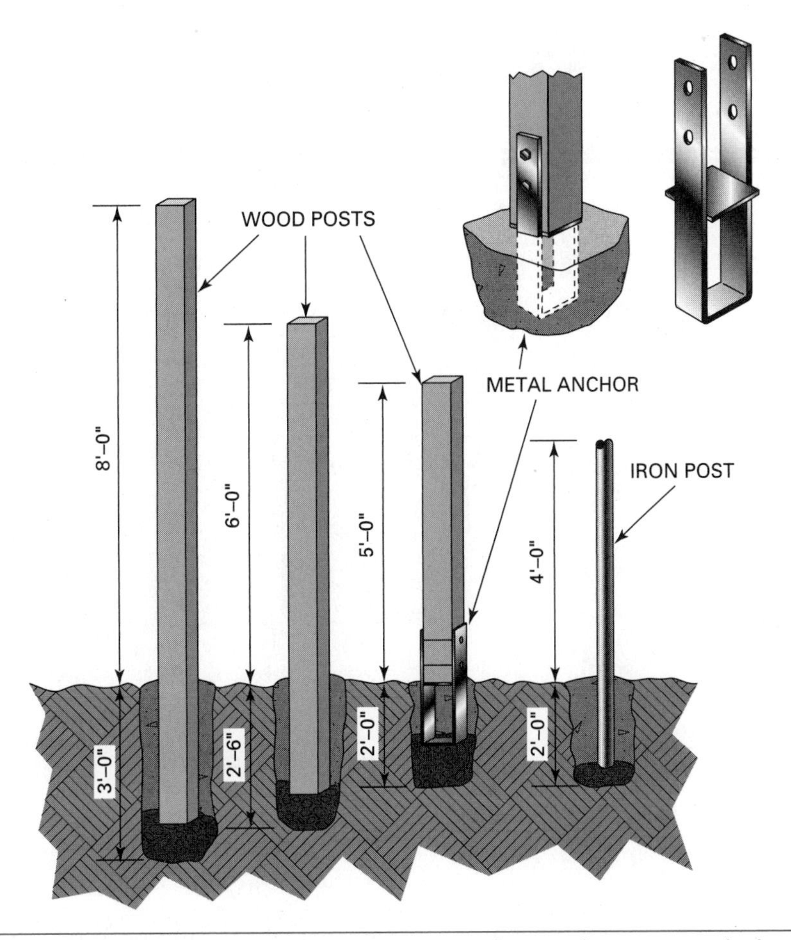

Fig. 67-12 A 3-foot hole is needed for an 8-foot fence. For shorter fences, post holes should be at least 2 feet deep.

Posts may be set in the earth. For the strongest fences, however, set the posts in concrete. Set the end posts first. Use a level to ensure that the posts are plumb in both directions. Brace them securely. The height of the post is determined by measuring from the ground, especially when the ground slopes. If top fence rails run across the tops of the posts, then the posts are left long. The tops are cut later in a straight or contoured line, as necessary. Make sure the face edges of the posts are aligned with the length of the fence. After the posts are braced, stretch lines between the end posts at the top and at the bottom.

Set and brace intermediate posts. The face edge of the posts are kept to the stretched line at the top and bottom. They are plumbed in the other direction with a level. Lag screws or spikes partially driven into the bottom of the post strengthen the set of fence posts in concrete.

Place concrete around the posts. The concrete may be placed in the hole dry as the moisture from the ground will provide enough water to hydrate the cement. Tamp it well into the holes. Form the top so the surface pitches down from the posts. Instead of setting the posts in concrete, metal anchors may be embedded. The posts are installed on them after the concrete has hardened (Fig. 67-13).

Installing Fence Rails

Usually two or three horizontal rails are used on most fences. However, the height or the design may require more. Rails may run across the face or be cut in between the fence posts. If rails are cut between posts, they are installed to a snapped chalk line across the edges of the posts. They are secured by toenailing or with metal framing connectors. The bottom rail should be kept at least 6 inches above the ground.

Very often, the faces of wood posts are not in the same line as the rails. Therefore, rail ends must be cut with other than square ends to fit between posts. Figure 67-14 shows a method of laying out rail ends to fit between twisted posts.

When iron fence posts are used, holes are bored in the rails. The rails are then installed over the previously set iron posts. Splices are made on the ends to continue the rail in a straight, unbroken

STEP BY STEP

PROCEDURES

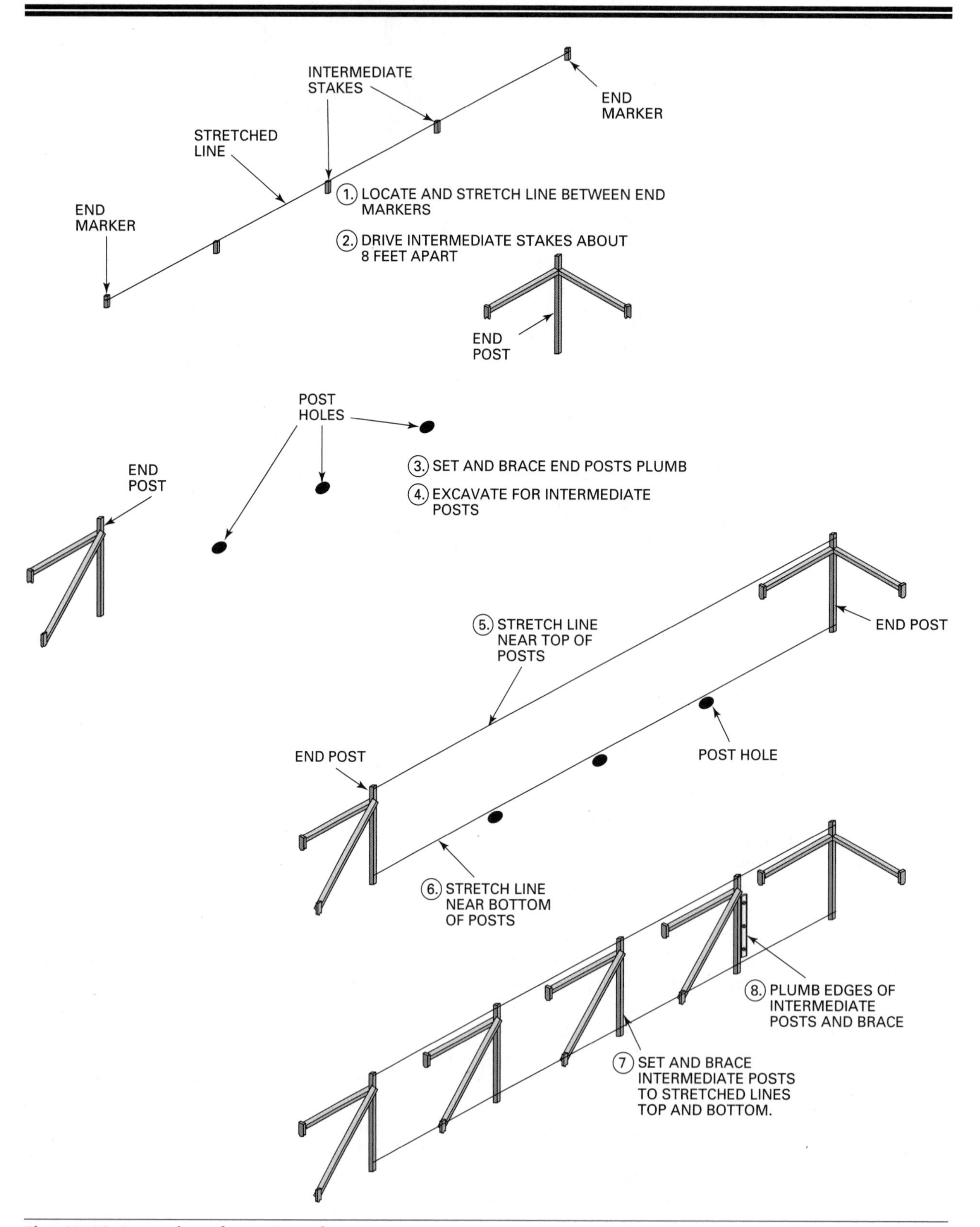

Fig. 67-13 Procedure for setting fence posts.

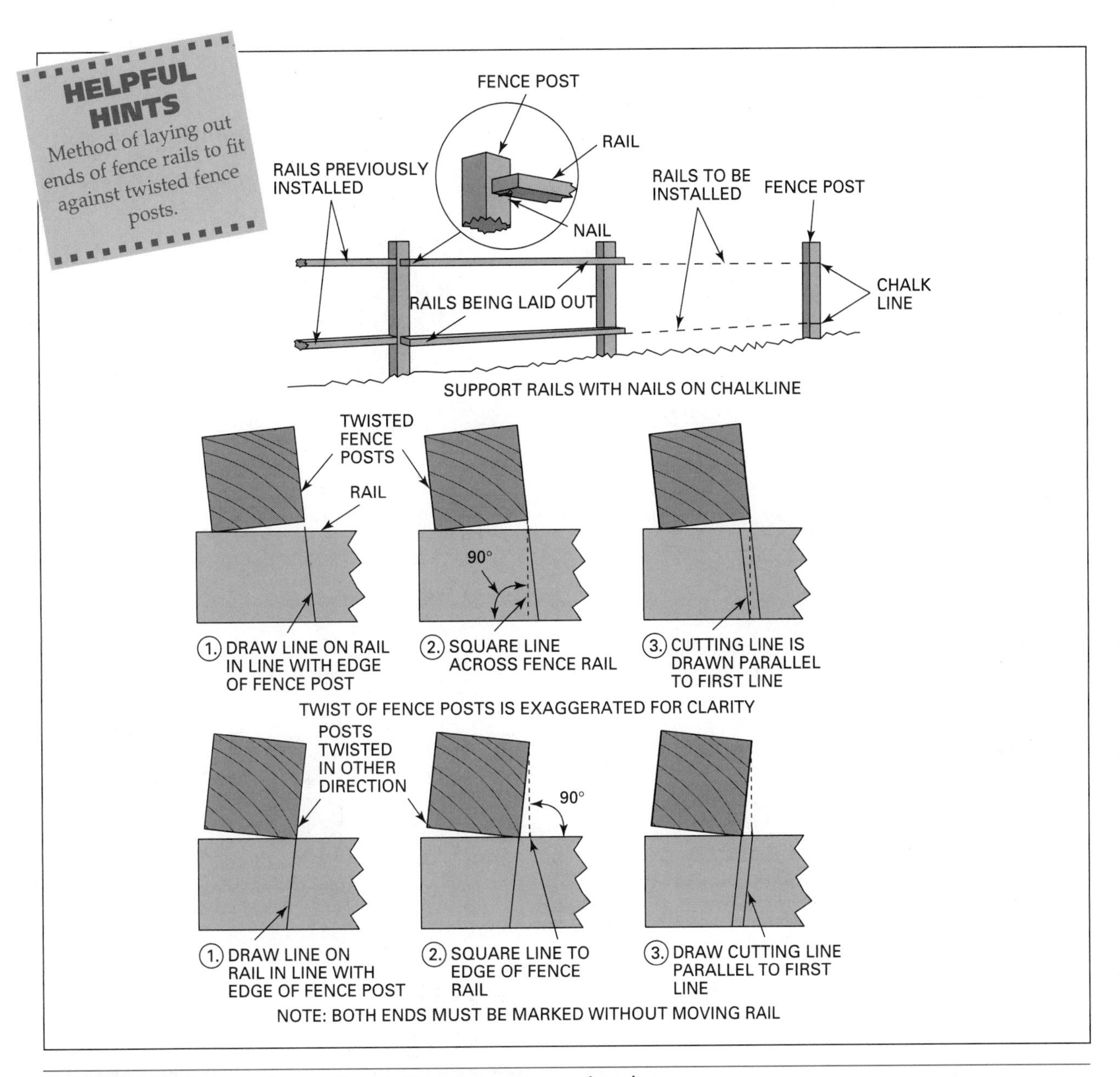

Fig. 67-14 Technique for cutting railing to meet a twisted post.

line. Special metal pipe grips are made to fasten fence rails to iron posts (Fig. 67-15). If iron posts are boxed with wood, then the rails are installed in the same manner as for wood posts.

On some rustic-style post and rail fences, the rails are doweled into the posts. In this case, post and rails must be installed together, one section at a time.

Applying Spaced Fenceboards and Pickets

Fasten pickets in plumb positions with their tops to the correct height at the starting and ending points. Stretch a line tightly between the tops of the two pickets. If the fence is long, temporarily install intermediate pickets to support the line from sagging. Sight the line by eye to see if it is straight. If not, make adjustments and add more support pickets if necessary. Use a picket for a spacer, and fasten pickets to the rails. If the spacing is different, rip a piece of lumber for use as a spacer.

Cut only the bottom end of the pickets when trimming their height. The bottom of pickets should not touch the ground. Place a 2-inch block on the ground. Turn the picket upside down with its top end on the block. Mark it at the chalk line. Fasten the picket with its top end to the stretched line (Fig. 67-16).

Continue cutting and fastening pickets using the spacer. Keep their tops to the line. Check the pickets for plumb frequently. If not plumb, bring back into plumb gradually with the installation of three or four pickets.

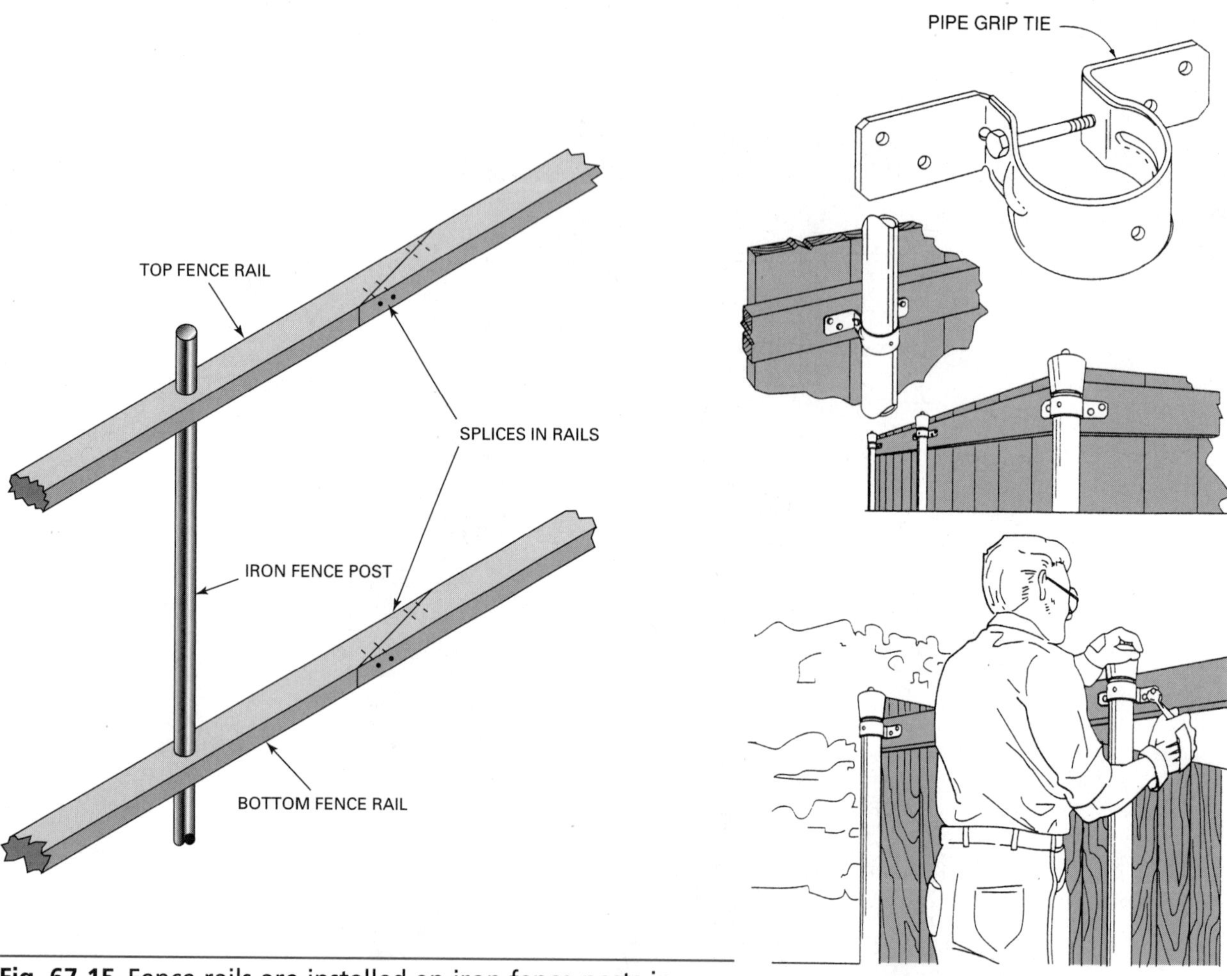

Fig. 67-15 Fence rails are installed on iron fence posts in several ways. (*Courtesy of Simpson Strong-Tie Company*)

STEP BY STEP

PROCEDURES

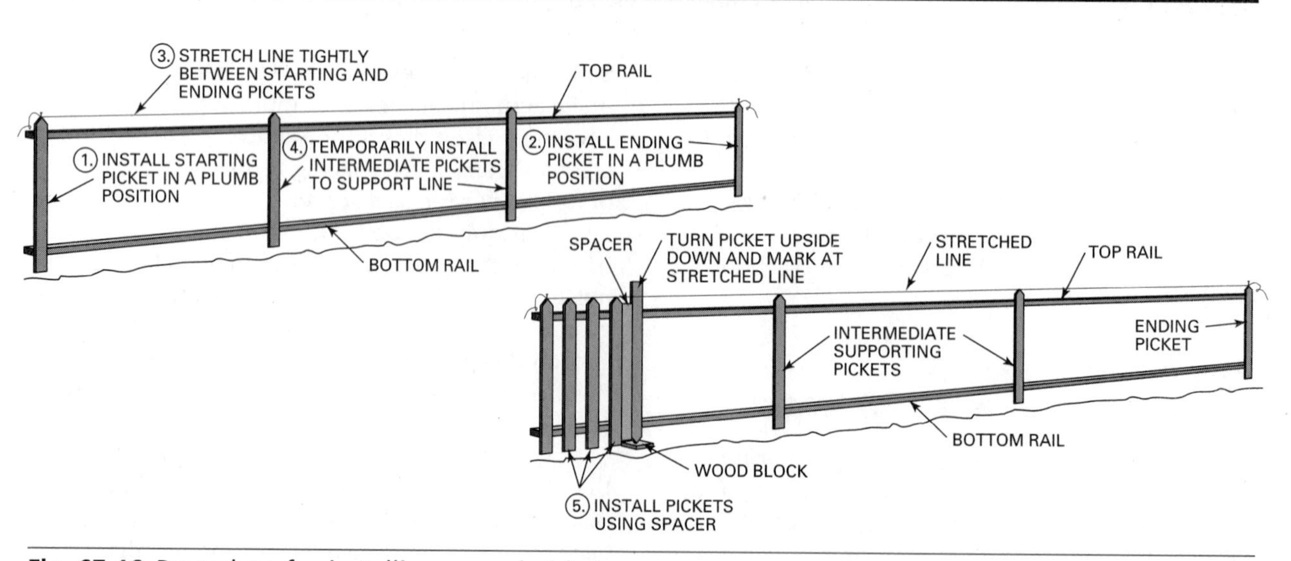

Fig. 67-16 Procedure for installing spaced pickets.

Stop 3 or 4 feet from the end. Check to see if the spacing will come out even. Usually the spacing has to be either increased or decreased slightly. Set the dividers for the width of a picket plus a space, increased or decreased slightly, whichever is appropriate. Space off the remaining distance. Adjust the dividers until the spacing comes out even. Any slight difference in a few spaces is not noticeable. This is much better than ending up with one narrow, conspicuous space (Fig. 67-17).

Installing Pickets in Concave Curves

In some cases, the pickets or other fenceboards are installed with their tops in concave curved lines between fence posts. If the fenceboards have shaped tops that cannot be cut, erect the fence using the following procedure.

Install a picket on each end at the high point of the curve. In the center, temporarily install a picket with its top to the low point of the curve. Fasten a flexible strip of wood in a curve to the top of the three pickets. Start from the center. Work both ways to install the remainder of the pickets with their tops to the curved strip. Space the pickets to each end (Fig. 67-18). Other fenceboards, such as board on board and louvers, are installed in a similar manner.

If fenceboard tops are to be cut in the shape of the curve, tack them in place with their tops above the curve. Bend the flexible strip to the curve. Mark all the fenceboard tops. Remove the fenceboards, if necessary. Cut the tops and replace them.

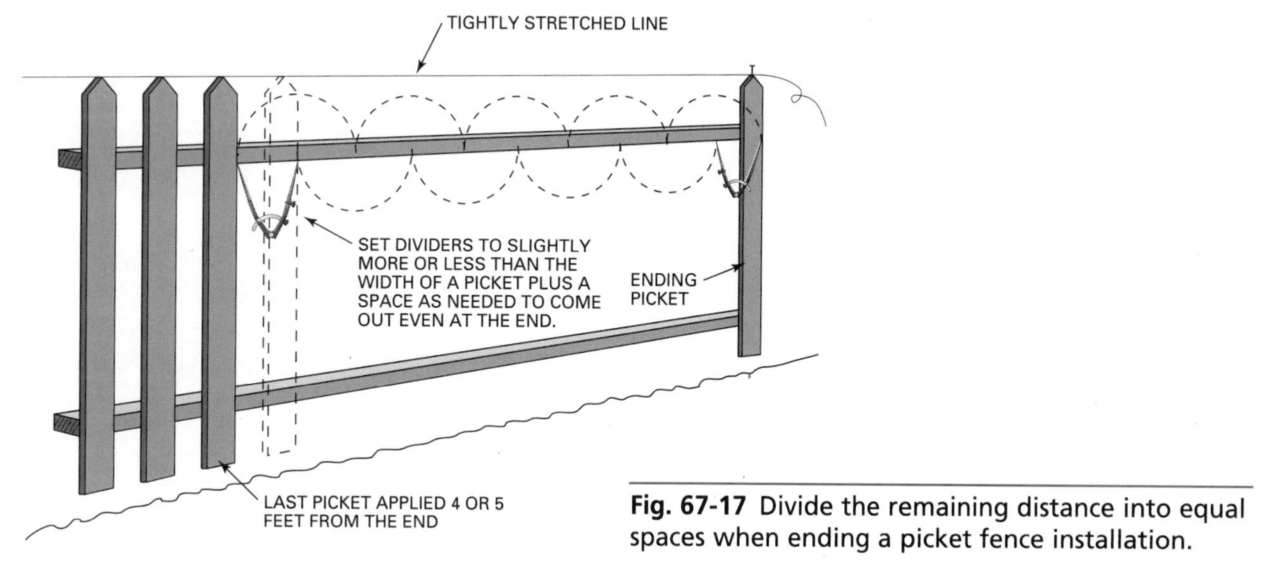

Fig. 67-17 Divide the remaining distance into equal spaces when ending a picket fence installation.

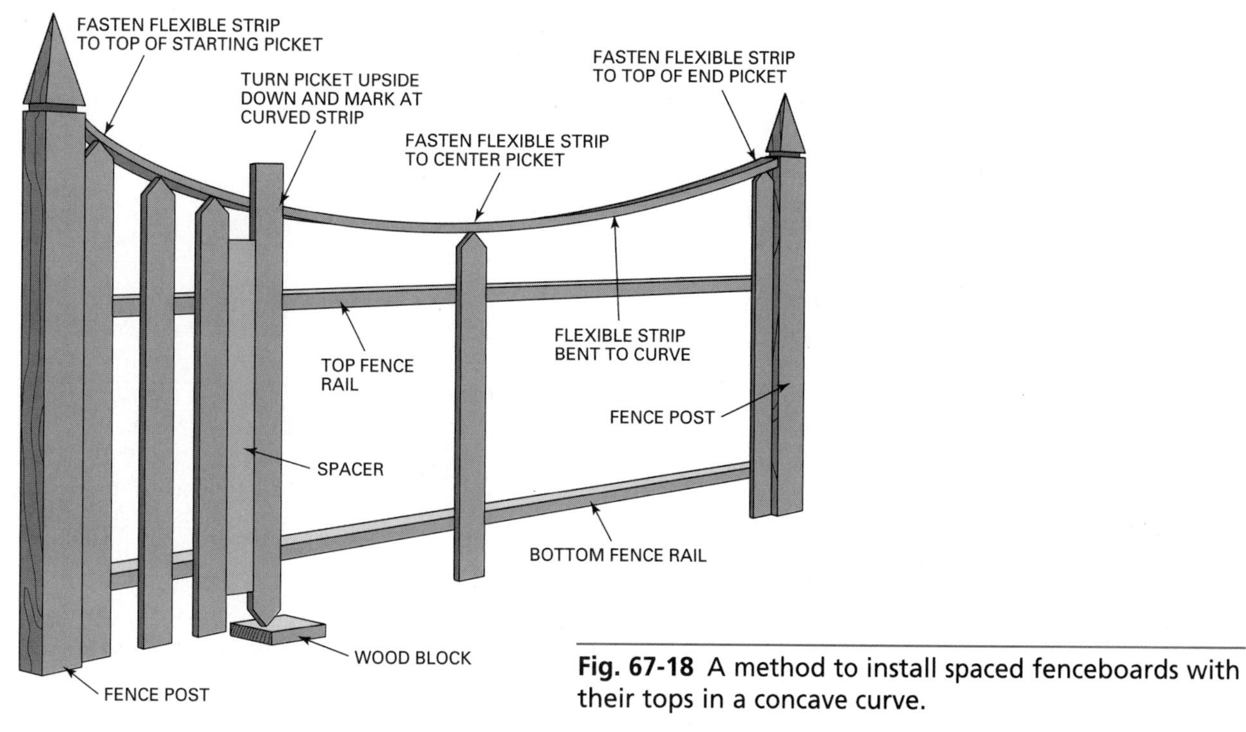

Fig. 67-18 A method to install spaced fenceboards with their tops in a concave curve.

Review Questions

Select the most appropriate answer.

1. No fasteners and hardware used on exposed decks and fences should be
   a. aluminum.
   b. hot-dipped galvanized.
   c. electroplated.
   d. stainless steel.

2. A ledger is a beam
   a. attached to the side of a building.
   b. supported by a girder.
   c. used to support girders.
   d. installed on supporting posts.

3. A footing for supporting posts must extend
   a. at least 12 inches below grade.
   b. 36 inches in depth.
   c. below the frost line.
   d. to stable soil.

4. Deck joists must be installed
   a. between girders.
   b. over girders.
   c. crowned edge up.
   d. using joist hangers.

5. A railing is required on deck stairs with more than
   a. 30 inches total rise.
   b. 2 risers.
   c. 3 feet total rise.
   d. 4 risers.

6. A railing is required on decks
   a. more than 3 stair risers above the ground.
   b. 30 or more inches above the ground.
   c. 4 feet or more above the ground.
   d. of any height.

7. The usual height of a bench seat without a cushion is
   a. 14 inches.
   b. 16 inches.
   c. 18 inches.
   d. 20 inches.

8. A high fence with picket edges applied tightly together is called a
   a. board on board fence.
   b. panel fence.
   c. post and rail fence.
   d. stockade fence.

9. The bottom rail of any type fence is installed above the ground by, at least,
   a. the thickness of a 2 × 4 block.
   b. the width of a 2 × 4 block.
   c. 6 inches.
   d. 8 inches.

10. When applying fence pickets to rails
   a. plumb each one before fastening.
   b. plumb them frequently.
   c. cut the top ends to the line.
   d. fasten bottom ends flush with bottom rail.

BUILDING FOR SUCCESS

Customer Relations in Construction

An important element affecting transactions that occur between the general public and a business is customer relations. Assuming that a primary goal of good customer relations is to gain customer satisfaction, some key considerations must be identified.

To set the stage for establishing good customer relations, businesspersons should follow certain practices. They must exhibit knowledge and credibility, demonstrate skill in their areas of expertise, and inspire customers. They must be aware of the customer's needs while possessing a good attitude. The information that goes out from the business must be accurate and verified.

Any mental image people have about a company or business will relate directly to that company's success. The target population of a business could be large or quite small, depending on the product and cost. The volume of sales will be reflected in the company's customer relations policies.

Another primary goal of customer relations is to create and maintain a positive image using sound business principles. This must be a high priority. As a builder conducts daily business with the public, his or her company is involved with maintaining a good image. The builder has the responsibility to conduct business in an ethical and proper manner if trust is to be established.

Without regular customers, a business will be forced to close its doors. The need for good customer relations must be recognized by every successful businessperson. This will involve developing good personal relations skills, sound business skills, and an awareness of customer satisfaction.

The customer is any person who purchases a commodity or service from another person or place of business. Any lasting business relationship that is created is established with mutual goals. The foundation of this relationship begins when the customer desires a quality service or product from a vendor for an agreed price. Once the transaction is initiated, a business relationship is formed. From that initial agreement, the process of "doing business" will involve the need to honor that agreement. In the process it is important how each party is treated. Basic responsibilities will be for the vendor to supply the service or product as agreed. The customer is to respond with payment once the goods are delivered.

In the more tangible terms of construction, we can use the home builder as an example. A potential client contacts a builder with a desire to build a new home. The builder sees this as an opportunity to "sell" a product or service to the client. The customer may have heard of the builder previously or may have seen one of the builder's products. This alone may have set in motion the foundation for any customer relations affecting the project. The customer will bring along any previous image of the builder based on personal observations and/or what others have said. This becomes the starting point of communications and relationships that will affect the success of any business that is conducted.

Good customer relations will have a major impact on the success of all business transactions. A builder's public image and reputation will be affected by the manner in which his or her employees view public relations. Mutual trust and respect must be established in the beginning and continue throughout the duration of the transaction. Sound business principles must be readily seen by the public and customers. Customer relation policies must be consistent in order to assure honesty and fairness. All future business activities between the builder and the clientele will be based on the previously established relationships. By establishing clear oral and written communications at the start of any business discussions, the future working relationships will almost always guarantee success. As builders, subcontractors, or other technicians interact with the clientele, they will need to establish good customer relations. The success of doing business with the public will depend on how well those relations are established.

FOCUS QUESTIONS: For individual or group discussion

1. How do you define customer relations?
2. What are five important considerations for establishing positive customer relations between builders and customers?
3. How can effective customer relations assure framing subcontractors that their perceived image is credible?
4. As you prepare to enter the construction world, how would you structure a plan to establish good customer relations for your personal business?

STUDENT ACTIVITIES for Section 3: Exterior Finish

Unit 18 Roofing

1. Demonstrate and explain the application of asphalt shingles, including felting, layout, nailing, starter course, cutting and spacing on the gable, hip, and valley. Include the installations of metal edge, gutter apron, valley flashing, roof vents, asphalt cement, step flashing, hip and ridge caps.
2. Demonstrate and explain the application of wood shingles and shakes using the same criteria as in activity #1.
3. Demonstrate the installation process of roll roofing with concealed nails and double coverage. Include felting and asphalt cement use.
4. Explain the available grades of asphalt shingles and wood roofing products, their advantages and disadvantages, costs, and how to calculate the quantities needed for the job.
5. Explain in detail any two typical roofing problems and how to prevent or properly repair those problems.

Unit 19 Windows

1. Using a major wood window manufacturer's catalog, explain how to read the catalog and obtain the following information: window type, sizes available, features, accessories, costs, and technical information for calculating heating, ventilating, & air conditioning data.
2. Using a blueprint and the window manufacturer's catalog, explain the process of selecting windows that will meet local building code requirements. List the information on a window schedule.
3. Duplicate activities #1 and #2 for metal windows.
4. Explain and demonstrate the manufacturer's installation process for a casement, double hung, or gliding window. Include a description of the specific features, such as cladding, hardware, weatherstripping, and sash venting action.

Unit 20 Exterior Doors

1. Describe the styles, sizes, features, and parts of an exterior door.
2. Assemble and set an exterior door frame. Hang a door in the frame. Install the lockset.
3. Rout hinge gains using a butt hinge template and router.
4. Explain the process of installing a sectional overhead garage door.

Unit 21 Siding and Cornice Construction

1. Make a product display board showing many samples of sidings, corner boards, accessories, and finishes available. Explain how the siding products are estimated, priced, and sold.
2. Demonstrate the application procedures for horizontal lap and vertical sidings, with the necessary trims.
3. Demonstrate the application procedures for the sheet type plywood, composite, and hardboard sidings, with the necessary trims.
4. Explain the installation process of metal and plastic sidings. Include the necessary accessories for each siding product.
5. Using a residential blueprint, estimate the quantity of siding needed to cover the side walls of the house with a $4' \times 8'$ hardboard product. Make no deductions for cutouts. Figure no waste.
6. Explain the theory of a story pole and demonstrate how to make one for horizontal lap siding.
7. Explain the use of a water table, where it is used, and how it is installed.
8. Using a typical wall section, explain the construction process and major parts of a box, snub, and open cornice.
9. Using aluminum soffit and fascia material, explain the installation process for a box cornice. Also demonstrate the required tools.

Unit 22 Decks, Porches, and Fences

1. Using a set of blueprints of a residence with no front porch, design a porch that adds character and aesthetic value to the house. Design a shed, gable, or hip roof into the plan. The open or enclosed wall design may be used. Include a complete list of construction materials and costs.
2. Using a blueprint, design a wood deck and backyard fence to provide an outside eating and recreation area for the home. Incorporate into the plan the use of pressure-treated and redwood or western red cedar products. The deck may have a sun roof or be open. Prepare a complete list of the building materials and costs.
3. Make a presentation of your design to others, stating how you arrived at the design, the reasons certain materials were selected, a time schedule, and how the project will be built. Exhibit examples of the materials and plans.

SECTION 4

Interior Finish

UNIT

23

Drywall Construction

The term *drywall construction* generally means the application of gypsum board to interior walls and ceilings. In contrast it does not refer to the application of wet gypsum plaster to a plaster base of gypsum or wire lath.

OBJECTIVES

After completing this unit, the student should be able to:

- describe various kinds, sizes, and uses of gypsum panels.
- describe the kinds and sizes of nails, screws, and adhesives used to attach gypsum panels.
- make single-ply and multi-ply gypsum board applications to interior walls and ceilings.
- conceal gypsum board fasteners and corner beads.
- reinforce and conceal joints with tape and compound.

UNIT CONTENTS

CHAPTER 68 GYPSUM BOARD

Gypsum board is sometimes called *wallboard, plasterboard,* **drywall,** or *Sheetrock®*. It is used extensively in construction (Fig. 68-1). The term *Sheetrock* is a brand name for gypsum panels made by the U.S. Gypsum Company. However, the brand name is in such popular use, it has become a generic name for gypsum panels. Gypsum board makes a strong, high-quality, fire-resistant wall and ceiling covering. It is readily available, easy to apply, decorate, or repair, and it is relatively inexpensive.

Gypsum Board

Many types of gypsum board are available for a variety of applications. The board or panel is composed of a gypsum core encased in a strong, smooth-finish paper on the **face** side and a natural-finish paper on the back side. The face paper is folded around the long edges. This reinforces and protects the core. The long edges are usually tapered. This allows the joints to be concealed with compound without any noticeable *crown joint* (Fig. 68-2). A crown joint is a buildup of the compound above the surface.

Types of Gypsum Panels

Most gypsum panels can be purchased, if desired, with an aluminum foil backing. The backing functions as a vapor retarder. It helps prevent the passage of interior water vapor into wall and ceiling spaces.

Fig. 68-1 The application of gypsum board to interior walls and ceilings is called drywall construction. (*Courtesy of U.S. Gypsum Company*)

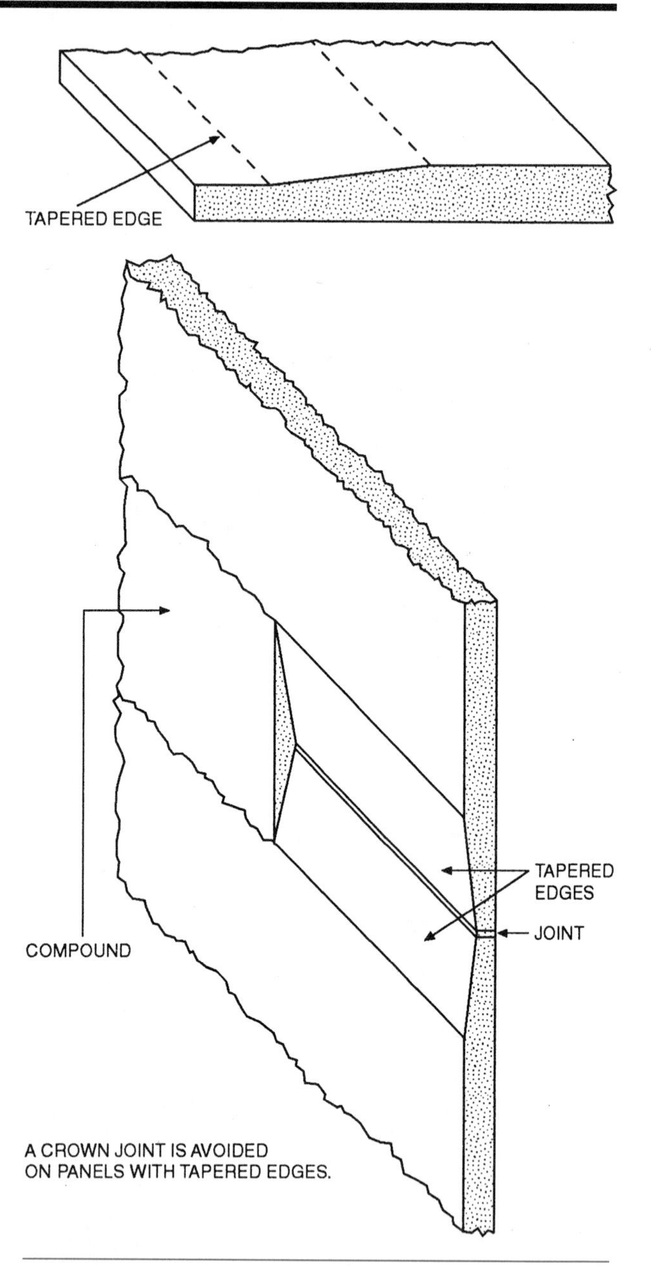

Fig. 68-2 The long edges of gypsum panels usually are tapered for effective joint concealment.

Regular. *Regular* gypsum panels are most commonly used for single- or multilayer application. They are applied to interior walls and ceilings in new construction and remodeling.

Eased Edge. An *eased edge* gypsum board has a special tapered, rounded edge. This produces a much stronger concealed joint than a tapered, square edge (Fig. 68-3).

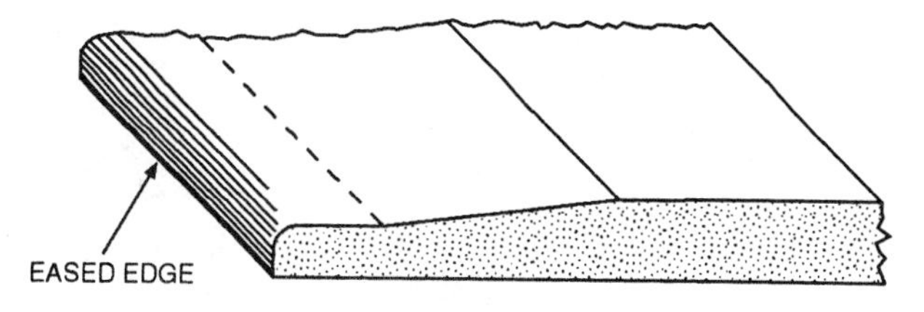

Fig. 68-3 An eased edge panel has a rounded corner that produces a stronger concealed joint.

Type X. *Type X* gypsum board is typically known as *firecode* board. It has greater resistance to fire because of special additives in the core. Type X gypsum board is manufactured in several degrees of resistance to fire. Type X looks the same as regular gypsum board. However, it is labeled Type X on the edge or on the back.

Water-Resistant. *Water-resistant* or *moisture resistant* (MR) gypsum board consists of a special moisture-resistant core and paper cover that is chemically treated to repel moisture. It is used frequently as a base for application of wall tile in bath, showers, and other areas subjected to considerable moisture. It is easily recognized by its distinctive green face. It is frequently called *green board* by workers in the field. Water-resistant panels are available with a Type X core for increased fire resistance.

Special Purpose. *Backing board* is designed to be used as a base layer in multilayer systems. It is available with regular or Type X cores. *Core-board* is available in 1-inch thickness. It is used for various applications, including the core of solid gypsum partitions. It comes in 24-inch widths with a variety of edge shapes. *Predecorated* panels have coated, printed, or overlay surfaces that require no further treatment. *Liner board* has a special fire-resistant core encased in a moisture-repellent paper. It is used to cover shaft walls, stairwells, chaseways, and similar areas.

Veneer Plaster Base. *Veneer plaster bases* are commonly called blue board. They are large, 4-foot wide gypsum board panels faced with a specially treated blue paper. This paper is designed to receive applications of *veneer plaster.* Conventional plaster is applied about 3/8 inch thick and takes considerable time to dry. In contrast, specially formulated veneer plaster is applied in one coat of about 1/16 inch, or two coats totaling about 1/8 inch. It takes only about forty-eight hours to dry.

Gypsum lath is used as a base to receive conventional plaster. (It is more fully described in Chapter 37.) Other gypsum panels, such as soffit board and sheathing, are manufactured for exterior use.

Sizes of Gypsum Panels

Widths and Lengths. Coreboards and liner boards come in 2-foot widths and from 8 to 12 feet long. Other gypsum panels are manufactured 4 feet wide and in lengths of 8, 9, 10, 12, 14, or 16 feet. Gypsum board is made in a number of thicknesses. Not all lengths are available in every thickness.

Thicknesses

- A 1/4-inch thickness is used as a base layer in multilayer applications. It is also used to cover existing walls and ceilings in remodeling work. It can be applied in several layers for forming curved surfaces with short radii.
- A 3/8-inch thickness is usually applied as a face layer in repair and remodeling work over existing surfaces. It is also used in multilayer applications in new construction.
- Both 1/2-inch and 5/8-inch are common used thicknesses of gypsum panels for single-layer wall and ceiling application in residential and commercial construction. The 5/8-inch thick panel is more rigid. It has greater resistance to impacts and fire.
- Coreboards and liner boards come in thicknesses of 3/4 and one inch.

Cement Board

Like gypsum board, *cement board,* or *wonder board* are panel products. However, they have a core of portland cement reinforced with a glass fiber mesh embedded in both sides (Fig. 68-4). The core resists water penetration and will not deteriorate when wet. It is designed for use in areas that may be subjected to high-moisture conditions. It is used extensively in bathtub, shower, kitchen, and laundry

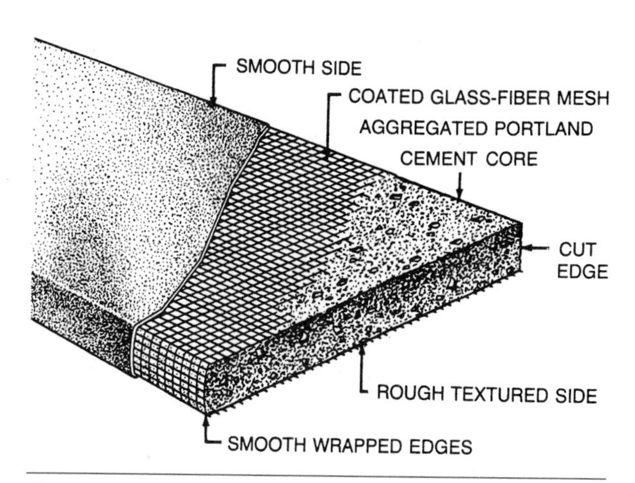

Fig. 68-4 Composition of cement board. *(Courtesy of U.S. Gypsum Company)*

areas as a base for ceramic tile. In fact, some building codes require its use in these areas.

Panels are manufactured in sizes designed for easy installation in tub and shower areas with a minimum of cutting. Standard cement board panels come in a thickness of 1/2 inch, in widths of 32 or 36 inches, and in 5-foot lengths. Custom panels are available in a thickness of 5/8 inch, widths of 32 or 48 inches, and lengths from 32 to 96 inches.

Cement board is also manufactured in a 5/16 inch thickness. It is used as an underlayment for floors and countertops. Exterior cement board is used primarily as a base for various finishes on building exteriors.

Drywall Fasteners

Specially designed nails and screws are used to fasten drywall panels. Ordinary nails or screws are not recommended. The heads of common nails are too small in relation to the shank. They are likely to cut the paper surface when driven. Staples may only be used to fasten the base layer in multilayer applications. They must penetrate at least 5/8 inch into supports. Using the correct fastener is extremely important for proper performance of the application. Fasteners with corrosion-resistant coatings must be used when applying water-resistant gypsum board or cement board. Care should be taken to drive the fasteners straight and at right angles to the wallboard to prevent the fastener head from breaking the face paper.

Nails

Gypsum board nails should have flat or concave heads that taper to thin edges at the rim. Nails should have relatively small diameter shanks with heads at least 1/4 inch, but no more than 5/16 inch in diameter. For greater holding power, nails with annular ring shanks are used (Fig. 68-5).

Smooth shank nails should penetrate at least 7/8 inch into framing members. Only 3/4-inch penetration is required when ring shank nails are used. Greater nail penetrations are required for fire-rated applications.

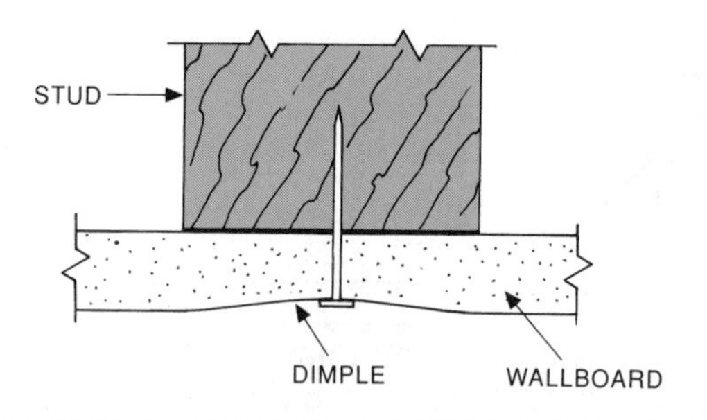

Fig. 68-6 Nails are driven with a convex-faced drywall hammer. This forms a dimple in the board. The dimple is filled with compound to conceal the fastener.

Nails should be driven with a drywall hammer that has a convex face. This hammer is designed to compress the gypsum panel face to form a dimple not more than 1/16 inch when the nail is driven home (Fig. 68-6). The dimple is made so the nail head can later be covered with compound.

Screws

Special drywall screws are used to fasten gypsum panels to steel or wood framing or to other panels. They are made with Phillips heads designed to be driven with a drywall power screwdriver (Fig. 68-7). A proper setting of the nosepiece on the power screwdriver assures correct countersinking of the screwhead. When driven correctly, the specially contoured bugle head makes a uniform depression in the panel surface without breaking the paper.

Different kinds of drywall screws are used for fastening into wood, metal, and gypsum panels. *Type W* screws are used for wood, *Type S* and Type *S-12* for metal framing, and *Type G* for fastening to gypsum backing boards (Fig. 68-8).

Type W screws have sharp points. They have specially designed threads for easy penetration and

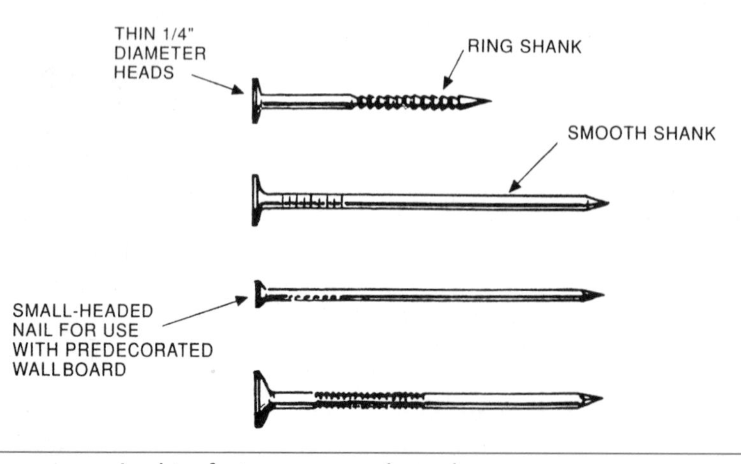

Fig. 68-5 Special nails are required to fasten gypsum board.

Fig. 68-7 Drywall screws are driven with a screwgun to the desired depth. (*Courtesy of U.S. Gypsum Company*)

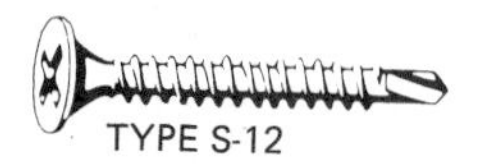

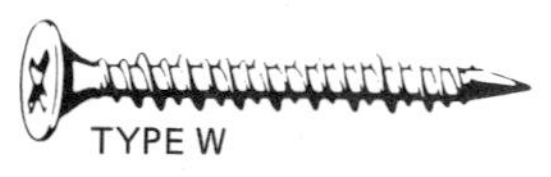

Fig. 68-8 Several types of screws are used to fasten gypsum panels. Selection of the proper type is important.

excellent holding power. The screw should penetrate the supporting wood frame by at least 5/8 inch.

Type S screws are self-drilling and self-tapping. The point is designed to penetrate sheet metal with little pressure. This is an important feature because thin metal studs have a tendency to bend away when driving screws. Type S-12 screws have a different drill point designed for heavier gauge metal framing.

Type G screws have a deeper, special thread design for effectively fastening gypsum panels together. These screws must penetrate into the supporting board by at least 1/2 inch. If the supporting board is not thick enough, longer fasteners should be used. Make sure there is sufficient penetration into framing members.

Adhesives

Drywall adhesives are used to bond single layers directly to supports or to laminate gypsum board to base layers. Adhesives used to apply gypsum board are classified as *stud adhesives* and *laminating adhesives.*

For bonding gypsum board directly to supports, special *drywall stud adhesive* or approved *construction adhesive* is used. Supplemental fasteners must be used with stud adhesives. Stud adhesives are available in large cartridges. They are applied to framing members with hand or powered adhesive guns (Fig. 68-9).

CAUTION **Some types of drywall adhesives may contain a flammable solvent. Do not use these types near an open flame or in poorly ventilated areas.**

For laminating gypsum boards to each other, *joint compound adhesives* and *contact adhesives* are used. Joint compound adhesives are applied over the entire board with a suitable spreader prior to lamination. Boards laminated with joint compound adhesive require supplemental fasteners.

When contact adhesives are used, no supplemental fasteners are necessary. However, the board cannot be moved after contact has been made. The adhesive is applied to both surfaces by brush, roller, or spray gun. It is allowed to dry before laminating. A *modified contact* adhesive is also used. It permits an open time of up to thirty minutes during which the board can be repositioned, if necessary.

Fig. 68-9 Applying drywall adhesive to stud edges. (*Courtesy of U.S. Gypsum Company*)

CHAPTER 69 SINGLE-LAYER AND MULTILAYER DRYWALL APPLICATION

Single-layer gypsum board applications are widely used in light commercial and residential construction. They adequately meet building code requirements. *Multilayer* applications are more often used in commercial construction. They have increased resistance to fire and sound transmission. Both systems provide a smooth, unbroken, quality surface if recommended application procedures are followed.

Single-Layer Application

Drywall should be not be delivered to the job site until shortly before installation begins. Boards stored on the job for long periods are subject to damage. The boards must be stored under cover and stacked flat on supports. Supports should be at least 4 inches wide and placed fairly close together (Fig. 69-1). Leaning boards against framing for long periods may cause the boards to warp. This makes application more difficult. To avoid damaging the edges, carry the boards. Do not drag them. Then, set the boards down gently. Be careful not to drop them.

Cutting and Fitting Gypsum Board

Take measurements accurately. Cut the board by first scoring the face side through the paper to the core. Use a utility knife. Guide it with a *drywall T-square,* if cutting a square end (Fig. 69-2). The board is then broken along the scored face. The back paper is scored along the fold. The sheet is then broken by snapping the board in the reverse direction (Fig. 69-3).

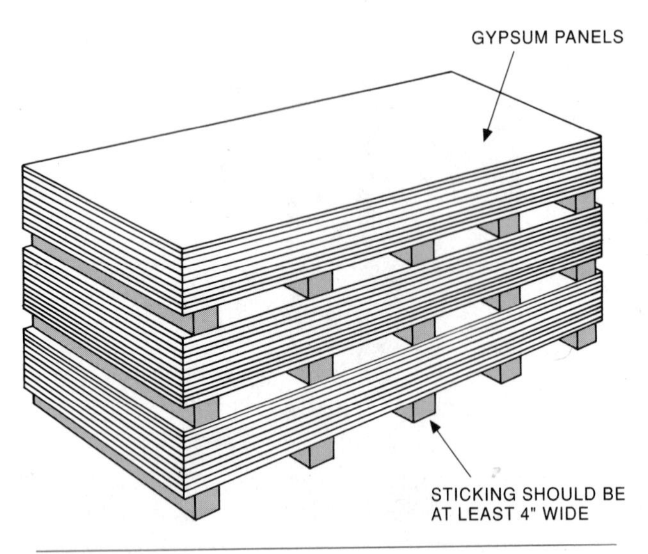

Fig. 69-1 Correct method of stacking gypsum board.

Fig. 69-2 Using a drywall T-square as a guide when scoring across the width of a board. *(Courtesy of U.S. Gypsum Company)*

To make cuts parallel to the long edges, the board is often gauged with a tape and scored with a utility knife. A *tape guide* and *tape tip* are sometimes used to aid the procedure. The tape guide permits more accurate gauging and protects the fingers. The tape tip contains a slot into which the knife is inserted. This prevents slipping off the end of the tape (Fig. 69-4).

When making parallel cuts close to long edges, it is usually necessary to score both sides of the board to obtain a clean break.

Smooth ragged edges with a drywall rasp, coarse sanding block, or knife. A job-made drywall rasp can be made by fastening a piece of metal lath to a wood block. Cut panels should fit easily into place without being forced. Forcing the panel may cause it to break.

Aligning Framing Members

Before applying the gypsum board, check the framing members for alignment. Stud edges should not be out of alignment more than 1/8 inch with adjacent studs. A wood stud that is out of alignment can be straightened by the procedure shown in Figure 69-5. Ceiling joists are sometimes brought into alignment with the installation of a *strongback* across

Fig. 69-3 When cutting gypsum board, start by cutting and breaking the board from the finished side. Then cut about 50 percent of the back paper, leaving the top and bottom intact to act as hinges. Finally, swing the board back the other way with a quick snap. (*Courtesy of U.S. Gypsum Company*)

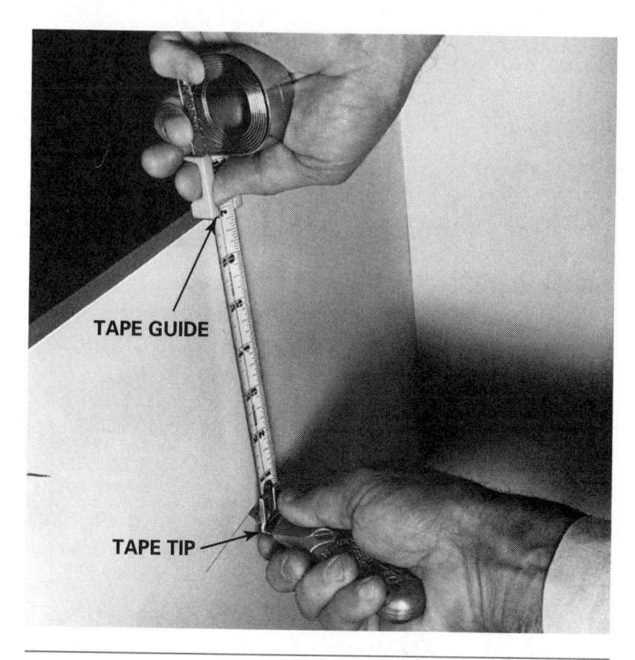

Fig. 69-4 Lines parallel to edges are scored by gauging with a tape. A tape guide and tape tip are useful accessories for gauging. (*Courtesy of U.S. Gypsum Company*)

the tops of the joists at about the center of the span (Fig. 69-6).

Fastening Gypsum Panels

Drywall is fastened to framing members with nails or screws. Hand pressure should be applied on the panel next to the fastener being driven. This ensures that the panel is in tight contact with the framing member. The use of adhesives reduces the number of nails or screws required. A single or double method of nailing may be used.

Single Nailing Method. With this method, nails are spaced a maximum of 7 inches OC on ceilings and 8 inches OC on walls into frame members. Nails

HELPFUL HINTS

How to straighten a crooked stud.

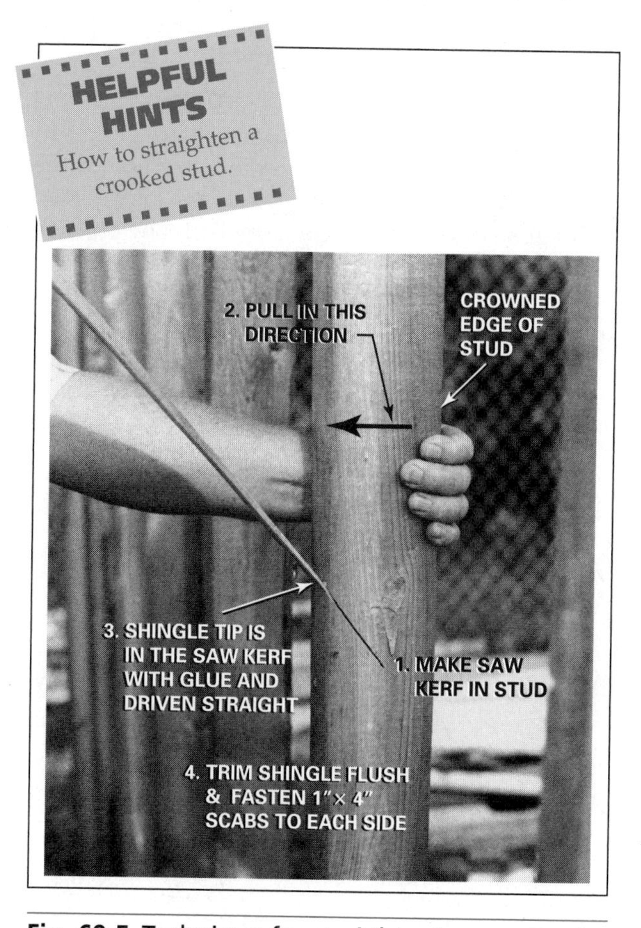

Fig. 69-5 Technique for straightening a severely crowned stud.

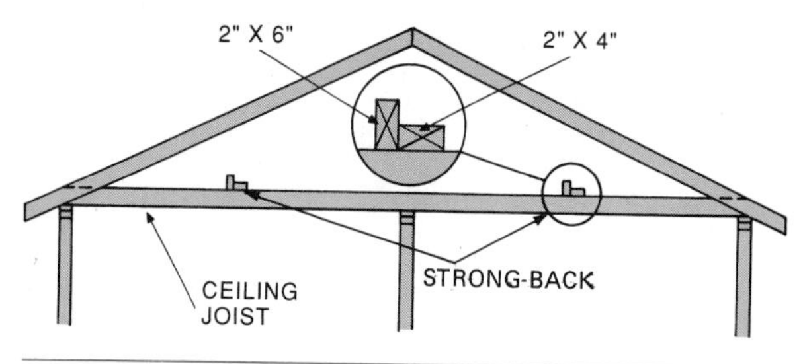

Fig. 69-6 A strongback sometimes is used to align ceiling joists or the bottom chord of roof trusses.

should be first driven in the center of the board and then outward toward the edges. Perimeter fasteners should be at least 3/8 inch, but not more than 1 inch from the edge.

Double Nailing Method. In double nailing, the perimeter fasteners are spaced as for single nailing. In the field of the panel, space a first set of nails 12 inches OC. Space a second set 2 to 2 1/2 inches from the first set. The first nail driven is reseated after driving the second nail of each set. This assures solid contact with framing members (Fig. 69-7).

Screw Attachment. Screws are spaced 12 inches OC on ceilings and 16 inches OC on walls when framing members are spaced 16 inches OC. If framing members are spaced 24 inches OC, then screws are spaced a maximum of 12 inches OC on both walls and ceilings.

Using Adhesives. Apply a straight bead about 1/4 inch in diameter to the centerline of the stud edge. On studs where panels are joined, two parallel beads of adhesive are applied, one on each side of the centerline. Zigzag beads should be avoided to prevent the adhesive from squeezing out at the joint.

On wall application, supplemental fasteners are used around the perimeter. Space them about 16 inches apart. On ceilings, in addition to perimeter

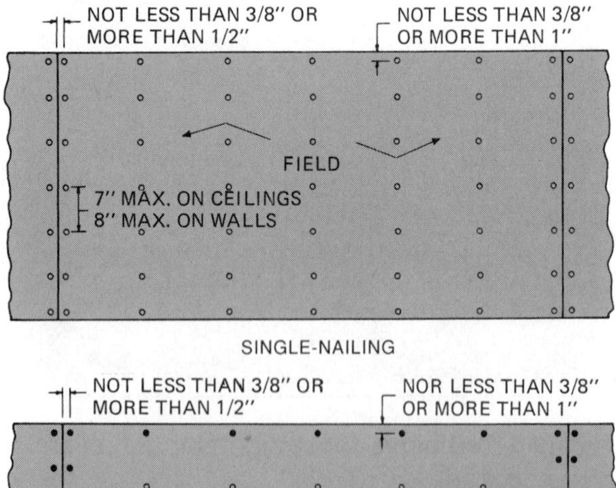

Fig. 69-7 Gypsum panels may be single-nailed or double-nailed.

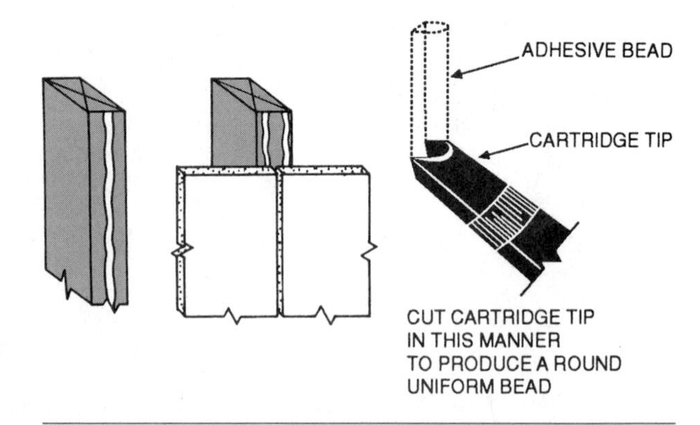

Fig. 69-8 Beads of stud adhesive are applied (left) straight under the field of a board, and (right) parallel beads under joints.

fastening, the field is fastened at about 24-inch intervals (Fig. 69-8).

Gypsum panels may be prebowed. This reduces the number of supplemental fasteners required. Prebow the panels by one of the methods shown in Figure 69-9. Make sure the finish side of the panel faces in the correct direction. Allow them to remain overnight or until the boards have a 2-inch permanent bow. Apply adhesive to the studs. Fasten the panel at top and bottom plates. The bow keeps the center of the board in tight contact with the adhesive until bonded.

Ceiling Application

Gypsum panels are applied first to ceilings and then to the walls. Panels may be applied parallel, or at right angles, to joists or furring. If applied parallel, edges and ends must bear completely on framing. If applied at right angles, the edges are fastened where they cross over each framing member. Ends must be fastened completely to joists or furring strips.

Carefully measure and cut the first board to width and length. Cut edges should be against the wall. Lay out lines on the panel face indicating the location of the framing in order to place fasteners accurately.

Gypsum board panels are heavy. At least two or more people are needed for ceiling application unless special equipment is available. Lift the panel overhead and place it in position (Fig. 69-10). Install two **deadmen** under the panel to hold it in position.

Deadmen are supports made in the form of a "T." They are easily made on the job using 1 × 3 lumber with short braces from the vertical member to the horizontal member. The leg of the support is made about 1/4 inch longer than the floor-to-ceiling height. The deadmen are wedged between the floor and the panel. They hold the panel in position while

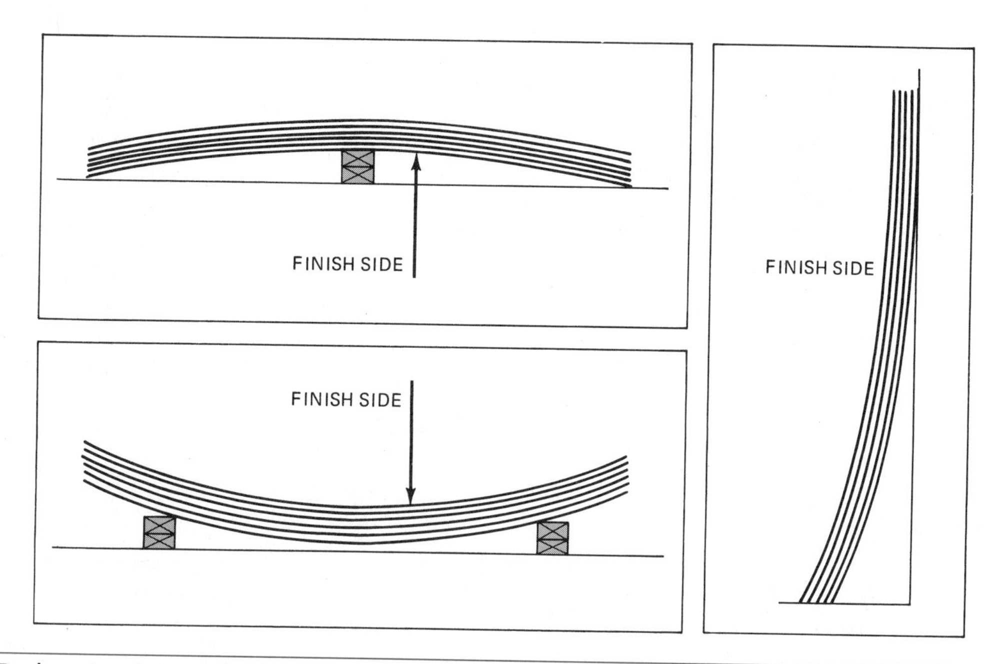

Fig. 69-9 Prebowing keeps the board in tight contact with the adhesive until bonded and reduces the number of fasteners required.

Fig. 69-10 Two or more persons are needed for ceiling application of gypsum board unless special lifting equipment is used.

it is being fastened (Fig. 69-11). Using deadmen is much easier than trying to hold the sheet in position and fasten it at the same time.

Fasten the sheet in one of the recommended manners. Hold the board firmly against framing to avoid nail pops or protrusions. Drive fasteners straight into the member. Fasteners that miss supports should be removed. The nail hole should be dimpled so that later it can be covered with joint compound (Fig. 69-12).

Continue applying sheets in this manner, staggering end joints, until the ceiling is covered.

To cut a corner out of a panel in case of a protrusion in the wall, make the shortest cut with a *drywall saw.* Then, score and snap the sheet in the other direction (Fig. 69-13). To cut a circular hole, mark the circle with pencil dividers, twist and push the drywall. Saw through the board. Cut out the hole, following the circular line.

Horizontal Application on Walls

When walls are less than 8′-1″ high, wallboard is usually installed horizontally, at right angles to the studs. If possible, use a board of sufficient length to go from corner to corner. Otherwise, use as long a board as possible to minimize end joints because they are difficult to conceal. Stagger end joints or center them over and below windows and over doors, if possible. That way not so much of the joint is visible. End joints should not fall on the same stud as those on the opposite side of the partition.

The top panel is installed first. Cut the board to length to fit easily into place without forcing it. Stand the board on edge against the wall. Start fasteners along the top edge opposite each stud. Raise the sheet so the top edge is firmly against the ceiling. Drive the fasteners home. Fasten the

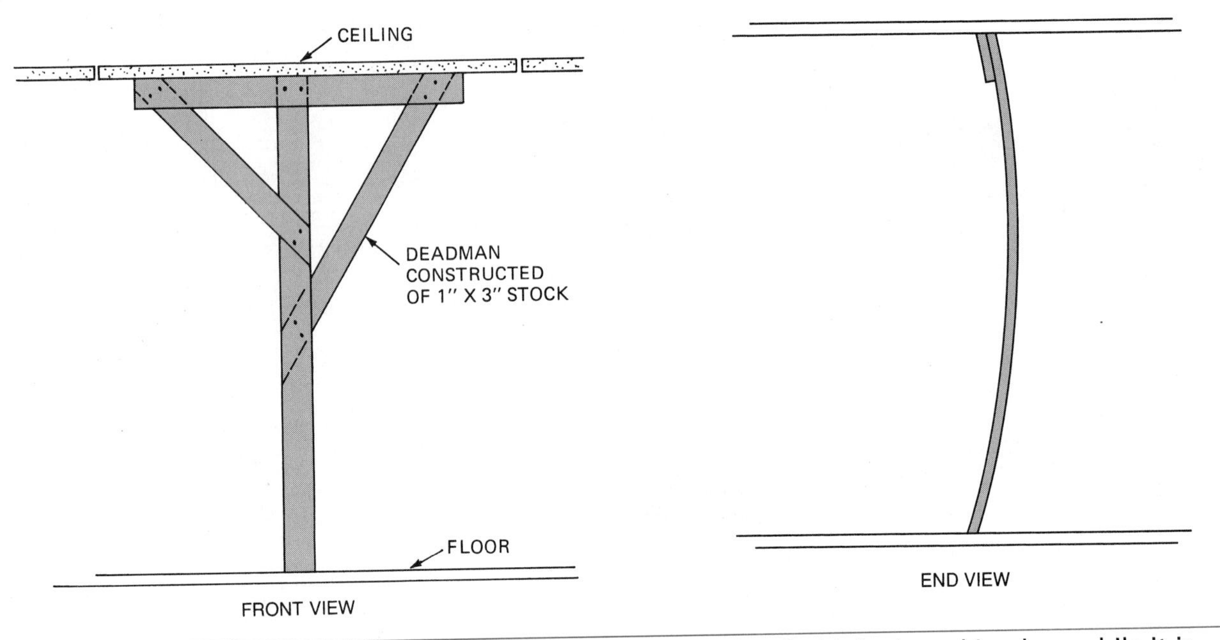

Fig. 69-11 Deadmen are wedged between floor and ceiling. They hold the board in place while it is being fastened.

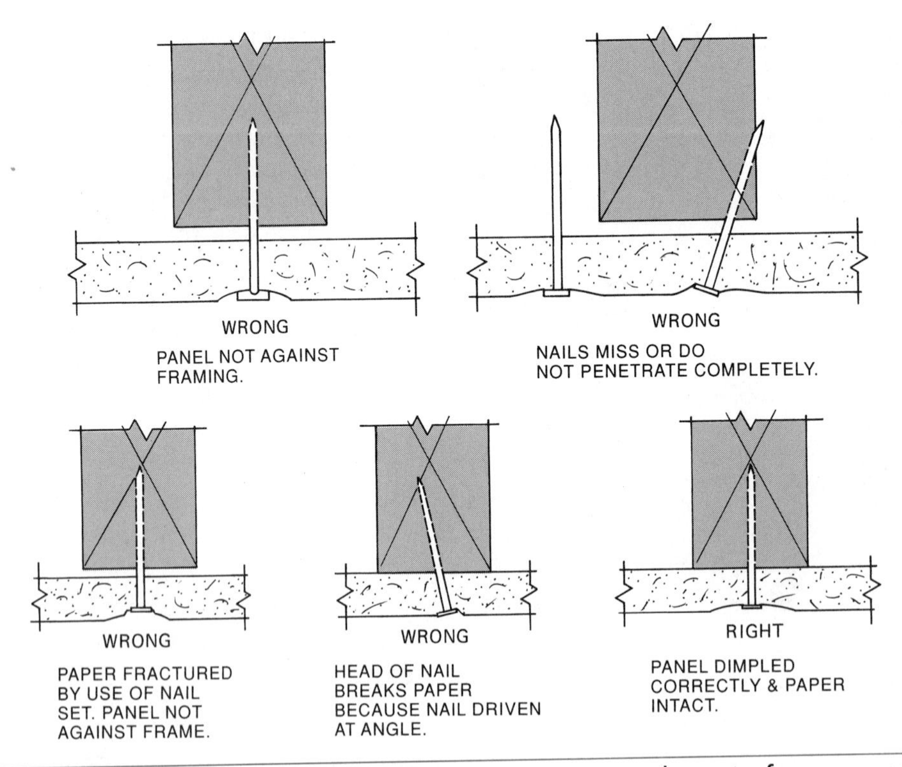

Fig. 69-12 It is important to drive fasteners correctly for secure attachment of gypsum panels.

rest of the sheet in the recommended manner (Fig. 69-14).

Measure and cut the bottom panel to width and length. Cut the width about 1/4 inch narrower than the distance measured. Lay the panel against the wall. Raise it with a *drywall foot lifter* against the bottom edge of the previously installed top panel. A drywall foot lifter is a tool especially designed for this purpose. However, one can be made on the job by tapering the end of a short piece of 1 × 3 or 1 × 4 lumber (Fig. 69-15). Fasten the sheet as recommended. Install all others in a similar manner.

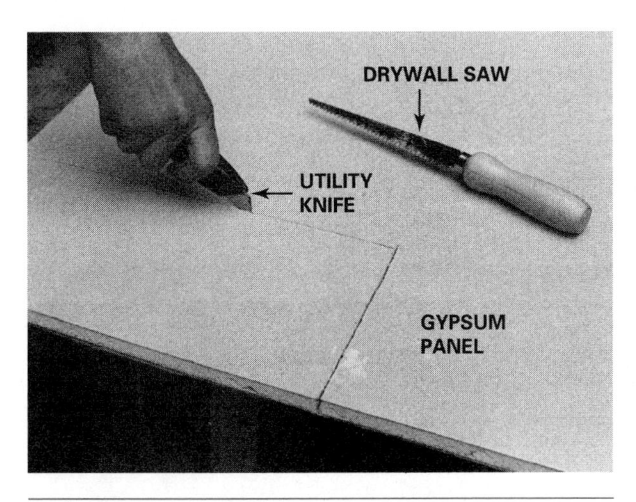

Fig. 69-13 Use a knife and drywall saw to cut the corner out of a gypsum panel.

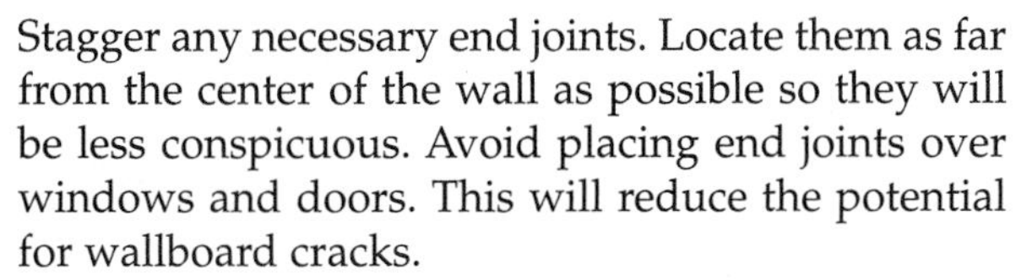

Stagger any necessary end joints. Locate them as far from the center of the wall as possible so they will be less conspicuous. Avoid placing end joints over windows and doors. This will reduce the potential for wallboard cracks.

Where end joints occur on steel studs, attach the end of the first panel to the open or unsupported edge of the stud. This holds the stud flange in a rigid position for the attachment of the end of the adjoining panel. Making end joints in the opposite manner usually causes the stud edge to deflect. This results in an uneven surface at the joint (Fig. 69-16).

Fig. 69-15 A foot lifter is used to lift gypsum panels in the bottom course up against the top panels. (*Courtesy of U.S. Gypsum Company*)

Making Cutouts in Wall Panels. There are several ways of making cutouts in wall panels for electrical outlet boxes, ducts, and similar objects. Care must be taken not to make the cutout much larger than the outlet. Most cover plates do not cover by much. If cut too large, much extra time has to be taken to patch up around the outlet, replace the panel, or install oversize outlet cover plates.

- Plumb the sides of the outlet box down to the floor, or up to the previously installed top

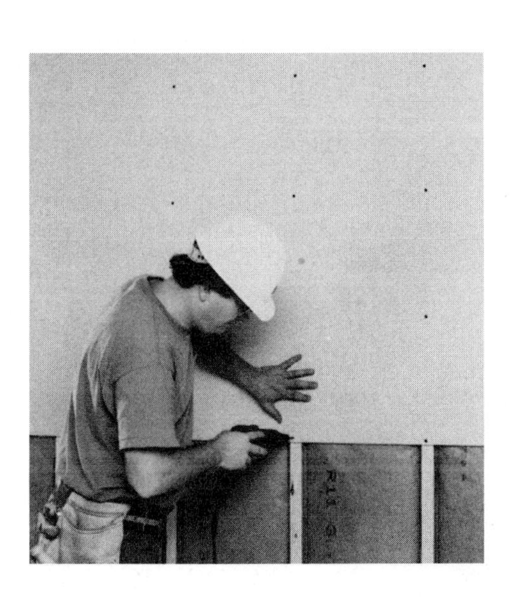

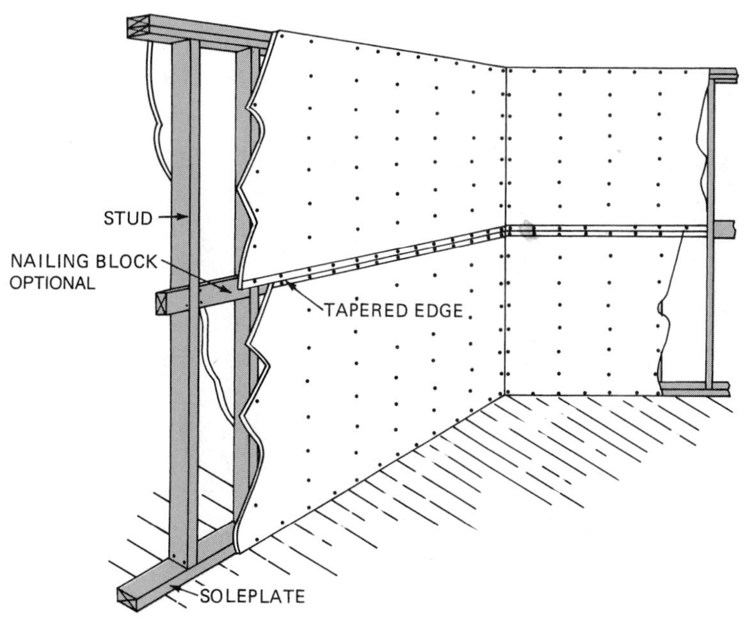

Fig. 69-14 Horizontal application of gypsum panels to walls. (*Part 1, Courtesy of U.S. Gypsum Company*)

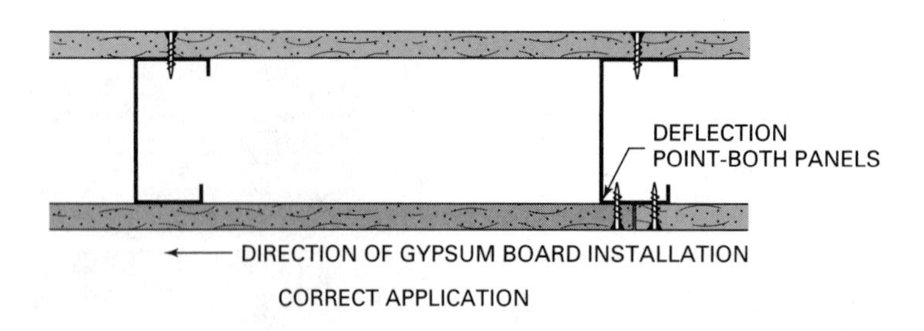

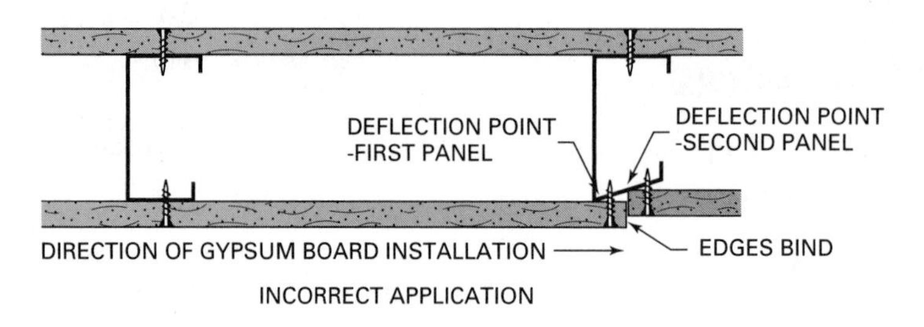

Fig. 69-16 Sequence of making end joints when attaching gypsum panels to steel studs. (*Courtesy of U.S. Gypsum Company*)

panel, whichever is more convenient. The panel is placed in position. Lines for the sides of the box are plumbed on it from the marks on the floor or on the panel. The top and bottom of the box are laid out by measuring down from the bottom edge of the top panel. With a saw or utility knife, cut the outline of the box. Take care not to injure the vapor retarder by pulling the lower end of the sheet away from the wall as you cut.

- A fast, easy, and accurate way of making cutouts is with the use of a portable electric *drywall cutout tool.* The approximate location of the center of the outlet box is determined and marked on the panel. The panel is then installed over the box. Using the cutout tool, a hole is plunged through the panel in the approximate center of the outlet box. Care must be taken not to make contact with wiring. The tool is not recommended for use around live wires. The tool is moved in any direction until the bit hits a side of the box. It is then withdrawn slightly to ride over the edge to the outside of the box. The tool is then moved so the bit rides around the outside of the box to make the cutout (Fig. 69-17). Usually cutouts are made for outlets after all the panels in a room have been installed.

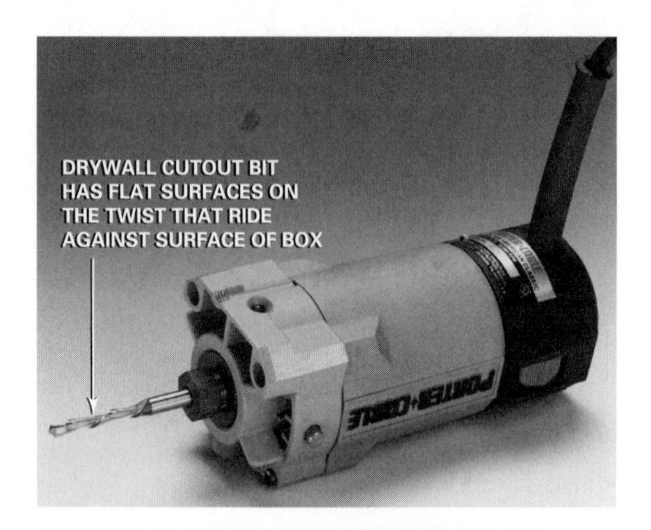

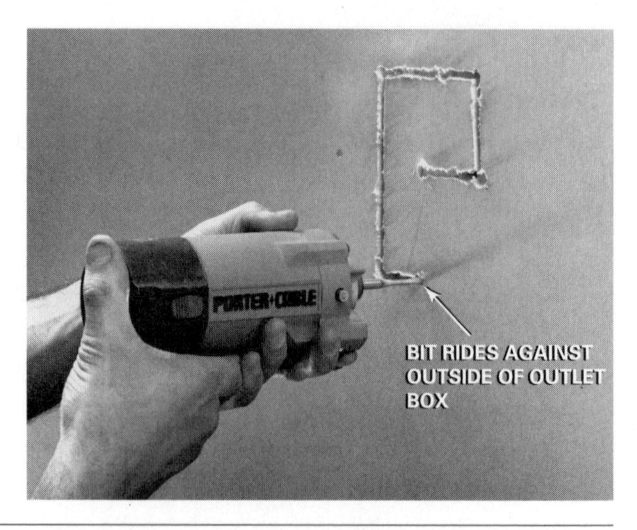

Fig. 69-17 A portable electric drywall cutout tool often is used to make cutouts for outlet boxes and similar objects. (*Courtesy of Porter-Cable*)

To make cutouts around door openings, either mark and cut out the panel before it is applied, or make the cutout after it is applied. To make the cutout after the panel is applied, use a saw to cut in one direction. Then score it flush with the opening on the back side in the other direction. Bend and score it on the face to make the cutout.

Vertical Application on Walls

Vertical application of gypsum panels on walls, with long edges parallel to studs, is more practical if the ceiling height is more than 8′-1″ or the wall is 4′-0″ wide or less. Cut the first board, in the corner, to length and width. Its length should be about 1/4 inch shorter than the height from floor to ceiling. It should be cut to width so the edge away from the corner falls on the center of a stud. All cut edges must be in the corners. None should butt edges of adjacent panels.

With a foot lifter, raise the sheet so it is snug against the ceiling. The tapered edge should be plumb and centered on the stud. Fasten it in the specified manner. Continue applying sheets around the room with tapered uncut edges against each other. There should be no horizontal joints between floor and ceiling (Fig. 69-18). Make any necessary cutouts as previously described.

Floating Interior Angle Construction

To help prevent nail popping and cracking due to structural stresses where walls and ceilings meet, the *floating angle* method of drywall application may be used. When joists or furring strips are at right angles to walls, fasteners in ceiling panels are located 7 inches from the wall for single nailing, and 11 to 12 inches for double nailing or screw attachment. When joists or furring are parallel to the wall, nailing is started at the intersection.

Gypsum panels applied on walls are fitted tightly against ceiling panels. This provides a firm support for the floating edges of the ceiling panels. The top fastener into each stud is located 8 inches down from the ceiling for single nailing and 11 to 12 inches down for double nailing or screw attachment.

At interior wall corners, the underlying wallboard is not fastened. The overlapping board is fitted snugly against the underlying board. This brings it in firm contact with the face of the stud. The overlapping panel is nailed or screwed into the interior corner stud (Fig. 69-19).

Applying Gypsum Panels to Curved Surfaces

Gypsum panels may be applied to curved surfaces. However, closer spacing of the frame members may be required to prevent flat areas from occurring on the face of the panel. If the paper and core of gypsum panels are moistened, they may be bent to curves with shorter radii than when dry. After the boards are thoroughly moistened, they should be stacked on a flat surface. They should be allowed to stand for at least one hour before bending. Moistened boards must be handled very carefully. They will regain their original hardness after drying. Wallboard marketed as *bendable* does not need to be wet before it is shaped. The minimum bending radii for dry and wet gypsum panels are shown in Figure 69-20.

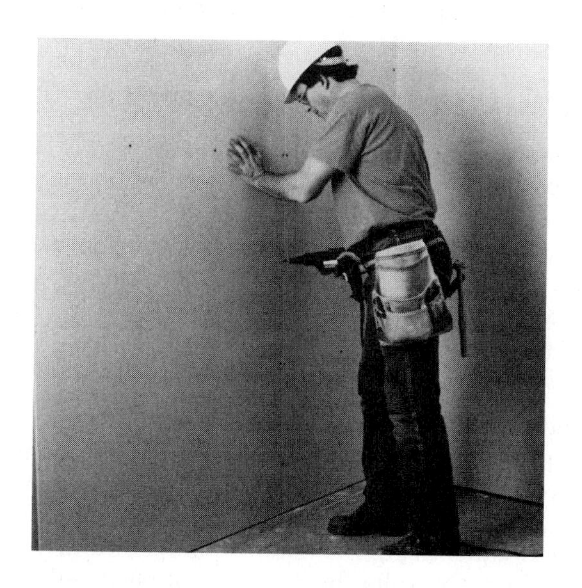

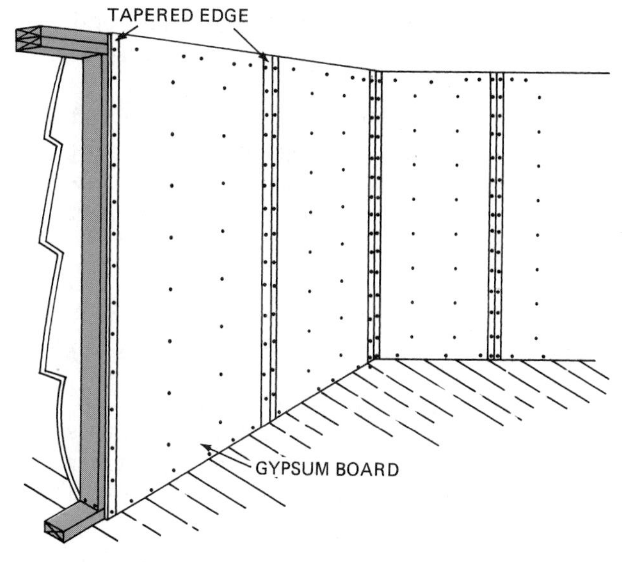

Fig. 69-18 Applying gypsum panels parallel to wall studs. (*Photo Courtesy of U.S. Gypsum Company*)

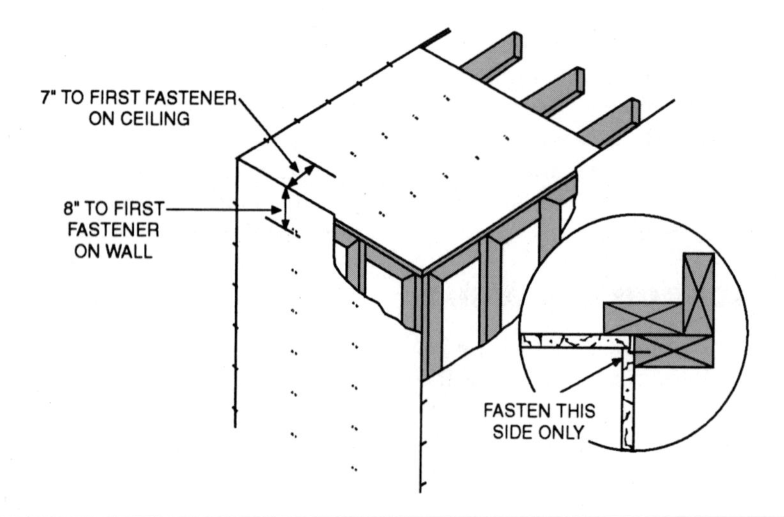

Fig. 69-19 Floating angle method of applying drywall. (*Courtesy of U.S. Gypsum Company*)

BOARD THICKNESS IN INCHES	DRY	WET
1/4	5 FT.	2 TO 2 1/2 FT.
3/8	7 1/2 FT.	3 TO 3 1/2 FT.
1/2	20 FT.	4 TO 4 1/2 FT.

Fig. 69-20 Minimum bending radii of gypsum panels.

To apply panels to a *convex* surface, fasten one end to the framing. Gradually work to the other end by gently pushing and fastening the panel progressively to each framing member. When applying panels to a *concave* curve, fasten a stop at one end. Carefully push on the other end to force the center of the panel against the framing. Work from the end against the stop. Fasten the panel successively to each framing member.

Gypsum board may be applied to the curved inner surfaces of arched openings. If the dry board cannot be bent to the desired curve, it may be moistened or parallel knife scores made about 1 inch apart across its width.

Drywall Application to Bath and Shower Areas

Water-resistant gypsum board and cement board panels are used in bath and shower areas as bases for the application of ceramic tile. Framing should be 16 inches OC. Steel framing should be at least 20-gauge thickness.

Apply panels horizontally with the bottom edge not less than 1/4 inch above the lip of the shower pan or tub. The bottom edges of gypsum panels should be uncut and paper-covered.

Check the alignment of the framing. If necessary, apply furring strips to bring the face of the board flush with the lip of the tub or shower pan (Fig. 69-21).

Provide blocking between studs about 1 inch above the top of the tub or shower pan. Install additional blocking between studs behind the horizontal joint of the panels above the tub or shower pan.

Cement board panels are cut in the same manner as gypsum board panels. Before attaching panels, apply thinned ceramic-tile adhesive to all cut edges around holes and other locations.

Attach panels with corrosion-resistant screws or nails spaced not more than 8 inches apart. When ceramic tile more than 3/8 inch thick is to be applied, the nail or screw spacing should not exceed 4 inches OC.

Multilayer Application

A multilayer application has one or more layers of gypsum board applied over a base layer. It provides greater strength, higher fire resistance, and better sound control. The base layer may be gypsum backing board, regular gypsum board, or other gypsum base material.

Base Layer

The base layer is fastened in the same manner as single-layer panels. However, double nailing is not necessary and staples may be used in wood framing. On ceilings, panels are applied with the long edges either at right angles or parallel to framing members. On walls, the panels are applied with the long edges parallel to the studs.

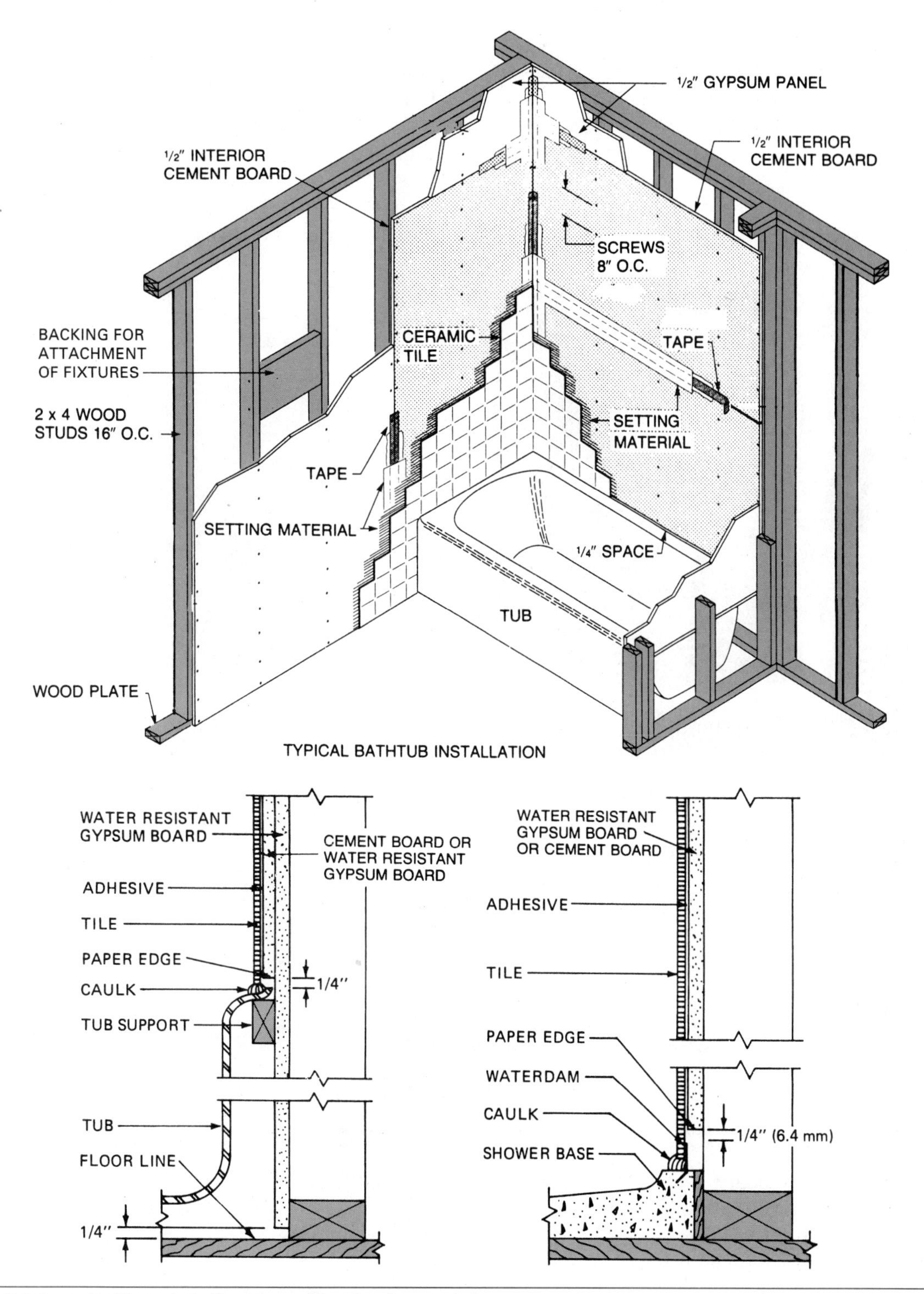

Fig. 69-21 Installation details around bathtubs and showers.

Face Layer

Joints in the face layer are offset at least 10 inches from joints in the base layer. The face layer is applied either parallel to or at right angles to framing, whichever minimizes end joints and results in the least amount of waste.

The face layer may be attached with nails, screws, or adhesives. If nails or screws are used without adhesive, the maximum spacing and minimum penetration into framing should be the same as for single-layer application.

Adhesive Attachment of the Face Layer

Panels of the face layer may be attached to the base layer by *sheet lamination, strip lamination,* or *spot lamination.* In sheet lamination, the entire back of the face layer is covered with laminating adhesive, using a notched spreader (Fig. 69-22).

In strip lamination, the adhesive is applied in ribbons with a special spreader, spaced 16 to 24 inches OC (Fig. 69-23). In *spot lamination,* plum-sized daubs of adhesive are applied to the panel in a regular pattern.

After the adhesive is applied, the face panel is held firmly against the base layer with supplemental fasteners or braces while the adhesive sets. The panel may be also be secured by temporarily nailing with duplex nails driven through wood or gypsum board scraps. The nail should penetrate into the framing by at least 3/4 inch. Type G screws may be used, but the base layer must be at least 1/2 inch thick.

Fig. 69-22 In sheet lamination, the entire back of the face layer is covered with adhesive. (*Courtesy of U.S. Gypsum Company*)

Fig. 69-23 In strip lamination, ribbons of the adhesive are spaced. (*Courtesy of U.S. Gypsum Company*)

CHAPTER 70 CONCEALING FASTENERS AND JOINTS

After the gypsum board is installed, it is necessary to conceal the fasteners and to reinforce and conceal the joints. One of several levels of finish may be specified for a gypsum board surface. The lowest level of finish may simply require the taping of wallboard joints and *spotting* of fastener heads on surfaces. This is done in warehouses and other areas where appearance is normally not critical. The level of finish depends, among other things, on the number of coats of compound applied to joints and fasteners (Fig. 70-1).

Description of Materials

Fasteners are concealed with *joint compound.* Joints are reinforced with *joint tape* and covered with *joint compound.* Exterior corners are reinforced with *corner bead.* Other kinds of drywall trim may be used around doors, windows, and other openings.

Joint Compounds

Drying type joint compounds for joint finishing and fastener spotting are made in both a dry powder form and a ready-mixed form in three general types. Drying type compounds provide smooth application and ample working time. A *taping compound* is used to embed and adhere tape to the board over the joint. A *topping compound* is used for second and third coats over taped joints. An *all-purpose compound* is used for both bedding the tape and finishing the joint. All-purpose compounds do

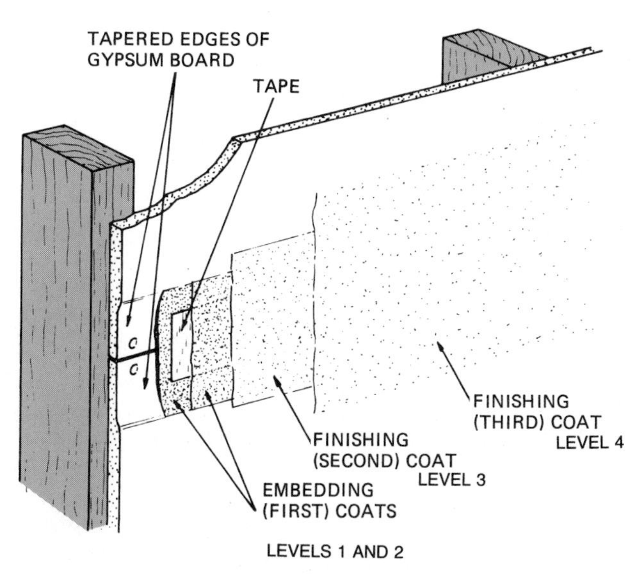

Fig. 70-1 The level of finish varies with the type of final decoration to be applied to drywall panels.

Fig. 70-2 Applying joint tape to an interior corner.

not possess the strength or workability of two-step taping and topping compound systems.

Setting type joint compounds are used when a faster setting time than that of drying types is desired. Drying type compounds harden through the loss of water by evaporation. They usually cannot be recoated until the next day. Setting type compounds harden through a chemical reaction when water is added to the dry powder. Therefore, they come only in a dry powder form and not ready-mixed. They are formulated in several different setting times. The fastest setting type will harden in as little as twenty to thirty minutes. The slowest type takes four to six hours to set up. Setting type joint compounds permit finishing of drywall interiors in a single day.

Joint Reinforcing Tape

Joint reinforcing tape is used to cover, strengthen, and provide crack resistance to drywall joints. One type is made of *high-strength fiber paper.* It is designed for use with joint compounds on gypsum panels. It is creased along its center to simplify folding for application in corners (Fig. 70-2).

Another type is made of *glass fiber mesh.* It is designed to reinforce joints on veneer plaster gypsum panels. It is not recommended for use with conventional compounds for general drywall joint finishing. It may be used with special high-strength setting compounds. Glass fiber mesh tape is available with a plain back or with an adhesive backing for quick application (Fig. 70-3). Joint tape is normally available 2 and 2 1/2 inches wide in 300-foot rolls.

Fig. 70-3 An adhesive-backed glass fiber mesh tape is quickly applied to drywall joints. (*Courtesy of U.S. Gypsum Company*)

Corner Bead and Other Drywall Trim

Corner beads are applied to protect exterior comers of drywall construction from damage by impact.

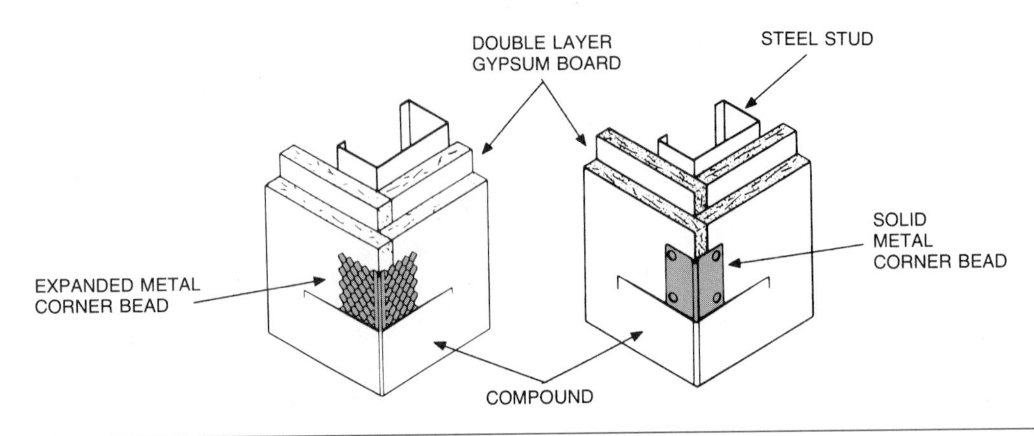

Fig. 70-4 Corner beads are used to finish and protect exterior corners of drywall panels. (*Courtesy of U.S. Gypsum Company*)

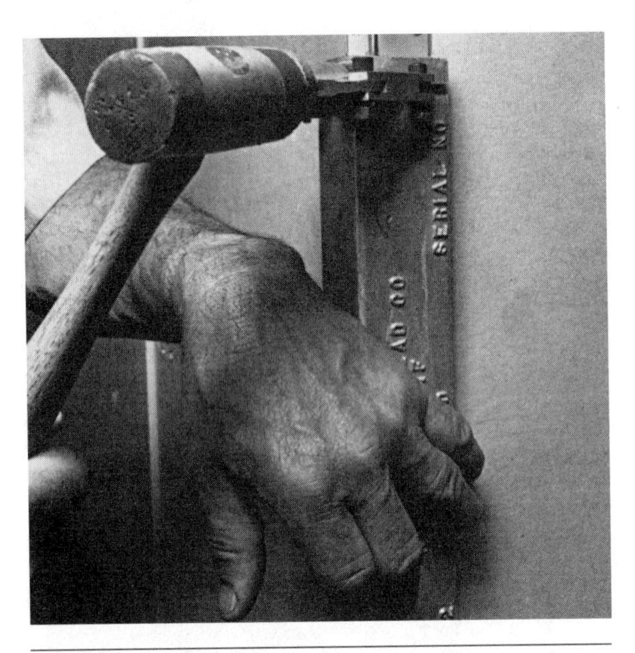

Fig. 70-5 A clinching tool is sometimes used to fasten corner beads to exterior corners. (*Courtesy of U.S. Gypsum Company*)

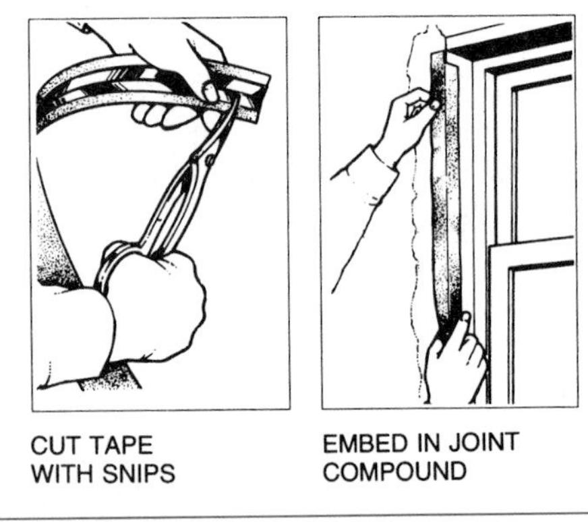

Fig. 70-6 Flexible metal corner tape is applied to exterior corners by embedding in compound. (*Courtesy of U.S. Gypsum Company*)

One type with solid metal flanges is widely used. Another type has flanges of expanded metal with a fine mesh. This provides excellent keying of the compound (Fig. 70-4).

Corner bead is fastened through the drywall panel into the framing with nails or staples. Instead of using fasteners, a *clinching tool* is sometimes used. It crimps the solid flanges and locks the bead to the corner (Fig. 70-5).

Metal corner tape is applied by embedding in joint compound. It is used for corner protection on arches, windows with no trim, and other locations (Fig. 70-6).

A variety of *metal trim* is used to provide protection and finished edges to drywall panels. They are used at windows, doors, inside corners, and intersections. They are fastened through their flanges into the framing (Fig. 70-7).

Control joints are metal strips with flanges on both sides of a 1/4-inch, V-shaped slot. Control joints are placed in large drywall areas. They relieve stresses induced by expansion or contraction. They are used from floor to ceiling in long partitions and from wall to wall in large ceiling areas (Fig. 70-8). The flanges are concealed with compound in a manner similar to corner beads and other trim.

A wide assortment of rigid vinyl drywall accessories is available (Fig. 70-9) including the metal trim previously discussed. They are designed for easy installation and workability to reduce installation time. Most have edges to guide the drywall knife, which allows for an even application of joint compound. Some have edges that are later torn away when the painting is done. This allows the finish to be applied more quickly and at the same time more uniformly. Vinyl accessories make it possible to create smooth joints easily, whether they are curved or straight.

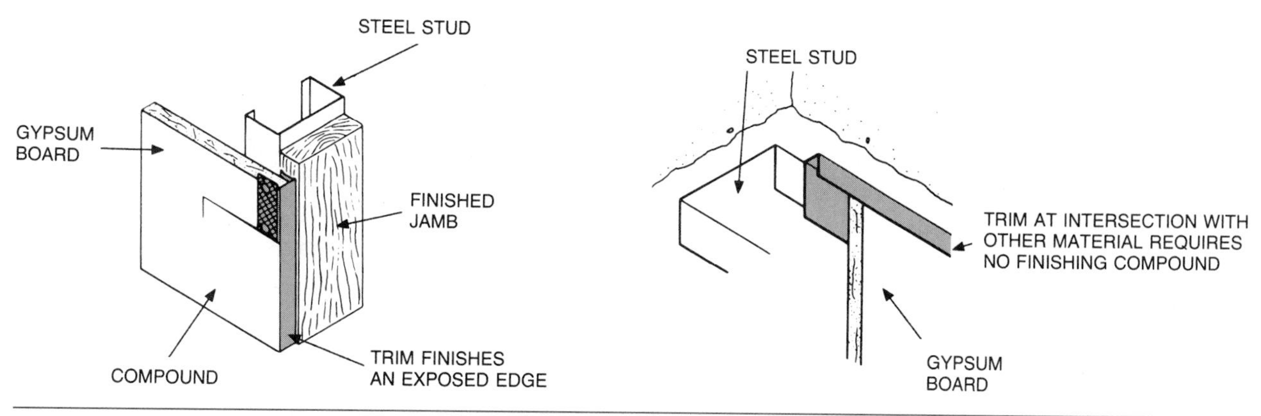

Fig. 70-7 Various metal trim is used to provide finished edges to gypsum panels. (*Courtesy of U.S. Gypsum Company*)

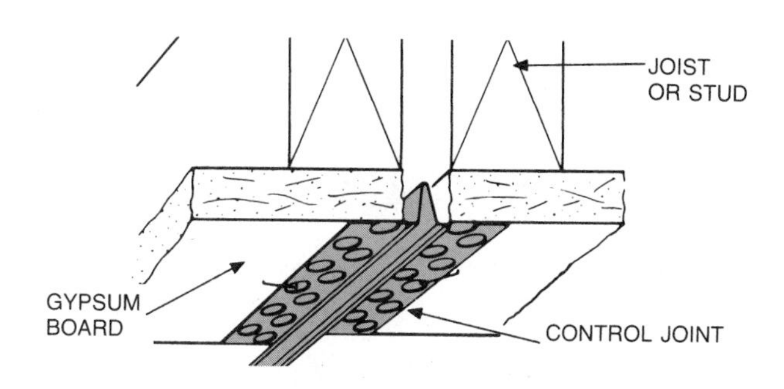

Fig. 70-8 Control joints are used in large wall and ceiling areas subject to movement by expansion and contraction. (*Courtesy of U.S. Gypsum Company*)

Applying Joint Compound and Tape

In cold weather, care should be taken to maintain the interior temperature at a minimum of 50 degrees Fahrenheit for twenty-four hours before and during application of joint compound, and for at least four days after application has been completed.

Care should also be taken to use clean tools. Avoid contamination of the compound by foreign material, such as sawdust, hardened, or different types of compounds.

Prefilling Joints

Before applying compound to drywall panels, check the surface for fasteners that have not been sufficiently recessed. Also look for other conditions that may affect the finishing. Prefill any joints between panels of 1/4 inch or more and all V-groove joints between **eased-edged** panels with compound. A twenty-four-hour drying period can be eliminated with the use of setting compounds for prefilling operations.

Embedding Joint Tape

Fill the recess formed by the tapered edges of the sheets with the specified type of joint compound. Use a joint knife (Fig. 70-10). Center the tape on the joint. Lightly press it into the compound. Draw the knife along the joint with sufficient pressure to *embed* the tape and remove excess compound (Fig. 70-11).

There should be enough compound under the tape for a proper bond, but not over 1/32 inch under the edges. Make sure there are no air bubbles under the tape. The tape edges should be well adhered to the compound. If not satisfactory, lift the portion. Add compound and embed the tape again. A *taping tool* sometimes is used. It applies the compound and embeds the tape at the same time (Fig. 70-12).

Immediately after embedding, apply a thin coat of joint compound over the tape. This helps prevent the edges from wrinkling. It also makes easier concealment of the tape with following coats. Draw the knife to bring the coat to feather edges on both sides of the joint. Make sure there is no excess compound left on the surface. After the compound has set up, but not completely hardened, wipe the surface with a *damp sponge.* This eliminates the need for sanding any excess after the compound has hardened.

Spotting Fasteners

Fasteners should be *spotted* immediately before or after embedding joint tape. *Spotting* is the application of compound to conceal fastener heads. Apply enough pressure on the taping knife to fill only the depression. Level the compound with the panel surface. Spotting is repeated each time additional coats of compound are applied to joints.

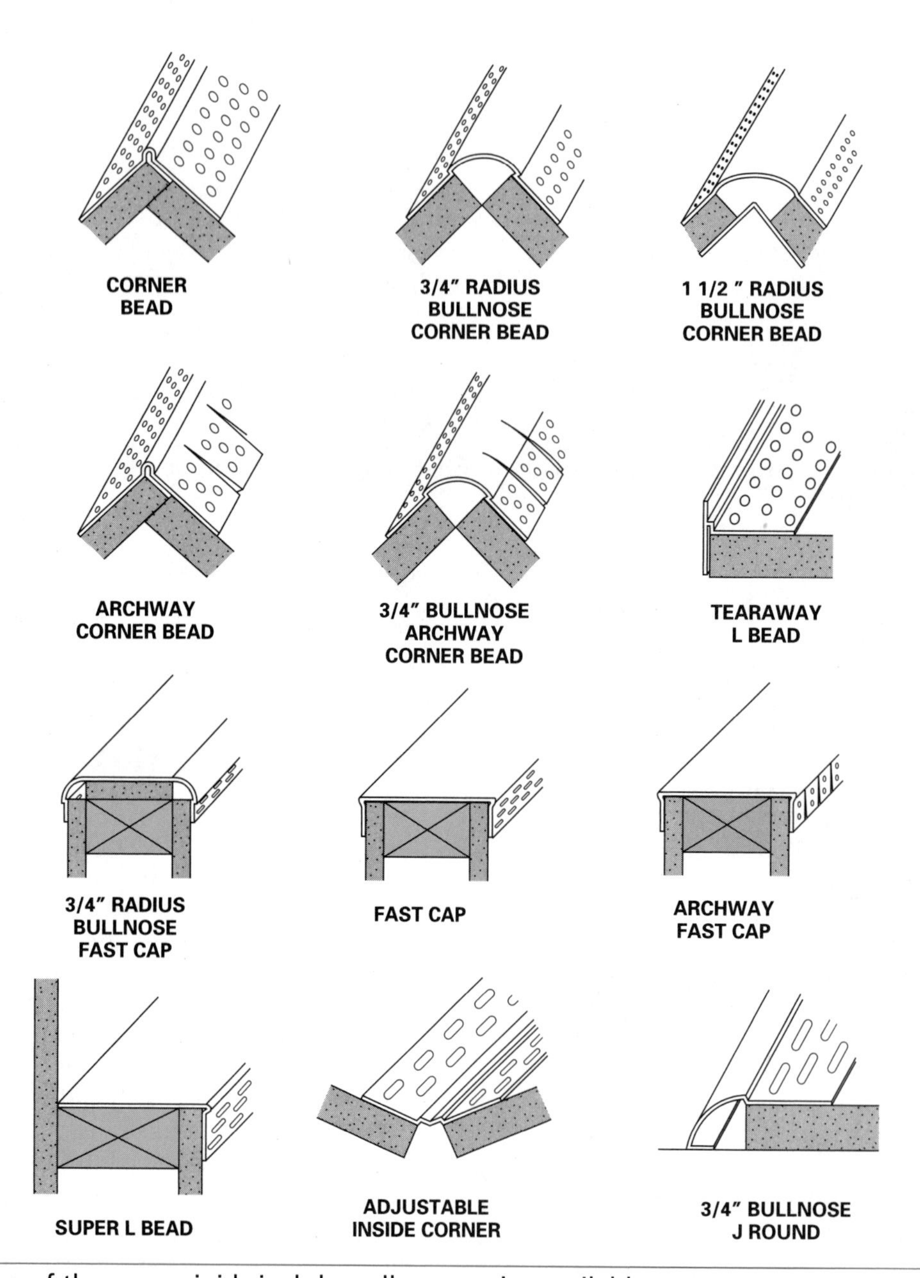

Fig. 70-9 Some of the many rigid vinyl drywall accessories available.

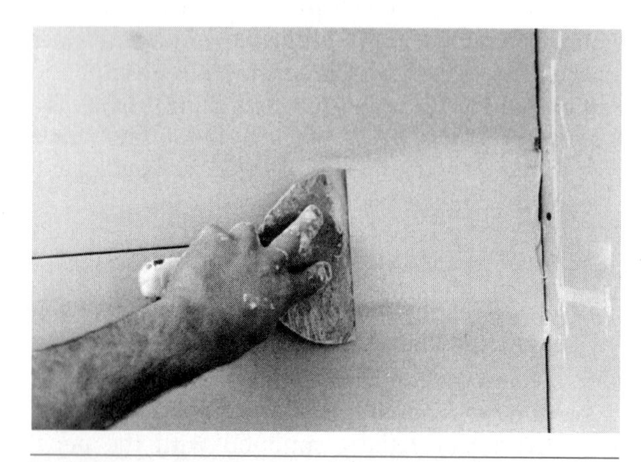

Fig. 70-10 Taping compound is first applied to the channel formed by the tapered edges between panels.

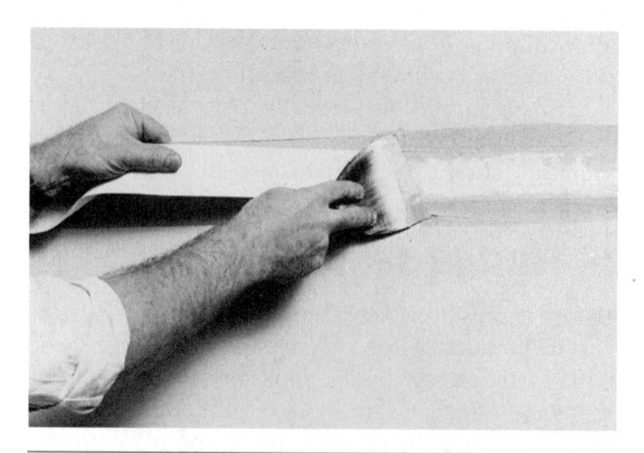

Fig. 70-11 The tape is embedded into the compound.

Fig. 70-12 A taping tool applies tape and compound at the same time. (*Courtesy of U.S. Gypsum Company*)

Applying Compound to Corner Beads and Other Trim

The first coat of compound is applied to corner beads and other metal trim when first coats are given to joints and fasteners. The nose of the bead or trim serves as a guide for applying the compound. The compound is applied about 6 inches wide from the nose of the bead to a feather edge on the wall. Each subsequent finishing coat is applied about 2 inches wider than the previous one.

Fill and Finishing Coats

Allow the first coat to dry thoroughly. This may take twenty-four hours or more depending on temperature and humidity unless a setting type compound has been used. It is common to use setting compounds for first coats and slower setting types for finishing coats. Feel the entire surface to see if any excess compound has hardened on the surface. Sand any excess, if necessary, to avoid interfering with the next coat of compound.

The second coat is sometimes called a *fill* coat. It is feathered out about 2 inches beyond the edges of all first coats, approximately 7 to 10 inches wide. Care must always be taken to remove all excess compound so that it does not harden on the surface. Professional drywall finishers rarely have to sand any excess in preparation for following coats. Remember, a damp sponge rubbed over the joint after the compound starts to set up will remove any small particles of excess. It will also help bring edges to a feather edge.

If the level of finish requires it, apply a third and *finishing* coat of compound over all fill coats. The edges of the finishing coat should be feathered out about 2 inches beyond the edges of the second coat (Fig. 70-13).

Interior Corners

Interior corners are finished in a similar way. However, the tape is folded in the center to fit in the corner. After the tape is embedded, drywall finishers usually apply a setting compound to one side only of each interior corner. By the time they have finished all interior corners in a room, the compound has set enough to finish the other side of the corners.

Estimating Drywall

To estimate the amount of drywall material needed, determine the area of the walls and ceilings to be covered. To find the ceiling area, multiply the length of the room by its width. To find the wall area, multiply the perimeter of the room by the height. Subtract all large wall openings. Combine all areas to find the total number of square feet of drywall required. Add about 5 percent of the total for waste. The number of drywall panels can then be determined by dividing the total area to be covered by the area of one panel.

About 1,000 screws are needed for each 1,000 square foot of drywall when applied to framing 16 inches OC, and 850 screws are needed for 24 inch OC framing. About 5 pounds of nails are required to fasten each 1,000 square foot of drywall.

Approximately 370 feet of joint tape and 135 pounds of conventional, ready-mixed joint compound are needed for every 1,000 square foot of drywall area.

Fig. 70-13 Applying a finishing coat of compound to a taped drywall joint.

Review Questions

Select the most appropriate answer.

1. Standard gypsum board width is
  a. 36 inches. c. 54 inches.
  b. 48 inches. d. 60 inches.

2. Standard gypsum board lengths are
  a. 8, 10, and 12 feet.
  b. 8, 10, 12, and 14 feet.
  c. 8, 9, 10, 12, and 14 feet.
  d. 8, 9, 10, 12, 14, and 16 feet.

3. When fastening drywall, minimum penetration of ring-shanked nails into the framing member is
  a. 1/2 inch. c. 7/8 inch.
  b. 3/4 inch. d. one inch.

4. Gypsum board is usually installed vertically on walls when the wall height is greater than
  a. 8′-0″. c. 8′-4″.
  b. 8′- 1″. d. 8′-6″.

5. Ceiling joists are sometimes aligned by the use of a
  a. deadman. c. strongback.
  b. dutchman. d. straightedge.

6. In the single nailing method, nails are spaced a maximum of
  a. 8 inches OC on walls; 7 inches OC on ceilings.
  b. 10 inches OC on walls; 8 inches OC on ceilings.
  c. 12 inches OC on walls; 10 inches OC on ceilings.
  d. 12 inches OC on walls and ceilings.

7. Screws are spaced
  a. 12 inches OC on walls; 10 inches OC on ceilings.
  b. 12 inches OC on walls and ceilings.
  c. 16 inches OC on walls and ceilings.
  d. 16 inches OC on walls; 12 inches OC on ceilings.

8. Joints in the face layer of a multilayer application are offset from joints in the base layer by at least
  a. 6 inches. c. 10 inches.
  b. 8 inches. d. 12 inches.

9. The paper-covered edge of water-resistant gypsum board is applied above the lip of the tub or shower pan not less than
  a. 1/4 inch. c. 1/2 inch.
  b. 3/8 inch. d. 3/4 inch.

10. When ceramic tile more than 3/8-inch thick is to be applied over water-resistant gypsum board, fasten the board with screws or nails spaced not more than
  a. 4 inches OC. c. 8 inches OC.
  b. 6 inches OC. d. 10 inches OC.

BUILDING FOR SUCCESS

A Formula for Success

The following information comes from an interview with Greg Shinaut, the owner of *Pride Homes,* a custom home builder in Lincoln, Nebraska. We will look at the events that led up to self-employment in the home building industry. Of primary importance are the basics that set the entire process of owning a business into motion. The initial dream, the goal setting, the plan to reach those goals, the experiences along the way, the skill building, the relationships formed, and the outcomes are presented. This example may appear to be unique, but it actually represents thousands of other success stories throughout the building industry today.

Shinaut's initial interest in building began during his elementary school years in

Fremont, Nebraska. When challenged by the question, "What do you want to do with your life?" he knew that he desired to work as a carpenter and build houses. His interest was spurred by outside influences because he was not reared in a construction-oriented family. With the assistance of his high school counselor and industrial technology instructor, Shinaut began to formulate ideas and goals for his career. He desired to learn the necessary carpentry skills that would allow him to enter the building field and experience a variety of jobs related to home building.

He set up a ladder with steps to obtain along the way. Knowing that he would eventually need a good working knowledge of construction systems and procedures, good carpentry skills, business skills, and people skills, he entered the building construction program of study at Southeast Community College. Upon graduation, he was employed by a builder in central Nebraska. He was given a chance to experience a variety of construction jobs. This led to later work as a framer for a builder in the Omaha area. Eventually he worked as a deck builder and siding applicator in the city of Lincoln. While working for a builder in Lincoln, Shinaut served as the "call-back" representative. A lot of the work involved customer relations. The total skills and attributes he gained during those years provided the field and business experiences he needed to reach another stage in the predetermined sequence of events that lead to self-employment.

In focusing on the necessary attributes and abilities needed for being a successful builder and businessperson, Shinaut identifies the following as important. Having an open mind, with the ability to look and listen to everything affecting the construction world, is the pace setter. "Doing what it takes to get it done" and meeting the client's demands are two ingredients that are necessary for success. Being open to learning, using quality people and their strengths, possessing a good attitude, and being able to take constructive criticism are extremely important along the way. Having a conscience, being honest with people in dealings, sharing information with others, and being dedicated to people and work must not be ignored. Shinaut feels that all these attributes and abilities are required to build the necessary trust in people needed for a successful business.

As he continued working toward the goal of long-term self-employment in the home building business, some typical unknowns were carried with him. Not knowing exactly what tomorrow would bring, whether there would be enough business to keep moving, or if success would displace failure, were daily questions,

When asked to share a few observations with others who may be considering a career in construction, Shinaut offered some good advice. He cautions interested persons to diligently work their way into their careers. They need to decide how they might meet their goals and build the necessary skills and attitudes that undergird success. Experience the field, pace yourself, and remain open to learning to allow for confidence building. Being careful not to act too quickly in considering going out on your own before all of your homework is completed is important. Disaster can set in when actions are not backed by experience.

Shinaut believes that education, both formal and informal, is the basis for learning the construction business. In reference to young people entering the construction business, he feels that each person needs to find the correct balance. Those reared in a construction environment need different experiences than those coming from other backgrounds. Eventually the majority of business personnel in the construction world will gain a composite of similar experiences to help focus on successful careers.

The day-to-day challenges center on spending quality time with his family, doing what it takes to get the job, and including the family in the business operations. Without a complete commitment from the family, the frustration and stress are increased. The benefits coming from being self-employed are centered on personal satisfaction. The gratification that comes from seeing a family enter a new house and convert it into a home is extremely satisfying. It is a good feeling to see what hard work and dedication can produce in the lives of people.

The opportunity to work with many community organizations produces other satisfactions. The home builder has the opportunity to provide an important service to a community that affects the lives of many people. Usually the contribution to the community comes back in the way of benefits. These benefits range from providing assistance in solving local problems to helping local children, elderly, and disadvantaged people reach a better standard of living. Local home builder associations are strong in community service due to the caring attitude of the associated members.

When asked if he felt successful and why, Shinaut responded that he did not feel established, but needed that feeling to stay competitive. The feeling of complacency or being overly assured of continual work without competition will bring failure. The quality home builder needs to be challenged each year in the business.

Shinaut advises those who are considering a career in the construction business to "do it." Begin with a dream, and start to set goals. Follow your heart and reinforce that feeling with a willingness to work and learn. Use your imagination. Go as far as you are willing to be challenged. Do not wait for something to happen, because it usually will not. You have to initiate the action. This listing of activities and considerations culminates in an exemplary "formula for success."

FOCUS QUESTIONS: For individual or group discussion

1. According to Shinaut, what appear to be some of the key considerations a person must think about when anticipating future self-employment? List six or more of the considerations.
2. As you look at the considerations made by Shinaut when concerning self-employment do you agree with his identified "formula for success"? Why or why not? How could you add to this formula to make it better?
3. How important is the planning process to successfully reaching your goals? What would be your recommendations to a person when goals seem out of reach?
4. How important is it to a person entering the business world to understand need for both technical and business skill building? Identify at least three important skills that you feel are necessary for eventual self-employment. Explain why they are necessary and how they can be obtained.

UNIT

24

Wall Paneling and Wall Tile

Plans and specifications often call for the installation of *wall paneling* in certain rooms of both residential and commercial construction. *Ceramic wall tile* is widely used in rest rooms, baths, showers, kitchens, and similar areas.

OBJECTIVES

After completing this unit, the student should be able to:

- describe and apply several kinds of sheet wall paneling.
- describe and apply various patterns of solid lumber wall paneling.
- describe and install ceramic wall tile to bathroom walls.
- estimate quantities of wall paneling and ceramic wall tile.

UNIT CONTENTS

CHAPTER 71 TYPES OF WALL PANELING

Two basic kinds of *wall paneling* are *sheets* of various prefinished material and solid *wood boards.* Many compositions, colors, textures, and patterns are available in *sheet* form. Solid *wood boards* of many species and shapes are used for both rustic and elegant interiors (Fig. 71-1).

Fig. 71-1 Solid wood board paneling provides warmth and beauty to interiors of residential and commercial buildings.

Description of Sheet Paneling

Sheets of *prefinished plywood,* **hardboard,** *particleboard,* **plastic laminate,** and other material are used to panel walls.

Plywood

Prefinished plywood is probably the most widely used sheet paneling. A tremendous variety is available in both hardwoods and **softwoods.** The more expensive types have a face veneer of real wood. The less expensive kinds of plywood paneling are prefinished with a printed wood grain or other design on a thin vinyl covering. Care must be taken not to scratch or scrape the surface when handling these types. Unfinished plywood panels are also available.

Some sheets are scored lengthwise at random intervals to imitate solid wood paneling. There is always a score 16, 24, and 32 inches from the edge. This facilitates fastening of the sheets and in case the sheet has to be ripped lengthwise to fit stud spacing.

Most commonly used panel thicknesses are 3/16 and 1/4 inch. Sheets are normally 4 feet wide and 7 to 10 feet long. An 8-foot length is most commonly used. Panels may be shaped with square, beveled, or shiplapped edges (Fig. 71-2). Matching molding is available to cover edges, corners, and

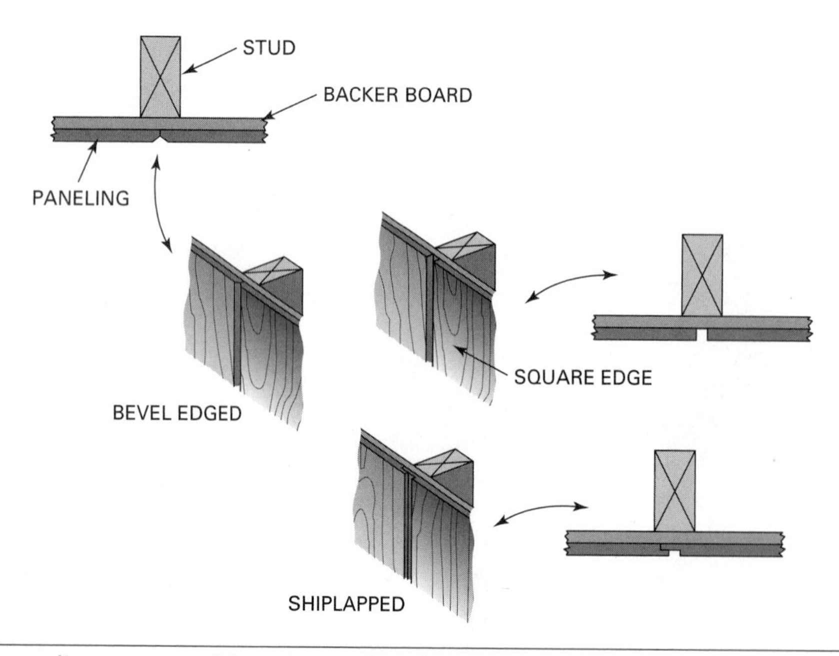

Fig. 71-2 Sheet paneling comes with various edge shapes.

joints. Thin ring-shanked nails, called *color pins,* are available in colors to match panels. They are used when exposed fastening is necessary.

Hardboard

Hardboard is available in many manmade surface colors, textures, and designs. Some of these designs simulate stone, brick, stucco, leather, weathered or smooth wood, and other materials. Unfinished hardboard is also used that has a smooth, dark brown surface suitable for painting and other decorating. Unfinished hardboard may be solid or perforated in a number of designs (Fig. 71-3).

Tileboard is a hardboard panel with a baked-on plastic finish. It is embossed to simulate ceramic wall tile. The sheets come in a variety of solid colors, marble, floral, and other patterns. Tileboard is designed for use in bathrooms, kitchens, and similar areas.

Hardboard paneling comes in widths of 4 feet and in lengths of from 8 to 12 feet. Commonly used thicknesses are from 1/8 to 1/4 inch. Color-coordinated molding and trim are available for use with hardboard paneling.

Particleboard

Panels of *particleboard* come with wood grain or other designs applied to the surface, similar to plywood and hardboard. Sheets are usually 1/4-inch thick, 4 feet wide, and 8 feet long. Prefinished particleboard is used when an inexpensive wall covering is desired. Because the sheets are brittle and break easily, care must be taken when handling. They must be applied only on a solid wall backing.

Unfinished particleboard is not usually used as an interior wall finish. One exception, made from aromatic cedar chips, is used to cover walls in closets to repel moths.

Plastic Laminates

Plastic laminates are commonly called *mica.* They are widely used for surfacing kitchen cabinets and countertops. They are also used to cover walls or parts of walls in kitchens, bathrooms, rest rooms, and similar areas where a durable, easy-to-clean surface is desired. Laminates can be scorched by an open flame. However, they resist heat, alcohol, acids, and stains. They clean easily with a mild detergent.

Laminates are known by such brand names as Formica, Pionite, Wilson Art, and others. They are manufactured in many colors and designs, including wood grain patterns. Surfaces are available in gloss, satin, and textured finishes, among others.

Laminates are ordinarily used in two thicknesses. *Vertical-type* laminate is relatively thin (about 1/32 inch). It is used for vertical surfaces, such as walls and cabinet sides. Vertical-type laminate is available only in widths of 4 feet or 8 feet. *Regular* or *standard* laminate is about 1/16-inch thick. It comes in widths of 24, 36, 48, and 60 inches and in lengths of 5, 6, 8, 10, and 12 feet. It is generally used on horizontal surfaces, such as countertops. It can be used on walls, if desired, or if the size required is not available in vertical type. Sheets are usually manufactured 1 inch wider and longer than the nominal size.

Laminates are difficult to apply to wall surfaces because they are so thin and brittle. Also, because a *contact bond* adhesive is used, the sheet cannot be moved once it makes contact with the surface. Thus, prefabricated panels, with sheets of laminate already bonded to a backer, are normally used to panel walls.

Description of Board Paneling

Board paneling is used on interior walls when the warmth and beauty of solid wood is desired. Wood

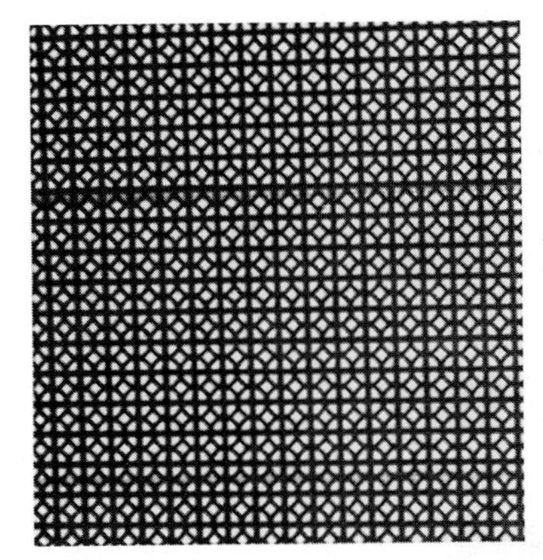

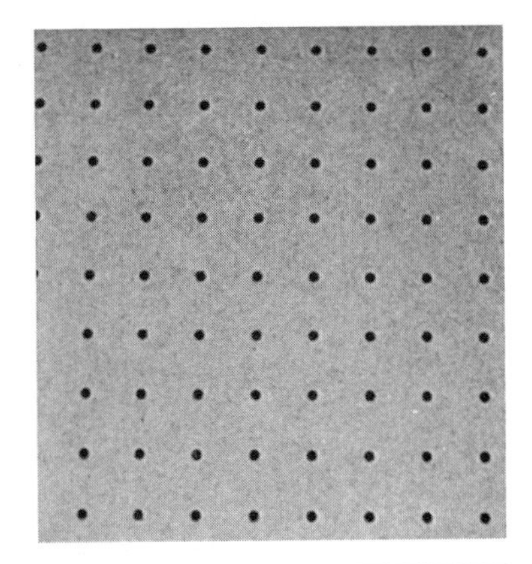

Fig. 71-3 Hardboard is also available in a number of perforated designs.

paneling is available in softwoods and hardwoods of many species. Each has its own distinctive appearance, unique grain, and knot pattern.

Wood Species

Woods may be described as light-, medium-, and dark-toned. Light tones include birch, maple, spruce, and white pine. Some medium tones are cherry, cypress, hemlock, oak, ponderosa pine, and fir. Among the darker-toned woods are cedar, mahogany, redwood, and walnut. For special effects, knotty pine, wormy chestnut, pecky cypress, and white-pocketed Douglas fir board paneling may be used.

Surface Textures and Patterns

Wood paneling is available in many shapes. It is either *planed* for smooth finishing, or *rough-sawn* for a rustic, informal effect. *Square-edge* boards may be joined edge to edge, spaced on a dark background, or applied in *board-and-batten* or *board-on-board* patterns. *Tongue-and-grooved* or *shiplapped* paneling comes in patterns, a few of which are illustrated in Figure 71-4.

Sizes

Most wood paneling comes in a 3/4-inch thickness and in nominal widths of 4, 6, 8, 10, and 12 inches. A few patterns are manufactured in a 9/16-inch thickness. *Aromatic cedar* paneling is used in clothes closets. It runs from 3/8 to 5/16 inch thick, depending on the mill. It is usually *edge-* and *end-matched* (tongue-and-grooved) for application to a backing surface.

Moisture Content

To avoid shrinkage, wood paneling, like all interior finish, should be dried to a *moisture equilibrium* content. That is, its moisture content should be about the same as the area in which it is to be used. Except for arid, desert areas and some coastal regions, the average moisture content of the air in the United States is about 8 percent (Fig. 71-5). Interior finish applied with an excessive moisture content will eventually shrink, causing open joints, warping, loose fasteners, and many other problems.

Other Paneling

The types of paneling described in this chapter are those that are most commonly used. In order to become acquainted with other types and methods of application, a study of manufacturers' catalogs dealing with sheet and board paneling, found in *Sweet's Architectural File,* is suggested.

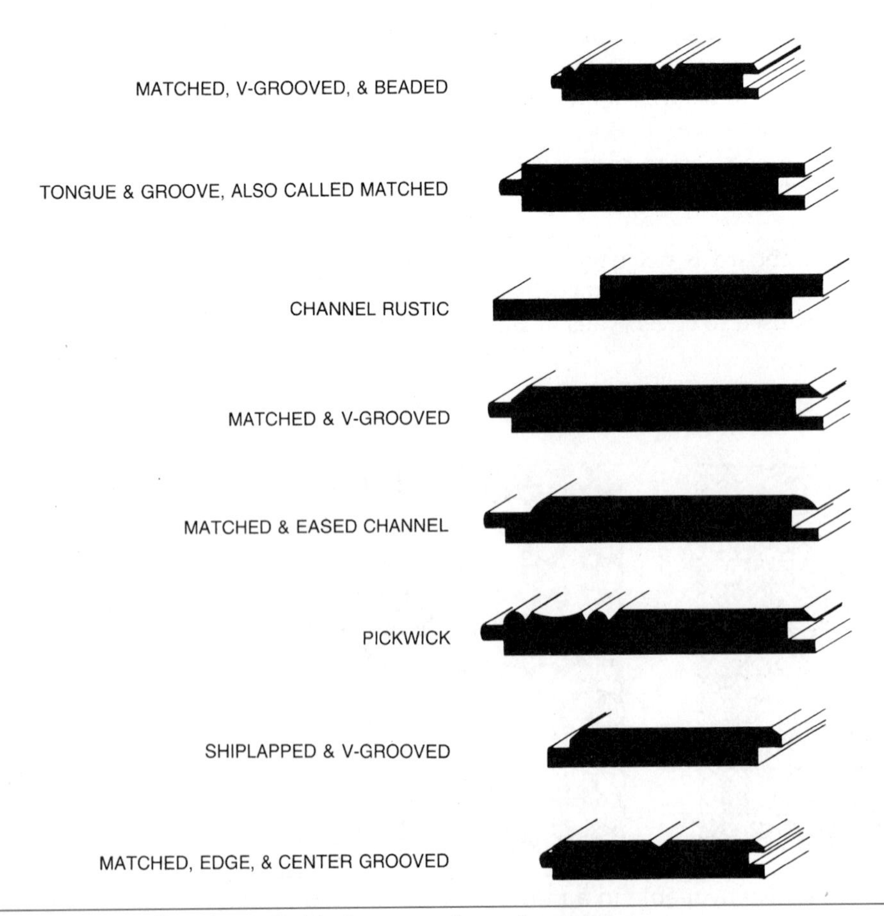

Fig. 71-4 Solid wood paneling is available in a number of patterns.

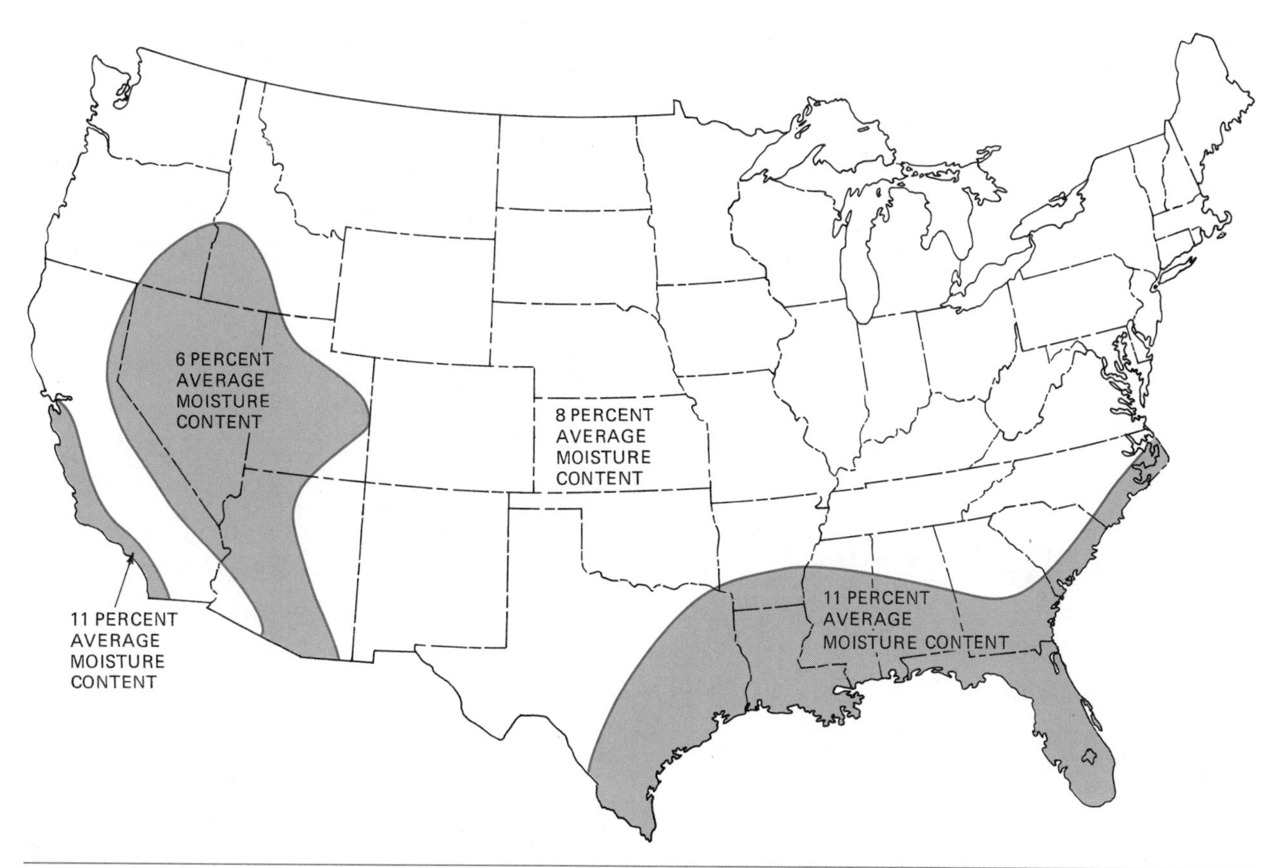

Fig. 71-5 Recommended average moisture content for interior finish woodwork in different parts of the United States.

CHAPTER 72 APPLICATION OF WALL PANELING

Sheet paneling, such as prefinished plywood and hardboard, is usually applied to walls with the long edges vertical. *Board* paneling may be installed vertically, horizontally, diagonally, or in many interesting patterns (Fig. 72-1).

Fig. 72-1 Wood paneling can be applied in many directions. *(Courtesy of California Redwood Association)*

Installation of Sheet Paneling

Some building codes require a base layer of gypsum board applied to studs or furring strips for the installation of sheet paneling. Even if not required, a backer board layer, at least 3/8-inch thick, should be installed on walls prior to the application of sheet paneling. The backing makes a stronger and more fire-resistant wall, helps block sound transmission, tends to bring studs in alignment, and provides a rigid finished surface for application of paneling. Without a base layer, depressions in the thin

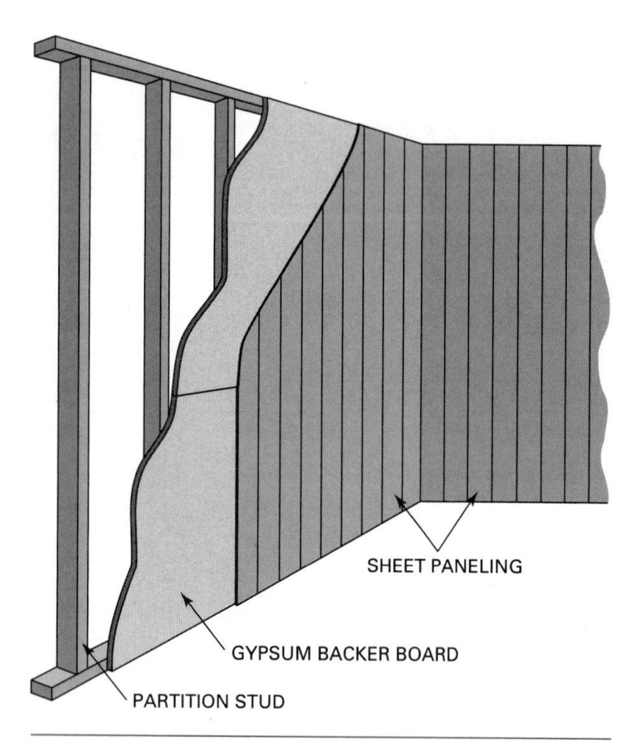

Fig. 72-2 Apply sheet paneling over a gypsum wallboard base.

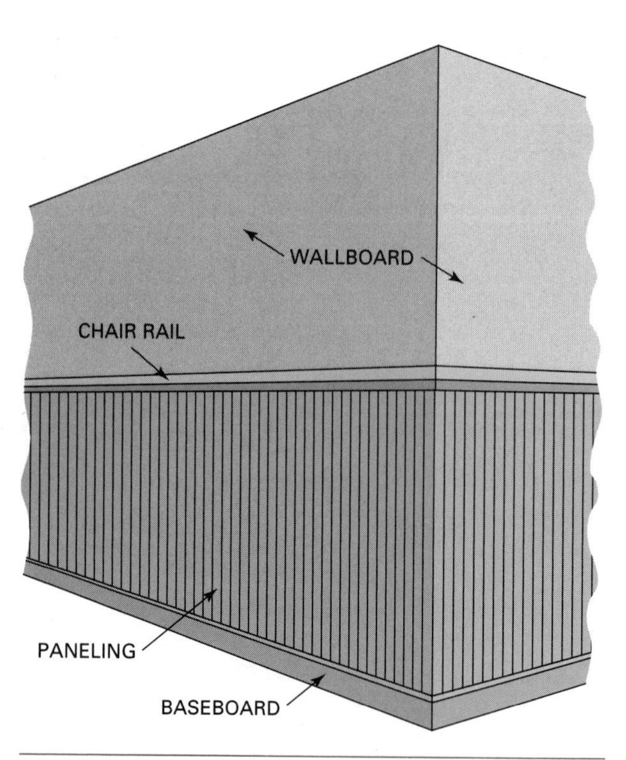

Fig. 72-3 Wainscoting is a wall finish, usually paneling, applied to the lower portion of the wall that is different from the upper portion.

sheets usually can be seen in the surface between the studs (Fig. 72-2).

Furring strips must be applied to masonry walls before the installation of paneling. The strips are usually applied by driving hardened nails with a pneumatic nailer. Instead of using furring strips, a freestanding wood wall close to the masonry wall can be built, if enough space is available.

Preparation for Application

Mark the location of each stud in the wall on the floor and ceiling. Paneling edges must fall on stud centers, even if applied with adhesive over a backer board, in case supplemental nailing of the edges is necessary. Panels are usually fastened with a combination of color pins and adhesive.

Apply narrow strips of black paint on the wall from floor to ceiling. Center them where joints between paneling will occur. If joints between sheets open slightly because of shrinkage during extended dry periods or heating seasons, it is not so noticeable with a black surface behind the joint.

Sometimes paneling does not extend to the ceiling but covers only the lower portion of the wall. This partial paneling is called **wainscoting**. It is usually installed about 3 feet above the floor (Fig. 72-3). If the wall is to be wainscoted, snap a horizontal line across the wall to indicate its height.

Stand panels on their long edge against the wall for at least forty-eight hours before installation. This allows them to adjust to room temperature and humidity (Fig. 72-4). Otherwise sheets may buckle

Fig. 72-4 Allow panels to stand in a room for forty-eight hours to adjust to temperature and humidity.

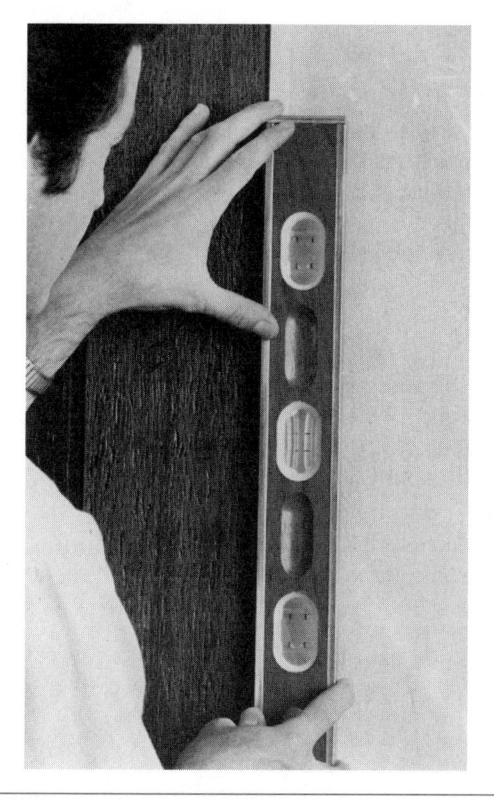

Fig. 72-5 The first sheet must be set plumb in the corner.

after installation, especially if they were very dry when installed.

Just prior to application, stand panels on end, side by side, around the room. Arrange them by matching the grain and color to obtain the most pleasing appearance.

Starting the Application

Start in a corner and continue installing consecutive sheets around the room. Select the starting corner, remembering that it will also be the ending point. This corner should be the least visible, such as behind an often-open door. Cut the first sheet to a length about 1/4-inch less than the wall height. Place the sheet in the corner. Plumb the outside edge and tack it in the plumb position. Set the distance between the points of the dividers the same as the amount the sheet overlaps the center of the stud (Fig. 72-5). Scribe this amount on the edge of the sheet butting the corner (Fig. 72-6).

Remove the sheet from the wall. Place it on saw horses on which a sheet of plywood, or two 8-foot lengths of 2 × 4s, have been placed for support. Cut to the scribed lines with a sharp, fine-toothed, hand crosscut saw. A handsaw is recommended for cutting sheets of prefinished paneling.

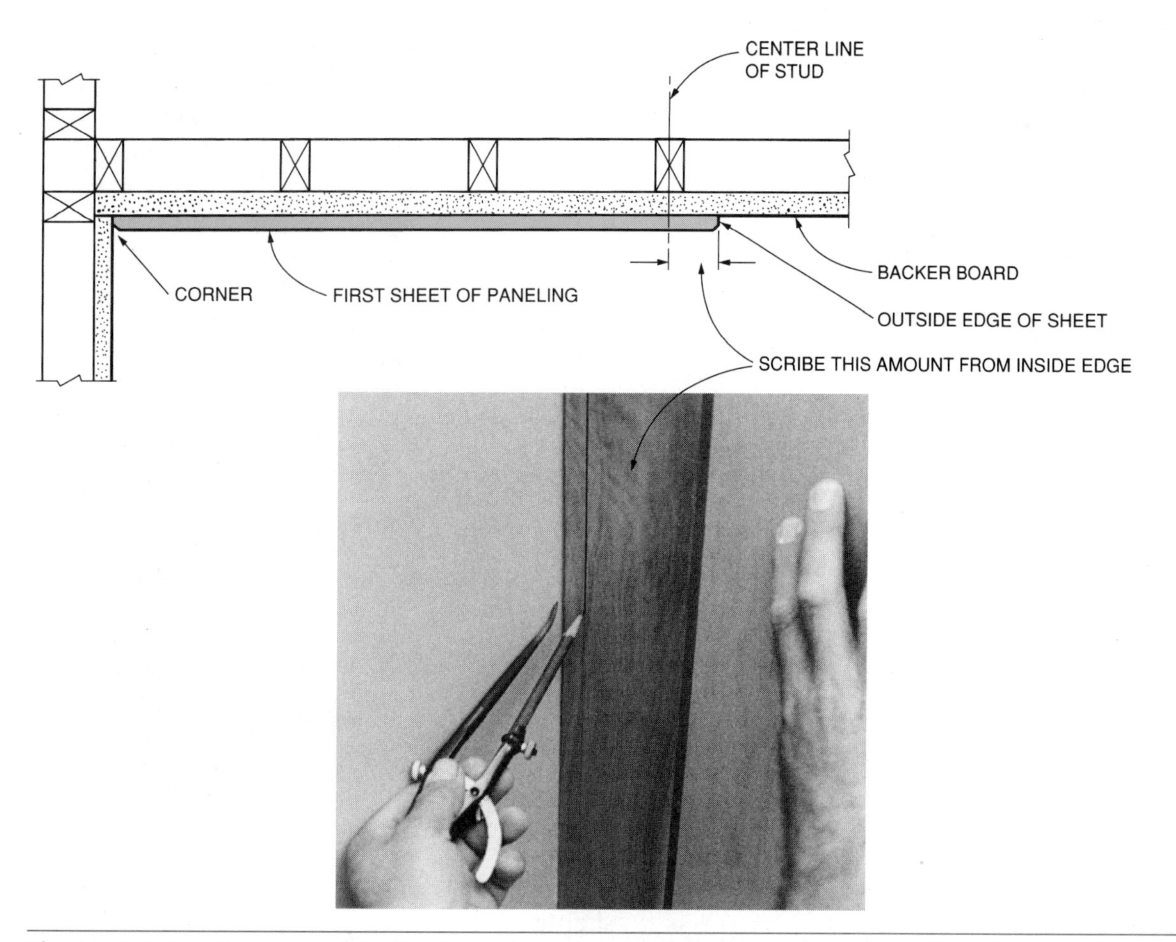

Fig. 72-6 Set scribers equal to the largest space and scribe the edge of the first sheet to the corner.

Power saws usually produce an undesirable splintered edge along the layout line. In addition, the irregularity of some scribed edges makes them difficult to cut with a portable circular saw.

Replace the sheet with the cut edge fitting snugly in the corner. The joint at the ceiling need not be fitted if a molding is to be used. If a tight fit between the panel and ceiling is desired, set the dividers and scribe a small amount at the ceiling line. Remove the sheet again. Cut to the scribed line. Replace the sheet, and raise it snugly against the ceiling. The space at the bottom will be covered later by a baseboard.

Fastening

If only nails are used, fasten about 6 inches apart along edges and about 12 inches apart on intermediate studs for 1/4-inch thick paneling. Nails may be spaced farther apart on thicker paneling. Drive nails at a slight angle for better holding power (Fig. 72-7).

If adhesives are used, apply a 1/8-inch continuous bead where panel edges and ends make contact. Apply beads 3 inches long and about 6 inches apart on all intermediate studs (Fig. 72-8). Put the panel in place. Tack it at the top. Be sure the panel is properly placed in position. Press on the panel surface to make contact with the adhesive. Use firm, uniform pressure to spread the adhesive beads evenly between the wall and the panel. Then, grasp the panel and slowly pull the bottom of the sheet a few inches away from the wall (Fig. 72-9). Press the sheet back into position after about two minutes. After about twenty minutes, recheck the panel. Apply pressure to assure thorough adhesion and to smooth the panel surface. Apply successive sheets

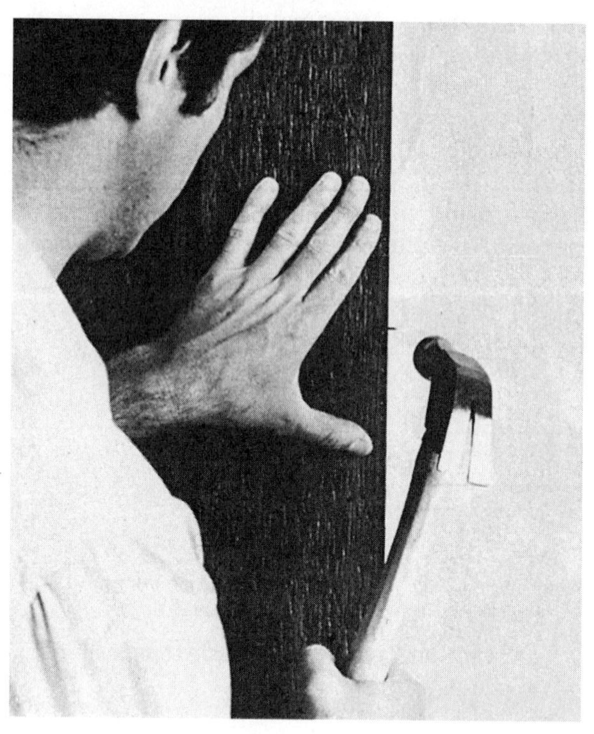

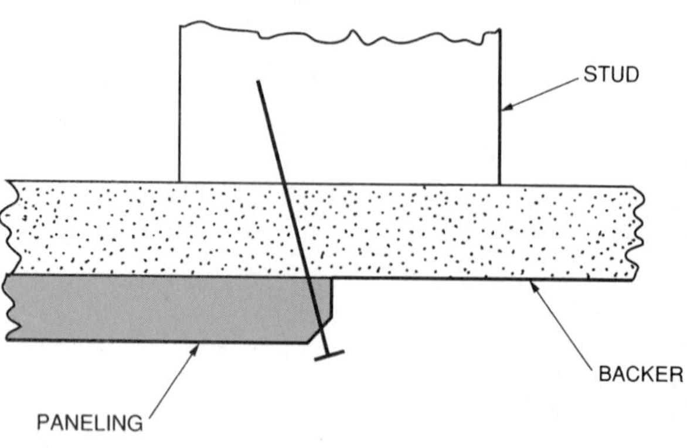

Fig. 72-7 Drive nails at a slight angle for better holding power. Color pins are used on prefinished sheet paneling.

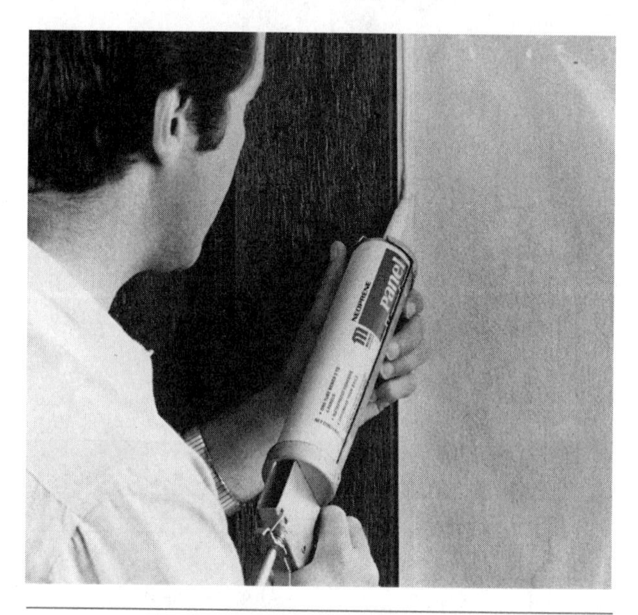

Fig. 72-8 Applying adhesive to a wall for the installation of sheet paneling.

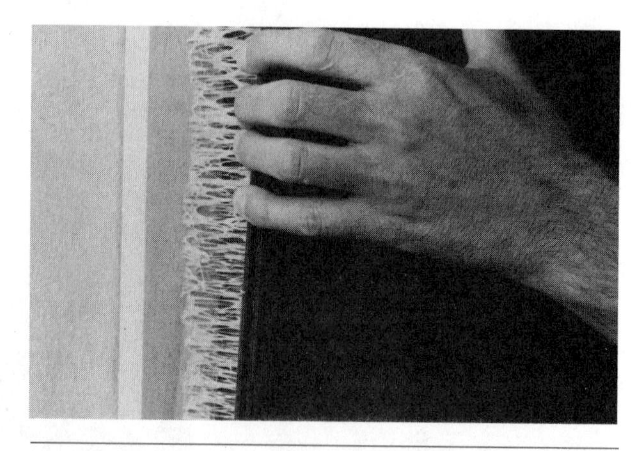

Fig. 72-9 The sheet is pulled a short distance away from the wall to allow the adhesive to dry slightly.

in the same manner. Do not force panels in position. Panels should touch very lightly at joints.

Wall Outlets

To lay out for wall outlets, plumb and mark both sides of the outlet to the floor. If the opening is close to the ceiling, plumb upward and mark lightly on the ceiling. Level the top and bottom of the outlet on the wall beyond the edge of the sheet to be installed. Or, level on the adjacent sheet, if closer. Cut, fit, and tack the sheet in position. Level and plumb marks from the wall and floor onto the sheet for the location of the opening (Fig. 72-10).

Another method is to rub a cake of carpenter's chalk on the edges of the outlet box. Fit and tack the sheet in position. Tap on the sheet directly over the outlet to transfer the chalked edges to the back of the sheet. Remove the sheet. Cut the opening for the outlet. Openings for wall outlets, such as electrical boxes must be cut fairly close to the location and size. The cover plate may not cover if the cutout is not made accurately. This could require replacement of the sheet. A saber saw may be used to cut these openings. When using the saber saw, cut from the back of the panel to avoid splintering the face (Fig. 72-11).

Ending the Application

The final sheet in the wall need not fit snugly in the corner if the adjacent wall is to be paneled or if interior corner molding is to be used. Take measurements at the top, center, and bottom. Cut the sheet to width, and install.

STEP BY STEP

PROCEDURES

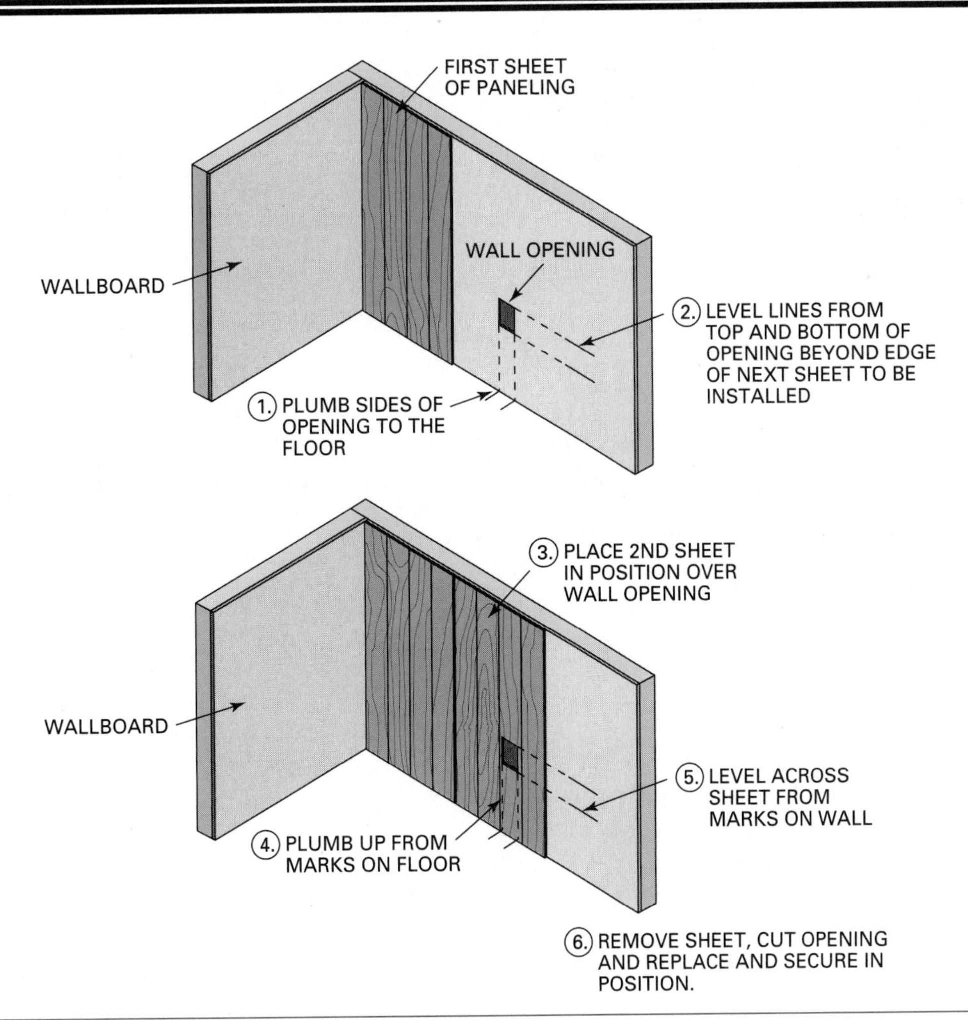

Fig. 72-10 Method of laying out for wall outlets.

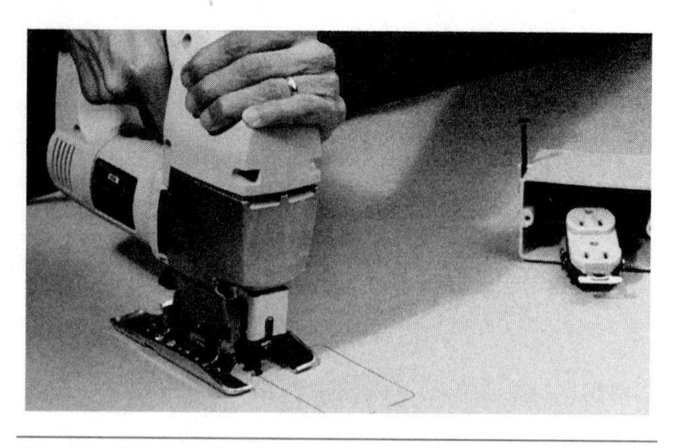

Fig. 72-11 A saber saw being used to cut an opening for an electrical outlet box.

If the last sheet butts against a finished wall and no corner molding is used, the sheet must be cut to fit snugly in the corner. To mark the sheet accurately, first measure the remaining space at the top, bottom, and about the center. Rip the panel about 1/2 inch wider than the greatest distance. Place the sheet with the cut edge in the corner and the other edge overlapping the last sheet installed. Tack the sheet in position. The amount of overlap should be exactly the same from top to bottom. Set the dividers for the amount of overlap. Scribe this amount on the edge in the corner (Fig. 72-12). Instead of dividers, it is sometimes more exact to use a small block of wood for scribing. The width is

STEP BY STEP PROCEDURES

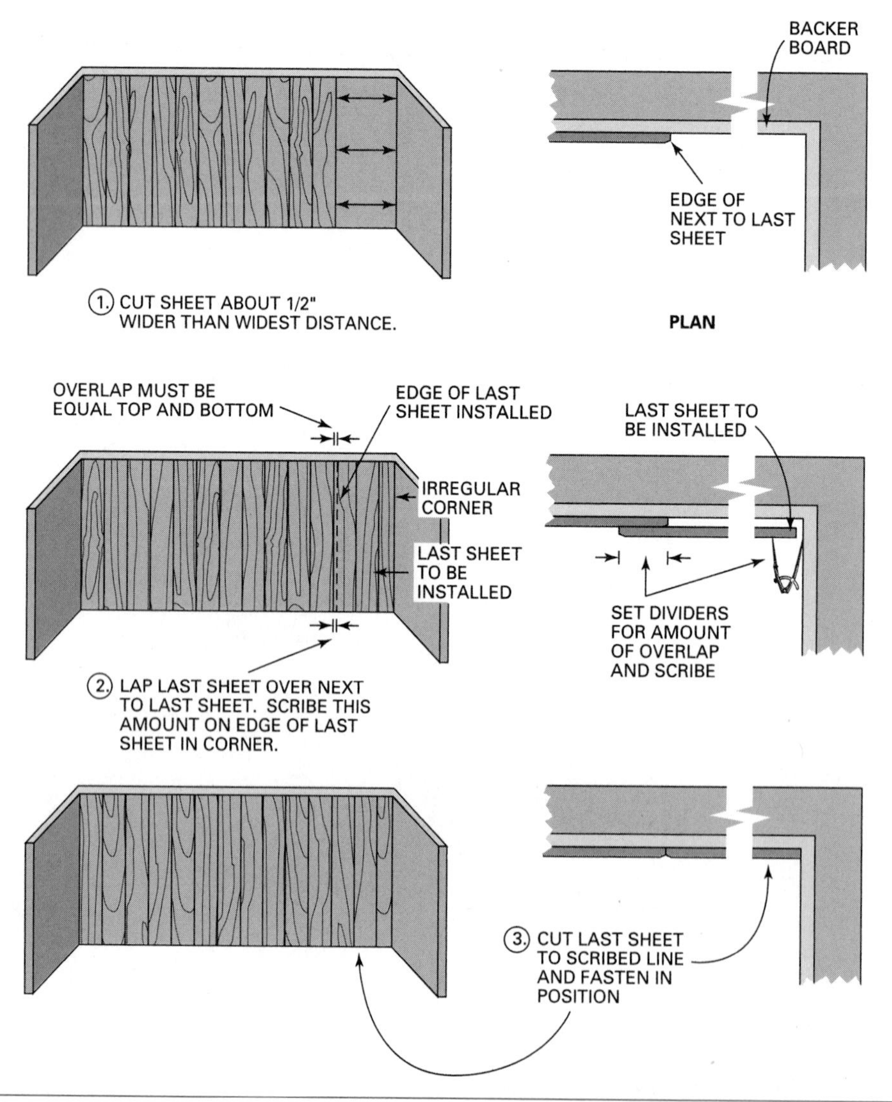

Fig. 72-12 Laying out the last sheet to fit against a finished corner without a molding.

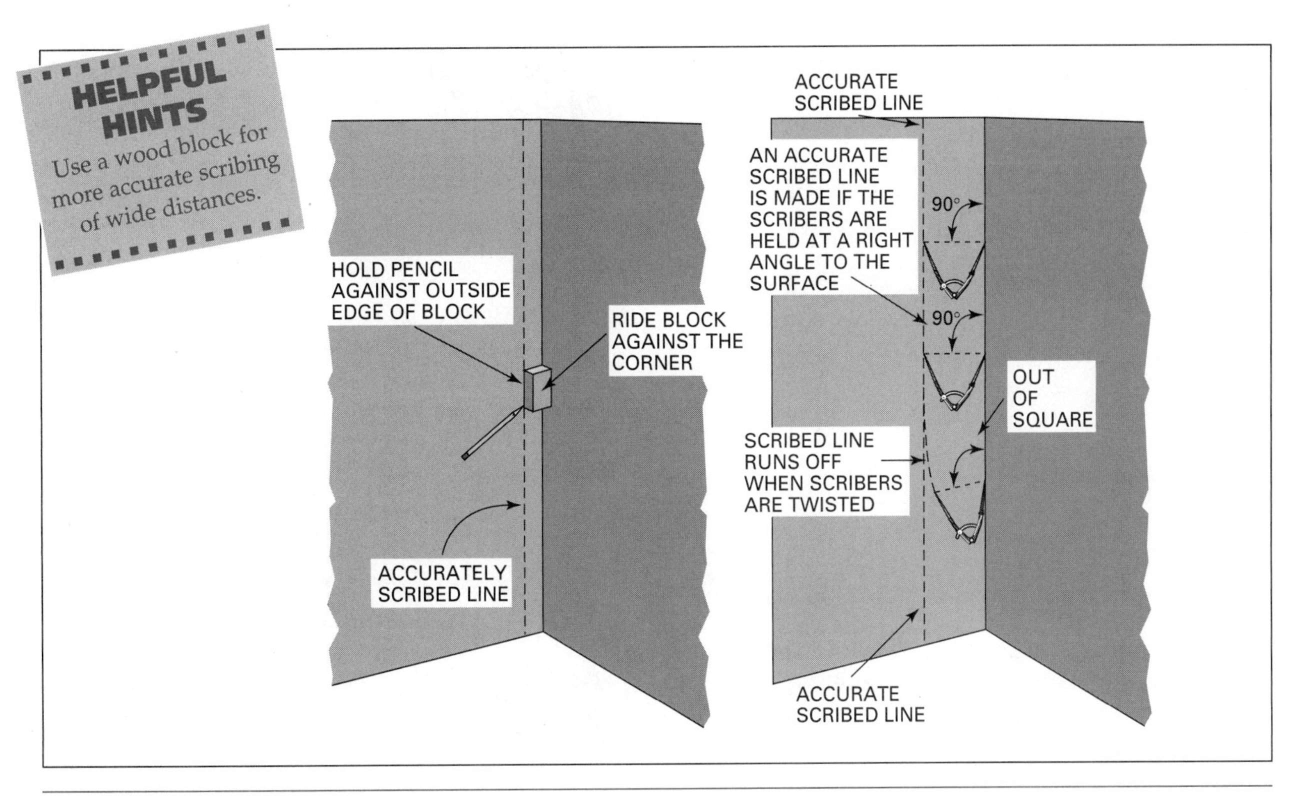

Fig. 72-13 Accurate scribing requires that the marked line be made perpendicular to the corner.

cut the same as the amount of overlap. Care must be taken to keep from turning the dividers while scribing along a surface (Fig. 72-13).

Cut to the scribed line. If the line is followed carefully, the sheet should fit snugly between the last sheet installed and the corner, regardless of any irregularities. On exterior corners, a quarter-round molding is sometimes installed against the edges of the sheets. Or, the joint may be covered with a wood, metal, or vinyl corner molding (Fig. 72-14).

Installing Solid Wood Board Paneling

Horizontal board paneling may be fastened to studs in new and existing walls (Fig. 72-15). For vertical application of board paneling in a frame wall, blocking must be provided between studs (Fig. 72-16). On existing and masonry walls, horizontal furring strips must be installed. Blocking or furring must be provided in appropriate locations for diagonal or pattern applications of board paneling.

Allow the boards to adjust to room temperature and humidity by standing them against the walls around the room. At the same time, put them in the order of application. Match them for grain and color (Fig. 72-17). If tongue-and-grooved boards are to be eventually stained or painted, apply the same finish to the tongues so that an unfinished surface is not exposed if the boards shrink after installation.

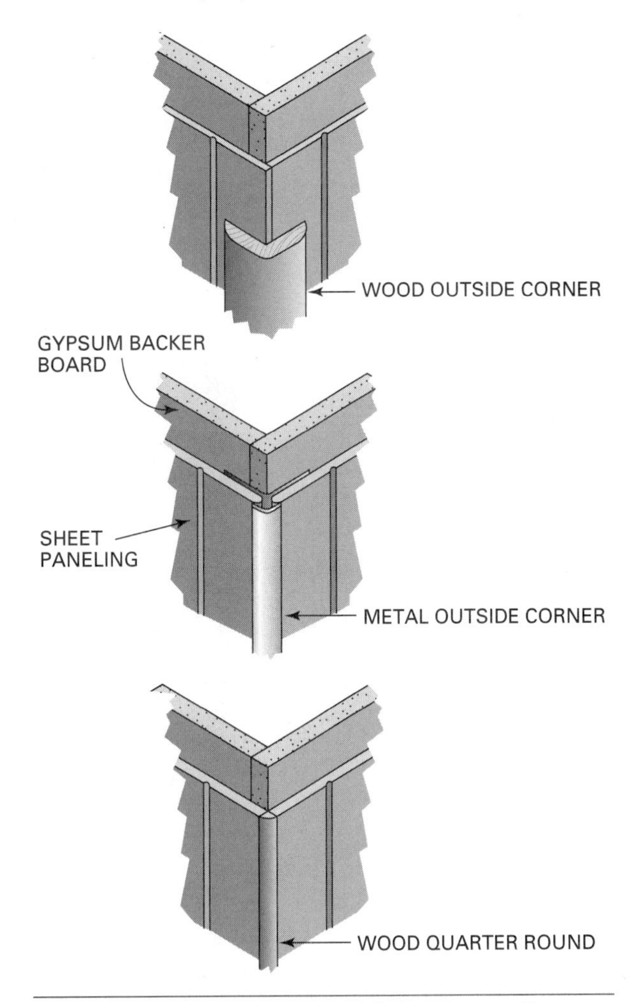

Fig. 72-14 Exterior corners of sheet paneling may be finished in several ways.

Fig. 72-15 No wall blocking is required for horizontal application of board paneling.

Fig. 72-17 Place boards around the room to adjust them to temperature and humidity and to match color and grain. (*Courtesy of California Redwood Association*)

Starting the Application

Select a straight board to start with. And cut it to length, about 1/4 inch less than the height of the wall. If tongue-and-grooved stock is used, tack it in a plumb position with the grooved edge in the corner. If a tight fit is desired, adjust the dividers to scribe an amount a little more than the depth of the groove. Rip to the scribed line. Face nail along the cut edge into the corner with finish nails about 16 inches apart. Blind nail the other edge through the tongue.

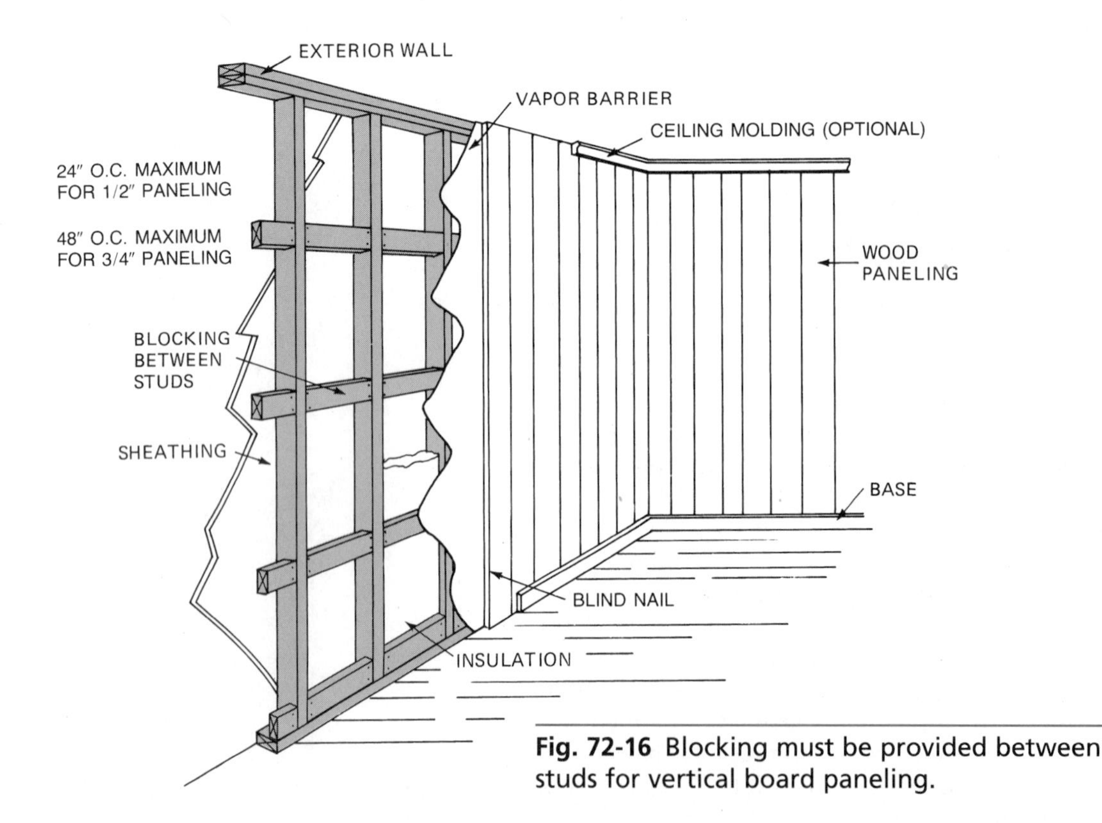

Fig. 72-16 Blocking must be provided between studs for vertical board paneling.

Continuing the Application

Apply succeeding boards by blind nailing into the tongue only (Fig. 72-18). Make sure the joints between boards come up tightly. See Figure 61-15 for methods of bringing edge joints of matched boards up tightly if they are slightly crooked. Severely warped boards should not be used. As installation progresses, check the paneling for plumb. If out of plumb, gradually bring back by driving one end of several boards a little tighter than the other end. Cut out openings in the same manner as described for sheet paneling.

Applying the Last Board

If the last board in the installation must fit snugly in the corner without a molding, the layout should be planned so that the last board will be as wide as possible. If boards are a uniform width, the width of the starting board must be planned to avoid ending with a narrow strip. If random widths are used, they can be arranged when nearing the end.

Cut and fit the next to the last board. Then remove it. Tack the last board in the place of the next to the last board. Cut a scrap block about 6 inches long and equal in width to the finished face of the next to the last board. Use this block to scribe the last board by running one edge along the corner and holding a pencil against the other edge (Fig. 72-19). Remove the board from the wall. Cut it to the scribed line. Fasten the next to the last board in position. Fasten the last board in position with the cut edge in the corner. Face nail the edge nearest the corner.

Horizontal application of wood paneling is done in a similar manner. However, blocking between studs on open walls or furring strips on existing walls are not necessary. On existing walls, locate and snap lines to indicate the position of stud centerlines. The thickness of wood paneling should be at least 3/8 inch for 16-inch spacing of frame members and 5/8 inch for 24-inch spacing. Diagonal and pattern application of board paneling is similar to vertical and horizontal applications. If wainscoting is applied to a wall, the joint between the different materials may treated in several ways (Fig. 72-20).

Applying Plastic Laminates

Plastic laminates are not usually applied to walls unless first prefabricated to sheets of plywood or similar material. They are then installed on walls in the same manner as sheet paneling. Special matching molding is used between sheets and on interior and exterior corners. (The application of plastic laminates is described in greater detail in a later unit on kitchen cabinets and countertops.)

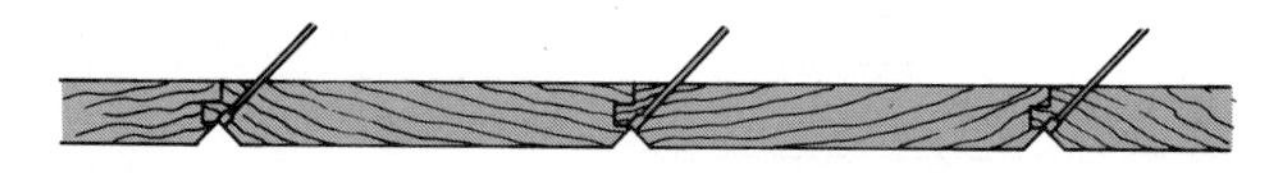

Fig. 72-18 Tongue-and-grooved paneling is blind nailed.

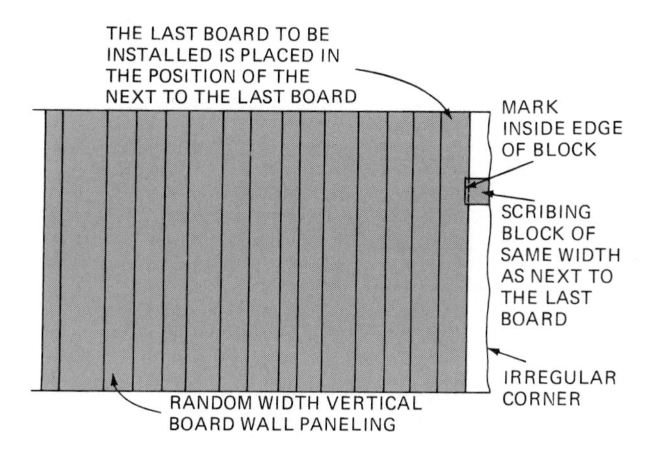

Fig. 72-19 Laying out the last board to fit against a finished corner.

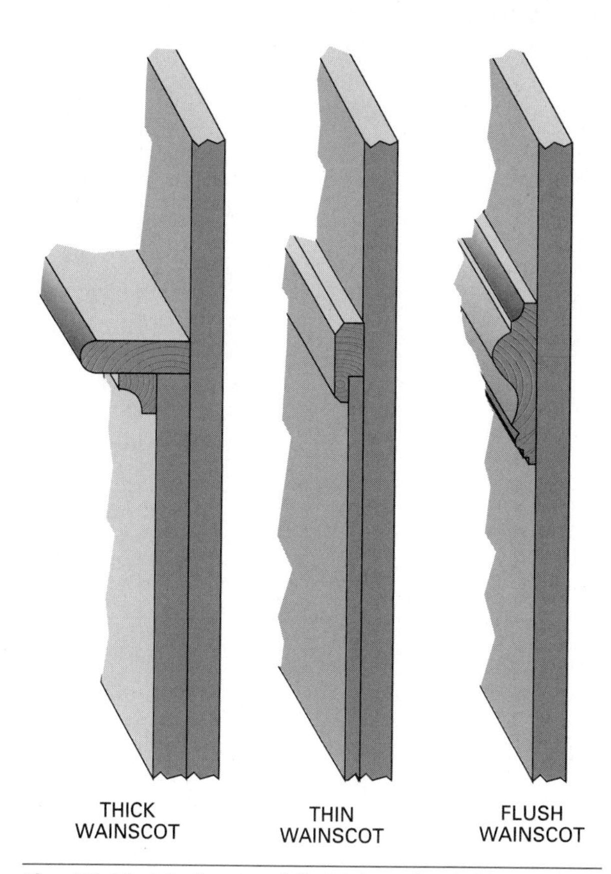

Fig. 72-20 Methods of finishing the joint at the top of wainscoting.

Estimating Paneling

Sheet Paneling

To determine the number of sheets of paneling needed, measure the perimeter of the room. Divide the perimeter by the width of the panels to be used. Deduct from this number any large openings such as doors, windows, or fireplaces. Deduct 2/3 of a panel for a door and 1/2 for a window or fireplace. Round off any remainder to the next highest number.

Estimating Board Paneling

Determine the square foot area to be covered by multiplying the perimeter by the height of each room. Deduct the area of any large openings. An additional percentage of the total area to be covered is needed because of the difference in the nominal size of lumber and its actual size.

Multiply the area to be covered by the area factor shown in Figure 72-21. Add 5 percent for waste in cutting. For example, the total area to be covered is 850 square feet, and 1 × 8 tongue and groove board paneling is to be used. Multiply 850 by 1.21, the sum of the coverage factor of 1.16 found in the table and .05 for waste in cutting. Round the answer of 1028.50 to 1029 for the number of board feet of paneling needed. To reduce waste in cutting, order suitable lengths.

Nominal Size	WIDTH		Area Factor*
	Dress	Face	
SHIPLAP			
1x6	5 1/2	5 1/8	1.17
1x8	7 1/4	6 7/8	1.16
1x10	9 1/4	8 7/8	1.13
1x12	11 1/4	10 7/8	1.10
TONGUE AND GROOVE			
1x4	3 3/8	3 1/8	1.28
1x6	5 3/8	5 1/8	1.17
1x8	7 1/8	6 7/8	1.16
1x10	9 1/8	8 7/8	1.13
1x12	11 1/8	10 7/8	1.10
S4S			
1x4	3 1/2	3 1/2	1.14
1x6	5 1/2	5 1/2	1.09
1x8	7 1/4	7 1/4	1.10
1x10	9 1/4	9 1/4	1.08
1x12	11 1/4	11 1/4	1.07
PANELING AND SIDING PATTERNS			
1x6	5 7/16	5 1/16	1.19
1x8	7 1/8	6 3/4	1.19
1x10	9 1/8	8 3/4	1.14
1x12	11 1/8	10 3/4	1.12

Fig. 72-21 Factors used to estimate amounts of board paneling.

CHAPTER 73 CERAMIC WALL TILE

Ceramic tile is used to cover floors and walls in rest rooms, baths, showers, and other high-moisture areas that need to be cleaned easily and frequently. On large jobs, the tile is usually applied by specialists. On small jobs, especially in rural or suburban communities, it is sometimes more expedient for the general carpenter to install tile. Ceramic tile was once set in *mortar.* Today *adhesive application* to a water-resistant backer board is the method commonly used (Fig. 73-1). This chapter describes the adhesive method of tile application to walls.

Fig. 73-1 Adhesive application of ceramic tile to shower walls. (*Courtesy of U.S. Gypsum Company*)

Description of Ceramic Wall Tile

Ceramic tiles are usually rectangular or square, but many geometric shapes, such as hexagons and octagons, are manufactured. Many solid colors, patterns, designs, and sizes give a wide choice to achieve the desired wall effect.

The most commonly used wall tile are nominal 4- and 6-inch squares, in 1/4-inch thickness. Special pieces such as *base, caps, inside corners,* and *outside corners* are used to trim the installation (Fig. 73-2).

Wall Preparation for Tile Application

Water-resistant gypsum board or cement board should be applied as a backing for ceramic wall tile (Fig. 73-3). (Application instructions for these products are given in Chapter 68.)

Minimum Backer and Tile Area in Baths and Showers

Tiles should overlap the lip and be applied down to the shower floor or top edge of the bathtub. On tubs without a showerhead, they should be installed to extend to a minimum of 6 inches above the rim. Around bathtubs with showerheads, tiles should extend a minimum of 5 feet above the rim or 6 inches above the showerhead, whichever is higher.

In shower stalls, tiles should be a minimum of 6 feet above the shower floor or 6 inches above the showerhead, whichever is higher. A 4-inch minimum extension of the full height is recommended beyond the outside face of the tub or shower (Fig. 73-4).

Calculating Border Tiles

Before beginning the application of ceramic wall tile, the width of the *border tiles* must be calculated. Border tiles are those that fit into the corner of the wall. The installation has a professional appearance if the border tiles are close to the same width and also as wide as possible on both sides of a wall (Fig. 73-5).

Measure the width of the wall from corner to corner. Change the wall measurement to inches. Divide it by the width of a tile. Measure the tile accurately. Sometimes, the actual size of a tile is different than its nominal size. Add the width of a tile to the remainder. Divide by 2 to find the width of border tiles.

For example: a wall measures 8′-6″ or 102 inches from corner to corner. If 4-inch, actual width, tiles are used, 102″ divided by 4″ equals 25 with a remainder of 2″. Add 4 to the remainder of 2, which equals 6. Divide 6 by 2 to get 3, the width of border tiles in inches.

Tile Layout

Measure out from the corner near the center of the wall. Mark a point that will be the edge of a full tile. Lay out a plumb line on the wall through this point from the floor to ceiling.

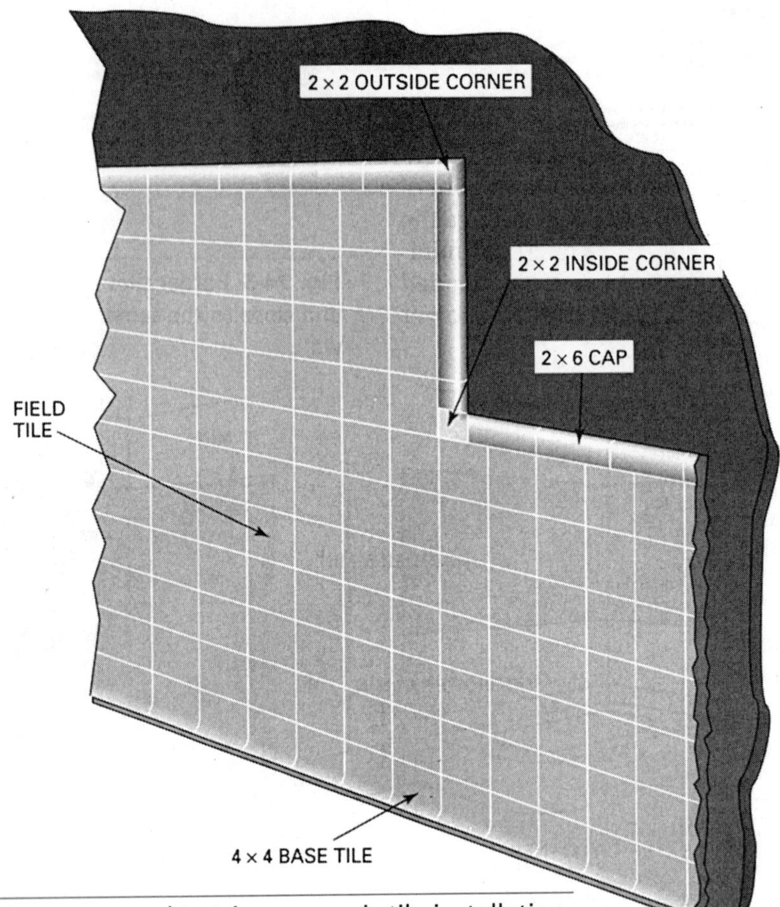

Fig. 73-2 Special pieces are used to trim a ceramic tile installation.

Fig. 73-3 Applying cement board on bathroom walls. (*Courtesy of U.S. Gypsum Company*)

If the floor is level, tiles may be applied by placing the bottom edge of the first row on the floor using plain tile or base tile. If the floor is not level, then base tile cannot be used. The bottom edge of the tiles must be cut to fit the floor while keeping the top edges level.

In this case, place a level on the floor and find the low point of the tile installation. From this point, measure up and mark on the wall the height of a tile. Draw a level line on the wall through the mark. Tack a straightedge on the wall with its top edge to the line. Tiles are then laid to the straightedge. When tiling is completed above, the straightedge is removed. Tiles are then cut and fitted to the floor.

Tile Application

It usually is best to install tile on the back wall first. This way, the joint in the corner is the least visible and the most watertight. Apply the recommended adhesive to the wall with a trowel. Use a flat trowel with grooved edges that allows the recommended amount of adhesive to remain on the surface. Too heavy a coat results in adhesive being squeezed out between the joints of applied tile, causing a mess. Too little adhesive results in failure of tiles to adhere to the wall. Follow the manufacturer's directions for

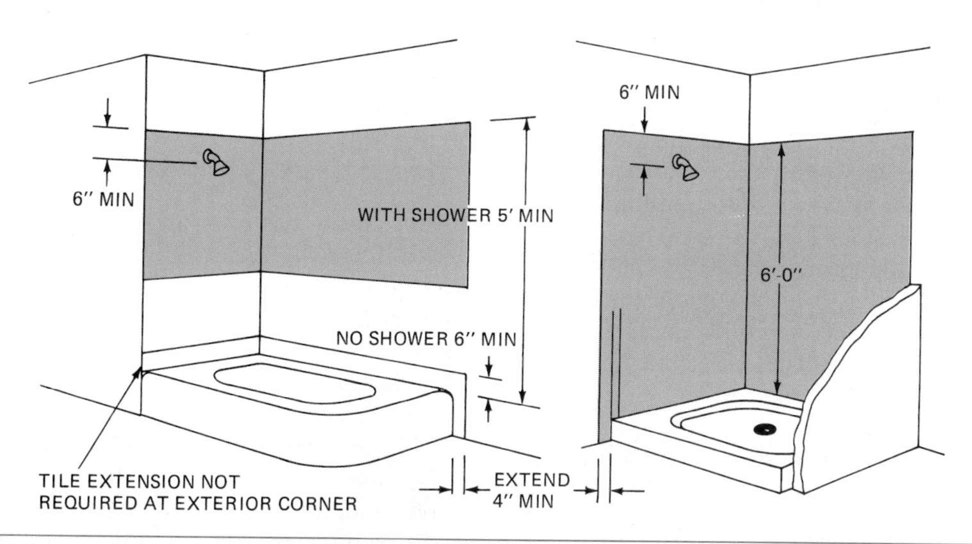

Fig. 73-4 Minimum areas recommended for the installation of ceramic tile around bathtubs and shower stalls.

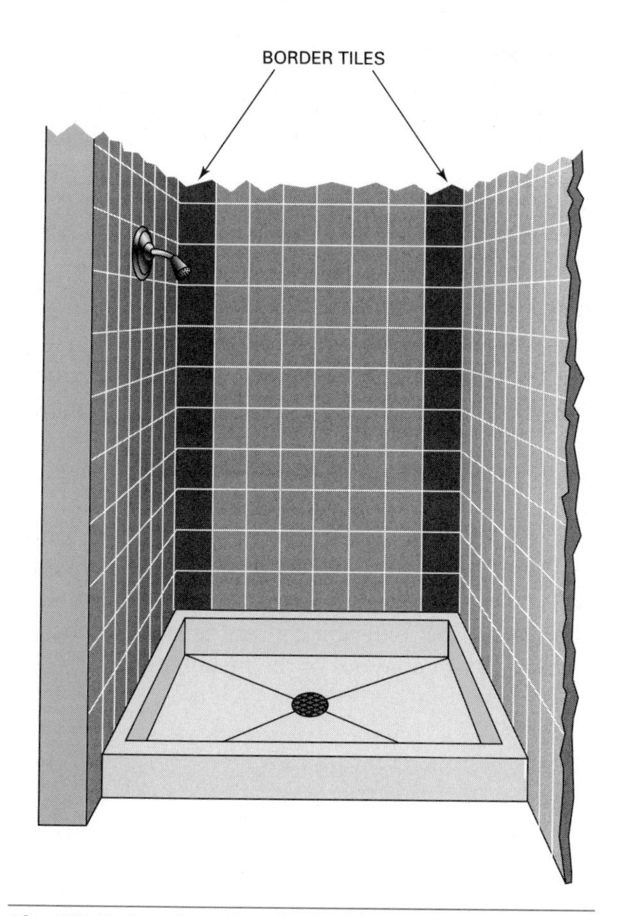

Fig. 73-5 Border tiles should be as wide as possible and close to the same width on both sides of the wall.

Fig. 73-6 Use a trowel that is properly grooved to allow the correct amount of adhesive to remain on the wall. (*Courtesy of U.S. Gypsum Company*)

the type of trowel to use and the amount of adhesive to be spread at any one time. Be careful not to spread adhesive beyond the area to be covered (Fig. 73-6).

Apply the first tile with its bottom edge to the straightedge or floor, as the case may be, and its vertical edge on the plumb line. Press the tile firmly into the adhesive. Apply other tiles in the same manner. Start from the center plumb line and pyramid upward and outward. As tiles are applied, slight adjustments may need to be made to keep them lined up. Keep fingers clean and adhesive off the face of the tiles. Clean tiles with a cloth dampened with the solvent recommended by the manufacturer to remove any adhesive from the faces of the tile as installation progresses.

Cutting Border Tiles

After all *field* tiles are applied, it is necessary to cut and apply border tiles. Field tiles are whole tiles that are applied in the center of the wall. Specialists on large jobs use a power tool for cutting ceramic tile (Fig. 73-7). On small jobs, the general carpenter may use a hand-operated ceramic tile cutter (Fig. 73-8). The hand cutter scores and snaps the tile in a way similar to that used in cutting glass. However, this tool cannot cut a small amount from the edge of a tile. For this purpose, a *nibbler,* shown in Figure 73-9, is used.

To finish the edges and ends of the installation, *caps* are used. Caps may be 2 × 6 or 4 × 4 pieces with one rounded finished edge. Special trim pieces are used to finish interior and exterior corners.

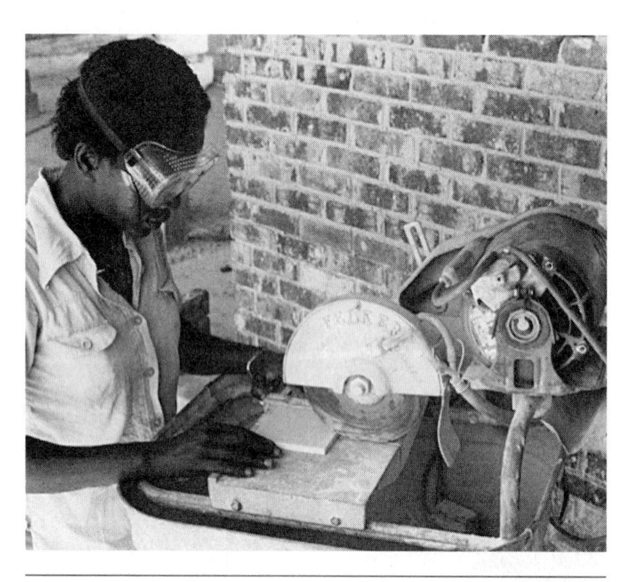

Fig. 73-7 Power tool for cutting ceramic tile.

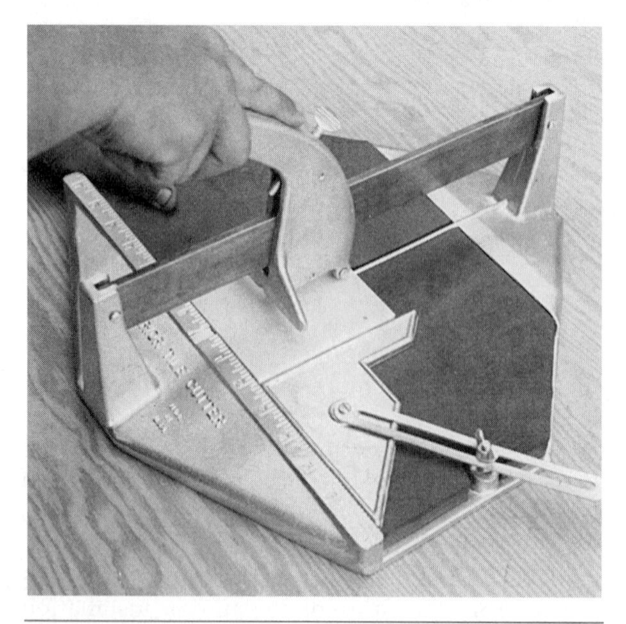

Fig. 73-8 Hand cutter for ceramic tile.

Grouting Tile Joints

After all tile has been applied, the joints are filled with tile *grout.* Grout comes in a powder form. It is mixed with water to form a paste of the desired consistency. It is spread over the face of the tile with a *rubber trowel* to fill the joints. The grout is worked into the joints. Then, the surface is wiped as clean as possible with the trowel.

The grout is allowed to set up, but not harden. The joints are then *pointed.* Removing excess grout and smoothing joints is called pointing. A small hardwood stick with a rounded end can be used as a pointing tool. The entire surface is wiped clean with a dry cloth after the grout has dried.

Estimating Tile

Determine the square foot area to be covered for the amount of tile to order. Divide the area to be covered by the area covered by one tile to find the number of tiles needed. For instance, if the area to be covered is 120 square feet and 4 × 4 tiles are being used, divide 120 by 1/9 to find that 1,080 tiles are needed.

Note: 1/9 is that part of a square foot that one 4 × 4 tile will cover. Because there are 144 square inches in 1 square foot, and there are 16 square inches in one 4 × 4 tile, the tile covers 16/144ths, or when reduced to its lowest term, 1/9 of a square foot.

The number of straight pieces of cap is found by determining the total linear feet to be covered and dividing by the length of the cap. The number of interior and exterior corners is determined by counting from a layout of the installation.

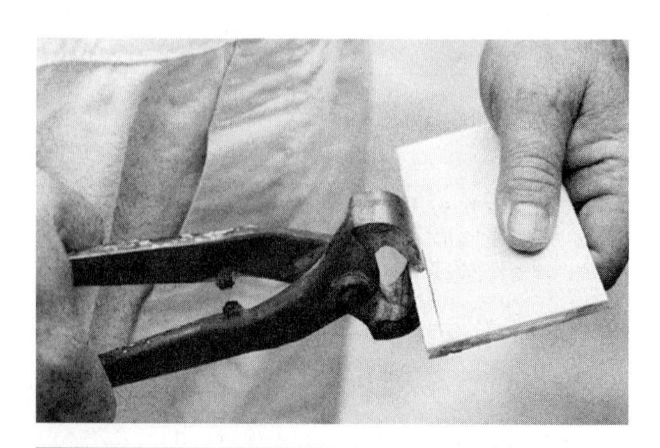

Fig. 73-9 Nibblers are used to cut a small amount from the tile.

Review Questions

Select the most appropriate answer.

1. On prefinished plywood paneling that is scored to simulate boards, some scores are always placed in from the edge
a. 12 and 16 inches.
b. 16 and 20 inches.
c. 12 and 24 inches.
d. 16 and 32 inches.

2. Tileboard is a type of
a. tile.
b. hardboard.
c. fiberboard.
d. plywood.

3. A wainscoting is a wall finish
a. applied diagonally.
b. applied partway up the wall.
c. used as a coating on prefinished wall panels.
d. used around tubs and showers.

4. For most parts of the country, wood used for interior finish should be dried to a moisture content of about
a. 8 percent.
b. 12 percent.
c. 15 percent.
d. 20 percent.

5. The thickest plastic laminate described in this unit is called
a. vertical type.
b. regular type.
c. backer type.
d. all-purpose type.

6. The tool recommended to cut thin plywood paneling is the
a. portable electric circular saw.
b. fine-tooth hand crosscut saw.
c. table saw.
d. saber saw.

7. If 3/4-inch thick wood paneling is to be applied vertically over open studs, wood blocking must be provided between studs for nailing at intervals of
a. 16 inches.
b. 24 inches.
c. 32 inches.
d. 48 inches.

8. When applying board paneling vertically to an existing wall
a. nail paneling to existing studs.
b. apply horizontal furring strips.
c. remove the wall covering and install blocking between studs.
d. use adhesives.

9. The number of 4 × 4 ceramic tiles needed to cover 150 square feet of wall area is:
a. 1,152.
b. 1,350.
c. 1,500.
d. 1,674.

10. Bathroom tile should extend over the tops of showerheads a minimum of
a. 4 inches.
b. 6 inches.
c. 8 inches.
d. 12 inches.

BUILDING FOR SUCCESS

Planning Is Critical

The old cliché "If you fail to plan, you plan to fail" is important in the construction world. Many builders, remodelers, and other skilled trade representatives will attest to the truth of this statement. The planning process must be at the forefront of all successful construction projects.

The construction industry has structured some ideal planning procedures through critical path processes and other scheduling plans. Why has so much emphasis been placed on the planning process? The following story is humorous, but conveys a lot of truth.

Poor Planning

(The following letter was written to clarify a health insurance claim made for injuries sustained on the job by a bricklayer. Author unknown.)

In response to your request for additional information in block number 3 of this accident report, I put "poor planning" as the cause of my accident. In your letter, you requested that I explain myself more fully. I trust the following details will be sufficient.

I am a bricklayer by trade. On the day of the accident, I was working alone on the roof of a new six-story building. When I completed my work, I discovered that I had about 500 pounds of brick left over. Rather than carry the bricks down by hand, I decided to lower them in a barrel using a pulley, which fortunately was attached to the side of the building, at the sixth floor.

Securing the rope at ground level, I went up to the roof, swung the barrel out, and loaded the brick into it. Then I went back to the ground and untied the rope, holding it tightly to ensure a slow descent of the 500 pounds of bricks. You will note in block number 11 of the accident report that I weigh 135 pounds.

Due to my surprise at being jerked off the ground so suddenly, I lost my presence of mind and forgot to let go of the rope. Needless to say, I proceeded at a rather rapid rate up the side of the building.

In the vicinity of the third floor, I met the barrel coming down. That explains the fractured skull and collarbone.

Slowed only slightly, I continued my rapid ascent, not stopping until the fingers of my right hand were two knuckles deep into the pulley.

Fortunately, by this time I had regained my presence of mind and was able to hold tightly to the rope despite my pain.

At approximately the same time, however, the barrel of bricks hit the ground and the bottom fell out of the barrel. Devoid of the weight of the bricks, the barrel now weighed approximately 50 pounds.

Again, I refer you to my weight in block number 11, As you might imagine I began a rapid descent down the side of the building.

In the vicinity of the third floor, I met the barrel coming up. This accounts for the two fractured ankles and the lacerations of my legs and lower body.

The encounter with the barrel slowed me enough to lessen my injuries when I fell into the pile of bricks. Fortunately, only three vertebrae were cracked.

I'm sorry to report, however, that as I lay there on the bricks—in pain, unable to stand, and watching the empty barrel six stories above me—again I lost my presence of mind. I let go of the rope!

As far fetched as this account may seem, there is something to be learned in planning a project. Arriving at the job site in the morning with no real advance thought about the project usually proves disastrous.

The purpose of planning is manifold. In order to begin and end a project on time, a plan must be formulated and carried out. Economics alone underscores the need to plan well. There are few builders, if any, who can afford to carry out their daily work with no plan to guide them.

In order to structure a realistic work flow, materials and job schedules have to become part of the overall plan. Experience becomes the best gauge when developing a project completion plan. For the beginning carpenter or other skilled craftsperson, planning skills can be learned daily on the job. Good record keeping is a good place to begin. Time cards can be used to accumulate information that will assist each person in defining how much work of each type can be accomplished in a day, week, or month. With this valuable information documented, a person can estimate the amount of time that will be necessary to complete the job under normal conditions. From this, cost estimates can be made.

Another criterion for good planning is to have an adequate understanding of the task at hand. The more that is understood about the building process, the more accurate the plan to build. Education can be interwoven with work experience to provide a big picture of the entire construction process involving all of the trades. With a broad knowledge of the various trade responsibilities and when all construction tasks are to occur, the plan can serve everyone well. Undoubtedly there will be interruptions, illnesses, weather delays, equipment breakdowns, and myriad other problems, but a good plan will include problem solving. "What if" situations should be discussed even if they are only remotely possible. Playing the odds and not looking for solutions to the "what if" problems can be a costly gamble.

Good communication among workers, crews, trades, and suppliers is essential for success. A major part of good planning is to communicate openly before, during, and after the construction process. With the complexity of the building industry today, it is essential to include all pertinent people in the planning process. General contractors should work consistently to improve communications within and outside their companies. Each person within the company should be a part of the planning process. Work incentives will inspire each employee to contribute valuable information to aid the planning process.

As the plans for a construction project are being formulated, the project coordinator needs to initiate the planning process well in advance of the actual construction. This allows for time to alter or rearrange any potential problem areas. One of the greatest planning problems involves extremes in weather conditions. Excessively hot or cold weather may necessitate altering a "typical" planning schedule. Characteristically, these elements will affect job and material scheduling. The best way to schedule a job under any condition is to recall all similar past experiences, and then use them as you plan the next project. Good records will help in this process.

Project planning and scheduling will be successful when the process is given full attention. As the bricklayer surely noticed when the brick removal process was completed, more planning could have saved time, money, and heartache.

Remember to plan your work and work your plan. It is critical.

FOCUS QUESTIONS: For individual or group discussion

1. Review any planning process that you have been exposed to in the classroom or on the job. List the main considerations in outline form. Create a detailed plan, from the earliest phases of a construction project through completion.
2. Explain how good planning can affect the profit/loss margin of a construction project. Include how planning is reflected in overhead, scheduling, time management, and cost concerns for time overruns.
3. Select a specific part of a construction project, such as framing. Create a list of the various considerations necessary for good planning by the crew leader.

UNIT

25

Ceiling Finish

Inexpensive and highly attractive ceilings may be created by installing suspended ceilings or ceiling tiles. They may be installed in new construction beneath exposed joists or in remodeling below existing ceilings.

OBJECTIVES

After completing this unit, the student should be able to:

- describe the sizes, kinds, and shapes of ceiling tile and lay in suspended ceiling panels.
- lay out and install suspended ceilings.
- lay out and install ceiling tile.
- estimate quantities of ceiling finish materials.

UNIT CONTENTS

CHAPTER 74 SUSPENDED CEILINGS

Suspended ceilings are widely used in commercial and residential construction as a ceiling finish. They also provide space for recessed lighting, ducts, pipes, and other necessary conduits (Fig. 74-1). Besides improving the appearance of a room, a suspended ceiling conserves energy by increasing the insulating value of the ceiling. It also aids in controlling sound transmission. In remodeling work, a suspended system can be easily installed beneath the existing ceiling. In basements, where overhead pipes and ducts may make other types of ceiling application difficult, a suspended type is easily installed. In addition, removable panels make pipes, ducts, and wiring accessible.

Fig. 74-1 Installing panels in a suspended ceiling grid. (*Courtesy of Armstrong World Industries*)

Suspended Ceiling Components

A *suspended ceiling* system consists of *panels* that are laid into a metal *grid*. The grid consists of *main runners, cross tees,* and *wall angles.* It is constructed in a 2 × 4 rectangular or 2 × 2 square pattern (Fig. 74-2).

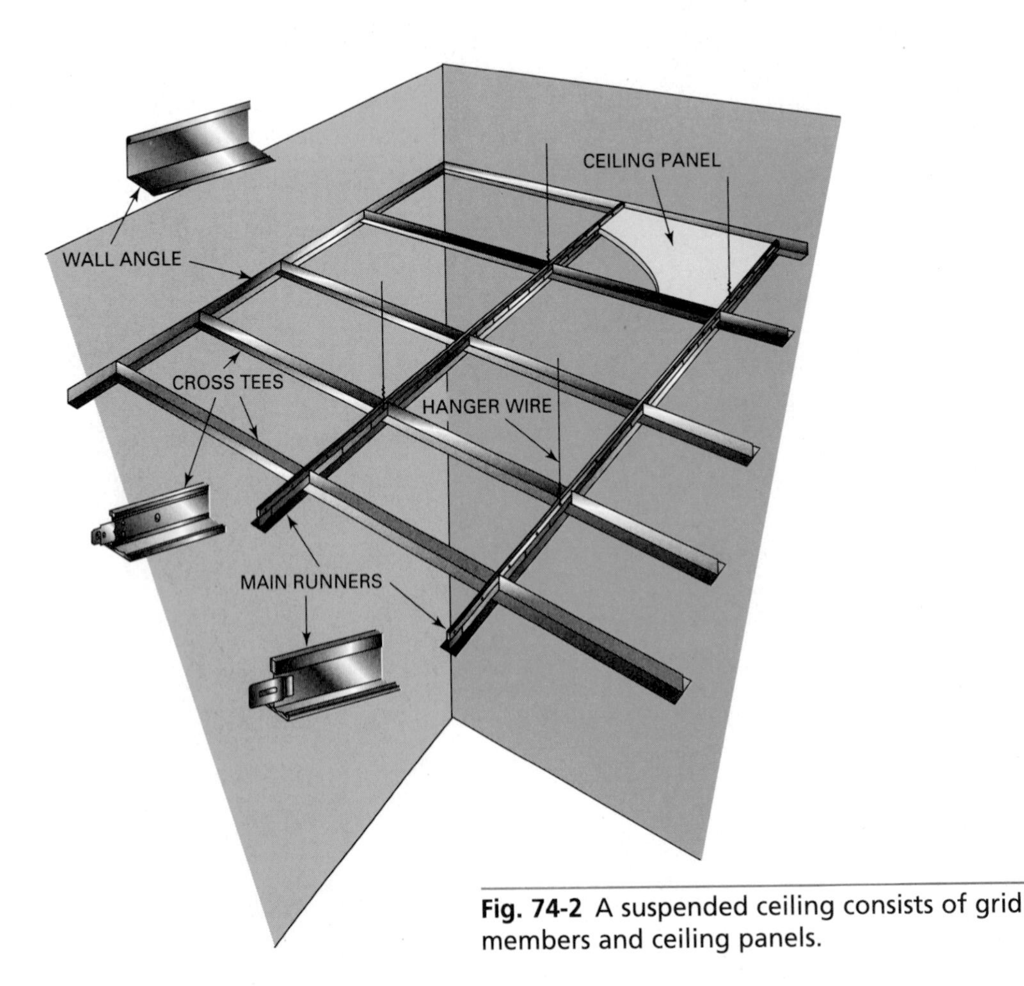

Fig. 74-2 A suspended ceiling consists of grid members and ceiling panels.

Grid members come prefinished in white, black, brass, chrome, and wood grain patterns, among others.

Wall Angles

Wall angles are L-shaped pieces that are fastened to the wall to support the ends of main runners and cross tees. They come in 10- and 12-foot lengths. They provide a continuous finished edge around the perimeter of the ceiling, where it meets the wall.

Main Runners

Main runners or *tees* are shaped in the form of an upside-down T. They come in 12-foot lengths. End splices make it possible to join lengths of main runners together. Slots are punched in the side of the runner at 12-inch intervals to receive cross tees. Along the top edge, punched holes are spaced at intervals for suspending main runners with *hanger* wire. Main runners extend from wall to wall. They are the primary support of a ceiling's weight.

Cross Tees

Cross tees come in 2- and 4-foot lengths. A slot, of similar shape and size as those in main runners, is centered on 4-foot cross tees for use when making a 2 × 2 grid pattern. They come with connecting tabs on each end. These tabs are inserted and locked into main runners and other cross tees.

Ceiling Panels

Ceiling panels are manufactured of many different kinds of material, such as gypsum, glass fibers, mineral fibers, and wood fibers. Panel selection is based on considerations such as fire resistance, sound control, thermal insulation, light reflectance, moisture resistance, maintenance, appearance, and cost. Panels are given a variety of surface textures, designs, and finishes. They are available in 2 × 2 and 2 × 4 sizes with square or rabbeted edges (Fig. 74-3).

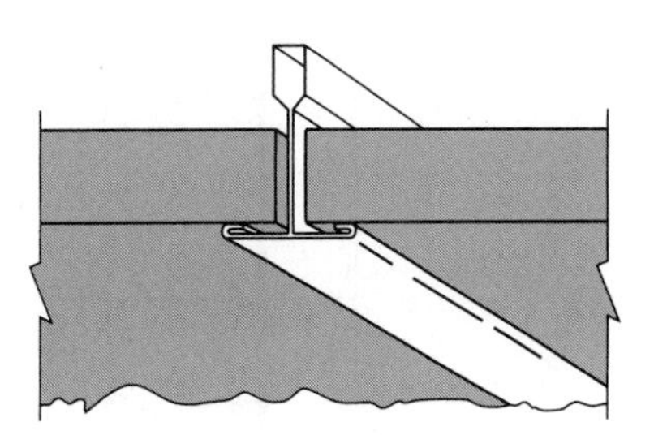

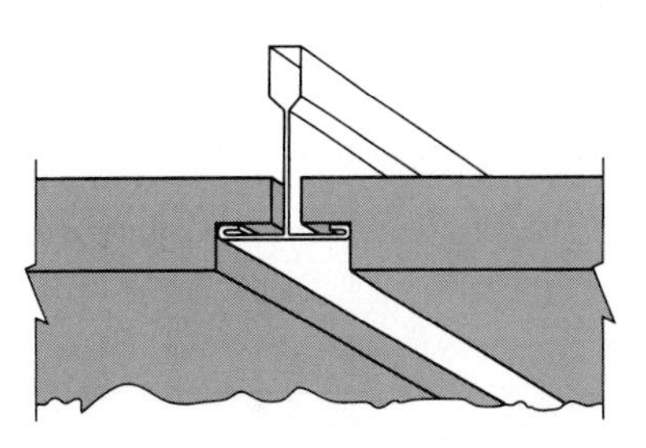

Fig. 74-3 Suspended ceiling panels may have square or rabbeted edges and ends.

Suspended Ceiling Layout

Before the actual installation of a suspended ceiling, a scaled sketch of the ceiling grid should be made. The sketch should indicate the direction and location of the main runners, cross tees, light panels, and border panels.

Main runners usually are spaced 4 feet apart. They usually run parallel with the long dimension of the room. For a standard 2 × 4 pattern, 4-foot cross tees are then spaced 2 feet apart between main runners. If a 2 × 2 pattern is used, 2-foot cross tees are installed between the midpoints of the 4-foot cross tees. Main runners and cross tees should be located in such a way that *border panels* on both sides of the room are equal and as large as possible (Fig. 74-4). Sketching the ceiling layout also helps when estimating materials.

Sketching the Layout

Sketch a grid plan by first drawing the overall size of the ceiling to a convenient scale. Use special care in measuring around irregular walls.

Locating Main Runners. To locate main runners, change the width of room to inches and divide by 48. Add 48 inches to any remainder. Divide the sum by 2 to find the distance from the wall to the first main runner. This distance is also the length of border panels.

For example, if the width of the room is 15'-8" changing to inches equals 188. Dividing 188 by 48 equals 3, with a remainder of 44. Adding 48 to 44 equals 92. Dividing 92 by 2 equals 46 inches, the distance from the wall to the first main runner.

Draw a main runner the calculated distance from, and parallel to, the long dimension of the ceiling. Draw the rest of the main runners parallel to the first, and at 4-foot intervals. The distance between the last main runner and the wall should be the same as the distance between the first main runner and the opposite wall (Fig. 74-5).

Locating Cross Tees. To locate 4-foot cross tees between main runners, first change the

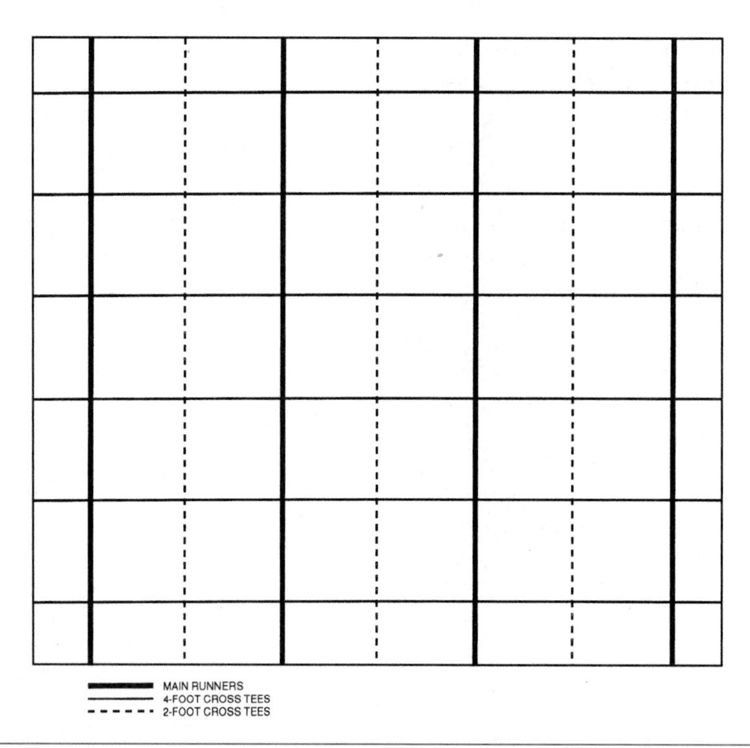

Fig. 74-4 A grid system with 4-foot cross tees spaced 2 feet apart along main runners is the recommended method of constructing a 2 × 4-foot grid. A 2 × 2-foot grid is made by installing 2-foot cross tees between the midpoints of the 4-foot cross tees.

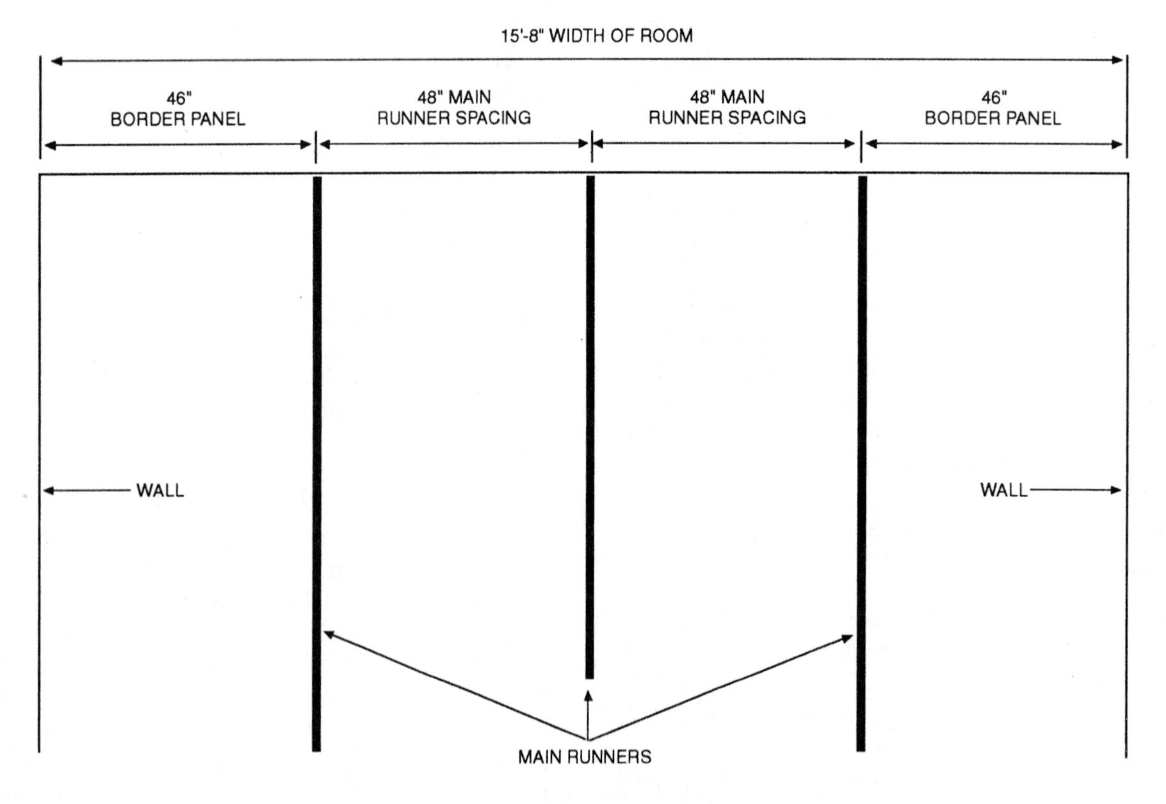

Fig. 74-5 Method of determining the location of main runners.

long dimension of the ceiling to inches. Divide by 24. Add 24 to the remainder. Divide the sum by 2 to find the width of the border panels on the other walls.

For example, if the long dimension of the room is 27′-10″, changing it to inches equals 334. Dividing 334 by 24 equals 13, with a remainder of 22. Adding 24 to 22 equals 46. Dividing 46 by 2 equals 23, the distance from the wall to the first row of cross tees.

Draw the first row of cross tees the calculated distance from, and parallel to the short wall. Draw the remaining rows of cross tees parallel to each other at 2-foot intervals. The distance from the last row of cross tees to the wall should be the same as the distance from the first row of cross tees to the opposite wall (Fig. 74-6).

Constructing the Ceiling Grid

The ceiling grid is constructed by first installing *wall angles,* then installing *suspended ceiling lags* and *hanger wires,* suspending the *main runners,* inserting full-length *cross tees,* and, finally, cutting and inserting *border* cross tees.

Installing Wall Angles

A suspended ceiling must be installed with at least 3 inches for clearance below the lowest air duct, pipe, or beam. This clearance provides enough room to insert ceiling panels in the grid. If recessed lighting is to be used, allow a minimum of 6 inches for clearance. The height of the ceiling may be located by measuring up from the floor. If the floor is rough or out of level, the ceiling line may be located with various leveling devices previously described. A combination of a hand level and straightedge, a water level, builders' level, transit-level, or laser level can be used. Snap chalk lines on all walls around the room to the height of the top edge of the wall angle (Fig. 74-7).

Fasten wall angles around the room with their top edge lined up with the chalk line. It may be easier to fasten the wall angle by prepunching holes

Fig. 74-7 Snap chalk lines on the walls for the location of wall angles.

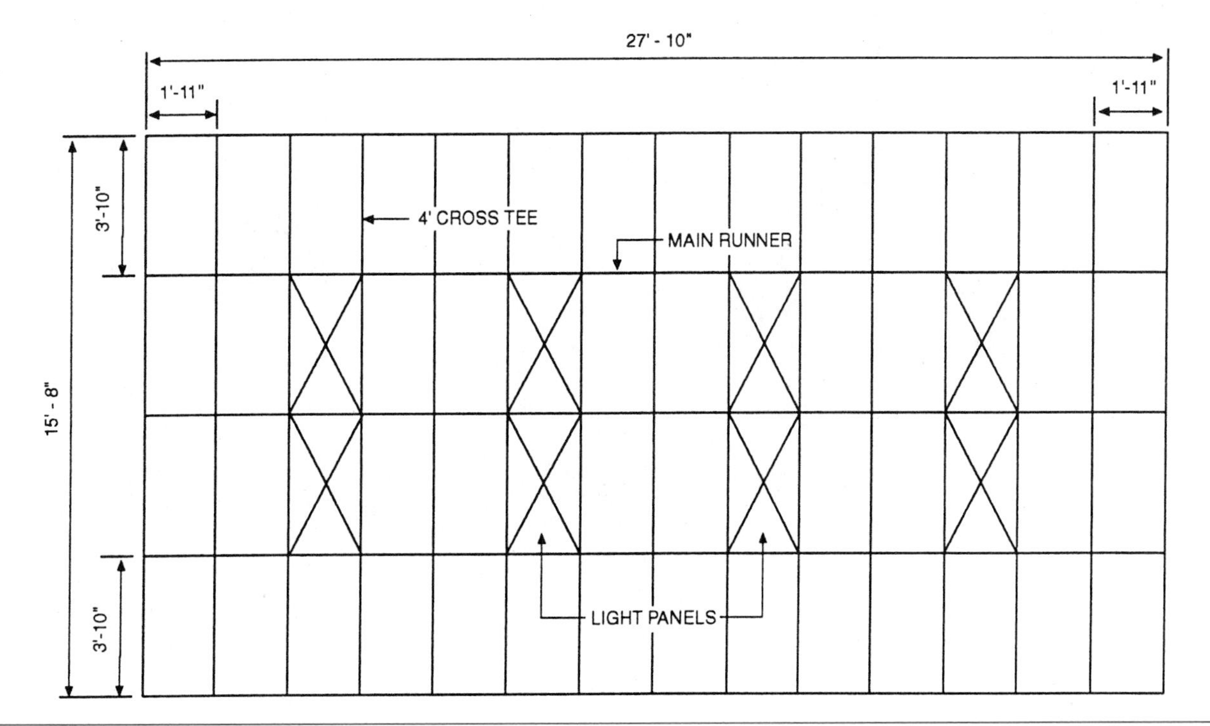

Fig. 74-6 Completed sketch of a suspended ceiling layout.

Fig. 74-8 Installing wall angle. (*Courtesy of Armstrong World Industries*)

with a center punch or spike. Fasten into framing wherever possible, not more than 24 inches apart (Fig. 74-8).

To fasten wall angles to concrete walls, short masonry nails sometimes are used. However, they are difficult to hold and drive. Use a small strip of cardboard to hold the nail while driving it with the hammer (Fig. 74-9). Lead or plastic inserts and screws may also be used to fasten the wall angles. Their use does require more time. If available, power nailers can be used for efficient fastening of wall angles to masonry walls.

Make miter joints on outside corners. Make butt joints in interior corners and between straight lengths of wall angle (Fig. 74-10). Use a combination square to layout and draw the square and angled lines. Cut carefully along the lines with snips.

CAUTION Use care in handling the cut ends of wall angle and other grid members. Cut metal ends are very sharp and can cause serious injury.

Installing Hanger Lags

From the ceiling sketch, determine the position of the first main runner. Stretch a line at this location across the room from the top edges of the wall angle. Stretch the line tightly on nails inserted between the wall and wall angle (Fig. 74-11). The line serves as a guide for installing *hanger lags* or *screw eyes* and *hanger wires* from which main runners are suspended.

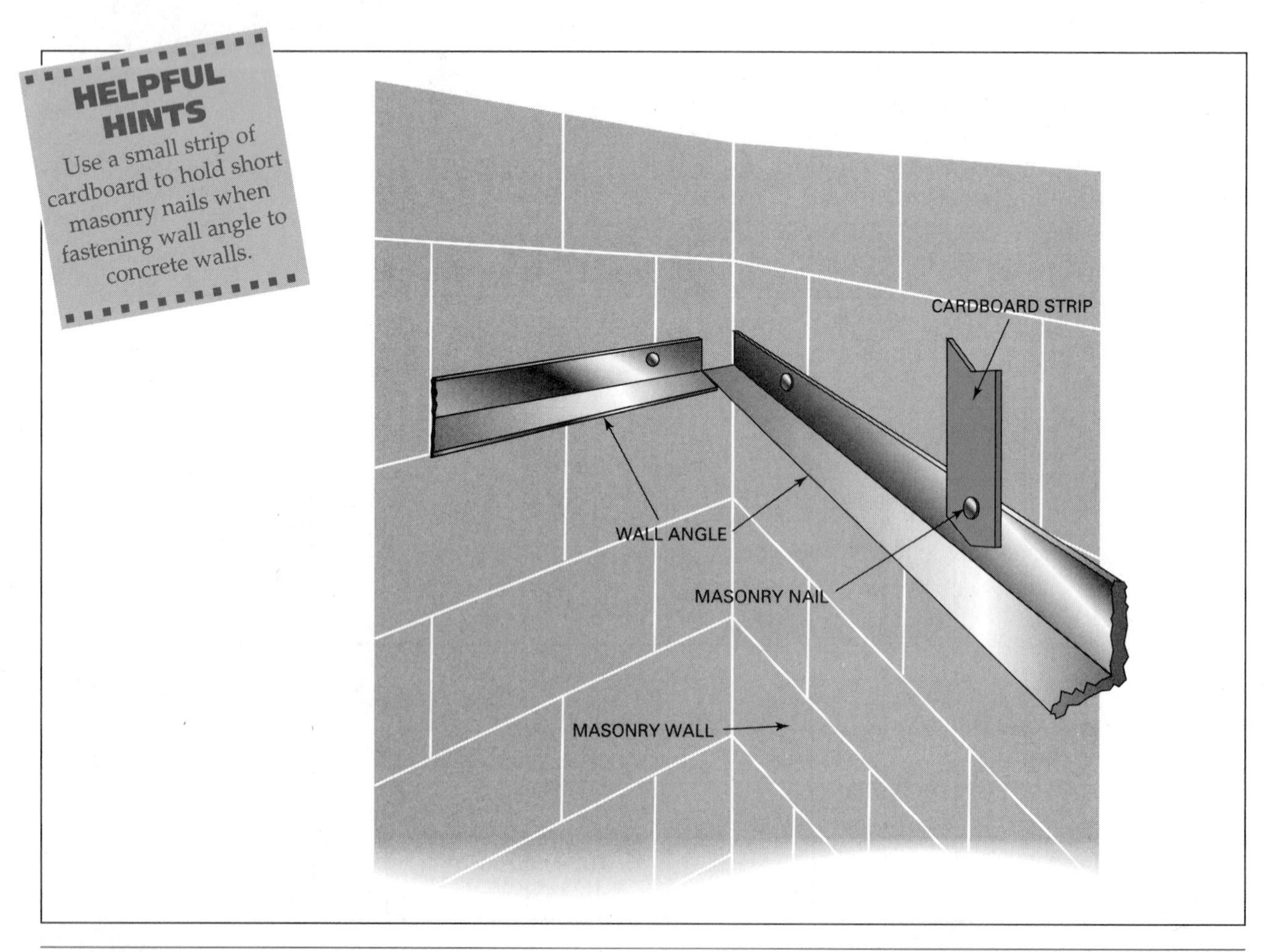

Fig. 74-9 Technique for driving short nails.

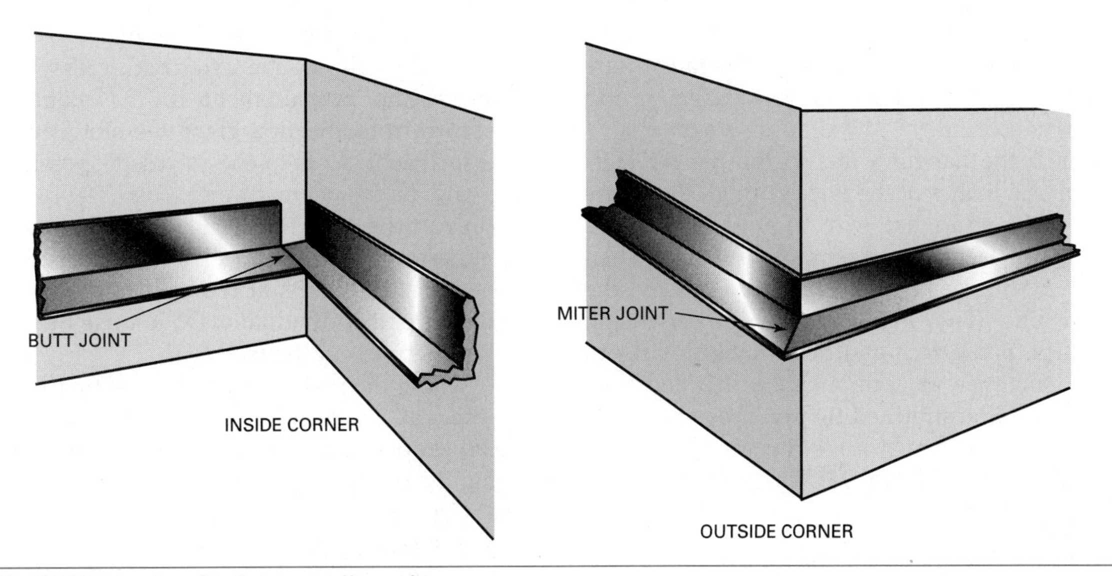

Fig. 74-10 Methods of joining wall angle.

Install hanger lags not over 4 feet apart and directly over the stretched line. Hanger lags should be of the type commonly used for suspended ceilings. They must be long enough to penetrate wood joists a minimum of 1 inch to provide strong support. *Eye pins* are driven into concrete with a *powder-actuated* fastening tool (Fig. 16-16). Hanger wires may also be attached directly around the lower chord of bar joists or trusses.

Installing Hanger Wire

Cut a number of hanger wires using wire cutters. The wires should be about 12 inches longer than the

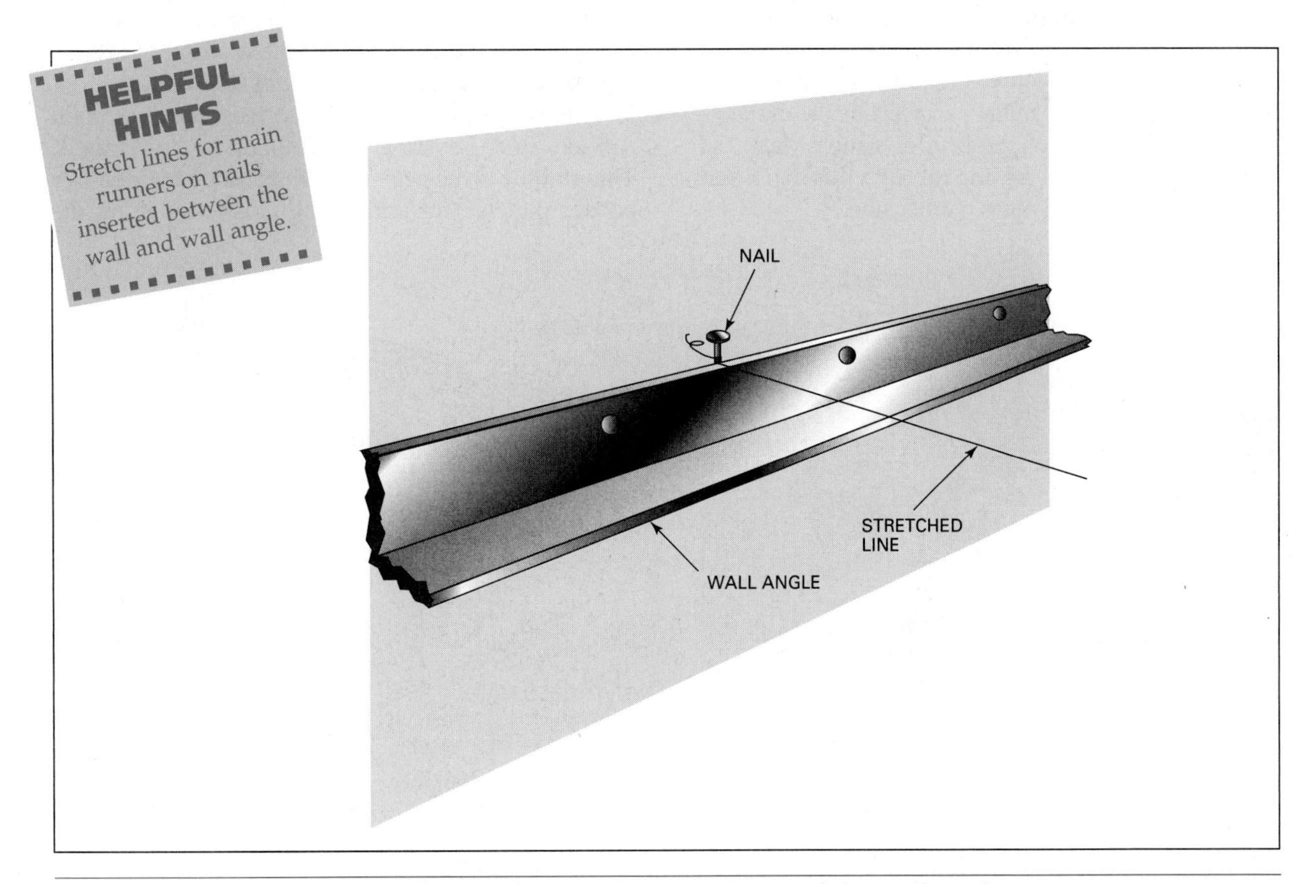

Fig. 74-11 Technique for stretching a line along the top edge of the wall angle.

distance between the overhead construction and the stretched line. For residential work, 16-gauge wire is usually used. For commercial work, 12-gauge and heavier wire is used.

Attach the hanger wires to the hanger lags. Insert about 6 inches of the wire through the screw eye. Securely wrap the wire around itself three times. Pull on each wire to remove any kinks. Then make a 90-degree bend where it crosses the stretched line (Fig. 74-12). Stretch lines, install hanger lags, and attach and bend hanger wires in the same manner at each main runner location. Leave the last line stretched tightly in position.

Installing Main Runners

The ends of the main runners rest on the wall angle. They must be cut so that a cross tee slot in the web of the runner lines up with the first row of cross tees. A cross tee line must be stretched, at wall angle height, across the short dimension of the room to line up the slots in the main runners. The line must run exactly at right angles to the main runner line and at a distance from the wall equal to the width of the border panels. If the walls are at right angles to each other, the location of the cross tee line can be determined by measuring out from both ends of the wall.

When the walls are not at right angles, the Pythagorean Theorem (Fig. 26-4) is used to square the grid system (Fig. 74-13). After the main runner line is installed, measure out from the short wall, along the stretched main runner line, a distance equal to the width of the border panel. Mark the line. Stretch the cross tee line through this mark and at right angles to the main runner line.

At each main runner location, measure from the short wall to the stretched cross tee line. Transfer the measurement to the main runner. Measure from the first cross tee slot beyond the measurement, so as to cut as little as possible from the end of the main runner (Fig. 74-14). Cut the main runners about 1/8 inch less to allow for the thickness of the wall angle. Backcut the web slightly for easier installation at the wall. Measure and cut main runners individually. Do not use the first one as a pattern to cut the rest.

Hang the main runners by resting the cut end on the wall angle and inserting suspension wires in the appropriate holes in the top of the main runner. Bring the runners up to the bend in the wires. Twist the wires with at least three turns to hold the main runners securely (Fig. 74-15).

More than one length of main runner may be needed to reach the opposite wall. Connect lengths of main runners together by inserting tabs into matching ends. Make sure end joints come up tight. The length of the last section is measured from the end of the last one installed to the opposite wall, allowing about 1/8-inch less for the thickness of the wall angle.

Installing Cross Tees

Cross tees are installed by inserting the tabs on the ends into the slots in the main runners. These fit into position easily, although the method of attaching varies from one manufacturer to another.

Install all full-length cross tees between main runners first. Lay in a few full-size ceiling panels. This stabilizes the grid while installing cross tees for border panels. Cut and install cross tees along the

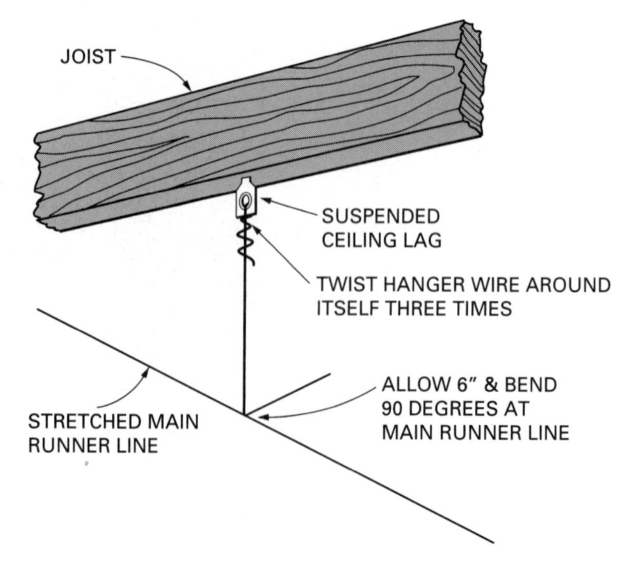

Fig. 74-12 Cut and fasten hanger wire to suspended ceiling lags. Bend the lower end at a 90-degree angle.

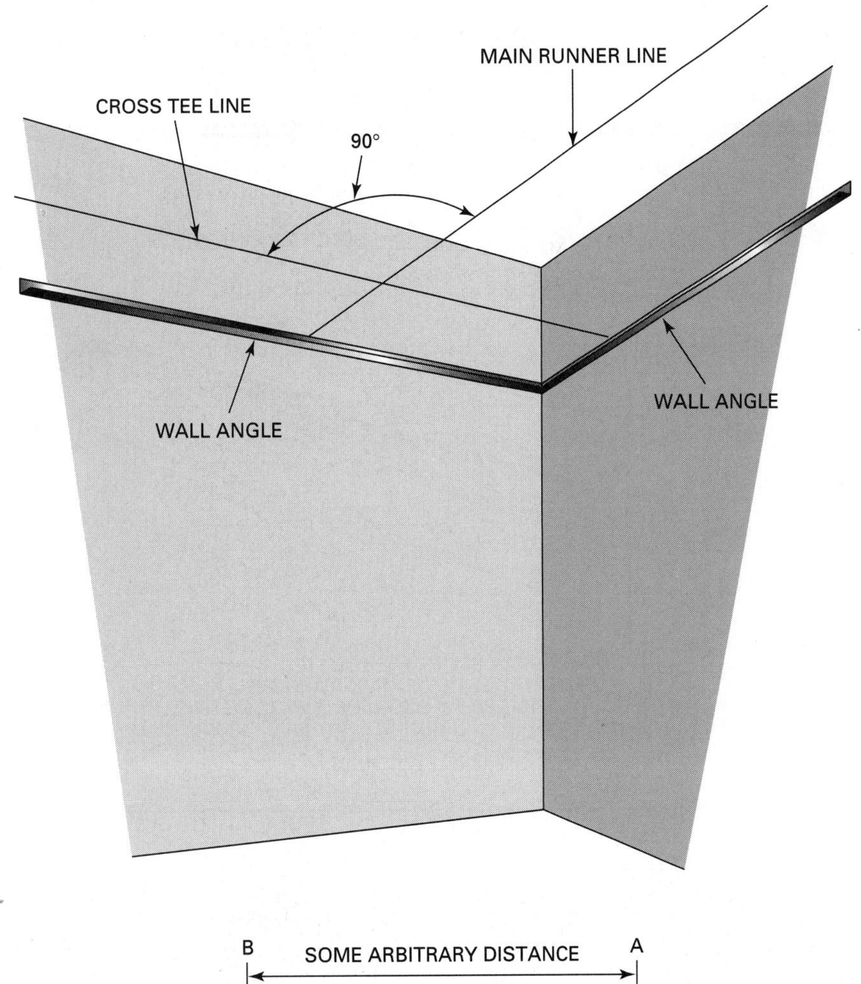

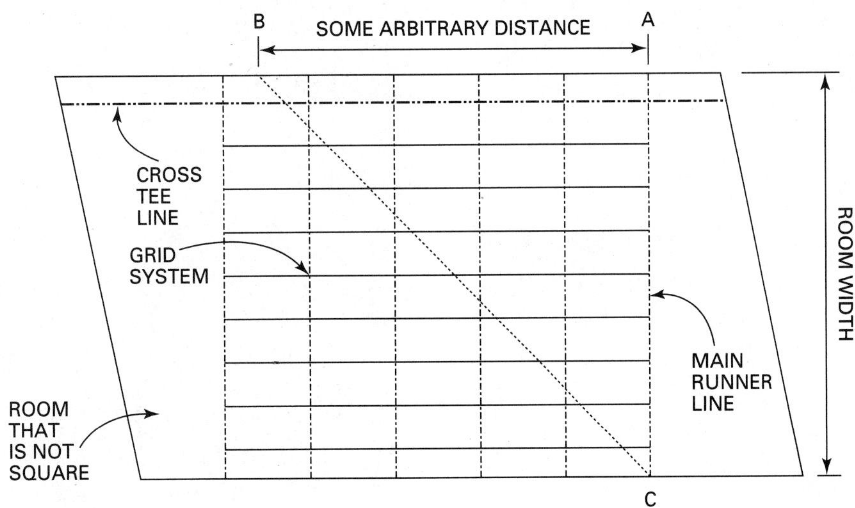

1) From A measure some arbitrary distance to B. The actual distance does not matter only that it should be large enough to make a big triangle, ABC.
2) Starting at point A, measure the width of the room to point C keeping your tape as square as possible with the first wall.
3) Use these numbers in the Pythagorean Theorem to determine the distance from B to C.
4) Measure and mark the distance from B to C. C is now square with point A.
5) Connect A and C with a string and measure each successive rows of main runner from it.

Fig. 74-13 Stretch a cross tee line at a right angle to the main runner line. Use the Pythagorean Theorem to determine square.

STEP BY STEP
PROCEDURES

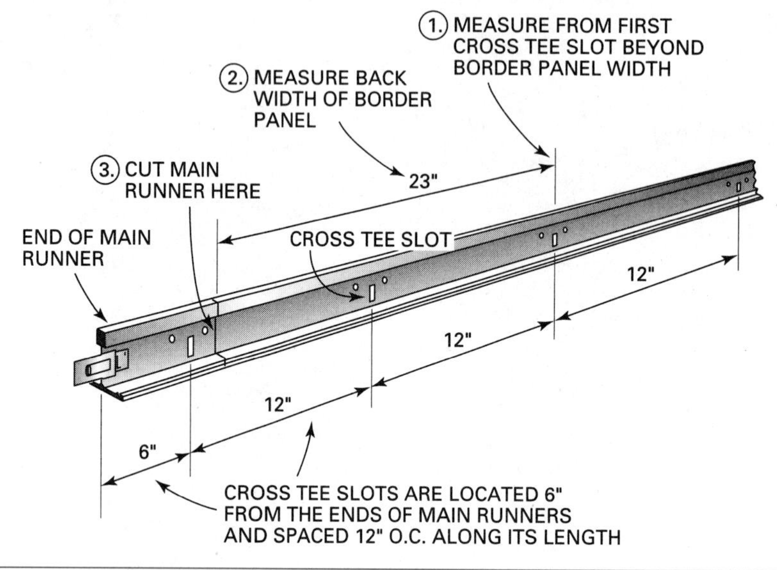

Fig. 74-14 Method of locating the end cut on main runners.

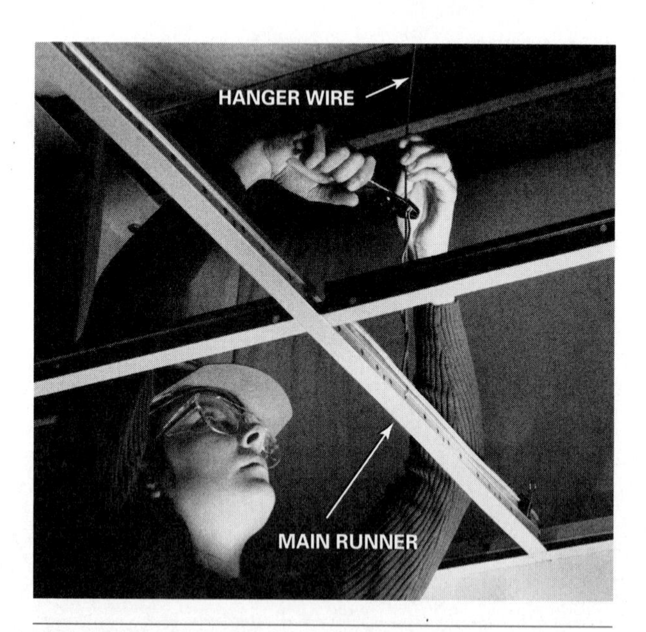

Fig. 74-15 Hanging the main runner. (*Courtesy of Armstrong World Industries*)

Fig. 74-16 Inserting a cross tee in the main runner. (*Courtesy of Armstrong World Industries*)

border. Insert the connecting tab of one end in the main runner and rest the cut end on the wall angle (Fig. 74-16). It may be necessary to measure and cut cross tees for border panels individually, if walls are not straight or square. For 2 × 2 panels, install 2-foot cross tees at the midpoints of the 4-foot cross tees. After the grid is complete, sight sections by eye. Straighten where necessary by making adjustments to border cross tees or hanger wires.

Installing Ceiling Panels

Ceiling panels are placed in position by tilting them slightly, lifting them above the grid, and letting

Fig. 74-17 Installing ceiling panels. (*Courtesy of Armstrong World Industries*)

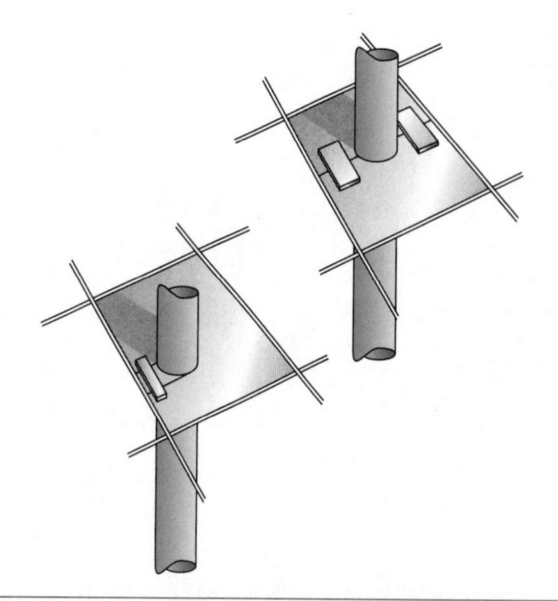

Fig. 74-18 Fitting ceiling panels around columns. (*Courtesy of Armstrong World Industries*)

them fall into place (Fig. 74-17). Be careful when handling panels to avoid marring the finished surface. Install all full-sized field panels first. Then cut and install border panels. Measure each border panel individually, if necessary. Cut them slightly smaller than measured so they can drop into place easily. Cut the panels with a sharp utility knife using a straightedge as a guide. A scrap piece of cross tee material can be used as a straightedge. Always cut with the finished side of the panel up.

Cutting Ceiling Panels Around Columns

When a column is near the center of a ceiling panel, cut the panel at the midpoint of the column. Cut semicircles from the cut edge to the size required for the panel pieces to fit snugly around the column. After the two pieces are rejoined around the column, glue scrap pieces of panel material to the back of the installed panel.

If the column is close to the edge or end of a panel, cut the panel from the nearest edge or end to fit around the column. The small piece is also fitted around the column and joined to the panel by gluing scrap pieces to its back side (Fig. 74-18).

Estimating Suspended Ceiling Materials

- Divide the perimeter of the room by 10, the length of a wall angle, to find the number needed.
- Find the number of main runners needed from the sketch. No more than two pieces can be cut from one 12-foot length.
- Count the number of 2- and 4-foot cross tees from the sketch. Border cross tees must be cut from full-length tees.
- From the sketch, count the number of hanger wires and screw eyes needed. Multiply the number needed by the length of each hanger wire to find the total linear feet of hanger wire needed.
- Count the number of ceiling panels from the sketch. Each border panel requires a full-size ceiling panel. There will be one less for each light fixture.

CHAPTER 75 CEILING TILE

Ceiling tile is usually stapled to furring strips that are fastened to exposed joists. They may also be cemented to existing ceilings provided the ceilings are solid and level. If the existing ceiling is not sound, furring strips should be installed and fastened through the ceiling into the joists above (Fig. 75-1).

Fig. 75-1 Installing ceiling tile to wood furring strips. (*Courtesy of Armstrong World Industries*)

Description of Ceiling Tiles

Most ceiling tiles are made of *wood fiber* or *mineral fiber*. Wood fiber tiles are lowest in cost and are adequate for many installations. Mineral fiber tiles are used when a more fire-resistant type is required.

Manufacture

Wood fibers are pressed into large sheets that are 7/16- to 3/4-inch thick. Mineral fiber tiles are made of rock that is heated to a molten state. The fibers are then sprayed into a sheet form. The surfaces of some sheets are **fissured** or **perforated** for sound absorption. The surfaces of other sheets are embossed with different designs or left smooth. Then, they are given a factory finish and cut into individual tiles. Most tiles are cut with *chamfered, tongue-and-grooved* edges with two adjacent *stapling flanges* for concealed fastening (Fig. 75-2).

Sizes

The most popular sizes of ceiling tile are a 12-inch square and a 12 × 24 inch rectangle in thicknesses of 1/2 inch. Tiles are also manufactured in squares of 16 inches and in rectangles of 16 × 32 inches.

Ceiling Tile Layout

Before installation begins, it is necessary to calculate the size of *border tiles* that run along the walls. It is desirable that border tiles be as wide as possible and of equal widths on opposite walls.

Calculating Border Tile Sizes

To find the width of the border tiles along the long walls of a room, first determine the dimension of the short wall. In most cases, the measurement will

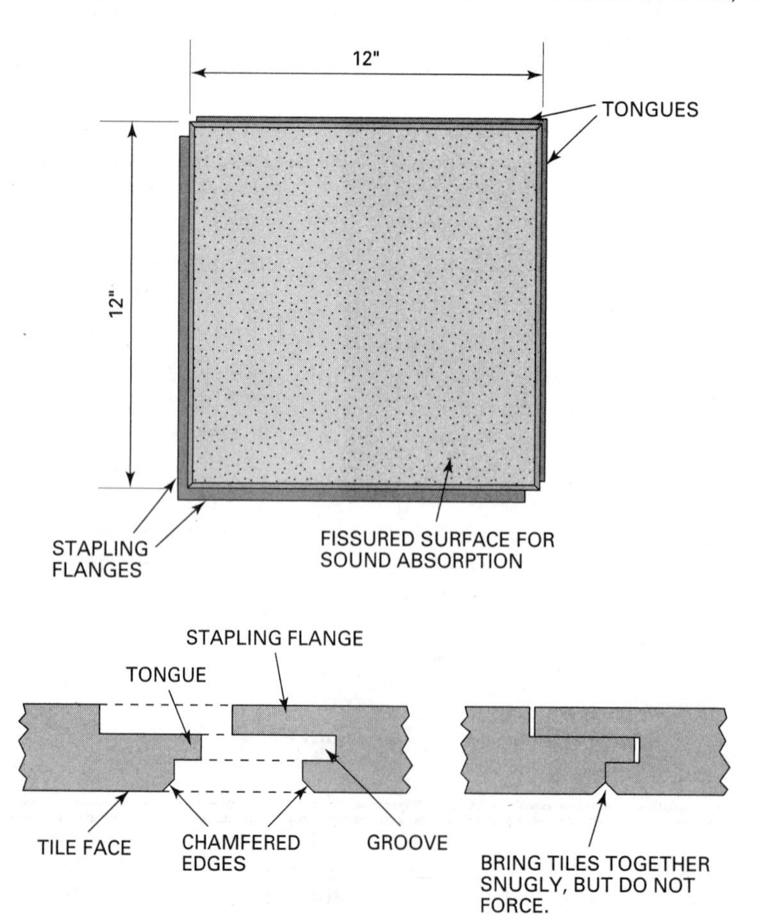

Fig. 75-2 A typical ceiling tile is tongue-and-grooved with stapling flanges.

be a number of full feet, plus a few inches. Not counting the foot measurement, add 12 more inches to the remaining inch measurement. Divide the sum by 2 to find the width of border tiles for the long wall. The following example is shown when the distance between the long walls is 10′-6″.

EXAMPLE

Room width	10′-6″
Add to inches	12″
Width of border tiles	18″ ÷ 2 = 9″

The width of border tiles along the short walls is calculated in the same manner. The following example is used when the distance between the short walls is 19′-8″.

EXAMPLE

Room length	19′-8″
Add to inches	12″
Width of border tiles	20″ ÷ 2 = 10″

Preparation for Ceiling Tile Application

Unless an adhesive application to an existing ceiling is to be made *furring strips* must be provided on which to fasten ceiling tiles. Furring strips are usually fastened directly to exposed joists. They are sometimes applied to an existing ceiling and fastened into the concealed joists above.

Locating Concealed Joists

If the joists are hidden by an existing ceiling, tap on the ceiling with a hammer. Drive a nail into the spot where a dull thud is heard, to find a concealed joist. Locate other joists by the same method or by measuring from the first location. Usually, ceiling joists are spaced 16 inches OC and run parallel to the short dimension of the room. When all joists are located, snap lines on the existing ceiling directly below and in line with the concealed joists.

Laying Out and Applying Furring Strips

For fastening 12-inch tiles, furring strips must be installed 12 inches OC. From the corner, measure out the width of the border tiles. This measurement is the center of the first furring strip away from the wall. To mark the edge, measure from the center, in either direction, half the width of the furring strip. Mark an *X* on the side of the mark toward the center of the furring strip. From the edge of the first furring strip, measure and mark, every 12 inches, across the room. Place Xs on the same side of the mark as the first one (Fig. 75-3).

Lay out the other end of the room in the same manner. Snap lines between the marks for the location of the furring strips. The strips are fastened by keeping one edge to the chalk line with the strip on the side of the line indicated by the X. Fasteners must penetrate at least 1 inch into the joist. Starting and ending furring strips are also installed against both walls.

Squaring the Room

First, snap a chalk line on a furring strip as a guide for installing border tiles against the long wall. The line is snapped parallel to, and the width of the border tiles, away from the wall.

A second chalk line must be snapped to guide the application of the short wall border tiles.

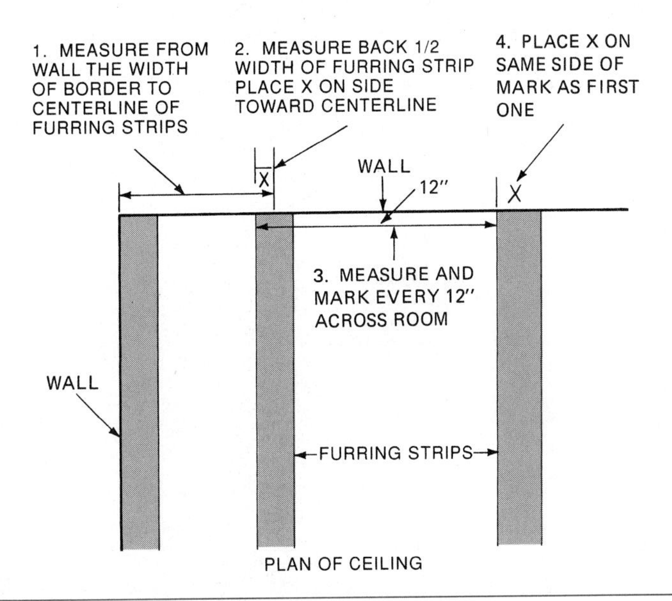

Fig. 75-3 Furring strip layout for ceiling tile.

The line must be snapped at exactly at 90 degrees to the first chalk line. Otherwise tiles will not line up properly. From the short wall, measure in along the first chalk line, the width of the short wall border tiles. From this point, use the Pythagorean Theorem and snap another line at a right angle to the first line. This method of squaring lines has been previously described in Figure 74-13 and Figure 26-4.

Installing Ceiling Tile

Ceiling tiles should be allowed to adjust to normal interior conditions for twenty-four hours before installation. Some carpenters sprinkle talcum powder or corn starch on their hands to keep them dry. This prevents fingerprints and smudges on the finished ceiling. Cut tiles face up with a sharp utility knife guided by a straightedge. All cut edges should be against a wall.

Starting the Installation

To start the installation, cut a tile to fit in the corner. The outside edges of the tile should line up exactly with both chalk lines. Because this tile fits in the corner, it must be cut to the size of border tiles on both long and short sides of the room. For example, if the border tiles on the long wall are 9 inches and those on the short wall are 10 inches, the corner tile should be cut twice to make it 9 × 10. Staple the tile in position. Be careful to line up the outside edges with both chalk lines (Fig. 75-4).

Completing the Installation

After the corner tile is in place, work across the ceiling. Install two or three border tiles at a time. Then fill in with full-sized field tiles (Fig. 75-5). Tiles are applied so they are snug to each other, but not jammed tightly. Fasten each tile with four 1/2- or 9/16-inch staples. Place two in each flange, using a hand stapler. Use six staples in 12 × 24 tiles. Continue applying tiles in this manner until the last row is reached.

When the last row is reached, measure and cut each border tile individually. Cut the tiles slightly less than measured for easy installation. Do not force the tiles in place. Face nail the last tile in the corner near the wall where the nailhead will be covered by the wall molding. After all tiles are in place, the ceiling is finished by applying molding between the wall and ceiling. (The application of wall molding is discussed in a later unit.)

Adhesive Application of Ceiling Tile

Ceiling tile is sometimes cemented to existing plaster and drywall ceilings. These ceilings must be completely dry, solid, level, and free of dust and dirt. If the existing ceiling is in poor condition or has loose paint, adhesive application is not recommended.

Special tile adhesive is used to cement tiles to ceilings. Four daubs of cement are used on 12 × 12 tile (Fig. 75-6). Six daubs are used on 12 × 24 tile. Before applying the daubs, prime each spot by using the putty knife to force a thin layer into the back

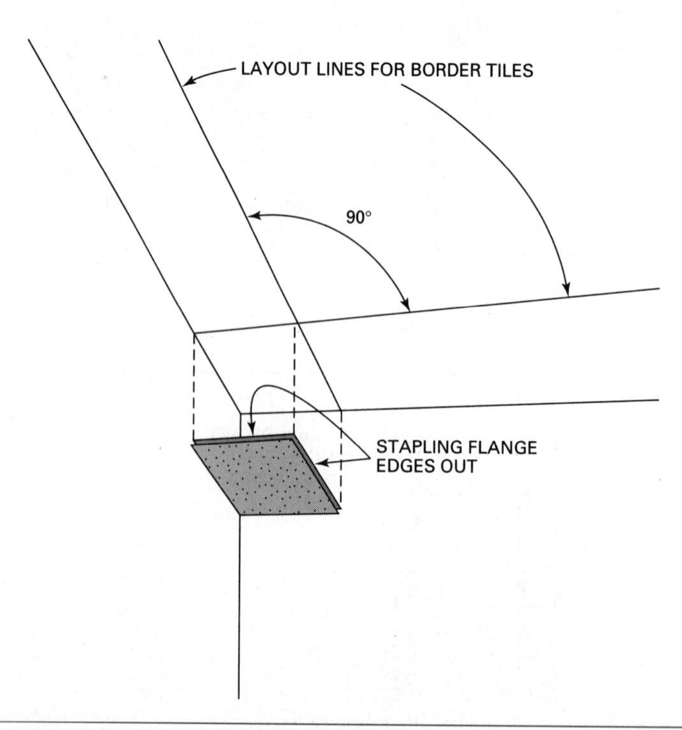

Fig. 75-4 Install the first tile in the corner with its outside edges to the layout lines.

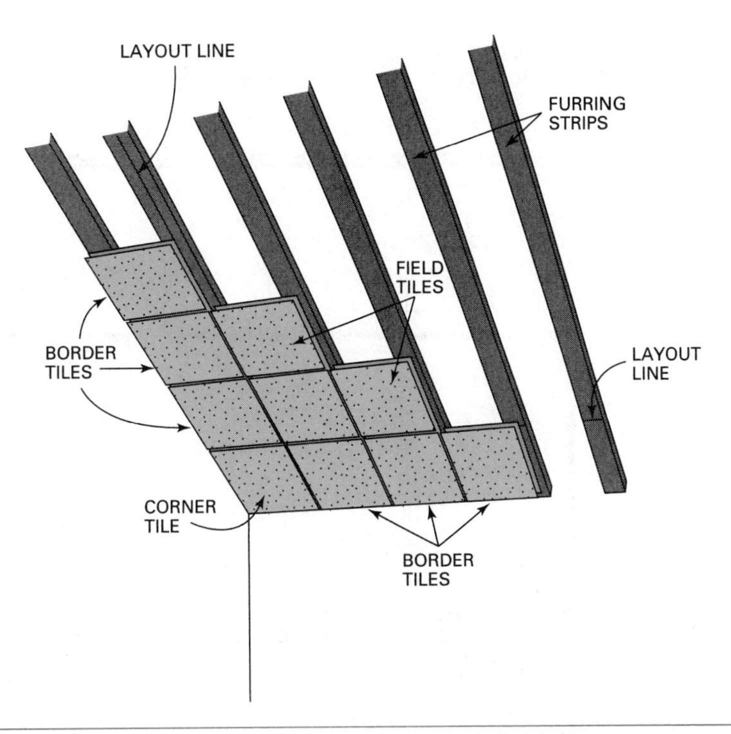

Fig. 75-5 Install a few border tiles. Then fill in with full-sized field tiles.

surface of the tile. Apply the daubs, about the size of a walnut. Press the tile into position. Keep the adhesive away from the edges to allow for spreading when the tile is pressed into position. No staples are required to hold the tile in place.

Easy Up® Installation System

Ceiling tile may also be installed using a patented system called Easy Up®. This system consists of specially shaped metal tracks and clips that are attached to the tracks to hold the ceiling tile (Fig. 75-7).

Fig. 75-6 Applying adhesive to the back of ceiling tiles.

Installing Tracks

Border tile sizes are determined in the same manner as previously described. If ceiling joists are hidden, they must be located. The tracks must be screwed or nailed at right angles to the joists. The procedure for installing the tracks is shown in Figure 75-8.

Installing Tile

Two lines must be stretched or snapped at right angles to each other to locate the inner edges of the border tiles. The installation is started in the corner of the room where the layout lines intersect. The procedure for installing tile with the Easy Up® system is described in Figure 75-9.

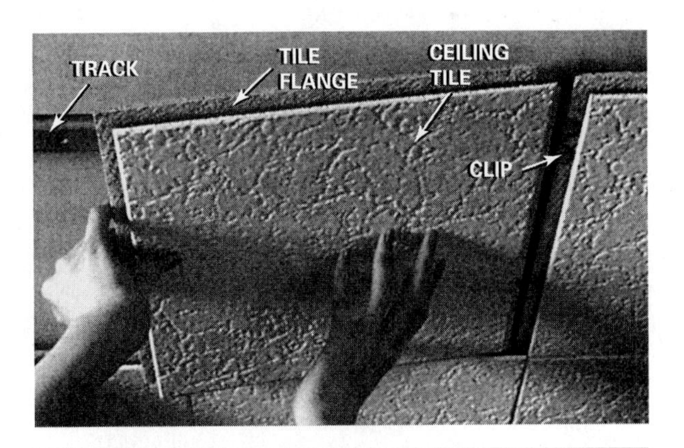

Fig. 75-7 Installing ceiling tile using the Easy Up® system. (*Courtesy of Armstrong World Industries*)

STEP BY STEP

PROCEDURES

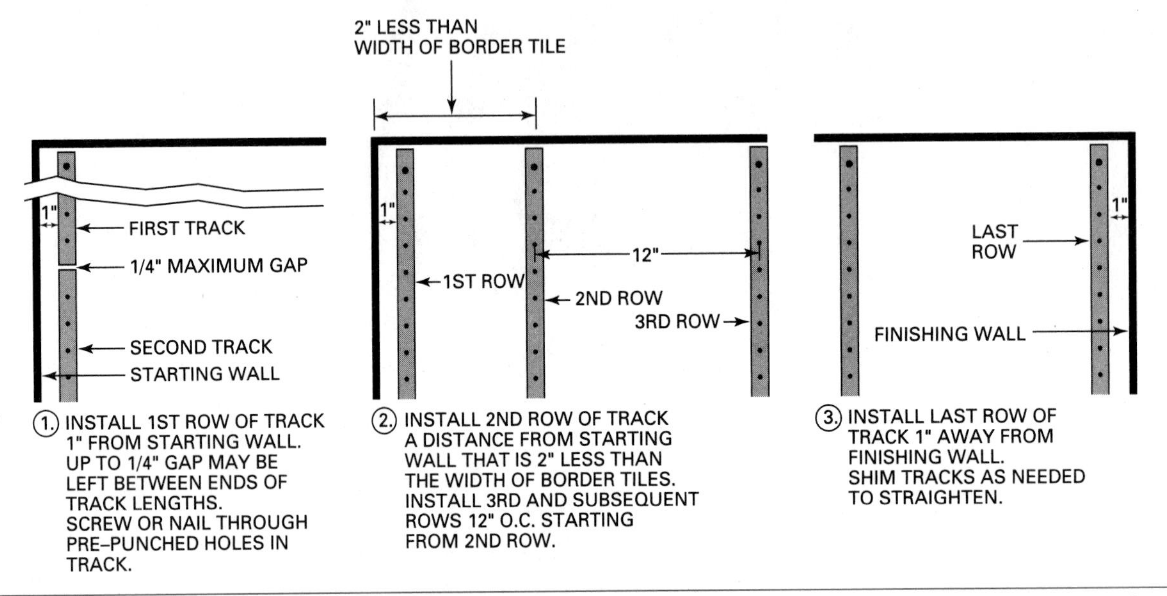

Fig. 75-8 Procedure for installing Easy Up® metal tracks. (*Courtesy of Armstrong World Industries*)

STEP BY STEP

PROCEDURES

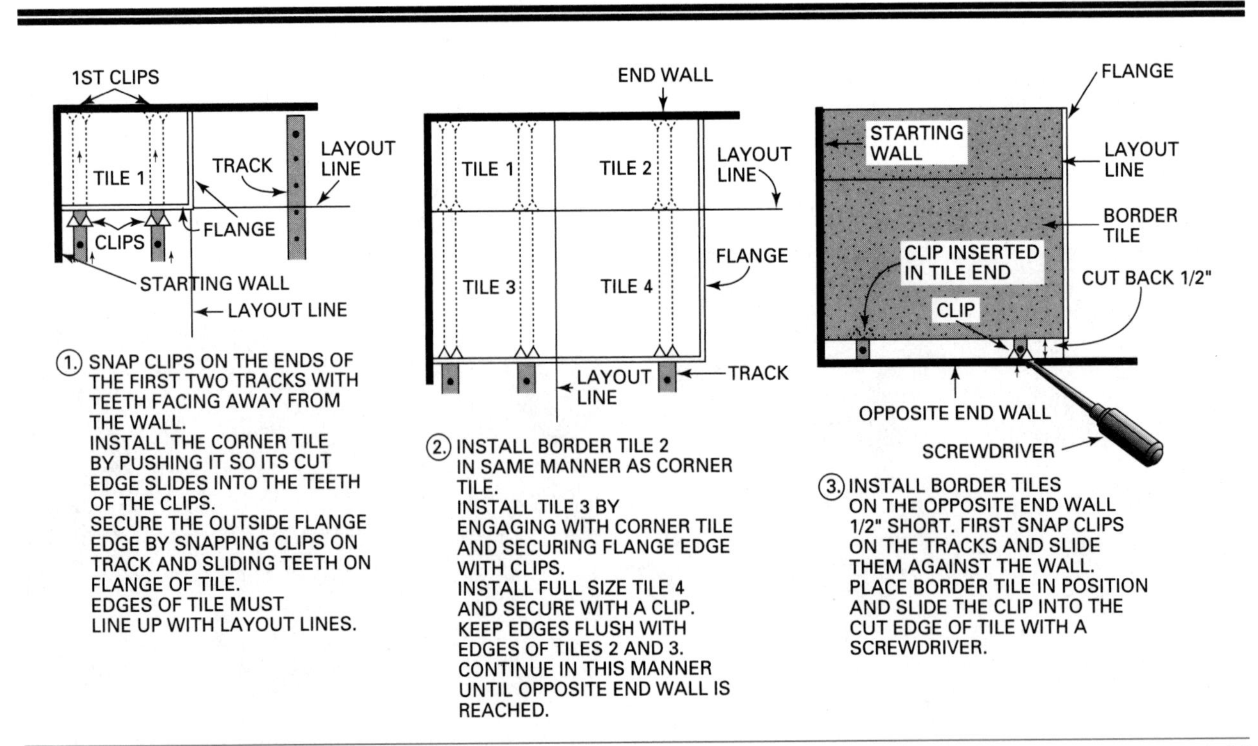

Fig. 75-9 Procedure for installing ceiling tile using the Easy Up® system. (*Courtesy of Armstrong World Industries*)

Estimating Materials for Tile Installation

To estimate ceiling tile, measure the width and length of the room to the next whole foot measurement. Multiply these figures together to find the area of the ceiling in square feet. Divide the ceiling area by the number of square feet contained in one of the ceiling tiles being used to find the number of ceiling tiles needed.

To estimate furring strips, measure to the next whole foot the length of the room in the direction that the furring strips are to run. Multiply this by the number of rows of furring strips to find the total number of linear feet of furring strip stock needed. To find the number of rows, divide the width of the room by the furring strip spacing and add one.

Review Questions

Select the most appropriate answer.

1. The most common sizes in inches of ceiling tile are
a. 8 × 12 and 12 × 12.
b. 12 × 12 and 16 × 16.
c. 12 × 12 and 12 × 24.
d. 14 × 24 and 24 × 48.

2. The most common sizes in inches of lay-in suspended ceiling panels are
a. 8 × 12 and 12 × 12.
b. 12 × 12 and 16 × 16.
c. 12 × 12 and 12 × 24.
d. 24 × 24 and 24 × 48.

3. The width in inches of border tiles for a room 14′-9″ × 22′-3″, when 12 × 12 inch tiles are used, is
a. 9 and 3.
b. 8 1/2 and 6 1/2.
c. 10 1/2 and 7 1/2.
d. 11 and 8.

4. The width in inches of border panel for a room 12′-6″ × 18′-8″, when 24″ × 24″ suspended ceiling panels are used, is
a. 6 and 8.
b. 10 and 12.
c. 14 1/2 and 18 1/2.
d. 15 and 16.

5. When squaring a room, use measurements of
a. 2, 3, and 4 feet.
b. 3, 4, and 5 feet.
c. 4, 5, and 6 feet.
d. 5, 6, and 7 feet.

6. To locate joists above an existing ceiling
a. determine the direction of the finished floor above.
b. tap on the ceiling with a hammer until a dull thud is heard.
c. look for lines of discoloration on the existing ceiling.
d. determine the location of studs in the wall below.

7. Nails used to fasten furring to ceiling joists must penetrate into the joists at least
a. 3/4 inch.
b. 1 inch.
c. 1 1/4 inches.
d. 1 1/2 inches.

8. Fastening 12″ × 12″ tiles requires
a. two staples.
b. three staples.
c. four staples.
d. six staples.

9. In a suspended ceiling, hanger wire is used to suspend
a. cross tees.
b. main runners.
c. wall angle.
d. furring strips.

10. The first step in installing a suspended ceiling is to
a. square the room.
b. make a sketch of the planned ceiling.
c. calculate border tiles.
d. install the wall angle.

BUILDING FOR SUCCESS

The Builder and Community Involvement

The professional builder has established an active, quality business in a community. This reputation stems from a desire to help people realize their dream of home ownership and continues on throughout an entire career. During their careers, builders become very important parts of their communities, as people benefit from their voluntary services.

The builder, like many others, is responsible for assuring that various populations are assisted and supported in their quest to improve their living standards. Builders represent a source of expertise and strength that can help solve both existing and potential roadblocks to community development. All construction-related businesses in each community should view their position as one of great potential. By helping those who lack the expertise, time, and financial resources to upgrade their living conditions, building trade representatives can make a valuable contribution toward community development.

With the acknowledged responsibility to assist in dealing with housing problems, health concerns, inadequate community facilities, needy children, and homeless families, the builder has a great opportunity to become involved in community concerns. One of the most accessible avenues to assist with community needs is through involvement with the local home builders' association.

Housing associations at the national, state, and local levels are but one of many organizations that assist people in solving housing-related problems. The builder, subcontractor, retiree, self-employed person, businessperson, and citizen who have the desire to offer assistance to the community may unite to tackle local problems that are obstacles to community improvement.

The person who has a construction-related business must be service-oriented. The desire to provide housing, remodeling, and products to a community must not stop with the issue of payment for each service rendered. The competitive builder will show great vision for the future by looking ahead to how the community will grow during the next one or two decades. This vision will involve participating in the work that attacks immediate and future constructed-related problems. With this investment of time and resources will come visibility, credibility, and integrity for those who participate.

Most local builders' associations have identified community needs that they can participate in and help provide solutions to. Habitat for Humanity projects and Christmas toy projects for local homeless or handicapped children are popular. Teaming up with other local organizations such as police and fire departments has assisted many diverse groups with health or housing needs. As these groups undertake united efforts to assist others, they identify community needs, form common goals, and work toward reasonable solutions in solving problems.

For example, the Home Builders' Association of Lincoln, Nebraska, annually works with Habitat for Humanity. This project assembles a wide variety of builders, materials suppliers, volunteer workers (including college National Association of Home Builders student chapters), businesses, lending agencies, and community organizations to assist families that cannot afford housing by traditional means. Builders and business representatives take time from their busy schedules to head up efforts of this type in order to help less fortunate people gain access to better housing and higher living standards.

Development projects of this magnitude are a high priority for each community. Local elected leaders such as the mayor, city council members, or other administrators should be visibly involved in all ongoing efforts to improve living conditions. There will always be a need to involve as many people as possible in each community for goals to be reached. The success of any national, state, or local development project can be evaluated by the involvement of those who assist others. There will be a continual need for builders to recognize their place in the community as quality living conditions are established and health hazards are eliminated. The element of charity has not lost its appeal or its place in a community. People will always need to assist others in the constant effort to raise living conditions. This brings into play the often neglected realm of responsibility to do all we can to help each other.

The access to better living conditions in the future will be based on the actions we take in our communities today. What initially appears to be an expense in time or finances up front usually turns out to be an intelligent investment. We must remember that potential clients make contact on the basis of what they know about each builder. The builder who invests time and finances in his or her community may be the recipient of unexpected business opportunities. Each builder and tradesperson has an opportunity to contribute something to his or her community. The decision to get involved can be made at any time.

FOCUS QUESTIONS: For individual or group discussion

1. What considerations must a person address before volunteering a service to the community? Are there advantages and disadvantages? Evaluate any.
2. As you prepare for a career in the construction industry, how do you see your future participation in volunteer community service? What is appealing and what is not?
3. What are some benefits to members of the community who are recipients of volunteer service by builders and others in the construction trades?
4. What are some local community housing problems that you as a construction student could help solve? How would you do this?

UNIT

26

Interior Doors and Door Frames

Interior doors used in residential and light commercial buildings are less dense than exterior doors. They are not ordinarily subjected to as much use and are not exposed to the weather. In commercial buildings, such as hospitals and schools, heavier and larger interior doors are specified that meet special conditions.

OBJECTIVES

After completing this unit, the student should be able to:

- describe the sizes and kinds of interior doors.
- make and set interior door frames.
- hang an interior swinging door.
- install locksets on interior swinging doors.
- set a prehung door and frame.
- install bypass, bifold, pocket, and folding doors.

UNIT CONTENTS

CHAPTER 76 DESCRIPTION OF INTERIOR DOORS

Interior doors are classified by style as *flush, panel, French, louver,* and *cafe* doors. Interior flush doors have a smooth surface, are usually less expensive, and are widely used when a plain appearance is desired. Some of the other styles have special uses. Doors are also classified by the way they operate, such as *swinging, sliding,* or *folding.*

Interior Door Sizes and Styles

For residential and light commercial use, most interior doors are manufactured in 1 3/8-inch thickness. Some, like cafe and bifold doors, are also made in 1 1/4- and 1 1/8-inch thicknesses. Most doors are manufactured in 6′-8″ heights. Some types may be obtained in heights of 6′-0″ and 6′-6″. Door widths range from 1′-0″ to 3′-0″ in increments of 2 inches. However, not all sizes are available in every style.

Flush Doors

Flush doors are made with *solid* or *hollow* cores. Solid core doors are generally used as entrance or fire-rated doors. (They have been previously described in Unit 20, Exterior Doors.) Hollow core doors are commonly used in the interior except when fire resistance or sound transmission is critical.

A hollow core door consists of a light perimeter frame. This frame encloses a mesh of thin wood or composition material supporting the faces of the door. Solid wood blocks are appropriately placed in the core for the installation of locksets. The frame and mesh are covered with a thin plywood called a *skin. Lauan* plywood is used extensively for flush door skins. Flush doors are also available with veneer faces of *birch, gum, oak,* and *mahogany,* among others (Fig. 76-1). When flush doors are to be painted, an overlay plywood or tempered hardboard may be used for the skin.

Panel Doors

Interior *panel doors* consist of a frame with usually one to eight wood panels in various designs (Fig. 76-2). They are similar in style to some exterior panel doors. (The construction of panel doors has been previously described in Unit 20, Exterior Doors.)

French Doors

French doors may contain from one to fifteen lights of glass. They are made in a 1 3/4-inch thickness for exterior doors and 1 3/8-inch thickness for interior doors (Fig. 76-3).

Louver Doors

Louver doors are made with spaced horizontal slats called louvers used in place of panels. The louvers

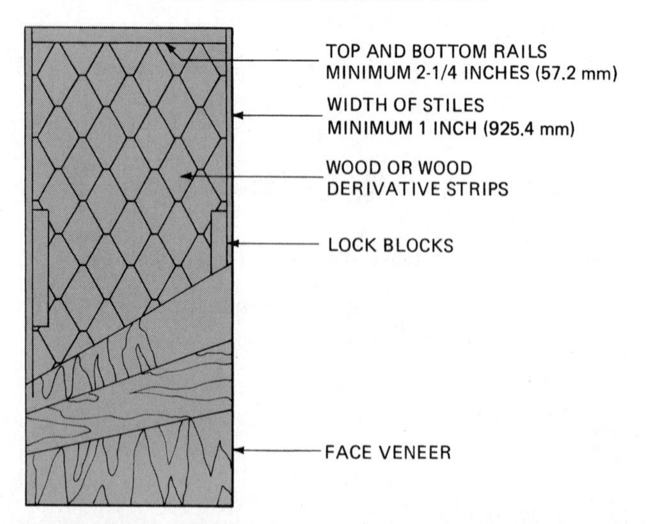

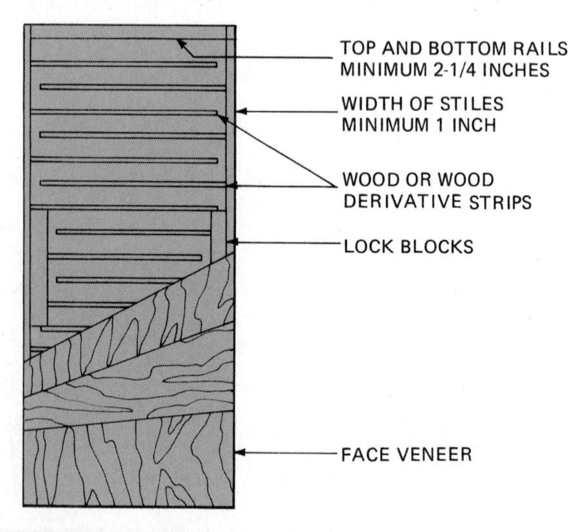

Fig. 76-1 The construction of hollow core flush doors.

Fig. 76-2 Styles of commonly used interior panel doors. (*Courtesy of Morgan Manufacturing*)

Fig. 76-3 French doors are used in the interior as well as for entrances. (*Courtesy of Morgan Manufacturing*)

are installed at an angle to obstruct vision but permit the flow of air through the door. Louvered doors are widely used on clothes closets (Fig. 76-4).

Cafe Doors

Cafe doors are short panel or louver doors. They are hung in pairs that swing in both directions. They are used to partially screen an area, yet allow easy and safe passage through the opening. The tops and bottoms of the doors are usually shaped in a pleasing design (Fig. 76-5).

Fig. 76-4 Louver doors obstruct vision, but permit the circulation of air. (*Courtesy of Morgan Manufacturing*)

Methods of Interior Door Operation

Doors are also identified by their method of operation. Doors either swing on hinges or slide on tracks. The choice of door operation depends on such factors as convenience, cost, safety, and space.

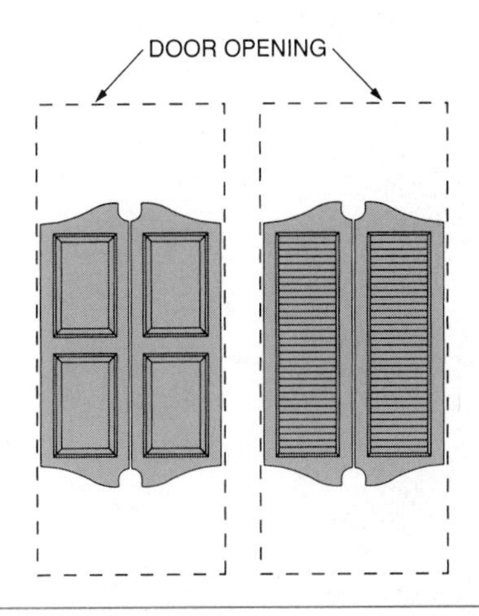

Fig. 76-5 Cafe doors usually are used between kitchens and dining areas. (*Courtesy of Morgan Manufacturing*)

Swinging Doors

Swinging doors are hinged on one edge. They swing out of the opening. When closed, they cover the total opening. Swinging doors that swing in one direction are called *single-acting*. With special hinges, they can swing in both directions. They are then called **double-acting** doors (Fig. 76-6). Swinging doors are the most commonly used type of door. They have the disadvantage of requiring space for the swing.

Bypass Doors

Bypass doors are commonly used on wide clothes closet openings. A double track is mounted on the header jamb of the door frame. Rollers that ride in the track are attached to the doors so that they slide by each other. A floor guide keeps the doors in alignment at the bottom (Fig. 76-7). Usually two doors are used in a single opening. Three or more doors may be used, depending on the situation.

The disadvantage of bypass doors is that although they do not project out into the room, access to the complete width of the opening is not possible. They are easy to install, but are not practical in openings less than 6 feet wide.

Fig. 76-7 Bypass doors are used on wide closet openings.

Pocket Doors

The *pocket door* is opened by sliding it sideways into the interior of the partition. When opened, only the lockedge of the door is visible (Fig. 76-8). Pocket doors may be installed as a single unit, sliding in one direction, or as a double unit sliding in opposite directions. When opened, the total width of the opening is obtained, and the door does not project out into the room. Pocket doors are used when these advantages are desired.

Fig. 76-6 A single-acting swinging door is the most widely used type of interior door.

Fig. 76-8 The pocket door slides into the interior of the partition.

The installation of pocket doors requires more time and material than other methods of door operation. A special pocket door frame unit and track must be installed during the rough framing stage (Fig. 76-9). The rough opening in the partition must be large enough for the door opening and the pocket.

Bifold Doors

Bifold doors are made in flush, panel, louver, or combination panel and louver styles. They are made in narrower widths than other doors. This allows them to operate in a folding fashion on closet and similar type openings (Fig. 76-10).

Bifold doors consist of pairs of doors hinged at their edges. The doors on the jamb side swing on pivots installed at the top and bottom. Other doors fold up against the jamb door as it is swung open. The end door has a guide pin installed at the top. The pin rides in a track to guide the set when opening or closing (Fig. 76-11). On very wide openings the guide pin is replaced by a combination guide and support to keep the doors from sagging.

Bifold doors may be installed in double sets, opening and closing from each side of the opening. They have the advantage of providing access to almost the total width of the opening. Yet they do not project out much into the room.

Fig. 76-9 A pocket door frame comes preassembled from the factory. It is installed when the interior partitions are framed.

Folding Doors

Folding doors are similar to bifold doors except they consist of many narrow panels. Each panel is about the same width as the door jamb. The entire door assembly is supported by rollers that slide in an overhead track. When opened, the panels stack against each other on side jambs. A folding door requires only a small amount of space and does not project out into the room (Fig. 76-12). Folding doors are used frequently for wardrobe closets, laundry rooms, and storage spaces.

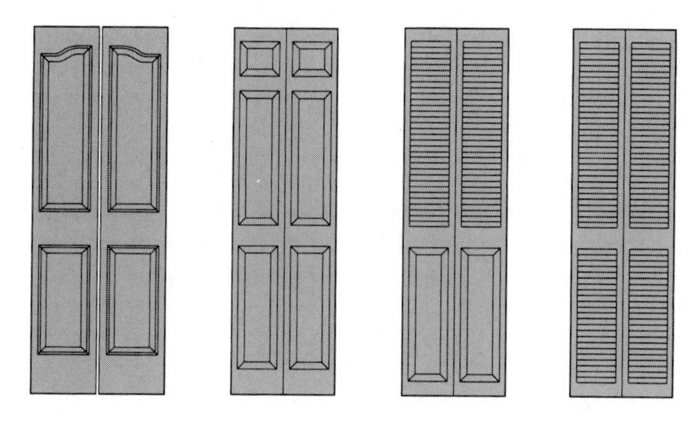

Fig. 76-10 Bifold doors are manufactured in many styles. (*Courtesy of Morgan Manufacturing*)

Fig. 76-11 Bifold doors provide access to almost the total width of the opening.

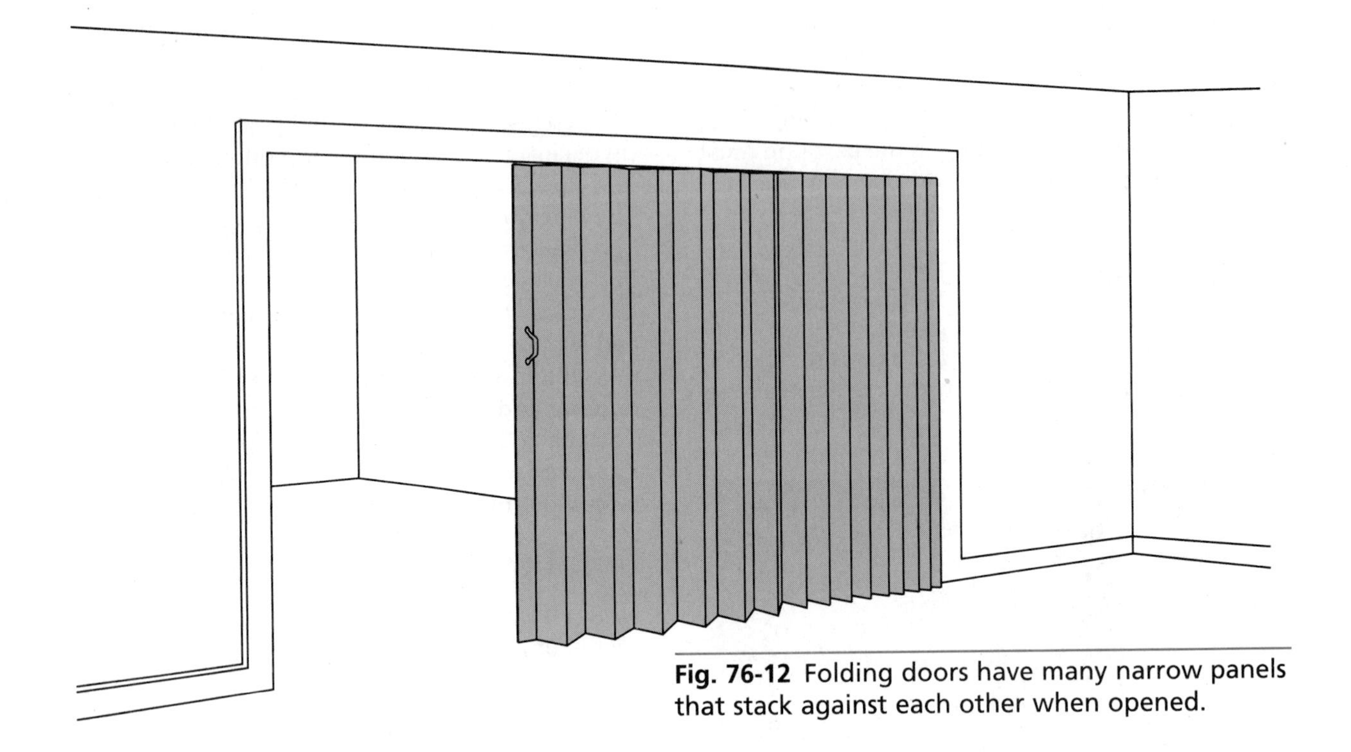

Fig. 76-12 Folding doors have many narrow panels that stack against each other when opened.

CHAPTER 77 INSTALLATION OF INTERIOR DOORS AND DOOR FRAMES

Many interior doors come *prehung* in their frames for easier and faster installation on the job. However, it is often necessary to build and set the door frame, hang the door, and install the locksets.

Interior Door Frames

Special *rabbeted* jamb stock or nominal 1-inch square-edge lumber is used to make interior *door frames.* If square-edge lumber is used, a separate *stop,* if needed, is applied to the inside faces of the door frame.

Checking Rough Openings

The first step in making an interior door frame is to measure the door opening to make sure it is the correct width and height. The rough opening width for single-acting swinging doors should be the width of the door plus twice the thickness of the side jamb, plus 1/2 inch for shimming between the door frame and the opening. For example, if the thickness of the side jamb beyond the door is 3/4 inch, the rough opening width is 2 inches more than the door width.

The rough opening height should be the height of the door, plus the thickness of the header jamb, plus 1/4 inch for clearance at the top, plus the thickness of the finished floor, plus a desired clearance under the door. An allowance of 1/2 inch is usually made for clearance between the finished floor and the door.

For example, if the header jamb and finished floor thickness are both 3/4 inch, the rough opening height should be 2 1/4 inches over the door height, if 1/2 inch is allowed for clearance under the door (Fig. 77-1).

The rough opening size for other than single-acting swinging doors, such as bypass and bifold doors, should be checked against the manufacturer's directions. The sizes of the doors and allowances for hardware may differ with the manufacturer.

Making an Interior Door Frame

Interior door frames are constructed like exterior door frames except they have no sill. Interior door frames usually are installed after the interior wall

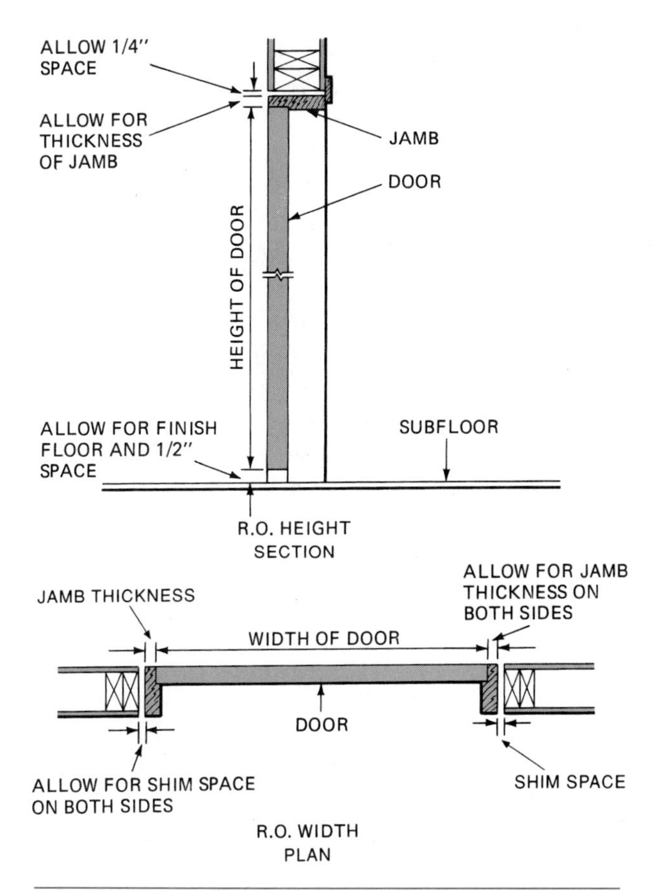

Fig. 77-1 Check the size of rough openings for swinging doors.

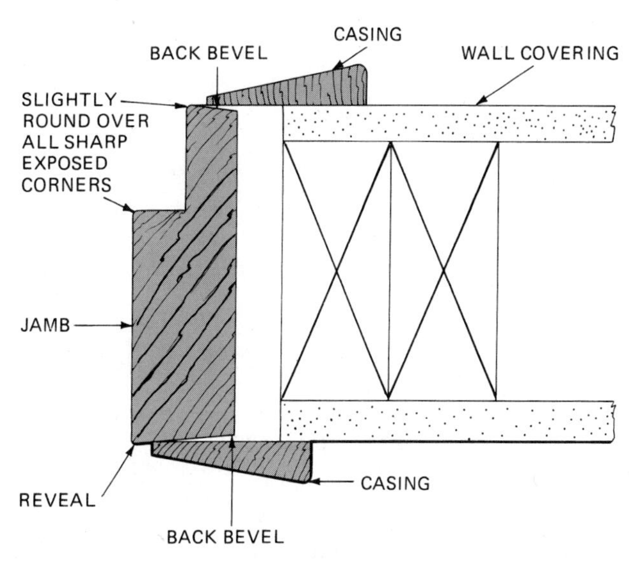

Fig. 77-2 Back-bevel jamb edges slightly to permit casings to fit snugly against them.

covering has been applied. Measure the total thickness of the wall, including the wall covering, to find the jamb width.

Cutting Jambs to Width

If necessary, rip the door jamb stock so its width is the same as the wall thickness. If rabbeted jamb stock is used, cut the edge opposite the rabbet. Plane and smooth both edges to a slight *back bevel.* The back bevel permits the door casings, when later applied, to fit tightly against the edges of the door frame in case there are irregularities in the wall (Fig. 77-2). *Ease* all sharp exposed corners.

Cutting Jambs to Length

On interior door frames, the *head jamb* is usually cut to fit between, and is dadoed into, the *side jambs.* The side jambs run the total height of the rough opening. Cut both side jambs to a length equal to the height of the opening.

Head Jambs of Door Frames for Hinged Doors. If rabbeted jambs are used, the length of the head jamb is the width of the door plus 3/16 inch. The extra 3/16 inch is for joints of 3/32 inch on each side, between the edges of the door and the side jambs. If square-edge lumber is used, its length is the same as a rabbeted head jamb. However, 1/2 inch is added for dadoing 1/4-inch deep into each side jamb (Fig. 77-3). Cut the head jamb to length with both ends square.

Side Jambs of Door Frames for Hinged Doors. Measure up from the bottom ends. Square lines across the side jambs to mark the location of the bottom side of the head jamb. This dimension is the sum of:

- the thickness of the finish floor, if the door frame rests on the subfloor
- an allowance of 1/2 inch minimum between the door and the finish floor
- the height of the door
- 3/32 inch for a joint between the door and the head jamb
- on rabbeted jambs, subtract 1/2 inch for the depth of the rabbet.

Hold a scrap piece of jamb stock to the squared lines. Mark its other side to lay out the width of the dado. Mark the depth of the dadoes on both edges of the side jambs. Cut the dadoes to receive the head jamb. On rabbeted jambs, dado depth is to the face of the rabbet. A dado depth of 1/4 inch is sufficient on plain jambs (Fig. 77-4).

Jamb Lengths for Other Types of Doors. The length of head and side jambs for other types of doors, such as bypass and bifold, must be determined from instructions provided by the manufacturer of the hardware and the door.

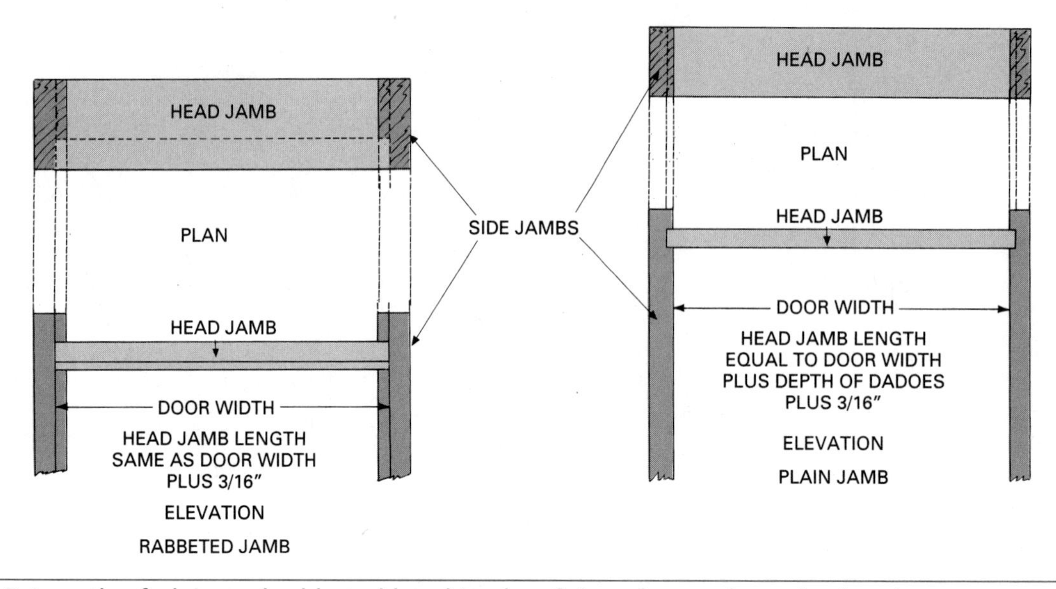

Fig. 77-3 Length of plain and rabbeted head jambs of door frames for swinging doors.

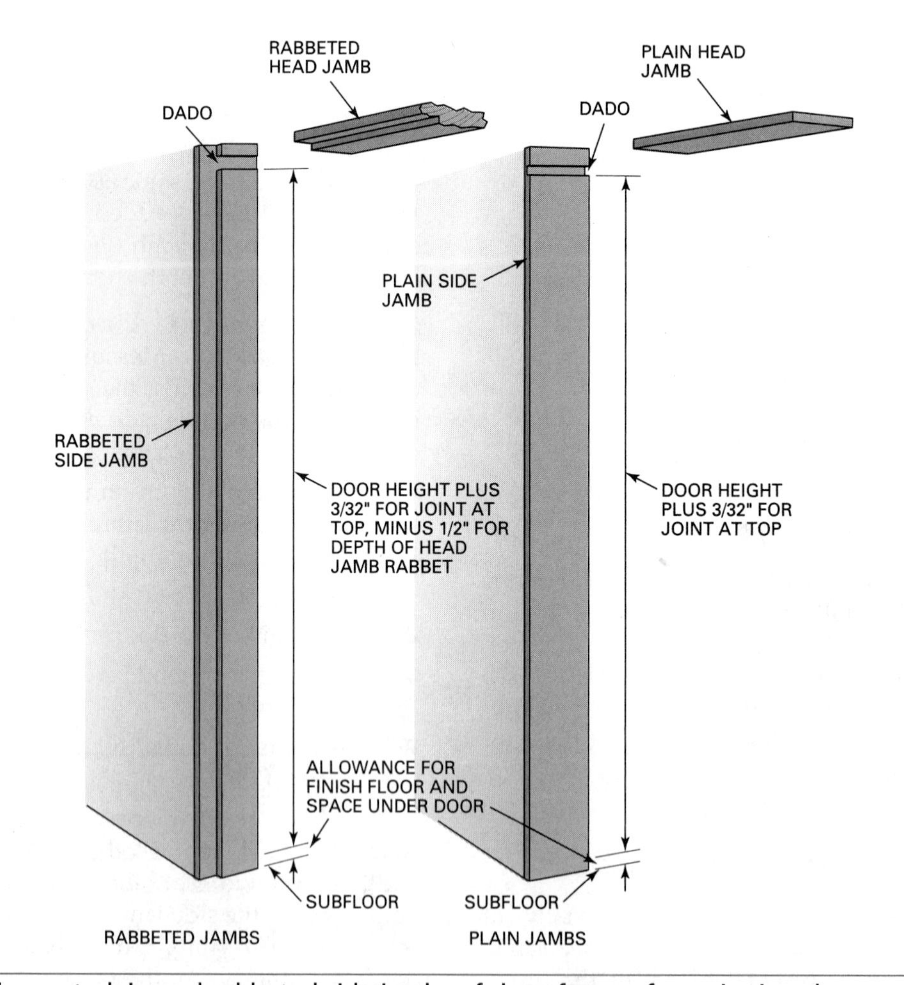

Fig. 77-4 Laying out plain and rabbeted side jambs of door frames for swinging doors.

Door hardware and door sizes differ with the manufacturer. This affects the length of the door jambs.

Plain, square-edge lumber jambs are used to make door frames for doors other than single-acting swinging doors. The rest of the procedures, such as checking rough openings, cutting jambs to width, assembling, and setting, is the same for all door frames.

Assembling the Door Frame

Fasten the side jambs to the head jamb, keeping the edges flush. If there is play in the dado, first wedge the head jamb with a chisel so the face side comes up tight against the dado shoulders before fastening.

Cut a narrow strip of wood. Tack it to the side jambs a few inches up from the bottom so that the frame width is the same at the bottom as it is at the top. This strip is commonly called a *spreader* (Fig. 77-5).

Setting the Door Frame

Door frames must be set so that the jambs are straight, level, and plumb. They are usually set before the finish floor is laid. If a rabbeted frame is used, determine the swing of the door so that the rabbet is facing toward the correct side.

Cut any *horns* from the top ends of side jambs. Place the frame in the opening. The horns are cut off in case the side jambs need to be shimmed to level the head jamb.

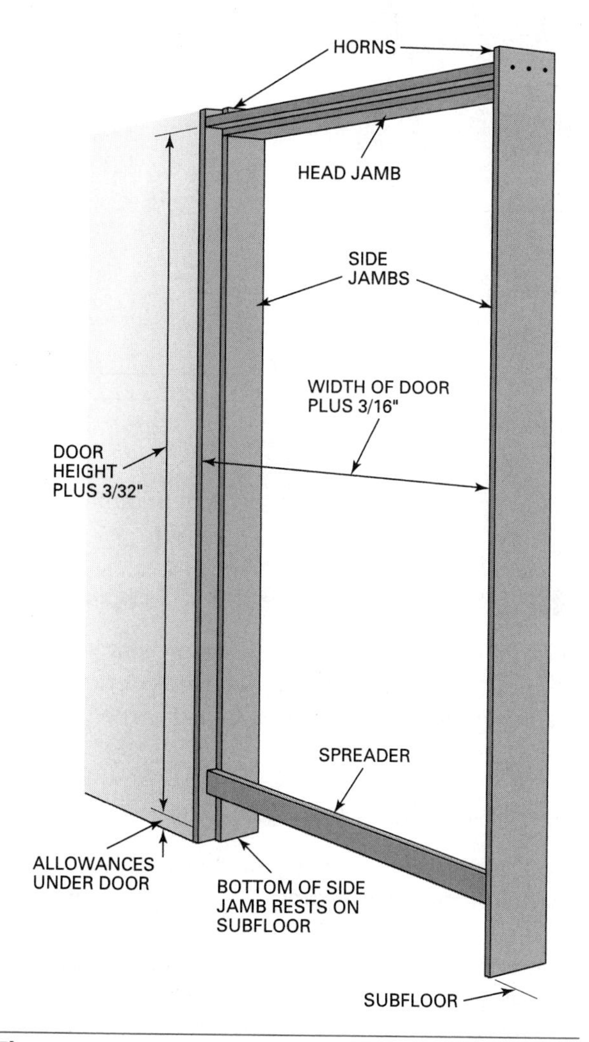

Fig. 77-5 An assembled interior rabbeted door frame.

Install shims directly opposite the ends of the header jamb between the opening and the side jambs. Shim an equal amount on both sides so that the frame is close to being centered at the top of the opening. Drive shims up snugly.

Leveling the Header Jamb

Keep the edges of the frame flush with the wall. With the bottom ends of both side jambs resting on the subfloor, check whether the head jamb is level. Level the head jamb, if necessary, by placing shims between the bottom end of the appropriate side jamb and the subfloor. When the header jamb is level, tack the frame in place on both sides, close to the top. Drive fasteners through side jambs and shims into the studs.

If the door frame rests on a finish floor, then a tight joint must be made between the bottom ends of the side jambs and the finish floor. If the floor is level, side jambs will fit the floor if their lengths are exactly the same and their bottom ends have been cut square. Once the frame is set, the head jamb should be level when the ends of the side jamb are resting on the floor.

If the floor is not quite level, or if the side jambs are of unequal length, level the head jamb by shimming under the bottom end of the side jamb on the low side. Set the dividers for the amount the jamb has been shimmed. Scribe that amount on the bottom of the opposite side jamb. Remove the frame from the opening. Cut to the scribed line. Replace the frame in the opening. The head jamb should be level. The bottom ends of both side jambs should fit snugly against the finish floor (Fig. 77-6).

Plumbing the Side Jambs

Several ways of plumbing door frames have been previously described. An accurate and fast method is with the use of a 6-foot level. When one side jamb is plumb, shim and tack its bottom end in place. Only one side needs to be plumbed. Locate the bottom end of the other side jamb by measuring across. The door frame width at the bottom should be the same as on the top. Shim and tack the bottom end of the other side jamb in place.

Straightening Jambs

Use a 6-foot straightedge against the side jambs. Straighten them by shimming at intermediate points. Besides other points, shims should be placed opposite hinge and lockset locations. Fasten the jambs by nailing through the shims. Header jambs on wide door frames are straightened, shimmed, and fastened in a similar way.

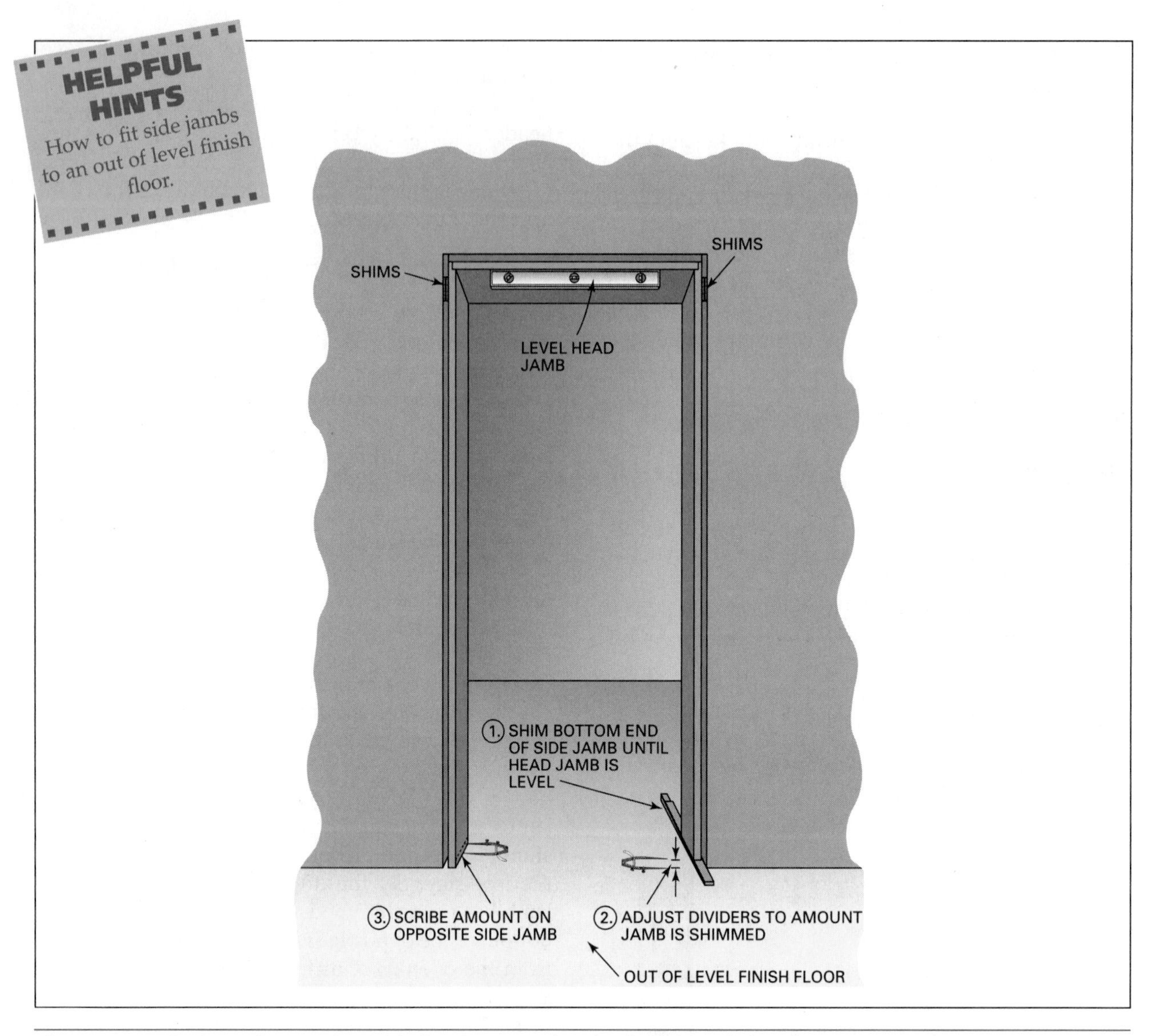

Fig. 77-6 Technique for cutting side jambs to make the head jamb level.

Sighting the Door Frame for a Wind

Before any nails are driven home, sight the door frame to see if it has a *wind.* The frame must be sighted by eye to make sure that side jambs line up vertically with each other and that the frame is not twisted. This is important when installing rabbeted jambs. The method of checking for a wind in door frames has been previously described (see Fig. 57-20).

If the frame has a wind, move the top or bottom ends of the side jambs slightly until they line up with each other. Fasten top and bottom ends of the side jambs securely. Set the nails (Fig. 77-7).

Applying Door Stops

At this time, *door stops* may be applied to plain jambs. The stops are not permanently fastened, in case they have to be adjusted when locksets are installed. A **back miter** joint is usually made between molded side and header stops. A *butt* joint is made between square-edge stops (Fig. 77-8).

Hanging Interior Doors

The method of fitting and hanging single-acting, hinged, interior doors is the same for exterior doors. See Chapter 57, Door Frame Construction and Installation.

Double-Acting Doors

Double-acting doors are installed with either special pivoting hardware installed on the floor and the head jamb or with spring-loaded double-acting hinges. Both types return the door to a closed position after being opened. When opened wide, the doors can be held in the open position. A different type of light-duty, double-acting hardware is used on *cafe* doors. To install double-acting door hardware, follow the manufacturer's directions.

STEP BY STEP
PROCEDURES

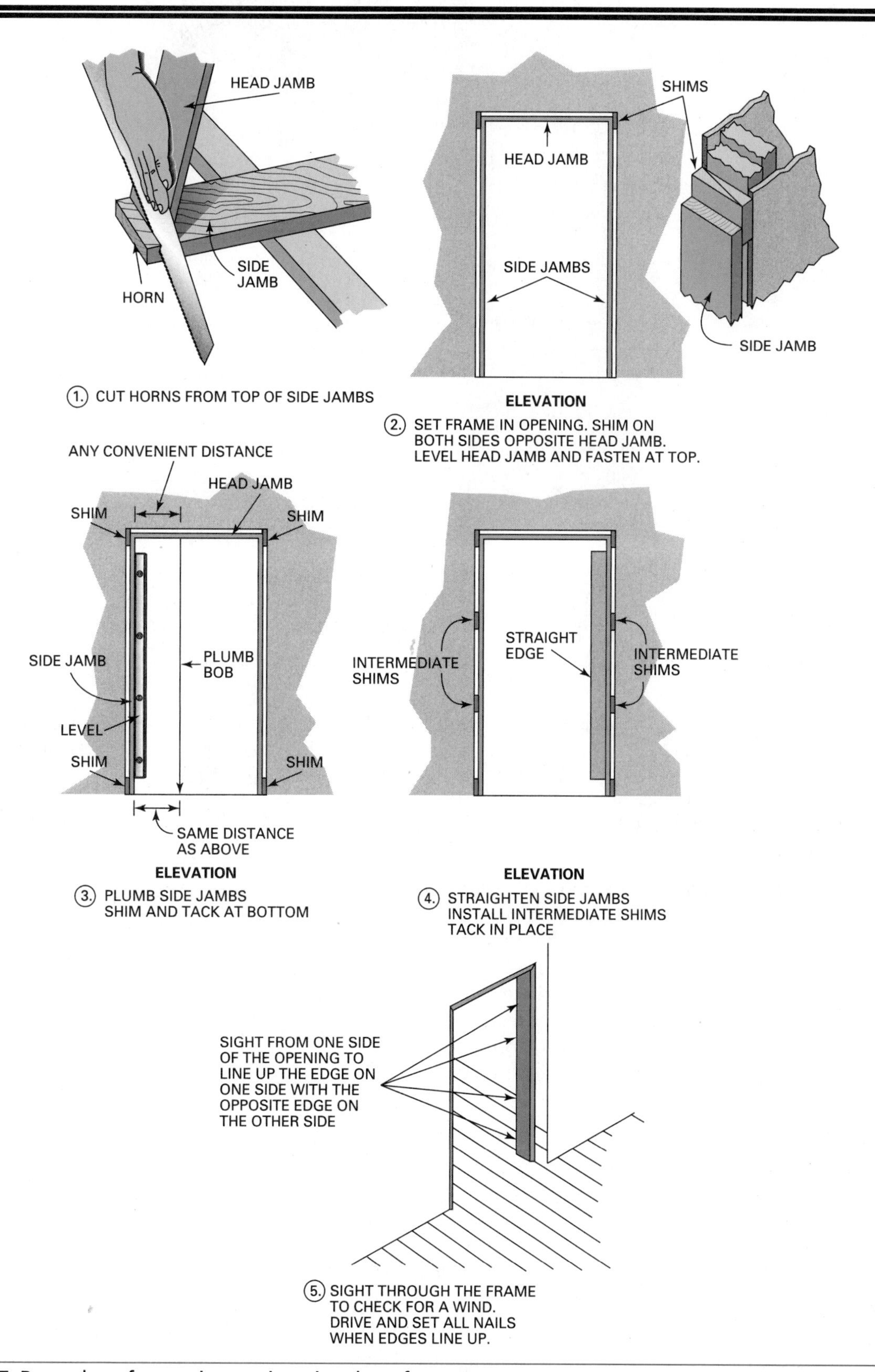

Fig. 77-7 Procedure for setting an interior door frame.

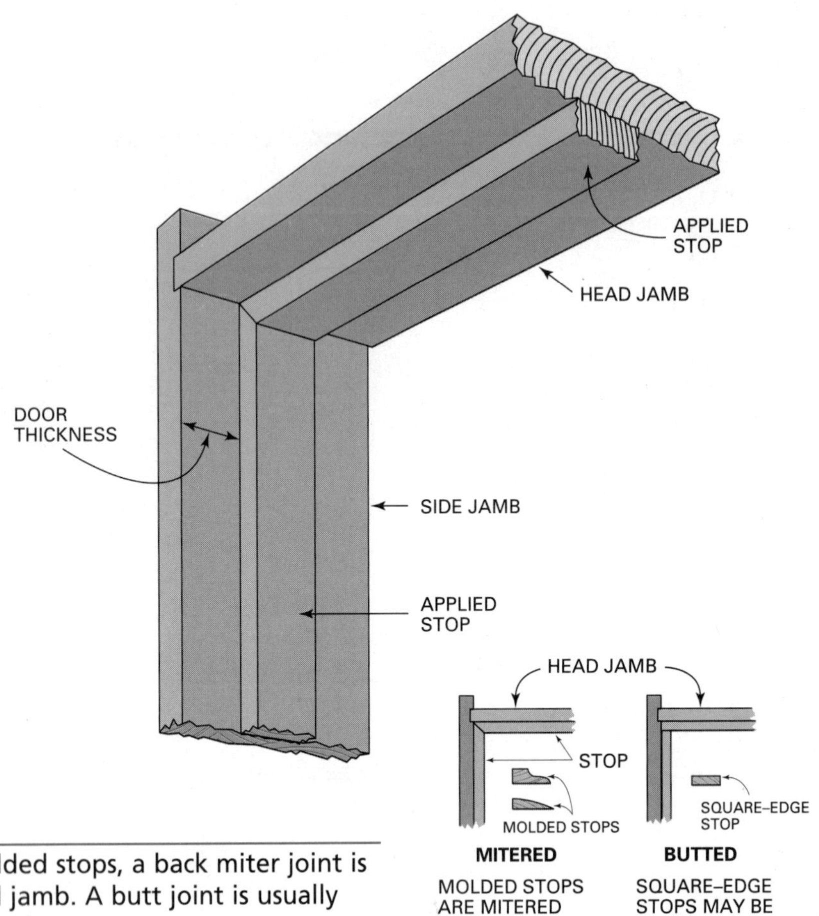

Fig. 77-8 On molded stops, a back miter joint is used at the head jamb. A butt joint is usually made on square-edge stops.

Installing Bypass Doors

Bypass doors are installed so they overlap each other by about 1 inch when closed. Cut the track to length. Install it on the header jamb according to the manufacturer's directions.

Installing Rollers. Install pairs of *roller hangers* on each door. The roller hangers may be offset a different amount for the door on the outside than the door on the inside. They are also offset differently for doors of various thicknesses. Make sure that rollers with the same and correct offset are used on each door (Fig. 77-9). The location of the rollers from the edge of the door is usually specified in the manufacturer's instruction sheet.

Installing Door Pulls. Mark the location and bore holes for *door pulls.* Flush pulls must be used so that bypassing is not obstructed (Fig. 77-10). The proper size hole is bored partway into the door. The pull is tapped into place with a hammer and wood block. The press fit holds the pull in place. Rectangular flush pulls, also used on bypass doors, are held in place with small recessed screws.

Hanging Doors. Hang the doors by holding the bottom outward. Insert the rollers in the overhead track. Then gently let the door come to a vertical position. Install the inside door first, then the outside door (Fig. 77-11).

Fitting Doors. Test the door operation and the fit against side jambs. Door edges must fit against side jambs evenly from top to bottom. If the top or bottom portion of the edge strikes the side jamb first, it may cause the door to jump from the track. The door rollers have adjustments for raising and lowering. Adjust one or the other to make the door edges fit against side jambs.

Installing Floor Guides. A *floor guide* is included with bypass door hardware to keep the doors in alignment. The guide is centered on the lap to steady the doors at the bottom. Mark the location of the guide. Remove the doors. Install the inside section of the guide. Replace the inside door. Replace the outside door. Install the rest of the guide (Fig. 77-12).

Installing Bifold Doors

Before installing *bifold* doors, make sure the opening size is as specified by the hardware or door manufacturer. Usually bifold doors come hinged together in pairs. The hardware consists of the track, pivot

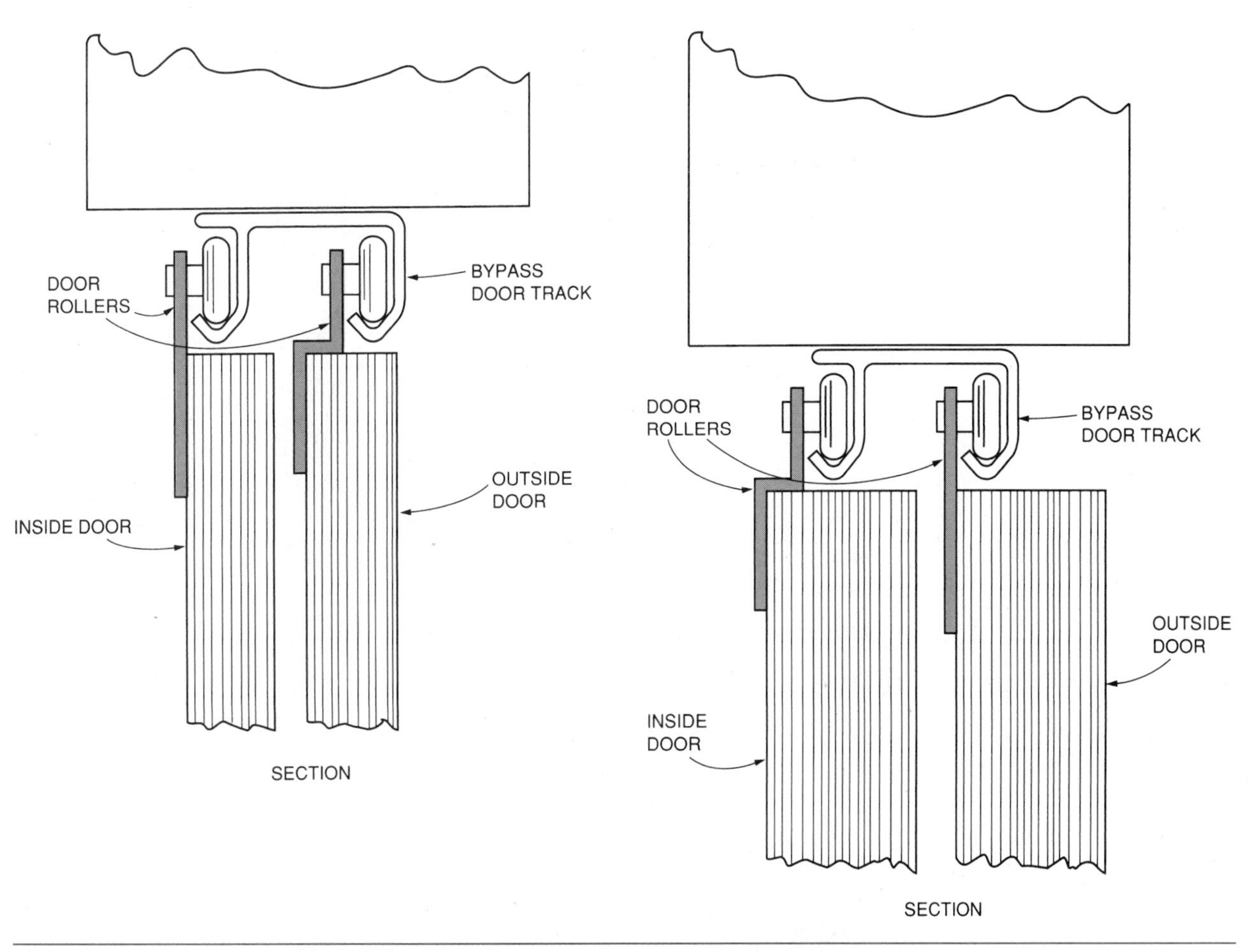

Fig. 77-9 Bypass door rollers are offset different distances for use on doors of various thicknesses.

Fig. 77-10 Bypass doors must have flush pulls.

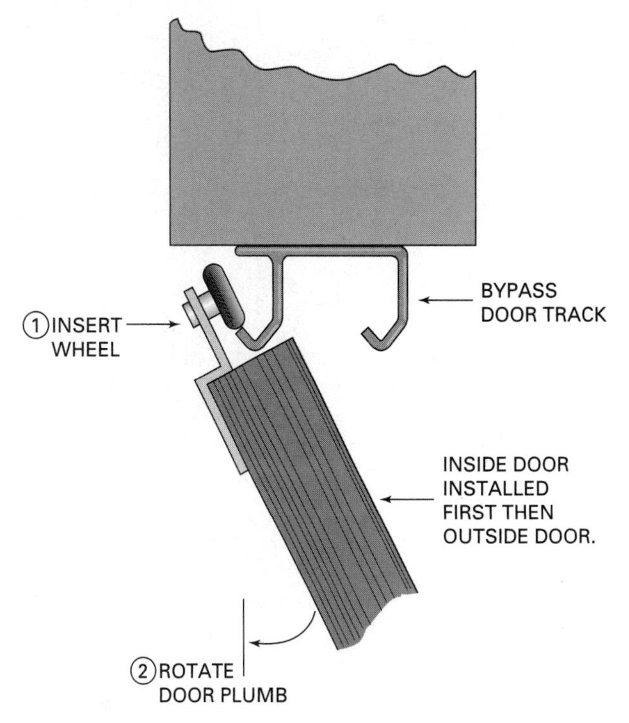

Fig. 77-11 Bypass doors are hung on the overhead track by holding the bottom of the door outward.

sockets, pivot pins and guides, door aligners, door pulls, and necessary fasteners (Fig. 77-13).

Installing the Track. Cut the track to length. Fasten it to the header jamb with screws provided in the kit. The track contains adjustable *sockets* for the door *pivot pins.* Make sure these are inserted before fastening the track in position. The

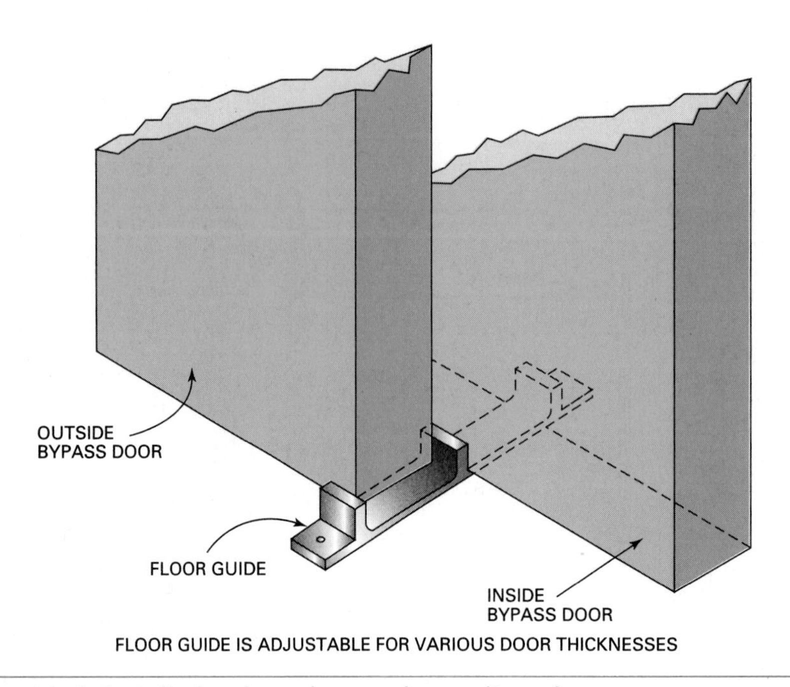

Fig. 77-12 A floor guide is installed to keep bypass doors aligned.

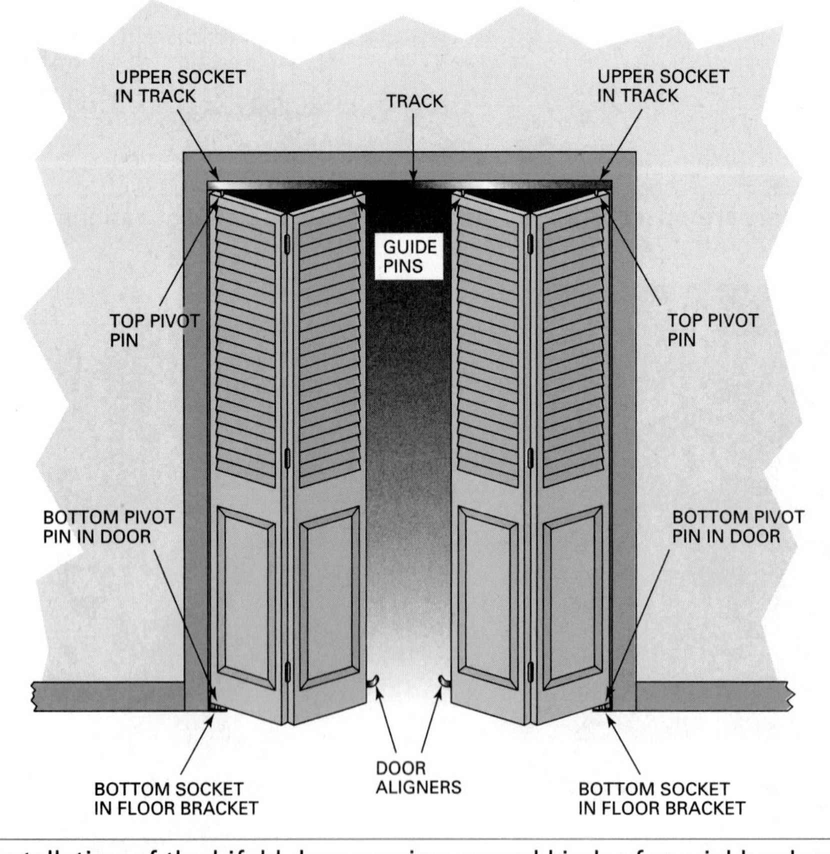

Fig. 77-13 Installation of the bifold door requires several kinds of special hardware.

position of the track on the header jamb is not critical. It may be positioned as desired (Fig. 77-14).

Installing Bottom Pivot Sockets. Locate the bottom pivot sockets. Fasten one on each side, at the bottom of the opening. The pivot socket bracket is L-shaped. It rests on the floor against the side jamb. It is centered on a plumb line from the center of the pivot sockets in the track on the header jamb above.

Installing Pivot and Guide Pins. In most cases, bifold doors come with prebored holes

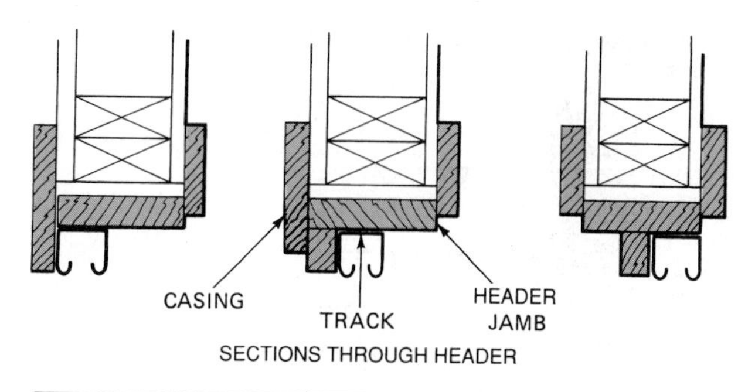

Fig. 77-14 The bifold door track may be located in any position on the header jamb in several ways. Trim conceals the track from view.

for *pivot* and *guide pins.* If not, it is necessary to bore them. Follow the manufacturer's directions as to size and location. Install pivot pins at the top and bottom ends of the door in the prebored holes closest to the jamb. Sometimes the top pivot pin is spring loaded. It can then be depressed for easier installation of the door. The bottom pivot pin is threaded and can be adjusted for height. The guide pin rides in the track. It is installed in the hole provided at the top end of the door farthest away from the jamb.

Hanging the Doors. After all the necessary hardware has been applied, the doors are ready for installation. Loosen the set screw in the top pivot socket. Slide it along the track toward the center of the opening about one foot away from the side jamb. Place the door in position by inserting the bottom pivot pin in the bottom pivot socket. Tilt the doors to an upright position. At the same time insert the top pivot pin in the top socket, and the guide pin in the track, while sliding the socket toward the jamb.

Adjusting the Doors. Adjust top and bottom pivot sockets in or out so the desired joint is obtained between the door and the jamb. Lock top and bottom pivot sockets in position. Adjust the bottom pivot pin to raise or lower the doors, if necessary.

If more than one set of bifold doors are to be installed in an opening, install the other set on the opposite side in the same manner. Install knobs in the manner and location recommended by the manufacturer.

Where sets of bifold doors meet at the middle of an opening, door aligners are installed, near the bottom, on the inside of each of the meeting doors. The door aligners keep the faces of the center doors lined up when closed (Fig. 77-15).

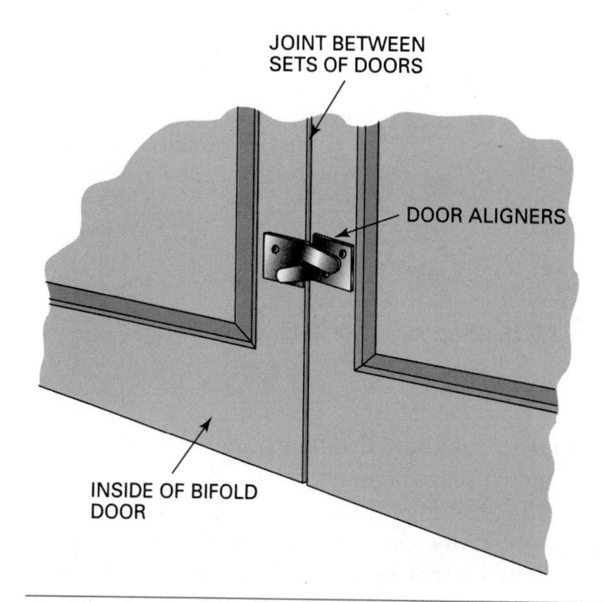

Fig. 77-15 Door aligners are used near the bottom where sets of bifold doors meet.

Installing Pocket Doors

The pocket door frame, complete with track, is installed when interior partitions are framed. The pocket consists of two ladder-like frames between which the door slides. A steel channel is fastened to the floor. The channel keeps the pocket opening spread the proper distance apart.

The frame, which is usually preassembled at the factory, is made of nominal 1-inch stock. The pocket is covered by the interior wall finish. Care must be taken when covering the pocket frame not to use fasteners that are so long that they penetrate the frame. If fasteners penetrate through the pocket door frame, they will probably scratch the side of the door as it is operated or stop its complete entrance into the pocket.

Installing Door Hardware. Attach rollers to the top of the door in the location specified by the manufacturer. Install pulls on the door. On pocket doors an edge pull is necessary, in addition to recessed pulls on the sides of the door. A special pocket door pull contains edge and side pulls. It is mortised in the edge of the door. In most cases, all the necessary hardware is supplied when the pocket door frame is purchased.

Hanging the Door. Engage the rollers in the track by holding the bottom of the door outward in a way similar to that used with bypass doors. Test the operation of the door to make sure it slides easily and butts against the side jamb evenly. Make adjustments to the rollers, if necessary. Stops are later applied to the jambs on both sides of the door. The stops serve as guides for the door. When the door is closed, the stops prevent it from being pushed out of the opening (Fig. 77-16).

Hanging Folding Doors

Because of many styles, manufacturer's instructions supplied with the *folding door* unit should be closely

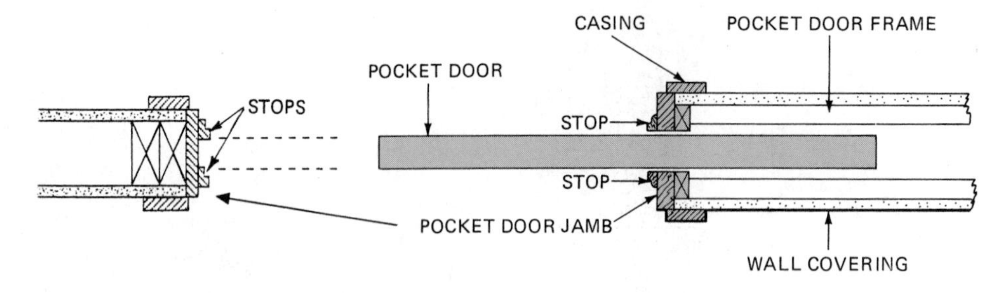

Fig. 77-16 Plan of a pocket door.

followed. Make sure the door frame is the size specified by the manufacturer.

The folding door track is fastened with screws, usually along the centerline, to the head jamb. The door is hung on the track with the *jamb panel* toward the side jamb on which the panels will be stacked. Fasten the jamb panel to the side jamb. Fasten the *stop,* against which the door closes, on the opposite side jamb. Usually, trim strips are fastened on both sides of track to conceal it (Fig. 77-17). Check the operation of the door and make any necessary adjustments.

Installing a Prehung Door

A prehung single-acting, hinged door unit consists of a door frame with the door hinged and casings installed. Holes are provided, if locksets have not already been installed. Small cardboard shims are stapled to the lock edge and top end of the door to maintain proper clearance between the door and frame.

Prehung units are available in several jamb widths to accommodate various wall thicknesses. Some prehung units have split jambs that are adjustable for varying wall thicknesses (Fig. 77-18).

A prehung door unit can be set in a matter of minutes. Many prehung units come without the casing attached, but if it is attached, remove the casings carefully from one side of solid jamb units.

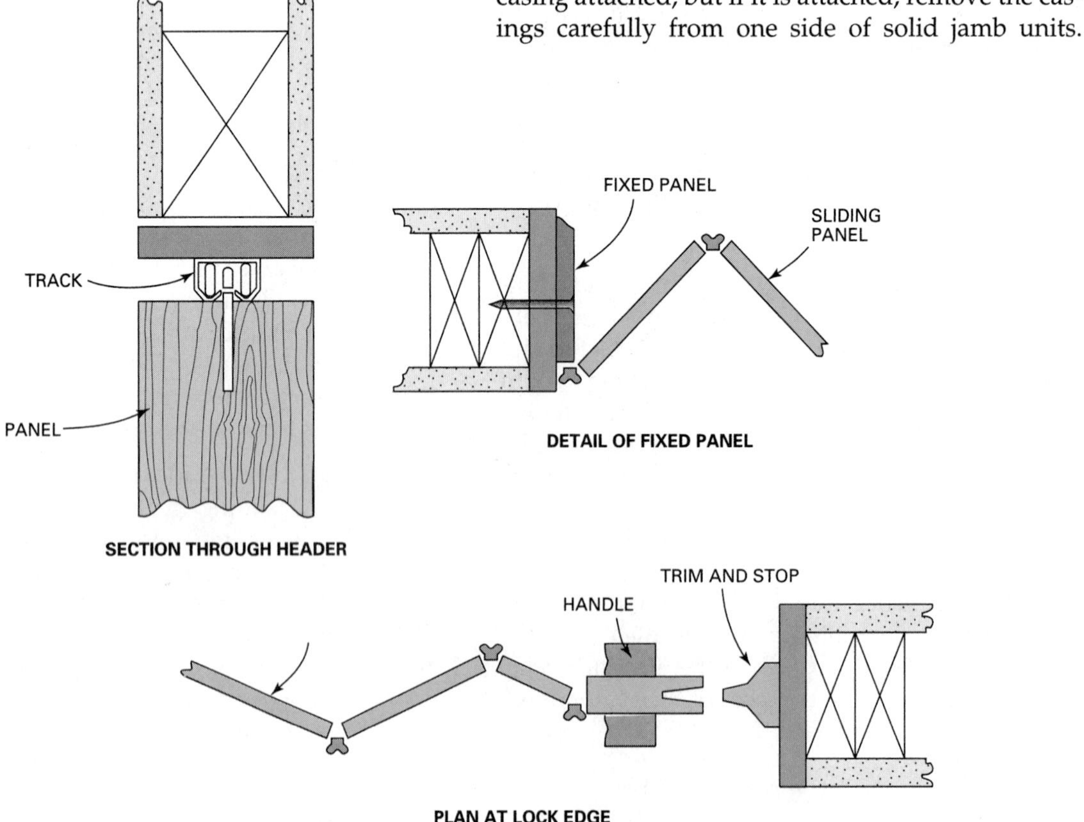

Fig. 77-17 Installation details for folding doors.

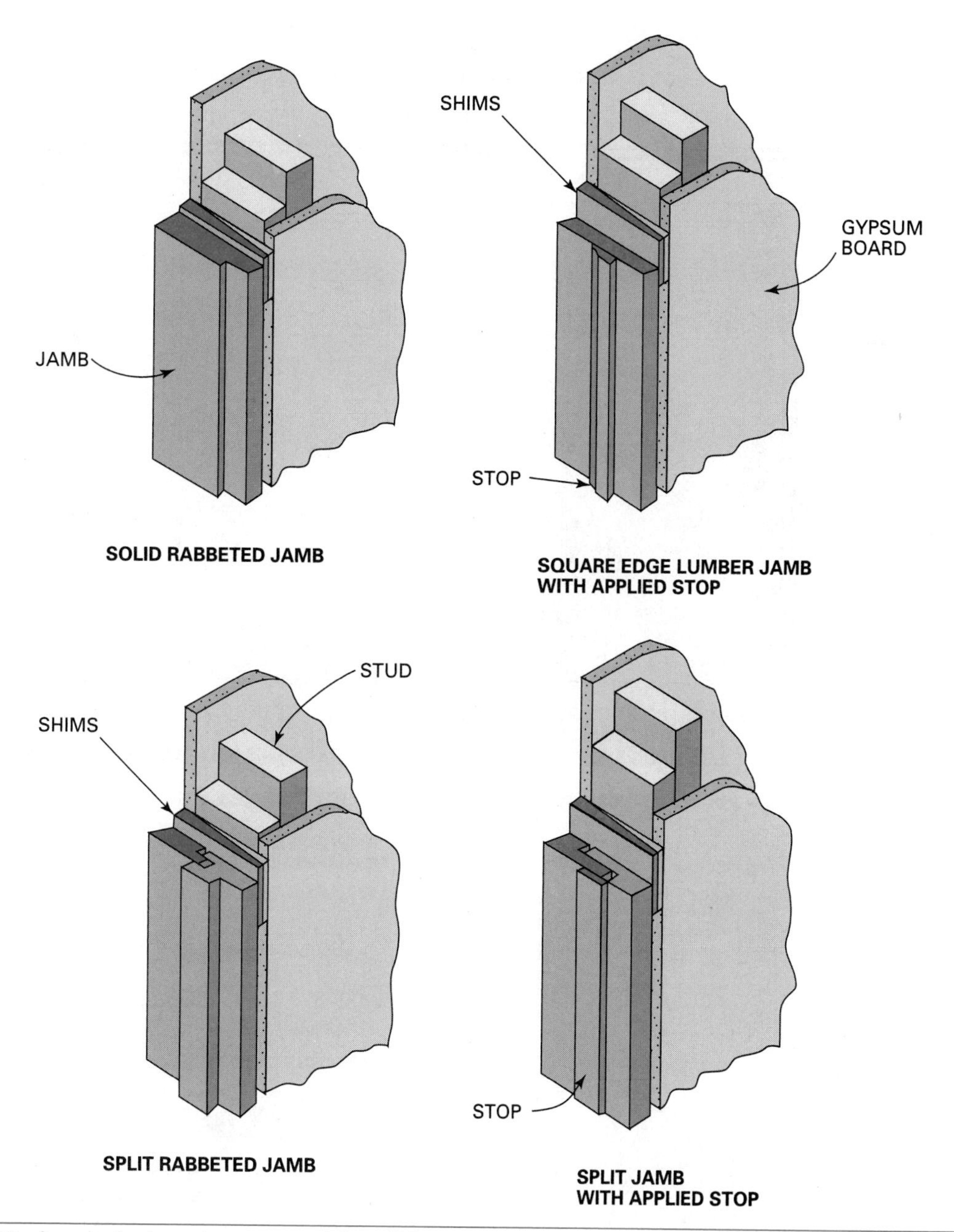

Fig. 77-18 Prehung door units come with solid or split jambs.

Center the unit in the opening, so the door will swing in the desired direction. Be sure the door is closed and spacer shims are in place between the jamb and door. Plumb the door unit. Tack it to the wall through the casing.

Open the door and move to the other side. Install shims between the side jambs and the rough opening at intermediate points, keeping side jambs straight. Nail through the side jambs and shims. Remove spacers. Check the operation of the door. Make any necessary adjustments. Replace the previously removed casings. Drive and set all nails (Fig. 77-19).

Prehung door units with split jambs are set in a similar manner. However, there is no need to remove the casings. One section is installed as described above. The remaining section is inserted into the one already in place.

Installing Locksets

Locksets are installed on interior doors in the same manner as for exterior doors and as described in Chapter 59, Installing Exterior Door Locksets. Although their installation is basically the same, some locks are used exclusively on interior doors.

STEP BY STEP PROCEDURES

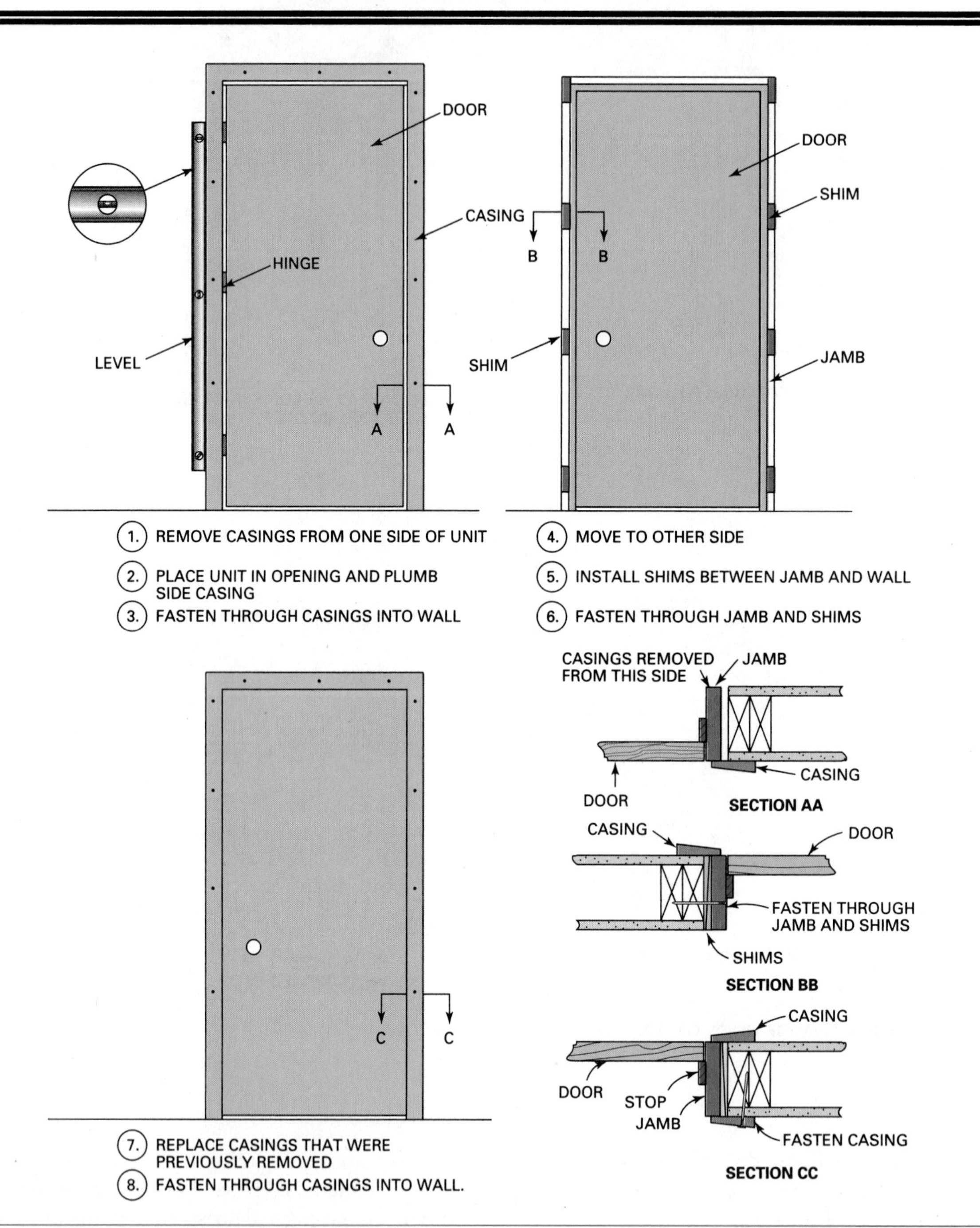

Fig. 77-19 Procedure for installing a prehung door unit.

The *privacy* lock is often used on bathroom and bedroom doors. It is locked by pushing or turning a button on the room side (Fig. 77-20). On most privacy locks, a turn of the knob on the room side unlocks the door. On the opposite side, the door can be unlocked by a pin or key inserted into a hole in the knob. The unlocking device should be kept close by, in a prominent location, in case the door needs to be opened quickly in an emergency.

The *passage* lockset has knobs on both sides that are turned to unlatch the door. This lockset is used when it is not desirable to lock the door.

Fig. 77-20 Privacy lockets for bathroom doors may be locked by pushing in the button in the center of the knob.

Review Questions

Select the most appropriate answer.

1. Most interior doors are manufactured in a thickness of
a. 1 inch. c. 1 1/2 inches.
b. 1 3/8 inches. d. 1 3/4 inches.

2. The height of most interior doors is
a. 6′-0″. c. 6′-8″.
b. 6′-6″. d. 7′-0″.

3. Interior door widths usually range from
a. 1′-6″ to 2′-6″.
b. 2′-2″ to 2′-8″.
c. 2′-6″ to 2′-8″.
d. 1′-0″ to 3′-0″.

4. Used extensively for flush door skins is
a. fir plywood. c. metal.
b. lauan plywood. d. plastic laminate.

5. The usual distance allowed between the finish floor and the bottom of swinging doors for clearance is
a. 1/4 inch. c. 3/4 inch.
b. 1/2 inch. d. 1 inch.

6. A disadvantage of bypass doors is that they
a. project out into the room.
b. cost more and require more time to install.
c. are difficult to operate.
d. do not provide total access to the opening.

7. If the jamb stock is 3/4-inch thick, the rough opening width for a swinging door should be the door width plus
a. 3/4 inches. c. 2 inches.
b. 1 1/2 inches. d. 2 1/2 inches.

8. If the jamb stock and the finished floor are both 3/4-inch thick, the rough opening height for a 6′-8″ swinging door should be
a. 7′-0″. c. 6′-10 1/4″.
b. 6′-11 1/2″. d. 6′-9 1/4″.

9. When a rabbeted door frame is made for a swinging door, the header jamb length is equal to the width of the door plus
a. 3/16 inches. c. 3/8 inches.
b. 3/32 inches. d. 1/2 inches.

10. An accurate and fast method of plumbing side jambs of a door frame is by the use of a
a. builder's level.
b. carpenter's 26-inch hand level.
c. plumb bob.
d. straightedge.

BUILDING FOR SUCCESS

Environmental Issues in Construction

Each year brings new environmental concerns to nations throughout the world. In certain areas of the United States, the availability of selected natural resources such as water, timber, and land suitable for building has become a front-page issue. Other issues, such as waste disposal and wetlands, will continue to affect the quality of land available for development. With larger populations to feed, clothe, house, and employ, every nation must conservatively utilize its natural resources. At the same time, it must maintain a strong economy in a well thought out fashion.

The United States has experienced confrontations involving construction material producers, environmentalists, and conservationists. The issues vary, but generally center on the consumption of natural resources or preservation of those resources and the creating or maintaining of natural habitat for wildlife. Directly and indirectly, many of these issues affect housing and construction. How should we, as people associated with the construction industry, look at these and other controversial issues? Do we consider ourselves only consumers of natural resources or only environmentalists? Most builders, producers of construction products, and material suppliers understand how critical both the preservation of the natural environment and economic growth are to our economy. How should members of the construction industry become informed and work with others as environmental, ecological, and economic disputes are resolved? How can they stay informed about the issues that have a direct bearing on construction-related industries?

Initially, all those earning their living in the construction industry should be as informed as possible about current environmental, ecological, and economic issues. They should participate in discussions and information exchanges that deal with local and national environmental concerns. Most national and local associations representing the construction industry will be prepared to take a stand on issues that affect their membership. The best decisions to be made about these issues will be made with the most information and concern for who and what are involved. The research process should involve data from all perspectives.

One large group with a great deal of concern for the preservation of our natural resources and habitats is elementary through postsecondary school students. Many young people are responding to the major concerns that will affect the environment of tomorrow. Many school-age children and young adults participate in awareness and problem resolution activities. They are willing to become involved in attempts to help others understand the urgency of responding. They realize that cooperation between opposing factions must be established and maintained. They are also involved with education and economic concerns that affect future job opportunities. In many ways, these concerned and motivated young people demonstrate how people can work together to find solutions. It is encouraging to know that other organizations and special interest groups have taken the same approach to solving controversial disputes.

Almost everyone has heard of the spotted owl issue and the desire of many to preserve the owl's natural habitat in the Pacific Northwest. Some opposition to this cause has come from the logging industry and lumber product producers. Industries' concerns to continue utilizing the forests in the production of building materials are also very important and require study. This dispute has involved people representing a large variety of associations, special interest groups, federal and state governments, private industry, and even the president of the United States.

This issue is representative of others that involve the future preservation and/or consumption of natural resources. Attempting to place blame in many cases should not be the goal. Realistically, the goal should be to look beyond selfish interests and focus on possible solutions that will preserve the environment and stimulate the economy at the same time. Left unaddressed, many controversial issues may prove to be devastating to our environment, our economy, and our ecological well-being. Few of these concerns, if any, can be solved on short notice. They will take time, understanding, flexibility, commitment, and money to resolve.

Historically, many solutions to issues of national concern have come about by compromise resulting from discussion. The goals on both sides of a controversial issue should be to bring solutions to the problems that benefit both people and the environment. The greatest challenge may be in arriving at a compromise that satisfies all opponents.

Economically, there is a great need to maintain growth in jobs affected by environmental issues. Stimulating the economy of a community that has been hit by a drastic reduction in producing construction materials and products is a tremendous challenge. The solutions need to be immediate, sound, and valuable long-term. Finding alternative industrial production and business potential must be an integral part of any discussion.

The construction industry as a whole must continue to be informed and desire to work with others in the process of working to save our natural environment and maintain economic stability. Each student of construction must understand the need to solve the ongoing environmental and economic problems that put our nation's workforce at risk. A continual desire to be a part of the solution and not the problem will be required. We have a good history of addressing concerns of all types in this country and finding workable solutions. That should continue to be a goal for everyone.

FOCUS QUESTIONS: For individual or group discussion

1. Identify one important environmental, ecological, or economic issue that directly affects a construction-related industry. Provide an overview of the issue.
2. How have issues such as the spotted owl controversy been addressed by the construction-related industries in your specific geographic area?
3. What solutions do you feel are workable in addressing the spotted owl issue where environmentalist, ecological, and industrial groups have not reached agreements? How can opposing parties begin to work toward "reasonable" solutions?
4. As a future participant in the nation's construction workforce, what controversial issues do you think will affect construction-related industries? What do you see as your personal obligation in following these types of issues?

UNIT

27

Interior Trim

Interior trim, also called *interior finish*, involves the application of *molding* around windows and doors; at the intersection of walls, floor, and ceilings; and to other inside surfaces. Moldings are strips of material, shaped in numerous patterns, for use in a specific location. Wood is used to make most molding, but some are made of plastic or metal.

OBJECTIVES

After completing this unit, the student should be able to:

- identify standard moldings and describe their use.
- apply ceiling and wall molding.
- apply interior door casings, baseboard, base cap, and base shoe.
- install window trim, including stools, aprons, jamb extensions, casings, and stop beads.
- install closet shelves and closet pole.
- install mantels.

UNIT CONTENTS

CHAPTER 78 DESCRIPTION AND APPLICATION OF MOLDING

Moldings are available in many *standard* types. Each type is manufactured in several sizes and patterns. Standard patterns are usually made only from softwood. When other kinds of wood, or special patterns, are desired, mills make *custom* moldings to order. All moldings must be applied with tight-fitting joints to present a suitable appearance.

Standard Molding Patterns

Standard moldings are designated as *bed, crown, cove, full round, half round, quarter round, base, base shoe, base cap, casing, chair rail, back band, apron, stool, stop,* and others (Fig. 78-1).

Molding usually comes in lengths of 8, 10, 12, 14, and 16 feet. Some moldings are available in odd lengths. Door casings, in particular, are available in lengths of 7 feet to reduce waste.

Finger-jointed lengths are made of short pieces joined together. These are used only when a paint finish is to be applied. The joints show through a stained or natural finish.

Molding Shape and Use

Some moldings are classified by the way they are shaped. Others are designated by location. For example, *beds, crowns, and coves* are terms related to shape. Although they may be placed in other locations, they are usually used at the intersection of walls and ceilings (Fig. 78-2). Also classified by their shape are *full rounds, half rounds,* and *quarter rounds.* They are used in many locations. Full rounds are used for such things as closet poles. Half rounds may be used to conceal joints between panels or to trim shelf edges. Quarter rounds may be used to trim outside corners of wall paneling and for many other purposes (Fig. 78-3).

Designated by location, *base, base shoe,* and *base cap* are moldings applied at the bottom of walls where they meet the floor. When square-edge base is used, a base cap is usually applied to its top edge. Base shoe is normally used to conceal the joint between the bottom of the base and the finish floor (Fig. 78-4).

Casings are used to trim around windows, doors, and other openings. They cover the space between the frame and the wall. *Back bands* are applied to the outside edges of casings for a more decorative appearance.

Aprons, stools, and *stops* are parts of window trim. Stops are also applied to door frames. On the same window, aprons should have the same molded shape as casings. Aprons, however, are not *backed out.* They have straight, smooth backs and sharp, square, top edges that butt against the bottom of the stool (Fig. 78-5).

Corner guards are also called *outside corners.* They are used to finish exterior corners of interior wall finish. *Caps* and *chair rail* trim the top edge of wainscoting. (These moldings have been previously described in Chapter 72, Application of Wall Paneling.) Others, such as *astragals, battens, panel,* and *picture* moldings are used for various purposes.

Making Joints on Molding

End joints between lengths of ceiling molding may be made square or at a 45 degree angle. Many carpenters prefer to make square joints between moldings because less joint line is shown. Also, the square end acts as a stop when bowing and snapping the last length of molding into place at a corner.

Usually, the last piece of molding along a wall is cut slightly longer. It is bowed outward in the center, then pressed into place when the ends are in position. This makes the joints come up tight. After the molding has been fastened, joints between lengths should be sanded flush, except on prefinished moldings. Failure to sand butted ends flush with each other results in a shadow being cast at the joint line. This gives the appearance of an open joint.

Joints on exterior corners are *mitered.* (Miter joints are defined in Chapter 12, Boring and Cutting Tools.) Joints on interior corners are usually *coped,* especially on large moldings. A coped joint is made by fitting the piece on one wall with a square end into the corner. The end of the molding on the other wall is cut to fit against the shaped face of the molding on the first wall (Fig. 78-6).

Methods of Mitering Using Miter Boxes

Moldings of all types may be mitered by using either hand or power *miter boxes.* A miter box is a tool that cuts a piece of material at an angle. The most common angle is 45 degrees. These boxes can be made from scraps of wood used to guide a handsaw or of metal (Fig. 78-7). These are effective tools in cutting a miter.

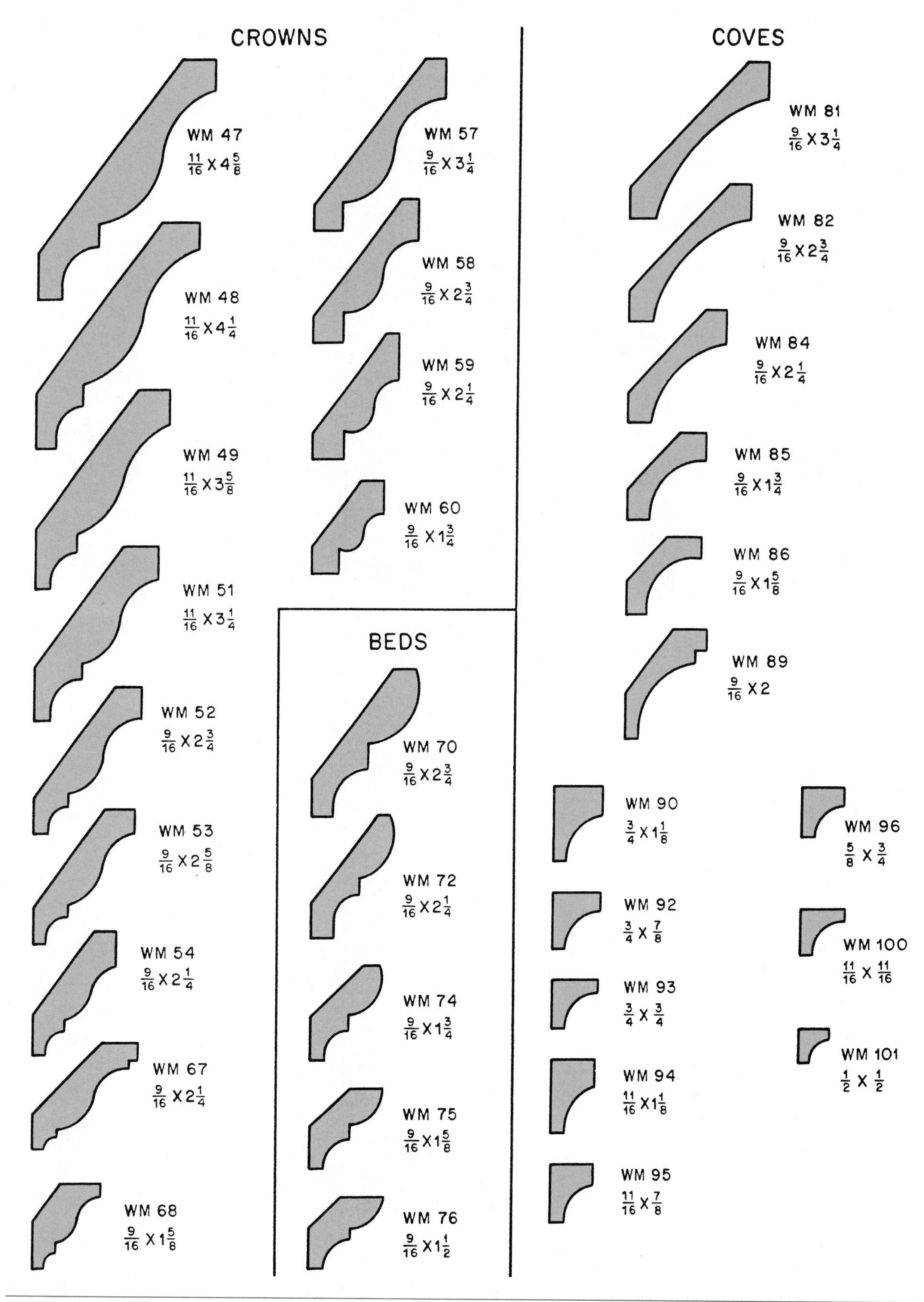

Fig. 78-1 Standard molding patterns (*Courtesy of Wood Molding and Millwork Producers, Inc., P.O. Box 25278, Portland, Oregon 97225*)

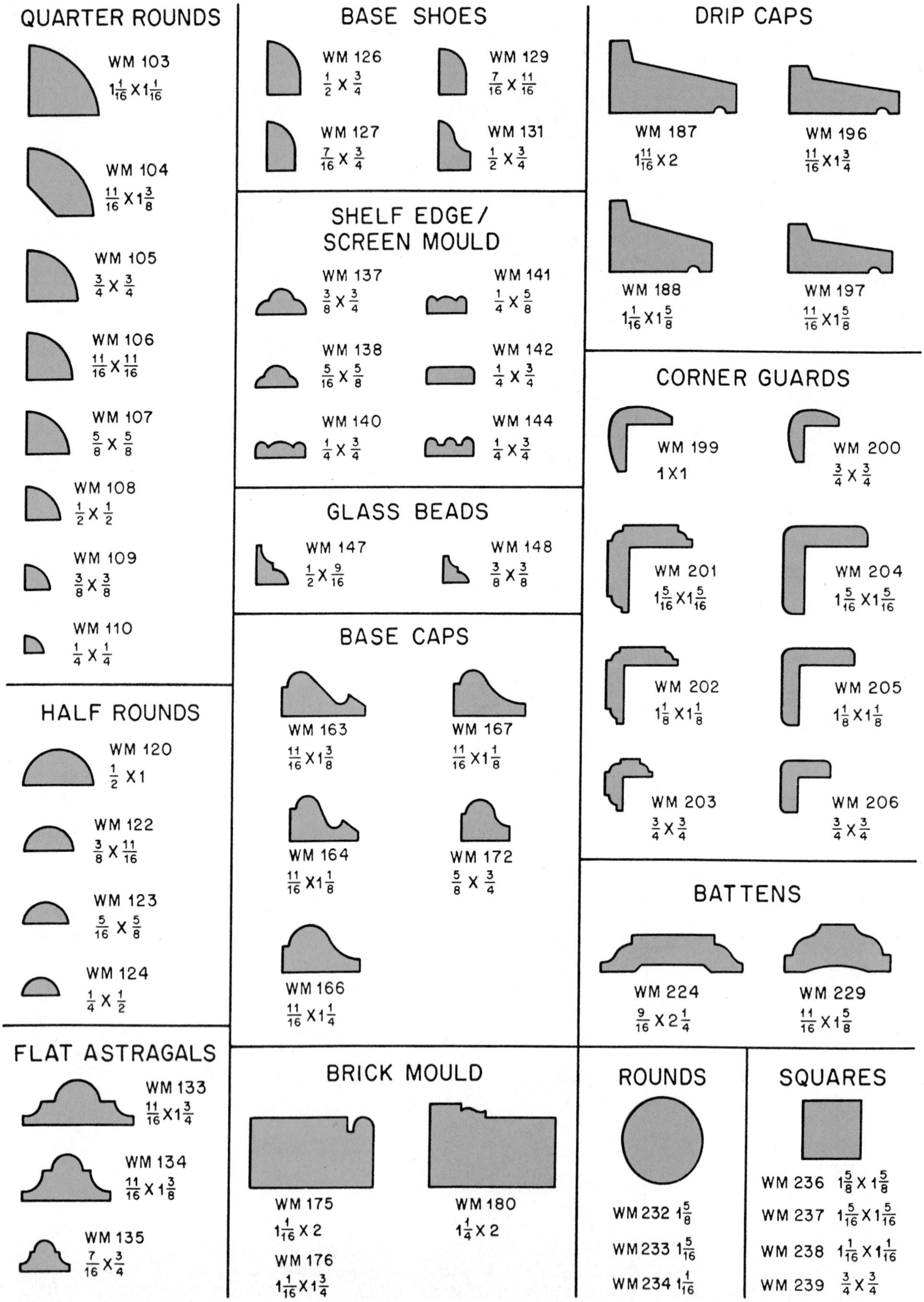

QUARTER ROUNDS

WM 103 $1\frac{1}{16} \times 1\frac{1}{16}$

WM 104 $\frac{11}{16} \times 1\frac{3}{8}$

WM 105 $\frac{3}{4} \times \frac{3}{4}$

WM 106 $\frac{11}{16} \times \frac{11}{16}$

WM 107 $\frac{5}{8} \times \frac{5}{8}$

WM 108 $\frac{1}{2} \times \frac{1}{2}$

WM 109 $\frac{3}{8} \times \frac{3}{8}$

WM 110 $\frac{1}{4} \times \frac{1}{4}$

HALF ROUNDS

WM 120 $\frac{1}{2} \times 1$

WM 122 $\frac{3}{8} \times \frac{11}{16}$

WM 123 $\frac{5}{16} \times \frac{5}{8}$

WM 124 $\frac{1}{4} \times \frac{1}{2}$

FLAT ASTRAGALS

WM 133 $\frac{11}{16} \times 1\frac{3}{4}$

WM 134 $\frac{11}{16} \times 1\frac{3}{8}$

WM 135 $\frac{7}{16} \times \frac{3}{4}$

BASE SHOES

WM 126 $\frac{1}{2} \times \frac{3}{4}$

WM 129 $\frac{7}{16} \times \frac{11}{16}$

WM 127 $\frac{7}{16} \times \frac{3}{4}$

WM 131 $\frac{1}{2} \times \frac{3}{4}$

SHELF EDGE/ SCREEN MOULD

WM 137 $\frac{3}{8} \times \frac{3}{4}$

WM 141 $\frac{1}{4} \times \frac{5}{8}$

WM 138 $\frac{5}{16} \times \frac{5}{8}$

WM 142 $\frac{1}{4} \times \frac{3}{4}$

WM 140 $\frac{1}{4} \times \frac{3}{4}$

WM 144 $\frac{1}{4} \times \frac{3}{4}$

GLASS BEADS

WM 147 $\frac{1}{2} \times \frac{9}{16}$

WM 148 $\frac{3}{8} \times \frac{3}{8}$

BASE CAPS

WM 163 $\frac{11}{16} \times 1\frac{3}{8}$

WM 167 $\frac{11}{16} \times 1\frac{1}{8}$

WM 164 $\frac{11}{16} \times 1\frac{1}{8}$

WM 172 $\frac{5}{8} \times \frac{3}{4}$

WM 166 $\frac{11}{16} \times 1\frac{1}{4}$

BRICK MOULD

WM 175 $1\frac{1}{16} \times 2$

WM 176 $1\frac{1}{16} \times 1\frac{3}{4}$

WM 180 $1\frac{1}{4} \times 2$

DRIP CAPS

WM 187 $1\frac{11}{16} \times 2$

WM 196 $\frac{11}{16} \times 1\frac{3}{4}$

WM 188 $1\frac{1}{16} \times 1\frac{5}{8}$

WM 197 $\frac{11}{16} \times 1\frac{5}{8}$

CORNER GUARDS

WM 199 1X1

WM 200 $\frac{3}{4} \times \frac{3}{4}$

WM 201 $1\frac{5}{16} \times 1\frac{5}{16}$

WM 204 $1\frac{5}{16} \times 1\frac{5}{16}$

WM 202 $1\frac{1}{8} \times 1\frac{1}{8}$

WM 205 $1\frac{1}{8} \times 1\frac{1}{8}$

WM 203 $\frac{3}{4} \times \frac{3}{4}$

WM 206 $\frac{3}{4} \times \frac{3}{4}$

BATTENS

WM 224 $\frac{9}{16} \times 2\frac{1}{4}$

WM 229 $\frac{11}{16} \times 1\frac{5}{8}$

ROUNDS

WM 232 $1\frac{5}{8}$

WM 233 $1\frac{5}{16}$

WM 234 $1\frac{1}{16}$

SQUARES

WM 236 $1\frac{5}{8} \times 1\frac{5}{8}$

WM 237 $1\frac{5}{16} \times 1\frac{5}{16}$

WM 238 $1\frac{1}{16} \times 1\frac{1}{16}$

WM 239 $\frac{3}{4} \times \frac{3}{4}$

Fig. 78-1 *(continued)*

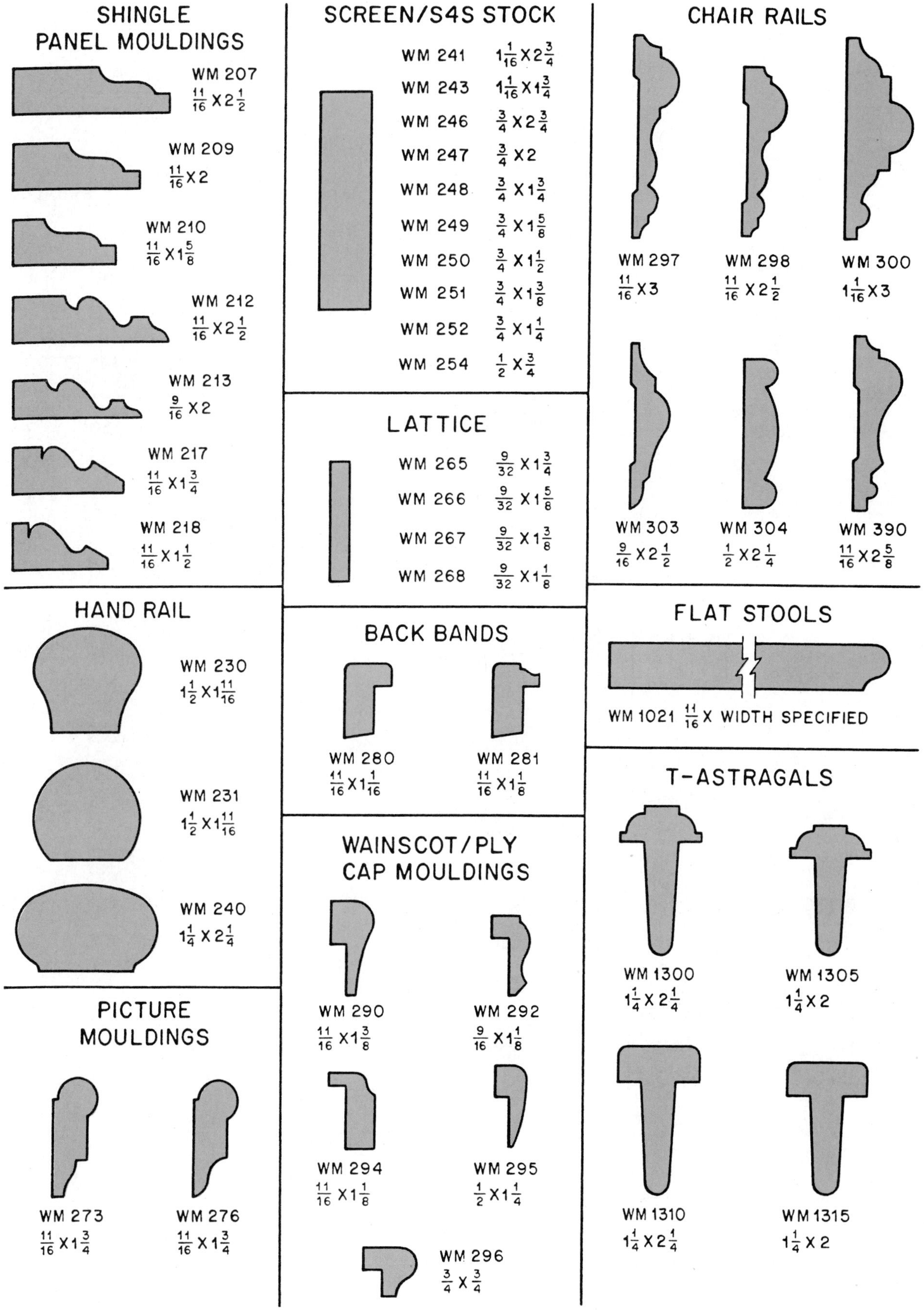

SHINGLE PANEL MOULDINGS

WM 207 $\frac{11}{16} \times 2\frac{1}{2}$

WM 209 $\frac{11}{16} \times 2$

WM 210 $\frac{11}{16} \times 1\frac{5}{8}$

WM 212 $\frac{11}{16} \times 2\frac{1}{2}$

WM 213 $\frac{9}{16} \times 2$

WM 217 $\frac{11}{16} \times 1\frac{3}{4}$

WM 218 $\frac{11}{16} \times 1\frac{1}{2}$

HAND RAIL

WM 230 $1\frac{1}{2} \times 1\frac{11}{16}$

WM 231 $1\frac{1}{2} \times 1\frac{11}{16}$

WM 240 $1\frac{1}{4} \times 2\frac{1}{4}$

PICTURE MOULDINGS

WM 273 $\frac{11}{16} \times 1\frac{3}{4}$

WM 276 $\frac{11}{16} \times 1\frac{3}{4}$

SCREEN/S4S STOCK

WM 241 $1\frac{1}{16} \times 2\frac{3}{4}$

WM 243 $1\frac{1}{16} \times 1\frac{3}{4}$

WM 246 $\frac{3}{4} \times 2\frac{3}{4}$

WM 247 $\frac{3}{4} \times 2$

WM 248 $\frac{3}{4} \times 1\frac{3}{4}$

WM 249 $\frac{3}{4} \times 1\frac{5}{8}$

WM 250 $\frac{3}{4} \times 1\frac{1}{2}$

WM 251 $\frac{3}{4} \times 1\frac{3}{8}$

WM 252 $\frac{3}{4} \times 1\frac{1}{4}$

WM 254 $\frac{1}{2} \times \frac{3}{4}$

LATTICE

WM 265 $\frac{9}{32} \times 1\frac{3}{4}$

WM 266 $\frac{9}{32} \times 1\frac{5}{8}$

WM 267 $\frac{9}{32} \times 1\frac{3}{8}$

WM 268 $\frac{9}{32} \times 1\frac{1}{8}$

BACK BANDS

WM 280 $\frac{11}{16} \times 1\frac{1}{16}$

WM 281 $\frac{11}{16} \times 1\frac{1}{8}$

WAINSCOT/PLY CAP MOULDINGS

WM 290 $\frac{11}{16} \times 1\frac{3}{8}$

WM 292 $\frac{9}{16} \times 1\frac{1}{8}$

WM 294 $\frac{11}{16} \times 1\frac{1}{8}$

WM 295 $\frac{1}{2} \times 1\frac{1}{4}$

WM 296 $\frac{3}{4} \times \frac{3}{4}$

CHAIR RAILS

WM 297 $\frac{11}{16} \times 3$

WM 298 $\frac{11}{16} \times 2\frac{1}{2}$

WM 300 $1\frac{1}{16} \times 3$

WM 303 $\frac{9}{16} \times 2\frac{1}{2}$

WM 304 $\frac{1}{2} \times 2\frac{1}{4}$

WM 390 $\frac{11}{16} \times 2\frac{5}{8}$

FLAT STOOLS

WM 1021 $\frac{11}{16}$ X WIDTH SPECIFIED

T-ASTRAGALS

WM 1300 $1\frac{1}{4} \times 2\frac{1}{4}$

WM 1305 $1\frac{1}{4} \times 2$

WM 1310 $1\frac{1}{4} \times 2\frac{1}{4}$

WM 1315 $1\frac{1}{4} \times 2$

Fig. 78-1 *(continued)*

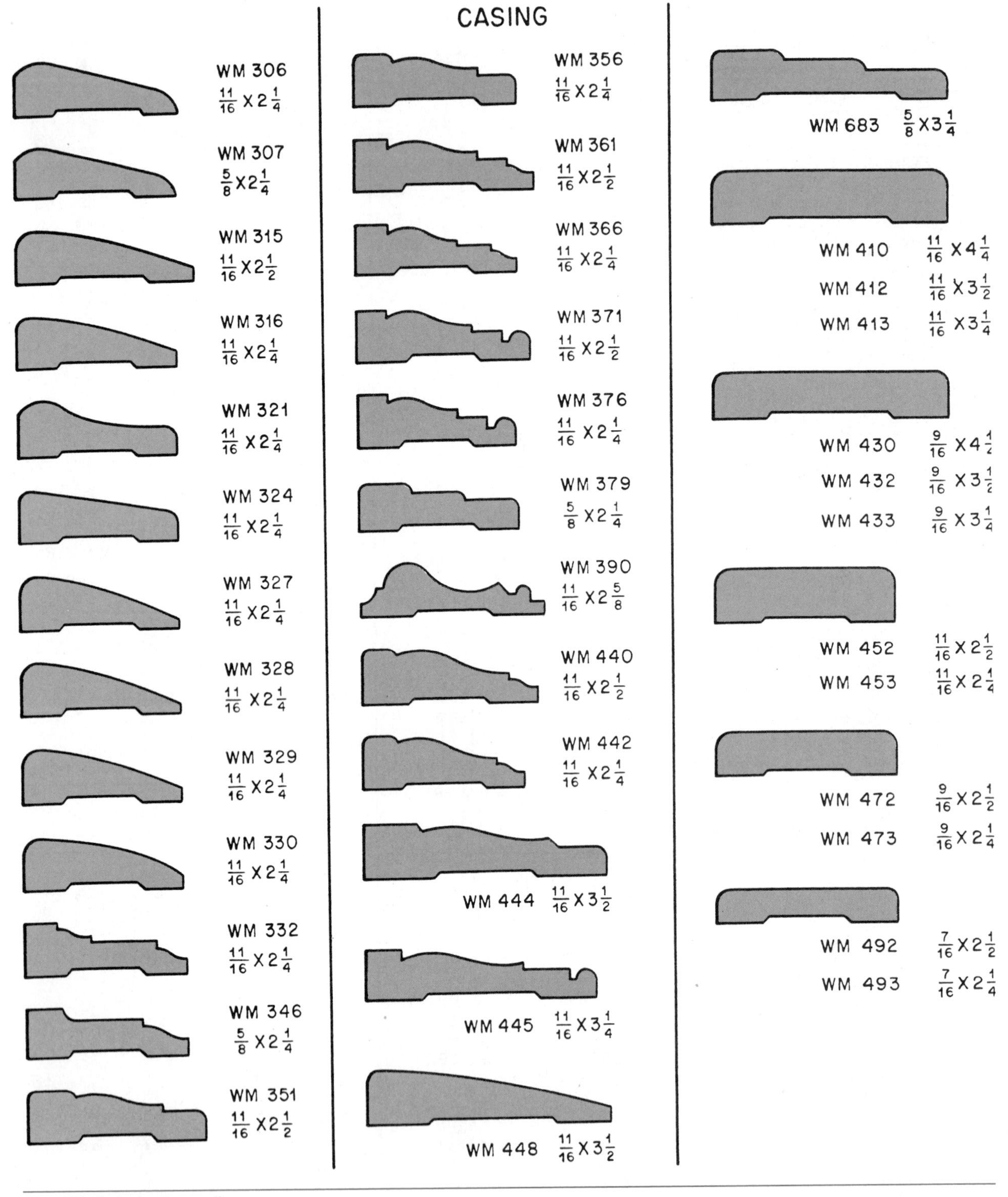

Fig. 78-1 *(continued)*

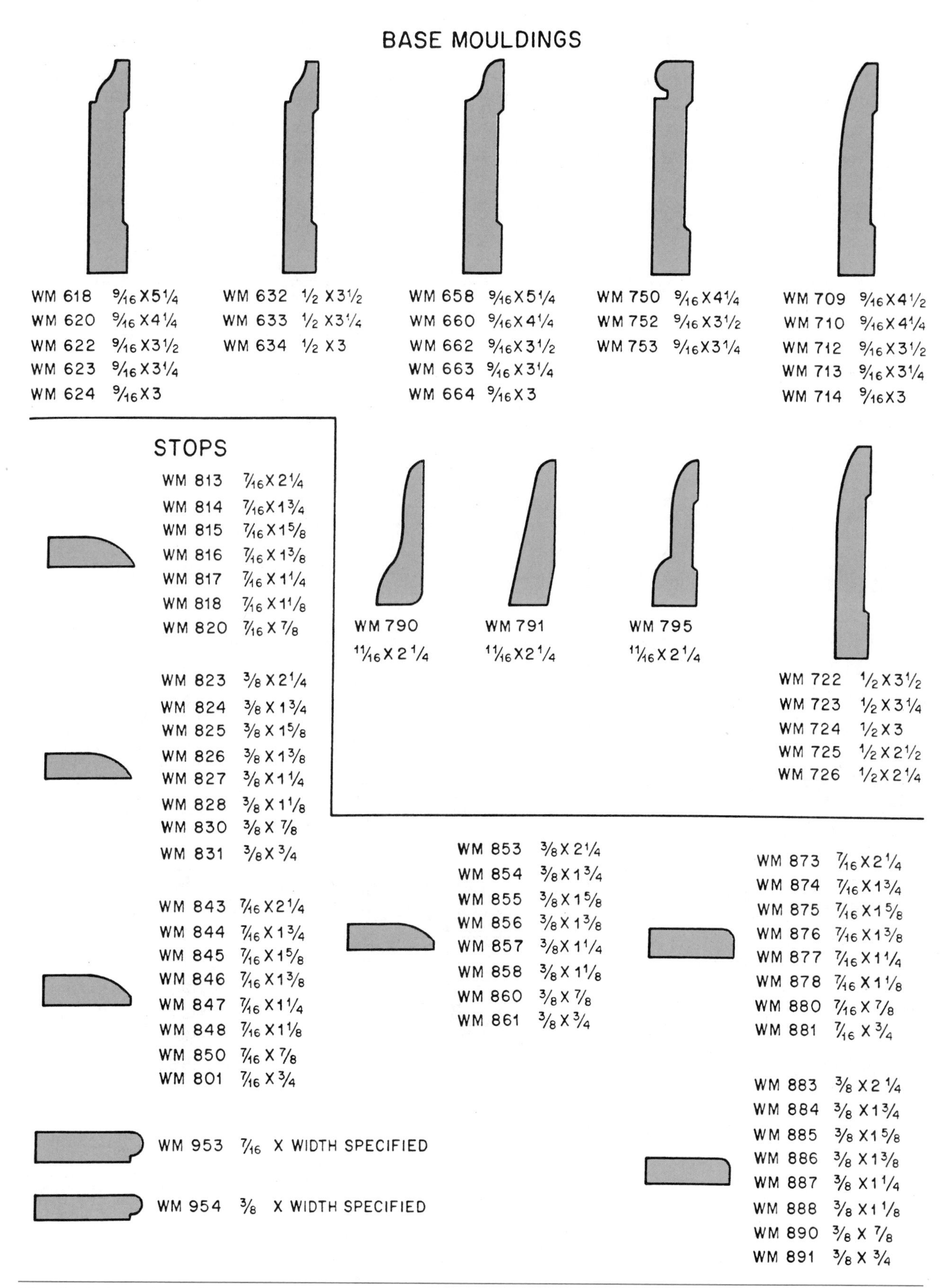

Fig. 78-1 (*continued*)

STOPS

WM 903 7/16 X 2 1/4	WM 913 3/8 X 2 1/4	WM 933 7/16 X 2 1/4	WM 943 3/8 X 2 1/4
WM 904 7/16 X 1 3/4	WM 914 3/8 X 1 3/4	WM 934 7/16 X 1 3/4	WM 944 3/8 X 1 3/4
WM 905 7/16 X 1 5/8	WM 915 3/8 X 1 5/8	WM 935 7/16 X 1 5/8	WM 945 3/8 X 1 5/8
WM 906 7/16 X 1 3/8	WM 916 3/8 X 1 3/8	WM 936 7/16 X 1 3/8	WM 946 3/8 X 1 3/8
WM 907 7/16 X 1 1/4	WM 917 3/8 X 1 1/4	WM 937 7/16 X 1 1/4	WM 947 3/8 X 1 1/4
WM 908 7/16 X 1 1/8	WM 918 3/8 X 1 1/8	WM 938 7/16 X 1 1/8	WM 948 3/8 X 1 1/8
WM 910 7/16 X 7/8	WM 920 3/8 X 7/8	WM 940 7/16 X 7/8	WM 950 3/8 X 7/8
WM 911 7/16 X 3/4	WM 921 3/8 X 3/4	WM 941 7/16 X 3/4	WM 951 3/8 X 3/4

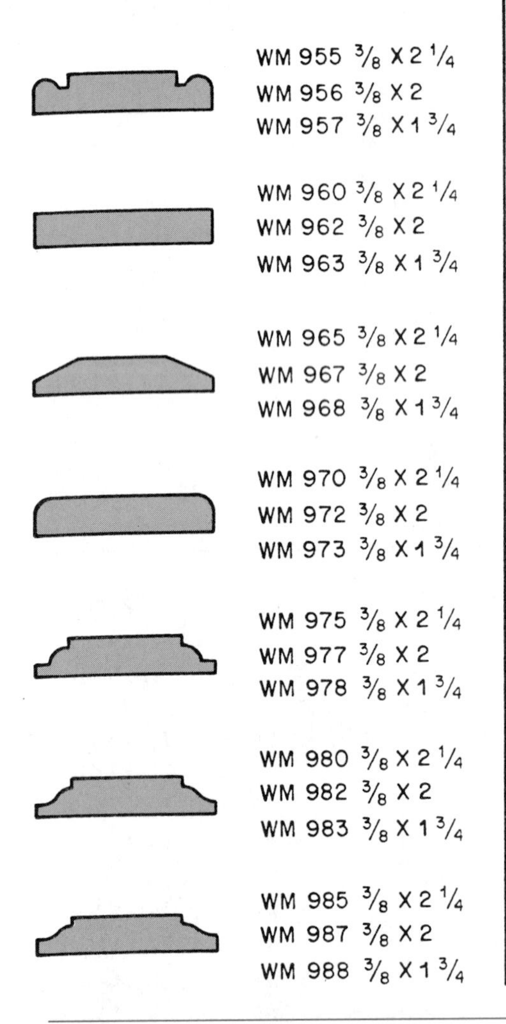

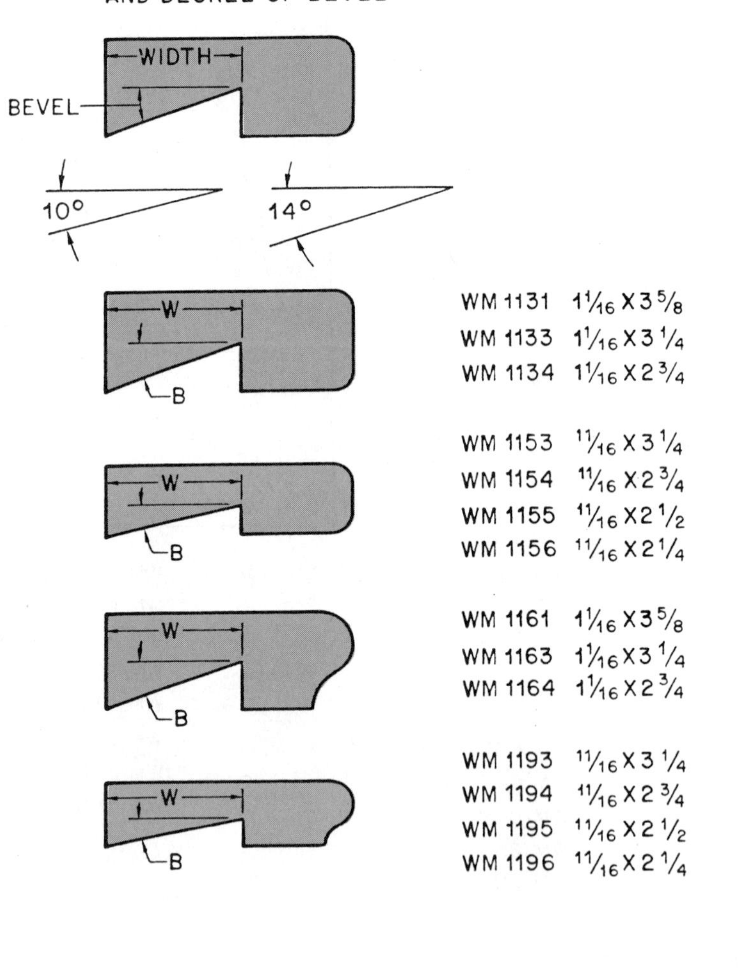

Fig. 78-1 *(continued)*

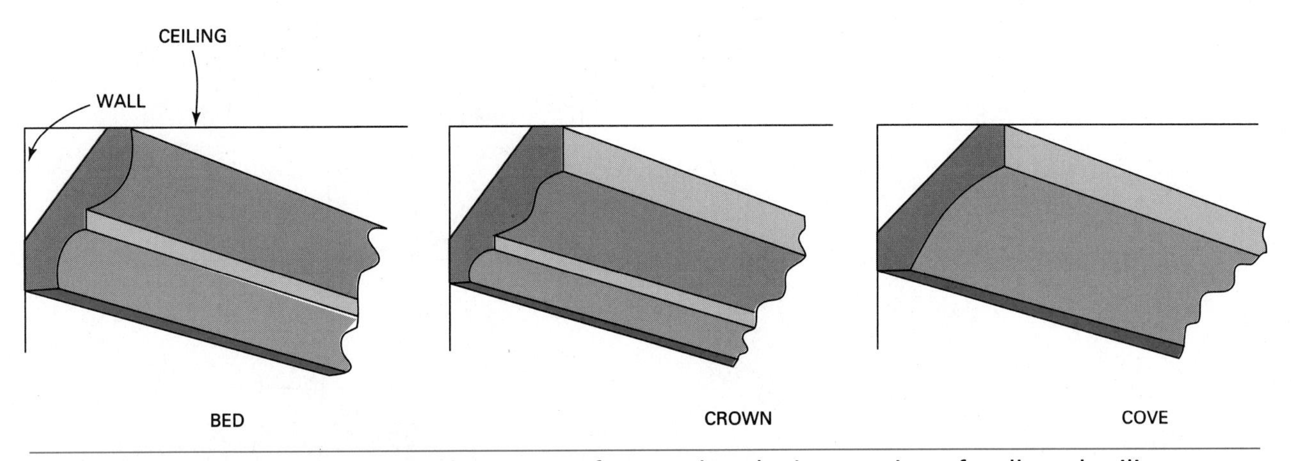

Fig. 78-2 Bed, crown, and cove moldings are often used at the intersection of walls and ceilings.

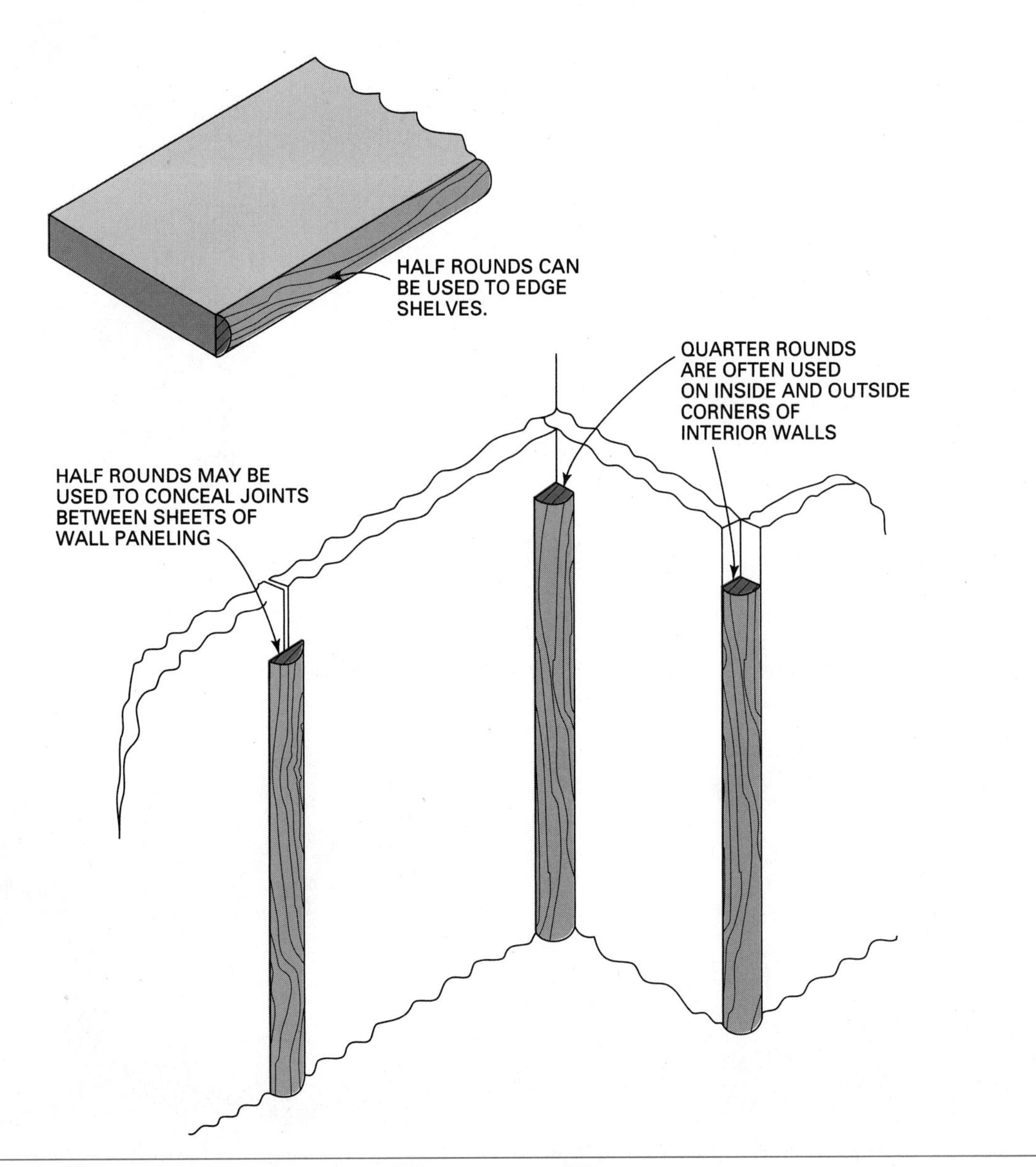

Fig. 78-3 Half round and quarter round moldings are used for many purposes.

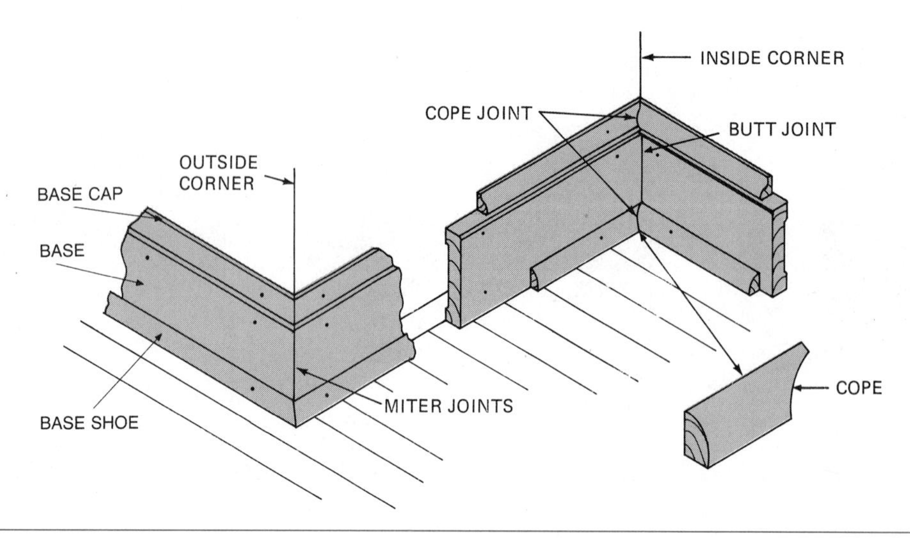

Fig. 78-4 Base, base shoe, and base cap are used to trim the bottom of the wall.

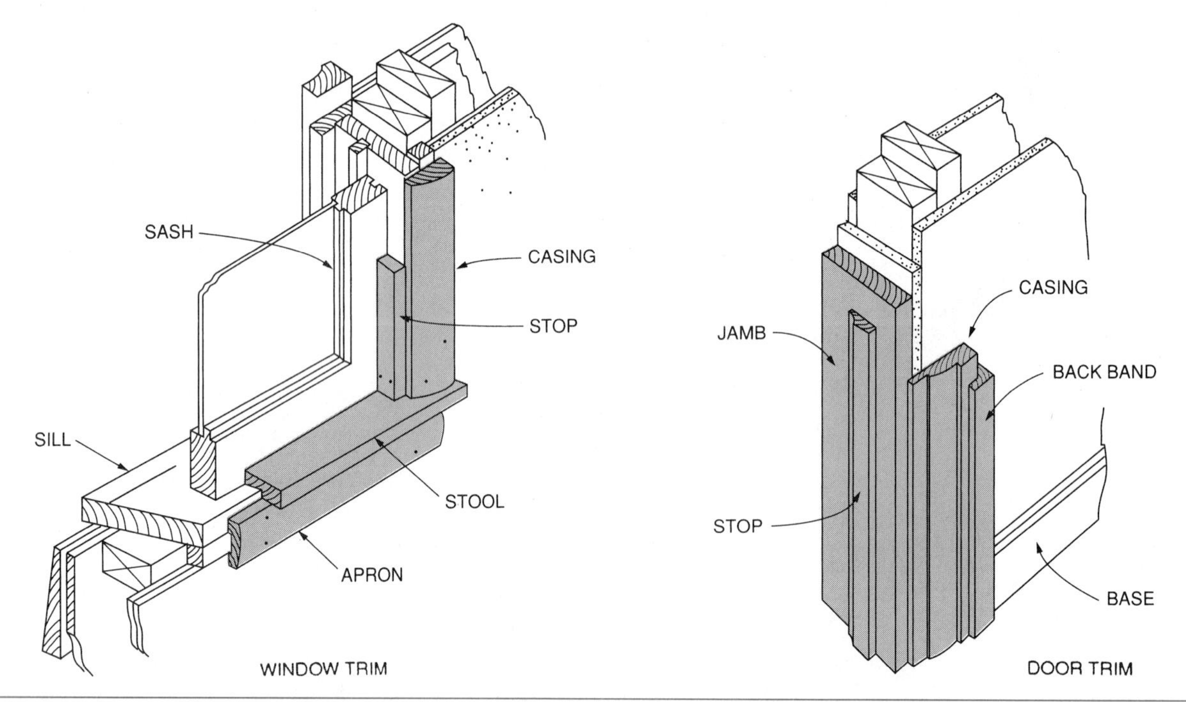

Fig. 78-5 Casing, back bands, and stops are used for window and door trim. Stools and aprons are part of window trim.

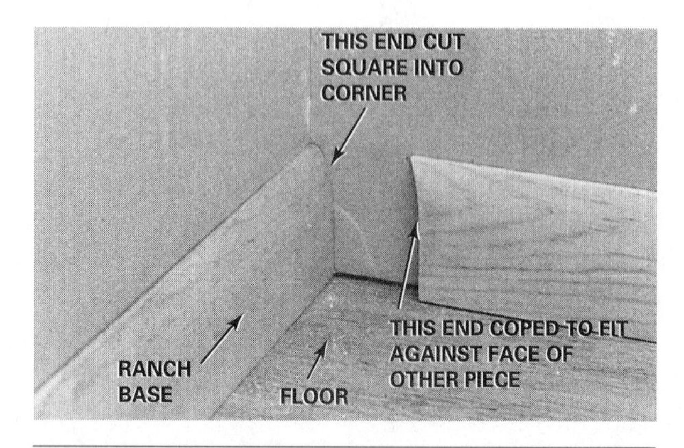

Fig. 78-6 A coped joint is made by fitting the end of one piece of molding against the shaped face of the other piece.

Fig. 78-7 Using a metal miter box.

Fig. 78-8 Using a power miter box. It is commonly called a chop saw by workers in the trade.

The most popular way to cut miters and other end cuts on trim is with a power miter box (Fig. 78-8). With this tool, a carpenter is able to cut virtually any angle with ease, whether it is a simple or a compound miter (one with two angles). Fine adjustments to a piece of trim, +/− 1/64 inch, can be made with great speed and accuracy. The power miter box is discussed in more depth in Chapter 18 and seen in figures 18-17, 18-18, and 18-19.

Positioning Molding in the Miter Box

Placing molding in the correct position in the miter box is essential for accurate *mitering*. Cut all moldings with their face sides or edges up or toward the operator so that the saw splinters out the back side, not the face side. Position the molding with one back side or edge against the bottom of the miter box and the other against the side of the miter box. On a wood miter box, which ordinarily has two sides, hold the molding against the side farthest away from the worker.

Flat miters are cut by holding the molding with its face side up and its thicker edge against the side of the miter box. Some moldings, such as base, base cap and shoe, and chair rail, are held right side up. Their bottom edge should be against the bottom of the miter box and their back against the side of the miter box (Fig. 78-9).

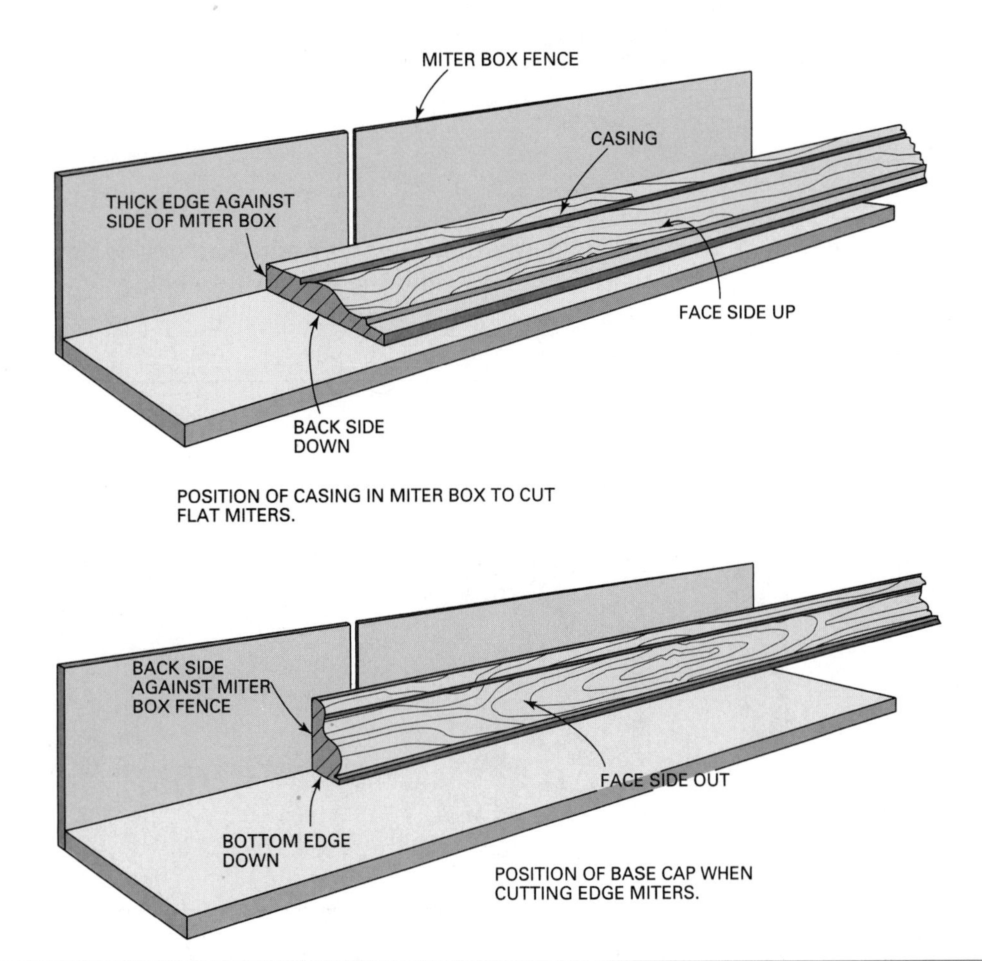

Fig. 78-9 Position of molding in a miter box to cut flat and edge miters.

Mitering Bed, Crown, and Cove molding. *Bed, crown,* and *cove* molding must be positioned upside down in the miter box. Their top edge is placed against the bottom of the miter box and their bottom edge against the far side (Fig. 78-10). This is the only way these types of moldings can be positioned accurately and held steady while being mitered. This position is also convenient because the bottom edge of this type of molding is the edge that is usually marked for cutting to length. With the bottom edge up, the mark, on which to start the cut, can be easily seen.

A thin, narrow strip of wood fastened to the bottom of the miter box and against the molding helps prevent the molding from moving when being mitered. The strip also assures that subsequent pieces of the same type of molding will be positioned at the same angle. Therefore, they will be mitered the same as the first piece (Fig. 78-11).

To position *corner guards* and other rabbeted molding, such as *back bands* and *caps,* for mitering, temporarily fasten a small strip of wood into the far corner of the miter box. The width and thickness of the strip should be slightly more than the rabbet of the molding. The molding is positioned on the strip and held steady while being mitered (Fig. 78-12).

Using a Miter Trimmer

Miters may also be made with a hand-operated *miter trimmer* (Fig. 78-13). All cuts on the trimmer must be made at or near the end of the molding. All softwood molding and trim are usually rough mitered with a chop cut. They are trimmed with one or two shaving

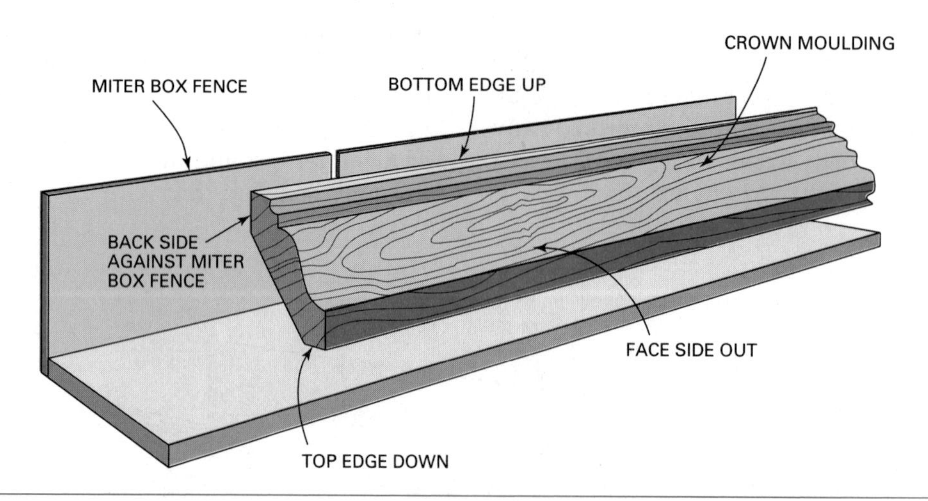

Fig. 78-10 To miter bed, crown, and cove moldings, they must be positioned upside down in the miter box. It may be helpful to think of the fence as the wall and the base as the ceiling.

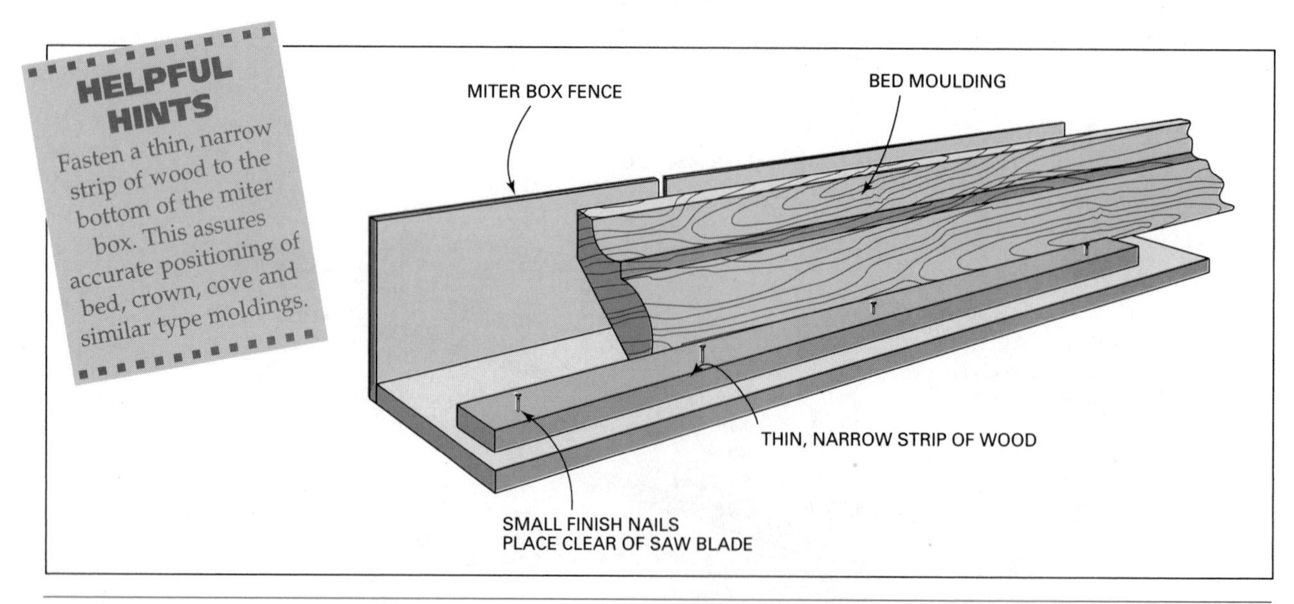

Fig. 78-11 Technique for holding a wide piece of molding in the proper position for cutting.

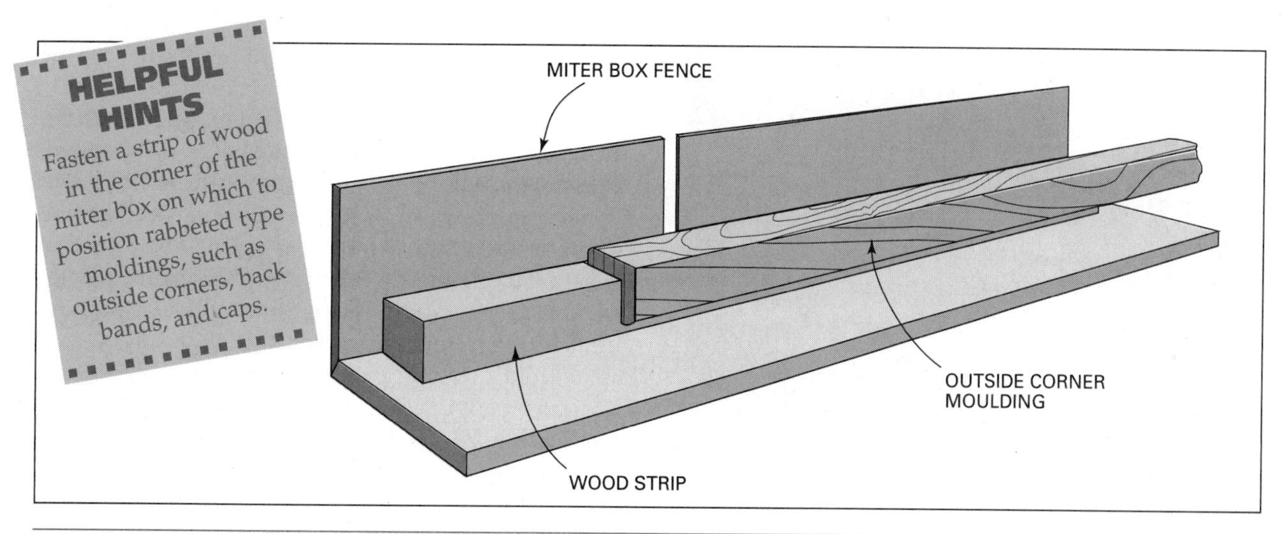

Fig. 78-12 Technique for holding an outside corner for cutting.

Fig. 78-13 The miter trimmer is widely used for fast and accurate mitering. (*Courtesy of Pootatuck Corporation*)

Fig. 78-14 With a mitering jig, left- and right-hand miters can be made quickly and easily without changing the setup (Guard has been removed for clarity).

slices. Hardwoods must be sawed at the approximate miter angle and then trimmed. The harder the wood, the thinner the cut must be made. When miter joints need fitting, paper-thin corrective cuts are made quickly and easily with the miter trimmer.

Mitering with the Table and Radial Arm Saw

Miters may also be made by using the table saw or the radial arm saw. The use of mitering jigs is helpful when making flat miters on window and door casings. The jigs allow both right- and left-hand miters to be cut quickly and easily without any changes in the setup (Fig. 78-14). (The construction and use of mitering jigs for table and radial arm saws are more fully described in Unit 7, Stationary Power Tools.)

Making a Coped Joint

To cope the end of molding, first make a *back miter* on the end. A back miter starts from the end and is cut back on the face of the molding (Fig. 78-15). The edge of the cut along the face forms the profile of the cope. Rub the side of a pencil point lightly along the profile to outline it more clearly.

Use a coping saw with a fine-tooth blade. Cut along the outlined profile with a slight undercut.

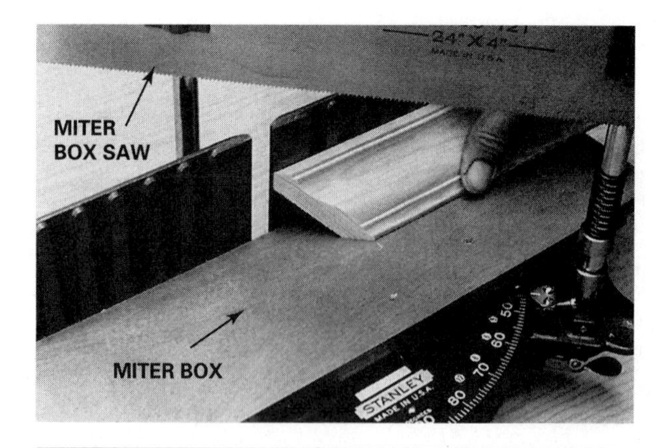

Fig. 78-15 Making a back miter on a piece of crown molding.

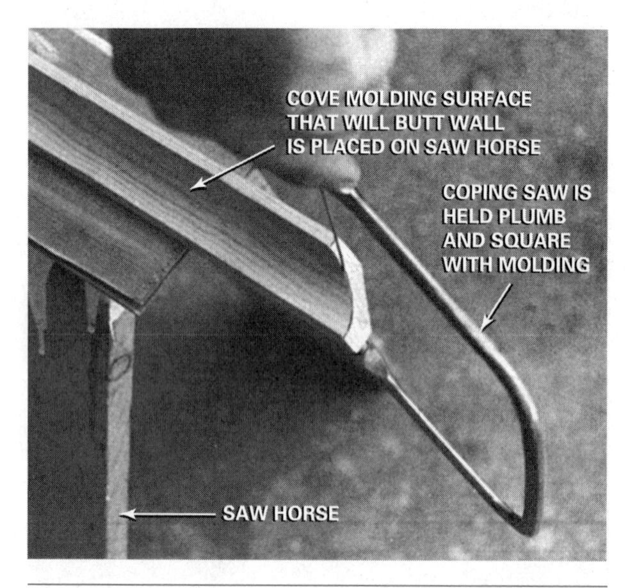

Fig. 78-16 Coping the end of a cove molding.

Cut with the handle of the coping saw above the work and the teeth of the blade pointing away from the handle (Fig. 78-16). Hold the molding so it is over the end of a sawhorse. The side of the molding that will butt the wall should be lying flat on the top of the sawhorse. It is important that the molding be held in this position. Holding it any other way makes it difficult to cut the cope with an undercut so that it will fit properly. It may be necessary to touch up the cut with a wood file or sharp utility knife.

Coped joints are used on interior wall corners. They will not open up when the molding is nailed in place, especially if the backing is not solid. Miter joints may open up in interior corners when the ends are fastened.

Applying Molding

To apply chair rail, caps, or some other type molding located on the wall, chalk lines should be snapped. This makes sure molding is applied in a straight line. No lines need to be snapped for base moldings or for small-size moldings applied at the intersection of walls and ceiling.

For large-size ceiling moldings, such as beds, crowns, and coves, a chalk line should be snapped. This assures straight application of the molding and easier joining of the pieces. Without a straight line to guide application, the molding may be forced at different angles along its length. This results in a noticeably crooked bottom edge and difficulty making tight-fitting miters and copes.

Hold a short scrap piece of the large-size molding at the proper angle at the wall and ceiling intersection (Fig. 78-17). Lightly mark the wall along the bottom edge of the molding. Measure the distance from the ceiling down to the mark. Measure and mark this same distance down from the ceiling on each end of each wall to which the molding is to be applied. Snap lines between the marks. Apply the molding so its bottom edge is to the chalk line.

Apply the molding to the first wall with square ends in both corners. If more than one piece is required to go from corner to corner, install the first piece with both ends square. If desired to make mitered joints between lengths, cut a square end into the corner and a back miter on the other end. On subsequent lengths, make matching end joints until the other corner is reached.

On some moldings, such as quarter rounds and small cove moldings, the straight, back surfaces should, but may not always, be of equal width. One of the back surfaces of these moldings should be marked with a pencil to assure positioning them in the miter box the same way each time. Mitering the molding with the same side down each time helps make fitting more accurate, faster, and easier (Fig. 78-18).

If a small-size molding is used, fasten it with finish nails in the center. Use nails of sufficient

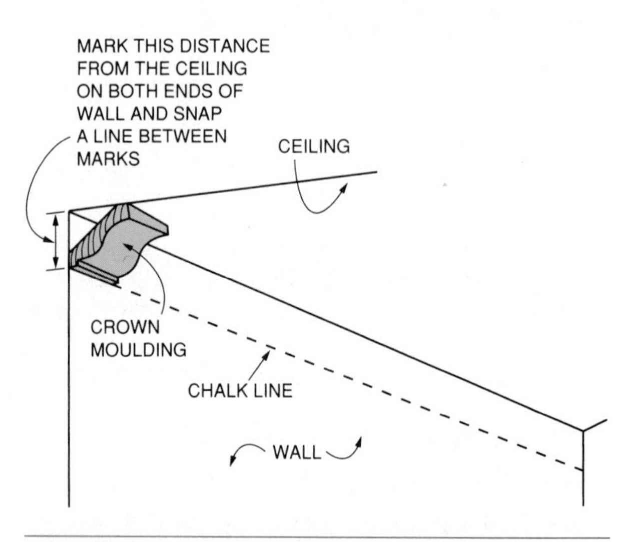

Fig. 78-17 Hold a scrap piece of molding against the ceiling and wall to determine the distance from the ceiling to its bottom edge.

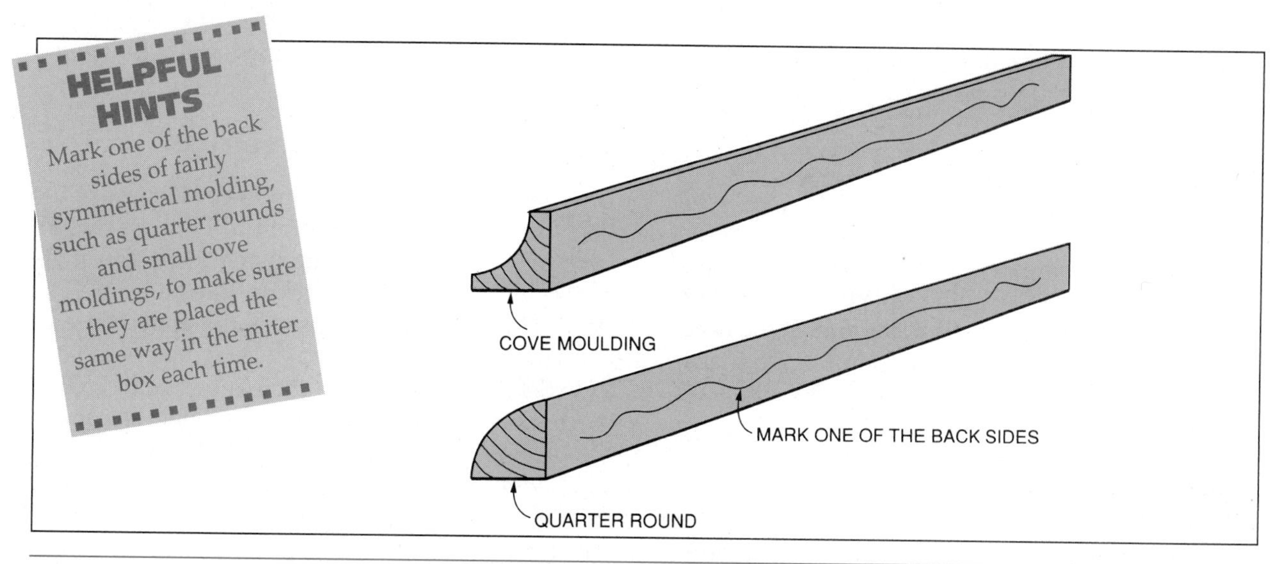

Fig. 78-18 Technique for reducing the confusion of working with molding that has a fairly symmetrical cross section.

length to penetrate into solid wood at least one inch. If large-size molding is used, fastening is required along both edges (Fig. 78-19).

Press the molding in against the wall or intersection with one hand while driving the nail almost home. Then set the nail below the surface. Nail at about 16-inch intervals and in other locations as necessary to bring the molding tight against the surface. End nails should be placed 2 to 3 inches from the end to keep from splitting the molding. If it is likely that the molding may split, blunt the pointed end of the nail. Or, drill a hole slightly smaller than the nail diameter.

Install the last piece on the first wall by first squaring one end. Place the square end in the corner. Let the other end overlap the first piece. Mark and cut it at the overlap. This method is more accurate than measuring and then transferring the measurement to the piece. Mark all pieces of interior trim for length in this manner whenever possible. Cut and fasten the last piece with its square end into the corner.

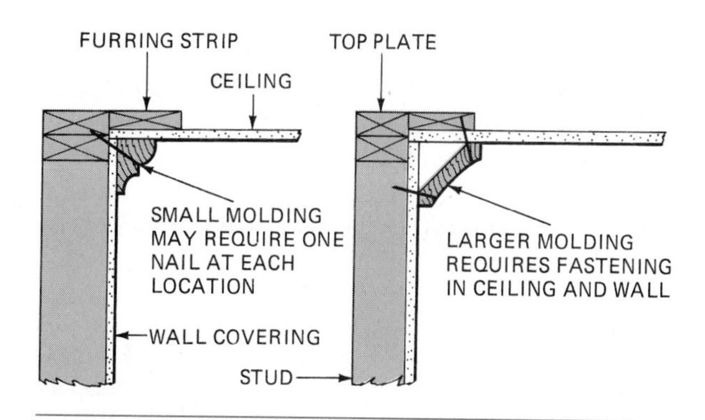

Fig. 78-19 Methods of fastening molding.

Cope the starting end of the first piece on each succeeding wall against the face of the last piece installed on the previous wall. Work around the room in one direction, either clockwise or counterclockwise. The end of the last piece installed must be coped to fit against the face of the first piece.

CHAPTER 79 APPLICATION OF DOOR CASINGS, BASE, AND WINDOW TRIM

In addition to wall and ceiling molding, the application of *door casings, base, base cap, base shoe,* and *window trim* is a major part of interior finish work. Care must be taken to avoid marring the work and to make neat, tight-fitting joints.

Door Casings

Door casings are moldings applied around the door opening. They trim and cover the space between the door frame and the wall. Casings must be applied before any base moldings because the base butts against the edge of door casing (Fig. 79-1). Door casings extend to the floor.

Design of Door Casings

Moldings or *S4S* stock may be used for door casings. S4S is the abbreviation for *surfaced four sides.* It is used to describe smooth, square-edge lumber. When molded casings are used, the joint at the head must be mitered unless butted against *plinth blocks.* Plinth blocks are small decorative blocks. They are thicker and wider than the door casing. They are used as part of the door trim at the base and at the head (Fig. 79-2).

When using S4S lumber, the joint may be mitered or butted. If a butt joint is used, the head casing overlaps the side casing. The appearance of S4S casings and some molded casings may be enhanced with the application of *back bands* (Fig. 79-3).

Molded casings usually have their back sides *backed* out. In cases where the jamb edges and the wall surfaces may not be exactly flush with each other, the backed out surfaces allow the casing to come up tight on both wall and jamb (Fig. 79-4). If S4S casings are used, they must be backed out on the job. (A method of backing out S4S lumber is described in Chapter 19, Table Saws, and is illustrated in Fig. 19-6.)

Applying Door Casings

Door casings are set back from the inside face of the door frame a distance of about 5/16 inch. This allows room for the door hinges and the striker plate of the door lock. This setback is called a *reveal* (Fig. 79-5). The reveal also improves the appearance of the door trim.

Set the blade of the combination square so that it extends 5/16 inch beyond the body of the square. Gauge lines at intervals along the side and head jamb edges by riding the square against the inside face of the jamb. Let the lines intersect where side and head jambs meet.

The following procedure applies to molded door casings mitered at the head. If several door openings are to be cased, cut the necessary number of *casings* to rough lengths with a miter cut on one end of each piece. Rough lengths are a few inches longer than actually needed. For each interior door opening, four *side casings* and two *head casings* are required. Cut side casings in pairs with right- and left-hand miters for use on both sides of the opening.

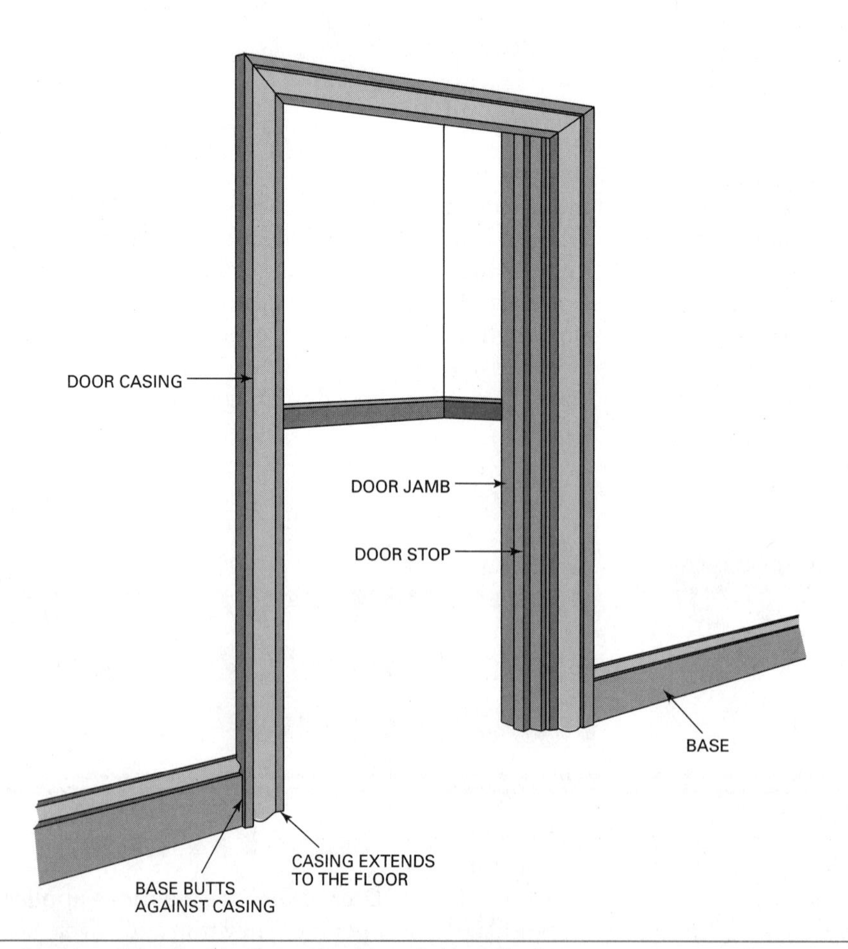

Fig. 79-1 Door casings are applied before the base is installed.

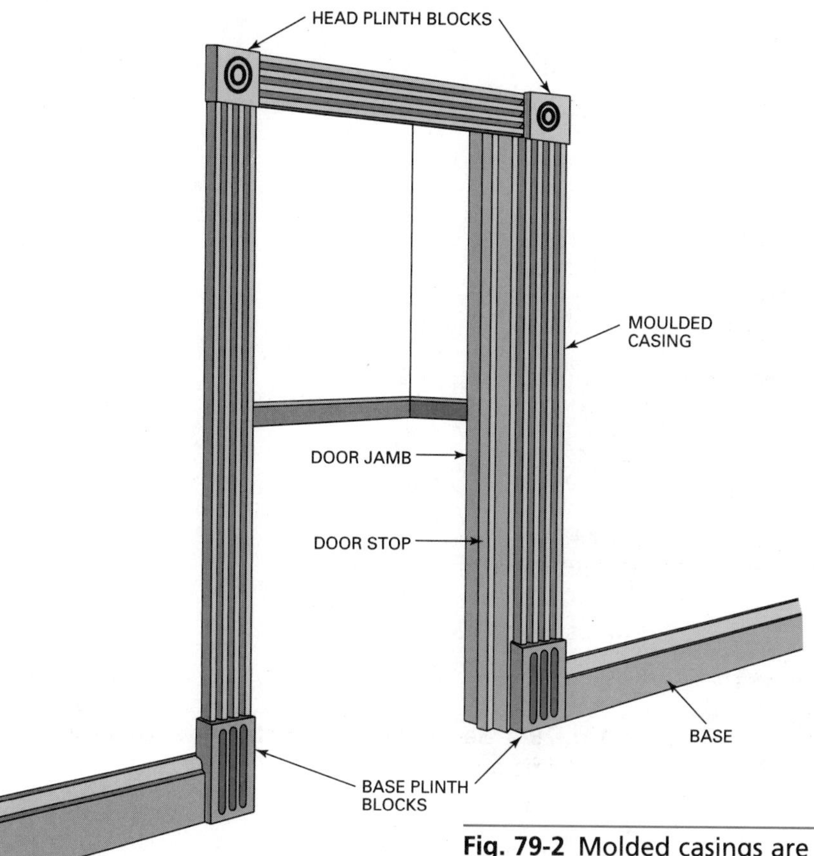

Fig. 79-2 Molded casings are mitered at the head unless plinth blocks are used.

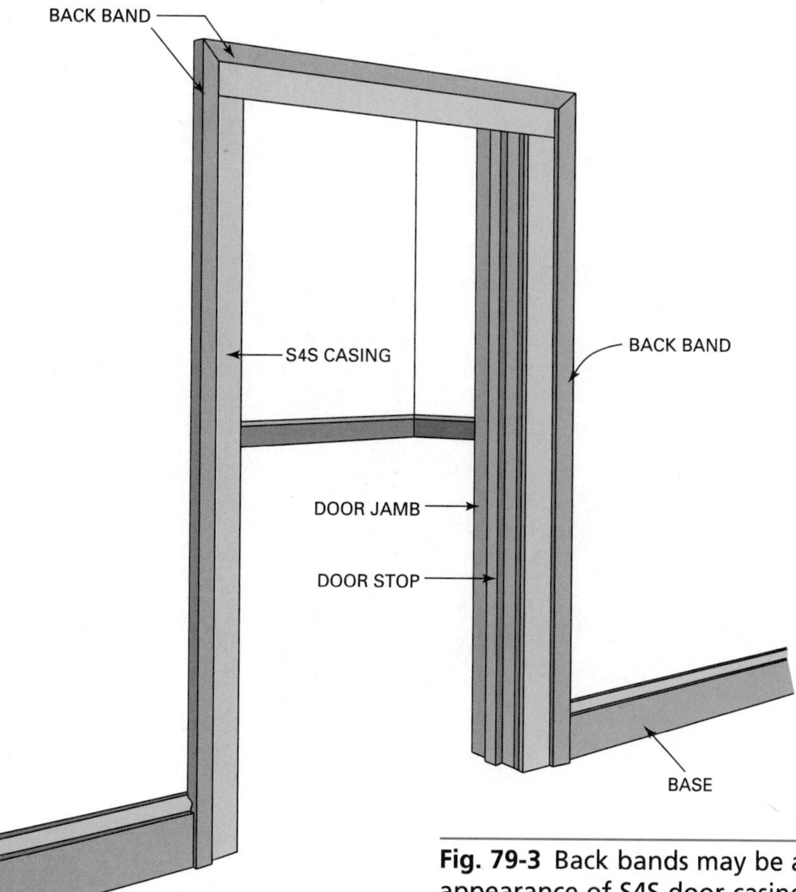

Fig. 79-3 Back bands may be applied to improve the appearance of S4S door casings. They may also be used on some types of molded casings.

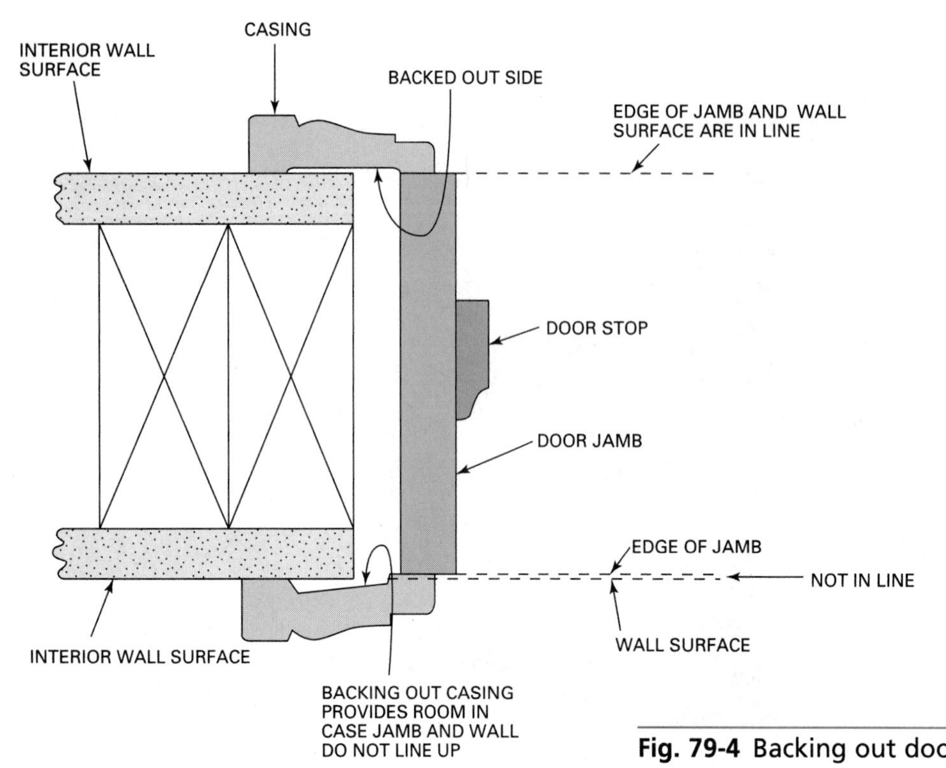

Fig. 79-4 Backing out door casings allows for a tight fit on wall and jamb.

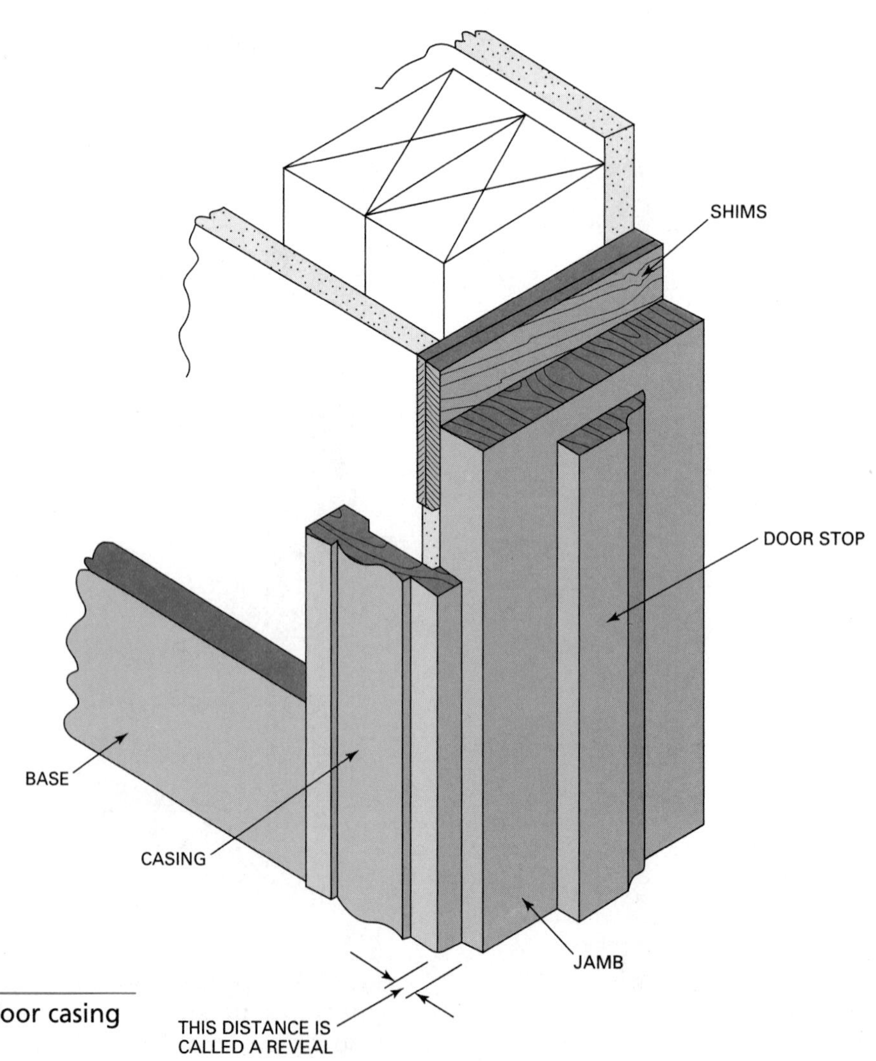

Fig. 79-5 The setback of the door casing on the jamb is called a reveal.

Applying the Head Casing

Miter one end of the head casing. Hold it against the head jamb of the door frame so that the miter is on the intersection of the gauged lines. Mark the length of the head casing at the intersection of the gauged lines on the opposite side of the door frame. Miter the casing to length at the mark.

Fasten the head casing in position. Its inside edge should be to the gauged lines on the head jamb. The mitered ends should be in line with the gauged lines on the side jambs. Use finish nails along the inside edge of the casing into the header jamb. If the casing edge is thin, use 3d or 4d finish nails spaced about 12 inches apart. Keep the edge of the casing to the gauged lines on the jamb. Straighten the casing as necessary as nailing progresses. Drive nails at the proper angle to keep them from coming through the face or back side of the jamb. Pneumatic finish nailers speed up the job of fastening interior trim.

Fasten the top edge of the casing into the framing. The outside edge is thicker, so longer nails are used, usually 6d or 8d finish nails. They are spaced farther apart, about 16 inches OC (Fig. 79-6). Do not drive end nails at this time. It may be necessary to move the ends slightly to fit the mitered joint between head and side casings.

Applying the Side Casings

Mark one of the previously mitered side casings by turning it upside down with the point of the miter touching the floor. If the finish floor has not been laid, hold the point of the miter on a scrap block of wood that is equal in thickness to the finish floor. Mark the side casing in line with the top edge of the head casing (Fig. 79-7). Make a square cut on the casing at the mark.

Place the side casing in position. Try the fit at the mitered joint. If the joint needs fitting, trim the mitered end of the side casing by planing thin shavings with a sharp block plane. The joint may also be fitted by shimming the casing away from the side of the chop saw and making a thin corrective cut. Shim either near or far from the saw blade as needed to hold the casing at the desired angle (Fig. 79-8). When fitted, apply a little glue to the joint. Nail the side casing in the same manner as the head casing.

Avoid sanding the joint to bring the casing faces flush. It is difficult to keep from sanding

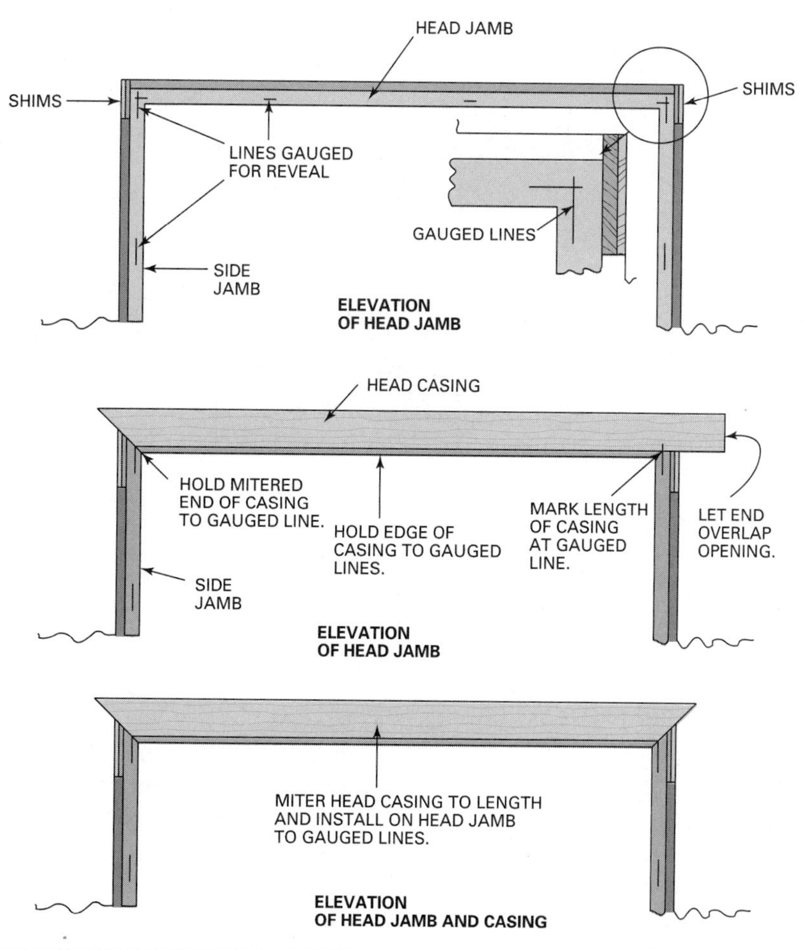

Fig. 79-6 The header casing is mitered on both ends and fastened in place.

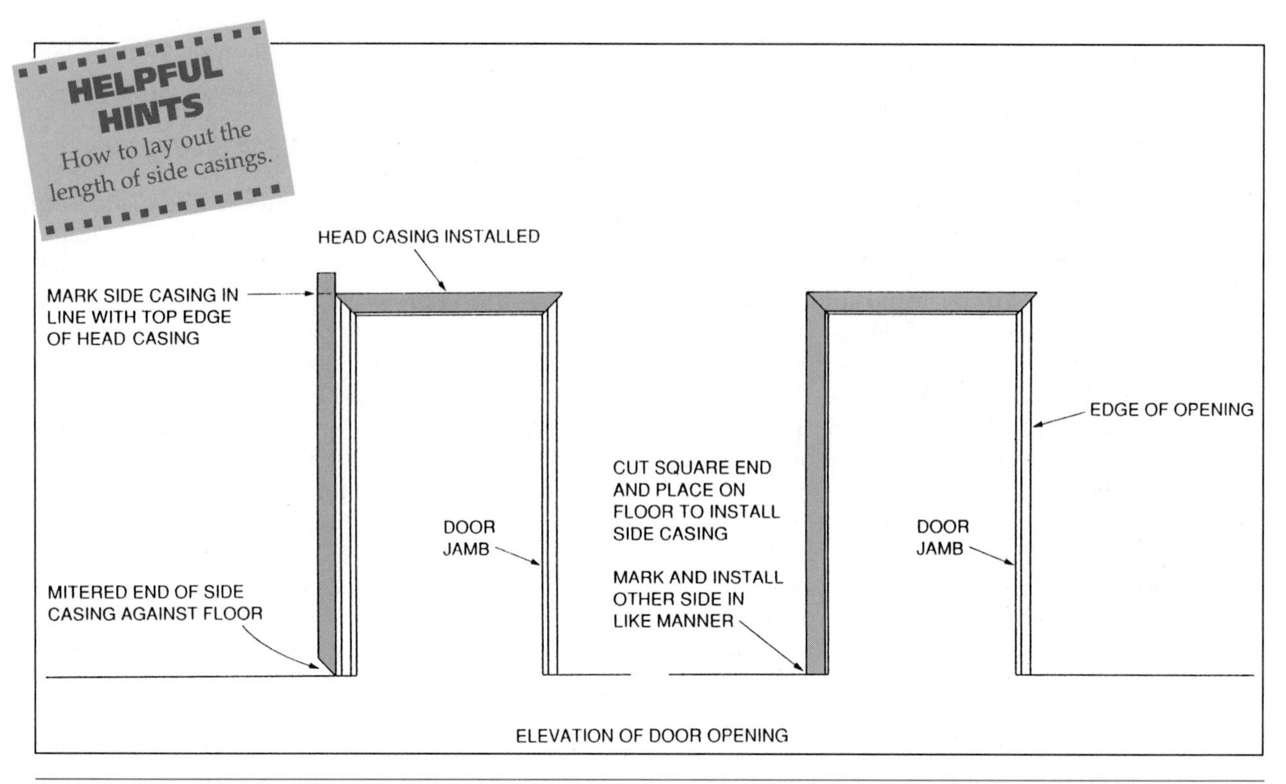

Fig. 79-7 Technique for cutting the casing side pieces.

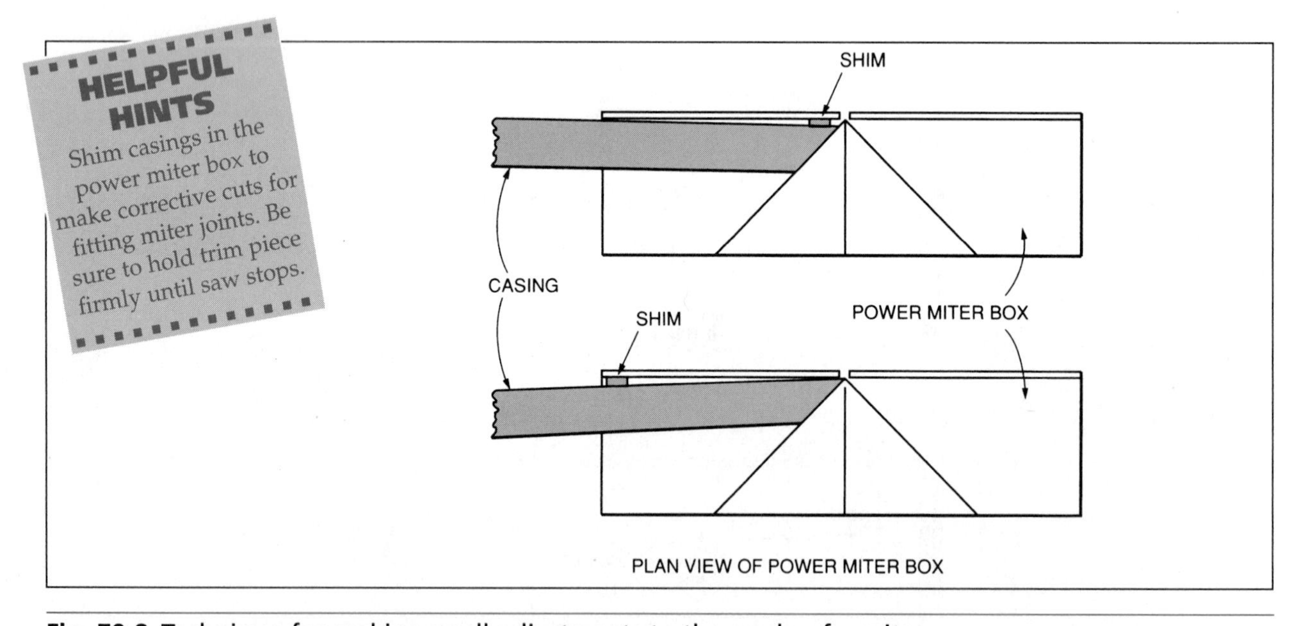

Fig. 79-8 Technique for making small adjustments to the angle of a miter.

across the grain on one or the other of the pieces. Cross-grain scratches will be very noticeable, especially if the trim is to have a stained finish. Bring the faces flush, if necessary, by shimming between the back of the casing and the wall. Usually, only very thin shims are needed. Any small space between the casing and the wall can be filled later with joint filling compound. Also, the backside of the thicker piece may be planed or chiseled thinner. Most carpenters prefer to do these rather than try to sand the joint.

Drive a 4d finish nail into the edge of the casing and through the mitered joint. Drive end nails. Then set all fasteners. Keep nails 2 or 3 inches from the end to avoid splitting the casing. If there is danger of splitting, blunt the pointed end or drill a hole slightly smaller in diameter than the nail.

Applying Base Moldings

Molded or *S4S* stock may be used for base. If S4S base is used, it should be backed out. A *base cap* should be applied to its top edge. The base cap conforms easily to the wall surface, resulting in a tight fit against the wall. The base trim should be thinner than the door casings against which it butts. This makes a more attractive appearance.

The base is applied in a manner similar to wall and ceiling molding. However, copes are laid out for joints in interior corners. Instead of back-mitering to outline the cope, it is usually more accurate to lay out the cope by *scribing* (Fig. 79-9). When placed against the wall, the face of the base may not always be square with the floor. Therefore, if the base is tilted slightly, back-mitering to obtain the outline of the cope will result in a poor fit against it.

Apply the base to the first wall with square ends in each corner. Drive and set two finishing nails, of sufficient length, at each stud location. Nailing blocks previously installed during framing provide solid wood for fastening the ends of the base in interior corners.

Cut the base to go on the next wall about an inch longer than required. Lay the base against the wall by bending it so the end to be scribed lies flat against the wall and against the first base. Set the dividers to scribe about 1/2 inch. Lay out the cope by riding the dividers along the face of the base on the first wall.

Hold the dividers while scribing so that a line between the two points is parallel to the floor. Twisting the dividers while making the scribe results in an inaccurate layout. Cut the end to the scribed line with a slight undercut. Bend the base back in position and try the fit. If scribed and cut accurately, no adjustments should be necessary. Its overall length must now be determined.

Cutting the Base to Length

The length of a baseboard that fits between two walls may be determined by measuring from corner to corner. Then, transfer the measurement to the baseboard. Another method of determining its length eliminates using the rule. It may be faster and more accurate.

With the base in the last position described above, place marks, near the center, on the top edge of the base and the wall so they line up with each other. Place the other end in the opposite corner. Press the base against the wall at the mark. The difference between the mark on the wall and the mark on the base is the amount to scribe off the end in the corner. Set the dividers to this distance. Scribe the end. Cut to the scribed line (Fig. 79-10). If a tighter fit is desired, set the dividers slightly less than the distance between the marks. This method of fitting long lengths between walls may be applied to other kinds of trim. However, this works especially well with the base.

Place one end in the corner, and bow out the center. Place the other end in the opposite corner, and press the center against the wall. Fasten in place. Continue in this manner around the room in a previously planned order. Make regular miter joints on outside corners.

If both ends of a single piece are to have regular miters for outside corners, it is imperative that it be fastened in the same position as it was marked. Tack the rough length in position with one finish nail in the center. Mark both ends. Remove, and cut the miters. Installing the piece by first fastening into the original nail hole assures that the piece is fastened in the same position as marked (Fig. 79-11).

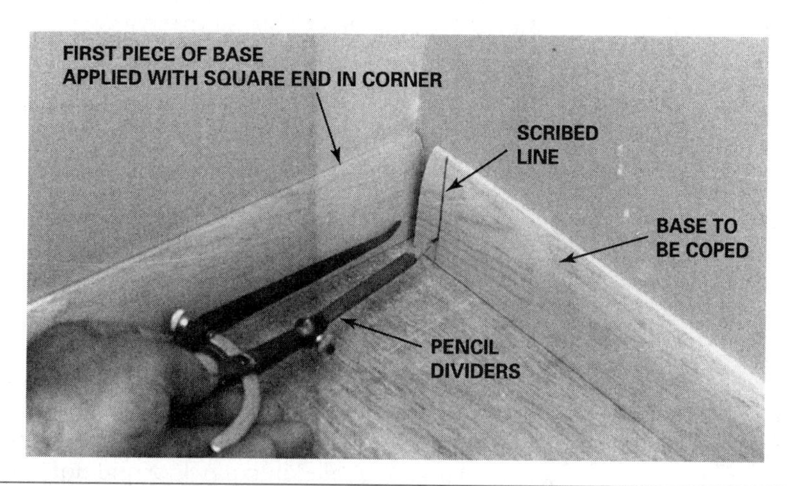

Fig. 79-9 Laying out a coped joint on base by scribing.

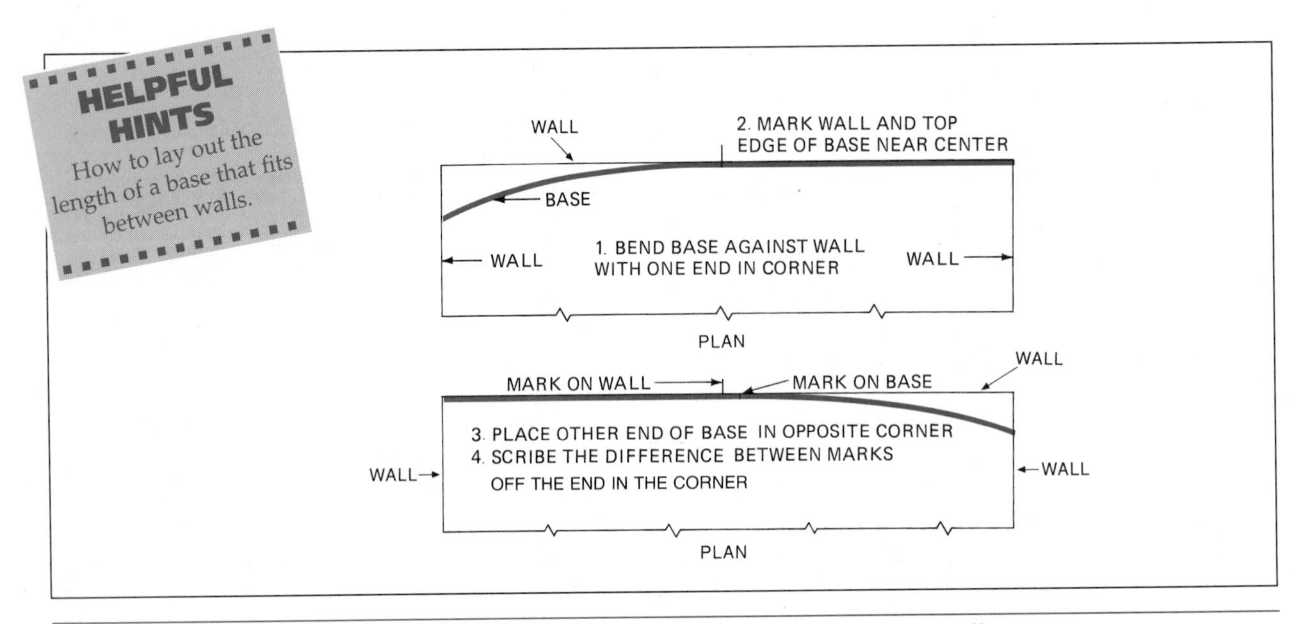

Fig. 79-10 Technique for cutting two ends of a trim board to fit between walls.

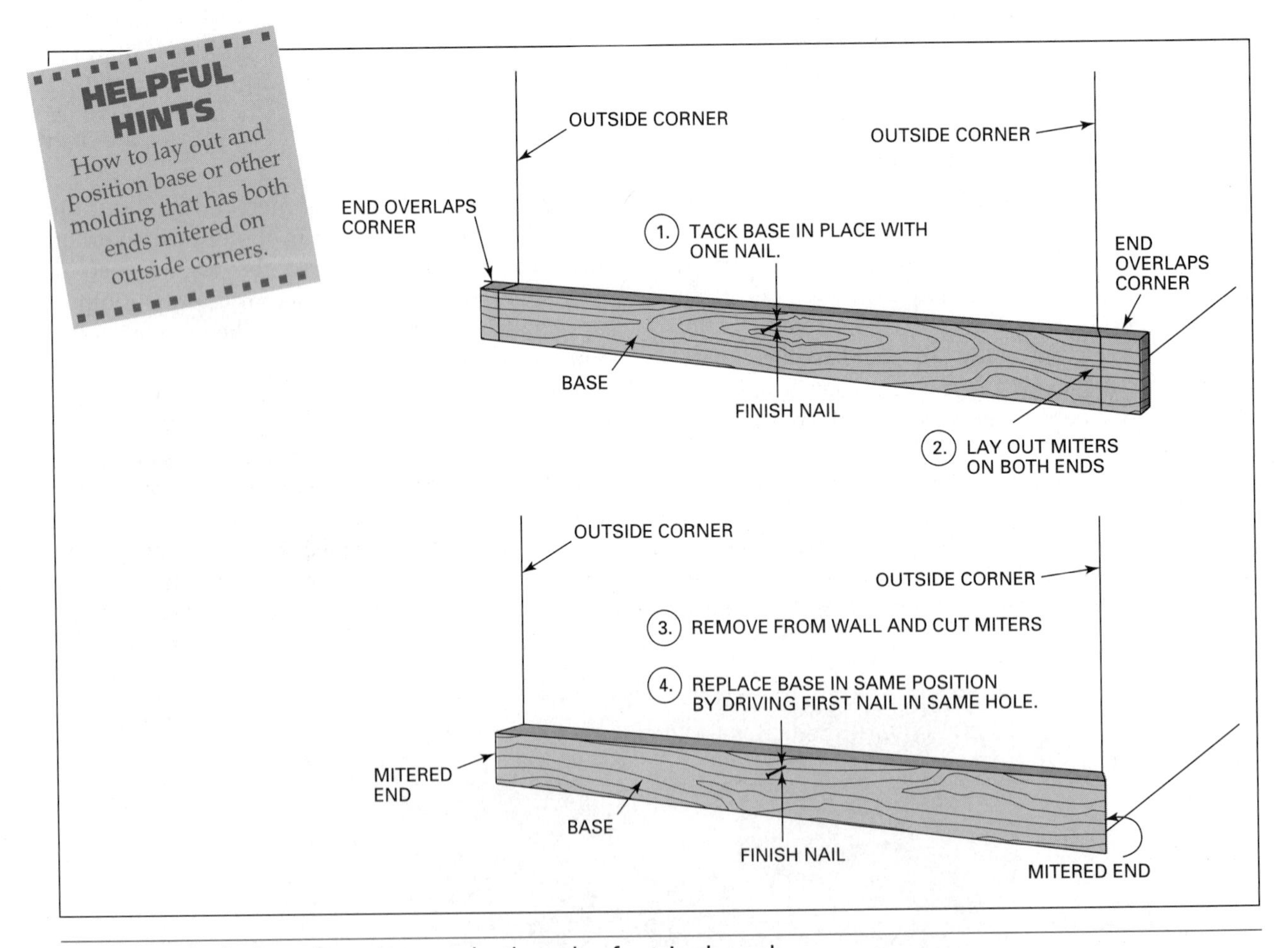

Fig. 79-11 Technique for miters on both ends of a trim board.

Applying the Base Cap and Base Shoe

The *base cap* is applied in the same manner as most wall or ceiling molding. Cope interior corners and miter exterior corners. The *base shoe* is also applied in a similar manner as other molding. However, it is ordinarily nailed into the floor and not into the baseboard. This prevents the joint under the shoe from opening should the shrinkage take place in the baseboard (Fig. 79-12).

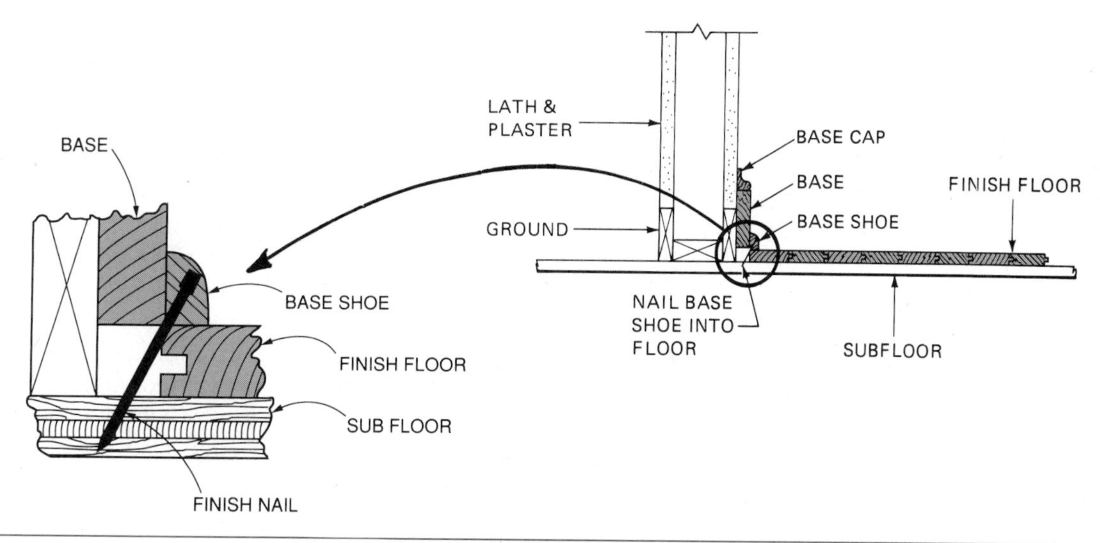

Fig. 79-12 The base shoe is fastened into the floor, not into the baseboard. This prevents the joint at the floor from opening due to any movement of the baseboard.

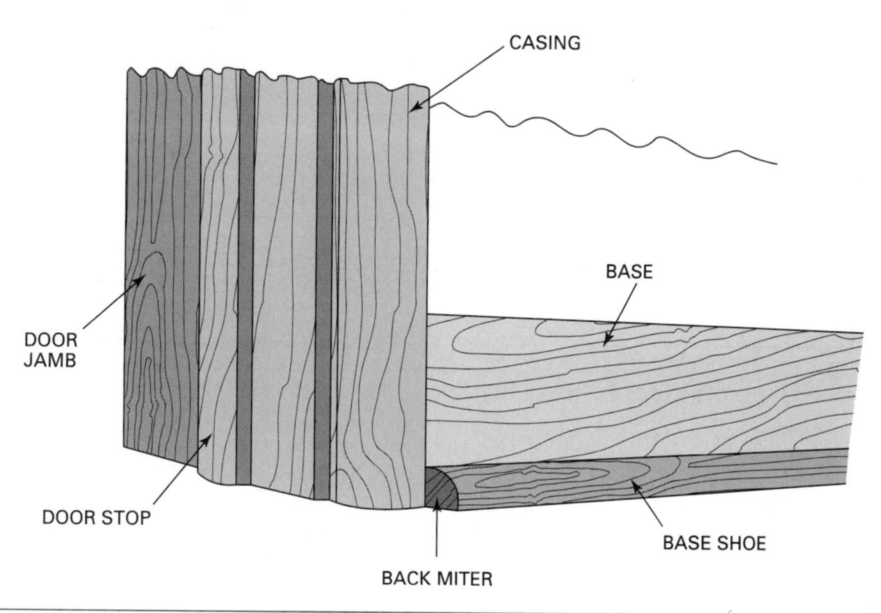

Fig. 79-13 The exposed ends of base shoe are usually back-mitered and sanded smooth.

Because the base shoe is a small-size molding and has solid backing, both interior and exterior corners are mitered. When the base shoe must be stopped at a door opening or other location, with nothing to butt against, its exposed end is generally *back-mitered* and sanded smooth (Fig. 79-13). No base shoe is required if carpeting is to be used as a finish floor.

Installing Window Trim

Interior window trim, in order of installation, consists of the *stool,* also called *stool cap, apron, jamb extensions, casings,* and *stops* or *stop bead.* Although the kind and amount of trim may differ, depending on the style of window, the application is basically the same. The procedure described in this chapter applies to most double hung windows.

Installing the Stool

The bottom side of the *stool* is rabbeted at an angle to fit on the sill of the window frame so its top side will be level. Its final position has the outside edge against the sash. Both ends are notched around the side jambs of the window frame. Each end projects beyond the casings by an amount equal to the casing thickness.

The stool length is equal to the distance between the outside edges of the vertical casing plus twice the casing thickness. On both sides of the window, just above the sill, hold a scrap piece of casing stock on the wall. Its inside edge should be flush with the inside face of the side jamb of the window frame. Draw a light line on the wall along the outside edge of the casing stock. Lay out a distance outward from each line equal to the thickness of the window casing. Cut a piece of stool stock to length equal to the distance between the outermost marks.

Raise the lower sash slightly. Place a short, thin strip of wood under it, on each side, which projects inward to support the stool while it is being laid out (Fig. 79-14). Place the stool on the strips. Raise or lower the sash slightly so the top of the stool is level. Position the stool with its outside edge against the wall. Its ends should be in line with the marks previously made on the wall.

Square lines, across the face of the stool, even with the inside face of each side jamb of the window frame. Set the pencil dividers or scribers so that, on both sides, an amount equal to twice the casing thickness will be left on the stool. Scribe the stool by riding the dividers along the wall on both sides and along the bottom rail of the window sash (Fig. 79-15).

Cut to the lines, using a handsaw. Smooth the sawed edge that will be nearest to the sash. Shape and smooth both ends of the stool the same as the inside edge. Apply a small amount of caulking compound to the bottom of the stool along its outside edge. Fasten the stool in position by driving finish nails along its outside edge into the sill. Set the nails.

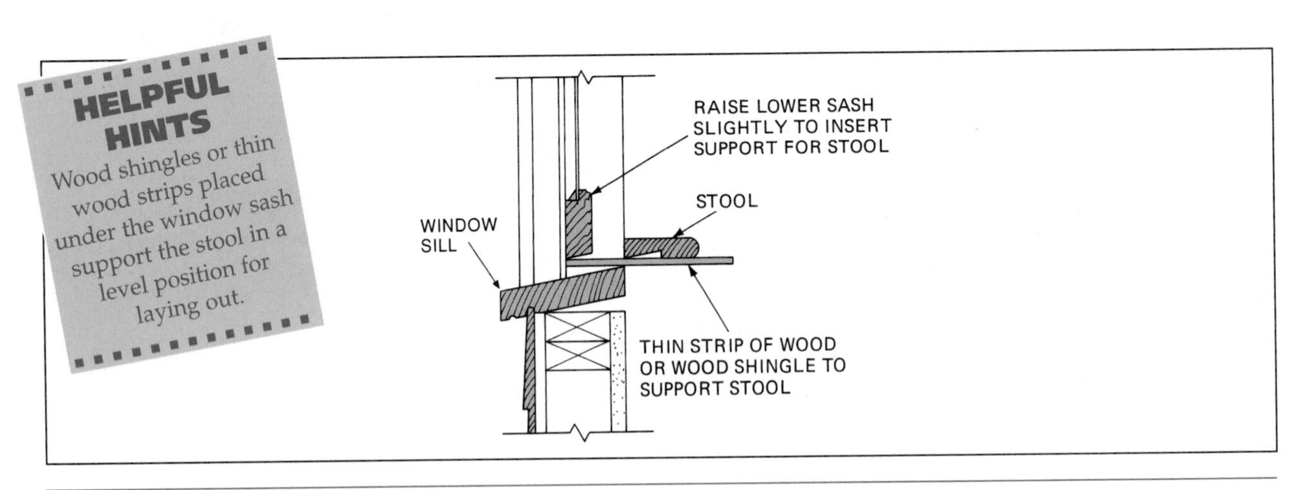

Fig. 79-14 Technique to hold a stool for easy layout.

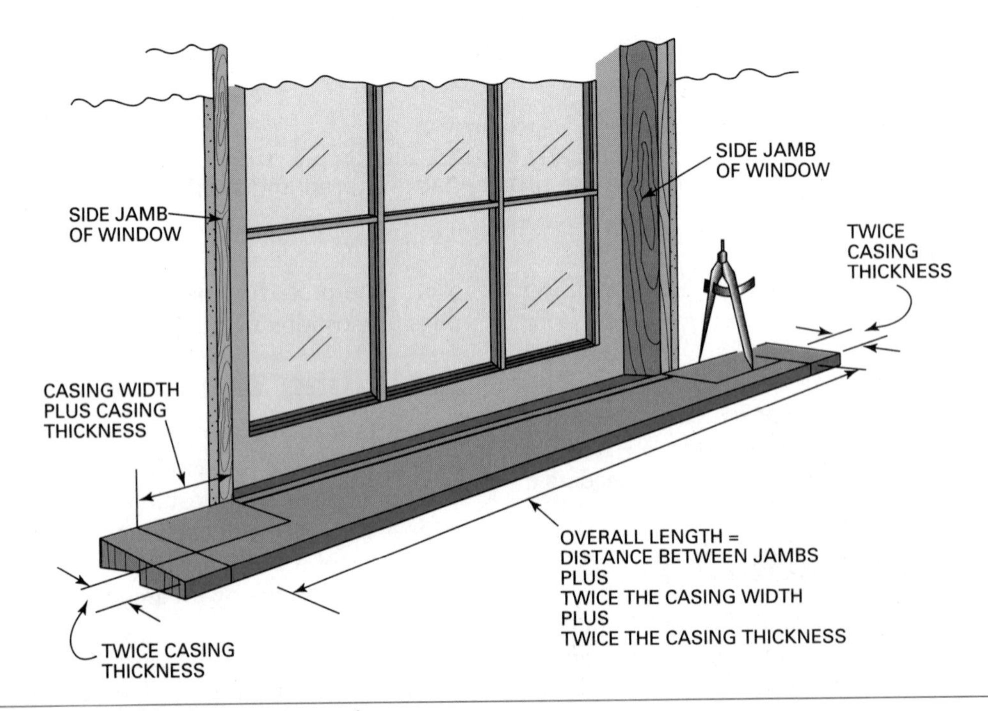

Fig. 79-15 Method of laying out a stool.

Applying the Apron

The apron covers the joint between the sill and the wall. It is applied with its ends in line with the outside edges of the window casing. Cut a length of apron stock equal to the distance between the outer edges of the window casings.

Each end of the apron is then *returned upon itself.* This means that the ends are shaped the same as its face. To return an end upon itself, hold a scrap piece on the apron. Draw its profile flush with the end. Cut to the line with a coping saw. Sand the cut end smooth (Fig. 79-16). Return the other end upon itself in the same manner.

Place the apron in position with its upper edge against the bottom of the stool. Be careful not to force the stool upward. Keep the top side of the stool level by holding a square between it and the edge of the side jamb. Fasten the apron along its bottom edge into the wall. Then drive nails through the stool into the top edge of the apron. When nailing through the stool, wedge a short length of 1 × 4 stock between the apron and the floor at each nail location. This supports the apron while nails are being driven. Failure to support the apron results in an open joint between it and the stool. Take care not to damage the bottom edge of the apron with the supporting piece (Fig. 79-17).

Installing Jamb Extensions

Windows are often installed with jambs that are narrower than the wall thickness. Strips must be fastened to these narrow jambs to bring the inside

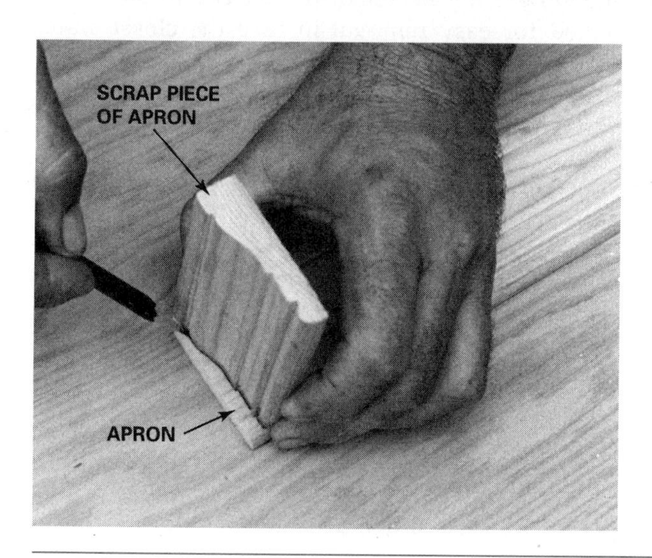

Fig. 79-16 Returning the end of an apron upon itself.

HELPFUL HINTS

Support the apron when fastening the stool to it.

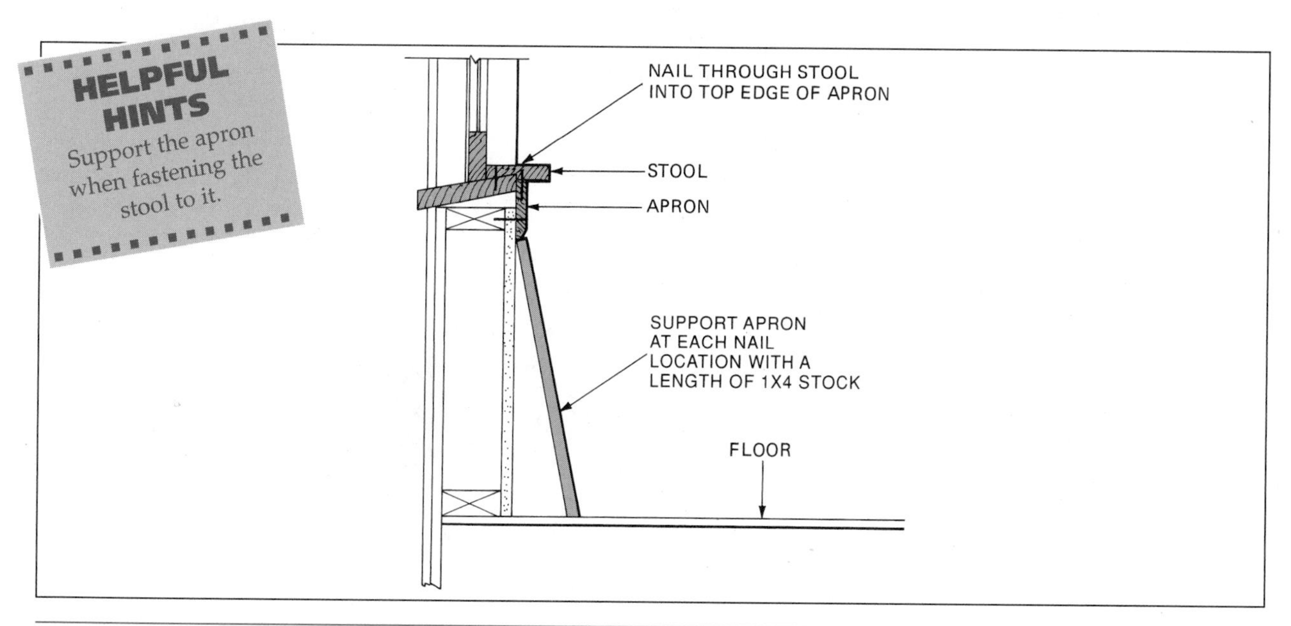

Fig. 79-17 Technique for holding an apron in place for nailing.

edges flush with inside wall surface. These strips are called *jamb extensions.*

Some manufacturers provide jamb extensions with the window unit. However, they are not applied when the window is installed, but when the window is trimmed. Therefore when windows are set, these pieces should be carefully stored and then retrieved when it is time to apply the trim. They are usually precut to length and need only to be cut to width.

Measure the distance from the inside edge of the jamb to the finished wall. Rip the jamb extensions to this width with a slight back-bevel on the inside edge. Cut the pieces to length, if necessary, and apply them to the header and side jambs. Drive finish nails through the edges into the edge of the jambs (Fig. 79-18).

Applying the Casings

Window casings usually are installed with a reveal similar to that of door casings. They also may be installed flush with the inside face. In either case, the bottom ends of the side casings rest on the stool. The window casing pattern is usually the same as the door casings. Window casings are applied in the same manner as door casings.

Cut the number of window casings needed to a rough length with a miter on one end. Cut side casings with left- and right-hand miters. Install the header casing first and then the side casings. Find the length of side casings by turning them upside down with the point of the miter on the stool in the same manner as door casings. Fasten casings with their inside edges flush with the inside face of the jamb. Make neat, tight-fitting joints at the stool and at the head.

Installing Closet Trim

A simple clothes closet is normally furnished with a *shelf* and a *rod* for hanging clothes. Usually a piece of 1 × 5 stock is installed around the walls of the closet to support the shelf and the rod. This piece is called a *cleat.* The shelf is installed on top of it. The closet pole is installed in the center of it. Shelves are not fastened to the cleat. Rods are installed for easy removal in case the closet walls need refinishing.

Shelves are usually 1 × 12 boards. Rods may be 3/4-inch steel pipe, 15/16-inch full round wood poles, or chrome plated rods manufactured for this purpose. On long spans, the rod may be supported in its center by special metal closet pole supports. On each end, the closet pole is supported by plastic or metal closet pole *sockets.* In place of sockets, holes and notches are made in the cleat to support the ends of the closet pole.

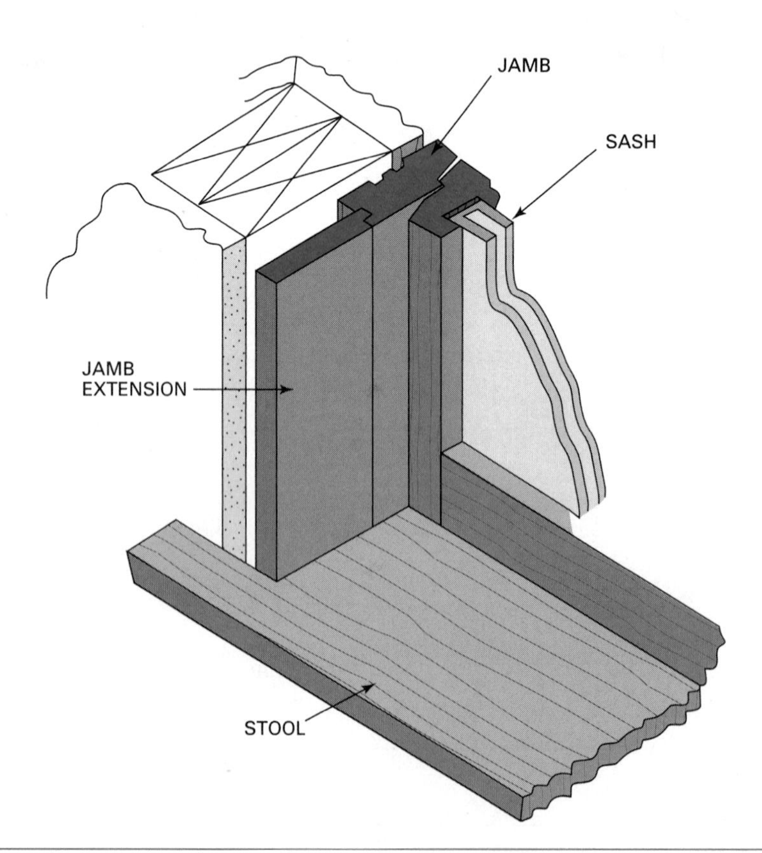

Fig. 79-18 Jamb extensions are used to widen the window jamb.

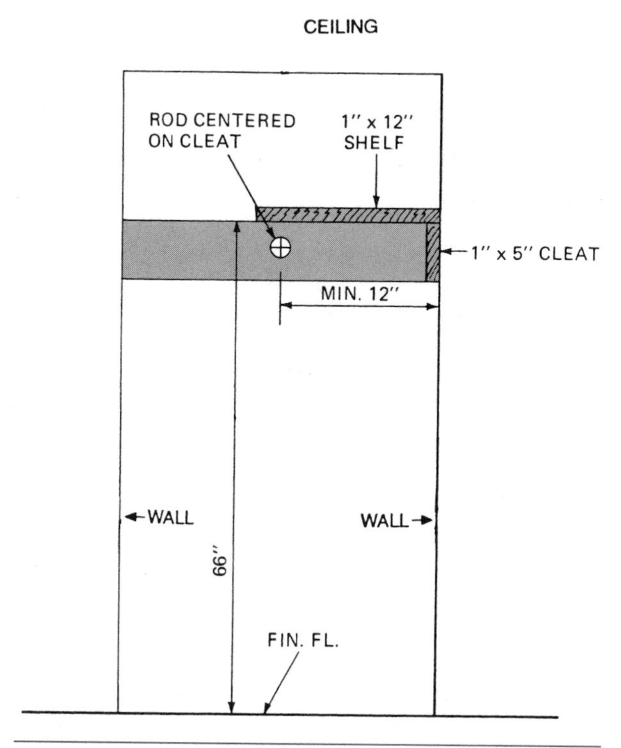

Fig. 79-19 Specifications for an ordinary clothes closet.

For ordinary clothes closets, the height from the floor to the top edge of the cleat is 66 inches (Fig. 79-19). Measure up from the floor this distance. Draw a level line on the back wall and two end walls of the closet. Ease the bottom outside corner and install the cleat so its top edges are to the line. The cleat is installed in the same manner as baseboard. Butt the interior corners. Fasten with two finish nails at each stud.

Install the closet pole sockets on the end cleats. The center of the socket should be at least 12 inches from the back wall and centered on the width of the cleat. Fasten the sockets through the predrilled holes with the screws provided.

Installing Closet Shelves

For a professional job, fit the ends and back edge of the shelf to the wall. Cut the shelf about 1/2 inch longer than the distance between end walls.

Place the shelf in position by laying one end on the cleat and tilting the other end up and resting against the wall. Scribe about 1/4 inch off the end resting on the cleat. Remove the shelf. Cut to the scribed line. Measure the distance between corners along the back wall. Transfer this measurement to the shelf, measuring from the scribed cut along the back edge of the shelf.

Place the shelf in position, tilted in the opposite direction. Set the dividers to scribe the distance from the wall to the mark on the shelf. Scribe and cut the other end of the shelf. Place the shelf into position, resting it on the cleats. Scribe the back edge to the wall to take off as little as possible. Cut to the scribed line. Ease the corners on the front edge of the shelf with a hand plane. Sand and place the shelf in position (Fig. 79-20).

Installing the Closet Pole

Measure the distance between pole sockets. Cut the pole to length. Install the pole on the sockets. One socket is closed. The opposite socket has an open top. Place one end of the pole in the closed socket. Then rest the other end on the opposite socket.

Linen Closets

Linen closets usually consist of a series of shelves spaced 12 to 16 inches apart. Cleats used to support shelves are $3/4 \times 1$ stock, chamfered on the bottom outside corner. A *chamfer* is a bevel on the edge of a board that extends only partway through the thickness of the stock.

Lay out level lines for the top edges of each set of cleats. Install the cleats and shelves in the same manner as described for clothes closets.

Mantels

Mantels are used to decorate fireplaces and to cover the joint between the fireplace and the wall. Most mantels come preassembled from the factory. They are available in a number of sizes and styles (Fig. 79-21).

Study the manufacturer's directions carefully. Place the mantel against the wall. Center it on the fireplace. Scribe it to the floor or wall as necessary. Carefully fasten the mantel in place and set all nails.

Conclusion

All pieces of interior trim should be sanded smooth after they have been cut and fitted, and before they are fastened. The sanding of interior finish provides a smooth base for the application of stains, paints, and clear coatings. Always sand with the grain, never across the grain.

All sharp, exposed corners of trim should be rounded over slightly. Use a block plane to make a slight chamfer. Then round over with sandpaper.

If the trim is to be stained, make sure every trace of glue is removed. Excess glue, allowed to dry, seals the surface. It does not allow the stain to penetrate, resulting in a blotchy finish.

Be careful not to make hammer marks in the finish. Occasionally rubbing the face of the hammer with sandpaper to clean it helps prevent it from glancing off the head of a nail.

STEP BY STEP
PROCEDURES

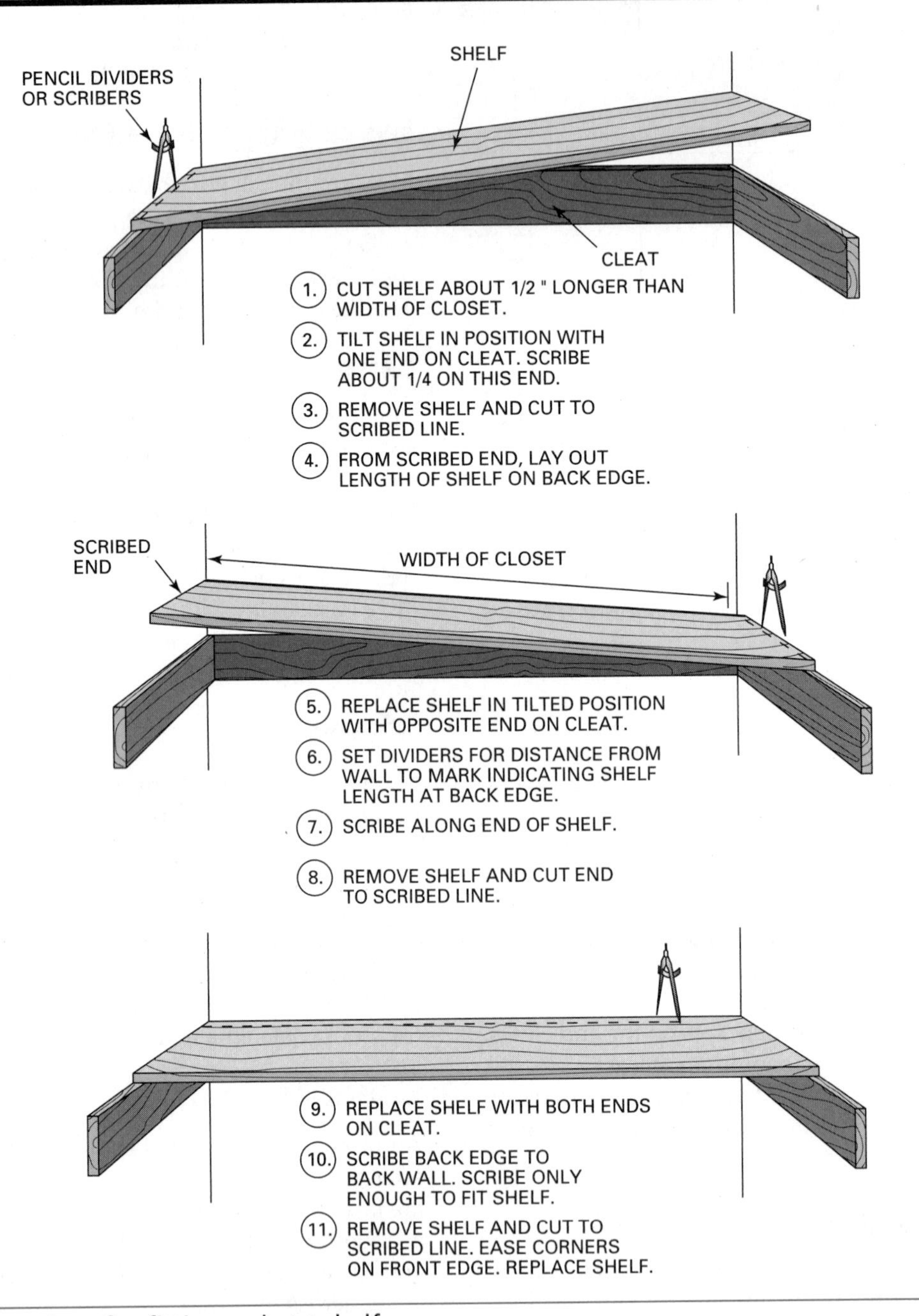

Fig. 79-20 Procedure for fitting a closet shelf.

Make sure any pencil lines left along the edge of a cut are removed before fastening the pieces. Pencil marks in interior corners are difficult to remove after the pieces are fastened in position. Pencil marks show through a stained or clear finish and make the joint appear open. When marking interior trim make light, fine pencil marks.

Note: Layout lines in the illustrations are purposely made dark and heavy only for the sake of clarity.

Make sure all joints are tight-fitting. Measure, mark, and cut carefully. Do not leave a poor fit. Do it over, if necessary!

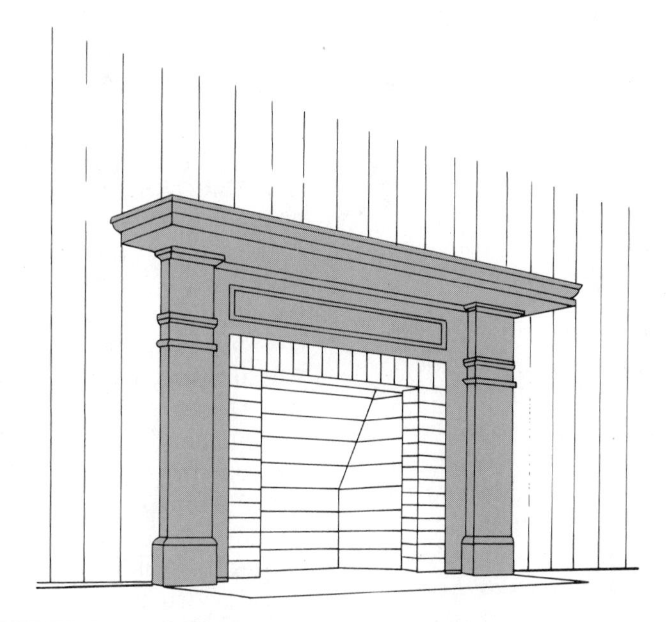

Fig. 79-21 Mantels come preassembled in a number of styles and sizes.

Review Questions

Select the most appropriate answer.

1. Bed, crown, and cove moldings are used frequently as
a. window trim.
b. ceiling molding.
c. part of the base.
d. door casings.

2. Back bands are applied to
a. wainscoting.
b. exterior corners.
c. casings.
d. interior corners.

3. A stool is part of the
a. soffit.
b. door trim.
c. base.
d. window trim.

4. The joint between moldings in interior corners is usually
a. coped.
b. mitered.
c. butted.
d. bisected.

5. The setback of door casings from the face of the jamb is often referred to as a
a. gain.
b. backset.
c. reveal.
d. quirk.

6. Find the length of side casings by
a. measuring the distance from floor to the header casing.
b. marking the length on a scrap strip and transferring it to the side casing.
c. turning the side casing upside down with the point of the miter against the floor.
d. holding the side casing with the right end up and marking the miter.

7. If the joint between side and head casings needs fitting
a. sand it.
b. fill it.
c. plane it.
d. nail it.

8. The cope on baseboard is laid out more accurately by
a. back mitering.
b. returning it.
c. a combination square.
d. scribing.

9. The base shoe is fastened
a. to the baseboard only.
b. to both the base and the floor.
c. to the floor only.
d. directly to the wall.

10. When the end of a molding has no material to butt against, its end is
a. back-mitered.
b. mitered.
c. returned upon itself.
d. coped.

BUILDING FOR SUCCESS

The Role of Subcontractors in Construction

Over the past four decades subcontractors representing various construction-related trades have served a vital role in successful building projects. This role has been altered occasionally by influences from various sectors. Today subcontracting is affected by economic factors, federal and state laws, management trends, and the financial world, to name a few. Therefore, reliance on subcontracting in new home construction and remodeling has fluctuated.

The need to streamline a business and cut unnecessary operational costs has caused many contractors to reduce their number of employees. In turn, they utilize more subs as a way to hold costs down, reduce overhead, and optimize on any potential profit. This need is a variable with each contractor in every section of the country. Once the decision is made to utilize more independent subcontractors, a basic plan must be set in motion to secure and work along with compatible subs who reflect the builder's concern for quality.

In selecting subs, a builder must be careful not to look at the low estimate first. The quality builder must be certain that the sub will also be concerned about quality customer service. The chances for a successful project are greatly increased by creating a workable plan with each sub in which contracts are understood, communications are established, priorities are set, schedules are agreed on, and expected outcomes are known.

Many times the best sources for locating a quality subcontractor may be other subs, contractors, clients, or material suppliers. It is advisable to have a firm knowledge of the sub's financial stability. This will help avoid dissatisfaction over when and how payments are to be made during construction. By providing all potential subs with a complete bid package, the builder will receive the most accurate and detailed bids. The subs are to be included in any necessary preconstruction meetings in which important information about the planned project is disseminated. At these meetings the builder's quality standards can be presented, written performances can be explained, and tolerances can be described. Anticipated construction schedules must be shared with all subs, material suppliers, and the client. During the basic information exchange among the builder, client, and any potential subs, communications can help assure that problems are solved as soon as possible, A preliminary "walk-thru" of the project can occur. Once the subs are selected and scheduled into the construction process, they should be introduced to the client. This begins the process of building positive relationships based on trust.

Flexibility in scheduling will help assure that the project is completed as planned. As both known and unknown obstacles are encountered during the planning and construction phases of the project, the builder and subs will better adapt and find solutions to problems.

By establishing a policy of being firm but fair to all subcontractors, the builder will increase the potential for success. Before the bottom line of generating a profit from a job can be realized, many other goals must be met. The goal of finding and utilizing the services of quality subcontractors begins with good planning. By keeping their word and establishing a good business relationship with all subs, builders will not become frustrated in the continual hunt for quality subs.

FOCUS QUESTIONS: For individual or group discussion

1. Explain the need to find good subcontractors who demonstrate quality work.
2. As a builder, how would you locate and utilize the services of a subcontractor?
3. What are some of the greatest challenges to be overcome in working with subcontractors?
4. As a subcontractor, how would you work with a general contractor or home builder? List at least six considerations for building successful relationships with the builder and the client.

UNIT 28

Stair Finish

The staircase is usually one of the most outstanding features of a building's interior. All stair finish work must be done in a first class manner. Joints between stair finish members must be tight-fitting for appearance. They must also be strong enough to provide support and prevent accidents.

OBJECTIVES

After completing this unit, the student should be able to:

- name various stair finish parts and describe their location and function.
- lay out, dado, and assemble a housed-stringer staircase.
- apply finish to the stair body of open and closed staircases.
- lay out treads for winding steps.
- install a post-to-post balustrade, without fittings, from floor to balcony on the open end of a staircase.
- install an over-the-post balustrade, with fittings, on an open staircase that runs from a starting step to an intermediate landing and, then, to a balcony.

UNIT CONTENTS

CHAPTER 80 DESCRIPTION OF STAIR FINISH

Many kinds of stair finish parts are manufactured in a wide variety of wood species, such as oak, beech, cherry, poplar, pine, and hemlock. It is important to identify each of the parts, know their location, and understand their function when learning to apply stair finish.

Types of Staircases

The stair finish may be separated in two parts: the *stair body* and the **balustrade.** Important components of the stair body finish are *treads, risers,* and *finish stringers.* The stair body may be constructed as an *open* or *closed* staircase. In an open staircase, the ends of the treads are exposed to view. In a closed staircase, they butt against the wall. Staircases may be open or closed on one or both sides.

Major parts of the balustrade include *handrails, newel posts,* and *balusters.* Balustrades are constructed in either a *post-to-post* or *over-the-post* method. In the post-to-post method, the handrail is fitted between the newel posts (Fig. 80-1). In the over-the-post method, the handrail runs continuously from top to bottom. It requires special curved sections, called *fittings,* where the handrail changes height or direction (Fig. 80-2).

Stair Body Parts

Many kinds of stair parts are required to finish the stair body.

Risers

Risers are vertical members that enclose the space between treads. They are manufactured in a thickness of 3/4 inch and in widths of 7 1/2 and 8 inches.

Treads

Treads are horizontal finish members upon which the feet are placed when ascending or descending stairs. High-quality treads are made from oak. Others are made from poplar or hard pine. They normally come in 3/4 and 1 1/32 inch thicknesses,

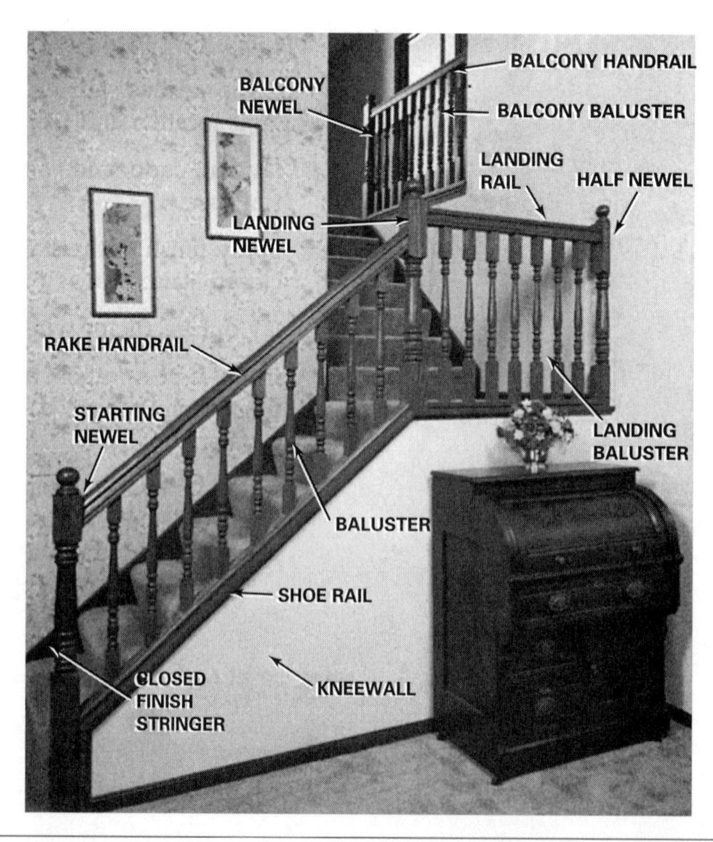

Fig. 80-1 A closed staircase with a post-to-post balustrade on a kneewall. (*Courtesy of L. J. Smith*)

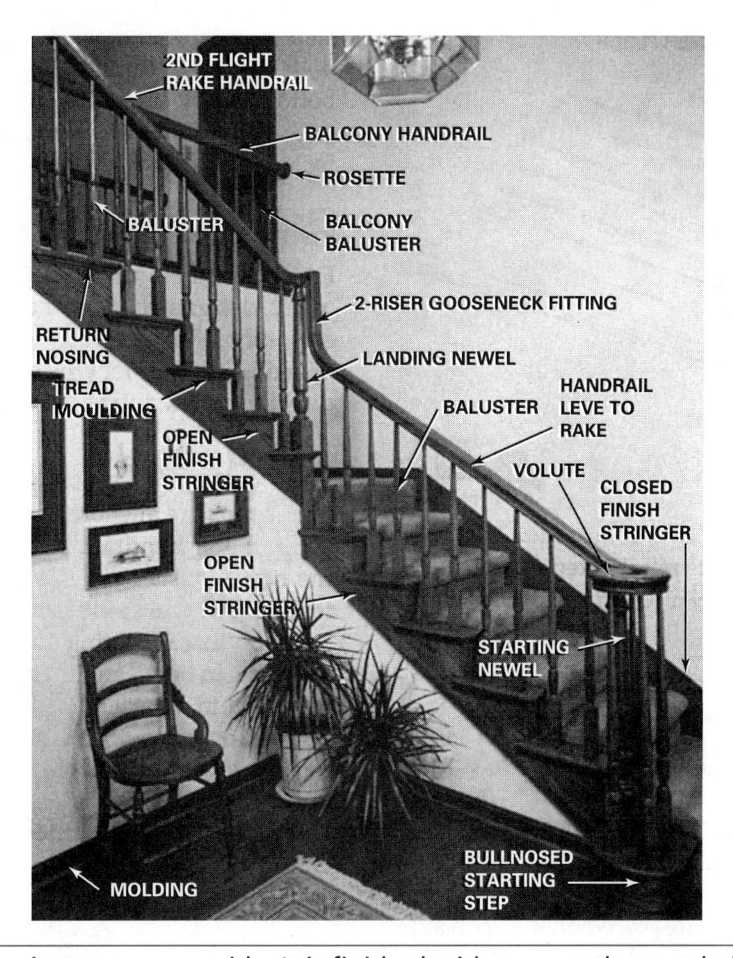

Fig. 80-2 This staircase is open on one side. It is finished with an over-the-post balustrade. (*Courtesy of L. J. Smith*)

in 10 1/2 and 11 1/2 inch widths, and in lengths from 36 up to 72 inches.

Nosings. The outside edge of the tread beyond the riser has a half round shape. It is called the *nosing.*

A *return nosing* is a separate piece mitered to the open end of a tread to conceal its end grain. Return nosings are available in the same thickness as treads and in 1 1/4 inch widths. Treads are available with the return nosing already applied to one end (Fig. 80-3).

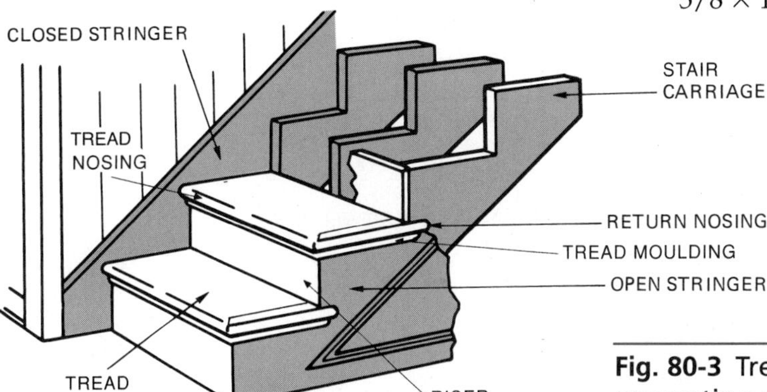

Fig. 80-3 Treads and risers are often fastened to supporting stair carriages.

Landing Treads. *Landing treads* are used at the edge of landings and balconies. They match the tread thickness at the nosing. However, they are rabbeted to match the finish floor thickness on the landing. They come in 3 1/2- and 5 1/2-inch widths. The wider landing tread is used when newel posts are more than 3 1/2 inches wide (Fig. 80-4).

Tread Molding. The *tread molding* is a small cove molding used to finish the joint under stair and landing treads. The molding should be the same kind of wood as the treads. Its usual size is 5/8 × 13/16.

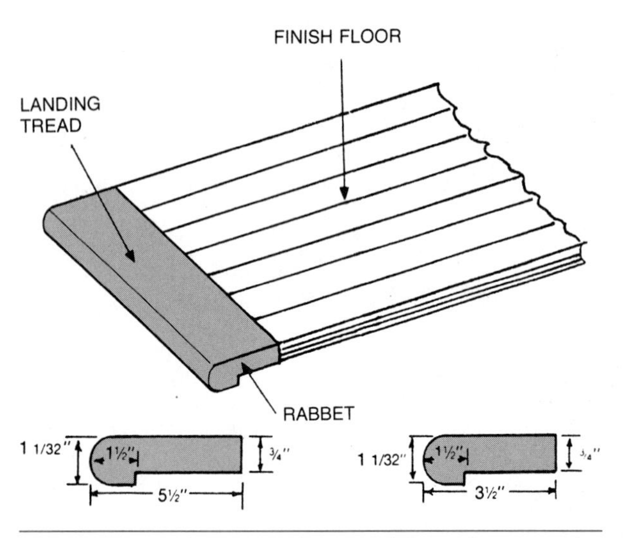

Fig. 80-4 Landing treads are rabbeted to match the thickness of the finish floor. (*Courtesy of L. J. Smith*)

Finish Stringers

Finish stringers, are sometimes called **skirt boards.** They are members of the stair body used to trim the intersection of risers and treads with the wall. They are called *closed finish stringers* when they are located above treads that butt the wall. They are termed *open finish stringers* when they are placed on the open side of a stairway below the treads (Fig. 80-5). Finish stringer lineal stock is available in a 3/4-inch thickness and in widths of 9 1/4 and 11 1/4 inches.

Starting Steps

A *starting step* or bull-nose step is the first tread-and-riser unit sometimes used at the bottom of a stairway. The starting step is used when the staircase is open on one or both sides and the handrail curves outward at the bottom. They are available in a number of styles with *bullnosed* ends, preassembled, and ready for installation (Fig. 80-6).

Balustrade Members

Finish members of the balustrade are available in many designs that are combined to complement each other. Various types of fittings are sometimes joined to straight lengths of handrail when turns in direction are required.

Newel Posts

Newel posts are anchored securely to the staircase to support the handrail. In post-to-post balustrades, the newel posts have flat square surfaces near the top, against which the handrails are fitted, and also at the bottom for fitting and securing the post to the staircase. In between the flat surfaces, the posts may be *turned* in a variety of designs (Fig. 80-7).

In over-the-post systems, a round pin at the top of newel posts fits into the underside of handrail fittings. The posts are tapered toward the top end in a number of turned designs (Fig. 80-8).

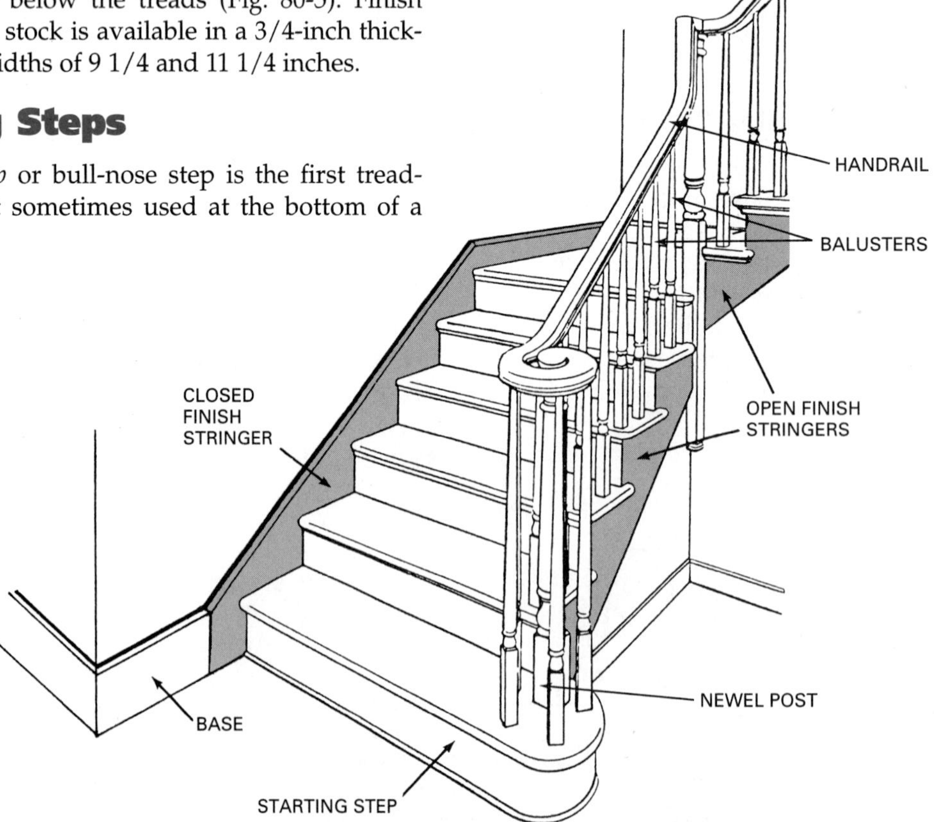

Fig. 80-5 Open and closed finish stringers trim the intersection of the stair body and the wall.

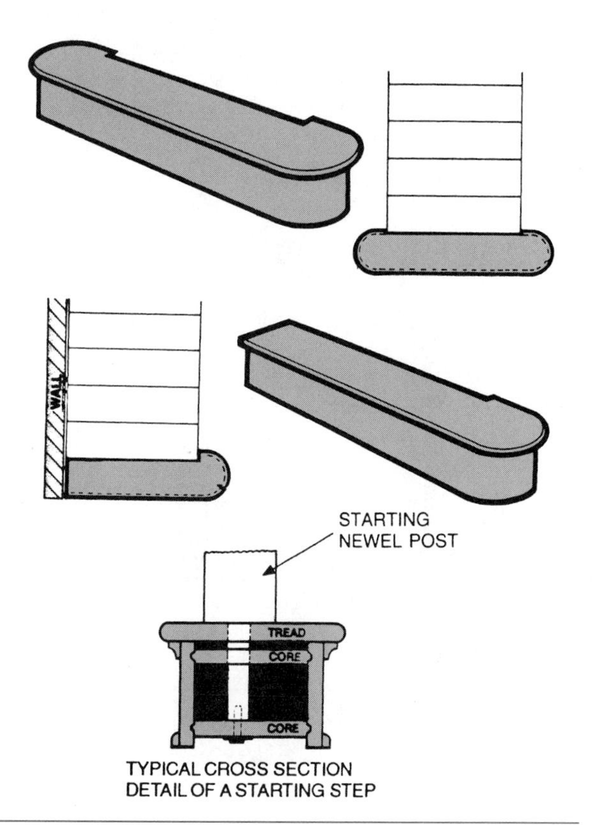

Fig. 80-6 Starting steps are available in a number of styles and sizes. Two commonly used types are shown. (*Courtesy of L. J. Smith*)

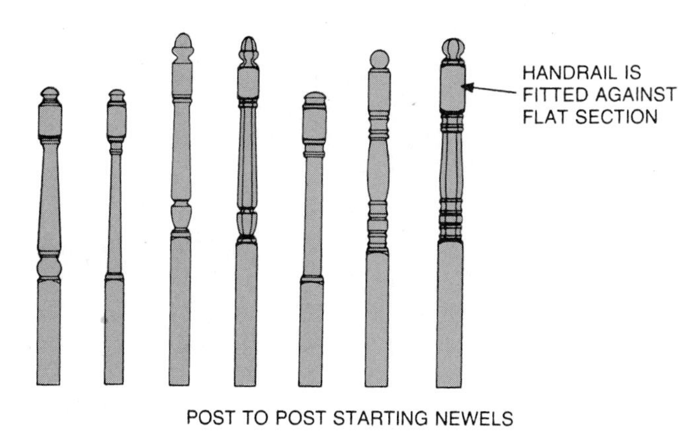

Fig. 80-7 Newels, in post-to-post balustrades, must have flat surfaces against which handrails are fitted. (*Courtesy of L. J. Smith*)

Three types of newel posts are used in a post-to-post balustrade. *Starting newels* are used at the bottom of a staircase. They are fitted against the first or second riser. If fitted against the second riser, the flat, square surface at the bottom must be longer. At the top of the staircase, *second floor newels* are used. *Intermediate landing newels* are also available. Because part of the bottom end of these newels are exposed, turned *buttons* are available to finish the end. The same design is used in the same staircase

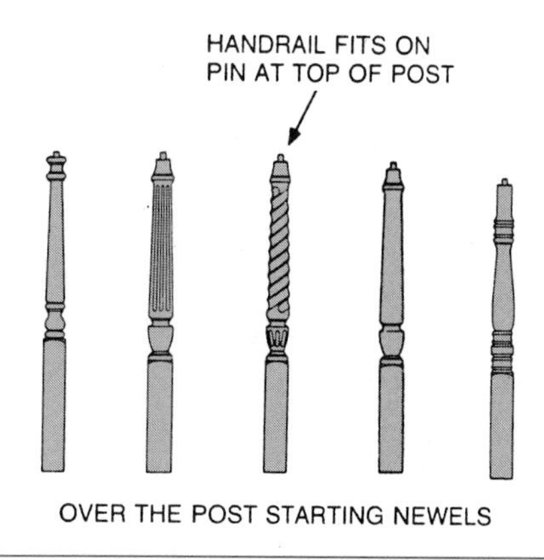

Fig. 80-8 Newels, in over-the-post balustrades, are made with a pin at the top. (*Courtesy of L. J. Smith*)

for each of the three types of posts. They differ only in their overall length and in the length of the flat surfaces (Fig. 80-9).

Four types of newel posts are used in an over-the-post balustrade. There are three types of *starting newels* depending on the type of handrail fitting used. If a *volute* or *turnout* is used, a newel post with a dowel at the bottom is installed on top of a required starting step. The fourth type is a longer newel for landings, where a gooseneck handrail fitting is used (Fig. 80-10).

When the balustrade ends against a wall, a *half newel* is sometimes fastened to the wall. The handrail is then butted to it. In place of a half newel, the handrail may butt against an oval or round *rosette* (Fig. 80-11).

Handrails

The *handrail* is the sloping finish member grasped by the hand of the person ascending or descending the stairs. It is installed horizontally when it runs along the edge of a balcony. Handrail heights are 30 to 38 inches vertically above the nosing edge of the tread. There should be a continuous 1-1/2-inch finger clearance between the rail and the wall. Several styles of handrails come in lineal lengths that are cut to fit on the job. Some handrails are *plowed* with a wide groove on the bottom side to hold square top balusters in place (Fig. 80-12).

On closed staircases, a balustrade may be installed on top of a *kneewall* or buttress. In relation to stairs, a kneewall is a short wall that projects a short distance above and on the same rake as the stair body. A *shoe rail* or buttress cap, which is plowed on the top side, is usually applied to the top of the kneewall on which the bottom end of balusters are fastened (Fig. 80-13). Narrow strips, called *fillets*, are used between balusters to fill the plowed groove on handrails and shoe rails.

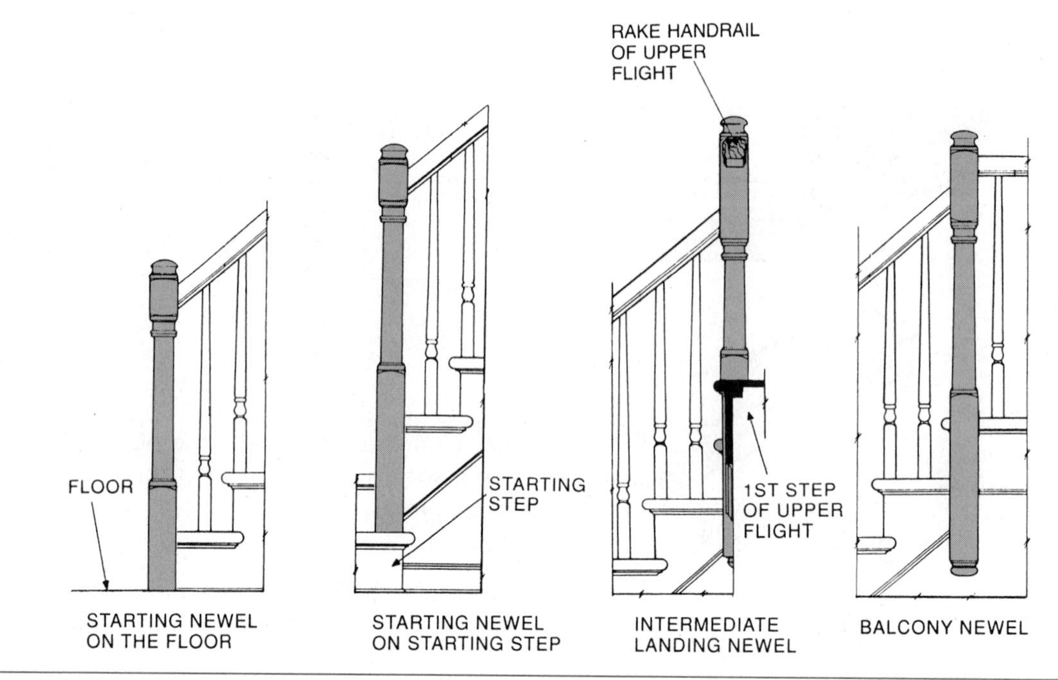

Fig. 80-9 Three types of newel posts are used in a post-to-post balustrade. (*Courtesy of L. J. Smith*)

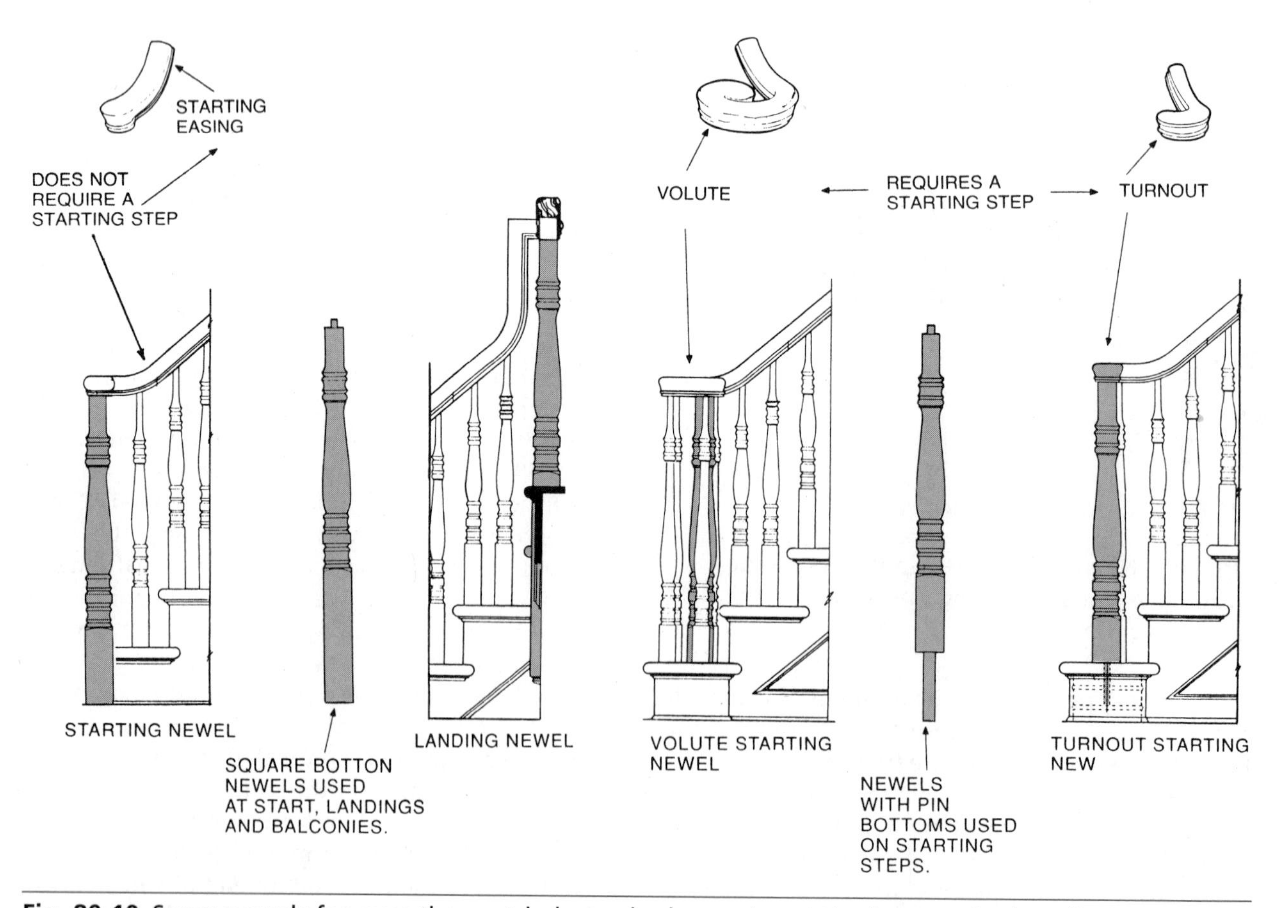

Fig. 80-10 Some newels for over-the-post balustrades have pins at the bottom for installation on a starting step. Others have square bottom ends for use in many locations. (*Courtesy of L. J. Smith*)

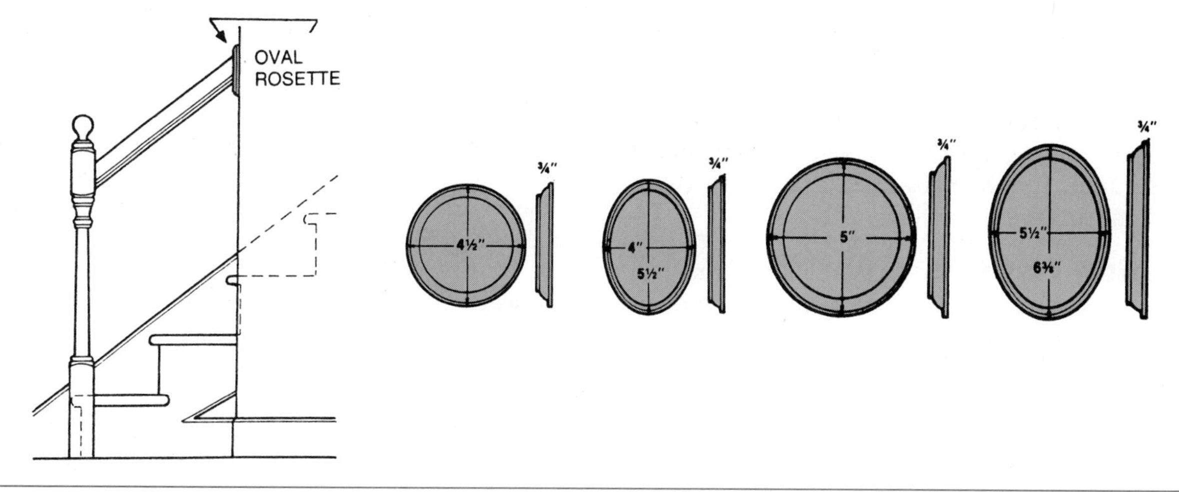

Fig. 80-11 Rosettes come in round and oval shapes. They are fastened to the wall to provide a surface on which to butt and fasten the end of a handrail. (*Courtesy of L. J. Smith*)

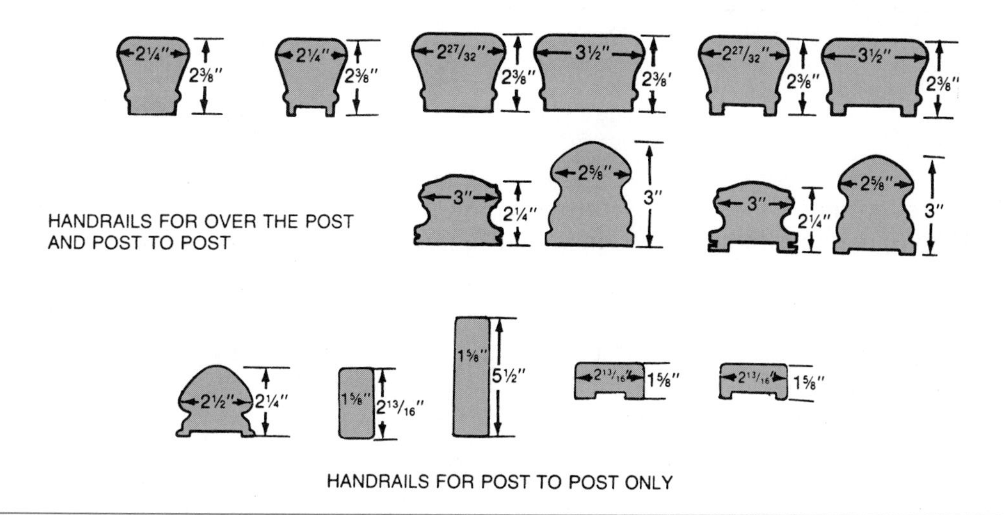

Fig. 80-12 Straight lengths of handrail are manufactured in many styles. (*Courtesy of L. J. Smith*)

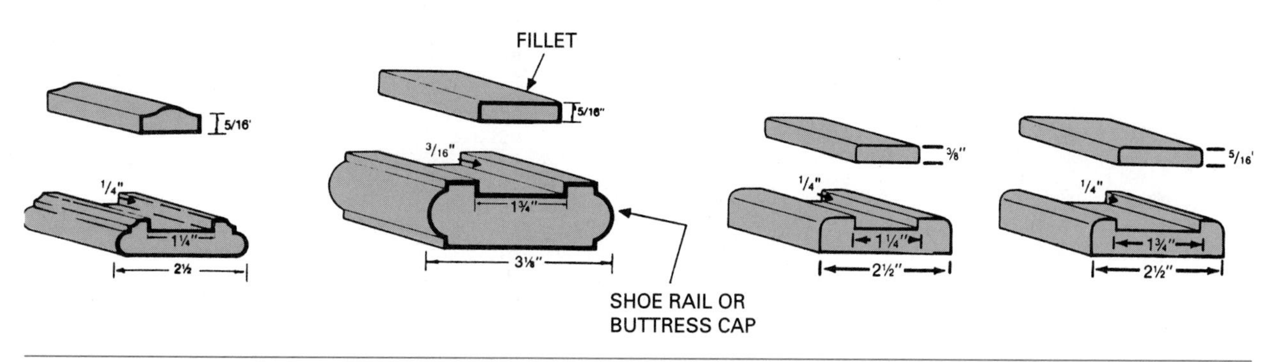

Fig. 80-13 A shoe rail is often used at the bottom of a balustrade that is constructed on a kneewall. (*Courtesy of L. J. Smith*)

Handrail Fittings

Short sections of specially curved handrail are called *fittings.* They are used at various locations, joined to straight sections, to change the direction of the handrail. They are classified as *starting, gooseneck,* and *miscellaneous* fittings.

Starting Fittings. To start an over-the-post handrail, starting fittings called *volutes,* **turnouts** or *starting easings* may be used. In a post-to-post system, a straight length of handrail may be used at the bottom. To start with a soft, graceful curve, an *upeasing* is used (Fig. 80-14).

Gooseneck Fittings. Over-the-post systems, in which the handrail is continuous, fittings called *goosenecks* are required at intermediate landings and at the top. This is because of changes in the handrail height or direction. In post-to-post systems, goosenecks are not required. However, their use makes the balustrade more attractive (Fig. 80-15).

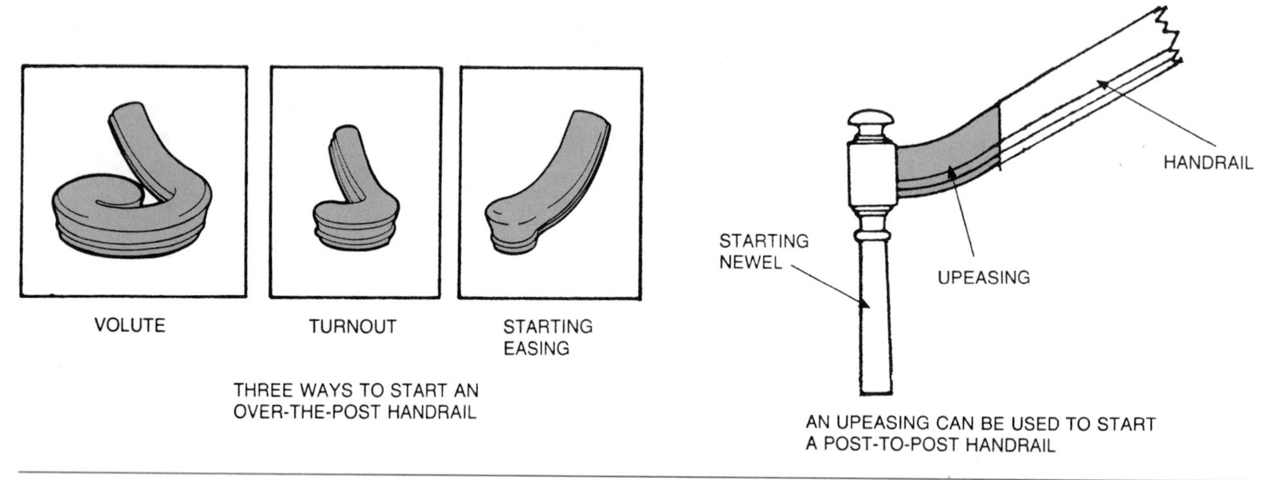

Fig. 80-14 Handrail fittings, called volutes, turnouts, and starting easings, are used to start an over-the-post handrail. An upeasing is sometimes used to start a post-to-post handrail. (Courtesy of L. J. Smith)

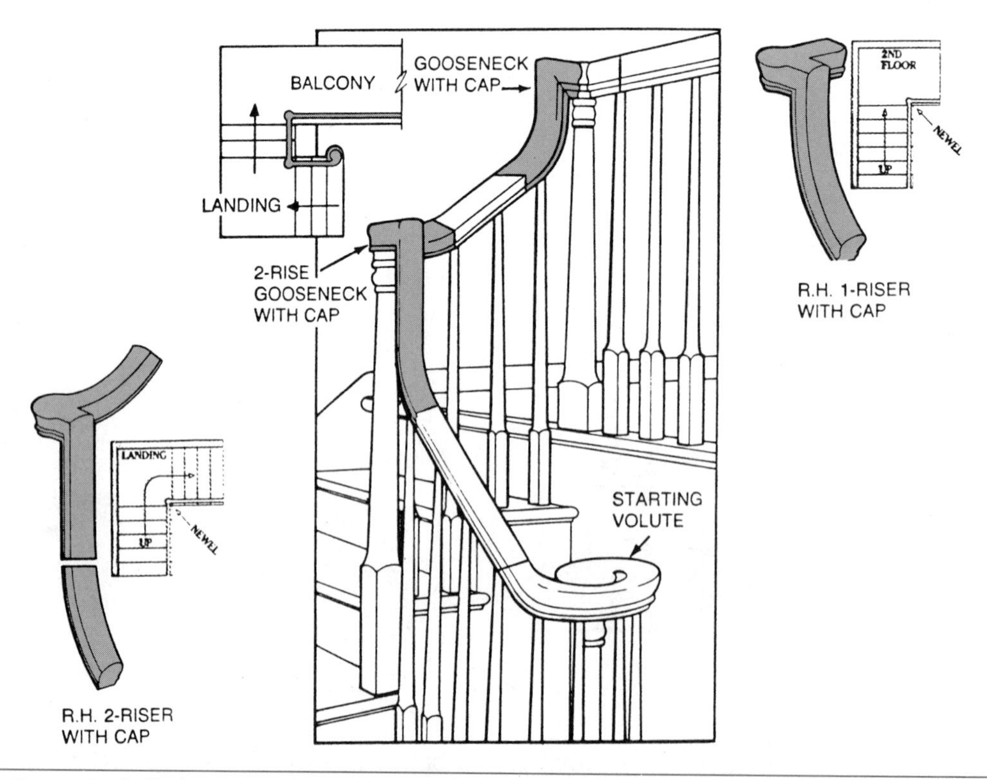

Fig. 80-15 Gooseneck fittings are used at landings when handrails change direction or height. Many shapes, with or without caps, are manufactured. (*Courtesy of L. J. Smith*)

Goosenecks are available for handrails that continue level or sloping or that turn 90 or 180 degrees right or left. They are made with or without caps for both types of handrail systems.

Miscellaneous Fittings. Among the miscellaneous handrail fittings are *easings* of various kinds, *coped* and *returned ends, quarterturns,* and *caps* (Fig. 80-16). They are used where necessary to meet the specifications for the staircase. All handrail fittings are ordered to match the straight lengths of handrail being used. When the handrail being used is plowed, a matching plow is specified for the handrail fittings. A special fillet, shaped to fit the plow, comes with the handrail fitting.

Balusters

Balusters are vertical, usually decorative pieces between newel posts. They are spaced close together and support the handrail. On a kneewall, they run from the handrail to the shoe rail. On an open staircase, they run from the handrail to the treads (see Figs. 80-1 and 80-2).

Balusters are manufactured in many styles. They should be selected to complement the newel posts being used (Fig. 80-17). For example, in a post-to-post staircase, balusters with square tops are usually used. In an over-the-post staircase, balusters that are tapered at the top complement the newel posts.

Most balusters are made in lengths of 31, 34, 36, 39, and 42 inches for use in any part of the stairway. Several lengths of the same style baluster are needed for each tread of the staircase because of the rake of the handrail.

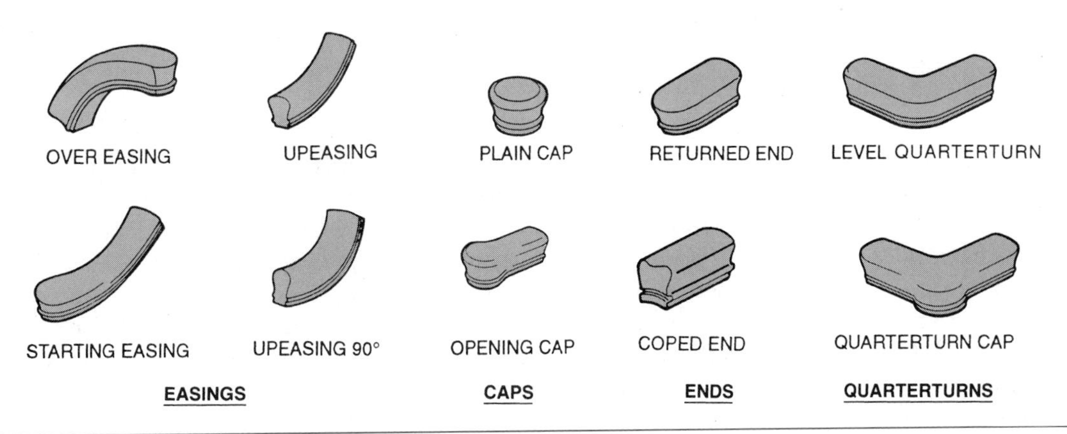

Fig. 80-16 Many miscellaneous fittings are available to meet the need of any type of handrail installation. (*Courtesy of L. J. Smith)*

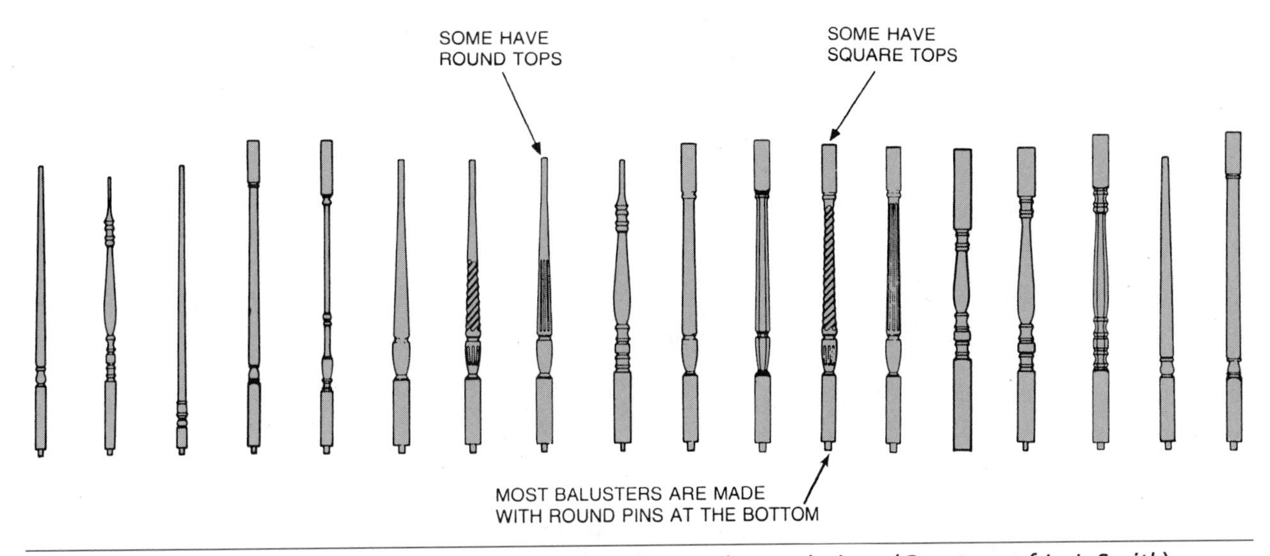

Fig. 80-17 Balusters are made in designs that match newel post design. (*Courtesy of L. J. Smith*)

CHAPTER 81 FINISHING THE STAIR BODY OF OPEN AND CLOSED STAIRCASES

Review the previous chapter for the description, location, and function of the finish members of the stair body. This chapter describes their installation.

Making a Housed Stringer

Because all the major components of a housed-stringer staircase are finish members, its construction is described in this chapter. The method of determining the riser height and tread width for a housed-stringer staircase is the same as for any staircase. (It is explained in Unit 16, Stair Framing.) After these dimensions have been determined, the layout is made on the stringer stock.

Using a Pitch Board

A *pitch board* is often used instead of a framing square for laying out a housed stringer. A pitch board is a piece of stock, usually 3/4 inch thick. It is cut to the rise and run of the stairs. A strip of wood, usually a 1 × 2, is fastened to the *rake* edge of the pitch board. This is used to hold the pitch board against the edge while laying out the stringer (Fig. 81-1).

On the face side of the stringer stock, draw a line parallel to and about 2 inches down from the top edge. This distance may vary, depending on the width of the stringer stock and the desired height of the top edge of the stringer above the stair treads. The line is the intersection of the tread and riser faces. Using the pitch board, lay out the risers and treads for each step of the staircase. These lines show the location of the face side of each riser and tread and are the outside edges of the housing (Fig. 81-2).

Cut and fit the bottom end of the stringer to the floor and the top end to the landing. Make end cuts that will join with the baseboard properly. The housed stringer and the baseboard should be joined in a professional manner to provide a continuous line of finish from one floor to the next. Since the stringer is usually S4S stock, the base should also be S4S. A base cap may then be used that continues from floor to floor on the raked top edge of the stringer (Fig. 81-3).

Housing the Stringer

After the stringer has been laid out, it is ready to be housed. One method of housing the stringer is to use a *template* to guide a router. The router must be equipped with a straight bit and a template guide of the correct size. Stair routing templates are manufactured that are adjustable for different rises and runs. Templates may also be made by cutting out thin plywood or hardboard to the shape of the housing. The template is shaped so the dadoes will be the exact width at the nosing and wider toward the inside. Then the treads and risers can be wedged tightly against the shoulders of the dadoes (Fig. 81-4). Clamp the template to the stringer. Rout the stringer, about 1/4 inch deep, for all treads and risers.

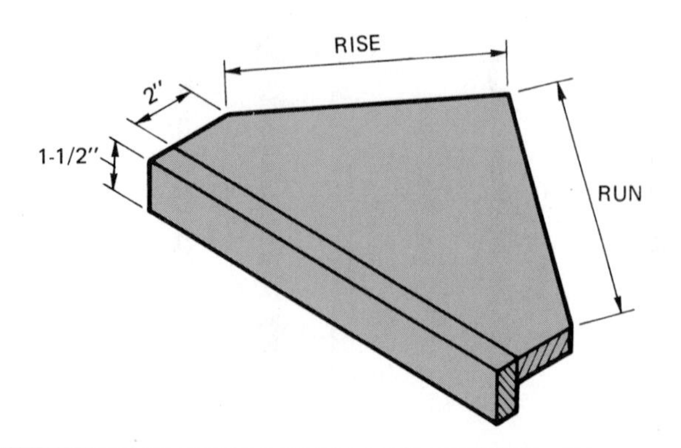

Fig. 81-1 A pitch board is often used for laying out a housed stringer.

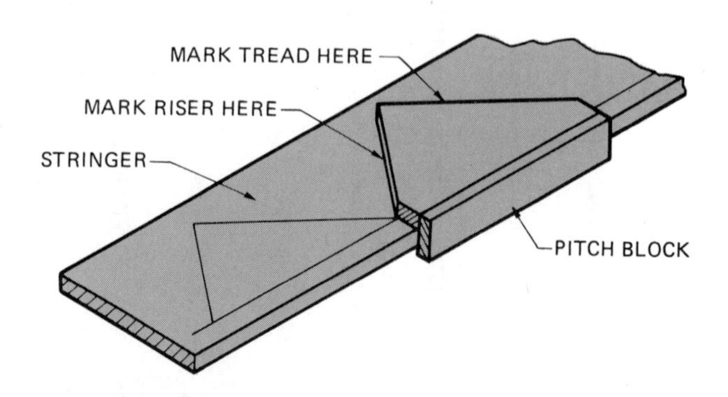

Fig. 81-2 Using a pitch board to lay out a housed finish stringer.

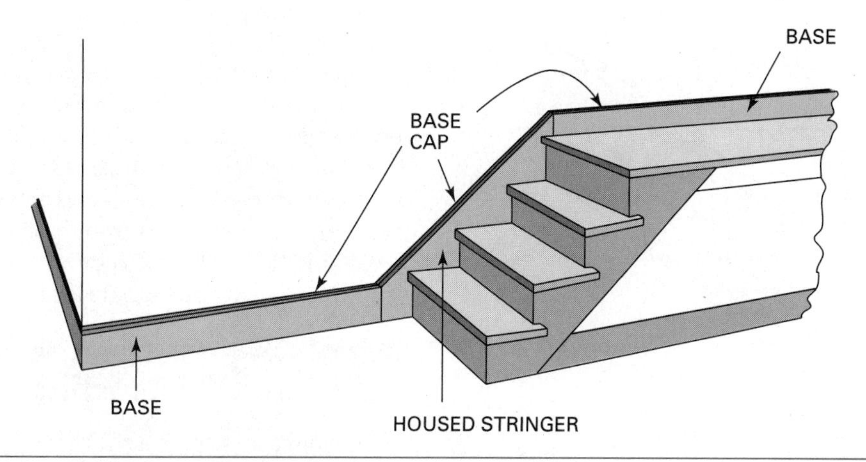

Fig. 81-3 The baseboard is joined to the housed stringer at the bottom and top of the staircase.

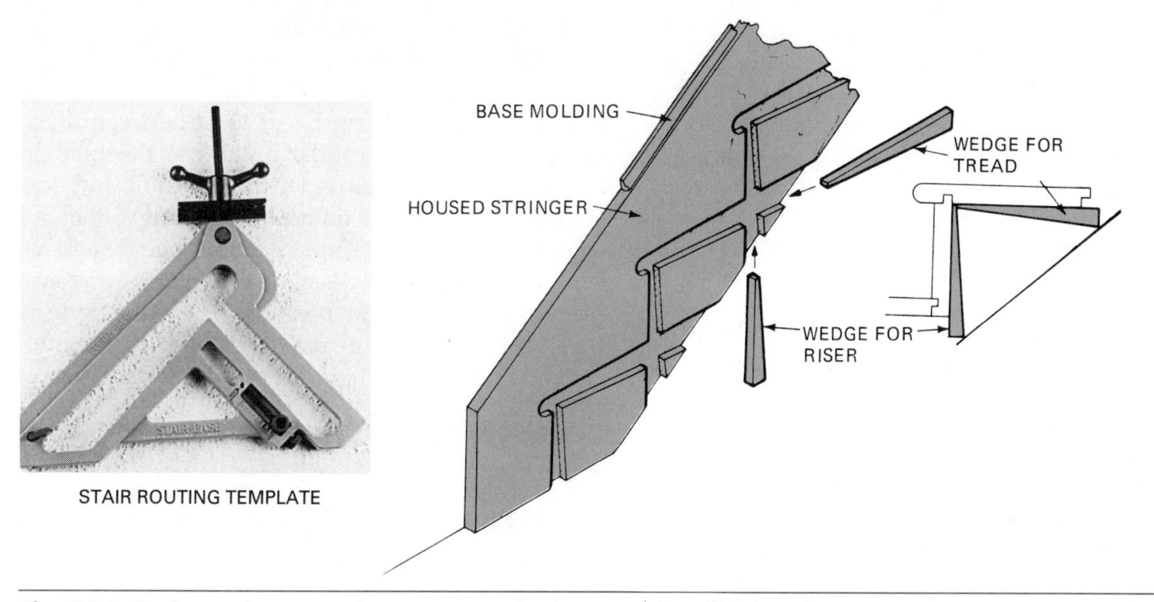

Fig. 81-4 Stair routing templates are used when routing finish stringers. Stringers are routed to allow treads and risers to be wedged and glued in place. (*Part 1, Courtesy of Porter-Cable*)

Laying Out an Open Stringer

The layout of an open (or *mitered*) stringer is similar to that of a housed stringer. However, riser and tread layout lines intersect at the top edge of the stringer, instead of against a line in from the edge. The riser layout line is the outside face of the riser. This layout line is mitered to fit the mitered end of the riser. The tread layout is to the face side of the tread. The risers and treads are marked lightly with a sharp pencil (Fig. 81-5).

To lay out the *miter cut* for the risers, measure in at right angles from the riser layout line a distance equal to the thickness of the riser stock. Draw another plumb line at this point. Square both lines across the top edge of the stringer stock. Draw a diagonal line on the top edge to mark the miter angle (Fig. 81-6).

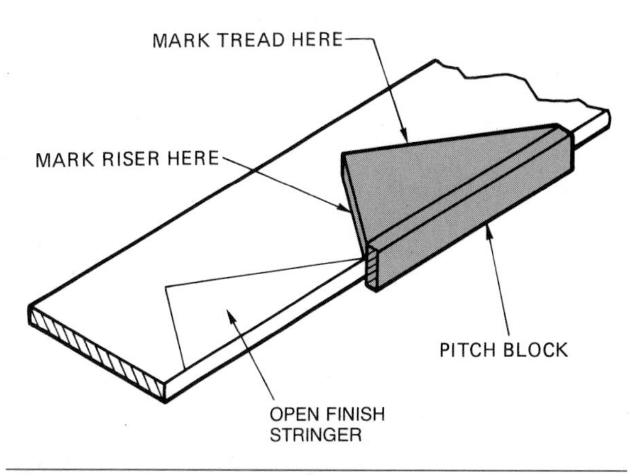

Fig. 81-5 Laying out the open finish stringer of a housed staircase.

To mark the tread cut on the stringer, measure down from the tread layout line a distance equal to the thickness of the tread stock. Draw a level line at

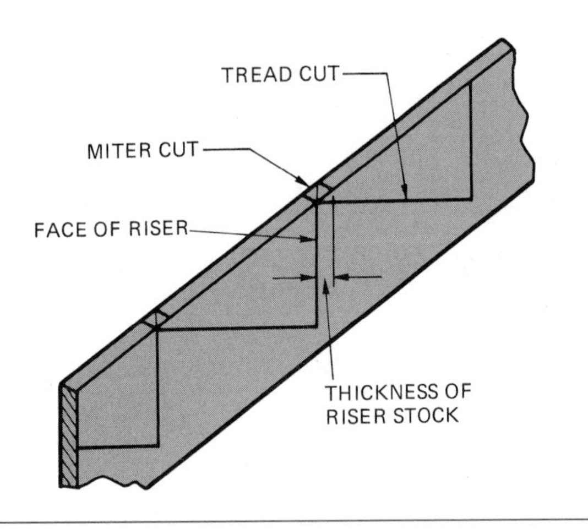

Fig. 81-6 Laying out the miter angle on an open finish stringer.

this point for the tread cut. The tread cut is square through the thickness of the stringer. Fit the bottom end to the floor. Fit the top end against the landing. Make the mitered plumb cuts for the risers and the square level cuts for the treads.

Installing Risers and Treads

Cut the required number of risers to a rough length. Determine the face side of each piece. Then cut a 3/8 × 3/8 groove near the bottom edge on the face side of all but the starting riser. The groove is located so its top is a distance from the bottom edge equal to the tread thickness. The groove is made for the rabbeted inner edge of the tread to fit into it (Fig. 81-7). Rip the risers to width by cutting the edge opposite the groove. Rip the treads to width. Rabbet their back edges to fit in the riser grooves. Cut the risers and tread to exact lengths.

On a closed staircase, the risers are installed with wedges, glue, and screws between housed stringers. On the open side of a staircase, where the riser and open stringer meet, a miter joint is made so no end grain is exposed (Fig. 81-8). The treads are then installed with wedges, glue, and screws on the closed side and with screws through screw blocks on the open side. Screw blocks reinforce interior corners on the underside at appropriate locations and intervals.

Applying Return Nosings and Tread Molding

If the staircase is open, a *return nosing* is mitered to the end of the tread. The back end of the return nosing projects past the riser the same amount as the tread overhangs the riser. The end is returned upon itself.

The tread molding is then applied under the overhang of the tread. If the staircase is closed on both sides, the molding is cut to fit between finish stringers. On the open end of a staircase, the molding is mitered around and under the return nosing. It is stopped and returned on itself at a point so the end assembly appears the same as at the edge (Fig. 81-9).

After the housed-stringer staircase is assembled, it is installed in position. The balustrade is then constructed in a manner similar to that described later in this chapter for framed staircases.

Finishing the Body of a Closed Staircase Supported by Stair Framing

The following section describes the installation of finish to a *closed* staircase in which the supporting

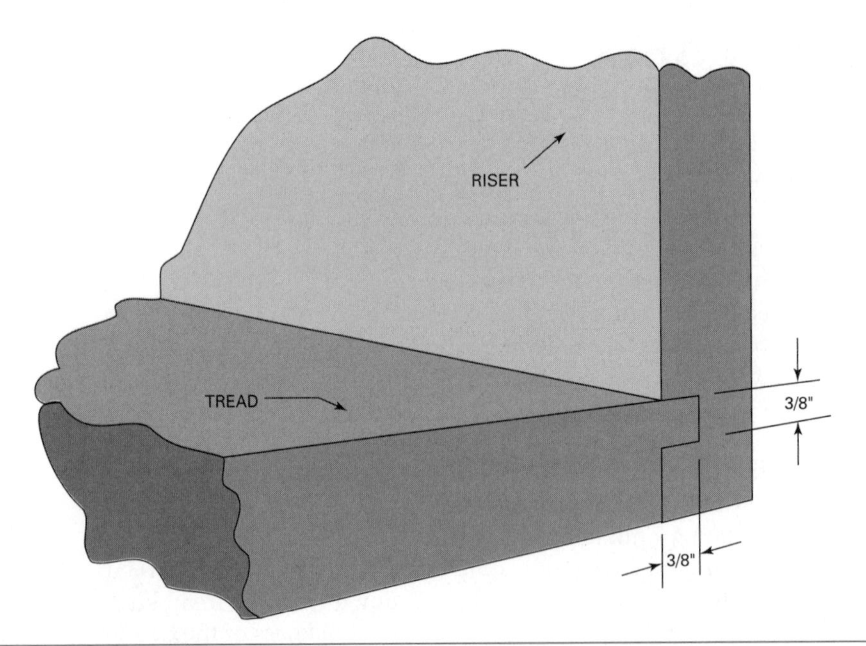

Fig. 81-7 A groove is made in the riser to receive the rabbeted inner edge of the tread.

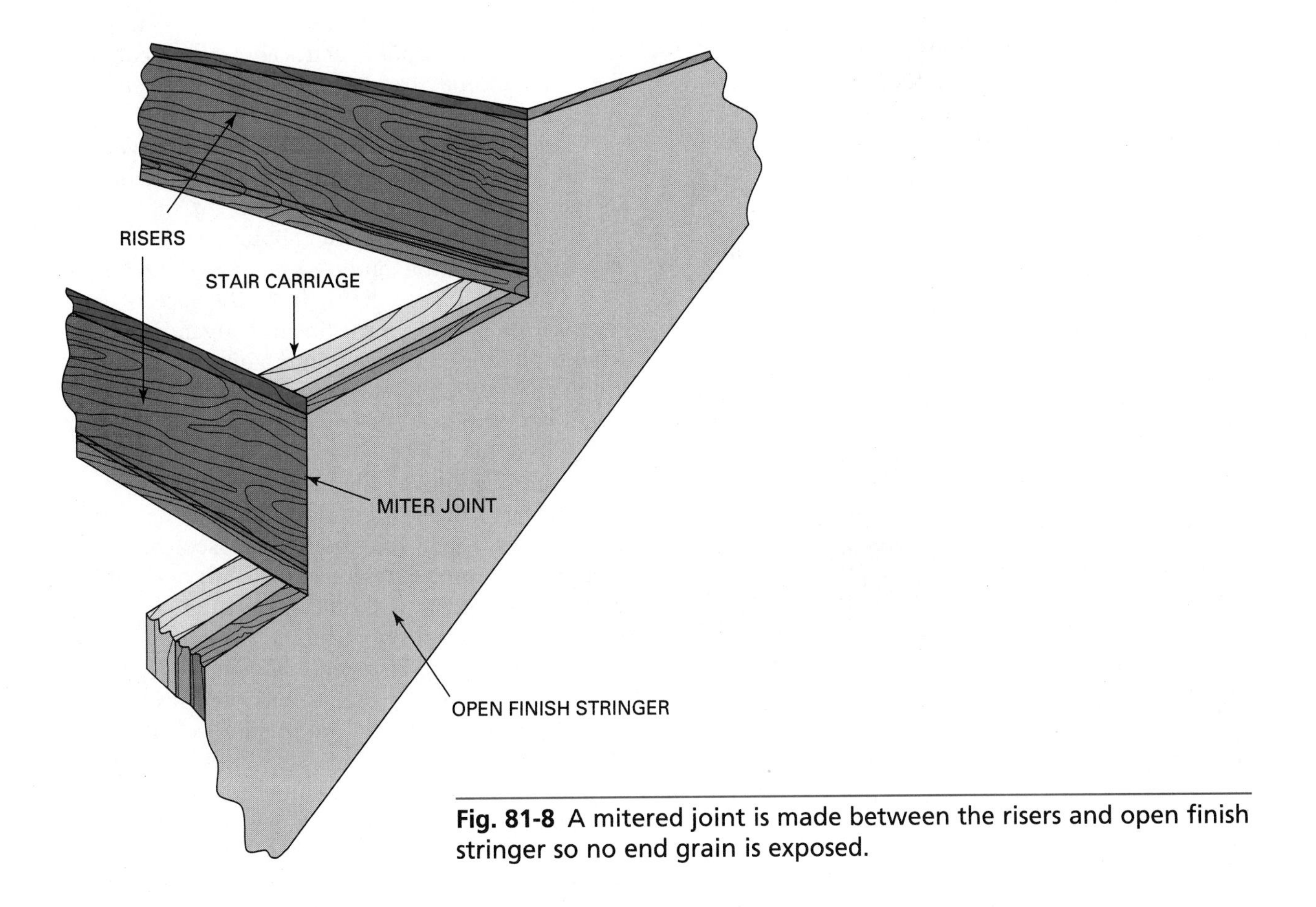

Fig. 81-8 A mitered joint is made between the risers and open finish stringer so no end grain is exposed.

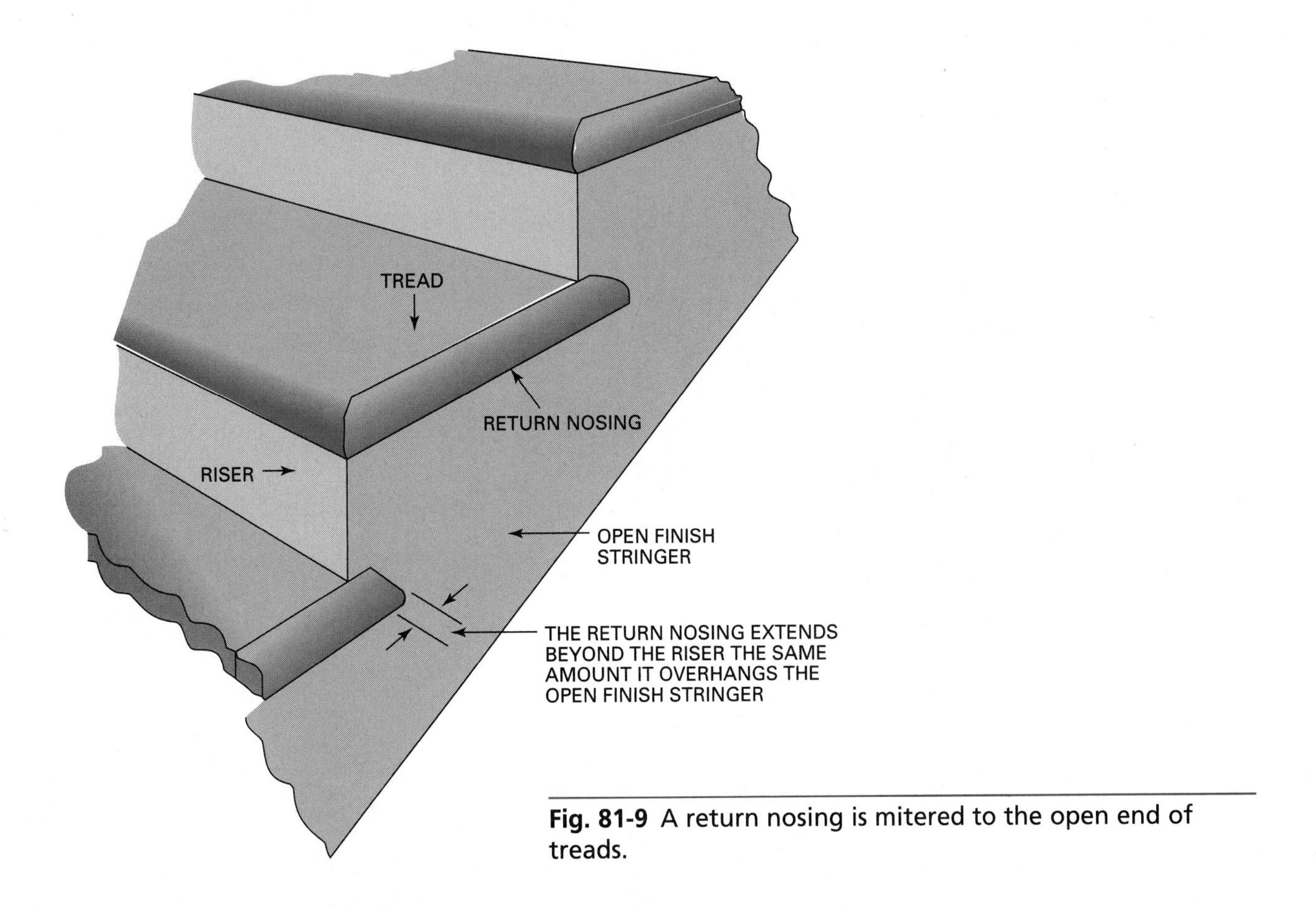

Fig. 81-9 A return nosing is mitered to the open end of treads.

stair carriages have already been installed between two walls that extend from floor to ceiling.

Applying Risers

The first trim applied to the stair carriages are the *risers.* Rip the riser stock to the proper width. Cut grooves as described previously for housed-stringer construction. Cut the risers to a length, with square ends, about 1/4 inch less than the distance between walls.

Fasten the risers in position with three 2 1/2-inch finish nails into each stair carriage. Start at the top and work down. Remove the temporary treads installed previously as work progresses downward (Fig. 81-10).

CAUTION Put up positive barriers at the top and bottom of the stairs so that the stairs cannot be used while the finish is applied. A serious accident can happen if a person who does not realize that the temporary treads have been removed uses the stairs.

Layout and Installation of the Closed Finish Stringer

After the risers have been installed, the *closed finish stringer* is cut around the previously installed risers. Usually 1 × 10 lumber is used. When installed, its top edge will be about 3 inches above the tread nosing. A 1 × 12 may be used if a wider finish stringer is desired.

Tack a length of stringer stock to the wall. Its bottom edge should rest on the top edges of the previously installed risers. Its bottom end should rest on the floor. The top end should extend about 6 inches beyond the landing.

Lay out plumb lines, from the face of each riser, across the face of the finish stringer. If the riser itself is out of plumb, then plumb upward from that part of the riser that projects farthest outward. Then, lay out level lines on the stringer, from each tread cut of the stair carriage and also from the floor of the landing above (Fig. 81-11).

Remove the stringer from the wall. Cut to the layout lines. Use a fine-toothed crosscut handsaw. Follow the plumb lines carefully with a slight undercut. Plumb cuts will butt against the face of the risers, so a careful cut needs to be made. Not as much care needs to be taken with level cuts because treads will later butt against and cover them.

After the cutouts are made in the finish stringer, tack it back in position. Fit it to the floor. Then, lay out top and bottom ends to join the base that will later be installed on the walls. Remove the stringer. Make the end cuts. Sand the board, and place it back in position. Fasten the stringer securely to the wall with finishing nails. Do not nail too low

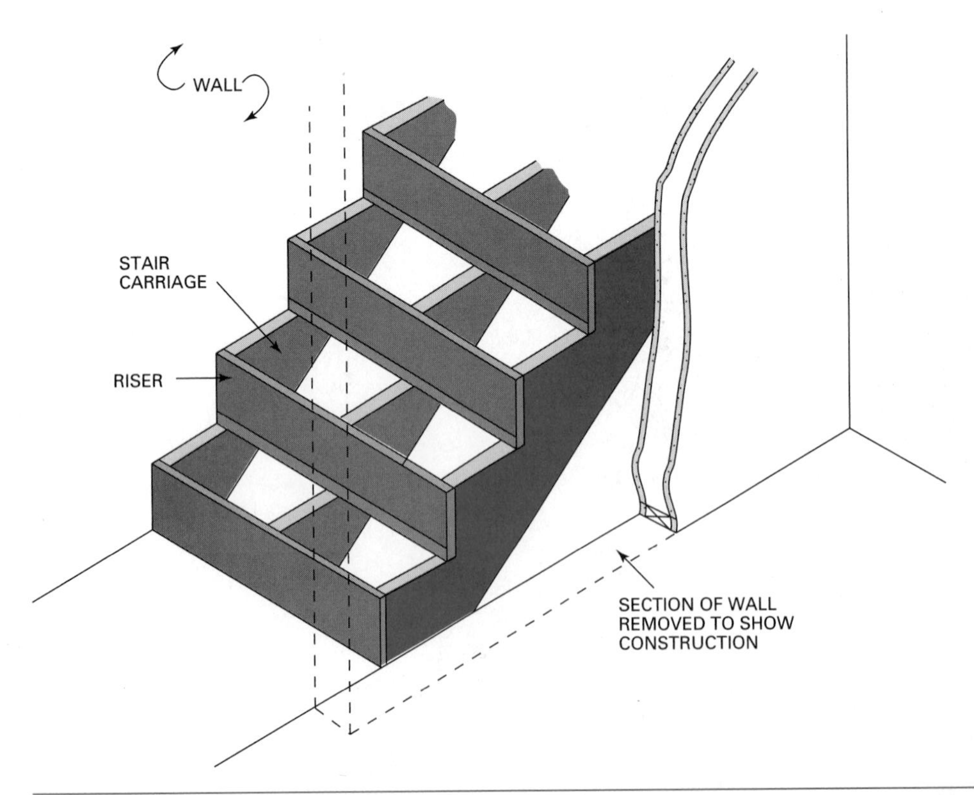

Fig. 81-10 Risers are ordinarily the first finish members applied to the stair carriage in a closed staircase.

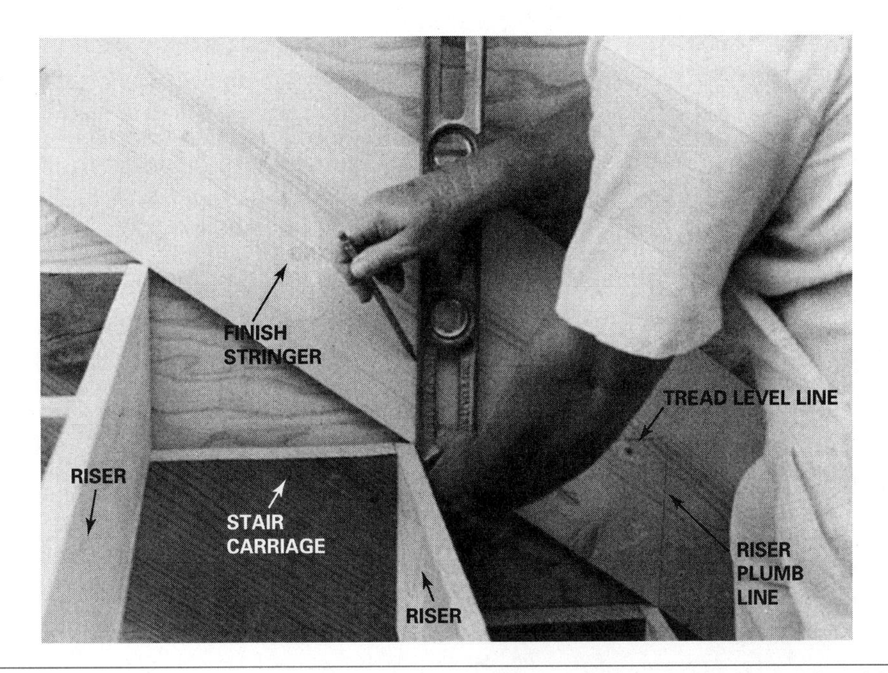

Fig. 81-11 The closed finish stringer is laid out using a level to extend plumb and level lines from the stair carriage.

to avoid splitting the lower end of the stringer. Install the finish stringer on the other wall in the same manner.

Drive shims, at each step, between the back side of the risers and the stair carriage. The shims force the risers tightly against the plumb cut of the finish stringer. Shim at intermediate stair carriages to straighten the risers, from end to end, between walls.

Installing Treads

Treads are cut on both ends to fit snugly between the finish stringers. The nosed edge of the tread projects beyond the face of the riser by 1 1/8 inches (Fig. 81-12).

Along the top edge of the riser, measure carefully the distance between finish stringers. Transfer the measurement and square lines across the tread.

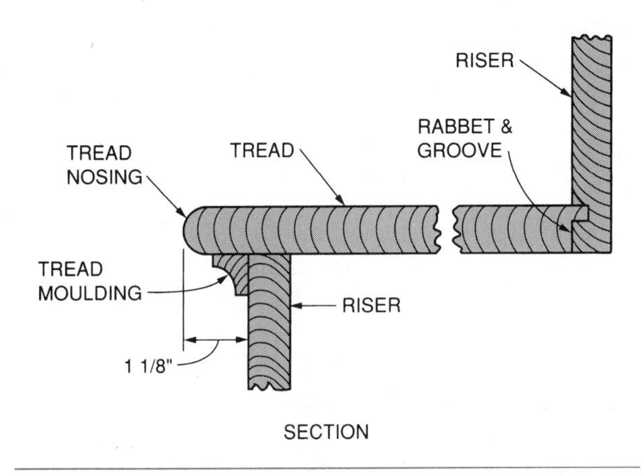

Fig. 81-12 Tread and riser detail.

Cut in from the nosed edge. Square through the thickness for a short distance. Then undercut slightly. Smooth the cut ends with a block plane. Rub one end with wax. Place the other end in position. Press on the waxed end until the tread lays flat on the stair carriages.

Place a short block on the nosed edge. Tap it until the inner rabbeted edge is firmly seated in the groove of the riser.

If it is possible to work from the underside, the tread may be fastened by the use of screw blocks at each stair carriage and at intermediate locations.

If it is not possible to work from the underside, the treads must be face nailed. Fasten each tread in place with three 8d finish nails into each stair carriage. It may be necessary to drill holes in hardwood treads to prevent splitting the tread or bending the nail. A little wax applied to the nail makes driving easier and helps keep the nail from bending.

Start from the bottom and work up, installing the treads in a similar manner. At the top of the stairs, install a landing tread. If 1 1/32-inch thick treads are used on the staircase, use a landing tread that is rabbeted to match the thickness of the finish floor (Fig. 81-13).

Tread Molding

The *tread molding* is installed under the overhang of the tread and against the riser. Cut the molding to the same length as the treads, using a miter box. Predrill holes. Fasten the molding in place with 4d finish nails spaced about 12 inches apart. Nails are driven at about a 45 degree angle through the center of the molding.

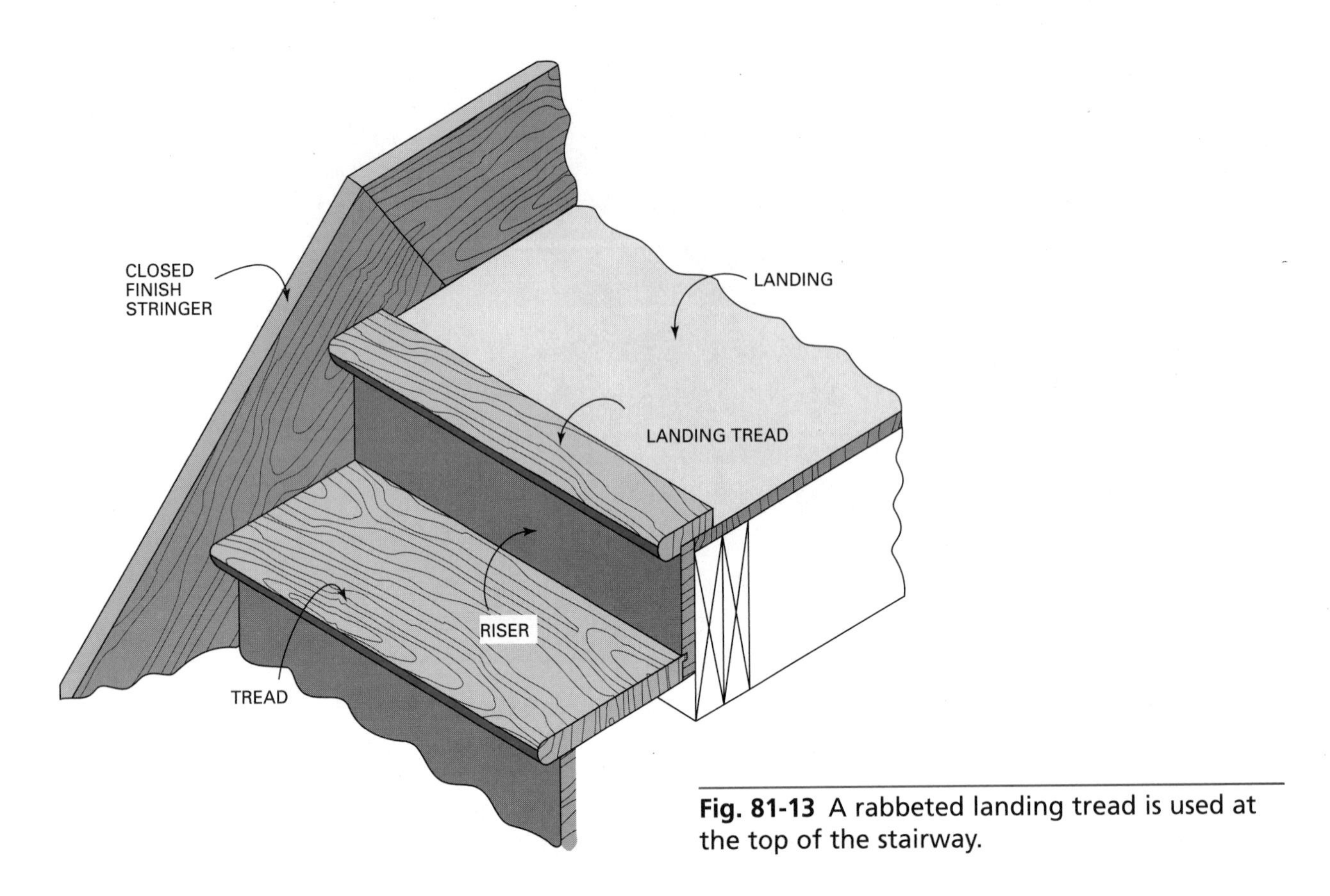

Fig. 81-13 A rabbeted landing tread is used at the top of the stairway.

Finishing the Body of an Open Staircase Supported by Stair Framing

The following section describes the installation of finish to the stair body of a staircase, supported by stair carriages, which is closed on one side and open on the other side.

Installing the Finish Stringers

The *open finish stringer* must be installed before the *risers* and the *closed finish stringer.* To lay out the *open finish stringer,* cut a length of finish stringer stock. Fit it to the floor and against the landing. Its top edge should be flush with the top edge of the stair carriage. Tack it in this position to keep it from moving while it is being laid out.

First, lay out level lines on the face of the stringer in line with the tread cut on the stair carriage. Next, plumb lines must be laid out on the face of the finish stringer for making miter joints with risers.

Using a Preacher to Lay Out Plumb Lines. Use a *preacher* to lay out the plumb lines on the open finish stringer. A preacher is made from a piece of nominal 4-inch stock about 12 inches long. Its thickness must be the same as the riser stock. The preacher is notched in the center. It should be wide enough to fit over the finish stringer. It should be long enough to allow the preacher to rest on the tread cut of the stair carriage when held against the rise cut.

Place the preacher over the stringer and against the rise cut of the stair carriage. Plumb the preacher with a hand level. Lay out the plumb cut on the stringer by marking along the side of the preacher that faces the bottom of the staircase (Fig. 81-14).

Mark the top edge of the stringer along the side of the preacher that faces the top of the staircase. Draw a diagonal line across the top edge of the stringer for the miter cut. Lay out all plumb lines on the stringer in this manner.

Remove the stringer. Cut to the layout lines. Make miter cuts along the plumb lines. Cut square through the thickness along the level lines. Sand the piece. Fasten it in position. To assure getting the piece in the same position as it was when laid out, fasten it first in the same holes where the piece was originally tacked.

Installing Risers

Cut *risers* to length by making a square cut on the end that goes against the wall. Make miters on the

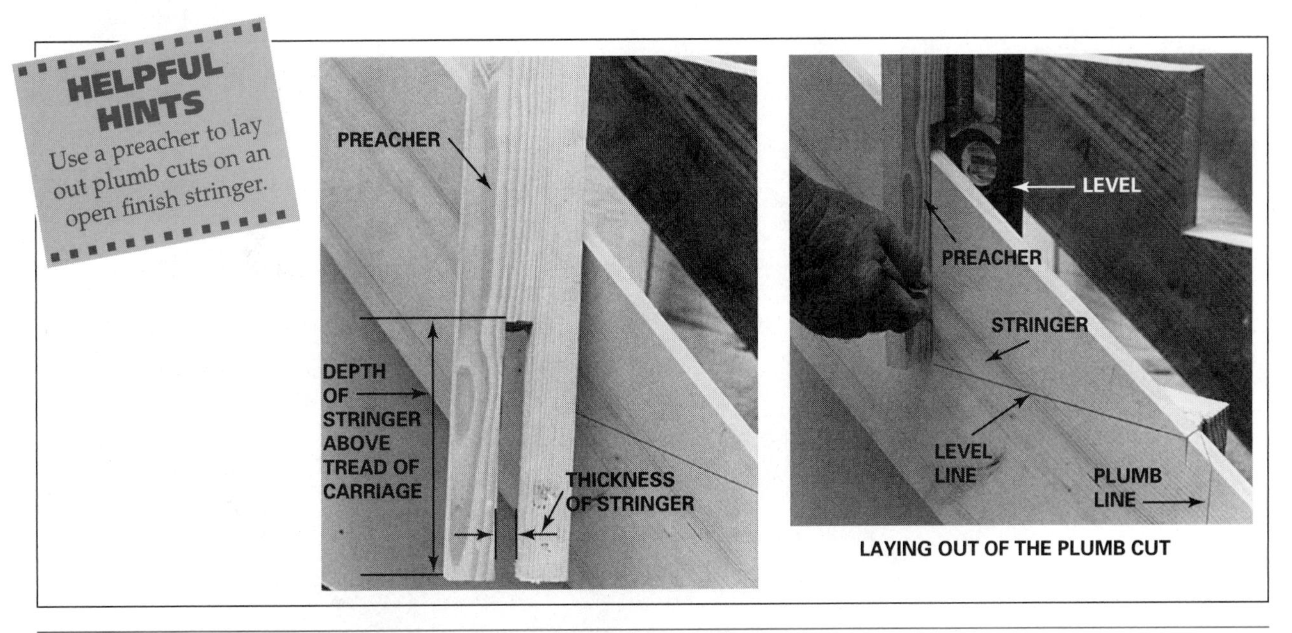

Fig. 81-14 Technique for easily marking a stringer on both faces.

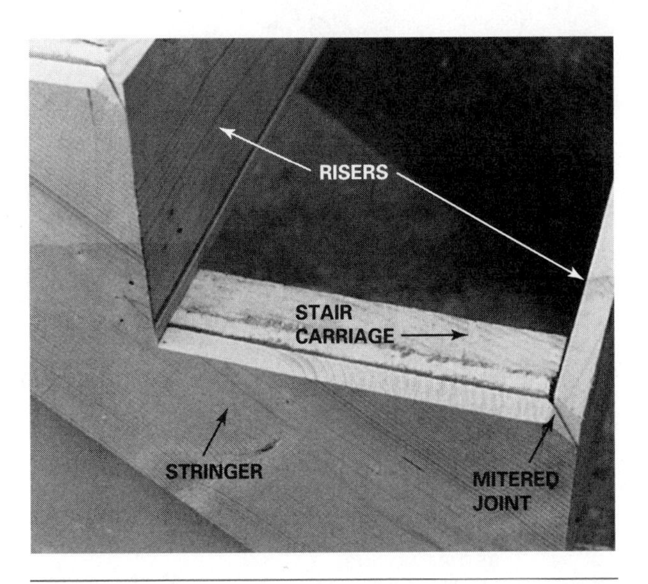

Fig. 81-15 The ends of the risers are mitered to fit against the miters of the open finish stringer.

other end to fit the mitered plumb cuts of the open finish stringer. Sand all pieces before installation. Apply a small amount of glue to the miters. Fasten them in position to each stair carriage. Drive finish nails both ways through the miter to hold the joint tight (Fig. 81-15). Wipe off any excess glue. Set all nails. Lay out and install the *closed finish stringer* in the same manner as described previously.

Installing the Treads

Rip the treads to width. Rabbet the back edges. Make allowance for the rabbet when ripping treads to width. Cut one end to fit against the closed finish stringer. Make a cut on the other end to receive the return nosing. This is a combination square and miter cut. The square cut is made flush with the outside face of the open finish stringer. The miter starts a distance equal to the width of the return nosing beyond the square cut as shown in Figure 82-9.

Applying the Return Nosings

The return nosings are applied to the open ends of the treads. Miter one end of the return nosing to fit against the miter on the tread. Cut the back end square. Return the end on itself. The end of the return nosing extends beyond the face of the riser, the same amount as its width.

Predrill pilot holes in the return nosing for nails. Locate the holes so they are not in line with any balusters that will later be installed on the treads. Holes must be bored in the treads to receive the balusters. Any nails in line with the holes will damage the boring tool (Fig. 81-16).

Apply glue to the joint. Fasten the return nosing to the end of the tread with three 8d finishing nails. Set all nails. Sand the joint flush. Apply all other return nosings in the same manner. Treads may be purchased with the return nosing applied in the factory. If used, the closed end of the tread is cut so the nosed end overhangs the finish stringer by the proper amount.

Applying the Tread Molding

The *tread molding* is applied in the same manner as for closed staircases. However, it is mitered on the open end and returned back onto the open stringer. The back end of the return molding is cut and returned upon itself at a point so the end assembly shows the same as at the edge (Fig. 81-17). Predrill pilot holes in the molding. Fasten it in place.

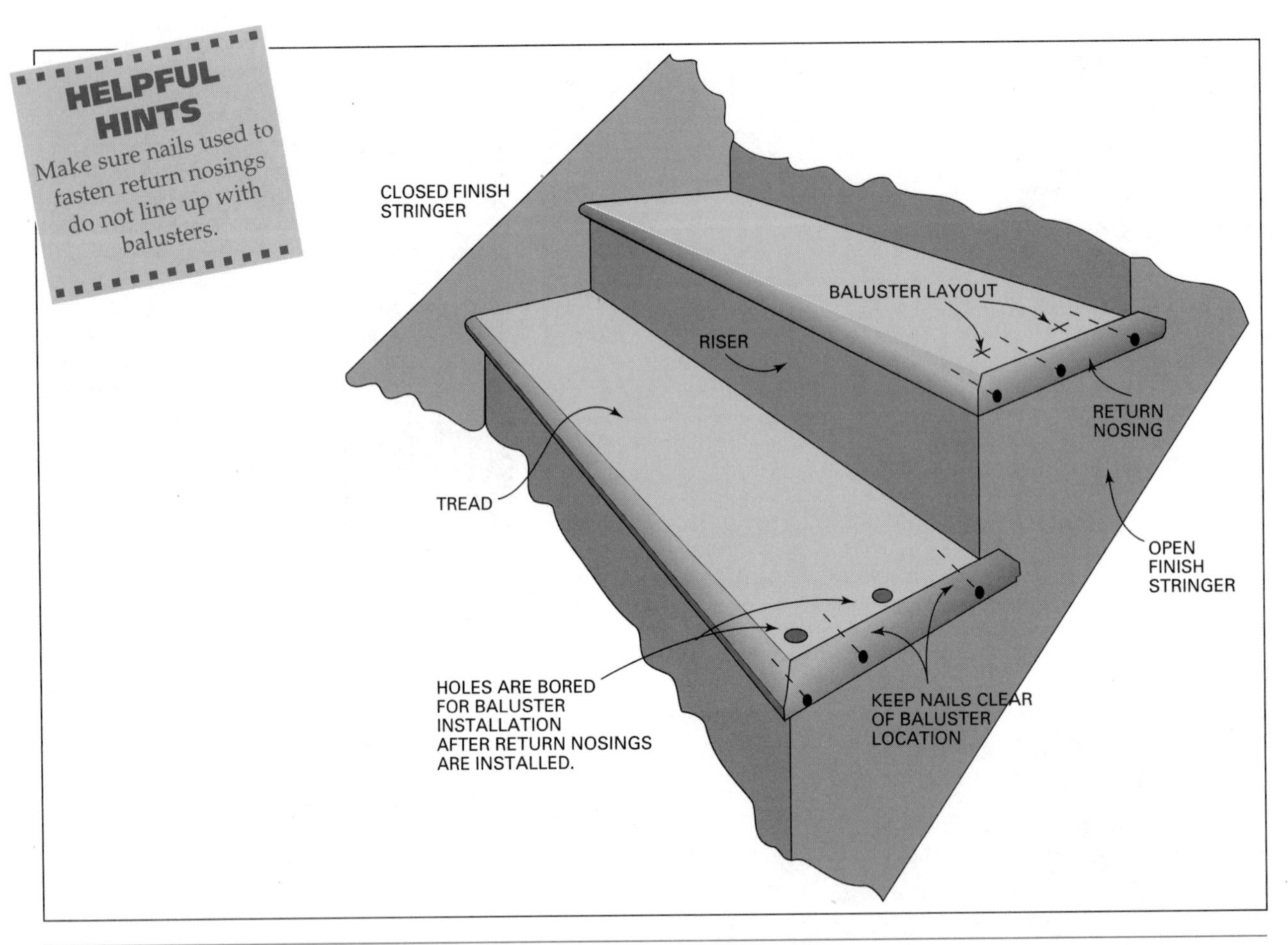

Fig. 81-16 Alignment of trim nails should be positioned to avoid future baluster holes.

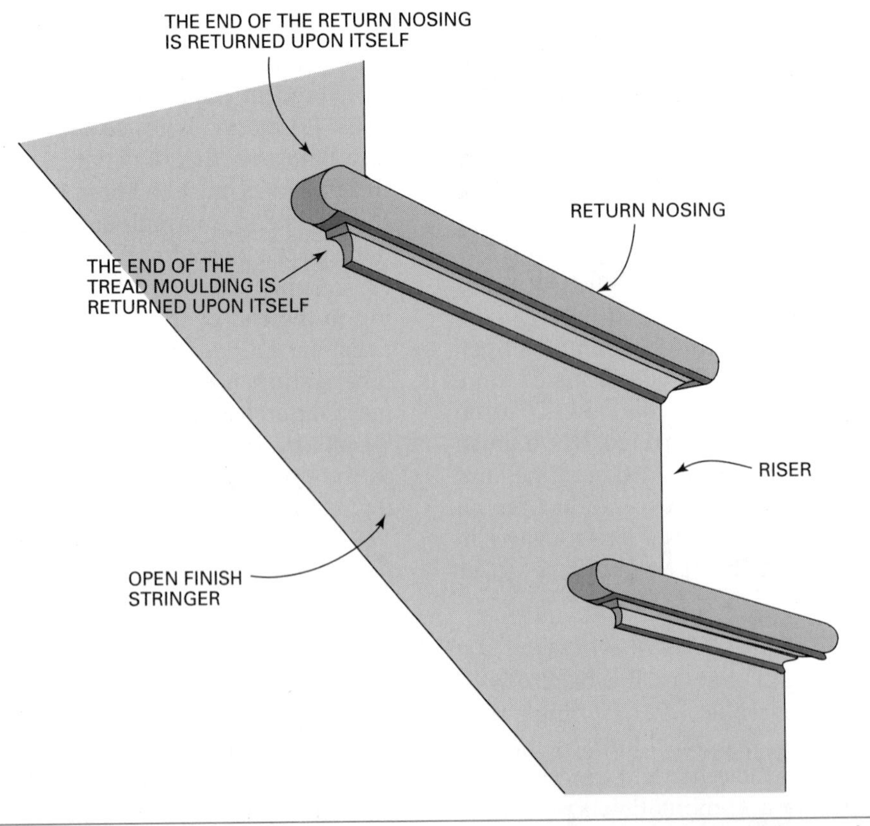

Fig. 81-17 The back ends of the return nosing and molding are returned upon themselves.

Molding on starting and landing treads should only be tacked in case it needs to be removed for fitting after newel posts have been installed.

Installing Treads on Winders

Treads on winding steps are especially difficult to fit because of the angles on both ends. A method used by many carpenters involves the use of a pattern and a scribing block. Cut a thin piece of plywood so it fits in the tread space within 1/2 inch on the ends and back edge. The outside edge should be straight and in line with the nosed edge of the tread when installed.

Tack the plywood pattern in position. Use a 3/4-inch block, rabbeted on one side by the thickness of the pattern, to scribe the ends and back edge. Scribe by riding the block against stringers, riser, and post while marking its inside edge on the pattern. Remove the pattern. Tack it on the tread stock. Place the block with its rabbeted side down and inside edge to the scribed lines on the pattern. Mark the tread stock on the outside edge of the block (Fig. 81-18).

Protecting the Finished Stairs

Protect the risers and treads by applying a width of building paper to them. Unroll a length down the stairway. Hold the paper in position by tacking thin strips of wood to the risers.

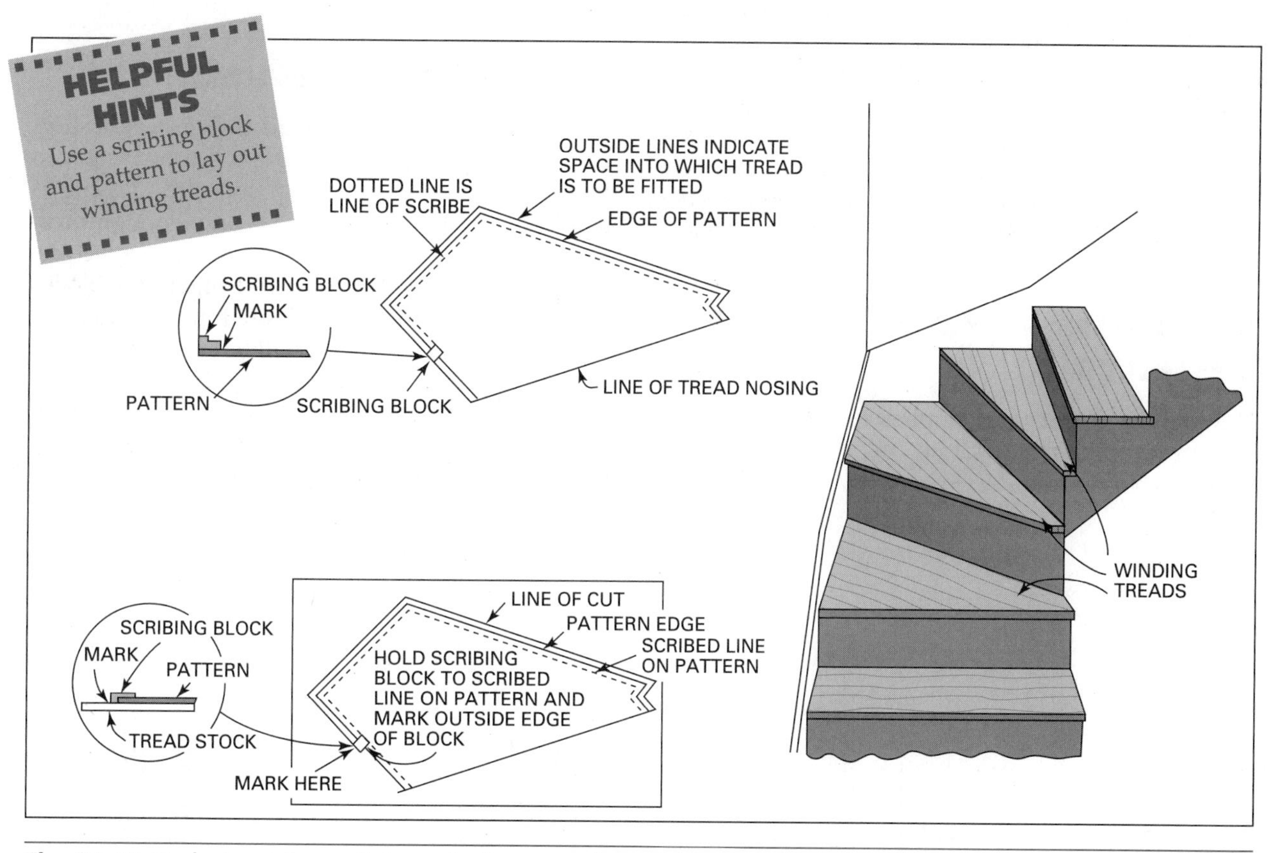

Fig. 81-18 Technique for scribing winding treads to the stringers, risers, and posts.

CHAPTER 82 BALUSTRADE INSTALLATION

Balustrades are the most visible and complex component of a staircase (see Fig. I-9). Installing them is one of the most intricate kinds of interior finish work. This chapter describes installation of post-to-post and over-the-post balustrades. Mastering the techniques described in this chapter will enable the student to install balustrades for practically any situation.

Laying Out Balustrade Centerlines

For the installation of any balustrade, its centerline is first laid out. On an open staircase, the centerline should be located a distance inward, from the face of the finish stringer, that is equal to half the baluster width. It is laid out on top of the treads. If the balustrade is constructed on a *kneewall,* it is centered and laid out on the top of the wall (Fig. 82-1).

Lay Out Baluster Centers

The next step is to lay out the baluster centers. Code requirements for maximum baluster spacing may vary. Check the local building code for allowable spacing. Most codes require that balusters be spaced so that no object 5 inches in diameter or greater can pass through.

On open treads, the center of the front baluster is located a distance equal to half its thickness back from the face of the riser. If two balusters can be used on each tread, the spacing is half the run. If codes require three balusters per tread, the spacing is one-third the run (Fig. 82-2).

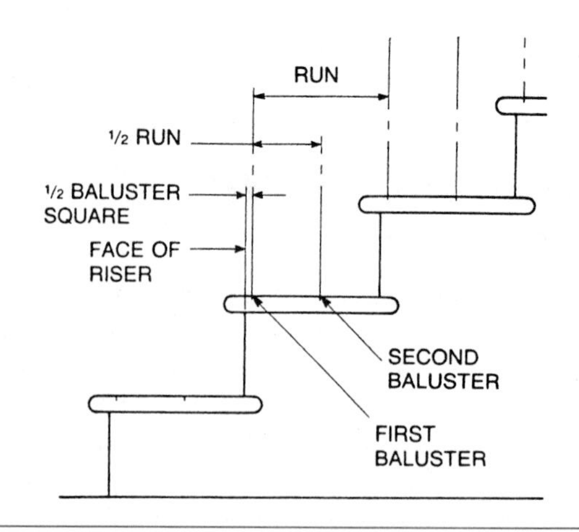

Fig. 82-2 Layout of baluster centers on open treads. (*Courtesy of L. J. Smith*)

Installing a Post-to-Post Balustrade

The following procedure applies to a post-to-post balustrade running, without interruption, from floor to floor.

Laying Out the Handrail

Clamp the handrail to the tread nosings. Use a short bar clamp from the bottom of the finish stringer to the top of the handrail. Clamp opposite a nosing to avoid bowing the handrail. Use only enough pressure to keep the handrail from moving. Protect the edges of the handrail and finish stringer with blocks to avoid marring the pieces. Use a framing square to mark the handrail where it will fit between starting and balcony newel posts (Fig. 82-3).

While the handrail is clamped in this position, use a framing square at the landing nosing to measure the *vertical thickness* of the rake handrail. Also,

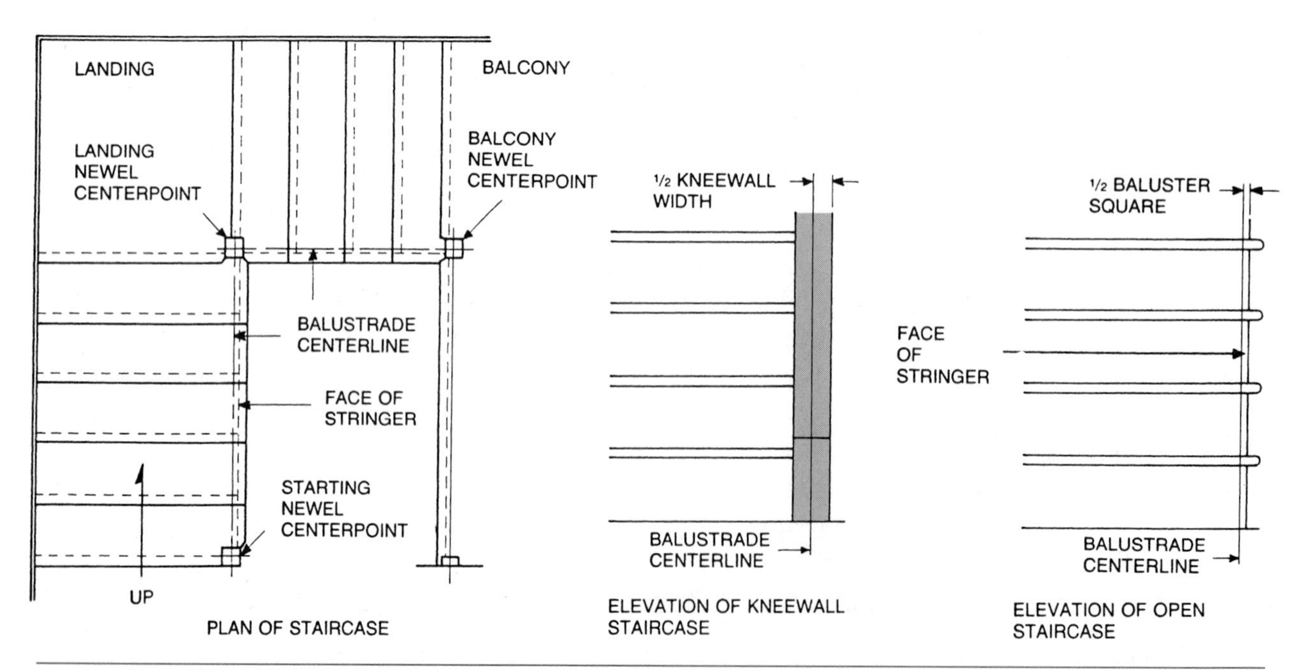

Fig. 82-1 The centerline of the balustrade is laid out on a kneewall or open treads. (*Courtesy of L. J. Smith*)

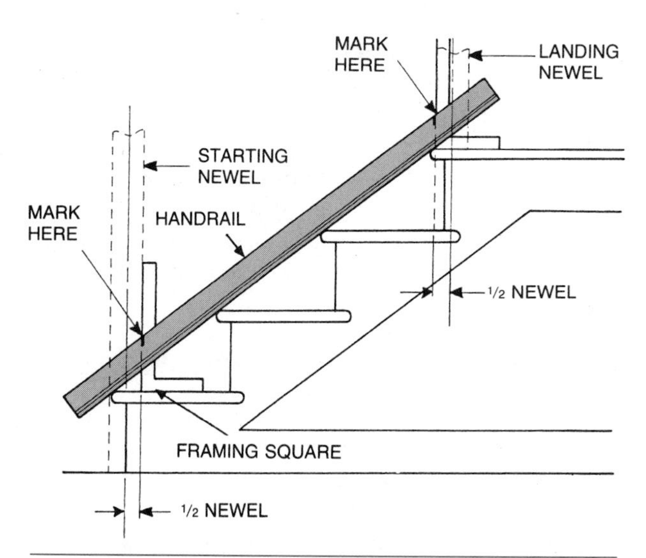

Fig. 82-3 The handrail is laid out to fit between starting and balcony newel posts. (*Courtesy of L. J. Smith*)

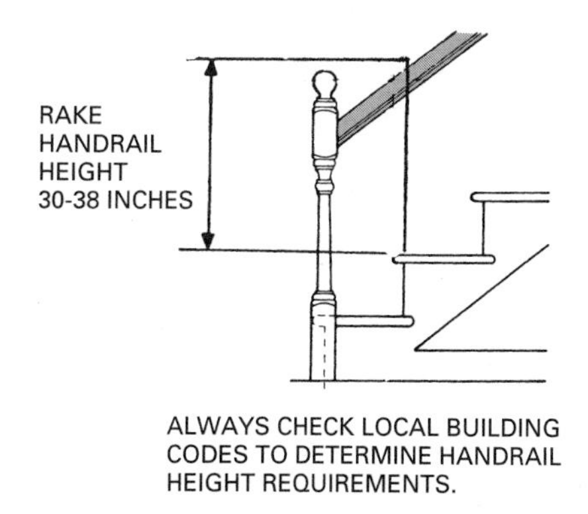

Fig. 82-5 The rake handrail height is the vertical distance from the tread nosing to the top of the handrail. (*Courtesy of L. J. Smith*)

at the bottom, measure the height from the first tread to the top of the handrail where it butts the newel post. Record and save the measurements for later use (Fig. 83-4).

Determining the Height of the Starting Newel

Most building codes state that the rake handrail height shall be no less than 30 inches or more than 34 inches. However, some codes require heights of no less than 34 inches or more than 38 inches. Verify the height requirement with local building codes.

The height of the stair handrail is taken from the top of the tread along a plumb line flush with the face of the riser (Fig. 82-5). Handrails are required only on one side in stairways of less than 44 inches in width. Stairways wider than 44 inches require a handrail on both sides. A center handrail must be provided in stairways more than 88 inches wide. Handrail heights are 30 to 38 inches vertically above the nosing edge of the tread. There should be a continuous 1 1/2-inch finger clearance between the rail and the wall.

If a *turned* starting newel post is used, add the *difference* between the two previously recorded measurements above to the required rake handrail height. Add 1 inch for a *block reveal.* The block reveal is the distance from the top of the handrail to the top of the square section of the post. This sum is the distance from the top of the first tread to the top of the upper block. To this measurement add the height of the turned top and the distance the newel extends below the top of the first tread to the floor (Fig. 82-6). Cut the starting newel to its total length.

Installing the Starting Newel

The starting newel is notched over the outside corner of the first step. One-half of its bottom thickness is left on from the front face of the post to the face of the riser. In the other direction, it is notched so its

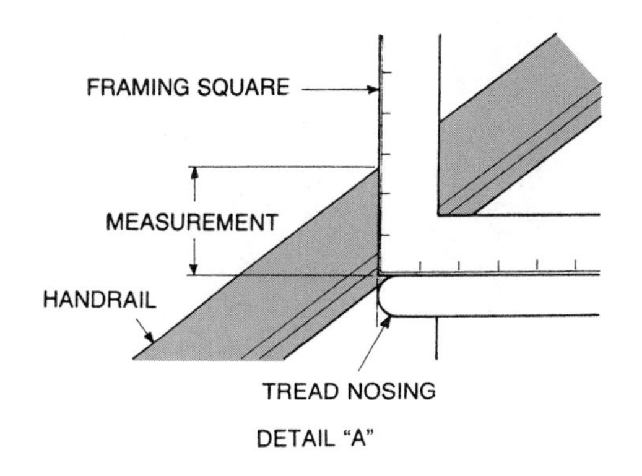

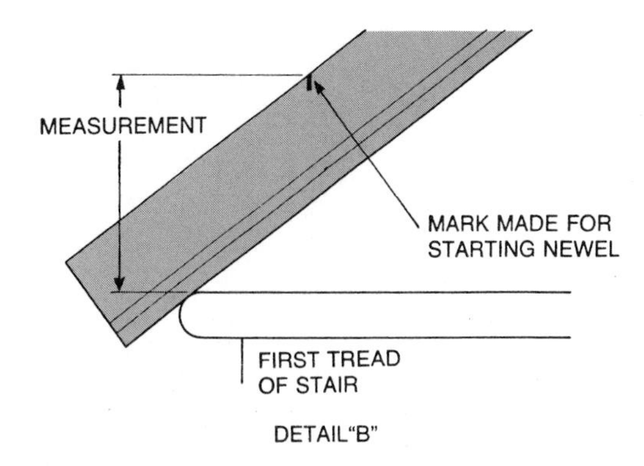

Fig. 82-4 Determine the two measurements shown above and record for future use. (*Courtesy of L. J. Smith*)

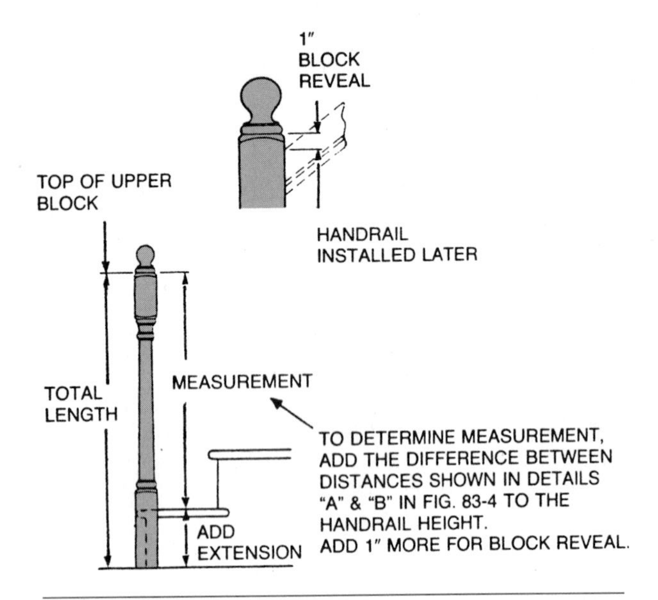

Fig. 82-6 Determining the height of the starting newel. (*Courtesy of L. J. Smith*)

centerline will be aligned with the handrail centerline (Fig. 82-7). The post is then fastened to the first step with lag screws. The lag screws are counterbored and later concealed with wood plugs.

Newel posts must be set plumb. They must be strong enough to resist lateral force applied by persons using the staircase. The post may be slightly out of plumb after it is fastened. If so, loosen the lag screws slightly. Install thin shims, between the post and riser or finish stringer, to plumb the post. On one or both sides, install the shims, near the bottom or top of the notch as necessary, to plumb the post. When plumb, retighten the lag screws.

Installing the Balcony Newel

Generally, codes require that *balcony rails* for homes be no less than 36 inches. For commercial or public structures, the rails are required to be no less than 42 inches. Check local codes for requirements.

The height of the *balcony newel* is determined by finding the sum of the required balcony handrail height, a block reveal of one inch, the height of the turned top, and the distance the newel extends below the balcony floor.

Trim the balcony newel to the calculated height. Notch and fit it over the top riser with its centerlines aligned with both the rake and balcony handrail centerlines. Plumb it in both directions. Fasten it in place with counterbored lag bolts (Fig. 82-8).

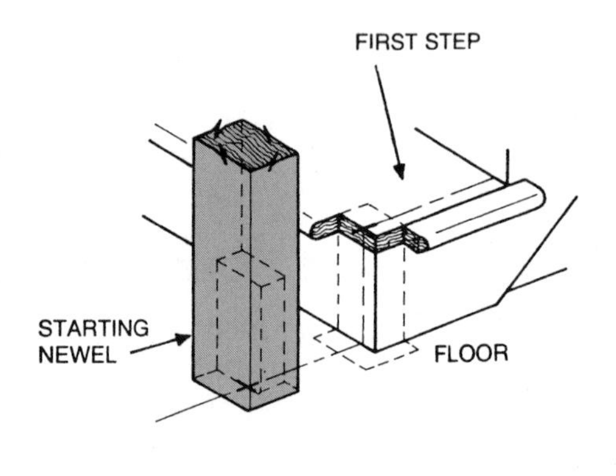

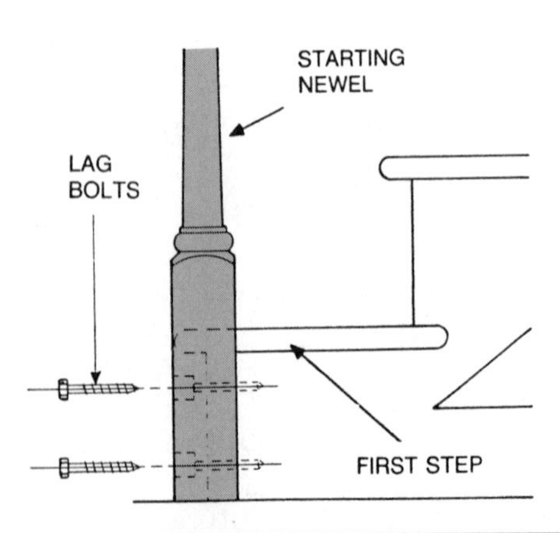

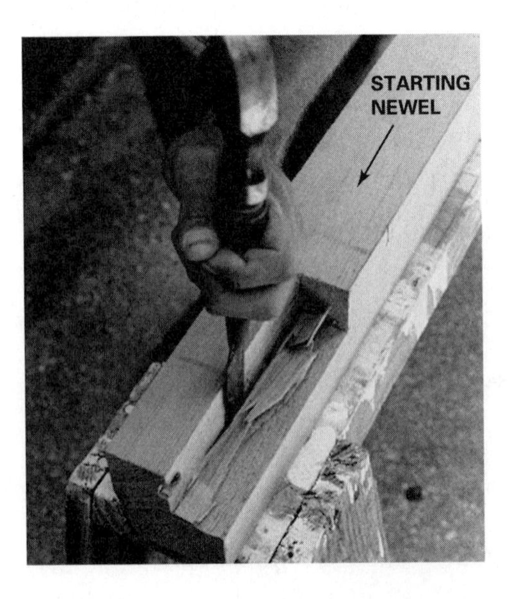

Fig. 82-7 The starting newel is notched to fit over the first step. It is then fastened in place. (*Courtesy of L. J. Smith*)

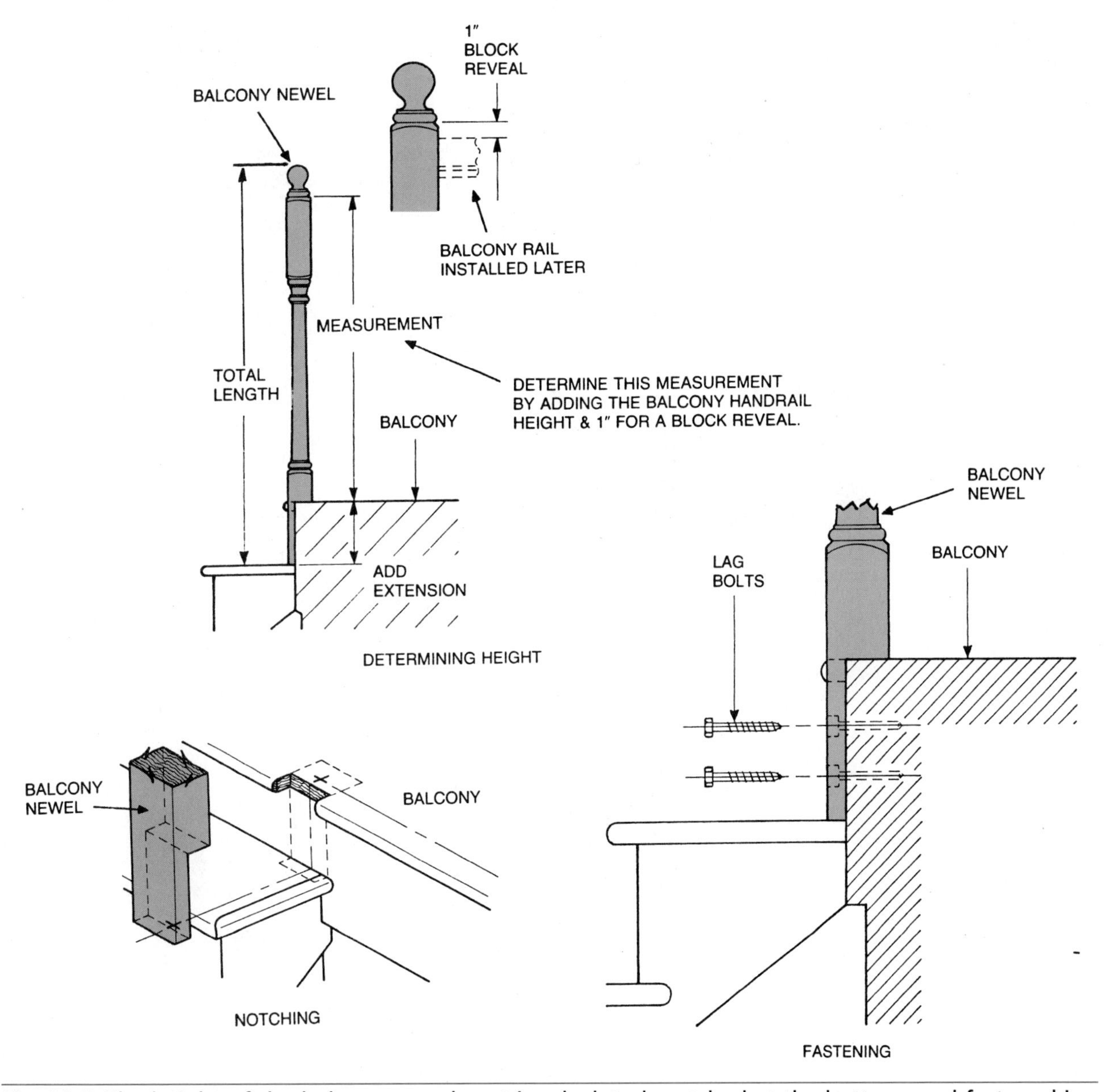

Fig. 82-8 The height of the balcony newel post is calculated, notched at the bottom, and fastened in place. (*Courtesy of L. J. Smith*)

Preparing Treads and Handrail for Baluster Installation

Bore holes in the treads at the center of each baluster. The diameter of the hole should be equal to the diameter of the pin at the bottom end of the baluster. The depth of the hole should be slightly more than the length of the pin (Fig. 82-9).

Cut the handrail to fit between starting and balcony newels. Lay it back on the tread nosings. The handrail can be cut with a handsaw, a radial arm saw, or a compound miter saw with the blade tilted. If using a power saw, make a practice cut to be sure the setup is correct, before cutting the handrail. Transfer the baluster centerlines from the treads to the handrail (Fig. 82-10).

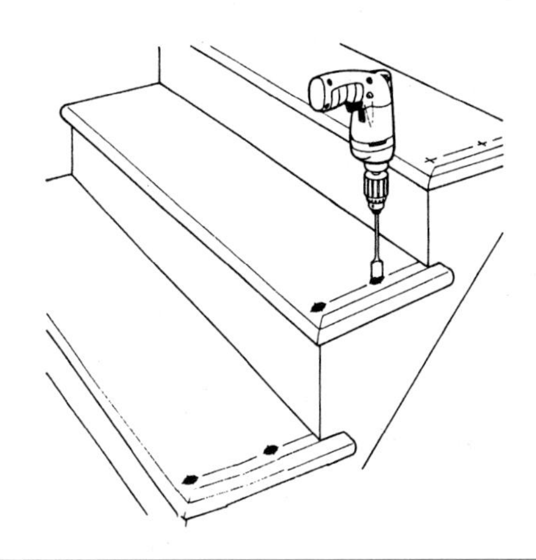

Fig. 82-9 Holes are bored in the top of the treads at each baluster center point. (*Courtesy of L. J. Smith*)

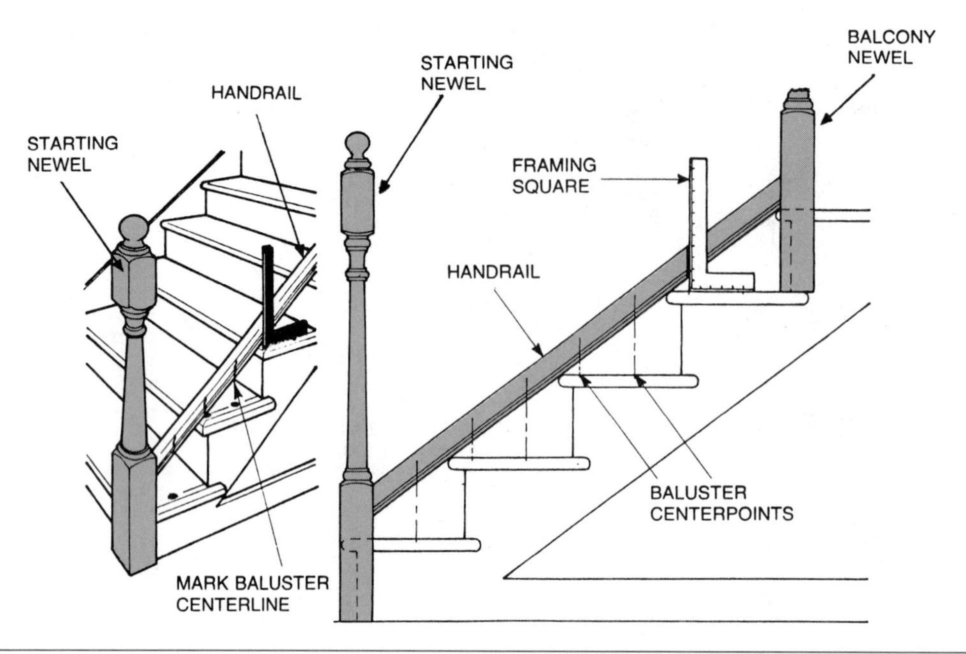

Fig. 82-10 The handrail is fitted between newel posts. The baluster centers are transferred to it. (*Courtesy of L. J. Smith*)

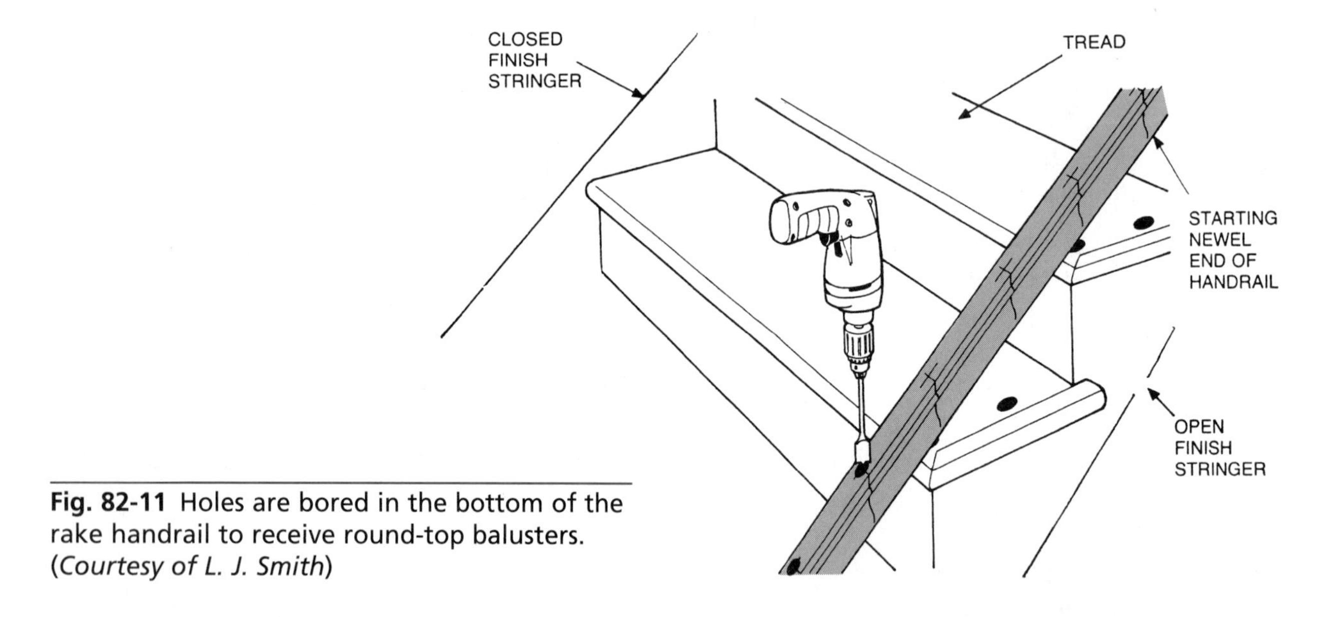

Fig. 82-11 Holes are bored in the bottom of the rake handrail to receive round-top balusters. (*Courtesy of L. J. Smith*)

Turn the handrail upside down and end for end. Set it back on the tread nosings with the starting newel end facing up the stairs. Bore holes at baluster centers at least 3/4 inch deep, if balusters with *round tops* are to be used (Fig. 82-11).

Installing Handrail and Balusters

Prepare the posts for fastening the handrail by counterboring and drilling **shank holes** for lag bolts through the posts. Place the handrail at the correct height between newel posts. Drill **pilot holes.** Temporarily fasten the handrail to the posts (Fig. 82-12).

Cut the balusters to length. Allow 3/4 inch for insertion in the hole in the bottom of the handrail. The handrail may have to be removed for baluster installation and then fastened permanently. The bottom pin is inserted in the holes in the treads. The top of the baluster is inserted in the holes in the handrail bottom.

If *square top balusters* are used, they are trimmed to length at the rake angle. They are inserted into a *plowed handrail.* The balusters are then fastened to the handrail with finish nails and glue (Fig. 82-13). Care must be taken to keep the handrail

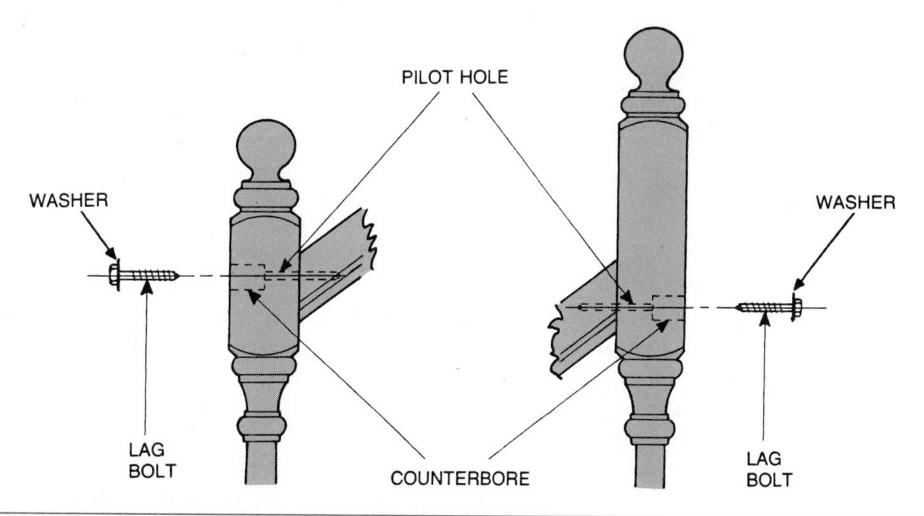

Fig. 82-12 The handrail is fastened temporarily to newel posts with lag bolts. The use of nails when constructing balustrades is discouraged. (*Courtesy of L. J. Smith*)

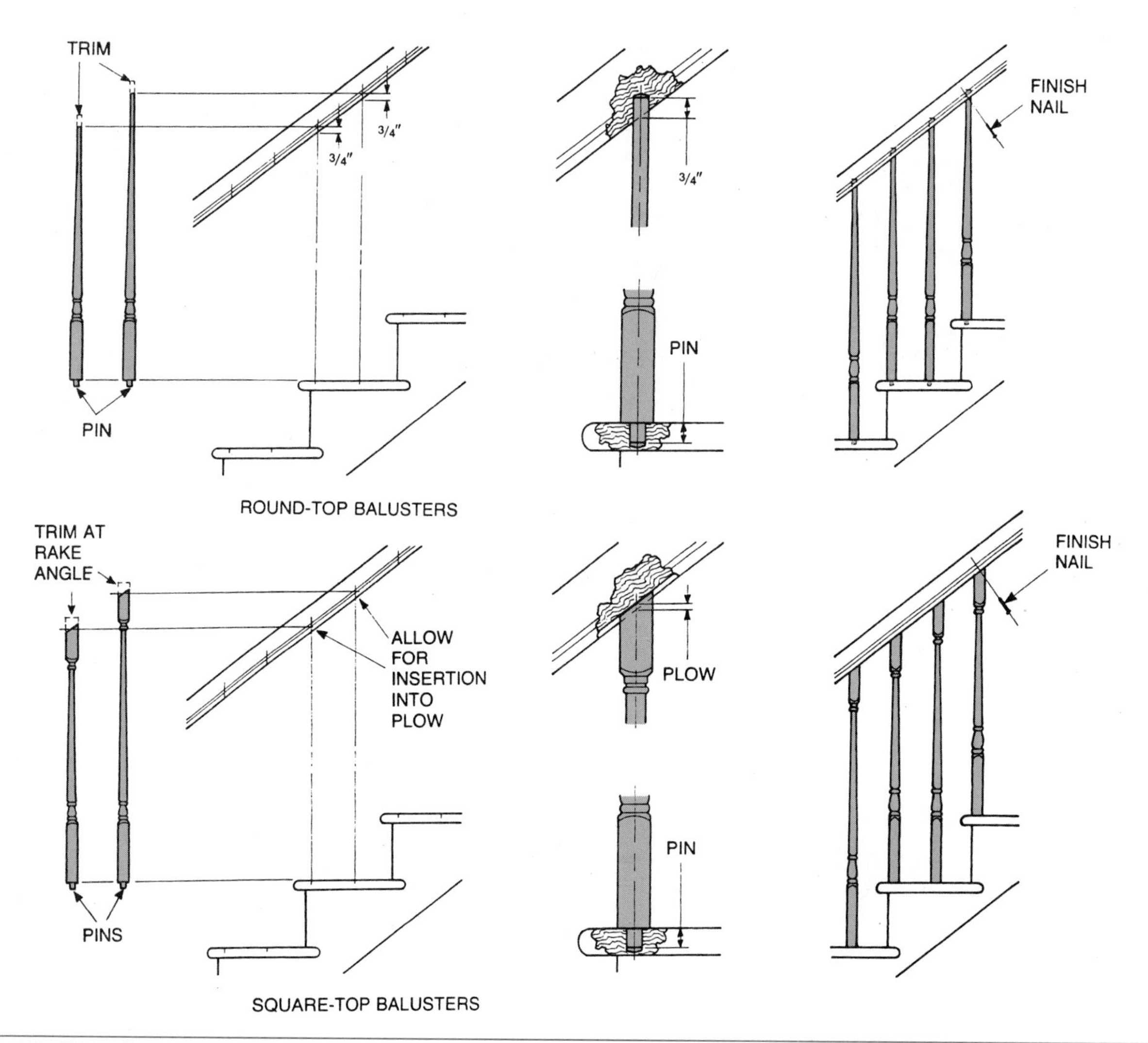

Fig. 82-13 Balusters are cut to length and installed between handrail and treads. (*Courtesy of L. J. Smith*)

in a straight line from top to bottom when fastening square top balusters. Care must also be taken to keep each baluster in a plumb line. Install *fillets* in the plow of the handrail, between the balusters.

Installing the Balcony Balustrade

Cut a *half newel* to the same height as the balcony newel. Temporarily place it against the wall. Mark the length of the balcony handrail (Fig. 82-14). Cut the handrail to length. Fasten the half newel to one end of it. Temporarily fasten the half newel to the wall and the other end of the handrail to the landing newel, if they must be removed to install the balcony balusters.

If the balcony handrail ends at the wall against a *rosette,* first fasten the rosette to the end of the handrail. Hold the rosette against the wall. Mark the length of the handrail at the landing newel. Cut the handrail to length. Temporarily fasten it in place (Fig. 82-15).

Spacing and Installing Balcony Balusters

The balcony balusters are spaced by adding the thickness of one baluster to the distance between the balcony newel and the half newel. The overall distance is then divided into spaces that equal, as close as possible, the spacing of the rake balusters (Fig. 82-16). The balcony balusters are then installed in a manner similar to the rake balusters.

Installing an Over-the-Post Balustrade

The following procedures apply to an over-the-post balustrade running from floor to floor with an in-

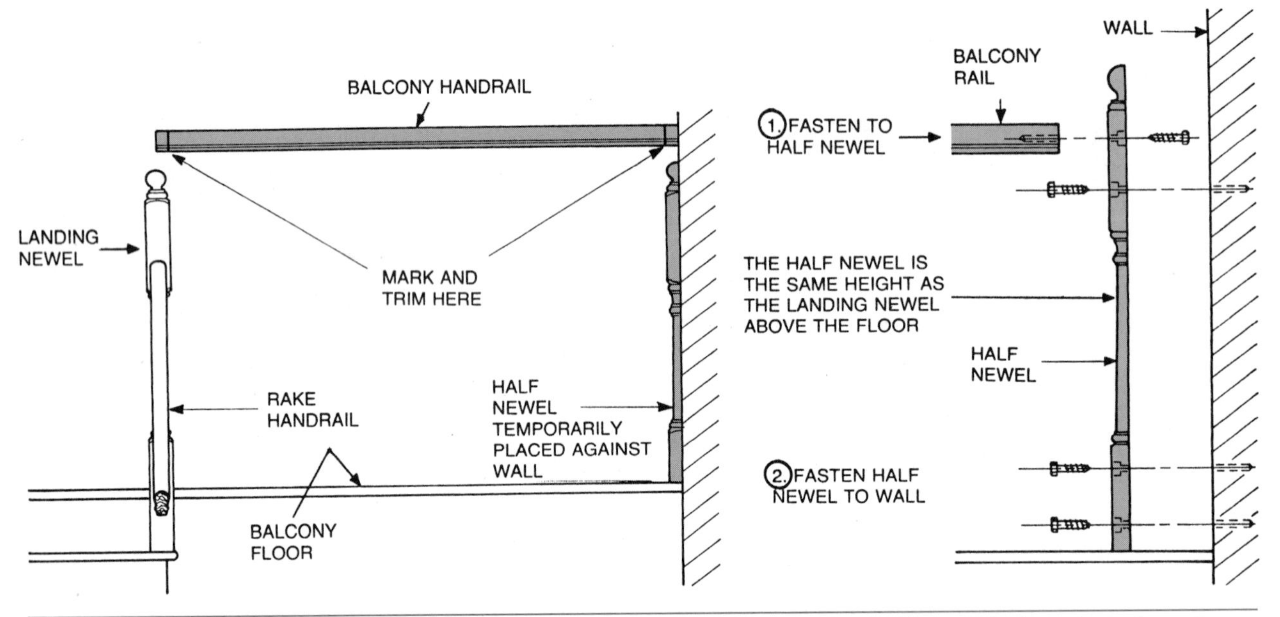

Fig. 82-14 The balcony rail is fitted between the landing newel and a half newel placed against the wall. (*Courtesy of L. J. Smith*)

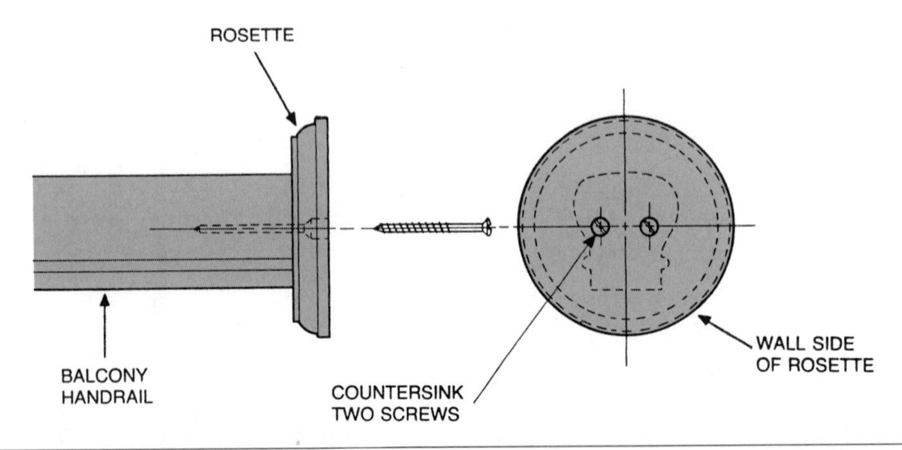

Fig. 82-15 A rosette is sometimes used to end the balcony handrail instead of a half newel. (*Courtesy of L. J. Smith*)

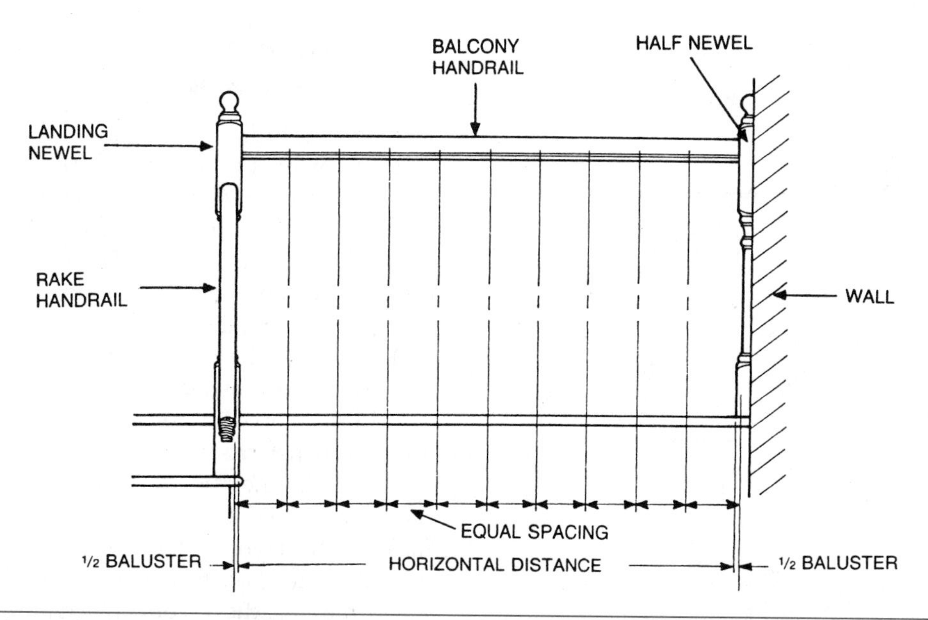

Fig. 82-16 Balcony balusters are installed as close to possible to the same spacing as the rake balusters. (*Courtesy of L. J. Smith*)

termediate landing (see Fig. 80-2). An over-the-post balustrade is more complicated. *Handrail fittings* are required to be joined to straight sections to construct a continuous handrail from start to end. The procedure consists of constructing the entire handrail first, then setting newel posts and installing balusters.

Constructing the Handrail

The first step is to lay out balustrade and baluster centerlines on the stair treads as described previously. If a *starting step* is used, lay out the baluster and starting newel centers using a template provided with the starting fitting (Fig. 82-17).

Laying Out the Starting Fitting

Make a *pitch block* by cutting a piece of wood in the shape of a right triangle whose sides are equal in length to the rise and tread run of the stairs. Hold the cap of the *starting fitting* and the run side of the pitch block on a flat surface. The rake edge of the pitch block should be against the bottom side of the fitting. Mark the fitting at the *tangent point,* where its curved surface touches the pitch block. A straight line, *tangent* to a curved line, touches the curved line at one point only.

Then turn the pitch block so its rise side is on the flat surface. Mark the fitting along the rake edge of the pitch block. Place the fitting in a power miter box, supported by the pitch block. Cut it to the layout line (Fig. 82-18).

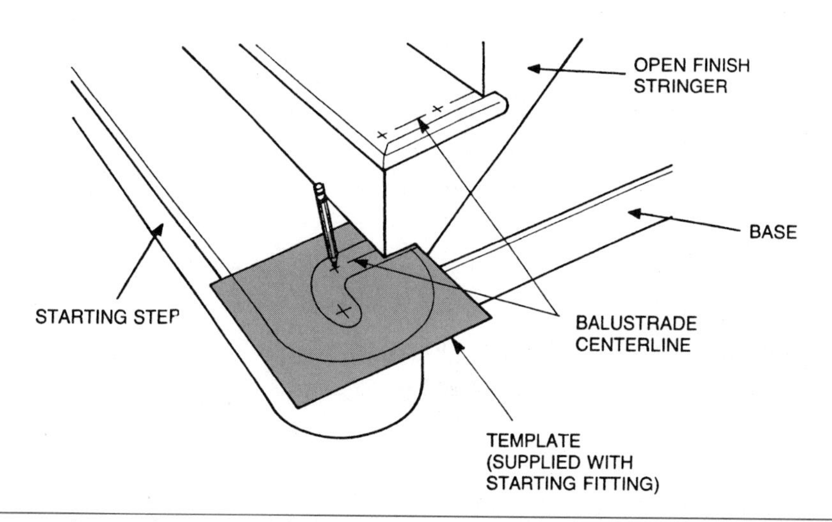

Fig. 82-17 Baluster and newel post centers are laid out on a starting step with a template. (*Courtesy of L. J. Smith*)

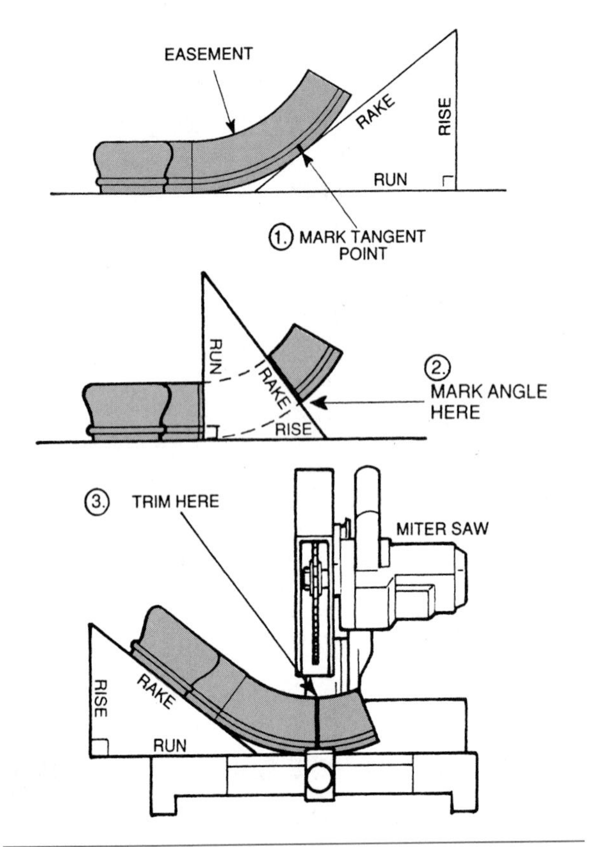

Fig. 82-18 Procedure for laying out and cutting the starting fitting. (*Courtesy of L. J. Smith*)

CAUTION When cutting handrail fittings, supported by a pitch block in a power miter box, clamp the pitch block and fitting securely to make sure they will not move when cut. If either one moves, during cutting, a serious injury could result. Even if no injury occurs, the fitting would probably be ruined.

Joining the Starting Fitting to the Handrail

The starting fitting is joined to a straight section of handrail by means of a special *handrail bolt.* The bolt has threads on one end designed for fastening into wood. The other end is threaded for a nut. Holes must be drilled in the end of both the fitting and the handrail in a manner that assures their alignment when joined together.

To mark the hole locations, make a template by cutting about a 1/8-inch piece of handrail. Drill a 1/16-inch hole centered on the template width and 15/16 inch from the bottom side. Mark one side *rail* and the other side *fitting.* This will ensure that the template is facing in the right direction when making the layout. If the hole in the template is drilled slightly off center and the template turned when making the layout, the handrail and fitting will not be in alignment when joined. Use the template to mark the location of the rail bolt on the end of the fitting and the handrail.

Drill all holes to the depth and diameter shown in Figure 82-19. Double nut the rail bolt and turn it into the fitting. Remove the nuts, place the handrail on the bolt. Install a washer and nut, and tighten to join the sections together. Clamp the assembly to the tread nosings. The newel center points on the fitting and the starting tread should be in a plumb line (Fig. 82-20). Note the measurement in Detail "C" of Figure 82-20.

Installing the Landing Fitting

The second flight of stairs turns at the landing, so a right hand, two-riser *gooseneck* fitting is used. It is laid out and joined to the bottom end of the handrail of the second flight in a similar manner as the starting fitting (Fig. 82-21). The assembled fit-

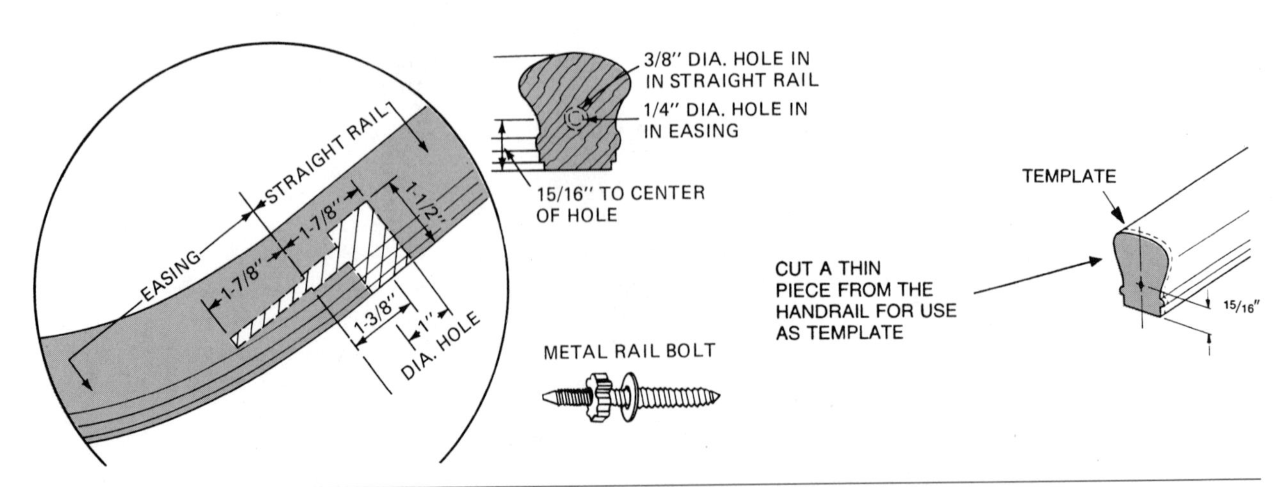

Fig. 82-19 A template is made to mark the ends of handrails and fittings for joining with rail bolts. The ends are drilled to specific depths and diameters.

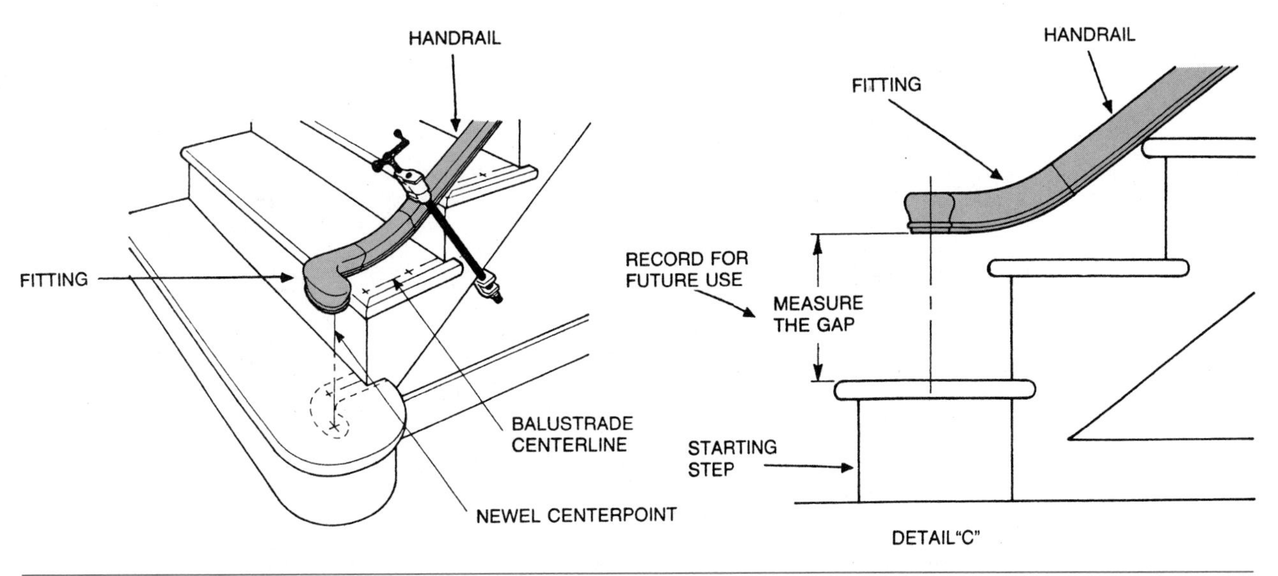

Fig. 82-20 The starting fitting and handrail assembly are clamped to the tread nosings of the first flight of stairs. (*Courtesy of L. J. Smith*)

ting and rail are then clamped to the nosings of the treads in the second flight. Position the rail so that the *newel cap* of the fitting is in a plumb line with the baluster centerlines of both flights (Fig. 82-22).

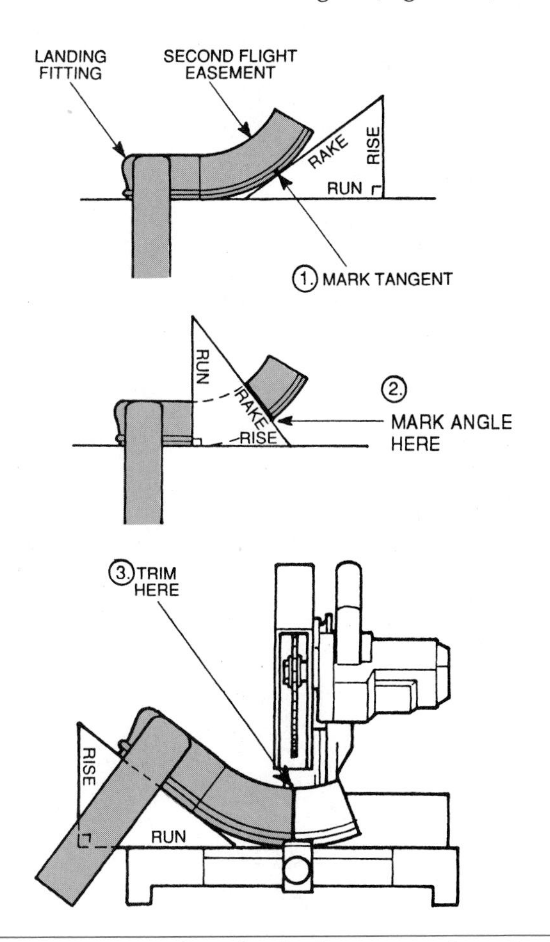

Fig. 82-21 Procedure for laying out and cutting the landing fitting for joining to the second flight handrail. (*Courtesy of L. J. Smith*)

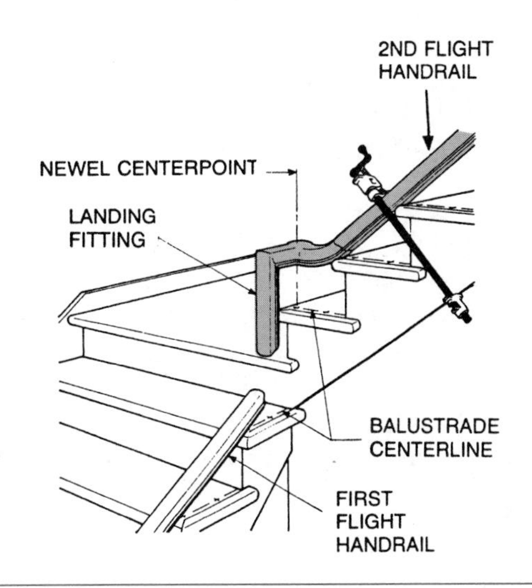

Fig. 82-22 The second flight handrail is clamped to the tread nosings after the landing fitting is joined to it. (*Courtesy of L. J. Smith*)

Joining First and Second Flight Handrails

In preparation for laying out the *easement* used to join the first and second flight handrails, tack a piece of plywood about 5 inches wide to the bottom side of the gooseneck fitting and the handrail of the first flight. These pieces are used to rest the connecting easement against when laying out the joint.

Clamp the handrails back on the treads. Make sure the newel cap centers are plumb with starting and landing newel post centers. Rest the connecting easement on the plywood blocks. Level its upper end. Mark the gooseneck in line with the upper end. Mark the lower end of the easement at a point *tangent* with the block under the handrail (Fig. 82-23). Note the measurement in Detail "D" of Figure 82-23.

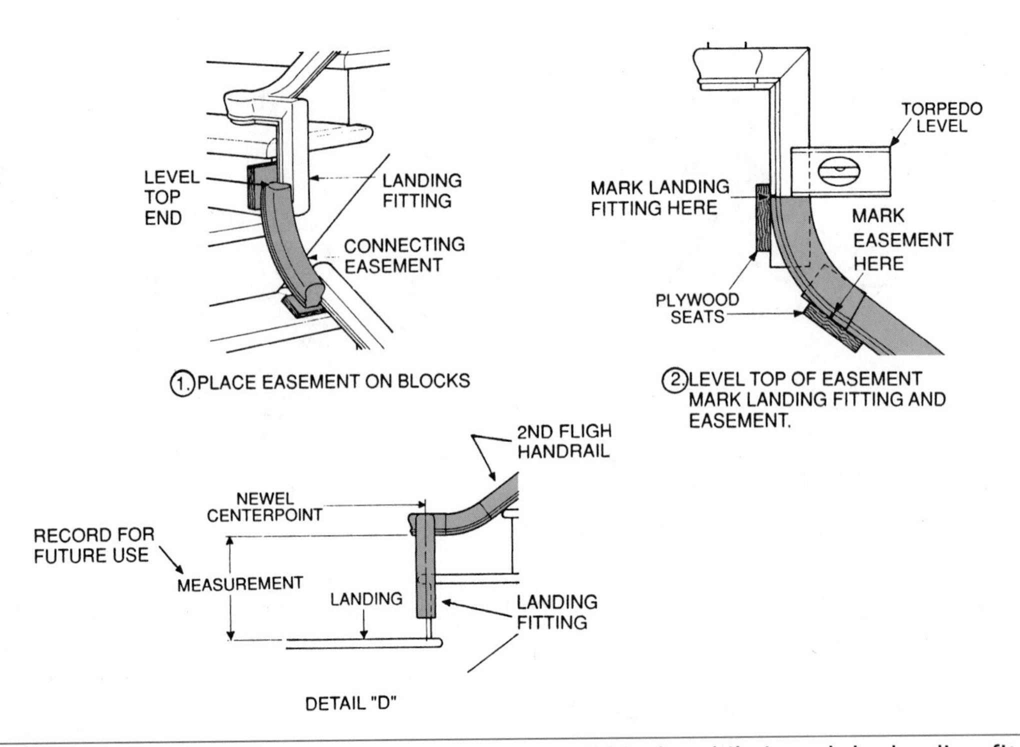

Fig. 82-23 The connecting easement is positioned on small blocks while it and the landing fitting are marked. (*Courtesy of L. J. Smith*)

Make a square cut at the mark laid out on the lower end of the gooseneck. Join the gooseneck and easement together with a rail bolt. Lay out the lower end of the easement with the pitch block. Cut it using the pitch block for support (Fig. 82-24).

Place the assembled landing fitting and handrail back on the stair nosings. Mark the handrail of the lower flight where it meets the end of the fitting. Cut the handrail square at the mark. Join it to the landing fitting with a rail bolt. Clamp the entire handrail assembly to the tread nosings with newel cap centers plumb with newel post centerlines (Fig. 82-25).

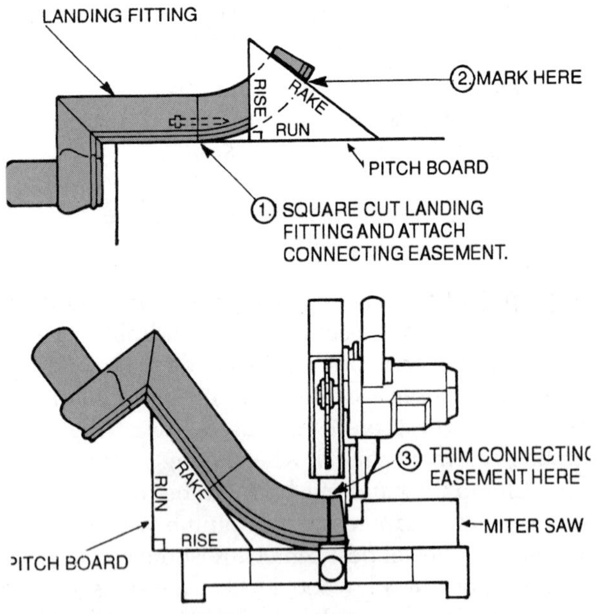

Fig. 82-24 The connecting easement and landing fitting are joined. Then the lower end is trimmed to form a two-riser gooseneck. (*Courtesy of L. J. Smith*)

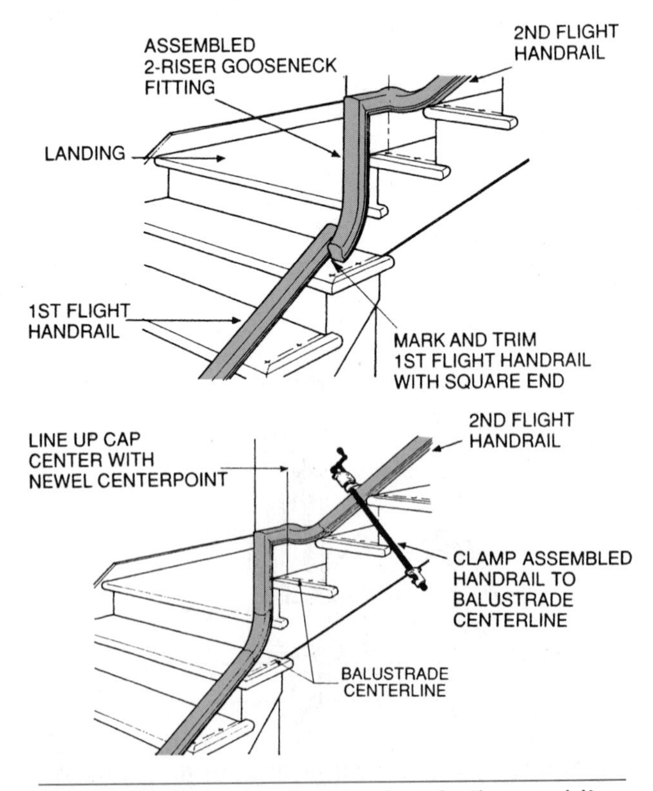

Fig. 82-25 Lower and upper handrail assemblies are joined. They are then clamped to the tread nosings. (*Courtesy of L. J. Smith*)

Installing the Balcony Gooseneck Fitting

A one-riser *balcony gooseneck* fitting is used when balcony rails are 36 inches high. Two-riser fittings are used for rails 42 inches high. In this case, a one-riser fitting is used at the balcony. Laying out and fitting a two-riser fitting at the landing has been previously described.

Hold the fitting so the center of its cap is directly above the balcony newel post centerline. Hold it against the handrail of the upper flight. Mark it and the handrail at the point of tangent (Fig. 82-26). Lay out and make the cut on the gooseneck with the use of a pitch block (Fig. 82-27). Cut the handrail square at the mark. Join the gooseneck fitting and the handrail with a rail bolt.

Clamp the entire rail assembly back on the nosing of the treads in line with the balustrade centerlines and the three newel post centers. Use a framing square to transfer the baluster centers from the treads to the side of the handrails. Remove the handrail assembly out of the way of newel post installation.

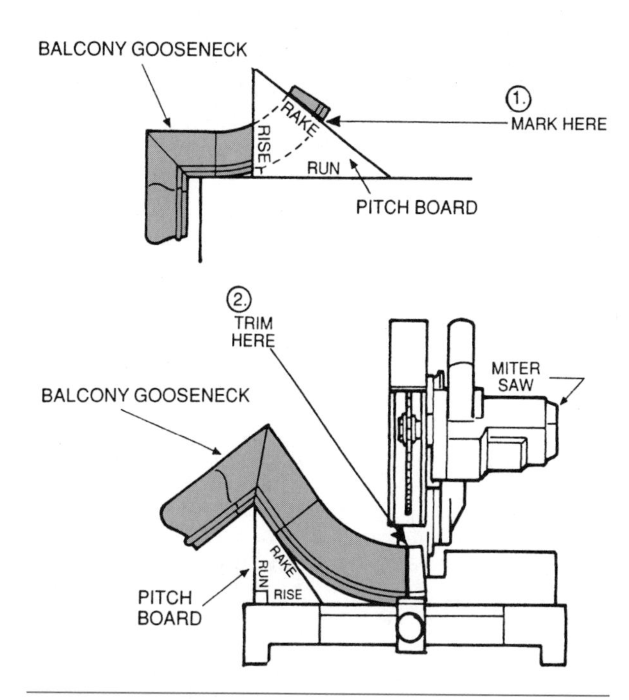

Fig. 82-27 Procedure for laying out and cutting the balcony gooseneck fitting. (*Courtesy of L. J. Smith*)

Installing Newel Posts

In this stair case, the starting newel is installed on a starting step. The height of the *rake handrail* is calculated, from the height of the starting newel, to make sure the handrail will conform to the height required by the building code.

From the height of the starting newel to be used, subtract the previously recorded distance between the starting fitting and the starting tread as shown in Detail "C" of Figure 82-20. Then, add the vertical thickness of the rake handrail shown in Figure 82-4, Detail "A". The result is the rake handrail height (Fig. 82-28). If the height does not conform to the building code, the starting newel post height must be changed.

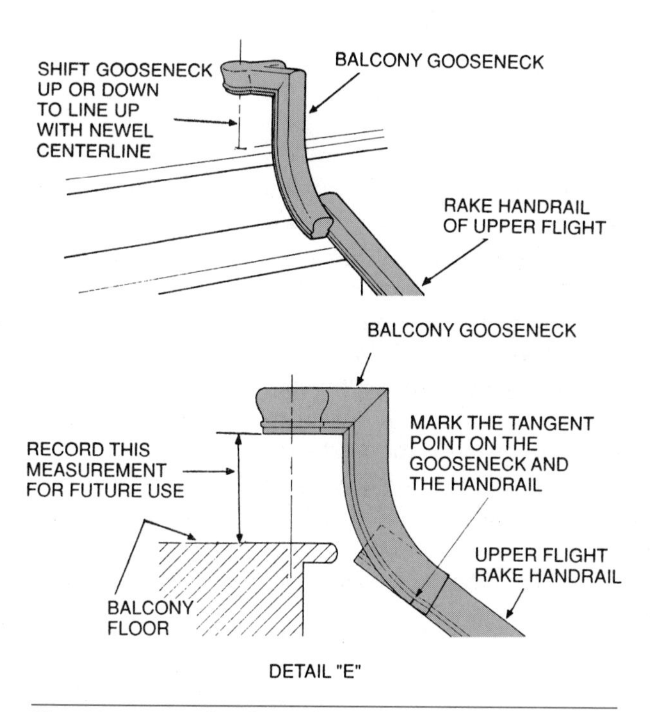

Fig. 82-26 Method used to mark the balcony gooseneck fitting and handrail of the upper flight. (*Courtesy of L. J. Smith*)

Installing the Starting Newel Post

Before installing the starting step, measure the diameter of the *dowel* at the bottom of the starting newel post. At the centerpoint of the newel post, bore a hole for the dowel through the tread and floor. Install the post. Wedge it under the floor. Wedges are driven in a *through mortise* cut in the doweled end of the post.

An alternate method is used when there is no access under the floor. Bore holes only through the tread and upper riser block of the starting step. Cut the dowel to fit against the lower riser block so the newel post rests snugly on the tread. Fasten the end of the dowel with a lag screw through the lower riser block (Fig. 82-29). Fasten the assembled starting newel and starting step in position.

Installing the Landing Newel Post

The height of *landing newel* above the landing is found by subtracting the previously recorded distance between the starting fitting and the starting tread (Fig. 82-20, Detail "C") from the height of the

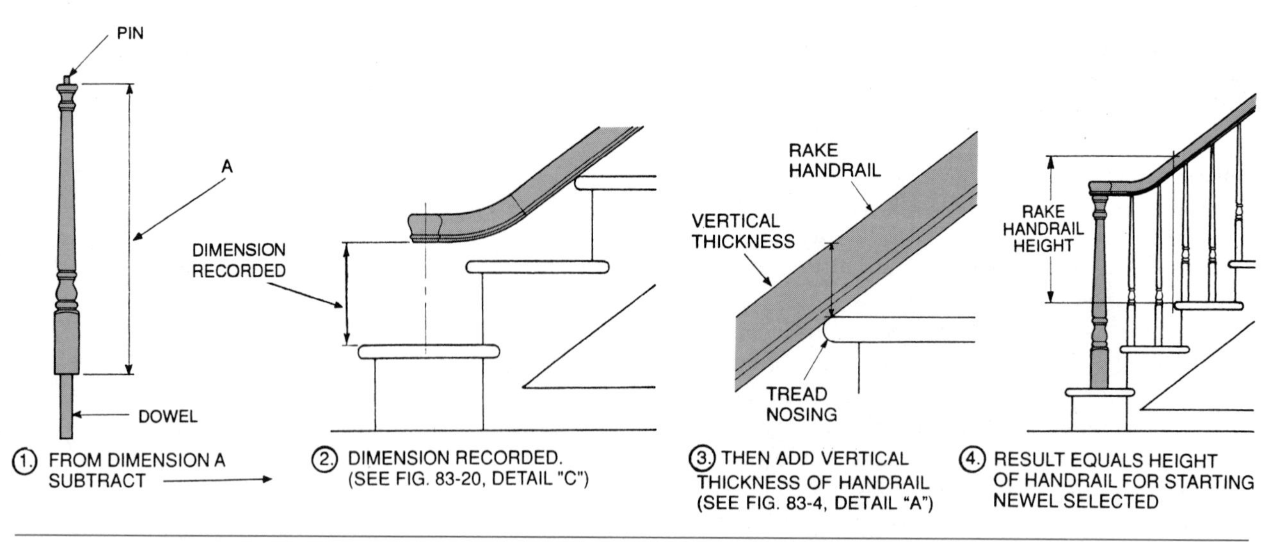

Fig. 82-28 The rake handrail height is calculated from the height of the starting newel post. (*Courtesy of L. J. Smith*)

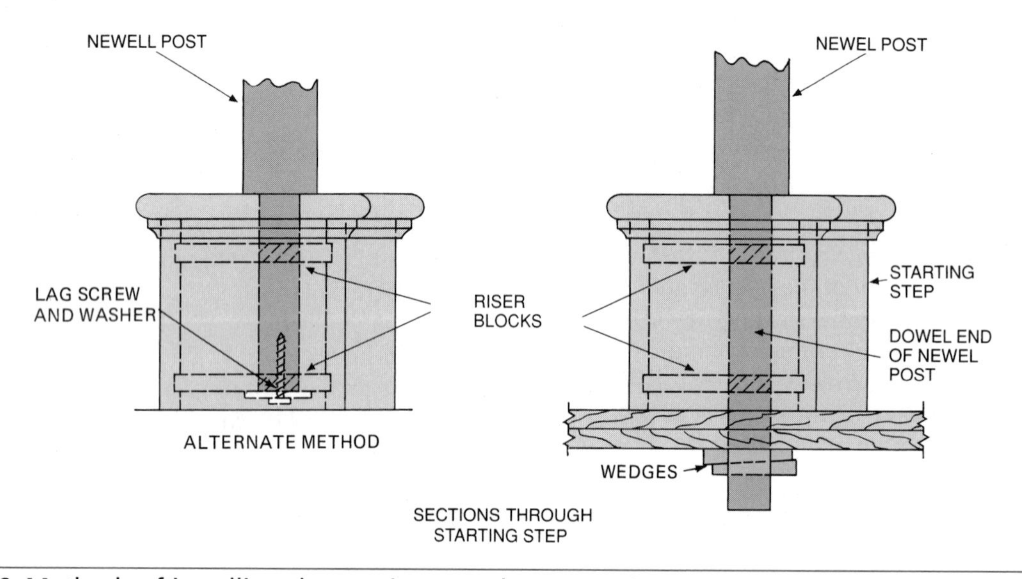

Fig. 82-29 Methods of installing the starting newel on a starting step.

starting newel. Then add the distance between the landing fitting and the landing as previously recorded and shown in Figure 83-23, Detail "D". To this length, add the distance that the landing newel extends below the landing.

Notch the landing newel to fit over the landing and the first step of the upper flight. Fasten the post in position with lag bolts in counterbored holes (Fig. 82-30).

Installing the Balcony Newel Post

The height of the *balcony handrail* must be calculated before the height of the *balcony newel* can be determined. The height of the balcony handrail is found by subtracting the previously recorded distance between the starting fitting and the starting tread (Fig. 82-20, Detail "C") from the height of the starting newel. Then add the previously recorded distance between the balcony gooseneck fitting and the landing (Fig. 82-27, Detail "E"). Then, add the thickness of the balcony handrail. The balcony handrail height must conform to the building code. If not, substitute a two-riser gooseneck fitting instead of a one-riser fitting.

The height of the balcony newel, above the balcony floor, is found by subtracting the handrail thickness from the handrail height. To this length, add the distance the post extends below the floor. Notch and install the post over the balcony riser in line with balustrade centerlines (Fig. 82-31).

Installing Balusters

Bore holes for balusters in the tread and bottom edge of the handrail. No holes are bored in the handrail if square top balusters are used. Install the handrail on the posts. Cut the balusters to length. Install them in the manner described previously for post-to-post balustrades.

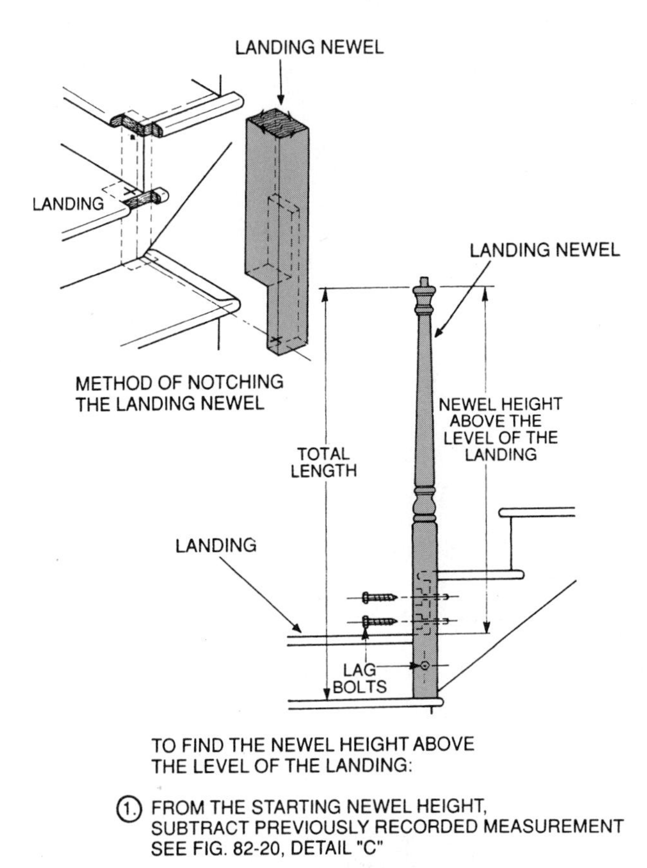

Fig. 82-30 The landing newel is notched to fit over the landing and first step of the upper flight. (*Courtesy of L. J. Smith*)

Installing the Balcony Balustrade

Cut a *half newel* to the same height as the balcony newel extends above the floor. Install it against the wall on the balustrade centerline. Cut an *opening cap* so it fits on top of the half newel. Join the cap to the end of the balcony handrail with a rail bolt. Place the cap on the half newel. Mark the length of the handrail at the balcony newel. Cut the handrail and join it, on one end, to the balcony newel post and, on the other end to the half newel (Fig. 82-32).

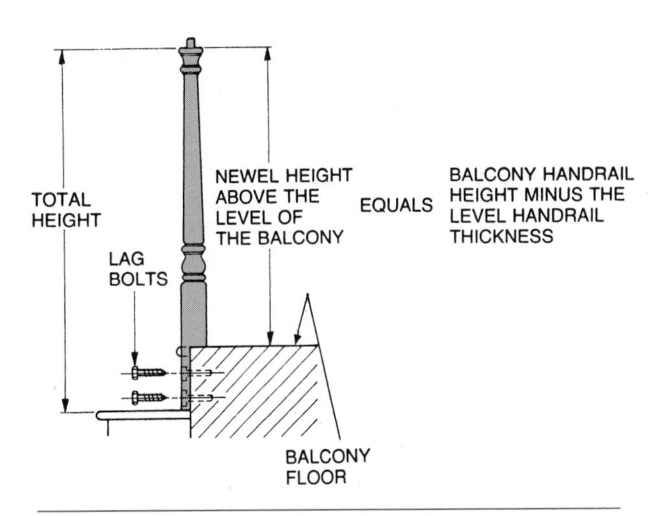

Fig. 82-31 The height of the balcony newel is determined. The post is then fastened in place. (*Courtesy of L. J. Smith*)

A rosette may be used against the wall instead of the half newel and opening cap. The procedure for installing a rosette has been previously described (see Fig. 82-15).

Installing Balcony Balusters

Balcony balusters are laid out and installed in the same manner as described previously for post-to-post balustrades. Balconies with a span of 10 feet or more should have intermediate balcony newels installed every 5 or 6 feet.

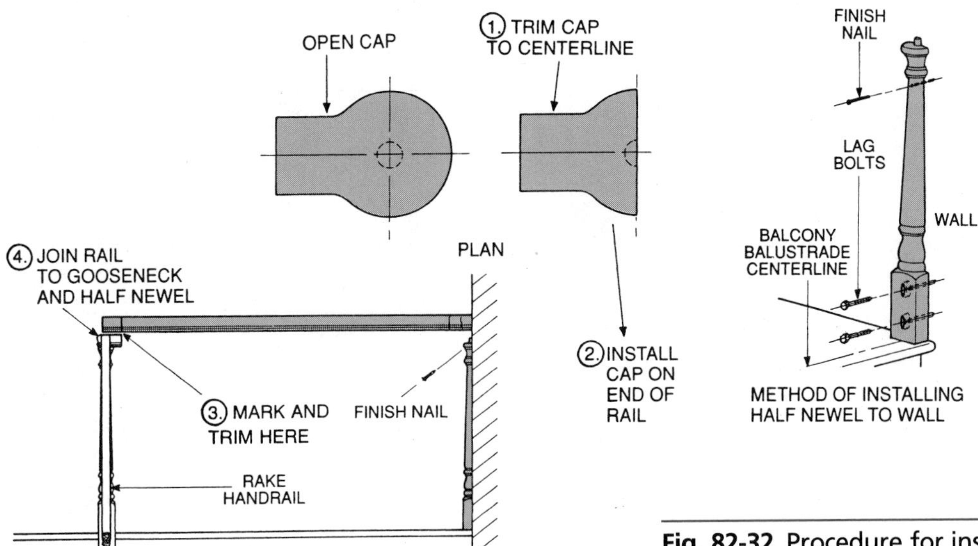

Fig. 82-32 Procedure for installing a half newel and the balcony handrail. (*Courtesy of L. J. Smith*)

Review Questions

Select the most appropriate answer.

1. The rounded outside edge of a tread that extends beyond the riser is called a
a. housing. c. coving.
b. turnout. d. nosing.

2. Finish boards between the stairway and the wall are called
a. returns. c. stringers.
b. balusters. d. casings.

3. Treads are rabbeted on their back edge to fit into
a. risers. c. return nosings.
b. housed stringers. d. newel posts.

4. An open stringer is
a. housed to receive risers.
b. mitered to receive risers.
c. housed to receive treads.
d. mitered to receive treads.

5. A volute is part of a
a. tread. c. newel post.
b. baluster. d. handrail.

6. The entire rail assembly on the open side of a stairway is called a
a. baluster.
b. balustrade.
c. guardrail.
d. finish stringer assembly.

7. In a framed staircase, the treads and risers are supported by
a. stair carriages. c. each other.
b. housed stringers. d. blocking.

8. One of the first things to do when trimming a staircase is
a. check the rough framing for rise and run.
b. block the staircase so no one can use it.
c. straighten the stair carriages.
d. install all the risers.

9. Treads usually project beyond the face of the riser
a. 3/4 inch. c. 1 1/4 inches.
b. 1 1/8 inches. d. 1 3/8 inches.

10. Newel posts are notched around the stairs so that their centerline lines up with the
a. centerline of the stair carriage.
b. centerline of the balustrade.
c. outside face of the open stringer.
d. outside face of the stair carriage.

BUILDING FOR SUCCESS

Selling Your Product

If you produce a quality product, does this mean that you will have no difficulty selling it? Not necessarily. The marketing of a product is not automatically assured. There are many challenges in convincing someone to purchase your product instead of a competitor's. How can a person be certain the market is out there? How can you know that your product is better than the others, or that it will sell?

The free enterprise system demands that quality be a part of every structure that is built. Assuming that quality is there, how do builders persuade the public to purchase their product? The builder can possess excellent technical skills and a good understanding of the construction industry, but fail to convince the public to buy. There are other pieces of the puzzle that must fall into place in order for a sale to be made. Let's see what criteria are necessary to convince someone to buy.

What do people want when they consider which builder or craftsperson to hire? Up front, potential buyers are searching for products that will meet their needs, are affordable, and are those with which they will feel satisfied. Many buyers who are considering a remodel or a new home are very consumer-smart. The single person, couple, or family that is researching the feasibility of a construction project probably will ask good questions of any builder, remodeler, or other construction technician. Builders must be knowledgeable and informed about all possible construction products and processes. They must be prepared to communicate with an informed public. With a large sum of money to be invested in a new home or home improvement, many buyers do their homework thoroughly.

Builders have an obligation to demonstrate that they are capable of successfully producing the desired product, by having the expertise, workforce, financial stability, and experience necessary to do the job.

Assuming that the need is there, the economy is good, the interest rates are low enough to encourage borrowing, and construction materials are affordable and available, many people will entertain the thought of building. They will be expecting the most for their money. The builder must thoroughly understand the potential client's needs and wants in a new home or remodel. In most cases, the money available may not be enough to purchase all of the amenities the buyer desires. The builder has to be able to discern what can be bought for the money at hand. Above all, the buyer must be satisfied that the proposed structure is a good value and investment. The builder can help assure this through good builder/client preconstruction discussions.

Potential clients may have looked at many homes or remodels in the community. Over a period of time, they will acquire many ideas that are incorporated into their desired structure. They will offer a multitude of pictures, examples, and suggestions for the builder to sort through in the design process. The challenge for the builder is how to put the puzzle pieces together and satisfy the customer within the prescribed price range.

On the surface, buyers want quality. They also want the largest possible amount of amenities, convenience, and comfort for the least amount of money. But there are equally important considerations that the builder must explain to the customer. Today many people may desire all of the above. But they must understand that the lowest bid may not be the best in all cases. What else is there? Part of the answer lies in the area of service, project coordination, and assistance with the client's needs during and after the construction process.

Explaining to the client how the remodel construction process will not disturb the family living situation will be of great value. Assuring the customer that all of the scheduling, coordination, and problem solving will be the responsibility of the builder will establish trust. People must understand that they are paying for things other than lumber, concrete, shingles, and paint when they engage a builder. This new home or remodel could be the largest single investment they will make in their life. They need to be sure it is the right thing to do and that they have employed the right person to do it.

In order to effectively sell a quality product, builders must demonstrate the ability to provide more than tangibles to the buyer. Builders have to legitimately convince the potential client that their company can provide more quality, service, and trustworthiness than competitors. Each builder or craftsperson will be challenged to do this on a regular basis. In order to provide this type of assurance and service to the customer, the builder will be looking for construction technicians that can meet these same standards. Each carpenter and other skilled technician must understand this and buy into the concept.

The construction business today has many marketing and selling demands placed on it that become everyday responsibilities for builders. Competition will dictate to each person in the industry a need to strive for excellence. Excellence will be visible not only in the durable product, but in the construction process that is used to build and market that product.

Before, during, and after the project is completed, clients will communicate to others how satisfied they were with the project and the builder. If they feel that the builder met their needs, provided quality, gained their trust, and satisfied their desires, the report will be positive. Satisfied customers will be one of the most effective marketing tools for the builder. That resource will prove to be invaluable in the quest to build and maintain a reputable business. By providing and maintaining customer satisfaction, many challenges in the selling process are lessened. It is a great advantage to any skilled worker to know what people want in a construction product and process, then meet the customer's desires. The selling process will be easier and the competition will be met.

FOCUS QUESTIONS: For individual or group discussion

1. As you begin to address the challenges of selling your construction projects on a regular basis, what do you see as considerations affecting a successful sale?
2. How should builders structure their sales strategies to develop trust with clients?
3. How could a good understanding of the competition's marketing/sales plan help your strategy as a builder? Is it possible to learn from others in the same profession?

UNIT 29

Finish Floors

A number of materials are used for finish floors. Wall to wall *carpet,* applied over an underlayment, is widely used. The carpenter usually applies the underlayment, but not the carpet. *Terrazo floors* are the responsibility of masons. Terrazo floors are made by grinding and polishing concrete in which small marble chips or colored stone are embedded. *Resilient sheet* and *tile* floors are widely used in bathrooms and kitchens. They are usually installed by specialists. On occasion, however, the carpenter may be required to install a resilient tile floor. Carpenters install *wood finish* floors. These floors are long-time favorites, because of their durability, beauty, and warmth.

OBJECTIVES

After completing this unit, the student should be able to:

- describe the kinds, sizes, and grades of hardwood finish flooring.
- apply strip, plank, and parquet finish flooring.
- estimate quantities of wood finish flooring required for various installations.
- apply underlayment and resilient tile flooring.
- estimate required amounts of underlayment and resilient tile for various installations.

UNIT CONTENTS

CHAPTER 83 DESCRIPTION OF WOOD FINISH FLOORS

Most hardwood finish flooring is made from white or red oak. Beech, birch, hard maple, and pecan finish flooring are also manufactured. For less expensive finish floors, some softwoods such as Douglas fir, hemlock, and southern yellow pine are used.

Kinds of Hardwood Flooring

The four basic types of solid wood finish flooring are *strip, plank, parquet strip,* and *parquet block. Laminated strip* wood flooring is a relatively new type that is gaining in popularity. *Laminated parquet blocks* are also manufactured.

Solid Wood Strip Flooring

Solid wood *strip* flooring is probably the most widely used type. Most strips are tongue-and-grooved on edges and ends to fit precisely together.

Unfinished strip flooring is milled with square, sharp corners at the intersections of the face and edges. After the floor is laid, any unevenness in the faces of adjoining pieces is removed by sanding the surface so strips are flush with each other.

Prefinished strips are sanded, finished, and waxed at the factory. They cannot be sanded after installation. A *chamfer* is machined between the face side and edges of the flooring prior to prefinishing. When installed, these chamfered edges form small V-grooves between adjoining pieces. This obscures any unevenness in the surface.

The most popular size of hardwood strip flooring is 3/4 inch thick with a *face width* of 2 1/4 inches. The face width is the width of the exposed surface between adjoining strips. It does not include the tongue. Other thicknesses and widths are manufactured (Fig. 83-1).

Laminated Wood Strip Flooring

Laminated strip flooring is a five-ply prefinished wood assembly. Each board is 9/16 inch thick, 7 1/2

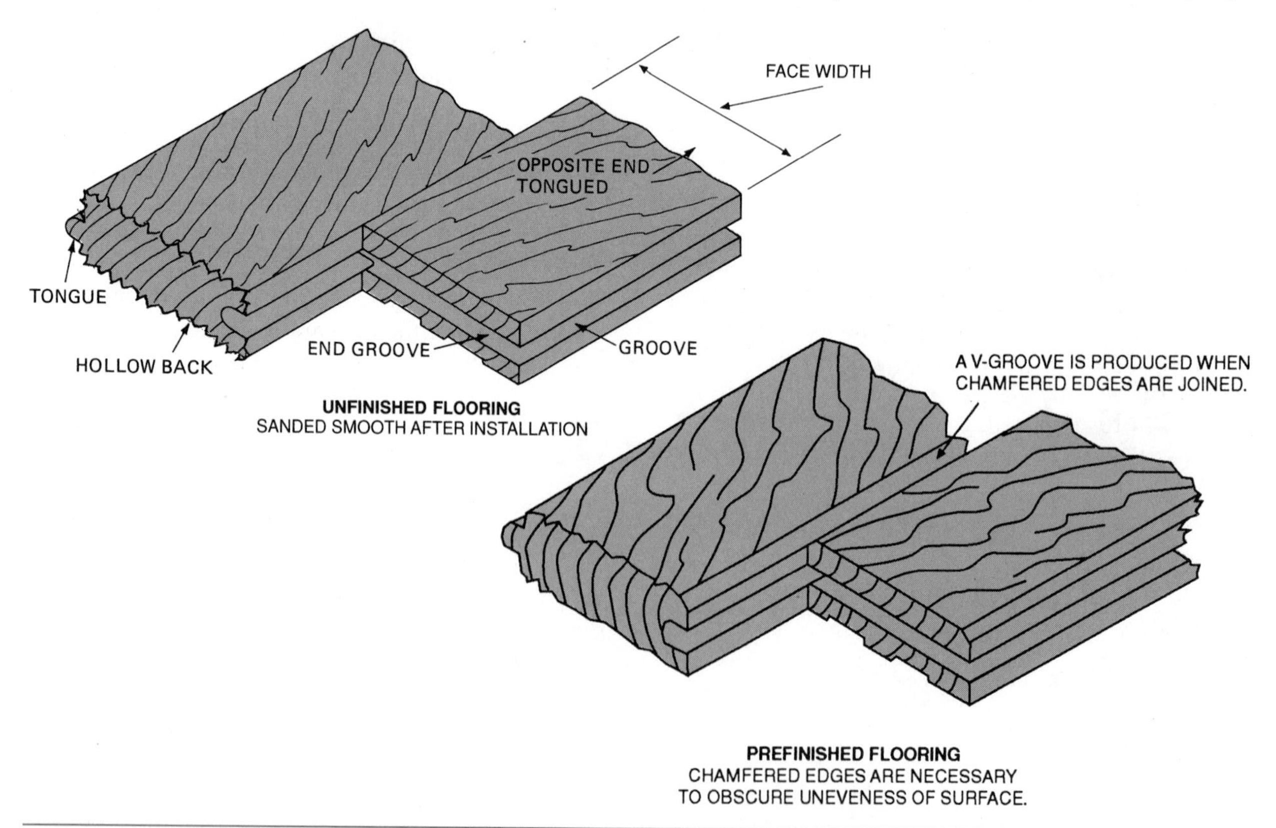

Fig. 83-1 Hardwood strip flooring is edge and end matched. The edges of prefinished flooring are chamfered.

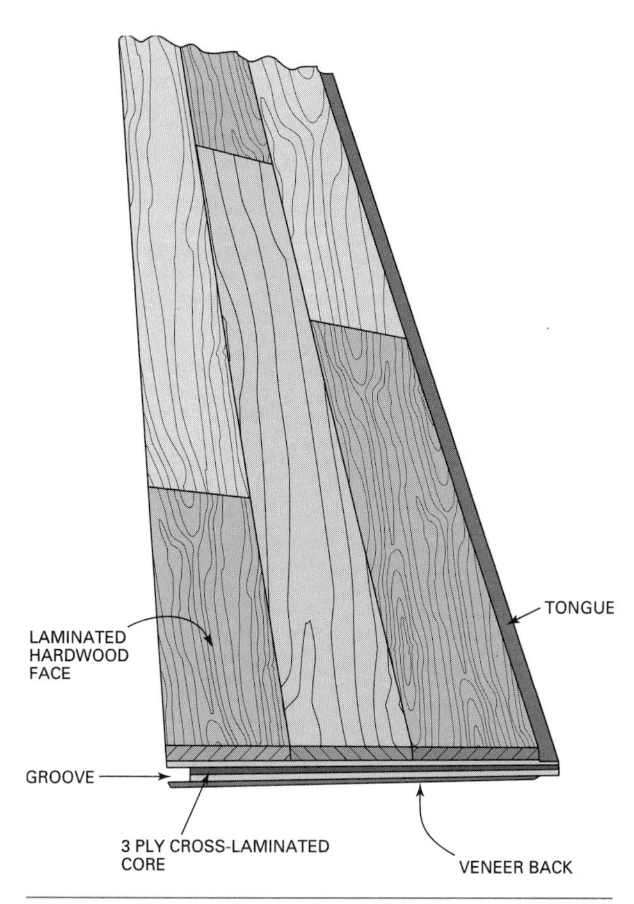

Fig. 83-2 Cross-section of a board of laminated strip flooring.

inches wide, and 7 feet, 11 1/2 inches long. The board consists of a bottom veneer, a three-ply cross-laminated core, and a face layer. The face layer consists of three rows of hardwood strips joined snugly edge to edge (Fig. 83-2).

The uniqueness of this flooring is in the exact milling of edge and end tongue and grooves. This precision allows the boards to be joined with no noticeable uneveness of the prefinished surface. This eliminates the need to chamfer the edges. Without the chamfers, a smooth continuous surface without V-grooves results when the floor is laid (Fig. 83-3).

Plank Flooring

Solid wood *plank* flooring is similar to strip flooring. However, it comes in various mixed combinations ranging from 3 to 8 inches in width. For instance, plank flooring may be laid with alternating widths of 3 and 4 inches; 3, 4, and 6 inches, 3, 5, and 7 inches, or any random width combination.

Like strips, planks are available unfinished or prefinished. The edges of some prefinished planks have deeper chamfers to accentuate the plank widths. The surface of some prefinished plank flooring may have plugs of contrasting color already installed to simulate screw fastening. One or more plugs, depending on the width of the plank, are used across the face at each end (Fig. 83-4).

Fig. 83-3 Completed installation of a laminated strip floor. (*Courtesy of Harris-Tarkett*)

Unfinished plank flooring comes with either square or chamfered edges and with or without plugs. The planks may be bored for plugs on the job, if desired.

Parquet Strips

Parquet strip flooring is a type in which short strips are laid to form various mosaic designs. The original parquet floors were laid by using short strips. Some, at the present time, are laid in the same manner. This type is manufactured in precise, short lengths, which are multiples of its width. For instance, 2 1/4 inch parquet strips come in lengths of 9, 11 1/4, 13 1/2, and 15 3/4 inches. Each piece is tongue-and-grooved on the edges and ends. Herringbone, basketweave, and other interesting patterns can be made using parquet strips (Fig. 83-5).

Parquet Blocks

Parquet block flooring consists of square or rectangular blocks, sometimes installed in combination with strips, to form mosaic designs. The three basic types are the *unit, laminated,* and *slat* block (Fig. 83-6).

Unit Blocks. The highest-quality parquet block is made with 3/4-inch thick, tongue-and-groove solid hardwood, usually oak. The widely used 9 × 9 *unit block* is made with six strips, 1 1/2 inches wide, or with four strips 2 1/4 inches wide. Unit blocks are laid with the direction of the strips

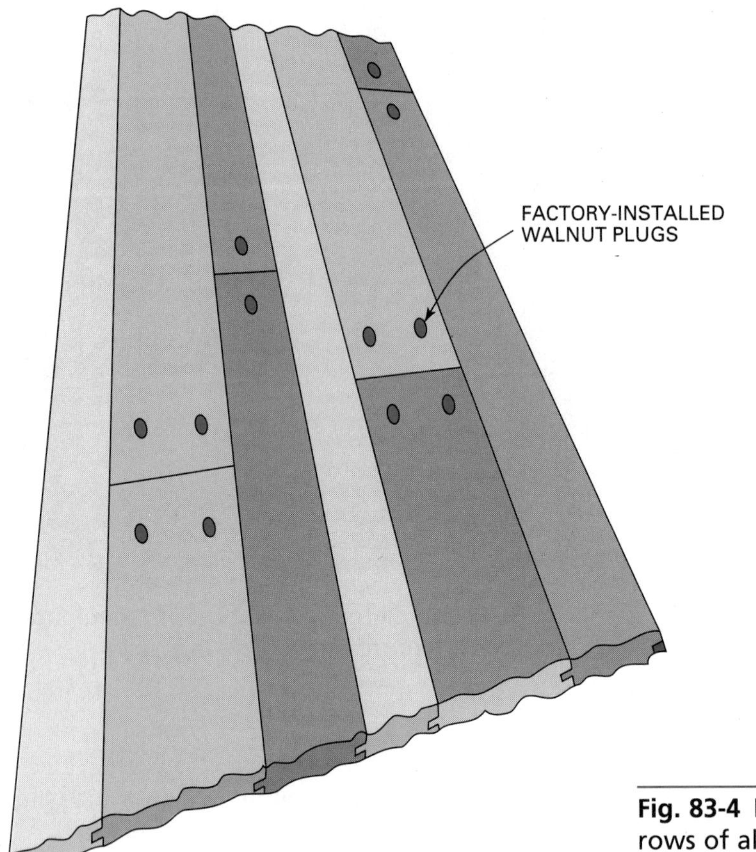

Fig. 83-4 Plank flooring usually is applied in rows of alternating widths. Plugs of contrasting color simulate screw fastening.

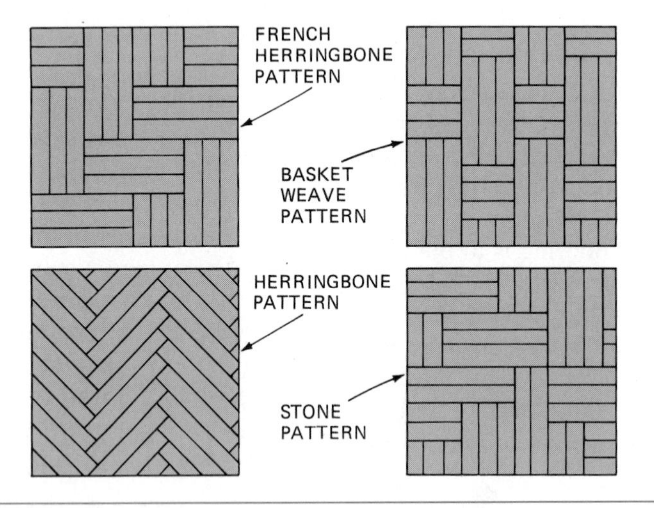

Fig. 83-5 Parquet strips are made in lengths that are multiples of its width. They are laid in a number of interesting patterns.

at right angles to adjacent blocks (Fig. 83-7). Unit blocks are made in other sizes and used in combination with parquet strips. Several patterns have gained popularity.

Monticello is the name of a parquet originally designed by Thomas Jefferson, the third president of the United States. The pattern consists of a 6 × 6 center unit block surrounded by 2 1/4-inch wide pointed *pickets.* Each center unit block is made of four 1 1/2-inch wide strips (Fig. 83-8). Each block comes with three pickets joined to it at the factory.

Another popular parquet, called the *Marie Antoinette,* is copied from part of the Versailles Palace in France. Square center unit blocks are enclosed by strips applied in a basketweave design (Fig. 84-9).

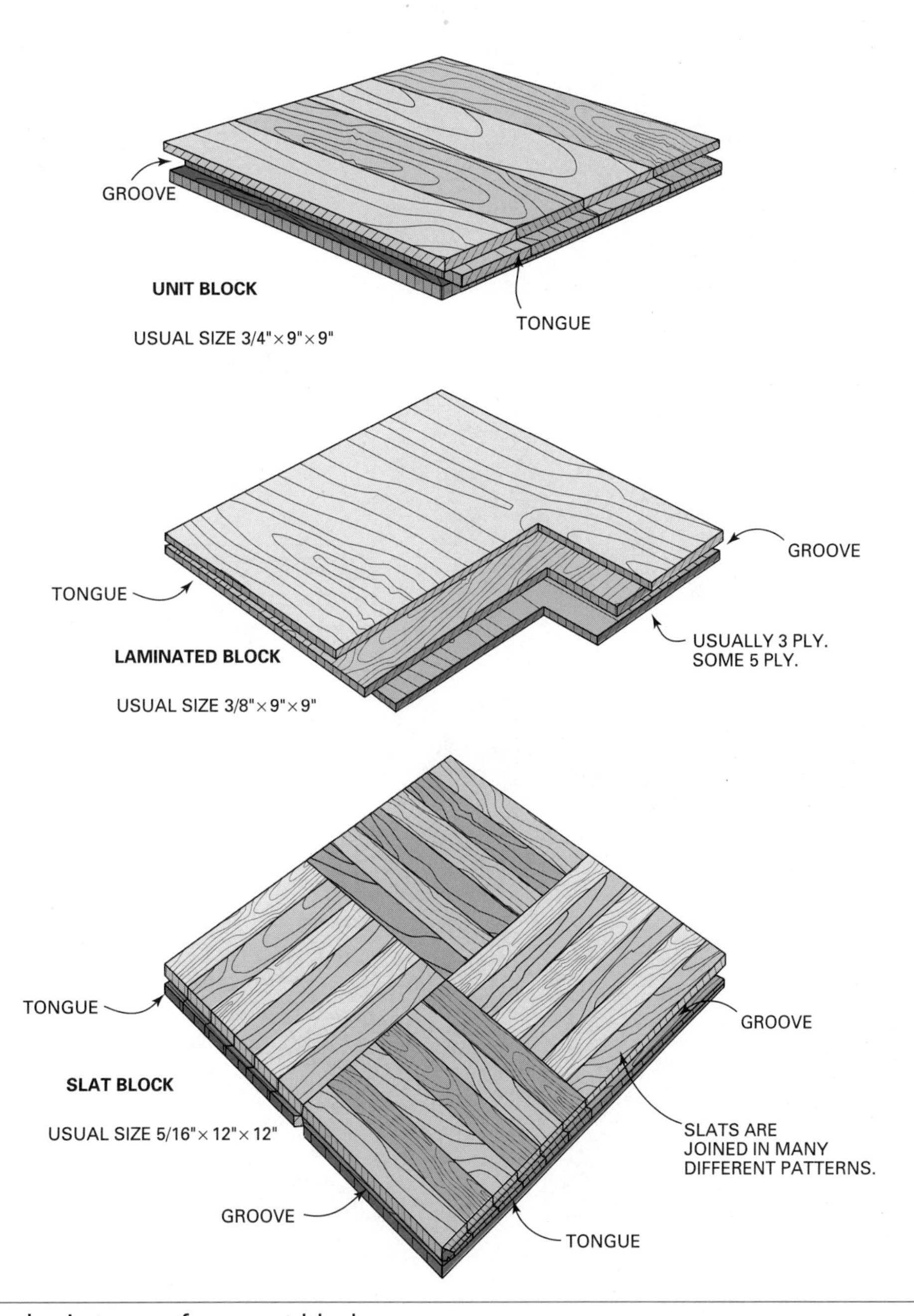

Fig. 83-6 Three basic types of parquet block.

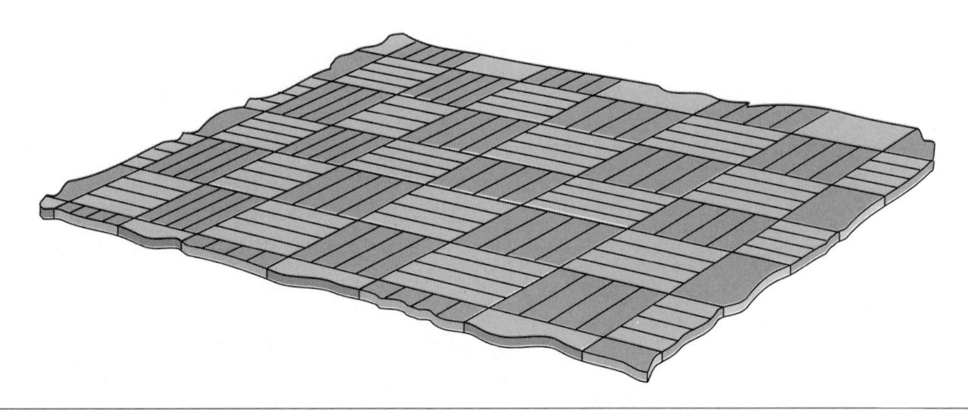

Fig. 83-7 Unit blocks are widely used in an alternating pattern.

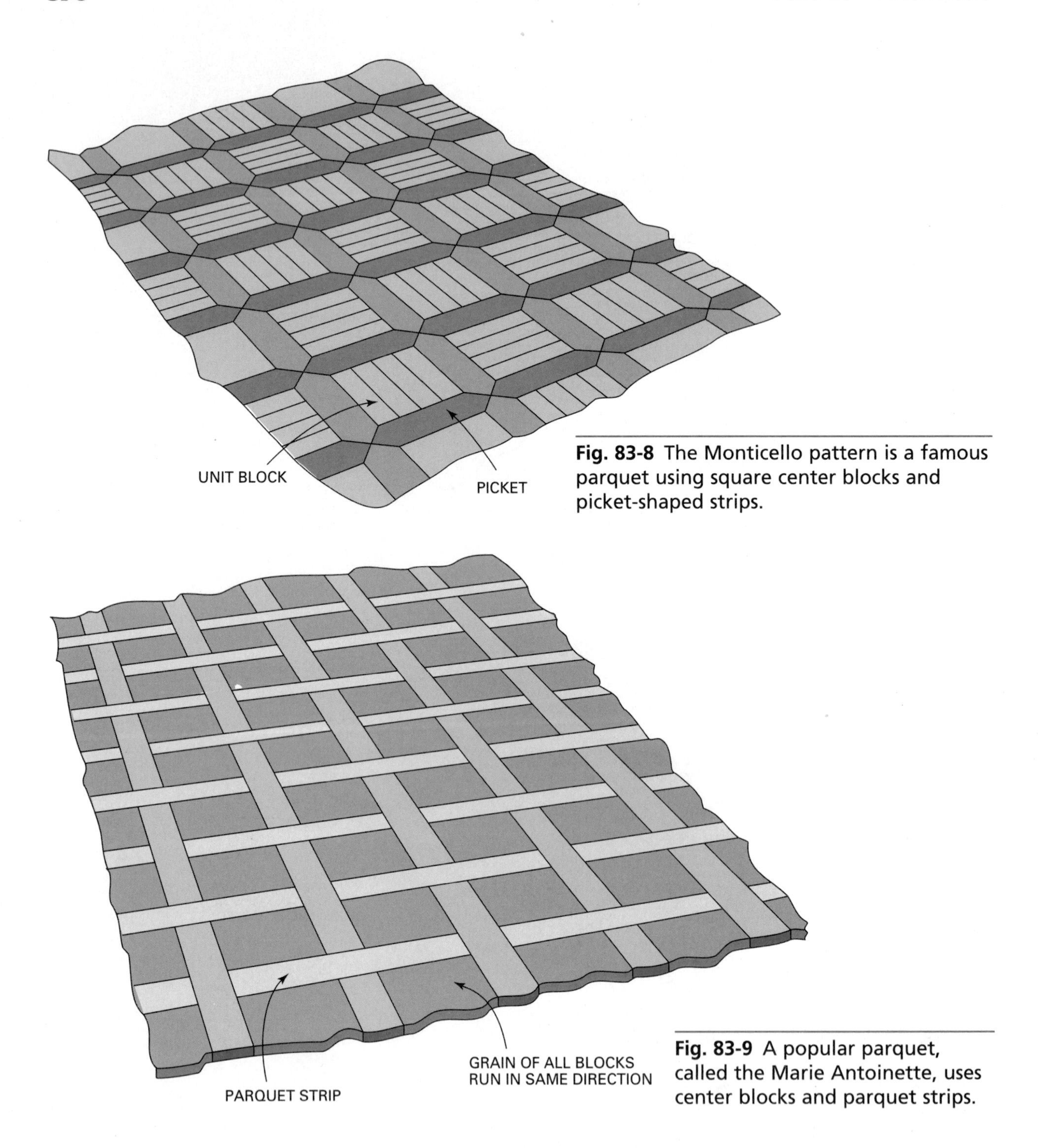

Fig. 83-8 The Monticello pattern is a famous parquet using square center blocks and picket-shaped strips.

Fig. 83-9 A popular parquet, called the Marie Antoinette, uses center blocks and parquet strips.

Rectangular parquet blocks are often used in a *herringbone* pattern. One commonly used block is 4 1/2 × 9 and made of three strips each 1 1/2 inches wide and 9 inches long (Fig. 83-10).

Laminated Blocks. *Laminated blocks* are generally made of three-ply laminated oak in a 3/8-inch thickness. Most blocks come in 8 × 8 or 9 × 9 sizes. A 2 × 12 laminated strip is manufactured for use in herringbone and similar patterns.

Slat Blocks. *Slat blocks* are also called *finger* blocks. They are made by joining many short, narrow strips together in small squares of various patterns. Some strips may be as narrow as 5/8 inch and as short as 2 inches or less. Several squares are assembled to make the block (Fig. 83-11). The squares are held together with a mesh backing or with a paper on the face side that is removed after the block is laid.

Presanded Flooring

Some strip, plank, and parquet finish flooring can be obtained *presanded* at the factory, but without

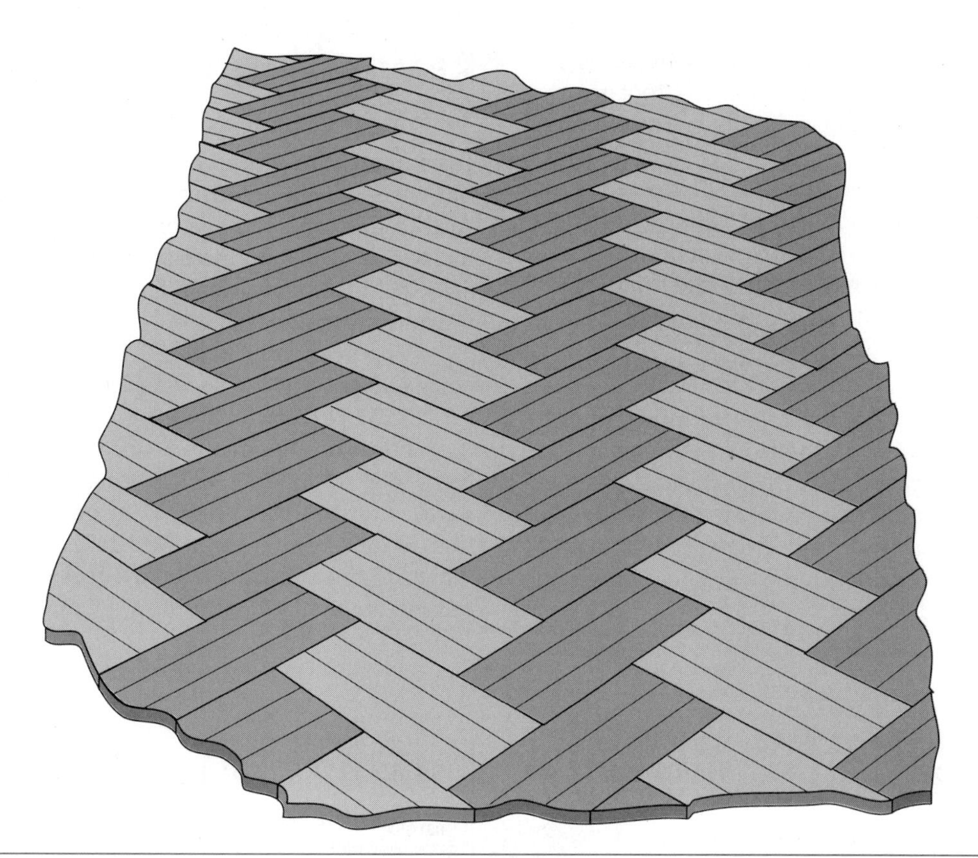

Fig. 83-10 Rectangular parquet blocks often are used to make herringbone and similar patterns.

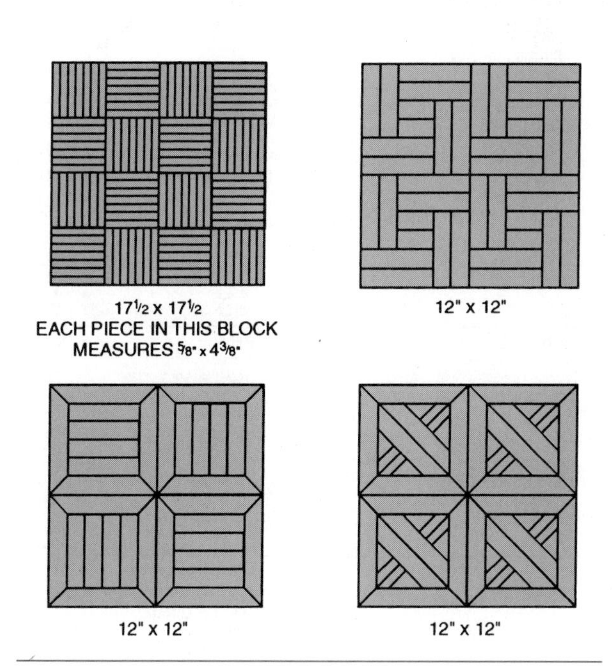

Fig. 83-11 Slat blocks are made of short, narrow strips joined together in various patterns. (*Courtesy of Bruce Hardwood Floors*)

the finish. The presanded surface eliminates the necessity for all but a touch-up sanding after installation to remove surface marks before the finish is applied.

Grades of Hardwood Flooring

Uniform grading rules have been established for strip and plank solid hardwood flooring by the National Oak Flooring Manufacturers Association. The association's trademark on flooring assures consumers that the flooring is manufactured and graded in compliance with established quality standards. Other types of wood finish flooring, such as parquet and laminated flooring, have no official grade rules.

Unfinished Flooring

Oak flooring is available *quarter-sawed* and *plain-sawed.* The grades for unfinished oak flooring, in declining order, are *clear, select, no. 1 common, no. 2 common,* and *1 1/4-foot shorts.* Quarter-sawed flooring is available in clear and select grades only.

Birch, beech, and hard maple flooring are graded in declining order as *first grade, second grade, third grade,* and *special grade.* Grades of pecan flooring are *first grade, first grade red, first grade white, second grade, second grade red,* and *third grade.* Red grades contain all heartwood. White grades are all bright sapwood.

In addition to appearance, grades are based on length. For instance, bundles of 1 1/4-foot shorts contain pieces from 9 to 18 inches long. The average length of clear bundles is 3 3/4 feet. The flooring comes in bundles in lengths of 1 1/4 feet and up. Pieces in each bundle are not of equal lengths. A bundle may include pieces from 6 inches under to 6 inches over the nominal length of the bundle. No pieces shorter than 9 inches are allowed.

Prefinished Flooring

Grades of *prefinished* flooring are determined after it has been sanded and finished. In declining order, they are *prime, standard and better, standard,* and *tavern.* Prefinished beech and pecan are furnished only in a combination grade called *tavern or better.* A guide to hardwood flooring grades is shown in Figure 83-12.

Estimating Hardwood Flooring

To estimate the amount of hardwood flooring material needed, first determine the area to be covered. Add to this a percentage of the area depending on the width of the flooring to be used. The percentages include an additional 5 percent for end matching and normal waste:

55 percent for flooring 1 1/2 inches wide
42.5 percent for flooring 2 inches wide
38.33 percent for flooring 2 1/4 inches wide.

For example, the area of a room 16 feet by 24 feet is 384 square feet. If 2 1/4-inch flooring is to be used, multiply 384 by .3833 to get 147.18. Round this off to 147 square feet. Add 147 to 384 to get 531, which is the number of board feet of flooring required. A calculator simplifies the mathematics.

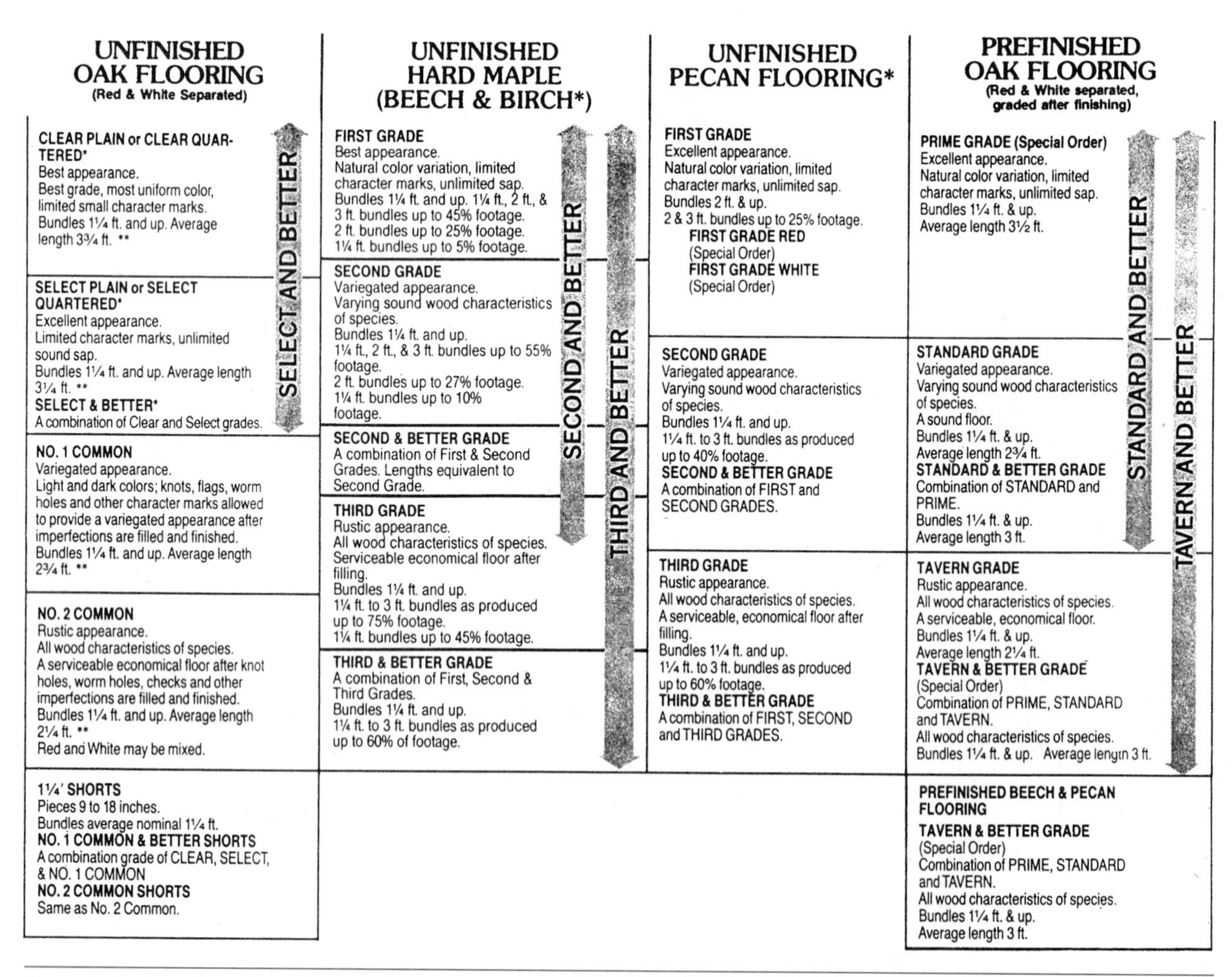

Fig. 83-12 Guide to hardwood flooring grades. (*Courtesy of National Oak Flooring Manufacturers Association*)

CHAPTER 84 LAYING WOOD FINISH FLOOR

Lumber used in the manufacture of hardwood flooring has been air-dried, kiln-dried, cooled, and then accurately machined to exacting standards. It is a fine product that should receive proper care during handling and installation.

Handling and Storage of Wood Finish Floor

Maintain moisture content of the flooring by observing recommended procedures. Flooring should not be delivered to the job site until the building has been closed in. Outside windows and doors should be in place. Cement work, plastering, and other materials must be thoroughly dry. In warm seasons, the building should be well ventilated. During cold months, the building should be heated, not exceeding 72 degrees Fahrenheit, for at least five days before delivery and until flooring is installed and finished.

Do not unload flooring in the rain. Stack the bundles in small lots in the rooms where the flooring is to be laid. Leave adequate space around the bundles for good air circulation. Let the flooring become acclimated to the atmosphere of the building for four or five days or more before installation.

Concrete Slab Preparation

Wood finish floors can be installed on an on-grade or above-grade *concrete slab.* Floors should not be installed on below-grade slabs. New slabs should be at least ninety days old. Flooring should not be installed when tests indicate excessive moisture in the slab.

Testing for Moisture

A test can be made by laying a smooth *rubber mat* on the slab. Put weight on it to prevent moisture from escaping. Allow the mat to remain in place for twenty-four hours. If moisture shows when the mat is removed, the slab is too wet. Another method of testing is by taping and sealing the edges of about a one-foot square of 6 mil clear *polyethylene film* to the slab. If moisture is present, it can be easily seen on the inside of the film after twenty-four hours.

If no moisture is present, prepare the slab. Grind off any high spots. Fill low spots with leveling compound. The slab must be free of grease, oil, or dust.

Applying a Moisture Barrier

A *moisture barrier* must be installed over all concrete slabs. This insures a trouble-free finish floor installation. Spread a skim coat of *mastic* with a straight trowel over the entire slab. Allow to dry at least two to three hours. Then, cover the slab with polyethylene film. Lap the edges of the film 4 to 6 inches and extending up all walls enough to be covered by the baseboard, when installed.

When the film is in place, *walk it in.* Step on the film, over every square inch of the floor, to make sure it is completely adhered to the cement. Small bubbles of trapped air may appear. The film may be punctured, without concern, to let the air escape.

Applying Plywood Subfloor

A *plywood subfloor* may be installed over the moisture barrier on which to fasten the finish floor. Exterior grade sheathing plywood of at least 3/4-inch thickness is used. The plywood is laid with staggered joints. Leave a 3/4-inch space at walls and 1/4- to 1/2-inch space between panel edges and ends. Fasten the plywood to the concrete with at least nine nails per panel (Fig. 84-1).

Instead of driving fasteners, the plywood may be cemented to the moisture barrier. Cut the plywood in 4 × 4 squares. Use a portable electric circular saw. Make scores 3/8 inch deep on the back of each panel that form a 12 × 12 grid. Lay the panel in asphalt cement spread with a 1/4-inch notched trowel.

Applying Sleepers

Finish flooring may also be fastened to *sleepers* installed on the slab. Sleepers are short lengths of lumber cemented to the slab. They must be pressure-treated and dried to a suitable moisture content. Usually, 2 × 4 lumber, from 18 to 48 inches long, is used.

Sleepers are laid on their side and cemented to the slab with mastic. They are staggered, with end laps of at least 4 inches, in rows 12 inches on center and at right angles to the direction of the finish floor. A polyethylene vapor barrier is then placed over the sleepers. The edges are lapped over the rows (Fig. 84-2). With end-matched flooring end joints need not meet over the sleepers.

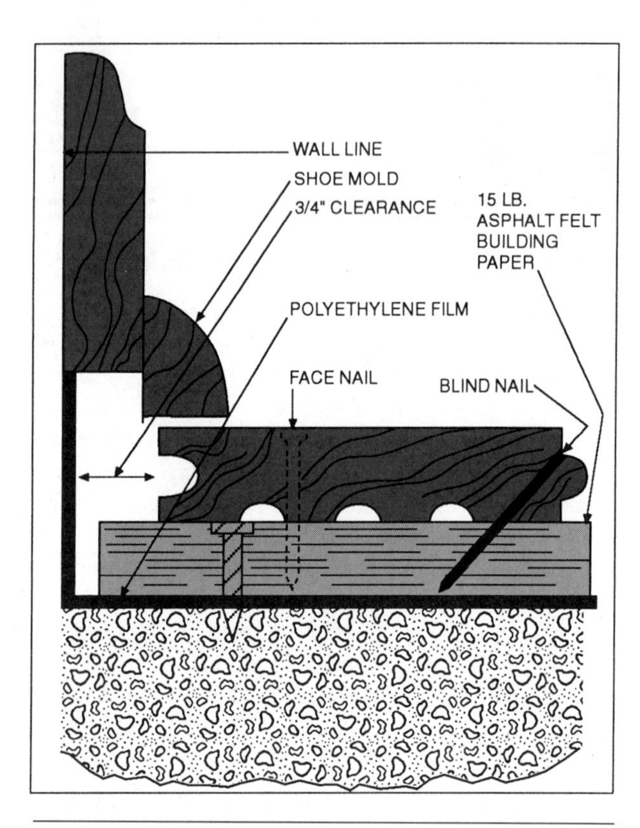

Fig. 84-1 Installation details of a plywood subfloor over a concrete slab.

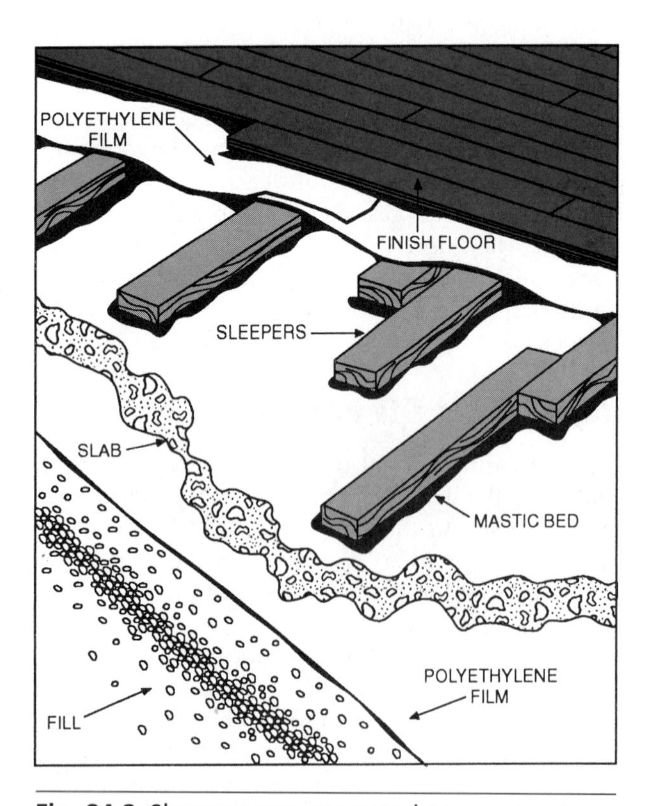

Fig. 84-2 Sleepers are cemented to a concrete slab. They provide fastening for strip or plank finish flooring.

Subfloors on Joists

Exterior plywood or boards are recommended for use as subfloors on joists when wood finish floors are installed on them. If plywood is used, a full 1/2 inch thickness is required. Thickness of 5/8 inch or more is preferred for 3/4-inch strip finish flooring. Use 3/4-inch thick subfloor for 1/2-inch strip flooring. The National Oak Flooring Manufacturers Association does not recommend fastening finish flooring to subfloors of nonveneered panels.

Laying Wood Finish Floor

In new construction, the base or door casings are not usually applied yet for easier application of the finish floor. In remodeling, the base and base shoe must be removed. Use a scrap piece of finish flooring as a guide on which to lay a handsaw. Cut the ends of any door casings that are extending below the finish floor surface.

Before laying any type of wood finish flooring, nail any loose areas. Sweep the subfloor clean. Scraping may be necessary to remove all plaster, taping compound, or other materials.

Laying Strip Flooring

Strip flooring laid in the direction of the longest dimension of the room gives the best appearance. The flooring may be laid in either direction on a plywood or diagonal board subfloor. It must be laid perpendicular to subfloor boards that run at right angles to the joists (Fig. 84-3).

When the subfloor is clean, cover it with building paper. Lap it 4 inches at the seams, and at right angles to the direction of the finish floor. The paper helps keep out dust, prevents squeaks in dry seasons, and retards moisture from below that could cause warping of the floor.

Snap chalk lines on the paper showing the centerline of floor joists so flooring can be nailed into them. For better holding power, fasten flooring through the subfloor and into the floor joists whenever possible. On 1/2-inch plywood subfloors, flooring fasteners must penetrate into the joists.

Starting Strip. The location and straight alignment of the first course is important. Place a strip of flooring on each end of the room, 3/4 inch from the starter wall with the groove side toward the wall. Mark along the edge of the flooring tongue. Snap a chalk line between the two points. The gap between the flooring and the wall is needed for expansion. It will eventually be covered by the base.

Hold the strip with its tongue edge to the chalk line. Face nail it with 8d finish nails, alternat-

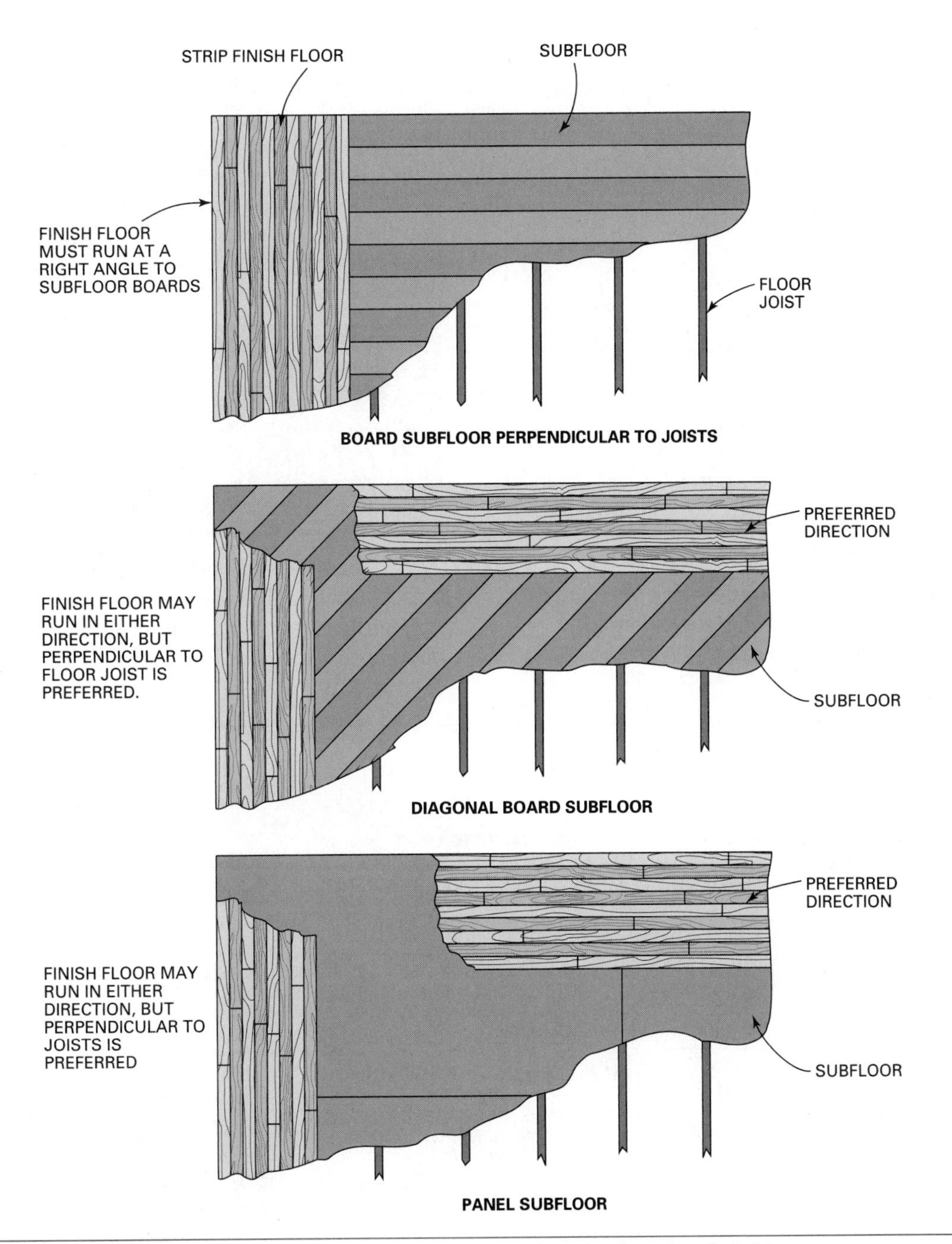

Fig. 84-3 Several factors determine the direction in which strip flooring is laid.

ing from one edge to the other, not less than 8 inches apart. Work from left to right with the grooved end of the first piece toward the wall. Left is determined by having the back of the person laying the floor to the wall where the starting strip is laid. Make sure end joints between strips are driven up tight (Fig. 84-4).

When necessary to cut a strip to fit to the right wall, use a strip long enough so that the cut-off piece is 8 inches or longer. Start the next course on the left wall with this piece.

Blind Nailing

Flooring is *blind nailed* by driving nails at about a 45 degree angle through the flooring. Start the nail in the corner at the top edge of the tongue. Usually 2 1/4-inch hardened cut or spiral screw nails are used. Recommendations for fastening are shown in Figure 84-5.

For the first two or three courses of flooring, a hammer must be used to drive the fasteners. For floor laying, a heavier than usual hammer, from 20

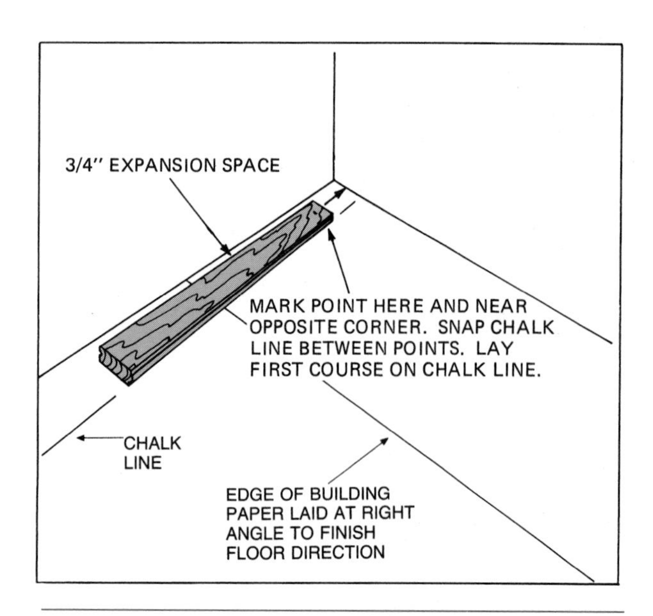

Fig. 84-4 A chalk line is snapped on the floor for alignment of the starting row of strip flooring. (*Courtesy of Chickasaw Hardwood Floors*)

to 28 ounces, is generally used for extra driving power. Care must be taken not to let the hammer glance off the nail. This may damage the edge of the flooring. Care also must be taken that, on the final blows, the hammer head does not hit the top corner of the flooring. To prevent this, raise the hammer handle slightly on the final blow so that the hammer head hits the nail head and the tongue, but not the corner of the flooring (Fig. 84-6).

CAUTION Eye protection should be worn when driving hardened steel nails with a hammer. A small piece of steel may break off the hammer or the nail, and fly out in any direction. This could cause serious injury to an unprotected eye.

After the nail is driven home, its head must be set slightly. This allows adjoining strips to come up tightly against each other. Floor layers use the head of the next nail to be driven to set the nail just driven.

STRIP T & G SIZE FLOORING	SIZE NAIL TO BE USED	BLIND NAIL SPACING ALONG THE LENGTH OF STRIPS. MINIMUM 2 NAILS PER PIECE NEAR THE ENDS. (1"-3")
3/4×1 1/2", 2 1/4", & 3 1/4"	2" SERRATED EDGE BARBED FASTENER, 2 1/4" OR 2 1/2" SCREW OR CUT NAIL, 2" 15 GAUGE STAPLES WITH 1/2" CROWN.	IN ADDITION-10-12" APART-8-10" PREFERRED
ON SLAB WITH 3/4" PLYWOOD SUBFLOOR USE 1 1/2" BARBED FASTENER, 1/2" PLYWOOD SUBFLOOR WITH JOISTS A MAXIMUM 16" O.C., FASTEN INTO EACH JOIST WITH ADDITIONAL FASTENING BETWEEN, OR 8" APART.		
MUST INSTALL ON A SUBFLOOR		
1/2×1 1/2" & 2"	1 1/2" SERRATED EDGE BARBED FASTENER, 1 1/2" SCREW, CUT STEEL, OR WIRE CASING NAIL.	10" APART 1/2" FLOORING MUST BE INSTALLED OVER A MINIMUM 5/8" THICK PLYWOOD SUBFLOOR.
3/8×1 1/4" & 2"	1 1/4" SERRATED EDGE BARBED FASTENER, 1 1/2" BRIGHT WIRE CASING NAIL.	8" APART
SQUARE-EDGE FLOORING		
5/16×1 1/2" & 2"	1" 15 GAUGE FULLY BARBED FLOORING BRAD.	2 NAILS EVERY 7"
5/16×1 1/3"	1" 15 GAUGE FULLY BARBED FLOORING BRAD.	1 NAIL EVERY 5" ON ALTERNATE SIDES OF STRIP
PLANK 3/4×3" TO 8"	2" SERRATED EDGE BARBED FASTENER, 2 1/4" OR 2 1/2" SCREW, OR CUT NAIL, USE 1 1/2" LENGTH WITH 3/4" PLYWOOD SUBFLOOR ON SLAB.	8" APART
FOLLOW MANUFACTURER'S INSTRUCTIONS FOR INSTALLING PLANK FLOORING		
WIDTHS 4" AND OVER MUST BE INSTALLED ON A SUBFLOOR OF 5/8" OR THICKER PLYWOOD OR 3/4" BOARDS. ON SLAB USE 3/4" OR THICKER PLYWOOD.		

Fig. 84-5 Nailing guide for strip and plank finish flooring. (*Courtesy of National Oak Flooring Manufacturers*)

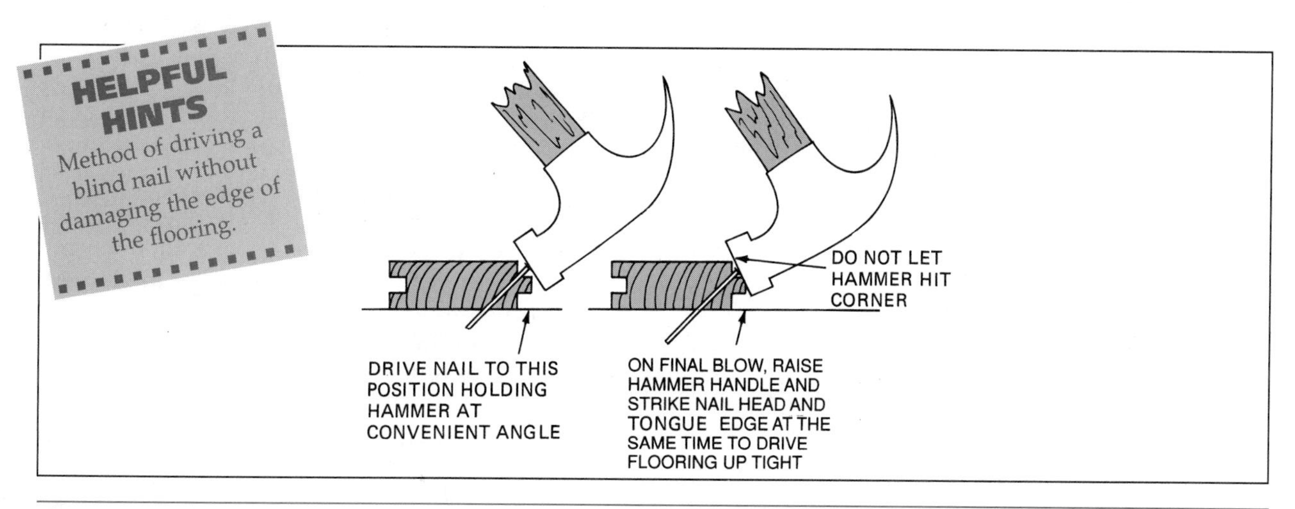

Fig. 84-6 Technique for driving a blind nail.

When fastening flooring, the floor layer holds a hammer in one hand and a number of nails in the other. While driving one nail, the floor layer *fingers* the next nail to be driven into position to be used as a set. When the nail being driven is home, the fingered nail is laid on edge with its head on the nail to be set. With one sharp blow, the nail is set (Fig. 84-7). The setting nail is then the next nail to be driven. In this manner, the floor layer maintains a smooth, continuous motion when fastening flooring.

Note: A nail set should not be used to set hardened flooring nails. If used, the tip of the nail set will be flattened, thus rendering the nail set useless. Do not lay the nail set flat along the tongue, on top of the nail head, and then set the nail by hitting the side of the nail set with a hammer. Not only is this method slower, but it invariably breaks the nail set, possibly causing an injury.

Racking the Floor

After the second course of flooring is fastened, lay out seven or eight loose rows of flooring, end to end. Lay out in a staggered pattern. End joints should be at least 6 inches apart. Find or cut pieces to fit within 1/2 inch of the end wall. Distribute long and short pieces evenly for the best appearance. Avoid clusters of short strips. Laying out loose flooring in this manner is called *racking the floor.* Racking is done to save time and material (Fig. 84-8).

Using the Power Nailer

At least two courses of flooring must be laid by hand to provide clearance from the wall before a *power nailer* can be used. The power nailer holds strips of special barbed fasteners. The fasteners are driven and set through the tongue of the flooring at the proper angle. Although it is called a power nailer, a heavy hammer is swung by the operator against a plunger to drive the fastener (Fig. 84-9).

The hammer is double-ended. One end is rubber and the other end is steel. The flooring strip is placed in position. The rubber end of the hammer is used to drive the edges and ends of the strips up tight. The steel end is used against the plunger of

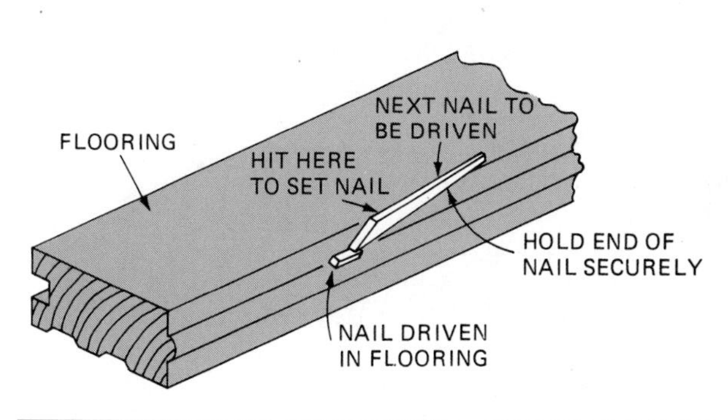

Fig. 84-7 Method used by floor layers to set nails driven by hand.

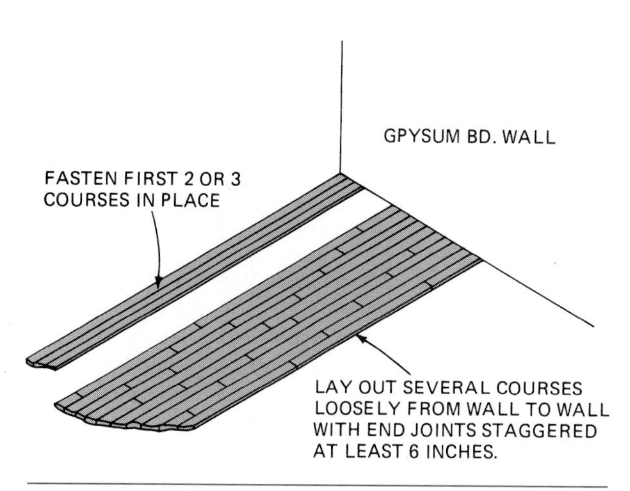

Fig. 84-8 Racking the floor places the strips in position for efficient installation.

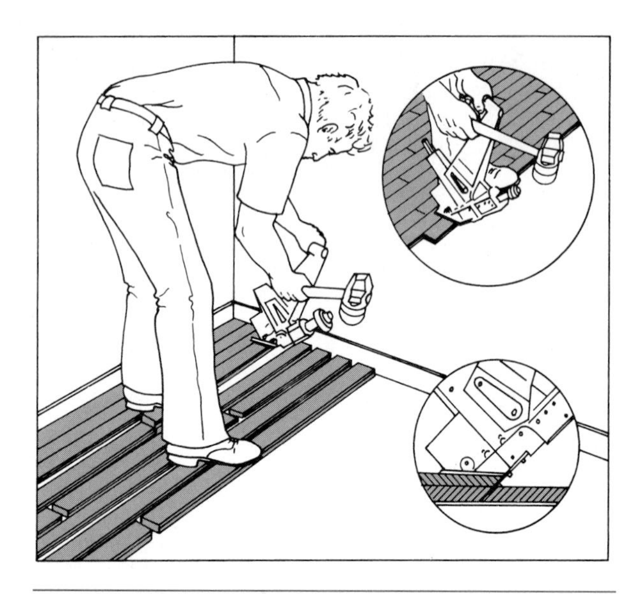

Fig. 84-9 A power nailer is widely used to fasten strip flooring. (*Courtesy of National Oak Flooring Manufacturers Association*)

the power nailer to drive the fasteners. Slide the power nailer along the tongue edge. Drive fasteners about 8 to 10 inches apart or as needed to bring the strip up tight against previously laid strips.

Note: When using the power nailer, one and only one heavy blow must be used to drive the fastener. After the first blow, another fastener drops into place ready to be driven. Make sure the first blow is powerful enough to drive the fastener.

Whether laying floor with a power nailer or driving nails with a hammer, the floor layer stands with heels on strips already fastened, and toes on the loose strip to be fastened. With weight applied to the joint, easier alignment of the tongue and groove is possible (Fig. 84-10). The weight of the worker also prevents the loose strip from bouncing when it is driven to make the edge joint tight. Avoid using a power nailer, pneumatic nailer, and hammer-driven fasteners on the same strip of flooring. Each method of fastening places the strips together with varying degrees of tightness. This variation, compounded over multiple strips, will cause waves in the straightness of the flooring.

Ending the Flooring

Continue across the room. Rack seven or eight courses as work progresses. The last three or four courses from the opposite wall must be nailed by hand. This is because of limited room to place the power nailer and swing the hammer. The next-to-the-last row can be blind nailed if care is taken. However, the flooring must be brought up tightly by prying between the flooring and the wall. Use a bar to pry the pieces tight at each nail location (Fig. 84-11).

The last course is installed in a similar manner. However, it must be face nailed. It may need to be ripped to the proper width. If it appears that the installation will end with an undesirable, difficult to apply, narrow strip, lay wider strips in the last row (Fig. 84-12).

HELPFUL HINTS

Step on the strip being fastened to align tongue-and-groove edges and prevent the strip from bouncing.

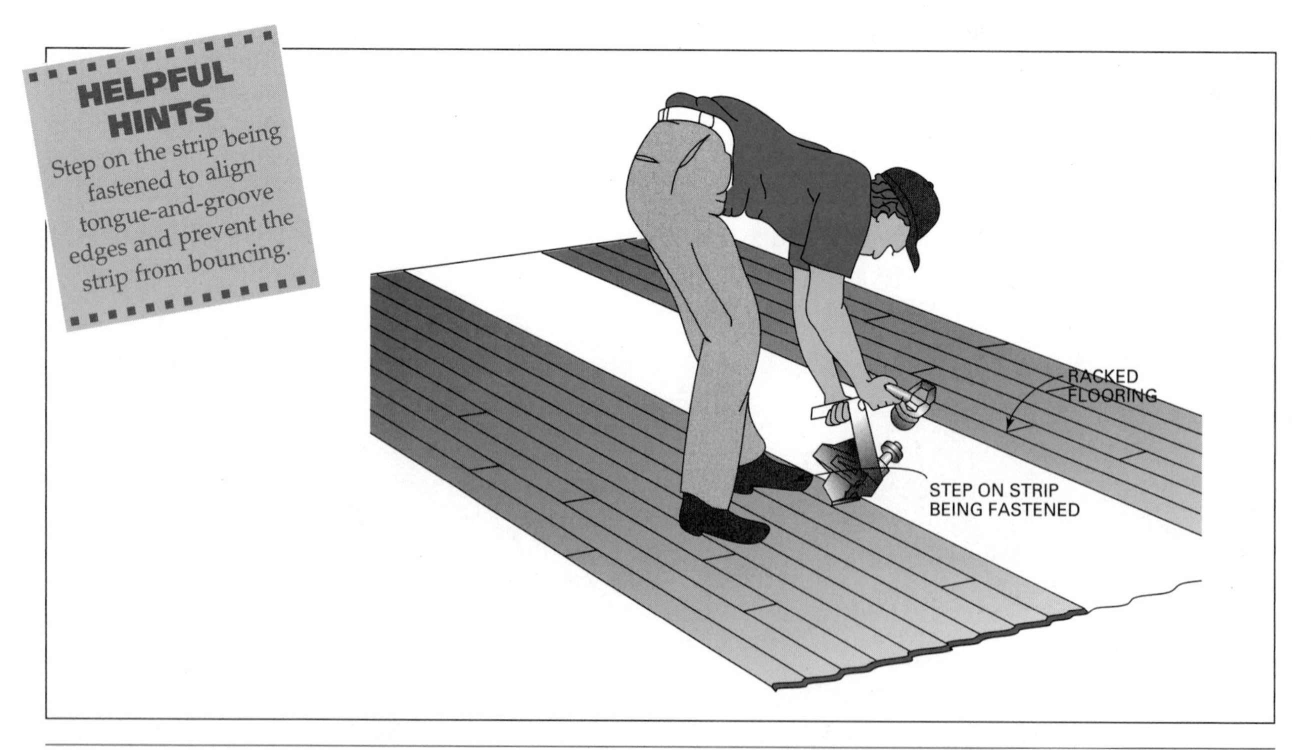

Fig. 84-10 It is important to push the board down and together tightly while nailing.

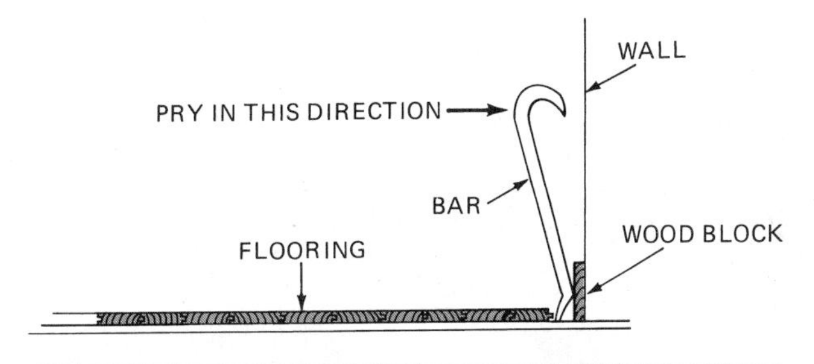

Fig. 84-11 The last two courses of strip flooring must be brought tight with a pry bar.

Framing Around Obstructions

A much more professional and finished look is given to a strip flooring installation if *hearths* and other floor obstructions are framed. Use flooring, with mitered joints at the corners, as framework around the obstructions (Fig. 84-13).

Changing Direction of Flooring

Sometimes it is necessary to change direction of flooring when it extends from a room into another room, hallway, or closet. To do this, face nail the extended piece to a chalk line. Change directions by joining groove edge to groove edge and inserting a *spline,* ordinarily supplied with the flooring (Fig. 84-14). For best appearance, avoid bunching short or long strips. Open extra bundles, if necessary, to get the right selection of lengths.

Laying Laminated Strip Flooring

Before laying *laminated strip* flooring, a 1/8-inch foam underlayment, supplied or approved by the manufacturer, is applied to the subfloor or slab. The flooring is not fastened or cemented to the floor. However, the boards must be glued to each other along the edges and ends. Apply glue on edges in

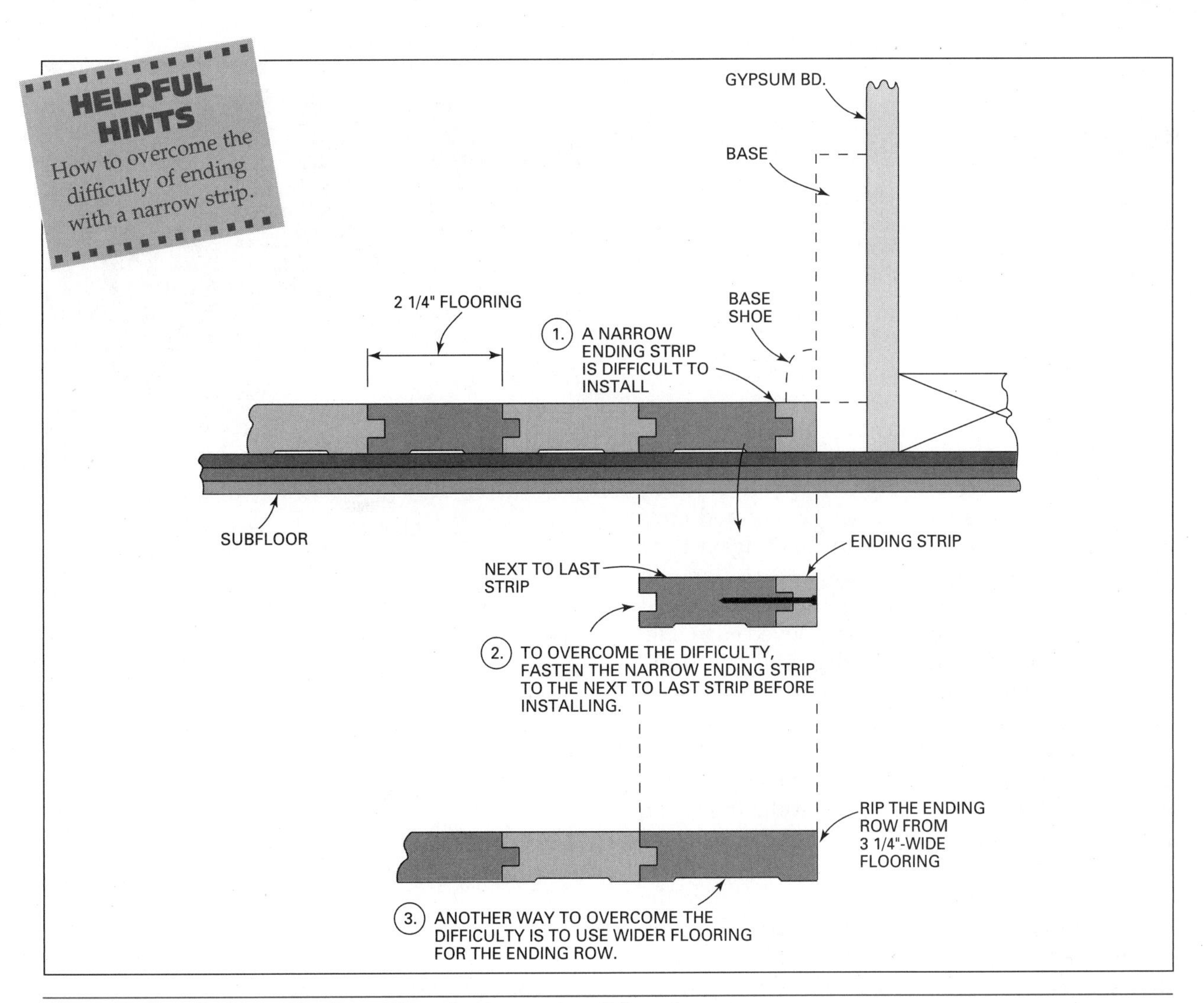

Fig. 84-12 Techniques for installing the last strip of flooring.

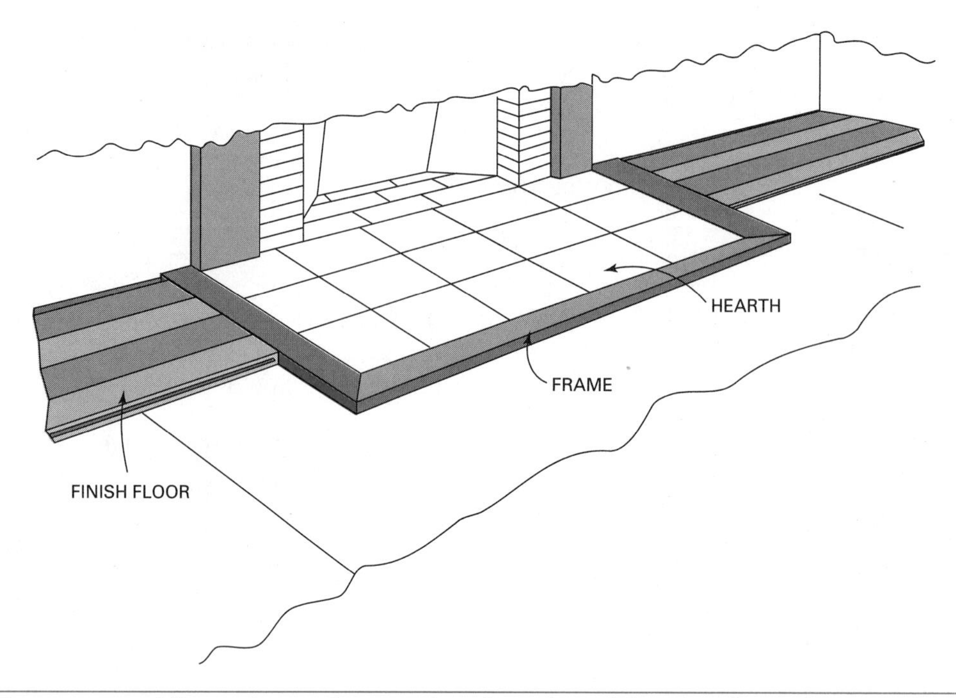

Fig. 84-13 Frame around floor obstructions, such as hearths, with strips that are mitered at the corners.

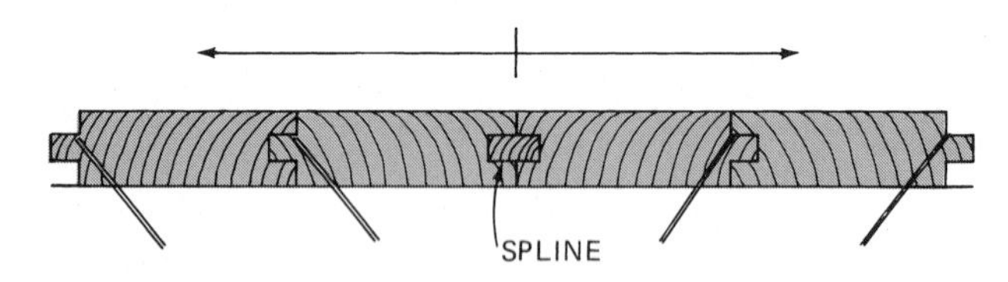

Fig. 84-14 The direction of strip flooring is changed by the use of a spline.

8-inch long beads with 12-inch spaces between them. Apply a full bead across the ends.

The first row is laid in a straight line with end joints glued. The groove edge is placed toward the wall, leaving about a 1/2-inch expansion space. Subsequent courses are installed. Edges and ends are glued. Each piece is brought tight against the other with a hammer and *tapping block.* The tapping block should be used only against the tongue. It should never be used against the grooved edge (Fig. 84-15). Tapping the grooved edge will damage it. Stagger end joints at least 2 feet from those in adjacent courses.

Usually the last course must be cut to width. To lay out the last course to fit, lay a complete row of boards, unglued, tongue toward wall, directly on top of the already installed next to last course. Cut a short piece of flooring for use as a *scribing* block. Hold the tongue edge against the wall. Move the block along the wall while holding a pencil against the other edge to lay out the flooring. The width of the tongue of the scribing block provides the necessary expansion space (Fig. 84-16). The last row, when cut, can be glued and wedged tightly in place with a pry bar.

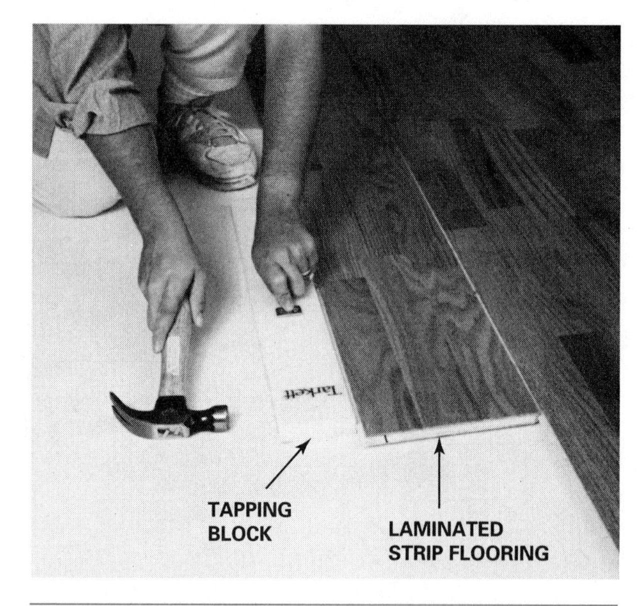

Fig. 84-15 A tapping block is used to drive laminated strip flooring boards tight against those already installed. (*Courtesy of Harris-Tarkett*)

Installing Plank Flooring

Plank flooring is installed like strip flooring. Alternate the courses by widths. Start with the narrowest pieces. Then use increasingly wider courses, and repeat the pattern. Stagger the joints in adjacent

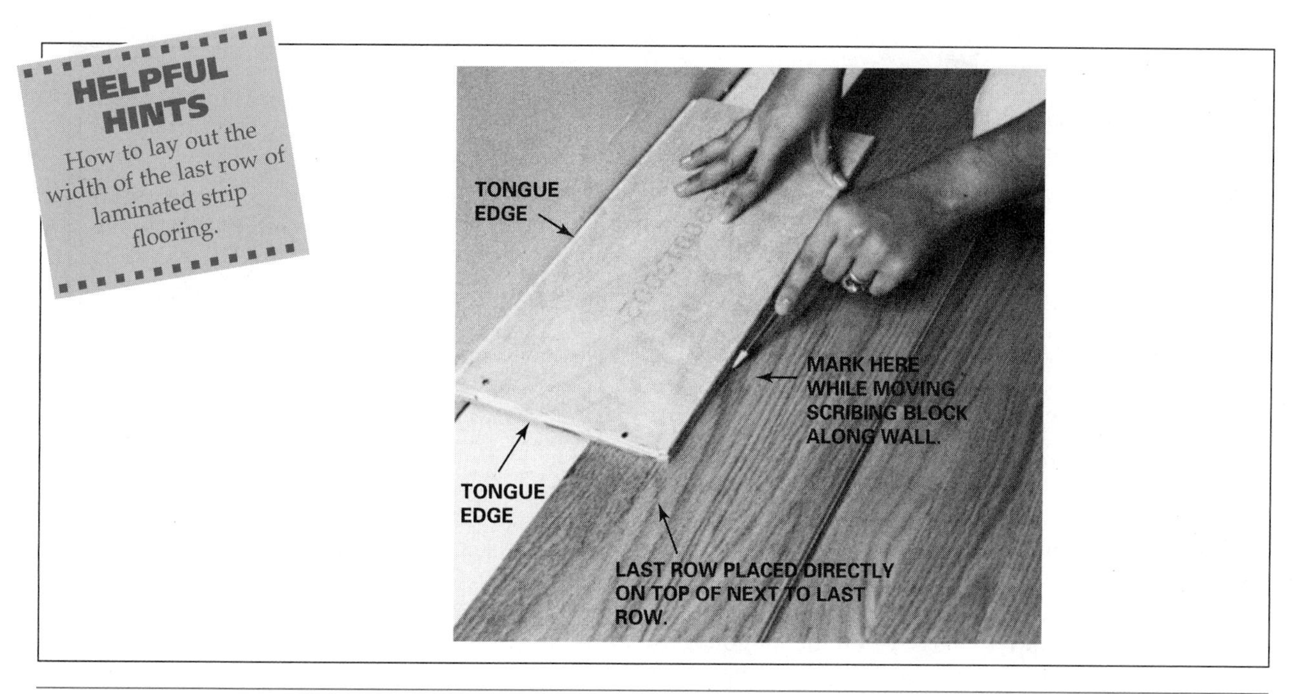

Fig. 84-16 Technique for scribing the last strip on flooring using a scrap piece of flooring.

courses. Use lengths so they present the best appearance.

Manufacturers' instructions for fastening the flooring vary and should be followed. Generally, the flooring is blind nailed through the tongue of the plank and at intervals along the plank in a manner similar to strip flooring.

Installing Parquet Flooring

Procedures for the application of parquet flooring vary with the style and the manufacturer. Detailed installation directions are usually provided with the flooring. Generally, both parquet blocks and strips are laid in *mastic.* Use the recommended type. Apply with a notched trowel. The depth and spacing of the notches are important to leave the correct amount of mastic on the floor. Parquet may be installed either square with the walls or diagonally.

Square Pattern

When laying *unit blocks* in a square pattern, two layout lines are snapped, at right angles to each other, and parallel to the walls. Blocks are laid with their edges to the lines. Lines are usually laid out so that rows of blocks are either centered on the floor or half the width of a block off center. This depends on which layout produces border blocks of equal and maximum widths against opposite walls.

To determine the location of the layout lines, measure the distance to the center of the room's width. Divide this distance by the width of a block. If the remainder is half or more, snap the layout line in the center. If the remainder is less than half, snap the layout line off center by half the width of the block. Find the location of the other layout line in the same way. It is possible that one of the layout lines will be centered, while the other must be snapped off center.

Other factors may determine the location of layout lines, such as ending with full blocks under a door, or where they meet another type of floor. Regardless of the location, two lines, at right angles to each other, must be snapped.

Place one unit at the intersection of the lines. Position the grooved edges exactly on the lines. Lay the next units ahead and to one side of the first one and along the lines. Install blocks in a pyramid. Work from the center outward toward the walls in all directions. Make adjustments as installation progresses to prevent misalignment (Fig. 84-17).

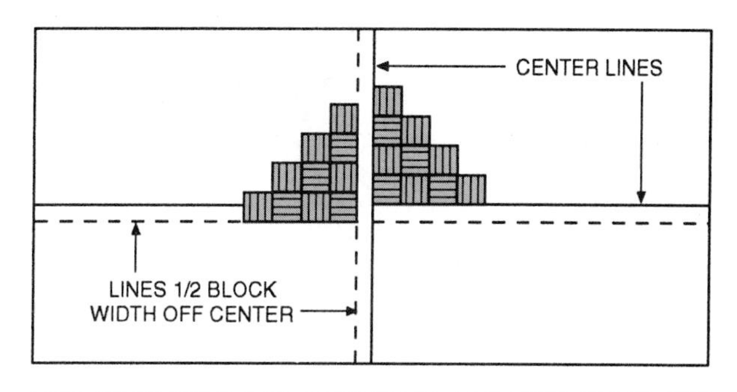

Fig. 84-17 Parquet blocks are laid to center or off-center lines. This depends on which produces the best size border blocks.

Diagonal Pattern

To lay unit blocks in a diagonal pattern, measure an equal distance from one corner of the room along both walls. Snap a starting line between the two marks. At the center of the starting line, snap another line at right angles to it. The location of both lines may need to be changed in order to end with border blocks of equal and largest possible size against opposite walls. The diagonal pattern is then laid in a manner similar to the square pattern (Fig. 84-18).

Special Patterns

Many parquet patterns can be laid out with square and diagonal layout lines. The *herringbone* pattern requires three layout lines. One will be the 90 degree line used for a square pattern. The other line crosses the intersection at a 45 degree angle. Align the first block or strip with its edge on the diagonal line. The corner of the piece should be lined up at the intersection. Continue the pattern in rows of three units wide, aligning the units with the layout lines (Fig. 84-19).

The *Monticello* pattern can be laid square or diagonally in the same way as unit blocks. For best results, lay the parquet in a pyramid pattern. Alternate the grain of the center blocks. Keep *picket* points in precise alignment (Fig. 84-20).

The *Marie Antoinette* pattern may also be laid square or diagonally. It is started by laying a *band* with its grooved edge and end aligned with the layout lines with the tongued edge to the right. Continue laying the pattern by placing center blocks and bands in a sequence so that bands appear woven (Fig. 84-21). Tongues of all mem-

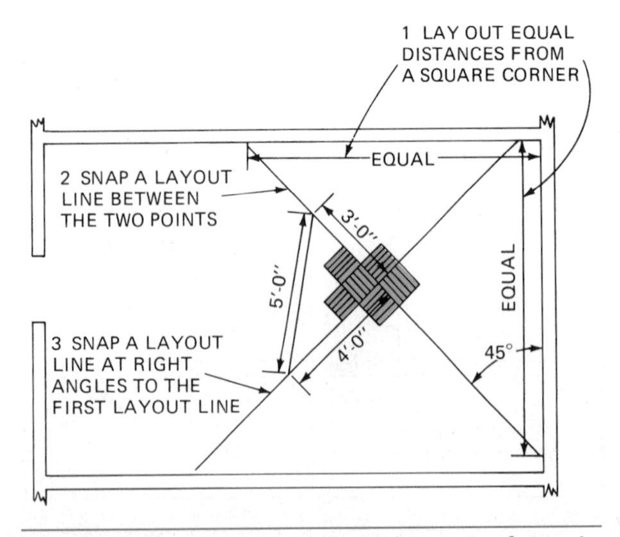

Fig. 84-18 Layout of a diagonal pattern for unit blocks using the 3-4-5 right triangle method. Make adjustments for satisfactory borders.

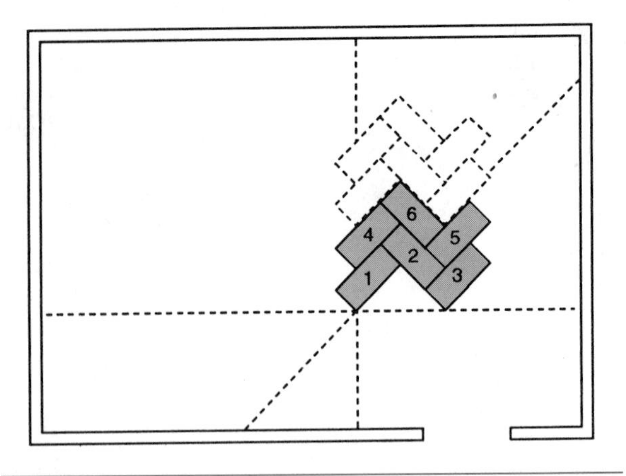

Fig. 84-19 The herringbone pattern requires 90- and 45-degree layout lines. (*Courtesy of Chickasaw Hardwood Floors*)

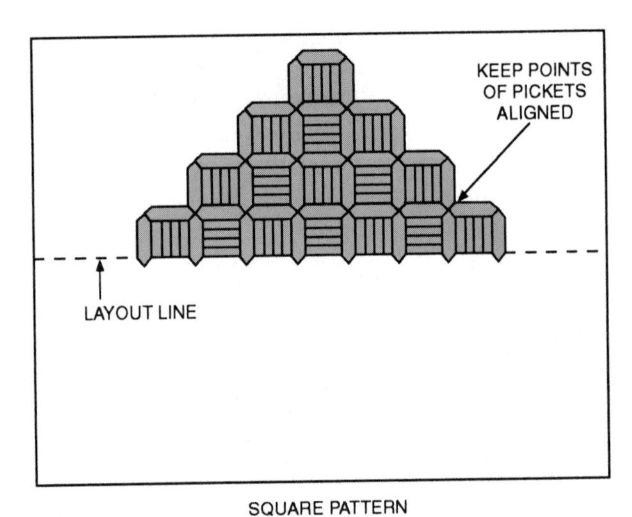

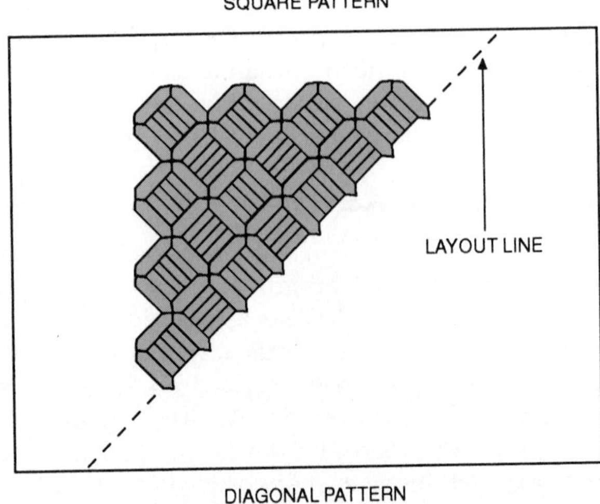

Fig. 84-20 Layout of the Monticello parquet pattern. (*Courtesy of Chickasaw Hardwood Floors*)

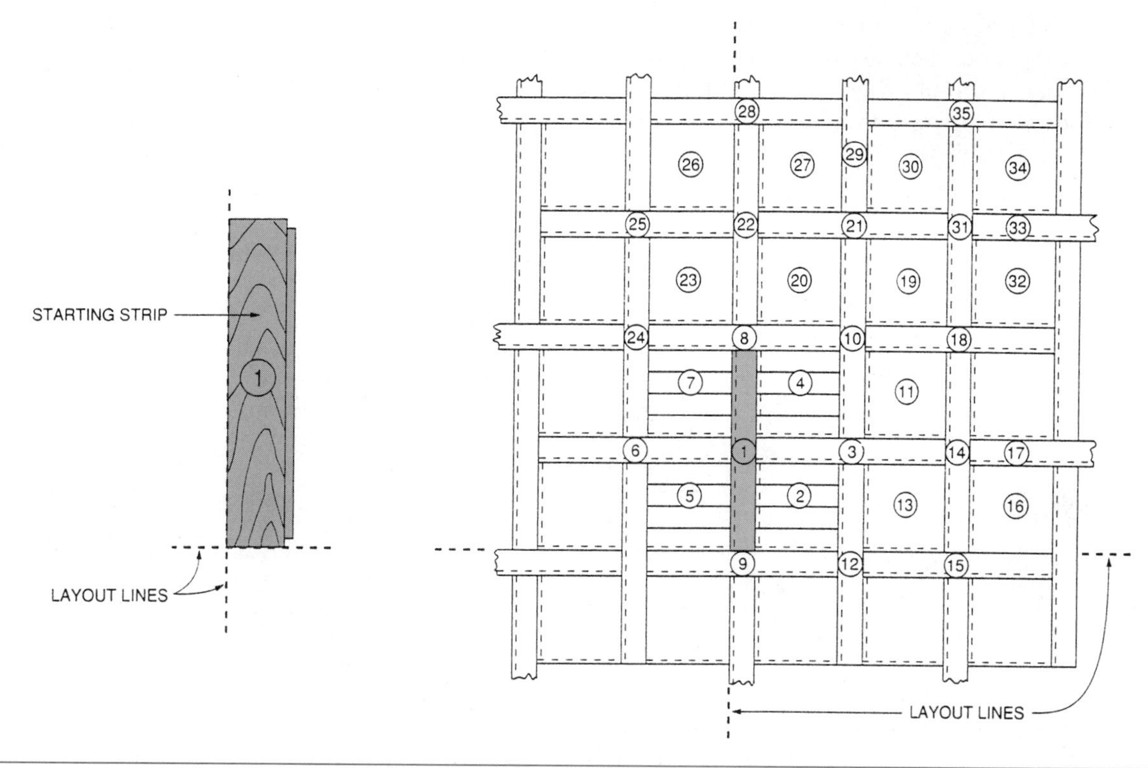

Fig. 84-21 Layout of the Marie Antoinette parquet pattern. (*Courtesy of Chickasaw Hardwood Floors*)

bers face toward the right or ahead. In this pattern, the grain of center blocks runs in the same direction.

Many other parquet patterns are manufactured. With the use of parquet blocks and strips, the design possibilities are almost endless. Competence in laying the popular patterns described in this chapter will enable the student to apply the principles for professional installations of many different parquet floors.

CHAPTER 85 UNDERLAYMENT AND RESILIENT TILE

Resilient flooring is widely used in residential and commercial buildings. It is a thin, flexible material that comes in *sheet* or *tile* form. It is applied on a smooth concrete slab or an *underlayment.*

Underlayment

Plywood, strandboard, hardboard, or particleboard may be used for *underlayment.* It is installed on top of the subfloor. This provides a base for the application of resilient sheet or tile flooring. The number of underlayment panels required is found by dividing the area to be covered by the area of the panel.

Underlayment thickness may range from 1/4 to 3/4 inch, depending on the material and job requirements. In many cases, where a finish wood floor meets a tile floor, the underlayment thickness is determined by the difference in the thickness of the two types of finish floor. Both floor surfaces should come flush with each other.

Installing Underlayment

Sweep the subfloor as clean as possible. Cover with asphalt felt lapped 4 inches at the edges. Stagger joints between the subfloor and underlayment. If installation is started in the same corner as the subfloor, the thickness of the walls will be enough to offset the joints. Leave about 1/32 inch between underlayment panels to allow for expansion.

If the underlayment is to go over a board subfloor, install the underlayment with its face grain across the boards. In all other cases, a stiffer and stronger floor is obtained if the face grain is across the floor joists (Fig. 85-1).

Fasten the first row in place with staples or nails. Underlayment requires more nails than subfloor to provide a squeak-free and stiff floor. Nail spacing and size are shown in Figure 85-2 for various thicknesses of plywood underlayment. Similar

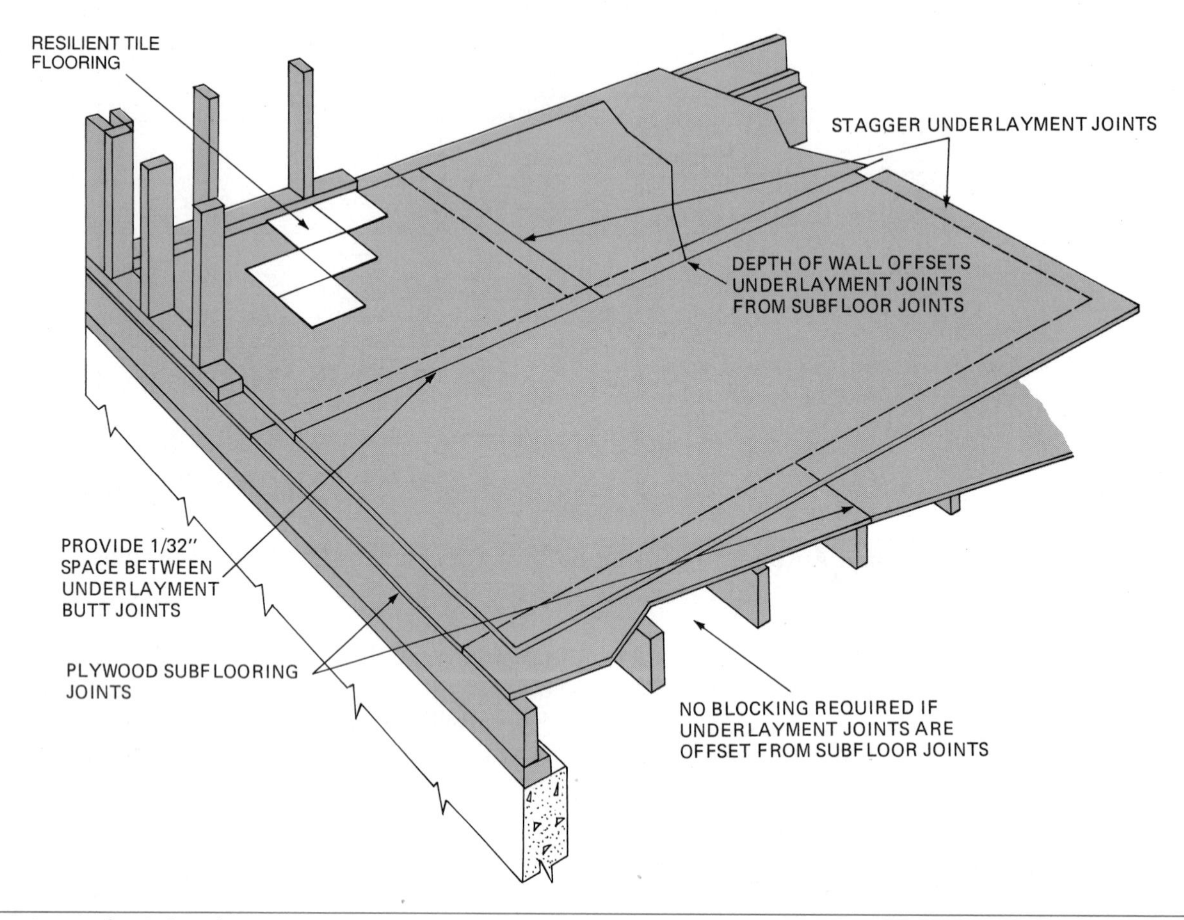

Fig. 85-1 The joints between underlayment and subfloor are offset. (*Courtesy of American Plywood Association*)

Plywood Thickness (inch)	Fastener Size (approx.) and Type	Fastener Spacing (inches)	
		Panel Edges	Panel Interior
1/4	1 1/4" ring-shank nails 18 gauge staples	3 3	6 each way 6 each way
3/8 or 1/2	1 1/4" ring-shank nails 16 gauge staples	6 3	8 each way 6 each way
5/8 or 3/4	1 1/2" ring-shank nails 16 gauge staples	6 3	8 each way 6 each way
1/4	1 1/4" ring-shank nails 18 gauge staples	3 3	6 each way 6 each way

Fig. 85-2 Nailing specifications for plywood underlayment.

nail spacing and size are appropriate for other types of underlayment.

Install remaining rows of underlayment with end joints staggered from the previous courses. The last row of panels is ripped to width to fit the remaining space.

If APA Sturd-I-Floor is used, it can double as a subfloor and underlayment. If square-edged panels are used, blocking must be provided under the joints to support the edges. If tongue-and-groove panels are used, blocking is not required (Fig. 85-3).

Resilient Tile Flooring

Most *resilient floor tiles* are made from asphalt, vinyl, rubber, or cork. Many different colors, textures, and patterns are available. Generally, tiles come in 9 × 9 and 12 × 12 squares. They are applied to the floor in a manner similar to that used for applying parquet blocks. The most common tile thicknesses are 1/16 inch and 1/8 inch. Thicker tiles are used in commercial applications subjected to considerable use.

Long strips of the same material are called *feature strips*. They are available from the manufacturer. The strips vary in width from 1/4 inch to 2 inches. They are used between tiles to create unique floor patterns.

Installing Resilient Tile

CAUTION If the application involves the removal of existing resilient floor covering, be aware that it, and the adhesive used, may contain asbestos. The presence of asbestos in the material is not easily determined. If there is any doubt, always assume that the existing flooring and adhesive do contain asbestos. Practices for removal of existing flooring or any other building material containing asbestos should comply with standards set by the U.S. Occupational Safety and Health Administration (OSHA) or corresponding authorities in other jurisdictions.

If the application is over an existing resilient floor covering, do not sand the existing surface unless absolutely sure it does not contain asbestos. Inhalation of asbestos dust can cause serious harm.

Before installing resilient tile, make sure underlayment fasteners are not projecting above the surface. Fill any open areas, such as splits, with a floor leveling compound. Use a portable disc

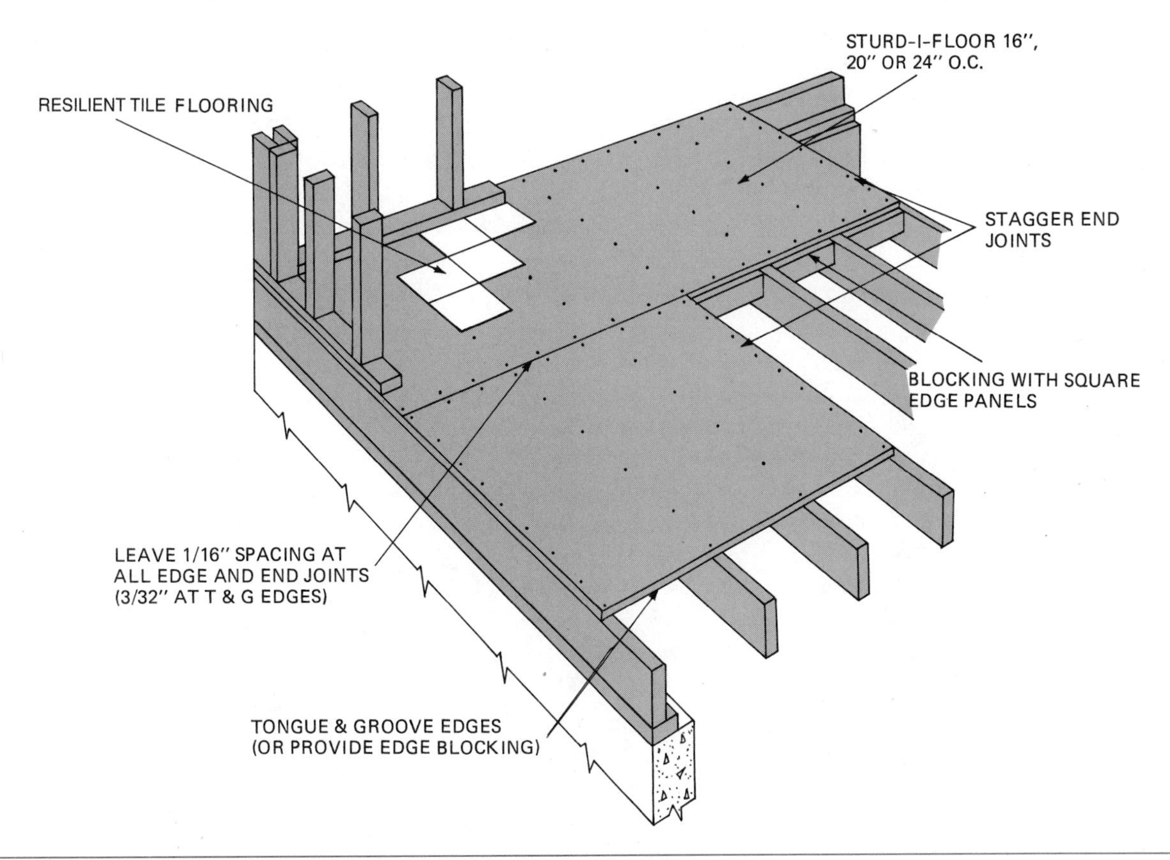

Fig. 85-3 APA Sturd-I-Floor requires no underlayment for the installation of resilient floors. (*Courtesy of American Plywood Association*)

sander to bring underlayment joints flush. Skim the entire surface with the sander to make sure the surface is smooth. Sweep the floor clean.

Layout

Snap layout lines across the floor at right angles to each other. The lines are centered if border tiles are a half width or more. If border tiles are less than a half width, layout lines are snapped half the width of a tile off center.

Applying Adhesive

Some floor tiles are manufactured with the adhesive applied at the factory. They are commonly known as *peel and stick* tiles. The adhesive is protected by a *release paper.* The paper is peeled off when ready to apply the tile. Professional floor tile installation normally includes the use of adhesives.

CAUTION The release paper is very smooth and slippery. When peeled from the tile, dispose of it in a container. If left on the floor, a serious accident could occur if a person stepped on it, slipped, and fell.

Adhesive usually is applied to one-quarter of the floor with a properly notched trowel, up to the layout lines (Fig. 85-4). Use an adhesive and notched trowel recommended by the manufacturer of the tile. If the notches in the trowel are too shallow, not enough adhesive will be applied. If the notches are too deep, more adhesive than necessary will be applied. This will result in the adhesive squeezing up through the joints onto the face of the tile. It will probably require removal of the application.

Laying Tiles

Start by applying a tile to intersection of the layout lines with the two adjacent edges on the lines. Lay tiles with edges tight. Work toward the walls (Fig. 85-5). Watch the grain pattern. It may be desired to alternate the run of the patterns or to lay the patterns, all in one direction. Some tiles are stamped with an arrow on the back. They are placed so the arrows on all tiles point in the same direction.

Lay tiles in place instead of sliding them into position. Sliding the tile pushes the adhesive through the joint. With some types of adhesives, it may be difficult or impossible to slide the tile. Apply all tiles in the quarter section except border tiles. Apply tiles to all other quarter sections in the same manner.

Applying Border Tiles

Border tiles may be cut by scoring with a sharp utility knife and bending or by the use of a special tile cutter. To lay out and score a border tile to fit snugly,

Fig. 85-5 Laying resilient floor tile.

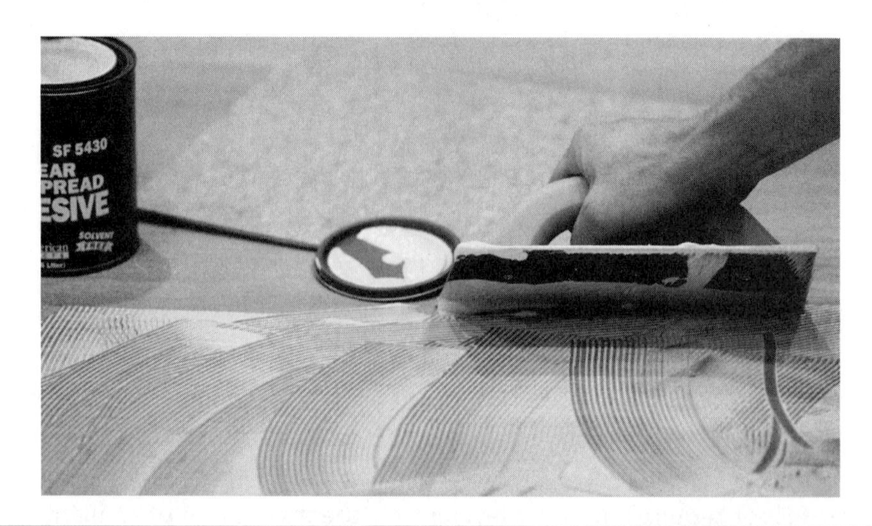

Fig. 85-4 Applying adhesive to the underlayment for the application of resilient floor tile.

first place the tile to be cut directly on top of the last tile installed. Make sure all edges are in line. Place a full tile with its edge against the wall and on top of the one to be fitted. Score the border tile along the outside edge of the top tile (Fig. 85-6). Bend the tile. Break it along the scored line, and fit it into place (Fig. 85-7). If a base has not yet been installed, then not so much care need be taken with the fit of border tiles. Simply measure and cut to an approximate width so the space can be covered by the base.

Applying a Vinyl Cove Base

Many times a vinyl cove base is used to trim a tile floor. A special vinyl base cement is applied to its back. The base is pressed into place (Fig. 85-8).

Estimating Resilient Tile

To estimate the amount of tile flooring needed, find the area of the room in square feet. To do this, measure the length and width of the room to the next whole foot. Multiply these figures together to find the area. For 12 × 12 tiles, the result is the number of pieces needed. Multiply the area by 1 1/3 to find the number of 9 × 9 tiles needed.

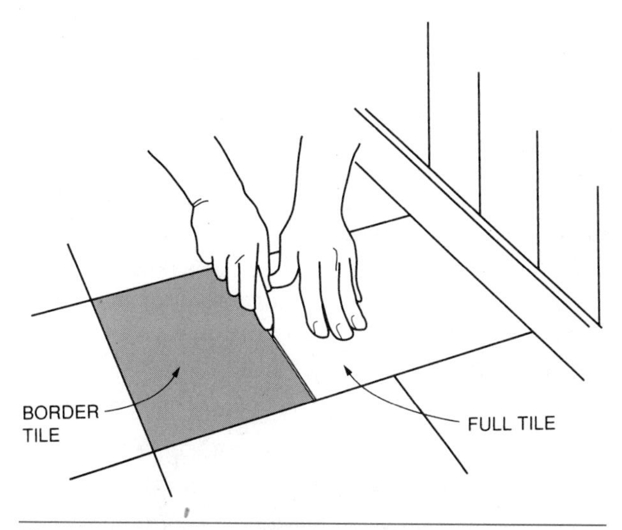

Fig. 85-6 Scoring a border tile to fit against the wall.

Fig. 85-7 Fitting a border tile in place

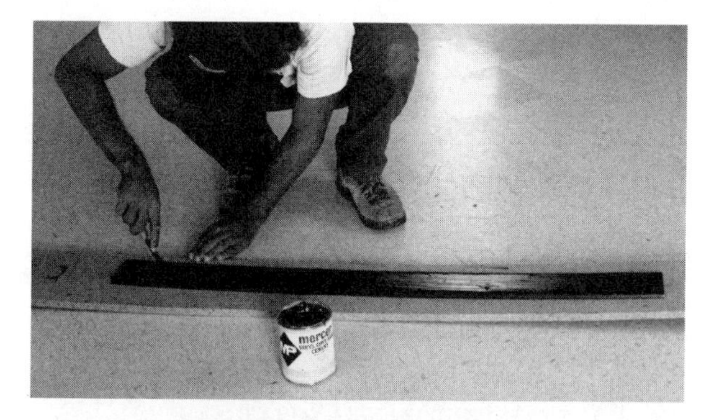

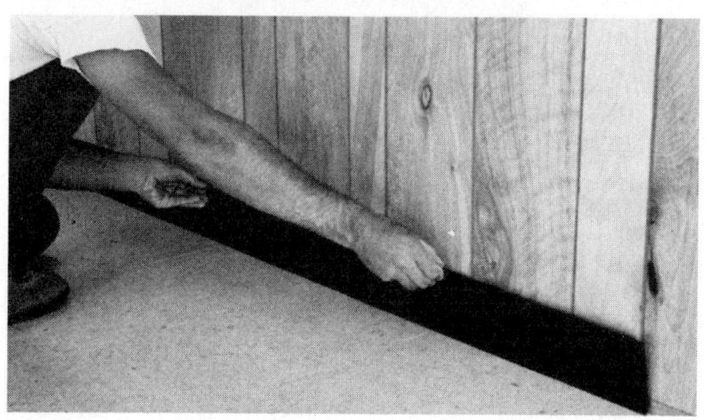

Fig. 85-8 Applying cement to the back of a vinyl cove base and pressing it into place.

Review Questions

Select the most appropriate answer.

1. If hardwood flooring is stored in a heated building, the temperature should not exceed
a. 72 degrees.
b. 78 degrees.
c. 85 degrees.
d. 90 degrees.

2. Most hardwood finish flooring is made from
a. Douglas fir.
b. hemlock.
c. southern pine.
d. oak.

3. Bundles of strip flooring may contain pieces over and under the nominal length of the bundle by
a. 4 inches. c. 8 inches.
b. 6 inches. d. 9 inches.

4. No pieces are allowed in bundles of hardwood strip flooring shorter than
a. 4 inches. c. 8 inches.
b. 6 inches. d. 9 inches.

5. The edges of prefinished strip flooring are chamfered to
a. prevent splitting.
b. apply the finish.
c. simulate cracks between adjoining pieces.
d. obscure any unevenness in the floor surface.

6. The best grade of unfinished oak strip flooring is
a. prime. c. select.
b. clear. d. quarter-sawed.

7. To estimate the amount of 2 1/4-inch face hardwood flooring, add to the area to be covered a percentage of the area of
a. 42.5. c. 29.
b. 38.33. d. 33.33.

8. When it is necessary to cut the last strip in a course of flooring, the waste is used to start the next course and should be at least
a. 8 inches long. c. 12 inches long.
b. 10 inches long. d. 16 inches long.

9. For floor laying, the hammer weight is generally
a. 13 to 16 ounces. c. 20 to 28 ounces.
b. 16 to 20 ounces. d. 25 to 30 ounces.

10. To change direction of strip flooring,
a. face nail both strips.
b. turn the extended strip around.
c. blind nail both strips.
d. use a spline.

BUILDING FOR SUCCESS

Communication Infrastructure

In any building project, communication must be a priority. The flow of pertinent information has to be continual if success is to be obtained. Each builder must create a personal communication infrastructure within the company and community that will leave no weak link for potential problems.

Successful, skilled technicians, subcontractors or builders will always work to improve their communication network within the industry. Experience usually proves to be the best teacher and will be the most valuable resource. By combining actual business communication experiences with observations in the field and continued education, builders can create the most meaningful communication system for their companies.

The following is a sample of a communication infrastructure that has proven to be successful for a home builder. Sam and Mike Manzitto, owners of Manzitto Brothers Custom Homes and Remodeling in Lincoln, Nebraska, have graciously provided this example. This planned communication infrastructure has the capacity for ongoing modification and improvement. Its flexibility makes it attractive for others who are addressing their own communications structure.

For seventeen years, goals have been met, volume in production has peaked, profit margins are high, and a good company team has been established. We have made many changes to keep up with the economy, competition, and technology.

Then comes a new problem: a breakdown in the communications infrastructure (inner office, supervisor to lead carpenters, production manager to client, production manager to job supervisor, lead carpenter to production manager, lead carpenter to client, and so on.)

To correct and counteract these problems, we built a system to accomplish the following goals:

1. Keep communication lines open with the customer throughout its construction process.
2. Improve communication between lead carpenters and subcontractors.
3. Improve communication between supervisor and lead carpenters.

4. Improve the scheduling and organization of the job.
5. Improve the quality of the finished product.
6. Improve our follow up with a punch list procedure.

This system consists of a series of ongoing weekly meetings, involving reports on jobs under construction, jobs to be completed, current clients, and marketing strategies by our sales and management team. Separate weekly meetings are held with supervisors and lead carpenters for job updates. Daily meetings begin with job supervisors. These activities involve completing written forms and procedures to be carried on throughout construction, from start to finish.

The following documents have been created and are used on every job in the following order.

PROJECT CHECKLIST: This list becomes part of each job file. These procedures must be met before the project can completed.

LEADPERSON CHECKLIST: This document is reviewed by the lead carpenter at the time the project is being estimated or at the signing of the contract.

PRECONSTRUCTION CONFERENCE: This document is used at the first meeting of the lead carpenter or supervisor, the client, and the salesperson.

SUBCONTRACTOR LETTER: This letter is sent out to all subcontractors working on the project.

GENERAL PROJECT PROCEDURES: This document is posted in poster form on the project before the work is to begin.

NEIGHBORHOOD LETTER: This document is sent to neighbors in the area of the project.

WALK-THRU PROCEDURES: This document helps the lead carpenter or supervisor with the final walk-thru at the completion of the job.

PUNCHLIST PROJECT: This list helps the final walk-thru of the project to run more smoothly. Any unfinished items can be agreed on at that point. The amounts of those items only are withheld from the final settlement of the project.

The following exhibit is of the *General Project Procedures* established by Manzitto Brothers Builders,

1. INTRODUCE YOURSELF TO HOME OWNER BY NAME AND COMPANY.
2. PLEASE AVOID USING PROJECT PROPERTY FOR BATHROOM FACILITIES.
3. PUNCTUALITY—MAKE A QUICK CALL IF YOU ARE RUNNING LATE.
4. USE DROP CLOTHS TO PROTECT FINISHED AREAS AT ALL TIMES.
5. PROPERLY DISPOSE OF ALL TRASH, INCLUDING PERSONAL TRASH.
6. DEMONSTRATE PROFESSIONAL ATTITUDE TOWARD FELLOW EMPLOYEES AND SUBCONTRACTORS.
7. BE JOB ACCESSIBLE AT ALL TIMES FOR HOME OWNER AND SUBCONTRACTORS.
8. NO PARKING IN DRIVEWAY. AVOID PARKING ON LOT.
9. WEAR APPROPRIATE ATTIRE (wear shirts in the house at all times).
10. NO PROFANITY.
11. NO SMOKING.
12. NO LOUD MUSIC.

CUSTOMER SERVICE DEPARTMENT

A customer service department has been created so that our follow up or call-back procedures run smoothly. We feel that following up on our product is one of the most important parts of our business.

CUSTOMER SERVICE DEPARTMENT: Explains how our customer service department is structured, and shows the steps that are taken throughout the year.

CUSTOMER EVALUATION: Each customer completes this after the project is completed so that our services can be evaluated.

SERVICE REQUEST LETTER: Each customer responds three months after the project is completed, requesting items that may need our attention.

WARRANTY SERVICE REQUEST FORM: This is sent out with the service letter in reference to the new home warranty.

SERVICE REQUEST LETTER #2: This is sent out after the client has been in the home for approximately nine months.

FOCUS QUESTIONS: For individual or group discussion

1. As you review this communication infrastructure concept, what may be the most productive uses for a plan of this type? Why?

2. What appears to be the best way this plan can be implemented? Who are the key people that need to participate in this plan for it to work?
3. As expectations of performance, scheduling, and quality are set forth in the communication process, what can it tell the prospective client, subcontractors, and public about the builder? Does this appear to be an advantage over a builder that does not communicate these important elements to others? Why?
4. How can the following "checklists" and service considerations be of benefit to the company's management employees and clients?
   a. Project checklist
   b. Leadman checklist
   c. Preconstruction conference
   d. General project procedures list
   e. Walk-thru procedures
   f. Punch list
5. What function does the customer service department serve?
6. How can the company follow-up process benefit the company?

UNIT

30

Cabinets and Countertops

Cabinets and countertops usually are purchased in preassembled units and may be installed by a carpenter. Manufactured cabinets are installed because of the great variety and shorter installation time than job-built cabinets. Cabinets can be custom-made to meet the specifications of most any job, but they are usually made in a cabinet shop. Countertops, cabinet doors, and drawers may be customized in a wide variety of styles and sizes.

OBJECTIVES

After completing this unit, the student should be able to:

- state the sizes and describe the construction of typical base and wall kitchen cabinet units.
- plan, order, and install manufactured kitchen cabinets.
- construct, laminate, and install a counter top.
- identify cabinet doors and drawers according to the type of construction and method of installation.
- identify overlay, lipped, and flush cabinet doors and proper drawer construction.
- apply cabinet hinges, pulls, and door catches.

UNIT CONTENTS

CHAPTER 86 DESCRIPTION AND INSTALLATION OF MANUFACTURED CABINETS

Manufactured kitchen and bath cabinets come in a wide variety of styles, materials, and finishes (Fig. 86-1). The carpenter must be familiar with the various kinds, sizes, uses, and construction of the cabinets to know how to plan, order, and install them.

Description of Manufactured Cabinets

For commercial buildings, many kinds of specialty cabinets are manufactured. They are designed for specific uses in offices, hospitals, laboratories, schools, libraries, and other buildings. Most cabinets used in residential construction are manufactured for the kitchen or bathroom. All cabinets, whether for commercial or residential use, consist of a case which is fitted with shelves, doors, and/or drawers. Cabinets are manufactured and installed in essentially the same way. Designs vary considerably with the manufacturer, but sizes are close to the same.

Kinds and Sizes

One method of cabinet construction utilizes a **face frame**. This frame provides openings for doors and drawers. Another method, called *European* or *frameless,* eliminates the face frame (Fig. 86-2). Face-framed cabinets usually give a traditional look. Frameless cabinets are used when a contemporary appearance is desired.

The two basic kinds of kitchen cabinets are the *wall* unit and the *base* unit. The surface of the countertop is usually about 36 inches from the floor. Wall units are installed about 18 inches above the countertop. This distance is enough to accommodate such articles as coffee makers, toasters, blenders, and mixers. Yet it keeps the top shelf within reach, not over 6 feet from the floor. The usual overall height of a kitchen cabinet installation is 7′-0″ (Fig. 86-3).

Wall Cabinets. Standard *wall* cabinets are 12 inches deep. They normally come in heights of 42, 30, 24, 18, 15, and 12 inches. The standard height is 30 inches. Shorter cabinets are used above sinks, refrigerators, and ranges. The 42-inch cabinets are for use in kitchens without soffits where more storage space is desired. A standard height wall unit usually contains two adjustable shelves.

Usual wall cabinet widths range from 9 to 48 inches in 3-inch increments. They come with single or double doors depending on their width.

Fig. 86-1 Manufactured kitchen cabinets are available in a wide variety of styles and sizes. (*Courtesy of Merillat Industries*)

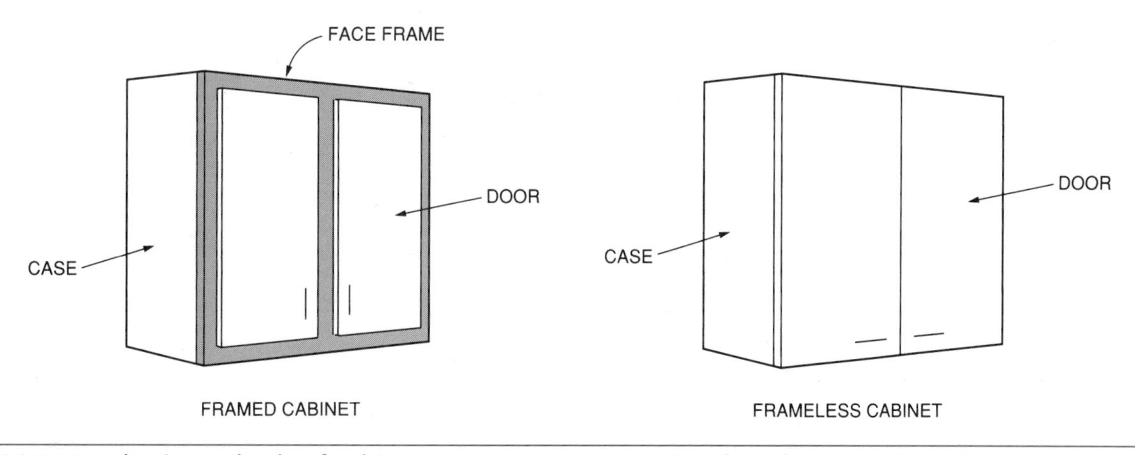

Fig. 86-2 Two basic methods of cabinet construction are with a face frame or frameless.

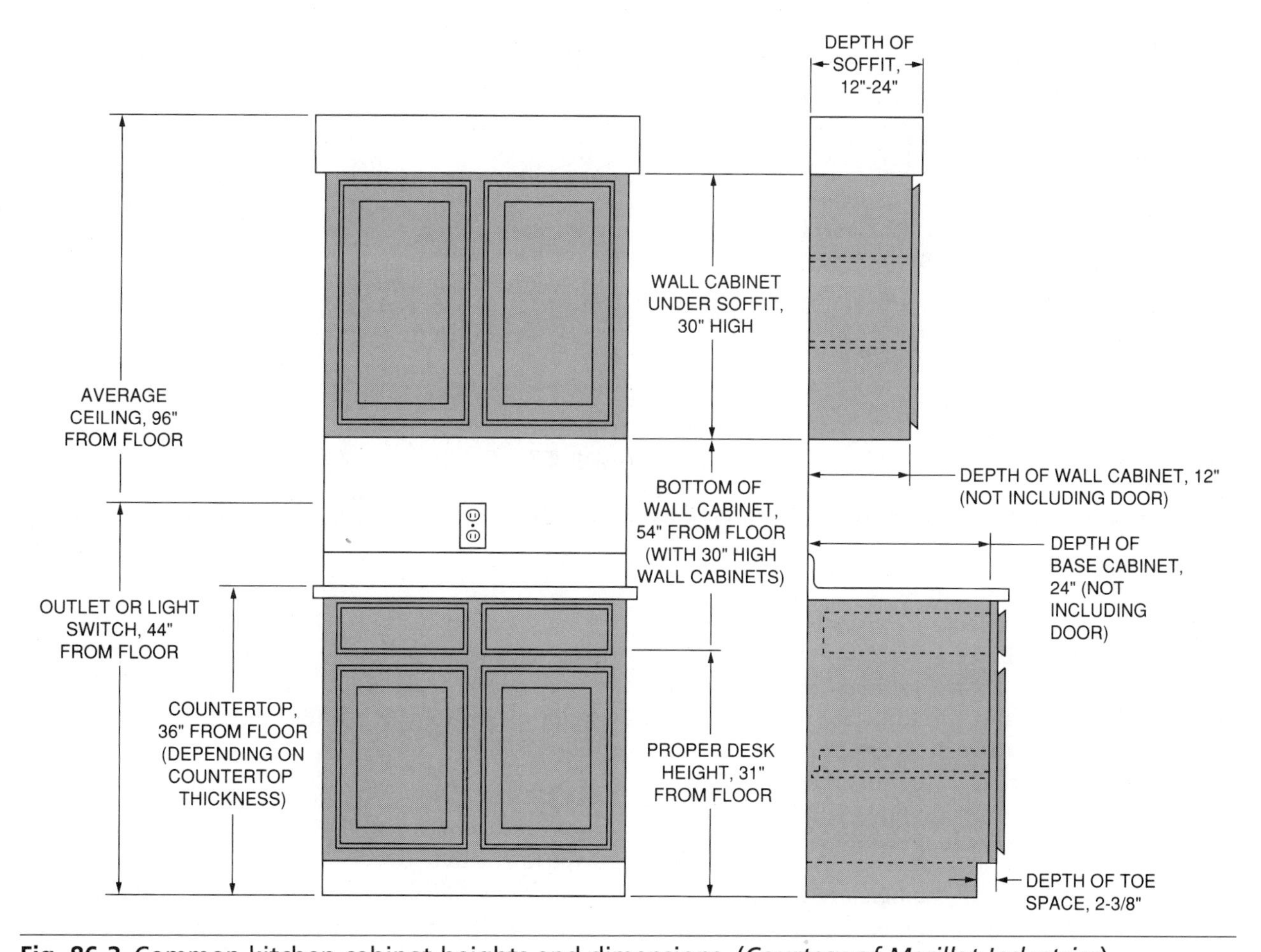

Fig. 86-3 Common kitchen cabinet heights and dimensions. (*Courtesy of Merillat Industries*)

Single-door cabinets can be hung so doors can swing in either direction.

Wall *corner* cabinets make access into corners easier. *Double-faced* cabinets have doors on both sides for use above island and peninsular bases. Some wall cabinets are made 24 inches deep for installation above refrigerators. A microwave oven case, with a 30-inch wide shelf, is available (Fig. 86-4).

Base Cabinets. Most *base* cabinets are manufactured 34 1/2 inches high and 24 inches deep. By adding the usual countertop thickness of 1 1/2 inches, the work surface is at the standard height of 36 inches from the floor. Base cabinets come in widths to match wall cabinets. Single-door cabinets are manufactured in widths from 9 to 24 inches. Double-door cabinets come in widths from 27 to

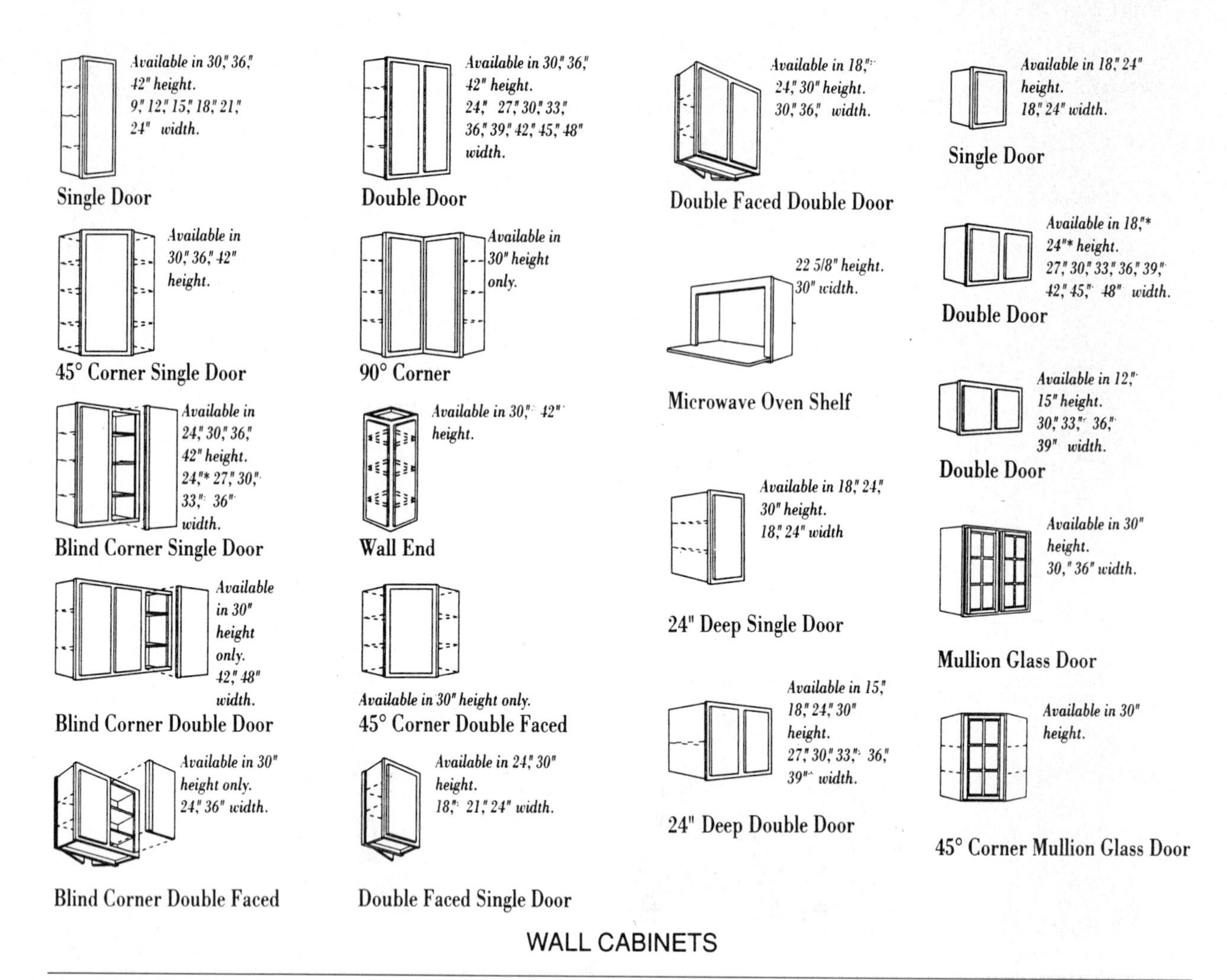

Fig. 86-4 Kinds and sizes of manufactured wall cabinets. (*Courtesy of Merillat Industries*)

48 inches. A recess called a *toe space* is provided at the bottom of the cabinet.

The standard base cabinet contains one drawer, one door, and an adjustable shelf. Some base units have no drawers; others contain all drawers. *Double-faced* cabinets provide access from both sides. *Corner* units, with round revolving shelves, make corner storage easily accessible (Fig. 86-5).

Tall Cabinets. *Tall* cabinets are usually manufactured 24 inches deep, the same depth as base cabinets. Some *utility* cabinets are 12 inches deep. They are made 66 inches high and in widths of 27, 30, and 33 inches for use as *oven* cabinets. Single-door *utility* cabinets are made 18 and 24 inches wide. Double-door *pantry* cabinets are made 36 inches wide (Fig. 86-6). Wall cabinets with a 24-inch depth are usually installed above tall cabinets.

Vanity Cabinets. Most *vanity* base cabinets are made 31 1/2 inches high and 21 inches deep. Some are made in depths of 16 and 18 inches. Usual widths range from 24 to 36 inches in increments of 3 inches, then 42, 48, and 60 inches. They are available with several combinations of doors and drawers depending on their width. Various sizes and styles of vanity wall cabinets are also manufactured (Fig. 86-7).

Accessories. *Accessories* are essential to or enhance a cabinet installation. *Filler* pieces fill small gaps in width between wall and base units when no combination of sizes can fill the existing space. They are cut to necessary widths on the job. Other accessories include cabinet end panels, face panels for dishwashers and refrigerators, open shelves for cabinet ends, and spice racks.

Laying Out Manufactured Kitchen Cabinets

The blueprints for a building contain plans, elevations, and details that show the cabinet layout. Architects may draw the layout. But they may not specify the size or the manufacturer's identification for each individual unit of the installation. In resi-

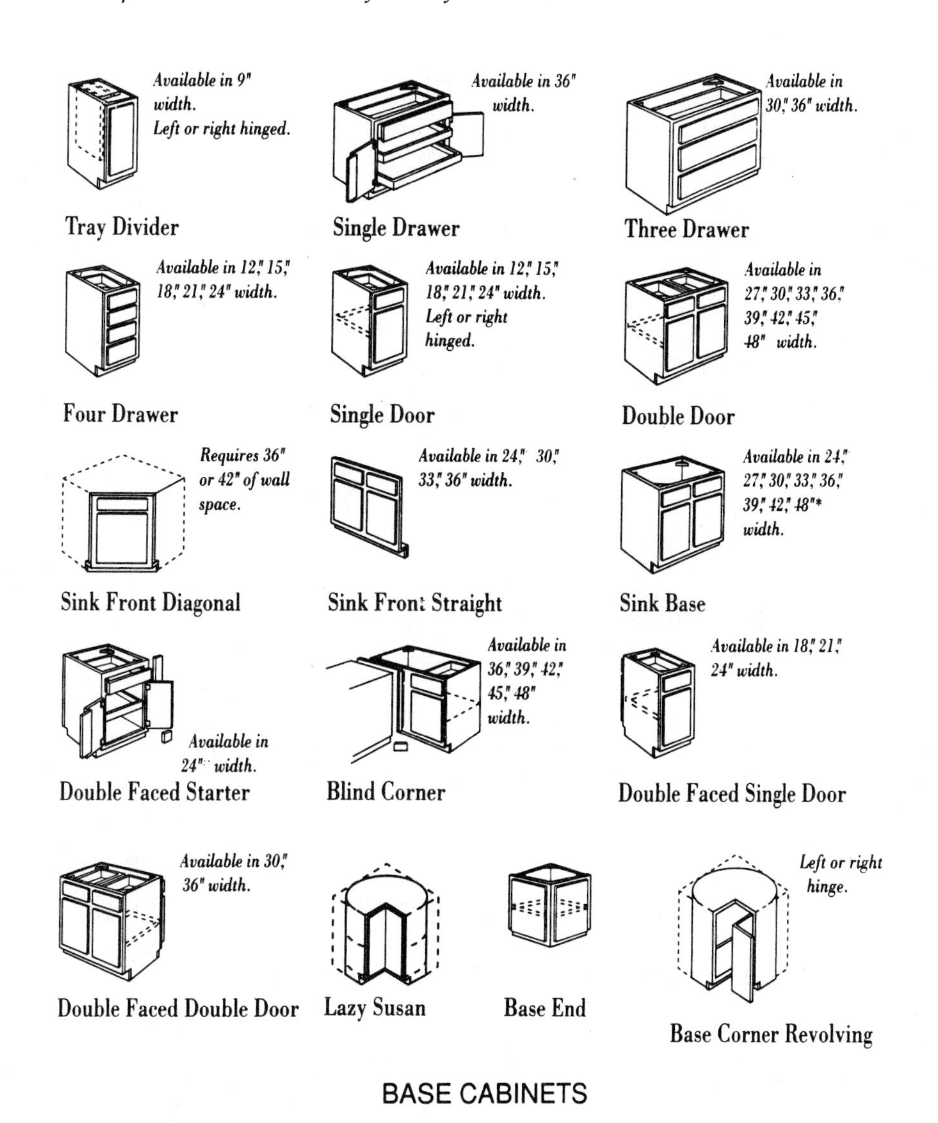

Fig. 86-5 Most base cabinets are manufactured to match wall units. (*Courtesy of Merillat Industries*)

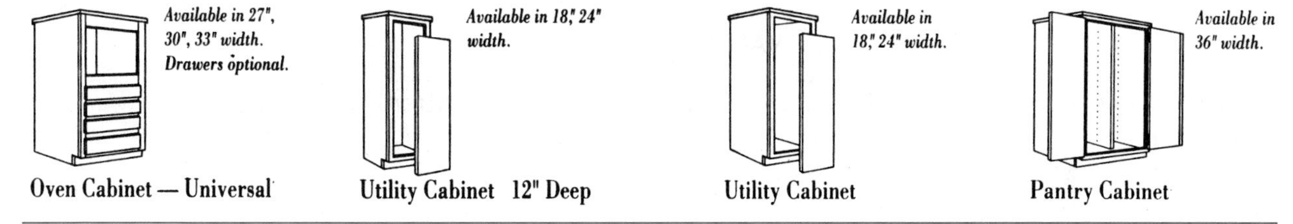

Fig. 86-6 Tall cabinets are manufactured as oven, utility, and pantry units. (*Courtesy of Merillat Industries*)

dential construction, particularly in remodeling, no plans are usually available to show the cabinet arrangement. In addition to installation, it becomes the responsibility of the carpentry contractor to plan, lay out, and order the cabinets, in accordance with the customer's specifications.

The first step is to measure carefully and accurately the length of the walls on which the cabinets are to be installed. A plan is then drawn to scale. It must show the location of all appliances, sinks, windows, and other necessary items (Fig. 86-8).

Next, draw elevations of the base cabinets, referring to the manufacturer's catalog for sizes. Always use the largest size cabinets available instead of two or three smaller ones. This reduces the cost and makes installation easier.

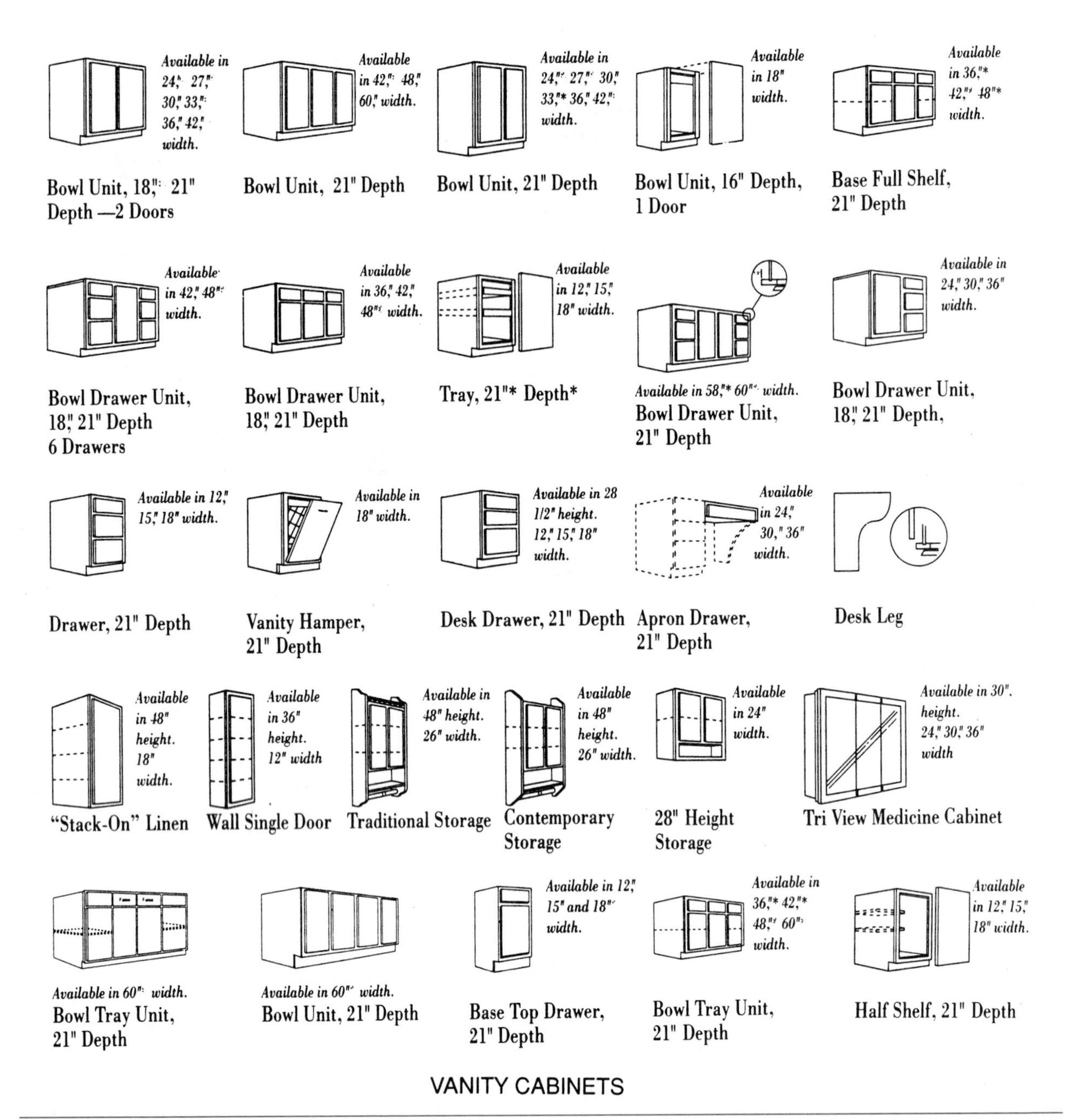

Fig. 86-7 Vanity cabinets are made similar to kitchen cabinets, but differ in size. (*Courtesy of Merillat Industries*)

Match up the wall cabinets with the base cabinets, where feasible. If filler strips are necessary, place them between a wall and a cabinet or between cabinets in the corner. Identify each unit on the elevations with the manufacturer's identification (Fig. 87-9). Make a list of the units in the layout. Order from the distributor.

Computer Layouts

Computer programs are available to help in laying out manufactured kitchen cabinets. When the required information is fed into the computer, a number of different layouts can be quickly made. When a acceptable layout is made, it can be printed with each of the cabinets in the layout identified and priced. Most large kitchen cabinet distributors will supply computerized layouts on request.

Installing Manufactured Cabinets

Cabinets must be installed level and plumb even though floors are not always level and walls not always plumb. Level lines are first drawn on the wall for base and wall cabinets. In order to level base

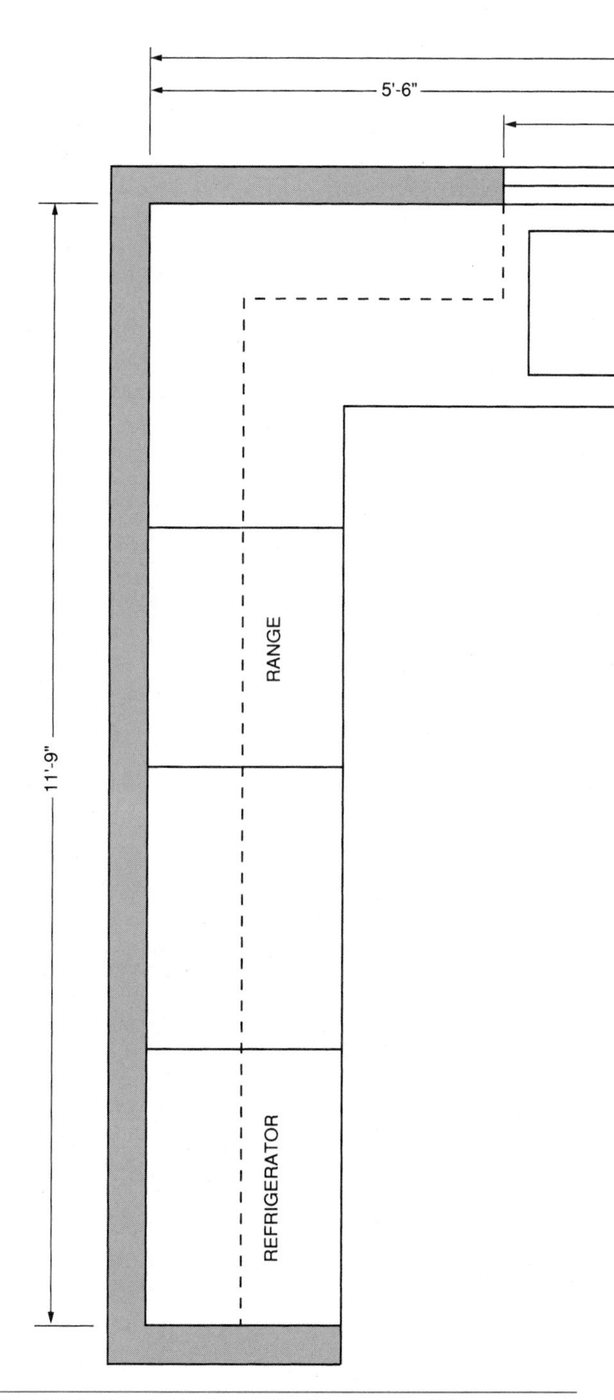

Fig. 86-8 Typical plan of a kitchen cabinet layout showing location of walls, windows, and appliances.

cabinets that set on an unlevel floor, either shim the cabinets from the high point of the floor or scribe and fit the cabinets to the floor from lowest point on the floor. Shimming the base cabinets leaves a space that must be later covered by a molding. Scribing and fitting the cabinets to the floor eliminate the need for a molding. The method used depends on various conditions of the job. If shimming base cabinets, lay out the level lines on the wall from the highest point on the floor where cabinets are to be installed. If fitting cabinets to the floor, measure up from the low point.

Laying Out the Wall

Measure 34 1/2 inches up the wall. Draw a level line to indicate the tops of the base cabinets. Use the most accurate method of leveling available (described in Chapter 25, Leveling and Layout Tools). Another level line must be made on the wall 54 inches from the floor. The bottoms of the wall units are installed to this line. It is more accurate to measure 19 1/2-inches up from the first level line and snap lines parallel to it than to level another line.

The next step is to mark the stud locations in a framed wall. (Cabinet mounting screws will be driven into the studs.) An electronic stud finder works well to locate framing. The other method is to lightly tap on and across a short distance of the wall with a hammer. Drive a finish nail in at the point where a solid sound is heard. Drive the nail where holes are later covered by a cabinet. If a stud is found, mark the location with a pencil. If no stud is found, try a little over to one side or the other.

Measure at 16-inch intervals in both directions from the first stud to locate other studs. Drive a finish nail to test for solid wood. Mark each stud location. If studs are not found at 16-inch centers, try 24-inch centers. At each stud location, draw plumb lines on the wall. Mark the outlines of all cabinets on the wall to visualize and check the cabinet locations against the layout (Fig. 86-10).

Installing Wall Units

Many installers prefer to mount the wall units first so that work does not have to be done over the base units. A *cabinet lift* may be used to hold the cabinets in position for fastening to the wall. If a lift is not available, the doors and shelves may be removed to make the cabinet lighter and easier to clamp together. If possible, screw a 1 × 3 strip of lumber so its top edge is on the level line for the bottom of the wall cabinets. This is used to support the wall units

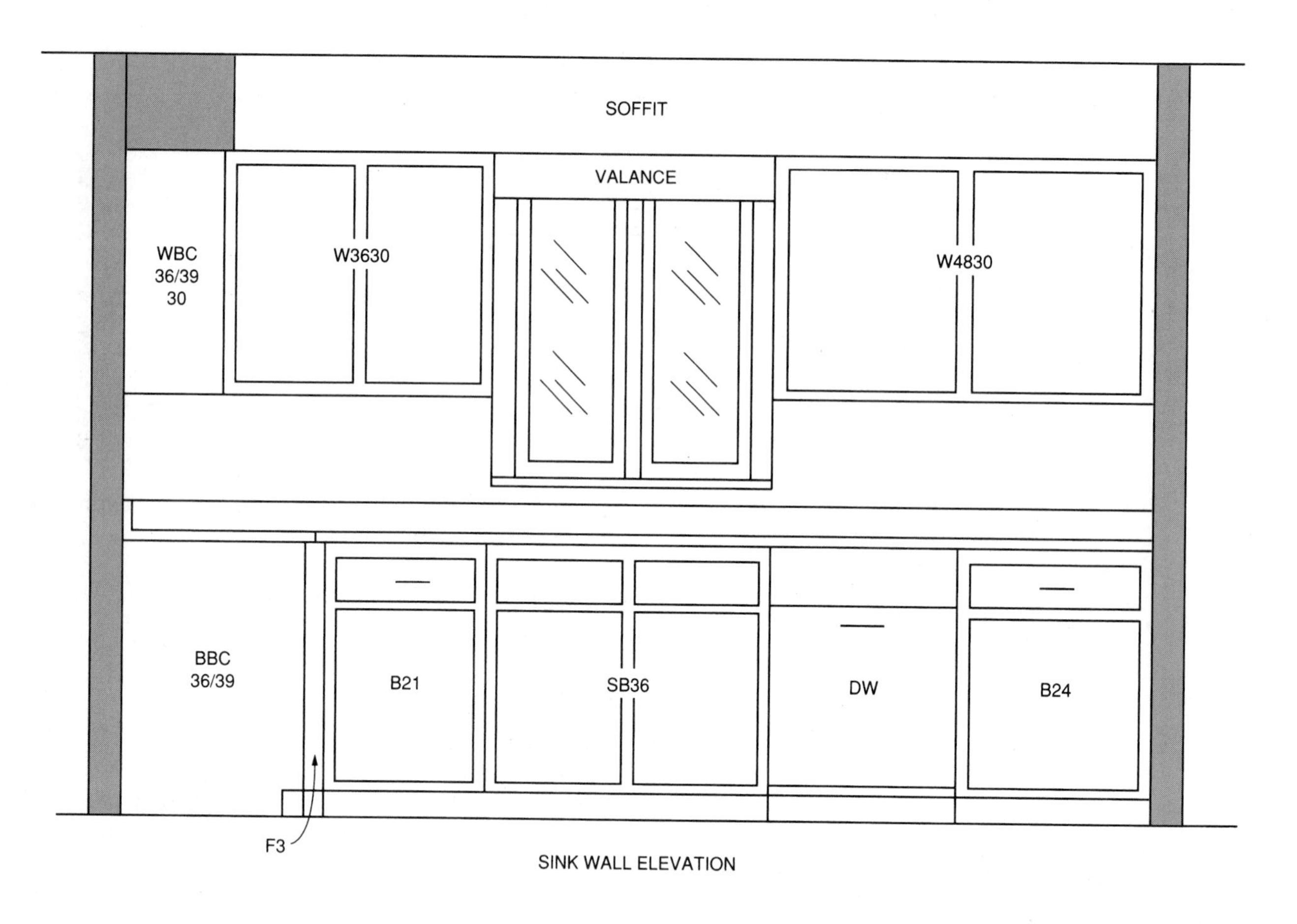

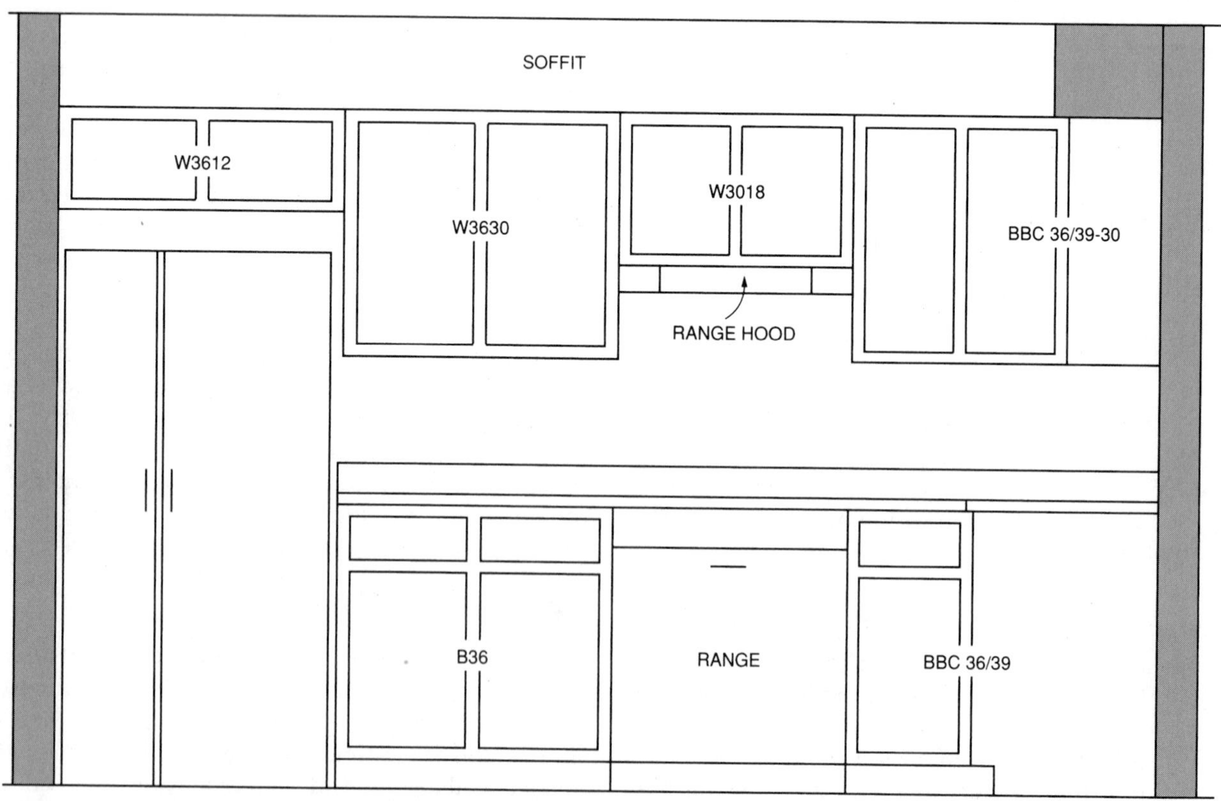

Fig. 86-9 Elevations of the installation are drawn and the cabinets identified.

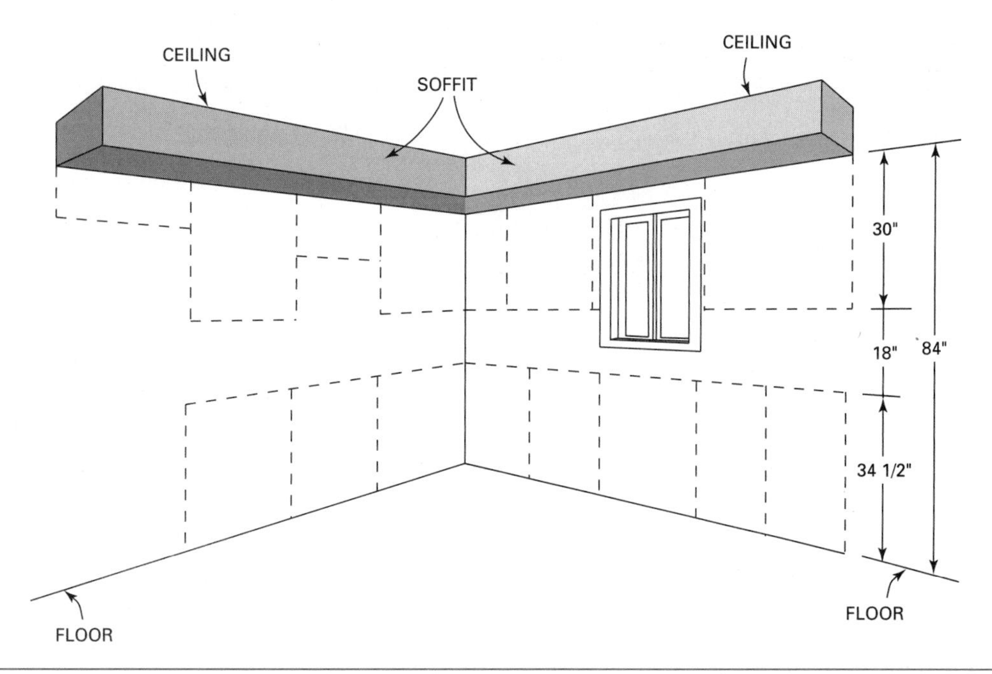

Fig. 86-10 The wall is laid out with outlines of the cabinets.

while they are being fastened. If it is not possible to screw to the wall, build a stand on which to support the unit near the line of installation (Fig. 86-11).

Start the installation of wall cabinets in a corner. On the wall, measure from the line representing the outside of the cabinet to the stud centers. Transfer the measurements to the cabinets. Drill shank holes for mounting screws through mounting rails usually installed at the top and bottom of the cabinet. Place the cabinet on the supporting strip or stand so its bottom is on the level layout line. Fasten the cabinet in place with mounting screws of sufficient length to hold the cabinet securely. Do not fully tighten the screws (Fig. 86-12).

On masonry walls, first drill holes through the mounting strips. Place the cabinet in position, and mark the location of the drilled holes on the wall. Remove the cabinet. Drill holes into the masonry wall for lead inserts. Replace the cabinet, and screw in place.

The next cabinet is installed in the same manner. Align the adjoining *stiles* so their faces are flush with each other. Clamp them together with C-clamps. Screw the stiles tightly together (Fig. 86-13). Continue this procedure around the room. Tighten all mounting screws.

If a *filler* needs to be used, it is better to add it next to a *blind corner* cabinet or at the end of a run. It may be necessary to scribe the filler to the wall. This procedure is shown in Figure 86-14.

Installing Base Cabinets

Start the installation of base cabinets in a corner. Shim the bottom until the cabinet top is on the layout line. Then level and shim the cabinet from back to front.

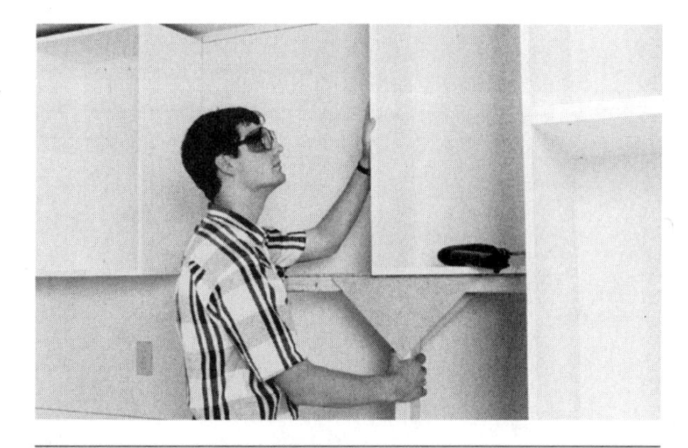

Fig. 86-11 Support the wall unit with a job-built stand. (*Courtesy of Merillat Industries*)

If cabinets are to be fitted to the floor, shim until their tops are level across width and depth. This will bring the tops above the layout line that was measured from the low point of the floor. Adjust the pencil dividers so the distance between the points is equal to the amount the top of the unit is above the layout line. Scribe this amount on the bottom end of the cabinets by running the dividers along the floor (Fig. 86-15).

Cut both ends and toeboard to the scribed lines. There is no need to cut the cabinet backs because they do not, ordinarily, extend to the floor.

Place the cabinet in position. The top ends should be on the layout line. Fasten it loosely to the wall.

The remaining base cabinets are installed in the same manner. Align and clamp the stiles of adjoining cabinets. Fasten them together. Finally, fasten all units securely to the wall (Fig. 86-16).

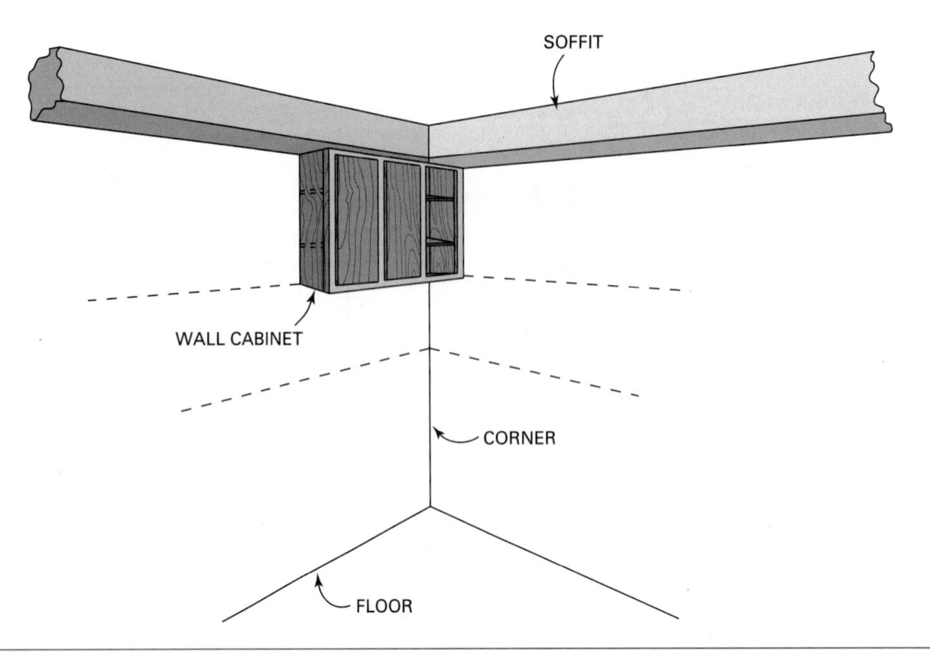

Fig. 86-12 Installation of wall cabinets is started in the corner.

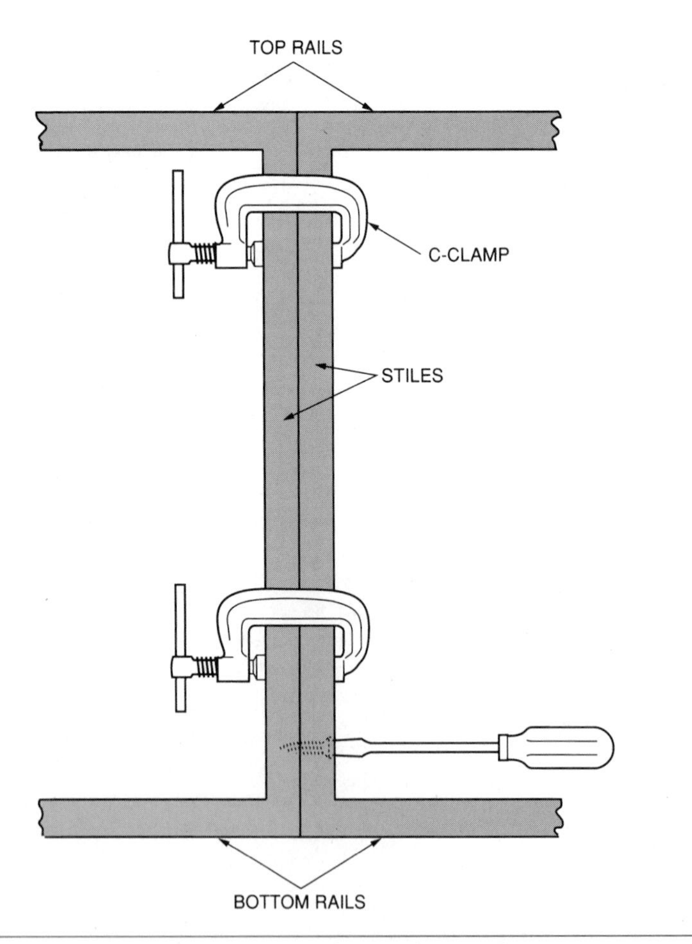

Fig. 86-13 The stiles of adjoining cabinets are joined together with screws.

Installing Manufactured Countertops

Countertops are manufactured in various standard lengths. They can be cut to fit any installation against walls. They are also available with one end precut at a 45 degree angle for joining with a similar one at corners. Special hardware is used to join the sections. The countertops are covered with a thin, tough *high-pressure plastic laminate.* This is

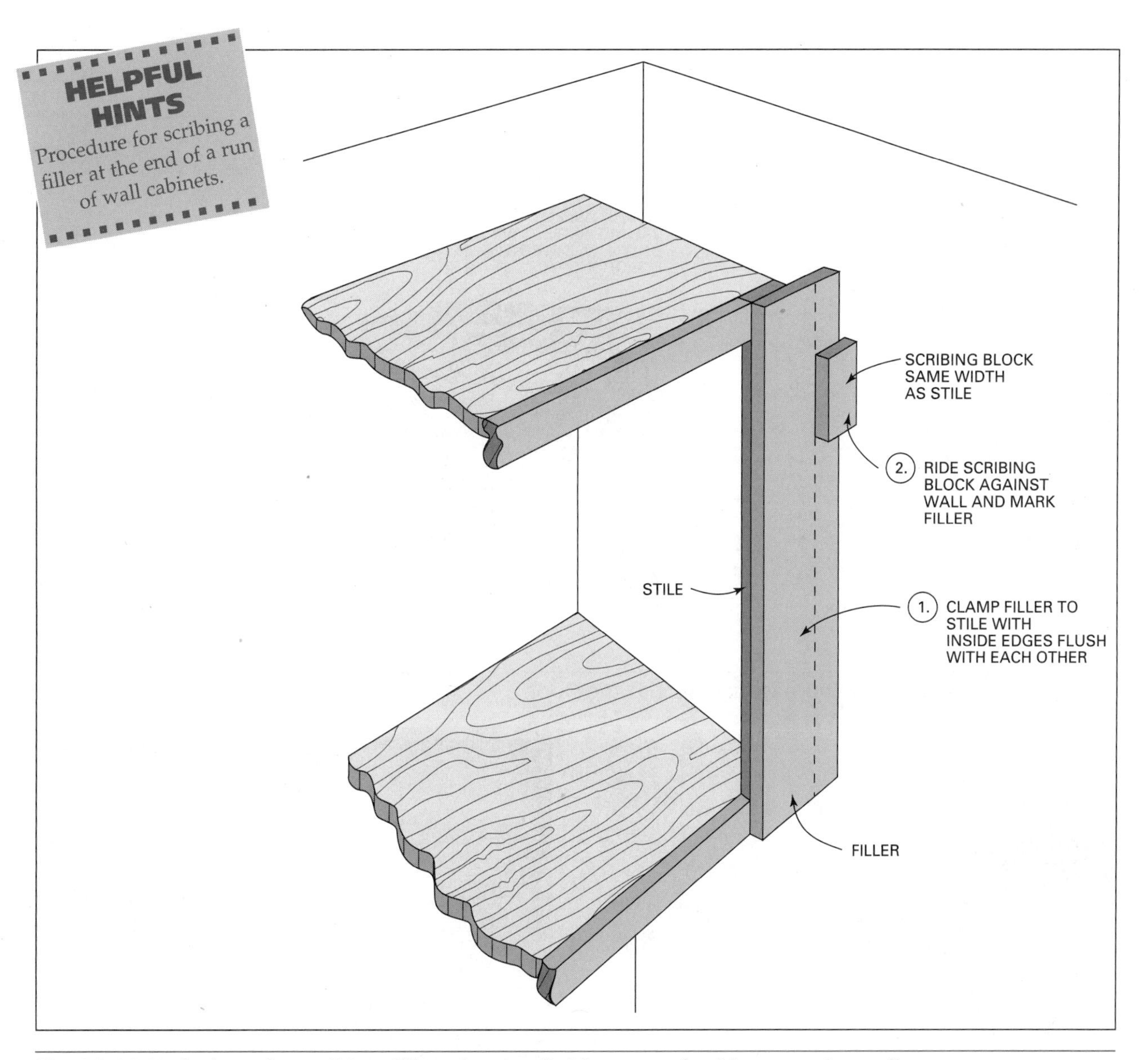

Fig. 86-14 Technique for scribing filler piece to finish a run of cabinets to the wall.

generally known as *mica.* It is available in many colors and patterns. The countertops are called *postformed countertops.* This term comes from the method of forming the mica to the rounded edges and corners of the countertop (Fig. 86-17). Postforming is bending the mica with heat to a radius of 3/4 inch or less. This can only be done with special equipment.

After the base units are fastened in position, the countertop is cut to length. It is fastened on top of the base units and against the wall. The backsplash can be scribed, limited by the thickness of its scribing strip, to an irregular wall surface. Use pencil dividers to scribe a line on the top edge of the backsplash. Then plane or belt sand to the scribed line.

Fasten the countertop to the base cabinets with screws up through triangular blocks usually installed in the top corners of base units. Use a stop on the drill bit. This prevents drilling through the countertop. Use screws of sufficient length, but not so long that they penetrate the countertop.

Exposed cut ends of **postformed** countertops are covered by specially shaped pieces of plastic laminate.

Sink cutouts are made by carefully outlining the cutout and cutting with a saber saw. The cutout pattern usually comes with the sink. Use a fine tooth blade to prevent chipping out the face of the mica beyond the sink. Some duct tape applied to the base of the saber saw will prevent scratching of the countertop when making the cutout.

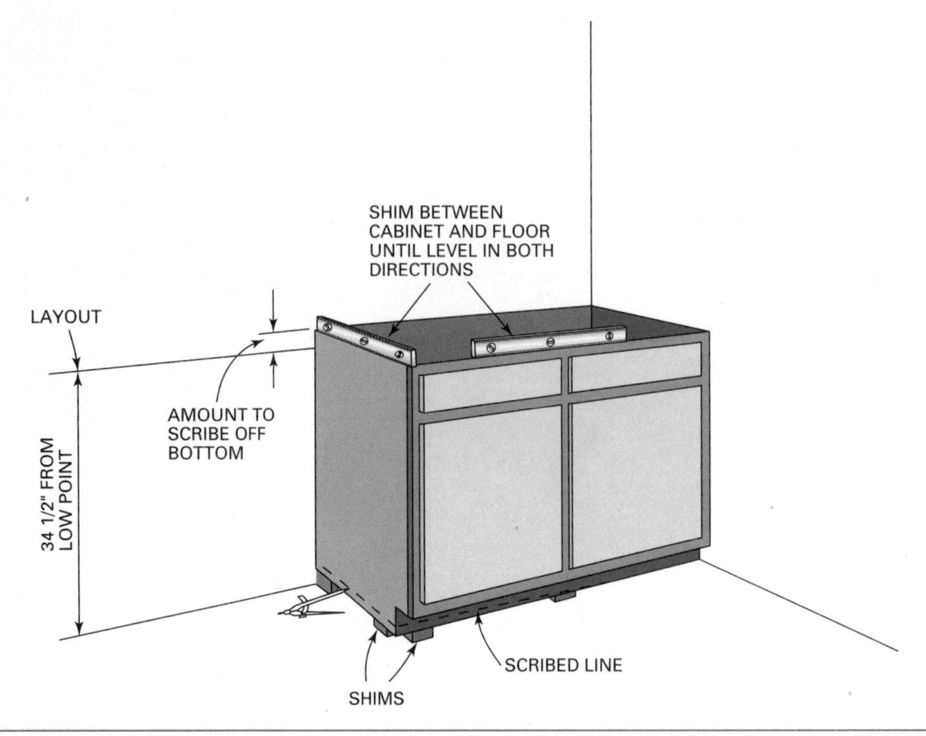

Fig. 86-15 Method of scribing base cabinets to the floor.

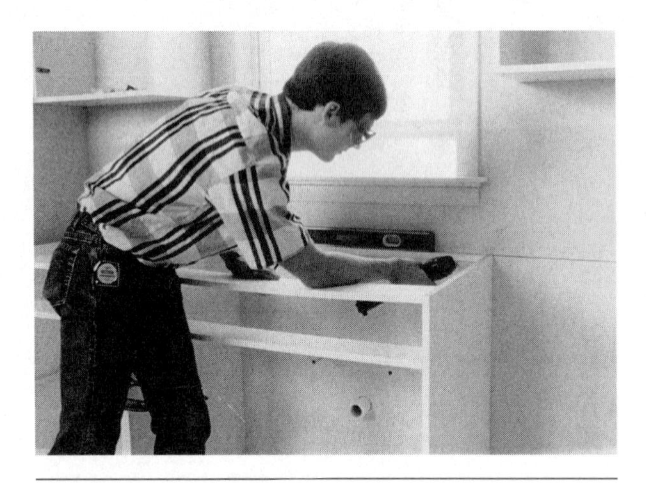

Fig. 86-16 Base cabinets are secured with screws to wall studs. (*Courtesy of Merillat Industries*)

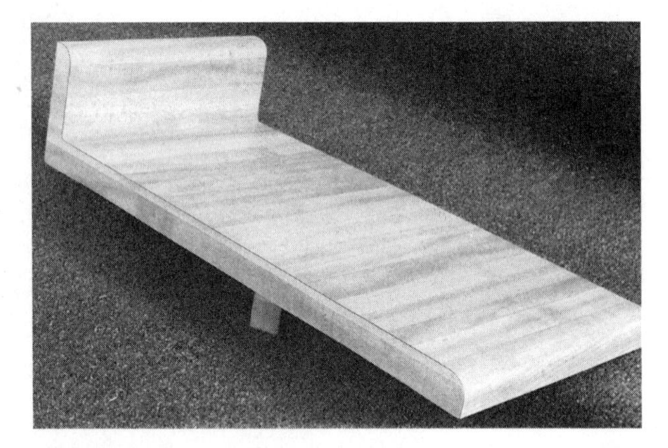

Fig. 86-17 A section of a manufactured postformed countertop. The edges and interior corner are rounded.

CHAPTER 87 COUNTERTOPS AND CABINET COMPONENTS

Custom cabinets usually are made at cabinet shops. Sometimes, though, counter tops, doors, and drawers are produced or modified on the job.

Making the Countertop

Use the pieces of 3/4 - or 5/8-inch panel material left from making the cabinet bottoms to make the *countertop.* If more than one length is required, join them with glue and screws to a short piece of backing plywood. The width of the pieces should be about 24 1/2 inches.

Fitting the Countertop

Place the countertop on the base cabinets, against the wall. Its outside edge should overhang the face frame the same amount along the entire length. Open the pencil dividers or scribers to the amount of overhang. Scribe the back edge of the countertop to the wall. Cut the countertop to the scribed line. Place it back on top of the base cabinets. The ends should be flush with the ends of the base cabinets. The front edge should be flush with the face of the face frame (Fig. 87-1). Install a 1 × 2 on the front edge and at the ends if an end overhang is desired. Keep the top edge flush with the top side of the countertop.

Applying the Backsplash

If a *backsplash* is used, rip a 4-inch wide length of 3/4-inch stock the same length as the countertop. Use lumber for the backsplash, if lengths over 8 feet are required, to eliminate joints. Fasten the backsplash on top of and flush with the back edge of the countertop by driving screws up through the countertop and into the bottom edge of the backsplash (Fig. 87-2). In corners, fasten the ends of the backsplash together with screws.

Laminating a Countertop

Most countertops are covered with *plastic laminate.* Before laminating a countertop, make sure all surfaces are flush. Check for protruding nailheads. Fill in all holes and open joints. Lightly hand or power sand the entire surface, making sure joints are sanded flush.

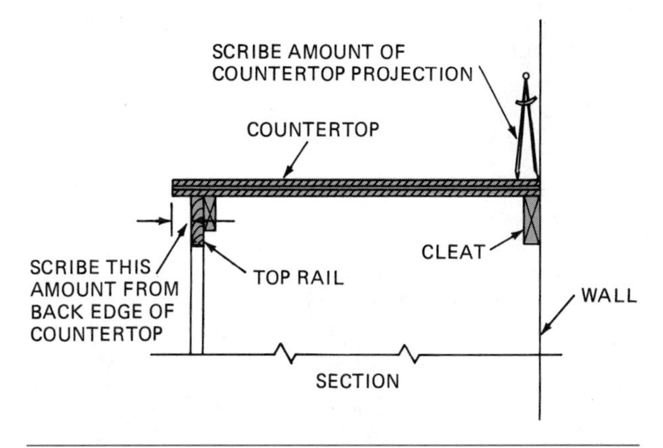

Fig. 87-1 Scribing the countertop to fit the wall with its outside edge flush with the face of the cabinet.

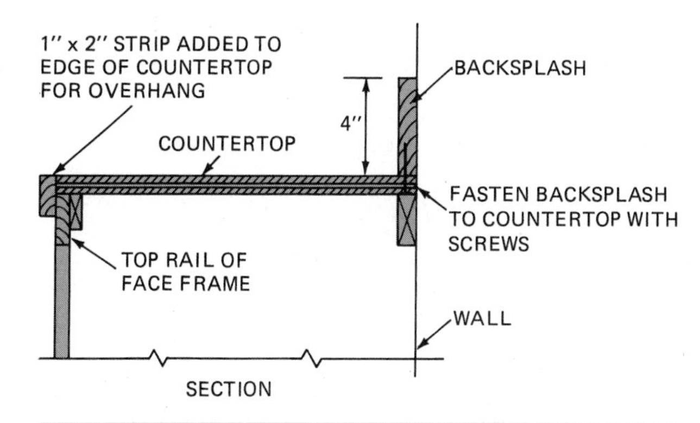

Fig. 87-2 Drive screws into the bottom edge of the backsplash to fasten it to the countertop.

Laminate Trimming Tools and Methods

Pieces of laminate are first cut to a *rough size,* about 1/4 to 1/2 inch wider and longer than the surface to be covered. A strip is then cemented to the edge of the countertop. Its edges are *flush-trimmed* even with the top and bottom surfaces. Laminate is then cemented to the top surface, overhanging the edge strip. The overhang is then *bevel-trimmed* even with the laminated edge. A *laminate trimmer* or a small *router* fitted with *laminate trimming bits* is used for rough cutting and flush and bevel trimming of the laminate (Fig. 87-3).

Cutting Laminate to Rough Sizes

Sheets of laminate are large, thin, and flexible. This makes them difficult to cut on a table saw. One method of cutting laminates to rough sizes is by clamping a straightedge to the sheet. Cut it by guiding a laminate trimmer with a flush trimming bit along the straightedge (Fig. 87-4). It is easier to run the trimmer across the sheet than to run the sheet across the table saw. Also, the router bit leaves a smooth, clean cut edge. Use a *solid carbide* trimming bit, which is smaller in diameter than one with ball bearings. It makes a narrower cut. It is easier to control and creates less waste. With this method, cut all the pieces of laminate needed to a rough width and length. Cut the narrow edge strips from the sheet first.

Using Contact Cement

Contact cement is used for bonding plastic laminates and other thin, flexible material to surfaces. A coat of cement is applied to the back side of the laminate and to the countertop surface. The cement must be dry before the laminate is bonded to the core. The bond is made on contact without the need of

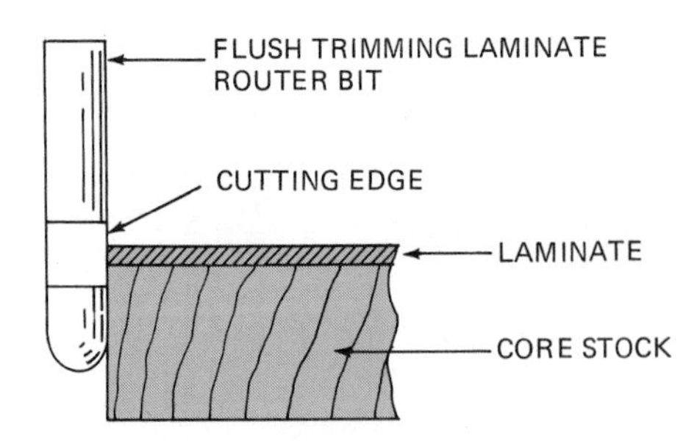

FLUSH TRIM

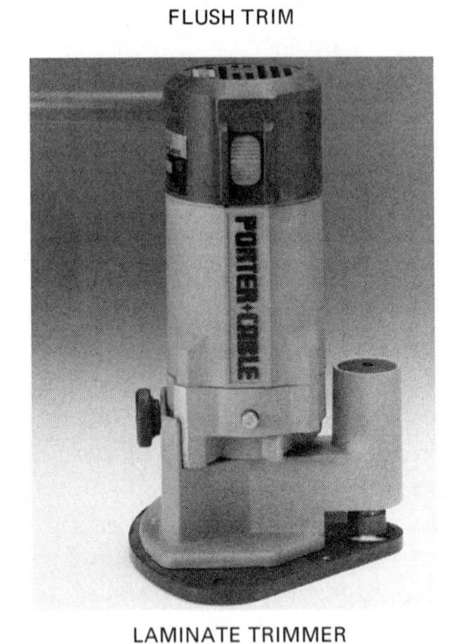

LAMINATE TRIMMER

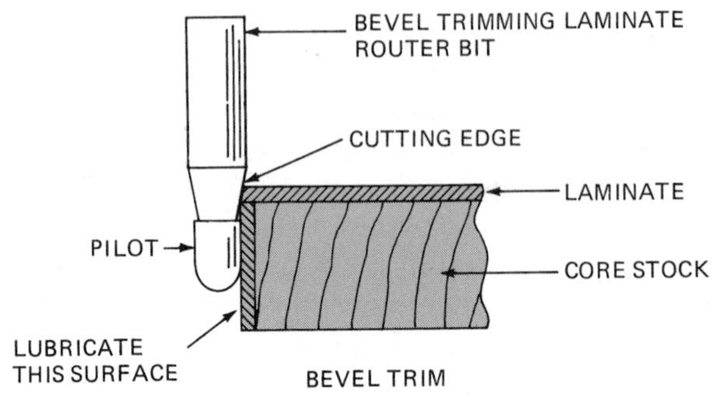

BEVEL TRIM

ADJUST BEVEL TRIMMING BIT TO CUT FLUSH WITH, BUT NOT INTO EDGE LAMINATE. THE BEVEL KEEPS THE CUTTING EDGE FROM GRAZING THE FIRST LAYER OF LAMINATE.

Fig. 87-3 The laminate trimmer is used with flush and bevel bits to trim overhanging edges of laminate.

HELPFUL HINTS

Clamp the laminate to a straightedge. Cut rough sizes with a laminate trimmer.

Fig. 87-4 Laminate trimmer being used to trim laminate back flush with the edge of the countertop.

clamps. There are several reasons why a contact cement bond may fail.

- Not enough cement is applied. If the material is porous, like the edge of particleboard or plywood, a second coat is required after the first coat dries. When enough cement has been applied, a glossy film appears over the entire surface when dry.
- Too little time is allowed for the cement to dry. Both surfaces must be dry before contact is made. To test for dryness, lightly press your finger on the surface. Although it may feel sticky, the cement is dry if no cement remains on the finger.
- The cement is allowed to dry too long. If contact cement dries too long (more than about 2 hours, depending on the humidity), it will not bond properly. To correct this condition, merely apply another coat of cement and let it dry.
- The surface is not rolled out or tapped after the bond is made. Pressure must be applied to the entire surface using a 3-inch **J-roller** or by tapping with a hammer on a small block of wood.

CAUTION Some contact cements are flammable. Apply only in a well-ventilated area around no open flame. Avoid inhaling the fumes.

Laminating the Countertop Edges

Remove the backsplash from the countertop. Apply coats of cement to the countertop edges and the back of the edge laminate with a narrow brush or small paint roller. After the cement is dry, apply the laminate to the front edge of the countertop (Fig. 87-5). Position it so the bottom edge, top edge, and ends overhang. A permanent bond is made when the two surfaces make contact. A mistake in positioning means removing the bonded piece—a time-consuming, frustrating, and difficult job. Roll out or tap the surface.

Apply the laminate to the ends in the same manner as to the front edge piece. Make sure that the square ends butt up firmly against the back side of the overhanging ends of the front edge piece to make a tight joint.

Trimming Laminated Edges

The overhanging ends of the edge laminate must be trimmed before the top and bottom edges. If the

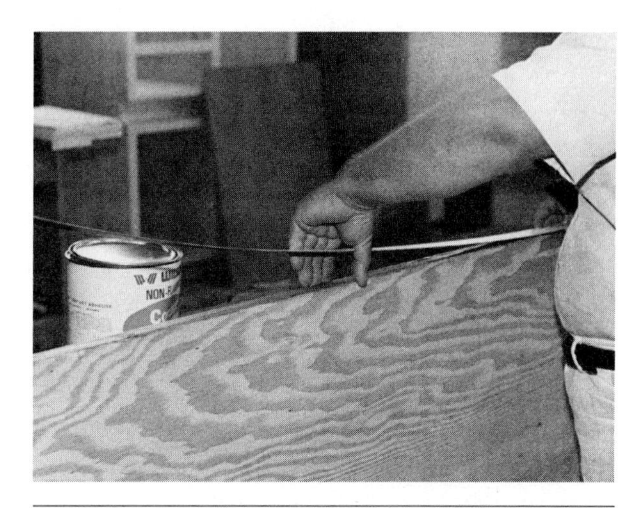

Fig. 87-5 Applying laminate to the edge of the countertop.

laminate has been applied to the ends, a bevel trimming bit must be used to trim the overhanging ends.

Using a Bevel Trim Bit.

When using a bevel trimming bit, the router base is gradually adjusted to expose the bit so that the laminate is trimmed flush with the first piece but not cutting into it. The bevel of the cutting edge allows the laminate to be trimmed without cutting into the adjacent piece (see Fig. 87-3). A flush trimming bit cannot be used when the pilot rides against another piece of laminate because the cutting edge may damage it.

Ball bearing trimming bits have *live pilots.* Solid carbide bits have *dead pilots* that turn with the bit. When using a trimming bit with a dead pilot, the laminate must be lubricated where the **pilot** will ride. Rub a short piece of white candle or some solid shortening on the laminate to prevent marring the laminate by the bit.

Using the bevel trimming bit, trim the overhanging ends of the edge laminate. Then, using the flush trimming bit, trim off the bottom and top edges of both front and end edge pieces (Fig. 87-6). To save the time required to change and adjust trimming bits, some installers use two laminate trimmers, one with a flush bit and the other with a bevel bit.

Use a belt sander or a file to smooth the top edge flush with the surface. Sand or file *flat* on the countertop core so a sharp square edge is made. This assures a tight joint with the countertop laminate. Sand or file *toward* the core to prevent chipping the laminate. Smooth the bottom edge. Ease the sharp outside corner with a sanding block.

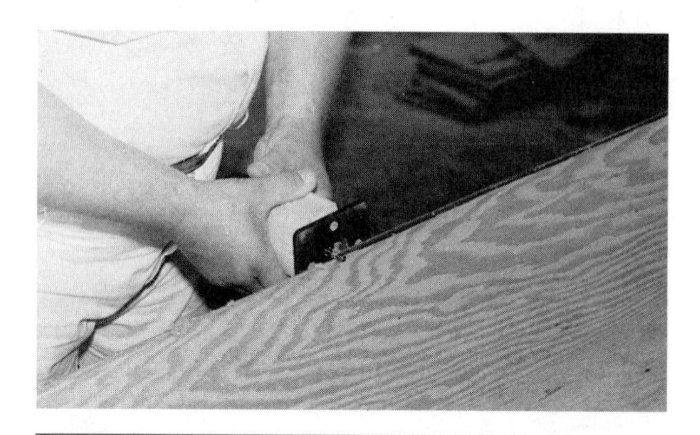

Fig. 87-6 Flush trimming the countertop edge laminate.

Laminating the Countertop Surface

Apply contact bond cement to the countertop and the back side of the laminate. Let dry. To position large pieces of countertop laminate, first place thin strips of wood or metal venetian blind slats about a foot apart on the surface. Lay the laminate to be bonded on the strips or slats. Then position the laminate correctly (Fig. 87-7).

Make contact on one end. Gradually remove the slats one by one until all are removed. The laminate should then be positioned correctly with no costly errors. Roll out the laminate (Fig. 87-8). Trim the overhanging back edge with a flush trimming bit. Trim the ends and front edge with a bevel trimming bit (Fig. 87-9). Use a flat file to smooth the trimmed edge. Slightly ease the sharp corner.

Laminating a Countertop with Two or More Pieces

When the countertop is laminated with two or more lengths, tight joints must be made between them. Tight joints can be made by clamping the two pieces

Fig. 87-7 Position the laminate on the countertop using venetian blind slats.

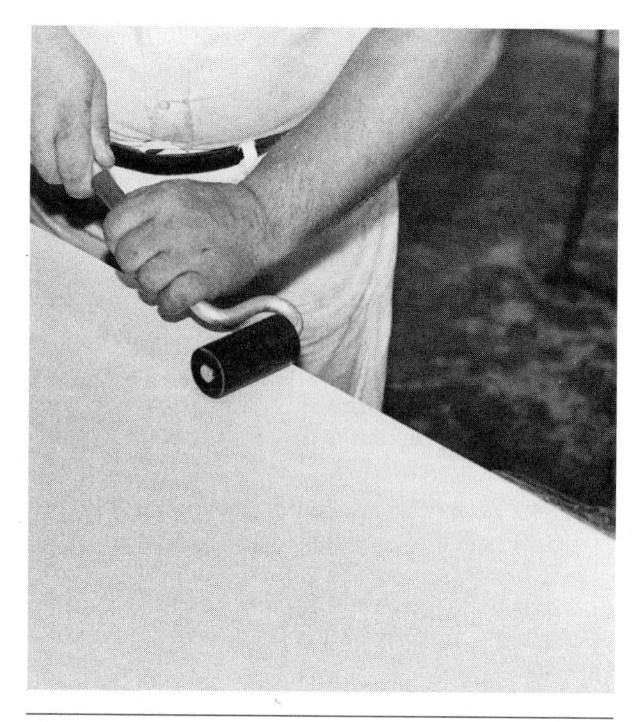

Fig. 87-8 Rolling out the laminate with a J-roller is required to assure a proper bond.

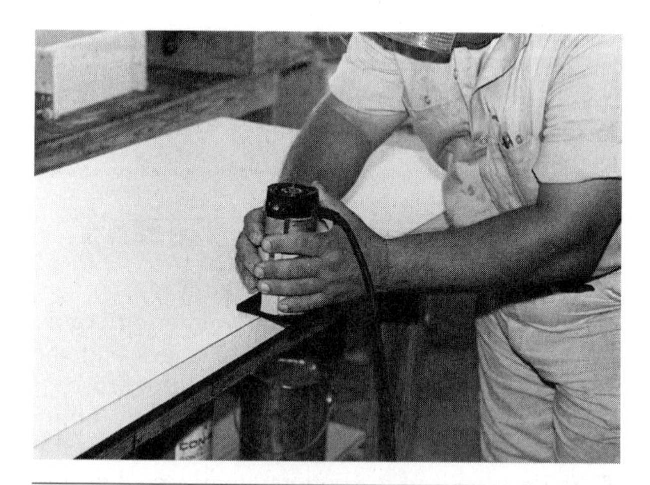

Fig. 87-9 The outside edge of the countertop laminate is bevel-trimmed.

of laminate in a straight line on some strips of 3/4-inch stock. Butt the ends together or leave a space less than 1/4 inch between them.

Using one of the strips as a guide, run the laminate trimmer, with a flush trimming bit installed, through the joint. Keep the pilot of the bit against the straightedge. Cut the ends of both pieces at the same time to assure making a tight joint (Fig. 87-10). Bond the sheets as previously described. Apply *seam-filling compound,* especially made for laminates, to make a practically invisible joint. Wipe off excess compound with the recommended solvent.

HELPFUL HINTS

How to make a tight joint between the ends of two lengths of laminate.

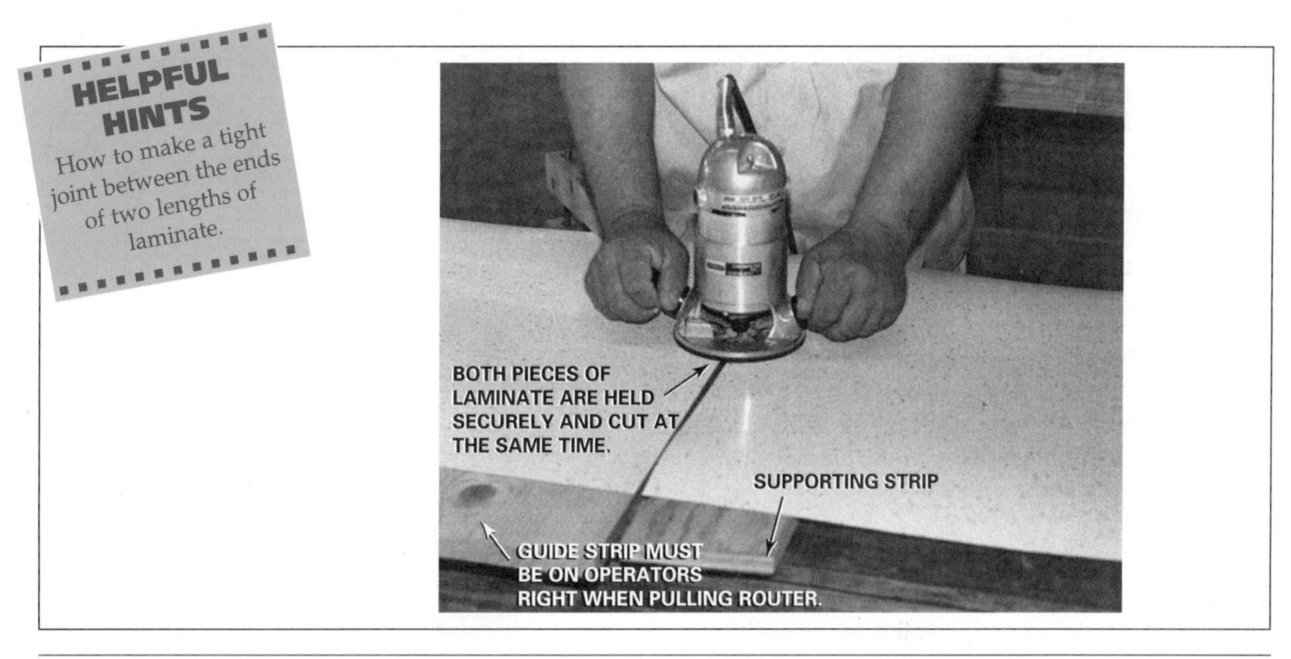

Fig. 87-10 Making a tight laminate butt seam.

Laminating Backsplashes

Backsplashes are laminated in the same manner as countertops. Laminate the backsplash. Then reattach it to the countertop with the same screws. Use a little caulking compound between the backsplash and countertop. This prevents any water from seeping through the joint (Fig. 87-11).

Laminating Rounded Corners

If the edge of a countertop has a rounded corner, the laminate can be bent. Strips of laminate can be cold bent to a minimum radius of about 6 inches. Heating the laminate to 325 degrees Fahrenheit uniformly over the entire bend will facilitate bending to a minimum radius of about 2 1/2 inches. Heat the laminate carefully with a *heat gun.* Bend it until the desired radius is obtained (Fig. 87-12). Experimentation may be necessary until success in bending is achieved.

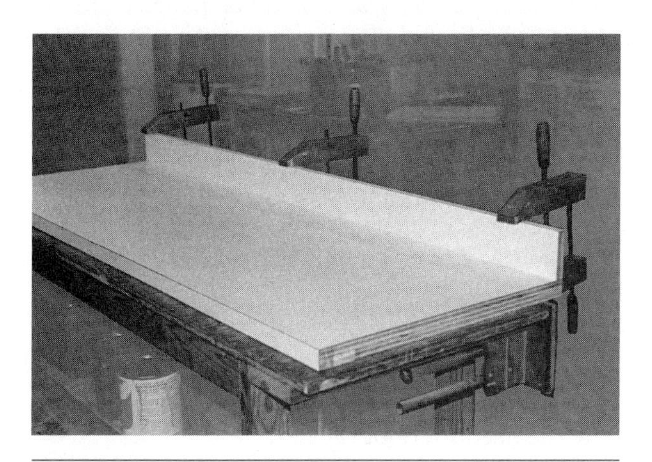

Fig. 87-11 Apply the laminate to the backsplash. Then fasten it to the laminated countertop by driving screws through the countertop into its bottom edge.

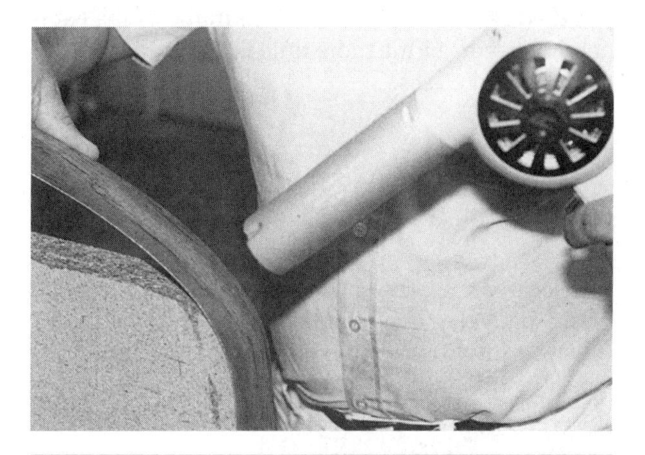

Fig. 87-12 Heating and bending laminate with a heat gun.

CAUTION Keep fingers away from the heated area of the laminate. Remember that the laminate retains heat for some time.

Kinds of Doors

Doors are classified by their construction and also by the method of installation. Sliding doors are occasionally installed, but most cabinets are fitted with hinged doors that swing.

Hinged cabinet doors are classified as *overlay, lipped,* and *flush,* based on the method of installation (Fig. 87-13). The overlay method of hanging cabinet doors is the most widely used.

Overlay Doors

The *overlay* type of door laps over the opening, usually 3/8 inch on all sides. However, it may overlay any amount. In many cases, it may cover the entire face frame. The overlay door is easy to install. it does not require fitting in the opening and the face frame of the cabinet acts as a stop for the door. *European-style* cabinets omit the face frame. Doors completely overlay the front edges of the cabinet (Fig. 87-14).

Lipped Doors

The *lipped* door has rabbeted edges that overlap the opening by about 3/8 inch on all sides. Usually the ends and edges are rounded over to give a more pleasing appearance. Lipped doors and drawers are easy to install. No fitting is required and the rabbeted edges stop against the face frame of the cabinet. However, a little more time is required to shape the rabbeted edges.

Flush Type

The *flush type* door fits into and flush with the face of the opening. They are a little more difficult to hang because they must be fitted in the opening. A fine joint, about the thickness of a dime, must be made between the opening and the door. Stops must be provided in the cabinet against which to close the door.

Door Construction

Doors are also classified, by their construction, as *solid* or *paneled.* Solid doors are made of plywood,

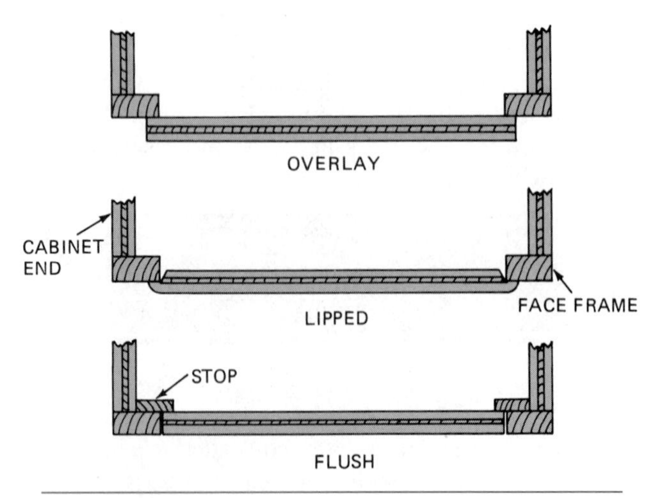

Fig. 87-13 Plan views of overlay, lipped, and flush doors.

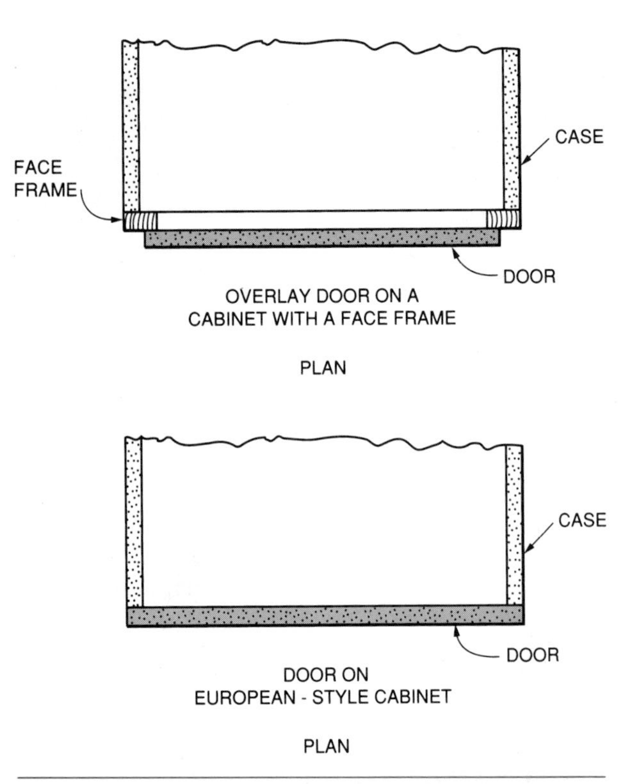

Fig. 87-14 Overlay doors lap the face frame by varying amounts. European-style doors are hinged to and completely overlay the case.

particleboard, or solid lumber. Particleboard doors are ordinarily covered with plastic laminate. Matched boards with V-grooves and other designs, such as used for wall paneling, are often used to make solid doors (Fig. 87-15). Designs may be grooved into the face of the door with a router. Small moldings may be applied for a more attractive appearance.

Paneled doors have an exterior framework of solid wood with panels of solid wood, plywood, hardboard, plastic, glass, or other panel material. Many complicated designs are manufactured by millworkers with specialized equipment. With the equipment available, carpenters can make paneled doors of simple design only (Fig. 87-16). Both solid doors and paneled doors may be hinged in overlay, lipped, or flush fashion.

Types of Hinges

Several types of cabinet hinges are *surface, offset, overlay, pivot,* and *butt.* For each type there are many styles and finishes (Fig. 87-17). Some types are *self-closing* hinges that hold the door closed and eliminate the need for door catches.

Surface Hinges

Surface hinges are applied to the exterior surface of the door and frame. The back side of the hinge

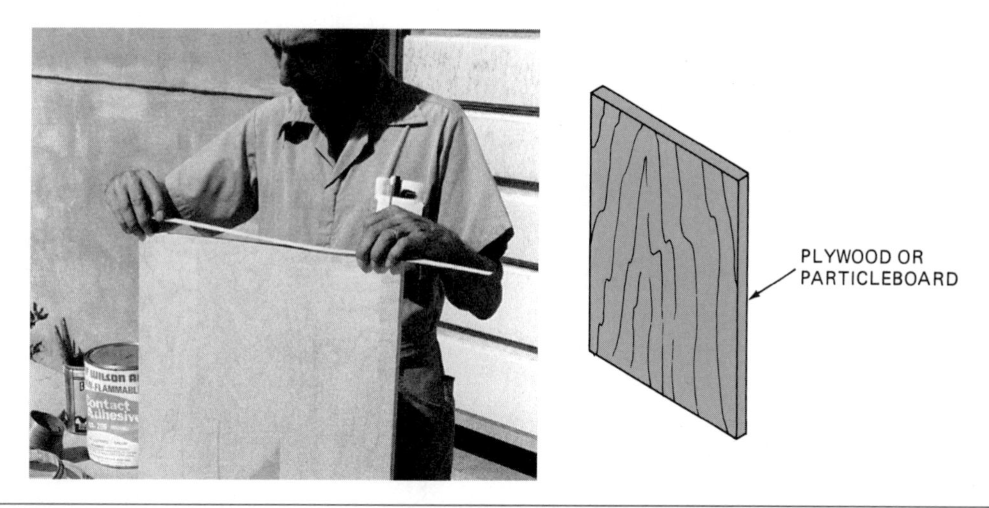

Fig. 87-15 Solid doors usually are made of plywood or particleboard.

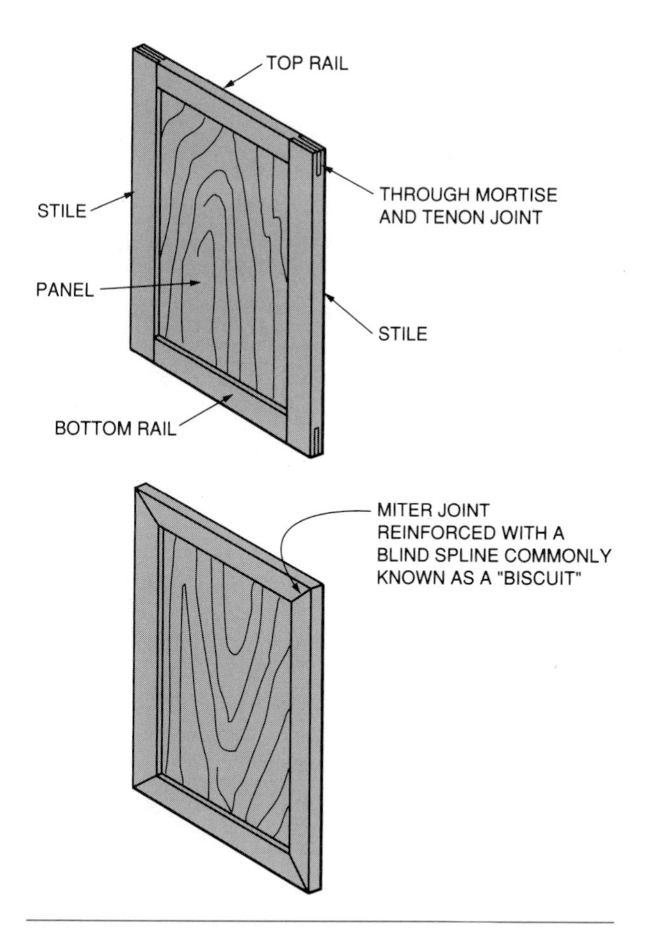

Fig. 87-16 Panel doors of simple design can be made on the job.

leaves may lie in a straight line for flush doors. One leaf may be *offset* for lipped doors (Fig. 87-18). The surface type is used when it is desired to expose the hardware, as in the case of wrought iron and other decorative hinges.

Offset Hinges

Offset hinges are used on lipped doors. They are called *offset surface* hinges when both leaves are fastened to outside surfaces. The *semiconcealed offset* hinge has one leaf bent to a 3/8-inch offset that is screwed to the back of the door. The other leaf screws to the exterior surface of the face frame. A *concealed offset* type is designed in which only the pin is exposed when the door is closed (Fig. 87-19).

Overlay Hinges

Overlay hinges are available in *semiconcealed* and *concealed* types. With semiconcealed types, the amount of overlay is variable. Certain concealed overlay hinges are made for a specific amount of overlay, such as 1/4, 5/16, 3/8, and 1/2 inch. European-style hinges are completely concealed. They are not usually installed by the carpenter because of the equipment needed to bore the holes to receive the hinge. Some overlay hinges, with one leaf bent at a 30-degree angle, are used on doors with reverse beveled edges (Fig. 87-20).

Pivot Hinges

Pivot hinges are usually used on overlay doors. They are fastened to the top and bottom of the door and to the inside of the case. They are frequently used when there is no face frame and the door completely covers the face of the case (Fig. 87-21).

Butt Hinges

Butt hinges are used on flush doors. Butt hinges for cabinet doors are a smaller version of those used on entrance doors. The leaves of the hinge are set into **gains** in the edges of the frame and the door, in the

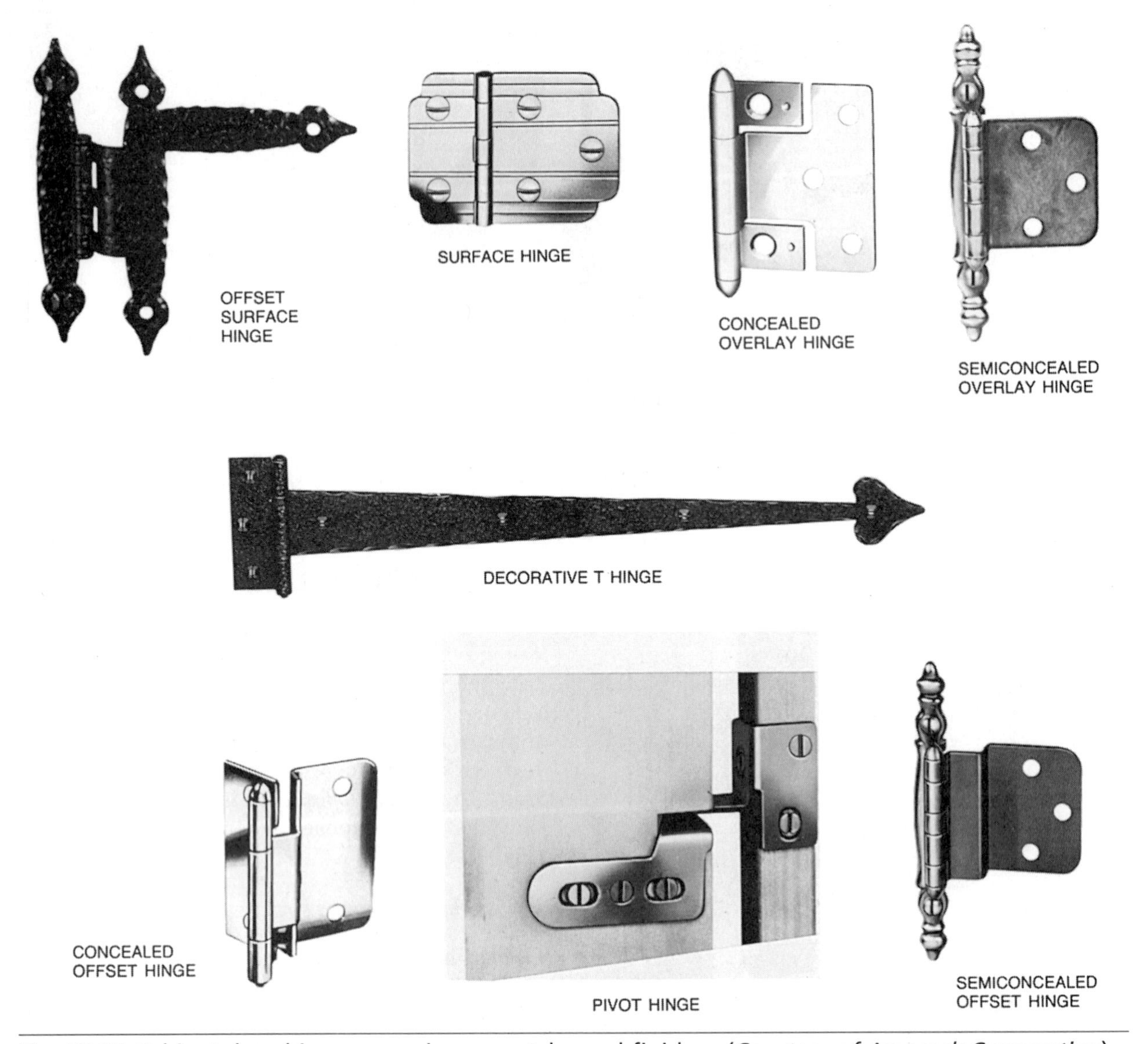

Fig. 87-17 Cabinet door hinges come in many styles and finishes. (*Courtesy of Amerock Corporation*)

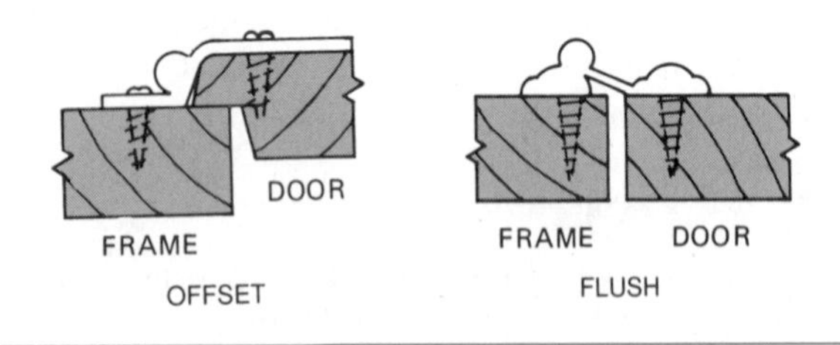

Fig. 87-18 Surface hinges.

same manner as for entrance doors. Butt hinges are used on flush doors when it is desired to conceal most of the hardware. They are not often used on cabinets because they take more time to install than other types (Fig. 87-22).

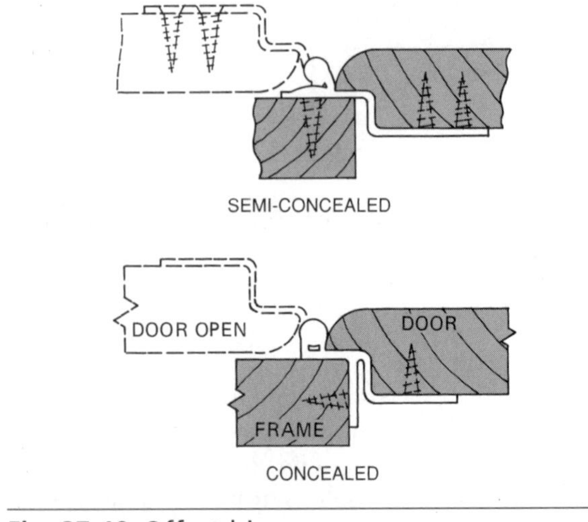

Fig. 87-19 Offset hinges.

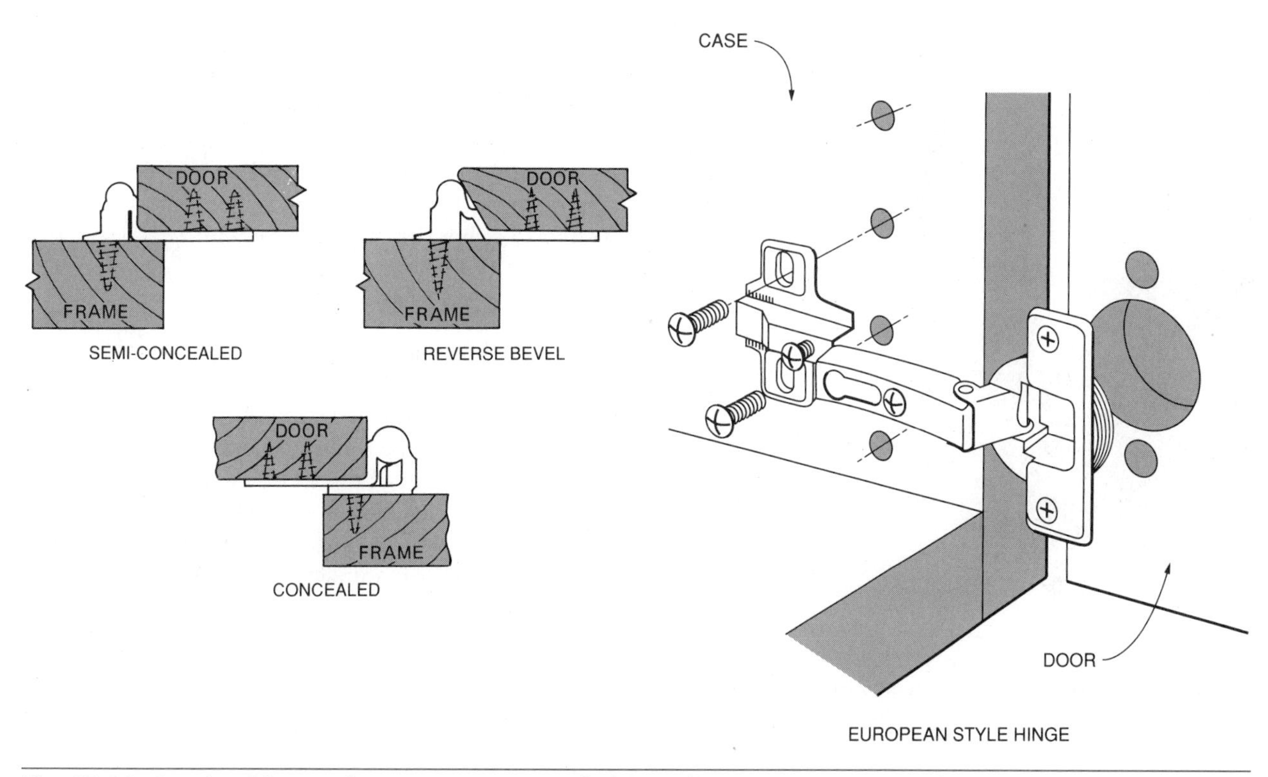

Fig. 87-20 Overlay hinges. (Note: European-style hinge line-art is Courtesy of Hettich America L.P.)

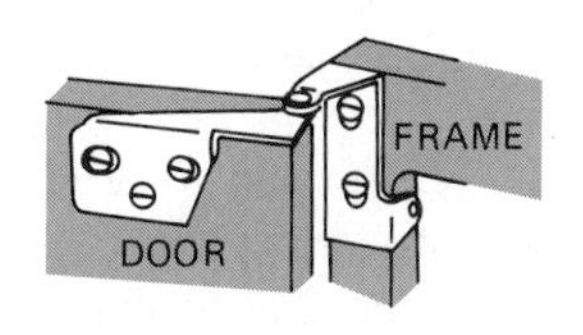

Fig. 87-21 Pivot hinges for an overlay door.

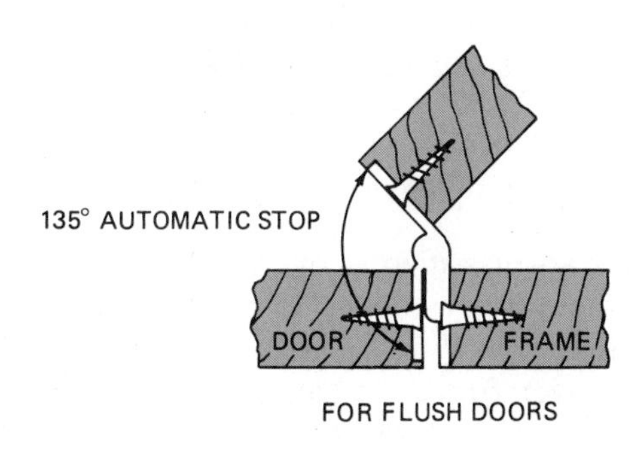

Fig. 87-22 Butt hinges.

Hanging Cabinet Doors

Surface Hinges

To hang cabinet doors with surface hinges, first apply the hinges to the door. Then shim the door in the opening so an even joint is obtained all around. Screw the hinges to the face frame.

Semiconcealed Hinges

For semiconcealed hinges, screw the hinges to the back of the door. Then center the door in the opening. Fasten the hinges to the face frame. When more than one door is to be installed side by side, clamp a straightedge to the face frame along the bottom of the openings for the full length of the cabinet. Rest the doors on the straightedge to keep them in line (Fig. 87-23).

Concealed Hinges

When installing concealed hinges, first screw the hinges on the door. Center the door in the opening.

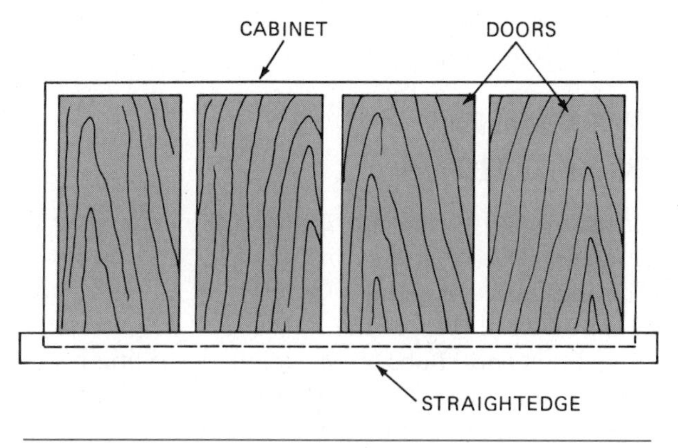

Fig. 87-23 When installing doors, use a straightedge to keep them in line.

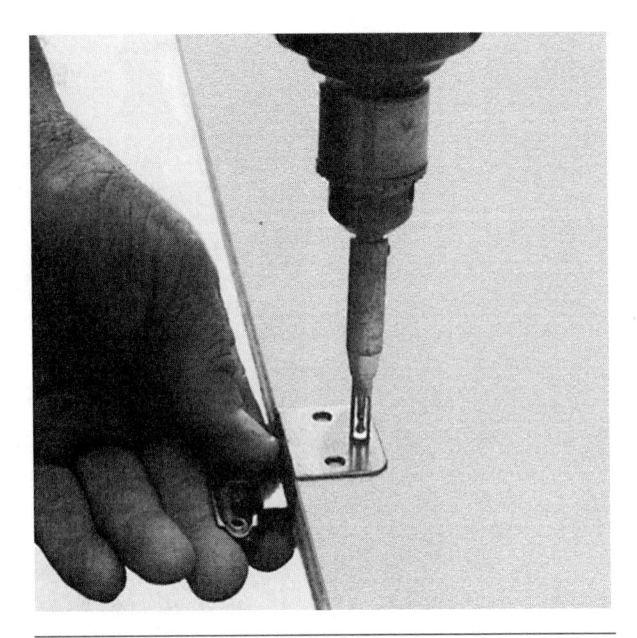

Fig. 87-24 The VIX bit is a self-centering drill stop used for drilling holes for cabinet hinges.

Press or tap on the hinge opposite the face frame. Small projections on the hinge make indentations to mark its location on the face frame. Open the door. Place the projections of the hinges into the indentations. Screw the hinges to the face frame.

Butt Hinges

Hanging flush cabinet doors with butt hinges is done in the same manner as hanging entrance doors. Drill pilot holes for all screws so they are centered on the holes in the hinge leaf. Drilling the holes off center throws the hinge to one side when the screws are driven. This usually causes the door to be out of alignment when hung. Many carpenters use a a self-centering tool, called a *VIX bit,* when drilling pilot holes for screw fastening of cabinet door hinges of all types (Fig. 87-24). The tool centers a twist drill on the hinge leaf screw hole. It also stops at a set depth to prevent drilling through the door or face frame.

Installing Pulls and Knobs

Cabinet *pulls* or *knobs* are used on cabinet doors and drawers. They come in many styles and designs. They are made of metal, plastic, wood, porcelain, or other material (Fig. 87-25).

Pulls and knobs are installed by drilling holes through the door. Then fasten them with machine screws from the inside. When two screws are used to fasten a pull, the holes are drilled slightly oversize in case they are a little off center. This allows the pulls to be fastened easily without cross-threading the screws. Usually 3/16-inch diameter holes are drilled for 1/8-inch machine screws. To drill holes quickly and accurately, make a *template* from scrap wood that fits over the door. The template can be made so that holes can be drilled for doors that swing in either direction (Fig. 87-26).

Door Catches

Doors without self-closing hinges need *catches* to hold them closed. There are many kinds of catches available (Fig. 87-27). Catches should be placed where they are not in the way, such as on the bottom of shelves, instead of the top.

Magnetic catches are widely used. They are available with single or double magnets of varying holding power. An adjustable magnet is attached to the inside of the case. A metal plate is attached to the door. First attach the magnet. Then place the plate on the magnet. Close the door and tap it opposite the plate. Projections on the plate mark its location on the door. Attach the plate to the door where marked. Try the door. Adjust the magnet, if necessary.

Friction catches are installed in a similar manner to that used for magnetic catches. Fasten the adjustable section to the case and the other section to the door.

Elbow catches are used to hold one door of a double set. They are released by reaching to the back side of the door. These catches are usually used when one of the doors is locked against the other.

Bullet catches are spring loaded. They fit into the edge of the door. When the door is closed, the catch fits into a recessed plate mounted on the frame.

Drawer Construction

Drawers are classified as overlay, lipped, and flush in the same way as doors. In a cabinet installation, the drawer type should match the door type.

Drawer fronts are generally made from the same material as the cabinet doors. Drawer sides and backs are generally 1/2 inch thick. They may be made of solid lumber, plywood, or particleboard.

Medium-density fiberboard with a printed wood grain is also manufactured for use as drawer sides and backs. The drawer bottom is usually made of 1/4-inch plywood or hardboard. Small drawers may have 1/8-inch hardboard bottoms.

Drawer Joints

Typical joints between the front and sides of drawers are the *dovetail, lock,* and *rabbet* joints. The dove-

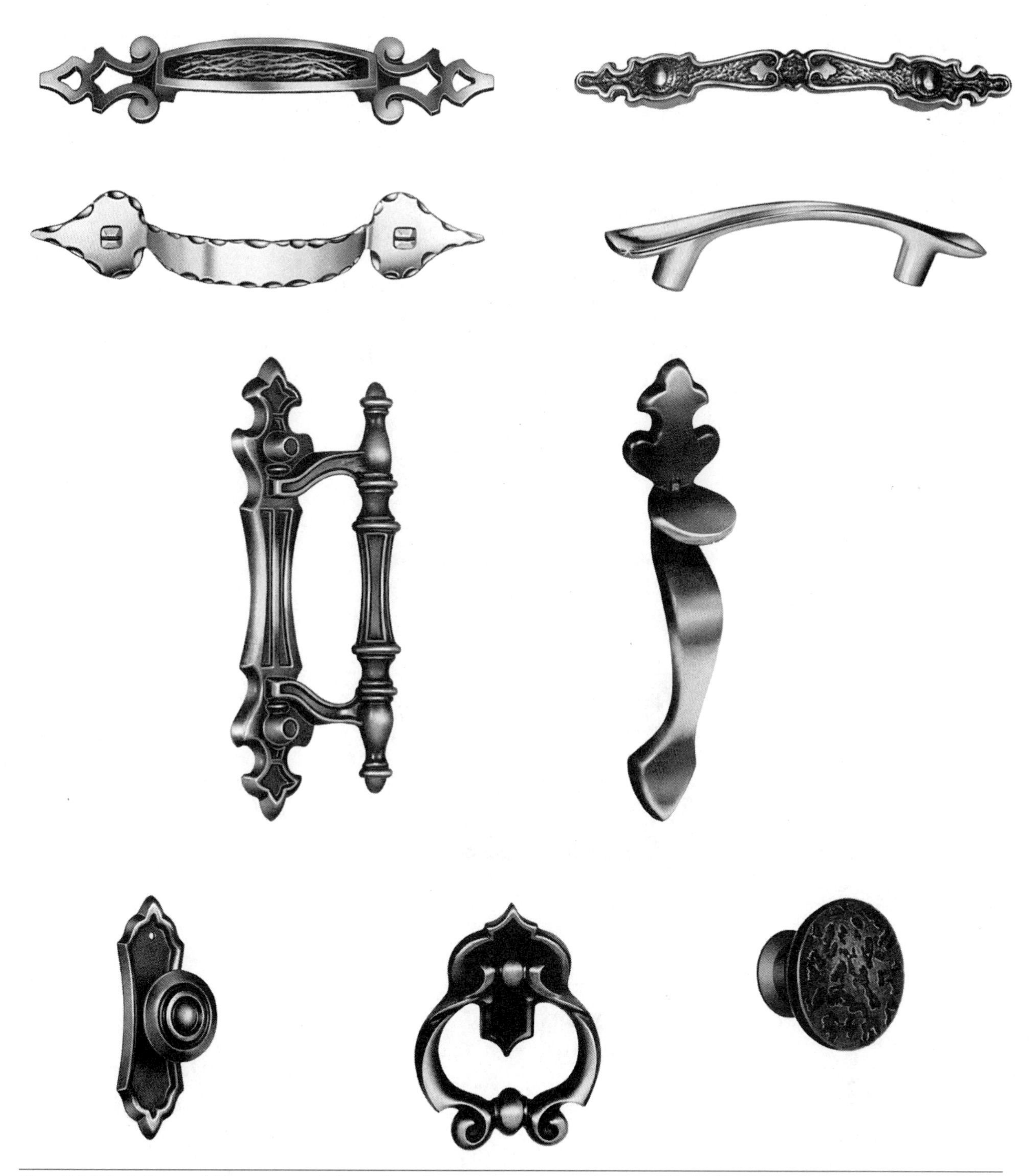

Fig. 87-25 A few of the many styles of pulls and knobs used on cabinet doors and drawers. (*Courtesy of Amerock Corporation*)

tail joint is used in higher-quality drawer construction. It takes a longer time to make, but is the strongest. Dovetail drawer joints may be made using a router and a dovetail template (Fig. 87-28). The lock joint is simpler. It can be easily made using a table saw. The rabbet joint is the easiest to make. However, it must be strengthened with fasteners in addition to glue (Fig. 87-29).

Joints normally used between the sides and back are the *dovetail, dado and rabbet, dado,* and *butt* joints. With the exception of the dovetail joint, the drawer back is usually set in at least 1/2 inch from the back ends of the sides to provide added strength. This helps prevent the drawer back from being pulled off if the contents get stuck while opening the drawer (Fig. 87-30).

HELPFUL HINTS
Use a template when drilling holes for cabinet door pulls.

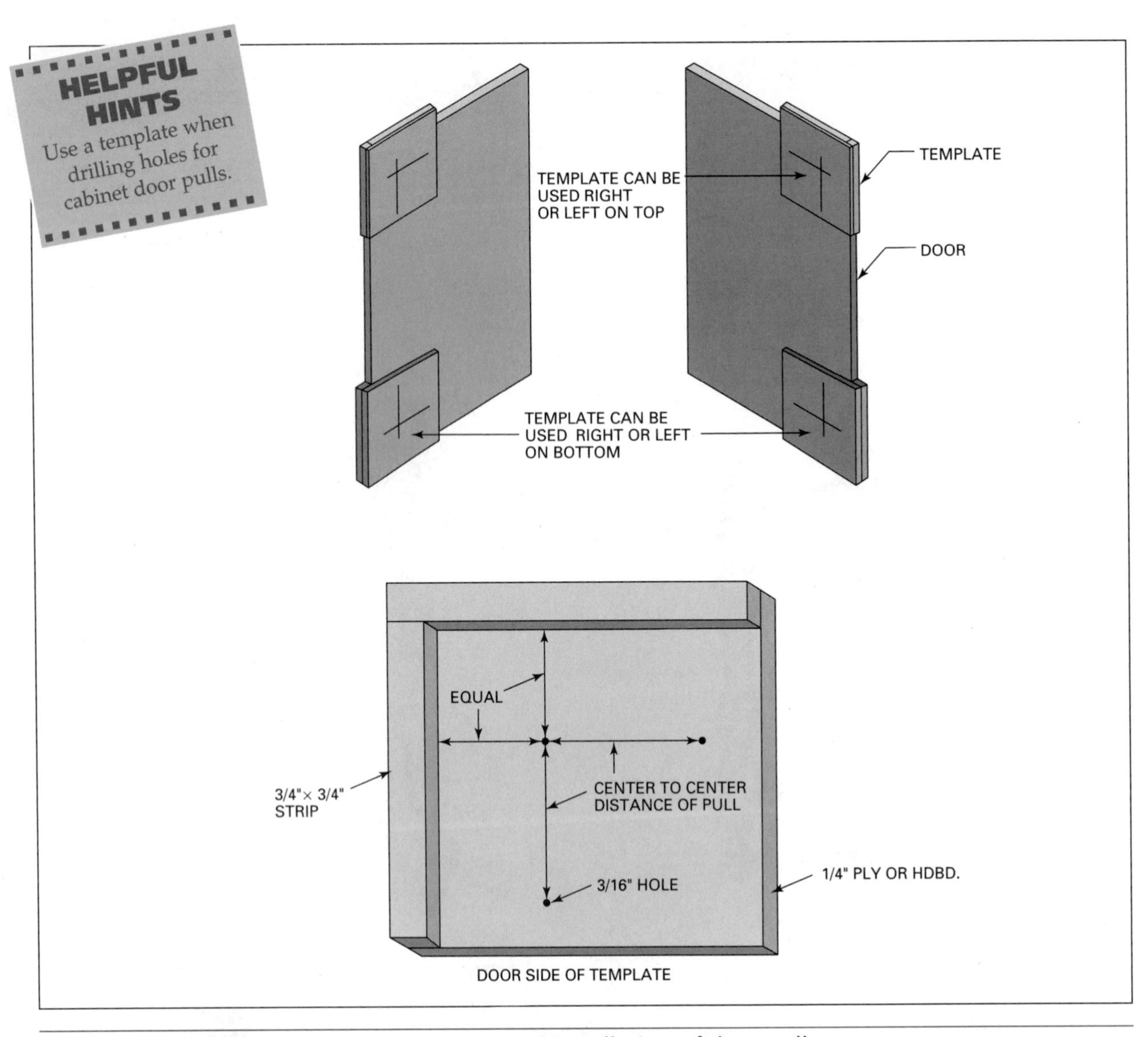

Fig. 87-26 Techniques for making a jig to speed installation of door pulls.

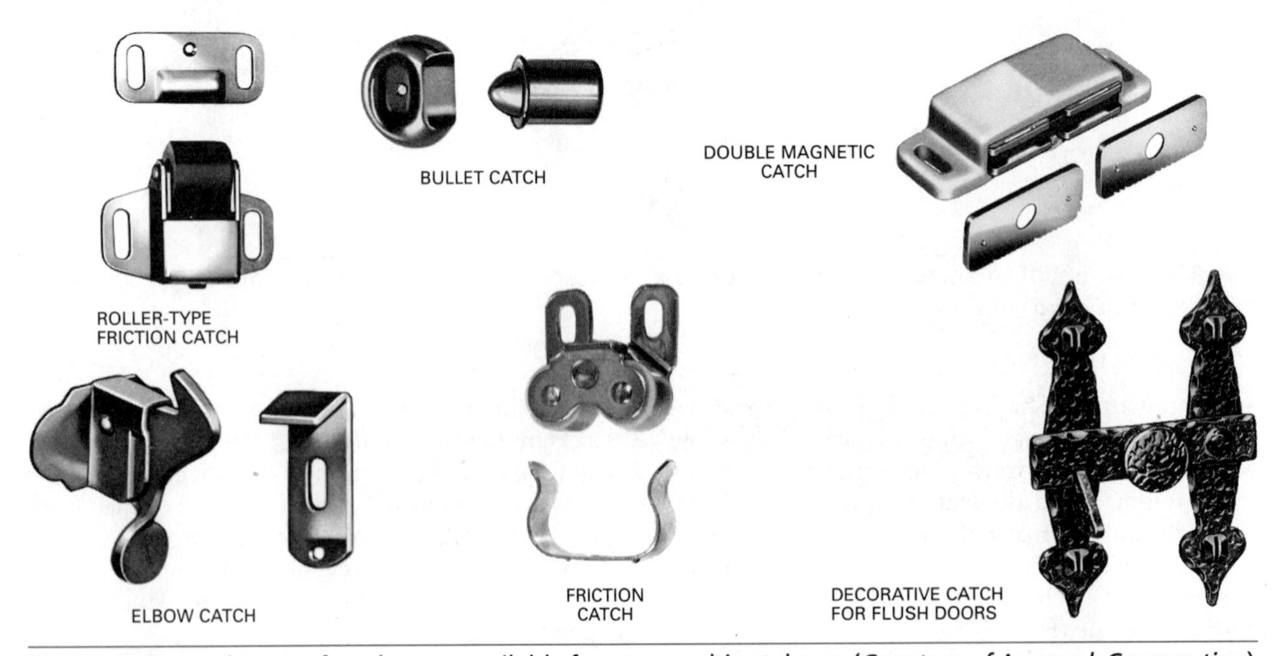

Fig. 87-27 Several types of catches are available for use on cabinet doors. (*Courtesy of Amerock Corporation*)

Fig. 87-28 Dovetail joints can be made with a router and a dovetail template.

Drawer Bottom Joints

The drawer bottom is fitted into a groove on all four sides of the drawer (Fig. 87-31). In some cases, the drawer back is made narrower, the four sides assembled, the bottom slipped in the groove, and its back edge fastened to the bottom edge of the drawer back (Fig. 87-32).

Drawer Guides

There are many ways of guiding drawers (Fig. 87-33). The type of drawer guide selected affects the size of the drawer. The drawer must be supported level and guided sideways. It must also be kept from tilting down when opened.

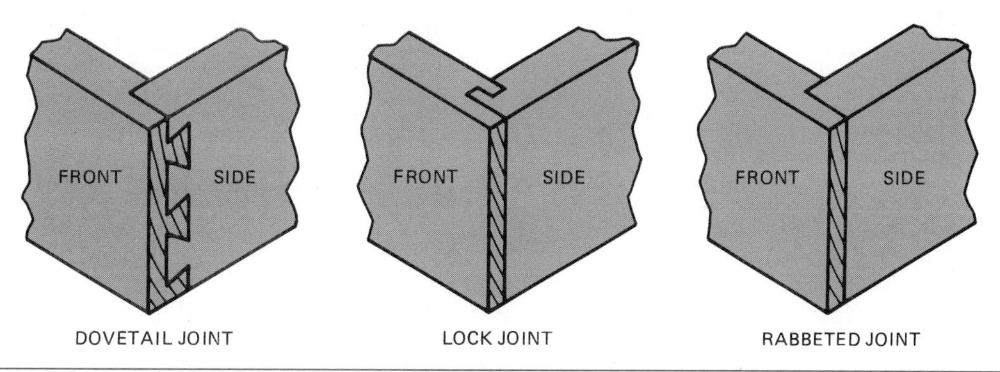

Fig. 87-29 Typical joints between drawer front and side.

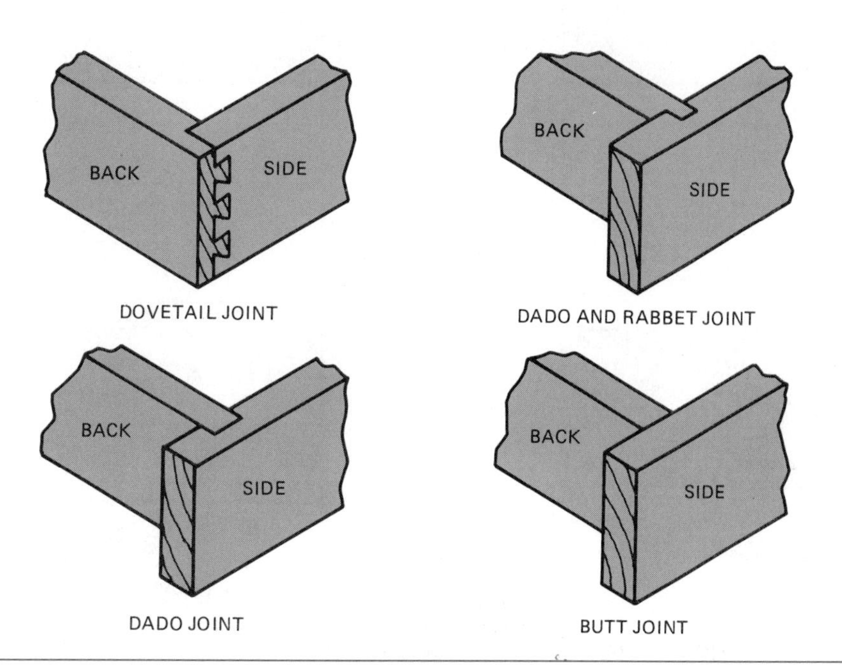

Fig. 87-30 Typical joints between drawer back and side.

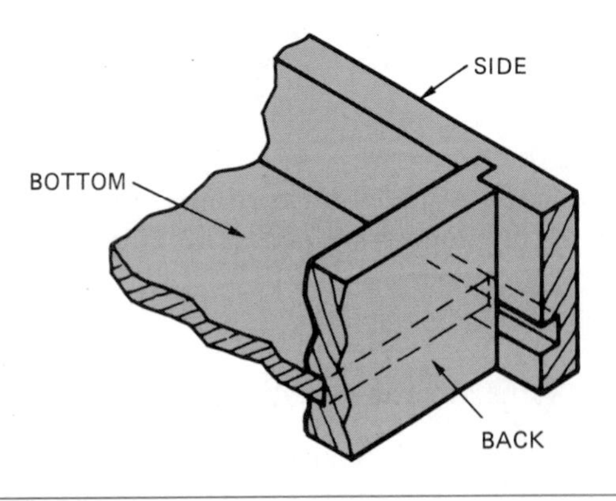

Fig. 87-31 Drawer bottom fitted in groove at drawer back.

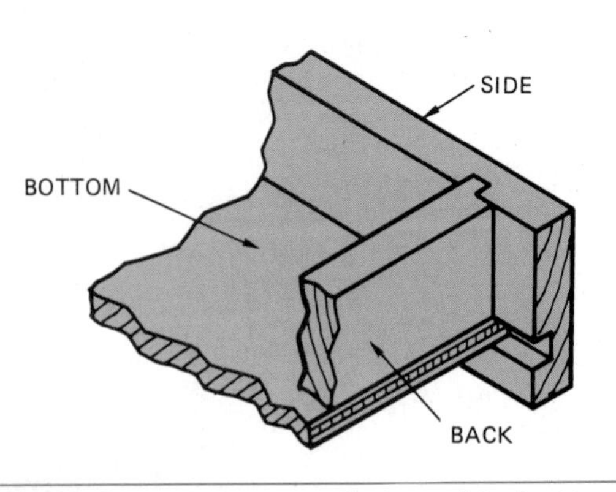

Fig. 87-32 Drawer bottom fastened to bottom edge of drawer back.

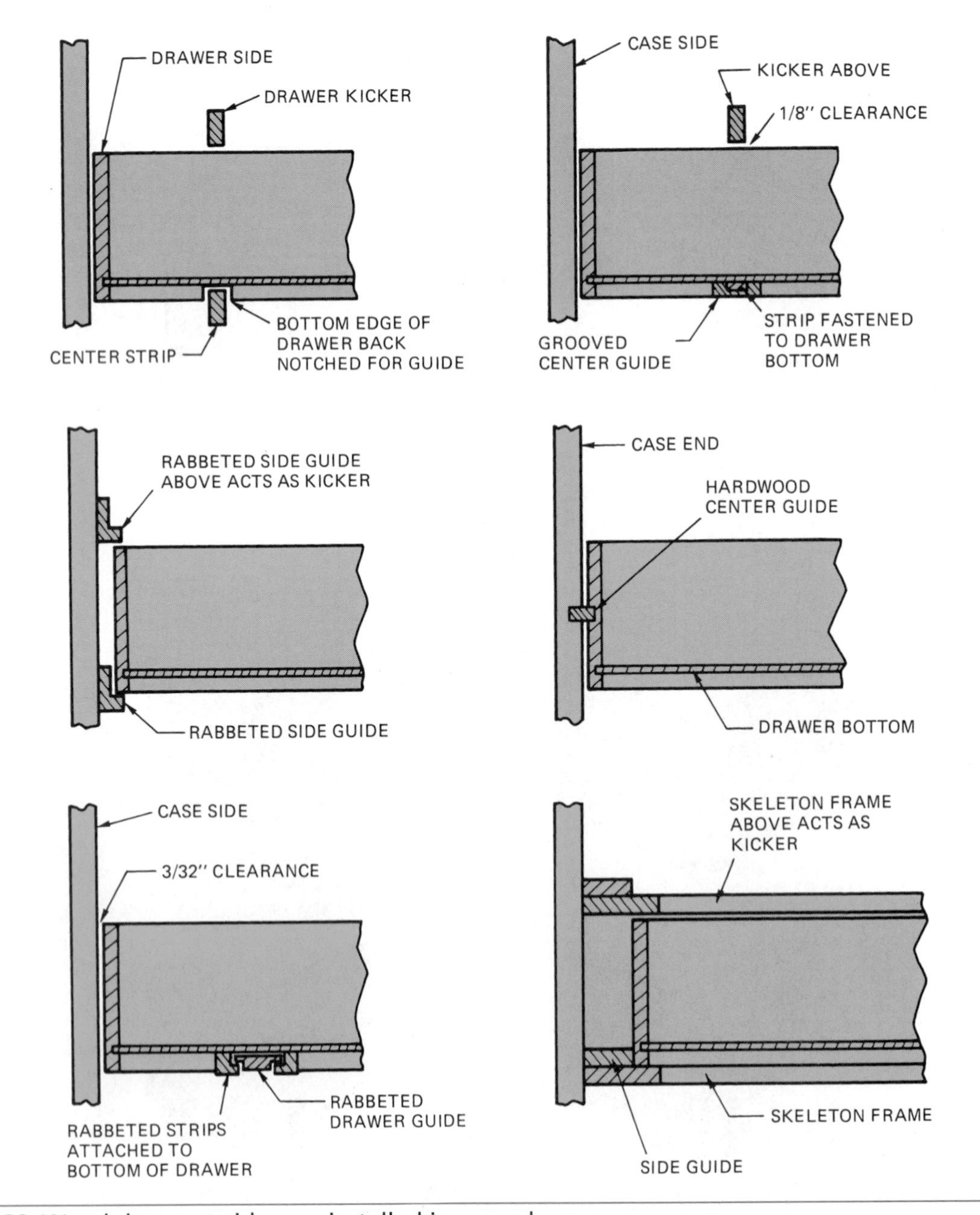

Fig. 87-33 Wood drawer guides are installed in several ways.

Wood Guides

Probably the simplest wood guide is the center strip. It is installed in the bottom center of the opening from front to back (Fig. 87-34). The strip projects above the bottom of the opening about 1/4 inch. The bottom edge of the drawer back is notched to ride in the guide. A *kicker* is installed. It is centered above the drawer to keep it from tilting downward when opened.

Another type of wood guide is the grooved center strip (Fig. 87-35). The strip is placed in the center of the opening from front to back. A matching strip is fastened to the drawer bottom. In addition to guiding the drawer, this system keeps it from tilting when opened, eliminating the need for drawer kickers.

Another type of wood guide is a rabbeted strip. Strips are used on each side of the drawer opening (Fig. 87-36). The drawer sides fit into and slide along the rabbeted pieces. Sometimes these guides are made up of two pieces instead of rabbeting one piece. A kicker above the drawer is necessary with this type guide.

Fig. 87-34 Simple center wood drawer guide. The back of the drawer is notched to run on the glide.

Fig. 87-35 The grooved center wood drawer guide eliminates the need of a kicker.

Metal Drawer Guides

There are many different types of metal drawer guides. Some have a single track mounted on the bottom center of the opening. Others may be centered above or on each side of the drawer. Nylon rollers mounted on the drawer ride in the track of the guide (Fig. 87-37).

Instructions for installation differ with each type and manufacturer. When using commercially made drawer guides, read the instructions first, before making the drawer, so proper allowances for the drawer guide can be made.

Fig. 87-36 Rabbeted wood guides are installed on each side of the drawer.

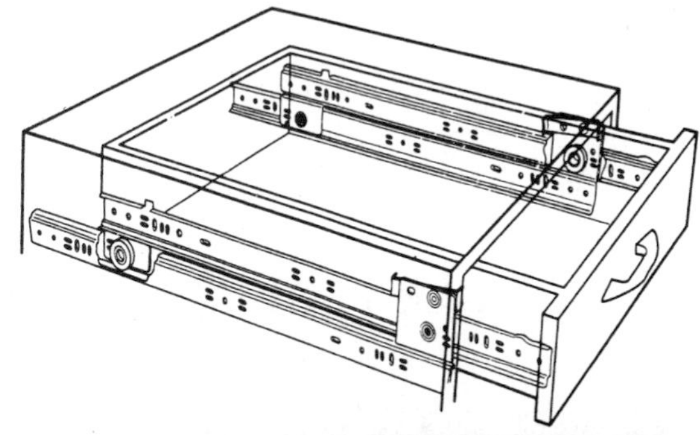

Fig. 87-37 Installing metal drawer guides. (*Courtesy of Knape and Vogt Mfg. Co.*)

Review Questions

Select the most appropriate answer.

1. The vertical distance between the base unit and a wall unit is usually
a. 12 inches. c. 18 inches.
b. 15 inches. d. 24 inches.

2. The distance from the floor to the surface of the countertop is
a. 30 inches. c. 36 inches.
b. 32 inches. d. 42 inches.

3. In order to accommodate sinks and provide adequate working space, the width of the countertop is
a. 25 inches. c. 30 inches.
b. 28 inches. d. 32 inches.

4. Standard wall cabinet height is
a. 24 inches. c. 32 inches.
b. 30 inches. d. 36 inches.

5. The height of most manufactured base kitchen cabinets is
a. 30 3/4 inches. c. 34 1/2 inches.
b. 32 1/2 inches. d. 35 1/4 inches.

6. A drawer front or door with its edges and ends rabbeted to fit over the opening is called
a. an overlay type. c. a lipped type.
b. a flush type. d. a rabbeted type.

7. The offset hinge is used on
a. paneled doors. c. lipped doors.
b. flush doors. d. overlay doors.

8. Butt hinges are used on
a. flush doors. c. overlay doors.
b. lipped doors. d. solid doors.

9. The joint used on high-quality drawers is the
a. dado joint.
b. dado and rabbet joint.
c. dovetail joint.
d. rabbeted joint.

10. A wood strip installed in the cabinet to prevent a drawer from tilting downward when opened is called a
a. top guide. c. sleeper.
b. kicker. d. tilt strip.

BUILDING FOR SUCCESS 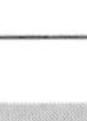

Demonstrating Quality Workmanship

The demonstration of quality workmanship begins with two basic premises: one involving need and one involving desire. First, each skilled worker must accept and understand the need to do quality work. Skill competencies cannot be reached without a commitment to quality. Second, each person has to want to do quality work. Without meeting these two requirements, quality workmanship cannot be accomplished.

Because customers expect quality materials and workmanship and are willing to pay for them, they will not settle for anything less. By looking at successful building and remodeling companies, quality work will surface as a trademark. With those companies quality will become a product. Therefore, all workers associated with those companies are expected to deliver quality. The mind-set for each person has to be a total buy-in to the quality concept.

How is quality to be defined? What are the characteristics of quality? In some cases, definitions of quality work will differ. Its definition is often based on expectations. At the beginning of each construction project, written statements of quality in materials, features, and workmanship must relate to expectations held by all parties. This in turn relates to satisfaction. Therefore it becomes imperative to agree on quality materials and workmanship before the construction process begins.

There are quality standard guidelines available for both new construction and remodeling. For example, *the Quality Standards for the Professional Remodeler*, published by the NAHB Remodeler's Council, has established good descriptions of industry quality standards. Other ideal publications by the NAHB Home Builder Bookstore are: *Production Checklist for Builders and Superintendents, Customer Service for Home Builders, The Positive Walk Through: Your Blueprint for Success, and Warranty Service for Builders and Remodelers.*

Typically, these standards manuals and guides present criteria, written specifications, and suggestions that help each builder or remodeler provide quality products, workmanship, and service. The quality standards presented in these publications define measureable criteria to be used throughout the construction process. These manuals serve well in resolving complaints and disputes between the builder or remodeler and the client. They are written in layperson's terms that are easily understood by everyone.

Written materials that explain quality workmanship and construction procedures set the stage for credibility in relationships in a construction project. Written standards that are presented at the beginning of the construction discussions reassure clients that they will receive the quality they are paying for. With that assurance, clients will be more trusting and willing to cooperate fully with the builder or remodeler.

Quality standards, along with construction building codes, can be used by the builder or remodeler as the very minimum in quality to be provided throughout the building process. The goal should be to meet or exceed these minimums during the entire construction process.

The minimum quality standards adopted by a company also become the basic guidelines for training employees in company expectations. Each prospective employee should be exposed to these quality standards before actual employment. This will give everyone a chance to see what is expected on the job. At that time the potential employee can see the company philosophy, the commitment to quality workmanship, and how it is provided to the customer. This preview of expectations will undoubtedly increase each employee's chances for success.

By involving all employees in developing and using quality standards, a company can better promote and defend its commitment to excellence. Each skilled craftsperson should desire to protect his or her image of doing quality work. Communicating in writing what quality workmanship is and what each client can expect from the skilled workers should be presented in the early stages of the project. The commitment by the builder or remodeler to meet or exceed those standards will be satisfying to the client.

Competition will never allow the skilled worker to abandon the need to demonstrate quality workmanship. Each worker must develop good technical skills and gain a good understanding of the construction processes used in building. Obtaining and maintaining quality tools will help install quality materials. People are willing to pay for these things and more if they are satisfied the results will be the best possible. Customer satisfaction will remain one of the primary gauges for determining the level of quality workmanship and service being provided by a construction company. The margin for error is small and the tolerance level for poor workmanship will continue to decrease. Each person entering the construction industry today will have to buy into the quality workmanship concept to be successful.

FOCUS QUESTIONS: For individual or group discussion

1. What criteria are necessary in order for a carpenter to demonstrate quality workmanship? Explain how these criteria will be observable to the client.
2. Is there a correlation between a carpenter's attitude toward quality workmanship and customer satisfaction? Explain your answer in detail.
3. What would be some advantages of starting the degree of quality materials and workmanship before a contract is agreed upon?
4. How could meeting or exceeding the quality standards and workmanship stated in the agreement prove to be a positive long-term strategy for the builder? Explain how this might affect customer satisfaction.

STUDENT ACTIVITIES for Section 4: Interior Finish

Unit 23 Drywall Construction

1. Explain the drywall construction process. Include the product types and descriptions, tools used, cutting, fasteners, mastics, taping procedures, trims, moldings, and accessories.
2. Demonstrate the estimating, measuring, cutting, hanging (nailing and screwing) process, and taping and sanding of drywall products.
3. Demonstrate the wall preparation and application of textures to drywall surfaces.
4. Using a set of blueprints, estimate the quantity and cost of the drywall needed for the interior walls and ceilings of the house.

Unit 24 Wall Paneling and Wall Tile

1. Create a display board that illustrates a variety of interior wall paneling, moldings, mastics, and fasteners.
2. Using a set of blueprints, estimate the necessary quantity of paneling needed to cover the walls of three rooms.
3. Explain and demonstrate the procedure for installing solid lumber paneling. Include furring strips, scribing, measuring, matching, cutting, and adhering with nails and mastic.
4. Explain and demonstrate the process of installing wainscoting on walls (less than full height). Include the use of moldings and trim.
5. Explain and demonstrate the installation of plastic laminate products. Include the measuring, cutting, fitting, and adhering processes.
6. Explain and demonstrate the process of installing ceramic tile and accessories on both walls and floors. Include the measuring, cutting, adhering, and grouting processes.
7. Using a blueprint, estimate the ceramic tile needed for the bathroom and kitchen.

Unit 25 Ceiling Finish

1. Explain the variety of ceiling tile and panel products used to finish interior ceilings. Include the product names, structure, accessories, sizes, fastening methods, and materials cost estimating.
2. Demonstrate the ceiling preparation, room squaring, applying furring strips, and the installation of acoustical ceiling tile.
3. Explain the process of installing a suspended ceiling in a room. Include the room preparation, squaring, leveling, installing wall angles, runners, cross tees, ceiling panels, and materials cost estimating.
4. Repeat the process with the concealed grid suspended ceilings.

Unit 26 Interior Doors and Door Frames

1. Describe the types of interior doors and jambs, their structure, operating functions, and the method of determining the hand of door.
2. Demonstrate installing a jamb, trimming and hanging a swinging door, and installing the hardware. Repeat the process for a bifold and bypass door. Use the hinge butt template, portable electric plane, and router.

Unit 27 Interior Trim

1. Identify the types of standard molding patterns and describe their use.
2. Demonstrate the installation of as many of these as possible, including the process of coping and mitering a joint. Use the hand miter box, the power miter box, the table saw, a coping saw, and basic hand tools. Include these moldings: beds, crowns, full rounds, half rounds, quarter rounds, base, base shoes, base caps, corner molds, back bands, aprons, stools, paneling caps, casings, and stops.

Unit 28 Stair Finish

1. Name and show the parts of stair finish. Describe their location and function.
2. Using a 4-foot piece of stringer material, lay out a housed stringer and a job-built (rough) stringer with given dimensions from the instructor.
3. Apply the following finish to a job-built staircase: risers, newel posts, open and closed finish stringers, treads, return nosings, tread moldings, handrail, and balusters.
4. Using a stair manufacturer's catalog, assemble a price list of the previously listed finish parts.

Unit 29 Finish Floors

1. Prepare and present a display of the various kinds and sizes of hardwood finish flooring. Include strip, plank, and parquet products. Explain the installation process for hardwood floors.
2. Prepare a wood or concrete floor to receive either strip, plank, or parquet flooring. Install a hardwood finish floor. Use both a hammer power nailer and mastics if possible. Install base trim.
3. Prepare a floor to receive resilient tiles or sheet products, and install resilient floor tile.

Unit 30 Kitchen Cabinets and Countertops

1. Describe in detail the construction process for the shop-built kitchen cabinet. Illustrate and use examples of the cabinet components, materials, and plans.
2. Using a cabinet plan, install a base and wall cabinet in the shop or on the job. Install and laminate a countertop.
3. Describe in detail the various styles of cabinet doors and drawers. Include the hardware that could be used for each.

GLOSSARY

Acoustical board—material used to control or deaden sound

Actual size—the size of lumber after it has been surfaced

Admixture—material used in concrete or mortar to produce special qualities

Aggregate—materials, such as sand, rock, and gravel, used to make concrete

Air-dried lumber—lumber that has been seasoned by drying in the air

Anchor—a device used to fasten structural members in place

Annular ring—the rings seen when viewing a cross-section of a tree trunk; each ring constitutes one year of tree growth

Apprentice—a beginner who serves for a stated period of time to learn a trade

Apron—a piece of the window trim used under the stool

Arbor—a shaft on which circular saw blades are inserted

Areaway—below-grade, walled area around basement windows

Asphalt felt—a building paper saturated with asphalt for waterproofing

Astragal—a semicircular molding often used to cover a joint between doors

Awning window—a type of window in which the sash are hinged at the top and swing outward

Back band—molding applied around the sides and tops of windows and doors

Back-bevel—a bevel on the edge or end of stock toward the back side

Backing—strips or blocks installed in walls or ceilings for the purpose of fastening or supporting trim or fixtures

Backing the hip—beveling the top edge of a hip rafter to line it up with adjacent roof surfaces

Back miter—an angle cut starting from the end and coming back on the face of the stock

Backset—the distance an object is set back from an edge, side, or end of stock, such as the distance a hinge is set back from the edge of a door

Backsplash—a raised portion on the back edge of a countertop to protect the wall

Balloon frame—a type of frame in which the studs are continuous from foundation to roof

Baluster—vertical members of a stair rail, usually decorative and spaced closely together

Balustrade—the entire stair rail assembly, including handrail, balusters, and posts

Baseboard—finish board used to cover the joint at the intersection of wall and floor

Base cap—a molding applied to the top edge of the baseboard

Base shoe—a molding applied between the baseboard and floor

Batten—a thin, narrow strip usually used to cover joints between vertical boards

Batter board—a temporary framework erected to hold the stretched lines of a building layout

Bay window—a window, usually three-sided, which projects out from the wall line

Bearer—horizontal members of a wood scaffold that support scaffold plank

Bearing partition—an interior wall that supports the floor above

Bedding—filling of mortar, putty, or similar substance used to secure a firm bearing, as in a bed of putty for glass panes

Benchmark—a reference point for determining elevations during the construction of a building

Bevel—the sloping edge or side of a piece with the angle between not a right angle

Bird's mouth—a notch cut in the underside of a rafter to fit on top of the wall plate

Blind joint—a type of joint in which the cuts do not go all the way through

Blind nail—a method of fastening that conceals the nails

Blind stop—part of a window finish applied just inside the exterior casing

Board—lumber usually 8 inches or more in width and less than 2 inches thick

Board foot—a board that measures 1 foot square and 1 inch thick or any equivalent lumber volume

Bow—a type of warp in which the side of lumber is curved from end to end

Box nail—a thin nail with a head, usually coated with a material to increase its holding power

Brad—a thin, short, finishing nail

Break joints—to stagger joints in adjacent rows of sheathing, siding, roofing, flooring, and similar materials

Bridging—diagonal braces or solid wood blocks between floor joists used to distribute the load imposed on the floor

Buck—a rough frame used to form openings in concrete walls

Bullnose—a starting step that has one or both ends rounded

Butt—the joint formed when one square cut piece is placed against another; also, a type of hinge

Cambium layer—a layer just inside the bark of a tree where new cells are formed

Cant strip—a thin strip of wood placed under a piece to tilt the piece at a slant or triangular-shaped ripping used to blunt an inside corner

Carbide-tipped—in reference to cutting tools that have small, extremely hard pieces of carbide steel welded to the tips

Casement window—a type of window in which the sash are hinged at the edge and usually swing outward

Casing—Molding used to trim around doors, windows, and other openings

Caulking—putty-like mastic used to seal cracks and crevices

Centering punch—a tool used to make an indentation at the centerline of holes

Chair rail—molding applied horizontally along the wall to prevent chair backs from marring the wall

Chamfer—an edge or end bevel that does not go all the way across the edge or end

Chase—a channel formed in buildings to run electrical, plumbing, or mechanical lines

Check—lengthwise split in the end or surface of lumber, usually resulting from more rapid drying of the end than the rest of the piece

Cheek cut—a compound miter cut on the end of certain roof rafters

Cleat—a small strip applied to support a shelf or similar piece

Closed grain—wood in which the pores are small and closely spaced

Closed valley—a roof valley in which the roof covering meets in the center of the valley, completely covering the valley

Collar tie—a horizontal member placed close to the ridge at right angles to the plate

Common rafter—extends from the wall plate to the ridge board, where its run is perpendicular to the plate

Compound miter—a bevel cut across the width and also through the thickness of a piece

Concrete—a building material made from portland cement, aggregates, and water

Condensation—when water, in a vapor form, changes to a liquid due to cooling of the air; the resulting drops of water that accumulate on the cool surface

Conductor—a vertical member used to carry water from the gutter downward to the ground; also called *downspout* or *leader*

Coniferous—cone-bearing tree; also known as *evergreen* tree

Contact cement—an adhesive used to bond plastic laminates or other thin material; so called because the bond is made on contact, eliminating the need for clamps

Contour line—lines on a drawing representing a certain elevation of the land

Coped joint—a type of joint between moldings in which the end of one piece is cut to fit the molded surface of the other

Corner bead—metal trim used on exterior corners of walls to trim and reinforce them

Corner boards—boards used to trim corners on the exterior walls of a building

Corner brace—diagonal member of the wall frame used at the corners to stiffen and strengthen the wall

Cornerite—metal lath, cut into strips and bent at right angles, used in interior corners of walls and ceilings, on top of lath to prevent cracks in plaster

Corner post—built-up stud used in the corner of a wall frame

Cornice—the entire finished assembly where the walls of a structure meet the roof; sometimes called the *eaves*

Counterbore—boring a larger hole partway through the stock so that the head of a fastener can be recessed

Countersink—making a flared depression around the top of a hole to receive the head of a flathead screw; also, the tool used to make the depression

Course—a continuous row of building material, such as brick, siding, roofing, or flooring

Cove—a concave-shaped molding

Crawl space—a shallow space below the living quarters of a structure without a basement

Cricket—a small, false roof built behind a chimney or other roof obstacle for the purpose of shedding water

Crook—a type of warp in which the edge of lumber is not straight

Crosscut—a cut made across the grain of lumber

Crown—usually referred to as the high point of the crooked edge of joists, rafters, and other framing members

Cup—a type of warp in which the side of a board is curved from edge to edge

Dado—a cut, partway through, and across the grain of lumber

Deadbolt—door-locking bolt operated by a key from the outside and by a handle or key from the inside

Deadman—a T-shaped wood device used to support ceiling drywall panels and other objects

Deciduous—trees that shed leaves each year

Dew point—temperature at which moisture in the air condenses into drops

Diagonal—at an angle, usually from corner to corner in a straight line

Door stop—molding fastened to door jambs for the door to stop against

Dormer—a structure that projects out from a sloping roof to form another roofed area to provide a surface for the installation of windows

Double-acting—doors that swing in both directions or the hinges used on these doors

Double-hung window—a window in which two sash slide vertically by each other

Dovetail—a type of interlocking joint resembling the shape of a dove's tail

Dowel—hardwood rods of various diameters

Downspout—see *conductor*

Drip—that part of a cornice or a course of horizontal siding that projects below another part; also, a channel cut in the underside of a windowsill that causes water to drop off instead of running back and down the wall

Drip cap—a molding placed on the top of exterior door and window casings for the purpose of shedding water away from the units

Drip edge—metal edging strips placed on roof edges to provide a support for the overhang of the roofing material

Dropping the hip—increasing the depth of the hip rafter seat cut so that the centerline of its top edge will lie in the plane of adjacent roof surfaces

Dry rot—dry, powdery residue of wood left after fungus destruction of wood due to excessive moisture

Drywall—a type of construction usually referred to as the installation of gypsum board

Dry well—gravel or stone-filled excavation for catching water so it can be absorbed into the earth

Duplex nail—a double-headed nail used for temporary fastening such as in the construction of wood scaffolds

Dutch hip—a roof consisting of a partial hip and partial gable roof

Dutchman—an odd-shaped piece usually used to fill or cover an opening

Eased edge—an edge of lumber whose sharp corners have been rounded

Easement—a curved member of a stair handrail. Also, an area of land that cannot be built upon because it provides access to a structure or utilities

Eaves—that part of a roof that extends beyond the sidewall

Edge—the narrow surface of lumber running with the grain

Edge grain—boards in which the annular rings are at or near perpendicular to the face; sometimes called *vertical grain*

Electrolysis—the decomposition of the softer of two unlike metals in contact with each other in the presence of water

Elevation—a drawing in which the height of the structure or object is shown; also, the height of a specific point in relation to another point

End—the extremities of a piece of lumber

Engineered Lumber Products (ELP)—manufactured lumber substitutes, such as wood I-beams, glue-laminated beams, laminated veneer lumber, parallel strand lumber, and laminated strand lumber

Equilibrium moisture content—the point at which the moisture content of wood is equal to the moisture content of the surrounding air

Escutcheon—protective plate covering the knob or key hole in doors

Exposure—the amount that courses of siding or roofing are exposed to the weather

Face—the best-appearing side of a piece of wood or the side that is exposed when installed, such as finish flooring

Face frame—a framework of narrow pieces on the face of a cabinet containing door and drawer openings

Fascia—a vertical member of the cornice finish installed on the tail end of rafters

Feather edge—the edge of material brought down in a long taper to a very thin edge, such as a wood shingle tip

Fence—a guide for ripping lumber on a table saw

Fiber saturation point—the moisture content of wood when the cell cavities are empty but the cell walls are still saturated

Fillet—small strips used to fill a space, such as between balusters in a plowed handrail

Finger joint—joints made in a mill used to join short lengths together to make long lengths

Finish carpentry—that part of the carpentry trade involved with the application of exterior and interior finish

Finish nail—a thin nail with a small head designed for setting below the surface of finish material

Finish stringer—the finish board running with the slope of the stairs and covering the joint between the stairs and the wall; also called a *skirt board*

Firecut—an angle cut made on the ends of floor joists bearing in a masonry wall designed to prevent the masonry wall from toppling in case the joists are burned through and collapse

Firestop—material used to fill air passages in a frame to prevent the spread of fire; in a wood frame, might consist of 2 × 4 blocking between studs

Firsts and Seconds—the best grade of hardwood lumber

Fissured—irregular shaped grooves made in material, such as ceiling tile,for acoustical purposes

Flashing—material used at intersections such as roof valleys, dormers, and above windows and doors to prevent the entrance of water

Flat grain—grain in which the annular rings of lumber lie close to parallel to the sides; opposite of edge grain

Flute—concave groove in lumber; usually a number are used closely spaced for decorative purposes as in a column, post, or pilaster

Footing—a foundation for a column, wall, or chimney made wider than the object it supports, to distribute the weight over a greater area

Foundation—that part of a wall on which the major portion of the structure is erected

Frieze—a part of the exterior finish applied at the intersection of a overhanging cornice and the wall

Frost line—the depth to which frost penetrates into the ground in a particular area; footings must be placed below this depth

Furring strip—strips of lumber spaced at desired intervals for the attachment of wall or ceiling covering

Gable end—the triangular-shaped section on the end of a building formed by the rafters in a common or gable roof and the top plate line

Gable roof—a type of roof that pitches in two directions

Gain—a cutout made in a piece to receive another piece, such as a cutout for a butt hinge

Galvanized—protected from rusting by a coating of zinc

Gambrel roof—a type of roof that has two slopes of different pitches on each side of center

Girder—a heavy timber or beam used to support vertical loads

Glass bead—small molding used to hold lights of glass in place

Glaze—to install glass in a frame

Glazier—a person who installs glass in a frame

Glazing—the act of installing glass in a frame

Glazing compound—a soft, plastic-type material, similar to putty, used for sealing lights of glass in a frame

Glazing points—small, triangular, or diamond-shaped pieces of metal used to secure and hold lights of glass in a frame

Glue-laminated lumber (glulam)—large beams or columns made by gluing smaller-dimension lumber together side to side

Gooseneck—a curved section of handrail used when approaching a landing; also, an outlet in a roof gutter

Grade—the level of the ground; also identifies the quality of lumber

Grain—in wood, the design on the surfaces caused by the contrast, spacing, and direction of the annular rings

Graphite—a mineral used as pencil lead and also as a lubricant for the working parts of locks and certain tools

Green lumber—lumber that has not been dried to a suitable moisture content

Groove—a cut, partway through, and running with the grain of lumber

Ground—strips of wood placed at the base of walls and around openings and used as a guide for the application of an even thickness of plaster; also, a system used for electrical safety

Ground Fault Circuit Interrupter (GFCI)—device used in electrical circuits for protection against electrical shock; it detects a short circuit instantly and shuts off the power automatically

Grout—a mixture of cement, fine aggregate, and water used to fill joints in masonry and tile

Gusset—a pad of wood or metal used over a joint to stiffen and strengthen it

Gutter—a wood or metal trough used at the roof edge to carry off rain water and water from melting snow

Gypsum—a chalky type rock that is the basic ingredient of plaster

Gypsum board—a sheet product made by encasing gypsum in a heavy paper wrapping

Half round—a molding with its end section in the shape of a semicircle

Handrail—a railing on a stairway intended to be grasped by the hand to serve as a support and guard

Hardboard—a building product made by compressing wood fibers into sheet form

Header—pieces placed at right angles to joists, studs, and rafters to form and support openings in a wood frame

Hearth—a section of the floor in front of a fireplace; usually covered with some type of fireproof material

Heartwood—the wood in the inner part of a tree, usually darker and containing inactive cells

Heel—the back end of objects, such as a handsaw or hand plane

Herringbone—a pattern used in parquet floors

Hexagon—a plane figure having six sides

Hip rafter—extends diagonally from the corner of the plate to the ridge at the intersection of two surfaces of a hip roof

Hip roof—a roof that slopes upward toward the ridge from four directions

Hip-valley cripple jack rafter—a short rafter running parallel to common rafters, cut between hip and valley rafters

Hopper window—a type of window in which the sash is hinged at the bottom and swings inward

Horn—an extension of the stiles of doors or the side jambs of window and door frames

Housed stringer—a finished stringer that is dadoed to receive treads and risers of a stairway

Housewrap—type of building paper with which the entire sidewalls of a building are covered

Insulated glass—multiple panes of glass fused together with an air space between them

Insulation—material used to restrict the passage of heat or sound

Intersecting roof—the roof of irregular shaped buildings; valleys are formed at the intersection of the roofs

Isometric—a drawing in which three surfaces of an object are seen in one view, with the base of each surface drawn at a 30° angle

J-roller—a 3-inch wide rubber roller used to apply pressure over the surface of contact cement-bonded plastic laminates

Jalousie window—a type of window containing movable, horizontal slats of glass

Jamb—the sides and top of window and door frames

Jamb extension—narrow strips of wood fastened to the edge of window jambs to increase their width

Jig—any type of fixture designed to hold pieces or guide tools while work is being performed

Joint—as a verb, denotes straightening the edge of lumber; as a noun, means the place where parts meet and unite

Joist—horizontal framing members used in a spaced pattern that provide support for the floor or ceiling system

Joist hanger—metal stirrups used to support the ends of joists

Journeyman—a tradesman who has completed an apprenticeship or who has gained enough experience to perform work without instruction

Juvenile wood—the portion of wood that contains the first seven to fifteen growth rings of a log. They are located in the pith.

Kerf—the width of a cut made with a saw

Keyway—a groove made in concrete footings for tying in the concrete foundation wall

Kicker—a member placed in such a manner to keep other members from moving

Kiln-dried—lumber dried by placing it in huge ovens called kilns

Knee wall—a wall of less than full height, also known as a pony wall

Knot—a defect in lumber caused by cutting through a branch or limb embedded in the log

Laminate—a thin layer of plastic often used as a finished surface for countertops

Laminated Strand Lumber (LSL)—lumber manufactured by bonding thin strands of wood, up to 12 inches long, with adhesive and pressure

Laminated Veneer Lumber (LVL)—lumber manufactured by laminating many veneers of plywood with the grain of all running in the same direction

Landing—an intermediate-level platform between flights

Laser—a concentrated, narrow beam of light; laser-equipped devices are used in building construction

Lath—a base for plaster; usually gypsum board or expanded metal sheets

Lattice—thin strips of wood, spaced apart and applied in two layers at angles to each layer resulting in a kind of grillwork

Lazy Susan—a set of revolving circular shelves; used in kitchen cabinets and other places

Ledger—a horizontal member of a wood scaffold that ties the scaffold posts together and supports the bearers; a temporary or permanent supporting member for joists or other members running at right angles

Level—horizontal or perpendicular to the force of gravity

Light—a pane of glass or an opening for a pane of glass

Linear measure—a measurement of length

Line length—the length of a rafter along a measuring line without consideration to the width or thickness of the rafter

Lintel—horizontal load-bearing member over an opening; also called a *header*

Lipped door—a cabinet door with rabbeted edges

Lookout—horizontal framing pieces in a cornice, installed to provide fastening for the soffit

Louver—an opening for ventilation consisting of horizontal slats installed at an angle to exclude rain, light, and vision, but to allow the passage of air

Low emissivity glass (LoE)—a coating on double-glazed windows designed to raise the insulating value by reflecting heat

Lumber—wood that is cut from the log to form boards, planks, and timbers

Magazine—a container in power nailers and staplers in which the fasteners to be ejected are placed

Mansard roof—a type of roof that has two different pitches on all sides of the building, with the lower slopes steeper than the upper

Mantel—the ornamental finish around a fireplace, including the shelf above the opening

Masonry—any construction of stone, brick, tile, concrete plaster, and similar materials

Mastic—a thick adhesive

Matched boards—boards that have been finished with tongue-and-grooved edges

Medullary ray—bands of cells radiating from the cambium layer to the pith of a tree to transport nourishment toward the center

Metes and Bounds—boundaries established by distances and compass directions

Millwork—any wood products that have been manufactured, such as moldings, doors, windows, and stairs for use in building construction; sometimes called *joinery*

Miter—the cutting of the end of a piece at any angle other than a right angle

Miter gauge—a guide used on the table saw for making miters and square ends

Miter joint—the joining of two pieces by cutting the end of each piece by bisecting the angle at which they are joined

Modular construction—a method of construction in which parts are preassembled in convenient-sized units

Moisture content—the amount of moisture in wood expressed as a percentage of the dry weight

Moisture meter—a device used to determine the moisture content of wood

Mortar—a mixture of portland cement, lime, sand, and water used to bond masonry units together

Mortise—a rectangular cavity cut in a piece of wood to receive a tongue or tenon projecting from another piece

Molding—decorative strips of wood used for finishing purposes

Mullion—a vertical division between windows or panels in a door

Muntin—slender strips of wood between lights of glass in windows or doors

Newel post—an upright post supporting the handrail in a flight of stairs

No. 1 common—a lower grade of hardwood lumber

Nominal size—the stated size of the thickness and width of lumber even though it differs from its actual size; the approximate size of rough lumber before it is surfaced

Nosing—the rounded edge of a stair tread projecting over the riser

OC ("on center")—the distance from the center of one structural member to the center of the next one

Octagon—a plane figure with eight sides

Ogee—a molding with an S-shaped curve

Open valley—a roof valley in which the roof covering is kept back from the centerline of the valley

Panel—a large sheet of building material

Panel door—a door in which panels are enclosed by a frame

Parallel Strand Lumber(PSL)— lumber manufactured by bonding strands of structural lumber, up to 8 feet long, with adhesive, heat, and pressure

Parquet—a floor made with strips or blocks to form intricate designs

Particleboard—a building product made by compressing wood chips and sawdust with adhesives to form sheets

Parting strip—a small strip of wood separating the upper and lower sash of a double-hung window

Partition—an interior wall separating one portion of a building from another

Penny (d)—a term used in designating nail sizes

Perforated—material that has closely spaced holes in a regular or irregular pattern

Perm—a measure of water vapor movement through a material

Phillips head—a type of screw head with a cross-slot

Pier—a column of masonry, usually rectangular in horizontal cross-section, used to support other structural members

Pilaster—column built within and usually projecting from a wall to reinforce the wall

Pile—concrete, metal, or wood pillar forced into the earth or cast in place as a foundation support

Pilot—a guide on the end of edge-forming router bits used to control the amount of cut

Pilot hole—a small hole drilled to receive the threaded portion of a wood screw

Pitch—the amount of slope to a roof expressed as ratio of the total rise to the span

Pitch block—a piece of wood cut in the shape of a right triangle; used as a pattern for laying out stair stringers, rafters, and handrail fittings

Pitch pocket—an opening in lumber between annular rings containing pitch in either liquid or solid form

Pith—the small, soft core at the center of a tree

Pivot—revolving around a point

Plain-sawed—a method of sawing lumber that produces flat-grain

Plan—in an architectural drawing, a object drawn as viewed from above

Plancier—the finish member on the underside of a box cornice; also called *soffit*

Plank—lumber that is 6 or more inches in width and from 1 1/2 to 6 inches in thickness

Plaster—a mixture of portland cement, sand, and water used for covering walls and ceilings of a building

Plastic laminate—a very tough, thin material in sheet form used to cover countertops; available in a wide variety of colors and designs

Plate—top or bottom horizontal member of a wall frame

Platform frame—method of wood frame construction in which walls are erected on a previously constructed floor deck or platform

Plinth block-a small, decorative block, thicker and wider than a door casing, used as part of the door trim at the base and at the head

Plot plan—a drawing showing a bird's-eye view of the lot, the position of the building, and other pertinent information; also called *site plan*

Plow—a wide groove cut with the grain of lumber

Plumb—vertical; at right angles to level

Plumb bob—a pointed weight attached to a line for testing plumb

Plunge cut—an interior cut made with a portable saw by a method that does not require boring holes before making the cut

Ply—one thickness of several layers of built-up material, such as one of the layers of plywood

Plywood—a building material in which thin sheets of wood are glued together with the grain of adjacent layers at right angles to each other

Pneumatic—powered by compressed air

Pocket—a recess in a wall to receive a piece, such as a recess in a concrete foundation wall to receive the end of a girder

Pocket door—a type of door that when opened slides into a recess in the wall

Polyethylene film—a thin plastic sheet used as a vapor barrier

Postforming—method used to bend plastic laminate to small radii

Preacher—a small piece of wood of the same thickness as the stair risers; it is notched in the center to fit over the finish stringer and rest on the tread cut of the stair carriage; it is used to lay out open finished stringers in a staircase

Preservative—a substance applied to wood to prevent decay

Pressure-treated—treatment given to lumber that applies preservative under pressure to penetrate the total piece

Primer—the first coat of paint applied to the surface or the paint used to prime a surface

Purlin—horizontal roof member used to support rafters between the plate and ridge

Quarter round—a type of molding, an end section of which is in the form of a quarter-circle

Quarter-sawed—a method of sawing lumber parallel to the medullary rays to produce edge-grain lumber. See *edge grain*

Rabbet—a cutout along the edge or end of lumber

Rafter—a sloping structural member of a roof frame that supports the roof sheathing and covering

Rafter tables—information found printed on the body of a framing square; used to calculate the lengths of various components of a roof system

Rail—the horizontal member of a frame

Rake—the sloping portion of the gable ends of a building

Rebar—steel reinforcing rod used in concrete

Reciprocating—a back-and-forth action, as in certain power tools

Return—a turn and continuation for a short distance of a molding, cornice, or other kind of finish

Return nosing—a separate piece mitered to the open end of a stair tread for the purpose of returning the tread nosing

Reveal—the amount of setback of the casing from the face side of window and door jambs or similar pieces

Ribbon—a narrow board let into studs of a balloon frame to support floor joists

Ridge—the highest point of a roof that has sloping sides

Ridgeboard—a horizontal member of a roof frame that is placed on edge at the ridge and into which the upper ends of rafters are fastened

Rip—sawing lumber in the direction of the grain

Rise—in stairs, the vertical distance of the Right; in roofs, the vertical distance from plate to ridge; may also be the vertical distance through which anything rises

Riser—the finish member in a stairway covering the space between treads

Rosette—a round or oval decorative wood piece used to end a handrail against a wall

Rough carpentry—that part of the trade involved with construction of the building frame or other work that will be dismantled or covered by the finish

Rough opening—opening in a wall in which windows or doors are placed

Rough stringer—cutout supports for the treads and risers in a staircase; also called *stair carriage* and *stair horse*

Run—the horizontal distance over which rafters, stairs, and other like members travel

R-value—a number given to a material to indicate its resistance to the passage of heat

S2S—surfaced two sides

S4S—surfaced four sides

Saddle—same as *cricket*

Sapwood—the outer part of a tree just beneath the bark containing active cells

Sash—that part of a window into which the glass is set

Sash balance—a device, usually operated by a spring or tensioned weatherstripping, designed to counterbalance double-hung window sash

Sawyer—a person whose job is to cut logs into lumber

S-beam—an I-shaped steel beam

Scab—a length of lumber applied over a joint to stiffen and strengthen it

Scaffold—an elevated, temporary working platform

Scratch coat—the first coat of plaster applied to metal lath

Screed—strips of wood, metal, or pipe secured in position and used as guides to level the top surface of concrete

Scribe—laying out woodwork to fit against an irregular surface

Scuttle—attic access or drain through a parapet wall

Seasoned lumber—lumber that has been dried to a suitable moisture content

Seat cut—see *bird's mouth*

Section—drawing showing a vertical cut through an object or part of an object

Selvage—the unexposed part of roll roofing covered by the course above

Set—alternate bending of saw teeth to provide clearance in the saw cut

Shake—a defect in lumber caused by a separation of the annular rings; also, a type of wood shingle

Shank hole—a hole drilled for the thicker portion of a wood screw

Sheathing—boards or sheet material that are fastened to roofs and exterior walls and on which the roof covering and siding are applied

Shed roof—a type of roof that slopes in one direction only

Shim—a thin, wedge-shaped piece of material used behind pieces for the purpose of straightening them, or for bringing their surfaces flush at a joint

Shingle tip—the thin end of a wood shingle

Shortened valley rafter—a valley rafter that runs from the plate to the supporting valley rafter

Side—the wide surfaces of a board, plank, or sheet

Sidelight—a framework containing small lights of glass placed on one or both sides of the entrance door

Siding—exterior sidewall finish covering

Sill—horizontal timbers resting on the foundation supporting the framework of a building; also, the lowest horizontal member in a window or door frame

Skirt board—another name for a finished stringer in a staircase

Skylight—a type of roof window

Sleeper—strips of wood laid over a concrete floor to which finish flooring is fastened

Slump test—a test given to concrete to determine its consistency

Snap tie—a metal device to hold concrete wall forms the desired distance apart

Soffit—the underside trim member of a cornice or any such overhanging assembly

Softwood—wood from coniferous (cone-bearing) trees

Soil stack—part of the plumbing; a vertical pipe extending up through the roof to vent the system

Soleplate—the bottom horizontal member of a wall frame

Specifications—written or printed directions of construction details for a building

Spike—a large nail 4 inches or longer

Spline—a thin, flat strip of wood inserted into the grooved edges of adjoining pieces

Spreader—a strip of wood used to keep other pieces a desired distance apart

Square—the amount of roof covering that will cover 100 square feet of roof area

Staging—see *scaffold*

Stair carriage—see *rough stringer*

Stair horse—see *rough stringer*

Stairwell—an opening in the floor for climbing or descending stairs or the space of a structure where the stairs are located

Standing cut—a cut made through the thickness of stock at more than a 90° angle between the side and edge or end

Staple—a U-shaped fastener

Starter course—usually used in reference to the first row of shingles applied to a roof or wall

Starting step—the first step in a flight of stairs

Sticker—machine that makes moldings or a thin strip placed between layers of lumber to create an air space for drying

Sticking—the molded inside edge of the frame of a panel door

Stile—the outside vertical members of a frame, such as in a paneled door

Stool—the bottom horizontal member of interior window trim

Stool cap—a horizontal finish piece covering the stool or sill of a window frame on the interior; also called *stool*

Stop bead—a vertical member of the interior finish of a window against which the sash butts or sides

Storm sash—an additional sash placed on the outside of a window to create a dead air space to prevent the loss of heat from the interior in cold weather

Story—the distance between the upper surface of any floor and the upper surface of the floor above

Story pole—a narrow strip of wood used to lay out the heights of members of a wall frame or courses of siding

Straightedge—a length of wood or metal having at least one straight edge to be used for testing straight surfaces

Strapping—application of furring strips at specified spacings for the purpose of attaching wall or ceiling finish; called *stripping* in some locations

Striated—finish material with random and finely spaced grooves running with the grain

Strongback—a member placed on edge and fastened to others to help support them

Stud—vertical framing member in a wall running between plates

Tack—to fasten temporarily in place; also, a short nail

Tail cut—a cut on the extreme lower end of a rafter

Tail joist—short joist running from an opening to a bearing

Tangent—straight line touching the circumference of a circle at one point

Taper—becoming thinner from one end to the other

Taper ground—the inner part of a saw blade ground thinner than the outside edge

Tempered—treated in a special way to be harder and stronger

Template—a pattern or a guide for cutting or drilling

Tenon—a tongue cut on the end of a piece usually to fit into a mortise

Termite shield—metal flashing plate over the foundation to protect wood members from termites

Threshold—a piece with chamfered edges placed on the floor under a door also called a *sill*

Tile—square or rectangular blocks placed side by side to cover an area

Timber—large pieces of lumber over 5 inches in thickness and width

Toe—the forward end of tools, such as a hand saw and hand plane

Toeboard—a strip of material located at the back of the toe space under a base cabinet; also the bottom horizontal member of a scaffold guardrail

Toenail—nail driven diagonally to fasten the end of framing

To the weather—a term used to indicate the exposure of roofing or siding

Transom—small sash above a door

Tread—horizontal finish members in a staircase upon which the feet of a person ascending or descending the stairs are placed

Trestle—similar to a sawhorse used to support scaffold plank

Trimmers—members of a frame placed at the sides of an opening running parallel to the main frame members

Truss—an engineered assembly of wood or wood and metal members used to support roofs or floors

Turnout—a type of handrail fitting

Undercut—a cut made through the thickness of finished material at slightly less than 90-degrees, such that butt joints in the material will fit tightly on the face sides

Underlayment—material placed on the subfloor to provide a smooth, even surface for the application of resilient finish floors

Unit length—the length of a stair stringer or rafter per unit of run

Unit rise—the amount a stair or rafter rises per unit of run

Unit run—a horizontal distance of a stair tread or horizontal segment of the total run of a rafter

Valley—the intersection of two roof slopes at interior corners

Valley cripple jack rafter—a rafter running between two valley rafters

Valley jack rafter—a rafter running between a valley rafter and the ridge

Valley rafter—the rafter placed at the intersection of two roof slopes in interior corners

Vapor retarder—a material used to prevent the passage of vapor

Veneer—a very thin sheet or layer of wood

Vermiculite—a mineral closely related to mica with the ability to expand on heating to form a lightweight material with insulating qualities

Volute—a spiral fitting at beginning of a handrail

Wainscoting—a wall finish applied partway up the wall from the floor

Waler—horizontal or vertical members of a concrete form used to brace and stiffen the form and to which ties are fastened

Wane—bark, or lack of wood, on the edge of lumber

Warp—any deviation from straightness in a piece of lumber

Water table—finish work applied just above the foundation that projects beyond it and sheds water away from it

W-beam—a wide-flanged, I-shaped steel beam

Weatherstripping—narrow strips of thin metal or other material applied to windows and doors to prevent the infiltration of air and moisture

Web—wood or metal members connecting top and bottom chords in trusses; also, the center section of a wood or steel I-beam

Whet—the sharpening of a tool on a sharpening stone by rubbing the tool on the stone

Wind—a defect in lumber caused by a twist in the stock from one end to the other

Winder—a tread in a stairway, wider on one end than the other, which changes the direction of travel

Index

The **bold** numbers are illustrations.